CHEMICAL ENGINEERING IN THE
PHARMACEUTICAL INDUSTRY

CHEMICAL ENGINEERING IN THE PHARMACEUTICAL INDUSTRY

Active Pharmaceutical Ingredients

Second Edition

Edited by

DAVID J. AM ENDE
Nalas Engineering Services, Inc., Centerbrook, CT, USA

MARY T. AM ENDE
Lyndra Therapeutics, Inc., Watertown, MA, USA

WILEY

AIChE
The Global Home of Chemical Engineers

This edition first published 2019
© 2019 John Wiley & Sons, Inc.

A Joint Publication of the American Institute of Chemical Engineers and John Wiley & Sons, Inc.

Edition History
Chemical Engineering in the Pharmaceutical Industry, First edition, John Wiley & Sons, 2011.

All rights reserved. No part of this publication may be reproduced, stored in a retrieval system, or transmitted, in any form or by any means, electronic, mechanical, photocopying, recording or otherwise, except as permitted by law. Advice on how to obtain permission to reuse material from this title is available at http://www.wiley.com/go/permissions.

The right of David J. am Ende and Mary T. am Ende to be identified as the authors of the editorial material in this work has been asserted in accordance with law.

Registered Office
John Wiley & Sons, Inc., 111 River Street, Hoboken, NJ 07030, USA

Editorial Office
111 River Street, Hoboken, NJ 07030, USA

For details of our global editorial offices, customer services, and more information about Wiley products visit us at www.wiley.com.

Wiley also publishes its books in a variety of electronic formats and by print-on-demand. Some content that appears in standard print versions of this book may not be available in other formats.

Limit of Liability/Disclaimer of Warranty
In view of ongoing research, equipment modifications, changes in governmental regulations, and the constant flow of information relating to the use of experimental reagents, equipment, and devices, the reader is urged to review and evaluate the information provided in the package insert or instructions for each chemical, piece of equipment, reagent, or device for, among other things, any changes in the instructions or indication of usage and for added warnings and precautions. While the publisher and authors have used their best efforts in preparing this work, they make no representations or warranties with respect to the accuracy or completeness of the contents of this work and specifically disclaim all warranties, including without limitation any implied warranties of merchantability or fitness for a particular purpose. No warranty may be created or extended by sales representatives, written sales materials or promotional statements for this work. The fact that an organization, website, or product is referred to in this work as a citation and/or potential source of further information does not mean that the publisher and authors endorse the information or services the organization, website, or product may provide or recommendations it may make. This work is sold with the understanding that the publisher is not engaged in rendering professional services. The advice and strategies contained herein may not be suitable for your situation. You should consult with a specialist where appropriate. Further, readers should be aware that websites listed in this work may have changed or disappeared between when this work was written and when it is read. Neither the publisher nor authors shall be liable for any loss of profit or any other commercial damages, including but not limited to special, incidental, consequential, or other damages.

Library of Congress Cataloging-in-Publication data applied for

Hardback ISBN: 9781119285861

Cover Design: Wiley
Cover Images: © Sigma surface for aspirin, COSMOBASE 2015, COSMOLogic Gmbh;
Stirred tank showing flow pattern courtesy of Francis X. McConville;
Stirred tank via computational fluid dynamics courtesy of Vivek Ranade;
Photomicrograph courtesy of Nalas Engineering Services, Inc.

Set in 10/12pt Times by SPi Global, Pondicherry, India

Printed and bound by CPI Group (UK) Ltd, Croydon, CR0 4YY

C9781119285861_231122

CONTENTS

LIST OF CONTRIBUTORS	xi
PREFACE	xv
UNIT CONVERSIONS	xvii

PART I INTRODUCTION — 1

1 Chemical Engineering in the Pharmaceutical Industry: An Introduction — 3
David J. am Ende and Mary T. am Ende

2 Current Challenges and Opportunities in the Pharmaceutical Industry — 19
Joseph L. Kukura and Michael P. Thien

PART II MASS AND ENERGY BALANCES — 27

3 Process Safety and Reaction Hazard Assessment — 29
Wim Dermaut

4 Calorimetric Approaches to Characterizing Undesired Reactions — 61
Megan Roth and Tom Vickery

5 Case Study of a Borane–THF Explosion — 91
David J. am Ende and Richard M. Davis

6 Analytical Aspects for Determination of Mass Balances — 115
Matthew Jorgensen

7 Quantitative Applications of NMR Spectroscopy — 133
Brian L. Marquez and R. Thomas Williamson

PART III REACTION KINETICS AND MIXING PROCESSES — 151

8 Reaction Kinetics and Characterization — 153
Utpal K. Singh, Brandon J. Reizman, Shujauddin M. Changi, Justin L. Burt, and Chuck Orella

9 Understanding Fundamental Processes in Catalytic Hydrogenation Reactions — 191
Yongkui Sun and Carl LeBlond

10 Characterization and First Principles Prediction of API Unit Operations — 203
Joe Hannon

11 Scale-Up of Mass Transfer-Limited Reactions: Fundamentals and a Case Study — 227
Ayman Allian and Seth Huggins

12 Scale-Up of Mixing Processes: A Primer — 241
Francis X. McConville and Stephen B. Kessler

13 Stirred Vessels: Computational Modeling of Multiphase Flows and Mixing — 261
Avinash R. Khopkar and Vivek V. Ranade

PART IV CONTINUOUS PROCESSING — 319

14 Process Development and Case Studies of Continuous Reactor Systems for Production of API and Pharmaceutical Intermediates — 321
Thomas L. LaPorte, Chenchi Wang, and G. Scott Jones

15 Development and Application of Continuous Processes for the Intermediates and Active Pharmaceutical Ingredients — 341
Flavien Susanne

16 Design and Selection of Continuous Reactors for Pharmaceutical Manufacturing — 367
Martin D. Johnson, Scott A. May, Michael E. Kopach, Jennifer McClary Groh, Timothy Braden, Vaidyaraman Shankarraman, and Jeremy Miles Merritt

PART V BIOLOGICS — 387

17 Chemical Engineering Principles in Biologics: Unique Challenges and Applications — 389
Sourav Kundu, Vivek Bhatnagar, Naveen Pathak, and Cenk Undey

PART VI THERMODYNAMICS — 417

18 Applications of Thermodynamics Toward Pharmaceutical Problem Solving — 419
Ahmad Y. Sheikh, Alessandra Mattei, Raimundo Ho, Moiz Diwan, Thomas Borchardt, Gerald Danzer, Nadine Ding, and Xinmin (Sam) Xu

19 A General Framework for Solid–Liquid Equilibria in Pharmaceutical Systems — 439
Thomas Lafitte, Vasileios Papaioannou, Simon Dufal, and Constantinos C. Pantelides

20	**Drug Solubility, Reaction Thermodynamics, and Co-Crystal Screening** *Karin Wichmann, Christoph Loschen, and Andreas Klamt*	467
21	**Thermodynamic Modeling of Aqueous and Mixed Solvent Electrolyte Systems** *Benjamin Caudle, Toni E. Kirkes, Cheng-Hsiu Yu, and Chau-Chyun Chen*	493
22	**Thermodynamics and Relative Solubility Prediction of Polymorphic Systems** *Yuriy A. Abramov and Klimentina Pencheva*	505
23	**Toward a Rational Solvent Selection for Conformational Polymorph Screening** *Yuriy A. Abramov, Mark Zell, and Joseph F. Krzyzaniak*	519

PART VII CRYSTALLIZATION AND FINAL FORM 533

24	**Crystallization Design and Scale-Up** *James Wertman, Robert McKeown, Lotfi Derdour, and Philip Dell'Orco*	535
25	**Introduction to Chiral Crystallization in Pharmaceutical Development and Manufacturing** *Jose E. Tabora, Shawn Brueggemeier, Michael Lovette, and Jason Sweeney*	569
26	**Measurement of Solubility and Estimation of Crystal Nucleation and Growth Kinetics** *Nandkishor K. Nere, Manish S. Kelkar, Ann M. Czyzewski, Kushal Sinha, and Evelina B. Kim*	591
27	**Case Studies On Crystallization Scale-Up** *Nandkishor K. Nere, Moiz Diwan, Ann M. Czyzewski, James C. Marek, Kushal Sinha, and Huayu Li*	617
28	**Population Balance-Enabled Model for Batch and Continuous Crystallization Processes** *Ajinkya Pandit, Rahul Bhambure, and Vivek V. Ranade*	635
29	**Solid Form Development for Poorly Soluble Compounds** *Alessandra Mattei, Shuang Chen, Jie Chen, and Ahmad Y. Sheikh*	665
30	**Multiscale Assessment of API Physical Properties in the Context of Materials Science Tetrahedron Concept** *Raimundo Ho, Yujin Shin, Yinshan Chen, Laura Poloni, Shuang Chen, and Ahmad Y. Sheikh*	689

PART VIII SEPARATIONS, FILTRATION, DRYING AND MILLING 713

31	**The Design and Economics of Large-Scale Chromatographic Separations** *Firoz D. Antia*	715
32	**Membrane Systems for Pharmaceutical Applications** *Dimitrios Zarkadas and Kamalesh K. Sirkar*	733
33	**Design of Distillation and Extraction Operations** *Eric M. Cordi*	751

viii CONTENTS

34 **Case Studies On the Use of Distillation in the Pharmaceutical Industry** — 787
    Laurie Mlinar, Kushal Sinha, Elie Chaaya, Subramanya Nayak, and Andrew Cosbie

35 **Design of Filtration and Drying Operations** — 799
    Praveen K. Sharma, Saravanababu Murugesan, and Jose E. Tabora

36 **Filtration Case Studies** — 833
    Seth Huggins, Andrew Cosbie, and John Gaertner

37 **Drying Case Studies** — 847
    John Gaertner, Nandkishor K. Nere, James C. Marek, Shailendra Bordawekar, Laurie Mlinar, Moiz Diwan, and Lei Cao

38 **Milling Operations in the Pharmaceutical Industry** — 861
    Kevin D. Seibert, Paul C. Collins, Carla V. Luciani, and Elizabeth S. Fisher

PART IX STATISTICAL MODELS, PAT, AND PROCESS MODELING APPLICATIONS — 881

39 **Experimental Design for Pharmaceutical Development** — 883
    Gregory S. Steeno

40 **Multivariate Analysis in API Development** — 909
    James C. Marek

41 **Probabilistic Models for Forecasting Process Robustness** — 919
    Jose E. Tabora, Jacob Albrecht, and Brendan Mack

42 **Use of Process Analytical Technology (PAT) in Small Molecule Drug Substance Reaction Development** — 937
    Dimitri Skliar, Jeffrey Nye, and Antonio Ramirez

43 **Process Modeling Applications Toward Enabling Development and Scale-Up: Chemical Reactions** — 957
    Anuj A. Verma, Steven Richter, Brian Kotecki, and Moiz Diwan

PART X MANUFACTURING — 971

44 **Process Scale-Up and Assessment** — 973
    Alan Braem, Jason Sweeney, and Jean Tom

45 **Scale-Up Do's and Don'ts** — 1001
    Francis X. McConville

46 **Kilo Lab and Pilot Plant Manufacturing** — 1011
    Matthew Casey, Jason Hamm, Melanie Miller, Tom Ramsey, Richard Schild, Andrew Stewart, and Jean Tom

47 **The Role of Simulation and Scheduling Tools in the Development and Manufacturing of Active Pharmaceutical Ingredients** — 1037
    Demetri Petrides, Doug Carmichael, Charles Siletti, Dimitris Vardalis, Alexandros Koulouris, and Pericles Lagonikos

PART XI QUALITY BY DESIGN AND REGULATORY   1067

48 Scientific Opportunities through Quality by Design   1069
Timothy J. Watson and Roger Nosal

49 Applications of Quality Risk Assessment in Quality by Design (QbD) Drug Substance Process Development   1073
Alan Braem and Gillian Turner

50 Development of Design Space for Reaction Steps: Approaches and Case Studies for Impurity Control   1091
Srinivas Tummala, Antonio Ramirez, Sushil Srivastava, and Daniel M. Hallow

INDEX   1123

LIST OF CONTRIBUTORS

Yuriy A. Abramov, Global R&D, Pharmaceutical Sciences, Pfizer, Inc., Groton, CT, USA

Jacob Albrecht, Product Development, Bristol-Myers Squibb, New Brunswick, NJ, USA

Ayman Allian, Synthetic Technologies and Engineering, Amgen Inc., Thousand Oaks, CA, USA

David J. am Ende, Nalas Engineering Services, Inc., Centerbrook, CT, USA

Mary T. am Ende, Lyndra Therapeutics, Inc., Watertown, MA, USA

Firoz D. Antia, Antisense Oligonucleotide Process Development and Manufacturing, Biogen Inc., Cambridge, MA USA

Rahul Bhambure, Chemical Engineering and Process Development Division, CSIR – National Chemical Laboratory, Pune, MH, India

Vivek Bhatnagar, Biologics R&D, Teva Pharmaceuticals, Inc., West Chester, PA, USA

Thomas Borchardt, Drug Product Development, AbbVie Inc., North Chicago, IL, USA

Shailendra Bordawekar, Process Research and Development, AbbVie Inc., North Chicago, IL, USA

Timothy Braden, Eli Lilly and Company, Indianapolis, IN, USA

Alan Braem, Product Development, Bristol-Myers Squibb, New Brunswick, NJ, USA

Shawn Brueggemeier, Product Development, Bristol-Myers Squibb, New Brunswick, NJ, USA

Justin L. Burt, Eli Lilly and Company, Indianapolis, IN, USA

Lei Cao, Operations Science and Technology, AbbVie Inc., North Chicago, IL, USA

Doug Carmichael, Intelligen, Inc., Scotch Plains, NJ, USA

Matthew Casey, Biogen Inc., Durham, NC, USA

Benjamin Caudle, Texas Tech University, Lubbock, TX, USA

Elie Chaaya, Process Research and Development, AbbVie Inc., North Chicago, IL, USA

Shujauddin M. Changi, Eli Lilly and Company, Indianapolis, IN, USA

Chau-Chyun Chen, Texas Tech University, Lubbock, TX, USA

Jie Chen, Solid State Chemistry, AbbVie Inc., North Chicago, IL, USA

Shuang Chen, Solid State Chemistry, AbbVie Inc., North Chicago, IL, USA

Yinshan Chen, Solid State Chemistry, AbbVie Inc., North Chicago, IL, USA

Paul C. Collins, Eli Lilly and Company, Indianapolis, IN, USA

LIST OF CONTRIBUTORS

Eric M. Cordi, Chemical R&D, Pfizer, Inc., Groton, CT, USA

Andrew Cosbie, Drug Substance Technologies and Engineering, Amgen Inc., Thousand Oaks, CA, USA

Ann M. Czyzewski, Process Research and Development, AbbVie Inc., North Chicago, IL, USA

Gerald Danzer, Drug Product Development, AbbVie Inc., North Chicago, IL, USA

Richard M. Davis, Global Environmental, Health and Safety, Pfizer, Inc., Groton, CT, USA

Philip Dell'Orco, Chemical Development, GlaxoSmithKline, King of Prussia, PA, USA

Lotfi Derdour, Chemical Development, GlaxoSmithKline, King of Prussia, PA, USA

Wim Dermaut, Chemical Process Development, Materials Technology Center, Agfa-Gevaert NV, Mortsel, Belgium

Nadine Ding, Abbott Vascular, Santa Clara, CA, USA

Moiz Diwan, Process Research and Development, AbbVie Inc., North Chicago, IL, USA

Simon Dufal, Process Systems Enterprise Ltd., London, UK

Elizabeth S. Fisher, Merck & Co., Inc., Rahway, NJ, USA

John Gaertner, Process Research and Development, AbbVie Inc., North Chicago, IL, USA

Jennifer McClary Groh, Eli Lilly and Company, Indianapolis, IN, USA

Daniel M. Hallow, Noramco, Athens, GA, USA

Jason Hamm, Product Development, Bristol-Myers Squibb, New Brunswick, NJ, USA

Joe Hannon, Scale-up Systems Limited, Dublin, Ireland

Raimundo Ho, Solid State Chemistry, AbbVie Inc., North Chicago, IL, USA

Seth Huggins, Drug Substance Technologies and Engineering, Amgen Inc., Thousand Oaks, CA, USA

Martin D. Johnson, Eli Lilly and Company, Indianapolis, IN, USA

G. Scott Jones, Bristol-Myers Squibb, New Brunswick, NJ, USA

Matthew Jorgensen, Nalas Engineering Services, Inc., Centerbrook, CT, USA

Manish S. Kelkar, Process Research and Development, AbbVie Inc., North Chicago, IL, USA

Stephen B. Kessler, Impact Technology Development, Lincoln, MA, USA

Avinash R. Khopkar, Reliance Industries Limited, Mumbai, MH, India

Evelina B. Kim, Process Research and Development, AbbVie Inc., North Chicago, IL, USA

Toni E. Kirkes, Texas Tech University, Lubbock, TX, USA

Andreas Klamt, COSMO logic GmbH & Co. KG, Leverkusen, Germany and Institute of Physical and Theoretical Chemistry, University of Regensburg, Regensburg, Germany

Michael E. Kopach, Eli Lilly and Company, Indianapolis, IN, USA

Brian Kotecki, Process Research and Development, AbbVie Inc., North Chicago, IL, USA

Alexandros Koulouris, Alexander Technological Education Institute of Thessaloniki, Thessaloniki, Greece

Joseph F. Krzyzaniak, Global R&D, Pharmaceutical Sciences, Pfizer, Inc., Groton, CT, USA

Joseph L. Kukura, Merck Research Laboratories, Merck & Co., Inc., Rahway, NJ, USA

Sourav Kundu, Biologics R&D, Teva Pharmaceuticals, Inc., West Chester, PA, USA

Thomas Lafitte, Process Systems Enterprise Ltd., London, UK

Pericles Lagonikos, Merck & Co., Inc., Singapore, Singapore

Thomas L. LaPorte, Product Development, Bristol-Myers Squibb, New Brunswick, NJ, USA

Carl LeBlond, Department of Chemistry, University of Pennsylvania, Indiana, PA, USA

Huayu Li, Material and Analytical Sciences, Boehringer Ingelheim Pharmaceuticals, Inc., Ridgefield, CT, USA

Christoph Loschen, COSMO logic GmbH & Co. KG, Leverkusen, Germany

Michael Lovette, Drug Substance Process Development, Amgen Inc., Thousand Oaks, CA, USA

Carla V. Luciani, Eli Lilly and Company, Indianapolis, IN, USA

Brendan Mack, Product Development, Bristol-Myers Squibb, New Brunswick, NJ, USA

James C. Marek, Process Research and Development, AbbVie Inc., North Chicago, IL, USA

Brian L. Marquez, Nalas Engineering Services, Inc., Centerbrook, CT, USA

Alessandra Mattei, Solid State Chemistry, AbbVie Inc., North Chicago, IL, USA

Scott A. May, Eli Lilly and Company, Indianapolis, IN, USA

Francis X. McConville, Impact Technology Development, Lincoln, MA, USA

Robert McKeown, Chemical Development, GlaxoSmithKline, King of Prussia, PA, USA

Jeremy Miles Merritt, Eli Lilly and Company, Indianapolis, IN, USA

Melanie Miller, Product Development, Bristol-Myers Squibb, New Brunswick, NJ, USA

Laurie Mlinar, Process Research and Development, AbbVie Inc., North Chicago, IL, USA

Saravanababu Murugesan, Product Development, Bristol-Myers Squibb, New Brunswick, NJ, USA

Subramanya Nayak, Process Research and Development, AbbVie Inc., North Chicago, IL, USA

Nandkishor K. Nere, Process Research and Development, AbbVie Inc., North Chicago, IL, USA

Roger Nosal, Global CMC, Pfizer, Inc., Groton, CT, USA

Jeffrey Nye, Product Development, Bristol-Myers Squibb, New Brunswick, NJ, USA

Chuck Orella, Merck & Co., Inc., Rahway, NJ, USA

Ajinkya Pandit, Chemical Engineering and Process Development Division, CSIR – National Chemical Laboratory, Pune, MH, India

Constantinos C. Pantelides, Process Systems Enterprise Ltd., London, UK

Vasileios Papaioannou, Process Systems Enterprise Ltd., London, UK

Naveen Pathak, Process Development and Technical Services, Shire plc, Cambridge, MA, USA

Klimentina Pencheva, Global R&D, Pharmaceutical Sciences, Pfizer, Inc., Sandwich, UK

Demetri Petrides, Intelligen, Inc., Scotch Plains, NJ, USA

Laura Poloni, Department of Chemical Engineering and Applied Chemistry, University of Toronto, Ontario, Canada

Antonio Ramirez, Product Development, Bristol-Myers Squibb, New Brunswick, NJ, USA

Tom Ramsey, Janssen, Raritan, NJ, USA

Vivek V. Ranade, School of Chemistry and Chemical Engineering, Queen's University of Belfast, Belfast, UK

Brandon J. Reizman, Eli Lilly and Company, Indianapolis, IN, USA

Steven Richter, Process Research and Development, AbbVie Inc., North Chicago, IL, USA

Megan Roth, Chemical Engineering R&D, Merck & Co., Inc., Rahway, NJ, USA

Richard Schild, TG Therapeutics, Inc. New York City, NY, USA

Kevin D. Seibert, Eli Lilly and Company, Indianapolis, IN, USA

Vaidyaraman Shankarraman, Eli Lilly and Company, Indianapolis, IN, USA

Praveen K. Sharma, Chemical Development, Tetraphase Pharmaceuticals, Inc., Watertown, MA, USA

Ahmad Y. Sheikh, Solid State Chemistry, AbbVie Inc., North Chicago, IL, USA

Yujin Shin, Solid State Chemistry, AbbVie Inc., North Chicago, IL, USA

Charles Siletti, Intelligen, Inc., Scotch Plains, NJ, USA

Utpal K. Singh, Eli Lilly and Company, Indianapolis, IN, USA

Kushal Sinha, Process Research and Development, AbbVie Inc., North Chicago, IL, USA

Kamalesh K. Sirkar, New Jersey Institute of Technology, Newark, NJ, USA

Dimitri Skliar, Product Development, Bristol-Myers Squibb, New Brunswick, NJ, USA

Sushil Srivastava, Bristol-Myers Squibb, New Brunswick, NJ, USA

Gregory S. Steeno, Worldwide Research and Development, Pfizer, Inc., Groton, CT, USA

Andrew Stewart, Allergan, Irvine, CA, USA

Yongkui Sun, Ionova Life Science Co., Ltd., Shenzhen, China

Flavien Susanne, Product and Process Engineering, GSK Medicines Research Centre, GlaxoSmithKline, Stevenage, UK

Jason Sweeney, Product Development, Bristol-Myers Squibb, New Brunswick, NJ, USA

Jose E. Tabora, Product Development, Bristol-Myers Squibb, New Brunswick, NJ, USA xiv   LIST OF CONTRIBUTORS

Michael P. Thien, Merck Manufacturing Division, Merck & Co., Inc., Whitehouse Station, NJ, USA

Jean Tom, Product Development, Bristol-Myers Squibb, New Brunswick, NJ, USA

Srinivas Tummala, Bristol-Myers Squibb, New Brunswick, NJ, USA

Gillian Turner, Product Development and Supply, GlaxoSmithKline, Stevenage, UK

Cenk Undey, Process Development, Amgen Inc., Thousand Oaks, CA, USA

Dimitris Vardalis, Intelligen, Inc., Scotch Plains, NJ, USA

Anuj A. Verma, Process Research and Development, AbbVie Inc., North Chicago, IL, USA

Tom Vickery, Chemical Engineering R&D, Merck & Co., Inc., Rahway, NJ, USA

Chenchi Wang, Manufacturing Science and Technology, Bristol-Myers Squibb, New Brunswick, NJ, USA

Timothy J. Watson, Global CMC, Pfizer, Inc., Groton, CT, USA

James Wertman, Technical Operations, Theravance Biopharma, South San Francisco, CA, USA

Karin Wichmann, COSMO logic GmbH & Co. KG, Leverkusen, Germany

R. Thomas Williamson, Department of Chemistry & Biochemistry, University of North Carolina Wilmington, Wilmington, NC, USA

Xinmin (Sam) Xu, Abbott Vascular, Santa Clara, CA, USA

Cheng-Hsiu Yu, Texas Tech University, Lubbock, TX, USA

Dimitrios Zarkadas, Merck & Co., Inc., Rahway, NJ, USA

Mark Zell, Takeda Oncology, Cambridge, MA, USA

PREFACE

Chemical Engineering in the Pharmaceutical Industry is unique in many ways as to what is traditionally taught in schools of chemical engineering. This book is thus intended to cover many concepts and applications of chemical engineering science that are particularly important to the pharmaceutical industry. Several excellent books have been written on the subjects of *Process Chemistry in the Pharmaceutical Industry* and separately on formulation development, but relatively little has been published specifically with a chemical engineering focus.

The intent of this book is to highlight the importance and value of chemical engineering to the development and commercialization of pharmaceuticals covering active pharmaceutical ingredients (APIs) and drug products (DPs). It should serve as a resource handbook to practicing chemical engineers as well as a resource for chemists, analysts, technologists, and operations and management team members – all those who partner to bring pharmaceuticals successfully to market. The latter will benefit through an exposure to the mathematical and predictive approach and the broader capabilities of chemical engineers as well as to illustrate chemical engineering science specifically to pharmaceutical problems. This book emphasizes the need for scientific integration of chemical engineers with synthetic organic chemists within process R&D, as well as the importance of the interface between R&D engineers and manufacturing engineers.

Although specific workflows for engineers in R&D depend on each company's specific organization, in general it is clear that, as part of a multidisciplinary team in R&D, chemical engineering practitioners offer value in many ways including API and DP process design, scale-up assessment from lab to plant, process modeling, process understanding, and general process development that ultimately reduces cost and ensures safe, robust, and environmentally friendly processes are transferred to manufacturing. How effective the teams leverage each of the various skill sets (i.e. via resource allocation) to arrive at an optimal process depends in part on the roles and responsibilities as determined within each organization and company. In general it is clear that with increased cost pressures facing the pharmaceutical industry, including R&D and manufacturing, opportunities to leverage the field of chemical engineering science continue to increase. The increased emphasis and broader implementation of continuous processing from R&D to manufacturing over the last 10 years is a good example.

In the first edition of this book, 44 chapters spanned API, drug product, and analytical. The second edition has expanded to 75 total chapters, and for this reason we divided the book into two discrete volumes to separately focus on API and drug product.

The second edition of this book is divided into following sections:

Volume 1: API/Drug Substance

- Introduction
- Mass and Energy Balances
- Reaction Kinetics and Mixing Processes
- Continuous Processing
- Biologics
- Thermodynamics
- Crystallization and Final Form
- Separations, Filtration, Drying, and Milling
- Statistical Models, PAT, and Process Modeling Applications

- Manufacturing
- Quality by Design and Regulatory

Volume 2: Drug Product

- Introduction
- Drug Product Design, Development, and Modeling
- Continuous Manufacturing
- Applied Statistics and Regulatory Environment

The second edition has many new chapters and significantly expanded previous chapters. We have 13 chapters devoted to applied thermodynamics, final form, and crystallization. Eight new chapters are case studies. New chapters were added on continuous processing and quality by design as well.

The contributors to these two volumes were encouraged to provide worked-out examples – so in most chapters a quantitative example is offered to illustrate key concepts, assumptions, and a problem solving approach. In this way, the chapters serve to help others solve similar problems.

There are many people to thank that made the original book project possible.

It was during my time at Pfizer from 1994 to 2013 in Chemical R&D where I began to truly appreciate my career choice and how chemical engineering science could add value to pharmaceutical projects and project teams. I was fortunate to get my start at Pfizer working with the Mettler RC1 and FTIR and having the opportunity to build the process safety, reaction engineering, and later engineering technologies group within Chemical R&D. It was there that I was inspired to take on this project for the first edition of this book. I am grateful to my Chemical R&D management for permitting me to fulfill that vision in 2010. In 2013, I partnered with Jerry Salan and joined Nalas Engineering.

Special thanks to my family (Mary, Nathan, Noah, Brianna) for their support during the preparation of this book. Special thanks to Mary, not only for contributing multiple chapters in this book but also for assisting in all phases of the project and as coeditor for the second edition. In addition, a special thanks to my parents for their encouragement to pursue chemical engineering in 1983 and their support ever since.

David J. am Ende, PhD

President
Nalas Engineering Services, Inc.
Centerbrook, CT, USA
December 2018

UNIT CONVERSIONS

Quantity	Equivalent Values
Length	1 m = 100 cm = 1000 mm = 10^6 μm = 10^{10} Å 1 m = 39.37 in = 3.2808 ft = 1.0936 yards = 0.0006214 mile 1 ft = 12 in = 0.3048 m = 1/3 yard = 30.48 cm
Area	$1 \text{ m}^2 = 10.76 \text{ ft}^2 = 1550 \text{ in}^2 = 10\,000 \text{ cm}^2$ $1 \text{ in}^2 = 6.4516 \text{ cm}^2 = 645.16 \text{ mm}^2 = 0.006\,94 \text{ ft}^2$ $1 \text{ ft}^2 = 929.03 \text{ cm}^2 = 0.092\,903 \text{ m}^2$ Example: cross sectional area of ¼″ ID tube: $\frac{\pi d^2}{4} = \frac{\pi 0.25^2}{4} = 0.0491 \text{ in}^2 = 0.3167 \text{ cm}^2$
Volume	$1 \text{ m}^3 = 1000 \text{ L} = 10^6 \text{ cm}^3 \text{ (ml)} = 1000 \text{ dm}^3$ $1 \text{ m}^3 = 35.3145 \text{ ft}^3 = 220.83 \text{ imperial gallons} = 264.17 \text{ gal (U)}$ $1 \text{ ft}^3 = 1728 \text{ in}^3 = 7.4805 \text{ gal (US)} = 0.028\,317 \text{ m}^3 = 28.317 \text{ L}$ 1 gal (US) = 3.785 L = 0.1337 ft^3 = 231 in^3 = 4 quart = 8 pints 1 L = 0.264 gal = 1.0567 quart = 2.113 pint = 4.2267 cup = 202.88 teaspoon (US) = 0.035 31 ft^3 = 61.02 in^3
Mass	1 kg = 1000 g = 0.001 metric ton (MT) = 2.204 62 lb_m = 35.273 92 oz 1 lb_m = 16 oz = 453.593 g = 0.453 593 kg 1 ton (metric) = 1000 kg = 2204.6 lb_m
Pressure	1 atm = 1.013 25 bar = 1.013 25 × $10^5 \frac{\text{N}}{\text{m}^2}$ (Pa) = 0.101 325 MPa = 101.325 kPa = 1.013 25 × $10^6 \frac{\text{dynes}}{\text{cm}^2}$ = 1.033 $\frac{\text{kg}_f}{\text{cm}^2}$ = 760 mm Hg at 0 °C (torr) = 10.333 m H_2O at 4 °C = 14.696 $\frac{\text{lb}_f}{\text{in}^2}$ (psi) = 33.9 ft H_2O at 4 °C = 2116 $\frac{\text{lb}_f}{\text{ft}^2}$ = 29.921 in Hg at 0 °C 1 MPa = 9.869 atm = 10 bar = 145.04 psi 1 psi = 2.31 ft H_2O = 0.0680 atm = 703.1 $\frac{\text{kg}_f}{\text{m}^2}$ = 0.070 31 $\frac{\text{kg}_f}{\text{cm}^2}$ = 51.71 mm Hg *Note:* $P_{\text{absolute}} = P_{\text{gauge}} + P_{\text{atmospheric}}$ For example, if a pressure gauge reads 30 psig, then the absolute pressure is 44.7 psia, *i.e.* $P_{\text{absolute}} = P_{\text{gauge}} + P_{\text{atmospheric}} = 30 + 14.7 = 44.7$ psia

Quantity	Equivalent Values
Pressure (*continued*)	Vacuum: A vacuum gauge may have range from 0 to −30 in *Hg*. So if the vacuum gauge is reading −25.0 in *Hg* then the absolute pressure is $P_{absolute} = P_{gauge} + P_{atmospheric} = -25.0 + 29.921 = +4.921\ in\ Hg = 0.16\ atm = 2.42\ psia$ *(assuming that the atmospheric pressure is taken at sea level equivalent to 1.0 atm = 29.921 in Hg)*
Temperature	°C = 5/9 (°F − 32) °F = 9/5 °C + 32 K = °C + 273.15 = 5/9 °R °R = °F + 459.67 Freezing point H_2O = 0 °C or 32 °F or 273.15 K Boiling point H_2O = 100 °C or 212 °F or 373.15 K
Density (ρ)	$1\ \frac{g}{cm^3} = 1\ \frac{g}{ml} = 1\ \frac{kg}{L} = 62.43\ \frac{lb_m}{ft^3} = 1000\ \frac{kg}{m^3} = 8.345\ \frac{lb_m}{U.S.gal}$ $100\ \frac{lb_m}{ft^3} = 1601.85\ \frac{kg}{m^3} = 1.602\ \frac{g}{cm^3}$ $\rho(H_2O,\ 20\ °C) = 998.2\ \frac{kg}{m^3} = 0.9982\ \frac{g}{cm^3}$ $\rho(H_2SO_4,\ 25\ °C,\ 95\ wt\ \%) = 1.84\ \frac{g}{cm^3}$
Force	$1\ N = 1\ \frac{kg \cdot m}{sec^2} = 10^5\ dynes = 10^5\ g\ \frac{g \cdot cm}{sec^2} = 0.224\ 81\ lb_f = 0.102\ kg_f$ $1\ lb_f = 32.174\ \frac{lb_m \cdot ft}{sec^2} = 4.4482\ N = 4.4482 \times 10^5\ dynes$ $1\ kg_f = 1\ kg \cdot 9.806\ 65\ \frac{m}{sec^2} = 9.806\ 65\ \frac{kg \cdot m}{sec^2} = 9.806\ 65\ N = 2.205\ lb_f$
Energy	1 W = J/sec 1 calorie = 4.184 J (thermochemical) $1\ J = 1\ N \cdot m = 1\ W \cdot sec = 0.239\ 01\ cal = 10^7\ ergs = 10^7\ dyne \cdot cm$ $1\ J = 2.778 \times 10^{-7}\ kW \cdot h = 0.7376\ ft\text{-}lb_f = 0.000\ 948\ 45\ Btu$ 1 Btu = 1054.35 J = 1.054 kJ = 251.996 cal = 0.2929 W·h = 1054.35 N·m 1 kWh = 3.6 MJ
Heat generation rate	$1\ \frac{Btu}{lb_m - hr} = 0.64612\ \frac{W}{kg}$
Heat transfer coefficient (U_o, h_o)	$1\ \frac{W}{(m^2 K)} = 0.1761\ \frac{Btu}{(hr\ ft^2\ °F)}$ $1\ \frac{Btu}{(hr\ ft^2\ °F)} = 5.678\ \frac{W}{(m^2\ K)} = 4.882\ \frac{kcal}{(hr\ m^2\ °C)}$
Nusselt Number (Nu)	$$Nu \equiv \frac{hD}{k} \equiv \frac{conduction + convection}{conduction}$$ where h is the heat transfer coefficient, D is the pipe diameter, and k is thermal conductivity. For purely *laminar* and fully developed pipe flow (conduction dominates) limiting cases: • *Case of uniform heat flux or constant temperature difference:* $$Nu_\infty = \frac{h_\infty D}{k} \approx 4.364$$ • *Case of constant wall temperature:* $$Nu_\infty = \frac{h_\infty D}{k} \approx 3.656$$

Quantity	Equivalent Values
Nusselt Number (Nu) (*continued*)	For *turbulent* pipe flow: *Dittus-Boelter Equation*: $$\mathrm{Nu} \equiv \frac{hD}{k} = 0.023\,\mathrm{Re}^{0.8}\,\mathrm{Pr}^n$$ $$n = \begin{pmatrix} 0.4 \text{ for heating} \\ 0.3 \text{ for cooling} \end{pmatrix}$$ Dittus-Boelter valid for: $10\,000 < \mathrm{Re} < 120\,000$, $0.7 < \mathrm{Pr} < 120$, $L/D > 60$ (ie fully developed) and *when the pipe temperature is within $10\,°F$ for liquids and $100\,°F$ for gases.* *Sieder-Tate Equation*: $$\mathrm{Nu} \equiv \frac{hD}{k} = 0.023\,\mathrm{Re}^{0.8}\,\mathrm{Pr}^{1/3} \left(\frac{\mu_b}{\mu_{\mathrm{wall}}}\right)^{0.14}$$ Valid for $\mathrm{Re} > 10\,000$, $L/D > 60$, higher Prandtl numbers $0.7 < \mathrm{Pr} < 16\,700$, and larger temperature differences between bulk and wall. Properties evaluated at bulk temperature except for μ_{wall} which is evaluated at the wall temperature. *Source*: From Pitts and Sissom [1].
Prandtl Number (Pr)	$$\mathrm{Pr} = \frac{C_p \mu}{k} = \frac{\text{viscous diffusion rate}}{\text{heat conduction rate}}$$ Prandtl number is a characteristic of the fluid. Liquids in general have high Prandtl numbers. • Ethylene glycol $0\,°C = 615$ • Water at $20\,°C = 7.02$ • Water at $80\,°C = 2.22$ • Steam $107\,°C \approx 1.06$ • Gases ≈ 0.7
Latent heat	$1\,\dfrac{\mathrm{Btu}}{\mathrm{lb_m}} = 2.326\,\dfrac{\mathrm{kJ}}{\mathrm{kg}}$ $1\,\dfrac{\mathrm{J}}{\mathrm{g}} = 0.239\,01\,\dfrac{\mathrm{cal}}{\mathrm{g}}$ H_2O: $\Delta H_{\mathrm{melting}} = \Delta H_{\mathrm{fusion}} = 6.01\,\mathrm{kJ/mol}$ or $334\,\mathrm{J/g}$ H_2O: $\Delta H_{\mathrm{vaporization}}\,H_2O = 2230\,\mathrm{J/g} = 40.65\,\mathrm{kJ/mol}$
Power	$1\,\mathrm{W} = 1\,\dfrac{\mathrm{J}}{\mathrm{sec}} = 1\,\dfrac{\mathrm{kg\cdot m^2}}{\mathrm{sec}^3} = 1\,\dfrac{\mathrm{Nm}}{\mathrm{sec}} = 0.23901\,\dfrac{\mathrm{cal}}{\mathrm{sec}}$ $\phantom{1\,\mathrm{W}} = 0.7376\,\dfrac{\mathrm{ft\cdot lb_f}}{\mathrm{sec}} = 0.0009485\,\dfrac{\mathrm{Btu}}{\mathrm{sec}}$ $\phantom{1\,\mathrm{W}} = 3.414\,\dfrac{\mathrm{Btu}}{\mathrm{hr}} = 0.001341\,\mathrm{hp}$
Power/volume	$1\,\dfrac{\mathrm{W}}{\mathrm{L}} = \dfrac{\mathrm{kW}}{\mathrm{m}^3} = 0.037\,98\,\dfrac{\mathrm{hp}}{\mathrm{ft}^3} = 96.67\,\dfrac{\mathrm{Btu}}{\mathrm{hr\text{-}ft}^3} = 12.9235\,\dfrac{\mathrm{Btu}}{\mathrm{hr\text{-}gal}}$
Specific heat (C_p)	$1\,\dfrac{\mathrm{kJ}}{(\mathrm{kg\cdot K})} = \dfrac{\mathrm{J}}{(\mathrm{g\cdot K})} = 0.239\,01\,\dfrac{\mathrm{kcal}}{(\mathrm{kg\cdot°C})} = 0.239\,01\,\dfrac{\mathrm{Btu}}{(\mathrm{lb_m\cdot°F})}$ $1\,\dfrac{\mathrm{Btu}}{(\mathrm{lb_m\cdot°F})} = 1\,\dfrac{\mathrm{cal}}{(\mathrm{g\cdot°C})} = 4184\,\dfrac{\mathrm{J}}{\mathrm{kg\cdot K}}$

Quantity	Equivalent Values
Specific heat (C_p) (continued)	For water (20 °C): $C_p = 4184 \dfrac{J}{kg \cdot K} = 1 \dfrac{cal}{gm \cdot °C} = 1 \dfrac{Btu}{(lb \cdot °F)}$ For air (20 °C): $C_p = 1013 \dfrac{J}{kg \cdot K} = 29.29 \dfrac{J}{mol \cdot K}$ $= 0.24 \dfrac{cal}{gm \cdot °C} = 7 \dfrac{cal}{mol \cdot °C}$
Thermal conductivity (k)	$1 \dfrac{Btu}{(hr \cdot ft \cdot °F)} = 1.7307 \dfrac{W}{(m \cdot K)} = 0.004\,13 \dfrac{cal}{(sec \cdot cm \cdot K)}$ $1 \dfrac{W}{(m \cdot K)} = 0.5779 \dfrac{Btu}{(hr \cdot ft \cdot °F)} = 0.859\,84 \dfrac{kcal}{(hr \cdot m \cdot °C)}$ $1 \dfrac{W}{(m \cdot K)} = 0.002\,39 \dfrac{cal}{sec \cdot cm \cdot °C} = 0.578 \dfrac{Btu}{hr \cdot ft \cdot °F}$ Thermal conductivity $k \cong$ independent of pressure For water (20 °C): $k = 0.597 \dfrac{W}{m \cdot K}$ For air (20 °C): $k = 0.0257 \dfrac{W}{m \cdot K}$ For ethanol (20 °C): $k = 0.17 \dfrac{W}{m \cdot K}$
Throughput (continuous at 365 days/yr)	$1\,yr = 365\,days = 8760\,hr = 5.256 \times 10^5\,min$ $1 \dfrac{kg}{hr} = 16.67 \dfrac{g}{min} = 24 \dfrac{kg}{day} = 8760 \dfrac{kg}{yr} = 8.76 \dfrac{MT}{yr}$ $10 \dfrac{MT}{yr} = 10\,000 \dfrac{kg}{yr} = 27.4 \dfrac{kg}{day} = 1.14 \dfrac{kg}{hr} = 19.03 \dfrac{g}{min}$ $1\,Billion \dfrac{tablets}{year} = 2.74 \times 10^6 \dfrac{tablets}{day}$ $= 114{,}155 \dfrac{tablets}{hr} = 31.7 \dfrac{tablets}{sec}$ $10 \dfrac{MT\,API}{yr} = 10\,000 \dfrac{kg \cdot API}{yr}$ of API formulated as a 10 mg dose $\dfrac{API}{tablet}$ $= 1.0\,Billion \dfrac{tablets}{yr}$
Thermal diffusivity	$\alpha = \dfrac{k}{\rho\,C_p} = \left[\dfrac{m^2}{s}\right]$ $1 \dfrac{m^2}{sec} = 10.76 \dfrac{ft^2}{sec} = 387\,49 \dfrac{ft^2}{hr}$ $1 \dfrac{ft^2}{sec} = 929.03 \dfrac{cm^2}{sec} = 0.092\,903 \dfrac{m^2}{sec}$ For air (20 °C): $\alpha = 2.12 \times 10^{-5} \dfrac{m^2}{s}$ For water (20 °C): $\alpha = 1.43 \times 10^{-7} \dfrac{m^2}{s}$
Viscosity	Dynamic viscosity (μ) $(1\,Pa \cdot sec) = \dfrac{1\,N \cdot sec}{m^2} = \left[\dfrac{1\,kg}{m \cdot sec}\right] = 1000\,cP\,(centipoise)$ $1\,cP = 0.01\,poise = 0.01 \dfrac{g}{(cm \cdot sec)} = 0.001\,Pa \cdot sec = 1\,mPa\,sec\,(milliPascal \cdot sec)$

Quantity	Equivalent Values
Viscosity (continued)	$1 \text{ cP} = 3.6 \dfrac{\text{kg}}{(\text{m}\cdot\text{hr})} = 0.001 \dfrac{\text{kg}}{(\text{m}\cdot\text{sec})} = 2.419 \dfrac{\text{lb}_m}{(\text{ft}\cdot\text{hr})}$ $1 \text{ poise} = 1 \dfrac{\text{g}}{(\text{cm}\cdot\text{sec})} = 100 \text{ cP}$ For liquid water (20 °C): $\mu = 1.002 \times 10^{-3}$ Pa·sec = 1.002 cP For gases (20 °C): $\mu \cong 10^{-5} \dfrac{\text{kg}}{\text{m sec}} = 0.01$ cP For air (20 °C): $\mu \cong 1.8 \times 10^{-5} \dfrac{\text{kg}}{\text{m sec}} = 0.018$ cP Kinematic viscosity (ν) $$\nu = \dfrac{\mu}{\rho} = \dfrac{\text{kg}}{\text{m sec}} \dfrac{\text{m}^3}{\text{kg}} = \left[\dfrac{\text{m}^2}{\text{s}}\right]$$ $\text{Stoke} = 1 \dfrac{\text{cm}^2}{\text{sec}} = 1 \text{ St}$ $\text{Centistoke} = 1 \times 10^{-6} \dfrac{\text{m}^2}{\text{sec}} = 0.01 \text{ stoke} = 0.01 \dfrac{\text{cm}^2}{\text{sec}} = 1 \text{ cSt}$ $= 0.0036 \dfrac{\text{m}^2}{\text{hr}} = 0.0388 \dfrac{\text{ft}^2}{\text{hr}}$ $1 \dfrac{\text{m}^2}{\text{s}} = 10^4 \dfrac{\text{cm}^2}{\text{s}} = 10^4 \text{ stoke} = 10^6 \text{ centistoke}$ $\nu(\text{H}_2\text{O } 20\,°C) = 1.004 \times 10^{-6} \dfrac{\text{m}^2}{\text{sec}} = 1.004 \text{ cSt}$
Gravitational force	$g = 9.8066 \dfrac{\text{m}}{\text{sec}^2} = 32.174 \dfrac{\text{ft}}{\text{sec}^2}$
Ideal Gas Law	$$PV = nRT \text{ and } R = \dfrac{PV}{nT}$$ $R = 8.314 \dfrac{\text{J}}{\text{mol K}} = 8.314 \dfrac{\text{m}^3\,\text{Pa}}{\text{mol K}} = 82.06 \times 10^{-6} \dfrac{\text{m}^3\,\text{atm}}{\text{mol K}}$ $R = 0.082\,06 \dfrac{\text{L}\cdot\text{atm}}{\text{mol K}} = 1.987 \dfrac{\text{cal}}{\text{mol K}}$ $R = 1.987 \dfrac{\text{Btu}}{\text{lb mol °R}} = 0.729 \dfrac{\text{ft}^3\,\text{atm}}{\text{lb mol °R}}$ $R = 82.057 \dfrac{\text{atm cm}^3}{(\text{mol K})} = 10.73 \dfrac{\text{psi ft}^3}{(\text{lb mol °R})} = 62.36 \dfrac{\text{L}\cdot\text{torr}}{\text{mol K}}$ At STP (Standard Temperature and Pressure), temperature is equal to 0 °C and pressure is equal to 1 atm. At STP, 1 mol of an ideal gas occupies 22.415 L.
Raoult's Law (approximation and generally valid for concentrated solutions when x_A is close to 1)	Raoult's Law: $$p_A = y_A P_T = x_A p_A^*(T)$$ where x_A is the mol fraction of A, y_A is the mol fraction in the vapor phase, P_T is the total pressure, $p_A^*(T)$ is the vapor pressure of A, and p_A is the partial pressure of A As x_A approaches 1, the partial pressure $p_A \approx p_A(T)$ approaches the vapor pressure of liquid A. Example: Drying 2-propanol (IPA) from a wetcake using a nitrogen stream (single-pass through the cake). How long will it take to remove 100 g of IPA from a wet-cake using nitrogen blow-through (the cake) at 1 L/min at 30 °C? Assume the drying cake remains isothermal at 30 °C and the nitrogen remains saturated and the system total pressure is 1 atm = 760 mmHg absolute. Solution: 1. Vapor pressure of IPA (30 °C) = 58.3 mmHg 2. Assume $x_A \approx 1$ for IPA 3. Ignoring solids and any mass transfer limitations, simply calculate the saturation condition for IPA in N_2. 4. Assume 30 °C isothermal wet-cake Mol fraction of saturated 2-PrOH in nitrogen: $$y_{\text{IPA}} = \dfrac{p_{\text{IPA}}^*(T)}{P_T} = \dfrac{58.3 \text{ mmHg}}{760 \text{ mmHg}} = 0.0767 \text{ mol frac IPA}$$

Quantity	Equivalent Values
Raoult's Law *(continued)*	Mol fraction of nitrogen:$$y_{N_2} = 1 - 0.0767 = 0.9233 \text{ mol frac } N_2$$5. Calculate mass of nitrogen per gram of IPA required to become fully saturated with 2PrOH (30 °C):$$\frac{0.9233 \times 28\,\text{g/mol}}{0.0767 \times 58.3\,\text{g/mol}} = 5.78 \text{ g of } N_2/(\text{g of 2PrOH})$$6. Flow of 1 L/min (N_2) at 30 °C: $\dfrac{1\,\text{L}}{\text{min}} \dfrac{(273\text{K})}{(303\text{K})} \dfrac{\text{mol}}{22.4\,\text{L}} \dfrac{28\,\text{g}}{\text{mol}} = 1.13\,\text{g}\,N_2/\text{min}$ 7. To saturate and remove 100 g of IPA using nitrogen at 30 °C and 1 L/min:$$100\,\text{g of 2PrOH}\left(\frac{\text{min}}{1.13\,\text{g}\,N_2}\right)\frac{(5.78\,\text{g}\,N_2)}{(\text{g 2PrOH})} = 511.5\,\text{min} = 8.5\,\text{hr}$$Note: This is only a rough estimate (order of magnitude). The assumptions are idealized since there is typically a significant drop in temperature due to evaporative cooling which will lower the vapor pressure and slow the rate of drying. For a comprehensive treatment of mass transfer during flow-through drying see Treybal [2].
Henry's Law *(generally valid for dilute solutions when x_A is close to 0; and commonly applied to solutions of noncondensable gases)*	Henry's Law: At a constant temperature, the amount of a gas dissolved in a specific type and volume of liquid is directly proportional to the partial pressure of that gas in equilibrium with that liquid.$$p_A = y_A P_T = x_A H'_A(T)$$where H_A is the Henry's Law constant with units of (pressure/mol fraction), x_A is the mol fraction of gas dissolved in liquid, y_A is the mol fraction in the vapor, P_T is the total pressure, p_A is the partial pressure of A. Commonly used variation for pure gas ($y_A \approx 1$) over liquid: (such as a hydrogenation)$$p_A = y_A P_T \approx (1.0) P_{sat} = C_{sat} \cdot H_A(T)$$where H_A is the Henry's Law constant with units of (pressure/(mol/L)), C_{sat} is the gas solubility in mol/L at saturation pressure P_{sat}. Example: Calculate the solubility of H_2 in methanol at 25 °C and 3.5 bar (absolute pressure of pure H_2): Solution: The Henry's constant for hydrogen in methanol at 25 °C:$$H_{H_2} = 268\,\frac{\text{bar L}_{soln}}{\text{mol}_{H_2}}$$Solubility $= C_{sat} = \dfrac{P_{sat}}{H_{H_2}} = \dfrac{3.5\,\text{bar}\,\text{mol}_{H_2}}{268\,\text{bar L}} = \dfrac{0.0131\,\text{mol}_{H_2}}{\text{L}} = 0.0131\,M_{H_2}$ at 3.5 bar and 25 °C Other commonly used forms of Henry's Law:$$p_A = y_A P_T = \frac{C_A}{k_H}$$where k_H is the Henry's Law constant, with units of $\dfrac{\text{mol}}{\text{L atm}}$: Example: a. Calculate the solubility of pure oxygen in equlibrium with water at 25 °C at 1 atm (absolute pressure of oxygen): Henry's constant, k_H for O_2 in water at 25 °C: $1.3 \times 10^{-3}\,\dfrac{\text{mol}}{\text{L atm}}$ (see table below)

Quantity	Equivalent Values
Henry's Law *(continued)*	$$C_{\text{sat},O_2} = p_{O_2} k_{H,O_2} = \frac{1.0 \text{ atm} \, 0.0013 \text{ mol}}{\text{L atm}} = \frac{0.0013 \text{ mol}_{O_2}}{\text{L}} = 0.0013 \, M_{O_2}$$ b. Instead of pure oxygen, calculate the solubility of oxygen in water while in equlibrium with air at 25 °C and 1 atm. Recognize the mol fractions in the gas phase: $y_{N_2} = 0.79$ and $y_{O_2} = 0.21$ $P_T = 1$ atm so $P_{O_2} = y_{O_2} P_T = (0.21)(1 \text{ atm}) = 0.21$ atm $$C_{\text{sat},O_2} = p_{O_2} k_{H,O_2} = \frac{0.21 \text{ atm} \, 0.0013 \text{ mol}}{\text{L atm}} = \frac{2.73 \times 10^{-4} \text{ mol}_{O_2}}{\text{L}}$$ **Forms of Henry's Law and Constants (Gases in Water at 298 K)**

Equation	$k_H = \dfrac{p_{\text{gas}}}{C_{\text{aq}}}$	$k_H = \dfrac{C_{\text{aq}}}{p_{\text{gas}}}$	$k_H = \dfrac{p_{\text{gas}}}{x_{\text{aq}}}$	$k_H = \dfrac{C_{\text{aq}}}{C_{\text{gas}}}$
Units	$\dfrac{L_{\text{soln}} \text{ atm}}{\text{mol}_{\text{gas}}}$	$\dfrac{\text{mol}_{\text{gas}}}{L_{\text{soln}} \text{ atm}}$	$\dfrac{\text{atm mol}_{\text{soln}}}{\text{mol}_{\text{gas}}}$	Dimensionless
O_2	769.23	0.001 3	42 590	0.031 80
H_2	1282.05	0.000 78	70 990	0.019 07
CO_2	29.41	0.034 0	1 630	0.831 7
N_2	1639.34	0.000 61	90 770	0.014 92
He	2702.7	0.000 37	149 700	0.009 051
Ne	2222.22	0.000 45	123 000	0.011 01
Ar	714.28	0.001 4	39 550	0.034 25
CO	1052.63	0.000 95	58 280	0.023 24

where:

C_{aq} = moles of gas per Liter of solution

p_{gas} = partial pressure of gas above the solution in atmospheres

x_{aq} = mole fraction of gas in solution

https://chemengineering.wikispaces.com/Henry%27s+Law

Reynold's Number (Re)	Reynolds Number for stirred vessel: $$\text{Re} = \frac{\rho N D^2}{\mu}$$ where ρ = density, N = stir speed, D = impeller diameter, and μ = viscosity For pipe or tube: $$\text{Re} = \frac{\rho u D_{\text{pipe}}}{\mu}$$ where u is fluid velocity, ρ = density, D = pipe inside diameter, and μ = viscosity Example (Stirred Tank): Calculate Re for a lab reactor containing water using the following parameters: Impeller diameter: $D = 5$ cm $= 0.05$ m

Quantity	Equivalent Values
Reynold's Number (Re) (*continued*)	Stirrer speed: $N = 600$ rpm/(60 sec/min) = $10^{\text{rotations}}/_{\text{sec}}$ $\rho = 1 \text{ g}/_{\text{cm}^3} = 1000 \text{ kg}/_{\text{m}^3}$ Viscosity of water: $\mu = 1$ cp = $0.001 \text{ kg}/_{\text{(m s)}}$ $$\text{Re} = \frac{1000 \text{ kg}/_{\text{m}^3} \times 10 \text{ rps} \times (0.05 \text{ m})^2}{0.001 \text{ kg}/_{\text{(m s)}}} = 25\,000$$ Example (pipe flow): What flow rate of water inside a ¼″ OD tube would provide a Reynolds number of 25 000: Assume inside diameter of tube, ID = 0.23 in = 0.584 cm Tube: cross-sectional area: $(3.14 \times (0.584)^2)/4 = 0.268 \text{ cm}^2$ $\rho = 1 \text{ g}/_{\text{cm}^3} = 1000 \text{ kg}/_{\text{m}^3}$ Viscosity of water: $\mu = 1$ cp = $0.001 \text{ kg}/_{\text{(m s)}}$ $$\text{Re} = \frac{(1000 \text{ kg}/_{\text{m}^3}) \cdot u \cdot (0.005\,84 \text{ m})}{0.001 \text{ kg}/_{\text{(m s)}}} = 25\,000$$ $$u = \frac{(25\,000)\left(0.001 \frac{\text{kg}}{\text{m s}}\right)}{\left(1000 \frac{\text{kg}}{\text{m}^3}\right)(0.005\,84 \text{ m})} = 4.28 \text{ m/s}$$ Volumetric flow rate = $u \cdot$area = $428 \text{ cm/sec} \times 0.268 \text{ cm}^2 = 114.7 \text{ cm}^3/\text{sec} = 6.88 \text{ L/min}$
Power Number (Np)	$$N_P = \frac{P}{\rho N^3 D^5}$$ where P = power, ρ = density, N = stir speed, D = impeller diameter For a given reactor + agitator configuration the mixing power of an impeller can be uniquely characterized by the power number: - Turbulent flow where Re > 2000: $N_P = N_{p,\text{turbulent}} = $ constant $\geq$ Power $\propto \rho N^3 D^5$ Example: If the measured power required to agitate a tank of liquid at 600 rpm with a 5 cm ID impeller is 1.0 W then what is the power at 1000 rpm (16.7 rps) in water? Solution: Check Reynolds number (Re = 25 000; see above for calculation); *Power Number*, N_P, will be a constant for this reactor + impeller when Re > 2000 $$N_P = \frac{P}{\rho N^3 D^5}$$ Calculate N_P: Note that $1.0 \text{ W} = 1.0 \frac{\text{kg m}^2}{\text{s}^3}$ Power Number, $N_P = \dfrac{1.0 \text{ kg m}^2/_{\text{s}^3}}{1000 \text{ kg}/_{\text{m}^3} \times (10 \text{ rps})^3 \times (0.05 \text{ m})^5} = 3.2$ will be constant for Re > 2000 Calculate the Power required for the higher stir speed of 1000 rpm $P = N_P \rho N^3 D^5 = 3.2$ $P = 3.2 \times 1000 \frac{\text{kg}}{\text{m}^3} \times (16.7 \text{ rps})^3 \times (0.05 \text{ m})^5$ $P = 4.7 \text{ W}$ Note the ~5× increase in power draw to increase stirring from 600 to 1000 rpm.
Surface tension (γ)	$$\left[\frac{\text{dyne}}{\text{cm}}\right]$$ Water–air interface: $\gamma_{\text{H2O-AIR}}$ (20 °C) = 72.75 dynes/cm CRC HB 62nd edition $$1 \frac{\text{dyne}}{\text{cm}} = 0.001 \frac{\text{N}}{\text{m}} = 1 \frac{\text{erg}}{\text{cm}^2}$$

Quantity	Equivalent Values
ppm	Percent (%) ppm 0.0001 1 0.001 10 0.01 100 0.1 1000
Moisture content	Moisture content (wet basis) Range = 0–100% $$\%M_{wet} = 100\, \frac{\text{mass}_{solvent}}{\text{mass}_{solvent} + \text{mass}_{dry\,solids}} = 100\, \frac{\text{mass}_{solvent}}{\text{mass}_{Total}}$$ Note wet basis is the same as % Loss on drying (%LoD) where: $$\%\text{LoD} = 100\, \frac{\text{mass}_{loss\,on\,drying}}{\text{mass}_{initial\,wet\,cake}}$$ Moisture content (dry basis) range = 0 to $\gg$100% $$\%M_{dry} = 100\, \frac{\text{mass}_{solvent}}{\text{mass}_{dry\,solids}}$$ To convert wet basis to dry basis: $\text{wt}\%M_{dry} = 100\, \dfrac{\%M_{wet}}{100\% - \%M_{wet}}$ To convert dry basis to wet basis: $\text{wt}\%M_{wet} = 100\, \dfrac{\%M_{dry}}{100\% + \%M_{dry}}$ Example: 100 g of wetcake contains 40 g of water and 60 g of dry API, compare the moisture contents (wet vs dry basis): $$\%M_{wet} = \%\text{LoD} = 100\, \frac{40\,g}{40\,g + 60\,g} = 40\%$$ $$\%M_{dry} = 100\, \frac{40\,g}{60\,g} = 67\%$$
Humidity	Absolute humidity for water in air: $h = \dfrac{kg_{water}}{kg_{dry\,Air}}$ Example: Calculate the absolute humidity of air saturated with water at 30 °C at 1 atm pressure. Solution: • The vapor pressure of water at 30 °C, is 31.8 torr = 0.0418 atm • Calculate the mole fraction water: $$y_{water} = \frac{p^*_{water}(T)}{P_T} = \frac{0.0418\,\text{atm}}{1\,\text{atm}} = 0.0418\,\text{mol frac water}$$ • Mole fraction of air $$y_{air} = 1 - 0.0418 = 0.9582\,\text{mol frac air}$$ • Calculate mass ratio of water to air to find the absolute humidity: $$\frac{0.0418 \times 18.02\,\text{g/mol}}{0.9582 \times 29\,\text{g/mol}} = 0.0271\, \frac{kg_{water}}{kg_{dry\,air}}$$ • This value can also be obtained be read from a pyschrometric chart for water/air; • Because the air is saturated, in this example, the relative humidity is 100%.

Quantity	Equivalent Values
Inertion of a tank via pressure cycles	$x_{final} = x_{initial} \left(\dfrac{P_{Low}}{P_{High}} \right)^k$ Example: Determine the number of pressure purges required to reduce the oxygen concentration in a reactor from 21 to 0.1 vol % O_2 using nitrogen. A nitrogen source is used to pressurize the reactor to 50 psig and is vented down to 5 psig, in several cycles. Calculate the approximate number of cycles required. $k = \dfrac{\ln\left(\dfrac{0.001}{0.21}\right)}{\ln\left(\dfrac{5\,\text{psig} + 14.7\,\text{psig}}{50\,\text{psig} + 14.7\,\text{psig}}\right)} = \dfrac{-5.347}{-1.189} = 4.5 \approx 5 \text{ cycles}$ *Source:* Adapted from Kinsley [3].

Polymath (6.10) program for Semi-Batch (i.e. Fed-Batch) with 1 hour Feed-Time. *A* Is Being Fed to *B*.
   Assume Isothermal Kinetics

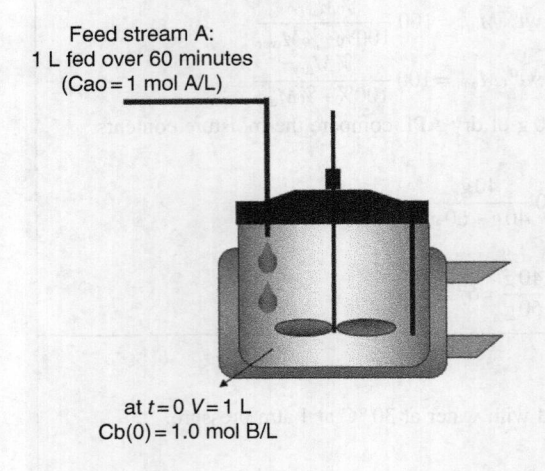

Feed stream A:
1 L fed over 60 minutes
(Cao = 1 mol A/L)

at $t = 0$ $V = 1$ L
Cb(0) = 1.0 mol B/L

$A + B \xrightarrow{k} C$

rate = $-kC_A C_B$ where $k \equiv \left(\dfrac{\text{L}}{\text{mol}\cdot\text{min}}\right)$
$\Delta H = -30$ kcal/mol
Initial conditions at $t = 0$

- Volume in the reactor, Vo = 1 L
- Concentration of *B* in the reactor, Cb(0) = 1 M
- Concentration of *A* and *C* in the reactor = 0

```
# A + B → C
# A is fed to B
d(Ca)/d(t) = if (t > dose) then ra else ra + Cao * vo/V - Ca*Vo/V # mols/(l·min)
d(Cb)/d(t) = if (t > dose) then ra else ra - vo*Cb/V #
d(Cc)/d(t) = if (t > dose) then -ra else -ra - vo * Cc/V #
Dose = 60 # minutes
V = if (t > dose) then Vo + vo * dose else Vo + vo * t #
```

molsAfed = if (t > dose) then Cao * vo * dose else Cao * vo * t #
molsB = Cb * V
Vo = 1 # liter (initial volume of the reactor)
vo = 1/60 # L/min (volumetric flow rate of the feed)
k = 0.1 # rate constant
Cao = 1 # mol/L (concentration of A in the feed stream)
Cbo = 1 # mol/L (initial concentration of B in the reactor)
ra = $-k$ * Ca * Cb # reaction rate expression
rate = $-$ra #

#Heat of Reaction
DeltaH = 30 × 1000/0.239 01 # Exothermic heat of reaction, (30 kcal/mol) × (1000 cal/kcal) × (J/0.23901 cal)
Q = DeltaH × rate × V/60# (J/mol) × (mol/(L·min) × (L) × (min/60 sec) = J/sec = W
WL = Q/V # W/L

#Yields
YC = if (t > 0) then (Cc * V)/(Cbo * Vo) else 0 #Yield of C
XB = if (t > 0) then (Cbo * Vo $-$ Cb * V)/(Cbo * Vo) else 0 #Conversion of B

#Initial Conditions
$t(0)$ = 0 #
Ca(0) = 0 #There is no A initially in the reactor
Cb(0) = 1 # initial concentration of B (mol/L) initially in the reactor
Cc(0) = 0
$t(f)$ = 240 # minutes

The plots below simulate concentration, heat, and yield profiles for rate constants of 0, 0.01, 0.05, 0.1, and 1 (L/(mol·min)) under Isothermal conditions.

xxviii UNIT CONVERSIONS

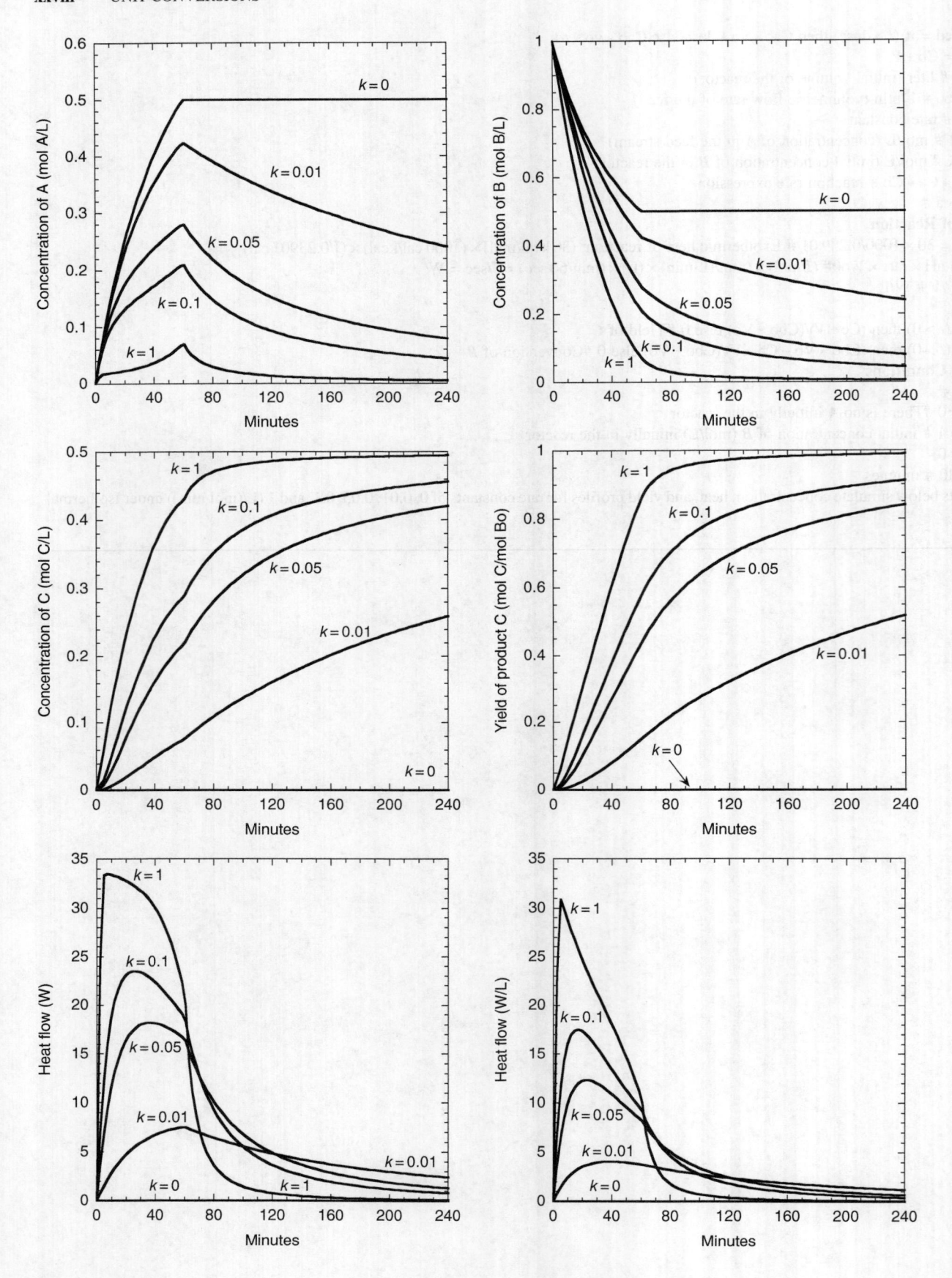

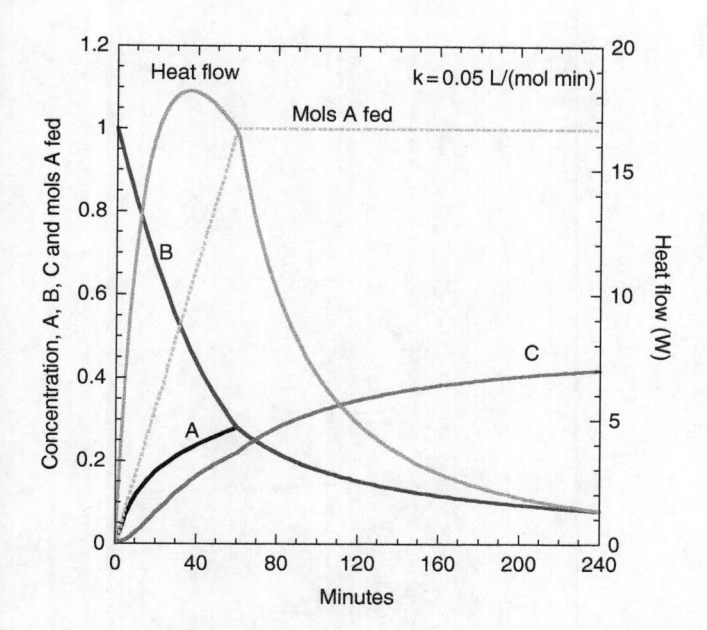

REFERENCES

1. Pitts, D.R. and Sissom, L.E. (1997). *Schaum's Outline of Heat Transfer*, Schaum's Outline Series. New York: McGraw-Hill.
2. Treybal, R. (1980). *Mass Transfer Operations*, 3rde, 684–686. New York: McGraw-Hill.
3. Kinsley, G.R. (2001). Properly purge and inert storage vessels. *Chemical Engineering Progress* 97: 57–61.

PART I

INTRODUCTION

1

CHEMICAL ENGINEERING IN THE PHARMACEUTICAL INDUSTRY: AN INTRODUCTION

DAVID J. AM ENDE
Nalas Engineering Services, Inc., Centerbrook, CT, USA

MARY T. AM ENDE*
Pfizer, Inc., Groton, CT, USA

Across the pharmaceutical industry chemical engineers are employed throughout research and development (R&D) to full-scale manufacturing and packaging in technical and managerial capacities. The chapters in these two volumes provide an emphasis on the application of chemical engineering science to process design, development, and scale-up for active pharmaceutical ingredients (APIs), drug products (DPs), and biologicals including sections on regulatory considerations such as design space, control strategies, process analytical technology (PAT), and quality by design (QbD). The focus of this introduction is to provide a high-level overview of bringing a drug to market and highlight industry trends, current challenges, and how chemical engineering skills are an exquisite match to address those challenges.

In general pharmaceuticals are drug delivery systems in which drug-containing products are designed and manufactured to deliver precise therapeutic responses [1]. The drug is considered the "active," i.e. active pharmaceutical ingredient (API) or "drug substance," and the formulated final dosage form is simply referred to as the drug product (DP).

This book focuses on API in volume 1 and DP in volume 2. The API and DP are designed and developed in R&D and then transferred to the commercial manufacturing sites by teams of organic chemists, analytical chemists, pharmaceutical scientists, and chemical engineers. Prior to the transition to the commercial site, co-development teams are formed with members from R&D and manufacturing working together to define the computational and experimental studies to conduct based on risk and scientific considerations. The outcome of this multidisciplinary team effort forms the regulatory filing strategy for the API and drug products.

Once the commercial API and DP have been established, the co-development teams support three major regulatory submissions for a global product. A New Drug Application (NDA) is submitted to the US Food and Drug Administration (FDA), whereas in the Europe Union a Marketing Authorization Application (MAA) is submitted to the European Medicines Agency (EMA), and in Japan a Japan New Drug Application (JNDA) is submitted to the Pharmaceuticals and Medical Devices Agency (PMDA). Subsequently, the rest of world regulatory filings are led by the commercial division with no significant involvement by R&D since more commercial experience is available at the site by that time.

In the United States, federal and state laws exist to control the manufacture and distribution of pharmaceuticals. Specifically, the FDA exists by the mandate of the US Congress with the Food, Drug, and Cosmetics Act as the principal law to enforce and constitutes the basis of the drug approval process [1]. Specifically in the United States, "The FDA is responsible for protecting the public health by assuring the safety, efficacy, and security of human and veterinary drugs, biological products, medical devices, our nation's food supply, cosmetics, and products that emit radiation. The FDA is also responsible for advancing the public health where possible by speeding innovations that make medicines and foods more effective, safer, and more affordable. They also serve the public by ensuring accurate,

*Current address: Lyndra Therapeutics, Inc., Watertown, MA, USA

Chemical Engineering in the Pharmaceutical Industry: Active Pharmaceutical Ingredients, Second Edition.
Edited by David J. am Ende and Mary T. am Ende.
© 2019 John Wiley & Sons, Inc. Published 2019 by John Wiley & Sons, Inc.

science-based information on medicines and foods to maintain and improve their health."[1] On 28 March 2018 the FDA announced organizational changes available on their website. Janet Woodcock remains the director of the small molecule division, referred to as Center for Drug Evaluation and Research (CDER).[2] Peter W. Marks is the director of the large molecule division, referred to as Center for Biologics Evaluation and Research (CBER).[3] Further information can also be easily obtained from the FDA website, including the overall drug review process, current good manufacturing practices (cGMP), International Council on Harmonization (ICH), and mechanisms to comment on draft guidances, recalls, safety alerts, and warning letters that have been issued to companies.[4]

EMA is a decentralized body of the European Union with headquarters in London whose main responsibility is the protection and promotion of public and animal health, through the evaluation and supervision of medicines for human and veterinary use.[5]

The Japan Pharmaceutical Affairs Law (JPAL) is a law intended to control and regulate the manufacturing, importation, sale of drugs, and medical devices.[6] It exists to assure the quality, efficacy and safety of drugs, cosmetics, and medical devices while improving public health and hygiene. The JPAL also provides guidance to pharmaceutical companies on how to translate their QbD control strategy, which was found to align well with the three levels of criticality initially used in early QbD filings for noncritical, key, and critical process parameters. Japan's Ministry of Health, Labour and Welfare (MHLW) has issued clear guidance in English for three key ministerial ordinances to assure compliance requirements for manufacturers.

Japan, Europe, and United States collaborate as the International Council on Harmonization – Quality (ICH) to establish greater expectations for science and risk-based approaches to transform the pharmaceutical industry over the past decade. Critical to that transformation were the QbD guidances, Q8, Q9, and Q10 [2–4]. The final versions of the guidances are readily available on the CDER website, including the more recent QbD guidance for drug substance composed in Q11.[7]

1.1 GLOBAL IMPACT OF THE INDUSTRY

The value of the pharmaceutical industry to the American economy is substantial. In 2016, the industry employed over 854 000 people with each job indirectly supporting an additional 4 jobs. Thus as an aggregate, the industry supported 4.4 million jobs and generated nearly $1.2 trillion in annual economic output when direct, indirect, and induced effects were considered for 2016.[8]

As an industry sector, the pharmaceutical industry is considered profitable, in spite of the high attrition rate for new chemical entities (NCEs).[9] For example, *Forbes* estimated the profit margin for the health-care technology industry in 2015 to be approximately 21%, clearly placing near the top for profitable industries.[10] The companies that are most profitable in this sector were major pharmaceutical and generics companies. As far as total revenues in pharmaceutical sales, the top 20 pharmaceutical companies are listed in Table 1.1.

Based on revenue, the pharmaceutical and biopharmaceutical companies are based in the following countries: 9 (United States), 2 (Switzerland), 2 (United Kingdom), 1 (France), 3 (Germany), 1 (Israel), 1(Denmark), and 1 (Republic of Ireland). Only 1 company in the top 20 revenue producing is privately held.

Global prescription drug sales are on the order of $800 billion in 2017. These drug sales are forecasted to grow at 6.3% compound annual growth rate (CAGR) between 2016 and 2022 to nearly $1.2 trillion (as shown in Figure 1.1),[11] while generic drugs account for approximately 10% of those sales figures.

There is considerable value in being the first company to deliver a new medicine that treats a new indication (e.g. breakthrough therapy designation from regulators) or uses a new mechanism of action to benefit patients. Therefore, new developments in pharmaceutical R&D that speed quality drug candidates to the market are important investments for the future.

1.2 INVESTMENTS IN PHARMACEUTICAL R&D

R&D is the engine that drives innovation of new drugs and therapies. Significant investment is required to discover and advance potential NCEs and new molecular entities (NMEs). For example, the pharmaceutical industry invested

---

[1] http://www.fda.gov/AboutFDA/WhatWeDo/default.htm
[2] https://www.fda.gov/AboutFDA/CentersOffices/OrganizationCharts/ucm350895.htm
[3] https://www.fda.gov/AboutFDA/CentersOffices/OrganizationCharts/ucm350556.htm
[4] https://www.fda.gov/default.htm
[5] http://www.ema.europa.eu/htms/aboutus/emeaoverview.htm
[6] http://www.jouhoukoukai.com/repositories/source/pal.htm
[7] https://www.fda.gov/Drugs/GuidanceComplianceRegulatoryInformation/Guidances/ucm065005.htm
[8] http://phrma.org/industryprofile
[9] http://money.cnn.com/magazines/fortune/global500/2009/performers/industries/profits
[10] https://www.fool.com/investing/2016/07/31/12-big-pharma-stats-that-will-blow-you-away.aspx
[11] http://info.evaluategroup.com/rs/607-YGS-364/images/wp16.pdf; some estimates are even higher, with 2017 global revenue of $1.05 trillion (see, for example, https://www.fool.com/investing/2016/07/31/12-big-pharma-stats-that-will-blow-you-away.aspx).

TABLE 1.1 Top 20 Pharmaceutical Companies Based on 2017 Revenue as Listed in Wikipedia

#	HQ	Company	2017 (USD Billions)	2016 (USD Billions)	2015 (USD Billions)	2014 (USD Billions)	2013 (USD Billions)	2012 (USD Billions)	2011 (USD Billions)
1		Johnson & Johnson NYSE: JNJ	76.50	71.89	70.10	74.30	71.31	67.20	65.00
2		Roche OTCQX: RHHBY	57.37	50.11	47.70	49.86	48.53	47.80	45.21
3		Pfizer NYSE: PFE	52.54	52.82	48.85	49.61	51.58	58.99	65.26
4		Novartis NYSE: NVS	49.11	48.52	49.41	58.00	57.36	56.67	58.57
5		Sanofi NYSE: SNY	42.91	36.57	36.73	43.07	42.08	46.41	44.34
6		GlaxoSmithKline LSE: GSK	42.05	34.79	29.84	37.96	41.61	39.93	41.39
7		Merck & Co. NYSE: MRK	40.10	39.80	39.50	42.24	44.03	47.27	48.05
8		AbbVie NYSE: ABBV	28.22	25.56	22.82	19.96	18.79	–	–
9		Bayer FWB: BAYN	27.76	25.27	24.09	25.47	24.17	24.30	23.11
10		Abbott Laboratories NASDAQ: ABT	27.39	20.85	20.4	20.25	21.85	39.87	38.85
11		Gilead Sciences NASDAQ: GILD	25.70	30.39	32.15	24.47	10.80	9.70	8.39
12		Eli Lilly & Co NYSE: LLY	22.90	21.22	20.00	19.62	23.11	22.60	24.29

(*continued*)

TABLE 1.1 (*Continued*)

#	HQ	Company	2017 (USD Billions)	2016 (USD Billions)	2015 (USD Billions)	2014 (USD Billions)	2013 (USD Billions)	2012 (USD Billions)	2011 (USD Billions)
13		Amgen NASDAQ: AMGN	22.80	22.99	21.66	20.06	18.68	17.30	15.58
14		AstraZeneca LSE: AZN	22.47	23.00	24.71	26.10	25.71	27.97	33.59
15		Teva Pharmaceutical Industries NASDAQ: TEVA	22.40	21.90	20.00	20.27	20.31	18.31	16.12
16		Boehringer Ingelheim Private	21.67	17.54	16.41	17.70	18.68	18.89	18.34
17		Bristol-Myers Squibb NASDAQ: BMY	20.80	19.43	16.56	15.88	16.39	17.62	21.24
18		Novo Nordisk NYSE: NVO	18.77	16.61	16.06	15.83	14.88	13.48	11.56
19		Merck Group ETR: MRK	15.32	15.80	13.95	14.99	14.77	13.02	12.83
20		Shire NASDAQ: SHPG	14.40	11.40	6.42	6.00	4.76	4.68	3.50

Source: From https://en.wikipedia.org/wiki/List_of_largest_pharmaceutical_companies_by_revenue#cite_note-28. Licensed under CC BY 3.0.

approximately $150 billion into R&D in 2015. Worldwide pharmaceutical R&D spending is expected to grow by 2.8% (CAGR) to $182 billion in 2022 (Figure 1.2).[12] The cost of advancing drug candidates and entire pharmaceutical portfolios in R&D is significant. In 2001 the average cost for an approved medicine was estimated to be $802 million, and by the end of 2014, the average cost escalated to $2.6 billion as reported by Tufts Center for the Study of Drug Development.[13] Although these figures clearly depend on the drug type, therapeutic area, and speed of development, the bottom line is that the up-front investments required to reach the market are massive especially when considering the uncertainty whether the up-front investment will payback.

Given there might be 10 or more years of R&D costs without any revenue generated on a NCE or NME, the gross margins of a successful drug need to cover prior R&D investments and candidate attrition and to cover the continuing marketing and production costs. Figure 1.3 shows the classic cash flow profile for a new drug developed and marketed. First there is a period of negative cash flow during the R&D phase. When the drug is approved and launched, only then are revenues generated, which have to be priced high enough to recoup the extensive R&D investment and provide a return on the investment.

The net present value (NPV) calculation is one way to assess return on investment with a discount rate of 10–12% generally chosen in the pharmaceutical industry as

[12] http://info.evaluategroup.com/rs/607-YGS-364/images/wp16.pdf
[13] Based on estimated average pre-tax industry cost per new prescription drug approval (inclusive of failures and capital costs: source, DiMasi et al. [5]).

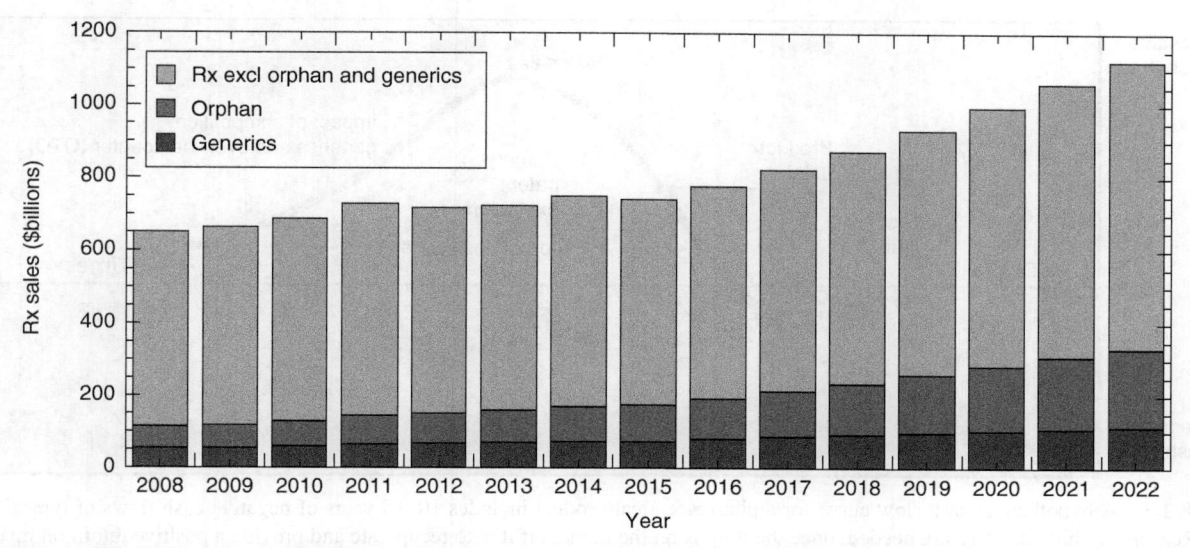

FIGURE 1.1 Global pharmaceutical prescription sales as a function of the type of drug. Global prescription drug sales were on the order of $800 billion in 2017. These drug sales are forecasted to grow at 6.3% compound annual growth rate (CAGR) between 2016 through 2022 to nearly $1.2 trillion while generic drugs account for approximately 10% of those sales figures. *Source*: From http://info.evaluategroup.com/rs/607-YGS-364/images/wp16.pdf.

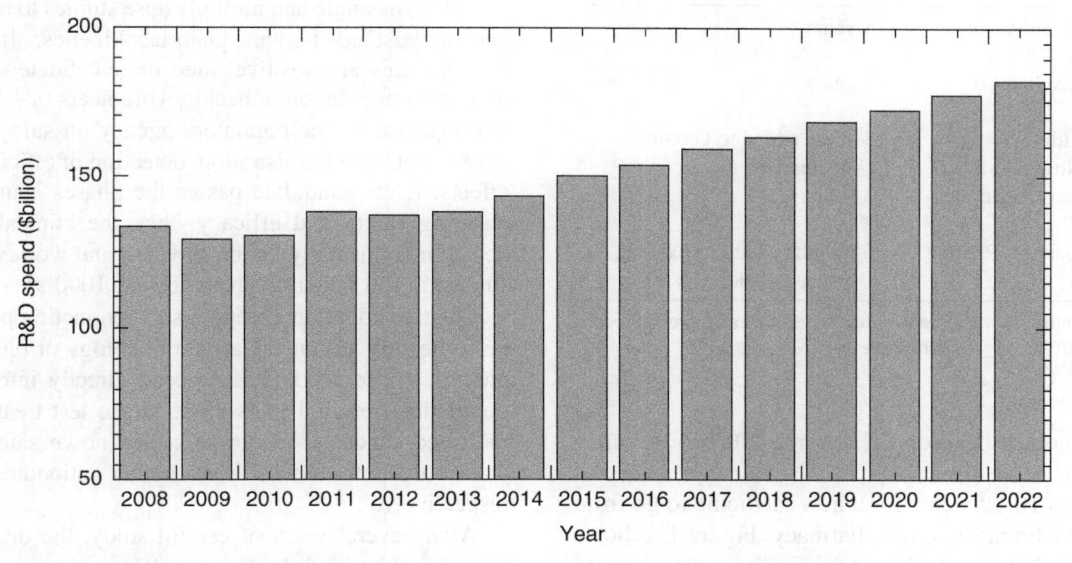

FIGURE 1.2 World-wide pharmaceutical R&D spend in 2015 was approximately $150 billion. Growth rate in R&D spend is projected to grow at a rate of 2.8%. *Source*: EvaluatePharma®.

the rate to value products or programs for investment decisions [6]. The highest revenues for a new drug are achieved during the period of market exclusivity (where no competitors can sell the same drug). So it is in the company's best interest to ensure the best patent protection strategy is in place to maximize the length of market exclusivity. Patents typically have a validity of 20 years from the earliest application grant date base on applications filed after 1995. In some cases the time of market exclusivity can be extended through new indications, new formulations, and devices, which may themselves be patent protected (see Table 1.2).

Once market exclusivity ends, generic competition is poised to immediately introduce a alternative cheaper option that will erode sales for the patent owner. A dramatic example of patent cliff can be seen in the sales of Lipitor (Figure 1.4). Peak sales occurred in 2006 with sales nearing $13 billion in revenue, but at the end of patent exclusivity in 2011, sales dropped off precipitously to less than $4 billion in 2012.

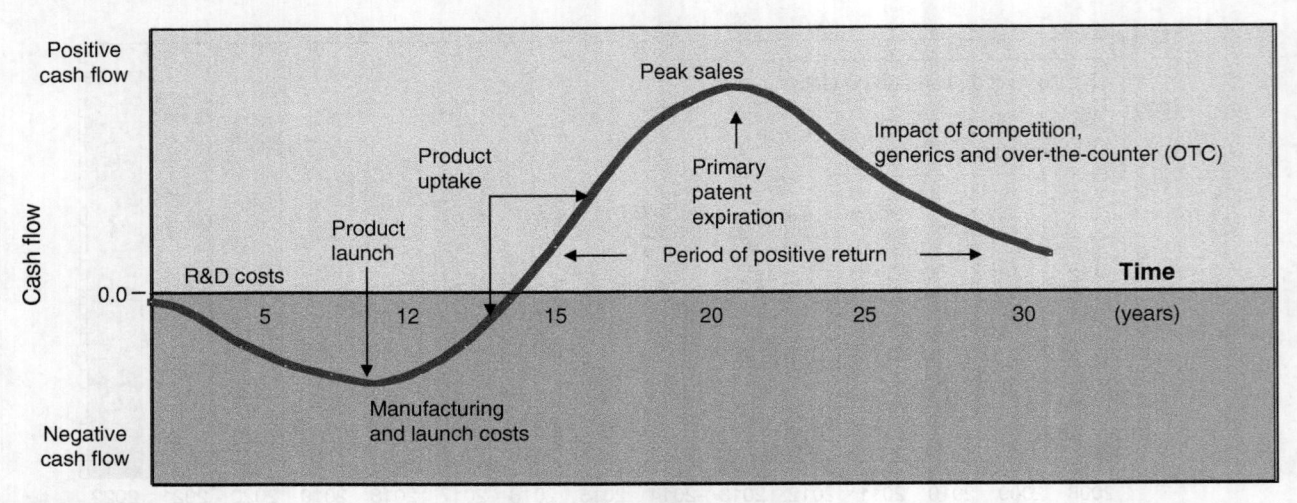

FIGURE 1.3 A hypothetical cash flow curve for a pharmaceutical product includes 10–15 years of negative cash flows of typically $1–3 billion. Reasonably high margins are needed, once the drug is on the market, if it is to recuperate and provide a positive return on investment (ROI) over its lifecycle.

TABLE 1.2 Periods of Exclusivity Granted by the FDA

Specific FDA Applications	Period of Exclusivity
New Chemical Entity Exclusivity (NCE)	5 years
Orphan Drug Exclusivity (ODE)	7 years
Generating Antibiotic Incentives Now (GAIN)	5 years added to certain exclusivities
New Clinical Investigation Exclusivity	3 years
Pediatric Exclusivity (PED)	6 months added to existing patents/exclusivity

Source: Form https://www.fda.gov/Drugs/DevelopmentApprovalProcess/ucm079031.htm#What_is_the_difference_between_patents_a

The trend continued to drop off through 2017 to less than $2 billion.

It now takes 10–15 years for a new medicine to go from the discovery laboratory to the pharmacy. Figure 1.5 shows the typical development activity timeline from discovery to launch. From thousands of compounds evaluated for potential therapeutic effect, very few will clear all the safety, efficacy, and clinical hurdles to make it to approval. Figure 1.5 also shows how a general range of volunteers, and clinical supplies, increases through phases I–III of clinical trials with clinical development typically lasting six years or more.

Before entering human clinical studies, the drug candidate is tested for safety and efficacy in preclinical studies. When the candidate looks promising for a targeted indication or potential therapeutic effect, the company files an Investigational New Drug Application (IND) for regulatory agency and clinical site approval. At this time, referred to as phase I, the drug candidate will be tested in a few healthy volunteers ($n \sim 10$'s) in single and multiple dose studies to test for safety and understand human pharmacokinetics. If the phase I evaluations are positive, then the candidate can progress to a larger population of healthy volunteers ($n \sim 100$'s) pending approval by the regulatory agency on study design, i.e. doses, route of administration, detection of efficacy, and side effects. If the candidate passes the phases I and II hurdles ensuring safety and efficacy, then the clinical teams will design incrementally larger, broader, and worldwide clinical studies in test patients (phase III, $n \sim 1000$'s).

The two common exceptions to conducting phase II studies in healthy volunteers are for oncology or biological candidates. These candidates proceed directly into the patient population, referred to as phase III, to test treatment of the indicated cancer or to progress the known safe and efficacious candidate derived from human antibodies or viruses, respectively.

After several years of careful study, the drug candidate may be submitted to the regulatory agency (e.g. FDA, EMA, PMDA) for approval. Depending on the type of API, the regulatory submission may need to be filed differently. For example, in the United States, a small molecule is submitted as an NDA, while a biologic is submitted as a Biologics Licensing Application (BLA).

As mentioned, the 2014 cost to advance a NCE or NME to market was estimated at $2.6 billion. The cost of product development that includes the cost to manufacture clinical supplies is estimated to be in the range of 30–35% of the total cost of bringing a NCE/NME to market with the following other cost contributors: discovery 20–25%, safety and toxicology 15–20%, and clinical trials 35–40% [7].

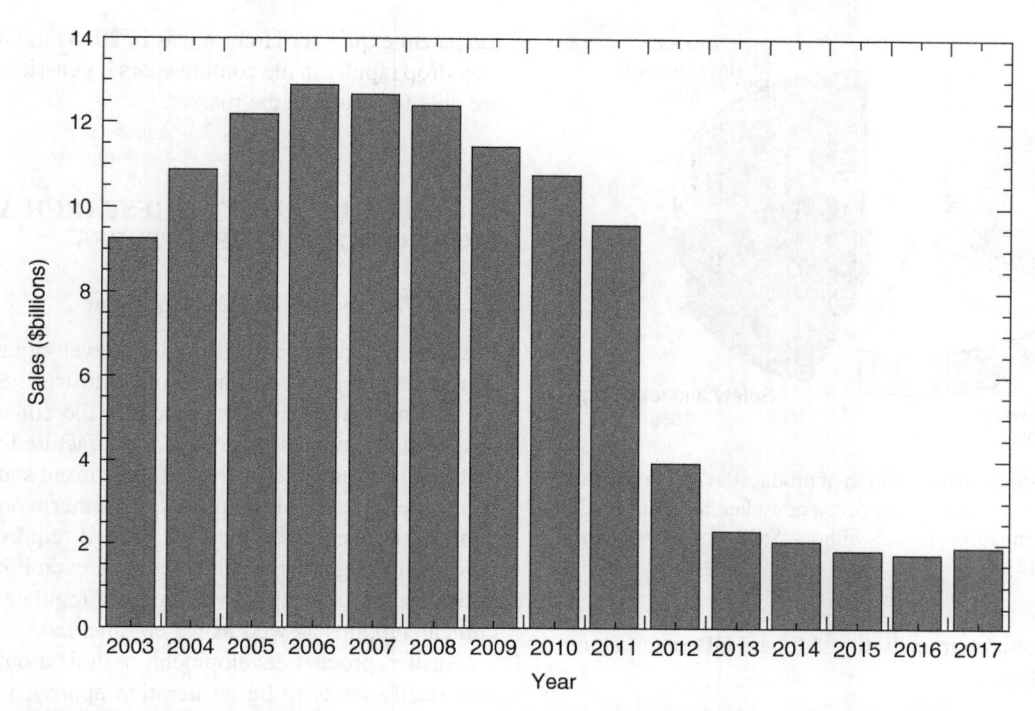

FIGURE 1.4 Sales of Pfizer's Lipitor (atorvastatin) between 2003 and 2017. In 2006 Lipitor generated nearly $13 billion in revenue. Patent exclusivity ended in 2011 and its impact was significant as seen by the significant drop on revenue in subsequent years (known as the "patent cliff"). *Source*: Data from www.statista.com.

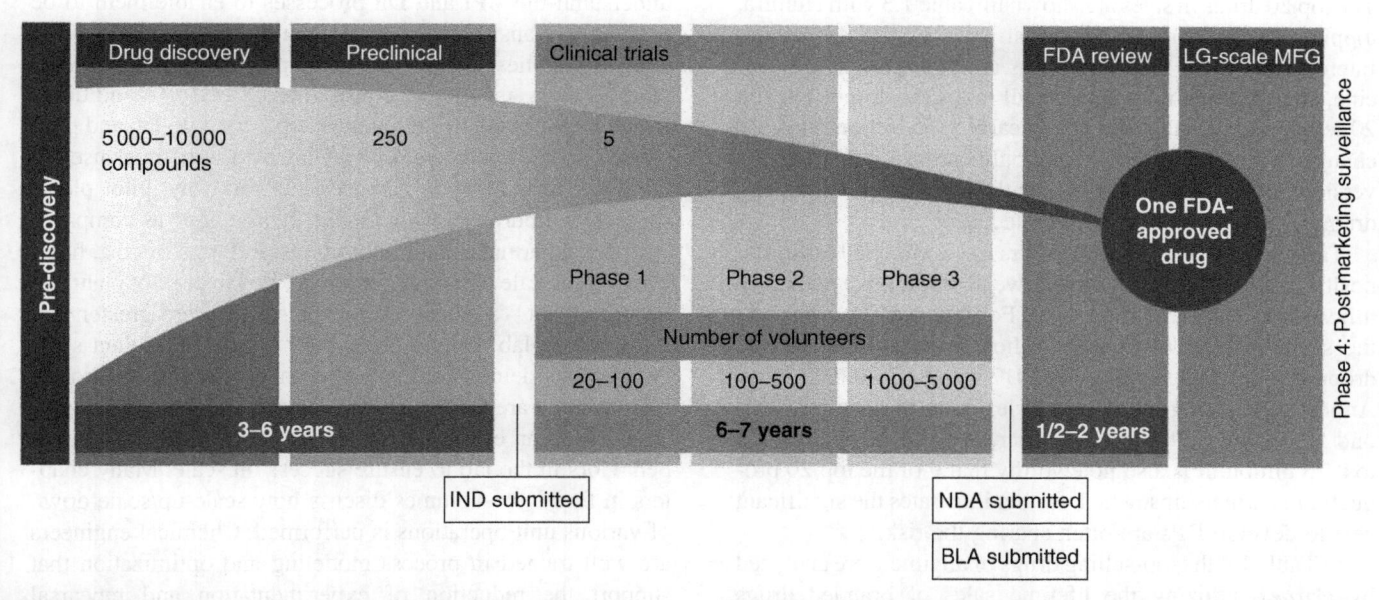

FIGURE 1.5 Drug research and development can take 10–15 years with one approval from 5 to 10 000 compounds in discovery. BLA, biologics license application; FDA, food and drug administration; IND, investigational new drug; NDA, new drug application. *Source*: Adapted from Pharmaceutical Research and Manufacturers of America (PhRMA), publication Pharmaceutical Industry Profile 2009 (www.phrma.org).

The distribution is graphically displayed in Figure 1.6. Clearly the distribution will depend on the specific drug, its therapeutic area, dose, and specific company.

Chemical engineers, chemists, biologists, pharmaceutical scientists, and others make up the diverse scientific disciplines of product development that include API and

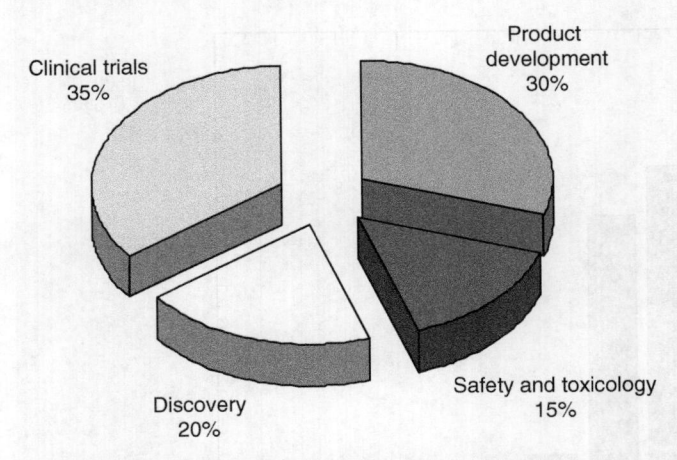

FIGURE 1.6 Estimated distribution of product development costs within R&D with the total cost to bring a new chemical entity NCE to market in the range of $1–3.5 billion. *Source*: Adapted from Suresh and Basu [7].

Pie chart values: Clinical trials 35%, Product development 30%, Discovery 20%, Safety and toxicology 15%.

formulation development including API and DP manufacture of clinical supplies.

1.3 BEST SELLERS

The top 20 drugs in sales are shown in Table 1.3 with Humira, topping the list with 2017 global sales of $18.43 billion. Interestingly 11 of these top drugs are biologics, 1 is a vaccine, and the remaining 8 are small molecule drugs. The top 20 selling drugs in that year total nearly $135 billion. This has changed significantly since the publication of the original version of this book in 2010 when the majority of top-selling drugs at that time were small molecules.

The majority of the 20 top sellers have remained in similar positions over the past 2 years; however a few have made significant moves in this short time. For instance, Harvoni was the second place with $9.08 billion in sales in 2016 and dropped to seventeenth place in 2017 with $4.37 billion sales. Another interesting move was Eylea from thirteenth to second place from 2016 to 2017 increasing sales from $5.05 to 8.23 billion. It is also noteworthy that 9 of the top 20 products are partnerships, which further illustrates the significant cost to develop DPs are often sharing the risk.

In Table 1.4 the top-selling drugs of all time were analyzed by *Forbes*, utilizing the lifetime sales of branded drugs between 1996 and 2012 and company reported sales data between 2013 and 2016. It is noteworthy that the number one in sales, Lipitor, at $148.7 billion is not even on the top 20 drug sales list for 2017 in Table 1.3. While there is a large gap between the top two selling drugs, amounting to $53 billion for Lipitor above Humira, if Humira annual sales continue at $18 billion, it will outperform Lipitor as the all-time best-selling drug in just under 3 years. However, the patent expiry for Humira was in 2016, and therefore sales may drop rapidly in the coming years if generics or biosimilars are able to penetrate the market.

1.4 PHARMACEUTICAL RESEARCH AND DEVELOPMENT EXPENDITURES

1.4.1 Pharmaceutical Development

In general, pharmaceutical product development is different than most other research intensive industries. Specifically in the pharmaceutical industry, there is the consistent need to ensure that clinical supplies are manufactured and delivered in a timely manner regardless of the current state of development or efficiency of the process. In other words, delivering clinical supplies when they are needed requires using technology that is good enough at the time even if it is not a fully optimized process. However, this is a regulated industry for clinical supplies as well as for commercial.

Further, process development, optimization, and scale-up historically tends to be an iterative approach [8] – clinical supply demands are met by scale-ups to kilo lab or pilot plant through phase I, phase II, and phase III, and it is through this period that R&D teams (including analysts/chemist/engineers, referred to as the ACE model) refine, optimize, and understand the API and DP processes to enable them to be eventually transferred to manufacturing. Manufacturing of clinical supplies in kilo lab, pilot plant, and solid dosage plants occurs under the constraints of cGMP conditions, which is discussed further in the chapter on kilo lab and pilot plant. The pilot plant and kilo lab are also sometimes used to "test" the scalability of a process. In this way, pilot plants serve a dual purpose, which make them unique as compared with non-pharmaceutical pilot plants. In terms of cost, however, large-scale experimentation in kilo lab or pilot plant can be significant – so there has been a shift toward greater predictability at lab scale to offset the need for pilot plant-scale "technology demonstration" experiments. Engineers through their training are well suited to scale-up and scale-down processes and can effectively model the chemical and physical behaviors in the lab to ensure success on scale. Many chapters in these two volumes discuss how scale-up/scale-down of various unit operations is performed. Chemical engineers are well trained in process modeling and optimization that support the reduction of experimentation and rehearsal batches prior to commercialization. This helps to reduce the number of larger-scale "experiments," thereby lowering costs during R&D. In this way, with the recent trend toward increasing efficiency and continuous improvement, the pilot plant and kilo labs are preferentially utilized to manufacture supplies for toxicological and clinical supplies rather than being used to "test" or verify that the chemistry or process will work on scale.

TABLE 1.3 Top 20 Global Pharmaceutical Products (2017 Sales)

Rank	Brand Name	API	Marketer	Indication	2017 Sales ($ Billion)
1 biologic	Humira	Adalimumab	AbbVie	Autoimmune diseases and rheumatoid arthritis	18.43
2 biologic	Eylea	Aflibercept	Regeneron Pharmaceuticals and Bayer	Macular degeneration	8.23
3	Revlimid	Lenalidomide	Celgene	Multiple myeloma	8.19
4 biologic	Rituxan	Rituximab (MabThera)	Roche and Biogen	Treatment of cancer	8.11
5 biologic	Enbrel	Etanercept	Amgen and Pfizer	Autoimmune diseases including rheumatoid arthritis, psoriasis, and other inflammatory conditions	7.98
6 biologic	Herceptin	Trastuzumab	Roche and Biogen	Treatment of cancer, mainly breast and gastric	7.55
7	Eliquis	Apixaban	BMS and Pfizer	Anticoagulant, mainly used to treat atrial fibrillation and deep vein thrombosis	7.40
8 biologic	Avastin	Bevacizumab	Roche and Biogen	Advanced colorectal, breast, lung, kidney, cervical, and ovarian cancer and relapsed glioblastoma	7.21
9 biologic	Remicade	Infliximab	Johnson & Johnson and Merck	Autoimmune diseases	7.16
10	Xarelto	Rivaroxaban	Bayer and Johnson & Johnson	Anticoagulant	6.54
11	Januvia/Janumet	Sitagliptin	Merck	Treatment of type 2 diabetes	5.90
12 biologic	Lantus	Insulin glargine	Sanofi	Long-acting human insulin analog for the treatment of diabetes	5.65
13 vaccine	Prevnar 13/Prevener	Pneumococcal 13-valent conjugate vaccine	Pfizer	Pneumococcal vaccine	5.60
14 biologic	Opdivo	Nivolumab	BMS	Melanoma	4.95
15 biologic	Neulasta/Peglasta/Neupogen	(Pegfilgrastim and Filgrastim)	Amgen and Kyowa Hakko Kirin	Neutropenia; decreases the incidence of infection during cancer treatment	4.56
16	Lyrica	Pregabalin	Pfizer	Anti-epileptic and neuropathic pain	4.51
17	Harvoni	Ledipasvir (sofosbuvir)	Gilead Sciences	HCV/HIV-1 infection	4.37
18	Advair	Fluticasone and Salmeterol	GlaxoSmithKline	Asthma	4.36
19	Tecfidera	Dimethyl fumarate	Biogen	Multiple sclerosis	4.21
20 biologic	Stelara	Ustekinumab	Johnson & Johnson	Plaque psoriasis	4.01

Source: From https://igeahub.com/2018/04/07/20-best-selling-drugs-2018 Shaded row indicates API is a new chemical entity; non-shaded row indicates API is a biologic.

TABLE 1.4 Fifteen Top-Selling Drugs (2013–2016) for Cumulative Sales Through 2016

	Drug/Drug Product Type[a]	API	Marketer	Approval	$ Billion
1	Lipitor Atorvastatin/film-coated tablet		Pfizer	1996	148.7
2	Humira Adalimumab/solution for injection	Antitumor necrosis factor (TNF) monoclonal antibody	AbbVie	2003	95.6
3	Advair (United States) Seretide(EU) fluticasone + salmeterol/dry powder inhaler		GlaxoSmithKline	2001	92.5
4	Remicade/lyophilized powder for constitution, solution injection	Anti-TNF chimeric monoclonal antibody	Janssen	1998	85.5
5	Plavix clopidogrel/film-coated tablet		Bristol-Myers Squibb	1997	82.3
6	Enbrel Etanercept/subcutaneous injection	Fusion protein produced by recombinant DNA	Amgen/Pfizer	1998	77.2
7	Rituxan Rituximab/solution injection	Chimeric monoclonal antibody	Roche Genentech	1997	75.9
8	Herceptin Trastuzumab/intravenous (IV) infusion	Monoclonal antibody	Roche Genentech	1999	65.2
9	Avastin Bevacizumab/IV infusion	Monoclonal antibody	Roche Genentech	2004	62.3
10	Nexium Esomeprazole/delayed release capsule; IV injection		AstraZeneca	2001	60.2

(continued)

TABLE 1.4 (Continued)

	Drug/Drug Product Type[a]	API	Marketer	Approval	$ Billion
11	Zyprexa Olanzapine/oral disintegrating tablets; injections		Eli Lilly	1996	60.2
12	Diovan Valsartan/film-coated tablet		Novartis	1997	60.1
13	Lantus Insulin glargine/ Subcutaneous Injection	Long-acting basal analog	Sanofi-Aventis	2001	58.3
14	Crestor Rosuvastatin/film-coated tablet		AstraZeneca	2003	55.2
15	Singulair Montelukast/ chewable tablet		Merck	1998	47.4

Source: Data from https://www.fool.com/amp/investing/2017/03/13/the-19-best-selling-prescription-drugs-of-all-time.aspx
[a] www.drugs.com source of dosage form type for originator drug.

A primary focus of process development is to drive down the cost contribution of the API to the final formulated pharmaceutical product cost while at the same time optimizing to ensure quality and process robustness. The impact of API costs on overall manufacturing costs is approximated in Figure 1.7. The cost contribution of API is expected to increase with increasing complexity of molecular structures of APIs, e.g. biologics. It is interesting to note that API molecular complexity can often impact API cost more than formulation or packaging costs. As Federsel points out that, "Given the importance of 'time to market' which remains one of the highest priorities of pharmaceutical companies, the need to meet increasingly stretched targets for speed to best route has come to the forefront in process R&D" [9]. In the not too distant past it was considered satisfactory to have a good-enough synthetic route that was fit for purpose (i.e. could support the quantities of material needed) but not one considered best or lowest cost ($/kg of API). The prevailing view was that the market would bear higher product pricing as compensation for higher cost of goods (COGs).

14 CHEMICAL ENGINEERING IN THE PHARMACEUTICAL INDUSTRY

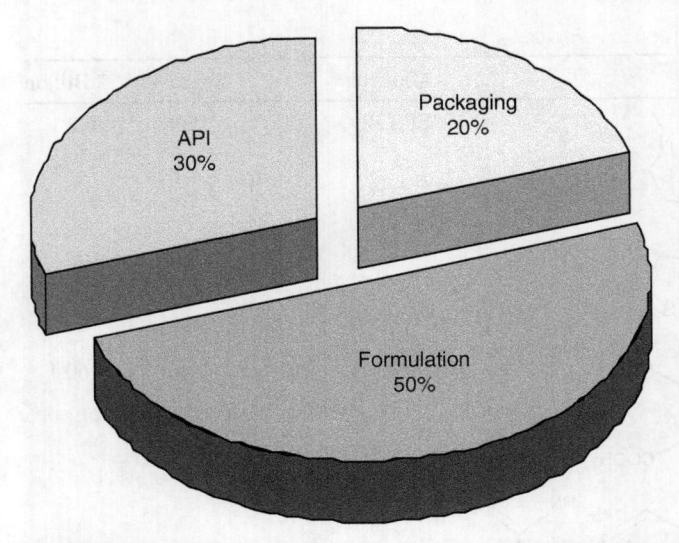

FIGURE 1.7 Average cost of goods (COG's) components in final dosage form across a large product portfolio – may vary widely for individual drugs (e.g. for API from 5 to 40%). *Source*: Reprinted with permission from Federsel [9]. Copyright (2006) Elsevier.

Further cost reduction through new routes could be and were pursued post-launch with savings realized later in the life cycle. According to Federsel, and evidenced frequently in contemporary R&D organizations, this approach is no longer viable, at least not as a default position. Instead the best synthetic route to API (i.e. route with ultimate lowest cost materials) coupled with best process design and engineering (process with lowest processing costs) must be worked out as early as possible in API process development [9]. The optimal API process developed by the time of launch is necessary to extract additional revenues and respond to reduced COG margins. Achieving this requires continuous improvement in scientific and technical tools as well as multidisciplinary skill sets in the R&D labs, including chemical engineering science. The implementation of process design principles, drawing on the right skill sets, both from chemistry and engineering perspectives during clinical phase II, is considered such an important step toward leaner more cost-effective processes readied for launch that several portions of this book will expand on this concept.

1.5 RECENT TRENDS FOR PHARMACEUTICAL DRUG AND MANUFACTURING

During the past decade, the pharmaceutical field has evolved to a science and risk-based industry. It is now commonplace for the regulatory dossier to contain scientifically rigorous information and descriptions of the risk management approach used for decision making. Now, the industry is undergoing significant changes in the API (from small molecule to biologics), manufacturing (from batch toward continuous), medicinal approach (generalized to personalized), and complexity of manufacturing (from simple dosage forms toward additive manufacturing or 3D printing).

1.5.1 Drug Substances Trend Toward Biologics

Biologic medicines are revolutionizing the treatment of cancer, autoimmune disorders, and rare illnesses and are therefore critical to the future of the pharmaceutical industry. Cancer immunotherapy includes monoclonal antibodies, checkpoint inhibitors, antibody-drug conjugates (ADCs), and kinase inhibitors, to name a few.

From the 2017 top-selling drugs shown in Table 1.3, there is a strong trend toward drugs derived from biological origins dominating the market than small molecules. In fact, the majority of best sellers are biologics, often monoclonal antibodies, which treat new diseases such as Crohn's and ulcerative colitis previously unmet medical needs by small molecule APIs.[14] It is also evident that the biologics retain their value even after patent exclusivity expires, e.g. Humira sales continue to grow post-patent expiry in 2016. The current generic industry is skilled in small molecule development but appears to be challenged to rapidly erode sales for biologics. In fact, in the coming years, it appears the first biologic medicine may take over as the all-time best seller from Lipitor.

Biological drug candidates include many different types of molecules including monoclonal antibodies, vaccines, therapeutic proteins, blood and blood components, and tissues.[15] In contrast to chemically synthesized drugs, which have a well-defined structure and can be thoroughly verified, biologics are derived from living material (human, animal, microorganism, or plant) and are vastly larger and more complex in structure. Biosimilars are versions of biologic products that reference the originator product in applications submitted for marketing approval to a regulatory body and are not exactly generic equivalents. However, biosimilar DPs are far more complex to gain regulatory approval in developed markets than for chemical generics and may involve costly clinical trials. Those that succeed will also have to compete with the originator companies who are unlikely to exit the market considering their expertise and investments. The biosimilars market is expected to increase significantly with the first FDA approval for Sandoz ZARXIO subcutaneous IV injection product in 2015 that helped establish a clear pathway for gaining regulatory approval [10]. Recently, Hospira, a Pfizer company, received FDA approval of their epoetin alfa biosimilar, Retacrit, in May 2018 [11].

---
[14] www.blog.crohnology.com
[15] https://www.selectusa.gov/pharmaceutical-and-biotech-industries-united-states

Biologic and biosimilar medicines are treating illnesses with unmet needs while retaining value even after post-exclusivity period. These are clear advantages for the originator, biopharmaceutical company developing biologic medicines, and are expected to continue to increase in the coming years. While the major disciplines making advancements in this area are biologists and chemists, there is a role for chemical and biochemical engineers in the design and development of the processing and purification steps. Chemical engineers are skilled at developing predictive models, and scale-up/scale-down principles, which make them a key contributor to this growing field. In fact, for biologics, scale-down predictive models of process steps were established and helped pave the way for biological products to use them for validation [12].

Chemical engineers that include biochemical engineering are well trained to impact the biotech industry, which utilizes cellular and biomolecular processes for new medicines [13]. Chemical engineers can also support the design of protein recovery, purification, and scaling up from lab to commercial production of the therapeutic proteins.

1.5.2 Lean Manufacturing

Pharmaceutical production of APIs and DPs can be generally characterized as primarily batch-operated multipurpose manufacturing plants. At these facilities commercial supplies of API intermediates, APIs, and DPs are manufactured before being packaged, labeled, and distributed to customers. Pharmaceutical production plants were typically designed to be flexible to allow a number of different products to be run in separate equipment trains, depending on the demand. Further, these facilities have various degrees of automation, relatively high levels of documentation, and change control to manage reconfigurations, with relatively long downtimes for cleanup and turnover of the plant between product changes [14]. These considerations are in part to meet regulatory requirements for commercial manufacturing. Manufacturing costs or COGs often account for approximately one-third of the total costs with expenses exceeding that of R&D [15]. *For this reason COG's have received considerable focus as an area of opportunity for potential savings* [7, 16].

It has been claimed that through adopting QbD principles and principles of lean manufacturing, pharmaceutical companies, as an aggregate, could save in the range of $20–50 billion/year by eliminating inefficiencies in current manufacturing [16]. This translates to 10–25% reduction in current COGs. An early QbD product approval of Chantix afforded an opportune chance to prove the benefits of these lean manufacturing and QbD principles. Chantix (varenicline tablet) was approved as an immediate release tablet commercially manufactured in the Pfizer Illertissen, Germany site. The OEB classification of this product required containment, available at small scale in this facility. Product demand increased dramatically in 1 year by 430%. The site employed lean manufacturing to eliminate process inefficiencies and wastes to increase production from 1 batch/day to 3 batches/day, delivering the desired 900 batches/year in the small-scale facility [17]. The lean manufacturing was indeed proven when the manufacturing maintained an inventory of only approximately one week lead ahead of demand.

The principles of lean manufacturing are often cited as an approach to reduce COGs in pharmaceutical development and manufacturing. Lean manufacturing describes a management philosophy concerned with improving profitability through the systematic elimination of activities that contribute to waste – thus the central theme to lean manufacturing is the elimination of waste where waste is considered the opposite of value. Based on the work of Taiichi Ohno, creator of the Toyota Production System, wastes are considered based on the following [18]:

- Overproduction
- Waiting
- Transportation
- Unnecessary processing
- Unnecessary inventory
- Unnecessary motion
- Defects

All of these wastes have the effect of increasing the proportion of non-value-added activities. Lean thinking is obviously applicable to many industries including pharmaceutical manufacturing as well as pharmaceutical development. Continuous processing, for pharmaceutical APIs and DPs, is one application of lean thinking applied to pharmaceutical manufacturing. The challenge is that batch processing inherently leads to overproduction and specifically the buildup of excess inventory of intermediates and DPs to supply the market. This leads to longer cycle times and is addressed through the concepts of continuous manufacturing (CM).

According to Ohno, "The greatest waste of all is excess inventory" where in simplest terms, excess inventory incurs cost associated with managing, transporting, and storing inventories adding to the waste. Large inventories also tie up large amounts of capital. Excess inventory represents an opportunity cost where capital is held up in the form of work in process (WIP), API finished goods, and formulated finished goods versus what could be invested elsewhere or back into R&D. Implementation of lean manufacturing principles can be used to develop workflows and infrastructures to reduce inventories. One way to reduce inventories is through continuous processing. Several chapters discuss the technical benefits of CM. A reliable steady delivery of product API and DP through small product-specific continuous plants could potentially reduce the level of inventory required in a dramatic way if the workflows were designed to ensure consistent delivery of product to packaging and distribution. *The facilities of continuous production trains tend to be significantly smaller.*

The costs of inventory holdings are significant and include both the carrying cost and the cash value of the inventory. The carrying costs of inventory include two main contributions – (i) weighted average cost of capital (WACC) and (ii) overhead [19].

Estimates for the combined carrying cost of WACC and overhead range from 14 to 25% that translates to approximately 20% return for every dollar of inventory eliminated [20]. Technology platforms and new workflows designed to minimize the need for stockpiling API and DP inventories across the industry therefore would seem to offer very rapid payback.

1.5.3 Continuous Manufacturing

For a large pharmaceutical company carrying $5 billion in inventories, the holding cost based on the combined WACC and overhead of 20% is approximately $1 billion/year. Considered another way, technologies that ensure a reliable and steady distribution of product with the result of eliminating the need to build and store massive inventories can return the company cost savings equivalent to a blockbuster drug (generating $billion/year). Indeed one of the three factors having the largest impact on the profitability of a manufacturing organization is inventory with the other two being throughput and operating expense according to Goldratt and Cox [21]. Continuous processing if designed for reliable operations essentially year-round or in other cases simply "on demand" could potentially eliminate the need to accumulate significant inventories above and beyond two to four weeks of critical safety stocks of finished goods.

CM across API and DP integrated under one roof as a platform technology is one long-term approach to transforming the way the industry manages their commercial supply chain.

As one reference cites, "Even for very small processes, continuous processes will prove to be less expensive in terms of equipment and operating costs. Dedicated continuous processes often put batch processes out of business" [22]. The real point here is that continuous is one approach to lean manufacturing and to reducing inventories and costs but certainly not the only approach. Other lean systems can be devised that utilize the existing batch facilities as well. Since the publication of the first edition of this book in 2010, there has been a significant wave of interest in considering continuous processing for pharmaceutical API and DP.

In July 2015 FDA granted Vertex Pharmaceuticals approval of the first DP, Orkambi, a cystic fibrosis (CF) drug, to be produced using a CM process. Vertex's second drug, Symdeko, for treating the underlying cause of CF occurred in February 2018. Janssen aims to manufacture 70% of their "highest volume" products using CM within eight years.[16] In addition, they intend to increase yield by reducing waste by 33% and reduce manufacturing and testing cycle times by 80% through the use of CM. Their claim is that CM can reduce operating costs by as much as 50%, gain higher throughputs, and significantly reduce waste [23]. Janssen's HIV drug Prezista is also manufactured via a continuous process after obtaining approval to convert from batch to continuous.[17] Pfizer and Eli Lilly have made investments in CM and recently submitted an NDA or gained approval for a product, respectively. Merck states that CM will help achieve their goals of well-controlled processes with flexible sizes to handle small-to-large volume products localized closer to the customer.[18]

Merck Manufacturing Division targets a total lead time of 90 days formulation to the patient, reducing the current timing by one-quarter. Multiple companies are teaming together to leverage CM, e.g. Novartis-MIT Center for Continuous Manufacturing. A critical component of CM is that PAT is embedded into the overall plan for monitoring and control of the process. As stated by Kevin Nepveux, "one of the best ways to go about implementing CM processes is to develop the analytics in-line with the application."[19] In summary, CM requires significant focus by chemical engineers as there is more attention on cost savings and cost efficiencies.[20]

1.5.4 Personalized Medicine

"One size doesn't fit all is a tenet of personalized medicine, also called precision medicine," states Lisa Esposito in a recent report [24]. In her article, she highlights the long-standing personalized medicine approach taken to treat cancer based on the individual patient's disease state and conditions. There is a growing expectation that the pharmaceutical industry should deliver DPs targeted to the individual, tailoring the amount of drug based on their mass, metabolism, genetic factors, and disease state. In this section, we discuss two approaches for manufacturing personalized/precision medicine through a pharm-on-demand concept for military personnel and for complex dosage forms using additive manufacturing (referred to as 3D printing).

The Defense Advanced Research Projects Agency (DARPA) Battlefield Medicine program is keenly interested in miniaturized, flexible platforms for end-to-end manufacturing of pharmaceuticals to support the troops on location. As discussed in other chapters within this book, advances

---

[16] http://www.pharmtech.com/fda-approves-tablet-production-janssen-continuous-manufacturing-line

[17] https://www.lifescienceleader.com/doc/merck-s-path-to-continuous-manufacturing-for-solid-oral-dose-products-what-stands-in-the-way-0001

[18] ibid.

[19] https://www.pharmaceuticalonline.com/doc/pfizer-s-hybrid-approach-to-implementing-continuous-manufacturing-processes-0001

[20] http://www.contractpharma.com/contents/view_features/2018-01-30/pharma-industry-outlook

in continuous flow synthesis, chemistry, biological engineering, and downstream processing, coupled with online analytics, automation, and enhanced process control measures, are proving that such capabilities are ready for implementation. The desire is to have a mobile, on-demand pharmacy located at the battlefield that could ensure readiness to treat threats of chemical, biological, radiological, and nuclear weapons [25].

1.5.5 Additive Manufacturing

Additive manufacturing, also referred to as three-dimensional (3D) printing, is an automated process of building layer by layer a complex dosage form personalized and manufactured on demand. FDA approved the first 3D printed DP in August 2015 for Aprecia Pharmaceuticals SPRITAM product as a disintegrating tablet [26].

The 3D printing in this case binds the powders while maintaining a porous structure (without the typical compression of a tablet press), providing a fast dissolving tablet. For example, 1000 mg of levetiracetam dissolves within seconds [27]. Extensions of 3D printing include printing extremely low dose APIs or highly potent APIs, but encapsulated with excipients, thus reducing potential exposure. Norman et al. [27] provide a thorough review of the different modalities of 3D printing for pharmaceutical manufacturing, which includes an analysis of the potential benefits of such products.

1.6 CHEMICAL ENGINEERS SKILLED TO IMPACT FUTURE OF PHARMACEUTICAL INDUSTRY

The fundamental principles taught in the chemical engineering curriculum ensure the chemical engineer is well poised to apply them to solve the coming challenging issues in the pharmaceutical industry. Chemical engineers are uniquely positioned to help address these needs in part derived from their ability to predict using mathematical models and their understanding of equipment and manufacturability. As Wu et al. highlighted, chemical engineers can help transform pharmaceutics from an industry focusing on inventing and testing to a process and product design industry [28]. Significant pressure exists on what used to be a historically high-margin nature of the pharmaceutical industry to deliver safe, environmentally friendly, and economic processes in increasingly shorter timelines. This means fewer scale-ups at kilo and pilot plant scale, with expectation that a synthesis or formulation can be designed in the lab to perform as expected (and right the first time) at the desired manufacturing scale.

Chemical engineers are also uniquely positioned to influence regulators by incorporating advancements such as continuous processing coupled with PAT into a highly regulated industry. From R&D through manufacturing within the pharmaceutical industry, chemical engineering can be leveraged to bring competitive advantage to their respective organizations through process and predictive modeling that lead to process understanding, improving speed of development, and developing new technology platforms and leaner manufacturing methods. The chapters in these two volumes are intended to provide examples of chemical engineering principles specifically applied toward relevant problems faced in the pharmaceutical sciences and manufacturing areas. Further the broader goal of this work is to promote the role of chemical engineering within our industry, to promote the breadth of skill sets therein, and to showcase the critical synergy between this discipline and the many scientific disciplines that combine to bring pharmaceutical drugs and therapies to patients in need around the world.

REFERENCES

1. Janusz Rzeszotarski, W. (2000). Encyclopedia of Chemical Technology.
2. FDA (2009). International conference on harmonisation: guidance on Q8(R1) pharmaceutical development. *Federal Register* 74 (109): 27325–27326.
3. FDA (2006). International conference on harmonisation: guidance on Q9 quality risk management. *Federal Register* 71 (106): 32105–32106.
4. FDA (2009). International conference on harmonisation: guidance on Q10 pharmaceutical quality system. *Federal Register* 74 (66): 15990–15991.
5. DiMasi, J.A., Grabowski, H.G., and Hansen, R.W. (2016). Innovation in the pharmaceutical industry: new estimates of R&D costs. *Journal of Health Economics* 47: 20–33.
6. Gregson, N., Sparrowhawk, K., Mauskopf, J., and Paul, J. (2005). Pricing medicines: theory and practice, challenges and opportunities. *Nature Reviews Drug Discovery* 4: 121–130.
7. Suresh, P. and Basu, P.K. (2008). Improving pharmaceutical product development and manufacturing: impact on cost of drug development and cost of goods sold of pharmaceuticals. *Journal of Pharmaceutical Innovation* 3: 175–187.
8. Dienemann, E. and Osifchin, R. (2000). The role of chemical engineering in process development and optimization. *Current Opinion in Drug Discovery and Development* 3 (6): 690–698.
9. Federsel, H.-J. (2006). In search of sustainability: process R&D in light of current pharmaceutical R&D challenges. *Drug Discovery Today* 11 (21/22): 966–974.
10. FDA (2015). FDA approves first biosimilar product Zarxio. *US Food and Drug Administration* (6 March 2015).
11. FDA (2018). FDA approves first epoetin alfa biosimilar for the treatment of anemia. *US Food and Drug Administration* (15 May 2018).
12. Li, F., Hashimura, Y., Pendleton, R. et al. (2006). A systematic approach for scale-down model development and characterization of commercial cell culture processes. *Biotechnology Progress* 22: 696–703.

13. Rovner, S.L. (2011). Process development shines in tough times. *Chemical and Engineering News* 89 (24): 41–44.
14. Behr, A., Brehme, V.A., Ewers, L.J. et al. (2004). New developments in chemical engineering for the production of drug substances. *Engineering in Life Sciences* 4 (1): 15–24.
15. Burns, L.R. (ed.) (2005). *The Business of Healthcare Innovation*. Cambridge: Cambridge University Press.
16. F-D-C Reports Inc. (2009). The Gold Sheet, Pharmaceutical & Biotechnology Quality Control, Attention turns to the business case of quality by design. F-D-C reports, January 2009.
17. am Ende, M.T., Bernhard, G., Lubczyk, V., Dressler, U. et al. (2008). Risk Management and Lifecycle Efficiency within Process Development. Leadership Forum 2008, Washington, DC, 29 May 2008.
18. Ohno, T. (1988). *Toyota Production System Beyond Large Scale Production*. New York: Productivity Press.
19. Cogdil, R.P., Knight, T.P., Anderson, C.A., and Drennan, J.K. (2007). The financial returns on investments in process analytical technology and lean manufacturing: benchmarks and case study. *Journal of Pharmaceutical Innovation* 2: 38–50.
20. Lewis, N.A. (2006). A tracking tool for lean solid-dose manufacturing. *Pharmaceutical Technology* 30 (10): 94–109.
21. Goldratt, E.M. and Cox, J. (2004). *The Goal: A Process for Ongoing Improvement*, 3e. London: Routledge.
22. Biegler, L.T., Grossmann, I.E., and Westerberg, A.W. (1997). *Systematic Methods of Chemical Process Design*. Upper Saddle River: Prentice Hall.
23. Kuehn, S. E. (2015). Janssen embraces continuous manufacturing for Prezista. *PharmaceuticalManufacturing.com* (8 October 2015), https://www.pharmamanufacturing.com/articles/2015/janssen-embraces-continuous-manufacturing-for-prezista/?show=all
24. Esposito, L. (2018). What does personalized medicine really mean? *US News & World Report* (26 January 2018).
25. Lewin, J.J., Choi, E.J., and Ling, G. (2016). Pharmacy on demand: new technologies to enable miniaturized and mobile drug manufacturing. *American Journal of Health-System Pharmacy* 73 (2): 45–54.
26. United States Food and Drug Administration (2015). Highlights of prescribing information: Spritam. https://www.accessdata.fda.gov/drugsatfda_docs/label/2015/207958s000lbl.pdf
27. Norman, J., Madurawe, R.D., Moore, C.M.V. et al. (2017). A new chapter in pharmaceutical manufacturing: 3D-printed drug products. *Advanced Drug Delivery Reviews* 108: 39–50.
28. Wu, H., Khan, M.A., and Hussain, A.S. (2007). Process control perspective for process analytical technology: integration of chemical engineering practice into semiconductor and pharmaceutical industries. *Chemical Engineering Communications* 194: 760–779.

2

CURRENT CHALLENGES AND OPPORTUNITIES IN THE PHARMACEUTICAL INDUSTRY

JOSEPH L. KUKURA
Merck Research Laboratories, Merck & Co., Inc., Rahway, NJ, USA

MICHAEL P. THIEN
Merck Manufacturing Division, Merck & Co., Inc., Whitehouse Station, NJ, USA

2.1 INTRODUCTION

The pharmaceutical industry bases its products, strategies, decisions, and actions and ultimately its very existence on the primary challenge of improving human health and the quality of life. The work of this industry uses a foundation of medical science to connect to the most basic struggle faced by all individuals and societies: the struggle for people to live healthy, productive lives. The industry has partnered with governments, health organizations, and society to achieve key successes in human history, including the eradication of smallpox, prevention of infection through the large-scale production and distribution of antibiotics such as penicillin, and significant reductions in cardiovascular events. Advances in pharmaceutics have contributed to lower infant mortality rates and longer life spans observed over the past century. When considering the landscape of the pharmaceutical industry, one should retain the perspective that a challenge or opportunity that relates to the improvement of human health is at the core of challenges and opportunities shared by pharmaceutical companies.

2.2 INDUSTRY-WIDE CHALLENGES

Just as life and the state of human health often undergo significant changes, the pharmaceutical industry is profoundly changing. Fundamental elements that molded past business models are dynamically moving into new realms in a manner that will challenge the continued vitality of pharmaceutical companies much in the same way that changes in environments can influence the health of people. The parallels between diseases that the industry seeks to address and the pharmaceutical business climate are distinctly apparent. Illnesses such as HIV/AIDS and cancer are more complex and evolve at a faster pace than many previous diseases, requiring new approaches and advances. Similarly, economic, societal, and scientific forces are rapidly driving changes to the industry's business models. The forces challenging the industry align into four categories: increased costs and risks, revenue/price constraints, globalization of activities, and increasing complexity of pharmaceutical science.

2.2.1 Increased Costs and Risks

Bringing a new medicine to market involves a long, complex process in a highly regulated industry. Estimating accurate or typical costs for successfully launching new pharmaceutical products is difficult. The estimates are a strong function of the success rates assumed for moving a program through various clinical trial stages. Estimates for the average cost to launch a pharmaceutical product ranged from $900 million to $1.7 billion in the early years of the twenty-first century and rose to $2.6 billion over the following decade [1, 2]. Across the industry, costs are trending upward in a manner that forces business practices to adjust.

Chemical Engineering in the Pharmaceutical Industry: Active Pharmaceutical Ingredients, Second Edition.
Edited by David J. am Ende and Mary T. am Ende.
© 2019 John Wiley & Sons, Inc. Published 2019 by John Wiley & Sons, Inc.

Developing new medicines for unmet medical needs always involved significant costs and risks. For every product brought to market, pharmaceutical companies have typically invested in several thousands of compounds during the drug discovery stage, hundreds of compounds in preclinical testing, and many (7–12) unsuccessful clinical trials over a period of 9–15 years, as depicted in Figure 2.1. Though the later-stage phase II and phase III clinical trials are performed on the fewest number of compounds, they are also the most costly stages of development since they can require testing hundreds of patients in phase II and thousands in phase III in order to get statistically meaningful results. The cost of developing new medicines is therefore particularly sensitive to success rates of late-stage clinical trials. The industry has generally experienced a decline in the fraction of compounds proceeding through phase II and phase III clinical trials to a successful regulatory approval and commercial launch of a new product. This decline translates to expending more resources on programs that do not return value on their investment and greater overall spending on research and development.

The lower clinical trial success rates are due in part from the fact that pharmaceutical companies are attempting to treat more complex therapeutic targets. Many diseases with straightforward cause–effect relationships and less sophisticated biological mechanisms have already been addressed, leaving more challenging and intricate problems for the future. The setbacks and frustrations relating to the development of treatments and vaccines for HIV infection serve as a case in point. Numerous articles and presentations have reported that the HIV virus mutates, adapts, and changes features faster than predecessors that were studied for vaccine development. Similarly, the causes of many forms of cancer being addressed in clinical trials have more complex physiological traits in comparison with successfully treated conditions such as high cholesterol and high blood pressure. Treating more complicated illnesses leads to higher risks for clinical trial evaluations.

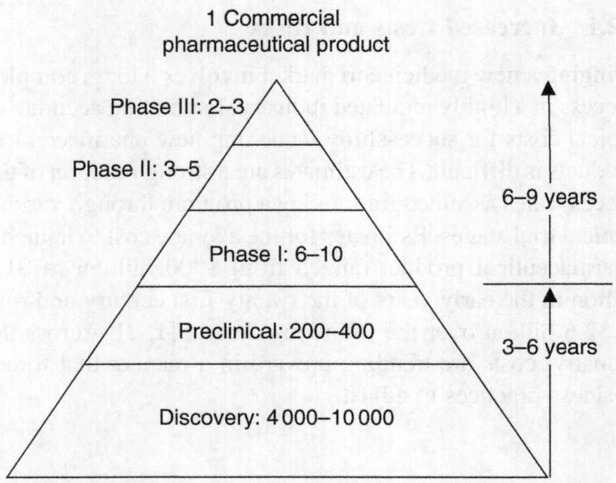
1 Commercial
pharmaceutical product
Phase III: 2–3
Phase II: 3–5
6–9 years
Phase I: 6–10
Preclinical: 200–400
3–6 years
Discovery: 4 000–10 000

FIGURE 2.1 Number of compounds in research and development for every successful launch of a pharmaceutical product.

The pharmaceutical industry research and development costs are also increasing due to greater regulatory hurdles for getting approval of new medication. Many agencies have raised their requirements for approval, leading to the need for larger and more comprehensive clinical trials and safety assessment testing. Government health agencies are showing a high level of caution with respect to side effects and risk–benefit assessments. This caution creates a need for outcome data and in turn longer running trials and longer review periods that can delay introduction of a product to market. A more conservative regulatory approach ultimately forces greater spending on development and testing of new medicines to collect the data needed for the higher standards.

2.2.2 Revenue/Price Constraints

Beyond the challenge of increasing costs, the pharmaceutical industry is also facing constraints to income and product pricing. The patents of large revenue "blockbuster" drugs are expiring faster than they are being replaced by a comparable portfolio of new highly profitable products. The challenges of addressing more complex therapeutic targets previously described in the context of increasing cost also directly affect revenue in the industry. The greater level of complexity not only makes the research and development process more expensive, but it also slows the realization of a return on investment in these areas. Many of these more sophisticated research efforts target a narrower patient base than preceding blockbusters. The largest sources of revenue for the industry in prior decades improved conditions that were widespread, such as depression, hypertension, and hypercholesterolemia. Far fewer people have conditions that many products currently in development aim to improve, such as specific forms of cancer. With a smaller base of potential patients, these new products can be expected to generate less revenue than broadly used products already on the market.

Another strong influence on pharmaceutical sales relates to the means by which patients pay for medicine. Organizations responsible to pay for prescriptions, the payers such as insurance companies and health maintenance organizations (HMOs), are influencing the medical options for their membership. The pharmaceutical companies used to be able to focus on physician–patient relationships when marketing products, but the decision-making process to select medicine now involves a more complex set of interactions between physicians, patients, and payers. The pharmaceutical industry must engage all three members of this collective to successfully bring products to those who need them. Payers acquired an increasingly important role in this process in the United States through consolidations that have allowed a few groups to represent larger numbers of people. Single payers can control access to millions of patients [3]. Payers can exert their influence on the pharmaceutical industry in several ways. They cannot directly specify which medications a patient may use, but they can make co-payments paid by patients much higher for some medications relative to others. If a payer wants to provide incentive for patients to request switching from a current treatment to a less expensive generic alternative, they can make the co-payment for the generic version significantly less expensive. Similarly, payers can also choose to reimburse pharmacists at a higher rate for supplying generics and drive policies at pharmacies to favor the generic options.

In addition to consolidation of private payers, other events elevate the importance of payers to the pharmaceutical industry. The US government became effectively the largest payer to the pharmaceutical industry in January 2006 with the implementation of Medicare Part D prescription plan, covering over 39 million people with that plan alone [4]. Even more people will be eligible for coverage benefits in the coming years. If the US government alters its current policy to not negotiate medicine covered by Medicare or reimportation policies, the changes will create significant challenges to the business models of the pharmaceutical industry. The issue is certainly not limited to the United States. The changing demographics of the world will dramatically affect social medicine policies. The patients themselves in the patient–physician–payer relationship are changing in ways that will challenge the business models of the pharmaceutical industries. Across the world, the fraction of people above age 65 is growing as life expectancy increases. Never will the world have had this many people this old. Along with economically challenged populations in emerging non-Western markets, this increasing fraction of the planet's population will generally have limited income available for health care, but they will have a disproportionally strong demand for pharmaceutical products. Ensuring access to medicine across the globe and across population sectors will require lower prices. The pharmaceutical industry must adapt to meet the needs of these large segments of customers.

Not only are the demographics of patients changing, but their behaviors and approaches to health care are differing from the past. Survey results show that health care is a diverse consumer market with people seeking greater access to information to make their own choices with respect to health-care needs [5]. With technology advances such as the internet, patients can get more information to play a larger role in selecting treatment options. In some parts of the world, direct advertising to consumers is prevalent and raising new levels of their awareness of options. Patients are also willing to explore innovative techniques or travel outside their area and even their country to find options that best suit their preferences. To face the revenue and price constraints introduced by patients' actions to the patient–physician–payer relationship, pharmaceutical companies need to understand the changing manner in which patient behaviors affect market demand and pricing.

2.2.3 Globalization of Activities

To address financial constraints and meet the global demands of emerging markets, the pharmaceutical industry is increasing the activity levels of its business in these regions, moving away from being primarily located and focused in the Unites States and Europe. Like many other industries, a greater fraction of manufacturing and research and development is shifting overseas from a western base into countries such as China and India. Numerous clinical trials are conducted in these regions to achieve cost savings and to more quickly enroll patients who are not already undergoing another therapy. Development activities such as medicinal chemistry and process scale-up are being performed there as well, leading to an expansion of sophisticated laboratories in these countries. Manufacturing is becoming increasingly well established in regions outside the United States and Europe, supplying global medical needs from truly global locations. Like international efforts in other industries, the globalization of pharmaceutical activities increases challenges associated with alignment of regulatory standards, logistics, language barriers, and cultural differences but pays dividends in cost reduction and increases in the size of the talent pool.

2.2.4 Increasing Complexity of Pharmaceutical Science

As already mentioned in the context of rising costs and constrained revenues, current research and development of new medicine is attempting to address afflictions and therapeutic categories that are more complex than their predecessors. In order to understand and treat these more complex targets, the industry must use more complex and difficult science. The pharmaceutical industry has always employed a highly talented collection of several scientific disciplines, ranging from biologists and chemists to engineers and statisticians. All of these professions now face harder problems down to the molecular level of their fields to bring forward the next generation of medicines. The scientific challenges take many forms. For example, a larger fraction of compounds in development have low solubility and low permeability in human tissue, making drug delivery within the body more difficult. Highly potent compounds dictate that the amount of drug in the formulation be very small, sometimes in the sub-milligram ranges, and this also adds to the challenges of formulation development. Innovative and novel delivery systems are required to ensure new medicines are effective. Advances in the academic understanding of the workings of human genetic code are creating especially challenging questions around how to translate this knowledge into practical improvements in human health. Employees of pharmaceutical companies must be prepared for a future with more difficult challenges.

2.3 OPPORTUNITIES FOR CHEMICAL ENGINEERS

The challenges faced by the pharmaceutical industry create several opportunities for its members, including chemical engineers. The pressures to reduce costs connect directly to engineering principles that seek economies of scale and the application of efficient technology. In a related manner, chemical engineers can also reduce the lead time required to manufacture pharmaceutical products. Reducing lead time can have the parallel benefits of lowering inventory costs and protecting revenue by increasing the responsiveness of the supply chain to ensure changes in patient demands for medicine can be met. Technological innovation and engineering analysis can enhance products to create meaningful differentiation for patients in a variety of ways including contributions to drug metabolism and pharmacokinetics, which provides value in the face of revenue constraints. The complexities of pharmaceutical science and constraints of approaches that need to be suitable for global use are interwoven with the application of chemical engineering tools to address cost issues and enable product value. Finally, the strategic management of technology used to meet industry challenges by chemical engineers is an additional overarching opportunity in the industry.

Cross-functional collaboration is vital to efforts to reduce costs, lower lead times, improve processes and products, and strategically manage new technology. Engineering alone cannot accomplish nearly as much as what can be achieved in combination with the organic chemists who design API synthetic routes, the analytical chemists who develop methods to measure the quality of the material, the formulators who design the final dosage form, the quality colleagues who help ensure processes are compliant, the operating staff who run the equipment, and the supply chain professionals who manage logistics. Chemical engineers are integrated into broad teams with a common goal to deliver the highest quality medicine to patients at the most affordable cost. To fully realize the more technical opportunities highlighted in this section, chemical engineers must engage in the opportunities to collaborate with a diverse group of professionals to learn from their perspectives and share innovative approaches to problem solving.

2.3.1 Reducing Costs with Engineering Principles

Owing to large margins, engineers in the nongeneric pharmaceutical industry historically did not have the same traditional focus on product cost as engineers in other businesses. With a renewed emphasis on cost, engineers are increasingly using a wide variety of engineering tools to improve costs and help maintain margins while maintaining high quality. These tools include modeling of unit operations, employment of efficient lab methods and design of experiments, combining the output of models and experiments to define advantageous processing options, and the use of standardized technology platforms. Some of these topics will be highlighted briefly here and discussed in greater detail in subsequent chapters.

Across many industries, engineers are employed to use process modeling and physical/chemical property estimation to maximize the yield and minimize the energy consumption and waste production associated with desired products. Using the broad applicability of this network of techniques is a continuing opportunity. Engineers in the pharmaceutical industry use computational tools originally created for oil refinery processes to optimize distillations and solvent recovery associated with the manufacture of active pharmaceutical ingredients (APIs) [6]. Similarly, thermodynamic solubility modeling can be applied to optimize crystallizations [7]. Computational fluid dynamics (CFD) has numerous applications to pharmaceutical flows [8]. The use of sound, fundamental chemical engineering science can eliminate bottlenecks, improve production, and unlock the full potential of biological, chemical, and formulation processes used to make medicine.

Chemical engineers can also use their training and expertise with technology to help reduce costs. In the R&D arena, the use of high-throughput screening tools and multi-reactor laboratory systems efficiently promotes the generation of data at faster rates. When modeling and estimation techniques cannot provide a complete picture, engineers can get the data they need quickly with high efficiency technology. It is important to recognize that not all of the advanced laboratory technology works universally well in all situations. A miniature reactor system suitable for homogeneous reactions may have insufficient mixing for heterogeneous chemistry. Selection of the appropriate laboratory technology and proper interpretation of results produced by these laboratory tools benefit from the perspective of combined chemical engineering principles such as mass transfer, heat transfer, reaction kinetics, and fluid mechanics. With the appropriate equipment in hand, engineers can utilize a statistically driven design of experiments to maximize the value of data generated via the experimental methodology.

The process understanding that comes from combining models and experimental data is a key opportunity for chemical engineers. Changes in the regulatory environment contribute to this opportunity. The advent of quality by design (QbD) principles (ICH Q8, Q9, and Q10) provides greater freedom after launching a product to modify operating parameters within a defined operating space. These changes can promote higher quality products and reduce process waste by applying the knowledge that comes with increased production experience once a medicine is commercialized. The modeling abilities of chemical engineers and their technology expertise will be able to provide crucial guidance to the definition and refinement of a QbD operating space. A thoughtful, well-conceived operating space will in turn lead to long-term gains in process efficiency and better results for consumers.

A primary method to achieve the benefits of chemical engineering principles at manufacturing scale comes from the development and application of technology platforms that can use a single set of equipment with common operating techniques across a portfolio of processes and products. The platforms not only reduce capital costs by allowing the purchase of a reduced amount of equipment for more applications, but development costs can also be lowered as well through a streamlined approach that comes from having a deep understanding and expertise with a technology platform. The familiarity and data obtained from running multiple projects in a single platform will translate into benefits for future projects that share common features. Broad uses of standardized platforms also make processes more portable for global applications. The key challenges of platforms are (i) knowing for which compounds the platform will be applicable and (ii) maintaining the knowledge gained about the platform and its underlying technology.

2.3.2 Reducing Lead Time

As the uncertainty and volatility of demand for pharmaceutical products increase, the value of reducing the lead time to manufacture medicine also increases. The lead time for ordering commodity raw materials for the first custom step in a synthesis to the release of an API can often take as long as 9–15 months, depending on the length and complexity of the synthesis, complexity of the supply chain, and availability of the required chemicals and manufacturing equipment. The lead time for the formulation of the API with excipients to make the drug product typically adds another two to six months, with a significant fraction of that time associated with the quality testing and release process.

The result of such long lead times is that companies will typically build various inventory levels of isolated intermediates, active ingredients, bulk drug product, and packaged products in order to fulfill unforecasted demand spikes that can arise during a replenishment lead time. The "safety stock" inventory level needs to be at least the replenishment lead time (LT) times the difference between the maximum demand rate (DR_{Max}) pulled by the consumer that a company wants to be capable of meeting and the forecasted demand rate (DR_{FC}) used for production planning:

$$\text{Minimum safety stock} = LT \times (DR_{Max} - DR_{FC})$$

By reducing lead time, the level of inventory to cover customer demand volatility can be reduced at least linearly. The gap between the maximum and forecasted demand rate may also decrease because forecasts are typically made with greater accuracy when the time between locking the forecast for production schedule and the final delivery is reduced [9]. Lead time reduction provides meaningful cost reduction because inventory costs include not only the warehousing costs to hold the material but also the interest costs of the money invested to manufacture the inventory. Money not spent on inventory builds can be invested elsewhere in the business. Further, if the actual demand rate is significantly lower than the forecasted demand, then material in the inventory can be vulnerable to expiring and needing to be retested, reprocessed, or discarded and replaced. A case study found that reducing the lead time from 10 months to less than 2 months for production of a vaccine can offset more than a 50% production unit cost increase, by reducing the vulnerability to discards stemming from uncertain demand [10]. Just as importantly, in cases where the actual maximum demand rate exceeds the level planned to be covered by inventory, having a shorter lead time can better meet the upside demands of the market and reduce the risks of stocking out of medicine needed by patients.

Engineers can reduce the lead time of production in a variety of ways. During development, process alternatives that improve reaction kinetics, reduce mass transfer restrictions, and enable more productive unit operations are fundamental to the engineering skill set and can lower lead times. However, the majority of the very long lead times for medicine do not involve the time required for material to change chemically or physically into a different form. Most of the lead time is commonly associated with holding material while waiting for it to be released for the next process step, transportation time between manufacturing sites, and/or waiting for the next available production window. Processes that follow multiple sequences of batch unit operations are especially vulnerable to such waiting periods. Engineers are strongly contributing to an increasing use of continuous manufacturing processes that inherently flow material from one production step to the next without building inventory of intermediates or pausing between transformations. Multiple types of processes and many unit operations are amenable to connecting steps that are segregated when run batchwise [11]. Continuous processes also lend themselves to integrating with process analytical technology (PAT) and other analytical tools to perform enhanced process control and real-time release of material at intermediate stages of production. By integrating engineering principles with advanced analytical measurements and fundamental understanding of the associated organic and physical chemistry mechanisms, processes of the future will be able to flow through hold points that were necessary in the past.

2.3.3 Improving Product Value

Chemical engineers also have opportunities to meet market demands for pharmaceutical products that deliver greater value to patients, payers, and physicians. These customers generally do not care about the manufacturing process, but they do care about product convenience, safety, and compliance. They are seeking meaningful differentiation in these areas among their options. The remainder of this section will discuss ways engineers can contribute to product value with two examples: drug delivery and diagnostics.

Contributions to improvements in drug delivery vehicles serve as an excellent example of how engineers can improve pharmaceutical product features. The application of particle engineering and convection modeling to inhalers can improve the consistency with which a dose is administered via the respiratory system independent of the strength of the patient's breath [12]. Greater consistency of delivery increases the associated compliance. In orally administered capsules and tablets, engineers can manipulate polymer properties and transport driving forces to afford a consistent extended release of an API [13]. A steady, slow release of medicine from a single delivery vehicle can reduce both the frequency with which the medicine needs to be taken and potential side effects, which can, in turn, improve conveniences for the patient and compliance with the dosing regimen. In order to realize the benefits of controlled release, the pure API particle size distribution usually must be kept consistent prior to formulation. A great deal of engineering effort has been applied to maintain control of crystal sizes during the crystallization, filtration, and drying unit operations for drug substances [14].

Engineering principles can also be used to improve diagnostic tools used to treat diseases. Diagnostics are especially important for payer organizations that want to utilize options that have the highest probability of success for the patient. A diagnostic tool that enables physicians to initially assign the best treatment without going through a trial-and-error approach reduces the costs charged to the payers through the preemptive elimination of ineffective options. The aforementioned chemical engineering skills that aid the process of making medicines also contribute to improvements in making diagnostic technology. The underlying governing equations that characterize the transport of medicine to a specific target in the body also have applications in the movement of a sample from a patient through a device to the analysis component. Beyond diagnostic effectiveness, the ability for chemical engineers to respond to patient preference and improve the convenience of diagnostics tools used in the home or other areas outside of hospitals and physician offices is a key opportunity with health care becoming an increasingly consumer-driven market [5]. Advances in polymer technology and manufacturing processes can lead to devices that are lighter, smaller, and more resilient to being dropped. Just like new models of an iPhone™ garner increased use over their heavier, larger, and more delicate predecessors, delivery vehicles and diagnostics serve as examples of significant opportunities for chemical engineers in the pharmaceutical industry to meet consumer demands for improved products.

2.3.4 Strategic Technology Management

In addition to direct scientific contributions that reduce costs and improve value for the pharmaceutical industry, engineers have the opportunity to help direct strategic investments in technology. Companies cannot afford to individually develop, implement, and advance all technologies required for their business. Several case studies show how good and poor strategies relating technology to business considerations have affected multiple industries, including computer companies and international distributors [15]. The availability of global development and supply options creates relatively new decisions for the pharmaceutical industry. Technology investment must be managed through a careful balance of internal capabilities, strategic partnerships, and reliance on external vendors. In order for this balance to be established and maintained, a holistic definition and view of technology must first be established. Is any laboratory or production device such as a granulator or a blender considered technology, or are they just pieces of equipment? In the context of strategic management, technology can be defined as a system composed of (i) technical knowledge, (ii) processes, and (iii) equipment that is used to accomplish a specific goal. The knowledge encompasses the understanding of fundamental principles and relationships that provide the foundation of the technology. The processes are the procedures, techniques, and best practices associated with the technology. The equipment is the physical manifestation of the technology such as devices, instruments, and machinery. The goal for strategic technology management is to make value-driven decisions around investments in the advancement, capacity, and capability with each of the technology components.

To make those investment decisions, industrially relevant technology can be assigned to categories. The concepts of core technology and noncore technology and associated subcategories are useful in this regard:

- *Core technology*: Core technology is sophisticated technology that makes critical contributions to the core business. As such, it justifies investment in *all three* components of the technology (knowledge, process, equipment) to afford a competitive advantage. Options for core technology include maintaining internal capability for all three components as a primary technology or managing the technology via external capacity as a partnered technology. The distinguishing feature of all core technology is that internal capability in the knowledge component of the technology is typically required to ensure the technology is adequately controlled to meet core business needs:
  - *Primary technology*: Primary technology is a category of core technology that offers competitive advantage by maintaining internal capability for

the complete technology system. The investment in internal capacity does not need to meet all business uses of the technology but ensures that adequate resources can be provided for critical projects.

- *Partnered technology:* Partnered technology is a technology that contributes to the core business, but the company can maintain a competitive advantage while relying on external partners to be primarily responsible for parts of the technology system. The company may invest in knowledge and process development for a partnered technology while utilizing external equipment capacity and potentially outside process expertise.

- *Noncore technology:* Noncore technology does not warrant investment or control in all three components of the technology system. The subcategories of emerging technology and commoditized technology characterize the most relevant noncore technology:

  - *Emerging technology:* Emerging technology has potential to contribute significant business value in the future but generally requires additional investment in the knowledge base before it can be applied in practice to the core business. Emerging technology is not necessarily brand new technology, but its application to the core business may be atypical or speculative.

  - *Commoditized technology:* Commoditized technology is mature technology that is reliable, well established within the industry, cost efficient, and available in the market such that little investment in the technology is required.

To determine if a technology of interest is core or noncore, the connection of the technology to business value must be assigned as well as risks associated with the technology's ability to meet business requirements. Business value (BV) can be determined by identifying the revenue enabled by products made via the technology. Risk (R) can be calculated or estimated by assessing the fraction of attempts that a technology fails to deliver intended results within predetermined specifications for both quality and efficiency. A minimum business value (BV_{min}) and minimum risk tolerance (R_{min}) for being a core technology should then be assigned based on a strategic business and financial perspective. If either the business value or risk associated with a technology is below the corresponding minimum, the technology should not be considered for core technology investment. When a technology meets the minimum risk and business value requirements, a threshold value (TV) can serve as the primary criteria for determining the core technology designation as follows:

Core technology: $(BV - BV_{min}) \times (R - R_{min}) > TV$

Noncore technology: $(BV - BV_{min}) \times (R - R_{min}) < TV$

The underlying principle of this approach is that the core technology investment ensures the value benefit of the technology to the core business and mitigates the risk of a severe failure in the application of the technology. This insurance and mitigation come from the direct investment and maintenance of expertise in all three components of the technology, whereas noncore technology takes more appropriate risks with lower investments. As technology evolves in importance and reliability, it can transition between the core and noncore regimes by regular assessment of the business value and risk associated with the technology.

Technology progression and the evolution of a concomitant investment approach can be illustrated graphically on a plot of business value versus risk. On such plots, core technologies fall into the upper right regions. Risk generally decreases as time progresses and experience with the technology increases. A life cycle thus moves from right to left on the value–risk plot, and two examples are shown in Figure 2.2. In both cases, the technologies start with a relatively high risk as emerging technologies and transition from noncore to core technology when the business value becomes sufficiently high. In case 1, the technology sustains business value long enough for the technology to become a low risk commoditized technology. The business value remains high as the risk with the technology decreases, and the reduced risk drives the transition from core to noncore technology in this case. Technologies that make sterilized vials serve as an example here. At one point, the pharmaceutical industry needed to invest internal resources to ensure vials for vaccines would be sterile, but they are now readily available as a commodity made from reliable, established technology managed by vendors. In contrast, case 2 illustrates a decreased business value driving the transition from core to noncore technology, perhaps due to the introduction of a better replacement. Obsolete open-top vacuum funnel

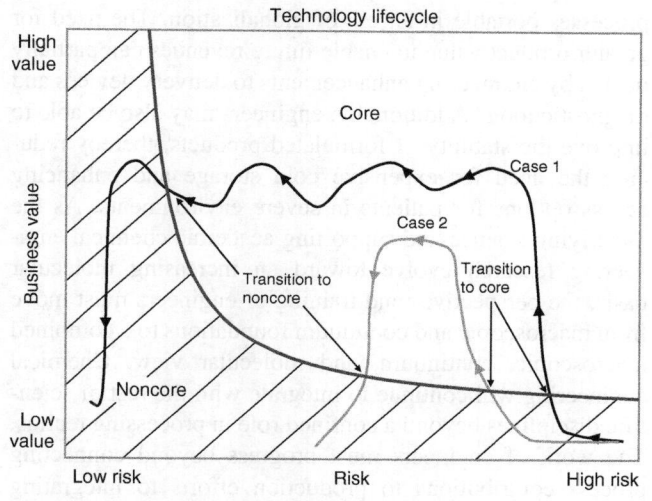

FIGURE 2.2 Examples of progression along a technology life cycle between core and noncore regimes.

filters that have been replaced by centrifuges and sealed filter dryers in manufacturing environments to improve industrial hygiene and efficiency provide an industrial example of case 2. Technologies relating to crystallization, spray drying, and roller compaction are representative of current chemical engineering core technology at several pharmaceutical companies.

Engineers have opportunities to substantially contribute to several facets of technology management as outlined above. Due to the complex nature of pharmaceutical processes, assigning a failure event to specific technology can be a challenging multivariant problem during risk assessment and risk management endeavors. Fundamental process understanding and technology expertise is vital to evaluating quantitative contributions to risk. Similar skills are useful for objectively determining whether value is enhanced through the use of internal capabilities versus external options. The understanding of a technology is important for determining the reliability of a prospective partner using that technology for critical business needs. Engineers also can contribute to investment choices among various emerging technologies with technical assessments of probabilities of success and potential applicability across a company's portfolio of products.

2.4 PROSPECTS FOR CHEMICAL ENGINEERS

Chemical engineers have made enabling contributions to health care, which serve as a strong foundation for future success. However, the pharmaceutical industry is profoundly changing, and the role of engineers must change with it. The industry challenges described in the second section of this chapter translate into the opportunities for chemical engineers in the third section. The use of modeling, standardized technology platforms, and a sound technology strategy will allow engineers to help reduce costs and lead times. Platform technologies also will assist with the challenges of making processes portable in an era of globalization. The need for greater product value to enable future revenues can partially be met by engineering enhancements to delivery devices and diagnostic tools. Additionally, engineers may also be able to improve the stability of formulated products, thereby reducing the need for expensive cold storage and enhancing access options for patients in severe environments. As the underlying science and supporting academic chemical engineering research evolve toward an increasing molecular basis, the perspectives and training of engineers must move from macroscopic and continuum foundations to a combined macroscopic, continuum, and molecular view. Chemical engineering will continue to integrate with the rest of scientific disciplines beyond a confined role in processing realms. The work of engineers must progress beyond connecting process contributions to production efforts to integrating the processes with the product itself. These efforts will be performed in the context of changing business models and pricing constraints on an increasingly global stage.

REFERENCES

1. Gilbert, J., Henske, P., and Singh, A. (2003). Rebuilding Big Pharma's business model. *In Vivo: The Business and Medicine Report* 21 (10): 73–80.
2. Mullin, R. (2014). Tufts study finds big rise in the cost of drug development. *Chemical and Engineering News* 92 (47): 6.
3. Steiner, M., Bugen, D., Kazanchy, B. et al. (2007). The continuing evolution of the pharmaceutical industry: career challenges and opportunities. Regent Atlantic Capital and Fiduciary Network, December 2007.
4. New Tech Media (2007). Medicare drug program hits 39 million, still open for advantage plans. *Senior Journal.com*, January 2007.
5. Deloitte Center for Health Solutions (2008). Opportunities for life science companies in a consumer-driven market. www.deloitte.com (accessed 2009).
6. Li, Y.-E., Yang, Y., Kathod, V., and Tyler, S. (2009). Optimization of solvent chasing in API manufacturing process: constant volume distillation. *Organic Process Research and Development* 13 (1): 73.
7. KoKitkar, P., Plocharczyk, E., and Chen, C.-C. (2008). Modeling drug molecule solubility to identify optimal solvent systems for crystallization. *Organic Process Research and Development* 12 (2): 249–256.
8. Kukura, J., Arratia, P., Szalai, E. et al. (2002). Understanding pharmaceutical flows. *Pharmaceutical Technology* 26 (1): 48–73.
9. de Treville, S., Schurhoff, N., Trigeorgis, L., and Avani, B. (2014). Optimal sourcing and lead-time reduction under evolutionary demand risk. *Production and Operations Management* 23: 2103–2117.
10. de Treville, S., Bicer, I., Chavez-Demoulin, V. et al. (2014). Valuing lead time. *Journal of Operations Management* 32: 337–346.
11. Cole, K., Groh, J.M., Johnson, M. et al. (2017). Kilogram scale prexasertib monolactate monohydrate synthesis under continuous-flow cGMP conditions. *Science* 356: 1144–1150.
12. Finley, W.H. (2001). *The Mechanics of Inhaled Pharmaceutical Aerosols: An Introduction*. New York: Academic Press.
13. Wise, D. (2000). *Handbook of Pharmaceutical Controlled Release Technology*. New York: CRC.
14. Tung, H.-H., Paul, E.L., Midler, M., and McCauley, J. (2009). *Crystallization of Organic Compounds: An Industrial Perspective*. Hoboken, NJ: Wiley.
15. Shimizu, T., Carvalho, M.M., and Laurindo, F.J.B. (eds.) Concepts and history of strategy in organizations. In: *Strategic Alignment Process and Decision Support Systems: Theory and Case Studies*, 2006. Hershey, PA: Idea Group Publishing.

PART II

MASS AND ENERGY BALANCES

PART II

MASS AND ENERGY BALANCES

3

PROCESS SAFETY AND REACTION HAZARD ASSESSMENT

Wim Dermaut
Chemical Process Development, Materials Technology Center, Agfa-Gevaert NV, Mortsel, Belgium

3.1 INTRODUCTION

When the issue of safety is raised in the context of pharmaceutical manufacturing, many might first think about issues of product and/or patient safety. There is another side of safety that might not get as much attention but that is also crucial to the production of pharmaceuticals: process safety. An often heard phrase in this context is "If the process can't be run safely, it shouldn't be run at all." Process safety should indeed be a concern, starting already in early development of a drug candidate. Running a small-scale synthesis in the lab only once is one thing, and running this process at metric ton scale on a routine basis in a chemical manufacturing plant is something completely different. Events such as exothermicity, gas generation, and stability of products might be relatively unimportant on a small scale, but they can pose tremendous challenges when this reaction is run at a larger scale.

This chapter provides an introduction to the field of process safety. The aim is to discuss some of the fundamentals of safety testing, in order to try to facilitate the communication between chemical engineers and development chemists. The focus will be on the interpretation and practical use of the different test results rather than on the tests itself. The discussion will focus mainly on (semi-) batch reactors, since this is still the most commonly used type of reactors. Some brief thoughts on the impact of flow chemistry on process safety will be added as well.

This chapter can be roughly divided in four main parts. We will start with a brief description of some general concepts like the runaway scenario and the criticality classes. After that, we will consider some safety aspects of the desired synthesis reaction and how it can be studied on the lab scale. The main focus will be on exothermicity (heat generation) and gas generation. We will then continue discussing how the data thus obtained can be used to scale-up the reaction safely, with a large emphasis on the heat transfer at a large scale. Finally we will take a closer look at the undesired decomposition reactions that can take place in the case of process deviations, how to study them at lab scale and how to minimize the associated risks.

In the next paragraphs, the reader is offered some first insights into the domain of process safety and safety testing. It is by no means the intention of the author to give an exhaustive overview of this field, but hopefully this introduction can provide some insight into the most common pitfalls of process safety. For a more in-depth review, the reader is referred to the widely available literature [1–3].

3.2 GENERAL CONCEPTS

3.2.1 Runaway Scenario

When discussing process safety, the cooling failure scenario is often used to illustrate the possibility of a runaway reaction in a reactor [4, 5]. In Figure 3.1, a possible cooling failure scenario is depicted. The normal process condition is indicated with the thin solid line: the reactants are being charged to the reactor (batch reaction), the reaction mixture is heated

---

Chemical Engineering in the Pharmaceutical Industry: Active Pharmaceutical Ingredients, Second Edition.
Edited by David J. am Ende and Mary T. am Ende.
© 2019 John Wiley & Sons, Inc. Published 2019 by John Wiley & Sons, Inc.

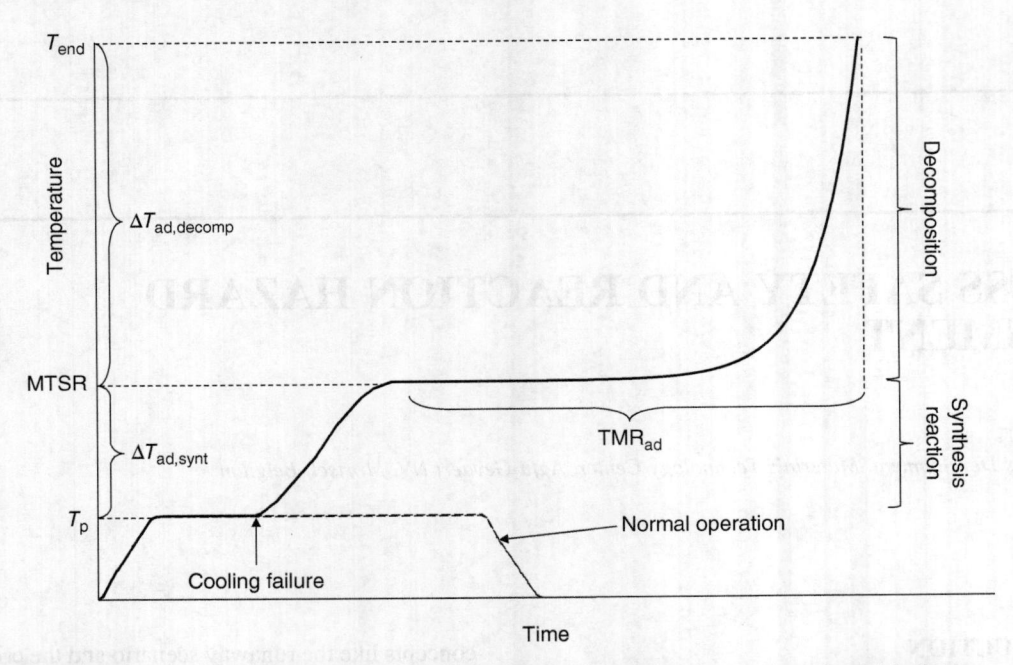

FIGURE 3.1 Cooling failure scenario. The thin line represents the normal mode of operation, the thick line represents the possible consequences of a cooling failure [3]. *Source*: Reproduced with permission from Stoessel [3]. Copyright 2008 Wiley-VCH Verlag GmbH & Co. KGaA.

to the desired process temperature (T_p), the mixture is then kept isothermally at this temperature with active jacket cooling (exothermic reaction), and when the reaction is finished, the mixture is brought back to room temperature for further workup. This is the process as it is intended to be run both at small scale in the lab and at larger scale in the plant.

Let us now consider the possible consequences of a loss of cooling. We will assume that the loss of cooling power occurs relatively short after the desired process temperature has been reached, as indicated in the graph. From this point on, the exothermic reaction will proceed, but since the reaction heat is no longer removed by the jacket, the temperature in the reaction mass will start to increase. There is no heat exchange between the reactor and the surroundings; the system is said to be adiabatic. After a certain time, the reaction has gone to completion, and hence a final temperature is reached, which is called the maximum temperature of the synthesis reaction (MTSR). The total temperature increase from the process temperature to the MTSR is called the adiabatic temperature rise of the synthesis reaction ($\Delta T_{ad,synt}$). Up to this point in the cooling failure scenario, we are only dealing with the desired synthesis reaction. The study of the desired reaction will therefore be discussed first in the following paragraph.

When the MTSR is reached, a secondary exothermic reaction may take place, i.e. a thermal decomposition of the reaction mixture or any of the ingredients. If such decomposition takes place, the temperature will increase further until the final temperature T_{end} has been reached. The time there is between reaching the MTSR, and the point of the maximum rate of the decomposition reaction (i.e. thermal explosion) is called the time to maximum rate (TMR) under adiabatic conditions (TMR$_{ad}$). It is generally accepted that a TMR$_{ad}$ of 24 hours or more can be considered as safe. The chance that a reactor would stay under adiabatic conditions for more than 24 hours is low. A cooling failure should be noticed quite rapidly, and this leaves ample time to take corrective measures like restoring the original cooling capacity, applying external emergency cooling, quenching the reaction mixture, or transferring it to another vessel or container with appropriate cooling. In analogy with the synthesis reaction, the total temperature increase from the MTSR to T_{end} is called the adiabatic temperature rise of the decomposition reaction ($\Delta T_{ad,decomp}$). How decomposition reactions are studied at lab scale and how they are dealt with during scale-up will be discussed in a later paragraph.

3.2.2 Criticality Classes

Starting from the cooling failure scenario, the criticality of any chemical process can be described in a relatively simple way by using the criticality classes as first introduced by Stoessel in 1993 [4]. In this method, four different temperatures need to be known in order to assess the possible consequences of a runaway reaction:

1. The process temperature under normal conditions (T_p).
2. The MTSR.

3. The temperature at which the TMR is 24 hours. In the description above, the TMR concept was introduced. Since the reaction rate is strongly dependent on the temperature, the TMR_{ad} will vary with temperature as well. The importance of a TMR that is longer than 24 hours was pointed out, and hence the third temperature we need to know is the temperature at which the TMR is 24 hours (we will denote this temperature as $TMR_{ad,24h}$).
4. The maximum temperature for technical reasons (MTT). In an open system, this is the boiling point of the reaction mixture, and in a closed system, it is the temperature that corresponds to the bursting pressure of the safety relief system. This is a temperature that cannot be surpassed under normal process conditions and can therefore act as a safety barrier. Only when dealing with very rapid temperature rise rates a risk of over pressurization or flooding of the condenser lines might occur. This will be discussed in Section 3.5.4.5.

When these four temperatures are known for a given process, the criticality class can be determined according to Figure 3.2. Five different criticality classes are defined, ranging from the intrinsically safe class 1 processes to the critical class 5 processes.

Let us consider a process that corresponds to the class 1 type. In this case, the process is run at the process temperature T_p, and when a cooling failure takes place, the temperature will increase to the MTSR. This temperature is below the $TMR_{ad,24h}$, meaning that even in case the reaction mixture would remain at this temperature (under adiabatic conditions) for 24 hours, there would be no serious consequences. Moreover, the MTT is situated between the MTSR and the $TMR_{ad,24h}$ giving an extra safety barrier for any possible further temperature increase. So even in case this process would run out of control due to a loss of cooling, there will be no real safety concerns.

The story is entirely different however when considering a class 5 process. In this case, a loss of cooling would raise the temperature inside the reactor to the MTSR, but here this temperature is higher than the $TMR_{ad,24h}$. This means that the secondary decomposition reaction will go to completion in less than 24 hours if the reaction mixture remains under adiabatic conditions for a prolonged period of time. The MTT is higher than $TMR_{ad,24h}$, so there is a possibility that it will not be sufficient to prevent a true thermal explosion. This type of reactions is truly critical from a safety point of view, and either a redesign of the process should be considered to bring it to a lower criticality class or appropriate safety measures should be taken.

The three other classes are intermediate cases and will not be described explicitly here, so the reader is referred to the original publication. The criticality index can be very useful to come to a unified risk assessment of a process. Some caution is needed however, as this classification does not take pressure increase into account. As will be discussed in a further paragraph, pressure effects are at least equally important as temperature effects in the assessment of process safety. This was addressed by the original author in a later publication [6], where a modified type of criticality index was proposed, indicating which gas generation data are needed for the different criticality classes in order to assure a safe scale-up. It is important to note that gas generation of both the desired (synthetic) and undesired (decomposition) reactions should be considered for each process, as is the case with heat generation.

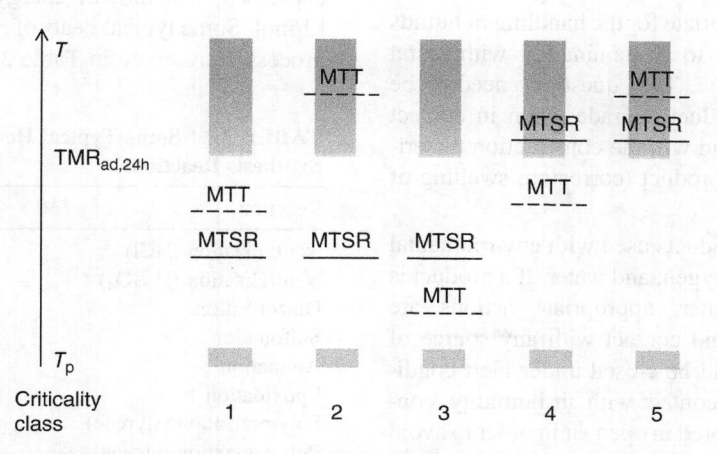

FIGURE 3.2 Criticality classes of a chemical process [3]. In this classification, processes are divided into five different criticality classes, ranging from class 1 (intrinsically safe) to class 5 (high risk). *Source*: Reproduced with permission from Stoessel [3]. Copyright 2008 Wiley-VCH Verlag GmbH & Co. KGaA.

3.3 STUDYING THE DESIRED SYNTHESIS REACTION AT LAB SCALE

3.3.1 Compatibility

Before starting with any further safety assessment of a chemical process, it is crucial to evaluate the compatibility of all reagents being used. Ideally, the reagents should show no reactivity other than that leading to the desired reaction. Some of the incompatibilities are very obvious: developing a chlorination reaction with thionyl chloride in an aqueous solution simply does not make sense. Some other incompatibilities might be less known but can also have very serious consequences. The stability of hydroxylamine, for instance, is catastrophically influenced by the presence of several metal ions [7, 8]; even in the parts-per-million range, this type of contamination can have severe consequences. A first starting point for any compatibility assessment should be *Bretherick's Handbook of Reactive Chemical Hazards* [9], a standard reference with a vast list of known stability and compatibility data on a wide range of chemicals.

Compatibility issues for several different conditions should be checked either from literature, or where the information is not available the data should be generated experimentally:

1. Compatibility of all reagents used in combination with the other reagents present.
2. Compatibility of the reagents with possible main contaminants in other reagents. Technical dichloromethane, for instance, is often stabilized with 0.1–0.3% of ethanol, which can turn out to be significant because of the large molar excess of the solvent in the reaction mixture.
3. Compatibility of the reagents with construction materials such as stainless steel (vessel wall), sealings (Kalrez, Teflon, etc.). E.g. the use of disposable Teflon dip tubes may be appropriate for the handling of liquids that are very sensitive to contamination with metal ions like hydroxylamine. Two questions need to be answered: will the product degrade when in contact with these materials, and will the construction materials be affected by the product (corrosion, swelling of gaskets, or sealings)?
4. Compatibility of all products used with environmental factors such as light, oxygen, and water. If a product is incompatible with water, appropriate actions are needed in order to avoid contact with any source of water: containers should be closed under inert conditions in order to avoid contact with air humidity, containers should not be stored in open air in order to avoid water ingression due to rain, reactions should be run in a reactor where the heat transfer media (such as jacket cooling and condenser cooling) are water free, etc.

A first indication of possible compatibility issues with oxygen can be obtained from two differential scanning calorimetry (DSC) experiments in an open crucible, once under nitrogen atmosphere and once under air. If there is a pronounced difference between the outcomes of both experiments, the product is very likely to show some degree of reactivity with oxygen.

3.3.2 Exothermicity

Most chemical processes run in pharmaceutical production plants are exothermic reactions. In general terms, a reaction is called exothermic when heat is being generated during the course of the reaction. Reactions that absorb heat during their course are called endothermic reactions. Chemical processes in pharmaceutical production are in most cases designed as isothermal processes, so the heat that is being generated during the course of reaction has to be removed effectively, usually through jacket cooling of the reactor. Intuitively one can understand that an effective heat removal will become increasingly difficult when the scale of the process is increased from milliliter (lab) scale to cubic meter (production) scale. Therefore, a correct assessment of the reaction heat becomes crucial when a process is being run at a larger scale.

From a thermodynamic point of view, the heat being released (or absorbed) by a reaction matches the difference in heat of formation between the reactants and the products. Hence, a first indication of the heat of reaction of any process can be obtained by making this calculation based on tabulated literature data [10]. By convention, reaction enthalpies for exothermic reactions are negative values; for endothermic reactions, they are positive values.

The heat of reaction of a chemical process is usually expressed in a unit of energy per mole, e.g. kcal/mol or kJ/mol. Some typical heats of reaction for common chemical processes are given in Table 3.1 [3].

TABLE 3.1 Some Typical Heat of Reactions for Common Synthesis Reactions

Reaction	ΔH_R (KJ/mol)
Neutralization (HCl)	−55
Neutralization (H_2SO_4)	−105
Diazotization	−65
Sulfonation	−150
Amination	−120
Epoxidation	−100
Polymerization (styrene)	−60
Polymerization (alkene)	−200
Hydrogenation (nitro)	−560
Nitration	−130

This table clearly shows that there is a big span in heats of reaction one can encounter in process chemistry, with the highest energies (and hence highest risks) being related to the usual suspects such as hydrogenations of nitro compounds and polymerizations. When developing this type of reactions, extra care should be taken, and a correct determination of the total reaction heat and the kinetics of the process by means of calorimetry is crucial.

The reaction heat of a chemical reaction can be determined by means of a reaction calorimeter. This is basically a small-scale reactor in which the reaction can be performed under controlled circumstances while recording any heat entering or leaving the system. Most used is the heat flow calorimeter, where the reaction heat is measured by continuously monitoring the temperature difference between the reaction mixture and the cooling/heating fluid in the jacket:

$$Q_{flow} = U \times A \times (T_R - T_J) \quad (3.1)$$

where

Q_{flow}: heat flowing in or out the reaction mixture (W)
U: heat transfer coefficient (W/m^{2}K)
A: heat exchange area (m^2)
T_R: reaction temperature (K)
T_J: jacket temperature (K)

There are different heat flow calorimeters available on the market such as the RC1 (Mettler Toledo), the Calo (Systag), and the Simular (combined with power compensation calorimetry, HEL). Other systems offer reaction calorimetry based on a more direct measurement of the heat flux, e.g. the Chemisens CPA (Peltier based). In our discussion, we will limit ourselves to heat flow calorimetry, since it is the most widespread technique to date, but the interpretation of the data obtained with other types of calorimeters will be very comparable.

In principle, a heat flow calorimeter can be considered as a scaled-down jacketed reactor (usually in the range from 100 ml to 2 L), with a very accurate temperature control. Usually such a calorimeter is run in isothermal mode, so the temperature of the reaction mixture is kept constant during the course of the reaction. If the reaction is exothermic, the jacket temperature will have to be lower than the reaction temperature in order to remove the reaction heat. As can be seen from Eq. (3.1), measuring T_J and T_R is not enough to obtain the reaction heat entering or leaving the reactor; we also need to know U and A. The heat transfer area A is usually easy to obtain: since the reactor geometry of the calorimeter is fixed, the heat exchange area as a function of the volume of the reaction mixture is known. The heat transfer coefficient U is most commonly obtained by the use of a calibration heater. During a certain period of time (typically 5 or 10 minutes), a calibration heater with a known heat output is switched on. The temperature of the jacket will be adjusted in such a way that the temperature in the reaction mixture remains unchanged. Since Q, A, T_R, and T_J from Eq. (3.1) are all known for this calibration period, U can be calculated. U is a function of a variety of factors such as viscosity, stirring speed, and temperature, as will be discussed further on in greater detail. This means that U will be different in each calorimetry experiment, and it will even be different before and after the reaction, according to the physical properties of the reaction mixture. Therefore, the calibration is performed once before the reaction takes place and once after the reaction is finished, yielding the appropriate U-values for the reaction mixture before and after the reaction.

An example of a semi-batch calorimetry experiment is shown in Figure 3.3. The reactor is filled with the appropriate reagents and brought to the reaction temperature. After the temperature of both the reactor (T_R) and the jacket (T_J) has reached stable values, the calibration procedure as described above is executed (not shown in the graph). After this calibration, the reaction is started by a gradual dosing of the desired reagent, as can be read from the mass signal. The response in heat profile is almost instantaneous, and a gradual increase in power can be observed until all the reagents have been dosed. At the end of the dosing, the heat signal does not drop to the baseline immediately; this phenomenon where heat is being released after the addition of the reagent has been stopped is called thermal accumulation. The thermal accumulation at the end of the dosing can be calculated according to Eq. (3.2).

$$\%\text{Thermal accumulation} = 100 \times \frac{\int_A^B q(t)dt}{\int_0^B q(t)dt} \quad (3.2)$$

with $t = 0$ representing the time at which dosing starts, $t = A$ the time at which dosing ends, and $t = B$ the time at which all exothermicity has faded away (Figure 3.3). If the heat signal dropped to zero immediately after the dosing had stopped, there would be no thermal accumulation. On the other hand, if the dosing was instant (which is the case in a batch reaction), there would be 100% thermal accumulation.

Let us now take a closer look at the key figures that can be extracted from a calorimetry experiment and how they should be interpreted.

3.3.2.1 Reaction Heat–Adiabatic Temperature Rise–MTSR
The integration of the heat signal versus time gives us the total reaction heat, usually expressed in kJ or kcal. From this reaction heat, the adiabatic temperature rise of the synthesis reaction can be calculated according to Eq. (3.3):

$$\Delta T_{ad} = \frac{-\Delta H_R}{c_p} \quad (3.3)$$

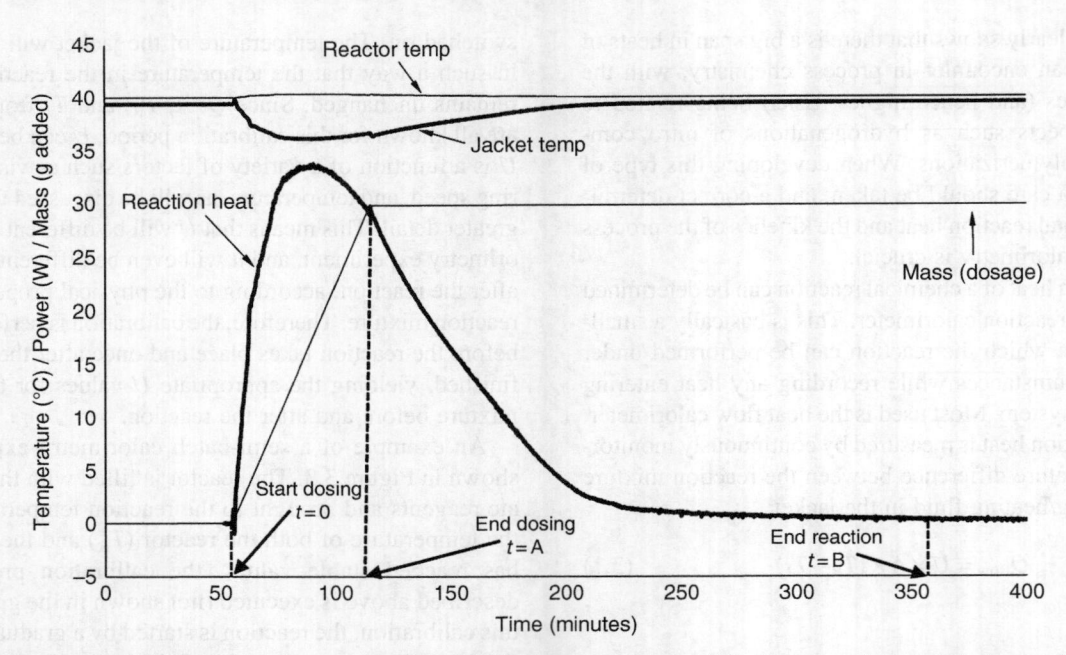

FIGURE 3.3 Example of a reaction calorimetry experiment, indicating time 0 (start of dosing), time A (end of dosing) end time B (end of reaction) as they are used in the definition of thermal accumulation (Eq. 3.2).

where

ΔH_R: reaction enthalpy (kJ/kg)

c_p: specific heat capacity of the reaction mixture (kJ/kgK)

The MTSR can be calculated merely by adding the adiabatic temperature rise to the reaction temperature (Eq. 3.4):

$$\text{MTSR} = T_{\text{Process}} + \Delta T_{\text{ad}} \quad (3.4)$$

Whereas the molar reaction enthalpy is an intrinsic property of a specific reaction, the adiabatic temperature rise is dependent on the reaction conditions. In the (hypothetical) example in Table 3.2, this difference is demonstrated.

This example clearly shows the importance of reaction conditions, with the adiabatic temperature rise being more than 10 times higher in case 1. This dramatic difference can be fully attributed to the effect of the solvent acting as a heat sink. Working at higher dilution in a solvent with a higher heat capacity can drastically reduce the possible consequences of reaction that runs out of control. Unfortunately, running a process at higher dilution has an impact on the overall economy, so both aspects should be considered.

The adiabatic temperature rise is often used as a measure for the severity of a runaway reaction. A process with an adiabatic temperature rise of less than 50 K is usually considered to pose no serious safety concerns, at least when there is no pressure increase associated with the reaction. When a process has an adiabatic temperature rise of more than 200 K, a runaway reaction would most probably result in a true thermal explosion, and hence such processes require a very thorough safety study.

TABLE 3.2 Theoretical Example of the Resulting Adiabatic Temperature Rise for the Hydrogenation of Nitrobenzene Under Different Reaction Conditions

	Case 1	Case 2
Reaction heat	−560 kJ/mol	−560 kJ/mol
Concentration	2 M	0.5 M
Solvent	Chlorobenzene	Water
Density solvent	1.11 kg/l	1 kg/l
Specific heat solvent	1.3 kJ/kg/K	4.2 kJ/kg/K
ΔT_{ad} (Eq. 3.3)	776 °C	67 °C

3.3.2.2 Thermal Accumulation According to Eq. (3.2), the thermal accumulation can be calculated by the partial integration of the heat signal. Thermal accumulation is an important parameter in the assessment of the safety of a process. If a problem occurs during a process (cooling failure, stirrer failure, etc.), it is common practice to stop the addition of chemicals immediately. In case there is no thermal accumulation, the reaction will also stop immediately, and there will be no further heat generation that can lead to a temperature increase in the reactor. A reaction with 0% thermal accumulation is therefore called dosing controlled. If there is thermal accumulation however, part of the reaction heat will still be set free after the dosing has been stopped, and hence the temperature in the reactor can increase.

TABLE 3.3 Example of the Effect of the Thermal Accumulation on the MTSR$_{semi\text{-}batch}$

	Case 1	Case 2
Reaction temperature (°C)	60	60
Total reaction enthalpy (kJ/kg)	−200	−200
Thermal accumulation (%)	2	60
Specific heat (J/gK)	2	2
$\Delta T_{ad,batch}$ (°C)	100	100
MTSR$_{batch}$ (°C)	160	160
$\Delta T_{ad,semi\text{-}batch}$ (°C)	2	60
MTSR$_{semi\text{-}batch}$ (°C)	62	120

Because of the importance of the thermal accumulation, the MTSR is often specified as either being MTSR$_{batch}$ or MTSR$_{semi\text{-}batch}$. For the calculation of the former, the total adiabatic temperature rise is added to the reaction temperature, whereas, for the latter, the adiabatic temperature rise is first multiplied by the percentage of thermal accumulation. An example is given in Table 3.3.

This example shows the big difference in intrinsic safety of the process between the two cases. Should a cooling failure occur in case 1, the temperature would never be able to rise significantly above the process temperature, provided of course that the dosing is stopped as soon as the failure occurs. In case 2, on the other hand, the temperature would increase to 120 °C without the possibility to cool, even when dosing is stopped immediately.

So, obviously, low thermal accumulation is to be preferred for any semi-batch process. A high degree of thermal accumulation is a sign that the reaction rate is low relatively to the dosing rate. Two possible measures can be taken to decrease the thermal accumulation of a given process:

1. Increase the reaction rate. This can be done by increasing the reaction temperature. Increasing the reaction rate means also increasing the heat rate of the reaction, so a calorimetry experiment at this new (higher) process temperature is required to make sure that the cooling of the reactor can cope with the heat generation under normal process conditions. Obviously, the increased temperature will lead to a smaller safety margin between reaction temperature and possible decomposition temperature, and this should be dealt with appropriately.
2. Decrease the dosing rate.

3.3.2.3 Heat Rate Whereas a correct determination of the total reaction heat is important, the rate at which this heat is being liberated is at least equally important for a proper safety study. Where a process is run under identical conditions both at large scale and in the calorimeter, the heat rate (in W/kg) is scale independent. It should be kept in mind, however, that the cooling capacity of a reaction calorimeter is in most cases several orders of magnitude higher than that of a large-scale production vessel. A reaction calorimeter might still be able to keep a constant heat rate of 200 W/kg under control, running this process at production scale will most certainly lead to a runaway reaction. In such a case, the process should be redesigned as to decrease the heat rate, and ideally the calorimetry experiment should be repeated under the new process conditions to make sure that no unwanted side effects occur (higher thermal accumulation, sudden crystallization, formation of extra impurities, etc.).

Not only the absolute value of the heat rate has to be considered, the duration of the heat evolution is also important. A peak in the heat evolution that surpasses the available cooling capacity but which only lasts for a short period of time is not necessarily problematic. If such a peak is observed, one should calculate the corresponding adiabatic temperature rise and evaluate its consequences. E.g. a heat rate of 200 W/kg for two minutes would give rise to a temperature increase of 12 °C under adiabatic conditions, assuming a specific heat of 2 J/gK. If a cooling capacity of 50 W/kg is available, this will only be 9 °C. The issues one might encounter when scaling up a reaction to meet the heat removal capacities of the production vessel will be discussed in more detail.

3.3.3 Gas Evolution

Up until now, we have only focused on the heat being generated by exothermic chemical processes. From a safety perspective, gas evolution and a resulting pressure buildup can have even more devastating consequences, so proper knowledge of any gaseous products being formed during a process is crucial to assuring a safe execution at production scale.

There are quite a few common reactions that do liberate considerable amounts of gas: chlorinations with thionyl chloride, BOC deprotections, quenching of excess hydride, and decarboxylations, to name but a few. The most appropriate way to quantify the gas evolution during a reaction is to couple any type of gas flow measurement device to a reaction calorimeter and run the process under the same conditions as it will be run on scale.

There are several possibilities for the measurement of gas evolution at small scale:

1. Thermal mass flow meters. This type of devices is probably the most widespread when a flow of gaseous products has to be measured. They are available in a large span of measuring ranges (from <1 ml/min to several thousand liters per minute), are relatively cheap, and deliver a signal that can be picked up easily as an input in the reaction calorimeter. However, this type of meters measures a mass flow (i.e. grams of gas per minute) and not a volumetric flow (milliliter per minute).

When dealing with one known single type of gas, this is no problem since the volumetric flow can be easily calculated from the mass flow signal. When the gas to be measured is a mixture of different components, or when the composition of the gas stream is entirely unknown, the volumetric flow cannot be obtained reliably.

2. Wet drum type flow meters. The gas is led through a drum that is half submerged in inert oil, causing this drum to rotate. This rotation is recorded and is proportional to the volumetric flow (as opposed to the thermal mass flow meters). When using a unit that is entirely made of an inert material (e.g. Teflon), a very broad range of gaseous products can be studied. However, the dynamic range of this type of instruments is only modest, accuracy at the low end of the flow ranges (0–20 ml/min) is rather limited, and the fact that the drum rotates in a chamber filled with inert oil makes it susceptible to mechanical wear. Although the oil in the gas meter is basically inert, most gasses are somewhat soluble in it, which can lead to a (small) underestimation of the gas flow in some cases.

3. Gas burette. This type of device measures the pressure increase in a burette that is filled with inert oil, releasing the overpressure at a predefined value, making an accurate determination of low gas flow rates possible. The signal is proportional to the volumetric flow, the setup is extremely simple without any moving parts, and it is fully corrosion resistant (only glass and silicon oil in contact with the gas). However, the output signal is difficult to integrate in any evaluation software (combination of pressure signal and count of the number of "trips"), and the flow range that can be measured is limited at the high end to approximately 50 ml/min (using the standard type of burette).

4. Rotameters, bubble flow meters, etc. There are different other types of laboratory gas flow meters that will not be discussed here since they give only a visual readout and not a signal that can be incorporated electronically.

5. Some labs have developed in-house solutions for the measurement of small gas streams. An example was published by Weisenburger et al. [11], where the gas stream from a small-scale experiment is collected inside a gastight bag that is suspended in a closed bottle filled with ambient air. When gas fills the bag, air is pushed out of the bottle and passes through a thermal mass flow meter. This way, only air is passing through the meter, making proper calibration possible while also ensuring a better protection of the instrument when corrosive gasses are being measured. This system is only suitable for the measurement of small gas streams (limited to the size of the bottle and bag), and a fair degree of noise on the signal should be tolerated.

In the explanation above, it has been emphasized that the determination of a volumetric gas flow is of interest rather than a mass flow. When scaling up the reaction to plant scale, we need to make sure that all gas that is being produced can be removed safely from the vessel to the exhaust. This means that all of this gas will have to flow through piping with a certain diameter, and the limiting factor for a gas flowing through a pipe without causing pressure buildup is its volume and not its mass. The maximum allowable gas rate for a specific process depends on the actual production plant layout, and this will be dealt with in the next paragraph.

Apart from the obvious importance of measuring the gas flow rate during a process, it might also be of interest to characterize the gas that is being emitted. Although there is no difference in possible pressure buildup, having a release of 50 m^3/h of carbon dioxide will obviously feel more comfortable for any chemist or operator than having a release of 50 m^3/h of hydrogen cyanide. When gas evolution comes into play, industrial hygiene, environmental emission limitations, and hazard classification (e.g. when hydrogen is being set free) should all be addressed appropriately.

Characterization of the gas being liberated during a process at lab scale is not an easy thing to do. Ideally, an online mass spectrometer can be used to quantify the exact composition of the gas stream at any time. Mass spectrometers with an appropriate measuring range (down to 28 Da when carbon monoxide is to be detected), low dead volume to eliminate unnecessary long holdup times, and high resolution (both nitrogen and carbon monoxide have a molecular weight of 28; very high resolving power is needed to discriminate between them) do not come cheaply. Collecting the gas leaving the reactor in a gas sampling bag, and subsequently injecting this gas into a regular mass spectrometer can be a viable alternative. Another widely used technique is to trap the gas in a wash bottle with an appropriate solvent in which the gas either dissolves or with which it reacts and then to analyze this solution in a traditional way. Which technique is being used is irrelevant, but one should always try to know the composition of the gas stream leaving the reaction mixture.

3.4 SCALE-UP OF THE DESIRED REACTION

3.4.1 Heat Removal

3.4.1.1 Film Theory When designing an exothermic reaction for scale-up, it is important to know what the heat removal capacity of the reactor at production scale is. Unfortunately this is easier said than done. The heat transfer between the heat transfer medium in the jacket and the reaction mixture is usually described in terms of a series of resistances, the so-called film theory. It considers three main factors governing the heat transfer in a stirred tank reactor:

the resistance of the inner film (boundary reaction mixture–vessel wall), the resistance of the vessel wall and the resistance of the outer film (boundary vessel wall–heat transfer fluid). This can be expressed numerically:

$$\frac{1}{U} = \frac{1}{h_r} + \frac{d}{\lambda} + \frac{1}{h_c} = \frac{1}{h_r} + \frac{1}{U_{max}} \qquad (3.5)$$

where

U: overall heat transfer coefficient (W/m²K)
h_r: inner film transfer coefficient (W/m²K)
d: thickness of the vessel wall (m)
λ: thermal conductivity of the vessel wall (W/mK)
h_c: inner film transfer coefficient (W/m²K)
U_{max}: maximum heat transfer coefficient (W/m²K)

From this equation, it can be understood that there are two main contributions to the overall heat transfer coefficient: one that is solely dependent on the characteristics of the reaction mixture and one that is solely dependent on the characteristics of the reactor. Indeed, the inner film transfer coefficient h_r is a measure for the resistance to heat transfer between the reaction mixture and the vessel wall and is strongly correlated to the physicochemical properties of the reaction mixture (viscosity, density, heat capacity, etc.) and the stirring speed. U_{max} on the other hand can be interpreted as the maximum obtainable heat transfer coefficient in a certain reactor in the hypothetical case when the inner film resistance would approach zero. This term comprises of two contributing parts: one that is due to the thermal conductivity of the vessel wall and one that is due to the outer film coefficient. These two terms are solely dependent on the characteristics of the reactor.

This explains why a correct scale-up of the heat transfer characteristics is so difficult: since U is dependent on both the process and the reactor, it ideally should be determined or calculated separately for each vessel–process combination. This is a rather time-consuming process, as can be seen from the list actions that should be undertaken:

1. Determine the U_{max} of the calorimeter at the process temperature.
2. Determine U for the reaction mixture under process conditions in the calorimeter.
3. From 1 and 2, calculate h_r for the reaction mass in the calorimeter.
4. Calculate h_r for the reaction mass in the production vessel (literature scale-up rules).
5. Determine the U_{max} of the production vessel at the intended jacket temperature.
6. From 4 and 5, calculate the according U-value for the reaction mixture in the production vessel.

For a thorough description of the theory behind this approach, the reader should refer to the literature [12, 13]. In the following part, we will briefly discuss some major issues related to the determination of the heat transfer coefficient at production scale.

3.4.1.2 Determination of **U** The most widely used approach to determine the heat transfer characteristics of a reactor is by means of the Wilson plot [14]. The heat transfer coefficient is determined experimentally (by means of a cooling curve) at several different stirring speeds. When plotting the reciprocal heat transfer coefficient at a certain jacket temperature versus the stirring speed to the power −2/3 (Eq. 3.6), a straight line is obtained, with the intercept being equal to U_{max}^{-1}.

$$\frac{1}{U} = c^{te} \times n^{-2/3} \qquad (3.6)$$

where

n = stirrer speed (in rpm)
c^{te} = constant.

An example of such a Wilson plot is shown in Figure 3.4.

This is intrinsically the most reliable way to determine the U_{max} for any reactor, since the result is independent of the solvent being used for the cooling experiment. When repeating the same experiments with a different solvent, the experimental points will differ, and the slope of the line will differ, but U_{max} will remain the same.

This method can be used for the characterization of a reaction calorimeter where a large number of either cooling curves or U determinations via the calibration heater can be run in an automated way. Since U_{max} is dependent on the filling degree and jacket temperature, a vast range of U determinations are necessary for a proper description of the heat transfer properties of the reaction calorimeter. It is our experience that isopropanol is the most suitable solvent for obtaining good Wilson plots in the reaction calorimeter.

This method can be applied to the reaction calorimeter in programmed mode, and it becomes rather cumbersome however for use at plant scale. Since each measuring point in the curve has to be obtained from 1 cooling curve at one stirring speed, it becomes very time-consuming to gather the data needed for this plot. Therefore, another approach can be used for the estimation of U_{max} at production scale. As mentioned above, constructing the Wilson plot with different solvents will alter the slope but not the intercept. If water is used for the construction of the Wilson plot, the slope turns out to be very low, i.e. the contribution of $1/h_r$ is low. This implies that determining the U-value with a reactor filled with water at the highest possible stirring speed will yield a value that is close to U_{max}. This way, a good

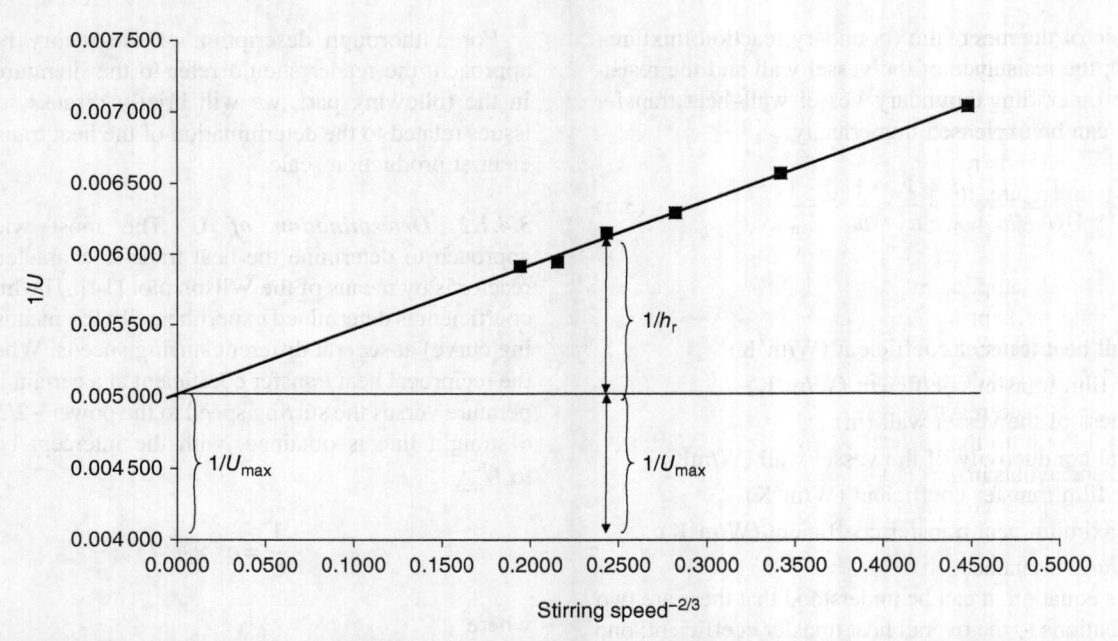

FIGURE 3.4 Example of a Wilson plot for the determination of the heat transfer characteristics of a reactor. The markers are experimentally obtained heat transfer coefficients at different stirring speeds. Through these points a straight line can be fitted that yields the contribution of both the maximum heat transfer coefficient and the inner film coefficient to the total heat transfer coefficient.

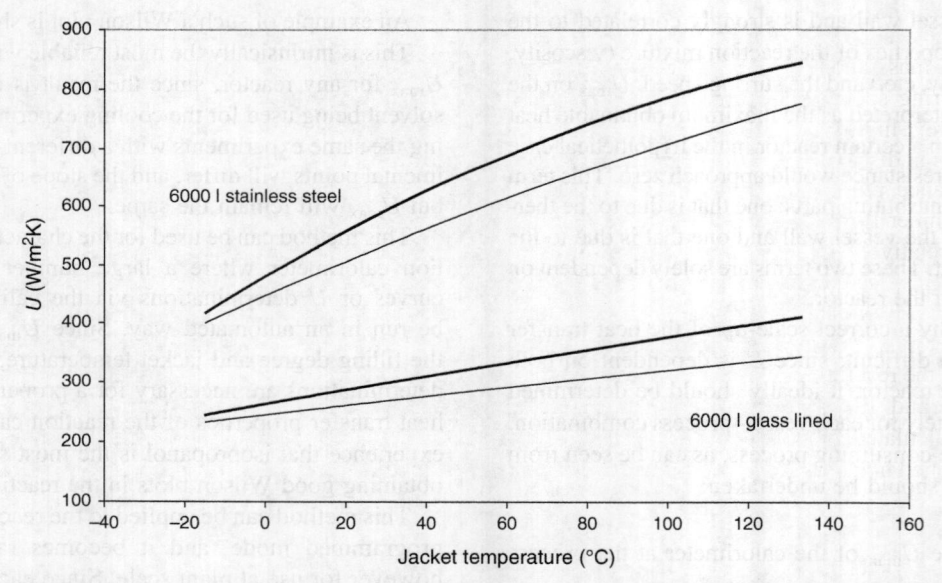

FIGURE 3.5 Heat transfer coefficients for a 6000 L stainless steel and a 6000 L glass lined reactor. The reactor was filled with water for 50% of the nominal volume (thick line) or 85% (thin line), heated to the boiling point and then cooled to room temperature with a constant temperature difference between jacket and reactor of 20 °C at a high stirring speed. The heat transfer coefficient was determined as a function of the jacket temperature, and a curve was fitted through these data points to allow for extrapolation to other temperatures.

approximation of the maximum heat transfer capacity of the vessel can be obtained from only one experiment. This approximation will obviously be less accurate, but the error made is always on the safe side: the U_{max} will be underestimated, and hence in reality there will be more cooling power available than anticipated. An example of this approach is being shown in Figure 3.5, where the heat transfer coefficients for a typical stainless steel reactor and a glass-lined reactor of 6000 L are shown as a function of jacket temperature for a filling degree of 50 and 85%. This shows clearly the large range of U-values that can be encountered in practice.

It is important to note that the U_{max} value is a function of the jacket temperature rather than the reactor temperature because it is linked to the properties of the jacket and vessel wall, which is always at approximately the same temperature as the jacket. This implies that cooling curves should be determined with a constant temperature difference between the jacket and reactor. If a cooling curve is recorded with the jacket constantly at its lowest temperature, the temperature dependence of the U_{max} is lost. Moreover, when calculating the final overall heat transfer coefficient of the reactor from its respective U_{max} value, the intended temperature offset between the reactor and jacket should be kept in mind. This is less of an issue in the reaction calorimeter, since the observed temperature differences between the jacket and reactor are usually a lot smaller than in a large-scale reactor. Obviously, the values from Figure 3.5 apply only to the specific reactors in this specific plant, as different layouts in the cooling system, different materials of construction, different heat transfer media, and different temperature control strategies can all have a large influence on the heat transfer characteristics of a vessel.

3.4.1.3 Influence of the Inner Film Coefficient
Now that the U_{max} has been determined at both lab scale and production scale, let us turn our attention to the other term in Eq. (3.5), i.e. the inner film coefficient h_r. This film coefficient is a measure for the resistance toward heat transfer between the reaction mass and the vessel wall. It is mainly governed by the stirring speed and the physical properties of the reaction mixture: a highly viscous reaction mixture will have more difficulties in dissipating reaction heat to the reactor wall than, for instance, pure water. Unfortunately enough, the influence of the inner film coefficient can be quite pronounced: if it were relatively small in comparison with the U_{max} term, it could be neglected, and one single heat transfer coefficient could be used for each vessel, irrespective of the reaction mixture.

When the U_{max} of the reaction calorimeter is known at the (jacket) temperature and fill degree used, the inner film coefficient can be calculated from the overall heat transfer coefficient as determined in the calibration procedure according to Eq. (3.5).

As the inner film coefficient is dependent on the mixing characteristics of the vessel used, it is scale dependent and should be scaled up accordingly. This is usually done according to Eq. (3.7):

$$h_{r_{(prod)}} = h_{r_{(lab)}} \times \left(\frac{D}{d}\right)^{1/3} \times \left(\frac{N}{n}\right)^{2/3} \times V_{i(prod)}^{0.14} \quad (3.7)$$

where

$h_{r_{(prod)}}$: the inner film coefficient at large scale

$h_{r_{(lab)}}$: the inner film coefficient at calorimeter scale

D: vessel diameter at large scale

d: vessel diameter at calorimeter scale

N: stirring speed at large scale

n: stirring speed at calorimeter scale

$V_{i_{(prod)}}$ = viscosity number at large scale

The viscosity number is the ratio of the viscosity of the reaction mixture at the reaction temperature and its viscosity at the jacket temperature. When considering standard organic reactions in solution, this ratio is quite close to unity, so this factor is usually neglected. When studying polymerization reactions, however, this effect should be taken into account.

Using this equation, the inner film coefficient at production scale can be calculated, and hence the overall heat transfer coefficient is now known, according to Eq. (3.5). In order to get a more quantitative feeling for the influence of the different parameters on the overall heat transfer coefficient, let us take a look at a realistic numerical example.

EXAMPLE PROBLEM 3.1

A reaction is run in the reaction calorimeter at a temperature of 60 °C. The reaction mixture is homogeneous, and the solvent is methanol. The details for both the reaction in the calorimeter and in the production vessel (6000 L stainless steel) are given in Table 3.4.

TABLE 3.4 Example Problem 3.1 to Show the Influence of the Inner Film Coefficient on the Overall Heat Transfer Coefficient

	Calorimeter	Production Vessel	Calorimeter	Production Vessel
Reactor temperature (°C)	60	60	60	60
Jacket temperature (°C)	±60	20	±60	20
Filling degree (%)	50	50	50	50
U_{max} at jacket temp (W/m²K)	215	500	215	500
Diameter reactor (m)	0.12	2	0.12	2
Stirring speed (rpm)	450	100	450	100
U experimental (W/m²K)	**180**		**120**	
U calculated (W/m²K)		**337**		**169**

The left part represents the case of a homogeneous nonviscous reaction mixture, the right part that of strongly heterogeneous reaction mixture. Bold values are the overall heat transfer coefficients for two different situations at lab and plant scale respectively.

The maximum heat transfer coefficient of the production vessel was estimated by running a cooling curve as follows: the reactor was filled with water up to 50% of its nominal volume, and the content was then heated to the boiling point and kept at that temperature for a while until both the reactor and jacket temperatures have reached stable values. The reactor is then cooled to room temperature with a constant temperature offset between the reactor and the jacket. During the entire cooling cycle, rapid stirring is applied. The reactor temperature and jacket temperature are recorded and put in a graph (temperature versus time). From the first derivative of this curve, the appropriate heat transfer coefficient is calculated. In this example, this yielded a value of 500 W/m² K at the intended jacket temperature (20 °C). As described above, the U-value for a reactor filled with water at high stirring speed is considered to be a good approximation of U_{max}.

The data from Table 3.4 are then inserted into Eqs. (3.5) and (3.7).

First, h_r for the reaction calorimeter is calculated (Eq. 3.5):

$$\frac{1}{h_{r_{lab}}} = \frac{1}{U_{lab}} - \frac{1}{U_{max_{lab}}} = \frac{1}{180} - \frac{1}{215} \approx \frac{1}{1106} \frac{m^2 K}{W}$$

From Eq. (3.7), we can now calculate h_r for the production vessel. We will assume that the viscosity of the reaction mixture at the reaction temperature is essentially equal to the viscosity of the reaction mixture at the jacket temperature.

$$h_{r_{prod}} = h_{r_{lab}} \times \left(\frac{D}{d}\right)^{1/3} \times \left(\frac{N}{n}\right)^{2/3} = 1106 \frac{W}{m^2 K} \times \left(\frac{2 m}{0.12 m}\right)^{1/3}$$

$$\times \left(\frac{100 \, rpm}{450 \, rpm}\right)^{2/3} = 1036 \frac{W}{m^2 K}$$

Inserting this again in Eq. (3.5) yields the final U-value for the production vessel.

$$\frac{1}{U_{prod}} = \frac{1}{h_{r_{prod}}} + \frac{1}{U_{max_{prod}}} = \frac{1}{1036} + \frac{1}{500} \approx 337 \frac{W}{m^2 K}$$

It is instructive to consider exactly the same reaction conditions, but this time with a strongly heterogeneous reaction mixture, where the experimentally obtained U-value in the calorimeter is only 120 W/m² K. If all other parameters are kept constant, a final U-value at production scale of 169 W/m² K can be calculated. This clearly shows the importance of the inner film coefficient on the finally observed heat transfer coefficient: going from the maximum value of 500 W/m² K for pure water at the maximum stirring speed (U_{max}) to as low as 169 W/m² K for a heterogeneous reaction mixture at moderate stirring speed.

3.4.1.4 Shortcuts to U-Value Determinations

When the procedure for the determination of correct heat transfer data as described above is out of reach, there are other possibilities to make a rough estimation of the U-values. If one chooses to go for these simplified estimation methods, care is needed to include a wide enough safety window when a batch is run for the first time in a certain reaction vessel.

A first possible estimation method is to simply use the heat transfer coefficient for a reactor filled with the neat solvent in which the reaction is to be performed. Cooling curves for some solvents are often readily available from cleaning campaigns. Since these cleaning cycles are usually repeated quite regularly, an indication of the evolution of the heat transfer characteristics of the reactor over time can also be obtained. It excludes the need for the separate determination of U_{max} with water, which is not often used as a cleaning solvent in a temperature cycle, and of the entire characterization of the U_{max} behavior of the reaction calorimeter. So if a reaction is to be run in a methanol solution at 30 °C, one could simply calculate the U-value at that temperature from a cooling curve with neat methanol. This approach will yield acceptable results, as long as the reaction mixture is not strongly heterogeneous or highly viscous.

Another possible approach is to use very conservative general heat transfer coefficients in the scale-up calculations. One could, for instance, record a cooling curve for methanol, calculate the U-values from this curve, and then base all calculations on 50% of the heat transfer coefficient found. Although this method does not consider any specific effect of the physical properties of the reaction mixture on the heat transfer coefficient and should therefore only be used with great care and large safety margins, it can be an easy tool to give some guidance in the scale-up calculations.

Again, it should be stressed that when the correct U-values are not known and can only be roughly estimated, broad safety margins should be incorporated in the process, and the reactor data from the first batch should be checked carefully for any inconsistencies.

3.4.1.5 Practical Use of U-Values

So now the U-value of the reaction mixture at production scale has been determined, but what can we do with it? The main use of heat transfer coefficients is to allow for a correct calculation of dosing times, making sure that all heat that is being generated during the reaction can be safely removed. This is illustrated in the example below for a dosing controlled reaction. When dealing with a non-dosing controlled reaction (i.e. with significant thermal accumulation), one should make sure that the available cooling capacity matches the heat release rate as observed in the calorimetry experiment at any time.

EXAMPLE PROBLEM 3.2

We will use the same reaction as in the previous example, with the U-value at lab scale being 180 W/m^2 K and at production scale 337 W/m^2 K. The reaction at 60 °C is dosing controlled and has a total reaction heat of 100 kJ/kg reaction mass. The density of the reaction mixture is 0.8 kg/l. What dosing time is needed for a jacket temperature of 20 °C? The reaction is performed in a 6000 L vessel with 3000 L of reaction mixture and the heat exchange area is 8 m^2. We assume that the change in volume (and hence in heat exchange area) due to dosing is small.

$$Q_{r \times n} = Q_{cooling} = UA(T_R - T_J)$$

Total heat to be removed = 3 000 L × 0.8 kg/l × 100 kJ/kg
= 240 000 kJ

Heat removal capacity (Eq. 3.1) = 337 W/m^2 K × 8 m^2 × (60 − 20) K = 108 kW = 108 kJ/s

Dosing time needed = 240 000 kJ/108 kJ/s = 2 222 seconds ≈ 37 minutes

If the same reaction was to be run in a glass-lined reactor at 25 °C, with a U-value of 150 W/m^2 K and a minimum obtainable jacket temperature of 5 °C (water-cooled reactor), what would the dosing time be?

Total heat to be removed = 3 000 L × 0.8 kg/l × 100 kJ/kg
= 240 000 kJ

Heat removal capacity = 150 W/m^2 K × 8 m^2 × (25 − 5) K
= 24 kW = 24 kJ/s

Dosing time needed = 240 000 kJ/24 kJ/s = 10 000 seconds ≈ 167 minutes

A graphical representation of this example is given in Figure 3.6. It clearly shows the big influence the heat transfer characteristics of a vessel can have on the way in which a process can be run at production scale. In this context, our first concern is safety: is our cooling capacity sufficient to guarantee a safe operation at production scale? But there might be other consequences as well. In the example above, the recommended dosing time for the same process but in different equipment varies between 37 minutes and 3 hours. Such a difference in dosing time can have serious consequences on the yield of the reaction, the impurity profile, thermal accumulation, etc. It is therefore advisable to perform this type of scale-up calculations at a relatively early stage in development in order to avoid unpleasant surprises later on.

To conclude, some general trends about heat transfer in stirred tanks can be listed:

U (W/m^{2}K)	337	P (kW)	108
A (m^2)	8	ΔH (kJ)	240 000
ΔT (K)	40	Dos time (sec)	2 222

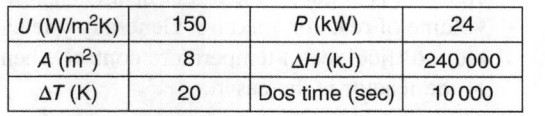

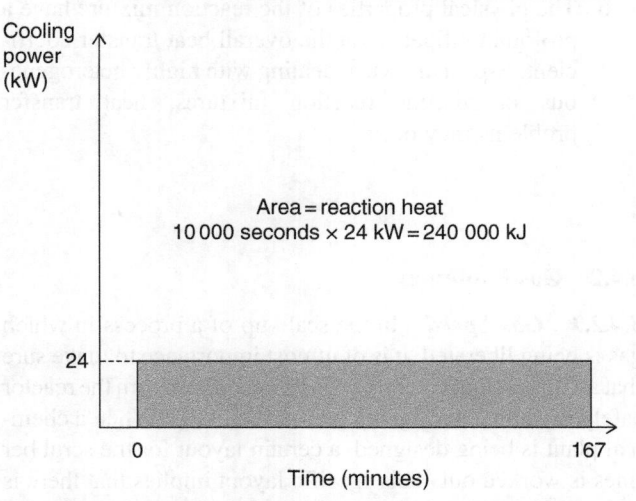

U (W/m^{2}K)	150	P (kW)	24
A (m^2)	8	ΔH (kJ)	240 000
ΔT (K)	20	Dos time (sec)	10 000

FIGURE 3.6 Influence of the heat transfer characteristics of a vessel on the dosing profile for an exothermic addition. The figure to the left shows the heat profile for a dosing controlled reaction in a vessel with a high heat transfer coefficient. The reaction mixture can be added relatively fast, since the cooling capacity is large. In the figure to the right, the same addition is shown in a reactor with a lower heat transfer coefficient. Less heat can be removed in the same period of time, hence the dosing has to be performed slower. The overall reaction heat (shaded area) is the same in both profiles.

TABLE 3.5 Influence of the Limiting Piping Diameter in the Scrubber Lines on the Maximum Allowable Gas Flow at Production Scale (Left) and the Corresponding Gas Flows at Lab- Scale (Right)

	Production Scale			Scale Down from 6000 L		
	Diameter (cm)	Max flow (l/min)	Max Flow (m^3/h)	2 L Reactor Max Flow (ml/min)	1 L Reactor Max Flow (ml/min)	100 ml Reactor Max Flow (ml/min)
DN25	2.5	295	18	98	49	5
DN50	5	1 178	71	392	196	20
DN100	10	4 712	283	1570	785	79
DN150	15	10 603	636	3534	1767	177

These values assume that the installation is designed for a maximum gas speed of 5 m/s.

1. Stainless steel reactors generally have better heat transfer characteristics than glass-lined reactors.
2. For the same temperature difference between the reactor and jacket, heat removal will be more efficient at higher reaction temperatures because of the higher U_{max} value at higher temperatures.
3. For the same heat transfer coefficient, increasing the temperature difference between the jacket and reactor will make the heat removal more efficient. Care has to be taken however when going too low in jacket temperature as to avoid crust formation.
4. The cooling circuit should be designed in accordance with the heat to be removed, in order to make sure that the heat transfer medium returning to the jacket is sufficiently cooled.
5. Smaller vessels usually have better heat transfer capacities. This is due to the larger heat exchange area per volume of reaction mixture. Generally one can expect more difficulties in temperature control when the size of the reactor is increased.
6. The physical properties of the reaction mixture have a profound influence on the overall heat transfer coefficient. Especially when dealing with highly heterogeneous or viscous reaction mixtures, heat transfer problems may occur.

3.4.2 Gas Evolution

3.4.2.1 Gas Speed In the scale-up of a process in which gas is being liberated, it is of utmost importance to make sure that all the gas that is set free can be evacuated from the reactor safely, without causing any pressure buildup. When a chemical plant is being designed, a certain layout for the scrubber lines is worked out. This specific layout implies that there is a maximum gas speed in the piping in order to avoid pressure buildup, entrainment of powders, or unintended changes in flow pattern or even flow direction. A maximum gas speed that is often used is 5 m/s. If the maximum design gas speed and the minimum diameter through which the gas has to pass are known, the resulting maximum gas flow can be calculated. Some typical piping sizes and the corresponding maximum gas flow to meet the 5 m/s criterion are given in Table 3.5.

These figures clearly demonstrate the pronounced effect of the diameter of the narrowest piping the gas has to pass on its maximum flow rate: doubling the diameter allows a fourfold increase in gas flow.

The part of Table 3.5 shows some interesting scale-down data. In these columns, we have calculated what the corresponding gas flow at lab scale is. For instance, if the narrowest piping in the scrubber line is a DN50, a maximum flow rate of 71 m^3/h can be allowed at production scale. Using this maximum gas flow and assuming it concerns a 6000 L reactor, we can calculate the gas flow at lab scale. In this case, the corresponding gas flow in a 2 L reaction calorimeter would be 392 ml/min. This flow can be detected easily, but if the reaction is run in a 100 ml calorimeter, the corresponding gas flow is only 20 ml/min. One can imagine that such a low gas flow rate can be overseen easily during the process development work. This illustrates the importance of an accurate gas flow measurement combined with each reaction calorimetry experiment. Especially when small-scale reactors are used (<1 L), care should be taken in choosing the appropriate gas flow measuring device.

These values for the maximum allowed gas flow apply mainly to the desired synthesis reaction. Exceeding this flow to a limited extent might result in operational problems such as slight pressure buildup, process gases entering a neighboring reactor that is connected to the same scrubber line, or suboptimal condenser and scrubber performance but will not necessarily lead to pronounced safety issues. When gas flow rates are considered an order of magnitude higher, vent sizing calculations come into play. This will be briefly discussed at the end of this chapter.

3.4.2.2 Reactive Gases In the previous discussion, the only parameter of concern was the gas flow rate. In many cases, however, the gas being emitted is reactive by itself, and this can cause particular safety problems. One example of having a very high yielding but unfortunately enough undesired synthetic reaction is depicted in Scheme 3.1.

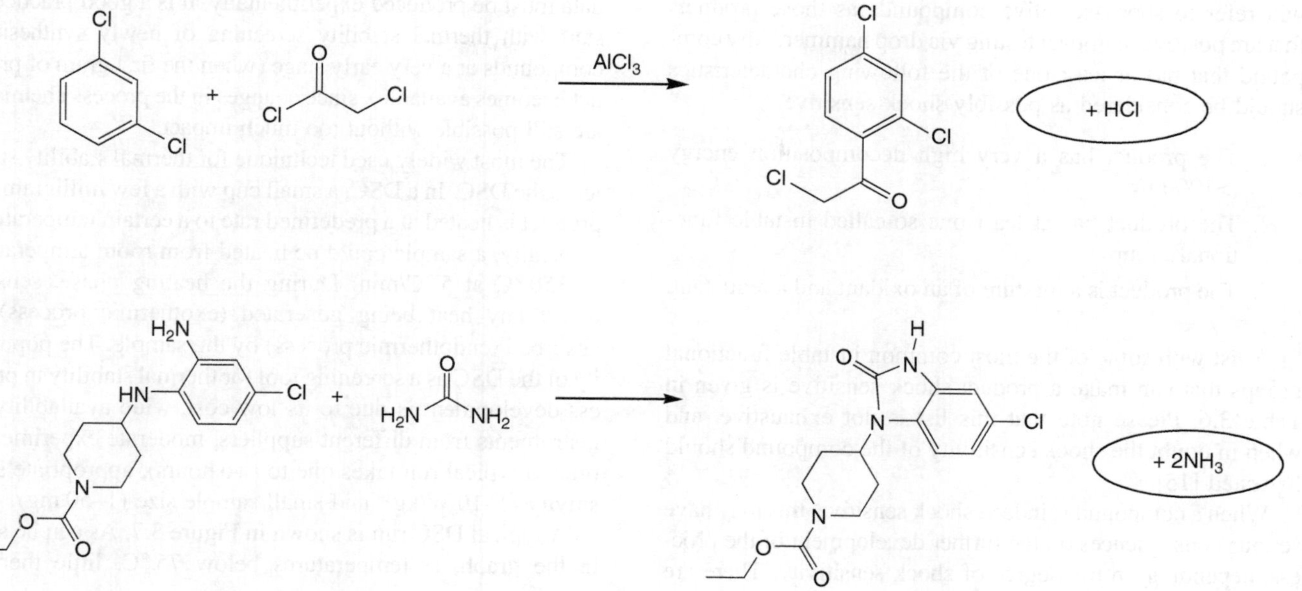

SCHEME 3.1 Example of two synthetic reactions that generate reactive gasses.

These two processes were both run in the same plant. By coincidence, they were being run at exactly the same time in two neighboring reactors. The reaction depicted at the top resulted in the emission of hydrogen chloride, while reaction at the bottom was releasing ammonia. Both reactors were connected to the same scrubber lines, and they inevitably reacted with each other, forming ammonium chloride in large amounts.

This solid material blocked the scrubber lines, and the reaction heat being evolved was large enough to partly melt the plastic scrubber lines. Fortunately enough there were no serious consequences, but this demonstrates the need for a broad safety overview in any chemical plant.

3.4.2.3 Environmental Issues Although this factor is not related to process safety in the strict sense, the importance of the gas flow rate for environmental compliance should be mentioned here as well. It is important to know the layout of the gas treatment facility of the plant where the process is going to be run. If a carbon absorption bed is used, it is important to keep an overview of what the capacity of this bed is for the process gas being emitted: some gasses are retained better than others, and some gasses might even lead to dangerous hot spot formation in the bed. If a catalytic oxidation installation is used, it is important to know that some compounds (like hydrogen and alkenes) will lead to overheating in the installation if the flow is too high. And when no air treatment facility is installed at all, one should always make sure that the gas streams being emitted are within all environmental requirements. This assessment needs to be made for each production plant separately.

3.5 STUDYING THE DECOMPOSITION REACTION AT LAB SCALE

Having dealt with the study of the desired synthesis reaction (the first part of the cooling failure scenario), let us now turn our attention to the study of the undesired decomposition reaction. Decomposition reactions are of extreme importance for safety studies: in most cases the energies being released in decomposition reactions are several orders of magnitude higher than those being released in the synthetic reaction, and hence the possible consequences of decomposition reactions can be catastrophic. Prerequisites for any compound being used in a process include that, firstly, it should be stable at the storage temperature for at least the time span anticipated for storage under normal operational conditions. Secondly, it should be at least sufficiently stable at the process temperature being used. And finally, it is important to assess its stability at the MTSR as well, since this is a temperature that can be attained in case of a cooling failure during the process (see Figure 3.1). In this paragraph we will discuss how the thermal stability of reagents and reaction mixtures can be studied at lab scale and how these data can be used to ensure a safe scale-up. But first we will start with another important characteristic of the stability of compounds, i.e. shock sensitivity.

3.5.1 Shock Sensitivity

Some compounds are known to be prone to explosive decomposition when subject to a sudden impact, and they are therefore called shock-sensitive compounds. In this chapter we will refer to shock-sensitive compounds as those products that are positive in impact testing via drop hammer. Any compound that has at least one of the following characteristics should be considered as possibly shock sensitive:

1. The product has a very high decomposition energy (>1000 J/g).
2. The product has at least one so-called instable functional group.
3. The product is a mixture of an oxidant and a reductant.

A list with some of the most common instable functional groups that can make a product shock sensitive is given in Table 3.6. Please note that this list is not exhaustive, and when in doubt the shock sensitivity of the compound should be tested [15].

When a compound is indeed shock sensitive, this may have serious consequences on the further development of the process, depending on the degree of shock sensitivity. There are restrictions for the transportation and storage of shock-sensitive compounds, so getting permission to purchase and store any of these products can be cumbersome. Therefore it is vital to be aware of shock sensitivity issues at an early stage.

When a reagent used in a synthesis is known to be shock sensitive, this does not necessarily exclude it from being used. For instance, hydroxybenzotriazole (HOBT) is known to be shock sensitive in its anhydrous form but not in its hydrate form. Making sure that the appropriate grade of the chemical is used from an early stage on can therefore avoid many practical problems later on.

3.5.2 Screening of Thermal Stability with DSC

When evaluating the stability and risk potential of commonly used reagents, common literature and references such as safety data sheets (SDS) can be a good starting point. More often than not in the development of active pharmaceutical ingredients, the compounds used are entirely new so the necessary safety data must be produced experimentally. It is a good practice to start with thermal stability screening of newly synthesized compounds at a very early stage (when the first gram of product becomes available), since changes in the process chemistry are still possible without too much impact.

The most widely used technique for thermal stability studies is the DSC. In a DSC, a small cup with a few milligrams of product is heated at a predefined rate to a certain temperature. Typically, a sample could be heated from room temperature to 350 °C at 5 °C/min. During the heating phase, sensors detect any heat being generated (exothermic process) or absorbed (endothermic process) by the sample. The popularity of the DSC as a screening tool for thermal stability in process development is due to its low cost, wide availability of instruments from different suppliers, moderate experimental time (a typical run takes one to two hours), appropriate sensitivity (1–10 W/kg), and small sample size (1–50 mg).

A typical DSC run is shown in Figure 3.7. As can be seen in the graph, at temperatures below 75 °C, little thermal

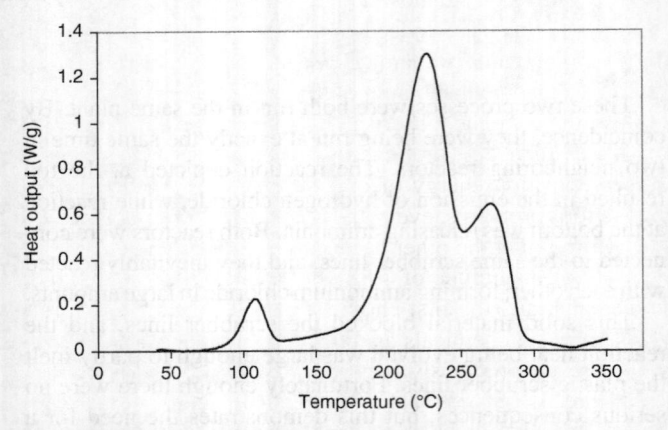

FIGURE 3.7 Example of a scanning DSC run of an unstable organic substance. The temperature was increased linearly from 30 to 350 °C and the subsequent heat signal was recorded (exotherm is shown as a positive, upward, signal).

TABLE 3.6 Non-Exhaustive List of Functional Groups That Can Be Shock Sensitive

Acetylenes	C≡C	Diazo	R=N=N
Nitroso	R—N=O	Nitro	R—NO$_2$
Nitrites	R—O—N=O	Nitrates	R—O—NO$_2$
Epoxides	△(O)	Fulminates	C≡N—O
N-Metal derivative	R—N—M	Dimercuryimmonium Salt	R—N=Hg=N—R
Nitroso	R—N—N=O	N-Nitro	N—NO$_2$
Azo	R—N=N—R	Triazene	R—N=N—N—R
Peroxyacid	R—O—OH	Peroxides	R—O—O—R
Peroxide salts	R—O—O—M	Azide	R—N=N=N
Halo-aryl metals	Ar—M—X	N-Halogen compounds	N—X
N—O compounds	N—O	X—O compounds	R—O—X

This list is extracted From Bretherick's [9] (version 4).

activity is observed. A first exothermic peak is observed at 110 °C, followed by a much larger exotherm exhibiting two peaks at around 225 and 270 °C. Integration of the entire exothermic signal yields a reaction enthalpy of more than 1000 J/g, indicating that this particular compound has a very high decomposition energy.

When evaluating a DSC run, there are three main parameters of interest.

3.5.2.1 Reaction Enthalpy
Some typical decomposition energies for the most common thermally unstable groups are given in Table 3.7.

This table clearly shows that merely looking at a molecular structure can already give an indication about the decomposition potential of a reagent. Note however that the data in this table are given in kJ/mol, whereas DSC data are usually reported in J/g. The latter gives in fact a better indication of the intrinsic energy potential of the product, since the influence of an instable group will obviously be much larger in a small molecule than in a very large one. It is therefore no surprise that many of the most dangerous reagents (in terms of thermal stability) are indeed small molecules: hydroxylamine, nitromethane, cyanamide, methylisocyanate, hydrogen peroxide, diazomethane, ammonium nitrate, etc.

According to the observed decomposition enthalpy, a first assessment of the energy potential can be made. A decomposition energy of only 50 J/g is very unlikely to pose serious problems, even when the product would decompose entirely. On the other hand, a decomposition energy in the order of magnitude of 1000 J/g should be considered as problematic, and the first choice should always be to avoid the use of such chemicals as much as possible.

As already previously mentioned (Eq. 3.3), the reaction heat is directly proportional to the adiabatic temperature rise. For ΔT_{ad} to be known, the c_p of the compound or reaction mixture is needed. The heat capacity can be determined separately in the DSC, but in order to get a first indication, an estimated value can be used as well. Usually a c_p of 2 J/gK can be used for organic solvents or dilute reaction mixtures, 3 J/gK for alcoholic reaction mixtures, 4 J/gK for aqueous solutions, and 1 J/gK for solids (conservative guess). A rough classification of severity of a decomposition reaction based on its reaction enthalpy and corresponding adiabatic temperature rise (in this case for $c_p = 2$ J/gK) is given in Table 3.8.

One final remark should be made here about endothermic decompositions. Some compounds decompose endothermically, and whereas one might consider them therefore to be harmless, special attention for these compounds is sometimes needed. Decomposition reactions in which gaseous products are formed (such as elimination reactions) are often endothermic. The intrinsic risk of this type of decompositions lies therefore not in the thermal consequences but in a possible pressure buildup. For this type of compounds, an evaluation of the possible pressure buildup associated with the decomposition is recommended.

3.5.2.2 Onset Temperature
Even when a compound has a very high decomposition energy, it can still be possible to use it safely in a process, provided that the difference between the process temperature and the decomposition temperature is sufficiently high. The term "onset temperature" is used quite frequently to denote the temperature at which a reaction or decomposition starts. In reality, however, there is no such thing as an onset temperature, since the temperature at which a reaction starts is strongly dependent on the experimental conditions. The point from which on a deviation from the baseline signal can be observed is determined by the sensitivity of the instrument, the sample size, and the heating rate of the experiment. This implies that great care is needed when comparing "onset temperatures" obtained with different methods, but it does not completely rule out the use of this parameter in early safety assessment.

It should be mentioned here that the description above holds for the way in which the term "onset temperature" is usually interpreted in the process safety field. In other fields of research, the term onset is defined as the point where the tangent to the rising curve at the inclination point crosses the baseline. This definition is, e.g. used in the calibration of a

TABLE 3.8 Classification of the Severity of Decomposition Reactions According to the Corresponding Adiabatic Temperature Rise

ΔH (J/g)	ΔT_{ad} (°C)	Severity
<−500	>250	High
−500 < ΔH < −50	25 < ΔT_{ad} < 250	Medium
>−50	<25	Low

Assuming a c_p of 2 J/gK.

TABLE 3.7 Typical Decomposition Energies for Some Common Instable Functional Groups [2]

Functional Group	ΔH (kJ/mol)	Functional Group	ΔH (kJ/mol)
Diazo —N=N—	−100 to −180	Nitro —NO$_2$	−310 to −360
Diazonium salt —N≡N$^+$	−160 to −180	N-hydroxide —N—OH	−180 to −240
Epoxide	−70 to −100	Nitrate —O—NO$_2$	−400 to −480
Isocyanate —N=C=O	−50 to −75	Peroxide —C—O—O—C	−350

DSC by measuring the melting point of a suitable metal (mostly indium). In this text, however, the term "onset" will always refer to the temperature at which a first deviation from the baseline signal is observed.

A rule of thumb that has been used quite extensively in the past is the so-called 100 °C rule. This rule states that a process can be run safely when the operating process temperature is at least 100 °C below the observed onset temperature. This rule has been shown to be invalid in certain cases, so it should certainly not be used as a basis of safety. This does not mean however that it is completely useless. If a process is to be run on a relatively small scale (50 L or less) in a well-stirred reaction vessel, natural heat losses to the environment are usually sufficiently large for the 100 °C rule to be valid. If the process is to be run at a larger scale, a proper and more detailed study of the decomposition kinetics (and especially of the TMR_{ad}) should be made, as will be discussed further on in this chapter. Here as well, it should be stressed that these remarks only apply to the thermal stability of the products, and extra care should be taken when gas evolution comes into play.

3.5.2.3 Reaction Type
When dealing with DSC data of compounds or reaction mixtures, merely looking at the shape of the peak can give some clues about the type of reaction taking place. The DSC run shown in Figure 3.7 consists of several different overlapping peaks, indicating a rather complex reaction type. Very sharp peaks are indicative for autocatalytic reactions, and extra care with this type of decompositions is needed [16].

Autocatalytic reactions are reactions in which the reaction product acts as a catalyst for the primary reaction. This implies that the reaction might run slowly at a certain temperature for a while, but as time passes more catalyst for the reaction is formed, and hence the reaction rate increases over time. Such reactions are therefore also called self-accelerating reactions. This is in contrast with the more classical behavior of reactions following Arrhenius kinetics where the reaction rate stays constant (zero order) or decreases (first order or higher) with time at a certain temperature. Having determined the decomposition reaction in such a case on a fresh sample will therefore always give the worst-case decomposition scenario, the initial rate measured will be the maximum rate for that sample at that particular temperature. This is not the case for autocatalytic reactions, where the reaction rate is strongly dependent on the thermal history of the sample. Measuring the reaction rate for a pristine sample might lead to an underestimation of the risk associated to the decomposition of this sample when it has been subject to a certain thermal history (e.g. prolonged residence time at higher temperature due to a process deviation). An example of an isothermal and a scanning DSC run for both an autocatalytic and a first-order reaction are shown in Figure 3.8.

As pointed out in the above, an autocatalytic reaction behavior is easily recognized by isothermal DSC experiments. If the temperature in the sample remains constant, the heat release over time will decrease in case of an nth-order reaction but will show a distinct maximum in case of an autocatalytic reaction. Although not always as easy as in an isothermal DSC run, autocatalytic reactions can also be recognized in scanning DSC experiments where the peak shape is notably sharper than for nth-order reactions. This is shown in Figure 3.8.

When a decomposition reaction is known to be autocatalytic, extra care is needed in order to avoid its triggering. For such compounds, any unnecessary residence time at elevated temperatures should be avoided. Extra testing might be appropriate to reflect the thermal history the product will experience at full production scale, since, e.g. heating and cooling phases may take considerably more time than in small-scale experiments. It should also be kept in mind that temperature alarms are not always an efficient basis of safety for this type of decomposition reactions: since the temperature rise can be very sudden, this type of alarm might simply be too slow to ensure sufficient time is available to take corrective actions.

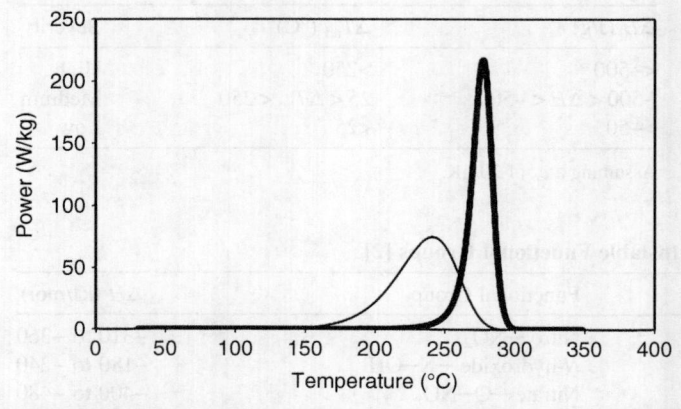

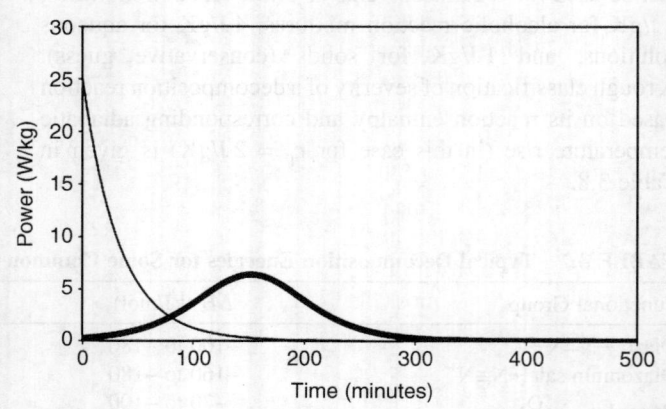

FIGURE 3.8 Example of a scanning DSC run (left) and an isothermal DSC run (right) of a reaction following nth order kinetics (thin line) and of an autocatalytic reaction (thick line).

3.5.3 Screening of Thermal Stability: Pressure Buildup

As already mentioned earlier, in many cases the gas being released during a (decomposition) reaction can have greater safety consequences than merely the exothermicity. A proper testing method to determine whether a decomposition reaction is accompanied by the formation of a permanent (i.e. non-condensable) gas is therefore very important.

There are several instruments commercially available that are suited for this type of testing. Generally speaking they consist of a sample cell of approximately 10 ml in which the sample is heated in a heating block or oven from ambient to around 300 °C. During this heating stage, the temperature inside the sample is measured, as well as the pressure inside the sample cell. The main criteria these instruments should meet are an appropriate temperature range (ideally from [sub] ambient to 300 °C), pressure range (up to 200 bar when measuring in metal test cells), and sample size (milligram to gram range). Some examples of such instruments available at the time of writing are the C80 from Setaram, the TSU from HEL, the RSD from THT, the mini-autoclave from Kuhner, the Carius tube from Chilworth, and the Radex from Systag.

The most difficult aspect of interpreting the pressure data from this type of experiments lies in the differentiation between a pressure increase that is due to the formation of a permanent gas and a pressure increase due to an increased vapor pressure of the compounds at higher temperatures. When dealing with reaction mixtures in a solvent, the vapor pressure as a function of temperature can be calculated easily by means of the Antoine coefficients. These coefficients are readily available in the literature for most common solvents. If a plot of the vapor pressure as a function of the sample temperature matches the observed pressure profile, one can conclude that the observed pressure increase is due to the increased vapor pressure only. When in doubt or when the Antoine coefficients of the product are not known, e.g. when dealing with a newly synthesized product that is an oil, it is advised to run an isothermal experiment with pressure measurement at a temperature at which the decomposition is known or believed to occur at a considerable rate. If the pressure during this experiment remains constant, the pressure is due to vapor pressure, and if the pressure increases over time, it is due to the formation of a gaseous product. Alternatively, one could run two scanning experiments, each with a different filling degree in the test cell. If the pressure profile in the two runs match each other, the pressure is most likely to be due to vapor pressure, since it is not dependent on the free headspace available. If the test with the higher filling degree leads to a higher pressure, it is most likely to be due to the formation of a non-condensable gas, since less headspace is available for a larger amount of gas, leading to higher pressure.

But why is it so important to differentiate between vapor pressure and the formation of a non-condensable gas? If a certain pressure at elevated temperatures is only due to vapor pressure, it is relatively unlikely to pose problems. In such a case, a very rapid temperature increase is needed before the amount of vapor produced surpasses the amount that can be removed through the vent lines. When dealing with vent sizing calculations for serious runaway reactions, the effect of vapor pressure should definitely be taken into account. However, when dealing with moderate temperature rise rates, or in isothermal operation, vapor pressure is unlikely to lead to major problems. The story is entirely different for the formation of a permanent gas due to a (decomposition) reaction. In this case, each process parameter that leads to an increased reaction rate will lead to an increased pressure rise rate, with possibly devastating effects. Obviously, a temperature rise will lead to an increased reaction rate, but other effects such as an increase in concentration due to the evaporation of the solvent (e.g. in case of a condenser failure) or sudden mixing of two previously separated layers (e.g. switching the stirrer back on after a failure) could also lead to an increased pressure rise rate in the vessel. This is also important for storage conditions: vapor pressure in a closed drum will reach an equilibrium at a certain pressure, whereas the formation of a permanent gas will lead to a pressure increase over time and the subsequent possibility of rupturing the drum.

EXAMPLE PROBLEM 3.3

In an isothermal stability test of a reaction mixture at 90 °C, a gradual (linear) pressure increase is observed from 1.5 bar at the start of the experiment to 20 bar after 10 hours. What is the gas release rate if the reaction is to be run at a 4000 L scale? We assume that the reaction behavior in an open system (production scale) is comparable with that in a closed system (lab test). In the test, 4 ml of reaction mixture was used in a system with an overall free headspace of 8 ml.

$$\frac{\left(\frac{\Delta P}{\text{time}} V\right) \text{lab}}{(\text{reaction volume})\text{lab}} = \frac{\left(P \frac{\Delta V}{\text{time}}\right) \text{vessel}}{(\text{reaction volume})\text{vessel}}$$

The pressure rise rate in the experiment is 18.5 bar in 10 hours, i.e. 0.031 bar/min.

The volume at lab scale is the headspace volume, in this case 8 ml, whereas the reaction volume is 4 ml. At vessel scale the pressure remains constant at 1 bar (open system), and the reaction volume is 4000 L. From this the volumetric gas rate can be calculated, and the result found equals to 248 L/min.

At production scale, a gas evolution rate of 248 L/min is expected. This corresponds to 14.9 m^3/h, which is only moderate (see Table 3.5).

3.5.4 Adiabatic Calorimetry

When discussing the cooling failure scenario earlier in this chapter, the concept of adiabaticity was introduced. A system is said to be adiabatic when there is no heat exchange with the surroundings. In a jacketed semi-batch reactor under normal process conditions, the reactor temperature is controlled by means of heat exchange between the reaction mixture and the heat transfer medium in the jacket. In case of a loss of cooling capacity (either because the heat transfer medium itself is no longer cooled or because it is no longer circulated), this heat exchange is no longer possible, and the reactor will behave adiabatically. This is considered to be the worst-case situation in a reactor apart from a constant heat input (e.g. through an external fire), which will not be considered here.

For a lab chemist working on small-scale experiments only, the concept of adiabatic behavior in a large vessel is often hard to imagine. "I did it in the lab and I didn't notice any exothermicity" is an often heard statement. However, heat losses at small scale are a lot higher than at large scale, so the heat generation should already be relatively high before it is noticed during normal synthesis work at lab scale. This can be seen in Table 3.9, where some heat losses for different types of equipment are listed.

This table shows the vast difference in heat losses between small scale and large scale and also the relevance of performing proper adiabatic tests. A 1 L Dewar calorimeter can be considered to be representative for other state-of-the-art adiabatic calorimeters, and it can be seen that its heat losses compare favorably to reactors in the m³ range.

Since this adiabatic behavior is considered to be the worst-case situation from a thermal point of view, it is of great interest to be able to mimic this situation in the lab under controlled conditions. An adiabatic calorimeter typically consists of a solid containment (to protect the operator against possible explosions that might take place inside the calorimeter) around a set of heaters in which the sample cell is placed. A thermocouple either inside the test cell or attached to the outside of the test cell records the sample temperature, and the heaters are kept at exactly the same temperature at any time in order to obtain fully adiabatic conditions. During the entire experiment, the pressure inside the test cell is recorded, as well as the sample temperature. The criteria an adiabatic calorimeter for safety studies should meet are obviously a high degree of adiabaticity (i.e. very low heat losses), an appropriate sample volume (typically between 5 and 50 ml), broad temperature range (ambient to 400 °C), high pressure resistance or a pressure compensation system (up to 200 bar), and high speed of temperature tracking (>20 °C/min). Some commercially available instruments are the ARC from Thermal Hazards Technologies, the Phi-TEC from HEL, the Dewar system from Chilworth, and the VSP from Fauske. Several pharmaceutical and chemical companies have developed their own adiabatic testing equipment, mainly based on a high pressure Dewar vessel.

3.5.4.1 Adiabatic Temperature Profile Let us consider the situation as being depicted in Figure 3.9. The reaction mixture is at a constant temperature of 120 °C when a cooling failure takes place. At this temperature, the reaction starts relatively slowly, and hence the temperature increases, albeit at a slow pace. Since most chemical reactions proceed faster at higher temperatures, the reaction rate (and thus the temperature rise rate) will increase as the reaction continues. This acceleration continues until finally the depletion of the reagents slows the reaction down again and a stable final temperature is achieved. The S-shaped temperature curve seen in the graph is very characteristic for an adiabatic runaway reaction. As indicated on the graph, there are three main parameters that describe the process of a runaway reaction, i.e. the adiabatic temperature rise (ΔT_{ad}), the TMR, and the maximum self-heat rate (max SHR). The first two have been discussed earlier in this chapter, and the max SHR is a measure for the maximum speed with which the reaction occurs and can be directly correlated to the power output of the reaction: a SHR of 1 °C/min corresponds to 33 W/kg reaction mixture for an organic medium with a c_p of 2 and to 70 W/kg in aqueous medium

$$1\frac{°C}{min} \times 2\frac{J}{gK} \times \frac{1}{60}\frac{min}{sec} \times 1000\frac{g}{kg} = 33\frac{W}{kg}$$

This is important for vent sizing calculations and the assessment of using the boiling point as a safety barrier, as will be discussed further on in this chapter.

3.5.4.2 Heat–Wait–Search Procedure The most commonly applied method for adiabatic testing is the so-called heat–wait–search procedure, as depicted in Figure 3.10. In this procedure, the sample is introduced in the instrument at room temperature and then heated (heating step) to the starting temperature of the test. The sample is then allowed to equilibrate at this temperature (waiting step), followed by the so-called search step. During this step (which usually takes between 5 and 30 minutes), the sample temperature is monitored to see if there is any sign of an exothermic reaction taking place. If the temperature rise rate under adiabatic conditions during this period is below the chosen detection threshold (typically 0.02 or 0.03 °C/min), the temperature is increased with a couple of degrees and the cycle starts

TABLE 3.9 Typical Heat Losses for Different Types of Equipment [1]

	Heat Loss (W/kg/K)	Time for 1 °C Loss at 80 °C
5000 L reactor	0.027	43 min
2500 L reactor	0.054	21 min
100 ml beaker	3.68	17 s
10 ml test tube	5.91	11 s
1 L Dewar	0.018	62 min

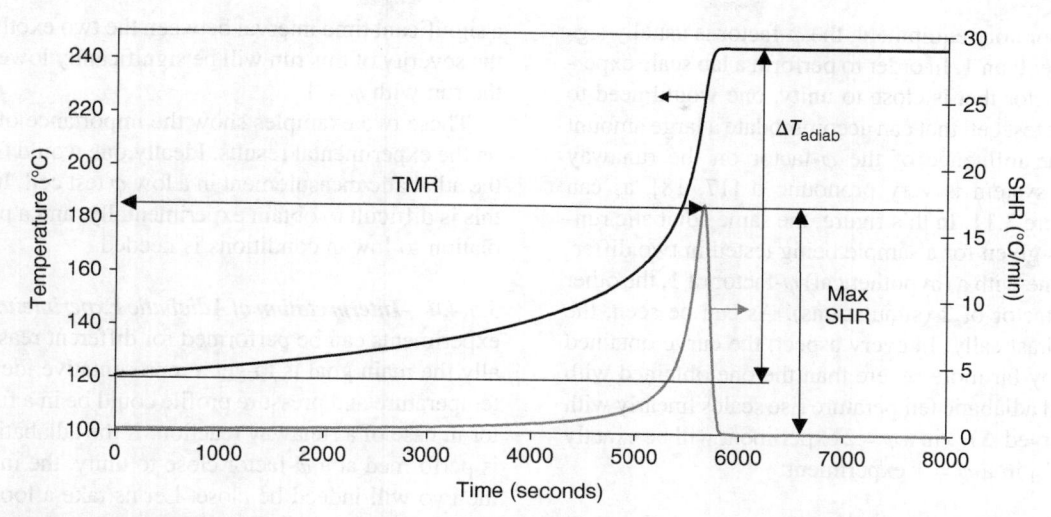

FIGURE 3.9 Typical adiabatic temperature versus time profile (left axis). The first derivative of the heat signal (self heat rate) is shown on the right axis. The TMR, adiabatic temperature rise (ΔT_{ad}) and the maximum selfheat rate (max SHR) are indicated.

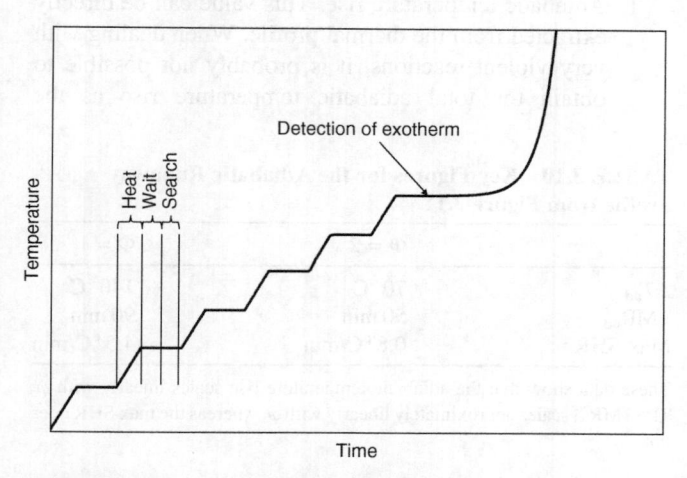

FIGURE 3.10 Typical representation of the heat-wait-search procedure in an adiabatic experiment. The sample is first heated to the desired temperature, then the temperature is allowed to stabilize during the wait period, and finally the temperature profile is checked for any sign of exothermicity during the search period. If exothermicity is detected, the temperature is adiabatically tracked until completion of the reaction or until the maximum experimental temperature has been reached. Otherwise the cycle is repeated.

all over again, until either an exotherm is detected or the preset final temperature has been reached. When an exotherm has been detected, the instrument will track the sample temperature adiabatically until the temperature rise rate drops below the threshold value (end of the reaction), after which the heat–wait–search cycle starts again. Alternatively, the run is aborted during an exotherm if the upper temperature limit of the experiment has been surpassed.

Although adiabatic calorimeters can usually operate in other thermal modes as well, the heat–wait–search procedure is still the most widely used because it allows determining the onset temperature of the exotherm with great accuracy while keeping the experimental time acceptably short.

3.5.4.3 Thermal Inertia: φ-Factor

As mentioned above, one of the main reasons why it is hard to extrapolate adiabatic behavior at large scale from small-scale lab work is the dramatic difference in heat losses between these two working environments. There is another reason as well, which plays a very important role in the interpretation of adiabatic calorimetry, i.e. the thermal inertia or φ-factor:

$$\varphi = \frac{m_c \times c_{p_c} + m_s \times c_{p_s}}{m_s \times c_{p_s}} \quad (3.8)$$

where

m_c: mass of the container (vessel or sample cell) (g)
m_s: mass of the reaction mass (g)
c_{pc}: heat capacity of the container (J/gK)
c_{ps}: heat capacity of the reaction mass (J/gK)

When heat is generated in the reaction mixture, this heat will be used to increase the temperature of not only the reaction mixture itself but also the container, being the vessel at large scale or the test cell at small scale. The φ-factor is therefore a measure of which fraction of the thermal mass of the entire system is due to the thermal mass of the reaction mixture and which part is due to the container.

In large-scale equipment, the φ-factor of a vessel during a runaway will be close to unity: i.e. the thermal mass of the vessel itself (mainly the jacket) will be low compared with the thermal mass of the reaction mixture (i.e. $\varphi = 1$). In small-scale laboratory equipment, the φ-factor is usually significantly higher than 1. In order to perform a lab scale experiment at a φ-factor that is close to unity, one would need to use a very light test cell that can accommodate a large amount of sample. The influence of the φ-factor on the runaway behavior of a system is very pronounced [17, 18], as can be seen in Figure 3.11. In this figure, the same adiabatic runaway profile is given for a sample being tested in two different test cells, one with a (hypothetical) φ-factor of 1, the other one with a φ-factor of 2 (simulations). As can be seen, the curves differ drastically. In every aspect, the curve obtained with $\varphi = 1$ is by far more severe than the one obtained with $\varphi = 2$. The total adiabatic temperature rise scales linearly with φ, i.e. the observed ΔT_{ad} in a $\varphi = 2$ experiment will be exactly half of the ΔT_{ad} in a $\varphi = 1$ experiment:

$$\Delta T_{ad, \varphi=1} = \Delta T_{ad, \exp} \times \varphi \qquad (3.9)$$

The TMR scales almost linearly with φ in most cases, but the max SHR scales far from linear with φ. The max SHR in an experiment with $\varphi = 1$ can easily be 10 times higher than in the $\varphi = 2$ experiment! The numerical data for the curves as depicted in Figure 3.11 are given in Table 3.10.

Figure 3.11 also gives the runaway behavior of one reaction in a test cell with a (hypothetical) φ-factor of 1 as compared with the same run in test cell with a φ-factor of 2. In this case, however, the difference between the two runs is even more pronounced. The reaction consists of two consecutive reactions, and running this reaction at $\varphi = 1$ will result in a temperature profile where the first exotherm continues into the second one, leading to a very rapid temperature rise. In the run with $\varphi = 2$, the temperature rise from the first exotherm will be far less pronounced, and this will lead to a significant time interval between the two exotherms. Hence the severity of this run will be significantly lower than that of the run with $\varphi = 1$.

These two examples show the importance of the φ-factor on the experimental results. Ideally one would try to perform the adiabatic measurement in a low φ test cell. In many cases this is difficult to obtain experimentally, and a proper extrapolation to low-φ conditions is needed.

3.5.4.4 Interpretation of Adiabatic Experiments
Adiabatic experiments can be performed for different reasons, but usually the main goal is to get a representative idea of what the temperature and pressure profile could be in a full-scale reactor in case of a runaway reaction. If the adiabatic experiment is performed at a φ-factor close to unity, the match between the two will indeed be close. Let us take a look at the most important parameters to analyze.

1. Adiabatic temperature rise. This value can be directly extracted from the thermal profile. When dealing with very violent reactions, it is probably not possible to obtain the total adiabatic temperature rise as the

TABLE 3.10 Key Figures for the Adiabatic Runaway Profile from Figure 3.11

	$\Phi = 2$	$\Phi = 1$
ΔT_{ad}	70 °C	140 °C
TMR_{ad}	50 min	90 min
Max SHR	0.8 °C/min	4.5 °C/min

These data show that the adiabatic temperature rise scales linearly with φ. The TMR_{ad} scales approximately linearly with φ, whereas the max SHR does not.

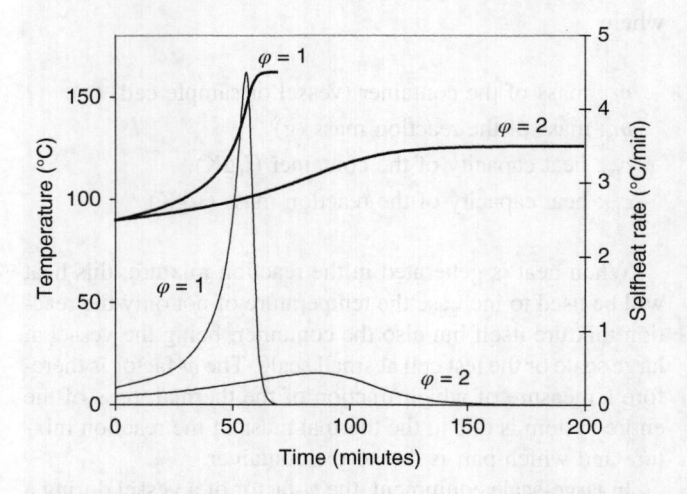

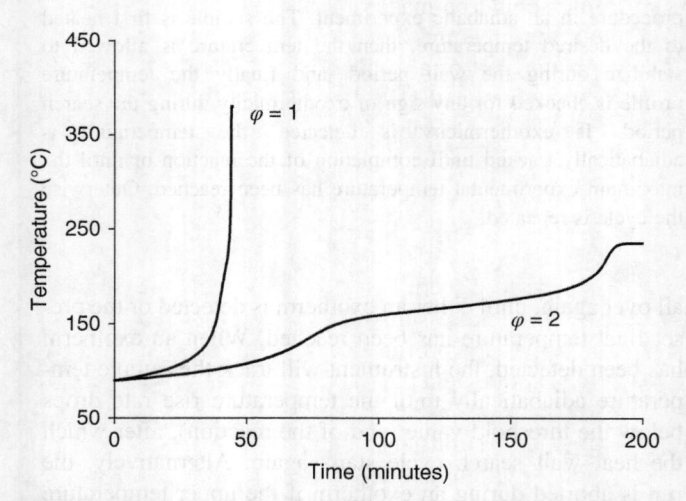

FIGURE 3.11 Influence of the φ-factor on the adiabatic temperature profile [16]. The figure to the left represents an adiabatic experiment on exactly the same sample, but once measured at a $\varphi = 1$ and once at $\varphi = 2$. The influence of the thermal inertia on the result is pronounced. This difference is even larger in the case of two consecutive reactions, as shown in the figure to the right. *Source*: Reprinted with permission from Dermaut [17]. Copyright 2006 American Chemical Society.

maximum safety temperature or pressure of the equipment will be surpassed and the experiment will be automatically stopped. This is not necessarily a problem since ΔT_{ad} can be obtained from other experiments as well (e.g. from DSC), and for this type of violent reactions, the temperature profile close to the onset temperature is far more important. Whether the final temperature would be 1000 or 700 °C is relatively unimportant, it will be a full-blown thermal explosion anyhow. It should be kept in mind that the observed ΔT_{ad} should be multiplied with the φ-factor in order to obtain the correct adiabatic temperature rise in case of a large-scale runaway (Eq. 3.9).

2. Onset of the exotherm. Here as well, the term onset refers to the point where deviation from the baseline can be observed and is hence instrument dependent. In adiabatic calorimetry a detection threshold of 0.02 °C/min is often used, which corresponds to a 1.4 W/kg in case of an aqueous reaction medium. Referring to Table 3.9, we know that the natural heat loss of a 5000 L reactor at 80 °C and an ambient temperature of 20 °C is 1.6 W/kg. These two figures match quite closely, so the temperature at which the exotherm is detected in the adiabatic calorimeter with this sensitivity is most likely to be the temperature at which exothermicity will be first noticed under adiabatic conditions at large scale (at least for temperatures higher than 80 °C). If the onset temperature is well above the MTSR, the decomposition is unlikely to be triggered, even in case of a cooling failure during the synthesis reaction. If the onset temperature is close to the MTSR, a calculation of the TMR_{ad} should be made, as will be discussed further on.

3. Pressure profile. A careful analysis of the pressure profile should be made after each experiment. Very often, the pressure signal will be more sensitive to detect the start of a decomposition reaction than the temperature signal [19]. If a slow pressure increase during any of the search periods is observed, a decomposition reaction with gas evolution should be suspected. Some software packages allow for a direct overlay of the vapor pressure of any chosen solvent related to the sample temperature. This can be very indicative to discern between permanent gas formation and vapor pressure. If this is not possible, looking at the pressure profile during the wait and search period should yield the same information: if the pressure remains constant during this stage (when the sample is at a constant temperature), vapor pressure is the most important contribution to the overall pressure. If the pressure rises during this stage, formation of a permanent gas is most likely to happen. If the pressure drops at a certain point, a leak of the test cell has most probably occurred.

4. Self-heat rate (temperature rise rate). The temperature rise rate of a runaway reaction can be calculated by taking the first derivative of the temperature versus time plot. Two things should be kept in mind when analyzing the SHR: firstly the dramatic influence of the φ-factor on the SHR, as discussed above, and secondly the fact that the associated pressure rise rate can have far more serious consequences. So in case the SHR is low at any time (e.g. below 2°C/min) in a low-φ experiment ($\varphi < 1.2$) and there is no strong pressure increase, the consequences of the runaway reaction are unlikely to be severe.

5. Pressure rise rate. The analysis of the pressure rise rate data is far from trivial. First thing to keep in mind is the influence the free headspace volume in the adiabatic calorimeter has on the finally observed pressure profile. If a permanent gas is being formed during the runaway reaction, the observed pressure increase will be considerably larger when the test has been performed with a small free headspace (e.g. a filling degree of 90% of the test cell) than in the case of a large free headspace (e.g. 50% filling degree). Therefore, it is advisable to use a ratio reaction mixture versus free headspace that is comparable with the situation at a large scale. As mentioned before, this is only relevant when dealing with the formation of a permanent gas, not when considering vapor pressure data. Moreover, as the pressure rise rate is in any case directly correlated to the temperature rise rate, the remarks made in the point above about the influence of the φ-factor hold here as well. The pressure rise rate can be calculated back to a gas evolution rate (see also the Example Problem 3.3), and if the thus obtained gas flow is below the design limits of the installation under normal process conditions, no problems are to be expected. If this limit is exceeded to a limited extent, some operational issues can be suspected (limited condenser capacity, disturbed flow patterns in the venting line, etc.) without serious safety consequences. If the gas flow rate is considerably above this design limit, proper vent sizing calculations are needed to make sure that the emergency relief system is sufficient to cope with a runaway reaction. This point will be briefly discussed in a later paragraph.

Let us now consider a real-life example of such an adiabatic experiment. In Figure 3.12, a heat–wait–search experiment of a relatively concentrated solution of dibenzoyl peroxide in chlorobenzene is shown. The experiment was conducted in glass test cell with a φ-factor of 1.5.

A close inspection of the temperature profile indicates that the thermal activity starts already at 45 °C, but the detection threshold of 0.02 °C/min is only reached at 58 °C. At this temperature the instrument goes into tracking mode, and a

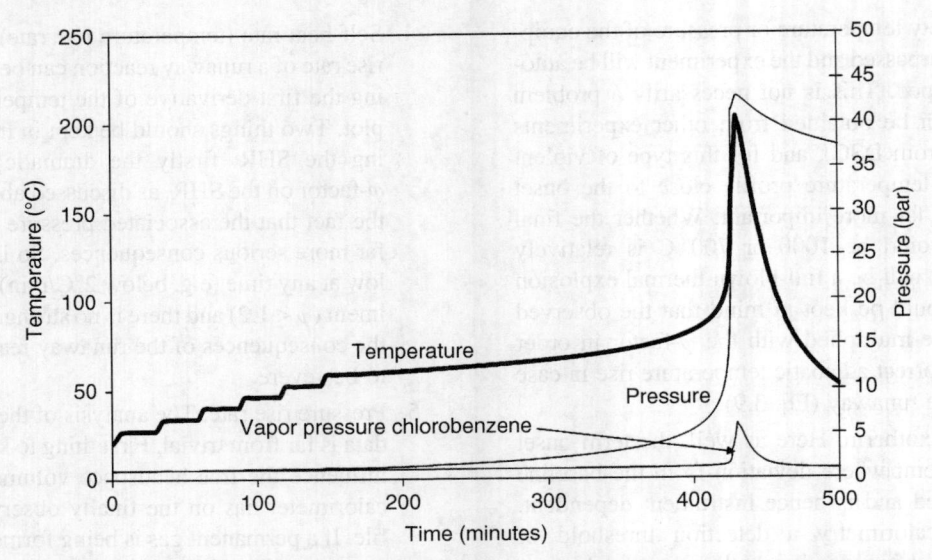

FIGURE 3.12 Adiabatic heat-wait-search experiment on a 0.75 M solution of dibenzoylperoxide in chlorobenzene. The test was conducted in a glass test cell with a φ-factor of 1.5. Exothermicity is observed at 58 °C, the run is aborted at 200 °C in order to prevent leaking of the silicone septa used to close the test cell [16].

max SHR of more than 40 °C/min is reached after 280 minutes. The observed adiabatic temperature rise is 140 °C, but in reality it will be higher since the run was aborted at 200 °C in order to prevent leakage of the silicone septa used to seal the glass test cell. An overlay of the vapor pressure curve with the pressure profile shows that a large part of the observed pressure is due to gas evolution of the decomposition reaction. The maximum pressure rise rate is very high, more than 200 bar/min. Keeping in mind that this experiment was conducted at a relatively high φ-factor, it is clear that the severity of this runaway is totally unacceptable for introduction at a large scale. An obvious safety advice would be to investigate the use of a more dilute solution of this compound, or to turn to other, more stable reagents.

3.5.4.5 Using the Boiling Point as a Safety Barrier
When a runaway reaction takes place in a reactor, the temperature of the reaction mixture can reach the boiling point. This can either be an extra risk that needs to be taken into account or it can act as an efficient safety barrier [20].

If the heat rate under adiabatic conditions at the boiling point is low, part of the solvent will be evaporated, but the temperature will remain constant, and this will temper the runaway reaction. When the detected onset temperature in the adiabatic calorimeter is close to the boiling point, this can be a very effective safety barrier.

If the heat rate at the boiling point is relatively high, other effects might come into play:

1. Evaporation of the solvent. If the boiling point is reached, part of the solvent will start to evaporate. This in itself will have a cooling effect, but when the vapor is no longer condensed and returned to the reactor, the reaction mixture will become more concentrated. This in turn will lead to an increased reaction rate and also to an increased boiling temperature. This should be taken into account when relying on the reflux barrier as a basis of safety.
2. Swelling of the reaction mass. If the reaction mixture starts to boil vigorously, a kind of "champagne effect" may take place, leading to an increase of the volume of the reaction mass. In such a case the reactor content might even be forced out of the reactor into the condenser and scrubber lines.
3. Flooding of the vapor line. This effect will occur particularly when countercurrent condensers are used (i.e. the vapor and the condensate flow in opposite direction through the condenser). If the vapor flow through the condenser is too high, this flow will prevent the condensed liquid to flow back into the reactor, and this liquid will be carried along with the vapor flow. Here as well, solvent will enter the scrubber lines with all possible problems associated with it.

Provided that all of the abovementioned factors are taken into account, the reflux barrier can be used as a very effective safety barrier. For a more quantitative description, the reader is referred to the literature [3].

3.5.5 TMRad Calculations

In the discussion of the cooling failure scenario, the importance of the temperature at which the TMR_{ad} is 24 hours was pointed out. It was stated that if the TMR_{ad} at the MTSR is more than 24 hours, the reaction can be considered to be safe. This obviously implies a correct determination of this important parameter. Several approaches can be followed to do this, each with their merits and shortcomings.

3.5.5.1 Determination from 1 DSC Run

A first approximation of the temperature at which the TMR_{ad} is 24 hours (we will call this temperature $TMR_{ad,24h}$) can be made from one single DSC run. A full discussion of the theory behind this approach is out of scope here, so the reader is referred to the original publication by Keller et al. [18]. We will however point out the basic concept with an example.

The idea behind this approach is quite simple: measure the heat release at one temperature, and assume that the reaction follows zero-order reaction kinetics with a low activation energy of 50 kJ/mol. From this, the heat release at any temperature and hence the $TMR_{ad,24h}$ can be calculated. The assumptions on which this approach is based and the practical calculations are discussed below.

1. Zero-order assumption. A "classical" behavior for a reaction is that it follows nth-order Arrhenius kinetics. This means that the reaction rate increases with concentration and temperature. This implies that the reaction rate will decrease over time when the reaction temperature is constant (see Figure 3.8) as the concentration of the reagents drops with increasing conversion. In a reaction that follows zero-order kinetics, however, the reaction rate is independent of the concentration. This means that the reaction rate remains constant at a given temperature from the start (0% conversion) until the end (full conversion) of the reaction. Assuming this type of reaction kinetics leaves out the concentration dependence and makes the calculations a lot easier. It should be kept in mind however that this is an assumption; in reality the reaction is most likely not going to follow these kinetics. It is however a "safe" assumption, since the reaction rate will be overestimated as the decrease in reaction rate with decreasing reagent concentration is neglected. The fact that this is indeed a "worst-case" assumption is discussed thoroughly in the original publication.

2. Low activation energy. The dependence of the reaction rate on the temperature is dictated by the activation energy: the decrease in reaction rate when lowering the reaction temperature is more pronounced for a reaction with high activation energy than for a reaction with a low activation energy. Since we will try to extrapolate the reaction rate at lower temperatures from one single measurement at one temperature, the activation energy needs to be known or estimated. A correct determination of the activation energy is possible by means of DSC, but not always straightforward. Therefore, the activation energy is assumed to be 50 kJ/mol, which is very low for organic reactions and decompositions. Choosing a low activation energy will again be on the safe side, since it will tend to overestimate the reaction rate at lower temperatures.

3. Determination of the reaction rate (heat rate) at one temperature. A correct determination of the heat rate at one temperature is needed, preferably in a relatively early stage of the reaction since this will minimize the error introduced by the zero-order assumption. The heat signal should be well separated from the baseline, however, in order to obtain an accurate signal. Keller et al. suggest to search for the temperature at which the heat rate is 20 W/kg. This is a sensitivity that is well within reach of any decent DSC apparatus. In the example shown in Figure 3.13, this heat rate is observed at 111 °C.

4. The heat rate at other temperatures can now be calculated according to Eq. (3.10):

$$q_0 = q_{onset} \times e^{\left(\frac{E_a}{R} \times \left(\frac{1}{T_{onset}} - \frac{1}{T_0}\right)\right)} \qquad (3.10)$$

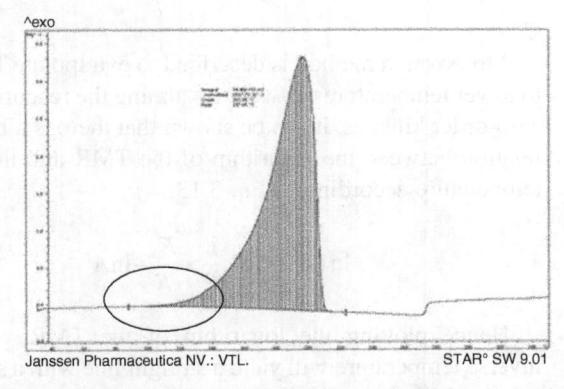

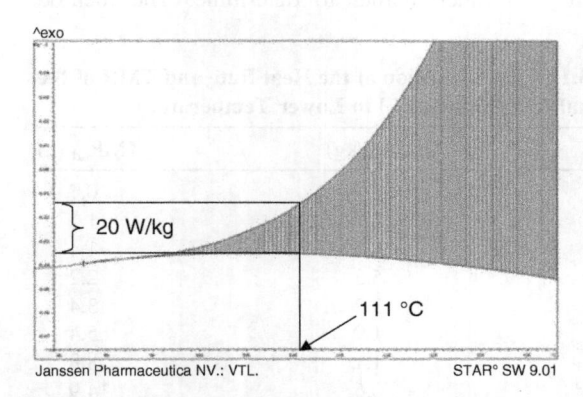

FIGURE 3.13 Example of a scanning DSC run of a highly unstable organic compound. The figure to the left shows the entire run, the figure to the right is zoomed in on the start of the exotherm.

where

q_0: heat rate at the new temperature

q_{onset}: heat rate at the onset temperature (in this case 20 W/kg)

E_a: activation energy (50 kJ/mol)

R: universal gas constant (8.31 J/mol K)

T_{onset}: onset temperature (in this case 111 °C or 384 K)

T_0: new temperature at which the heat rate is to be calculated

5. From this, the TMR_{ad} can be calculated for any temperature according to

Eq. (3.11):

$$TMR_{ad} = \frac{c_p \times R \times T_0^2}{q_{T_0} \times E_a} \qquad (3.11)$$

If this calculation is performed for a number of temperatures, $TMR_{ad,24h}$ can be determined, as shown in Table 3.11.

In this example (using a c_p of 2 kJ/kgK), the $TMR_{ad,24h}$ is estimated to be 31 °C. Another interesting point we can learn from this table is that a heat release of only 10 W/kg corresponds to a TRM_{ad} of less than one hour!

This example shows how the $TMR_{ad,24h}$ can be extrapolated from only one DSC experiment. Because of the assumptions made, this will only be a rough estimate that can differ considerably from the true $TMR_{ad,24h}$. The merit of this method however lies in the fact that all assumptions are on the safe (conservative) side. If the thus obtained TMR_{ad} at the MTSR is longer than 24 hours, no further testing is needed. If it is shorter than 24 hours, a more accurate determination of the $TMR_{ad,24h}$ might be needed, as will be discussed below.

One final remark is needed about autocatalytic reactions. Since the thermal history of a sample is so important in the characterization of autocatalytic (decomposition) reactions, their TMR_{ad} is much harder to determine. The method described here should therefore not be used for this type of reactions, and more elaborate adiabatic testing will be needed.

3.5.5.2 TMR_{ad} from 1 Adiabatic Experiment Probably the best way to determine the $TMR_{ad,24h}$ accurately is to perform a number of adiabatic experiments in a low-φ test cell, each at a different starting temperature, and then determine the TMR for each of these experiments until the temperature at which this TMR is 24 hours is found. Obviously, this will be a very time-consuming procedure, and better alternatives are to be sought for. It would be beneficial if we could extract a reliable $TMR_{ad,24h}$ from one single adiabatic heat–wait–search experiment at a somewhat higher φ-factor. This way, we would run an experiment under adiabatic conditions, which is closer to the real situation in a vessel during a runaway reaction than a scanning DSC experiment. Also, if we can run in HWS mode, the experimental time will be reduced significantly, and running at a higher φ-factor (e.g. 1.5) is experimentally easier than at a φ-factor close to unity. The question is how to extrapolate the data from this single adiabatic experiment to other temperatures and other φ-factors.

In one of the classical papers about adiabatic calorimetry, Townsend and Tou [21] evaluated in great detail the analysis of experimental adiabatic data. In this article, they present a method to extrapolate the experimental TMR to other temperatures and other φ-factors. A full description of the kinetic evaluation made in the original paper is out of scope here, but the general concepts and the practical use of this approach are discussed below.

We start from an adiabatic experiment, where we can determine the TMR at the onset temperature directly from the temperature versus time plot. The first extrapolation to be made is from the experimental φ-factor (i.e. >1) to the "ideal" case of $\varphi = 1$. Townsend and Tou state that for most relevant decomposition reaction with a high activation energy, the TMR scales linearly with φ (Eq. 3.12):

$$TMR_{\varphi=1} = \frac{TMR_{exp}}{\varphi} \qquad (3.12)$$

Moreover, a method is described to extrapolate TMR data to lower temperatures as well. Assuming the reaction follows zero-order kinetics, it can be shown that there is a linear correlation between the logarithm of the TMR and the inverse temperature according to Eq. 3.13:

$$\ln(TMR) = \frac{1}{T} \times \frac{E_a}{R} - \ln A \qquad (3.13)$$

Hence plotting the logarithm of the TMR versus the inverse temperature will yield a straight line with a slope proportional to the activation energy and the intercept being equal to the logarithm of the frequency factor.

TABLE 3.11 Extrapolation of the Heat Rate and TMR of the DSC Signal from Figure 3.13 to Lower Temperatures

T_0 (°C)	q_0 (W/kg)	TMR_{ad} (h)
111	20	0.7
101	13.2	1
91	8.5	1.4
81	5.3	2.2
71	3.2	3.4
61	1.9	5.4
51	1.1	8.8
41	0.6	14.9
31	0.3	26.2

Using these two equations, the $TMR_{ad,24h}$ can be determined if the TMR_{ad} is known at a number of different temperatures. Since the approach is partly based on zero-order assumptions, it is important to focus on the early part of the exotherm, since the influence of a decrease in concentration due to conversion can be neglected there. This way, the experimentally obtained TMR_{ad} at the onset temperature and at a couple of temperatures that are slightly higher are determined. These values are then corrected for the experimental φ-factor according to Eq. (3.12). Finally a straight line is fitted through the plot of $\ln(TMR_{ad})$ versus $1/T$, and the point at which the TMR_{ad} is equal to 24 hours can be read from the graph. This approach is illustrated in the example below.

Let us consider the adiabatic experiment as represented in Figure 3.12. The φ-factor used in this experiment was 1.5. The onset of the exothermicity is detected at 58 °C, 146 minutes after the start of the experiment. The maximum rate is reached after 429 minutes; hence the TMR_{ad} at this temperature is 283 minutes. We will consider the TMR at five different points in the early part of the exotherm, i.e. at 58, 60, 62, 64, and 66 °C. The corresponding TMR_{ad} at the experimental φ-factor and at $\varphi = 1$ are calculated, as shown in Table 3.12.

The according plot based on these data is shown in Figure 3.14. It can be seen that the correlation is indeed linear for both experimental and φ-corrected data points. From the graph, the $TMR_{ad,24h}$ can be obtained directly. In this case, the TMR will be 24 hours at 38.3 °C.

The advantage of this method over the above method based on one DSC measurement is the increased accuracy. The reason for this can be found in different aspects of this approach:

1. This method is based on an adiabatic experiment that will be by definition more representative for the situation in a large-scale reactor during a runaway reaction than a DSC experiment.
2. The temperature range over which the extrapolation takes place is fairly limited: in our example the difference between the experimental onset temperature and the finally obtained $TMR_{ad,24h}$ is only 20 °C (as compared with 80 °C in the DSC example).
3. The extrapolation for the φ-correction is also limited (from $\varphi = 1.5$ to $\varphi = 1$).
4. By using only data points in the early part of the reaction, the error introduced by assuming zero-order kinetics is limited. Indeed, in this early stage with low conversion, the concentration of the reagents can be assumed to be constant.

Also for this method of determining $TMR_{ad,24h}$, a word of caution is needed. Only the data points in the early part of the

TABLE 3.12 TMR at Different Temperatures from the Experimental Adiabatic Run as Shown in Figure 3.12

		TMR_{ad} (min)	
Temperature (°C)	Time (min)	$\Phi = 1.5$	$\Phi = 1$
58	146	283	189
60	189	240	160
62	230	199	133
64	265	164	109
66	294	134	90

The TMR_{ad} at $\varphi = 1$ is calculated according to Eq. (3.12).

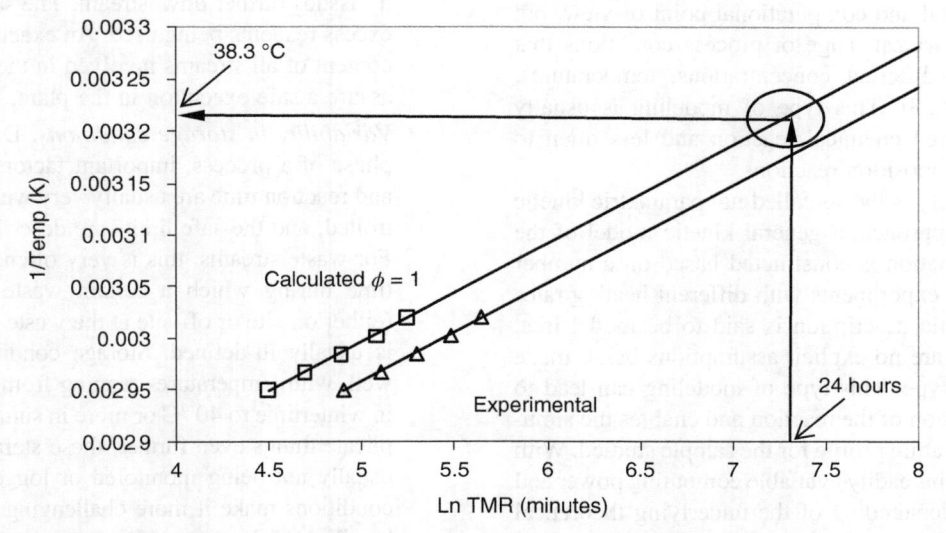

FIGURE 3.14 Extrapolation of the experimentally obtained TMR to lower temperatures and to $\varphi = 1$. The triangles represent the experimentally obtained TMR data at $\varphi = 1.5$, the squares are the corresponding calculated data at $\varphi = 1$, and the $TMR_{ad,24h}$ can be read from the graph as indicated by the arrows.

exotherm are used; consequently we are dealing with very low heat rates at that moment. This poses high demands on the quality of the experimental data: a small amount of drift in the temperature stability of the instrument (either positive or negative drift) can have a profound effect on the final result. It is our experience that a well-operated adiabatic calorimeter should be able to deliver reliable results, provided that regular drift checks on empty test cells are performed to confirm the stability of the instrument.

Here as well, extra attention is needed when dealing with strongly autocatalytic reactions, since this method might overestimate the TMR_{ad} for that kind of reactions, leading to unsafe extrapolations. When a very sudden and sharp temperature increase is noticed in an adiabatic experiment, autocatalysis should be suspected, and more testing will be appropriate. When in doubt, an iso-aging experiment at a temperature close to the calculated $TMR_{ad,24h}$ should be conducted to check the validity of the calculations.

3.5.5.3 Kinetic Modeling Both of the abovementioned methods make it possible to extract the $TMR_{ad,24h}$ from one single experiment but with a limited accuracy due to the assumptions made. There are more advanced methods available as well, which ask for a larger number of experiments and more advanced mathematical models, but which lead to more accurate description of the reaction and allow for a broader range of simulations.

One possible approach to kinetic modeling is the fully mechanistic description of the reaction, which implies a complete understanding of the (decomposition) reaction at a molecular level. The reaction is therefore split up into its elementary reactions, and for each of those the frequency factor, reaction order and activation energy are determined. It goes without saying that not only is this method quite elaborate from an experimental and computational point of view, but it also enables the widest range of process conditions that can be simulated (different concentrations, temperatures, φ-factor, etc.) [22, 23]. This type of modeling is usually applied to the desired chemical reaction and less often to the undesired decomposition reactions.

Another possibility is the so-called nonparametric kinetic modeling. In this approach, a general kinetic model of the (decomposition) reaction is constructed based on a number (usually 5) of DSC experiments with different heating rates [24, 25]. This kinetic description is said to be model free, meaning that there are no explicit assumptions being made about the reaction type. This type of modeling can lead to an accurate description of the reaction and enables the simulation of any temperature profile for the sample studied. With advancements both in readily available computing power and in fundamental understanding of the underlying theoretical kinetic models, this type of modeling is becoming more widespread and has even reached a point where some large companies tend to prefer modeling over adiabatic testing.

It is worth pointing out however that, in order to guarantee a safe process development, the combination of data from different sources (modeling and experimental isothermal, non-isothermal, and adiabatic data) is the preferred way to get to a fundamental understanding of a process during intended, non-intended, and storage conditions.

The reader is referred to the literature for a more detailed discussion of the different possibilities of kinetic modeling for both safety studies and process development [23, 24, 26–28].

3.6 REACTIVE WASTE

In the past decades, knowledge and awareness with regard to reactive chemistry hazards have increased considerably both in industry and academia. One aspect of chemical reactive hazards that is still often overlooked and causes incidents on a regular basis is related to waste streams. When designing a chemical process, be it in the lab or in a production plant, one has to keep the reactivity of all chemicals in mind "from cradle to grave." When using excess reagents, reactivity does not magically disappear once poured into the waste container.

There are several reasons why waste streams can cause unexpected safety issues:

- *Incomplete understanding of the mass balance of the process and waste streams*. When considering process safety, one has to keep in mind the entire mass balance of all products, including waste streams. By-products that are formed should always be kept in mind (e.g. triethylammonium hydrochloride when capturing HCl formed in the process) since they may cause compatibility issues further downstream. The same holds for any excess reagents being used. An exact knowledge of the content of all streams involved in the process is key to assure a safe execution in the plant.
- *Variability in storage conditions*. During the reactive phase of a process, important factors like temperature and reaction time are usually very well known and controlled, and the safe limits are described and analyzed. For waste streams, this is very often not the case. The time during which a certain waste stream is stored (either on-site or off-site at the waste treatment facility) is usually ill-defined. Storage conditions can vary as well, with temperatures ranging from as low as −10 °C in wintertime to 40 °C or more in summertime. To complicate things even further, these storage conditions are usually not being monitored or logged. These varying conditions make it more challenging to assure the safe handling of these waste streams.
- *Mixing with other streams*. In many cases, waste streams from one process are mixed with those of other processes.

Generally speaking, it is very hard to know in advance which other chemicals a certain waste stream may be exposed to. Therefore, one has to consider all reactivity left in a certain waste stream and neutralize those where it is appropriate.
- *Transfer between sites.* Most companies send their waste streams to other companies for final disposal and/or treatment. Therefore the knowledge of the exact composition is very limited at the receiving end of this process, making it difficult to anticipate any reactivity hazards.

It is clear that the points above should be a warning to have a closer look at the waste streams involved in each process. Some examples of possible reactive waste incidents are briefly discussed:

- *Water reactive chemicals.* As a basic rule, all waste streams containing water reactive reagents should be decomposed in a controlled manner before further mixing. Certainly when the reaction with water gives rise to gas generation (which is often the case, like with thionyl chloride, oxalyl chloride, isocyanates, hydrides, etc.), since the presence of water in mixed waste streams can never be ruled out completely. Even a stoichiometric excess of as little as a couple of mole percent should always be decomposed under controlled conditions, since relatively small and slow gas release rates can lead to dangerous pressure buildup in a closed container. Also when the reaction with water does not give rise to gas generation, care is needed. When using acetic anhydride as a solvent, one has to keep in mind that it will react in an autocatalytic manner with water and alcohols, releasing a significant amount of heat during this process. Therefore even this relatively common solvent deserves further attention when dealing with waste streams. It is also noteworthy that most water-reactive chemicals also react with alcohols, adding further to the risk of encountering unwanted side reactions when poured down the waste.
- *Strong acids.* Many synthetic reactions are acid catalyzed, so a multitude of unwanted reactions can be expected when adding strong acids to a waste stream. Most chemical waste treatment facilities do not accept strongly acidic waste streams to start with, so in many cases they should be neutralized before mixing with other streams or be collected and disposed of separately.
- *Nitric acid.* Nitric acid deserves a special place, as the description in Bretherick's justifies [9]: "…Nitric acid is the common chemical most frequently involved in reactive incidents, and this is a reflection of its exceptional ability to function as an effective oxidant even under fairly dilute conditions (unlike sulfuric acid) or an ambient temperature (unlike perchloric acid). Its other notable ability to oxidize most organic compounds to gaseous carbon dioxide, coupled with its own reduction to gaseous 'nitrous fume' has been involved in many incidents in which closed, or nearly closed reaction vessels or storage cabinets have failed from internal gas pressure." Chemical reactivity incidents involving nitric acid still occur all too frequently, in a study from Etchells et al., and it was shown that from the 142 incidents reported to HSE between 1989 and 2005, 14% of them were related to nitric acid reactivity [29].
- *Intrinsically unstable reagents.* Some reagents are inherently unstable and can therefore pose a safety risk when ending up in waste streams. Typical examples are hydrogen peroxide (gas formation upon thermal decomposition) or hydroxylamines (exothermic decomposition, strongly influenced by contamination with metal ions). When such reagents are used in a process, a neutralization procedure is needed to ensure a complete removal of the reactivity before addition to the general waste stream. Also in this case, extra care is needed when the decomposition of the reagent is accompanied by gas generation. Very slow decomposition rates can lead to significant pressure buildup in drums, so attention is needed to follow up on the shelf life of these reagents under storage conditions in order to avoid bulging drums or even rupture and spills.
- *Catalysts.* Special care is needed when catalysts are used in chemical processes. By nature, these catalysts tend to accelerate reactions not necessarily limited to the desired synthetic reaction. Catalysts used for hydrogenations may be pyrophoric as well, so extra attention is needed for the treatment of the spent catalyst. An example of reactive waste incident involving platinum black is described in the literature [30].

These are just some examples of reactive hazards that may cause problems in waste streams. In any case, any excess reagents or reactive and/or unstable products should be accounted for in the entire process, including the waste streams.

3.7 CONTINUOUS PROCESSING

In the description above, we have always considered batch processing, since it is still the most widespread method of chemical manufacturing in the pharmaceutical industry. In recent years, however, continuous processing (or flow chemistry) has come to a point where the first GMP-approved processes are being introduced at a commercial scale. One of the main drivers to consider continuous processing rather than batch processing is the inherent safety benefits of the former. There are several reasons why continuous processing can lead to an intrinsically safer process; some of them are briefly touched upon underneath. For a more in-depth overview, some elaborate review articles about the benefits of flow chemistry can be found in the literature [31, 32]:

- *Better temperature control.* For highly exothermic reactions, batch reactors are fairly limited with regard to temperature control. This has mainly to do with the modest surface to volume ratio of large-scale reactors: in a reactor of 6000 L this ratio is approximately $2\,m^2/m^3$. In a microreactor with 1 mm channels, this ratio is approximately $4000\,m^2/m^3$. This ensures a much better heat transfer, making it easier to control even highly exothermic reactions. This was demonstrated by DSM, who introduced a continuous nitration reaction at full production scale under GMP conditions in 2009 [31]. Since then, many reports have been made in the literature on the successful transfer of highly exothermic reactions from batch to continuous on the lab scale [33]. Because of intellectual property reasons, the number of disclosed large-scale commercial continuous processes remains limited however.
- *Short reaction times.* Related to the previous point, the very efficient heat transfer implies that very rapid heating, and cooling is feasible when using flow chemistry. This way, a short reaction time at a high temperature can be used, making it possible to deal with unstable compounds at higher temperatures for very short time periods.
- *Small reactor volume.* Continuous flow reactors are typically a lot smaller than batch reactors; usually the reactive volume is limited to a few liters at most. This means that even in case a runaway reaction would take place inside the reactor, the consequences will be far less severe. Another consequence of this is that working at more extreme reaction conditions (high or low temperature, high pressure, etc.) is easier and requires less engineering effort.
- *No free headspace.* In flow reactors (apart from continuously stirred tank reactors), there is no free headspace above the reaction mixture. E.g. in the case of azide chemistry, this is a significant safety advantage. When running a batch process using azides, one of the many safety issues during scale-up involves the control of shock sensitive and explosive hydrazoic acid that may accumulate in the headspace above the reaction mixture [34]. When running this type of chemistry in flow, all hydrazoic acid being formed will remain in the solution phase, with an immediate quench with sodium nitrite at the end. This way, a significant gain in process safety can be achieved [35].
- In situ *generation of reagents.* Many highly reactive chemicals are very attractive to use in production processes because of their intrinsic reactivity and atom efficiency but are used only reluctantly because of the severe safety precautions they require. Examples of such reagents are hydrazoic acid, diazomethane, hydrogen cyanide, and phosgene. Several flow processes have been developed where these reagents are generated in situ [36, 37], starting from more benign chemicals. This way of processing can add significantly not only to the overall process safety but also to the sustainability of a process since more atom efficient routes can be explored.

These points clearly illustrate that many processes can benefit from continuous processing when it comes to process safety, but making such a shift in technology implies that other factors should be accounted for instead.

- *Accuracy of pumping.* In batch chemistry, the actual stoichiometry of the reagents is mostly controlled gravimetrically at the endpoint. The total amount of all reagents being fed to the reactor is usually very well known. In flow chemistry, however, the actual stoichiometry of the process will depend on the accuracy of the flow rates being used. This is less straightforward to control with sufficient precision, and care is needed to make sure that there is no unwanted accumulation of excess reagents due to fluctuations in the flow speeds used. If one of the streams stops entirely (e.g. because of an empty feed vessel), the entire process should stop automatically in order to avoid the addition of pure unreacted reagent to the receiving vessel.
- *Flow direction.* Probably the worst-case scenario for a continuous process is to end up in a situation where one of the reagent streams is pumped through the pump of another stream into the feed vessel of this latter stream. Proper flow control and non-return valves should be in place in order to avoid this type of situations.
- *Temperature control and hotspots.* It was mentioned above that temperature control is usually better in continuous processing than in batch processing, but a word of caution is in its place here. In continuous processing, often the temperature of the circulation fluid is measured but not the temperature inside the reactor (which is virtually impossible when using true microreactors). This may be acceptable in many cases, since the heat transfer in continuous processing is very efficient, but one should always be aware that the actual reaction temperature may be different from the intended temperature. Certainly when dealing with strongly exothermic reactions, the formation of hotspots is very likely, and one should therefore always have a good understanding of the thermal stability of the process stream.

- *Consider the workup as well.* In many cases, the process stream leaving the continuous reactor will be collected in a tank or vessel prior to further workup. If excess reagents are being used, the stability and reactivity of this final mixture over a time period that is relevant for the process should be taken into account when assessing the overall process safety.
- *Reactor clogging/blocking.* One of the main process disturbances that can be expected in flow chemistry is reactor clogging. Proper procedures should be in place to deal with this, making sure that the flow of all reagents is either immediately shut off or diverted. Certainly when a process is being run unattended for a long period of time, these precautions become more and more important.

To conclude, one could state that continuous processing has the possibility to increase process safety significantly, but a higher degree of engineering and controlling standards will be needed.

3.8 OTHER POINTS TO CONSIDER

In the paragraphs above, we have discussed the fundamentals of process safety testing. Most of these techniques should be at least considered when developing any chemical process for scale-up. Some other techniques or practices should be brought to the attention of the reader, but, for lack of space, we will only touch upon them very briefly and refer to the literature for more details.

3.8.1 Flammability–Explosivity

Probably the single largest source of hazards in any chemical production plant is not due to intrinsic process safety, but to the risk of fire, especially when working with highly flammable organic solvents and reagents.

For solvents and liquid reagents, the flash point should be known in order to make sure that the instrumentation being used is suited for the job (hazardous area classification, zones, and Ex protection types). The flash point usually needs to be known for regulatory reasons for storage and transportation as well.

For solids, it might be necessary to determine the dust explosion characteristics. If a finely dispersed cloud of an organic solid finds an ignition source, a dust explosion can occur. Some products are more prone to this type of explosion hazards than others, so experimental testing is often needed. This will especially be the case in situations where a cloud of finely divided product can be formed in a non-inerted atmosphere (e.g. in fluidized bed driers, dry mills, or a reactor during charging of a solid). Since these tests require relatively large amounts of product, they are usually only conducted in a late stage of development.

3.8.2 Static Electricity

Since a flammable atmosphere can be present in a chemical production plant, it is important to exclude any type of ignition source at any time. Especially when dealing with organic compounds, static electricity discharges can become very relevant. Many organic solvents and also a lot of the organic solids have a very low conductivity and can be charged easily. If the equipment being used is insufficiently grounded, a sudden discharge might occur, leading to a spark. This spark can in turn act as an ignition source for any flammable atmosphere, either of a vapor cloud or a dust cloud. Therefore a proper understanding of static electricity is very important, and proper testing of the solvents and solids used might be needed (electrical conductivity measurements, charge decay measurements, etc.). The most obvious preventive measures against incidents related to static electricity are a proper grounding of any equipment being used (including the operator) and an appropriate inertization of reactors, driers, etc. whenever possible.

3.8.3 Vent Sizing

When a runaway reaction takes place in a reactor, the amount of gas (and vapor) being liberated can surpass the amount that can be removed through the conventional way, i.e. through the condenser and scrubber lines. Therefore, vents are placed on the reactor. They usually consist of either a bursting disc or a pressure relief valve connected to a vent line of an appropriate diameter. In case of a serious runaway reaction or an external fire leading to an overpressure inside the reactor, the vents will open, and the vent line will allow for a safe depressurization of the reactor. This obviously implies a proper design of the vent system. The amount of gas that can be removed by such a system will obviously depend on the diameter of vent lines and also on the back pressure being generated by these lines. This back pressure is mainly a function of the amount and type of bends in the vent line and its total length. Another important factor one should consider is the composition of the gas flowing through the vent lines: if there is only gas flowing through the vent line, the minimum diameter needed will differ from the case when a mixture of liquid and gas is leaving the reactor. Proper vent sizing calculations are very complex and should only be undertaken by experts with the proper experience [38–40].

3.8.4 Safety Culture and Managerial Issues

Merely having a proper technical understanding of the process hazards present is not enough to guarantee the safety in a chemical production plant. The safety culture of the entire company, from the highest management level down to the shop floor, is of utmost importance. "Nothing is that important or urgent that it should not be done safely" should not be a hollow phrase but a natural part of the everyday work.

There are also managerial systems that are very important for reaching a high level of safety. Management of change should be taken very seriously: even minor changes made to a process can turn it from a safe process into an unsafe one, so safety should be kept in mind with every change made, and a proper hazard reevaluation might be needed. Adequate systems should be in place to establish the roles and responsibilities of all involved in process safety and to guarantee a standardized framework for safety assessments like HAZOP, HAZAN, and PHA.

REFERENCES

1. Barton, J. and Rogers, R. (1993). *Chemical Reaction Hazards*. Rugby: Institution of Chemical Engineers.
2. Grewer, T. (1994). *Thermal Hazards of Chemical Reactions*, Industrial Safety Series, vol. 4. Amsterdam, New York: Elsevier.
3. Stoessel, F. (2008). *Thermal Safety of Chemical Processes, Risk Assessment and Process Design*. Weinheim: Wiley.
4. Stoessel, F. (1993). *Chemical Engineering Progress* 89 (10): 68.
5. Gygax, R. and IUPAC (1991). *Safety in Chemical Production*. Oxford: Blackwell Scientific.
6. Stoessel, F. (2007). *Loss Prevention and Safety Performance in the Process Industries Symposium*, Edinburgh (22–24 May 2007), IChemE Symposium Series no. 153, IChemE.
7. Cisneros, L., Rogers, W.J., and Mannan, M.S. (2004). *Thermochimica Acta* 414: 177.
8. Chervin, S., Bodman, G.T., and Barnhart, R.W. (2006). *Journal of Hazardous Materials* 130: 48–52.
9. Urben, P.G. (ed.) (2007). *Bretherick's Handbook of Reactive Chemical Hazards*, 7e. Oxford: Academic Press.
10. Weisenburger, G., Barnhart, R.W., Clark, J.D. et al. (2007). *Organic Process Research and Development* 11 (6): 1112–1125.
11. Weisenburger, G., Barnhart, R.W., and Steeno, G.S. (2008). *Organic Process Research and Development* 12 (6): 1299–1304.
12. Burli, M. (1979). Ueberpruefung einer neuen Methode zur voraussage des Waermeueberganges in Ruehrkesseln. Dissertation ETHZ, 6479.
13. Zufferey, B. (2006). Scale-down approach: chemical process optimisation using reaction calorimetry for the experimental simulation of industrial reactors dynamics. Dissertation No 3464, Ecole Polytechnique Federale de Lausanne.
14. Wilson, E.E. (1915). *The American Society of Mechanical Engineers* 37: 47.
15. Yosida, T. (1987). *Safety of Reactive Chemicals*. Amsterdam: Elsevier.
16. Dien, J.M., Fierz, H., Stoessel, S., and Kille, G. (1994). *Chimia* 48: 542–550.
17. Dermaut, W. (2006). *Organic Process Research and Development* 10: 1251–1257.
18. Keller, A., Stark, D., Fierz, H. et al. (1997). *Journal of Loss Prevention in the Process Industries* 10 (1): 31–41.
19. McIntosh, R.D. and Waldram, S.P. (2003). *Journal of Thermal Analysis and Calorimetry* 73: 35–52.
20. Wiss, J., Stoessel, F., and Kille, G. (1993). *Chimia* 47 (11): 417–423.
21. Townsend, D.I. and Tou, J.C. (1980). *Thermochimica Acta* 37: 1–30.
22. Gigante, L., Lunghi, A., Martinelli, S. et al. (2003). *Organic Process Research and Development* 7: 1079–1082.
23. Remy, B., Brueggemeier, S., Marchut, A. et al. (2008). *Organic Process Research and Development* 12: 381–391.
24. Dermaut, W., Fannes, C., and Van Thienen, J. (2007). *Organic Process Research and Development* 11: 1126–1130.
25. Roduit, B., Dermaut, W., Lunghi, A. et al. (2008). *Journal of Thermal Analysis and Calorimetry* 93 (1): 163–173.
26. Roduit, B., Hartmann, M., Folly, P. et al. (2015). *Thermochimica Acta* 621: 6–24.
27. Abele, S., Schwaninger, M., Fierz, H. et al. (2012). *Organic Process Research and Development* 16 (12): 2015–2020.
28. Waser, M., Obermuller, R., Wiegand, J. et al. (2010). *Organic Process Research and Development* 14 (3): 562–567.
29. Etchells, J., James, H., Jones, M., and Summerfield, A. (2010). *Loss Prevention Bulletin* 210: 18–23.
30. Okada, K., Akiyoshi, M., Ishizaki, K. et al. (2014). *Journal of Hazardous Materials* 278: 75–81.
31. Braune, S., Pochlauer, P., Reintjens, R. et al. (2009). *Chemistry Today* 27 (1): 26.
32. Gutmann, B., Cantillo, D., and Kappe, O. (2015). *Angewandte Chemie International Edition* 54: 6688–6728.
33. Plutschack, M., Pieber, B., Gilmore, K., and Seeberger, P. (2017). *Chemical Reviews* 117 (18): 11796–11893.
34. Gonzales-Bobes, F., Kopp, N., Li, L. et al. (2012). *Organic Process Research and Development* 16 (12): 2051–2057.
35. Gutmann, B., Roduit, J.-P., Roberge, D., and Kappe, O. (2010). *Angewandte Chemie International Edition* 49: 7101–7105.
36. Fuse, S., Tanabe, N., and Takahashi, T. (2011). *Chemical Communications* 47: 12661.
37. Poechlauer, P., Braune, S., Dielemans, B. et al. *Chimica Oggi* 30 (4): 51–54.
38. CCPS (1998). *Guidelines for Pressure Relief and Effluent Handling Systems*. Hoboken, NJ: Wiley.
39. Fisher, H.G., Forrest, H.S., Grossel, S.S. et al. (1992). *Emergency Relief System Design Using DIERS Technology, the Design Institute for Emergency Relief Systems (DIERS) Project Manual*. New York: AICHE.
40. Etchells, J. and Wilday, J. (1998). *Workbook for Chemical Reactor Relief System Sizing*. Norwich: HSE.

4

CALORIMETRIC APPROACHES TO CHARACTERIZING UNDESIRED REACTIONS

Megan Roth and Tom Vickery
Chemical Engineering R&D, Merck & Co., Inc., Rahway, NJ, USA

4.1 INTRODUCTION

Identification and characterization of the thermal and pressure consequences of both desired and undesired reactions are critical components of assessing the safety of a process. Although the desired reactions are typically studied in depth, undesired reactions are frequently neither identified nor characterized. An undesired reaction is the reaction that occurs when as few as one process deviation such as a single human or equipment error occurs, causing the process to not run as intended. It is important to identify these hazards and understand the associated consequences in order to take the appropriate corrective actions to either mitigate or avoid the risks. This data may influence the design of the process or equipment as it feeds into associated hazard assessments (e.g. process hazard analysis (PHA), hazard and operability study (HAZOP), failure mode and effects analysis (FMEA)) [1]. The end goal is to protect people, protect equipment/facilities, and prevent disruption to the supply of medicines.

Even though it is uncommon to reach an undesired reaction that has severe consequences associated with it, when these reactions are present in a process, it is important that inherent protection is provided. A process may run for years without an incident before one day an error leads to a runaway reaction and vessel explosion. Below are excerpts from a relevant incident investigation by the Chemical Safety Board (CSB) [2] that highlights the importance of characterizing undesired reactions:

In April 1998, "an explosion and fire occurred during the production of Automate Yellow 96 Dye at the Morton International, Inc. (now Rohm & Haas) Plant in Paterson, New Jersey. The explosion and fire were the consequence of a runaway reaction, which over-pressured a 2000-gallon capacity chemical reactor vessel (or kettle) and released flammable material that ignited." "Nine employees were injured, including two seriously, and potentially hazardous materials were released into the community." "The reaction accelerated beyond the heat-removal capability of the kettle. The resulting high temperature led to a secondary runaway decomposition reaction causing an explosion, which blew the hatch off the kettle and allowed the release of the kettle contents."

Morton's initial research and development for the Yellow 96 process identified the existence of both a desired and undesired exothermic reaction. "The Paterson facility was not aware of the decomposition reaction. The Process Safety Information (PSI) package, which was used at the Paterson plant to design the Yellow 96 production process in 1990 and which served as the basis for a Process Hazard Analysis (PHA) conducted in 1995, noted the desired exothermic reaction, but did not include information on the decomposition reaction." Also, there was a discrepancy in the MSDS for Yellow 96, where the boiling point was listed at 100C when the actual boiling point is approximately 330C, greater than the decomposition temperature. In addition, no action had been taken on a recommendation from 1989 to perform additional studies to characterize the reaction rate under worst-case conditions and obtain pressure rise data to size emergency venting equipment. These findings are among various other key findings by the CSB.

Chemical Engineering in the Pharmaceutical Industry: Active Pharmaceutical Ingredients, Second Edition.
Edited by David J. am Ende and Mary T. am Ende.
© 2019 John Wiley & Sons, Inc. Published 2019 by John Wiley & Sons, Inc.

Several other relevant incident investigations can also be found on the CSB website (www.csb.gov).

This chapter will build upon the introduction to process safety (Chapter 3) to provide further details about the use of calorimetric approaches for studying the undesired decomposition reaction. The approaches include the use of thermal and pressure screening tools in addition to the use of more advanced instrumentation and techniques. Instruments highlighted include differential scanning calorimetry (DSC), multiple module calorimeter (MMC), accelerating rate calorimeter (ARC), and advanced reactive systems screening tool (ARSST). Several case studies will then be presented to demonstrate the use of these tools and highlight the importance of identifying and characterizing undesired reactions.

4.2 BACKGROUND

In order to accurately assess the safety of a process, it is important to understand the consequences of heat and pressure and the importance of scale on the process risks. The selection of samples for evaluation can also have a serious impact on the effectiveness of the hazards evaluation.

4.2.1 Heat and Pressure Effects

The generation of heat by desired reaction or decomposition requires one of two things to occur: either the heat must be removed via cooling or phase change, or the batch temperature must increase. For example, a 2 J/g energy release could raise the temperature of an organic material approximately 1 K, depending on the heat capacity (typically ~2 J/g·K for an organic liquid):

$$\Delta T_{ad} = \frac{Q}{Cp} = \frac{2\,(\text{J/g})}{2\,(\text{J/g}\cdot\text{K})} = 1\,\text{K}$$

Acid–base reactions typically have energies on the order of 40–60 kJ/mol. Therefore to produce a 1 K temperature increase would require only a modest reaction concentration of approximately 0.04 mol/l:

$$C = \frac{Cp \times \Delta T \times \rho}{\Delta H_{rxn}} = \frac{2\,(\text{J/g*K}) \times 1\,\text{K} \times 1000\,(\text{g/l})}{50\,000\,(\text{J/mol})} = 0.04\,(\text{mol/l})$$

The amount of heat generated by either the desired chemistry or decomposition, and the associated temperature rise, should be used in combination with operating parameters such as batch temperature, maximum desirable temperature, vessel cooling capability, and maximum jacket temperature to determine the risk. For example, if an exothermic reaction is being heated to temperature by an electrical heating mantle,

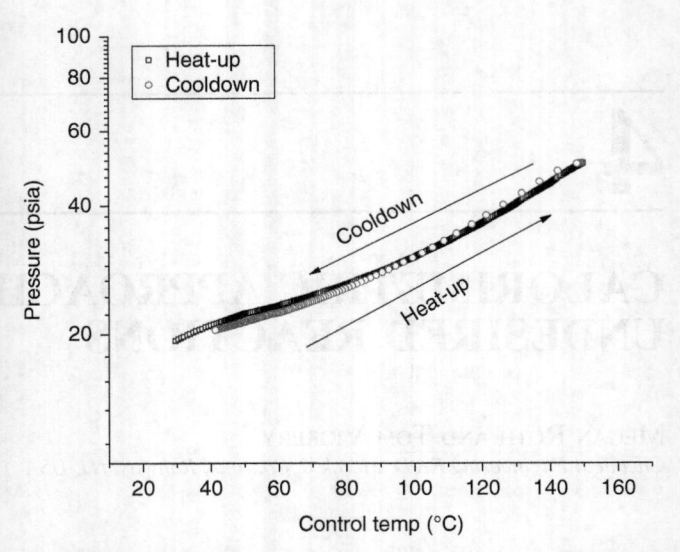

FIGURE 4.1 Pressure–temperature curve of a reaction mixture without permanent gas generation.

the mantle can serve as insulation, preventing the removal of heat as the temperature rises as a result of the reaction.

As the temperature increases, the rates of most reactions increase, with a general approximation that the rate doubles for every 10 K temperature increase; the exact rate of increase in temperature will depend on the activation energy. Therefore failing to remove the heat of the reaction or decomposition can lead to a rapid increase in the rate that heat is generated, which can further increase the batch temperature. This phenomenon is known as a runaway reaction and is implicated in many serious chemical incidents, such as the one in the introduction.

The total pressure present in a closed reactor has three contributors: the initial permanent gas in the vessel, which undergoes thermal expansion as a result of heating, the vapor pressure of any solvents, and the generation of additional permanent gas. The first two terms are a function primarily of the vessel temperature and are reversible. In the absence of permanent gas generation, the heat-up and cooldown curves should be nearly identical on a plot of pressure versus temperature. First the pressure increases as the temperature increases, and the pressure decreases as temperature is cooled back to the starting temperature. One exception is when the solvent is involved in the reaction, which may shift the curve up or down. See Figure 4.1.

If the reaction generates a permanent gas[1] such as CO_2, N_2, or HCl, the pressure change becomes a function of both temperature and time and is irreversible. When the heat-up and cooldown pressure–temperature curves are plotted, the pressure during cooldown is much higher than when it was heated. The difference in pressure following heat-up and cooldown is from the permanent gas that was generated. See Figure 4.2.

---
[1]Permanent gases are sometimes also referred to as non-condensable gases.

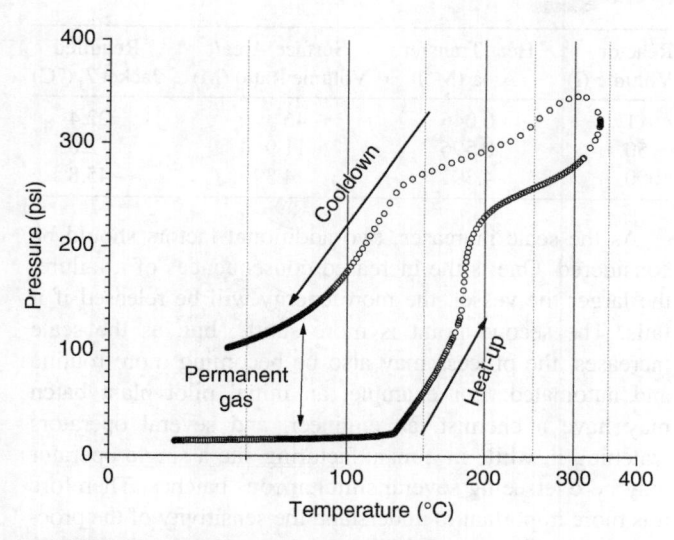

FIGURE 4.2 Pressure–temperature curve of a reaction mixture with permanent gas generation.

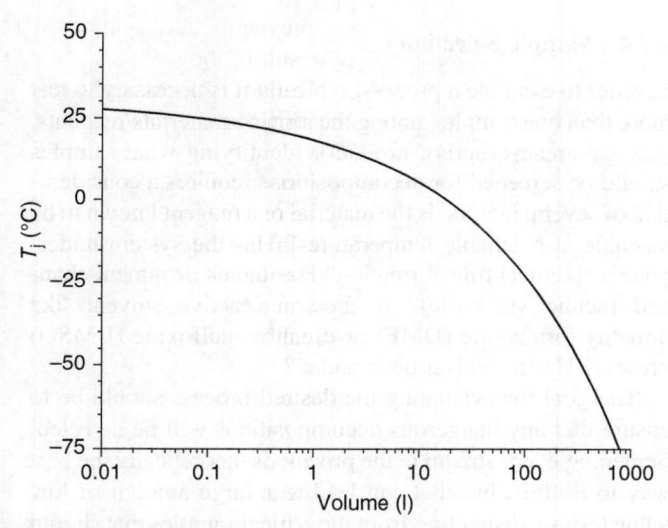

FIGURE 4.3 Jacket temperature vs. reactor size for the reaction modeled in Example Problem 4.2.

The total pressure generated, in addition to the rate of pressure generation, should be compared with operating parameters such as the vessel maximum allowable working pressure (MAWP) and available venting capacity to determine the associated level of risk.

By comparing two versions of a reaction, one of which generates gas, and one of which does not:

1. $A(l) + B(l) \rightarrow C(l) + D(l)$
2. $A(l) + B(l) \rightarrow C(l) + N_2(g)$

we can show that even in fairly dilute conditions, the formation of gas leads to a major increase in the potential maximum pressure with associated risk increase.

EXAMPLE PROBLEM 4.1 Calculation of Pressure Generation from a Reaction

ΔH_{rxn} (J/mol) = −100 000 = generic heat of reaction
C (mol/l) = 0.1
V (l) = 0.5 = reaction mixture volume
ρ (g/l) = 785
Cp (J/g·K) = 1.9
P_0 (psia) = 14.7 = initial permanent gas pressure
VP (psi) = $f(T)$ = vapor pressure
T_0 (K) = 323.1 = temperature before reaction
T_f (K) = 329.8 = temperature after reaction
V_h (l) = 0.5 = headspace volume
R (psi·l)/(mol·K) = 1.206 = ideal gas constant
n (mol) = 0.05 = moles involved in the reaction

A reaction is run at 50 °C in acetone as the solvent:

$$\Delta T_{ad} = \frac{\Delta H_{rxn} \times C}{\rho Cp} = \frac{100\,000 \left(\frac{J}{mol}\right) \times 0.1\,(mol/l)}{1.9 \left(\frac{J}{g \cdot K}\right) \times 785\,(g/l)}$$

$= 6.7\,K, T_f = T_0 + \Delta T_{ad} = 323.1\,K + 6.7\,K = 329.8\,K$

Thermal expansion = $\Delta P = \left(\frac{T_f}{T_0} - 1\right) P_0 = \left(\frac{329.8\,K}{323.1\,K} - 1\right)$
$\times 14.7\,psi = 0.3\,psi$

Change in acetone vapor pressure: $\Delta P = VP_{329.8K} - VP_{323.1K} = 15.0\,psi - 11.9\,psi = 3.1\,psi^2$

Gas generation $n = C\,V = 0.1\,(mol/l) \times 0.5\,l = 0.05\,mol$

$$\Delta P = \frac{nRT_f}{V_h} = \frac{0.05\,mol \times 1.206 \frac{l\text{-psi}}{mol \cdot K} \times 329.8\,K}{0.5\,L} = 39.8\,psi$$

Factor	Reaction with No Gas	Reaction with Gas
Thermal expansion (psi)	0.3	0.3
Change in VP (psi)	3.1	3.1
Gas generation (psi)	0.0	39.8
Total ΔP (psi)	3.4	43.2

[2] Calculated from parameters for acetone in Perry's [3].

4.2.2 Onset Temperature Versus Exotherm Initiation Temperature (EIT)

Throughout this chapter two key temperatures will be discussed. One is the onset temperature, which is used in discussing DSC results. This concept is discussed in Section 3.5.2; typically about a 10 mW/g heat generation rate can be detected. The other temperature is the exotherm initiation temperature (EIT), which is defined as a temperature rise rate of 0.02 K/min at a phi factor of 1.0, typically measured in an adiabatic calorimeter such as an ARC or vent sizing package (VSP), and corresponds to a heat generation rate of 0.7 mW/g for a typical organic sample [4], compared with the typical 10 mW/g threshold for DSC onset. The EIT, which is normally considered to be the temperature at which the reaction becomes hazardous, can be from 25 to 50 K or more lower than the onset temperature.

4.2.3 Scale Dependence

The scale at which a reaction is run can have a dramatic effect on the safety of process. This is commonly referred to as the "square-cube effect" or "surface-area-to-volume ratio," where the hazard (heat generation, gas generation) increases as the cube of the "radius" (proportional to volume), while the protection available (jacket cooling, gas removal) increases as the square of the "radius" (proportional to surface area).

For example, a reaction that generates 50 W/l (~0.05 W/g) of heat might require a jacket temperature of 22 °C to maintain the 30 °C temperature at a 1 L scale, but 2 °C at a 50 L prep lab scale, and −45 °C at a 1000 L manufacturing scale (See Figure 4.3).

EXAMPLE PROBLEM 4.2 Calculation of Cooling Requirements

Rearranging the energy balance equation $UA(T_j - T_r) + \dot{Q} = mCp\frac{dT_r}{dt}$, assuming a cylindrical vessel with diameter = height, and assuming that the objective is to hold the temperature constant ($dT_r/dt = 0$):

U (W/M^2·K) = 150 = overall heat transfer coefficient
T_r (°C) = 30 = desired reaction temperature
$\dot{Q}(\frac{W}{l})$ = 50 = rate of heat generation by the reaction
T_j (°C) = required jacket temperature

$$T_r - \frac{\dot{Q}(\pi r^2 h)}{U(\pi r^2 + 2\pi rh)} = T_j$$

Reactor Volume (l)	Heat Transfer Area (M^2)	Surface Area/ Volume Ratio (M)	Required Jacket T_j (°C)
1	0.046	46	22.4
50	0.596	11.9	2.0
1000	4.39	4.39	−45.8

As the scale increases, two additional factors should be considered. One is the increased consequences of a failure; the larger the vessel, the more energy will be released if it fails. The second point is more subtle, but, as the scale increases, the process may also be becoming more routine and automated. For example, an initial pilot plant batch may have a chemist, an engineer, and several operators watching it, while in a manufacturing site a single operator may be overseeing several simultaneous batches. Therefore it is more important to understand the sensitivity of the process to deviations, and the consequences of the deviations, which may trigger increased and more sensitive testing, using adiabatic calorimetry, for example.

4.2.4 Sample Selection

In order to evaluate a process, typically it is necessary to test more than one sample (among the starting materials/reagents, process streams, and/or product). Identifying what samples should be screened for decompositions requires a consideration of several factors. Is the material or a reagent known to be unstable at reachable temperatures? Has the system undergone a chemical transformation? Has the environment changed (acidic vs. basic)? Is there a reactive solvent like dimethylformamide (DMF) or dimethyl sulfoxide (DMSO) present? Has a catalyst been added?

The goal for evaluating the desired process should be to ensure that any dangerous decompositions will be detected. Screening every stream in the process is theoretically the best way to do this, but that can lead to a large amount of low value testing, distracting from the critical samples that should be carefully evaluated. Therefore testing of select streams may be sufficient to evaluate the process, if the streams that present the highest likelihood of being process safety risks can be systematically identified.

For example, a nitro compound is dissolved in ethanol and reduced with a catalyst and hydrogen to an amine and then isolated as an HCl salt. Possible samples might include the nitro compound itself (known instability from the literature), the solution plus catalyst before hydrogen is added (does the catalyst increase instability), the solution after the hydrogenation (is the amine stable, and are unstable impurities such as hydroxylamines and nitroso compounds present), the mother liquors (contain all process impurities and may be held for an extended period of time), and the isolated amine salt, which is a new species with a new environment (acid vs. base).

TABLE 4.1 Comparison of Calorimetric Techniques for Studying Decompositions

Instrument	Sample Size (g)	Typical Run Time (h)	Lowest Detectable Heat Flow (mW)	Sensitivity (W/kg)	Maximum Temperature (°C)	Maximum Pressure (psi)
DSC	0.005	0.5–5	0.02	4	450	400
Small-scale reaction calorimeter	5	5–24	1	0.2	150	500
Large-scale reaction calorimeter	100	1–24	50	0.5	200	150
Adiabatic calorimeter	2	4–72	2	0.7	500	1500

Beyond the streams from the desired process, if a particular deviation from a single point of failure is considered to either be likely, or of serious consequences, then it may be desirable to assess the stream or streams that would result from the deviation. For example, a hazardous reagent such as hydrogen peroxide might be overcharged, or 5 equiv. of sodium azide instead of 1 may be used, or the hydrogenation discussed earlier might stall at 50% completion, leaving a dangerous combination of intermediates in the batch. Once the appropriate samples to test have been identified, thermal screening for heat and gas generation should be carried out.

4.2.5 Techniques for Evaluating the Thermal Behavior of Decompositions

DSC is the most common but not the only tool for evaluating the thermal behavior of decomposition reactions. Table 4.1 compares several techniques used to study them. The DSC's main advantages are small sample size and short run times. The main disadvantage is that an exotherm rate must exceed a certain threshold before it is detected, which typically means that the reported onset temperature is higher than the actual EIT, so if only DSC data is available, a conservative safety margin may need to be applied. This is the origin of the traditional "100-degree rule" referred to in Section 3.5.2, where the exothermic activity is considered to be avoided 100 °C below the exotherm onset. As noted, in some cases this rule may not be adequate, in particular if autocatalytic activity or low activation energies are involved. A more detailed discussion of onset temperature issues can be found in Hofelich [5].

As the most common method for detecting the presence of decomposition exotherms, DSC will be discussed below.

4.3 THERMAL STABILITY SCREENING: DIFFERENTIAL SCANNING CALORIMETRY (DSC)

The basics of DSC screening for hazards evaluation are covered in Section 3.5.2. This section discusses some additional points to consider in using DSC testing for evaluating decompositions.

4.3.1 Cell Selection

There are several types of DSC cells and seals in common use. They differ mainly in weight, material of construction (MOC), sealing, and pressure rating.[3] The cell weight affects the sensitivity and the quality of the baseline: lighter cells generally produce better signal-to-noise ratios. The MOC of the cell is important if a decomposition is being studied. In some cases certain materials can catalyze the decomposition (hydrazine in Hastelloy, for example). In other cases either the stream contains a corrosive material (HCl, for example) or generates a corrosive material. Corrosion reactions are highly exothermic ($1/3\ Al(s) + HCl(aq) \rightarrow 1/3\ Al^{+3} + 3Cl^- + 3/2\ H_2\ \Delta H = -177\ kJ/mol$[4]) and can appear as false positives.

Sealing and pressure rating of the cells are related; if a DSC cell is not hermetically sealed, the evaporation of either the solution solvent or the solvate solvent (commonly water) will occur. For studying decompositions, this is undesirable for several reasons: (i) the heat loss from evaporation can mask exotherms that may occur near the boiling point. (ii) The sample no longer reflects what will be in the process equipment – it will either desolvate or turn into a neat solid or liquid, both of which can have very different onset temperatures or energies than the original sample. If the seal fails during a decomposition, all of the volatile material and any confined gases will escape, both providing evaporative cooling and decompressive cooling, thus preventing accurate evaluation of the exotherm size.

For studying decompositions up to high temperatures (~400 °C), stainless steel, Hastelloy, and gold- and tantalum-lined cells with gold or gold-plated or PTFE high pressure seals are commonly used, combining corrosion resistance, a high pressure rating, and good sealing in one package.[5]

---

[3] A good overview of the variety of cells available is available at: Ref. [6].
[4] Based on heat of formation values from: Ref. [7].
[5] Pans are available both directly from DSC vendors (TA Instruments, Netzsch, Perkin-Elmer, Mettler-Toledo and others) and aftermarket suppliers.

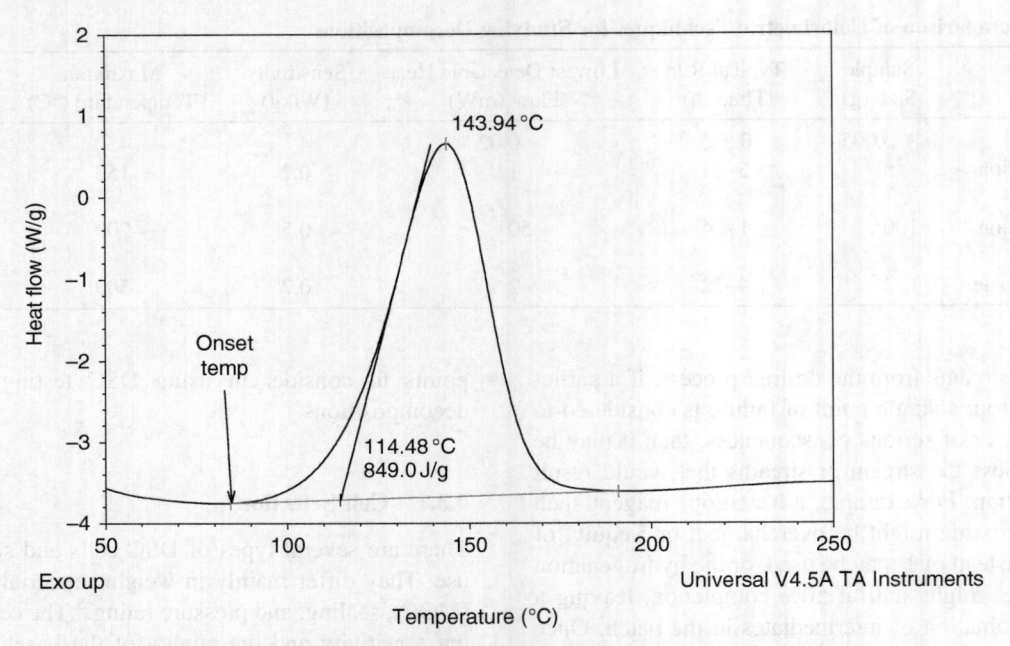

FIGURE 4.4 Typical scanning DSC trace of an exothermic decomposition.

Decompositions cannot normally be studied in standard DSC aluminum cells, even if crimp sealed. The cells cannot withstand the high pressures associated with either gas formation or solvent vapor pressures, and if the cell begins to leak, much of the information on the decomposition will be lost. In addition, aluminum has poor resistance to acids and bases, and the highly exothermic corrosion reaction may completely mask the desired signal. For this reason, DSCs for process safety testing are typically run in gold-lined or tantalum-lined cells, equipped with pressure-resistant closures that are either crimped or threaded.

Although decomposition reactions typically follow Arrhenius-type kinetics, $k = A \exp(E_a/RT)$, there are frequently several reactions involved, the exact reaction pathway is seldom known, and the contribution of each reaction to the overall heat generation is difficult to determine. Therefore DSC data is mainly used to get overall data on the decomposition: the onset temperature, the onset time, the total energy of decomposition, and the peak rate of decomposition.

4.3.2 Scanning

A typical DSC screening scan starts at room temperature and continues to a temperature of 250–400 °C to ensure all exotherms of interest have at least been detected if not completed. A normal scan rate is 5–10 K/min to minimize run time.[6]

A typical scanning DSC trace of an exothermic decomposition is shown in Figure 4.4.

[6] http://www.tainstruments.com/pdf/literature/TA297.pdf

The onset is the point at which the signal deviates from baseline (typically estimated visually) or temperature at which the signal has risen to a certain level above baseline (typically 0.02–0.1 W/g). In Figure 4.4 the DSC onset was called at 84.5 °C.

Determining the onset can be further complicated by the presence of an endothermic melt, as exothermic activity could have initiated either during or immediately after the melt. In the DSC shown in Figure 4.5, the exothermic onset could be anywhere between 73 and 81 °C.

If an exotherm is suspected to initiate below 50–75 °C, pre-cooling the DSC and the sample and starting the scan at a lower temperature and/or scan rate will allow more accurate evaluation of the exotherm.

Sometimes rerunning the sample at a lower scan rate (0.5–5 K/min) is desirable. The most common reasons for doing so are (i) to allow the exotherm to reach completion, (ii) to improve the separation between the melt and a follow-up exotherm to see if the exotherm truly initiates in the melt, and (iii) to improve separation of overlapping peaks.

The basics of evaluating DSC results are discussed in Section 3.5.2. Some additional useful information that can be obtained from the DSC tests are discussed in Section 4.3.4.

4.3.2.1 Isothermal Aging
The other common DSC technique used for decomposition analysis is isothermal aging, where a sample is rapidly heated to a target temperature and then aged for a period of time (6–24 hours is typical in the DSC). The most straightforward use of isothermal aging technique is to detect or confirm the presence of autocatalytic activity, where

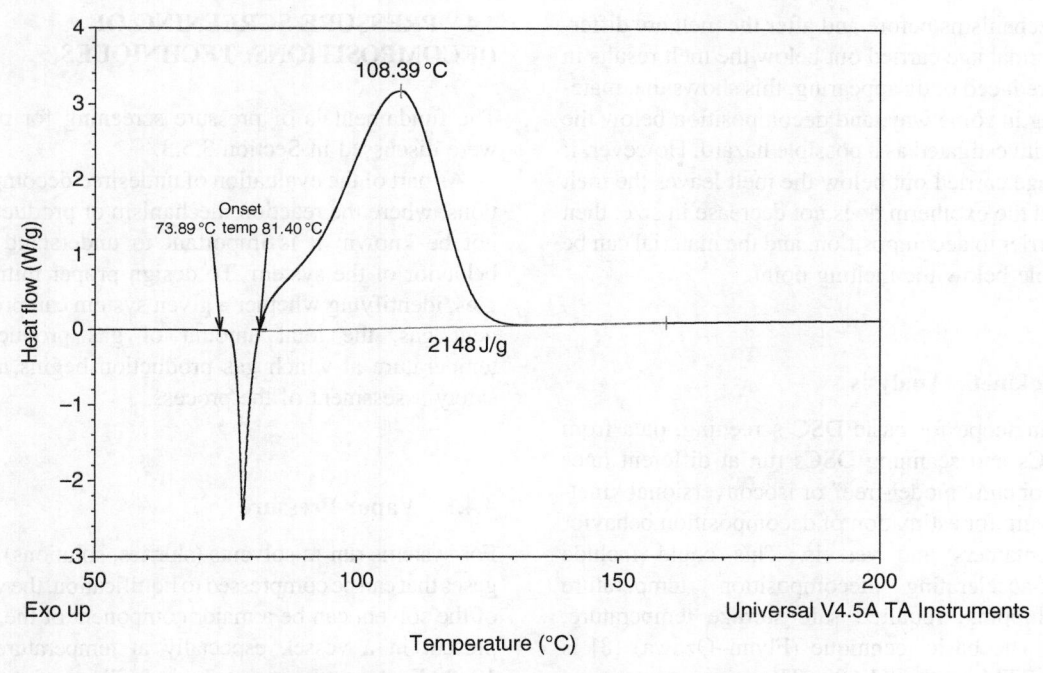

FIGURE 4.5 DSC trace of an exotherm with a melt.

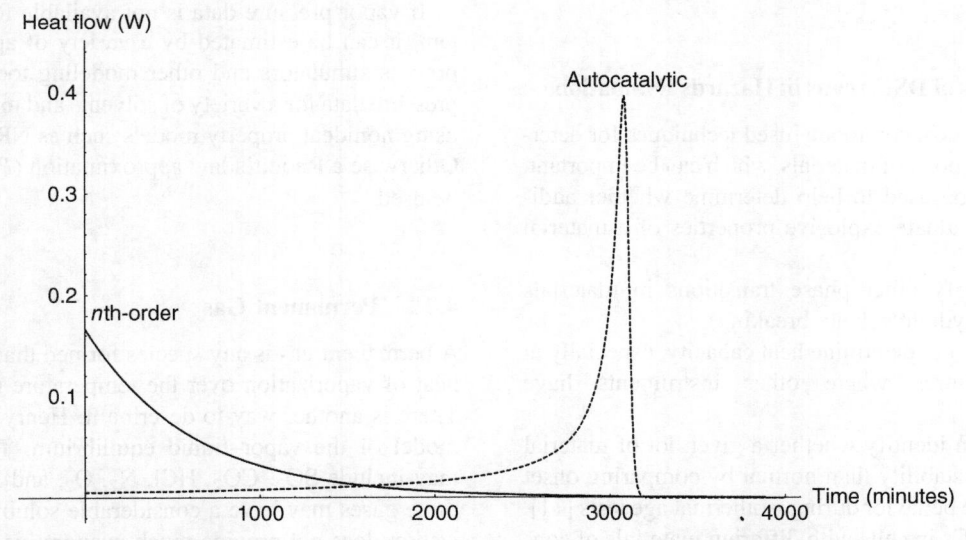

FIGURE 4.6 Comparison of heat flow during an isothermal age of an autocatalytic decomposition (dashed line) vs. an nth-order decomposition (solid line).

an nth-order reaction will show either an effectively constant or steadily decreasing heat flow, whereas an autocatalytic reaction will show a distinct maximum in the heat flow as the sample is aged at constant temperature (see Figure 4.6).

The second use for isothermal aging is to estimate a decomposition rate from the decrease in the exotherm peak size. By holding a sample at constant temperature for a period of time and then rescanning the sample and comparing the new exotherm size to the original, an estimate of the decomposition rate can be made. Typically an exotherm decrease of 10% can be reproducibly obtained, and assuming a 24-hour (86 400 seconds) age, and an exotherm size of 500 J/g, a decomposition rate of $(0.1 \times 500$ J/g$)/86\,400$ seconds $= 0.000\,58$ W/g $= 0.58$ mW/g can be studied, which is an order of magnitude increase in the sensitivity vs. a scanning DSC.

The final use of isothermal aging is to determine whether a melt as in Figure 4.5 presents a thermokinetic barrier, where the reaction mechanisms before and after the melt are different. If an isothermal age carried out below the melt results in the melt being reduced or disappearing, this shows that material is degrading in some way, and decomposition below the melt should be investigated as a possible hazard. However, if an isothermal age carried out below the melt leaves the melt unchanged, and the exotherm does not decrease in size, then the melt is a barrier to decomposition, and the material can be considered stable below the melting point.

4.3.3 Thermokinetic Analysis

Although not in scope for basic DSC screening, data from isothermal DSCs and scanning DSCs run at different rates can be used to obtain "model-free" or isoconversional kinetics [8, 9], allowing for estimation of decomposition behavior in real-life containers and vessels. This could include TMR24, self-accelerating decomposition temperature (SADT) for shipping, required safe storage temperature, and shelf life. The basic technique (Flynn–Ozawa) [8] is described in ASTM method E698. There are commercial packages to carry out this analysis from Mettler Toledo, AKTS, Netzsch, and others.

4.3.4 Other Uses of DSC Scans in Hazards Evaluation

DSC is one of the most commonly used techniques for determining the melting point of materials, which can be important.

DSC data can be used to help determine whether additional testing to evaluate explosive properties of a material is needed [10].

DSC can identify other phase transitions in materials (glass transition, hydrate/solvate breaking).

DSC can be used to determine heat capacity, especially at elevated temperatures where other instruments have challenges.

DSC testing can identify whether a given lot of material has lower thermal stability than normal by comparing onset temperatures or the behavior during isothermal age tests [11].

By running DSCs in cells with differing materials of construction, possible catalytic effects can be identified.

4.3.5 Summary of DSC for Decomposition Evaluation

DSC testing is a simple way to screen for hazardous decompositions, giving good data on exotherm size and onset temperature, using only 1–10 mg of sample. DSC screening can identify where more accurate techniques (ARC, for example) are needed, but DSC screening alone should not be used where accurate rate vs. temperature data is needed.

4.4 PRESSURE SCREENING OF DECOMPOSITIONS: TECHNIQUES

The fundamentals of pressure screening for process safety were discussed in Section 3.5.3.

As part of the evaluation of undesired decomposition reactions, where the reaction mechanism or product species may not be known, it is important to understand the pressure behavior of the system. To design proper mitigation strategies, identifying whether a given system can produce permanent gas, the total amount of gas produced, and the temperature at which gas production begins all impact the safety assessment of the process.

4.4.1 Vapor Pressure

For systems run in solvents (slurries, solutions) and for some gases that can be compressed to liquification, the vapor pressure of the solvent can be a major component of the total pressure present in a vessel, especially at temperatures more than 10–20 K above the atmospheric boiling point. For example, acetone at 373 K has a vapor pressure of 54 psi and at 383 K 69 psi [3].

If vapor pressure data is not available for the actual system, it can be estimated by a variety of approaches. Many process simulators and other modeling tools include vapor pressure data for a variety of solvents and mixture calculators using nonideal property models such as NRTL or UNIFAC. Otherwise a Raoult's law approximation ($P_{tot} = \sum_i x_i P_i$) can be used.

4.4.2 Permanent Gas

A permanent gas is any species formed that does not incur a heat of vaporization over the temperature range of interest. There is another way to describe it: Henry's law[7] is a good model of the vapor–liquid equilibrium. Typical examples may include SO_2, CO_2, HCl, N_2, O_2, and H_2. Even though these gases may have a considerable solubility, their vaporization does not provide much evaporative cooling.

The formation of permanent gas is a major complexity in evaluating the safety of a material or stream, as unlike vapor pressure and Gay-Lussac[8] pressure, which are only a function of temperature, permanent gas formation will cause the pressure to be a function of both liquid fill fraction and temperature. Permanent gas formation is of importance for both desired and undesired chemistries.

---

[7]Henry's law: $C_{gas} = K_H \times p_{gas}$ where K_H has units of (mol/l·bar).
[8]$P = nRT/V$.

4.4.3 Measurement Techniques

All of the following tests are carried out in closed equipment cells capable of withstanding pressure. Although metal cells (e.g. Hastelloy C-276, 316 L stainless steel, tantalum) are most common, glass or glass-lined cells are available for certain pieces of equipment.

4.4.3.1 Isothermal
This is the simplest method of obtaining pressure data. A sample is heated to a target temperature and held until the pressure reaches either a constant value (showing that there is no gas generation or that it is complete) or a time limit is reached without the pressure stabilizing (showing that gas generation is continuing). The sample would then be cooled down to the starting temperature, and final pressure recorded. Using the ideal gas law, the amount and rate of permanent gas generation can be calculated. If the pressure reached a constant value, all of the gas-forming reaction that will occur at the temperature has taken place. This technique produces the best kinetic information on the gas-forming reaction but requires a separate run for each temperature of interest.

4.4.3.2 Stepwise Isothermal
Similar to the isothermal method, this involves heating a sample up in a series of isothermal steps of predetermined length, typically 1–4 hours. This allows development of the kinetics of gas formation by plotting the rate vs. temperature.

There are two possible complications. One is that if the gas generation does not go to completion, the total gas-generating potential of the system cannot be accurately estimated. For the data used to generate Figures 4.7 and 4.8, there is an assumption that the reagent concentration is approximately constant, which is reasonable for the first 10% of the reaction. The second possible complication is that once more than 10% of the reagent is consumed, the rate regressions may have significant errors.

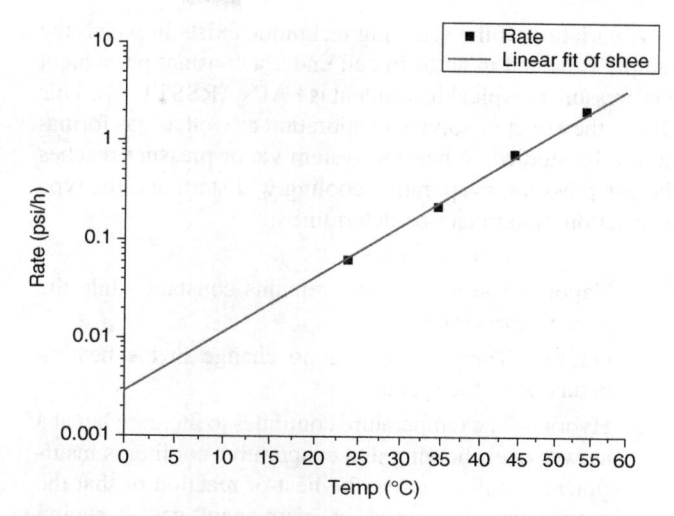

FIGURE 4.8 Plot of the $\log_{10}$ of the pressure rate vs. inverse temperature to show Arrhenius-type behavior.

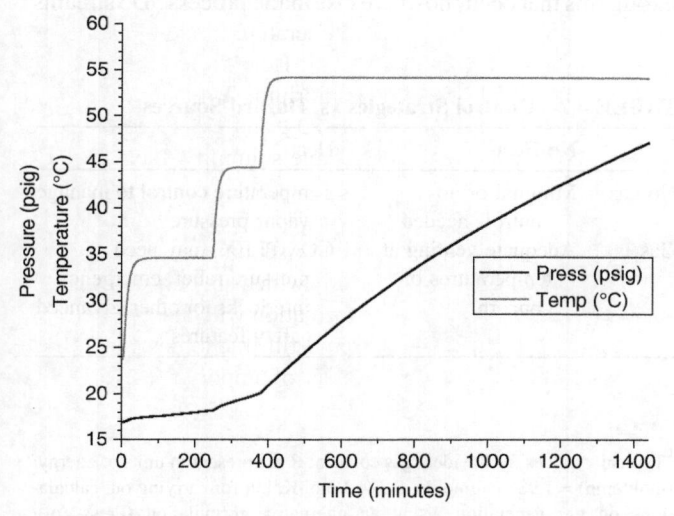

FIGURE 4.7 Pressure vs. time plot showing the generation of gas during a sequential aging run.

4.4.3.3 Scanning/Continuous
The fastest way to screen for gas generation is by heating the sample continuously at a ramped rate. There are several commercially available instruments for carrying this out: HEL's TSu [12], THT's RSD [13], and Netzsch's MMC [14] are some of the more commonly encountered. The basics are discussed in Chapter 3. The environment can be heated at a controlled rate (typically 1–5 K/min), the sample can be heated at a controlled rate (typically 1–5 K/min), or the environment can be heated rapidly to a fixed temperature, and the sample allowed to equilibrate. The sample would then be cooled down to the starting temperature, and final pressure recorded. Using the ideal gas law, the amount of permanent gas generated can be calculated.

Both scanning and scan and hold techniques can be used to generate useful data on the amount and rate of gas formation. The main limitation is in systems with a significant vapor pressure, where separating the rate of permanent gas formation from the vapor pressure increase can be complicated. However, if using the total rate of pressure increase is adequate to predict whether the process is safe to scale up, then deconvoluting the sources of pressure may not be necessary. The other limitation is ensuring that the gas generation is complete, if containment of the decomposition is the basis of safety, as it can sometimes be a challenge to see if gas formation has ended in a scanning experiment.

A variation of the scanning technique exists in which the sample is heated in an open cell under a constant permanent gas pressure (a typical instrument is FAI's ARSST [15]). This allows the effect of solvent evaporation as well as gas formation to be studied. When the system vapor pressure reaches the gas pressure, evaporative cooling will start, and the type of reaction system can be determined:

1. Vapor – The temperature remains constant while the solvent evaporates.
2. Gassy – There is little or no change in the heating behavior of the system.
3. Hybrid – The temperature continues to increase but at a slower rate, showing that evaporative cooling is insufficient to fully remove the heat of reaction or that the evaporation is caused by permanent gas sweeping out solvent.

Knowledge of the reaction type is important in determining how to perform vent sizing [16].

4.4.3.4 Adiabatic The most sophisticated technique for obtaining pressure data is adiabatic calorimetry (such as Fauske's VSP or one of several manufacturers' ARC). These typically start out as stepwise isothermal (also referred to as heat-wait-search), but once the sample reaches the EIT, it begins to generate heat above a certain threshold and switches into pseudoadiabatic mode and tracks the reaction [4]. This gives a gas generation profile that corresponds to the adiabatic decomposition, such as might occur in a pile of hot solids or a vessel during a jacket cooling failure. The technique does suffer from the same drawback of not deconvoluting vapor pressure from gas generation in solvent systems, but the lower initial rates of temperature increase involved do make the start of gas formation easier to detect. If the reaction is run with a sufficiently low phi (1.2 or less), the data can be used directly for vent sizing; otherwise a run with a lower phi to ensure accurate pressure rise rates is needed to size emergency relief devices.

4.4.4 Pressure Hazard Assessment of Permanent Gas Formation

Once screening tests have determined the amount of permanent gas formed (typically in mole gas per liter, or kilogram, of batch), a preliminary assessment of the risk can be made. Assuming a certain maximum vessel fill (90%, for example) at the operating temperature, the predicted pressure rise and gas formation rate can be calculated. Comparing those values with the vessel's MAWP and/or available venting capacity, the need for more detailed evaluation of the actual hazard scenario and/or gas generation rate is warranted. For example

(assuming 0.1 mol gas generated/l batch at 90% fill and 50 °C in a 100 L container), the permanent gas pressure increase in that container would be [9]

V (l) = 90 = reaction mixture volume
T_f (K) = 323.1 = temperature
V_h (l) = 10 = headspace volume
R (psi·l)/(mol·K) = 1.206 = ideal gas constant
n (mol/l) = 0.1 = gas formation

$$\Delta P = \frac{nVRT_f}{V_h} = \frac{0.1 \frac{\text{mol}}{\text{l}} \times 90\,\text{l} \times 1.206 \frac{\text{L-psi}}{\text{mol·K}} \times 323.1\,\text{K}}{10\,\text{l}} = 351\,\text{psi}$$

For a typical glass-lined vessel with an MAWP of 100 psi, the above scenario is considered a potential hazard and may require a reduction in the maximum vessel fill or other mitigation such as emergency relief.

4.5 HAZARD SCENARIOS

The goal of collecting undesired decomposition reaction heat generation and gas formation data on molecules and process streams is to determine how credible it is that they pose a risk during processing. Then the consequences of risks can be assessed in an informed manner. A grid can be constructed of hazards present and the means needed to control them. See Table 4.2.

4.5.1 Whole Process Overview

Once data is available on all of the key streams, it is desirable to take a holistic look at the whole process. Do any streams have exotherms that occur nowhere else in the process? Do streams

TABLE 4.2 Control Strategies vs. Hazard Sources

	No Heat	Heat
No gas	Minimal or no controls needed	Temperature control to manage vapor pressure
Gas	Adequate venting at temperatures of concern	COMPLEX, may need pressure relief, emergency interlocks, or other advanced safety features

---

[9]Typically values for the ideal gas constant R are present in units of energy/(mol·temp) = 1.987 cal/mol·K, 8.314 J/mol·K, but for carrying out calculations on gas generation, using an alternative formulation (Press·Vol)/(mol·temp) makes the calculation easier. Common R's are 1.206 l·psi/mol·K, and 0.082 l·atm/mol·K.

before the reaction generate gas but after the reaction they do not? Does the chemistry proceed through a known unstable intermediate that might not have been captured by testing done thus far? The purpose of this exercise is to ensure that all of the possible hazards associated with a process have at least been considered and if necessary identified and evaluated.

The safety data collected will then be used to feed a hazard assessment of the process. This could be as simple as reviewing the data with the chemist running the reaction in the lab, and making them aware of any hazards, to a full PHA or HAZOP. The goal is to match both the desired process and potential process deviations with the potential consequences. Common process deviations that are considered include the following:

1. Overheating (either by a jacket high temperature failure or electrical heating malfunction).
2. Overcharge or undercharge of a reagent.
3. Inadequate cooling during the reaction.
4. Incorrect addition rate (too fast or too slow).
5. Overconcentration during a distillation.
6. Charging the wrong reagent or reagents in the wrong order.
7. Equipment failure (controller, valve, pump, agitator).

An example for (1) might be that the vessel jacket fluid is heated via a heat exchanger that uses 180 °C steam. If there is a leak or other loss of control, the jacket fluid could reach the steam temperature of 180 °C.

How a company assesses and addresses the safety data and these deviations will be a function of a company's risk tolerance and safety posture, but there are several published papers with guidance on how to deal with these hazards [17]. The Center for Chemical Process Safety (CCPS), a division of the American Institute for Chemical Engineers (AIChE), has a good overview and extensive links to relevant material.[10]

The following case studies will discuss several cases where hazards were identified and evaluated against the proposed operating conditions and available safety features of the equipment. Measures were taken to either avoid encountering the hazard or make sure that the protections in place were adequate to protect the equipment and the personnel.

4.6 CASE STUDY 1: REACTION IN DIMETHYL SULFOXIDE (DMSO)

A batch process for making a pharmaceutical ingredient involved the addition of base to a pre-reaction solution composed of a brominated starting material, DMSO, and acid.

[10]https://www.aiche.org/ccps/topics/process-safety-technical-areas/chemical-reactivity-hazards/reactive-material-hazards

The base addition was exothermic with an adiabatic temperature rise of greater than 40 K. The batch was then heated and aged to drive the reaction forward. Gas generation was not expected, based on the stoichiometry of the desired reactions.

Review of the chemistry raised the following process safety concerns:

- The use of DMSO; a reactive solvent with a high boiling point and large exothermic decomposition at elevated temperatures [18].
- The use of DMSO in the presence of a halogen (bromine) and base [19].

Thermal screening of all process streams was carried out by DSC, while initial pressure screening of select streams [20] (starting with the end of reaction) was carried out in scanning mode using the MMC. The screen showed that there were no major process safety risks identified with the desired reactions. There was no gas generation and no exothermic decompositions of concern at the normal operating temperatures, and the exothermic base addition could be controlled by addition rate. However, consistent with the known high temperature decomposition of DMSO, DSC thermal scanning of all process streams prior to the end of reaction showed a significant exothermic decomposition (undesired reaction). See Figure 4.9. The decomposition occurs at temperatures above the normal operating temperature but is potentially reachable by a reactor jacket loss of control (overheating) process deviation, which is a single point of failure. In this case, the reactor targeted for use has a silicone-based heat transfer fluid-filled jacket with a maximum jacket temperature of 170 °C versus the exothermic decompositions of approximately 400–700 J/g with a DSC onset temperature of 175–225 °C. Note that the general guidance is that the EIT can be up to 50 K lower than the DSC onset temperature.

As shown by the shape of the exotherm peaks in Figure 4.9, prior to base addition the rate of the exotherm was rapid, while after the base addition the rate was much slower. Once base is charged, the presence of water in the base solution creates tempering and slows the decomposition reaction. Tempering occurs when the heat of evaporation of the solvent removes some or all of the heat generation from the decomposition.

Due to the presence of DMSO, with a freezing point of approximately 16–19 °C, DSC testing was also used to measure the potential freezing point of the stream prior to base addition (pre-reaction stream) in order to determine if a potential deviation involves the batch freezing. Batch freezing could lead to agitation issues and lack of thermocouple reliability. A DSC scan starting at sub-ambient temperatures confirmed that under normal operating temperatures, the batch would not freeze, as the DSC freezing point of

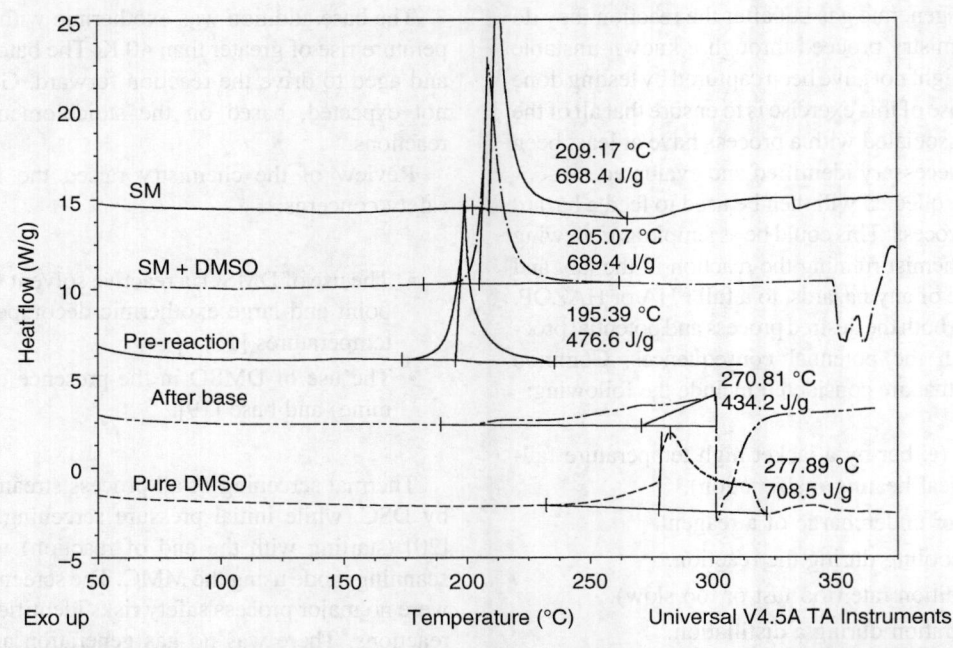

72 CHEMICAL ENGINEERING IN THE PHARMACEUTICAL INDUSTRY

FIGURE 4.9 DSC overlay plot of process streams prior to the end of reaction. SM, starting material. The temperature was increased linearly from room temperature to 350 °C (exotherm is shown upward). Tantalum-lined DSC cells with high pressure seals were used.

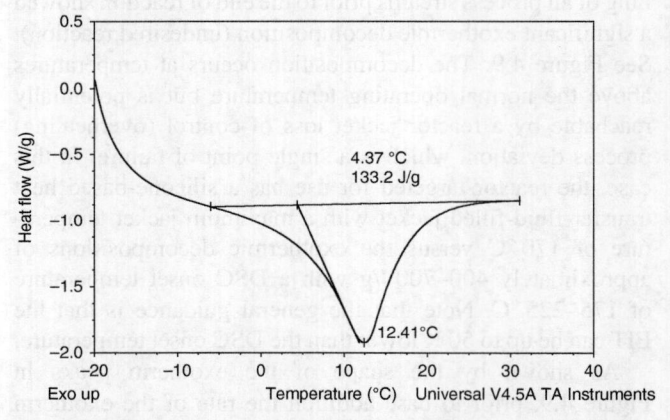

FIGURE 4.10 Pre-reaction stream freezing point determination using sub-ambient DSC.

approximately 13 °C was lower than the minimum allowable batch temperature. See Figure 4.10.

In addition, due to the presence of acid, bromine species, and DMSO, a DSC was used to determine if there was a MOC impact on the undesired decomposition reaction onset, size, or rate. DSC scans of the various process streams up through base addition were performed in tantalum, Hastelloy C, and stainless steel test cells. These results were used both to assist with reactor MOC evaluation and to confirm that test results across instruments were not impacted by different MOC test cells. The DSC results using tantalum and Hastelloy C were comparable, while results for streams in stainless steel showed comparable exotherm size and onset temperature, but had a slower exotherm rate. See Figure 4.11. Therefore, data generated in tantalum and Hastelloy C test cells was considered conservative. All of the MOCs assessed are considered acceptable, but based on the slower degradation rate, stainless steel would be the preferred reactor MOC.

With thermal screening complete, the next step was to determine the pressure consequences associated with the exothermic decompositions. It was known from testing of the prior step that the exothermic decomposition in the brominated starting material had significant and rapid gas generation associated with it. ARC testing of the brominated starting material showed an EIT of 158 °C. However, the undesired activity in the brominated starting material is considered a low risk since a double upset (two deviations) would need to occur in order for this activity to be reached, a mischarge of the starting material first to the reactor with a jacket overheating deviation at that time, prior to solvent addition.

Due to the onset temperature, large size, and rapid rate of the exothermic decomposition seen in the process streams prior to base addition, the screening step for pressure consequences of these streams was skipped, and the decomposition evaluation was performed using a more advanced instrument (ARC). ARC testing was first performed on the stream prior to base addition (pre-reaction stream), as it was likely to be

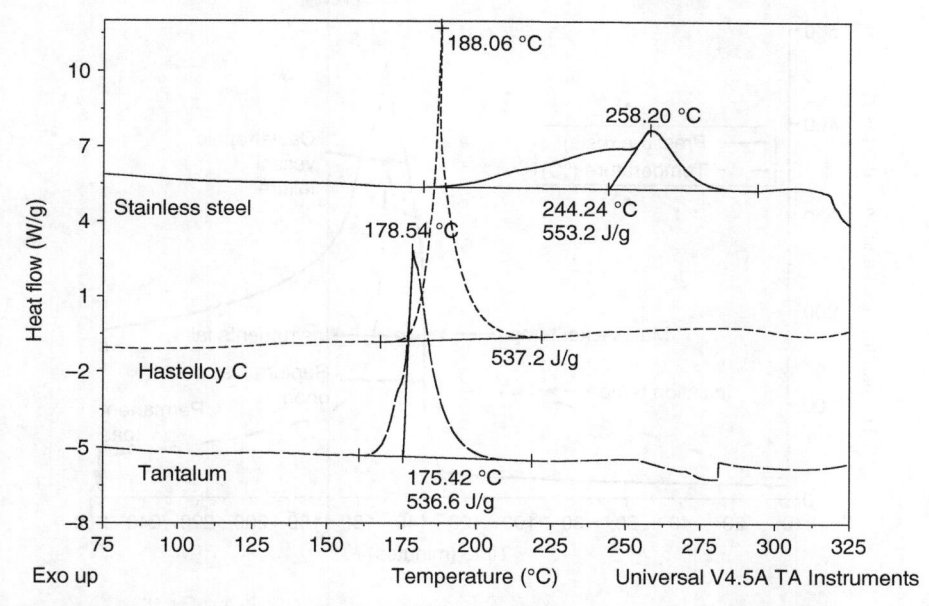

FIGURE 4.11 Material of construction (MOC) evaluation of pre-reaction stream using DSC.

the worst-case stream for this part of the process. This was suspected to be the pre-reaction stream based on DSC results showing that it had the lowest exotherm onset temperature, with a rapid exotherm rate. The worst-case stream must be identified as this contains the undesired reaction for which corrective actions must be taken to either mitigate against or avoid the risks for this part of the process.

Figure 4.12 shows the plots from an ARC heat-wait-search conducted on the pre-reaction stream. The heat-wait-search was conducted from 110 °C, at which point a series of exotherms were immediately observed, which continued to approximately 350 °C before the test was ended. Once the exotherm initiated, gas generation occurred and the pressure immediately increased. The ARC test showed that within minutes the pressure would exceed the 100 psig MAWP of the reactor, and the rupture disc would open (by 1.16x MAWP) [22]. If the rupture disc does not open or the venting capacity is inadequate then at approximately 170 °C, the maximum jacket temperature, the instruments would start failing (at 1.5x MAWP). Ten minutes later, the pressure is such that vessel failure would occur (at 3.5x MAWP). The adiabatic temperature rise associated with the first exotherm was 420 °C, and the total residual pressure (permanent gas) from the ARC test was 145 psi (at an 8 vol % test cell fill versus approximate 40 vol % vessel fill upon scale-up). Note that DMSO boils at 189 °C, so most of the activity would be over before the temperature reaches the boiling point of the solvent.

This first ARC heat-wait-search was started at 110 °C based on the DSC onset of approximately 175 °C and the general guidance that the ARC EIT can be up to 50 K lower than the DSC onset. The fact that the exotherm initiated immediately at 110 °C indicates that the actual EIT may be lower; therefore a second ARC was run starting at a lower temperature to determine the EIT of 84 °C. See Figure 4.13. In this case, the ARC EIT was greater than 50 °C from the DSC onset and did not follow the general rule of thumb; this is likely due to the fact that DMSO is a reactive solvent and the decomposition is suspected to be autocatalytic. The ARC data confirms that this exotherm is reachable by a reactor jacket loss of control (overheating) process deviation.

ARC testing of additional process streams, before and after the pre-reaction stream, was completed and confirmed that the pre-reaction stream was the worst-case stream with the lowest EIT (see Table 4.3) and most severe pressure consequences. The higher EIT seen after the base addition is likely attributed to tempering from the presence of water.

Because this testing was for a multi-kilo scale process, in addition to evaluating the thermal and pressure consequences from a jacket runaway deviation, the consequences of several other credible single point of failure process deviations, identified in a hazard assessment, were investigated. None of the process deviations resulted in an exothermic decomposition worse than overheating of the pre-reaction stream. Examples of process deviations investigated included the following:

- Missed charge of the brominated starting material.
- Undercharge of DMSO/overcharge of starting material.
- Pre-reaction stream temperature lower than the normal operating temperature.

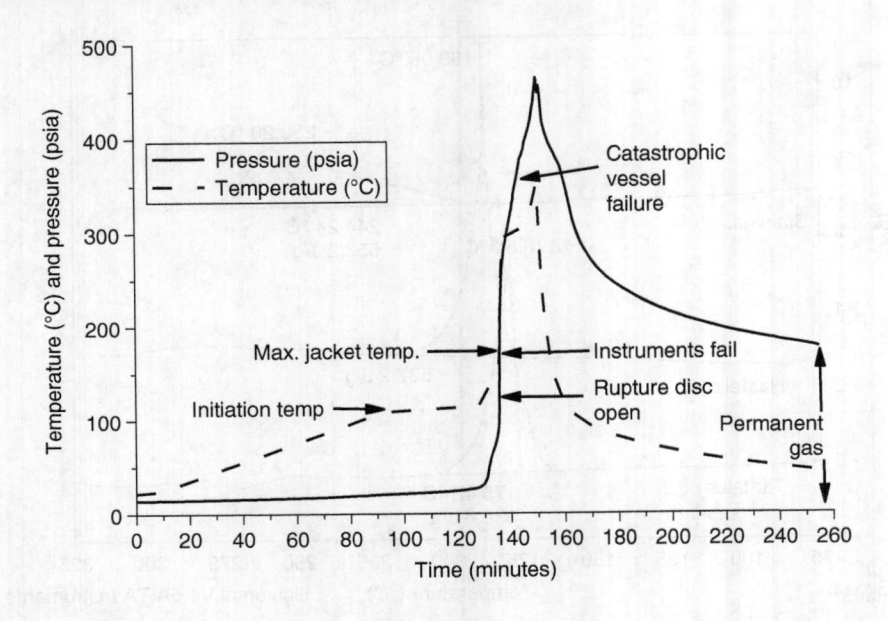

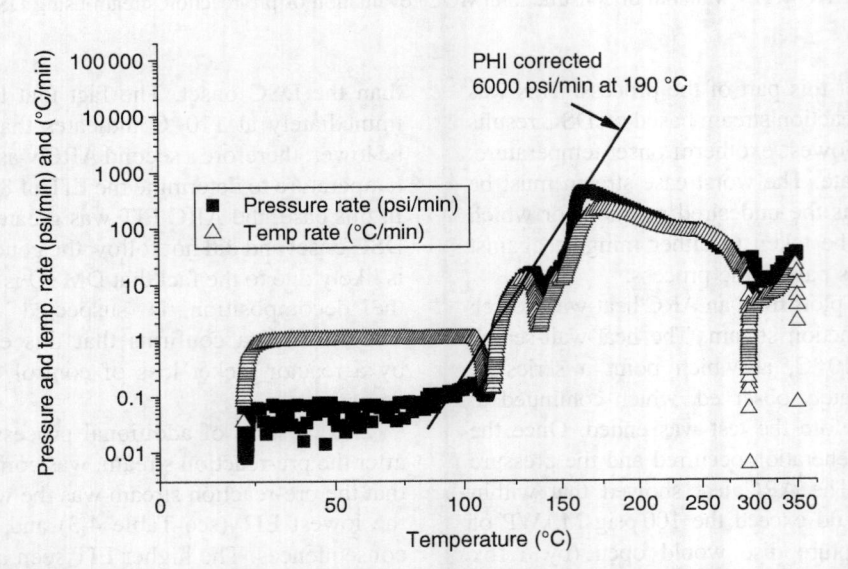

FIGURE 4.12 Plots of temperature and pressure vs. time and pressure and temperature rate vs. temperature from an ARC heat-wait-search on the pre-reaction stream conducted from 110 °C with 10°C steps and an onset threshold of 0.04 °C. A Hastelloy C ARC cell was used with a phi = 2.2. This figure is not corrected for phi; therefore the actual situation would likely be worse with a shorter time frame [21].

- Pre-reaction stream aged for several days.
- Rapid addition of base.
- Mischarge of a more concentrated base solution.

Now that the worst-case process stream, the pre-reaction stream, has been identified, further testing on this stream was needed to fully characterize the exothermic decomposition. First, an ARSST experiment was run to accurately identify vent sizing parameters, as the ARSST is an open system and most closely represents the deviation at scale. Vent sizing was needed since the ARC pressure results rapidly exceed the MAWP of the vessel and indicate that the decomposition has a high likelihood of overpressurizing the vessel. From the ARSST experiment it was concluded that the system is gassy (no tempering). The ARSST also provided the temperature at which the peak pressure rate occurs, 190 °C. As shown in Figure 4.12, this was then used to predict the peak pressure rise rate of 6000 psi/min from the ARC data. In order to predict (extrapolate) the peak pressure rise rate under adiabatic conditions (phi = 1.0), a straight line was drawn to describe the effect of temperature on the pressure rise rate.

It was suspected that the undesired reaction was autocatalytic due to the sharp peak observed in the DSC and the low EIT seen in the ARC (see Section 3.5.2.3). Due to this suspicion,

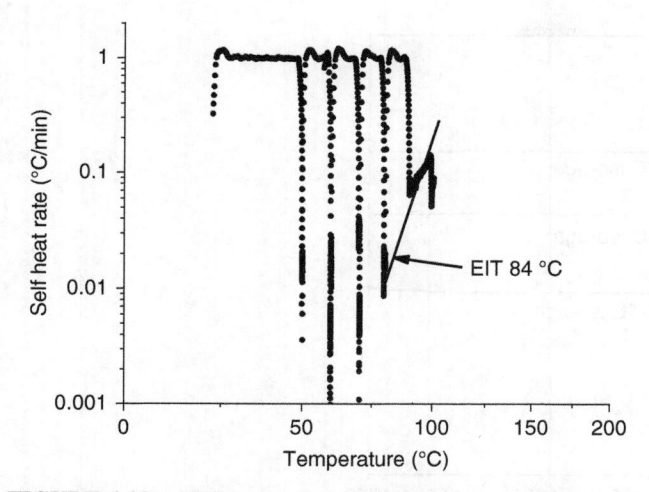

FIGURE 4.13 ARC heat-wait-search repeat run on pre-reaction stream starting at a lower temperature to capture the EIT. A Hastelloy C ARC test cell was used with a phi = 1.4. The EIT was obtained from extrapolating to a self-heat rate of 0.02 °C/min.

TABLE 4.3 Exotherm Initiation Temperature (EIT) of Process Streams Prior to the End of Reaction

Process Stream	EIT (°C)
DMSO + starting material	105
Pre-reaction	84
After base addition	144

several 24+ hour isothermal ages were performed in the DSC in order to gain further confidence in the ARC EIT. As shown in Figure 4.14, the isothermal age at 85 °C showed an exothermic event approximately 440 minutes into the age, in alignment with the ARC EIT of 84 °C. The 65 °C and 75 °C iso-ages showed no decomposition after 30 and 72 hours, respectively. This confirmed the temperature of concern as 84 °C.

With the pre-reaction stream fully characterized, vent sizing was performed to show the consequences of reaching the undesired activity. Using the ARC/ARSST results and Design Institute for Emergency Relief Systems (DIERS) [16] vent sizing two-phase model for a gassy system, it was determined that a vessel vent diameter of 133 cm would be needed to prevent overpressurizing the vessel, an 11 300 L reactor. This determination was based on an L/D[11] of 200 versus the actual L/D of greater than 900; therefore the actual vent requirement would be even larger. As a final confirmation of the worst-case pre-reaction stream, vent sizing was also performed on the "DMSO+starting material" and "after base addition" streams,

in addition to a rough vent sizing analysis on select process deviations; in all cases the vent sizing showed a smaller vent requirement. Since the 11 300 L reactor only had a 20 cm diameter nozzle, installing a 133 cm vent presented a major challenge; it would have required more than forty 20 cm relief vents to provide the relief area equivalent to a 133 cm vent. As this size vent is unreasonably large, a secondary vent could not be provided to protect against a vessel failure.

The above assessment identified a significant undesired reaction that could lead to vessel failure as a result of a single process deviation, reactor jacket loss of control (overheating). With all of this information in hand, a HAZOP and layers of protection analysis (LOPA) were used to determine the appropriate processing safeguards needed to mitigate the consequences and reduce the likelihood of reaching this activity to an acceptable level. In order to completely eliminate the risk, either a change in chemistry or how the process was run would be necessary.

This case study demonstrates the successful use of various calorimetric approaches to identify and characterize the thermal and pressure consequences of an undesired reaction:

- DSC for initial screening for identification of exothermic decompositions, freezing point determination, and MOC impact evaluation.
- DSC isothermal age for confirmation of the ARC EIT and verifying the suspected autocatalytic decomposition.
- ARC for EIT determination and quantification of thermal/pressure consequences to enable vent sizing.
- ARSST for determination of system type (e.g. gassy vs. tempering) and quantification of thermal/pressure consequences to enable vent sizing.

This case study also shows why reactive solvent use is discouraged, especially high boiling point solvents, as the solvent participates in the decomposition activity rather than helping to absorb heat from the decomposition. In addition, the case study shows that small changes in stream composition can greatly impact the EIT, size, and rate; for example, the pre-reaction stream containing acid had an EIT approximately 20 °C lower than the prior stream without the acid.

4.7 CASE STUDY 2: CONTINUOUS GAS GENERATION IN A WASTE STREAM CONTAINING CARBONATE

A batch process for making an active pharmaceutical ingredient (API) involves a salt break via the addition of an aqueous 1 M potassium carbonate solution to a tosylate salt intermediate in ethyl acetate at 15–25 °C. After cutting away the first aqueous layer containing potassium tosylate, two

---

[11]L/D is the ratio of pipe length to pipe diameter, and correlates to pressure drop during flow; the higher the L/D the lower the flow at a given pressure, or the more pressure is required for a given flow.

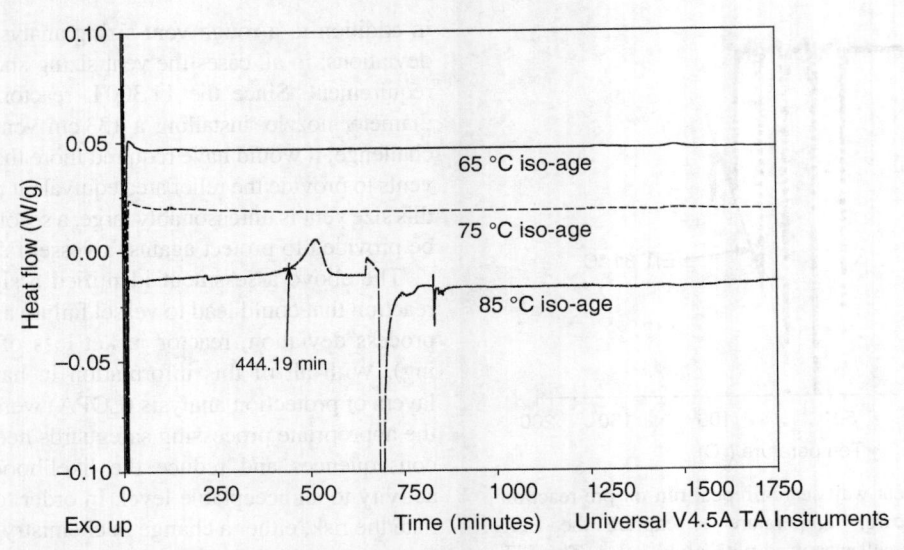

FIGURE 4.14 DSC isothermal age overlay plot of the pre-reaction stream aged at 65, 75, and 85 °C to gain further confidence in the ARC EIT.

additional washes are performed: a second aqueous potassium carbonate wash and a water wash.

Preparation of the aqueous 1 M potassium carbonate solution involves the dissolution of potassium carbonate solids, which was expected to be exothermic (~9 K) with potential minor gas generation (carbon dioxide) during initial mixing, until the pH is above approximately 8. Based on the pK_a buffer curve of carbonate and carbonic acid, at pH > 8 little carbon dioxide is expected to evolve since the main component is carbonate [23].

Addition of the aqueous potassium carbonate solution to the batch was also observed to be exothermic ($\Delta T_{ad} < 5$ K). Carbon dioxide may be generated from the potassium carbonate, but any carbon dioxide generated was expected to be dissolved in the batch stream. Since more than one mole of carbonate was present per mole of salt being broken, the main by-products were projected to be potassium tosylate and potassium bicarbonate. However, due to the potential for carbon dioxide generation from the presence of carbonate, testing was performed to confirm if gas was being generated.

The first stream assessed for gas generation was the first aqueous potassium carbonate layer, as this stream was the most likely to generate gas and could be stored in a closed nonpressure-rated container (e.g. drum or tank) for off-site waste disposal. With a stream pH of approximately 8.5–9.3, carbon dioxide was not expected to evolve based on the pK_a buffer curve of carbonate and carbonic acid. Using a MMC, an isothermal age (iso-age) was performed on the first aqueous layer at 50 °C for approximately 1200 minutes (~20 hours). This temperature represents the maximum potential ambient temperature a container may be exposed to if temperature control is not provided. As shown in Figure 4.15, the iso-age showed continuous slow gas generation.

Due to the results on the first aqueous layer, an MMC iso-age was also performed on the second carbonate aqueous layer. This stream had a pH of approximately 9.9–10.9. As shown in Figure 4.16, the iso-age exhibited initial minor gas generation, but the gas generation was not sustained.

With the potential need to combine the three aqueous layers together in a vessel, an MMC iso-age was performed on the combined stream. The combined stream had a pH of approximately 10.0. The stream was first degassed (with multiple vacuum-pressure purges) to try to remove any potential dissolved gas. As shown in Figure 4.17, the combined stream after degassing showed continual slow gas generation, persisting from the first aqueous layer.

Due to the continuous gas generation from the first aqueous layer and the potential need to also store the organic layers in a closed nonpressure-rated container, MMC iso-age testing was also performed on the first organic layer (the worst-case stream from the two carbonate washes) and the final organic stream after the water wash. As shown in Figure 4.18, the iso-ages each showed no gas generation.

In summary, the MMC iso-ages on the various aqueous and organic layers showed that there is a potential pressure concern with storing either the first aqueous layer or combined aqueous layers in a closed nonpressure-rated container, while the remaining streams do not present a pressure hazard.

The rate of gas formation from the MMC iso-age was then used to determine whether a 55 gal (208 L) poly drum containing the combined aqueous stream, which had a lower gas generation rate of 3.5 psi/day versus the first aqueous layer at 8.2 psi/day, could be safely transported off-site for

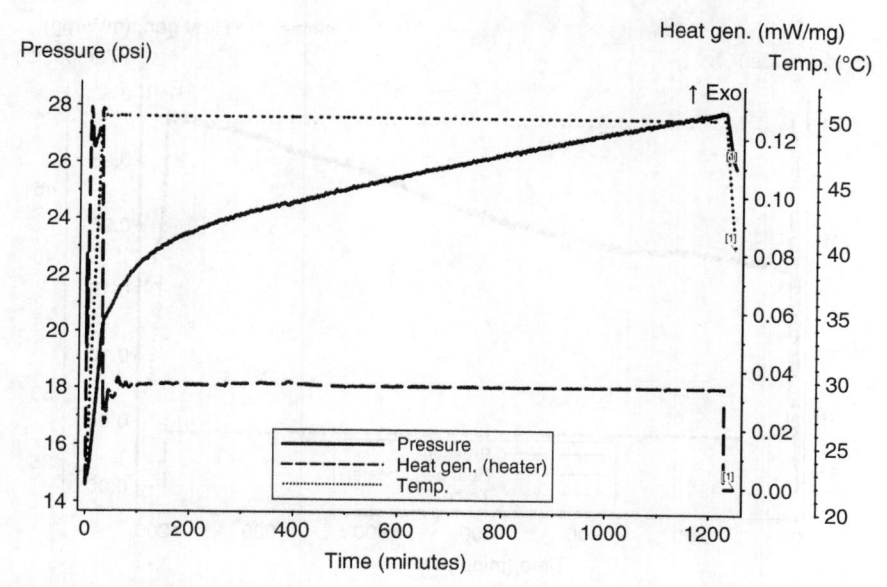

FIGURE 4.15 MMC iso-age of first carbonate aqueous layer at 50 °C showing continuous slow gas generation.

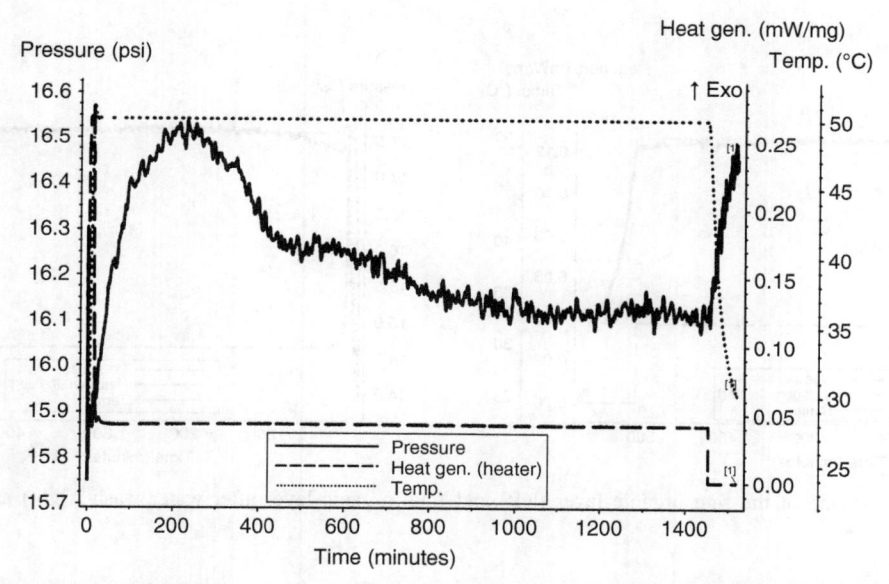

FIGURE 4.16 MMC iso-age of second carbonate aqueous layer at 50 °C showing initial minor gas generation. But the gas generation was not sustained.

disposal. The drum was assumed to bulge if the pressure exceeded 3 psig.[12]

Gas generation rate calculated from MMC iso-age (Figures 4.15 and 4.17):

$$\text{First aqueous layer}: \frac{6.4\,\text{psi}}{1120\,\text{min}} = \frac{X\,\text{psi}}{1440\,\text{min}}$$

$$X = 8.2\,\text{psi}$$

$$\text{Combined aqueous layer}: \frac{2.2\,\text{psi}}{900\,\text{min}} = \frac{X\,\text{psi}}{1440\,\text{min}}$$

$$X = 3.5\,\text{psi}$$

Calculations showed that at a reasonable drum fill (50 vol %), the combined stream could only be stored in a closed drum for less than 2 days before the pressure reaches 3 psig

---

[12] 3 psig is a conservative assumption for when drum bulging may occur in a 55 gal poly drum, compared to the typical hydrostatic test pressure of approximately 21.7 psi at which point the drum will leak. https://phmsa.dot.gov/staticfiles/PHMSA/DownloadableFiles/Files/LOGSA%20Policies,%20Procedures%20and%20Standards.pdfhttps://www.grainger.com/product/GRAINGER-APPROVED-Transport-Drum-19H059https://www.grainger.com/ec/pdf/19H053_1.pdf

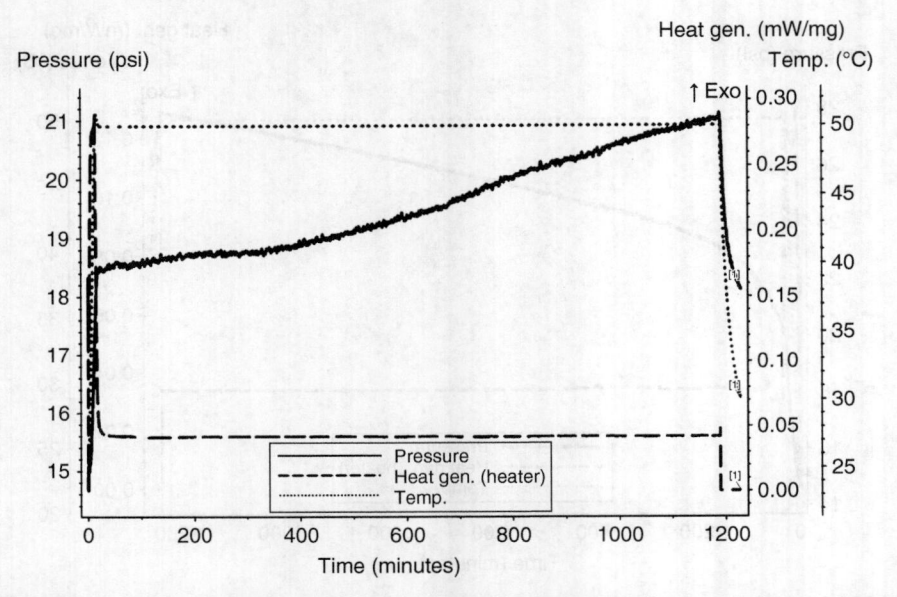

FIGURE 4.17 MMC iso-age of the three aqueous layers combined at 50 °C showing continuous slow gas generation.

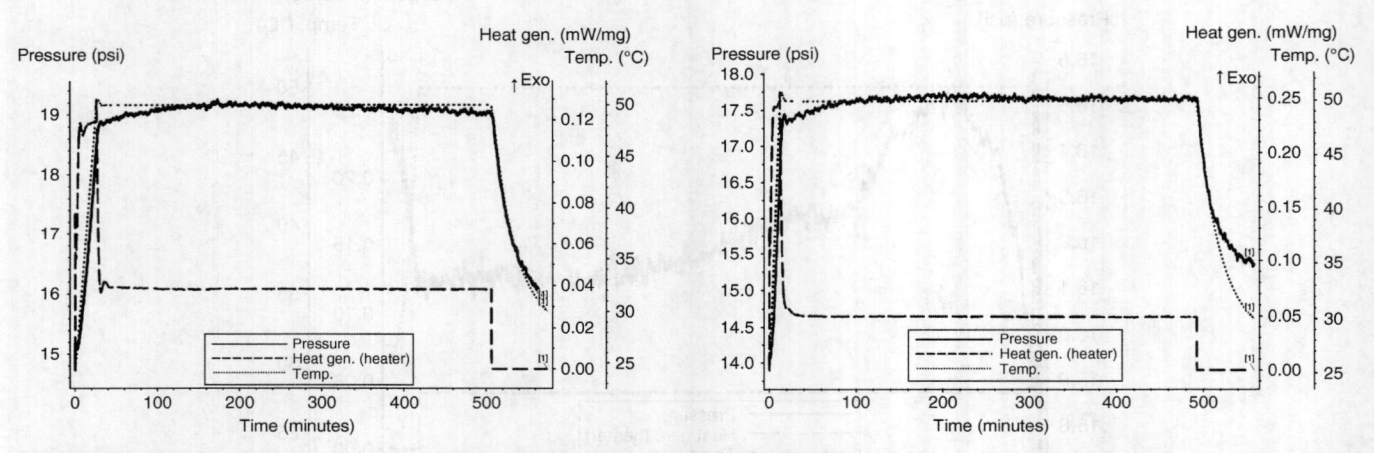

FIGURE 4.18 MMC iso-age of the first organic layer (left) and final organic layer after water wash (right) at 50 °C showing no gas generation.

and the drum begins to bulge. In order to maintain the pressure below 3 psig in the drum for 30 days, the drum fill would have to be 6 vol % (12.5 L). Due to the low drum fill requirement and short length of time leading to drum bulging, it was concluded that the combined aqueous stream cannot be stored safely in a closed drum for transport off-site for disposal. The same is true for the first aqueous layer, as it would require even lower drum fills and shorter storage times due to the increased gas generation rate. However, at a drum fill of less than 75 vol % (157 L), the combined stream can be temporarily stored in a drum for up to 15 hours without presenting a bulging hazard (to enable necessary short-term transportation within a site). See calculations in Appendix 4.1 at the end of this case study. A similar calculation can also be done to determine short-term storage of the first aqueous layer.

Note that the gas generation rate does not pose a hazard for a vented vessel. This was confirmed by calculating the volumetric gas flow rate, Q, upon scale-up in a vessel, using the MMC iso-age gas generation rate per unit volume, $\dot{G}$. See Appendix 4.1 at the end of this case study for equation input values, except that in this case V_{batch} (l) is the 10 000 L batch stream volume in the vessel:

$$\dot{Q}\left(\frac{1}{min}\right) = (\dot{G} \times V_{batch}) \times 22.4 \frac{\text{liters gas}}{\text{mole gas}} \times \left(\frac{T_{IA}}{273.15\,K}\right)$$

$$\dot{Q} = \sim 1 \frac{1}{min}$$

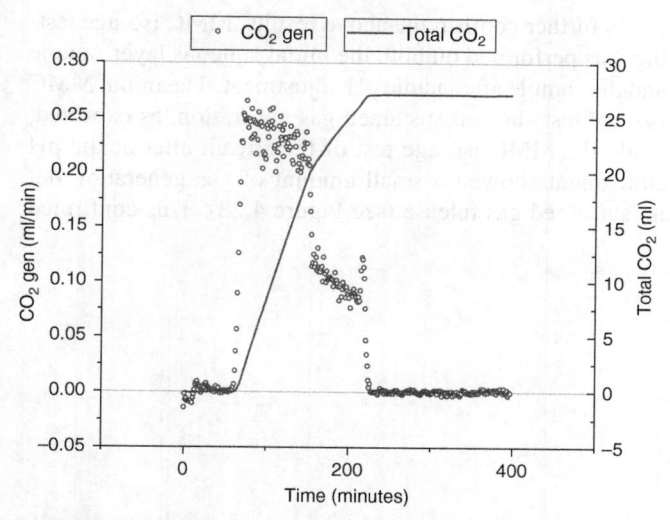

FIGURE 4.19 Experiment results showing sharp carbon dioxide peak at 60 minutes once the aqueous potassium carbonate solution was added to the intermediate in ethyl acetate. At approximately 150 minutes the batch was settled, which resulted in a decrease in carbon dioxide level. Then at 210 minutes, the first aqueous layer was removed, and the second potassium carbonate wash was charged, at which point the carbon dioxide generation stopped.

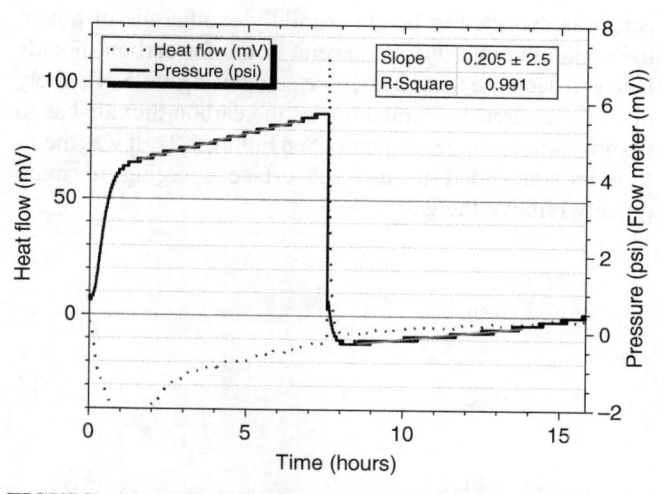

FIGURE 4.20 SuperCRC experiment of pTSA with aqueous potassium carbonate and ethyl acetate showing slow continuous carbon dioxide generation at a rate of 0.20 psi/h.

This calculated volumetric gas flow rate is lower than a typical vessel vent capacity.

An EasyMax (by Mettler Toledo) was then used to more accurately determine the rate of gas formation during the carbonate washes. Fourier-transform infrared (FTIR) spectroscopy was also used to confirm the gas identity, since at this time the root cause of the continuous gas generation was unknown. The experiment showed that carbon dioxide, shown by a sharp peak in the FTIR trend at 2312 cm^{-1}, was slowly generated from the time the first charge of aqueous potassium carbonate solution is made until the first aqueous layer is cut away. The gas formation slowed down when the first aqueous layer was being settled (agitation turned off) and stopped when the second aqueous potassium carbonate charge was carried out. See Figure 4.19. In alignment with the prior MMC iso-ages, minor carbon dioxide generation may be seen from the second aqueous layer, although the gas generation is not sustained and does not present a hazard.

As stated previously, up until this point the root cause of the continuous gas generation in the first aqueous layer was unclear. Since only the first aqueous layer continually generated gas, a closer look was taken at the tosylate salt intermediate. It was discovered that the tosylate salt intermediate may contain residual *p*-toluenesulfonic acid (pTSA) from the prior processing step. Experiments were then performed to determine if the presence of pTSA with aqueous potassium carbonate and ethyl acetate would cause continuous carbon dioxide off-gassing. A SuperCRC (by Omnical) experiment confirmed that this mixture did slowly generate pressure at a rate of 0.20 psi/h. See Figure 4.20. It was therefore concluded that the presence of pTSA with aqueous potassium carbonate and ethyl acetate does cause a slow continuous gas formation reaction.

In addition to not storing the first aqueous layer in a closed nonpressure-rated container for off-site transport, the first aqueous layer also cannot be sewered directly due to the contents exceeding local sewering limits for total organic carbon (TOC) [24]. Another storage option evaluated was to rent or purchase a pressure-rated container for transport off-site for disposal. This option was not pursued due to the high cost, long lead time, and complicated logistics. Therefore, further calorimetry experiments were carried out using an ARC, a SuperCRC, and an EasyMax (with FTIR) to determine if there was a way to stop the gas generation and enable safe storage of the first aqueous waste stream in a closed nonpressure-rated container. Previous work already showed that degassing the layers was insufficient to remove all of the gas. Additional experiments performed included thermal treatment, raising the pH, and lowering the pH of the stream.

First it was concluded that raising the temperature of the first aqueous layer for an extended period of time did not stop the gas generation. There was still a low level of pressure generation seen in the ARC after performing a heat-wait-search up to 115 °C over a total exposure time of 15 hours. Then, caustic (e.g. 5 M NaOH) addition to the first aqueous layer in a SuperCRC slowed the pressure generation rate but did not stop the gas generation. Finally, addition of a strong acid (e.g. 1 M H2SO4) to an EasyMax (with FTIR) to adjust the pH of the first aqueous layer to below 4.5 (required 1 mol of H$_2$SO$_4$ per mole K$_2$CO$_3$) did successfully stop the gas generation; all carbonate was converted to carbon dioxide. The gas generation rate during the pH adjustment was directly proportional to the rate of addition of the acid. Also, the addition of acid was expected to be either mildly exothermic or potentially endothermic. It took several hours for carbon dioxide levels to decrease to zero after the acid dosing at 25 °C, likely due to mass transfer limitations as the carbon dioxide has to desorb from the aqueous phase. See Figure 4.21. It was therefore recommended to use subsurface sparging to more quickly remove the gas.

To further confirm the above results, MMC iso-age testing was performed on both the initial aqueous layer sample and the sample after acidic pH adjustment. The initial MMC iso-age test showed sustained gas generation, as expected, while the MMC iso-age test of the stream after acidic pH adjustment showed a small amount of gas generation but no sustained gas release (see Figure 4.22). This confirmed

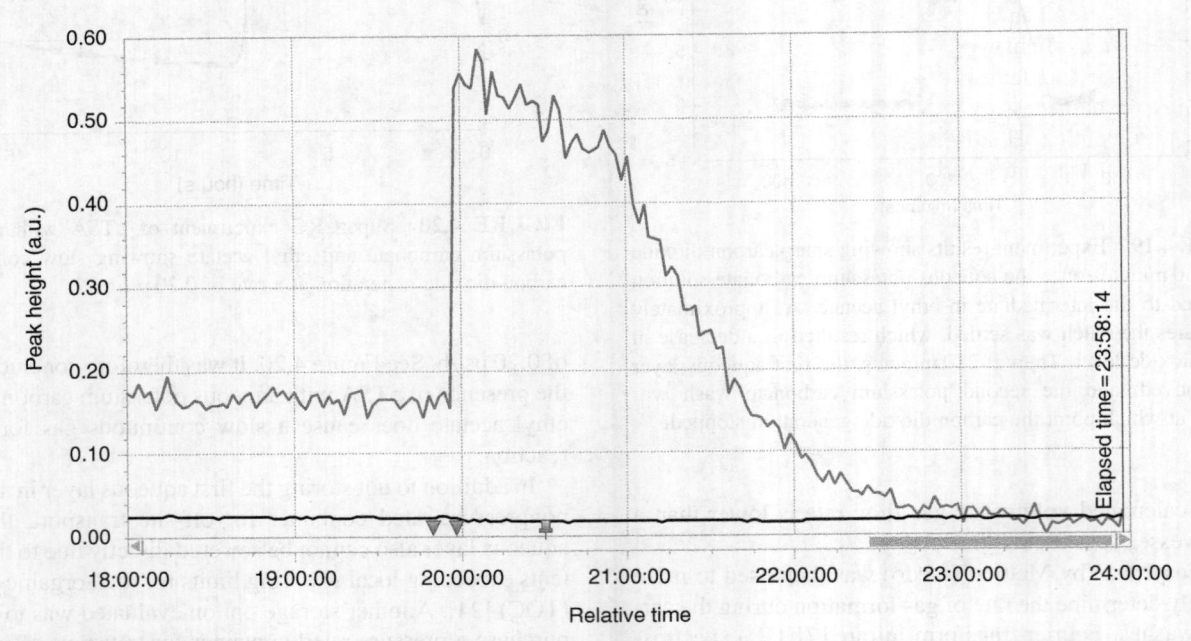

FIGURE 4.21 FTIR trend showing sharp carbon dioxide peak (at 2386 cm^{-1}) once the strong acid is added to the first aqueous layer (shortly before 20 : 00 hours). The carbon dioxide level then dropped to zero by 24 : 00 hours.

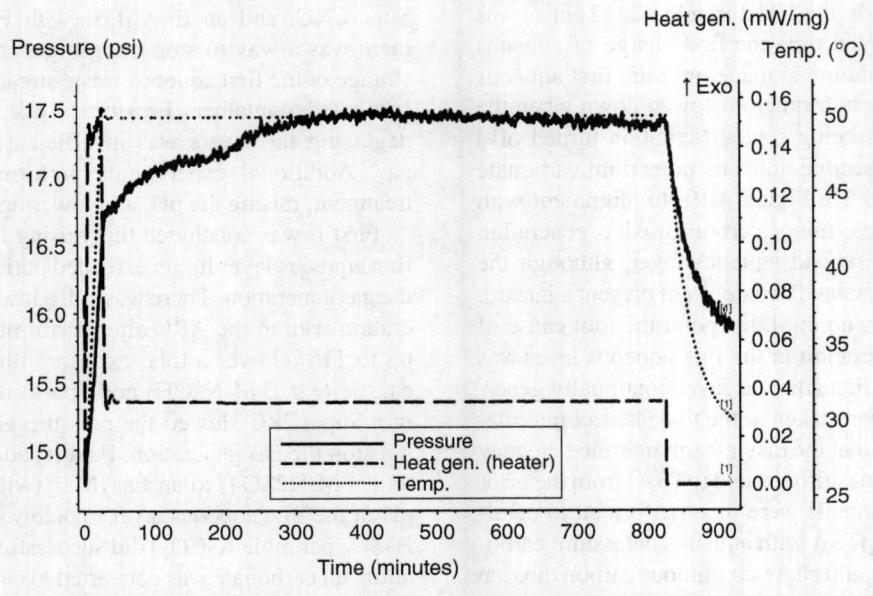

FIGURE 4.22 MMC iso-age at 50 °C of first carbonate aqueous layer after acidic pH adjustment to less than 4.5 showing a small amount of gas generation but no sustained gas release.

that lowering the pH of the first aqueous layer to less than 4.5 with a strong acid removes the carbon dioxide from the system and stops the gas generation in this stream, therefore making it safe to store in a closed nonpressure-rated container.

This case study demonstrates the successful use of various calorimetric approaches to identify and characterize off-gassing of carbonate (or bicarbonate) containing streams due to the potential for gas generation, particularly if the stream may be stored in a closed nonpressure-rated container, as this assessment will greatly reduce the potential for drum bulging incidents. It also provides an example of when it may be necessary (or desired) to perform additional operations (e.g. pH adjustment) in order to eliminate gas generation in a stream. Lastly, the case study highlights that residual quantities of materials from a prior processing step can impact downstream processing; in this case the presence of residual pTSA and its previously unknown interaction with aqueous potassium carbonate slowly generated carbon dioxide.

APPENDIX 4.1 Drum Calculations

V_{sample} (ml) = 1.8 = sample volume

$V_{headspace}$ (ml) = 0.8 = MMC cell headspace volume

P_{MMC} (psi) = 2.2 = pressure generated during MMC iso-age

T_{IA} (K) = 323.2 = MMC iso-age temperature

R (psi·ml)/(mmol·K) = 1.206 = gas constant

t_{IA} (minutes) = 900 = MMC iso-age pressure generation time

V_{drum} (l) = 208 = total volume of drum

V_{batch} (l) = see below = batch stream volume in drum

To determine the length of time that the combined stream can be stored in a closed drum at a 50 vol % (104 L) fill, first calculate the gas generation per unit volume in the MMC iso-age, G:

$$G\left(\frac{\text{mmoles gas}}{\text{ml sample}}\right) = \frac{P_{MMC} \times V_{headspace}}{R \times T_{IA} \times V_{sample}}$$

Then calculate the gas generation rate per unit volume, $\dot{G}$:

$$\dot{G}\left(\frac{\text{mmoles gas}}{\text{min} \times \text{ml sample}}\right) = \frac{G}{t_{IA}}$$

Using a V_{batch} of 104 L, solve for the number of days, X, which will result in a drum pressure of 3.0 psi by plugging in X values in the below equation until the pressure generated in the drum, P_{drum}, is equal to 3.0 psi:

$$P_{drum}(\text{psi}) = \frac{\left(\dot{G} \times X\,\text{days} \times 1440\frac{\text{min}}{\text{day}} \times V_{batch}\right)RT_{IA}}{(V_{drum} - V_{batch})} = 3.0$$

$$X = 1.9\,\text{days}$$

Therefore, at a drum fill of 50 vol % (104 L), the drums should not be stored for more than 1.9 days.

To determine the maximum fill level for 30-day storage of the combined stream in a closed drum, solve for the batch stream volume in the drum, V_{batch}, which will result in a drum pressure of 3.0 psi by plugging in V_{batch} values in the below equation until the pressure generated in the drum, P_{drum}, is equal to 3.0 psi:

$$P_{drum}(\text{psi}) = \frac{\left(\dot{G} \times 30\,\text{days} \times 1440\frac{\text{min}}{\text{day}} \times V_{batch}\right)RT_{IA}}{(V_{drum} - V_{batch})} = 3.0$$

$$V_{batch} = 12.51$$

Therefore, at a drum fill of 6 vol % (12.5 L), the drums should not be stored for more than 30 days.

To determine the maximum fill level for short-term storage of the combined stream in a closed drum, solve for the batch stream volume in the drum, V_{batch}, which will result in a drum pressure of 3.0 psi by plugging in V_{batch} values in the below equation until the pressure generated in the drum, P_{drum}, is equal to 3.0 psi:

$$P_{drum}(\text{psi}) = \frac{(G \times V_{batch})RT_{IA}}{(V_{drum} - V_{batch})} = 3.0$$

$$V_{batch} = 1571$$

Note that this calculation uses the quantity of gas generated over 900 minutes of the iso-age. Therefore, at a drum fill of 75 vol % (157 L), the drums should not be stored for more than 900 minutes or 15 hours.

4.8 CASE STUDY 3: EVALUATING A FUNCTIONAL GROUP

A research group wanted to prepare a variety of molecules containing a potentially unstable functional group (diazirine: R₁–C(R₂)–N=N ring). This type of molecule can upon irradiation with UV light readily form carbenes by elimination of nitrogen, making it useful in a variety of biological labeling studies [25]. Because of the potential instability from the ring strain, the research group asked for an investigation of the decomposition properties of this class of molecules.

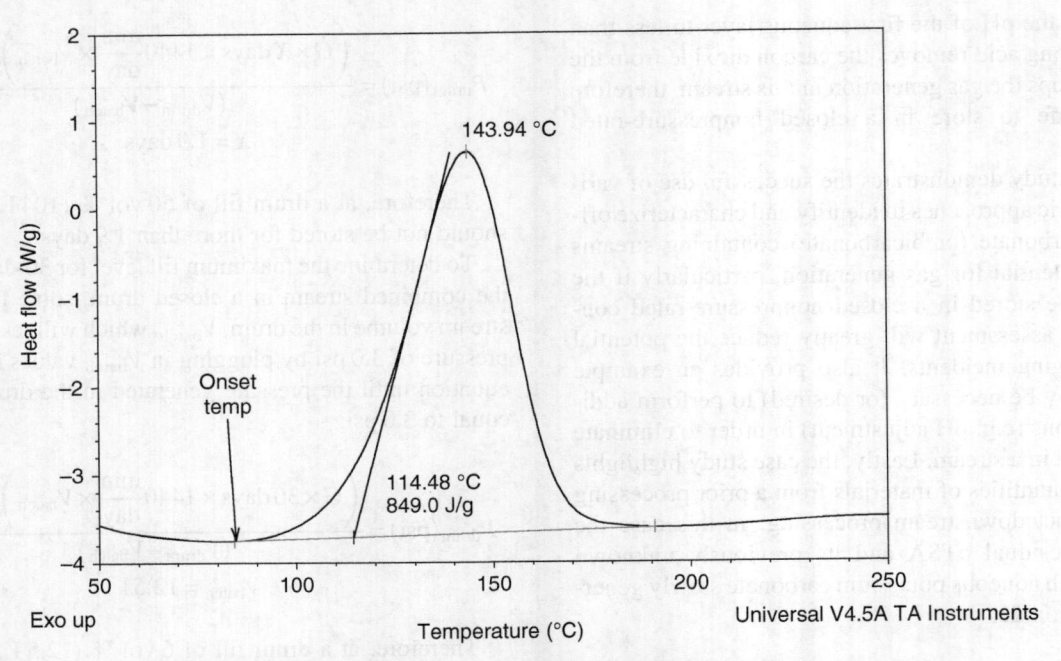

FIGURE 4.23 DSC scan on compound 1 (CAS 92367-11-8) diazirine.

The first compound studied was a commercially available liquid (compound 1, CAS 92367-11-8): [structure].[13] The DSC on this compound (see Figure 4.23) showed an exothermic decomposition with low onset temperature (84 °C) and a large amount of energy (849 J/g).

Based on this data, several follow-up questions were raised that needed to be evaluated before making handling and shipping recommendations:

1. Is the heat of decomposition due to the functional group only, or is there an interaction with the rest of the molecule?
2. Is the decomposition different in the presence of solvents that could be used to carry out any reactions or dilutions?
3. How is the decomposition affected by dilution?
4. Are these compounds detonable by impact?

It was assumed that the decomposition would release 1 mol of nitrogen per mole of compound. For the compound above, for example, this corresponds to 87 ml of gas per gram of compound at 25 °C and 1 atm, which could easily produce dangerous pressures if it decomposed. This increased the importance of studying the decomposition behavior of this class of compounds.

4.8.1 Heat of Decomposition of Multiple Chemicals Sharing a Common Functional Group

From a compound collection maintained by the research group, nine different molecules, containing a wide variety of R1 and R2 non-energetic substituents,[14] were analyzed by DSC. They all gave a single decomposition peak with onset temperatures ranging from 66 to 120 °C and energies ranging from 454 to 3000 J/g. To evaluate the trend, the decomposition energy was plotted versus molecular weight (MW) (see Figure 4.24).

It was shown that a curve drawn at 320 000 (J/g) divided by the MW (g/mol) bounded the energies of decomposition. Therefore $\frac{320\,000}{MW}$ can be used as a conservative estimate of the energy of decomposition.

When the curves were analyzed further, in all of the decompositions where R1 and R2 were aryl or alkyl substituents, the peaks were broad like the sample compound. One compound where R1 and R2 were both part of the same ring gave a sharp peak, more typical of a potentially explosive material. As a result, any generalizations that would be drawn are limited to R1 and R2 being separate chains.

---

[13]CAS 92367-11-8, 4-[3-(Trifluoromethyl)-3H-diazirin-3-yl]benzyl Bromide.

[14]No –ONO$_2$, –NO$_2$, –N=O, cyclopropyl, –N$_3$ groups, for examples. A more comprehensive list can be found in Appendix 6, "Screening Procedures" in Ref. [26].

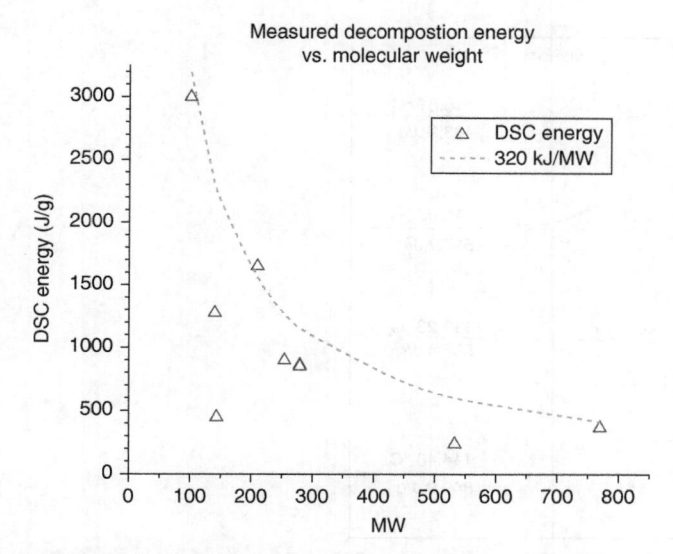

FIGURE 4.24 DSC energy vs. molecular weight (MW) for multiple diazirine compounds.

4.8.2 The Effect of Solvents

In order to make and use these compounds, they will need to be dissolved in an organic solvent, and it might be desirable to store them as dilute solutions. To see what effect several common solvents had on the diazirine functionality, DSC testing was carried out on 50% (volume/volume) solutions of compound 1 in isopropyl acetate, DMSO, and dichloromethane. (See Figure 4.25.)

The results from Figure 4.25 show that the molar energy of decomposition and onset temperature were unchanged by the presence of solvent.

4.8.3 Does the Decomposition Decrease Linearly with Concentration

To support the idea that the decomposition mechanism is similar over the entire range of concentration, the commercial diazirine was tested by DSC from concentrated (100% compound) to dilute (~30 wt %) in IPAc. If there was no significant change in mechanism, the peak shape should remain the same; if the molar heat of decomposition is constant, a plot of energy per mass vs. mass fraction should give a linear plot. Based on Figures 4.26 and 4.27, the decomposition energy and mechanism is independent of concentration.

4.8.4 Is Compound 1 Detonable by Impact?

In order to ship compound 1, or any diazirine, it must not be an explosive (class 1 dangerous goods), as defined by UN regulations [26]. DSC screening of the compound can identify whether a material needs impact sensitivity testing and can help guide interpretation of the results.

As can be seen in Figure 4.23, the peak shape of the decomposition is fairly Gaussian, indicating a smooth decomposition. For comparison, the DSC of benzoyl peroxide (98%) is shown in Figure 4.28 and shows a very sharp and sudden increase in heat flow at 103 °C. This sort of change is an indicator of possible explosive properties.

Drop-weight testing[15] was carried out on both the benzoyl peroxide and compound 1 at 5 J of impact energy. The benzoyl peroxide gave a series of strong positive indicators of detonation at that energy level, and that particular material does have serious shipping restrictions on it (no more than 1 Lb per container for a start). Drop-weight testing on compound 1 gave the appearance of a strong positive but closer examination of the appearance of the test cells after the runs showed that the material was not detonating, but merely decomposing. Signs included the lack of damage to a witness plate in the holder and the sample O-ring, and the fact that the plunger was jammed in the up position, and the sample holder was hard to disassemble. When evaluated in conjunction with DSC shape, what happened was a decomposition being triggered that was slow enough that the gas formed could escape around the witness plate and lift up the plunger and wedge it in. This was verified when the holder was finally unscrewed, and large amounts of gas and liquid escaped. If a detonation had occurred, the "explosion" would have punched a hole through the witness plate, and the gas escaped through vent holes. Without the DSC evidence of a smooth decomposition, the material would have had to be classified as potentially explosive.

This difference is important, as detonations produce much more severe consequences, and can readily propagate across multiple containers. A gassy decomposition is much less severe in terms of consequences.

4.8.4.1 Thermokinetics and Safe Handling Parameters

With the DSC onset of the diazirines being between 66 and 120 °C, some questions about proper handling should be asked. Should these compounds be stored refrigerated? Are they safe to leave on the benchtop? Do they need refrigeration during shipping? With the advent of thermokinetic analysis, a series of DSC runs at different rates can be used to estimate the kinetics of decomposition without a detailed understanding of the mechanism(s). In classical kinetics it is assumed that a decomposition can be modeled by a network of reactions such as

$$A \xrightarrow{R1} B \xrightarrow{R2} C \xrightarrow{R3} D$$

and that each individual reaction rate equation is of the form $r = A\exp\left(-\frac{E_a}{RT}\right)(1-x)^n$. By carrying out a series of runs at

---

[15]Drop weight testing is a standard test for potential decomposition or detonation initiation by impact and is described in detail in ASTM E680-79(2011)e1 [27].

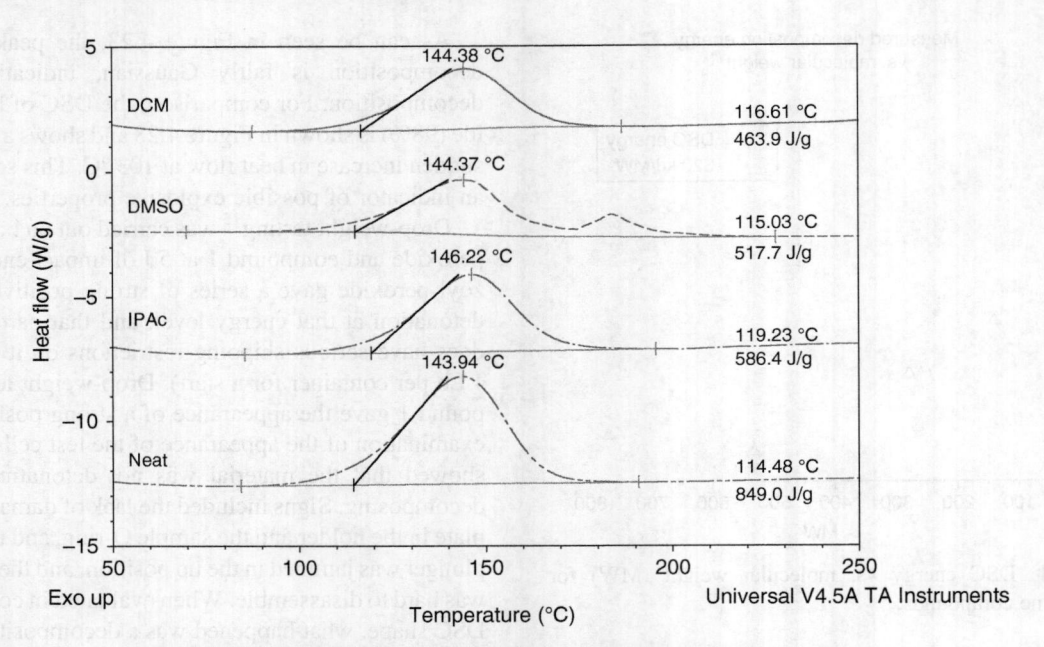

FIGURE 4.25 Decomposition of compound 1 diazirine in solvents.

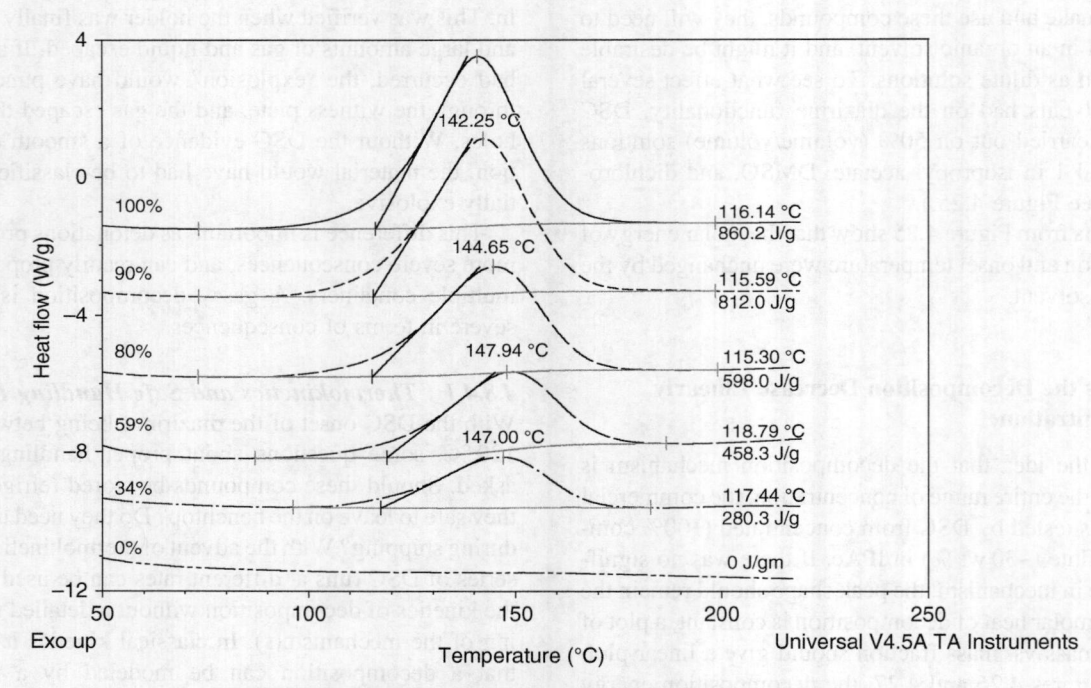

FIGURE 4.26 DSC peaks vs. concentration for compound 1.

different temperatures and concentrations, the rate parameters can be determined. By using a technique called "model-free kinetics," DSC runs at different rates can be used to estimate the rate parameters as a function merely of thermal conversion, where the model is simply $r = A(\alpha)\left(-\frac{E_a(\alpha)}{RT}\right)(1-\alpha)$ and the decomposition is treated as a single lumped reaction with the parameters a function of α, the fractional conversion.

A new diazirine was being prepared, and it was suspected to be less stable than the benchmark compound 1. Three DSC runs at three different rates for the new diazirine as shown in Figure 4.29 demonstrated the complicated effect of changing the scan rate, and thereby the temperature–time relationship

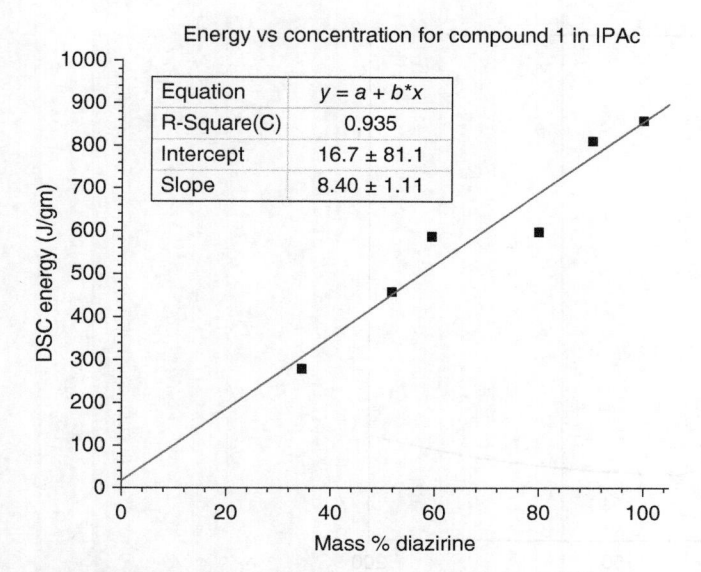

FIGURE 4.27 Plot of energy vs. concentration of compound 1 in IPAc.

of a sample is exposed. Similar runs had been previously carried out on compound 1.

Commercial software packages are available[16,17,18] to convert these DSC curves, often in conjunction with isothermal age DSC studies and ARC runs, into an empirical model of the decomposition (and its heat generation). Once this model is available, a variety of behaviors can be predicted. For this new compound, three parameters were of particular interest. The maximum safe temperature for long-term storage is the temperature at which 1% decomposition would take at least one month to occur. For working with the material on the benchtop, if the TMR24 was greater than 25 °C, the material would be considered sufficiently stable to handle at room temperature. The predicted SADT for a 50 kg package of the material was used to evaluate if the material needed to be classified as self-reactive for shipping purposes and if it should be shipped cold. For this diazirine, with a 50 kg package SADT below 65 °C, UN shipping guidance [26] would state the material is a class 4.1 self-reactive substance, and the proper shipping conditions for the actual package would need to be determined. If a 50 kg package needed to be shipped, it would need to ship under temperature-controlled conditions with a control temperature of −12 °C [28].

Table 4.4 shows the researchers' suspicion that the new compound was less stable was correct, and it is recommended that it be held in a freezer rather than a refrigerator.

This case study illustrates why DSC is a valuable tool for studying decompositions and understanding their behavior while requiring the preparation and use of less than 200 mg to carry out a safety assessment and guiding handling, storage, and synthesis. As a result of the DSC screening program, the research organization is now comfortable working with these highly energetic compounds.

Two additional case studies where the thermal stability of an intermediate was of practical importance for shipping and handling are shown in these articles Zhao et al. [11] and Duh et al. [29].

4.9 CASE STUDY 4: USE OF THERMAL AND PRESSURE SCREENING TOOLS TO ASSESS AN mCPBA SOLUTION

A pharmaceutical batch process involves the use of a 109 g/l solution of 3-chloroperoxybenzoic acid (mCPBA) in a mixture of toluene and ethanol. Due to the known instability of mCPBA [30] at low temperatures and the exothermic reaction between mCPBA and ethanol, process safety testing was performed to quantify the thermal and pressure hazards associated with the solution.

Thermal screening was performed first using a DSC scan. As shown in Figure 4.30, there is a 240 J/g exotherm with an onset temperature of approximately 35 °C. This exotherm was potentially of concern because of both the large exotherm size and low onset temperature. If the exotherm was reached, decomposition would occur, and the temperature would increase approximately 120 K:

$$\Delta T_{ad}(K) = \frac{Q}{Cp} = \frac{240 \frac{J}{g}}{2 \frac{J}{g \cdot K}} = 120 K$$

A screening test was then performed using an MMC instrument to determine the pressure consequences associated with the exotherm. As shown in Figure 4.31a, in alignment with the DSC result, the MMC scan showed a 179 J/g exotherm with an onset of approximately 56 °C. There was minor permanent gas formation associated with the decomposition that resulted in a residual pressure of 2.7 psi after cooldown. See Figure 4.31b. Therefore the majority of the pressure increase during the scan was attributed to vapor pressure. Since the MMC scan did not show an obvious gas onset temperature, it was conservatively assumed that the gas formation was associated with the exothermic decomposition and had the same onset temperature as the exotherm.

The pressure data from the MMC scan was then scaled-up to determine the consequences at large scale. Calculations were performed assuming an 80% vessel fill and a maximum temperature of concern of 413.15 K (140 °C), the maximum jacket service fluid temperature, which was also used as the worst-case batch temperature, T_{batch}:

---

[16] AKTS Thermokinetic Software.
[17] Netzsch Thermokinetics®.
[18] Mettler Advanced Model Free Kinetics.

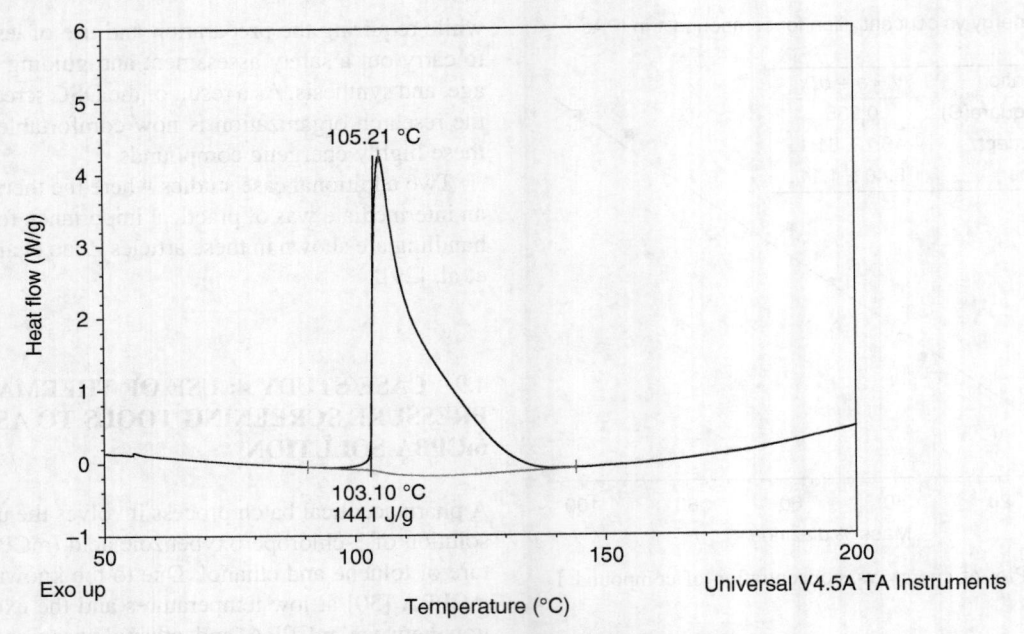

FIGURE 4.28 Decomposition of 98% benzoyl peroxide.

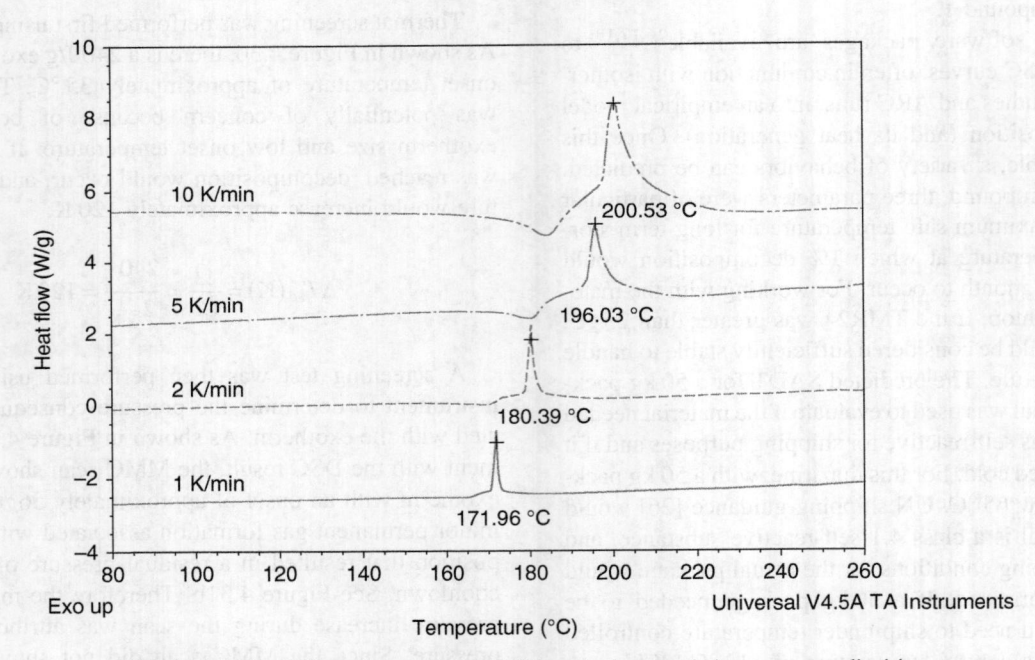

FIGURE 4.29 Effect of scan rate on DSC peak shape and size on a new diazirine.

TABLE 4.4 Comparison of Properties of the Two Diazirines

Parameter	New Diazirine	Compound 1
ΔH_{decomp}	−590 J/g ± 50	−890 J/g ± 30
50 kg SADT	8 °C	18 °C
t_{mr} (25 °C)	37.5–40 h	57.5–60 h
1% decomposition (30 days)	−10 to −5 °C	5–10 °C

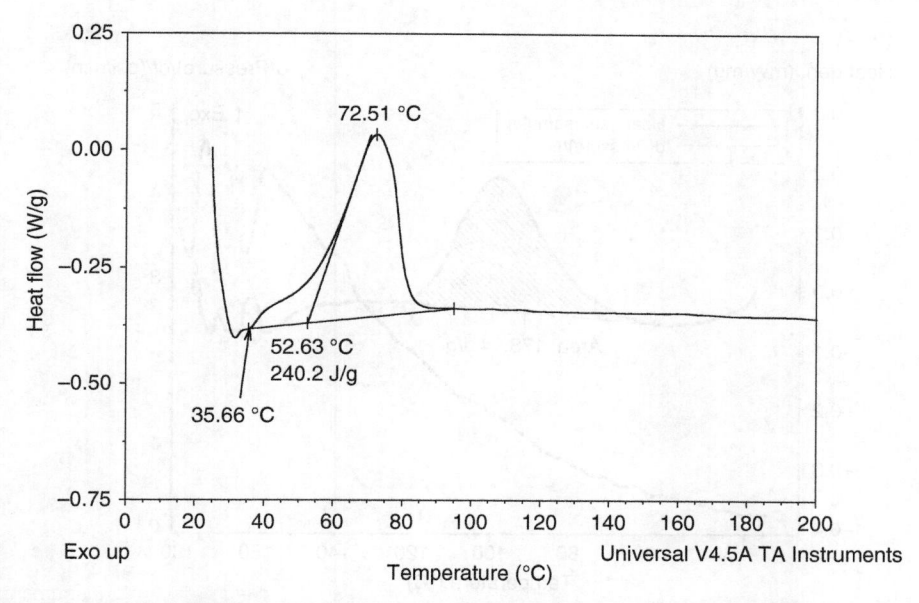

FIGURE 4.30 DSC plot of a 109 g/l mCPBA solution showing the exotherm of concern (exotherm is shown upward). The temperature was increased linearly from room temperature to 350 °C. The above plot is magnified to only show up to 200 °C; no other exothermic activity was observed in the scan to 350 °C.

V_{sample} (ml) = 1.0 = sample volume
$V_{headspace}$ (ml) = 1.6 = MMC cell headspace volume
P_{MMC} (psi) = 2.7 = residual pressure after cooldown
T_{max} (K) = 413.15 = maximum temperature of concern
T_f (K) = 298.15 = final MMC temperature
dP/dt_{max} (psi/min) = 3.3 = maximum MMC pressure rate up to T_{max}
T_{batch} (K) = 413.15 = batch temperature
R (psi·ml)/(mmol·K) = 1.206 = gas constant
V_{vessel} (l) = 3785 L = total vessel volume
V_{batch} (l) = 3028 L = batch volume in vessel

First, the pressure observed in the MMC at the maximum temperature of concern was compared with the vessel MAWP to determine if there was a potential pressure hazard that may result in the safety valve/rupture disc lifting. In this case, if the mCPBA solution were to heat up to 140 °C, the pressure would be 59 psig (see Figure 4.31b and calculation below). Therefore the pressure at the maximum temperature did not pose a potential hazard when compared with a vessel MAWP of 100 psig:

$$59\,\text{psig} = 96\,\text{psia} - \left(\frac{16\,\text{psia pad gas} \times (140 + 273.15\,\text{K})}{298.15\,\text{K}}\right) - 14.7\frac{\text{psig}}{\text{psia}}$$

The residual pressure in the vessel was then determined and compared with the vessel MAWP. This was done by first calculating the gas generation per unit volume, G, in the MMC and then using this value to calculate the residual pressure upon scale-up, $P_{scale\text{-}up}$, as follows:

$$G\left(\frac{\text{mmoles gas}}{\text{ml sample}}\right) = \frac{P_{MMC} \times V_{headspace}}{R \times T_f \times V_{sample}}$$

$$P_{scale\text{-}up}(\text{psi}) = \frac{(G \times V_{batch})RT_{batch}}{(V_{vessel} - V_{batch})}$$

These calculations showed that if the mCPBA solution were to heat up to 140 °C at the specified vessel fill, the 24 psig residual pressure generated did not present a potential hazard when compared with a vessel MAWP of 100 psig.

Lastly, the pressure generation rate in the vessel was compared with the vessel safety valve vent capacity to determine if there was a potential pressure hazard that may result in the safety valve lifting or equipment damage to occur. The pressure generation rate per unit volume in the MMC, $\dot{G}$, was calculated and used to determine the volumetric gas flow rate upon scale-up, Q, as follows:

$$\dot{G}\left(\frac{\text{mmoles gas}}{\text{min} \cdot \text{ml sample}}\right) = \frac{\frac{dP}{dt_{max}} \times V_{headspace}}{R \times T_{max} \times V_{sample}}$$

$$\dot{Q}\left(\frac{1}{\text{min}}\right) = (\dot{G} \times V_{batch}) \times 22.4\frac{\text{liters gas}}{\text{mole gas}} \times \left(\frac{T_{batch}}{273.15\,\text{K}}\right)$$

These calculations showed that if the mCPBA solution were to heat up to 140 °C at the specified batch volume, the predicted volumetric gas flow rate of 1084 L/min did not present a potential hazard when compared with a typical vessel safety valve vent capacity of greater than 30 000 L/min.

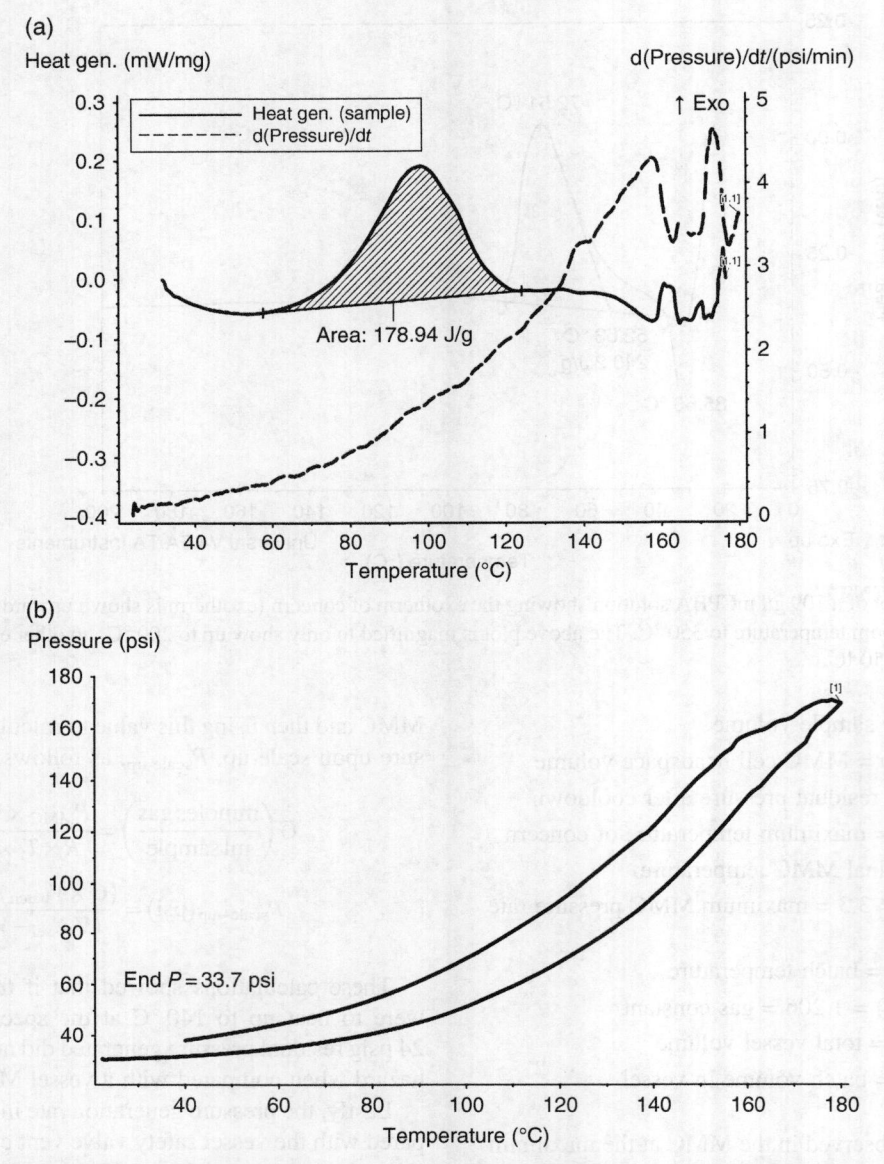

FIGURE 4.31 (a) MMC scan heat generation and pressure rate vs. temperature plot for 109 g/l mCPBA solution (exotherm is shown upward). The temperature was increased linearly from room temperature to 180 °C. This plot shows the exotherm onset and size, along with the pressure generation rate. (b) MMC scan pressure vs. temperature plot for 109 g/l mCPBA solution. This plot was used to determine the residual pressure after cooldown and the pressure at maximum temperature of concern. The MMC scan was run under 16 psi of nitrogen pad gas to prevent refluxing.

Note that this calculation used the maximum dP/dt, which conservatively assumes all gas generation when in this case it is mostly vapor pressure, thus neglecting any evaporative cooling benefits.

In conclusion, pressure screening using an MMC showed that there are no potential pressure hazards associated with reaching the 240 J/g exotherm observed starting at approximately 35 °C in the DSC.

This case study demonstrates the use of a thermal screening tool (DSC) to identify a potential exotherm of concern and the use of a pressure screening tool (MMC) to conclude that the exotherm was not a hazard due to only minimal permanent pressure generation associated with the exotherm.

4.10 NOTATIONS

The common symbols used are as follows:

Q = heat release (J/g)
ΔH = enthalpy change (J/mol)
Cp = heat capacity (J/g·K)

ρ = density (g/l)

ΔT = temperature change (K)

ΔT_{ad} = adiabatic temperature change (K)

T = temperature (°C)

C = concentration (mol/l)

ϕ (phi) = thermal inertia = $1 + \frac{m_{cell} Cp_{cell}}{m_{sample} Cp_{sample}}$ (dimensionless)

where m = mass

TMR24 = temperature at which the adiabatic time to maximum rate is 24 hours (°C)

EIT = exotherm initiation temperature, typically measured in an adiabatic calorimeter such as an ARC or VSP, which is defined as a temperature rise rate of 0.02 K/min at a phi factor of 1.0. In other sources this may be referred to as the exothermic onset temperature.

ACKNOWLEDGMENTS

The author would like to thank Dan Muzzio and Don Bachert from Merck & Co., Inc., Chemical Engineering R&D Rahway, NJ 07065 USA.

REFERENCES

1. (a) Center for Chemical Process Safety (2008) Selection of hazard evaluation techniques, in *Guidelines for Hazard Evaluation Procedures*, 3rd Edition, Wiley, Hoboken, NJ. doi: https://doi.org/10.1002/9780470924891.ch6;
   (b) Center for Chemical Process Safety (1995) References, in *Guidelines for Chemical Reactivity Evaluation and Application to Process Design*, Wiley, Hoboken, NJ. doi: https://doi.org/10.1002/9780470938058.refs;
   (c) Center for Chemical Process Safety (2010) Overview of LOPA, in *Layer of Protection Analysis: Simplified Process Risk Assessment*, Wiley, Hoboken, NJ. doi:https://doi.org/10.1002/9780470935590.ch2;
   (d) Stoessel, F. (2008) Introduction to risk analysis of fine chemical processes, in *Thermal Safety of Chemical Processes: Risk Assessment and Process Design*, Wiley-VCH Verlag GmbH & Co. KGaA, Weinheim, Germany. doi: https://doi.org/10.1002/9783527621606.ch1.

2. Poje, G.V., Taylor, A.K., and Rosenthal, I. (2000). *US Chemical Safety and Hazard Investigation*. Report No. 1998-06-I-NJ, Chemical Manufacturing Incident at Morton International, Inc., 16 August 2000. https://www.csb.gov/assets/1/20/morton_report.pdf?13798 (accessed 16 October 2018).

3. Perry, R.H., Green, D.W., and Maloney, J.O. (1999). *Perry's Chemical Engineering Handbook*, 7the. New York: McGraw-Hill.

4. ASTM E1981-98(2012)e2 *Standard Guide for Assessing Thermal Stability of Materials by Methods of Accelerating Rate Calorimetry*, ASTM International, West Conshohocken, PA, 2012. www.astm.org (accessed 17 November 2017).

5. Hofelich, T.C. and LaBarge, M.S. (2002). On the use and misuse of detected onset temperature of calorimetric experiments for reactive chemicals. *Journal of Loss Prevention in the Process Industries* 15 (3): 163–168.

6. Crucible Brochure. Crucibles overview: DSC and TGA/SDTA. https://www.mt.com/mt_ext_files/editorial/generic/4/crucible_brochure_0x00024947000255120 0068070_files/51724175.pdf (accessed 29 September 2017).

7. Weast, R.C. (ed.) (1987). *CRC Handbook of Chemistry and Physics*, 68the. Boca Raton, FL: CRC Press.

8. ASTM E698-16, *Standard Test Method for Kinetic Parameters for Thermally Unstable Materials Using Differential Scanning Calorimetry and the Flynn/Wall/Ozawa Method*, ASTM International, West Conshohocken, PA, 2016. www.astm.org (accessed 17 November 2017).

9. AKTS (2017). Thermokinetic Software. www.akts.com (accessed 29 September 2017).

10. Yoshida, T., Yoshizawa, F., Itoh, M. et al. (1987). Prediction of fire and explosion hazards of reactive chemicals. I. Estimation of explosive properties of self-reactive chemicals from SC-DSC data. *Kogyo Kayaku* 48 (5): 311–316.

11. Zhao, R., Muzzio, D., Vickery, T. et al. (2018). Thermal hazard investigation of a pharmaceutical intermediate. *Process Safety Progress* 37 (2): 263–267. https://doi.org/10.1002/prs.11917.

12. Singh, J. and Simms, C. (2001). The thermal screening unit (TSU) a tool for reactive chemical screening. In: *Hazards XVI*, IChemE Symposium Series, vol. 148 (ed. Institution of Chemical Engineers). Rugby: Institution of Chemical Engineers.

13. Thermal Hazard Technology (THT), Rapid Screening Device (RSD) A cost-effective calorimeter for screening samples for exothermic reaction hazards. www.thtuk.com (accessed 29 September 2017).

14. NETZSCH. Multiple mode calorimetry (MMC). https://www.netzsch-thermal-analysis.com/en/products-solutions/multiple-mode-calorimetry (accessed 29 September 2017).

15. Burelbach, J.P. (2000). Advanced reactive system screening tool (ARSST). Presented at the 28th Annual North American Thermal Analysis Society (NATAS) Conference, Orlando, FL, (4–6 October 2000).

16. Fisher, H. G., Forrest, H. S., Grossel, S. S., Huff, J. E., Muller, A. R., Noronha, J. A., Shaw, D. A. and Tilley, B. J. (1993) Front matter, in *Emergency Relief System Design Using DIERS Technology: The Design Institute for Emergency Relief Systems (DIERS) Project Manual*, Wiley, Hoboken, NJ. doi:https://doi.org/10.1002/9780470938317.fmatter.

17. Fauske, H.K. (2006). Managing chemical reactivity: minimum best practice. *Process Safety Progress* 25: 120–129. https://doi.org/10.1002/prs.10126.

18. (a) Lam, T.T., Vickery, T., and Tuma, L. (2006). Thermal hazards and safe scale-up of reactions containing dimethyl sulfoxide. *Journal of Thermal Analysis and Calorimetry* 85 (1): 25–30.
    (b) Wang, Z., Richter, S.M., Gates, B.D., and Grieme, T.A. (2012). Safety concerns in a pharmaceutical manufacturing process using dimethyl sulfoxide (DMSO) as a solvent.

Organic Process Research & Development 16 (12): 1994–2000. https://doi.org/10.1021/op300016m.

(c) Brandes, B.T. and Smith, D.K. (2016). Calorimetric study of the exothermic decomposition of dimethyl sulfoxide. *Process Safety Progress* 35: 374–391. https://doi.org/10.1002/prs.11802.

19. Bassan, E., Ruck, R.T., Dienemann, E. et al. (2013). Merck's reaction review policy: an exercise in process safety. *Organic Process Research & Development* 17 (12): 1611–1616. https://doi.org/10.1021/op4002033.

20. Frurip, D.J. (2008). Selection of the proper calorimetric test strategy in reactive chemicals hazard evaluation. *Organic Process Research & Development* 12 (6): 1287–1292. https://doi.org/10.1021/op800121x.

21. Townsend, D.I. and Tou, J.C. (1980). Thermal hazard evaluation by an accelerating rate calorimeter. *Thermochimica Acta* 37 (1): 1–30. https://doi.org/10.1016/0040-6031(80)85001-5.

22. (a) ASME *ASME Boiler and Pressure Vessel Code, Section VIII Pressure Vessels, Division 1*. New York, NY: American Society of Mechanical Engineers http://www.asme.org/shop/standards/new-releases/boiler-pressure-vessel-code-2013.

(b) Crowl, D.A. and Tippler, S.A. (2013). Sizing pressure-relief devices. *Chemical Engineering Progress* 109: 68–76.

(c) Sandler, H.J. and Luckiewicz, E.T. (1993). *Practical Process Engineering*. New York: McGrew-Hill.

23. International Atomic Energy Agency (IAEA). Environmental isotopes in the hydrological cycle: Vol. 1, Ch. 9 pg 94, Figure 9.6. http://www-naweb.iaea.org/napc/ih/documents/global_cycle/Environmental%20Isotopes%20in%20the%20Hydrological%20Cycle%20Vol%201.pdf (Accessed 18 October 2018).

24. Environmental Protection Agency (EPA) (1990). Environmental protection agency land disposal restrictions for the third third scheduled wastes. *Federal Register* 55 (106), 1 June 1990. https://www.epa.gov/sites/production/files/2016-02/documents/55_fr_22519_to_22720_june_1_1990_thirdthird.pdf (accessed 17 November 2017).

25. Dubinsky, L., Krom, B.P., and Meijler, M.M. (2012). Diazirine based photoaffinity labeling. *Bioorganic & Medicinal Chemistry* 20 (2): 554–570. https://doi.org/10.1016/j.bmc.2011.06.066.

26. United Nations (1999). *Recommendations on the Transport of Dangerous Goods*, Third Revised. New York: United Nation.

27. ASTM E680-79(2011)e1 (2011). *Standard Test Method for Drop Weight Impact Sensitivity of Solid-Phase Hazardous Materials*. West Conshohocken, PA: ASTM International.

28. US Department of Transportation Title 49 (2011). Code of Federal Regulations on Forbidden materials and packages. https://www.gpo.gov/fdsys/pkg/CFR-2011-title49-vol2/pdf/CFR-2011-title49-vol2-sec173-21.pdf (accessed 17 November 2017).

29. Duh, Y.S., Yo, J.M., Lee, W.L. et al. (2014). Thermal decompositions of dialkyl peroxides studied by DSC. *J. Therm. Anal. Calorim.* 118: 339.

30. Sigma Aldrich Corporation (2018). Safety Data Sheet (SDS) for mCPBA. http://www.sigmaaldrich.com/MSDS/MSDS/DisplayMSDSPage.do?country=US&language=en&productNumber=273031 (accessed 2 March 2018).

5

CASE STUDY OF A BORANE–THF EXPLOSION

DAVID J. AM ENDE*
Nalas Engineering Services, Inc., Centerbrook, CT, USA

RICHARD M. DAVIS
Global Environmental, Health and Safety, Pfizer, Inc., Groton, CT, USA

5.1 INTRODUCTION

Borane–THF (BTHF) is a specialty chemical supplied as a solution in tetrahydrofuran (THF) often used in pharmaceutical research and industrial applications as a reducing agent for aldehydes, ketones, amides, and other functional groups [1]. In 2002, a large cylinder of BTHF experienced a boiling liquid expanding vapor explosion (BLEVE) upon storage at Pfizer's R&D site in Groton, CT. This chapter provides a case study specific to the thermochemical aspects of BTHF, its storage condition and how the analysis ultimately provided understanding into the cause of the explosion.

5.2 BACKGROUND

The original lot of six cylinders of BTHF, each with a capacity of 400 L (350 kg) was purchased with the intention to be used in a pilot plant campaign to perform a chemical reduction step in the synthesis of a drug candidate. The cylinders were shipped from the manufacturer on 4 March 2002 and delivered to Pfizer R&D (Groton, CT) on 6 March 2002 and placed in one of Pfizer's flammable materials storage buildings external to the site's pilot plant. A little less than four months after receiving the cylinders, on 25 June at 8:00 a.m., one of the six cylinders exploded, resulting in a large fire ball and significant structural damage to the warehouse and adjacent buildings. Multiple injuries were suffered as well [2]. The bottom of the exploded drum is shown in Figure 5.1. The type of explosion that occurred was ultimately classified as a BLEVE. A BLEVE occurs when a tank containing a liquid held above its normal boiling point ruptures, resulting in the explosive vaporization of the tank contents, and when the vaporized liquid is flammable, can result in a vapor cloud explosion. [3] The focus of this case study is on the technical details and analysis with worked out example problems and calculations used to help understand the cause of the thermal runaway reaction. It is intended as a process safety educational tool to highlight the challenges and approaches to thermal characterization of reactive materials, with respect to added stabilizers, and their packaging configurations and storage conditions.

BTHF solution is typically prepared industrially by dissolving diborane gas into THF. A small amount of additive such as sodium borohydride, $NaBH_4$, is typically added to the solution as a stabilizer [4]. The stabilizer is used to slow the rate of decomposition that occurs by ether cleavage, leading to loss of purity of BTHF as shown in Scheme 5.1.[1]

The decomposition pathway occurs through ring opening of the THF and is assumed to proceed through monobutoxyborane (M). Since [M] is not actually observed, it is assumed to be short-lived, rapidly disproportionating, so

---

*Previous address: Chemical R&D, Pfizer, Inc., Groton, CT, USA

[1]Borane-Tetrahydrofuran 1 and 2 M Solutions, manufactured by Callery Chemical Company, Division of Mine Safety Appliances Co., MSDS dated 22 March 2001.

Chemical Engineering in the Pharmaceutical Industry: Active Pharmaceutical Ingredients, Second Edition.
Edited by David J. am Ende and Mary T. am Ende.
© 2019 John Wiley & Sons, Inc. Published 2019 by John Wiley & Sons, Inc.

FIGURE 5.1 One of the unbreached 400 L cylinders of BTHF remaining after the explosion (left). The base of the cylinder impacted by the BLEVE recovered after the explosion (right).

SCHEME 5.1 BTHF decomposition via ether cleavage: borane–THF (BTHF) [A], mono-butoxyborane [M], dibutoxyborane [B], and tributyl borate [C].

the rate to form [B] from [M] is considered fast. More detailed kinetic analysis and modeling, based on a slightly more complex mechanism, is described later in this case study. As the material ages, especially at ambient temperatures, this ring-opening decomposition leads to a loss of potency of BTHF with accumulation of species B and C as shown in Scheme 5.1.

The stabilizer slows the ring-opening decomposition, but the exact mechanism of how the sodium borohydride inhibits the ring opening is not well understood. More recently new stabilizers have been developed, claiming improved thermal properties and stability of BTHF solutions [5]. Flanagan and coworkers have recently published on safe handling of 1 M solutions [6]. For this case study, however, our focus is on sodium borohydride-stabilized solutions of BTHF with concentrations between 1 and 2 M solution.

Chemical stability data obtained from *isothermal* studies indicated that the material loses potency at a rate that depends on temperature. Thus the storage condition recommended by the manufacturer was to store cold (at 0 °C or lower) to protect product purity. There was no adiabatic test data or self-accelerating decomposition temperature (SADT[2]) determinations specific to the packaging configuration (e.g. 400 L cylinders containing 2M BTHF).

At the time of the incident, there were six cylinders of BTHF stored inside the flammable materials warehouse adjacent to the pilot plant. On the day of the incident, one cylinder exploded, and a second one was breached (by the explosion of the first cylinder), while the remaining four cylinders remained intact. After the explosion, the remaining four cylinders were measured to have elevated temperatures and pressures. As part of the emergency response, the cylinders were sprayed with water and packed with ice. Pressure gauges were installed on the cylinders to monitor any further

---

[2]SADT refers to self-accelerating decomposition temperature.

buildup of pressure. Technical representatives from the manufacturer arrived on-site to provide consultation for handling the remaining cylinders.

In addition to injuries to personnel and damage to buildings, the incident resulted in closure of the R&D site, employing approxiamtely 5000, for approximately 5 days until the remaining cylinders could be safely transported off-site to the manufacturer.

5.3 INVESTIGATION

One facet of the investigation focused on the thermochemical stability of the material.

Once it was determined that a cylinder of BTHF exploded, there were many questions as to why it happened. Was this material actually BTHF? Was it contaminated? Why did only one cylinder explode? What was the cause?

One of the first items to be established was the chemical composition of the BTHF in the unbreached cylinders. It was important to understand whether it was typical or atypical and if there was any suspected type of contamination that may have altered its stability.

A sample was obtained from one of the four surviving cylinders. Boron NMR (^{11}BNMR) was one of the tests used for confirming the identity and composition of the BTHF in the cylinder. The ^{11}BNMR spectrum of the sample confirmed the identity of the material (i.e. the material was indeed BTHF), and the spectrum is shown in Figure 5.2.

EXAMPLE PROBLEM 5.1

A liquid sample was withdrawn from one of the surviving cylinders recovered post-incident. It was found to have a ^{11}BNMR spectrum with the integral data for the three major components as shown in Figure 5.2.

Assuming the original potency was 2.0 M BTHF (in solution with THF) and using the NMR data:

(a) Calculate the molarity of BH_3-THF complex remaining in the cylinder.
(b) Calculate the molarity of active hydride "B-H" species.
(c) A hydrogen evolution test showed that a 5.0 ml of sample of the BTHF sample injected into a large excess of methanol liberated 535 ml (at 25 °C) of hydrogen (via an inverted graduated cylinder and measuring the displacement of water). Calculate the active hydride and compare the result with the molarity calculated from NMR results in part (b).

Solution

(a) There is only one boron in each of the species, so Integral% = mol % in this case; using the data from Figure 5.2, calculate the relative mol %:

$$\frac{3.6751}{1.000 + 3.6751 + 0.9385} = \frac{3.6751}{5.6136} = 65.5 \, mol\% BH_3-THF$$

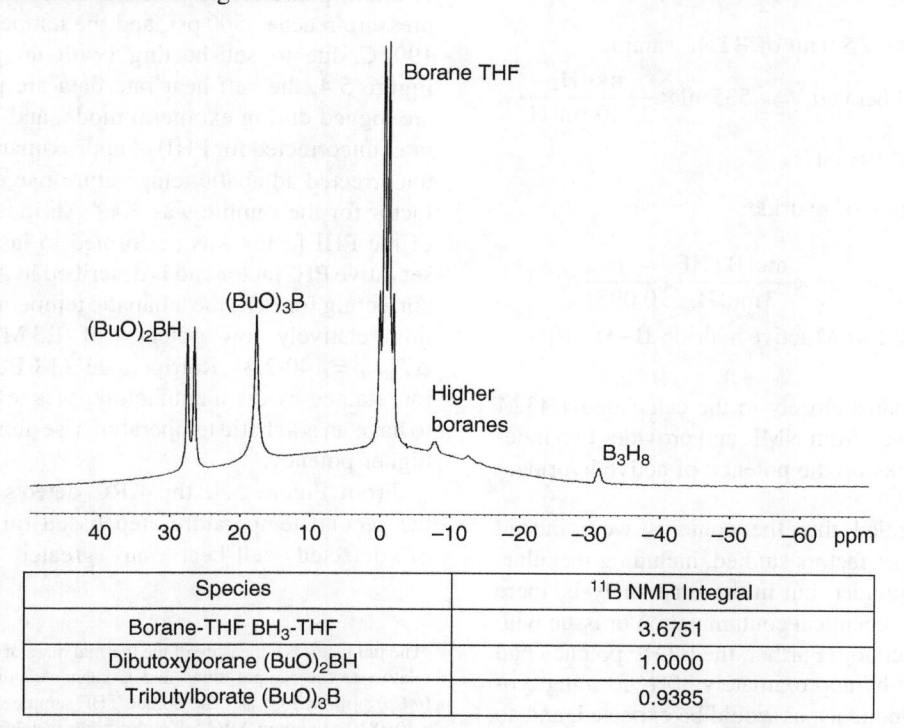

Species	^{11}B NMR Integral
Borane-THF BH_3-THF	3.6751
Dibutoxyborane $(BuO)_2BH$	1.0000
Tributylborate $(BuO)_3B$	0.9385

FIGURE 5.2 ^{11}BNMR spectrum obtained from cylinder 24 893 tested showed partial decomposition of BTHF to dibutoxyborane and tributoxyborane. Insets are the respective integrals for the three major components.

- Dibutoxyborane = $\dfrac{1.000}{5.6136} = 17.8\,\text{mol\%}\,(\text{BuO})_2\text{BH}$
- Tributyl borate = 16.7 mol % $(\text{BuO})_3\text{B}$
- If the original potency was 2.0 M, the new potency is 65.5% × 2 M = 1.31 M BH3-THF. Similarly, the concentration of dibutoxyborane is 0.36 M, and tributyl borate is 0.33 M.

(b) BH$_3$ contributes three hydrides, while dibutoxyborane only contributes one. Tributyl borate contributes no active hydride.
- Mols of active "B-H" hydrides:

$$\dfrac{[(3)\times 65.5\% + (1)\times 17.8\% + (0)\times(16.7)]}{100} = 2.143$$

- Theoretical mols "B-H" hydride per mol of BH3-THF = 3.0
- Active hydride:

$$\% = \dfrac{2.143}{3.0} = 71.4\,\text{mol\%}$$

- Remaining molarity of active hydride:

$$2\,\text{M} = 2.0 \times 71.4\% = 1.43\,\text{M active hydride}$$

(c) The methanol rapidly quenches BTHF releasing hydrogen gas.

$$\text{THF}\cdot\text{BH}_3 + 3\text{MeOH} \rightarrow 3\text{H}_2 + (\text{OMe})_3\text{B} + \text{THF}$$

- Liquid sample = 5.0 ml of BTHF sample
- Mols gas liberated = $535\,\text{ml} \times \dfrac{\text{mol H}_2}{22\,400\,\text{ml H}_2} \times \dfrac{273\,\text{K}}{298\,\text{K}} = 0.0219\,\text{mol H}_2$
- Overall potency of hydride:

$$0.0219\,\text{mol H}_2\,\text{liberate} \times \dfrac{\text{mol BTHF}}{3\,\text{mol H}_2} * \dfrac{1}{0.0051}$$
$$= 1.46\,\text{M active hydride"B-H"}$$

- 1.46 M compares closely to the calculated 1.43 M hydride potency from NMR and provides two independent checks on the potency of active hydride.

These results revealed that the contents were indeed BTHF. There were other factors studied, including metallurgical studies on the cylinder, but in the final analysis, there was nothing to suggest chemical contamination or issue with the materials of construction. Further, the BTHF potency had diminished in potency by approximately 30%, forming typical ring-opening components, as would be expected, and so its concentration did not appear to be atypical. The BTHF manufacturer also tested the contents of each of the returned cylinders, and their data also supported this conclusion that the material, although aged with some loss of potency, otherwise appeared normal.

5.4 ARC DATA OF BTHF FROM AN ADJACENT CYLINDER

Accelerating rate calorimetry (ARC) is able to provide information on thermal runaway behavior of substances and reaction mixtures. It is considered an industry standard test for the determination of thermal stability and specifically, self-heating characteristics of materials [7]. The ARC uses a *heat–wait–search (HWS)* cycle where it first *heats* the sample to the starting temperature and then *waits* for a set period and then *searches* for an exotherm. If the exotherm in any cycle exceeds a certain threshold during the search period, an "exotherm" is detected. If an exotherm does not exceed the threshold, the ARC will heat the sample incrementally (typically 5 °C) to the next plateau.

A sample obtained from one of the adjacent cylinders post-incident was tested using ARC. The composition of this sample was 1.3 M BTHF and 1.46 M active hydride (71% of original), having lost 29% potency of active hydride due to the extended storage time under ambient conditions.

A photo of the ARC system is shown in Figure 5.3. The ARC results are shown in Figures 5.3 and 5.4 from recovered sample cylinder #24893. In Figure 5.3 both temperature and pressure are profiled as a function of time. The pressure reaches 500 psi, and the temperature reaches over 190 °C due to self-heating (with no phi correction). In Figure 5.4, the self-heat rate data are plotted. These data are logged during exotherm mode, and it shows a self-heat rate (uncorrected for PHI) of approximately 4 °C/min and an uncorrected adiabatic temperature rise of 140 °C. The PHI factor for the sample was 2.4.[3] An independent validation of the PHI factor was performed to justify this more conservative PHI factor and is described in Appendix 5.A. After correcting for PHI, the adiabatic temperature rise, even from this relatively low potency of 1.3 M, was 336 °C (i.e. $\Delta T_{ad}\cdot\varphi = 140\cdot 2.4$). Refrigerated 2 M BTHF from the same lot retained by the manufacturer was tested and was shown to have an adiabatic temperature rise above 500 °C due to its higher potency.

From Figure 5.4, the ARC detects an "exotherm" on the second temperature step based on the strict criterion of detected self-heat rate greater than 0.02 °C/min.

---

[3] The phi factor is a measure of the thermal mass of the sample and container relative to the sample alone and is calculated from $\varphi = [(mCp)_{sample} + (mCp)_{cell})]/(mCp)_{sample}$. The borane THF sample was 4.357 g; the cell was a 10.3810 g titanium ARC cell with fittings weighting 10.7730 g. The calculated phi factor for the sample and test cell (no fittings) was 1.57. The calculated PHI for the sample and test cell with fittings was 2.40.

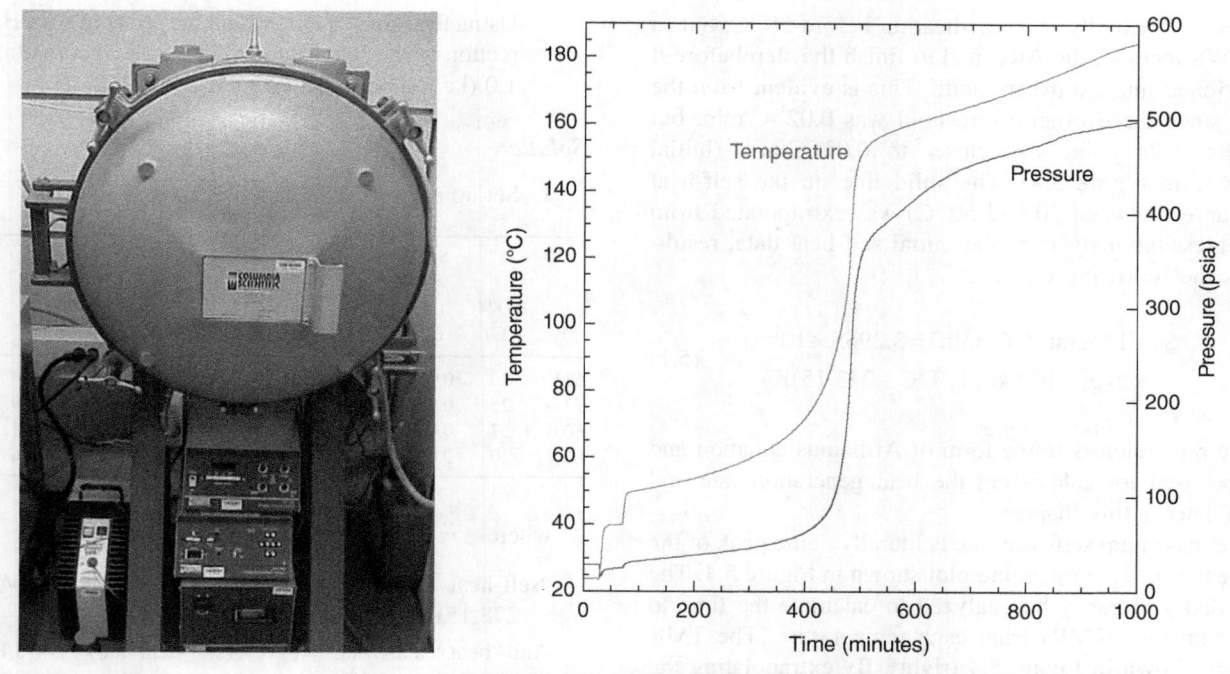

FIGURE 5.3 Left: accelerating rate calorimeter (Columbia Scientific). Right: heat–wait–search profile ARC experiment on "aged" BTHF recovered from a surviving cylinder. The composition was 1.46 M "B-H" active hydride B-H and 1.3 M BH_3-THF.

Self heat rate (°C/min)=3.2965×10^{12} exp(−10 446 (1/T°C+273.15)K)

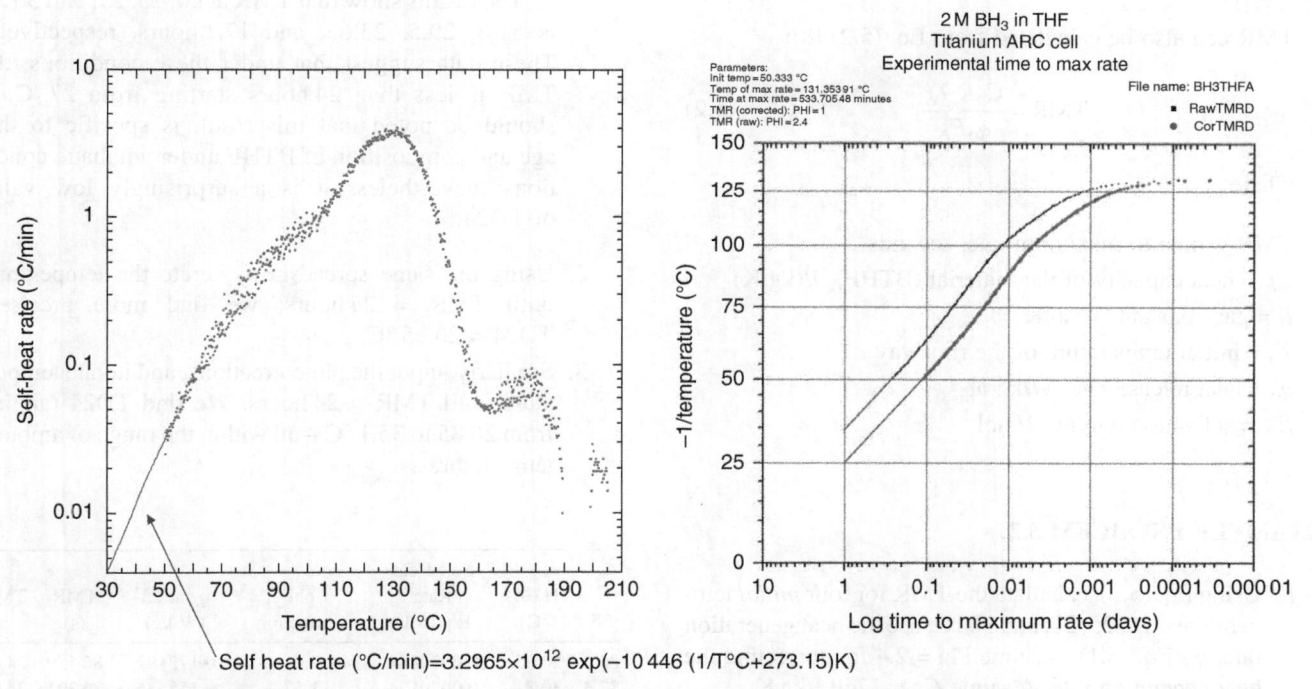

FIGURE 5.4 Left: self-heat rate curve of aged BTHF. Right: time to maximum rate plot for aged BTHF (1.3 M BH_3-THF).

However, it actually was self-heating before 50 °C, but in the HWS method, the ARC had to finish the step before it transitioned into exotherm mode. This is evident from the curve since the exotherm threshold was 0.02 °C/min, but the first data point was closer to 0.03 °C/min (initial points from Figure 5.4). The solid line on the self-heat rate curve (between 30 and 50 °C) was extrapolated from the regression analysis of the initial self-heat data, resulting in the following equation:

$$\text{Self-heat rate (°C/min)} = 3.2965 \times 10^{12} \times \exp(-10\,446\,(1/T°C + 273.15)\,K) \quad (5.1)$$

The regression is in the form of Arrhenius equation and will be used for calculating the heat generation rate and SADT later in this chapter.

The maximum self-heat rate is literally at the peak of the self-heat rate vs. temperature plot shown in Figure 5.4. The ARC data can easily be analyzed to calculate the time to maximum rate (TMR) from each temperature. The TMR data are shown in Figure 5.4 (right). By extrapolating the TMR curve, an estimate for TD24 can be made where TD24 is the temperature where the TMR is 24 hours when held under adiabatic conditions. Graphically the phi-corrected TD24 is near 25 °C for this age and potency of BTHF.

TMR can also be calculated from Eq. (5.2) [8]:

$$\text{TMR} = \frac{C_p R T_o^2}{q_o E_A} \quad (5.2)$$

where

TMR = time to maximum rate, seconds
C_p = heat capacity of the material (BTHF), J/(kg·K)
R = gas constant, J/molK
T_o = initial temperature of the runaway
q_o = heat release rate, W/kg at T_o
E_A = activation energy, J/mol

EXAMPLE PROBLEM 5.2

1. Using Eq. (5.2), calculate the TMR for four *initial* temperatures: 20, 25, 27, and 30 °C; use the heat generation rate, q_o, Eq. (5.1). Assume Phi = 2.4 for correcting the heat generation rate. Assume C_p = 1740 J/kg K.
2. Now find the TD24 (hint: find the temperature where TMR = 24 hours).
3. Using the same equation, find the effect of the PHI correction on the calculated TD24's for PHI corrections of 1.0 (i.e. no correction), 1.57, 2.0, and 2.4.

Solution

1. Set up a spreadsheet and calculate as follows:

°F	TD24 (°C)	Self Heat Rate (°C/min) Eq. (5.1)	phi Correction	Heat Release q_o (W/kg)	TMR (s)	TMR (h)
68	20	0.001103	2.4	0.076766443	186 469.4	51.80
77	25	0.0020049	2.4	0.124190015	106 113.9	29.48
80.6	27	0.0025321	2.4	0.156846833	85 151.1	23.65
86	30	0.0035732	2.4	0.221335451	61 553.6	17.10

where

"self-heat rate" = $(3.2965 \times 10^{12}) \times \exp(-10\,446/(°C + 273.15)) \equiv °C/min$

And "heat release rate" = PHI × (self-heat rate)(1740 J/kg K)/(60 s/min) ≡ W/kg

And TMR = $(1740 \times (°C + 273.15)^2)/(\text{heat release} \times 10\,446) \equiv$ seconds.

Note 10 446 = E/R.

The results show that TMR at 20, 25, 27, and 30 °C is 51.8, 29.5, 23.65, and 17.1 hours, respectively. These data suggest that under these conditions, the TMR is less than 24 hours starting from 27 °C. It should be noted that this result is specific to this age and composition of BTHF under adiabatic conditions; nevertheless it is a surprisingly low value of TD24.

2. Using the same spreadsheet, iterate the temperature until TMR = 24 hours; we find more precisely TD24 = 26.85 °C.

3. Similarly, input the phi corrections, and iterate temperature until TMR = 24 hours. We find TD24 ranging from 26.85 to 35.1 °C – all within the range of ambient temperatures.

°F	TD24 (°C)	Self Heat Rate Eq. (5.1)	phi Correction	q_o at 25 (W/kg)	TMR (s)	TMR (h)
80.3	26.85	0.0024884	2.4	0.173192	86 559	24.04
87.4	30.8	0.0039124	1.57	0.158536	86 391	24.00
95.2	35.1	0.006319	1	0.163093	86 369	23.99
83.3	28.5	0.0030105	2	0.155401	86 804	24.11

Based on the calculations and the graphical data, the TD24, under adiabatic conditions, was in the range of 25–27 °C for this potency. These TD24's were surprising low as it suggested runaway behavior is expected at ambient temperatures under adiabatic (well-insulated or low heat loss) conditions. This was a significant finding in the investigation.

Reiterating, the TD24 is the temperature under adiabatic conditions where the maximum rate is achieved over a 24 hour period. For designing safe processes in the kilo lab or pilot plant, it is good practice to characterize TD24 and to ensure temperatures are typically controlled to well below TD24. TD24 is useful because it provides a 24 hour "window" to respond if a rise in temperature is detected. Although since these cylinders were not continuously monitored, with the information developed as part of the investigation, it should be obvious to the reader that a TD24 anywhere near ambient temperatures for these cylinders turned out to be a serious safety concern. Values this low, if previously known, would have signaled the need for refrigeration for safety reasons. Specifically the TD24 values were in the range of 25–27 °C for this age and relatively low potency of BTHF. It is likely the TD24 is even lower for higher potency BTHF that have become unstabilized during aging. The ambient temperature the day prior to explosion reached 29–30 °C.

Under US Department of Transportation (DOT) regulations, shipment of a package at ambient temperature containing self-reactive materials with an SADT of less than or equal to 50 °C is prohibited unless the material is sufficiently stabilized or inhibited (Potyen et al. [6]). The SADT also depends on the specific packaging, in this case a 400 L cylinder, so information about heat transfer from a filled cylinder is necessary to properly assess the SADT. This will be discussed in a later section.

Based simply on this ARC experiment, the extrapolated data suggested there was a previously unknown but significant thermal runaway potential for this material even at ambient temperatures. In addition, the kinetics of decomposition, based on self-heat rate vs. temperature, appear to follow first-order kinetics. As such there appears to be little or no effect of stabilizer remaining. Further it was suggested that aged BTHF solutions, like those held under ambient conditions in the flammable materials warehouse, eventually became effectively unstabilized. Confirmation was made by preparing laboratory scale BTHF from diborane gas and THF, without stabilizer, and comparing the self-heat rates with those samples pulled from the cylinders remaining after the incident. This is shown in Figure 5.5. The concentrations of BTHF are slightly different (1.3 vs. 1.5 M potency), but the initial rates were similar, indicating that *aged* BTHF essentially behaves as *unstabilized* BTHF. The effect of time–temperature history on loss of stabilizer and temperature stability will be discussed in more detail in later sections.

5.5 TEMPERATURE HISTORY OF THE BORANE–THF CYLINDERS

A batch of stabilized BTHF (Lot # 385765) was manufactured on 1 March 2002, and 6 cylinders each containing 350 kg of 2 M BTHF solution (~400 L per cylinder) were

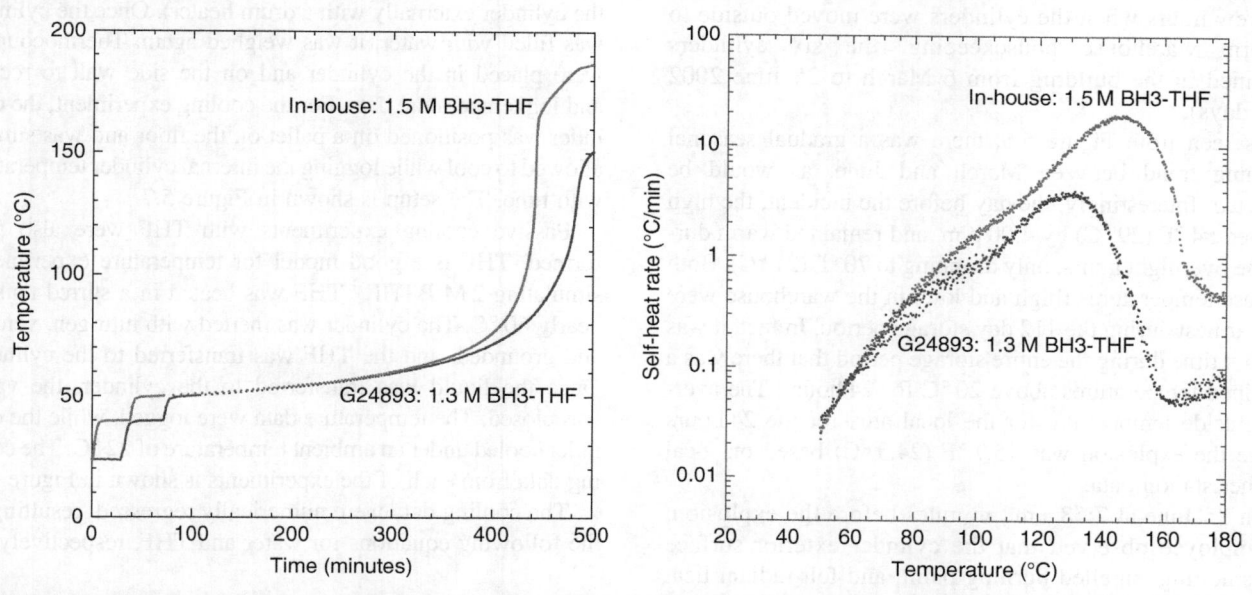

FIGURE 5.5 ARC data comparing aged BTHF G24893 to laboratory-prepared *unstabilized* BH3-THF of similar potencies. The unstabilized BTHF was prepared at laboratory scale by bubbling diborane gas into neat THF.

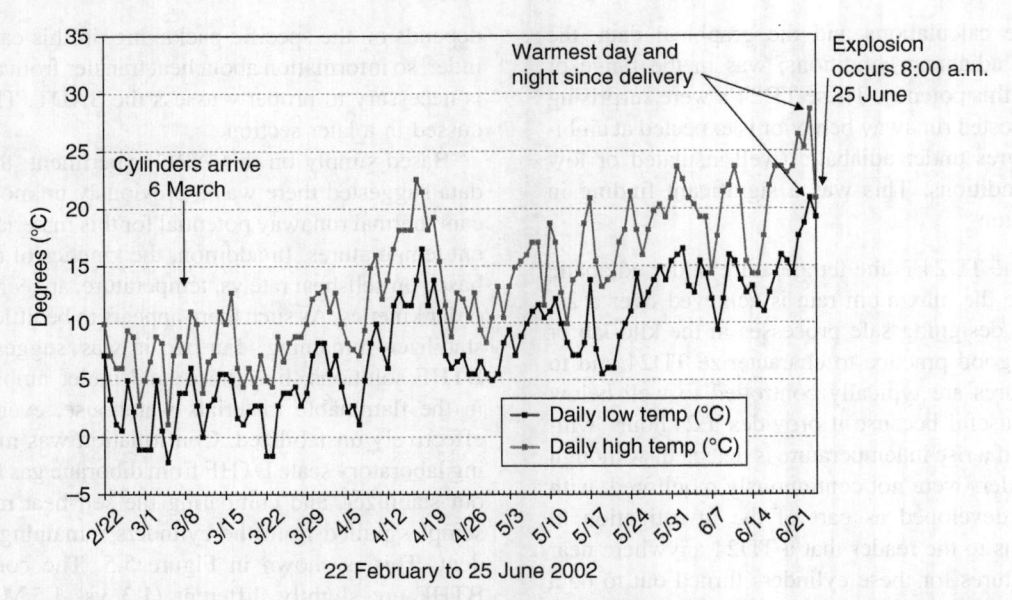

FIGURE 5.6 Daily high and low temperature readings during storage of BTHF cylinders in the warehouse.

filled at the manufacturer. The six cylinders were transported by truck (unrefrigerated) to Groton, CT. Several other cylinders were filled on the same day from the same lot and were retained at the manufacturer. The six cylinders arrived at Pfizer's, R&D site in Groton, CT, on 6 March 2002 and placed into the flammable materials warehouse. Temperatures in the warehouse were maintained at ambient temperatures using ventilation fans, and interior temperatures were recorded via a chart recorder. With the exception of a few hours when the cylinders were moved outside to perform warehouse housekeeping, the six cylinders remained in the building from 6 March to 25 June 2002 (112 days).

As seen from Figure 5.6, there was a gradual seasonal warming trend between March and June, as would be expected. Interestingly, the day before the incident, the high reached 84 °F (29 °C) by 4:00 p.m. and remained warm during the overnight hours, only dropping to 70 °F (21 °C). Both of these temperatures (high and low) in the warehouse were the warmest during the 112 day storage period. In fact, it was the first time during the entire storage period that there was a sustained temperatures above 20 °C for 24 hours. The average outside temperature for the local area for the 24 hours before the explosion was 75.7 °F (24.3 °C) based on local weather station data.

On 25 June at 7:58 a.m., minutes before the explosion, an employee observed that the cylinder exterior surface was smoking, smelled burning paint, and felt radiant heat from the cylinder. The employee also observed that the cylinder had become misshapen, described as football shaped. Two minutes later, at 8:00 a.m., the BLEVE occurred.

5.6 HEAT LOSS MEASUREMENTS

To understand the role of the storage container on the storage and shipping of two molar BTHF, Pfizer obtained a nearly identical, new, empty 400 L cylinder from a cylinder manufacturer to perform heat loss experiments. As a first trial, the cylinder was filled with hot water from a separate tank. Since the liquid was uniformly heated in the separate vessel, initial temperature gradients could be avoided (compared with heating the cylinder externally with a drum heater). Once the cylinder was filled with water, it was weighed again. Thermocouples were placed in the cylinder and on the side wall to record and log temperature data. For the cooling experiment, the cylinder was positioned on a pallet on the floor and was simply allowed to cool while logging the internal cylinder temperature with time. The setup is shown in Figure 5.7.

Passive cooling experiments with THF were also performed. THF is a good model for temperature experiments simulating 2 M BTHF.[4] THF was heated in a stirred tank to nearly 60 °C. The cylinder was inerted with nitrogen, vented, and grounded, and the THF was transferred to the cylinder. Once the liquid was transferred to the cylinder, the valve was closed. The temperature data were logged, while the cylinder cooled under an ambient temperature of 22 °C. The cooling data from each of the experiments is shown in Figure 5.8.

The cooling data were numerically regressed, resulting in the following equations for water and THF, respectively:

---

[4]Since BTHF is a solution in THF, the properties of THF are a good assumption – thus the heat capacity and thermal conductivity are expected to be very similar.

FIGURE 5.7 An identical, new, empty 400 L carbon steel cylinder. The cylinder was equipped with internal thermocouples and filled to perform a passive cooling experiment to determine the rate of heat loss from a typical 400 L steel cylinder.

$$\text{Water}: T(t) = 22°C + 37.919 \exp(-0.0010436 \times t) \quad (5.3)$$

$$\text{THF}: T(t) = 22°C + 29.082 \exp(-0.0023253 \times t) \quad (5.4)$$

where t is in minutes. These fits of course are based on Newton's law of cooling equation:

$$T(t) = T_{\text{final}} + (T_{\text{initial}} - T_{\text{final}}) \exp((-hA/mCp) \times t) \quad (5.5)$$

The curve fits are described in more detail in Example Problem 5.3.

EXAMPLE PROBLEM 5.3

Two passive cooling experiments were performed in a 400 L steel cylinder. The temperature data are plotted from each cooling experiment as shown in Figure 5.8 below. The data were regressed via nonlinear curve fit of the form shown in Eqs. (5.3), (5.4), and (5.5):

1. Derive Eq. (5.5) from Newton's law of cooling.
2. Using the regression coefficients from the curve fits, estimate a lumped parameter heat transfer coefficient for the convective cooling of the 400 L cylinder for both water and THF. Assume:
   - Cylinder containing water: 350 kg of water; Cp Water = 4.18 J/(g·K)
   - Cylinder containing THF: 333 kg of THF; Cp THF = 1.765 J/(g·K)
3. Estimate the PHI factor for the 400 L cylinder filled with BTHF.

Additional tank specifications:

1000# DOT cylinder
Working pressure = 240 psig
Outside diameter OD = 30″ length = 45.6″
Outside surface area = 32.7 ft^2
Volume = 16.1 ft^3 = 456 L
Empty weight = 340 lb
Note: 1000# refers to 1000 lb water capacity.

The internal thermocouple (temperature) data from each experiment is plotted in Figure 5.8 below:

Solution

1. Derive the lumped parameter regression equation from Newton's law of cooling (assuming no reaction).

   The change in temperature as a function of time assuming a lumped parameter approach is described by an energy balance (with no reaction):

   Rate of heat change in the cylinder = rate of heat change to the surroundings

$$m_{\text{liquid}} Cp_{\text{liquid}} \frac{dT}{dt} = h_o A (T(t) - T_{\text{amb}}) \quad (5.6)$$

where

m_{liquid} is the mass of the liquid in cylinder
Cp_{liquid} is the specific heat of the liquid
h_o is the overall convective heat transfer coefficient for the cylinder
A is surface area for heat exchange of the cylinder.

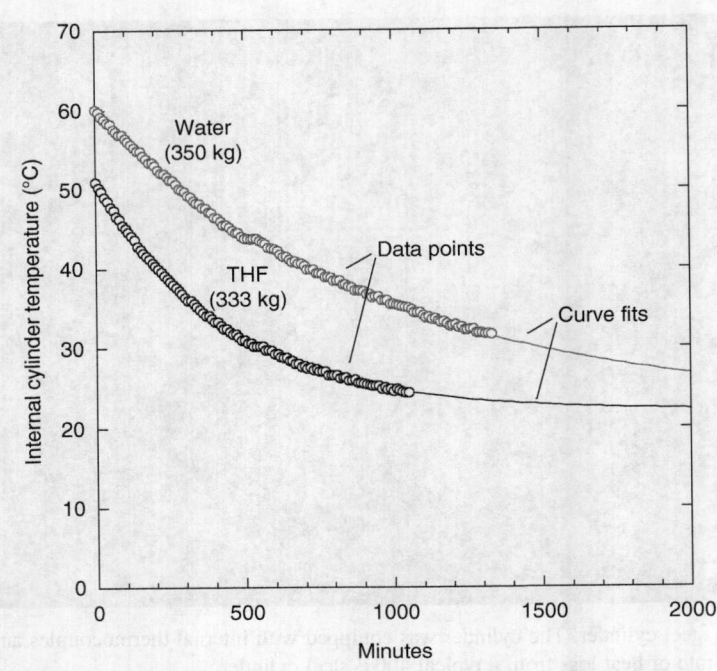

100 CHEMICAL ENGINEERING IN THE PHARMACEUTICAL INDUSTRY

Water cooling curve

Water	Value	Error
m_2	37.919	0.056 659
m_3	0.001 043 6	2.958 8e-6
Chisq	7.950 2	NA
R	0.999 55	NA

$y = 22 + m_2 \times \exp(-m_3 \times m_0)$

THF cooling curve

THF	Value	Error
m_2	29.082	0.040 717
m_3	0.002 325 3	4.985e-6
Chisq	1.812 5	NA
R	0.999 84	NA

$y = 22 + m_2 \times \exp(-m_3 \times m_0)$

FIGURE 5.8 Cooling curve profiles for two separate experiments – cooling of 400 L cylinder. Ambient temperature is 22 °C. The points are experimental, and the line through the data is a regression model. The fitted parameters for the regression are shown inset as m_2 and m_3 where m_0 is time (minutes).

Note that the surface area, A, can be defined as the total surface area of the cylinder or wetted surface area based on the fill volume as long as it is used consistently.

By integrating Eq. (5.6), rearrangement, and solving for $T(t)$:

$$T(t) = T_{\text{ambient}} + (T_i - T_{\text{amb}})\exp\left(-\frac{h_o A}{m_{\text{liquid}} C p_{\text{liquid}}} t\right) \quad (5.7)$$

where

T_{amb} is the ambient temperature
T_i is the initial temperature.

The parameters in Eq. (5.7) were regressed shown as inset in Figure 5.8.

Notice that the term $h_o A/mCp$ has units of (1/time), and thus the reciprocal $mCp/h_o A$ has units of time and can be considered a thermal time constant for the vessel. This constant characterizes the temperature dynamics of the vessel:

$$\tau = \left(\frac{m_{\text{THF}} C p_{\text{THF}}}{h_o A}\right) \equiv \text{units of time}$$

The half-life $t_{1/2}$ for cooling can be defined as

$$t_{1/2} = \ln 2 \cdot \tau = 0.693\tau$$

2. From the regression fits (Eqs. 5.3 and 5.4), the individual cooling curves were described as:
   - Cooling curve for 350 kg of water in 400 L cylinder:

$$T(t) = 22°C + 37.919\exp(-0.001\ 043\ 6t)$$

   - Cooling curve for 333 kg of THF in 400 L cylinder:

$$T(t) = 22°C + 29.082\exp(-0.002\ 325\ 3t)$$

where t is in minutes. From the regression equations, we find that the thermal time constants for THF and water are 7.2 and 16 hours, respectively:

$$\tau_{\text{THF}} = \left(\frac{m_{\text{THF}} C p_{\text{THF}}}{h_o A}\right) = \frac{1}{0.002\ 325\ 3\ \text{minutes}^{-1}}$$
$$= 430\ \text{minutes} = 7.2\ \text{hours}$$

$$t_{1/2,\text{THF}} = 0.693 \times 7.2\ \text{hours} = 5\ \text{hours}$$

$$\tau_{\text{water}} = \left(\frac{m_{\text{water}} C p_{\text{water}}}{h_o A}\right) = \frac{1}{0.001\ 043\ 6\ \text{minutes}^{-1}}$$
$$= 958\ \text{minutes} = 16\ \text{hours}$$

$$t_{1/2,\text{water}} = 0.693 \times 16\ \text{hours} = 11\ \text{hours}$$

and so the thermal half-life constants for cooling are 5 and 11 hours for THF and water, respectively. Given the much higher heat capacity of water compared with THF, it takes longer to cool the tank of water for a similar mass.

Knowing the mass and heat capacity used in the cooling experiments, we can solve for the individual heat transfer coefficients, $h_o A$

$$h_o A = mCp/\tau$$

- Water:

$$h_o A = \frac{350 \text{ kg} \times 4180 \text{ J/kgK}}{958 \text{ minutes}} \left(\frac{\text{minutes}}{60 \text{ seconds}}\right) = 25.45 \text{ W/K}$$

- THF:

$$h_o A = \frac{333 \text{ kg} \times 1765 \text{ J/kgK}}{430 \text{ minutes}} \left(\frac{\text{minutes}}{60 \text{ seconds}}\right) = 22.8 \text{ W/K}$$

For simplicity if A is assumed to be the outside surface area (32.7 ft² = 3.04 m²) of the tank, then the individual heat transfer coefficient, h_o, can be calculated:

Water: $h_o = hA/A = (25.45 \text{ W/K})/3.04 \text{ m}^2 = 8.4 \text{ W/m}^2\text{K}$

THF: $h_o = hA/A = (22.8 \text{ W/K})/3.04 \text{ m}^2 = 7.5 \text{ W/m}^2\text{K}$

Using a per mass basis (of liquid) eliminates the uncertainty in the area term:

Water: $h_o A/m = 25.46 \text{ W/K} \times (1/350 \text{ kg of H}_2\text{O})$
$\times 1000 \text{ mW/W} = 72.7 \text{ mW/kgK}$ water

THF: $h_o A/m = 22.8 \text{ W/K} \times (1/333 \text{ kg of THF})$
$\times 1000 \text{ mW/W} = 68.5 \text{ mW/kgK}$ THF

These data show good consistency for two separate cooling experiments for two different liquids with dramatically different specific heats, both resulting in an outside convective heat transfer coefficient of approximately 8 W/m²K or 70 mW/kgK for a 400 L cylinder. Table 5.1 summarizes the calculated time constants.

3. Calculate the PHI factor for the 400 L cylinder containing 350 kg of THF:

- Assume a specific heat of carbon steel of 0.47 J/(g·°C).
- Assume a specific heat for THF: 1765 J/(kg·K).
- Assume mass of empty cylinder: 340 lb or 154 kg.

Solution

$$\varnothing = 1 + \frac{(mCp)_{\text{tank}}}{(mCp)_{\text{THF}}}$$

$$\phi = 1 + \frac{(154 \text{ kg})\left(0.473 \frac{\text{kJ}}{\text{kg} \cdot \text{K}}\right)}{(350 \text{ kg})\left(1.765 \frac{\text{kJ}}{\text{kg} \cdot \text{K}}\right)} = 1 + \frac{72.84 \frac{\text{kJ}}{\text{K}}}{617.8 \frac{\text{kJ}}{\text{K}}} = 1.12$$

SADT

The SADT of a material depends on the reactivity of the material, its packaging configuration, and the heat loss characteristics of that packaging. In the case of BTHF, the packaging is the particular size cylinder and the surface area to volume for heat transfer that influences the heat loss characteristics. The SADT value is important for characterizing self-reactive chemicals and establishing whether the material requires refrigeration for storage and shipping. UN transportation requirements, for example, specify that a material to be shipped must be stable at 55 °C for 7 days, while US DOT regulations require stability at 130 °F (54.4 °C) for the duration of the shipment, which could potentially be 1 week to 6 months [7].

The SADT can be determined by several different methods, one of which is via the classic Semenov plot. This is done by plotting the heat generation rate and heat loss rate as a function of temperature. The rate of heat loss of the container depends on the difference between the internal cylinder temperature and the ambient temperature (i.e. $T_r - T_{\text{amb}}$). The Semenov diagram is one approach used to estimate the critical ambient temperature where the heat generation rate exceeds the heat loss. This critical temperature is the SADT, typically rounded up to the nearest 5 °C interval.

TABLE 5.1 Heat Transfer/Parameters Obtained for the Cooling Curves (Shown in Figure 5.8)

Test	Mass Used (kg)	Overall A (m²)	mCp/hA (hr)	Half-Life $t_{1/2}$ (h)	$h_o A$ (W/K)	$h_o A$/mass (mW/kgK)	h_o (W/m²K)
Water	350	3.04	16	11	25.46	72.7	8.4
THF	333	3.04	7.2	5	22.8	68.5	7.5
Average					24.1	70.6	8.0

Parameter h_o is the heat transfer coefficient between the tank and ambient air.

EXAMPLE PROBLEM 5.4

1. Construct a Semenov diagram based on the ARC data shown in Figure 5.4 and the cooling curve data for THF from Table 5.1. Assume the tank contains 350 kg of BTHF in the packaging configuration of a 400 L cylinder. Assume a PHI correction of 2.4 for the ARC data and a specific heat (Cp) of 1740 J/kgK for the BTHF sample.
2. Estimate the SADT (i.e. the critical ambient temperature).
3. Estimate the temperature of no return (TNR) (i.e. temperature inside the cylinder).

Solution

1. To construct the Semenov diagram, first start with the heat generation rate:
   The ARC data shown in Figure 5.4 was used to regress parameters from a plot of self-heat rate vs. $1/T$ where T is in Kelvin:
   From ARC data (Figure 5.4)

$$\text{Self-heat rate}(°C/\min) = \frac{dT}{dt} = 3.2965 \times 10^{12} \exp\left(-10446\left(\frac{1}{(T°C + 273.15)}\right)\right)$$

To convert the self-heat rate data from the ARC to heat generation rate, we need to correct for phi and convert the units to watts per liter:

$$\text{Heat Generation Rate (HGR)} = \emptyset \cdot \rho \cdot Cp \frac{dT}{dt}$$

$$= 2.4 \times 1740 \frac{J}{kg \cdot K} \cdot \frac{0.89\,kg}{L}$$

$$\left(3.2965 \times 10^{12} \exp\left(-10446\left(\frac{1}{T}\right)\right)\frac{K}{\min}\right) \cdot \left(\frac{\min}{60\,s}\right)$$

$$\equiv W/L$$

You should recognize that 10 446 K = E_a/R (E_a = 86.8 kJ/mol = 20.7 kcal/mol) where E_a is the activation energy and R is the gas constant.

The heat loss rate for the 400 L cylinder containing 350 kg of BTHF inside the storage cylinder and using h_oA obtained from Table 5.1:

$$Q_{\text{Heat loss}} = h_oA(T - T_{\text{amb}}) \equiv W$$

$$\text{Heat loss rate} = Q_{\text{Heat loss}} \cdot \frac{1}{\text{mass}} \cdot \rho \equiv W/L$$

$$= \frac{22.8\,W}{K} \cdot \frac{1}{350\,kg} \cdot \frac{0.89\,kg}{L}(T - T_{\text{amb}}) = 0.058 \frac{W/K}{L}(T - T_{\text{amb}}) \equiv W/L$$

Next, using a spreadsheet, generate points for the heat generation rate and heat loss functions (shaded in red indicate when HGR exceeds heat loss).

Temp (°C)	Heat Gen. Rate (W/L)	Heat Loss			
		T_{amb}=32 °C (W/L)	T_{amb}=29 °C (W/L)	T_{amb}=28 °C (W/L)	T_{amb}=27 °C (W/L)
25	0.124	−0.406	−0.232	−0.174	−0.116
30	0.221	−0.116	0.058	0.116	0.174
35	0.387	0.174	0.348	0.406	0.464
40	0.665	0.464	0.638	0.696	0.754
43	0.913	0.638	0.812	0.870	0.928
45	1.124	0.754	0.928	0.986	1.044
50	1.867	1.044	1.218	1.276	1.334
55	3.056	1.334	1.508	1.566	1.624
60	4.928	1.624	1.798	1.856	1.914

Keep in mind that the x-axis of the Semenov plot is the internal temperature of the cylinder. So when the internal temperature is exactly equal to the outside temperature, there is no heat exchange taking place. So for the heat loss curve, plot lines at various ambient temperatures. For example, at T_{ambient} = 29 °C starts at 29 °C on the x-axis where y = 0 and with a slope of 0.058 (W/K)/L as shown in Figure 5.9. Plot the heat generation and heat loss equations between 25 and 50 °C using T_{ambient} of 29 °C. As shown in Figure 5.9, the straight heat loss curve intersects the heat generation curve very close to a single point. The point of

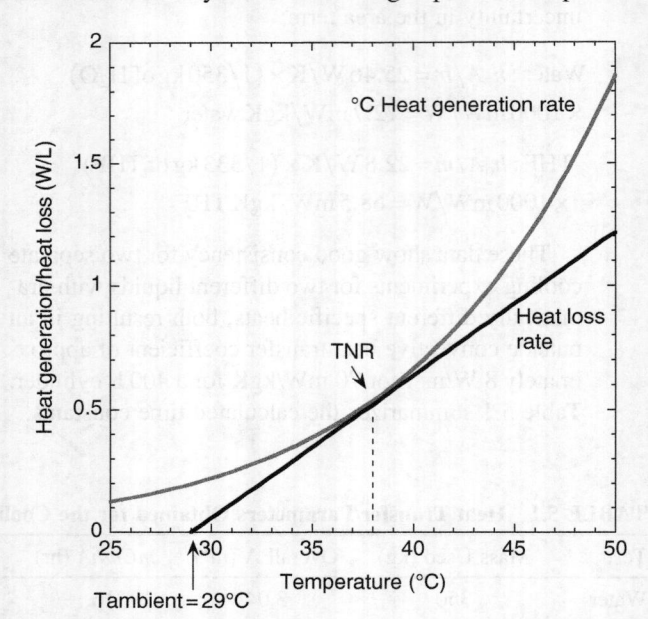

FIGURE 5.9 Semenov plot to determine SADT for aged BTHF. Red curve is the heat generation based on ARC data. The blue curve is the heat loss from the 400 L cylinder when the ambient temperature is 29 °C. The intersection at the tangent is the critical point or temperature of no return (TNR).

intersection is the critical point where heat generation and heat loss are equal and is considered unstable because a slight shift will cause the heat generation to exceed heat loss, leading to a thermal runaway.

2. The SADT is the ambient temperature at which this intersection is found, in this case near 29 °C. Solve in Excel or trial and error at 28.8 °C.
3. From the same graph, the TNR inside the cylinder is found near 38 °C. Above this temperature thermal equilibrium is no longer possible, resulting in thermal runaway. Note that T_{NR} can also be calculated from

$$\frac{E_a}{R} = \frac{T_{NR}^2}{(T_{NR} - T_{ambient})} = 10\,446\,K$$

where E_a/R was obtained from the ARC data. (Note: keeping all temperatures in Kelvin, and setting $T_{ambient} = 29$ °C, the T_{NR} is found to be 38.285 K.)

From the initial ARC tests of the aged BTHF, we showed that the TD24 was approximately 27 °C. Thus, under purely *adiabatic* (no heat loss) conditions, this composition of BTHF can self-heat and have a runaway reaction from 27 °C (81 °F) such that the maximum rate will be reached in 24 hours. By constructing the Semenov plot, we added a new constraint of heat loss occurring between the tank and the surrounding air. With the heat loss considered, the data showed the heat generation exceeds the rate of heat loss when the outside air temperature reaches 29 °C (84 °F). The explosion occurred on 25 June where the peak temperature the previous day was (29 °C) (84 °F) by 4:00 p.m., 16 hours prior to the explosion. As the ambient temperatures increased, the internal tank temperature was even warmer than 29 °C due to the internal self-heating, and after several hours (due to the time constant for heating/cooling), it was likely closer to 38 °C inside the tank. At this critical TNR point, the internal temperature can keep rising even if the ambient temperature starts to cool (e.g. evening/nighttime).

What is not immediately evident is how long must the peak ambient temperatures need to be sustained for the runaway reaction to occur. The peak daytime temperature was maintained for a short time before it began decreasing again. This required additional kinetic analysis, which will be discussed in more detail in a subsequent section. However, what we can say from the Semenov analysis is that the SADT was determined to be surprisingly low (below 30 °C) for aged BTHF, certainly well below the limits necessary to ship without refrigeration per US DOT transportation guidelines for self-reactive materials.

5.7 THERMAL STABILITY OF FRESH 2M BORANE–THF

Up to now we have discussed ARC data and analysis for "aged BTHF" recovered from the surviving cylinders where the potency of BTHF was approximately 1.3 M (originally ~1.9 M) BTHF and the remaining active hydride was approximately 1.4 M. After the incident, the manufacturer provided samples from the original lot that had been retained at their site under refrigeration. Those retained samples were also studied by ARC.

ARC testing of the retained refrigerated (fresh) 2M BTHF, from the same originating lot, using the standard HWS method is shown in Figure 5.10. The effect of stabilizer is clearly

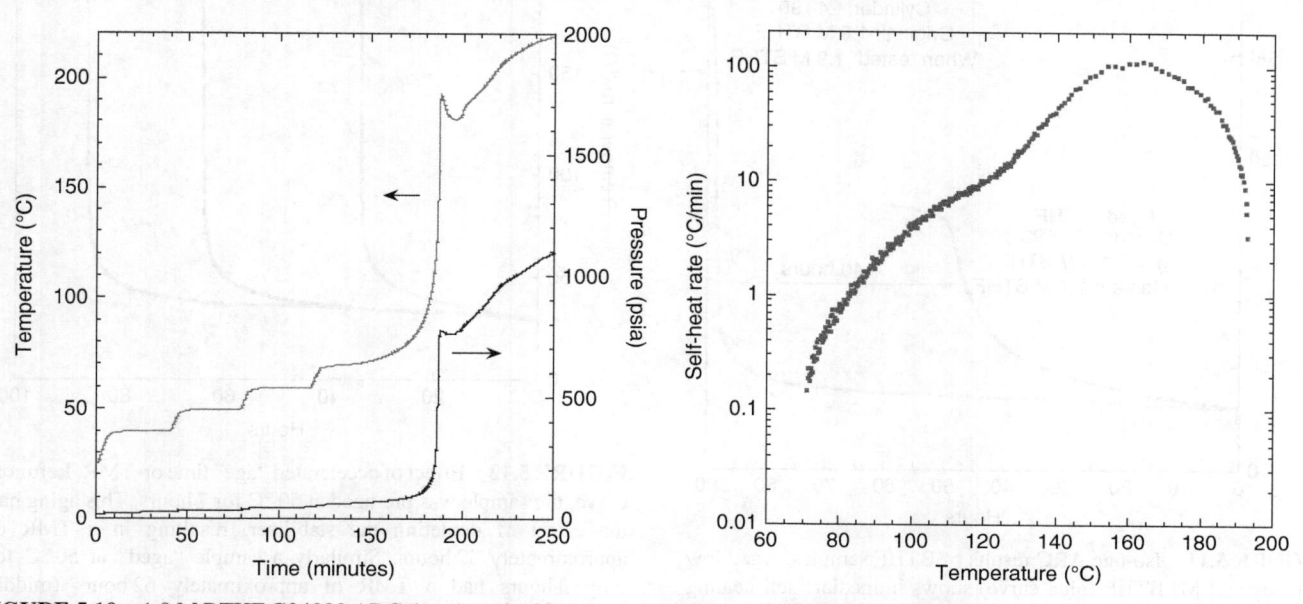

FIGURE 5.10 1.9 M BTHF G24890 ARC data (standard heat-wait-search method (HWS) in a titanium test cell; phi = 2.4). Temperature vs. time (left) and self-heat rate (SHR) (right) vs. time. The effect of stabilizer in refrigerated BTHF is apparent by the delayed onset temperature and curvature in SHR starting at 70 °C.

evident as the onset temperature is delayed as compared with the aged sample (70 °C vs. 50 °C). In the standard HWS method, the exotherm criterion is met when the self-heating rate exceeds 0.02 °C/min. And there is a significant difference in the self-heat profiles between the refrigerated and aged samples. The adiabatic temperature rise when corrected for PHI was 288 °C (120 × 2.4) with pressure rise exceeding 1000 psi.

The role of the stabilizer is to slow the ring-opening decomposition kinetics, but as the stabilizer's effectiveness diminishes, the rate of decomposition increases. This explains the *observed* pseudo-autocatalytic behavior in the ARC data where the apparent change in mechanism is due to the loss of effectiveness of the stabilizer. This was not observed in the aged samples because the effect of stabilizer was fully depleted and effectively absent.

To study the effect of aging, isothermal aging studies were performed in the ARC from 40 °C, as well as 30 °C, initial temperatures. In all cases, the BTHF samples were observed to self-heat from the lowest test temperatures of 30 °C in the ARC. For refrigerated 2M BTHF, several days at 30 °C were required to deplete the stabilizer. Refrigerated BTHF with stabilizer has a longer TMR than aged samples. The TMR depends on the remaining level of stabilization, which depends on the temperature–time history of the sample. The dramatic difference between the refrigerated and aged samples is reflected in the iso-age ARC study shown in Figure 5.11.

The aged sample shows no stabilizer effect remaining, while the refrigerated material required 3.5 days when starting the experiment at 30 °C before thermal runaway.

We also performed accelerated aging studies on refrigerated BTHF. In this case we took small samples of previously refrigerated BTHF and held them isothermally at 50 °C for 3 and 7 hours, respectively, and compared them without any accelerated aging. The results are shown in Figure 5.12.

The stabilization effect of the sodium borohydride on the decomposition complicates the thermal analysis. For example, if we only looked at the HWS profile for refrigerated 2M BTHF, it would appear to have much higher stability (onset at 70 °C). Further with the apparent pseudocatalytic behavior due to the stabilizer, we cannot reliably extrapolate the self-heat curve to lower temperatures. To adequately address the influence of stabilizer, iso-aging studies were required as was shown in Figure 5.12.

What we can conclude from the refrigerated BTHF thermal testing is that the TMR for thermal decomposition depends on the level of stabilizer effectiveness, which in turn depends on the time–temperature history of the samples.

All 2M BTHF samples (whether refrigerated or aged) were found to self-heat in less than 4 days from our lowest starting ARC temperature of 30 °C.

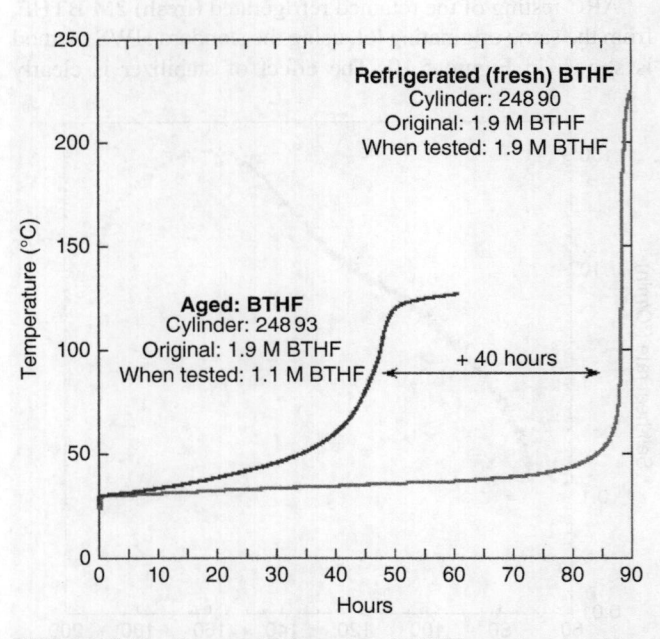

FIGURE 5.11 Iso-age ARC results of BTHF samples. Aged low potency (1.1 M) BTHF (blue curve) shows immediate self-heating from 30 °C. Refrigerated 2M BTHF (1.9 M) requires 87 hours (3.5 days) before runaway (red curve). These data illustrate the effect of loss of stabilization after aging.

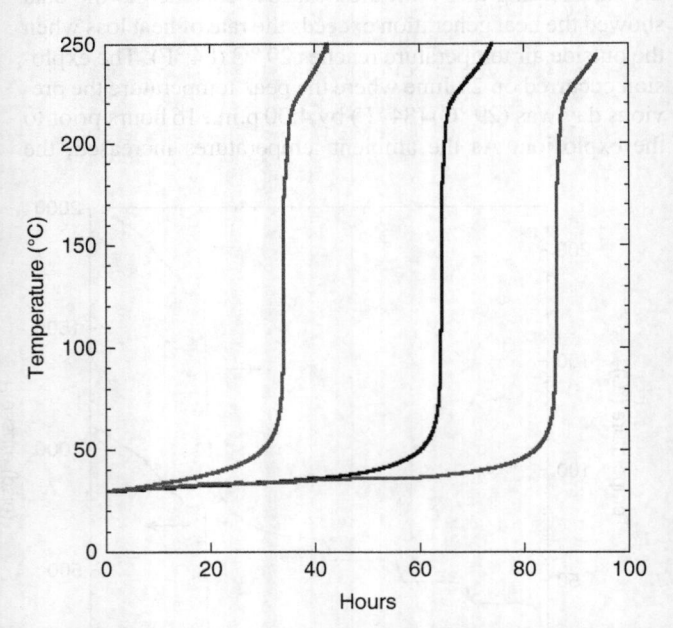

FIGURE 5.12 Effect of accelerated "age" time on TMR. Leftmost curve, the sample was pre-aged at 50 °C for 7 hours. This aging had the effect of depleting the stabilizer, resulting in a TMR of approximately 32 hours. Similarly a sample "aged" at 50 °C for only 3 hours had a TMR of approximately 62 hours (middle curve). No pre-aging sample showed a TMR of over 80 hours (rightmost curve). Regardless of the "aging" time, all samples eventually self-heated from 30 °C.

5.8 KINETICS OF DECOMPOSITION

It is known that BTHF solutions slowly degrade to tributyl borate. In a patent granted to the manufacturers of BTHF, the authors wrote that the "exact mechanism of thermal decomposition has not been determined. The overall decomposition does not follow first- or second-order kinetics. Several possible mechanistic routes can be envisioned" (Burkhardt and Corella [4]).

As part of the incident investigation, a kinetic model was developed to help understand the kinetics of decomposition, including the role of the stabilizer and its depletion with time and temperature. Temporal compositional data were collected at several temperatures, and plausible mechanisms were tested by regressing their respective kinetic models to the actual data. A good fit of the model does not necessarily prove the mechanism, but it demonstrates that the proposed mechanism is at least consistent with kinetic data within the range of temperatures and compositions tested.

One plausible mechanism that was highly consistent with the kinetic data is shown in Scheme 5.2:

The kinetic profiles based on isothermal NMR data are shown in Figure 5.13. These data were obtained by placing refrigerated (fresh) 2M BTHF in quartz NMR tubes and holding them at a constant temperature in temperature baths for days. NMR scans were taken periodically to obtain the temporal compositions.

Notice the induction times, especially for the lower temperatures, where the effect of the stabilizer is evident. After this initial induction time, the stabilizer effectiveness essentially becomes depleted, and the decomposition kinetics transitions to what appears to be unstabilized kinetics. The data points were obtained from ^{11}BNMR measurements and converted to molar concentrations. The kinetics (line fits through the data) were mathematically described by the respective differential rate equations shown in Schemes 5.3 and 5.4.

A kinetic inhibition function was used to describe the level of stabilizer $[S]$ present where $[S]$ has a value between 0 and 1. Given $[S]$, the conversion (or loss) of stabilizer is defined as $X_s = 1 - [S]$. For example, new BTHF is fully stabilized, so $S = 1$ and the loss of stabilizer $X_s = 0$, while fully aged BTHF has effectively no stabilizer, so $S = 0$ and $X_s = 1$. The rate at which the stabilizer is depleted is temperature dependent and is well captured by the model. This is evident by the line through the data points of Figure 5.13, which is based on the regression fit of the kinetic model.

When the stabilizer is new, the first reaction (A → M) is fully inhibited so mathematically k_1 is inhibited by the initial level of stabilizer where $S = 1$ and the "loss" of stabilizer $X_s = 0$. As the stabilizer degrades with time and temperature $[S]$ tends toward 0 and $X_s = 1$. So to accurately describe the effect of stabilizer in the model, we multiply k_1 by X_s to slow the rate of decomposition. As the stabilizer is depleted, X_s approaches 1 (i.e. unstabilized), having the effect of accelerating the rate, so the detailed kinetic model describes both stabilized and unstabilized decompositions of BTHF. The detailed model is shown in Scheme 5.4 where Eqs. (5.1)–(5.6) describe the compositional changes in the liquid phase.

Each data set (Figure 5.13) was regressed using the detailed model shown in Scheme 5.4. Since [M] is never observed, suggesting k_2 must be fast, we assigned k_2 to be

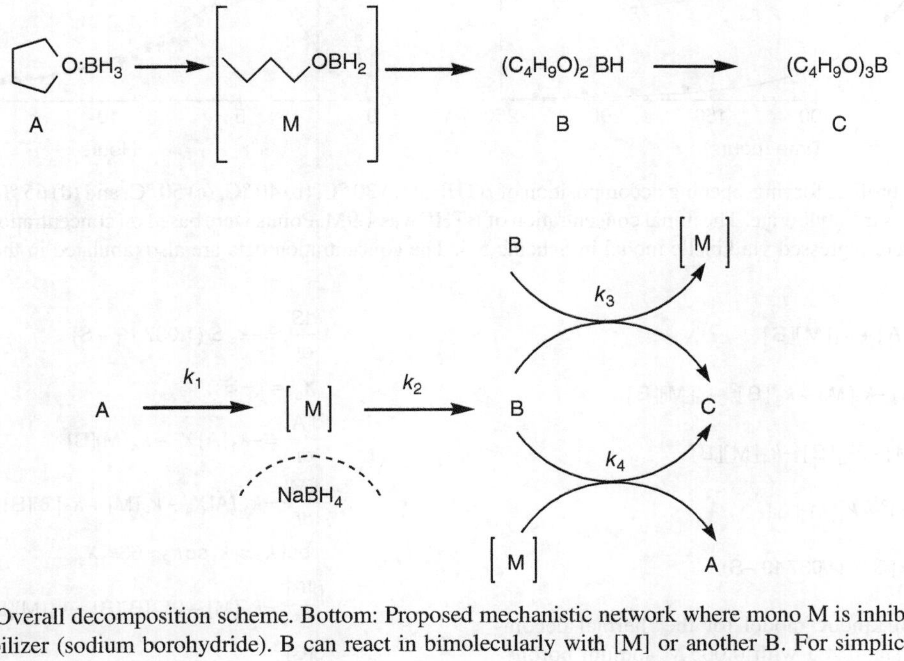

SCHEME 5.2 Top: Overall decomposition scheme. Bottom: Proposed mechanistic network where mono M is inhibited by the presence of active inhibitor or stabilizer (sodium borohydride). B can react in bimolecularly with [M] or another B. For simplicity, these reactions are considered irreversible.

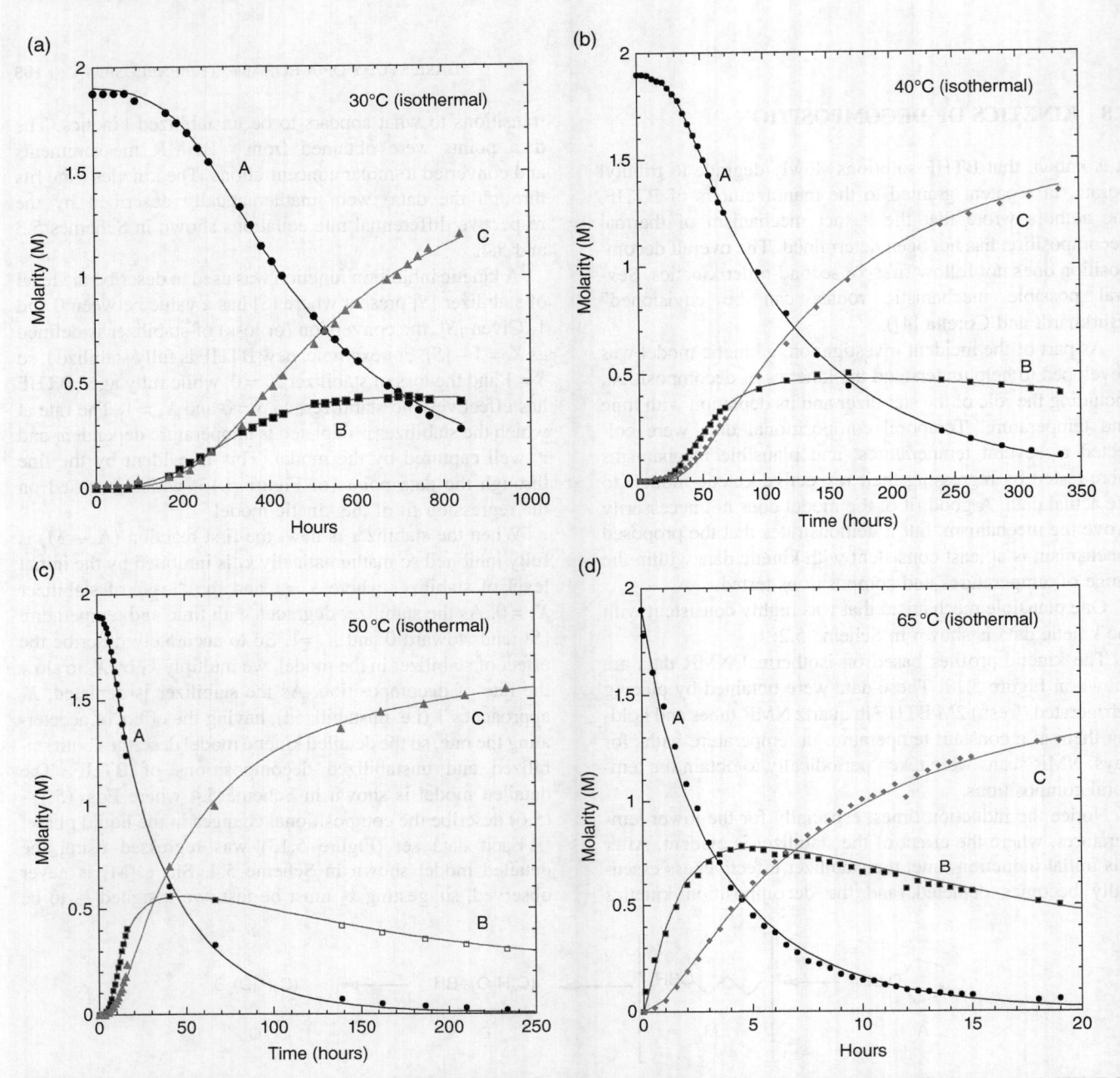

FIGURE 5.13 Kinetic profiles for ring-opening decomposition of BTHF at (a) 30 °C, (b) 40 °C, (c) 50 °C, and (d) 65 °C, where A is BTHF, B is dibutoxyborane, and C is tributylborate. The initial concentration of BTHF was 1.9M. Points were based on concentrations from boron ^{11}NMR and the curve fit lines were regressed via kinetic model in Scheme 5.4. The concentration data are also tabulated in the Appendix 5.A.

$$\frac{dA}{dt} = -k_1[A] + k_4[M][B]$$

$$\frac{dM}{dt} = k_1[A] - k_2[M] + k_3[B]^2 - k_4[M][B]$$

$$\frac{dB}{dt} = k_2[M] - 2k_3[B]^2 - k_4[M][B]$$

$$\frac{dC}{dt} = k_3[B]^2 + k_4[M][B]$$

$$\frac{dS}{dt} = -k_s \times [S] \times (1.03719 - S)$$

SCHEME 5.3 General kinetic model for the thermal decomposition of 2 M BTHF stabilized with 0.005 M sodium borohydride and where the initial conditions: A = 1.9 M, B = 0, C = 0, M = 0, S = 1.

$$\frac{dS}{dt} = -k_s S (1.03719 - S)$$

$$X_s = 1 - S$$

$$\frac{d[A]}{dt} = -k_1[A] X_s + k_4[M][B]$$

$$\frac{d[M]}{dt} = k_1[A] X_s - k_2[M] + k_3[B][B] - k_4[M][B]$$

but $k_2 \gg k_1$ so $k_2 \approx 63 k_1 X_s$

$$\frac{d[B]}{dt} = k_2[M] - 2k_3[B][B] - k_4[M][B]$$

$$\frac{d[C]}{dt} = k_3[B][B] + k_4[M][B]$$

SCHEME 5.4 Detailed kinetics used for parameter estimation and fitting of the data shown in Figure 5.13.

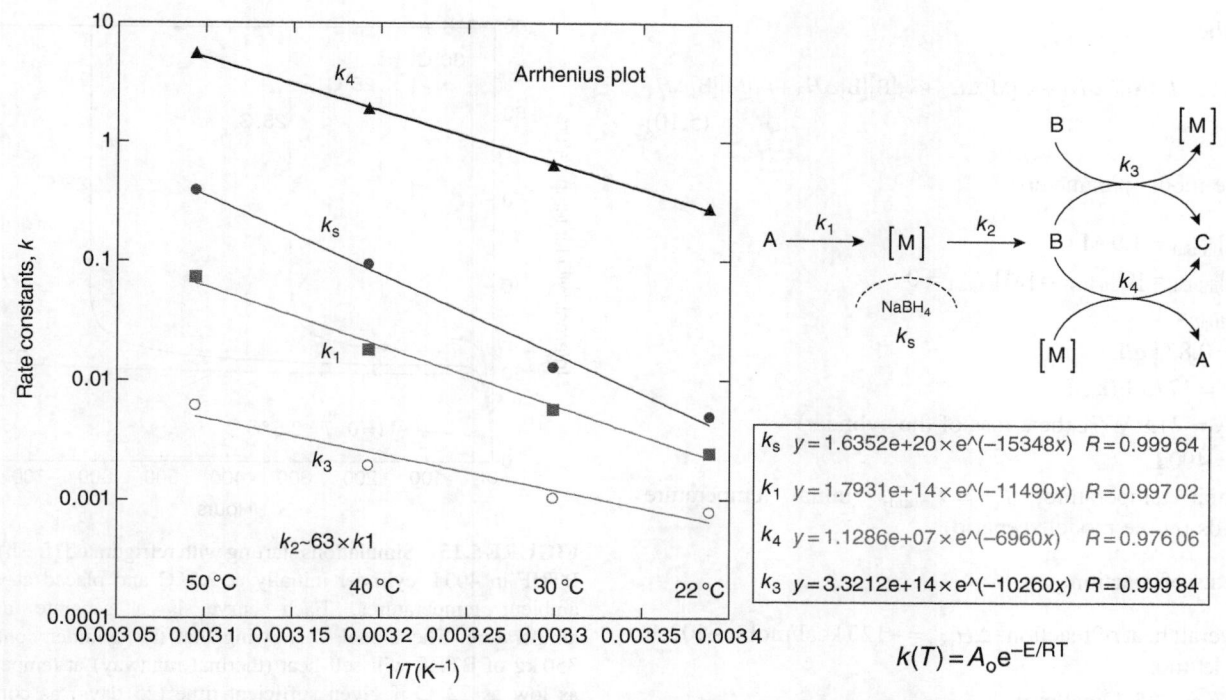

FIGURE 5.14 Arrhenius plot based on the best fits from each of the isothermal kinetic data sets.

$$k_s \quad y = 1.6352e+20 \times e^{\wedge}(-15348x) \quad R = 0.99964$$
$$k_1 \quad y = 1.7931e+14 \times e^{\wedge}(-11490x) \quad R = 0.99702$$
$$k_4 \quad y = 1.1286e+07 \times e^{\wedge}(-6960x) \quad R = 0.97606$$
$$k_3 \quad y = 3.3212e+14 \times e^{\wedge}(-10260x) \quad R = 0.99984$$

$$k(T) = A_o e^{-E/RT}$$

proportional to k_1. After regression analysis at all temperatures, we determined $k_2 \approx 63 k_1$. After each set of concentration data was regressed at each temperature, an Arrhenius plot provided activation parameters as shown in Figure 5.14. The highest activation energy was for the loss of stabilization, k_s at 30.5 kcal/mol. And the larger the activation energy, the more sensitive the reaction rate is to temperature. It follows that the loss of stabilization is the most sensitive to temperature where, for example, BTHF is stable for months below 0 °C, while only a few days at 30 °C.

The parameter estimates were obtained from each set of kinetic data. k_2 is considered much faster than k_1 and was thus fixed at $63 \times k_1$.

T(°C)	1/T (K)	k_s (h^{-1})	k_1 (h^{-1})	k_3 (h^{-1}M^{-1})	k_4 (h^{-1}M^{-1})
22	0.0033881	0.004982	0.0024720	0.0007846	0.2750
30	0.0032987	0.012854	0.0057092	0.0010380	0.6417
40	0.0031934	0.093740	0.0180450	0.0019403	1.9213
50	0.0030945	0.392750	0.0731300	0.0062100	5.5474

In order to couple the reaction kinetics to the storage conditions, the model was extended to include the thermochemistry and energy balances. Equation (5.8) is the basic energy balance for the cylinder, which describes how the temperature changes, depending in the rate of heat generation from BTHF decomposition and the rate of heat removed from the storage container. Equation (5.9) describes the rate of heat transferred through the wall of the tank or storage vessel. Equation (5.10) describes the rate of heat generation from the BTHF decomposition pathway where

$$H_1 = -35.4 \text{ kcal/mol}$$
$$H_2 = -39.4 \text{ kcal/mol}$$
$$H_3 = -5.8 \text{ kcal/mol}$$
$$H_4 = -9.8 \text{ kcal/mol}$$

The overall heat of reaction was measured by CRC90 calorimeter to be approximately $\Delta H = -120$ kcal/mol of BTHF. This was consistent with thermodynamic calculations based on heats of formation of BTHF and tributyl borate, resulting in a calculated $\Delta H - 126$ kcal/mol.

Energy balance:

$$\frac{dT}{dt} = \frac{Q_{generation} - Q_{loss}}{Cp \cdot \rho} \quad (5.8)$$

where

$$Q_{loss} = \frac{U \cdot A \cdot (T_r - T_{ambient})}{V} \quad (5.9)$$

and where

$$Q_{generation} = k_1[A]X_s\Delta H_1 + k_2[M]\Delta H_2 + k_3[B][B]\Delta H_3 + k_4[M][B]\Delta H_4 \quad (5.10)$$

The model parameters:

$[A]_{initial} = 1.9$ M
$[B]_{initial} = [C]_{initial} = [M]_{initial} = 0$
$S_{initial} = 1$
$\rho = 0.87$ kg/L
$Cp = 1775$ J/(kg·K)
UA = 21.9 W/K (heat loss of the cylinder)
$V = 400$ L

Initial temperature, $T_r = T_{initial}$, (initial temperature depends on the model scenario)

Heats of reaction:

Overall heat of reaction: $\Delta H_{rxn} = -120$ kcal/mol $= -502.1$ kJ/mol

$\Delta H_1 = -35.4$ kcal/mol
$\Delta H_2 = -39.4$ kcal/mol
$\Delta H_3 = -5.8$ kcal/mol
$\Delta H_4 = -9.8$ kcal/mol

Note that
$\Delta H_3 = \Delta H_{rxn} - \Delta H_1 - 2\Delta H_2$
$\Delta H_4 = \Delta H_{rxn} - 2\Delta H_1 - \Delta H_2$

Arrhenius parameters from (Figure 5.14):

$k_s = 1.635 \times 10^{20}\exp(-15\,348/(T_r + 273.15)) \equiv \text{hour}^{-1}$
($E_s = 30.5$ kcal/mol)

$k_1 = 1.79 \times 10^{14}\exp(-11\,500/(T_r + 273.15)) \equiv \text{hour}^{-1}$
($E_1 = 22.9$ kcal/mol)

$k_3 = 1.13 \times 10^7\exp(-6\,960/(T_r + 273.15)) \equiv \text{hour}^{-1}$ ($E_3 = 13.8$ kcal/mol)

$k_5 = 3.32 \times 10^{14}\exp(-10\,300/(T_r + 273.15)) \equiv \text{hour}^{-1}$
($E_5 = 20.5$ kcal/mol)

The model (based on Scientist/Micromath software) enabled dynamic temperature and composition simulations. With this working model, we were able to simulate composition and temperature of the 400 L cylinder based on different storage conditions. For example, in Figure 5.15 we calculate the internal temperature profiles for various ambient temperatures. The simulation shown in Figure 5.15 assumes the cylinder is removed from a cold storage of 2.5 °C and held at constant ambient temperature. We show that sustained ambient temperature of 22 °C for the 400 L cylinder would likely result in thermal runaway starting with new 2M BTHF, requiring about 1 month to destabilize and for thermal runaway to occur.

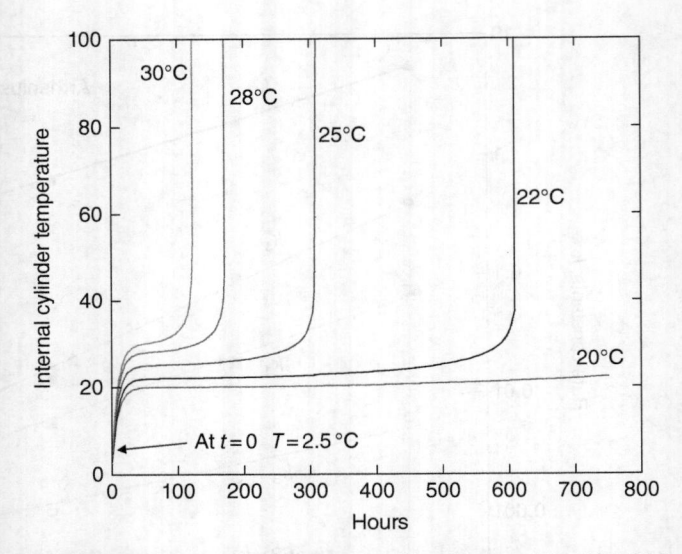

FIGURE 5.15 Simulations starting with refrigerated (fresh) 1.9 M BTHF in 400 L cylinder initially at 2.5 °C and placed at various ambient temperatures. Each curve is a separate ambient temperature. The model is showing that the cylinder containing 350 kg of BTHF will self-heat (thermal runaway) at temperatures as low as 22 °C if given sufficient time (25 days) as compared with approximately 5 days at 30 °C.

In a similar way we simulate the effect of initial concentration in Figure 5.16. It shows that concentrations above 1 M are most susceptible to thermal runaway in this size cylinder. We took the analysis a step further by modeling the nonlinear ambient temperature profile for the full duration of storage in the flammable materials warehouse as shown in Figure 5.17.

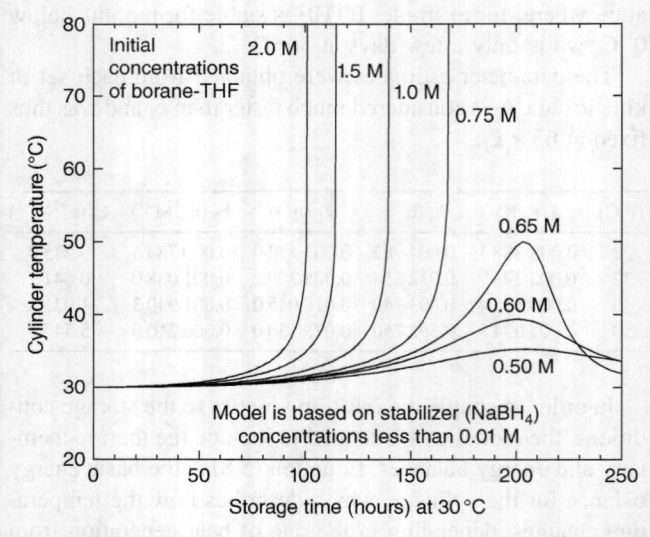

FIGURE 5.16 Simulations of different starting concentrations of new BTHF in 400 L cylinder stored at 30 °C. Each curve assumes a different initial concentration of BTHF.

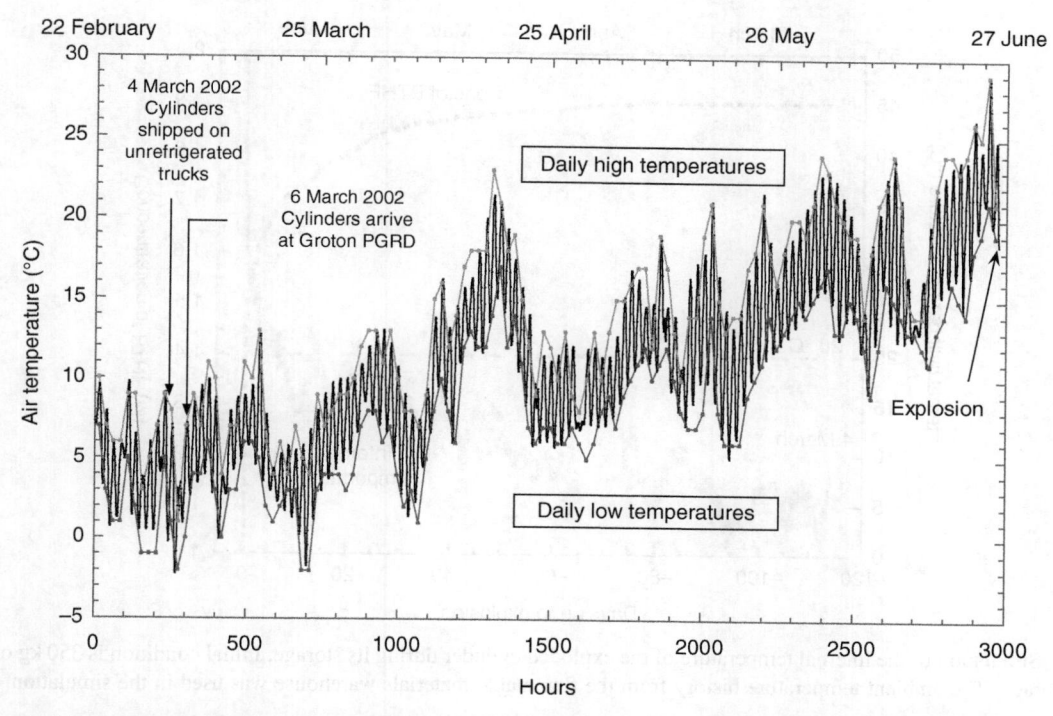

FIGURE 5.17 Ambient temperature history based on chart recorder in the flammable materials warehouse where the cylinders were stored. The high points are the daily high temperatures. The low points are the daily low temperatures. The black line is the numerical function describing the temperature history as a sinusoidal $T_{ambient} = f(\sin(x))$, which was used in the kinetic model to describe the external temperature.

Each daytime high and low temperature was fit to a sinusoidal temperature function to approximate the general daily temperature trend. Figure 5.17 shows an accurate ambient temperature history that the six cylinders were exposed to during the duration of storage in B196 warehouse. By applying this nonlinear ambient temperature profile to the thermokinetic model, the internal cylinder temperature can be modeled as shown in Figures 5.18 and 5.19.

From Figure 5.18 we see the internal temperature climb during the spring months. The stabilizer has been depleted by the end of May based on the potency starting to drop (near 1.8 M). For the month of June, the BTHF cylinders are essentially unstabilized. The last five days are shown in Figure 5.19.

What we see from Figure 5.19 is the temperature oscillation is very apparent (due to the daytime heating and cooling). For 23 and 24 June, however, was the first time that the nighttime temperatures remained high (above 20 °C), pushing the internal temperature to over 25 °C. The peak temperature on June 24th, the day before the explosion, was the hottest day of the year to that date, reaching 29 °C. This appears to be where the internal temperature in the cylinder slipped past the critical TNR point consistent with the Semenov analysis. It took 16 more hours, as can be seen in Figure 5.19, for the cylinder to approach its maximum rate, and by 8:00 a.m. on 25 June, the cylinder was observed to be misshapen and smoking, moments before it exploded.

5.9 CONCLUSIONS

This case study focused on understanding the cause of a BLEVE involving a cylinder of 2M BTHF. The cause of the explosion was a thermal runaway reaction of a self-reactive material (BTHF) in a package configuration that was unable to adequately dissipate the heat-generated reaction. Upon the prolonged storage, between 6 March and 25 June 2002, especially as the temperatures became warmer, the stabilizer in the 2M BTHF eventually became depleted. As a result, the rate of the ring-opening decomposition reaction increased, leading to a corresponding drop in potency of BTHF. The rate of decomposition of BTHF is fairly slow, but the reactions are highly exothermic (~120 kcal/mol). In fact, there is enough exothermic potential to raise the temperature to approximately 600 °C based on the initial concentration of BTHF.

Further, the SADT for this packaging configuration had not been characterized before it arrived at the Pfizer

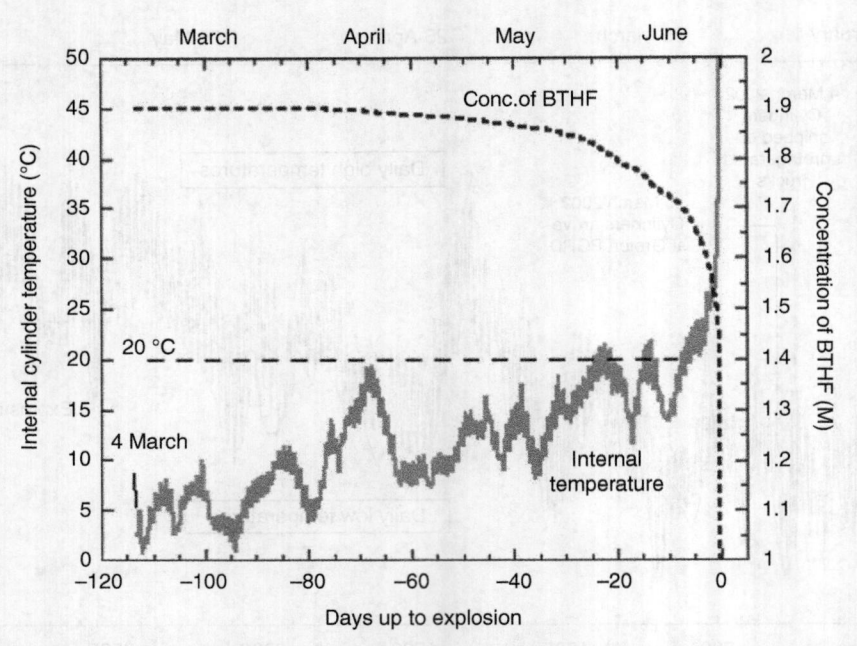

FIGURE 5.18 Simulation of the internal temperature of the exploded cylinder during its storage. Initial condition is 350 kg of 1.9 M BTHF and placed in storage. The ambient temperature history from the flammable materials warehouse was used in the simulation.

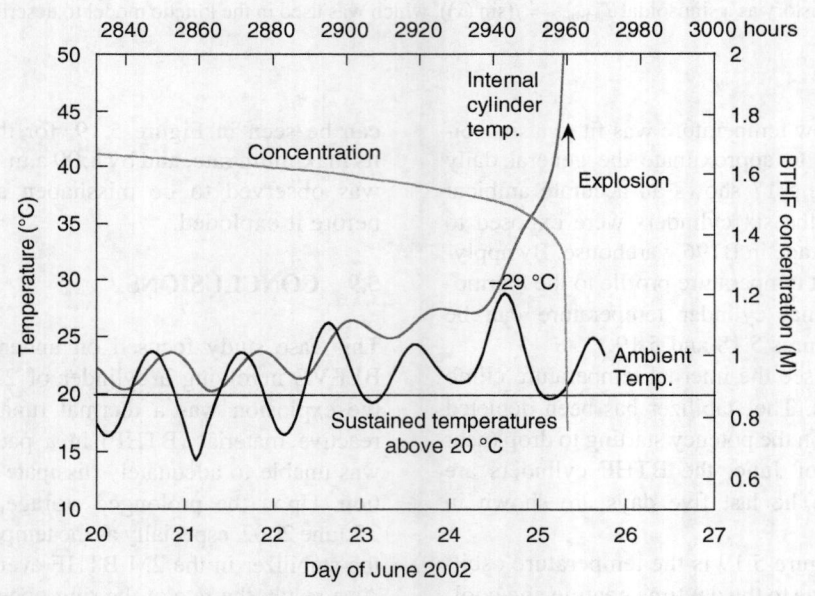

FIGURE 5.19 Simulation for the internal temperature (400 L cylinder), concentration of remaining BTHF, and ambient temperature for the five days, leading up to the explosion.

site. As part of the incident investigation, the SADT was determined to be well below 50 °C, which would have required refrigeration during transportation per DOT guidelines. The investigation also showed through ARC studies that even refrigerated, fully stabilized 2M BTHF can self-heat from the lowest test temperature of 30 °C. Performing an energy balance (Semenov plot) and using the thermal data from the aged low potency BTHF recovered from one of the surviving cylinders showed the SADT of BTHF to be 29 °C despite its lower potency.

Kinetic modeling suggests that this SADT could be even lower, when starting with new 2M BTHF and allowing it to be held at isothermal (ambient conditions) for extended periods. SADT values this low represent a serious safety concern especially considering the potential heat that can be liberated.

The investigation also showed that sodium borohydride stabilizer provides only temporary inhibitory benefit and, in fact, becomes depleted with time and temperature. Further it was shown that BTHF that is retained at room temperature will eventually behave as if it is unstabilized. Aged BTHF recovered after the incident behaved similarly to *unstabilized* laboratory-prepared BTHF of similar concentration as both exhibited very similar self-heating rates by ARC testing.

A comprehensive kinetic model was developed that was found to be consistent with extensive decomposition kinetic data over the temperature range of interest. The thermodynamics of the BTHF decomposition and thermal characteristics of the packaging configuration (400 L cylinder) were also incorporated into a complete thermokinetic model. This modeling effectively described how the internal temperature of a 400 L cylinder of BTHF responds with time and the variable and oscillating daily temperatures experienced during the three months of storage. This showed that the stabilizer became depleted and the BTHF was effectively unstabilized during the month of June 2002. The model has been provided as part of this case study.

Two important points are noteworthy regarding the ambient temperature:

1. The day before the explosion was the hottest day recorded during the entire storage time reaching 29 °C.
2. The two days preceding the explosion (23 and 24 June) represented the only window of time during the nearly 3000 hours of storage where the ambient temperatures remained above 20 °C for a duration of 48 hours. All previous days saw significantly cooler nights. It was during this 48 hours window that the internal cylinder temperature exceeded the critical point of no return.

It is also important to note that the potency of the cylinders was slowly diminishing. The kinetic model suggests that most of the decline in potency of the BTHF occurred in the month of June. In fact, the model suggests that the potency was dropping from 1.65 to 1.5 M during those few days prior to the explosion. This is also consistent with the measured potencies of the remaining cylinders. Had the cylinders been delivered a week or two later than they were, the concentrations of BTHF would have been even higher (e.g. 1.7 M) during those critically warm days of 23 to 24 June. In that case, with slightly higher potencies, and higher rates of decomposition, the slightly lower SADT's may have resulted in thermal runaway in multiple cylinders.

Self-reactive materials require thorough safety evaluations. Materials that require stabilization (added stabilizers and inhibitors) present additional complexities to fully characterize. With BTHF, the thermal stability depends on both temperature and age of the sample. Thus a single scanning experiment by DSC, ARC, or adiabatic dewar is not sufficient to adequately characterize the SADT of a stabilized self-reactive material.

Aging studies and iso-aging studies are required to assess the impact of age and temperature history on the self-heating rates and SADT. With BTHF stabilized by $NaBH_4$, the effectiveness of the stabilizer diminishes, eventually behaving as unstabilized BTHF when stored at ambient temperatures. In general for self-reactives, thorough characterization and understanding of the energy potential, self-heat rates, and effect of aging are required to properly assign SADTs and storage conditions.

ACKNOWLEDGMENTS

There were many contributors to the technical data presented in this chapter. Extensive ARC testing and analysis including the validation of phi experiments described in the appendix, David R. Bill; heat of reaction estimation and data fitting to provide sinusoidal function for ambient temperature as function of storage time, Ray Bemish; additional heat of reaction estimates, Don Knoechel; a heat of reaction measurement from CRC90, Matt Jorgensen; heat transfer (cooling curves) measurements (in various cylinders), Steve Brenek, Sandeep Kedia, and Brian Morgan; NMR analysis and studies, Andy Jensen and Linda Lohr; and additional BTHF studies, Eric Dias.

5.A BORON NMR KINETIC DATA FROM FIGURE 5.13

room temperature ≈22 °C NMR data				30 °C NMR data				40 °C NMR data				50 °C NMR data				65 °C NMR data			
hrs	BTHF	DI	TRI	hrs	BTHF	DI	TRI	hrs	BTHF	DI	TR	hrs	BTHF	DI	TR	hrs	BTHF	DI	TR
0	1.9	0	0	0	1.875	0	0.01875	0	1.8988	0	0	0	1.8971	0	0	0	1.899164	0	0
24.0	1.893688	0	0	24	1.875	0	0.01875	4	1.8978	0	0	0	1.9	0	0	0.1	1.780204	0.095063	0
48.0	1.881188	0	0.018812	48	1.875	0	0.01875	8	1.8949	0	0	1	1.9	0	0	0.5	1.608217	0.242358	0.018495
72.5	1.875	0	0.01875	72.5	1.875	0	0.01875	12	1.8801	0	0.012784	2	1.9	0	0	1	1.448723	0.370728	0.05143
96.0	1.875	0	0.01875	96	1.83871	0.018387	0.036774	16	1.8686	0.006914	0.015696	3	1.9	0.000378	0	1.5	1.258851	0.493721	0.114681
168.0	1.856678	0.018567	0.018567	168	1.779582	0.053387	0.053387	20	1.863	0.009501	0.017698	4	1.9	0.017307	0.009491	2	1.08086	0.596743	0.188178
192.0	1.856678	0.018567	0.018567	192	1.730944	0.086547	0.069238	24	1.835	0.02991	0.023671	5	1.8	0.030771	0.011793	2.5	0.966352	0.640305	0.246613
216.0	1.856678	0.018567	0.018567	216	1.684895	0.117943	0.084245	28	1.8078	0.04863	0.031818	6	1.8	0.052336	0.016355	3	0.804037	0.709562	0.334801
240.0	1.856678	0.018567	0.018567	240	1.61336	0.145202	0.129069	32	1.7755	0.07173	0.041724	7	1.8	0.075044	0.032187	3.5	0.706051	0.746084	0.389175
265.0	1.856678	0.018567	0.018567	265	1.546392	0.170103	0.170103	36	1.7167	0.10867	0.060599	8	1.7	0.11346	0.041856	4	0.582424	0.769324	0.483936
336.0	1.821086	0.036422	0.036422	336	1.328981	0.265796	0.292376	40	1.6668	0.13951	0.079007	9	1.7	0.14391	0.053145	4.5	0.5271	0.785116	0.54091
360.0	1.832797	0.036656	0.018328	360	1.242101	0.285683	0.360209	44	1.6143	0.16692	0.10461	10	1.6	0.18159	0.066418	5	0.449712	0.773011	0.603199
384.5	1.815287	0.036306	0.036306	384.5	1.144119	0.308912	0.434765	48	1.5642	0.19803	0.12357	11	1.6	0.22175	0.08223	5.5	0.401736	0.764463	0.653865
408.5	1.798107	0.053943	0.035962	408.5	1.071026	0.332018	0.481962	52	1.4981	0.23221	0.15581	12	1.5	0.24955	0.10675	6	0.371826	0.743279	0.705279
432.0	1.78125	0.07125	0.035625	432	1.002639	0.350924	0.531398	56	1.4264	0.26531	0.19342	13	1.5	0.30233	0.12198	6.5	0.312744	0.743267	0.753431
504.0	1.715317	0.102919	0.068613	504	0.809199	0.396508	0.679727	60	1.3591	0.29887	0.22724	14	1.4	0.33772	0.1538	7	0.276929	0.732864	0.807192
528.0	1.69997	0.101998	0.084999	528	0.738629	0.39886	0.746015	64	1.2984	0.32628	0.25734	15	1.3	0.37604	0.17834	7.5	0.233751	0.733208	0.835755
552.0	1.670085	0.116906	0.100205	552	0.69973	0.405843	0.7767	68	1.2507	0.3427	0.28992	16	1.2	0.40794	0.23452	8	0.221483	0.718404	0.864693
576.0	1.654092	0.132327	0.099246	576	0.638799	0.408831	0.836826	95	0.97109	0.42806	0.48661	40.5	0.6	0.55898	0.70411	8.5	0.199541	0.704578	0.904438
605.0	1.625784	0.146321	0.113805	605	0.596484	0.423504	0.864902	117	0.77782	0.46662	0.63704	66.5	0.3	0.54039	0.99807	9	0.172455	0.691545	0.945554
672.0	1.533907	0.184069	0.16873	672	0.478951	0.421477	0.98185	141	0.58532	0.49308	0.80365	139	0.1	0.4187	1.3637	9.5	0.156922	0.677481	0.97408
696.5	1.521623	0.182595	0.182595	696.5	0.450344	0.427827	1.004266	166	0.44092	0.50128	0.9325	162	0.0	0.38486	1.4417	10	0.144323	0.654474	1.005856
720.0	1.474392	0.206415	0.206415	720	0.397878	0.429708	1.054377	239.5	0.22666	0.46527	1.1878	186.5	0.0	0.35638	1.476	10.5	0.140115	0.659396	1.01987
744.0	1.440849	0.216127	0.230536	744	0.365338	0.423792	1.09236	262.5	0.1887	0.45152	1.2329	210	0.0	0.32888	1.5176	11	0.122155	0.650195	1.039112
768.0	1.408799	0.225408	0.253584	768	0.337178	0.414729	1.129547	287	0.154	0.43936	1.2794	233.5	0.0	0.30638	1.55	11.5	0.112254	0.64188	1.073587
840.0	1.31975	0.250752	0.31674	840	0.275017	0.407025	1.199074	310.5	0.13017	0.42046	1.3128					12	0.099745	0.576427	1.014906
								334	0.11148	0.40399	1.3458					12.5	0.092192	0.6093	1.119216
																13	0.09158	0.593533	1.139352
																13.5	0.084069	0.590248	1.160908
																14	0.078753	0.578516	1.176957
																14.5	0.07564	0.564274	1.192615
																15	0.073839	0.562728	1.202968
																18	0.052361	0.515543	1.272468
																19	0.057331	0.500561	1.277575
																45	0.013921	0.315718	1.523463
																68	0	0.244438	1.613451
																92	0	0.204585	1.657897

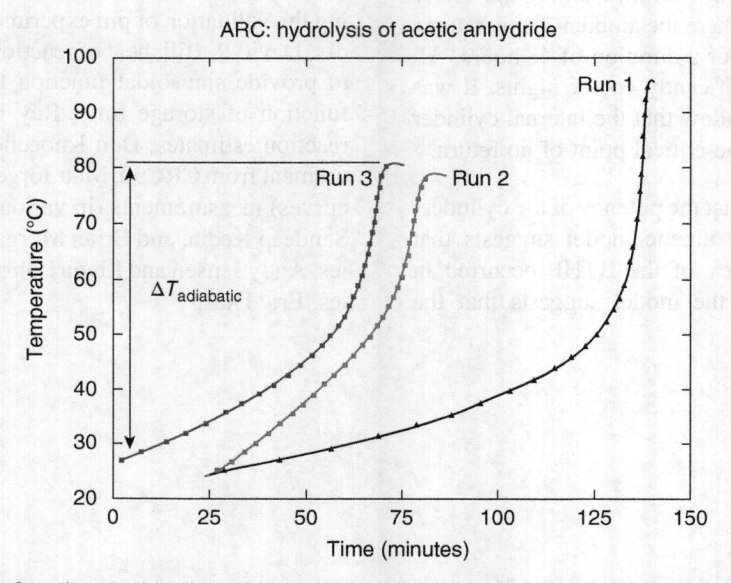

FIGURE 5.B.1 ARC profiles of acetic anhydride injected into water. Given the known heat of hydrolysis of acetic anhydride, the adiabatic temperature rise data was used to validate the PHI.

5.B CALIBRATION FOR PHI FOR ARC ANALYSIS

The PHI factor, ϕ, has been defined as a correction factor applied to time and temperature differences observed in exothermic reactions, which accounts for the sensible heat absorbed by the sample container that would otherwise underestimate ΔH, ΔT_{ad} and result in longer TMR [8].

Mathematically,

$$\phi = 1 + \frac{(mCp)_{container}}{(mCp)_{sample}} \quad (5.B.1)$$

Using ϕ, ARC data can generally be used to calculate ΔH for a process according to Eq. (5.B.2):

$$\Delta H = \phi (mCp)_{sample} \cdot \Delta T_{exp} \quad (5.B.2)$$

where

ΔH is the heat of reaction, decomposition, etc.

ΔT_{exp} is the observed adiabatic temperature rise from experiment

On the other hand, by using a well-characterized reaction with a well-known heat of reaction over a limited temperature range, Eq. (5.B.2) can be rearranged and used to validate ϕ experimentally.

5.B.1 Experimental Discussion

The hydrolysis of acetic anhydride was chosen as a standard reaction for experimentally determining ϕ in the ARC, as it has a well-known ΔH_r^o of 14.01 kcal/mol [9].

Three tests were conducted to evaluate ϕ. In each case a quantity of water was placed into the ARC bomb. The bomb was heated to 25 °C and allowed to equilibrate. Once thermal equilibrium had been established, a quantity of acetic anhydride was injected into the bomb via a syringe through the pressure tube. After the addition, the system was manually placed into EXOTHERM mode to allow for the tracking of the reaction under adiabatic conditions. Using the measure ΔT_{ad}, and averaged specific heat values from the start and end of the experiment, and assuming 100% conversion, ϕ was calculated according from

$$\phi = \frac{\Delta H_{Literature}}{(mCp)\Delta T}$$

5.B.2 Results and Discussion

The results from these experiments are summarized in Table 5.B.1.

In Table 5.B.2 the experimentally determined values of PHI, ϕ, from Table 5.B.1 are compared with values calculated both with and without the Swagelok fittings. Looking at the results in Table 5.B.2, there are three points worth noting:

1. The calculated values for PHI with the fittings included in the mass of the bomb *are more consistent* with the experimentally determined values especially when the sample mass was below 5 g.
2. When the mass of the fittings was omitted, the calculated value of PHI was consistently low, which is undesirable for thermal hazard evaluation as it will lead to an erroneously higher onset temperature and lower heat generation rate.

TABLE 5.B.1 Experiment Summary (ΔH Calculated Based on 14.01 kcal/mol of Anhydride or 58.62 kJ/mol)

Expt. No.	Empty Bomb (g)	Fittings (g)	H_2O (g)	Acetic Anhydride (g)	Mols of Anhydride	Calculated ΔH (J)	Avg Cp of Sample (J/g·°C)	ΔT_{ad} exp (from ARC)	ϕ Calculated from: $\Delta H/mCp\Delta T$
1	10.3955	10.6950	5.2581	4.94	0.0484	2836	3.09	71.43	1.260
2	10.2028	10.7054	2.4794	2.38	0.0233	1366	3.08	54.42	1.678
3	10.4089	10.6520	2.5172	2.66	0.0261	1527	3.02	54.77	1.781

TABLE 5.B.2 Comparison of Experimental ϕ and Calculated PHI with and Without Fittings

Experiment	ϕ Experimental from Table 5.B.1	ϕ Without Fittings[a]	ϕ With Fittings[a]	Mass of Titanium Bomb (g)	Mass of SS Swagelok Fittings (g)
1	1.260	1.166 (7%)	1.336 (−6%)	10.3955	10.6950
2	1.678	1.342 (20%)	1.702 (−1%)	10.2028	10.7054
3	1.781	1.334 (25%)	1.675 (6%)	10.4089	10.6520

[a] Numbers in parentheses indicate deviation from the experimentally determined value. The heat capacity values of both T_i and SS equal 0.12 cal/(g·°C).

3. For the larger sample size experiment (~10 g used in experiment 1), either PHI results in deviations of 6–7%. An average of the two would be closest. But as the sample size is reduced (~5 g), there is a greater deviation and greater error if the fittings are not included in the phi calculation. By including the fittings, PHI was most consistent with the acetic anhydride hydrolysis validations, within 1% for experiment 2 and 6% for experiment 3.

In this work we developed a procedure for estimating PHI directly in the ARC to compare and validate with the two methods of calculating PHI. Based upon these results, it can be generally concluded that the mass and heat capacity of the fittings should be included for low sample masses (<5 g). For this reason PHI values including the fittings were employed in the assessment of ARC data described in this case study since the sample sizes of BTHF used for our ARC studies were typically approximately 4.5 g, and the specific heats were relatively low (~1.7 J/g·°C).

REFERENCES

1. Nettles, S.M., Matos, K., Burkhardt, E.R. et al. (2002). Role of NaBH$_4$ stabilizer in the oxazaborolidine-catalyzed asymmetric reduction of ketones with BH$_3$-THF. *The Journal of Organic Chemistry* 67: 2970–2976.
2. (a) MARC REISCH (2002). Two seriously injured at Pfizer research site. *Chemical and Engineering News* 80 (27): 11–12. (b) am Ende, D.J. and Vogt, P.F. (2003). Safety notables: information from the literature. *Organic Process Research and Development* 7: 1029–1033.
3. Crowl, D.A. and Louvar, J.F. (2002). *Chemical Process Safety*, 2e. Upper Saddle River, NJ: Prentice Hall.
4. Burkhardt, E.R., and Corella, J.A. (2000). Borane-THF Complex Method of Storing and Reacting Borane-THF Complex, Mine Safety Appliances Co. US Patent 6,048,985, 11 April 2000.
5. (a) Potyen, M., Josyula, K., Schuck, M. et al. (2007). Borane-THF: new solutions with improved thermal properties and stability. *Organic Process Research and Development* 11: 210–214. (b) Aldrich Technical Bulletin AL-218 (2004). New, safer, amine-stabilized borane-tetrahydrofuran solutions for hydroboration and reduction. Note: 1.0 M BH3-THF, whether stabilized with amines PMP (1,2,2,6,6-pentamethyl-piperidine) or NIMBA (*N*-isopropyl-*N*-methyl-*tert*-butylamine) or with NaBH$_4$, should be stored at 2–8 °C.
6. Monteiro, A.M. and Flanagan, R.C. (2017). Process safety considerations for the use of 1 M borane tetrahydrofuran complex under general purpose plant conditions. *Organic Process Research and Development* 21 (2): 241–246.
7. CCPS (1995). *Guidelines for Chemical Reactivity Evaluation and Application to Process Design*. New York: Center for Chemical Process Safety, AICHE.
8. Townsend, D.I. and Tou, J.C. (1980). Thermal hazard evaluation by an accelerating rate calorimeter. *Thermochimica Acta* 37: 1–30.
9. Wadso, I. (1962). Heats of aminolysis and hydrolysis of some N-acetyl compounds and of acetic anhydride. *Acta Chemica Scandinavica* 16: 471–478.

6

ANALYTICAL ASPECTS FOR DETERMINATION OF MASS BALANCES

MATTHEW JORGENSEN*

Chemical R&D, Pfizer, Inc., Groton, CT, USA

6.1 INTRODUCTION

The role of the analytical chemist in API process development is critically important in the pharmaceutical industry. The analyst and the analytical data they provide are the "eyes" on the process. Without accurate analytical results, the process would be running blind. Often the process engineer and chemist know what to expect. But without reliable analytical data, it is impossible to know if the processes have quantitatively met expectations.

The level of importance placed on the analytical data highlights how critical it is that the data be sound and truly representative of the process.

Occasionally the analytical results may be confounded with unquantified or unseparated components or simply may be nonrepresentative due to oversight on the part of the chemist, engineer, analyst, or a combination of the three. This breakdown in the quality of the analytical results is traced back to a breakdown in the communication between the parties involved. Information that one or all parties are unaware of can directly impact the quality of the analytical results. The entire process team needs to be cognizant of information such as the stability of reaction components, composition of samples (in addition to starting materials intermediates and products), and what level of precision is required of the results.

This chapter will deal directly with what a process engineer should know about the analytical data. This includes information around what is required to ensure that the data that are produced, be it by an analyst or engineer, is of the highest quality needed for a particular study. Details about what each analytical technique is tracking and what are its limitations, common mistakes that may confound analytical results, and coupling analytical methods to overcome these limitations will all be covered in this chapter. Finally, it will be shown through examples how this level of understanding of the analytical techniques can be leveraged by the engineer to solve the problems of mass balance and estimating kinetic parameters.

High quality analytical data are paramount if one wishes to accurately know how a process is truly performing. In most cases certain assumptions are made during the application of the analytical data, and understanding the validity of the assumptions is important. Information in this chapter will help the engineer be aware of these typical assumptions and their applicability.

6.2 THE USE OF ANALYTICAL METHODS APPLIED TO ENGINEERING

Occasionally the analyst and the engineer can sometimes feel that the other is speaking different languages. For example, the terms potency and purity are commonly used and can be a source of confusion without clarification around what these numbers mean and how their values were arrived at. Both potency and purity refer to a measure of the active or desired ingredient relative to the sample. The details of how purity and potency are actually determined are important to understand and are the subject of the next section.

---

*Current address: Nalas Engineering Services, Inc., Centerbrook, CT, USA

Chemical Engineering in the Pharmaceutical Industry: Active Pharmaceutical Ingredients, Second Edition.
Edited by David J. am Ende and Mary T. am Ende.
© 2019 John Wiley & Sons, Inc. Published 2019 by John Wiley & Sons, Inc.

6.2.1 Purity

A strict definition of percent purity would require qualifying what the purity basis is, i.e. purity percent by weight or purity percent by high pressure/performance liquid chromatography (HPLC) area at 254 nm. Often in the pharmaceutical industry, purity percent by HPLC area is shortened to just purity, and when the more rigorous definition is applied, purity percent is stated as purity by wt %.

The term purity typically is based on area% values alone:

$$\text{Purity} = \left[\frac{\text{Active}_{\text{area}}}{\text{Active}_{\text{area}} + \text{Other}_{\text{area}}}\right] \times 100\% \quad (6.1)$$

where $\text{Other}_{\text{area}}$ refers to the peak areas of all the other peaks in the chromatogram.

Thus any impurity is assumed to have the same response factor as that of the main component. The reason area% purity is reported is one of timing. In early development of a new chemical entity, there are usually no standards. As area% purity is something that can be reported from the first injection, a meaningful metric can be generated without a lot of work to develop standards. The area% purity values can be used to compare the historical samples with each other to compare differing chemical approaches to the project. Later on in the project, when standards have been made and characterized, percent purity by mass values can also be reported. This percent purity by mass relative to the standard (also referred to as potency) taken with the percent purity by area value is a good indicator of how well the standards are characterized. For the remainder of this chapter, the percent purity by area will be referred to as just purity.

6.2.2 Potency

$$\text{Potency of } A^s = \left[\frac{\text{Mass}_A^S}{\text{Mass}_{\text{total}}^S}\right] \times 100\% \quad (6.2)$$

$$\text{where Mass}_A^S = \text{Area}_A^S R_{fA} \quad (6.3)$$

$$\text{where } R_{fA} = \frac{\text{Mass}_A^{\text{STD}}}{\text{Area}_A^{\text{STD}}} \quad (6.4)$$

Superscript S denotes sample.
Superscript STD denotes standard.
Subscript A denotes material A.

The term potency in Eq. (6.2) is a bit deceiving at first glance as it appears that when samples are reported at a given percent potency, then this is a percent by mass intrinsic to the test sample. This is not the case. Actually this value is a percent active compared to an external standard as shown in Eq. (6.3) through the use of a proportionality constant called a response factor designated by R_f. This response factor is generated from a reference standard as shown in Eq. (6.4) by taking the ratio of the response, HPLC area in this case, to the mass of the sample. The reference standard is typically a well-characterized purified sample of the desired material used to calibrate HPLC peak area to mass of the sample. In most cases in the pharmaceutical industry, the reference standard is not commercially available and has to be purified by thorough crystallization or preparative-scale chromatography.

Response factors can be simplified as the ratio of the output response to the input material and is used in most analytical technique with linear responses such as mid-IR spectroscopy, mass spectroscopy, or in this case ultraviolet (UV) spectroscopy. After close inspection of Eqs. (6.2)–(6.4), it becomes apparent that the accuracy of the potency value hinges on the quality of the standard used to generate the response factor. As all reported values are relative to the standard, it is possible to have a test result indicating a potency greater than 100%, indicating the sample is more potent than the standard.

Early in development, this standard may be nothing more than the most pure obtained sample to date. The limited characterization of the standard consists of analysis for residual solvents including water as well as residue on ignition testing (ash). Anything that is not ash or solvent is then attributed to the material of interest. So to reiterate, potency refers to the % active component in a given sample relative to an external reference standard for that active component.

EXAMPLE PROBLEM 6.1

(A) An isolated sample is submitted for the typical purity and potency analysis. The results reported back were the following:

Purity: 99.5%

Potency: 97.3%

What do these results tell us about the sample?

Solution

A typical mass balance for the reference standard is the following:

$$\text{Mass}_{\text{total}} = \text{Mass}_A^\# + \text{Mass}_{\text{residualsolvent}}^\# + \text{Mass}_{\text{ASH}}^\# + \text{Mass}_{\text{impurities}}^\# \quad (6.5)$$

where

Superscript # = either STD or S if the mass balance is for the standard or the sample, respectively.

$\text{Mass}_A^\#$ = the mass of the desired compound of interest in the standard or sample.

Writing Eq. (6.5) in terms of the sample and solving for Mass_A^s results in Eq. (6.6):

$$\text{Mass}_A^s = \text{Mass}_{total}^s - \text{Mass}_{residual\ solvent}^s - \text{Mass}_{ASH}^s - \text{Mass}_{impurities}^s \quad (6.6)$$

Substituting Eq. (6.6) into Eq. (6.2) is shown in Eq. (6.7):

Potency of A^s

$$= \left[\frac{\text{Mass}_{total}^s - \text{Mass}_{residual\ solvent}^s - \text{Mass}_{ASH}^s - \text{Mass}_{impurities}^s}{\text{Mass}_{total}^s}\right] \times 100\% = 97.3\% \quad (6.7)$$

From Eq. (6.7) it becomes apparent that the terms for residual solvent, ash, and impurities are why the potency is less than 100%. At first glance it would be easy to assume that because the purity value is 99.5%, then the $\text{Mass}_{impurities}^s$ term is low. Remember though that the percent purity is actually percent purity by area%. If there are any impurities that have a drastically larger response factor than the desired material, then they will be under reported. At best the purity value of 99.5% infers what are the dominate terms that are reducing the potency of the sample. Submitting the sample for further analysis for residual solvents and ash is the only way to identify if the missing 2.7% is due to residual solvents and ash or underreported impurities.

(B) Consider the example where the magnitudes of purity and potency are reversed, i.e. the sample has a higher potency than purity:

Purity: 95.3%

Potency: 103.2%

What does this tell us about the product and more importantly about the standard?

Solution
The potency value of 103.2% means that the sample is more potent than the standard. Because potency is a relative activity of a sample compared with the standard, it is possible to have values greater than 100%. What this indicates is that the mass balance around the standard is not fully closed. From Eq. (6.5) the mass of the active material in the standard is determined by difference, so

$$\text{Mass}_A^{STD} = \text{Mass}_{total}^{STD} - \text{Mass}_{residual\ solvent}^{STD} - \text{Mass}_{ASH}^{STD} - \text{Mass}_{impurities}^{STD} \quad (6.8)$$

Substituting Eqs. (6.2)–(6.4) and (6.8) into Eq. (6.2) is shown in Eq. (6.9):

Potency of A^s

$$= \left[\text{Area}_A^s \times \frac{\text{Mass}_{total}^{STD} - \text{Mass}_{residual\ solvent}^{STD} - \text{Mass}_{ASH}^{STD} - \text{Mass}_{impurities}^{STD}}{\text{Area}_A^{STD}} \Big/ \text{Mass}_{total}^s\right]$$

$$= 103.2\% \quad (6.9)$$

If the areas and total mass of both the sample and standard are accurate than from Eq. (6.8), the only way to have potency greater than 100% is if the characterization of the reference standard around ash, solvent, or impurities is off. The source of the failure to close the mass balance of the standard is most likely due to the impurities not being fully characterized, as residual solvent and ash are standard analysis. If the purity of the standard (UV area%) was near 100%, then there may be impurities that are not showing up at the wavelength that the detector is set at, or they may not be UV active. In such a case, further purification of the standard by chromatography or recrystallization is needed to better close the mass balance of the standard and gain an accurate R_f.

To further complicate the issue, a sample purity value of 95.3% indicates that not only is the R_f too high due to a poorly characterized standard, but the samples' total impurities of 4.7% indicate that some of these impurities have higher molar absorption coefficient relative to the desired and thus will appear to be present in higher concentration. The assumption with area% values is that everything has the same R_f as the main peak. If any of the impurities have a lower R_f than the main peak, then the impurities will be over reported by the area% value. The various scenarios discussed above have been summarized in Table 6.1.

For comparing processes with each other based solely on isolated yield and relative potency, a less than fully characterized standard still allows for relative comparison, i.e. 103% potent material is better than 95% potent material. For work that would require a more stringent mass balance, kinetics, or process understanding, the mass balance should be closed by utilizing a combination of complementary analytical techniques such as quantitative H^1-NMR and HPLC.

Two areas where analytical data are most frequently needed by the API process development engineer are data to close the mass balance and data to develop kinetic models. These two utilizations of the data are not independent of each other, as it is necessary to have a reasonable mass balance before attempting to develop a kinetic model. As such it is imperative to have analytical techniques available that can both "see" what needs to be tracked and give values of concentrations that are needed for both the mass balance and the kinetic model.

6.3 METHODS USED AND BACKGROUND

What follows is a brief overview of the most common analytical techniques used and some concepts that need to be kept in mind when attempting to analyze the data generated. A more thorough discussion about each technique can be found elsewhere [1].

All methods of column chromatography rely on the same basic principles. First there is a sample that is made up of a mixture of components. This mixture is loaded onto a column that separates the individual components as they partition

TABLE 6.1 Possible Scenarios of Purity and Potency Values

$$\text{Purity: Purity(area\%)} = \frac{(A_{\text{area}})}{\left(\sum \text{UVactive}_{\text{area}}\right)} \cdot 100 \qquad \text{Potency: Potency(wt\%)} = \frac{(A_{\text{area}})(R_{f_A})}{(\text{Sample}_{\text{mass}})} \cdot 100$$

Purity = potency	Can occur if the response factors of all the UV-active components including impurities are very similar so that an area% of A is equivalent to a wt %. It also requires that the reference standard for the potency determination be highly accurate
Purity < potency	If we assume the reference standard is accurate, then this situation can arise if the impurity peaks have a higher extinction coefficient and higher absorbance than the desired component A. This translates to artificially high impurity count and lower purity by area%
Purity > potency	If we assume the reference standard is accurate, then purity will exceed potency when there are non-UV components present that are not being detected by HPLC. This will contribute to lower potency values and higher HPLC area% – for example, if the sample has high salt content (ash). This will look pure by HPLC because the ash is not detected
Potency > 100%	Reference standard likely not well characterized with respect to wt % ash, residual solvent, or impurities

TABLE 6.2 Properties of Most Common Detector Types

	Chromatography Method	Detection of	Sensitivity	Notes
Ultraviolet (UV)	LC	Absorption of UV light by pi–pi bonds, i.e. conjugation	Dependent on ε of the analyte	Variation of ε on the order of 100- to 1000-fold is possible
Mass spectrometer (MS)	LC/GC	Charged particles	μM concentrations	Does not detect mass but rather mass to charge m/z ratio
Flame ionization detector (FID)	GC	Ionized particles from combustion of organic compounds	ppm–ppb	Signal proportional to number of carbon atoms, i.e. signal proportional to mass not concentration
Conductivity	LC	Charged ions by measuring resistance in detection cell	5×10^{-9} g/ml	Most often used for ion exchange chromatography
Electrochemical	LC	Current generated by oxidation or reduction of sample	Order of magnitude more sensitive then UV	More selective and sensitive then UV, but detector not as rugged as UV
Refractive index (RI)	LC	Variations in refractive index	0.1×10^{-7} g/ml	Universal detector, poor detection limit, sensitivity to external condition (temperature, dissolved gas, etc.) limit practical use
Evaporative light scattering detector (ELSD)	LC	Nonvolatile particles of analyte scattering light	0.1×10^{-7} g/ml	Analyte needs to be nonvolatile were mobile phase needs to be volatile
Fluorescence	LC		Low ng/ml range	Typically require derivatization with fluorophore reagents

Source: From Ref. [2].

between two phases, the mobile and the stationary phases. In liquid chromatography (LC), the partitioning is driven by the polarity of the components and the differing polarity of the mobile phase versus the stationary phase, absorbing and de-absorbing onto the stationary phase down the length of the column. In gas chromatography (GC), the partitioning is driven by the relative volatility of the components as it alternates between the gas phase and dissolution into the stationary phase. The net effect of any chromatographic system is to separate the components of the sample mixture. It is the detector that is attached to the outlet of the chromatographic system that allows one to see the relative concentrations of each species in the sample. As such, the type of detector used will dictate what is "seen" by the analytical method. Table 6.2 lists the types of detectors available, what type of chromatographic system they are most often paired with, and what they are capable of detecting.

The underling similarity in all these methods of detection excluding FID is that the resulting signal is proportional to the concentration. The important thing to remember is that for every component of a sample that is being analyzed, there is a proportionality constant that is unique to that compound. So in the example of the UV detector, the most common detector for different LC methods, this proportionality constant is the molar absorptivity ε. The relationship of ε to concentration and absorption is described by Beer–Lambert law shown in Eq. (6.10):

$$A = \varepsilon b c \quad (6.10)$$

where

A = absorption, dimensionless

ε = molar absorptivity, L/mol/cm

c = concentration, mol/L

b = detector path length, cm

In the case of mass spectrum (MS) detectors, the proportionality constant is the ionization potential; in electrochemical detection, it is the RedOx potential. Even in the case of non-chromatographic methods, the idea of proportionality constants should always be remembered. As an example quantitative NMR has relaxation times that can be thought of as proportionality factors. So for every sample analyzed, be it with chromatography or not, the individual components each will have a unique proportionality constant that may or may not be similar to other components in that sample. This is why most detectors are not universal detectors; all species that are chemically different will have different proportionality constants. In many instances if the components are all structurally similar, then their proportionality constants may be very similar as well, but this is not always the case. This is why taking area% values as direct replacements for concentration can lead to erroneous results. At best these area% values can be used to indicate relative abundances, but care around the possibility of different response factors must be taken if area% values are used as replacements for concentration values for calculating mass balances or kinetic profiles. The following example illustrates this point.

EXAMPLE PROBLEM 6.2 Response Factors vs. Area%

A high temperature coupling reaction was being evaluated with potassium hydroxide in a high boiling solvent. Initial reaction completion HPLC looked promising with apparent conversion of greater than 80% although long reaction times were required (Figure 6.1). However the isolated yields were low (<50%), but this was attributed to a laborious workup. The workup involved two extractions followed by distillation and then crystallization. An extensive amount of time was spent trying to optimize the reaction with an eye toward fixing the workup, and increasing isolated yield after the reaction was optimized.

The "conversion" was being calculated in the lab as

$$\text{Conversion} = \frac{\text{Area\%product}}{(\text{area\%starting material} + \text{area\%product})} \quad (6.11)$$

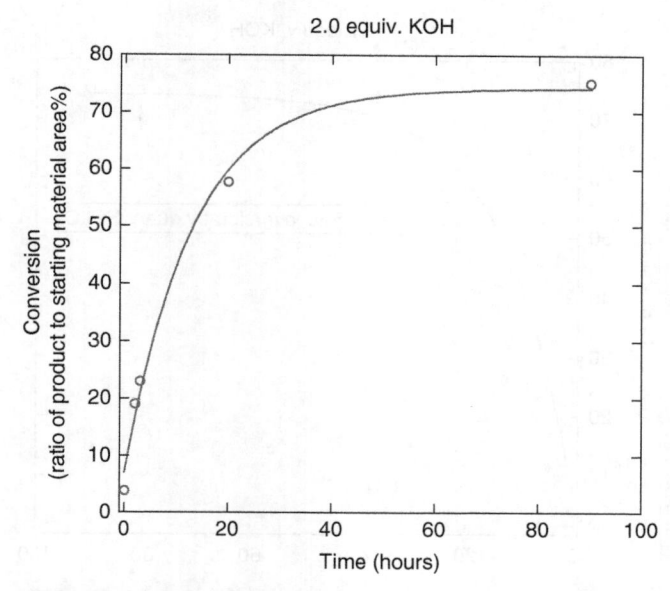

FIGURE 6.1 Conversion as calculated by the chemist in the lab by Eq. (13.6) only taking the ratio of starting material and product area% into account.

There was some concern around this approach, but the argument against pulling samples for quantitative HPLC was the reaction was very thick and heterogeneous, making it hard to sample representatively. Provide a solution to the approach.

Solution

To get accurate quantitative HPLC data and potency values, the sampling limitation was avoided by not sampling. A reaction was run and the quantitative HPLC sample made by using the entire reaction to make up the HPLC sample in a volumetric flask. By doing this there would be no sampling error as the entire reaction would be used. In this instance the quantitative HPLC conversion was calculated as

$$\text{Conversion} = \frac{\text{moles product}}{(\text{moles starting material at } T_0)} \quad (6.12)$$

where moles product is calculated as

$$\text{moles product} = \text{Area}_{\text{product}} \times R_{\text{f}}^{\text{mole}} \text{product} \quad (6.13)$$

and $R_{\text{f}}^{\text{mole}}$ is calculated from Eq. (6.4) on a mole basis.

Quantitative HPLC showed there to be much less product after 45 hours than originally assumed (Figure 6.2). Low yield was not due to product loss in workup, but rather was never formed to begin with. The starting material was reacting/degrading to something other than product. Forcing the mass balance on area% (between starting material and product), it appeared higher than it actually was. Quantization resulted in the decision to discontinue further development on these conditions.

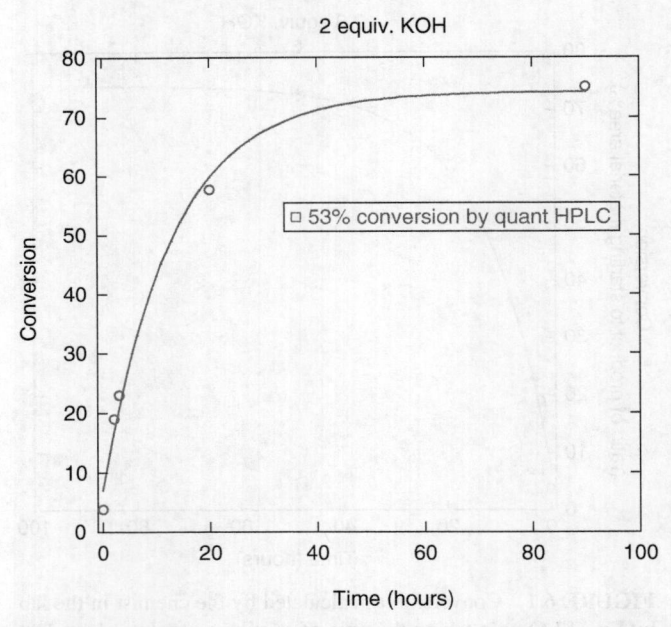

FIGURE 6.2 Conversion calculated by Eq. (13.6) shown as solid line compared with conversion calculated by Eq. (13.7) shown as solid box. This discrepancy between the two values at the 48 hours time point indicates that the forced mass balance of Eq. (13.6) elevated the product concentration as by not taking into account a possible side reaction of the starting material that did not result in product.

6.3.1 Mechanics of HPLC and UPLC

HPLC and the more recent ultra-pressure/performance liquid chromatography (UPLC) are considered the standard lab equipment when it comes to understanding what is going on in a synthesis or process. The difference between these two techniques lies in the size of the solid-phase packing in the columns as well as the pressures that are employed, hence the high/ultra descriptors. For HPLC the solid-phase packing is between 5 μM–3 μm and 200–400 bar pressure, where UPLC solid phase is below 3 μm in size and pressures above 1000 bar.

A quick aside about the equation that governs the efficacy of both techniques, as well as any other column chromatography, the van Deemter equation is in its simplified version (Eq. 6.14) [3].

$$H = A + \frac{B}{u} + Cu \quad (6.14)$$

where

H (sometimes shown as HETP) is the variance per unit length, also referred to as height equivalent to a theoretical plate.

u is the volumetric flow rate.

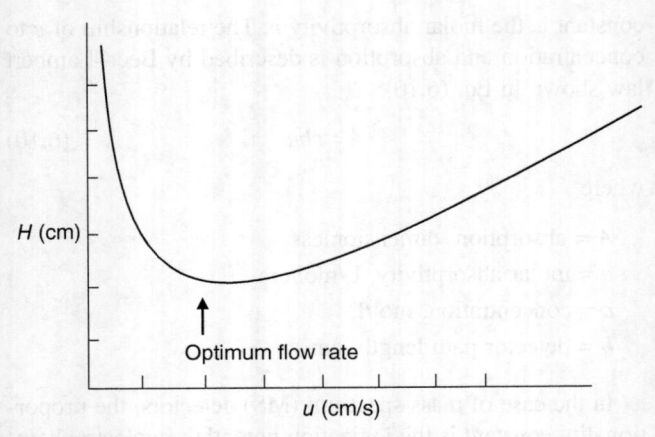

FIGURE 6.3 Characteristic van Deemter plot shape illustrating the presence of an optimum flow rate to maximize column efficiency.

A is the term describing the multipaths in the packed bed.
B is the term describing longitudinal diffusion.
C is the term describing resistance to mass transfer.

This hyperbolic function relates the variance per unit length to particle size, mass transfer between the stationary and mobile phases, and the linear velocity of the mobile phase. This relationship was the first result from applying rate theory to the chromatography process and was originally developed to describe GC. It has been extended to describe LC as well with modifications to the lumped parameters terms A, B, and C in Eq. (6.14). A typical van Deemter plot is shown in Figure 6.3.

The van Deemter equation is useful in describing the theory and mechanism of the chromatography process, not only for the small analytical chromatography used in analysis but also for large-scale separations done on large pilot plant and commercial scale. This equation explains why the problem of unresolved peaks cannot be solved by just going to a longer column at the same flow rate. The increased resolving power of more packing in a longer column is lost to the increase of the B term in Eq. (6.14) (increase of eddy and longitudinal diffusion) due to increased time spent on the column. Thus the number of theoretical plates is less for the longer column (when held at the same flow rate) even though it is longer with more packing because the height of the plates, the H term in Eq. (6.14), is larger as well. This is where UPLC comes into its own. By decreasing the packing size and increasing the pressure, the linear velocity is kept high, and the increased resolving power of more packing is not lost to increased diffusion, thus giving higher plate counts for a given time on the column. The net result is increased resolution and throughput for the analyses.

Both methods, HPLC and UPLC, are only a tool for separating individual components from a mixture and feeding

6.4 THINGS TO WATCH OUT FOR IN LC AND GC

6.4.1 Injections Have Everything in Them Not Just the Desired Reactants

The first thing that must be communicated to the analyst, or kept in mind for those that are acquiring their own data, is to account for what is in the reaction mixture. The most common mistake in LC that everyone makes once and hopefully only once is the toluene mistake. This is what happens when people forget that toluene unlike most organic solvent has both a chromophore and is retained on most LC columns. I cannot tell you how many bright analytical chemists have come running down in a panic telling everyone that there is this major new impurity only to find out that the project has switched to toluene in the process and had not notified the analyst. Worse yet is the chemist that reports that they have excellent in situ yield only to be looking at a nonexistent reaction because they assumed that large peak that was not starting material was product when in reality it was the toluene peak. This can lead to wasted development time chasing a nonexistent reaction.

In GC the major concern is nonvolatiles, i.e. salts. If a lot of reaction mixtures that contain large percentages of salts are injected, then the injector may become plugged, necessitating in cleaning the injector before accurate analysis can resume. What is more complicated is when the product or reactants are salts, i.e. charged species. These will not "fly" on the GC and will require some sort of quench to run on the GC. Most often this is a neutralization of the reaction mixture to quench the charge on the desired compounds so as to facilitate GC analysis.

6.4.2 Unplanned/Planned Modification of Stationary Phase

If running one's own analysis, one must be cognizant of possible changes to the HPLC column due to history. Depending upon the nature of the mobile phase being used, "conditioning" of the column may take place such that the results may not be repeatable, or representative of differing HPLC system. This opens the possibility of analysis that cannot be duplicated, leading to confusion around what results are accurate. A prime example of conditioning of the column is ion-pairing mobile phase such as sodium dodecyl sulfate, or a weak ion-pairing agent such as perchloric acid. In the case of ion-pairing mobile phases, the stationary phase is modified or conditioned over time to be more retentive of polar species such as primary amines due to the stationary phase being modified by the mobile phase containing the ion-pairing agents. There is a memory effect now for this column that will still maintain the effect even if the mobile phase is switched to a more traditional acidic mobile phase. This has the greatest impact when someone develops a method with such a conditioned column as it will be impossible to replicate these results without this preconditioned column.

The other extreme is when the column is conditioned negatively or destroyed by running samples of reaction mixtures that destroy the resolving power of the stationary phase. This is most often seen with samples from reactions such as hydrogenations that contain metal species that bind to the stationary phase, resulting in reduced resolving power. If a column is suspected of being conditioned either negatively or positivity, then the only option available is to replace the column and see if the previous analysis is replicated. Thus it is always good to periodically run a system suitability test to check analytics with a reference mixture to confirm retention times/peak shapes.

6.4.3 Product Stability/Compatibility with Analysis Method

Stability of the reaction mixture or products to the chromatographic conditions is another major concern that needs to be addressed before a strategy for analyses can be agreed upon. The majority of aqueous mobile phases utilized in UPLC and HPLC are acidic. This regularity of acid mobile phases is due to two major factors. The first factor is that until recently the silicon support for the column mobile phase was not stable to high pH values as silicon is soluble at pH levels above 11. The second factor is that if the mobile phase pH is near the pK_a of any of the sample components, then slight variations in the pH of the mobile phase can change the polarity of the components. This change in polarity will then change retention time and order of elution of the components.

Because of these two factors, the majority, near 80% of the mobile phases, is acidic (pH < 1) to both maintain the stability of the mobile phase as well as to prevent any change in the analysis due to pH variations. The idea is to protonate everything and prevent pH gradients from forming on the column that may cause chromatographic artifacts.

This proclivity of mobile phases to be acidic makes stability to acid aqueous conditions one of the main stability concerns. As the amount of sample that will be loaded on the column for each injection is so miniscule, in the order of microliters, the sample gets swamped by the mobile phase. If the sample is not stable to aqueous/acidic conditions, then degradation will be taking place as the sample travels and elutes from the column. The net effect is that the sample that was representative of the reaction at a given time or point in the process is scrambled by the analysis method, rendering the results no longer representative. This is unfortunate if one lab-scale reaction is lost because of this, and devastating if three weeks of DOE experimentation is rendered useless also because of this; both have happened.

In GC the major issue with stability of the reaction mixture is that of thermal stability. Remembering that the standard injector temperature for GC is 280 °C, this is the temperature that the reaction samples have to "endure" just to get on the column.

If stability is a problem in LC or GC, then quenching the reaction samples (if this improves stability) or some sort of derivatization method may be required.

6.4.4 Derivatization

Derivatization is the process by which a reaction sample is further reacted to form a new compound as part of the sample preparation. This may be done for various reasons such as increasing reactant stability to the analysis method, or modifying the components of a sample to make them detectable such as attaching a chromophore, or to increase volatility for GC analysis [4]. The major issue with derivatization is that this is a second reaction that is in series with the desired reaction. The net effect of this is that if the reaction of interest is to be accurately characterized, then the derivatization reaction needs to be quantitative in reaction completion and at a reaction rate that is orders of magnitude faster than the desired so as to not skew the analysis. For a system that a kinetic model is being developed and derivatization is required, a quench of the reaction should be used before derivatization or a derivatization that also quenches the reaction. This is to ensure that the analysis is representative of the time the sample was taken, not the time the sample was analyzed.

6.5 USE OF MULTIPLE ANALYTICAL TECHNIQUES

Oftentimes when trying to understand a process by developing a mass balance as well as a kinetic model, it is best to start at the beginning. In most cases the beginning is a full characterization of the feedstocks going into the process. It will be very difficult to close the mass balance if one is not aware of what is going into the process. The use of two or more complementary analytical techniques can greatly aid in fully understanding the inputs for a process. The following Case Study 6.1 illustrates this point.

CASE STUDY 6.1: MASS BALANCE AROUND STARTING MATERIALS TO DEVELOP A KINETIC MODEL CASE STUDY

A process involves coupling secondary aniline with a volatile chiral epoxide utilizing an ytterbium catalyst in isopropyl acetate at 60 °C. The secondary aniline was synthesized from the primary aniline as shown in Scheme 6.1.

The secondary aniline was telescoped into the reaction with the chiral epoxide with some residual primary aniline present. The observation was that when the process to coupling the secondary aniline and epoxide was first scaled up in a kilo scale facility, a second charge of the epoxide reagent was required to drive the reaction to completion. The reaction had been run in a sealed reactor so as to limit losses of the epoxide with its very high vapor pressure. Owing to the sealed reactor configuration, the time for the reaction to complete was believed to be 16 hours but had not been confirmed as reaction completion samples were not taken during this initial 16 hours over fears of venting the epoxide during sampling. In addition the incoming secondary aniline reagent was in a solution of isopropyl acetate, which was known to have residual primary aniline from the first reaction. The question was what the impact of residual primary aniline was on the desired reaction of the secondary aniline with the epoxide.

It was decided to undertake a kinetic study of this reaction to identify the answers to the following questions:

1. How long does the reaction take?
2. Why did the first scale-up require the second charge of epoxide to drive the reaction to completion?
3. What is the impact of residual primary aniline?

The first step to developing a kinetic model was to fully characterize the two incoming reagent streams. The aniline reagent stream had an unknown potency as standards were not available for quantitative HPLC. The epoxide reagent had a certificate of analysis (COA) from the vendor, but its potency would be reevaluated to confirm these numbers.

To gain a handle on the composition of the aniline reagent stream, an H^1-NMR was taken that resolved the isopropyl acetate from the aniline compounds. Figure 6.4 shows the H^1-NMR with the peaks assigned to the structure. From Figure 6.4 it becomes apparent that there was not enough resolution between the primary and secondary aniline compounds with NMR to

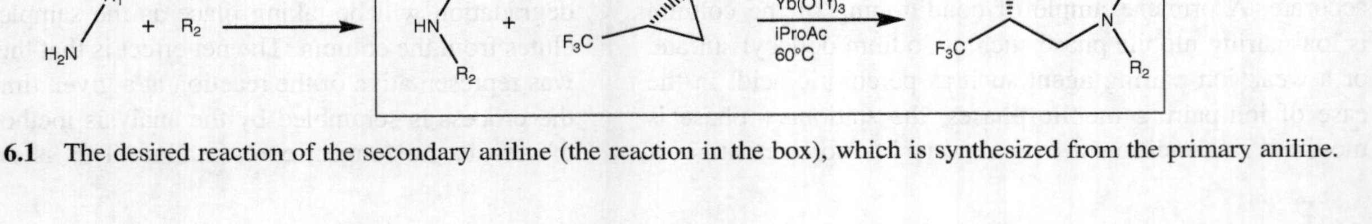

SCHEME 6.1 The desired reaction of the secondary aniline (the reaction in the box), which is synthesized from the primary aniline.

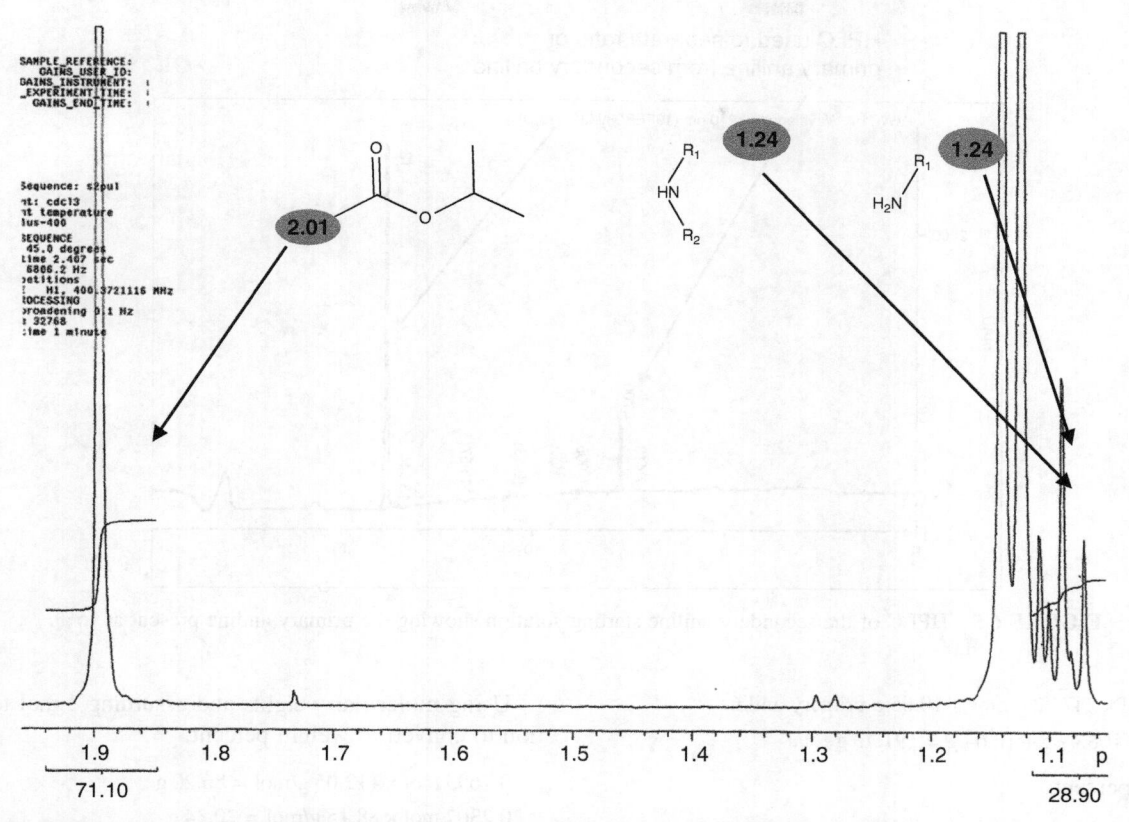

FIGURE 6.4 The NMR scan of the secondary aniline solution with the methyl group protons integrated for both the primary and secondary aniline compounds as well as the methyl protons for the isopropyl acetate.

decouple their individual concentrations. Using the HPLC area %, the concentration of the different anilines was calculated.

The assumption in this approach was that the only other components in the stream besides the secondary aniline were isopropyl acetate and the primary aniline. The second assumption was that the NMR relaxation times for all the components were on the same time scale. The third assumption was that the response factors for the two anilines were similar enough to be able to use the area% values directly. The calculation for the composition of the starting material is shown below:

From Figure 6.4 the ratio of isopropyl acetate to aniline compounds is as follows:

$$\text{Isopropyl acetate } \frac{71.1}{3H} = 23.7$$

where

H = proton

$$\text{Aniline compounds } \frac{28.9}{3H} = 9.63$$

23.7 and 9.63 represent the relative number of moles of IPAC and aniline compounds. So the mol % isopropyl acetate is easily calculated from

$$\frac{23.7}{23.7 + 9.63} = 71.1\%$$

And the mol % aniline compounds is

$$\frac{9.63}{23.7 + 9.63} = 28.2\%$$

The step of taking the ratio of the area to the number of protons in this case is redundant as both peak in the NMR being compared are for methyl groups (three protons), but this is an important step that can often be missed.

The area% values from the HPLC in Figure 6.5 were used in order to decouple the aniline compounds concentration.

Secondary aniline 84.6%

Primary aniline 15.4%

So on a mol %, the composition is

0.846 × 0.282 = 0.239 × 100 = 23.9% secondary aniline

0.154 × 0.282 = 0.043 × 100 = 4.3% primary aniline

Using molecular weights and assuming a basis of 1 mol,

0.711 mol × 102.13 g/mol = 72.61 g isopropyl acetate

0.239 mol × 453.86 g/mol = 108.47 g secondary aniline

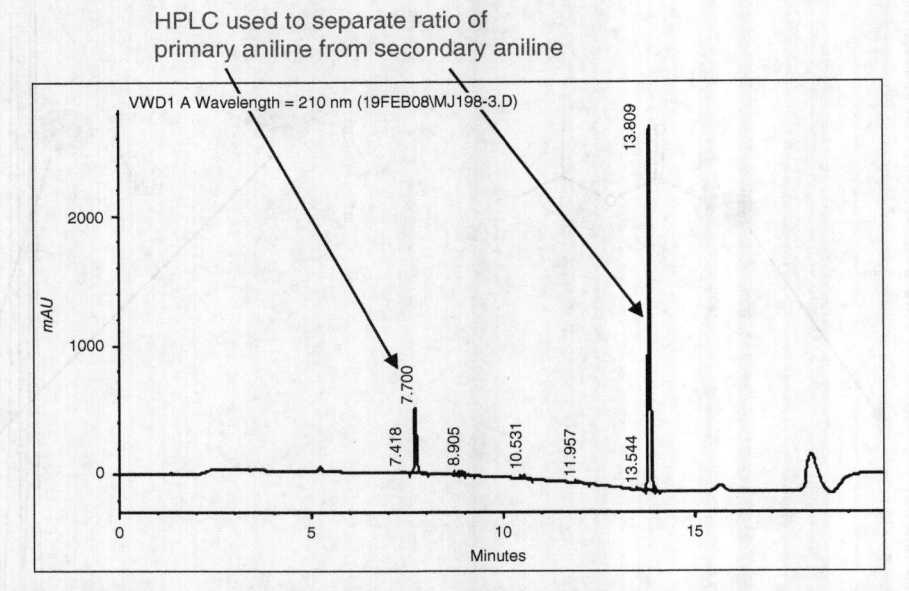

FIGURE 6.5 HPLC of the secondary aniline starting solution showing the primary aniline present at 15%.

0.043 mol × 247.72 g/mol = 10.65 g primary aniline

72.61 g + 108.47 g + 10.64 g = 191.73 g total

Weight percent:

$$\frac{72.61\,g}{191.73\,g} \times 100 = 37.87\,wt\%\ \text{isopropyl acetate}$$

$$\frac{108.47\,g}{191.73\,g} \times 100 = 56.57\,wt\%\ \text{secondary aniline}$$

$$\frac{10.65\,g}{191.73\,g} \times 100 = 5.55\,wt\%\ \text{primary aniline}$$

This characterizes the incoming aniline reagent stream; now the same process is repeated with the epoxide reagent stream. By the COA from the vendor, the epoxide, which is a liquid, it is known to have methyl *tert*-butyl ether present at 12% by weight. As this compound cannot be detected with HPLC with a UV detector, H^1-NMR was again used to characterize the material as shown in Figure 6.6. Taking the area of the MTBE compared with the area of the epoxide peaks, we are able to calculate the mol % of each of the following:

$$\frac{12.99 + 40.54}{12H} = 4.46\,\%/H$$

$$\frac{14.79 + 29.82}{3H} = 14.87\,\%/H$$

$$\frac{14.87}{4.46 + 14.87} \times 100 = 76.93\%\ \text{epoxide}$$

$$\frac{4.46}{4.46 + 14.87} \times 100 = 23.07\%\ \text{MTBE}$$

Using molecular weights and assuming 1 mol total solution to convert to weight percent,

0.7693 mol × 112.05 g/mol = 86.20 g

0.2307 mol × 88.15 g/mol = 20.34 g

20.34 g + 86.20 g = 106.54 g total

Weight percent:

$$\frac{86.20\,g}{106.54\,g} \times 100 = 80.91\%\ \text{epoxide}$$

$$\frac{20.34\,g}{106.54\,g} \times 100 = 19.09\%\ \text{MTBE}$$

Remembering that from the COA at the time the epoxide was received, it was 12 wt % MTBE, but due to the volatile nature of the epoxide every time the container was opened, it had been concentrating the MTBE to nearly 20% by evaporation of the epoxide. This is most likely why when it was first scaled up, an additional amount of epoxide had to be charged as the first charge was effectively an undercharge due to lower than expected potency of the epoxide.

With these characterized reagent streams, a kinetic model could now be developed for the system. Reactions were set up in small septum-capped vials at three temperatures and two catalysts loadings. The septum caps allowed for sampling of the reaction mixtures without venting the epoxide. The HPLC of these IPC samples showed the fate of the aniline species when reacted with the chiral epoxide. Over time the secondary aniline reacts with the epoxide as the desired reaction but so does the primary aniline according to Scheme 6.2.

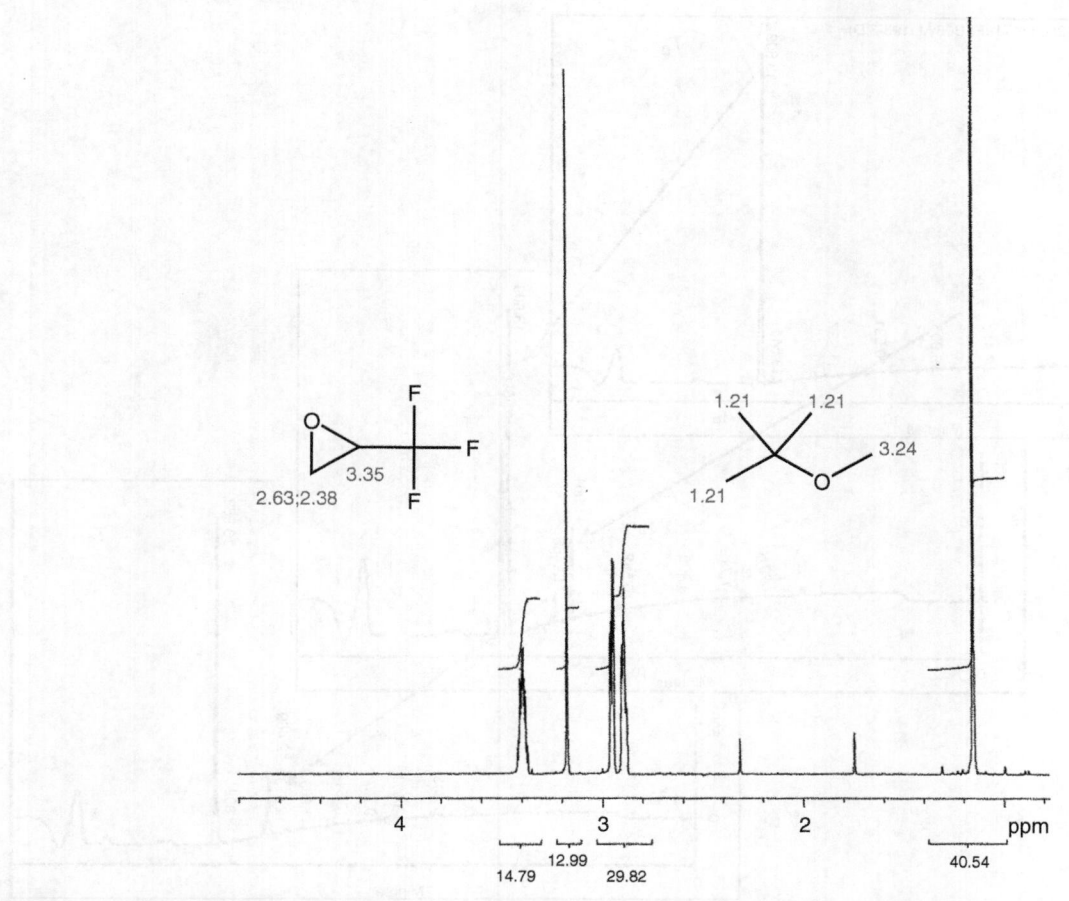

FIGURE 6.6 H¹-NMR of epoxide starting material with three separate peaks each representing 1 proton, while the two singlets at 1.21 ppms (9 protons) and 3.24 (3 protons) indicate the presence of MTBE.

SCHEME 6.2 Undesired reaction pathway for primary aniline with epoxide.

The primary aniline reacts with the epoxide to form impurity 1, which then reacts with a second mole of the epoxide to form impurity 2. This reaction progression is shown in the HPLC traces in Figure 6.7. The shoulder peak on impurity 2 is in fact the diastereomer that is formed when the second chiral epoxide is added. Standard reverse phase HPLC column packing is not capable of separating enantiomers but can separate diastereomers.

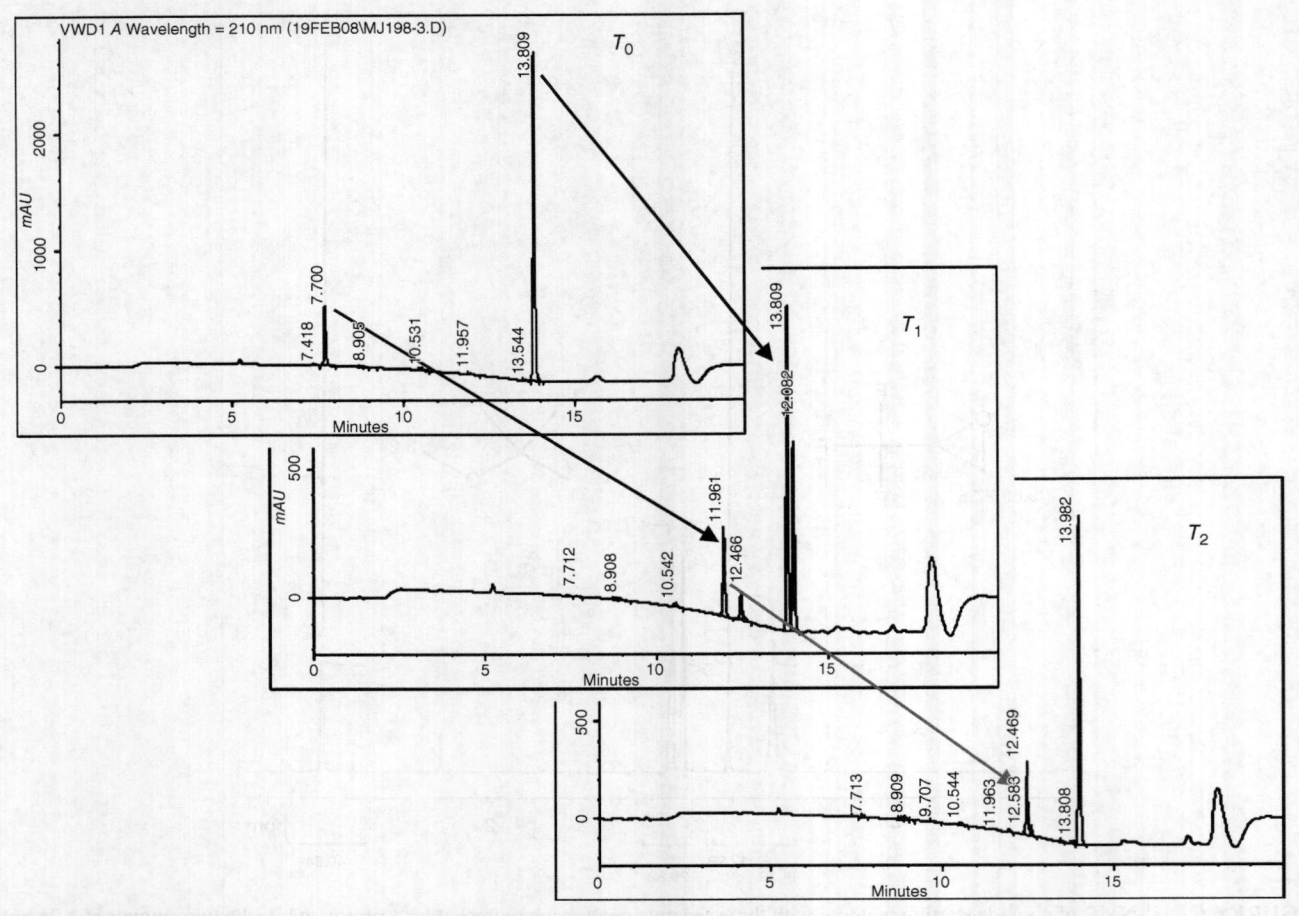

FIGURE 6.7 HPLCs over time showing both desired and undesired reactions. The desired reaction is secondary aniline at 13.8 minutes going to product at 13.9, while the undesired reactions are primary aniline at 7.7 minutes going to impurity 1 at 11.9 minutes which further reacts to impurity 2 at 12.4 minutes.

TABLE 6.3 HPLC Area and Normalized to Each Reaction System

Time (s)	Secondary Aniline Area	Desired Product Area	Secondary Aniline Area	Imp#1 Area	Imp#2 Area	Normalized Area for System Scheme 6.1		Normalized Area for System Scheme 6.2		
0	11 763.66	0.00	2138.86	0.00	0.00	1.00	0.00	1.00	0.00	0.00
720	7 701.32	755.84	636.18	1281.50	40.90	0.91	0.09	0.32	0.65	0.02
4 680	3 884.43	4631.85	23.46	1202.17	864.00	0.46	0.54	0.01	0.58	0.41
9 420	1 771.80	6652.65	15.72	639.57	1432.61	0.21	0.79	0.01	0.31	0.69
16 380	619.75	7895.18	25.21	289.47	1830.73	0.07	0.93	0.01	0.13	0.85
19 380	292.27	6106.68	25.17	160.02	1451.73	0.05	0.95	0.02	0.10	0.89
78 240	15.82	6365.85	0.00	0.00	1626.31	0.00	1.00	0.00	0.00	1.00

In order to calculate the concentration over time, the area% values from the HPLC were used. Each reaction system (Schemes 6.1 and 6.2) had the area% values normalized only for that system. The normalized values were then used to calculate the concentration at that time point by multiplying by the initial starting concentrations. To illustrate this process, Table 6.3 lists the area and normalized values for each reaction system for the 60 °C reaction using the standard catalyst loading.

TABLE 6.4 Concentrations vs. Time for All Components in Schemes 6.1 and 6.2

Time (s)	Secondary Aniline (mol/L)	Desired Product (mol/L)	Secondary Aniline (mol/L)	Imp#1 (mol/L)	Imp#2 (mol/L)
0	7.97E-01	0.00E+00	1.45E-01	0.00E+00	0.00E+00
720	7.25E-01	7.12E-02	4.71E-02	9.48E-02	3.02E-03
4 680	3.63E-01	4.33E-01	1.63E-03	8.33E-02	5.99E-02
9 420	1.68E-01	6.29E-01	1.09E-03	4.44E-02	9.94E-02
16 380	5.80E-02	7.39E-01	1.70E-03	1.95E-02	1.24E-01
19 380	3.64E-02	7.60E-01	2.23E-03	1.42E-02	1.28E-01
78 240	1.97E-03	7.95E-01	0.00E+00	0.00E+00	1.45E-01

TABLE 6.5 The Output Values for the Kinetic Model with the Confidence Interval (C.I.)

	Final Value	Units	C.I. (%)
k_{1ref}	0.0031	$l^2/mol^2 \cdot s$	11.157
k_{2ref}	0.0173	$l^2/mol^2 \cdot s$	7.745
k_{1ref}	0.0029	$l^2/mol^2 \cdot s$	9.346
Ea_1	59.554	kJ/mol	10.953
Ea_2	57.586	kJ/mol	7.521
Ea_3	56.99	kJ/mol	9.932

The data from Table 6.3 was then transformed into concentration data by multiplying the normalized area values for each reaction system to the starting concentrations. The secondary aniline starting concentration of 0.797 M was used for the desired reaction system of Scheme 6.1 and the concentration of the primary aniline of 0.145 M for the undesired reaction system of Scheme 6.2. This resulted in the concentration versus time data shown in Table 6.4.

With this understanding of the reactions involved, the temperature-dependent kinetic model could be developed using DynoChem software as shown in Eq. (6.15):

$$Yb(OTf)_3 + \text{secondary analine} + \text{epoxide} \xrightarrow{k_1} \text{product} + Yb(OTf)_3$$

$$Yb(OTf)_3 + \text{primary analine} + \text{epoxide} \xrightarrow{k_2} \text{imp1} + Yb(OTf)_3$$

$$Yb(OTf)_3 + \text{imp1} + \text{epoxide} \xrightarrow{k_3} \text{imp2} + Yb(OTf)_3 \quad (6.15)$$

where the temperature-dependent rate constants $k_\#$ are defined in Eq. (6.16):

$$k = k_{ref} \times \exp^{-Ea \times R}\left(\frac{1}{T} - \frac{1}{T_{ref}}\right) \quad (6.16)$$

The fit in DynoChem resulted in the values in Table 6.5.

The equations in Eq. (6.15) with the values from Table 6.5 result in the predicted vs. actual plots of reaction progression shown in Figure 6.8. The first reaction of the primary aniline with the epoxide was found to have a rate constant (k_{2ref}) that was an order of magnitude faster than the desired reaction rate constant (k_{1ref}), further explaining why additional epoxide was needed to consume all the of the secondary aniline starting material.

From Case Study 6.1, we see that using the complimentary analytical techniques of NMR and HPLC-UV, it was possible to fully characterize the starting materials. Once this was done, the area% values were used to calculate the concentration over time, which was then used to develop the kinetic model that gave the necessary process understanding. The assumption in this case was that all the species in a reacting system, i.e. secondary aniline to the desired product for one system and primary aniline to impurity 1 onto impurity 2 for the other reactive system, had the same response factor and that area% could be used without response factors. This is a reasonable assumption to make as the epoxide that was being added to the molecules did not have a chromophore, and the electronics of the UV chromophore between starting materials and products were not changing much, i.e. the response of primary aniline differs slightly from that of impurity 1 and 2, as did the response of secondary aniline to the desired product. But what if this assumption about starting materials and products having the same response cannot be made, what is the course of action? This problem is explored in Case Study 6.2.

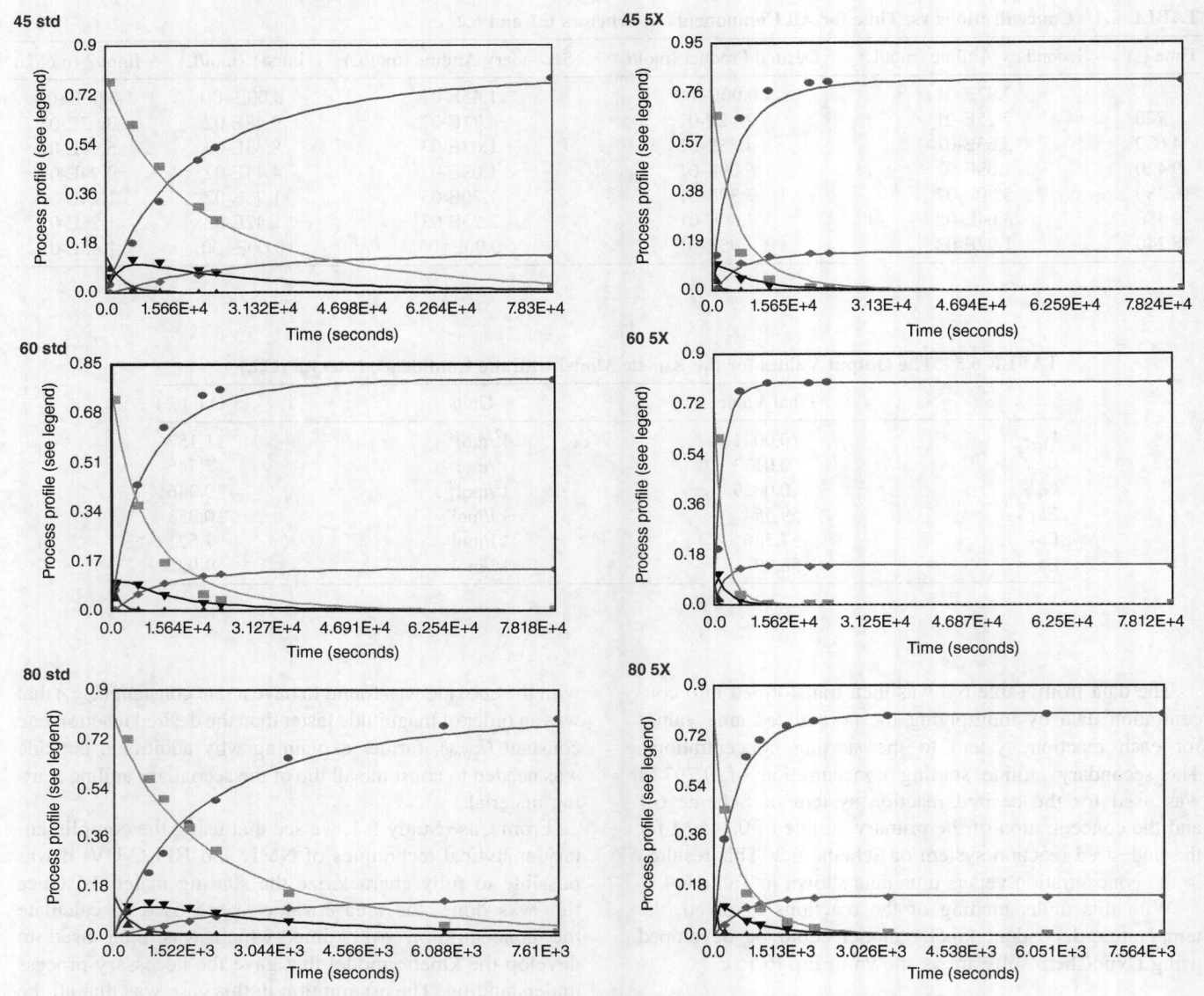

FIGURE 6.8 Predicted vs. actual values for reaction progression from kinetic model.

CASE STUDY 6.2: PROCESS UNDERSTANDING FOR DEVELOPMENT OF CONTINUOUS PROCESS

Two reactions in series need to have CSTR reactors sized for a given annual throughput. In order to do this reactor sizing, absolute reaction rates as a function of temperature are needed. These reaction rates have to track the impurity levels throughout the process, not only the desired reaction, so as to arrive at an optimum reactor configuration. In the first reaction, referred to as reaction A, "feed" reacts with the "starting material" forming "product" and a series of impurities. This system is further complicated as the reagent "feed" can exist as two different tautomers, with only one of which is reactive. The six reactions in this system are shown in Eq. (6.17):

$$\begin{aligned}
&\text{"Feed"} + \text{TEA} = \text{"Feed"}^* + \text{TEA} \\
&(\text{Feed and Feed}^* \text{ are tautomers}) \\
&\text{Starting material} + \text{"Feed"}^* \rightarrow \text{Product} \\
&\text{"Feed"}^* + \text{TEA} \rightarrow \text{decomp} \\
&\text{Product} + \text{"Feed"}^* \rightarrow \text{ImpA} \\
&\text{ImpA} \rightarrow \text{ImpB} \\
&\text{Product} + \text{H}_2\text{O} \rightarrow \text{Hydrolysis product}
\end{aligned} \qquad (6.17)$$

In this case HPLC data was available with comparison to external standard response factors to arrive at wt % of each species. In order to get this wt % data, all reaction samples

ANALYTICAL ASPECTS FOR DETERMINATION OF MASS BALANCES

were made up as quantitative samples, i.e. mg/mg reaction in samples in volumetric flasks.

With these data it is possible to now close the mass balance around the incoming limiting reagent "starting material." This mass balance at time t was calculated by comparing with initial starting material (SM_0) with Eq. (6.18):

$$\frac{SM_t + prod_t + impA_t + impB_t + Hydrolysis\_prod_t}{SM_0} \times 100 = \% \text{measured mass balance} \quad (6.18)$$

Calculations using Eq. (6.18) were performed for every sample taken from the reaction over the course of the reaction to give the mass balance. Step A mass balance around the starting material including the impurities that are being tracked is shown in Figure 6.9. This mass balance illustrates that for three reactions at three different temperatures and over the course of each reaction, the mass balance fluctuates near 100%. In this case the mass balance is not being forced to 100% by taking ratios, but is calculated from external standards. What we can conclude from this data is that there is not an unaccounted for reaction as this would cause a systematic drain on the system as a function of temperature or over the course of the reaction. The variability of the mass balance around the 100% point is most likely due to variability in sampling.

The data can be smoothed by reprocessing with relative molar response factors, resulting in smoothed data that has had the sampling error removed. This is done by setting one of the compounds, usually the starting material or product, as the reference and having a relative response factor of one. The other compounds then have their response factors calculated as a fraction of this reference response factor by Eq. (6.19):

$$\frac{R_{f_{comp}}}{R_{f_{product}}} = R_{f_{relative}} \quad (6.19)$$

The relative response factors can now be used to adjust the individual components area values. These areas are then summed and used to calculate new response-corrected area % values. These area% values are then used in conjunction with the starting material concentration at time zero to calculate individual component concentrations.

An example of these calculations for one time point is shown below for the reaction where B and C are reacted together, with C in excess to give product A and impurities D and E shown in Table 6.6.

The fraction of T_0 concentration is calculated by taking the relative response factor-corrected areas summed together, resulting in 1 747 786.85 total area counts. The subtle point now is to make sure that the reagent in excess is not double counted. So product and impurities fraction of T_0 concentration values are calculated as shown in Eq. (6.20):

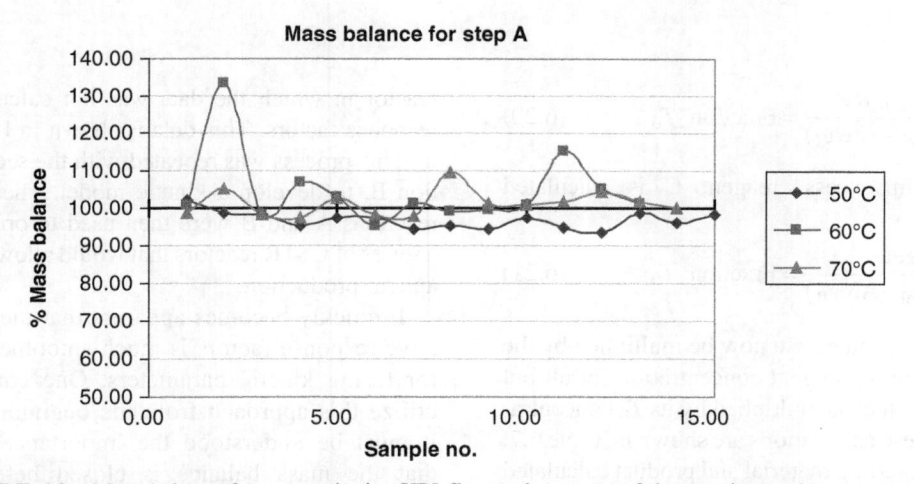

FIGURE 6.9 Mass balance from quantitative HPLC over the coarse of the reaction at three temperatures.

TABLE 6.6 Relative Response Factors Used to Smooth the Data by Converting Area into Fraction T_0 Concentration

Component	A	B	C	D	E
Relative response factor	1	0.441 5	0.262 8	0.4173	1.9011
Area	980 582.93	138 507.25	118 161.23	1125.98	2110.25
RRF corrected area	980 582.93	313 730.71	449 665.07	2698.10	1110.04
Fraction of T_0 [ ]	0.755	0.242	0.314	0.002	0.001

TABLE 6.7 Final Conversion of T_0 Concentration to Concentration at This Time Point

Component	A	B	C	D	E
T_0 [ ]	0	0.512	0.603	0	0
Fraction of T_0 [ ]	0.755	0.242	0.314	0.002	0.001
[ ] at this time point	3.869E-01	1.238E-01	1.890E-01	1.065E-03	4.380E-04

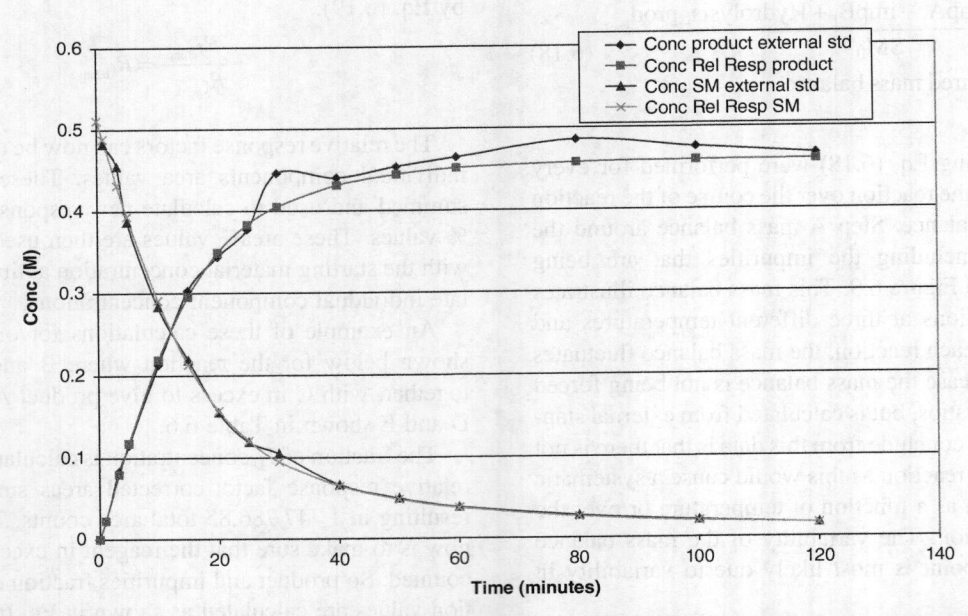

FIGURE 6.10 Starting material and product profiles comparing external standards (triangle) and relative responds factors (square). The effect of smoothing the data and indicating where possible errors in sampling may have occurred is relatively straightforward once displayed graphically.

$$\frac{Area_{A,B,D,E}}{(Area_{total} - Area_C)} = Fraction\_T_0 \quad (6.20)$$

where the reagent in excess, reagent C is calculated by Eq. (6.21):

$$\frac{Area_C}{(Area_{total} - Area_B)} = Fraction\_T_0 \quad (6.21)$$

These fractions of T_0 values can now be multiplied by the T_0 concentrations (limiting reagent concentration for all but the reagent in excess, which is multiplied by its T_0 concentration). The results of these calculations are shown in Table 6.7.

The profiles of the starting material and product calculated with external standards versus calculated with relative response factors are shown in Figure 6.10. This smoothed data was then used to develop the kinetic model with Dyno-Chem software package. The model versus predicted from this model is shown in Figure 6.11.

The model for step A was then validated by comparing the model to a semi-batch reaction done in a Mettler 0.5L RC-1 reactor in which the data was not calculated with relative response factors. This data is shown in Figure 6.12.

This process was repeated with the second reaction, reaction B, to develop a kinetic model. These models for both reactions A and B were then used to optimize a design for a series of CSTR reactors that would allow for the appropriate annual production.

It quickly becomes apparent that the data from the relative response factors is much smoother and better suited for fitting kinetic parameters. One could be tempted to utilize this approach from the beginning of the analysis. It must be understood the importance of first verifying that the mass balance is closed before using relative response factors, as the use of relative response factors is a normalization of the data that forces the closure of the mass balance. If there had been a secondary reaction pathway that was not accounted for, then the kinetic model would have not represented the process, and any reactor configuration that was designed would not have performed as expected.

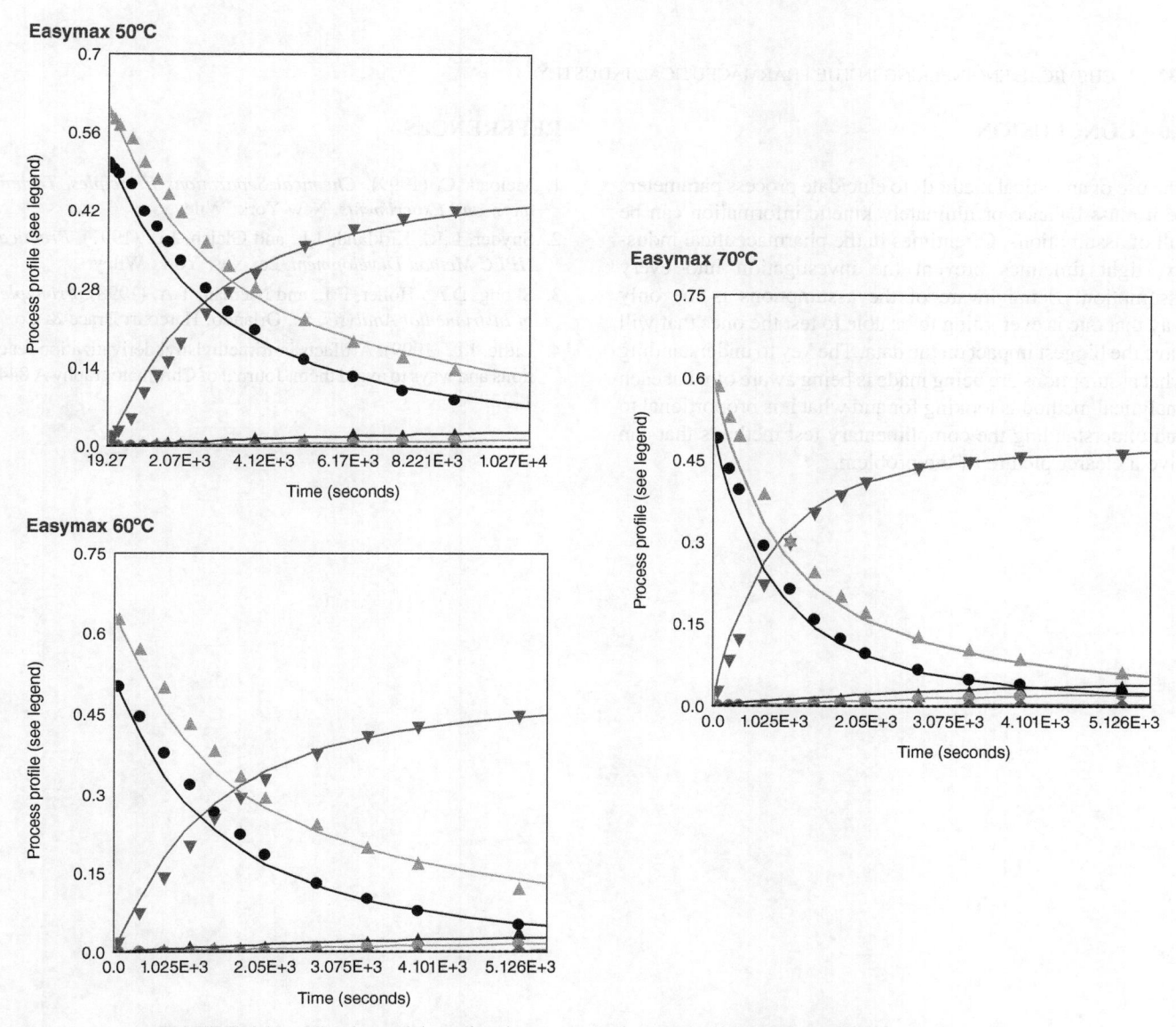

FIGURE 6.11 Kinetic model predicted versus actual concentrations at three different temperatures.

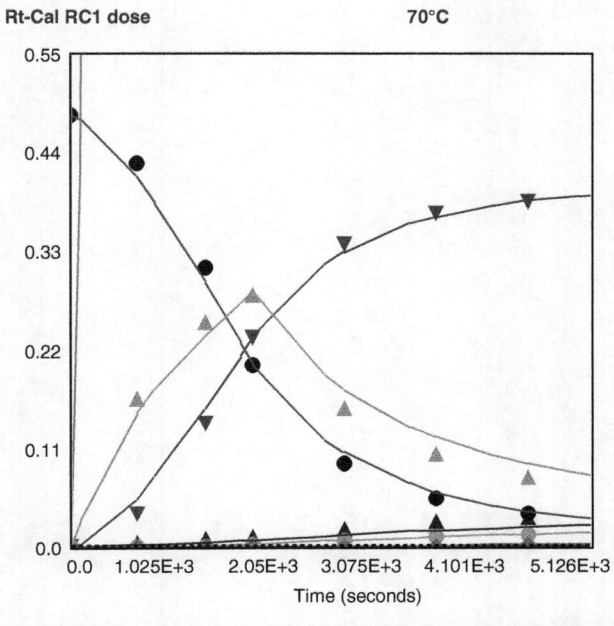

FIGURE 6.12 Validation semi batch reaction with 30 addition done in Mettler RC-1 reactor.

6.6 CONCLUSION

The use of analytical methods to elucidate process parameters be it mass balance or ultimately kinetic information can be full of assumptions. Oftentimes in the pharmaceutical industry, tight timelines prevent the investigation into every assumption. Being aware of the assumptions is the only way that one is ever going to be able to test the ones that will have the biggest impact on the data. The key to understanding what assumptions are being made is being aware of what each analytical method is looking for and what it is proportional to and understanding the complimentary test methods that can give a clearer picture of the problem.

REFERENCES

1. Meloan, C. (1999). *Chemical Separations Principles, Techniques, and Experiments*. New York: Wiley.
2. Snyder, L.R., Kirkland, J.J., and Glajch, J.L. (1997). *Practical HPLC Method Development*, 2e. New York: Wiley.
3. Skoog, D.A., Holler, F.J., and Nieman, T.A. (1998). *Principles of Instrumental Analysis*, 5e. Orlando: Harcourt Brace & Co.
4. Little, J.L. (1999). Artifacts in trimethylsilyl derivatization reactions and ways to avoid them. Journal of Chromatography A 844: 1–22.

7

QUANTITATIVE APPLICATIONS OF NMR SPECTROSCOPY

BRIAN L. MARQUEZ*
Pfizer, Inc., Groton, CT, USA

R. THOMAS WILLIAMSON**
Roche Carolina, Inc., Florence, SC, USA

7.1 INTRODUCTION

7.1.1 General Principles of NMR

Nuclear magnetic resonance (NMR) is an analytical method that takes advantage of the magnetic properties of certain atomic nuclei. This approach is similar to other types of spectroscopy in that the absorption or emission of electromagnetic energy at characteristic frequencies provides analytical information. However, NMR differs from other types of spectroscopy in that the discrete energy levels and the transitions between them are created by placing the samples in a strong magnetic field (B_0).

When an atom is placed in a magnetic field, its electrons circulate about the direction of the applied magnetic field. The circulation of these nuclei generates a very small magnetic field, which is generally on the order of 1–20 ppm of the total applied magnetic field for ^1H and 1–200 ppm for ^{13}C. This field opposes the applied magnetic field and can be detected through the same R_f coil that is used to excite the nuclei of interest. When nuclei spin about the axis of this externally applied magnetic field, they possess an angular momentum. This angular momentum can be expressed as a function of a proportionality constant, I, which can be either an integer or half-integer. I is referred to as the spin quantum number or more simply as the nuclear spin. It is possible for some isotopes to have a spin quantum number $I = 0$. These nuclei are not considered magnetic and cannot be detected

*Current address: Nalas Engineering Services, Inc., Centerbrook, CT, USA
**Current address: Department of Chemistry & Biochemistry, University of North Carolina Wilmington, Wilmington, NC, USA

by NMR. In order for a nuclei to have a spin quantum $I = 0$, it must have an even atomic number and even mass. Commonly occurring non-NMR active nuclei include ^{12}C, ^{16}O, and ^{32}S. The group of nuclei most commonly observed by NMR methods is nuclei with a spin of ½. These include ^1H, ^{13}C, ^{19}F, ^{31}P, and ^{15}N. Other spins with $I = 1$ or $I > 1$ nuclei can be observed with slightly more difficulty. These nuclei include ^2H and ^{14}N ($I = 1$) and ^{10}B, ^{11}B, ^{17}O, and ^{23}Na ($I > 1$) (Figure 7.1).

The rate at which a particular nuclei spins in a particular magnetic field is known as its precession frequency. This frequency is both a function of the externally applied field, the nucleus of interest, and the environment in which it resides. For a proton (^1H), in an applied 2.35 T magnetic field, the reference precession frequency is approximately 100 MHz. In the same externally applied field, other nuclei have different gyromagnetic ratios. For example, ^{13}C with a gyromagnetic ratio of 4 will precess at 25 MHz in the same magnetic field. This characteristic precession frequency is known as the Larmor frequency of the nucleus. An NMR sample may contain many different magnetization components, each with its own Larmor frequency. Therefore, an NMR spectrum may be made up of many different frequency lines.

Nuclei aligned with the axis of the externally applied magnetic field will be in the lowest possible energy state. Thermal processes oppose this tendency, such that there are two populations of nuclei in an externally applied magnetic field. One is aligned with the axis of the field, and another, which is only slightly smaller, is aligned opposite to the direction of the applied field. The distribution of spins between these two energy levels is referred to as the Boltzmann distribution,

Chemical Engineering in the Pharmaceutical Industry: Active Pharmaceutical Ingredients, Second Edition.
Edited by David J. am Ende and Mary T. am Ende.
© 2019 John Wiley & Sons, Inc. Published 2019 by John Wiley & Sons, Inc.

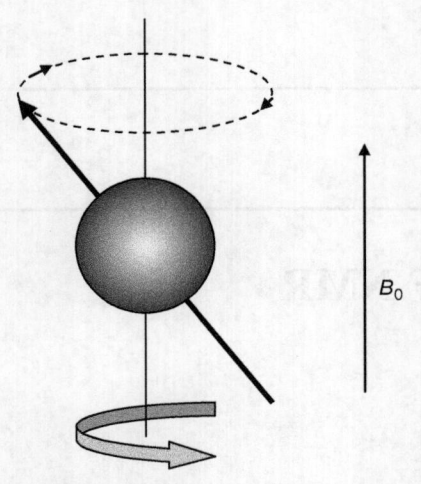

FIGURE 7.1 A spinning magnetic nuclei in an externally applied magnetic field with its axis precessing around the direction of the applied field (B_0) analogous to a gyroscope.

STRUCTURE 7.1 Santonin.

and it is this population difference between the two levels that provides the observable collection of spins in an NMR experiment. This difference is very small compared to the total number of spins present, which leads to the comparative inherent insensitivity of NMR.

In the modern NMR experiment, pulsed radio frequency (R_f) energy is used to excite all frequencies at once. In a simplistic sense, a certain amount of energy from an R_f pulse is absorbed by each nuclei. As these nuclei relax back from the excited state to the ground state, a corresponding amount of R_f energy is emitted. The frequency and the amplitude of this emitted energy contain important information about the nucleus from where it originated. In other words, the application of a radio frequency (R_f) pulse of energy orthogonal to the axis of the applied magnetic field perturbs the Boltzmann distribution, thereby producing an observable event that is governed by the Bloch equations. When the application of that pulse is completed, the vector that has been rotated into the x,y plane will continue to precess about the axis of the externally applied field (z-axis), generating an oscillating signal in a receiver coil as the vector rotates about the z-axis at its characteristic Larmor frequency. The signal from the magnetization vector will decay back to an equilibrium condition along the B_0 axis as a function of two time constants, the spin–spin or transverse relaxation time, T_2, and the spin–lattice relaxation time, T_1. Once equilibrium is reestablished after a short delay, the process can be repeated, and multiple acquisitions can be added to increase S/N. Immediately after the original R_f pulse, a receiver is turned on, and a signal known as a time-domain interferogram is acquired via an R_f receiver. With the help of an analogue to digital convertor, this data is saved to a computer. This so-called interferogram contains information on all signals emitted by the sample at various NMR frequencies. This information is present as a sum of all damped sinusoid signals emitted by the sample at various nuclear resonance frequencies. The specific resonance frequencies of these signals vary by the strength of the corresponding magnetic field. For instance, at 11.7 Tesla (T) the ^{1}H resonance frequency is approximately 500 MHz, and at 18.8 T it is 800 MHz. The collected interferogram is called a free induction decay (FID). An example of FID can be found in Figure 7.2.

Once acquired and saved, these data are converted from the time domain to frequency data to make them more meaningful and easier to interpret for the end user. This conversion is typically done through a mathematical manipulation known as the Fourier transform named for the mathematician Jean-Baptiste Joseph Fourier. This mathematical transformation also provides signal intensity information that is a key to the usefulness of NMR data. Typically, in common practice, a derivation of the Fourier transform is done by computer using the Cooley–Tukey fast Fourier algorithm otherwise known as the fast Fourier transform (FFT).

The frequency of the R_f energy absorbed by a particular nucleus is strongly affected by its chemical environment. These variables in the chemical environment can lead to changes in its so-called chemical shift. Three basic components of an NMR spectrum reveal its extreme usefulness. These include (i) the chemical shift, (ii) the amplitude of the signal at that chemical shift, and (iii) the splitting of the signal in response to its interaction with neighboring nuclei.

The location of an NMR signal in a spectrum is known as the signal's chemical shift. The location of this chemical shift is a function of the chemical environment of the sampled nuclei and is designated with a scale value referenced to an internal standard. For solution-state ^{1}H and ^{13}C NMR experiments, the reference standard is tetramethylsilane, or TMS, which has an accepted chemical shift of 0.00 ppm Most proton signals appear to the left or "downfield" of TMS. Aliphatic hydrocarbon signals will generally be grouped nearer the position of TMS and are said to be "shielded" relative to vinyl or aromatic signals that are as a group referred to as "deshielded" (Figure 7.3). Chemical moieties involving heteroatoms, e.g. –OCH$_3$, –NCH$_2$–, typically will be located in a region of the NMR spectrum between the aliphatic and vinyl/aromatic signals.

In addition to chemical shift information, an NMR spectrum may also contain scalar coupling information. For

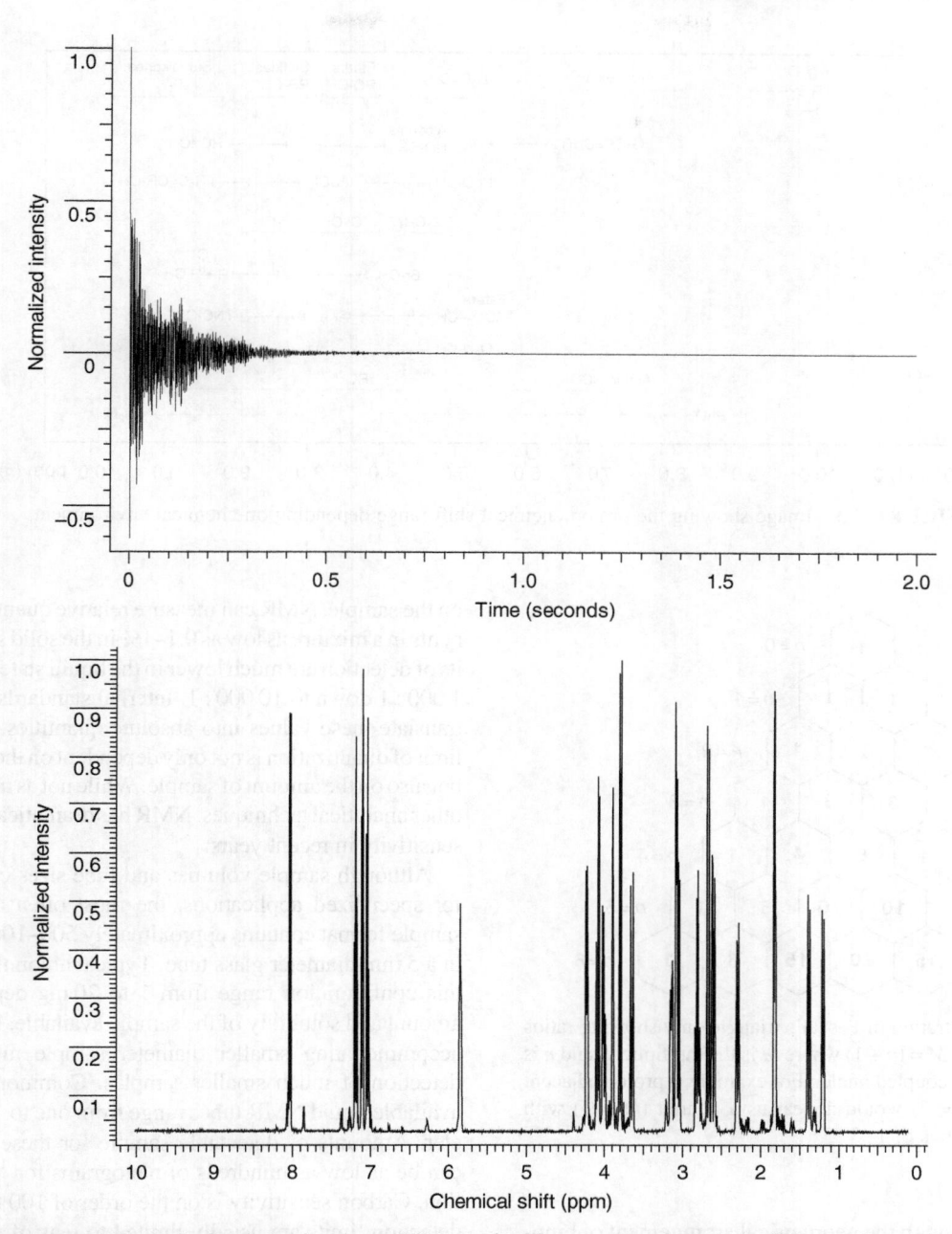

FIGURE 7.2 The top panel shows the free induction decay (FID) acquired for a sample of strychnine [1] at an observation frequency of 500 MHz. The spectrum was digitized with 16 K points and an acquisition time of approximately 2 seconds. Fourier transforming the data from the time domain to the frequency domain yields the spectrum of strychnine presented as intensity vs. frequency shown in the bottom panel.

STRUCTURE 7.2 Astemizole.

NMR of liquids, scalar (J) couplings in proton spectra provide information about the local chemical environment of a given proton resonance. Proton resonances are split into multiplets related to the number of neighboring protons. For example, an ethyl fragment will be represented by a triplet with relative peak intensities of 1 : 2 : 1 for the methyl group, the splitting due to the two neighboring methylene protons, and a 1 : 3 : 3 : 1 quartet for the methylene group, with the splitting due to the 3 equiv. methyl protons. This concept

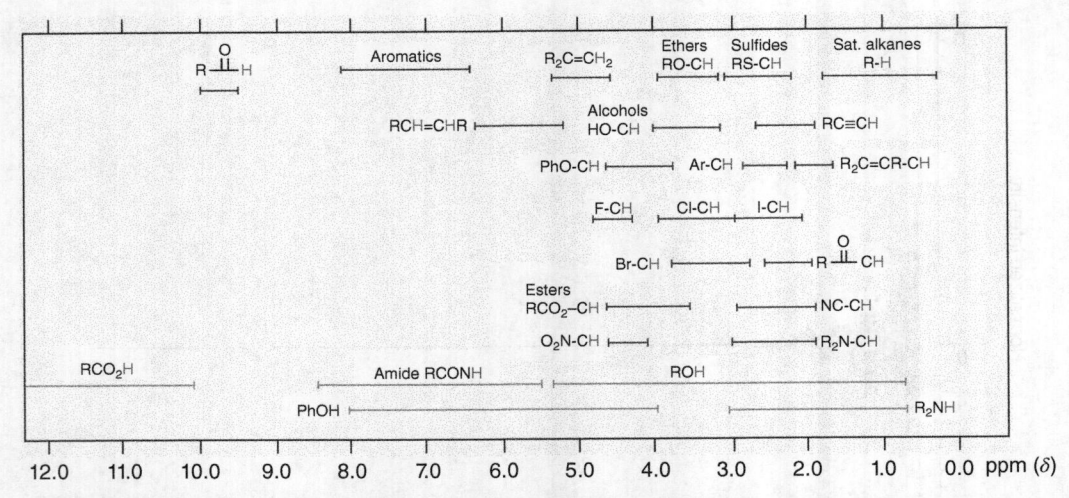

FIGURE 7.3 Image showing the proton chemical shift range depending on chemical environment.

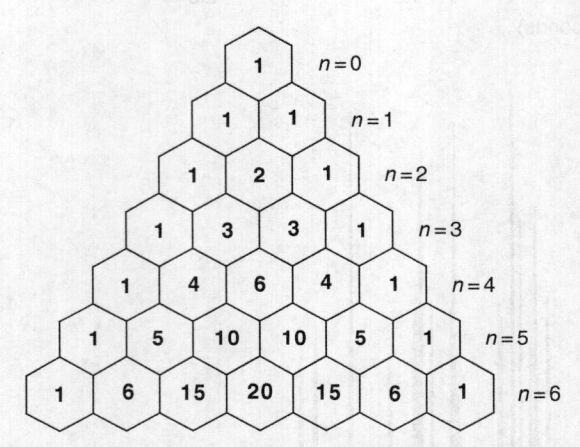

FIGURE 7.4 Illustration of Pascal's triangle only showing ratios to $n = 6$ according to $M = (n + 1)$ where M is the multiplicity and n is the number of scalar coupled nuclei. For example, a proton adjacent to three protons ($n = 3$) would appear as a quartet ($M = 4$) with relative peak intensities of $1 : 3 : 3 : 1$.

is easily illustrated with the geometrical arrangement of binomial coefficients known as Pascal's triangle shown in Figure 7.4. More complex molecules, of course, lead to considerably more complicated spin-coupling patterns as shown in Figure 7.5.

Prior to the advent of homonuclear two-dimensional (2D) NMR experiments, it was necessary to rigorously interpret a proton NMR experiment and identify all of the homonuclear couplings to assemble the structure. Alternatively, there are multidimensional NMR experiments that provide similar information in a more readily interpretable way. These techniques will be discussed in more detail later.

Beyond the qualitative molecular information afforded by NMR, one can also obtain quantitative information. Depending on the sample, NMR can measure relative quantities of components in a mixture as low as 0.1–1% in the solid state. NMR limits of detection are much lower in the liquid state, often as low as 1 000 : 1 down to 10 000 : 1. Internal standards can be used to translate these values into absolute quantities. Of course, the limit of quantization is not only dependent on the type of sample but also on the amount of sample. While not as mass sensitive as other analytical techniques, NMR has dramatically improved in sensitivity in recent years.

Although sample volumes and tube sizes can vary greatly for specialized applications, the most often analyzed NMR sample format contains approximately 500–1000 µl of solvent in a 5 mm diameter glass tube. Typical amounts of sample for this configuration range from 1 to 20 mg depending on the amount and solubility of the sample available. NMR hardware accommodating smaller diameter sample tubes enable the detection of much smaller samples. Common commercially available liquid NMR tubes range from one to 11 mm in diameter. Amounts of detectable sample for these configurations can be as low as hundreds of nanograms for ^{1}H NMR detection. Carbon sensitivity is on the order of 100 times worse, so detection limits are usually limited to tens of micrograms.

Essentially all magnetic nuclei can be observed by NMR, but the sensitivity of each is determined by the relative sensitivity and the natural abundance of the nuclei. For example, not only is ^{1}H (with a spin of ½) one of the most sensitive nuclei, but it also has a natural abundance of 99.985%. Alternatively, ^{13}C is fairly sensitive but only has a natural abundance of 1.108%. Other nuclei like ^{15}N (spin $\mathbf{I} = -1/2$) have both low natural abundance (0.37%) and low sensitivity, making them doubly difficult to detect. These types of nuclei are very difficult to observe directly and are most commonly detected using what is known as "inverse detection" through a more sensitive nucleus. This technique can dramatically increase sensitivity for insensitive nuclei and can make

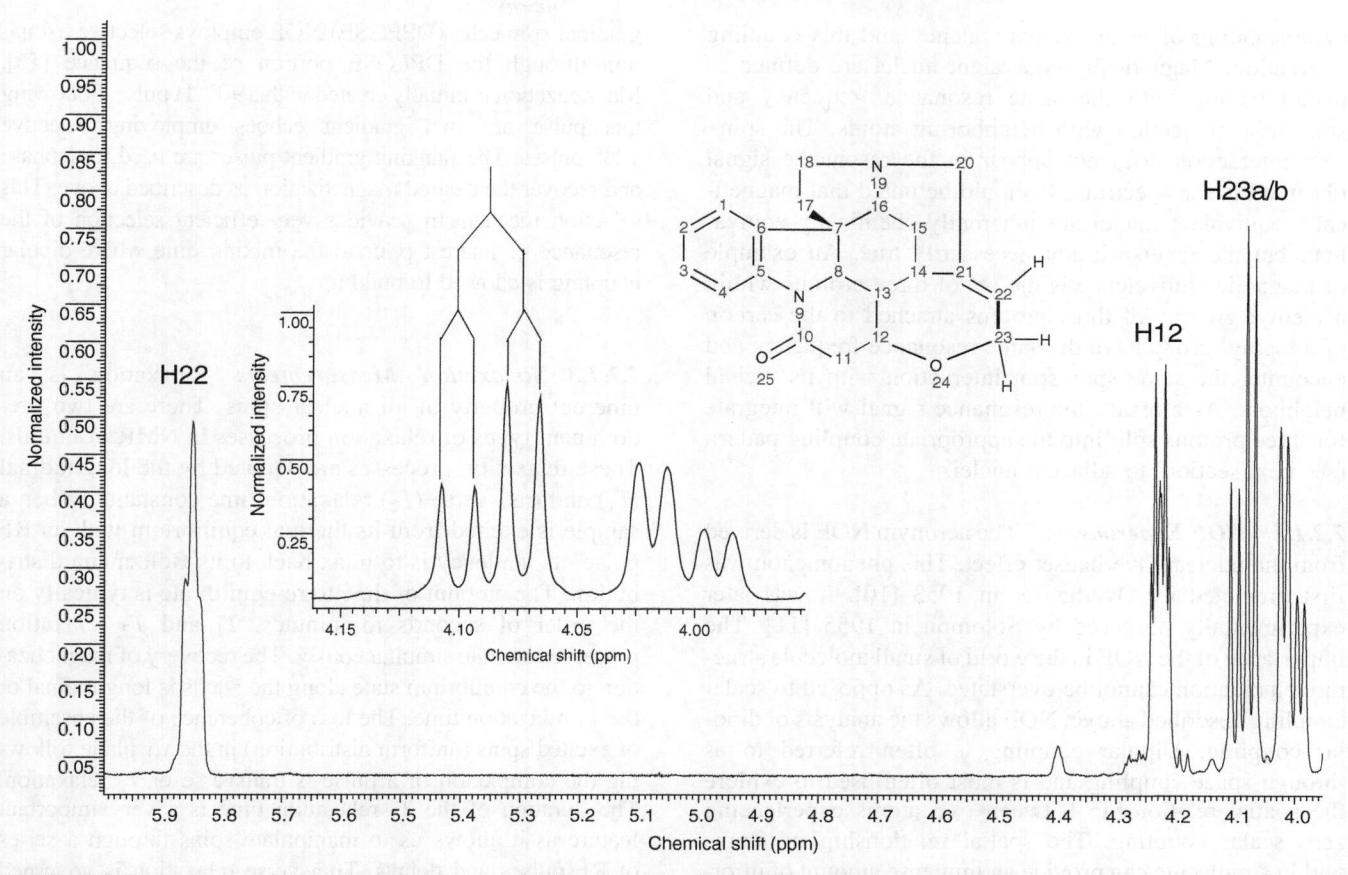

FIGURE 7.5 Expansion of a portion of the proton NMR spectrum of strychnine (inset structure). The full proton spectrum is shown in Figure 7.1. The resonances for the H22 vinyl proton and the H12 and H23 oxygen-bearing methine and methylene resonances, respectively, are shown. The inset expansion of the H23 methylene protons shows a splitting diagram for this resonance. The larger of the two couplings is the geminal coupling to the other H23 resonance and the smaller coupling is the vicinal coupling to the H22 vinyl proton.

available chemical shift and coupling information available that otherwise would have never been possible with existing technologies.

For the NMR analysis of solid samples, the amount of sample required is much greater in part because the apparent signal-to-noise ratio is significantly reduced. This is due to much broader line shapes, easily an order of magnitude wider than those observed in equivalent spectra of liquids. A standard solid NMR sample is a powder packed tightly into a small zirconia rotor and sealed with end caps. As for liquids, solid sample configurations are described in terms of their diameter, in this case the diameter of the sample rotor. Common commercially available rotors range from 2.5 to 11 mm in diameter. Amounts of sample for these configurations depend on the sample and its density and typically range from 30 to 500 mg.

In common practice, most ^{1}H NMR spectra are recorded in solutions prepared with deuterated solvents. The advantages of this customary practice of sample preparation are twofold. Firstly, the lack of protons in the solvent enables observation of the protons of interest without interference from solute protons. Secondly, the presence of deuterium in the sample allows the NMR instrument to "lock" onto a reference frequency outside of those frequencies being observed. The ability to lock on a signal as the magnetic field drifts slightly over time can substantially improve the quality of spectra in experiments with long acquisitions. This can be very important when observing small amounts of sample or when the highest quality spectra with narrow lines are required. References [1–9] are listed as suggested reading for a deeper understanding of the theory, application, and experimental setup of many commonplace experiments [1–9].

7.2 ONE-DIMENSIONAL NMR METHODS

7.2.1 1D Proton NMR Methods

7.2.1.1 Magnetically Equivalent Nuclei A critical element to one of the fundamental phenomenon that is used for structure characterization is that of *J*-coupling (spin–spin coupling, scalar coupling, or often referred to as simply "coupling"). Of particular importance is the understanding of magnetic equivalency and this coupling interaction. Magnetically equivalent nuclei are defined as nuclei having both the same resonance frequency and spin–spin interaction with neighboring atoms. The spin–spin interaction does not appear in the resonance signal observed in the spectrum. It should be noted that magnetically equivalent nuclei are inherently chemically equivalent, but the reverse is not necessarily true. An example of magnetic equivalence is the set of three protons within a methyl group. All three protons attached to the carbon of a methyl group have the same resonance frequency and encounter the same spin–spin interaction with its vicinal neighbors. As a result, the resonance signal will integrate for three protons split into the appropriate coupling pattern (see next section) to adjacent nuclei.

7.2.1.2 NOE Experiments

The acronym NOE is derived from the nuclear Overhauser effect. This phenomenon was first predicted by Overhauser in 1953 [10]. It was later experimentally observed by Solomon in 1955 [11]. The importance of the NOE in the world of small molecule structure elucidation cannot be overstated. As opposed to scalar coupling described above, NOE allows the analysis of dipolar coupling. Dipolar coupling is often referred to as through-space coupling and is most often used to explore the spatial relationship between two atoms experiencing zero scalar coupling. The spatial relationship of atoms within a molecule can provide an immense amount of information about a molecule, ranging from the regiochemistry of an olefin to the three-dimensional solution structure.

A simplified explanation of the NOE is the magnetic perturbations induced on a neighboring atom, resulting in a change in intensity. This is usually an increase in intensity but may alternatively be zero or even negative. One can envision saturating a particular resonance of interest for a time t. During the saturation process a population transfer occurs to all spins that feel an induced magnetization.

There are two types of NOE experiments that can be performed. These are referred to as the steady-state NOE and the transient NOE. The steady-state NOE experiment is exemplified by the classic NOE difference experiment [12]. Steady-state NOE experiments allow one to quantitate relative atomic distances. However, there are many issues that can complicate their measurement, and a qualitative interpretation is more reliable [13]. Spectral artifacts can be observed from imperfect subtraction of spectra. In addition, this experiment is extremely susceptible to inhomogeneity issues and temperature fluctuations.

One-dimensional (1D) transient NOE experiments employing gradient selection are more robust and therefore are more reliable for measuring dipolar coupling interactions [14]. Shaka et al. published one such 1D transient NOE experiment that has, in most cases, replaced the traditional NOE difference experiment [15]. The sequence dubbed the double pulsed field gradient spin echo (DPFGSE) NOE employs selective excitation through the DPFGSE portion of the sequence [16]. Magnetization is initially created with a 90° ^{1}H pulse. Following this pulse are two gradient echoes employing selective 180° pulses. The flanking gradient pulses are used to dephase and recover the desired magnetization as described above. This selection mechanism provides very efficient selection of the resonance of interest prior to the mixing time where dipolar coupling is allowed to build up.

7.2.1.3 Relaxation Measurements

Relaxation is an inherent property of all nuclear spins. There are two predominant types of relaxation processes in NMR of liquids. These relaxation processes are denoted by the longitudinal (T_1) and transverse (T_2) relaxation time constants. When a sample is excited from its thermal equilibrium with an RF pulse, its tendency is to relax back to its Boltzmann distribution. The amount of time to re-equilibrate is typically on the order of seconds to minutes. T_1 and T_2 relaxation processes operate simultaneously. The recovery of magnetization to the equilibrium state along the z-axis is longitudinal or the T_1 relaxation time. The loss of coherence of the ensemble of excited spins (uniform distribution) in the x,y plane following the completion of a pulse is transverse or T_2 relaxation. The duration of the T_1 relaxation time is a very important feature as it allows us to manipulate spins through a series of RF pulses and delays. Transverse relaxation is governed by the loss of phase coherence of the precessing spins when removed from thermal equilibrium (e.g. an RF pulse). The transverse or T_2 relaxation time is visibly manifested in an NMR spectrum in the line width of resonances; the line width at half height is the reciprocal of the T_2 relaxation time. In liquids NMR many of the same physical mechanisms dictate these two rate constants and in most cases are equal. These two relaxation mechanisms can provide very important information concerning the physical properties of the molecule under study, tumbling in solution, binding or interaction with other molecules, etc.

T_1 relaxation measurements provide information concerning the time constant for the return of excited spins to thermal equilibrium. For spins to fully relax, it is necessary to wait a period of five times T_1. To accelerate data collection, in most cases one can perform smaller flip angles than 90° and wait a shorter time before repeating the pulse sequence. Knowing the value of T_1 proves to be very useful in some instances, and it is quite simple to measure. The pulse sequence used to perform this measurement is an inversion recovery sequence [17]. The basic linear sequence of RF pulses (an NMR pulse sequence) consists of a 180-τ-90-acquire. Knowledge of the T_1 is of paramount importance when looking to increase the accuracy of quantitative NMR (qNMR) experiments. This will be described in greater detail below in the section on qNMR.

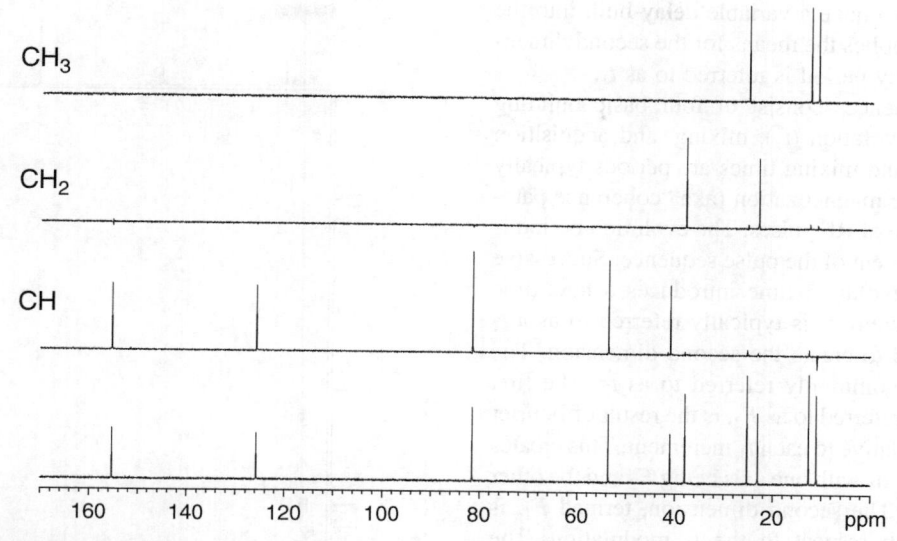

FIGURE 7.6 Multiplicity edited DEPT traces for the methine, methylene and methyl resonances of santonin [1]. Quaternary carbons are excluded in the DEPT experiment and must be observed in the ^{13}C reference spectrum or through the use of another multiplicity editing experiment such as APT.

7.2.2 1D Carbon NMR Methods

7.2.2.1 Proton Decoupling
Carbon-13, or ^{13}C, is a rare isotope of carbon with a natural abundance of 1.13% and a gyromagnetic ratio, γ_C, that is approximately one quarter that of ^1H. Early efforts to observe ^{13}C NMR signals were hampered by several factors. First, the 100% abundance of ^1H and the heteronuclear spin coupling, $^nJ_{CH}$ where $n = 1$–4, split the ^{13}C signals into multiplets, thereby making them more difficult to observe. The original efforts to observe ^{13}C spectra were further hampered by attempts to record them in the swept mode, necessitating long acquisition times and computer averaging of scans. These limitations were circumvented, however, with the advent of pulsed Fourier transform NMR spectrometers with broadband proton decoupling capabilities [18].

Broadband ^1H decoupling, in which the entire proton spectral window is irradiated, collapses all of the ^{13}C multiplets to singlets, vastly simplifying the ^{13}C spectrum. An added benefit of broadband proton decoupling is NOE enhancement of protonated ^{13}C signals by as much as a factor of three.

Early broadband proton decoupling was accomplished by noise modulation that required considerable power, typically 10 W or more, and thus caused significant sample heating. Over the years since the advent of broadband proton decoupling methods, more efficient decoupling methods have been developed including GARP, WURST, and others [19]. The net result is that ^{13}C spectra can now be acquired when needed with low power pulsed decoupling methods and almost no sample heating.

7.2.2.2 Standard 1D Experiments
Of the multitude of 1D ^{13}C NMR experiments that can be performed, the two most common experiments are a simple broadband proton-decoupled ^{13}C reference spectrum, and a DEPT sequence of experiments [20]. The latter, through addition and subtraction of data subsets, allows the presentation of the data as a series of "edited" experiments containing only methine, methylene, and methyl resonances as separate subspectra. Quaternary carbons are excluded in the DEPT experiment and can only be observed in the ^{13}C reference spectrum or by using another editing sequence such as APT [21]. The individual DEPT subspectra for CH, CH_2, and CH_3 resonances of santonin (**1**) are presented in Figure 7.6.

7.3 TWO-DIMENSIONAL NMR METHODS

7.3.1 Basic Principles of 2D

2D NMR methods are highly useful for structure elucidation. Jeener described the first 2D NMR experiment in 1971 [22]. In standard NMR nomenclature, a data set is referred to by one less than the total number of actual dimensions, since the intensity dimension is implied. The 2D data matrix therefore can be described as a plot containing two frequency dimensions. The inherent third dimension is the intensity of the correlations within the data matrix. This is the case in "1D" NMR data as well. The implied second dimension actually reflects the intensity of the peaks of a certain resonance frequency (which is the first dimension). The basic 2D experiment consists of a series of 1D spectra. These data are run in sequence and have a variable delay built into the pulse program that supplies the means for the second dimension. The variable delay period is referred to as t_1.

All 2D pulse sequences consist of four basic building blocks: preparation, evolution (t_1), mixing, and acquisition (t_2). The preparation and mixing times are periods typically used to manipulate the magnetization (aka "coherence pathways") through the use of RF pulses. The evolution period is a variable time component of the pulse sequence. Successive incrementing of the evolution time introduces a new time domain. This time increment is typically referred to as a t_1 increment and is used to create the second dimension. The acquisition period is commonly referred to as t_2. The first dimension, generally referred to as F_2, is the result of Fourier transformation of t_2 relative to each t_1 increment. This creates a series of interferograms with one axis being F_2 and the other the modulation in t_1. The second dimension, termed F_1, is then transformed with respect to the t_1 modulation. The resultant is the two frequency dimensions correlating the desired magnetization interaction, most typically scalar or dipolar coupling. Also keep in mind that there is the "third dimension" that shows the intensity of the correlations.

When performing 2D NMR experiments, one must keep in mind that the second frequency dimension (F_1) is digitized by the number of t_1 increments. Therefore, it is important to consider the amount of spectral resolution that is needed to resolve the correlations of interest. In the first dimension (F_2), the resolution is independent of time relative to F_1. The only requirement for F_2 is that the necessary number of scans is obtained to allow appropriate signal averaging to obtain the desired S/N. These two parameters, the number of scans acquired per t_1 increment and the total number of t_1 increments, are what dictate the amount of time required to acquire the full 2D data matrix. 2D homonuclear spectroscopy can be summarized by three different interactions, namely, scalar coupling, dipolar coupling, and exchange processes.

7.3.2 Homonuclear 2D Methods

7.3.2.1 Scalar-Coupled Experiments: COSY and TOCSY

The correlated spectroscopy (COSY) experiment is one of the simplest 2D NMR pulse sequences in terms of the number of RF pulses it requires [23]. Once the time-domain data are collected and Fourier transformed, the data appear as a diagonal in the spectrum that consists of the ^{1}H chemical shift centered at each proton's resonance frequency. The off-diagonal peaks are a result of scalar coupling evolution during t_1 between neighboring protons. The data allow one to visualize contiguous spin systems within the molecule under study.

In addition to the basic COSY experiment, there are phase-sensitive variants that allow one to discriminate the active from the passive couplings, allowing clearer

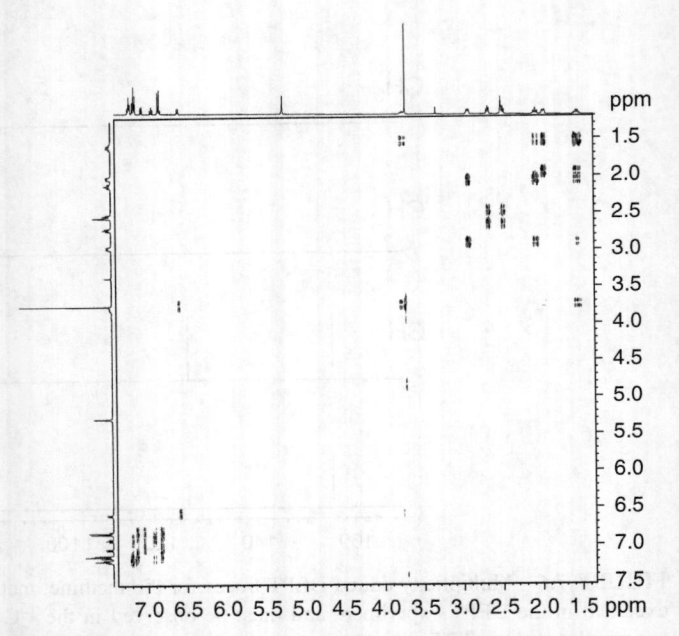

FIGURE 7.7 DQF COSY data of astemizole [2]. The black bars indicate the contiguous spin-system drawn with arrows on the relevant portion of astemizole.

measurement of the former. Active couplings give rise to the off-diagonal cross-peak. However, the multiplicity of the correlation has couplings inherent to additional coupled spins. These additional couplings are referred to as passive couplings. One such experiment is the double quantum-filtered (DQF) COSY experiment [24]. Homonuclear couplings can be measured in this experiment between two protons isolated in a single spin system. Additional experiments have been developed that allow the measurement of more complicated spin systems involving multiple protons in the same spin system [25]. The 2D representation of the scalar coupled experiment is useful when identifying coupled spins that are overlapped or are in a crowded region of the spectrum. An example of a DQF COSY spectrum is shown in Figure 7.7. This data set was collected on astemizole (**2**).

Total correlation spectroscopy (TOCSY) is similar to the COSY sequence in that it allows observation of contiguous spin systems [26]. However, the TOCSY experiment additionally will allow observation of many coupled spins simultaneously (contiguous spin system). The basic sequence is similar to the COSY sequence with the exception of the last pulse, which is a "spin-lock" pulse train.

7.3.2.2 Scalar Coupled Experiments: INADEQUATE

2D homonuclear correlation experiments are typically run using ^{1}H as the nucleus in both dimensions. This is advantageous, as the sensitivity for proton is quite high. However, there are 2D homonuclear techniques that detect other nuclei. One such experiment is the INADEQUATE experiment [27].

Its insensitive nature arises from the low natural abundance of ^{13}C and the fact that one is trying to detect an interaction between two adjacent ^{13}C atoms. The chance of observing this correlation is 1 in every 10 000 molecules. Given sufficient time or isotope-enriched molecules, the INADEQUATE provides highly valuable information. The data generated from this experiment allow one to map out the entire carbon skeleton of the molecular structure. The only missing structural features occur where there are intervening heteroatoms (e.g. O, N).

7.3.2.3 Dipolar Coupled Experiments: NOESY

The 2D experiments described thus far rely solely on the presence of scalar coupling. There are other sequences that allow one to capitalize on the chemical shift dispersion gained with the second dimension for dipolar coupling experiments as well. One example in this category is the 2D NOESY pulse sequence.[1] The pulse sequence for the 2D NOESY experiment is essentially identical to that of the DQF COSY experiment with the exception of an element that allows the buildup of dipolar couplings. The processed data is reminiscent of a COSY spectrum in that there is a diagonal represented by the 1D ^1H spectrum in both frequency dimensions. In sharp contrast to the scalar coupled cross-peaks in the COSY experiment, NOESY provides off-diagonal responses that correlate spins through space. This sequence is used extensively in the structure characterization of small molecules for the same reason as its 1D counterparts. The spatial relationship of ^1H atoms is an invaluable tool.

The similarity of the NOESY to the COSY also causes some artifacts to arise in the 2D data matrix of a NOESY spectrum. The artifacts arise from residual scalar coupling contributions that survive throughout the NOESY pulse sequence. These artifacts are usually quite straightforward to identify, as they have a similar antiphase behavior as can be seen for the DQF COSY data, Figure 7.8.

7.3.3 Heteronuclear 2D Methods

7.3.3.1 Direct Heteronuclear Chemical Shift Correlation

There are many numbers of heteronuclear correlation experiments that reach into the spin systems of many different chemical environments. This chapter will focus on the two major types of experiments that are used for structure elucidation. These are the $^1J_{CH}$ scalar coupled experiments and the long-range ($^nJ_{CH}$, where n is >1) scalar coupled experiments. The two predominant experiments for $^1J_{CH}$ are the heteronuclear multiple quantum coherence (HMQC) and the heteronuclear single quantum coherence (HSQC) methods. Both

---
[1]Jeener et al. [28]. Please note this is the first appearance of the NOSY sequence; however, this publication deals with chemical exchange. For a quite comprehensive text on NOE, please see the above reference by Neuhaus and Williamson [13].

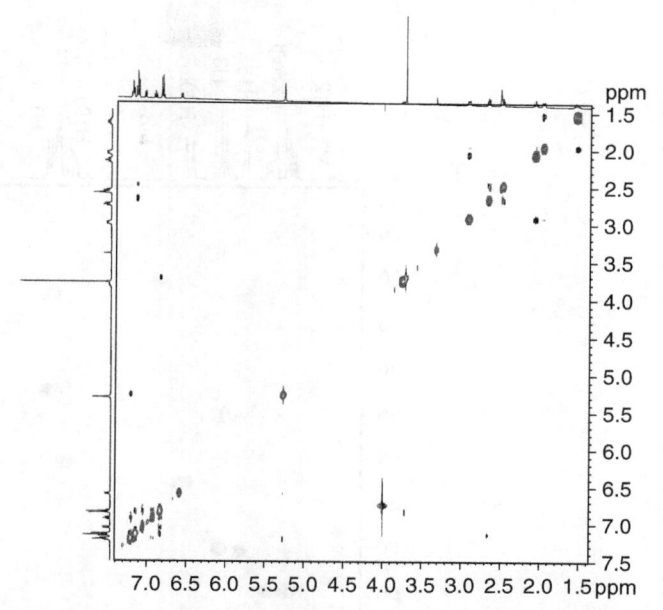

FIGURE 7.8 2D NOESY data of astemizole [2]. The mixed phase correlations in the aromatic region (boxed) are examples of the artifacts described in the text.

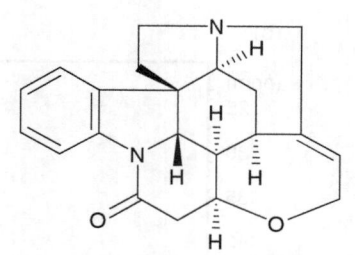

STRUCTURE 7.3 Strychnine.

of these methods rely on the sensitive and time-efficient proton, or so-called "inverse"-detected heteronuclear chemical shift correlation experiments are preferable [29]. For molecules with highly congested ^{13}C spectra, ^{13}C rather than ^1H detection is desirable due to high resolution in the F_2 dimension [30].

Using strychnine (Structure 7.3) as a model compound, a pair of HSQC spectra is shown in Figure 7.9. The top panel shows the HSQC spectrum of strychnine without multiplicity editing. All resonances have positive phase. In contrast, in the opinion of the authors, the much more useful multiplicity-edited variant of the experiment is shown in the bottom panel. The multiplicity-editing feature allows one to phase the data so that the correlations representing methyl and methine groups are in the same direction and methylenes are the opposite phase. Other less common direct heteronuclear shift correlation experiments have been described in the literature [31].

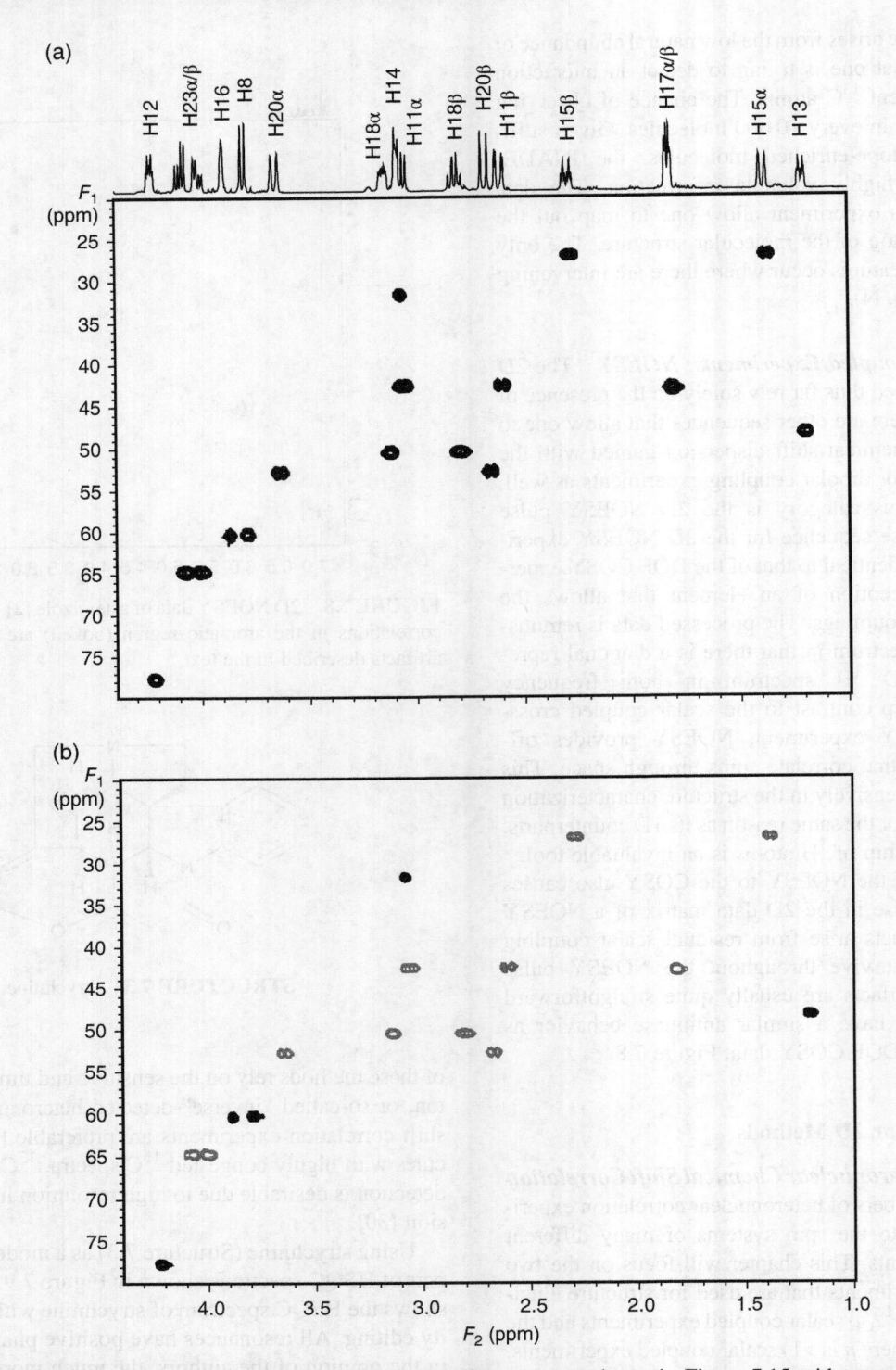

FIGURE 7.9 (a) GHSQC spectrum of strychnine [3] using the pulse sequence shown in Figure 7.15 without multiplicity editing. (b) Multiplicity-edited GHSQC spectrum of strychnine showing methylene resonances (gray contours) inverted with methine resonances (black contours) with positive phase. (Strychnine has no methyl resonances.)

7.3.3.2 Long-Range Heteronuclear Shift Correlation Methods

There are numerous ^{13}C-detected long-range heteronuclear shift correlation methods developed [32]. The primary reason that these methods have largely fallen into disuse is because of the heteronuclear multiple bond correlation (HMBC) experiment [33]. The proton-detected HMBC experiment and its gradient-enhanced variant offer considerably greater sensitivity than the original heteronucleus-detected methods.

7.4 QUANTITATIVE NMR SPECTROSCOPY (qNMR)

7.4.1 The Basics

One of the tremendous qualities of NMR spectroscopy is its intrinsic quantitative attributes. A few recent publications have done an excellent job of detailing the basic principles and examples of the use of qNMR as well as pertinent parameters and reference standards [34–36]. The ability to quantitatively determine relative concentrations of multiple components in a mixture, determine potency, and product recovery, to name a few, has led this technique to gain tremendous popularity in the last 15 years. As with all things, popularity can lend itself to usage of the technique without carefully considering the physical phenomenon that allow these properties to be measured. This chapter will deal with the typical uses of qNMR, including the key parameters to consider and examples of applications.

With respect to qNMR the most relevant relationship is that between the integrated signals to the number of nuclei responsible for the resonance:

$$I_X = K_S * N_X \tag{7.1}$$

where

I = the integrated resonance that is directly correlated to the number of nuclei responsible for the integrated resonance

K = a spectrometer constant.

The factor K cancels as the substances are undergoing the same experimental conditions, assuming proper care is taken to ensure appropriate parameters are in place [34]. The utilization of NMR for quantitative explanations is also given in the USP-NF 28, Chapter 761.[2] There are two primary ways in which qNMR is used. These are a relative approach and an absolute method.

7.4.2 Relative and Absolute Methods

The most frequently used method, often in a pseudo-qualitative mode (not optimal acquisition parameters), is the relative approach. For a quick analysis of two components of similar chemical properties (e.g. similar molecular weight and atomic composition), one can select two resonances of similar hybridization and integrate and get a rough estimate of molar ratios $\left(\frac{n_x}{n_y}\right)$ using Eq. (7.2). One can clearly solve this equation to provide the mole fraction of either of the two components:

[2] USP.

$$\frac{n_x}{n_y} = \frac{I_x N_y}{I_y N_x} \tag{7.2}$$

If the chemical compositions of both components are known, then the weight percent is inherent. However, one must be careful when using this "crude" method with components of an unknown chemical composition as there are many factors that can lead to ever increasingly erroneous results. One of the most important of these is the intrinsic relaxation rates of the compounds under study. If little is known of the compound being measured, then large differences in this chemical property can lead to a poor result. To make a rudimentary comparison, it would be equivalent to measuring the relative response of two compounds by UV at a single wavelength with no knowledge of the extinction coefficient/relative response factors.

The above description provides a very fast approach, yet it is not a rigorous quantitative method. To make this a rigorous relative quantitative method, one needs to measure a few critical NMR parameters that are intrinsic to the compounds in the mixture. These parameters, especially relaxation times, are well defined in the suggested reading list. In addition, there have been a few papers that have been published that detail general sets of parameters to be used to achieve a particular level of confidence in the measurement [35, 36]. Publications by Malz et al. go into a very good level of detail concerning the validation of quantitative methods [34, 35]. Utilizing careful experiment parameterization, one can get approximately 0.5–2% accuracies in their quantitative measurements. A "validated" parameter setup is shown below. Table 7.1 has been taken directly from the paper published by Malz et al. [35]. The reference by Malz et al. also deals a great deal with attributes related to specificity and selectivity as well as accuracy, precision, measurement uncertainty, and sensitivity. It is the authors opinion that this reference should serve as the primary reference to begin ones exploration into the use of NMR as a quantitative tool.

The use of the absolute method traditionally involves the use of internal standards that are added into the NMR sample and used as an internal reference to use the relative method described above. The amount of care one takes in both sample preparation as well as experiment setup will dictate the level of accuracies that are obtainable. The primary literature also provides several good reviews that include a vast number of potential internal standards to utilize. Standards should be selected for the sample at hand based on several factors including chemical shift, similar molecular weight, relaxation rates, etc. One should also consider the use of solvents that allow for reasonable manipulation without evaporation as this can also lead to inaccurate measurements (Figure 7.10).

One thing that has not been discussed is the criticality of processing of the data once it is required. Several steps should be taken to ensure proper data processing. Three of the most

TABLE 7.1 Summary of the Universal Spectrometer Parameters for qNMR

Parameter	Bruker	JEOL	Varian	Value
90° pulse strength	Pl1	x_atn	tpwr	Instrument specific
90° pulse length	P1	X_90_width	pw90	Instrument specific
Spin rotation		Spin_set	spin	Optional
Measurement temperature	TE	Temp_get	temp	300 K
Frequency of excitation	o1	Irr_freq	tof	Middle of spectrum
Pulse angle		X_angle	pw	30°
Preacquisition delay	DE	Initial_wait	alfa	5 μs
Acquisition time	AQ	X_acq_time	at	3.41 s
Relaxation delay	D1	Relaxation_delay	D1	$\geq(7/3)x$ longest T_1
Sweep width	SW	X_sweep	sw	16 ppm
Filter width	FW	Filter_width	fb	≥ 20 ppm
Number of FID points	TD	X_points	np	32 k
Number of scans	ns	Scans	nt	Declined of reached S/N
Signal-to-noise ratio	S/N	Sn_ratio	dsn	≥ 150
Line broadening (em)	lb	Width	lb	0.3 Hz
Number of frequency points	SI	X_points	fn	64 k

Source: Taken from Ref. [35].

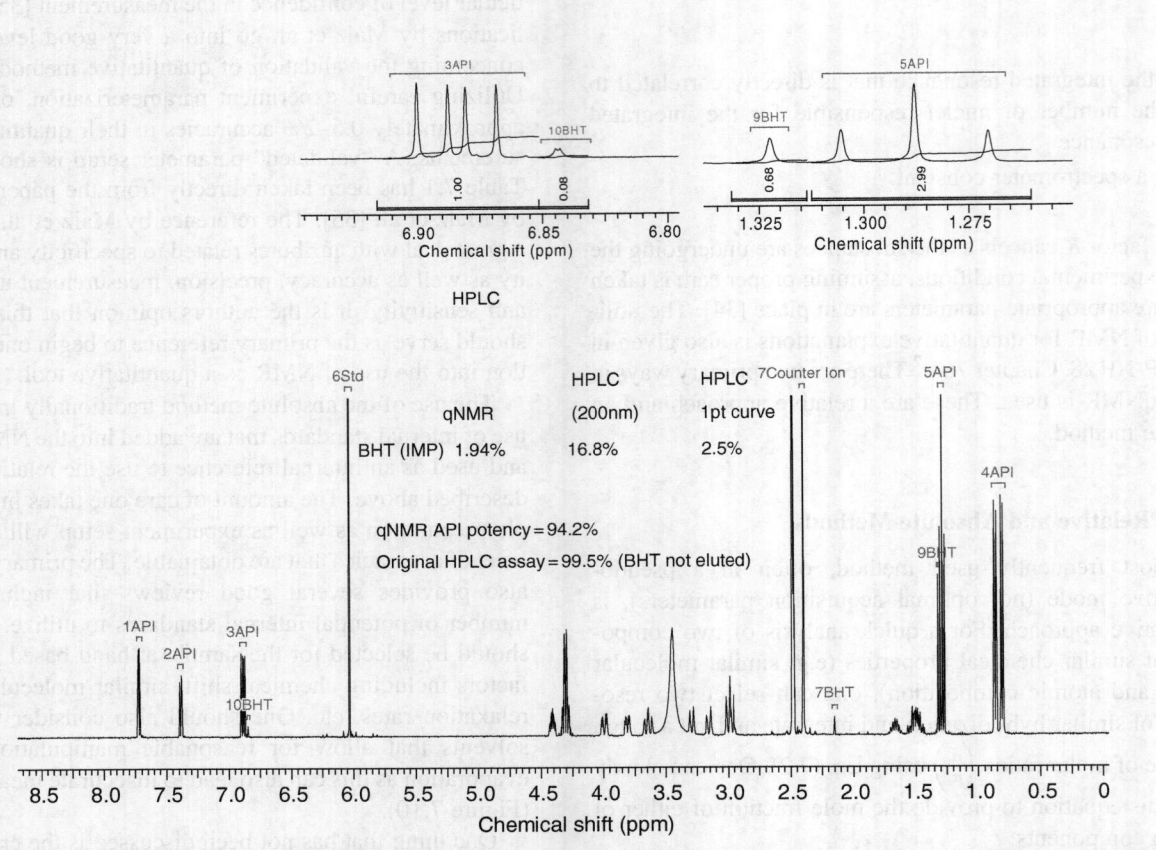

FIGURE 7.10 Example of the "absolute" method for qNMR experiments. In this case BHT was used as an internal reference with a know active pharmaceutical (API). See Table 7.2 to observe the impact of using NMR as opposed to UV for quantitation.

TABLE 7.2 Data Showing the Variation of Potency Determination with Variable Wavelength and NMR

Method	% BHT	% API
HPLC-UV Area% (200 nm)	16.8	83.2
HPLC-UV Area% (214 nm)	11.2	88.8
HPLC-UV Area% (224 nm)	2.7	97.3
HPLC-UV Area% (282 nm)	0.98	99.0
HPLC-UV RF wt/wt %	2.5	
qNMR wt/wt %	1.94	94.2

important features involve the use of baseline correction, proper phasing of the resultant Fourier transformed spectrum, and choosing appropriate integration ranges to ensure the majority of the signal is encompassed in the integration [34–36].

7.4.3 Electronic Referencing in qNMR

The use of internal reference materials has been the method of choice for many years to do quantitative measurements. In the pharmaceutical industry this is particularly true as it relates to potency measurements, mass balance, etc. An alternative approach has been gaining traction for those that work in this area. This approach utilizes an electronic signal that is generated artificially and is calibrated to a particular value and subsequently inserted into the collected NMR spectrum and used as a "standard" for comparison. The method was published under the acronym of ERETIC [37]. The original application was developed for imaging. It was later elaborated for use in high resolution NMR [38]. In effect, an electronic signal is routed through a "spare" coil in the probe. The amplitude of the signal is set by the operator, and this signal can be inserted at a user-defined resonance frequency so as to not interfere with any of the signals related to the compound under study. This signal can then be used with a reference standard independently and then related back to the sample of interest using the same amplitude.

Subsequent to the ERETIC method, there have been alternate approaches developed. One such method uses software to simulate a signal that is incorporated into a reference spectrum [39]. The simulated signal is then integrated relative to the reference. As with the ERETIC the signal is then placed in the spectrum containing the compound/chemistry under investigation as a reference. The simulated signal can be placed anywhere in the spectrum and can be crafted to have similar lineshape and intensity to the resonance/species of interest.

7.4.4 Quantitation in Flow NMR

The use of quantitation under flow NMR conditions requires a bit more consideration than for a static NMR tube

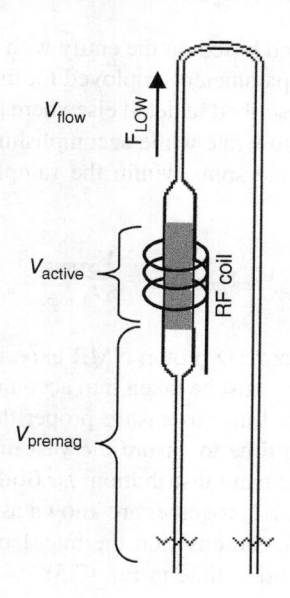

FIGURE 7.11 Diagram showing the key components within the flow path of the flow cell located in the NMR probe. This diagram is showing the key elements along the flow path that must be considered when designing flow-NMR experiments under quantitative condition.

arrangement. The theory and practicality of flow NMR has been described thoroughly in the literature, including several comprehensive reviews.[3,4,5,6] Unlike static tube-based experiments where all spins in the sample are uniformly experiencing B_0 and are therefore at Boltzmann distribution prior to excitation with a radio frequency pulse, flow experiments require the scientist to adjust the flow rate in such a manner to ensure that the sample flowing through has enough residence time within the B_0 field that they reach Boltzmann distribution prior to the active region of the flow cell for excitation. To ensure quantitation this condition must be met (Figure 7.11).

$$\frac{1}{T_{i,\text{flow}}} = \frac{1}{T_{i,\text{static}}} + \frac{1}{\tau} \text{ with } \tau = \frac{V_{\text{active}}}{V_{\text{flow}}} \quad (7.3)$$

Equation (7.3) shows the interdependency of the flow of the system to the intrinsic relaxation processes as well as the geometry of the probe ensuring quantitative conditions, where T_i represents either T_1 or T_2^* and τ is flow dependent [34]. Ensuring the sample is at or near Boltzmann's conditions prior to excitation is critical. The typical amount of time that the sample must reside in the magnetic field prior to irradiation is about five times T_1. It should be noted that this time

[3] Albert Book.
[4] Jones, D.W.
[5] Dorn.
[6] keifer.

should be calculated based on the entity with the slowest T_1. The experimental parameters employed for the measurement of T_1 values are described in detail elsewhere [9]. Calculation of the maximum flow rate while accomplishing optimal premagnetization of the spins within the sample can be done through Eq. (7.4):

$$V_{\text{flow, maximum}} = \frac{V_{\text{premag}}}{5T_{1,\text{max}}} \quad (7.4)$$

Within a standard 1D proton NMR experiment, there are several factors that must be taken into account. Two of these are the acquisition time (to ensure proper digitization), t_{aq}, and the relaxation time to ensure the vast majority of spins "re-achieve" Boltzmann distribution, t_{d}. Both of these parameters (see Table 7.1) together are known as the pulse repetition time t_{p}. One can envision the interdependency of the flow and the repetition time in Eq. (7.5):

$$t_{\text{p}} = t_{\text{aq}} + t_{\text{d}} = \frac{V_{\text{active}}}{V_{\text{flow, max}}} \quad (7.5)$$

An example would be a compound with a T_1 of 5 seconds, and a 100 µl premagnetization volume would have a maximum flow rate ($V_{\text{flow, maximum}}$) of approximately 0.24 ml/min.

7.4.5 Reaction Kinetics

One important and sometimes overlooked application of qNMR is the ability to monitor the progress and kinetics of a chemical reaction directly by NMR. A recent publication does a very good job of reviewing this topic and provides many references to the primary literature. This section will give highlights of the methodology and a few examples [40]. A simplistic example of the use of NMR to monitor reactions is shown for the hydrolysis of acetic anhydride to acetic acid in the presence of D_2O. Figure 7.12 shows the reaction progression under two different experimental conditions. One is static in an NMR tube, and the other is measured in real-time through the application of flow NMR. Both of these methods give good accuracy in the measurement of this first-order rate constant.

The practice of monitoring reactions in NMR tubes has been used since the inception of NMR and involves preparing reactions on a small scale and initiating them in an NMR tube with a suitable deuterated solvent. Although there are certain and obvious advantages to following this practice, the use of solvents lacking protons ($CDCl_3$, D_2O, CCl_4, etc.) is by no means a necessity. Recently, with the advent of more stable magnets and improved instrument electronics, it has become routine practice to monitor reactions by collecting NMR spectra in non-deuterated solvents [42]. The applications of

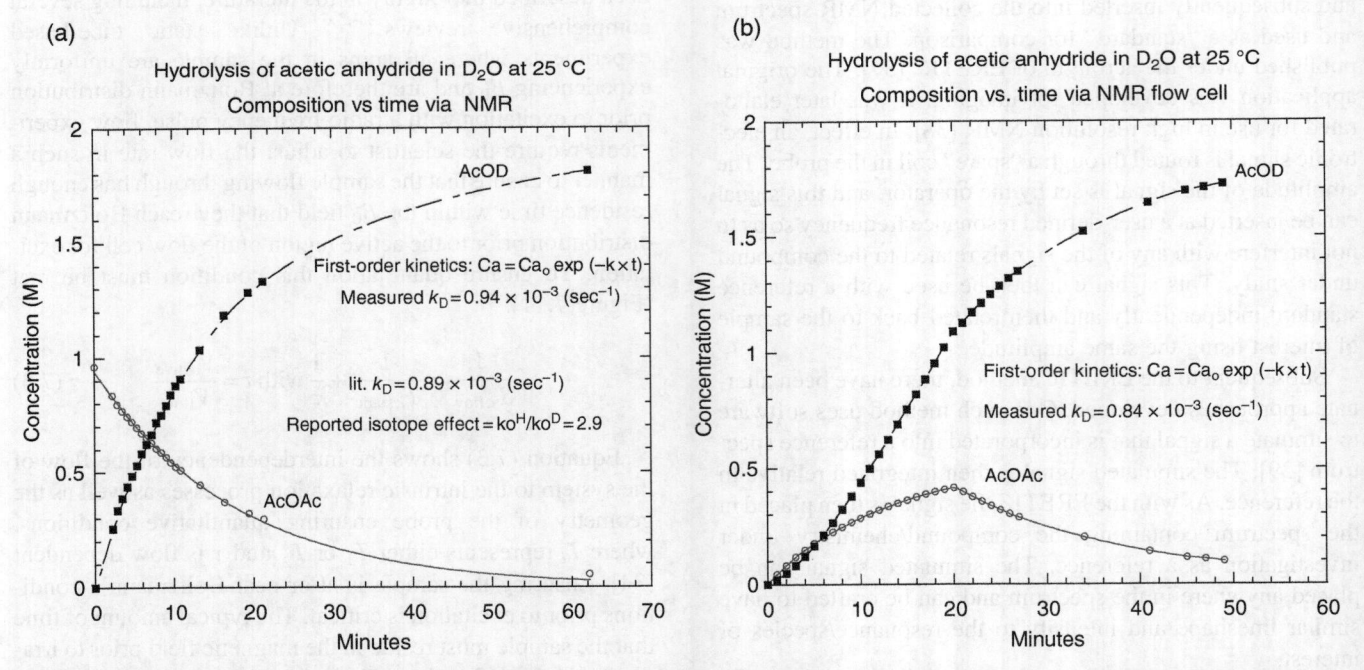

FIGURE 7.12 (a) NMR Tube Kinetics the reagent is injected, NMR tube shaken and placed in the magnet and spectra are recorded. (b) Flow NMR Kinetics, the reactor is integrated to the NMR and run in SemiBatch mode. The conditions of the reaction were 20 ml acetic anyhydride (AcOAc) dosed over 20 minutes into 200 ml of D_2O. The flow rate through the probe was 3 ml/min. *Source*: am Ende, et. al. AIChE 2010 [41].

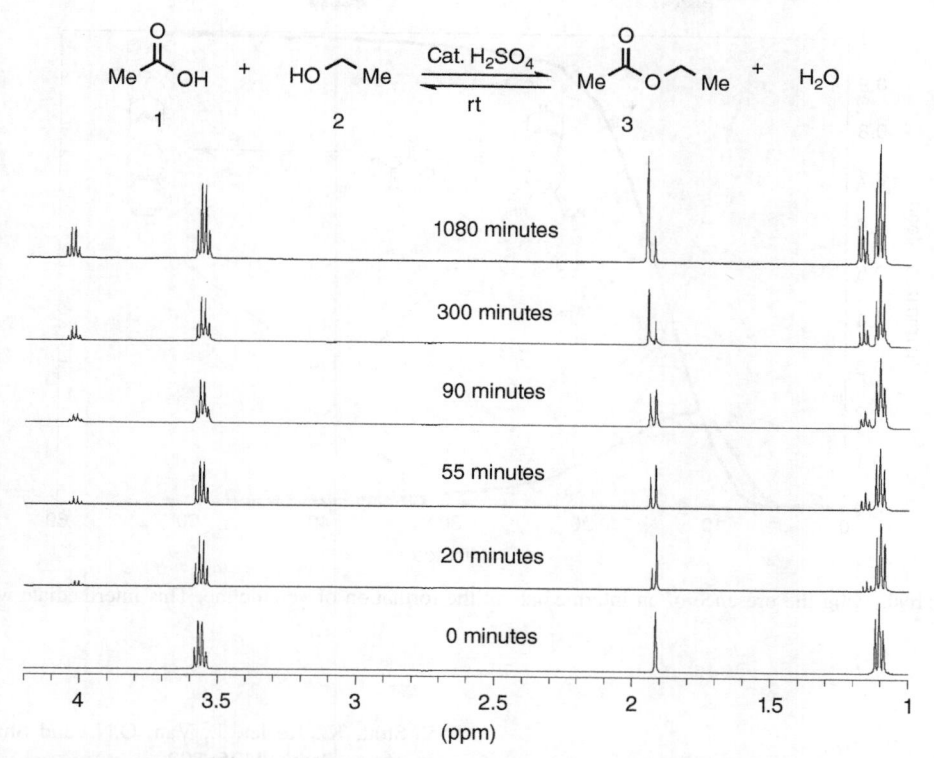

FIGURE 7.13 Example of the No-D NMR method Fischer esterification of acetic acid in ethanol (1 : 4 M ratio). *Source*: Reprinted with permission from Hoye et al. [42]. Copyright (2004) American Chemical Society.

the technique are wide ranging and can be used to directly monitor or assay most any reaction mixture or reagent solution. As mentioned previously, no deuterium solvents are used so the spectra are recorded in an "unlocked mode." One factor that facilitates the acquisition of NMR data in this way is that the concentration of most commonly used neat organic solvents is usually somewhere around 10 M. The concentration of the reactants in most reaction solutions is in the range of 0.1–1 M (and of solutions of most commercial reagents ca. 0.5–2.5 M). Thus, the ratio of solvent to solute/analyte molecules is usually between 100 : 1 and 10 : 1 in most solutions of interest. It is a straightforward matter for most NMR spectrometer hardware to handle dynamic ranges of proton intensities of greater than 4 orders of magnitude. If needed, techniques can be applied that use special "pulse sequences" to suppress unwanted signals from the solvent [34]. By doing so, the dynamic range and spectral quality of the reactions species can be improved (Figure 7.13).

Other more advanced techniques such as flow NMR can be utilized to monitor larger-scale reactions in real time. The use of flow NMR has been around for many years and has been used to monitor everything from reaction completion, kinetics, combinatorial chemistry, etc. [5]. Of late there has been a great deal of activity in the area of monitoring process chemistry using flow NMR [40, 42–45].[7,8]

There have been many different designs (instrument setup) to accomplish real-time reaction monitoring. However, one that works very well is that published by Maiwald et al. and is used by many others in the field [40]. The basic premise is to have a fast loop that carries the reaction mixture from a reactor to ensure that the loop is just a "real-time" extension of the reactor. From this loop there is a split (many ways to accomplish this) that allows a slow loop to flow into the NMR flow cell. One must keep in mind that the criteria must be met as described in Section 7.4.5 to be under quantitative conditions if quantitation is desired. Obviously if one is interested in strictly observing the reaction for gross features (e.g. reaction completion and/or reaction optimization), the quantitative conditions are not needed, and a less rigorous approach can be taken. A general schematic is described in Ref. [40} (Figure 7.14).

[7]Andreas at 2006 SMASH.
[8]SMASH Workshop 2008.

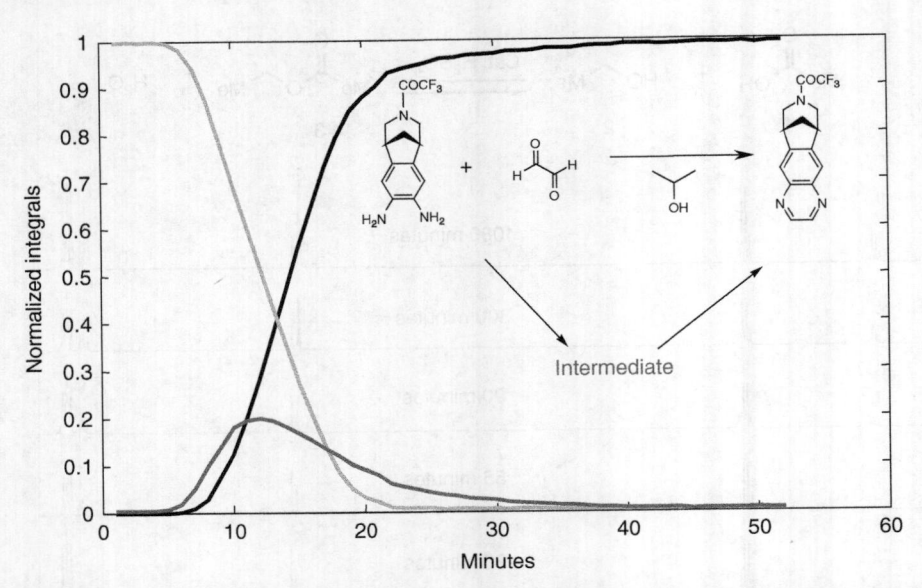

FIGURE 7.14 A graph showing the presence of an intermediate in the formation of veranicline. This intermediate was readily observed by NMR.

REFERENCES

1. Friebolin, H. (1993). *Basic One- and Two-Dimensional NMR Spectroscopy*. New York: VCH Publishers.
2. Gunther, H. (1995). *NMR Spectroscopy*, 2e. New York: Wiley.
3. Bloch, F., Hensen, W.W., and Packard, M. (1946). *Phys. Rev.* 69: 127.
4. Derome, A.E. (1987). *Modern NMR Techniques for Chemistry Research*. New York: Pergamon Press.
5. Claridge, T.D.W. (1999). *High Resolution NMR Techniques in Organic Chemistry*. New York: Pergamon Press.
6. Keeler, J. (2006). *Understanding NMR Spectroscopy*. New York: Wiley.
7. Levitt, M.H. (2002). *Spin Dynamics: Basics of Magnetic Resonance*. New York: Wiley.
8. Croasmun, W.R. and Carlson, R.M.K. (1994). *Two-Dimensional NMR Spectroscopy: Applications for Chemists and Biochemists*, 2e. New York: VCH Publishers.
9. Berger, S. and Braun, S. (2004). *200 and More NMR Experiments: A Practical Course*. Weinheim: Wiley-VCH Verlag GmbH and Co. KGaA.
10. (a) Overhauser, A.W. (1953). *Phys. Rev.* 89: 689.
    (b) Overhauser, A.W. (1953). *Phys. Rev.* 92: 411.
11. Solomon, I. (1955). *Phys. Rev.* 99: 559.
12. Richarz, R. and Wuthrich, K. (1978). *J. Magn. Reson.* 30: 147.
13. Neuhaus, D. and Williamson, M. (2000). *The Nuclear Overhauser Effect in Structural and Conformational Analysis, P.*, 2e. New York: Wiley.
14. Claridge, T.D.W. (1999). *High-Resolution NMR Techniques in Organic Chemistry*, Tetrahedron Organic Chemistry Series, vol. 19. Oxford: Elsevier Science.
15. Stott, K., Keeler, J., Van, Q.N., and Shaka, A.J. (1997). *J. Magn. Reson.* 1125: 302.
16. Stott, K., Stonehouse, J., Keeler, J. et al. (1995). *J. Am. Chem. Soc.* 117: 4199.
17. Hahn, E.L. (1949). *Phys. Rev.* 76: 145.
18. (a) Levy, G.C. and Nelson, G.L. (1972). *Carbon-13 Nuclear Magnetic Resonance for Organic Chemists*. New York: Wiley-Interscience.
    (b) Stothers, J.B. (1972). *Carbon-13 NMR Spectroscopy*. New York: Academic Press.
19. Shaka, A.J. and Keeler, J. (1987). *Prog. Nucl. Magn. Reson. Spectrosc.* 19: 47–129.
20. (a) Doddrell, D.M., Pegg, D.T., and Bendall, M.R. (1982). *J. Magn. Reson.* 48: 323–327.
    (b) Doddrell, D.M., Pegg, D.T., and Bendall, M.R. (1982). *J. Chem. Phys.* 77: 2745–2752.
21. Patt, S.L. and Shoolery, J.N. (1982). *J. Magn. Reson.* 46: 535–539.
22. Jeener, J. (1971). Ampere International Summer School, Basko Polje, (proposal).
23. (a) Aue, W.P., Bartholdi, E., and Ernst, R.R. (1976). *J. Chem. Phys.* 64: 2229–2246.
    (b) A. Bax, R. Freeman, and G. A. Morris, *J. Magn. Reson.*, 42, 164–168 (1981).
24. Piantini, U., Sorensen, O.W., and Ernst, R.R. (1982). *J. Am. Chem. Soc.* 104: 6800.
25. Griesinger, C., Sorensen, O.W., and Ernst, R.R. (1985). *J. Am. Chem. Soc.* 107: 6394.
26. Braunschweiler, L. and Ernst, R.R. (1983). *J. Magn. Reson.* 53: 521.
27. Bax, A., Freeman, R., and Frenkiel, T.A. (2102). *J. Am. Chem. Soc.* 1981: 103.

28. Jeener, J., Meier, B.H., Bachmann, P., and Ernst, R.R. (1979). *J. Chem. Phys.* 71: 4546.
29. (a) Müller, L. (1979). *J. Am. Chem. Soc.* 101: 4481.
    (b) Bodenhausen, G. and Ruben, D.J. (1980). *Chem. Phys. Lett.* 69: 185–189.
30. Reynolds, W.F., MacLean, S., Jacobs, H., and Harding, W.W. (1999). *Can. J. Chem.* 77: 1922–1930.
31. Martin, G.E. (2002). Qualitative and quantitative exploitation of heteronuclear coupling constants. In: *Annual Report NMR Spectros*, vol. 46 (ed. G.A. Webb), 37–100. New York: Academic Press.
32. Martin, G.E. and Zektzer, A.S. (1988). *Magn. Reson. Chem.* 26: 631.
33. Bax, A. and Summers, M.F. (1986). *J. Am. Chem. Soc.* 108: 2093–2094.
34. Malz, F. (2008). Quantitative NMR in the solution state. In: *NMR Spectroscopy in Pharmaceutical Analysis*, 1e (ed. U. Holzgrabe, I. Wawer and B. Diehl), 43–60. Amsterdam, The Netherlands: Elsevier.
35. Malz, F. and Jancke, H. (2005). *J. Pharm. Biomed. Anal.* 38: 813.
36. Pauli, G.F., Jaki, B.U., and Lankin, D.C. (2005). *J. Nat. Prod.* 68: 133.
37. Barantin, L., Akoka, S., and LePape, A. (1995). French Patent CNRS no.95 07651, 26 June 1995.
38. Akoka, S., Barantin, L., and Trierweiler, M. (1999). *Anal. Chem.* 71: 2554–2557.
39. Wider, G. and Drefer, L. (2006). *J. Am. Chem. Soc.* 128 (8): 2571–2576.
40. Maiwald, M., Steinhof, O., Sleigh, C. et al. (2008). Quantitative NMR in the solution state. In: *NMR Spectroscopy in Pharmaceutical Analysis*, 1e (ed. U. Holzgrabe, I. Wawer and B. Diehl), 471–491. Amsterdam, The Netherlands: Elsevier.
41. am Ende, D., Dube, P., Gorman, E., Marquez, B., and Mark Zell and R. Krull, D. Piroli, M. Fey and K. Colson, Reaction NMR: A Quantitative Kinetic Analysis "Probe" for Process Development, presented at AIChE National Meeting, Salt-Lake City November 6-10, 2010; The reported kinetic isotope ratio is from Batts, B.D. and Gold, V. (1969). *J. Chem. Soc. A* (6): 984.
42. Hoye, T.R., Eklov, B.M., Ryba, T.D. et al. (2004). *Org. Lett.* 6: 953–956.
43. Maiwald, M., Fischer, H.H., Kim, Y.K. et al. (2004). *J. Mag. Reson.* 166: 135–146.
44. Albert, K., Hasse, H. et al. (2005). *Chem. Eng. Process.* 44: 653–660.
45. Horvath, I.T. and Millar, J.M. (1991). *Chem. Rev.* 91: 1339–1351.

PART III

REACTION KINETICS AND MIXING PROCESSES

PART II

REACTION KINETICS AND MIXING PROCESSES

8

REACTION KINETICS AND CHARACTERIZATION

UTPAL K. SINGH, BRANDON J. REIZMAN, SHUJAUDDIN M. CHANGI, AND JUSTIN L. BURT
Eli Lilly and Company, Indianapolis, IN, USA

CHUCK ORELLA
Merck & Co., Inc., Rahway, NJ, USA

8.1 INTRODUCTION

The ability to effectively characterize and interpret reaction kinetic data is of fundamental importance for chemical development. In the pharmaceutical industry, reaction kinetics factor prominently in the areas of process optimization, process safety evaluation, understanding of scale sensitivity, and assessment of process robustness. Beyond these specific applications, the role that reaction kinetic understanding can have at all stages of process development is difficult to overstate. Examples in which detailed kinetic understanding can impact the viability of a process include instances:

- Where mechanistic insights are exploited in the mitigation of a key impurity, degradant, or other process failure mode.
- Where a successful process scale-up is contingent upon understanding the competition between kinetic and mass transfer effects.
- Where online measurements enable process adjustments for safer operation, improved product quality, or increased yield.
- Or where a reaction pathway model is leveraged to direct experimental work.

Above all else, the investigation of reaction kinetics affords scientists and engineers a means to quantifiably impact a chemical system throughout development from the molecular scale to commercial-scale manufacturing.

The literature has many excellent texts and articles devoted to a variety of perspectives on chemical reaction engineering and kinetics. Generally, these fall into the categories of mechanistic chemistry, reaction kinetics, and reactor design and operation. We encourage readers to sample texts from all of these perspectives [1–6]. In this chapter, we focus on those aspects of reaction characterization and scale-up that are of greater relevance to chemists and engineers working in the pharmaceutical industry and highlight those methods that have the potential to significantly improve the efficiency and effectiveness of development activities.

Two aspects in particular have historically made pharmaceutical processing unique from commodity and (to a lesser degree) specialty chemicals. The first differentiating feature is the complexity and richness of the chemistry, with multiple reactive moieties present in molecules available for desired and, often, undesired reaction pathways. Second, while bulk chemical manufacturers have embraced continuous processing as a means of enabling smaller process footprints and more sophisticated engineering controls, pharmaceutical processing has mostly remained steadfast in carrying out sequential batch operations in stirred tank reactors. Transitioning away from this batch mindset requires increased integration of process chemistry and chemical engineering in the design of continuous manufacturing processes for pharmaceuticals, a compelling area of development that will be referenced in examples throughout this chapter.

Chemical Engineering in the Pharmaceutical Industry: Active Pharmaceutical Ingredients, Second Edition.
Edited by David J. am Ende and Mary T. am Ende.
© 2019 John Wiley & Sons, Inc. Published 2019 by John Wiley & Sons, Inc.

The diversity in chemistry practiced in the pharmaceutical industry can be viewed from two competing perspectives. From a molecule-to-molecule standpoint, each chemical synthesis brings new challenges, ranging from transformations that disobey accepted heuristics, to the control of competing impurities and degradants, to difficult-to-quantify mass transfer effects. To address this complexity, it is necessary to design and execute the proper experiments that help uncover the underlying mechanisms, reaction pathways, and rates. The application of kinetic models in describing each of these phenomena offers a quantitative approach that can help with testing mechanistic hypotheses or with transitioning to a new target reaction condition or processing scale.

From a macroscopic perspective, although active pharmaceutical ingredients vary greatly in structure and function, the variety of chemical transformations employed by pharmaceutical chemists to construct molecules is hardly as diverse as the molecules themselves. For example, a survey of the most commonly employed chemical transformations at companies such as GlaxoSmithKline (GSK), AstraZeneca, and Pfizer revealed that more than 80% of developed reactions can be categorized into one of eight different classes: heteroatom alkylations and arylations, acylations, C—C bond formations, aromatic heterocycle formations, deprotections, protections, reductions, and oxidations [7]. The majority of the remaining reactions were functional group interconversions or additions. Efforts to minimize the time and labor involved in process development should begin with standardizing the knowledge of mechanistic and scale-dependent factors encountered in each of these common reaction classes.

In addition to mechanistic insight, characterization of reaction kinetics requires an understanding of the interplay between the rate of chemical transformation and the physics of the system. Scale sensitivity is exhibited when the rate of a physical transformation (i.e. rate of mixing, heat input, or heat removal) becomes nearly equal to or slower than the rate of chemical transformation. A different study of the pharmaceutical industry surveyed 22 different processes and classified the 86 reactions used in those processes according to overall kinetics and the physical nature of the reaction mixture [8]. Nearly 75% of the reactions were classified as having a potential for scale sensitivity. A vast majority of the surveyed reactions were heterogeneous. The combination of the multiphasic nature of the reaction mixture with the intrinsic speed of the reaction results in the potential for scale sensitivity. This chapter will present some of the techniques for quantitative assessment when there is a scale-dependent competition between the rates of chemical transformation and physical transformation, i.e. limitations due to the rates of mass and heat transfer.

A unique historical aspect of the pharmaceutical industry has been the use of batch processing for the majority of operations. The demand placed on the development scientist has then been the scale-up of reactions from the milliliter scale to the cubic meter scale (6 orders of magnitude). Over such a wide range of scale, changes in the observed reaction behavior owing to the competition between reaction rate and mass transfer are often observed. Such changes can lead to suboptimal cycle time, or even compromised product quality upon scale-up for manufacturing if not managed appropriately. To successfully scale-up, it is necessary to take both kinetic and equipment design considerations into account such that the perceived rate-limiting step is understood at every scale.

Pharmaceutical companies have begun to leverage the operation of chemical reactions in continuous mode to decrease the overall magnitude of scale-up, thus lessening the severity of scale sensitivity observed upon manufacturing [9–11]. The ratio of surface area to volume is generally much greater for continuous reactors than for industrial batch reactors. This affords heat and mass transfer rates that are more consistent with laboratory-scale experiments. Classical examples of continuous reactors offering an advantage over batch operating conditions involve fast reactions that risk "overreaction" of the desired product [5] or cases where safety is enhanced by decreasing the total amount of material in the reactor at a given time. Examples of safety enhancement in continuous reactors include cases where there is a risk of thermal runaway (e.g. Grignard reactions, lithiations, or nitrations) or where hazardous reagents are employed (e.g. syntheses involving azides, peroxides, strong reductants, or reactive gases such as hydrogen, oxygen, fluorine, or chlorine). For the purpose of the ensuing discussion, it is important to emphasize that in the absence of mixing or thermal effects, the fundamental kinetic data obtained from a plug flow continuous reactor will be identical to the kinetics measured in a stirred tank batch reactor [5]. Continuous processing also creates opportunities for more in-line and at-line kinetic measurements, as will be discussed later in this chapter, and can lead to higher plant productivity than a comparable batch operation [12]. A few of the advantages that continuous reactors offer in comparison with traditional batch processing are summarized in Table 8.1.

This chapter is divided into four sections. We begin by reviewing the common factors contributing to observed reaction kinetic behavior: kinetic effects, mass transfer effects, thermal (energy-related) effects, and dispersion. Subsequent sections discuss strategies for the characterization of reaction behavior in the presence of these different phenomena and lay the framework for transforming the data collected from these characterization experiments into a kinetic model. Both empirical and mechanistically based approaches are highlighted. Finally, we discuss some of the more recent developments in the field and the impact that new methodologies are having both in the laboratory and in the manufacturing environment. A number of academic and industrial examples are provided throughout the chapter to illustrate the implementation of the discussed techniques. We hope the reader will come

TABLE 8.1 Advantages of Continuous Reactors in Comparison to Batch Reactors

Characteristic	Batch (8000 L Stirred Vessel)	Continuous (1 in pipe reactor)	Advantage
Mixing time	>10 s (bulk blending)	>0.1 s	Rapid blending of reagents for fast reactions
Surface to volume available for heat transfer	~2 m^{-1}	~200 m^{-1}	Superior temperature uniformity with fewer "hot spots"
Typical temperature and pressure limit (in the absence of special reactor designs)	150 °C 10 bar	200–250 °C 30–150 bar	Ease of running reactions above the normal boiling point of solvents. Higher concentrations of gaseous reagents dissolved
Instantaneous amount reacting	100–1000 kg	1–5 kg	Lower energy potential and impact of runaway reaction

to appreciate the complexity of problems solvable using these characterization and numerical tools in concert with an applied understanding of the competition between reaction kinetic and physical phenomena.

8.2 FUNDAMENTALS OF CHEMICAL REACTION KINETICS

Fundamental understanding of the relevant rates of reactions is crucial in order to optimize process performance criteria such as yield and selectivity and to build quantitative relationships between input attributes, process parameters, and the desired outcomes of a reaction. Reactive chemical systems can be divided into two classes: homogeneous reactions, involving only a single phase, or heterogeneous reactions, involving two or more phases (e.g. solid–liquid, liquid–liquid, gas–liquid, or gas–solid–liquid). A small subset of homogenous reactions are elementary, i.e. these reactions occur in a single step with a rate that scales proportionally with the concentrations of the species in the reaction. The majority of reactions comprise multiple steps occurring in series or parallel [4, 13], passing through both detectable and undetectable intermediates to ultimately afford the desired product and (undesired) reaction impurities. In general, a robust process chemistry and reproducible reaction performance require an understanding of the competing rates of reactions, together with mixing, mass transfer, distribution among phases, and heat transfer considerations. This section provides considerations for determining the rates of reaction and mixing for various categories of reactions encountered in pharmaceutical processes. Understanding of these rates can then be used in conjunction with the physics and energetics of the system to select the most appropriate reactor type and mode of operation.

8.2.1 Reaction Rate and Mass Balance

For a reactive single-phase system,

$$aA + bB \rightarrow cC + dD$$

there exists a reaction time scale defined by the kinetics and independent of mass transfer limits (e.g. mixing, diffusion). The reaction rate (r_A) for disappearance of moles of limiting reactant A (N_A) can be written in a differential form as shown in Eq. (8.1), with V being the overall volume of the system:

$$r_A = \frac{1}{V}\frac{dN_A}{dt} \text{ (mol/m}^3\text{s)} \qquad (8.1)$$

The rates of reaction for each component (r_i) are interrelated based on the mass balance for the system, as shown in Eq. (8.2):

$$\frac{-r_A}{a} = \frac{-r_B}{b} = \frac{r_C}{c} = \frac{r_D}{d} \qquad (8.2)$$

The rate of a chemical reaction is generally determined empirically as a function of the concentration of reactive species and the temperature. In a limited number of cases, the reaction is elementary in that no reactive intermediate is formed in the transition from reactants to products. For elementary reactions, the reaction rate is then exactly proportional to the concentrations of reacting species. Many reacting systems, however, consist of a sequence of mechanistic steps (in series, in parallel, or both) proceeding at unique rates that in combination afford an observed rate law. Depending upon the relative rates of the mechanistic steps, the measured rate law may still simplify to a power law along the lines of Eq. (8.3):

$$r_{\text{power law}} = -\frac{dC_A}{dt} = kC_A^{\alpha}C_B^{\beta} \qquad (8.3)$$

where

k is the reaction rate constant.
C_i are concentrations of the ith species.
α, β are the reaction orders.

A rate law may even simplify to the same expression that would be obtained was the reaction assumed to be elementary ($\alpha = \beta = 1$ in Eq. (8.3)). Methods are discussed in this chapter for testing whether a sequence of proposed elementary or nonelementary reactions consistently describes the experimental data.

Though not a universal statement, reaction kinetics that do not adhere to a simple power law often indicate the presence of competing processes within the reaction system. Examples of such behavior may be the result of a reversible or inhibitory reaction involving one or more of the products, a reaction cycle in which a reactive species or catalyst must be regenerated in order for the reaction to proceed, or chemistry involving multiple phases where the overall rate is dependent on the intrinsic reaction rate and the rate at which the species can transition between phases. This last example is of particular emphasis in this chapter, as the rate of mass transfer of reagents between phases is a contribution that often appears only when examining the chemistry across different scales. In light of this, it is important to keep in mind that the converse of this paragraph's opening sentence should not be assumed, i.e. reaction kinetics that adhere to a simple power law do not necessarily imply simplicity of the underlying chemical system. Several examples showing a change in apparent kinetics with reaction conditions and/or process scale can be found throughout this chapter.

8.2.2 Kinetic Considerations

The act of measuring a rate law – power law or otherwise – is a simple activity in comparison with the act of interpreting kinetic results with the goal of optimizing for yield or selectivity, scaling up a process, or understanding the reaction mechanism. When embarking on any of these tasks, it is important to have awareness of all factors that may contribute to the observed reaction rate, including contributions from the reaction medium (solvent), mass transfer, and heat transfer. Above all else, it is of utmost importance to have an understanding of the factors that govern the reaction itself.

In addition to the concentration of species, the rates of elementary reaction steps are governed by a rate constant, assumed to be of the Arrhenius form

$$k = Ae^{-\frac{E_A}{RT}} \quad (8.4)$$

where

A is the pre-exponential factor.
E_A is the activation energy.
R is the gas constant.
T is the temperature.

Whereas A and E_A are commonly found in practice by regression to experimental data (see Section 8.4), quantum mechanical tools can in principle be used for the estimation of both parameters in an effort to confirm mechanistic hypotheses or to predict reactivity [6, 14]. A physical interpretation of A and E_A applicable to many reactive systems can be extracted from transition state theory, which postulates that reactants must traverse a barrier of higher-energy states before conversion to products. At the minimum height of this barrier is the transition state, which resides in energy above the starting reactants at a difference equal to the activation energy, E_A. The pre-exponential factor, A, reflects the number of degrees of freedom available at that transition state; a lesser value of A indicates a more constrained transition state and would imply a lower probability of proper alignment of the reactant orbitals for a reaction to occur.

In the simplest terms, the art of designing chemical reactions to be faster or more selective reduces to the identification of reagents or methods that impact either or both of the coefficients in Eq. (8.4). Most notably, catalysts are employed to reduce the height of the activation energy barrier. Transition metal catalysts are prevalent throughout chemical processing because of their ability to adopt different oxidation states in support of what would otherwise be much more energetically unfavorable intermediates. For reactions directed at a particular site of a molecule, heterogeneous reactions occurring at the surface of a catalyst can also constrain reactive species such that a targeted transition state becomes more favorable. Among the attributes of the reactants themselves that contribute to the reaction rate are the electronics of the molecules, i.e. the ability to donate or withdraw charge and hence minimize the energy burden of a transition state, and steric factors that impact the ability of the molecules to conform to the proper orientation for the reaction to occur. The reaction medium (solvent) has the potential to significantly impact both of these attributes.

The solvent in which solution-phase chemistry occurs can impact the rate, selectivity, and mechanism of the desired reaction [15]. Factors impacting solvent selection include compatibility with the mechanism of the proposed reaction, the ability to dissolve substrate, reagent, and/or product at minimal dilution, physical compatibility with the intended process conditions (e.g. reaction temperatures above the freezing point and below the boiling point of the solvent), occupational health, process safety, and environmental considerations, as well as solvent cost and commercial availability. In practice, solvent selection is often driven by empirical knowledge of solvents that have worked in the past for a given type of reaction, derived from institutional history, accumulated literature, or personal experience.

An industrial example of solvent selection impacting reaction selectivity was reported by researchers at Eli Lilly [16].

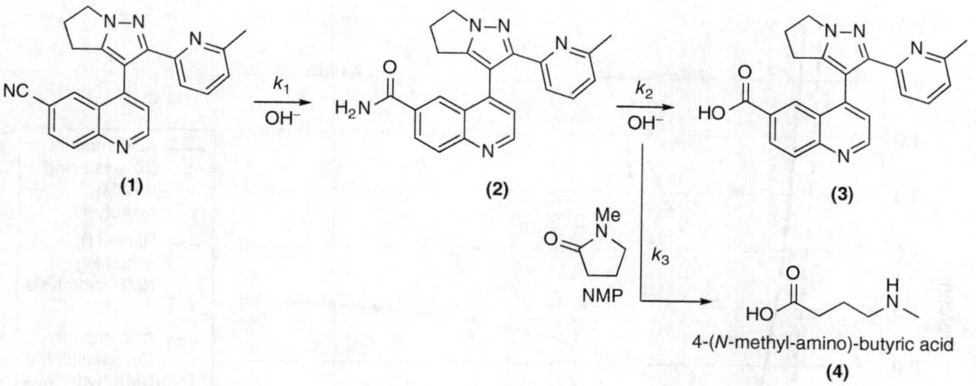

SCHEME 8.1 Hydroxide-catalyzed hydrolysis of 1 in the presence of NMP.

Referring to Scheme 8.1, hydroxide-catalyzed conversion of nitrile **1** to the desired amide **2** was susceptible to over-hydrolysis, affording a carboxylic acid impurity **3**. N-Methyl-2-pyrrolidone (NMP) was identified as a solvent that, in addition to affording favorable solubility properties, exhibited a rate of hydrolysis comparable with the amide. Sacrificial hydrolysis of NMP to 4-(methylamino)butyric acid **4** consumed excess hydroxide, protecting the amide and limiting formation of the carboxylic acid impurity.

A kinetic model was developed based on the elementary reactions:

$$1 + OH^- \rightarrow RC(OH)=N^-$$
$$RC(OH)=N^- + H_2O \rightarrow 2 + OH^-$$
$$2 + OH^- \rightarrow RCO_2^- + NH_3$$
$$NMP + OH^- \rightarrow MeNH(CH_2)_3 CO_2^-$$

Kinetic parameters were regressed, and the resultant kinetic model was employed to probe the robustness of the process chemistry *in silico*. Figure 8.1 illustrates the enhanced process robustness realized by employing NMP as a sacrificial solvent. The design requirements were less than 0.3% residual nitrile and less than 2.5% carboxylic acid impurity at the reaction endpoint. Referring to the dashed reaction profiles in Figure 8.1, in the absence of NMP hydrolysis, there was only an approximately 30 minute window (from 90 to 120 minutes) in which to stop the reaction while satisfying both design requirements. With the sacrificial hydrolysis of NMP (Figure 8.1, solid reaction profiles), the design requirement of less than 0.3% residual substrate was satisfied after 3 hours of reaction, and the level of **3** remained within its target range of less than 2.5% for several hours, affording a robust window in which to stop the reaction. The acceptable time intervals for reaction control without (shorter duration) and with (longer duration) NMP are included for comparison in Figure 8.1.

8.2.3 Mass Transfer Considerations

An understanding of mixing rates is important to the characterization and scale-up of heterogeneous reactions. Mass transfer-limited processes can give an erroneous sense of kinetics when scaled up, impacting the desired outcome. Detailed reviews can be found in literature that describe mixing dynamics at molecular and larger scales [17, 18].

The Damköhler number can be used to assess the effect of scale on reaction kinetics:

$$Da = \frac{\text{rate of chemical transformation}}{\text{rate of physical processes}} \quad (8.5)$$

Here, the rate of physical processes can include any rate of mixing, including those associated with mass transfer such as liquid–liquid mixing, gas absorption, gas desorption, and solid suspension. In general, no scale sensitivities would be expected when the rate of the chemical transformation is slower than that of the relevant physical process. In contrast, scale sensitivities are observed when the rate of the physical process is slower than the rate of the chemical transformation. This statement holds not only for the desired reaction pathway but also for other chemical pathways that may result in impurity formation.

The rate of chemical transformation generally takes the form of Eq. (8.1). The rate of mass transfer is expressed generally as a function of the mass transfer coefficient ($k_s a_s$) and the difference in concentration of a given species between the two phases:

$$\text{Rate of mass transfer} = k_s a_s \left(C_{A,\text{phase 1}} - C_{A,\text{phase 2}} \right) \quad (8.6)$$

Numerous correlations have been reported in the literature describing functional relationships between the nondimensional groups of Reynolds number (Re), Schmidt number

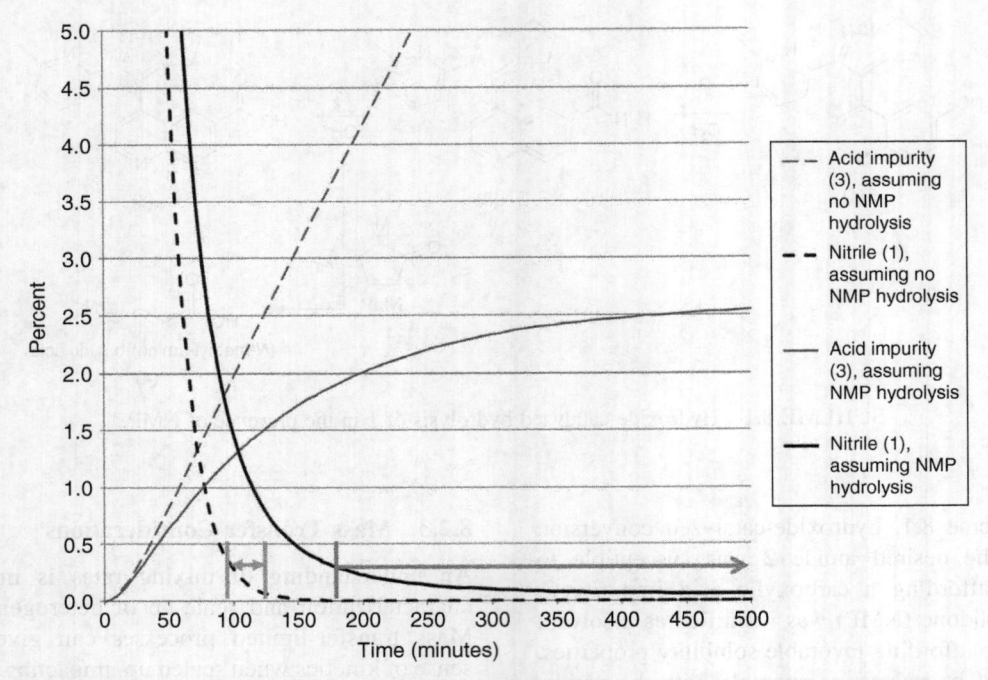

FIGURE 8.1 Model-predicted reaction profiles with and without use of NMP as a sacrificial solvent, assessed at target reaction conditions (75 °C and 0.25 equiv. NaOH). *Source*: Reprinted with permission from Niemeier et al. [16]. Copyright 2014, American Chemical Society.

(Sc), and Sherwood number (Sh). These dimensionless parameters are defined by Eqs. (8.7)–(8.9):

$$\mathrm{Re} = \frac{\rho u d}{\mu} \quad (8.7)$$

$$\mathrm{Sc} = \frac{\mu/\rho}{D_m} \quad (8.8)$$

$$\mathrm{Sh} = \frac{k_s a_s d}{D_m} \quad (8.9)$$

where

ρ = the fluid density
u = the velocity
μ = the fluid viscosity
D_m = the diffusion coefficient
d = characteristic linear length traveled by the fluid

The characteristic length d is system-dependent and may represent an average particle diameter, the diameter of an impeller, or a pipe diameter, depending on the application. In general, these correlations have the functional form

$$\mathrm{Sh} = z\mathrm{Re}^x \mathrm{Sc}^y \quad (8.10)$$

where the constants x, y, and z vary depending on the system under consideration. While it is difficult to make broad generalizations regarding how to measure competing reaction kinetic and mass transfer effects over a diverse range of chemical environments, generalizations can be made for specific physical processes. To that extent, relevant mass transfer regimes will be discussed as follows.

8.2.3.1 Solid–Liquid Transfer Many industrial applications involve insoluble reagents, catalysts, or intermediates. In these cases, solid–liquid transfer effects need to be characterized and understood, but the deconvolution of mass transfer rates from reaction kinetics may be complex. A number of different mass transfer correlations for solid–liquid systems are available in the literature [19]. One issue that arises when utilizing correlations in the form of Eq. (8.10) is the formulation of the Reynolds number. A number of different modified particle Reynolds number expressions are presented in the literature [20]. Many studies have been published on the mass transfer to particles in both stirred tanks and pipes [21–27]. It should be noted that a variety of substrates (e.g. lead sulfate, barium sulfate, silver chloride) have been used for dissolution measurements, with some systems being especially susceptible to agglomeration and/or subjected to additional surface resistances. There is always some uncertainty when applying these correlations to a new system.

Understanding the limitations of correlations, it is often preferable to explicitly measure the mass transfer constant across the solid–liquid interface using dissolution measurements [28]. This approach allows a direct measurement of the mass transfer rate constant for comparison with the corresponding reaction rate constant. Alternatively, one could combine the rate constant for solid–liquid transfer with the intrinsic reaction rate constant and then use the relative activation energies as a means to deconvolute mass transfer and reaction-limited regimes. For a first-order reaction with rate constant k_r, the rate expression that combines the reaction rate and the rate of mass transfer can be written as Eq. (8.11) [29, 30]:

$$r_i = \frac{k_r k_s a_s}{k_r + k_s a_s} C_i \quad (8.11)$$

where $k_s a_s$ is the rate constants for mass transfer across the solid–liquid interface. In cases where the mass transfer across the solid-to-liquid interface is rapid ($k_s a_s \gg k_r$), the rate expression simplifies to $r_i = k_r C_i$, and chemical kinetics are rate controlling. In cases where the mass transfer across the solid–liquid interface is slow ($k_s a_s \ll k_r$), the rate expression simplifies to $r_i = k_s a_s C_i$, and mass transfer across the boundary layer is the rate-limiting step.

The temperature dependence of the rate constant k_r and $k_s a_s$ allows for deconvolution of the chemical kinetics from the mass transfer kinetics. The influence of temperature on the mass transfer rate is primarily through its influence on viscosity and/or diffusion coefficients. There is only a modest effect of temperature on these variables, and as a result, the mass transfer rates typically exhibit a weak dependence with temperature. As a general heuristic, activation energies for mass transfer-limited processes are typically on the order of 10–20 kJ/mol, compared with 40–60 kJ/mol for reaction-limited processes [31].

Calculations to estimate the transport from particles in heterogeneous reactions have been outlined by Zwietering [32]. These mass transfer rates are influenced by agitation speed, up to a certain point (called the *just-suspended* speed) beyond which the particles no longer form a layer at the bottom of the vessel. Experimental data and mathematical correlations indicate that the rate of mass transfer in solid–liquid systems changes appreciably up to the just-suspended speed for particles. Further increases in mixing intensity once solids have already been suspended give only marginal increases in mass transfer [20]. Changi and Wong illustrate one such example using a Grignard reaction that accounts for mass transfer effects and the corresponding intrinsic kinetic rate constant [33]. Although the presented model does not capture the actual mechanism for Grignard reagent formation and is purely phenomenological, it is able to capture the physical and chemical aspects of the process across various scales by using the Zwietering criterion to estimate the just-suspended speed for the particles.

8.2.3.2 Liquid–Liquid Transfer There have been several reviews documenting the effect of liquid–liquid mixing in pharmaceutical applications [34]. Among the powerful tools in characterizing liquid–liquid mixing are Bourne reactions; these are pairs of competing reactions designed such that the selectivity toward a slower-forming by-product is characteristic of the rate of mixing. The known rate constants of the Bourne reaction can be used to quantify mixing times and thus understand the interplay of mixing and chemical kinetics. Prudhomme and Johnson [35], Mahajan and Kirwan [36], and Singh and coworkers [37] have used such reaction systems to characterize different mixing geometries to enhance mixing efficiency and reduce mixing times.

For miscible liquid–liquid systems, the impact of mixing can be determined by conducting one of a number of diagnostic tests, depending on the reaction kinetics and the mode of mixing: micromixing or macromixing. Micromixing is associated with molecular diffusion and stretching of the fluid under small-length-scale laminar flow conditions, under which viscous forces dominate over inertial forces. Macromixing occurs in conventional batch or continuous stirred tank reactors due to mechanical agitation. This is typically in the turbulent regime. Correlations for mixing times for macromixing and micromixing regimes have been articulated in the literature [38]. Use of the Damköhler number offers guidance on determining the effect of mixing on reaction performance.

In lieu of correlations, the impact of the order of addition upon reaction performance is a simple metric for assessing mixing-limited regimes. Consider a case of parallel reactions, in which a stream of A is added to a vessel containing B to yield product C and an impurity D.

$$A + B \rightarrow C$$
$$A + B \rightarrow D$$

If the reaction is conducted in the reverse order of addition, i.e. a stream of B is added to a vessel containing A and no effect on rate of impurity formation is observed, then it can be concluded that mixing effects are negligible. This is because the two addition modes mimic conditions of highly segregated concentrations of either A or B. In contrast, if the order in B for formation of species C is greater than that for formation of D, then the two addition modes would afford different ratios of species C and D, hence mixing sensitivity could be pronounced upon scale-up.

There are several ways to manage mixing sensitivity. For example, static mixers or auxiliary mixing devices such as mixing elbows or vortex mixers can be used to enhance mixing while leaving the reaction kinetics unaffected. Alternatively, the reaction kinetics can be modified by leveraging differences in reaction rates between the different chemical pathways. This may involve running the reaction at a different concentration (to exploit differences in reaction order) or at a different temperature (to exploit differences in activation energy).

SCHEME 8.2 Debenzylation and corresponding fumaric acid reduction.

8.2.3.3 Gas–Liquid Transfer

Gas–liquid mixing plays a central role in a number of commercialized synthetic processes. Transport of gas into and out of solutions can drive reaction rates and selectivity. A procedure for measuring the rate of mass transfer from the gas to liquid phases has been detailed previously [39]. The integral approach for measuring the vapor–liquid mass transfer coefficient $k_L a$ is shown in Eq. (8.12):

$$k_L a \times t = \frac{P_f - P_o}{P_i - P_o} \ln \frac{P_i - P_f}{P(t) - P_f} \qquad (8.12)$$

where

- P_o is the solvent vapor pressure.
- P_f is the final system pressure.
- P_i is the initial system pressure.
- $P(t)$ is the system pressure measured during the course of the experiment.

Plotting the left-hand side of Eq. (8.12) versus time yields a slope with units of 1/time and represents the mass transfer constant from gas phase to liquid phase. Alternatively, the initial slope of the pressure drop at the start of an uptake experiment to estimate the value of $k_L a$ is given by Eq. (8.13):

$$k_L a \approx -\frac{dP(t)}{dt} \frac{1}{P_i - P_f} \qquad (8.13)$$

For both large- and small-scale measurements, it is important to understand the ramp-up time for an agitator to reach full power. Experimental details for measuring $k_L a$ and factors that affect gas–liquid mixing efficiency have been captured elsewhere [40].

As in the case of solid–liquid and liquid–liquid systems, the convolution of reaction rate with mass transfer from gas phase to liquid phase can be described using the Damköhler number:

$$\mathrm{Da} = \frac{r_{rxn}}{r_{MT}^{max}} = \frac{r_{rxn}}{k_L a \times C_{H_2}^{sat}} \qquad (8.14)$$

where

- r_{rxn} is the intrinsic reaction rate.

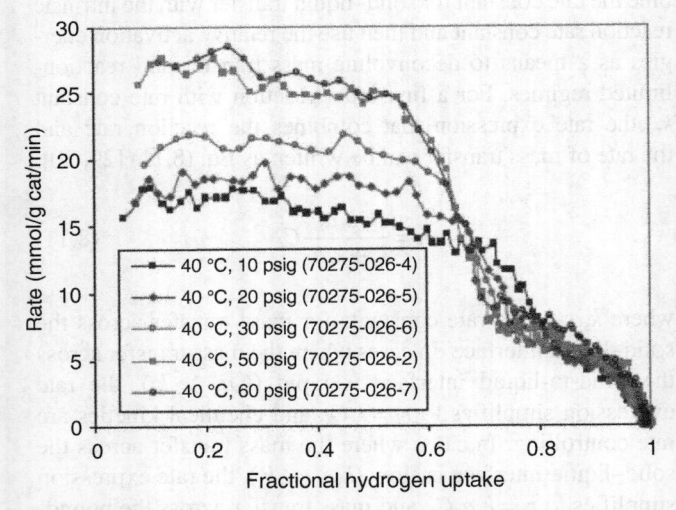

FIGURE 8.2 Rate profile for concomitant debenzylation and fumaric acid reduction over Pd/C for reaction in Scheme 8.2.

- r_{MT}^{max} is the maximum rate of transfer from the gas phase to the liquid phase.

A ratio of Da > 1 is indicative of mass transfer limitations whereas Da < 0.1 is indicative of a regime free of mass transport limitations. An example of the utility of the Damköhler number arises for debenzylation of a fumurate salt of an amine to give the corresponding succinate salt of the secondary amine (see Scheme 8.2). The hydrogenation process initially involves reduction of the fumaric acid to succinic acid followed by debenzylation to form the corresponding secondary amine succinate salt. The reaction rate profile as a function of hydrogen pressure is shown in Figure 8.2. The results indicate a positive-order dependence of the rate of fumaric acid reduction on hydrogen pressure compared with zero-order dependence for debenzylation. Hydrogen starvation resulted in significant decrease in the rate of fumaric acid reduction with little or no effect on the rate of debenzylation, resulting in the accumulation of the fumaric acid in the presence of a secondary amine, thereby increasing the propensity for the formation of the Michael adduct **7** (Scheme 8.3).

The Damköhler number for this process is defined by Eq. (8.15):

$$\mathrm{Da} = \frac{\text{rate of fumaric acid hydrogenation}}{\text{rate of hydrogen transfer from gas phase to liquid phase}} \qquad (8.15)$$

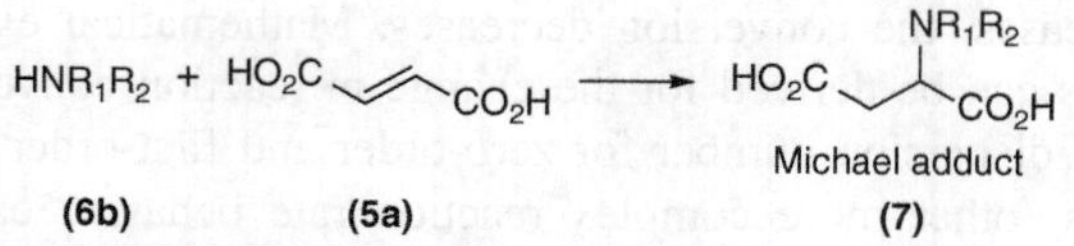

SCHEME 8.3 Michael adduct formation reaction.

When Da < 1, the rate of hydrogen transfer from the gas phase to liquid phase is rapid compared with fumaric acid reduction, and as a result the hydrogenation proceeds rapidly. When Da > 1, the rate of hydrogen transfer is slower than the rate for fumaric acid reduction; as a result, the rate of fumaric acid reduction is slowed to the point that subsequent debenzylation can occur simultaneously, thereby allowing the deprotected secondary amine to react with the fumaric acid to form the Michael adduct.

In considering gas–liquid reactions, one may also need to account for cases when gas is desorbed from liquid phase to gas phase. This is routinely encountered during oxygen-sensitive reactions such as asymmetric hydrogenations, coupling reactions in which trace concentrations of oxygen can poison catalysts, or decarboxylation reactions in which effective desorption of CO_2 is necessary prior to forward processing. The fundamental rate expression that describes this driving force is similar to that for the rate of transfer from gas phase to liquid phase. Specifically, the rate can be described by

$$\text{Desorption rate} = \frac{dC}{dt} = k_L a(C - C^*) \quad (8.16)$$

where

- $k_L a$ is the mass transfer coefficient of the system.
- C is the solution-phase concentration of the gas at a given time.
- C^* is the equilibrium concentration of the gas described by Henry's law.

It must be noted that the $k_L a$ describing the desorption rate constant is different from that for absorption processes. Depending on the measurement approach, the value of C^* may vary during the measurement process, and an additional mass balance in the gas phase would be necessary.

The reactor design and configuration will influence the mass transfer rate of gas–liquid reactions. Johnson et al. illustrated a comparison of three different designs of continuous reactor types (coiled tubes, horizontal pipes in series, and vertical pipes in series) for a direct asymmetric reductive amination reaction [41]. For all three continuous reactors analyzed, it was shown quantitatively that sufficient mass transfer rates in terms of $k_L a$ were obtained for reaction residence times on the order of hours, with the reaction kinetics being rate limiting. A comparison of the flow reactors showed that the vertical pipes-in-series reactor had the highest $k_L a$, followed by horizontal pipes in series, and lastly the coiled tubes. On account of the higher $k_L a$, only 3 equiv. of hydrogen were needed for the production scale. The use of a continuous reactor in production led to a substantial reduction in process volume and enhanced process safety.

8.2.3.4 Gas–Liquid–Solid Transfer
Certain pharmaceutical catalytic reaction systems involve three phases (e.g. solid, liquid, and gas phases for a catalytic reduction in a trickle bed reactor [42]) or even four phases (e.g. hydrogenation in a gas–liquid–liquid–solid system of nitrobenzene to p-aminophenol, an intermediate for paracetamol [43]). The complexity of such situations generally warrants a comprehensive assessment of several factors such as competing reaction rates, solubility changes, and changes in adsorption and desorption rates due to evolving hydrodynamic profiles in the reactor. From a reaction engineering perspective, the following considerations must be made:

- Identification of the physical and chemical processes for the different phases under consideration.
- Identification of any reactions occurring in an interfacial boundary layer.
- Formulation of rate equations to account for the kinetics and mass transfer rates in the various phases.
- Identification of the rate-limiting regime under the operating conditions, using the available correlations in literature and experimental measurements.

Mills and Chaudhari [17] have reviewed extensively different kinetics rate models in literature, while several textbook chapters discuss the performance equations for scale-up of multiphasic reactions in detail [4, 13].

8.2.4 Thermal Considerations

Just as the overall rate of a reaction can be hindered by the transport of molecules, the rate at which energy is delivered or removed from a chemical system can also affect the rate at which the reaction proceeds. Reactions that are either exothermic or endothermic will create an imbalance of heat within a reactor, which may impact the conversion and selectivity of a process if the reaction temperature is not controlled adequately [44]. Similar to the Damköhler expression, Eq. (8.17) defines a dimensionless number (β) to express the rate of heat formation to the heat removal by jacket services [18]:

$$\beta = \frac{\text{heat generated}}{\text{heat removed}} = \frac{-r\Delta H_{rxn} d_B}{6\Delta T_{ad} h} \quad (8.17)$$

where

r is the reaction rate.
ΔH_{rxn} is the heat of reaction.
d_b is the vessel diameter.
ΔT_{ad} is the adiabatic temperature rise.
h is the convective heat transfer coefficient.

When $\beta < 1$, physical heat removal from a system is not a concern, and the outcome of the reaction is predictable based on intrinsic kinetics. Hartman et al. [18] have reviewed the heat transfer considerations for reactions carried out in different operation modes (microreactor, batch, and continuous flow), including several pharmaceutical examples, showing the interplay of heat transfer impacting the reaction outcome. Classic examples of chemical systems where energy considerations are important include lithiations, diazotizations, Grignard reactions, reductions, and oxidations.

8.2.5 Axial Dispersion

All continuous processes are impacted by probabilistic variations in the amount of time each of the reacting components spends inside the reactor. These variations are a result of the extent of mixing that occurs in the reaction process and will be different depending upon whether the reaction is performed in a plug flow reactor (ideally no mixing) or in a continuously stirred tank reactor (ideally complete mixing). For a tubular reactor, two physical phenomena – diffusion and convection – contribute to the degree of mixing observed within the reactor. This tubular reactor mixing is observed experimentally as a diffusion-like spreading of material and is called dispersion. For a complete evaluation of reaction performance, it is important to understand the impact of dispersion upon observed reaction kinetics and to recognize cases where assumptions of ideal mixing do or do not apply. An excellent discussion can be found in the book by Levenspiel [4].

For flow through a tube, the dispersion D (in m²/s) is a function of the molecular diffusion D_m, the velocity u, and the tube diameter d_t:

$$D = D_m + \frac{u^2 d_t^2}{192 D_m} \qquad (8.18)$$

The extent to which dispersion influences reactor performance is determined primarily by the dispersion number, expressed as D/uL, where L is the reactor length. For values of $D/uL \ll 1$, the assumption of plug flow behavior in the reactor is reasonable. As the dispersion number increases, the significance of mixing in the reactor also increases. For any nth-order isothermal reaction, the conversion obtained in a mixed-flow reactor is always less than that obtained in the ideal case of plug flow. Hence, as the rate of dispersion increases, the conversion decreases. Mathematical expressions can be derived for the change in reaction conversion with dispersion number for zero-order and first-order reactions (other more complex reaction rate behavior can be captured numerically) [13]. Figure 8.3 illustrates the impact the dispersion number will have upon conversion for the case of first-order reaction kinetics as a function of the Damköhler number ($k\tau$) for any single-input, single-output continuous reactor. For a reactor where $D/uL = 0.1$, the reaction time required to achieve 90% conversion is more than 20% greater than the time required to achieve the same conversion in an ideal, unmixed plug flow reactor. Likewise for $D/uL \geq \sim 10$, the reaction effectively progresses as slowly as it would progress in a well-mixed stirred tank reactor.

Packed bed columns require careful consideration of axial dispersion for successful scale-up. Delgado has critically reviewed the phenomenon of dispersion for packed beds and presented several empirical correlations for prediction of dispersion coefficient over different flow regimes [45].

For direct asymmetric reductive amination, Changi et al. showed that the combination of reaction kinetics and dispersion understanding can be used to simulate the performance of a plug flow reactor at manufacturing scale [46]. The impact of variation in catalyst pump flow rate was considered for a complex reaction network comprising eleven species. The model predicted acceptable product quality along the length of the reactor, consistent with the manufacturing results.

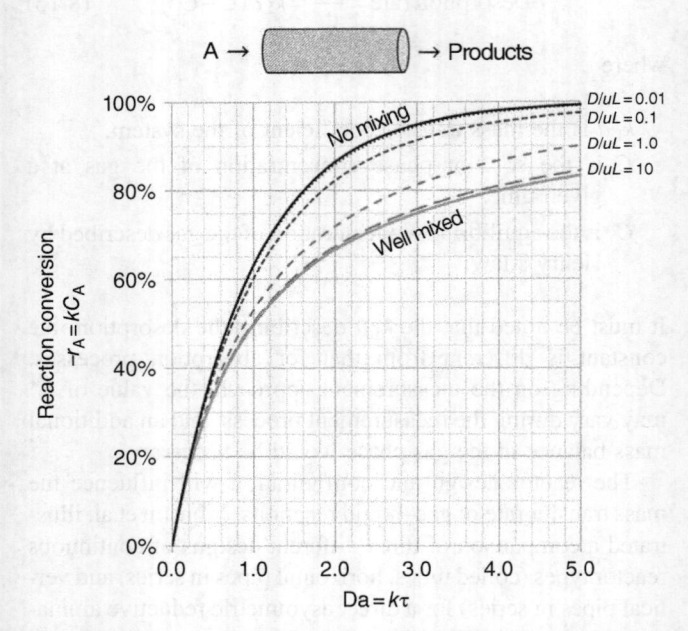

FIGURE 8.3 Reaction conversion as a function of Damköhler number (Da) and dispersion number (D/uL) for the case of first-order reaction kinetics. Profile assumes a single-input, single-output continuous reactor and is derived analytically in Ref. [13]. "No mixing" case is provided for the limit of $D/uL \to 0$; "well mixed" is in the limit of $D/uL \to \infty$.

8.3 METHODS FOR THE CHARACTERIZATION OF CHEMICAL KINETICS

The previous section introduced the fundamental considerations for assessing a chemical system in terms of kinetics, mass transfer, and heat transfer. Keeping this background in mind, this section focuses upon the experimental and analytical tools available for understanding competing rates, which will be needed in order to optimize a chemical system for yield and selectivity. Several different instruments and technologies are available to aid in reaction kinetics measurements. This area is constantly evolving as the levels of automation and analyzer sophistication increase.

8.3.1 Calorimetry

Reaction calorimetry is a versatile and highly effective tool for reaction characterization in the pharmaceutical industry. The technique requires conducting an energy balance around the batch reactor, yielding the following:

$$MC_p \frac{dT_r}{dt} = UA(T_j - T_r) + r_{rxn}\Delta H_{rxn} + MC_p(T_{addn} - T_r) \quad (8.19)$$

where

M is the reaction mixture mass.
C_p is the heat capacity of reaction mixture.
UA is the heat transfer coefficient.
T_j is the jacket temperature.
T_r is the reactor temperature.
r_{rxn} is the reaction rate.
ΔH_{rxn} is the heat of reaction.
T_{addn} is the temperature of added stream.

The measurement can be conducted in an isothermal or nonisothermal mode, which changes the relevant terms in Eq. (8.19). Since this technique measures the total heat of reaction, it convolutes the heat associated with several chemical processes including heats of mixing, dissolution, and crystallization, as well as heats associated with all reactions including the desired reaction and side reactions. For safety testing, this is ideal since such a measurement allows a lumped measurement of heat associated with all relevant chemical events in the process. For measurement of detailed reaction kinetics requiring deconvolution of different processes, reaction calorimetry offers the advantage that subtle changes in concentration profiles are magnified in heat flow measurements, since the heat flow is directly proportional to the reaction rate. This methodology has been routinely highlighted in the work of Blackmond and coworkers for the example of cross-coupling reactions [47–50]. A systematic use of reaction progress kinetic analysis using an *in situ* reaction calorimeter has also been documented by Blackmond and coworkers, and several review articles articulate this approach in great detail [47, 48].

One important caveat when measuring rapid reaction kinetics, especially when the process kinetics are of the same scale or faster than the equipment time constant, is that the measured rate constant can vary significantly. Table 8.2 shows a comparison of the rate constant for acetic anhydride hydrolysis found by calorimetry with that from literature. As the reaction half-life is shortened to less than one minute, the difference between the measured reaction rate and the literature value increases. A number of different algorithms are available for deconvoluting the equipment time constant from the measured kinetics [52]; however, this process can be a black box. Nevertheless, these results indicate that reaction calorimetry can adequately measure reaction rates under synthetically relevant conditions with half-lives greater than one minute.

Results from reaction calorimetry are further enhanced when orthogonal techniques are utilized in parallel. One such example of using orthogonal techniques is in the kinetic investigation of heterogeneous catalytic hydrogenation of nitro compounds shown in Scheme 8.4 [53]. Hydrogen uptake and reaction calorimetry data are shown in Figure 8.4 [54]; similar temporal profiles are observed with both hydrogen uptake and reaction calorimetry. Concomitant LC sampling indicated that the zero-order kinetics observed during the first 120 minutes, as evidenced by a flat temporal hydrogen uptake profile, are attributed to hydrogenation of the nitro moiety to the corresponding hydroxyl amine, as shown in Scheme 8.5.

Taking the ratio of the two curves shown in Figure 8.4 yields the plot in Figure 8.5, which allows for deconvolution of the energetics of hydroxylamine formation from those of

TABLE 8.2 Comparison of Reaction Kinetics for Acetic Anhydride Hydrolysis in the Presence of Acetic Acid Using an Omnical Z3 Calorimeter to Values from Literature

Temperature (°C)	k_{obs} (s^{-1})	k_{lit} (s^{-1}) Ref. [51]	Measured Half-Life (s)	Expected Half-Life from k_{lit} (s)
55	0.017	0.024	41	29
45	0.012	0.011	58	63
35	0.005 85	0.005 25	118	132

SCHEME 8.4 Hydrogenation of 1-(4-nitrobenzyl)-1,2,4-triazole.

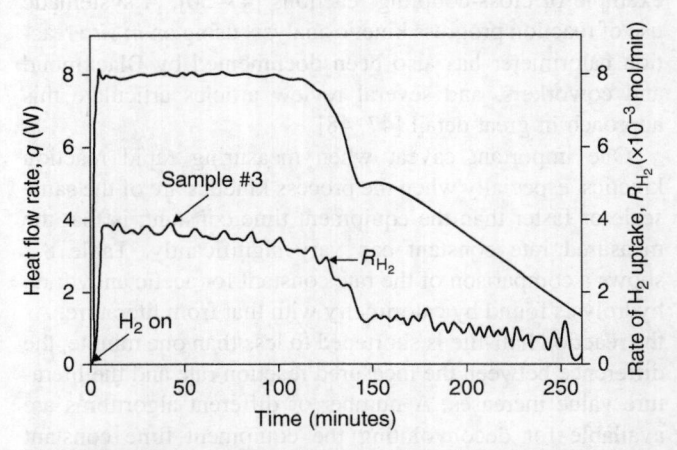

FIGURE 8.4 Temporal hydrogen uptake and reaction calorimetry for hydrogenation shown in Scheme 8.4. *Source*: Reprinted with permission from LeBlond et al. [53]. Copyright 1998, Wiley-Blackwell.

$$Ph\text{-}NO_2 + 2H_2 \longrightarrow Ph\text{-}NHOH + H_2O$$
$$Ph\text{-}NHOH + H_2 \longrightarrow Ph\text{-}NH_2 + H_2O$$

SCHEME 8.5 Stepwise reduction of the nitro moiety.

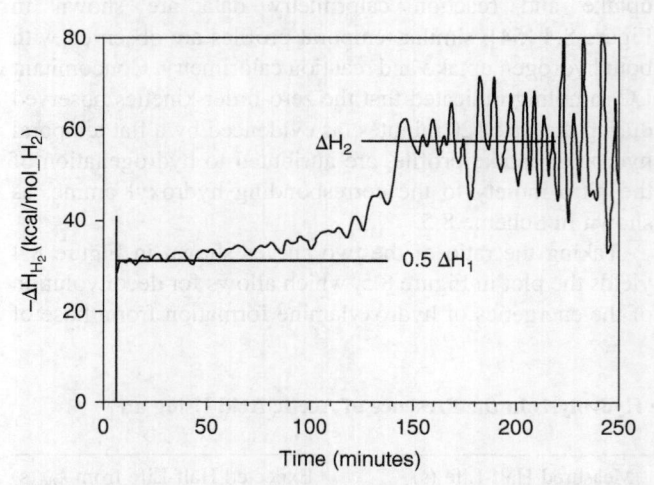

FIGURE 8.5 Ratio of temporal hydrogen uptake and calorimetry to elucidate the energetic of stepwise hydrogenation kinetics. *Source*: Reprinted with permission from LeBlond et al. [53]. Copyright 1998, Wiley-Blackwell.

amine formation. The corresponding energetics extracted from the graph were found to be −65 and −58 kcal/mol for the first and second reductions, respectively. Such information and characterization is useful for safety assessment as well as reaction optimization. Understanding of reaction orders and energetics for each pathway in the reaction can be used to understand the operating design space. This example highlights the power of using orthogonal techniques to characterize reaction kinetics. Clearly the use of any one of the analytical techniques alone was not as powerful as the synergy of leveraging hydrogen uptake and calorimetry with off-line LC measurements. This theme of employing simultaneous orthogonal analytical techniques to probe reaction kinetics is elaborated in a review [54], and we will revisit this theme in the ensuing section on process analytical technology (PAT).

Other calorimetry types, especially accelerated rate calorimetry (ARC), are frequently used for process safety evaluation. Several other reviews have been written discussing the details of ARC testing and analysis [55].

8.3.2 Parametric Measurements

Physical measurements taken during the process can also serve as a means to track reaction progress and characterize reaction kinetics. These physical measurements can take many forms; however, temperature, gas flow, and pH are three more common measurements to characterize reactions. As mentioned with calorimetry, such measurements lump several different chemical events; hence caution must be exercised for complex reaction systems.

Gas uptake measurements are particularly useful for multiphasic reactions such as hydrogenations, as outlined in the preceding example. As with calorimetry, care must be taken to ensure that the observed gas uptake measurement is correlated with the desired chemical transformation that is being tracked. Side reactions such as over-reduction of desired products or catalyst reduction often mask the details of the chemical transformation that is to be tracked. Conversely, gas evolution measurements can also be used to track progress. This is frequently the case for decarboxylation reactions in which CO_2 evolution can be used to monitor and characterize decarboxylation kinetics.

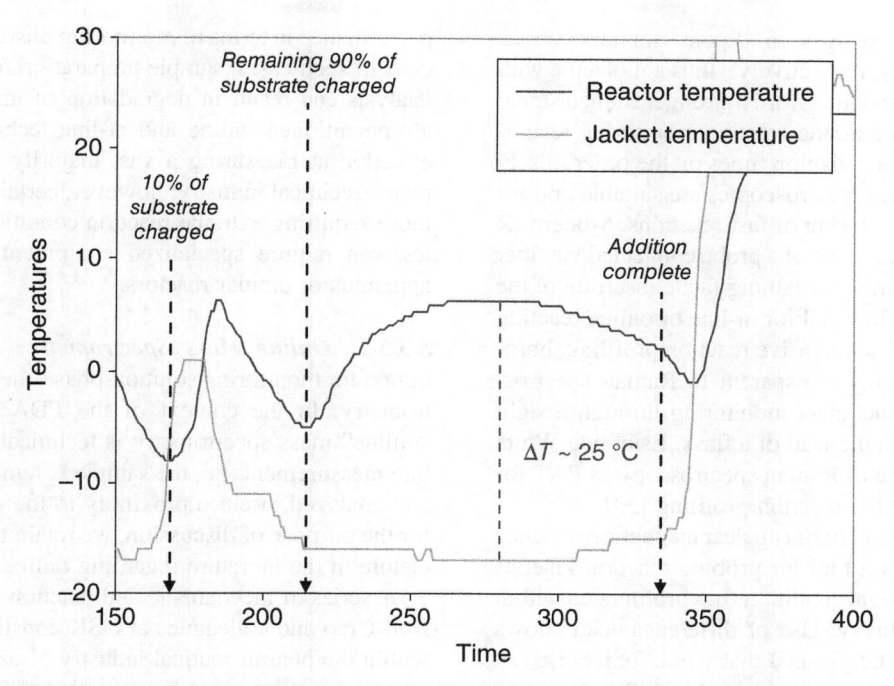

FIGURE 8.6 Reactor and jacket temperature profiles during the formation of a Grignard in a 200 gal reactor. Both the initiation and post-initiation reactive regimes are indicated.

Temperature has been used for decades to track reaction progress and is sometimes mistakenly neglected in favor of more complicated online sensors. Tracking reaction progress with temperature, especially for exothermic reactions such as Grignard reactions, is effective. Figure 8.6 shows the tracking of reaction progress at 200 gal scale during a benzyl Grignard formation. Initiation is evident during the time span of 150–200 minutes, followed by formation of the Grignard reagent in a feed-limited manner up to approximately 330 minutes. The use of these physical measurements allows characterization and estimation of reaction rate constants both on laboratory scale and pilot plant scale, which, in turn, can be used to understand scale sensitivity.

8.3.3 Process Analytical Technology

Particularly since the issuance of formal FDA guidance on the topic of PAT in 2004, the pharmaceutical industry has witnessed broad adoption of online and in-line technologies that have proven effective for reaction characterization and measurement of reaction kinetics. While a detailed review of the various types of PAT employed in reaction kinetic studies is beyond the scope of this chapter, readers are referred to a comprehensive, multiauthor review of PAT applications within the industry [56]. Another recommended resource is a review from members of the IQ Consortium on the topic of PAT applications in drug substance process development [57].

Per FDA guidance, the following nomenclature holds with respect to modes of PAT implementation [58]:

- *At-line measurement*: The sample is removed, isolated from, and analyzed in close proximity to the process stream.
- *Online measurement*: The sample is diverted from the manufacturing process and may be returned to the process stream.
- *In-line measurement*: The sample is not removed from the process stream.

In keeping with the emerging PAT paradigm, a general trend within pharmaceutical process development in recent years has involved movement from off-line and at-line analyses toward newly developed online and in-line alternatives. The following sections relate this general trend toward online and in-line analyses to the specific areas of spectroscopy, mass spectrometry, and high performance liquid chromatography (HPLC).

8.3.3.1 Online Spectroscopy
While several online spectroscopic techniques are available, infrared (IR) and Raman spectroscopies are the two techniques that have been most commonly used by practicing chemists and engineers to extract detailed reaction kinetics and mechanistic information. IR and Raman spectroscopies are complementary techniques, but selection rules for IR- and Raman-active

vibrations differ (net changes in dipole moment versus changes in polarizability, respectively); thus a molecule with weak IR signal can potentially afford a stronger Raman signal and vice versa. Both are nondestructive monitoring techniques, and with spectral acquisition times on the order of seconds, both IR and Raman spectroscopies are suitable options for online or in-line monitoring of fast reactions. Modern IR and Raman instruments consist of a probe connected via fiber optic cable to a spectrometer, enabling facile insertion of the probe into a reactor or flow cell for in-line or online reaction monitoring. In terms of noninvasive reaction profiling, borosilicate glass is essentially transparent to Raman spectroscopy, thus enabling noncontact monitoring through a sight glass or directly through the wall of a flask. Esmonde-White et al. reviewed the scope of Raman spectroscopy as PAT for pharmaceuticals, including reaction profiling [59].

In recent years, the use of *in situ* nuclear magnetic resonance (NMR) spectroscopy as a tool for probing reaction kinetics under synthetically relevant conditions has proliferated within the pharmaceutical industry. Use of different nuclei allows specific information to be gleaned that would otherwise not have been possible by conventional methods. Reaction profiles are obtained via analysis of peak integrals from sequentially acquired NMR spectra. An important development in recent years has been the advent of spectroscopy-grade compact magnets, which has facilitated development of low-field compact NMR spectrometers, enabling direct deployment of online NMR reaction monitoring capabilities to laboratory chemists and engineers. A review of low-field NMR spectroscopy, including several examples of reaction profiling via benchtop NMR, can be found in the literature [60].

One challenge associated with the use of low-field benchtop NMR spectrometers is loss of spectral resolution, often resulting in significant peak overlap. 2D NMR can afford enhanced spectral resolution, but the long acquisition duration makes this a suboptimal solution for time-sensitive applications such as reaction profiling. Gouilleux et al. report the application of ultrafast NMR methodology to a compact NMR spectrometer, affording ultrafast 2D NMR at low field [61]. A reaction monitoring case study is presented, in which ultrafast 2D NMR spectroscopy affords spectral acquisition every 2.6 minutes.

A report from Foley and colleagues at Pfizer demonstrates significant differences in kinetic data obtained from static NMR tube experiments versus online NMR (wherein a continuous flow of process solution is withdrawn from a well-mixed reaction vessel, subjected to NMR analysis, and then returned to the reaction vessel) [62]. These differences were attributed to the lack of mixing in the NMR tube experiments. For studies intended to extract detailed kinetic data, online NMR is the preferred configuration, as it allows the bulk solution to be maintained in the reaction vessel and with adequate agitation for the duration of the experiment.

A general advantage of in-line and online spectroscopic techniques over at-line or off-line methods is their superior performance in terms of monitoring unstable or transient species; in such cases, sample preparation for at-line or off-line analysis can result in degradation of unstable species. The aforementioned online and at-line technologies are highly effective at measuring a vast majority of processes in the pharmaceutical industry; however, certain applications, such those requiring extreme reaction conditions and rapid kinetics, can require specialized equipment such as stop-flow apparatus or tubular reactors.

8.3.3.2 Online Mass Spectrometry An emerging PAT option for monitoring solution-phase chemistry is mass spectrometry. In the context of the FDA guidance on PAT, "online" mass spectrometry is technically an automated at-line measurement (i.e. the sample is removed, isolated from, and analyzed in close proximity to the process stream), but for the purpose of discussion, we retain the common nomenclature in the literature regarding online mass spectrometry.

A series of mechanistic and reaction kinetic studies from Dell'Orco and colleagues at GSK constitute early examples within the pharmaceutical industry of time-resolved electrospray ionization mass spectrometry (ESI-MS) as an online monitoring tool for solution-phase chemistry [63–65]. A notable application of this work was realized in a publication describing the optimization of a semi-batch reaction protocol for production of an intermediate in the commercial manufacture of eprosartan, wherein the reaction mechanism was elucidated and experimental data for kinetic parameter fitting was obtained via online ESI-MS. [66] Analogous to recent developments in compact NMR spectrometers, recently developed small footprint mass spectrometers have facilitated broader use of online MS, as the analyzer can be more easily staged in close proximity to the process chemistry [67]. A review of recent advances in reaction monitoring by online MS is available in the literature [68].

Revisiting the theme of employing simultaneous orthogonal analytical techniques to enable robust kinetic and mechanistic studies, a recent investigation of the hydroacylation reaction of 2-(methylthio)benzaldehyde with 1-octyne, catalyzed by a cationic rhodium catalyst, demonstrates the power of coupling online IR with online MS [69]. Conversion of the aldehyde substrate to the ketone product was monitored by online IR, while online ESI-MS was simultaneously employed to probe for catalytically relevant species, some of which were detected at concentrations five orders of magnitude less than the initial substrate concentration.

Figure 8.7 presents online IR spectra of aldehyde substrate and ketone product, as well as online ESI-MS spectra of the precatalyst and resting state, catalyst impurities, a reaction intermediate, and catalyst decomposition products. The species identified in Figure 8.7b–e were detected by ESI-MS at concentrations of approximately (b) 1/20th, (c) 1/8 000th, (d) 1/50 000th, and (e) 1/100 000th the initial substrate concentration. The proposed catalytic cycle based on this mechanistic study is presented in Scheme 8.6. Conversion of substrate

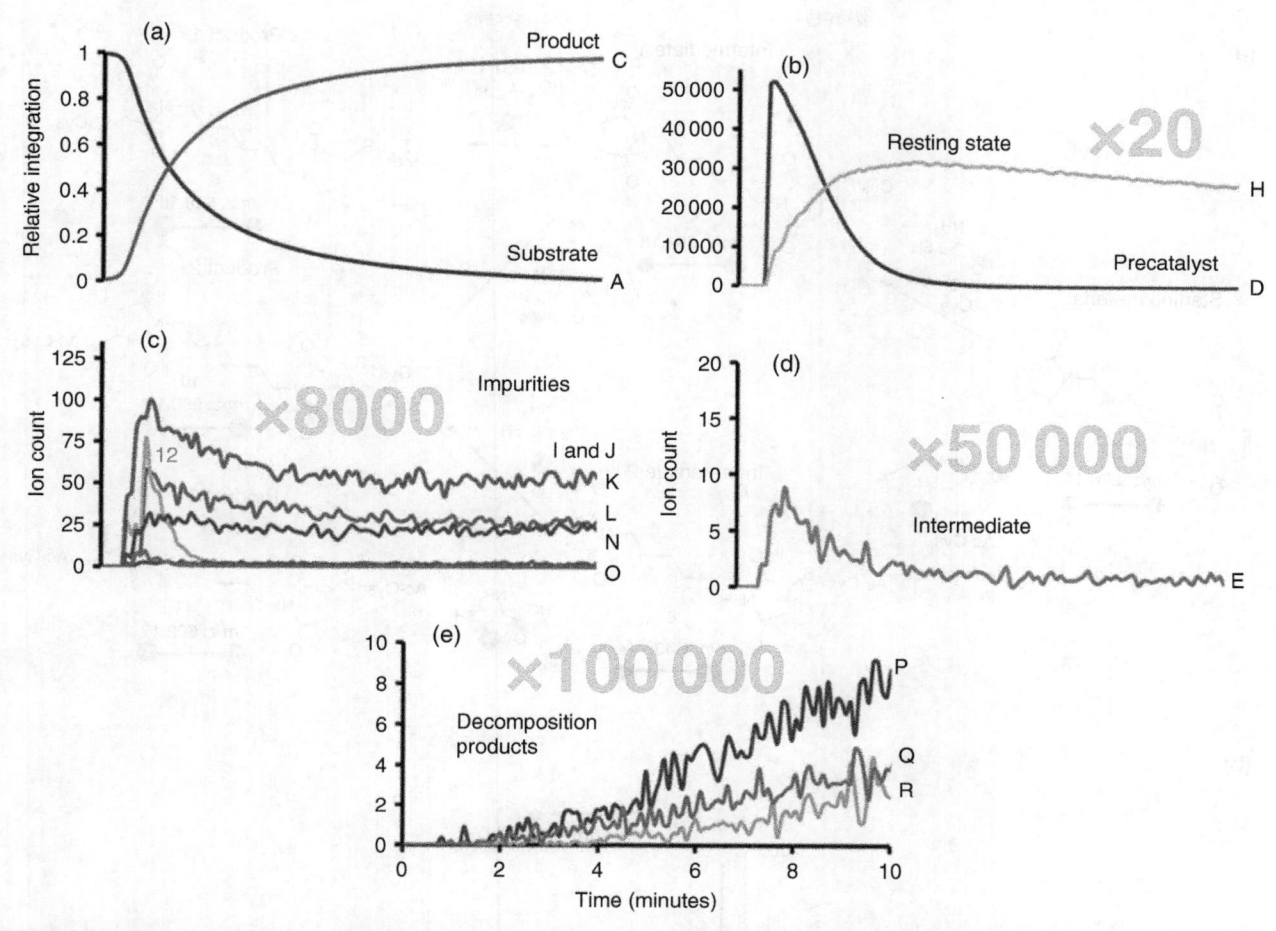

FIGURE 8.7 IR spectra of (a) substrate and product, and ESI-MS spectra of (b) precatalyst and catalyst resting state, (c) catalyst impurities, (d) reaction intermediate, and (e) catalyst decomposition products. *Source*: Reprinted with permission from Theron et al. [69]. Copyright 2016, American Chemical Society.

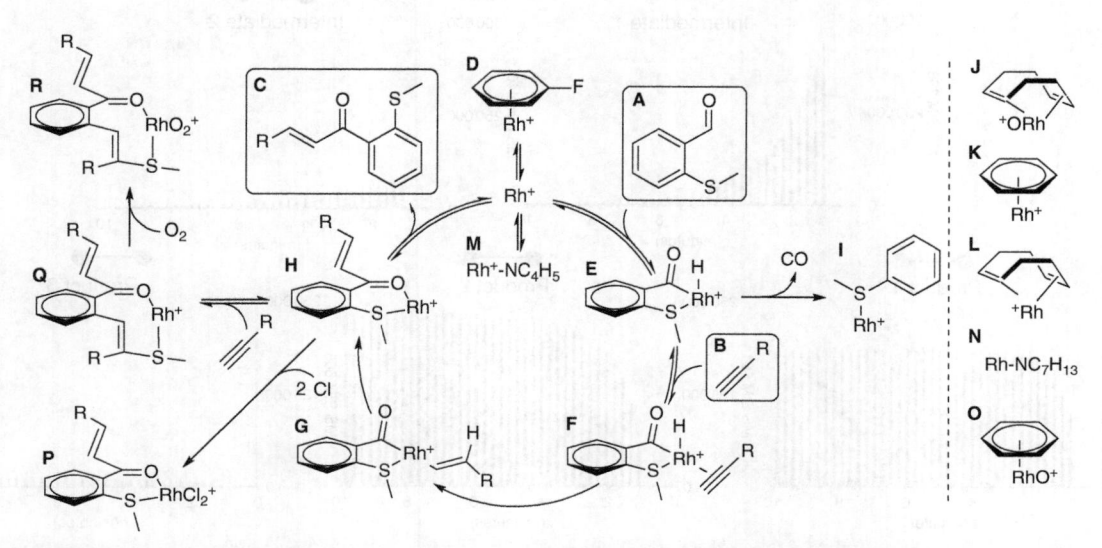

SCHEME 8.6 Catalytic cycle, proposed based on online reaction studies in Figure 8.7. Labels correlate with those in Figure 8.7. *Source*: Reprinted with permission from Theron et al. [69]. Copyright 2016, American Chemical Society.

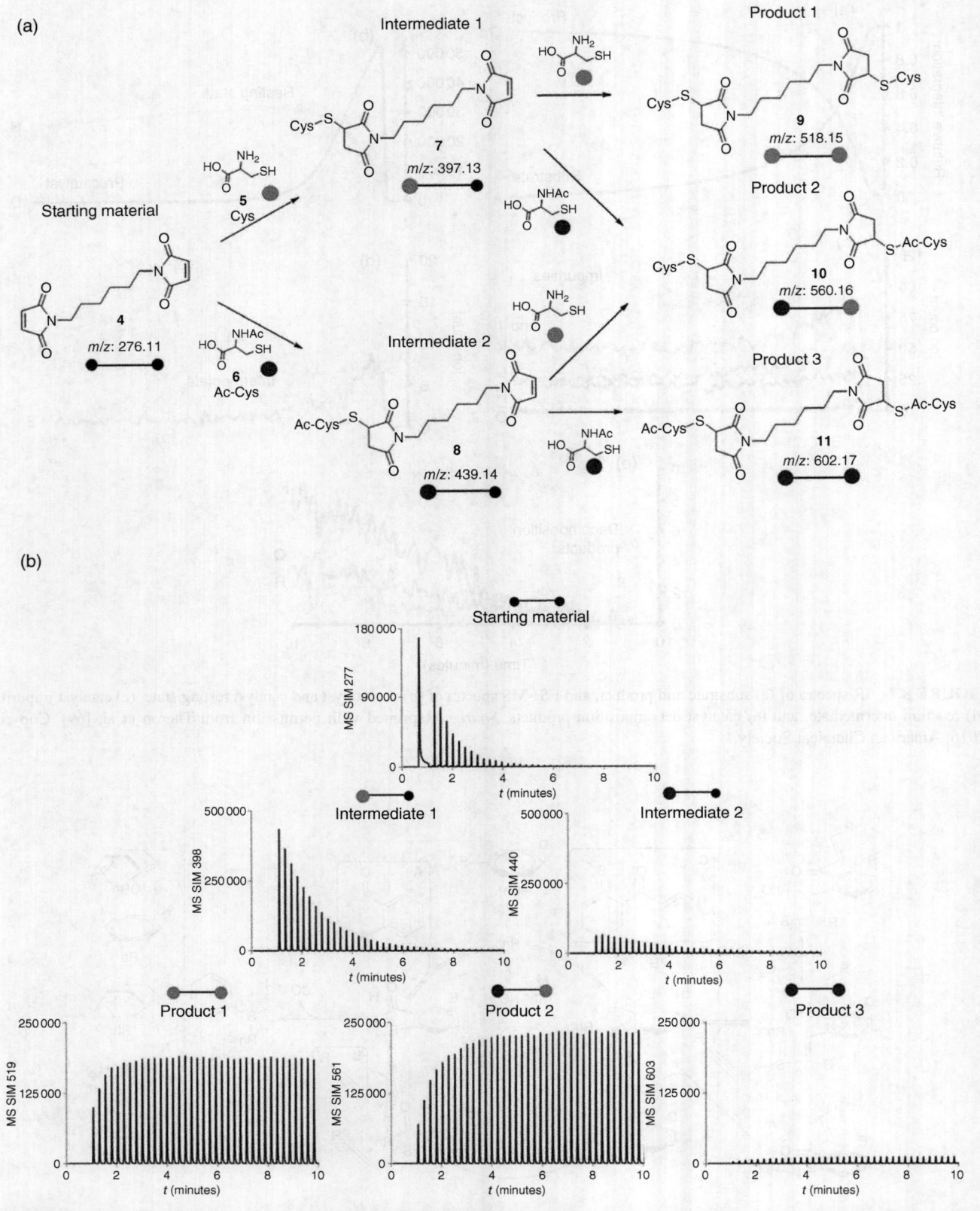

FIGURE 8.8 Kinetic profiling of the reaction of bismaleimidohexane with a 1:1 mixture of cysteine and *N*-acetylcysteine, analyzed by MISER LC-MS at an injection frequency of 14 seconds. (a) Reaction pathway. (b) MISER LC-MS profiles of reaction species. *Source*: Reprinted with permission from Zawatzky et al. [72]. Copyright 2017, Elsevier.

to product is first order, $k = 0.011 \pm 0.001$ s^{-1}; thus the reaction time scale is too fast for robust monitoring via NMR. As mentioned in the discussion of online spectroscopy, IR and Raman spectroscopies are complementary techniques, and in the event that the substrate and/or product in this study had not been IR active, one could envision the substitution of Raman spectroscopy to enable an analogous dual-monitoring strategy (i.e. IR or Raman spectroscopy to monitor the main reaction and ESI-MS to probe for catalytically relevant species down to part per million concentrations).

8.3.3.3 Online HPLC
Concentration measurements by HPLC provide a powerful means to track reaction progress, especially with complex reaction networks and when tracking impurities at levels less than 0.5%. Automated sampling and online HPLC measurements can significantly decrease the time required of a process development scientist to profile a reaction, relative to manual sampling and off-line HPLC analysis. Additionally, recent advances have allowed researchers to innovatively use a system of online HPLCs in series to sample and track the *in situ* progress of high pressure reactions, identify impurities, and gain better understanding of the reaction system with increased safety and circumventing error associated with manual sampling [46]. As mentioned in the case of online MS, in the context of the FDA guidance on PAT, "online" HPLC is an automated at-line measurement (i.e. the sample is removed, isolated from, and analyzed in close proximity to the process stream), but in the ensuing discussion we retain the common nomenclature in the literature regarding online HPLC.

Sampling a minimum of 5–10 points across the reaction gives qualitative data regarding overall reaction kinetics. Because of the separation capability and sensitivity of HPLC analysis, the kinetics of minor and major pathways leading to low-level impurities as well as desired intermediates and products can be followed in this manner. To generate a richer set of data for quantitative analysis, more frequent sampling is required. This can be accomplished by means of integral data for concentrations of the major species using IR or Raman spectroscopy or online HPLC with samples taken at intervals of about 2–4% conversion.

The advent of "multiple injections in a single experimental run" (MISER) chromatography by Welch and colleagues has pushed the boundaries of LC as mobile tool for online reaction monitoring [70]. Standard HPLC reaction profiling (both online and off-line) involves a series of time point samples, each of which is analyzed as an individual chromatogram, post-processed, and then compiled and analyzed to afford a reaction profile for a single experiment. In the MISER approach, multiple injections occur over the course of a single isocratic chromatography run. For a reaction profiling experiment, the end product is a graph derived from sequential injections, the shape of which is directly correlated to the reaction profile [71].

An example from Welch and colleagues illustrates the MISER approach to reaction profiling via LC-MS [72].

Bismaleimidohexane was subjected to competitive coupling with a 1 : 1 mixture of cysteine and *N*-acetylcysteine (Figure 8.8a). Selected ion monitoring (SIM) of substrate, intermediates, and products at an injection frequency of 14 seconds (Figure 8.8b) afforded rich kinetic information, including relative rates of formation and consumption of the two monoadduct intermediates. Quantitative interpretation of MISER LC-MS chromatograms can be complicated by nonlinear MS responses, as well as ion suppression and enhancement effects. Nonetheless, the SIM reaction profiles afforded by the MISER LC-MS approach can provide key mechanistic insights and allow for direct assessment of relative reaction rates.

To extract quantitative data from online HPLC chromatograms, relative response factors must be applied for the analytes of interest. Obtaining reference standards for direct assessment of relative response factors can be challenging in the earlier stages of process development or for unstable intermediates. Revisiting once more the theme of employing simultaneous orthogonal analytical techniques to enable robust kinetic and mechanistic studies, a report from Foley and colleagues at Pfizer describes simultaneous reaction monitoring via online NMR and online HPLC as an efficient means to establish relative response factors [73]. The approach was demonstrated for the reaction of aniline and 4-fluorobenzaldehyde to afford *N*-(4-fluorobenzylidene)aniline (**12**) (Scheme 8.7). The imine product undergoes hydrolysis due to the water generated as a reaction by-product, establishing an equilibrium between substrate and product.

Consumption of the 4-fluorobenzaldehyde substrate and formation of the imine product were simultaneously monitored by online ^{19}F NMR and online HPLC. Figure 8.9a presents the mole percent of the aldehyde substrate and imine product as determined by ^{19}F NMR (open squares), overlaid with the corresponding HPLC area percent (solid lines) over the time course of the reaction. Comparison with the quantitative NMR data suggested a UV under-response for the aldehyde substrate along with UV over-response for the imine product. The online NMR data were employed directly to determine the HPLC relative response factors for the substrate and product, the results of which are shown in Figure 8.9b. Combinations of approaches such as NMR-derived HPLC relative response factors and MISER online HPLC could provide an efficient route to highly automated reaction profiling experiments that provide a wealth of quantitative data to support kinetic model development while circumventing the need for extensive preparation of reference standards for impurities and (potentially unstable) process intermediates.

8.4 TRANSFORMING EXPERIMENTAL DATA INTO A KINETIC MODEL

By nature, development of a kinetic model requires simplification of what is often a complex reaction system into a refined set of pathways or elementary steps. Depending on

SCHEME 8.7 Condensation of aniline with 4-fluorobenzaldehyde to afford imine product N-(4-fluorobenzylidene)aniline.

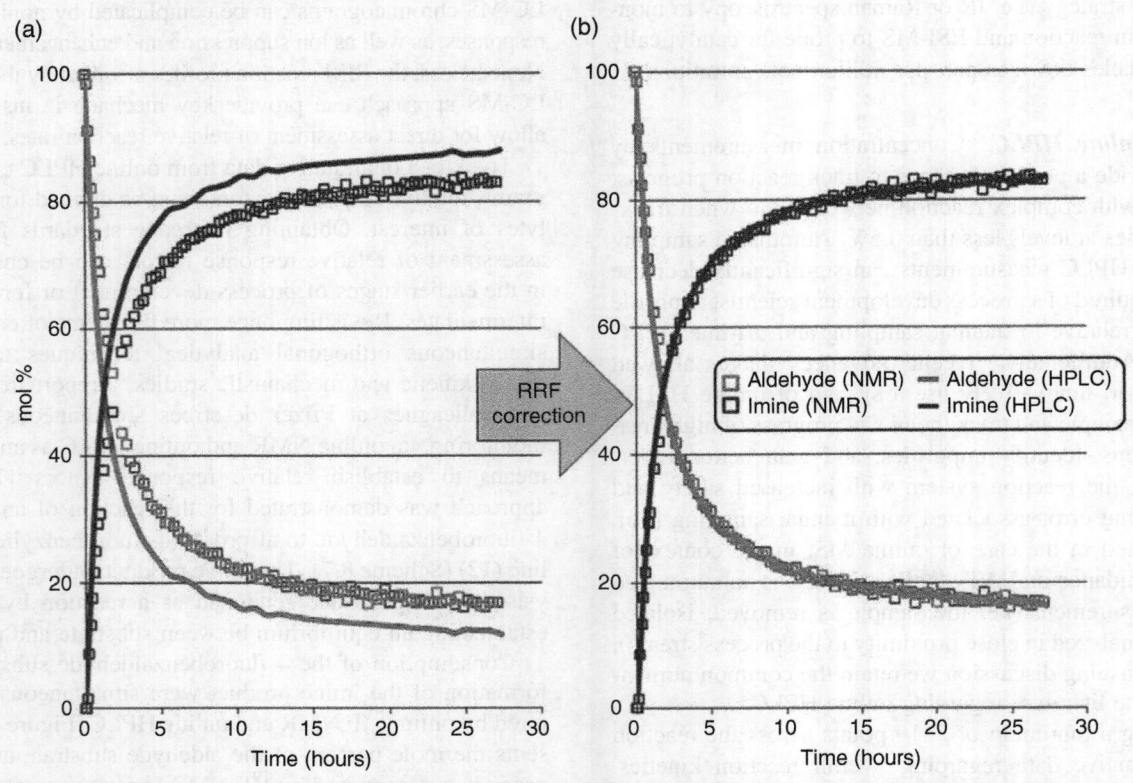

FIGURE 8.9 Kinetic profiling of the reaction shown in Scheme 8.7. (a) Profiles of the aldehyde substrate and imine product as monitored by online HPLC area percent (solid lines) and online ^{19}F NMR (open squares). (b) Quantitative reaction profiles, with HPLC relative response factors established based on the quantitative online NMR data. *Source*: Reprinted with permission from Foley et al. [73]. Copyright 2013, American Chemical Society.

the application, a model may only be needed to provide correlation between process inputs and outputs (e.g. rates or yields). With more insight into the chemistry and the mode of operation, apparent reaction behavior may instead be quantified in terms of kinetic, mass transfer, and heat transfer expressions that can be used to predict performance under a broader scope of process conditions. It is important to recognize in modeling that the accuracy and generality of predictions often go hand in hand with the amount of experimental data and mechanistic understanding put forth in compiling the model. Underlying reaction and transport effects unobservable at one set of operating conditions may greatly influence the chemistry at a different set of reaction conditions or processing scale. Without thorough exploration of the process operating space, one may find that a proposed model overly simplifies the complexity of the reaction system.

That caution notwithstanding, it should be the goal of the researcher to identify the model that captures a high degree of complexity without creating undue complexity in the model itself. Models are developed to simplify and rationalize an often complicated system into its most meaningful components. Experimental data are generated to test whether the simplified description captures the observed physical behaviors of the system. In such a way, the model guides the experiments, and the experimental data are used to support or refute the model. This allows a refining process for the model, which reflects refinement in the underlying process knowledge. A comprehensive model is an end goal, but smaller and simpler models are also helpful to the development of knowledge. As soon as a first draft model exists, it can be challenged with experimental data that help improve the model and thereby enhance process understanding.

8.4.1 Univariate Methods for Model Development

Often the easiest way to start into development of a reaction model is to profile the reaction evolution from start to final conversion. Sampling as the reaction progresses in time affords a richer set of information for analysis than simply analyzing the final conversion and yield. To further enhance the utility of the data, reaction profiles should capture the concentrations of reagents, intermediates, products, and by-products when possible.

The ability to recognize the distinguishing characteristics of the basic power law rate expressions and low-order dependence in one or more variables is of great utility in model construction. The reason for this is that even complex chemical reaction systems over a limited range of experimental conditions may appear to follow well-behaved and low-order kinetics. By understanding the situations in which a particular rate governs the reaction, valuable insights may be gleaned into the overall reaction rate law or into the reaction mechanism itself. For extrapolation of rates in apparently zero-, first-, or second-order systems, concentration-dependent time course data can be extracted using methods such as those described in Section 8.3. These data are then transformed into linear functions of the starting material concentration(s) (C_A and/or C_B) and time (t). Table 8.3 shows the appropriate linear equation for the commonly encountered zero-, first-, and second-order reaction systems. Excluding the effect of temperature, conformance to each of these models can be tested based on linearity for plots of $f(C_A)$ vs. t, with the proportionality factor being the rate constant k. Assuming an Arrhenius relationship for k, estimates for both the pre-exponential factor A and the activation energy E_A are found by graphing $\ln[f(C_A)/(t - t_0)]$ vs. $1/T$. The slope of such a plot will correspond to $-E_A/R$ and the intercept to $\ln(A)$.

Particularly in model construction, there are cases where it is preferable to measure the reaction rate independent of time. This can make the relationship between the reaction order in each species and the rate easier to interpret, as changes in reaction order become more pronounced when using differential methods – e.g. calorimetry and differential reactor analysis – as opposed to concentration-based methods [13, 74]. Examples of the rate-based approach in model construction can be found throughout chemical development. In particular, such examples appear regularly in catalytic systems, where the rate of catalyst turnover is regulated by one or a few key steps in the catalyst cycle.

In a relatively benign example, one can consider the catalytic cycle for the hydrogenation of alkenes using the rhodium catalyst shown in Scheme 8.8, first presented by Wilkinson and coworkers [75]. Like many other transition metal-catalyzed homogeneous reactions, the Wilkinson hydrogenation initiates with activation of the transition metal (rhodium) catalyst to form the complex I. The olefin then enters the catalytic cycle by coordinating to the rhodium complex, creating complex II, which undergoes migratory insertion to complex III and affords a vacancy for the oxidative addition of hydrogen. Finally the reduced alkane is eliminated from the complex IV, and the initial catalyst complex I is regenerated.

O'Connor and Wilkinson explored several contributing factors to the rate of the hydrogenation reaction using the substrates hex-1-ene and dec-1-ene. The reaction rate was found to scale linearly with hydrogen pressure and nearly linearly with catalyst charge. Subtle deviations in the rate with reduced catalyst charge were attributed to the loss of PPh_3 from the active species I. The dependence of the reaction rate

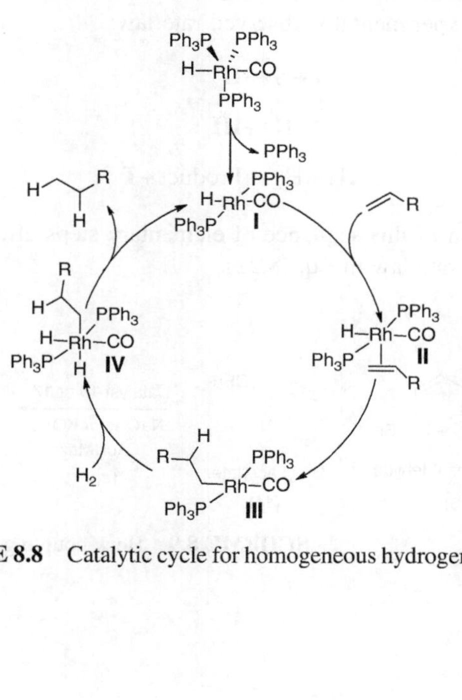

TABLE 8.3 For Reaction of A (+ B) → Products, Linear Solutions for Zero-, First-, and Second-Order Reaction Kinetics, Assuming Arrhenius Dependence on Temperature for the Rate Constant k

Apparent Zero Order

	Linearized Solution ($y = mx + b$)
$-\frac{dC_A}{dt} = k$	$C_{A0} - C_A = k(t - t_0)$
	$\ln\left(\frac{C_{A0} - C_A}{t - t_0}\right) = -\frac{E_A}{R}\left(\frac{1}{T}\right) + \ln(A)$

Apparent First Order

	Linearized Solution ($y = mx + b$)
$-\frac{dC_A}{dt} = kC_A$	$\ln\left(\frac{C_{A0}}{C_A}\right) = k(t - t_0)$
	$\ln\left[\frac{\ln(C_{A0}/C_A)}{t - t_0}\right] = -\frac{E_A}{R}\left(\frac{1}{T}\right) + \ln(A)$

Apparent Second Order

	Linearized Solution ($y = mx + b$)
$-\frac{dC_A}{dt} = kC_A^2$ or kC_AC_B with $C_{A0} = C_{B0}$	$\frac{1}{C_A} - \frac{1}{C_{A0}} = k(t - t_0)$
	$\ln\left[\frac{1/C_A - 1/C_{A0}}{t - t_0}\right] = -\frac{E_A}{R}\left(\frac{1}{T}\right) + \ln(A)$
$-\frac{dC_A}{dt} = kC_AC_B$ with $C_{A0} \neq C_{B0}$	$\frac{1}{C_{B0} - C_{A0}}\ln\left(\frac{C_{A0}C_B}{C_{B0}C_A}\right) = k(t - t_0)$
	$\ln\left[\frac{\ln\left(\frac{C_{A0}C_B}{C_{B0}C_A}\right)}{(C_{B0} - C_{A0})(t - t_0)}\right] = -\frac{E_A}{R}\left(\frac{1}{T}\right) + \ln(A)$

SCHEME 8.8 Catalytic cycle for homogeneous hydrogenation of alkenes.

upon alkene concentration was found to approach an asymptotic limit with increasing alkene concentration. Comparison of the inverse of the alkene concentration ($1/C_S$) to the inverse of the reaction rate ($-1/r_S$) resulted in an excellent linear fit of the data in the form

$$\frac{1}{-r_S} = \alpha\left(\frac{1}{C_S}\right) + \beta \quad (8.20)$$

From compilation of these observations, the rate law in Eq. (8.21) was proposed to describe the hydrogenation system independent of catalyst degradation:

$$-r_S = \frac{kKC_S C_{H_2} C_{Cat}}{1 + KC_S} \quad (8.21)$$

where

- k and K are constants.
- C_{H_2} is the concentration of dissolved hydrogen.
- C_{Cat} is the catalyst concentration (constant for a given experiment).

To interpret this empirical rate law in the context of the reaction mechanism (Scheme 8.8), one can examine Eq. (8.21) and identify two limiting cases for the reaction rate. Considering first the case where $KC_S \gg 1$, the reaction rate simplifies to a first-order rate dependence upon the hydrogen concentration, implying that the insertion of hydrogen determines the rate for the catalytic cycle under these conditions. As the alkene concentration is reduced ($KC_S \ll 1$), the rate law effectively becomes first order in both C_S and C_{H_2}. This would be indicative of a limiting dependence of the rate upon the alkene coordination, with a relatively fast equilibrium established for the molecular rearrangement of II and III. The following steps can therefore be written out to support the experimentally observed rate law:

$$I + S \rightleftharpoons II$$

$$II \rightleftharpoons III$$

$$III + H_2 \rightarrow Product + I$$

Evaluation of this sequence of elementary steps affords the proposed rate law in Eq. (8.21).

In general, were the reaction mechanism (or mass transfer effects) to be well known at the outset of a study, the need for experimentation would be minimal and would amount only to an exercise of fitting rate coefficients. In practice, the approach of monitoring for rate-limiting behavior is important in that new mechanistic insights are often gained by observing the chemistry at conditions where apparently "simple" kinetics transition to new regimes. An example of such an occurrence was demonstrated by Blackmond and coworkers for the Heck coupling reaction shown in Scheme 8.9 [74]. The catalytic cycle is nominally consistent with the general catalytic cycle for a cross-coupling reaction and begins with oxidative addition of p-bromobenzaldehyde (**13**) to the dimeric palladacycle catalyst **16**, followed by the addition of the olefin butyl acrylate (**14**) and finally reductive elimination to generate the desired product and regenerate the catalyst. Using reaction calorimetry, Blackmond and coworkers examined this reaction and observed multiple irregularities in the reaction rate in comparison with conventional Heck reaction kinetics. Whereas kinetic analyses of the Heck reaction are usually performed with a large excess of olefin (making the kinetics pseudo-zero order in olefin), the researchers in this case found that the reaction rate both had a dependence on olefin concentration and was sublinear when the catalyst concentration increased. A profile of the reaction rate with time (shown in Figure 8.10a) revealed the reaction to undergo a transition in rate behavior after about 90% conversion of the aryl bromide.

Given these observations, the researchers hypothesized the modified catalytic cycle illustrated in Scheme 8.10. The perceived inhibitory effect of operating with increased catalyst was attributed to the equilibrium formation of a [PdLArX]$_2$ dimer after the catalyst had undergone oxidative addition. Equilibrium for this reaction was theorized to significantly favor formation of the dimer. In the limiting case of complete inhibition because of dimer formation, the rate law was proposed to scale with the square root of the total palladium concentration (C_{Pd}):

$$r_{\text{dimer limited}} = \frac{k_2 k_3}{\sqrt{2K_R(k_{-2} + k_3)}} C_{14} C_{Pd}^{1/2} \quad (8.22)$$

This expression was then incorporated into a comprehensive rate law [74], generating a rate law that would simplify to

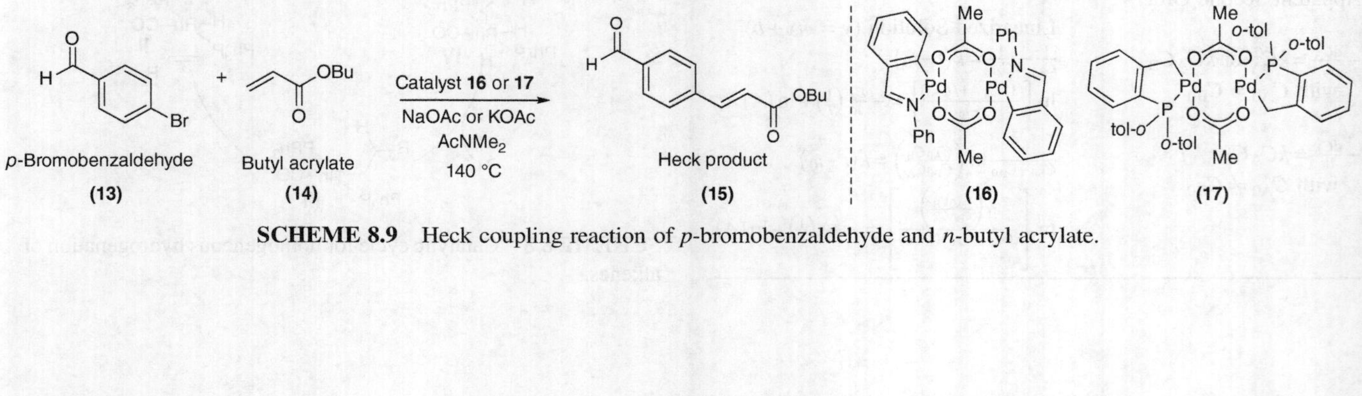

SCHEME 8.9 Heck coupling reaction of p-bromobenzaldehyde and n-butyl acrylate.

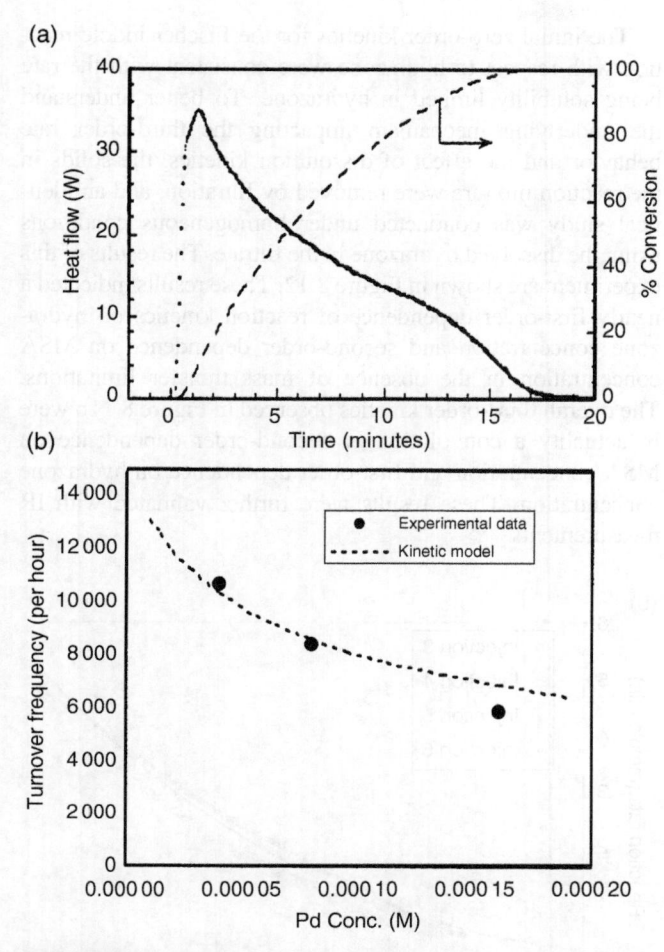

FIGURE 8.10 (a) Reaction heat flow as a function of time for Heck reaction in Scheme 8.9 with catalyst **16** and calculated conversion. (b) At 50% conversion, dependence of reaction rate on catalyst **16** and fit to proposed half-order catalyst dependence (Eq. 8.22). *Source*: Reprinted with permission from Rosner et al. [74]. Copyright 2001, American Chemical Society.

a first-order dependence on the kinetics in C_{Pd} when K_R was small and transition to a half-order dependence on C_{Pd} when K_R was large. The model fit is illustrated in Figure 8.10b. The modified rate law suggested that non-first-order dependence of the rate on the catalyst would only be observed when oxidative addition of the aryl halide was not limiting. Beyond

SCHEME 8.10 Modified catalytic cycle for Heck reaction including formation of [PdLArX]$_2$ dimer.

90% conversion, the olefin concentration was depleted sufficiently such that the oxidative addition step became rate controlling. Qualitatively similar behavior was observed using the dimeric catalyst **17**.

From the discussions in Section 8.2, it should be apparent that in addition to relative reaction rates, other factors such as mass transfer or heat transfer may also contribute to effective rate-limiting behaviors. An example illustrating the convolution of reaction rate with mass transfer can be found in the case of the Fischer indole reaction shown in Scheme 8.11. The Fischer indole reaction is expected to proceed through a hydrazone intermediate (Scheme 8.12), which exists as a slurry before strong acid drives the cyclization to close the pyrrole ring and form the bicyclic indole [76].

The kinetics of this reaction were studied using reaction calorimetry and off-line HPLC. Multiple small (0.15 equiv.) injections of methane sulfonic acid (MSA) were introduced into a slurry of hydrazone in the calorimeter. Figure 8.11a shows modified calorimetry data indicating the reaction rate data as a function of hydrazone concentration. Each injection of MSA can be thought of and analyzed as an individual batch reaction. Each peak observed in Figure 8.11a is due to an identical spike of MSA, which is nearly depleted after each reaction. Changes in the initial hydrazone concentration allow for calculation of the reaction order in hydrazone during the

SCHEME 8.11 Fischer indole reaction.

SCHEME 8.12 Proposed pathway for the Fischer indole reaction of Scheme 8.11.

course of the reaction. Superposition of the reaction profiles from the four MSA injections revealed that the reaction rate was zero order in initial hydrazone concentration and exhibited an overall third-order behavior when plotted as a function of MSA concentration, as shown in Figure 8.11b (solid line represents the rate profile expected from third-order kinetics).

The initial zero-order kinetics for the Fischer indole reaction with respect to hydrazone were consistent with the rate being solubility limited in hydrazone. To better understand the underlying mechanism impacting the third-order rate behavior and the effect of dissolution kinetics, the solids in the reaction mixture were removed by filtration, and an identical study was conducted under homogeneous conditions using the dissolved hydrazone in the filtrate. The results of this experiment are shown in Figure 8.12. These results indicated a nearly first-order dependence of reaction kinetics on hydrazone concentration and second-order dependence on MSA concentration in the absence of mass transfer limitations. The overall third-order kinetics observed in Figure 8.11b were in actuality a convolution of second-order dependence on MSA concentration and first-order dependence on hydrazone concentration. These results were further validated with IR measurements.

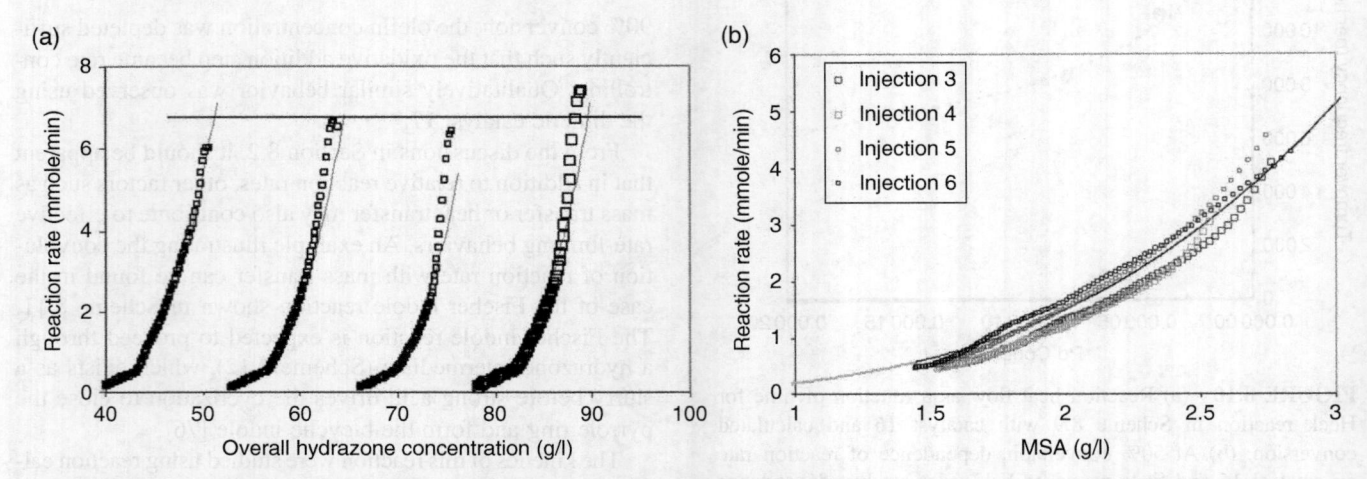

FIGURE 8.11 Rate of Fischer indole reaction as a function of (a) overall hydrazone concentration and (b) MSA concentration.

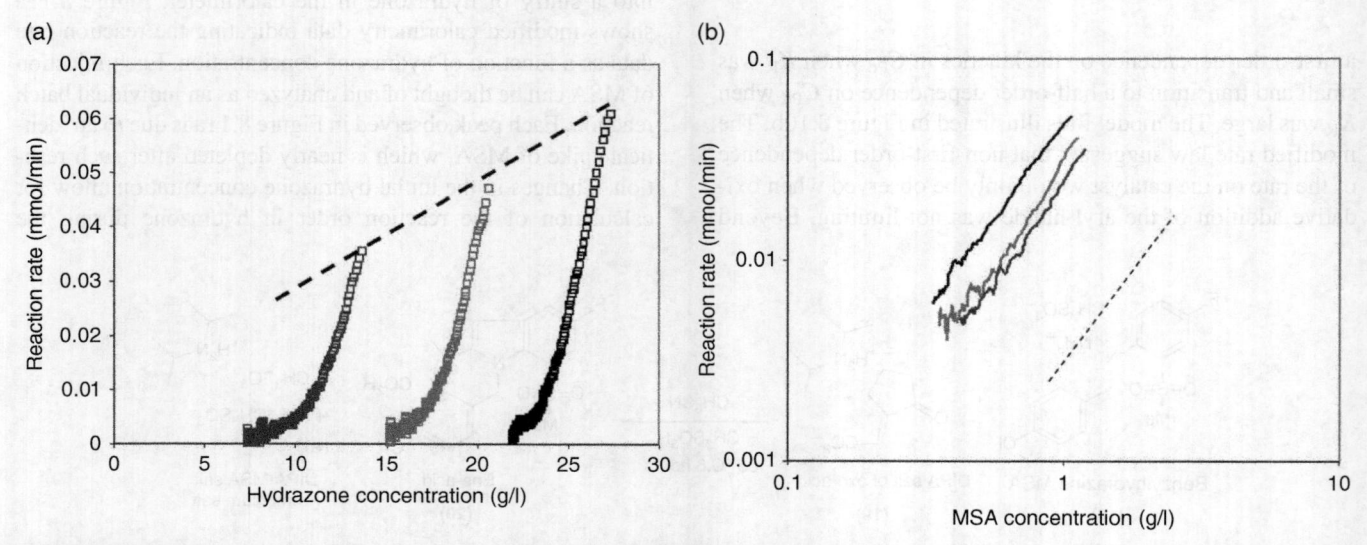

FIGURE 8.12 Plot of Fischer indole reaction rate under homogenous conditions as a function of (a) hydazone and (b) MSA concentration. Dashed lines in (a) and (b) represent first- and second-order curves, respectively.

SCHEME 8.13 Scheme for the Paal–Knorr reaction.

The methods of reaction profiling need not be reserved to batch kinetic studies only. An example of the analogous approach to model generation in flow was demonstrated by Moore and Jensen [77], who examined the Paal–Knorr reaction summarized in Scheme 8.13 using an automated microreactor system. In the experiment, the two reactants, 2,5-hexanedione (**21**) and ethanolamine (**22**), were delivered independently to a heated microreactor, and the reaction conversion was measured at the exit of the reactor with an in-line Mettler Toledo ReactIR.

Taking advantage of the short mixing and heat transfer times in the microreactor system, the authors allowed the reaction to reach steady state in the flow device and then imposed a gradual reduction in the overall flow rate such that the instantaneous residence time in the reactor increased as the experiment progressed, with the reaction stoichiometry held constant. Reaction kinetics were monitored for 60 minutes using IR spectroscopy, then the reactor temperature set point was increased by 20 °C, and the ramp of flow rate was repeated. This process repeated until all experimental conditions of interest had been evaluated. Temperatures in the range of 50–170 °C and reaction times in the range of 0.5–40 minutes were examined comprehensively and automatically in the span of 8 hours. The resulting product concentration and time profiles are provided in Figure 8.13. These results were transferred to a least squares regression algorithm, which estimated activation energies for the two slow Paal–Knorr reaction steps as $E_{A1} = 12.2 \pm 0.4$ kJ/mol and $E_{A2} = 20.0 \pm 0.9$ kJ/mol, respectively.

This method was later revisited by the same authors in the case of the continuous flow aminocarbonylation of p-bromobenzonitrile, shown in Scheme 8.14 [78]. With the goal of modeling the factors influencing the formation of either the mono-amide (**26**) or the di-α-keto amide (**27**), the authors examined the reaction in the context of both gas pressure and reaction temperature. From preliminary screening results in the microreactor system, the gas–liquid mass transfer rate was found to be non-limiting in the experimental condition range of interest. Based on this result, the rate-limiting kinetics were simplified to

$$C_{Pd} = \frac{(C_{Pd})_0}{1 + KC_{CO}} \quad (8.23)$$

$$-\frac{dC_{24}}{dt} = k_1 C_{Pd} C_{24} = -k_{obs} C_{24} \quad (8.24)$$

where

- k, k_1 and k_{obs} are constants
- C_{Pd} represented the molar concentration of the active palladium species (relative to initial palladium $(C_{Pd})_0$).
- C_{24} represented concentration of the aryl halide.
- C_{CO} represented the concentration of carbon monoxide

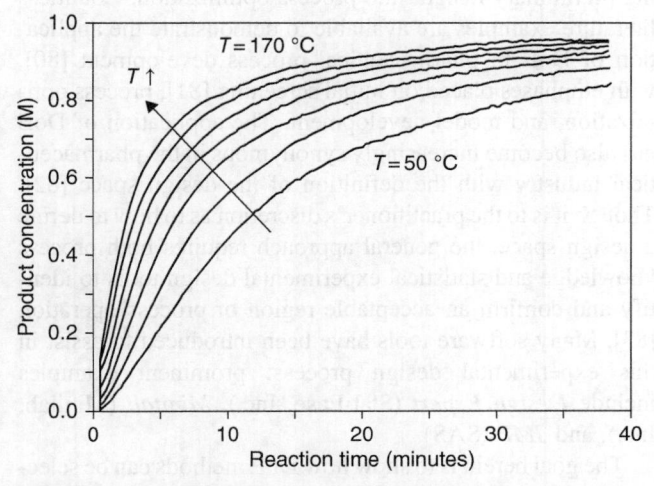

FIGURE 8.13 Kinetic profiling of the Paal–Knorr reaction in flow with IR spectroscopy. *Source*: Reprinted with permission from Moore and Jensen [77]. Copyright 2014, John Wiley and Sons.

The reaction was then conducted in a tubular reactor containing stainless steel beads for mixing and was heated in 1 °C increments every 2 minutes from 80 to 160 °C. The reaction effluent was continuously monitored by IR spectroscopy, and principle component analysis (PCA) was used two distinguish the two

SCHEME 8.14 Continuous flow aminocarbonylation reaction of *p*-bromobenzonitrile.

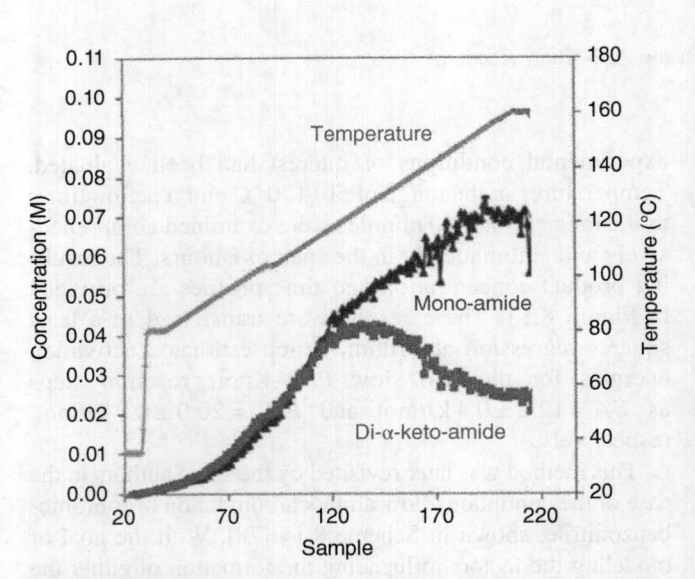

FIGURE 8.14 IR-measured concentration profile for aminocarbonylation products in Scheme 8.14 with increasing tubular reactor temperature (8.3 minutes residence time; 8.3 bar CO). *Source*: Reprinted with permission from Moore et al. [78]. Copyright 2016, The Royal Society of Chemistry.

products from the starting aryl halide. Automated operation resulted in generation of the reaction profile in Figure 8.14, which showed the reaction to afford nearly identical selectivity toward the mono- and di-α-keto amide products at temperatures less than 120 °C. However, once the temperature climbed to above 120 °C, the selectivity appeared to shift to as much as 2 : 1 in favor of the mono-amide **26**.

In evaluating the observed results, the researchers first considered the independence of the reaction selectivity at below 120 °C, attributing this observation to the originally proposed limiting oxidative addition kinetics from Eqs. (8.23) to (8.24). For this limiting case, it was found through further experimentation that the rate was linear with the inverse of the CO pressure (due to competition from the reversible reaction of the Pd–ligand catalyst with CO to form [PdLCO]). The observed rate constant from Eq. (8.24) was then simplified to the ratio $k_1(C_{Pd})_0/KC_{CO}$ (a function of temperature only with C_{CO} in excess), and an effective activation energy for the reaction was estimated deriving from the first-order dependence of the rate upon C_{24}. Based upon the results at higher temperature conditions, the researchers proposed an alternative mechanistic route that drew upon literature precedent [79], suggesting two competing routes to the formation of the mono-amide (Scheme 8.15). The "inner" cycle via k_2 and k_4, by virtue of forming a more entropically favorable intermediate, was hypothesized to become more favorable with increasing temperature. Such a mechanism was shown numerically to support both the enhanced selectivity toward the mono-amidation product at high temperatures and the pseudo-first-order dependence upon the aryl halide concentration at low temperatures upon regression to experimental data. This example cited a two to three times speed improvement compared with conventional flow experimentation and highlights the impact that automation and PAT are having upon kinetic modeling in next-generation development.

8.4.2 Design of Experiments

For cases where a well-characterized kinetic model may simply be too complex to compile or large volumes of data are too expensive to generate, response surface models constructed from design of experiments (DoE) principles can be invaluable in providing the researcher with basic guidance on the interactions of multiple process inputs and for providing preliminary insight into process optimization. Countless literature examples are available to demonstrate the application of DoE in pharmaceutical process development [80], with emphases placed on initial screening [81], process optimization, and model development. The application of DoE has also become increasingly synonymous in the pharmaceutical industry with the definition of the design space [82]. Though it is to the practitioner's discretion as to how to define a design space, the general approach requires both process knowledge and statistical experimental design tools to identify and confirm an acceptable region of process operation [83]. Many software tools have been introduced to assist in this experimental design process; prominent examples include *Design Expert* (Stat-Ease, Inc.), *Minitab* (Minitab, Inc.), and *JMP* (SAS).

The goal herein is to show how DoE methods can be selectively applied to model development in the spirit of other case studies already highlighted in this chapter. If the DoE methodology has a fault, it is that those who leverage it can get lost in overestimating the value of statistical models generated

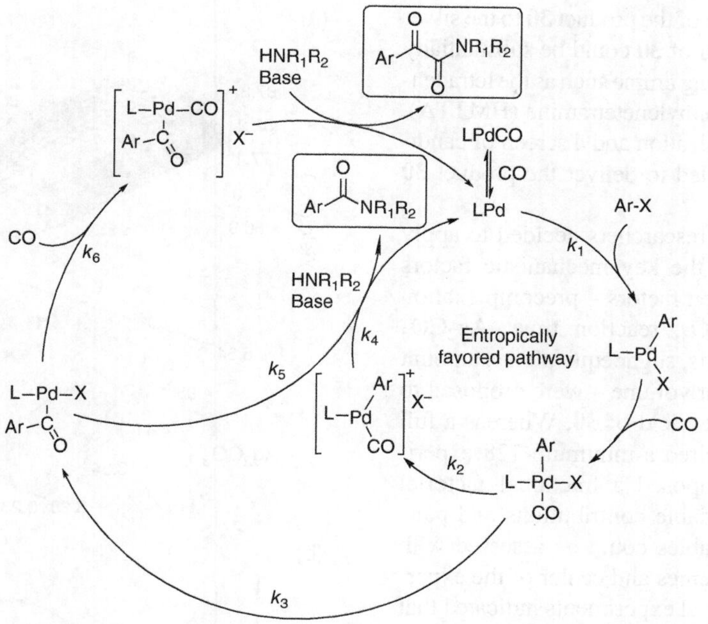

SCHEME 8.15 Catalytic cycle for aminocarbonylation reaction in Scheme 8.14.

from a small sample of experiments. To avoid this trap (particularly for reaction kinetics, which could be otherwise deduced through first principles mechanistic understanding), it is essential that empirical results be examined critically in the context of any simplifications made in generating the original experimental design. An effective use of DoE may be to review the initial experimental data generated and apply the knowledge gained either to a new experimental design or to construction of a model that derives from mechanistic principles.

The DoE process generally begins with the identification of one or more responses (e.g. yields, quality attributes, material properties) to control and the identification of factors/variables (e.g. temperature, time, amount) believed to have the most significant influence upon those responses. Depending upon the number of variables under consideration, it may be prudent to conduct an initial set of screening experiments to prioritize factors and eliminate those with little or no impact upon the targeted outcome. Such a screen could be a full factorial design, where all factors k are considered in combination at high and low levels to afford a total of 2^k experimental conditions.

Alternatively one could use a fractional factorial design, where the number of high- and low-level experiments is divided by 2^n to save experimental time, at the expense of losing some information as to the higher-order interactions among variables. After refining the list of candidate variables, a more comprehensive set of experiments is then prescribed. Examples could include, but are not limited to, full factorial or higher-order fractional factorial designs, central composite designs, or Box–Behnken designs. These latter two designs include experiments not situated at the vertices of the experimental design region, which are necessary to detect nonlinearity in individual factors. The prescribed experiments are then executed, and the results are compiled to afford a response surface model that relates the response(s) to the variables in combination. Various statistical tools can be used to quantify the accuracy of the empirical model.

A general example showing the merits of the DoE procedure can be found in Stazi et al. [84], who examined the Koenigs–Knorr glucuronidation of the nitrophenol 28 as shown in Scheme 8.16. Initial efforts to react 28 in the presence of the sugar 29, Ag_2O, and 4 Å molecular sieves resulted in very low

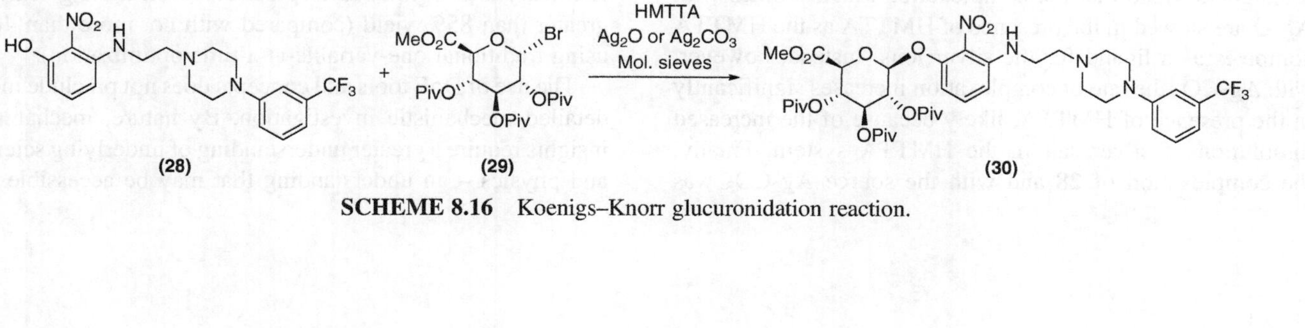

SCHEME 8.16 Koenigs–Knorr glucuronidation reaction.

yields on account of the chelation of the product **30** to the silver ions. It was learned that the yield of **30** could be substantially increased with the use of a chelating amine such as the tetradentate 1,1,4,7,10,10-hexamethyltriethylenetetramine (HMTTA), but one-variable-at-a-time optimization and a screen of candidate Ag sources and amines failed to deliver the product **30** in better than 40% yield.

In light of these results, the researchers decided to apply DoE to identify and optimize the key mechanistic factors in the HMTTA chemistry. Seven factors – precomplexation time for HMTTA and Ag_2CO_3, reaction time, Ag_2CO_3 equivalents, HMTTA equivalents, sugar equivalents, amount of molecular sieves, and solvent volume – were proposed to contribute in the reaction to the yield of **30**. Whereas a full factorial DoE would have required a minimum 128 experiments, the authors instead proposed a fractional factorial design in which individual variable contributions and pairwise interactions between variables could be assessed with only 11 experiments at the extremes and center of the experimental design space. These initial experiments indicated that the yield depended most significantly upon the precomplexation time and upon the amounts of Ag_2CO_3, HMTTA, and sugar **29** used in the experiment. A consequence of using such a reduced fractional factorial design, the interaction of Ag_2CO_3, and the amount of **29** was strongly correlated to the precomplexation time in the preliminary experiments. Because of this and in order to limit the overall number of experiments, the authors elected in subsequent response surface modeling to only consider the three reagents, Ag_2CO_3, HMTTA, and sugar **29**, and to maintain a constant precomplexation time of 60 minutes. A 20-experiment central composite design was conducted, which sampled from points on both the edges and interior of the experimental space.

Following response surface generation from the results of the three-factor central composite design, Stazi et al. noted the abnormality of the bimodal dependence of the product **30** upon the additive HMTTA, shown in Figure 8.15a. Such an effect was captured empirically with a cubic polynomial model and was postulated to arise from any of several factors, including the involvement of HMTTA in competing as a ligand for silver ions, in activating the silver source, in deprotonating the phenol **28**, and in contributing to the dehydrohalogenation of sugar **29** when used at high concentration. To discern and quantify each of these effects, the researchers conducted additional experiments comparing the complexation of reactant **28** with silver from either Ag_2O or Ag_2CO_3 in the presence or absence of HMTTA. From these studies, the researchers found that the complexation kinetics of **28** with Ag_2O are slowed in the presence of HMTTA as the HMTTA competes as a ligand for the silver ions. Instead, however, with Ag_2CO_3 the rate of complexation increases significantly in the presence of HMTTA, likely because of the increased dissolution of silver salt in the HMTTA system. Finally, the complexation of **28** and with the source Ag_2CO_3 was

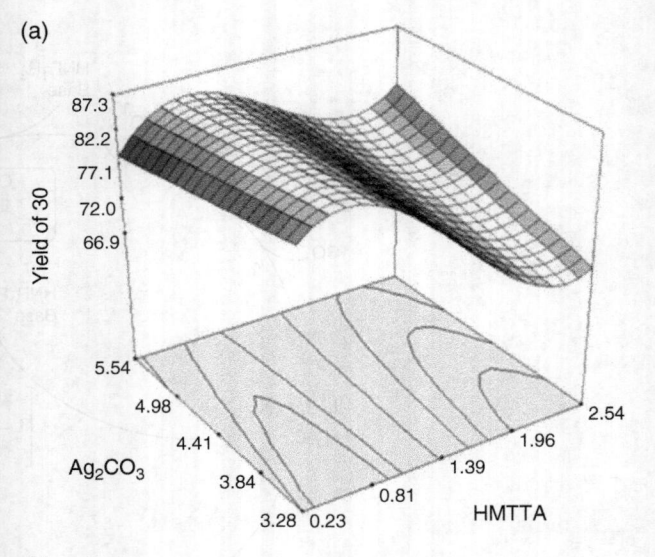

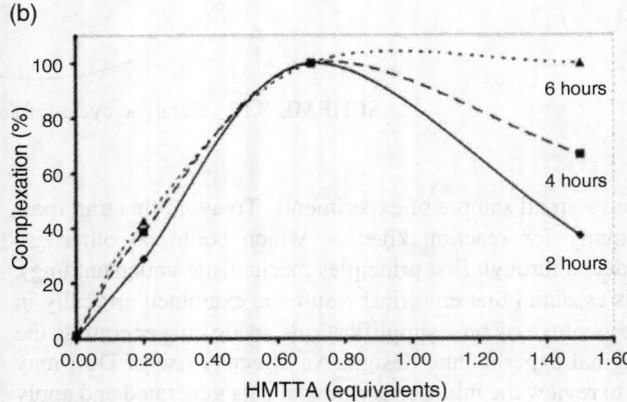

FIGURE 8.15 (a) Influence of stoichiometry of Ag_2CO_3 and HMTTA on the *in situ* yield of **30** for 2.42 equiv. sugar **29**. (b) Effect of different amounts of HMTTA on complexation equilibria between **28** and Ag_2CO_3 at reaction times of 2, 4, and 6 hours. *Source*: Reprinted with permission from Stazi et al. [84]. Copyright 2004, American Chemical Society.

assessed as a function of HMTTA equivalents and found to exhibit a maximum as shown in Figure 8.15b. At low loadings of HMTTA, the complexation rate increased with increasing HMTTA charge on account of increased solubility of Ag^+ in solution. However, an increase of HMTTA to above approximately 0.7 equiv. led to a decrease in the complexation of **28** as the increased amine competed for the binding to the silver ions. With this added knowledge, the chemistry was optimized to produce **30** on a 3.5 g scale in greater than 85% yield (compared with not more than 40% using traditional one-variable-at-a-time optimization).

The use of DoE tools and concepts does not preclude more detailed mechanistic investigation. By nature, mechanistic insights require a greater understanding of underlying science and physics – an understanding that may be accessible for primary reaction pathways but may be difficult to discern for low-level impurities. In this case, it would seem reasonable that a control strategy could be based on both mechanistic considerations (to account for all that is well-understood in the reaction system) and empirical considerations (to capture the impact of other, less-readily characterized factors upon a proposed design space). In utilizing this approach, the question becomes how to define, and then verify, these empirical relationships while accounting for the framework provided by the established mechanistic knowledge. Researchers at Bristol-Myers Squibb (BMS) have produced several papers [85–87] illustrating an approach to these problems that is consistent with the FDA's guidelines for quality by design [83]. Given that the problem involves determining design space robustness, it should not be a surprise that the methodology relies heavily upon the application of DoE.

An example of this approach was provided by Burt et al. in a case study involving the base-mediated conversion of an ester to a primary amide [86, 88]. The postulated network is shown in Scheme 8.17. Formation of the drug substance 34 occurs in a multistep sequence: deprotonation of the amidation reagent 31, followed by reaction of the active amide species 32 and amine 31 with the input ester 33 to afford the drug substance 34 and an imide by-product 35. Both reactions are equilibrium reactions. As hydrolysis with water can cause degradation of the input ester 33 in the presence of base, an orthoester 37 is used to promote chemical removal of water prior to initiation of the amidation reaction, yielding formate 38. However, 38 can react with the amide 32 to produce the imide 35, shifting the equilibrium of the desired pathway away from the formation of the drug substance. The imide 35 can also form via side reactions originating with either the drug substance or a diester impurity 39 introduced in the feed. These last two side reactions also afford bis-amide species 40 that was the key impurity in the drug substance manufacturing process.

Implementation of a multistaged DoE began with the identification of factors most likely to impact product purity; those factors were identified through a process risk assessment as reaction temperature, equivalents of base, initial water content of the feed solution (i.e. prior to chemical water removal), and the amount of diester impurity 39 in the feed. For the initial stage DoE, three of these factors – reaction temperature, base amount, and water amount – were screened in a 2^3 factorial design. The results of these experiments were applied to estimate temperature-dependent rate coefficients for the formation of drug substance 34, and its degradation to bis-amide impurity 40. For this initial stage DoE, input-related diester impurity 39 was maintained at a uniformly low level to allow for clean regression of kinetic parameters for drug substance degradation to the bis-amide impurity.

With a preliminary mechanistic model in place, a second stage DoE was undertaken to delineate the limits of the multivariate parameter space (encompassing both process parameter ranges and input quality attributes), affording in-specification drug substance. A 2^{4-1} fractional factorial design was considered (including the amount of orthoester 37 as a fourth factor), and the diester impurity 39 was included in the reaction at its specification limit in the input ester 33. In a third stage DoE, axial points were assessed for the initial three factors (DoE stage 2 had confirmed that variation of orthoester amount across its univariate PAR was not quality impacting). For DoE stages 2 and 3, key responses included levels of residual input ester 33 and bis-amide impurity 40, both in process and in the isolated product, and the isolated drug substance yield. The combined results of DoE stages 2 and 3 gave indication of the curvature of the system and delineated two regions of the multivariate parameter space corresponding to process failure modes: a combination of high reaction temperature and high base for formation of bis-amide impurity 40 and a combination of high reaction temperature and low base for unreacted ester 33.

In addition to identifying a subset of the multivariate parameter space that afforded in-specification drug substance, the BMS team recognized that the compilation of results from their multistaged DoE could support the development of a hybrid mechanistic and empirical process chemistry model. Elementary rate expressions were assumed for the formation of drug substance 34, the imide 35, and the degradation of drug substance to bis-amide 40. The rate of bis-amide impurity generation from input-related diester 39 was quite fast, and the level of the diester itself was small; hence the model parameters for the elementary rate expressions were first regressed, and then an empirical component was applied to the mechanistic modeling results to account for the varying magnitude of bis-amide impurity resulting from the initial presence of the diester in the feed. Once a hybrid model was compiled, the BMS team was able to verify the parameter estimates with additional experimentation.

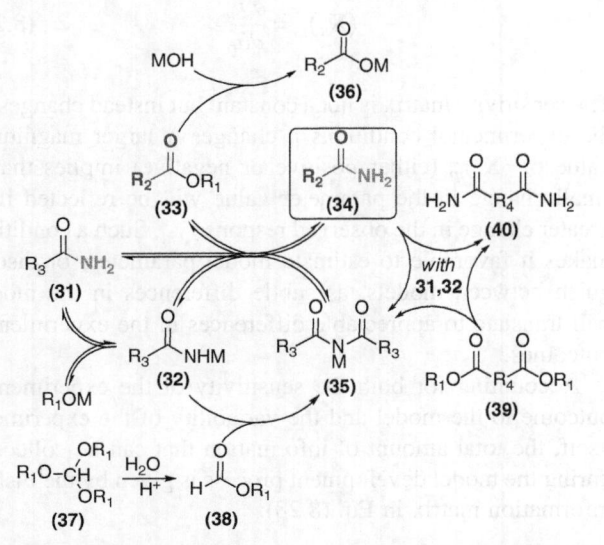

SCHEME 8.17 Network of reactions for base-mediated conversion of an ester to a primary amide and competing by-product formation.

The resultant hybrid process chemistry model was leveraged to guide the selection and verification of a design space for the drug substance manufacturing process.

8.4.3 Parameter Estimation and Model Discrimination

The preceding examples illustrate both the advantages and disadvantages of leveraging DoE in process development. DoE is an efficient tool for surveying the impact of several (perhaps interacting) factors upon specified responses. As exemplified, the DoE process can often reveal new regions of the experimental space that are overlooked when changing only a single variable at a time, making it more likely that a better process optimum is found and that the empirical kinetics more comprehensively represent system behavior across a robust design space. A negative of the approach is that without proper vetting of candidate variables, the number of experiments needed to complete the DoE can quickly become unmanageable. Choosing all or the majority of these experiments at the outset can be costly as well, as experiments will inevitably be chosen that either (i) are non-informative in that they do not generate knowledge beyond what has already been learned from previous experiments or (ii) are nonoptimal and hence unlikely to be relevant for process optimization or design space considerations. To address scalability concerns, there is no substitute for having access to a kinetic model describing the process in detail.

The question may be asked then as to how to bridge the more general knowledge gained through the DoE approach with established methods for kinetic model development, enabling greater process understanding to be gained in less experimental time. One route to achieving this is through the use of feedback optimization, wherein one or a handful of selected experiments are executed, then the data are analyzed, and a new experiment is proposed, which should – given the prior data collected – produce the most information to support or refute the proposed model. The selected experiment is then conducted and the procedure repeats. Though exceptions exist [89–91], such a sequential improvement methodology is usually too tedious to implement when developing processes in batch when factoring time for sample preparation/analysis and off-line process modeling. With continuous flow, however, the integration of equipment control, analytical instrumentation, and computational tools can greatly facilitate the process of iterative optimization or model construction, shortening process development lead times from weeks to a few hours or days [92, 93].

Justification of the feedback optimization approach requires a more generalizable understanding of the method for model regression. Such a method to nonlinear model parameter estimation is documented, for instance, in Ref. [5] and involves minimization of the sum of squared residuals between observed (measured) responses, y_i, and predicted responses, $\hat{y}_i$, for a set of conducted experiments, indexed here by u:

$$\theta_{opt} = \underset{\theta}{\text{argmin}} \sum_{i=1}^{N_{resp}} \sum_{u=1}^{N_{expt}} (y_{iu} - \hat{y}_{iu})(\mathbf{V}_{\varepsilon u})_{ij}^{-1}(y_{iu} - \hat{y}_{iu}) \quad (8.25)$$

Here, θ is a group of parameters (e.g. activation energies and pre-exponential factors) relating the adjustable variables in the reaction to the predicted responses. The term $\mathbf{V}_{\varepsilon u}$ is a weighting matrix that approximates the relative uncertainty in each measured response for the experimental conditions u. Estimation of the elements of $\mathbf{V}_{\varepsilon u}$ comes from taking the variance or covariance expected for each pair of responses upon repeated experimentation at the set of experimental conditions. For cases where it is impractical to conduct many replicate experiments, an estimate for $\mathbf{V}_{\varepsilon u}$ can sometimes be employed by calculating a general covariance matrix $\mathbf{V}_\varepsilon$ with elements $(\mathbf{V}_\varepsilon)_{ij}$ as follows:

$$(\mathbf{V}_\varepsilon)_{ij} \approx s_{ij}^2 = \frac{1}{N_{expt} - N_{param}} \sum_{u=1}^{N_{expt}} (y_{iu} - \hat{y}_{iu})(y_{ju} - \hat{y}_{ju}) \quad (8.26)$$

Because this calculation assumes the best-fit parameters in the models describing $\hat{y}_i$ and $\hat{y}_j$, Eq. (8.26) must be evaluated iteratively with the optimization in Eq. (8.25).

Independent of the variance from experiment to experiment, precise estimation of kinetic parameters is also contingent upon selection of the correct experiments to distinguish different elements of the model from one another. This should be intuitive given the many examples in this chapter of changes in an apparent reaction model that are only observable at particular processing conditions. The sensitivity of a model ($\mathbf{X}_u$) is a quantitative measurement of how much each model prediction changes as its defining parameters are varied:

$$(\mathbf{X}_u)_{ip} = \frac{\partial \hat{y}_{iu}}{\partial \theta_p} \quad (8.27)$$

The sensitivity matrix is not a constant but instead changes as the experimental conditions u change. A larger magnitude value of $(\mathbf{X}_u)_{ip}$ (either positive or negative) implies that a small change in the parameter value will be reflected in a greater change in the observed response $\hat{y}_{iu}$. Such a condition makes it favorable to estimate model parameters or distinguish between models, as subtle differences in the model will translate to appreciable differences in the experimental outcomes.

Accounting for both the sensitivity of the experimental outcome to the model and the variability of the experiment itself, the total amount of information that can be collected during the model development process is given by the Fisher information matrix in Eq. (8.28):

$$\mathbf{Z} = \sum_{u=1}^{N_{expt}} \mathbf{X}_u^T \mathbf{V}_{\varepsilon u}^{-1} \mathbf{X}_u \quad (8.28)$$

Among the important properties of this matrix is that its inverse $\mathbf{Z}^{-1}$ corresponds to the covariance matrix of the estimated parameter values, meaning that the elements of $\mathbf{Z}$ give a quantifiable measure as to with how much certainty the proposed model and kinetic coefficients describe the experimental data. In sequential experimental design, the goal then becomes selection of new experiments that can lead to the smallest uncertainty in the model, i.e. by selecting experiments that minimize the size of elements in $\mathbf{Z}^{-1}$. A widely used metric is D-optimality, which aims to shrink total volume of the region of uncertainty for all model parameters as shown in Eq. (8.29):

$$u = \underset{u}{\mathrm{argmin}}\ \det\left[\left(\mathbf{Z} + \mathbf{X}_u^T \mathbf{V}_{\varepsilon u}^{-1} \mathbf{X}_u\right)^{-1}\right] \quad (8.29)$$

The addition of experimental data to the optimization in Eq. (8.29) serves to (hopefully) lessen $|\mathbf{Z}^{-1}|$. By lessening $|\mathbf{Z}^{-1}|$, the variance and covariance estimates for the model parameters are reduced, and hence the model is supported to greater precision. With further refinement of the model parameters, the ability of the model to predict sensitivity coefficients in future experiments should improve, meaning that the quality of experimental conditions chosen to inform model development should improve as well.

An example employing the D-optimal design method for kinetic model development was demonstrated by Reizman and Jensen in consideration of the series–parallel nucleophilic aromatic substitution reaction pathway shown in Scheme 8.18 [94]. Reagents 2,4-dichloropyrimidine (**41**) and morpholine (**25**) were reacted in an automated flow system. At the outlet of the flow reactor, the concentrations of pyrimidine derivatives **41**, **42**, **43**, and **44** were measured by online HPLC. Results of this analysis were supplied to a computer algorithm, which estimated kinetic parameters assuming elementary second-order reaction kinetics of the reactions of morpholine with **41**, **42**, and **43**. After an initial set of factorial experiments, the computer algorithm selected a new set of experimental conditions (flow rates and temperature) that would most greatly reduce the uncertainty in the kinetic model parameters. These experimental conditions were automatically provided to the continuous flow system, which executed the new experiment to allow the feedback loop to repeat.

The initial factorial design afforded reasonable estimates for the pre-exponential and activation energy parameters associated with k_1 and k_2, with reported uncertainties in parameter values not exceeding ±12%. This same factorial design, however, proved to be unacceptable for resolving the rate of formation of the disubstituted product **44** and in the estimation of parameters associated with k_3 and k_4. Not surprisingly, the majority of ensuing feedback experiments were selected at higher temperatures and longer reaction times, where a greater yield of the product **44** was expected. After an additional 12 optimal experiments, the relative uncertainty on the activation energies for k_1 and k_2 had been reduced to at most ±6% and for k_3 and k_4 to at most ±14%, and qualitatively reasonable model predictions were obtained. For further resolution of the kinetic parameters, Reizman and Jensen reverted to the more "traditional" method of kinetic estimation by isolating each of the intermediate monosubstituted products and then using the automated flow system to estimate the rate constant parameters associated with the one-step conversion of these intermediates to the disubstituted product **44**. The rate parameters for k_3 and k_4 and their uncertainties were then supplied to a final all-at-once automated parameter fitting exercise that afforded rate coefficients with no more than ±4% error. In total, 78 automated experiments were needed to fit the 8-parameter model.

The parameter estimation approach can be expanded to model discrimination using the goodness-of-fit metric χ^2. To compare two or more candidate models, a reduced χ^2 value is calculated for each model using sum-of-squares residuals as follows:

$$\chi^2 = \frac{1}{N_{\mathrm{expt}} - N_{\mathrm{param}}} \sum_{i=1}^{N_{\mathrm{resp}}} \sum_{u=1}^{N_{\mathrm{expt}}} (y_{iu} - \hat{y}_{iu})(\mathbf{V}_{\varepsilon u})_{ij}^{-1} (y_{iu} - \hat{y}_{iu}) \quad (8.30)$$

The reduced χ^2 is scaled by the available degrees of freedom and therefore is reduced both for a better model fit and the use of fewer model parameters. In this regard it

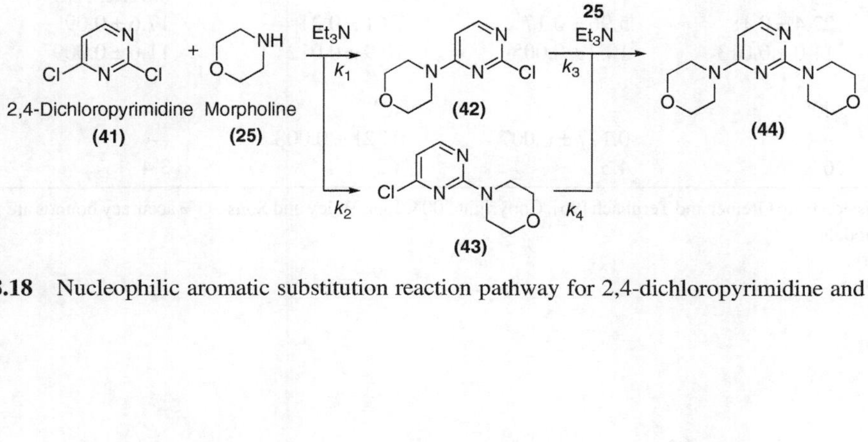

SCHEME 8.18 Nucleophilic aromatic substitution reaction pathway for 2,4-dichloropyrimidine and morpholine.

is unfavorable to add more parameters to a model without improving upon the goodness of fit.

The use of the goodness-of-fit approach was demonstrated by Greiner and Ternbach for the hydrogenation chemistry in Scheme 8.19 [95]. The homogeneous rhodium catalyst complex {[Rh(PyrPhos)-(COD)]BF$_4$;**45-COD**} was used in the study. The catalyst **45** is liberated from the precatalyst complex before entering into the catalytic cycle and facilitating the reduction of *N*-acetylaminocinnamic acid (**46**) to **47**. The authors proposed inclusion of first-order kinetics describing both the activation and deactivation of **45** in the model:

$$\frac{dC_{45}}{dt} = k_{act}C_{45-COD} - k_{deact}C_{45} \quad (8.31)$$

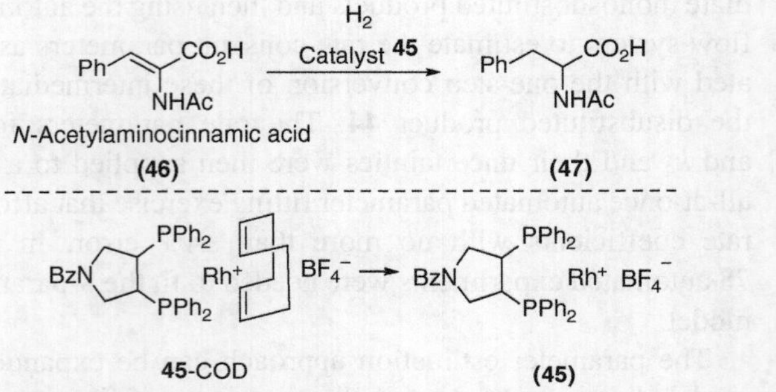

SCHEME 8.19 Rhodium-catalyzed hydrogenation of *N*-acetylaminocinnamic acid.

Likewise the consumption of **46** was hypothesized to be dependent on the equilibria of intermediates of the catalytic cycle. A general form of the rate expression was assumed to be:

$$\frac{dC_{46}}{dt} = -\frac{v_{max}C_{46}C_{47}}{R} \quad (8.32)$$

Six candidate expressions for R were proposed, as summarized in Table 8.4. All expressions included an equilibrium binding constant for **46** to the catalyst, K, and a few included additional inhibitive equilibrium constants for binding to the reactant **46** ($K_{i,46}$) or the product **47** ($K_{i,47}$). Best-fit parameters and reduced χ^2 values were then calculated for each proposed model (also Table 8.4). Overall minimal χ^2 values were found in cases where catalyst deactivation was included in the model, namely, in models 5 and 6. Model 6, which included the maximum six parameters, gave a slightly improved model fit in comparison with model 5, but the authors questioned the legitimacy of the predicted inhibition kinetics for the more advanced model. Based on simulations of the optimized models, the authors suggested that increasing the concentration of the substrate **46** would be necessary in order to better discriminate between models 5 and 6. Such experiments would be limited by the solubility of **46** in the solvent methanol.

The observation that additional experiments are needed in untested regions of the experimental space is almost ubiquitous to kinetic model discrimination and is reflective of the challenges of scale-up. Statistical tools for model discrimination can help in this regard, provided that the candidate models incorporate the relevant information needed for scale-up.

TABLE 8.4 Best-Fit Parameters for Candidate Rate Models Describing Hydrogenation of *N*-Acetylaminocinnamic Acid

Model	1	2	3	4	5	6
Description	Michaelis–Menten	Substrate Surplus Inhibition	Competitive Product Inhibition	Both Inhibition Types	Model 1 with Deactivation	Model 4 with Deactivation
R in Eq. (8.32)	$K + C_{46}$	$K + \frac{C_{46}}{1 + C_{46}/K_{i,46}}$	$K\left(1 + \frac{C_{47}}{K_{i,47}}\right) + C_{46}$	$K(1 + C_{47}/K_{i,47}) + C_{46}(1 + C_{46}/K_{i,46})$	$K + C_{46}$	$K(1 + C_{47}/K_{i,47}) + C_{46}(1 + C_{46}/K_{i,46})$
k_{act} (h^{-1})	1.541 ± 0.012	1.566 ± 0.012	1.403 ± 0.010	1.404 ± 0.010	1.045 ± 0.007	1.103 ± 0.008
k_{deact} (h^{-1})	—	—	—	—	0.012 ± 9E-5	0.012 ± 13E-5
K (mM)	25.3 ± 0.11	27.4 ± 0.18	5.96 ± 0.17	7.61 ± 0.21	17.6 ± 0.09	20.5 ± 0.23
v_{max} (min^{-1})	10.8 ± 0.006	11.0 ± 0.013	10.7 ± 0.005	10.9 ± 0.012	11.6 ± 0.009	12.0 ± 0.019
$K_{i,46}$ (M)	—	40[a]	—	40[a]	—	14.8 ± 0.45
$K_{i,47}$ (M)	—	—	0.177 ± 0.007	0.221 ± 0.008	—	11.8 ± 2.6
χ^2	7.0	6.8	4.3	4.2	3.4	3.0

Source: Reprinted with permission from Greiner and Ternbach [95]. Copyright 2004, John Wiley and Sons. 95% accuracy bounds are provided for parameters and χ^2 values are provided for models.
[a] Values are at upper boundary.

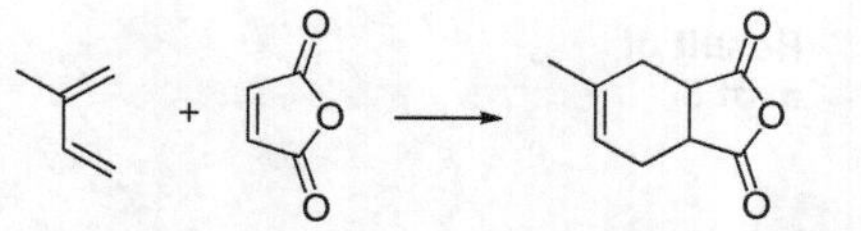

SCHEME 8.20 Diels–Alder reaction of isoprene and maleic anhydride.

Consider as an example the simplified case of the Diels–Alder reaction of isoprene (**48**) and maleic anhydride (**49**) studied by McMullen and Jensen [96]. The researchers were interested in using an automated flow system to discriminate among four candidate reaction rate laws:

$$r_{48,\text{I}} = -k_\text{I} C_{48} C_{49} \qquad (8.33)$$

$$r_{48,\text{II}} = -k_\text{II} C_{48}^2 C_{49} \qquad (8.34)$$

$$r_{48,\text{III}} = -k_\text{III} C_{48} C_{49}^2 \qquad (8.35)$$

$$r_{48,\text{IV}} = -k_{\text{IV},f} C_{48} C_{49} + k_{\text{IV},r} C_{50} \qquad (8.36)$$

The reactants and the solvent DMF were automatically delivered to a microreactor and analyzed online by HPLC. The result of the analysis was returned to a computer algorithm that determined best-fit parameters for the four candidate models and then selected new experimental conditions that would most greatly accentuate the differences between the models. These new reaction conditions were transferred to the automated system, and the process repeated.

To compare candidate models, McMullen and Jensen chose to use a Bayesian approach first introduced by Box and Hill [97]. A half-factorial, four-experiment DoE was prescribed to calculate these variances and generate initial model predictions; after this design, only two optimized experiments were needed to confirm with greater than 99% certainty that the second-order model given by Eq. (8.33) was the most representative of the four candidate models. (Note that even though Eq. (8.36) is the reversible form of Eq. (8.33), the Bayesian approach was able to discriminate this model by demonstrating $k_{\text{IV},r}$ to be effectively zero.) Figure 8.16 shows how, entering the fifth experiment (i.e. the first optimized experiment), the probability distributions for the isoprene concentration at the reactor outlet for the first three models under consideration were nearly overlapped, indicating that any of these models could reasonably describe the data. However, given the result observed in the fifth experiment, a much greater discrepancy in the probability distributions among models was observed in the ensuing experiment. The result of this experiment was consistent only with the second-order overall model, Eq. (8.33). Once this model was selected, the activation energy and pre-exponential parameters for k_I were further refined by conducting iterative experiments with feedback.

Following the kinetic study, the researchers engaged in a 500-fold scale-up of the Diels–Alder reaction to the 60 ml Corning Advanced Flow Reactor (AFR) system. To accurately scale to the larger continuous reactor, both the dispersion in the reactor and the heat transfer rate for the AFR needed to be estimated. The dispersion coefficient, D, was measured by measuring the tracer signal following a pulse injection in the AFR. This value was then used as an input into the dispersion model for a laminar flow reactor. To account for the rate of heat transfer, the heat of reaction for the Diels–Alder chemistry and the heat transfer coefficient for the AFR were both estimated from literature values. Using these approximations, the product concentrations at four different experimental conditions were evaluated numerically and found to be in good agreement (within 3%) with the experimental results found in the AFR. Overall, this example demonstrates the efficiency of scale-up that can be achieved with limited experimental effort and through the applications of automation, optimal experimental design, and continuous flow technology.

8.5 EMERGING AREAS FOR INNOVATION AND IMPLEMENTATION

Notwithstanding the innovative approaches to reaction kinetic characterization already discussed in this chapter, it is recognized that best-practice approaches will continue to evolve as new technologies enter into the process development space. This section considers a few of the emerging challenges impacting the pharmaceutical industry and a select group of the disruptive technologies expected to grow in utilization in response to these challenges.

8.5.1 Computational and Statistical Tools for Model Development

As chemical systems increase in complexity – both in terms of molecular structures and in the modes of implementation – the roles of computational modeling and data mining will become more prominent in the description of kinetic phenomena. As of 2018, complete *in silico* modeling of a chemical reaction system relevant to the pharmaceutical industry has yet to be realized [98]; however progress by way of computational chemistry and in particular with the use of density

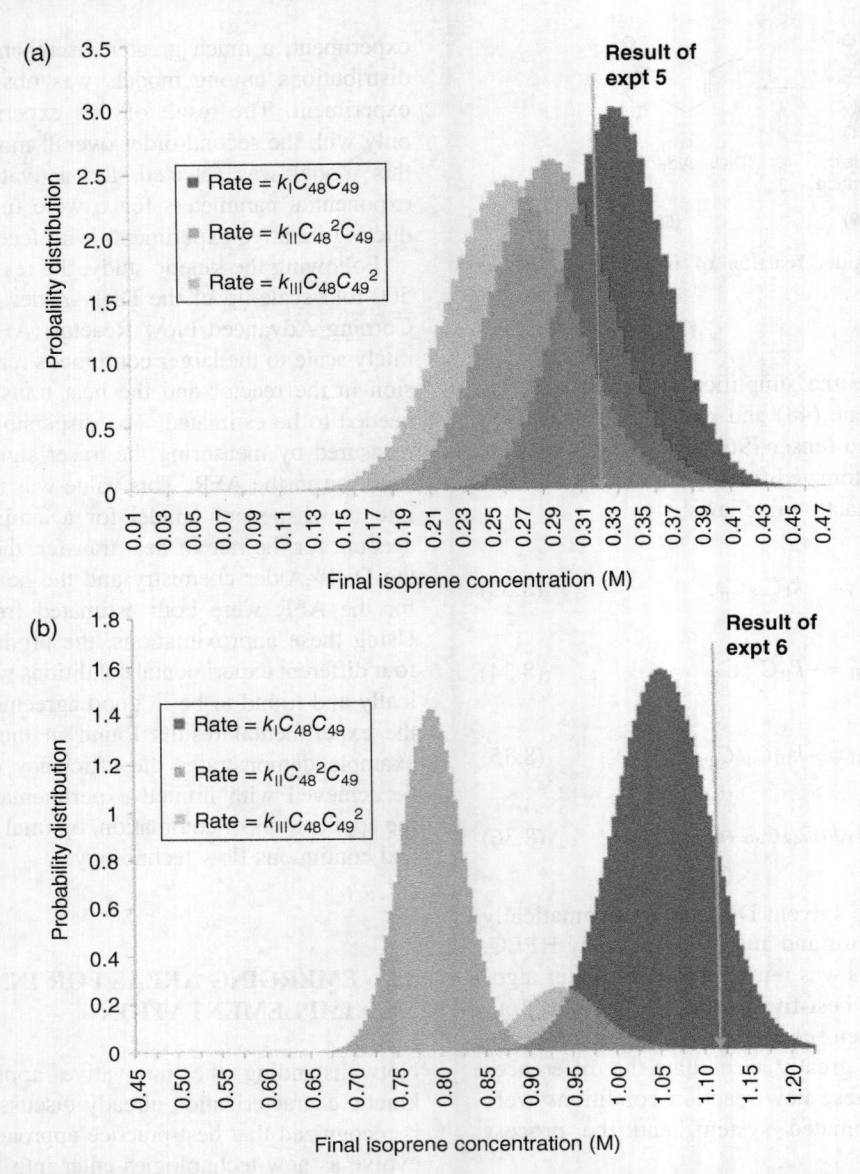

FIGURE 8.16 Probability distribution maps showing likelihood of output isoprene concentration for a given kinetic model after (a) fifth and (b) sixth automated model reduction experiments for the Diels–Alder reaction. Fifth experiment selected conditions: $\tau = 1$ minutes, $C_{48} = C_{49} = 0.5$ M. Sixth experiment selected conditions: $\tau = 1$ minutes, $C_{48} = 2.0$ M, $C_{49} = 1.5$ M. *Source*: Reprinted with permission from McMullen and Jensen [96]. Copyright 2011, American Chemical Society.

functional theory (DFT) [99] has led to breakthroughs in optimizing reaction selectivity and substrate selection, predicting impurity formation, and deciphering mechanisms in both organocatalysis and organometallic catalysis [100]. Among the more notable DFT tools for predicting reaction mechanisms are the artificial force-induced reaction method [101] and the *ab initio* nanoreactor [102]. For faster elucidation of reaction mechanisms and/or evaluation of larger molecular systems, more heuristic graph-based methods [103] and machine learning-based methods [104] have been introduced. Regarding solvent selection for chemical reactions, Adjiman and colleagues presented a quantum mechanical computer-aided molecular design (QM-CAMD) methodology for identification of rate-accelerating solvents in application to a Menschutkin reaction [105]. A virtual screen of over 1300 candidate solvents and afforded a 40% improvement in reaction rate compared with an initial six-solvent screen. Despite these advancements, uptake of quantum mechanical simulation technologies into industrial process development remains slow on account of lack of speed of the simulation in comparison with lab experimentation, restrictions on maximum molecule size and computing power, and simply a lack of accessibility to the non-computational specialist. As the accessibility and accuracy of molecular-scale simulation improve, the field of process development should experience an increased impact of these tools in areas such as reaction optimization, solubility prediction, and prediction of mass transfer effects.

Apart from more sophisticated first principles and computation-rich models, research in high-throughput reaction development is progressing in the use of statistical tools to augment the information extracted from DoE methodologies. In the optimization of new chemical reactions, the use of PCA has already been demonstrated as an efficient tool for quantifying shared properties of discrete factors, such as solvent attributes or ligand properties [106–108]. The PCA technique reduces a range of physical properties for these discrete variables to a few (perhaps 1–3) nonphysical descriptors that can be treated continuously like the factors normally incorporated into DoE optimization. Carlson was the first to propose the combination of PCA with DoE for the purpose of optimizing solvent selection for chemical reactions [107, 109]. This approach enables a discrete parameter, namely, solvent identity, to be treated as a continuous parameter on the basis of principal component scores that are derived from loading vectors encompassing a multivariate range of physicochemical solvent properties. Exploration of a solvent system described by two or three principal components can be achieved by investigating five or nine solvents, respectively (with solvents selected based on their PCA scores to best approximate a 2^2 or 2^3 factorial design, respectively, plus the inclusion of a center point).

A publication from AstraZeneca describes an updated corporate solvent selection guide consisting of a PCA-based interactive tool [110]. In contrast to previously published pharmaceutical solvent selection guides that focused predominantly upon occupational health, safety, and environmental considerations [111], this interactive solvent selection tool allows for consideration of a broader range of selection criteria. The PCA solvent map was constructed on the basis of 272 solvents and 30 solvent properties. Three principal components captured 70% of the variation in the solvent property data. Literature-derived examples of reactions exhibiting significant variation in rate or selectivity as a function of solvent were retrospectively applied to the PCA solvent map. In one example, the rate of the S_NAr reaction of 4-fluoronitrobenzene with azide anion was increased by six orders of magnitude upon moving from methanol to hexamethylphosphoramide (HMPA) as solvent [112]. Figure 8.17a shows how the principal component score plots reflect the observed trend in relative reaction rate as the PCA solvent space is traversed. A second example assessed the selectivity of O- versus

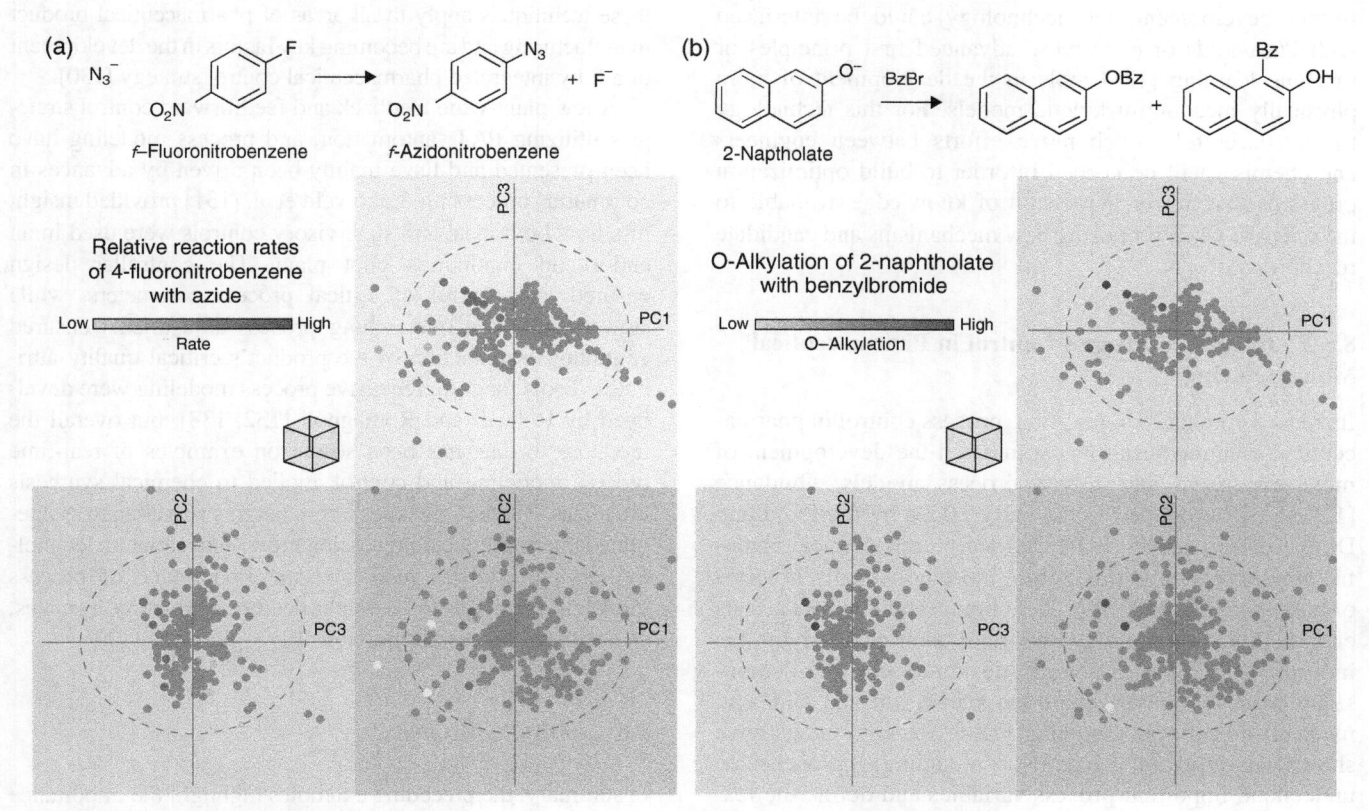

FIGURE 8.17 (a) Reaction rate of 4-fluoronitrobenzene with azide anion. (b) O-Alkylation selectivity. *Source*: Reprinted with permission from Diorazio et al. [110]. Copyright 2016, American Chemical Society.

C-alkylation of 2-naphtholate by benzyl bromide in a range of solvents (Figure 8.17b) [113]. In this case the O-alkylation selectivity was adjusted from 7 to 97% upon moving from 2,2,2-trifluoroethanol to N,N-dimethylformamide (DMF).

In addition to solvent selection, the coupling of PCA with DoE has accelerated the understanding of how solvent and ligand properties contribute to optimal performance in industrially relevant examples such as Suzuki–Miyaura, Heck, Buchwald–Hartwig, Ullman, S_NAr, and borrowing hydrogen reactions [109, 114, 115]. Researchers at GSK have also demonstrated how PCA and projection to latent structure (PLS) models can used to diagnose regions of the experimental design space where impurity formation is most prevalent, through the augmentation of response surface models with selected spiking studies [116].

The merger of discrete and continuous factors into automated kinetic modeling is an intuitive next step for reaction optimization platforms. Though a one-stop machine for chemical optimization and scale-up is still far from reach, automated on-demand synthesis [117, 118] and automated reaction optimization for both continuous and discrete factors [119, 120] are emerging technologies. In the latter, even black box optimal experimental design approaches to reaction optimization have shown a propensity to identify shared response surface model attributes among solvents and catalysts with known physical property similarities. With further development, this technology could be interfaced with PCA tools or even more advanced first principles or machine learning [121] tools in the development of more physically meaningful kinetic models. For this technology to be successful, collaborative efforts between engineers and chemists will be needed in order to build optimization programs that mirror the wealth of knowledge available to the scientist when proposing new mechanisms and candidate reaction pathways.

8.5.2 Integrated Process Control in Pharmaceutical Manufacturing

Increased emphasis on real-time process control in pharmaceutical manufacturing has stimulated the development of more expansive and robust process models. Guidance [122, 123] introduced in the early 2000s by the Food and Drug Administration (FDA) has led pharmaceutical manufacturers to examine production more holistically in terms of how the variability of process inputs and controls affects end quality. Among the important metrics used by the industry to propose and verify design spaces and demonstrate process robustness are the statistics-based DoE and mechanistic process modeling [124]. Recent studies have shown an influx of mechanistic modeling approaches to understand important process variables and define the feasible region by constructing probability maps (see, for example, Garcia et al. [84] and Changi et al. [46]). Factors supporting the use of mechanistic models in these exercises include the ability to quantify the risk associated with different process variations (using sensitivity analyses as presented in Section 8.4) and the ability to guide decision making during processing with the use of higher-level control architectures incorporating data derived from real-time (or near real-time) PAT.

The FDA has recognized the importance of models as a tool to assist regulatory submission. The implementation of integrated process diagnostic and control systems in manufacturing will be among the most important drivers for mechanistic modeling in the pharmaceutical industry. Applicable for all manufacturing – but particularly in the case of continuous manufacturing – the coupling of multiple reaction, separation, and/or crystallization steps in a synthetic route dictates the need for predictive understanding of the effects that changes in materials and upstream processing can have on downstream processes and the final product. These changes, when diagnosed through PAT, can be fed to the process model to induce automated process manipulations utilizing feedforward control (changing a downstream operation on account of a change in an upstream material property), feedback control (a corrective change made to the upstream process), or combinations thereof. Examples of how modeling can be used to steer the process toward optimality can be found prominently in crystallization [125, 126] and tablet manufacturing [127–129] literature, but these techniques apply to all areas of pharmaceutical product manufacturing and are becoming key factors in the development of a fully integrated pharmaceutical control strategy [130].

A few plant-wide feedback and feedforward control strategies utilizing PAT, automation, and process modeling have been presented and have mainly been driven by advances in continuous processing. Lakerveld et al. [131] provided insight into how both local and supervisory controls were used in an end-to-end continuous pilot plant. The controller design ensured fast control of critical process parameters, while slower and more overarching process adjustments ensured continuous satisfaction of the product's critical quality attributes. Tools for comprehensive process modeling were developed by Rolandi and Romagnoli [132, 133], but overall the literature to date has been scarce on examples of real-time process modeling and control applied to chemical synthesis processes. There is an expectation that this will change as integrated pharmaceutical processing grows into a mature technology in the coming years and the importance of process modeling continues to be emphasized by regulatory agencies, as has been the case in previous FDA guidance [134].

8.6 CONCLUSIONS

In summary, the preceding examples highlight the importance of understanding the interplay of chemical transformations with physical rate processes in the context of successful process development, commercial scale-up, and manufacturing. Although every chemical system introduces new challenges, continued use of the characterization and numerical tools presented herein should enable a more systematic approach to the diagnosis of competing rate effects. Teamwork between chemists and engineers is invaluable in these efforts.

8.7 QUESTIONS

1. Derive the expression for $k_L a$ (gas–liquid).

The mass transfer from the gas phase to the liquid phase can be described as shown:

$$\text{Rate} = \frac{dC}{dt} = k_L a (C_t - C_{sat})$$

where

C_t is the solution-phase concentration of the gas at a given time.

C_{sat} is the equilibrium concentration of the gas in solution.

Integrating the above expression yields the following expression:

$$\ln\left[\frac{(C_t - C_{sat})}{(C_o - C_{sat})}\right] = \ln\left[\frac{(n_t - n_{sat})}{(n_o - n_{sat})}\right] = k_L a \times t$$

where

C_o is the concentration of the gas in solution at $t = 0$.
n_o is the moles of gas initially in solution.
n_t is the moles of gas in solution at a given time t.
n_{sat} is the moles of gas in solution at the saturation point.

Mass balance for the gas yields the following:

$$n_t = n_o + (P_o - P_t)\frac{V_g}{RT}$$

$$n_{sat} = n_o + (P_o - P_f)\frac{V_g}{RT}$$

Substitution of the mass balance equation above yields the following expression for $k_L a$:

$$k_L a \times t = \ln \frac{P_f - P_t}{P_f - P_o}$$

2. For a reaction with a first-order rate law, develop a rate expression that integrates the mass transfer constant for transport across the solid–liquid interface and the intrinsic reaction kinetics.

The rate of diffusion across the boundary layer is defined as follows:

$$R_{diff} = k_s a_s (C_\infty - C_I)$$

where C_∞ and C_I are the bulk and interface concentrations, respectively. The reaction rate for a power law rate model can be described as follows:

$$R_{rxn} = k_r C_I$$

Equating the above two expressions and solving for C_I yields the following expression:

$$C_I = \frac{k_r k_s a_s}{k_r + k_s a_s} C_\infty$$

Substitution of this expression into the equation for R_{rxn} yields the following:

$$R_{rxn} = \frac{k_r k_s a_s}{k_r + k_s a_s} C_\infty$$

3. Show that for the elementary reaction steps below, the overall reaction rate law can be simplified to Eq. (8.21):

$$\mathbf{I} + S \rightleftharpoons \mathbf{II}$$

$$\mathbf{II} \rightleftharpoons \mathbf{III}$$

$$\mathbf{III} + H_2 \rightarrow \text{Product} + \mathbf{I}$$

(Hint: The total amount of catalytic species **I**, **II**, and **III** is conserved.)

For the catalytic cycle above, all intermediate species must be generated and consumed at the same rate in order to afford the overall reaction:

$$S + H_2 \rightarrow \text{Product}$$

The reactions to produce **II** and **III** are both in equilibrium; hence we know that

$$C_{II} = K_1 C_I C_S$$
$$C_{III} = K_2 C_{II} = K_1 K_2 C_I C_S$$

For convenience, we will use the final elementary step to define the overall rate:

$$r_P = -r_S = k_3 C_{III} C_{H_2}$$

Of course C_{III} is not easily measured before or during the reaction. We do know, however, the total amount of catalyst

(C_{cat}) in the reactor. With some rearranging, an expression for C_{cat} can be substituted into the rate expression above in place of C_{III}:

$$C_{cat} = C_I + C_{II} + C_{III}$$

$$C_{cat} = C_{III}\left(\frac{1}{K_1 K_2 C_S} + \frac{1}{K_2} + 1\right)$$

$$\Rightarrow -r_S = \frac{k_3 C_{cat} C_{H_2}}{\frac{1}{K_1 K_2 C_S} + \frac{1}{K_2} + 1} = \frac{k_3 K_1 K_2 C_{cat} C_{H_2} C_S}{1 + K_1 C_S + K_1 K_2 C_S}$$

This expression can be simplified further by assuming that the equilibrium between species **II** and **III** is shifted toward the intermediate **III**, such that $K_2 \gg 1$ and $K_1 K_2 C_S \gg K_1 C_S$. For simplicity, define $k = k_3$ and $K = K_1 K_2$ to afford the rate law:

$$-r_S = \frac{kKC_{cat} C_{H_2} C_S}{1 + KC_S}.$$

REFERENCES

1. Smith, M.B. (2013). *March's Advanced Organic Chemistry: Reactions, Mechanisms, and Structure*, 7e. Hoboken, NJ: Wiley.
2. Astarita, G. (1967). *Mass Transfer with Chemical Reaction*. Amsterdam: Elsevier.
3. Dankwerts, P.V. (1970). *Gas-Liquid Reactions*. New York: McGraw-Hill.
4. Levenspiel, O. (1999). *Chemical Reaction Engineering*, 3e. Hoboken, NJ: Wiley.
5. Froment, G.F. and Bischoff, K.B. (1990). *Chemical Reactor Analysis and Design*, 2e. New York: Wiley.
6. Masel, R.I. (2001). *Chemical Kinetics and Catalysis*. New York: Wiley.
7. Carey, J.S., Laffan, D., Thomson, C., and Williams, M.T. (2006). *Org. Biomol. Chem.* 4: 2337–2347.
8. Roberge, D.M. (2004). *Org. Process Res. Dev.* 8: 1049–1053.
9. Mascia, M., Heider, P.L., Zhang, H. et al. (2013). *Angew. Chem. Int. Ed.* 52: 12359–12363.
10. Cole, K.P., Groh, J.M., Johnson, M.D. et al. (2017). *Science* 356: 1144–1150.
11. Lee, S.L., O'Connor, T.F., Yang, X. et al. (2015). *J. Pharm. Innov.* 10: 191–199.
12. Schaber, S.D., Gerogiorgis, D.I., Ramachandran, R. et al. (2011). *Ind. Eng. Chem. Res.* 50: 10083–10092.
13. Fogler, H.S. (2006). *Elements of Chemical Reaction Engineering*, 4e. Upper Saddle River, NJ: Pearson Education.
14. Steinfeld, J.I., Francisco, J.S., and Hase, W.L. (1999). *Chemical Kinetics and Dynamics*, 2e. Upper Saddle River, NJ: Pearson Education.
15. Reichardt, C. and Welton, T. (2011). *Solvents and Solvent Effects in Organic Chemistry*, 4e. Weinheim: Wiley-VCH.
16. Niemeier, J.K., Rothhaar, R.R., Vicenzi, J.T., and Werner, J.A. (2014). *Org. Process Res. Dev.* 18: 410–416.
17. Mills, P.L. and Chaudhari, R.V. (1997). *Catal. Today* 37: 367–404.
18. Hartman, R.L., McMullen, J.P., and Jensen, K.F. (2011). *Angew. Chem. Int. Ed.* 50: 7502–7519.
19. Nienow, A.W. (1975). *Chem. Eng. J.* 9: 153–160.
20. Paul, E.L., Atiemo-Obeng, V.A., and Kresta, S.M. (2004). *Handbook of Industrial Mixing: Science and Practice*. Hoboken, NJ: Wiley.
21. Barker, J.J. and Treybel, R.E. (1960). *AIChE J.* 6: 289–295.
22. Harriott, P. (1962). *AIChE J.* 8: 93–101.
23. Harriott, P. (1962). *AIChE J.* 8: 101–102.
24. Brian, P.L.T., Hales, H.B., and Sherwood, T.K. (1969). *AIChE J.* 15: 727–733.
25. Davies, J.T. (1986). *Chem. Eng. Process.* 20: 175–181.
26. Marrone, G.M. and Kirwan, D.J. (1986). *AIChE J.* 32: 523–525.
27. Armenante, P.M. and Kirwan, D.J. (1989). *Chem. Eng. Sci.* 44: 2781–2796.
28. Tamas, A., Martagiu, R., and Minea, R. (2007). *Chem. Bull. "POLITEHNICA" Univ. (Timisoara)* 52: 133–138.
29. Garside, J., Mersmann, A., and Nyvlt, J. (2002). *Measurement of Crystal Growth and Nucleation Rates*, 2e. Rugby: Institution of Chemical Engineers.
30. Green, D.W. and Perry, R.H. (2008). *Perry's Chemical Engineering Handbook*, 8e, 7–19. New York: McGraw-Hill.
31. Singh, U.K., Pietz, M.A., and Kopach, M. (2009). *Org. Process Res. Dev.* 13: 276–279.
32. Zwietering, T.N. (1958). *Chem. Eng. Sci.* 8: 244–253.
33. Changi, S. and Wong, S.-W. (2016). *Org. Process Res. Dev.* 20: 525–539.
34. Bourne, J.R. (2003). *Org. Process Res. Dev.* 7: 471–508.
35. Johnson, B.K. and Prud'homme, R.K. (2003). *AIChE J.* 49: 2264–2282.
36. Mahajan, A.J. and Kirwan, D.J. (1996). *AIChE J.* 42: 1801–1814.
37. Singh, U.K., Spencer, G., Osifchin, R. et al. (2005). *Ind. Eng. Chem. Res.* 44: 4068–4074.
38. Bockhorn, H., Mewes, D., Peukert, W., and Warnecke, H.-J. (eds.) (2010). *Micro and Macro Mixing: Analysis, Simulation and Numerical Calculation*. Berlin, Heidelberg: Springer-Verlag.
39. Deimling, A., Karandikar, B.M., Shah, Y.T., and Carr, N.L. (1984). *Chem. Eng. J.* 29: 127–140.
40. Oldshue, J.Y. (1980). *Chem. Eng. Prog.* 76: 60–64.
41. Johnson, M.D., May, S.A., Haeberle, B. et al. (2016). *Org. Process Res. Dev.* 20: 1305–1320.
42. Hickman, D.A., Holbrook, M.T., Mistretta, S., and Rozeveld, S.J. (2013). *Ind. Eng. Chem. Res.* 52: 15287–15292.
43. Rode, C.V., Vaidya, M.J., and Chaudhari, R.V. (1999). *Org. Process Res. Dev.* 3: 465–470.
44. Smith, J.M., Van Ness, H.C., and Abbott, M.M. (2005). *Introduction to Chemical Engineering Thermodynamics*, 7e. New York: McGraw-Hill.
45. Delgado, J.M.P.Q. (2006). *Heat Mass Transfer.* 42: 279–310.
46. Changi, S.M., Yokozawa, T., Yamamoto, T. et al. (2017). *React. Chem. Eng.* 2: 720–739.

47. Blackmond, D.G. (2005). *Angew. Chem. Int. Ed.* 44: 4302–4320.
48. Mathew, J.S., Klussmann, M., Iwamura, H. et al. (2005). *J. Org. Chem.* 71: 4711–4722.
49. Shekhar, S., Ryberg, P., Hartwig, J.F. et al. (2006). *J. Am. Chem. Soc.* 128: 3584–3591.
50. Singh, U.K., Strieter, E.R., Blackmond, D.G., and Buchwald, S.L. (2002). *J. Am. Chem. Soc.* 124: 14104–14114.
51. Zogg, A., Fischer, U., and Hungerbuhler, K. (2003). *Ind. Eng. Chem. Res.* 42: 767–776.
52. Boddington, T., Chia, H.A., Halford-Maw, P. et al. (1992). *Thermochim. Acta.* 195: 365–372.
53. LeBlond, C., Wang, J., Larsen, R. et al. (1998). *Top. Catal.* 5: 149–158.
54. Chung, R. and Hein, J.E. (2017). *Top. Catal.* 60: 594–608.
55. Stoessel, F. (2008). *Thermal Safety of Chemical Processes: Risk Assessment and Process Design*. Weinheim: Wiley-VCH.
56. Simon, L.L., Pataki, H., Marosi, G. et al. (2015). *Org. Process Res. Dev.* 19: 3–62.
57. Chanda, A., Daly, A.M., Foley, D.A. et al. (2015). *Org. Process Res. Dev.* 19: 63–83.
58. Center for Drug Evaluation and Research (U.S.); Center for Veterinary Medicine (U.S.); United States. Food and Drug Administration. Office of Regulatory Affairs (2004). *Guidance for Industry: PAT – A Framework for Innovative Pharmaceutical Development, Manufacturing, and Quality Assurance*. Rockville, MD: Food and Drug Administration.
59. Esmonde-White, K.A., Cuellar, M., Uerpmann, C. et al. (2017). *Anal. Bioanal. Chem.* 409: 637–649.
60. Singh, K. and Blümich, B. (2016). *Trends Anal. Chem.* 83: 12–26.
61. Gouilleux, B., Charrier, B., Akoka, S. et al. (2016). *Trends Anal. Chem.* 83: 65–75.
62. Foley, D.A. and Dunn, A.L. (2016). Zell MT. *Magn. Reson. Chem.* 54: 451–456.
63. Brum, J. and Dell'Orco, P. (1998). *Rapid Commun. Mass Spectrom.* 12: 741–745.
64. Dell'Orco, P., Brum, J., Matuoka, R. et al. (1999). *Anal. Chem.* 71: 5165–5170.
65. Brum, J., Dell'Orco, P., Lapka, S. et al. (2001). *Rapid Commun. Mass Spetrom.* 15: 1548–1553.
66. Muske, K.R., Badlani, M., Dell'Orco, P.C., and Brum, J. (2004). *Chem. Eng. Sci.* 59: 1167–1180.
67. Snyder, D.T., Pulliam, C.J., Ouyang, Z., and Cooks, R.G. (2016). *Anal. Chem.* 88: 2–29.
68. Ray, A., Bristow, T., Whitmore, C., and Mosely, J. (2018). *Mass Spec. Rev.* 37: 565–579.
69. Theron, R., Wu, Y., Yunker, L.P.E. et al. (2016). *ACS Catal.* 6: 6911–6917.
70. Schafer, W.A., Hobbs, S., Rehm, J. et al. (2007). *Org. Process Res. Dev.* 11: 870–876.
71. Welch, C.J., Gong, X., Schafer, W. et al. (2010). *Tetrahedron Asymmetr.* 21: 1674–1681.
72. Zawatzky, K., Grosser, S., and Welch, C.J. (2017). *Tetrahedron* 73: 5048–5053.
73. Foley, D.A., Wang, J., Maranzano, B. et al. (2013). *Anal. Chem.* 85: 8928–8932.
74. Rosner, T., Le Bars, J., Pfaltz, A., and Blackmond, D.G. (2001). *J. Am. Chem. Soc.* 123: 1848–1855.
75. O'Connor, C. and Wilkinson, G. (1968). *J. Chem. Soc. A.* 2665–2671.
76. Hughes, D. and Zhao, D. (1993). *J. Org. Chem.* 58: 228–233.
77. Moore, J.S. and Jensen, K.F. (2014). *Angew. Chem. Int. Ed.* 53: 470–473.
78. Moore, J.S., Smith, C.D., and Jensen, K.F. (2016). *React. Chem. Eng.* 1: 272–279.
79. Lin, Y.-S. and Yamamoto, A. (1998). *Organometallics* 17: 3466–3478.
80. Weissman, S.A. and Anderson, N.G. (2015). *Org. Process Res. Dev.* 19: 1605–1633.
81. Mateos, C., Mendiola, J., Carpintero, M., and Minguez, J.M. (2010). *Org. Lett.* 12: 4924–4927.
82. ICH (2009). ICH harmonised tripartite guideline: pharmaceutical development Q8 (R2). *Proceedings of the International Conference on Harmonisation of Technical Requirements for Registration of Pharmaceuticals for Human Use*, Geneva (August 2009).
83. Garcia-Munoz, S., Luciani, C.V., Vaidyaraman, S., and Seibert, K.D. *Org. Process Res. Dev.* 205 19: 1012–1023.
84. Stazi, F., Palmisano, G., Turconi, M. et al. (2004). *J. Org. Chem.* 69: 1097–1103.
85. Burt, J.L., Braem, A.D., Ramirez, A. et al. (2011). *J. Pharm. Innov.* 6: 181–192.
86. Hallow, D.M., Mudryk, B.M., Braem, A.D. et al. (2010). *J. Pharm. Innov.* 5: 193–203.
87. Ramirez, A., Hallow, D.M., Fenster, M.D.B. et al. (2016). *Org. Process Res. Dev.* 20: 1781–1791.
88. Ramirez, A., Mudryk, B., Rossano, L., and Tummala, S. (2012). *J. Org. Chem.* 77: 775–779.
89. Issanchou, S., Cognet, P., and Cabassud, M. (2005). *AIChE J.* 51: 1773–1781.
90. Lindsey, J.S. (1997). Automated approaches to reaction optimization. In: *A Practical Guide to Combinatorial Chemistry* (ed. A. Czarnik and S. Hobbs-Dewitt), 309–326. Washington, DC: American Chemical Society.
91. Wagner, R.W., Li, F., Du, H., and Lindsey, J.S. (1999). *Org. Process Res. Dev.* 3: 28–37.
92. McMullen, J.P. and Jensen, K.F. (2010). *Annu. Rev. Anal. Chem.* 3: 19–42.
93. Reizman, B.J. and Jensen, K.F. (2016). *Acc. Chem. Res.* 49: 1786–1796.
94. Reizman, B.J. and Jensen, K.F. (2012). *Org. Process Res. Dev.* 16: 1770–1782.
95. Greiner, L. and Ternbach, M.B. (2004). *Adv. Synth. Cat.* 346: 1392–1396.
96. McMullen, J.P. and Jensen, K.F. (2011). *Org. Process Res. Dev.* 15: 398–407.
97. Box, G.E.P. and Hill, W.J. (1967). *Technometrics* 9: 57–71.

98. Houk, K.N. and Liu, F. (2017). *Acc. Chem. Res.* 50: 539–543.
99. Hohenberg, P. and Kohn, W. (1964). *Phys. Rev.* 136: B864.
100. For examples, see (a) QNN, N. and Tantillo, D.J. (2014). *Chem. Asian. J.* 9: 674–680.
    (b) Lam, Y.-H., Grayson, M.N., Holland, M.C. et al. (2016). *Acc. Chem. Res.* 49: 750–762.
101. Sameera, W.M.C., Maeda, S., and Morokuma, K. (2016). *Acc. Chem. Res.* 49: 763–773.
102. Wang, L.P., Titov, A., McGibbon, R. et al. (2014). *Nat. Chem.* 6: 1044–1048.
103. Dewyer, A.L. and Zimmerman, P.M. (2017). *Org. Biomol. Chem.* 15: 501–504.
104. Kayala, M.A., Azencott, C.-A., Chen, J.H., and Baldi, P. (2011). *J. Chem. Inf. Model.* 51: 2209–2222.
105. Struebing, H., Ganase, Z., Karamertzanis, P.G. et al. (2013). *Nat. Chem.* 5: 952–957.
106. Carlson, R. and Carlson, J.E. (2005). *Org. Process Res. Dev.* 9: 680–689.
107. Maldonado, A.G. and Rothenberg, G. (2010). *Chem. Soc. Rev.* 39: 1891–1902.
108. Murray, P.M., Tyler, S.N.G., and Moseley, J.D. (2013). *Org. Process Res. Dev.* 17: 40–46.
109. Carlson, R., Lundstedt, T., and Albano, C. (1985). *Acta Chem. Scand. Ser. B* 39: 79–91.
110. Diorazio, L.J., Hose, R.J., and Adlington, N.K. (2016). *Org. Process Res. Dev.* 20: 760–773.
111. Prat, D., Halyer, J., and Wells, A. (2014). *Green Chem.* 16: 4546–4551.
112. Cox, B.G. and Parker, A.J. (1973). *J. Am. Chem. Soc.* 95: 408–410.
113. Kornblum, N., Seltzer, R., and Haberfield, P. (1963). *J. Am. Chem. Soc.* 85: 1148–1154.
114. Murray, P.M., Bellany, F., Benhamou, L. et al. (2016). *Org. Biomol. Chem.* 14: 2373–2384.
115. Moseley, J.D. and Murray, P.M. (2014). *J. Chem. Technol. Biotechnol.* 89: 623–632.
116. Wang, H., Goodman, S.N., Dai, Q. et al. (2008). *Org. Process Res. Dev.* 12: 226–234.
117. Li, J., Ballmer, S.G., Gillis, E.P. et al. (2015). *Science* 347: 1221–1226.
118. Adamo, A., Beingessner, R.L., Behnam, M. et al. (2016). *Science* 352: 61–67.
119. Reizman, B.J. and Jensen, K.F. (2015). *Chem. Commun.* 51: 13290–13293.
120. Reizman, B.J., Wang, Y.-M., Buchwald, S.L., and Jensen, K.F. (2016). *React. Chem. Eng.* 1: 658–666.
121. Coley, C.W., Barzilay, R., Jaakkola, T.S. et al. (2017). *ACS Cent. Sci.* 3: 434–443.
122. ICH (2000). *Guidance on Q6A Specifications: Test Procedures and Acceptance Criteria for New Drug Substances and New Drug Products: Chemical Substances*. Washington, DC: Food and Drug Administration.
123. US Food and Drug Administration (2006). *Guidance for Industry: Quality Systems Approach to Pharmaceutical CGMP Regulations*. Washington, DC: Food and Drug Administration.
124. Rantanen, J. and Khinast, J. (2015). *J. Pharm. Sci.* 104: 3612–3638.
125. Fujiwara, M., Nagy, Z.K., Chew, J.W., and Braatz, R.D. (2005). *J. Process Control.* 15: 493–504.
126. Zhou, G., Moment, A., Yaung, S. et al. (2013). *Org. Process Res. Dev.* 17: 1320–1329.
127. Singh, R., Roman-Osprino, A.D., Romanach, R.J. et al. (2015). *Int. J. Pharm.* 495: 612–625.
128. Muteki, K., Swaminathan, V., Sekulic, S.S., and Reid, G.L. (2011). *AAPS PharmSciTech.* 12: 1324–1334.
129. Townsend Haas, N., Ierapetritou, M., and Singh, R. (2017). *J. Pharm. Innov.* 12: 110–123.
130. Myerson, A.S., Krumme, M., Nasr, M. et al. (2015). *J. Pharm. Sci.* 104: 832–839.
131. Lakerveld, R., Benyahia, B., Heider, P.L. et al. (2015). *Org. Process Res. Dev.* 19: 1088–1100.
132. Rolandi, P.A. and Romagnoli, J.A. (2005). *Comp. Aided Chem. Eng.* 20: 1315–1320.
133. Rolandi, P.A. and Romagnoli, J.A. (2010). *Comp. Chem. Eng.* 34: 17–35.
134. FDA Guidance for Industry (2012). *ICH Q8, Q9, & Q10 Questions and Answers—Appendix: Q&As from Training Sessions (Q8, Q9, & Q10 Points to Consider)*. Silver Spring, MD: Food and Drug Administration.

9

UNDERSTANDING FUNDAMENTAL PROCESSES IN CATALYTIC HYDROGENATION REACTIONS

YONGKUI SUN*
Department of Process Research, Merck & Co., Inc., Rahway, NJ, USA

CARL LEBLOND
Department of Chemistry, University of Pennsylvania, Indiana, PA, USA

9.1 INTRODUCTION

Hydrogenation is a powerful methodology in synthetic organic chemistry and has been broadly employed by organic chemists in drug synthesis. Heterogeneous-catalyzed hydrogenation has been very popular traditionally and continues to play a critical role in modern organic synthesis [1, 2]. Hydrogenation by homogeneous catalysts, particularly asymmetric hydrogenation, is gaining momentum in applications in the pharmaceutical industry. Since the pioneering work in the development of chiral catalysts for asymmetric hydrogenation by William Knowles in late 1960s and the first commercial application of asymmetric hydrogenation in the early 1970s by Monsanto in the production of the anti-Parkinsonian drug L-DOPA [3, 4], asymmetric hydrogenation has developed into a powerful chemical transformation, achieving enantioselectivities matching those previously seen only in enzymatic processes. Over the ensuing two to three decades, there has been rapid development in the science and technology of asymmetric hydrogenation. One of the key milestones in the development of this chiral technology was the discovery of BINAP in the 1980s by Ryoji Noyori and coworkers [5], which significantly broadened the scope of utility of asymmetric hydrogenation [6]. A recent special issue of the *Accounts of Chemical Research* documented the growing application of asymmetric hydrogenation in the pharmaceutical industry [7].

Hydrogenation reactions in the liquid phase are complex processes. Even for a homogeneously catalyzed hydrogenation for which the catalyst and the substrate are fully soluble in the solvent, the hydrogenation process is heterogeneous in nature since the dihydrogen reducing reagent H_2 is in a different phase. In addition to the catalytic reaction occurring on the catalyst, there are a number of mass transfer processes that can exert direct influence on the outcome of the catalytic reaction itself.

In the case of heterogeneous catalysis, the mass transfer processes include H_2 transport across the gas–liquid and the liquid–solid interfaces before the molecular H_2 chemisorbs dissociatively on the metal catalyst, as well as H_2 diffusion inside the pore structure of the catalyst particles. A schematic depicting the mass transfer processes (pore diffusion not displayed) is shown in Figure 9.1, along with the corresponding H_2 concentration profile. Among the mass transfer processes, the gas–liquid mass transfer needs significant attention in designing a hydrogenation process because it can have profound impact on the performance of the hydrogenation reaction [8], *and* it is greatly affected by agitation, reactor design and configuration, solvent properties such as viscosity, and solvent fill level. Mass transfer of H_2 across the liquid–solid interface and through the pore structure in the catalyst particles is often dominated by the physical characteristics of the catalyst, i.e. particle size, shape and support material, and density and pore structures. While agitation intensity and reactor design can influence their kinetics, for catalytic system in which the diameter of the catalyst particle is less than 50 μm, the effects become

*Current address: Ionova Life Science Co., Ltd, Shenzhen, China

---

Chemical Engineering in the Pharmaceutical Industry: Active Pharmaceutical Ingredients, Second Edition.
Edited by David J. am Ende and Mary T. am Ende.
© 2019 John Wiley & Sons, Inc. Published 2019 by John Wiley & Sons, Inc.

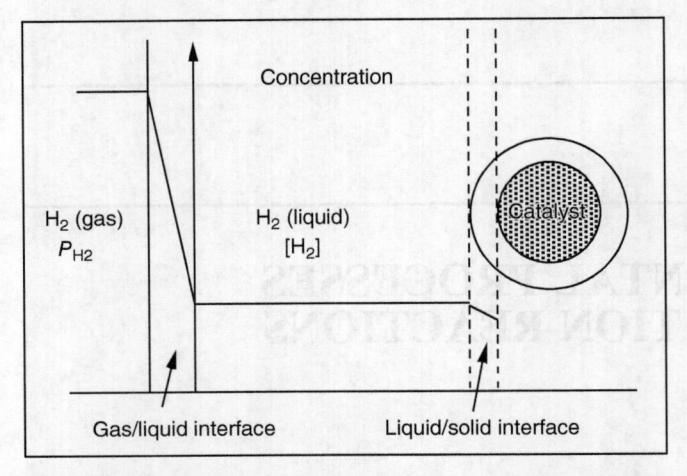

FIGURE 9.1 A schematic of hydrogen mass transfer processes across the gas/liquid and liquid/solid interfaces during a heterogeneously catalyzed hydrogenation reaction. Also plotted is a schematic of the hydrogen concentration profile.

minimal once a uniform catalyst suspension is achieved. It is not the intention of this chapter to review all the mass transfer issues encountered in hydrogenation reactions carried out in slurry reactors [9]. Instead, this chapter focuses only on the simple but crucial hydrogen gas–liquid mass transfer issue and its impact on the development of hydrogenation processes.

9.2 SOLUTION HYDROGEN CONCENTRATION DURING HYDROGENATION REACTIONS, [H$_2$]

When gas–liquid delivery is the dominant mass transfer step, the pathway followed by H$_2$ from the gas phase to its incorporation into the product in catalytic hydrogenation reactions using either a homogeneous or a heterogeneous catalyst may be simply described as follows:

$$H_2(g) \xrightarrow{k_L a} H_2(l) \xrightarrow{k_r} \text{Hydrogenation product} \quad (9.1)$$

where $k_L a$ is the mass transfer coefficient for the H$_2$ transfer across the gas–liquid interface and k_r is the rate coefficient of the catalytic reaction. The intrinsic kinetics in the catalytic hydrogenation is a function of concentrations of the substrate, the catalyst, and the dissolved H$_2$, in addition to other factors such as the temperature. In developing a chemical process, one is normally demanding in the knowledge of the concentration of the substrate and tries to follow [substrate] via various means during the course of the reaction. One is frequently, however, less demanding in the knowledge of the solution concentration of hydrogen, [H$_2$], due, in most part, to an assumption that [H$_2$] equals the equilibrium solubility of hydrogen, [H$_2$]$_{sat}$, at the temperature and pressure of the reaction.

The assumption holds, however, only when the rate of H$_2$ mass transfer is far greater than that of hydrogenation reaction itself. While [H$_2$]$_{sat}$ is fixed at a constant gas-phase H$_2$ pressure, [H$_2$] may vary widely from nearly zero to nearly saturation, depending critically upon the relative magnitude of the rate of H$_2$ mass transfer from the gas to the liquid phase and the rate of reactive H$_2$ removal from the liquid phase due to the hydrogenation reaction. This is readily seen from a mass balance of the dissolved H$_2$ below:

$$\frac{d[H_2]}{dt} = k_L a \left([H_2]_{sat} - [H_2] \right) - k_r f([H_2], [\text{catalyst}], [\text{substrate}]) \quad (9.2)$$

The first term to the right of Eq. (9.2) is the rate of H$_2$ mass transfer from the gas to the liquid phase, and the second term the rate of reactive removal of the dissolved H$_2$ from the liquid phase by the catalytic hydrogenation.

The kinetics of H$_2$ mass transfer resembles that of the familiar first-order chemical reaction and may be characterized by a single parameter, the mass transfer coefficient $k_L a$. A comparison of the characteristics, including the kinetic expression and the maximum rate, between these two processes is given in Table 9.1. The significance of the mass transfer coefficient $k_L a$ is that it is the kinetic factor that dictates the maximum rate of mass transfer of hydrogen across a gas–liquid interface, i.e.

$$R^{max}_{H_2, g/l} = k_L a [H_2]_{sat} \quad (9.3)$$

in conjunction with the thermodynamic factor [H$_2$]$_{sat}$.

While k_r is an intrinsic kinetic property of the catalytic system, $k_L a$ is strongly affected by characteristics of the reactor vessel, including reactor type, configuration, liquid fill level, viscosity, and particularly agitation speed. At a constant H$_2$ pressure, the magnitude of [H$_2$] during the reaction depends critically upon the relative magnitude of these two rate coefficients. In a situation where the kinetics of the mass transfer is much slower than the intrinsic reaction kinetics, i.e.

TABLE 9.1 A Comparison of the Characteristics of Gas–Liquid Mass Transfer and a First-Order Chemical Reaction

	Gas–Liquid H$_2$ Mass Transfer	First-Order Reaction
Kinetic expression	$-\frac{d[H_2]}{dt} = k_L a \left([H_2] - [H_2]_{sat} \right)$	$-\frac{d[C]}{dt} = k_r [C]$
First-order rate coefficient (s^{-1})	$k_L a$	k_r
Maximum rate	$R^{max}_{H_2, g/l} = k_L a [H_2]_{sat}$	$R^{max}_{rxn} = k_r [C]_0$

$$R^{max}_{H_2,g/l} \ll R^{max}_{H_2,rxn}$$

or

$$k_L a[H_2]_{sat} \ll k_r f([H_2]_{sat},[catalyst],[substrate]), \quad (9.4)$$

$[H_2]$ deviates significantly from $[H_2]_{sat}$, and the solution is starved for H_2. In other words, the effective H_2 pressure, i.e. the pressure that the catalyst experiences, is lower than the H_2 pressure in the gas phase. Under extreme hydrogen-starved conditions, $[H_2]$ or the effective H_2 pressure may approach zero. In this case, the reaction becomes entirely limited by the mass transfer instead of by the catalytic processes on the catalyst, and the observed reaction rate equals the maximum rate of H_2 mass transfer. On the other hand, if the kinetics of mass transfer is much faster than the intrinsic reaction kinetics, i.e.

$$R^{max}_{H_2,g/l} \gg R^{max}_{H_2,rxn},$$

or

$$k_L a[H_2]_{sat} \gg k_r f([H_2]_{sat},[catalyst],[substrate]), \quad (9.5)$$

$[H_2]$ approaches $[H_2]_{sat}$, and the effective H_2 pressure approaches the pressure in the gas phase.

Assuming that $[H_2]$ in Eq. (9.2) varies slowly with time, the lowest value of $[H_2]$ may be expressed as

$$[H_2] \approx [H_2]_{sat}\left(1 - R^{max}_{H_2,rxn}/R^{max}_{H_2,g/l}\right) \quad (9.6)$$

To ensure that the solution is nearly saturated with H_2 throughout the entire course of the reaction and the observed rate is representative of the kinetics intrinsic to the catalytic system under the specified hydrogen pressure, a rule of thumb is that the maximum intrinsic reaction rate should be less than 10% of the maximum H_2 delivery rate:

$$\frac{R^{max}_{H_2,rxn}}{R^{max}_{H_2,g/l}} \leq 10\% \quad (9.7)$$

Equation (9.6) shows

$$[H_2] \geq 90\%[H_2]_{sat} \quad (9.8)$$

when the 10% rule of thumb is satisfied.

9.3 IMPACT OF $\kappa_L a$ ON REACTION KINETICS AND SELECTIVITY

As shown in Section 9.2, even when the pressure in the gas phase is specified, the reaction conditions could in actuality be unspecified due to uncharacterized deviation of $[H_2]$ from $[H_2]_{sat}$ as a result of a lack of knowledge of the hydrogen mass transfer coefficient $k_L a$. Under the unspecified conditions, rate measured reflects kinetics at an unknown $[H_2]$ instead of the intended constant $[H_2]_{sat}$. The kinetic data obtained under such conditions are not helpful and can even be harmful to the development of scalable processes. Irreproducibility in rate from reactor to reactor may be observed under seemingly identical reaction conditions due to different mass transfer coefficients in different reactors. A process developed under such conditions may pose safety as well as selectivity problems in scale-up. In addition, the deviation of $[H_2]$ from $[H_2]_{sat}$ may cause irreversible alteration of the catalyst properties. The catalyst may be deactivated due to lack of hydrogen atoms on the catalytic surface, which not only reduces the catalytic activity but also may change the selectivity of the catalyst in undesirable manners.

In addition, selectivity of the hydrogenation reactions may be strongly influenced by mass transfer by virtue of the intrinsic dependence of the selectivity on $[H_2]$, the effective pressure that the catalyst experiences. Many reactions exhibit strong dependence of selectivity on hydrogen pressure. Examples include the [Rh(dipamp)]$^+$-catalyzed asymmetric hydrogenation of α-acylaminoacrylic acid derivatives [10], asymmetric hydrogenation of γ-geraniol and geraniol catalyzed by BINAP-Ru(II) [11], and enantioselective hydrogenation of ethyl pyruvate over cinchonidine-modified Pt [12]. In the case of asymmetric hydrogenation of γ-geraniol catalyzed by (S)-BINAP-Ru(II), the enantioselectivity to (R)-β-citronellol decreases precipitously from 90%ee to nearly racemic with increasing effective H_2 pressure from nearly zero to 100 psia. The case of asymmetric hydrogenation of geraniol is even more dramatic. The enantioselectivity flips from an ee of 93%(R) to 91%(S) when the H_2 effective pressure changes from 100 psia to nearly zero due to the presence of the competitive isomerization of geraniol to γ-geraniol under the hydrogenation conditions (Figure 9.2) [13]. *In all these cases, even at constant H_2 gas-phase pressure, uncharacterized deviation of $[H_2]$ from $[H_2]_{sat}$ as a result of H_2 mass transfer limitations translates into unpredictable selectivity.*

It is interesting to note that the mass transfer limitations may work for or against the desired selectivity, depending on how the selectivity is related to $[H_2]$. For instance, at a constant gas-phase pressure, mass transfer limitations help to enhance the enantioselectivity of the (S)-BINAP-Ru(II)-catalyzed asymmetric hydrogenation of γ-geraniol to (S)-citronellol, in which case the selectivity increases with decreasing pressure.

The effect of the interplay between the mass transfer and the intrinsic rate processes on kinetics and selectivity for reactions carried out at constant H_2 pressure is further demonstrated in Figure 9.3, using as an example the enantioselective hydrogenation of ethyl pyruvate over cinchonidine-modified Pt [8b]. The effects of the two extreme situations discussed

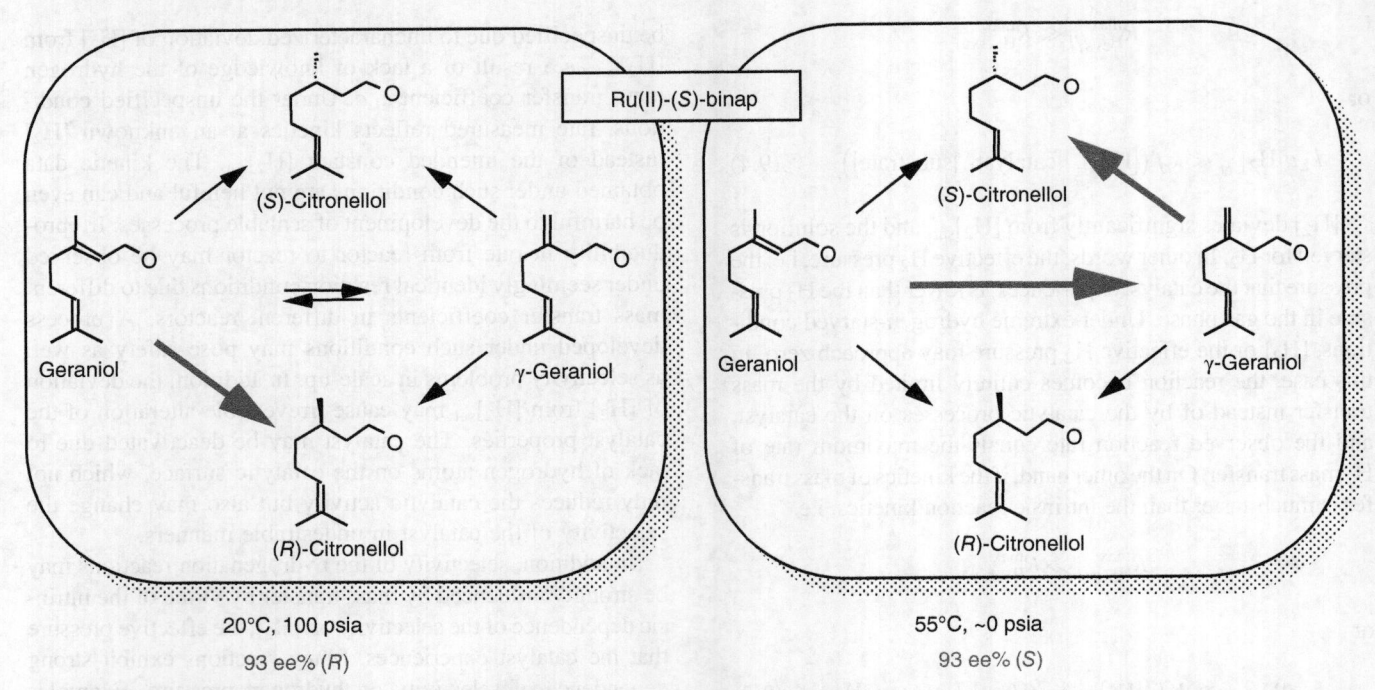

FIGURE 9.2 Striking dependence of enantioselectivity in the asymmetric hydrogenation of geraniol as a result of the interplay of rate processes in the isomerization/hydrogenation network.

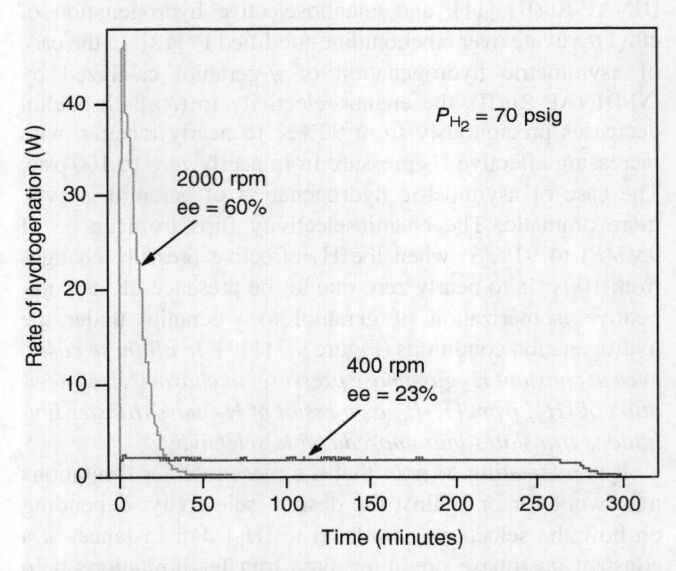

FIGURE 9.3 Effect of hydrogen mass transfer on kinetics and enantioselectivity in the chiral hydrogenation of ethyl pyruvate over cinchonidine-modified Pt. Except for the different stir rates, the two experiments were carried out under otherwise identical conditions (30 °C, solvent: 1-propanol).

above are graphically illustrated by the difference in the observed kinetics and enantioselectivity. At 400 rpm, the H_2 mass transfer is much slower than the intrinsic hydrogenation rate ($k_L a = 4 \times 10^{-3}$ s^{-1}). [H_2] is virtually zero. The kinetics, being completely limited by the rate of the H_2 delivery across the gas–liquid interface instead of by the catalytic hydrogenation of ethyl pyruvate on the Pt surface, is independent of the substrate concentration and therefore exhibits zero-order kinetic behavior. A process developed under these conditions would potentially run into safety issues when scaled up as the hydrogenation rate would change greatly with differences in the mass transfer coefficient $k_L a$ associated with large hydrogenators. At 2000 rpm, the mass transfer is no longer limiting the rate ($k_L a = 0.7$ s^{-1}); the solution is saturated with H_2 at all times during the reaction, i.e. [H_2] ≅ [H_2]$_{sat}$; and the rate is limited by the catalytic hydrogenation of ethyl pyruvate over Pt. Further increases in the mass transfer coefficient no longer change the rate profile. As a result, a picture of the intrinsic kinetics emerges. In addition to the kinetics, the mass transfer also influences selectivity by virtue of changing the availability of H_2 that the catalyst experiences. Figure 9.3 shows that the enantioselectivity increases from 23 to 60 %ee due to a change in [H_2] from starvation to saturation upon increasing the agitation speed from 400 to 2000 rpm.

In developing a hydrogenation process, it is the intrinsic catalyst activity and selectivity, i.e. the catalytic behaviors under well-defined conditions, which are of foremost concern, not those convoluted with the hydrogen mass transfer process. Obviously, it is not sufficient to specify H_2 pressure alone in order to unravel kinetics and selectivity of a catalytic hydrogenation reaction intrinsic to the catalytic system. It is necessary to characterize the mass transfer properties of the hydrogenator and conduct the hydrogenation experiments according to the 10% "rule of thumb" as described by Eq. (9.7).

9.4 CHARACTERIZATION OF GAS–LIQUID MASS TRANSFER PROCESS

Section 9.3 described how the gas–liquid mass transfer process, characterized by the simple parameter $k_L a$, can exert significant influence on the outcome and robustness of the hydrogenation process. It is not uncommon in the pharmaceutical industry, however, to see chemists and chemical engineers employ hydrogenation vessels for screening, scaling up, and commercialization of hydrogenation processes without characterizing their mass transfer coefficients. In this section, a simple practical procedure for measuring $k_L a$ is reviewed, and the examples are given for measuring $k_L a$ in a variety of hydrogenators at scales ranging from 100 ml to 760 gal in volume. Given the importance of the mass transfer coefficient, it is recommended that all hydrogenators, including those used for screening, for process development and scaling up, and for commercialization, be characterized in term of their mass transfer coefficients.

Among the methodologies available, the most straightforward one for measuring the gas–liquid mass transfer coefficient $k_L a$ is to measure directly the kinetics of nonreactive hydrogen uptake by the solution at various agitation speeds [14]. In this method, the solution is first degassed thoroughly by vacuum with the agitation on. The agitator is then turned off; hydrogen gas is introduced to the headspace of the hydrogenator to a pressure close to the hydrogenation pressure at which point the hydrogen line valve to the hydrogenator is closed. When the agitation commences, the pressure in the reactor is recorded as a function of time using a fast-response pressure transducer. A schematic depicting the pressure drop in the headspace of the hydrogenator and the corresponding concentration rise of hydrogen in the liquid phase is shown in Figure 9.4. The rate at which the pressure decreases is directly related to the gas–liquid transfer rate, whereas the extent to which the pressure drops is related to the solubility of the gas in the liquid. The dependence of pressure P as a function of time t is governed by the simple first-order rate equation below:

$$\frac{P_f - P_0}{P_i - P_0} \ln\left(\frac{P_i - P_f}{P - P_f}\right) = (k_L a) t \quad (9.9)$$

where

P_i: the initial pressure
P_f: the final pressure
P_0: the solvent vapor pressure.

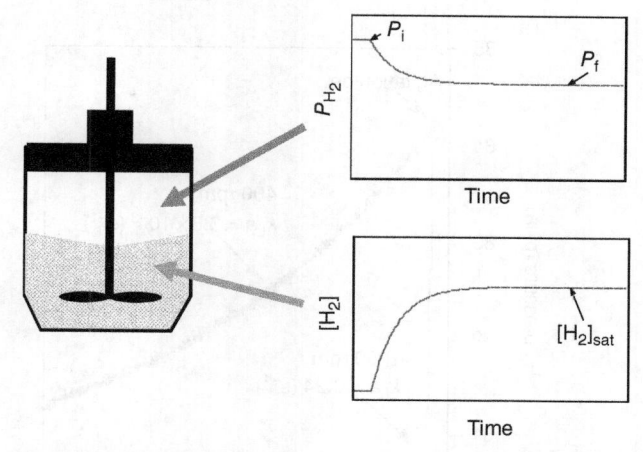

FIGURE 9.4 Schematics of the pressure drop and the corresponding concentration rise in a $k_L a$ measurement experiment.

Typical pressure–time curves in the Metter Toledo's 1 L RC1 MP10 reactor are displayed in Figure 9.5a. They graphically illustrate the difference in mass transfer rates at agitation rates of 400 and 1000 rpm. Figure 9.5b shows the pressure plot obtained using the data in Figure 9.5a according to Eq. (9.9). The slope of the pressure plot yields $k_L a$, which is 3.8×10^{-3} s^{-1} at 400 rpm and 0.24 s^{-1} at 1000 rpm. Another way to look at the difference in the mass transfer rates is a comparison of the "half-life," i.e. $t_{1/2} = \ln2/k_L a$. While reaching 50% of the hydrogen saturation concentration takes only approximately 3 seconds at 1000 rpm, it takes approximately 3 minutes at 400 rpm.

The mass transfer coefficient can be strongly influenced by a number of parameters including the agitation rate, the type of reactor, reactor configuration such as reactor size and shape, the type of agitator and its position in the reactor, fill level, temperature, solvent and solute, the use of baffle, and subsurface sparge line. The methodology described here is a convenient and fast way to measure the mass transfer characteristics of the hydrogenator relevant to the specific conditions of the hydrogenation process.

Using the simple methodology described in this section, the full mass transfer characteristics of three different laboratory hydrogenators, the 1 L Mettler Toledo's RC1 MP10 reactor, the 250 ml Parr Shaker, and a 5 gal stainless steel hydrogenator with mechanical agitation, have been conveniently and rapidly measured. The mass transfer coefficients and their dependence on agitation speed are shown in Figure 9.6.

Taking advantage of its design precision, e.g. the reproducible agitation ramp from zero to the desired agitation rate in less than 0.2 seconds, we first used the 1 L Mettler Toledo's RC1 MP10 system to study the effect of the hydrogen pressure on $k_L a$ using methanol as the solvent. The results show that the $k_L a$ value is rather independent of the hydrogen pressure over the range of 15–100 psig. Given this

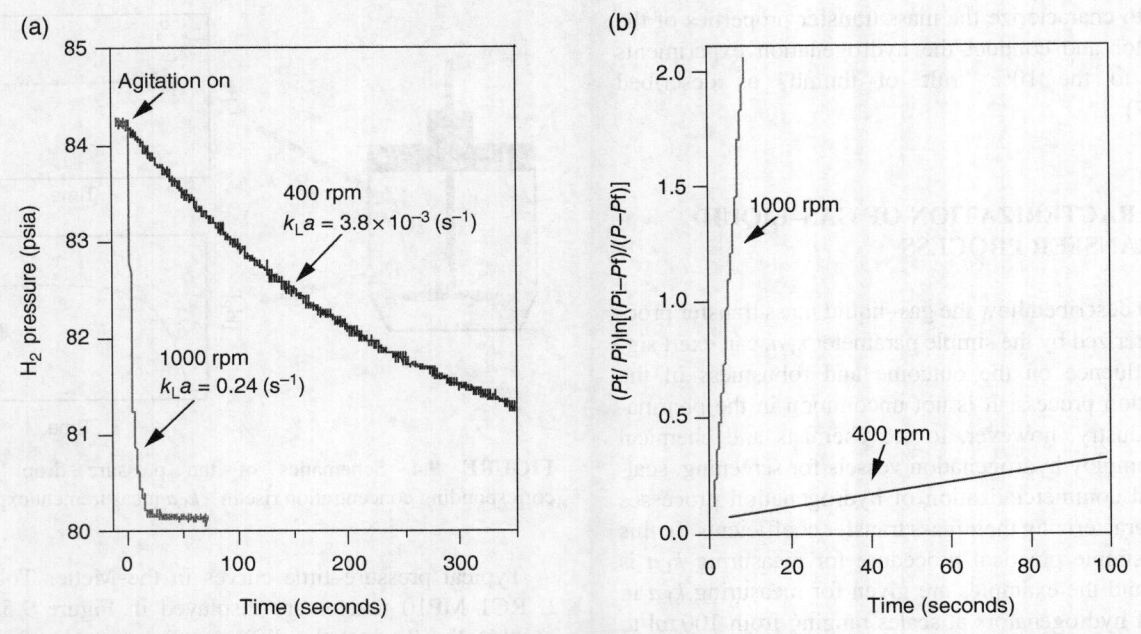

FIGURE 9.5 Measurements of the H_2 uptake kinetics in 1-propanol (0.5 L) in the Mettler Toledo's 1 L RC1 MP10 reactor at 30 °C. (a) Hydrogen pressure drop as a function of time. (b) Corresponding plots of the pressure function in Eq. (9.9) vs. time. The slope is k_La.

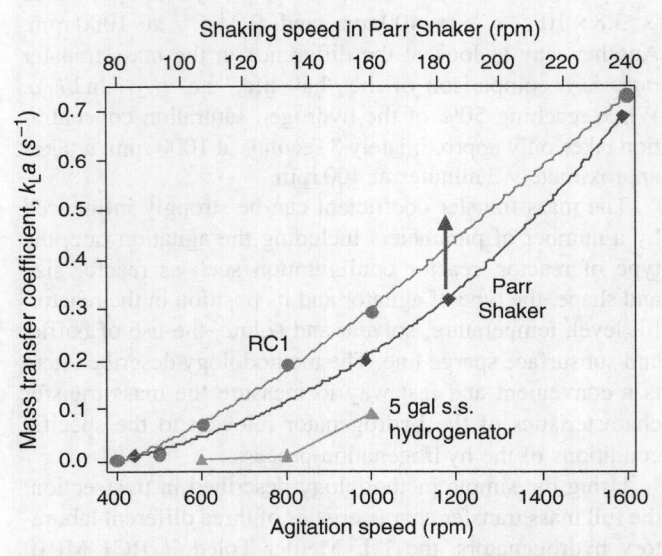

FIGURE 9.6 Hydrogen mass transfer coefficient k_La as a function of the agitation intensity for three types of hydrogenators: the 1 L Mettler Toledo's RC1 MP10 reactor (500 ml MeOH), a 5 gal stainless steel hydrogenator with mechanical agitation (2.5 gal MeOH) and the 250 ml Parr shaker (90 ml MeOH).

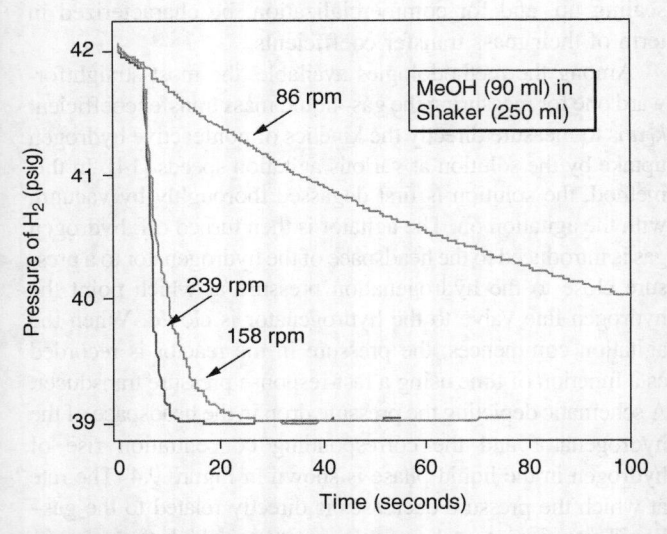

FIGURE 9.7 Decay of pressure in the head space due to the non-reactive hydrogen uptake by methanol (90 ml) in the Parr Shaker (250 ml), parametric in the shaking frequency.

observation, we simply choose any convenient hydrogen pressure in this pressure range for the k_La measurement.

To the best of our knowledge, mass transfer characteristics of the Parr Shaker system, traditionally and frequently employed for development of hydrogenation processes, have not been reported. Figure 9.6 shows that the Parr Shaker possesses surprisingly good mass transfer capability rivaling the best achievable in the RC1 system that is known for its good mass transfer capability. At the shaking frequency typically used for process development, the value of k_La is moderate, 0.1 s^{-1} at 130 rpm. Higher k_La values are achievable at higher shaking frequencies. Figure 9.7 shows several pressure curves measured in the Parr Shaker. Figure 9.6 also shows that the mass transfer capability of the specific 5 gal hydrogenator studied even at the highest allowable rpm (1000 rpm) is somewhat limited ($k_L a = 0.1 \text{ s}^{-1}$).

The simple methodology is applicable to $k_L a$ measurement in large hydrogenators in manufacturing facilities. The mass transfer capability of a factory glass-lined hydrogenator with a nominal 750 gal volume (900 gal actual, with a retreat blade impeller, one baffle, subsurface sparging) was characterized. The results revealed a limited gas–liquid mass transfer capability of this specific hydrogenator even with 100% agitation power. For instance, at a 460 gal fill and with the full agitation power, the $k_L a$ value is 0.035 s^{-1}, a mass transfer coefficient equivalent to that in the RC1 reactor at 500 rpm only (Figure 9.6). The low mass transfer ability in this hydrogenator makes it not suitable for running fast hydrogenation reactions that require $k_L a > 0.035 \text{ s}^{-1}$ at the existing configuration.

9.5 CHARACTERIZATION OF CATALYST REDUCTION PROCESS

For hydrogenation processes in a solvent with a solid supported metal catalyst, an important process involved is the reduction of the catalyst itself to its metallic state. For example, the Pd in the typical Pd/C catalysts is either in the form of palladium hydroxide or in the form of metallic palladium particles with its surface layers oxidized. Reduction of the surface Pd to its metallic state is necessary for the catalytic hydrogenation processes. While reduction of Pd/C catalysts used in the gas–solid reactions can be conveniently studied, to the best of our knowledge, the kinetics of catalyst reduction under the gas–liquid–solid slurry hydrogenation conditions has not been reported. How long does it take to reduce the catalyst under the slurry hydrogenation conditions? What does the kinetics look like? Is it instantaneous upon pressurization of the hydrogenator? How do the solvent, the additives, and the substrate influence the kinetics of the catalyst reduction? How do the transient properties of the catalyst during the catalyst reduction process influence reactivity and selectivity of the hydrogenation of the substrate? These questions remain unanswered due to a lack of research tools to characterize the kinetics of the catalyst reduction *in situ* under the slurry hydrogenation conditions. In this section, a simple procedure is described that allows one to measure the characteristics of the catalyst reduction process.

The procedure is an extension of the $k_L a$ measurement protocol described in Section 9.4. First, profile of the nonreactive uptake by the solvent is measured. Profile of the sum of the nonreactive and reactive hydrogen uptake due to the catalyst reduction is subsequently measured by repeating the same procedure (under the identical conditions) after addition of the heterogeneous catalysts to the solvent in the batch. The uptake profile due to the catalyst reduction can be extracted by taking the difference of the two uptake profiles.

The methodology is demonstrated in Figure 9.8 for measuring the reduction rate of a Pd/C catalyst in methanol at $-10\,°\text{C}$. The hydrogen uptake curves at 40 psig H_2 and at $-10\,°\text{C}$ upon agitation (1000 rpm) for MeOH (500 ml) only and for MeOH (500 ml) plus the 5%Pd/C catalyst (10 g) were measured consecutively, and the results are shown in Figure 9.8a. Because the two uptake curves were measured under the identical conditions except that one is with the Pd/C catalyst added, the nonreactive uptake curve can be directly subtracted from the sum curve to generate the uptake curve associated only with the Pd/C reduction. By subtracting the two curves in Figure 9.8a, the reduction profile of the Pd/C catalyst emerges and is shown in Figure 9.8b.

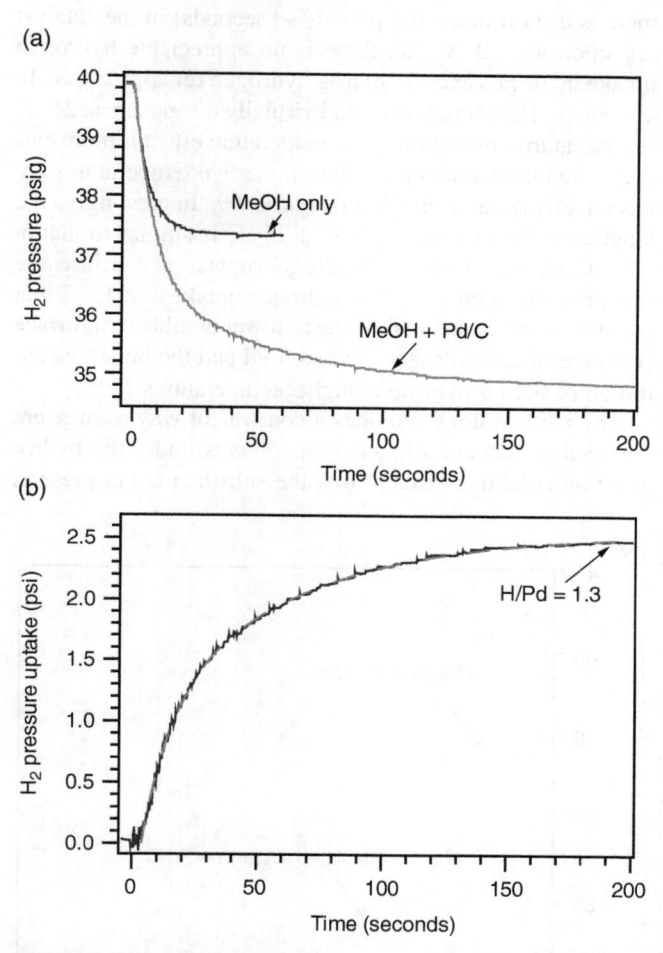

FIGURE 9.8 (a) Hydrogen uptake curves at 40 psig H_2 and at $-10\,°\text{C}$ upon agitation (1000 rpm) for MeOH (500 ml) only and for MeOH (500 ml) plus 5%Pd/C (10 g, Type 21, JM), following the standard $k_L a$ measurement procedure described in Section 9.4. Apparatus: Mettler-Toledo's RC1 with an MP10 reactor (b) Difference between the two uptake curves in (a), representing the reactive hydrogen uptake by the reduction of the Pd/C catalyst only. The dash line is a fit to a double exponential function Uptake $= 2.55 - 1.65 e^{-0.0177 t} - 1.34 e^{-0.0864 t}$.

A few properties of the reduction process become apparent from Figure 9.8b. The reduction is a relatively fast process – it is nearly completed in 2 minutes at 40 psig H_2 and at $-10\,°C$. The reduction kinetics exhibits two distinct regimes and can be fitted nearly perfectly by a double exponential function, i.e. Uptake = $2.55 - 1.65e^{-0.0177t} - 1.34e^{-0.0864t}$. The fast rate process has a half-life of ca. 8 seconds and is likely associated with the reduction of the outside layers of Pd catalyst particles, whereas the slower rate process has a half-life of ca. 39 seconds and is likely associated with the reduction of the bulk of the Pd catalyst particles and with the formation of bulk palladium hydrides. Figure 9.8b also shows that the overall stoichiometry of the reactive hydrogen update for this catalyst is H/Pd = 1.3.

A closer look at the uptake profile in Figure 9.8b shows there is a short induction period (~4 seconds) in the catalyst reduction at $-10\,°C$, i.e. there is no appreciable hydrogen uptake in the presence of 40 psig hydrogen for approximately 4 seconds. The induction period virtually disappears at $25\,°C$ for the catalyst reduction. This temperature effect is more evident when the reduction profiles at the two temperatures are placed in the same graph in Figure 9.9. Interestingly, the kinetics in the fast rate regime at $25\,°C$ is similar to that at $-10\,°C$ (similar slopes of hydrogen uptake at the fast rate regimes). The total reduction hydrogen uptake is at $25\,°C$ that is however less, presumably due to lower equilibrium surface coverage of the hydrogen atoms on Pd and the bulk concentration of the Pd hydride at higher temperatures.

The methodology provides a convenient way to measure the catalyst reduction kinetics in solvents under the hydrogenation conditions except that the substrate is not present.

In the presence of the substrate, the reactive hydrogen uptake would originate from the catalyst reduction and from the substrate hydrogenation. The methodology described here cannot deconvolute the rate of the catalyst reduction from the rate of the substrate hydrogenation. How the presence of the substrate effect the reduction kinetics of the heterogeneous catalysts remains unclear. One thing clear is that the kinetics of the catalyst reduction can be altered by the substrate in the hydrogenation reaction. Figure 9.10 shows the effect of the addition of ammonia to the methanol solvent on the catalyst reduction at $-10\,°C$. In this experiment, a

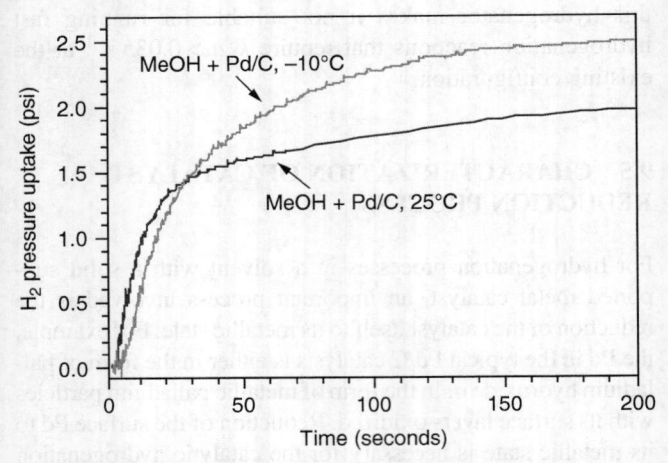

FIGURE 9.9 A comparison of the reduction kinetics of the 5% Pd/C catalyst (Type 21, JM) at -10 and $25\,°C$. The experimental conditions are identical to those described in Figure 9.8.

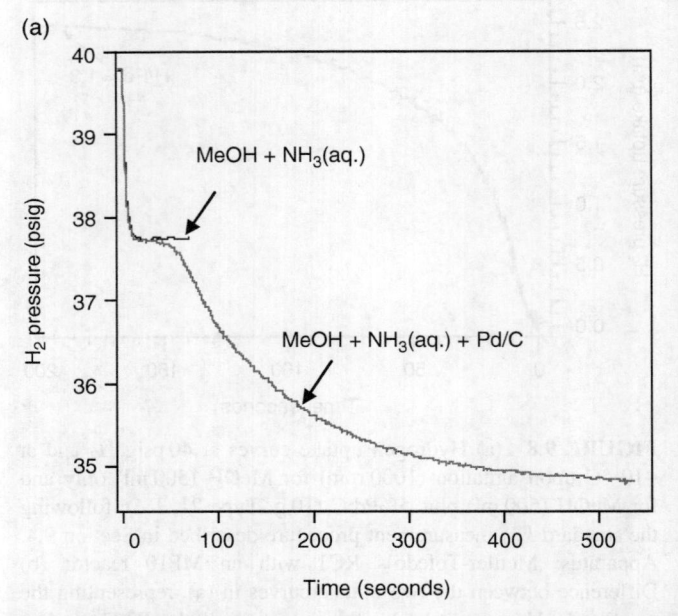

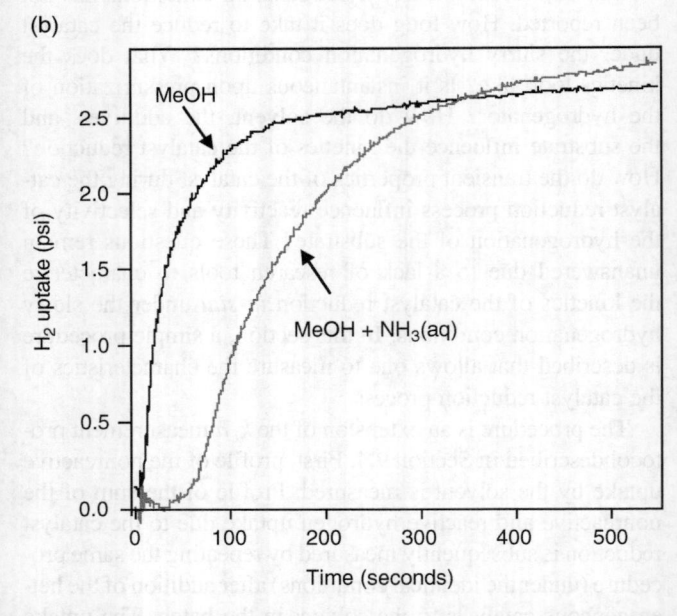

FIGURE 9.10 (a) Hydrogen uptake curves at 40 psig H_2, $-10\,°C$, 1000 rpm for MeOH + NH_3(aq.) (500 ml) and for MeOH + NH_3(aq.) (500 ml) plus 5%Pd/C (10 g, Type 21, JM), following the standard $k_L a$ measurement procedure described in Section 9.4. (b) Kinetics of the Pd/C catalyst reduction in MeOH + NH_3(aq.), in comparison with the kinetics in MeOH only. *Source*: Reproduced from Figure 9.8b.

small amount of aqueous ammonia (0.78 mol) was added into methanol to a total volume of 500 ml, and the catalyst reduction kinetics was studied following the standard procedure. The two hydrogen uptake curves measured are displayed in Figure 9.10a. The catalyst reduction kinetic profile derived from Figure 9.10a is plotted in Figure 9.10b along with the kinetic profile of the catalyst reduction in methanol alone as a reference.

The most striking feature of the reduction kinetics in the presence of ammonia is that there is a significant induction period, approximately 60 seconds, much longer than the 4 seconds induction period in MeOH without the ammonia addition. Over this long induction period, there is virtually no hydrogen uptake. At the end of the induction period, the rate of the Pd catalyst reduction accelerates. This induction period may result from strong chemisorption of ammonia on the surface of the Pd catalyst, inhibiting dissociative chemisorption of dihydrogen on Pd. Adsorption of a small amount of hydrogen on the catalyst surface through competitive adsorption process reduces a small fraction of the surface Pd, which conversely facilitates dissociative chemisorption of additional hydrogen, leading to the rate acceleration. The maximum rate of the reduction, however, is slower than that in the absence of ammonia (Figure 9.10b). The long induction period is attributed to the presence of ammonia in the hydrogenation batch. Independent experiments showed that water as a result of the aqueous ammonia addition does not alter the catalyst reduction. Overall, the addition of ammonia extended the catalyst reduction time scale from ca. 2 to 7 minutes, suggesting that additives used in hydrogenation reactions and the substrate itself can influence the catalyst reduction kinetics.

9.6 BASIC SCALE-UP STRATEGY FOR HYDROGENATION PROCESSES

When a hydrogenation process is translated from the laboratory to the plant, chemists and chemical engineers sometimes encounter scale-up issues such as problems with reactivity and/or selectivity. Often the scale-up issues can be traced to a lack of characterization or understanding of the various rate processes and an ensemble of key factors that affect the reproducibility and robustness of the hydrogenation process. A successful scale-up necessitates that these factors be identified, measured, and controlled. This section discusses some of the fundamental factors that need to be considered for a successful scale-up. A basic strategy for scaling up hydrogenation processes is described in terms of several quantitative criteria.

The most basic requirement for a successful scale-up is for the chemists and engineers to have at the laboratory development stage a true understanding of reaction kinetics that is intrinsic to the catalytic system, i.e. one that is not masked by H_2 mass transfer limitations but is obtained under well-defined conditions, in particular, with a known $[H_2]_{lab}$. The best strategy to achieve this is to characterize the mass transfer capability of the hydrogenation reactor and use the 10% rule of thumb (as described by Eq. (9.7)) in the laboratory developmental work so that the condition

$$[H_2]_{lab} \approx [H_2]_{sat} \qquad (9.10)$$

is also satisfied at all times. Availability of the intrinsic kinetic information allows one to make an intelligent choice of reactors and reaction conditions for reproducible and robust scale-ups.

The hydrogen mass transfer coefficient $k_L a$ of the hydrogenation reactor is one of the primary factors to consider in scale-up because it can have direct impact on kinetics, process safety, and selectivity. It is not uncommon, however, to find hydrogenation process descriptions specifying only the gas-phase H_2 pressure and agitation speed without specifying the requirement for the mass transfer coefficient $k_L a$ of the hydrogenator. To scale up a process in this manner subjects the process to the risk of running under ill-defined conditions, i.e. undefined $[H_2]$ that is disengaged from the gas-phase hydrogen pressure. To reproduce in the plant runs the rate and selectivity observed in the laboratory, it is important that $[H_2]$ in the plant runs be the same as the $[H_2]$ used in the laboratory development work, i.e.

$$[H_2]_{plant} = [H_2]_{lab} \approx [H_2]_{sat} \qquad (9.11)$$

Using the same pressure as that employed in the laboratory development without specifying $k_L a$ does not necessarily guarantee that this condition (Eq. 9.11) is satisfied.

When one runs the plant process far from the hydrogen mass transfer limited situations, the hydrogen concentration $[H_2]$ is known and constant (as long as the gas-phase pressure remains unchanged) over the entire course of reaction. This is particularly important for hydrogenation reactions with $[H_2]$-dependent selectivity. If this type of reactions is operated under hydrogen diffusion limitations, $[H_2]$ deviates from $[H_2]_{sat}$ and becomes disengaged from the gas-phase hydrogen pressure. The solution goes from hydrogen starved at the beginning stage of reaction when the intrinsic hydrogenation rate is fast to hydrogen saturated at the later stage of the reaction when the intrinsic hydrogenation slows down due to depletion of the starting material. The selectivity would vary throughout the course of the reaction as a result. In addition, rate of a process running under hydrogen mass transfer control is sensitive to slight changes in reaction conditions. For example, the rate can change significantly with changes in rpm due to the commonly observed exponential dependence of $k_L a$ on the agitation rate. The rate would not be affected by changes in the agitation rate, however, for processes running away from hydrogen mass transfer limitations.

To satisfy the scale-up requirement described by Eq. (9.11), the 10% rule of thumb (Eq. 9.7) provides a good general guideline for matching the mass transfer capability of the hydrogenation reactor to the kinetics of a process. Given a process with known intrinsic kinetics, the chosen reactor for scale-up needs to have a minimum mass transfer coefficient of

$$k_L a \geq 10 \frac{R_{H_2,rxn}^{max}}{[H_2]_{sat}} \quad (9.12)$$

In addition, heat transfer issue also needs to be considered for safe operations in scale-up. The following criteria need to be met:

$$R_{H_2,rxn}^{max} \leq \frac{q_r^{max}}{\Delta H_{H_2}} \quad (9.13)$$

where q_r^{max} is the maximum heat removal capability of the reactor and ΔH_{H_2} is the heat of hydrogenation per mole of H_2 reacted.

The inherent reaction kinetics and the criteria described by Eqs. (9.12) and (9.13) form some of the basic requirements that need to be considered for successful scale-ups of hydrogenation processes.

9.7 SUMMARY

Dissolution of H_2 into the liquid phase is the first rate process along the H_2 pathway in catalytic hydrogenation reactions. The dissolution kinetics can play a key role in the kinetics and selectivity of catalytic reactions and the outcome of process scale-up. At a given gas-phase H_2 pressure, while thermodynamics determines the solubility $[H_2]_{sat}$, it is *kinetics* that determines the actual solution H_2 concentration $[H_2]$ or the "effective hydrogen pressure" that the catalyst experiences *during* hydrogenation reactions. Depending upon the relative magnitude of the rate of H_2 mass transfer across the gas–liquid interface vs. the rate of the reactive removal of the dissolved H_2 from the liquid phase by hydrogenation, *$[H_2]$ may vary greatly from saturation to nearly zero, even at a constant pressure in the gas phase*. The influence of mass transfer on $[H_2]$ exerts a direct impact not only on rate but also on selectivity for reactions whose selectivity depends on $[H_2]$. It is also often the fundamental cause of irreproducibility in rate and selectivity observed from reactor to reactor (e.g. when scaling up a lab process in the manufacturing facility) when the same reaction is carried out under seemingly identical conditions. This is because different mass transfer capabilities of reactors of different types and scales can lead to different "effective pressures" even at a constant gas-phase pressure.

A catalytic hydrogenation process should be carried out under conditions where the intrinsic hydrogenation rate is at least 10 times lower than the maximum rate of H_2 mass transfer across the gas–liquid interface. This ensures that the observed kinetics and catalytic behaviors are not masked by the mass transfer limitations but are intrinsic to the catalytic system under well-defined conditions. This requirement may serve as a general guideline for designing scalable processes, since as long as the 10% rule of thumb is satisfied, the hydrogenation process is "portable," i.e. the same kinetics and catalytic behavior will be reproduced from reactor to reactor at any scale. Strategy for a successful scale-up of hydrogenation processes is formulated in terms of a set of quantitative criteria.

The process of reduction of a Pd/C catalyst under the hydrogenation conditions is characterized using a novel methodology. The results show that the catalyst reduction process can have an induction period, is relatively fast (on the order of <5 minutes) under typical hydrogenation conditions, and can be significantly influenced by the nature of the additives or the hydrogenation substrates.

ACKNOWLEDGMENTS

We thank Prof. Donna Blackmond for helpful discussions and Andy Newell, Charlie Bazaral, and Steve Conway for assistance in experiments.

REFERENCES

1. (a) Rylander, P. (1990). *Hydrogenation Methods*. London: Academic Press.
   (b) Rylander, P. (1979). *Catalytic Hydrogenation in Organic Synthesis*. New York: Academic Press.
2. Augustine, R.L. (1995). *Heterogeneous Catalysis for the Synthetic Organic Chemists*. New York: CRC Press.
3. Knowles, W.S. (2002). Asymmetric hydrogenations. *Angewandte Chemie International Edition* 41: 1998.
4. Knowles, W.S. and Noyori, R. (2007). Pioneering perspectives on asymmetric hydrogenation. *Accounts of Chemical Research* 40: 1238.
5. Noyori, R. (2002). Asymmetric catalysis: science and opportunities. *Angewandte Chemie International Edition* 41: 2008.
6. Noyori, R. (1994). *Asymmetric Catalysis in Organic Synthesis*. New York: Wiley, and references therein.
7. Krische, M.J. and Sun, Y.K. (2007). Hydrogenation and transfer hydrogenation. *Accounts of Chemical Research* 40: 1237–1237.
8. See for example(a)Sun, Y.K., Landau, R.N., Wang, J. et al. (1996). A re-examination of pressure effects on enantioselectivity in asymmetric catalytic hydrogenation. *Journal of the American Chemical Society* 118: 1348.

(b) Sun, Y.K., Wang, J., Landau, R.N. et al. (1996). Asymmetric hydrogenation of ethyl pyruvate: diffusion effects on enantioselectivity. *Journal of Catalysis* 161: 759.

9. (a) Ramachandran, P.A. and Chaudhari, R.V. (1983). *Three-Phase Catalytic Reactors, Topics in Chemical Engineering*, vol. 2, Chap. 9. New York: Gordon and Breach Science Publishers.
(b) Beenackers, A.A.C.M. and Van Swaaij, W.P.M. (1993). Mass transfer in gas-liquid slurry reactors. *Chemical Engineering Science* 48: 3109.
(c) Hines, A.L. and Maddox, R.N. (1985). *Mass Transfer: Fundamentals and Applications*. Englewood Cliffs, NJ: Prentice-Hall.

10. (a) Landis, C.R. and Halpern, J. (1987). Asymmetric hydrogenation of methyl (Z)-.alpha.-acetamidocinnamate catalyzed by [1,2-bis((phenyl-o-anisoyl)phosphino)ethane]rhodium(I): kinetics, mechanism and origin of enantioselection. *Journal of the American Chemical Society* 109: 1746.
(b) Sun, Y.K., Landau, R.N., Wang, J. et al. (1996). A reexamination of pressure effects on enantioselectivity in asymmetric catalytic-hydrogenation. *Journal of the American Chemical Society* 118: 1348.

11. Sun, Y.K., LeBlond, C., Wang, J. et al. (1995). Observation of a $[RuCl_2((S)-(-)-tol-binap)]_2 \cdot N(C_2H_5)_3$-catalyzed Isomerization-Hydrogenation Network. *Journal of the American Chemical Society* 117: 12647.

12. Wang, J., Sun, Y.K., LeBlond, C. et al. (1996). Asymmetric hydrogenation of ethyl pyruvate: relationship between conversion and enantioselectivity. *Journal of Catalysis* 161: 752.

13. Sun, Y.K., Wang, J., LeBlond, C. et al. (1997). Kinetic influences on enantioselectivity in asymmetric catalytic hydrogenation. *Journal of Organometallic Chemistry* 581: 65.

14. (a) Matsumara, M., Masunaga, H., and Kobayashi, J. (1979). Gas absorption in an aerated stirred tank at high power input. *Journal of Fermentation Technology* 57: 107.
(b) Deimling, A., Karandikar, B.M., Shah, Y.T., and Carr, N.L. (1984). Solubility and mass transfer of CO and H_2 in Fischer-Tropsch liquids and slurries. *Chemical Engineering Journal* 29: 140.
(c) Blaser, H.U., Garland, M., and Jallet, H.P. (1993). Enantioselective hydrogenation of ethyl pyruvate: kinetic modeling of the modification of Pt catalysts by cinchona alkaloids. *Journal of Catalysis* 144: 569.

10

CHARACTERIZATION AND FIRST PRINCIPLES PREDICTION OF API UNIT OPERATIONS

JOE HANNON

Scale-up Systems Limited, Dublin, Ireland

10.1 INTRODUCTION

This chapter deals with how unit operations in API synthesis can be described, predicted, and scaled using classical chemical engineering principles. This allows design to be completed more quickly and with greater success than adopting a trial-and-error approach. Several common applications are worked through in detail for illustration. Examples and references to industry projects utilizing these concepts are included.

10.1.1 Rate Processes in API Unit Operations

Table 10.1 lists the rate processes involved in common unit operations in API synthesis. Characterization of each rate process is often feasible; the apparent complexity of unit operations arises from the combination of several rate processes in a single operation. In many cases, one rate process is limiting and dominates the others.

Other chapters in this book also address several of these rate processes and operations, including "REACTION KINETICS AND CHARACTERIZATION," "ANALYTICAL METHODS FOR MASS BALANCE AND REACTION KINETIC ANALYSES," "DESIGN OF DISTILLATION AND EXTRACTION OPERATIONS," "DESIGN AND SCALE-UP OF PHARMACEUTICAL CRYSTALLIZATIONS," "DESIGN OF FILTRATION AND DRYING OPERATIONS," "KILO LAB AND PILOT PLANT MANUFACTURING," "QUALITY BY DESIGN APPROACHES FOR API," and "QbD CASE STUDY: API CASE STUDY."

10.1.2 Scale Dependence and Scale Independence

The intrinsic rate of a chemical reaction is independent of scale; in other words, if the reaction could be conducted without limitation by any other rate process, it would run at the same rate (per unit volume) at all scales; the reaction time, starting from the same initial composition and held at the same temperature, would be the same at all scales. This is true for reactions whose kinetics are slow compared with other rate processes. The choice of solvent, reagents, and/or catalyst combination for reactions has been largely the domain of development chemists, and these variables are taken here to be fixed already; in any case, those effects, including reagent solubility, are also independent of scale.

All other rate processes listed in Table 10.1 are scale dependent. For example, the rate of heat transfer depends strongly on the ratio of surface area to volume, which reduces on scale-up. This means that heat removal on scale will be slower than in the laboratory, unless specific measures are taken to provide additional surface area or an increased temperature driving force. Therefore the time required to heat or cool a batch of material tends to increase on scale. Similarly, the rate of mass (or phase) transfer is scale dependent; the agitation conditions inside the vessel determine the time required to reach equilibrium between the phases; it is easier to make this time short in the lab than on scale.

Chemical Engineering in the Pharmaceutical Industry: Active Pharmaceutical Ingredients, Second Edition.
Edited by David J. am Ende and Mary T. am Ende.
© 2019 John Wiley & Sons, Inc. Published 2019 by John Wiley & Sons, Inc.

TABLE 10.1 Typical Incidence of Classical Chemical Engineering Rate Processes in Unit Operations

Operation Rate Process	Reaction/Quench	Distillation	Extraction	Crystallization	Filtration	Drying
Chemical kinetics	☑			☑		
Mass transfer	☑	☑	☑	☑		☑
Heat transfer	☑	☑	☑	☑		☑
Addition	☑	☑	☑	☑	☑	☑
Removal	☑	☑	☑		☑	☑

The rates of addition and removal of material also change with scale, normally taking longer at larger scales. For example, the rate of disengagement of gas evolved from a reaction depends again on the surface-area-to-volume ratio, so reactions such as decarboxylation have to be conducted more slowly on scale to avoid partial loss of the reactor contents due to swelling of the batch.

The impact of scale dependence is to lengthen processing times on scale, reducing productivity somewhat but more importantly potentially allowing additional phenomena to occur that were not observed to the same extent in the laboratory, such as impurity-forming reactions, product degradation, catalyst poisoning or deactivation, unwanted crystal forms, nucleation, agglomeration, or breakage. Fortunately, each of the rate processes that contribute to such problems can be characterized or estimated using classical chemical engineering methods, allowing scale-up problems to be anticipated and resolved in advance or at least resolved quickly once they occur.

10.1.3 Scale-Up and Achievement of Similarity

Unit operations need to produce similar results and a similar quality product at each scale; historically the pharmaceutical industry has demonstrated this through process validation at each facility and operation thereafter with rigid limits on process parameters. Recent regulatory guidance [1] encourages adoption of "quality by design" (QbD) and development of a "design space" in which the process may operate, with flexibility to move operating conditions around in this space without needing to obtain additional regulatory approval. The design space is proposed by the applicant and is intended to be applicable to the process at any scale. This implies a greater investment in developing process understanding during the development phase, balanced by flexibility, a reduced regulatory burden, and opportunities for continuous improvement once the process is in full-scale manufacturing.

There is some debate about how best to apply these principles while also satisfying business, process safety, and environmental requirements, and the underlying benefit of improved process understanding is underlined in this chapter. The operating window in which acceptable or similar results are obtained can be determined, demonstrated, and justified in terms of the chemical engineering rate processes involved in unit operations. For example, the design space for a chemical reaction whose intrinsic kinetics have been shown to be slow (compared with other rate processes involved) can be demonstrated to be scale independent, giving a flexible operating window that can be used at any scale. On the other hand, where specific equipment performance requirements exist, these can be framed in terms of the chemical engineering "rate constants" or time constants required by the process.

In the next section, we will see that these rate constants arise naturally in the chemical engineering rate equations, which provide a first principles basis for defining a scale-independent design space. The application of this approach is illustrated for several common types of chemical reaction, and the same general principles and methods apply to other API synthesis operations. Examining the rate equations also leads to familiar scale-up rules.

10.1.4 Characterization of Reaction Systems

Guidance specific to certain classes of reaction is given below, but to avoid repetition for each class, the following general guidelines on reaction characterization are provided here. Guidelines on equipment characterization are given toward the end of the chapter.

Assuming that the solvent and reagents/catalyst have been selected, reactions may be characterized by a sequence of initial screening experiments to identify scale-dependent physical rate process limitations, followed by a more detailed kinetics study if necessary.

In all experiments aimed at characterizing a reaction, progress should be followed by either taking multiple samples or using an *in situ* analytical technique, not by relying on a single endpoint sample. If composition is analyzed using HPLC, determination of response factors will become more important as characterization proceeds. Other possibilities include infrared (IR), ultraviolet (UV), and Raman spectroscopy or heat flow (Q_r).

Samples should be analyzed to indicate both reaction progress (e.g. product level, conversion, or current yield) and product quality (e.g. impurity level). If either of these variables is sensitive to physical rates (e.g. agitation), the reaction is not limited by chemical kinetics, at least under typical processing conditions. There may be some scope to reduce the rates of the chemical reactions (relative to mixing) by operating at, e.g. lower concentrations and/or temperatures; otherwise, if the process persists in its current form, the focus of characterization should be the physical capabilities of lab and larger-scale equipment (see Section 10.6).

Under conditions where there is no effect of physical rate processes, a kinetics study may follow in which temperature, equivalents, and/or other case-specific variables affecting the scale-independent chemical rates are varied experimentally. The choice of conditions and the sampling program for these experiments should be informed by the results of the previous experiments and an initial kinetic model [2], and in any event, multiple samples should be taken again to capture when the reactions are occurring as well as the final result.

Where possible, the parameters in Eq. (10.3) such as rate constant and activation energy should be fitted using a kinetic model to reliable results of experiments. Model development in this manner tends to be iterative, and the proposed reaction scheme may change a number of times before the model fits the data to an acceptable degree. Chemical knowledge that certain intermediate species is unlikely to exist for long may allow the rate constants of certain reactions to be set arbitrarily high so that the remaining parameters can be fitted with greater confidence. Model development should take place at the same time as experimentation and the design of remaining experiments may change as a result of indications from the model.

It is difficult to overemphasize the importance of conducting a few well-designed, well-monitored experiments versus a large set of experiments with scant data recording. Isothermal experiments are preferred, though not absolutely necessary as long as the temperature versus time is recorded during the heat-up and the location of "time zero" is clear. HPLC remains the most powerful method to follow product and impurity levels, and it is preferable to convert directly from HPLC area to moles rather than HPLC area percent. When starting materials are sufficiently pure, enough peaks are captured by HPLC, several response factors are known, and (preferably) an internal standard is also used, assumptions about the reaction scheme proposal can be checked directly using HPLC area. This can reveal the existence of undetected intermediates or by-products and allow the reaction scheme to be revised accordingly. An excellent illustration of the procedure is available [3].

At the end of this phase, such a model may be used to find optimum conditions inside or outside the experimental region and to assist with definition of a scale-independent design space. Some excellent examples of the success of this approach are available [4–8]. Further experimentation should be focused on model verification, especially at the most forcing conditions over which it may be used.

10.2 BATCH PROCESSES WITH HOMOGENEOUS REACTIONS

Figure 10.1 shows a "process scheme" for a batch reaction. This consists of a simple representation of the (i) phases and (ii) rates involved in the operation. The phase names and components initially present are indicated together with the rate processes. More or less detail may be added to such a schematic as required, and this representation will be used below to introduce other operations.

10.2.1 Rate Equations

If, for example, the reaction involved is simply $A + B \rightarrow P$, the rate equations for this system may be written as follows. The rate of change of concentration of A or B is

$$\frac{dC}{dt} = -r \qquad (10.1)$$

where the rate expression on the right-hand side is of this or similar form:

$$r = kC_A^\alpha C_B^\beta \qquad (10.2)$$

and the kinetic rate constant follows the Arrhenius relationship

$$k = k_0 \exp\left(-\frac{E_A}{RT}\right) \qquad (10.3)$$

More generally than (Eq. 10.1), and when several reactions are occurring in the same mixture, the rate of change of concentration of any species in the mixture is given by

$$\frac{dC_i}{dt} = \sum_j -\nu_{ij} r_j \qquad (10.4)$$

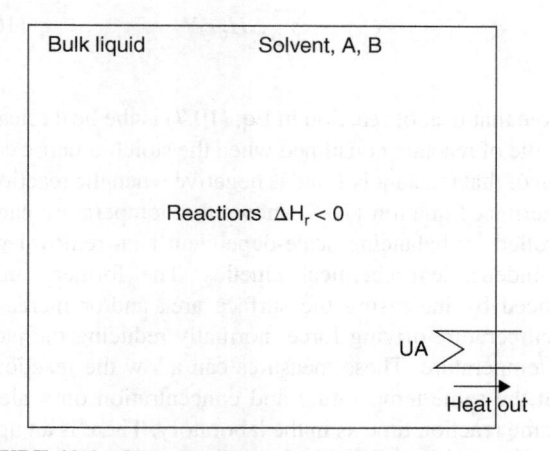

FIGURE 10.1 Process scheme for batch homogeneous reaction.

From Eqs. (10.1) to (10.4), the rate constant depends on temperature, and the rates of reaction depend on concentrations. This is the chemical engineering basis for the classical approach by chemists to study temperature and "equivalents" as two of the primary variables affecting the outcome of a reaction. In the above example, with 1 mol of A reacting with 1 mol of B, 1 equiv. of A would mean 1 mol of A/mol of B, 2 equiv. of A would be 2 mol of A/mol of B, etc.

Integrating Eq. (10.4) in time, the final outcome (composition) of reaction will depend on the concentrations and temperature profiles followed during reaction and the time allowed for reaction. Temperature is often (but not always) held constant or nearly constant, in which case final concentrations at completion depend only on initial concentrations and initial temperature. Example Problem 10.4 follows this behavior.

Temperature is an important influence and can change the relative rates of the reactions, depending on their activation energies. From a safety standpoint, temperature must be controlled to avoid thermal runaway, and the batch reactor is not ideal for this purpose. The rate of change of reactor temperature is given by

$$\frac{dT}{dt} = -\frac{UA}{\rho V C_p}\Delta T_{LM} + \frac{Q_r}{\rho V C_p} \quad (10.5)$$

In which the log mean temperature difference is

$$\Delta T_{LM} = \frac{T_{jout} - T_{jin}}{\ln\frac{T_{jout} - T}{T_{jin} - T}} \cong (T - T_j) \quad (10.6)$$

The approximation on the right of Eq. (10.6) applies when the heat transfer fluid flow rate is high, i.e. the heat transfer fluid temperature is almost constant between the jacket inlet and outlet.

The rate of heat evolution summed over all chemical reactions in the mixture is

$$Q_r = -\sum_j \Delta H_{rj} r_j V \quad (10.7)$$

Note that heat of reaction in Eq. (10.7) is the heat released per mole of reactant consumed when the stoichiometric coefficient of that reactant is 1 and is negative when the reaction is exothermic. Equation (10.5) shows that temperature can be controlled by balancing scale-dependent heat removal with scale-independent chemical kinetics. The former can be enhanced by increasing the surface area and/or increasing the temperature driving force, normally reducing the jacket inlet temperature. These measures can allow the reaction to run at the same temperature and concentration on scale, in the same reaction time as in the laboratory. There is an upper limit on the scale at which this will be feasible. On the other hand, reaction rate can be reduced to balance reduced heat removal by operating at lower concentration or lower temperature; the former reduces productivity; the latter could change the balance between desired and undesired reactions, lengthens reaction times, and also reduces the available temperature driving force for cooling.

For the common case where the temperature is held constant during reaction and the initial concentrations are the same at each scale, Eq. (10.5) with the left-hand side set to zero leads to the scale-up basis

$$\frac{UA}{V}\Delta T_{LM} = \text{const} \quad (10.8)$$

Equations (10.4) and (10.8) taken together indicate that for scaling batch homogeneous reactions that are limited by chemical kinetics, a scale-independent design space can be expressed using temperature and equivalents as factors, as long as there is sufficient assurance that adequate heat transfer to maintain temperature control will be available at all scales. The latter condition amounts to a statement about the equipment capability at each location, specifically the heat removal capacity (in W/l) of the equipment operating at the required temperature.

All of the above effects can be calculated using mechanistic models in which Eqs. (10.2)–(10.7) are solved by integration in time. In practice, batch homogeneous reactions may not be allowed to run to completion, because stopping the reaction earlier produces higher yield (yield peaks during the batch) and less impurity than waiting until the end. Similarly, batch homogeneous reactions may be run using temperature profiles, e.g. charge at low temperature, heat to reaction temperature, then hold, and possibly heat again. In these cases, the temperature profile followed by the reaction may be scale dependent for the reasons given above, with, e.g. longer heating ramps on scale; this will impact the relative rates of the reactions and could affect quality as well as rate.

If Eq. (10.3) is known for each reaction, these effects can be predicted easily, and operating conditions can be optimized to a high degree as shown in Example Problem 10.4; good examples of this approach covering a variety of reaction types are available [9–13].

10.2.2 Characterization Tests for Batch Homogeneous Reactions

The general guidelines described in Introduction should be followed with regard to experimental design.

The reaction should first be run in 2–3 otherwise identical (replicate) experiments in which stirrer speed is varied between 200 and 1000 rpm; four or more samples should be taken, biased toward the beginning of the reaction, e.g. after 10, 30, 60 minutes, and at the end.

Under conditions where there is no effect of physical rate processes, a kinetics study may follow in which temperature and equivalents are varied experimentally and the parameters in Eq. (10.3) fitted.

If agitation conditions affected the result of the initial screening experiments, the result of the reaction is influenced by the rate of mixing or more specifically the rate of bulk blending of the vessel contents, known as "macromixing." This is unusual for a batch homogeneous reaction, but if it occurs, the scale-dependent macromixing time constant will be important as the outcome of reaction is influenced by this characteristic and not just chemical kinetics; see the section below on equipment characterization. A reaction such as this might be better run in fed-batch mode, with the fed reagent maintained at a low concentration. Alternatively, there may be some scope to reduce the rates of the chemical reactions (relative to mixing) by operating at lower concentrations and/or temperatures.

10.2.3 Achieving Similarity on Scale-Up

Achieving similarity on scale-up relies on having characterized both the reaction and the intended scale-up facility. Equations (10.2)–(10.7) provide the basis for a scale-independent design space.

When kinetics are rate limiting, other than agitation that is necessary to promote heat transfer, no particular additional agitation requirement exists for homogeneous reactions. As noted above, there may be an optimum temperature profile to adopt during reaction and/or an optimum time at which to stop the reaction. These may be determined from a kinetic model, as in Example Problem 10.4. A scale-independent design space may include factors such as equivalents, temperature, and reaction time.

When mixing is rate limiting, unusual for batch homogeneous reactions, agitation influences the outcome, and similar macromixing time constants will be required at each scale; if the process persists in this form, the macromixing time constant should be a factor in design space definition.

10.3 MULTIPHASE BATCH PROCESSES WITH REACTIONS

Figure 10.2 shows a schematic of a multiphase batch reaction, in this case a hydrogenation.

These reactions are also referred to as "heterogeneous," "biphasic," or "slurry phase." Figure 10.2 shows the main liquid phase initially containing substrate, another phase (the headspace) containing reagent (hydrogen), reaction in the liquid phase between substrate and dissolving reagent (hydrogen) and removal of heat. (In Figure 10.2, the solid catalyst phase that is normally present for hydrogenation is omitted, and the catalyst particles are assumed to follow

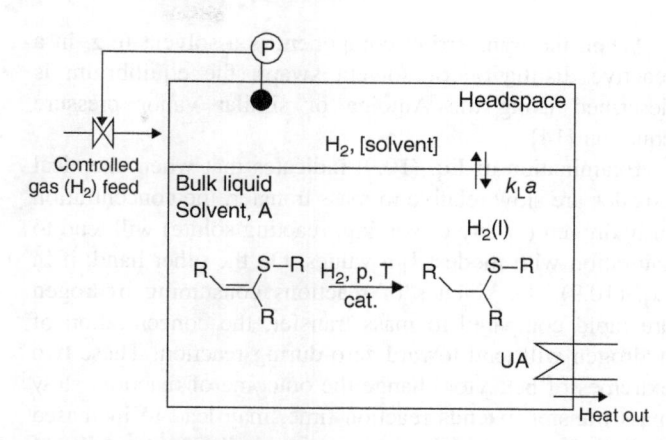

FIGURE 10.2 Process scheme for a hydrogenation, an example of batch multiphase reaction.

the liquid. The headspace is continuously replenished with hydrogen as reaction proceeds.)

10.3.1 Rate Equations

Although relatively complex at first sight, this multiphase, chemically reacting, time-dependent problem can be broken into classical chemical engineering elements and rate processes just like the batch homogeneous reaction above. The rate equation for chemical species concentration in the liquid phase is given by combination of chemical kinetics with the film theory of scale-dependent mass transfer:

$$\frac{dC_i}{dt} = \sum_j -\nu_{ij} r_j + k_L a (C_i^* - C_i) \quad (10.9)$$

Only the equations for components that transfer between phases contain the second term; for dissolving gases, solubility is given by Henry's law:

$$C_i^* = \frac{p_i}{RTH_i} \quad (10.10)$$

In other heterogeneous reactions, the solubility expression differs; for liquid–liquid (aqueous–organic) systems,

$$C_i^* = \frac{C_{id}}{S_i} \quad (10.11)$$

where S_i is the partition coefficient for component i between the phases. For solid–liquid systems,

$$C_i^* \approx f(T, \text{composition}) \quad (10.12)$$

where $f()$ is a function determined from phase equilibrium fundamentals and/or by experimental measurement (e.g. using NRTL equations or similar).

When the transferring component is a solvent (e.g. in a reactive distillation or solvent swap), the equilibrium is described using the Antoine or similar vapor pressure equation [14].

Examination of Eq. (10.9) indicates that when chemical kinetics are slow relative to mass transfer, the concentration of hydrogen (or any dissolving, reacting solute) will tend to saturation with modest $k_L a$ values. On the other hand, if in Eq. (10.9), the kinetics of reactions consuming hydrogen are rapid compared to mass transfer, the concentration of hydrogen will tend toward zero during reaction. These two extremes of behavior change the outcome of reaction; slow mass transfer extends reaction time, may lead to increased impurities, and can cause reaction to stall. The outcome of a batch multiphase reaction at constant temperature therefore depends on temperature, concentration, equivalents (including catalyst loading), pressure (if the solute is in the gas phase), and potentially $k_L a$.

In reactions with three phases present, such as the additional solid catalyst phase in hydrogenation, that phase should be dispersed, and its surface area made available for mass transfer in a similar way to the gas phase in the present example. Complete suspension of a solid phase is usually sufficient to prevent mass transfer to or from that phase being rate limiting; when the solid particles have the same characteristics at both scales, complete suspension is approximately equivalent to maintaining constant $k_L a$ for those particles (see the Section 10.6).

For multiphase reactions, the rate equation for solution phase temperature is

$$\frac{dT}{dt} = -\frac{UA}{\rho V C_p}\Delta T_{LM} + \frac{Q_r}{\rho V C_p} - \frac{k_L a V (C_i^* - C_i)\Delta H_{transfer}}{\rho V C_p} \quad (10.13)$$

Equation (10.13) contains one additional term compared with Eq. (10.5) to account for any heat effect associated with mass transfer of dissolving material between phases; this effect is often negligible but can be significant when solvent is vaporized (in distillation) or when a large quantity of solute is transferred quickly (e.g. when crystals come out of solution suddenly). The heat flow signature from the chemical reactions will be determined by kinetics when mass transfer is rapid and by mass transfer when kinetics are rapid. Neglecting the final term in Eq. (10.13), the scale-up relationship for maintaining constant temperature when kinetics limit is identical to the corresponding equation for homogeneous reactions:

$$\frac{UA}{V}\Delta T_{LM} = \text{const} \quad (10.14)$$

Equation (10.14) when applied to pressure reactions assumes constant pressure, i.e. constant solubility. The scale-up relationship when kinetics are rapid relative to mass transfer can be obtained by noting that in this case, the rate of dissolution of the limiting solute dominates the rates of the main reactions:

$$Q_r \cong -\sum_j \Delta H_{rj} k_L a C^*_{solute} V \quad (10.15)$$

Substituting Eq. (10.15) into Eq. (10.13) leads to

$$\frac{UA}{k_L a V}\Delta T_{LM} = \text{const} \quad (10.16)$$

The above result may also be obtained by writing Eq. (10.13) in dimensionless form using $t^* = k_L a \cdot t$ and $T^* = T/T_0$, neglecting the last term and substituting Eq. (10.15) for Q_r:

$$\frac{dT^*}{dt^*} = -\frac{UA\Delta T_{LM}}{k_L a V \rho C_p T_0} + \frac{-\sum_j \Delta H_{rj} C^*_{solute}}{\rho C_p T_0} \quad (10.17)$$

The second term on the right-hand side is constant when pressure and initial temperature are fixed, and when the left-hand side is set equal to zero (constant temperature), Eq. (10.16) is obtained.

Equations (10.16) and (10.17) indicate that when the solution concentration of solute is close to zero, the required rate of heat removal is directly proportional to the rate of mass transfer, i.e. higher $k_L a$ implies the need for greater heat removal.

An additional constraint exists for pressure reactions, such as hydrogenation, which cannot be taken for granted in practice. The ability to maintain a desired or constant pressure in the headspace depends on the balance between supply of fresh gas to the headspace and removal of gas by mass transfer:

$$\frac{dp}{dt} = \frac{RT}{V_{head}}(N_{H_2} - k_L a V(C_i^* - C_i)) \quad (10.18)$$

Equation (10.18) shows that rapid mass transfer may cause the headspace pressure to reduce, while a temperature rise may cause it to increase. To maintain a constant pressure on scale-up requires adequate temperature control and adequate gas supply:

$$N_{H_2,max} > k_L a V(C_i^* - C_i) \quad (10.19)$$

The right-hand side of Eq. (10.19) requires knowledge of the solution concentration of the dissolved gas. A conservative estimate may be made by setting the dissolved concentration to zero, leading to

$$\frac{N_{H_2,max}}{k_L a V C_i^*} > 1 \quad (10.20)$$

A crude estimate of the required flow rate to maintain constant pressure may also be made from the number of moles of gas required for reaction and the intended reaction time, i.e.

$$N_{H_2, max} \gg \frac{moles\_required}{reaction\_time} \quad (10.21)$$

Similar to homogeneous batch reactions, Eqs. (10.9)–(10.13) and (10.18) can be easily incorporated into dynamic mechanistic models for multiphase reactions and the unknown parameters (e.g. rate constants, activation energies) regressed against experimental data. The effects of changing concentrations and equivalents (including catalyst loading), temperatures, and pressures and the effects of mass transfer, heat transfer, and gas supply limitations can then be predicted. In many hydrogenations, for example, neither pressure nor temperature is maintained constant, and the profiles they follow may be scale dependent; even the sequencing of nitrogen and hydrogen purges before reaction can affect the result. The impact of these changes can be predicted using classical chemical engineering rate equations with appropriate reaction and equipment characterization. Example Problems 10.1 and 10.2 illustrate this approach.

10.3.2 Characterization Tests for Multiphase Reactions

The general guidelines described in Introduction should be followed with regard to experimental design.

For gas–liquid, gas–liquid–solid, liquid–liquid, and liquid–liquid–solid systems, the reaction should first be run in 2–3 otherwise identical (replicate) experiments in which stirrer speed is varied between 200 and 1000 rpm; 4 or more samples of the reacting phase should be taken during reaction, biased toward the beginning of the reaction, e.g. after 10, 30, 60 minutes, and at the end. With some reactions, additional profiling may be possible by following variables such as hydrogen uptake, pressure, and both pot and jacket temperature. The purpose of changing stirrer speed is to change the mass transfer contact area between the phases. In each of these experiments, any solids present should be well suspended.

To check for mass transfer limitation due to dissolving solids, experiments should be run at an impeller speed, which guarantees suspension, but either the particle size or the mass of solid reagent should be varied.

Under conditions where there is no effect of physical rate processes, a kinetics study may follow in which any of temperature, equivalents, pressure, catalyst loading, and/or other case-specific variables affecting the scale-independent chemical rates are varied experimentally.

If agitation conditions affected the result of the initial screening experiments, the outcome of the reaction is influenced by the rate of mass transfer. This is frequently the case for heterogeneous reactions, and characterization of the scale-dependent equipment characteristics will be important. There may be some scope to reduce the rates of the chemical reactions (relative to mixing) by operating at lower concentrations, pressures, catalyst loadings, and/or temperatures, but this will also reduce volumetric productivity.

10.3.3 Achieving Similarity on Scale-Up

For batch multiphase reactions, achieving similarity on scale-up relies on having characterized both the reaction and the intended scale-up facility. Equations (10.9)–(10.13) and (10.18) provide the basis for a scale-independent design space.

When kinetics are slow (relative to mass transfer), adequate agitation is necessary to create a dispersion and to promote heat transfer; adequate gas supply is required for pressure reactions. Justifying the scale independence of the design space in this case reduces to demonstrating that pressure and temperature can be maintained in the same range on scale as in the laboratory and that agitation is sufficient to disperse the phases.

When kinetics are fast, $k_L a$ should be included explicitly to make the design space scale independent.

For hydrogenations and other catalyzed pressure reactions, a design space could therefore contain concentrations, equivalents, temperature, pressure, catalyst loading, and time as factors; when kinetics are fast, $k_L a$ should also be a factor.

As noted above, there may be an optimum temperature or pressure profile to adopt during the reaction and/or an optimum time at which to stop the reaction. These may be determined from a kinetic model, as used in the example described below [15]. Many examples of kinetic modeling of multiphase reactions are available, including a methanethiol reaction [16] and for other hydrogenations [17].

When kinetic models are used in reverse, to work backward from a desired end result to determine the possible operating conditions (or factors) that would produce this result, multiple acceptable combinations of factor settings may be found. For example, very poor mass transfer (low $k_L a$) can be partially compensated for by operating at higher pressure [18]; the effects of low $k_L a$ may also be partially mitigated by operating at lower concentrations of starting materials or catalyst. These combinations may be difficult to express in a design space definition but can be found easily using a mechanistic model and justified if the model has been properly verified against experimental data. This has led to discussion of the use of models to more flexibly capture the definition of a design space and verify that a given set of operating conditions lie within it [19].

EXAMPLE PROBLEM 10.1

Figures 10.3 and 10.4 show examples of hydrogenation reactions with mass transfer limitations [9]. Figure 10.3 illustrates hydrogen heat flow and thermal conversion for a nitro reduction in a lab reactor, with reaction exhibiting apparent zeroth-order kinetics (constant rate, linear conversion profile) and taking over six hours. The mass transfer capability ($k_L a$) of the lab reactor was known from previous equipment characterization to be rate limiting. The results of scale-up to the pilot plant are shown in Figure 10.4 (temperature, hydrogen uptake, and substrate level). The reaction time is reduced to 1.5 hours due to superior mass transfer in the pilot plant reactor. This result was predicted and expected based on the reactor characteristics.

EXAMPLE PROBLEM 10.2

Figure 10.5 shows the reaction profile (hydrogen uptake) for a hydrogenation with dissolving solid substrate. On scale, the reaction had a long "tail" due to the bimodal size distribution of the substrate coming from the previous step. Larger particles dissolve more slowly than smaller particles, and this caused the extended reaction time.

10.4 FED-BATCH PROCESSES WITH REACTIONS

Figure 10.6 shows a process scheme for a homogeneous fed-batch reaction.

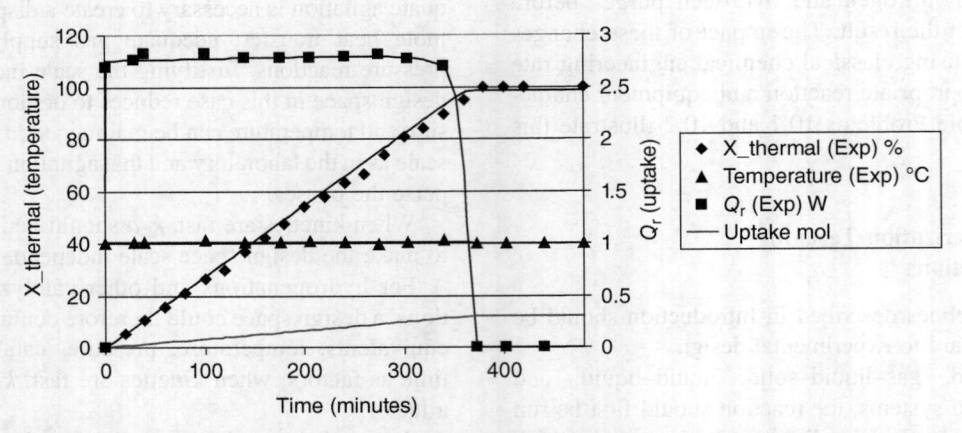

FIGURE 10.3 Lab results (discrete points) of this hydrogenation (reduction of a nitro group, Example Problem 10.1) indicated a six hour reaction time. Modeling (continuous curves) and vessel characterization revealed this was due to severe gas–liquid mass transport limitations ($k_L a$) in the lab equipment and that the process could perform much better.

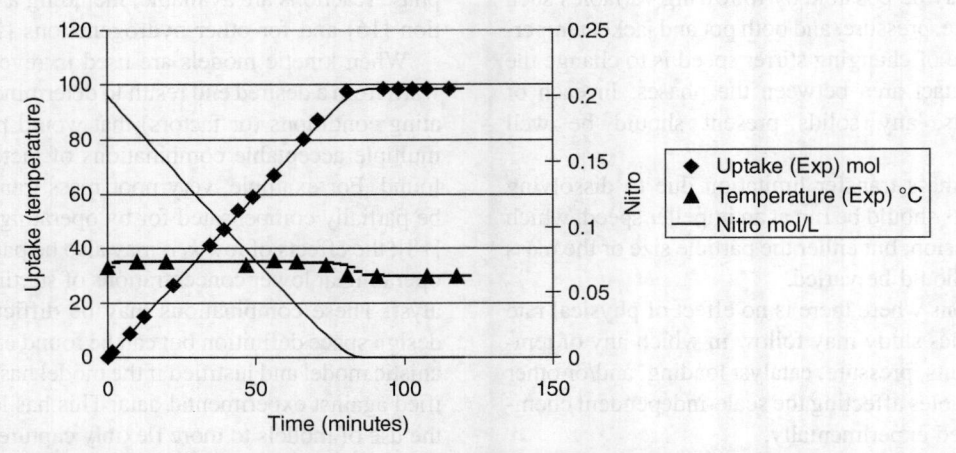

FIGURE 10.4 A reaction time of 1.5 hours was predicted (curves) and achieved (symbols) in the pilot plant for the same reaction as in Figure 10.3 (Example Problem 10.1). The mass transfer characteristics of the pilot plant reactor were superior to those of the lab reactor.

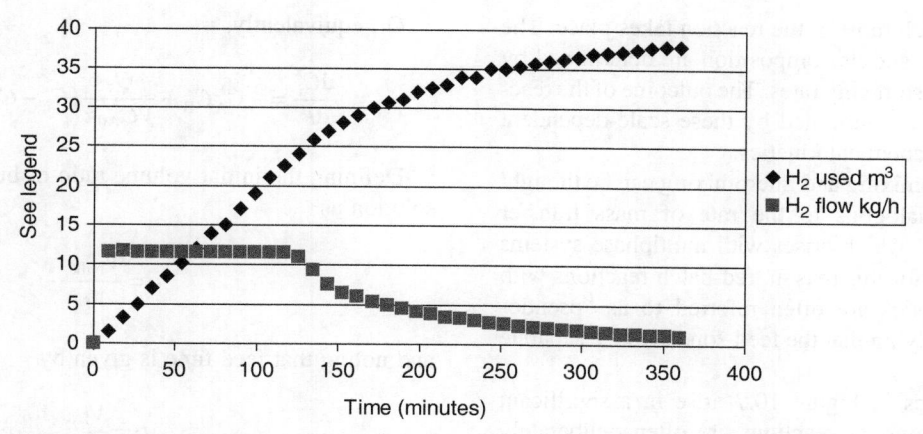

FIGURE 10.5 On scale-up of another hydrogenation (Example Problem 10.2), reaction times extended to over six hours, due to the appearance of a bimodal particle size distribution in the solid substrate, with about 50% of the solids mass having a much larger particle size; this reduces the rate of reaction, especially once the smaller solids have dissolved.

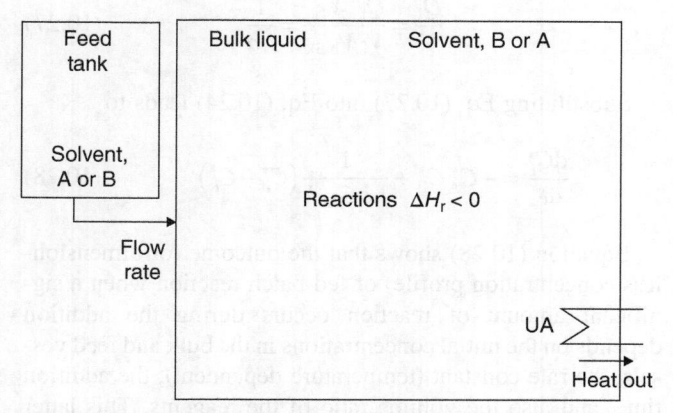

FIGURE 10.6 Process scheme for a homogeneous fed batch reaction system.

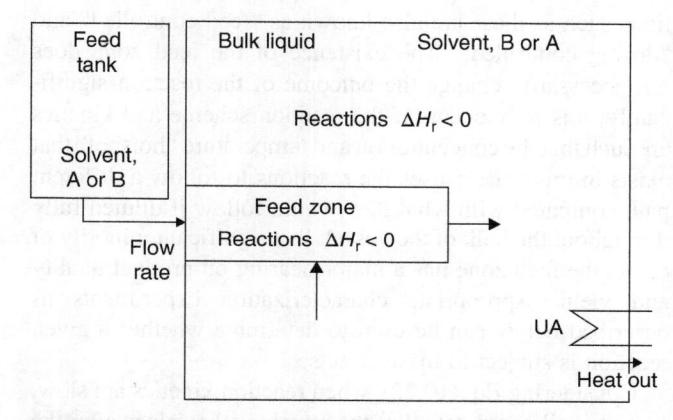

FIGURE 10.7 Process scheme required for a fed batch reaction with rapid chemical kinetics.

These operations are common both in the laboratory and on scale and are also known as "exothermic additions" or "semi-batch reactions." Figure 10.6 shows the main reaction liquid phase initially containing solvent and one or more reactants, with another reactant initially in the feed tank. Liquid is added from the feed tank, reaction begins, and the resulting heat is removed.

10.4.1 Rate Equations

The rate of change of concentration of each species in a homogeneous fed-batch reaction is

$$\frac{dC_i}{dt} = \sum_j -\nu_{ij} r_j + \frac{Q_f}{V}(C_{if} - C_i) \qquad (10.22)$$

Compared with Eq. (10.4), the additional term on the right-hand side represents addition of feed and the accompanying dilution effect.

When the kinetic rates of the reactions are slow compared with the rate of addition, reaction takes place after the addition, and the result is very similar to a homogeneous batch reaction, described above. When the kinetic rates are comparable with the addition rate, a significant amount of reaction occurs during the feed, which normally changes the outcome of reaction compared with the batch case. This is because the concentration of the fed reactant remains low during the addition.

When the kinetic rates are much faster than the addition rate, the concentration of the fed reactant tends to zero, and in this case the reaction is highly localized around the feed region. Equation (10.22) in this case no longer fully describes the rates of change, and a more detailed analysis involving phenomena known as mesomixing and micromixing is required [20, 21] for an accurate mathematical description. A process scheme for this situation is shown in Figure 10.7, with the notable addition of a "feed zone" or reaction zone in which most of the reaction takes place. The size of the feed zone and its composition are determined by micromixing and mesomixing rates. The outcome of the reaction is predominantly determined by these scale-dependent rates rather than by chemical kinetics.

The rates of mesomixing and micromixing can be thought of as somewhat analogous to the rate of mass transfer between two phases, which arises with multiphase systems as described above; for this reason, fed-batch reactions with rapid chemical kinetics are often referred to as "pseudo-homogeneous," implying that the feed zone is like a separate phase.

The characteristics in Figure 10.7 arise in a significant fraction of applications, as reactions are often deliberately engineered to run with fast kinetics (e.g. by operating at high concentration, catalyst loading, and temperatures) because the heat output from such systems can then be halted in the event of a cooling failure by stopping the addition. Reactions such as these are also known as "feed controlled" and "dosing controlled." The existence of the feed zone does not necessarily change the outcome of the reaction significantly; this only occurs if the reaction scheme and kinetics are such that the concentration and temperature "hotspot" that exists in this zone causes the reactions to follow a different path compared with what they would follow if diluted fully throughout the bulk of the vessel. In a significant minority of cases, the feed zone has a major bearing on product quality and yield. Appropriate characterization experiments as described below can be used to determine whether a given reaction is subject to these effects.

Considering Eq. (10.22), when reaction kinetics are slow, such that all or almost all of the reaction takes place after the addition is complete, the solution tends toward that for a batch reaction with the same volume, as described by Eq. (10.4). Integrating Eq. (10.22) in time, the final outcome (composition) of the reaction will then depend on the concentrations and temperature profiles followed during the reaction and the time allowed for the reaction. Temperature is often (but not always) held constant or nearly constant, so final concentrations at completion depend only on initial concentrations and initial temperature. Example Problem 10.4 follows this behavior.

When a significant amount, but not all, of the reaction occurs during feeding, this indicates that some of the chemical kinetic rates are comparable with the addition rate. Equation (10.22) is more informative for this case when examined in dimensionless form; for a second-order reaction between A and B, using $C_i^* = C_i/C_{A0}$ and $t^* = kC_{A0} \cdot t$, the rate equation for the dimensionless A or B concentration is, from Eq. (10.22),

$$\frac{dC_i^*}{dt^*} = -\frac{C_A C_B}{C_{A0}^2} + \frac{Q_f}{VC_{A0}k}\frac{(C_{if}-C_i)}{C_{A0}} \quad (10.23)$$

Or, equivalently,

$$\frac{dC_i^*}{dt^*} = -C_A^* C_B^* + \frac{Q_f}{VC_{A0}k}\left(C_{if}^* - C_i^*\right) \quad (10.24)$$

Defining the initial volume ratio of bulk solution to feed solution as

$$\alpha = \frac{V_{\text{bulk}}}{V_f} \quad (10.25)$$

and noting that feed time is given by

$$t_f = \frac{V_f}{Q_f} \quad (10.26)$$

then, at the start of the addition,[1]

$$\frac{Q_f}{V} = \frac{Q_f}{V_f}\frac{V_f}{V_{\text{bulk}}} = \frac{1}{t_f\alpha} \quad (10.27)$$

Substituting Eq. (10.27) into Eq. (10.24) leads to

$$\frac{dC_i^*}{dt^*} = -C_A^* C_B^* + \frac{1}{t_f\alpha C_{A0}k}\left(C_{if}^* - C_i^*\right) \quad (10.28)$$

Equation (10.28) shows that the outcome (or dimensionless concentration profile) of fed-batch reaction when a significant amount of reaction occurs during the addition depends on the initial concentrations in the bulk and feed vessels, the rate constant (temperature dependent), the addition time, and also the volume ratio of the reagents. This latter point is important as the volume ratio is independent of the number of equivalents, i.e. reactions run at the same equivalents, temperature, and addition time may produce different results when the volume ratios mixed are different.

Apart from volume ratio, a further degree of freedom exists in fed-batch reactions that is not present in the batch case: the order of addition; when the order of addition is reversed, this can radically change the solution concentrations compared with the forward addition. Reverse addition is worth considering and testing when feasible and safe. From a quality point of view, order of addition should follow from the reaction scheme (expressed in Eq. (10.4)) using the stoichiometric matrix ν_{ij}. For example, in the following case, to maximize product it is more favorable to add A to B:

A + B → product
A + product → impurity

---

[1] By the end of the addition, the denominator contains $\alpha + 1$ instead of α.

In the next case, it is more favorable to add B to A:

A + B → product
B + product → impurity

When essentially the entire reaction occurs during the addition, Eqs. (10.22) and (10.28) are no longer directly applicable, as the reaction zone will be localized and some reactions will run to completion before the feed has been diluted into the bulk contents. In this case, the dominant effect on the composition of the reaction mixture is the rate of localized mixing near the addition point, a physical phenomenon somewhat analogous to mass transfer in the multiphase example above. The degree to which the reaction exhibits this characteristic may depend on the order of addition. The final outcome of this type of reaction will depend on the time constants for mesomixing or micromixing, whichever is slower and therefore rate determining; these influence the size and residence time of the feed zone, which in turn determines how much of the chemistry takes place in the concentration and temperature hotspot created by the feed. Both meso- and micromixing time constants are affected by the addition rate, the addition location, the diameter of the addition nozzle, the agitator configuration, and the agitator rotational speed [14]. Factors directly affecting the intrinsic kinetics have less influence in this scenario, such as reaction temperature.

The rate of change of temperature in fed-batch reaction systems is given by

$$\frac{dT}{dt} = -\frac{UA}{\rho V C_p}\Delta T_{LM} + \frac{Q_r}{\rho V C_p} + \frac{\rho_f C_{pf} Q_f}{\rho V C_p}(T_f - T) - \frac{\rho_f Q_f \Delta H_m}{\rho V C_p} \quad (10.29)$$

Compared with Eq. (10.5) for batch reactions, the two additional terms on the right-hand side represent contribution of sensible heat by the fed material (e.g. when warmer than the bulk) and any associated thermodynamic heat of mixing that also accompanies the addition. The heat flow signature from the chemical reactions will be determined by kinetics (kinetically controlled) when reaction occurs after the addition, by the rate of addition when the reactions occur entirely during the addition, and by both kinetics and the rate of addition when a significant amount of reaction, but not all, occurs during the addition.

From Eq. (10.29) with $Q_f = 0$, the scale-up relationship for maintaining constant temperature when kinetics limit is identical to the corresponding equation for homogeneous reactions:

$$\frac{UA}{V}\Delta T_{LM} = \text{const} \quad (10.30)$$

At the other extreme, when kinetics are instantaneous relative to addition, the rate of heat output is primarily dependent on the rate of addition of limiting fed reactant:

$$Q_r \cong -\sum_j \Delta H_{rj} Q_f C_{fed} \quad (10.31)$$

Substitution of Eq. (10.31) into Eq. (10.29) and setting the right-hand side equal to zero leads to the widely used scale-up relationship

$$\frac{UA}{Q_f}\Delta T_{LM} = \text{const} \quad (10.32)$$

Noting that the addition volumetric flow rate is the addition volume divided by the addition time, Eq. (10.32) may be rearranged in terms of feed time as

$$t_f = \text{const}\frac{V_f}{UA\Delta T_{LM}} \quad (10.33)$$

A more complete version of Eq. (10.33) may be obtained by writing Eq. (10.29) in dimensionless form using $T^* = T/T_0$ and $t^* = UAt/\rho V C_p$ and then substituting Eq. (10.31) for Q_r and Eq. (10.27) for Q_f, giving at the start of the addition

$$\frac{dT^*}{dt^*} = -\Delta T^*_{LM} + \frac{V}{UA}\frac{1}{\alpha t_f}\frac{-\sum_j \Delta H_{rj} C_{fed}}{T_0} \quad (10.34)$$

Rearranging Eq. (10.34) with the right-hand side set equal to zero leads to

$$\frac{\alpha t_f UA\Delta T_{LM}}{-\sum_j \Delta H_{rj} C_{fed} V} = 1 \quad (10.35)$$

From Eq. (10.35), a longer feed time implies that a lower UA or ΔT may be tolerated; a higher volume ratio is most likely to be combined with a more concentrated feed, and these effects cancel out.

From Eqs. (10.33) and (10.35), it is clear that the addition time on larger scale will often need to be longer than on smaller scale in order to operate the reaction at the same constant temperature as at lab scale; the magnitude of the increase in addition time will depend on the degree to which the temperature driving force can be increased on scale-up.

Similarly to homogeneous batch reactions, the equations above can be easily incorporated into dynamic mechanistic models for fed-batch reactions and unknown parameters (e.g. rate constants, activation energies) regressed against experimental data. The effects of changing order of addition, concentrations (including catalyst loading), temperatures, volume ratio, and addition time and the effects of mixing and heat transfer can be predicted. In some fed-batch reactions, for example, temperature is deliberately profiled (e.g. using a heating ramp to drive reaction to completion), and the profile may be scale dependent. The impact of these changes can be predicted using classical chemical engineering rate equations. There may be an optimum time at which to stop the reactions, e.g. after product concentration has peaked and before impurities are able to form. To understand these effects and take advantage of the opportunities, a detailed mechanistic model of the reaction is valuable. Several examples illustrating this approach are given below, and more are included in the Refs. [7, 11].

10.4.2 Characterization Tests for Fed-Batch Reactions

For fed-batch homogeneous reactions, if feasible the reaction should first be run using forward and reverse additions. If the desired process involves addition of A to B, then run this case followed by an otherwise identical case with B added to A. If the results of these experiments differ significantly, reaction occurs during the addition and is at least in the intermediate regime described above. Temperature or heat flow profiles when monitored during these experiments may indicate the degree of addition-controlled behavior and the amount of reaction happening during the feed versus after the feed. Likewise, *in situ* or sample data following the concentration of the bulk reactant both during and after the addition will be informative in this regard.

When a dosing-controlled reaction is suspected, a further 2 replicate experiments in which only stirrer speed is varied between 200 and 1000 rpm will further indicate the influence of agitation on the rate and outcome of reaction.

In each of the above initial screening experiments, four or more samples should be taken, e.g. after 25, 50, 75, and 100% of the feed have been added. With some reactions, additional profiling may be possible by following variables such as the rate of gas evolution and both pot and jacket temperatures.

Under conditions where there is no effect of agitation on rate or quality, a kinetics study may follow in which addition time, volume ratio, temperature, equivalents, and catalyst loading are varied experimentally. The choice of conditions and the sampling program for these experiments should be informed by the results of the previous experiments and an initial kinetic model [2], and in any event, multiple samples should be taken again to capture when the reactions are occurring as well as the final result.

If rate or quality is sensitive to agitation, the reaction is not limited by chemical kinetics at current conditions but by localized mixing, at least under typical agitation conditions. The main focus of process design and characterization should be the mixing capabilities of lab and larger-scale equipment (see Section 10.6). There may be some scope to reduce the rates of the chemical reactions (relative to mixing) by operating at lower concentrations and/or temperatures.

10.4.3 Achieving Similarity on Scale-Up

Achieving similarity on scale-up relies on having characterized both the reaction and the intended scale-up facility. Equations (10.22) and (10.29) provide the foundation for a scale-independent design space.

When kinetics are rate limiting, other than agitation that is necessary to promote heat transfer, no particular additional agitation requirement exists. As noted above, there may be an optimum temperature profile to adopt during reaction and/or an optimum time at which to stop the reaction. These may be determined from a kinetic model, as in Example Problem 10.4. A scale-independent design space may include factors such as equivalents, temperature, and reaction time.

When mixing is rate limiting (reaction is dosing controlled and mixing affects quality), agitation influences the outcome, and similar mesomixing or micromixing time constants will be required at each scale; in order to achieve a scale-independent design space, the mixing time constants should be factors in design space definition. Additional factors may include order of addition, addition time, volume ratio, equivalents, and temperature.

In the intermediate regime where a significant portion but not all of reaction occurs during the addition, reaction will tend to become more dosing controlled on scale-up, as the addition time is increased to compensate for lower heat transfer area per unit volume. A scale-independent design space may include factors such as order of addition, addition time, volume ratio, equivalents, temperature (profile), and reaction time.

EXAMPLE PROBLEM 10.3

A reaction whose heat flow profile showed typical very different behavior when run with forward and reverse additions is shown [22] in Figure 10.8. In this case the forward addition produced the correct material but had an undesirable sudden exotherm at the end of the addition. The reverse addition was dosing controlled, as shown in Figure 10.9.

EXAMPLE PROBLEM 10.4

Many examples of design space mapping using chemical engineering rate equation based models are available [5, 13, 23], and the results of one case are presented below [24, 25].

The production of epi-pleuromutilin in a homogeneous liquid-phase reaction was believed to follow the pathway shown in Figure 10.10.

To characterize and scale this reaction, six fed-batch experiments were carried out, in which acid was added to the other reactants and two factors were varied, temperature and equivalents of acid. The progress of reaction was

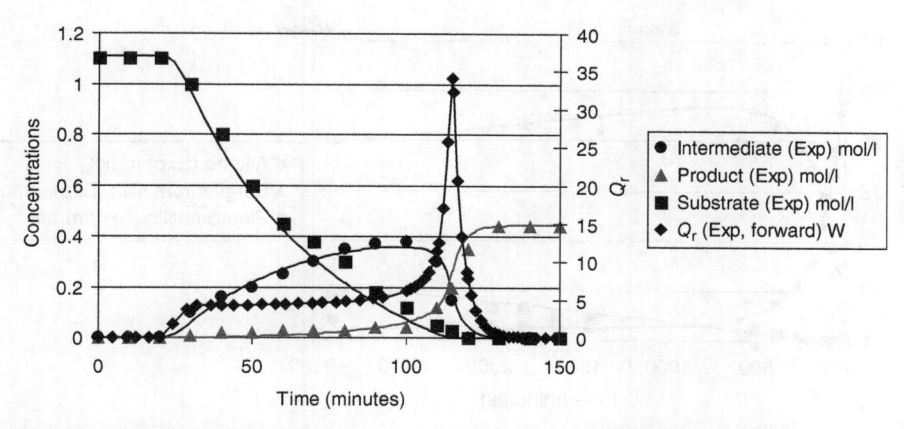

FIGURE 10.8 Concentration and heat flow profiles for Example Problem 10.3: symbols are measured data and curves are model predictions. A large spike in heat flow occurred at the end of the forward addition.

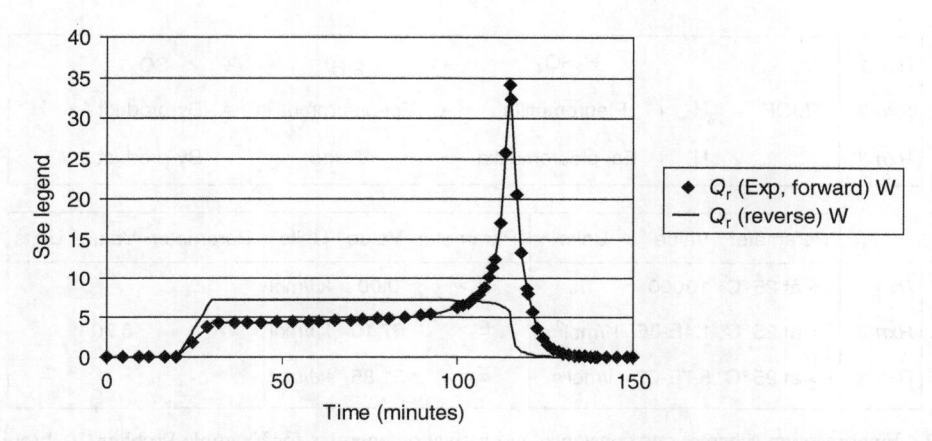

FIGURE 10.9 Heat flow profiles for Example Problem 10.3, comparing forward and reverse addition: symbols are measured data (forward addition) and curves are model predictions (forward and reverse addition). The heat flow curve indicates dosing-controlled conditions when the addition is reversed.

FIGURE 10.10 Overall reaction scheme for Example Problem 10.4 (epi-pleuromutilin).

followed by taking multiple samples; a typical reaction profile is shown below (Figure 10.11).

There is a small temperature rise during the addition of acid, and the majority of the reaction takes place after the addition is complete. The latter signals that the results of these experiments are determined by chemical kinetics. The fact that product increases to a peak and then reduces while the alkene increases lends support to the overall reaction scheme shown above. The raw HPLC area percent data for each sample were converted with the aid of relative response factors to molar quantities for modeling. Kinetic fitting to the results of all six experiments led to the set of reactions and fitted parameters in Figure 10.12; the dissociation rate constant for sulfuric acid was not fitted but set to a high value.

During the kinetic fitting process, close attention was paid to the statistics relating to each parameter, each measured

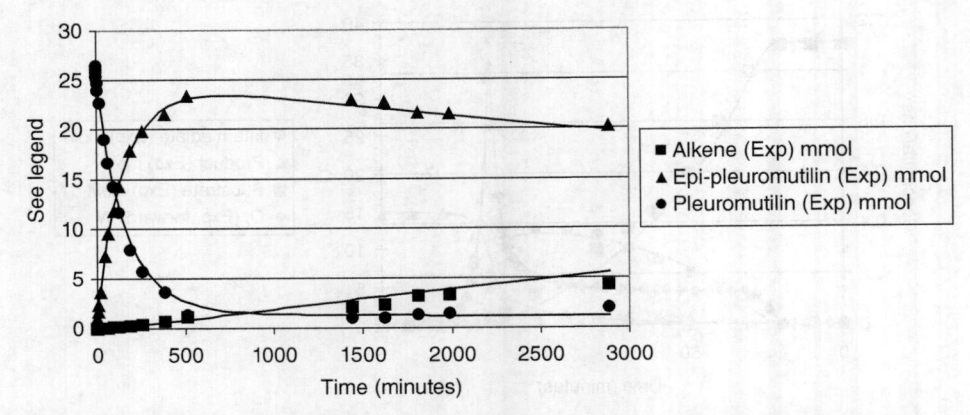

FIGURE 10.11 Typical reaction profiles for Example Problem 10.4. Symbols are measured data (from HPLC area percent) and curves are model predictions.

Rxn 1			H_2SO_4	→	$2H^+$	+	SO_4^-
Rxn 2	TMOF +	H^+ +	Pleuromutilin	→	Epi-pleuromutilin	+ By-product 1 +	H^+
Rxn 3		H^+ +	Epi-pleuromutilin	→	Alkene	+ By-product 2	

	Parameter	Value	Units	Parameter	Value	Units	Parameter	Value	Units
Rxn 1	k > at 25 °C	10000	1/s	E_a >	0.00	kJ/mol			
Rxn 2	k > at 25 °C	1.4E-05	$l^2/mol^2 \cdot s$	E_a >	67.10	kJ/mol	K_{eq}	6.70	-
Rxn 3	k > at 25 °C	5.7E-07	$l/mol \cdot s$	E_a >	51.85	kJ/mol			

FIGURE 10.12 Final reaction scheme, rate constants, and activation energies for Example Problem 10.4 (epi-pleuromutilin).

response, and the model as a whole. In particular, confidence intervals on parameters of tens of percent or less were targeted. When comparing possible alternative reaction schemes/hypotheses, a low final sum of squares (quantifying the lack of fit) combined with a high value of the model discrimination statistic (the Akaike information criterion [26]) was achieved using the parameter values in Figure 10.12.

To leverage the model for process design and scale-up, two additional responses were defined:

- ProductMaxTime, i.e. the time when the product concentration reaches its peak.
- QuenchWindow, i.e. the time after that until 1% of product is lost to impurity (alkene).

The specification on these responses was that ProductMaxTime should be less than eight hours, for productivity and operational reasons, while QuenchWindow should be more than two hours, to allow adequate time for sampling, analysis, and quenching the reaction before more than 1% of product was lost to alkene.

The design space was defined as the region of temperature and acid levels that simultaneously met both criteria. A series of 440 virtual experiments were carried out using the model to produce the responses surface shown in Figures 10.13 and 10.14. The region of overlap of these responses is shown in Figure 10.15.

The original publication in 2007 [18] was based on 30 virtual simulations to which a polynomial equation was fitted for plotting; the above figures were reproduced in 2009 by running 440 virtual experiments in the same factor space with no interpolation.

Because kinetics limit this reaction at or near current operating conditions and because most of the reaction occurs after the addition, a scale-independent design space for these responses may be constructed using two variables: temperature and equivalents of acid. When all other input material quantities are held constant and adequate temperature control is maintained on scale, the result of this reaction at any point in this space would be independent of scale.

Figure 10.15 is quite typical in that there appear to be two "edges of failure" when the responses are overlapped [27].

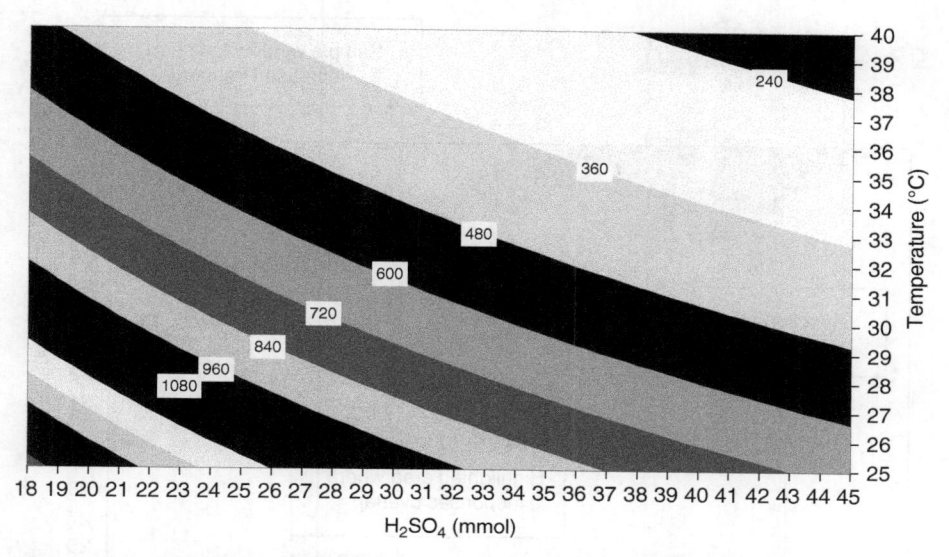

FIGURE 10.13 Response surface of ProductMaxTime (minutes) versus initial amount of acid and temperature (1 equiv. = 26.5 mmol) for Example Problem 10.4. Results produced using mechanistic model to run 440 virtual experiments.

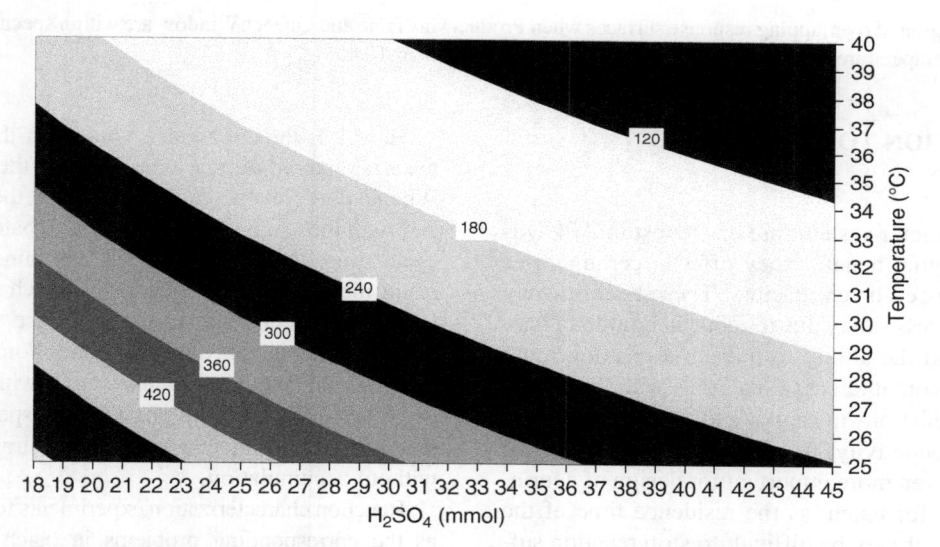

FIGURE 10.14 Response surface of QuenchWindow (minutes) versus initial amount of acid and temperature (1 equiv. = 26.5 mmol) for Example Problem 10.4. Results produced using mechanistic model to run 440 virtual experiments.

One response relates to the quality attribute (amount of alkene in this case), and another relates in this case to a business attribute (reaction time). In a strictly QbD context, Figure 10.15 therefore has only one edge of failure: if the quench window is too short, the alkene level will exceed its limit; if the reaction time is too long, this will not directly impact quality but will slow productivity.

The design space shown in Figure 10.15 is approximately 10 mmol wide and 5° high. However if a rectangular region of proven acceptable ranges had to be defined inside this space, the maximum width would be 7 mmol, and the maximum height 2–3°. This illustrates how the design space has the potential to offer greater flexibility than rigidly defined proven acceptable ranges.

On the other hand, Figure 10.15 remains quite restrictive, in that all other factors (such as substrate concentration) have to be held constant for it to apply. This is one of the reasons why a more dynamic design space definition based on the full mechanistic model, rather than one set of response surfaces at otherwise fixed conditions, has been advocated by industry [13]. The model verification statistics associated with this example are discussed at the end of this chapter.

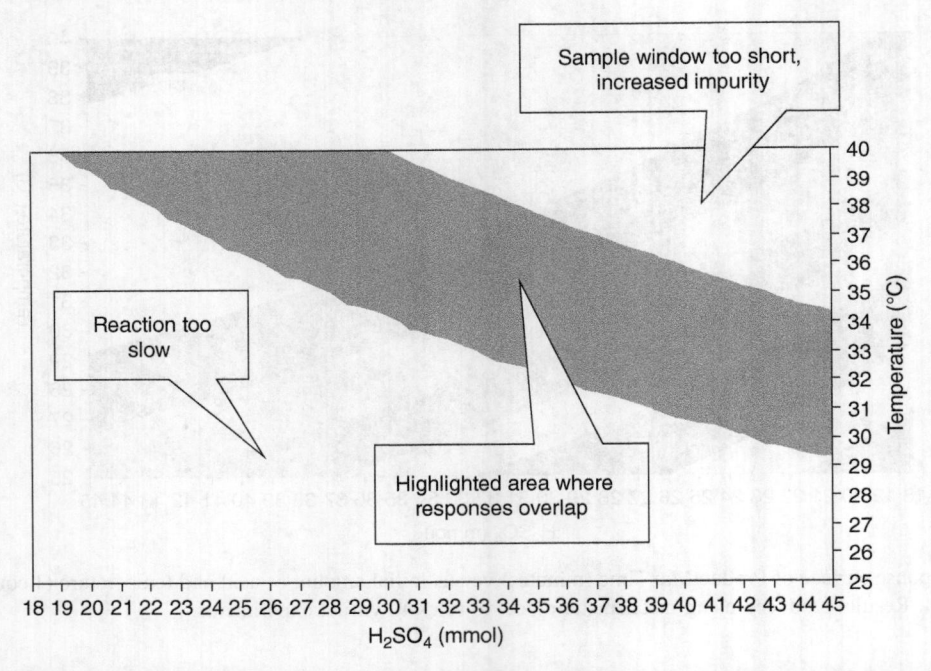

FIGURE 10.15 Region of overlapping response surfaces when ProductMaxTime and QuenchWindow are within specification, versus initial amount of acid and temperature (1 equiv. = 26.5 mmol) for Example Problem 10.4.

10.5 APPLICATION TO CONTINUOUS FLOW SYSTEMS

Continuous flow reactor systems are of interest in API synthesis for the potential benefits they offer in certain cases relative to batch or fed-batch reactors. These benefits may include greater process safety due to reduced holdup of hazardous material and the ability to quench a reaction more rapidly, improved containment of materials with low exposure limits, and reductions in capital and/or operating costs and volumetric productivity. In general, knowledge of the rate of reaction is even more important for design of continuous systems than for batch, as the residence time of the reactor is finite and it may be difficult to stop reaction sufficiently rapidly if problems arise. The economics of low volume production in a dedicated unit may limit the scope for continuous operation. Process issues specific to each application may also present challenges when reactors are operated for longer than typical batch cycle times, such as catalyst deactivation and difficulties handling slurries.

10.5.1 Plug Flow

Equations presented above for batch reactions may be applied directly to plug flow reactions after noting that the independent time variable now applies to position along the reactor [28]:

$$t = \frac{V}{Q} \quad (10.36)$$

Here V is the cumulative volume of the reactor since the material entered at time zero and Q is the volume flow rate. When a plug flow reactor is operated at the same temperature and with the same residence time as a batch reactor with that cycle time, the end results are the same. This convenient result means that data collected in batch mode may be used for design in continuous mode and vice versa.

Plug flow reactors may be used for homogeneous systems, liquid–liquid and gas–liquid systems, with appropriate attention to ensuring that the phases are dispersed and separated when required and that both phases travel along the tube without accumulation.

Reaction characterization experiments follow the same logic as the corresponding problems in batch systems described above [14]. Equipment performance characterization in terms of heat transfer is always required; mixing and mass transfer characterization will be required when kinetics are rapid and in multiphase systems [14, 29]. Design spaces are expressed by taking into account similar factors to batch reactors.

Plug flow reactors with multiple addition points along the reactor length are analogous to fed-batch systems with multiple sequential additions and likewise may be characterized and predicted using a similar approach to fed-batch systems.

Useful additional information on application of continuous flow systems for API is available in the Refs. [30–32].

EXAMPLE PROBLEM 10.5

Figure 10.16 shows heat flow profiles measured in fed-batch mode for the oxidation conversion of a tertiary alcohol to a

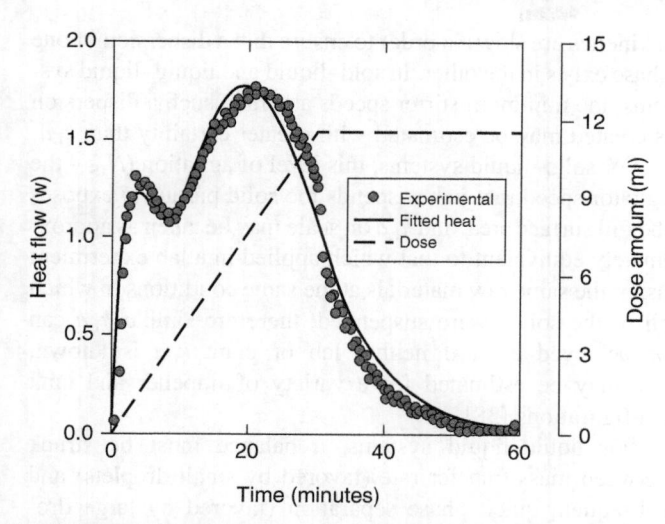

FIGURE 10.16 Fed batch heat flow data (discrete points) for peroxide oxidation reaction in Example Problem 10.5, showing double peak in heat output. Kinetic model results (curves) were used to design an intrinsically safer continuous reactor system.

primary alcohol using excess hydrogen peroxide catalyzed in the presence of an acidic environment [33]. It was known from accelerating rate calorimetry (ARC) experiments that hydrogen peroxide in the presence of the acidic solution medium was thermally unstable at temperatures above 20 °C. Carefully controlled fed-batch experiments were performed at two temperatures (5 and 15 °C) with the same addition time; a reaction scheme and chemical kinetic were fitted to the resulting heat flow data. Although not shown here, there was generally good agreement between model and experiments for both the heat and independently measured concentration profiles. An intrinsically safer continuous flow reactor was designed using these kinetics to produce kilograms of material at both pilot and manufacturing scales.

10.5.2 CSTRs in Series

Plug flow behavior may be approximated using a cascade of stirred reactors in series [21]; this can also ease the problem of running slurry reactions in continuous mode. The approach to reaction characterization and design for these systems is described above. Equipment characterization in terms of heat transfer is always required, and batch characterization tests may be used for this purpose (see below).

Models developed for process prediction in batch or fed-batch systems may be reused to design continuous stirred tank reactor systems by adding appropriate feeding and removal between reactors in the cascade [34]. When the number of reactors is large, a simpler plug flow model will give almost equivalent results.

10.6 EQUIPMENT CHARACTERIZATION AND ASSESSMENT

Equipment performance characteristics play a major role in most operations in API synthesis including each of the reaction types described above. This section reviews methods to evaluate some of the key characteristics for reactions, quenches, extractions, distillation/solvent swap, and crystallization.

Equipment performance can in general be quantified using either or both of the following methods:

- Characterization tests, in which purpose-designed tests are carried out on the equipment and responses such as pressure, temperature, or concentration, are monitored versus time.
- Assessment calculations, in which empirical correlations, often based on dimensionless groups, are used to estimate equipment performance as a function of dimensions, geometry, fluid properties, and the intended operating volume and recipe.

Pharma API development laboratories, kilo labs, pilot plants, and full-scale manufacturing plants are dominated by relatively standard, multipurpose equipment into which each new process is fitted; this means that performance data can be reused many times over once generated. To support QbD, equipment performance characteristics are stored in equipment databases that allow users at any location to retrieve performance data for equipment at both their own and other locations.

10.6.1 Heat Transfer

The product of overall heat transfer coefficient U and wetted area A appears throughout this chapter as equipment characteristics required for process scale-up.

The product UA is best evaluated using a solvent test in the intended process vessel, to which solvent is charged and the fill level and agitator speed are set to those of the intended process. The batch is heated and/or cooled over the range of pot and jacket temperatures required by the process. Pot and jacket inlet temperatures are monitored continuously; jacket outlet temperature is monitored if available. In the absence of reaction and any other heat effects, Eq. (10.5) may be used to fit UA for both heating and cooling.

UA may also be estimated from chemical engineering correlations developed from measurements in similar types of equipment [35]. The coefficients in such heat transfer correlations may vary depending on the precise configuration, and in some cases, new coefficients may be required to describe unusual vessel configurations. In mature applications of equipment characterization, those coefficients are stored with the equipment configuration data in an equipment database for future reuse.

10.6.2 Mass Transfer

The product of mass transfer coefficient k_L and interfacial area a (per unit volume) appears in this chapter whenever interphase transfer is involved.

The product $k_L a$ is best evaluated using a test in the intended process vessel, in which mass transfer either is the only phenomenon occurring or is the rate-limiting phenomenon.

In one such test for headspace fed gas–liquid reactions, typically hydrogenations [36], solvent is charged, and the fill level set to that of the intended process. The headspace is evacuated and then pressurized with gas, and the agitator is turned on, with the speed increasing quickly to the intended process speed. Headspace pressure and both headspace and liquid temperature are monitored continuously. In the absence of reaction and any other phase transfer effects, Eq. (10.17) may be used to fit both gas–liquid (surface gassing) $k_L a$ and solubility. Alternatively, a reaction that consumes the dissolving gas may be run under conditions of high catalyst loading, such that mass transfer is rate limiting; in this case, Eq. (10.9) may be used to fit $k_L a$ to an indicator of reaction progress, such as hydrogen uptake. Note that (i) the $k_L a$ for surface gassing is very sensitive to the submergence of the top impeller [37] and (ii) sparging gas may be of little value in a "dead-end" system (i.e. unless gas is continuously removed from the headspace) and mass transfer due to surface gassing is more important.

The depressurization test described above must be done carefully in order to provide useful data; for example, the headspace and liquid must be at the same temperatures before the agitator is turned on, to avoid a pressure recovery due to heating of the gas by the liquid.

For solid–liquid and liquid–liquid reactions, analogous characterization tests in which the uptake of solute is monitored (e.g. by sampling the liquid) may be used, or a known fast chemical reaction may be run under conditions in which dissolution of solute is rate limiting. Equation (10.9) again provides the basis for fitting $k_L a$ to the monitored profiles.

Alternatively, empirical estimates may be made using chemical engineering correlations; for these, the molecular diffusion coefficient of the solute in the solvent is required, which limits applicability. A feature that makes solid–liquid systems somewhat easier to predict is that the wetted area does not change with stirrer speed once the solids are suspended, that is, unless the particles are broken as a result of agitation. For liquid–liquid systems, the effect of minor components on the droplet size and the resulting surface area can be very significant, making accurate estimates difficult.

In all operations involving contact between multiple phases, a certain minimum level of agitation is required (even if kinetics are slow) in order to ensure that a dispersion of one phase exists in the other. In sold–liquid and liquid–liquid systems, the minimum stirrer speeds at which such a dispersion is created may be estimated with greater certainty than $k_L a$.

For solid–liquid systems, this level of agitation (N_{JS} – the agitator speed that just suspends the solid particles) exposes the full surface area, and $k_L a$ on scale may be taken as approximately equivalent to that which applied in a lab experiment using the same raw materials at the same conditions in which all of the solids were suspended; therefore similar $k_L a$ can be achieved even if neither lab or plant $k_L a$ is known. N_{JS} may be estimated for a variety of impeller and tank configurations [38].

For liquid–liquid systems, a balance must be struck between mass transfer rate (favored by small droplets) and subsequent quick phase separation (favored by large droplets). Operating at N_{JD} – the agitator speed that just disperses the liquid droplets of one phase in the other – exposes significant surface area while reducing the likelihood of forming a stable emulsion; once again the $k_L a$ on scale may be taken as similar to that which applied in a lab experiment with the same raw material at the same conditions in which the liquids were just dispersed; therefore similar $k_L a$ can be achieved even if neither lab or plant $k_L a$ is known. N_{JD} may be estimated for a variety of impeller and tank configurations [34, 39].

If solid–liquid or liquid–liquid $k_L a$ is a factor in definition of a design space, the proximity of agitator speed to N_{JS} or N_{JD}, respectively, is a reasonable surrogate variable for $k_L a$; for example, a dimensionless agitator speed that is scale independent could be defined as a factor:

$$N^* = \frac{N}{N_{JS}} \quad (10.37)$$

N^* in Eq. (10.37) might, for example, in a particular application need to be above 1.0 in order to avoid mass transfer limitations.

10.6.3 Liquid Mixing

In batch homogeneous reactions there is the possibility of reaction rate limitation by macromixing, at very low agitation rates. A macromixing time constant 30–60 seconds should eliminate this dependence, and correlations are available to estimate this factor [40]. The liquid mixing characteristics relevant for fed-batch reactions are meso- and micromixing time constants; these appear in equations that describe the rate of localized mixing near the addition point in fed-batch and continuous reactors and may be rate determining when kinetics are fast, i.e. for pseudo-homogeneous, dosing-controlled reactions as described above.

The time constants for meso- and micromixing may be characterized by running tests reactions with known kinetics in the target equipment [14, 41]. The outcome of these reactions is mixing sensitive under certain conditions and varies with factors such as impeller speed, addition rate, and addition location. When the product distribution or selectivity from each experiment is combined with a mathematical model based on Figure 10.7, the time constants for meso- or micromixing for each set of conditions may be obtained.

Alternatively, formulas are available for these time constants [14], and if the spatial distribution of the local rate of energy dissipation, ε, is known, the time constants may be obtained from these. Specialized tools such as computational fluid dynamics or laser anemometry may be used to estimate the spatial distribution of ε. As a general rule, ε is a multiple (e.g. 10–100) of the vessel-averaged power input per unit mass (W/kg) near the impeller and a fraction (e.g. 1–10%) of the average near the liquid surface.

In continuous plug flow reactor systems, the extent of "axial dispersion" or longitudinal mixing along the axis of the flow may be characterized using pulse or step response experiments. Given the small scale that often applies in continuous pharmaceutical production/flow chemistry, the flow regime is not always turbulent (e.g. small diameter tubes with concentrated materials flowing at low velocities). In general, long, thin tubes, whether coiled or straight, lead to a low level of axial dispersion, and inserts such as static mixers further enhance the degree of plug flow and reduce axial dispersion to the point where assuming idealized plug flow behavior is a good approximation. An extensive discussion on this topic is available [42] containing further detail of measurement techniques and results to support process development and scale-up. Application of PFRs in "flow chemistry"/continuous manufacturing also allows operation at higher temperature and pressure than typical for batch reactors, and in this case, fluid properties may differ significantly from those at ambient conditions, resulting in typically lower residence times for a given mass flow rate, due to thermal expansion.

10.6.4 Phase Separation

Phase separation times, e.g. after a liquid–liquid reaction or extraction, are longer on scale that in the laboratory [43]. As a rough indicator, the separation time scales with the liquid depth, so a separation time of one minute in the laboratory can easily extend to one hour on scale. Excessive separation times can be avoided by avoiding the formation of very fine droplets or stable emulsions; operating a liquid–liquid reaction at or near N_{JD} as described above represents a good balance between high mass transfer area and short phase separation time.

10.6.5 Gas Disengagement

When reactions evolve gas, the rate of gas evolution scales with the reaction volume, but the ability of the liquid surface to allow the gas to escape scales with the cross-sectional area of the vessel. The maximum velocity of gas escape through the liquid surface is approximately 0.1 m/s [34], allowing the maximum volume flow rate of gas evolution to be estimated:

$$Q_{gas} = 0.1 \frac{\pi T^2}{4} \quad (10.38)$$

This can be converted into a molar rate using the ideal gas law

$$N_{gas} = 0.1 \frac{\pi T^2}{4} \frac{p}{RT} \quad (10.39)$$

If the rate of reaction exceeds the maximum rate of gas evolution, significant foaming will occur, and in some cases material will be lost from the reaction vessel. This problem becomes more likely on scale, as the maximum rate of gas evolution per unit volume reduces:

$$\frac{N_{gas}}{V} \approx 0.1 \frac{1}{H} \frac{p}{RT} \quad (10.40)$$

Equation (10.40) indicates that the volumetric rate of reactions evolving gases may need to be reduced on scale (e.g. by slower feeding or otherwise) in order not to exceed the limitations imposed by the equipment.

For example, in a vessel of 2 L nominal volume with diameter 0.115 m, filled to a level of 1 L at 20 °C, the maximum volumetric rate of a reaction evolving 1 mol of gas (e.g. CO_2) per mole of product without batch swelling is $0.1 \times (1/0.104) \times (1.013\,25 \times 10^5/8.314 \times (273.15 + 20)) = 39.97$ mol/m^3·s ≈ 0.04 mol/l·s. The same process running in a 2000 L vessel with diameter 1.5 m at 1600 L batch size is limited to a rate of $0.1 \times (1/1)(1.013\,25 \times 10^5/8.314 \times (273.15 + 20)) = 4.16$ mol/m^3·s ≈ 0.004 mol/l·s, i.e. 10 times slower, to avoid batch swelling.

10.7 MODEL VERIFICATION STATISTICS

10.7.1 Parameter Uncertainty

All models or equations with parameters that are derived originally from experimental data have a degree of uncertainty associated with their calculations or predictions. This is true for calibration curves, regression lines, response surfaces generated by design of experiments, chemical engineering correlations for estimating equipment performance, models based on ordinary differential equations (such as many of the examples shown above), and those using partial differential equations (such as computational fluid dynamics, in which turbulence models, "wall laws," and other model parameters are ultimately based on experimental data). Unless these latter models involving differential equations are integrated over a sufficiently fine "grid," their uncertainty is further increased by lack of precision.

Once accepted as useful, models are often used without taking this uncertainty into account, but experienced practitioners will always allow a factor of safety to compensate for a margin of error, even if the size of that error is not known. When the original data and the model are both available, it is possible to use statistical methods to quantify the uncertainty level. This makes the potential deficiencies of the model more evident and explicit and may also focus further experimentation on reducing those uncertainties.

More details about how to calculate uncertainty are given elsewhere [44], and Figure 10.17 illustrates the typical situation for a linear model in which the slope and intercept have been fitted.

Confidence bands define an envelope within which there is a certain confidence level (typically 95%) of the true location of the best-fit line. Prediction bands (or intervals) define a wider envelope within which there is a certain confidence level (e.g. 95%) that all datapoints/observations will lie. The width of prediction bands relative to the model prediction indicates the likely relative error of the model predictions; this also reflects underlying variability or lack of reproducibility in the experimental data. For typical linear models such as that in Figure 10.17, the band widths are at a minimum at the center of the experimental data and increase in either direction from the center. This tallies with the belief that models are most applicable near the conditions where the experiments were run and become less reliable at more extreme conditions.

Any such model should in the first instance be based on reliable, reproducible data; there should also be sufficient data to avoid "overfitting," i.e. the number of observations should be significantly greater than the number of model parameters fitted. The degree to which such a model can be said to be verified depends on the width of its prediction bands compared with the accuracy needed in the intended application.

For example, if a model predicts an impurity level of 0.1%, with 95% prediction band widths of 0.05%, one could state with a high degree of confidence that the impurity level will be less than 1%; more formally, the probability of impurity exceeding 1% is almost zero, $p(\text{Impurity} > 1\%) \approx 0$. The same model is not sufficiently accurate to state with the same degree of confidence that the impurity level will be less than 0.2%, i.e. $p(\text{Impurity} > 0.2\%) > 0$. However if the 95% prediction band width was 0.01%, the model would be suitable for more confident predictions at lower impurity levels. There is therefore an element of fitness for purpose when judging whether a model has been verified sufficiently to apply it in a given situation.

Similarly, if the correlation used to calculate the stirrer speed required to suspend solids has a stated accuracy of 20% and the predicted $N_{JS} = 80$ rpm, this level of verification is sufficient to say that 40 rpm is too little – $p(\text{Suspended}) \approx 0$ – and 120 rpm is too much – $p(\text{Suspended}) \approx 1$, but not whether 75 rpm is sufficient – $0 < p(\text{Suspended}) < 1$.

Returning to Example Problem 10.4, confidence and prediction bands may be calculated for a dynamic model based on chemical engineering rate equations, and typical results are shown in Figures 10.18 and 10.19.

Comparing Figures 10.18 and 10.19, the relative width of prediction bands is greater for alkene than for product. This reflects greater uncertainty in the alkene measurements and predictions.

In QbD work, the criticality of factors is sometimes judged according to their proximity to factor settings that define the "edge of failure," e.g. crossing an impurity limit. Prediction bands should be taken into account when judging criticality in this way, as these will show that failure will occasionally occur at factor settings that are nearer to intended operating conditions.

10.7.2 Implications for Design Space Definition

The existence of uncertainty means that formal probability statements may be required to properly define a design space; while this implies some additional work, it has the benefit of quantifying the degree of assurance (or risk) that exists in relation to how the process will perform. In general, this approach will lead to design spaces that are smaller and more conservative than when uncertainty is ignored and that maximize the probability that the product quality will be in specification. Most popular statistical software packages at present do not take uncertainty into account in this way [45].

The relevant probability can be calculated in a variety of ways, and Figure 10.20 illustrates application to Example Problem 10.4. Recall from Figure 10.19 that the relative uncertainty of alkene is greater than that for product in this

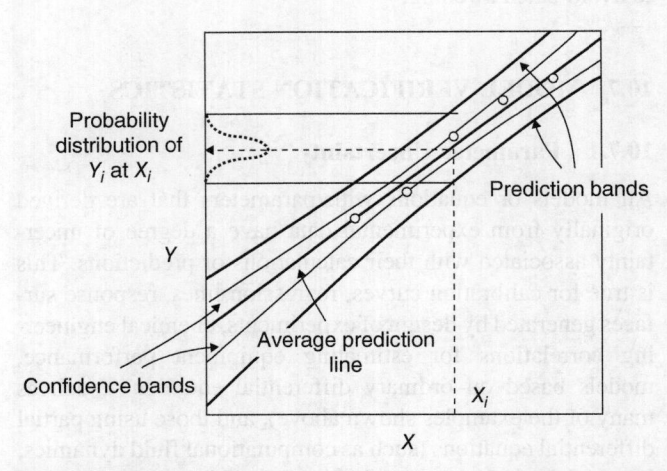

FIGURE 10.17 Schematic of confidence and prediction bands for a linear model in which the slope and intercept of the best fit line have been fitted to data (symbols).

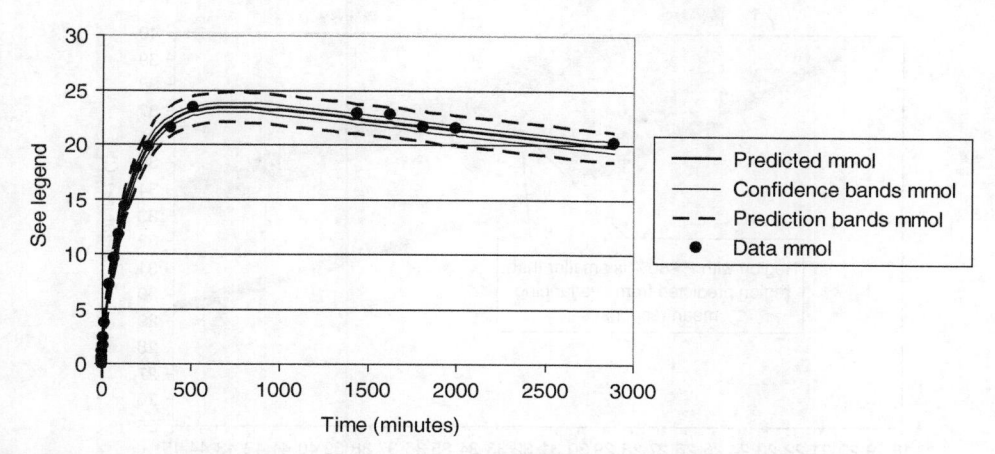

FIGURE 10.18 Model predictions with confidence and prediction bands/limits for the product profile in Example Problem 10.4, compared with experimental measurements of that profile. All measured data lie within the envelope defined by the 95% prediction bands.

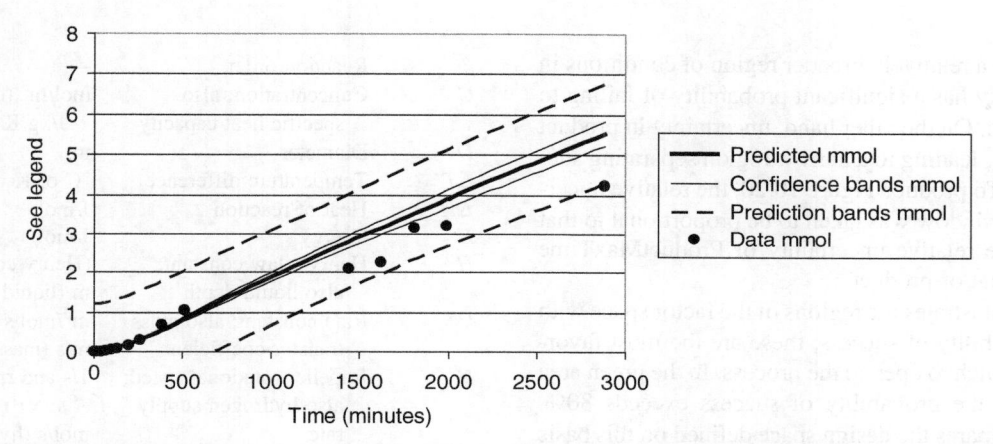

FIGURE 10.19 Model predictions with confidence and prediction bands/limits for the alkene profile in Example Problem 10.4, compared with experimental measurements of that profile. All measured data lie within the envelope defined by the 95% prediction bands.

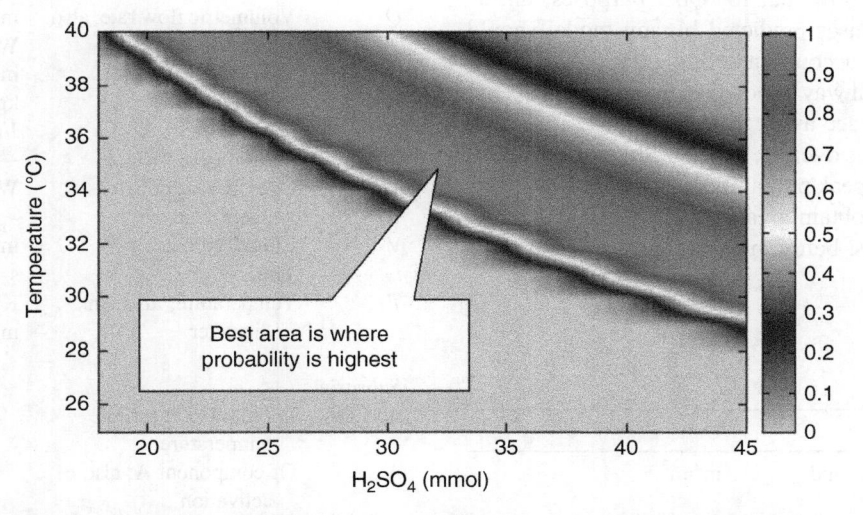

FIGURE 10.20 Response surface showing joint probability that responses ProductMaxTime and QuenchWindow will be in specification for Example Problem 10.4.

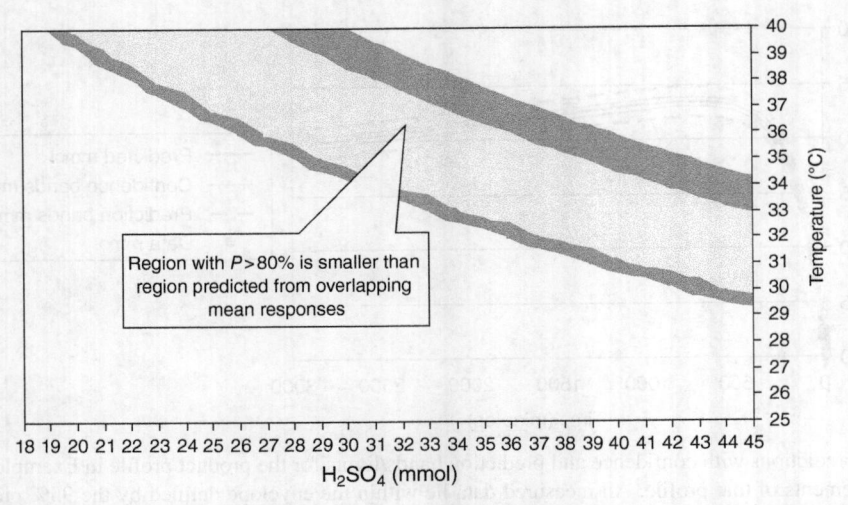

FIGURE 10.21 Comparison of design space for Example Problem 10.4 defined using average responses and that defined using probability of success greater than 80%.

Region with P>80% is smaller than region predicted from overlapping mean responses

case; this leads to a relatively broader region of conditions in which the impurity has a significant probability of failing to meet specification. On the other hand, uncertainty in product predictions is low, leading to a narrow region separating success and failure. To produce Figure 10.20, the relative uncertainty in quench window was taken to be proportional to that of alkene and the relative uncertainty of ProductMaxTime proportional to that of product.

Figure 10.20 illustrates the regions of the factor space with the highest probability of success; these are the most favorable regions in which to operate the process. In the green area of Figure 10.20, the probability of success exceeds 80%. Figure 10.21 compares the design space defined on this basis ($p > 80\%$) with that obtained by overlapping the average responses; as expected, taking account of uncertainty reduces the size of the design space.

The above results show that for QbD purposes, uncertainty in data or responses predicted by any model should be explicitly taken into account in defining the design space; probability is the natural way to do this. Uncertainty will tend to shrink the design space away from the edges of the area where average responses overlap, in line with good engineering practice. When the peak probability is far from 100%, this highlights the need to obtain greater process understanding, or an improved process, before proceeding.

10.8 NOTATIONS

Symbol	Meaning	Units
a	Area per unit liquid volume	m^2/m^3
α	Reaction order; also volume ratio	—
A	Area for heat transfer	m^2
β	Reaction order	—
C	Concentration; also specific heat capacity	mol/m^3 (concentration) J/kg K (heat capacity)
D	Diameter	m
ΔT	Temperature difference	°C or K
ΔH	Heat of reaction	J/mol
E	Energy	J/mol
H	Henry's law constant; also liquid depth	(Henry constant) m (liquid depth)
k	Rate constant; also mass transfer coefficient	$m^3/mol \cdot s$ (kinetics) m/s (mass transfer)
N	Impeller rotational speed; also hydrogen supply rate	1/s and rpm (impeller speed) mol/s (hydrogen supply rate)
ν	Stoichiometric coefficient	—
p	Pressure; also probability	Pa (pressure) -(probability)
Q	Volumetric flow rate; also heat flow rate	m^3/s (flow rate) W (heat flow rate)
r	Rate of reaction	$mol/m^3 s$
ρ	Density	kg/m^3
R	Gas constant	$J/mol \cdot K$
S	Partition coefficient	—
U	Overall heat transfer coefficient	$W/m^2 K$
V	Liquid volume	m^3
t	Time	s
T	Temperature; also tank diameter	K or °C (temperature) m (tank diameter)

Subscripts

0	Initial; also at infinite temperature	
A	Of component A; also of activation	
B	Of component B	
bulk	Of the bulk solution	
d	Of the dispersed phase	

H$_2$	Of hydrogen
f	Of the feed
fed	Of limiting fed reactant
gas	Of gas
head	Of the headspace
i	Of the ith component
ij	Of the ith component in the jth reaction
j	At the inlet to the cooling jacket or coil; also of the jth reaction
JS	Just suspended
JD	Just dispersed
L	In the liquid phase
LM	Logarithmic mean
m	Of mixing
max	Maximum
p	At constant pressure
r	Of reaction
solute	Of limiting dissolving solute
transfer	Of reaction

Superscripts

*	At saturation, i.e. equilibrium between the phases; also dimensionless

REFERENCES

1. ICH. (2012). International Conference on Harmonization (ICH) Q8, Q9, Q10 guidance documents. https://www.ich.org/products/guidelines/quality/article/quality-guidelines.html (accessed 24 October 2018).
2. Place, D. (2009). Using DynoChem to determine a suitable sampling endpoint for reaction analysis in a DoE. *DynoChem User Meeting*, Philadelphia (14–15 May 2009). https://dcresources.scale-up.com/?t=pu&id=223 (accessed 24 October 2018).
3. Hoffmann, W. (2015). Obtain reaction scheme proposal from chemical structures and HPLC data, training exercise published 1. https://dcresources.scale-up.com/?t=tr&id=5306&pid=17 (accessed 24 October 2018).
4. Weires, N.A., Caspi, D.D., and Garg, N.K. (2017). Kinetic modeling of the nickel-catalyzed esterification of amides. *ACS Catal.* 7 (7): 4381–4385.
5. Niemeier, J.K., Rothhaar, R.R., Vicenzi, J.T., and Werner, J.A. (2014). Application of kinetic modeling and competitive solvent hydrolysis in the development of a highly selective hydrolysis of a nitrile to an amide. *Org. Process Res. Dev.* 18 (3): 410–416.
6. Shujauddin, M. (2016). Changi and Sze-Wing Wong, kinetics model for designing Grignard reactions in batch or flow operations. *Org. Process Res. Dev.* 20 (2): 525–539.
7. Mandrelli, F., Buco, A., Piccioni, L. et al. (2017). The scale-up of continuous biphasic liquid/liquid reactions under superheating conditions: methodology and reactor design. *Green Chem.* 19: 1425–1430.
8. Ramirez, A., Hallow, D.M., Fenster, M.D.B. et al. (2016). Development of a control strategy for a final intermediate to enable impurities control. *Org. Process Res. Dev.* 20 (10): 1781–1791.
9. Vickery, T. (2007). Scale-up from RC1 and ARC safety tests using DynoChem. *DynoChem User Meeting*, Philadelphia (15–16 May 2007). https://dcresources.scale-up.com/?t=pu&id=264 (accessed 24 October 2018).
10. Bright, R., Dale, D.J., Dunn, P.J. et al. (2004). Identification of new catalysts to promote imidazolide couplings and optimisation of reaction conditions using kinetic modelling. *Org. Proc. Res. Dev.* 8 (6): 1054–1058.
11. Jorgensen, M. (2009). Modeling is the Easy Part!: getting the right data and getting the data right is the challenging part!. *DynoChem User Meeting*, Philadelphia (14–15 May 2009). https://dcresources.scale-up.com/?t=pu&id=212 (accessed 24 October 2018).
12. Eyley, S. (2009). Why study a synthetically useless reaction? Unravelling sulphonate ester formation using DynoChem. *DynoChem User Meeting*, Philadelphia (14–15 May 2009). https://dcresources.scale-up.com/?t=pu&id=212 (accessed 24 October 2018).
13. Niemeier, J. (2009). Using DynoChem to scale up data from various calorimeters. *DynoChem User Meeting*, Philadelphia (14–15 May 2009). https://dcresources.scale-up.com/?t=pu&id=217 (accessed 24 October 2018).
14. Nyrop, J. (2009). Development of a high performance, company specific DynoChem front-end. *DynoChem User Meeting*, Philadelphia (14–15 May 2009). https://dcresources.scale-up.com/?t=pu&id=218 (accessed 24 October 2018).
15. Hannon, J., Hearn, S., and Brechtelsbauer, C. (2002). Characterisation of the scalability of hydrogenation reactions. *Scientific Update Scale-up Conference*, St Helier, Jersey (23–26 September 2002). https://dcresources.scale-up.com/?t=pu&id=281 (accessed 24 October 2018).
16. Remy, B., Brueggemeier, S., Marchut, A. et al. (2008). Modeling-based approach towards on-scale implementation of a methanethiol-emitting reaction. *Org. Process Res. Dev.* 12: 381–391.
17. Richter, S. and Allian, A. (2008). Process safety testing and process modeling in the PSL using DynoChem. *3rd US Pharmaceutical Process Safety Forum*, Pearl River, New York (14 October 2008). https://dcresources.scale-up.com/?t=pu&id=235 (accessed 24 October 2018).
18. Hannon, J. (2008). Design space for a synthesis reaction, Part 3. http://blog.scale-up.com/2008/04/design-space-for-synthesis-reaction.html (accessed 24 October 2018).
19. Stonestreet, P., Hodnett, N., Squires, B., and Escott, R. (2009). Roles of mechanistic and empirical modeling/DOE in achieving Quality by Design. *DynoChem User Meeting*, Philadelphia (14–15 May 2009). https://dcresources.scale-up.com/?t=pu&id=216 (accessed 24 October 2018).

20. Baldyga, J., Bourne, J.R., and Hearn, S.J. (1997). Interaction between chemical reactions and mixing on various scales. *Chem. Eng. Sci.* 52 (4): 457–466.
21. Hoffmann, W. (2007). DynoChem and homogeneous mixing: an example. *DynoChem user Meeting*, Philadelphia (15–16 May 2007). https://dcresources.scale-up.com/?t=pu&id=266 (accessed 24 October 2018).
22. Hoffmann, W. (2001). Workshop on Basics of Kinetics/Application of Software Tools, Introduction to the determination of kinetic parameters. Mettler Toledo RXE Forum, Lucerne 2001.
23. Hallow, D., Mudryk, B., Braem, A. et al. (2009). Application of DynoChem® reaction modeling to quality by design. *DynoChem User Meeting*, Philadelphia (15–21 November 2008). https://aiche.confex.com/aiche/2008/techprogram/P135622.htm (accessed 24 October 2018).
24. Wertman, J. (2007). GSK approach to enhancing process understanding using DynoChem: reaction kinetics examples. *DynoChem User Meeting*, Philadelphia (15–16 May 2007). https://dcresources.scale-up.com/?t=pu&id=258 (accessed 24 October 2018).
25. Hannon, J. (2009). Quality by design for drug substance scale-up. *Presented at Scientific Update Conference on Scale-up of Chemical Processes*, Vancouver, Canada (7–10 July 2009). https://dcresources.scale-up.com/?t=pu&id=205 (accessed 24 October 2018).
26. Akaike, H. (1974). A new look at the statistical model identification. *IEEE Trans. Autom. Control* 19 (6): 716–723.
27. am Ende, D., Bronk, K.S., Mustakis, J. et al. (2007). API quality by design example from the torcetrapib manufacturing process. *J. Pharm. Innov.* 2: 71–86.
28. Levenspiel, O. (1999). *Chemical Reaction Engineering*, 3e. New York: Wiley.
29. Zhu, Z.M., Hannon, J., and Green, A. (1992). Use of high intensity gas-liquid mixers as reactors. *Chem. Eng. Sci.* 47: 2847–2852.
30. am Ende, D. (2009). Lean and green, the value of API process design. *DynoChem User Meeting*, Philadelphia (14–15 May 2009). https://dcresources.scale-up.com/?t=pu&id=220 (accessed 24 October 2018).
31. Roberge, D., Gottsponer, M., Eyholzer, M., and Kockmann, N. (2009). Industrial design, scale-up, and use of microreactors. *Chemistry Today* 27 (4): 8–11.
32. Ford, D. (2015). Using data-rich experimentation to enable the development of continuous processes. *Mettler Toledo Symposium*, Cambridge (20 May 2015). https://dcresources.scale-up.com/?t=pu&id=6063 (accessed 24 October 2018).
33. Chan, S.H., Wang, S.S.Y., and Kiang, S. (2005). Modeling and alternative reactor design for a highly exothermic reactive system. *AIChE Annual Meeting*, Cincinnati, OH (30 October to 4 November 2005).
34. Erdman, D. (2009). DynoChem modelling of 3 continuous stirred tank reactors. *DynoChem User Meeting*, Philadelphia (14–15 May 2009). https://dcresources.scale-up.com/?t=pu&id=221 (accessed 24 October 2018).
35. Kayode Coker, A. (2001). *Modeling of Chemical Kinetics and Reactor Design*. Houston, TX: Gulf Professional Publishing.
36. Machado, R. (1994). Fundamentals of mass transfer and kinetics for the hydrogenation of nitrobenzene to aniline. *Presented at 7th RC User Forum USA*, St Petersburg Beach (23–26 October 1994).
37. Lines, P.C. (2000). Gas-liquid mass transfer using surface-aeration in stirred vessels, with dual impellers. *Chem. Eng. Res. Des.* 78: 342–347.
38. Zwietering, T.N. (1958). Suspending of solid particles in liquid by agitators. *Chem. Eng. Sci.* 8: 244.
39. Lines, P.C. and Carpenter, K.J. (1990). The effect of physical property ranges on correlations for minimum impeller speeds to disperse immiscible liquid mixtures. *IChemE Symp. Series* 121: 167–182.
40. Paul, E.L., Atiemo-Obeng, V.A., and Kresta, S.M. (2004). *Handbook of Industrial Mixing*. Hoboken, NJ: Wiley.
41. Bourne, J.R. and Yu, S. (1994). Investigation of micromixing in stirred tank reactors using parallel reactions. *Ind. Eng. Chem. Res.* 33 (1): 41–55.
42. May, S.A., Johnson, M.D., Braden, T.M. et al. (2012). Rapid development and scale-up of a 1H-4-substituted imidazole intermediate enabled by chemistry in continuous plug flow reactors. *Org. Process Res. Dev.* 16: 982–1002.
43. Atherton, J.H. and Carpenter, K. (1999). *Process Development: Physiochemical Concepts*. Oxford: Oxford Science.
44. Box, G.E.P., Hunter, W.G., and Stuart Hunter, J. (1978). *Statistics for Experimenters: An Introduction to Design, Data Analysis, and Model Building*. Hoboken, NJ: Wiley.
45. Peterson, J.J. (2008). A Bayesian approach to the ICH Q8 definition of design space. *J. Biopharm. Stat.* 18 (5): 959–975.

11

SCALE-UP OF MASS TRANSFER-LIMITED REACTIONS: FUNDAMENTALS AND A CASE STUDY

SETH HUGGINS
Drug Substance Technologies and Engineering, Amgen Inc., Thousand Oaks, CA, USA

AYMAN ALLIAN
Synthetic Technologies and Engineering, Amgen Inc., Thousand Oaks, CA, USA

11.1 INTRODUCTION

In the pharmaceutical industry, chemical engineers are challenged to reduce the cost and time to market in drug development; this motivates seamless process scale-up – that is, translating the process, which has been designed and developed in the lab, to production scale, while maintaining the same process efficiency and product quality. This type of seamless scale-up, however, is not always easy to achieve and may require *a priori* understanding of when, and to what extent, an operation might be scale sensitive.

Indeed, unexpected results, such as the formation of new impurities and yield losses, during scale-up are frequent occurrence. The latter can introduce delays and often require significant resources to rectify. Such scale-up surprises are not uncommon even when biopharmaceutical companies use their internal pilot plants for scale-up. To make matters more challenging, in the last decade, low cost manufacturing around the globe particularly in emerging markets like India and China has pushed biopharmaceutical companies to externalize portions, if not all, of their scale-up operations around the globe. This explosion of outsourcing across pharmaceutical companies has compounded the challenge of scale-up due to the simple fact that the staff that developed the process in the lab and those in charge of scale-up are in different geographical locations, with probably different time zones. Consequently, when scale-up issues do arise in these scenarios, the cost can be even higher, as it often requires development staff to travel and oversee the scale-up operations in person, resulting in delays and high travel costs. In this current manufacturing paradigm, scientist and staff involved in scale-up and tech transfer simply cannot afford to fail at scale. This has given rise to heavy emphasis on the concept of right first time [1].

This chapter provides guidance for successful scale-up of reaction operations: first showing when these operations are likely to be scale sensitive – namely, when there is a strong mass transfer dependence – and then providing a brief review of the fundamentals of mass transfer to illuminate why scale sensitivities exist when mass transfer plays a large role in the overall system dynamics. At the end, an illustrative example of a mass transfer-limited reaction system is given.

11.1.1 When a Reaction Operation Is Likely Scale Sensitive

The root cause of scale sensitivity is transport phenomena, and not, generally, chemical transformation. Transport phenomena can take a different shape based on the system at hand as outline in the next examples. In processes where gas is required, transport refers to the adsorption of gas to liquid, such as the use of hydrogen to reduce chemical species or oxygen to enable cell growth. For the same system, gas–liquid, there are also operations where inertion of the bulk liquid medium is required to remove undesirable gases such as oxygen and carbon dioxide. For example, removal of oxygen may be required to prevent catalyst/chemistry poisoning.

Chemical Engineering in the Pharmaceutical Industry: Active Pharmaceutical Ingredients, Second Edition.
Edited by David J. am Ende and Mary T. am Ende.
© 2019 John Wiley & Sons, Inc. Published 2019 by John Wiley & Sons, Inc.

This transport phenomenon commonly referred as gas desorption studies the transport of dissolved gas into bubbles and headspace. In liquid–liquid mixing, transport refers to species transport between micelles or between immiscible phases – as in liquid extraction, a ubiquitous unit operation in pharmaceutical processing. Solid–liquid transport is also very critical to pharmaceutical operations describing the dissolution of suspended solids into solution and is often a dominant mechanism to the startup of a chemical. The transport of dissolved species to solid is the key mechanism in critically important operations like crystallization utilized for the separation and purification of active pharmaceutical ingredients (APIs).

These transport phenomena can be coupled with a reaction, which can complicate understanding the unit operation at hand. For example, in hydrogenation reactions, the hydrogen transport and reaction can take place simultaneously. Solid dissolution can also commonly be coupled with a reaction of the dissolved substance in the bulk liquid. For these various unit operations, the overall observed reaction rate can depend on a number of competing mass transfer and chemical kinetic rate processes that chemical engineers need to understand in order to evaluate these unit operations and how their performance will be impacted by scaling up and/or changing equipment. A common methodology for the assessment of scale sensitivity is carried out by evaluating the rate of mass transfer relative to the rate of chemical transformation using the classical Damköhler number:

$$\mathrm{Da} \equiv \frac{[\text{Reaction rate}]}{[\text{Mass transport rate}]} \quad (11.1)$$

In the case where the Damköhler number is high (e.g. Da > 100), the rate of mass transfer is slow compared with the rate of reaction. The overall observed rate is therefore dictated by the slow rate of mass transfer and very likely to be scale sensitive. On the other hand, when mass transfer is very fast and reaction rate is slower (e.g. Da < 0.01), the overall system dynamics are dictated by the reaction and likely to be scale insensitive. In other words, if the Damköhler number is low, the system dynamics observed at the lab scale are likely to be accurate as the reaction is scaled up, but if Damköhler number is high, scale effects should be anticipated.

11.1.2 Survey of Mass Transfer Challenges in the Pharmaceutical Industry

Understanding mass transfer early on in the process is critical in order to avoid costly delays during scale-up or tech transfer to pilot plants. Indeed, in a recent survey of 240 cases in which yield or quality was negatively impacted during scale-up, 30% of these failures were attributed to mass

TABLE 11.1 Breakdown of Reported Mass Transfer-Related Challenges in Scale-Up

Systems	Percentage
Solid–liquid	50
Gas–liquid	20
Liquid–liquid	20
Solid–gas–liquid	10

transfer [1]. If we look closer at these cases, Table 11.1, we also see that 50% of those mass transfer challenges are related to solid–liquid systems.

The fact that the vast majority of mass transfer challenges are related to solid–liquid systems is not surprising, as most APIs and their intermediates are typically generated in agitated batch reactors that involve solid starting materials, solid reagents, and/or solid catalysts mixed in a liquid solution. If any of the latter species is not completely soluble during the process, i.e. the mixture is heterogeneous, then mass transfer can play a role in overall observed reaction rates with a potential to introduce challenges to scale-up. While this chapter is focused on solid–liquid mass transfer, the reader should come away with an understanding that is general: mass transfer, and transport phenomena more broadly, depends on the shape, size, and configuration of the reactor and is therefore affected by scale.

11.2 MASS TRANSFER IN A SOLID–LIQUID SYSTEM WITHOUT REACTION

11.2.1 Background

As mentioned earlier, many pharmaceutical unit operations involve a solid phase dispersed in a liquid solution, as shown in Figure 11.1a. For simplicity, we assume the solids are spheres, and agitation is sufficient to ensure particle suspension. Progressively zooming in on the solid–liquid interface, we can highlight the key components that dictate mass transfer. Zooming in one level (Figure 11.1b), we see that the fluid inside the vessel is moving at a certain velocity; however, the fluid velocity directly in contact with the solid sphere is approaching zero due to drag forces. This difference in velocity is referred to as the slip velocity. As we move away from the sphere, drag forces are weakened, and fluid velocity will gradually increase until it reaches the bulk liquid velocity. The distance from the solid surface to the point where the fluid is 99% of the bulk velocity is a critical parameter in solid dissolution and is referred to as the boundary layer thickness, δ [2]. It is generally accepted that all resistance to mass transfer is within the boundary layer. As we will see, the boundary

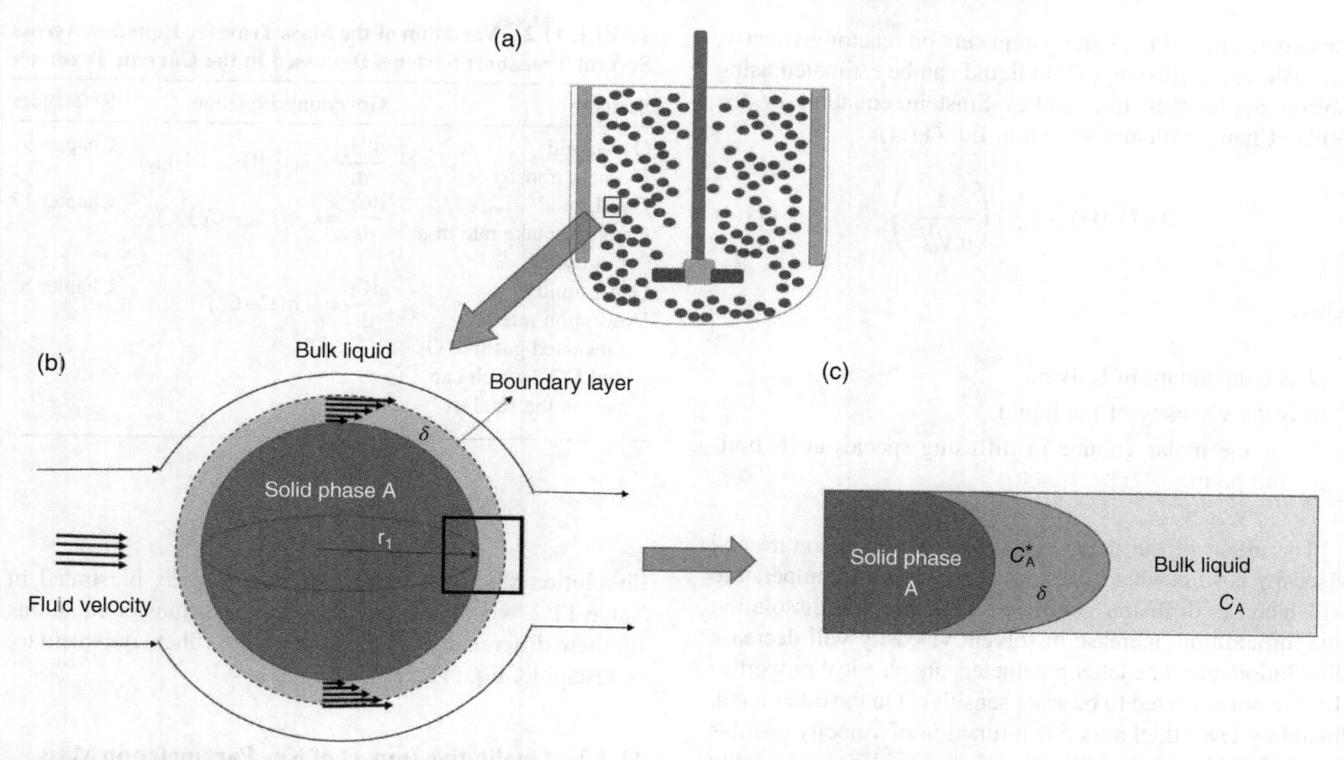

FIGURE 11.1 Illustration of transport in a solid-liquid system showing (a) solid phase, black spheres suspended in a agitated liquid solution, (b) a closer look at a suspended sphere showing the fluid velocity profiles around it and (c) a closer look at the solid-liquid interphase showing the diffusion boundary layer.

layer thickness is inversely proportional to agitation speed, and, thus, agitation speed can impact mass transfer rate. Zooming in one more level, Figure 11.1c, we illustrate the second key component of mass transfer: a concentration gradient. In our illustration, the solid is assumed to be made up of pure species A, while the solution is a mixture that contains A. In this scenario, the concentration of A is high on the surface of solid (and can generally be assumed to be saturation concentration C_A^*), while the concentration of A is lower in the bulk liquid (at least initially). Consequently, there is a concentration gradient that will drive transport in the direction of decreasing concentration through the boundary layer. Eventually, the concentration in the liquid bulk C_A will reach saturation, i.e. solubility limit, and the diffusion will cease.

Rate of transport in solid–liquid transport is proportional to the concentration gradient and inversely proportional to the boundary layer thickness. Taking the latter parameters, one can calculate the time-dependent concentration profile of solid-phase A using the classical chemical engineering equation, also known as the Noyes–Whitney equation, below:

$$\frac{dC_A}{dt} = k_L a (C_A^* - C_A) \qquad (11.2)$$

Rate of mass transfer across the boundary layers $\frac{dC_A}{dt}$	=	Term 1: k_L mass transfer coefficient	Term 2: a interfacial area	Term 3: $(C_A^* - C_A)$ gradient between the bulk phase concentration C_A and the concentration at the boundary layer C_A^*

Term 1 Mass Transfer Coefficient k_L: It is worth mentioning that Eq. (11.2) can be derived from Fick's law, which consists the boundary layer thickness δ whose magnitude is unknown. Therefore, and for convenience, mass transfer coefficient is normally used to replace the diffusivity coefficient D and the boundary layer δ as shown in Eq. (11.3):

$$k_L = \frac{D}{\delta} \qquad (11.3)$$

where D is the solute diffusivity in the bulk liquid in liquid solution across the boundary layer δ. Next, we will assess the dependence of the latter parameters on reactor geometry, i.e. scale-up. Diffusivity, D, in liquid can be estimated using several approaches like Stokes–Einstein equation or the Wilke–Chang estimates shown in Eq. (11.4):

$$D = (1.05)(10^{-9})\left(\frac{T}{\mu V_m^{1/3}}\right) \quad (11.4)$$

where

T is temperature in Kelvin.
μ is the viscosity of the liquid.
V_m is the molar volume of diffusing species at its boiling point.

The effect of the latter parameters like temperature and viscosity is somewhat expected, i.e. increase in temperature will increase diffusion coefficient and thus the dissolution rate. In addition, increase in solvent viscosity will decrease dissolution rate. The latter parameters are physical properties that are not expected to be scale sensitive. On the other hand, boundary layer thickness δ is a function of velocity profiles within the agitated vessel and therefore very dependent on the reactor geometrical parameters. This includes, but is not limited to, impeller type, size, and speed. Furthermore, baffling and geometry of the system can also impact δ. Changes in these parameters, due to scale or change in equipment, can result in significant changes in δ and thus the overall observed rate that explains the large number of cases where surprises are encountered during scale-up.

Term 2 Interfacial Area a: The rate of the mass transfer is directly impacted by the interfacial area per unit volume, a, as shown in Eq. (11.2). An increase in the area will directly increase the overall mass transfer rates. Interfacial area per unit volume of the same solid increases by decreasing particle size that can directly increase the overall mass transfer rate, $k_L a$, assuming k_L remains constant.

Term 3 Concentration Gradient $(C_A^* - C_A)$: The concentration gradient across the boundary layer, Figure 11.1c, is the driving force for the transport. Typically, the concentration at the surface is high, and diffusion takes place until the concentration in the bulk phase reaches saturation, C_A^*.

While the focus has been on solid–liquid mass transfer, the author wants to point out that the dynamics of mass transfer in other multiphase systems follow similar equations. The concepts discussed here can be, in a general sense, extrapolated to other applications such as hydrogenation, gas solubility in a bioreactor, oxygen and CO_2 desorption, and even drug dissolution studies in aqueous phase. This is illustrated in Table 11.2, where we show that the mass transfer equations for these different two-phase systems are either equivalent to, or resemble, Eq. (11.2).

TABLE 11.2 Variation of the Mass Transfer Equation Across Several Transport Systems Discussed in the Current Textbook

Scenario	Governing Equation	References
Gas–liquid H_2 mass transfer	$\dfrac{d[H_2]}{dt} = k_L a([H_2] - [H_2]_{sat})$	Chapter 9
Gas–Liquid Oxygen uptake rate in a bioreactor	$\dfrac{dC_L}{dt} = k_L a(C_{sat} - C_L)$	Chapter 17
Gas–Liquid Desorption rate of unwanted gas like O_2 and CO_2, which can poison the catalyst	$\dfrac{dC_L}{dt} = k_L a(C - C^*)$	Chapter 8

11.2.2 Qualitative Impact of Key Parameter on Mass Transfer Coefficient k_L

There are several geometrical parameters that can impact mass transfer coefficient k_L and interfacial area a. In this section, more light is shed on the degree of the impact of two key parameters on the overall mass transfer coefficient, $k_L a$, namely, particle size and agitation rate.

11.2.2.1 Effect of Particle Size on the Overall Mass Transfer Coefficient Particle size has a significant effect on the interfacial area a. In this section we want to highlight the impact of particle size on k_L. In contrast to the effect on interfacial area, decreasing the particle size interestingly has a negative effect on k_L. However, the effect on k_L is very minimal. A study showed that reducing particle size from 10 000 to 40 mm decreased k_L by 33% and this reduction is expected to enormously increase the area interfacial area a [3].

11.2.2.2 Agitation Speed and State of Solid Suspension Impact on the Overall Mass Transfer Coefficient Understating the state of solid suspension in your vessel is very critical to mass transfer. In an agitated vessel, the degree of solid suspension is classified into three states, namely, partial suspension, complete suspension, and uniform suspension, as shown in Figure 11.2 [4]. In the partial suspension, due to no or little stirring, much of the solid is settled at the bottom of reactor. The surface area of the settled particles will not be available, as they are obscured by other particles; therefore, interfacial area in this scenario is low. Once sufficient agitation is reached, we move

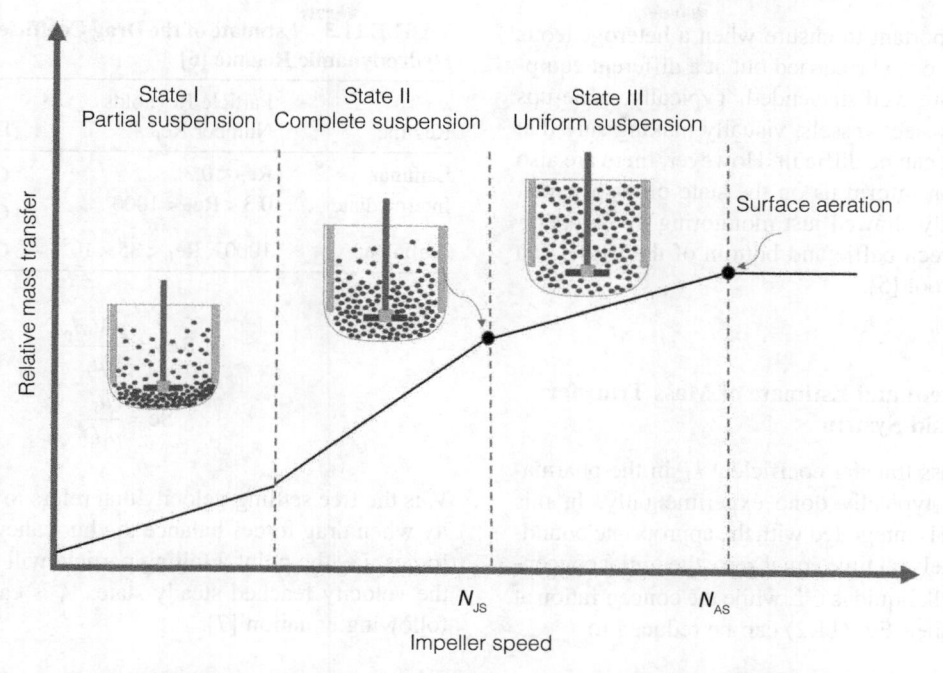

FIGURE 11.2 Mass transfer coefficient over a wide range of stirring speed.

into State II where solids begin to suspend until complete suspension is realized and no particles remaining at the base of the agitated reactor for more than one to two seconds. At this stage, the entire surface area of the particles is exposed for mass transfer or reaction. The minimum impeller speed required to achieve full suspension is termed the just-suspended speed and is denoted N_{js}, as shown Figure 11.2. Uniform suspension occurs when agitation is high enough beyond N_{js} and at a state where particle size distribution is uniform throughout the agitated vessel (Figure 11.2).

In the partial suspension (State I), agitation impact is limited to rolling particles at the bottom of the reactor. As we enter into State II, increasing impeller speed has significant positive effect on the relative mass transfer. Specifically, (i) increasing impeller speed will increase the velocity profile in the agitated vessel and consequently decrease the boundary layer thickness, and therefore increase mass transfer, and (ii) increase the portion of the solid particle that are suspended, which will increase the effective interfacial area, and again increase the mass transfer. Indeed, in this region, the overall transfer coefficient $k_L a$ will rapidly increase with increasing agitation until reaction complete suspension state. However, increasing impeller speed beyond N_{js} (State III) improves the homogeneity of the suspension but does not greatly improve the mass transfer characteristics. It is worth noting that increasing agitation speed any further, beyond State III, can bring about surface aeration, N_{SA}, where ingested air due to high agitation blanket the particles surface.

It is clear from the upper discussion that for solid–liquid mass transfer, it is very desirable to operate at an impeller speed $\geq N_{js}$, which also means that chemical engineers need to understand and estimate N_{js} for their vessel wherein scale-up will take place and ultimately carry out the development in appropriate scale-down models. In practice, however, this is easier said than done, as the development teams are typically not informed of the site where the process will be scaled until late in development. To estimate N_{js}, a number of correlations are available. The most widely used correlations for N_{js} is the Zwietering correlation shown below:

$$N_{js} = S \left(\frac{\mu_l}{\rho_l}\right)^{0.1} \left[\frac{g_c(\rho_s - \rho_l)}{\rho_l}\right]^{0.45} X^{0.13} d_p^{0.2} D_{imp}^{-0.85} \quad (11.5)$$

where

S is a dimensionless number that is a function of the impeller type.
μ_l is liquid viscosity.
ρ_s is the solid density.
ρ_l is bulk liquid density.
g_c is the gravitation constant (9.81 m/s^2).
D_{imp} is the impeller diameter.
X is the mass ratio of suspended solids to liquid × 100.

It is critically important to ensure when a heterogeneous reaction is scaled up or to be carried out at a different equipment that particles are well suspended. Typically, scale-ups are done in stainless steel vessels, visually making sure that solids are suspended can be difficult. However, there are also several tools that can inform us on the state of suspension. Indeed, a recent study showed just monitoring the temperature difference between baffle and bottom of the vessel can be a very effective tool [5].

11.2.3 Measurement and Estimate of Mass Transfer k_L for a Solid–Liquid System

Measurement of mass transfer coefficient k_L in the pharmaceutical industry is typically done experimentally. In this approach, Eq. (11.2) is integrated with the appropriate boundary conditions. Namely, at time equal zero, the initial concentration of A in the bulk liquid is C_A^0, while the concentration at time (t) is C_A, and then Eq. (11.2) can be reduced to

$$\ln\left[\frac{(C_A^* - C_A^0)}{(C_A^* - C_A)}\right] = k_L a \cdot t \qquad (11.6)$$

Equation (11.6) resembles a straight line, $y = mx$; therefore, the common practice is to collect concentration profile date over time and then plot the experimental value on the left hand of equation of 1.4 versus time. The results will follow a straight line with a slope of $k_L a$. The latter equation can be used when the bulk solute concentration can be easily measured, which is the case when solids are dissolved in liquid, or when the oxygen concentration can be measured using oxygen sensors (see Chapter 9).

In case experimental methods are not available for solid–liquid, the mass transfer coefficient k_L can be estimated using available correlations. However, the reader is warned that these correlations do not take into account the effect of geometry, number and shape of impellers, and location and their impact on mass transfer coefficient and thus can produce uncertainty in the outcome. Despite the earlier warning on using correlation, the widely accepted correlation referred to as Froessling equation (Eq. 11.7) will be discussed. The latter equation has proven useful for estimating k_L for a wide range of configuration and bulk flow regimes, i.e. from laminar and turbulent flow [6]:

$$k_L = \frac{D}{d_p}\left[2 + 0.44 \mathrm{Re}_P^{1/2} \mathrm{Sc}^{1/3}\right] \qquad (11.7)$$

where Re_P is the particle Reynolds number and Sc is Schmidt number, which are defined below:

TABLE 11.3 Estimate of the Drag Coefficient C_D Based on the Hydrodynamic Regime [6]

Regime	Particle Reynolds Number Re_P	Drag Coefficient C_D
Laminar	$\mathrm{Re}_P < 0.3$	$C_D = 24/\mathrm{Re}_P$
Intermediate	$0.3 < \mathrm{Re}_P < 1000$	$C_D = 18.5/\mathrm{Re}_P^{3/5}$
Turbulent	$1000 < \mathrm{Re}_P < 35 \times 10^4$	$C_D = 0.445$

$$\mathrm{Re}_P = \frac{\rho_1 V_t d_p}{\mu_1} \qquad (11.8)$$

$$\mathrm{Sc} = \frac{\mu_1}{\rho_1 D} \qquad (11.9)$$

V_t is the free settling velocity that refers to the particle velocity when drag forces balance the buoyancy and gravitational forces, i.e. the point a falling particle will not accelerate and the velocity reached steady state. V_t is calculated using the following equation [7]:

$$V_t = \left(\frac{4 g_c d_p (\rho_s - \rho_1)}{3 C_D \rho_1}\right)^{1/2} \qquad (11.10)$$

C_D is the drag coefficient that can be estimated based on the hydrodynamic regimes, namely, laminar, intermediate, and turbulent, as shown in Table 11.3.

11.3 MASS TRANSFER WITH CHEMICAL REACTION

In the previous section, we have discussed the mass transfer in the absence of reaction. In this section, we will introduce the complexity of chemical reaction. In Figure 11.1, we assumed dissolution of a solid-phase A that is dispersed in a liquid solution. Now, let us assume bulk solution contain B at a certain concentration, C_B, which reacts with A to form product P as shown in the Figure 11.3.

The overall reaction kinetics in the bulk phase are given by the following:

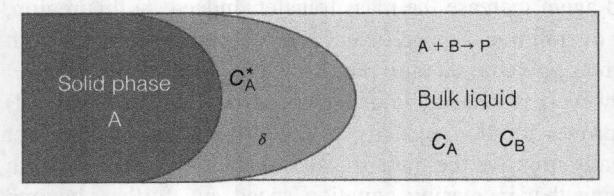

FIGURE 11.3 Mass transfer and reaction in a solid–liquid system.

$$\text{Rate} = \frac{dC_A}{dt} = -k_r[C_A][C_B] \quad (11.11)$$

where

- k_r is the reaction rate constant.
- C_A and C_B are the concentration of A and B in the bulk solution, respectively.

The reaction rate constant can be described by Arrhenius expression:

$$k_r = k_0 \exp\left(-\frac{E_a}{RT}\right) \quad (11.12)$$

where k_0, E_a, R, T are the Arrhenius pre-exponential factor, activation energy, universal gas constant, and temperature in Kelvin, respectively.

An important point to note here is that none of the parameters in the equations describing reaction kinetics are related to reactor geometry. This is in contrast to what was observed earlier, for mass transfer dynamics, and the key reason why the Damköhler number can be used as metric relating the likelihood of scale sensitivity. We can see this, with greater clarity, coupling the reaction kinetic and mass transfer equations.

If we couple reaction kinetic in Eq. (11.11) with mass transfer in Eq. (11.2), we come to the following equation for the *overall* dynamics:

$$\frac{dC_A}{dt} = k_L a\left(C_A^* - C_A\right) - k_r C_A C_B \quad (11.13)$$

If we assume that B concentration is very high, we can assume its concentration is almost constant through the reaction we can replace $k_R = k_r C_B$, and Eq. (11.13) is further simplified to

$$\frac{dC_A}{dt} = k_L a\left(C_A^* - C_A\right) - k_R C_A \quad (11.14)$$

At steady state $dC_A/dt = 0$, which means the transport rate equals the reaction rate:

$$0 = k_L a\left(C_A^*\right) - k_L a(C_A) - k_R(C_A) \quad (11.15)$$

Then, rearrange Eq. (11.15) and solve for C_A:

$$C_A(k_L a + k_R) = k_L a\left(C_A^*\right) \quad (11.16)$$

$$C_A = \frac{k_L a\left(C_A^*\right)}{(k_L a + k_R)} \quad (11.17)$$

Then, substituted Eq. (11.17) from steady-state approximation in Eq. (11.11), the following classical equation is obtained:

$$\frac{dC_A}{dt} = k_R C_A = k_R \frac{k_L a\left(C_A^*\right)}{(k_L a + k_R)} \quad (11.18)$$

$$\frac{dC_A}{dt} = k_R \frac{k_L a\left(C_A^*\right)}{(k_L a + k_R)} = k_{obs}\left(C_A^*\right) \quad (11.19)$$

where k_{obs} is the overall observable rate

$$k_{obs} = \frac{k_R k_L a}{(k_L a + k_R)} \quad (11.20)$$

Next, we will use Eq. (11.18) to understand the governing dynamics when mass transfer is coupled with a chemical reaction.

11.3.1 Very Fast Chemical Reaction: Regime 1

In the case where chemical reaction is very fast compared with mass transfer, i.e. $k_R \gg k_L$, the denominator in the upper equation $(k_R + k_L a) \cong k_R$,

$$\text{Rate} = k_R \frac{k_L a\left(C_A^*\right)}{(k_R)} \quad (11.21)$$

$$\text{Rate} = k_L a\left(C_A^*\right) \quad (11.22)$$

The upper equation shows that for mass transfer-limited step coupled with a very fast chemical reaction, the overall observed rate is constant. A plot of concentration versus time will produce a straight line with a slope of $k_l a\left(C_A^*\right)$. The latter correlation will be utilized in the case study (Section 11.4). In regime 1, the Damköhler number is high (Da > 100), and as discussed earlier, the observed rate is expected to be scale sensitive and dominated by mass transfer dynamics. Actually, Eq. (11.22) is strikingly similar to Eq. (11.2) with the exception that the solute concentration in the bulk liquid is assumed to be negligible, $C_A \cong 0$, which is expected under the current assumption of fast reaction. In other words, solute A never accumulates in the bulk and reacts as soon as it diffuses through the boundary layer. The overall system dynamics are dictated by the mass transfer dynamics, which are, in turn, influenced by the parameters that change with scale. Specifically, this is the scenario where the overall observed dynamics will be most sensitive to agitation, reaction geometry, and particle size. Therefore, special attention should be given to the particle size and to assurance that the stir speed is at or above the just-suspended speed, see Section 11.2.2.2.

11.3.2 Slow Chemical Reaction: Regime 2

In case where reaction rate is much slower than mass transfer $k_R \ll k_L a$, Eq. (11.18) can be reduced to

$$\text{Rate} = k_R \left(C_A^* \right) \quad (11.23)$$

The overall observed rate is predicted to be linear. A plot of overall observed rate will generate a straight line with a slope of $k_R \left(C_A^* \right)$. In regime 2, chemical reaction controls the overall observed system dynamics. The term $k_L a$ does not even appear on observed rate. In regime 2 Damköhler number is low. In this regime, increase in the agitation rate will not produce increase in the overall observed reaction rate. Indeed, for extremely slow reaction even complete suspension of the particles might not be necessary. Only gentle turnover for solids to prevent stagnant pockets inside the reaction is all what is needed. Reaction in this regime is not very sensitive to scale and tends to follow first principle reaction kinetics. In this regime with reaction kinetic control, the reaction is expected to be dependent on temperature based on Arrhenius equation (11.12) but not sensitive to scale assuming concentration and reagent equivalence were kept constant.

11.3.3 Mass Transfer and Moderate Chemical Reaction: Regime 3

In the case where mass transfer and chemical reaction are occurring at a comparable rate, Eq. (11.18) cannot be further simplified as both terms contribute to the overall dynamics. In other words, one cannot decouple the rate of reaction from rate of diffusion.

11.3.3.1 Effect of Particle Size and Agitation Speed on the Overall Observed Rate in Regime 3
The effect of certain parameters such as agitation, particle size, and temperature is harder to predict in regime 3 when compared with regimes 1 and 2. However, we can look at the equation to understand the impact of several of the latter variables in these scenarios. Equation (11.18) shows that increase in the interfacial area, by reducing particle size, will increase the overall rate of reaction especially when mass transfer is contributing to the overall observed reaction. However, the increase will not be indefinite. In other words, at some point the particle size will be so small that mass transfer is much faster and reaction dynamics dominate, which is not a function of particle size, and thus the overall observed rate will plateau as the system moves from regime 1 to regime 2. Indeed, the latter conclusion is in agreement with experimental data, as shown in Figure 11.4a [8], where reducing particle size increased the overall rate. However, further reduction beyond approximately 2 mm did not improve the overall observed rate. In regime 2, particle size has a very little influence on the rate of reaction in the bulk.

In this regime, the impact of impeller speed is similar to that of particle size discussed above where increasing stirring speed is expected to increase the overall rate by increasing the fluid velocity and subsequently reducing the boundary layer film thickness; see Figure 11.3. Again, as with agitation, the increase will not be infinite according to Eq. (11.20) and at some point the reactions will not be mass transfer limited and dominated by reaction dynamics that is not a strong function of agitation rate; therefore, any further increase is predicted not to impact the overall observed rate. Indeed, as

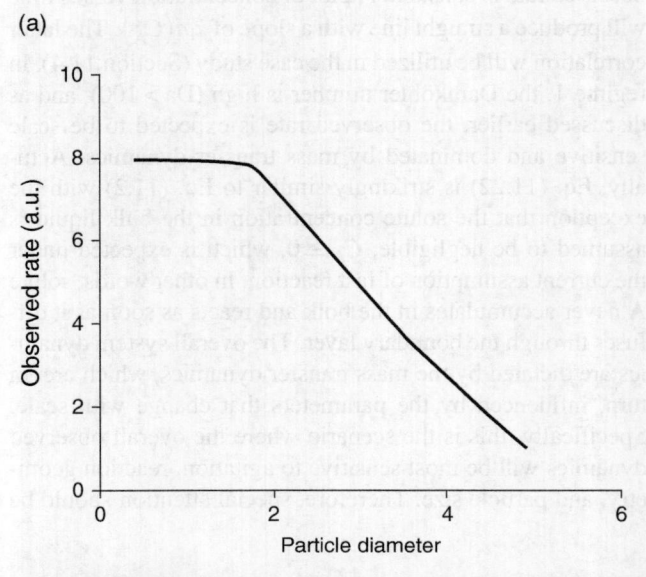

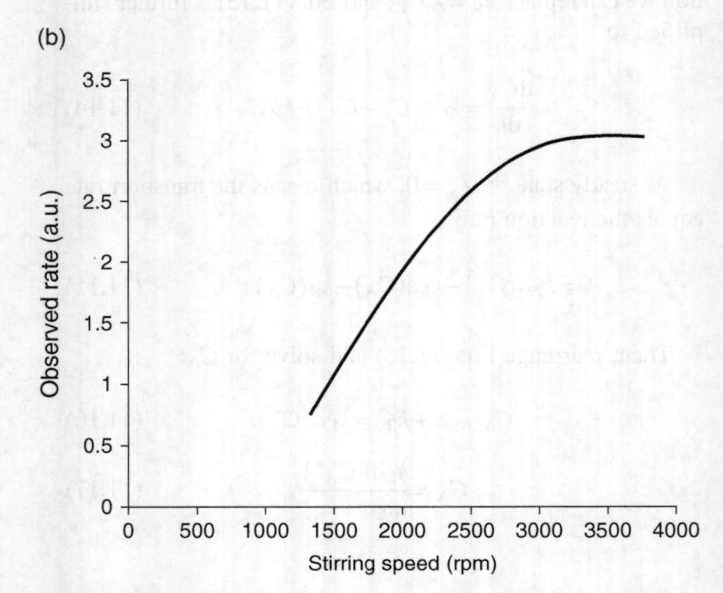

FIGURE 11.4 Effect of (a) particle size and (b) agitation speed size on the overall observed rate. *Source*: Adapted from Ref. [8].

shown in Figure 11.4b, experimental data support the latter hypothesis. In the latter experiment, stirring speed increased the overall rate; however, at stirring speeds greater than 3600 rpm, the overall rates measured were essentially free of mass transfer limitations and independent of stirring speed.

11.3.3.2 Effect of Temperature on the Overall Observed Rate: Moving from Regime 2 to Regime 1
The temperature dependency of k_R has been discussed using Arrhenius; see Eq. (11.12). Mass transfer coefficient k_L can be explained by the fact that it is directly related to the solute diffusivity, D, Eq. (11.3). D can also be described by the Arrhenius equation as shown below:

$$D = D_0 \exp\left(-\frac{E_D}{RT}\right) \quad (11.24)$$

where D_0, E_D, R, T are the Arrhenius pre-exponential factor, activation energy of diffusion, universal gas constant, and temperature in Kelvin, respectively.

If one plots the logarithmic of the observable rate versus temperature, one would obtain a straight line whose slope is the reaction activation energy of the reaction, E_a, in regime 2, i.e. free of mass transfer limitation. On the other hand, in regime 1 where mass transfer dominates, the slope will be activation energy of diffusion, E_D. However, the effect of temperature is typically modest, compared with reaction dependence. This week influence of temperature translates to a low/lower activation energy, E_D. Activation energies for mass transfer limited are 10–20 kJ/mol compared with 40–60 kJ/mol for reaction processes. Therefore, an overall reaction can be slow, regime 2; however, with increasing temperature, the rate of reaction will accelerate, and eventually that reaction rate will be very fast and will be regime 1, i.e. mass transfer limited in which diffusion control, as shown in Figure 11.5.

11.4 CASE STUDY: SCALING OF A MASS TRANSFER-LIMITED REACTION

An early phase, API intermediate was planned to be run at a 250 L scale. The desired diaryl ether was formed via a $S_N Ar$ reaction between the phenol (B) and fluorobenzonitrile (D) starting materials that were used in stoichiometric proportions with an excess of potassium carbonate base (A). The reaction was run in DMSO, where the starting materials were fully soluble and the potassium carbonate had sparing solubility. This heterogeneous reaction, depicted in Figure 11.6, had been developed on the bench scale to achieve yields of greater than or equal to 99% with a high purity profile with a reaction time of less than 12 hours.

Upon transfer of the process to the kilo scale facilities, the reaction was observed to take on the order of 4–5 times longer to achieve completion, prompting a root cause investigation. As discussed earlier, scale sensitivities are typically not associated with the chemical transformation, but the transport processes. In this case the chemical transformations were maintained but at a slower rate; there were no observations of an increase in the extent of competitive side reactions and no observable decomposition of the starting materials. In addition, the reaction could be accelerated to completion by additional charges of potassium carbonate. Further inquiry revealed that the grade of potassium carbonate was different than that used during development. The grade for the kilo-scale run was purchased because it was more readily available and was a higher purity. However, upon inspection the particle size was noted to be larger, as seen in Figure 11.7.

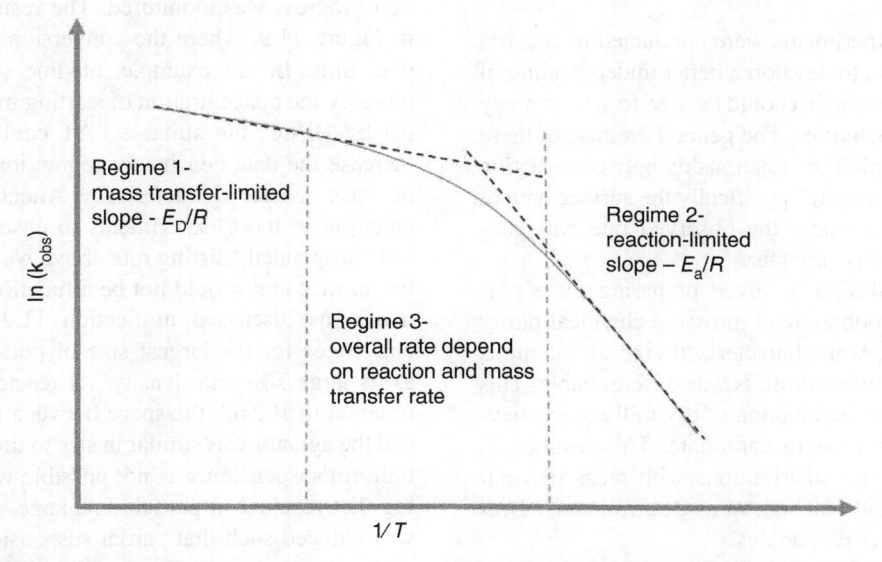

FIGURE 11.5 Effect of temperature on the overall observed rate. *Source*: Adapted from Ref. [2].

FIGURE 11.6 Heterogeneous coupling reaction to form the diaryl ether, API intermediate.

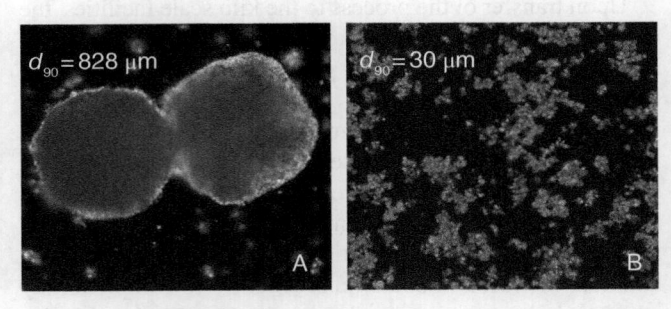

FIGURE 11.7 Polarized microscope images of potassium carbonate. Reaction completion was observed at 77 hours for material A and 12 hours for material B.

As discussed in Section 11.2, mass transfer dynamics is a very strong function of the interfacial area and thus the observed scale sensitivity pointed to mass transfer being the rate-limiting step.

A short series of experiments were conducted to confirm the suspected cause and to develop a better understanding of the mass transfer limitation that could be used to set a strategy for future production activities. The general premise of these experiments was to exploit the relationship between reaction rate and particle size, or more specifically the surface area of potassium carbonate, to show the observed rate was controlled by the rate of mass transfer.

The experimental design involved preparing three portions of potassium carbonate with the same chemical purity but each with unique, well-characterized size distributions such that the total area per unit mass was differentiable. This was accomplished by a combination of dry milling and sieving from one lot of potassium carbonate. This resulted in three distinct particle size distributions with areas per unit mass of 4902, 2387, and 1201 cm^2/g as determined by Brunauer–Emmett–Teller (BET) analysis.

In instances where BET is not available, the surface area may sometimes be approximated from the particle size distribution through calculation within the particle size analyzer software using the appropriate moments of the distribution. Provided the particles are not of an irregular shape or highly aggregated, the estimates from the particle size distribution can be used in the mass transfer calculations. As seen within the table of Figure 11.8, this system has good agreement for the materials with the large and medium specific surface areas. However, the larger particles exhibited both agglomeration and irregular characteristics, leading to a deviation from the actual specific surface area and the estimate from the particle size distribution.

The coupling reactions were performed in parallel, where all conditions were kept equivalent with the exception of the particle size of potassium carbonate. Using three different particle size distributions of potassium carbonate, the reaction progress was monitored. The results of this are shown in Figure 11.9, where the conversion to product is plotted over time. In this example, off-line samples were used to quantify the concentration of starting materials and the product by HPLC, but suitable PAT could have been used to increase the data density for regression of the kinetics and/or mass transfer characteristics. Attention was given to the agitation in these experiments to ensure the particles were fully suspended, stirring rate above N_{js}, such that the available surface area would not be a function of agitation performance as discussed in Section 11.2.2.2. This presented challenges for the largest size of potassium carbonate due to its large size and density. At reactor scales of less than or equal to 100 ml, the space between the reactor inner wall and the agitator was similar in size to the largest particle; thus uniform suspension was not possible without vigorous mixing that resulted in particle breakage. As a result, agitation was reduced such that partial suspension was achieved and

SCALE-UP OF MASS TRANSFER-LIMITED REACTIONS: FUNDAMENTALS AND A CASE STUDY 237

Particle size distribution: statistics, moments, and calculations

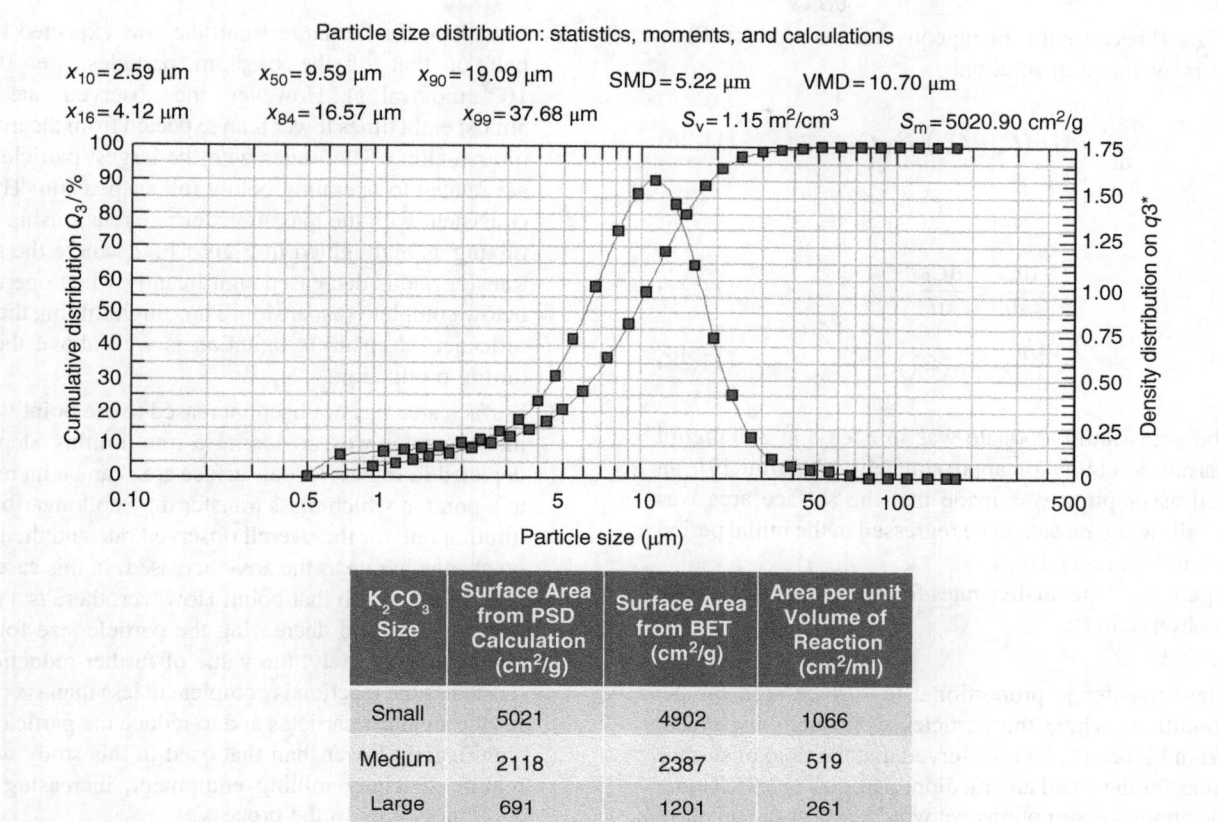

$x_{10} = 2.59\ \mu m$ $x_{50} = 9.59\ \mu m$ $x_{90} = 19.09\ \mu m$ $SMD = 5.22\ \mu m$ $VMD = 10.70\ \mu m$

$x_{16} = 4.12\ \mu m$ $x_{84} = 16.57\ \mu m$ $x_{99} = 37.68\ \mu m$ $S_v = 1.15\ m^2/cm^3$ $S_m = 5020.90\ cm^2/g$

K_2CO_3 Size	Surface Area from PSD Calculation (cm^2/g)	Surface Area from BET (cm^2/g)	Area per unit Volume of Reaction (cm^2/ml)
Small	5021	4902	1066
Medium	2118	2387	519
Large	691	1201	261

FIGURE 11.8 Particle size distribution for the smallest potassium carbonate and comparison of surface areas of the three portions using the PSD calculations and from BET measurement.

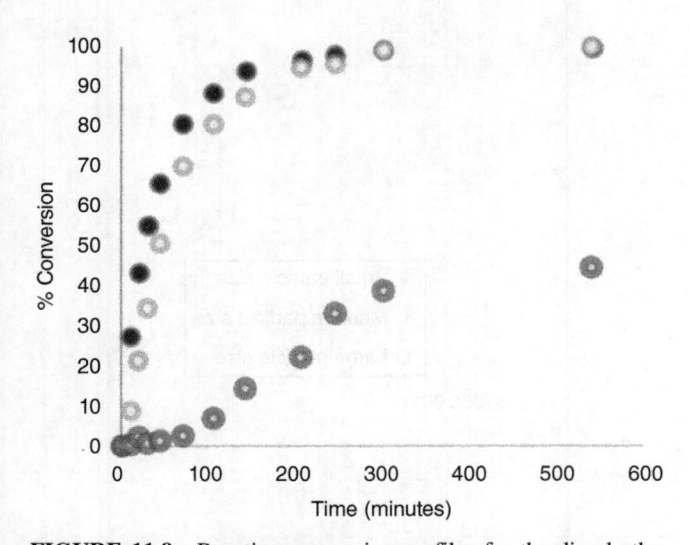

FIGURE 11.9 Reaction conversion profiles for the diaryl ether formation using potassium carbonate with three different particle size distributions. Conversion on the y-axis is equal to the moles of product at the sampling point divided by the moles of limiting substrate multiplied by 100.

onset of particle breakage was delayed. An inflection in the reaction profile was observed after approximately two hours as a result of particle breakage, which was conferment via microscopy.

This case study resembles the scenario described in Figure 11.1 where K_2CO_3 is (A) suspended in an agitated vessel. Starting from general Eq. (11.13) to describe the role of K_2CO_3,

$$\frac{dC_A}{dt} = k_{L1}a(C_A^* - C_A) - k_{R2}C_A C_B \qquad (11.25)$$

Based on the magnitude that conversion was impacted by the particle size of the potassium carbonate, it was decided to start the regression from assumption that the mass transfer rate was the rate-limiting step for the reaction. More specifically the rate of reactions (R2 and R3) was significantly faster than the rate of mass transfer (R1). Initially, while substrates are in high concentration, the reaction is assumed to be fast, leading to the approximation that in the initial period $C_A \approx 0$ because it is being converted to product (P) as fast as it is available in solution:

$$\frac{dC_A}{dt} = k_L a \left(C_A^* - C_A\right) - k_R C_A C_B \quad (11.26)$$

and

$$\frac{dC_A}{dt} \approx \frac{dC_P}{dt} \quad (11.27)$$

$$\frac{dC_P}{dt} = k_L a \left(C_A^*\right) \quad (11.28)$$

As the potassium carbonate was in excess at 304 mg/ml with sparing solubility of approximately 47 mg/ml [9], an additional assumption was made that the surface area was constant, allowing the data to be regressed in the initial period as shown in Figure 11.10.

Comparison of the relative rates for the initial period leads to three observations:

1. Mass transfer is proportional to surface area for the conditions where the particles were well suspended. From Figure 11.8 it is observed that the ratio of surface areas for the small and medium particles is 2.05. Under the applied assumptions we would expect the relationship of the mass transfer rates to be the same ratio. From Figure 11.10 we see the rate for the small particles is 2.84×10^{-7} mol/(ml s) and that of the medium particles is 1.39×10^{-7} mol/(ml s), giving a ratio of 2.04.

2. The slope for the largest particle was expected to be half of that of the medium particles, i.e. 0.7×10^{-7} mol/(ml s). However, the observed rate was almost eight times lower than expected from the impact of area alone. The issues with the largest particle size are related to operating below full suspension. This is consistent with the mass transfer rate relationship with mixing as highlighted in Figure 11.2, where the mass transfer rate is decreased significantly when operating below complete suspension. Thus, highlighting the significance of ensuring agitation is well above the N_{js} during production.

3. Surface area has not been increased to the point where mass transfer was no longer a rate-limiting step. As depicted in Figure 11.4a, surface area can be increased to a point at which mass transfer may no longer be the limiting rate for the overall observed rate and therefore no change in rate as the area increased. In this case, the area did not reach that point. However, there is a practical limitation to decreasing the particle size for this application. Namely, the value of further reduction is limited as the reaction is complete in less than six hours for the smallest particles and to reduce the particle size significantly lower than that used in this study would require specialty milling equipment, increasing cost and complexity to the process.

To ensure that these results were not appreciably impacted by mixing above the power necessary to fully suspend the solids, additional data was gathered at high and low mixing speeds and at a scale change greater than one order of

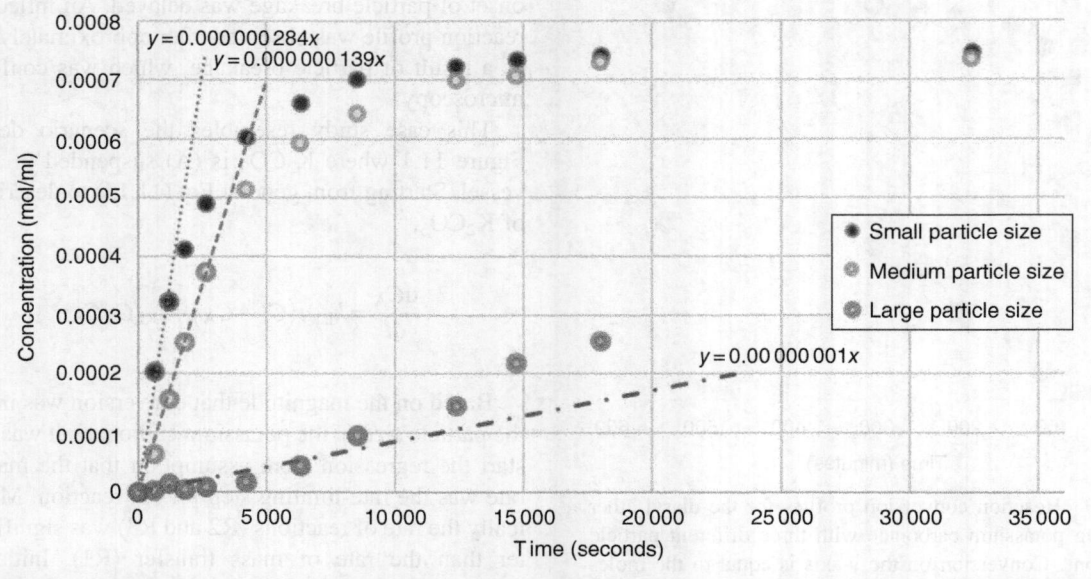

FIGURE 11.10 Observed rates of reaction for the initial period where the rate is controlled by mass transfer for the three different sizes of potassium carbonate.

magnitude. In all instances the results are consistent with the observed rates on Figure 11.10.

This knowledge was used to introduce a milling step for potassium carbonate in the kilo-scale run and define a raw material specification for the pilot plant that balance the reaction time with commercially available materials. In the case of the kilo scale runs, the resulting particle size from milling was used to predict a reaction time of 6.5 hours, in agreement with the observed reaction times of 6.5–7 hours. The particle size specified for the pilot scale predicted a reaction time of 12 hours, identical to the time observed on scale.

11.5 NOTATIONS

a	Interfacial area	m^2/m^3
C_A	Concentration of solute A	mol/m^3
C_B	Concentration of solute B	mol/m^3
C_A^*	Saturation concentration of A, solubility limit	mol/m^3
C_D	Drag coefficient	n/a
D	Solute diffusivity	m^2/s
D_0	Diffusivity Arrhenius pre-exponential factor	$1/s$
Da	Damköhler number	n/a
D_{imp}	Impeller diameter	m
d_p	Particle diameter	m
E_a	Activation energy	kJ/mol
E_D	Activation energy of diffusion	kJ/mol
g_c	The gravitation constant	m/s^2
k_0	Arrhenius pre-exponential factor – first order	$1/s$
k_L	Mass transfer coefficient	m/s
k_{obs}	The overall observable rate	$1/s$
k_r	The reaction rate constant	$m^3/(s \cdot mol)$
k_R	Observed reaction rate $k_r[C_B]$, assuming C_B is in excess and not changing during the course of reaction	$1/s$
δ	Boundary layer	m
N_{js}	Just-suspended speed	rps
μ_l	Viscosity of the bulk liquid, respectively	$Pa \cdot s$
ρ_l	Bulk liquid density	kg/m^3
ρ_s	Solid density	kg/m^3
Re_P	Particle Reynolds number	n/a
S	Dimensionless number that is a function of impeller type	n/a
Sc	Schmidt number	n/a
T	Temperature	Kelvin
V_m	Molar volume of diffusing species at its boiling point	m^3/mol
V_t	Free settling velocity	m/s
X	The mass ratio of suspended solids to liquid time 100	n/a

ACKNOWLEDGMENTS

The authors of this chapter would like to acknowledge and thank the following for their contributions to the preceding text: Daniel Griffin, Kenneth McRae, Jacqueline Milne, Scott Roberts, Shawn Walker, and Margaret Faul.

REFERENCES

1. Hulshof, L.A. (2013). *Right First Time in Fine-Chemical Process Scale-Up: Avoiding Scale-Up Problems: The Key to Rapid Success*. Mayfield, U.K.: Scientific Update LLP.

2. Fogler, H.S. (2000). *Elements of Chemical Reaction Engineering*, Prentice Hall International Series, 4the. Upper Saddle River, NJ; London: Prentice Hall PTR.

3. Harnby, N., Edwards, M.F., and Nienow, A.W. (1992). *Mixing in the Process Industries*, 2nde. Oxford: Butterworth-Heinemann.

4. Houson, I. (2011). *Process Understanding: For Scale-Up and Manufacture of Active Ingredients*, 1ste. Weinheim, Germany: Wiley-VCH.

5. Mohan, A.E., Kukura, J., Spencer, G., and Ulis, J. (2015). Solid–Liquid Suspension in Pilot Plants: Using Engineering Tools to Understand At-Scale Capabilities. Organic Process Research & Development 19 (9): 1128–1137.

6. Paul, E.L., Atiemo-Obeng, V., and Kresta, S.M. (2003). *Handbook of Industrial Mixing: Science and Practice*. New York: Wiley.

7. Perry, R.H. and Green, D. (1984). *Chemical Engineer's Handbook*. New York: McGraw-Hill.

8. Davis, M.E. and Davis, R.J. (2003). *Fundamentals of Chemical Reaction Engineering*. New York, NY: McGraw-Hill Higher Education.

9. Cella, J.A. and Bacon, S.W. (1984). Preparation of Dialkyl Carbonates via the Phase-Transfer-Catalyzed Alkylation of Alkali Metal Carbonate and Bicarbonate Salts. The Journal of Organic Chemistry 49 (6): 1122–1125. https://doi.org/10.1021/jo00180a033.

12

SCALE-UP OF MIXING PROCESSES: A PRIMER

Francis X. McConville and Stephen B. Kessler

Impact Technology Development, Lincoln, MA, USA

12.1 INTRODUCTION

The problems associated with the scale-up of mixing processes are universal. This is because the dynamics and mechanics of liquid agitation and blending are often poorly understood, and yet these operations play a fundamental role in many aspects of the chemical and pharmaceutical industries. The success of homogeneous and heterogeneous chemical reactions, crystallizations, liquid–liquid extractions, and so many other operations is critically dependent on effective mixing and appropriately designed mixing systems. Unfortunately, as we shall see below, duplicating the energy and quality of mixing available in the laboratory at commercial scale can prove extremely difficult.

For example, the motor power required to turn agitators increases exponentially as the diameter of the agitators increases, making it prohibitively expensive to match, one to one, the mixing power input of small-scale reactors in large commercial vessels. This results in batch blend times, the time it takes for the contents of a batch reactor to become homogenized, sometimes orders of magnitude longer in commercial reactors than in the laboratory. This can have severe consequences for the results of many chemical operations.

Frequently, heterogeneous reactions such as catalytic hydrogenations fail to achieve expected reaction rates upon scale-up because there is insufficient mixing to fully suspend the catalyst particles. The catalyst settles to the bottom of the vessel where it is inaccessible to the reactants in solution and therefore cannot effectively catalyze the reaction.

Differences in local and average shear conditions due to differences in impeller diameter and impeller tip speeds in commercial vessels can have unexpected consequences for shear-sensitive processes such as fermentations using living cells or crystallization of materials that require a specific particle size distribution. High shear can also cause severe emulsification at large scale that might not have been experienced in the lab.

These are just a few of the types of problems often encountered at large scale due to the fact that mixing conditions differ so much from those available in the laboratory. Mixing scale-up often proves to be a compromise between cost and performance and between achieving the desired result and minimizing unexpected negative effects. The better the understanding of the fundamental principles of mixing and of the specific requirements of the process involved, the better the results of this compromise will be.

12.2 BASIC APPROACHES TO MIXING SCALE-UP

Over the years, scientists and engineers have considered many approaches to scaling up mixing processes, with the ultimate goal of successfully matching laboratory results at commercial scale at a reasonable cost. As a result, numerous scale-up parameters, equations, and principles have been developed, some of which work better or are more reliable than others depending on the specific application. No single method has been successful for all situations, and the characteristics of the system must be understood as well as possible to maximize the chances for success.

Chemical Engineering in the Pharmaceutical Industry: Active Pharmaceutical Ingredients, Second Edition.
Edited by David J. am Ende and Mary T. am Ende.
© 2019 John Wiley & Sons, Inc. Published 2019 by John Wiley & Sons, Inc.

12.2.1 Principles of Similarity

Modeling theory considers two processes similar if they possess geometric, kinematic, and dynamic similarity. Geometric similarity requires that linear dimensions of two systems are scaled by the same ratios at different scales. Kinematic similarity requires geometric similarity and also that characteristic velocities scale by the same ratio. Dynamic similarity requires both geometric and kinematic similarity and adds the requirement that characteristic forces scale by the same ratio.

Rigorous application of modeling theory is rarely applied to scale-up of industrial mixing processes. One reason for this is that when more than two force properties are important in a mixing process, full dynamic similarity cannot be achieved. Since most mixing processes involve three or more force properties, a choice must be made among the possible properties to select one as a scaling factor. This choice is made by considering the nature of the process at hand and applying scaling factors that have been proven to work in similar processes. Some commonly used approaches to mixing scale-up and their utility in specific situations are described in the following sections.

12.2.2 Geometric Similarity

The concept of geometric similarity is illustrated in Figure 12.1. Adhering to geometric similarity can be extremely important in designing systems for scale-up or for building small-scale experimental vessels designed to mimic the behavior of a larger system for research purposes. This latter approach, called scaling down or modeling, is an important aspect of mixing engineering and widely used to study the mixing behavior of commercial systems at a more convenient scale.

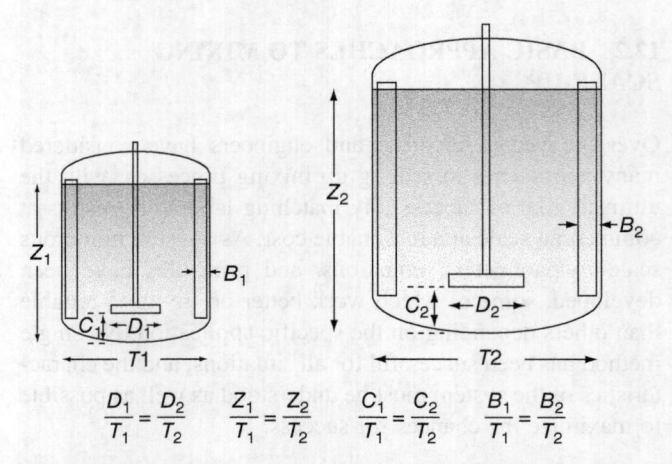

FIGURE 12.1 The principle of geometric similarity for stirred tanks. Key ratios (D/T, C/T, B/T, Z/T) are held equal at both scales. (B, baffle width; C, impeller bottom clearance; D, impeller diameter; T, tank diameter; Z, liquid height.)

Figure 12.1 shows how certain key ratios would be held equal in two geometrically similar vessels of different sizes. Thus the ratio of impeller diameter to tank diameter (D/T) is identical in both cases, as are the ratios of the liquid level (Z), the impeller bottom clearance (C), and the baffle width (B) to the tank diameter.

A number of practical issues limit the usefulness of this technique alone as a primary scale-up method. Firstly, mechanical limitations may limit its utility in some cases. For example, marine impellers are often used in laboratory systems. However, in large-scale industrial mixing applications, these impellers are impractically heavy if scaled up by geometric similarity. Also, the shape of the vessel heads is usually not limited by mechanical considerations in the laboratory, but in most industrial applications, vessel head design is defined by codes that take mechanical stresses into account. These types of limitations can sometimes be overcome by anticipating large-scale design issues and creating scaled-down laboratory vessels that match the large-scale geometry.

In addition to such limitations in the application of geometric similarity, there are limitations in what can be achieved when it is applied. Due to the rules of geometry, as a vessel doubles in diameter, its volume increases by a factor of 8 (2^3). Thus, when scaling up by a factor of 2, it is not possible to maintain certain key ratios such as surface area per unit volume or the volume/diameter ratio. It also proves impossible to operate these two systems in such a way that the intensity of mixing (as measured by power input per unit volume, P/V, for example) and the velocity of fluid circulation are *both* identical. It is possible to design and operate two systems of different sizes at an identical P/V, but the fluid circulation patterns, fluid velocities, degree of turbulence, etc. would likely be very different. If a successful process result relies on a particular fluid motion, it might not be achieved upon scale-up by simply maintaining geometric similarity and matching P/V. Such limitations are the source of much confusion and difficulty. In most cases, geometric similarity proves to be useful as a starting point for scale-up, but several other factors must be considered to ensure success.

Consequently there are situations where deliberate deviation from geometric similarity is the best approach to scale down. Oldshue [1] uses the term "nongeometric similarity" to describe a situation where conventional concepts of similarity must be sacrificed so that certain factors can be controlled to achieve successful scale-up. For example, with geometric similarity observed, a scaled-down vessel could be operated at the same tip speed as its full-scale counterpart, but this requires that the impeller in the scaled-down vessel be operated at higher rpm. In this example, the maximum shear rate in the two vessels is the same, but the average shear rate in the impeller region is higher in the small vessel. If shear rate is one of the key process variables being modeled, the mismatch in maximum and average shear rates can be reduced by increasing the diameter of the impeller in the scaled-down vessel relative to the vessel diameter (*D/T*). Tip speeds will now match at lower rpm in the small vessel, which would correlate with a smaller difference in impeller average shear rate.

More detailed information on system geometry and the application of geometric similarity to mixing processes can be found in Uhl and Von Essen [2] and Johnstone and Thring [3].

Figure 12.2 shows some typical "shape factors" – geometric ratios that have historically proven effective in systems designed for mixing processes and can be used as a general guide to vessel design. For example, many mixing vessels employ agitators with diameters approximately 1/3 of the vessel diameter, located one impeller diameter off the bottom. Again, these values are typical but can vary significantly in equipment designed for specific applications.

12.2.3 Rate of Turbulent Energy Dissipation and *P/V*

A particularly useful and widely used approach to mixing scale-up involves maintaining a constant rate of turbulent energy dissipation, ε, across the various scales. ε, which is defined by Eq. (12.1), is usually expressed in units of W/kg:

$$\varepsilon = \frac{P}{\rho V} \quad (12.1)$$

where

P is power input (W).
ρ is liquid density (kg/m^3).
V is liquid volume (m^3).

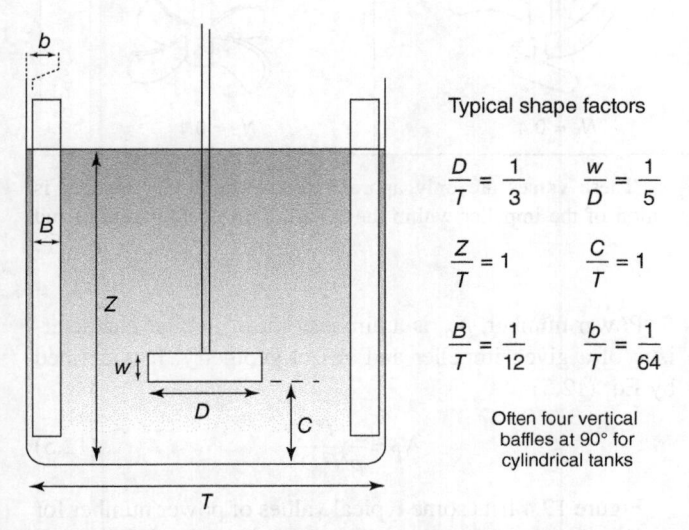

FIGURE 12.2 Typical shape factors, or geometric ratios, found useful for general mixing applications in stirred tanks.

Typical shape factors:

$\frac{D}{T} = \frac{1}{3}$ $\frac{w}{D} = \frac{1}{5}$

$\frac{Z}{T} = 1$ $\frac{C}{T} = 1$

$\frac{B}{T} = \frac{1}{12}$ $\frac{b}{T} = \frac{1}{64}$

Often four vertical baffles at 90° for cylindrical tanks

ε is fundamental in describing the interrelationship between turbulence and mass transfer in mixing operations. This statement is illustrated by the Kolmogorov length scale, which characterizes the smallest eddies associated with turbulent mixing. The Kolmogorov eddy length η is defined by Eq. (12.2), where ν is kinematic viscosity:

$$\eta = \left(\frac{\nu^3}{\varepsilon}\right)^{1/4} \quad (12.2)$$

At the length scale represented by η, viscous forces in the eddy are equal to inertial forces due to turbulent velocity fluctuations. The Kolmogorov eddy length underlies and informs the use of ε as a scaling parameter. Kinematic viscosity, ν, is a liquid property that is scale independent; thus constant ε is sufficient to fix a value for η over a range of scales. However, it remains to define the region for which ε is applicable. A mean value of ε can be calculated from the total power input and mass of liquid in the vessel. This overall mean value of ε is useful where an operation is governed by bulk flow characteristics. For geometrically similar vessels, it is sometimes assumed that holding overall mean ε constant is sufficient to provide accurate scaling in the impeller region. For more accurate scaling of local characteristics, a better approach is to define a volume based on the swept volume of the impeller instead of the total batch volume. A method for calculating impeller swept volume is given in Kresta and Brodkey [4].

Another parameter that is used to represent average mixing intensity in a vessel is power/unit volume (*P/V*). *P/V* is sometimes called power intensity and is usually expressed in either W/l or HP/1000 gal. In scaling equations that involve ratios of ε or *P/V* to represent different sizes of equipment, either ε or *P/V* works equally well. However, because of its units, *P/V* cannot be applied in fundamental equations that define turbulence and mass transfer in mixing systems. Also keep in mind that in large vessels, local values of ε may vary widely in different regions of the vessel.

When mean values of either ε or *P/V* are used for scale-up, it is important to also maintain geometric similarity. This is because in some mixing applications, a local value of ε may be of greater importance than the vessel average value. This point is well made in Figure 12.3, which shows three vessels all operated at the same *P/V*, but the fact that their geometries are very different (specifically impeller size) results in very different results in the suspension of solids. The effects shown are the result of calculations made with the commercial computational fluid dynamics (CFD) program VisiMix® [5].

As shown in Eqs. (12.3) and (12.4), ε and *P/V* can be expressed in terms of the impeller diameter *D*, its rotational speed *N*, liquid volume *V*, batch density ρ, and a parameter

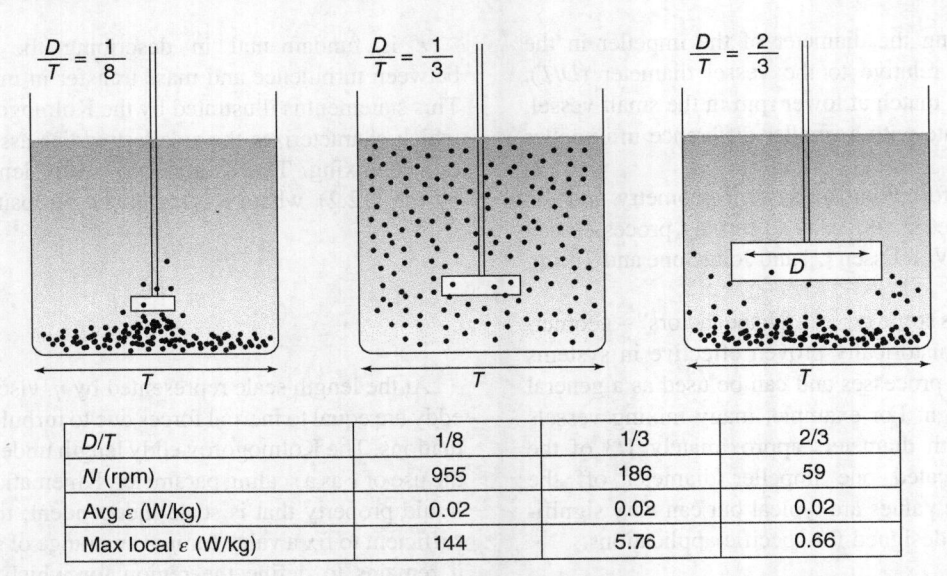

244 CHEMICAL ENGINEERING IN THE PHARMACEUTICAL INDUSTRY

D/T	1/8	1/3	2/3
N (rpm)	955	186	59
Avg ε (W/kg)	0.02	0.02	0.02
Max local ε (W/kg)	144	5.76	0.66

FIGURE 12.3 These three cases illustrate the importance of system geometry and the distinction between mean rate of energy dissipation (ε) and maximum local rate of energy dissipation. While all three vessels are operating at the same average value of ε, differences in geometry result in very different fluid motion and mixing behavior, in this case manifested by differences in the suspension of solids as predicted by VisiMix [5].

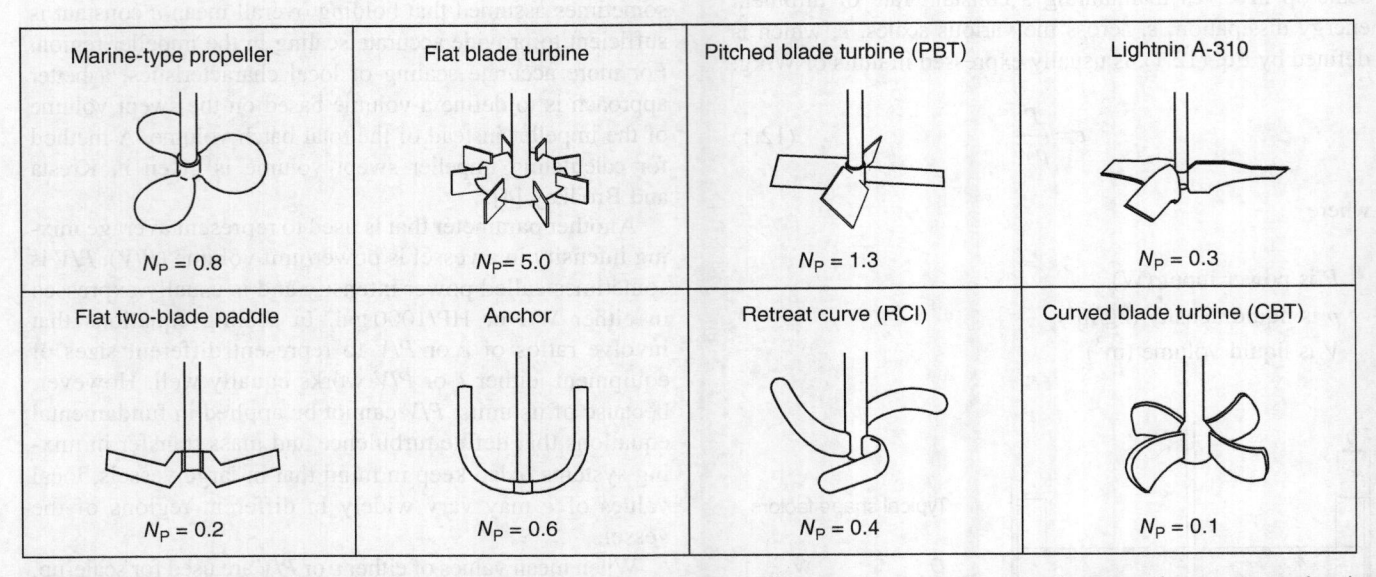

Marine-type propeller	Flat blade turbine	Pitched blade turbine (PBT)	Lightnin A-310
$N_P = 0.8$	$N_P = 5.0$	$N_P = 1.3$	$N_P = 0.3$
Flat two-blade paddle	Anchor	Retreat curve (RCI)	Curved blade turbine (CBT)
$N_P = 0.2$	$N_P = 0.6$	$N_P = 0.4$	$N_P = 0.1$

FIGURE 12.4 Typical power numbers (N_P) for various impeller types. These values are only approximate as the power number is significantly affected by number and pitch of blades, tip chord angle, position of the impeller within the vessel, baffle configuration, and other geometric factors.

called the power number N_P that is explained in more detail below:

$$\varepsilon = \frac{N_P N^3 D^5}{V} \quad (12.3)$$

$$\frac{P}{V} = \frac{N_P \rho N^3 D^5}{V} \quad (12.4)$$

Power number, N_P, is a dimensionless number characteristic of a given impeller and vessel geometry. It is defined by Eq. (12.5):

$$N_P = \frac{P}{\rho N^3 D^5} \quad (12.5)$$

Figure 12.4 lists some typical values of power number for various types of impellers, but geometric factors such as impeller tip chord angle, number of blades, position of the impeller within the vessel, and the number and dimensions of baffles all affect the value of the power number. For this reason, accurate power number values for a particular system can only be obtained experimentally, by measuring power draw via a wattmeter, or, more accurately, by directly measuring torque on the impeller shaft, under well-defined experimental conditions.

As in any type of fluid flow, fluid motion in mixing can be generally classified as either turbulent or laminar, depending on the velocity and other physical parameters. A common term for quantifying this is the impeller Reynolds number N_{Re}, a dimensionless parameter defined by Eq. (12.6). Values of N_{Re} can range from single digits for highly viscous flow to hundreds of thousands for very turbulent flow:

$$N_{Re} = \frac{\rho D^2 N}{\mu} \qquad (12.6)$$

At impeller Reynolds numbers greater than about 10^4, fluid motion is considered turbulent, and under such conditions, the power number N_P assumes a constant value. Under laminar mixing conditions ($N_{Re} < 100$) and in the transitional regime between laminar and turbulent mixing, power number varies, typically increasing with decreasing Reynolds number as shown by the curves in Figure 12.5. The values of N_{Re} that delineate the transitional region are only approximate and will vary depending on the system.

Note that the fluid viscosity term does not appear in the equations that define ε or P/V, but is captured indirectly in this relationship between power number and Reynolds number. It should be assumed that published values such as those given in Figure 12.4 represent turbulent power numbers. It is usually necessary to use empirical relationships such as those shown in Figure 12.5 to estimate N_P values under nonturbulent conditions.

The importance of the power number N_P and its application in typical mixing calculations are illustrated in Example Problem 12.1.

EXAMPLE PROBLEM 12.1

Determine what size of motor will be required to turn a 0.33 m diameter A310 hydrofoil impeller at 120 rpm in a crystallizer with a working volume of 500 L. The process fluid has a density of 1150 kg/m^3 and a viscosity similar to water (~0.001 Pa-s).

The A310 has a published turbulent power number of 0.3. To use this number we must ensure that we are operating in the turbulent mixing regime ($N_{Re} > 10^4$). Apply Eq. (12.6) to calculate the Reynolds number. Note that the rotational speed must be expressed in rev/s:

$$N_{Re} = \frac{\rho D^2 N}{\mu} = \frac{1\,150 \text{ kg}}{\text{m}^3} \times (0.33 \text{ m})^2 \times \frac{2}{\text{s}} \times \frac{\text{m} \cdot \text{s}}{0.001 \text{ kg}} = 250\,470$$

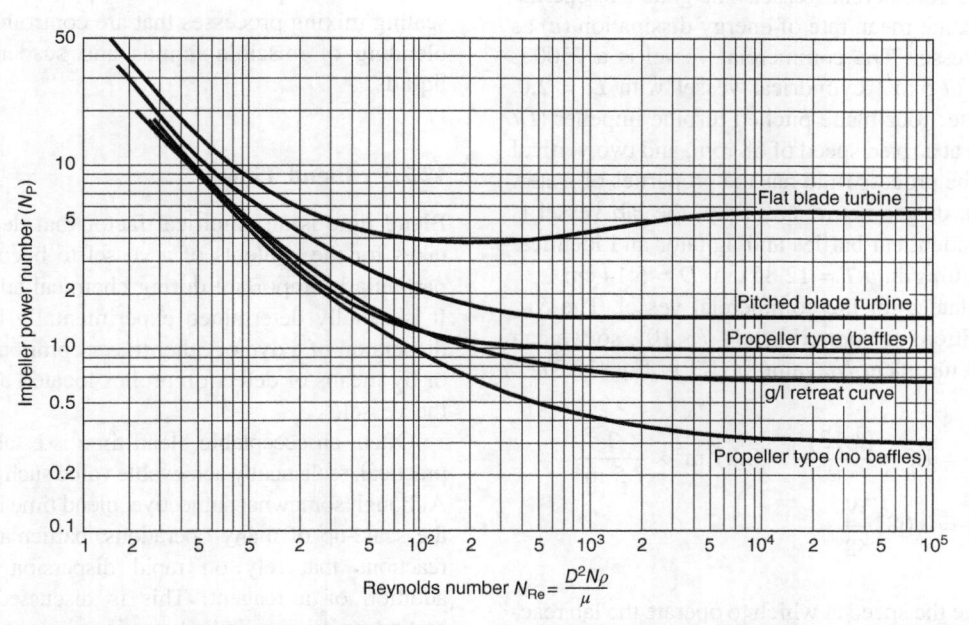

FIGURE 12.5 Relationship between impeller power number N_P and impeller Reynolds number N_{Re} for some typical impeller types. At $N_{Re} > 10^4$, flow is turbulent and N_P reaches a constant value. The power number generally increases under laminar conditions ($N_{Re} < 10$) and in the transitional regime between laminar and turbulent. *Source*: Adapted from Hemrajani and Tatterson [6] and Rushton [7].

This indicates that the mixing flow is clearly in the turbulent regime, so it is appropriate to use the published value of $N_P = 0.3$ for power draw estimation.

The power requirement is calculated by rearranging Eq. (12.5) as shown:

$$P = N_P \rho N^3 D^5 = 0.3 \times \frac{1150 \text{ kg}}{\text{m}^3} \times \left(\frac{2}{\text{s}}\right)^3 \times (0.33 \text{ m})^5$$

$$= 10.8 \frac{\text{kgm}^2}{\text{s}^3} = 10.8 \text{ W}$$

Estimating that frictional losses amount to roughly 20%, the total power requirement would be approximately 13 W. It is common practice to add an additional 15% safety margin at the design stage and then select the next commercially available motor size above that.

In some cases, the agitator is designed with multiple impellers. If, for example, the agitator in this example were designed with two identical impellers mounted on the same shaft, the power requirements would approximately double.

Example Problem 12.2 illustrates the basic approach for scaling up by maintaining constant mean rate of energy dissipation, ε, in two vessels of different scales.

EXAMPLE PROBLEM 12.2

A 2 L laboratory system is being designed to study the mixing characteristics of a commercial vessel. The goal is to operate the model at the same mean rate of energy dissipation (ε) as the commercial vessel. The commercial vessel is a 7500 L working volume (7.5 m³) cylindrical vessel with $D = 2.0$ m, a 0.8 m diameter four-blade pitched turbine impeller ($D/T = 0.4$) that turns at a fixed speed of 68 rpm, and two vertical baffles. Assume the process fluid has the properties of water.

In the interests of geometric similarity, the lab vessel is designed to have identical baffles and agitator and identical D/T and Z/T, resulting in a $T = 12.85$ cm, $D = 5.14$ cm.

First we calculate ε for the commercial vessel. The N_{Re} under these conditions is approximately 7×10^5, so we can use the published turbulent N_P value of 1.3 in Eq. (12.3):

$$\varepsilon = \frac{N_P N^3 D^5}{V} = 1.3 \times \frac{(1.13)^3}{\text{s}^3} \times (0.8)^5 \text{m}^5 \times \frac{1}{7.5 \text{ m}^3}$$

$$= 0.082 \frac{\text{m}^2}{\text{s}^3} = 0.082 \frac{\text{W}}{\text{kg}}$$

Now, determine the speed at which to operate the lab reactor to achieve the same mean value of ε by rearranging Eq. (12.3) and solving for N:

$$N^3 = \frac{\varepsilon V}{N_P D^5} = 0.082 \frac{\text{m}^2}{\text{s}^3} \times 0.002 \text{ m}^3 \times \frac{1}{1.3} \times \frac{1}{(0.0514)^5 \text{m}^5}$$

$$N = \frac{7.06}{\text{s}} = 423 \text{ rpm}$$

Thus we can match the commercial-scale mean rate of energy dissipation in the laboratory by operating the 5.14 cm impeller at 423 rpm.

12.2.4 Tip Speed

Tip speed is simply tangential velocity of the impeller at its maximum diameter and is calculated according to Eq. (12.7):

$$S_t = \pi D N \qquad (12.7)$$

Tip speed is related to maximum shear rate in stirred vessels. For this reason, tip speed is often applied as a scaling parameter for operations where maximum shear is a critical determinant of the process outcome. This includes those processes for which shear can be either beneficial or detrimental. This issue is discussed in more detail in the section on shear below. When vessels are scaled according to geometric similarity and at constant mean energy dissipation rate, tip speed will be higher in the larger vessel, a fact supported by Eq. (12.7).

In addition to its relationship to maximum impeller shear, in geometrically similar vessels, tip speed scaling corresponds exactly to scaling at constant torque per unit volume. In fully turbulent flow, i.e. above $N_{Re} = 10^4$, all velocities scale with tip speed regardless of viscosity. Because of these relationships, tip speed or torque per unit volume is useful in scaling mixing processes that are controlled by flow such as blending of miscible liquids and suspension of solids in liquids.

12.2.5 Blend Time

Blend time is an empirical factor that describes the time it takes for the contents of a vessel to become homogenized, particularly important during chemical additions to a batch. It is usually determined experimentally by monitoring the dispersion of a dye or other tracer compound, either visually or by means of detection probes located at various points in the vessel.

Often, an acceptable blend time is established based on a practical, realistically achievable value such as 99% uniformity. Although somewhat subjective, blend time is a critical factor in the scale-up of many operations, particularly rapid chemical reactions that rely on rapid dispersion during controlled addition of a reagent. This is discussed in detail in the section on mixing-limited reactions below.

Figure 12.6 illustrates that blend time increases rapidly when P/V is held constant but vessel size increases. Holding blend time constant with increasing vessel size requires maintaining constant impeller speed in geometrically similar

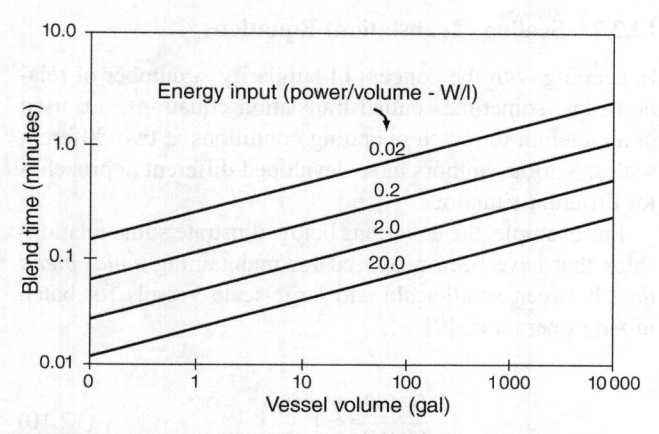

FIGURE 12.6 Relationship between blend time and vessel volume for various levels of power input (P/V).

vessels. This approach leads to increasing P/V with vessel size and, ultimately, to unrealistically high power requirements. Values much higher than 1–2 W/l are difficult to achieve in standard stirred tanks at large scale as beyond that, motors would become impractically large.

Impeller design and number of impellers will also have a significant effect on blend time. Some types of impellers, such as standard anchor blade impellers, which are not designed for good bulk mixing, generally result in very long blend times, whereas a pitched blade turbine operated at typical speeds in the same vessel would result in much shorter blend times.

Various correlations have been developed to help maintain constant blend time at different scales, such as the translation equations introduced below, but their success depends heavily on impeller design and other geometric factors. For standard vessel and impeller geometries, correlations are available that estimate blend times for the turbulent, transitional, and laminar regimes [8].

In mixing calculations, it is common to see a variable called "dimensionless blend time," which is essentially the product of the actual blend time and the impeller rotational speed, although often other geometric factors are included in equations used to calculate it.

12.2.6 Shear

As mentioned earlier, shear in a batch mixing operation can have desirable or undesirable effects, depending on the intended result of the operation. For example, maintaining sufficiently high shear rate in the impeller region may be required to rapidly disperse a reactant being fed into a vessel during a chemical reaction. However, the product of this same reaction may be a solid precipitate whose particles are shear sensitive and would suffer attrition, creating fines and complicating downstream recovery if high shear rates are maintained for too long.

Controlling shear rates when such a process is scaled up can become a complex undertaking. Various correlations presented in the literature to estimate shear rates in mixing vessels predict a broad range of shear rate values. Moreover, shear rates may be predicted to increase, decrease, or remain constant on scale-up, depending on the shear correlation and scale-up approach that are chosen. Some examples of the available correlations are described below.

One widely used correlation, the Metzner–Otto relationship, predicts average shear rate in the impeller region. This relationship, defined by Eq. (12.8), is valid for laminar, transitional, and moderately turbulent conditions:

$$\dot{\gamma} = k'N \quad (12.8)$$

where

$\dot{\gamma}$ is shear rate in per second.
k' is a dimensionless Metzner–Otto coefficient characteristic of the impeller.
N is impeller speed in rev/s.

For a Lightnin® A310 hydrofoil impeller (see Figure 12.4), the value of k' is 8.6. Thus the estimated average shear rate in the impeller region for an A310 running at 100 rpm is

$$\dot{\gamma} = 8.6 \times 100/60 = 14 \, \text{s}^{-1}$$

The value of shear rate predicted by the Metzner–Otto relationship depends only on impeller type and speed and is independent of impeller and vessel dimensions.

To estimate maximum shear rates produced in the flow near the impeller tip, an approach analogous to the Metzner–Otto relationship is used. In this case, a single value of the coefficient, $k' = 150$, is applied regardless of the impeller type. For estimating maximum shear on the impeller surface, a value of $k' = 2000$ is sometimes applied.

As mentioned in the section on tip speed, this factor can be related to maximum shear rate near the impeller tip. While the $k' = 150$ rule described above applies for moderate Reynolds number (laminar and transitional) conditions, tip speed is recommended for higher N_{Re} conditions as a means of scaling on the basis maximum shear rate. There is no general rule found in the literature that correlates tip speed with shear rate. When used for scaling purposes, tip speed is held constant as scale increases, which is assumed to provide constant maximum shear rate.

For estimates of shear rate averaged throughout a vessel under turbulent conditions, a vessel average shear rate can be calculated based on total energy dissipation. Eq. (12.9) defines this correlation:

$$\dot{\gamma} = \frac{P}{V}\mu^{0.5} \quad (12.9)$$

As described above, the various shear rate correlations provide widely divergent values on scale-up. To illustrate this point, Figure 12.7 compares the shear correlations that are presented above. The graph presented in Figure 12.7 covers a range of 100 : 1 scale-up of impeller diameter under the condition of geometric similarity, which corresponds to a range of 10^6 : 1 in vessel volume. ε and P/V are held constant. Under these conditions, shear rates in the impeller region, in the flow near the tip, and on the tip surface, which are each defined by a constant multiplied by rpm, all decrease with increasing impeller diameter. Vessel average turbulent shear rate remains constant when P/V is held constant. Tip speed as an indicator of shear rate increases with increasing impeller diameter at constant P/V.

General guidelines in the literature indicate that Metzner–Otto-type correlations are best applied over the laminar and transitional Reynolds number ranges. Vessel average shear applies only under fully turbulent conditions. Tip speed can be used as a scaling factor for maximum shear under fully turbulent conditions. However, these guidelines should not be relied upon if a process is to be scaled up in which shear is an important consideration. In this case, lab- and pilot-scale experiments should be conducted to evaluate the effects of shear over a range of scales whereby a correlation can be selected for commercial scale-up.

12.2.7 Scaling (Translation) Equations

In keeping with the concept of similarity, a number of relationships, sometimes called translation equations, are used in an attempt to match operating conditions at two different scales. Various authors have developed different approaches for different situations.

For example, the equations below illustrate some relationships that have been proposed for maintaining *equal blend time* between small-scale and large-scale vessels for batch mixing operations [9]:

$$\frac{(P/V)_2}{(P/V)_1} \approx \left(\frac{D_2}{D_1}\right)^2 \quad (12.10)$$

$$\frac{(T_Q/V)_2}{(T_Q/V)_1} \approx \left(\frac{D_2}{D_1}\right)^2 \quad (12.11)$$

The application of translation equations such as these depends very much on the specific application and system geometry, and as always experimental validation at two different scales is strongly recommended when applying them to predict performance in commercial-scale operations. A wide array of translation equations used for various

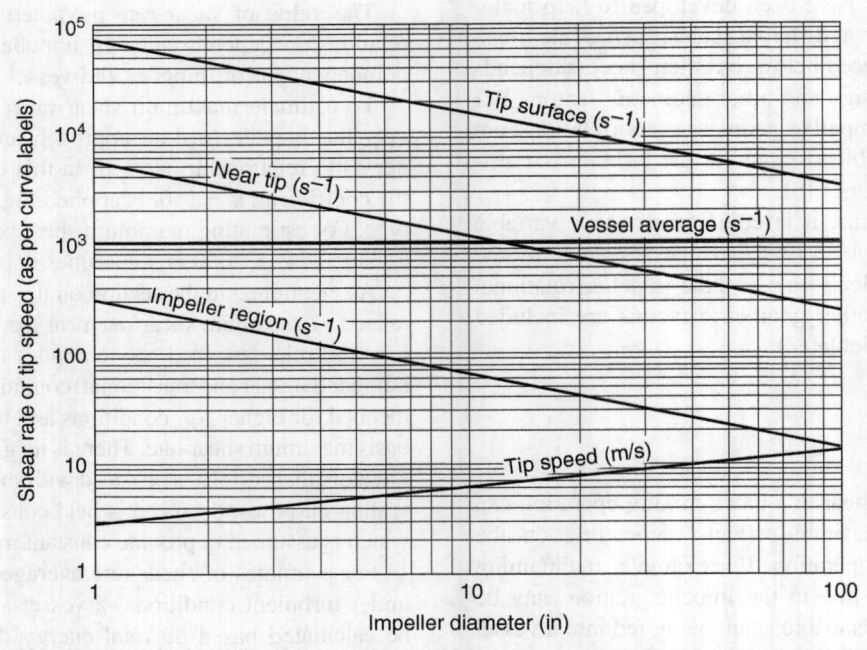

FIGURE 12.7 Theoretical effect of scale on various manifestations of shear in a mixing vessel under conditions of geometric similarity and constant ε (and P/V). Shear rates in the region swept by the impeller, in the flow near the tip, and on the tip surface, which are each defined by a constant multiplied by rpm, all decrease with increasing impeller diameter. Vessel average turbulent shear rate remains constant under these conditions. Tip speed as an indicator of shear rate increases with increasing impeller diameter.

purposes under various conditions is examined by Uhl and Von Essen [2].

12.3 OTHER CONSIDERATIONS IN MIXING SCALE-UP

12.3.1 Importance of Fluid Rheology

While Reynolds number is seldom used as a mixing scale-up correlation per se, knowing whether a mixing operation is conducted under laminar, transitional, or turbulent flow conditions is vital to successful scale-up. For example, if the commercial-scale operation will be fully turbulent, then lab- and pilot-scale experiments must be designed to operate under turbulent conditions as well. Liquid viscosity is a primary determinant of the value of impeller Reynolds number, defined by Eq. (12.6), so knowledge of the viscosity that is characteristic of a given mixing operation is vital as well.

If an operation comprises blending of Newtonian liquids or suspending an immiscible solid in a Newtonian liquid, then obtaining the required viscosity data is straightforward. The liquids may have well-known viscosities that can be found in literature references. If not, then measurement of Newtonian viscosity is a simple matter that can be performed with inexpensive instruments. However, if the liquid being mixed contains macromolecular solutes or colloidal-size particles, it may exhibit non-Newtonian characteristics. In this case, defining its rheology requires more than a single coefficient, and measurements may require more sophisticated instruments and techniques.

The flow properties of a Newtonian liquid are defined by Eq. (12.12):

$$\tau = \mu \dot{\gamma} \qquad (12.12)$$

where

- τ is shear stress.
- μ is coefficient of viscosity.
- $\dot{\gamma}$ is shear rate.

Shear-thinning fluids are often encountered when dealing with macromolecular solutes or colloidal suspensions. Such fluids can be effectively modeled by a power law, shown in Eq. (12.13):

$$\tau = K \dot{\gamma}^n \qquad (12.13)$$

where

- K is a consistency index.
- n is a behavior index.

If a shear-thinning fluid also exhibits a yield stress (i.e. there exists a shear stress below which no flow occurs), then the Herschel–Bulkley model, shown in Eq. (12.14), can be applied:

$$\tau = \tau_0 + K \dot{\gamma}^n \qquad (12.14)$$

where

τ_0 is yield stress.

Mixing of yield stress fluids can prove particularly challenging to scale-up. If the yield stress is of sufficient magnitude, then use of a conventional turbine-style impeller may result in a well-mixed cavern of liquid surrounding the impeller with little or no liquid motion closer to the vessel walls. In this case, a close-clearance impeller that sweeps close to the walls of the vessel, such as an anchor or helical ribbon, may be required. Empirical correlations are available to assist with these scale-up problems. These correlations are specific to impeller geometrical factors, and their efficacy will depend on choosing an appropriate rheological model and thorough characterization of the fluid.

Rheological models exist for many known types of fluid behavior. In addition to the non-Newtonian behaviors discussed above, additional levels of complexity such as time dependency and viscoelasticity can also be modeled. To support such models, measurement of the properties of non-Newtonian fluids requires the use of a rheometer. Rheometers are capable of controlling either shear stress or shear rate applied to a sample and are adaptable to multiple test geometries. A detailed discussion of non-Newtonian rheometry is beyond the scope of this chapter. A recommended reference in this regard is Macosko [10]. A particularly powerful combination of techniques for scaling of mixing operations for non-Newtonian fluids is the use of appropriate rheological models in conjunction with CFD as described in a later section.

12.3.2 The Role of Mixing in Heat Transfer

Heating and cooling batch vessels is a fundamental operation in any chemical processing endeavor. The rate and efficiency of heat transfer in and out of such vessels depend on many things, including the intrinsic heat transfer coefficient of the system, the temperature difference between the batch contents and the heat transfer medium (such as the fluid in the vessel heating jacket), and certain key properties of the batch itself (such as density, thermal conductivity, and heat capacity).

However, mixing also plays a major role in determining heat transfer efficiency. While a full treatment of heat transfer in agitated vessels is beyond the scope of this chapter, it is worth pointing out some fundamental principles.

A common dimensionless group that characterizes process-side heat transfer in stirred tanks is the Nusselt number Nu_L, which is a measure of the ratio of convective heat transfer to conductive heat transfer. It is defined in Eq. (12.15):

$$Nu_L = \frac{h_i D}{k} \quad (12.15)$$

where

h_i is the process-side heat transfer coefficient,
D is the vessel diameter, and
k is the thermal conductivity of the batch.

The value of Nu_L is strongly dependent on the mixing Reynolds number. Nu_L values close to unity indicate sluggish motion and heat transfer driven primarily by thermal conduction. Under highly turbulent conditions, Nu_L values can range from 100 to 1000, which indicates highly efficient convective heat transfer. Thus providing a sufficient degree of mixing is an important factor in designing vessels that will be used for heating and cooling. Many very comprehensive texts on process heat transfer are available for additional information, such as Serth [11].

12.3.3 Continuous Mixing Scale-Up

While a majority of mixing operations in the pharmaceutical industry are still performed in batch mixing vessels, there is a trend toward instituting continuous processing. This trend is being fostered by the FDA through elimination of regulatory constraints that previously limited most pharmaceutical process to the batch approach. The advantages of continuous mixing include potentially much higher productivity and improved heat transfer, mass transfer, and mixing. The latter three advantages are attributable primarily to reduced mixing volume.

Current examples of continuous mixing processes in the pharmaceutical industry mainly involve mixing-sensitive chemical reactions, that is, fast consecutive reactions that occur on a time scale that is short compared with practical blend times for commercial-scale batch mixing vessels. Most of the reactors used for these operations are of tubular configuration – for example, in-line static mixers. The continuous stirred tank reactor (CSTR) is also used and may find new applications with the current regulatory environment favoring continuous processes.

The typical objective of batch mixing operations is to achieve a spatially homogeneous mixture within a fixed process volume, within a specified blend time. Continuous mixing operations are designed to produce a spatially and temporally homogeneous effluent stream within a specified residence time. While blend time is the key parameter characterizing batch mixing operations, residence time distribution (RTD) is the key parameter characterizing continuous operations. Figure 12.8 plots the response of a sensor at the outlet of a mixing vessel to a step input of tracer at the inlet. The two RTD curves illustrate ideal flow patterns that establish the bounds within which real stirred tanks operate: the plug flow reactor (PFR) and the CSTR. A PFR represents the unmixed limit or complete segregation, while the ideal CSTR represents perfect macromixing.

The y-axis parameter in Figure 12.8, $F_{CSTR} = A/A_0$, is the concentration of tracer measured at the outlet (A) divided by the concentration applied at the inlet (A_0). The parameter θ in Figure 12.8 is dimensionless residence time, defined by Eq. (12.16):

$$\theta = \frac{t}{\bar{t}} \quad (12.16)$$

where

$\bar{t}$ is the mean residence time (vessel volume/flow rate).
t is time elapsed following application of tracer.

In the case of plug flow (dotted curve), no tracer is detected at the outlet until $\theta = 1$, at which point the tracer concentration jumps to the value at the inlet. For the CSTR (solid curve), when tracer enters the vessel, some is detected instantly at the outlet, and its concentration continues to rise

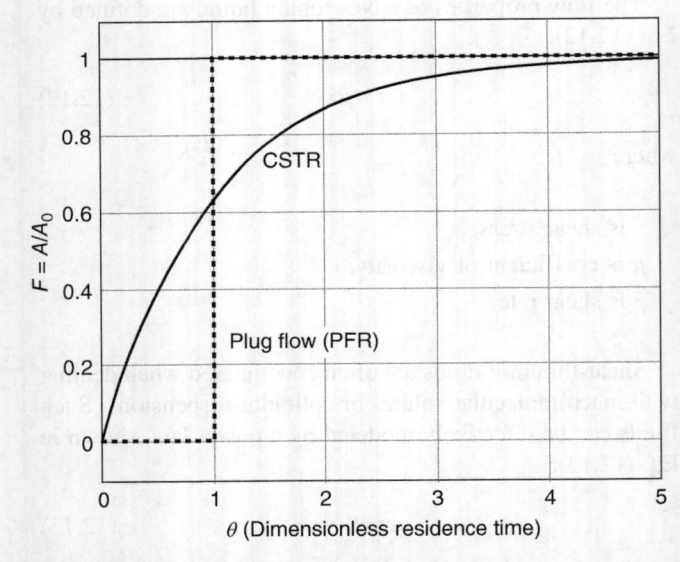

FIGURE 12.8 The response of a sensor at the outlet of a mixing vessel to a step input of tracer at the inlet for both plug flow reactors (PFR) and continuous stirred tank reactors (CSTR). The two ideal retention time distribution (RTD) curves illustrate the bounds within which real stirred tanks operate.

exponentially, approaching asymptotically the inlet concentration (see Eq. (12.17)):

$$F_{CSTR}(\theta) = 1 - e^{-\theta} \quad (12.17)$$

Real stirred tanks are often assumed to behave as ideal CSTRs. However, some degree of nonideal flow is likely to occur due to channeling, recycling, stagnant regions, or a combination of these effects. In many cases a real stirred tank may approximate the ideal CSTR closely enough that deviations from ideal flow have a negligible effect on the process. However, such deviations must be considered when scaling up. To quote Levenspiel [12], "The problems of nonideal flow are intimately tied to those of scale-up...Often the uncontrolled factor in scale-up is the magnitude of the nonideality of flow, and unfortunately this very often differs widely between large and small units. Therefore ignoring this factor may lead to gross errors in design."

For purposes of design and scale-up of continuous mixing operations, the ratio of the mean residence time in a CSTR divided by the batch blend time for the same vessel is defined by Eq. (12.18):

$$\alpha = \frac{V}{Q\Theta} \quad (12.18)$$

where

V is mixed volume.

Q is flow rate through the vessel.

Θ is the batch blend time for the vessel, which is either measured or estimated.

To ensure continuous mixing that is near-ideal CSTR in character, a rule of thumb states that the ratio α should be greater than 10. The basis for this ratio is discussed by Roussinova and Kresta [13].

In addition, the location of inlet and outlet ports must be considered. The rule to be followed in this regard is that a straight line drawn from the inlet port to the outlet port should pass through the impeller(s).

Scaling of the ratio of inlet flow to the impeller flow must also be considered. The simplest approach is to limit the average velocity of the liquid in the inlet port to be less than the tip speed of the impeller. Recommended ranges for this ratio can be found in Hemrajani and Tatterson [6]. Ratios of momentum and specific energy dissipation between the entering liquid jet and the impeller flow are sometimes used as scaling factors instead of a velocity ratio. These scaling approaches are also frequently applied in semi-batch mixing, which is discussed in Section 12.2.3.

Scale-up of static mixers for use in continuous mixing processes is beyond the scope of this chapter. See Etchells and Meyer [14] for further reading regarding this topic.

12.4 COMMON MIXING EQUIPMENT

Because of the wide variety of mixing processes encountered in the industry, a great number of mixing types and geometries have been developed, including fluidized beds, jet nozzles, and gas sparging. Here however, we will focus on mechanically stirred tanks and examine the typical impeller types used in this application. Such stirred vessels are used for batch production of the vast majority of specialty chemicals and pharmaceuticals, for blending and homogenization, for creating dispersions, and for running chemical reactions.

12.4.1 Major Impeller Types Used in Batch Mixing

Batch vessels may employ a broad range of impeller designs, each optimized for a particular type of process duty. The impellers shown in Figure 12.4 are among the more common types used in agitated vessels in chemical processing. Some vessels use multiple impellers of different types on a single shaft to obtain better mixing results. For example, it is common to utilize a high-shear flat-blade turbine at the bottom and a high-flow pitched blade impeller higher up the shaft in certain blending and dispersion operations.

Based on their design, impellers can be broadly categorized as generating an axial flow pattern or a radial flow pattern. In the case of a stirred tank, the axial flow pattern results in a pumping action, usually downward, that is very useful for preventing the settling of solids and generating good cross-mixing. Examples of axial flow impellers are marine propellers, pitched blade turbines, and hydrofoils such as the Lightnin® A310. These impellers will be found in crystallizers, solid suspension applications, and the like. Radial flow impellers do not tend to generate a vertical flow field, but tend more to push the fluid outward radially from the impeller. Most high-shear impellers, such as flat-blade turbines or paddles, are radial flow styles. Figure 12.9 illustrates axial and radial flow patterns.

One design commonly seen in the industry is the retreat curve impeller (RCI), sometimes called a "crowfoot" impeller. Originally designed to prevent flexing and cracking of the enamel coating when mixing viscous polymers, this impeller has been ubiquitous in glass-lined chemical reactors for decades. Nowadays, it is being largely replaced in glass-lined reactors by the curved blade turbine (CBT) for general process mixing.

Close-clearance impellers, the so-called anchor styles, serve a rather specialized need in mixing highly viscous or non-Newtonian fluids, since a high-speed center-shaft impeller might simply rotate in the liquid without generating any movement at the vessel wall. The anchor provides this action near the wall that is critically important when heating or cooling the batch in a jacketed vessel. Often an anchor will be combined with a center-mounted high-speed turbine to achieve sufficient heat transfer and good bulk mixing.

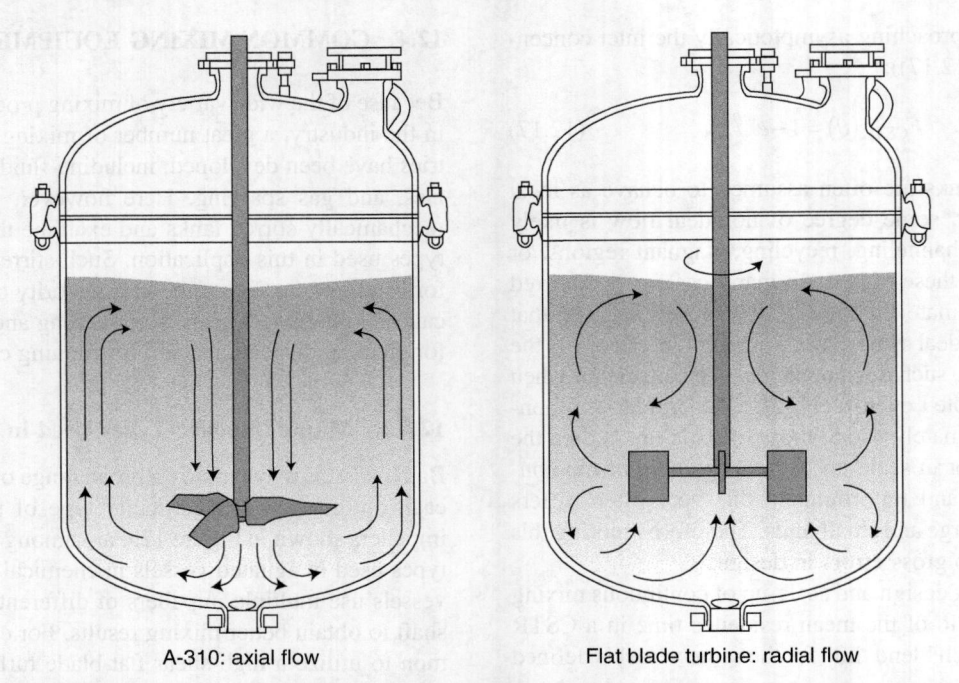

FIGURE 12.9 Typical stirred tank flow patterns. This figure shows an A310 hydrofoil generating axial flow and a flat paddle impeller generating radial flow. Many impellers or combinations of impellers exhibit components of both types of flow patterns.

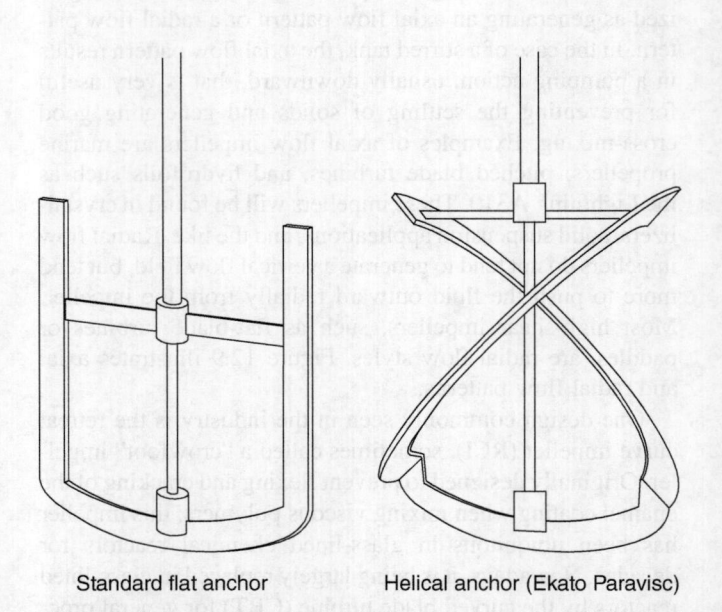

FIGURE 12.10 Anchor-type impellers. The flat anchor generates motion at the wall that is critical for heat transfer in mixing viscous liquids. Pitched anchor or helical styles (such as the "Paravisc" model designed by Ekato, Inc.) generate this motion at the wall and better bulk mixing throughout the rest of the vessel.

Figure 12.10 illustrates two types of anchor blades. Pitched anchors and other helical-type designs, albeit more expensive to construct than a flat anchor, can provide both wall motion and good bulk mixing.

The very fact that so many types of impellers are used industrially further illustrates that there are many types of mixing duties and no one impeller type is suitable for all. This is another source of difficulty in properly scaling up from the chemistry lab, where flat PTFE paddles are used almost exclusively for all overhead mixing service.

12.4.2 Mixing Baffles

Mixing baffles play a critical role in achieving efficient mixing in cylindrical vessels by preventing swirling and vortexing, increasing turbulence and cross-mixing, and providing better distribution of kinetic energy, especially for low viscosity fluids (viscosity < 5000 cP). Their use is limited to high-speed impeller applications and would not be found in vessels utilizing close-clearance impellers such as anchor or helical impellers.

Many baffle designs exist, including those shown in Figure 12.11. The majority of those shown can be found in various glass-lined vessels and are designed to be suspended from the vessel head. The rightmost baffle illustrated would be more typically used in stainless steel or other metal vessels where bolting directly to the wall is feasible. A space is normally left between the vessel wall and the baffle to allow flow and prevent collection of material there and simplify cleaning.

The introduction of baffles can actually have unwanted effects in certain cases, for example, tanks used for the dissolution of solids. Solids that are difficult to wet or that tend to float on the surface of the liquid may require the presence of a strong vortex to draw the material under the liquid surface. Baffles tend to reduce or eliminate this vortex and can actually make this sometimes problematic processing step more difficult.

12.4.3 High-Shear Impellers

As part of the discussion of conventional impeller types in Section 12.2.2, impellers were described as axial flow (e.g. A310), mixed flow (e.g. pitched blade turbine), or radial flow (e.g. flat-blade turbine). With respect to shear in stirred tanks, axial, mixed, and radial flow impellers are considered to be low, medium, and high shear, respectively. As discussed in Section 12.2, there are various definitions for impeller shear. Average shear in the impeller region as defined by the Metzner–Otto relationship is one definition of shear that supports the categorization of impeller types given above.

While a radial flow impeller is considered high shear among conventional impellers, operations that are intended to create dispersions may require higher shear than can be produced by standard radial flow impellers. While gas–liquid dispersions are often created with flat-blade turbines, liquid–liquid and solid–liquid dispersions usually require higher shear to reduce droplet or particle sizes to desired levels. For dispersions that must be stable or settle slowly on standing, particle sizes of less than 10 µm are usually required. When solid particles require de-agglomeration or attrition to achieve the desired size, intense shear stresses must be generated at the length scale of single particles.

For the kinds of applications described above, the preferred dispersion devices are high-speed disperser blades or rotor–stator homogenizers. High-speed dispersers are simple devices that can be used in a stirred tank configuration but are operated at much higher tip speeds than convention impellers. Figure 12.12 shows two high-speed disperser blades of differing design. The blade on the right has a smaller number of teeth, but the teeth are larger than the standard Cowles design on the left. The blade with fewer, larger teeth will generate more flow than the Cowles blade but sacrifices some shear to achieve this. Scale-up of high-speed dispersers is typically done by tip speed. Commonly used tip speeds for these devices range from 2 to 25 m/s. The low end of this range is adequate for delumping of solids being introduced into a mixing tank, while the high end of the range is typical for producing fine particle dispersions.

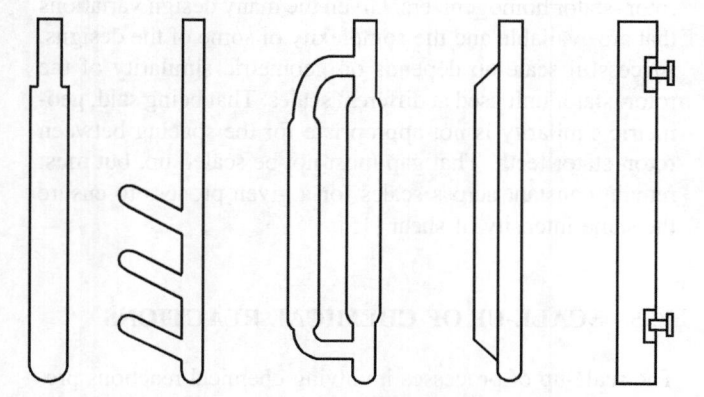

FIGURE 12.11 Various mixing baffle designs found in industrial tanks. From left to right: beaver tail, finger baffle, D-type baffle, fin baffle (all of which can be found in glass-lined vessels and do no attach to vessel side), and flat rectangular style for bolting directly to inner side wall of vessel.

FIGURE 12.12 Two high-speed disperser blades of different design. The blade on the right has a smaller number of teeth, but the teeth are larger than the standard Cowles design on the left. The blade with fewer, larger teeth will generate more flow than the Cowles blade but sacrifices some shear to achieve this.

Rotor–stator homogenizers provide a higher range of shear and energy dissipation than can be achieved by high-speed dispersers. A typical rotor–stator homogenizer is shown in Figure 12.13. While the usual tip speeds (5–50 m/s) are not that much higher than high-speed dispersers, much of the energy dissipation occurs within a small volume of liquid near the rotor and stator. This results in energy dissipation rates from 10^3 to 10^5 W/kg. This intense shear field results in very high shear stresses being transmitted to particles as they pass through the rotor–stator. Figure 12.14 shows the distribution of turbulent kinetic energy in a rotor–stator as predicted by CFD [15].

FIGURE 12.13 A typical rotor–stator homogenizer.

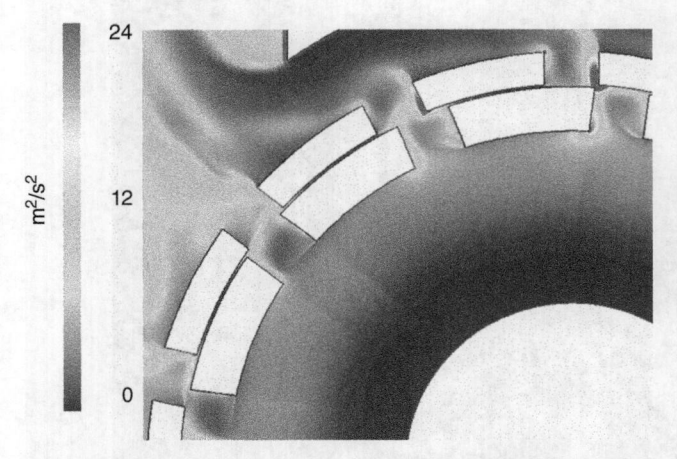

FIGURE 12.14 The distribution of turbulent kinetic energy in a rotor–stator as predicted by CFD. *Source*: From Atiemo-Obeng and Calabrese [15]. Copyright (2004) Wiley. Reprinted with permission of John Wiley & Sons, Inc.

There are many design variations of rotor–stator units, which alter the balance between pumping and shear. Rotor–stator homogenizers can be used in batch mode within a stirred tank or in-line. In a stirred tank, it is best to provide an additional impeller to provide circulation and rely on the rotor–stator unit only to produce shear. In this way, the two effects can be decoupled, providing better control. This approach is essential when the fluid being mixed has significant yield stress. For yield stress fluids, multi-shaft mixers can offer both close-clearance impellers and rotor–stator homogenizers. While in-line rotor–stator homogenizers provide some pumping, it is best to use a separate pump so that flow and shear can be controlled independently.

Tip speed is the most common approach for scaling up rotor–stator homogenizers. Given the many design variations that are available and the complexity of some of the designs, successful scale-up depends on geometric similarity of the rotor–stator unit used at different scales. That being said, geometric similarity is not appropriate for the spacing between rotor–stator teeth. That gap must not be scaled up, but must remain constant across scales for a given process to ensure the same intensity of shear [15].

12.5 SCALE-UP OF CHEMICAL REACTIONS

The scale-up of processes involving chemical reactions presents a special set of challenges, particularly in nonhomogeneous systems or in semi-batch reactions involving the controlled addition of reactive chemical reagents to a stirred vessel. Most chemical reactions are not 100% selective, that is to say that unwanted side reactions often accompany the main reaction. These reactions can reduce yield by consuming valuable starting materials, and the products of these side reactions can accumulate as contaminants that may be difficult or impossible to remove from the final product. These contaminants can also alter the crystal structure of some products, resulting in unexpected polymorphic crystal forms with poor solubility or other undesirable physical characteristics.

This can be a particularly vexing issue in an industry such as pharmaceutical manufacture, in which product quality is highly regulated and the presence of mere tenths of a percent of an unwanted impurity can result in an entire batch being rejected. Unfortunately, scaling up certain classes of reactions from a laboratory to commercial scale almost inevitably results in changes in selectivity, and often not for the better.

When a reaction is optimized in a laboratory setting, mixing is usually not an issue that comes into serious consideration, because laboratory stirrers provide very vigorous mixing and blend times in the one to two second range or less. However, upon scale-up, the reaction will be run in a system in which blend time may be on the order of 30 seconds or longer (see Figure 12.6).

Consequently, as the reactive material is added to the reactor, it may swirl around in a highly concentrated plume for some time before it becomes dispersed throughout the reaction mixture. This localized zone of high concentration can cause an increase in side reactions that may not have been an issue in the lab, resulting in poor reaction selectivity and low batch quality. This is an extremely common problem in reaction scale-up, and below we discuss some possible solutions. First, some examples of reactions that are affected by this phenomenon (so-called mixing-limited reactions) are in order.

12.5.1 Examples of Mixing-Limited Reactions

Consider the so-called Bourne [16] reaction, in which trimethoxybenzene (TMB) is treated with bromine to produce monobromotrimethoxybenzene (see Scheme 12.1).

SCHEME 12.1 The reaction between trimethoxybenzene and and bromine, and the consecutive competing reaction resulting in the formation of the di-bromated side product.

This reaction suffers from a consecutive competing reaction, in which the mono-Br product reacts with a second Br to form the di-Br product. The rate of the primary reaction (k_1) is about 1000× faster than that of the secondary reaction (k_2), so one would expect little of the di-Br to form. However, the rate of the secondary reaction is still fast enough that under typical mixing conditions, the mono-Br product is not swept away from the site of the reaction quickly enough and undergoes the second reaction. Figure 12.15 shows that reaction selectivity can be somewhat improved by increasing the intensity of agitation.

Another excellent example of the effect of mixing on reaction selectivity is the stereoselective enzymatic hydrolysis of a chiral organic ester (Scheme 12.2).

This is a biphasic reaction in which the enzyme is dissolved in the aqueous phase and preferentially hydrolyzes only one enantiomer of the chiral ester (an insoluble organic liquid) as it slowly enters the aqueous phase by diffusion. Base is added to maintain a constant pH as the acid product is formed.

Figure 12.16 shows the results of the reaction under conditions of good mixing and poor mixing. Note that under conditions of rapid mixing, the product purity is on the order of 99%, a result of the intrinsic selectivity of the enzyme, until the conversion reaches roughly 50%, at which point the preferred enantiomer is essentially all consumed and the enzyme begins to hydrolyze the other enantiomer. This results in reduced product purity at high conversions. However with poor mixing, the addition of the base causes nonselective chemical hydrolysis at the point of addition, resulting in lower product purity even at very low conversions.

A final example will illustrate the importance of mixing in heterogeneous reacting systems due to its effect on mass

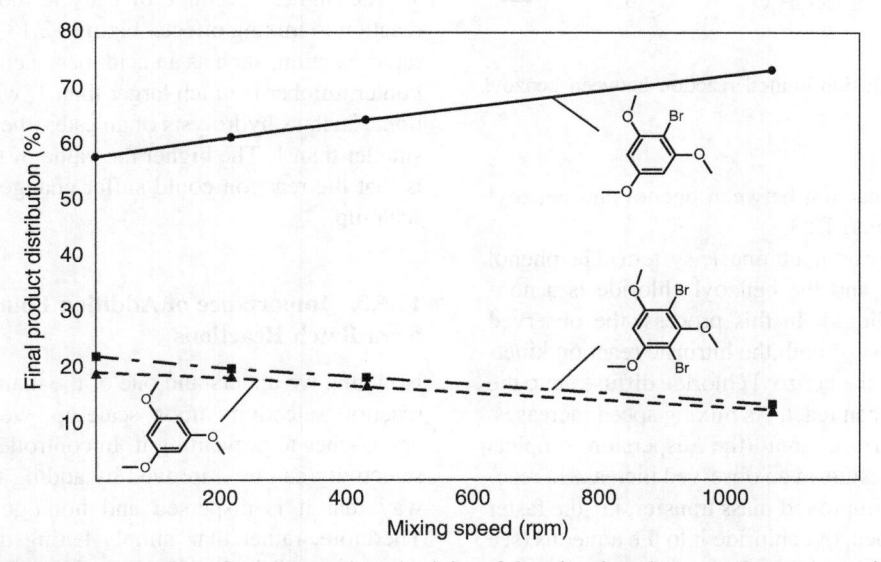

FIGURE 12.15 The effect of mixing on the selectivity of the trimethoxybenzene (Bourne) reaction [16].

SCHEME 12.2 The stereoselective enzymatic hydrolysis of a chiral ester.

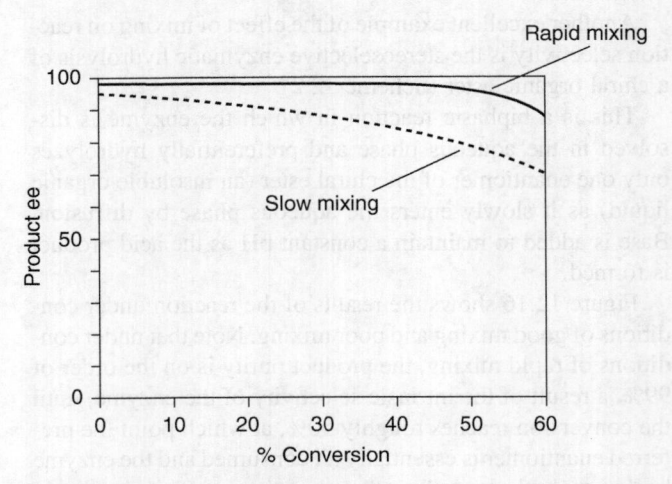

FIGURE 12.16 The effect of mixing intensity on stereoselectivity of the enzymatic hydrolysis of a chiral carboxylate ester. y-axis is enantiomeric excess (ee) of the product, a measure of enantiomeric purity. x-axis is degree of enzymatic conversion.

SCHEME 12.3 The diffusion-limited reaction between benzoyl chloride and phenol.

transfer. Consider the reaction between phenol and benzoyl chloride shown in Scheme 12.3.

This reaction can be run in a biphasic system. The phenol is in aqueous solution, and the benzoyl chloride is a non-water-soluble organic liquid. In this process, the observed reaction rate is a function of both the intrinsic reaction kinetics and the rate at which the benzoyl chloride diffuses into the aqueous phase where it can react. As mixing speed increases, so does interfacial surface area (the dispersion droplets become smaller). This results in an observed increase in reaction rate because of the improved mass transfer, i.e. the faster rate of diffusion of the benzoyl chloride into the aqueous (see Figure 12.17).

12.5.2 Identifying Mixing-Limited Reactions

The types of reactions most likely to be affected by mixing upon scale-up are highly rapid reactions, such as acid–base neutralizations. It is wise to try to identify any mixing-dependent behavior of reactions in the laboratory prior to scale-up to minimize surprises and failed batches. Sometimes it is simply a matter of running the chemistry in the laboratory under conditions of intense rapid mixing and of slow poor mixing. For example, for a controlled addition reaction, one could set up two side-by-side experiments. In one, the reagent is added slowly to a well-mixed flask; in the other the reagent is added quickly to a flask with poor or no mixing. If there is a significant difference in product purity, then this system will likely experience issues at scale, and measures can be taken to try to minimize these effects prior to scale-up.

A more theoretical approach is to calculate the Damkohler number for the reacting system. The Damkohler number (Da) is a dimensionless reaction time that represents the dependence of a given chemical reaction on mixing. Da is a function of reaction rate constant, reaction order, and reactant concentrations, but in simple terms, for semi-batch reactions, Da is generally defined as a ratio between some characteristic mixing time scale and the time scale of the reaction.

The higher the value of Da, the more susceptible is the reaction to mixing effects. Figure 12.18 shows that for a very rapid reaction, such as an acid–base neutralization, the Damkohler number is much larger than 1, whereas for slow reactions, such as hydrolysis of an ester, the value of Da is much smaller than 1. The higher the value of Da, the more likely it is that the reaction could suffer changes in selectivity upon scale-up.

12.5.3 Importance of Addition Point Design in Semi-Batch Reactions

Now that we understand one of the main causes behind poor reaction selectivity upon scale-up, we can examine some approaches to preventing it. In controlled addition reactions, selectivity can be improved by adding the reagent in such a way that it is dispersed and homogenized more rapidly. Therefore, rather than simply letting the reagent drip onto the surface of the batch or run down the side of the reactor,

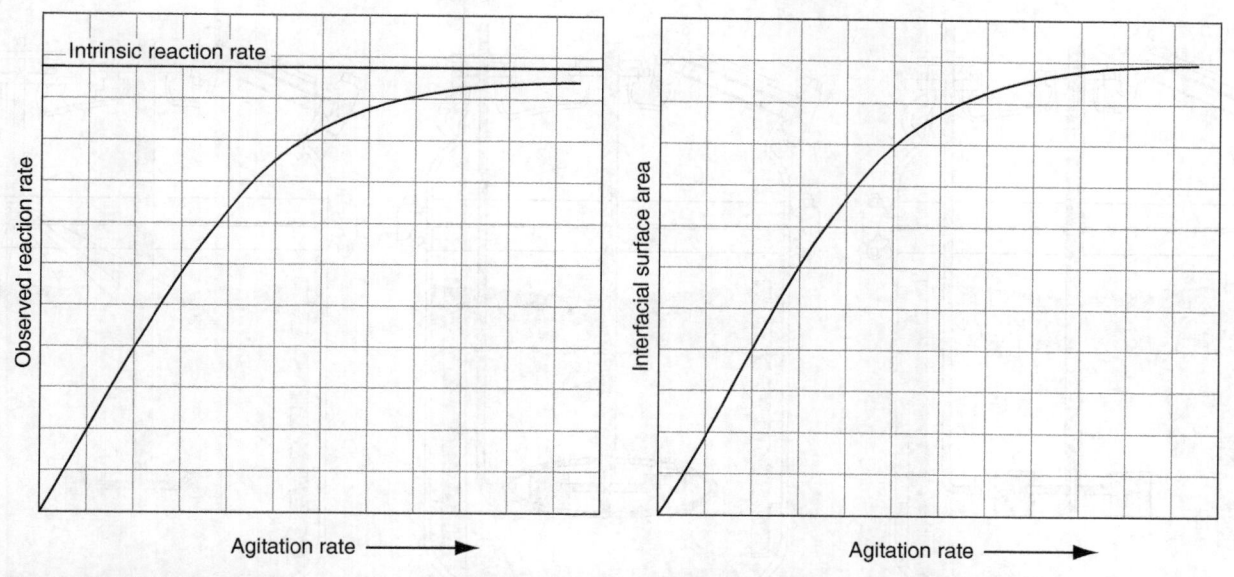

FIGURE 12.17 The effect of mixing speed on the reaction between benzoyl chloride and an aqueous solution of phenol. As agitation rate increases, so does interfacial surface area and diffusion rate and consequently the observed reaction rate. *Source*: Adapted from Atherton and Carpenter [17].

$$Da = \frac{\text{Characteristic time scale}}{\text{Reaction time scale}}$$

Acid–base neutralization
(rapid reaction)
$k \sim 10^{11}$ l/mol-s
rxn time scale $\sim 10^{-9}$ seconds

$$Da = \frac{10 \text{ seconds}}{10^{-9} \text{ seconds}} = 10^8$$

Base hydrolysis of ester
(slow reaction)
$k \sim 10^{-1}$ l/mol-s
rxn time scale $\sim 10^3$ seconds

$$Da = \frac{10 \text{ seconds}}{10^3 \text{ seconds}} = 0.01$$

FIGURE 12.18 An example of Damkohler number (Da) calculations for a rapid reaction and a slow reaction. The higher the value of Da, the more susceptible the reaction is to changes in selectivity due to mixing effects. The characteristic mixing time scale in this example is the blend time, here set to a typical value of 10 seconds.

it should be injected at a zone of very high shear, such as right at the periphery of the rotating impeller using a delivery or "dip" tube.

For the best results, the tube must be properly sized (i.e. small enough diameter and high enough flow velocity) to prevent back mixing in the tube, which can lead to the same selectivity issues. Numerous setups are used to achieve rapid dispersion during chemical additions, some of which are shown in Figure 12.19. The perforated dispersion ring can be particularly useful in controlling pH by acid or base addition in biological or enzymatic systems that may be sensitive to high concentrations of these reagents. Some reactions have been significantly improved by spraying the reagent onto the surface of the batch by means of a "shower-head"-type arrangement.

A number of other techniques are available for scaling up mixing-sensitive reactions. One common approach is to install a static or mechanically agitated mixer in a forced recirculation loop. The reactive chemical reagent is then injected in a controlled fashion into the recirculation line just upstream of the in-line mixer. This can speed dispersion and minimize the likelihood that a zone of very high concentration will exist in the vessel for any significant length of time. The static-type mixers, of which there are many designs, are particularly useful because they are generally well characterized and have no moving parts that minimize maintenance.

12.6 CFD AND OTHER MODELING TECHNIQUES

One of the major tools for studying mixing and predicting mixing behavior in process equipment is CFD modeling. The advent of high-speed personal computers has made CFD widely available, and it is finding use in many areas of technology, from plasma physics to the relatively simple liquid agitation we are concerned with here. Nonetheless, accurate modeling of fluid behavior requires the simultaneous calculation of huge numbers of equations, and even the simplest of problems consumes considerable CPU time.

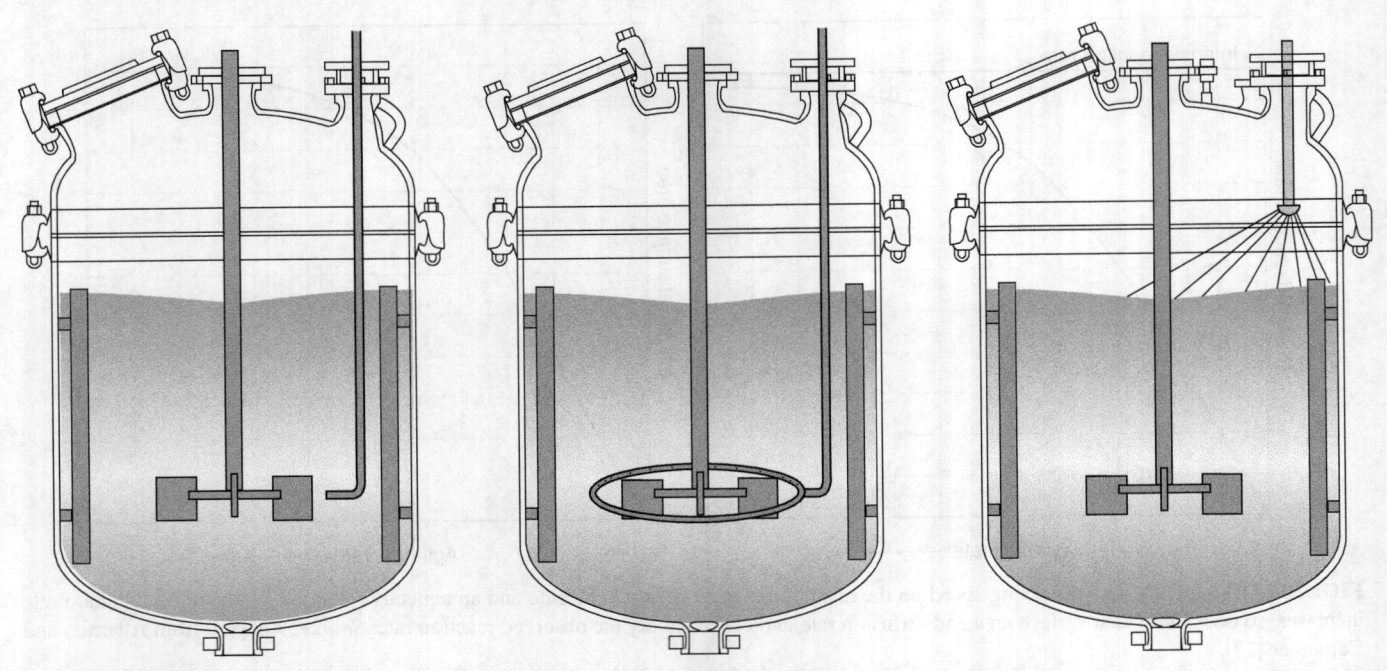

FIGURE 12.19 Some systems for improving performance of semi-batch or controlled addition reactions (from left to right: addition tube, dispersion ring, spray nozzle).

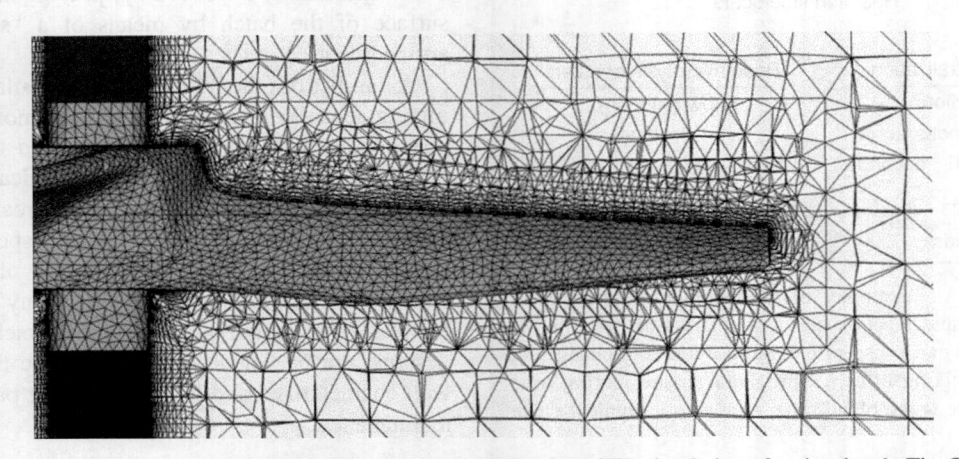

FIGURE 12.20 Showing the "wire mesh" for a portion of an A310 impeller for a CFD simulation of a stirred tank. The CFD mesh can consist of millions of three-dimensional cells, usually with a finer grid size in the vicinity of the impeller where highest velocity and shear occur and coarser in the bulk fluid to save CPU time.

Basically a mathematical model is constructed of the system of interest by dividing the fluid volume into hundreds of thousands or perhaps millions of contiguous cells (the model mesh; see, for example, Figure 12.20). The CFD software then tries to simultaneously solve the numerous momentum, velocity, force, heat transfer, and reaction mass balance equations associated with each of these cells in an attempt to converge on a single solution. When successful, these programs can accurately predict torque and mixing power requirements and can generate visual images or animations of fluid motion and circulation patterns that aid in identifying zones of high shear, stagnation, or other nonideal mixing behavior. Figure 12.14 shows an example of this type of image.

Several commercial software platforms are available for CFD modeling, but they are all quite expensive and require considerable expertise to properly code the model, create the mesh, and run the simulations. Needless to say, the success of the model depends on the accuracy of the rheological, chemical, and geometric data that is used to build it.

Some uncertainty is inevitable when conducting mixing experiments or modeling studies using computer simulations. For this reason most experts agree that for critical work, the CFD model should be validated by comparing predicted results with experimental measurements at least two different scales. In a typical scenario in which an industrial mixing system is to be designed based on the results of CFD modeling, the best results will be obtained if experiments are conducted first at some laboratory scale and then at a small pilot scale. CFD simulations of these two smaller-scale operations would then be carried out, and once fine-tuned to the point where predictions agree well with experiments, the model can be used for simulations to support the full-scale design.

12.7 NOTATIONS

A	tracer concentration (outlet)
A_0	tracer concentration (inlet)
B	tank baffle width
C	impeller bottom clearance
D	impeller diameter
Da	Damkohler number
F_{CSTR}	concentration ratio in CSTR
Gal	US Gal
h_i	internal (process-side) heat transfer coefficient
HP	horsepower
k	thermal conductivity
k'	dimensionless Metzner–Otto constant
K	rheological consistency index
n	rheological behavior index
N	impeller rotational speed
N_P	impeller power number
N_Q	impeller flow number
N_{Re}	impeller Reynolds number
Nu_L	Nusselt number
P	mixing power
Q	flow rate
S_t	tip speed
t	time
$\bar{t}$	mean residence time
T	tank diameter
T_Q	torque applied to a mixer shaft
V	batch liquid volume
W	watt
Z	liquid height in batch vessel
α	ratio mean residence time/batch blend time
ε	rate of turbulent energy dissipation
ρ	density
η	Kolmogorov eddy length
ν	kinematic viscosity
$\dot{\gamma}$	shear rate
μ	viscosity
τ	shear stress
τ_0	yield stress
θ	dimensionless residence time
Θ	batch blend time

REFERENCES

1. Oldshue, J.Y. (1985). Current trends in mixer scale-up techniques. In: *Mixing of Liquids by Mechanical Agitation* (ed. J.J. Ulbrecht and G.K. Patterson). New York: Gordon and Breach.
2. Uhl, V.W. and Von Essen, J.A. (1986). Scale-up of fluid mixing equipment. In: *Mixing: Theory and Practice Vol III* (ed. V.W. Uhl and J.B. Gray), 155–167. Orlando, FL: Academic Press.
3. Johnstone, R.E. and Thring, M.W. (1957). *Pilot Plants, Models, and Scale-Up Methods*. New York: McGraw Hill.
4. Kresta, S.M. and Brodkey, R.S. (2004). Turbulence in mixing applications. In: *Handbook of Industrial Mixing* (ed. E.L. Paul, V.A. Atiemo-Obeng and S.M. Kresta). Hoboken, NJ: Wiley-Interscience.
5. VisiMix (2006). Mixing Simulation Software, www.visimix.com
6. Hemrajani, R.R. and Tatterson, G.B. (2004). Mechanically stirred vessels. In: *Handbook of Industrial Mixing* (ed. E.L. Paul, V.A. Atiemo-Obeng and S.M. Kresta). Hoboken, NJ: Wiley-Interscience.
7. Rushton, J.H. et al. (1950). Power Characteristics of Mixing Impellers. *Chemical Engineering Progress* 46: 8 p395.
8. Grenville, R.K. and Nienow, A.W. (2004). Blending of miscible liquids. In: *Handbook of Industrial Mixing* (ed. E.L. Paul, V.A. Atiemo-Obeng and S.M. Kresta). Hoboken, NJ: Wiley-Interscience.
9. Penney, W.R. (1971). Recent trends in mixing equipment. *Chem Eng* 78 (7): 86.
10. Macosko, C.W. (1994). *Rheology: Principles, Measurements, and Applications*. New York: Wiley-VCH.
11. Serth, R.W. (2007). *Process Heat Transfer: Principles and Applications*. Amsterdam: Elsevier.
12. Levenspiel, O. (1999). *Chemical Reaction Engineering*, 3e. Hoboken, NJ: Wiley.
13. Roussinova, V.T. and Kresta, S.M. (2008). Comparison of continuous blend time and residence time distribution models for a stirred tank. *Industrial and Engineering Chemistry Research* 47 (10): 3532–3529.
14. Etchells, A.W. and Meyer, C.F. (2004). Mixing in pipelines. In: *Handbook of Industrial Mixing* (ed. E.L. Paul, V.A. Atiemo-Obeng and S.M. Kresta). Hoboken, NJ: Wiley-Interscience.
15. Atiemo-Obeng, V.A. and Calabrese, R.V. (2004). Rotor-Stator mixing devices. In: *Handbook of Industrial Mixing* (ed. E.L. Paul, V.A. Atiemo-Obeng and S.M. Kresta). Hoboken, NJ: Wiley-Interscience.
16. Bourne, J. (2003). Mixing and the selectivity of chemical reactions. *Organic Process Research & Development* 7 (4): 471–508.
17. Atherton, J.H. and Carpenter, K.J. (1999). *Process Development: Physicochemical Concepts*. Oxford: Oxford University Press.

13

STIRRED VESSELS: COMPUTATIONAL MODELING OF MULTIPHASE FLOWS AND MIXING*

AVINASH R. KHOPKAR
Reliance Industries Limited, Mumbai, MH, India

VIVEK V. RANADE
School of Chemistry and Chemical Engineering, Queen's University of Belfast, Belfast, UK

Stirred vessels are widely used in pharmaceutical industry to carry out a large number of multiphase applications (reactions, precipitations, emulsions, etc.) and recipes. They offer unmatched flexibility in operation to manipulate the performance of the vessel. A skilled reactor engineer can use the offered flexibility to tailor the fluid dynamics and therefore performance of a reactor by appropriately adjusting the reactor hardware and operating parameters. Performance of stirred vessels is influenced by a variety of parameters such as the number, type, location and size of impellers, degree of baffling, sparger type, inlet/outlet locations, aspect ratio, and reactor shape. It is therefore essential to first translate the "wish list" of the reactor performance into a "wish list" of desired fluid dynamics. Despite the widespread use of these stirred vessels, the fluid dynamics in them, essentially for multiphase flows, is not well understood. This lack of understanding and the knowledge of the underlying fluid dynamics have caused reliance on empirical information [1–3]. Available empirical information is usually described in an overall/global parametric form. This practice conceals detailed localized information, which may be crucial in the successful design of the process equipment. Reliability of such empirical information and, in particular, extrapolation beyond the range of parameters studied often remains questionable. It is therefore, essential to develop and apply new tools to enhance our understanding of the fluid dynamics prevailing in stirred vessels. Such understanding will be useful in devising cost-effective and reliable scale-up of stirred vessels.

In last two decades, with improvement in the knowledge of numerical techniques, turbulence models and the availability of fast computational resources have made it possible to develop models based on computational fluid dynamics (CFD) and use them for "a priori" prediction of the flow field in chemical process equipment [4–6]. However, unlike single-phase flow, which is possible to predict with reasonable confidence [4], the computational models capable of predicting real-life turbulent multiphase flows involving complex geometries and with a wide range of space and time scales are yet to be established. The development of such models will be a significant step toward the prediction of local fluid dynamics. Such models will be useful for exploring the possibilities for performance enhancement of existing reactors, for evolving better reactor configurations, and for reliable scale-up. In this chapter, we have critically reviewed the state of the art of computational modeling of multiphase flows in stirred vessels and discussed application of computational models to address a wide range of industrially relevant processes.

13.1 ENGINEERING OF MULTIPHASE STIRRED REACTORS

Multiphase stirred vessels are ubiquitous in pharmaceutical industry right from R&D to manufacturing. In many

---

* Submitted for consideration as a chapter in a book "Chemical Engineering in the Pharmaceutical Industry: R&D to Manufacturing".

Chemical Engineering in the Pharmaceutical Industry: Active Pharmaceutical Ingredients, Second Edition.
Edited by David J. am Ende and Mary T. am Ende.
© 2019 John Wiley & Sons, Inc. Published 2019 by John Wiley & Sons, Inc.

situations of practical interest, more than one phase need to be contacted in a stirred vessel. In several cases, phase transitions such as generation of vapors by evaporation of volatile components, precipitation of solid particles via reactions, or solidification and generation of liquid droplets via melting of solids or phase inversion occur in stirred vessels. Some examples of industrial multiphase processes carried out in stirred reactors are listed in Table 13.1. Engineering of these reactors begins with the analysis of the process requirements and evolving a preliminary configuration of the reactor. More often than not the reactor has to carry out several functions like bringing reactants into intimate contact (to allow chemical reactions to occur), providing an appropriate environment (temperature and concentration fields by facilitating mixing, heat transfer, and mass transfer) for adequate time and allowing for removal of products. Naturally, successful reactor engineering requires expertise from various fields ranging from thermodynamics, chemistry, and catalysis to reaction engineering, fluid dynamics, mixing, and heat and mass transfer. Reactor engineer has to interact with chemists to understand intricacies of the considered chemistry. Based on such understanding and proposed performance targets, reactor engineer has to abstract the information relevant for identifying the characteristics of desired fluid dynamics of the reactor. Reactor engineer has to then conceive suitable reactor hardware and operating protocol to realize this desired fluid dynamics in practice.

The laboratory study and reactor engineering models, based on idealized fluid dynamics and mixing, help in this step. This step helps in defining performance targets of the reactor. The reactor engineer faces a major difficulty in translating the preliminary reactor configuration (lab or pilot scale) to the industrial reactor. Transformation of a preliminary reactor configuration to an industrial reactor proceeds through several steps. Some of these scale-up steps that are discussed in other chapters are highlighted here:

- *Scale-down/scale-up analysis*: It is essential to analyze the possible effects of scale of the reactor on the prevailing fluid dynamics and reactor performance. Conventionally, such an analysis is carried out with certain empirical rules (for example, equal power per unit volume, equal tip speed, and so on) and prior experience. However, it was observed that these rules do not guarantee the identical performance of reactor at two different scales. This can be explained by using the case of gas–liquid stirred reactor. A small-scale reactor provides a higher shear rate and more rapid circulations compared with a large-scale reactor. Gas dispersion, therefore, is often breakage (dispersion) controlled in a small-scale reactor but coalescence controlled in a large-scale reactor. The interfacial area per unit volume of reactor for gas–liquid interphase mass transfer decreases as the scale of the reactor increases.
- *Presence of conflicting process requirements*: Presence of conflicting process requirements is also a major issue a reactor engineer needs to tackle in a multiphase stirred reactor. For example, the fluid dynamic characteristics required for better blending and heat transfer (flow-controlled operations) are quite different from those required for better dispersion of a secondary phase and better mass transfer (shear-controlled operations). Such conflicting process requirements make the task of evolving a "wish list" of desired fluid dynamics difficult. The reactor engineer needs to achieve a compromise between conflicting processes to obtain the best performance. It is therefore necessary to have a good understanding of the prevailing fluid dynamics and its relation with design parameters on one hand and with the processes of interest on the other hand.
- *Designing new reactor concepts:* Development of reactor technologies relies on prior experience. Testing of new reactor concepts/designs is often sidelined due to lack of resources (experimental facilities, time, funding, and so on). Experimental studies have obvious limitations regarding the extent of parameter space that can be studied and regarding the extrapolation beyond the studied parameter space.

This brief review of the modeling of multiphase stirred reactors indicates that the detailed knowledge of the prevailing fluid dynamics will allow a reactor engineer to exploit the available degrees of freedom of stirred reactors. However, obtaining the detailed information on fluid dynamics in stirred reactors for multiphase flow is challenging. The

TABLE 13.1 Some Industrial Applications of Multiphase Stirred Reactor

Phases Handled	Applications
Gas–liquid	Chlorination, oxidation, carbonylation, manufacture of adipic acid and oxamide, and so on
Gas–liquid–solid	Fermentation, hydrogenation, oxidation (*p*-xylene), wastewater treatment, and so on
Liquid–liquid	Suspension and emulsion polymerization, oximations, methanolysis, extraction, and so on
Liquid–solid	Calcium hydroxide (from calcium oxide), anaerobic fermentation, regeneration of ion-exchange resins, leaching, and so on
Gas–liquid–liquid	Biphasic hydroformylation, carbonylation

Source: Reprinted with permission from Khopkar and Ranade [7]. Copyright 2011 John Wiley and Sons Inc.

complexity in modeling the fluid dynamics increases significantly for multiphase flows. Till recent past, the complexity of fluid dynamics and multiphase processes occurring in stirred vessels was too overwhelming, and most of the practical engineering decisions were based on empirical and semiempirical analysis. Several excellent reviews and books on such design procedures are available (for example, Refs. [1, 3, 8] and so on). However the information obtainable from these methods is usually described in an overall/global parametric form. This practice conceals detailed local information about turbulence and mixing, which may ultimately determine overall performance. The conventional approach essentially relies on prior experience and trial-and-error method to evolve suitable reactor hardware. These tools, therefore, are being increasingly perceived to be expensive and time consuming for developing better reactor technologies. It is necessary to adapt and develop better techniques and tools for relating reactor hardware with fluid dynamics and resultant transport processes.

In recent years, chemical engineers have started using the power of CFD models to address some of these reactor engineering issues. CFD is a body of knowledge and techniques to solve mathematical models of fluid dynamics on digital computers. Considering the central role of stirred vessels in pharmaceutical industry, there is tremendous potential for applying these tools for better engineering of stirred vessels. Computational flow modeling (CFM) can make substantial contributions to scale-up by providing quantitative information about the fluid dynamics at different scales. The computational model may offer a unique advantage for understanding the requirements of conflicting processes and their subsequent prioritization. The CFD model will allow a reactor engineer to switch on and off various processes and study interactions between different processes. Such numerical experiments can help to reduce and to resolve some of the challenges posed by conflicting demands made by different processes. CFD models can make valuable contributions to developing new reactor technologies by allowing "a priori" prediction of fluid dynamics for any configuration with just knowledge of reactor geometry and operating parameters. These simulations allow detailed analysis, at an earlier stage in the design cycle, with less cost, with lower risk, and in less time than experimental testing. It sounds almost too good to be true. Indeed, these advantages of CFD are conditional and may be realized only when the fluid dynamic equations are being solved accurately, which is quite difficult for most of the engineering flows of interest. It must be remembered that numerical simulations will always be approximate. There can be various reasons for differences between computed results and "reality" such as errors associated with fluid dynamic equations being solved, input data and boundary conditions, numerical methods and convergence, computational constraints, interpretation of results, and so on.

It is indeed necessary to develop an appropriate methodology to harness the potential of CFD for better reactor engineering, design, and scale-up despite some of the limitations. This chapter is written with an intention of assisting practicing engineers and researchers to develop such methodology and approach.

Various aspects of CFM and its application to multiphase stirred vessels are discussed and related in a coherent way. The emphasis is not on providing a complete review but is on equipping the reader with adequate information and tips to undertake a complex flow modeling project. Modeling of single-phase flows and mixing in stirred vessels are not discussed, and the reader is referred to chapter 10 of Ranade [9]. While CFD simulations for single-phase systems have been widely used for designing and optimizing operation and control of existing processes, their use is limited for systems containing reactive and/or multiphase flows. The efficient design and operation of multiphase flow systems is currently limited by a number of factors. In the technology roadmap vision 2020 listed several reasons (see Table 13.2) associated with the design, operation and control, and process-related issues for efficient design of multiphase system. Some of these are due to the lack of accurate modeling tools for multiphase flow regimes, and others result from problems inherent to specific chemical processes.

The scope of this chapter is restricted to multiphase flows and mixing. Significant efforts were made in last 15 years for improving the computational modeling of multiphase flow in process equipment. Here we critically analyze role of turbulence, multiphase flow, interphase interactions (drag, lift, virtual mass, coalescence and breakup, and so on), and flow regimes for multiphase flows in stirred vessels. The presented computational models were found to capture key features of two-phase flows in stirred vessels reasonably well. The present work highlighted the limited applicability of direct extension of gas–liquid and solid–liquid modeling approaches for simulating three-phase flow. The computational models were found to predict the implications of reactor hardware, flow regimes, and suspension quality on the transport and mixing process. In some conditions, the steady-state approach may not be appropriate, and a full unsteady-state approach may be necessary. This is especially crucial when fast reactive mixing and interaction of nozzle and impeller stream are important or in formulation processes where the rheology constantly changes either due to reaction or due to physical change in the dispersed phase.

It is indeed necessary to develop an appropriate methodology to harness the potential of CFD for better reactor engineering, design, and scale-up despite some of the limitations. This chapter is written with an intention of

TABLE 13.2 Problems Limiting the Efficient Design and Operation of Multiphase Systems

Design	Operation and Control	Process-Related Issues
Current designs are artificially constraints	Lack of means for controlling product attributes	Inefficient pneumatic handling of solids (feeds and products) resulting from poor design and reliability
Current designs are based on precedence and empirical methods	Poor utilization of existing process vessels	Problems associated with chemical containment and safety
Data at the macroscopic rather than microscopic level is used in current designs	Excessive downtimes due to corrosion and erosion	General process inefficiencies leading to unnecessary energy consumption, production of waste, and emission of pollutants
Not all design alternatives are explored – limited possibility thinking is the rule	Limited ability to optimize existing reactors and separation units, leading to low yields and poor performance	Limited availability of designs and computational tools that target specific production processes or plants
Steady state is often used to explain transient and segregated flows	High cost of experimenting in full-scale production facilities	
Limited ability to do real reactor design	Poor visualization of process phenomenon	Mass transfer-controlled operations
Current design simulations are based on idealized conditions (often quite different from actual)		Multiphase flow in channels
		Viscous and non-Newtonian mass transfer operations
Current codes provide inaccurate predictions when extended to other flow conditions		Safety pressure relief multiphase discharge designs

assisting practicing engineers and researchers to develop such methodology and approach. Various aspects of CFM and its application to multiphase stirred vessels are discussed and related in a coherent way. The emphasis is not on providing a complete review but is on equipping the reader with adequate information and tips to undertake a complex flow modeling project. Some aspects of single-phase mixing are also discussed for the sake of completeness. Adequate attention is provided to addressing key issues in solid–liquid flows like solid suspension in viscous liquids, solid drawdown, and solid dissolution considering the importance of crystallization in pharmaceutical industry. The basics of computational modeling and the extent of its applicability to simulating multiphase stirred reactors are discussed with various examples. After describing these, possible applications to practical problems relevant to pharmaceutical industry are briefly discussed. Overall discussion is organized in two parts: the first part deals with computational modeling of multiphase flows and the second with applications to engineering of stirred vessels. Key conclusions and some suggestions for further work are outlined at the end.

PART I: COMPUTATIONAL MODELING OF MULTIPHASE FLOWS IN STIRRED VESSELS

13.2 COMPUTATIONAL MODELING OF MULTIPHASE STIRRED REACTOR

The subject of modeling of multiphase flow processes is quite vast and covers a wide range of subtopics. It is virtually impossible to treat all the relevant issues in a single book, let alone in a single chapter. The scope here is restricted to modeling of dispersed multiphase flows in stirred reactors where continuous phase is a liquid phase and a dispersed phase may gas, liquid, or solid. There are mainly three approaches for modeling such dispersed multiphase flows:

- VOF: Volume of fluid approach (Eulerian framework for both the phases with reformulation of interface forces on volumetric basis).
- EL: Eulerian framework for the continuous phase and Lagrangian framework for all the dispersed phases.
- EE: Eulerian framework for all the phases (without explicit accounting of interface between phases).

If the shape and flow processes occurring near the interface are of interest, VOF approach should be used. This approach is, however, naturally limited to modeling the motion of only a few dispersed phase particles. The EL approach is suitable for simulating dispersed multiphase flows containing low (motion of dispersed particles is not influenced by collisions) volume fraction of the dispersed phases. For denser dispersed phase flows, it is usually necessary to use the EE approach. Considering that most of the pharmaceutical applications will involve dense dispersions, the scope here is restricted to EE approach. More information on modeling of other approaches may be found in Ranade [9] and references cited therein.

In the EE approach, the dispersed phases are also treated as continuum. All the phases "share" the domain, and they may interpenetrate as they move within it. A concept of volume fraction of phase q, α_q, is used while deriving governing equations. Various averaging methods have been proposed (see Ranade [9] for more details). In this section, we will present a general form of governing equations for dispersed multiphase flows, which will be suitable for further numerical solution, without going into details of their derivation.

13.2.1 Model Equations

For most of the operating regimes used in practice, flows in multiphase stirred vessels are turbulent. Therefore, the mass and momentum balance equations governing such flows can be written as (Favre averaged equations for each phase without considering mass transfer)

$$\frac{\partial(\alpha_q \rho_q)}{\partial t} + \nabla \cdot (\alpha_q \rho_q \vec{U}_{q,i}) = 0 \quad (13.1)$$

$$\frac{\partial(\alpha_q \rho_q \vec{U}_{q,i})}{\partial t} + \nabla \cdot (\alpha_q \rho_q \vec{U}_{q,i} \times \vec{U}_{q,i})$$
$$= -\alpha_q \nabla \vec{p} - \nabla \cdot (\alpha_q \overline{\overline{\tau}}_{q,ij}^{(\text{lam})}) - \nabla \cdot (\alpha_q \overline{\overline{\tau}}_{q,ij}^{(t)})$$
$$+ \alpha_q \rho_q g_i + \vec{T}_{\text{fl}} + \vec{F}_{12,i} \quad (13.2)$$

where

$\vec{T}_{\text{fl}}$ is the turbulent dispersion force accounting for the turbulent fluctuation in the phase volume fraction.

It is modeled as

$$\vec{T}_{\text{fl}} = K V_{\text{dr}} \quad \text{where} \quad V_{\text{dr}} = -\left(\frac{D_p}{\sigma_{pq} \alpha_p} \nabla \alpha_p - \frac{D_q}{\sigma_{pq} \alpha_q} \nabla \alpha_p\right) \quad (13.3)$$

Here, V_{dr} is the drift velocity, D_p and D_q are the diffusivities of the continuous and dispersed phase, respectively, and σ_{pq} is the turbulent Prandtl number. The diffusivities D_p and D_q can be calculated from the turbulent quantities following the work of Simonin and Viollet [10]. The turbulent Prandtl number σ_{pq} is usually set to 0.75–1.0. $\overline{\overline{\tau}}_{q,ij}^{(\text{lam})}$ is the stress tensor in the phase q due to molecular viscosity, and $\overline{\overline{\tau}}_{q,ij}^{(t)}$ is the Reynolds stress tensor of phase q (representing contributions of correlation of fluctuating velocities in momentum transfer). Boussinesq's eddy viscosity hypothesis is usually used to relate the Reynolds stresses with gradients of time-averaged velocity as

$$\overline{\overline{\tau}}_{q,ij}^{(t)} = \mu_{tq}\left(\left(\nabla \vec{U}_{q,i} + (\nabla \vec{U}_{q,i})^T\right) - \frac{2}{3} I (\nabla \vec{U}_{q,i})\right) \quad (13.4)$$

Here, μ_{tq} is the turbulent viscosity of the phase q and I is the unit tensor.

The turbulent viscosity may be related to the characteristic velocity and length scales of turbulence. Several turbulence models have been proposed to devise suitable methods/equations to estimate these characteristic length and velocity scales in order to close the set of equations. Despite the known deficiencies, the overall performance of the standard k–ε turbulence model for simulating flows in stirred vessels is adequate for many engineering applications [9]. Most of the modeling attempts of complex turbulent multiphase flows mainly rely on the practices followed for the single-phase flows, with some ad hoc modifications to account for the presence of dispersed phase particles. Here we present the standard k–ε turbulence model to estimate the turbulent viscosity of the liquid phase without going into critical review of different models and approaches. Additional details may be found in Ranade [9]. The governing equations for turbulent kinetic energy, k, and turbulent energy dissipation rate, ε, are listed below:

$$\frac{\partial}{\partial t}(\alpha_l \rho_l \phi_l) + \nabla \cdot (\alpha_l \rho_l \vec{U}_{l,i} \phi_l) = -\nabla \cdot \left(\alpha_l \frac{\mu_{tl}}{\sigma_{\phi l}} \nabla \phi_l\right) + S_{\phi l} \quad (13.5)$$

where

ϕ_l is the turbulent kinetic energy or turbulent energy dissipation rate in the liquid phase.

The symbol $\sigma_{\phi l}$ denotes the turbulent Prandtl number for variable ϕ. $S_{\phi l}$ is the corresponding source term for ϕ in liquid phase. Note that the turbulence equations are solved only for the continuous liquid phase. Source terms for turbulent kinetic energy and dissipation can be written as

$$S_{kl} = \alpha_l[(G_l + G_{el}) - \rho_l \varepsilon_l]$$
$$S_{el} = \alpha_l \frac{\varepsilon_l}{k_l}[C_1(G_l + G_{el}) - C_2 \rho_l \varepsilon_l] \quad (13.6)$$

where

G_l is generation in the liquid phase.

G_{el} is extra generation (or dissipation) of turbulence in the liquid phase.

Generation due to mean velocity gradients, G_l and μ_{tl}, turbulent viscosity was calculated as

$$G_l = \frac{1}{2}\mu_{tl}\left(\nabla \vec{U}_{l,i} + (\nabla \vec{U}_{l,i})^T\right)^2 \quad \mu_{tl} = \frac{\rho_l C_\varpi k_l^2}{\varepsilon_l} \quad (13.7)$$

Extra generation or damping of turbulence due to the presence of dispersed phase particles is represented by G_{el}. Kataoka et al. [11] have analyzed the influence of the gas bubbles on liquid-phase turbulence. Motion of larger bubbles generates extra turbulence. However, their analysis indicates that the extra dissipation due to small-scale interfacial structures almost compensates for the extra generation of turbulence due to large bubbles. Numerical experiments on bubble columns also indicate that one may neglect the contribution of extra turbulence generation (see Ref. [12]

for more details). Therefore, for stirred vessels where impeller rotation generates significantly higher turbulence than that observed in bubble columns, the contribution of the additional turbulence generation due to bubbles can be neglected.

Following the general practice, the same values of parameters proposed for single-phase flow ($C_{1\varepsilon} = 1.44$, $C_{2\varepsilon} = 1.92$, $C_{3\varepsilon} = 1.3$, $C_\mu = 0.09$, $\sigma_k = 1.0$, and $\sigma_\varepsilon = 1.3$) may be used to simulate the turbulence in two-phase flow. In the dispersed k–ε turbulence model, no extra transport equations were solved for estimating the turbulent quantities for dispersed phase. Instead, a set of algebraic relations can be used to couple the dispersed phase turbulence to continuous phase turbulence using Tchen's theory [10, 13]. The turbulence of dispersed phase depends mainly on three important time scales, characteristic time of turbulent eddy τ_1^t, bubble relaxation time τ_{12}^b, and eddy–bubble interaction time τ_{12}^t (see Ref. [14] for more details). This approach of modeling turbulent dispersed phase is computationally less expensive and can adequately simulate the turbulence in two-phase flow with low dispersed phase holdup (<10%). In the case of higher dispersed phase holdup, simulating turbulence equations for individual phases may be required.

Interphase coupling terms make multiphase flows fundamentally different from single-phase flows. The formulation of time-averaged $\vec{F}_{12,i}$, therefore, must proceed carefully. The interphase momentum exchange term consists of four different interphase forces: Basset history force, lift force, virtual mass force, and drag force [15]. Basset force arises due to the development of a boundary layer around bubbles and is relevant only for unsteady flows. The Basset force involves a history integral, which is time consuming to evaluate, and in most cases, its magnitude is much smaller than the interphase drag force. Considering this, the Basset history force is usually not considered for simulating dispersed multiphase flows in stirred vessels. The interphase momentum exchange term that included the lift, virtual mass, and drag force terms is written as

$$\vec{F}_{12,i} = \vec{F}_{D,i} + \vec{F}_{VM,i} + \vec{F}_{\text{lift},i} \quad (13.8)$$

The virtual mass term in i direction is given as

$$\vec{F}_{VM,i} = \alpha_2 \rho_1 C_{VM} \left(\frac{D\vec{U}_{2,i}}{Dt} - \frac{D\vec{U}_{1,i}}{Dt} \right) \quad (13.9)$$

where

C_{VM} is virtual mass coefficient.

In the present work, the value of C_{VM} was set to 0.5. The lift force in i direction is given as

$$\vec{F}_{L,i} = -0.5 \rho_1 \alpha_1 \alpha_g (V_{1,i} - V_{g,i})(\nabla \times V_{1,i}) \quad (13.10)$$

The interphase drag force exerted on phase 2 in i direction is given by

$$\vec{F}_{D,i} = -\frac{3 \alpha_1 \alpha_2 \rho_1 C_D \left(\sum (\vec{U}_{2,i} - \vec{U}_{1,i})^2 \right)^{0.5} (\vec{U}_{2,i} - \vec{U}_{1,i})}{4 d_b} \quad (13.11)$$

where

C_D is a drag coefficient.

This expression can be generalized to more than one dispersed phases in a straightforward way. It is necessary to correct the estimation of drag coefficient to account for the particle size distribution and nonspherical shapes of the particles, for the presence of other particles, and for the presence of prevailing turbulence. Specific discussion of these as well as formulation of boundary conditions related to simulations of multiphase flows in stirred vessels is included in the following sections.

Denser suspensions of gas or liquid phases within continuous liquid phase lead to issues like coalescence and breakup. It is possible to extend the approach to incorporate population balance models to account for such processes. However, this may require significantly larger computational resources as well as input data on model parameters of coalescence and breakup kernels. Presence of dense solid suspension exhibits various additional complexities and requires substantial modifications in the governing equations. The governing equations for such cases are not included here for the sake of brevity and may be found in Ranade [9]. Other conservation equations (enthalpy and species) for multiphase flows that can be written following the similar general format are also not included here. Application of these governing equations to simulate multiphase flows in stirred vessels is discussed in the following.

13.2.2 Application to Simulate Gas–Liquid Flow in Stirred Reactor

The most important step in the application of model equations to simulate a gas–liquid stirred reactor is the appropriate selection of interphase force formulations. They play a very important role while simulating gas dispersion [16]. Lane et al. [16] carried out order-of-magnitude analysis of all interphase forces. They observed that in the bulk region of the stirred reactor, interphase drag force dominates the total magnitude of interphase forces and hence can determine the gas dispersion pattern. There are few studies available in the literature highlighting the influence of interphase drag force on the predicted gas holdup distribution (for example,

Refs. [17, 18]). However, not much information is available in the literature on the virtual mass force and lift force and their effect on the predicted gas holdup distribution. To explain the influence of different interphase forces, we reproduce some of the results obtained by Khopkar and Ranade [17]. They have carried out simulations of gas–liquid flow in a stirred vessel in an experimental setup used by Bombac et al. [19]. All the relevant dimensions like impeller diameter, impeller off-bottom clearance, reactor height and diameter, sparger location and diameter, and so on were the same as used by Bombac et al. [19]. Considering the symmetry of the geometry, half of the reactor was considered as a solution domain (see Figure 13.1). The solution domain and details of the finite volume grid used were similar to those used by Khopkar and Ranade [17]. A QUICK discretization scheme with SUPERBEE limiter function (to avoid nonphysical oscillations) was used. Standard wall functions were used to specify wall boundary conditions. The computational results are discussed in the following section.

13.2.2.1 Interphase Forces

13.2.2.1.1 Interphase Drag Force In stirred reactors, bubbles experience significantly higher turbulence generated by impellers. Unless the influence of this prevailing turbulence on bubble drag coefficient is accounted, the CFD model was not found to predict the pattern of gas holdup distribution adequately. Relatively few attempts (experimental as well as numerical) have been made to understand the influence of prevailing turbulence on drag coefficient (see, for example, Refs. [18, 20–23]). Bakker and van den Akker [20], Brucato et al. [21], and Lane et al. [18] have attempted to relate the influence of turbulence on drag coefficient to the characteristic spatiotemporal scales of prevailing turbulence and therefore seem to be promising. Khopkar and Ranade [17] evaluated the three alternative proposals using a two-dimensional (2D) CFD-based model problem. They have observed that the predicted results deviate from the trends estimated by correlation of Lane et al. [18]. However, the predicted results show reasonable agreement with estimation based on correlation Bakker and van den Akker [20] (Eq. 13.12) and Brucato et al. [21] (Eq. 13.13), with 100 times lower correlation constant ($K = 6.5 \times 10^{-6}$). Interestingly, in both of these correlations, they have used volume-averaged values of the turbulent viscosity and Kolmogorov scale, respectively:

$$C_D = \frac{24}{Re^*}\left[1 + 0.15(Re^*)^{0.687}\right] \quad \therefore Re^* = \frac{\rho_l U_{slip} d_b}{\mu_l + \frac{2}{9}\mu_t} \quad (13.12)$$

$$\frac{C_D - C_{D0}}{C_{D0}} = K\left(\frac{d_b}{\lambda}\right)^3$$

$$C_{D0} = \max\left\{\left(\frac{2.667 \times Eo}{Eo + 4.0}\right), \left(\frac{24}{Re_b}\left(1 + 0.15 \times Re_b^{0.687}\right)\right)\right\} \quad (13.13)$$

where

C_D is the drag coefficient in turbulent liquid.
C_{D0} is the drag coefficient in a stagnant liquid.
d_b is bubble/particle diameter.
λ is the Kolmogorov length scale (based on volume-averaged energy dissipation rate).

The gas–liquid flow in stirred reactor was simulated using the drag coefficients estimated with volume average values of

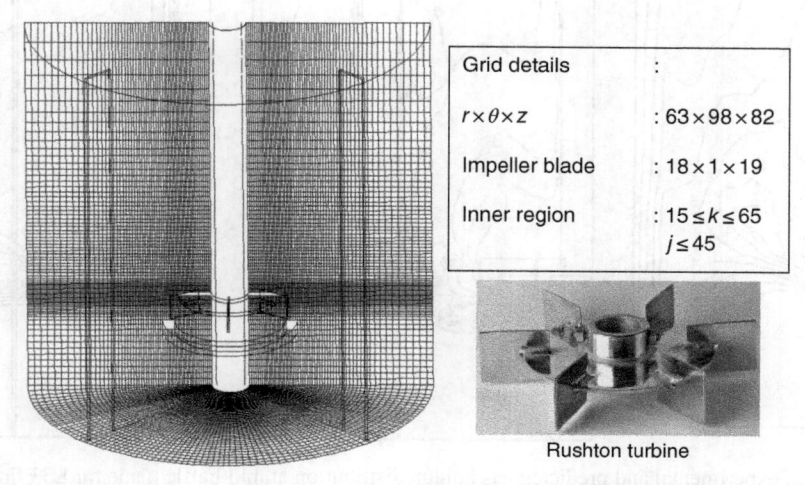

FIGURE 13.1 Computational grid and solution domain of stirred reactor. *Source*: Reprinted with permission from Khopkar and Ranade [7], Copyright 2011 John Wiley and Sons Inc.

Kolmogorov scale (Eq. 13.13) and turbulent viscosity (Eq. 13.12) for operating conditions of Fl = 0.1114 and Fr = 0.3005. This operating condition represents L33 flow regime (large 33 cavities) in stirred reactor. The quantitative comparison of the predicted gas holdup distribution with the experimental data [19] is shown in Figure 13.2. It can be seen from Figure 13.2a and b that the gas holdup distribution predicted based on Eq. (13.12) shows fairly different gas distribution from the experimental data (shown in Figure 13.2a). The major disagreement was observed in the region below the impeller. The impeller-generated flow was not sufficient to circulate gas in a lower circulation loop. The computational model has underpredicted total gas holdup (predicted holdup was 2.55% compared to the experimental measurement of 3.3%). The predicted results based on Eq. (13.13) are closer to the experimental data (see Figure 13.2a and c). This model resulted in overprediction of total gas holdup (predicted holdup was 3.97% compared to the experimental measurement of 3.3%). Despite the overprediction, the predicted gas holdup distribution showed better agreement with the data than predicted by Eq. (13.12). Equation (13.13) can therefore be recommended for carrying out gas–liquid flow simulations in stirred tanks.

13.2.2.1.2 Virtual Mass and Lift Force The other two important interphase forces are virtual mass force and lift force. The virtual mass effect is significant when the secondary phase density is much smaller than the primary phase density. The effect of the virtual mass force was first studied. The predicted gas holdup distributions obtained with and without considering virtual mass force are shown in Figure 13.3a and b. It can be seen from Figure 13.3 that the influence of the virtual mass force on the predicted pattern of gas distribution was significant only in the impeller discharge stream. However, the influence of virtual mass force was not found to be significant in the bulk of the reactor. It should be noted that the value of virtual mass coefficient used in the present study (0.5) is valid for spherical bubble and may not be appropriate for wobbling bubbles. The reported value of virtual mass coefficient is somewhat higher than 0.5 (see, for example, Ref. [24]). However, it should be noted that the predicted results are not very sensitive to the consideration of virtual mass terms. A comparison of the predicted results obtained with values of virtual mass coefficients as 0 and 0.5 did not show any significant differences (see Figure 13.3). Considering this, no specific effort was made to obtain accurate value of virtual mass coefficient.

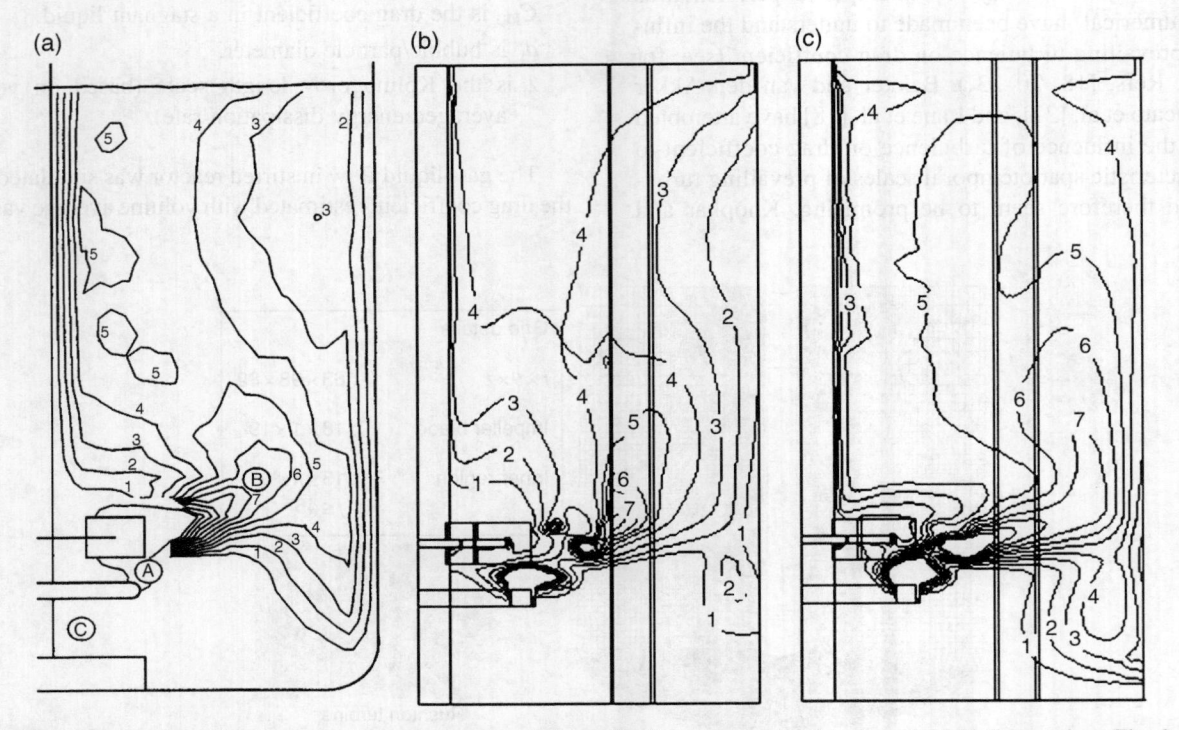

FIGURE 13.2 Comparison of experimental and predicted gas holdup distribution at mid-baffle plane for L33 flow regime, Fl = 0.1114 and Fr = 0.3005. (a) Experimental [19]. (b) Predicted results with Bakker and van den Akker correlation (Eq. 13.12). (c) Predicted results with modified Brucato et al. correlation, (Eq. 13.13). (Contour labels denote the actual values of gas holdup in percentage). *Source*: Reprinted with permission from Khopkar and Ranade [17]. Copyright 2006 American Institute of Chemical Engineers.

Similarly, the simulations were carried out with and without considering lift force. The predicted gas holdup distribution obtained with considering lift force is shown in Figure 13.3c. It can be seen from Figure 13.3 that the influence of lift force on the predicted pattern of gas distribution was significant in the impeller discharge stream and below impeller region. The predicted results with lift force predict a lower gas holdup in the region below impeller. In the upper impeller region of the reactor, the influence of lift force was not found to be significant. Therefore, it can be said that the modeling of lift force and virtual mass force may not be essential while simulating gas–liquid flow in stirred vessels.

13.2.2.2 Modeling Bubble Size Distribution

In a gas–liquid stirred reactor, gas bubbles of different sizes coexist. Very fine bubbles are observed in the impeller discharge stream (<1 mm), whereas bubbles of the size of few mm (~5 mm) are observed in the region away from the impeller [25]. The width of the bubble size distribution (BSD) depends upon the turbulence level and prevailing flow regime. Appropriate selection of bubble sizes is very important for the correct prediction of the slip velocity and mass transfer area. Both the slip velocity and mass transfer area can be more accurately estimated by modeling with local BSDs. Local bubble size or gas–liquid mass transfer can be estimated more accurately from local BSDs based on either a population balance [26] or by modeling the bubble number density function [18]. The bubble density function approach [18] is computationally less intensive and requires one additional equation to solve along with the two-fluid model. This approach predicts the Sauter mean diameter at every grid node point. The predicted results of Lane et al. [18] show reasonable agreement with the experimental data of Barigou and Greaves [25]. Laakkonen et al. [26] simulated the gas–liquid flows in a stirred reactor using population balance modeling. Their simulated results are discussed here to explain the need for modeling the BSD while simulating gas–liquid flow in stirred reactors. The details of the population balance formulation, bubble breakage, and coalescence model are not discussed here and can be found in Laakkonen et al. [26]. The influence of the prevailing turbulence on the interphase drag force was modeled with a slightly modified Bakker and van den Akker [20] correlation. The comparison of the predicted BSD and the mean bubble diameter with the experimental data is shown in Figure 13.4. The following conclusions can be drawn from the comparison between predicted results and experimental data:

- The parameters of the coalescence and breakage models were tuned to fit the experimental measurements. This

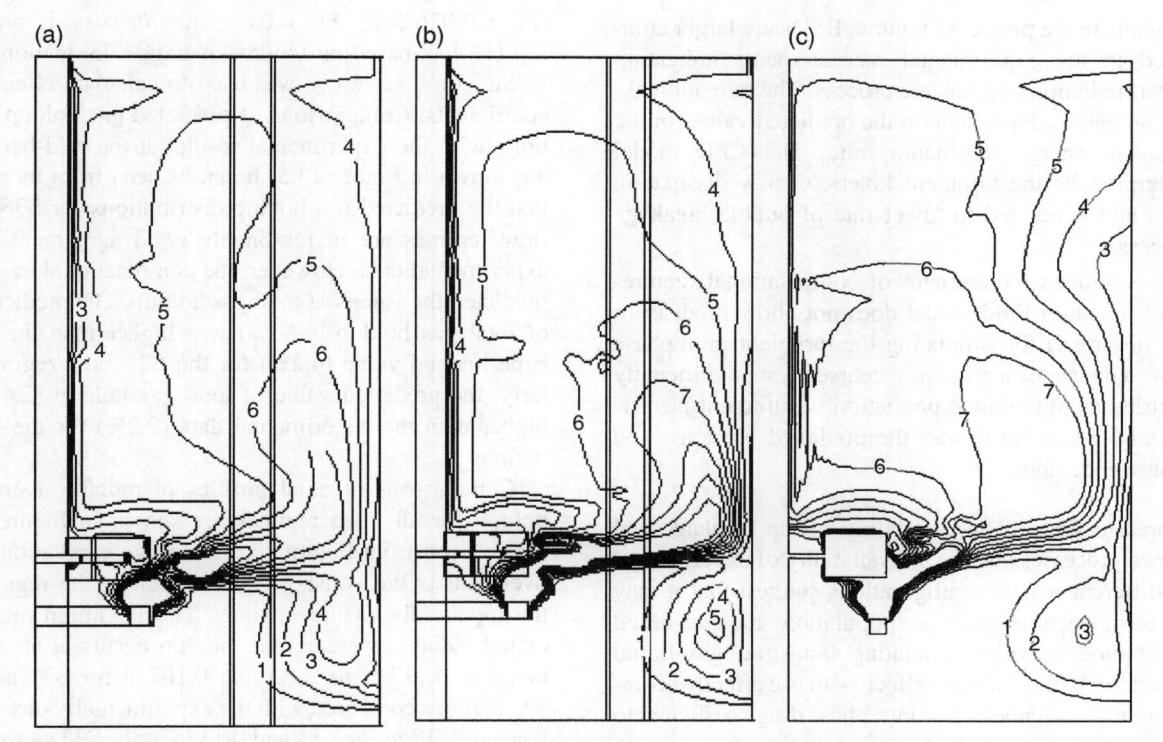

FIGURE 13.3 Comparison of predicted gas holdup profiles for with and without virtual mass and lift force effect for L33 flow regime, Fl = 0.1114 and Fr = 0.3005. (a) Predicted, without virtual mass and lift force effect: mid-baffle plane. (b) Predicted, with virtual mass effect: mid-baffle plane. (c) Predicted, with lift force effect: mid-baffle plane. (Contour labels denote the actual values of gas holdup in percentage). *Source*: Reprinted with permission from Khopkar and Ranade [17]. Copyright 2006 American Institute of Chemical Engineers.

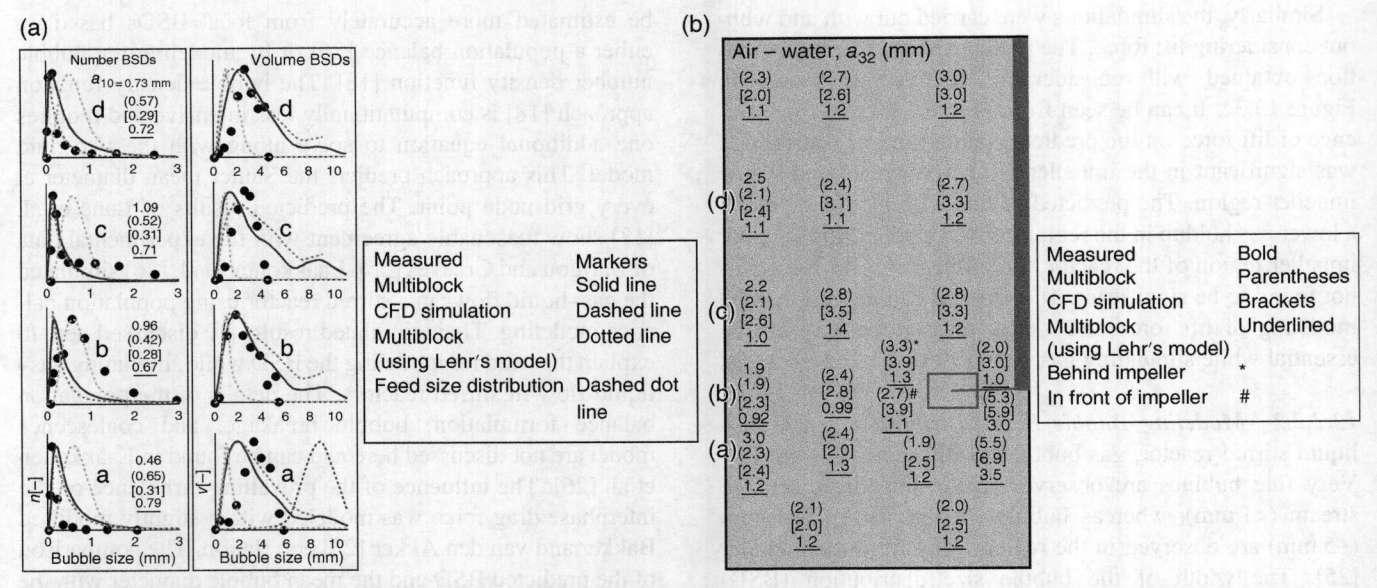

FIGURE 13.4 Comparison of predicted bubble size distribution with experimental data. (a) Local bubble size distributions in the air–water dispersion, 14 L tank, $N = 700$ rpm, and $Q = 0.7$ vvm. (b) Mean diameters (mm). *Source*: Reprinted with permission from Laakkonen et al. [26]. Copyright 2006 Elsevier Ltd.

limits the applicability of the model for different configurations of stirred reactor.

- The tails in the predicted volume BSDs are larger compared to the experimental measurements indicating underprediction of breakage process. The rate of breakage process is dependent on the predicted values of the turbulent energy dissipation rate. The CFD model underpredicts the turbulent kinetic energy dissipation rates and hence led to lower rate of bubble breakage process.
- The enormous requirement of computational requirement for multi-fluid model does not allow modeler to use fine mesh for simulating the turbulent multiphase flow. The use of a relatively coarse mesh significantly contributes to the underprediction of turbulent properties and hence influences the predicted breakage and coalescence rates.

The present state of understanding of the breakage and coalescence processes and the unavailability of experimental data for different reactor configurations suggest that it may not be advantageous to use population balance-based multi-fluid models while simulating industrial gas–liquid stirred reactors. It may be more effective to use effective combination of bubble diameter and interphase drag coefficient to get realistic results.

13.2.2.3 Gas Holdup Distribution in L33, S33, and VC Flow Regimes

Gas–liquid flows generated by the Rushton turbine in a stirred vessel were simulated for other two flow regimes representing S33 (Fl = 0.0788; Fr = 0.6) and VC (Fl = 0.026 267; Fr = 0.6). As discussed previously, Eq. (13.13), based on volume-averaged dissipation rate and Kolmogorov scale (λ), was used to calculate effective drag coefficients. Comparisons of predicted gas holdup distributions with the experimental results at the mid-baffle plane are shown in Figure 13.5. It can be seen from these figures that the predicted gas holdup distributions for S33 and VC flow regimes are in reasonably good agreement with the experimental data. However, the computational model overpredicted the values of total gas holdup. The predicted value of total gas holdup (4.85%) was higher than the reported experimental value (4.2%) for the S33 flow regime. Similarly, the predicted value of total gas holdup (2.63%) was higher than the experimental data (2.2%) for the VC flow regime.

Comparisons of axial profiles of radially averaged gas holdup for all three regimes are shown in Figure 13.6. It can be seen from Figure 13.6 that the computational model overpredicts the values of gas holdup in the region above the impeller for all three regimes. The maximum value of predicted radially averaged gas holdup occurs at an axial distance of 0.117 m for L33 and 0.107 m for S33 as well as VC regimes compared with the experimentally observed distance of 0.13 m for L33 and 0.1125 m for S33 as well as VC regimes. The predicted values of gas holdups at this maximum are underpredicted (7.3% for L33, 7.94% for S33, and 3.82 for VC) compared with the experimental value

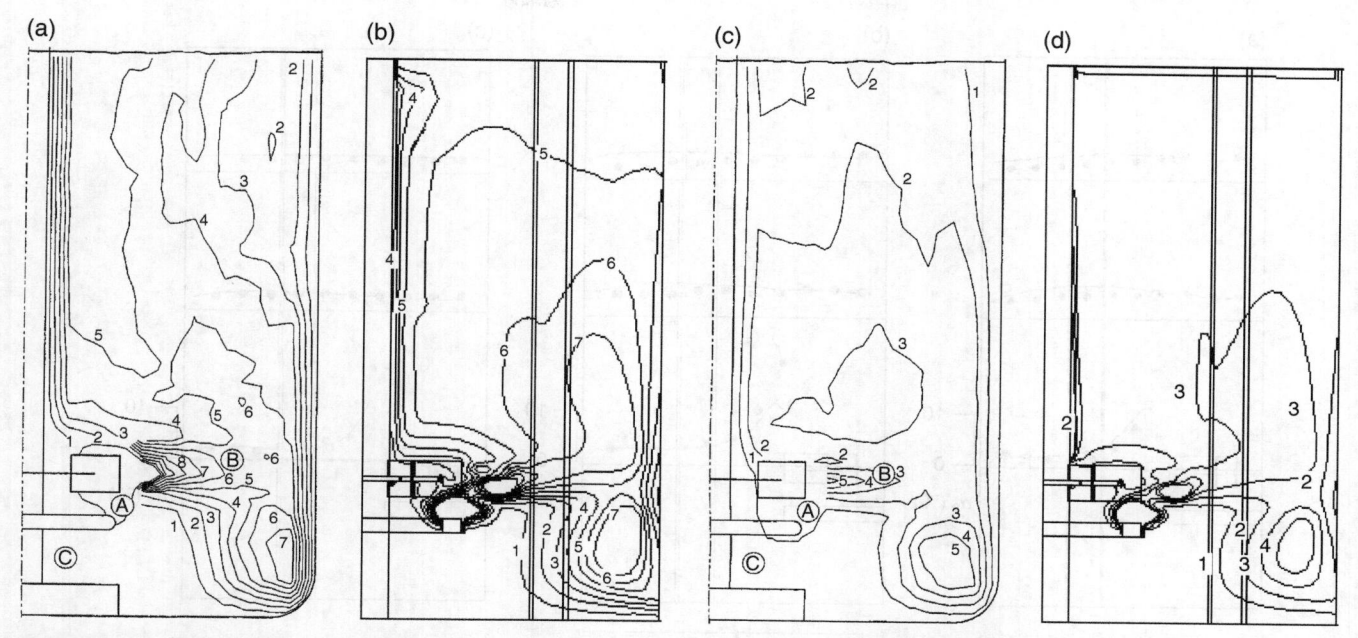

FIGURE 13.5 Comparison of experimental and predicted gas holdup distribution for S33 and VC flow regimes (experimental data of Bombac et al. [19]). (a) Experimental, S33 flow regime, Fl = 0.0788 and Fr = 0.3005 (mid-baffle). (b) Predicted, S33 flow regime, Fl = 0.0788 and Fr = 0.6 (mid-baffle). (c) Experimental, VC flow regime, Fl = 0.026 267 and Fr = 0.6 (mid-baffle). (d) Predicted, VC flow regime, Fl = 0.026 267 and Fr = 0.6 (mid-baffle). (Contour labels denote the actual values of gas holdup in percentage). *Source*: Reprinted with permission from Khopkar and Ranade [17]. Copyright 2006 American Institute of Chemical Engineers.

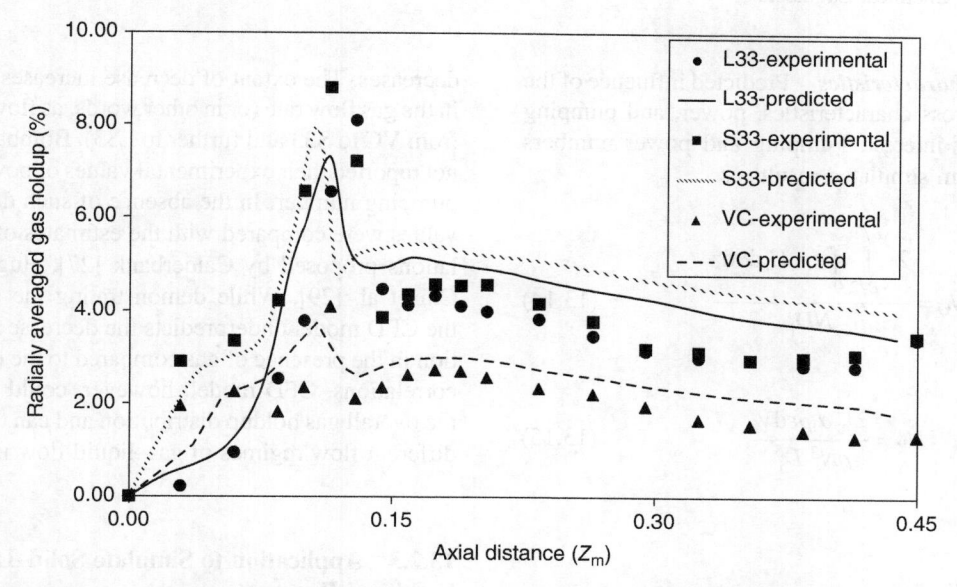

FIGURE 13.6 Comparison of predicted axial profile of radially averaged gas holdup with experimental data for L33, S33, and VC flow regimes (symbol denotes the experimental data of Bombac et al. [19]). *Source*: Reprinted with permission from Khopkar and Ranade [17]. Copyright 2006 American Institute of Chemical Engineers.

(8.1% for L33, 8.8% for S33, and 4.1% for VC). Quantitative comparisons of angle-averaged values of predicted gas holdup and experimental data at three different axial locations for all three regimes are shown in Figure 13.7. It can be seen from Figure 13.7 that comparisons of the predicted values of gas holdup and experimental data are reasonably good for all three regimes. The computational model was thus able to simulate all three regimes reasonably well.

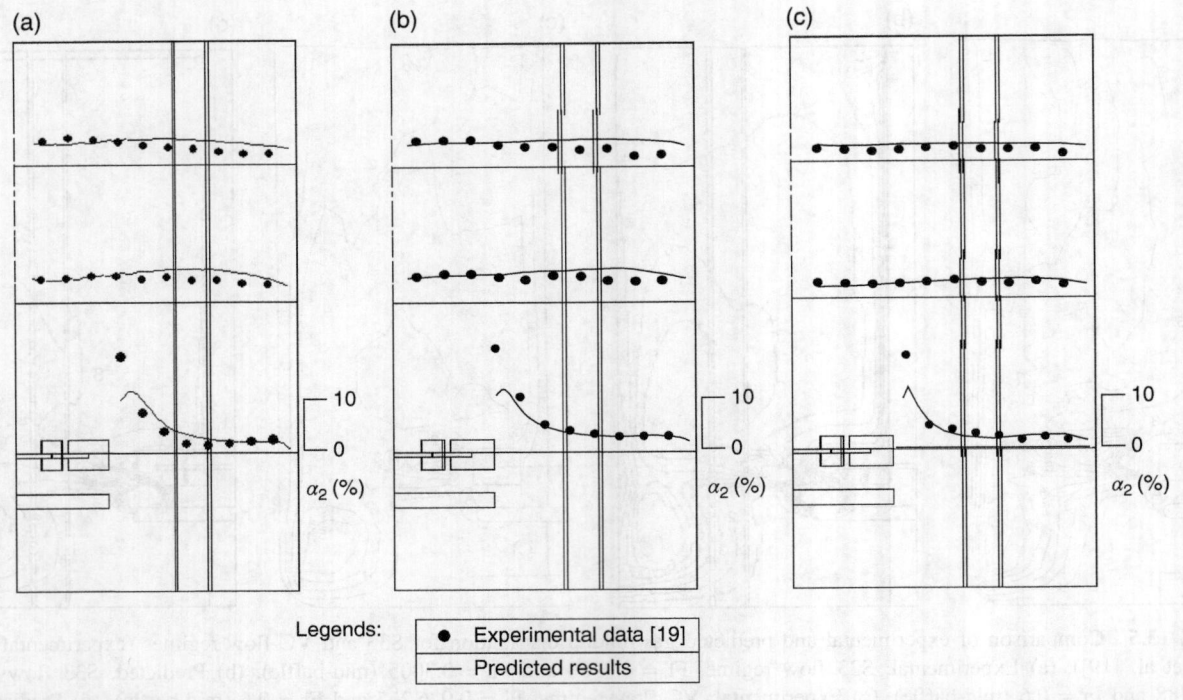

FIGURE 13.7 Comparison of predicted angle-averaged values of gas holdup (α_2) with experimental data for L33, S33, and VC flow regimes. (a) L33 flow regime, Fl = 0.1114 and Fr = 0.3005. (b) S33 flow regime, Fl = 0.0788 and Fr = 0.6. (c) VC flow regime, Fl = 0.026267 and Fr = 0.6. ● Experimental data [19]; — Predicted results. *Source*: Reprinted with permission from Khopkar and Ranade [17]. Copyright 2006 American Institute of Chemical Engineers.

13.2.2.4 Gross Characteristics

Predicted influence of the gas flow rate on gross characteristics, power, and pumping numbers is also of interest. Pumping and power numbers were calculated from simulated results as

$$N_Q = \frac{2\int_{-B/2}^{B/2}\int_0^{\pi}\alpha_1 r_i U_r \,d\theta \,dz}{ND_i^3} \quad (13.14)$$

$$N_P = \frac{2\int_V \alpha_1 \rho \varepsilon \,dV}{\rho N^3 D_i^5} \quad (13.15)$$

where

- B is blade height.
- D_i is impeller diameter.
- N is impeller speed.
- r_i is impeller radius.
- U_r is radial velocity.

The calculated values of pumping and power number from the simulated results are listed in Table 13.3. As the gas flow rate increases, impeller pumping as well as power dissipation decreases. The extent of decrease increases with an increase in the gas flow rate (or in other words, as flow regime changes from VC to S33 and further to L33). Bombac et al. [19] have not reported their experimental values of power dissipation or pumping number. In the absence of such data, the predicted values were compared with the estimates of empirical correlations proposed by Calderbank [27], Hughmark [28], and Cui et al. [29]. While demonstrating the qualitative trend, the CFD model underpredicts the decrease in power dissipation in the presence of gas compared to the estimates of these correlations. CFD model, however, could correctly capture the overall gas holdup distribution and can therefore simulate different flow regimes of gas–liquid flow in stirred reactors.

13.2.3 Application to Simulate Solid–Liquid Flow in Stirred Reactor

Suspension of solid particles in a stirred reactor either in presence or in absence of gas is commonly encountered in chemical process industry (refer to Table 13.1). All these processes involve mass transfer between the solid and liquid phases. There are various studies reported in literature, such as Nienow [30], Nienow and Miles [31], Chaudhari [32], and Conti and Sicardi [33], explaining the effect of agitation on the mass transfer coefficient, k_{SL}. They observed that the

TABLE 13.3 Gross Characteristics of an Aerated Stirred Reactor

Operating Conditions	Total Gas Holdup (%)		Predicted Results		Influence of Gas on Power Number, N_{Pg}/N_P				Influence of Gas on Pumping Number, Predicted N_{Qg}/N_Q
	Predicted	Experimental [19]	N_{Pg}	N_{Qg}	Predicted by CFD	Predicted by Empirical Correlations			
						Calderbank [27]	Hughmark [28]	Cui et al. [29]	
Single-phase flow	—	—	4.15	0.66	—	—	—	—	—
VC flow regime (Fl = 0.026267 and Fr = 0.6)	2.63	2.20	2.76	0.615	0.66	0.67	0.64	0.61	0.93
S33 flow regime (Fl = 0.0788 and Fr = 0.6)	4.85	4.20	2.196	0.6	0.53	0.47	0.49	0.41	0.9
L33 flow regime (Fl = 0.1114 and Fr = 0.3005)	3.97	3.30	1.66	0.49	0.4	0.41	0.51	0.41	0.74

Source: Reprinted with permission from Khopkar and Ranade [17]. Copyright 2006 American Institute of Chemical Engineers.

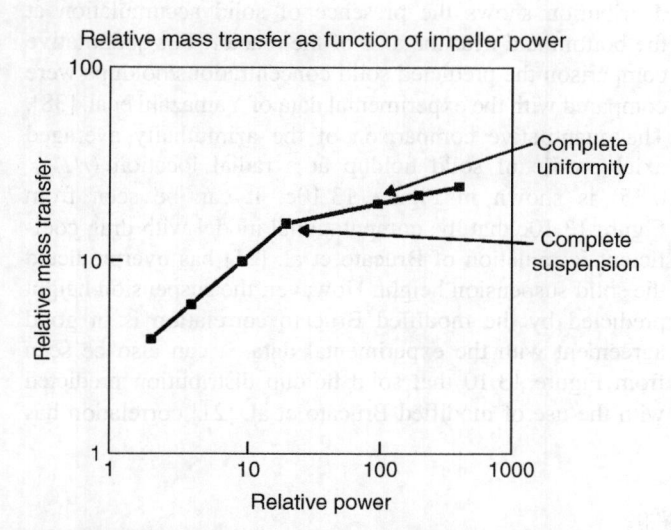

FIGURE 13.8 Influence of suspension quality on the mass transfer coefficient. Source: Reprinted with permission from Atiemo-Obeng et al. [34]. Copyright 2004 John Wiley and Sons Inc.

agitation speed influences the mass transfer coefficient (see Figure 13.8). This phenomenon was explained through the two important parameters, viz. mesoscopic availability of solids in bulk vessel volume or suspension quality and the rate of renewal of the diffusional boundary layer around the solid particle. The mass transfer curve clearly explains that before complete suspension, the mass transfer coefficient linearly increases with the impeller speed and after that the rate of increase drops. This suggests that before complete suspension, both the parameters positively increase with an increase in the impeller rotational speed. However, the mesoscopic availability of solids in bulk vessel volume does not change much after complete suspension condition, and hence the rate of increase in mass transfer rate with an increase in impeller rotational speed drops after the complete suspension condition. Additional energy dissipation does not yield much benefit in mass transfer after it.

Despite significant research efforts, prediction of design parameters to ensure an adequate solid suspension is still an open problem for design engineers. Design of stirred slurry reactors relies on empirical correlations obtained from the experimental data. These correlations are prone to great uncertainty as one departs from the limited database that supports them. Moreover, for higher values of solid concentration, very few experimental data on local solid concentration is available because of the difficulties in the measurement techniques. Considering this, it would be most useful to develop computational models, which will allow "a priori" estimation of the solid concentration over the reactor volume.

The discussed two-fluid model is applied to simulate solid–liquid flow in a stirred reactor. In addition to interphase drag force, turbulent dispersion force plays an important role while simulating solid–liquid flows. There are few studies available in the literature highlighting the influence of interphase drag force on the predicted solid holdup distribution (for example, Refs. [35–37]). To explain the influence of different interphase forces, we reproduce some of the results obtained by Khopkar et al. [37]. They have carried out simulations of solid–liquid flow in a stirred vessel in an experimental setup used by Yamazaki et al. [38]. The system investigated consists of a cylindrical flat-bottomed reactor (of diameter, $T = 0.3$ m; liquid height, $H = T$). Four baffles of width $0.1T$ were mounted perpendicular to the reactor wall. The shaft of the impeller was concentric with the axis of the reactor. A standard Rushton turbine with diameter

$D = T/3$ has been used. The impeller off-bottom clearance has been set equal to $C = T/3$, measured from the bottom of the reactor to the center of the impeller blade height. Water as liquid phase and glass beads (having density equal to 2470 kg/m^3 and particle diameter, $d_p = 264$ µm) as solid phase were used in the simulations.

Considering geometrical symmetry, half of the reactor was considered as a solution domain. It is very important to use an adequate number of computational cells while numerically solving the governing equations over the solution domain. The prediction of the turbulence quantities is especially sensitive to the number of grid nodes and grid distribution within the solution domain. In the present work, the numerical simulations for solid–liquid flows in stirred reactor have been carried out with grid size of 298 905 ($r \times \theta \times z$: $57 \times 93 \times 57$). The details of computational grid used in the present work are shown in Figure 13.9. In the present work, the standard wall functions were used to specify wall boundary conditions.

13.2.3.1 Interphase Drag Force
In stirred reactors, particles experience significantly higher turbulence generated by impellers. Similar to the discussion included in previous subsection, unless the influence of this prevailing turbulence on particle drag coefficient is accounted, the CFD model will not predict the solid suspension adequately. Khopkar et al. [37] evaluated the two alternative proposals [21, 39] using a 2D CFD-based model problem. They have observed that the predicted results deviate from the trends estimated by correlation of Pinelli et al. [39]. However, the predicted results show reasonable agreement with estimation based on correlation by Brucato et al. [21]. They correlated the predicted results by considering the sole dependence on d_p/λ for a range of solid holdup values ($5 < \alpha < 25\%$). They observed that the predicted results require ten times lower proportionality constant ($K = 8.76 \times 10^{-5}$) in Eq. (13.13) as compared with that proposed by Brucato et al. [21].

Solid–liquid flow generated by the Rushton turbine has been simulated for a solid volume fraction equal to 10.0%, $d_p = 264$ µm, and at an impeller rotation speed $N = 20$ rps. Both the formulation of drag coefficient and actual Brucato et al. [21] ($K = 8.76 \times 10^{-4}$) and the modified Brucato correlation ($K = 8.76 \times 10^{-5}$) were used for the evaluation of the interphase drag force formulation. The value of dispersion Prandtl number, σ_{pq}, has been set to the default value of 0.75. The predicted solid holdup distributions by using both drag coefficient formulation at the mid-baffle plane are shown in Figure 13.10a and b. It can be seen from Figure 13.10a that with actual Brucato et al. [21] correlation, the computational model has predicted almost complete suspension of the solid particles. However, the simulated solid holdup distribution using modified Brucato et al. [21] did not capture the complete suspension of solid particles in stirred reactor (see Figure 13.10b). The simulated solid holdup distribution shows the presence of solid accumulation at the bottom and near the axis of the reactor. For quantitative comparison the predicted solid concentrations/holdups were compared with the experimental data of Yamazaki et al. [38]. The quantitative comparison of the azimuthally averaged axial profile of solid holdup at a radial location ($r/T = 0.35$) is shown in Figure 13.10c. It can be seen from Figure 13.10c that the computational model with drag coefficient formulation of Brucato et al. [21] has overpredicted the solid suspension height. However, the suspension height predicted by the modified Brucato correlation is in good agreement with the experimental data. It can also be seen from Figure 13.10 that solid holdup distribution predicted with the use of modified Brucato et al. [21] correlation has

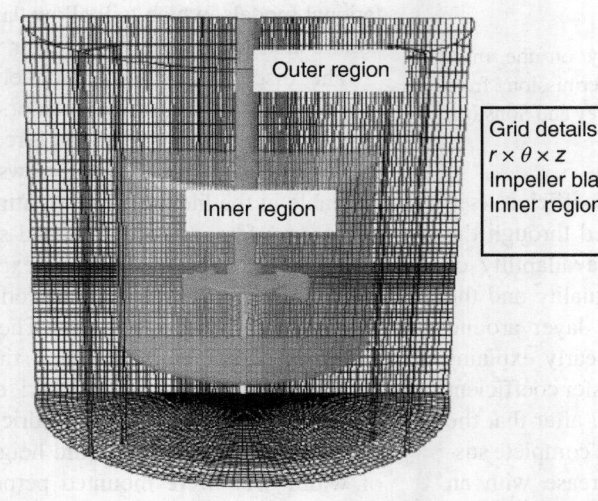

FIGURE 13.9 Computational grid and solution domain of stirred reactor. *Source*: Reprinted with permission from Khopkar et al. [37]. Copyright 2006 American Chemical Society.

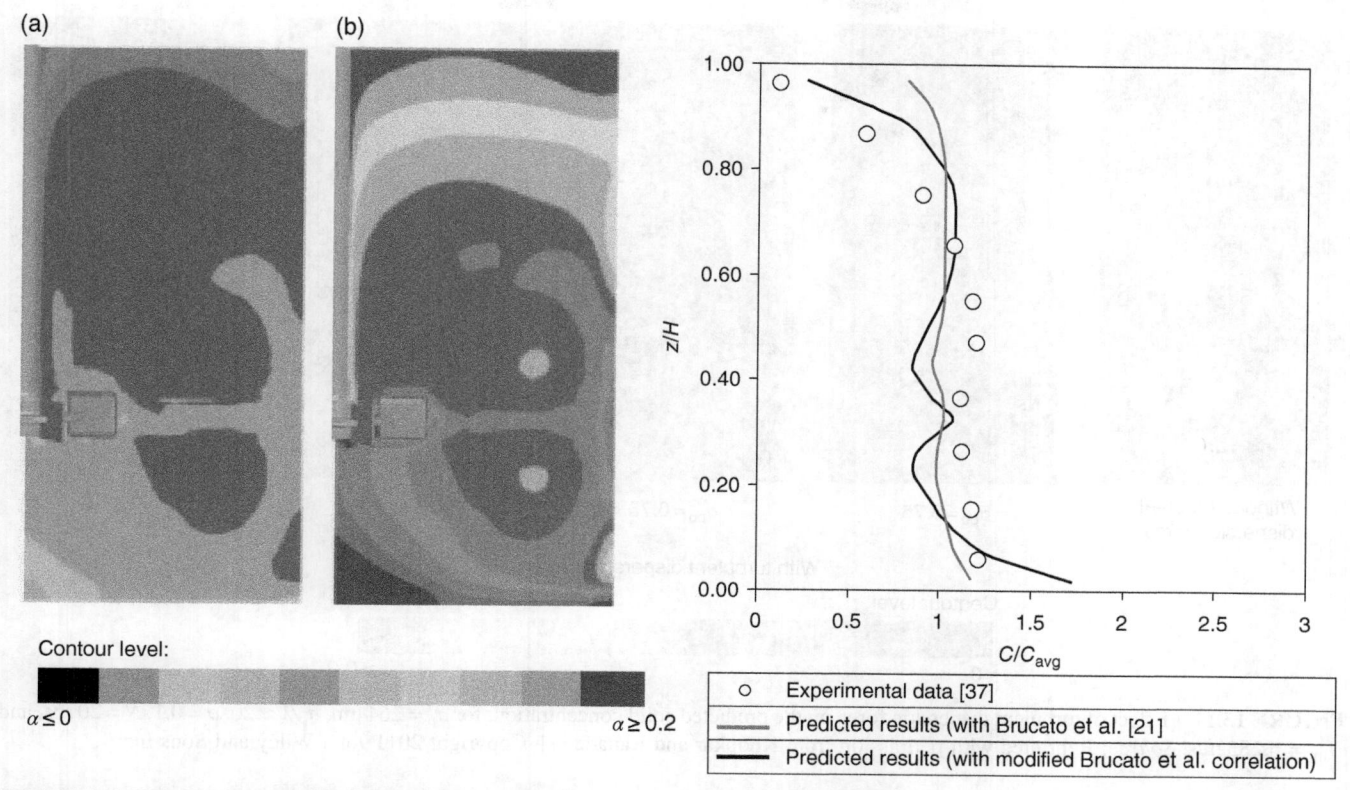

FIGURE 13.10 Simulated solid holdup distribution at mid-baffle plane, for $d_p = 264$ μm, $d_p/\lambda \approx 20$, $\alpha = 0.1$, $N = 20.0$ rps, and $U_{tip} = 6.283$ m/s. (a) $K = 8.76 \times 10^{-4}$. (b) $K = 8.76 \times 10^{-5}$. (c) Comparison of predicted results with experimental data. *Source*: Reprinted with permission from Khopkar et al. [37]. Copyright 2006 American Chemical Society.

captured the presence of higher solid concentration in the impeller discharge stream (a bell shaped in the concentration profile), which is a characteristic of the solid–liquid flow generated by Rushton turbine. However the prediction with Brucato et al. [21] correlation does not show any such characteristics. Overall, it can be said that the modified Brucato et al. [21] correlation predicted solid holdup distribution in stirred vessel more accurately.

13.2.3.2 Turbulent Dispersion Force

The developed computational model was then extended to study the influence of the turbulent dispersion force on the suspension quality in the stirred reactor. The magnitude of the turbulent dispersion force was varied by varying the value of the dispersion Prandtl number, σ_{pq}, in the range of 0.0375–3.75. Figure 13.11 shows the comparison of the predicted solid concentration distribution in the stirred reactor with the experimental data of Yamazaki et al. [38] with and without turbulent dispersion force. It can be seen from Figure 13.11 that the turbulent dispersion has a significant effect on the predicted suspension quality in the stirred reactor. The computational model predicted a more uniform suspension with a decrease in the value of dispersion Prandtl number. This is expected as the drift velocity (or turbulent dispersion force) is inversely proportional to the dispersion Prandtl number (see Eq. (13.3)). Decreasing the latter means increase in the turbulent dispersion force, which consequently results in more dispersion of the particles, resulting in more uniform suspension. Overall, it can be said that the simulations carried out with $\sigma_{pq} = 0.375$ and 0.0375 have overpredicted the suspension quality. However, for $\sigma_{pq} = 3.75$ the computational model has underpredicted the suspension quality. Therefore, CFD simulation of solid–liquid stirred reactor needs to be carried out with $\sigma_{pq} = 0.75$ for adequate prediction of suspension quality.

13.2.4 Application to Simulate Gas–Solid–Liquid Flow in Stirred Reactor

Suspension of solid particles in the presence of gas has various applications in the process industry. These applications include catalytic hydrogenations, oxidations, fermentations, evaporative crystallizations, and froth flotation. In a gas–liquid–solid system, the impeller plays a dual role of keeping the solids suspended in the liquid while dispersing the gas bubbles. Dyalag and Talaga [40] have found that in a gas–liquid–solid stirred reactor, the gas phase is always uniformly dispersed before the solids are completely suspended.

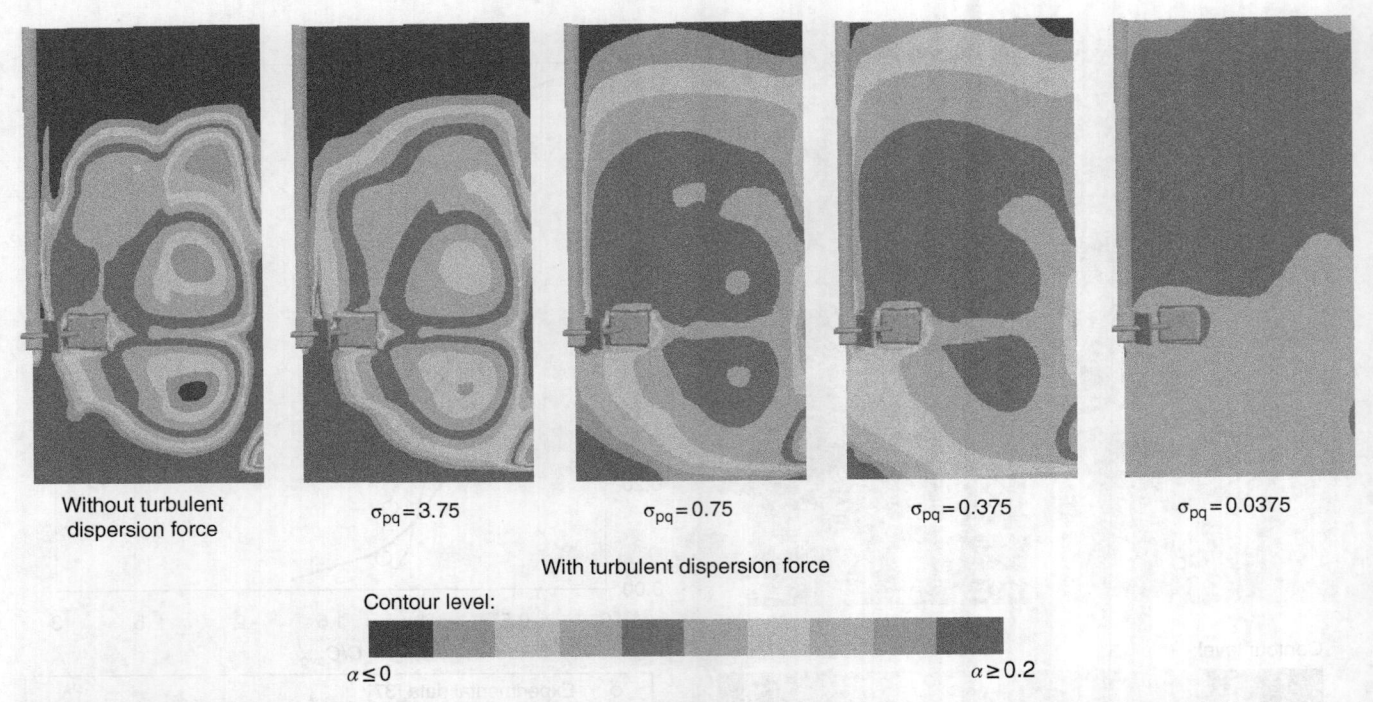

FIGURE 13.11 Effect of turbulent dispersion force on the predicted solid concentration, for $d_p = 264$ μm, $d_p/\lambda \approx 20$, $\alpha = 0.1$, $N = 20$ rps, and $U_{tip} = 6.283$ m/s. *Source*: Reprinted with permission from Khopkar and Ranade [7]. Copyright 2011 John Wiley and Sons Inc.

Therefore, the formation of a completely dispersed three-phase system depends on the condition under which the solids are suspended by the impeller action. Identification of these operating conditions is very important for operating a stirred reactor in an energy-efficient mode. Some mass transfer studies (for example, Ref. [31]) in two-phase solid–liquid mixing in stirred tanks have also shown that the particle–fluid mass transfer rate is comparable at the just off-bottom suspension (JS) point, irrespective of the power input level. Any incremental power input beyond this point for improving the mass transfer coefficient is often uneconomical. Attempts to extend the above hypothesis to three-phase systems introduce an additional complexity, as the impeller pumping efficiency changes in the presence of gas. It is also observed that the tank, impeller, and sparger geometry variations that have been proposed [41–43] are highly system specific with respect to gas–liquid and solid–liquid systems and may not be possible to directly extend to gas–liquid–solid systems. It is therefore necessary to develop tools to examine the role of reactor hardware in meeting the demands associated with the simultaneous gas dispersion and solid suspension.

Critical analysis of available literature suggests that practically no information is available in the literature on the CFD simulation of three-phase gas–liquid–solid stirred reactor. Complex interactions between the solid particles, gas bubbles, and the liquid phase make the fluid dynamics of three-phase stirred reactor very complex. Recently, Murthy et al. [44] made an attempt to simulate a three-phase stirred reactor. They used the approach proposed by Khopkar and Ranade [17] for modeling gas–liquid flow and approach of Pinelli et al. [39] for simulating solid–liquid flow. Murthy et al. [44] were able to predict the critical impeller speed required for solid suspension. However, their study was limited to very low solids loading (maximum solids loading is <10 wt %). The applicability of the same computational model to simulate solid suspension at higher solids loading (>20 wt % or 10% by volume fraction) is not known. In this work, a CFD model was developed to simulate solid suspension in a three-phase stirred reactor. The approaches discussed in the last two subsections were used to model the gas–liquid and solid–liquid interactions in the three-phase stirred reactor.

Experimental setup of Pantula and Ahmed [45] was used to simulate gas–liquid–solid flows in a stirred reactor. The system investigated consists of a cylindrical flat-bottomed reactor (of diameter, $T = 0.4$ m; liquid height, $H = T$). Four baffles of width $0.1T$ were mounted perpendicular to the reactor wall. The shaft of the impeller was concentric with the axis of the reactor. A standard Rushton turbine with diameter $D = T/3$ has been used. The impeller off-bottom clearance has been set equal to $C = T/4$, measured from the bottom of the reactor to the center of the impeller blade height. A ring sparger of diameter $D_s = 2D/3$ with evenly spaced holes at a clearance (C_s) of $T/6$ was provided for gas input. Water as liquid phase, air as gas phase, and glass beads (having density equal to 2500 kg/m³ and particle diameter $d_p = 174$ μm) as solid phase were used in the simulations. Simulations were carried out with solid-phase volume fraction equal to 12% (i.e. 30 wt %).

Considering the geometrical symmetry, half of the reactor was considered as a solution domain. In the present work, the numerical simulations for gas–liquid–solid flows in stirred reactor have been carried out with grid size of 436 170 ($r \times \theta \times z$: 70 × 93 × 67). The details of computational grid used in the present work are shown in Figure 13.12. In the present work, the standard wall functions were used to specify wall boundary conditions.

13.2.4.1 Solid Suspension in an Aerated Stirred Vessel

Minimum speed for JS is a very important hydrodynamic parameter for designing gas–liquid–solid stirred reactor. Experimental studies so far on gas–liquid–solid suspensions have clearly indicated the requirement of increased suspension speed, thereby more power input, on the introduction of gas [40–43]. This is because of a decrease in impeller pumping efficiency and power draw due to the formation of ventilated cavities behind the impeller blades on gassing [46]. Recently, Zhu and Wu [47] carried out experimental measurements in a three-phase stirred reactor to determine the JS speed for a variety of solid sizes, solids loading, impeller sizes, and tank sizes. They suggested the possibility of relating relative just off-bottom suspension speed ($RJSS$) with just suspension aeration number (based on just suspension speed for solid–liquid system). They also observed that the proposed relation was independent of impeller size, solid size, solids loading, and tank size and can be used to scale up laboratory data to full-scale vessel. The same definition (Eq. 13.16) was used in the present study to identify the JS speed for different gas flow rates:

$$RJSS = 1 + mNa_{js}^n$$
$$RJSS = \frac{N_{jsg}}{N_{js}} \quad \text{and} \quad Na_{js} = \frac{Qg}{N_{js}D^3} \quad (13.16)$$

where

m and n are constants.

For the Rushton turbine, the values of m and n are 2.6 and 0.7, respectively. The simulations were carried out for three just suspension aeration numbers: 0, 0.025, and 0.05. The impeller rotational speeds (N_{jsg}) for the three aeration numbers are 9.30, 11.16, and 12.27 rps, respectively.

The predicted solid holdup distribution at mid-baffle plane for all three aeration numbers is shown in Figure 13.13. It can be seen from Figure 13.13 that for three-phase system the computational model has predicted more accumulation of solids at the bottom of reactor near the central axis in comparison with the two-phase system. The predicted cloud height values were also found to drop in the presence of gas. The predicted solid volume fraction values were then used to describe the suspension quality in the reactor. The criterion based on the standard deviation value, calculated using Eq. (13.17), was used to describe suspension quality for all three cases. It was observed that the computational model predicted standard deviation value (σ) equal to 0.45 for two-phase flow. However, for three-phase flow, computational model predicted σ equal to 0.82 and 0.90 for Na_{js} equal to 0.025 and 0.05, respectively. Overall, it can be said that the computational model has predicted JS condition for two-phase flow ($\sigma < 0.8$), but incomplete suspension for three-phase system ($\sigma > 0.8$):

$$\sigma = \sqrt{\frac{1}{n}\sum_{i=1}^{n}\left(\frac{\alpha_i}{\alpha_{avg}} - 1\right)^2} \quad (13.17)$$

The predicted gas holdup distribution at mid-baffle plane for two suspension aeration numbers is shown in Figure 13.14. It can be seen from Figure 13.14 that for both conditions, the computational model has predicted higher values of gas holdup in both circulation loops of flow. This indicates that the computational model predicted complete dispersion condition of gas phase in the vessel. These

FIGURE 13.12 Computational grid and solution domain of stirred reactor. *Source*: Reprinted with permission from Khopkar and Ranade [7]. Copyright 2011 John Wiley and Sons Inc.

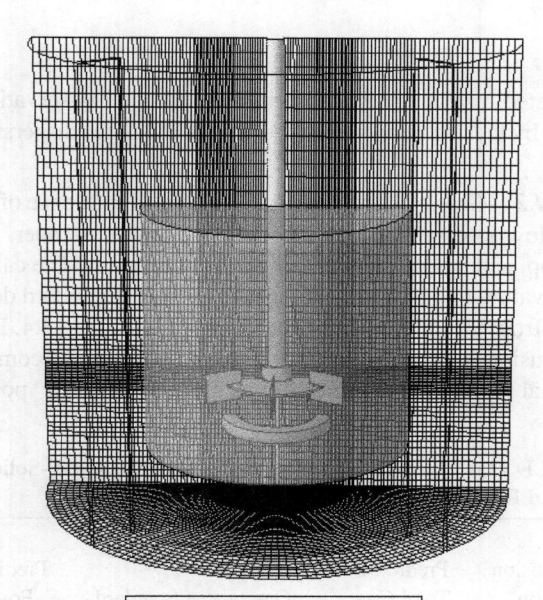

Grid details :
$r \times \theta \times z$: 70 × 93 × 67
Impeller blade : 20 × 1 × 15
Inner region : 8 ≤ k ≤ 55
  j ≤ 51

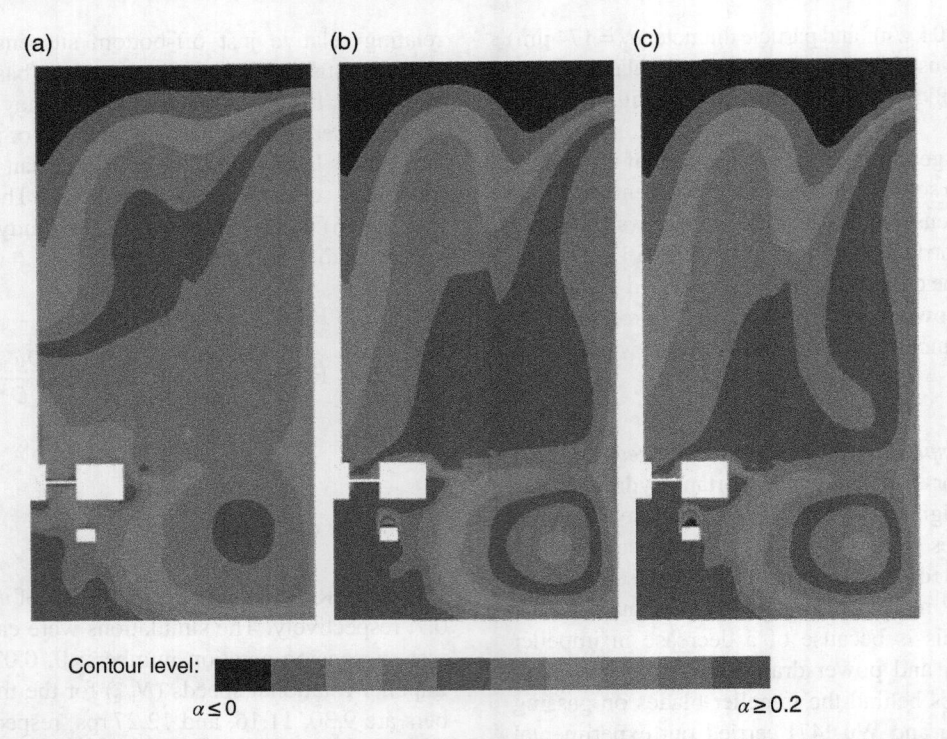

FIGURE 13.13 Simulated solid holdup distribution at mid-baffle plane, for $d_p = 174\,\mu m$ and $\alpha_s = 0.12$. (a) $Na_{js} = 0$ and $N = 9.3$ rps. (b) $Na_{js} = 0.025$ and $N = 11.16$ rps. (c) $Na_{js} = 0.05$ and $N = 12.27$ rps. *Source*: Reprinted with permission from Khopkar and Ranade [7]. Copyright 2011 John Wiley and Sons Inc.

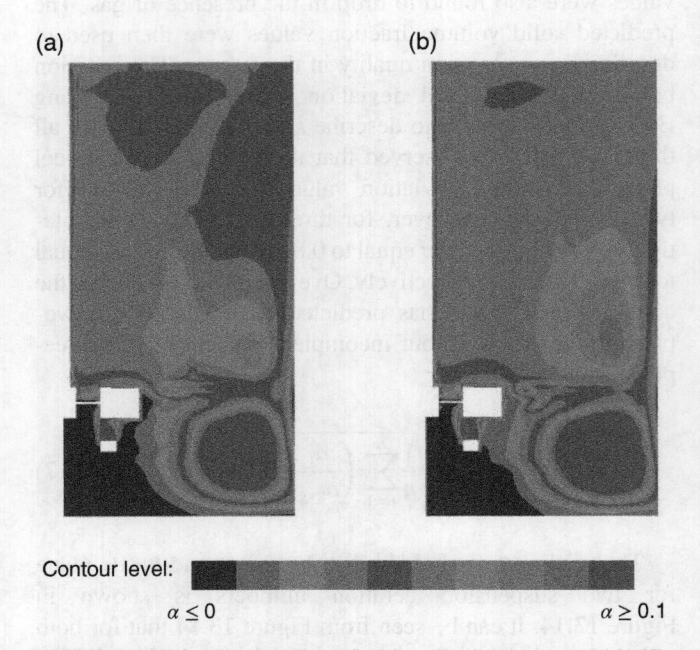

FIGURE 13.14 Simulated gas holdup distribution at mid-baffle plane, for $d_p = 174\,\mu m$ and $\alpha_s = 0.12$. (a) Najs = 0.025. (b) Najs = 0.05. *Source*: Reprinted with permission from Khopkar and Ranade [7]. Copyright 2011 John Wiley and Sons Inc.

predicted results also support the experimental observations made by Dyalag and Talaga [40] on quality of gas dispersion.

13.2.4.2 Gross Characteristics

Predicted influence of the gas flow rate on gross characteristics, power number, gas holdup, and suspension quality is also of interest. The calculated values of power number, gas holdup, and standard deviation from the simulated results are listed in Table 13.4. Few conclusions can be drawn from Table 13.4. First, the computational model has predicted the drop in impeller power

TABLE 13.4 Gross Characteristics of a Gas–Liquid–Solid Stirred Reactor

Just Suspension Aeration Number, Na_{js}	Predicted Total Gas Holdup (%), ε_g	Estimated Gas Holdup (%), [45]	Standard Deviation, σ	Predicted Power Number, N_P
0	—	—	0.45	3.95
0.025	5.33	4.80	0.81	2.61
0.05	6.45	9.55	0.90	2.54

Source: Reprinted with permission from Khopkar and Ranade [7]. Copyright 2011 John Wiley and Sons Inc.

number value in the presence of gas. While demonstrating the qualitative trend, the CFD model has underpredicted the actual power number value (predicted value of power number for single-phase flow equal to 3.85). Second, the CFD model was able to predict just suspension condition for solid–liquid flows. However, the model failed to predict just suspension condition in the presence of gas. The standard deviation value (describing suspension quality) increases with an increase in gas flow rate. For lower aeration rate the model has predicted the total gas holdup values reasonably well. However, model has underpredicted total gas holdup value for higher aeration rate. Overall, CFD model with the presented modeling approach could reasonably predict gas–liquid–solid flow in a stirred reactor at low aeration rates. Further work is needed to develop adequately accurate model capable of simulating gas–liquid–solid flow in stirred reactor at higher aeration rates.

13.2.5 Application to Simulate Liquid–Liquid Flows in Stirred Reactor

Stirred tanks represent the most popular reactors and mixers that are widely used in carrying out operations involving liquid–liquid dispersions. Drop size distributions and dynamics of their evolution are important characteristics of such dispersions as they are related to the rate of mass transfer and chemical reactions that may occur in a process. In some cases the drops are stabilized against coalescence by the addition of stabilizers to have drops sized by agitation before chemical reaction begins (suspension polymerization). In other areas the breakup and coalescence processes can affect directly a reaction in the dispersed phase. It is well known that other than physical chemistry, fluid dynamic interaction between the two phases plays a significant role in determining the features of the dispersion, but it is far from being fully understood. Average properties over the whole vessel are usually considered for system description and for scale-up. The following main aspects have been studied: minimum agitation speed for complete liquid–liquid dispersion [48], correlation of mean drop size and DSD to energy dissipation rate and mixer geometric parameters [49] as well as to energy dissipation rate and flow in the vessel [50], the influence of various impellers on the dispersion features [50–52], and description of the interaction between the liquid phases in terms of intermittent turbulence [53–56].

CFD modeling of these systems has also been attempted in recent years by using the Eulerian–Eulerian approach coupled with breakup and coalescence models (see, for example, [57–61]). All these efforts are analogous to the efforts made for simulating gas–liquid stirred reactor. In spite of the highly complex system and significant simplifications, the first results are encouraging [57]. To explain the CFD modeling of liquid–liquid stirred reactor, we have reproduced some of the results obtained by Alopaeus et al. [57] here.

Alopaeus et al. [57] simulated liquid–liquid dispersion in a stirred vessel coupled with population balance equations (PBE). Readers are requested to refer to Alopaeus et al. [57] for the working equation of population balance simulation. The general PBE call for the drop rate functions and convection terms before it can be used for simulating drop size distributions. In liquid–liquid dispersion the dispersed drops first deform and then break. The magnitude of deformation and breakage depends on the flow pattern around the drop. Most often the systems characterized were having low dispersed phase viscosity. Such drops break up, provided that the local instantaneous turbulent stresses exceed the stabilizing forces due to the interfacial tension. Therefore, the earlier drop breakage models were only function of turbulence present in the continuous phase. In most of the practical applications, dispersion of high viscosity drops is commonly encountered. In such situations, contribution of the local turbulence on the drop breakage is not sufficient for modeling drop breakage. A viscous drop exposed to the pressure fluctuations causing its deformation tries to return to spherical shape by the action of stabilizing stresses. Therefore, stabilizing effect is found to be dominant in high viscosity dispersed phase. One can assume that for the breakage of drop, normal turbulent stress outside the drop has to be greater than the sum of viscous stresses developed within the drop due to deformation and stress due to interfacial tension. Narsimhan et al. [62] used both viscous and interfacial forces for estimating breakage frequency. Alopaeus et al. [57] used the same model for simulating breakage frequency in PBE.

Coalescence of two drops depends on two subprocesses, viz. collision between two drops and drainage of film between two drops. Alopaeus et al. [57] used frequency of both these processes to estimate coalescence efficiency. They carried out preliminary simulations with multi-block model (see Ref. [57]) for fitting the parameters of breakage and coalescence efficiency for dense dispersion. Alopaeus et al. [57] simulated dispersion of Exxsol in water in a 50 L stirred reactor equipped with Rushton turbine. Twenty drop size groups were used in the population balance model, with constant viscosity and density for both of the phases. Each group was introduced as mass fraction using user-defined scalars. The conservation equations for user-defined scalars are solved for each cell. Thus, only the source terms for drop breakage and coalescence had to be introduced. Alopaeus et al. [57] did not model the effects of drop size distribution and the volume fraction of the dispersed phase on the prevailing turbulence. They only modeled the effect of the population balance model on velocity, and turbulence calculation is through density. The predicted distribution of Sauter mean drop diameter and turbulent kinetic energy distribution is shown in Figure 13.15. The comparison of the predicted local Sauter mean drop diameter with experimentally measured data at three different locations is shown in Table 13.5. It can be seen from Table 13.5 that the CFD model was able to predict local

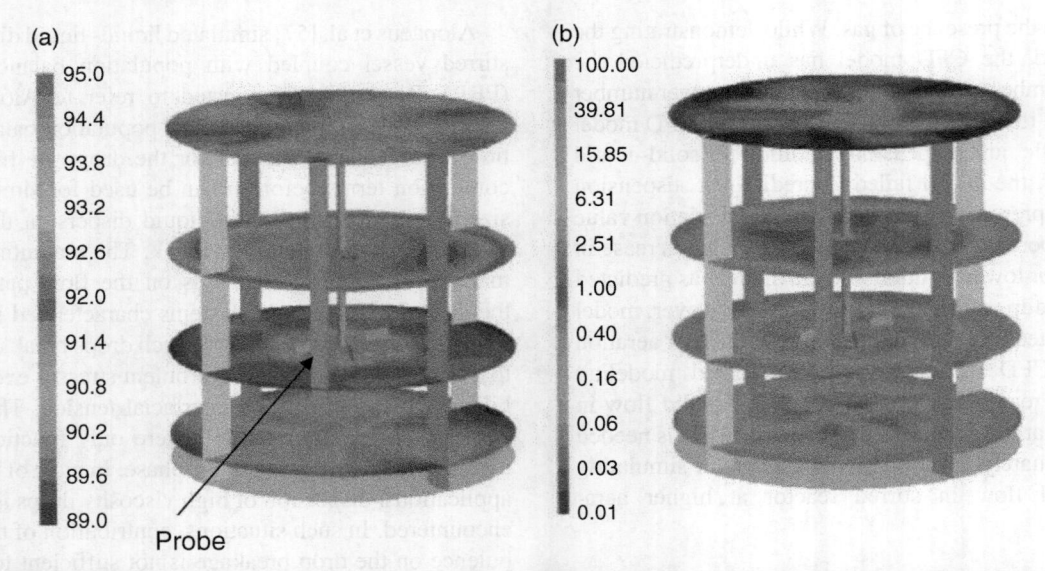

FIGURE 13.15 Predicted distribution of Sauter mean diameter and turbulent kinetic energy dissipation rate, at heights of 0.03, 0.133, 0.25, and 0.4 m. (a) Sauter mean diameter, μm. (b) Turbulent kinetic energy dissipation rate (W/kg). *Source*: Reprinted with permission from Alopaeus et al. [57]. Copyright 2002 Elsevier Science Ltd.

TABLE 13.5 Comparison of Predicted Values of Sauter Mean Diameter with Experimental Data

Point	Measured Value (μm)	Predicted Tangential Distribution (μm)	Predicted Value (μm)
1	93.2	93.3–94.3	93.8
2	91.1	93.2–94.1	94.1
3	88.7	89.7–91.8	90.5

Source: Reprinted with Permission from Alopaeus et al. [57]. Copyright 2002 Elsevier Science Ltd.

Sauter mean drop diameters reasonably well. Overall, CFD model with discussed modeling approach could reasonably predict dense liquid–liquid flow in a stirred reactor. Further work is needed to evaluate CFD model for simulating dispersion of high viscosity drops in a stirred reactor.

PART II: APPLICATION TO MIXING IN STIRRED VESSELS

13.3 APPLICATION TO ENGINEERING OF STIRRED VESSELS

Engineering of stirred vessels involves designing of vessel configuration and operating protocols to realize desired chemical and physical transformations. A reactor engineer has to ensure that the evolved reactor hardware and operating protocol satisfies various process demands without compromising safety, environment, and economics. Engineering of stirred reactors essentially begins with the analysis of process requirements. This step is usually based on laboratory study and on reactor models based on idealized fluid dynamics and mixing. In most of the industrial cases, this step itself may involve several iterations, especially for multiphase systems. Converting this understanding of process requirements to configuration and operating protocols for industrial reactor proceeds through several steps, such as examining sensitivity of reactor performance with various flow and mixing related issues (short circuiting, bypass, residence/circulation time distribution) and resolving conflicting process requirements and scale-up.

Not much progress can be made without better understanding of the underlying fluid dynamics of stirred reactors and its relation with the variety of design parameters on one hand and with the processes of interest on the other. Experimental investigations have contributed significantly to the better understanding of the complex hydrodynamics of stirred vessels in the recent years. However, computational models offer unique advantages for understanding conflicting requirements of different processes and their subsequent prioritization. Using a computation model, one can switch on and off various processes, which otherwise is not possible while carrying out experiments. Such numerical experiments can give useful insight into interactions between different processes and can help to resolve the conflicting requirements.

It is essential to analyze possible influence of scale of reactor on its fluid dynamics and performance. It should be noted that small-scale reactor would invariably have higher shear and more rapid circulation than large-scale reactor.

Multiphase processes, therefore, are often dispersion controlled in small scale and are coalescence controlled in large-scale reactor. The interfacial area per unit volume of reactor normally reduces as the scale of reactor increases. Scale-up/scale-down analysis is important to plan useful laboratory and pilot plant tests. It may be often necessary to use pilot reactor configuration, which is not geometrically similar to the large-scale reactor in order to maintain the similarity of the desired process. Conventionally such analysis is carried out based on certain empirical scaling rules and prior experience. CFM can make substantial contributions to this step by providing quantitative information about the fluid dynamics. Computational flow models, which allow "a priori" predictions of the flow generated in a stirred reactor of any configuration (impellers of any shape), with just the knowledge of geometry and operating parameters, can make valuable contributions in evolving optimum reactor designs.

Recent advances in physics of flows, numerical methods, and computing resources open up new avenues of harnessing power of CFM for engineering of stirred vessels. It is however important to use this power judiciously. Conventional reaction engineering models and accumulated empirical knowledge about the hydrodynamics of stirred vessels must be used to get whatever useful information that can be obtained before undertaking rigorous CFD modeling. Distinguishing the "simple" (keeping the essential aspects intact and ignoring nonessential aspects) and "simpler" (ignoring some of the crucial issues along with the nonessential issues) formulations is a very important step toward finding useful solutions to practical problems. More often than not CFM projects are likely to overrun the budget (of time and other resources) due to inadequate attention paid to this initial step of the overall project.

Another important point is that it is beneficial and more efficient to develop computational flow models in several stages rather than directly working with and developing a one-stage comprehensive model. For example, even if the objective is to simulate non-isothermal reactive multiphase flows, it is always useful to undertake a stagewise development. Such stages could be like (i) simulating isothermal single-phase flow; (ii) evaluating isothermal turbulent simulations, verifying existence of key flow features, and using the simulations to extract useful quantities such as circulation time distributions; (iii) including non-isothermal effects (without reactions); (iv) including multiphase models; and (v) including reactive mixing models. Such a multistage development process also greatly reduces various numerical problems, as the results from each stage serve as a convenient starting point for the next stage. The stagewise process also provides insight about relative importance of different processes, which helps to make judicious choice between "simple" and "simpler" representations.

More often than not, in many practical situations, models and results obtained at intermediate stages of such a stepwise process can provide useful support for decision making and continuous improvements without waiting for complete development of a comprehensive model. In this section, we illustrate application of computational flow models discussed in previous section to obtain useful information to some of the industrially relevant cases. It may not be possible to present actual case studies for various reasons. The presented examples may however be useful to indicate power and methodology of applying CFM to address industrially relevant multiphase mixing issues.

There are various practical multiphase problems whose solutions can be obtained by solving them with simplification of single-phase flow. For example, in slurry polymerization, residence time of catalyst particle in stirred vessel determines the final product particle size distribution. It is not necessary to include whole complex multiphase reactive system for simulating the particle size distribution. A smart process engineer would simulate the residence time distributions (RTD) of the feed stream for various operating conditions and/or reactor design modifications and then relate them to the product particle size distribution. Right definition of the problem with correct simplification will definitely help process engineer getting solution in a very short time. These simulations can help in optimizing mesoscale flow properties essential for determining convective transport processes and its implication on the process performance. In the following subsection, some examples on single-phase mixing are discussed, which has potential to be extended to multiphase flow.

13.3.1 Single-Phase Mixing

13.3.1.1 Shortstop of Runaway Reaction Through Reaction Inhibition Classical reactor analysis and design engineering commonly assume one of two idealized flow patterns: plug flow or completely back-mixed flow. Real reactors may approach one of these; however, it is often the nonidealities and their interaction with chemical kinetics that lead to poor reactor design and performance [63]. For well-designed batch stirred tank with simple reaction schemes and kinetics that are slow relative to the mixing time, the completely mixed flow (CMF) works well. But in many practical situations, the kinetics is relatively faster than the mixing process. In such situations, addition location of reactant or additive and prevailing fluid dynamics plays very important role. Validated computational models play a very important role not only in elucidating the implications of reaction kinetics, mixing time, and additive addition or inlet/outlet locations but also in designing effective operating protocols or desired design modifications for meeting the process requirement. To explain the influence of addition location and amount of additive on the mixing-controlled reaction performance, we reproduce some results obtained by Dakshinamoorthy et al. [64].

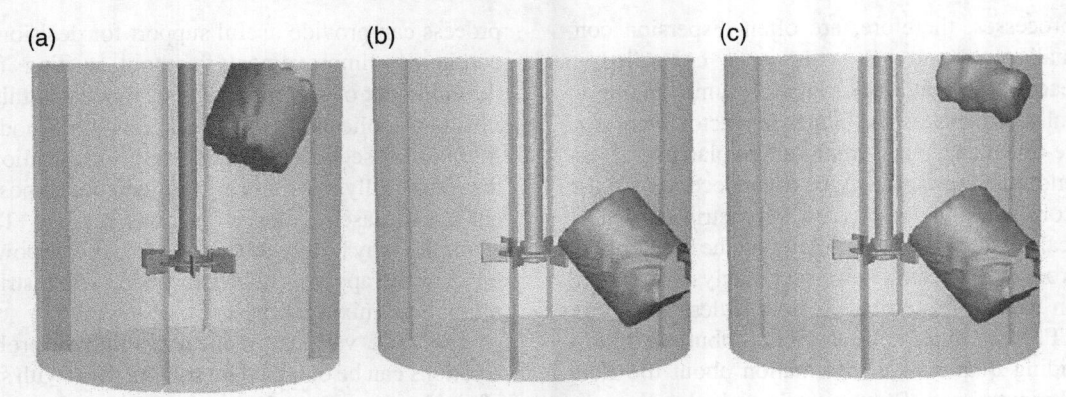

FIGURE 13.16 Different inhibitor addition locations. (a) First addition location. (b) Second addition location. (c) Third addition location. *Source*: Reprinted with permission from Dakshinamoorthy et al. [64]. Copyright 2004 Elsevier Ltd.

Dakshinamoorthy et al. [64] used CFD-based model to understand the role of imperfect mixing on shortstopping of a runaway reaction in a fully baffled stirred reactor. The propylene oxide (PO) polymerization runaway reaction in a stirred vessel was considered as a model problem. The polymerization of PO is catalyzed by potassium hydroxide. The reaction rate is proportional to temperature and concentrations of the monomer and catalyst. Adding an acid to neutralize the basic catalyst inhibits the runaway reaction.

A CMF model is conventionally used in practice to study the runaway and inhibition of runaway reactions. The application of a CMF model for developing operating protocols may be adequate when the characteristic runaway time is much greater than the mixing time of the reactor. When the characteristic runaway time is smaller or comparable to the mixing time, the CMF model cannot give reliable and useful information on the shortstopping process. To demonstrate this, they simulated a case with an operational constraint such as late detection of the runaway and hence delayed addition of inhibitor. The predicted temperatures from the CMF model for this case indicated that the late detection of the runaway at an average reactor temperature 450 K allows a design engineer only 60 seconds to control the reactor by inhibiting the runaway reaction. The time available (60 seconds) compared with the vessel mixing time of 450 seconds is very small and makes it more difficult and challenging to shortstop, due to imperfect mixing of the inhibitor present in the reactor. In such situations, the addition location of inhibitor and its quantity play important role.

Dakshinamoorthy et al. [64] investigated the influence of three different addition locations on the performance. Figure 13.16 shows the inhibitor addition locations considered in their study. For the first inhibitor addition location, the inhibitor was added just below the top surface of the liquid (see Figure 13.16a). For second addition location, the inhibitor was added in the impeller discharge stream to facilitate faster mixing of the inhibitor (Figure 13.16b). The selection of the third location was chosen after the analysis of the predicted temperature distributions of the first two addition locations. In this case, the inhibitor is distributed to both of the addition location: 75% of the inhibitor is added in the impeller discharge stream, and the remaining 25% is added at the top surface (see Figure 13.16c). The total quantity of inhibitor added in the reactor was first kept constant for all three additions and was added in the reactor when the reactor temperature reached to 450 K (delayed addition) due to runaway reaction.

The computational model was then used to simulate the inhibition of the runaway reaction for all three addition locations. They recorded the average reactor temperature for all the three cases and compared with the average reactor temperature predicted using CMF model (see Figure 13.17). They observed that the addition location has significant impact on the predicted reactor temperature and the average reactor temperature approaches to the CMF model predictions for the third addition location.

The simulated contour plot of temperature for all the three cases was then used for identifying the implications of addition location. Figure 13.18 shows the comparison of the predicted temperature distributions. They observed quite different distributions of the high temperature zones for the first and second inhibitor addition locations. The first location showed a high temperature zone in the bulk volume of the vessel, but no hotspots were predicted near the impeller shaft. Also the highest temperature predicted near the top surface was found to be lower than the temperature predicted for the second location. For the third inhibitor addition location, the final reactor temperature (475 K) was found to be slightly higher than the CMF results (470 K). The predicted temperature distribution at the mid-baffle plane shows the advantage of adding the inhibitor at both locations with substantial reduction in the hotspot volumes compared with the addition at a single location.

The performance of the inhibition process was substantially improved for the third addition location. However, some part of reactor was still showing high temperature

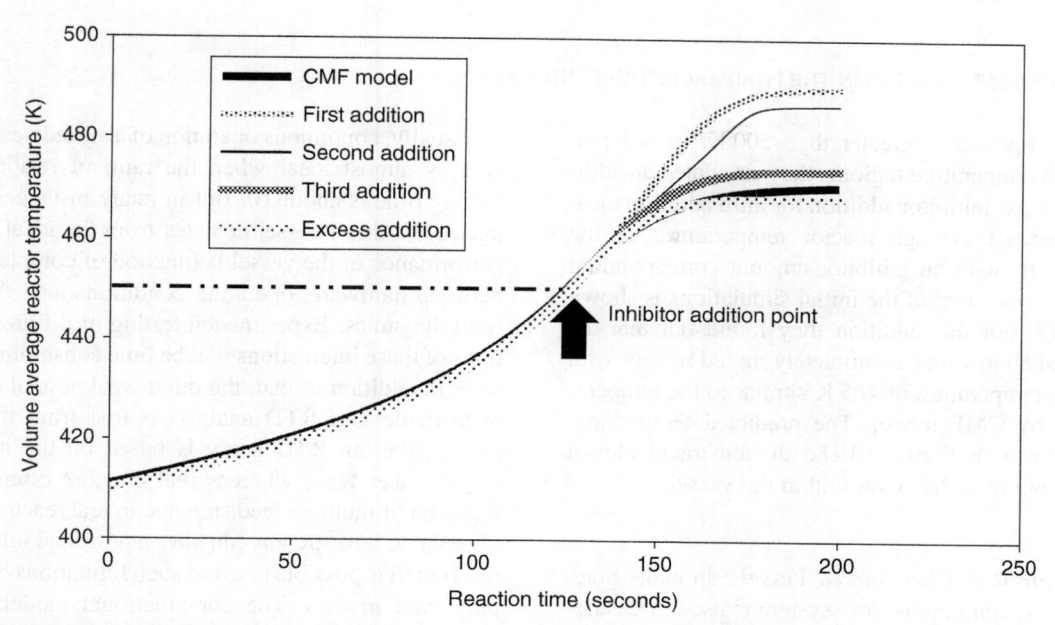

FIGURE 13.17 Predicted evolution of average reactor temperature for different inhibitor addition locations. *Source*: Reprinted with permission from Dakshinamoorthy et al. [64]. Copyright 2004 Elsevier Ltd.

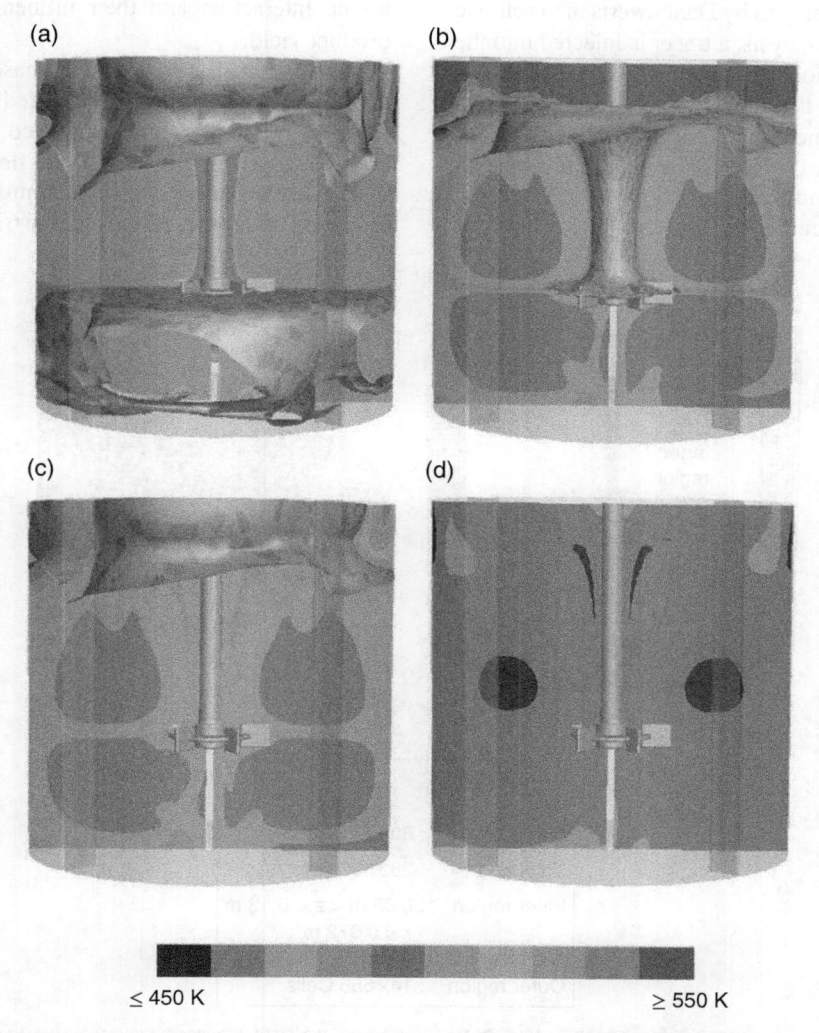

FIGURE 13.18 Predicted temperature distribution after 200 seconds for different inhibitor addition locations. (a) First addition location. (b) Second addition location. (c) Third addition location. (d) Two and half times of inhibitor (10 uniform contour levels between 450 and 550 K and iso-surface of temperature: iso-value = 500 K). *Source*: Reprinted with permission from Dakshinamoorthy et al. [64]. Copyright 2004 Elsevier Ltd.

regions with temperatures greater than 500 K. In order to avoid such high temperature regions, they simulated an additional case of excess inhibitor addition for third addition location. The predicted average reactor temperature, as the reaction proceeds, with an inhibitor amount corresponding to two and half times that of the initial simulations is shown in Figure 13.17. For this addition they found out that the reactor behaved almost like a completely mixed reactor with a final average temperature of 465 K similar to the temperature predicted by CMF model. The predicted temperature distribution shown in Figure 13.18d demonstrated almost uniform temperature distribution within the vessel.

13.3.1.2 Continuous Flow Stirred Vessel

In many practical applications, continuous flow system is used. Measurement of RTD is the only mixing tool covered in the undergraduate studies. It is a well-known method for assessing the nonideality of continuous process equipment. RTD is a concept first developed by Danckwerts in his classic 1953 paper [65]. In RTD analysis, a tracer is injected into the flow, and the concentration of tracer in the outlet line is recorded over time. From the concentration history, the distribution of fluid residence times in the vessel can be extracted. Information on the mixing nonidealities such as channeling, bypassing, and dead zones are then extracted from the measured RTD curve.

Usually, continuous operation of a stirred vessel is considered as almost ideal when the ratio of residence time to mixing time is about 10. But in many instances, the performance of stirred vessel deviates from the ideal. The mixing performance of the vessel is function of complex interaction between hardware, operating conditions, and the prevailing fluid dynamics. Experimental testing of different configurations of these interactions will be time consuming and expensive. In addition to that, the other weakness of experimental measurement of RTD analysis is that from the diagnostic perspective, an RTD study is based on the injection of a single tracer feed, whereas real reactors often employ the injection of multiple feed streams. In real reactors the mixing of separate feed streams can have a profound influence on the reaction. It is possible to avoid such limitations by using computational models. The computational models provide an opportunity to mathematically simulate mixing of multiple species entering into the reactor through multiple inlets. Such simulations can also help in simulating inlet plume interaction and their influence on the process and product yield.

Khopkar et al. [66] used CFD-based model for simulating RTD in a continuous flow stirred vessel equipped with Mixel TT impeller. The simulated geometry is shown in Figure 13.19. They simulated the flow in a continuous flow reactor having residence time to mixing time ratio equal to 9.6 ($N = 360$ rpm and $Q_l = 2.01667 \times 10^{-4}$ m^3/s) closer to

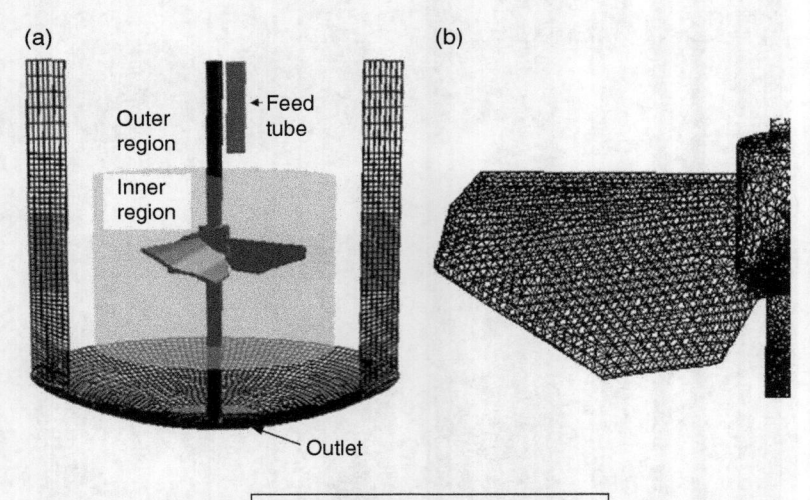

FIGURE 13.19 Solution domain and computational grid. (a) Solution domain. (b) Grid distribution on impeller blade. *Source*: Reprinted with permission from Khopkar et al. [66]. Copyright 2004 The Institution of Chemical Engineers, Published by Elsevier B.V.

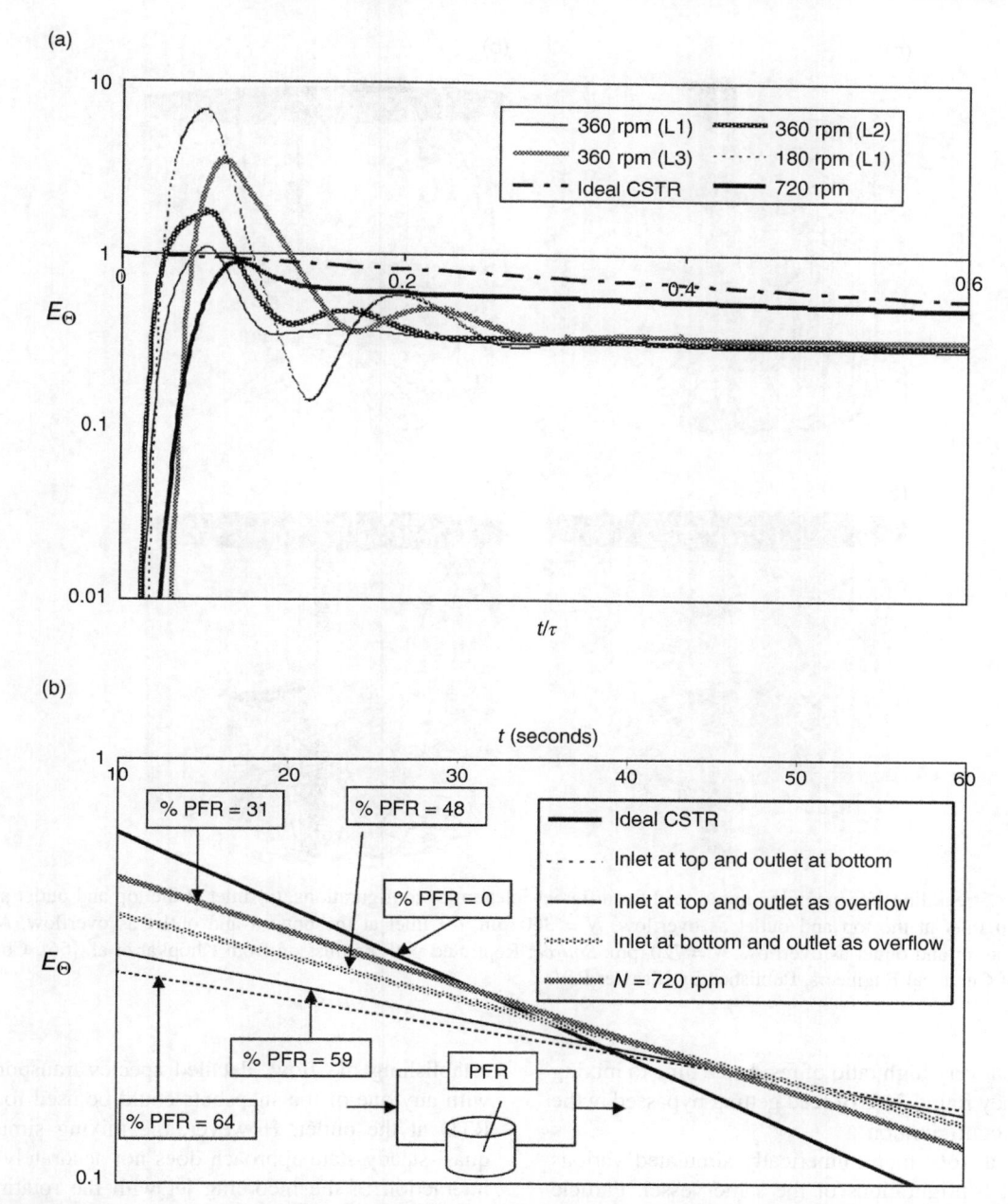

FIGURE 13.20 Comparison of predicted exit age distribution with ideal CSTR. *Source*: Reprinted with permission from Khopkar et al. [66]. Copyright 2004 The Institution of Chemical Engineers, Published by Elsevier B.V.

the standard value of 10 commonly used in practice for determining the ideal performance behavior. The predicted exit age distribution is shown in Figure 13.20. From the overshoot in tracer concentration observed at the outlet, it appears that the high velocity inlet jet may be interacting directly with the outlet. This was further confirmed by the lower slope of predicted RTD curve compared with that of ideal CSTR. The combination of overshoot at the beginning and lower slope at later stage indicates that part of the incoming fluid bypasses stirred vessel and flows effectively through a small volume plug flow reactor and the remaining part of the incoming fluid flows through a stirred vessel with much larger effective residence time than that calculated from the total incoming flow (Figure 13.20b). The nature of predicted exit age distribution they modeled was a combination of ideal stirred reactor and plug flow reactor operating in parallel. The analysis indicated that the effective residence time of the ideal CSTR part is about 68 seconds (this means only about 36% of the incoming liquid flows through a vessel and about 64% of the incoming fluid short-circuits to the outlet). They further studied the influence of impeller speed on the exit age distribution. They found significant influence of impeller speed on

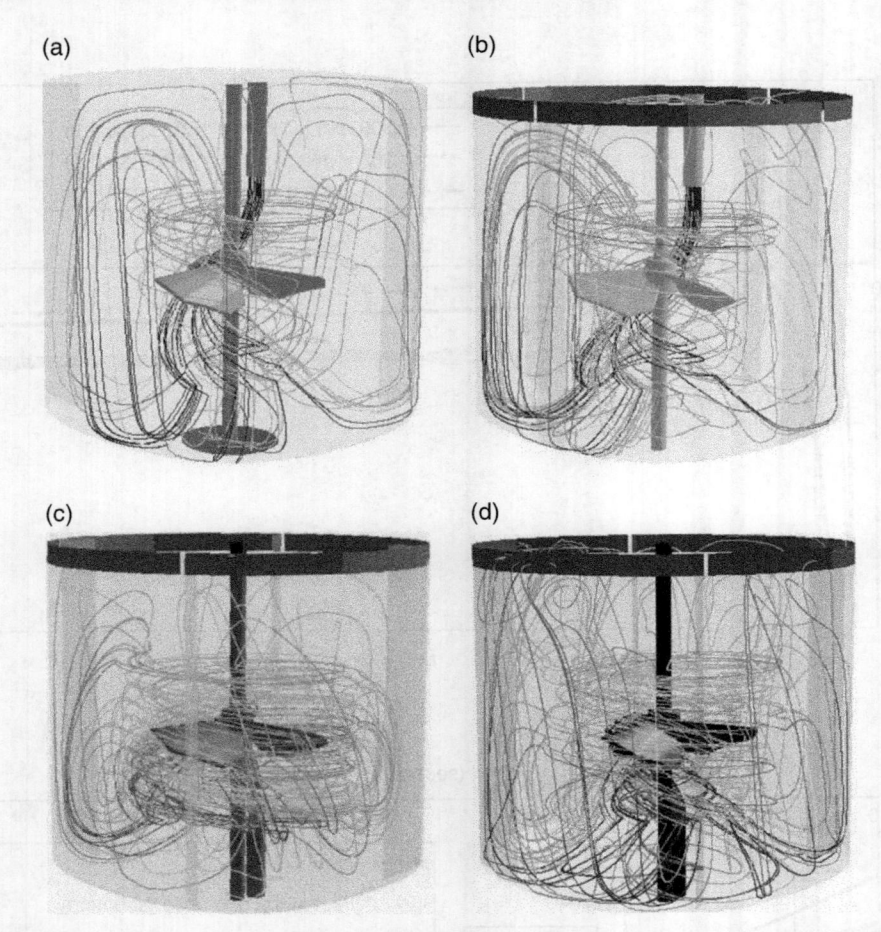

FIGURE 13.21 Streak lines for incoming liquid feed for different inlet/outlet configurations. (a) Inlet at the top and outlet at the bottom, $N = 360$ rpm. (b) Inlet at the top and outlet as overflow, $N = 360$ rpm. (c) Inlet at the bottom and outlet as overflow, $N = 360$ rpm. (d) Inlet at the bottom and outlet as overflow, $N = 720$ rpm. *Source*: Reprinted with permission from Khopkar et al. [66]. Copyright 2004 The Institution of Chemical Engineers, Published by Elsevier B.V.

the RTD. Even at very high ratio of residence time to mixing time (=19.2), they found 31% of feed getting bypassed in the selected vessel configuration.

Khopkar et al. [66] then numerically simulated various inlet and outlet configurations of the same vessel. Particle streak lines were simulated by releasing neutrally buoyant tracer particles from the inlet pipe. The simulated particle streak lines (for a flow time of five seconds) for different configurations are shown in Figure 13.21. Particle streak lines shown in Figure 13.21 help in understanding the complex interaction of incoming feed with impeller-generated flow and outlet.

Khopkar et al. [66] used quasi-steady-state approach with multiple snapshots for simulating the flow in an industrial continuous flow stirred tank reactor. They observed that multiple numbers of snapshots are needed to adequately represent the flow generated by an impeller and the interaction of the incoming liquid jet with the impeller would depend on the design of the impeller. They further recommended that after establishing the flow, detailed species transport equations with any one of the snapshots could be used for simulating RTD at the outlet. However, the mixing simulation with quasi-steady-state approach does not accurately capture the interaction of the incoming jet with the rotating impeller, but the obtained results are found to be sufficient for the "a priori" suggestion for the evaluation of the different design configuration.

13.3.1.3 Viscous Mixing Formulations are widely used in pharmaceutical, paint, food, polymer, and personal care products. Some of the most difficult mixing problems do get encountered in making these formulations. In many situations these formulations exhibit highly viscous, non-Newtonian, and viscoelastic fluids. With all non-Newtonian fluids, the potential exists that a portion of a tank will remain unmixed because of inadequate fluid motion. Because of both the high viscosity and non-Newtonian behavior, special equipment is often required for mixing.

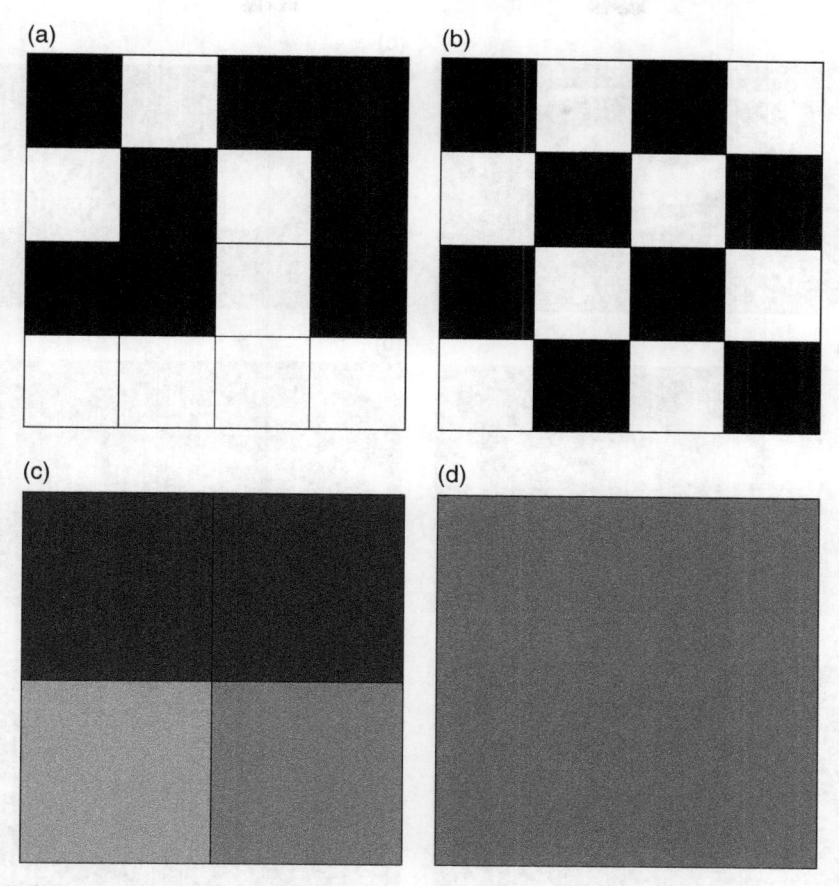

FIGURE 13.22 Schematic illustration of distributive and dispersive mixing. (a) Bad dispersion and bad distribution. (b) Bad dispersion and good distribution. (c) Good dispersion and bad distribution. (d) Good dispersion and good distribution.

Mixing can be brought about in viscous systems only by mechanical action or by the forced shear or by elongational flow of the matrix. Mixing achieved through the history of deformation imparted to the fluid is called distributive mixing. In distributive mixing the homogeneity of the mixture can be quantified through the scale of segregation [67], whereas in dispersive mixing a consequence of the history of the fluid mechanical stresses is imposed on the mixture for breakup of agglomerates and drops. A dispersive mixing index can be quantified through strength of the pure elongational flow. The interrelationship between dispersive and distributive mixing is illustrated in Figure 13.22. In general, viscous mixing operations require some combination of dispersive and distributive actions. Before explaining some examples of viscous mixing, basic definitions of distributive and dispersive mixing are explained in the following subsection. These definitions will help in quantifying mixing in the process equipment.

13.3.1.3.1 Measures of Distributive and Dispersive Mixing As said earlier the purpose of the distributive mixing is to create uniform macrostructure. It is created due to the presence of chaotic motion in the deterministic laminar flows [68]. A variety of tools have been developed to examine and characterize chaotic flows including Poincare sections, periodic point analysis [69], and stretching distributions [70]. These tools do provide rich insight into the nature of chaotic flows and reveal the mechanism of chaotic mixing, but do not provide tools to quantify the distributive mixing behavior. Tucker III and Peters [71] numerically simulated mixing in a 2D time-periodic flows in a rectangular cavity with upper and lower moving walls (see Figure 13.23). The mixing pattern was simulated using mapping method, which has a finite spatial resolution and certain amount of numerical diffusion. A variety of protocols involving time-periodic sliding motions of the upper and lower cavity surfaces were studied. For quantifying distributive mixing, two measures of distributive mixing were proposed and examined: the standard deviation among samples σ and the maximum sample error E. Both these measures are closely related to the distributive mixing and small value of either means that the cell concentrations are nearly uniform:

$$\langle C \rangle_i = \frac{1}{V_{avg}} \int_{V_i} C dV \qquad (13.18)$$

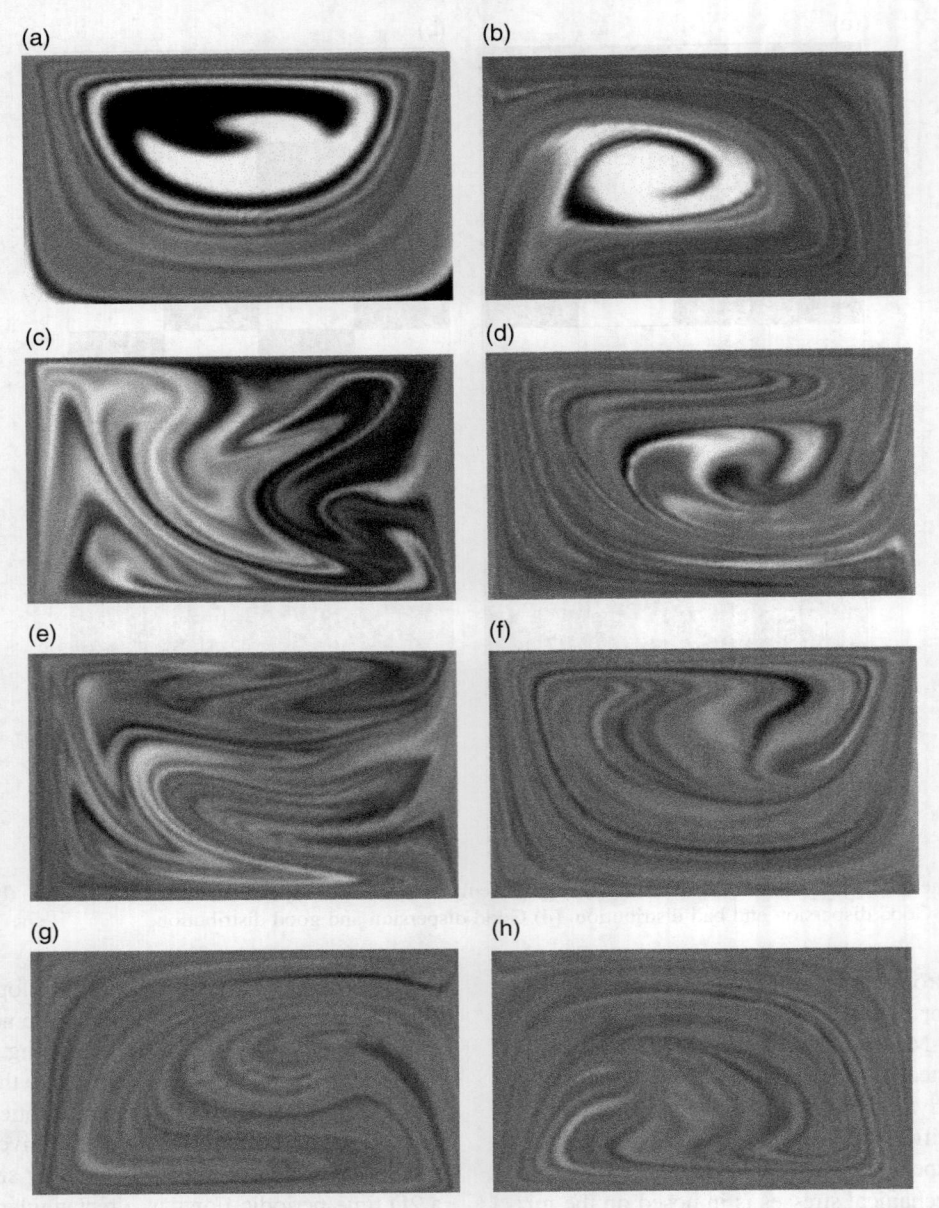

FIGURE 13.23 Eight example mixtures simulated by Tucker III and Peters [71] using mapping method. (a) A8. (b) B8. (c) D2. (d) C8. (e) D4. (f) C16. (g) E8. (h) B16. *Source*: Reprinted with permission from Tucker and Peters [71]. Copyright 2003 Korean Society of Rheology and Australian Society of Rheology.

$$\sigma(L) = \sqrt{\frac{1}{M}\sum_{i=1}^{M}\left(\langle C \rangle_i - c_{\text{avg}}\right)^2} \quad (13.19)$$

$$E(L) \equiv \max\left|\langle C \rangle_i - c_{\text{avg}}\right| \quad (13.20)$$

where

- V_{avg} is the averaging volume.
- c_{avg} is the final uniform concentration.
- L is the length scale of the averaging volume.

M is the number of points in the mixer separated by the distance L.

Tucker III and Peters [71] observed that E and σ both provide the ability to compare the mixture patterns that may be quite different in appearance and to say which mixture is better distributed.

Dispersive mixing is usually more difficult to achieve than the distributive mixing. Dispersive mixing involves breakup of agglomerates and droplets in flow, caused by stresses large enough to overcome the cohesive or interfacial forces that

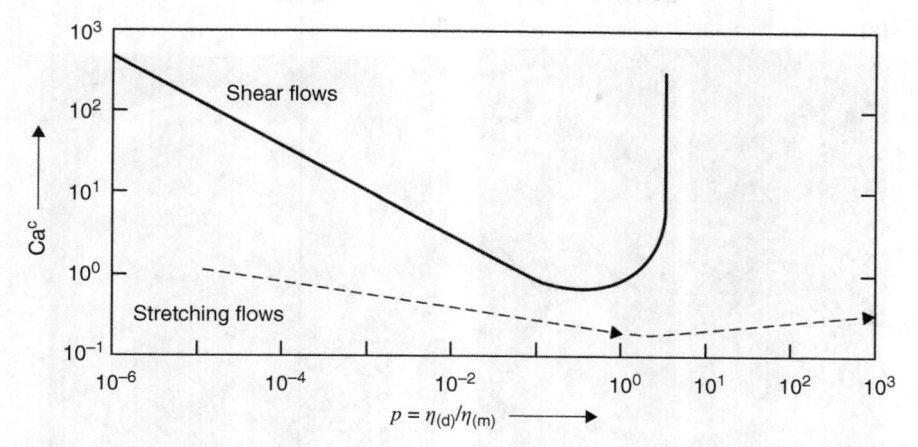

FIGURE 13.24 Critical Weber number versus viscosity ratio. *Source*: Reprinted with permission from Grace [72]. Copyright 1982 Gordon and Breach Science Publisher Inc.

tend to keep the agglomerates or the droplets intact. Two main mechanisms are responsible for dispersive mixing, viz. shear flow and elongational or extensional flow. Quantitative studies of droplet breakup in simple shear and pure elongational flows have shown that the elongational flow is more effective than the simple shear flow especially in the case of high viscosity ratios and low interfacial tensions. This is been clearly explained in Figure 13.24 where the Weber or capillary number is plotted again the viscosity ratio. Based on Figure 13.24, the minimum dispersed phase drop radius can be achieved where the viscosity p is close to unity, but the dispersion by shear flow is not possible if p exceeds 4. This limit could be different for viscoelastic fluids.

Accurate estimation of dispersive mixing efficiency would involve tracking of the dispersed phase during their entire residence time in the equipment and following the dynamics of their breakup and coalescence. However, such approach is numerically very expensive. A simpler global approach is required, which will help in discriminating between various designs and processing conditions for effective mixing equipment selection. A flow field characteristic relevant to dispersive mixing is known as flow strength [73, 74]. The flow strength is function of rate of deformation tensor and Jaumann time derivative of rate of deformation tensor. The flow strength parameter ranges from zero for pure rotational fluid to infinity for pure elongational flow; its value is unity for simple shear flow. However, numerically determining flow strength for complex geometry is difficult. Yang and Manas-Zloczower [75] defined dispersive mixing index λ that quantifies the relative strength of the pure elongational flow component:

$$\lambda = \frac{|D|}{(|D| + |\omega|)} \quad (13.21)$$

where

|D| and |ω| are the magnitudes of the rate of strain and vorticity tensors, respectively.

The above parameter assumes values between 0 for pure rotation and 1 for pure elongation, with a value of 0.5 for simple shear. This mixing index is not frame invariant, but still can be used as first approximation to discriminate between various equipment designs and processing conditions [75, 76].

13.3.1.3.2 Examples of Viscous Mixing It has been well established that the chaos is necessary for achieving efficient mixing in laminar stirred vessel [77]. Chaotic mixing is characterized by the exponential rate of stretching and folding of fluid elements. To illustrate this, Arratia et al. [78] used the planar laser-induced fluorescence (PLIF) snapshot through the axis of the mixing tank filled with Newtonian fluid and stirred with concentric disks and Rushton turbines. Figure 13.25a demonstrates that the fluid does not mix down to a very small length scale. No sign of chaotic mixing was observed. Flow was characterized by a linear rate of stretching. Later, they induced chaos in the system by replacing the disk by Rushton turbines. They found out that the passage of blades periodically perturbed the flow and triggered the stretching and folding of fluid material, which helped in achieving efficient mixing (Figure 13.25b).

Zalc et al. [79, 80] used the computational model for simulating the chaotic mixing in the stirred vessel equipped with multiple Rushton turbines. They computed flow field using the ORCA software (Fujitsu, Campbell, CA). Extensive comparison of simulated results was done with the PIV and PLIF measurements. Figure 13.26 shows the comparison of experimentally measured flow field with numerical

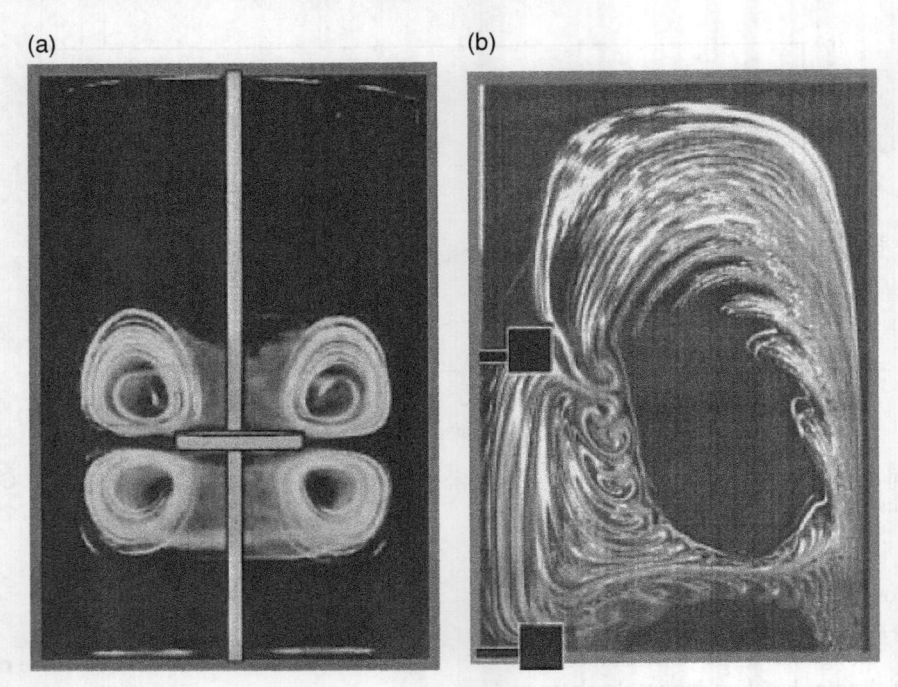

FIGURE 13.25 Chaotic mixing in stirred tank. (a) Flow generated by axisymmetric disk. (b) Flow generated by Rushton turbine. *Source*: Reprinted with permission from Arratia et al. [78]. Copyright 2006 American Institute of Chemical Engineers.

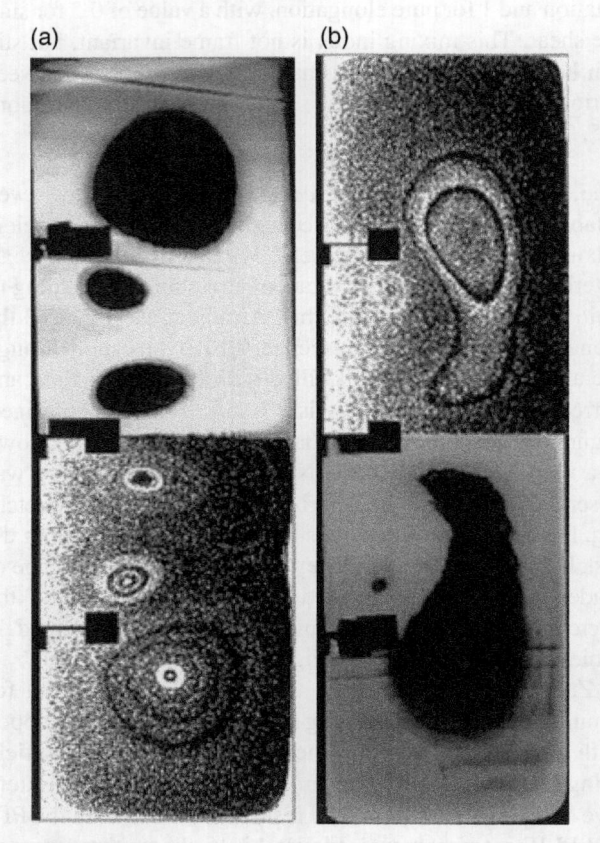

FIGURE 13.26 Comparison of experimental and simulated flow field results in stirred vessel equipped with three Rushton turbines. (a) Re = 20. (b) Re = 40. *Source*: Reprinted with permission from Zalc et al. [80]. Copyright 2002 American Institute of Chemical Engineers.

results. CFD results accurately predicted size and location of the poorly mixed regions in the stirred vessel. Zalc et al. [79, 80] carried out quantitative measure of mixing intensities in chaotic flows by computing the accumulated stretching of small fluid filaments. These simulations were performed by placing small vectors in the flow. The deformation of each infinitesimal vector by the instantaneous velocity gradient along its trajectory while being convected throughout the flow domain was calculated for quantifying stretching value (λ). Figure 13.27 illustrates the stretching field for the different Reynolds number values 20, 40, and 160 after 20 revolutions. The contour maps reveal the heterogeneity of stretching and subsequently help in identifying poor mixing regions in the stirred vessel. The contour plot also shows the change in the poor mixing region structure with impeller Reynolds number. A priori knowledge of such stretching distribution pattern not only helps in identifying the poor mixing regions but also provides guideline in deciding the injection point for additives to achieve optimum distribution.

In order to improve the chaotic mixing, few approaches were suggested in literature, viz. using variable agitation speed instead of constant agitation by Szalai et al. [81], using eccentrically mounted impellers [77, 78, 82], and using dual shaft mixers [83]. Cabaret et al. [83] experimentally studied the mixing kinetics in a laminar flow stirred vessel equipped with multiple Rushton turbines. They used color discoloration method for quantifying the mixing in vessel. They observed that the mixing in a dual shaft stirred vessel is more efficient in comparison with centrally mounted shaft and off-centered shaft systems (see Figure 13.28 for details).

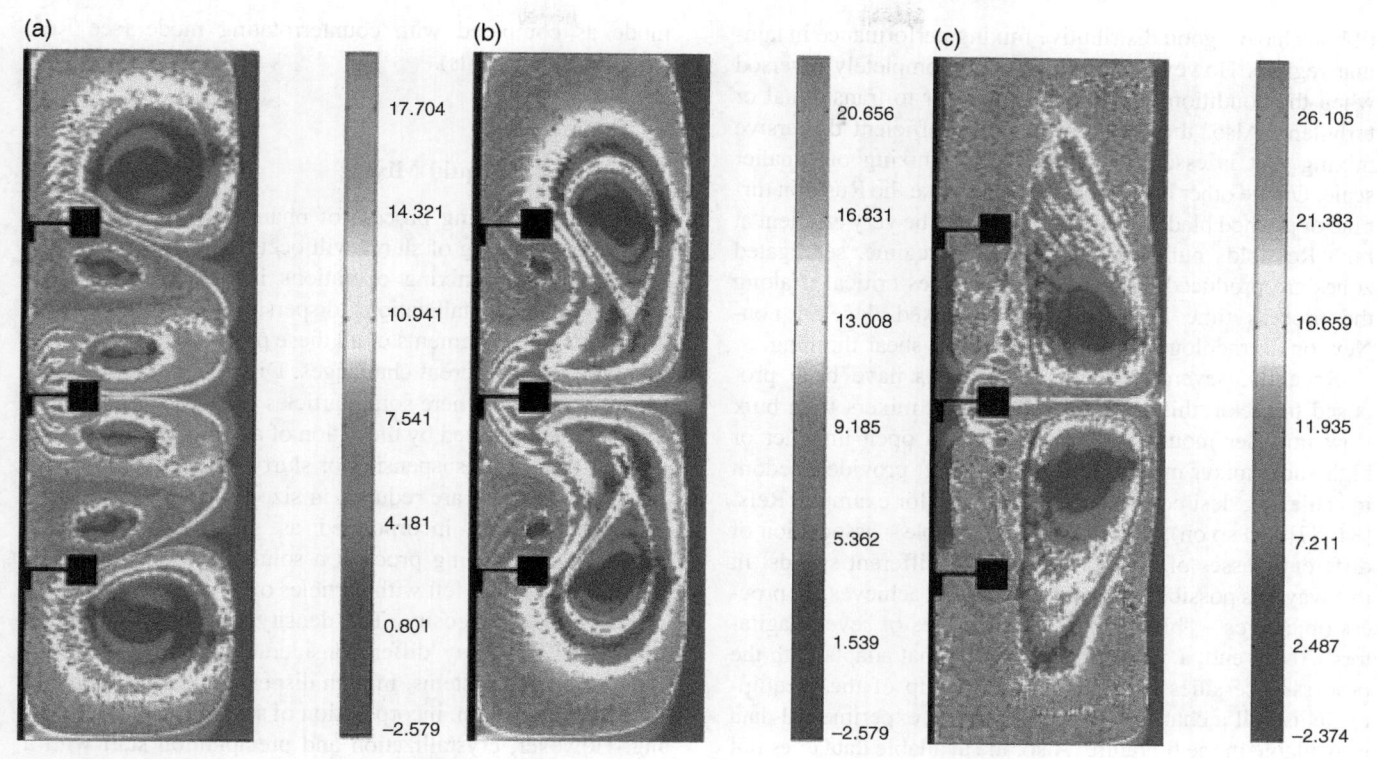

FIGURE 13.27 Simulated stretching field at 20 seconds for different Reynolds numbers. (a) Re = 20. (b) Re = 40. (c) Re = 160. *Source*: Reprinted with permission from Zalc et al. [80]. Copyright 2002 American Institute of Chemical Engineers.

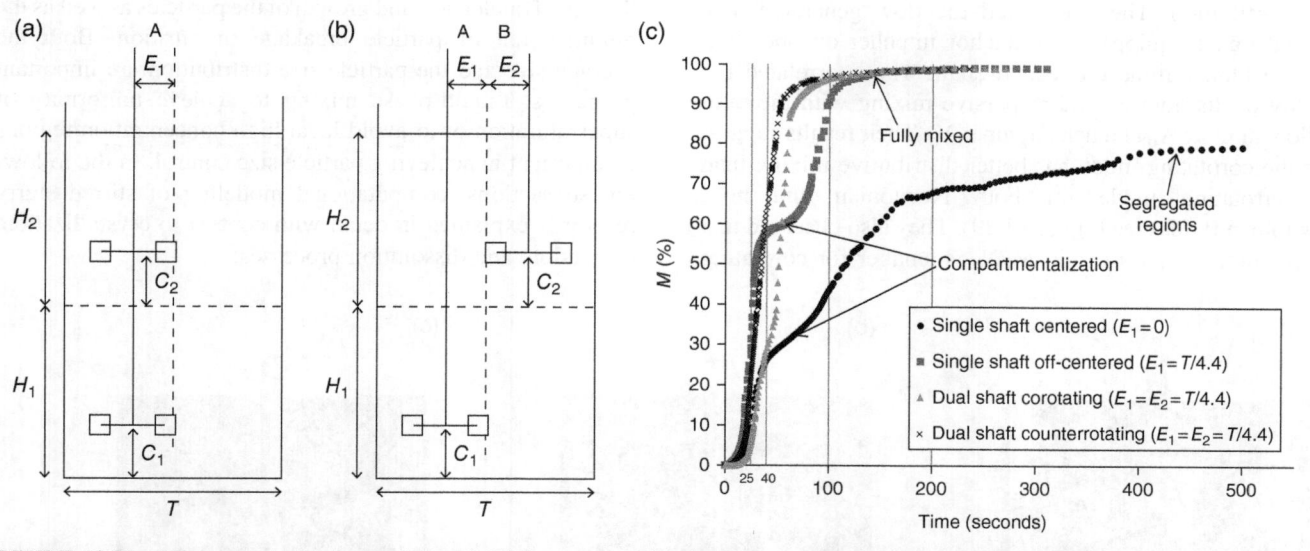

FIGURE 13.28 Mixing analysis of stirred vessel. Mixing systems (a) single shaft configuration. (b) Dual shaft configuration. (c) Mixing analysis for various configurations. *Source*: Reprinted with permission from Cabaret et al. [83]. Copyright 2007 The Institution of Chemical Engineers, Published by Elsevier B.V.

Dual shaft system also provides an additional degree of freedom, i.e. direction of rotation for influencing prevailing fluid dynamics. Cabaret et al. [83] observed better mixing in counterrotating mode as compared with corotating mode of operation.

Nowadays, the industry needs impellers that can work in laminar, transitional, or turbulent regimes with minimum modifications. Standard agitators like close-clearance and open impellers exhibit some limitations with this aspect. On one hand, close-clearance impellers such as helical ribbons have a good distributive mixing performance in laminar regime. However, this situation is completely reversed when the condition changes from laminar to transitional or turbulent. Also, they do not generate sufficient dispersive mixing that is essential for improving mixing on smaller scale. On the other hand, open impellers like the Rushton turbine or pitched blade turbine are known to be very efficient at high Reynolds number, but in laminar regime, segregated zones are produced. The situation becomes critical if along the process time the phases to be mixed develop non-Newtonian rheological properties such as shear thinning.

Recently, several innovative strategies have been proposed to tackle this problem. Multi-shaft mixers with bulk flow impeller mounted on one shaft and open impeller or high-shear mixer mounted on second shaft provide freedom in achieving desired level of mixing (see, for example, Refs. [84–87] and so on). The main idea is simple – association of different classes of agitators rotating at different speeds. In this way it is possible to create a mixer that achieves the process objectives – "blending" the capabilities of several agitators. At the end, a dynamic mixing unit that adapts with the process necessities is obtained. Scaling up of these equipments is still a challenge as not sufficient experimental data is available in the literature. Also, the available data does not cover all the possible degrees of freedom available with the designs of the multi-shaft mixers. Validated CFD model can help us in filling these gaps with the experiments. Rivera et al. [86] simulated flow in a coaxial mixer using POLY3D™ (Rheosoft, Inc.). They simulated the flow generated in a coaxial mixer equipped with anchor impeller on one shaft and Rushton turbine on second shaft. They correlated the quality of distributive and dispersive mixing with the ratio of flow number N_q and head number N_h. Their results suggest that the corotating mode has better distributive mixing than counterrotating mode for both Newtonian and non-Newtonian fluids (see Figure 13.29). They also observed that the predicted value of ratio N_h/N_q is smaller for corotating mode as compared with counterrotating mode (see Ref. [86] for more details).

13.3.2 Solid–Liquid Mixing

In the manufacturing process of pharmaceutical intermediates, often handling of slurry will occur at some point. Typical solid–liquid mixing operations in the pharmaceutical industry are crystallization, dispersion, and dissolution. The process requirements of all these processes are different, and they have different challenges. Dispersion of solids is a physical process where solid particles or aggregates are suspended and dispersed by the action of an agitator in a fluid to achieve a uniform suspension or slurry. In dissolution process, solid particles are reduced in size and ultimately disappear as they are incorporated as solute in the liquid. However, in leaching process, a soluble component of the solid dissolves and left with particles of having different size, density, and/or porosity. The density and viscosity of the resulting liquid may differ considerably from the original liquid for some systems. In both dispersion and in some dissolution application, incorporation of solids is very challenging. However, crystallization and precipitation start with a solid-free liquid phase if unseeded, and the solid particles form during the operation. The solids grow in size as well as in population. The viscosity and density of the slurry thus formed usually changes. The process goals include control of the rate of nucleation and growth of the particles as well as the minimization of particle breakage or attrition. Both the average size and the particle size distribution are important properties. Liquid-phase mixing to achieve uniformity of supersaturation or to avoid local high concentration regions is important in achieving particle size control. In the following subsections, computational modeling of stirred slurry reactor is explained in detail with context to crystallization, dispersion, and dissolution processes.

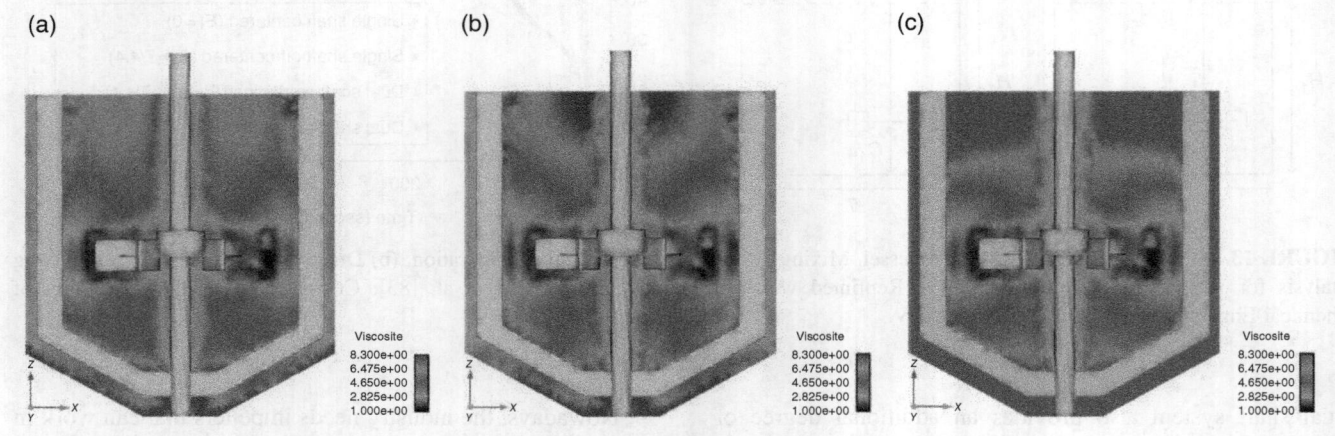

FIGURE 13.29 Simulated viscosity profile for non-Newtonian fluid in coaxial mixer. (a) Corotating mode. (b) Counterrotating mode. (c) Single Rushton turbine. *Source*: Reprinted with permission from Rivera et al. [86]. Copyright 2005 Elsevier Ltd.

13.3.2.1 Crystallization in Stirred Vessel

In pharmaceutical industry, crystallization and precipitation are widely used processes in manufacturing of various drug molecules. Precipitation and crystallization refer to unit operations that generate a solid from a supersaturated solution. The nonequilibrium supersaturated condition can be induced in a variety of ways such as removal of solvent by evaporation, addition of another solvent, changes of temperature or pressure, addition of other solutes, oxidation–reduction reactions, or even combinations of these. The distinction between precipitation and crystallization is quite often based on the speed of the process and the size of the solid particles produced. The term precipitation commonly refers to a process that results in rapid solid formation that can give small crystals that may not appear crystalline to the eye, but still may give very distinct X-ray diffraction peaks. Amorphous solids (at least as indicated by X-ray diffraction) may also be produced through precipitation. The term precipitation also tends to be applied to a relatively irreversible reaction between an added reagent and other species in solution, whereas crystallization products can usually be redissolved using simple means such as heating or dilution. Precipitation processes usually begin at high supersaturation where rapid nucleation and growth of solid phases occur. In both precipitation and crystallization processes, the same basic steps occur: supersaturation, nucleation, and growth.

Supersaturation affects both crystal growth and nucleation rates, which in turn impact the particle size distribution. A higher level of nucleation leads to smaller particles and vice versa. Also, a high degree of nucleation rate over crystal growth rate due to a high degree of supersaturation can lead to poorer rejection of impurities [88]. Nucleation does not necessarily begin immediately on reaching a supersaturated condition, except at very high supersaturation, and there may be an induction period before detection of the first crystals or solid particles.

Tung [88] used these time scales with the liquid-phase mixing time for explaining complexity of the process. He defined two dimensionless numbers, viz. $Da_{nucleation}$ and $Da_{crystallization}$:

$$Da_{nucleation} = \frac{\tau_{mixing}}{\tau_{induction}} \quad (13.22)$$

$$Da_{crystallization} = \frac{\tau_{mixing}}{\tau_{release\ of\ supersaturation}} \quad (13.23)$$

where

τ_{mixing} is the mixing time.

$\tau_{induction}$ is the induction time for primary nucleation.

$\tau_{release\ of\ supersaturation}$ is the time required to release the supersaturation.

$Da_{nucleation} \ll 1$ means a complete mixing before the nucleation is achieved. Similarly, $Da_{crystallization} \ll 1$ means a complete mixing is achieved before the supersaturation is released. This would be the case for crystallization with relatively slow crystallization kinetics in releasing the supersaturation. Since the order of mixing time in a crystallizer is generally available, it is straightforward to learn if the crystallization system could be sensitive to mixing by comparing the induction time for nucleation or time for release of supersaturation. This suggests that the liquid-phase mixing time plays a very important role in the crystallization.

It is important to recognize the influence of mixing on the product characteristics. In general mixing is defined at three scales, viz. macromixing, mesomixing, and micromixing. Macromixing is defined on the scale of the vessel, mesomixing is defined in context of dispersion of the antisolvent plume at the feed point, and micromixing determines the mixing at the molecular level and influences the induction time. Micromixing is influenced by the impeller type and speed plus relative location of the antisolvent feed pipe with respect to impeller. Thus, feed pipe location, pipe diameter, and antisolvent flow can impact both micromixing and mesomixing times. For example, when the antisolvent feed rate is faster than the local mixing rate, resulting in a plume of highly concentrated antisolvent that is not mixed at the molecular level. This can yield a high localized nucleation rate. This will present scale-up difficulties, requiring a thorough engineering analysis for success.

The modeling of well-mixed crystallizers involves the computation of the PBE together with the material balance equations for each species in solution. Numerous numerical techniques that compute the full crystal size distribution (CSD) have been used to model well-mixed batch, semibatch, or continuous crystallizers. In most of the studies, the actual prevailing mixing condition was not simulated and is approximated with ideal mixing behaviors. For a more realistic simulation, it is necessary to solve the standard momentum and mass transport equations together with the PBE coupled flow simulation. Rielly and Marquis [89] provided an explanation on the pivotal role of fluid dynamics on the kinetics of crystallization and resulting CSD. Figure 13.30 explains how different scales of crystallizer fluid dynamics influence the crystallization process. Rielly and Marquis [89] built a computational model where particle or crystal motion was simulated in Lagrangian framework. They observed that the crystal particle experiences region with very different micro-flow properties and it can be explained through the variation in the distribution of instantaneous slip velocities. They also concluded that the distribution of the variation in slip velocities experienced by the crystals is strongly dependent on the particle microscale and macroscale Stokes number defined by the two flow time scales, viz. lifetime of turbulent eddy and circulation time or mixing time. They also observed that neglecting the effects of microscale flow properties significantly reduces the variance of the macroscale time scale.

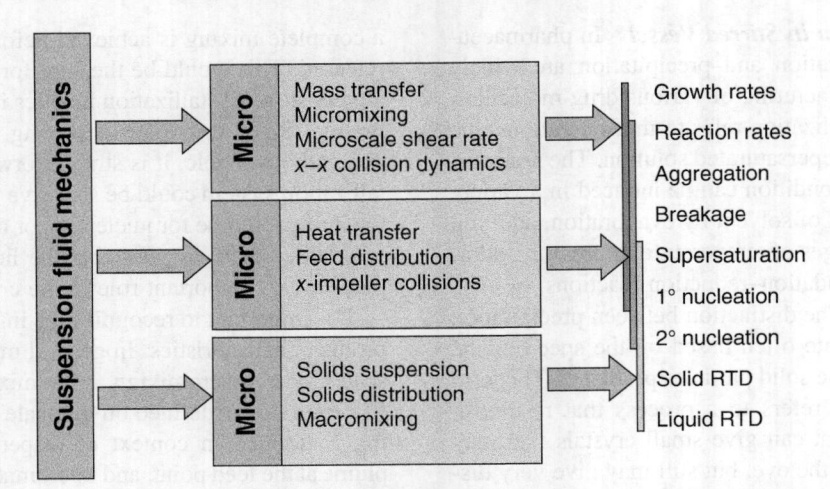

FIGURE 13.30 Suspension fluid dynamics effect on the crystallizer kinetics. *Source*: Reprinted with permission from Rielly and Marquis [89]. Copyright 2001 Elsevier Science Ltd.

In all the studies available on the CFD modeling of crystallizer, the presence of solids is modeled by treating the slurry as a pseudo-homogeneous fluid with a spatial distribution of effective viscosity that depends on the local solid fraction. Thus, they do not exactly simulate the flow essential for estimating accurate microscale flow properties that will further influence the macroscale flow time scales. There are multiple reasons for not yet development of a comprehensive computational model for simulating whole crystallization process. First, the computational model for simulating hydrodynamics of slurry is not yet fully validated. Second, although certain progress has been made in simulating micromixing effect on the crystallization kinetics, other mechanisms such as secondary nucleation, agglomeration, and breakage are yet to be modeled and validated. Third, most important reason is huge computational requirement for coupling algorithms of multiphase flow and crystallization kinetics. It is therefore necessary to validate these models separately and develop an innovative approach for combining the results of these models for effective design of crystallization process.

13.3.2.2 CFD Simulation of Solid Suspension in Stirred Vessel

In the previous section a detailed discussion on the computational modeling of stirred slurry reactor is presented. The developed model was evaluated by comparing results for solid volume fraction profiles averaged over reactor cross section. A close look at the prevailing fluid dynamics in the crystallizer shows very complex behavior. The average density and viscosity continuously changes due to formation and growth of solids. There is a possibility of solid suspension mechanism demonstrating various transient phenomena in crystallizer. These transients will definitely influence the transport processes. It is therefore necessary that the computational model must predict the transients associated with the slurry suspension and its implications on the liquid-phase mixing. Sardeshpande et al. [90] observed the hysteresis in the suspension quality with respect to impeller speed. Figure 13.31 shows the visual observation of hysteresis observed by Sardeshpande et al. [90]. Their data also suggest that the observed hysteresis is more profound at higher solids loading.

Sardeshpande et al. [90] used computational model with different interphase drag force formulations for simulating the sudden hysteresis observed in cloud or suspension height. Figure 13.32 shows the comparison of the cloud height predicted using Brucato et al. [21] drag law and modified Brucato's drag law recommended by Khopkar et al. [38]. They observed that for both increasing and decreasing impeller rotational speed, the computational model with Brucato's drag law overpredicted the cloud height. However, the computational model recommended by Khopkar et al. [37] reasonably captured the cloud height for both increasing and decreasing impeller rotational speed conditions. Ability of computational model to predict hysteresis enhances the capability of process engineer for a priori controlling the performance of slurry reactor, where small deviation in the suspension quality could influence the final product properties through different rates of transport and mixing rates as well as different particle–particle interaction.

13.3.2.3 Solid Suspension and Mixing in Stirred Reactor

Liquid-phase mixing is quite important in many solid–liquid reactions as well. It not only affects the selectivity of reactions but also controls the temperature distribution inside the reactor for exothermic reactions. In many cases, stirred reactors are operated with higher solids loading (solid volume fraction >5.0%). In such situations, the liquid-phase mixing process was found to show a complex interaction with the suspension quality (for example, Ref. [91]).

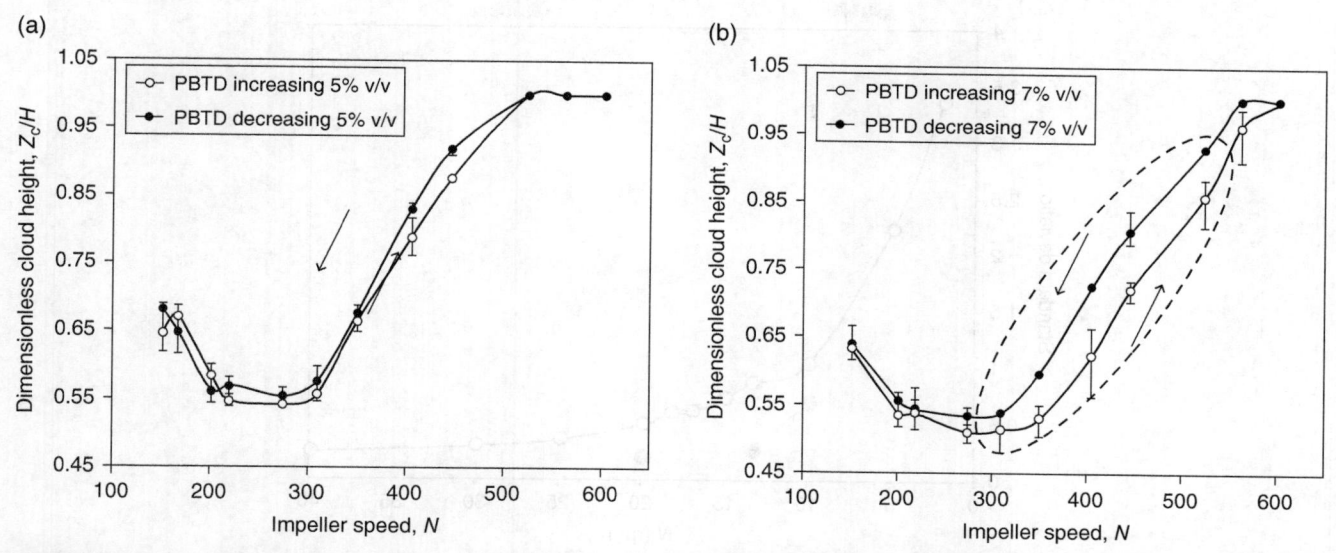

FIGURE 13.31 Visually observed cloud height hysteresis. *Source*: Reprinted with permission from Sardeshpande et al. [90]. Copyright 2010 American Institute of Chemical Engineers.

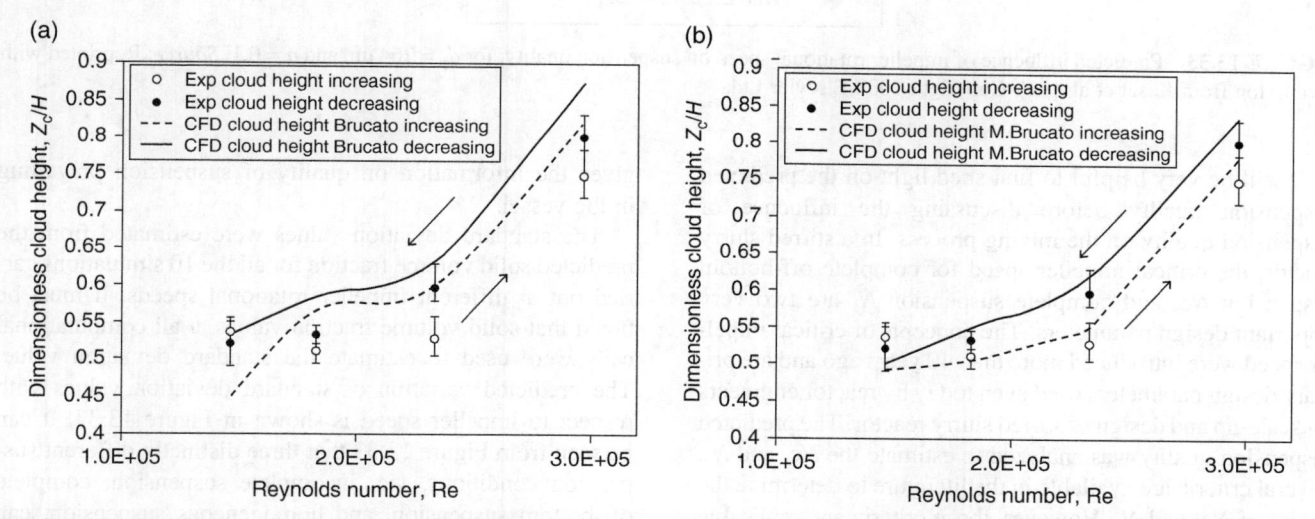

FIGURE 13.32 Predicted cloud height hysteresis using two drag formulations. (a) Brucato's drag coefficient. (b) Modified Brucato's drag coefficient suggested by Khopkar et al. *Source*: Reprinted with permission from Sardeshpande et al. [90]. Copyright 2010 American Institute of Chemical Engineers.

A computational model, which is able to predict suspension quality and its influence of liquid-phase mixing, will definitely help reactor engineers to obtain optimum performance of stirred slurry reactors. Recently, Kasat et al. [92] simulated liquid-phase mixing in a stirred reactor for different operating conditions. Their simulations are reproduced here to explain the liquid-phase mixing in a stirred slurry reactor.

The simulations are carried out in the experimental setup of Yamazaki et al. [38] with solid volume fraction equal to 10.0% and particle diameter equal to 264 μm. The simulations are carried out for 10 different impeller rotational speeds starting from 2 to 40 rps. The completely converged solid–liquid flow simulations were used to simulate liquid-phase mixing. Mixing simulations were carried out with 1.0% (by volume) of tracer, having same physical properties of liquid in the vessel. The tracer history was recorded at eight different locations. In stirred slurry reactor, delayed mixing was usually observed near the top surface of the liquid. Therefore, tracer history was recorded at four different locations close to top surface (Ref. [92] for more details). The mixing time in the present work is defined as the time required for the tracer concentration at these locations to lie within ±5.0% of the final concentration (C_∞).

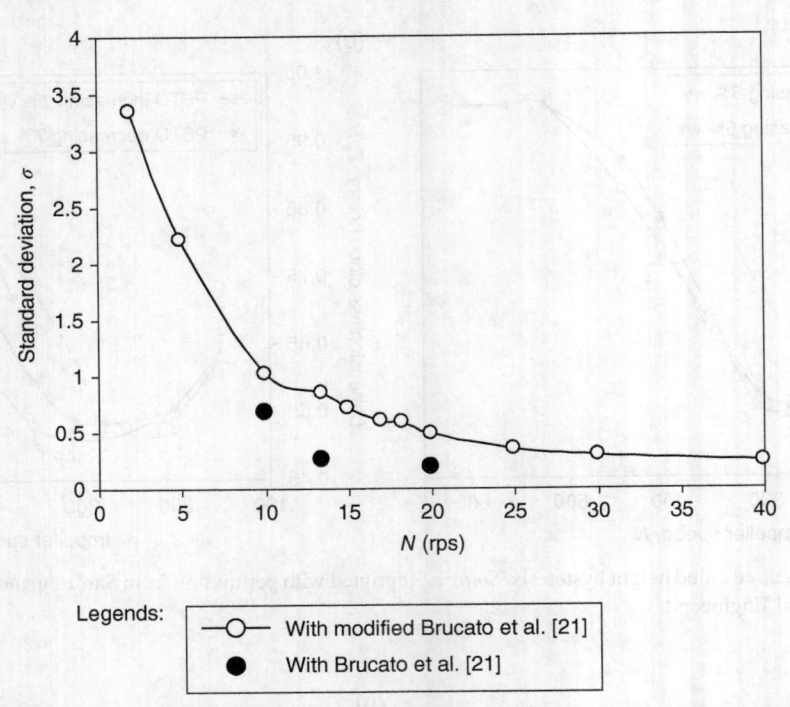

FIGURE 13.33 Predicted influence of impeller rotational speed on suspension quality, for $d_p = 264$ μm and $\alpha = 0.1$. *Source*: Reprinted with permission from Kasat et al. [92]. Copyright 2008 Elsevier Ltd.

It will be very helpful to first shed light on the predicted suspension quality before discussing the influence of suspension quality on the mixing process. In a stirred slurry reactor, the critical impeller speed for complete off-bottom suspension Nc_s and complete suspension N_s are two very important design parameters. The concepts of critical impeller speed were introduced more than 40 years ago and are primary design parameters used even today by reactor engineers for scale-up and design of stirred slurry reactor. The predicted suspension quality was analyzed to estimate the Nc_s and N_s. Several criteria are available in the literature to determine the values of Nc_s and N_s. However, those criteria are applicable for experimental measurements and cannot be extended directly to the CFD simulations with the EE approach. Bohnet and Niesmak [93] have proposed alternative criteria based on the standard deviation σ of solid concentration to describe the suspension quality (see Eq. (13.17)). The same criterion is used in the present work to describe the suspension quality. The decrease in standard deviation value is manifested as an increase in the quality of the suspension. Based on the range of the standard deviation, the quality of the suspension is broadly divided into three regimes: homogeneous suspension, where the value of the standard deviation is smaller than 0.2 ($\sigma < 0.2$); complete off-bottom suspension, where the value of the standard deviation lies between 0.2 and 0.8 ($0.2 < \sigma < 0.8$); and incomplete suspension, where the standard deviation value was found to be higher than 0.8 ($\sigma > 0.8$). This criterion enables the prediction of Nc_s and N_s and also gives the information on quality of suspension prevailing in the vessel.

The standard deviation values were estimated from the predicted solid volume fraction for all the 10 simulations carried out at different impeller rotational speeds. It must be noted that solid volume fraction values at all computational cells were used to estimate the standard deviation value. The predicted variation of standard deviation values with respect to impeller speed is shown in Figure 13.33. It can be seen from Figure 13.33 that three distinctly different suspension conditions, viz. incomplete suspension, complete off-bottom suspension, and homogeneous suspension, can be identified in the vessel. At a lower impeller speed, the computational model predicted very high values of the standard deviation ($\sigma > 0.8$), indicating incomplete suspension in the vessel. It is also observed that the standard deviation values drop sharply with an increase in the impeller rotational speed until complete off-bottom condition is achieved. The computational model predicted standard deviation value equal to 0.7 for 15 rps. This indicates the presence of a critical impeller speed for complete off-bottom suspension ($\sigma = 0.8$) close to 15 rps. This is in good agreement with the Nc_s (= 13.4 rps) estimated using correlation proposed by Zwietering [94] for the experimental setup of Yamazaki et al. [38]. With further increase in the impeller rotational speed, the values of standard deviation drop slowly till the system achieves homogeneous suspension condition. The predicted results suggest that the homogeneous suspension condition

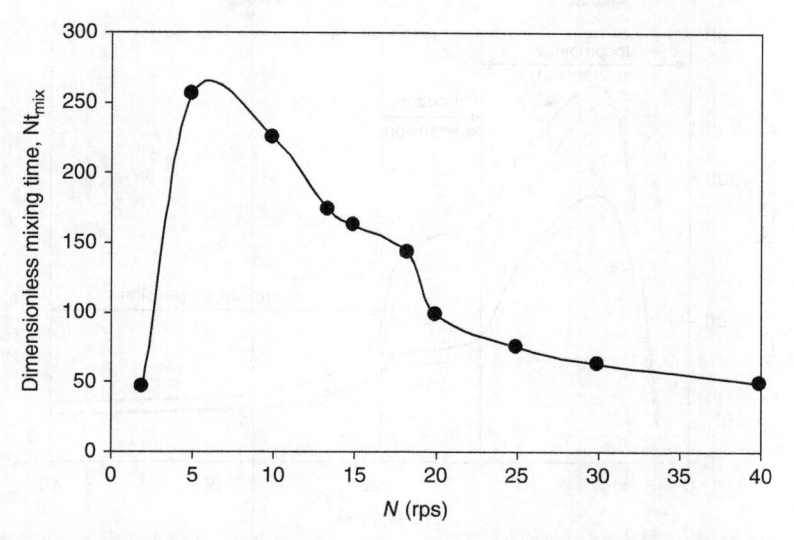

FIGURE 13.34 Predicted influence of impeller rotational speed on the dimensionless mixing time, for $d_p = 264\,\mu m$ and $\alpha = 0.1$. *Source*: Reprinted with permission from Kasat et al. [92]. Copyright 2008 Elsevier Ltd.

for the experimental condition of Yamazaki et al. [38] is achieved at impeller speed N_s equal to 40 rps ($\sigma = 0.17$).

The species transport simulations are then carried out to understand the mixing process in the experimental setup of Yamazaki et al. [38]. The variation of predicted dimensionless mixing time (Nt_{mix}) with impeller rotational speed is shown in Figure 13.34. It can be seen from Figure 13.34 that the dimensionless mixing time first increases sharply with increase in the impeller rotational speed and then drops slowly with further increase in impeller speed. Figure 13.34 shows minimum value of dimensionless mixing time for lowest impeller speed (2 rps). The predicted liquid velocity vector plot was studied to understand the possible reason behind the observance of a minimum mixing time. The predicted flow characteristics for an impeller rotational speed equal to 2 rps are shown in Figure 13.35. It can be seen from Figure 13.35a that the computational model predicted single-loop velocity pattern in the vessel. It is possible that for such a low impeller speed, impeller action is not sufficient to lift the particles from the vessel bottom (see Figure 13.35b). The solid bed present at the bottom of the reactor might offer apparent low off-bottom clearance to the impeller-generated flow and therefore lead to single-loop flow pattern for Rushton turbine. In such a scenario, all the energy dissipated by the impeller becomes available for generating liquid circulations in the vessel and for fluid mixing. Therefore, it is possible to have faster mixing in the reactor at low impeller rotational speed.

With increase in impeller rotational speed, the dimensionless mixing time increases, reaches maxima, and then drops slowly (see Figure 13.34). In the present work, the maximum in the mixing time was found to happen at around 5 rps. At 5 rps the impeller-generated flow becomes sufficient to start

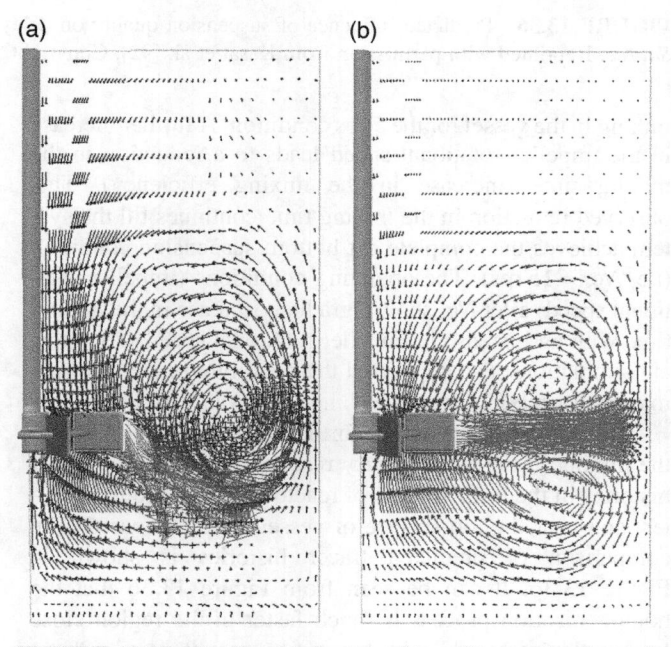

FIGURE 13.35 Predicted influence of impeller rotational speed on the liquid-phase flow field, for $d_p = 264\,\mu m$ and $\alpha = 0.1$. (a) $N = 2$ rps. (b) $N = 5$ rps. *Source*: Reprinted with permission from Kasat et al. [92]. Copyright 2008 Elsevier Ltd.

suspending solids into the bulk volume of the reactor. The energy dissipated by the impeller is now distributed for generating liquid circulations and fluid mixing and suspending the solids. The single-loop flow pattern changes to the classical two-loop structure for the Rushton turbine. The part of the energy dissipation for solid suspension and the rate of exchange between the two loops contribute to the slower

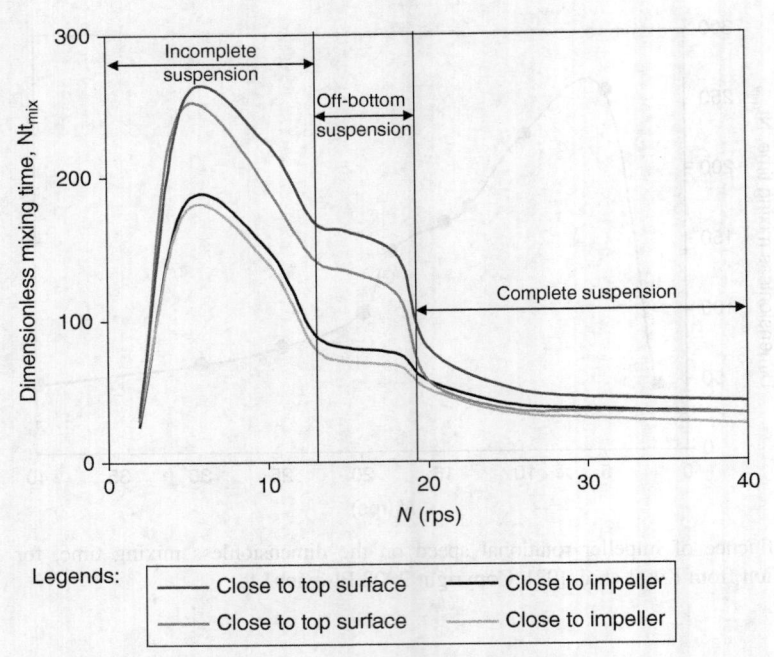

FIGURE 13.36 Predicted influence of suspension quality on delayed mixing in the top clear liquid layer, for $d_p = 264$ μm and $\alpha = 0.1$. *Source*: Reprinted with permission from Kasat et al. [92]. Copyright 2008 Elsevier Ltd.

mixing in the vessel for the 5 rps condition. A further increase in the impeller rotational speed leads to a reduction in the mixing time (increase in the mixing efficiency). This observed reduction in the mixing time continues till the system achieves the complete off-bottom suspension condition (i.e. $Nc_s = 15$ rps). The operating conditions (impeller rotational speed) after Nc_s show a gradual decrease in the mixing time with an increase in impeller rotational speed. The present simulations also supported the operating range at which maxima of mixing time occur, i.e. $N = Nc_s/3$ [91].

The simulated results are further analyzed to understand the mixing in the stirred slurry reactor. The predicted tracer histories in the bulk volume of reactor (more close to impeller) and near the top surface of vessel were compared. The comparison of predicted tracer histories is shown in Figure 13.36. It can be seen from Figure 13.36 that the homogenization process is much faster in the region close to impeller compared with the region near the top surface. It was also observed that the difference between the top region and the impeller region is strongly dependent on the suspension quality present in the vessel. Figure 13.36 shows that in incomplete suspension and in complete off-bottom suspension conditions, the time required for the homogenization near the top surface is significantly high compared with the time required for homogenization close to the impeller. The difference decreases as the system approaches the homogeneous suspension condition. Kasat et al. [92] showed that the lower liquid velocities present in the clear liquid layer above the solid cloud is responsible for the slower mixing process.

Preceding subsections discuss mixing issues in gas–liquid and solid–liquid systems. Similar approach can be used for addressing issues in other systems like liquid–liquid or gas–liquid–solid stirred vessels. Apart from predicting dispersion or suspension quality and mixing time, CFD models allow estimation of circulation time distribution, different zones in stirred vessels with different prevailing shear rates, interaction of impeller stream with inlet and outlet nozzles, and so on. It is possible to creatively use information obtainable from CFD to gain better insight and support engineering decision making. For example, information on different shear zones and RTD in these different zones often provides used clues for quantifying influence of scale on "breakup" and "coalescence" dominated zones. It is not possible to discuss actual industrial cases here for the sake of protecting confidential information. It is however hoped that information provided here will allow resourceful engineer to develop appropriate computational flow models and use the simulated results for addressing practical design and scale-up issues.

13.3.2.4 CFD Simulation of Antisolvent Crystallization
Pharmaceutical and fine chemical makers frequently rely on antisolvent crystallization, also known as precipitation for generating solid from a solution in which the product has high solubility. This technique is used for a variety of applications such as polymorph control, purification from a reaction mixture, and yield improvement. Antisolvent crystallization achieves supersaturation and solidification by exposing a solution of the product to another solvent (or multiple ones) in which the product is sparingly soluble.

Although this technique has the potential to achieve a controlled and scalable size distribution, it is not without problems. The product requires purification or separation steps to remove the antisolvent(s).

Woo et al. [95] discussed the model development in the framework of CFD for simulating the effect of mixing on the antisolvent crystallization process. In the CFD model they included PBE for the evolution of the CSD and the PDF of the local turbulent fluctuations. In addition to that, they used separate sub-models for nucleation and growth kinetics and Einstein equation for modeling effect of solid concentration on the rheology. For detail model equations and its implementation, please refer to Woo et al. [95]. Computational model was then used to predict the evolution of volume-averaged antisolvent, mass percent, supersaturation, nucleation rate, and mean growth rate. While comparing with the earlier published experimental observations, Woo et al. [95] found that the computational model was able to simulate temporal evolution of nucleation and growth rates at the inlet. They also predicted higher growth rates at the impeller region due to higher turbulence (Figure 13.37). They attributed this to improved mass transfer. This improved mass transfer at higher impeller rotational speed consequently led to faster desupersaturation, which resulted in lowering overall nucleation rate. Its impact on the overall crystallization process is explained through predicted final CSD. Figure 13.38 shows the predicted CSD at the end of one hour for three different rotational speeds. The predicted results show that fewer and slightly larger crystals were obtained for higher agitation rate. These results are qualitatively in agreement with the experimental findings.

Woo et al. [95] then used the computational model for simulating antisolvent crystallization with reverse mode of addition. Figure 13.39a shows the predicted final volume-averaged CSD. As anticipated, the computational model predicted finer particle size with reverse mode of addition in comparison with the normal mode of addition. They also used computational model for predicting influence of scale of the equipment on the CSD. Figure 13.39b shows the comparison of the predicted CSD for different scale and different scale-up criteria. The computational model predicted nonsignificant impact of scale of operation on the predicted CSD for constant power per unit volume that is consistent with the experimental findings by Torbacke and Rasmuson [96].

13.3.2.5 Process Innovation in Crystallization

It is been known that by reducing drug particle size to the absolute minimum may lead to significantly improving drugs' wettability and bioavailability. This will require a small mean crystal size and a narrow size distribution. Normally, pharmaceutical powders are polydisperse, i.e. consisting particles of different sizes. Polydisperse powders create considerable difficulties in the production of dosage forms. Particles of monosize (equal size) may be ideal for pharmaceutical purposes. In practice, powders with narrow range of size distribution can obviate the problems in processing them further. Making such products demand continuous processing either through an in-line mixer or via conducting milling on finished product [97]. At times milling technique can cause negative results such as dusting, caking, electrostatic charges, and in some times a polymorphic transformation. Process intensification by combining impinging jet reactor with stirred vessel (see Figure 13.40) provides an opportunity to control the final product particle size. It is been observed that uniform or nearly homogeneous nucleation happens in the impinging jet reactor [99]. These uniformly formed nucleates then feed into a stirred vessel where crystal growth or ripening will happen. Validated computational model not only will help in designing new innovative such systems but also will help in optimizing the operating parameters and mixing and transport processes for achieving desired process and product requirements.

13.3.2.6 Solid Suspension in Viscous Medium

There are several pharmaceutical suspensions available in the market. In these suspensions therapeutically active ingredients are uniformly dispersed in the medium. The solid particles used in these suspensions are smaller than 5 μm but have a tendency to settle. These suspensions are broadly differentiated into dilute suspensions (2–10% w/w) or concentrated suspensions (up to 50% w/w). It is imperative from the perspective of product specification to have uniform suspension in order to have the administration of a measured dose. Formulation experts avoid or minimize the settling either by reducing particle size or by increasing the viscosity of the continuous phase. Increasing the viscosity of the continuous phase is more predominantly used in formulations of these suspensions. The influence of rheology on solid suspension has received relatively less attention in the literature. Ibrahim and Nienow [100] showed that at higher viscosity, the Zwietering equation is likely to fail, with an error as large as 90%. The possible reason could be noninclusion of complex rheology behavior in the equation. It is known that in laminar condition, the presence of fine solids in the liquid produces shear-thinning non-Newtonian behavior [101].

Wu et al. [102] experimentally studied the influence of non-Newtonian rheology on the characteristics of solid suspension in a mechanically agitated vessel. They studied the suspension of glass beads in water and Carbopol solution. Their experimental results are presented in Figure 13.41. Very interesting observations were made by Wu et al. [102]. They observed that adding 0.04% Carbopol in water resulted in the reduction in the just suspension speed of impeller (N_{js}) by 15%. They attributed this drop to the reduction in the settling velocity of the solids. They also observed that above 0.09% of Carbopol in water, the N_{js} sharply reduced to zero (Figure 13.41b). To explain this, the variation

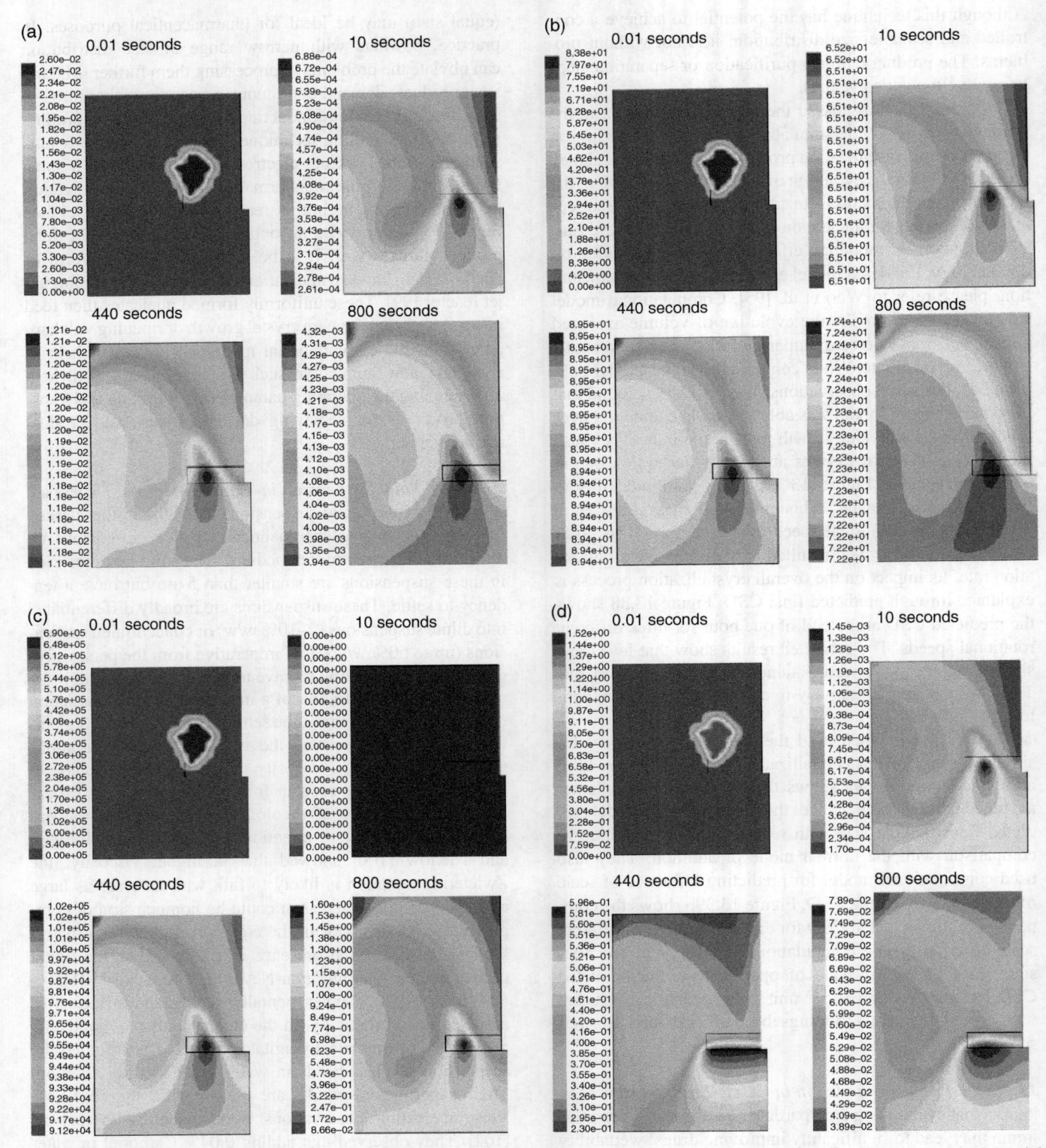

FIGURE 13.37 Simulated spatial distributions of process parameters for antisolvent crystallization at different time intervals. (a) Spatial distribution of the antisolvent mass fraction. (b) Spatial distribution of the supersaturation. (c) Spatial distribution of the nucleation rate. (d) Spatial distribution of the mean growth rate. *Source*: Reprinted with permission from Woo et al. [95]. Copyright 2006 American Chemical Society.

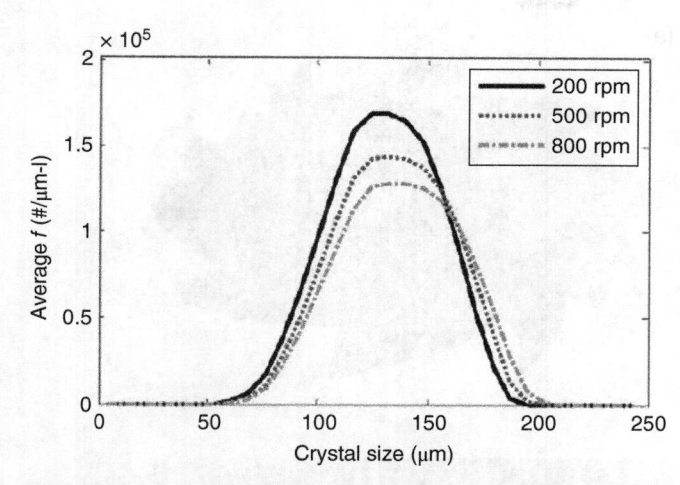

FIGURE 13.38 Simulated volume-averaged crystal size distribution for different impeller rotational speed. *Source*: Reprinted with permission from Woo et al. [95]. Copyright 2006 American Chemical Society.

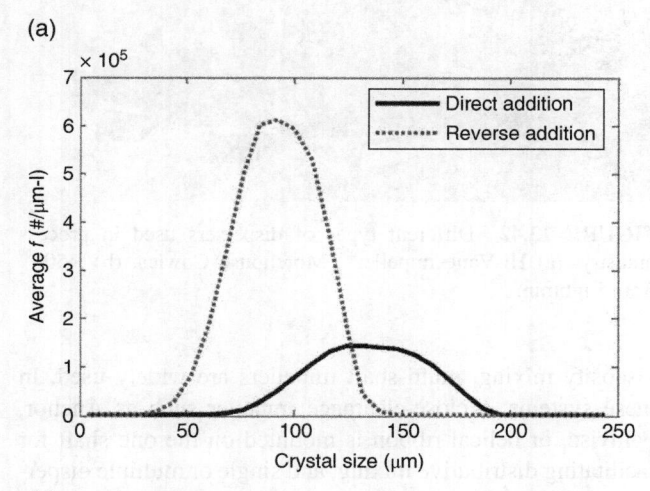

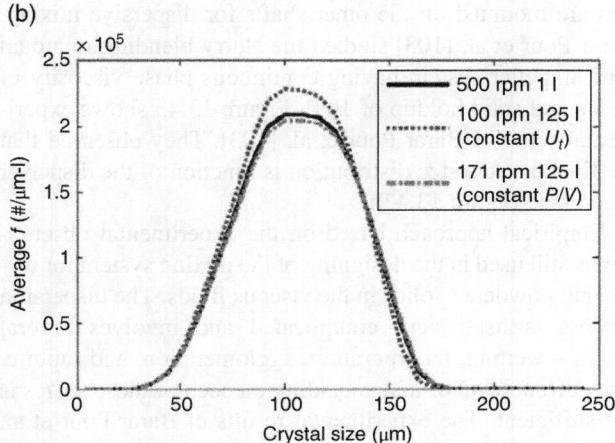

FIGURE 13.39 Simulated volume-averaged crystal size distribution. (a) Predicted CSD for different addition mode. (b) Predicted influence of scale-up on the CSD. *Source*: Reprinted with permission from Woo et al. [95]. Copyright 2006 American Chemical Society.

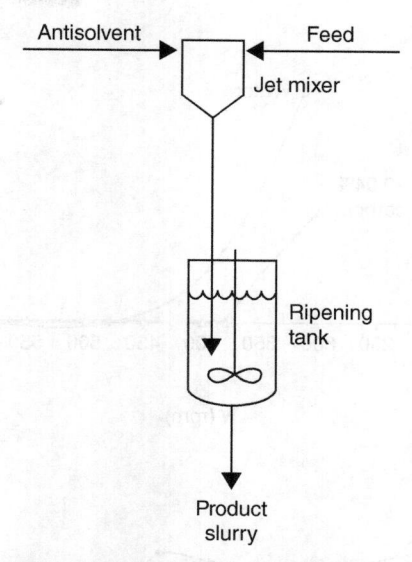

FIGURE 13.40 Conceptual process intensification of antisolvent crystallization. *Source*: Reprinted with permission from Genck [98]. Copyright Genck International.

of settling velocity is plotted in Figure 13.41c for different viscosity values for glass beads having 100 μm diameter. It is seen that the settling velocity approaches to the few tens of μm/s for 0.09% Carbopol solution and hence will not see settling for longer time. In such situations, the definition of N_{js} becomes irrelevant in defining impeller operating conditions, and definition of effective dispersion becomes more relevant.

Handling solid dispersion in viscous medium is not an easy task. A complication that arises is that the fine particles (microscopic in size) maintain electrical and molecular attraction. These fine particles tend to lump together and form agglomerates that no amount of mixing will break. An aggregate (or agglomerate) is composed of a group of particles that are strongly adherent and can be broken down only by the application of relatively strong mechanical forces. In the days before the advancement of disperser, different mills were used in practice for reducing the size of the product. However, the processing was very time consuming and led to very long batch time. With the advent of the disperser, the de-agglomeration process could be accomplished much more rapidly within the same vessel while mixing, resulting in a smoother, more uniform end product. The disperser is an impeller having thin disk with carefully designed teeth distributed radially about the circumference. Its action tears particles apart and disperses them uniformly throughout the product. This work is done with two actions: firstly, particles hitting the impeller are broken apart, or de-agglomerated, and secondly, the intense turbulence surrounding the impeller causes particles to hit each other with great momentum and inertia. The energy of this impact physically breaks apart agglomerates. Figure 13.42 shows the commonly used disperser in the industry.

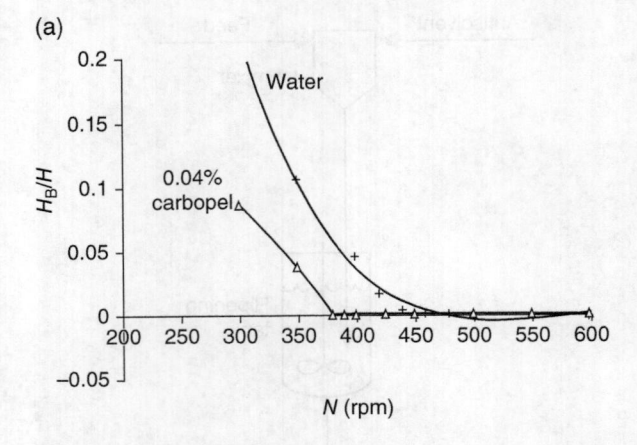

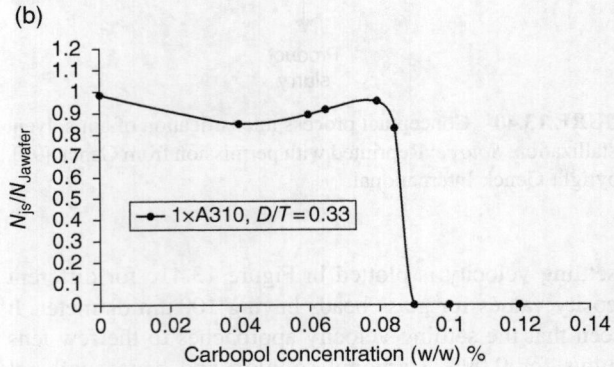

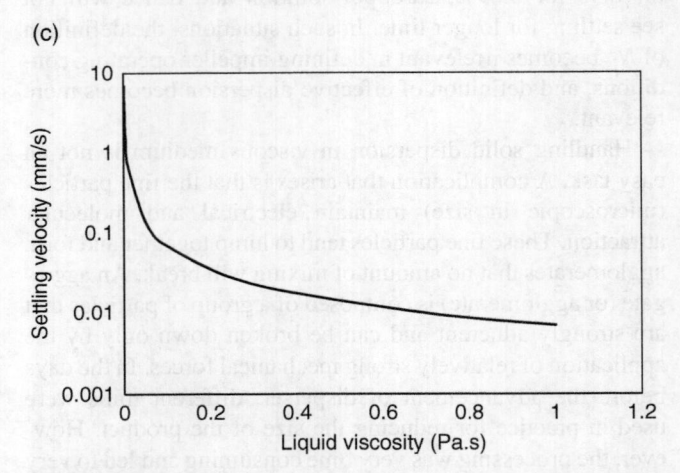

FIGURE 13.41 Influence of liquid-phase viscosity on the suspension. (a) Bed height versus impeller rotational speed for water and 0.04% Carbopol solution. (b) Bed height versus impeller rotational speed for water and 0.04% Carbopol solution. (c) Effect of continuous phase viscosity on the settling velocity of glass particle, $\rho = 2250\,kg/m^3$ and $dp = 100\,\mu m$. *Source*: Reprinted with permission from Wu et al. [102]. Copyright 2001 John Wiley and Sons.

Disperser has very limited pumping capacity, and hence in a larger mixing vessel, it is imperative to use other flow impeller to continuously feed disperser for achieving better product results. In many industrial applications having high viscosity mixing, multi-shaft impellers are widely used. In these systems, a close-clearance impeller such as Anchor, Paravisc, or helical ribbon is mounted on the one shaft for facilitating distributive mixing, and single or multiple dispersers are mounted on the other shafts for dispersive mixing. Barar Pour et al. [103] studied the slurry blending in a dual shaft impeller system having continuous phase viscosity of 1 Pa·s and solid holdup of 10%. Figure 13.43 shows experimental setup of Barar Pour et al. [103]. They observed that the final particle size distribution is function of the disperser speed (see Figure 13.43b).

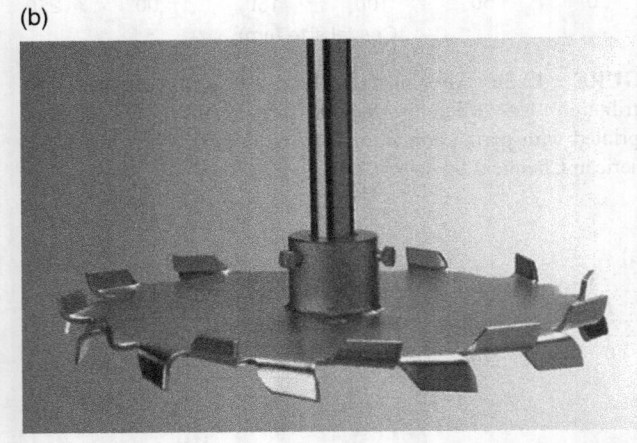

FIGURE 13.42 Different types of dispersers used in process industry. (a) Hi-Vane impeller®, Morehouse Cowles. (b) R500® from Lightnin.

Empirical approach based on the experimental observation is still used in the designing of the mixing system for dispersing powdered solids in the viscous fluids. The dispersion process is itself very complicated and involves several stages – wetting, incorporation, agglomeration, and rupture. The current level of understanding on each of these stages is not sufficient. The experimental results of Barar Pour et al. [103] show different behavior of torque experienced by Paravisc impeller for different solids. The actual reason for this variation is not very clear. Also, the current level of understanding on agglomerate formation and its breakup in a

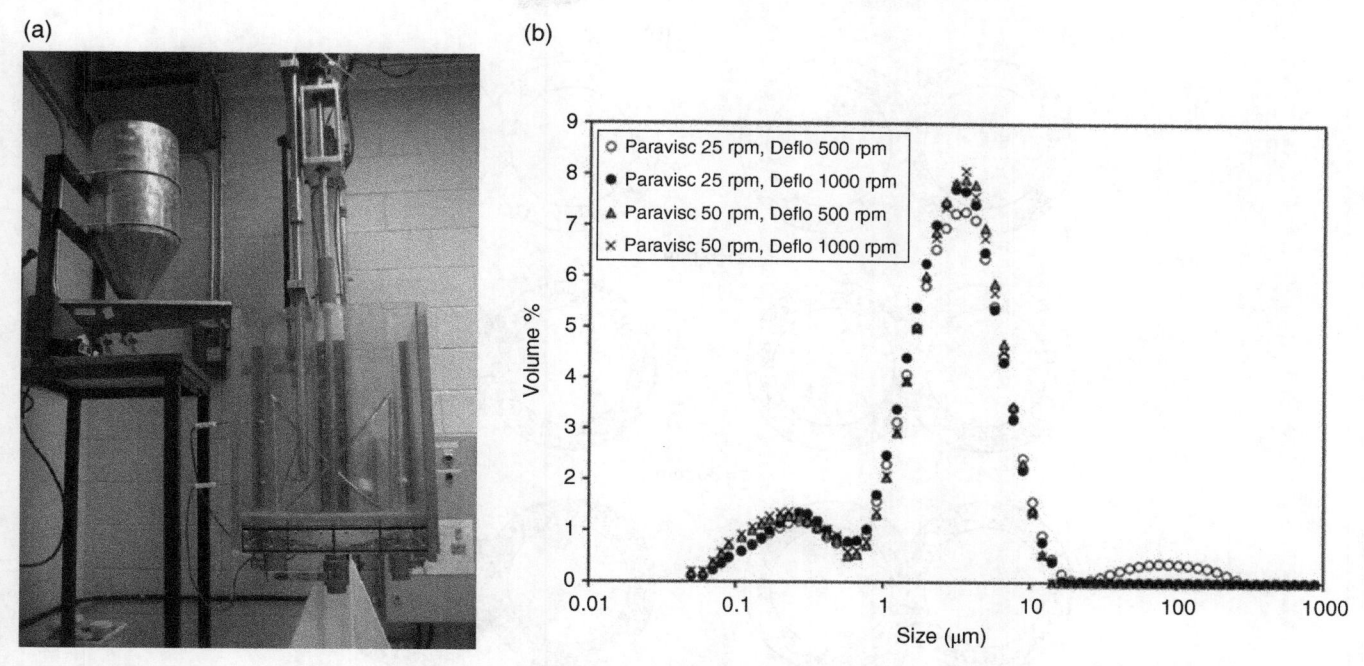

FIGURE 13.43 Slurry blending in a dual shaft stirred vessel. (a) Experimental setup. (b) Influence of deflo speed on the particle size distribution. *Source*: Reprinted with permission from Barar Pour et al. [103]. Copyright 2007 The Institution of Chemical Engineers, Published by Elsevier B.V.

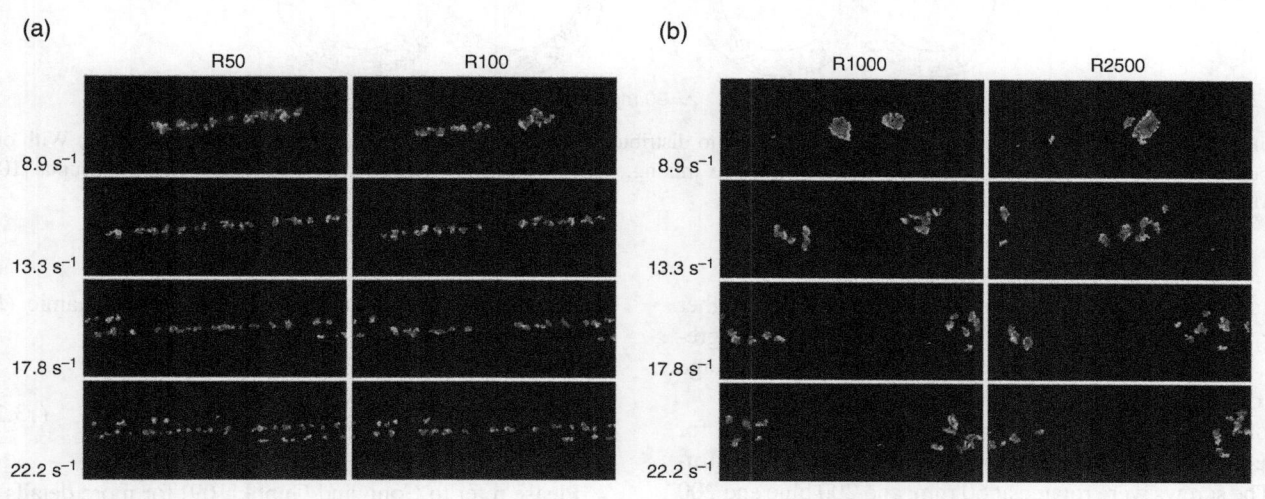

FIGURE 13.44 Simulated aggregate rupture for different cohesive strengths. (a) Rupture of weak aggregates. (b) Erosion of strong aggregates. *Source*: Reprinted with permission from Zeidan et al. [108]. Copyright 2007 The Institution of Chemical Engineers, Published by Elsevier B.V.

complex flow is not up to the level of understanding of bubble and/or droplet coalescence and breakage. There are few studies reported in literature mainly by Manas-Zloczower and coworkers [104–106], and Hansen et al. [107] presented models for simulating agglomerate breakup. Zeidan et al. [108] simulated the evolution of aggregates subjected to simple shear flow using a combined continuum and discrete model (CCDM). They solved the motion of discrete particles using Newton's second law of motion (DEM), and the continuum fluid flow was solved using locally averaged Navier–Stokes equations (CFD). This allowed Zeidan et al. [108] to study fully coupled aggregate deformation and breakup in a simple shear flow. They studied the breakage for aggregate having different cohesive strength. Figure 13.44 shows the simulated aggregate breakage for two different cohesive strengths. It can be seen that two different aggregate

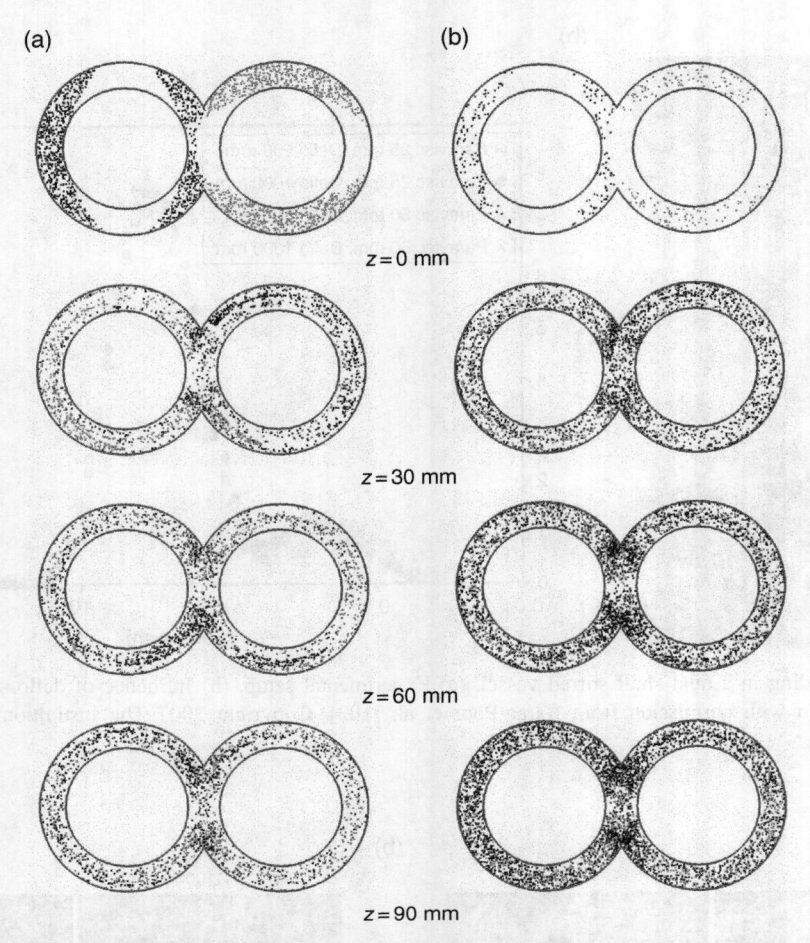

FIGURE 13.45 Change in spatial distribution of solids due to distributive and dispersive mixing in a twin-screw extruder. (a) With only distributive mixing. (b) Combined distributive and dispersive mixing. *Source*: Reprinted with permission from Cong and Gupta [109]. Copyright 2008 Society of Plastics Engineers.

mechanisms exist. Weak aggregates rupture and at higher shear rate produce smaller and more uniform-sized aggregates. However, erosion mechanism is dominant in breakage of strong aggregates, and it results in wider distribution.

Cong and Gupta [109] used PELDOM™ software for simulating solid dispersion in a corotating twin-screw extruder. The screws were rotated at 60 rpm, and 200 blue and 200 red segregated agglomerates were placed in the two halves of the twin-screw extruder entrance. Quality of distributive mixing in a corotating twin-screw extruder was evaluated by finding the change in the spatial distribution of initially segregated particles. Shannon entropy of mixture was first estimated, and then it was used for estimating color homogeneity index (CHI). CHI value lies between 0 and 1. Zero corresponds to completely segregated system, whereas one corresponds to completely mixed system. For the same twin-screw extruder, quality of dispersive mixing was determined by using the erosion model of Scurati et al. [105]. In this model the rate of agglomerate size reduction is proportional to the shear rate and the difference between hydrodynamic (F_h) and cohesive (F_c) forces:

$$-\frac{dR}{dt} = K(F_h - F_c)\frac{\dot{\gamma}}{2} \qquad (13.24)$$

Please refer to Cong and Gupta [109] for more details on the definition of hydrodynamic and cohesive forces. Particle tracing scheme is used for the simulating particle motion. The change in the spatial distribution was determined by following the particle path lines to the desired axial location of the extruder. The predicted CHI for distributed and dispersive mixing is shown in Figure 13.45. It can be seen from Figure 13.45a that starting with completely segregated red and blue particles in each lobes, at $z = 30$ mm, the distribution of the particles in both the lobes is nonhomogeneous and the homogeneity improves with distance. At $z = 90$ mm, the particle distribution in the twin-screw extruder is quite uniform. Initial guess of agglomerate and fragment radii determines the number of fragments erodes from the single agglomerate. For 0.5 and 0.11 mm for agglomerate and fragment radii, respectively, Alemaskin et al. [110] observed 93 fragments from a single agglomerate. Very high number of particles negatively influences the computational efficiency, and hence Cong and Gupta [109] in their simulations used 0.3 and 0.1 mm as the agglomerate and fragment radii, respectively. This led to 26 number of fragments resulting from erosion of single agglomerate. Figure 13.45b shows the spatial distribution of agglomerates and eroded fragments at three different locations of twin-screw extruder. It shows that with only 400 agglomerates at entrance, the population of the fragments increases substantially with extruder length. They also observed that the CHI, which determines the mixing efficiency, improves significantly due to dispersive mixing. This suggests that alone simulating distributive mixing will not be sufficient for quantifying mixing efficiency in such applications.

13.3.2.7 Solid Drawdown in a Stirred Vessel In many practical applications, low density solids, more particularly powders, are added into the liquid. Powder addition is usually accompanied with a variety of problems, irrespective of whether powder is soluble or insoluble. Typical changes do happen during hydration of solids in liquid and are explained schematically in Figure 13.46. The first case is simpler case compared with rest of the three. The challenge in first case is to wet the surface of solids and incorporate them in the liquid if required without entraining air. However, in the remaining three cases, rheological challenges do crop up. For example, if the rate of dissolution is faster than the rate of surface wetting, then there is a possibility of having a high viscosity layer of liquid near the top surface. This layer will eventually cover the dry solids and may delay the further dissolution. If the liquid develops Bingham behavior, i.e. a yield stress, wetting will stop completely. Swelling of particles always results in a slower rate of wetting, which may even approach zero. Many food ingredients like starch and some proteins tend to swell in water; in general the particles may swell and/or dissolve in the liquid to various degrees (Figure 13.46). It is impossible to predict in general whether or not such behavior leads to a faster or slower rate of wetting, as compared with an unchanged bulk material.

The rate of addition and surface motion can either worsen or improve powder addition. Many powders need to be added slowly enough that they do get sufficient time for wetting their surface and their incorporation into the liquid. Some hydrating thickeners such as cellulosic polymers need to be added quickly, while the fluid is still low viscosity and prevailing turbulence is there to provide aid in the addition and dispersion of the powder. This suggests that there is a specific rate of addition exists based on the wetting characteristics of the powders. Different mechanisms are recommended in the open literature for incorporating low density solids into the

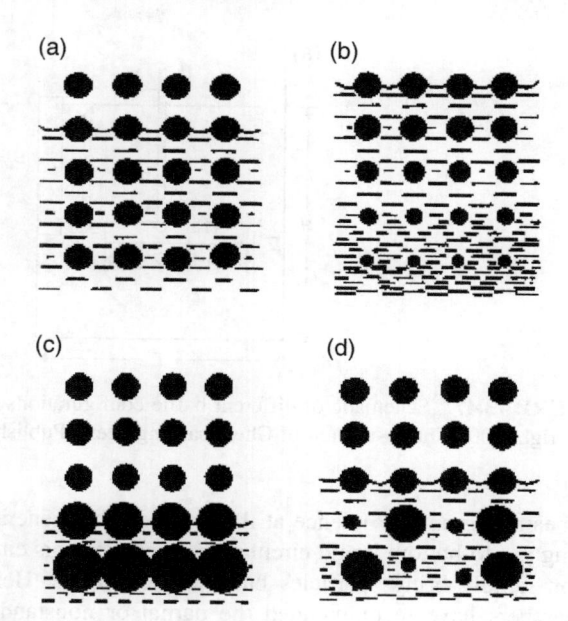

FIGURE 13.46 Schematic illustration of typical changes taking place during hydration of powders. (a) Unchanged. (b) Dissolving. (c) Swelling. (d) Dissolving and swelling. *Source*: Reprinted with permission from Schubert [111]. Copyright 1990 VCH Verlagsgesellschaft GmbH.

liquid in turbulent condition. Khazam and Kresta [112] identified three mechanisms of solid drawdown in stirred tanks: (1) Formation of stable single vortex (with no baffles or single baffle system) causes downward axial velocities at the surface responsible for drawdown. (2) Turbulent fluctuations form mesoscale eddies/vortices on the surface, which intermittently pull particles in the liquid. (3) Mean drag produced by the liquid circulation loops draw particles into the liquid where the downward axial velocities are greater than the particle slip velocity.

Most commonly used mechanism is generating a strong single vortex for incorporating solids (mechanism 1). The strong vortex can be generated using either no or partial baffling. Joosten et al. [113] used single baffle, Hemrajani et al. [114] used four baffles of width 1/50 tank diameter, and Siddiqui [115] recommended three partially immersed baffles 90° apart. Edwards and Ellis [116] found three-blade marine propeller without any baffles to be the most energy-efficient design. In this mechanism, during operation, headspace gas/air also gets entrained in the liquid. It is possible that in many practical applications, gas entrainment in the liquid needs to be avoided. This definitely puts challenge on the process engineer in designing the vessel and its operating conditions for efficient drawdown of the solids.

Several works have been reported in the past explaining the different baffling concepts and its implications on the drawdown of solids. It is well known that baffling suppresses the stable surface vortex formation and increases the intensity

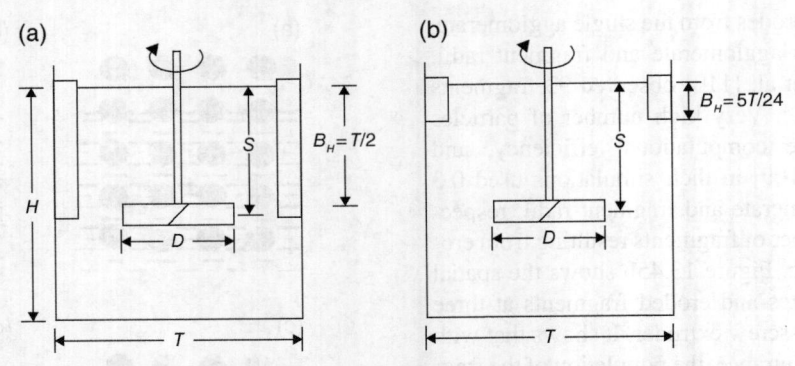

FIGURE 13.47 Schematic of different baffle configurations studied. *Source*: Reprinted with permission from Khazam and Kresta [112]. Copyright 2008 The Institution of Chemical Engineers, Published by Elsevier B.V.

of mean drag and turbulence at the surface. This generates strong top-to-bottom liquid circulation. However, this circulation rapidly brings particles back to the surface. Hence researchers have recommended the partial or nonstandard baffling. Khazam and Kresta [112] experimentally and computationally studied the drawdown of the floating solids in a stirred vessel for two different baffle designs, viz. half baffles and surface baffles. Schematics of the different baffling configurations used by them are shown in Figure 13.47. The objective was to maintain a high level of turbulence at the surface while reducing the return circulation from the bottom of the tank. The performance of the baffle configurations was compared using the just drawdown speed, N_{jd}, and cloud depth, CD, for the PBTD, PBTU, and A340 impellers. CFD simulations and measurements of the power number for the fully baffled and the surface baffled configurations were also reported. They found out that the drawdown speed for all the impellers was very similar for full and half baffles (see Figure 13.48). But with surface baffles, they found advantage in significant reduction in the drawdown speed N_{jd}. They also observed that a more robust performance at large submergences and a better distribution of solids was obtained for the surface baffles.

Atibeni et al. [117] used the PIV technique for elucidating the effect of baffle design on the drawdown of the floating particles. They observed that for the standard baffle system, just drawdown speed reduces by placing impeller off-center. They observed more than 50% reduction in impeller power by this modification. They also found that around 50% impeller power reduces by modifying standard baffles with down triangle baffles. Hsu et al. [118] used both visual observations and CFD model for simulating the effect of impeller clearance and baffle design on the drawdown of the floating particles. They also observed positive impact of new baffle designs on the drawdown process over the standard baffles.

13.3.2.8 Solid Dissolution in Stirred Vessel

Solid dissolution is a common process unit operation during liquid mixing in the pharmaceutical industry. Incomplete dissolution can result in subpotent batches, resulting in batch failure. Competing elements such as the time required for dissolution versus the potential of microbial contamination over an extended mixing period may also be in play in a manufacturing setting. Dissolution in a stirred tank is influenced by many parameters. Khinast and coworkers [119, 120] provided an Ishikawa diagram (Figure 13.49) for explaining the various parameters that influence the operation. Many of these parameters cannot be changed in a pharmaceutical manufacturing setting. For example, most of the physicochemical properties of a raw material are fixed, although variations do occur due to the batch-to-batch variability of APIs. The process engineer has to work with process parameters, human environment, and equipment for improving the dissolution rate or time required for achieving complete dissolution. Other than the particle properties, the dissolution kinetics depend on the local mass transfer coefficient (which in turn is a function of suspension state and the local turbulence level), on the thermodynamics of the crystal, and on the equilibrium solubility at the fluid temperature.

A key fluid dynamic parameter that will govern the dissolution time is the solid–liquid mass transfer coefficient. The rate of mass transfer is defined as

$$\text{Rate of mass transfer} = ka(C^* - C) \quad (13.25)$$

where

k is the solid–liquid mass transfer coefficient.

a is interfacial area.

C^* is the concentration at the interface that is saturation concentration.

C is the concentration in the bulk liquid phase.

Thus by definition, achieving a constant mass transfer coefficient across different scales, one should achieve similar dissolution rates and hence similar dissolution times. In order

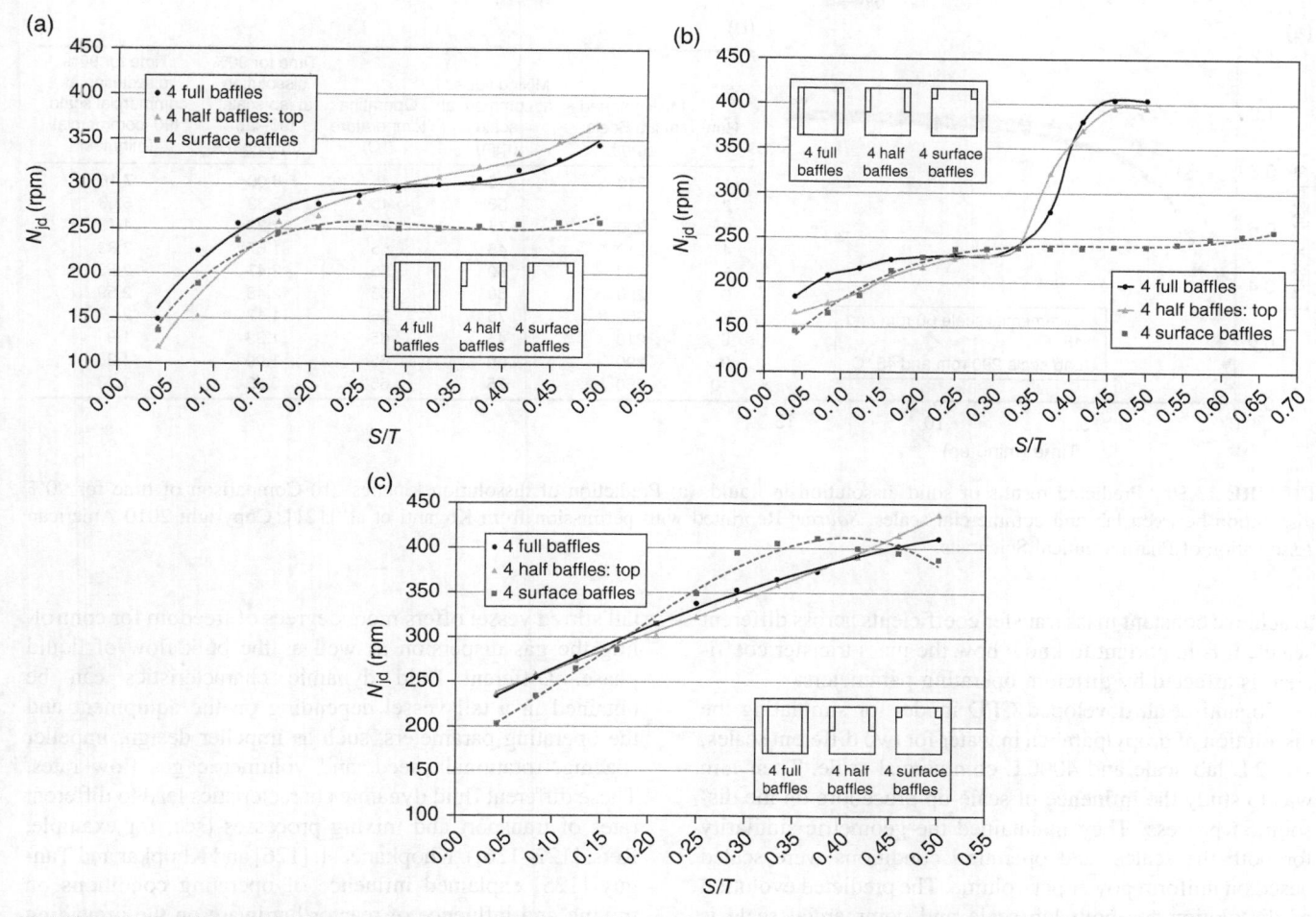

FIGURE 13.48 Effect of baffle configuration on the N_{jd}. (a) Up-pumping pitched blade turbine. (b) Down-pumping pitched blade turbine. (c) A340. *Source*: Reprinted with permission from Khazam and Kresta [112]. Copyright 2008 The Institution of Chemical Engineers, Published by Elsevier B.V.

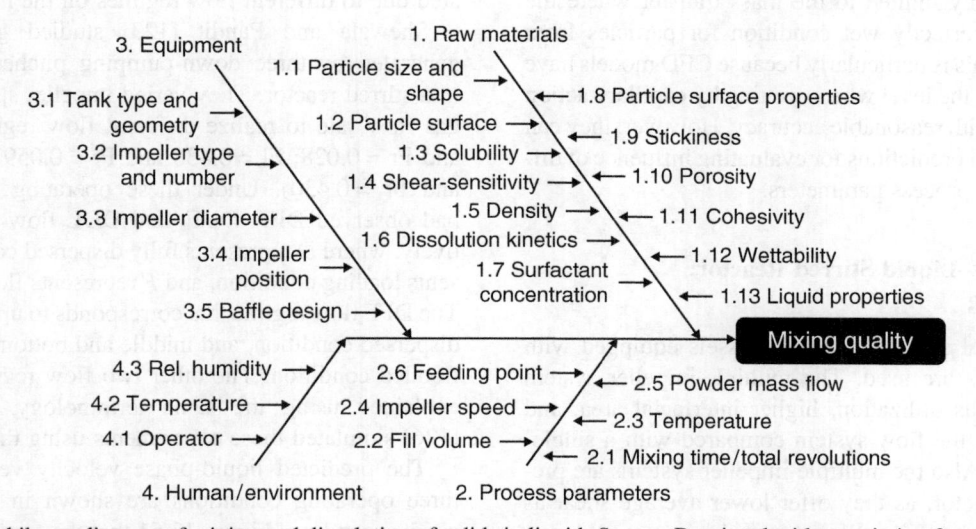

FIGURE 13.49 Ishikawa diagram of mixing and dissolution of solids in liquid. *Source*: Reprinted with permission from Adam et al. [119]. Copyright 2010 Elsevier B.V.

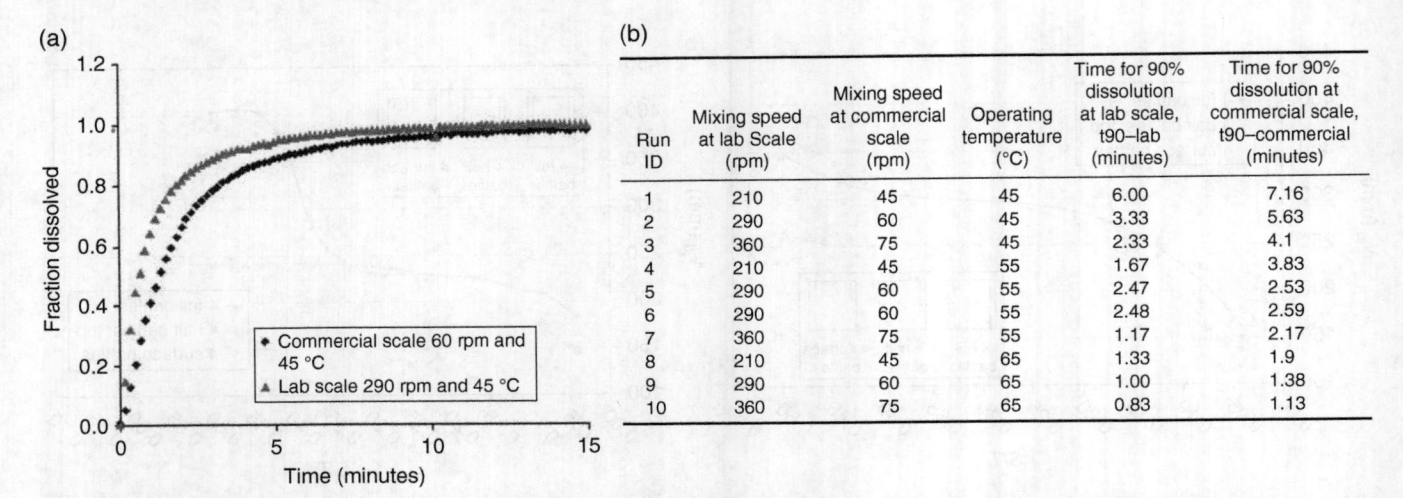

FIGURE 13.50 Predicted results of solid dissolution in liquid. (a) Prediction of dissolution kinetics. (b) Comparison of time for 90% dissolution between lab and commercial scales. *Source*: Reprinted with permission from Koganti et al. [121]. Copyright 2010 American Association of Pharmaceutical Scientists.

Run ID	Mixing speed at lab Scale (rpm)	Mixing speed at commercial scale (rpm)	Operating temperature (°C)	Time for 90% dissolution at lab scale, t90–lab (minutes)	Time for 90% dissolution at commercial scale, t90–commercial (minutes)
1	210	45	45	6.00	7.16
2	290	60	45	3.33	5.63
3	360	75	45	2.33	4.1
4	210	45	55	1.67	3.83
5	290	60	55	2.47	2.53
6	290	60	55	2.48	2.59
7	360	75	55	1.17	2.17
8	210	45	65	1.33	1.9
9	290	60	65	1.00	1.38
10	360	75	65	0.83	1.13

to achieve constant mass transfer coefficients across different scales, it is important to know how the mass transfer coefficient is affected by different operating parameters.

Koganti et al. developed CFD model for simulating the dissolution of propylparaben in water for two different scales, viz. 2 L lab scale and 4000 L commercial scale. Their aim was to study the influence of scale-up procedure on the dissolution process. They maintained the geometric similarity for both the scales, and operating conditions were scaled based on uniform power per volume. The predicted evolution of dissolution for both lab scale and commercial scale is shown in Figure 13.50. It shows that the difference between the laboratory scale and commercial scale is reasonably small and in order of 75–88% confidence for various operating conditions. It must be noted that the CFD simulation of dissolution process is only limited to the mass transfer where the model assumes perfectly wet condition for particles from the initial time. This is particularly because CFD models have not yet reached to the level where wetting by capillary action can be modeled with reasonable accuracy. However, they can be used for a priori predictions for evaluating influence of different design and process parameters.

13.3.3 Tall Gas–Liquid Stirred Reactor: Flow and Mixing

In many industrial applications, tall vessels equipped with multiple impellers are used. The multiple-impeller system provides better gas utilization, higher interfacial area, and narrower RTD in the flow system compared with a single-impeller system. Also the multiple-impeller systems are preferred in a bioreactor, as they offer lower average shear as compared with a single-impeller system due to overall lower operational speed with nearly same power input. Overall, the tall stirred vessel offers more degrees of freedom for controlling the gas dispersion as well as the bulk flow of liquid phase. Different fluid dynamic characteristics can be obtained in a tall vessel depending on the equipment and the operating parameters, such as impeller design, impeller spacing, rotational speed, and volumetric gas flow rates. These different fluid dynamic characteristics lead to different rates of transport and mixing processes (see, for example, Refs. [122–126]). Khopkar et al. [126] and Khopkar and Tanguy [125] explained influence of operating conditions on mixing and influence of reactor hardware on the prevailing local fluid dynamics, respectively. In this chapter, the case of gas–liquid flow generated by three down-pumping pitched blade turbines studied by Khopkar et al. [126] was considered to explain the implications of prevailing flow patterns generated due to different flow regimes on the mixing process.

Shewale and Pandit [123] studied gas–liquid flows generated by three down-pumping pitched blade turbines in a stirred reactor. They varied impeller speed at a specific gas flow rate to realize different flow regimes (Fl = 0.638 and Fr = 0.028; Fl = 0.438 and Fr = 0.0597 and Fl = 0.163 and Fr = 0.430). Under these operating conditions, they had observed DFF, DDF, and DDL flow regimes, respectively, where D represents fully dispersed condition, L represents loading condition, and F represents flooding condition. The DFF flow regime that corresponds to upper impeller is in dispersed condition, and middle and bottom impellers are in flooded condition. The other two flow regimes can also be explained using the same terminology. Khopkar et al. [126] simulated these experiments using the EE approach.

The predicted liquid-phase velocity vectors for all the three operating conditions are shown in Figure 13.51. It can be seen from Figure 13.51 that the computational model captured the significantly different flow fields for all the three

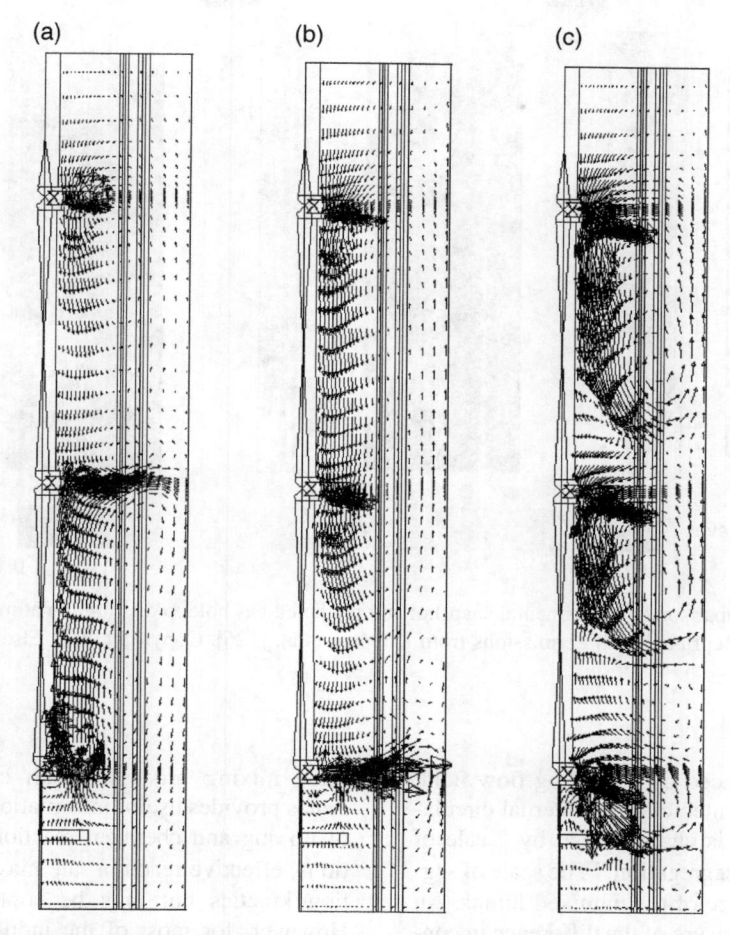

FIGURE 13.51 Predicted mean liquid velocity field at mid-baffle plane for DFF, DDF, and DDL flow regimes. (a) DFF flow regime (Fl = 0.678 and Fr = 0.028). (b) DDF flow regime (Fl = 0.438 and Fr = 0.0597). (c) DDL flow regime (Fl = 0.163 and Fr = 0.430). *Source*: Reprinted with Permissions from Khopkar et al. [126]. Copyright 2005 Elsevier Ltd.

conditions. For DFF (Fl = 0.638 and Fr = 0.028) flow regime, the predicted velocity field shows the presence of two-loop structure. The predicted liquid-phase velocity field for DDF flow regime (Fl = 0.438 and Fr = 0.0597) also shows the two-loop structure (Figure 13.51b). However, the predicted two-loop structure for DDF flow regime was significantly different from the two-loop structure predicted for DFF flow regime. Along with these two primary circulation loops, the computational model has also captured a secondary circulation loop, present between both circulation loops. For the DDL flow regime (Fl = 0.163 and Fr = 0.430), simulated results show (Figure 13.51c) three separate circulation loops for each impeller. The predicted velocity field for DDL condition also captured two secondary circulation loops, one at the bottom of the reactor and another between the lower and middle impeller circulation loops. The complex interaction between the impeller-generated flow and gas-generated flow was responsible for the formation of these two secondary circulation loops in the reactor.

The qualitative comparison of predicted gas holdup distributions for all the three operating conditions [Fl = 0.638 and Fr = 0.028 (DFF); Fl = 0.438 and Fr = 0.0597 (DDF) and Fl = 0.163 and Fr = 0.430 (DDL)] with experimental snapshots is shown in Figure 13.52. It can be seen from Figure 13.52a that similar to experimental condition, the simulation has captured the inefficient dispersion of gas at the bottom and middle impellers and dispersed condition of gas at the upper impeller for DFF flow regime. It can be seen from Figure 13.52b that the simulation has correctly captured the inefficient dispersion of gas by the bottom impeller and the complete dispersed conditions by the middle as well as upper impeller as observed in the case of DDF flow regime. For the DDL flow regime (Figure 13.52c), the predicted gas holdup distribution shows the fully dispersed condition for upper and middle impeller and loading condition for the bottom impeller.

One of the major interests in developing such complex flow models is to gain insight into mixing. Mixing in the

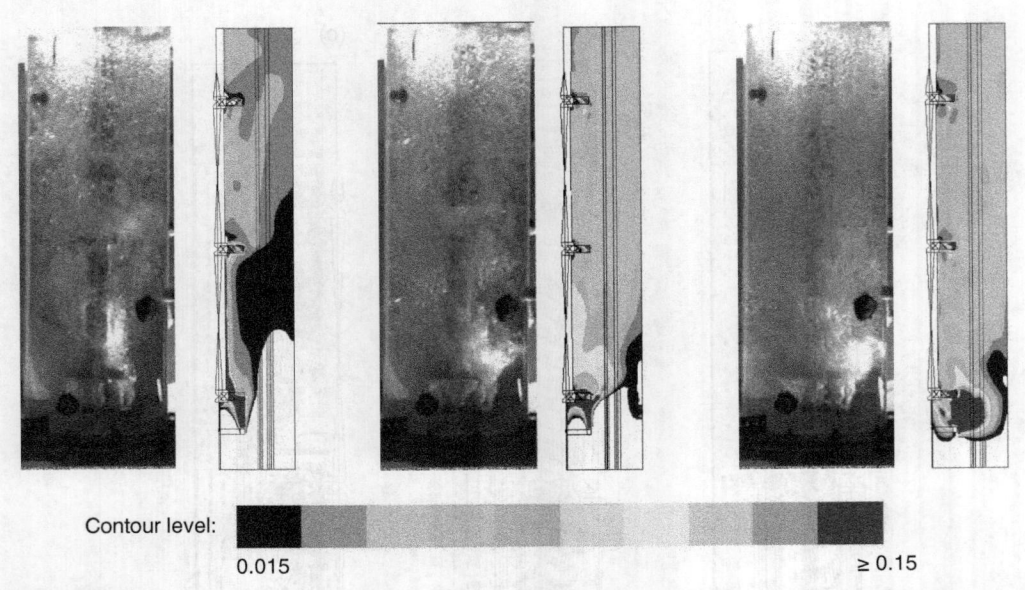

FIGURE 13.52 Qualitative comparison of experimental snapshot and predicted gas holdup distribution at mid-baffle plane for DFF, DDF, and DDL flow regimes. *Source*: Reprinted with Permissions from Khopkar et al. [126]. Copyright 2005 Elsevier Ltd.

reactors is significantly influenced by prevailing flow field, particularly flow regimes and interaction of internal circulation loops. Generally, mixing is characterized by "scale of segregation" and "intensity of segregation." The scale of segregation is a measure of the size of the unmixed lumps. An intensity of segregation is a measure of the difference in concentration between neighboring lumps of fluid. The lower the intensity of segregation, the more is the extent of molecular mixing (see Ref. [9] and references cited therein for more detailed discussion). Since most of the multiphase flows in industrial reactors will be turbulent, we will limit our discussion here to turbulent mixing. The convection and turbulent dispersion by large eddies lead to macroscale mixing and do not cause any small-scale mixing. Fluid motions in the inertial subrange reduce the scale of segregation via vortex stretching. Such a reduction in scale increases interfacial area between segregated lumps of tracer fluid and the base fluid, which increases the rate of mixing by molecular diffusion. However, increase in interfacial area by the inertial subrange eddies may not be substantial. The mixing caused by this step is typically called as "mesomixing." Mesomixing reduces the scale of mixing substantially but does not affect intensity of mixing much. Engulfment and viscous stretching by Kolmogorov scale eddies lead to substantial increase in the interfacial area for molecular diffusion and therefore contribute significantly to molecular mixing. The last step is diffusion process through such interfacial area between layers of different fluids accompanied by chemical reactions, if any. Molecular diffusion leads to complete mixing and dissipates concentration fluctuations. Comparison of time scales of these mixing processes with characteristic reaction time scales provides useful information about possible interaction of mixing and chemical reactions. For fast chemical reactions, effective reaction rate may not be controlled by reaction kinetics but may be controlled by rate of mixing. However, for most of the industrially relevant multiphase flow processes, fast reactions may often be controlled by interphase mass transfer rather than liquid-phase mixing. It is however often important to quantify characteristic time scale of "mixing" to understand interaction of interphase transport and mixing as well as possibility of short circuiting and channeling. Usually "mixing time" and "circulation time," which essentially characterize macromixing in stirred tanks, are used for this purpose.

Mixing time is the time required to achieve a certain degree of homogeneity [127]. However, circulation time is the time necessary for a fluid element to complete a one circulation within the vessel (time difference between an event of fluid element exiting from the impeller swept volume and an event of its reentry into impeller swept volume). The circulation time distributions provide useful insight about possible short circuiting and channeling. The mixing time is also usually related to mean circulation time [8]. In this example of tall gas–liquid stirred tanks, computational flow models are used to estimate mean circulation time to gain better understanding of macromixing process.

Using the Eulerian flow field obtained as discussed in the previous subsection, the particle trajectories were simulated for all the three operating conditions (DFF, DDF, DDL). Based on the study of Rammohan et al. [128], neutrally buoyant particles of size less than 0.25 mm were released into the liquid at 10 randomly selected positions in the solution domain. The motion of particles in the liquid phase was simulated using the Lagrangian framework. The simulated particle trajectories were used to calculate the circulation time distribution.

The simulated circulation time distributions for all the three operating conditions are shown in Figure 13.53. It can be seen from Figure 13.53 that for DFF flow regime, significant fraction of (18%) show circulation time higher than 16 seconds. These circulations were for particles following the upper circulation loop and may lead to slower mixing in the reactor. Almost no circulations (<1%) with circulation times less than four seconds were found in the simulated circulation time distribution. For the DDF flow regime, significant fraction (~60%) show circulation time less than six seconds, indicating faster mixing. For the DDL regime, not insignificant fraction (9%) show circulation times more than 30 seconds, indicating slower mixing despite increase in the impeller speed. The predicted values of average circulation time and the experimental data are listed in Table 13.6. Figure 13.54 shows the variation in the mixing time with impeller speed as reported by Shewale and Pandit [123] and the time required for a fixed number of circulations as per the simulations in this work. It can be seen from Table 13.6 and Figure 13.54 that the predicted values of average circulation times have captured the apparently counterintuitive trend (increase in mixing time with increase in impeller speed) observed in the experimental study of Shewale and Pandit [123]. The developed computational model can thus be creatively used to address industrially important issues.

13.4 SUMMARY AND PATH FORWARD

In this chapter, we have demonstrated the extent of applicability of computational models for simulating multiphase flows in stirred vessels with some examples. Role of turbulence, multiphase flow, interphase interactions (drag, lift, virtual mass, coalescence and breakup, and so on), and flow regimes are critically analyzed for gas–liquid and solid–liquid flows. The presented computational models were found to capture key features of two-phase flows in stirred tank reasonably well. The present work highlighted the limited applicability of direct extension of gas–liquid and solid–liquid modeling approaches for simulating three-phase flow. Despite some of the limitations, computational models were shown to provide useful information on important flow characteristics around the impeller blades as well as in the bulk. The computational models were able to predict the implications of reactor hardware, flow regimes, and suspension quality on the transport and mixing process. Careful numerical experiments using these CFD models can be used for better understanding of the characteristics of existing reactors to enhance their performance, assess different configurations, and greatly assist the engineering decision-making process. The approach, models, and the results discussed here will provide useful basis for practical applications as well as for further developments.

Though the models discussed here are capable of providing valuable and new insights that hitherto were unavailable, there is still significant scope to improve fidelity of these multiphase flow models. Some of the ways for improving the discussed models are listed in the following:

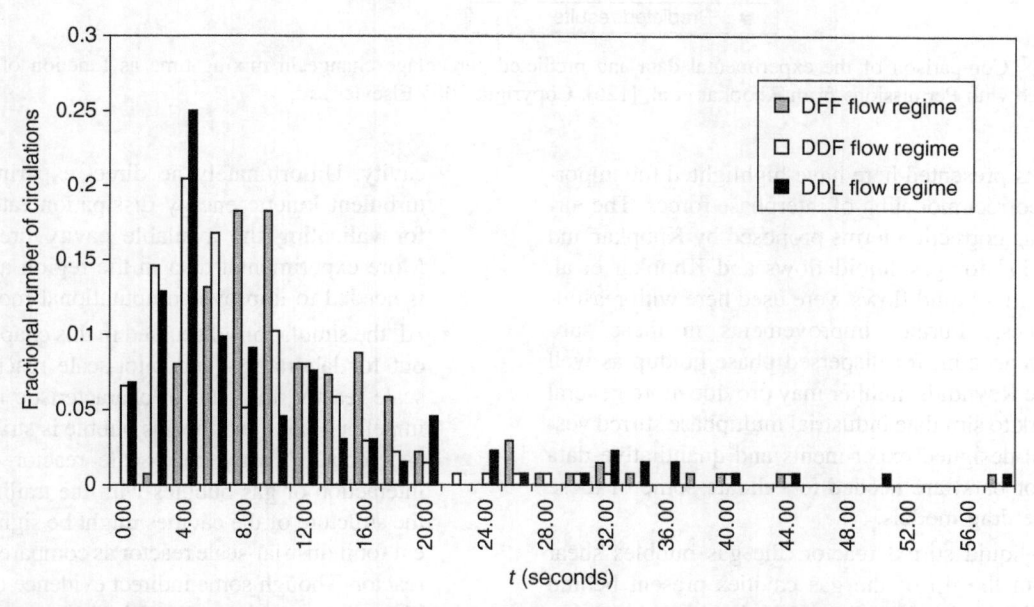

FIGURE 13.53 Predicted circulation time distribution for DFF, DDF, and DDL flow regimes. *Source*: Reprinted with Permissions from Khopkar et al. [126]. Copyright 2005 Elsevier Ltd.

TABLE 13.6 Gross Characteristics of a Tall Gas–Liquid Stirred Reactor

Flow Regime	Total Gas Holdup (%)		Power Number, N_{Pg}		Average Circulation Time, t_c (Predicted)	Mixing Time, t_m (Experimental)	Percentage Change	
	Predicted	Experimental	Predicted	Experimental			$t_c/t_{c,min}$	$t_m/t_{m,min}$
DFF (Fl = 0.6328 and Fr = 0.028)	2.99	2.47	2.64	2.2	13.851	59	1.493	1.553
DDF (Fl = 0.438 and Fr = 0.0597)	3.43	2.79	2.98	2.55	9.277	38	1	1
DDL (Fl = 0.163 and Fr = 0.430)	5.58	3.65	4.05	3.45	11.234	45	1.211	1.184

Source: Reprinted with permissions from Khopkar et al. [126]. Copyright 2005 Elsevier Ltd. Experimental data from Shewale and Pandit [123].

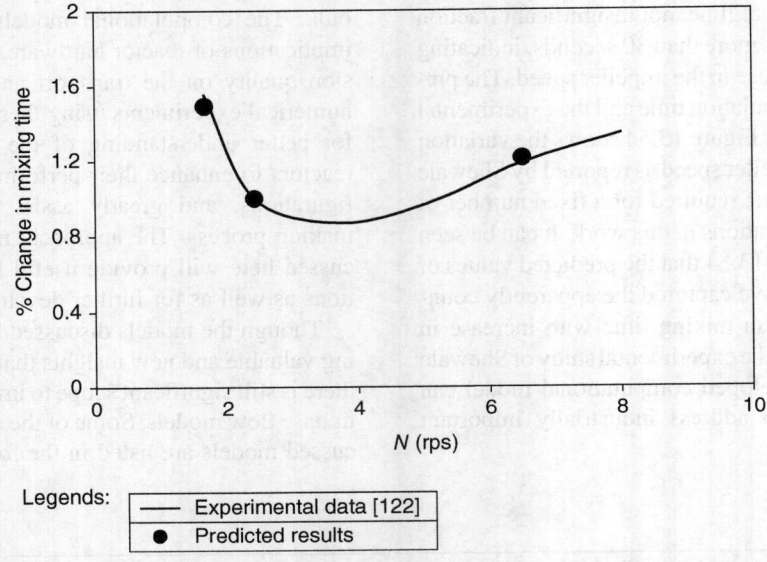

FIGURE 13.54 Comparison of the experimental data and predicted percentage change in mixing time as function of impeller speed. *Source*: Reprinted with Permissions from Khopkar et al. [126]. Copyright 2005 Elsevier Ltd.

- The results presented here have highlighted the importance of correct modeling of interphase forces. The turbulent drag correction terms proposed by Khopkar and Ranade [17] for gas–liquid flows and Khopkar et al. [38] for solid–liquid flows were used here with reasonable success. Further improvements in these submodels to account for dispersed phase holdup as well as particle Reynolds number may provide more general framework to simulate industrial multiphase stirred vessels. Well-designed experiments and quantitative data (with error bars) are needed to validate some of these interphase drag models.
- In a gas–liquid stirred reactor, the gas bubbles shear away from the tip of the gas cavities present behind the impeller blades. The size of the bubbles emanating from the cavity tip is controlled by the size of the cavity, breakage of cavity, and the turbulence level around the cavity. Unfortunately no direct experimental data for turbulent kinetic energy dissipation rate are available for validating the available cavity breakage models. More experimental data in the region around impeller is needed to improve computational models.
- All the simulations discussed in this chapter were carried out for laboratory- and pilot-scale reactors. For large-scale reactor, the ratio of characteristic length scales of impeller blades and the gas bubble is strikingly different as compared with small-scale reactor. Therefore, the interaction of gas bubbles with the trailing vortices and the structure of the cavities might be significantly different for industrial-scale reactor as compared to small-scale reactor. Though some indirect evidence of this is available, no systematic study of the influence of the scale on relative performance of different newly proposed impellers for dispersing secondary phase is available.

- Solid–liquid systems with polydispersed solid phase are encountered in process industry. However, there are no reports in the literature on the experimental measurements of the concentration profiles for the polydispersed system. Experimental and computational efforts are therefore needed to study hydrodynamics of the solid–liquid stirred reactor with polydispersed solid phase.
- In a three-phase stirred reactor, suspended solid particles will interact with the wake of gas bubble. This interaction will not only influence the drag experienced by solid particles as well as gas bubble but also influence the lift experienced by solid particles. This might be a possible reason for limited applicability of direct extensions of gas–liquid and solid–liquid modeling approaches for simulating gas–liquid–solid stirred reactor. Well-designed experiments and computational efforts need to be undertaken for the estimation of bubble and particle interaction.

Complexity of reactive flows may greatly expand the list of issues on which further research is required. Another area that deserves mention here is modeling of unsteady flows in stirred vessels. Most of the examples discussed in this chapter used a steady-state modeling approach for simulating flows in stirred vessels. In some conditions, the steady-state approach may not be appropriate, and a full unsteady-state approach may be necessary. This is especially crucial when fast reactive mixing and interaction of nozzle and impeller stream are important. Fortunately, for many multiphase stirred vessels, the overall performance is dominated by interphase transport rather than micromixing, and therefore full unsteady simulations may not be necessary.

Adequate attention to key issues mentioned in this chapter and creative use of CFM will hopefully make useful contributions to reactor engineering of multiphase stirred reactors. New advances made in modeling of multiphase flows in stirred vessels may be assimilated using the framework discussed in this chapter. We hope that this work will stimulate applications of CFM to reactor engineering in pharmaceutical industry.

13.5 NOTATIONS

Roman Symbols

C	impeller off-bottom clearance, m
C_1, C_2	model parameters (Eq. 13.5)
C_D	drag coefficient
C_{D0}	drag coefficient in stagnant water
C_{VM}	virtual mass coefficient
C_ω	model parameter (Eq. 13.6)
D_{12}	turbulent diffusivity, m^2/s
d_b	bubble diameter, m
d_p	diameter of particle, m
d_s	impeller shaft diameter, m
D_i	impeller diameter, m
d_{sp}	outer diameter of ring sparger, m
F_D	interphase drag force, N/m^3
F_L	lift force, N/m^3
F_q	interphase momentum exchange term
F_{VM}	virtual mass force, N/m^3
g	acceleration due to gravity, m/s^2
H	vessel height, m
k	turbulent kinetic energy, m^2/s^2
K	constant (Eq. 13.13)
N	impeller rotational speed, rps
p	pressure, N/m^2
Q_g	volumetric gas flow rate, m^3/s
r	radial coordinate, m
T	vessel diameter, m
t	time, s
T_{fl}	turbulent dispersion force, N/m^3
T_L	integral time scale of turbulence, s
t_{mix}	mixing time, s
U	velocity, m/s
U_{slip}	slip velocity, m/s
V	volume of vessel, m^3
V_{dr}	drift velocity, m/s

Greek Symbols

α	secondary phase volume fraction
ε	turbulent kinetic energy dissipation rate, m^2/s^3
τ	shear stress, N/m^2
τ_p	particle relaxation time, s
λ	Kolmogorov length scale, m
ρ	density, kg/m^3
$\sigma_{\phi,1}$	model parameter (Eq. 13.4)
σ	standard deviation
σ_{pq}	dispersion Prandtl number
θ	tangential coordinate
μ	viscosity, kg/ms
ϕ	variable

Dimensionless Numbers

Eo	Eötvös number
Fl	gas flow number
Fr	Froude number
N_P	power number
N_Q	pumping number
Re	impeller Reynolds number
Re$_b$	bubble Reynolds number
Re$_p$	particle Reynolds number

Subscripts

1	liquid
2	secondary phase
g	gas
i	direction
l	liquid
s	solid particle
q	phase number
t	turbulent

Superscript

‾	time-averaged value
`	rms value

REFERENCES

1. Oldshue, J.Y. (1983). *Fluid Mixing Technology*. New York: McGraw Hill.
2. Smith, J.M. (1985). *Dispersion of gases in liquids*. In: *Mixing of Liquids by Mechanical Agitation* (ed. J.J. Ulbrecht and G.K. Patterson). London: Gordon and Breach.
3. Tatterson, G.B. (1991). *Fluid Mixing and Gas Dispersion in Agitated Tanks*. London: McGraw Hill.
4. Joshi, J.B. and Ranade, V.V. (2003). Computational fluid dynamics for designing process equipment: expectations, current status and path forward. *Ind. Eng. Chem. Res.* 42: 1115–1128.
5. Kuipers, J.A.M. and van Swaij, W.P.M. (1997). Application of computational fluid dynamics to chemical reaction engineering. *Rev. Chem. Eng.* 13: 1.
6. Ranade, V.V. (1995). Computational fluid dynamics for reactor engineering. *Rev. Chem. Eng.* 11: 225.
7. Khopkar, A.R. and Ranade, V.V. (2010). Stirred vessels: computational modeling of multiphase flows and mixing. In: *Chemical Engineering in Pharmaceutical Industry: R&D to Manufacturing* (ed. D.J. am Ende), 269–297. Hoboken: Wiley.
8. Joshi, J.B., Pandit, A.B., and Sharma, M.M. (1982). Mechanically agitated gas-liquid reactors. *Chem. Eng. Sci.* 37: 813–844.
9. Ranade, V.V. (2002). *Computational Flow Modelling for Chemical Reactor Engineering*. New York: Academic Press.
10. Simonin, C. and Viollet, P.L. (1998). Prediction of an oxygen droplet pulverization in a compressible subsonic conflowing hydrogen flow. *Numer. Methods Multi. Flows* 1: 65–82.
11. Kataoka, I., Besnard, D.C., and Serizawa, A. (1992). Basic equation of turbulence and modelling of interfacial terms in gas-liquid two phase flows. *Chem. Eng. Commun.* 118: 221.
12. Ranade, V.V. (1997). Modeling of turbulent flow in a bubble column reactor. *Chem. Eng. Res. Des.* 75: 14.
13. Mudde, R. and Simonin, O. (1999). Two- and three-dimensional simulations of a bubble plume using a two-fluid model. *Chem. Eng. Sci.* 54: 5061–5069.
14. Oye, R.S., Mudde, R., and van den Akker, H.E.A. (2003). Sensitivity study on interfacial closure laws in two fluid bubbly flow simulations. *AICHE J.* 49: 1621–1636.
15. Ranade, V.V. (1992). Numerical simulation of dispersed gas-liquid flows. *Sadhana* 17: 237–273.
16. Lane, G.L., Schwarz, M.P., and Evans, G.M. (2000). Modelling of the interaction between gas and liquid in stirred vessels. *Paper presented at Proceedings of 10th European Conference on Mixing*, Delft, The Netherlands (2–5 July 2000, pp. 197–204).
17. Khopkar, A.R. and Ranade, V.V. (2006). CFD simulation of gas-liquid flow in stirred vessels: VC, S33 and L33 flow regimes. *AICHE J.* 52: 1654–1671.
18. Lane, G.L., Schwarz, M.P., and Evans, G.M. (2005). Computational modelling of gas-liquid flow in mechanically stirred tanks. *Chem. Eng. Sci.* 60: 2203–2214.
19. Bombac, A., Zun, I., Filipic, B., and Zumer, M. (1997). Gas-filled cavity structure and local void fraction distribution in aerated stirred vessel. *AICHE J.* 43 (11): 2921–2931.
20. Bakker, A. and van den Akker, H.E.A. (1994). A computational model for the gas-liquid flow in stirred reactors. *Trans. Inst. Chem. Eng.* 72: 594–606.
21. Brucato, A., Grisafi, F., and Montante, G. (1998). Particle drag coefficient in turbulent fluids. *Chem. Eng. Sci.* 45: 3295–3314.
22. Clift, R. and Gauvin, W.H. (1971). Motion of entrained particles in gas streams. *Can. J. Chem. Eng.* 49: 439–448.
23. Pinelli, D., Montante, G., and Magelli, F. (2004). Dispersion coefficients and settling velocities of solids in slurry vessels stirred with different types of multiple impellers. *Chem. Eng. Sci.* 59: 3081–3089.
24. Tomiyama, A. (2004). Drag lift and virtual mass forces acting on a single bubble. *Proceedings of the 3rd International Symposium on Two-Phase Flow Modeling and Experimentation, Pisa, Italy* (22–24 September).
25. Barigou, M. and Greaves, A. (1992). Bubble size distribution in a mechanically agitated gas-liquid contactor. *Chem. Eng. Sci.* 47 (8): 2009–2025.
26. Laakkonen, M., Moilanen, P., Alopeaus, V., and Aittamaa, J. (2007). Modeling local bubble size distributions in agitated vessel. *Chem. Eng. Sci.* 62: 721–740.
27. Calderbank, P.H. (1958). Physical rate processes in industrial fermentation: part I, the interfacial area in gas-liquid contacting with mechanical agitation. *Trans. Inst. Chem. Eng.* 36: 443.
28. Hughmark, G. (1980). Power requirements and interfacial area in gas-liquid turbine agitated systems. *Ind. Eng. Chem. Proc. Des. Dev.* 19: 641–646.
29. Cui, Y.Q., van der Lans, R.G.J.M., and Luben, K.C.A.M. (1996). Local power uptake in gas-liquid systems with single and multiple Rushton turbines. *Chem. Eng. Sci.* 51: 2631–2636.
30. Nienow, A.W. (1975). Agitated vessel particle–liquid mass transfer: a comparison between theories and data. *Chem. Eng. J.* 9: 153.
31. Nienow, A.W. and Miles, D. (1978). The effect of impeller/tank configurations on fluid-particle mass transfer. *Chem. Eng. J.* 15: 13–24.
32. Chaudhari, R.V. (1980). Three phase slurry reactors. *AICHE J.* 26: 179.
33. Conti, R. and Sicardi, S. (1982). Mass transfer from freely-suspended particles in stirred tanks. *Chem. Eng. Commun.* 14: 91.
34. Atiemo-Obeng, V.A., Penney, W.R., and Armenante, P. (2004). Solid-liquid mixing. In: *Handbook of Industrial Mixing: Science and Practice* (ed. E.L. Paul, V.A. Atiemo-Obeng and S.M. Kresta), Chapter 10, 543–584. Hoboken: Wiley Interscience.
35. Angst, R., Harnack, E., Singh, M., and Kraume, M. (2003). Grid and model dependency of the solid/liquid two-phase flow CFD simulation of stirred reactors. *Proceedings of 11th European Conference of Mixing*, Bamberg, Germany (14–17 October 2003).
36. Montante, G. and Magelli, F. (2005). Modeling of solids distribution in stirred tanks: analysis of simulation strategies and comparison with experimental data. *Int. J. Comp. Fluid Dyn.* 19: 253–262.
37. Khopkar, A.R., Kasat, G.R., Pandit, A.B., and Ranade, V.V. (2006b). CFD simulation of solid suspension in stirred slurry reactor. *Ind. Eng. Chem. Res.* 45: 4416–4428.

38. Yamazaki, H., Tojo, K., and Miyanami, K. (1986). Concentration profiles of solids suspended in a stirred tank. *Powder Technol.* 48: 205–216.

39. Pinelli, D., Nocentini, M., and Magelli, F. (2001). Solids distribution in stirred slurry reactors: influence of some mixer configurations and limits to the applicability of a simple model for predictions. *Chem. Eng. Commun.* 188: 91–107.

40. Dyalag, M. and Talaga, J. (1994). Hydrodynamics of mechanical mixing in a three-phase liquid-gas-solid system. *Int. Chem. Eng.* 34 (4): 539–551.

41. Chapman, C.M., Nienow, A.W., Cook, M., and Middleton, J.C. (1983). Particle-gas-liquid mixing in stirred vessel. Part III: Three phase mixing. *Chem. Eng. Res. Des.* 61: 167–181.

42. Frijlink, J.J., Bakker, A., and Smith, J.M. (1990). Suspension of solid particles with gassed impellers. *Chem. Eng. Sci.* 45 (7): 1703–1718.

43. Rewatkar, V.B., Raghava Rao, K.S.M.S., and Joshi, J.B. (1991). Critical impeller speed for solid suspension in mechanically agitated three-phase reactors: experimental part. *Ind. Eng. Chem. Res.* 30: 1770–1784.

44. Murthy, B.N., Ghadge, R.S., and Joshi, J.B. (2007). CFD Simulations of gas-liquid-solid stirred reactor: prediction of critical impeller speed for solid suspension. *Chem. Eng. Sci.* 62: 7184–7195.

45. Pantula, P. R. K. and Ahmed, N. (1998). Solid suspension and gas holdup in three phase mechanically agitated reactors. *Presented at Chemeca 98, Paper No. 132*, Port Douglas, Queensland, Australia (28–30 September 1998).

46. Warmoeskerken, M.M.C.G. and Smith, J.M. (1985). Flooding of disk turbines in gas-liquid dispersions: a new description of the phenomenon. *Chem. Eng. Sci.* 40: 2063.

47. Zhu, Y. and Wu, J. (2002). Critical impeller speed for suspending solids in aerated agitation tanks. *Can. J. Chem. Eng.* 80: 1–6.

48. Armenante, P.M. and Huang, Y.T. (1992). Experimental determination of the minimum agitation speeds for complete liquid–liquid dispersion in mechanically agitated vessels. *Ind. Eng. Chem. Res.* 31: 1398–1406.

49. Calabrese, R.V., Chang, T.P.K., and Dang, P.T. (1986). Drop breakup in turbulent stirred-tank contactors, Part I: effect of dispersed-phase viscosity. *AICHE J.* 32: 657–666.

50. Zhou, G. and Kresta, S.M. (1998). Correlation of mean drop size and minimum drop size with the turbulence energy dissipation and the flow in an agitated tank. *Chem. Eng. Sci.* 53: 2063–2079.

51. Giapos, A., Pachatouridis, C., and Stamatoudis, M. (2005). Effect of the number of impeller blades on the drop sizes in agitated dispersions. *Chem. Eng. Res. Des.* 83 (A12): 1425–1430.

52. Pacek, A.W., Man, C.C., and Nienow, A.W. (1998). On the Sauter mean diameter and size distributions in turbulent liquid/liquid dispersions in a stirred vessel. *Chem. Eng. Sci.* 53: 2005–2011.

53. Bałdyga, J. and Bourne, J.R. (1995). Interpretation of turbulent mixing using fractals and multi-fractals. *Chem. Eng. Sci.* 50: 381–400.

54. Bałdyga, J. and Bourne, J.R. (1999). *Turbulent Mixing and Chemical Reactions*. Chichester: Wiley.

55. Bałdyga, J. and Podgorska, W. (1998). Drop break-up in intermittent turbulence: maximum stable and transient sizes of drops. *Can. J. Chem. Eng.* 76: 456–470.

56. Bałdyga, J., Bourne, J.R., Pacek, A.W. et al. (2001). Effects of agitation and scale-up on drop size in turbulent dispersions: allowance for intermittency. *Chem. Eng. Sci.* 56: 3377–3385.

57. Alopaeus, V., Koskinen, J., Keskinen, K., and Majander, J. (2002). Simulation of the population balances for liquid-liquid systems in a non-ideal stirred tank. Part-2: Parameter fitting and the use of the multiblock model for dense dispersions. *Chem. Eng. Sci.* 57: 1815–1825.

58. Derksen, J. and van den Akker, H.E.A. (2007). Multi-scale simulations of stirred liquid-liquid dispersions. *Chem. Eng. Res. Des.* 85 (A2): 169–179.

59. Laurenzi, F., Coroneo, M., Montante, G. et al. (2009). Experimental and computational analysis of immiscible liquid-liquid dispersions in stirred vessels. *Chem. Eng. Res. Des.* 87: 507–514.

60. Wang, F., Mao, Z., Wang, Y., and Yang, C. (2006). Measurement of phase holdups in liquid–liquid–solid three-phase stirred tanks and CFD simulations. *Chem. Eng. Sci.* 61: 7535–7550.

61. Zaccone, A., Gabler, A., Maaβ, S. et al. (2007). Drop breakup in liquid-liquid dispersions: modeling of single drop breakage. *Chem. Eng. Sci.* 62: 6297–6307.

62. Narsimhan, G., Gupta, J.P., and Ramkrishna, D. (1979). A model for transitional breakage probability of droplets in agitated lean liquid-liquid dispersions. *Chem. Eng. Sci.* 34: 257–265.

63. Levenspiel, O. (1998). *Chemical Reaction Engineering*. Hoboken, NJ: Wiley.

64. Dakshinamoorthy, D., Khopkar, A.R., Louvar, J.F., and Ranade, V.V. (2004). CFD simulations to study shortstopping of runaway reactions in a stirred vessel. *J. Loss Prev. Process Ind.* 17 (5): 355–364.

65. Danckwerts, P.V. (1953). Continuous flow systems: distributions of residence times. *Chem. Eng. Sci.* 2: 1–13.

66. Khopkar, A.R., Mavros, P., Ranade, V.V., and Bertrand, J. (2004). Simulation of flow generated by an axial flow impeller: batch and continuous operation. *Chem. Eng. Res. Des.* 82 (A6): 737–751.

67. Vyakaranam, K. and Kokini, J.L. (2011). Advances in 3D numerical simulation of viscous and viscoelastic mixing flows. In: *Food Engineering Interfaces* (ed. J.M. Aguilera, R. Simpson, J. Welti-Chanes, et al.). New York: Springer.

68. Aref, H. (1984). Stirring by chaotic advection. *J. Fluid Mech.* 143: 1–21.

69. Meleshko, V.V. and Peters, G.W.M. (1996). Periodic points for two dimensional Stokes flow in a rectangular cavity. *Phys. Lett. A* 216: 87–96.

70. Muzzio, F.J., Swanson, P.D., and Ottino, J.M. (1991). The statistics of stretching and stirring in chaotic flows. *Phys. Fluids A* 3: 822–834.

71. Tucker, C.L. III and Peters, G.W.M. (2003). Global measures of distributive mixing and their behavior in chaotic flows. *Korea-Australia Rheo. J.* 15 (4): 197–208.

72. Grace, H.P. (1982). Dispersion phenomena in high viscosity immiscible fluid systems and application of static mixers as dispersion devices in such systems. *Chem. Eng. Commun.* 14: 225–277.
73. Larsen, R.G. (1985). Constitutive relationships for polymeric materials with power law distributions of relaxation times. *Rheol. Acta* 24 (4): 327–334.
74. Tanner, R.I. and Huigol, R.R. (1975). On a classification scheme for flow fields. *Rheol. Acta* 14 (11): 959–962.
75. Yang, H.H. and Manas-Zloczower, I. (1992). Flow field analysis of the kneading disc region in a co-rotating twin screw extruder. *Polym. Eng. Sci.* 32 (19): 1411–1417.
76. Li, T. and Manas-Zloczower, I. (1995). A study of distributive mixing in counter-rotating twin screw extruders. *Int. Polym. Process.* 10: 314–320.
77. Alvarez, M.M., Zalc, J.M., Shinbrot, T. et al. (2002). Mechanisms of mixing and creation of structure in laminar stirred tanks. *AIChE J.* 48: 2135–2148.
78. Arratia, P.E., Kukura, J., Lacombe, J., and Muzzio, F.J. (2006). Mixing of shear thinning fluids with yield stress in stirred tanks. *AIChE J.* 52: 2310–2322.
79. Zalc, J.M., Alvarez, M.M., Muzzio, F.J., and Arik, B.E. (2001). Extensive validation of computed laminar flow in a stirred tank with three Rushton turbines. *AIChE J.* 47: 2144–2154.
80. Zalc, J.M., Szalai, E.S., Alvarez, M.M., and Muzzio, F.J. (2002). Using CFD to understand chaotic mixing in laminar stirred tanks. *AIChE J.* 48: 2124–2134.
81. Szalai, E.S., Arratia, P., Johnson, K., and Muzzio, F.J. (2004). Mixing analysis in a tank stirred with Ekato Intermig impellers. *Chem. Eng. Sci.* 59: 3793–3805.
82. Luan, D., Zhang, S., Wei, X., and Duan, Z. (2017). Effect of the 6 PBT stirrer eccentricity and off-bottom clearance on mixing of pseudoplastic fluid in a stirred tank. *Results Phys.* 7: 1079–1085.
83. Cabaret, F., Rivera, C., Fradette, L. et al. (2007). Hydrodynamics performance of a dual shaft mixer with viscous Newtonian liquids. *Chem. Eng. Res. Des.* 85: 583–590.
84. Farhat, M., Rivera, C., Fradette, L. et al. (2007). Numerical and experimental study of a dual shaft coaxial mixer with viscous fluids. *Ind. Eng. Chem. Res.* 46: 5021–5031.
85. Khopkar, A.R., Fradette, L., and Tanguy, P.A. (2007). Hydrodynamic of a dual shaft mixer in the laminar regime. *Chem. Eng. Res. Des.* 85 (6): 863–871.
86. Rivera, C., Foucault, S., Heniche, M. et al. (2006). Mixing analysis in a coaxial mixer. *Chem. Eng. Sci.* 61: 2895–2907.
87. Rudolph, L., Schafer, M., Atiemo-Obeng, V., and Kraume, M. (2007). Experimental and numerical analysis of power consumption for mixing of high viscosity fluids with coaxial mixer. *Chem. Eng. Res. Des.* 85 (5): 568–575.
88. Tung, H.-H. (2013). Industrial perspectives of pharmaceutical crystallization. *Org. Process Res. Dev.* 17: 445–454.
89. Rielly, C.D. and Marquis, A.J. (2001). A particle's eye view of crystallizer fluid mechanics. *Chem. Eng. Sci.* 56: 2475–2493.
90. Sardeshpande, M.V., Juvekar, V.A., and Ranade, V.V. (2010). Hysteresis in a cloud heights during solid suspension in stirred tank reactor: experiments and CFD simulations. *AIChE J.* 56: 2795–2804.
91. Michelletti, M., Nikiforaki, L., Lee, K.C., and Yianneskis, M. (2003). Particle concentration and mixing characteristics of moderate-to-dense solid-liquid suspensions. *Ind. Eng. Chem. Res.* 42: 6236–6249.
92. Kasat, G.R., Khopkar, A.R., Ranade, V.V., and Pandit, A.B. (2008). CFD simulation of liquid-phase mixing in solid–liquid stirred reactor. *Chem. Eng. Sci.* 63: 3877–3885.
93. Bohnet, M. and Niesmak, G. (1980). Distribution of solids in stirred suspensions. *Ger. Chem. Eng.* 3: 57–65.
94. Zwietering, T.N. (1958). Suspending of solid particles in liquid by agitation. *Chem. Eng. Sci.* 8: 244–253.
95. Woo, X.Y., Tan, R.B.H., Chow, P.S., and Braatz, R.D. (2006). Simulation of mixing effects in antisolvent crystallization using a coupled CFD-PDF-PBE approach. *Cryst. Growth Des.* 6: 1291–1303.
96. Torbacke, M. and Rasmuson, A.C. (2004). Mesomixing in a semi-batch reaction crystallization and influence of reactor size. *AIChE J.* 50: 3107–3119.
97. Loh, Z.H., Samanta, A.K., and Heng, P.W.S. (2015). Overview of milling techniques for improving the solubility of poorly water soluble drugs. *Asian J. Pharm. Sci.* 10: 255–274.
98. Genck, W. (2010). Make the most of antisolvent crystallization. *Chem. Process.* 73 (12): 21–25.
99. Gavi, E., Rivautella, L., Marchisio, D.L. et al. (2007). CFD modelling of nanoparticle precipitation in confined impinging jet reactors. *Chem. Eng. Res. Des.* 85 (5): 735–744.
100. Ibrahim, S. and Nienow, A.W. (1994). The effect of viscosity on mixing pattern and solids. *Proceedings of 8th European Conference on Mixing, IChemE Symposium Series*, 136 (21–23 September 1994, pp. 25–32).
101. Phillips, R.J., Armstrong, R.C., Brown, R.A. et al. (1992). A constitutive equation for concentrated suspensions that accounts for shear-induced particle migration. *Phys. Fluids A* 4: 30.
102. Wu, J., Zhu, Y., and Pullum, L. (2001). The effect of impeller pumping and fluid rheology on solid suspension in a stirred vessel. *Can. J. Chem. Eng.* 79: 177–186.
103. Barar Pour, S., Fradette, L., and Tanguy, P.A. (2007). Laminar and slurry blending characteristics of a dual shaft impeller system. *Chem. Eng. Res. Des.* 85 (9): 1305–1313.
104. Manas-Zloczower, I. and Feke, D.L. (2009), Dispersive mixing of solid additives, *Mixing and Compounding of Polymers: Theory and Practice*, Manas-Zloczower I., Hanser Publication, Munich.
105. Scurati, A., Feke, D.L., and Manas-Zloczower, I. (2005). Analysis of the kinetics of agglomerate erosion in a simple shear. *Chem. Eng. Sci.* 60: 6564–6573.
106. Wang, W. and Manas-Zloczower, I. (2001). Dispersive and distributive mixing characterization in extrusion equipment. *Technical Papers of the Annual Technical Conference- Society of Plastics Engineers Incorporated (ANTEC 2001)*, Dallas (6–10 May 2001).
107. Hansen, S., Khakhar, D.V., and Ottino, J.M. (1998). Dispersion of solids in nonhomogeneous viscous flows. *Chem. Eng. Sci.* 53: 1803–1817.

108. Zeidan, M., Xu, B.H., Jia, X., and Williams, R.A. (2007). Simulation of aggregate deformation and breakup in simple shear flows using a combined continuum and discrete model. *Chem. Eng. Res. Des.* 85 (12): 1645–1654.
109. Cong, L. and Gupta, M. (2008). Simulation of distributive and dispersive mixing in a co-rotating twin-screw extruder. *ANTEC 2008*, Philadelphia (4–8 May 2008).
110. Alemaskin, K., Manas-Zloczower, I., and Kaufman, M. (2003). Simultaneous characterization of dispersive and distributive mixing in a single screw extruder. *ANTEC Proceedings*, Nashville (4–8 May 2003).
111. Schubert, H. (1990). Instantisieren pulverförmiger lebensmittel. *Chem. Ing. Tech.* 62 (11): 892–906.
112. Khazam, O. and Kresta, S.M. (2009). A novel geometry for solids drawdown in stirred tanks. *Chem. Eng. Res. Des.* 87: 280–290.
113. Joosten, G.E.H., Schilder, J.G.M., and Broere, A.M. (1977). The suspension of floating solids in stirred vessels. *Trans. IChemE* 55: 220–222.
114. Hemrajani, R. R., Smith, D. L., Koros, R. M., and Tarmy, B. L. (1988). Suspending floating solids in stirred tanks: mixer design, scale-up and optimization. *Presented at 6th European Conference on Mixing*, Pavia, Italy (24–26 May 1988).
115. Siddiqui, H. (1993). Mixing technology for buoyant solids in a nonstandard vessel. *AICHE J.* 39: 505–509.
116. Edwards, M.F. and Ellis, D.I. (1984), The drawdown of floating solids into mechanically agitated vessels. In M F Edwards; N Harnby; J C Middleton; *Fluid Mixing II: A Symposium Organised by the Yorkshire Branch and the Fluid Mixing Processes Subject Group of the Institution of Chemical Engineers and Held at Bradford University, 3–5 April 1984* Symposium Series (Institution of Chemical Engineers (Great Britain)) 89, Burlington: The Institution/Elsevier Science. 1–13, Rugby.
117. Atibeni, R., Gao, Z., and Bao, Y. (2013). Effect of baffles on fluid flow field in stirred tank with floating particles by using PIV. *Can. J. Chem. Eng.* 91: 570–578.
118. Hsu, R. C., Chiu, C. K., and Lin, S. C. (2016). The study of suspending floating solids in stirred vessel by using CFD. *Proceedings of the 5th Asian Conference of Mixing*, Tendo, Japan (September 2016).
119. Adam, S., Suzzi, D., Radeke, C., and Khinast, J.G. (2011). An integrated quality by design (QbD) approach towards design space definition of a blending unit operation by discrete element method (DEM) simulation. *Eur. J. Pharm. Sci.* 42: 106–115.
120. Hormann, T., Suzzi, D., and Khinast, J.G. (2011). Mixing and dissolution processes of pharmaceutical bulk materials in stirred tanks: experimental and numerical investigation. *Ind. Eng. Chem. Res.* 50: 2011–2025.
121. Koganti, V., Carrol, F., Ferraina, R. et al. (2010). Application of modeling to scale-up dissolution in pharmaceutical manufacturing. *AAPSPharmSciTech* 4: 1541–1548.
122. Hudcova, V., Machon, V., and Nienow, A.W. (1989). Gas-liquid dispersion with dual Rushton turbine impellers. *Biotechnol. Bioeng.* 34: 617–628.
123. Shewale, S.D. and Pandit, A.B. (2006). Studies in multiple impeller agitated gas-liquid contactors. *Chem. Eng. Sci.* 61: 489–504.
124. Kerdouss, F., Bannari, A., and Proulx, P. (2006). CFD modeling of gas dispersion and bubble size in double stirred tank. *Chem. Eng. Sci.* 61: 3313–3322.
125. Khopkar, A.R. and Tanguy, P.A. (2008). CFD simulation of gas-liquid flows in a stirred vessel equipped with dual Rushton turbines: influence of parallel, merging and diverging flow configurations. *Chem. Eng. Sci.* 63: 3810–3820.
126. Khopkar, A.R., Kasat, G.R., Pandit, A.B., and Ranade, V.V. (2006a). CFD simulation of mixing in tall gas-liquid stirred vessel: role of local flow pattern. *Chem. Eng. Sci.* 61: 2921–2929.
127. Ranade, V.V., Bourne, J.R., and Joshi, J.B. (1991). Fluid mechanics and mixing in agitated tanks. *Chem. Eng. Sci.* 46: 1883–1893.
128. Rammohan, A.R., Dudukovic, M.P., and Ranade, V.V. (2003). Eulerian flow field estimation from particle trajectories: numerical experiments for stirred tank type flows. *Ind. Eng. Chem. Res.* 42: 2589–2601.

PART IV

CONTINUOUS PROCESSING

14

PROCESS DEVELOPMENT AND CASE STUDIES OF CONTINUOUS REACTOR SYSTEMS FOR PRODUCTION OF API AND PHARMACEUTICAL INTERMEDIATES

Thomas L. LaPorte, Chenchi Wang, and G. Scott Jones
Bristol-Myers Squibb, New Brunswick, NJ, USA

14.1 INTRODUCTION

Batch processing in stirred tank reactors is the default mode of operation for production of process intermediates and active pharmaceutical ingredients in the pharmaceutical industry. This is true of both homogeneous and heterogeneous reactions, as well as the subsequent workup unit operations and final crystallization. While commonplace in the commodity chemical industry, continuous processes are somewhat rare in the pharmaceutical industry. However, the potential advantages of organic synthesis reactions operated via a continuous mode include enhanced safety, improved quality, reduced energy costs, and greater cycle efficiencies [1]. These benefits are largely the result of smaller active reaction volumes and superior mass and heat transfer. Recently in the pharmaceutical industry, there has been renewed emphasis on holistic continuous processing where not only the reaction but also downstream extractions, solvent exchanges, and crystallizations are performed continuously as well [2]. However, most examples of continuous processing in the pharmaceutical industry are reaction only at this point, and this chapter will primarily focus on implementation of continuous reactions.

Part of the appeal of stirred tank batch reactors is their general versatility and the fact that a single piece of capital equipment can serve as a reactor, an extractor, a still, or a crystallizer depending on the needs of the process. This versatility enables a wide array of unit operation combinations, and therefore, a single plant with multiple stirred tank reactors can manufacture a large number of products, with different processes. Additionally, batch processing on scale is similar to how a process chemist typically works in the laboratory. For example, charge ingredients, heat to reaction conditions, react for a specified time, and sample for reaction completion. This systematic approach affords a simple and reproducible methodology for processing. However, laboratory- and manufacturing-scale batch reactors typically have vastly different heat and mass transfer characteristics. For chemistries in which heat or mass transfer controls selectivity, a direct scale-up of a laboratory batch process may be problematic in manufacturing. Similarly, limitations in heat transfer in stirred tank reactors may render some energetic lab processes unsafe at manufacturing scale. Finally, the versatility provided by general purpose stirred tanks comes at an efficiency cost when compared with continuous equipment designed for a specific unit operation. In each of these instances, continuous processing can offer advantages over traditional batch processing.

This chapter will discuss opportunities for continuous processing of pharmaceutical intermediates and API, review some considerations for developing and implementing continuous processes, present two brief case studies from the authors' experience, and consider some of the barriers to widespread use of continuous processes. Since the engineering design equations for continuous reactors are covered extensively in undergraduate chemical engineering curricula, that level of detail is not presented here.

Chemical Engineering in the Pharmaceutical Industry: Active Pharmaceutical Ingredients, Second Edition.
Edited by David J. am Ende and Mary T. am Ende.
© 2019 John Wiley & Sons, Inc. Published 2019 by John Wiley & Sons, Inc.

14.2 BENEFITS OF CONTINUOUS PROCESSING

14.2.1 Safety

Process safety is probably the greatest driver for development of continuous processes within the pharmaceutical industry. The two attributes of continuous processes that facilitate improved safety are a reduced inventory of reactive species and improved heat transfer. For a given throughput, continuous reactors are relatively small when compared with batch reactors. Additionally, continuous reactors are often operated at higher temperatures than batch reactors, resulting in higher rates of conversion. Both of these factors reduce the potential heat release contained within the reactor volume, by reducing the inventory of reactive species. The reduced chemical inventory greatly reduces the severity of failure and also allows for a rapid emergency quench of the entire reactor contents in the case of potential runaway reaction. The improved heat transfer rates of continuous reactors also help to reduce safety concerns when scaling exothermic reactions. This characteristic results in dramatically improved temperature control and enables operation within a safe operating window. In some cases, continuous processing may be the only practical means of scaling a highly exothermic process. Some examples employing continuous processing to mitigate safety concerns are given below.

Many pharmaceutical syntheses involve reactions with short half-lives and high heats of reaction and thereby pose thermal runaway potential. Some examples include nitrations, oxidations, and other reactions involving energetic compounds such as peroxides, azides, and diazo compounds [3, 4]. Nitrations are highly exothermic and involve explosive or hazardous nitrating agents, and continuous processes have been developed to implement this chemistry more safely. In one example, the nitration of a pharmaceutical intermediate utilized a continuous reactor to enable high chemoselectivity while mitigating temperature control and decomposition concerns that existed in the batch process [4]. The continuous process operated at 90 °C with a 35 minute residence time in a microreactor. In contrast, the batch process operated for 8 hours at 50 °C and required very precise addition control for nitrating reagents.

14.2.2 Product Quality

The selectivity of organic reactions is determined by the amount of time molecules are exposed to a given set of conditions, i.e. stoichiometry and temperature. In batch processing, spatial gradients exist for temperature and reactant concentration due to the mixing times achievable with conventional batch reactors. Restated, in batch reactors, the reaction conditions vary with location in the reactor. That nonuniform reaction environment can lead to undesirable side products, and the extent of their formation depends upon the mixing characteristics of the reactor and the rate laws for both desired and undesired reactions. The increased heat and mass transfer capability of continuous reactors can result in improved reaction impurity profiles since conditions can be controlled more uniformly than with batch reactors. Improvements in impurity profiles at the reaction stage lessen the burden of downstream unit operations designed to remove impurities. This can allow for yield improvements due to optimization of downstream workup and crystallization. The improved control of reaction conditions should also help to minimize batch-to-batch variability that sometimes exists with batch processes.

There are additional consequences of the inferior heat and mass transfer properties of conventional batch reactors. Often reagents must be added over extended periods of time, and this means that there is a wide distribution in the amount of time that substrate molecules, starting material or product, are exposed to reaction conditions. While this increases cycle times, it also affects product quality and choice of operating conditions. These temporal gradients necessitate that conditions are defined to accommodate those molecules exposed to process conditions for the longest periods of time. Mean residence times are reduced in continuous reactors, and molecules experience reaction conditions for more uniform periods of time. Additionally, the increased heat and mass transfer rates also mean that reaction conditions can be manipulated more rapidly than with batch reactors. The minimization of temporal gradients, coupled with the ability to rapidly manipulate reaction conditions, allows the process development engineer to consider operating conditions that would lead to unacceptable impurity profiles in batch processes. One example of this benefit is in the case where a relatively unstable intermediate is produced. Consider the time–temperature stability envelope displayed for a hypothetical first-order decomposition in Figure 14.1. The stability of a chemical intermediate increases at lower temperature and decreases with time, and these parameters are coupled. This fact means that batch reactions requiring low temperatures and long reaction times for stability reasons can possibly be converted to high temperature continuous processes when operated for much shorter periods of time. This same concept applies to all reactions, desired and undesired, and by understanding the rate laws governing them, continuous processing conditions that improve reaction selectivity can often be identified.

Many examples exist where continuous processing led to improved product quality [5, 6]. For example, the biphasic BOC protection of an amine was investigated with continuous flow reactors due to its high heat of reaction, −213 kJ/mol, and the propensity to form dimeric impurities [5]. The dimeric impurities were reduced, and the overall selectivity was improved from 97 to 99.9% in the continuous process. The improvements were attributed to the reduction of spatial and temporal gradients in reaction conditions.

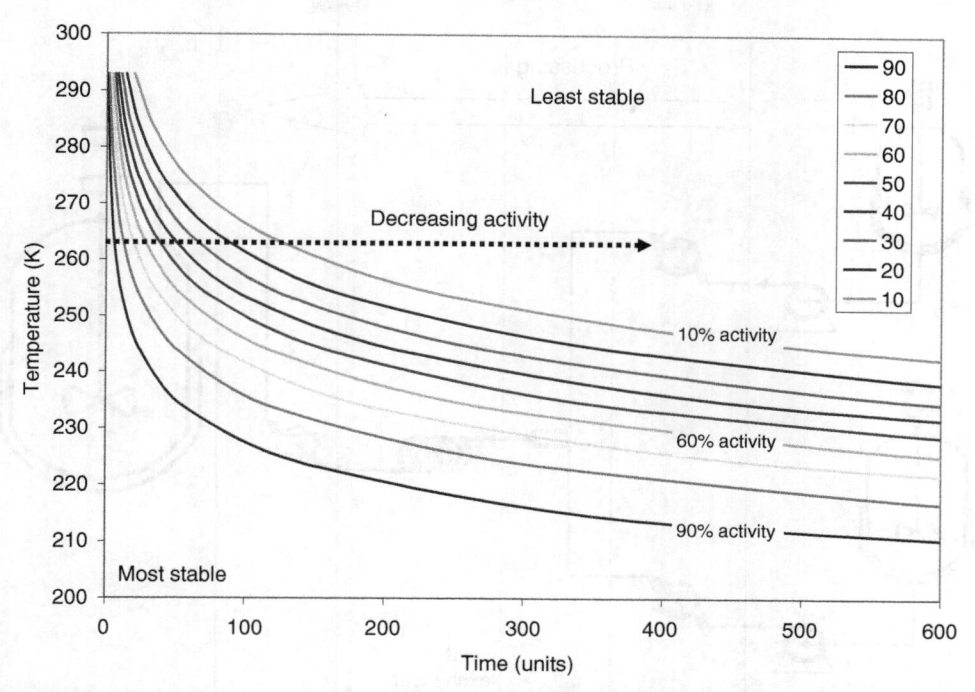

FIGURE 14.1 Operating chemistry envelope for a hypothetical first-order degradation mechanism. The stability (activity) of an intermediate is a function of time and temperature. The rate of degradation increases with higher temperatures.

14.3 CONTINUOUS REACTOR AND ANCILLARY SYSTEMS CONSIDERATIONS

The three main components of a continuous reaction process include the feed solutions, the reactor, and the quench [7] (Figure 14.2). We will first consider the reactor followed by the ancillary systems for the feed solutions and quench.

14.3.1 Continuous Reactors

14.3.1.1 Plug Flow Reactors (PFRs) Ideal plug flow reactors (PFRs) have flow with minimal backmixing along the flow path, no radial concentration or temperature gradients, and a precise residence time for all flowing material. In the case of laminar flow, radial gradients may exist, and there may not be plug flow in the truest sense. However, for the rest of this chapter, the term PFR will refer to all tubular flow reactors, regardless of the degree of turbulence and radial gradients. PFRs are composed of mixing zones for mass transfer and heat exchangers for heat transfer, and often these components are present in a single device. The static mixers and heat exchangers commonly utilized for these purposes in the pharmaceutical industry are described below and can be used in any combination required to meet the demands of the particular process.

In-line static mixers are commonly utilized in PFR systems to efficiently mix multiple feed streams. Tubular static mixers have characteristic mixing times of a few seconds or less depending on the degree of turbulence and provide efficient mixing even in the case of laminar flow. Static mixers consist of sequential static, often helical, mixing elements housed in a tube. The mixing elements typically alternate between left- and right-handed torsion and simultaneously produce flow division and efficient radial mixing, minimizing radial gradients in velocity, temperature, and concentration. A cartoon demonstrating the operating concept of a static mixer is given in Figure 14.3. In the case of the common Kenics® static mixer, each mixing element divides the process fluid in half. Each fluid division is further divided by subsequent mixing elements, and the number of fluid striations is theoretically equal to 2^N, where N is the number of individual elements. In this simple way, miscible fluids can be thoroughly mixed within a very short length of tubing, even under laminar flow conditions. Selection of the mixing inserts and number of elements depends on the fluid properties and the specific processing application [8]. A variety of vendors (Kenics®, Komax, Sulzer, etc.) manufacture static mixers and can aid in the selection of the most appropriate mixing elements. Static mixers are typically jacketed to control temperature when used as a reactor, and in one example, the mixing elements are made of heat transfer tubes for improved temperature control [9]. Figure 14.4 shows an example of a lab static mixer where 27 helical mixing elements are contained within 7 inch of ¼″ tubing, equating to a theoretical 134 million striations. Static mixers offer advantages over mechanical agitators such as more rapid mixing, ease of maintenance, and lower operating costs.

Heat removal and temperature control in PFRs are achieved with heat exchangers. The versatility of heat

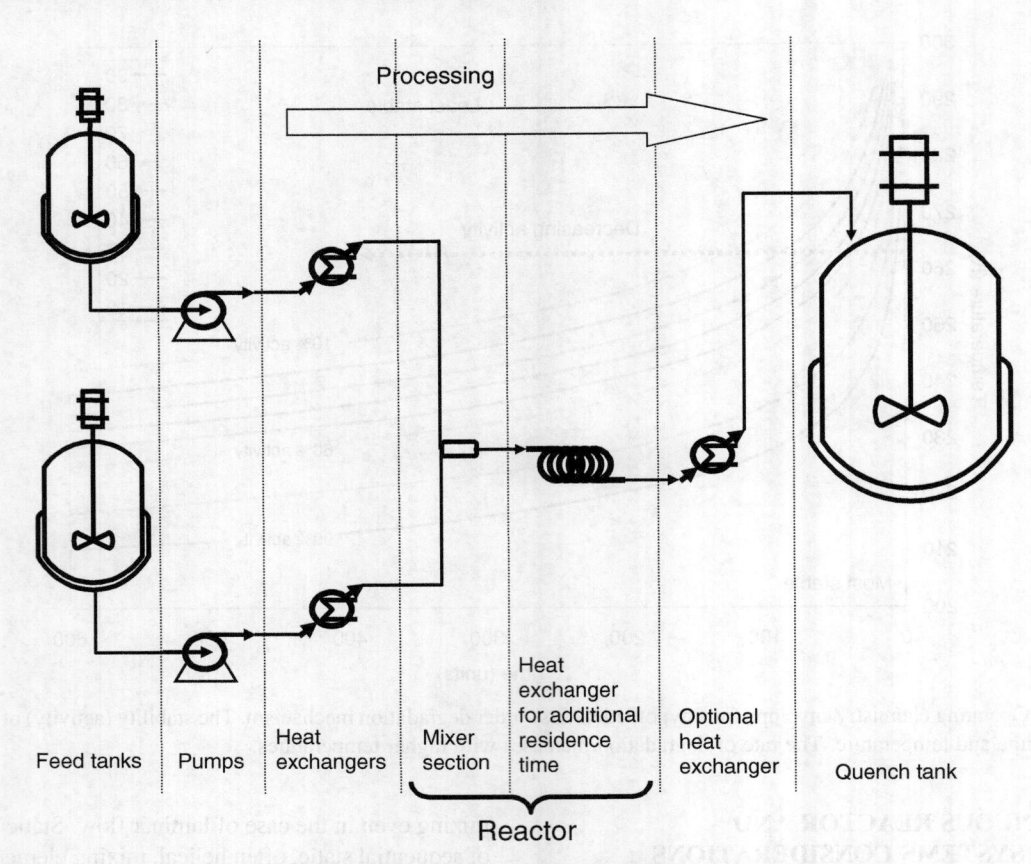

FIGURE 14.2 Typical continuous processing scheme.

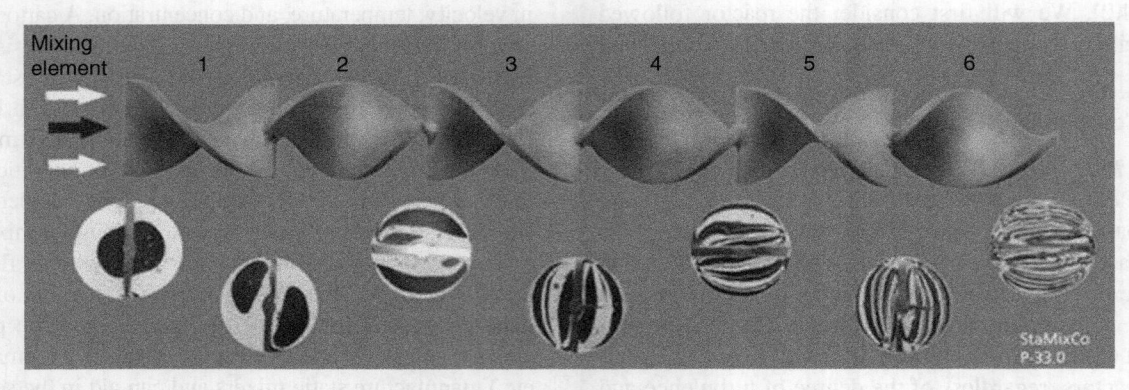

FIGURE 14.3 Example of static mixer elements demonstrating the operating principle (increased mixing with increased distance along the mixer element). Courtesy of StaMixCo LLC, www.stamixco-usa.com.

exchangers is demonstrated in the case studies of the latter sections of this chapter, where heat exchangers are utilized to adjust the temperature of feedstocks prior to reaction, to control the temperature in the mixing zone of the reactor, to provide additional residence time for complete conversion of the reaction, and to thermally quench reactions. They are an essential component of plug flow systems. The key advantage of heat exchangers, versus conventional stirred tanks, is their improved heat transfer rates that result from much higher surface area to volume ratios. Most commonly used by the authors are concentric tube and shell and tube heat exchangers that are readily available from vendors and easily constructed in-house as well. Schematics of several variants of these are shown in Figure 14.5. These heat exchangers are readily available at low prices, have reasonable pressure drops, and meet the heat transfer requirements

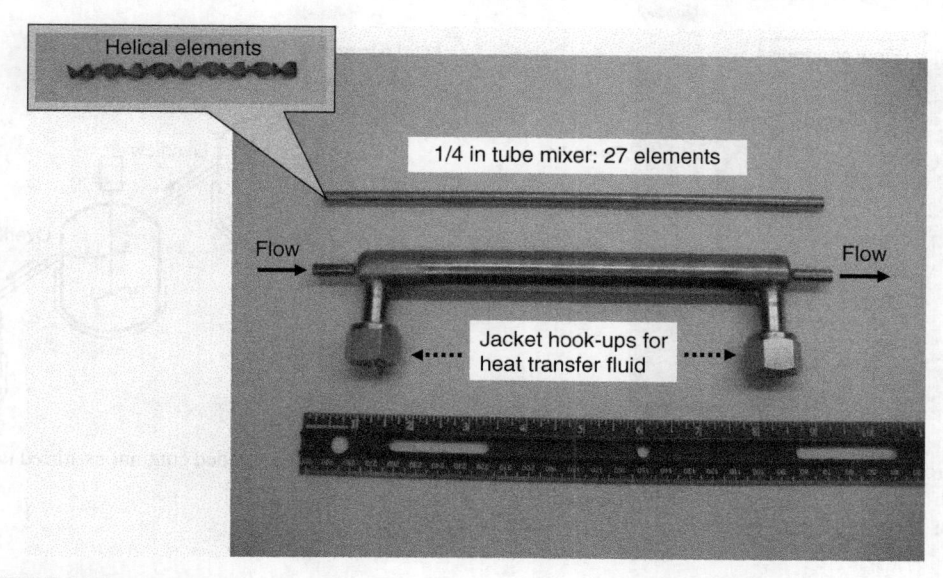

FIGURE 14.4 ¼ inch tube mixer with 27 elements and a ½ inch jacket for added temperature control.

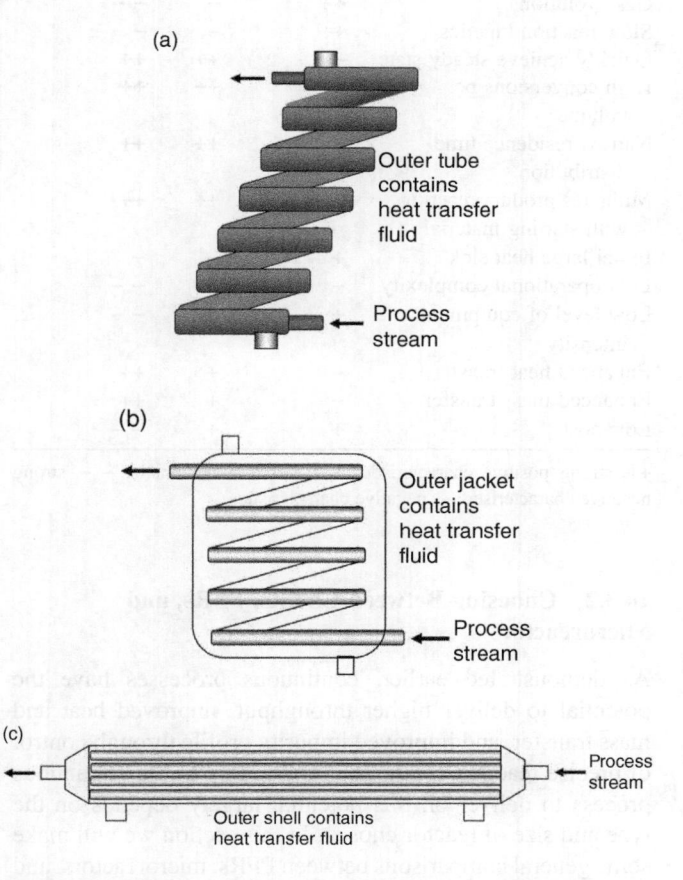

FIGURE 14.5 Heat exchangers commonly employed in continuous processing: (a) concentric tube, (b) jacketed coil, and (c) shell and tube.

for most reactions. By inserting mixing elements into one of these heat exchangers (Figure 14.5), a PFR can be constructed that provides good heat and mass transfer. Due to the simplicity of construction and lack of moving parts and associated seals, a PFR provides a cost-efficient reactor that is easy to construct and operate. Although they have better heat transfer properties than batch tank reactors, PFRs often operate non-isothermally for exothermic reactions.

14.3.1.2 Microreactors Microreactors are another type of flow reactor that have been increasingly studied and applied as laboratory tools for process screening and scale-up studies. The term microreactor typically implies a single unit integrating a static micromixer and heat exchanger combined with an additional heat exchanger that provides time for reaction conversion beyond the mixing zone. Other more specialized reactors such as spinning tube-in-tube [10] and spinning disk reactors [11] are less widespread and will not be discussed here. Laboratory microreactors fabricated by glass or metals are available with an internal volume of less than 1 ml. As an example, a standard microreactor from Micronit Microfluidics [12] includes a preheating section for each input stream, a mixing section, and a quenching section from a third input. The total volume of this borosilicate reactor is 3.4 ml of which the mixing zone is 2.4 ml.

Microreactors are suitable tools to employ with fast reactions that require extremely efficient mixing, and the reaction requires only low flow rates. Micromixers have internal microchannels that typically lie in the range of 50–500 μm. In these microfluidic devices, molecular diffusion is the

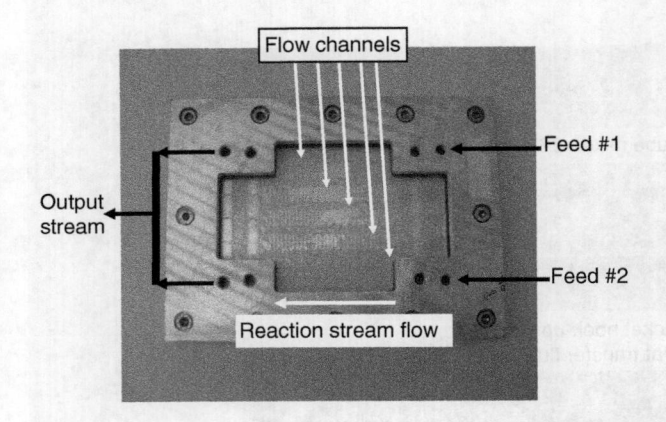

FIGURE 14.6 Microreactor from mikroglas chemtech GmbH: Interdigital mixer with heat exchanger, 5 channels with a width of 500 μm and a depth of 250 μm.

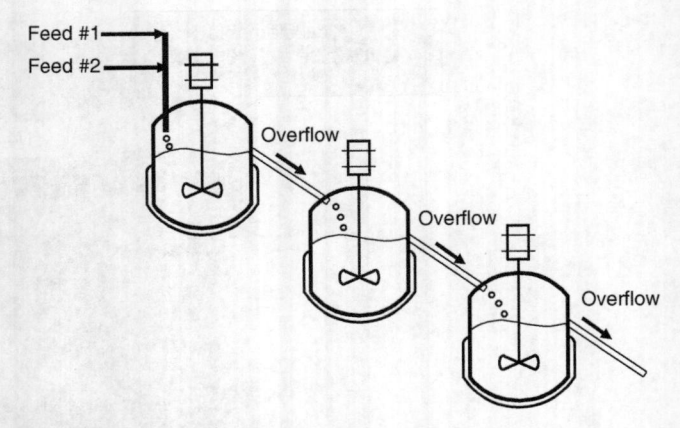

FIGURE 14.7 Cascaded continuous stirred tank reactors in series.

governing mixing mechanism within the laminar flow domain, unlike turbulent mixing created in a pipe or mechanically agitated vessels. Many micromixers can maximize the interfacial surface contact of fluid lamination and efficiently minimize concentration gradients. The internal microstructures promote multiple flow divisions and recombinations and are designed for specific flow arrangements and fluid types [13]. For example, T-mixers and interdigital mixers (Figure 14.6) are routinely used in microreactors for obtaining efficient liquid–liquid mixing. Microreactors also possess extremely high surface to volume ratios for enhanced heat transfer and can therefore operate isothermally even with exothermic reactions. This expands processing opportunities for managing hazardous or highly energetic chemistries with enhanced safety.

14.3.1.3 Continuous Stirred Tank Reactors (CSTRs)

Continuous stirred tank reactors (CSTRs) are presented last because they are less commonly used in the pharmaceutical industry. They are essentially batch tank reactors that are operated continuously by simultaneously flowing reactants in and product out. Since they are tank reactors, their heat and mass transfer characteristics are equivalent to similarly sized batch tank reactors. Additionally, single CSTRs have broad residence time distributions and low conversion rates per unit volume. Some of these characteristics of CSTRs can be improved by using a series of smaller reactors cascaded together as shown in Figure 14.7. CSTRs cascaded in this way have been used for several processes at AMPAC Fine Chemicals LLC for the production of hazardous or energetic chemicals [14]. In one facility, they utilize up to 7 cascaded reactors from 0.25 to 1 L in volume for a continuous process [15].

TABLE 14.1 Attributes of CSTRs, PFR, and Microreactors

Reactor Mode	Multiple CSTRs	PFR	Multiple Microreactors
Handling of solids	++	–	– –
Gas evolution	++	–	– –
Slow reaction kinetics	++	–	– –
Quickly achieve steady state	–	++	++
High conversions per volume	–	++	++
Narrow residence time distribution	–	++	++
Mitigates product reacting with starting material	–	++	++
Initial large heat sink	+	–	–
Low operational complexity	–	++	– –
Low level of equipment intensity	+	++	– –
Enhanced heat transfer	–	+	++
Enhanced mass transfer	–	+	++
Low cost	+	+	– –

++, strong positive characteristic; +, positive characteristic; – –, strong negative characteristic; –, negative characteristic.

14.3.2 Choosing Between CSTRs, PFRs, and Microreactors

As demonstrated earlier, continuous processes have the potential to deliver higher throughput, improved heat and mass transfer, and improved impurity profile through control of precise reaction conditions. The ability of the continuous process to deliver on this potential largely depends on the type and size of reactor chosen. In this section we will make some general comparisons between PFRs, microreactors, and CSTRs. A qualitative comparison of key attributes for these reactors is given in Table 14.1. The case will be made that

TABLE 14.2 Categorization of Reactions for Continuous Process Fit [16]

	Reaction Rate	Characteristics
1	Very fast	Reaction ½ life of <1 s. Reaction is mixing sensitive since rate is faster than mixing. Most of the reaction occurs in the mixing stage for a continuous reaction. Heat management can be an issue
2	Rapid	Reaction ½ life is 1 s to 10 min. Reaction may be mixing sensitive but typically kinetically controlled. Heat management may be an issue
3	Slow	Reaction ½ life >10 min. Implemented in a continuous process mainly for hazardous chemistries

TABLE 14.3 The High Surface Area to Volume Ratios for Continuous Reactors Are Due to the Small Characteristic Reactor Dimension

Reactor	Characteristic Dimension (mm)	Surface Area/ Volume (cm^{-1})
2000 L tank	680	2.9×10^{-2}
50 L CSTR	200	0.10
Tubular PFR (500 ml)	3.2	3.1
Microreactor[a]	0.05	400

[a]50+ units would be needed to meet throughput requirements.

PFRs and microreactors are generally preferred over CSTRs. PFRs are preferred over microreactors when they are capable of meeting the heat and mass transfer demands of the reaction of interest.

As shown in Table 14.2, reactions can be grouped into three general kinetic categories: (i) very fast with a half-life of less than 1 second, (ii) rapid reactions, typically 1 second to 10 minutes, and (iii) slow reactions greater than 10 minutes [16]. The rate of heat and mass transfer required by the process varies between these categories and in large part determines the choice of reactor. Since the rate of reaction depends upon the conditions chosen, it is sometimes possible for categorization of a reaction to change based upon reaction conditions.

The mass transfer requirements for a reaction depends upon the reaction categorization. For reactions with a half-life less than 1 second, microreactors may be the only practical choice due to mass transfer limitations of PFRs and especially CSTRs. Even though static mixers can greatly enhance mixing in PFRs, they pale in comparison with the millisecond mixing times that are characteristics of micromixers. The degree to which PFRs may be acceptable for these reactions depends in part on the extent to which concentration gradients influence reaction selectivity. Reactions in the second category have less stringent mass transfer requirements. With PFRs or microreactors, they are likely kinetically controlled, but concentration gradients may influence selectivity if conducted in CSTRs. Reactions in the third category have even less stringent mass transfer requirements, and either PFRs or CSTRs may be appropriate depending upon specific process needs.

Another major factor in reactor selection is heat transfer requirements of the process. Since the heat generated by a reaction is proportional to reactor volume and the heat removal is proportional to reactor surface area, the ratio of surface area to volume provides an easy means of comparing a reactor's ability to remove heat. Table 14.3 shows this ratio for a 2000 L batch reactor compared with a smaller CSTR, tubular PFR, and a system of microreactors capable of similar throughputs. Obviously conversion of an existing batch reactor to a CSTR does nothing to improve the heat transfer characteristics. Although a smaller CSTR represents a great improvement over batch reactors in terms of surface to volume ratios, it cannot compete with tubular PFRs or integrated microreactors. Reactions in category 1 are often highly energetic, and the rapid generation of heat may require a microreactor if a high degree of temperature control is required. PFRs however meet the heat removal requirements of many common reactions and can be used for more energetic reactions if non-isothermal operation is acceptable. The non-isothermal characterization refers primarily to temperature gradients in the axial rather than radial direction.

In order to realize all of the benefits described earlier, a narrow residence time distribution is often required. For an ideal PFR, all molecules have the same residence time, and the distribution is represented by a Dirac delta function. While real PFRs are less perfect, they have very narrow residence time distributions. Microreactors operate in laminar flow, and axial dispersion models have been used to model the residence time distribution [17]. This deviation from ideal plug flow is less important for microreactors since they are typically operated with shorter mean residence times, and it is the absolute value of the residence time at the upper boundary of the distribution that impacts impurity profiles, not the percent deviation. CSTRs have a broad residence time distribution, and some molecules spend considerably longer in the reactor than others. In fact, the standard deviation of the CSTR residence time distribution is equal to the mean residence time. For processes requiring exposure to reaction conditions for a precise period of time, PFRs are preferred. Furthermore, the residence time distribution in CSTRs results in longer transient periods, typically 3–4 residence times, prior to reaching steady state. The longer transient periods for CSTRs result in larger amounts of wasted product and are an additional drawback of CSTRs.

For processes that are not constrained by heat and mass transfer or residence time distribution considerations, throughput considerations may be important when choosing between CSTRs and PFRs. Continuous reactors will always offer a higher throughput than batch processing. Conceptually this is quite simple since in continuous processing reactants are constantly fed to the reaction and reactors operate at a constant volume, usually full. For positive-order reactions with simple kinetic rate laws, the design equations for CSTRs and PFRs dictate that PFRs deliver a given conversion with smaller reactor volumes than CSTRs. The extent of divergence between CSTR and PFR volumes depends upon the rate law and the conversion required in the reaction, and it is especially pronounced when high conversions are required. Conversions in the pharmaceutical industry are nearly always greater than 95% and quite frequently are greater than 99%. Table 14.4 compares the CSTR reactor volume, relative to a PFR, required for a first- or second-order reaction to achieve 99% conversion. For a single CSTR to reach 99% for a second-order reaction, it would need to be one hundred times larger than its PFR counterpart. The conversion efficiency of CSTRs is improved by cascading several in series, and in the limit of an infinite number of CSTRs in series, performance equals that of a PFR. Based upon conversion, or throughput per unit volume, PFRs are clearly superior to CSTRs.

Considerations for ease of operation and reactor costs may factor into choice of reactor as well. Based upon the authors' experience, PFRs represent a good balance between cost, heat and mass transfer efficiency, and ease of operation, and they are preferred when they meet the demands of the process. To the extent possible, attempts are made to modify reaction conditions to allow the use of PFRs. While microreactors have superior heat and mass transfer rates, they are significantly more expensive than PFRs due to the fine machining required to construct their microchannel flow paths. PFRs on the other hand are simple jacketed tubes with mixing elements, and their cost reflects this simplicity. Another practical drawback of microreactors is their relative inability to handle even small amounts of solids. Individual particles may be sufficient to block flow and interrupt operation. By comparison, the larger diameters of most PFRs allow slurries with low solid density to flow. Slurries with higher solid loading likely require CSTRs for operation or may not lend themselves to continuous processing at all. Additionally, CSTRs can be better suited for handling reactions that involve large amounts of gas evolution. In microreactors, generation of large amounts of gas can serve to reduce the residence time by forcing the process stream through the reactor more rapidly than intended. Non-CSTR reactors can be designed to handle gases, and one of the authors has developed and implemented a continuous trickle bed oxidation column for production of a pharmaceutical intermediate [18]. A final instance where CSTRs might be preferred is the case of slower reactions requiring longer residence times for complete conversions. In such circumstances the length of a PFR required to accommodate the longer residence time may result in impractical pressure drops.

14.3.3 Ancillary Systems

14.3.3.1 Feed Solutions All of the reactants for a continuous process must be in a form that is easily transported by pumps or pressure transfer. Since a higher number of feed solutions require a proportional number of tanks and feed control systems, it is generally desirable to combine several solvents and reactants when possible. Of course, species that react with one another should be prepared in separate feed streams. Typically, one feed solution will contain the starting material and the bulk solvent, while a second feed contains the reagent. In some cases, a third feed may contain a second reagent, a catalyst, or possibly a second compound in the case of a coupling reaction. Ideally the feed solutions should be homogeneous to avoid reactor plugging or fouling, and knowledge of substrate solubility in all process streams is desirable. Additionally, knowledge of the chemical stability of each feedstock is imperative for successful operation.

14.3.3.2 Quench The quench brings the reaction mixture to a nonreactive and stable condition for downstream processing. The quench can be chemical or thermal in nature, and the choice depends on the reactivity of the processing stream and downstream processing needs. Examples of both are given in the sections on two case studies. The three predominant quench modes utilized in continuous processing are demonstrated in Figure 14.8. While these modes are depicted for chemical quenches, slight variants can be envisioned for thermal quenches as well. The first is a reverse batch quench where the reaction stream flows into a reactor containing the quench material. Depending upon processing needs, parallel

TABLE 14.4 Comparison of Reactor Volume for Multiple Stirred Tank Reactors in Series Versus a Plug Flow Reactor Based on 99% Conversion of Starting Material

No. of Stirred Tank Reactors in Series	Volume Relative to a Plug Flow Reactor	
	First-Order Kinetics	Second-Order Kinetics
1	22	100
2	4	8
3	2.4	4
4	1.8	2.6
6	1.5	2
∞	1	1

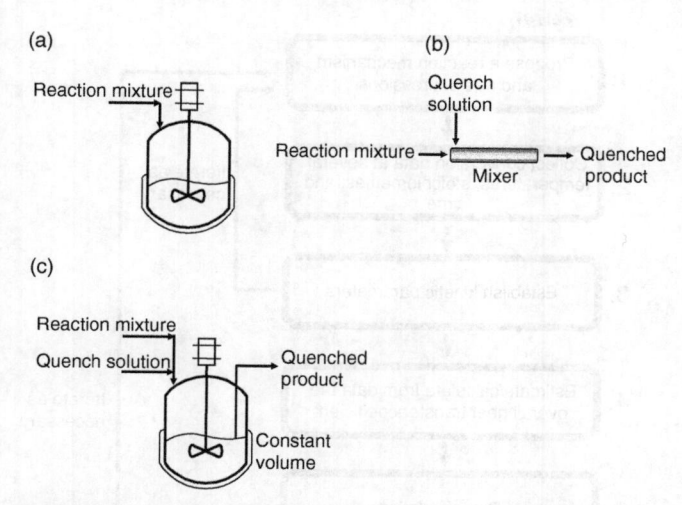

FIGURE 14.8 Continuous reaction stream quench scenarios. (a) Reverse quench, (b) continuous quench with in-line mixer, and (c) continuous quench with CSTR.

quench vessels can be set up to alternately receive the continuous reaction stream and allow uninterrupted operation of the reactor. Under this scenario, the contents of the off-line quench tank are worked-up, while the second quench tank continues to receive the reaction stream. This operating mode is the least complex, ensures an excess of quench solution, may provide a good heat sink for exothermic quenches, and provides a well-defined delineation of batches from a GMP perspective. The second mode of quenching utilizes a static mixer, typically jacketed, to introduce the quench solution. Assuming downstream processing is conducted batchwise, the quenched solution would be collected in stirred tank reactors. The third mode of quenching is similar to the second but uses a CSTR for the continuous quench. Since this quench mode would require an additional reactor as a collection vessel, it is relatively impractical unless subsequent processing is conducted continuously.

14.4 PROCESS DEVELOPMENT OF THE CONTINUOUS REACTION

14.4.1 Reaction Kinetics

A prerequisite to designing a continuous process is to understand the rate laws governing the kinetics of the desired and undesired reactions. The level of knowledge required depends upon the complexity of the process, but where possible a complex reaction should be broken down into its elementary reaction steps and the rate laws for each step established. In some cases, the development of an overall apparent rate law may be sufficient. In either case, the activation energy and effects of reactant concentration should be established for both desired and undesired reactions. It should be noted that the kinetic experiments need not be conducted in a continuous reactor since the reaction kinetics do not depend upon mode of operation. However, in some instances such as fast reactions, flow reactors may offer a practical means of studying reaction rates. With rate laws established, an overall kinetic model can be constructed to help identify operating conditions – temperature, concentration, and time – that promote high rates of conversion and selectivity toward the desired product. In this way, the process development engineer can realize the full potential of the improved heat and mass transfer rates and precise residence times of plug flow and microreactors.

EXAMPLE PROBLEM 14.1

Show how residence time varies with conversion in a PFR for a constant density first-order reaction. Generate a table of conversion versus residence time. Assume the first-order rate constant is $0.01\ s^{-1}$. Compare the residence times required to reach 90% conversion, 99, 99.9, and 99.99%.

Solution
Starting with the design equation for a PFR:

$$\frac{V}{F_{A0}} = \int_0^X \frac{dX}{-r_A} \tag{14.1}$$

Substituting the rate equation:

$$-r_A = kC_A \tag{14.2}$$

$$C_A = C_{A0}(1-X) \tag{14.3}$$

$$\frac{V}{F_{A0}} = \frac{1}{kC_{A0}} \int_0^X \frac{dX}{1-X} \tag{14.4}$$

where

F_{A0} is entering molar flow rate
C_{A0} is initial molar concentration.

Integrate and substitute the residence time, τ, relation to obtain

$$\tau = \frac{VC_{A0}}{F_{A0}} = -\frac{1}{k}\ln(1-X) \tag{14.5}$$

Create a table for X vs. τ using $k = 0.01\ s^{-1}$.

Conversion	Res. Time (s)	Difference
X	τ	Δ
0	0	
0.1	10.5	
0.2	22.3	
0.3	35.7	
0.4	51.1	
0.5	69.3	
0.6	91.6	
0.7	120.4	
0.8	160.9	
0.9	230.3	
0.99	460.5	230.3
0.999	690.8	230.3
0.9999	921.0	230.3

Thus the residence time required to achieve 90% conversion is 230.3 seconds. It takes another 230 seconds to convert from 90 to 99% and another 230 seconds to convert from 99 to 99.9%.

14.4.2 Reaction Engineering

In addition to the kinetics, an understanding of the heat generated by the process needs to be understood in order to design an appropriate reactor for the process. The two main sources of heat generation are the heat of mixing and heat of reaction. The heat of mixing refers to heat generated upon mixing of the feed streams, including heats of dilution. The heat of reaction is proportional to reaction conversion and is distributed across the length of the PFR based upon the extent of conversion. For PFRs and microreactors, the bulk of the heat is generated in the first part of the reactor and may primarily be in the mixing stage. This is in part due to the localization of the heat of mixing, but is primarily due to the distribution of reaction conversion, and thus heat in the axial direction. For a first-order reaction in an isothermal PFR, the length of reactor required to reach 90% conversion is the same length required to go from 90 to 99% conversion. The first half of the reactor would need to dissipate an order of magnitude more heat than the second half. The amount of heat generated in the early part of the reactor is even greater for higher-order reactions and in non-isothermal operation where the heat of reaction increases the temperature of the process stream early in the reactor, thus increasing the reaction rate and heat generated. It is therefore important to understand the intended reactor's overall heat transfer coefficient or to design a reactor that meets the process requirements.

Combining the kinetic rate laws, heats of reaction, and knowledge of the reactor's heat transfer coefficients provides a powerful means to model expected outcomes. A combined experimental and modeling approach is essential for rapid process development since so many parameters depend on one another. Figure 14.9 shows a generic workflow for continuous reaction development from a reaction engineering perspective. An understanding of the factors influencing heat generation is established in steps 1 through 3 by combining the kinetic rate laws and heats of reaction. The reactor's heat transfer properties and ability to remove heat are established in step 4. Simulations can then be conducted to determine reaction conditions at different positions along the reactor and ultimately product quality. Such simulations can be used to evaluate different reactor types and configurations, as well as changes in flow rates, stoichiometry, and temperature. With a good model, much of the process development can be facilitated by virtual experiments. The final step is to experimentally verify the optimized conditions or redesign the reactor.

Examples of this type of methodology exist in the literature [3, 19], and an example is also given in the sections on case studies of this chapter. Bogaert-Alvarez et al. [3] undertook a good example of this approach. They solved the rate laws and non-isothermal heat transfer equations as two ordinary differential equations for a PFR. They assumed a constant temperature for the heat transfer medium although an energy balance of it could also be included. The model enabled them to evaluate the effects of various parameters including jacket temperature, reactor length, flow rate, and heat transfer coefficients on reaction conversion and peak reaction stream temperatures.

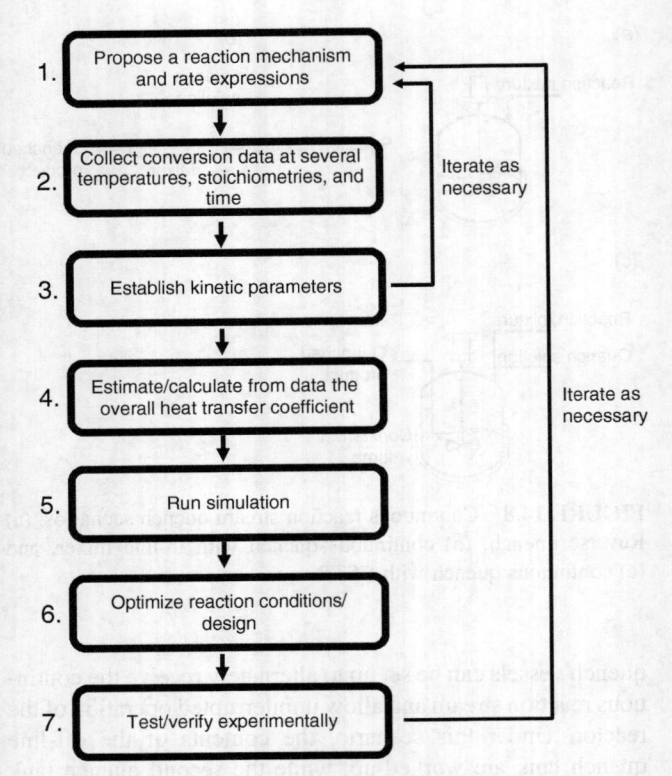

FIGURE 14.9 Workflow for process modeling.

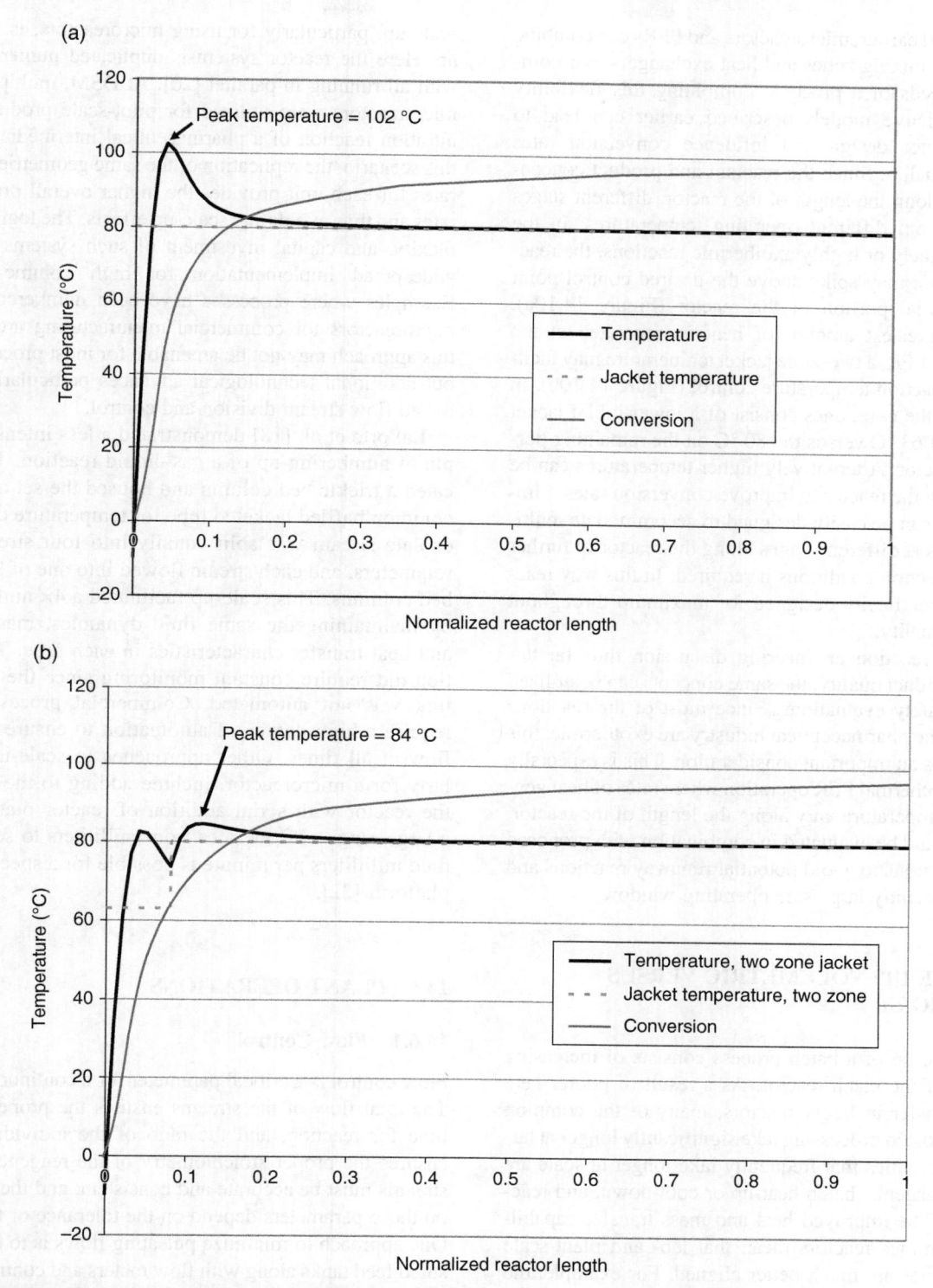

FIGURE 14.10 Simulated reaction temperature profile for a second-order reaction in a plug flow reactor. A constant jacket temperature is assumed for a single-zone jacket control (a) and a two-zone jacket control (b).

332 CHEMICAL ENGINEERING IN THE PHARMACEUTICAL INDUSTRY

As discussed earlier, microreactors and PFRs can combine any number of mixing zones and heat exchangers to accommodate the needs of a process. Combining this flexibility with the predictive models described earlier can lead to improved reactor design and influence conversion rates and product quality. Since the reactant and product concentrations vary along the length of the reactor, different stages may benefit from different operating temperatures. In the case of moderately or highly exothermic reactions, the reaction temperature may spike above the desired control point during the initial portion of the reactor (Figure 14.10a). Because the greatest amount of heat is generated at the entrance of the PFR, a two-zone jacket temperature may facilitate greater reaction temperature control (Figure 14.10b). In this example, the two zones consist of a lower initial jacket temperature of 63 °C versus the 80 °C on the remaining portion of the reactor. Alternatively higher temperatures can be utilized later in the reactor to improve conversion rates. Similarly reactors can be easily designed to accommodate multiple feed points at different stages along the reactor to further manipulate reaction conditions if required. In this way reactors can be specifically designed for maximum throughput and product quality.

While the reaction engineering discussion thus far has focused on product quality, the same concepts can be utilized for process safety evaluations. Since most of the reactions employed in the pharmaceutical industry are exothermic, this safety aspect is an important consideration. This is especially true of non-isothermal PFR operation where rates of heat generation and temperature vary along the length of the reactor. Reactions should be evaluated in combination with proposed reaction conditions to avoid potential runaway reactions and ensure a sufficiently large safe operating window.

14.5 SCALE-UP: VOLUMETRIC VERSUS NUMBERING UP

Classical scale-up of a batch process consists of increasing the volume of the batch reactor. As a result of poorer heat and mass transfer in larger reactors, many of the common operations of batch processing take significantly longer at larger scales. Activities that frequently take longer at scale are charging of reagents, batch heat-up or cool-down, and reaction quench. The improved heat and mass transfer capabilities of continuous reactors mean that lab- and plant-scale processing times are much better aligned. For example, the reaction time does not change with scale since it is the design criterion for the continuous process. Additionally, the reaction stream is quenched immediately upon completion of the reaction at any scale.

Scale-up of most continuous processes occurs by increasing the total reactor volume and the flow rate to maintain the same residence time established during development. However, an alternative approach in continuous processing scale-up, particularly for using microreactors, is to number up. Here the reactor system is duplicated numerous times, with all running in parallel [20]. At DSM, multiple parallel microreactors were utilized for pilot-scale production of a nitration reaction of a pharmaceutical intermediate [21]. In this scenario, the replication of the same geometries and flow rates for each unit provides the higher overall process flow rates and thus avoids any scale-up effects. The logistics, complexity, and capital investment of such systems may limit widespread implementation for high volume products. Examples where processes have been numbered up using microreactors for commercial manufacturing are rare, and this approach may not be amenable for most processes without additional technological advances particularly in automated flow stream division and control.

LaPorte et al. [18] demonstrated a less intensive example of numbering-up of a gas–liquid reaction. They replicated a trickle bed column and housed the set of four in a common baffled jacketed tube for temperature control. An enolate stream was split equally into four streams using rotameters, and each stream flowed into one of four trickle bed columns. This scale-up facilitated a 4× numbering-up by maintaining the same fluid dynamics, mass transfer, and heat transfer characteristics in each tube. This operation did require constant monitoring since the flow splitting was not automated. Commercial processes would require a high degree of automation to ensure the proper flow at all times. Other approaches to scale-up, particularly for a microreactor, include adding to the volume of the reactor with serial addition of reactor plates. Hence, a large range of flow rates from milliliters to several hundred milliliters per minute is possible for a specific reactor platform [22].

14.6 PLANT OPERATIONS

14.6.1 Flow Control

Flow control is a critical parameter for a continuous process. The total flow of the streams ensures the proper residence time for reaction, and the ratio of the individual streams ensures the proper stoichiometry of the reagents. The feed streams must be accurate and consistent, and the constraints on those parameters depend on the tolerance of the process. One approach to minimize pulsating flows is to utilize pressured feed tanks along with flow meters and control valves to control the flow rate. Another possibility is to use metering pumps for each of the feed streams. For pumps that pulsate (piston, diaphragm, etc.), synchronization, dampening devices, or multiple pistons that are sequenced and positioned on one pump may be required. Pulseless pumps with an integrated mass flow meter and feedback control system are ideal. These systems provide precise metering for processes with tight flow tolerances.

14.6.2 Process Analytical Technology (PAT)

Process analytical technology (PAT) is an important part of most continuous processes as it provides a useful means of monitoring the state of the reaction. Indeed, one of the stated goals of the FDA's PAT initiative is "Facilitating continuous processing to improve efficiency and manage variability" [23]. Typical PAT tools include Raman, FTIR, NIR, and UV–Vis spectroscopy [24, 25] or other noninvasive monitoring techniques that can be adapted to a flow cell or tube reactor. These tools can be used during both transient and steady-state operations. As an example, FTIR was used to determine the proper ratio of reagent feed rate to starting material feed rate during the startup of a continuous process to make an active pharmaceutical ingredient [18]. For this particular process, the same PAT equipment could have been used to monitor the product quality. In a well-defined continuous process, it is envisioned that feedback controllers could adjust operating parameters based on input signals from PAT analyzers. Even in the absence of feedback control, PAT can provide valuable information to plant operators who can modify the operation if necessary. If PAT analyzers indicate that product quality is suspect, flow can be diverted to alternative holding tanks for further analysis. In the absence of spectroscopic analyzers, simple temperature measurements at various reactor positions can provide a wealth of information regarding reaction performance.

14.7 CASE STUDY: CONTINUOUS DEPROTECTION REACTION – LAB TO KILO LAB SCALE-UP

A batch process to carry out an acidolysis and deprotection chemistry for a pharmaceutical intermediate involved adding the substrate solution to triflouroacetic acid (TFA) at approximately 0 °C. The complete reaction mixture was immediately quenched into a biphasic mixture of aqueous base and ethyl acetate. An amide impurity was formed at high levels of greater than 2%. The longer quench times anticipated upon scale-up were expected to further increase the level of the amide impurity.

A continuous processing approach was undertaken to minimize impurity formation through improved control of reaction time and reduced quenching time. The continuous reaction was assessed in the laboratory by mixing two feed streams, one for the substrate and the other TFA, in a glass microreactor with an overall volume less than 10 ml. Experiments varying temperature and residence time identified process conditions, 25 °C and a minimum residence time of 4 minutes, which provided complete conversion and a significantly lower level of amide impurity, approximately 1%.

The preliminary reaction kinetics obtained from the small-scale continuous reactions paved the way for a rapid process scale-up. A 100-fold increase in flow rate in the substrate and TFA streams was used to process approximately 5 kg of starting material using the setup shown in Figure 14.11. The starting material solution was not stable at room temperature and required storage at −10 °C. A preconditioning heat exchanger was used to continuously heat up the starting material feed stream to the reaction temperature, 25 °C, just prior to reaction. A PFR, constructed of a jacketed static mixer for mixing the two feeds and three sequential concentric tube heat exchangers, operated with an overall residence time of five minutes. The reaction stream was continuously quenched in a jacketed static mixer, and the quenched mixture flowed into a receiver for subsequent processing. The use of a static mixer for the quench ensured effective mixing of the biphasic process stream while rapidly quenching the reactive species.

This particular batch process was relatively simple to convert to a continuous process. However, it is a good example to demonstrate the key components and strategies behind the

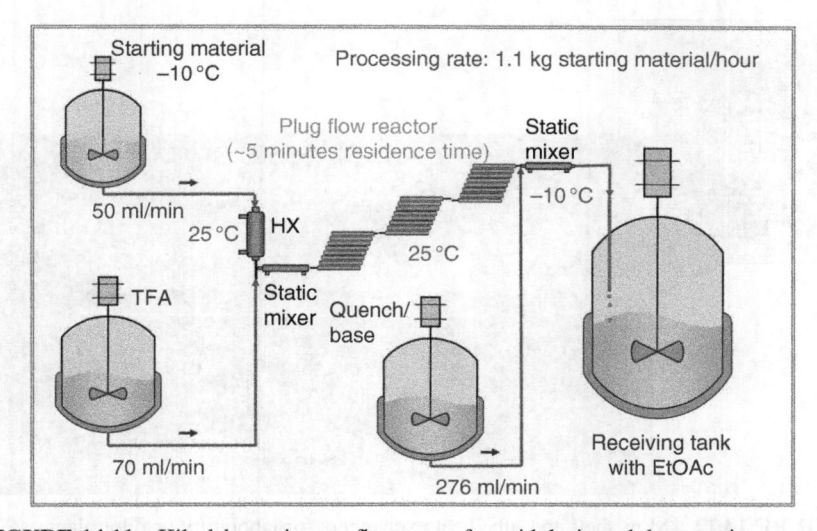

FIGURE 14.11 Kilo lab continuous flow setup for acidolysis and deprotection process.

development process. This includes the use of stable feeds, preconditioning of a feed, and combined in-line jacketed static mixer and heat exchangers as the reactor.

14.8 CASE STUDY: CONTINUOUS PRODUCTION OF A CYCLOPROPONATING REAGENT

14.8.1 Introduction

The Simmons–Smith cyclopropanation is a well-known reaction to form cyclopropanes from olefins utilizing zinc and an alkyl iodide. The structure of the reactive zinc carbenoid species is the subject of numerous papers [26, 27]. Formation of the active species is relatively exothermic with an adiabatic temperature rise above 120 °C. Additionally, the complexes are known to be unstable for extended periods of time above 0 °C. The exothermic nature of the reaction, combined with the complexity and incomplete understanding of the mechanism, and relative instability of the active species made scale-up very challenging in a batch process. One solution to the scale-up was the development of a continuous process for formation of the cyclopropanating reagent. The process was demonstrated at laboratory scale, scaled up to pilot plant scale, and was used to make launch supplies for the starting material of a commercial API. The development and implementation of this process are discussed here.

14.8.2 Process Development

The strategy for developing a continuous process was to operate a PFR with a short residence time and higher temperatures, followed by a rapid thermal quench. A short reaction time was required to minimize the size of the PFR as well as reagent degradation. Initial screening work utilized a coiled 1/8 inch stainless steel jacketed tube as the reactor. Later in development, multiple 26 ml shell and tube heat exchangers containing up to 19 1/8 inch stainless steel tubes (Figure 14.12) were utilized. The heat exchangers were sequenced end to end to form the PFR. The reactor was operated with a short 50 second residence time. Due to the short residence time and the large amount of heat generated early in the reactor, isothermal operation was not possible. Details of the process are described below.

The laboratory setup used for development of the continuous process is shown in Figure 14.13. Two feed streams, one containing diethyl zinc (13.5 wt %) and dimethoxyethane in toluene and the other containing diiodomethane in dichloromethane, were held at ambient temperature. These streams were fed to the reactor with gear pumps and mass flow meters to ensure proper stoichiometry and residence time. Both streams passed through independent heat exchangers with a 30 °C jacket temperature prior to mixing in a non-jacketed static mixer containing 27 helical mixing elements. The feeds entered the static mixer at about 29 °C and exited at about 48 °C. The reaction mixture flowed through a series of three shell and tube heat exchangers with 30 °C jacket temperature to facilitate formation of the reagent. The process stream exit temperature was 52 °C after the first heat exchanger and 32 °C after the third. The 2 °C temperature difference observed between process and jacket sides of the third heat exchanger indicates that the reaction was nearly complete by that stage. Additional calorimetric laboratory experiments that evaluated residual heat generation of the reaction mixture confirmed this observation. Finally, the reaction was thermally quenched to less than −10 °C, again with a shell and tube heat exchanger. The process attained a steady state within three residence times based on multiple temperature measurements at various points along the PFR.

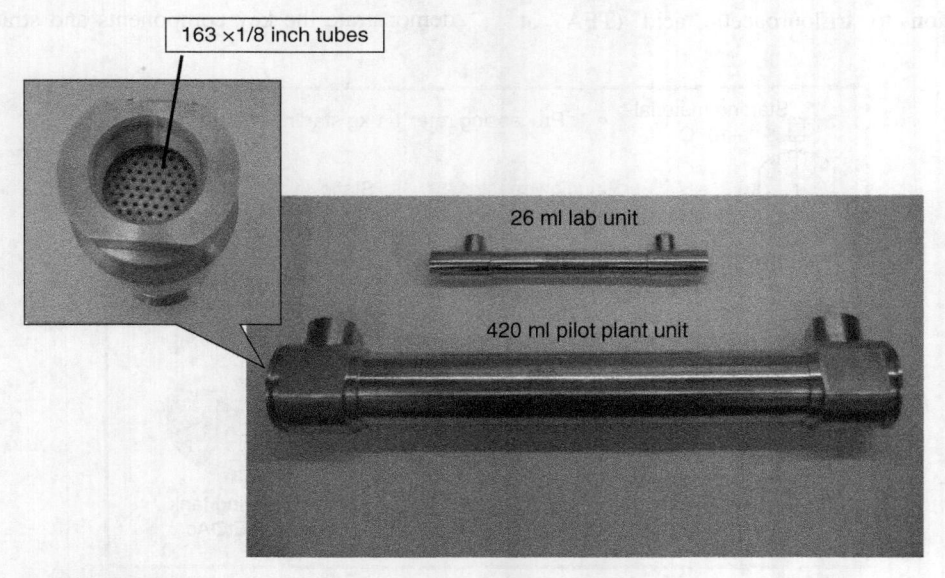

FIGURE 14.12 Mini shell and tube heat exchangers for laboratory or pilot plant use.

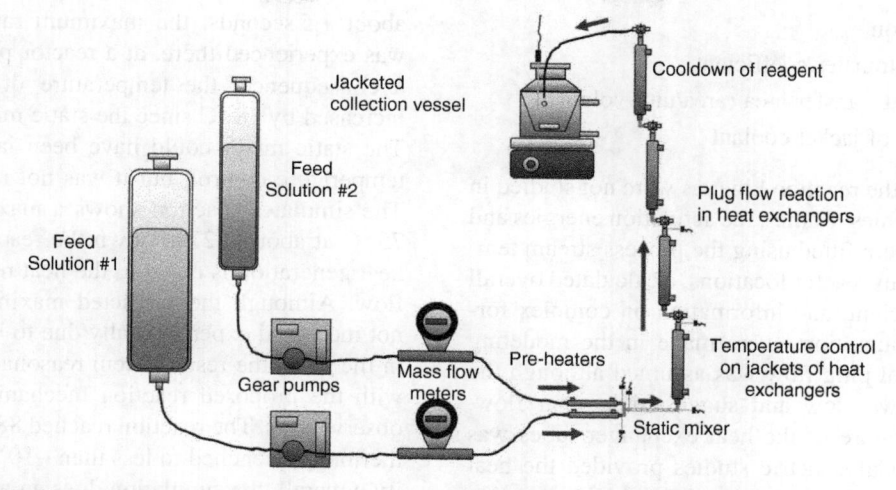

PROCESS DEVELOPMENT AND CASE STUDIES OF CONTINUOUS REACTOR SYSTEMS FOR PRODUCTION OF API

FIGURE 14.13 Laboratory setup for development of the Simmons–Smith reagent continuous process.

$$Et_2Zn + DME + CH_2I_2 \longrightarrow EtZnCH_2I * DME + EtI \quad \text{Furukawa}$$

$$EtZnCH_2I * DME + CH_2I_2 \longrightarrow Zn(CH_2I)_2 * DME + EtI \quad \text{Wittig}$$

FIGURE 14.14 Modeled formation of complexes for the cyclopropanating reagent.

14.8.3 Modeling and Simulation

A reaction engineering approach similar to that described earlier was employed here to gain further insight into the continuous process. The proposed reactions for the model are formation of the Furukawa complex and Wittig complex and are shown in Figure 14.14. A proposed kinetic model and energy balance equation governing the reaction are shown as follows:

Non-isothermal Plug Flow Reaction Model

Assumptions

- Completely mixed in radial direction.
- No diffusion in flow direction (axial).
- Constant shell side temperature.
- Constant stream density.
- A two-step reaction mechanism producing first the Furukawa complex followed by the Wittig complex.
- Modeled on a per tube basis in the heat exchanger.
- Overall heat transfer coefficient is independent of the number of tubes in the heat exchanger (constant shell side temperature)

Reaction mechanism:

$$A + B \rightarrow F \quad \Delta H_1 = -99 \text{ kJ/mol} \quad (14.6)$$

$$F + B \rightarrow W \quad \Delta H_2 = -94 \text{ kJ/mol} \quad (14.7)$$

A = Diethyl zinc/dimethoxyethane
B = Diiodomethane
F = Furukawa complex
W = Wittig complex

Rate expressions:

$$k = Ae^{-E/RT} \quad (14.8)$$

$$r_1 = -k_1 C_A C_B \quad (14.9)$$

$$r_2 = -k_2 C_F C_B \quad (14.10)$$

Simplified mass/energy balance:

$$u\rho C_p \frac{dT}{dz} = \Delta H_1 r_1 + \Delta H_2 r_2 - UA_V(T - T_c) \quad (14.11)$$

u = reaction stream velocity
ρ = reaction stream density
C_P = reaction stream heat capacity
T = reaction stream temperature
z = axial position in PFR
ΔH_i = heat of reaction

r_i = rate of reaction
U = overall heat transfer coefficient
A_V = specific heat transfer area (area/unit volume)
T_c = temperature of jacket coolant

In this example, the reaction kinetics were not studied in separate detailed studies. Rather, the activation energies and frequency factors were fitted using the process stream temperatures at numerous reactor locations, a calculated overall heat transfer coefficient, and information on complex formation. Several assumptions were made in the modeling of this process. Ideal plug flow was assumed although the Reynolds number was low and suggested laminar flow. The surface temperature of the heat exchanger tubes was assumed constant. Calorimetric studies provided the heat of reaction data. The kinetic model was fitted to the temperature profile with an estimated overall heat transfer coefficient [28]. Using the kinetic parameters, the reaction was simulated as it progressed through the reactor. Several local minima were determined during the model fitting exercise, requiring additional data to refine the model. The simulation results are shown in Figure 14.15, which plots temperature and heat generated versus axial position in the reactor. Although the residence time in the static mixer was only about 1.5 seconds, the maximum rate of heat generation was experienced there, at a reactor position of 0.18 m. As a consequence, the temperature of the reaction stream increased by 18 °C since the static mixer was not jacketed. The static mixer could have been jacketed for additional temperature control, but it was not necessary in this case. The simulated reaction shows a maximum temperature of 75 °C at about 0.27 m down the reactor. At this point, the heat generation is equal to the heat removal by the coolant flow. Although the predicted maximum temperature was not measured experimentally due to limited thermocouples in the PFR, the results seem reasonable and are consistent with the proposed reaction mechanism and experimental observations. The reaction reached 88% yield prior to being thermally quenched to less than −10 °C for complex stability. Overall, the simulation does an adequate job in modeling the observed behavior and results from the laboratory. Despite limited knowledge of the reaction kinetics prior to modeling, the exercise demonstrates the utility of simulating a non-isothermal PFR. This type of process knowledge could be used to modify reaction conditions if necessary but in this case was primarily used to guide the design and operation of pilot plant and commercial manufacturing reactors.

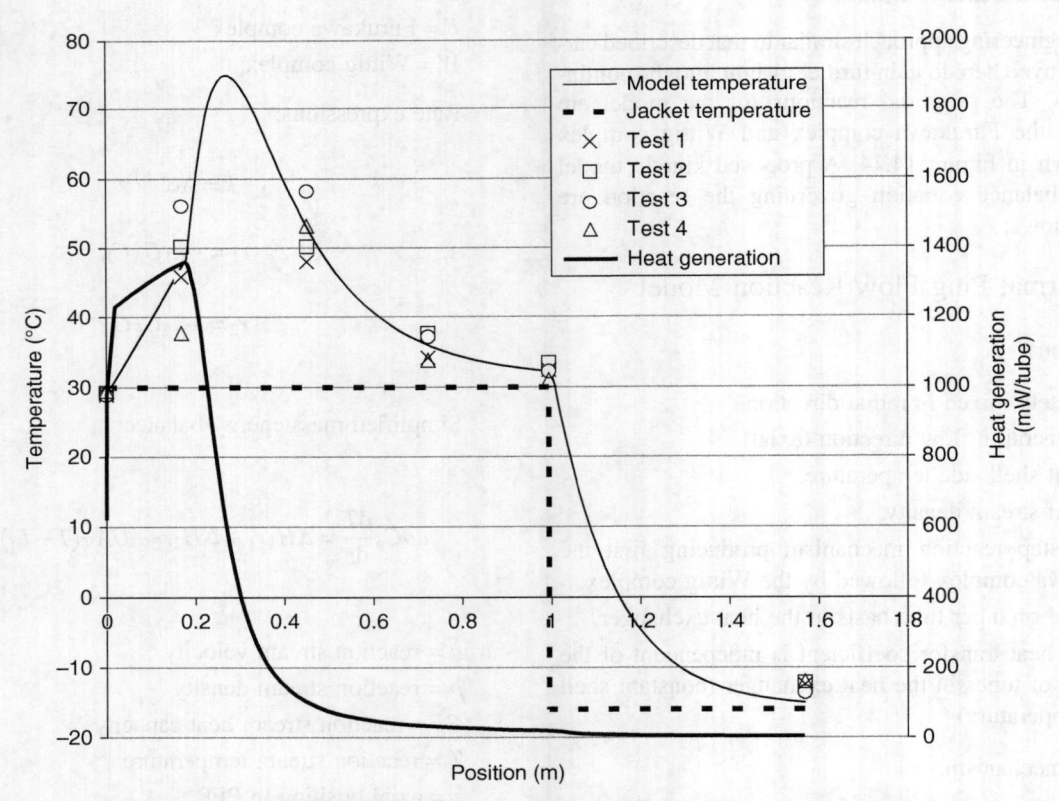

FIGURE 14.15 Test and model results for lab-scale plug flow reactor. Note that the axial velocity is higher in the static mixer due to a lower cross-sectional area relative to the heat exchangers. As a result, the shape of the temperature curve is influenced when plotted against position.

14.8.4 Process Scale-Up to Pilot Plant

With little additional development work, the process described in the previous sections was scaled up in a pilot plant to generate 700 kg of the cyclopropanating reagent solution. The feed tanks were pressurized, and an actuated diaphragm valve coupled with a mass flow meter controlled the flow rate of each feed. The process was scaled by maintaining a similar residence time as the lab, and essentially numbering up the lab setup by having a larger number of tubes in each heat exchanger, while maintaining the tube diameter. Unlike the batch process, this approach ensured similar heat and mass transfer characteristics and little change in reaction conditions when moving from the lab to the pilot plant. The PFR was constructed from a static mixer and five shell and tube heat exchangers, each containing 163 1/8 inch tubes. The total residence time was 65 seconds, similar to the 51 second residence time utilized in the lab. In the pilot plant, a spiral heat exchanger was used to facilitate the thermal quench. The design of the pilot plant system was otherwise similar to the previously described laboratory system.

Table 14.5 shows the different specifications for the laboratory and the pilot plant setups. The operation on pilot scale was similar to the laboratory process with slight differences in measured peak process temperatures, most likely due to differences in heat transfer characteristics. The maximum reaction stream temperature, measured after the first heat exchanger in the PFR, was 56–62 °C. As a result, the PFR was operated with a higher jacket temperature relative to the laboratory reactor. After exiting the PFR, the reaction stream was thermally quenched and collected in a jacketed 2000 L reactor for later use. The process ran until all the feed solutions were consumed. The implemented continuous process facilitated production of 700 kg of reagent of consistent quality under reproducible conditions.

TABLE 14.5 Laboratory to Pilot Plant PFR Specifications

	Lab	Pilot Plant
DME/DEZ (g/min)	70	1358
Diiodomethane/DCM (g/min)	48	940
Total mass flow rate (g/min)	118	2298
Volumetric flow rate (ml/min)	94	1868
PFR residence time (s)	50	65
Fluid velocity (cm/s)	1.7	3.85
Number of tubes (per shell)	19	163
Tube size (ID, cm)	0.254	0.254
Re #	53	120
Reactor length (m)	1.0	2.55
Heat load (W)	240	4700

14.9 INTEGRATED CONTINUOUS PROCESSING IN PHARMA

While implementing continuous reactions can lead to improved safety and product quality, the increased manufacturing efficiencies experienced in the commodity chemical industry are largely unrealized when the downstream processing is conducting in a semi-batch fashion. This is a result of equipment downtime in such a scenario. Coupling multiple unit operations into a continuous process train has the potential to accelerate introduction of new drugs through more efficient production processes and decrease the costs of production with smaller facilities, minimization of waste, lower energy consumption, and decreased raw material use [29]. Post-reaction processing in the pharmaceutical industry typically involves extractions, solvent exchanges, crystallizations, and drying, and technologies currently exist to perform many of these unit operations continuously. For example, traditional chemical processing equipment such as Podbielniak centrifugal extractors, wiped film evaporators, and continuous crystallizers can perform extractions, solvent exchanges, and crystallizations continuously. These devices offer not only higher throughput but also increase efficiencies as well, resulting in yield improvements and less waste. Additionally, parallel drying trains can be setup to alternately receive material from upstream continuous process trains. While integrated continuous processing of API is in its infancy, some companies have efforts underway [30], and others are collaborating with academia to develop new technologies for such purposes [2]. An integrated continuous processing plant may become more common in the pharmaceutical industry as technologies develop and as cost pressures rise. Whether these exist as smaller plants dedicated to a single product or modular multiproduct plants remains to be seen. Either way, the evolution will likely be slow given the entrenchment of existing batch processing plants and the real, or perceived, barriers to widespread acceptance of continuous processes.

14.10 BARRIERS TO IMPLEMENTATION OF CONTINUOUS PROCESSING IN PHARMA

The barriers to continuous processing in pharma are largely historical and involve GMP documentation concerns, lack of experience and understanding, and an existing infrastructure designed for batch processing. The pharmaceutical industry has traditionally preferred batch processing largely because of GMP documentation and traceability purposes. By having obviously defined discrete batches or lots of material, it is straightforward to meet GMP requirements to document and verify the processing activities, parameters, and raw materials that go into each batch. Since continuous processes do not have well-defined and frequent beginning points and endpoints, there is a perception that the definition of a batch is less obvious. This is a misperception however, since the FDA's own guidance states, "In the case of continuous production, a batch may correspond to a defined fraction of the production" [31]. Clearly, the FDA is willing to work with industry to adapt traditional GMP approaches to work with continuous processing. The prolonged absence of continuous processing in pharma leads to a dearth of continuous processing know-how, both in development and manufacturing. That barrier has largely been reduced, especially on the development side, over the last decade as the regulatory hurdles to continuous processing have lessened and chemical engineers bring their skill sets to bear on the industry. With regulatory acceptance and development capabilities in place, the question then becomes one of economics. Where continuous processes enable improvements in safety or product quality, they are currently being utilized on a case-by-case basis. The large investments that the pharmaceutical industry has in existing batch plants represent a significant hurdle to widespread adoption of continuous processing. Furthermore, given the high rates of attrition during development, companies may be hesitant to invest in less familiar processing technologies. Transitioning to a continuous process post NDA also represents a significant cost and regulatory burden. While many scientists and engineers recognize the benefits of continuous processing, the transition from a batch industry to a continuous industry will likely be very slow.

14.11 SUMMARY

In this chapter, we have discussed how to implement a continuous processing paradigm for organic synthesis reactions. Converting processes from batch to continuous has the advantages of improved intermediate stability, enhanced safety, greater risk management, and enhanced mass and heat transfer. A wide range of reactors, including traditional PFRs and CSTRs, as well as novel microreactors are available for continuous processing. Knowledge of the kinetics and heats of reactions is a prerequisite for the development of a continuous process, and modeling helps to guide reactor choice and identify operating conditions. Continuous operation may not be appropriate for all processes. When looking for development opportunities, the initial focus should be on processes that have safety issues, followed by issues of quality, and lastly economics. The economic considerations are difficult to realize during process development and may not be substantial in manufacturing in the absence of integrated continuous processing. Chemical engineers can take the lead in helping the pharmaceutical industry realize all the benefits of continuous processing.

REFERENCES

1. Higgins, S. (1998). Are more fine specialty chemicals being moved into continuous process plants? *Chimica Offi/Chemistry Today* 16 (9): 38–41.
2. Pellek, A. and Van Arnum, P. (2008)). Continuous processing: moving with or against the manufacturing flow. *Pharmaceutical Technology* 9 (32): 52–58.
3. Bogaert-Alvarez, R.J., Demena, P., Kodersha, G. et al. (2001). Continuous processing to control a potentially hazardous process: conversion of aryl 1,1-dimethylpropargyl ethers to 2,2-dimethylchromens (2,2-dimethyl-2H-1-benzopyrans). *Organic Process Research and Development* 5: 636.
4. Panke, G., Schwalbe, T., Stirner, W. et al. (2003). A practical approach of continuous processing to high energetic nitration reactions in microreactors. *Synthesis* 18: 2827.
5. Brechtelsbauer, C. and Ricard, F. (2001). Reaction engineering evaluation and utilization of static mixer technology for the synthesis of pharmaceuticals. *Organic Process Research and Development* 5: 646.
6. Roberge, D.M., Bieler, N., and Thalmann, M. (2008). Microreactor technology and continuous processes in the fine chemical and pharmaceutical industries: is the revolution underway? *Organic Process Research and Development* 12 (5): 905–910.
7. LaPorte, T.L. and Wang, C. (2007). Continuous processes for the production of pharmaceutical intermediates and active pharmaceutical ingredients. *Current Opinion in Drug Discovery and Development* 10 (6): 738–745.
8. Thakur, R.K., Vial, C., Nigam, K.D.P. et al. (2003). Static mixers in the process industries: a review. *Chemical Engineering Research and Design* 81 (7): 787–826.
9. Stankiewicz, A. and Drinkenburg, A.A.H. (2004). Process intensification: history, philosophy, principles. *Chemical Industries (Dekker)* 98: 1–32.
10. Ritter, S.K. (2002). A new spin on reactor design refined rotor-stator system has the potential to boost chemical process intensification efforts. *Chemical and Engineering News Archive* 80: 26–27.
11. Brechtelsbauer, C., Lewis, N., Oxley, P., and Richard, F. (2001). Evaluation of a spinning disc reactor for continuous processing. *Organic Process Research and Development* 5: 65–68.
12. Micronit Microfluidics BV (2009). Lab-on-a-chip and MEMS solutions. www.micronit.com.
13. Ehrfeld, W., Hessel, V., and Löwe, H. (2000). *Microreactors, New Technology for Modern Chemistry*, Chapter 3, 1e. Weinheim: Wiley-VCH.
14. Dapremont, O., Zeagler, L., DuBay, W. (2007). Reducing costs through continuous processing. *Sp2* June, 22–24.
15. AMPAC Fine Chemicals (2009). Reuters: AMPAC Fine Chemicals Announces the Inauguration of the New cGMP Continuous Processing Development Facility, 9 March 2009.
16. Roberge, D.M., Ducry, L., Bieler, N. et al. (2005). Microreactor Technology: a revolution for the fine chemical and

pharmaceutical industries? *Chemical Engineering and Technology* 28 (3): 318–323.

17. Günther, M., Scheider, S., Wagner, J. et al. (2004). Characterisation of residence time and residence time distribution in chip reactors with modular arrangements by integrated optical detection. *Chemical Engineering Journal* 101: 373–378.

18. LaPorte, T.L., Hamedi, M., DePue, J.S. et al. (2008). Development and scale-up of three consecutive continuous reactions for production of 6-hydroxybuspirone. *Organic Process Research and Development* 12: 956–966.

19. Choe, J., Kim, Y., and Song, K.H. (2003). Continuous Synthesis of an intermediate of quinolone antibiotic drug using static mixers. *Organic Process Research and Development* 7 (2): 187–190.

20. Schwalbe, T., Kursawe, A., and Sommer, J. (2005). Application report on operating Cellular Process Chemistry plants in fine chemical and contract manufacturing industries. *Chemical Engineering Technology* 28 (4): 408–419.

21. Thayer, A. (2009). Handle with care. *Chemical and Engineering News* 87 (11): 17–19.

22. Lonza News Release (2009). Lonza secures important manufacturing contract using its proprietary microreactor technology. *Lonza News Release*, 26 May.

23. US Food and Drug Administration (2004). *Guidance for Industry PAT A Framework for Innovative Pharmaceutical Manufacturing and Quality Assurance*. Rockville, MD: U.S. Department of Health and Human Services, Food and Drug Administration, Center for Drug Evaluation and Research.

24. Lobbecke, S., Ferstl, W., Panic, S., and Turcke, T. (2005). Concepts for modularization and automation of microreaction technology. *Chemical Engineering and Technology* 28 (4): 484–493.

25. Ferstl, W., Klahn, T., Schweikert, W. et al. (2007). Inline analysis in microreaction technology: a suitable tool for process screening and optimization. *Chemical Engineering and Technology* 30 (3): 370–378.

26. Charette, A. and Marcoux, J.F. (1996). Spectroscopic characterization of (iodomethyl)zinc reagents involved in stereoselective reactions: spectroscopic evidence that $IZnCH_2I$ is not $An(CH_2I)_2 + ZnI_2$ in the presence of an ether. *Journal of the American Chemical Society* 118: 4539–4549.

27. Davies, S., Ling, K., Roberts, P. et al. (2007). Diastereoselective Simmons-Smith cyclopropanation of allylic amines and carbamates. *Chemical Communications* 39: 4029–4031.

28. Pitts, D.R. and Sissom, L.E. (1977). *Heat Transfer*. New York: McGraw-Hill.

29. Mullin, R. (2007). Cover Story – Breaking Down Barriers: Drugmakers are paving the way to more streamlined manufacturing via culture change in R&D. *Chemical and Engineering News* 85 (4): 11–17.

30. Berry, M. (2009). After a Century of Batch Manufacturing API, What Does Continuous Processing Offer? PhRMA API Workshop.

31. ICH (2000). ICH Harmonised Tripartite Guideline Q7: Good Manufacturing Practice Guideline for Active Pharmaceutical Ingredients. International Conference on Harmonisation of Technical Requirements for Registration of Pharmaceuticals for Human Use, Geneva, Switzerland (November 2000).

15

DEVELOPMENT AND APPLICATION OF CONTINUOUS PROCESSES FOR THE INTERMEDIATES AND ACTIVE PHARMACEUTICAL INGREDIENTS

FLAVIEN SUSANNE
Product and Process Engineering, GSK Medicines Research Centre, GlaxoSmithKline, Stevenage, UK

15.1 INTRODUCTION

Continuous process operation is a manufacturing methodology that dominates the production of bulk chemicals. It is seen as the method of choice for delivering products associated with the oil and gas, food, and polymer industries. The process operations of such industries are driven by allowing consistency of the quality, enabling high demands and predominantly lowering the cost of goods (CoGs); positive margins can be achieved from low value raw materials and products. In contrast, the pharmaceutical industry has not placed much focus on the manufacturing cost of their final products, drug substance (DS) and drug product (DP) as, until recently, these were considered high value products with an extremely high margin. However, in the last 10–15 years, the emergence of generic companies,[1] manufacturing off-patent drugs in parts of the world where there are lower labor and overhead costs, has reduced the revenues of the pharmaceutical companies' overall.

Active pharmaceutical ingredient (API) manufacture involves low volumes, where demand is uncertain. Therefore, API manufacture had traditionally been in the domain of versatile and multipurpose batch manufacture, which remains the mainstay of the pharmaceutical industry. Pharmaceutical companies are now being forced to adjust their business models to react to generic competition. Enhanced value brought by new technologies is seen as the direction forward. We now see among process transformation the emergence of enzymatic transformations, specific catalytic transformations, and high value continuous processing. In the past decade, there has been a growing interest in the application of continuous processing in the pharmaceutical industry for two reasons. Firstly, flow chemistry enables a diversity of reactions from which to synthesize a molecule [1, 2]. Such approaches facilitate the synthetic route to a molecule that may in turn produce a simpler manufacturing process. Secondly, a rationale has emerged based around factory or supply chain benefit. This approach argues that telescoping multiple synthetic steps together in a continuous supply chain leads to benefits including reducing inventories and allowing more responsiveness to product demand [3].

Only recently have continuous processes started being implemented into pilot plant and small-scale manufacturing facilities. Most of these recent examples are predominantly single-stage processing for chemistry-enabling applications where a mass and heat transfer problem needs to be solved or controlled. With progressive innovation and development of new technologies, drug development and production is expected to move toward lower volume, customized, or patient-specific drugs. This trend of manufacturing many new (and low volume) products will force the pharmaceutical industry to improve drug quality yet at reduced manufacturing cost, which accounts for 36% of total industry cost [4]. In the recent years, given the pressure of cost reduction, improved consistency in DS or DP quality, and enhanced

---
[1] https://www.pharmacompass.com/radio-compass-blog/top-drugsby-sales-in-2016-who-sold-the-blockbuster-drugs (accessed 10 November 2018).

Chemical Engineering in the Pharmaceutical Industry: Active Pharmaceutical Ingredients, Second Edition.
Edited by David J. am Ende and Mary T. am Ende.
© 2019 John Wiley & Sons, Inc. Published 2019 by John Wiley & Sons, Inc.

manufacturability, the design of multistep continuous manufacturing has emerged as a promising solution to address most of these constraints.

The response from the pharmaceutical industry cannot be a simple transfer of technology used in other industries. While oil and gas or polymer product volumes represent extremely high tonnage per annum, API volumes rarely exceed 100 metric tons per annum, with an obvious trend leading toward smaller volumes in the future due to high efficiency and potency drugs.

Implementing continuous processing in the pharmaceutical industry is not trivial. It is often the answer when solving complex problems. Due to the high complexity of the chemical transformations, including intricate pathways from the starting materials to the product and a high number and diversity of impurities to be controlled to precise level, continuous process development requires the solving of interlinked chemical and engineering problems. To enable the efficient implementation of continuous process into the pharmaceutical world, the approach of process development and scale-up, first, has to be reassessed. Until recently, the pharmaceutical industry has relied on chemistry knowledge to bring their drugs from candidate selection to manufacture. For the development of continuous processing, the merging of multiple skill sets is necessary, complementing chemistry expertise with engineering application as part of the process design. Continuous process development involves generating deep process understanding across each unit operation. Together, kinetic and thermodynamic process models of the unit operations are generated including physical rates and performance of the process equipment. Understanding the intricate mechanisms taking place between the chemistry and the technology is key to the design, as it enables optimization and process transfer to manufacture in the most efficient way.

15.2 VALUE OF CONTINUOUS PROCESSES FOR THE PHARMACEUTICAL INDUSTRY

15.2.1 Access to a New Range of Operational Conditions and Transformations

Traditionally, due to the batch technology being used, process operation in the pharmaceutical industry is limited to a narrow range of pressure, temperature, and reactivity. Despite the benefit of such vessels in terms of flexibility and versatility to perform a recipe, they are limited in the range of conditions, which they could achieve in terms of pressure and temperature. In addition, due to their size, geometry, design, and poor surface to volume ratio, such vessels provide limited efficiency of heat and mass transfer. Therefore, the way processes could be operated becomes restricted, and generally processes do not achieve full optimization.

In contrast, continuous process technologies are made to enhance all physical rates such as heat and mass transfer. Smaller size technology enables an increase of the heat transfer coefficient, surface to volume ratio, and high efficiency of fluid dynamic. By increasing the operating space, less conventional chemistry can now be explored within a wider range of conditions. Process optimization can be extended toward new boundaries no longer limited by a narrow range of operations dictated by the large manufacturing vessels. The process conditions such as pressure, temperature, reaction time, or stoichiometry can be tuned more easily to reach optimum process output. In addition, transformations requiring high pressure and high temperature, such as Claisen, Cope, or Fisher indole rearrangement, can now be industrialized for manufacturing demand [2].

New types of transformation are now being explored with some success, among them photochemistry and electrochemistry. Photochemistry, despite its great potential, was limited until now by the light penetration depth and the photon exposition achieved in a big batch vessel. Therefore, photochemistry never broke through, as no scale-up pathway was identified. With the development of continuous technology with high surface to volume ratio see-through reactors and efficient penetration of the photon, simplified access and scale-up of photochemistry have become possible.

15.2.2 Safety of Operation

Operation in pharmaceutical manufacture has often been limited by the efficiency of the technology in place. We can find two classes of reaction where, to enable operations to be performed in a safe manner, processes are suboptimized and efficiency is needed to be compromised.

The first class of reaction is fast and/or exothermic; the poor mass and heat transfer achieved in big batch vessels forced these processes to be operated as dose-controlled addition to keep the heat release under control. From literature, a large range of examples such as nitro reduction, organometallic, or oxidation transformations could be found where suboptimum processes must be implemented in manufacture to keep them safe [5, 6]. These transformations have the specificity to generate extremely high exotherm, up to 1000s of kJ/mol and an extremely high adiabatic temperature raise, above 300–400 °C in some cases.

The second class of reaction can be defined by unstable and/or explosive intermediates, triggering the possibility of generating uncontrolled runaway reactions. The slow dynamic of large batch vessels forces these processes to be operated, predominantly, as dose-controlled addition, conditions far from their kinetic optimum to prevent accumulation. In the pharmaceutical industry, a safety margin of 50 °C is often applied from any onset of decomposition, to ensure safe operation. Reactions that are commonly found in API synthetic routes such as Curtius, nitration, or oxidation reactions are then underperformed [7, 8].

By enhancement of the heat and mass transfer, the use of continuous technology enables a more precise control of the process parameters and a much faster response to changes ensuring safe operation. The heat released by these reactions, or thermal instability, becomes easier to control by the high heat transfer coefficient and surface to volume ratio of the continuous technology. In some cases, such as peroxide oxidation, the process is operated in flow significantly above the onset of the reagent decomposition as the safety is achieved through an understanding of the relative rates of decomposition versus main reaction.

15.2.3 Quality Control

In the pharmaceutical industry, high quality of the finished product is mandatory. DS and DP are defined with extremely high quality specifications. However, the inaccuracy and lack of reproducibility of large batch processing generates out of specification material on a regular basis. This material can either be reprocessed or sometimes completely discarded from the chain of supply. Such defects, due to the high cost of raw materials and labor and overhead, have a significant negative impact on the overall cost of a drug.

Continuous processing by implementing tight controls of the process parameters and highly responsive process control and by minimizing manual intervention removes a significant number of the variables triggering poor quality material. A more consistent high quality end product can be achieved during state of control operations. The process is often monitored using various soft sensors and process analytical technologies. The quality output of the material can be achieved using two approaches. The first, closely related to batch philosophy, links the process input reading to a predefined output using design spaces. When disturbances are observed and the process parameters exceed the acceptance limit, the material is diverted and quarantined. The second one, more aligned with the philosophy applied in other industries such as bulk chemicals or oil and gas, looks at implementation of model predictive control (MPC). The process inputs are processed through a series of linear or nonlinear models to correct disturbances by applying feedback and feedforward counteractive measures.

15.2.4 Reduction of Cost

Most pharmaceutical industries are still focusing on the technical value that continuously can provide to their company. However, few pharmaceutical companies are now moving toward developing and integrating continuous multistep processes into their manufacturing facilities. For those processes, the implementation of new technology can be aligned with the integration of a concept-based recipe with a high degree of automation of the process operations. In addition, as defined in the quality control section above, process control can be achieved using more advanced process control such as MPC. The manual operation becomes minimal. The direct impact is the reduction of labor and overhead costs. However, as most process variation of material quality output is mainly related to human intervention or error, automated procedures allow significant opportunity for improvements in robustness and reproducibility. Less defects are then observed leading to less reprocessing or rejection of the final product.

Such an approach of automating and controlling processes using feedback and feedforward is well established in other industries such as the oil and gas or the tobacco manufactures. Those industries operate at a much greater sigma level than the pharmaceutical industry, enabling a significant overall CoGs reduction.

15.3 CONTINUOUS PROCESS DEVELOPMENT WORKFLOW

Continuous process development and scale-up into manufacturing facilities are not necessarily trivial. As described in Section 15.2, continuous processing is implemented to tackle considerations of quality control and safety, for highly demanding mass and heat transfer during chemical transformations. Therefore, continuous processing requires a detailed process understanding to ensure appropriate operating condition definitions, equipment design, and process control. Efficient continuous process development and scale-up are based on a workflow, where process understanding and equipment performance are combined through process modeling (Figure 15.1). Process simulation acts as a platform centralizing all process information, enabling holistic process definition and optimization.

The process understanding workflow can be applied similarly to all unit operations; however, the methodology will be exemplified using reactions in the following sections.

15.3.1 Process Understanding

Regardless of the type of unit operation considered, a process is defined by its energy and mass balance. The process understanding described below is focusing on providing all critical information; enabling the definition, the optimization of the process, and further linking with equipment performance.

15.3.1.1 Energy Balance Often for batch processes, the energy balance is neglected during the early process development activities and only integrated toward the end, to ensure safe operation during scale-up. The reverse is true for continuous process development; the understanding of the energy balance is considered critical to process design. The transfer of highly exothermic processes to continuous operation and the optimization can be challenging due to the overall heat and rate of release. Therefore, a tight control of the process parameters is required to ensure safe operation and quality attribute of the product. Being able to map out the enthalpy

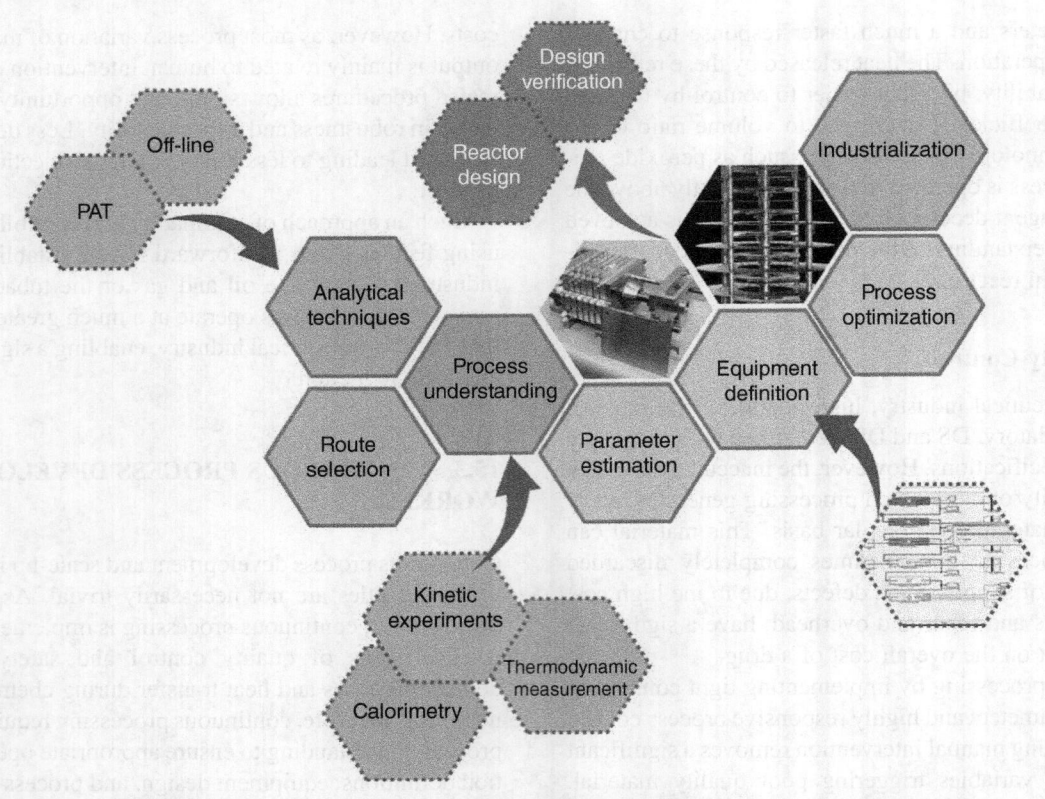

FIGURE 15.1 Schematic of the various phases of continuous process development.

related to each major aspects of the process is critical. Across the process, the enthalpy of reactions, dissolution, crystallization, mixing, and evaporation are summed up as a function of their rates to provide a dynamic picture of the process energy.

Techniques to measure enthalpies are not specific to batch or flow; classic batch calorimeters such as RC1, adiabatic accelerating rate calorimetry (ARC), and differential scanning calorimetry (DCS) [9–12] where the enthalpy is calculated from the integration of the overall heat flow can equally be used for flow. Heat flow traces of any investigated reactions can be gathered by measurements of non-isothermal or isothermal conditions. Despite their mode of operation and slow dynamic, they provide crucial information regarding the enthalpy of each single operation, even for flow operation. Batch calorimeters are limited in their function when fast and energetic reactions need to be assessed. In those cases, rates are impossible to regress as the heat released is limited by physical rates such as dosage. However, when the operation analyzed is not limited by the dynamic of the batch calorimeter, in addition to the heat flow, estimation of rate constants (k) and activation energy (E_a) can be achieved.

The reaction considered in the example illustrated in Figure 15.2, where A reacts with B forming product C, is followed by a by-product formation D and decomposition to form E. The heat flow measurements allow the overall heat released to be obtained where each reaction contributes in a certain specific ratio to the total observed heat flow. From the integration of the overall heat of reaction (Q_r), the enthalpy of reactions (ΔH_r), the rate constant (k), and activation energy (E_a) of each transformation can be calculated using expression (15.1) and the methods described below [10]:

$$Q_r = \left(\sum r_i \Delta H r_i\right) V = UA(Tr - Tj) + mCp\frac{dT}{dt} - \frac{dm}{dt} Cpdos(Tr - Tdos) \quad (15.1)$$

Two methods have been developed and commonly used to regress kinetic parameters from heat flow. The first method, the Borchardt and Daniels kinetics approach, allows the calculation of activation energy (E_a), pre-exponential factor ($k_{0,r}$), heat of reaction (ΔH_r), and reaction order (n) from a single DSC scan [12]. The kinetic approach is based on Eq. (15.2):

$$\frac{\partial [A]}{\partial t} = \ln k_{0,r} - \frac{E_a}{RT} + n\ln[1-\alpha] \quad (15.2)$$

The equation above can be solved with a multiple linear regression of the general form $k_{0,r}$ where the two basic parameters, $\partial[A]/\partial t$ and α, are determined from the DSC exotherm as shown in the DSC (Figure 15.2). This approach requires

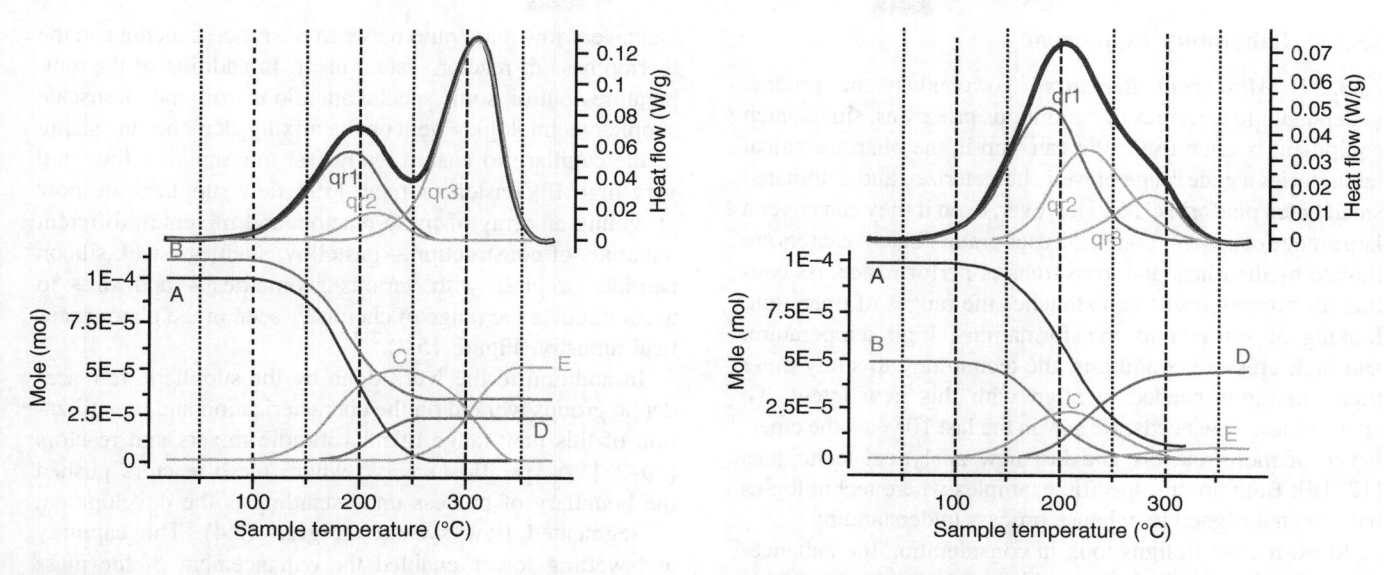

FIGURE 15.2 Heat flow signals and reagent profiles obtained by DSC under non-isothermal conditions, heating rate of 2 K/min; stoichiometry excess of 1.1 equiv. of B over A (left), sub-stoichiometric of 0.5 equiv. of B over A (right).

only a single temperature programmed experiment. This makes the approach highly attractive.

The second method, the ASTM E698 approach, is based on the variable program rate method of Ozawa, which requires a minimum of three DSC scans at different heating rates, usually between 1 and 10 °C/min [13]. The method assumes that the conversion at the peak exotherm is constant and independent of heating rate. By plotting ln k versus $1/T$, calculation of the activation energy (E_a) and pre-exponential factor ($k_{0,r}$) can be achieved. The ASTM E698 shows great accuracy on the regression of the activation energy, but the pre-exponential factor ($k_{0,r}$) may not be valid since the calculation of k assumes a reaction order predefined. However, the ASTM method is often the only means to analyze reactions with multiple exotherms [14].

Following the successful parameter estimation of the reactions using the methods described above, the evolution of the process can be expressed as molar time course profile where depletion and formation of the components A, B, C, D, and E are shown. Such model can be used for the prediction of any conditions and scenarios: isothermal, isoperibolic (constant temperature of the heat exchanger), adiabatic (cooling failure), and runway.

In the cases of fast reactions with unstable intermediates or competitive pathways, the use of traditional batch calorimeters to provide time-relevant information becomes limited. To prevent approximation in the energy balance calculation between the reaction studied and various parallel decompositions, some universities have been working on the design of flow calorimeters, allowing a better measurement of the heat flow for sub-minute reaction time. So far, the design of these new devices remains within the hand of academic groups with no real commercial equipment being available soon.

15.3.1.2 Mass Balance It was shown on the section above that calorimetry data could be used not only for calculation of enthalpy of reactions but also to regress other kinetic parameters such as rate constants and activation energies. However, these techniques show limited sensitivity, and the possibility to decouple all the phenomena taking place at the same time can be difficult to achieve. Therefore, for the process understanding of reactions, in addition to the energy balance and calorimetry data, molar time course profiles are often generated. Conventionally, they are generated at various concentrations and temperature mostly under isothermal conditions; however, non-isothermal profile can be applied [15]. They comprehend the major components of the mass balance such as starting materials, reagents, products, and major side products. The molar time course profile is sometime extended to critical impurities called critical quality attributes (CQAs) when the control of their formation is deemed to be important to the API.

With the difference from other industries where kinetic understanding is often limited to the understanding of a single, even complex, transformation, e.g. continuous hydrogenation in the oil and gas industry or polymerization in the polymer industry, the pharmaceutical industry is often working on the synthesis of new molecules involving reaction of building blocks never seen before. The transformation pathway is not always well defined, and a large range of new impurities appear. The control of those impurities to a low level, often below 0.1% w/w, is critical to the attribute of the final product and its commercialization. Generating mechanistic understanding of such complex transformations is challenging but highly valuable as it enables the definition and the optimization of the process in respect of the control of those impurities.

15.3.2 Laboratory Equipment

15.3.2.1 Mesoscale Reactors

To conduct the process understanding activities of continuous processes, small batch equipment is often used. We can find in the pharmaceutical laboratories a wide range of well characterized and automated small batch platforms [16]. However, even if they can cover a large range of process conditions, laboratory batch reactors are limited by their heat and mass transfer performance. As continuous process opens and stretches the ranges of operation, looking at sub-second transformations, high temperature, and high pressure conditions, the technology to study these transformations needed to align with this new intent. To answer these new needs, we saw in the last 10 years the emergence of meso-reactors and fast new analytical techniques [17, 18]. Both are the specific examples where technologies have been designed to enhance process understanding.

Meso-reactor designs took in consideration the enhancement of both aspects of mass and heat transfer. Such systems demonstrate the capability to achieve very high heat transfer coefficient due to an astonishing high surface to volume ratio and a very low mass ratio between the process media and the thermomass of reactor system itself. In addition to the temperature control, some special attentions from the mesoscale supplier technologies went on the mixing elements and sizing of the capillary to ensure really fast micromixing time and very high Bodenstein number [19]. Few suppliers are now providing an array of meso-reactors and mixers in different materials of construction – hastelloy, stainless steel, silicon carbide, or glass with various geometries – and sizes to accommodate the range of chemistry seen in the pharmaceutical industry (Figure 15.3).[2-5]

In addition to the work done by the suppliers, few academic groups worked on the characterization and the utilization of this new range of microfluidic mixers and reactors [20–23]. A specific example where meso-reactors pushed the boundary of process understanding is the development of segmented flow systems (Figure 15.4). The capillary and wetting forces enabled the enhancement of the mass transfer. In these systems, through an extensive organic film deposition on the walls of the reactor, a large specific surface area for the dissolution of the gas into the organic phase is

FIGURE 15.3 Examples of various meso-reactors available off the shelves.

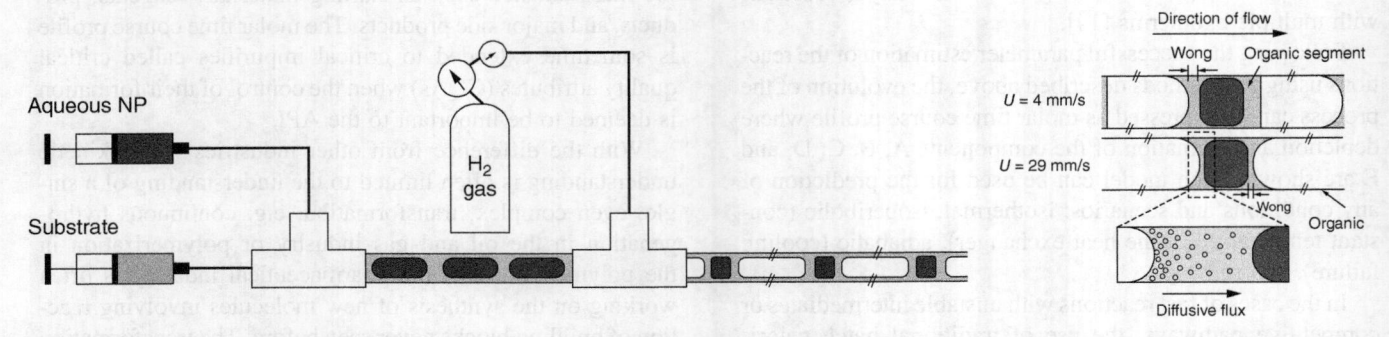

FIGURE 15.4 Schematic illustrating the mass transfer of dissolved reactants in the triphasic segmented flow. *Source*: Courtesy of Saif Khan, National University of Singapore.

---

[2] https://ehrfeld.com/en/home.html (accessed 10 November 2018).
[3] www.chemtrix.com (accessed 10 November 2018).
[4] https://www.corning.com/emea/en/innovation/corning-emerging-innovations/advanced-flow-reactors/Corning-Advanced-Flowreactors-take-continuous-flow-process-production-from-labto-full-scale-manufacturing.html (accessed 10 November 2018).
[5] www.ltf-gmbh.com (accessed 10 November 2018).

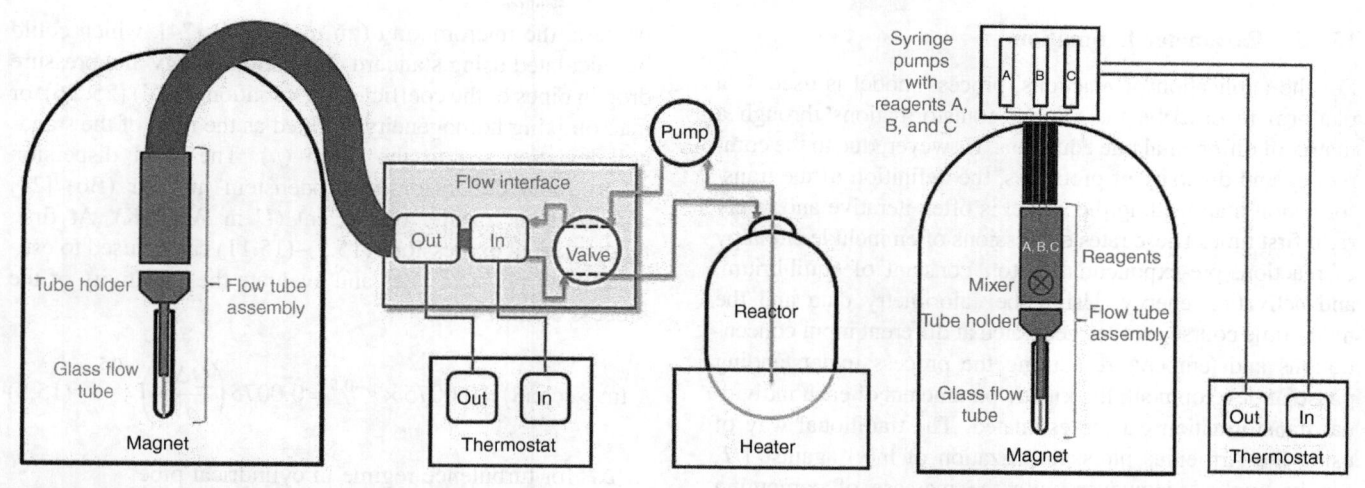

FIGURE 15.5 Schematic illustrating the application of online and in-line NMR spectrometer for process understanding. *Source*: Courtesy of Anna Codina, Bruker.

generated. The concentration of dissolved gas at the gas–liquid interface can be assumed to be equal to the gas solubility. In addition, as the width of the organic segment is significantly smaller than the inner diameter (ID) of the reactor, internal convection could be considered inexistent, and mass transfer is driven by diffusion only. Therefore, a simple film theory-based treatment of transport and reaction is applicable. Despite the nature of the multiphase system studied in this type of reactor, a tremendous acceleration of the mass transfer phenomenon is observed enabling opportunity to study intrinsic kinetics, which could not be achieved previously.

15.3.2.2 In-line, Online PAT

The pharmaceutical industry is already using a large array of process analytical technology (PAT) to measure in situ evolution of processes. Mid-IR, Raman, or UV instruments are now commonly spread across development and manufacturing. However, the type of probe used and their response time are not appropriate to continuous processing applications. New and fast analytical technologies have been developed in the recent years to support the understanding of continuous processes. Extremely fast reactions can be a challenge for their monitoring using traditional off-line method; therefore, generating intrinsic understanding of these reactions is limited. Suppliers invested time in the development of new PAT and flow cells, enabling the process understanding and development of continuous processes by collecting compositional information.[6,7]

Equally, NMR suppliers adapted their technology to respond to the new requirement of continuous process understanding. Complex and new design NMR flow cells and low-field NMR benchtop are now available.[8,9] The reworking done on the NMR spectrometer now allows minimum calibration and the use of non-deuterated solvents. Two modes of usage are currently possible (Figure 15.5), either by circulating reacting materials from an external reactor through the magnet for rapid spectral acquisition, thereby enabling online monitoring, or by in situ measurements where reagents are mixed directly inside the magnet. This enables measurement of the kinetics of fast reactions.

More standard technologies, traditionally used as off-line method are now transitioning to in-line or online PAT. With the support of suppliers, pharmaceutical research and development and academic groups are repurposing LC, MS, and GC instruments to online systems where major and low-level components could be quantified. Not only are those PATs of interest for process understanding, but they are also finding their way to pilot and manufacturing facilities to ensure compliance of the product, preventing long and costly traditional in-process control (IPC) analysis and batch defect.

Using some of the technology described above, either standard batch or mesoscale reactor, combined with new PAT technologies, molar time course profiles can easily and reliably be generated. Process understanding of complex and fast reactions that remained in the hands of specialized academic groups in the past can now be studied in the pharmaceutical laboratories.

---

[6] https://www.bruker.com/products/infrared-near-infrared-andraman-spectroscopy/ft-irnir-for-process.html (accessed 10 November 2018).

[7] https://www.mt.com/id/en/home/products/L1_AutochemProducts/ReactIR/flow-ir-chemis.html (accessed 10 November 2018).

[8] http://go.magritek.com/spinsolve-80-benchtop-nmr-brochure?utm_campaign=Adwords%20-%20Spinsolve%20Fam%20-%2080&utm_source=ppc&gclid=EAIaIQobChMI-qTdmMmY2-QIVRbobCh1Ddg-MeEAAYASAAEgJWbvD_BwE (accessed 10 November 2018).

[9] https://www.oxford-instruments.com/products/spectrometers/nuclear-magnetic-resonance-nmr/pulsar (accessed 10 November 2018).

15.3.3 Parameter Estimation

For the application of reactions, process model is used as a platform to describe the various transformations through a matrix of differential rate equations. However, due to the complexity and diversity of processes, the definition of the transformation map feeding the model is often iterative and rarely right first time. These rates expressions often include enthalpy of reaction, pre-exponential factor, constant of equilibrium, and activation energy. Using the calorimetry data and the molar time course profiles generated at different input concentrations and temperatures during the process understanding phase of development, the kinetic parameters of each individual transformation can be estimated. The traditional way of using the Arrhenius plots, linearization of $\ln(k)$ against $1/T$, can be applied. However, with the increase of computing power and emergence of simulation software, the estimation of the kinetic parameters can be simplified using the expressions (15.3 and 15.4) in dynamic kinetic models by performing the regression of all parameters simultaneously.

$$r = k_{0,r}\left(\frac{T}{T_{ref}}\right)^n e^{\left(-\frac{E_{a,r}}{R}\left(\frac{1}{T}-\frac{1}{T_{ref}}\right)\right)} \prod_{i=1}^{N} C_i^{b_{r,i}} \quad (15.3)$$

$$r = -\frac{d[A]}{dt} = k_{T_{ref}}[A][B] \quad (15.4)$$

Comparing the experimental data with the model predictions allows a constant refinement of the mechanistic proposal until acceptable agreement between the model prediction and the data (Figure 15.6).

When satisfactory prediction has finally been achieved, verification of the model using an additional set of experiments is usually performed. These experiments must be within the same ranges of conditions used for the parameter estimation as the model should not be used for extrapolation. The set of experiments used to verify the model is defined, using sensitivity analysis tools, as a function of the process sensitivity against the input parameters. It often happens that the verification experiments are grouped within a specific part of the process parameter ranges as the response is greater in this region (Figure 15.7).

When the verification is complete and successful, the model can be used within the range of conditions defined as a predictive tool to evaluate impact of process parameters to yield and process quality.

15.3.4 Equipment Performance

Having a detailed understanding of the equipment performance is critical for continuous processing as it allows to properly understand the process output in respect to the technology. Two physical rates are often seen important to the process: the micromixing (tm in second) [24], which could be calculated using standard dissipation energy and pressure drop in pipes or the coefficient of variation (CoV) [25, 26], or macromixing homogeneity, defined as the ratio of the standard deviation (σ) to the mean (μ). The axial dispersion referred the dimensionless Bodenstein number (Bo) [27] and the heat transfer coefficient (U in W/m$^2 \cdot$K). At first, mathematical expressions (15.5)–(15.11) can be used to estimate these physical rates and evaluate the sensitivity of the process against them:

$$\text{tm (second)} = 0.0075 \times \varepsilon^{-0.5} = 0.0075 \left(\frac{Q\Delta P}{\rho V}\right)^{-0.5} \quad (15.5)$$

ΔP for turbulence regime in cylindrical pipe:

$$\left(\frac{\Delta P}{L}\right) = fD\frac{\rho(v)^2}{2\;D} \quad (15.6)$$

ΔP for laminar regime in cylindrical pipe:

$$\left(\frac{\Delta P}{L}\right) = \frac{128}{\Pi} \cdot \frac{\mu Q}{D^4} \quad (15.7)$$

$$\text{CoV} = \frac{\sigma}{\mu} = \left(\frac{1-q/Q}{q/Q}\right)^{-0.5} \quad (15.8)$$

where q is the minor volumetric flow rate and Q the major volumetric flow rate:

$$\text{Bo} = \frac{VL}{D} = \text{Re} \cdot \text{Sc} \, D = Dv + \frac{V^2 d^2}{4\beta Dv} \quad (15.9)$$

$$\frac{1}{U} = \frac{1}{h_{in}} + \frac{1}{h_{out}} + \frac{1}{h_{wall}} + \frac{1}{h_{fin}} + \frac{1}{h_{fout}} \quad (15.10)$$

with $h_{fluid} = \text{Nu}\frac{Cp}{Dh}$ and $h_{wall} = \frac{1}{ri}\ln\left(\frac{ri}{ro}\right) \quad (15.11)$

For a low-energy release processes, an estimation of the heat transfer coefficient and wetted area (UA) based on the dimensionless number of Reynolds (Re), Nusselt (Nu), and Prandtl (Pr), reflecting the flow velocity, the physical properties of the process media, and the heat transfer fluid (HTF), the thickness, and conductivity of the material of construction can be estimated with reasonable accuracy. A similar approach with the mass transfer, e.g. macromixing and the axial dispersion, should be applied. Initial calculations based on the design of the reactor allow calculations of the blending

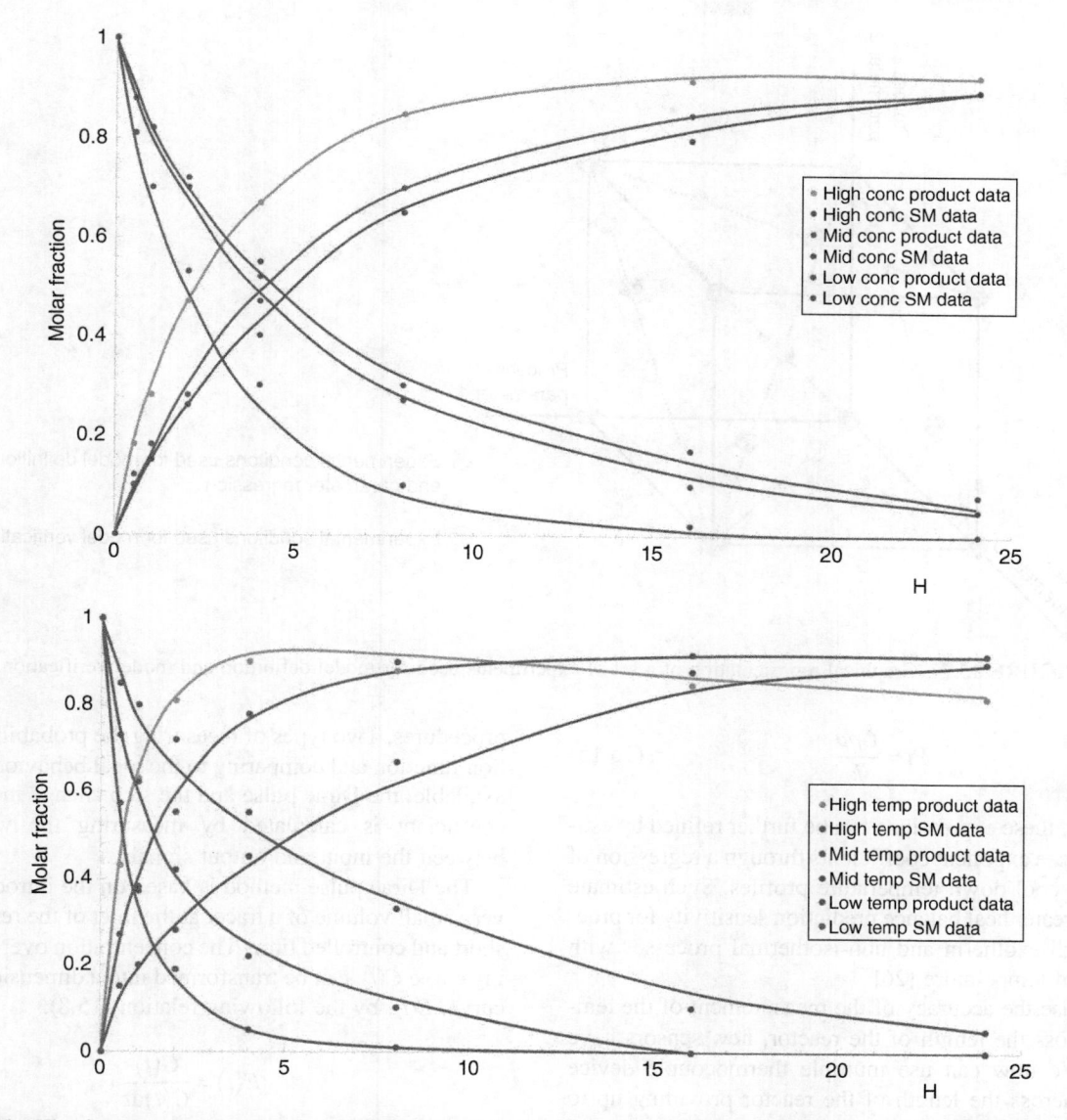

FIGURE 15.6 Kinetic parameter estimation at various temperatures and concentrations. Dots are experimental data as the trends are the predictions provided by the software after fitting of the pre-exponential factors and the activation energy.

time between two streams using the CoV or tm and the calculation of equivalent number of reactor using Bo mathematical expressions. If further accuracy is required as the sensitivity of the process is high in respect of any of these physical rates, they could be accurately measured using well-defined procedures.

A comprehensive definition of UA using the Sieder–Tate correlation can be conducted using Eqs. (15.11)–(15.15).

Nu for laminar flow:

$$Nu = C_l Re^\alpha Pr^\beta \left(\frac{D}{L}\right)^\sigma \quad (15.12)$$

where α, β, and σ could be estimated to 0.33 and C_l to 1.86

Nu for turbulent flow:

$$Nu = C_t Re^\alpha Pr^\beta \quad (15.13)$$

where α and β could be estimated, respectively, to 0.8 and 0.4 and C_t to 0.023:

$$Re = \frac{QDh}{vA} \quad (15.14)$$

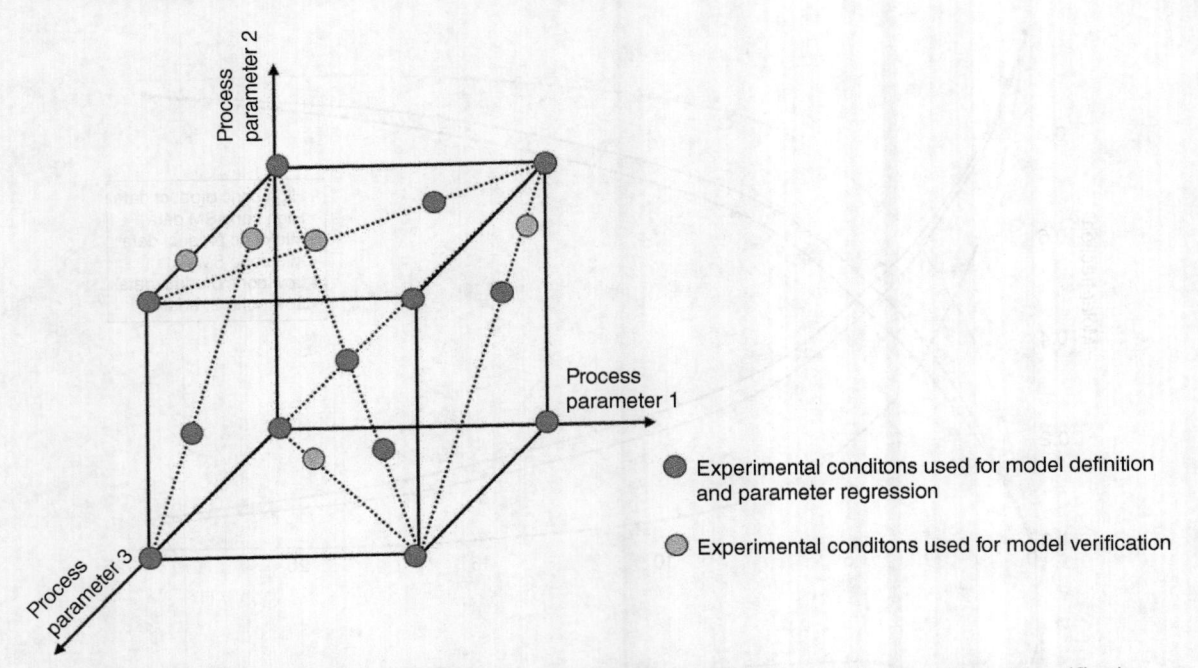

FIGURE 15.7 Graphical representation of a set of experiments used for model definition and model verification.

$$\Pr = \frac{Cp\mu}{k} \qquad (15.15)$$

However, these expressions can be further refined by estimation of these exponent coefficients through a regression of heat up and cool down temperature profiles. Such estimate will allow greater heat balance prediction sensitivity for process with high exotherm and non-isothermal processes with large delta of temperature [26].

To increase the accuracy of the measurement of the temperature across the length of the reactor, new sensors have emerged. We now can use multiple thermocouple device positioned across the length of the reactor providing up to one measurement every centimeter. However, more recently, fiber optics have been used for the same application, and a measurement every 20 nm is now achievable. The generation of such temperature profiles provides valuable and reliable information to regress the U value.

For meso-reactor, where direct measurement of the temperature is not feasible, a new innovative methodology has been tested with success. For see-through reactors, for example, procedures using quant dot where the dot emits a different color depending of their temperature have been implemented to characterize the U value of those highly efficient reactors [28]. For metal reactors, procedures using reactions with high activation energy can be used as the quality output is directly related to the temperature profile into the reactor and therefore the heat transfers. Among them, we can find a library of Diels–Alder reaction [12].

The Bodenstein number or axial dispersion coefficient can be easily measured using residence time distribution (RTD) procedures. Two types of measuring the probability distribution function and comparing to the ideal behavior are readily available, the Dirac pulse and the step change methods. The coefficient is calculated by measuring the variance (δ) between the input and output signals.

The Dirac pulse method is based on the introduction of a very small volume of a tracer at the inlet of the reactor over a short and controlled time. The concentration over time resulting curve $C(t)$, can be transformed into a dimensionless RTD curve, $E(t)$, by the following relation (15.8):

$$E(t) = \frac{C(t)}{C(t)\mathrm{d}t} \qquad (15.16)$$

The step method is based on the introduction of an abrupt change from 0 to C_0 of a tracer at the inlet of the reactor. The concentration of tracer at the outlet is measured and normalized to the concentration C_0 to obtain the nondimensional curve $F(t)$, which goes from 0 to 1 and then transformed into a dimensionless RTD curve $E(t)$ (Eq. 15.17):

$$F(t) = \frac{C(t)}{C_0} \quad E(t) = \frac{\mathrm{d}F(t)}{\mathrm{d}t} \qquad (15.17)$$

An ideal plug flow reactor (PFR), where each component enters and leaves in the exact same order, has a variance (δ) of zero, $E(t) = \delta(t-\tau)$. As the behavior of the reactor deviates from ideal PFR, the variance increases toward the behavior of a continuous stirred tank reactor (CSTR) system $E(t) = 1/\tau e(-t/\tau)$ with a variance (δ) of ∞. The variance can

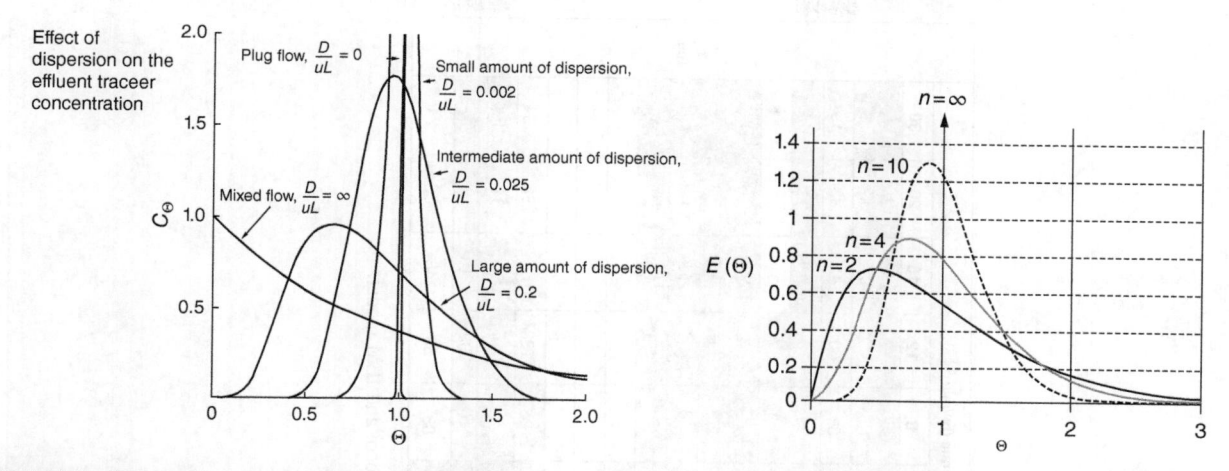

FIGURE 15.8 Effect of the concentration of a pulse based on the axial dispersion into a reactor. *Source*: From Fogler [29]. Reproduced with permission of John Wiley & Sons.

be equally referred as equivalence of reactor n, from $n = \infty$ for ideal PFR to $n = 1$ for ideal CSTR [29] (Figure 15.8).

The RTD of reactors can be used in association to kinetic models to evaluate the performance of a process during state of control and when disturbances occur. In addition, RTD is often used during the execution of the control strategy for traceability of materials.

Over the years, several methods have been developed to measure the micromixing time. Among them, two are well established, Bourne IV and Villermaux–Dushman [30, 31]. Villermaux–Dushman can be seen as the method of choice as it provides a protocol associated with the mean (IEM) mixing model to quantify the micromixing time based on the selectivity between two reactions, an estimated instantaneous reaction and a really fast reaction with their rates constant well defined. The concentration ratio between the two products could directly be linked with the mixing time of the equipment and the mixer (Figure 15.9).

15.3.5 Process Simulation

Statistical models such as design of experiments are commonly used in the pharmaceutical industry: first, during process development to support process knowledge by mapping out yield and quality of the product as a function of process parameters input and then, during the registration of the DS and the DP to demonstrate the control of the process during manufacture. More recently, supported by the regulatory agencies, the pharmaceutical industry has been seeking not only for process knowledge but also for process understanding to feed into the quality by design (QbD) approach. Originally developed for the oil and gas and the bulk chemical industry, various commercial simulation companies have adjusted their products to be more aligned with the pharmaceutical industry requirement and to respond to the change of strategy. Continuous processing in the pharmaceutical industry covers reaction and workup; however, the most significant part remains chemical transformations. Therefore, emphasis was made to the kinetic aspect of the model to respond to the complexity of the transformations. In addition to the kinetic and thermodynamic expressions, the physical rates of heat and mass transfer were added to allow prediction of complex nonuniform behaviors. As described in Section 15.3.4, following the characterization of the equipment, heat transfer, mixing rate, and axial dispersion can be added to a kinetic model. The development and industrialization of continuous processes for pharmaceutical application is a difficult task due to the intrinsic interaction between the process rates, of often fast and complex reactions, and the physical rates associated with the equipment. These models play a key role in a successful continuous process development and implementation into manufacturing facilities.

15.3.5.1 Process Optimization and Design Space Definition
Following the definition of the model, including the verified kinetic and thermodynamic expressions of the process and the relevant expression of the physical rates of heat and mass balance, the model can be used to execute a comprehensive optimization of the process. A list of process parameters must be selected with initial guessed values and boundaries. Usually, the boundaries are the one used for the verification of the model. Then, an objective function must be defined, which is usually relative to yield, process performance, and product quality. However, process robustness and process CoGs can be defined as objectives too. The objective function is usually defined either against a set target, for example, a yield greater than 95% w/w and impurities lower than 0.1% w/w or against a maximize/minimize function, for example, minimize impurity level and CoGs. To achieve the optimization, a large list of algorithms is

Micromixer supplier		Mixing principle Mixer information
Cascade 06, 15 **0216-3** Ehrfeld mikrotechnik BTS	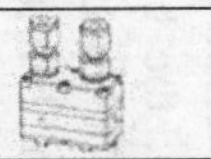	**Split and recombine** Internal volume: 103 μl (06) and 173 μl (15) Channel width: 0.6 mm (06) and 1.5 mm (15)
Lonza TG large **1701-2380** Ehrfeld mikrotechnik BTS	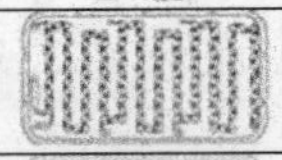	**Flow obstacles** Internal volume: 1.2 ml Mixer nominal width :05 mm
Lonza multiinjection SZ **1701-1642** Ehrfeld mikrotechnik BTS	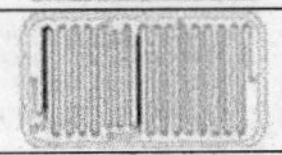	**Flow obstacles** Internal volume: 400 μl Nominal width: 200–600 μm
Slit plate LH 2 **0113-4** Ehrfeld mikrotechnik BTS	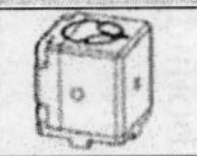	**Multilamination** Internal volume: 40 μl Narrowest diameter: 25–100 μm depending on the mixing and aperture plate
Micromixer **2101411** Syrris asia	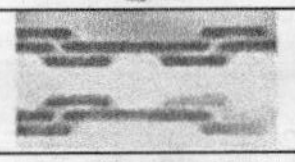	**Split and recombine** Internal volume: 6.25 μl Mixing channel dimension:50 × 125 μm
62.5 μl and 250 μl microreactors **2100141, 2100143** Syrris asia	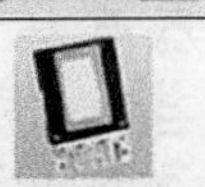	**Flow obstacles** Internal volume: 62.5 μl and 250 μl Mixing channel dimension: 85 × 220 μm (62.5 μl): 250 × 300 μm (250 μl)
T-mixer **SS-200-3, SS-100-3** Swagelok	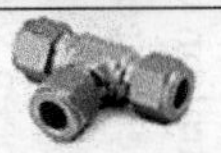	**Simple contacting** 1/8" T-mixer 1/6" T-mixer Narrowest diameter : 2.3 mm (SS-200-3) ; 1.3 mm (SS-100-3)

		Total flow rate (ml/min)[b]											
		0.1	**0.2**	**0.5**	**1**	**2**	**4**	**8**	**12**	**16**	**20**	**30**	
Ehrfeld cascade 06			65 ± 5	41 ± 5	37 ± 4	19 ± 3	7.8 ± 1.4	4.5 ± 1.5	3.4 ± 1.0	2.4 ± 0.6	1.8 ± 0.2	0.7 ± 0.1	t_m (ms)
Ehrfeld cascade 15					177	133	87	81	50	42	33	19	
Ehrfeld lonza TG large							94	35	19	7.4	5.5	3.1	
Ehrfeld lonza multiinjection SZ			82	52	40	11	3.6	1.1	0.7				
Ehrfeld slit plate	**25/25 μm mixing; 25 μm aperture**			117	98	91	5.1	0.9	0.6	0.5	0.3	0.03	
	25/25 μm mixing; 50 μm aperture				114	70	26	2.5	1.0	0.6	0.4	0.2	
	100/25 μm mixing; 25 μm aperture					30	4.6	1.2	0.9	1.4	1.3	0.6	
	100/25 μm mixing; 50 μm aperture				121	98	29	6.9	1.9	1.2	0.8	0.4	
Syrris asia micromixer		45	38	25	10	4.9	3.4						
Syrris asia 62.5 μl microreactor		33	22	5.4									
Syrris asia 250 μl microreactor		45	33	25	25	12	15						
1/8" T-piece	**1/16" tubing**				565	448	211	128	128	160	96	34	
	1/8" tubing				970	896	766	744	645	670	319	102	
1/16" T-piece	**1/16" tubing**				1220	1100	645	216	232	201	191	164	
	1/8" tubing				1910	1300	941	88	59	43	33	24	

FIGURE 15.9 Table of characterized meso-reactors and mixers using the Villermaux–Dushman protocol and IEM model.

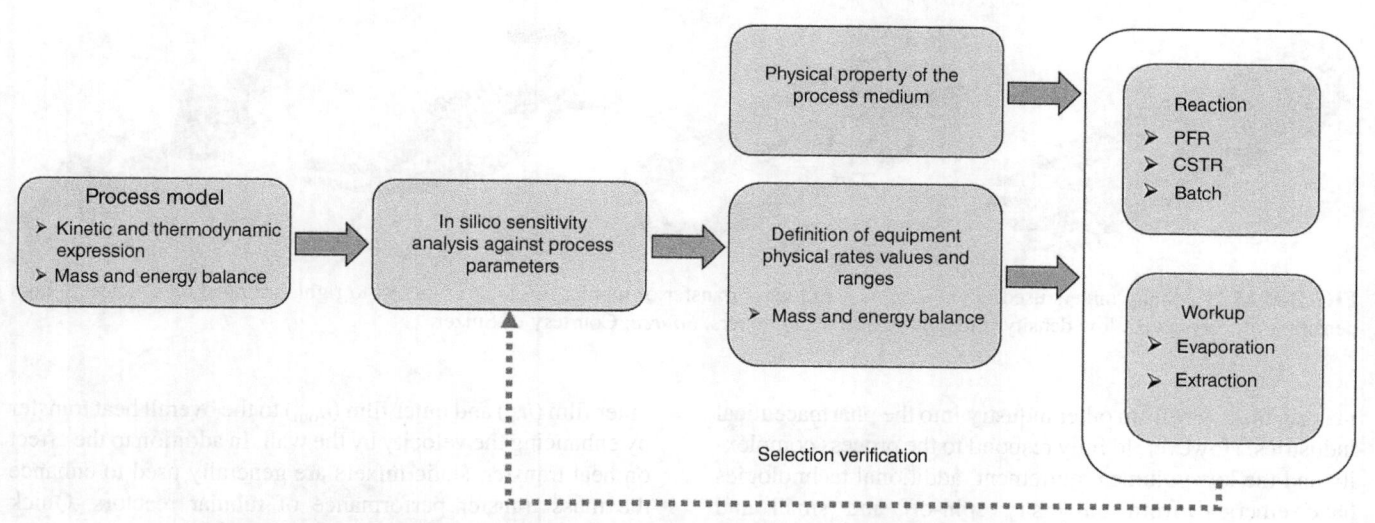

FIGURE 15.10 Workflow for equipment selection using process modeling and sensitivity analysis.

available; the most commonly used are simplex linear algorithms [32] or least square methods [33].

Simplex algorithm, for linear programing, is operating by defining a vector in a geometrical figure. It proceeds by performing successive pivot operations, the least optimal point is discarded, and a new vector is defined. It usually operates by following one of the four following actions: reflection, reflection with expansion, contraction, and multiple contractions.

The least square method minimizes the squared residuals between the data generated and the fitted values provided by a model for every single equation. It exists as linear and nonlinear. For the linear least square method, the model includes a linear combination of the parameters, and it has a closed-form solution. For the nonlinear method, the problem is solved by iterative refinement. At each iteration, the system is approximated by a linear solution. It is important to notice that in the linear least square method, the solution is unique. However, in the nonlinear least square method, there may be multiple minima in the sum of squares (SSQ).

Process optimization is not limited to the definition of the operating set point where the optimum function is converging to a global minimum. Process optimization can equally include the definition of the design space capturing ranges of conditions where the objective function is still valid. This design space represents the boundaries where the process fails or passes the specification defined. In addition, depending on the definition of the objective function, the process design space can include risk and robustness of the process for manufacture.

15.3.5.2 Reactor Definition and Selection The process development phase supported by a deep process understanding of the kinetic and thermodynamic processes can be linked to the identification of manufacturing technology. Ground rules are applicable for the selection of the equipment; however, process simulation with the opportunity to solve complex problem becomes an efficient design and selection tool. Sensitivity studies of the process output against the process parameters can be performed. The impact of each process parameter can be analyzed independently or globally within verified range of the model. Process simulation allows a more complex sensitivity analysis coupling multiple process parameters together.

From this in silico exercise, the impact of the input parameters to the process can be predicted. The required range of physical rates of heat and mass transfers are calculated to achieve the yield, quality, CoGs, and robustness predefined for the process. The information generated with the support of the process model can be incorporated in a reactor user requirement specification (URS) or reactor datasheet to be shared with equipment suppliers. The purpose of this exercise is to identify the optimum technology to run the process. From the nature of the chemistry studied, flow equipment is most likely to be selected. However, in some cases, batch or semi-batch type of process still provides the desired yield and quality output with the addition of flexibility. Using process simulation as schematized in Figure 15.10 allows to only consider the process benefits.

15.4 CONTINUOUS TECHNOLOGY

Continuous operation in the pharmaceutical industry is often focusing on reaction as the value is mostly seen in the chemical transformation rather than in the process and unit operation enhancement. However, continuous technology covers a wide array of unit operations demonstrating benefit across reaction, workup, and isolation. Tubular reactor, continuous extraction, and continuous evaporation technologies are now

FIGURE 15.11 Static mixers used to enhance heat and mass transfer of tubular reactors. From left to right, arranged by efficiency, high density grid, corrugated, low density grid, and helical static mixers. *Source*: Courtesy of Sulzer.

making their way from other industry into the pharmaceutical industries. However, to fully respond to the process complexity and the low volume requirement, additional technologies have emerged from university spin-off and small and medium enterprises (SMEs). Pharmaceutical continuous technology is the combination of the well-established and the new state-of-the-art technologies.

15.4.1 Chemical Reactors

15.4.1.1 Plug Flow Reactors and Static Mixers PFRs are the most common type of reactors used in pharmaceutical industry as a replacement of batch technology. The equivalence of batch reaction time is residence time under ideal plug flow conditions simplifying their implementation. Their enhanced performance in heat and mass transfer is the perfect response of the initial intent, enabling fast and temperature-sensitive chemical transformations. PFR is a versatile technology; as demonstrated across various applications in the oil and gas and the polymer industries, when operating at high flow velocity and high Reynolds numbers, they are efficient in adding or removing heat to a process stream due to their high heat transfer coefficient and surface to volume ratio. They are very efficient in mixing miscible fluid streams quickly and homogeneously in a compact volume both for laminar or turbulent flow regimes and for streams with similar or widely different viscosities. To maximize the performance of PFRs, tubular reactors often integrate static mixers.[10–13] Multiple types of insert are currently available from high to low mixing capacity (Figure 15.11).

Such inserts can be added on the shell and process media side as they minimize the impact of the contribution of the inner film (h_{in}) and outer film (h_{out}) to the overall heat transfer by enhancing the velocity by the wall. In addition to the effect on heat transfer, static mixers are generally used to enhance the mass transfer performance of tubular reactors. Quick micromixing time or blending time is usually achieved by using dense elements with high pressure drop associated. The efficiency often relies on the capacity to slip and recombine layers of the process stream until full homogenization of the process flow is achieved (Figure 15.12).

In addition, static mixers have the effect to minimize the RTD in the reactor, with Bodenstein numbers of up to three L/D (L and D *b*eing length and diameter of the reactor tube, respectively).

Static mixer can also be used for multiphase systems to generate droplets of one immiscible liquid stream into another or gas bubble in a liquid stream (Figure 15.13). In this case, the mixers are usually operated in the mist flow regime, where films form on the surfaces of the static mixer and at the same time droplets are present in the gas flow. Such installations are often used to enhance mass transfer of a dispersed liquid or gas in a liquid stream. Such biphasic application could equally be used for cleaning procedure, scrubbing residuals out of an equipment, or for the evaporation of a liquid in a hot gas stream.

15.4.1.2 Plate Reactors and Meso Channels Reactors The enhancement of heat transfer has been achieved in flow processing by the significant increase of surface to volume ratio of the reactors. As described in Section 15.4.1.1, PFR can provide significantly greater heat transfer coefficient than standard batch vessels. However, alternative designs to standard PFRs can even further improve the heat transfer. Among these designs, we can find plate or meso channel reactors (Figure 15.14). Their surface to volume ratio can be increased by two orders of magnitude, resulting in examples where $U > 3000$ W/m^2·K. can be achieved. In addition, their small ID maximizes the efficiency of micromixing by enabling a high flow velocity.

However, the design of these reactors provides some limitations to their use. Due to the small channel cross sections, the pressure drop reaches a high value even at low flow rate,

---

[10] https://www.sulzer.com/en/products/static-mixers (accessed 10 November 2018).

[11] https://www.primix.com/en/products.html (accessed 10 November 2018).

[12] http://www.stamixco.com/Static-Mixers-Overview-Injection-Molding-Extrusion-Melt-Blender-X-Grid-Plastic-Disposable-Double-Roof-Disk (accessed 10 November 2018).

[13] http://www.fluitec.ch/static-mixing.html?gclid=EAIaIQobCh-MI_fyw-griw2QIVrp3tCh3u8gK0EAAYAiAAEgI0PfD_BwE (accessed 10 November 2018).

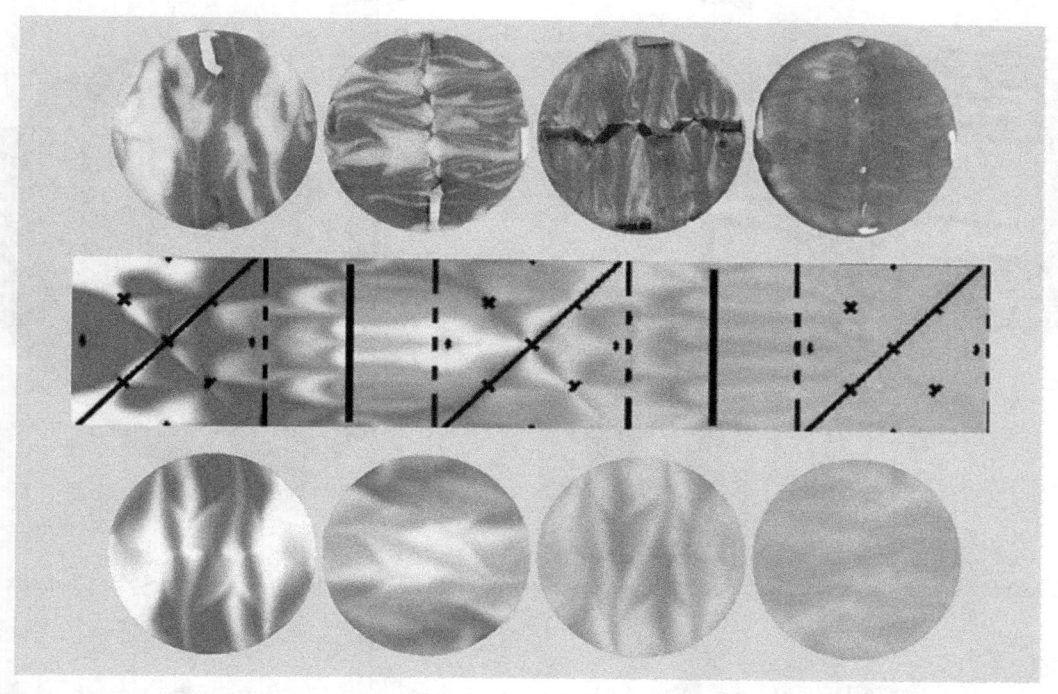

FIGURE 15.12 Cross section of an intensive high density grid static mixer during mixing of two streams. *Source*: Courtesy of Fluitec.

FIGURE 15.13 Generation of droplet in high intensity grid static mixer. *Source*: Courtesy of Sulzer.

lowering their potential productivity rate. In addition, their capability of handling solid remains limited and the reactor becomes subject to fouling.

Tubular and plate reactors have been, so far, the preferred technology of the pharmaceutical industry, as the high plug flow behaviors minimize the axial dispersion and allow a tight control of the genealogy of the product, minimizing the risk of product loss associated with deviations and contaminations.

15.4.1.3 Continuous Stirred Tank Reactors

Continuous stirred tank reactors (CSTRs) are often underutilized in the pharmaceutical industry to the favor of shell and tube reactors and high plug flow behavior reactors. However, the simplicity and versatility of these reactors make them extremely attractive to continuous operation. Their design is similar to standard batch reactors with a height to diameter ratio (L/D) around 1 and an elliptical bottom. The flow through the reactor is provided by constant feed using pumping or pressure, and the flow out is enabled by controlled pumping or valve systems connected to level controllers or in some cases by a simple gravity connection.

The mixing is provided by the mechanical agitation of an impeller, and the heat transfer is delivered by standard outer wall jacket. CSTRs rarely exceed a heat transfer coefficient of 300–400 $W/m^2 \cdot K$ and their macromixing time is often, depending on size and design, at least one order of magnitude lower than the blending time achieved between miscible streams in PFRs at high flow velocity. Despite these obvious limitations, CSTRs provide some specific advantages. First, as the mixing is decoupled to the flow velocity, good mixing efficiency can be achieved at even low flow rate. This mixing efficiency can be even further enhanced by the implementation of the baffle system, by tailoring the impeller type to the physical property[14] (Figure 15.15) or by the implementation of jet mixing through recirculation.

In contrary to most of the overflow technology, CSTRs have the ability to handle solids up to 10–15 w/w mass fraction. As chemical transformations in the pharmaceutical industry often contain solid base reagents or precipitate

[14] http://www.ekato.com/en/products/agitator-components/impellers (accessed 10 November 2018).

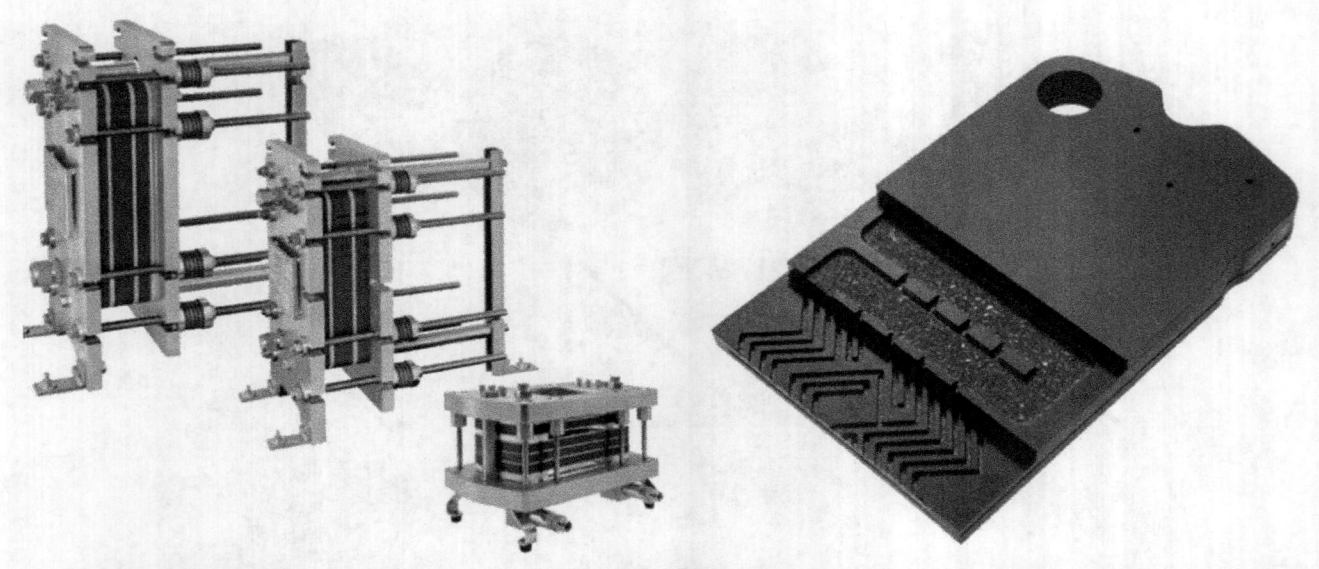

FIGURE 15.14 Schematic of plate reactor design and assembly of multiple elements. *Source*: Courtesy of Chemtrix catalog.

FIGURE 15.15 Type of impellers providing different mixing properties and flow directions from single to multiphase processes and different viscosities. *Source*: KMPS.

(product or by-product), CSTRs offer an alternative enabling continuous operation where other technology cannot be applied.

CSTRs can be operated as high dilution system. Therefore, the concentration profile achieved at state of control is significantly different to PFRs. This could offer a significant advantage to the control of side product formations and second-order decomposition of starting materials.

15.4.1.4 Packed Bed and Trickle Bed Reactors Continuous packed bed and trickle bed reactors have been the expertise of the bulk chemical and petrochemical industry for decades. In these industries, the implementation of such technology targeted improvement of the process efficiency, resulting to substantial reduction of process cost. Currently, pharmaceutical portfolio analysis reveals that up to 20% of the chemical transformation to the DS contain catalytic gas/liquid transformations. Therefore, pharmaceutical companies are now devoting significant efforts, shifting traditional batch solid–gas–liquid operation to flow processes. Packed bed and trickle bed reactors are the key technology to unleash these cost benefits. Through decades of development, the oil and gas industry generated a significant amount of knowledge for the development of three-phase reactions, in particular, the hydrosulfurization or hydrocracking of heavy oils and the hydrotreating of lubricating oils. However, the tonnage of pharmaceutical processes is significantly lower, and the variety of transformation on which three-phase reactions are applied is knowingly more complex and subject of significant number of by-product formations. For these reasons, not only a transfer of the knowledge is required, but also new sized equipment associated with new type of catalysts must be developed. For existing suspended catalytic transformations in batch reactor, over hundreds of different catalysts are referenced in catalogues from various suppliers to address the high requirement of selectivity of the pharmaceutical chemical transformations.

Despite the significant change required for the catalyst, the reactor fundamentally remains similar. This three-phase reactor flows downward gas and liquid concurrently over a fixed bed of catalyst. Liquid trickles down, while gas phase is continuous. In a trickle bed reactor, various flow regimes can be

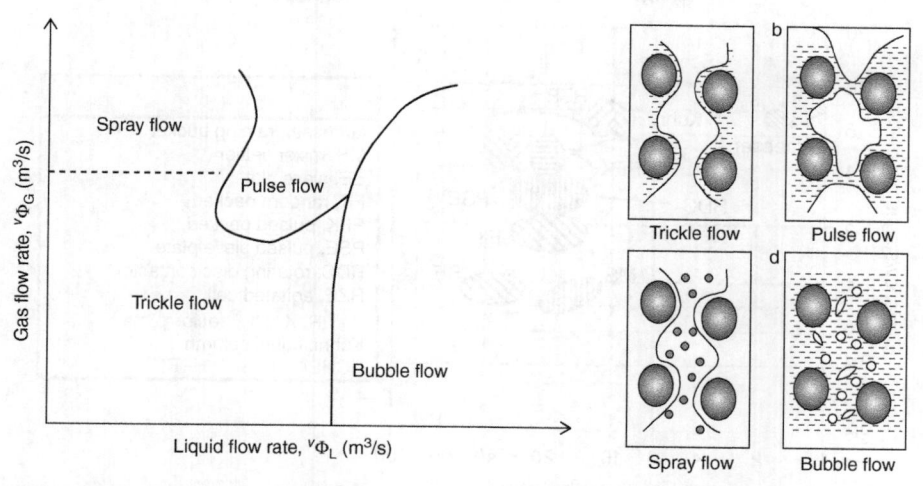

FIGURE 15.16 Schematic of the various flow regimes observed in trickle bed reactor due to the gas and liquid velocity. *Source*: Reprinted from Nguyen et al. [34] Reproduced with permission of Elsevier.

observed, from trickle flow (low interaction) to pulse flow (high interaction), spray flow, and bubble flow regimes (Figure 15.16). The effect of bed geometry, flow velocity, and fluid phase properties are the key parameters impacting the reactor regime.

Despite the multiple advantages demonstrated across processes in over industries, trickle bed reactors have complex fluid dynamic and therefore difficult to industrialize. Nonideal bed packing and nonuniform particle size distribution could lead to large deviation in flow behaviors, and bypassing and inefficient contacting can be observed. Despite the low shear applied by the flow, catalytic bed can be eroded, leading toward degradation of the bed behaviors and change of pressure drop in the system over time. Therefore, to enable efficient operation, the analysis and design of the multiphase reactor become a critical part of the process development. It is recommended to follow a three-phase strategy [35] for the selection of the type of multiphase reactor. The first level focuses on the catalyst design such as activity, shape, particle size distribution, and strength. The second level is about the phase injection and dispersion onto the catalytic bed. The final level is the understanding of the hydrodynamic of the multiphase flowing onto the bed.

15.4.1.5 Extraction Columns with Trays and Packings

Static extraction units are generally used for simple extraction applications that do not have significant change in the volumetric flow rate for the carrier feed stream and where there is little change in the fluid densities, viscosities, and interfacial surface tension. Several static extractors are available with different relative capacities and efficiencies. Spray columns are the simplest and oldest extractors. Dispersed-phase droplets simply rise (or fall) through the continuous phase in an empty column. Therefore, these units are susceptible to axial mixing, which generally prevents them from achieving more than one nominal theoretical stage of separation.

Structured or random packing can be added to spray columns to reduce axial mixing. Movement of the droplets across the packing surfaces induces turbulence inside and outside of the droplets, which accelerates the mass transfer. Structured packing surface is perforated and usually supplied with smooth surface. The periphery of the packing elements is equipped with a specialized wiper band to reduce wall effects. The corrugated sheets of the packing element guide the direction of flow within each layer. Some aspects of the structured packing material and wetting by the extraction fluids need to be considered. In general, the metal packing surface is wetted by the aqueous continuous phase, and plastic packings are wetted by the organic continuous phase. The dispersed droplets coalesce together when they reach the next sieve tray deck, and droplets form as the dispersed phase flows through the holes.

Pulsation can be added to static extraction units to enhance mass transfer. Such columns are equipped with static internals like structured packing or dual-flow sieve trays. Bellows or a pulsation pump provides oscillating pulses to the continuous phase. This improves the mass transfer efficiency but reduces the hydrodynamic capacity of the column. The main drawbacks of pulsation are high costs and energy consumption.

15.4.1.6 Agitated Extraction Columns

In the range of extraction columns with energy input, two main principles apply, pulsation and agitation. An overview about agitated extraction columns can be found in various engineering books [36–38]. The basics of this technology are still the same as in the rotating disc contactor (RDC). The column is vertically divided into compartments, and each compartment has a mixing element mounted on a common shaft.

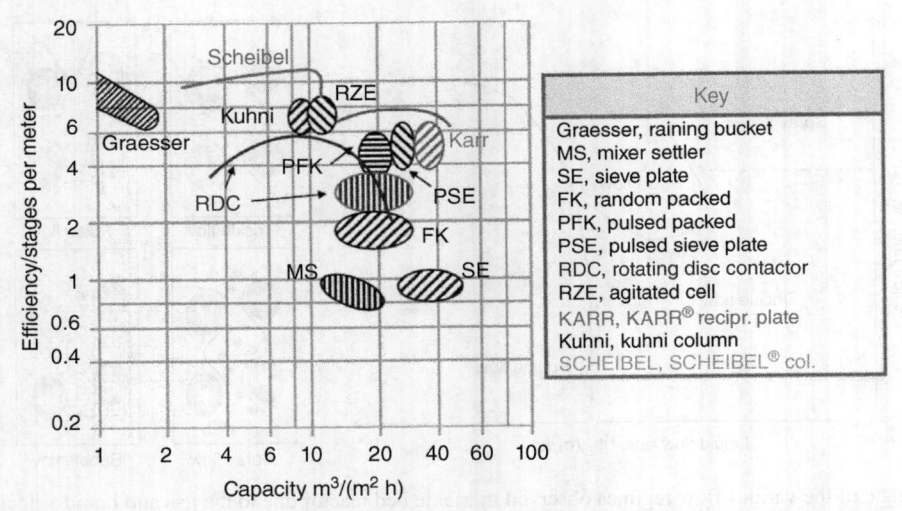

FIGURE 15.17 Examples of different types of extraction column and their efficiencies. *Source*: Courtesy of KMPS.

The mixing element was improved starting from the simple disc in a RDC to paddle and turbine agitators as applied in the Scheibel and Kuhni as shown in Figure 15.17. The aim of all developments is to improve the separation performance of the column and increase the throughput. Mainly this is achieved by improving the agitation in the column compartments to intensify the phase contact and to reduce the axial back mixing with optimized partition plates between the individual compartments.

The main benefit of the agitated columns is the wider flexibility and high separation performance as shown by Kumar and Hartland [16] compared with other column technologies. The agitated column type ECR covers the entire operating of RDC, simple spray columns, and pulsed sieve plate columns while providing a significant higher separation performance at high throughput. These columns are also well known for adapting the internals to cope with high mass transfer and changing physical properties over the column height.

15.4.1.7 Annular Centrifugal Extractors An alternative to gravity separation is centrifugal extractor (Figure 15.18). The technology was initially used to enhance the separation of hydrocarbon and water, solvent recovery, and waste treatment in the nuclear industry. Compared with the gravity separator, centrifugal extractors have a really short contacting time and residence time preventing degradation.

The two immiscible liquids are fed to the extractor and are rapidly mixed in the annular space between the spinning rotor and stationary housing. The mixed phases are directed toward the center of the rotor by radial vanes in the housing base. As the liquids enter the central opening of the rotor, the mixed phases are rapidly accelerated to rotor speed and separated based on their density difference. Annular centrifugal contactors are relatively low revolutions-per-minute (rpm), moderate-gravity-enhancing (100–2000 G) machines and can therefore be powered by a direct drive, variable speed motor. The effectiveness of a centrifugal separation can be easily described as proportional to the product of the force exerted in multiples of gravity (g) and the residence time in seconds or g-seconds. Achieving a particular g-seconds value in a liquid–liquid centrifuge can be obtained in two ways: increasing the multiples of gravity or increasing the residence time.

This technology has seen its number increasing in pharmaceutical industry as it provides a solution for implementation of continuous extraction with a linear and easy path from laboratory to large-scale facilities.

15.5 PROCESS DESIGN METHODOLOGY FOR ORGANOMETALLIC CHEMISTRY IN CONTINUOUS: FROM R&D TO MANUFACTURE

15.5.1 Introduction

Organometallic reagents, such as Grignard's and organolithium compounds, are frequently used within the pharmaceutical industry. Process design associated with organometallic chemistry is often governed by traditional batch controls of low temperatures and long hold durations to accommodate the poor heat transfer and limit process degradation [39]. Process development and industrialization of challenging organometallic chemistries in flow is greatly relevant as it is part of a large number of pharmaceutical assets.

This case study summarizes the process methodology described in the sections above to continuous process development of the synthesis of 3-borono-5-fluoro-4-methylbenzoic acid (compound A; Figure 15.19), a proposed registered starting material for the synthesis of Losmapimod, a p38 kinase candidate.

DEVELOPMENT AND APPLICATION OF CONTINUOUS PROCESSES FOR THE INTERMEDIATES

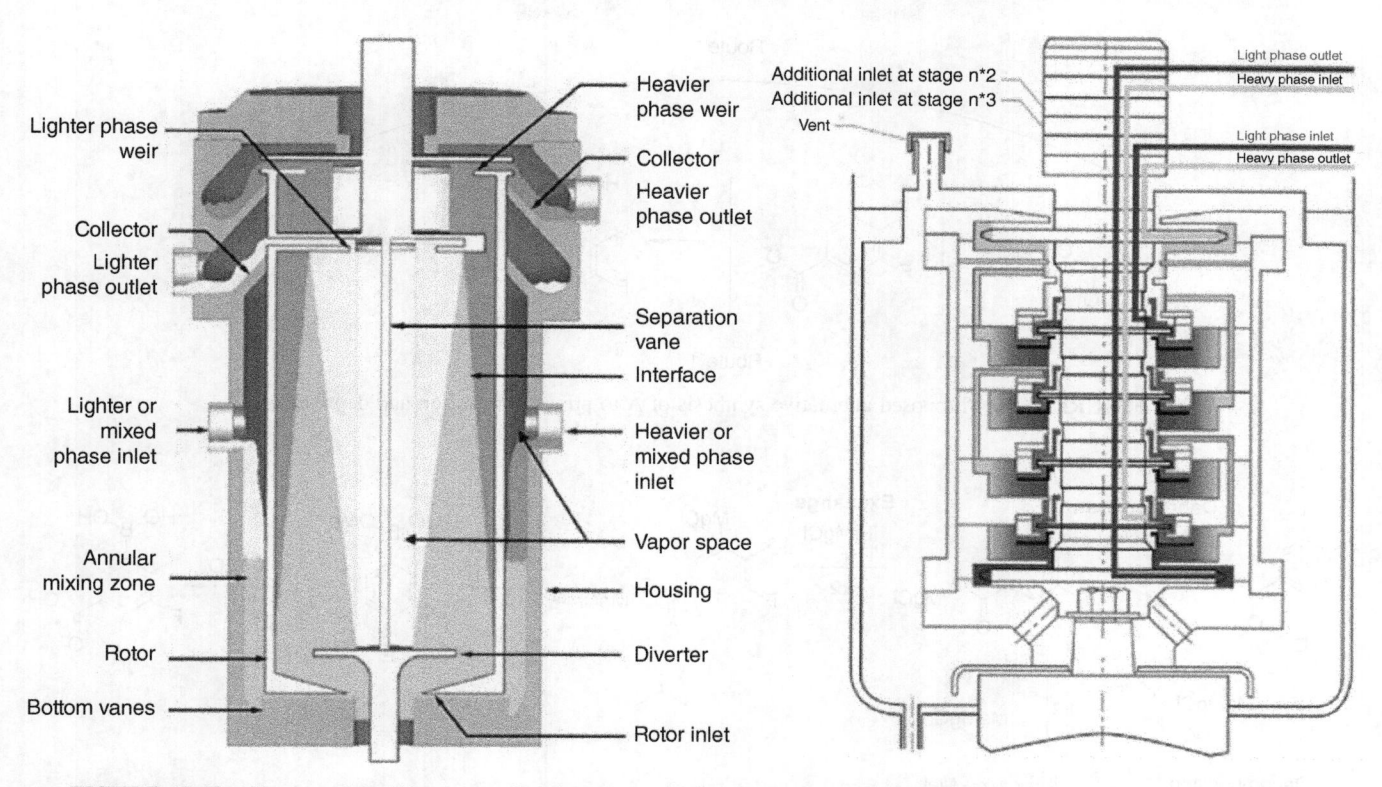

FIGURE 15.18 Single and multistage centrifugal extractor with design and flow routes. *Source*: Courtesy of Rousselet Robatel.

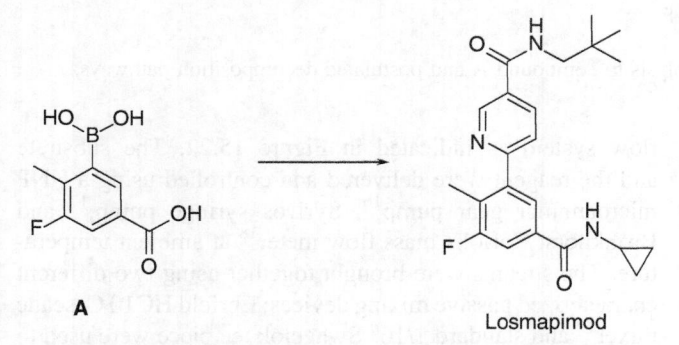

FIGURE 15.19 Proposed starting material (A) in the formation of Losmapimod.

The design approach focuses on fundamental mechanistic understanding and kinetics of main reactions and degradation pathways. Combined with calorimetric data, this allows the mass and energy balance of the process to be simulated in silico.

Compound A had initially been prepared from 3-fluoro-5-iodo-4-methylbenzoic acid (compound B; Figure 15.20) by protection as the methyl ester, conversion to the Grignard reagent, quenching with triisopropyl borate followed finally by hydrolysis of the methyl ester. A transformation directly from the carboxylic acid B, with in situ protection by deprotonation, had been proposed following the work of P. Knochel [40, 41].

It was shown that methyl magnesium chloride showed good selectivity for deprotonation of compound B over halogen–magnesium exchange. Excess methyl magnesium chloride does undergo the exchange reaction to generate the desired aryl Grignard (compound D). Unfortunately the methyl iodide side product was found to quench the intermediate generating 3-fluoro-4,5-dimethylbenzoic acid (compound F). This degradation reaction of the aryl Grignard was not observed when isopropyl magnesium chloride was used for the exchange reaction, giving 2-iodo-propane as the side product. Therefore, a sequential deprotonation with methyl magnesium chloride then halogen magnesium exchange with isopropyl magnesium chloride was required to generate Grignard reagent D, which could be quenched with trimethyl borate (introduced for improved atom efficiency) giving compound A after hydrolytic workup. The intermediates are unstable and moisture sensitive and decompose over time with strong temperature dependence. This sequential process is dependent on precise reagent charges and efficient mixing, making it an ideal candidate for flow operation. The intent was to enable tight control of the critical process parameters by utilizing flow chemistry to the organometallic transformation. The borate quench, aqueous workup, and final isolation were completed as batch operations.

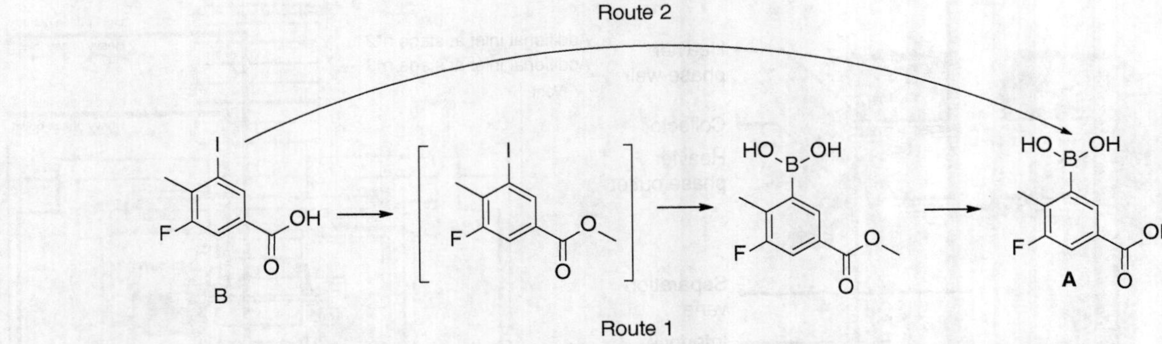

FIGURE 15.20 Proposed alternative synthesis of A to prevent protection and deprotection.

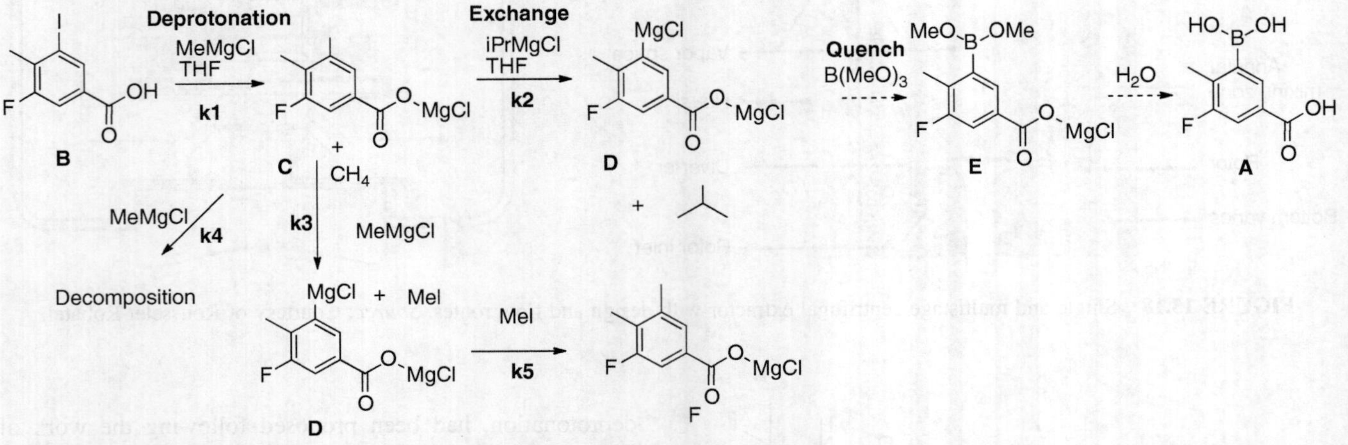

FIGURE 15.21 Deprotonation metalation borylation and hydrolysis to compound A and postulated decomposition pathways.

15.5.1.1 Methodology of Development

To enhance process understanding, a detailed study of the reactions was undertaken. The objectives of the study was to measure reaction calorimetry, estimate kinetic parameters for both the main process reactions, deprotonation and exchange, and degradation pathways as indicated in Figure 15.21 and to understand the impact of physical rate effects to yield and quality. Experimentation was completed in the most appropriate technology as discussed below. The equipment selected for the study were fully characterized (mass and heat transfer) to ensure accurate interpretation of the data.

To generate the reaction process understanding, two systems were used in response of the dynamic of the various reactions. The degradation kinetics, being slow enough, were monitored using small automated batch reactors[15] to evaluate impurity formation against process parameters and time. The main reactions were studied at steady state using a bespoke flow system as indicated in Figure 15.22. The substrate and the reagent were delivered and controlled using a HNP microannular gear pump[16], SyrDos syringe pump,[17] and Bronkhorst Coriolis mass flow meter[18] at ambient temperature. The streams were brought together using two different characterized passive mixing devices; Ehrfeld HC10 Cascade mixer[19] and standard 1/16″ Swagelok tee piece were used to evaluate the impact of mixing. The output mixture was then passed through variable lengths of polytetrafluoroethylene (PTFE) pipe with an internal diameter of 1/32″. Length of the pipe was used to adjust system residence time to 0.4, 0.3, 0.2, and 0.14 second. The mixture was then analyzed

---

[15] https://www.mt.com/gb/en/home/products/L1_AutochemProducts/Chemical-Synthesis-and-Process-Development-Lab-Reactors/Synthesis-Reactor-Systems.html (accessed 10 November 2018).

[16] https://www.hnp-mikrosysteme.de/en/products/micro-annulargear-pumps/high-performance-pump-series.html (accessed 10 November 2018).

[17] http://www.hitec-zang.de/en/laboratory-devices/syringe-doser.html (accessed 10 November 2018).

[18] https://www.bronkhorst.co.uk/en/products/coriolis_mass_flow_meters_-mass_flow_controllers_for_liquids_and_gases (accessed 10 November 2018).

[19] https://ehrfeld.com/en/products/mmrs.html (accessed 10 November 2018).

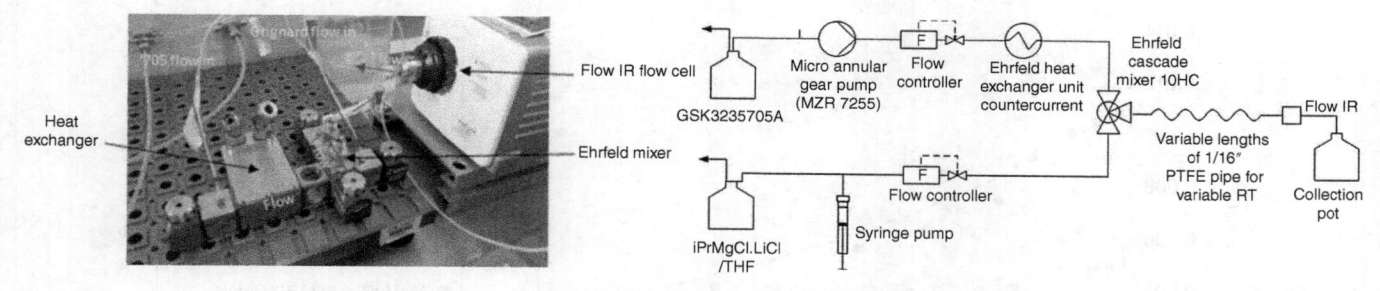

FIGURE 15.22 Photograph of fast kinetics rig and equipment line diagram for fast kinetics rig.

TABLE 15.1 Description of Experiences Used to Generate the Process Understanding and Regressing the Kinetic Parameters

Reaction	SM Concentration (M)	MeI Concentration (M)	MeMgCl Concentration (M)	Temperature (°C)
1	0.15	0.165	0.195	30
2	0.15	0.165	0.195	30
3	0.15	0.165	0.195	30
4	0.15	0.165	0.195	40
5	0.15	0.165	0.195	50
6	0.15	0.135	0.195	50
7	0.15	0.225	0.195	50
8	0.15	0.165	0.1575	50
9	0.15	0.165	0.1575	20
10	0.15	0.165	0.225	50
11	0.15	0.165	0.225	20

in-line using a Mettler Toledo flow IR[20], and additional samples were collected for off-line HPLC analysis.

To keep the methane generated during the deprotonation in solution, the pressure in the system was controlled with a back pressure Equilibar membrane device at 7 bar. This value was calculated using Henry coefficient for pure solvent with an extra precautionary allowance factored in to compensate for nonideal behaviors.

15.5.1.2 Kinetic Model
A kinetic model was developed using both DynoChem and MatLab to describe the main transformations and side reactions described in Figure 15.21. All reactions were modeled as second-order reactions; therefore, the rate equation can be expressed as shown in Eq. (15.18) where the rate constant, k, is expressed in terms of the activation energy (E_A) by the Arrhenius equation:

$$r = -\frac{d[A]}{dt} = k_{T_{ref}}[A][B] \quad k = k_o e^{-\frac{E_A}{RT}} \quad (15.18)$$

Time course profiles were generated for the degradation of intermediate C and D at various temperature and concentration as shown in Table 15.1.

[20] https://www.mt.com/gb/en/home/products/L1_AutochemProducts/ReactIR/flow-ir-chemis.html (accessed 10 November 2018).

Those batch kinetic data generated for the degradation of the intermediates were used, in a standard batch mechanistic model, where the mechanism described in Figure 15.21 was converted into differential rate equations. The activation energies and rate constants were regressed as shown in Figure 15.23 by using the weighted total SSQ of the deviations between the data and model results to ensure an accurate fitting of the various profiles regardless of the respective contribution to the mass balance and the number of points.

The estimation of the kinetic parameters achieved using the Levenberg–Marquardt algorithm and the weighted SSQ as objective function were reported in Table 15.2.

Due to extremely fast reaction and significant exotherm ($\Delta H_r = -220$ kJ/mol, $\Delta T_{add} = 77$ °C for the deprotonation reaction, $\Delta H_r = -210$ kJ/mol, $\Delta T_{add} = 64$ °C for the exchange reaction), generating valuable kinetic data for the main process reaction was achieved with great difficulty, and non-isothermal conditions were observed with a significant dependence to mixing. To address the interdependence between the kinetic and the physical rates, the performance of each part of the set up was characterized. The heat transfer and area (UA) was calculated through heat loss measurement experiments; the micro-mixing time was calculated as a function of flow rates using the Villermaux–Dushman procedure, and the axial dispersion was estimated using standard mathematical expression for the Bodenstein number as described in Section 15.3.3.

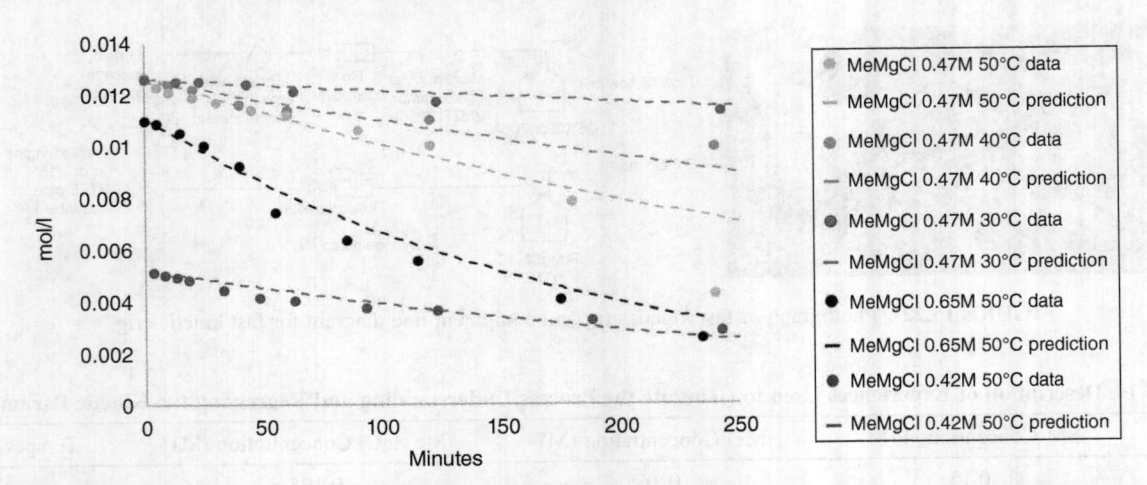

FIGURE 15.23 Examples of mol % versus time profiles for degradation of compound C (left) and compound D (right) generated at various temperatures and concentrations.

TABLE 15.2 Kinetic Parameters for Each Transformation Considered as Relevant to the Process

	K_{Tref}		E_A	
Main Reaction				
B + MeMgCl → C + Methane	1×10^2 min^{-1}	(Observed)	60 kg/mol	(Observed)
C + iPrMgCl → D + iPrI	1×10^2 min^{-1}	(Observed)	60 kg/mol	(Observed)
Degradation				
C + MeMgCl → D + MeI	0.03 min^{-1}		39 kg/mol	
C + MeMgCl → Degradent	4×10^{-3} min^{-1}		54.4 kg/mol	
D + MeI → F + MgClI	0.03 min^{-1}		45 kg/mol	

The flow steady-state system used for generating kinetic data resulted in conclusion that the desired reaction steps were complete in less than 140 ms, faster than it was possible to quantify using the setup. This was the case across a range of dilutions and temperatures. Based on this knowledge, observed rate constants and activation energies were applied, which satisfied these conditions (Table 15.2).

The partial differential equations for each transformations and side reactions were integrated into a series of PFR models in both DynoChem and MatLab to be the exact in silico representation of the existing setup described in Figure 15.21. In addition to the mechanistic description, the physical rates calculated and specific to the capillary PFR were included into the model.

15.5.2 Process Design

Some documented studies of transformations of organic molecules using Grignard reagents at room temperature are recorded in the literature, ranging from the micro- to mesoscale [42–44] with reduced or no heat transfer; however little supported understanding of the process kinetics is documented. This kinetic understanding is crucial for process scale-up where different temperature profiles due to change of heat transfer and heat loss reduced mixing efficiencies and more challenging control of residence times.

The PFR model including a full kinetic description of the process with the physical rates associated as described in the section above allows the evaluation of the process parameters and their impact to the product quality output. Impact of temperature was investigated in silico by enabling various heat transfer performance of the reactor from high heat transfer coefficient to adiabatic conditions. Impact of mixing time and nonuniformity of the concentration gradient across the length of the reactor was assessed from instantaneous to slow mixing. Impact of axial dispersion was equally studied from ideal plug flow to high axial distribution and small reactor equivalence.

Following a global sensitivity analysis, it demonstrated that axial dispersion and micromixing did not have limited impact as long as the mixing time was faster than the residence time and the equivalence of reactors was greater than 5 units. Equally, up to 0.5 second residence time and adiabatic conditions could be tolerated with acceptable impact on the product quality. It negates the need to reduce the system to cryogenic temperatures and simplifies the process control requirement and the reactor design.

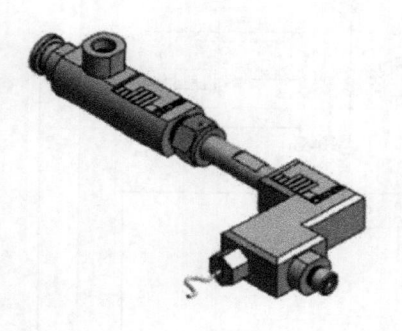

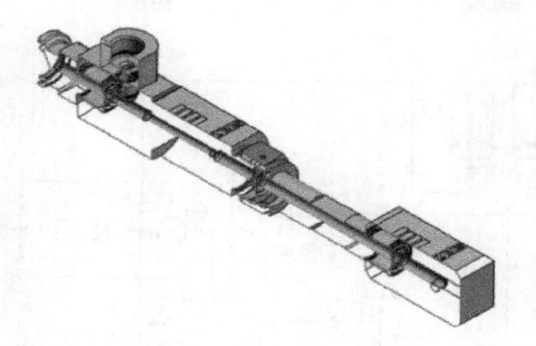

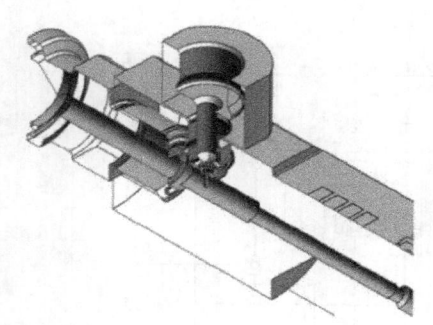

FIGURE 15.24 3D drawing of the reactor design.

15.5.3 Equipment Design

After completion of the process design, the various performance attributes of the reactor, enabling high yield and high quality, were captured in a user specific requirement document to support the design and construction of the equipment. Working alongside technology specialist, the design of the tubular reactor, injector, and static mixer was completed (Figure 15.24).

Specific attention was brought into the design of the injector to ensure robustness during operation as solid was observed. The injection system was designed to prevent back mixing and local concentration excess of organometallic. The side stream was injected at the point of highest turbulence to ensure the most efficient blending of both streams into each other. In addition, the design of the static mixer was reviewed with the supplier to ensure maximum efficiency for this process, and SMX-plus static mixers[21] were selected. The main inner body of the reactor was 4.8 mm diameter as the injector was 0.3 mm. The overall length of the reactor was only 300 mm before reaching an elbow connector where a thermocouple type K was inserted to measure the temperature of operation with an accuracy of ±0.5 °C. The same exact design was used for both transformations.

Following the construction of the reactor, a series of tests, as part of the site acceptance test (SAT), were completed to ensure the performance of the unit against the initial requirements. RTD, micromixing time, and heat loss were measured through a series of experiments (Figure 15.25).

Using the in-process model including the kinetic of the desired reaction, the side reactions, and the equipment performance, the optimum conditions were identified as described on the contour plots below (Figure 15.26).

The resulting operating conditions are described below.

A solution of compound B in tetrahydrofuran (THF) (4.4 wt %, 0.14 M) was combined with methyl magnesium chloride (3 M, 22% w/w 1.15 equiv.) in a PFR matching all the above performance characteristics (Bodenstein number, mixing time, and heat loss) for a residence time of 0.50 seconds. The resulting solution was then combined with isopropyl magnesium chloride (2 M, 20% w/w 1.1 equiv.) using similar PFR for an additional 0.50 second residence time. The aryl Grignard solution was then cooled to −15 °C and quenched by addition of trimethyl borate (2.5 equiv.) in a batch reactor. The resultant component E solution was worked up by adding 2 vol of 15% w/w sodium chloride aqueous solution and a 50% w/w aqueous citric acid (4 vol, 3.6 equiv.) followed by phase separation. The organic phase was then washed three times with sodium hydroxide (4 M, 5 vol), and the basic aqueous streams are combined. The combined aqueous base stream is then overlaid with cyclopentyl methyl ether (CPME) (6 vol) and then acidified by the addition of hydrochloric acid. The organic phase containing component A is washed with brine (26% w/w), concentrated to 6 vol (0.6 M) by vacuum distillation, and then the product is precipitated by the addition of 2,2,4-trimethylpentane (10 vol). The slurry is then filtered, and the solid is washed with CPME:2,2,4-trimethylpentane (3 vol).

Prior to the clinical campaign, the process was successfully verified in a kilogram continuous demonstration. Several batches of 1 kg of compound B were processed, collected, and isolated in 500 g batch of compound A yielding 70% as described in Table 15.3.

Following that demonstration, the process was transferred to pilot plant GMP facility to synthesize 200 kg of component B for a phase three clinical make.

15.5.4 Conclusion

A systematic approach using intrinsic process understanding and in silico enabled design has been shown to be an effective tool in the development of organometallic flow processes and industrialization. Low volume, residence time restricted systems allowed all processes to operate well outside of the usual cryogenic controls. By generating process understanding, such as quantification of kinetic constants and characterization of equipment, systems can be engineered to handle instable or hazardous chemistry while inherently increasing the process safety and sustainability.

---

[21] https://www.sulzer.com/en/products/static-mixers (accessed 10 November 2018).

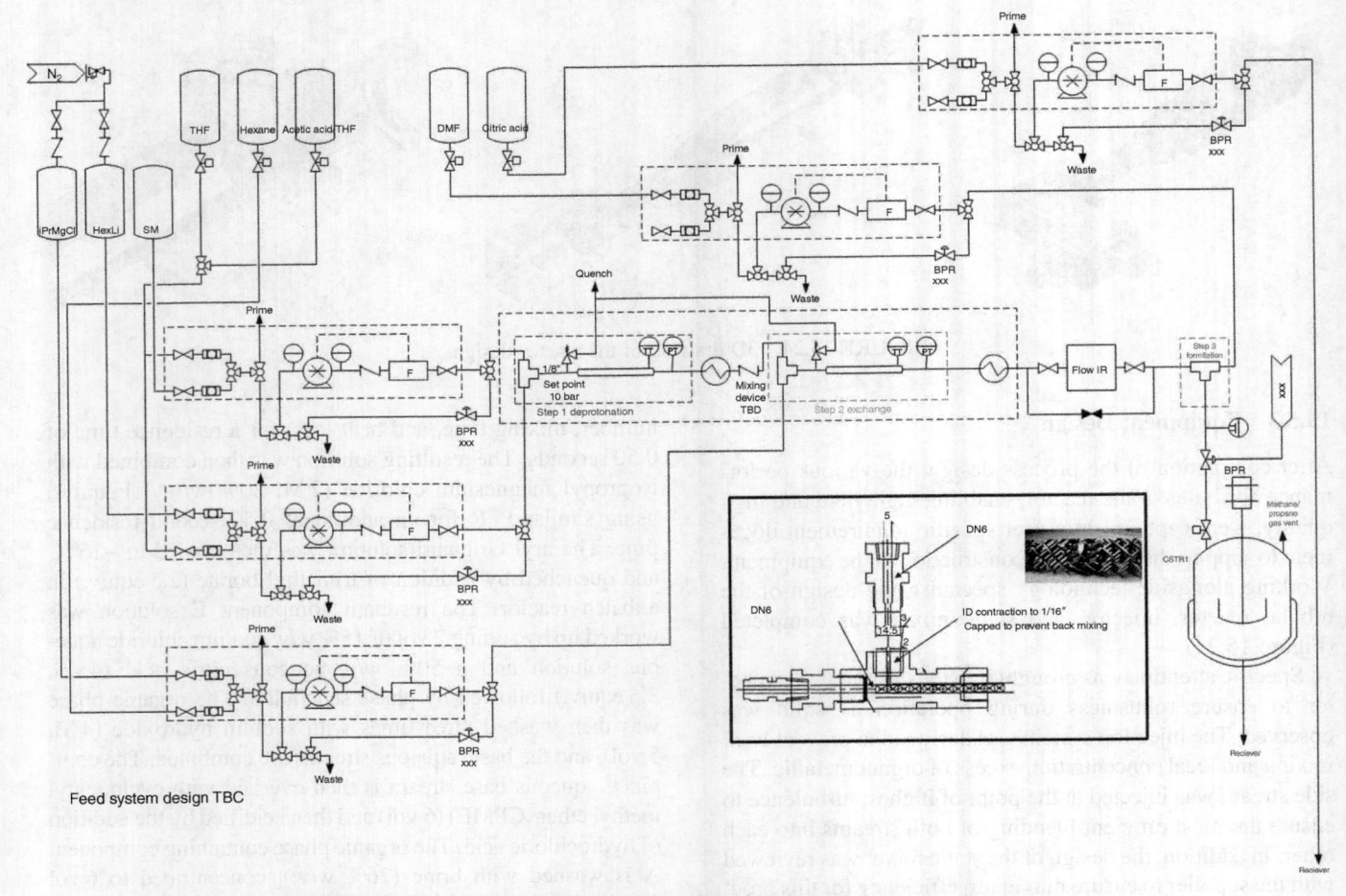

FIGURE 15.25 Process and instrumentation diagram for the two-stage flow process.

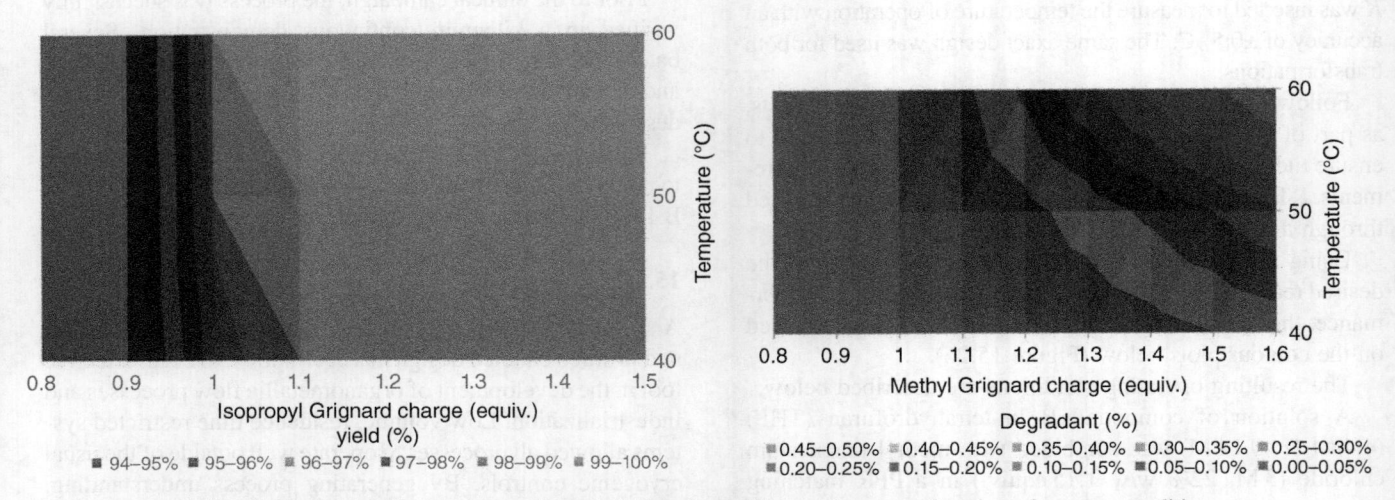

FIGURE 15.26 In silico representation of yield and overall degradants as a function of process conditions.

TABLE 15.3 Experimental verification of the process design

Reaction	SM Flow Rate (g/min)	MeMgCl Flow Rate (g/min)	IsoPrMgCl Flow Rate (g/min)
1	150	59	35
2	130	51	30
3	130	51	30
4	130	52	28.8
5	130	52	28.8
6	130	49	28.8
7	130	49	28.8
8	130	52	27.3
9	130	49	30.3
10	130	52	24.5
11	170	63	38.5
12	215	79	48

The workflow employed successfully demonstrated the efficiency of integration of process simulation into the development and scale-up of continuous processing. This offers significant advantages in terms of time and material saving, efficiency of scale-up, and process robustness. Future application of this workflow could be used to reduce scale-up risk, by formulating an in silico design space and verifying model predication across scales.

REFERENCES

1. Wiles, C. and Watts, P. (2012). Continuous flow reactors: a perspective. *Green Chem.* 14 (1): 38–54.
2. Malet-Sanz, L. and Susanne, F. (2012). Continuous flow synthesis. A pharma perspective. *J. Med. Chem.* 55 (9): 4062–4098.
3. Strang, D. Pharma 2020: Supplying the Future. Which Path Will You Take? https://www.pwc.com/gx/en/pharma-life-sciences/pdf/pharma-2020-supplying-the-future.pdf (accessed 5 May 2017).
4. Abboud, L. and Hensley, S. (2003). Factory shift: new prescription for drug makers: update the plants. *Wall Street Journal* (3 September 2003).
5. IChemE (2004). *Chemical Reaction Hazards – A Guide to Safety*, 2e (ed. J. Barton and R. Rogers). Institution of Chemical Engineers.
6. Barton, K. and Rogers, R. (1997). *Chemical Reaction Hazards*. Elsevier.
7. Di Miceli Raimondi, N., Olivier Maget, N., Gabas, N. et al. (2015). Safety enhancement by transposition of the nitration of toluene from semi-batch reactor to continuous intensified heat exchanger reactor. *Chem. Eng. Res.Des.* 94: 182–193.
8. Barton, J.A. and Nolan, P.F. Runaway reactions in batch reactors. I. CHEM. E. Symposium Series No. 85.
9. Willey, R.J. (ed.) *Process Saf. Prog.* 16 (2): 94–100.
10. Capellos, C. and Bielski, B.H.J. (1980). *Kinetic Systems: Mathematical Description of Chemical Kinetics in Solution*. R. E. Krieger Pub. Co.
11. Swarin, S.J. and Wims, A.M. (1976). A method for determining reaction kinetics by differential scanning calorimetry. *Anal. Calorim.* 4: 155.
12. Borchardt, H.J. and Daniels, F.J. (1956). The application of differential thermal analysis to the study of reaction kinetics. *Am. Chem. Soc.* 79: 41.
13. Ozawa, T.J. (1970). Kinetic analysis of derivative curves in thermal. *Thermal Anal.* 2: 301.
14. Mass, T.A.M.M. (1978). Optimization of processing conditions for thermosetting polymers by determination of the degree of curing with a differential scanning calorimeter. *Polym. Eng. Sci.* 18: 29.
15. Hoffmann, W. (2007). Kinetic data by nonisothermal reaction calorimetry: a model-assisted calorimetric evaluation. *Org. Process Res. Dev.* 11: 25–29.
16. Kumar, A. and Hartland, S. (1988). Mass transfer in a Kühni extraction column. *Ind. Eng. Chem. Res.* 27 (7): 1198–1203.
17. Keles, H., Susanne, F., Livingstone, H. et al. (2017). Development of a robust and reusable microreactor employing laser based mid-IR chemical imaging for the automated quantification of reaction kinetics. *Org. Process Res. Dev.* 21 (11): 1761–1768.
18. Sarrazin, F., Salmon, J.-B., Talaga, D., and Servant, L. (2008). Chemical reaction imaging within microfluidic devices using confocal Raman spectroscopy: the case of water and deuterium oxide as a model system. *Anal. Chem.* 80 (5): 1689–1695.
19. Schwolow, S., Hollmann, J., Schenkel, B., and Röder, T. (2012). Application-oriented analysis of mixing performance in microreactors, OPRD. *Org. Process Res. Dev.* 16 (9): 1513–1522.
20. Khan, S.A., Günther, A., Schmidt, M.A., and Jensen, K.F. (2004). Microfluidic synthesis of colloidal silica. *Langmuir* 20 (20): 8604–8611.
21. Günther, A., Khan, S.A., Thalmann, M. et al. (2004). Transport and reaction in microscale segmented gas–liquid flow. *Lab Chip* 4: 278–286.
22. McMullen, J.P. and Jensen, K.F. (2010). An automated microfluidic system for online optimization in chemical synthesis. *Org. Process Res. Dev.* 14 (5): 1169–1176.
23. Hwang, Y.-J., Coley, C.W., Abolhasani, M. et al. (2017). A segmented flow platform for on-demand medicinal chemistry and compound synthesis in oscillating droplets. *Chem. Commun.* 53: 6649.
24. Falk, L. and Commenge, J. (2010). Mixing performance of micromixers by Villermaux-Dushman reaction protocol at low Reynolds Number. *Chem. Eng. Sci.* 65: 405–411.
25. Giorges, A.T.G., Forney, L.J., and Wang, X. (2001). Numerical study of multi-jet mixing. *Chem. Eng. Res. Design* 79 (5): 515–522.

26. Paul, E.L., Atiemo-Obeng, V., and Kresta, S.M. (eds.) (2003). *Handbook of Industrial Mixing: Science and Practice*. Wiley-Blackwell.
27. Nagy, K.D., Shen, B., Jamison, T.F., and Jensen, K.F. (2012). Mixing and dispersion in small-scale flow systems. *Org. Process Res. Dev.* 16 (5): 976–981.
28. Li, S., Zhang, K., Yang, J.-M. et al. (2007). Single quantum dots as local temperature markers. *Nano Lett.* 7 (10): 3102–3105.
29. Scott Fogler, H. (2012). Distributions of residence times for chemical reactors. In: *Elements of Chemical Reaction Engineering*, 4e, 867–944.
30. Pinot, J., Commenge, J.-M., Portha, J.-F., and Falk, L. New protocol of the Villermaux–Dushman reaction system to characterize micromixing effect in viscous media. *Chem. Eng. Sci.* 118 (18): 94–101.
31. Reckamp, J.M., Bindels, A., Duffield, S. et al. Mixing performance evaluation for commercially available micromixers using Villermaux–Dushman reaction scheme with the interaction by exchange with the mean model. *Org. Process Res. Dev.*.
32. Stone, R.E. and Tovey, C.A. (1991). The simplex and projective scaling algorithms as iteratively reweighted least squares methods. *SIAM Rev.* 33 (2): 220–237.
33. Björck, Å. (1996). *Numerical Methods for Least Squares Problems*. Society for Industrial and Applied Mathematics.
34. Nguyen, N.L., van Buren, V., von Garnier, A. et al. (2005). Application of magnetic resonance imaging (MRI) for investigation of fluid dynamics in trickle bed reactors and of droplet separation kinetics in packed beds. *Chem. Eng. Sci.* 60 (22): 6289–6297.
35. Krishna, R. and Sie, S.T. (1994). Strategies for multiphase reactor selection. *Science* 49 (24A): 4029–4065.
36. Thornton, J.D. (1992). *Science and Practise of Liquid-Liquid-Extraction*. Oxford.
37. Lo, T.C., Baird, M.H.I., and Hanson, C. (1983). *Handbook of Solvent Extraction*. New York: Wiley.
38. Godfrey, J.C. and Slater, M.J. (1994). *Liquid-Liquid Extraction Equipment*. New York: Wiley.
39. Laue, S., Haverkamp, V., and Mleczko, L. (2016). Experience with scale-up of low-temperature organometallic reactions in continuous flow. *Org. Process Res. Dev.* 20: 480–486.
40. Castello-Mico, A., Herbert, S.A., Leon, T. et al. (2016). Functionalizations of mixtures of regioisomeric aryllithium compounds by selective trapping with dichlorozirconocene. *Angew. Chem. Int. Ed.* 55: 401–404.
41. Gutmann, B., Cantillo, D., and Kappe, O. (2015). Continuous-flow technology – a tool for the safe manufacturing of active pharmaceutical ingredients. *Angew. Chem.* 54 (23): 6688–6728.
42. Murray, P.R.D., Browne, D.L., Pastre, J.C. et al. (2013). Continuous flow-processing of organometallic reagents using an advanced peristaltic pumping system and the telescoped flow synthesis of (E/Z)-tamoxifen. *Org. Process Res. Dev.* 17 (9): 1192–1208.
43. Riva, E. et al. (2010). Reaction of Grignard reagents with carbonyl compounds under continuous flow conditions. *Tetrahedron* 66: 3242–3247.
44. Odille, F.G.J., Stenemyr, A., and Pontén, F. (2014). Development of a Grignard-type reaction for manufacturing in a continuous-flow reactor. *Org. Process Res. Dev.* 18: 1545–1549.

16

DESIGN AND SELECTION OF CONTINUOUS REACTORS FOR PHARMACEUTICAL MANUFACTURING

Martin D. Johnson, Scott A. May, Michael E. Kopach, Jennifer McClary Groh, Timothy Braden, Vaidyaraman Shankarraman, and Jeremy Miles Merritt

Eli Lilly and Company, Indianapolis, IN, USA

16.1 DRIVERS FOR CONTINUOUS REACTIONS IN DRUG SUBSTANCE MANUFACTURING

Continuous processing applications in the pharmaceutical industry have been increasing in recent years [1]. There are many reasons why an API manufacturer may choose to run chemical reactions in continuous mode instead of batch. Typical drivers for choosing continuous instead of batch reaction are fast kinetics, highly exothermic heats of reaction, hazardous reagents, extreme temperatures and pressures, and high throughput. Yield and selectivity may be better for fast reactions in series with unstable intermediates [2]. Safety may be improved for highly exothermic reactions [3]. Operating ranges may be wider, and safety may be improved for high pressure reactions with hazardous gas reagents such as hydrogenations, hydroformylations, aerobic oxidations, and carbonylations. Operating ranges may be larger, and safety may be improved for reactions at extreme temperatures less than −40 °C or greater than 200 °C or superheated reactions with vapor pressure greater than 50 bar at reaction temperature [4]. There are safety advantages of minimizing volumes and maximizing heat transfer rates for reactions with hazardous reagents, potential for thermal runaway or large exotherms [5]. Safety is improved for reactions with hazardous reagents like reactions with hydrazine, bromonitromethane, diazomethane [6], sodium cyanide, methanesulfonyl cyanide, and cyanogen chloride [7]. If there is an advantage to running with no headspace, such as organic azide or tetrazole formations where hydrazoic acid would partition into the headspace of the batch reactor, then a plug flow reactors (PFR) may be superior to batch because it operates 100% liquid filled [8]. If the impurity profile is improved by all-at-once addition, coaddition, fast heat-up, and/or fast cooldown, then a PFR or a continuous stirred tank reactor (CSTR) may be superior to batch. This depends on the kinetics of the side reactions relative to the main desired reaction. Continuous can provide throughput advantages and debottlenecking for high volume products [9]. End-to-end fully continuous processes offer advantages of quality control, extreme operating conditions, telescoping, and introducing excipients further upstream to improve processing API to drug product [10]. The use of continuous reactors in the pharmaceutical industry is rapidly increasing [11]. Product consistency advantages can be achieved via continuous processing because of steady-state operation, feedback control, and real-time product quality information by online process analytical technology (PAT). It aligns with quality by design (QbD) principles, and implementation of continuous processes is encouraged and supported by the FDA [12].

There are several additional criteria to look for in the batch reaction that may cause one to think about continuous. Is the yield or impurity profile sensitive to reaction time? Are there significant late-forming impurities? Is the product unstable to end-of-reaction conditions? If the answer is yes to any of these questions, then consider reaction in a plug flow tube reactor (PFR), because reaction time is more scalable, the late-forming impurities can be avoided by precise control of time in the reactor, and the unstable product can be quenched in-line immediately after the precisely controlled

---

Chemical Engineering in the Pharmaceutical Industry: Active Pharmaceutical Ingredients, Second Edition.
Edited by David J. am Ende and Mary T. am Ende.
© 2019 John Wiley & Sons, Inc. Published 2019 by John Wiley & Sons, Inc.

reaction time. Is the yield or impurity profile sensitive to mixing rate, heat-up rate, or cooldown rate? Reagent addition time is more scalable in flow than in batch because of in-line mixing and because the production-scale reactors are not as large as batch. Heat-up and cooldown rates are more scalable in continuous than batch. This is because process liquids flow through heat exchangers at the reactor inlet and outlet. Does the yield or impurity profile benefit from all-at-once mixing of reagents? If so, then a PFR is a better reactor choice than batch for all-at-once stoichiometric reagent mixing because of the ability to control (or remove) heat from an exothermic reaction. Is there a lag time or delayed catalyst initiation time in the batch reactor? In this case, the advantage of a CSTR is that the reaction is always operating at the end-of-reaction conditions. Is the product cytotoxic or highly potent? Using continuous processing, cytotoxic API production may be achieved in inexpensive, dedicated, and "disposable" equipment sets for production of low volume (<1000 kg/year) cytotoxic APIs in laboratory fume hoods [13]. Is it difficult to recover a batch in the event of operational problems? For example, if the catalyst deactivates because of something like poor inertion and if it is difficult to recover from this event by simply adding more catalyst, then the flow advantage of the PFR is that there is less material *at risk* due to the small reactor volume compared with the total production volume (one to one in batch operations).

16.2 TWO MAIN CATEGORIES OF CONTINUOUS REACTORS: PFRs AND CSTRs

Continuous reactors fall into two main categories, CSTR and plug flow with dispersion reactor (PFDR) [14]. They are shown schematically in Figure 16.1.

The design equations are derived by solving the material balance for conversion versus mean residence time (τ) or distance after making simplifying assumptions [15]. The design equations relate changes in concentration with time to flow of reacting species in and out of reactor and species formation/disappearance by chemical reaction. The design equation for the CSTR assumes no spatial changes within the reactor and no change with time at steady state. If the mixing time is sufficiently fast compared with the mean residence time, then the entire contents of the reactor remain at outlet concentrations. The simplified design equation for the PFDR assumes no change with time for a given location in the reactor, but change in conversion with distance along the axial direction from reactor inlet to outlet. The PFDRs are commonly called PFRs, but PFR implies perfect mixing in the radial direction and no mixing in the axial direction, which is not true for real reactors. All real reactors have some degree of axial dispersion, especially with laminar flow homogeneous liquids. If the axial dispersion number is small enough so that it has negligible impact on conversion versus τ, then the PFDRs are called PFRs. This chapter contains an example of quantifying axial dispersion and then modeling whether or not it has significant impact on conversion versus time in real rectors. There are many types of PFRs such as:

- Microreactors.
- Static mix tube reactors.
- Open tube or pipe reactors.
- Oscillatory baffled tube reactors.
- Flow tube or structured channel reactors that have alternative energy input like microwave, photo, electro, or sono energy input.

With proper engineering design, each of these variations can be designed to minimize axial dispersion such that highly plug flow profiles result.

Reaction conversion versus τ is different in a CSTR than in a PFR, and the respective residence time distributions (RTD) are also very different. Much longer residence times are required for full conversion in a CSTR compared with a PFR, unless the reaction rate is zero order. Example calculations are shown in this chapter to illustrate this point.

Judging by the number of published examples and the number of commercially available continuous reactor systems, PFRs are more commonly applied than CSTRs. PFRs maximize heat transfer A/V compared with stirred tanks, especially if they are microreactors, i.e. characteristic dimension less than 1 mm.

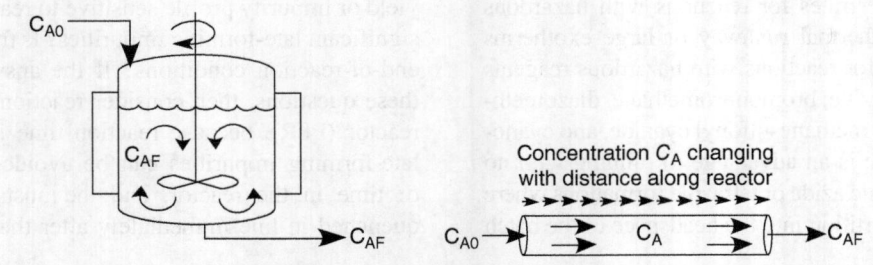

FIGURE 16.1 Schematic drawings of CSTR and PFDR. C_{A0} is initial concentration of reagent A entering the reactor, C_A is concentration of reagent A in PFR, and C_{AF} is final concentration of reagent A exiting the reactors, which is the same as the concentration of reagent A everywhere in the ideal CSTR.

A PFR minimizes total reactor volume and thus τ in the reactor for a given conversion. A PFR minimizes transition time and volume to reach a new steady state after a step change in process parameter because of narrow RTD. The PFR can run 100% liquid filled and as a consequence is the best reactor choice to eliminate headspace. This also helps to minimize equivalents of gas reagent for two-phase vapor–liquid reactions. The working volume of research-scale PFRs can be less than 0.1 ml; therefore they minimize materials required for research and development. Compared with CSTRs in series or intermittent flow CSTRs, the PFRs can be lower in complexity and lower in cost. One of the biggest challenges is to maintain solubility of reagents and reaction products in PFRs. Most PFRs are not capable of solids in flow because of fouling or blockage. Solids in packed columns are acceptable, for example, packed catalyst bed reactors, but the particle size of the solids should be greater than 100 μm to minimize pressure drop across the bed.

On the other hand, CSTRs, CSTRs in series, or intermittent flow CSTRs may be selected instead of PFRs in certain circumstances. These reactors are more suitable for heterogeneous systems, including slurry flow such as reactive crystallizations, reaction with solid precipitate, or reaction with solid reagent feeds like Grignard formation reactions with solid magnesium reagent. CSTRs tolerate and buffer out fluctuations and/or pulsing of reagent feeds. In fact, a useful reactor configuration is a small CSTR followed by a PFR, where the CSTR dampens out fluctuations in stoichiometry of two reagents combining prior to entering the PFR. While PFRs need precise flow control especially for multiple reagents, CSTRs are more forgiving of short oscillations in flow control. If stoichiometry is critical and the pump flow rates oscillate, then consider CSTRs. CSTRs can be stopped and restarted without loss of mixing during flow stoppage. Also, mixing is independent of flow rate in CSTRs. CSTRs achieve fast liquid–liquid two-phase mixing over a wider range of scales and reaction times, for example, a one hour reaction time in production-scale reactor. CSTRs help to overcome induction time or lag time in reaction kinetics, e.g. autocatalytic reaction, because they operate at "end-of-reaction" conditions at steady state. Also, because CSTRs do not run completely liquid filled, it is more feasible to adjust residence time in the reactor without changing liquid flow rate.

Furthermore, the choice of PFR versus CSTR versus batch depends on the best method of reagent addition. Reaction kinetics will determine the best reactor configuration for maximizing yield and minimizing side reactions and impurities. In any of the reactor configurations (batch, plug flow, continuous CSTRs, or intermittent flow CSTR), one can easily use kinetics from the batch reactions to design and size the continuous reactor for full conversion. The bigger challenge is understanding and predicting the impurity profile for the different reactor configurations. For a reaction $A + B \rightarrow C$, if the impurity profile and yield are most favorable when reagent B is added in a slow and controlled manner to a vessel containing reagent A, then batch and intermittent flow CSTR are better reactor choices than a PFR or true CSTR. A PFR with multiple addition points of reagent B along the length of the reactor is an alternative, but it requires more pumps and mass flow controllers. A mathematical example is given in Section 16.5. If impurity profile and yield are best for an all-at-once addition, then a PFR is the best choice. Batch is usually not practical for all-at-once stoichiometric addition because of safety concerns and heat transfer limitations. If yield and selectivity are most favorable as a result of slow coaddition of both A and B at stoichiometric ratios (i.e. coaddition) while keeping the conversion high in the reactor at all times, then a CSTR is the best reactor choice.

16.3 EXAMPLE OF COILED TUBE PFR FOR TWO-PHASE GAS–LIQUID REACTIONS

An asymmetric hydrogenation reaction operating at 68 bar hydrogen produced 144 kg of penultimate using a 73 L coiled tube PFR [16]. The reactor and the downstream separations and purification unit operations were run in laboratory facilities at 13 kg/day throughput. The hydrogenation was in a special laboratory facility with all explosion proof equipment, utilities, and lighting, high air volume turnovers, specialized doors, and blowout walls. The reaction scheme is shown below in Scheme 16.1. It was a challenging hydrogenation of a tetrasubstituted enone.

The rhodium and Josiphos homogeneous catalyst ligand system was very oxygen sensitive. Running the reaction continuous rather than batch provided advantages for excluding oxygen, because the reactor only needed to be inerted once at the beginning of the campaign. The catalyst system was also very expensive. To be economically viable, this reaction required a 2000 : 1 substrate to catalyst ratio ($S : C$) and high pressure hydrogen (68 bar) to maintain a high dissolved H_2 concentration. The enone substrate was 0.15 M in 2.5 : 1 ethyl acetate/methanol, and the dissolved H_2 concentration was expected to be on the same order or magnitude at the elevated pressure. Batch processing is normally the default option, but a 68 bar rated autoclave was not available at the selected manufacturing facility for this product. A capital spend was needed to support either a batch or continuous process.

There are several good references for determining kinetics in continuous reactors [17]. Reaction kinetics were measured

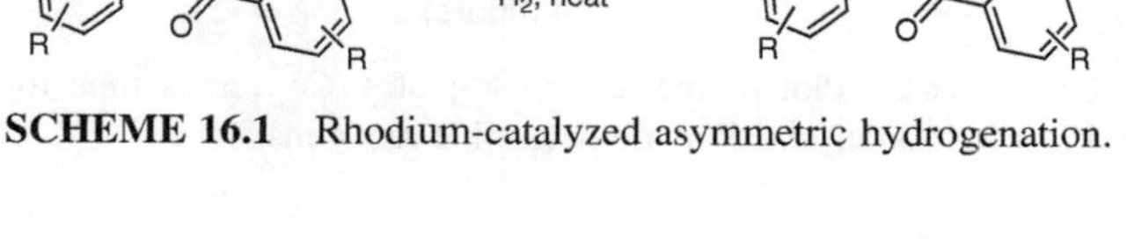

SCHEME 16.1 Rhodium-catalyzed asymmetric hydrogenation.

in a well-mixed batch PARR autoclave at research scale. Conversion versus time was measured in the autoclave, and the results are listed in Table 16.1.

Plotting the natural log of C/C_o versus time gives a straight line with slope equal to the reaction rate constant for a first-order reaction, where C is concentration of starting reagent at time t and C_o is initial concentration.

The rate expression is

$$\frac{dC}{dt} = -kC \qquad (16.1)$$

where

k is the first-order rate constant.

By separating variables and integrating, this becomes

$$\int_0^t -k \, dt = \int_{C_o}^C \frac{dC}{C} \qquad (16.2)$$

$$-kt = \ln\left(\frac{C}{C_o}\right) \qquad (16.3)$$

A plot of the natural log of C/C_o versus time is shown in Figure 16.2.

In this example, the product concentration is log linear; therefore the reaction is first order or pseudo first order with $k = 0.63$/h within the time period investigated. The first-order reaction rate model with $k = 0.63$/h is plotted with the actual conversion versus time data in Figure 16.3 and Table 16.1.

Overall gas–liquid mass transfer coefficient, $k_L a$, was the order of 0.01–0.1/s; therefore conversion versus time was likely not mass transfer rate limited. The kinetic data informs the choice of reactor type: batch versus PFR versus CSTR. The rate data was used to predict how long the reaction must run to reach a desired 99.9% conversion level in batch mode and what is the required reaction time to reach the same conversion in the PFR or CSTR.

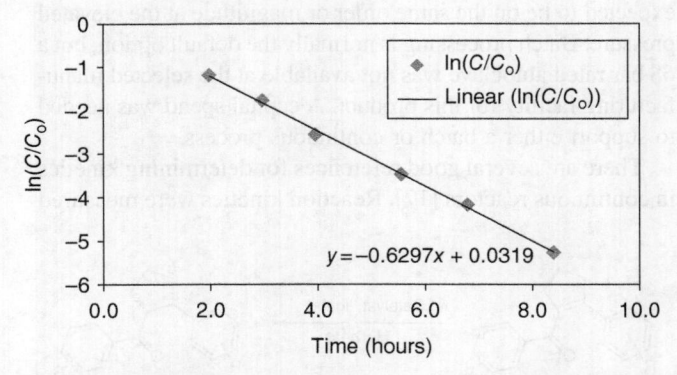

FIGURE 16.2 Plot of the natural log of C/C_o versus time to confirm first-order reaction and determine rate constant.

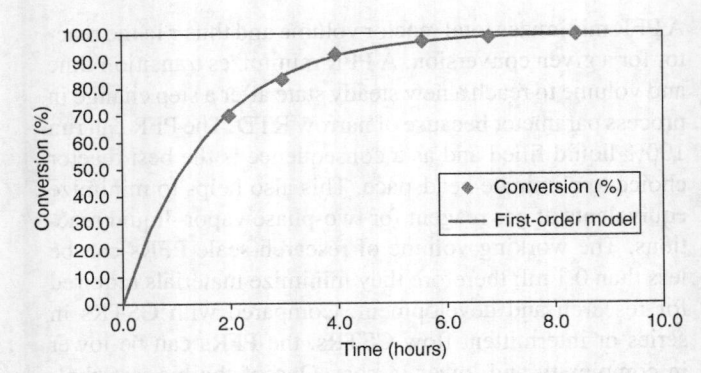

FIGURE 16.3 Conversion versus time and first-order reaction rate model fit to the data for asymmetric hydrogenation. Reaction at 70 °C and 68 bar.

TABLE 16.1 Hydrogenation Reaction Experimental Rate Data

Time (h)	Conversion (%)
2.0	69.7
3.0	83.1
4.0	92.3
5.6	96.9
6.8	98.5
8.4	99.5

Temperature 70 °C and pressure 68 bar.

CASE STUDY 16.1: BATCH REACTOR

In any reactor, the material balance equation is

$$[\text{in} - \text{out}] + \text{reaction} = \text{accumulation} \qquad (16.4)$$

"Reaction" is the difference between generation and consumption. In a batch reactor, there is no flow in or out, so the material balance simplifies to

$$\text{reaction} = \text{accumulation}$$

For a reaction that is first order in concentration of reagent A, this becomes the same as Eq. (16.1):

$$-kC_A = \frac{dC_A}{dt}$$

C_A = concentration of reagent A in reactor
k = first-order rate constant
t = reaction time.

Upon separating variables and integrating, this becomes the same as Eq. (16.2), except that we are integrating to C_{AF}:

$$\int_0^t -k\,dt = \int_{C_{A0}}^{C_{AF}} \frac{dC_A}{C_A}$$

C_{A0} = initial concentration of reagent A.
C_{AF} = final concentration of reagent A.

Integrating, this becomes the same as Eq. (16.3):

$$-kt = \ln\left(\frac{C_{AF}}{C_{A0}}\right)$$

Raising each side of the equation to e^x, this becomes

$$\frac{C_{AF}}{C_{A0}} = e^{-kt} \tag{16.5}$$

99.9% conversion requires

$$\frac{C_{AF}}{C_{A0}} = 0.001$$

For this case it is known that $k = 0.63/h$. Thus, $t = 11$ hours. This is the time required to reach 99.9% conversion in a batch reactor.

CASE STUDY 16.2: IDEAL PFR

Reaction time equals total reactor length (L) divided by linear velocity (v_x) in an ideal PFR:

$$\tau = \frac{L}{v_x} \tag{16.6}$$

At steady state there is no accumulation, so the material balance Eq. (16.4) simplifies to

$$[\text{in} - \text{out}] + \text{reaction} = 0$$

The material balance equation including effects of dispersion is [15]

$$\frac{D}{uL}\frac{d^2 C_A}{dx^2} - v_x \frac{dC_A}{dx} - kC_A = 0 \tag{16.7}$$

D/uL is axial dispersion number, where

D is the Taylor longitudinal dispersion coefficient that incorporates the effect of both diffusion and convection.

u is average flow velocity.

L is PFR length.

The first term represents flux in and out of a differential volume by dispersion, the second term represents flux in and out of a differential volume by bulk flow, and the third term represents reaction. This is described in detail in chemical engineering text books [15]. If we assume ideal plug flow, then $D/uL = 0$:

$$-v_x \frac{dC_A}{dx} - kC_A = 0$$

Separating variables and integrating gives the following. The steady-state reactor is integrated over the length, in contrast to the batch reactor that is integrated over time:

$$\int_{C_{A0}}^{C_{AF}} \frac{dC_A}{C_A} = \int_0^L \frac{-k}{v_x} dx \tag{16.8}$$

Solving the integrals and substituting τ for L/v_x gives an equation that is similar to Eq. (16.3) except that it uses mean residence time τ instead of time t:

$$\ln\left(\frac{C_{AF}}{C_{A0}}\right) = -\frac{kL}{v_x} = -k\tau$$

Likewise, taking the exponential of both sides leads to

$$\frac{C_{AF}}{C_{A0}} = e^{-k\tau}$$

Therefore, just like in a batch reactor, 99.9% conversion requires

$$\frac{C_{AF}}{C_{A0}} = 0.001$$

Again, $k = 0.63/h$. Thus, we calculate $\tau = 11$ hours. Notice that conversion versus time was the same in batch and in an ideal PFR. This reaction was not mass transfer rate limited. It was catalyst turnover rate limited. Speeding up the reaction by increasing catalyst loading was not a viable option because of the cost of the catalyst and ligand. In addition, increasing catalyst loading would increase the burden of removing of the metal from the drug substance during the purification steps. Increasing the reaction rate (and decreasing the required mean residence time in a PFR) by increasing the temperature was not a viable option because the ee was lower at higher temperatures. Therefore, a PFR would need to be capable at least an 11 hours τ.

CASE STUDY 16.3: CSTR

In a CSTR, reaction time equals volume of process fluid in reactor (V) divided by volumetric flow rate corrected for thermal expansion in the reactor (Q):

$$\tau = \frac{V}{Q} \tag{16.9}$$

The material balance Eq. (16.4) is simplified for steady state with no accumulation:

$$[\text{in} - \text{out}] + \text{reaction} = 0$$

Substituting in variables for the terms, rearranging, and substituting τ for V/Q,

$$QC_{A0} - QC_{AF} - kC_{AF}V = 0 \tag{16.10}$$

Rearranging gives the following equation:

$$\frac{C_{AF}}{C_{A0}} = \frac{Q}{Q + kV} \tag{16.11}$$

Dividing by Q and substituting Eq. (16.9) gives

$$\frac{C_{AF}}{C_{A0}} = \frac{1}{1 + k\tau} \tag{16.12}$$

If we use the same $k = 0.63 \frac{1}{h}$ and $\frac{C_{AF}}{C_{A0}} = 0.001$, then $\tau = 1586$ hours $= 66$ days to achieve 99.9% conversion in a CSTR.

A τ of 66 days is obviously not a practical reaction time. In general, pharmaceutical processing frequently requires very high conversion, for example, 99.9% or greater. Conversion greater than 99% is desired from an economics standpoint, and greater than 99.9% conversion can be required if the control strategy is dependent upon starting material rejection and if the unreacted starting material is not well rejected in the purification step, for example, crystallization. Figure 16.4 shows how much larger the CSTR volume must be compared with PFR volume to achieve the same conversion for a simple elementary irreversible first-order reaction. The ratio increases steeply at high conversions. (CSTR volume)/(PFR volume) = 21 for 99% conversion and 145 for 99.9% conversion.

For a positive-order reaction, CSTRs in series reduces the total residence time to achieve a target conversion compared with a single CSTR. Three equal volume CSTRs in series would bring total combined reactor residence time down to only 1.8 days compared with 6 days for a single CSTR to achieve 99.9% conversion. The design equation for three equal-sized CSTRs in series for irreversible first-order reaction is [15]

$$\frac{C_{AF}}{C_{A0}} = \left(\frac{1}{1 + k\tau}\right)^3 \tag{16.13}$$

Calculating for 99.9% conversion $\frac{C_{AF}}{C_{A0}} = 0.001$ and a rate constant $k = 0.63 \frac{1}{h}$, $\tau = 14.3$ hours.

With three CSTRs in series, τ is 14.3 hours in each CSTR, and thus $\tau = 42.9$ hours total in all three combined. As the number of CSTRs in series increases, the behavior of conversion versus time approaches that of an ideal PFR [15]. This is shown in Figure 16.5. Here, C is concentration at time t, C_f is concentration of nonreacting tracer that started flowing into the reactor at time zero. A system with 20 equal-sized CSTRs in series has about the same RTD as a PFDR with axial dispersion number 0.024, and one with 100 equal-sized CSTRs in series has about the same RTD as a PFDR with axial dispersion number 0.005. Obviously, it is more practical to design a PFDR with axial dispersion number 0.005 if the goal is to achieve plug flow behavior.

In the PFR, the reagent solution, catalyst solution, and reagent gas flowed cocurrently through the coiled tubes. A simplified schematic of the reactor system is shown in Figure 16.6.

Photographs of the 73 L reactor are shown in Figure 16.7. The overall height and diameter of the cylindrical coiled tubing assembly was designed to fit inside an existing 0.91 m diameter jacketed single plate filter that was used as a constant temperature bath, also shown in Figure 16.7.

The PFR was constructed of 316 L stainless steel tubing with outer diameter (o.d.) = 19.1 mm, inner diameter (i.d.) = 16.5 mm, and $L = 340$ m. Overall L/d was about 20 600, which is unusually large for tubular reactor designs. This is because the reactor was designed for 12 hours τ and flow was in the laminar regime. Reynolds number was about

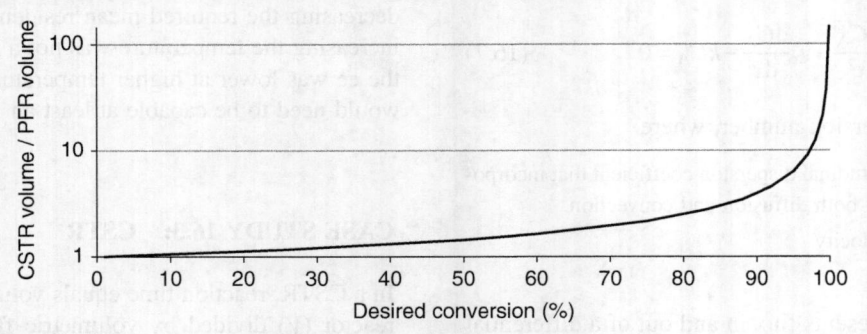

FIGURE 16.4 Volume comparison for a single CSTR versus a PFR to achieve the same reaction conversion for a first-order reaction, up to 99.9% conversion.

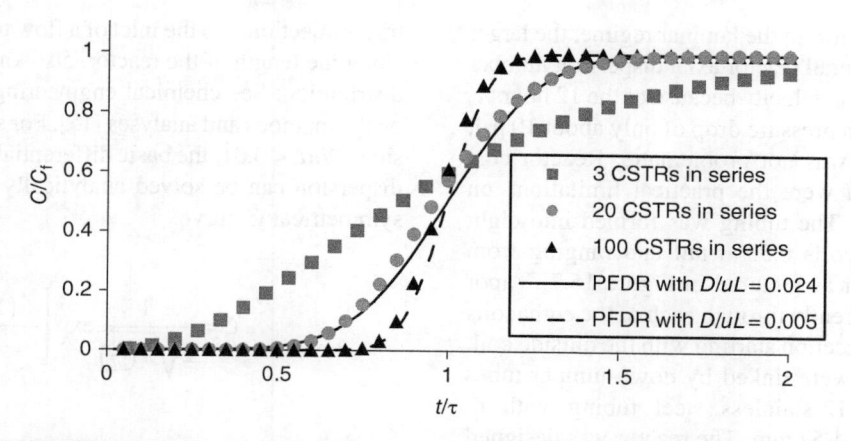

FIGURE 16.5 F-curve transitions comparing RTDs in CSTRs in series versus PFDR.

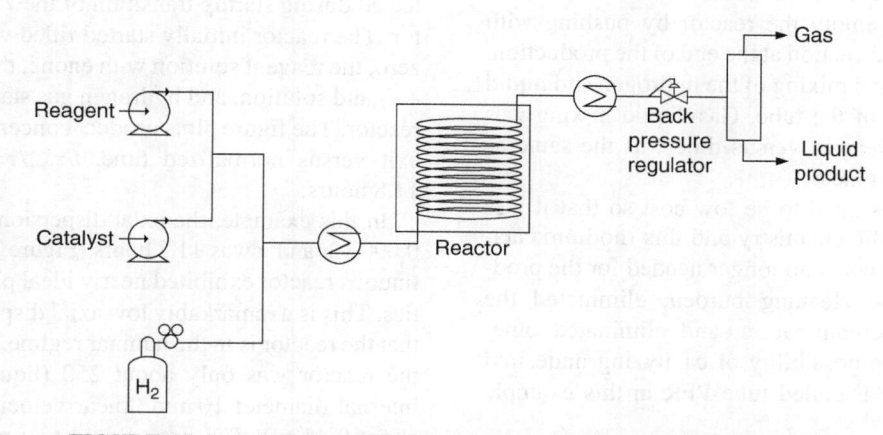

FIGURE 16.6 Schematic of PFR for two-phase reaction with hydrogen gas.

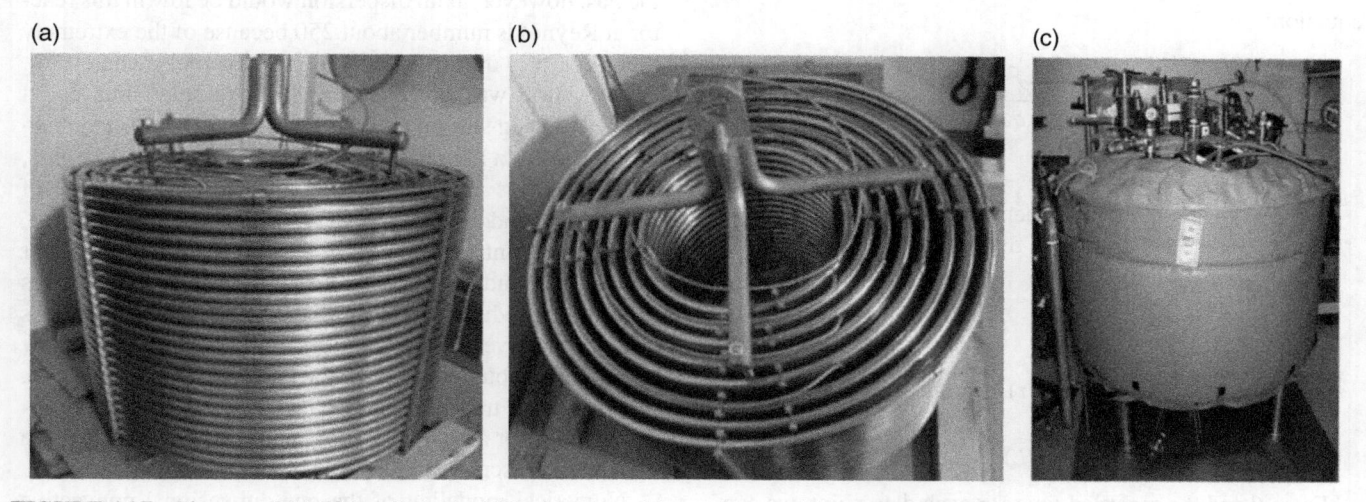

FIGURE 16.7 Pictures of (a) 73 L coiled tube PFR side view, (b) top view, and (c) constant temperature heating bath for submerging the reactor.

250 in the flow tube reactor. In the laminar regime, the larger the L/d, the lower the overall reactor axial dispersion number [18]. Due to the low linear velocity because of the 12 hours τ, the reactor operated with pressure drop of only about 10 psi; thus the pressure drop was not problematic. Reactor cost and ease of fabrication were the practical limitations on higher L/d in this case. The tubing was formed into eight concentric cylindrical coils 0.53 m tall and ranging from 0.36 to 0.80 m diameter as pictured in Figure 16.7. Vapor and liquid flowed cocurrently through each of the eight coils in series in the uphill direction starting with the outside coil. These individual coils were linked by down-jumper tubes constructed from 316 L stainless steel tubing with o.d. = 6.35 mm and i.d. = 4.57 mm. The reactor was designed with these narrow diameter down-jumpers to maintain greater than 99% of the total reactor volume in the uphill flow direction, which helped the reactor run more liquid filled in the forward direction. This also made it possible to almost completely empty the reactor by pushing with nitrogen in the reverse direction at the end of the production. There was no mechanical mixing of the hydrogen and liquid throughout the length of the tube. Gas–liquid mixing was sufficient so that conversion versus time was the same as in a well-mixed batch reactor.

The reactor was designed to be low cost so that it was dedicated to this specific chemistry and this rhodium catalyst and then disposed when no longer needed for the product. This reduced the cleaning burden, eliminated the possibility of cross-contamination, and eliminated other metal catalyst from the possibility of catalyzing undesired side reactions. The 73 L coiled tube PFR in this example cost $16 000.

Using the PFDR model, vessel dispersion number D/uL can be calculated by fitting experimental F-curve data to the basic differential equation representing dispersion by numerically solving for D/uL and θ as shown in the following equation:

$$\frac{\partial C}{\partial \theta} = \left(\frac{D}{uL}\right)\frac{\partial^2 C}{\partial z^2} - \frac{\partial C}{\partial z} \quad (16.14)$$

D = Taylor longitudinal dispersion coefficient that incorporates the effect of both diffusion and convection
u = average flow velocity
L = vessel length
C = concentration
θ = dimensionless time (t/τ), where τ is mean residence time
Z = dimensionless length (x/L)

From θ and the recorded time for each data point, we can calculate τ. The C-curve characterizes spreading of a pulse tracer injection into the inlet of a flow tube reactor as it travels along the length of the reactor, also known as residence time distribution. See chemical engineering textbooks for details on the methods and analyses [15]. For small extents of dispersion, $D/uL < 0.01$, the basic differential equation representing dispersion can be solved analytically to give the following symmetrical C-curve:

$$C = \frac{1}{2\sqrt{\pi\left(\frac{D}{uL}\right)}} \exp\left[-\frac{(1-\theta)^2}{4\left(\frac{D}{uL}\right)}\right] \quad (16.15)$$

Given that the F-curve is the integral of the C-curve, this equation was used to quantify D/uL and θ for the experimental F-curve using a simple Excel® spreadsheet.

Figure 16.8 shows the experimental F-curve data collected during startup transition of the 73 L coiled tube reactor. The reactor initially started filled with solvent. At time zero, the reagent solution with enone, the dissolved catalyst + ligand solution, and hydrogen gas started flowing into the reactor. The figure plots product concentration at the reactor exit versus normalized time, t/τ. $t/\tau = 1$ corresponds to 11.8 hours.

In this example, the axial dispersion number, D/uL, was 0.000 12, and τ was 11.8 hours (Figure 16.8). Thus, the continuous reactor exhibited nearly ideal plug flow characteristics. This is a remarkably low axial dispersion number given that the reactor is in the laminar regime. Reynolds number in the reactor was only about 250 (liquid density 0.9 g/ml, internal diameter 16 mm, linear velocity 8 mm/s, viscosity about 0.45 cp). The gas + liquid two-phase flow decreases axial dispersion compared with liquid-only flow because the gas bubbles disrupt the parabolic velocity profile and cause some mixing in the radial direction [16]. Even without the gas, however, axial dispersion would be low in this reactor at Reynolds number about 250 because of the extremely high length to diameter ratio, as explained elsewhere [16]. Steady state was achieved in this example after about $1.03 \times \tau$ during startup transition from a solvent-filled reactor, as shown in Figure 16.8. Target reaction τ was 12 hours, but actual was 11.8 hours. This emphasizes the value of measuring and modeling the startup transition curve, because it quantifies the actual τ. In addition to many other things, τ depends on % vapor space in the reactor, which is difficult to predict because it depends on relative gas versus liquid linear velocities. This makes it difficult to know actual τ without measuring the startup transition curve. A nonreactive tracer could also be used, but this was a production run for a pharmaceutical intermediate, so any extra additives are scrutinized.

Numerical modeling of the one-dimensional convective diffusion equation was performed to illustrate the effects of

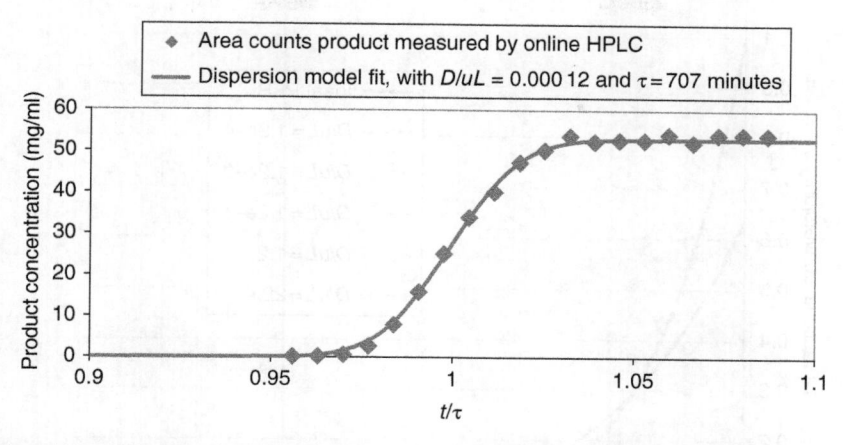

FIGURE 16.8 Experimental F-curve data and model fit for startup transition of a 73 L continuous hydrogenation reactor.

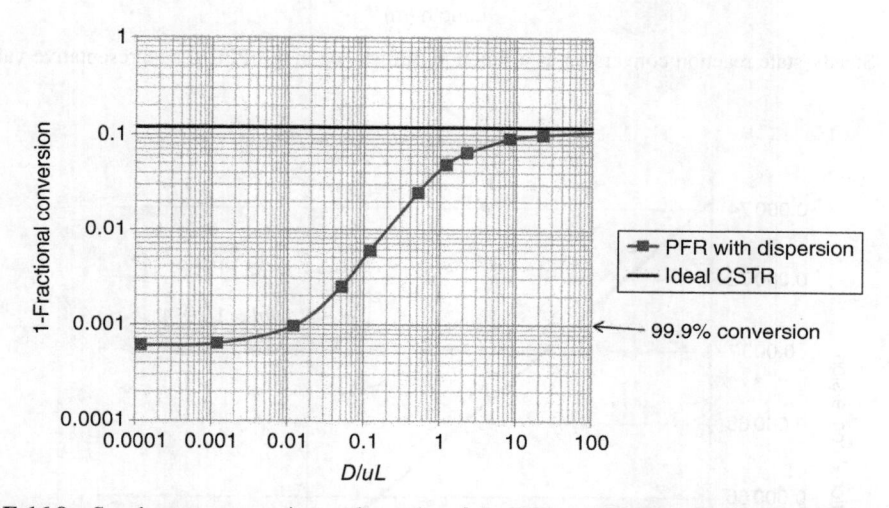

FIGURE 16.9 Steady-state conversion at the outlet of the PFR as a function of the dispersion coefficient.

dispersion in the 73 L PFR. The partial differential equation was solved using gPROMS custom modeler [19] discretizating along the length of the tube reactor. Backward finite differences were used, and a Danckwerts boundary condition was employed [20]. Numerical diffusion dictated the number of discretization points needed in order to simulate very low D/uL scenarios. Approximately 10 000 discretization grid points were needed to ensure numerical diffusion was small relative to the diffusion expected for a D/uL of 1.2e-4. Analytical expressions were used for the simulation of ideal PFR and CSTR cases.

First the effect of dispersion on the steady-state conversion at the exit of the tube reactor was examined. The numerical results of the model are shown in Figure 16.9. The limiting conversion at low D/uL is due to the residence time needed to achieve the design criteria of 99.9% conversion. At high D/uL the limiting case for an ideal CSTR is obtained. The conversion is insensitive to dispersion for $D/uL < 1\text{e-}3$.

Next the steady-state fractional conversion was simulated as a function of distance along the PFR for several representative values of D/uL, and the results are shown in Figure 16.10. At a D/uL of 1e-4, the concentration profile is virtually indistinguishable from the ideal PFR curve. For an ideal CSTR (infinitely high D/uL), the concentration at $L = 0$ drops (instantaneously) to the steady-state conversion of a CSTR with 11.8 hours τ.

Figure 16.11 shows the comparison of ideal PFR and the real PFR with $D/uL = 0.000\,12$ for the last 10 m of the reactor. The difference in conversion versus distance along the reactor is quantifiable but negligible for this low value of axial dispersion.

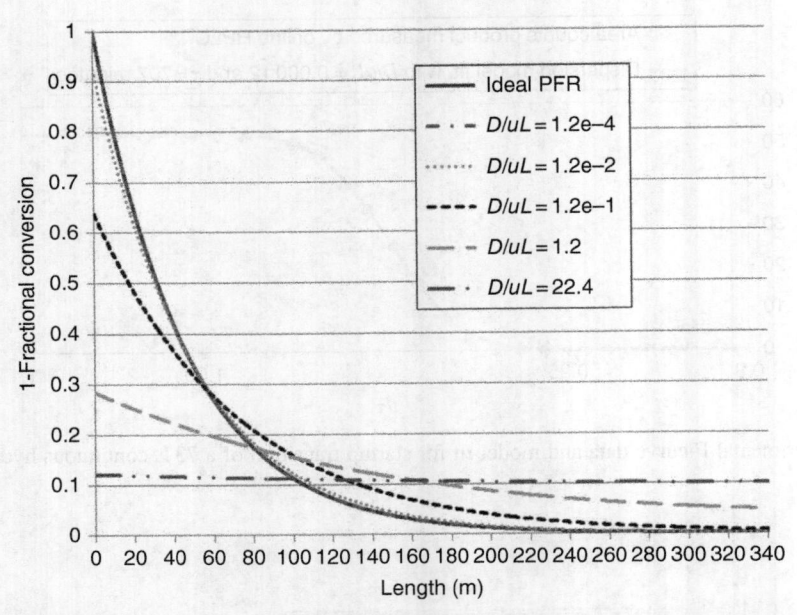

FIGURE 16.10 Steady-state reaction conversion as a function of distance in the PFR for representative values of D/uL.

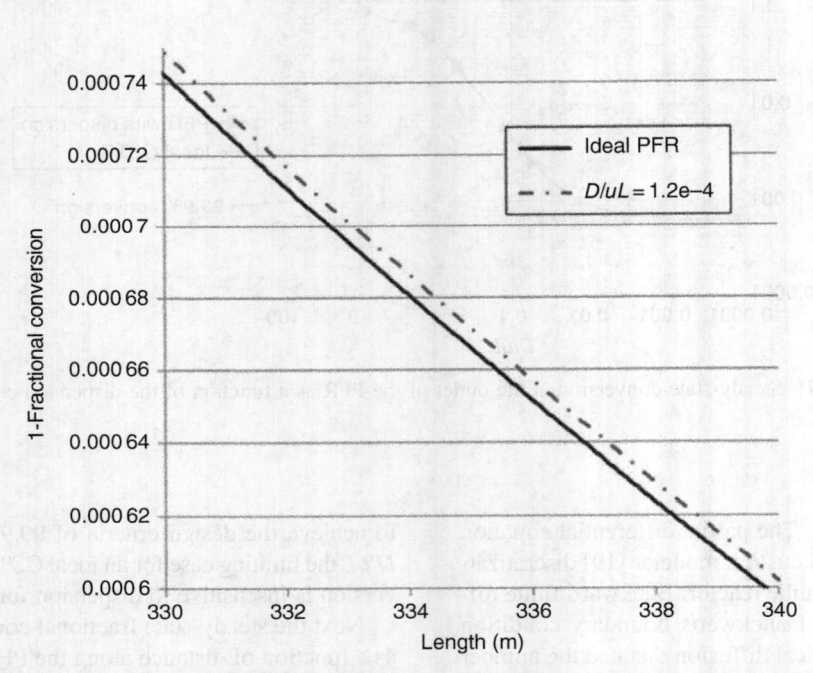

FIGURE 16.11 Steady-state reaction fractional conversion as a function of distance in the PFR for the final 10 m and comparison to ideal plug flow.

The ideal PFR curve was done analytically, and the plug flow with dispersion curve was done using the gPROMS numerical solver and 10 000 grid points. The simulations show that conversion versus distance is slightly less for the reactor with dispersion but clearly negligible for practical purposes.

To illustrate the effect of dispersion when not at steady state, concentration profiles along the length of the PFR are shown in Figure 16.12 for time = 1 hour after switch to the reactive feeds. For an ideal PFR the step change is infinitely narrow; however for larger values of dispersion, the step change gradually broadens out.

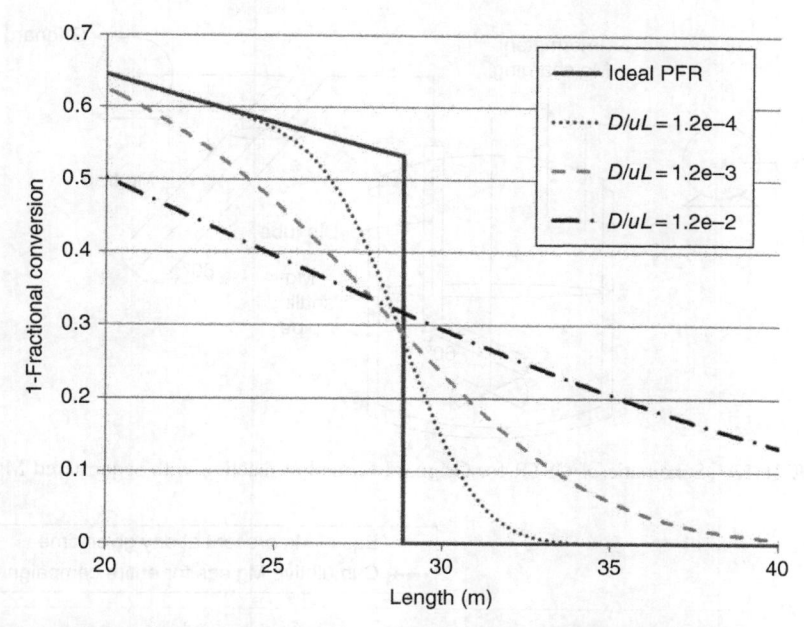

FIGURE 16.12 Reaction conversion as a function of distance in the PFR for different values of D/uL and time = 1 hour after switchover of the feeds from solvent to reagents in the startup transition.

16.4 EXAMPLE OF CSTR FOR GRIGNARD FORMATION REACTION WITH SEQUESTERED Mg SOLIDS

A continuous Grignard formation reaction was run in a CSTR with sequestered magnesium solids for reasons of safety and minimizing key impurities. The reaction is shown below in the Scheme 16.2.

The key impurities are shown below.

Wurtz Coupling   **Proteo**   **Phenol**

SCHEME 16.2 Grignard reaction in a CSTR.

The Wurtz coupling impurity was minimized by maintaining a high stoichiometric ratio of magnesium to aryl bromide starting material, the proteo impurity was reduced by minimizing water, and phenol was decreased by minimizing oxygen. Control of each of these impurities was easier to accomplish in a continuous reactor compared with batch. Oxygen and water are easier to minimize in a continuous reactor versus batch because the reactor is inerted and dried once and then remains inert and dry at steady state. The explanation for why the CSTR maintained high stoichiometric ratio of magnesium to aryl bromide starting material is given in the following discussion.

Liquid solutions were continuously pumped into the reactor containing a large molar excess of a solid activated magnesium metal, while liquid product solution was continuously pumped out, with the solid magnesium being sequestered in the reactor. Solid magnesium particles were periodically added once every four hours at an average rate equal to the molar feed rate of the starting material. In other words, Mg solids were added at a rate to match consumption. The magnesium was added in a very simple way. In a 6 L pilot-scale reactor, a 2 in diameter cap was removed from a nozzle on the top of a glass reactor, and the Mg was poured into the vessel from a beaker through a funnel. In a 100 L manufacturing reactor, the port on the Hastelloy reactor was removed, and Mg was poured into the vessel from 1 gal wide-mouth containers inside a glove bag. The freshly added metal rapidly activated when it mixed with the existing Grignard solution without the need for any additional activating agent. The kinetics of the Grignard formation reaction were extremely fast, achieving greater than 99% conversion in a single CSTR with a 1 hour mean residence time. This corresponds to a batch reaction with 99% conversion in 26 seconds, as described later in this section. A simplified schematic of the reactor system is shown in Figure 16.13.

Keeping solid magnesium particles in the CSTR was a key operational challenge. The two main methods/devices used to prevent magnesium from exiting the system were the settling pipe in the CSTR and the magnesium settling trap immediately downstream from the CSTR. These are described in more detail in the literature [21]. A 6 L pilot-scale reactor operated at the 4.5 L fill level and a 100 L manufacturing-scale

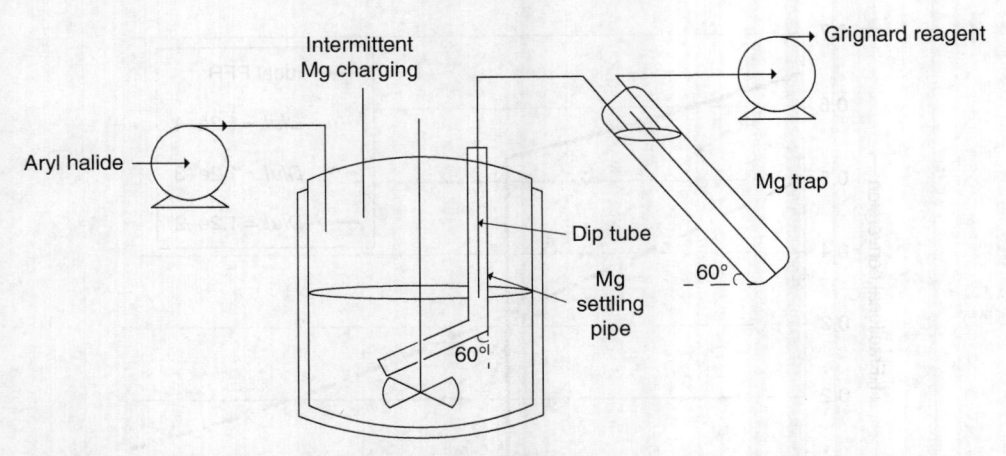

FIGURE 16.13 Schematic of CSTR for Grignard formation reaction with sequestered Mg solids.

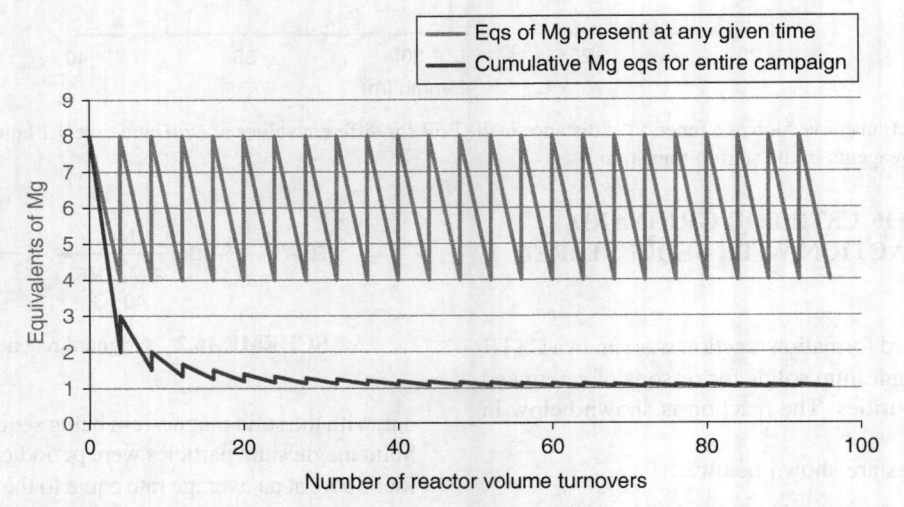

FIGURE 16.14 Instantaneous molar equivalents of magnesium in the reactor and also the overall cumulative molar equivalents of magnesium added for the 96 turnover manufacturing production campaign.

reactor operated at the 45 L fill level. The 100 L CSTR was used to generate 4000 L of 0.85 M Grignard reagent solution for an API starting material. Forward processing of the Grignard reagent into a coupling reaction and subsequent workup and isolation were done in batch 8000 L vessels. There were significant safety advantages of running the reaction in a 100 L CSTR instead of 8000 L batch reactor. The Grignard formation reaction was highly exothermic, initiation was difficult, and the reaction had runaway potential; therefore minimizing the size of the reactor minimized the safety risks. The Grignard initiation was confirmed in the CSTR by stopping flow of aryl bromide after five minutes and quantifying the exotherm before continuing. The Grignard was first initiated batch but only at 250 ml reactor scale. A 250 ml CSTR produced enough active Grignard reagent to initiate a 6 L CSTR, and the 6 L CSTR produced enough active Grignard reagent to initiate a 100 L CSTR.

The number of molar equivalents of magnesium in the reactor relative to substrate oscillated. The goal was to keep at least 4 M equiv. Mg in the CSTR relative to substrate, which was greater than 99% Grignard reagent rather than aryl bromide at any time. Therefore, the CSTR started with 8 M equiv., and the first Mg recharge was done after four reactor volume turnovers. Plotting the instantaneous molar equivalents of magnesium in the reactor results in a sawtooth plot over the 96 turnover manufacturing campaign (Figure 16.14). Observe that the Grignard reaction maintained high instantaneous magnesium equivalents (4–8) while achieving usage of 1.04 equiv. of magnesium overall for the manufacturing campaign. This can be compared with 1.5 equiv. of magnesium that would have been used in the batch campaign. This is probably the main reason why the CSTR approach gave less Wurtz coupled impurity compared with the batch reaction approach. In addition, a batch reaction with 1.5 M equiv. of Mg would have resulted in more magnesium to quench at the end, compared with the CSTR manufacturing that used 1.04 M equiv. overall. The magnesium quench liberates hydrogen; therefore the continuous process generated less hydrogen than the batch counterpart would have generated, which was a safety advantage of the CSTR approach.

The reaction achieved 99.9% conversion with a 60 minute residence time in a single CSTR operating at 41 °C. Assuming pseudo first-order reaction process, the rate constant can be calculated:

$$\frac{C_{AF}}{C_{A0}} = \frac{1}{1+k\tau} \quad (16.12)$$

$$0.001 = \frac{1}{1+k \times 60}$$

$k = 16/\text{minute}$

Conversion for different τ can be predicted once the rate constant is known. For example, if the residence time is reduced to 30 minutes, then conversion would drop to 99.8%:

$$\frac{C_{AF}}{C_{A0}} = \frac{1}{1+16 \times 30} = 0.998$$

One can calculate what the reaction time would be for 99.9% conversion if the reaction was run isothermally batch. Of course, it is not possible to run this reaction isothermally batch because of the large exotherm and the unrealistically high heat removal rate that would be required:

$$\frac{C_{AF}}{C_{A0}} = e^{-kt} \quad (16.5)$$

$$0.001 = e^{-16 \times t}$$

$t = 0.43$ minutes

It is important to note that it is not practical to measure this rate constant by running a batch reaction. The adiabatic temperature rise was 130 °C; therefore it would not be possible to keep the reactor temperature at 40 °C during a 26 second reaction. Performing this reaction in batch would also have mass transfer limitations. Therefore running this reaction in a CSTR is a more effective way to experimentally quantify the true kinetics for this fast exothermic reaction.

At pilot scale, the automation system monitored the process with a real-time energy balance in the 6 L CSTR campaign, in order to prove the reaction was operating at a high conversion. At steady state, the heat removal rate by the cooling jacket was 188 W:

$$Q_{\text{jacket}} = \dot{m}_{\text{jacket}} \cdot Cp_{\text{chiller}} \cdot \Delta T_{\text{jacket}}$$

$\dot{m}_{\text{jacket}}$: flow rate of chiller fluid $= 2.20 \dfrac{\text{kg}}{\text{min}}$

Cp_{chiller} : specific heat of chiller fluid $= 1.77 \dfrac{\text{kJ}}{\text{kg} \cdot \text{K}}$

$\Delta T_{\text{jacket}} : (T_{\text{out,jacket}} - T_{\text{in,jacket}}) = (21.8\,°\text{C} - 18.9\,°\text{C}) = 2.90\,°\text{C} = 2.90\,\text{K}$

$$Q_{\text{jacket}} = \left(2.20 \frac{\text{kg}}{\text{min}}\right)\left(1.77 \frac{\text{kJ}}{\text{kg} \cdot \text{K}}\right)(2.90\,\text{K})\left(1000 \frac{\text{J}}{\text{kJ}}\right)$$
$$\left(\frac{1\,\text{minute}}{60\,\text{seconds}}\right) = 188 \frac{\text{J}}{\text{s}} \quad (16.16)$$

Increasing the incoming feed from room temperature to the reaction temperature consumed 26.4 W:

$$Q_{\text{feed}} = \dot{m}_{\text{feed,in}} \cdot Cp_{\text{solution}} \cdot \Delta T$$

$\dot{m}_{\text{feed,in}}$: flow rate of feed $= 45.8\,\text{g/min}$

Cp_{solution} : specific heat of solution $= 1.85 \dfrac{\text{J}}{\text{g} \cdot \text{K}}$

$\Delta T : (T_{\text{reactor}} - T_{\text{feed}}) = (41.7\,°\text{C} - 23\,°\text{C}) = 18.7\,°\text{C} = 18.7\,\text{K}$

$$Q_{\text{feed}} = \left(45.8 \frac{\text{g}}{\text{min}}\right) \cdot \left(1.85 \frac{\text{J}}{\text{g} \cdot \text{K}}\right) \cdot (18.7\,\text{K}) \cdot \left(\frac{1\,\text{minute}}{60\,\text{seconds}}\right) = 26.4 \frac{\text{J}}{\text{s}}$$

The steady-state heat generation rate from reaction was 208 W:

Energy generation rate $= \Delta H_{\text{reaction}} \cdot C_{\text{feed}} \cdot V_{\text{feed}} \cdot X_a$

$\Delta H_{\text{reaction}}$: molar enthalpy of reaction, measured by calorimeter
$= 292\,000 \dfrac{\text{J}}{\text{mol}}$

C_{feed} : concentration of feed $= 0.85\,\text{mol/l}$

V_{feed} : volumetric flow rate $0.0504\,\text{l/min}$

X_a : conversion of component A (Grignard) $= 0.999$

$$\left(292\,330 \frac{\text{J}}{\text{mol}}\right)\left(0.85 \frac{\text{mol}}{\text{l}}\right)\left(0.0504 \frac{1}{\text{min}}\right)(0.999)$$
$$\left(\frac{1\,\text{minute}}{60\,\text{seconds}}\right) = 208 \frac{\text{J}}{\text{s}} \quad (16.17)$$

The heat loss to surroundings was assumed zero since the reactor jacket was room temperature.

Real-time energy balance was 103%, because total heat removed was 214 W and heat generated was 208 W. This real-time energy balance was an important safety feature of the process, and it was constantly calculated and monitored by the distributed control system (DCS). Automated shutoffs were in place to stop the reagent feeds if the real-time energy balance was below a limit, preventing the possibility of unreacted aryl bromide reagent accumulating to unsafe levels in the Grignard reactor.

16.5 NUMERICAL MODELING TO SELECT THE BEST REACTOR TYPE FOR MINIMIZING IMPURITIES

A simple hypothetical series–parallel reaction is used next to illustrate that there can be significant differences in impurity profiles for the same reaction depending on the reactor choice (PFR vs. batch with controlled addition vs. CSTR). The reaction and the kinetics (at 25 °C) in this example are

$$A + B \rightarrow P \quad k_1 = 2 \times 10^{-3} \frac{1}{\text{mol} \cdot \text{s}}$$

$$P + 2B \rightarrow \text{Imp} \quad k_2 = 3 \times 10^{-5} \frac{l^2}{\text{mol}^2 \cdot \text{s}}$$

where

- A and B are reactants.
- P is the product.
- Imp is an undesired impurity.

The details of the governing equations and the parameters used in this illustration are listed in Table 16.2. The governing equation for PFR is the same as batch with all-at-once addition, because time variable in batch translates to residence time along the reactor for a PFR.

The concentrations after all reagents are mixed (if there was no reaction) would be the same in all three cases in the table. For the batch with controlled addition over 30 minutes, c_A after mixing would be $2 \times 3/(3 + 4 \times 0.5) = 1.2$ M, and c_B after mixing would be $4 \times 4 \times 0.5/(3 + 4 \times 0.5) = 1.6$ M.

The model for this example was implemented using DynoChem [22]. The governing equations are different between PFR, batch with controlled addition, and CSTRs in series. Even though the kinetics are the same, the impurity profiles are different for the three different reactors. In this example, the PFR with residence time of 1 hour results in 96% *in situ* yield and 2.5% Imp. Controlled addition of B over 30 minutes, followed by 30 minutes of additional reaction mixing time, results in 94% *in situ* yield and 1.8% Imp. If this reaction is run using three CSTRs each with 1 hour residence time, an *in situ* yield of 94% can be achieved, but the impurity level at steady state is 5%. Figure 16.15 shows the impurity level versus conversion of A for (i) a PFR, (ii) batch with controlled addition, and (iii) three equal volume CSTRs in series.

In this case, a controlled addition (semi-batch) gives a lower impurity level than a PFR because it keeps the concentration of B lower. CSTRs give the highest level of impurities in this case because they are operating at steady-state conditions with P and B together for longer times. B is at low levels, but P is always at high concentration levels, and contact time is 3 hours in the three CSTRs in series. The figure

TABLE 16.2 Governing Equations and Parameters Used to Illustrate Difference in Impurity Profile Between PFR, Batch with Controlled Addition (Semi-batch), and CSTR

PFR (Batch with All-at-Once Addition)	Controlled Addition (Semi-batch)	Three CSTRs in Series
$\frac{dc_j}{dt} = r_j$	$\frac{dc_j}{dt} = \left(c_j^f - c_j\right) v^f + r_j$	$\frac{c_j^1 - c_j^0}{\tau} = r_j(c^1)$
$c_j(t=0) = c_j^0$	$c_j(t=0) = c_j^0$	$\frac{c_j^2 - c_j^1}{\tau} = r_j(c^2)$
c_j^0: Inlet concentration of species j to reactor	c_j^0: Initial concentration of species j	$\frac{c_j - c_j^2}{\tau} = r_j(c)$
c_j: Concentration of species j along the reactor (at residence time t)	c_j: Concentration of species j at time t	c_j^0: Inlet concentration of species j to CSTR 1
r_j: rate of formation per unit volume of species j	v^f: Ratio of volumetric flow rate of addition stream to volume of reactor	c_j^1, c_j^2, c_j: Concentration of species j in CSTR 1, 2, and 3, respectively
	c_j^f: Species j concentration in addition stream	τ: Residence time in each CSTR
	r_j: Rate of formation per unit volume of species j	r_j: Rate of formation per unit volume of species j

Reaction Rate Expressions

$$r_A = -k_1 c_A c_B$$
$$r_B = -k_1 c_A c_B - 2k_2 c_P c_B^2$$
$$r_P = k_1 c_A c_B - k_2 c_P c_B^2$$
$$r_{\text{Imp}} = k_2 c_P c_B^2$$

PFR
Inlet concentration: $c_A^0 = 1.2$ M, $c_B^0 = 1.6$ M
Residence time: 1 h

Controlled Addition (Semi-batch)
Initial concentration in reactor: $c_A^0 = 2$ M, $c_B^0 = 0$ M
Addition stream concentration: $c_A^f = 0$ M, $c_B^f = 4$ M
Addition stream flow rate: 4 L/h, initial reactor volume: 3 L, and dosing time: 30 min
Reaction time: 1 h

CSTR: (Three CSTRs in Series with Equal Residence Time)
Inlet concentration to CSTR 1: $c_A^0 = 1.2$ M, $c_B^0 = 1.6$ M
Residence time of each CSTR: 1 h

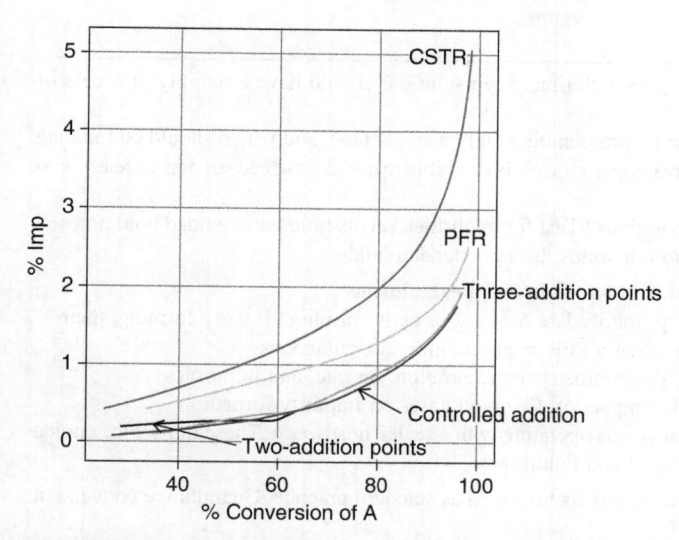

FIGURE 16.15 Impurity level vs. conversion for PFR, batch with controlled addition, 3 CSTRs with equal residence times, and PFR with multiple addition points along the length.

also shows two additional scenarios that begin to mimic batch with controlled addition in a PFR. Reagent B is added at multiple points along the PFR. The normal PFR has one addition point, which is at the inlet to the tube. The PFR with one side stream has two addition points, one at the inlet and one 1/3 of the way down the reactor length. Half of the reagent B feed solution was added at each location. The PFR with two side streams has three addition points. One is at the inlet, the second is at a distance 1/6 of the way down the reactor length, and the third is at a distance 1/3 of the way down the reactor length. Reagent B feed solution was added in three equal portions at the three locations. Interestingly, the impurity level in the PFR with multiple addition points matched batch reactor with controlled addition, even with only one side stream. Practically speaking, the cost of running a PFR with side streams is that another pump or control valve with mass flow control is needed for each additional side stream. This may be low cost compared to the benefit in yield and selectivity. Clearly, the numerical modeling is useful for reactor design and selection provided that the kinetics of the main reaction and of the unwanted side and series reactions are known. It is especially important when the impurities are difficult to reject and/or could be reactive downstream.

16.6 WHAT CAN BE DONE BATCH BEFORE RUNNING CONTINUOUS REACTION EXPERIMENTS?

Much of the experimental work to generate the data needed to develop a continuous flow process can actually be accomplished more quickly and efficiently in batch equipment. The impact of different reagents, solvents, and catalysts can often be run using less material in batch. In addition, batch experiments do not have the transitional waste produced in flow experiments to achieve steady state after changing an operational parameter. One may observe reaction characteristics in batch (such as solid precipitation) that will influence the eventual choice of flow reactor type or even advise when a flow process undesirable. Table 16.3 is a guide to what can be done batch to help design the continuous reaction before running continuous flow experiments.

16.7 WHAT IS DIFFICULT TO PREDICT FROM BATCH EXPERIMENTS?

Table 16.4 lists information that is best obtained in the actual flow chemistry experiments because it is difficult to predict from batch experiments.

TABLE 16.3 Batch Experimental Work That Can Inform Design for Flow

Solubility	Try to make all reagent solutions in addition funnels or syringe pumps, i.e. no solid charges if possible. Solid precipitation in the continuous reactor is much easier to handle than a continuous slurry feed to the reactor. Establish homogeneous solution feeds
	Measure solubility of reagents, products, intermediates, and by-products
	Visual observation over time. If possible, observe changes in physical state throughout Monitor for solid precipitation or gas evolution at all extents of conversion during reaction. Solids sticking to the walls of the batch reactor indicate that fouling will likely occur if the reaction were run in flow
	If there are solids, are they sticky and clumpy or more like a well-behaved crystallization? Use of a CSTR or slurry flow PFR is OK for well-behaved solids but not for sticky clumpy solids
CSTRs	If considering a CSTR, then batch-on-batch reactions can be run to mimic what would be expected in a CSTR. Run a batch reaction at minimum stir volume; then run another batch reaction on top of it. Is impurity profile better or worse for the second reaction? An intermittent flow CSTR can be mimicked by running a batch reaction, followed by a decant leaving a heel, followed by another batch, then decant with a heel, and so on. Test different heel volumes
	Test the impact of order of addition. Coaddition of A and B versus controlled addition of A to B versus all-at-once addition of A and B (although all-at-once addition may be difficult to test batch). This will help to select between a CSTR (coaddition), intermittent flow CSTR (A to B), and PFR (all-at-once addition)

(continued)

TABLE 16.3 (Continued)

Stability	Test the chemical stability of reagent solutions with time. Feed solutions should have a stability in excess of one week
	What reagents, catalysts, and additives can be pre-combined into a single feed, and which should be separate feeds for stability in the feed tank at room temperature? It is desirable to use as few feed streams as feasible to simplify the flow process
	Is the reaction stable to end-of-reaction conditions? Test for impurities versus time for extended hold times in the batch reactor. Watch for precipitation of solids during extended hold
Impurities	Identify impurities and develop analytical methods from batch experiments
	Monitor impurities versus time. Are they primarily late forming or early forming? If early forming, then consider using a CSTR. If late forming, then a PFR might be more advantageous
	Test the impact of temperature, pressure, stoichiometry, concentration, on rate, and impurities
	Test the impact of level of inertion and the impact of dissolved gases on impurity formation
	Test the impact of stripping during reaction versus operating with a sealed headspace. These impact the choice between a PFR designed for gas and liquid and liquid only
Kinetics	Measure rates. Collect samples for conversion versus time data as standard practice. Quantitative conversion data is much more valuable than area %
	Quantify kinetics of secondary reactions especially with respect to impurity profile
	Monitor concentration of intermediates over time
	Use of PAT for reaction monitoring and for model fitting to the rate profile curve. This provides the reaction rate and also proves a PAT tool to use when you run flow chemistry. Flow NMR is a valuable screening tool for kinetics because it gives quantitative conversion versus time with direct molar ratios
Safety	Test materials of construction for corrosion and also impact on reaction
	Do chemical reaction safety testing for heat of reaction: ARC, DSC, and RC1
Screening	If you plan to use a stirred tank because of solid precipitation, then try to find conditions that achieve complete conversion in 1 h or less in the batch reactor
	In a PFR, if the reaction is a homogeneous solution at reaction temperature but not after cooling to ambient temperature, then you may be able to add in a diluting solvent at the exit of the PFR before cooling to keep solids from precipitating downstream from the reactor. Determine this in batch experiments first
	For reactions involving a homogeneous solution in the reactor or that are gas–liquid with only one liquid phase, then 12 h reaction time is acceptable in a PFR. In other words, find conditions that achieve complete conversion in about 12 h or less in the batch reactor
	Discreet variables (solvent, reagent, catalyst, ligand) are best scouted in batch; continuous variables (T, P, τ, and in some cases stoichiometry and concentration) may be best screened in flow
	Screen at high and low temperatures and at high pressures in batch reactions. Go more extreme in these process parameters than you would if you were developing a batch reaction for standard manufacturing equipment
	Think more broadly about solvent selection. One can operate far above the normal solvent boiling temperature by operating in a high pressure flow reactor. Solvents can be selected based on simplifying workup, greenness, and cost, in addition to yield and selectivity, even if the normal boiling point of the solvent is much lower than the desired reaction temperature, because PFRs enable running the reaction in superheated solvent
	You may be able to more easily perform the reaction in a high boiling solvent and then isolate the product from a low boiling solvent. Solvent exchange from a high to a low boiling solvent can be more efficient continuous than batch
	Initial route scouting should be done batch
	If you are running a gas–liquid reaction with a mixture of gases, like H_2/CO, or O_2/N_2, test the impact of the feed gas stoichiometry
	Investigate large particle size catalyst. Reuse catalyst several times. Does activity decrease over time? Does metal leach? If not then consider a packed catalyst bed flow reactor
	Use the breathing autoclaves first if the reaction is an aerobic oxidation with dilute oxygen in nitrogen
	Test the impact of the reactor headspace volume. If there is advantage to zero headspace, then the reaction is a good candidate to utilize a PFR
	Investigate the effect of mixing rate
Thermodynamics and fluid properties	Measure phase transition points for the reaction solution, i.e. boiling point, freezing point, and precipitation point
	Measure the thermal expansion of the reaction mixture and the product mixture
	Measure densities of all feed and product solutions at room temperature. This is especially important if you will be pumping with a volumetric pump like a syringe or peristaltic

TABLE 16.4 In Batch Experiments It Would Be Difficult to Predict Information About the Following

Packed catalyst bed	If designing for a packed catalyst bed PFR, then catalyst life, i.e. number of reactor volume turnovers achievable with single catalyst charge, is best measured in flow
	Localized equivalents of catalyst relative to dissolved reagents in a packed bed reactor can be orders of magnitude higher than with solids suspended in a stirred tank. The impact of this high catalyst loading is best tested in flow. The extremely high gas–liquid mass transfer rate of a trickle bed reactor is not feasible to duplicate in batch reaction experiments
	The following should be probed in flow:
	• Catalyst wetting efficiency
	• The impact of adsorption/desorption on the startup and shutdown transition and on the difference between substrate residence time and solvent residence time
	• The procedures for preconditioning, pre-wetting, and pretreating a new packed bed
	• The pressure drop across the packed catalyst bed
	• The liquid fill fraction in the packed catalyst bed
	Pressure, temperature, concentration, vapor–liquid flow conditions, and residence time to minimize problematic impurities and minimize catalyst deactivation over time are better tested in flow because it is difficult for batch reactions to predict impurity profiles in a continuous packed catalyst bed
	Flow regime should be tested in the continuous reactor (trickle, pulse, spray, or bubble). What are the gas and liquid linear velocities at the transition points from one flow regime to another? For example, if increasing flows in a trickle bed reactor, at what flow rates does the transition from trickle flow to bubble flow occur?
Recycle	Effect of recycle on yield and impurity profile and impurity buildup in recycle loop over time
Fouling	Plugging and fouling issues and thus reliability over time for the flow process
Heat and mass transfer rates	Heat and mass transfer coefficients in tube reactors
	Effect of fast heat-up and cooldown times (on order of seconds) on yield, selectivity. Are yield and impurity profile sensitive to heat-up time, cooldown time, and time at reaction temperature? These are easier to test flow than batch
	Extreme temperatures and reactions in supercritical fluids with short residence time, for example, Newman–Kwart thermal rearrangement in supercritical DME [23] and thermal BOC deprotection in supercritical THF [24]
	All-at-once stoichiometric reagent addition, i.e. PFR operating mode. It is especially difficult to measure batch for fast exothermic reactions, for example, cryogenic lithiation and coupling, because it could be difficult or impossible to maintain nearly isothermal reaction conditions with all-at-once addition batch
PFR	Yield and impurity profile of organic azide formation reactions because of the need to run 100% liquid filled to avoid hydrazoic acid partition into the headspace [8]
	Impurity profile for fast reactions in series with unstable intermediates, like cryogenic lithiation, coupling, and quench
CSTR	Steady-state yield and selectivity for Grignard formation reactions in a CSTR. The batch startup event with initiating agent like DIBAL or iodine may not be representative of steady-state CSTR operation [25]
	Reaction rates for extremely fast exothermic reaction like a Grignard formation. The reaction rate constants are more accurately measured by running a CSTR at steady state with incomplete conversion and calculating the rate constant from the CSTR design equation

ACKNOWLEDGMENTS

Mark Kerr was the process chemist for the continuous Grignard formation reaction in the 100 L CSTR, and Sylvia Nwosu was the engineer on the scale-up project. The third-party CMO who did the scale-up manufacturing work did an outstanding job and are largely responsible for the success of the project. Ed Deweese, Paul Milenbaugh, and Rick Spears of D & M Continuous Solutions constructed and operated the hydrogenation PFR system. Jake Remacle, Miguel Gonzales, Wei-Ming Sun, and Nick Zaborenko developed the 73 L coiled tube PFR for hydrogenation, and the chemistry was designed and developed by Joel Calvin. We thank Bret Huff for initiating, leading, and sponsoring the continuous reaction design and development work at Eli Lilly and Company.

REFERENCES

1. (a) Malet-Sanz, L. and Susanne, F. (2012). Continuous flow synthesis. A pharma perspective. *J Med Chem* 55 (9): 4062–4098.
   (b) LaPorte, T.L. and Wang, C. (2007). Continuous processes for the production of pharmaceutical intermediates and active pharmaceutical ingredients. *Curr Opin Drug Discov Devel* 10 (6): 738–745.
   (c) Wegner, J., Ceylan, S., and Kirschning, A. (2012). Flow chemistry: a key enabling technology for (multistep) organic synthesis. *Adv Synth Catal* 354 (1): 17–57.
   (d) Mullin, R. (2007). Novartis and MIT study drug production. *Chemical & Engineering News* (8 October 2007), p. 10.
   (e) Pellek, A. and Van Arnum, P. (2008). Continuous processing: moving with or against the manufacturing flow. *Pharm Technol* 9 (32): 52–58.
   (f) Gutmann, B., Cantillo, D., and Kappe, C.O. (2015). Continuous-flow technology: a tool for the safe manufacturing of active pharmaceutical ingredients. *Angew Chemie Int Ed* 54 (23): 6688–6728.
   (g) Anderson, N.G. (2012). Using continuous processes to increase production. *Org Process Res Dev* 16 (5): 852–869.
   (h) Hessel, V., Kralisch, D., Kockmann, N. et al. (2013). Novel process windows for enabling, accelerating, and uplifting flow chemistry. *ChemSusChem* 6 (5): 746–789.
   (i) Baxendale, I.R. (2013). The integration of flow reactors into synthetic organic chemistry. *J Chem Technol Biotechnol* 88 (4): 519–552.
   (j) Baxendale, I.R., Braatz, R.D., Hodnett, B.K. et al. (2016). Achieving continuous manufacturing: technologies and approaches for synthesis, workup, and isolation of drug substance May 20-21, 2014 continuous manufacturing symposium. *JPharmSci* 104(3):781–791.
   (k) Hartman, R.L., McMullen, J.P., and Jensen, K.F. (2011). Deciding whether to go with the flow: evaluating the merits of flow reactors for synthesis. *Angew Chem Int Ed Engl* 50 (33): 7502–7519.
   (l) Mascia, S., Heider, P.L., Zhang, H. et al. (2013). End-to-end continuous manufacturing of pharmaceuticals: integrated synthesis, purification, and final dosage formation. *Angew Chem Int Ed Engl* 52 (47): 12359–12363.
   (m) Scott, A. (2013). Pfizer adds biologics capacity, continuous processing in Ireland. *Chem Eng News* 91 (29): 9.
2. (a) Webb, D. and Jamison, T.F. (2010). Continuous flow multistep organic synthesis. *Chem Sci* 1 (6): 675.
   (b) McQuade, D.T. and Seeberger, P.H. (2013). Applying flow chemistry: methods, materials, and multistep synthesis. *J Org Chem* 78 (13): 6384–6389.
   (c) Ahmed-Omer, B., Barrow, D.A., and Wirth, T. (2011). Multistep reactions using microreactor chemistry. *Arkivoc* 4: 26–36.
   (d) Grongsaard, P., Bulger, P.G., Wallace, D.J. et al. (2012). Convergent, kilogram scale synthesis of an Akt kinase inhibitor. *Org Process Res Dev* 16 (5): 1069–1081.
   (e) Browne, D.L., Baumann, M., Harji, B.H. et al. (2011). A new enabling technology for convenient laboratory scale continuous flow processing at low temperatures. *Org Lett* 13 (13): 3312–3315.
   (f) Fandrick, D.R., Roschangar, F., Kim, C. et al. (2011). Preparative synthesis via continuous flow of 4, 4, 5, 5-tetramethyl-2-(3-trimethylsilyl-2-propynyl)-1, 3, 2-dioxaborolane: a general propargylation reagent. *Org Process Res Dev* 16 (5): 1131–1140.
   (g) Desai, A.A. (2012). Overcoming the limitations of lithiation chemistry for organoboron compounds with continuous processing. *Angew Chem Int Ed* 51 (37): 9223–9225.
3. Kockmann, N. and Roberge, D.M. (2009). Harsh reaction conditions in continuous-flow microreactors for pharmaceutical production. *Chem Eng Technol* 32 (11): 1682–1694.
4. (a) Baxendale, I.R., Braatz, R.D., Hodnett, B.K. et al. (2015). Achieving continuous manufacturing: technologies and approaches for synthesis, workup, and isolation of drug substance. May 20–21, 2014 continuous manufacturing symposium. *JPharmSci* 104(3):781–791.
   (b) Anderson, N.G. (2012). Using continuous processes to increase production. *Org Process Res Dev* 16 (5): 852–869.
   (c) Hessel, V., Kralisch, D., Kockmann, N. et al. (2013). Novel process windows for enabling, accelerating, and uplifting flow chemistry. *ChemSusChem* 6 (5): 746–789.
   (d) Mascia, S., Heider, P.L., Zhang, H. et al. (2013). End-to-end continuous manufacturing of pharmaceuticals: integrated synthesis, purification, and final dosage formation. *Angew Chem Int Ed* 52 (47): 12359–12363.
5. (a) Wegner, J., Ceylan, S., and Kirschning, A. (2012). Flow chemistry–a key enabling technology for (multistep) organic synthesis. *Adv Synth Catal* 354 (1): 17–57.
   (b) Brocklehurst, C.E., Lehmann, H., and La Vecchia, L. (2011). Nitration chemistry in continuous flow using fuming nitric acid in a commercially available flow reactor. *Org Process Res Dev* 15 (6): 1447–1453.
   (c) Brandt, J.C., Elmore, S.C., Robinson, R.I., and Wirth, T. (2010). Safe and efficient Ritter reactions in flow. *Synlett* (20): 3099–3103.
   (d) Delville, M.M., Nieuwland, P.J., Janssen, P. et al. (2011). Continuous flow azide formation: optimization and scale-up. *Chem Eng J* 167 (2): 556–559.
   (e) Palde, P.B. and Jamison, T.F. (2011). Safe and efficient tetrazole synthesis in a continuous-flow microreactor. *Angew Chem Int Ed* 50 (15): 3525–3528.
6. Mastronardi, F., Gutmann, B., and Kappe, C.O. (2013). Continuous flow generation and reactions of anhydrous diazomethane using a Teflon AF-2400 tube-in-tube reactor. *Org Lett* 15 (21): 5590–5593.
7. Dunn, P.J., Wells, A., and Williams, M.T. (eds.) (2010). *Green chemistry in the pharmaceutical industry*. Weinheim: Wiley.
8. Kopach, M.E., Murray, M.M., Braden, T.M. et al. (2009). Improved synthesis of 1-(azidomethyl)-3, 5-bis-(trifluoromethyl) benzene: development of batch and microflow azide processes. *Org Process Res Dev* 13 (2): 152–160.
9. Poechlauer, P., Colberg, J., Fisher, E. et al. (2013). Pharmaceutical roundtable study demonstrates the value of continuous manufacturing in the design of greener processes. *Org Process Res Dev* 17 (12): 1472–1478.
10. Mascia, S., Heider, P.L., Zhang, H. et al. (2013). End-to-end continuous manufacturing of pharmaceuticals: integrated

synthesis, purification, and final dosage formation. *Angew Chem Int Ed* 52 (47): 12359–12363.

11. (a) Gutmann, B., Cantillo, D., and Kappe, C.O. (2015). Continuous-flow technology: a tool for the safe manufacturing of active pharmaceutical ingredients. *Angew Chem Int Ed* 54 (23): 6688–6728.
    (b) Anderson, N.G. (2012). Using continuous processes to increase production. *Org Process Res Dev* 16 (5): 852–869.
    (c) Hessel, V., Kralisch, D., Kockmann, N. et al. (2013). Novel process windows for enabling, accelerating, and uplifting flow chemistry. *ChemSusChem* 6 (5): 746–789.
    (d) Baxendale, I.R. (2013). The integration of flow reactors into synthetic organic chemistry. *J Chem Technol Biotechnol* 88 (4): 519–552.
    (e) Baxendale, I.R., Braatz, R.D., Hodnett, B.K. et al. (2015). Achieving continuous manufacturing: technologies and approaches for synthesis, workup, and isolation of drug substance. May 20–21, 2014 continuous manufacturing symposium. *J Pharm Sci* 104 (3): 781–791.

12. (a) Lee, S.L., O'Connor, T.F., Yang, X. et al. (2015). Modernizing pharmaceutical manufacturing: from batch to continuous production. *J Pharm Innov* 10 (3): 191–199.
    (b) Allison, G., Cain, Y.T., Cooney, C. et al. (2015). Regulatory and quality considerations for continuous manufacturing. May 20–21, 2014 continuous manufacturing symposium. *J Pharm Sci* 104 (3): 803–812.

13. White, T.D., Berglund, K.D., Groh, J.M. et al. (2012). Development of a continuous Schotten–Baumann route to an acyl sulfonamide. *Org Process Res Dev* 16 (5): 939–957.

14. May, S.A. and Johnson, M.D. (2013). Continuous processing to enable more efficient synthetic routes and improved process safety. *Pharm Outsourcing* 14 (5 API Supplement): 10–16.

15. (a) Levenspiel Octave, S. (1979). *The Chemical Reactor Omnibook*. Corvallis, OR: Oregon State University Book Stores.
    (b) Levenspiel Octave, S. (1962). *Chemical Reaction Engineering. An Introduction to the Design of Chemical Reactors*. New York: Wiley.
    (c) Fogler, H.S. (1999). *Elements of Chemical Reaction Engineering*, 3e. New Delhi: Prentice Hall PTR.
    (d) Weber, W.J. and DiGiano, F.A. (1996). *Process Dynamics in Environmental Systems*. New York: Wiley.

16. Johnson, M.D., May, S.A., Calvin, J.R. et al. (2012). Development and scale-up of a continuous, high-pressure, asymmetric hydrogenation reaction, workup, and isolation. *Org Process Res Dev* 16 (5): 1017–1038.

17. (a) McMullen, J.P. and Jensen, K.F. (2011). Rapid determination of reaction kinetics with an automated microfluidic system. *Org Process Res Dev* 15 (2): 398–407.
    (b) Jensen, K.F., Reizman, B.J., and Newman, S.G. (2014). Tools for chemical synthesis in microsystems. *Lab Chip* 14 (17): 3206–3212.
    (c) Chaudhari, R.V., Parande, M.G., Ramachandran, P.A. et al. (1985). Hydrogenation of butynediol to cis-butenediol catalyzed by Pd-Zn-Caco3: reaction-kinetics and modeling of a batch slurry reactor. *AIChE J* 31: 1891–1903.
    (d) Makiarvela, P., Salmi, T., and Paatero, E. (1994). Kinetics of the chlorination of acetic-acid with chlorine in the presence of chlorosulfonic acid and thionyl chloride. *Ind Eng Chem Res* 33: 2073–2083.
    (e) Zaldivar, J.M., Molga, E., Alos, M.A. et al. (1996). Aromatic nitrations by mixed acid. Fast liquid-liquid reaction regime. *Chem Eng Process* 35: 91–105.

18. (a) May, S.A., Johnson, M.D., Braden, T.M. et al. (2012). Rapid development and scale-up of a 1 H -4-substituted imidazole intermediate enabled by chemistry in continuous plug flow reactors. *Org Process Res Dev* 16 (5): 982–1002.
    (b) Levenspiel Octave, S. (1962). *Chemical Reaction Engineering. An Introduction to the Design of Chemical Reactors*. New York: Wiley.

19. Process Systems Enterprise (1997–2016). gPROMS. www.psenterprise.com (accessed 14 October 2018).

20. Wehner, J.F. and Wilhelm, R.H. (1956). Boundary conditions of flow reactor. *Chem Eng Sci* 6 (2): 89–93.

21. (a) Braden, T.M., Johnson, M.D., Kopach, M.E. et al. (2017). Development of a commercial flow Barbier process for a pharmaceutical intermediate. *Org Process Res Dev* 21 (3): 317–326.
    (b) Wong, S.-W., Changi, S.M., Shields, R. et al. (2016). Operation strategy development for Grignard reaction in a continuous stirred tank reactor. *Org Process Res Dev* 20 (2): 540–550.

22. Scale-up Systems Limited (2011). Dynochem 2011 (4.0.0.0), Ireland. www.scale-up.com (Accessed June 2015).

23. Tilstam, U., Defrance, T., Giard, T., and Johnson, M.D. (2009). The Newman–Kwart rearrangement revisited: continuous process under supercritical conditions. *Org Process Res Dev* 13 (2): 321–323.

24. May, S.A., Johnson, M.D., Braden, T.M. et al. (2012). Rapid development and scale-up of a 1 H-4-substituted imidazole intermediate enabled by chemistry in continuous plug flow reactors. *Org Process Res Dev* 16 (5): 982–1002.

25. Kopach, M.E., Roberts, D.J., Johnson, M.D. et al. (2012). The continuous flow Barbier reaction: an improved environmental alternative to the Grignard reaction? *Green Chem* 14 (5): 1524–1536.

PART V

BIOLOGICS

17

CHEMICAL ENGINEERING PRINCIPLES IN BIOLOGICS: UNIQUE CHALLENGES AND APPLICATIONS

SOURAV KUNDU AND VIVEK BHATNAGAR
Biologics R&D, Teva Pharmaceuticals, Inc., West Chester, PA, USA

NAVEEN PATHAK
Process Development and Technical Services, Shire plc, Cambridge, MA, USA

CENK UNDEY
Process Development, Amgen Inc., Thousand Oaks, CA, USA

Therapeutic proteins now represent a major class of pharmaceuticals. These macromolecules, generally referred to as biologics, are either natural (e.g. protein fractions from donor human plasma) or are commonly made in bacterial or mammalian host systems by recombinant DNA technology. In the case of the latter, through modern molecular biology techniques, genes representing human proteins, protein fragments, or antibodies to therapeutic targets can be inserted into a host system such as the bacterium *Escherichia coli* or a mammalian host cell such as Chinese hamster ovary (CHO) cells. The host cells are then grown in fermenters or bioreactors using nutrient-rich media and highly controlled conditions. The host cells are internally programmed to synthesize the protein of interest. The synthesized protein may reside in high concentrations within the cell in inclusion bodies or secreted into the extracellular environment. The protein then can be harvested in large quantities and purified to the final therapeutic dosage form. This is of course an oversimplification of a very complex biological production process that brings a protein therapeutic to patients.

Chemical engineering principles apply widely in development and production of biologics. During discovery, biologic drugs are first made in the laboratory in small scale (less than a milliliter to a few liters), quantities sufficient for testing *in vitro* or in an animal model. As the feasibility is proven, the process is refined and scaled up to manufacture quantities needed for extensive pharmacological, toxicological, and clinical testing. At this stage, the culture volume is typically increased to a few hundred to a few thousand liters. After the clinical safety and efficacy are proven and regulatory approval for commercialization is obtained, the drug may be manufactured at even larger scale such as 10–20 000 L. At each of these stages of product development life cycle, chemical engineering principles such as mass transfer, heat transfer, fluid mechanics, chemical reaction kinetics, etc. are used. As the production process is scaled up during the product development, many of the process characteristics defined by dimensionless numbers widely used in chemical engineering (e.g. Reynolds number, Nusselt number, Newton number [also known as Power number], etc.) are kept similar between scales. In fact, many of these dimensionless numbers provide the basis for scale continuum that is extremely important for consistency and comparability.

In the cell culture process for mammalian systems, chemical engineering principles are used to optimize transport of nutrients, supply of oxygen to the cells, and facilitate mixing for homogeneity, removal of undesired metabolites and toxins, and collection of the protein of interest. In the downstream purification process, chemical engineering principles are used for separation of the protein drug from associated impurities, concentration of the protein drug, removal of undesired microorganisms, and stabilization of the protein drug to a state appropriate for storage, transport, and administration. Later in the chapter we will describe in greater detail various unit operations that are typically involved in production of biologics and the specific fundamental principles associated with these unit operations.

Chemical Engineering in the Pharmaceutical Industry: Active Pharmaceutical Ingredients, Second Edition.
Edited by David J. am Ende and Mary T. am Ende.
© 2019 John Wiley & Sons, Inc. Published 2019 by John Wiley & Sons, Inc.

17.1 WHY ARE BIOLOGICS UNIQUE FROM A MASS, HEAT, AND MOMENTUM TRANSFER STANDPOINT?

In order to fully appreciate why biologics are unique, it is important to discuss the complexity of the structure of these macromolecules. As stated earlier, biologics are protein molecules of therapeutic value most commonly administered as aqueous solution. The molecular weight of these macromolecules can range from 5–30 kDa (e.g. recombinant insulin, molecular weight ~6 kDa; recombinant human growth hormone, molecular weight ~22 kDa; etc.) to hundreds of kDa (e.g. recombinant factor VIII, molecular weight ~280 kDa). More complex and larger macromolecules are also being produced in recombinant forms such as recombinant von Willebrand factor, which is made as a functional multimer with an approximate molecular weight reaching in the millions of daltons. Monoclonal antibodies represent an important class of protein drugs. Chimeric or humanized recombinant monoclonal antibodies have found wide applications as receptor blockers, receptor stimulators, delivery vehicles, or absorbents for undesirable antigens in various therapeutic, diagnostic, or imaging areas.

Protein molecules are composed of amino acid sequences determined by the genetic code residing in the DNA. Protein synthesis takes place on ribosomes. The amino acids are connected to each other by peptide bonds, forming polypeptide chains that provide the backbone of a protein molecule. Additional cross-links may be formed within a polypeptide chain or between adjacent branches of polypeptide chains through the disulfide bridges, creating complex folds and a three-dimensional geometry. Protein architecture comprises four levels of structure. The primary structure describes the amino acid sequence and the location of any disulfide bridges within the polypeptide chain and branches. The secondary structure describes the spatial arrangement of neighboring amino acids within the linear sequence. The tertiary structure of a protein describes the spatial arrangement of amino acids that are far apart in the linear sequence. If the protein molecule contains more than one polypeptide chain (also called the subunits), the spatial arrangement of these subunits are described in the quaternary structure. Additional complexity may arise when polypeptide chains fold into two or more compact globular regions joined by flexible portions. These domains resemble each other at times and vary in function at other times [1].

In addition to the characteristic three-dimensional peptide structures, many protein macromolecules also have covalently linked carbohydrate branches. These are known as glycoproteins and represent an important class of therapeutic molecules. The carbohydrate structure takes its initial shape in the endoplasmic reticulum during synthesis and modified into its final form in the Golgi complex. The carbohydrates (also known as oligosaccharides) can be linked to the polypeptide chain of the glycoprotein through the side-chain oxygen atom on the amino acids serine or threonine residues by *O*-glycosidic linkages or to the side-chain nitrogen of asparagine residues by *N*-glycosidic linkages. In addition to these carbohydrate cores, there can be other polysaccharides attached to these cores in a variety of configuration causing a diverse and defining carbohydrate structure [1].

The three-dimensional structure of a protein is extremely important for its functionality. Many interactions between protein molecules or between a protein molecule and a receptor on a cell require a specific spatial geometry of the polypeptide chain. Any disturbance in the three-dimensional structure bears the risk of loss of the protein's biological activity, bioavailability, or its therapeutic value. Environmental conditions that can alter the three-dimensional structure include pH, temperature, physical forces such as shear, specific concentration of chemicals such as chaotropic agents (chemicals that disrupt the protein three-dimensional structure such as urea or guanidine hydrochloride at high concentrations) and organic solvents, etc. The alteration of structure can be reversible and can actually be used for recovery of proteins. For example, highly aggregated recombinant protein housed in the inclusion bodies of *E. coli* can be solubilized in high concentration of urea or guanidine hydrochloride followed by a controlled removal of the chemicals. The process allows the protein to disaggregate and refold into its native desirable state. Under certain conditions the alteration of structure becomes irreversible. The common environmental conditions causing permanent structural alteration of a protein are exposure to extreme heat, pH, or chemical concentration. In response to environmental conditions, such as heat, protein unfolds into a denatured state from its native state. Subsequent aggregation of the denatured molecules results in irreversible denaturation. Denaturation and aggregation during processing are of particular concern to a manufacturer of biologics. Thermal denaturation and unfolding of proteins can be studied by using differential scanning calorimetry (DSC) where heat flux is measured in a sample as the temperature is gradually increased. The protein sample provides a characteristic curve representative of glass transition and phase changes. Denaturation of a protein drug can lead to the loss of its biological activity. In one study [2], natural and recombinant Protein C evaluated by DSC was observed to undergo rapid decline in activity as temperature was increased from 20 °C with a complete loss of activity at temperatures above 70 °C. Narhi et al. observed irreversible thermal unfolding of recombinant human megakaryocyte growth factor upon heating, leading to formation of soluble aggregates ranging in size from tetramer to 14-mer [3]. Proteins can be affected by cold temperatures as well. Protein solutions may be exposed to temperatures lower than −50 °C during freeze-drying in order to prepare a protein drug dosage form. Although less concerning than heat-induced structural damage, cold inactivation has been seen with enzymes such as phosphofructokinase [4] or with proteins such as β-lactoglobulin [5].

In addition to loss of biological activity, when administered to a patient, the structural change from the native state can trigger adverse cellular and immunological response, causing serious medical consequence. Therefore, manufacturing processes for biologic drugs are designed to have appropriate controls such that the protein's active state is maintained. The likelihood of irreversible denaturation is high during operations involving elevated temperature, contact with excessively high or low pH, vigorous fluid movement (e.g. vortex formation), or contact with air interface (e.g. foaming). Thermal denaturation can occur during processing if the temperature is not maintained at the facility or the equipment level or if unintended local heat buildup occurs in the processing equipment such as in a rotary lobe pump, a commonly used component of a bioprocess skid. Most biopharmaceutical plants have sophisticated environmental controls with high-efficiency HVAC systems to maintain the ambient temperature at a level that is far away from the denaturation temperature of the protein. Temperature is also controlled at the equipment level through water or ethylene glycol heat exchange fluids circulating through the jacket of a process vessel. When heating of a chilled protein solution (typically in between process steps) is needed, the jacket temperature is capped at 45–50 °C to prevent excessive vessel skin temperature that may cause denaturation. Exposure to excessively high or low pH can occur during titration of a protein solution with acid or alkali during processing. It can be avoided by ensuring rapid mixing in the vessel through appropriately designed agitators, by controlling the flow rate of the reagents, and by using lower titrant concentrations.

Appropriate handling of a protein solution during processing is extremely important. Abrupt and vigorous fluid movement such as those encountered during movement of tanks during shipping of liquid protein solutions can be detrimental and result in denaturation. Air entrapment and exposure of protein to an air interface create the bulk of the problem in these cases [6]. Process equipment and piping are designed such that the protein solution is not subjected to highly turbulent flow regions during transport. Effect of shear on proteins remains a controversial topic. Thomas provides an excellent overview of the issue of shear in bioprocessing [7]. While loss of activity has been seen with enzymes when exposed to high shear forces in a viscometer, denaturation of globular proteins is less likely just from high shear fields. As mentioned earlier, the primary mechanism of denaturation appears to be from the gas–liquid interface, especially if high velocity gradients are present. Lower concentration solutions are more susceptible to the damage. Design of equipment to minimize air entrapment (e.g. avoidance of pump cavitation) allows control of inactivation and protein denaturation in bioprocessing.

Cell culture fluids pose unique chemical engineering challenges as well. The cells require oxygen and respire carbon dioxide. Oxygen is sparged into the culture that produces bubbles and foam. Care needs to be taken to control the bubble size, velocity, and amount since the bubbles can carry cells to the surface and cause cell death upon bursting. Other chemical engineering challenges include the heat input (mammalian cells) or removal (microbial cells) requirements, mixing requirement, maintenance of appropriate pH conditions, nutrient delivery, and metabolite removal for large industrial-scale cell culture systems. Further details of bioreactor design and operation are provided in Section 17.3.1.

The complexity of proteins drugs and their biological production systems pose extraordinary challenges to the chemical engineers with responsibility for industrial-scale manufacturing of these products. Design, development, and operation of these processes require complete understanding and appreciation for the intricacies of proteins and living cells. In this chapter we will examine how concepts taught in traditional chemical engineering are applied to design equipment and develop processes that allow modern-day mass manufacture of biotechnology products.

17.2 SCALE-UP APPROACHES AND ASSOCIATED CHALLENGES IN BIOLOGICS MANUFACTURING

With the introduction of recombinant human insulin by Genentech/Roche (later licensed to Eli Lilly & Co.) in 1978, the commercial industrial biotechnology was born. It was clear that products created by recombinant DNA technology hold an immense potential for curing "uncurable" diseases such as cancer. The biotechnology products no longer were just research tools in the laboratory made in milligram quantities, but they needed to be manufactured in bulk in commercial-scale manufacturing facilities. Scale-up and optimization of biologics manufacturing processes rapidly became critical for the success and growth of the industry.

Scale-up of biologics manufacturing processes utilizes similar engineering principles as scale-up of chemical manufacturing processes. In the laboratory, the cell culture may be performed in shaker flasks, roller bottles, or small glass stirred tank bioreactors. Flask- or bottle-based cultures can be scaled up by simply increasing the number of flasks or bottles with some increase in the size. Bioreactors can be successfully scaled up linearly from 1 to 25 000 L or above by preserving the aspect ratios, impeller sizing ratios, impeller spacing ratios, and baffle geometries. Design of spargers for oxygen delivery to the cell culture may differ considerably between small-scale and industrial-scale bioreactors. While a small sparge stone may be adequate for oxygen delivery in a small-scale bioreactor, much more elaborate sparging systems with one or more drilled pipes may be necessary to deliver the quantity of oxygen needed in a large-scale bioreactor. The oxygen concentration is maintained at the same level between the small and large scales by controlling to a specified dissolved oxygen (DO) setting determined during process development and preserved through scale-up. In order to achieve the level of DO during cell culture, an adequate amount of oxygen is fed to the bioreactor typically through a flow controller. For large-scale bioreactors, the piping and delivery system for gases can become rather massive. The delivery of oxygen to the cells is dependent on the mass transfer of oxygen from bubbles in the stirred tank to the liquid where the cells are situated. The bubble size is governed by the sparger type and whether the bubbles are broken up further with a Rushton (turbine)-type impeller at the bottom of the bioreactor. The mass transfer coefficient $k_L a$ is an important parameter that determines if adequate transport of oxygen to the cells can be achieved and must be calculated for each bioreactor where scale-up is being performed. The oxygen uptake rate (OUR) is the oxygen required for the optimum culture performance that must be matched by the oxygen transfer rate (OTR) that is determined by the mass transfer coefficient. The relationship between OUR and OTR at a constant DO level is shown by the following formula:

$$OUR = \frac{\mu X}{Y_{X/O_2}} = k_L a (C_{sat} - C_L) = OTR$$

where

μ is the specific growth rate.

X is the measured cell density

Y_{X/O_2} is the calculated cell yield per unit oxygen consumption.

C_{sat} and C_L are the DO concentrations at saturation and at any given time in the liquid phase, respectively.

Scaling up a suspension cell culture requires a thorough evaluation of mixing. Adequate mixing is necessary to deliver nutrients and oxygen to cells, removal of metabolites from the microenvironment, and dispersion of additives such as shear protectants and antifoam. Scale-up of mixing is often performed by maintaining similar power per unit volume (P/V) between the small-scale and large-scale bioreactors. P/V is calculated from the impeller geometry, agitation rate, and working volume of the stirred tank according to the following formula:

$$\frac{P}{V} = \frac{\rho n N_p N^3 D_i^5}{V}$$

where

ρ is density.

n is the number of impellers.

N_p is the impeller power number.

N is the agitation rate.

D_i is the impeller diameter.

V is the working volume of the tank.

When P/V is preserved during scale-up, it is expected that with geometric similarity between the small- and large-scale bioreactors, circulation time, mixing time, and impeller tip speed increase, but the size of the eddies do not change, hence ensuring mixing and mass transport [8]. P/V can be reported as an ungassed value or a gassed value. Typically, the measured ungassed value is slightly higher than the measured gassed value as a result of loss of power upon introduction of gas sparging in the bioreactor. Impeller tip speed is determined by the following formula:

$$\text{Impeller tip speed} = \pi N D_i$$

Maintaining impeller tip speed between scales can ensure similar shear-induced damage of cells between scales. It is important to minimize shear-induced cell damage in order to maintain cell viability and prevent release of unwanted enzymes, host cell DNA, and other impurities into the cell culture. The ability of cells to withstand shear occurring at the tip of the impeller blade is dependent on the type of cells. Typically microbial cells can withstand more shear and therefore can be agitated more aggressively than mammalian cells. It is helpful to scale up in stages when transferring a process from a laboratory to a commercial manufacturing facility. The stepwise approach reduces the probability of failure and allows troubleshooting and understanding of scale-dependent product quality characteristics that are not uncommon. Beyond mixing, mass transfer, and shear characteristics, insight is needed with respect to bioreactor control parameters such as pH, temperature, DO, dissolved carbon dioxide, overlay pressure, and feed addition requirements for a successful bioreactor scale-up [8–11].

Scale-up of the downstream purification steps utilizes similar concepts as the upstream cell culture process. In the scale-up of the chromatography columns, the bed height and linear flow rate are generally preserved between the small-scale and large-scale columns. The volumetric output needed for commercial manufacturing is achieved by increased diameter and bed volume of the large-scale column. Modern chromatography resins provide high protein loading capacity often exceeding 30 g/L. This helps reduce the total bed volume and size of the column necessary. In addition, modern chromatography resins can withstand higher pressures without getting crushed or deformed. This allows operation of the column at higher flow rates, shortening the overall processing time. Column packing methods may differ significantly between the small columns used in the laboratory and large columns used in a commercial manufacturing facility. The complexity of packing increases significantly in larger columns just because of the size. In addition, column hardware components such as the screen, flow distribution system, and head plate play much more important role in large-scale columns. Section 17.3.3 in this chapter provides more information on column packing.

Scale-up of tangential flow filtration (TFF) systems have become easier due to the availability of membrane cassettes of various sizes. For example, Pellicon™ 2 TFF cassettes manufactured by Millipore Corporation (Billerica, MA) are available in 0.1, 0.5, and 2.5 m^2 membrane area configurations. Multiple 2.5 m^2 cassettes can be stacked together to provide the necessary membrane area for large-scale ultrafiltration operations. Other manufacturers also offer similar choice of sizes and molecular weight cutoffs to match scale and process needs. TFF can be scaled up by maintaining filtrate volume to membrane surface area ratios the same between laboratory-scale and commercial-scale systems. In addition, membrane material, molecular weight cutoff (determines retention characteristics), channel height and flow path type, and retentate and filtrate pressures are kept similar between the two scales [12, 13]. Other considerations for scale-up of a TFF system include the type of pump used, number of pump passes that the protein solution experiences during the operation, configuration and sizing of the piping, and the process time. Rotary lobe pumps are commonly used in large scale, while a small-scale TFF operation may be performed with an air-driven diaphragm pump or a peristaltic pump. The number of pump passes is an important consideration for thermally labile proteins. The retentate protein solution may experience a rise in temperature if the concentration factor is high. Some large-scale retentate tanks are equipped with cooling jackets to prevent heat buildup and potential denaturation of the protein. The size of the piping and configuration of the flow path in the large-scale equipment are carefully selected to ensure that excessive pressure drop does not occur or frictional forces do not become too large. Generally these are not major concerns in the laboratory-scale equipment. It is a good idea to maintain similar process time between the small-scale and large-scale systems. This allows for process consistency across scale and ensures predictable product quality during scale-up.

Microporous membrane filtration is another important type of unit operation employed in biopharmaceutical manufacturing. Hollow fiber cartridges can be used to produce filtrate in microfiltration. In this case linear scaling can be readily accomplished by keeping the fiber length similar between scales. The size of the fiber bundle may increase depending on the volume of material to be processed.

Other unit operations that are typically used for bioprocessing are normal flow filtration (NFF), centrifugation, and various types of mixing. While we will not discuss scale-up of each one separately, it is clear from the discussion so far that similar approaches utilizing basic principles of chemical engineering can be successfully used in all of these cases.

17.3 CHALLENGES IN LARGE-SCALE PROTEIN MANUFACTURING

Large-scale protein manufacturing utilizes a series of unit operations to grow cells, produce product, and isolate and purify product from the cell culture. In the following sections, we will examine these steps in more detail and discuss chemical engineering challenges associated with these steps.

17.3.1 Bioreactor

A bioreactor provides a well-controlled artificial environment to the protein-producing host cells that promote cell growth and product synthesis. Bioprocesses have multiple seed bioreactor steps to sequentially increase culture volume that finally culminates in a production bioreactor where product expression occurs. The design criteria for production bioreactor and the seed bioreactors could differ substantially as the goals in the two systems are different. However, the concepts presented here are applicable for either system.

Mammalian cells lack an outer cell wall and are sensitive to shear forces and damage by bursting bubbles. Lack of a cell wall directly exposes a cell's plasma membrane to environmental stresses. Plasma membrane contains enzymes and structural proteins that play a key role in the communication between cell and its environment [14]. Microbial cells are enveloped by a cell wall that imparts much higher tolerance to shear. Growth rates for the microorganisms are also much higher. Thus, the design criteria for fermenters are geared toward providing adequate nutrients to the culture and less concerned with the shear introduced through agitation. Due to these conflicting requirements, very few dual-purpose vessels designed for both mammalian and microbial cell cultivation are in use.

Environmental conditions in the bioreactor such as osmolality, pH, and nutrient concentrations could significantly affect quantity of protein expressed as well as product quality attributes. By product quality attributes we mean properties of the protein in terms of its structure, function, and stability. Complexity of maintaining live cells under optimal conditions dictates the design of the bioreactors. Over the years different designs of bioreactors have been developed. Most common is the stirred tank system, which is widely used for free suspension cell culture. Other designs that are also used commercially are hollow fiber bioreactors, air-lift bioreactors, and, more recently developed, various types of disposable plastic bag-based bioreactors. Each design provides some advantages as well as unique challenges. We will limit the scope of this discussion to the stirred tank bioreactor system.

Stirred tank bioreactors have emerged as a clear winner in the twenty-first century for culturing mammalian cells. One of the key reasons is that the mammalian cells have proved to be sturdier than initially thought of [15]. Stirred tank bioreactors are available in the widest capacity range (1–30 000 L) and commercially used in cultivating different cell lines (e.g. CHO, MDCK, BHK, NS/0, etc.). Fermenters used in commercial microbial and yeast fermentation processes are also primarily stirred tank vessels. Fermentation processes have been successfully scaled up to 250 000 L capacity.

Bioreactor and fermenter systems operate under aseptic conditions, which require exclusion of undesired contaminating organisms. Bioreactors and associated piping are designed as pressure vessels such that these can withstand heat sterilization with saturated steam. All entry points to the vessel for adding and removing gases or liquids are designed to maintain aseptic conditions. The requirement for aseptic processing and heat sterilization of the system is specific to bioprocesses. Factors that determine success of any sterilization regimen are exposure time, temperature, and system conditions. Equipment is designed to avoid buildup of condensate that forms as the saturated steam transfers heat to the vessel interior or to the piping. Buildup of condensate can cause "cold spots," resulting in insufficient microorganism "kill" and failed sterilization. Piping connected to the bioreactor is equipped with steam traps to ensure aseptic removal of the condensate. Exposure to saturated steam causes coagulation of proteins in the microorganisms, rendering them nonviable. The thermal death of microorganisms is a first-order process [16] and can be described by the following equation:

$$\frac{dC_v}{dt} = -kC_v$$

where

C_v is the concentration of viable microorganisms.
t is the time.
k is the thermal death rate constant.

Integration of this equation from a time 0 (t_0) when the viable microorganism concentration is C_{v0} to a finite time t provides the following expression:

$$\ln\left[\frac{C_{v0}}{C_v}\right] = k(t - t_0)$$

The equation can be used to calculate the time required to achieve a log reduction of microorganisms when the thermal death rate constant for a type of microorganism is known.

Bioreactor operation can be considered to be a two-step chemical reaction. The first step is the cell growth and the final step is product synthesis by the cells. For growth-associated products such as monoclonal antibodies, the two steps are not mutually exclusive and thus have to be considered as concurrent reactions. Protein expressed from genetically engineered mammalian cells is constitutive, i.e. protein expression is independent of the growth phase of the culture. For non-growth-associated products like expression of viral antigens, the two steps could be treated as different phases and may even require different operating conditions such as temperature shifts, pH shifts, etc. Induction or infection of the culture for expression of these non-growth-associated products may even trigger cell destruction, e.g. seen in production of interferons. Similar to the chemical reaction kinetics, stoichiometric and kinetic equations for cell growth and product formation have been developed [17]. Modern laboratory systems are used to determine concentrations of cells, nutrients, metabolites, and product at different stages of the process. These key experimental data generated in the laboratory are used to develop temporal correlations for cell growth, nutrient uptake, and product synthesis. These correlations are used to calculate kinetic parameters such as specific growth rates or specific nutrient uptake rate. Table 17.1 lists the typical cell culture parameters related to growth and product formation along with their definitions.

TABLE 17.1 Parameters Commonly Used for Measurement of Cell Culture Growth and Product Formation

Parameter	Definition	Unit	Mathematical Expression
Growth rate (G)	Rate of change in number of cells	No. of cells/L-h	$G = \frac{dx}{dt}$ x, viable cell density (cells/mL) at time t
Specific growth rate	Normalized growth rate	h^{-1}	$\mu = \left(\frac{1}{x}\right)\frac{dx}{dt}$
Specific nutrient consumption rate	Normalized uptake rate of a nutrient (e.g. glucose, glutamine, oxygen, etc.)	g nutrient/g cell-h	$q_s = -\left(\frac{1}{x}\right)\frac{ds}{dt}$ s, nutrient concentration
Specific product formation rate	Normalized rate of product synthesis	g product/g cell-h	$q_p = \left(\frac{1}{x}\right)\frac{dp}{dt}$ p, product concentration
Specific lactate formation	Lactate formed per unit glucose consumed	g lactate produced/g glucose consumed	$q_{lac} = \frac{d[lac]}{d[glu]}$ [lac], lactate concentration [glu], glucose concentration

For other design aspects of a bioreactor, most chemical engineering principles for chemical reactor design remain applicable. Chemical engineers play an important role in establishing the design criteria for the bioreactors. These design criteria are established with the goal of providing the organism with optimal conditions for product synthesis. To prevent formation of localized environment, homogeneous mixing of the culture is essential. Further, transfer of oxygen to the culture and removal of carbon dioxide from the system have to be achieved. The critical role for the process engineer is the identification of the design criteria that can meet these requirements. The process engineer analyzes the data generated in the laboratory using small-scale bioreactors and identifies key process operating parameters and their appropriate settings. These often include pH, DO, agitator speed, dissolved carbon dioxide, and initial viable cell density (VCD). The set points and normal ranges of these operating parameters are selected such that consistent and acceptable quality of the protein can be obtained run after run. Typically the cell culture is optimized to yield the highest product concentration, commonly called titer. However, in many instances, as more titer is obtained from the culture, the amount of unwanted impurities such as fragmented or misfolded proteins, host cell DNA, and cellular proteins, cell debris, etc. starts to increase. Therefore, a careful balance has to be achieved between titer and amount of impurities. The capacity of the downstream purification steps for impurity removal is a major consideration. If the downstream steps are capable of removing large quantities of impurities coming from the production bioreactor, the cell culture can be forced to produce high titers accompanied by higher amounts of impurities that can be reproducibly cleared. During process development, process is scaled up from a small-scale to an intermediate-scale pilot laboratory bioreactor operated at the same set points as the small-scale bioreactor. Acceptable quality of the pilot-scale material confirms the validity of the scale-up. If successful, then the same criteria are used to scale up to the commercial-scale bioreactors. Otherwise, further characterization work is undertaken to better understand the cell culture process requirements and the process fit to the equipment, and the cycle is repeated.

Key characteristics of stirred tank bioreactors used for production of therapeutic protein molecules are described in the following subsections.

17.3.1.1 Vessel Geometry
The tanks used for mammalian cell culture were "short and fat." Original bioreactor systems had an aspect ratio of vessel height to inside diameter of 1:1. This geometry allows proper mixing at low agitation speed. Currently, use of an aspect ratio of 2:1 is widely accepted, while use of tanks with 3:1 aspect ratio has also been reported. Matching aspect ratios between the bench-scale and the commercial-scale bioreactors facilitates process scale-up but is seldom used as a strict scale-up criterion. Lower aspect ratio for small-scale bioreactors is acceptable, but for large-scale bioreactors lower aspect ratio creates a larger footprint that becomes cost prohibitive to implement. Figure 17.1 shows a simplified schematic diagram of a bioreactor showing important internal components of the vessel. There are many possible variations for the configurations of the impellers, sparger assembly, and the baffle system.

17.3.1.2 Mechanical Agitation
Gentle mixing in the bioreactor is essential to maintain homogeneity, facilitate mass transfer, and support heat transfer. Three-blade down-pumping axial flow impellers are the most common design used in bioreactors. Number of impellers used is governed by the aspect ratio of the vessel. A bioreactor with an aspect ratio of 2:1 is typically equipped with two axial flow impellers of diameter equal to one-half of the tank diameter. If the impellers are too closely placed, then maximum power transfer is not achieved. For optimal results, space between the two impellers is between one and two impeller diameters. Actual placement of the impeller within this range is also governed by the operating volume in the tank so as to avoid the liquid levels where the rotating impeller is not fully immersed (also known as "splash zone") and may entrap air, causing unnecessary and detrimental foaming.

Commercial-scale bioreactors have bottom mount drives. This is preferred because it allows for a relatively shorter shaft. A shorter shaft provides structural integrity and absence of wobbling when the agitator is running. Removing the impellers for maintenance becomes easier with a bottom mount impeller system. In addition, head room requirement within the production suite is also less compared to a top mount drive for which sufficient ceiling height is required to be able to remove the impeller without having to open the ceiling hatch and exposing the bioreactor suite to the outside environment. Shaft seal design for a bottom mount drive bioreactor poses significant challenges. Typically, a double mechanical seal is used that is constantly lubricated by clean steam condensate fed to the seal interface, and a differential pressure is maintained across the seal so that bioreactor fluid does not enter the seal space, reducing possibility of contamination. For smaller-sized seed bioreactors, a top mount drive could be considered to eliminate mechanical seal design issues. Design of the seal is often not given the same level of attention as some of the other parts of the system. Apart from the absolute requirement of maintaining aseptic conditions, the ease of maintenance of the seal should also be considered.

The Reynolds number (Re) in a bioreactor is expressed by the following equation:

$$\mathrm{Re} = \frac{\rho_L N D^2}{\mu}$$

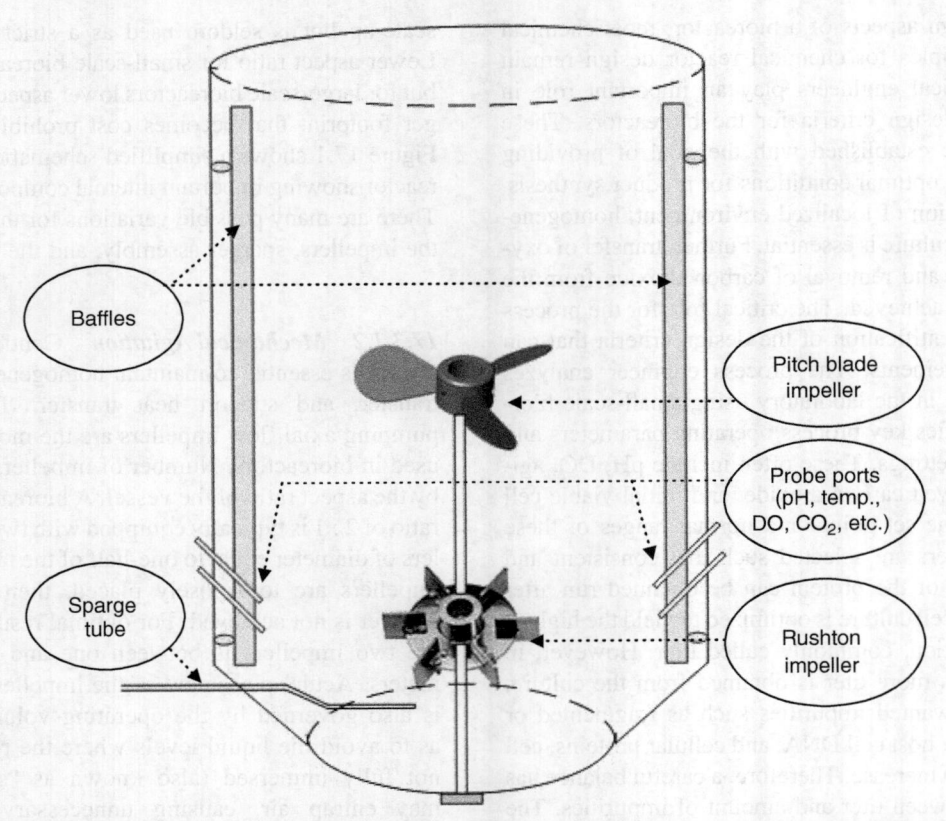

FIGURE 17.1 A simplistic schematic diagram of a cell culture bioreactor showing internal components. Many variations of sparge tube, impeller, and baffle configuration are possible.

where

- ρ_L is the density of the media (kg/m^3).
- μ is the viscosity of the media (Pa·s).
- D is the impeller diameter (m).
- N is the impeller speed (revolution/second).

The viscosity of the cell culture fluid is close to that of water. Thus, Reynolds number $> 10^4$ is often experienced in the bioreactor, which allows utilization of turbulent flow theories to analyze the fluid mechanics in the bioreactor.

17.3.1.3 Mass Transfer

Rate of oxygen transfer to the liquid in a bioreactor is governed by the difference between the oxygen concentration in the gas phase and the concentration of oxygen in the liquid culture, fluid properties, and the contact area between the gas and the liquid. Pressurized air–oxygen mixture is supplied to the sparger, which is either a ring or a tube with open end or with holes on the sidewall. Two spargers may be required in a production bioreactor, while the seed bioreactors may have only one sparger. The size of the bubbles and their distribution are key to the efficiency of mass transfer to the liquid. For a given air–oxygen mix, the higher the concentration of DO levels in the cell culture fluid, the lower the efficiency of oxygen transfer to the fluid as the concentration difference is the driving force. To increase mass transfer in the existing equipment, composition of the gas (oxygen enrichment) may be altered, or the gas flow rate may be increased. Higher sparge gas flows result in increased carbon dioxide stripping [5]. Thus, as a result of the selected strategy, carbon dioxide levels in the culture may vary. Above a certain concentration (usually partial pressure of >140 mmHg), carbon dioxide has a toxic effect on the cell culture, resulting in product quality issues. Therefore, careful control of carbon dioxide concentration in the culture is essential for cell health and product quality.

Respiratory quotient (RQ) is defined as the rate of carbon dioxide formation divided by the rate of oxygen consumption. RQ for mammalian cells is close to 1.0. This implies that for each mole of oxygen consumed, one mole of carbon dioxide is produced. Generally, bioreactors are maintained at DO levels of greater than 30%.

In the steady state, the OTR from the gas bubbles to the cells is matched by the OUR by the cells. The relationship between these two quantities has been described earlier.

The mass transfer coefficient $k_L a$ in s^{-1} can be determined using the correlation developed by Cooper et al. [18] for a bioreactor with multiple impellers:

$$k_L a = K * \left(\frac{P_g}{V}\right)^a (v_s)^\beta$$

where

P_g is the gassed power input to the bioreactor agitator.
V is the volume of liquid in the tank.
v_s is the superficial velocity.
K is a function of number of impellers (N_i) used in the bioreactor given by the formula

$$K = A + B * N_i$$

where

A and B are positive constants.

EXAMPLE PROBLEM 17.1

Mass transfer coefficient $k_L a$ needs to be determined experimentally in a 10 000 L bioreactor in order to ensure that the oxygen transfer rate (OTR) can match the oxygen uptake rate (OUR) in a cell culture process.

Solution

OTR is dependent on the design of the bioreactor, impellers, sparging system, and operational parameters such as agitation speed. OUR is determined by the oxygen requirement of the cells for growth and functionality. For a successful cell culture operation, OUR must be, at minimum, met and preferably exceeded by OTR. This will ensure that adequate oxygen is available to the cells for biological functionality. OUR is generally determined from small-scale cell culture experiments. OTR of a large-scale bioreactor can be determined using the following relationship:

$$k_L a (C_{sat} - C_L) = OTR$$

where

k_L is the mass transfer coefficient that can be determined experimentally.
a is the liquid–gas contact area or interfacial area of gas bubbles from the sparger.
$(C_{sat} - C_L)$ is the concentration gradient of oxygen between gas bubbles and the liquid.

To experimentally determine $k_L a$ for our 10 000 L bioreactor, the dynamic gassing technique is utilized. In this technique, the bioreactor is first filled to the working volume with a suitable pseudo-medium or water, and oxygen is purged out of the liquid by equilibrating with nitrogen. Thereafter, oxygen is sparged in the bioreactor, and the concentration of oxygen in the liquid (C_L) is measured as % DO at regular intervals using a fast-response calibrated dissolved oxygen probe. Table 17.2 shows simulated data from such an experiment at a specific oxygen sparge rate and agitation speed.

The rate of change in oxygen concentration can be described by the following equation:

$$\frac{dC_L}{dt} = k_L a (C_{sat} - C_L)$$

This equation can be integrated to yield

$$\ln\left[\frac{(C_{sat} - C_0)}{(C_{sat} - C_t)}\right] = k_L a t$$

where

C_0 is the initial concentration of dissolved oxygen.
C_t is the concentration of dissolved oxygen at time t.

From Table 17.2, % DO at $t = 0$ (or C_0) is 18.61, and % DO at saturation (C_{sat}) is 100.25. Substituting these values in the equation above and plotting over time, we get a straight line with a slope of 3.43. Therefore, $k_L a$ for our 10 000 L bioreactor at the experimental sparge rate and agitation speed settings is 3.43 h^{-1}. Substituting $k_L a$ in the earlier equation and assuming C_L to be a % DO set point typically between 20 and 60%, OTR can be calculated. If the OTR is not sufficient to match OUR, $k_L a$ can be raised by increasing the sparge flow rate and/or the agitation speed until sufficient OTR is obtained.

17.3.1.4 Heat Transfer In design of bioreactors for mammalian cell culture, heat transfer considerations are not as significant as in design of fermenters for microbial culture. Mammalian cells have lower metabolic activity and generate less heat that needs to be removed during normal processing. The large-scale bioreactors are jacketed tanks, and the bioreactor temperature is maintained by a temperature control module using chilled or hot water. The temperature of the culture generally remains between 30 and 40 °C. However, the vessel is exposed to high temperatures during the sterilization by steam-in-place and requires an appropriate design to withstand thermal stresses.

17.3.1.5 Bioreactor Control Commercial-scale bioreactors are hooked up to a distributed control system (DCS) or, at the very least, have their own stand-alone programmable logic controllers (PLC). Parameters such as pH, DO,

TABLE 17.2 Simulated Dissolved Oxygen Measurements over Time at a Specific Oxygen Sparge Rate and Agitation Speed for a 10 000 L Bioreactor

Time (min)	% DO
0.0	18.61
1.0	19.82
2.0	24.22
3.0	29.12
3.5	31.35
4.0	33.49
4.5	36.12
5.0	39.5
5.5	41.6
6.0	43.26
6.5	45.14
7.0	48.16
7.5	49.94
8.0	51.52
8.5	53.0
9.0	55.21
9.5	57.29
10.0	58.82
10.5	60.59
11.0	61.94
11.5	63.9
12.0	65.15
12.5	66.45
13.0	67.97
13.5	69.04
14.0	70.68
14.5	71.67
15.0	73.3
15.5	74.3
16.0	75.53
16.5	76.21
17.0	77.51
17.5	79.31
18.0	80.18
18.5	80.43
19.0	81.03
20.0	83.47
21.0	84.72
22.0	86.14
23.0	87.8
24.0	89.24
25.0	90.5
26.0	91.52
27.0	92.95
28.0	93.65
29.0	94.81
30.0	95.7
31.0	96.6
32.0	97.08
33.0	98.33
34.0	98.5
35.0	99.84
36.0	100.25

temperature, agitation, and gas flow rates are controlled in real time. The control of pH is achieved by adding carbon dioxide to lower pH and sodium carbonate solution to raise pH. A potentiometric pH probe measures the pH of the cell culture; the signal is sent to a pH control module that then determines the necessity of the additions. Culture pH is maintained within a predefined range for optimum cell culture performance and typically controlled within a "deadband" around a set point. Use of a deadband prevents the constant interaction of the pH control loops and overuse of the titrants. Similarly, the DO is typically measured by a polarimetric DO sensor. The control system determines the need for activation of gas flow to the sparger of the bioreactor through flow control devices. Feed solutions can be added to the bioreactor to supply enough nutrients to the growing and productive cells according to a predetermined regimen. In fed-batch cell cultures, product and metabolites remain in the culture until harvest. In perfusion cell cultures, the supernatant is periodically taken out of the culture to remove product and metabolites, while the cells are returned to the culture. Periodic samples are drawn from the bioreactor for optical examination of the culture to assess culture health. VCD, viability, nutrient and metabolite concentrations (e.g. glucose, glutamine, lactate, etc.), osmolality, offline pH, and carbon dioxide concentration are determined at regular intervals for confirmation of culture health and process controls. Figure 17.2 provides an example of the time profile of some of these parameters. In a fed-batch culture, feed is added at regular interval, stabilizing the nutrient supply to the cells. Significant effort has been made to mathematically model the mass transfer of oxygen and carbon dioxide in cell culture. This is described in greater detail in Section 17.4.1.

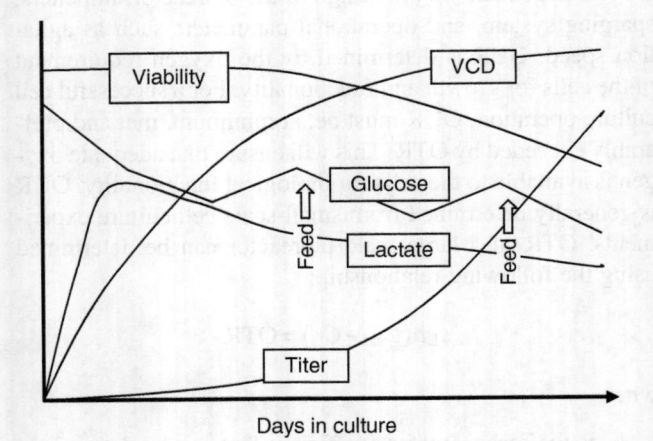

FIGURE 17.2 Example of cell culture parameter profiles that may be observed during the operation of a bioreactor in fed-batch mode (VCD is viable cell density). Many other parameters not shown here may also be monitored. The profiles will change based on cell line, culture system, product, and selection of operating parameters.

17.3.2 Centrifugation

Initial challenge upon completion of the cell culture is to separate the cellular mass from the product stream. The protein of interest is generally soluble, while the cells are suspended in the liquid culture media. In some cases with microbial fermentation, the protein of interest can be present at very high concentration within the cells in structures known as "inclusion bodies." Retrieval of the protein from these structures requires not only separation of the cells but also rupture of cells and processing of the inclusion bodies. We will not discuss processing of inclusion bodies in this chapter. The cells also contain DNA, host cell proteins, and proteolytic enzymes that have the potential to damage the protein of interest if released. Thus, to simplify the purification process, it is essential to remove the cellular mass without rupturing the cell wall in order to keep the intracellular contents out of the process stream. To achieve this goal, various solid–liquid separation techniques have been used. These techniques include centrifugation, microfiltration, depth filtration, membrane filtration, flocculation, and expanded bed chromatography [19]. These unit operations are used in a combination that is dependent on the process stream properties. Expanded bed chromatography has been developed as an integrated unit operation that combines harvest with product capture, but to date, practical limitations have kept this technique from being adopted in large-scale operations [19].

The initial step of the harvest is either centrifugation or microfiltration for a large-scale operation. While not very common, processes can also have only a depth filtration step to separate cells from the culture fluid. This is known as the primary recovery step and is followed by a secondary clarification step that typically consists of a sequence of depth filtrations. Final secondary clarification step uses a sterilizing grade membrane filter. It provides a bioburden-free process stream for the downstream purification operation.

In the quest for higher titers, the cell densities in the production bioreactor have been increased consistently along with processes being run for longer durations. This combination results in lower cell viability and increased cell debris at the harvest stage. In this scenario, centrifugation for primary recovery is becoming the method of choice as it can handle higher concentrations of insoluble material compared with the alternative techniques. Centrifugation uses the difference in densities of the suspended particles and the suspension medium to cause separation. A centrifuge utilizes the centrifugal force for accelerating the settling process of the insoluble particles. Cells in the harvest fluid can be approximated to be spherical. During centrifugation of this fluid, the cells are accelerated by the centrifugal force and at the same time experience the drag force that retards them.

From Stokes' law, a single spherical particle in a dilute solution experiences a drag force that is directly proportional to its velocity. A cell in suspension accelerates to a velocity where the force exerted by the centrifugal force is balanced by the drag force. Thus, based on the applied centrifugal force, cells achieve a terminal velocity in the centrifuge bowl. This property is exploited to affect the desired separation of the insoluble cellular mass from the process fluid. The terminal velocity of the cell (or other small insoluble spherical particles) can be calculated by the following equation [20]:

$$v_\omega = \frac{d^2}{18\mu}(\rho_s - \rho)\omega^2 r$$

where

- v_ω is the settling velocity.
- d is the diameter of the insoluble particle (cell).
- μ is the viscosity of the fluid.
- ρ and ρ_s are the densities of the fluid and the solids, respectively.
- ω is the angular rotation in rad/s.
- r is the radial distance from the center of the centrifuge to the sphere.

In commercial applications, continuous disk-stack centrifuges are used. The design of these units is based on the above principle. Continuous disk-stack centrifuges have multiple settling surfaces (disks) that yield high-throughput and consistent cell-free filtrate (centrate). Settling velocity (v_s) is correlated to the operation and scale-up of the centrifuge by using the Σ (sigma) factor. The Σ factor relates the liquid flow rate through the centrifuge, Q, to the settling velocity of a particle according to the following equation:

$$v_s = \frac{Q}{\Sigma}$$

The ratio of flow rate to the Σ factor is held constant during scale-up of a centrifugal separation [20]. This ratio can be used as a scale-up factor even if two different models of centrifuges are involved. For a disk-stack centrifuge, Σ factor is given by

$$\Sigma = \frac{2\pi n \omega^2}{3g}(R_0^3 - R_1^3)\cot\theta$$

where

- n is the number of disks.
- ω is the angular velocity.
- R_0 and R_1 are the distance from the center to the outer and inner edges of the disks, respectively.
- θ is the angle at which the disks are tilted from the vertical.

The Σ factor is expressed in the unit of (length)^2. The calculation of the Σ factor varies for different types of centrifuges, while the measurement unit remains the same.

Disk-stack centrifuges have a high up-front cost. Also, because of the complexity of the design, availability of equipment suitable for various scales of operation is very limited. Therefore, often harvesting at small scale is performed with a completely different type of centrifuge or sometimes without a centrifuge. Even at pilot scale, representative equipment is not always available. This brings additional risk to the scale-up of the harvest process. Over the past decade, the two dominant continuous disk-stack centrifuge manufacturers, Alfa Laval (Lund, Sweden) and GEA Westfalia (Oelde, Germany), have developed specific products for the biopharmaceutical application and have attempted making comparable pilot-scale equipment to address scale-up concerns. In addition to improvements in centrifuge design, depth filtration systems have also improved with filtration media capable of reliably removing cell debris, impurities, and other process contaminants. The combination has greatly improved the reliability and robustness of the harvest operation despite confronting more and more challenging source material. With the recent improvements, harvest yields of greater than 98% are being reported [19]. Harvest operation has been made robust to absorb the variability in the feed stream and yield a consistent output stream for the downstream chromatographic steps.

17.3.3 Chromatography

Liquid chromatography has been used in biotechnology in all phases of product development. In research, small-scale columns and systems have been the workhorse for purifying proteins that are used in several facets of drug development including high-throughput screening, elucidation of the three-dimensional structure using X-ray crystallography and NMR, and nonclinical studies. Although chemical engineering concepts apply universally for chromatography, they are neither as relevant nor as necessary to adhere to at the research stage. This is primarily because the cost of producing the protein and the amount of protein required are both small and success is primarily measured by being able to achieve the required purity in the shortest time frame. The situation is quite different when a therapeutic protein is approaching commercialization. This now requires chromatography to be used as a key unit operation in the manufacturing process for the protein. Therapeutic proteins, including monoclonal antibodies, antibody fragments, fusion proteins, and hormones, derived from bacterial and mammalian cell culture processes represent a sizeable portion of commercially available therapeutic proteins. It is typical for the purification process for such molecules to include two to three chromatography steps. The column used in these processes can be up to two meters in diameter and can weigh several tons. To use chromatography successfully in large scale, it is important to ensure that the concepts of chemical engineering are incorporated early during process development and are carried forward through the development process and subsequently during routine manufacturing.

Fundamental to success of using chromatography is selecting the right ligand chemistry irrespective of scale or phase of development. This results in the required "selectivity." Selectivity is a measure of relative retention of two components on the chromatographic media. Most of the chromatography media used across scales consists of porous beads with an appropriate ligand (for the desired selectivity) immobilized throughout the surface area available. The beads may be compressible or rigid and can be made of a number of substances such as carbohydrate, methacrylate, porous glass, mineral, etc. Although most beads are spherical, there are asymmetrical ones that are commercially available (e.g. ProSep® media from Millipore Corp., Billerica, MA). For the porous beads, due to high porosity, most of the ligand is immobilized in the area that is "internal" to the bead. The average bead diameter for most commercially used resins is between 40 and 90 μm. In order for the chemical interaction to occur, the protein of interest present in the mobile phase needs to be transported to the internal area of the beads via the pore structure and subsequently from the mobile phase to the ligands immobilized within the pores. Mass transfer theory, a key chemical engineering concept, plays a major role in understanding and predicting the behavior of such transport. Since the chromatography media are packed in relatively large-diameter columns, appropriate packed bed stability and ensuring that the mobile phase is equally distributed across the column are critically important.

17.3.3.1 Mass Transfer in Chromatography Columns

In the simplest case of a chromatography process, a mixture of a protein of interest (*A*) and an undesirable contaminant (*B*), referred to as the "load," is applied to a packed bed. As a result of convective and diffusive mass transfer, the load injected into the column as a bolus starts to assume a broader shape. As the load traverses the packed bed length, the peaks corresponding to the components *A* and *B* begin to separate based on the selectivity of the resin. At the same time, the peaks start to get broader as they move through the bed (Figure 17.3).

The peak broadening, which is a result of mass transfer resistances and diffusion, may result in the overlap of the peaks of components *A* and *B*. The best separation can be achieved if there is no overlap between the peaks of components *A* and *B* by the time the whole packed bed length is traversed. Hence, the overlap caused in part by mass transfer plays a key role in achieving the desired level of purification. The characteristics of mass transfer in a packed bed under flow have been extensively studied [21]. The effect of mass transfer on separation of components and purification

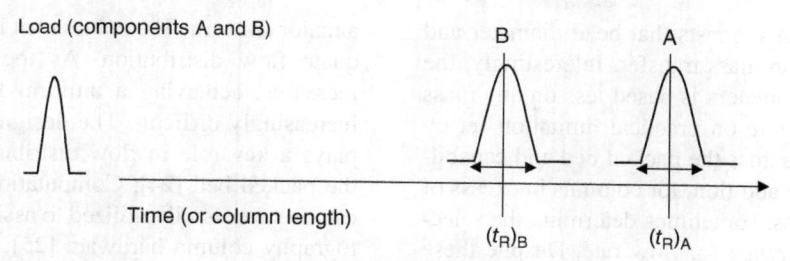

FIGURE 17.3 Schematic illustrating band broadening phenomenon.

efficiency is described by the term resolution. Resolution is described by the following equation:

$$R = \frac{2\left[(t_R)_B - (t_R)_A\right]}{W_A + W_B}$$

where

$(t_R)_A$ is the retention time of early eluting peak.
$(t_R)_B$ is the retention time of late eluting peak.
W_A is the width of early eluting peak.
W_B is the width of late eluting peak.

The higher the separation efficiency, the better the resolution of the column. Concepts of height equivalent to a theoretical plate (HETP) and residence time distribution have been successfully applied to quantify efficiency of a chromatography column. The van Deemter equation describes various mass transfer phenomena ongoing during chromatography, their impact on efficiency (as quantified by HETP), and the operating parameters that impact mass transfer. The resistance to mass transfer results in what is commonly referred to as "band broadening," i.e. broadening of the peaks as they travel through the column as mentioned earlier. Higher HETP represents more "band broadening" and less separation efficiency. According to the van Deemter equation [22], the HETP is composed of three terms that describe three different mass transfer mechanisms:

$$H = A + \frac{B}{u} + Cu$$

where

A represents eddy diffusion resulting from flow path inequality.
B represents molecular diffusion.
C represents all other resistances to mass transfer [21, 23].
u is the interstitial velocity defined by

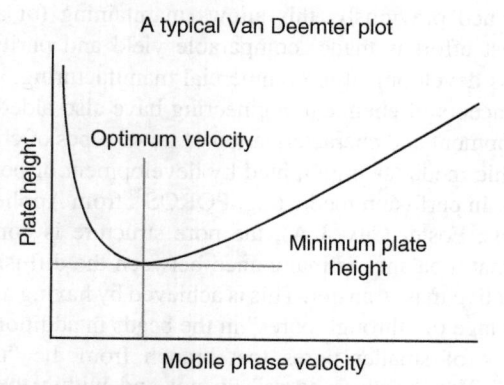

FIGURE 17.4 Van Deemter equation is the most commonly used model for "band broadening" in chromatography. For conventional media (plot shown), the resolving power decreases as the flow rate increases.

$$u = \frac{\text{Mobile phase flow rate}}{\text{Porosity} \times \text{column cross-sectional area}}$$

HETP is expressed in the unit of length. As the flow rate increases beyond an optimum, the van Deemter equation predicts that for conventional media the efficiency of the column decreases (leading to increased plate height) (Figure 17.4). In addition to the van Deemter equation, many other mathematical relationships have been derived to model the band broadening phenomenon [21].

Having an understanding of how various parameters impact the efficiency of the column is critical for large-scale operation. A compromise in efficiency may lead to loss in yield, effective binding capacity, and product quality attributes such as purity. Acceptance criteria are generally established during process development and applied during commercial manufacturing to ensure that a chromatography column has the requisite separation efficiency (as measured by HETP) and appropriate peak shape during use. More recently, transition analysis is being used to get a more comprehensive understanding of column efficiency and packed bed integrity and will be discussed later in this chapter.

The van Deemter equation suggests that bead diameter and flow rate play a key role in mass transfer. Interestingly, the selection of these two parameters is based less on any mass transfer calculation and more on practical limitation set by acceptable pressure drop across the packed bed and capability of equipment in use. In addition, for columns in excess of 1 m diameter, time and cost sometimes determine the selection of normal operating range for flow rate. Despite these economic considerations, the scale-up principles are indeed based on the engineering fundamentals. Linear flow rate and bed height are generally held constant during scale-up to ensure that the mass transfer conditions under which the process was developed remain consistent across scale. As mentioned previously, this allows maintaining (or at least the best effort is made) comparable yield and purity from process development to commercial manufacturing.

Concepts of chemical engineering have also aided in the development and characterization of novel types of chromatographic media as highlighted by development of perfusion media. In perfusion media (e.g. POROS® from Applied Biosystems, Foster City, CA), the pore structure is controlled such that a balance is maintained between the diffusive and convective mass transfer. This is achieved by having a certain percentage of "through pores" in the beads in addition to the network of smaller pores that branch from the "through pores." The "through pores" do not end within the beads and thus serve as transportation highways for solutes through the beads. The "through pores" with pore diameter greater than 5 nm maintain high degree of intra-particle mass transfer compared to diffusive (non-perfusive) media, whereas the smaller pores branching from them (diameters in the range of 300–700 Å) provide the adequate binding capacity [23]. Thus, such media can retain efficiency at significantly higher flow rates compared with conventional diffusive media, such as agarose-based beads. Tolerance to high flow rates and pressures enables the perfusion media to be the media of choice for high-throughput analytical chromatography.

17.3.3.2 Flow Distribution in Chromatography Columns

A chromatography unit operation consists of process steps such as equilibration, loading, washing, elution, and regeneration. The aim of these steps is to achieve the right condition for ligand binding, facilitate target/ligand interaction, collect the purified target protein, or restore the ligand to the initial state for the next batch. Each of these steps is carried out by flow of predetermined volumes of fluid through the column. For acceptable performance of these individual steps and the chromatography operation as a whole, uniformity of flow distribution across the column cross section is critical. Uniform flow distribution is a prerequisite for ensuring comparable mass transfer throughout a large commercial size column. Little attention is paid during bench-scale process development to flow distribution, since for small columns, typically of 1–3 cm in diameter, this is not a major concern. The column frit is sufficient to achieve adequate flow distribution. As the diameter of the column increases, achieving a uniform flow distribution becomes increasingly difficult. The design of the column head plate plays a key role in flow distribution and the efficiency of the packed bed [24]. Computational fluid dynamics (CFD) can be successfully utilized to assist in the design of chromatography column hardware [25]. CFD utilizes fundamental fluid mechanical and mass transport relationships, representative boundary conditions, and mathematical algorithms to predict fluid properties in a variety of fluid flow scenarios. Recently, the pharmaceutical industry has shown increasing interest in CFD for providing insight into flow characteristics and related phenomena that can help mitigate risks associated with scale-up as well as in troubleshooting [25]. Further CFD discussion can be found in Chapters 12 and 13. Frontal analysis techniques such as dye testing can be used to study the flow distribution. In brief, dye is injected into a packed and equilibrated column and allowed to flow for a predetermined period of time. Subsequently, the column is dismantled to expose the resin bed. Upon excavation, the dye profile in the bed reveals the quality of flow distribution. Alternatively, one can also look at the washout of the dye front from a column under flow and evaluate the effectiveness and uniformity of the column cleaning conditions. An appropriate cleaning procedure would be indicated when little or no residual dye is left, which can be confirmed by excavation of the resin bed after fluid flow through the column simulating the cleaning conditions.

It is ideal to use a complementary approach where CFD is used early in the hardware design process to ensure that predicted flow distribution is appropriate. Subsequent to the fabrication, dye testing can be performed to confirm the uniformity of flow distribution and flow conditions. This approach was utilized for characterization of flow dynamics in a process chromatography column [25]. A few highlights of this study are shown here. Figure 17.5 shows liquid velocity profile predicted by CFD (light gray color is low velocity and dark gray color is high velocity on the CFD tracer model picture) within a packed bed under specific flow rates with clear areas of stagnation. Subsequently, when dye testing was performed, as shown in the lower panel of Figure 17.5, the same areas failed to show dye washout demonstrating zones of stagnation and confirmed the predictions from CFD modeling.

Why is flow distribution so critical for the success of large-scale chromatography? In addition to the issue of mass transfer and its relationship to purity and yield, there are many other factors that come into play for manufacture of biopharmaceuticals using chromatography. Without appropriate flow distribution, part of the packed bed may remain unreachable for the process fluids. This reduces the total binding capacity of the bed and underutilizes expensive chromatography media such as protein A media. The lack of flow uniformity and pockets of stagnation within the bed also

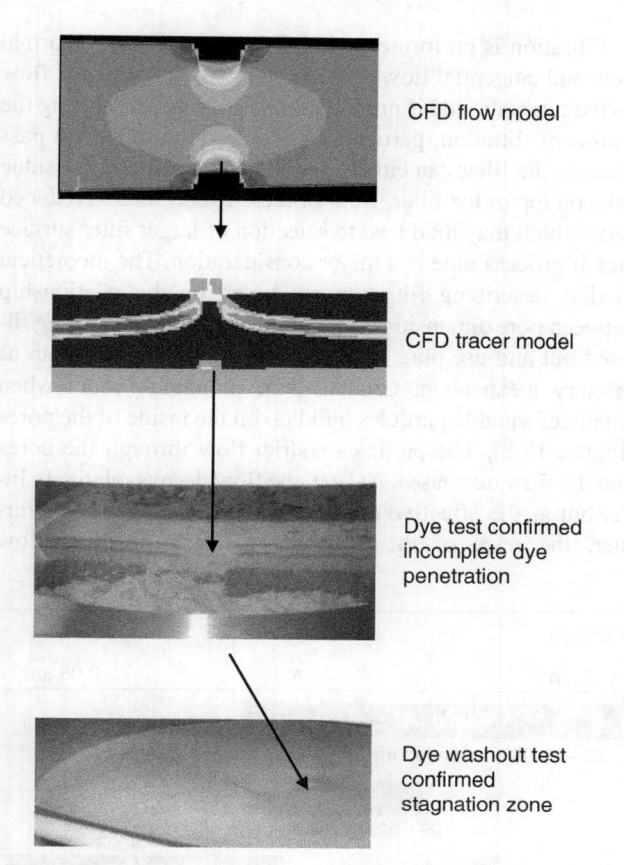

FIGURE 17.5 CFD flow and tracer modeling predicted stagnation zone beneath the chromatography resin introduction port that was confirmed by dye test.

increases the possibility of suboptimal cleaning and elevates the risk of microbial growth and protein carryover from one manufacturing batch to the next, creating significant safety and compliance concerns.

The chromatography system or "skid" design should ensure that appropriate flow can be delivered without excessive pressure drop from system components. Skid components (e.g. pumps) and piping should be selected appropriately to meet this requirement. The chromatography media used for biotechnology products are generally compressible and hence cannot withstand pressures in excess of three bars, thus putting practical limitation on flow rates and bed heights.

17.3.3.3 Column Packing and Packed Bed Stability

Column packing for compressible resins is generally achieved by delivering a predetermined amount of media slurry to the column followed by packing the bed to a predetermined bed height. The slurry amount and the bed height are related such that the resin bed, upon achieving the target bed height, is under a target compression. The compression factor can be recommended by the manufacturer of the chromatography media or can be determined experimentally. In either case it should be confirmed during packing development. Several engineering considerations are relevant during scale-up and commercial manufacturing to ensure that packed bed is fit for use. As mentioned earlier, due to the compressible nature of most of the chromatography media, operations are limited to lower pressures only. In a small-scale column with a smaller diameter, a significant part of the packed bed is supported by the frictional forces between the bed and the column wall. This is termed as the "wall effect." As the diameter of the column increases, the extent of the "wall effect" decreases. As a result, for the same bed height and flow rate, the pressure drop increases as the column diameter increases. Work performed by Stickel and Fotopoulos [26] can be used to predict pressure drop for larger columns once the data from the small scale is available. Additional work has been performed recently in an effort to improve the prediction models [27].

Column packing is time and resource intensive. For commercial manufacturing, the columns, once packed, are used for many cycles. This mandates that the packed bed remain stable and integral during multiple uses over long periods of time. Tools have been developed to measure and monitor bed stability. Qualification tests can be performed using these tools upon completion of packing. Traditionally, this has been done by injecting a tracer solution (e.g. salt solution or acetone, usually 1–2% of the column volume) and recording the output signal (either solution conductivity or absorbance of an ultraviolet light beam) at the column exit. Peak attributes are then used to calculate HETP and asymmetry (A_f) and compared to predetermined acceptance criteria. It should be noted that due to the significant resistance to protein mass transfer inside a bead, estimated plate heights for protein solutes are much greater than those obtained using a salt or an acetone solution. Hence, the results of column qualification are primarily indicative of packing consistency and integrity rather than the extent of protein separation [28]. Transition analysis, a noninvasive technique for monitoring the packed bed, is finding increasing utilization in process chromatography. Transition analysis is a quantitative evaluation of a chromatographic response at the column outlet to a step change at the column inlet [29]. It utilizes routine process data to calculate derived parameters that are indicative of the quality of the packed bed. All, or a subset of these parameters, can be used to qualify the column after packing and subsequently during the lifetime of the column to assess bed integrity.

Moreover, a robust column packing procedure needs to ensure that the amount of resin packed in the bed is accurate and reproducible. This is achieved by effective mixing of media slurry, accurate measurement of media fraction in the slurry, and the slurry volume. Application of chemical engineering concepts can help ensure success of these measurements.

17.3.4 Filtration

Filtration is used across pharmaceutical and biopharmaceutical industry spanning all phases of drug development and commercialization. It provides a "quick" option to achieve a size-based separation. This is especially true when the difference in the molecular weight/size is significant (one order of magnitude or more). For this reason research laboratories have used filtration as a workhorse in a variety of applications ranging from buffer exchange (dialysis) to removal of particulates. Similar to discussions presented in the chromatography section, generally, yield is not a major performance parameter at the laboratory scale. Also, the filter and system sizing are not rigorously performed. It is generally based on picking an "off-the-shelf" item that is judged to best suit the needs. In large-scale operation, yield is a major consideration, especially when the protein is expensive to produce. Appropriate filter sizing is also important as it determines the filtrate quality, time of operation, and cost.

At the heart of achieving separation by filtration are the membranes and filter media that provide the pore structure and size required to meet the separation performance requirement. Even though the primary mechanism of separation is based on size, in many cases, that alone is not sufficient, and charge interaction and/or adsorption is also exploited. Manufacturers of filter membranes do not always follow the same methodology to rate the membranes for pore size. Therefore, the rating provides only a first approximation of the retention capability of the membrane. Based on pore size and filter media structure, three broad categories exist as described in Figure 17.6. These categories and their application will be discussed in this chapter.

Filtration is performed primarily in two modes – normal flow and tangential flow (Figure 17.7). In NFF the bulk flow on the retentate side is normal to the filter surface. During the course of filtration, particulates that are not allowed to pass through the filter can either plug the pores or build a residue cake on top of the filter. Both of these events lead to reduced flux, which may then lead to selection of larger filter surface area if process time is a major consideration. The theoretical models describing filtration are based on the relationship between pore dimensions and nature of components being filtered out and use pore plugging or cake buildup, or both as primary mechanism. Gradual pore plugging occurs when small deformable particles build up on the inside of the pores (Figure 17.8). The particles restrict flow through the pores and the flow decreases. At first, the flow decays relatively little, but as the effective diameter of the pores decreases further, the pace of the blockage and the resulting flow

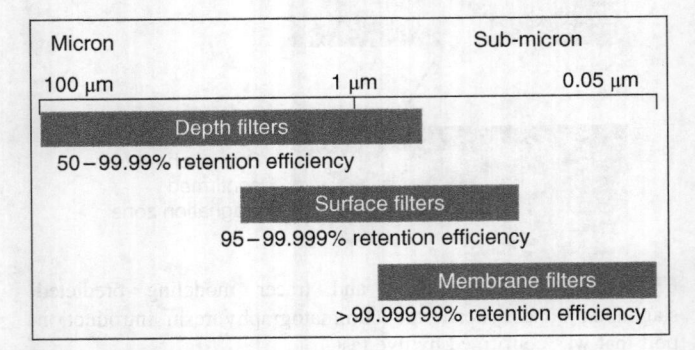

FIGURE 17.6 A comparison of filter media with respect to retention efficiency and pore size.

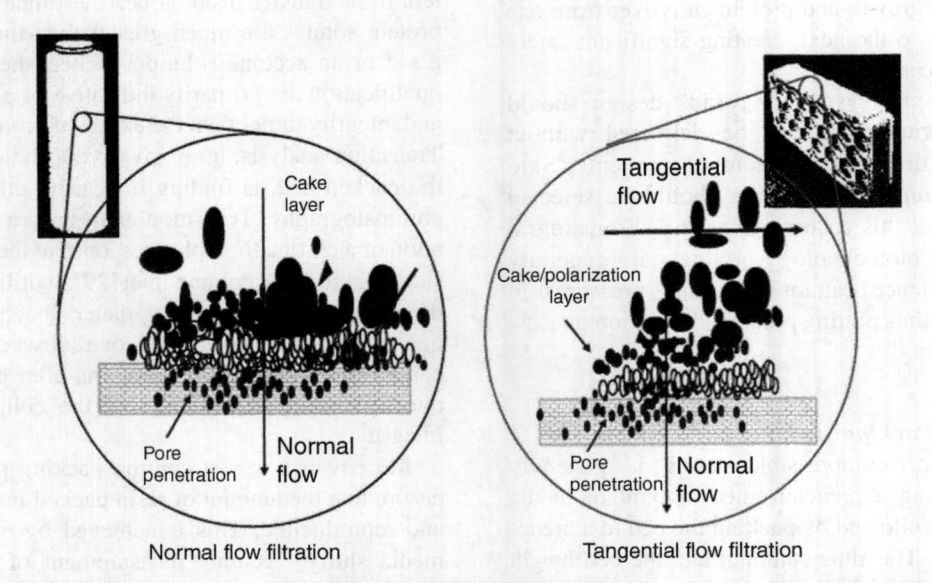

FIGURE 17.7 An illustration of resistance to flux during the two modes of filtration. Components that penetrate the filter result in fouling. The components that build up on the surface result in the formation of cake and concentration polarization layer. Concentration polarization can be controlled by tangential flow.

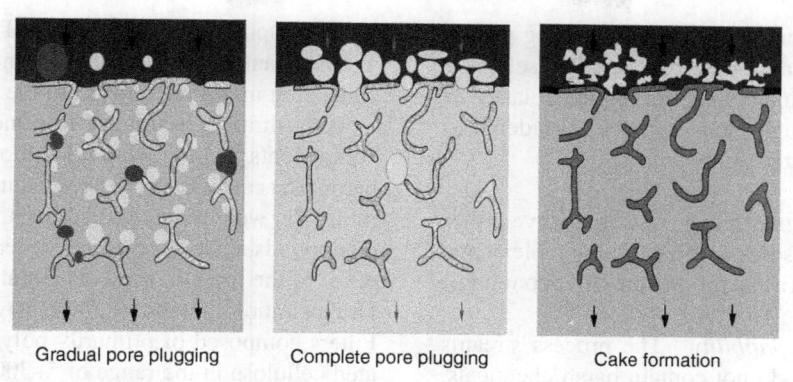

FIGURE 17.8 Mechanism and models for sizing normal flow filtration. Gradual pore plugging model is recognized as most applicable to biological processes.

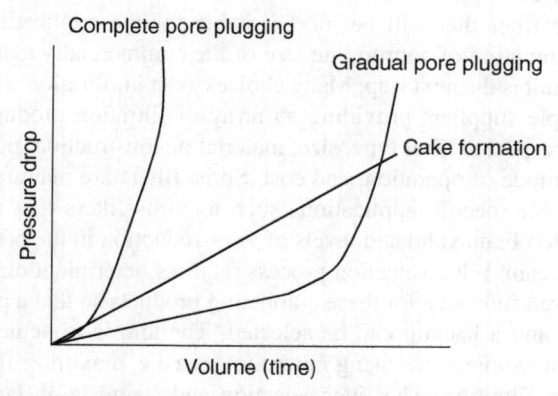

FIGURE 17.9 Retention mechanism and progression of filtration.

substantially decays (Figure 17.9). The gradual pore plugging model is recognized as most applicable to biological process streams.

The buildup on the surface, referred to as formation of concentration polarization or gel layer, results in increased resistance to flow, which is undesirable. The TFF mode of operation attempts to mitigate this undesirable situation by having the flow parallel or tangential to the filter surface (Figure 17.7). This tangential flow, also called cross-flow, creates a "sweeping" action, resulting in less gel layer and increased filtrate flux. Chemical engineering principles and empirical modeling have been extensively used to model membrane fouling and gel layer formation and understand their relationship to operating parameters. This understanding leads to successful process development and scale-up of filtration. The flux through the filter can be expressed as the ratio of driving force (i.e. transmembrane pressure) and the net resistance as shown below:

$$J = \frac{TMP}{(R_m + R_g + R_f)}$$

where

J is the flux through the membrane (commonly in L/m^2-h).

TMP is the transmembrane pressure defined as the difference of average pressures on the retentate and permeate side of the membrane.

R_m is the resistance to flow through the membrane.

R_g is the resistance to flow through the gel layer.

R_f is the resistance to flow due to membrane fouling.

Empirical correlations relating flux to tangential flow rate under laminar and turbulent flow conditions have been developed [30].

Normal flow filters can be further classified as depth or membrane filters. Depth filters are used for clarification and prefiltration with higher solids content. They remove particles via size exclusion, inertial impaction, and adsorption. They have high capacity for particulate matter and lower retention predictability. Membrane filters are generally used for prefiltration and sterilization. In membrane filtration, particles are removed via size exclusion. Membrane filters have very high retention predictability. Tangential flow filters are also membrane filters but with capability to retain various molecular sizes. Microfiltration devices are made of membranes that can retain cells, cell debris, and particulate matter of comparable size. Microfiltration can be utilized for separation of cells from the cell culture fluid at harvest. Ultrafiltration is used for concentration of protein solutions or for buffer removal or exchange. Most ultrafiltration operations in large scale are performed in the tangential flow mode to maximize throughput and minimize process time.

In order to successfully use filtration in a biopharmaceutical manufacturing process, the following three steps should be followed: (i) Correct filter type and mode of operation should be selected. (ii) Filter size should be optimized for the type and quantity of product. (iii) System components (e.g. pumps, piping, housing, etc.) and size should be optimized for the application. Chemical engineering principles related to filtration play a key role in filter selection and determination of optimal size. Principles related to pumps and fluid flow through pipes must be considered to ensure that the system is sized appropriately.

17.3.4.1 Filter Selection and Sizing
At a high level, filter selection and sizing is based on three primary elements: (i) chemical compatibility, (ii) retention, and (iii) economics.

17.3.4.1.1 Chemical Compatibility The process streams in biotechnology generally do not contain harsh chemicals, and therefore, chemical resistance to process streams is not a major concern. The filters operated in NFF mode are typically not reused and do not require cleaning. The filters operated in TFF mode are typically reused and are cleaned after every use. These membranes need to be compatible with cleaning agents such as sodium hydroxide, sodium hypochlorite, or detergent. Given that many biopharmaceuticals are injectables, the material of construction of a filter should be such that it does not add significant amounts of leachables and extractables into the product. In addition, material chosen should not result in high levels of adsorption of the product as it may lead to significant yield loss. Chemical engineers working in biopharmaceutical industry are responsible for selecting the appropriate material of construction for the filters used. Knowledge of materials sciences and polymer sciences are important in successful filter selection.

17.3.4.1.2 Retention At the initiation of the selection and sizing process, a well-defined design requirement is finalized. This requirement, at minimum, consists of (i) nature of the process fluid and components desired to be retained/not retained during the filtration step, (ii) volume of process fluid that will undergo filtration, and (iii) total time allowed for execution of the filtration operation. The required retention characteristic of a specified process fluid will generally steer the selection in the target range of pore size and membrane type as shown in Figure 17.6. Generally, for primary recovery steps such as separation of cells from the cell culture fluid or removal of large particulates from the process stream, depth filters are used. Filtration mechanisms in depth filters include size-based exclusion, charged interaction, and impaction. Complete depth of the filter is utilized for catching particulates. Generally for complex mixtures, a filter train is used where different types of filters may be placed in sequence. Alternatively, newer filtration products that have multiple layers of different filtration media assembled in one unit can also be utilized. Membrane filters are typically placed at the end of depth filters to obtain particulate-free process fluid. These can be nominally rated, or absolute rated, or a combination of the two. Absolute rated filters follow a strict cutoff that allows retention of particles above a certain size. Unlike depth filters, only the topmost layer of a membrane filter is responsible for separation. The rest of the filter serves as a mechanical support that prevents the filtering layer from collapsing under pressure from the process fluid.

Ultrafiltration is utilized to remove or exchange soluble components from a protein solution. Its widespread use in biopharmaceutical processes includes removal of smaller molecular weight process additives, removal of small organic compounds, change of buffer system to impart special properties to the protein, and concentration of a protein solution. Ultrafiltration membrane filters have the smallest pore size. Filters composed of primarily polyether sulfone or regenerated cellulose in the range of 5–300 kDa nominal molecular weight cutoff are commercially available.

17.3.4.1.3 Filtration Economics Upon deciding the type of the filter that will be most suitable for the application, determination of appropriate size of the commercially available unit is the next step. Many choices exist in filtration with multiple suppliers providing an array of filtration products with respect to filter type, size, material of construction, pore size, mode of operation, and cost. Some filters are manufactured for specific applications such as virus filters that are used to obtain validated levels of virus reduction in the process stream. Filter selection process requires determination of required filter size for these competing products so that a primary and a backup can be selected. The aim is to achieve design requirements using minimum area, i.e. maximize flux during filtration. The filter selection and sizing is, in large part, an experimental process. However, chemical engineering principles, as they apply to filtration, have been extensively used to understand and model flow through the filters so that the experimental design is optimal. Sizing of filter is often done experimentally by determination of a filter's maximum "filterable" fluid volume before plugging (V_{max}) using the actual process fluid. A characteristic curve generated by such experimentation is shown in Figure 17.10. The V_{max} can be subsequently used to determine the minimum filter area required per unit volume of process fluid as shown in the following equation:

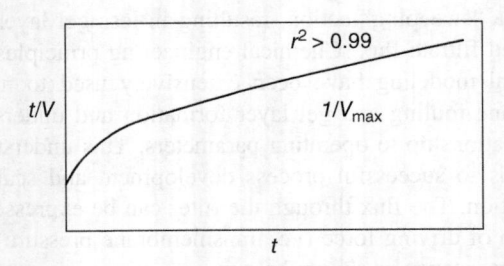

FIGURE 17.10 Filter sizing experiment using gradual pore plugging model. Process solution is filtered at a constant pressure at lab scale. Volume of solution filtered (V) with respect to time (t) is recorded and plotted as shown above. Filter with the highest V_{max} will result in least area requirement.

$$A_{\min} = V_B \left(\frac{1}{V_{\max}} + \frac{1}{J_i \times t_B} \right)$$

where

- $A_{\min}$ is the minimum filter area.
- V_B is the batch volume.
- t_B is the batch processing time.
- J_i is the initial filtrate flux.

The actual filter area recommended is always greater than the minimum area but generally is based on economic factors like filter cost and processing time.

EXAMPLE PROBLEM 17.2

For a cell culture operation, 10 000 L of culture media is to be sterilized through a 0.2 μm absolute grade filter. The operational limitations in the plant require that the filtration be completed in about two hours. The filter selected can withstand a maximum pressure drop of 20 psid. In order to design the appropriate filtration system for the plant, the filtration area needs to be calculated.

Solution

Filtration is a physical process of removing insoluble solids from a fluid stream by placing a porous filter in the flow path, which retains the insoluble components and allows the fluid to pass through. As fluid passes through the filter, the filter medium becomes clogged, and the resistance to the flow increases. Thus, either the driving force determined by the pressure differential across the filter is increased to maintain constant flow, or a constant pressure is maintained across the filter, and the flow rate of the fluid is allowed to recede as the pores in the filter medium become plugged. We will assume that the differential pressure across the filter is kept constant for our application.

In order to estimate the minimum filter area and the filtration time, a small-scale flow decay study is executed using a capsule filter with 40 cm^2 (0.004 m^2) filtration area that is a scale-down version of the large-scale filter. Media prepared in the laboratory are passed through the capsule filter, and the filtrate volume is measured over time. The inlet and outlet pressures are monitored so not to exceed the rated pressure drop. Table 17.3 provides simulated data the study conducted at ambient temperature.

Since constant pressure differential is maintained in this case, the partial pore plugging model can be utilized. A relationship can be derived by plotting t/V against t and then performing a linear fit of the data. The plot resembles the curve shown in Figure 17.10. The slope of the line is $1/V_{\max}$, where $V_{\max}$ is the maximum volume of media that

TABLE 17.3 Simulated Data from a Bench-Scale Filtration Study

Elapsed Time, t (min)	Volume Passed, V (ml)
0	0
3	91
6	272
12	496.6
17	683.5
22	607.5
27	896
32	964
37	1034
42	1123
47	1195.1
52	1246
57	1295
62	1346
67	1402
69	1416

can be filtered through the capsule. A linear fit of t/V against t from Table 17.3 (plot not shown) provides a slope of 0.0004 m/L with an R^2 of 0.93 indicating a good fit to the model. $V_{\max}$ is then calculated to be 1/0.0004 or 2500 mL (2.5 L). The minimum filter area $A_{\min}$ can then be calculated as follows:

$$A_{\min} = \frac{10\,000\,L}{2.5\,L} \times 0.004\,m^2 = 16\,m^2$$

Finally, we check for the processing time requirement. Fluid flux is calculated for each data point, and an average flux of 294 L/m^2-h is calculated from the individual fluxes. Using the calculated filter area above and the average flux determined from the data, we can calculate the estimated processing time to be 10 000 L/(16 m^2 × 294 L/m^2-h) = 2.12 h, close to our processing time requirement of 2 h. For the large-scale filter sizing, we would include a factor of safety of 50% over the calculated minimum filtration area to account for any scale-up issues and lot-to-lot variability in media components. Therefore, applying the safety factor, recommended filter area for this application would be 16 × 1.5 = 24 m^2.

EXAMPLE PROBLEM 17.3

For a cell culture operation, a depth filter is required to separate the culture broth from the cells and the cell debris. The protein is expressed in the culture broth during production culture and harvested for additional purification and processing to the final dosage form. The filter manufacturer offers two different depth filters with slightly different physical properties, and a study needs to be performed to select the most optimal depth filter.

Solution

The study evaluated two different types of depth filters offered by the filter manufacturer. Depth filter 1 had a larger average pore size than depth filter 2 and produced filtrate with slightly higher turbidity. The study was performed with laboratory-scale filters of 25 mm diameter placed in appropriate filter holders. Cell culture broth was filtered through these laboratory-scale filters at a constant flow rate of 4.5 mL/min, and the filtrate was collected for evaluation of turbidity. The weight of the filtrate and the pressure were monitored during the filtration. The filtration was stopped when a pressure of 20 psig was reached, which was set as the upper limit of pressure based on manufacturer's recommendation. The filter throughput was calculated from the filtrate weight (hence volume) and filter surface area for each filter. Table 17.4 shows the filtration data from the study.

The depth filter 1 produced a throughput of 112.32 L/m^2 compared with 96.08 L/m^2 with depth filter 2. The filtrate turbidity was 16.2 NTU (NTU is a measure of turbidity) for depth filter 1 compared with 10.2 NTU for depth filter 2. Both depth filters provided filtrates within the acceptable turbidity level of 20 NTU. Figure 17.11 shows the plot of the pressure vs. filter throughput. The higher throughput achieved with depth filter 1 allows reducing the filter size and is beneficial for scale-up. Smaller depth filter size provides a better fit with the existing manufacturing equipment. Based on these considerations, the depth filter 1 was selected to be the filter of choice.

The filtration cost also plays a role in determining if the mode of filtration will be normal flow filtration (NFF) or tangential flow filtration (TFF). Microfiltration operations can be mostly run successfully in an NFF mode even though TFF has been used in cell culture harvest operation. The ultrafiltration step, however, is exclusively run in TFF mode for commercial operations. This is primarily due to the extent of concentration polarization layer on the filter surface under normal flow operations. The layer consisting of high concentrations of retained proteins acts as an additional source of resistance and significantly impedes the flux. In fact, in some cases the gel layer can also change the retention characteristics such that

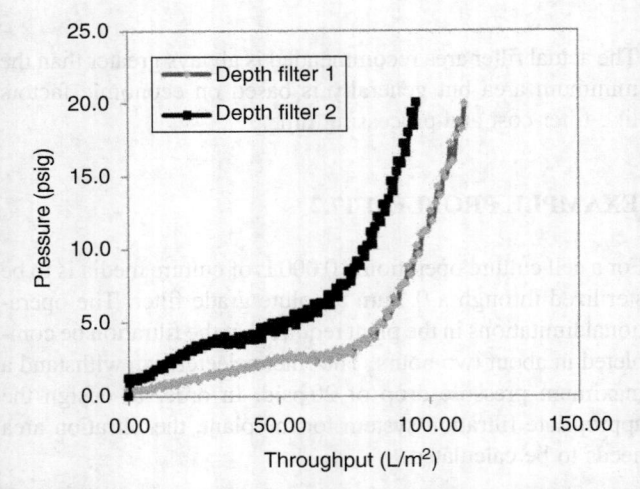

FIGURE 17.11 A plot of pressure vs. filter throughput for filter selection problem in Example Problem 17.3. Once the pores start to plug from cells and cell debris, the pressure rapidly goes up. In this case, a limit of 20 psig was recommended by the filter manufacturer. Depth filter 1 provides a higher throughput without a process-relevant increase in the turbidity.

TABLE 17.4 Laboratory-Scale Filtration Study Results for Comparison of Two Depth Filters

Elapsed Time (min)	Depth Filter 1			Depth Filter 2		
	Weight (g)	Throughput (L/m^2)	Pressure (psig)	Weight (g)	Throughput (L/m^2)	Pressure (psig)
0	0	0	0	0	0	0
5	20.7	8.28	0.5	18.1	7.24	1.4
10	44	17.6	1.1	40.1	16.04	2.5
15	67.5	22.0	1.5	62.2	24.88	3.6
20	90.8	36.32	1.8	84.6	33.84	3.8
25	114.2	45.68	2.2	106.7	42.68	4.2
30	137.6	55.04	2.5	128.8	51.52	4.5
35	160.7	64.28	2.7	151.1	60.44	5.3
40	183.9	73.56	3.3	172.8	69.12	7.3
45	207.1	82.84	4.7	194.7	77.88	9.9
50	230.2	92.08	7.6	216.7	86.68	13.4
55	253.2	101.28	11.9	238	95.2	18.4
55.5	255.6	102.24	13.1	240.2	96.08	20
60	276.2	110.48	18.7	—	—	—
61	**280.8**	**112.32**	**19.9**	—	—	—

Note: The throughput and pressure data were collected at 30 second intervals during the filtration study. For space considerations for the data table, the data are shown in five minute intervals.

components smaller than the membrane pore size are also retained. Operation in a tangential flow mode helps reduce the buildup of the gel layer due to the sweeping action caused by the flow tangentially to the membrane. Concepts of chemical engineering have also shaped the understanding of how the gel layer impacts the flux. It has led to the understanding that the flow rate across the membrane and the pressure differential between the retentate and the permeate side can be varied to get two filtration regimes – one where pressure differential controls the flow and the other where the gel layer controls the flow. This understanding helps form a strategy for operation of an ultrafiltration unit in a manner that minimizes the gel layer and maximizes the flux.

17.3.4.2 System Sizing and Selection

Similar to the chromatography system, the filtration system design should ensure that appropriate flow can be delivered without generating excessive pressure from system components. The process control is designed such that the normal flow through the membrane (flux) and tangential flow (retentate) can be monitored and controlled by the system. Flux in NFF is generally controlled by the differential pressure across the filter. In the case of TFF, both transmembrane pressure, a measure of differential pressure across the membrane, and tangential flow are used to control the flux. The control system should ensure that maximum allowable filter pressure is not exceeded. The filters used in NFF mode are placed in housings generally of stainless steel construction. All stainless steel components are designed to be able to withstand clean-in-place and steam-in-place conditions. The housing design ensures that all entrapped air can be removed so that complete filter area comes in contact with process fluid. Many filters are self-contained capsules that do not require housings. These types of self-contained and single-use filter systems are gaining popularity rapidly since they provide significant advantages by eliminating complicated setup and dismantling, shortening process time, and eliminating cleaning and sterilization.

17.4 SPECIALIZED APPLICATIONS OF CHEMICAL ENGINEERING CONCEPTS IN BIOLOGICS MANUFACTURING

In the preceding sections, we discussed how fundamental chemical engineering principles are used in process and equipment design for biologics manufacturing. In this section, we will examine how the area of process systems engineering in chemical engineering discipline offers unique solutions in biologics process development and manufacturing. Similar to a chemical plant, a biopharmaceutical plant utilizes the basic principles of mass, heat, and momentum transport to cultivate cells and produce product from raw materials and nutrients supplied to the cells and then to purify the protein in the final dosage form. Also similar to a chemical plant, a biopharmaceutical plant consumes energy and uses utilities such as water, steam, and compressed air. Because of these similarities, techniques for modeling of chemical plants can also be adapted to model a biopharmaceutical plant and the production processes.

Development of process models allows obtaining greater process understanding and predicting process behavior and is the first step toward process monitoring and control. Various levels of complexities can be incorporated in a process model. In biopharmaceutical manufacturing processes, it can vary between simple empirical models (data driven) to more sophisticated mathematical means such as metabolic flux and pathway models [31–33]. Comprehensive process understanding is imperative in achieving industry-wide guidelines set by regulatory agencies to promote quality-by-design (QbD), which is building quality in the process and product design. This can be achieved via correlative, causal, or mechanistic knowledge and at the highest level via first principles models [34].

First principles modeling involves in-depth understanding of the process dynamics that is typically defined in a set of material and energy balances via differential and algebraic equations and, depending on the modeling objectives, partial differential equations. As they demonstrate the process understanding at a highest level and provide opportunities for advanced process control, first principles models are desirable, and industry is encouraged by regulatory agencies as mentioned in various guidelines from US Food and Drug Administration [35, 36] and International Conference on Harmonization [37].

In this section, we will briefly review correlative (via multivariate modeling) and first principles-based approaches by providing case studies to demonstrate their benefits and practical use.

17.4.1 A Case Study Using First Principles Modeling: Mass Transfer Models in Cell Culture

Mathematical modeling of cell culture in bioreactors has received significant attention and has been successfully utilized for design and characterization. Most of the models developed included unstructured and unsegregated modeling approaches to describe the process at high level. In this section we will demonstrate how first principles modeling is used to develop representation of dissolved carbon dioxide (dCO_2) mass transfer in a bioreactor and practical use of the model in large-scale setting.

Excess accumulation of dCO_2 at high viable cell concentrations is known to result in adverse effects to cellular growth and specific productivity in large-scale mammalian cell culture bioreactors. The accumulation can occur as a result of reduced surface to volume ratios and low CO_2 removal rates [38–40]. It is also known that high dCO_2 might be detrimental to protein structure and function (due to the alteration of glycosylation pattern of therapeutic protein). On the other hand, excess stripping of CO_2 can alter bioreactor pH profile. Therefore, it is imperative to control dCO_2 levels.

There are three main sources of dCO_2 in cell culture: (i) CO_2 produced by cells during respiration, (ii) CO_2 addition to control pH at its desired level, and (iii) CO_2 produced by dissociation of sodium bicarbonate added to culture for pH control. Dissolved CO_2 mass balance in culture broth in relation to abovementioned sources are provided in the following equations [40]:

$$CO_2(gas) \xleftrightarrow{k_La(CO_2)} CO_2(liq) [\text{also from cells}] \xleftrightarrow{K_1} H_2CO_3 \xleftrightarrow{K_2} H^+$$

$$+ HCO_3^- \leftrightarrow 2H^+ + CO_3^{2-}$$

and

$$\frac{dCO_2}{dt} = \frac{10^{-pH}}{10^{-pH} + K_1}\{CER + k_La([CO_2]^* - [CO_2])\}$$

where

k_La is mass transfer coefficient.
K_1 and K_2 are equilibrium constants.
CER is CO_2 evolution rate.
$[CO_2]^*$ is the CO_2 concentration at the equilibrium.

Each of these sources can be mathematically described given the cell growth and lactate generation curves as well as the bioreactor configuration and operating conditions as depicted in Figure 17.12. While more detailed generic mathematical description of the model can be found in the literature [38–40], we will focus on the industrial use of the model in this section.

The key inputs for model are the VCD and lactate concentration sampled over the time course of the bioreactor operation. In Figure 17.12, t_d designates discrete measurements performed, e.g. daily measurements, and t designate continuous outputs. The model receives a number of other values such as gas flow rates and base addition just for comparison purposes against model predictions for those. In other words, the model predicts those within its differential and algebraic equation set for every time instance that it solves and compares its outputs against the daily observations from the plant results. Model also allows studying the effect of pH control loop influence on the oxygen composition of the sparge gas so that controlling the levels of dCO_2 can be improved. Once a model is developed, it should be tested with existing process data, and the output should be compared against the process outcome. In this case, the model was tested against large-scale production bioreactor data and found to be predictive of known performance. The model was able to provide a predicted relationship between CO_2 flow rate and dissolved CO_2 when the cell growth (VCD) and lactate concentration ([Lac]) time courses are provided and pH and aeration are at control set points. The predictive model can be utilized in a variety of applications. Perhaps the most beneficial application of such a model is in its ability to predict process behavior in a scale-up. During process development, a large body of small-scale process data is generated. These data can be used in the development of the model. Typically, the process is then scaled up to a pilot scale with perhaps a handful of process runs at that scale. Further refinement of the model can be performed with this data, and scale comparability can also be assessed. The model can then be utilized to make predictions of large-scale process conditions.

In our specific scenario, the model was being utilized to evaluate process behavior of a follow-on (second-generation) process when a current process was already being operated at large scale. Therefore, we had access to process data at small, pilot, and large scales for the current (legacy) process. The follow-on process had small-scale and some pilot-scale process data. The scenario can be pictorially represented in Figure 17.13. The follow-on process was expected to have much higher cell densities, and we were interested to know how we would select certain design criteria for large-scale bioreactors. One such criterion was the size of the mass flow

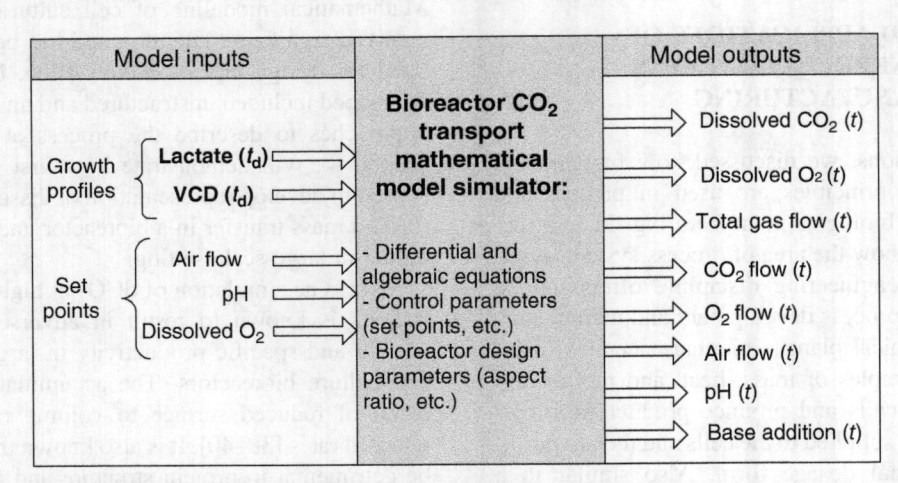

FIGURE 17.12 Schematic diagram representing the main input and output structure of CO_2 transport model.

controller that determines the amount of gas input to the culture. As can be seen from Figure 17.13, there were no process data available from a large-scale bioreactor (since no follow-on process had been run at this scale, yet, at that time). The approach undertaken was to first utilize the CO_2 first principles model for the current process to ensure that the predictions are comparable to what was observed at both pilot and the large scales from historical runs. Once that was confirmed, the follow-on process conditions were used to generate predicted values at pilot scale. Since we had pilot-scale data from the follow-on process, we could verify that the model was still predictive. The final step was then to run the model using the large-scale bioreactor conditions with expected follow-on process metabolic response (VCD and lactate concentrations) to predict the dCO_2 levels that might be seen at large scale. This then allowed us to do appropriate calculations and size selection of large-scale sparge equipment.

In another application of the first principles CO_2 model, we evaluated the impact of reducing agitator tip speed in our process. As mentioned earlier, the agitator speed is linked to mass transport in the culture but also can be a concern for shear or for the effect of air interface on the cells. Using our model we were able to predict the culture properties at a lower agitator tip speed and determine whether our existing systems would be adequate for supporting the process requirement. Figure 17.14 shows some of the output of the

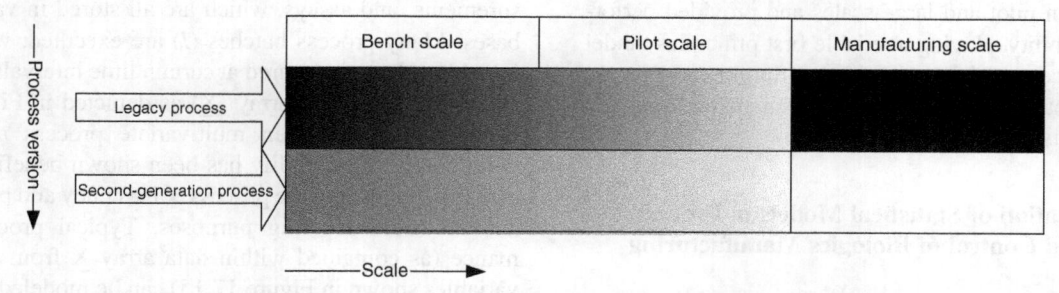

FIGURE 17.13 Color map representation of the available runs and data across scales (darker color means more manufacturing runs are available).

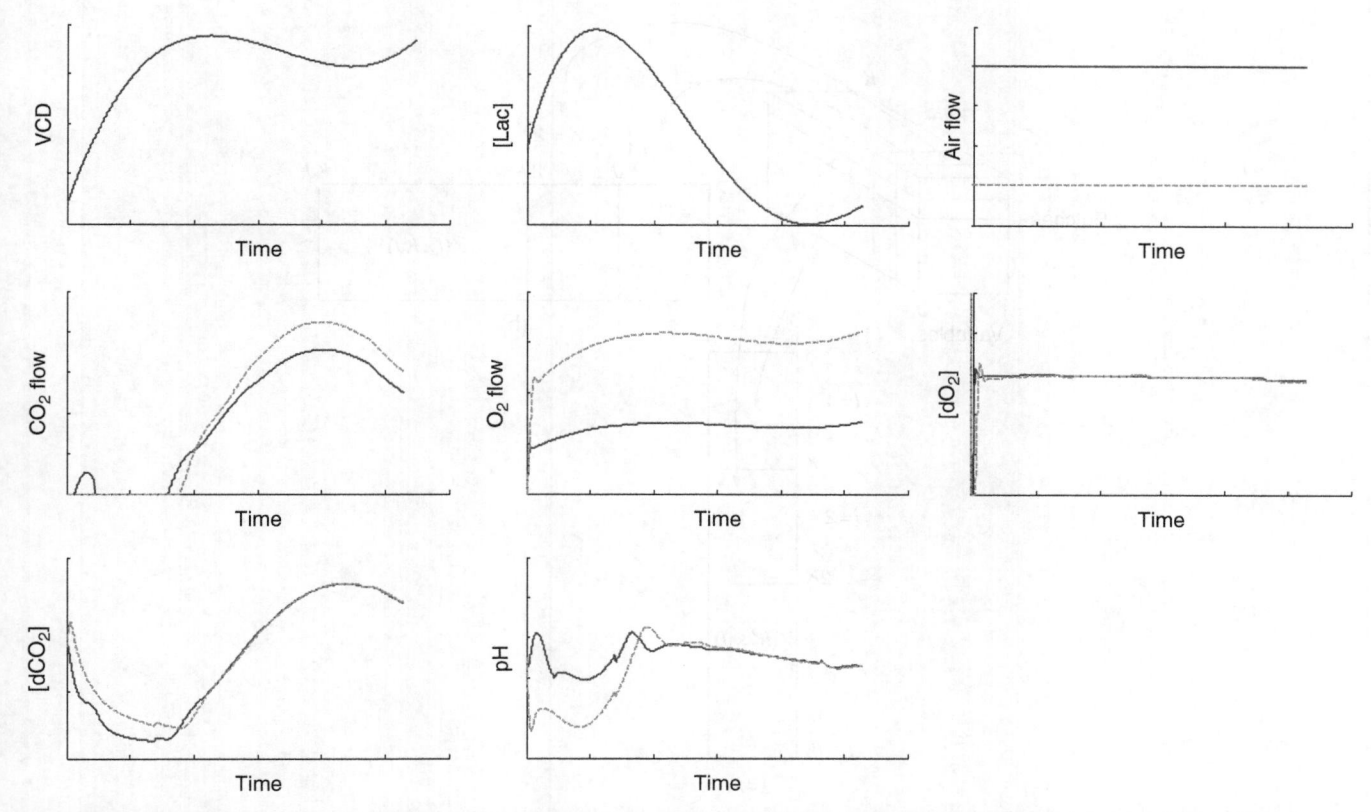

FIGURE 17.14 Simulation results for agitator tip speed reduction, effect on CO_2 and O_2 requirements. Solid line (—), high agitation speed and high aeration rate; dashed line (- -), reduced agitation speed and aeration rate.

model and demonstrates that for the most part key parameters such as pCO_2, pO_2, pH, etc. are comparable at different tip speeds. Despite the reduced agitation speed and aeration, DO was maintained at the required set point (mass transfer from gas to liquid is not rate limiting with respect to its consumption). Increase in sparged O_2 has allowed to attain the DO set point. Predicted increase in the demand on the sparged CO_2 under these conditions was well within the delivery capability of the existing mass flow controllers, and therefore, no resizing was required.

In these examples, use of a first principles model has shown its benefits by providing guidance for scale-up decisions without having to run actual experiments at scale, which is a costly activity. It has also helped increase process understanding of scale-up between pilot and large scales and provided performance comparability. Having a reliable first principles model provides bioprocess engineers ample opportunities for process optimization, troubleshooting, and improvement and supports engineering decision-making process.

17.4.2 Application of Statistical Models in Process Monitoring and Control of Biologics Manufacturing Processes

There are challenges in modeling, monitoring, and control of batch biopharmaceutical processes due to their inherently complex biological and biochemical mechanisms and nonlinear time-variant process dynamics. There are also many variables measured during the course of a batch either offline or online and at variable frequencies depending on the measurement system used. It is important to efficiently monitor and diagnose deviations from the in-control space for troubleshooting and process improvement purposes. One of the solutions successfully applied in chemical industry (both batch and continuous processing) is multivariate modeling and real-time statistical process monitoring [41–43]. These applications have been also successfully extended to pharmaceutical and biopharmaceutical cases [44–47].

In a typical industrial setting, data is generated by the process via various equipment online controls, offline measurements, and assays, which are all stored in various databases. Many process batches (I) are executed, where many variables (J) are measured at certain time intervals (K), forming a three-way data array ($\underline{X}$) as depicted in Figure 17.15. Developing data-driven multivariate process models that define process variability has been shown beneficial in proactively monitoring the process consistency and performance and in troubleshooting purposes. Typical process performance (as contained within data array $\underline{X}$ from the process variables shown in Figure 17.15) can be modeled using multivariate techniques such as principal component analysis (PCA) and partial least squares (PLS). PCA is used when

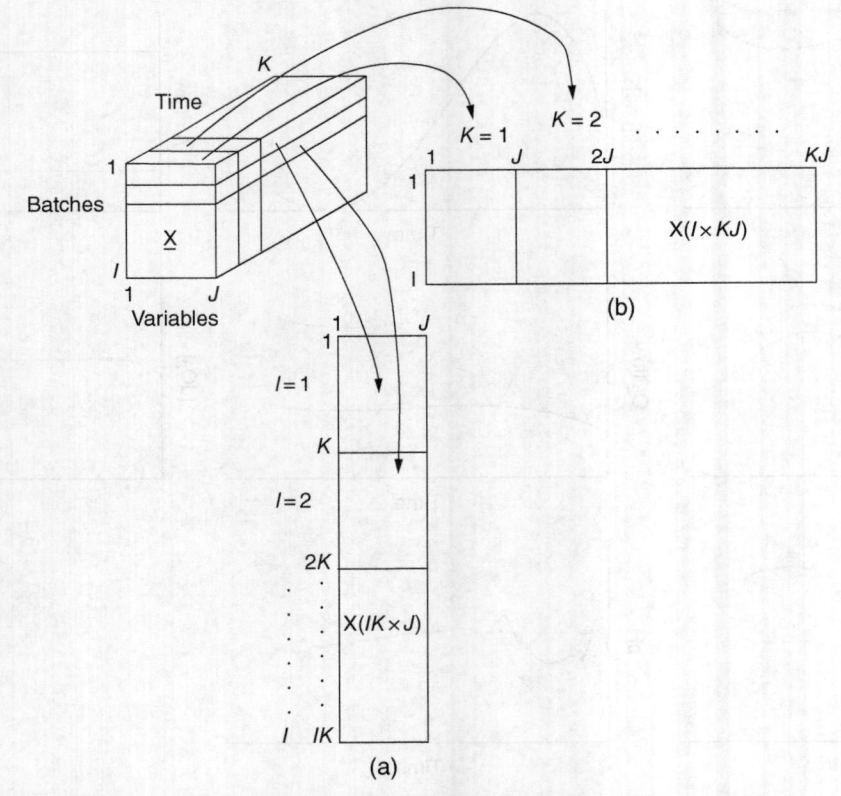

FIGURE 17.15 Unfolding of three-way batch data array: (a) observation level, which preserves variable direction, and (b) batch level, which preserves batch direction.

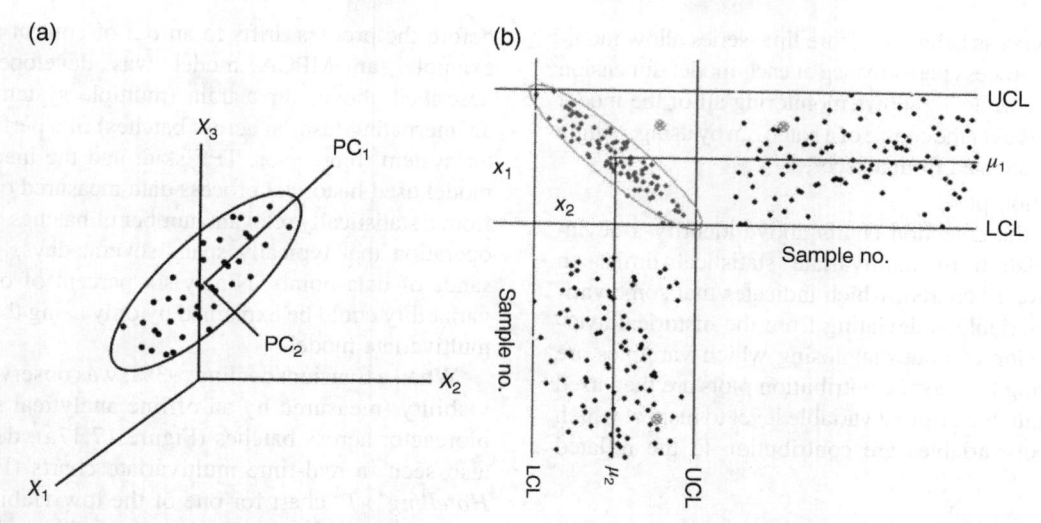

FIGURE 17.16 (a) Dimensionality reduction of a three-variable (x_1, x_2, and x_3) process. Overall variability of the process can be explained by using two principal components (Cinar et al. [33]). (b) Multivariate SPC vs. univariate SPC (MacGregor and Kourti [41]), a batch observation (depicted with a circled-cross sign) that is in control in univariate charts is actually out of control in bivariate in control region defined by the ellipsis (95% confidence region).

overall variability is to be modeled to find major variability dimension in a process (as shown for three variables in Figure 17.16a). This way the dimensionality problem is solved by reducing from many raw process variables to a few derived variables called principal components (PCs) that have linear contributions from raw variables. If the objective is also to correlate how changes in process variables impacting, say, a performance variable defining an endpoint, then PLS is the preferred choice. In that case, the multivariate regression model (PLS) is fitted to data from process variables (inputs) in such a way that it maximizes the correlation to response variable (process output or endpoint). These techniques are extremely useful including (i) reducing the dimensionality problem (summarizing the overall process variability into a few variables from many), (ii) explaining the correlation structure of the variables, (iii) handling missing data, (iv) reducing the noise inherent to measurements so the actual signal can be extracted, and (v) providing means for multivariate statistical process monitoring (MSPM) of the entire process (Figure 17.16).

MSPM for batch (bio) processes can be achieved in real time. Setting the framework up involves the following steps:

1. Collect historical batch data (from typically 20–30 batches of many process variables that are measured online) that define normal operating region.
2. Detect and remove outliers in the data and apply preconditioning (autoscaling).
3. If the objective is only fault detection and diagnosis, develop a PCA model, and if the objective is also to predict process performance variables or critical quality attributes in real time, develop multivariate regression models such as PLS.
4. Construct multivariate charts for real-time monitoring. Monitor a new batch by projecting its data (applying the same data conditioning) onto the model spaces that are developed for comparing to nominal performance.

PCA and PLS are performed on the unfolded array (X) in either direction as shown in Figure 17.15. Observation level models are suggested to avoid estimating the future portion of new batch trajectories [48]. These models are used in real-time monitoring of a new batch, whereas batch-level models are used at the end of the batch for analyzing across batch trends, also known as "batch fingerprinting." Details and mathematical formulation of the modeling in PCA and PLS for batch processes (when it is performed for batch processes, they are usually referred to as multiway PCA or PLS models) can be found in the literature [46, 48].

A number of multivariate statistics and charts are used for monitoring new batches in real time. These include:

1. Squared prediction error (SPE) (also known as Q-residuals or DModX) chart
   SPE is used for process deviation detection where events are not necessarily explained by the model. When SPE control limit violation is observed, it is likely that a new event is observed that is not captured by the reference model (this can be triggered by a normal event that is part of inherent process variability that is not captured or a process upset).
2. Score time series and Hotelling's T^2 charts
   These charts are also used for process deviation detection. They allow detecting deviations that are explained by the process model and within the overall variability but represent unusually high variation compared to the

average process behavior. Score time series allow monitoring the process performance at each model dimension separately, while T^2 allows monitoring all of the model dimensions over the course of a batch run by using a single statistic calculated from all scores.

3. Contribution plots

When the detection charts above identify a deviation (violation of multivariate statistical limits on SPE and/or T^2 charts), which indicates that some variables or variable is deviating from the historical average behavior without diagnosing which variables are contributing the most. Contribution plots are then used to delve into the original variable level to inspect which variable or variables are contribution to the inflated statistic.

17.4.2.1 Utilization of Real-Time Multivariate Statistical Process Monitoring in Bioprocessing MSPM tools provide an important means to rapidly detect process anomalies as the process is running and take action to correct issues before the process drifts to an out-of-control region. In our example, an MPCA model was developed (steps are described above) for a train (multiple systems operated in an alternating fashion across batches) of a perfusion bioreactor system (bioreactor, TFF skid, and the media tank). The model used historical process data measured on 21 variables from a statistically relevant number of batches throughout the operation that typically spans several days, yielding thousands of data points. Ninety-six percent of overall process variability could be explained by only using three PCs in this multivariate model.

When a transient decline (~3%) was observed in final day viability (measured by an offline analytical system) in the bioreactor across batches (Figure 17.17a), deviations were also seen in real-time multivariate charts (Figure 17.17b; *Hotelling's* T^2 chart for one of the low viability batches is shown). Variable contribution plots (Figure 17.16c) identified that higher perfusion feed and retentate temperature conditions occurred in the equipment in comparison to the normal process behavior demonstrated by the historical batches. Further investigation revealed that an equipment mismatch in the system caused the higher than normal

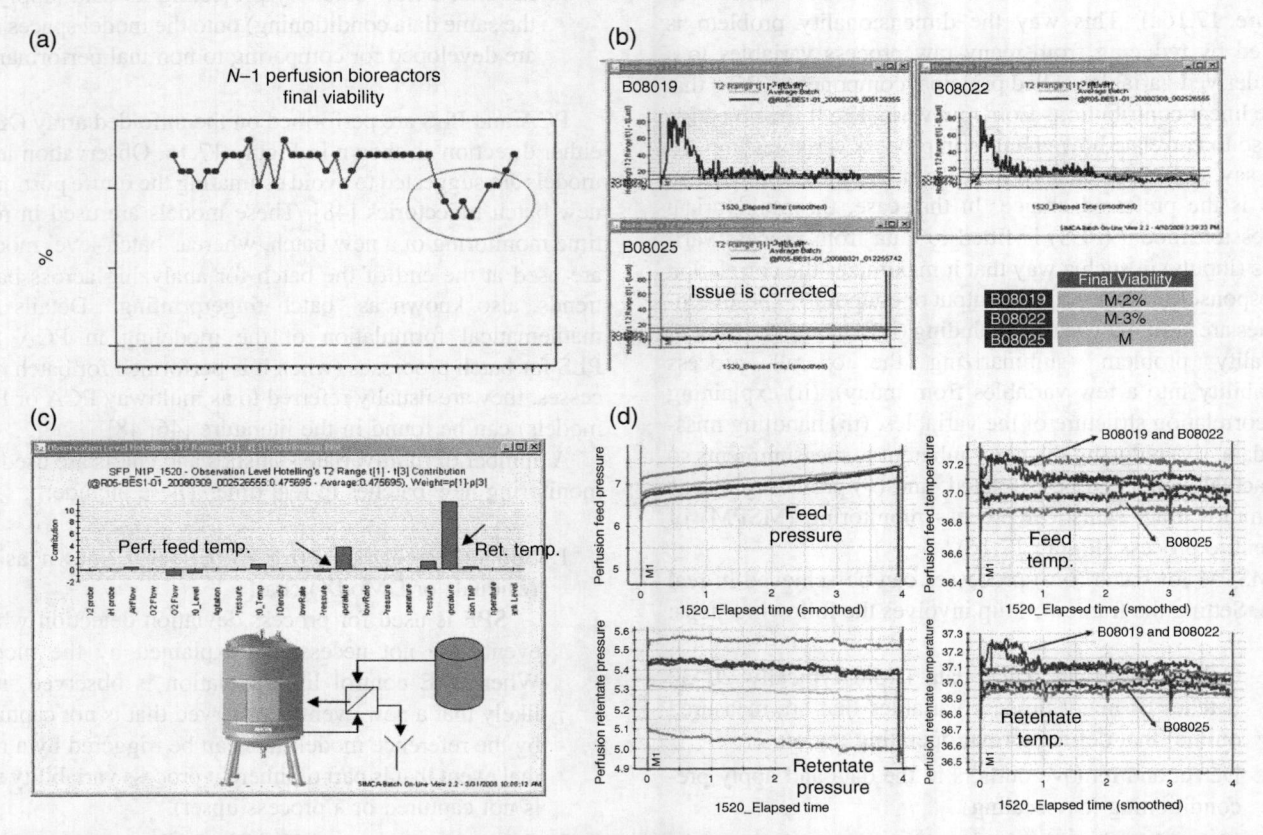

FIGURE 17.17 Real-time multivariate statistical process monitoring (RT-MSPM) steps involved in troubleshooting of a perfusion bioreactor system (shown in c): (a) offline observation of viability decline, (b) detection via T^2 charts, (c) contribution plot for diagnosing the potential root cause, and (d) inspection of the univariate variable profiles.

temperature. The issue was corrected prior to the next batch, and the temperature profile and final day viability returned to their normal ranges (lower left corner of Figure 17.16b).

The example provided here demonstrates how MSPM can detect an abnormal process behavior, assist in determination of root cause, and confirm when a resolution is implemented and the process has returned to normal. The technology has the potential to increase operational success and reduce production costs in highly complex biopharmaceutical processing systems [49].

17.5 CONCLUSIONS

Since the time of synthesizing penicillin in milk bottles in the 1940s, bioprocesses have come a long way. Today, bioprocesses are used for commercial manufacturing of biologics that include therapeutic proteins, vaccines, enzymes, diagnostic reagents, and nonprotein complex macromolecules such as polysaccharides. Biologics are expected to be a major sector within the life sciences industry. The global market for biologics, currently at about $48 billion, is expected to rapidly grow to about $100 billion in the next few years. Currently 6 out of 10 best-selling drugs are biotechnology derived, and over a third of all pipeline products in active development are biologics. The pipeline amounts to approximately 12 000 drug candidates in various stages of preclinical and clinical development covering about 150 disease states and promises to bring better lifesaving treatment to patients. These staggering statistics demonstrate that the fundamentals of biologics and bioprocess industry are strong.

Industrial biotechnology requires blending three key disciplines – biology, biochemistry, and chemical engineering. While our focus in this chapter has been on application of chemical engineering principles to biotechnology, we urge the readers to recognize that the principles of biology and biochemistry play equally (or more) important role in biologics and bioprocessing. Innovation is directly linked to the success of biotechnology products and processes. Innovation in protein design, genetic engineering, expression systems, and manufacturing technologies has allowed rapid commercialization of highly profitable biopharmaceutical products despite the presence of strict regulatory requirements. The innovation, however, has come at a considerable cost. Biologics are the most expensive among all medicinal products. As the biopharmaceutical industry proceeds through its evolution to maturity, it will be forced to undergo significant reinvention to include sound economic principles in addition to sound scientific and engineering principles that provide most cost-effective and value-appropriate therapies to patients.

REFERENCES

1. Stryer, L. (1988). Protein structure and function. In: *Biochemistry*, 3e, 15–42. New York: W. H. Freeman & Co.
2. Medved, L.V., Orthner, C.L., Lubon, H. et al. (1995). Thermal stability and domain-domain interactions in natural and recombinant Protein C. *J. Biol. Chem.* 270: 13652–13659.
3. Narhi, L., Philo, J., Sun, B. et al. (1999). Reversibility of heat-induced denaturation of the recombinant human megakaryocyte growth and development factor. *Pharm. Res.* 16: 799–807.
4. Franks, F. (1988). Conformational stability – denaturation and renaturation. In: *Characterization of Proteins*, 1e (ed. F. Franks), 104–107. New York: Humana Press.
5. Tang, X. and Pikal, M. (2005). The effect of stabilizers and denaturants on the cold denaturation temperatures of proteins and implications for freeze-drying. *Pharm. Res.* 22: 1167–1175.
6. Maa, Y.-F. and Hsu, C.C. (1997). Protein denaturation by combined effect of shear and air-liquid interface. *Biotechnol. Bioeng.* 54: 503–512.
7. Thomas, C.R. (1990). Problems of shear in biotechnology. In: *Chemical Engineering Problems in Biotechnology*, 1e (ed. M. A. Winkler), 23–93. New York: Elsevier.
8. Junker, B.H. (2004). Scale-up methodologies for *E. coli* and yeast fermentation processes. *J. Biosci. Bioeng.* 97: 347–364.
9. Xing, Z., Kenty, B., Jian, Z., and Lee, S.S. (2009). Scale-up analysis for a CHO cell culture process in large-scale bioreactors. *Biotechnol. Bioeng.* 103: 733–746.
10. Yang, J., Lu, C., Stasny, B. et al. (2007). Fed-batch bioreactor process scale-up from 3-L to 2500-L scale for monoclonal antibody production from cell culture. *Biotechnol. Bioeng.* 98: 141–154.
11. Ju, L.-K. and Chase, G.G. (1992). Improved scale-up strategies of bioreactors. *Bioprocess Eng.* 8: 49–53.
12. Dosmar, M., Meyeroltmanns, F., and Gohs, M. (2005). Factors influencing ultrafiltration scale-up. *Bioprocess Int.* 3: 40–50.
13. Van Reiss, R., Goodrich, E.M., Yson, C.L. et al. (1997). Linear scale ultrafiltration. *Biotechnol. Bioeng.* 55: 737–746.
14. Prokop, A. (1991). Implications of cell biology in animal cell biotechnology. In: *Animal Cell Bioreactors* (ed. C.S. Ho and D.I.C. Wang), 21–58. Boston, MA: Butterworth-Heinemann.
15. Nienow, A.W. (2006). Reactor engineering in large scale animal cell culture. *Cytotechnology* 50: 9–33.
16. Blanch, H.W. and Clark, D.S. (1997). *Biochemical Engineering*, 415–426. New York: Marcel Dekker Inc.
17. Xie, L. and Wang, D.I.C. (1994). Stoichiometric analysis of animal cell growth and its application in medium design. *Biotechnol. Bioeng.* 43: 1164–1174.
18. Cooper, C.K., Fernstrom, G.A., and Miller, S.A. (1944). Performance of agitated gas-liquid contactors. *Ind. Eng. Chem.* 36: 504–509.
19. Shukla, A.A. and Kandula, J.R. (2009). Harvest and recovery of monoclonal antibodies: cell removal and clarification. In: *Process Scale Purification of Antibodies* (ed. U. Gottschalk), 53–78. Hoboken, NJ: Wiley.

20. Blanch, H.W. and Clark, D.S. (1997). *Biochemical Engineering*, 461–467. New York: Marcel Dekker Inc.
21. Usher, K.M., Simmons, C.R., and Dorsey, J.G. (2008). Modeling chromatographic dispersion: a comparison of popular equations. *J. Chromatogr.* 1200: 122–128.
22. Van Deemter, J.J., Zuiderweg, F.J., and Klinkenberg, A. (1956). Longitudinal diffusion and resistance to mass transfer as causes of nonideality in chromatography. *Chem. Eng. Sci.* 5: 271–289.
23. McCoy, M., Kalghatgi, K., Regnier, F.E., and Afeyan, N. (1996). Perfusion chromatography – characterization of column packings for chromatography of proteins. *J. Chromatogr.* 743: 221–229.
24. Moscariello, J., Purdom, G., Coffman, J. et al. (2001). Characterizing the performance of industrial-scale columns. *J. Chromatogr.* 908: 131–141.
25. Pathak, N., Norman, C., Kundu, S. et al. (2008). Modeling flow distribution in large-scale chromatographic columns with computational fluid dynamics. *Bioprocess Int.* 6: 72–81.
26. Stickel, J.J. and Fotopoulos, A. (2001). Pressure-flow relationships for packed beds of compressible chromatography media at laboratory and production scale. *Biotechnol. Prog.* 17: 744–751.
27. McCue, J.T., Cecchini, D., Chu, C. et al. (2007). Application of a two-dimensional model for predicting the pressure-flow and compression properties during column packing scale-up. *J. Chromatogr.* 1145: 89–101.
28. Teeters, M.A. and Quinones-Garcia, I. (2005). Evaluating and monitoring the packing behavior of process-scale chromatography columns. *J. Chromatogr.* 1069: 53–64.
29. Larson, T.M., Davis, J., Lam, H., and Cacia, J. (2003). Use of process data to assess chromatographic performance in production-scale purification columns. *Biotechnol. Prog.* 19 (2): 485–492.
30. Lyderson, B.K., D'elia, N.A., and Nelson, K.L. (1994). *Bioprocess Engineering Systems, Equipment and Facilities*. New York: Wiley.
31. Bailey, J.E. (1998). Mathematical modeling and analysis in biochemical engineering: past accomplishments and future opportunities. *Biotechnol. Prog.* 14: 8–20.
32. Stephanopoulos, G., Aristidou, A.A., and Nielsen, J. (1998). Chapter 8: Metabolic flux analysis; Chapter 9: Method for experimental determination of metabolic fluxes by radioisotope labeling; Chapter 10: Applications of metabolic flux analysis. In: *Metabolic Engineering: Principles and Methodologies*, 309–459. London, UK: Academic Press.
33. Cinar, A., Parulekar, S.J., Undey, C., and Birol, G. (eds.) (2003). *Batch Fermentation: Modeling, Monitoring and Control*, 21–58. New York: CRC Press.
34. Rathore, A.S. and Winkle, H. (2009). Quality by design for biopharmaceuticals. *Nat. Biotechnol.* 27: 26–34.
35. US Food and Drug Administration (2002). *Pharmaceutical cGMPs for the 21st Century: A Risk-based Approach* (August 2002). Rockville, MD: FDA. https://www.fda.gov/downloads/drugs/guidances/ucm070305.pdf (accessed 23 July 2009).
36. US Food and Drug Administration (2004). *Guidance for Industry: PAT – A Framework for Innovative Pharmaceutical Development, Manufacturing, and Quality Assurance*. Rockville, MD: FDA. http://www.fda.gov/Cder/guidance/6419fnl.pdf (accessed 23 July 2009).
37. International Conference on Harmonisation (ICH) (2008). *Pharmaceutical Development Q8(R1)*. Geneva: ICH. https://www.ich.org/fileadmin/Public_Web_Site/ICH_Products/Guidelines/Quality/Q8_R1/Step4/Q8_R2_Guideline.pdf (accessed 23 July 2009).
38. Gray, D.R., Chen, S., Howarth, W. et al. (1996). CO_2 in large-scale and high-density CHO cell perfusion culture. *Cytotechnology* 22: 65–78.
39. Zupke, C. and Green, J. (1998). Modeling of CO_2 concentration in small and large scale bioreactors. Presented at Cell Culture Engineering VI, San Diego, CA (1998).
40. Mostafa, S. and Gu, X. (2003). Strategies for improved dCO_2 removal in large-scale fed-batch cultures. *Biotechnol. Prog.* 19: 45–51.
41. MacGregor, J.F. and Kourti, T. (1995). Statistical process control of multivariate processes. *Control Eng. Pract.* 3: 403–414.
42. Neogi, D. and Schlags, C.E. (1998). Multivariate statistical analysis of an emulsion batch process. *Ind. Eng. Chem. Res.* 37: 3971–3979.
43. Undey, C., Ertunc, S., Tatara, E. et al. (2004). Batch process monitoring and its applications in polymerization systems. *Macromol. Symp.* 206: 121–134.
44. Westerhuis, J.A., Coenegracht, P.M.J., and Lerk, C.F. (1997). Multivariate modeling of the tablet manufacturing process with wet granulation for tablet optimization and in-process control. *Int. J. Pharm.* 156: 109–117.
45. Albert, S. and Kinley, R.D. (2001). Multivariate statistical monitoring of batch processes: an industrial case study in fed-batch fermentation supervision. *Trends Biotechnol.* 19: 53–62.
46. Undey, C., Ertunc, S., and Cinar, A. (2003). Online batch/fed-batch process performance monitoring, quality prediction and variable contributions analysis for diagnosis. *Ind. Eng. Chem. Res.* 42: 4645–4658.
47. Undey, C., Tatara, E., and Cinar, A. (2004). Intelligent real-time performance monitoring and quality prediction for batch/fed-batch cultivations. *J. Biotechnol.* 108: 61–77.
48. Wold, S., Kettaneh, N., Fridén, H., and Holmberg, A. (1998). Modelling and diagnostics of batch processes and analogous kinetic experiments. *Chemom. Intel. Lab. Syst.* 44: 331–340.
49. Undey, C., Ertunc, S., Mistretta, T., and Pathak, M. (2009). Applied advanced process analytics in biopharmaceutical manufacturing: challenges and prospects in real-time monitoring and control. *IFAC ADCHEM Proceedings, Istanbul, Turkey* (13–15 July 2009).

PART VI

THERMODYNAMICS

18

APPLICATIONS OF THERMODYNAMICS TOWARD PHARMACEUTICAL PROBLEM SOLVING

AHMAD Y. SHEIKH, ALESSANDRA MATTEI, AND RAIMUNDO HO
Solid State Chemistry, AbbVie Inc., North Chicago, IL, USA

MOIZ DIWAN
Process Research and Development, AbbVie Inc., North Chicago, IL, USA

THOMAS BORCHARDT AND GERALD DANZER
Drug Product Development, AbbVie Inc., North Chicago, IL, USA

NADINE DING AND XINMIN (SAM) XU
Abbott Vascular, Santa Clara, CA, USA

18.1 INTRODUCTION

Interplay of thermodynamics and kinetics underpins many aspects of pharmaceutical development. Understanding of the underlying thermodynamics helps conceive, develop, and control a range of processes and the resulting properties/quality attributes. Quality and regulatory considerations also require thorough understanding and robust control of the processes to ensure patient safety. An obvious application of thermodynamics within the context of chemistry, manufacturing, and control (CMC) development relates to the importance of crystal form in solid dosage development. Polymorphism is prevalent for molecular crystals [1, 2], and a recent detailed study showed that well over a third of all pharmaceutical compounds exhibit polymorphism [3]. Polymorphism affects a range of physical properties including melting point, intrinsic solubility, dissolution rate, and stability. These differences are related to thermodynamics and in particular free energy differences often caused by differences in molecular packing configurations. The utility of co-crystals to modulate biopharmaceutical properties and processability [4–7] also stems from the underlying differences in free energies. Additionally, interactions between components in common multicomponent/multiphase pharmaceutical systems are governed by thermodynamics of mixing.

In this chapter, five distinct case studies are presented to demonstrate a range of approaches to establish and apply foundational thermodynamics in the design and control of a variety of different pharmaceutical processes. The case studies cover (i) desolvation behavior of a polymorphic system, (ii) solid form control during crystallization, (iii) scalable solution crystallization of co-crystals, (iv) coating process for drug-eluting bioresorbable vascular scaffold (BVS), and (v) polymer–plasticizer mixing performance in the presence of water.

While the most consequential assumptions inherent in various thermodynamic models are highlighted, a detailed discussion on the actual mathematical constructs is outside of the scope of this chapter.

18.2 DESOLVATION OF PARECOXIB SODIUM

18.2.1 Introduction

Parecoxib sodium is a water-soluble prodrug belonging to COX-2 class of drugs. It is administered intravenously [8] and marketed in Europe as Dynastat for short-term perioperative pain control. The chemical structure is shown in Figure 18.1. Polymorphism studies conducted at Pharmacia Corp. during development in the early 2000s identified three anhydrous (Forms I (A), II (B), and III (E)) and various

---
Chemical Engineering in the Pharmaceutical Industry: Active Pharmaceutical Ingredients, Second Edition.
Edited by David J. am Ende and Mary T. am Ende.
© 2019 John Wiley & Sons, Inc. Published 2019 by John Wiley & Sons, Inc.

solvated and hydrated forms [9]. The development of a more soluble (metastable) form (preferably mostly Form I) was desirable from a manufacturing perspective for the IV formulation. None of known anhydrous forms at the time could be isolated directly from the crystallization process and had to be made by desolvation of solvates. Form conversions in the drying process represented significant challenge for robust design, scale-up, and technology transfer of the API manufacturing process.

18.2.2 Semiquantitative Phase Diagrams

Free energy–temperature and composition–temperature diagrams were considered necessary to understand crystal form

FIGURE 18.1 Chemical structure of parecoxib sodium.

landscape and desolvation pathways for parecoxib sodium. For a system comprising "n" solvent-free forms, "$2n$" properties need to be measured or determined to construct a sufficiently detailed free energy–temperature diagram [10]. These properties can include but are not limited to melting temperature, heat of fusion, heat of solution [11], solubility, and intrinsic dissolution rate [12].

Thermal data was collected on the nonsolvated forms using Mettler Toledo differential scanning calorimetry (DSC). Form I displayed a single melting endotherm with an onset at about 273.1 °C ($\Delta H_t = 23.8$ kJ/mol). Form II displayed an endotherm with an onset at about 195.9 °C ($\Delta H_t = 20.71$ kJ/mol) representing transition to Form I, followed by a sharp melting endotherm for Form I at 273.7 °C. Form III displayed a broad endotherm with an onset at about 206.6 °C ($\Delta H_t = 18.35$ kJ/mol) representing transition to Form I, followed by a sharp melting endotherm for Form I at 273.2 °C. Qualitative thermal data for the three forms is shown in Figure 18.2.

The transitions for Forms II and III to Form I prior to melting were verified by hot-stage microscopy.

Based on the heat of transition rule [13], both Forms II and III can be considered enantiotropically related to Form I. Since, melting and recrystallization events were not sufficiently resolved; calculation of the transition temperature using the standard DSC [14] data was not possible. Method of eutectic melting [15] was therefore employed. It allows

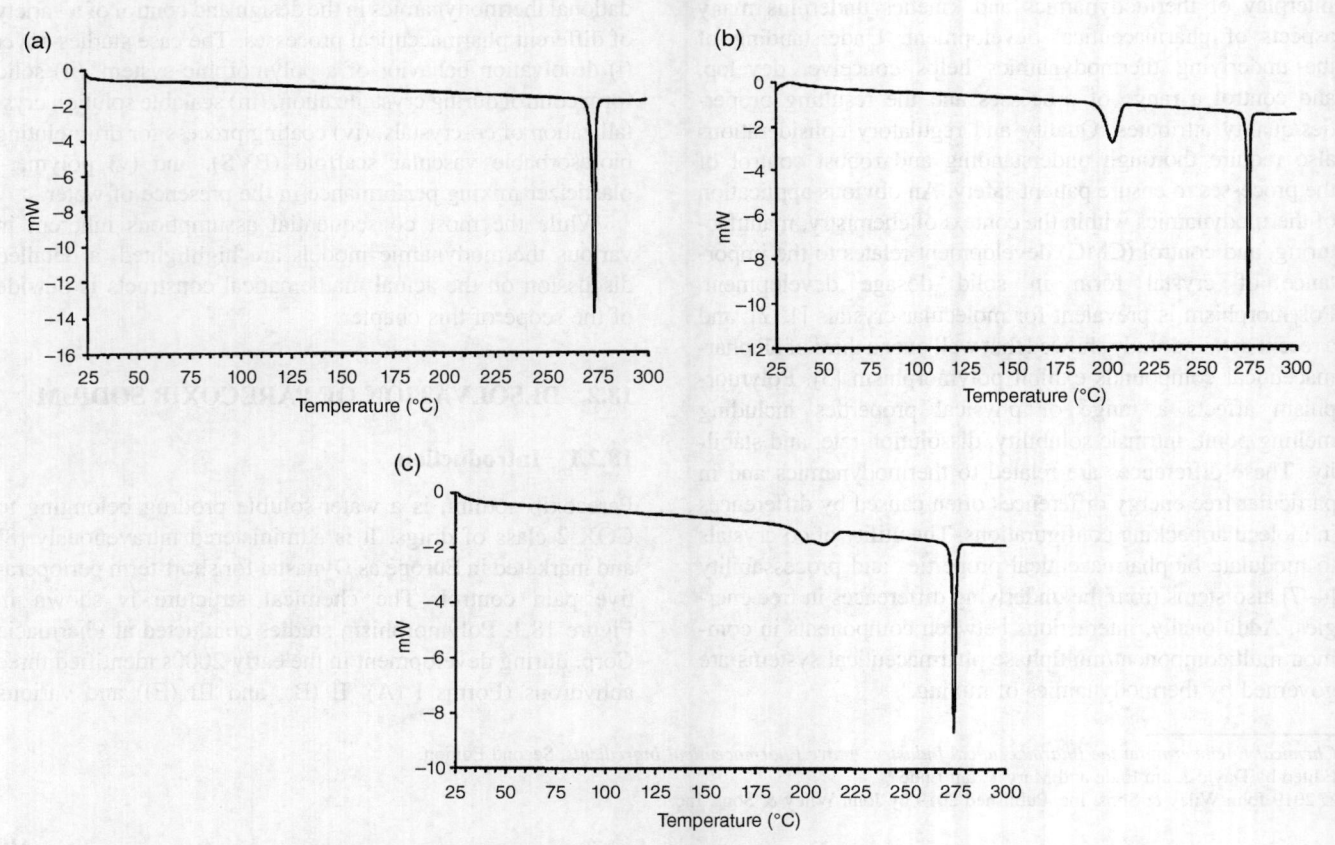

FIGURE 18.2 Qualitative DSC data for Forms I (a), II (b), and III (c).

modulation of melting points by using reference compounds at eutectic compositions to help isolate melting and crystallization events and accurately determine heat of fusion data. Eutectic compositions and corresponding heat of fusion data can then be used to derive the free energy differences between crystal forms using Eq. (18.1):

$$x_{ej}(G_j - G_i)_{T_{ei}} = \frac{\Delta H_{mej}(T_{ei} - T_{ej})}{T_{ej}} + \Delta Cp_{ej}\left[T_{ei} - T_{ej} - \left(T_{ei}\ln\frac{T_{ei}}{T_{ej}}\right)\right] + RT_{ei}\left\{\left[x_{ej}\ln\frac{x_{ej}}{x_{ei}}\right] + \left[(1-x_{ej})\ln\frac{1-x_{ej}}{1-x_{ei}}\right]\right\}$$
(18.1)

wherein

x_{ej} and x_{ei} are the mole fraction of crystal forms j and i, respectively, in the eutectic.

$(G_j - G_i)$ is the free energy difference between crystal forms i and j at T_{ei}.

ΔH_{mej} is the enthalpy of eutectic melting of crystal forms j.

T_{ei} and T_{ej} are the temperatures of eutectic melting of crystal forms i and j, respectively.

ΔCp_{ej} is the heat capacity change across the eutectic melt.

R is the ideal gas constant.

Plots of $(G_j - G_i)$ vs. T and $(G_j - G_i)/T$ vs. $1/T$ can then be used to determine two thermodynamic parameters (ΔH and ΔS) essential for determining the transition temperatures (T_t) using Eq. (18.2):

$$T_t = \frac{\Delta H}{\Delta S}$$
(18.2)

Phenacetin, benzanilide, and salophen were used as the reference compounds to ensure that a broad range of eutectic melting temperatures could be achieved for each of the three anhydrous forms. Physical mixtures comprising varying mole fractions of anhydrous form and reference compounds were prepared, and DSC data was collected. Compositions that led to detection of a single melting endotherm were recorded as eutectic compositions. Thermodynamic properties obtained at eutectic compositions for Forms I, II, and III with each reference compound are given in Table 18.1.

Plots of ΔG vs. T and $\Delta G/T$ vs. $1/T$ were used (not shown here) to determine enthalpic and entropic differences between pairs of parecoxib forms, and key thermodynamic parameters are shown in Table 18.2.

The data confirmed an enantiotropic relationship between Forms I and either II or III. Forms II and III were found to be very close in energy, whereas Form I was found to be the highest energy form. The rank order of stability correlated

TABLE 18.1 Eutectic Melting Data for Forms I, II and III

Crystal Form	Form I	Form II	Form III
T_m (°C)	274–276	Phase conversion	Phase conversion
RC = phenacetin			
X_e	0.25	0.25	0.25
T_e (°C) (mean)	118.2	124.7	124.7
ΔH_{me} (kJ/mol)	24.64	25.99	27.08
RC = benzanilide			
X_e	0.17	0.18	0.18
T_e (°C) (mean)	155.6	156.6	156.2
ΔH_{me} (kJ/mol)	28.32	31.95	31.42
RC = salophen			
X_e	0.42	0.42	0.42
T_e (°C) (mean)	171.7	170.1	170.1
ΔH_{me} (kJ/mol)	25.82	36.83	34.62

X_e is the eutectic composition on molar basis, T_e is eutectic melting point, ΔH_{me} is enthalpy of fusion for the eutectic melt, and T_m is the melting point of parecoxib sodium crystal forms.

TABLE 18.2 Thermodynamic Parameters for Interconversion of Forms

Form/Transition	ΔH (kJ/mol)	ΔS (J/[K mol])	Transition Temperature (°C)
II to I	16.63	38.1	163.3
III to I	17.15	39.2	163.9

with true density data as measured by helium pycnometry (Form II, 1.46 g/cm^3; Form III, 1.42 g/cm^3; Form I, 1.34 g/cm^3). Similar transition temperatures for Form III/Form I (163.9 °C) and Form II/Form I (163.3 °C) were obtained due to the narrow energy difference between Forms II and III. In fact, the similarity of free energies of Forms II and III makes it difficult to conclusively ascertain the thermodynamically most stable form at ambient temperature. The data were used to construct a semiquantitative G-T diagram shown in Figure 18.3.

While the G-T diagram is highly informative in understanding the phase relationships and difficulty of thermodynamically controlling Form II or III, it does not help understand the desolvation behavior of Form IV (ethanol solvate) that was isolated from the crystallization process. Desolvation or decomposition of a crystalline solvate can happen in two distinct environments: under constant composition or steady loss of solvent. For the latter, kinetics of conversion can be key determinant of the outcome, whereas for the former thermodynamics also plays an important role. To understand the thermodynamics of desolvation, experiments need to be conducted to understand the melting behavior of the solvate and determine if it exhibits eutectic or peritectic

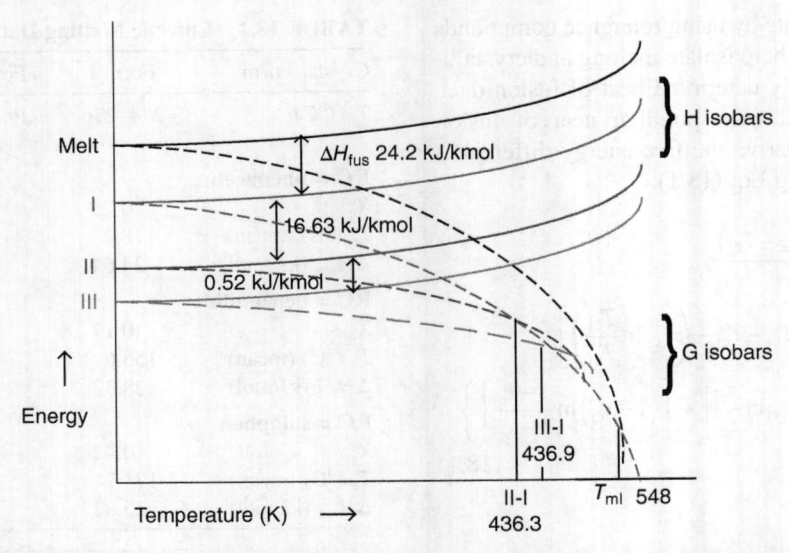

FIGURE 18.3 Semiquantitative G-T diagram for three anhydrous forms of parecoxib sodium (temperatures in K).

melting. Eutectic melting occurs when the two components of the solvate melt into each other and remain liquid with no recrystallization. Peritectic melting occurs when a solvate becomes unstable and phase converts in the presence of either saturated headspace or in a solution that has solvent activity of one for solvent of solvation. Phase conversions just above the peritectic temperature tend to be fast, and the resulting solids are usually solvent-free and most stable at that temperature. Carefully executed sealed pan and open pan DSC and sealed and unsealed hot-stage microscopy (with appropriate mineral oils) [16] can be used to determine the melting behavior of solvates. Peritectic melting behavior is detected in the DSC by delayed "melting/desolvation" endotherm, immediately followed by a crystallization exotherm and eventual melting endotherm of the anhydrous form. The crystallization exotherm can be sharp or broad depending on scan rate and ability to hold the solvent vapors in the DSC pan. Verification of peritectic melting behavior can be performed by comparing DSC thermal events under unsealed conditions. In this case, earlier onset of desolvation followed by very slow conversion to the desolvated phase or kinetically favored anhydrous phase is often observed.

When such experiments were conducted under sealed conditions with Form IV of parecoxib, peritectic melting to Form III was observed. The peritectic melting point was determined to be approximately 72 °C. Analogous experiments under unsealed conditions resulted in either Form I or II depending on the heating rates and starting solvent content of the wet Form IV. Solid form was confirmed by opening DSC pan after the desolvation event and analyzing the solids by powder X-ray diffraction.

Melting data on the anhydrous forms, transition temperatures (between anhydrous forms), and peritectic melting point were used to construct a semiquantitative composition–temperature or x-T diagram (Figure 18.4). It is worth emphasizing that the horizontal axis is exaggerated at both ends of the composition range to clearly illustrate the key features of the system. It is also noteworthy that the left-hand side of the diagram is purely conceptual and no efforts were made to experimentally determine eutectic composition through solubility measurements.

The x-T phase diagram helps visualize two distinct drying/desolvation pathways (ABC and ADE), which can lead to different polymorphic outcomes. In pathway ABC, the solvent content of solids is at or above mole fraction corresponding to the composition of Form IV solvate when drying temperature approaches/exceeds the peritectic melting point. As a consequence Form IV readily converts to Form III. In pathway ADE on the other hand, most of the solvent is removed below the peritectic melting point, and as a consequence saturation conditions with respect to the solvent are not achieved when the drying temperature approaches/exceeds the peritectic melting point. This results in relatively slow conversion of partially desolvated Form IV to either Form I or II. Kinetics of drying/desolvation play a leading role in the outcome of ADE pathway.

18.2.3 Implications for Process Design and Control

As previously mentioned during development for the IV formulation of parecoxib sodium, Form I or II was more desirable than Form III. Drying process design and scale-up efforts were therefore focused on efficient removal of the solvent from wet cake below 72 °C. Appropriate washing solvents, heating ramps, mixing protocols, and hold times were established. Only when the established low levels of residual solvents

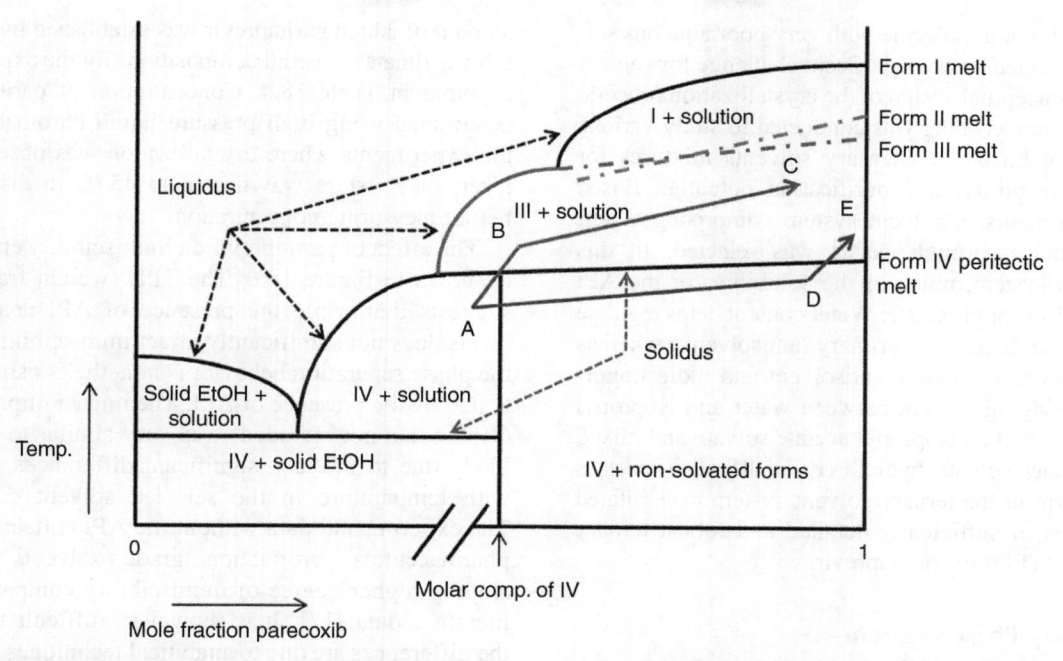

FIGURE 18.4 Semiquantitative x-T diagram for the ethanol solvate and three anhydrous forms of parecoxib sodium.

had been achieved was the drying temperature increased to 80 °C to complete form conversion and fully dry the cake. Flexibility was also built into the control strategy with respect to solid-phase purity of Form I. The process and control strategy was successfully filed with global regulatory authorities and implemented at the commercial manufacturing site.

18.3 SOLID FORM CONTROL OF PARITAPREVIR

18.3.1 Introduction

Paritaprevir is a protease inhibitor [17] that has been developed as part of treatment for hepatitis C virus (HCV). A regimen of three direct-acting antiviral drugs, with the NS5A inhibitor ombitasvir, and the non-nucleoside polymerase inhibitor dasabuvir, is marketed in the United States as Viekira. Due to very low solubility of paritaprevir, hot melt extrusion technology [18] is used to manufacture the enabling formulation for drug product.

In the extrusion process, the drug substance is converted from a crystalline solid form to an amorphous state and dissolved in a polymer/surfactant matrix to form an amorphous solid dispersion (ASD) [19]. For HME formulations, low melting crystalline anhydrous forms or hydrates with low dehydration temperatures are preferred to facilitate conversion to the ASD at relatively low temperatures (<200 °C). Given the complexity and structural flexibility of the paritaprevir molecule (see Figure 18.5), numerous crystalline solid

FIGURE 18.5 Chemical structure of paritaprevir.

forms were discovered during development [20]. Many isomorphic forms (with varying chemical compositions [water or solvents]) were also identified. Class I [21] (or Form II) structures were found to be most suitable for HME formulation. This class represented variable stoichiometry hydrates with similar crystallographic parameters that dehydrated to an amorphous state below a relatively low temperature of 140 °C. Form I, a crystallographically distinct structure, was the most stable under ambient temperature and did not loose crystallinity until 200 °C. Solid-state conversions between Form I and Form II were, however not observed under ICH stability conditions.

18.3.2 Solvent Selection and Phase Stability

While Form II had ideal solid-state properties for downstream processing, slurry competition studies had established that it was only thermodynamically stable at very high water activities (>0.8). For a molecule with very poor aqueous solubility, this presented a very significant challenge for solvent selection and conceptual design of the crystallization process. Extensive solvent screening was conducted to study various combinations of binary and ternary solvents mixtures for solubility, phase purity, and purification potential. Based on these experiments, a solvent system comprising water, isopropanol, and isopropyl acetate was selected. In this ternary solvent system, much of the solubility for the API is attributed to isopropyl acetate. Water content helps regulate water activity and acts as primary antisolvent, whereas isopropanol acts as secondary antisolvent and more importantly as the bridging solvent between water and isopropyl acetate. The fact that an isopropyl acetate solvate and mixed isopropanol water solvate hydrate (Form III) also exist as stable solid form in the ternary solvent system necessitated the construction of sufficiently detailed and robust ternary phase diagram (TPD) for paritaprevir.

18.3.3 Ternary Phase Diagram

Experiments for the construction of four-component (paritaprevir, water, isopropanol, isopropyl acetate) TPD at 25 and 60 °C was divided into three segments. First the immiscibility boundary for water, isopropanol, and isopropyl acetate was experimentally determined using manufacturing grade solvents. The effect of paritaprevir on the immiscibility boundary was then established. Finally, relative stability of process relevant solid forms was determined for the relevant domains of the solvent space. Immiscibility data from the literature [22] was used to define nine unique experiments in the biphasic space. Details of the overall solvent compositions can be found in Table 18.3. Each experiment was allowed to phase separate in a separation funnel for one hour, before the layers were separated. The composition of the layers was determined through gas chromatography for isopropanol and isopropyl acetate and Karl Fischer titration for water. Mass fraction of each phase was also recorded.

Nine additional experiments were performed to assess impact of paritaprevir on the immiscibility boundary. The amount of added paritaprevir was established based on solubility estimates. Overall compositions for the experiments are captured in Table 18.4. Concentration of paritaprevir was determined using high pressure liquid chromatography. In the experiments where crystallization was observed in any layer, the mixture was heated to 45 °C to dissolve solids before measuring concentration.

The effect of paritaprevir on immiscibility curve at 25 °C is shown in Figure 18.6. The TPD (weight fraction basis) suggests that while the presence of API near saturation levels does not significantly affect immiscibility boundary, the phase separation behavior is nonetheless slightly accentuated by the presence of API. The impact of paritaprevir at 60 °C (data not included) was very similar to 25 °C, most likely due to the lack significant differences in solubility with temperature in the selected solvent compositions. The experimental data without the API collected here with pharmaceutical production grade solvents also show slightly higher degree of immiscibility compared with the literature data [17]. It is, however, difficult to discern if the differences are due to analytical techniques or the subtle differences in actual solvents. Slight quantities of water in isopropyl acetate can, for instance, have significant effect on phase compositions.

Finally solid-phase stability was determined via slurry competition experiments in the key domains of ternary solvent space. Physical mixtures of Form II, Form III, isopropyl acetate solvate, and Form I (most stable form under ambient conditions) were used and up to four weeks were afforded for crystal form conversion at 25 and 60 °C. After equilibration, slurries were filtered and solids characterized by powder X-ray diffraction. Form I was found to be metastable in the entire solvent space, and the isopropyl acetate solvate was only found to be stable in 100% isopropyl acetate. Studies at 60 °C did not result in any changes in the relative stability of the forms. However, the conversion rates were significantly faster at the higher temperature. All the data was used to construct complete phase diagrams. TPD at 25 °C diagram is shown in Figure 18.7 for illustration purposes.

TABLE 18.3 Solvent Ratios for Immiscibility Boundary Determination

Experiment	Water (g)	IPA (g)	IPAc (g)
1	15.0	3.0	12.0
2	10.5	9.0	9.0
3	15.0	4.8	9.7
4	5.0	5.0	9.0
5	10.0	4.0	2.0
6	10.0	4.0	6.0
7	4.0	5.0	7.0
8	5.0	0.0	5.0
9	3.0	4.0	13.0

TABLE 18.4 Solvent Ratios for Immiscibility Boundary Determination with Paritaprevir

Experiment	Water (g)	IPA (g)	Total IPAc (g)	Paritaprevir (g)
1	15.0	3.0	12.0	0.6
2	10.5	9.0	9.0	0.6
3	15.0	4.8	9.7	0.6
4	5.0	5.0	9.0	0.4
5	10.0	4.0	2.0	0.3
6	10.0	4.0	6.0	0.4
7	4.0	5.0	7.0	0.3
8	5.0	0.0	5.0	0.2
9	3.0	4.0	13.0	0.4

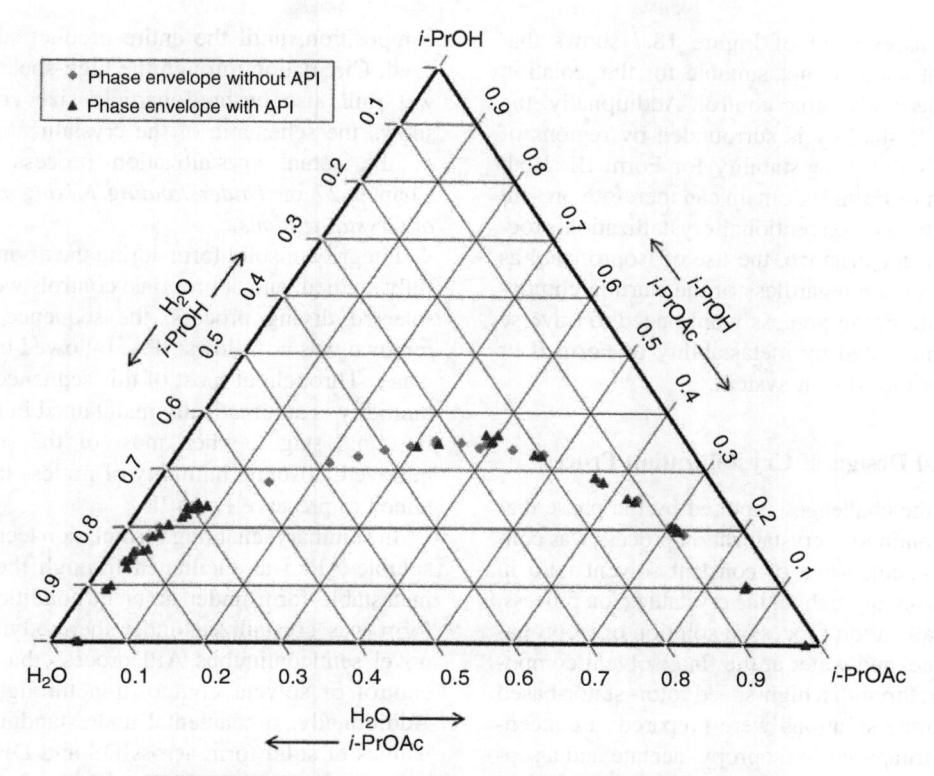

FIGURE 18.6 Impact of paritaprevir on immiscibility boundary.

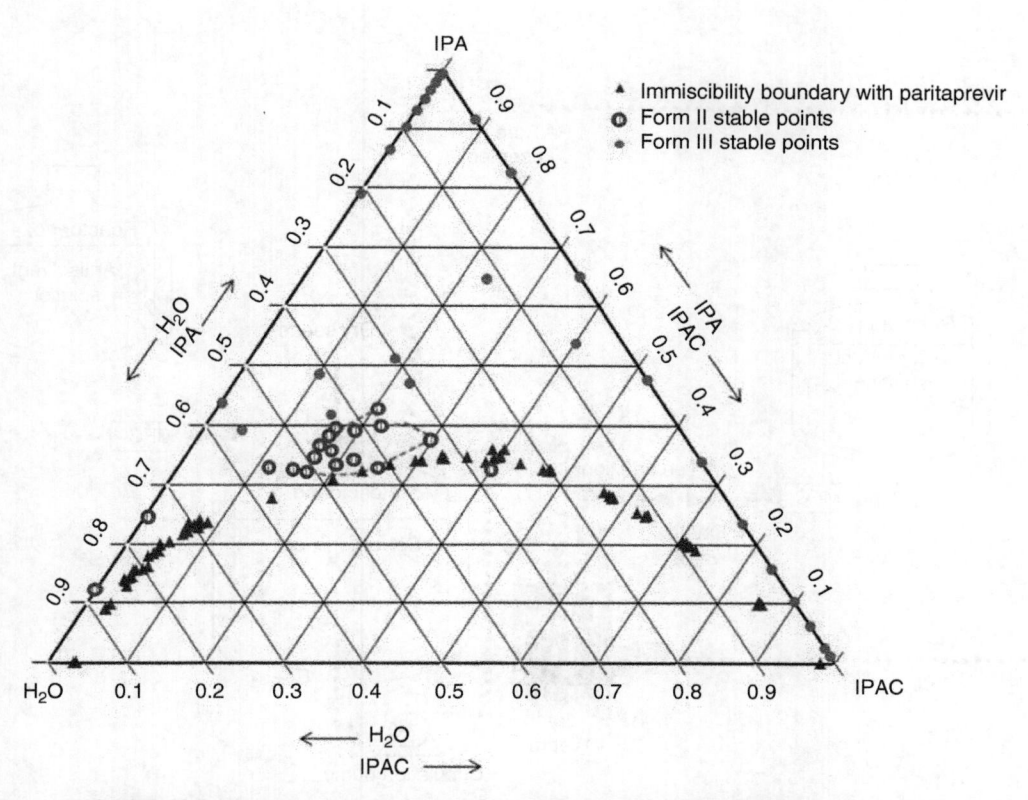

FIGURE 18.7 Complete phase diagram from paritaprevir in crystallization solvent system.

A very cursory assessment of Figure 18.7 shows that much of the solvent space is not suitable for the isolation of Form II under thermodynamic control. Additionally, the narrow domain for II stability is surrounded by regions of either solvent immiscibility or stability for Form III. Both the size and location of Form II domain can therefore present significant challenges for conventional crystallization process design based on, for instance, the use of isopropanol as the antisolvent. In essence regardless of the starting composition for crystallization, the process would need to traverse through regions represented by metastability of Form II or partial miscibility of the solvent system.

18.3.4 Conceptual Design of Crystallization Process

In order to address the challenges captured by the phase diagram, a novel semicontinuous crystallization process was conceived. It afforded maintenance of constant solvent ratio in Form II stability domain throughout the crystallization process.

During the crystallization process, a solution of isopropanol, isopropyl acetate, and water at the final solvent composition was circulated through a high-speed rotor–stator-based wet mill. Two separate solutions were prepared: a concentrated solution of paritaprevir in isopropyl acetate and a separate solution with appropriate composition of isopropanol and water. These solutions were added to the circulating solution to initiate and continue crystallization at fixed solvent composition, until the entire product solution had crystallized. Circulation through the high-speed rotor–stator-based wet mill also ensured particle size control. Figure 18.8 shows the schematic of the crystallization process. Details of the actual crystallization process can be found in Chapter 27 on *Understanding Mixing Effects on Scale-Up of Crystallizations*.

Integrity of solid form during the drying process was carefully studied, and appropriate controls were instituted. In the selected drying process, the sequence of "bulk" solvent removal was as follows: IPA, followed by IPAC, and finally water. Throughout most of this sequence, very high relative humidity is automatically maintained in the headspace. During final stages, when most of the solvents are already removed, absolute humidity of no less than 20% was maintained to preserve Form II.

In summary, enabling formulation technology for a poorly soluble API was facilitated through the development of a metastable form under ambient conditions. The metastable form was crystallized under thermodynamic control via a novel semicontinuous API process that ensured very tight control of solvent composition throughout crystallization. Additionally, fundamental understanding of the thermodynamics of solid form across DS and DP manufacture along with the understanding of the physical stability of the metastable form underpinned sound control strategy and robust commercial manufacturing.

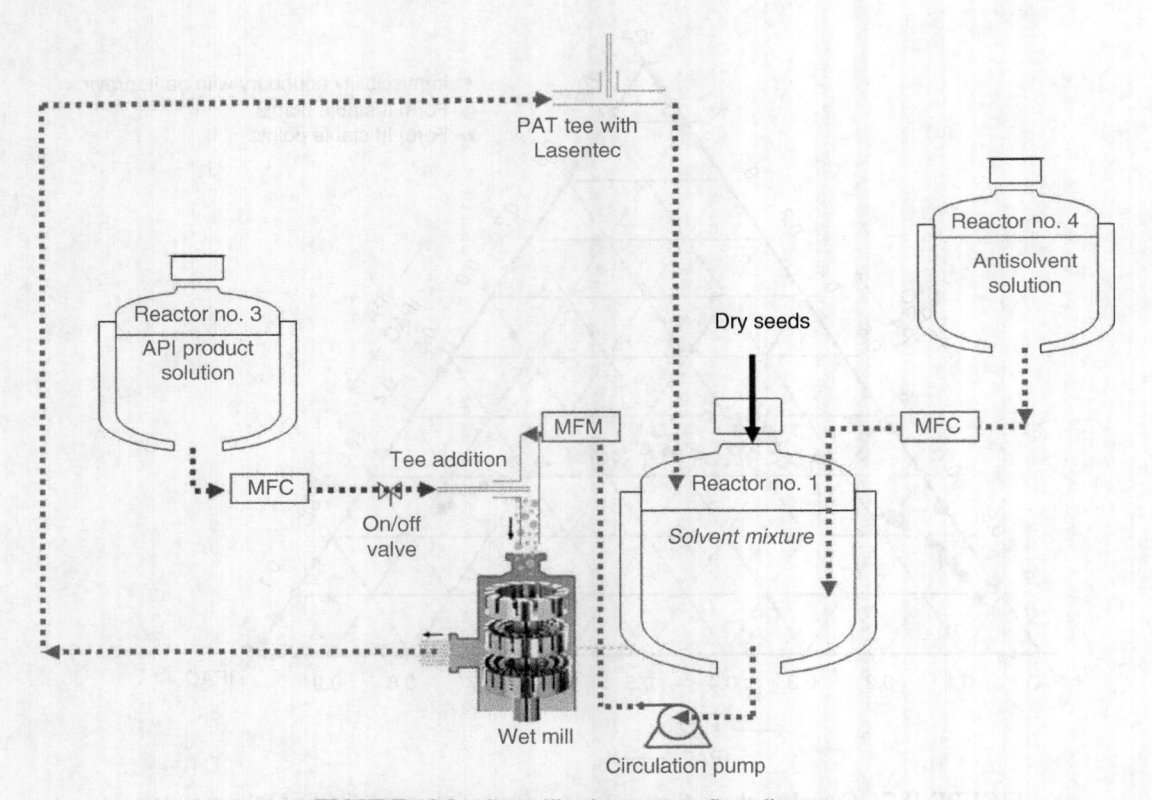

FIGURE 18.8 Crystallization process flow diagram.

18.4 SCALABLE SOLUTION CYRYSTALLIZATION OF CO-CRYSTALS

18.4.1 Introduction

Co-crystals are made by a wide variety of techniques including slow solvent evaporation [23], solution crystallization [24], solid-state grinding (dry grinding and grinding with solvent-drop addition) [25–27], sublimation and growth from melts [28], and slurry conversions [29]. Slow evaporation and grinding seem to be the most commonly used techniques for isolating co-crystals. Their prevalence primarily stems from the ability to quickly and simply screen for co-crystals. Large-scale production of crystals is, however, usually achieved most robustly through solution (cooling/antisolvent) crystallization [30]. Solution crystallization also offers opportunity for purification and particle size and shape control. Successful outcome of a solution co-crystallization experiment is predicated on sound understanding of domains of thermodynamic stability in multicomponent solid–liquid-phase equilibrium diagrams. Building on this essential requirement, we proposed a co-crystallization process design concept [31], wherein three essential features of the overall approach are (i) rational solvent selection, (ii) demarcation of domains of thermodynamic stability in the multicomponent solid–liquid-phase equilibrium diagram, and (iii) control of desupersaturation kinetics for desired process performance. Some of these concepts are illustrated in the following case study using carbamazepine–nicotinamide (CBZ–NIC) Form I co-crystals (see Figure 18.9).

18.4.2 Solvent Selection and Ternary Phase Diagram

TPDs are essential to understand thermodynamic stability in multicomponent solid–liquid systems [32, 33]. These diagrams depict the impact of solvent on domains of phase stability and hence provide primary criteria for solvent selection to design the crystallization process. A simple and sufficiently detailed phase diagram can be constructed from solubility data on coformer as a function of API concentration (solid line dividing regions 1 and 6 in Figure 18.10) and API and co-crystal as a function of coformer concentration (solid lines dividing regions 1 and 2 and 1 and 4, respectively). For a non-polymorphic single stoichiometry co-crystal system, three different conceptual TPDs can be drawn to capture the impact of differences in the relative solubility of the API and coformer in different solvents. Figure 18.10 shows scenarios where (a) solubility of the API and coformer are similar, (b) solubility of API is significantly lower than the coformer, and (c) solubility of API is significantly higher than the coformer.

Very significant practical insights can be gleaned from Figure 18.10. In scenarios (b) and (c), co-crystal exhibits incongruent dissolution behavior wherein phase change to either coformer or API accompanies a nonstoichiometric solution, whereas congruent dissolution occurs for scenario (a). As a consequence evaporative crystallization, for instance, could only be supported for scenario (a). In scenarios (a) and (c), the concentration of coformer corresponding to its critical activity to stabilize co-crystal (intersection of

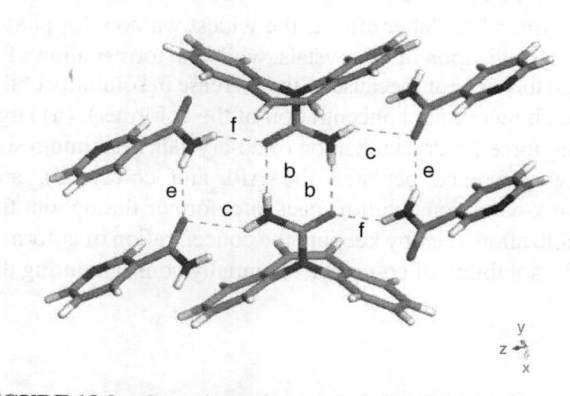

FIGURE 18.9 Crystal structure of CBZ–NIC I highlighting interring hydrogen bonding interaction for NIC–NIC (bond e) and CBZ–CBZ (bond b) [31]. *Source*: Reproduced with permission of Royal Society of Chemistry.

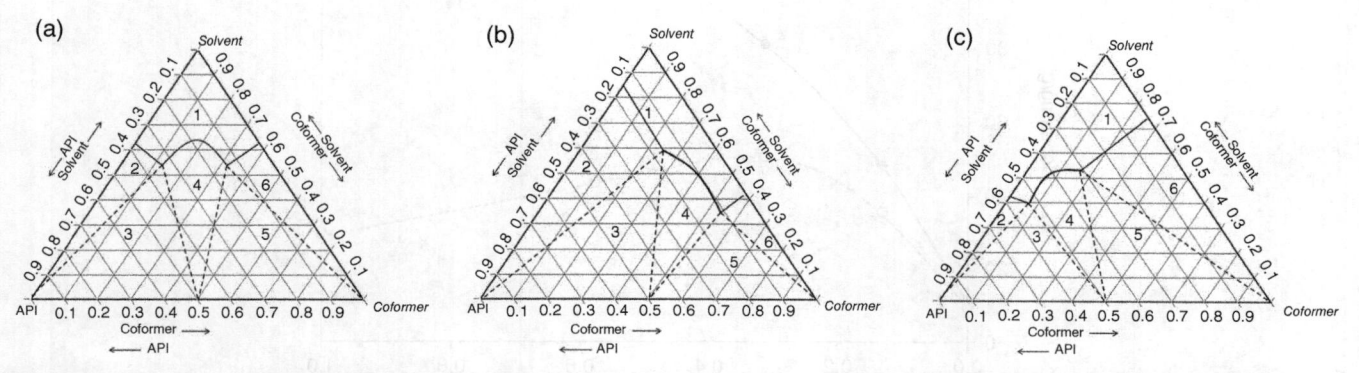

FIGURE 18.10 (a–c) TPDs for single polymorph single composition co-crystal system. 1, represent liquidus; 2, API and solution; 3, API, co-crystal, and solution; 4, co-crystal and solution; 5, coformer, co-crystal, and solution; 6, coformer and solution.

regions 1, 4, 5, and 6) is significantly higher than for scenario (b). This could for instance, be detrimental for solution crystallization, where relatively low coformer concentration to achieve co-crystal stability is more desirable. Higher area of the liquidus region (1) would also have negative consequences for throughput and potentially yield in solution co-crystallization.

Deeper insights can be gathered by the slopes and lengths of the lines/curve that define the boundaries of liquidus region. The slopes are related to the order of speciation in solution for the API and coformer and provide insights into the consequences of solution-phase equilibrium. Ability to influence the "length" of the lines by appropriate choice of solvent can have a direct effect on the region of co-crystal stability. Construction of TPDs as a function of temperature can help identify crystallization trajectories for cooling crystallization that ensure thermodynamic stability throughout the entire process.

As a general principle, the solvent selection criterion is derived from TPD and looks for essential features including (i) high solubility of the coformer compared with the co-crystal and API and (ii) a very significant difference between the critical concentration of coformer and solubility of coformer. The latter affords the widest window for phase-pure crystallization of co-crystals, while the former allows for (i) high throughput (because of the increase in solubility of the API with increasing concentration of the coformer), (ii) large driving force for crystallization of co-crystals (maximum solubility difference between the API and co-crystal), and (iii) sink conditions with respect to coformer throughout the crystallization, thereby keeping the concentration of coformer and the solubility of co-crystal essentially constant during the course of crystallization. Other solvent/solution properties such as deviations from ideal behavior, solvent complexation K_c, solubility product K_{sp}, and their temperature dependence can be studied to further refine solvent selection.

18.4.3 Co-crystallization of CBZ–NIC I

18.4.3.1 Solvent Selection
Ethanol, ethyl acetate, and mixtures thereof were used in our study because they provide extremes in solubility of CBZ–NIC I and NIC. Solubility data of NIC and CBZ in ethanol/ethyl acetate mixtures is shown in Figure 18.11. NIC solubility seems to show a very similar trend to nonideal mixing between ethanol and ethyl acetate. The solubility of CBZ on the other hand shows maxima at 0.3 mol fraction ethanol. Using both CBZ and NIC solubility data, it seems that solvent compositions with an ethanol content greater than 0.3 mol fraction (>50% v/v) could best meet the essential elements of the solvent selection criterion (higher differential between NIC and CBZ solubility) described in Section 18.4.2.

The solubility of CBZ–NIC I was measured at 25 °C in four solvent compositions (25, 50, 75, and 100% ethanol in ethyl acetate v/v) as a function of NIC concentration (denoted by [NIC] from hereon). The data captured in Figure 18.12 show that at lower [NIC], solubility of CBZ–NIC I does not follow the general trend of a reduction in solubility with increasing ethyl acetate concentration. Solubility in the low [NIC] region is in fact higher in 75% ethanol than in 100% ethanol. The data also seem to show that a [NIC]$_{critical}$ of greater than 0.97 M required for CBZ–NIC I stability does not vary significantly with solvent composition.

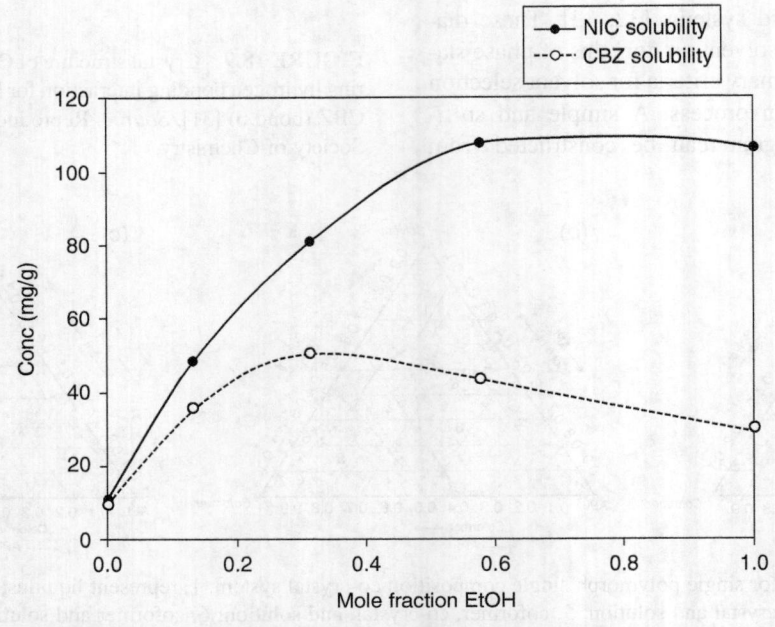

FIGURE 18.11 Solubility of CBZ and NIC in ethanol–ethyl acetate solvent system at 25 °C.

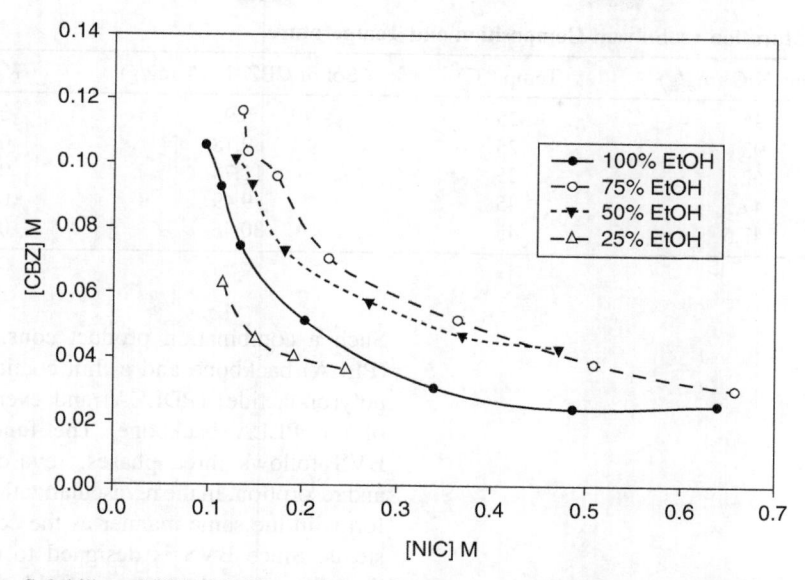

FIGURE 18.12 Solubility of CBZ–NIC I vs. NIC on CBZ basis in ethanol–ethyl acetate solvent system at 25 °C.

TABLE 18.5 K_{sp} and K_c Values at 25 °C in Ethanol/Ethyl Acetate

Volume % EtOH	K_{sp} (M^2)	K_c (M^{-1})	R^2
100	0.0096	0.521	0.99
75	0.0139	0.0776	0.98
50	0.0105	1.7809	0.99
25	0.0056	6.3428	0.95

The solubility behavior of co-crystals in solutions of coformers is best described by solubility product (K_{sp}) and solution complexation (K_c) [24]. Representation of solubility diagrams in these terms is especially useful when comparing the relative merits of different solvent systems. The data shown in Figure 18.12 were transformed and plotted according to Eq. (18.3) to calculate K_{sp} and K_c values for the four solvent mixture compositions (see Table 18.5):

$$[CBZ]_T = \frac{K_{sp}}{[NIC]_T} + K_c K_{sp} \quad (18.3)$$

where [CBT]$_T$ and [NIC]$_T$ are the total concentrations of two components on molar basis.

The intercepts of plots used to calculate K_c values in 100 and 75% ethanol are not statistically different from zero, and therefore the values are not expected to be very accurate. Suffice to say that complexation at these two compositions is negligible. K_{sp} values in the two solvent compositions reflect the greater solubility of CBZ–NIC I in 75% ethanol. As the ethanol content is further reduced from 75%, the expected trend of a reduction in K_{sp} and increase in K_c is observed. Both the K_{sp} and K_c values in 100% ethanol compare favorably with the previously reported values [24]. Although very low K_{sp} value in 25% ethanol helps eliminate this composition from further consideration, the similarity for remaining three compositions does not help in further constraining solvent composition.

Ease of nucleation was used to further refine the solvent selection. Solubility data for CBZ–NIC as a function of temperature and [NIC] in 50, 75, and 100% ethanol was used to determine concentrations of CBZ–NIC I to generate different levels of supersaturation (C/C^*) via undercooling. Eight levels of supersaturation were studied to identify onset of co-crystal nucleation from a clear solution. Table 18.6, summarizing the results, shows that, while the onset supersaturation for 100 and 50% ethanol systems under NIC saturated conditions is similar (ss 1.73 and 1.70, respectively), a significantly higher supersaturation (3.35) is required for nucleation in 75% ethanol.

These data along with the high NIC solubility distinguish 100% ethanol as the optimum solvent choice based on both the solution properties and nucleation kinetics. The solubility of CBZ–NIC I in a saturated solution of NIC/EtOH (at 25 °C) was measured as a function of temperature up to 65 °C. A linear van 't Hoff plot (not shown) was obtained indicating that at the selected [NIC], CBZ–NIC I is stable across the temperature range of interest.

18.4.3.2 Process Design Based on the understanding of solid–liquid-phase equilibrium, a conceptual design of solution crystallization was developed. Figure 18.13 shows the block flow diagram.

Milled seeds were employed at 2 wt % (CBZ–NIC I basis) to induce nucleation at low supersaturation and to desupersaturate in a more controlled fashion. Post filtration crystals were washed with a 5 mg/g NIC solution in ethyl acetate. This concentration is below the solubility of NIC in ethyl acetate (>8 mg/g) but above the critical NIC concentration for

TABLE 18.6 Onset Supersaturation vs. Solvent Composition and Temperature

Volume % EtOH	Conc NIC (mg/g)	Temp (°C)	Sol of CBZ–NIC (mg/g)	K_{sp} (M^2)	Onset SS
100	95	25	5.92	0.0096	1.73
75	93	25	9.18	0.0139	3.35
50	62	25	11.74	0.0105	1.70
100	47.5	45	9.49	0.0094	2.68
100	47.5	45	30.46	0.0521	1.48

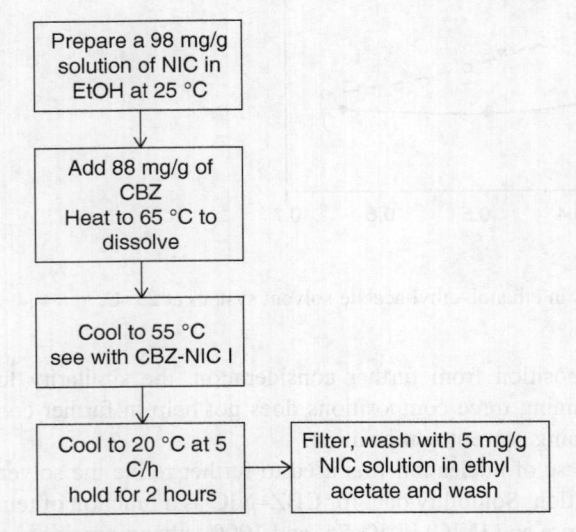

FIGURE 18.13 Block flow diagram of co-crystallization process for CBZ–NIC I.

the stability of CBZ–NIC I co-crystals and therefore allowed for successful removal of the NIC-rich mother liquor from the wet cake without affecting the stability of CBZ–NIC I during washing. The solids did not have any detectable excess NIC post drying when NIC levels were checked by quantitative HPLC. Yields in excess of 90% were reproducibly obtained for the process.

In summary we have demonstrated a generic approach for designing solution co-crystallization processes that uses solid–liquid-phase equilibrium as the foundation of optimum solvent selection and processing trajectories. The methodology has incorporated well-established techniques and procedures also commonly used in single-component crystallization to manipulate and control the process for desired performance and product attributes.

18.5 THERMODYNAMICS OF COATING PROCESS DURING THE MANUFACTURE OF DRUG-ELUTING BIORESORBABLE VASCULAR SCAFFOLD

18.5.1 Introduction

BVSs such as ABSORB™ [34] have the potential to significantly improve treatment of coronary artery disease [35]. Such a combination product consists of a poly(L-lactide) (PLLA) backbone and a thin coating of 1 : 1 w/w ratio of poly(DL-lactide) (PDLLA) and everolimus, applied on top of the PLLA backbone. The functional performance of BVS follows three phases: revascularization, restoration, and resorption. In the revascularization phase, the system performs in the same manner as the conventional drug-eluting stents. Since BVS is designed to disappear over time, in the restoration phase the scaffold starts losing its structural integrity to allow for gradual return of vessel functions. In the resorption phase, the implant is resorbed in a benign fashion, only leaving behind a restored vessel [36, 37].

The overall performance of BVS is influenced by the microstructure of PDLLA/everolimus coating and its relationship to the film coating process. These two aspects affect mechanical property of the coating and *in vivo* drug release profile. As an extended drug release product with only sub-milligram quantities of the drug in the device, control of BVS drug release profile is obviously critical for the product performance. Therefore, much scientific, quality and regulatory impetus exists for developing a thorough understanding of the film and processing factors affecting its properties.

The PDLLA/everolimus coating film is typically 1–2 μm in inner diameter and 3–4 μm in outer diameter of PLLA scaffold. It is amorphous in nature and spray-coated from a 4 wt % solution of 1 : 1 w/w everolimus and PDLLA in acetone [38]. The process involves spraying atomized form of the solution onto the PLLA scaffold backbone, followed by drying to remove the solvent. The resulting film is phase separated in nature, with everolimus-rich domains dispersed within the PDLLA-rich continuous phase [39]. Approximately 80% of everolimus within the film is released within 28 days (*in vivo*), and the elution is fully complete within 120 days [40]. In order to better understand the film-forming process and properties thereof, an experimental TPD between acetone, everolimus, and PDLLA (solution) was constructed. High-level details and key findings are summarized in this case study.

18.5.2 Construction of the TPD for Everolimus/Acetone and PDLLA System

Two well-established aspects of PDLLA/acetone/everolimus include high solubility of both everolimus and PDLLA in acetone (10 wt % each solution in acetone is easily made) and limited miscibility of PDLLA and everolimus.

These attributes provide the basic framework to develop a fully descriptive TPD by defining (i) liquidus lines describing change in solubility of PDLLA in acetone as a function of everolimus and vice versa, (ii) solubility/miscibility of PDLLA in everolimus and vice versa, and (iii) immiscibility boundaries between single-phase and other multiphase regions.

18.5.2.1 Solubility of Everolimus in PDLLA and PDLLA in Everolimus

As a first pass estimate, Flory–Fox [41] equation was used to estimate the solubility/miscibility of PDLLA in everolimus and vice versa. Solubility of PDLLA in amorphous everolimus was estimated to be approximately 8% by weight, whereas the solubility of everolimus in PDLLA was estimated to be approximately 5% by weight. Semiquantitative Raman mapping experiments were conducted to confirm that the estimates are sufficiently accurate. A direct thermal scanning method by Sun et al. [19] was also attempted; however relatively low solubility of the two components did not allow for an accurate determination of the endset temperature from the relevant thermal events in the DSC data. In the end, Flory–Fox estimates were used for the TPD.

18.5.2.2 Solubility of Everolimus and PDLLA in Acetone

Amorphous everolimus can convert to a crystalline anhydrous form in acetone. Hence dissolution and recrystallization profiles were determined for slurries of amorphous everolimus in acetone by measuring concentration every hour. Maximum concentration of 1.57 g/g was observed and recorded as the solubility of the amorphous phase [42]. Full conversion was achieved well within 24 hours, and final equilibrium concentration of 0.8 mg/g was measured for the crystalline anhydrous phase.

Slurry method showed that even at concentrations as high as 3 g/g, PDLLA and acetone remained miscible. To assess miscibility beyond this concentration, a 90/10 w/w PDLLA/acetone mixture was prepared and equilibrated for one month to achieve homogenization. The solids were isolated and analyzed for thermal behavior using DSC. A single, significantly reduced T_g of 9.7 °C was observed compared with approximately 48 °C for pure PDLLA [43]. Since single T_g indicated existence of a solid solution, it can be concluded that miscibility of PDLLA/acetone system extends up to 90/10 w/w PDLLA/acetone. Even more refined estimate can be obtained by extrapolating the liquidus line representing change in solubility of PDLLA as a function of everolimus.

18.5.2.3 Solubility of Everolimus as a Function of PDLLA and PDLLA as a Function of Everolimus in Acetone

A total of five different ratios of everolimus and PDLLA were studied (3.5/96.5, 7/93, 85/15, 89/11, and 93/7 w/w PDLLA/everolimus). Incremental amount of acetone was added to these physical mixtures (some were prepared through solvent evaporation and vacuum drying), and sufficient time was afforded for equilibration (one to six hours depending on the state of phase separation) to determine the limit of miscibility. The latter was established by high resolution polarized light microscopy on a customized sealed and temperature-controlled stage.

18.5.2.4 Immiscibility Boundaries Between Single- and Multiphase Regions

For the purposes of this subsection, immiscibility boundary is defined as the boundary between a single-phase homogeneous solution and multiphase mixtures that may or may not contain solids. The immiscibility boundary also connects two liquidus lines that can be constructed with the data described in Sections 18.5.2.1 through 18.5.2.3. Six different ratios of PDLLA/everolimus (15/85, 25/75, 33/66, 50/50, 66/33, and 75/25 on w/w basis) were equilibrated with different amounts of acetone, and the nature of the resulting phases was assessed by high resolution polarized light microscopy on a customized sealed and temperature-controlled stage.

Contours of a "concave-up" phase diagram become apparent (see Figure 18.14), when all the data from the experiments described above is plotted as a TPD. For such a system, construction of thermodynamically appropriate tie lines inadvertently results in a void in the middle of the immiscibility region representing a 3-phase region that borders with three 2-phase regions adjacent to the three binary axes [44]. The vertices of the 3-phase triangle can, in principle, be defined by (i) experimental determination of all the tie lines or (ii) experimental discrimination of the 2-phase region from the 3-phase region within the immiscible region to establish the boundaries. The latter is a more practical approach for the current system, since the existence of three-phase region can easily be captured by the techniques used herein.

Five additional experiments (at high acetone levels) were conducted to help construct the boundaries of the 3-phase triangle. In general, the data points in 3-phase region were characterized by "biphasic" dispersed phase or the presence of solids in two liquid phases (see Figure 18.15). The complete TPD shown in Figure 18.16 is constructed by incorporating the five data points to the data generated in Sections 18.5.2.1 through 18.5.2.4.

It is worth reiterating that the accurate determination of boundaries between 3- and 2-phase regions is affected by the experimental constraints imposed by the system behavior, wherein everolimus precipitates out at lower acetone content. As a consequence, well-spread-out points could not be accessed to establish the vertices of the triangle, and hence, a higher level of uncertainty exists in the boundaries of the triangle.

18.5.3 Conclusions

The miscibility behavior of acetone with PDLLA and everolimus and free energy of mixing for the three components is such that a "concave-up"-shaped phase diagram is obtained.

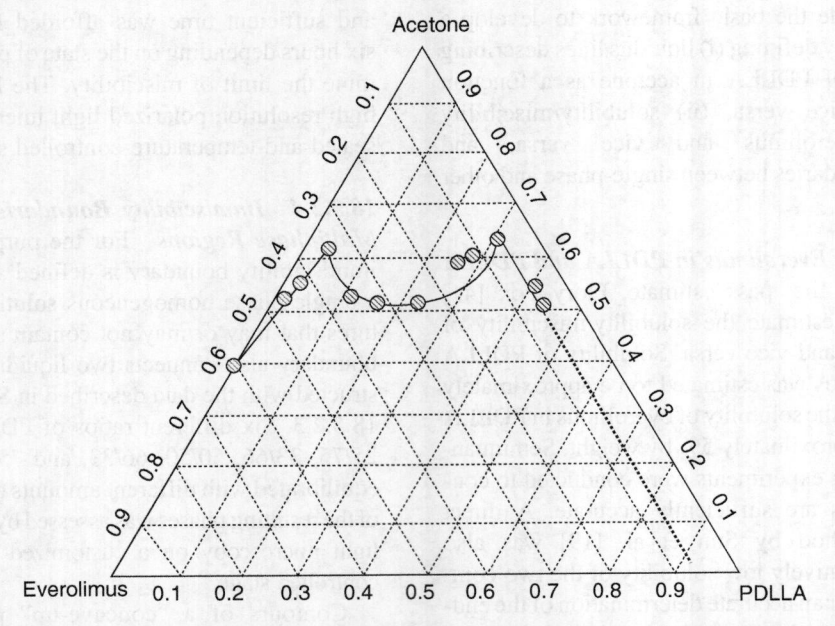

FIGURE 18.14 Liquidus and immiscibility in the concave-up phase diagram for everolimus/PDLLA/acetone, where (◉) points represent everolimus liquidus line as a function of PDLLA concentration, (○) points represent PDLLA liquidus line as a function of Everolimus, and (◍) points represent the immiscibility boundary connecting the two liquidus.

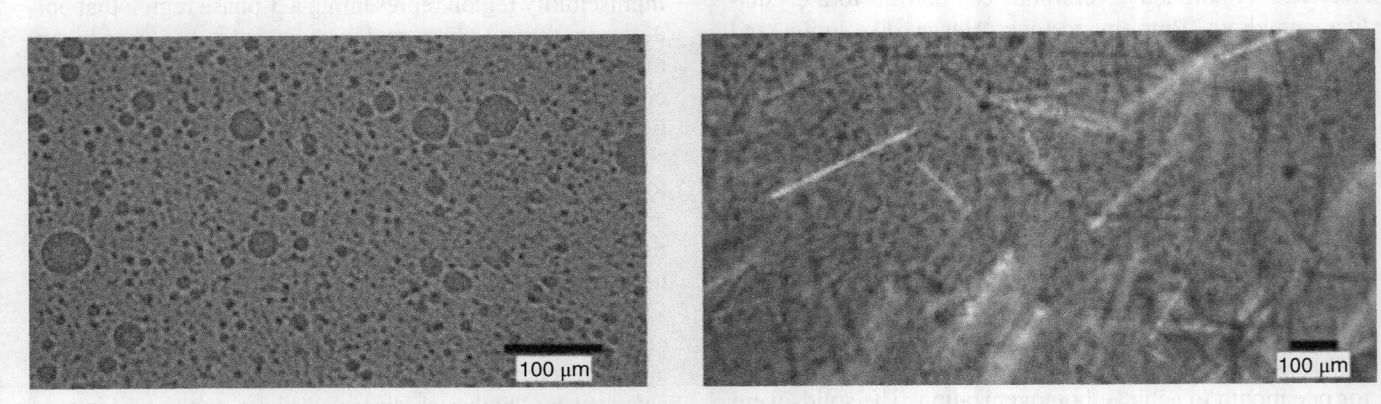

FIGURE 18.15 Illustration of data points in three-phase triangular region.

The diagram presents a complex phase landscape, characterized by two 1-phase regions, three biphasic regions, and most importantly one 3-phase region.

The manufacturing process for spraying the film onto the BVS starts with an equal weight % solution of everolimus and PDLLA in acetone, which is subsequently "dried" to remove acetone. A vertical line along the vertex of the TPD can be drawn to represent the trajectory of this film-forming process. The trajectory suggests that between the start as a single-phase, homogeneous solution and the end as a phase-separated biphasic film; the system goes through at least two additional transitions. As the immiscibility curve is approached through the removal of acetone, a biphasic mixture composing of a PDLLA-rich phase and an acetone-rich phase is formed. Continued removal of acetone results in the formation of an additional everolimus-rich phase as the system enters the 3-phase triangle. Further removal of acetone results in continuous depletion of acetone-rich phase with redistribution of its contents in the PDLLA- and everolimus-rich phases. When almost all the acetone is removed, the acetone-rich phase disappears, leaving behind a biphasic system composing of polymer-rich and everolimus-rich phases at the base of the phase diagram. Consistent properties of the film could be best achieved if the kinetics of the process is well controlled or the process is operated close to equilibrium during the key transitions to the biphasic film.

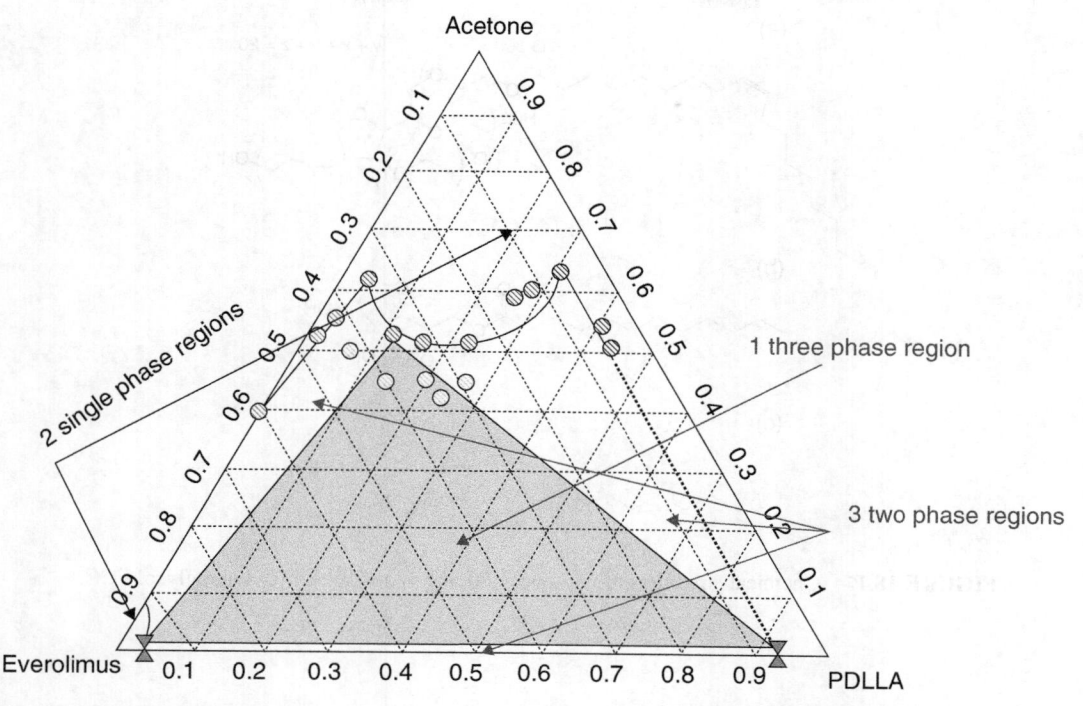

FIGURE 18.16 Complete experimental ternary phase diagram for PDLLA/everolimus/acetone at 25 °C. (◎) points represent the boundary between two-phase and three-phase regions.

18.6 POLYMER–PLASTICIZER MIXING PERFORMANCE IN THE PRESENCE OF WATER

18.6.1 Introduction

The ability for polymers and additives to physically mix in many industrial applications is dictated by a combination of kinetic and thermodynamic factors. In the manufacture of ASD via HME, the API, polymer, and other excipients, for instance, need to be homogenously mixed to achieve content uniformity and complete solubility/dissolution of the API in the ASD excipient matrix. The presence of moisture can complicate the mixing performance depending on the hydrophilicity of the materials. Water may also be intentionally added as part of the process. Polymer–plasticizer mixing performance can therefore be altered by the interfacial interactions and wetting behavior between the components due to moisture. In this case study [45], a model polymer and various plasticizers were evaluated for their mixing performance in the presence of water.

18.6.2 Physical Mixing

A ternary system consisting of a polymer (copovidone, a copolymer of vinyl pyrrolidone [poly(vinyl pyrrolidone)] and vinyl acetate), a plasticizer, and water was examined for physical mixing performance. Three different liquid plasticizers (Figure 18.17) representing a range of hydrophilic–lipophilic properties characterized by their hydrophilic–lipophilic balance (HLB) and viscosities were studied (Figure 18.18). Copovidone is a hygroscopic copolymer, with a water uptake of more than 41% wt/wt at 90% RH (Figure 18.19), when measured by dynamic vapor sorption (DVS) from Surface Measurement Systems, United Kingdom. Among the plasticizers, Tween 80 (PEG-20 sorbitan monooleate) is more hygroscopic than Span 20 (sorbitan monolaurate), and Lauroglycol FCC (propylene glycol monolaurate type I) is least hygroscopic. Span 20 is the most viscous with viscosity at 25 °C approximately 7 times higher than Tween 80. On the other hand, Lauroglycol FCC has a viscosity approximately 20 times lower than Tween 80.

Copovidone was equilibrated to different moisture contents and then physically mixed with each individual plasticizer in a ratio of 10 : 1 w/w at 25 °C to achieve a consistent mixture. The polymer–plasticizer mixing performance was directly quantified through characterization of the surface of each moisture-equilibrated primary component (polymer and plasticizers) and polymer–plasticizer binary mixtures using inverse gas chromatography. The studies identified the following mixing behavior:

- For Lauroglycol FCC, mixing performance was independent of moisture level in copovidone.
- For Span 20, lower moisture level promoted mixing performance.
- For Tween 80, higher moisture level promoted mixing performance.

FIGURE 18.17 Chemical structures of (a) Tween 80, (b) Span 20, and (c) Lauroglycol FCC.

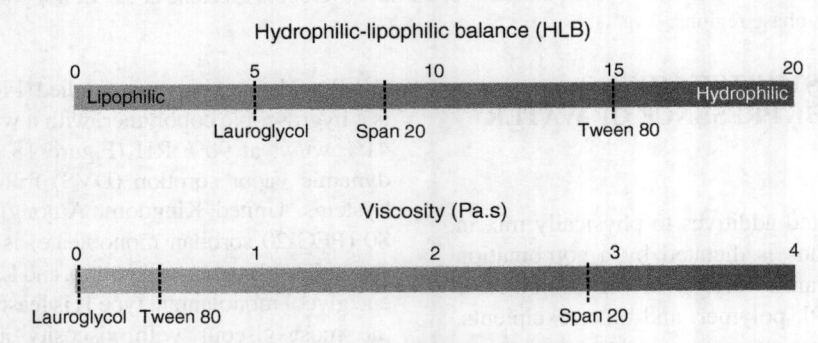

FIGURE 18.18 HLB and viscosity values at 25 °C of the model plasticizers.

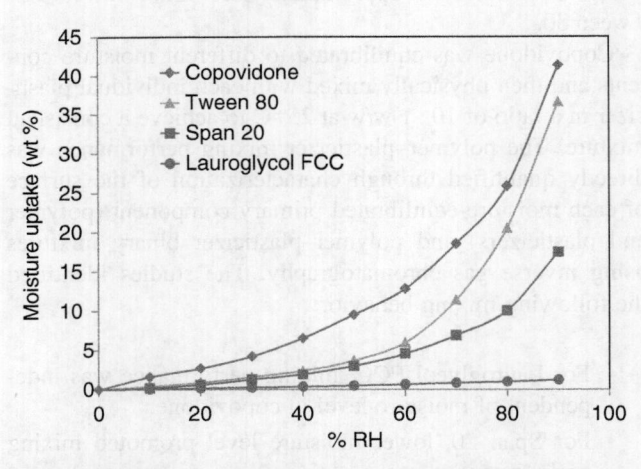

FIGURE 18.19 Water sorption isotherms for copovidone, Tween 80, Span 20, and Lauroglycol FCC at 25 °C.

Gibbs free energy of mixing and other thermodynamic and kinetics factors were evaluated to understand these findings.

18.6.3 Gibbs Free Energy of Mixing

The thermodynamic mixing behavior of the polymer–plasticizer–water ternary system was estimated by applying the well-known Flory–Huggins model. In the Flory–Huggins model for ternary systems, ΔG_{mix} is made up of entropic and enthalpic terms as given by Eq. (18.4):

$$\frac{\Delta G_{mix}}{RT} = n_1 \ln \varphi_1 + n_2 \ln \varphi_2 + n_3 \ln \varphi_3 + n_1 \varphi_2 \chi_{12} + n_1 \varphi_3 \chi_{13} + n_2 \varphi_3 \chi_{23} \quad (18.4)$$

where

n is mole fraction.

φ is volume fraction.

χ_{xy} is binary Flory–Huggins interaction parameter between components x and y.

R is universal gas constant.

T is temperature.

The subscripts 1, 2, and 3 denote water, plasticizer, and copovidone, respectively. Since Gibbs free energy of mixing, ΔG_{mix}, is a quantitative measure of mixing tendency between components, a negative ΔG_{mix} implies thermodynamic favorability, leading to spontaneous mixing and system stability. Some of the relevant parameters for the three components are captured in Table 18.7.

Binary Flory–Huggins interaction parameters χ_{xy} between water–polymer, water–plasticizer, and polymer–plasticizer can be determined by comparing experimental and calculated water sorption isotherms of the individual components and mixtures [49, 50]. The calculated moisture sorption can be constructed using Eq. (18.5) [51, 52], with the assumption that absorption of water into the polymer or plasticizer can be treated as a dissolution process:

$$\ln\left(\frac{p}{p_0}\right) = \ln\varphi_1 + \left(1 - \frac{1}{x_{1k}}\right)\varphi_k + \chi_{1k}\varphi_k^2 \quad (18.5)$$

Here, subscripts 1 and k refer to water and the individual component (copovidone, Span 20, Tween 80, or Lauroglycol FCC), respectively. φ_k is the volume fraction of component k. The term χ_{1k} is the binary Flory–Huggins interaction parameter dictating the strength of the interaction between water and component k. The term p/p_0 is ratio of the partial vapor pressure of moisture to the saturated vapor pressure (i.e. RH%). x_{1k} is the relative molecular volumes between water and component k.

For the "dissolution" assumption to hold, application of the above model is most appropriate at high relative humidities where the moisture is able to sufficiently plasticize the

TABLE 18.7 Material Parameters for Input into the Flory–Huggins Model

Material	Density (g/cm³)	Molecular Weight (g/mol)	Molecular Volume (cm³/mol)
Span 20 [46]	1.03	346.5	5.6×10^{-22}
Tween 80 [47]	1.08	1310	2.0×10^{-21}
Lauroglycol FCC [48]	0.93	258.4	4.6×10^{-22}
Copovidone	1.21	39 800	5.5×10^{-20}
Water	1.00	18.0	3.0×10^{-23}

sample such that its glass transition temperature (T_g) is near or below the experimental temperature (25 °C). As a reasonable assumption, focus was therefore placed on the moisture sorption results at 80 and 90% RH to estimate the interaction parameter. Using these two data points, a value of around 0.8 was obtained for water–copovidone interaction parameter. The value, being greater than the critical value of 0.5 for polymer solutions [47], suggests that the mixing of copovidone with water is somewhat unfavorable. The unfavorability is most likely due to the hydrophobic (polyvinyl acetate) component of copovidone, because for the pure hydrophilic component (polyvinylpyrrolidone [PVP]), the interaction parameter has been reported as 0.36 [49].

The aqueous interaction parameters for the individual plasticizers were estimated as follows: the weight change due to water uptake from the moisture sorption isotherm in Figure 18.18 is first converted to water volume fraction φ_1. The interaction parameter between water and component k, χ_{1k}, is then estimated by fitting the DVS moisture sorption profile to Eq. (18.5). Results for binary interactions are tabulated in Table 18.8. With increasing hydrophobicity (in the order of Tween 80, Span 20, and Lauroglycol FCC), the interaction parameter increases due to the decreasing hygroscopicity. Additionally, positive interaction parameters for all the plasticizers indicate unfavorable mixing tendency with water.

For a ternary system, assuming that the water–copovidone and water–plasticizer interactions in the ternary system are the same as in the binary systems, Eq. (18.5) can be extended [49], whereby the moisture sorption profile of the plasticizer–copovidone mixture can be described by

$$\ln\left(\frac{p}{p_0}\right) = \ln\varphi_1 + (\varphi_2 + \varphi_3) - \frac{\varphi_2}{x_{12}} - \frac{\varphi_3}{x_{13}} \\ + (\chi_{12}\varphi_2 + \chi_{13}\varphi_3)(\varphi_2 + \varphi_3) - \chi_{23}\frac{\varphi_2\varphi_3}{x_{12}} \quad (18.6)$$

The subscripts 1, 2, and 3 refer to water, plasticizer (Span 20, Tween 80, or Lauroglycol FCC), and copovidone, respectively, as defined above. χ_{23} is the binary Flory–Huggins interaction parameter dictating the strength of the interaction between the plasticizer and copovidone. If the binary interaction parameters between water and individual components are known, then the interaction parameter between the plasticizer and copovidone, χ_{23}, can be predicted from the moisture sorption profile of the binary plasticizer–copovidone mixture using Eq. (18.6).

Experimental moisture sorption data of the copovidone–plasticizer mixtures was used to (i) confirm the presence of the copovidone–plasticizer interactions and (ii) determine if Eq. (18.6) is required to determine χ_{23}. First, a theoretical isotherm was calculated using weighted sum of the individual components isotherm. Second, the calculated isotherm was compared against the experimental DVS moisture sorption isotherm. The presence of any copovidone–plasticizer

TABLE 18.8 Flory–Huggins Interaction Parameters Determined from Moisture Sorption Isotherms

System	Interaction Parameter (Estimated)
Copovidone–water	0.8
Tween 80–water	1.0
Span 20–water	1.5
Lauroglycol FCC–water	3.6

interactions would lead to deviations between calculated and experimental profiles. However, it was found that the water uptake properties of the individual components were not altered in the physical mixtures, and therefore, no significant interaction existed between the copovidone and plasticizer (i.e. $\chi_{23} \approx 0$). In essence, the last enthalpic term of Gibbs free energy (Eq. 18.4) is reduced to zero. Furthermore the binary interaction parameters (χ_{12}) obtained from moisture sorption isotherms (Table 18.8) are sufficient to calculate the free energy of mixing as a function of copovidone, plasticizer, and water composition using Eq. (18.4).

18.6.4 Thermodynamic and Kinetic Considerations to Mixing Behavior

The Gibbs free energy of mixing can be calculated for all copovidone, plasticizer, and water compositions at 25 °C using Eq. (18.4), from the binary interaction parameters provided in Table 18.8 and assuming no interaction between the copovidone and plasticizer (i.e. $\chi_{23} \approx 0$). The resulting Gibbs free energy of mixing is shown as ternary diagrams in Figure 18.20. The lowest free energies are achieved at low water content, indicating that mixing between copovidone and plasticizer is more favorable when moisture is minimized. Free energy contour lines in the system involving Tween 80 are less dependent on the copovidone and plasticizer compositions than in the systems involving Span 20 and, especially, Lauroglycol FCC. This is due to relatively similar interaction parameters between water–copovidone ($\chi_{13} = 0.8$) and water–Tween 80 ($\chi_{12} = 1.0$). In other words, physical mixing is most sensitive to copovidone and plasticizer compositions for Lauroglycol FCC and least sensitive for Tween 80. These results also show that for a more hydrophobic plasticizer (with lower HLB value), increasing water results in decreased mixing favorability.

Comparison of this predicted mixing performance to the experimental results, however, shows that physical mixing behavior was not solely influenced by thermodynamic factors, since Lauroglycol FFC displayed the best mixing performance regardless of water content. Kinetic barrier could also be very important. Lauroglycol FCC has low kinetic barrier because its viscosity is one and two orders of magnitude lower than Tween 80 and Span 20, respectively. The relatively low viscosity of this plasticizer would allow it to be easily incorporated into the copovidone regardless of the

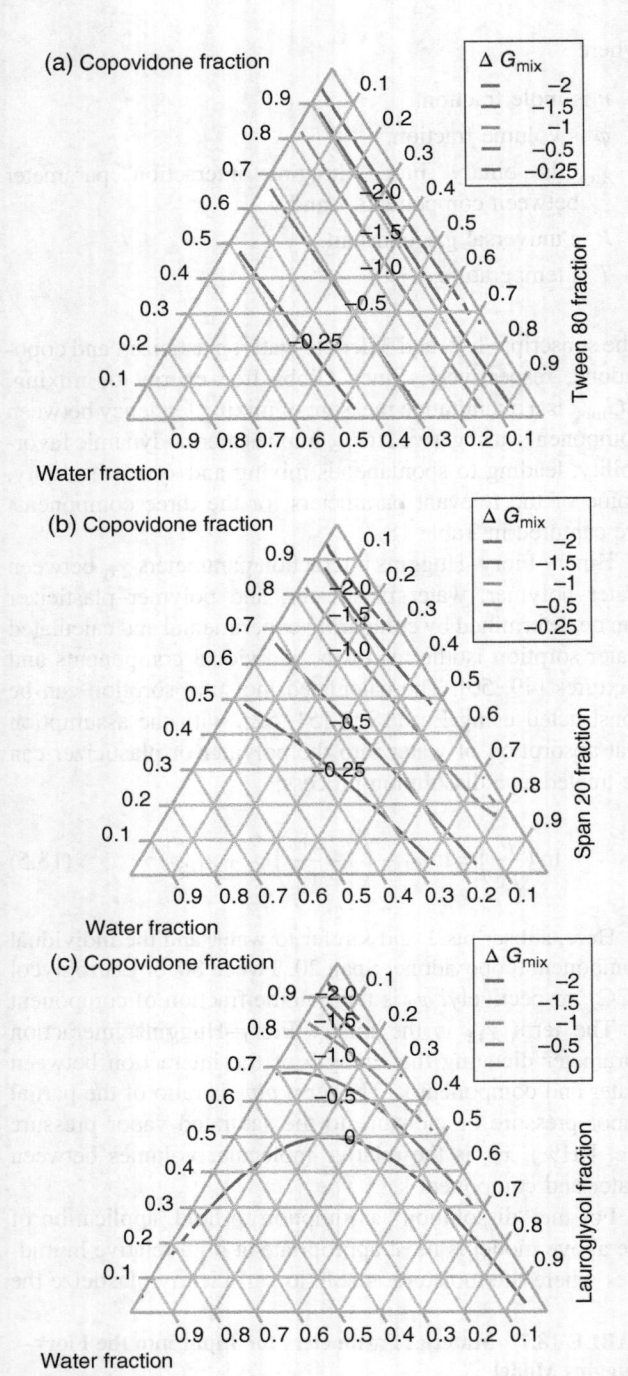

FIGURE 18.20 Contour lines of the Gibbs free energy of mixing for the ternary systems of copovidone, plasticizer, and water. Plasticizers are (a) Tween 80, (b) Span 20, and (c) Lauroglycol FCC.

moisture content. It is also interesting to note that when kinetic barriers are high due to high viscosity such as for Tween 80, mixing performance is dictated by more favorable interactions with water. For viscous polymers, the study therefore shows that underlying thermodynamic characteristics become more dominant for mixing favorability.

18.6.5 Conclusions

The mixing of polymers and additives in the presence of water is dictated by a combination of kinetic and thermodynamic factors; this arises from material physical properties, such as viscosity, hydrophobicity (surface properties), and moisture uptake behavior. As the viscosity of the components decreased in the system, the influence of the thermodynamic characteristics became less important. In summary, the Flory–Huggins model and ΔG_{mix} calculations provide a quick and quantitative estimation of performance for industrial mixing applications between polymer and surfactant at different water contents.

REFERENCES

1. Haleblian, J. and McCrone, W. (1969). Pharmaceutical applications of polymorphism. *J. Pharm. Sci.* 58: 911–929.
2. Rodriguez-Spong, B., Price, C.P., Jayasankar, A. et al. (2004). General principles of pharmaceutical polymorphism: a supramolecular perspective. *Adv. Drug Deliv. Rev.* 56: 241–274.
3. Cruz-Cabeza, A., Reutzel-Edens, S.M., and Bernstein, J. (2015). Facts and fictions about polymorphism. *Chem. Soc. Rev.* 44: 8619–8635.
4. Almarsson, Ö. and Zaworotko, M.J. (2004). Crystal engineering of the composition of pharmaceutical phases. Do pharmaceutical co-crystals represent a new path to improved medicines? *Chem. Commun.* 1889–1896.
5. Vishweshwar, P., McMahon, J.A., Bis, J.A., and Zaworotko, M.J. (2006). Pharmaceutical cocrystals. *J. Pharm. Sci.* 95: 499–516.
6. Friscic, T. and Jones, W. (2010). Benefits of co-crystallization in pharmaceutical material science – an update. *J. Pharm. Pharmacol.* 62: 1547–1559.
7. Thakuria, R., Delori, A., Jones, W. et al. (2013). Pharmaceutical co-crystals and poorly soluble drugs. *Int. J. Pharm.* 453: 101–125.
8. Crane, I.M., Mulhern, M.G., and Nema, S. (2003). Stability of reconstituted parecoxib for injection with commonly used diluents. *J. Clin. Pharm. Therap.* 28: 363–369.
9. Sheikh, A.Y., Borchardt, T.B., Ferro, L.J., and Danzer, G.D. (2003). Crystalline parecoxib sodium. US Patent US20030232871.
10. Morris, K.R., Griesser, U.J., Eckhardt, C.J., and Stowell, J.G. (2001). Theoretical approaches to physical transformations of active pharmaceutical ingredients during manufacturing processes. *Adv. Drug Deliv. Rev.* 48: 91–114.
11. Oliveira, M.A., Peterson, M.L., and Davey, R.J. (2011). Relative enthalpy of formation for co-crystals of small organic molecules. *Cryst. Growth Des.* 11: 449–457.
12. Grant, D.J.W. and Higuchi, T. (eds.) (1990). Dissolution rates of solids. In: *Solubility Behavior of Organic Compounds*, Techniques of Chemistry, vol. 21, 474–551. New York: Wiley.
13. Bernstein, J., Davey, R.J., and Henck, J. (1999). Concomitant polymorphs. *Angew. Chem. Int. Ed.* 38: 3440–3461.
14. Yu, L. (1995). Inferring thermodynamic stability relationship of polymorphs from melting data. *J. Pharm. Sci.* 84: 966–974.
15. Yu, L., Huang, J., and Jones, K.J. (2005). Measuring free-energy differences between crystal polymorphs through eutectic melting. *J. Phys. Chem. B* 109: 19915–19922.
16. McCrone, W. (1957). *Fusion Methods in Chemical Microscopy: A Textbook and Laboratory Manual*. New York: Interscience Publishers.
17. Ku, Y., McDaniel, K.F., Chen, H.-J. et al. (2010). Preparation of heterocyclic macrocyclic peptides as hepatitis C serine protease inhibitors. International Patent WO 2010/030359 A2.
18. Liepold, B., Moosmann, A., Pauli, M. et al. (2015). Solid pharmaceutical compositions useful in HCV treatment. International Patent WO 2015/071488 A1.
19. Sun, Y., Tao, Z., Zhang, G.Z.Z., and Yu, L. (2010). Solubilities of crystalline drugs in polymers: an improved analytical method and comparison of solubilities of indomethacin and nifedipine in PVP, PVP/VA and PVAc. *J. Pharm. Sci.* 99: 4023–4031.
20. Sheikh, A.Y., Diwan, M., Pal, A.E. et al. (2014). Crystalline forms of an HCV protease inhibitor. International Patent WO 2014/011840 A1.
21. Brackemeyer, P.J., Diwan, M., Gong, Y. et al. (2015). Crystal form of ABT 450. International Patent WO 2015/084953 A1.
22. Hong, G.-B., Lee, M., and Lin, H. (2002). Liquid-liquid equilibrium of ternary mixtures of water-2-propanol with ethyl acetate and isopropyl acetate or ethyl caproate. *Fluid Phase Equilib.* 202: 239–252.
23. Walsh, R., Bradner, M.W., Fleischman, S.G. et al. (2003). Crystal engineering of the composition of pharmaceutical phases. *Chem. Commun.* 186–187.
24. Nehm, S.J., Rodriguez-Spong, B., and Rodriguez-Hornedo, N. (2006). Phase solubility diagrams of cocrystals are explained by solubility product and solution complexation. *Cryst. Growth Des.* 6: 592–600.
25. Trask, A.V., Streek, J.V.D., Motherwell, W.D.S., and Jones, W. (2005). Achieving polymorphic and stoichiometric diversity in cocrystal formation: importance of solid-state grinding, powder X-ray structure determination, and seeding. *Cryst. Growth Des.* 5: 2233–2241.
26. Etter, M.C. and Reutzel, S.M. (1991). Hydrogen bond directed cocrystallization and molecular recognition properties of acyclic imides. *J. Am. Chem. Soc.* 113: 2586–2598.
27. Chadwick, K., Davey, R.J., and Cross, W. (2007). How does grinding produce co-crystals? Insights from the case of benzophenone and diphenylamine. *CrystEngComm* 9: 732–734.
28. Seefeldt, K., Miller, J., Alvarez-Nunez, F., and Rodriguez-Hornedo, N. (2007). Crystallization pathways and kinetics of carbamazepine–nicotinamide cocrystals from the amorphous state by in situ thermomicroscopy, spectroscopy, and calorimetry studies. *J. Pharm. Sci.* 96: 1147–1158.
29. Zhang, G.G.Z., Henry, R., Borchardt, T.B., and Lou, X.C. (2007). Efficient co-crystal screening using solution-mediated phase transformation. *J. Pharm. Sci.* 96: 990–995.

30. Mullin, J.W. (ed.) (2001). Crystallizer design and operation. In: *Crystallization*, 4e, 315–402. Oxford: Butterworth-Heinemann.
31. Sheikh, A.Y., Abd, R.S., Hammond, R.B., and Roberts, K.J. (2009). Scalable solution cocrystallization: case of carbamazepine-nicotinamide I. *CrystEngComm* 11: 501–509.
32. Chiarella, R.A., Davey, R.J., and Peterson, M.L. (2005). Making co-crystals – the utility of ternary phase diagrams. *Cryst. Growth Des.* 5: 2233–2241.
33. Lange, L. and Sadowski, G. (2015). Thermodynamic modeling of efficient co-crystal formation. *Cryst. Growth Des.* 15: 4406–4416.
34. Rapoza, R., Veldhof, S., Oberhauser, J., and Hossainy, S.F.A. (2015). Assessment of a drug eluting bioresorbable vascular scaffold. US Patent 2015/0073536 A1.
35. Tarantini, G., Masiero, G., Granada, J.F., and Rapoza, R.J. (2016). The BVS concept, from the chemical structure to vascular biology; the bases for a change in interventional cardiology. *Minerva Cardioangiol.* 64: 419–441.
36. Kossuth, M.B., Perkins, L.E.L., and Rapoza, R.J. (2016). Design principles of bioresorbable polymeric scaffolds. *Interv. Cardiol. Clin.* 5: 349–355.
37. Rapoza, R., Ding, N., Wang, Y. et al. (2014). Bioresorbable implants for transmyocardial revascularization. US Patent 2014/0336747 A1, 13 November 2014.
38. Chen, Y., Van Sciver, J., Hossainy, S.F.A., and Pacetti, S.D. (2010). Drying bioresorbable coating over stents. US Patent 20100323093 A1.
39. Wu, M., Kleiner, L., Tang, F.W. et al. (2010). Surface characterization of poly(lactic acid)/everolimus and poly(ethylene vinyl alcohol)/everolimus stents. *Drug Deliv.* 17: 376–384.
40. Hossainy, S.F.A. (2015). Phase separated block co-polymer coatings for implantable medical devices. US Patent US 2015/9028859 B2, 12 May 2015.
41. Fox, T.G. and Flory, P.J. (1950). Second-order transition temperatures and related properties of polystyrene. *J. Appl. Phys.* 21: 581–591.
42. Shefter, E. and Higuchi, T. (1963). Dissolution behavior of crystalline solvated and non-solvated forms of some pharmaceuticals. *J. Pharm. Sci.* 52: 781–791.
43. Nakafuku, C. and Takeshia, S. (2004). Glass transition temperature and mechanical properties of PLLA and PDLLA-PGA co-polymer blends. *J. App. Polym. Sci.* 93: 2164–2173.
44. Duffy, J.D., Stidham, H.D., Hsu, S.L. et al. (2002). Effect of polyester structure on the interaction parameters and morphology development of ternary blends: model for high performance adhesives and coatings. *J. Mater. Sci.* 37: 4801–4809.
45. Ho, R., Sun, Y., and Chen, B. (2015). Impact of moisture and plasticizer properties on polymer–plasticizer physical mixing performance. *J. Appl. Polym. Sci.* 132: 41679–41688.
46. Gangolli, S. (1999). *The Dictionary of Substances and Their Effects: Volume 6 O-S*, 2e. Cambridge, UK: Royal Society of Chemistry.
47. Ravve, A. (2012). *Principles of Polymer Chemistry*, 3e. New York: Springer.
48. Gattefosse (2010). Technical data sheet: lauroglycol FCC, Specification number 3219/5.
49. Rumondor, A.C.F. (2010). Analysis of the moisture sorption behavior of amorphous drug–polymer blends. *J. Appl. Polym. Sci.* 117: 1055–1063.
50. Crowley, K.J. and Zografi, G. (2002). Water vapor absorption into amorphous hydrophobic drug/poly(vinylpyrrolidone) dispersions. *J. Pharm. Sci.* 91: 2150–2165.
51. Flory, P.J. (1942). Thermodynamics of high polymer solutions. *J. Chem. Phys.* 10: 51–61.
52. Hancock, B. and Zografi, G. (1993). The use of solution theories for predicting water vapor absorption by amorphous pharmaceutical solids: a test of the Flory-Huggins and Vrentas models. *Pharm. Res.* 10: 1262–1267.

19

A GENERAL FRAMEWORK FOR SOLID–LIQUID EQUILIBRIA IN PHARMACEUTICAL SYSTEMS

THOMAS LAFITTE, VASILEIOS PAPAIOANNOU, SIMON DUFAL, AND CONSTANTINOS C. PANTELIDES
Process Systems Enterprise Ltd., London, UK

19.1 INTRODUCTION

A large proportion of active pharmaceutical ingredients (APIs) are produced and administered in the solid form. As a result, the thermodynamic equilibrium of systems involving solid and liquid phases is of central importance to API manufacturing where unit operations such as crystallization, particularly in organic solvents, play a prominent role in the separation and purification of APIs and their intermediates during chemical synthesis. Equilibrium compositions determine the driving forces for crystal formation and growth, and are important factors for the selection of an appropriate solvent for each separation step.

Solid–liquid equilibrium in aqueous environments also plays a central role in the oral absorption of pharmaceuticals [1], this time in determining the driving forces for dissolution of solid APIs in the lower gastrointestinal (GI) tract, consequently affecting drug bioavailability, particularly in the case of very low solubility drugs. The variation of pH along the GI tract and the presence of a variety of ions may also cause the reprecipitation of the API in various forms (e.g. salts) at different sections of the GI tract, which may have an inhibitory effect on bioavailability.

Recent years have witnessed a significant increase in the use of sophisticated mathematical models both in the design of pharmaceutical manufacturing operations and for the understanding and enhancement of oral absorption via appropriate formulations. The accurate computation of solid–liquid equilibria in mixtures of given temperature and composition is particularly important in this context as it can greatly affect the accuracy of predictions of the overall model. However, this poses a number of challenges. Some of these arise because of the complexity of the solid forms themselves. For example, acidic or basic APIs may combine with cations or anions present in the system to form salts. Crystallization in organic or aqueous media may result in solvent molecules being incorporated in the crystal structure, resulting in the formation of solvates or hydrates of fixed composition. There is also increasing interest in the use of co-crystals [2, 3] as a means of improving the solubility of poorly soluble APIs. Moreover, in any particular system, multiple solid forms relating to the same API may exist either simultaneously or in different regions of the composition and temperature domains; and multiple crystal structures (polymorphs [4]) having the same molecular composition may exist under different conditions.

Another set of challenges arises from the potential complexity of the solvents and their interactions with the API in the liquid phase. This is particularly pronounced not only in the case of aqueous media but also in other strongly associating solvents (e.g. alcohols) involving hydrogen bonding. The use of mixed solvents, either for enhancing the solubility of a target solute or for causing its precipitation via antisolvent effects, is also widespread. Moreover, under certain conditions, multiple liquid phases may form as is the case, for example, with undesirable "oiling out" effects that result in API-rich and API-lean phases [5, 6]. In aqueous mixtures, further complications arise because of the presence of ions and liquid-phase reactions such as acid/base dissociation.

Chemical Engineering in the Pharmaceutical Industry: Active Pharmaceutical Ingredients, Second Edition.
Edited by David J. am Ende and Mary T. am Ende.
© 2019 John Wiley & Sons, Inc. Published 2019 by John Wiley & Sons, Inc.

This chapter presents a unified framework for the determination of solid–liquid equilibria in complex systems of the types outlined above. The framework is based on a fundamental thermodynamics formulation of the combined phase and reaction equilibrium problem (see Section 19.2), supported by the emergence, over the past decade, of new equations of state (EoS) that are capable of accurately predicting liquid-phase behavior of mixtures involving complex intermolecular interactions (see Section 19.3).

An important consideration in the context of any modeling framework is the amount of experimental data required for the characterization of any particular system of interest in terms of the parameters required by the model. This issue is considered in detail in Section 19.4, with particular focus on the characterization of solid-phase behavior with minimal experimental solubility data.

Section 19.5 presents several examples illustrating key aspects of the proposed framework, and Section 19.6 draws some general perspectives from these examples. Finally, Section 19.7 concludes with a summary of the presented approach.

19.2 THERMODYNAMIC FUNDAMENTALS FOR SOLUBILITY CALCULATIONS

In this section, we review the fundamental thermodynamics underpinning the computation of solubilities, to the extent necessary for understanding the overall approach presented in this chapter.

Section 19.2.1 reviews the general conditions for thermodynamic equilibrium in a multiphase multi-reaction system. Section 19.2.2 then discusses how these general conditions can be applied to a wide class of systems of interest to the pharmaceutical industry, including those involving solid phases with API salts, hydrates or solvates, or co-crystals. It will be seen that the only physical properties required in all cases are chemical potentials in liquid and solid phases; these are covered in Sections 19.2.3 and 19.2.4, respectively.

19.2.1 General Conditions for Phase and Reaction Equilibrium

We consider a system comprising NC chemical compounds $i = 1, \ldots, NC$ that may be neutral or charged; the charge on compound i is denoted by q_i. The compounds may take part in a number NR of chemical reactions $j = 1, \ldots, NR$; the stoichiometric coefficient of compound i in reaction j is denoted by ν_{ij}.

The compounds are distributed over a number NP of phases $k = 1, \ldots, NP$. If the initial molar amount of compound i in the system is denoted by n_i^0, then the amounts $n_i^{[k]}$ of each compound i in each phase k at a given temperature T and pressure P are such that the total Gibbs free energy of the system is minimized. This leads to the following equilibrium conditions:

- For each neutral compound i ($q_i = 0$):

$$\mu_i^{[k]}(T, P, \boldsymbol{n}^{[k]}) = \mu_i^{[1]}(T, P, \boldsymbol{n}^{[1]}), \quad k = 2, \ldots, NP \quad (19.1)$$

- For each pair of charged compounds i, i' ($q_i, q_{i'} \neq 0$):

$$\frac{\mu_i^{[k]}(T, P, \boldsymbol{n}^{[k]}) - \mu_i^{[1]}(T, P, \boldsymbol{n}^{[1]})}{q_i} = \frac{\mu_{i'}^{[k]}(T, P, \boldsymbol{n}^{[k]}) - \mu_{i'}^{[1]}(T, P, \boldsymbol{n}^{[1]})}{q_{i'}},$$
$$k = 2, \ldots, NP \quad (19.2)$$

- For each chemical reaction j:

$$\sum_{i=1}^{NC} \nu_{ij} \mu_i^{[1]}(T, P, \boldsymbol{n}^{[1]}) = 0, \quad j = 1, \ldots, NR \quad (19.3)$$

- For each compound i:

$$\sum_{k=1}^{NP} n_i^{[k]} = n_i^o + \sum_{j=1}^{NR} \nu_{ij} \xi_j, \quad i = 1, \ldots, NC \quad (19.4)$$

- For each phase k:

$$\sum_{i=1}^{NC} n_i^{[k]} q_i = 0, \quad k = 2, \ldots, NP \quad (19.5)$$

where

$\mu_i^{[k]}$ is the chemical potential of compound i in phase k; as indicated above, it is generally a function of temperature, pressure and the vector $\boldsymbol{n}^{[k]}$ of molar amounts of the compounds in phase k.

ξ_j is the extent of reaction $j = 1, \ldots, NR$.

We note that the chemical reaction equilibrium condition (19.3) is written arbitrarily in terms of the chemical potentials in phase 1. It can be shown that, because of the phase equilibrium conditions (19.1) and (19.2), this also holds for all other phases. Similarly, the electroneutrality condition (19.5) is omitted for phase 1 as it is implied by the material balance Eq. (19.4) and the fact that the overall system is neutral.

Conditions (19.1)–(19.5) constitute a system of NC × NP + NR equations that can be solved to determine the phase compositions $n_i^{[k]}$ and reaction extents ξ_j.

The above mathematical formulation assumes that the number NP of phases in the system is known. In reality, this needs to be determined in an iterative manner that postulates a certain number of phases, solves the phase and reaction equilibrium problem, and then considers the stability of each individual phase at the equilibrium point. We will consider the phase stability problem in Section 19.2.5.

19.2.2 Fundamental Conditions for Solid–Liquid Equilibria

The conditions presented above are completely general and apply to any set of phases of any kind, with all compounds under consideration appearing in each and every phase. For solubility calculations in a pharmaceutical context, our interest is focused primarily on systems comprising only solid and liquid phases. Moreover, in most cases of interest, including those of simple APIs, API salts, hydrates and solvates, and co-crystals, the solid phases have fixed composition, with a subset of the compounds in the system being present in fixed proportions. In fact, for the purposes of solubility calculations, it is convenient to think of these combinations as separate "compounds." The latter may not exist in the liquid phases; instead, upon dissolution, they dissociate to their constituent compounds in a manner that can be described by an equilibrium chemical reaction. For example, in the case of a (2 : 1) co-crystal $A_2 \cdot B$ between an API A and a coformer B, we have a dissociation reaction of the form

$$A_2 \cdot B(s) \leftrightarrow 2\,A(l) + B(l)$$

while a sodium salt of an acidic API may dissociate in an aqueous environment according to

$$NaA(s) \leftrightarrow Na^+(l) + A^-(l)$$

We note that, although each solid phase comprises a single compound, the same compound may appear in multiple solid phases corresponding to different polymorphs. In general, at most one of these phases will appear in a nonzero amount at equilibrium, the others corresponding to metastable phases.[1] Furthermore, in addition to the above solid–liquid dissociation reactions, the system may also involve reactions taking place entirely within the liquid phases.

In view of the above discussion, the general phase equilibrium conditions presented in Section 19.2.1 may be amended to take account of the specific features of the solubility calculation problem. The main modification required is to the phase equilibrium condition (19.1). Assuming, for simplicity, that phase 1 is always a liquid:

- For compounds i that exist both in solid and liquid phases (e.g. non-dissociating or weakly dissociating APIs), Eq. (19.1) is modified to remove the composition dependence of the chemical potential in all solid phases

$$\mu_i^{[k]}(T,P) = \mu_i^{[1]}\left(T,P,\boldsymbol{n}^{[1]}\right) \quad (19.6)$$

for any solid phase k comprising compound i. Note that Eq. (19.1) remains unchanged for other liquid phases k.

- For compounds i that exist exclusively in solid phases (e.g. API salts, hydrates or solvates, or co-crystals), Eq. (19.1) is replaced by the equilibrium conditions for the corresponding dissociation reaction. These are a special case of Eq. (19.3), and can be written as

$$\mu_i^{[k]}(T,P) = \sum_{i'=1}^{NC} \sigma_{ii'} \mu_{i'}^{[1]}\left(T,P,\boldsymbol{n}^{[1]}\right) \quad (19.7)$$

for any solid phase k comprising compound i. Here $\sigma_{ii'}$ denotes the stoichiometry of composite compound i in terms of other compounds i' in the system (i.e. one mole of compound i comprises $\sigma_{ii'}$ moles of compound i').

As an illustration, consider again the (2 : 1) co-crystal and salt examples introduced earlier in this section in a system comprising one solid and one liquid phase. For these cases, Eq. (19.7) will take the form

$$\mu_{A_2 \cdot B}^{[s]}(T,P) = 2\mu_A^{[l]}\left(T,P,\boldsymbol{n}^{[1]}\right) + \mu_B^{[l]}\left(T,P,\boldsymbol{n}^{[1]}\right)$$

and

$$\mu_{NaA}^{[s]}(T,P) = \mu_{Na^+}^{[l]}\left(T,P,\boldsymbol{n}^{[1]}\right) + \mu_{A^-}^{[l]}\left(T,P,\boldsymbol{n}^{[1]}\right)$$

respectively. Here $\boldsymbol{n}^{[l]}$ denotes the entire vector of composition including the solvent(s), other ions (e.g. H^+ and OH^- for the salt example), etc.

On the other hand, for an API A that can also exist in the liquid phase, the relevant equation is (19.6), taking the simple form

$$\mu_A^{[s]}(T,P) = \mu_A^{[l]}\left(T,P,\boldsymbol{n}^{[1]}\right)$$

19.2.3 Liquid-Phase Chemical Potentials

The discussion in Sections 19.2.1 and 19.2.2 demonstrates that the only thermodynamic property required for solubility calculations is the chemical potential of each compound present in the system. In particular, there is no need for introducing concepts such as "reaction equilibrium constants" (or related quantities such as pK_a) or "solubility products". In this section, we consider the computation of liquid-phase chemical potentials, while Section 19.2.4 focuses on solid-phase chemical potentials.

Since in general we need to be able to handle chemical transformations (e.g. dissociation of composite compounds from solid to liquid phases; liquid-phase reactions), we need to adopt a reference datum of chemical elements at a reference temperature $T^{\ominus}$ and pressure $P^{\ominus}$. Thus, the chemical potential of a compound i in a liquid phase l can be computed by

$$\mu_i^{[l]}(T,P,\boldsymbol{n}) = \mu_i^{ig}(T,P,\boldsymbol{n}) + \mu_i^{res}(T,P,\boldsymbol{n}) \quad (19.8)$$

---

[1] The determination of the solid phases that are actually present in the system is discussed in Section 19.2.5.

where μ_i^{ig} is the ideal gas chemical potential of compound i given by

$$\mu_i^{ig}(T,P,n) = \Delta h_i^{F,ig} - T\Delta s_i^{F,ig}$$
$$+ \left[\int_{T^\ominus}^{T} c_{p,i}^{ig}(T')\,dT' - T\int_{T^\ominus}^{T} \frac{c_{p,i}^{ig}(T')}{T'}\,dT'\right] \quad (19.9)$$
$$+ RT\ln\left(\frac{P}{P^\ominus}\right) + RT\ln(x_i)$$

where

- $\Delta h_i^{F,ig}$ and $\Delta s_i^{F,ig}$ are, respectively, the enthalpy and entropy of formation of compound i in the ideal gas standard state at temperature $T^\ominus$ and pressure $P^\ominus$.
- $c_{p,i}^{ig}(T)$ is the ideal gas specific heat capacity of compound i.
- x_i is the molar fraction of compound i given by $x_i = n_i / \sum_{i'} n_{i'}$.

The quantity μ_i^{res} is the residual chemical potential of compound i, representing the difference between the real chemical potential and the ideal gas one under identical conditions of temperature, pressure, and composition.

The liquid-phase chemical potential may also be expressed in terms of reference states other than the ideal gas, such as the pure liquid state or a variety of infinite dilution states. In such cases, the residual term $\mu_i^{res}(T,P,n)$ is replaced by a term based on the corresponding activity coefficient $\gamma_i(T, P, n)$, which may be computed via a wide range of empirical and semiempirical models [7, 8]. However, as we shall see in Section 19.3, a significant development over the last decade has been the development of EoS that can accurately compute $\mu_i^{res}(T,P,n)$ over wide ranges of conditions for general liquid mixtures involving different types of molecules (e.g. neutral compounds, ions, polymers) and complex intermolecular interactions including hydrogen bonding or other forms of association and electrostatic (coulombic) forces. Moreover, the group contribution basis of these EoS implies that they can be applied to systems for which little experimental data are available.

In view of the above, our approach to computing solubilities is based on Eqs. (19.8) and (19.9). It is worth noting that, contrary to what is sometimes stated in the literature, these equations are applicable to *any* type of compound. In cases where experimentally derived values for the enthalpy and entropy of formation are available for reference states other than the ideal gas, they can easily be converted to the quantities $\Delta h_i^{F,ig}$ and $\Delta s_i^{F,ig}$ required by Eq. (19.9) using the EoS itself. For example, the enthalpy $\Delta h_i^{F,1}$ and entropy of formation $\Delta s_i^{F,1}$ of a pure liquid at temperature $T^\ominus$ and pressure $P^\ominus$ are related to the corresponding ideal gas quantities via

$$\Delta h_i^{F,1} = \Delta h_i^{F,ig} + h_i^{res}(T^\ominus, P^\ominus) \quad (19.10a)$$
$$\Delta s_i^{F,1} = \Delta s_i^{F,ig} + s_i^{res}(T^\ominus, P^\ominus) \quad (19.10b)$$

where the residual enthalpy $h_i^{res}(T^\ominus,P^\ominus)$ and entropy $s_i^{res}(T^\ominus,P^\ominus)$ of the pure liquid can be obtained from the corresponding $\mu_i^{res}(T^\ominus,P^\ominus)$ computed by the EoS via the standard thermodynamic relations:

$$s_i^{res}(T^\ominus,P^\ominus) = -\frac{\partial}{\partial T}\mu_i^{res}(T^\ominus,P^\ominus) \quad (19.11a)$$
$$h_i^{res}(T^\ominus,P^\ominus) = \mu_i^{res}(T^\ominus,P^\ominus) + T^\ominus s_i^{res}(T^\ominus,P^\ominus) \quad (19.11b)$$

Similar relations can be derived for enthalpies and entropies of formation in infinite dilution reference states and are presented in Appendix 19.A.

Figure 19.1 summarizes the different approaches to computing liquid-phase chemical potentials in the form of thermodynamic paths from chemical elements at standard conditions ($T^\ominus$, $P^\ominus$) to the mixture under the conditions (T, P, n) of interest. Different paths make use of formation properties corresponding to different reference states. All these quantities can be interconverted to each other via the use of an EoS that can accurately compute residual properties of pure compounds. This allows the use of whatever experimentally determined values of formation properties happen to be available, even if these are with respect to different reference states for different compounds in the system under consideration.

All calculations reported in this chapter make use of the thermodynamic path going through the ideal gas reference state as shown by the solid arrows in Figure 19.1. For completeness, this diagram also shows paths based on activity coefficient models as dashed arrows.

19.2.4 Solid-Phase Chemical Potentials

We now turn our attention to the computation of solid-state chemical potentials of the kind that appear in the phase equilibrium conditions (19.6) and (19.7).

Below we consider two alternative approaches that differ in their requirements for solid-state properties. In particular, the approach described in Section 19.2.4.1 makes use of enthalpy and entropy of formation, while that described in Section 19.2.4.2 relies on properties related to the melting of the solid phase.

19.2.4.1 Solid-Phase Potentials in Terms of Formation Properties
Although, in principle, Eq. (19.8) holds for solid phases too, there is currently no generally applicable way of

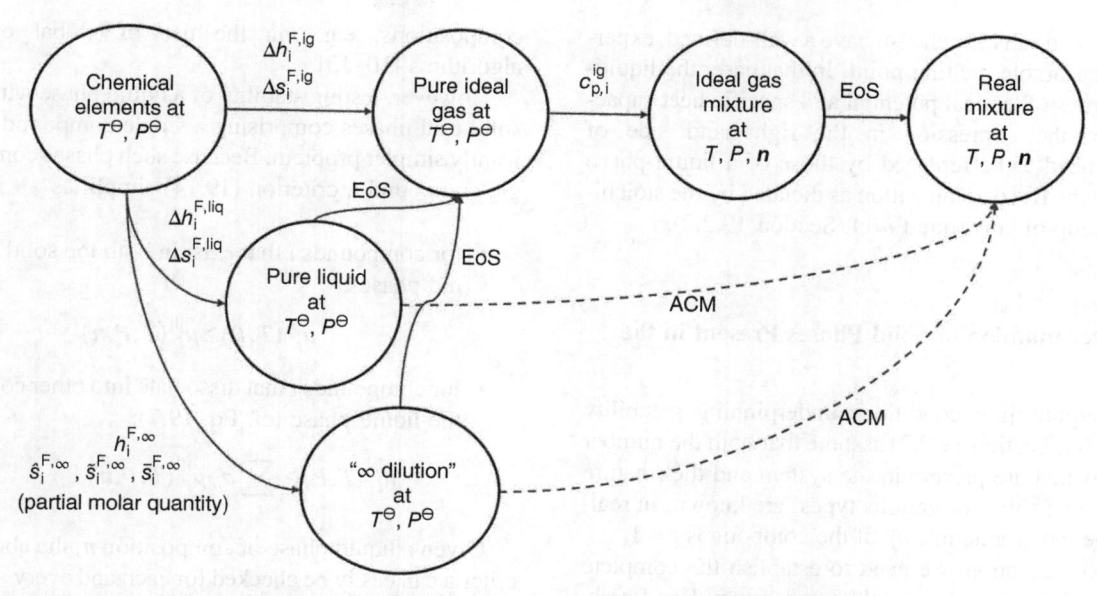

FIGURE 19.1 Summary of the various possible thermodynamic paths for the computation of chemical potential of compound i in liquid phases $\mu_i^l(T,P,\mathbf{n})$ using formation properties at different standard states. The thermodynamic integration path depends on whether an equation of state (EoS) or activity coefficient model (ACM) is used to model the mixture nonidealities.

computing the residual term μ_i^{res}. In particular, both EoS and activity coefficient models are restricted to phases exhibiting liquid-like structures only, and cannot describe the behavior of phases where the underlying microscopic structure follows a well-defined geometric pattern such as crystalline solid phases.[2]

In view of the above, we express the chemical potential of a compound i in a solid phase corresponding to a particular crystal structure as

$$\mu_i^{[s]}(T) = \Delta h_i^{F,s} - T\Delta s_i^{F,s} + \int_{T^\ominus}^{T} c_{p,i}^s(T')dT' - T\int_{T^\ominus}^{T} \frac{c_{p,i}^s(T')}{T'}dT' \quad (19.12)$$

where

- $\Delta h_i^{F,s}$ and $\Delta s_i^{F,s}$ are, respectively, the enthalpy and entropy of formation of the solid phase of the same crystal structure at standard temperature $T^\ominus$ and pressure $P^\ominus$.
- $c_{p,i}^s(T)$ is the specific heat capacity of the solid at temperature T and pressure $P^\ominus$.

The above equation omits the pressure dependence of $\mu_i^{[s]}(T)$ as this is typically very small at the pressures of interest to pharmaceutical applications.

---
[2] Note that *amorphous* solid phases often exhibit liquid-like structures and may, therefore, be amenable to description by fluid-phase EoS. For example, Ref. [59] reports the application of the SAFT-VR SW EoS to predict the solubility of gas in amorphous polyethylene.

19.2.4.2 Solid-Phase Potentials in Terms of Melting Properties

The melting properties approach takes advantage of the fact that at the melting point temperature $T_{m,i}$ of pure compound i, its liquid- and solid-phase chemical potentials (i.e. free energy) are equal. Neglecting pressure effects and integrating along the thermodynamic path [3]

Liquid at $T \to$ Liquid at $T_{m,i} \to$ Solid at $T_{m,i} \to$ Solid at T

we obtain the following expression for the solid chemical potential:

$$\mu_i^{[s]}(T) = \mu_i^{[l]}(T,P) + \Delta h_i^{[s\to l]}\left(\frac{T}{T_{m,i}} - 1\right) + \int_T^{T_{m,i}} \Delta c_{p,i}^{[s\to l]}(T')dT' - T\int_T^{T_{m,i}} \frac{\Delta c_{p,i}^{[s\to l]}(T')}{T'}dT' \quad (19.13)$$

where

- $\Delta h_i^{[s\to l]}$ is the enthalpy of melting at $T_{m,i}$.
- $\Delta c_{p,i}^{[s\to l]}$ is the difference between liquid- and solid-specific heat capacities.

Since $T < T_{m,i}$, the chemical potential $\mu_i^{[l]}(T,P)$ corresponds to a metastable (subcooled) liquid phase but can nevertheless be computed from an EoS.

This approach is usually applied to APIs that are stable up to their melting points. In principle, it can be extended to co-crystals and salts which also have a well-defined, experimentally measurable melting point. In that case, the liquid-phase properties (chemical potential and specific heat capacity) used in the expression in the right-hand side of Eq. (19.13) need to be replaced by those of a liquid-phase mixture with the fixed composition as dictated by the stoichiometric makeup of compound i (cf. Section 19.2.2).

19.2.5 Determination of Solid Phases Present in the System

The phase equilibrium conditions underpinning solubility calculations (cf. Section 19.2.2) assume that both the number NP of phases that are present in the system and their nature (liquid mixtures; solids of various types) are known. In reality, given the molar amounts n_i^o of the compounds $i = 1, \ldots,$ NC under consideration, we need to establish the complete set of phases that are present at the temperature T and pressure P of interest. One approach for achieving this aim is to start with a single liquid phase comprising all of the specified amounts and then test its stability with respect to all solid phases that may potentially form given the set of compounds under consideration.[3] For the general systems considered here, any nonzero amounts n_i^o of compounds i that exist only in solid phases (e.g. co-crystals) can be converted to the equivalent amounts $\sigma_{ii'}$ of the compounds i' formed by their dissociation (cf. Eq. 19.7). The resulting liquid phase can then be equilibrated with respect to any chemical reactions taking place in it.

A widely applied stability criterion is that based on the Gibbs surface tangent plane [7]. Given a fluid phase of composition n, it tests whether the formation of an infinitesimally small amount of a new "trial" phase would result in an overall reduction of the Gibbs free energy. If no such trial phase can be found, the original fluid phase is deemed to be stable. Mathematically, this is equivalent to requiring that the quantity

$$\Omega_G(w) = \sum_{i=1}^{NC} w_i \left[\mu_i^{\text{trial}}(T,P,w) - \mu_i(T,P,n) \right] \quad (19.14)$$

is nonnegative for any trial composition w.

The application of the above criterion in cases where the trial phase is a fluid is a complex problem that has been studied extensively in the literature. The main challenge is to ensure that $\Omega_G(w) \geq 0$ for *all* possible nonzero w, something that can be formally guaranteed only via the use of techniques that systematically search over the entire range of trial compositions, e.g. via the use of global optimization algorithms [10–13].

However, testing stability of a liquid phase with respect to solid trial phases comprising a single compound is a significantly simpler problem. Because such phases comprise a single compound i, criterion (19.14) simplifies

- for compounds i that exist in both the solid and the liquid phases:

$$\mu_i^{[s]}(T,P) \geq \mu_i^{[l]}(T,P,n) \quad (19.15a)$$

- for compounds i that dissociate into other compounds in the liquid phase (cf. Eq. 19.7):

$$\mu_i^{[s]}(T,P) \geq \sum_{i'} \sigma_{ii'} \mu_{i'}^{[l]}(T,P,n) \quad (19.15b)$$

Given a liquid phase of composition n, the above stability criteria can easily be checked for each and every solid phase s under consideration using the values of the liquid- and solid-phase chemical potentials (cf. Sections 19.2.3 and 19.2.4, respectively). If the criterion is found to be violated for any phase s, then that phase may be present in the system and therefore must be added to the set of phases included in the formulation of the phase equilibrium conditions presented in Section 19.2.2. The solution of these equations will then yield a new liquid-phase composition n, which again can be tested for stability against potential existence of more solid phases. The above procedure is therefore repeated until a stable liquid phase is obtained.

It should be noted that for a given liquid-phase composition, more than one trial solid phase may be found to violate the stability criterion (19.15). Our approach in such cases is to tentatively add each such solid phase s separately to the system, perform the associated solubility calculation, and then use the resulting equilibrium compositions to compute the total Gibbs free energy of the system:

$$G_s = \sum_{k=1}^{NP} \sum_i n_i^{[k]} \mu_i^{[k]} \left(T,P,n^{[k]} \right) \quad (19.16)$$

The trial solid phase s that results in the lowest free energy G_s is selected to be added to the system, and the corresponding liquid-phase composition is subjected to further stability tests as described above.

19.3 THE SAFT-γ MIE GROUP CONTRIBUTION EoS

In Section 19.2 we showed that the general problem of computing the solubility (and stability) for any type of pharmaceutical system relies on solving for a set of equations that are expressed entirely in terms of liquid- and solid-phase

---

[3] As already mentioned, in general such solids may include both pure API phases and API salts, solvates, hydrates, and/or co-crystals, some of which may exist in multiple polymorphic forms.

chemical potential. The complexity of the solubility calculation is mainly due to the composition dependence of the liquid-phase chemical potentials, whereas the solid phase is always a function of temperature only.

Therefore a prerequisite for the successful prediction of solubility is the choice of a suitable approach to compute the liquid-phase chemical potentials of complex multifunctional molecular structures, which are ubiquitous in pharmaceutical applications. The general framework of the statistical associating fluid theory (SAFT) [14, 15] has been shown to be particularly promising in this context, having been successfully applied to a wide range of systems; see the reviews in Refs. [16–18] for a more extensive discussion. SAFT takes explicit account of intermolecular forces, incorporating the description of nonidealities due to asymmetry in molecular shape and anisotropic interactions. The fact that this theoretical framework is based on a physically sound representation of the APIs makes it one of the most promising approaches to date for the solubility prediction in solvent mixtures.

In this contribution, all the calculations are performed using the SAFT-γ Mie group contribution approach [19, 20], a variant of the general SAFT framework based on a generalized Lennard-Jones (or so-called Mie) potential. As in other group contribution approaches, molecules are described in terms of the functional (chemical) groups they comprise. The functional groups are assumed to behave independently of the molecular structure on which they are found. An example of the decomposition of a molecular structure into the functional groups is given for the particular case of ibuprofen in Figure 19.2.

19.3.1 Intermolecular Potential Model

Within the SAFT-γ Mie EoS, each functional group k is characterized by a shape factor parameter S_k, describing the group non-sphericity as well as a set of parameters that relate to various group-group interactions, as described below.

Repulsive and dispersive interactions are modeled via the Mie potential and are considered for all group pairs. On the other hand, hydrogen bonding interactions are included only where necessary (i.e. for groups that exhibit association interactions) and are modeled using short-ranged square-well sites. It is also possible to formally take into account the presence of permanent charges on functional groups. This allows for the rigorous (and explicit) modeling of the speciation of APIs in aqueous mixtures as will be shown in Section 19.5. This does not require any additional parameters besides the charge itself (positive or negative), which is always known a priori, thus making the extension to charged systems straightforward.

The parameters that describe the Mie intermolecular potential are the diameter σ, the repulsive and attractive exponents λ^{rep} and λ^{att}, respectively, and the potential depth ϵ. The association interactions between pair of sites (denoted, for example, as a and b) are, if present, described by two additional parameters, namely, the energy of interaction, $\epsilon_{ab}^{\text{HB}}$, and the volume available for bonding, K_{ab}^{HB}. Note that no additional parameters are required for describing the formation of the molecular chains; this contribution is inferred by the knowledge of the distinct types of functional groups that comprise each molecule and their multiplicity on the molecular chains. Finally, the type and number of different association sites on a given group need to be defined a priori based on chemical understanding of hydrogen bonding donors and acceptors.

The interaction parameters described above relate to all pair interactions. A schematic representation of the physical meaning of each parameter is given in Figure 19.3. In practice, when developing the group pair interaction parameters, we commonly separate the treatment of the groups that are of the same type (self-interactions) and the interactions between groups of different types (unlike or cross-interactions). The self-interaction parameters are almost always obtained by regression to experimental data, whereas the parameters that describe unlike interaction can be approximated by combining rules. Overall, the application of the SAFT-γ Mie approach for pharmaceutical system relies on the development of databases of group interaction parameters for

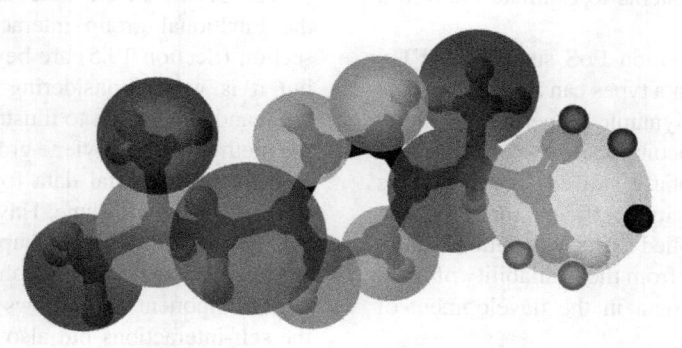

FIGURE 19.2 Decomposition of ibuprofen into functional groups ($3 \times CH_3$, $1 \times CH$, $1 \times aCCH_2$, $4 \times aCH$, $1 \times aCCH$, $1 \times COOH$).

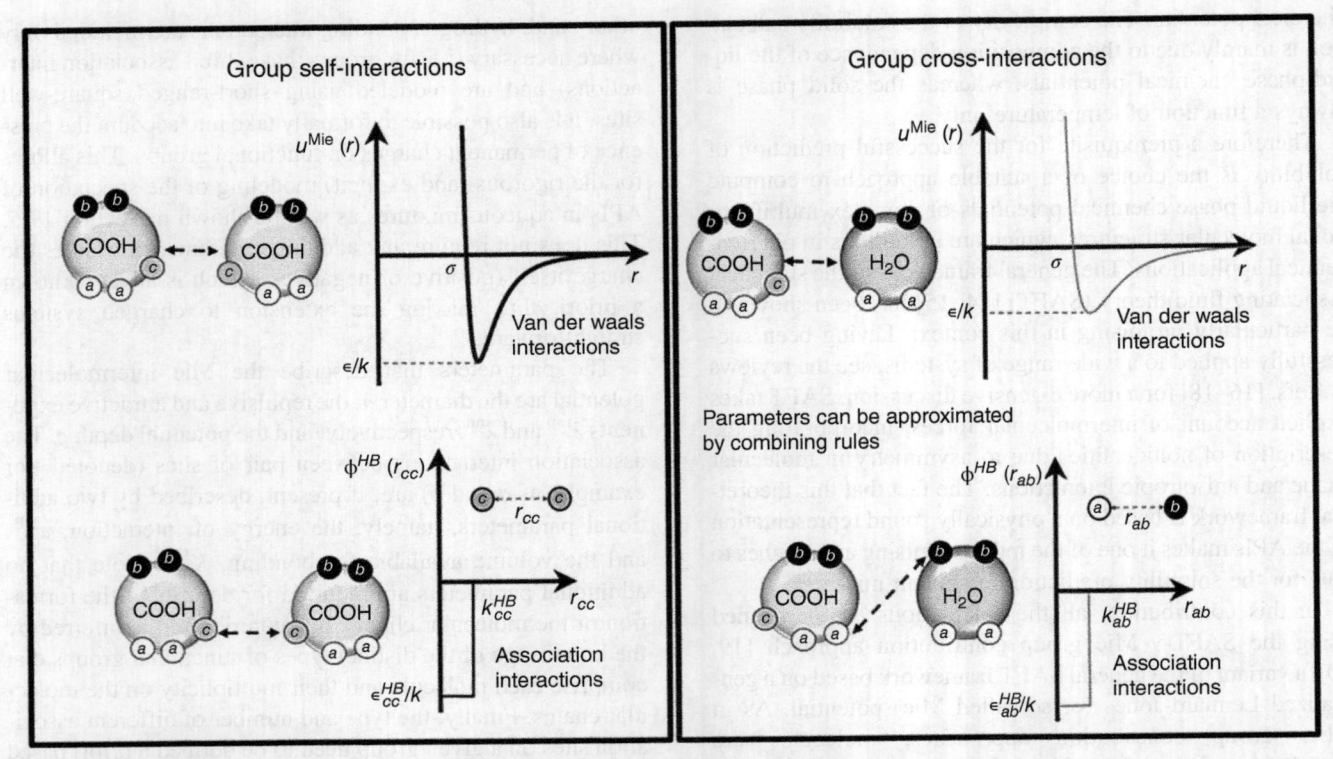

FIGURE 19.3 Schematic representation of the SAFT-γ Mie parameters needed to describe the group self- and cross-interactions. The choice of the association scheme for the COOH and H$_2$O groups is discussed in detail in Ref. [21]. *Source*: Reproduced with permission of Elsevier.

relevant chemical groups typically found in APIs. Section 19.3.2 explains the procedure for developing such database based on experimental data.

19.3.2 Estimation of Group Parameters from Experimental Data

As we have seen in Section 19.3.1, most of the parameters in the SAFT-γ Mie EoS relate to interactions between functional groups. In practice, one can make use of available experimental data on simple systems to estimate these in a sequential manner.

When using a group contribution EoS such as SAFT-γ Mie, a wide range of different data types can be used, including not only data on the thermodynamic properties and phase behavior of binary and/or multicomponent systems but also pure component data (single phase and equilibrium). This is a significant benefit over activity coefficient models, since only EoS approaches can be applied to the description of pure components and, hence, benefit from the availability of readily available pure component data in the development of group parameters.

In order to obtain reliable parameter estimates, a commonly used strategy is to include various types of experimental data in the regression procedure, including, for instance, densities, vapor pressure, and caloric data (heat capacity and/or vaporization enthalpy). The key aspect of the SAFT-γ Mie group contribution methodology is that experimental data on any molecule containing the group of interest can be used in the regression procedure. In principle, a database of group parameters relevant to pharmaceutical applications can be developed without the use of API-specific data.

The details on the exact sequence of development of all the functional group interactions used in the application section (Section 19.5) are beyond the scope of this chapter, but it is worth considering the simple example of CH$_3$, CH$_2$, and OH groups to illustrate the general workflow. First the methyl and methylene groups can be obtained by regression to experimental data for a series of (pure) n-alkanes, from ethane to n-decane. Having determined the parameters for the CH$_3$ and CH$_2$ groups, the OH group can then be obtained using primary alcohol data. Note that the usage of pure component data allows the determination of not only the self-interactions but also the cross-interactions between each group pair. A similar sequence can then be used to determine any other group, by selecting either pure component or mixture data.

Finally, it is worth noting that the SAFT-γ Mie EoS can be used in a "molecular" mode, with an entire compound being represented as a single functional group. This is typically the case for the first member of each chemical family – e.g. methane for the *n*-alkanes; methanol for the alcohols; etc. – but can also be employed in the study of more complex molecules. Note that in such cases compound-specific experimental data are required for the development of the (single) group parameters.

19.4 SYSTEM CHARACTERIZATION FOR SOLUBILITY CALCULATIONS

As described in Section 19.2.2, the computation of solid–liquid equilibria requires accurate values of the chemical potential of the individual compounds in the liquid and solid phases using the approaches presented in Sections 19.2.3 and 19.2.4, respectively, and, in particular, Eqs. (19.8), (19.9), (19.12), and (19.13).

This section focuses on the parameters that are required by these equations, and the way in which they can be obtained from experimental data. Section 19.4.1 provides a summary and categorization of the relevant parameters.

Because of the fundamental thermodynamic approach adopted in this chapter, these parameters are not specific to solubility calculations; in fact, most of them can be obtained or estimated without access to experimental solubility data. However, even limited availability of such data for the specific API of interest can substantially reduce the effort required for accurate solubility prediction. Interestingly, this may be true even if the available data pertain to completely different solvents or solvent mixtures to the ones involved in the system of interest. This is possible because of the use of predictive EoS such as the SAFT-γ Mie (cf. Section 19.3).

More specifically, Section 19.4.2 considers the use of experimental solubility data for the characterization of the solid state of solid-forming compounds that also exist in the liquid phase. Section 19.4.3 covers the same topic for compounds such as API salts, solvates, hydrates, and co-crystals that only exist in the solid phase, dissociating into other compounds in the liquid phase. Finally, Section 19.4.4 considers the characterization of liquid-phase chemical reactions using data from any system that contains a liquid phase in which these reactions occur. The use and power of all of these techniques will be demonstrated extensively in the examples considered in Section 19.5.

19.4.1 Parameters Involved in Solubility Calculations

Equations (19.8), (19.9), (19.12), and (19.13) involve a number of parameters:

- Pure compound parameters
  - Enthalpies and entropies of formation in the ideal gas reference state at standard temperature $T^{\ominus}$ and pressure $P^{\ominus}$.
  - Parameters describing the temperature dependence of specific heat capacities in the ideal gas state.
  - Either
    - Enthalpies and entropies of formation of crystalline phases at standard temperature $T^{\ominus}$ and pressure $P^{\ominus}$.
    - Parameters describing the temperature dependence of solid-phase specific heat capacities.
  - or
    - Melting point temperature and enthalpy of melting at that temperature.
    - Parameters describing the temperature dependence of the difference between liquid- and solid-phase specific heat capacities.
- EoS parameters used for the computation of liquid-phase chemical potentials and related quantities, including:
  - Residual chemical potentials used in Eq. (19.8).
  - Pure compound potential used in Eq. (19.13).
  - (Optionally) the terms required for obtaining the ideal gas formation properties from values of formation properties in other reference states as shown, for example, in Eqs. (19.10) and (19.11).

Some of these parameters can be estimated using group contribution methods, such as the Joback and Reid [22] method for ideal gas specific heat capacities. Others may be determined experimentally either via direct measurement (e.g. melting parameters using DSC techniques) or indirect estimation from experimental data (e.g. formation enthalpies and entropies). As already mentioned in Section 19.3, in the case of the SAFT-γ Mie EoS parameters for the various functional groups, these experimental data could relate to fluid-phase measurements (e.g. vapor–liquid or liquid–liquid equilibria) obtained from systems that do not actually involve any of the compounds occurring in the system of interest.

19.4.2 Solid-Forming Compounds That Also Exist in the Liquid Phase

We start by revisiting the equilibrium condition (19.6) for compounds that exist in both the liquid and solid phases.

Using Eqs. (19.8) and (19.9), ignoring the pressure dependence of the solid-phase potential and applying some minor rearrangement, we obtain an expression of the form

$$\Delta g_i(T) = RT\ln\left(\frac{P}{P^{\ominus}}\right) + RT\ln\left(x_i^{[1]}\right) + \mu_i^{\text{res}}\left(T, P, \boldsymbol{n}^{[1]}\right) \quad (19.17)$$

where we have introduced a new quantity Δg_i defined as

$$\Delta g_i(T) \equiv \mu_i^{[k]}(T) - \Delta h_i^{F,\text{ig}} + T\Delta s_i^{F,\text{ig}}$$
$$- \left[\int_{T^{\ominus}}^{T} c_{p,i}^{\text{ig}}(T')\mathrm{d}T' - T\int_{T^{\ominus}}^{T} \frac{c_{p,i}^{\text{ig}}(T')}{T'}\mathrm{d}T'\right] \quad (19.18)$$

and have retained the phase labeling conventions of the original Eq. (19.6), i.e. k denotes the solid phase formed by compound i and phase "1" is a liquid phase.

We note that the definition (19.18) implies that $\Delta g_i(T)$ is purely a characteristic of the solid-forming compound i and does not depend on any other compound in the system, and that it is solely a function of temperature. Therefore the value of $\Delta g_i(T)$ for a given solid API i at a given temperature T will always be the same, irrespective of the system in which this API occurs. Moreover, turning to Eq. (19.17), we can see that we can compute this $\Delta g_i(T)$ from any available experimental solubility data for this API by using the measured liquid-phase composition to replace the quantities $x_i^{[1]}$ and $\boldsymbol{n}^{[1]}$ appearing in the right-hand side – provided, of course, we have an EoS that can compute the residual term $\mu_i^{\text{res}}\left(T, P, \boldsymbol{n}^{[1]}\right)$.

It is worth noting that the original equilibrium condition (19.6) between solid phase k and liquid phase 1 is valid irrespective of the existence of other solid and/or liquid phases. Therefore, this type of calculation is correct even if the experimental data were derived in systems with multiple solid and/or liquid phases. It is sufficient to have the measured composition of one of the liquid phases.

As an illustration, Figure 19.4 shows the function $\Delta g_i(T)$ for ibuprofen. The values shown are computed from the right-hand side of Eq. (19.17) using published experimental solubility data [23–26] from a range of solvents over a range of temperatures, with the $\mu_i^{\text{res}}\left(T, P, \boldsymbol{n}^{[1]}\right)$ term being computed via the SAFT-γ Mie EoS. It is clear that all the points define a single function, irrespective of the solvent being considered and the actual solubility values. Moreover, at least in this case, this function can be well approximated by a straight line empirically determined to be of the form $\Delta g_{\text{ibuprofen}}(T) = 238.78T - 117\,460$ ($\Delta g_{\text{ibuprofen}}$ in J/mol, T in K).

Once the function $\Delta g_i(T)$ for a given API i has been determined, it can be used directly for the computation of its solubility within any solvent or solvent mixture simply by using Eq. (19.17) to replace the original condition (19.6) in the solubility calculations. Importantly, the values of the various pure compound parameters appearing on the right-hand side of Eq. (19.18) are not necessary. Effectively, the use of an advanced EoS allows transferability of solubility information from one system (or systems) to another.

As mentioned in Section 19.2.4.2, an alternative to the use of solid-phase chemical potentials based on formation energies (Eq. 19.12) is to employ an expression based on melting properties as indicated by Eq. (19.13). The latter requires

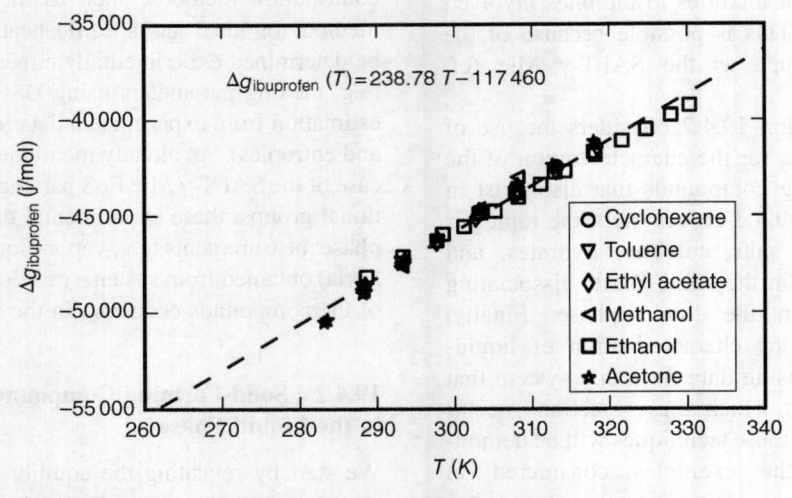

FIGURE 19.4 Calculation of empirical model for $\Delta g_i(T)$ for ibuprofen.

knowledge of the pure liquid chemical potential $\mu_i^{[l]}(T,P)$, which can itself be calculated using Eqs. (19.8) and (19.9):

$$\mu_i^{[l]}(T,P) = \Delta h_i^{F,ig} - T\Delta s_i^{F,ig}$$
$$+ \left[\int_{T^\ominus}^T c_{p,i}^{ig}(T')dT' - T\int_{T^\ominus}^T \frac{c_{p,i}^{ig}(T')}{T'}dT'\right]$$
$$+ RT\ln\left(\frac{P}{P^\ominus}\right) + \mu_i^{res}(T,P) \qquad (19.19)$$

where the last term on the right-hand side is the residual chemical potential for the pure liquid that can be calculated using the EoS. It will be noted that the above does not include the $+RT\ln(x_i)$ term that appears on the right-hand side of Eq. (19.9) as $x_i = 1$ for the case of pure liquid.

The expression for $\mu_i^{[l]}(T,P)$ given by Eq. (19.19) can then be substituted in Eq. (19.13) to yield a complete expression for the solid-phase chemical potential $\mu_i^{[s]}(T,P)$. The latter may then be used on the left-hand side of equilibrium condition (19.6), with Eqs. (19.8) and (19.9) being used to compute the liquid-phase chemical potential $\mu_i^{[l]}(T,P,\boldsymbol{n})$ on the right-hand side. At this point, all terms arising from the right-hand side of Eq. (19.19) except the last one will appear identically on both sides and will cancel out, resulting in an equation of the form

$$\Delta g_i^{[s\to l]}(T) = RT\ln(x_i) + \mu_i^{res}(T,P,\boldsymbol{n}) - \mu_i^{res}(T,P) \qquad (19.20)$$

where the quantity $\Delta g_i^{[s\to l]}(T)$ is defined as

$$\Delta g_i^{[s\to l]}(T) \equiv \Delta h_i^{[s\to l]}\left(\frac{T}{T_{m,i}} - 1\right)$$
$$+ \int_T^{T_{m,i}} \Delta c_{p,i}^{[s\to l]}(T')dT' \qquad (19.21)$$
$$- T\int_T^{T_{m,i}} \Delta \frac{c_{p,i}^{[s\to l]}(T')}{T'}dT'$$

Analogously to the quantity $\Delta g_i(T)$ defined by Eq. (19.18), $\Delta g_i^{[s\to l]}(T)$ is purely a characteristic of compound i and is solely a function of temperature. It can, therefore, also be described by an empirical relationship derived by fitting values obtained by computing the right-hand side of Eq. (19.20) at available experimental data points. On the other hand, if the melting properties $T_{m,i}, \Delta h_i^{[s\to l]}$ and $\Delta c_{p,i}^{[s\to l]}(T)$ of compound i are known, then they can be used directly for the computation of $\Delta g_i^{[s\to l]}(T)$ using Eq. (19.21) without the need to resort to experimental solubility data. In either case, there is no need to know the ideal gas properties $\Delta h_i^{F,ig}, \Delta s_i^{F,ig}$, and $c_{p,i}^{ig}(T)$.

19.4.3 Solid-Forming Compounds That Exist Only in the Solid Phase

As already mentioned in Section 19.2.2, API salts, solvates, hydrates, and co-crystals are assumed to exist only in the solid phase, dissociating fully upon entering the liquid phase. In such cases, the relevant equilibrium condition is Eq. (19.7). By following an approach similar to that used for the simpler case considered in Section 19.4.2 above, we can derive the equation

$$\Delta g_i(T) = \left(\sum_{i'}\sigma_{ii'}\right)RT\ln\left(\frac{P}{P^\ominus}\right)$$
$$+ \sum_{i'}\sigma_{ii'}\left[RT\ln\left(x_{i'}^{[l]}\right) + \mu_{i'}^{res}\left(T,P,\boldsymbol{n}^{[l]}\right)\right] \qquad (19.22)$$

where the quantity Δg_i is now defined as

$$\Delta g_i(T) \equiv \mu_i^{[k]}(T) - \sum_{i'}\sigma_{ii'}$$
$$\left(\Delta h_{i'}^{F,ig} - T\Delta s_{i'}^{F,ig} + \left[\int_{T^\ominus}^T c_{p,i'}^{ig}(T')dT' - T\int_{T^\ominus}^T \frac{c_{p,i'}^{ig}(T')}{T'}dT'\right]\right) \qquad (19.23)$$

As in the previous case, Eq. (19.23) shows that $\Delta g_i(T)$ is a characteristic of the compound i under consideration (e.g. a particular co-crystal) and is solely a function of temperature. Inserting any available experimental solubility data for this compound on the right-hand side of Eq. (19.22) allows us to compute the corresponding $\Delta g_i(T)$ and fit an empirical function of temperature to it. The latter can then be used in conjunction with Eq. (19.22) instead of the original condition (19.7) to compute solubilities of compound i in other systems of interest. Once again, we do not actually need to know any of the pure compound quantities appearing on the right-hand side of Eq. (19.23).

19.4.4 Characterization of Liquid-Phase Chemical Reactions

Liquid-phase chemical reactions are described by the chemical equilibrium condition (19.3). Following an approach similar to that presented in Sections 19.4.2 and 19.4.3, the latter can be reformulated to an equation of the form

$$\Delta g_j^R(T) = \left(\sum_i \nu_{ij}\right)RT\ln\left(\frac{P}{P^\ominus}\right)$$
$$+ \sum_i \nu_{ij}\left[RT\ln\left(x_i^{[l]}\right) + \mu_i^{res}\left(T,P,\boldsymbol{n}^{[l]}\right)\right] \qquad (19.24)$$

where the quantity Δg_j^R is now defined as

$$\Delta g_j^R(T) \equiv -\sum_i \nu_{ij} \left(\Delta h_i^{F,ig} - T\Delta s_i^{F,ig} + \left[\int_{T^\ominus}^T c_{p,i}^{ig}(T')dT' - T\int_{T^\ominus}^T \frac{c_{p,i}^{ig}(T')}{T'}dT' \right] \right) \quad (19.25)$$

From Eq. (19.25), we note that the free energy quantity Δg_j^R is a characteristic of the particular chemical reaction j and depends only on temperature. As before, we can follow a two-step procedure:

1. Determine the values of this quantity at different temperatures by computing the expression on the right-hand side of Eq. (19.24) using experimentally measured liquid-phase compositions for any system[4] in which this reaction occurs.
2. Use the values computed above to fit an empirical expression (e.g. a linear or quadratic relation) to describe the temperature variation of Δg_j^R.

Once we have the empirical expression $\Delta g_j^R(T)$, we can use it for characterizing reaction j in any system of interest where it may occur by replacing the original equilibrium condition (19.3) by Eq. (19.24).

It is worth noting that it is not always possible to measure directly the complete liquid-phase composition. For instance,[5] in aqueous ionic systems what is typically measured is the pH (i.e. the activity of the H$^+$ ion) and the API *total* molar fraction, which comprises contributions from both the undissociated API and its ion(s). In such cases, step 1 of the above procedure needs to be amended as it is no longer possible to explicitly compute $\Delta g_j^R(T)$ at the experimental temperature T simply by evaluating the right-hand side of Eq. (19.24). Instead, the latter needs to be considered simultaneously with the material balance Eq. (19.4) to determine the value of $\Delta g_j^R(T)$ that leads to the best fit of the available measurements.

19.5 ILLUSTRATIVE EXAMPLES

This section presents a number of examples illustrating the key aspects of the general framework for solid–liquid equilibria presented in this chapter and demonstrating its applicability to systems of relevance to pharmaceutical applications.

Section 19.5.1 provides a brief overview of a computer implementation of this framework. Sections 19.5.2–19.5.7 present solid–liquid equilibrium calculations for six different systems involving:

- One or more different types of solid phases (unionizable or weakly acidic APIs, API salts, hydrates, co-crystals, and polymers).
- Pure organic solvents and their mixtures.
- Aqueous systems with pH-modifying agents.

Particular emphasis is paid to the characterization of liquid- and solid-phase behavior from practically available experimental data.

Finally, Section 19.6 attempts to draw some more general conclusions from these examples.

19.5.1 The gSAFT Physical Properties Code

All calculations presented in this section were performed using gSAFT [27], a physical properties code developed by Process Systems Enterprise Ltd. for use within its gPROMS® [28] process modeling tool.

Designed to support pharmaceutical applications in the context of drug substance and drug product manufacturing and also oral absorption, gSAFT incorporates several elements of relevance to the topic of this chapter:

- An algorithm for the reliable computation of equilibria in multiple vapor–liquid–solid phases in the presence of chemical reactions, based on the general equilibrium conditions presented in Section 19.2.1 and including a comprehensive phase stability analysis (cf. Section 19.2.5).
- A robust implementation of the SAFT-γ Mie EoS (cf. Section 19.3) that can be applied efficiently to mixtures with large numbers of neutral and/or charged compounds in the presence of association and/or coulombic interactions.
- A database of SAFT-γ Mie parameters for functional groups commonly occurring in organic and aqueous systems, including a range of (mainly inorganic) ions.
- A tool for the reliable[6] estimation of SAFT-γ Mie parameters from multiple data sets derived from a range of standardized experiments such as pure compound vapor pressures and saturated liquid densities, vapor–liquid or liquid–liquid phase equilibria for binary mixtures, solubility measurements in single solvents or solvent mixtures, titration curves, etc.

---

[4] It should be emphasized that this system may be different to the one under consideration, and may not even involve any solid phases. The required composition measurements can relate to *any* liquid phase irrespective of the presence or absence of any other phases; for example, they can be obtained from experiments involving a single liquid phase provided it is equilibrated with respect to the chemical reaction j of interest.

[5] See, for example, the application considered in Section 19.5.5.

[6] A particularly important consideration in this context is the determination of *globally* optimal solutions of the mathematical optimization problem underpinning the estimation of the SAFT parameters.

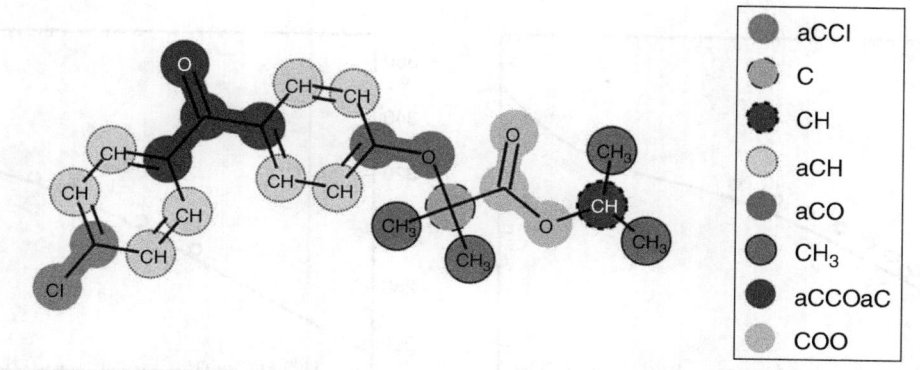

FIGURE 19.5 Molecular structure of fenofibrate (propan-2-yl 2-{4-[(4-chlorophenyl)carbonyl]phenoxy}-2-methylpropanoate) and its decomposition into functional groups in SAFT-γ Mie.

- A tool for the derivation of empirical models for the temperature dependence of quantities:

  - $\Delta g_i(T)$ and $\Delta g_i^{[s \to l]}(T)$ from experimental solubility measurements (cf. Sections 19.4.2 and 19.4.3).
  - $\Delta g_j^R(T)$ from experimental liquid-phase composition measurements (cf. Section 19.4.4).

In addition to SAFT-γ Mie, gSAFT also incorporates implementations of the SAFT-VR SW [29, 30] and PC-SAFT [31, 32] EoS and associated databanks. In principle, these EoS can also be used to perform solid–liquid equilibrium calculations within the framework described in this chapter (see, for example, [33–36]). However, they require parameters to be estimated specifically for each individual API, typically using experimental solubility data. This is in contrast to SAFT-γ Mie EoS, which, by virtue of its functional group basis, is capable of making use of experimental information from systems involving different compounds and/or phases. This advantage of SAFT-γ Mie EoS is particularly pronounced in systems involving complex molecules (such as pharmaceutical APIs) for which relatively little experimental data may be available.

19.5.2 Example 1: Solubility of Fenofibrate in Alcohols

The molecular structure of fenofibrate and its decomposition into functional groups is presented in Figure 19.5. The parameters for all the functional groups are already available within the gSAFT databank, having been estimated entirely from published experimental data derived from fluid-phase experiments; no solubility data of any kind were used for this purpose.

The solid-phase chemical potential is computed using the melting point approach described in Section 19.2.4.2, with the experimentally measured values of melting temperature and melting enthalpy reported in [37]. The specific heat capacity difference $\Delta c_{p,i}^{[s \to l]}(T)$ was set to zero.

The approach described in this chapter is now used to predict the solubility of fenofibrate in ethanol and 1-propanol.[7] A comparison between the predicted values and experimental data reported in the literature is shown in Figure 19.6. The solubility is presented as the liquid-phase molar fraction of fenofibrate, with a logarithmic scale being employed to allow a clearer inspection of the low solubility values. It can be seen that the predictions are in good agreement with the experimental data, the slight deviations being attributed to the fact that some of the unlike interactions between functional groups of the solute and the solvents are approximated by means of combining rules, and have not been regressed using experimental data.

In order to assess further the predictive capability of SAFT-γ Mie, we present in Figure 19.7 the computed solubilities in a binary mixture of water + ethanol. In general the computation of solubilities of APIs in a mixture of solvents is straightforward in gSAFT, the accuracy of the prediction depending on whether the majority of the group cross-interactions available in the gSAFT database have been regressed using experimental data, instead of being approximated by combining rules.

19.5.3 Example 2: Solubility of Ibuprofen in PEG Polymers

This example considers the solubility of ibuprofen in polyethylene glycol polymer of given molecular weight (MW = 10 000 g/mol), an application of relevance to the solubilization of APIs with poor aqueous solubility via the preparation of solid dispersions. The molecular structure of

---

[7] The functional group decomposition employed by gSAFT all compounds relevant to the examples of this section is presented in Appendix 19.B.

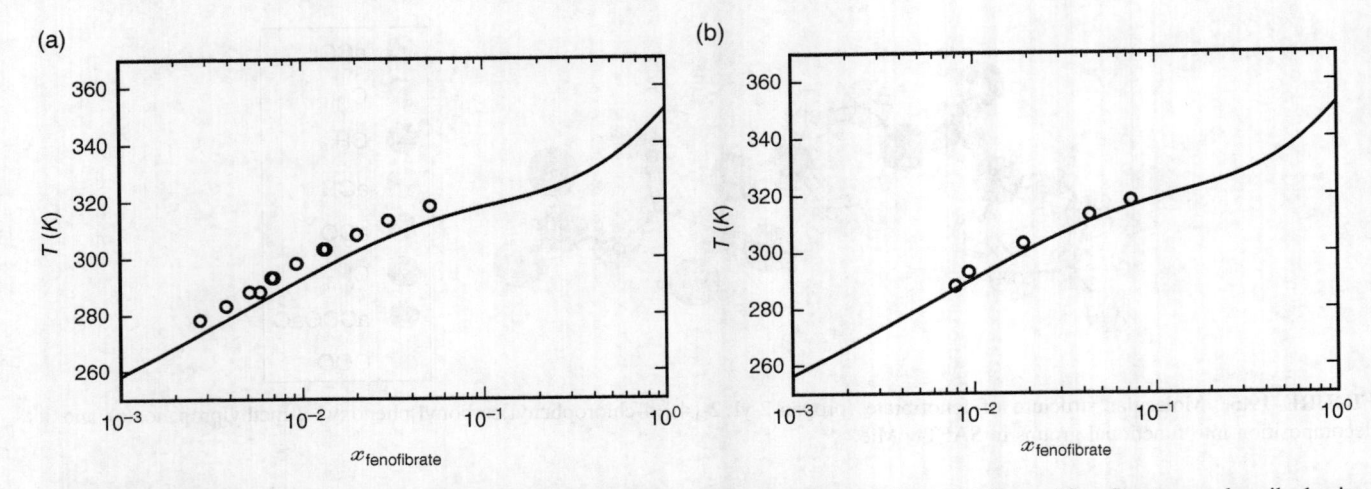

FIGURE 19.6 Mole fraction solubility of fenofibrate in (a) ethanol and (b) 1-propanol. The solid form of fenofibrate was described using melting properties (T_m = 352.05 K, $\Delta h^{[s \to l]}(T_m)$ = 33.53 kJ/mol [37]). The circles represent experimental data [37, 38] and the lines are pure predictions by gSAFT; no experimental solubility data of any kind were used in fitting model parameters. *Source*: Adapted from Watterson et al. [37] and Sun et al. [38].

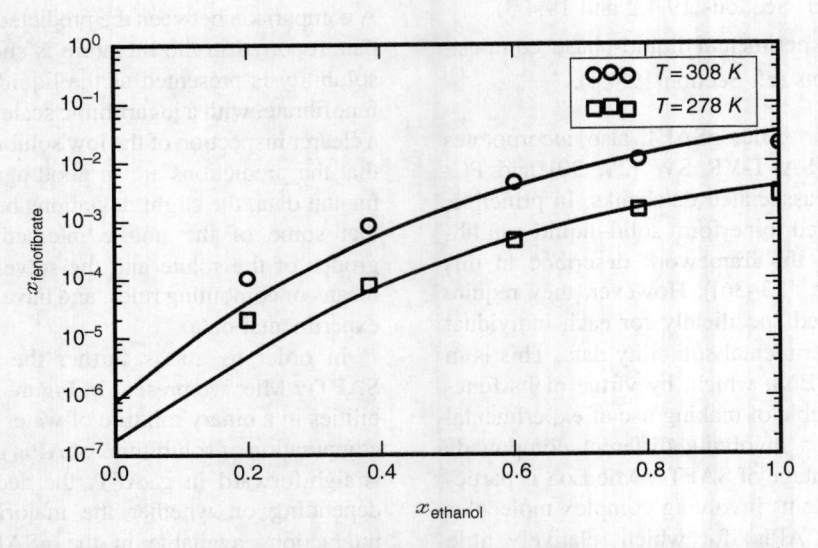

FIGURE 19.7 Mole fraction solubility of fenofibrate in a binary mixture of water + ethanol. The circles and squares represent experimental data [38] and the lines are pure predictions by gSAFT. *Source*: Reproduced with permission of American Chemical Society.

ibuprofen and its decomposition into functional groups in gSAFT are presented in Figure 19.8.

The functional group decomposition of the PEG polymer is shown in Figure 19.9. The group contribution basis of SAFT-γ Mie provides a natural mechanism for modeling complex polymers of arbitrary molecular weight. For example, PEG of molecular weight 10 000 g/mol is described as 2 hydroxyl groups combined with 454 methylene groups and 226 ether groups.

The melting point approach was used for the characterization of the solid phases of both ibuprofen and PEG, with the corresponding melting temperatures and enthalpies shown in Table 19.1. The specific heat capacity differences $\Delta c_{p,i}^{[s \to l]}(T)$ were set to zero.

Experimental data for this particular system [39] suggest the potential appearance of a co-crystal with a 5 : 1 ibuprofen-to-PEG ratio, reflecting the experimental composition at which the solid–liquid equilibrium temperature exhibits a maximum. Within our general framework for solid–liquid equilibria, the co-crystal is considered as a compound that appears only in a solid phase, dissociating to its constituents in the liquid phase:

$$IB_5 \cdot PEG\,(s) \leftrightarrow 5\,IB\,(l) + PEG\,(l)$$

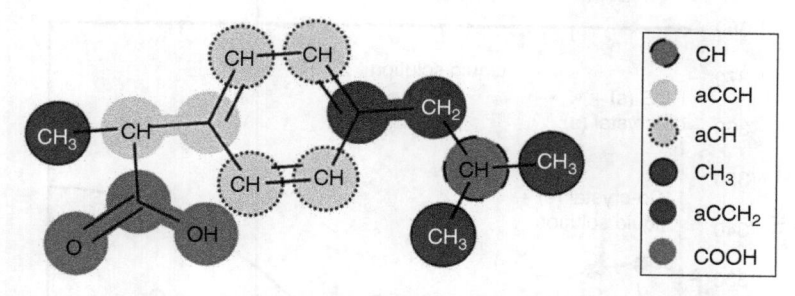

FIGURE 19.8 Molecular structure of ibuprofen (2-(4-(2-methylpropyl)phenyl)propanoic acid) and its decomposition into functional groups in SAFT-γ Mie.

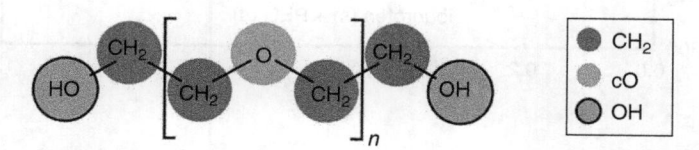

FIGURE 19.9 Molecular structure of PEG and its decomposition into functional groups in SAFT-γ Mie.

TABLE 19.1 Solid-Phase Characterization for Ibuprofen and PEG 10000 System

Compound	Solid-Phase Characterization
Ibuprofen	$T_m = 347.15$ K [23]
	$\Delta h^{[s \to l]}(T_m) = 25.50$ kJ/mol [23]
PEG 10000	$T_m = 331.05$ K [39]
	$\Delta h^{[s \to l]}(T_m) = 3360.0$ kJ/mol [39]
IB$_5$·PEG	$\Delta g(T) = -381.344 - 0.80695\, T$ (in kJ/mol; fitted)

Moreover, in view of the lack of the necessary solid-phase data, we characterize this compound using the approach described in Section 19.4.3 to derive an empirical representation for the function $\Delta g_i(T)$. In this case, we assume that the latter has a simple linear form obtained by drawing a straight line through two experimental solubility data points (shown by black circles in Figure 19.10). The resulting empirical model is also shown in Table 19.1.

Based on this information, we can now trace the entire solid–liquid phase diagram for this system over a range of temperatures T and ibuprofen mass fractions w. This is shown in Figure 19.10, where the lines represent calculations obtained using SAFT-γ Mie and the symbols the experimental solubility points. It can be seen that starting off with a liquid solution at a given temperature, a range of different solid forms can be obtained upon cooling depending on the initial composition.

19.5.4 Example 3: Solubility of Ketoprofen in PEG Polymers

This example examines the solubility of ketoprofen in PEG of molecular weight 6000 g/mol. The molecular structure and functional group decomposition of ketoprofen are shown in Figure 19.11.

Experimental data [40] suggest the potential formation of a co-crystal with a 10 : 1 ratio of ketoprofen to PEG. As in the previous example, the co-crystal is considered as a compound that appears only in a solid phase, dissociating to its constituents in the liquid phase:

$$KT_{10} \cdot PEG\,(s) \leftrightarrow 10\,KT\,(l) + PEG\,(l)$$

The approach used for the characterization of the solid-phase behavior of ketoprofen, PEG 6000, and the KT$_{10}$·PEG co-crystal is identical to that adopted for the ibuprofen/PEG 10000 system presented in the previous Section 19.5.3. The parameters used for the solid–liquid equilibrium calculations are summarized in Table 19.2.

A comparison between the predictions obtained using the SAFT-γ Mie approach and the experimental solubility data is presented in Figure 19.12. The latter illustrates both the complexity of the phase diagram and the accuracy of the proposed approach to the computation of solid–liquid equilibria.

19.5.5 Example 4: Aqueous Solubility of Ibuprofen as Function of pH

An application of great theoretical and practical interest is the study of the solubility of an ionizable API at different values of pH. Here we consider the aqueous solubility of ibuprofen in the presence of the pH-modifying agents HCl and NaOH at a constant temperature of 298.15 K.

Figure 19.13 shows the results of experimental investigations [42, 43] conducted on this system. As expected for a weakly acidic compound such as ibuprofen, an increase in the pH generally results in increased solubility. However, above a pH of about 9, the observed solubility remains

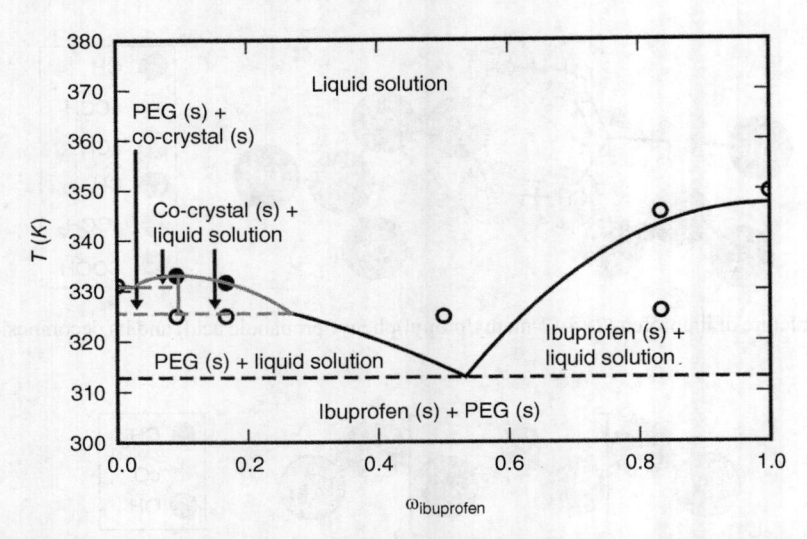

FIGURE 19.10 Solid–liquid phase diagram of ibuprofen + PEG 10000 g/mol assuming the formation of a co-crystal $IB_5 \cdot PEG$. Circles represent experimental data [39] and lines are gSAFT predictions. The latter made use of a linear empirical model of the co-crystal free energy function $\Delta g_i(T)$ (cf. Eq. 19.23) derived from only two experimental solubility data points shown as filled black circles.

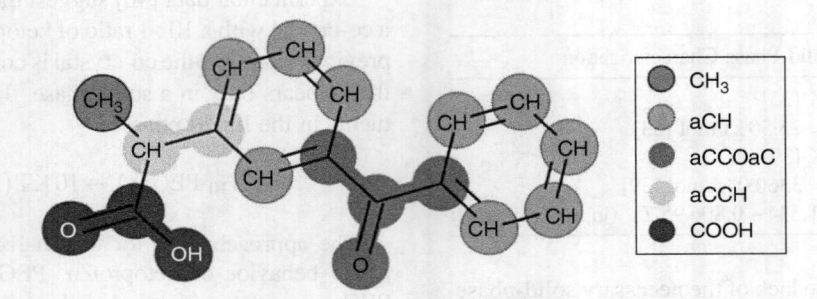

FIGURE 19.11 Molecular structure of ketoprofen (2-(3-benzoylphenyl)propanoic acid) and its decomposition into functional groups in SAFT-γ Mie.

TABLE 19.2 Solid-Phase Characterization for Ketoprofen and PEG 6000 System

Compound	Solid-Phase Characterization
Ketoprofen	$T_m = 368.00$ K [40]
	$\Delta h^{[s \to l]}(T_m) = 37.30$ kJ/mol [40]
PEG 6000	$T_m = 336.78$ K [41]
	$\Delta h^{[s \to l]}(T_m) = 1182.0$ kJ/mol [41]
$KT_{10} \cdot PEG$	$\Delta g(T) = -305.767 - 0.425399\, T$ (in kJ/mol; fitted)

essentially constant. This is because further addition of NaOH causes precipitation of ibuprofen in the form of sodium ibuprofen dihydrate [44].

19.5.5.1 System Modeling and Characterization

Both water and ibuprofen dissociate partially in the liquid phase according to the reactions

$$H_2O\,(l) \leftrightarrow H^+\,(aq) + OH^-\,(aq)$$

and

$$IB\,(l) \leftrightarrow H^+\,(aq) + IB^-\,(aq).$$

On the other hand, the pH-modifying agents are assumed to dissociate fully in the aqueous environment.

Ibuprofen may form a crystalline solid form. Sodium ibuprofen dihydrate is assumed to exist only in the solid phase, dissociating fully into its constituents in the liquid phase according to the reaction

$$NaIB \cdot (H_2O)_2\,(s) \leftrightarrow Na^+\,(aq) + IB^-\,(aq) + 2H_2O\,(l)$$

The functional group decomposition of ibuprofen employed by gSAFT has already been shown in Figure 19.8. The gSAFT databank also includes groups corresponding to the water molecule (see Appendix 19.B) and to the ionic species H^+, OH^-, Na^+, and Cl^-. The SAFT-γ Mie parameters for all these compounds were estimated from experimental data sets for simple inorganic systems.

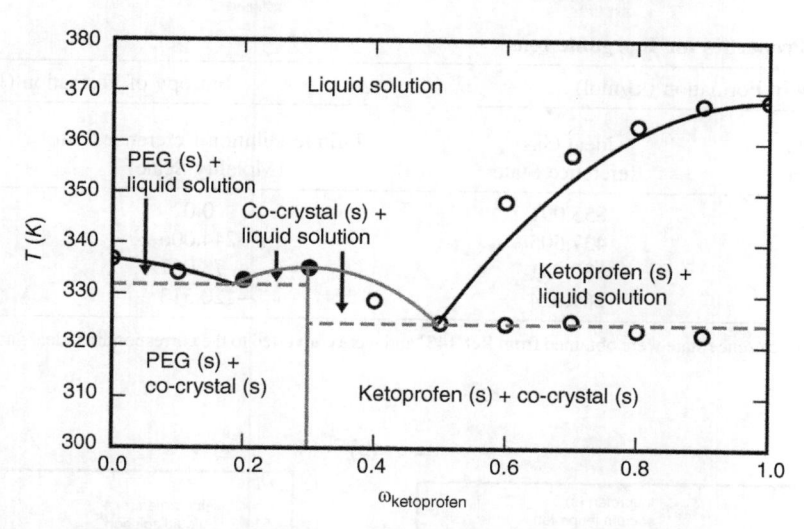

FIGURE 19.12 Solid–liquid phase diagram of ketoprofen + PEG 6000 g/mol assuming the formation of a co-crystal KT_{10}.PEG. Circles represent experimental data [41] and lines are gSAFT predictions. The latter made use of a linear empirical model of the co-crystal free energy function $\Delta g_i(T)$ (cf. Eq. 19.23) derived from the two experimental solubility data points shown as filled black circles. *Source*: Data from Margarit et al. [41].

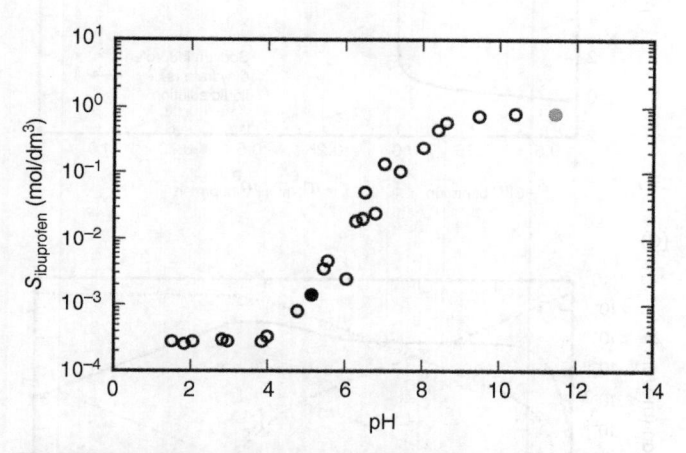

FIGURE 19.13 Experimentally determined [42, 43] effects of pH on the aqueous solubility of ibuprofen at 298.15 K. The pH modification was effected using aqueous solutions of HCl and NaOH. The two shaded points were used in our calculations for characterizing the free energies associated with the liquid-phase dissociation of ibuprofen (black point) and the solubility of sodium ibuprofen dihydrate (gray point); see text for more details. *Source*: Adapted from Avdeef et al. [42] and Shaw et al. [43].

The ideal gas formation enthalpy $\Delta h_i^{F,ig}$ and entropy $\Delta s_i^{F,ig}$ at standard temperature $T^\ominus$ and pressure $P^\ominus$ for water were obtained directly from Ref. [45]. The enthalpies $h_i^{F,\infty}$ and entropies $\tilde{s}_i^{F,\infty}$ of ion formation are tabulated in Ref. [45] for the infinite dilution reference state under the molality scale and were converted to the corresponding ideal gas formation properties via Eqs. (19.A.3b) and (19.A.4) in Appendix 19.A; the results are shown in Table 19.3.

The ibuprofen anion, IB^-, produced by the dissociation reaction of ibuprofen, is modeled using the same functional groups as for ibuprofen, simply replacing the carboxylic acid group (COOH) with a carboxylate group (COO^-), with the latter featuring a negative charge. As its enthalpy and entropy of formation are unknown, we employ the procedure described in Section 19.4.4 to estimate the free energy quantity Δg^R associated with the liquid-phase dissociation of ibuprofen using a single experimental data point, namely, the one shaded in black in Figure 19.13. In this context, we need to take account of the fact that the reported experimental ibuprofen solubility is actually the *combined* molar concentration of ibuprofen and ibuprofen ion and that the pH value is a measurement of the activity (and not the molarity) of the H^+ ion. The value of $\Delta g_{IB\ dissociation}^R (T = 298.15\,K)$ was estimated to be 29.231 kJ/mol.

The solid-phase characterization of crystalline pure ibuprofen has already been shown in Table 19.1. For the characterization of sodium ibuprofen dihydrate, we apply the procedure described in Section 19.4.3 for compounds that exist only in the solid phase. A single experimental data point, that is, the one shaded in gray in Figure 19.13, is used to estimate the free energy quantity $\Delta g_{NaIB \cdot (H_2O)_2}$, yielding a value of −698.18 kJ/mol.

19.5.5.2 Results Starting with a system containing water, an excess of ibuprofen, and a given quantity of the pH-modifying agent, the combined phase and reaction equilibrium calculation is performed at $T = 298.15\,K$ to determine the liquid-phase composition in terms of the molar fractions of water, ibuprofen, and the ions IB^-, H^+, OH^-, Na^+, and Cl^-. From these, other derived quantities can be computed, including the apparent solubility of ibuprofen (i.e. the combined molar concentrations of ibuprofen and ibuprofen

TABLE 19.3 Formation Properties for Inorganic Ions

Ion	Enthalpy of Formation (kJ/mol)		Entropy of Formation (J/mol·K)	
	Infinite Dilution Reference State	Ideal Gas Reference State	Infinite Dilution Reference State (Molality Scale)	Ideal Gas Reference State
H^+	0.0	855.091	0.0	421.584
OH^-	−229.994	437.505	−244.006	794.196
Na^+	−240.120	82.051	73.067	409.285
Cl^-	−167.159	269.024	−120.513	111.431

The values for the infinite dilution reference state were obtained from Ref. [45] and were converted to the corresponding ideal gas properties via Eqs. (19.A.4) and (19.A.3b).

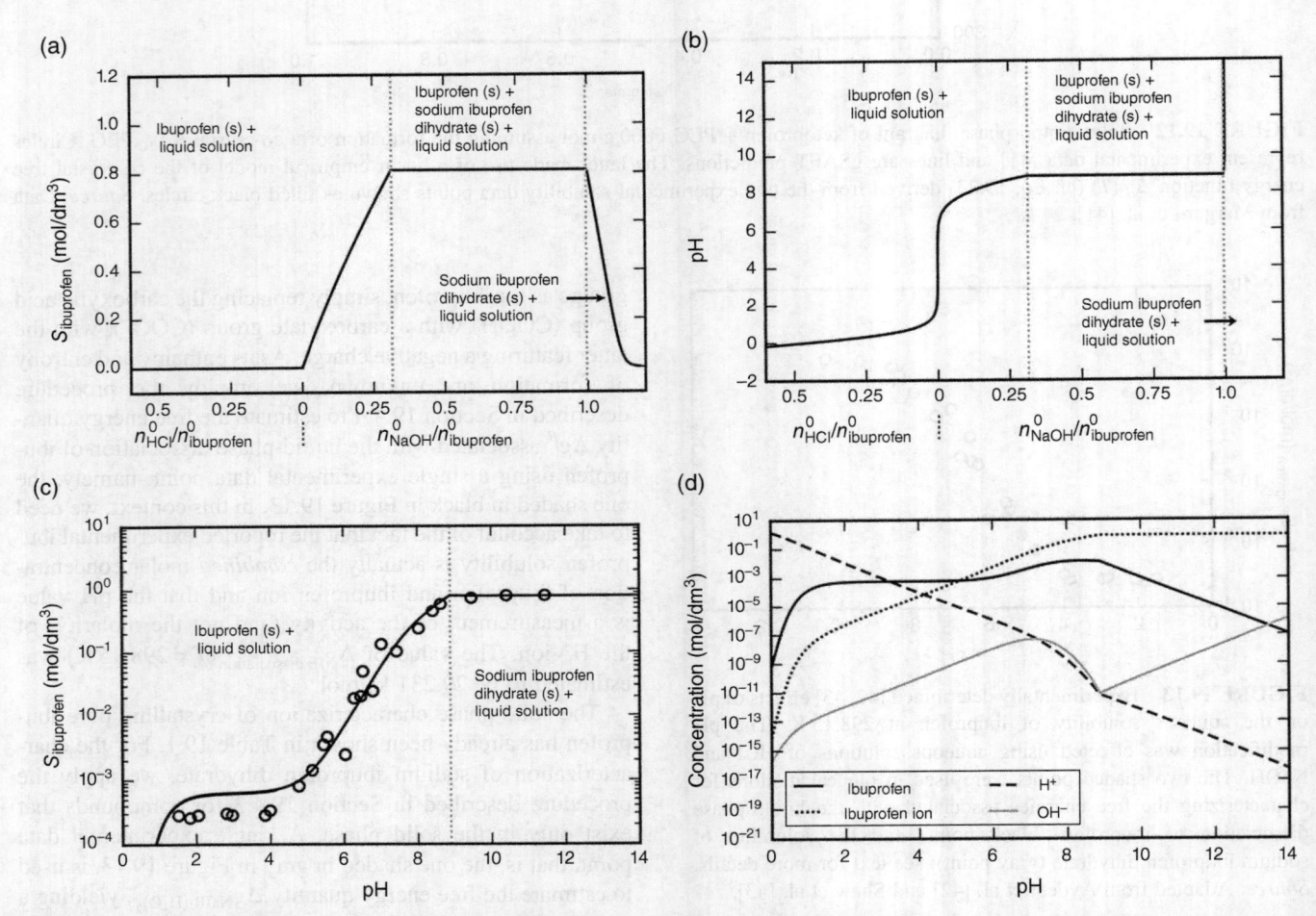

FIGURE 19.14 Solid–liquid phase equilibria of ibuprofen in aqueous solutions at 298.15 K, 1 atm using either NaOH or HCl as pH-modifying agents. (a) Solubility as a function of the initial amount of pH-modifying agent added; (b) pH of the saturated solution as a function of the initial amount of pH-modifying agent added; (c) solubility as a function of pH; (d) molar concentrations of H^+, OH^-, ibuprofen, and ibuprofen ion as functions of pH. Vertical dotted gray lines delineate the regions of precipitation of different solid forms. Symbols represent experimental data [42, 43] and lines SAFT-γ Mie predictions. *Source*: Adapted from Avdeef et al. [42] and Shaw et al. [43].

anion) and the H^+ activity (and hence the pH) illustrated in Figure 19.14a and b, respectively.

As can be seen from the left parts of Figure 19.14a and b, the solubility of ibuprofen in acidic environments, induced by the addition of HCl, is very low. On the other hand, the addition of NaOH causes a significant increase in the solubility of ibuprofen until solid sodium ibuprofen dihydrate starts forming. Thereafter, both crystalline molecular ibuprofen and sodium ibuprofen dihydrate coexist with the liquid solution. The solubility remains constant until the initial amount of ibuprofen has been fully converted to ions $\left(n^o_{\text{NaOH}} \approx n^o_{\text{Ibuprofen}}\right)$, after which point the solubility decreases as expected for a basic salt in a basic solution. The pH increases upon initial addition of NaOH, remains constant while the two solid phases coexist, and then rises sharply.

Combining the results of Figure 19.14a and b, the solubility of ibuprofen can be plotted against pH, as shown in Figure 19.14c, which employs a logarithmic solubility scale in order to allow the very low solubility values to be seen more clearly. We note that in this figure, the region of coexistence of the two solids is reduced to a single point corresponding to the constant solubility and constant pH regions in Figure 19.14a and b, respectively. In the acidic region, ibuprofen solubility remains constant until the pH drops below 1, beyond which point it decreases rapidly.

Figure 19.14c also shows the experimental data of Figure 19.13 for comparison. Agreement between predictions and experiments is generally good despite the fact that, as described in Section 19.5.5.1, only two experimental data points were used to characterize this system.

Figure 19.14d shows the concentrations of ibuprofen, ibuprofen ion, H^+, and OH^- as functions of pH along the solubility boundary. It can be seen that the phase and reaction equilibrium algorithms implemented in gSAFT are capable of resolving reliably very small ionic concentrations across the entire pH spectrum.

Finally, it is common practice in the pharmaceutical industry to describe the behavior of weakly dissociating acids in aqueous environments in terms of pK_a constants. The values of the latter are published in the literature and are used widely for the calculation of reaction equilibria in important applications such as the determination of bioavailability in oral absorption [1]. In the specific case of ibuprofen, pK_a corresponds to the quantity $-\log_{10}(C_{H^+} C_{IB^-})$, with the molar concentrations C_i being expressed in mol/dm^3. Figure 19.15 shows the variation of this quantity over the pH range based on C_i values computed from the liquid-phase compositions established by our solid–liquid phase equilibrium calculations. For comparison, the dashed line in Figure 19.15 indicates a published value [25] of the pK_a. It is clear that, especially in the alkaline region, the quantity $-\log_{10}(C_{H^+} C_{IB^-})$ is neither constant nor even approximately equal to the published pK_a values of around 5; in fact, the latter may overestimate the value of the concentration product $C_{H^+} C_{IB^-}$ by 3–4 orders of magnitude.

19.5.6 Example 5: Polymorphic Transition and Hydrate Formation of Caffeine in Water

Caffeine (CFN) is known to exhibit a polymorphic transition between two different crystalline forms (here denoted as forms I and II). Moreover, a separate solid phase consisting of a hydrate CFN$_5 \cdot$(H$_2$O)$_4$ with a ratio of caffeine to water of

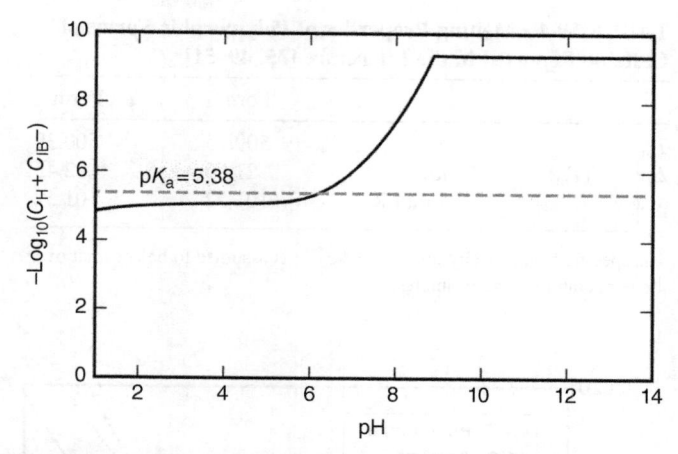

FIGURE 19.15 Apparent pK_a of ibuprofen at different pH values. The dashed line corresponds to the constant pK_a of 5.38 value reported in literature [25]. The solid line shows the value of the quantity $-\log_{10}(C_{H^+} C_{IB^-})$ using molar concentrations (in mol/dm^3) computed from the liquid-phase compositions determined by our solid–liquid equilibrium calculations. *Source*: Reproduced with permission of American Chemical Society.

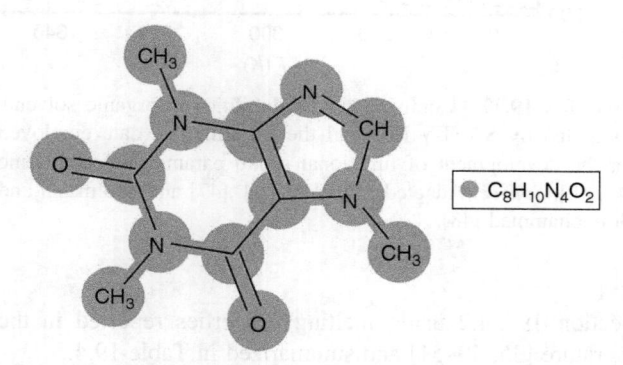

FIGURE 19.16 Molecular structure of caffeine (1,3,7-trimethylpurine-2,6-dione). In SAFT-γ Mie, the entire molecule is modeled as a single functional group.

5 : 4 may also form [45, 46]. These factors lead to a complex phase diagram describing the behavior of caffeine in aqueous solutions.

19.5.6.1 System Modeling and Characterization The molecular structure of caffeine is shown in Figure 19.16. For the purposes of describing the liquid-phase behavior of this compound in the SAFT-γ Mie framework, we choose to represent the entire molecule as a single functional group.

The necessary SAFT parameters were estimated from published solubility data [47, 48] for caffeine in simple organic solvents. This type of parameter estimation also requires a description of the solid-phase behavior of caffeine. This was based on the melting point approach described in

TABLE 19.4 Melting Properties of Polymorphic Forms of Caffeine Reported in the Literature [35, 49–51]

		Form I	Form II
T_m	K	509.15	500.39
$\Delta h^{[s \to l]}(T_m)$	kJ/mol	21.60	22.45
$\Delta c_p^{[s \to l]}$	J/mol·K	101.52	101.52

The specific heat capacity difference $\Delta c_p^{[s \to l]}$ is assumed to be constant over the temperature range of interest.

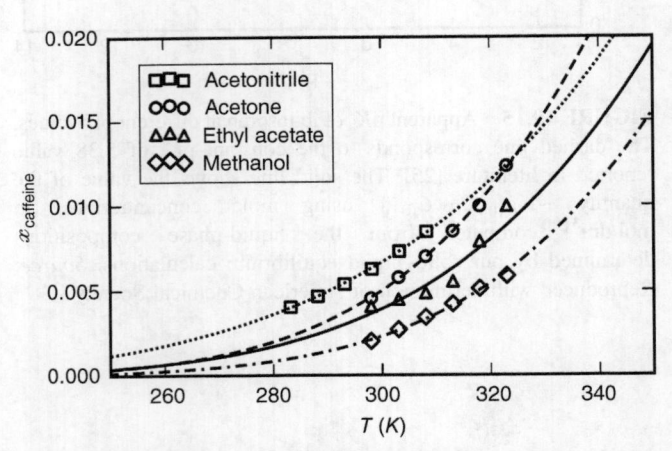

FIGURE 19.17 Comparison of solubilities in organic solvents computed by SAFT-γ Mie and the experimental data employed for the development of functional group parameters for caffeine [47, 48]. *Source*: Adapted from Yu et al. [47] and Shalmashi and Golmohammad [48].

Section 19.2.4.2 using melting properties reported in the literature [35, 49–51] and summarized in Table 19.4.

Figure 19.17 compares the solubilities computed by gSAFT using the estimated SAFT-γ Mie parameters with the experimental solubility data used for the estimation. As can be seen, a single set of SAFT-γ Mie parameters leads to a good fit for all four organic solvents, which provides a degree of confidence in employing this molecular model for describing the behavior of caffeine in other systems, such as the aqueous one of interest to this example.

The standard SAFT-γ Mie representation of the water molecule as a single functional group with two electron donor and two electron acceptor association sites was employed. The SAFT-γ Mie parameters describing the cross-interactions between the caffeine and water functional groups were estimated from the experimental solubility data for the high temperature polymorph (form I) shown in Figure 19.18a.

Since caffeine is a non-ionizable compound, no liquid-phase reactions need to be considered in this system.

The caffeine hydrate $CFN_5 \cdot (H_2O)_4$ is assumed to exist only in the solid phase, dissociating into its constituents in the liquid phase:

$$CFN_5 \cdot (H_2O)_4 \, (s) \leftrightarrow 5\,CFN\,(l) + 4\,H_2O\,(l)$$

The solid-phase free energy of the hydrate was characterized via the procedure described in Section 19.4.3 for compounds that exist only in the solid phase. Two experimental data points (shown as filled black circles in Figure 19.18b) were used to determine the value of the free energy quantity $\Delta g_{CFN_5 \cdot (H_2O)_4}$ at two different temperatures, which allows the derivation of the empirical model:
$\Delta g_{CFN_5 \cdot (H_2O)_4}(T) = -1777.229 - 0.16185\,T$ (in kJ/mol).

19.5.6.2 Calculation of Solid–Liquid Phase Diagram

Figure 19.18 shows the solid–liquid phase diagram for caffeine in water computed by gSAFT and compares it with experimental data reported in the literature [35, 52].

Overall, good agreement is observed across the entire phase diagram involving five distinct regions delineated by six phase boundaries. It is worth noting that only a relatively small subset of the experimental data points shown here were used for the characterization of the system (cf. Section 19.5.6.1). All other results shown represent predictions.

Figure 19.18b employs a logarithmic scale for the liquid-phase molar fraction of caffeine in order to show the low solubility region more clearly. Again, the computed solubilities are in good agreement with the experimental data.

19.5.7 Example 6: Co-Crystal Solubility of Nicotinamide/Succinic Acid in Mixed Solvents

The system of nicotinamide and succinic acid is an example of the general category of API + coformer systems, where alongside the crystalline forms of the two pure components, a co-crystal can be formed. In this specific case, the co-crystal, denoted as $NIC_2 \cdot SUCC$, involves a ratio of nicotinamide to succinic acid of 2:1.

The study presented here focuses on the solid–liquid phase diagram of nicotinamide and succinic acid in a 50:50 (by weight) mixture of ethanol and water at 298.15 K.

19.5.7.1 System Modeling and Characterization
The molecular structures of nicotinamide and succinic acid are shown in Figure 19.19.

As in the previous example (cf. Section 19.5.6.1), each solute molecule is modeled as a single functional group in the SAFT-γ Mie framework, and the corresponding

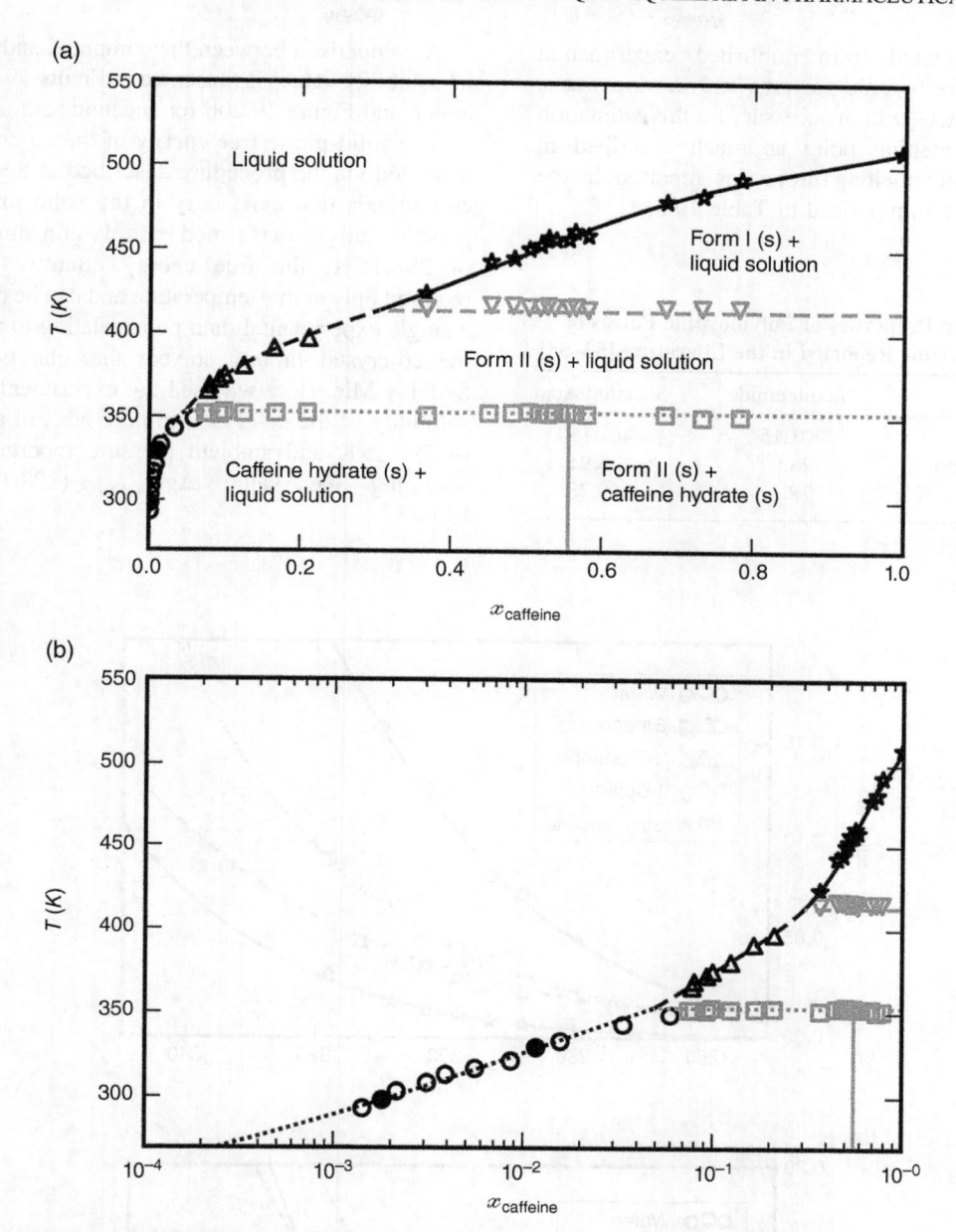

FIGURE 19.18 Solid–liquid phase diagram of caffeine with water. Symbols represent experimental data [35, 52] and lines results obtained with gSAFT. (a) Overall phase diagram comprising five distinct regions and six phase boundaries. The Form I solubility data shown as stars were used for fitting the SAFT-γ Mie unlike interaction parameters between water and caffeine. (b) Low solubility region shown using logarithmic molar fraction scale. The two experimental data points used for the derivation of the empirical free energy function $\Delta g_{CFN_5 \cdot (H_2O)_4}(T)$ are shown as filled black circles. *Source*: Adapted from Lange et al. [35] and Suzuki et al. [52].

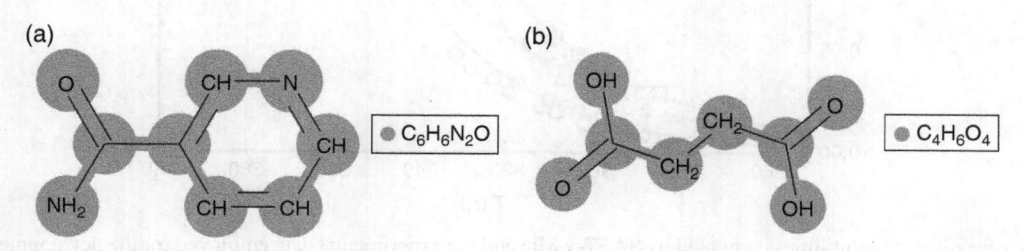

FIGURE 19.19 Molecular structure of (a) nicotinamide (pyridine-3-carboxamide) and (b) succinic acid (1,4-butanedioic acid). In SAFT-γ Mie, each of these molecules is modeled as a single functional group.

parameters are estimated from published experimental solubility data in pure organic solvents and in pure water. The solid-phase characterization necessary for this estimation was based on the melting point approach described in Section 19.2.4.2 using melting properties reported in the literature [53–55] and summarized in Table 19.5.

TABLE 19.5 Melting Properties of Polymorphic Forms of Nicotinamide and Succinic Reported in the Literature [53–55]

		Nicotinamide	Succinic Acid
T_m	K	401.15	461.15
$\Delta h^{[s \to l]}(T_m)$	kJ/mol	28.0	38.91
$\Delta c_p^{[s \to l]}$	J/mol·K	78.12	69.78

A comparison between the computed and the experimental solubility values is presented in Figure 19.20a for nicotinamide and Figure 19.20b for succinic acid.

The solid-phase free energy of the co-crystal was characterized via the procedure described in Section 19.4.3 for compounds that exist only in the solid phase. Since our present study is performed entirely at a single temperature of 298.15 K, the free energy quantity $\Delta g_{NIC_2 \cdot SUCC}$ is required only at this temperature and can be determined from a single experimental data point relating to the solubility of the co-crystal in *any* solvent that can be described by SAFT-γ Mie. Here we used the experimentally determined solubility of the co-crystal in pure ethanol at a temperature of 298.15 K and ambient pressure reported in Ref. [46], obtaining the value $\Delta g_{NIC_2 \cdot SUCC}(298.15\,K) = -36.404$ kJ/mol.

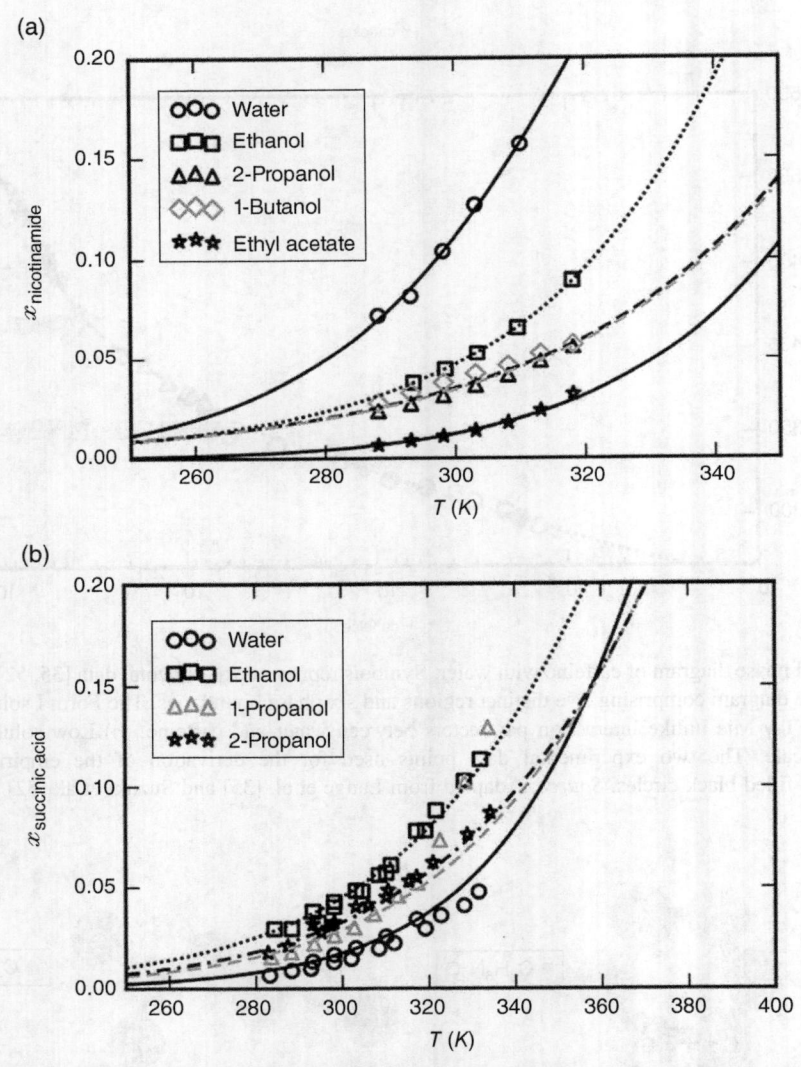

FIGURE 19.20 Comparison of solubilities computed by SAFT-γ Mie and the experimental data employed for the development of functional group parameters for (a) nicotinamide [36, 56] and (b) succinic acid [36, 57, 58].

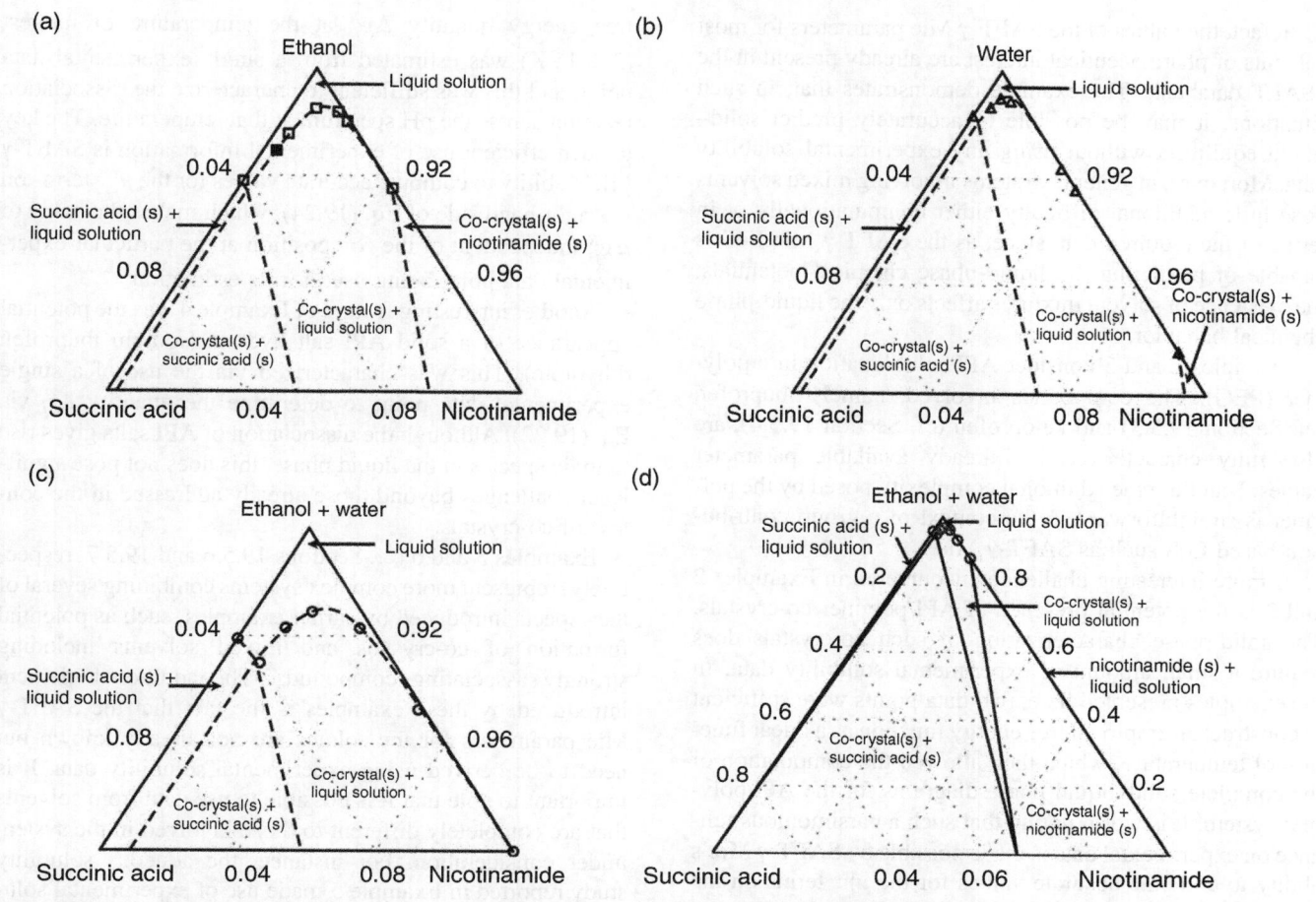

FIGURE 19.21 Solid–liquid phase diagrams at 298.15 K and 1 atm of nicotinamide. Succinic acid and their (2 : 1) co-crystal in (a) pure ethanol, (b) pure water, and (c), (d) a 50 : 50 (by weight) mixture of ethanol and water. Symbols represent experimental data [46] and lines results obtained with gSAFT. The single experimental data point used for the estimation of the empirical free energy $\Delta g_{\text{NIC}_2\cdot\text{SUCC}}(298.15K)$ is shown as a filled black square in (a). *Source*: Lange et al. [46]. http://www.mdpi.com/1420-3049/21/5/593. Licensed under CC BY 3.0.

No liquid-phase reactions were taken into account in this study.

19.5.7.2 Calculation of Solid–Liquid Phase Diagrams

Figure 19.21 shows solid–liquid phase diagrams of nicotinamide and succinic acid in pure ethanol, pure water, and a 50 : 50 (by weight) mixture of water with ethanol, all computed by gSAFT at a temperature of 298.15 K. Experimental data reported in Ref. [46] are also shown on these diagrams. It is noted that only the single data point shown as a filled black square in Figure 19.21a was used for the characterization of this system (cf. Section 19.5.7.1); all other gSAFT results reported represent predictions.

Overall, there is good agreement between gSAFT predictions and experimental data. In particular, the results in Figure 19.21c confirm the co-crystal's significantly increased solubility in the mixed solvent compared to that in pure water or pure ethanol.

19.6 DISCUSSION

The six applications considered in Sections 19.5.2–19.5.7 were selected to illustrate the key aspects of the general framework and methodology for solid–liquid equilibria presented in this chapter. In this section, we revisit each one of these applications and attempt to identify the more general aspects relating to computation of solid–liquid equilibria.

The main compound (fenofibrate) in Example 1 (cf. Section 19.5.2) was fully characterized in terms of already known values of all necessary parameters:

- The SAFT-γ Mie parameters for its functional groups had already been estimated using fluid-phase experimental data for unrelated compounds.
- The melting point properties had already been reported in the literature.

In fact, the values of the SAFT-γ Mie parameters for most solvents of pharmaceutical interest are already present in the gSAFT databank. The example demonstrates that, in such situations, it may be possible to accurately predict solid–liquid equilibria without using any experimental solubility data. Moreover, in general, systems involving mixed solvents pose little additional difficulty either computationally or in terms of the required data since, as the SAFT-γ Mie EoS is capable of predicting the liquid-phase chemical potentials, the existence of solvent mixtures affects only the liquid-phase chemical behavior.

Examples 2 and 3 consider API solubilization in a polymer (PEG). The compounds involved, namely ibuprofen (cf. Section 19.5.3) and ketoprofen (cf. Section 19.5.4), are also fully characterized by already available parameter values. Handling the additional complexity posed by the polymer is straightforward in the context of a group contribution-based EoS such as SAFT-γ Mie.

A more interesting challenge encountered in Examples 2 and 3 is the potential presence of API/polymer co-crystals. The solid-phase characterization of such co-crystals does require a small amount of experimental solubility data. In the examples presented here, two data points were sufficient to construct an empirical free energy function as a linear function of temperature, which then allowed the computation of the complete solid–liquid phase diagrams for the API/polymer system. It is worth noting that such a parsimonious reliance on experimental data is made possible by SAFT-γ Mie's ability to compute accurate values for the μ_i^{res} terms on the right-hand side of Eq. (19.22); this then ensures that the quantity Δg_i computed by Eq. (19.22) is composition independent despite the use of experimentally determined compositions for its evaluation.

Example 4 (cf. Section 19.5.5) involved an aqueous system, focusing on API solubility across the entire pH range from strongly acid to strongly alkaline environments. The nonideal behavior of water itself can already be handled well by the ability of SAFT-γ Mie to describe strongly associating compounds. Also important in this context is SAFT-γ Mie's handling of ionic species, both on the theoretical level and in terms of the ability to accurately convert formation enthalpies and entropies reported in the literature from one reference state to another. The SAFT-γ Mie parameterization for the API ion was derived directly from the corresponding parameters of the API itself.

A more significant challenge in such systems is the characterization of the liquid-phase dissociation of the API. As discussed in Section 19.5.5.2, the common practice of relying on pK_a "constants" may be unreliable as the corresponding product of API ion and H$^+$ concentrations is far from constant in at least some parts of the pH spectrum. Instead, the dissociation reaction may be characterized from a small amount of experimental measurements. In this example, the required free energy quantity Δg_j^R at the temperature of interest (298.15 K) was estimated from a single experimental data point, and this was sufficient to characterize the dissociation reaction across the pH spectrum at that temperature. The key to such efficient use of experimental information is SAFT-γ Mie's ability to compute accurate values for the μ_i^{res} terms on the right-hand side of Eq. (19.24), which makes the value of Δg_j^R independent of the composition at the particular experimental data point being used for its estimation.

Another interesting feature of Example 4 was the potential appearance of a solid API salt hydrate (sodium ibuprofen dihydrate). This was characterized via the use of a single experimental data point to determine the quantity Δg_i via Eq. (19.22). Although the dissociation of API salts gives rise to ionic species in the liquid phase, this does not pose significant challenges beyond those already addressed in the context of co-crystals.

Examples 5 and 6 (cf. Sections 19.5.6 and 19.5.7, respectively) represent more complex systems combining several of the aspects introduced by earlier examples, such as potential formation of co-crystals and mixed solvents including strongly associating compounds. The additional element introduced by these examples is the fact that the SAFT-γ Mie parameters for the solutes are not already known but need to be derived using experimental solubility data. It is important to note that it is possible to use data from solvents that are completely different to those involved in the system under consideration. For instance, the aqueous solubility study reported in Example 5 made use of experimental solubility data in organic solvents. The accuracy of free energy calculations by SAFT-γ Mie is key to ensuring reliable transferability of information between different systems.

An interesting feature of Example 5 is that, in addition to a hydrate form, there exist two different "neat" polymorphs of the solute. Form II is stable up to the transition temperature of 414.15 K [52], above which Form I becomes stable up to its melting temperature of 509.15 K. In this example, both polymorphs were characterized using melting properties quoted in Ref. [29] that, from the experimental point of view, would imply that the low temperature polymorph, Form II, can be heated up to its melting point without undergoing a solid-state polymorphic transition to Form I. For systems in which such polymorphic transitions cannot be avoided, the solid-phase behavior of some polymorphs will need to be characterized using alternative methods, such as those making use of experimental solubility data as described in Section 19.4.2. It is worth noting that these data do not need to relate to the specific solvent(s) for the system under consideration; Eq. (19.17) coupled with SAFT-γ Mie for the computation of the μ_i^{res} term allows an empirical form of the free energy function $\Delta g_i(T)$ to be determined from available solubility data for that particular polymorph in other solvents.

From a computational perspective, the examples demonstrate the ability of gSAFT to determine complex solid–liquid phase diagrams (cf. Figures 19.10, 19.12, 19.18, and 19.21) and also to handle very small concentrations such as those corresponding to ionic species (cf. Figure 19.14d). This is primarily a result of the reliable numerical solution of the phase equilibrium conditions (cf. Section 19.2.2) and the application of the correct stability criteria (cf. Section 19.2.5).

19.7 CONCLUSIONS

This chapter has presented a general framework for solid–liquid equilibrium calculations that covers many of the systems of practical interest to the pharmaceutical industry including those involving diverse types of:

- Solid forms such as simple APIs, API salts, solvates, hydrates, and co-crystals, as well as multiple polymorphs of these solid forms.
- Solvents, including both water and organic molecules and their mixtures, and polymers.

The framework can describe important aspects such as the effects of pH on solubility, the partial ionization of weakly acidic or basic APIs, and the simultaneous existence of multiple solid and/or liquid phases.

By adopting an approach based on fundamental thermodynamics, all of the above can be handled in a unified manner without the need for diverse concepts such as activity coefficients with respect to multiple reference states, solubility products, acid or base dissociation "constants," etc. The only thermodynamic properties required are the liquid- and solid-phase chemical potentials of the compounds that potentially exist in the system. In this context, an important development over the last decade has been the emergence of EoS, such as SAFT-γ Mie, which can describe the fluid-phase behavior of complex mixtures by taking account of different types of intermolecular interactions. On the other hand, solid-phase potentials can be evaluated using rigorous thermodynamic expressions involving either the enthalpy and entropy of formation of solid forms or their melting properties.

As demonstrated by the examples presented in Section 19.5, it is now possible to combine the various theoretical elements mentioned above with reliable numerical solvers for phase equilibrium and tests for phase stability, to produce computer codes for the accurate computation of entire solid–liquid phase diagrams for complex systems of pharmaceutical interest.

A key objective of any methodology for the prediction of solid–liquid equilibria is the minimization of the amount of new experimental data that are necessary for characterizing a particular system of interest. Because of its group contribution basis, SAFT-γ Mie can use information derived from already existing experimental data to be used for the accurate prediction of liquid-phase behavior, even if the system under consideration involves a completely different set of compounds to those used in the original experiments. On a practical level, the SAFT-γ Mie parameter databank in gSAFT currently (January 2019) includes parameters for the functional groups describing almost all FDA-approved solvents [59] for API manufacturing and an increasing number of functional groups occurring in APIs. With the continual addition of more functional groups to this databank, the approach described here becomes applicable to more and more APIs without the need for experimental data for developing descriptions of their liquid-phase behavior.

Unfortunately, at present there is no general theoretical approach that will allow similarly accurate cross-compound transferability of information for solid phases. However, as explained in Section 19.4, by combining small amounts of experimental solubility data for the solid forms of interest with SAFT-γ Mie liquid-phase predictions, it is possible to predict solid–liquid equilibria for these forms over wide ranges of conditions and/or in conjunction with different solvents. The practical examples presented in Section 19.5 provide some evidence for the applicability and effectiveness of this approach.

19.A USING FORMATION PROPERTIES IN THE INFINITE DILUTION STATE

Ionic compounds in aqueous systems cannot exist in pure form. Accordingly, from the perspective of relating thermodynamic theory to experimentally observable quantities, it is preferable to consider their properties with respect to an "infinite dilution" reference state corresponding to the limit of their presence tending to zero (or, equivalently, the molar fraction of water, x_w, tending to 1) in an aqueous solution at standard $T^\ominus$, $P^\ominus$.

The chemical potential of ionic compound i generally depends on the logarithm of the corresponding molar fraction x_i (cf. the term $+RT\ln(x_i)$ in Eq. 19.9). This dependence needs to be removed in order to ensure that the partial molar Gibbs free energy g_i^∞ at the reference state has a finite value. Historically, three different approaches have been proposed to achieve this, each making use of a different measure of ionic concentration:

- Molar fraction-based reference state:

$$\hat{g}_i^\infty \equiv \lim_{x_w \to 1} \left(\mu_i^{[l]}(T^\ominus, P^\ominus, \boldsymbol{n}) - RT\ln x_i \right) \quad (19.A.1a)$$

- Molality-based reference state:

$$\tilde{g}_i^\infty \equiv \lim_{x_w \to 1} \left(\mu_i^{[l]}(T^\ominus, P^\ominus, \boldsymbol{n}) - RT\ln \frac{m_i}{m_0} \right) \quad (19.A.1b)$$

where $m_i \equiv x_i/x_w M_w$ denotes the molality, with M_w being the molecular weight of water and $m_0 = 1$ mol/kg.

- Molarity-based reference state:

$$\bar{g}_i^\infty \equiv \lim_{x_w \to 1}\left(\mu_i^{[l]}(T^\ominus,P^\ominus,\boldsymbol{n}) - RT\ln\frac{C_i}{C_0}\right) \quad (19.A.1c)$$

where $C_i \equiv x_i\rho_w/M_w$ denotes the molarity (molar concentration), with ρ_w being the mass density (in kg/dm^3) of pure water at $T^\ominus$, $P^\ominus$ and $C_0 = 1$ mol/dm^3.

The free energy $g_i^{F,\infty}$ and enthalpy $h_i^{F,\infty}$ of formation in the infinite dilution reference state that are typically reported in the literature are related to s_i^∞ via the standard thermodynamic relations:

$$s_i^{F,\infty} = \frac{h_i^{F,\infty} - g_i^{F,\infty}}{T^\ominus} \quad (19.A.2)$$

It can be shown that, while each of the three infinite dilution reference states leads to a different value for $s_i^{F,\infty}$, they all result in the same value for $h_i^{F,\infty}$.

Using Eqs. (19.8) and (19.9) for the chemical potential $\mu_i^{[l]}(T^\ominus,P^\ominus,\boldsymbol{n})$ appearing in Eqs. (19.A.1a)–(19.A.1c), we can derive the following relations between the ideal gas formation entropy and the formation entropies in the various infinite dilution reference states:

- Molar fraction-based reference state:

$$\Delta s_i^{F,ig} = \hat{s}_i^{F,\infty} + \frac{\partial}{\partial T}\mu_i^{res}(T^\ominus,P^\ominus,\boldsymbol{n}^\infty) \quad (19.A.3a)$$

- Molality-based reference state:

$$\Delta s_i^{F,ig} = \tilde{s}_i^{F,\infty} + \frac{\partial}{\partial T}\mu_i^{res}(T^\ominus,P^\ominus,\boldsymbol{n}^\infty) + R\ln m_0 M_w \quad (19.A.3b)$$

- Molarity-based reference state:

$$\Delta s_i^{F,ig} = \bar{s}_i^{F,\infty} + \frac{\partial}{\partial T}\mu_i^{res}(T^\ominus,P^\ominus,\boldsymbol{n}^\infty) + R\ln C_0 M_w/\rho_w \quad (19.A.3c)$$

where $\mu_i^{res}(T^\ominus,P^\ominus,\boldsymbol{n}^\infty)$ is the residual chemical potential of compound i in an infinitely diluted composition. In practical terms, this can be computed by applying an EoS such as SAFT-γ Mie to a binary mixture comprising water and compound i and setting the corresponding molar fractions to $x_w = 1$ and $x_i = 0$, respectively.

In all three cases, the ideal gas enthalpy of formation $\Delta h_i^{F,ig}$ can be obtained from the corresponding infinite dilution values $h_i^{F,\infty}$ via the equation

$$\Delta h_i^{F,ig} = h_i^{F,\infty} - \mu_i^{res}(T^\ominus,P^\ominus,\boldsymbol{n}^\infty) + T^\ominus\frac{\partial}{\partial T}\mu_i^{res}(T^\ominus,P^\ominus,\boldsymbol{n}^\infty) \quad (19.A.4)$$

In summary, Eqs. (19.A.3a)–(19.A.3c) and (19.A.4), coupled with an EoS to compute the residual term $\mu_i^{res}(T^\ominus,P^\ominus,\boldsymbol{n}^\infty)$, allow the determination of the ideal gas formation properties $\Delta h_i^{F,ig}$ and $\Delta s_i^{F,ig}$ from experimentally derived values $h_i^{F,\infty}$ and $s_i^{F,\infty}$ in any one of the commonly used infinite dilution reference states.

19.B SAFT-γ MIE FUNCTIONAL GROUP DECOMPOSITION

The functional group decomposition of the compounds involved in the examples of Section 19.5 within the SAFT-γ Mie framework is shown in the table below.

Species	Functional Group Decomposition														
1-Butanol	1	CH_3	3	CH_2	1	OH									
1-Propanol	1	CH_3	2	CH_2	1	OH									
2-Propanol	2	CH_3	1	CH	1	OH_Sec									
Ethanol	1	CH_3	1	CH_2	1	OH									
Ethyl acetate	2	CH_3	1	CH_2	1	COO									
Toluene	1	$aCCH_3$	5	aCH											
Acetone	1	CH_3COCH_3													
Acetonitrile	1	CH_3CN													
Cyclohexane	6	$cyCH_2$													
Methanol	1	CH_3OH													
Water	1	H_2O													
Caffeine	1	$C_8H_{10}N_4O_2$													
Fenofibrate	1	aCCOaC	8	aCH	1	aCCl	1	aCO	1	COOH	1	COO	4	CH_3	1	CH
Ibuprofen	3	CH_3	1	CH	1	$aCCH_2$	1	aCCH	4	aCH	1	COOH			
Ibuprofen ion	3	CH_3	1	CH	1	$aCCH_2$	1	aCCH	4	aCH	1	COO–			
Ketoprofen	1	aCCOaC	1	CH_3	1	aCCH	9	aCH	1	COOH					
Nicotinamide	1	$C_6H_6N_2O$													
PEG 10000	454	CH_2	226	cO	2	OH									
PEG 6000	272	CH_2	135	cO	2	OH									
Succinic acid	1	$C_4H_6O_4$													
H^+	1	H^+													
OH^-	1	OH^-													
Na^+	1	Na^+													
Cl^-	1	Cl^-													

REFERENCES

1. Sugano, K. (2012). *Biopharmaceutics Modeling and Simulations: Theory, Practice, Methods and Applications*. Hoboken: Wiley.
2. Shan, N. and Zaworotko, M.J. (2008). The role of cocrystals in pharmaceutical science. *Drug Discov. Today* 13: 440–446.
3. Qiao, N., Li, M., Schlindwein, W. et al. (2011). Pharmaceutical cocrystals: an overview. *Int. J. Pharm.* 419: 1–11.
4. Lu, J. and Rohani, S. (2009). Polymorphism and crystallization of active pharmaceutical ingredients (APIs). *Curr. Med. Chem.* 16: 884–905.
5. Kiesow, K., Tumaka, F., and Sadowski, G. (2008). Experimental investigation and prediction of oiling out during crystallization process. *J. Cryst. Growth* 310: 4163–4168.
6. Lu, J., Li, Y.-P., Wang, J. et al. (2012). Crystallization of an active pharmaceutical ingredient that oils out. *Sep. Purif. Technol.* 96: 1–6.
7. Michelsen, M.L. and Mollerup, J.M. (2007). *Thermodynamic Models: Fundamentals & Computational Aspects*. Holte, Denmark: Tie-Line Publications.
8. Chen, C.-C. (2011). Molecular thermodynamics for pharmaceutical process modeling and simulation. In: *Chemical Engineering in the Pharmaceutical Industry* (ed. D.J. am Ende), 505–514. Hoboken: Wiley.
9. Paricaud, P., Galindo, A., and Jackson, G. (2004). Modeling the cloud curves and the solubility of gases in amorphous and semicrystalline polyethylene with the SAFT-VR approach and Flory theory of crystallization. *Ind. Eng. Chem. Res.* 43: 6871–6889.
10. Michelsen, M.L. (1982). The isothermal flash problem. Part I. Stability. *Fluid Phase Equilib.* 9: 1–19.
11. Pereira, F.E., Jackson, G., Galindo, A., and Adjiman, C.S. (2010). A duality-based optimisation approach for the reliable solution of (P, T) phase equilibrium in volume-composition space. *Fluid Phase Equilib.* 299 (1): 1–23.
12. Pereira, F.E., Jackson, G., Galindo, A., and Adjiman, C.S. (2012). The HELD algorithm for multicomponent, multiphase equilibrium calculations with generic equations of state. *Comput. Chem. Eng.* 36: 99–118.
13. McDonald, C.M. and Floudas, C.A. (1995). Global optimization for the phase stability problem. *AIChE J.* 41 (7): 1798–1814.
14. Jackson, G., Chapman, W.G., and Gubbins, K.E. (1988). Phase equilibria of associating fluids – spherical molecules with multiple bonding sites. *Mol. Phys.* 65 (1): 1–31.
15. Chapman, W.G., Gubbins, K.E., Jackson, G., and Radosz, M. (1990). New reference equation of state for associating liquids. *Ind. Eng. Chem. Res.* 29 (8): 1079–1721.
16. Economou, I. (2002). Statistical associating fluid theory: a successful model for the calculation of thermodynamic and phase equilibrium properties of complex fluid mixtures. *Ind. Eng. Chem. Res.* 41: 953–962.
17. Müller, E.A. and Gubbins, K.E. (2001). Molecular-based equations of state for associating fluids: a review of SAFT and related approaches. *Ind. Eng. Chem. Res.* 40: 22193–22211.
18. McCabe, C. and Galindo, A. (2011). SAFT associating fluids and fluid mixtures. In: *Applied Thermodynamics of Fluids* (ed. A.R.H. Goodwin, J.V. Sengers and C.J. Peters), 215–279. London: Royal Society of Chemistry.
19. Papaioannou, V., Lafitte, T., Avendaño, C. et al. (2014). Group contribution methodology based on the statistical associating fluid theory for heteronuclear molecules formed from Mie segments. *J. Chem. Phys.* 140: 054107.
20. Dufal, S., Lafitte, T., Haslam, A.J. et al. (2015). The A in SAFT: developing the contribution of association to the Helmholtz free energy within a Wertheim TPT1 treatment of generic Mie fluids. *Mol. Phys.* 113: 948–984.
21. Sadeqzadeh, M., Papaioannou, V., Dufal, S. et al. (2016). The development of unlike induced association-site models to study the phase behaviour of aqueous mixtures comprising acetone, alkanes and alkyl carboxylic acids with the SAFT-γ Mie group contribution methodology. *Fluid Phase Equilib.* 407: 39–57.
22. Joback, K.G. and Reid, R.C. (1987). Estimation of pure-component properties from group contributions. *Chem. Eng. Commun.* 57: 233–243.
23. Gracin, S. and Rasmuson, Å.C. (2002). Solubility of phenylacetic acid, p-hydroxyphenylacetic acid, p-aminophenylacetic acid, p-hydroxybenzoic acid, and ibuprofen in pure solvents. *J. Chem. Eng. Data* 47 (6): 1379–1383.
24. Wang, S., Song, Z., Wang, J. et al. (2010). Solubilities of ibuprofen in different pure solvents. *J. Chem. Eng. Data* 55 (11): 5283–5285.
25. Domanska, U., Pobudkowska, A., Pelczarska, A., and Gierycz, P. (2009). pK_a and solubility of drugs in water, ethanol, and 1-octanol. *J. Phys. Chem. B* 113 (26): 8941–8947.
26. Garzon, L.C. and Martinez, F. (2004). Temperature dependence of solubility for ibuprofen in some organic and aqueous solvents. *J. Solut. Chem.* 33 (11): 1379–1395.
27. Process Systems Enterprise Ltd. (2018). gPROMS advanced thermodynamics [Online]. https://www.psenterprise.com/products/gsaft (accessed 5 January 2018).
28. Process Systems Enterprise Ltd. (2018). gPROMS platform [Online]. https://www.psenterprise.com/products/gproms (accessed 5 January 2018).
29. Gill-Villegas, A., Galindo, A., Whitehead, P.J. et al. (1997). Statistical associating fluid theory for chain molecules with attractive potentials of variable range. *J. Chem. Phys.* 106: 4168.
30. Galindo, A., Davies, L.A., Gil-Villegas, A., and Jackson, G. (1998). The thermodynamics of mixtures and the corresponding mixing rules in the SAFT-VR approach for potentials of variable range. *Mol. Phys.* 93 (2): 241–252.
31. Gross, J. and Sadowski, G. (2001). Perturbed-chain SAFT: an equation of state based on a perturbation theory for chain molecules. *Ind. Eng. Chem. Res.* 40: 1244–1260.
32. Gross, J. and Sadowski, G. (2002). Application of the perturbed-chain SAFT equation of state to associating systems. *Ind. Eng. Chem. Res.* 41: 5510–5515.
33. Cuevas, J., Llovell, F., Galindo, A. et al. (2011). Solid–liquid equilibrium using the SAFT-VR equation of state: solubility

of naphthalene and acetic acid in binary mixtures and calculation of phase diagrams. *Fluid Phase Equilib.* 306: 137–147.

34. Prudic, A., Ji, Y., and Sadowski, G. (2014). Thermodynamic phase behavior of API/polymer solid dispersions. *Mol. Pharm.* 11: 2294–2304.

35. Lange, L., Schleinitz, M., and Sadowski, G. (2016). Predicting the effect of pH on stability and solubility of polymorphs, hydrates, and cocrystals. *Cryst. Growth Des.* 16 (7): 4136–4147.

36. Lange, L., Lehmkemper, K., and Sadowski, G. (2016). Predicting the aqueous solubility of pharmaceutical cocrystals as function of pH and temperature. *Cryst. Growth Des.* 16 (5): 2726–2740.

37. Watterson, S., Hudson, S., Svärd, M., and Rasmuson, Å.C. (2014). Thermodynamics of fenofibrate and solubility in pure organic solvents. *Fluid Phase Equilib.* 367: 143–150.

38. Sun, H., Liu, B., Liu, P. et al. (2016). Solubility of fenofibrate in different binary solvents: experimental data and results of thermodynamic modeling. *J. Chem. Eng. Data* 61 (9): 3177–3183.

39. Khan, G.M. and Jiabi, Z. (1998). Preparation, characterization, and dissolution studies of ibuprofen solid dispersions using polyethylene glycol (PEG), talc, and PEG-talc as dispersion carriers. *Drug Dev. Ind. Pharm.* 24 (5): 455–462.

40. Wassvik, C.M., Holmén, A.G., Bergström, C.A. et al. (2006). Contribution of solid-state properties to the aqueous solubility of drugs. *Eur. J. Pharm. Sci.* 29 (3–4): 294–305.

41. Margarit, M.V., Rodriguez, I.C., and Cerezo, A. (1994). Physical characteristics and dissolution kinetics of solid dispersions of ketoprofen and polyethylene glycol 6000. *Int. J. Pharm.* 108: 101–107.

42. Avdeef, A., Berger, C.M., and Brownell, C. (2000). pH-metric solubility. 2: correlation between the acid–base titration and the saturation shake-flask solubility-pH methods. *Pharm. Res.* 17 (1): 85–89.

43. Shaw, L.R., Irwin, W.J., Grattan, T.J., and Conway, B.R. (2005). The effect of selected water-soluble excipients on the dissolution of paracetamol and ibuprofen. *Drug Dev. Ind. Pharm.* 31: 515–525.

44. Censi, R., Martena, V., Hoti, E. et al. (2013). Sodium ibuprofen dihydrate and anhydrous: study of the dehydration and hydration mechanisms. *J. Therm. Anal. Calorim.* 111: 2009–2018.

45. Wagman, D.D., Evans, W.H., Parker, V.B. et al. (1982). The NBS tables of chemical thermodynamic properties. Selected values for inorganic and C_1 and C_2 organic substances in SI units. National Standard Reference Data System.

46. Lange, L., Heisel, S., and Sadowski, G. (2016). Predicting the solubility of pharmaceutical cocrystals in solvent/anti-solvent mixtures. *Molecules* 21: 593.

47. Yu, Z.Q., Chow, P.S., and Tan, R.B.H. (2010). Operating regions in cooling cocrystallization of caffeine and glutaric acid in acetonitrile. *Cryst. Growth Des.* 10 (5): 2382–2387.

48. Shalmashi, A. and Golmohammad, F. (2010). Solubility of caffeine in water, ethyl acetate, ethanol, carbon tetrachloride, dichloromethane, and acetone between 298 and 323 K. *Lat. Am. Appl. Res.* 40: 283–285.

49. Cesaro, A. and Starec, G. (1980). Thermodynamic properties of caffeine crystal forms. *J. Chem. Phys.* 84: 1345–1346.

50. Bothe, H. and Cammenga, H.K. (1979). Phase transitions and thermodynamic properties of anhydrous caffeine. *J. Thermal Anal.* 16: 267–275.

51. Kolska, Z., Kukal, J., Zabransky, M., and Ruzicka, V. (2008). Estimation of the heat capacity of organic liquids as a function of temperature by a three-level group contribution method. *Ind. Eng. Chem. Res.* 47: 2075–2085.

52. Suzuki, E., Shirotani, K.-I., Tsuda, Y., and Sekiguchi, K. (1985). Water content and dehydration behavior of crystalline caffeine hydrate. *Chem. Pharm. Bull.* 33 (11): 5028–5035.

53. Laube, F.S., Klein, T., and Sadowski, G. (2015). Partition coefficients of pharmaceuticals as functions of temperature and pH. *Ind. Eng. Chem. Res.* 54: 3968–3975.

54. Riipinen, I., Svenningsson, B., Bilde, M. et al. (2006). A method for determining thermophysical properties of organic material in aqueous solutions: succinic acid. *Atmos. Res.* 82: 579–590.

55. Marrero, J. and Abildskov, J. (2003). *Solubility and Related Properties of Large Complex Chemicals*, vol. 15. DECHEMA.

56. Wu, H., Dang, L., and Wei, H. (2014). Solid-liquid phase equilibrium of nicotinamide in different pure solvents: measurements and thermodynamic modeling. *Ind. Eng. Chem. Res.* 53 (4): 1707–1711.

57. Zhang, H., Yin, Q., Liu, Z. et al. (2014). An odd-even effect on solubility of dicarboxylic acids in organic solvents. *J. Chem. Thermodyn.* 77: 91–97.

58. Yu, Q., Black, S., and Wei, H. (2009). Solubility of butanedioic acid in different solvents at temperatures between 283 K and 333 K. *J. Chem. Eng. Data* 54 (7): 2123–2125.

59. International Council for Harmonisation of Technical Requirements for Pharmaceuticals for Human Use (2016). *Impurities: Guideline for Residual Solvents – Q3C(R6)*.

20

DRUG SOLUBILITY, REACTION THERMODYNAMICS, AND CO-CRYSTAL SCREENING

KARIN WICHMANN AND CHRISTOPH LOSCHEN
COSMO logic GmbH & Co. KG, Leverkusen, Germany

ANDREAS KLAMT
COSMO logic GmbH & Co. KG, Leverkusen, Germany and Institute of Physical and Theoretical Chemistry, University of Regensburg, Regensburg, Germany

20.1 INTRODUCTION

20.1.1 Methods for Compounds in Solution

There is a variety of computational methods for the treatment of compounds in solution. The scope of this chapter is not to give an overview of them, but to concentrate on applications of COSMO-RS, a very efficient method for the a priori prediction of thermophysical data of liquids. COSMO-RS combines unimolecular quantum chemical calculations that provide the necessary information for the evaluation of molecular interaction in the fluid phase, with a very fast and accurate statistical thermodynamic procedure. It has established itself as an alternative to structure-based group contribution methods (GCMs) on the one hand and to force field-based simulation methods on the other hand. Because of its special approach, COSMO-RS is a generally applicable method for compounds in solution. It has been applied successfully in such diverse areas as solvent screening, partitioning behavior, liquid–liquid and vapor–liquid equilibria, and ADME property prediction and for such diverse compound types as drugs, pesticides, common organic compounds, halocarbons, and ionic liquids. COSMO-RS is used in chemical, pharmaceutical, agrochemical, and petrochemical industry.

In this contribution, three application fields important in drug development and drug production will be considered: solubility prediction and prediction of free energy of reaction in solution and co-crystal screening. Solubility prediction methods are important during the drug design and development process, because in the early drug design phase compounds are often only virtually considered by computational drug design methods or the synthesized amount of substance is insufficient for experiments. In both cases the only tools for the selection of promising drug candidates with adequate solubility are computational methods that predict the solubility with sufficient accuracy just from the chemical structure of the compound. A method requiring experimental data for solubility prediction is unfeasible in this situation, since such data will not be available.

Prediction of thermodynamic properties of compounds in solution is also important in industrial process development. Here, specifically reaction energies and equilibrium constants of reaction in solution are of particular interest when a new process is developed or alternative pathways for existing processes are explored. The Gibbs free energy of reaction varies with the choice of the solvent or solvent mixture, and hence the chosen solvent system can strongly influence the process in solution. Generally, experimental data for the equilibrium constant or the free energy of reaction in solution are rare but are available relatively straightforward from a computational approach. Prediction of thermochemical data like heat of reaction and heat of vaporization furthermore helps in designing a chemical process such that process hazards can be prevented.

Finally, the prediction of compounds that form co-crystals together with drugs is of high interest for pharmaceutical research and development, because co-crystalline materials often show modified properties such as improved solubility or dissolution behavior.

Chemical Engineering in the Pharmaceutical Industry: Active Pharmaceutical Ingredients, Second Edition.
Edited by David J. am Ende and Mary T. am Ende.
© 2019 John Wiley & Sons, Inc. Published 2019 by John Wiley & Sons, Inc.

In order to identify new co-crystals, usually large sets of excipients are screened experimentally, whereas selection of compounds is done by trial and error or according to some empirical rules. Computational methods such as COSMO-RS allow for a rational design and a more focused search for new co-crystal candidates, and hence valuable resources can be saved.

20.1.2 COSMO

In conventional quantum chemistry, molecules are treated as isolated particles at a temperature of 0 K. In physical reality however, the major part of reactions takes place in solution and at higher temperatures. Since direct treatment of a large number of molecules is computationally very demanding, solvent effects are often treated indirectly by continuum solvation models, where the solute is embedded in a dielectric continuum and the solvent is represented by a mean interaction with a surrounding dielectric medium. The interaction of the solute with such a dielectric solvent is taken into account in the quantum chemical calculation by polarization charges that arise from the dielectric boundary condition.

The "COnductor-like Screening MOdel" (COSMO) is an efficient variant of dielectric continuum solvation methods [1]. In quantum chemical COSMO calculations, the solute molecules are calculated in a virtual scaled conductor environment, i.e. the scaled boundary condition of a conducting medium is used, where the molecule is ideally screened and not the exact dielectric boundary condition. In such a conducting environment, the solute molecule induces a polarization charge density σ on the interface between the molecule and the conductor, i.e. on the molecular surface. These charges act back on the solute and generate a more polarized electron density than in vacuum. During the quantum chemical self-consistency algorithm, the solute molecule is thus converged to its energetically optimal state in a conductor with respect to electron density. Due to the analytic gradients available for the COSMO energy contributions, the molecular geometry can be optimized using conventional methods for calculations in vacuum. The quantum chemical calculation has to be performed once for each molecule of interest.

20.1.3 COSMO-RS

As discussed in more detail elsewhere, the simple dielectric continuum models suffer from a number of insufficiencies [2, 3]. The polarization charge density σ resulting from unscaled COSMO calculations (also called screening charge density σ), which is a good local descriptor of the molecular surface polarity, is used to extent the model toward "real solvents" (COSMO-RS).

In COSMO-RS, a liquid is considered to be an ensemble of closely packed ideally screened molecules, as shown in Figure 20.1. In this picture, each piece of surface has one direct contact partner but is still separated from its partner by a thin film of conductor. Since the conducting medium that was assumed to surround the molecules in the COSMO calculation is not existent in reality, the energy difference between the pairwise contacts and the ideally screened situation has to be defined as a local electrostatic interaction energy that results from the removal of the conductor film between the molecules. Considering a contact on a region of molecular surface of area a_{eff} (effective contact area), and considering that the two contacting pieces of molecular surface have average ideal screening charge densities σ and σ' in the conductor, it is possible to calculate this interaction energy as the energy that is necessary to remove the residual screening charge density $\sigma + \sigma'$ from the contact. In the special case of $\sigma = -\sigma'$, the contact is an "ideal electrostatic contact," and the interaction energy is zero, because the two molecules screen each other as well as the conductor. In the general case however, $\sigma + \sigma'$ does not vanish, and the arising electrostatic interaction energy is

$$E_{\text{misfit}}(\sigma,\sigma') = a_{\text{eff}} e_{\text{misfit}}(\sigma,\sigma') = a_{\text{eff}} \frac{\alpha'}{2}(\sigma+\sigma')^2, \quad (20.1)$$

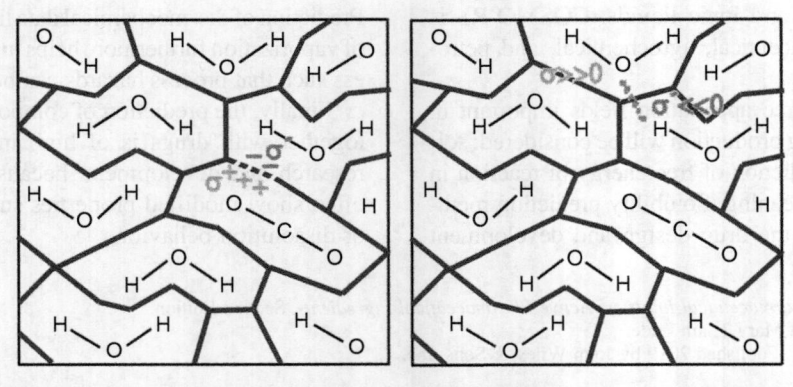

FIGURE 20.1 Schematic illustration of contacting molecular cavities and contact interactions.

where

- $e_{\text{misfit}}(\sigma, \sigma')$ is the misfit energy density on the contact surface.
- α' is a general constant that can be calculated approximately but in COSMO-RS is fitted to experimental data as fine-tuning.

The misfit term (Eq. 20.1) subsumes the polarization response of the molecules to the electrostatic misfit quite well [2, 4, 5].

Hydrogen bonding interactions are to some extent already covered by the description of electrostatic interactions, but we still have to parameterize the additional hydrogen bonding energy resulting from interpenetration of the atomic electron densities in some reasonable way. This energy should only be relevant if two sufficiently polar pieces of surface of opposite polarity are in contact, and its strength increases with increasing polarity of both surface pieces. Taking the screening charge density σ as a local measure of polarity, the following function realizes such behavior:

$$E_{\text{hb}}(\sigma, \sigma') = a_{\text{eff}} e_{\text{hb}}(\sigma, \sigma') = a_{\text{eff}} c_{\text{hb}} \min \{0, \min(0, \sigma_{\text{don}} + \sigma_{\text{hb}}) \max(0, \sigma_{\text{acc}} - \sigma_{\text{hb}})\} \quad (20.2)$$

with $\sigma_{\text{don}} = \min(\sigma, \sigma')$ and $\sigma_{\text{acc}} = \max(\sigma, \sigma')$. General parameters are σ_{hb}, the threshold for hydrogen bonding, and c_{hb}, the coefficient for the hydrogen bond strength. Both parameters have to be adjusted to experimental data. With Eq. (20.2), the hydrogen bond interaction energy is zero, unless the more negative of the two screening charge densities is less than the threshold $-\sigma_{\text{hb}}$ and the more positive exceeds σ_{hb}. Because positive molecular regions have negative screening charge, the negative σ now is the donor part of the hydrogen bond and the positive is the acceptor. In this case, the hydrogen bonding energy is proportional to the product of the excess screening charge densities $(\sigma_{\text{don}} + \sigma_{\text{hb}})(\sigma_{\text{acc}} - \sigma_{\text{hb}})$.

Van der Waals (vdW) interactions are described by element-specific parameters τ in COSMO-RS. The τ parameters have to be fitted to experimental data. Then, the vdW energy gain of a molecule X during the transfer from the gas phase to any solvent is given by

$$E_{\text{vdW}}^X = \sum_{\alpha \in X} a_\alpha^X \tau_{\text{vdW}}(e(\alpha)) \quad (20.3)$$

The vdW term is spatially nonspecific. Because E_{vdW} is independent of any neighboring relations, it is not really an interaction energy, but may be considered as an additional contribution to the energy of the reference state in solution. Currently nine of the vdW parameters (for elements H, C, N, O, F, S, Cl, Br, and I) have been optimized. For the majority of the remaining elements, reasonable estimates are available. Non-additive vdW corrections are used for a few element pairs, but they are of minor importance for the topics of this contribution.

The transition from microscopic surface interaction energies to macroscopic thermodynamic properties of a liquid is possible via a statistical thermodynamic procedure. The exact solution of the thermodynamic problem would require sampling of all different arrangements of all molecules of the systems, weighting the contribution of each arrangement by its Boltzmann factor. This direct approach, which is used in the molecular dynamics and Monte Carlo-type methods, is very time consuming and requires compromises regarding sampling and regarding the accuracy of the energy evaluations. COSMO-RS follows a different concept. The basic approximation is that the ensemble of interacting molecules may be replaced by the corresponding ensemble of independent pairwise interacting surface pieces. This approximation implies the neglect of any neighborhood information of surface pieces on the molecular surface and the loss of steric information. The advantage of this approximation is the extreme reduction of the complexity of the problem, which allows for a fast and exact solution. It should be noted that GCMs as UNIFAC are also based on the assumption of independent pairwise interacting surfaces.

Since the screening charge density σ is the only descriptor determining the interaction energy terms in Eqs. (20.1) and (20.2), the ensemble of surface pieces characterizing an ensemble S is sufficiently described by its composition with respect to σ. For this purpose we introduce the molecular σ-profile $p^X(\sigma)$, which is a histogram of the screening charge densities σ on the surface of a molecule X (Figure 20.2). The σ-profile can easily be derived from the COSMO files produced as output of the quantum chemical COSMO calculation for molecule X, applying a local averaging algorithm in order to take into account that only screening charge densities σ averaged over an effective contact area are of physical meaning in COSMO-RS [2].

The σ-profile for the entire solvent of interest S, which might be a mixture of several compounds, $p_S(\sigma)$, is given by the weighted surface area normalized sum of the σ-profiles of the components X_i:

$$p'_S(\sigma) = \frac{p_S(\sigma)}{A_S} = \frac{\sum_{i \in S} x^i p^{X_i}(\sigma)}{\sum_{i \in S} x^i A^{X_i}} \quad (20.4)$$

where A^{X_i} is the COSMO surface of a compound X_i in the system.

Under the condition that there is no free surface in the bulk of the liquid, i.e. each piece of molecular surface has a direct contact partner, the statistical thermodynamics of the system can be solved using the exact equation:

$$\mu_S(\sigma) = -RT \ln \left\{ \int p'_S(\sigma') \exp\left(\frac{\mu_S(\sigma') - a_{\text{eff}} e(\sigma, \sigma')}{RT}\right) d\sigma' \right\} \quad (20.5)$$

In this equation, $\mu_S(\sigma)$ is the chemical potential of an average molecular contact segment of size a_{eff} in the ensemble S

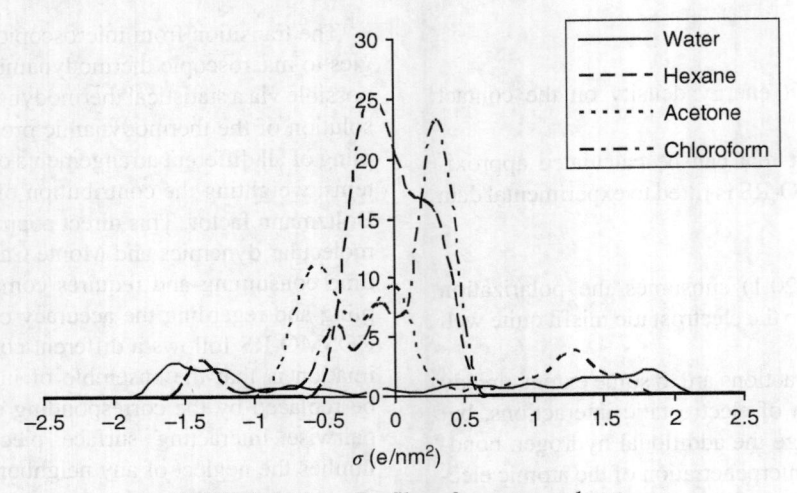

FIGURE 20.2 σ-Profiles of common solvents.

at temperature T, and $e(\sigma,\sigma')$ is the interaction energy functional $e(\sigma,\sigma') = e_{\text{misfit}}(\sigma,\sigma') + e_{\text{hb}}(\sigma,\sigma')$.

Since $\mu_S(\sigma)$ appears on both sides of Eq. (20.5), it must be solved by iteration, starting with $\mu_S(\sigma') = 0$ on the right-hand side. Fortunately, the solution converges rapidly, and $\mu_S(\sigma)$ can be computed up to numerical precision within milliseconds on a personal computer. For a formal derivation of Eq. (20.5), we refer to Ref. [4].

Now it is straightforward to define the chemical potential of a solute X in the ensemble S by

$$\mu_S^X = \mu_{\text{res},S}^X + \mu_{\text{comb},S}^X = a_{\text{eff}}^{-1} \int p^X(\sigma)\mu_S(\sigma)\mathrm{d}\sigma + \mu_{\text{comb},S}^X \quad (20.6)$$

where the residual part, i.e. the part resulting from the interactions of the surfaces in the liquid, is given by the surface integral of function $\mu_S(\sigma)$ over the solute surface, which is expressed using the σ-profile of the solute in Eq. (20.6). The second part is the combinatorial contribution, which arises from the different shapes and sizes of the solute and solvent molecules. Expressions based on the surface areas and volume ratios of solvents and solutes, similar to standard chemical engineering expressions as Staverman–Guggenheim, are used in the context of COSMO-RS [6]. COSMO surface areas and volumes are used for the evaluation of the combinatorial term. Hence Eq. (20.6) can be completely evaluated based on the information resulting from the COSMO calculations of the individual compounds.

The chemical potential of Eq. (20.6) is a pseudo-chemical potential [7], i.e. the standard chemical potential without the concentration term $RT \ln x_i$. We will shortly use the term chemical potential for the pseudo-chemical potential from Eq. (20.6) throughout this contribution. Providing the chemical potential of an arbitrary compound X in almost arbitrary solvents and mixtures as a function of temperature and concentration, Eq. (20.6) allows for the prediction of almost all thermodynamic properties of compounds or mixtures, such as activity coefficients, partition coefficients, or solubility, as shown in the flowchart for a COSMO-RS property prediction in Figure 20.3.

As mentioned above, the COSMO-RS method depends on a small number of adjustable parameters. Some of the parameters are predetermined from physics, while others are determined from selected properties of mixtures. The parameters are not specific to functional groups or types of molecule. As a result, COSMO-RS is the least parameterized of all quantitative methods for the prediction of chemical properties in the liquid phase [5].

20.1.4 Treatment of Conformers in COSMO-RS

Many molecules can adopt more than one conformation, and different conformers of one molecule can have different σ-profiles. The chemical potentials of the individual conformers and hence the conformer distribution as well as the chemical potential of the compound represented by an ensemble of conformers depend on the composition of the system and the temperature. Thus, it is essential for property prediction with COSMO-RS to take conformers with different σ-profiles into account, each described by individual quantum chemical COSMO calculations. The relative contributions of the conformers are determined by an iterative procedure using the Boltzmann weight of the free energies of the conformers in the liquid phase. This results in a thermodynamically fully consistent treatment of multiple molecular conformations.

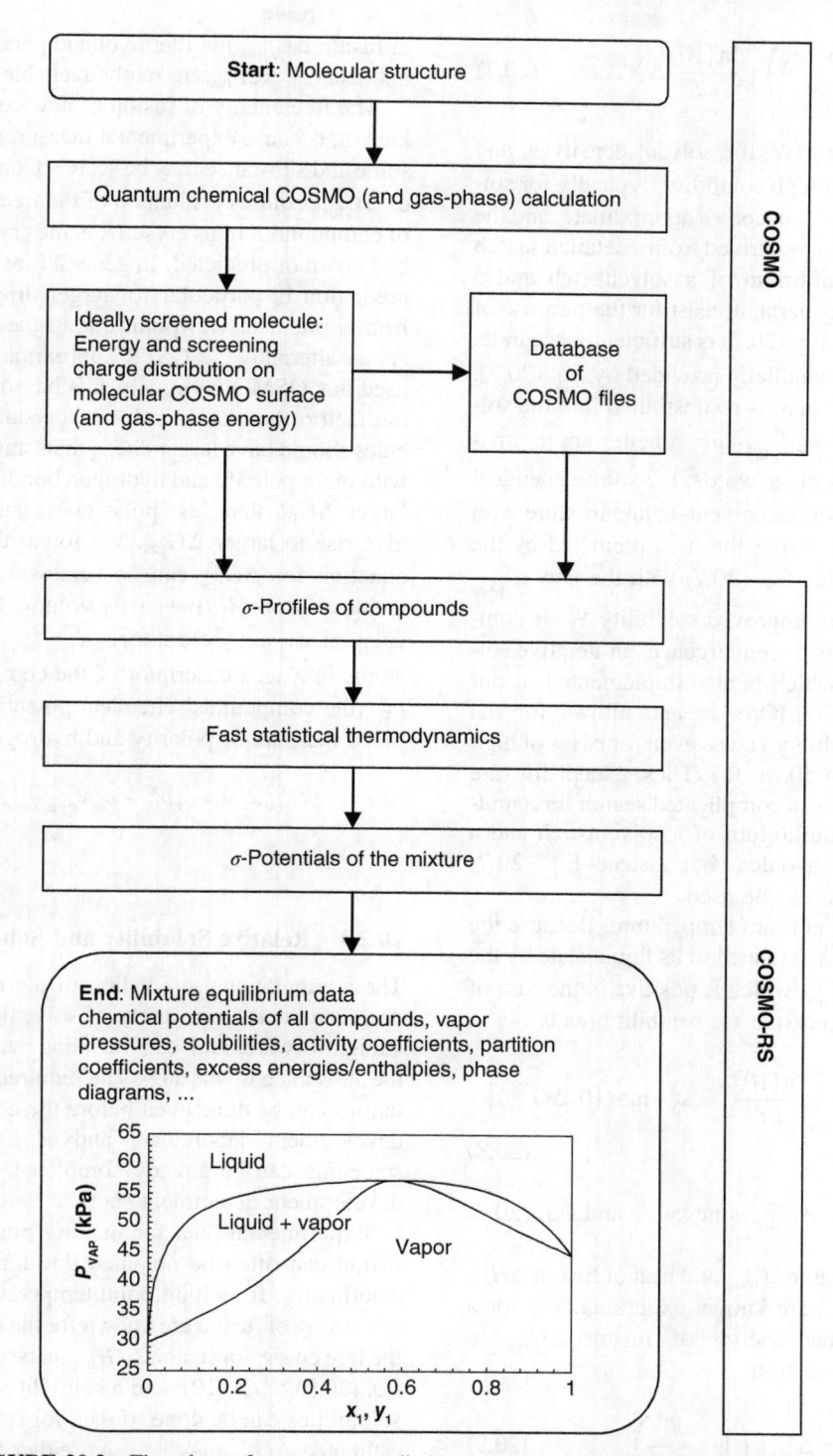

FIGURE 20.3 Flowchart of a property prediction procedure with COSMO and COSMO-RS.

20.2 SOLUBILITY PREDICTION WITH COSMO-RS

For the calculation of the solubility S_S^X of a liquid compound X in a solvent S, we require the chemical potentials of X in S and in its pure liquid state, μ_S^X and μ_X^X. If S_S^X is sufficiently small, so that the solvent behavior of the X-saturated solvent S is not significantly influenced by the solute X, then the decadic logarithm of the solubility is given by

$$\text{Log} \, S_S^X = \log\left(\frac{MW^X \rho_S}{MW_S}\right) - \frac{\ln(10)}{kT}\Delta_S^X \quad (20.7)$$

with the molecular weight MW, the solvent density ρ, and $\Delta_S^X = \mu_S^X - \mu_X^X$. In the case of high solubility (typically for solubility >10 wt. %), Eq. (20.7) becomes approximate, and the true solubility would have to be derived from a detailed search for a thermodynamic equilibrium of a solvent-rich and a solute-rich phase. But, in general, at least for the purpose of estimating drug solubility, Eq. (20.7) is sufficiently accurate.

If the zeroth-order $S_S^{X_0}$ as initially provided by Eq. (20.7), using infinite dilution of X in S, is resubstituted into the solubility calculation via $\Delta_S^{X_1} = \mu_{S(X_0)}^X - \mu_X^X$, a better approximation for S_S^X is achieved. In other words, the solute chemical potential μ_S^X is computed for the solvent–solute mixture with the finite mole fraction of X in S that was predicted by the zeroth-order S_S^X. Then, using Eq. (20.7) with the new $\mu_{S(X_0)}^X$ and the resulting values, an improved solubility $S_S^{X_1}$ is computed. Iterating this process to convergence, an iterative solubility can be achieved, which is also implemented in our COSMO-RS program COSMO*therm* and allows for the accurate prediction of solubility values even for cases of high solubility (solubility up to 50 wt. %). Thus, except for rare cases of very high solubility, a complicated search for a multiphase thermodynamic equilibrium of a solvent-rich and a solute-rich phase can be avoided, but instead Eq. (20.7) and its iterative refinement can be used.

Drugs are mostly solid at room temperature. Because the solid state of a compound X is related to its liquid state by the free energy difference ΔG_{fus}^X, which is positive in the case of solids, a more general expression for solubility reads

$$\text{Log} \, S_S^X = \log\left(\frac{MW^X \rho_S}{MW_S}\right) + \frac{\ln(10)}{kT}\left[-\Delta_S^X - \max\left(0, \Delta G_{\text{fus}}^X\right)\right] \quad (20.8)$$

For liquid compounds, ΔG_{fus}^X is negative and Eq. (20.8) reduces to Eq. (20.7).

If melting point temperature T_{melt} and heat of fusion ΔH_{fus} or entropy of fusion ΔS_{fus} are known experimentally for a solid compound, the free energy of fusion ΔG_{fus} in Eq. (20.8) can be estimated from

$$\Delta G_{\text{fus}}(T) = \Delta H_{\text{fus}}\left(1 - \frac{T}{T_{\text{melt}}}\right) \quad (20.9)$$

or

$$\Delta G_{\text{fus}}(T) = \Delta S_{\text{fus}}(T_{\text{melt}} - T) \quad (20.10)$$

Equations (20.9) and (20.10) can be complemented by an additional temperature-dependent term using the heat capacity of fusion $\Delta C p_{\text{fus}}$ in order to obtain good absolute predictions, but data for $\Delta C p_{\text{fus}}$ are rarely available from experiment.

The free energy of fusion of new compounds is often not known, because experimental measurements can be cumbersome and substance may be scarce. Computational prediction of ΔG_{fus}^X requires evaluation of the free energy of a molecule of compound X in its crystal, i.e. the crystal structure needs to be known or predicted. In general however, crystal structure prediction in particular for larger drugs is not yet feasible from a practical viewpoint due to the involved complexity. As an alternative, a QSPR approximation for ΔG_{fus} can be used in COSMO*therm*, which is based on a few rather obvious factors that should influence crystallization. Larger molecules should have larger ΔG_{fus} than smaller ones, compounds with more polarity and hydrogen bonding ability should have larger ΔG_{fus} than less polar ones, and also rigidity should give rise to larger ΔG_{fus}. We found that a good regression equation for ΔG_{fus}^X can be achieved by a combination of the descriptors V^X (the cavity volume from the COSMO calculation as size descriptor), $N_{\text{ring atom}}^X$ (the number of ring atoms in X as a descriptor of the compounds' rigidity), and μ_W^X (the compounds' chemical potential in water as a combined measure of polarity and hydrogen bonding) [8]:

$$\Delta G_{\text{fus}}^X = c_0 + c_1 \mu_W^X + c_2 N_{\text{ring atom}}^X + c_3 V^X \quad (20.11)$$

20.2.1 Relative Solubility and Solubility Screening

The computational prediction of the relative solubility of a drug candidate in a variety of solvents with COSMO-RS is straightforward and can be done without wasting any of the substance in this step. The required DFT/COSMO calculations can be done even before the compound comes to the development laboratory, and a COSMO-RS solubility screening can be already completed when the work in the development department starts.

Experimental data for melting point and free energy of fusion can often be obtained through differential scanning calorimetry. If melting point temperature and heat of fusion or entropy of fusion are known for the compound in question, the free energy of fusion ΔG_{fus} can be calculated according to Eq. (20.9) or (20.10), and a solubility screening for absolute solubilities can be done. If data for ΔG_{fus} are not known, an estimated ΔG_{fus} may be used, either from the QSPR model implemented in COSMO*therm* or from an external model.

EXAMPLE PROBLEM 20.1 Relative and Absolute Solubility of Acetaminophen in Pure Solvents at 30 °C

Experimental data for the solubility of acetaminophen in pure solvents were reported by Granberg and Rasmuson [9]. We use this data set to validate the calculated acetaminophen solubilities. Furthermore, a melting point temperature of $T_{melt} = 441.2$ K, a heat of fusion of $\Delta H_{fus} = 26.0$ kJ/mol, and an entropy of fusion of $\Delta S_{fus} = 59.0$ J/mol have been reported for acetaminophen [10]. These data will be used to compute absolute solubility predictions for acetaminophen in the solvent data set.

With the case study of acetaminophen solubility, we want to show first the prediction of relative solubility. For relative solubility calculations, we do not make use of any experimental data like melting point temperature or enthalpy of fusion, as that kind of data is usually not available in the early drug design phase.

Solute and solvents for this example were calculated on the BP86/TZVP level of theory, which corresponds to the higher quantum chemical level for which the COSMO*therm* program is parameterized. Different conformations of the compounds were taken into account where the conformers showed different σ-profiles and COSMO energies. All compounds of the data set, including conformers, are available from the COSMO*base*, a collection of validated COSMO files for common compounds and solvents (Figure 20.4).

For the calculation of relative solubility in pure solvents, we do not use the iterative refinement procedure, since the assumed value of $\Delta G_{fus} = 0$ kcal/mol will influence computed zeroth-order solubility and hence the iterative refinement. Therefore, the relative solubilities are calculated in infinite dilution in the respective solvent at 30 °C (Figure 20.5).

Calculation results can be read from the output and table files (Figure 20.6). The relative solubility of a compound can be calculated from the chemical potentials of the compound in the solvent $\mu_i^{(j)}$ and in its pure state $\mu_i^{(P)}$ as

$$\text{Log} S_{rel}(x) = \frac{\mu_i^{(P)} - \mu_i^{(j)}}{RT \ln(10)} \quad (20.12)$$

Since a logarithmic solubility value larger than 0 indicates only that the two compounds are miscible, there is a cutoff at 0 in the COSMO*therm* results for the logarithmic solubility in Figure 20.6. However, in order to provide insight in the whole range of the solvent data set independent of potential misestimates of ΔG_{fus}, positive values for log $S_{rel}(x)$ are allowed here for both relative and absolute solubility predictions. Relative solubility data in Table 20.1 were calculated directly from the chemical potential differences as described in Eq. (20.12).

The relative solubility predictions correlate well with the experimental data, revealing an overall shift of 1.8 log units, which arises mainly from the neglect of ΔG_{fus} and is therefore irrelevant for real solubility considerations. The predicted relative solubility data apparently fall into two groups, as can be seen from Figure 20.7: Solubility data in alcoholic solvents are grouped together on a rather straight line with a slope of 0.7 and a relative shift of 1.4 log units compared with the experimental data, while the second, more scattered group has a larger shift (≥ 2 log units), but a similar slope. There is one severe outlier in the data set, carbon tetrachloride, where the acetaminophen solubility is, in contrast to the trend observed in the other solvents, underestimated by 1.5 log units. Since the experimental value for the solubility in carbon tetrachloride comes from a single measurement, and the other solvents appear to be described reasonably

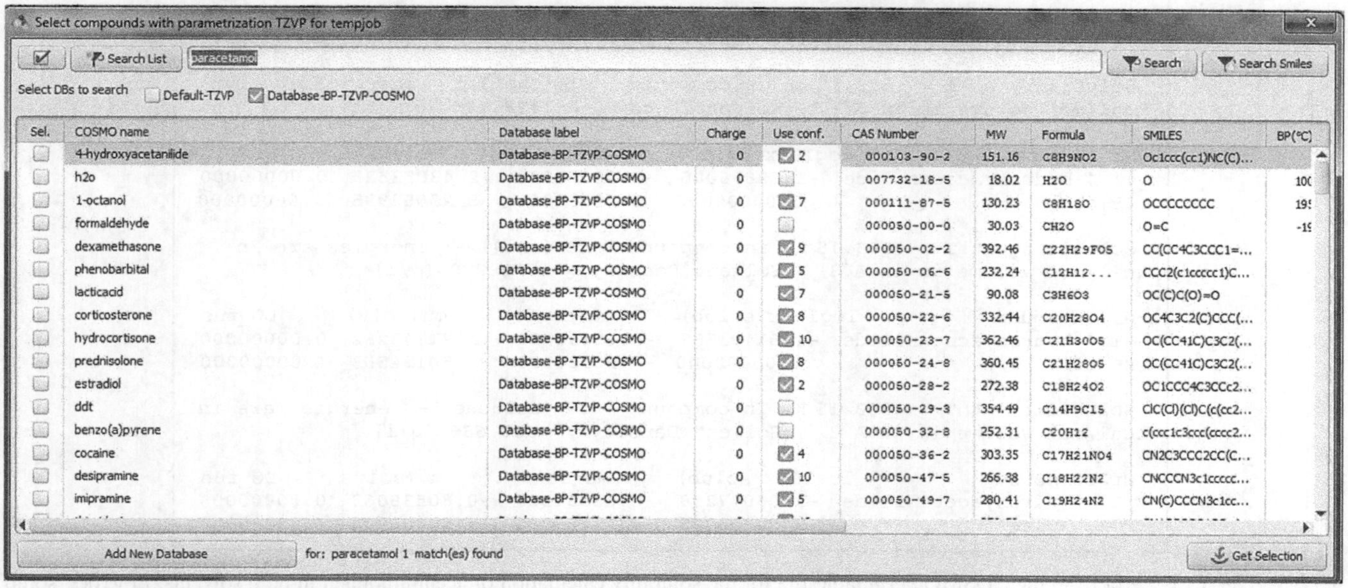

FIGURE 20.4 Database view in COSMO*therm*X. Databases can be searched and columns are sortable.

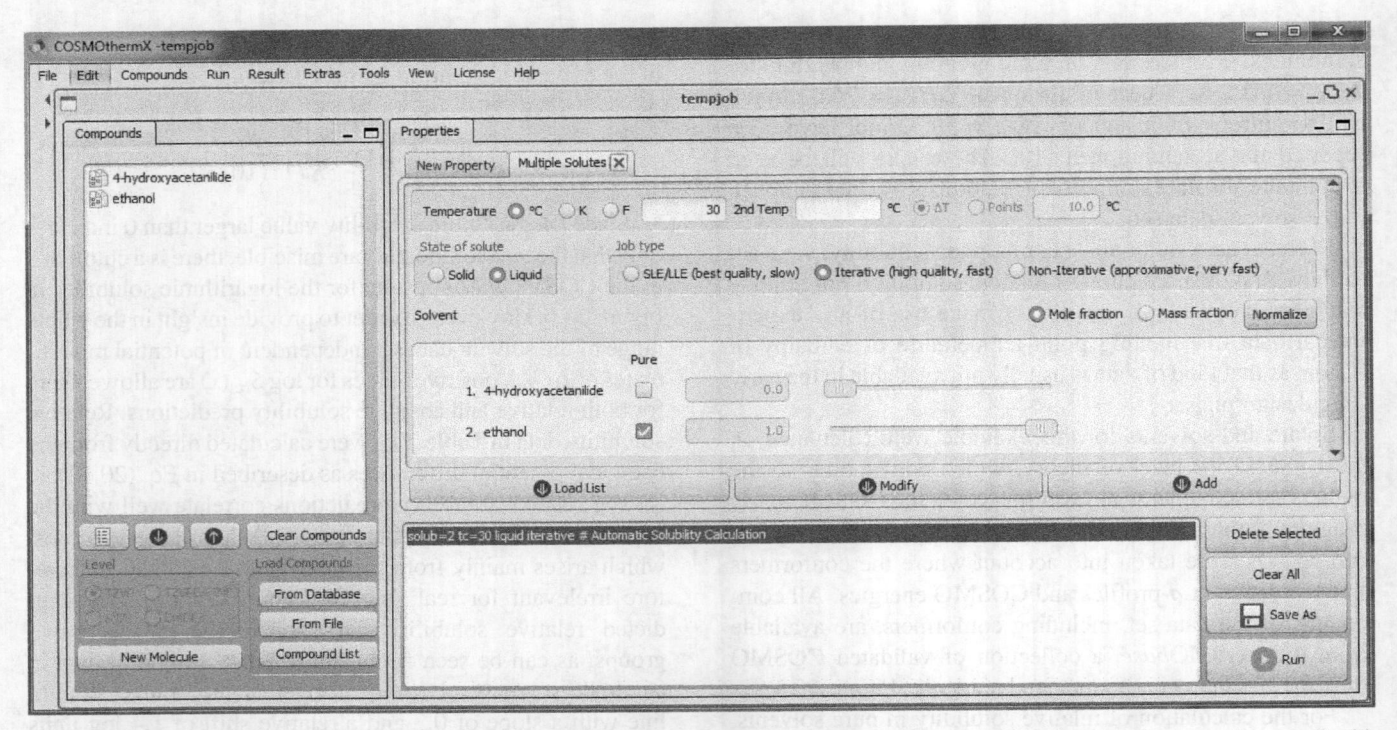

FIGURE 20.5 Overview of COSMO*therm*X with compound list, solubility panel, and input section. When the solute state is set to liquid, $\Delta G_{fus} = 0$ is used, while with solid solute state, given or estimated values for ΔG_{fus} are used. The iterative refinement can also be set in the solubility panel. In the solvent frame, the solvent composition is set to pure for the respective solvent. Pictured here are settings for absolute solubility using the iterative refinement procedure.

```
 Solubility at T =  303.15 K  in compound   2 (h2o) - energies are in
kcal/mol volume is in A^3 - Solvent Density =  995.363 [g/l]

Nr Compound            log10(x_solub)      mu(self)      mu(solv)      DG_fus
 1 4-hydroxyacetanilide  -1.68194563     -5.05034328   -2.71690337   0.00000000
 2 h2o                    0.00000000     -2.90487525   -2.90487525   0.00000000
...
 Solubility at T =  303.15 K  in compound   25 (chcl3) - energies are in
kcal/mol volume is in A^3 - Solvent Density = 1478.286 [g/l]

Nr Compound            log10(x_solub)      mu(self)      mu(solv)      DG_fus
 1 4-hydroxyacetanilide  -1.88836005     -5.05034328   -2.43053523   0.00000000
25 chcl3                  0.00000000     -5.29052995   -5.29052995   0.00000000
...
 Solubility at T =  303.15 K  in compound   26 (ccl4) - energies are in
kcal/mol volume is in A^3 - Solvent Density = 1572.230 [g/l]

Nr Compound            log10(x_solub)      mu(self)      mu(solv)      DG_fus
 1 4-hydroxyacetanilide  -4.51493557     -5.05034328    1.21343282   0.00000000
26 ccl4                   0.00000000     -7.50124553   -7.50124553   0.00000000
...
 Solubility at T =  303.15 K  in compound   27 (toluene) - energies are in
kcal/mol volume is in A^3 - Solvent Density =  860.666 [g/l]

Nr Compound            log10(x_solub)      mu(self)      mu(solv)      DG_fus
 1 4-hydroxyacetanilide  -3.05977298     -5.05034328   -0.80538062   0.00000000
27 toluene                0.00000000     -4.62709909   -4.62709909   0.00000000
```

FIGURE 20.6 Excerpt from the COSMO*therm* table file for the solubility calculation of acetaminophen in pure solvents. Log10(x_solub) indicates the logarithmic solubility in mole fractions.

TABLE 20.1 Experimental, Predicted Relative, and Predicted Absolute Solubilities of Acetaminophen in Pure Solvents

Solvent	Experimental		Predicted Relative			Predicted Absolute			Error
	c_S (g/kg)	Log S (mg/g)	Log $S(x)$	S (mg/g)	Log S (mg/g)	Log $S(x)$	S (mg/g)	Log S (mg/g)	
Water	17.39	1.24	−1.6819	174.53	2.24	−3.0745	7.07	0.85	−0.39
Methanol	371.61	2.57	0.6155	19 465.33	4.29	−1.0119	459.02	2.66	0.09
Ethanol	232.75	2.37	0.4809	9929.62	4.00	−1.0687	280.11	2.45	0.08
1,2-Ethanediol	144.3	2.16	0.1829	3711.29	3.57	−1.2535	135.86	2.13	−0.03
1-Propanol	132.77	2.12	0.2182	4157.36	3.62	−1.2441	143.37	2.16	0.03
2-Propanol	135.01	2.13	0.3858	6115.50	3.79	−1.1268	187.86	2.27	0.14
1-Butanol	93.64	1.97	0.0690	2390.35	3.38	−1.3675	87.50	1.94	−0.03
1-Pentanol	67.82	1.83	−0.0639	1480.10	3.17	−1.4866	55.92	1.75	−0.08
1-Hexanol	49.71	1.70	−0.1756	987.32	2.99	−1.5932	37.75	1.58	−0.12
1-Heptanol	37.43	1.57	−0.2630	709.92	2.85	−1.6779	27.31	1.44	−0.14
1-Octanol	27.47	1.44	−0.3589	507.95	2.71	−1.7716	19.64	1.29	−0.15
Acetone	111.65	2.05	0.9328	22 297.15	4.35	−0.8016	411.01	2.61	0.57
2-Butanone	69.99	1.85	0.6508	9380.65	3.97	−0.9153	254.79	2.41	0.56
4-Methyl-2-pentanone	17.81	1.25	0.1126	1956.04	3.29	−1.2844	78.41	1.89	0.64
Tetrahydrofuran	155.37	2.19	1.6842	101 306.53	5.01	0.0527	2366.72	3.37	1.18
1,4-Dioxane	17.08	1.23	1.0644	19 898.20	4.30	−0.7332	317.14	2.50	1.27
Ethyl acetate	10.73	1.03	0.0872	2097.21	3.32	−1.2415	98.38	1.99	0.96
Acetonitrile	32.83	1.52	0.1079	4720.92	3.67	−1.2689	198.25	2.30	0.78
Diethylamine	1316.9	3.12	3.5201	6 845 852.30	6.84	1.4457	57 683.45	4.76	1.64
N,N-Dimethylformamide	1012.02	3.01	2.2018	276 105.51	5.44	0.4567	4965.64	3.70	0.69
Dimethyl sulfoxide	1132.56	3.05	3.3062	3 915 389.90	6.59	0.2699	3601.53	3.56	0.50
Acetic acid	82.72	1.92	0.3232	5298.62	3.72	−1.1632	172.87	2.24	0.32
Dichloromethane	0.32	−0.49	−1.8354	26.00	1.41	−3.1820	1.17	0.07	0.56
Chloroform	1.54	0.19	−1.8884	16.37	1.21	−3.2647	0.69	−0.16	−0.35
Carbon tetrachloride	0.89	−0.05	−4.5149	0.03	−1.52	−5.9289	0.00	−2.94	−2.89
Toluene	0.34	−0.47	−3.0598	1.43	0.16	−4.4573	0.06	−1.24	−0.77

Log $S(x)$ indicates the logarithmic solubility in mole fractions.
Source: Data from Granberg and Rasmuson [9].

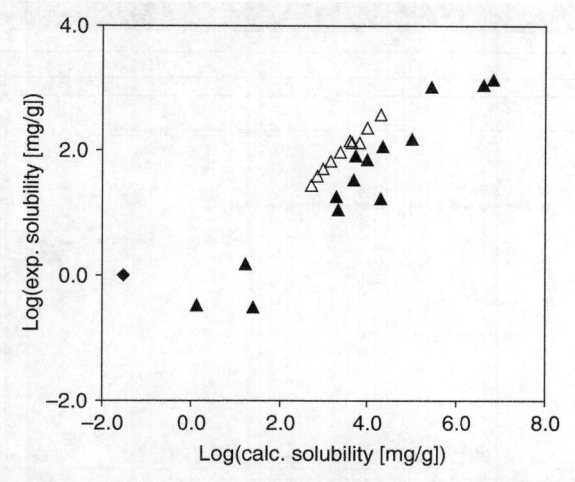

FIGURE 20.7 Predicted relative solubility of acetaminophen versus experimental data in pure solvents at 303.15 K. Triangles represent relative solubility data, with empty triangles (△) representing alcoholic solvents and water and solid triangles (▲) representing the remainder of the solvent data set. One outlier (carbon tetrachloride) is represented by a solid diamond (♦).

by the model prediction apart from a general overestimation due to the missing free energy of fusion term, we tend to consider this experimental value as questionable.

Indeed, in a personal communication, the original authors of Ref. [9] stated that a more recently determined value for the solubility of acetaminophen in CCl$_4$ of log S (mg/g) = −2.5 may be more accurate [12]. This would also perfectly correspond with the respective COSMO-RS prediction.

Using the experimental data for T_{melt} and ΔH_{fus}, absolute solubilities of acetaminophen in the solvent data set are also computed. The absolute predictions are in good quantitative agreement with experimental data, as shown in Figure 20.8. Of the 26 solvents, 4 are predicted with a positive log $S(x)$: tetrahydrofuran, diethylamine, N,N-diethylformamide, and dimethyl sulfoxide. The rmse for all solvents is 0.77 log units. There are, however, four significant outliers in the data set; one of them, carbon tetrachloride, deviates by almost 3 log units from the flawed experimental solubility as has already been discussed above. The other outliers are overpredicted solubilities. For diethylamine, the predicted solubility is 1.64 log units too high. We attribute this error to the known systematic error of COSMO and COSMO-RS for secondary and tertiary aliphatic amines [5]. Another outlier is the predicted solubility of acetaminophen in 1,4-dioxane (317.14 mg/g), which deviates by 1.27 log units from the value reported by Granberg and Rasmuson (17.1 mg/g). It is noteworthy that the deviation is much less (0.56 log units) if compared with the experimental solubility reported by Romero et al. (86.9 mg/g) [13]. The fourth outlier in the data set is in tetrahydrofuran, where the predicted solubility deviates by 1.13 log units from the experimental data. Here, we do not have an explanation for the error of the model prediction. However, since tetrahydrofuran and 1,4-dioxane are rather similar solvents, the question arises whether the uncertainty of the experimental data for tetrahydrofuran might be comparable with the case of 1,4-dioxane, where there is a deviation between the published experimental data from the different sources.

With the four outliers removed, the overall rmse reduces to 0.46 log units. While the rmse is very small for alcohols (0.10 log units), the calculated solubilities in polar aprotic solvents like acetone or hydrophobic solvents like toluene are systematically overpredicted. The rmse for the remainder of the data set without the alcoholic solvents is 0.86 log units. It should be noted that the estimate of ΔG_{fus} based on ΔH_{fus} and T_{melt} itself may have an error of the order of 0.5 log units, making these absolute deviations uncertain.

EXAMPLE PROBLEM 20.2 Solubility of Acetaminophen in Binary Mixtures of Water–Acetone and Toluene–Acetone at 25 °C

Experimental data for acetaminophen solubility in water–acetone and acetone–toluene binary solvents were also reported by Granberg and Rasmuson [14] and are used here for comparison with the model prediction.

As in the previous example, the COSMO files of the compounds were calculated on the BP86/TZVP level of theory. Absolute solubility predictions are calculated using the experimental data for ΔG_{fus} of acetaminophen and employing the iterative refinement procedure for the solubility. The calculations are done for the compositions that were measured by Granberg and Rasmuson (Figure 20.9).

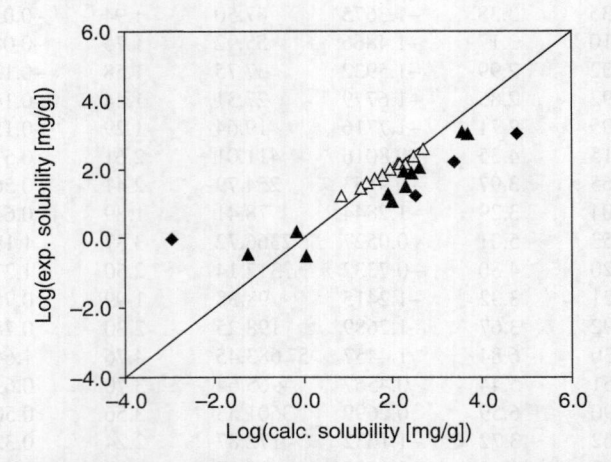

FIGURE 20.8 Predicted absolute solubility of acetaminophen versus experimental data in pure solvents at 303.15 K. Empty triangles (△) represent absolute solubility data of alcoholic solvents and water; solid triangles (▲) represent the remainder of the solvent data set. Four outliers (carbon tetrachloride, diethylamine, 1,4-dioxane, tetrahydrofuran) represented by solid diamonds (◆).

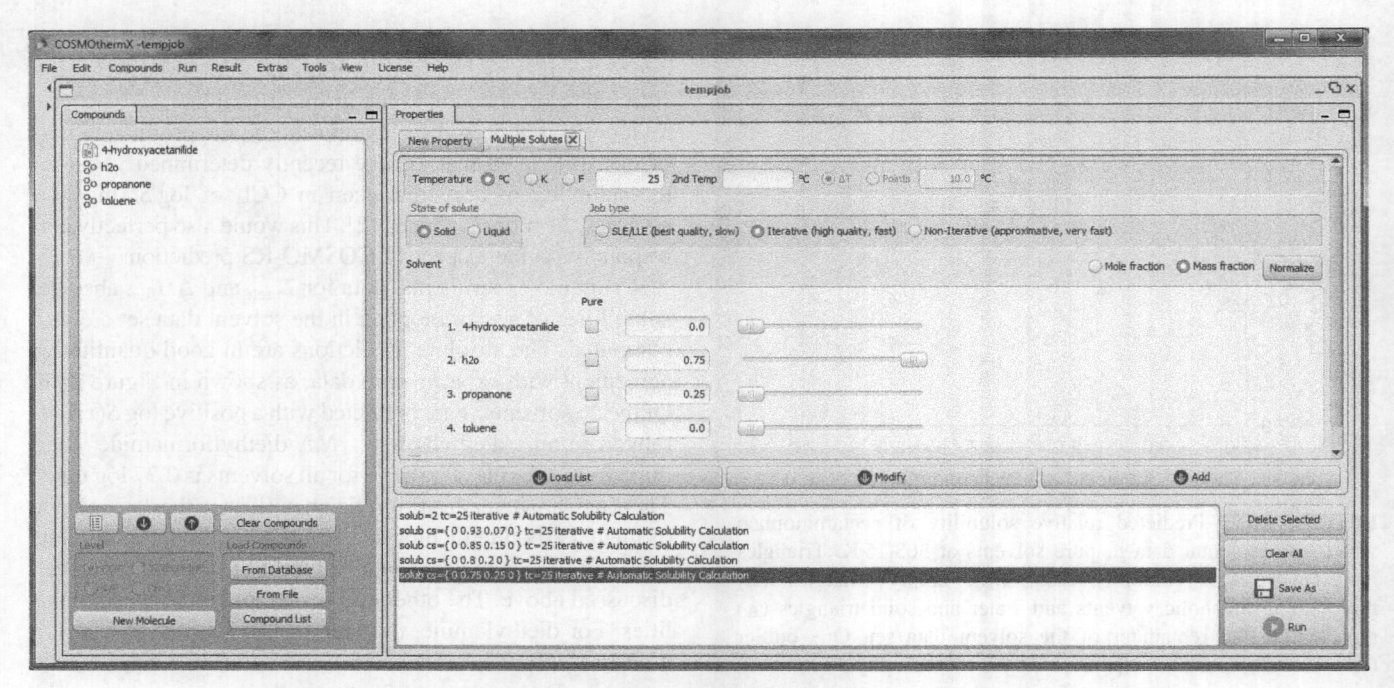

FIGURE 20.9 Input section of COSMO*therm*X with entries for a list of solvent compositions.

The predicted solubilities of acetaminophen in the binary solvent system can be extracted from the COSMO*therm* table file shown in Figure 20.10. Table 20.2 lists the predicted solubilities together with the experimental data.

Figures 20.11 and 20.12 show the prediction results for the acetaminophen solubility in water–acetone and acetone–toluene binary solvent mixtures. The solubility of acetaminophen in the water–acetone binary mixture is nonideal, with a solubility peak at approximately 70% mass fraction of acetone. The nonideal solubility behavior is also captured by the model prediction, with the solubility peak slightly shifted to higher acetone content of the binary solvent. Since the solubility in pure acetone is overpredicted by 0.59 log units, the model prediction for the binary mixture does not show the strong decrease in solubility for very high acetone content of the solvent mixture that is found in the experiment.

The prediction results for the acetone–toluene system (Figure 20.12) are consistent with the trends exhibited by the experimental data. Again, we see effects of the overpredicted solubility in pure acetone and the lower predicted

```
Solubility at T =  298.15 K  at given concentration CS={ 0.0 85 15 0 } -
energies are in kcal/mol volume is in A^3

Nr Compound              log10(x_solub)      mu(self)      mu(solv)       DG_fus
 1 4-hydroxyacetanilide     -2.39941309    -5.22146393   -3.98117592    2.03362348
 2 h2o                       0.00000000    -2.98262140   -2.98262140    0.00000000
 3 propanone                 0.00000000    -1.72077455   -1.72077628    0.00000000
 4 toluene                  -3.08354307    -4.65191005   -0.44452822    0.00000000
```

FIGURE 20.10 Excerpt from the COSMO*therm* table file for the solubility calculation of acetaminophen in a binary solvent system.

TABLE 20.2 Experimental and Predicted Solubilities of Acetaminophen in Water–Acetone and Acetone–Toluene Binary Solvent Mixtures

% Mass Fraction			Experimental		Predicted			
Water	Acetone	Toluene	c_S (g/kg)	Log S (mg/g)	Log $S(x)$	S (mg/g)	Log S (mg/g)	Error
100	0	0	14.90	1.17	−3.1508	5.93	0.77	−0.40
93	7	0	28.18	1.45	−2.7658	13.69	1.14	−0.31
85	15	0	53.0	1.72	−2.3994	29.99	1.48	−0.25
80	20	0			−2.1978	45.87	1.66	1.66
75	25	0			−2.0125	67.47	1.83	1.83
70	30	0	150.0	2.18	−1.8424	95.66	1.98	−0.20
65	35	0			−1.6879	130.58	2.12	2.12
60	40	0			−1.5500	171.24	2.23	2.23
55	45	0			−1.4290	215.47	2.33	2.33
50	50	0	327.0	2.51	−1.3242	260.58	2.42	−0.10
30	70	0	454.6	2.66	−1.0260	408.68	2.61	−0.05
15	85	0	420.3	2.62	−0.8851	452.18	2.66	0.03
7	93	0	302.2	2.48	−0.8364	438.41	2.64	0.16
3	97	0	197.1	2.29	−0.8247	415.72	2.62	0.32
0	100	0	99.8	2.00	−0.8285	386.32	2.59	0.59
0	95	5	91.7	1.96	−0.8517	359.46	2.56	0.59
0	90	10	82.4	1.92	−0.8774	332.39	2.52	0.61
0	85	15	75.7	1.88	−0.9062	305.14	2.48	0.61
0	80	20	66.4	1.82	−0.9386	277.63	2.44	0.62
0	70	30	52.8	1.72	−1.0165	222.79	2.35	0.63
0	60	40	37.08	1.57	−1.1186	168.79	2.23	0.66
0	50	50	26.56	1.42	−1.2563	117.58	2.07	0.65
0	30	70	8.55	0.93	−1.7149	37.19	1.57	0.64
0	20	80	3.39	0.53	−2.1018	14.50	1.16	0.63
0	15	85	2.12	0.33	−2.3663	7.68	0.89	0.56
0	7	93	0.78	−0.11	−2.9938	1.73	0.24	0.35
0	0	100	0.37	−0.43	−4.6734	0.03	−1.46	−1.03

Log $S(x)$ indicates the logarithmic solubility in mole fractions.
Source: Data from Granberg and Rasmuson [14].

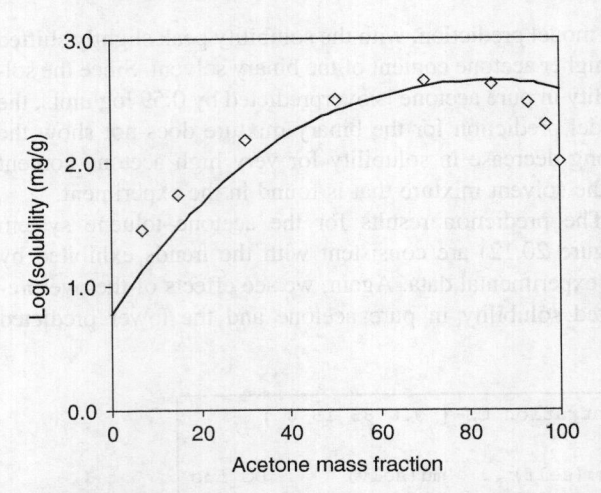

FIGURE 20.11 Experimental and predicted solubility of acetaminophen in acetone–water binary mixtures at 298.15 K. Diamonds ($\Diamond$) are experimental data; the solid line is from the model prediction.

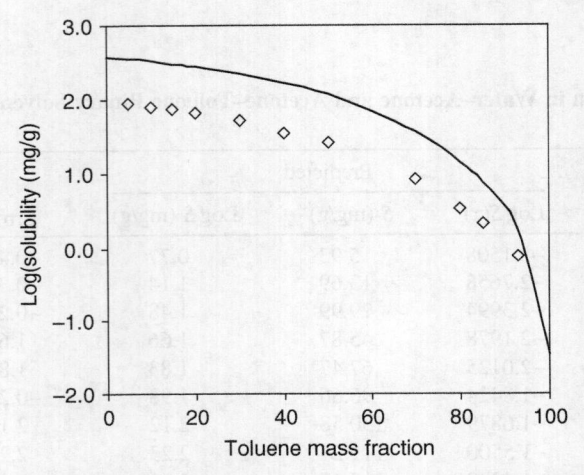

FIGURE 20.12 Experimental and predicted solubility of acetaminophen in toluene–acetone binary mixtures at 298.15 K. Diamonds ($\Diamond$) are experimental data; the solid line is from the model prediction.

solubility in pure toluene, but the relatively ideal solubility behavior of acetaminophen in the acetone–toluene binary mixture is found also by the model prediction.

20.3 CHEMICAL REACTIONS IN SOLUTION

Calculation of reaction energies in the gas phase is a standard application in quantum chemistry. The computational prediction of free energies of reaction in solution is more involved, but still a rather straightforward procedure. Generally, the free energy of reaction is the difference of the total free energies of the reactants and the products of the reaction. For a reaction

$$aA + bB \rightarrow cC + dD,$$

where A and B are the reactants with stoichiometric coefficients a and b and C and D are the reaction products with stoichiometric coefficients c and d, the free energy of reaction can be calculated from the difference of the sums of free energies on both sides of the reaction:

$$\Delta G_r = [c \cdot G(C) + d \cdot G(D)] - [a \cdot G(A) + b \cdot G(B)] \quad (20.13)$$

The free energies of the reactants and products, and thus the free energy of reaction, depend on the conditions under which the reaction takes place. The free energy of reaction in the gas phase differs from the free energy of reaction in solution, and it is different in each specific solvent.

In solution, the Gibbs free energy of a species is

$$G(i) = E_{gas}(i) + \Delta G_{solv}(i) \quad (20.14)$$

Here $E_{gas}(i)$ is the gas-phase energy of the compound, and for computational predictions of the reaction free energy, it should be taken from an adequate quantum chemical (DFT or post-Hartree–Fock) level. $\Delta G_{solv}(i)$, the free energy of solvation, describes the change of the free energy that occurs when the compound is dissolved from the gas phase into the liquid phase. This contribution to the total free energy of a compound can be computed using COSMO-RS.

Using the gas-phase energies of the compounds and the free energies of solvation of the compounds, the free energy of reaction in solution can be calculated according to a thermodynamic cycle as depicted in Figure 20.13.

In order to compute the lower horizontal leg of the cycle, corresponding to the reaction in solution, we have to take the appropriate sums and differences of the upper horizontal leg, i.e. the gas-phase reaction, and the vertical legs, i.e. the solvation energies of the compounds:

$$\Delta G_r(\text{sol}) = \Delta G_r(\text{gas}) + [c\Delta G_{solv}(C) + d\Delta G_{solv}(D)] \\ - [a\Delta G_{solv}(A) + b\Delta G_{solv}(B)] \quad (20.15)$$

$\Delta G_r(\text{gas})$ can be calculated from the chemical potential of the compounds in the gas phase. As already mentioned, the quantum chemical gas-phase energy E_{gas} of a compound is computed in vacuum at absolute zero. Furthermore, E_{gas} does not account for vibrational motion that is present even at $T = 0$ K. The so-called zero-point energy or zero-point vibrational energy (ZPE) can be computed quantum chemically from the vibrational frequencies of the compound and is a standard correction to E_{gas}. Using the ZPE, the free energy of a compound can be calculated as

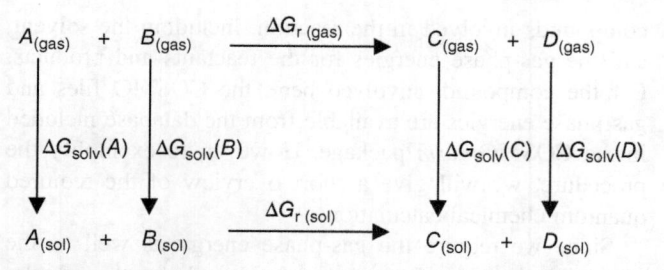

FIGURE 20.13 Cycle for computation of a free energy change in solution.

$$G(i) = E_{gas}(i) + ZPE(i) + \Delta G_{solv}(i) \quad (20.16)$$

As a further refinement for the gas-phase energies of the compounds and the resulting reaction energy, the temperature-dependent thermal contributions to the free energy μ_{vib} of the molecule can be calculated. From vibrational frequencies, the molecular translational, rotational, and vibrational partition functions, q_{trans}, q_{rot}, and q_{vib}, can be calculated, thus enabling prediction of thermodynamic functions at temperatures other than 0 K and finite pressure:

$$G(i) = E_{gas}(i) + ZPE(i) - RT\ln(q_{trans} \cdot q_{rot} \cdot q_{vib}) + \Delta G_{solv}(i)$$
$$= E_{gas}(i) + \mu_{vib} + \Delta G_{solv} \quad (20.17)$$

The other terms required for Eq. (20.15), the free energies of solvation ΔG_{solv}, can be obtained from a COSMO*therm* prediction of the reverse process, i.e. from a vapor pressure prediction. The partial vapor pressure $P(i)$ that is calculated by COSMO*therm* corresponds to the pure compound vapor pressure times the activity coefficient and is related to ΔG_{solv} by

$$\Delta G_{solv}(i) = RT\ln(10)[\log_{10}(P(i)) - \log_{10}(P)] \quad (20.18)$$

with P being the reference pressure at which the reaction takes place.

Note that if the reactant or product compounds are present in the mixture at a finite concentration with a mole fraction $x(i)$ (e.g. if the reaction takes place in bulk reactant liquid), an entropic contribution $RT\ln(x(i))$ of the compound has to be added to the compounds' free energy $G(i)$.

20.3.1 Heat of Reaction

The heat of reaction or reaction enthalpy in solution can be calculated by a procedure similar to the free energy of reaction. Instead of the free energy of solvation of the compounds, we make use of the heat of vaporization ΔH_{vap}. Since ΔH_{vap} is the enthalpy that is needed to transfer the compound from the liquid phase to the gas phase, it has to be subtracted from the gas-phase energy to obtain the enthalpy of the compound in solution:

$$H(i) = E_{gas}(i) - \Delta H_{vap}(i) \quad (20.19)$$

ZPE corrections or thermal correction terms for the enthalpy in the gas phase can also be used as corrections to the gas-phase energies of the compounds for the calculation of the heat of reaction.

Similarly to Eq. (20.13) for the free energy of reaction ΔG_r, the heat of reaction ΔH_r can be calculated from the enthalpy of the compounds:

$$\Delta H_r = [c \cdot H(C) + d \cdot H(D)] - [a \cdot H(A) + b \cdot H(B)] \quad (20.20)$$

20.3.2 Equilibrium Constants

The equilibrium constant K of a reaction is related to the free energy of reaction by

$$\ln K = -\frac{\Delta G^0}{RT} \quad (20.21)$$

For a reaction in an ideal solution, i.e. in infinite dilution, the equilibrium constant can be calculated using the reaction free energy in solution according to Eq. (20.15). The free energies of the individual compounds can be computed using different quantum chemical correction terms as described above.

20.3.3 Accuracy

In the described procedure the free energies of the compounds are calculated from two main contributions, the quantum chemical gas-phase energy and the free energy of solvation. The accuracy of the resulting reaction energy is determined mainly by the accuracy of the underlying quantum chemical method. With DFT methods like the BP86 functional, errors of the absolute reaction energy can be in the range of 10 kcal/mol or more [15, 16]. However, for relative reaction energies of one reaction in different solvents in a solvent screening application, i.e. if we are looking at the variation of the solvation energy only, calculated as the COSMO*therm* contribution in the liquid phase, the accuracy is much higher. For such relative reaction energy predictions considering ΔG_{solv} or ΔH_{vap} from the COSMO*therm* vapor pressure prediction only, an accuracy of 0.5 kcal/mol can be expected.

For a higher accuracy of absolute predictions of the reaction energy or enthalpy, it follows that a more accurate quantum chemical method should be applied for the calculation of the gas-phase energy, e.g. the MP2 or coupled-cluster methods combined with adequate basis sets. Quantitative improvement of the total free energy of a compound $G(i)$ can also be achieved by the ZPE and thermal corrections to the gas-phase energy of the compound.

20.3.4 Calculation of the Free Energy of Reaction and Heat of Reaction

A procedure for the computational prediction of the free energy of reaction ΔG_r or heat of reaction ΔH_r using quantum chemical gas-phase energies and free energies of solvation or heats of vaporization from COSMO-RS is described in the following:

- Compute the reactant and product molecules using the DFT methods for which COSMO*therm* is parameterized. Both COSMO and gas-phase quantum chemical calculations are required. Then use COSMO*therm* to obtain the ΔG_{solv} and/or ΔH_{vap} values.
- Compute gas-phase energies E_{gas} of the reactant and product molecules with a high-level ab initio method, e.g. the coupled-cluster method.
- Compute vibrational frequencies for reactant and product molecules to obtain the ZPE correction or the thermal corrections to the gas-phase energies of the compounds. Vibrational frequency calculations at a DFT level of theory are usually sufficiently accurate.
- Combine E_{gas}, ZPE or thermal corrections, and ΔG_{solv} to compute the compounds $G(i)$ according to Eqs. (20.16) and (20.17) or E_{gas}, ZPE or thermal corrections, and ΔH_{vap} to compute the compounds $H(i)$.
- Calculate the free energy of reaction ΔG_r or the heat of reaction ΔH_r from the difference of the sums of free energies or enthalpies on both sides of the reaction as described by Eqs. (20.13) and (20.20).

EXAMPLE PROBLEM 20.3 Estimate the Heat of Reaction ΔH_r for the Reduction of Nitrobenzene to Aniline in the Liquid Phase in THF at 25 °C:

$$PhNO_2\ (\mathbf{1}) + 3H_2\ (\mathbf{2}) \rightarrow PhNH_2\ (\mathbf{3}) + 2H_2O\ (\mathbf{4})$$

Following the procedure described above, we calculate the gas-phase energies and the heats of vaporization. In this example, we calculate the gas-phase energies of the compounds on the MP2/TZVPP quantum chemical level. Furthermore, we employ the zero-point vibrational energy (ZPE) and thermal corrections to the enthalpy in the gas phase to refine the calculated heat of reaction.

For the calculation of the heat of vaporization with COSMO*therm*, we have to provide COSMO files for all compounds involved in the reaction, including the solvent, and the gas-phase energies for the reactants and products. For the compounds involved here, the COSMO files and gas-phase energies are available from the database included in the COSMO*therm* package. However, to exemplify the procedure, we will give a short overview of the required quantum chemical calculations.

Since we require the gas-phase energy as well as the screening charge surface of the compounds for the calculation of the heat of vaporization, we need to do geometry optimizations both in the gas phase and in the conductor using the COSMO model with ideal screening. COSMO*therm* is parameterized for the BP86/TZVP and the BP86/SVP//AM1 quantum chemical levels. Here, we will use the higher one of the two levels, i.e. the BP86 functional and the TZVP basis set. The QC calculations are performed using the TURBOMOLE [17] quantum chemical program suite.

For the QC calculation we have to provide 3D structures of the compounds, which can be generated by an external tool or built with the molecular builder of the TURBOMOLE graphical user interface. With the starting geometries, the following QC calculations are performed for each compound:

- A gas-phase geometry optimization on the BP86/TZVP quantum chemical level.
- A geometry optimization on the BP86/TZVP/COSMO quantum chemical level.
- A gas-phase geometry optimization on the MP2/TZVPP quantum chemical level.
- A vibrational frequency calculation with the optimized BP86/TZVP molecular gas-phase structure to obtain the ZPE.
- Using the results from the vibrational frequency calculations, we also compute the thermal contributions with the corresponding interactive tool of the TURBOMOLE suite.

These steps are described in more detail in the Appendix. Further information about how the quantum chemical calculations are carried out can also be found in the documentation of the TURBOMOLE program suite and the TmoleX documentation.

The heats of vaporization H_{vap} are calculated with COSMO*therm* at a temperature of 25 °C and a solvent composition of pure tetrahydrofuran (THF). For the COSMO-RS vapor pressure prediction, quantum chemical gas-phase energies from the BP/TZVP level are used to calculate the chemical potential of the compound in the gas phase.

The calculated data for the heats of vaporization of the compounds can be extracted from the COSMO*therm* output file (Figure 20.14). In Table 20.3 the results for the individual energy contributions of the compounds from both the QC calculations and the COSMO*therm* calculation are tabulated

```
Results for mixture    1
-----------------------
Temperature            :       298.150 K

Compound Nr.           :         1         2         3         4         5
Compound               :    nitrobenz  aniline      h2       h2o       thf
Mole Fraction          :     0.0000    0.0000    0.0000    0.0000    1.0000

Compound:   1   (nitrobenzene)
Chemical potential of the compound in the mixture :        -3.54480 kcal/mol
Log10(partial pressure [mbar])                    :        -0.32317
Free energy of molecule in mix (E_COSMO+dE+Mu)    :   -274197.04529 kcal/mol
Total mean interaction energy in the mix (H_int)  :        -5.23132 kcal/mol
Misfit interaction energy in the mix (H_MF)       :         2.46378 kcal/mol
H-Bond interaction energy in the mix (H_HB)       :        -0.27976 kcal/mol
VdW interaction energy in the mix (H_vdW)         :        -7.41534 kcal/mol
Ring correction                                   :        -1.14821 kcal/mol
Vapor pressure of compound over the mixture       :         0.00000 mbar
Chemical potential of compound in the gas phase   :         0.98955 kcal/mol
Heat of vaporization                              :        12.02087 kcal/mol
```

FIGURE 20.14 Excerpt from the mixture output section of the COSMO*therm* output file for a vapor pressure calculation of compounds in pure tetrahydrofuran at 25 °C.

TABLE 20.3 Gas-Phase Energies, Heat of Vaporization, Zero-Point Vibrational Energies, Thermal Corrections, Total Enthalpies of Compounds, and Heat of Reaction for the Hydrogenation of Nitrobenzene

	1	2	3	4	ΔH_r MP2/TZVPP	(BP86/TZVP)
H_{vap}	12.02	1.35	14.54	10.24		
E_{gas} (BP86/TZVP) (Hartree)	−436.944 122	−1.177 446	−287.715 215	−76.465 165		
E_{gas} (MP2/TZVPP) (Hartree)	−435.958 485	−1.164 647	−287.000 283	−76.323 461		
ZPE (BP86/TZVP) (Hartree)	0.099 295	0.009 850	0.113 236	0.020 640		
ΔH_T (thermal corrections)	67.36	8.26	75.45	15.32		
$H = E_{gas} - H_{vap}$	−273 580.11	−732.17	−180 109.95	−47 903.93	−141.19	(−125.06)
$H = E_{gas} + ZPE - H_{vap}$	−273 517.80	−725.99	−180 038.89	−47 890.98	−125.08	(−108.95)
$H = E_{gas} + \Delta H_T - H_{vap}$	−273 512.75	−723.92	−180 034.50	−47 888.61	−127.22	(−111.09)

Enthalpy terms are in kcal/mol; quantum chemical gas-phase energies and ZPE are in Hartree.

together with the resulting data for the heat of reaction. The data for the heat of reaction were calculated from $\Delta H_r = [H(3) + 2 \cdot H(4)] - [H(1) + 3 \cdot H(2)]$, taking into account the stoichiometry of the reaction (Eq. 20.20). For the enthalpy values of the compounds $H(i)$, different correction terms for the quantum chemical gas-phase energies were employed. For comparison, heats of reaction using the quantum chemical gas-phase energies from the MP2/TZVPP level and the BP86/TZVP level are tabulated.

The catalytic reduction of aromatic nitro compounds in the gas phase is known to be a highly exothermic process. For the gas-phase reduction of nitrobenzene, a heat of reaction of $\Delta H_r = -131 \pm 3$ kcal/mol was published [18]. Absolute values for the heat of reaction of the reduction of nitrobenzene to aniline in THF solution could not be found in the literature, but heats of reaction for the reduction of other R-NO_2 compounds in solution have been found independent of R or solvent to be in the range of −125 to −130 kcal/mol.[1]

Using MP2/TZVPP gas-phase energies for the compounds, thermal corrections, and the heat of vaporization from the COSMO-RS prediction, the calculated heat of reaction in solution is $\Delta H_r = -127$ kcal/mol. With ZPE correction only, the calculated heat of reaction in solution is $\Delta H_r = -125$ kcal/mol. Both predicted values are well inside the range of the experimental data for comparable reactions. In contrast, when ZPE and thermal corrections are ignored, the heat of reaction is overestimated by several kcal/mol with a value of $\Delta H_r = -141$ kcal/mol. Table 20.3 also shows heats of reaction calculated from the BP86/TZVP gas-phase energies. The heat of reaction without ZPE or thermal corrections is $\Delta H_r = -125$ kcal/mol, which agrees well with experimental data, but when the correction terms, which should in general lead to a better prediction, are included, the heat of reaction in solution is significantly underestimated with predicted values of $\Delta H_r = -109$ kcal/mol and $\Delta H_r = -111$ kcal/mol, respectively. However, it should be noted that absolute errors in the range of 10–20 kcal/mol are not unusual for pure DFT functionals like BP86 [14, 15].

[1] am Ende, D.J., private communication, 2009.

EXAMPLE PROBLEM 20.4 Estimate Reaction Free Energy, Equilibrium Constant, and Equilibrium Composition for the Reaction of 1-Methoxy-2-Propanone 1 and Isopropylamine 2

First, the gas-phase energies of the reactants and products of the transamination reaction (Figure 20.15) are calculated quantum chemically. In this example, we also employ the MP2 level of theory and the TZVPP basis set. As starting structures for the gas-phase geometry optimizations, the 3D structures from the BP86/TZVP level are used, which are available from the COSMO*base*, a database of validated COSMO files and gas-phase structures. With these structures we perform gas-phase geometry optimizations using the MP2/TZVPP method and basis set combination.

As a further refinement for the gas-phase energies, vibrational frequency calculations for the compounds are carried out, and the thermal contributions from the vibrational frequencies to the gas-phase energy are computed using the corresponding tool of the TURBOMOLE suite. These calculations are carried out on the BP86/TZVP level.

In our next step for the calculation of the reaction free energy, the free energy of solvation of each compound is calculated from Eq. (20.18) using the partial pressures from the vapor pressure prediction of the COSMO*therm* program.

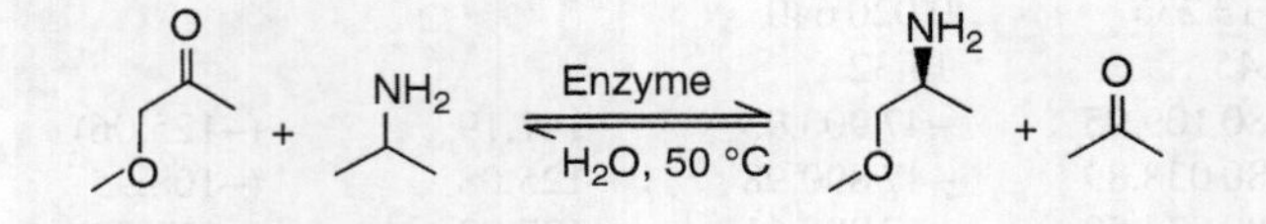

Reactants, products, and solvent for the reaction are taken from the COSMO*base*. The conditions for the vapor pressure calculation are set to a temperature of 50 °C, and the solvent composition is set to pure water. Quantum chemical gas-phase energies of the compounds are used to calculate the chemical potentials of the compounds in the gas phase. The calculated partial pressures can be extracted from the COSMO*therm* output file (Figure 20.16).

Energy terms and corrections for the reactants and products from this procedure are tabulated in Table 20.4. Calculated data for the free energy of reaction ΔG_r are also tabulated using different QC correction terms. ΔG_r is calculated according to the stoichiometry of the reaction as $\Delta G_r = [G(3) + G(4)] - [G(1) + G(2)]$.

With the MP2/TZVPP gas-phase energies and the solvation free energies of the compound, the free energy of reaction in solution is $\Delta G_r = -3.27$ kcal/mol, corresponding to an equilibrium constant of $\ln K = 5.10$. The zero-point vibrational energy corrections for the reactants and products have very little influence on the overall reaction energy ($\Delta G_r = -3.29$ kcal/mol, $\ln K = 5.12$ kcal/mol), but when thermal corrections are included in the free energies $G(i)$ of the compounds, the free energy of reaction decreases to $\Delta G_r = -2.20$ kcal/mol, and the equilibrium constant decreases to $\ln K = 3.43$. The equilibrium constant indicates that the equilibrium position of the reaction lies on the right-hand side of the reaction equation. The relative amount of reactants at equilibrium is 0.08 each, and the relative amount of products at equilibrium is 0.42 each. The experimental equilibrium constant for the reaction of 1-methoxy-2-propanone with isopropylamine is $K = 7.8$ ($\ln K = 2.05$) [11]. Thus, the calculated free energy of reaction and equilibrium constant are in excellent agreement with the experimental data.

FIGURE 20.15 Catalytic transamination of 1-methoxy-2-propanone and isopropylamine.

```
Results for mixture    1
-----------------------
Temperature               :      323.150 K

Compound Nr.              :        1         2         3         4         5
Compound                  :    isopropyl 1-methoxy 1-methoxy propanone     h2o
Mole Fraction             :      0.0000    0.0000    0.0000    0.0000    1.0000

Compound: 1  (isopropylamine)
Chemical potential of the compound in the mixture   :        -1.17237 kcal/mol
Log10(partial pressure [mbar])                      :         4.11202
Free energy of molecule in mix (E_COSMO+dE+Mu)      :   -109542.55539 kcal/mol
Total mean interaction energy in the mix (H_int)    :        -7.42349 kcal/mol
Misfit interaction energy in the mix (H_MF)         :         2.15679 kcal/mol
H-Bond interaction energy in the mix (H_HB)         :        -4.65492 kcal/mol
VdW interaction energy in the mix (H_vdW)           :        -4.92535 kcal/mol
Ring correction                                     :         0.00000 kcal/mol
Vapor pressure of compound over the mixture         :         0.00000 mbar
Chemical potential of compound in the gas phase     :        -2.81690 kcal/mol
Heat of vaporization                                :        10.89292 kcal/mol
```

FIGURE 20.16 Excerpt from the mixture output section of the COSMO*therm* output file for a vapor pressure calculation of compounds in pure water at 50 °C.

TABLE 20.4 Energy Terms and Energy Corrections for the Reactants and Products for Reaction 2 and Free Energies of Reaction with the Various Correction Terms

	1	2	3	4	ΔG_r	ln K
$\text{Log}_{10}(P(i)\text{ [mbar]})$	2.720 32	4.112 02	3.123 39	3.755 18		
G_{solv}	−0.41	1.64	0.18	1.12		
E_{gas} (MP2/TZVPP) (Hartree)	−307.100 753	−174.113 883	−288.440 644	−192.779 317		
ZPE (BP86/TZVP) (Hartree)	0.112 968	0.117 222	0.149 297	0.080 872		
μ_{vib} (thermal corrections)	54.44	227.770	71.54	32.03		
$G = E_{\text{gas}}$ (MP2) + G_{solv}	−192 709.05	−109 256.47	−180 999.06	−120 969.74	−3.27	5.10
$G = E_{\text{gas}}$ (MP2) + G_{solv} + ZPE	−192 638.16	−109 182.91	−180 905.38	−120 918.99	−3.29	5.12
$G = E_{\text{gas}}$ (MP2) + G_{solv} + μ_{vib}	−192 660.99	−109 202.03	−180 927.52	−120 937.70	−2.20	3.43
Relative amount of compound in equilibrium	0.08	0.08	0.42	0.42		

The relative amount of the compounds in equilibrium has been calculated from the equilibrium constant ln K = 3.43. Free energy terms and chemical potential are in kcal/mol; quantum chemical gas-phase energies and ZPE are in Hartree.

20.4 SCREENING OF CO-CRYSTALS

Co-crystalline materials consisting of an active pharmaceutical ingredient (API) and an excipient, usually termed coformer (co-crystal former), have attracted strong interest in recent years by pharmaceutical research and development. This is mostly due to the fact that co-crystals provide a leverage to modify the physicochemical properties of a drug substance, in particular its solubility and dissolution behavior and hence its bioavailability. In addition, they open up new interesting possibilities from an intellectual property and patent perspective. This has generated the need for the rational design of such systems and for approaches that go beyond a simple trial-and-error process.

The application of COSMO-RS theory to this field emerged rather recently when it was demonstrated that the excess enthalpy (or mixing enthalpy) of the supercooled co-crystal melt as computed by COSMO*therm* gives rather good correlation with the materials' tendency to co-crystallize [19].

The excess enthalpy H_{ex} of a mixture is defined as the enthalpy difference between the mixture (H_{mix}) and the pure state (H_{pure}) for each component i with mole fraction x_i:

$$H_{\text{ex}} = \sum_i x_i H_{\text{mix}}^i - \sum_i x_i H_{\text{pure}}^i \quad (20.22)$$

The excess enthalpy H_{ex} corresponds to the mixing enthalpy ΔH_{mix} of the system, whereas the latter usually refers to integer moles of the reactants. The more negative the excess enthalpy, the more the compounds prefer the mixture over their pure states. In this respect H_{ex} of a binary system may also be understood as a quantifier of complementarity of two structures. Figure 20.17 shows a histogram of the computed H_{ex} for about 500 molecule pairs including experimentally confirmed co-crystals (blue) and failed co-crystallization attempts (red). It is quite apparent that there is a significant correlation with the tendency to co-crystallize, in particular at regions with strongly negative H_{ex}. For positive H_{ex} co-crystal formation becomes significantly less probable, whereas in the area slightly below 0 kcal/mol, the discrimination between co-crystals and non-co-crystals is more difficult. A similar picture as in 4.1 would be obtained using the Gibbs free energy of mixing ΔG_{mix}, which is related to H_{ex}:

$$\Delta G_{\text{mix}} = G_{\text{ex}} + RT \sum_i x_i \ln(x_i) = H_{\text{ex}} - TS_{\text{ex}} \quad (20.23)$$

Here, G_{ex} is the excess free energy and S_{ex} the excess entropy in the mixture. In retrospective, the relation of liquid-phase properties with co-crystal formation can be better understood by the examination of the thermodynamic cycle for co-crystallization (Figure 20.18) that links the free energy of co-crystal formation $\Delta_r G_{\text{cocrystal}}$ to the free energy of mixing ΔG_{mix}.

In this thought experiment the drug A and the coformer B are transferred into their respective supercooled liquid state by expending the free energy of fusion ΔG_{fus}. Mixing of A and B yields the co-crystal $A_m B_n$ in its supercooled liquid state (ΔG_{mix}), which then upon solidification gives the final co-crystal. The correlation of the mixing or excess energies ($H_{\text{ex}}, \Delta G_{\text{mix}}$) with co-crystal formation can be explained by the assumption that the remaining terms, the sum of $\Delta G_{\text{fus}}(A)$ and $\Delta G_{\text{fus}}(B)$ of reactants A and B, and the $-\Delta G_{\text{fus}}(AB)$ of the co-crystal are of similar magnitude and hence cancel each other partially when going along the thermodynamic cycle:

$$\Delta_r G_{\text{cocrystal}} = \Delta G_{\text{mix}} + \Delta\Delta G_{\text{fus}} \quad (20.24)$$

Hence, the term $\Delta\Delta G_{\text{fus}}$ (or at least its variance) needs to be small compared with ΔG_{mix} in order to have a good prediction (ranking). Furthermore, the overall entropic changes related to the conversion of *solids A* and *B* to *solid* $A_m B_n$ will be comparably small; thus the enthalpy, i.e. H_{ex}, and not the

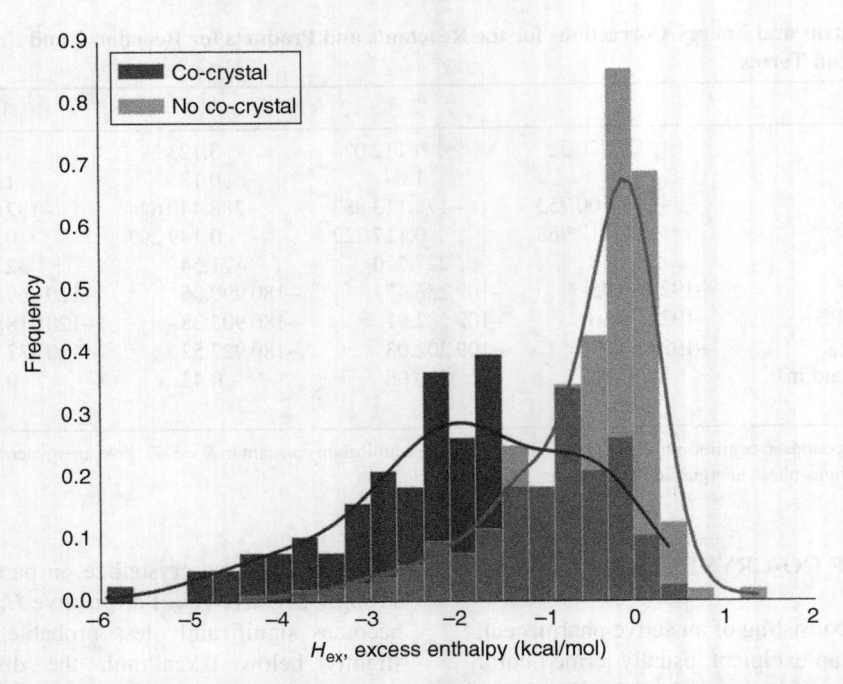

FIGURE 20.17 Histogram and normalized probability distribution for the computed excess enthalpy of a set of about 500 co-crystal experiments.

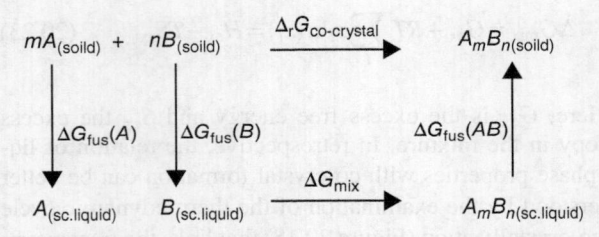

FIGURE 20.18 Thermodynamic cycle for the formation of co-crystals A_mB_n out of components A and B via a hypothetical supercooled liquid state. The free energy of fusion ΔG_{fus} is the energy that is needed to convert the material from the solid state into the supercooled liquid state (sc. liquid) at a specific temperature T.

free energy should be the prevailing quantity in this process. Although the drug–coformer interaction in co-crystals is in many cases dominated by intermediate to strong hydrogen bonding, electrostatics and vdW interaction also play an important role in the condensed state, all of which are properly taken into account by COSMO-RS theory.

So far only thermodynamic effects have been considered; kinetic effects however may also play a role for crystallization. Therefore, for rather large molecules, it is recommended to introduce an additional term, i.e. a penalty, that takes into account the molecular flexibility that poses a major barrier for crystallization. It was found that using this number of rotatable bonds as a simple additional linear term gives an improvement in predictivity for flexible systems. This can be rationalized by the assumption that the probability of crystallization is an exponentially decreasing function of the molecular flexibility that becomes an essentially linear term when being converted to the energy domain.

Table 20.5 shows the result of the excess enthalpy-based screening for the anti-malaria drug artemisinin against 67 potential coformers. Experimentally only two co-crystals, namely, resorcinol and orcinol, have been found. The data from the experimental screening has been taken from the study of Karki et al. [20].

Both experimentally confirmed co-crystals are placed at the fifth and seventh positions on the whole list of 67 tested coformers. This highlights the basic idea of a COSMO-RS-based prescreening that is not to obtain absolute predictions but to improve the chances to successfully find a coformer for a follow-up experimental search. As expected it shows some disagreement with experiment, which has mainly two reasons. Firstly, several approximations have been made from a model point of view, the severest one certainly the total neglect of crystalline long-range order. Secondly, from an experimental point of view, it can never be ruled out that some of the low excess enthalpy pairs indeed form a co-crystal that just has not been found in the experiment. Hence, it is possible that some of the false-positive low excess pairs may still form a co-crystal at the right experimental conditions. In this context a study by Corner et al. has to be mentioned where COSMO*therm* was used to prescreen coformers for the highly polymorphic drug precursor ROY [21]. Here, one of the top predicted coformers (pyrogallol) did not form a

TABLE 20.5 Results of the Co-crystal Screening for the Anti-malaria Drug Artemisinin and 67 Coformers, Where for the Sake of Conciseness Only the Top and the Bottom of the Whole Coformer Set Are Presented

Rank	Coformer	Co-crystal	H_{ex} (kcal/mol)
1	Ethanedisulfonic acid	No	−3.86
2	1,3-Dihydroxynaphthalene	No	−2.93
3	Phloroglucinol	No	−2.53
4	Benzenesulfonic acid	No	−2.01
5	*Resorcinol*	*Yes*	*−1.98*
6	Olivetol	No	−1.94
7	*Orcinol*	*Yes*	*−1.84*
8	Gentisic acid	No	−1.80
9	Cyanuric acid	No	−1.78
10–63	...	No	...
64	Lysine	No	0.78
65	Arginine	No	0.82
66	L-Serine	No	0.82
67	Alanine	No	0.92

Coformers are sorted due to increasing excess enthalpy.
Source: Experimental data taken from Ref. [19].

co-crystal but instead an amorphous complex with ROY. In other words, ROY could be stabilized in its amorphous form due to a COSMO*therm*-predicted strong intermolecular interaction with pyrogallol. This is an important finding considering that usually the amorphous form shows a higher solubility/bioavailability than its crystalline counterparts.

So far, the excess enthalpy is merely a quantity that *ranks* the potential coformers relatively and does not give absolute prediction or 1/0 classification of co-crystal formation. However, according to Figure 20.17, the probability to form a co-crystal decreases strongly close above 0 kcal/mol, and pairs having a positive excess enthalpy can be ruled out as co-crystal formers with high confidence. The introduction of a fixed optimal threshold for a co-crystal vs. non-co-crystal classification is possible, but the exact size of the threshold may vary with the specific targeted drug under consideration.

A thorough statistical analysis of a whole array of drug and coformer sets has been carried out in Reference [19] and in [23] using the area under the curve (AUC) metric from a receiver operating characteristic (ROC) curve. Using an up-to-date COSMO*therm* parameterization, the overall performance including the penalty term amounts to an AUC of about 0.83 (AUC = 0.81 without the penalty term) on a set of 20 drugs. In other words, the approach ranks any drug–excipient pair that forms a co-crystal before any non-co-crystal pair by a probability of about 83%.

An interesting aspect is related to the original objective of improving solubility via transferring the drug into a co-crystal: apart from kinetic effects during dissolution, the thermodynamic solubility can only be improved as compared with the pure drug if there is significant interaction between drug and coformer in solution, thereby reducing the activity of the drug as compared with its state without the coformer. This will however strongly correlate with the strength of the excess enthalpy. Hence screening for low excess enthalpy mixtures will simultaneously improve the chance of finding coformers that improve the overall solubility.

Solvates are strongly related to co-crystal as they differ only by the fact that one of the components is liquid at room temperature. Consequently, some attempts have been undertaken to correlate the excess enthalpy with the tendency of a drug to form a solvate [19, 24, 25]. Again, the correlation is significant, however in many cases less pronounced than for co-crystals. Usually interactions between solvents and drug are smaller than for coformer and drug, and packing effects are more important. In particular solvents having the right shape to fit into voids or channels of the crystalline may form inclusion compounds in the crystalline state that will show a H_{ex} close to zero and hence are out of the scope of the current approach. However, it was shown recently that overall predictions can be improved by considering the shape of the solvents and the topology of the drug [25].

A special subset of solvates are hydrates where the solvent that crystallizes together with the drug is water. Abramov tested the excess enthalpy approach for a set of 61 diverse hydrate- and non-hydrate-forming drugs [22]. The aim of this study was to predict not only the propensity of the pharmaceutical compounds to form a hydrate but also the resistance of co-crystals to hydration at high humidity conditions. Besides G_{ex} and H_{ex}, also other descriptors were taken into account such as the octanol–water partition coefficient (*c*log *P*), topological surface area (TPSA), and different donor–acceptor counts.

As a result it was demonstrated that the excess energy approach based on COSMO-RS offers the most efficient way to probe the tendency of a drug to form a hydrate. Moreover, for the investigated caffeine and theophylline co-crystals, the size of the excess enthalpy correlated well with the improved resistance toward hydration at 98% RH conditions.

EXAMPLE PROBLEM 20.5 Compute the Excess Enthalpy for Bicalutamide and a Set of Potential Coformers

The experimental data for this data set is taken from the work of Zaworotko and coworkers [26] where the drug bicalutamide and several coformers where experimentally screened for their ability to form co-crystals. The target molecule bicalutamide is a nonsteroidal antiandrogen and mainly used in the treatment of prostate cancer. The coformers used in this example are shown in Table 20.6. For a larger list of potential coformers, it is recommended to provide COSMO*thermX* with a list of compound names, for example, in the form of a text file. A subsequent database request is then carried out, and the compounds that are available are automatically added to the compound list. Once all the compounds are selected, the further procedure just consists in selecting the co-crystal module, choosing the right target API, and then adding the potential coformers. After submitting the job the excess enthalpies and mixing free energies are computed according to Eqs. (20.22) and (20.23).

The results of the screening are shown in Table 20.6. As all coformers are rather small and stiff, it was not necessary to take the molecular flexibility into account here, and only the excess enthalpy is shown. Table entries that have formed a co-crystal/non-co-crystal in the laboratory are labeled with a 1 or 0, respectively. The table is sorted according to increasing excess enthalpy, and except for 1,2-bis(4-pyridyl)ethane, the top entries with the lowest excess enthalpy form indeed a co-crystal.

Although co-crystals are most often dominated by hydrogen bonds, this case demonstrates that it is important to consider all intermolecular interactions in a screening. For example, taking only hydrogen bonding into account may lead to the false high ranking of essentially hydrophobic compounds such as 1-hexadecanol that are barely miscible with bicalutamide [27].

20.5 CONCLUSION AND OUTLOOK

In this chapter we presented an overview of three different applications of COSMO-RS in the drug development process. The power and main benefit of the COSMO-RS model is that properties in solution can be obtained from ab initio calculation without any experimental input. It does not require external data for modeling and can also be used when empirical models are not parameterized. Complex multifunctional molecules and new chemical functionalities are treated on the same footing as simple organic molecules.

Prediction of relative drug solubility with COSMO-RS is based on a consistent thermodynamic modeling of interactions in the solvent and the supercooled state of the drug. Solvent mixtures can be treated in the same way as pure solvents and with similar accuracy. Absolute solubility prediction is limited by the availability of free energy of fusion data.

Although COSMO-RS in its present state cannot be proven to be more accurate than more empirical models with many adjusted parameters, its strength is the essential independency from experimental data. This allows for independent modeling and avoids errors resulting from erroneous experimental data on which empirical models rely. Potential improvement to the current COSMO-RS solubility prediction model includes a more accurate fusion term for absolute solubility prediction, and improvement of the COSMO-RS interaction terms themselves, and requires more reliable experimental data as are available at present.

With COSMO-RS, solubility prediction is also possible for salts. This is important as many drugs are formulated as salts. Solvent systems involving ionic liquids can also be treated with very good accuracy [28]. Furthermore, different conformational forms of molecules can be used for solubility screening, and the relative weight of conformers in different solvents can be determined. This feature basically allows for examination of conditions influencing the crystallization process and solvent screening for conformational polymorphism and pseudopolymorphism.[2]

Another application of COSMO-RS frequently used in pharmaceutical and agrochemical industry deals with reaction modeling. In principle, reaction equilibrium, mechanism, rate,

TABLE 20.6 COSMO-RS-Based Co-crystal Screening for Bicalutamide (Example Problem 20.5)

Coformer	H_{ex} (kcal/mol)	Co-crystal
1,2-Bis(4-pyridyl)ethane	−1.57	0
trans-1,2-Bis(4-dipyridyl)ethylene	−1.21	1
4,4′-Bipyridine	−1.13	1
1-Naphthol	−0.99	0
1,3,5-Trihydroxybenzene	−0.89	0
4,4-Biphenol	−0.71	0
4-Phenylpyridine	−0.68	0
1,3-Dihydroxybenzene	−0.62	0
4-Cyanopyridine	−0.24	0
3-Cyanopyridine	−0.19	0
3-Cyanophenol	0.01	0
1-Naphtalenecarbonitrile	0.01	0
4-Cyanophenol	0.10	0
1,3-Benzenedicarbonitrile	0.13	0
1,4-Benzenedicarbonitrile	0.13	0
3-Hydroxypyridine	0.30	0
5-Hydroxyisoquinoline	0.31	0
1-Hexadecanol	1.28	0

Entries are ordered by increasing excess enthalpy H_{ex}.

---

[2] Abramov, Y., Mustakis, J., and am Ende, D.J., private communication, 2009; Geertman, R., private communication and presentation at COSMO-RS symposium, 2009. http://www.cosmologic.de/index.php?cosId=1901&crId=5.

and by-product formation may be solvent dependent. Here we investigated the influence of solvation on the free energy and heat of reaction and the reaction equilibrium only. A straightforward procedure for the computational prediction of the free energy of reaction and heat of reaction has been shown, and the effect of the employed quantum chemical level on the absolute heat of reaction has been demonstrated. Elsewhere, it has been shown that although, depending on the quantum chemical level, absolute values for the free energy of reaction may differ substantially from experimental data, general trends are predicted correctly [29].

Finally, the prediction of co-crystal formation can be carried out using COSMO-RS. Although not being a mere solution process, it is however strongly related to it. In fact, the problem of co-crystal formation that cannot be treated yet rigorously in all its complexity is approximated by assuming the co-crystal to be a supercooled mixture of API and coformer. This is of course a strong simplification, with a larger error bar than the usual COSMO-RS applications. Nevertheless, as it was demonstrated in this chapter, it allows to guide experimental work quite efficiently.

All applications of COSMO-RS require quantum chemical calculations of compounds, taking into account the various molecular conformations of the compounds. This constitutes the computationally most demanding part of the procedure but has to be done only once per compound. The COSMO files of the involved compounds can be reused for other projects and all types of properties. Thus, if combined with a database of precalculated COSMO files for common compounds, thermodynamic property calculations with COSMO*therm* can be carried out quite fast and efficiently.

20.A DETAILS OF COSMO AND GAS- PHASE CALCULATIONS

For later use of the COSMO files in COSMO*therm*, the details of the quantum chemical COSMO calculation should be consistent with one of the parameterizations of COSMO*therm*.

There are three levels of different qualities mainly used in COSMO*therm*: a lower, computationally less expensive level; an intermediate level, which is sufficiently accurate for most problems; and a higher level, which is computationally more time consuming and also resolves some particular problem cases. As the intermediate level is a good compromise between accuracy and efficiency, it has been used throughout in this contribution and requires the following:

- BP86 DFT geometry optimization with a TZVP quality basis set and the RI approximation applied in the gas phase and in the conductor.
- For COSMO calculations only: COSMO applied in the conductor limit ($\varepsilon = \infty$) using optimized element-specific COSMO radii for the cavity construction.
- If more than one conformation is considered to be potentially relevant for a compound, QC calculations have to be done for all conformations.

20.B STEPS FOR CALCULATING THE FREE ENERGY OR ENTHALPY OF A COMPOUND

Apart from the heat of vaporization or free energy of solvation that is calculated by COSMO*therm*, the free energies and enthalpies of compounds involved in the reaction examples are composed from several quantum chemical contributions.

20.B.1 Gas-Phase Energy

The BP86/TZVP gas-phase energy is obtained from a gas-phase geometry optimization of the compound using the settings described in Appendix 20.A. The resulting energy value can be found in output files of the TURBOMOLE program suite. The protocol file `job.last` comprises the output of the last complete cycle and information about the settings that were applied in the program run, e.g. convergence criteria. The gas-phase energy can be read from "`total energy`" line in the output section displayed in Figure 20.B.1. If the graphical user interface TmoleX is used, the gas-phase energy can also be taken from the Energy block of the Job Results panel.

```
******************************************************************
nitrobenzene
******************************************************************

       | total energy       =    -436.94412249634  |

       : kinetic energy     =     435.14545720492  :
       : potential energy   =    -872.08957970126  :
       : virial theorem     =       1.99588353476  :
       : wavefunction norm  =       1.00000000000  :
       ..........................................
```

FIGURE 20.B.1 Excerpt from the TURBOMOLE output file `job.last` from a BP86 gas-phase geometry optimization of nitrobenzene.

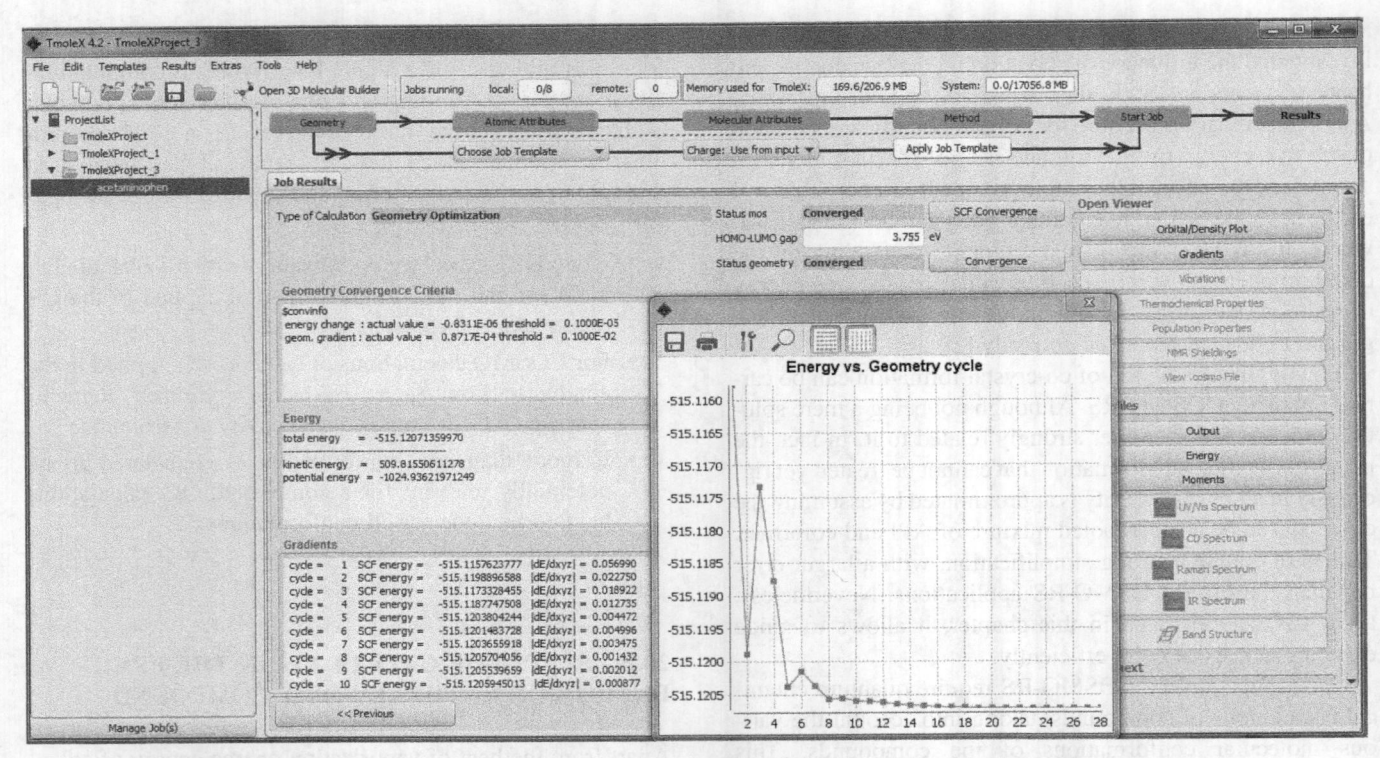

FIGURE 20.B.2 Job Results panel of the graphical user interface TmoleX of the TURBOMOLE program suite.

```
*****************************************************************
*                                                                 *
*<<<<<<<<<<   GROUND STATE FIRST-ORDER PROPERTIES   >>>>>>>>>>>*
*                                                                 *
*****************************************************************


    ---------------------------------------------
     Method       :   MP2
     Total Energy :   -435.9584853137
    ---------------------------------------------
```

FIGURE 20.B.3 Excerpt from the TURBOMOLE output file `job.last` from an MP2 gas-phase geometry optimization of nitrobenzene using the `ricc2` module of TURBOMOLE.

This panel also displays information about the run of the job (Figure 20.B.2).

The MP2/TZVPP gas-phase energy is obtained from an MP2 geometry optimization employing the TZVPP basis set (also called def-TZVPP) and the RI approximation. The module used for this type of calculation is `ricc2` (Figure 20.B.3). In TmoleX, the method has to be set to RI-MP2 in the "Level" section of the "Level of Theory" panel. In the MP2 calculations for this contribution, the 1s orbitals of elements C, N, and O were kept frozen. Details about settings for frozen core orbitals can be found in the TURBOMOLE documentation.

20.B.2 Correction Terms from Vibrational Frequency Calculations

For the ZPE contribution, a vibrational frequency calculation is required. This type of calculation can be done with the `aoforce` module of TURBOMOLE. The ZPE can be taken either directly from the `aoforce.out` file or from the interactive module `freeh`, which also allows for calculation of the molecular partition functions at temperatures other than 0 K and finite pressure. Results of the `freeh` module are printed to standard I/O (Figure 20.B.4). Note that vibrational frequency calculations are based on the assumption of a harmonic oscillator and partition functions are computed within

```
enter the range of temperatures (K) and pressures (MPa)
at which you want to calculate partition sums and free enthalpies :
tstart=(real) tend=(real) numt=(integer) pstart=(real) pend=(real) nump=(integer)
default values are  :
tstart=298.15 tend298.15 numt=1 pstart=0.1 pend=0.1 nump=1
or enter q or * to quit

         ------------------
         your wishes are :
         ------------------
 pstart=   0.1000      pend=    0.1000     nump=    1
 tstart=    298.1      tend=     298.1     numt=    1
         zero point vibrational energy
         -----------------------------
         zpe=   260.7    kJ/mol
   T        p       ln(qtrans) ln(qrot) ln(qvib)  chem.pot.    energy      entropy
  (K)     (MPa)                                   (kJ/mol)    (kJ/mol)   (kJ/mol/K)
 298.15  0.1000000     17.82     12.90      3.49   175.88      279.36      0.35539
   T        P              Cv            Cp
  (K)     (MPa)        (kJ/mol-K)    (kJ/mol-K)
 298.15  0.1000000      0.1128303     0.1211446
```

FIGURE 20.B.4 Excerpt from the interactive output of the `freeh` module of TURBOMOLE.

the assumption of an ideal gas and no coupling between degrees of freedom. Vibrational frequency calculations can also be started from TmoleX. Start the job as a single-point calculation with the "Frequency Analysis" radio button ticked in the "Job Selection" section of the single-point calculation panel.

The thermal enthalpy contribution ΔH_T can be calculated from the thermal energy contribution printed in the `freeh` output. This is done via $\Delta H_T = \Delta G_T + RT$, where ΔG_T is the value in the "energy" column of the `freeh` output, T is the temperature, and R is the gas constant $R = 8.314\,472$ J/K·mol.

For the calculation of the thermal contribution to the free energy of a compound, μ_{vib}, the `freeh` value for "chem. pot." should be used and not the value from the "energy" column.

After converting the contributions from the individual steps to consistent units, the total enthalpy H of a compound can be calculated as the sum of the individual contributions. For compound nitrobenzene on the MP2/TZVPP level of theory, this involves the sum $H = E_{gas}(\text{MP2}) + \Delta H_T - H_{vap}$ consisting of the following contributions listed in Table 20.3:

- $E_{gas}(\text{MP2}) = -435.958\,485$ Hartree $= -273\,568.09$ kcal/mol from the MP2 gas-phase geometry optimization.
- $\Delta H_T = \Delta G_T + RT = 279.36$ kJ/mol $+ 8.314\,472$ J/(K·mol) $\cdot 298.15$ K $= 281.84$ kJ/mol $= 67.36$ kcal/mol from the vibrational frequency calculation and subsequent computation of partition functions with the `freeh` module.
- $H_{vap} = 12.02$ kcal/mol from the COSMO-RS vapor pressure prediction.

Note that quantum chemical energies are usually expressed in Hartree, which is the atomic unit of energy. The conversion factor to the kcal/mol unit system is 627.5095 kcal/mol.

The free energy G of a compound can similarly be calculated from

$$G = E_{gas}(\text{MP2}) + G_{solv} + \mu_{vib}$$

For compound isopropylamine, this requires the following energy terms, listed in Table 20.4:

- $E_{gas}(\text{MP2}) = -174.113\,883$ Hartree $= -109\,258.16$ kcal/mol from the MP2 gas-phase geometry optimization.
- $\mu_{vib} = 227.77$ kJ/mol, the chemical potential value ("chem. pot.") from a vibrational frequency calculation and subsequent computation of partition functions with the `freeh` module.
- $G_{solv} = 1.64$ kcal/mol, calculated from $\Delta G_{solv}(i) = RT \ln(10)[\log_{10}(P(i)) - \log_{10}(P)]$ (Eq. 20.18). The partial pressure of isopropylamine, $\log_{10}(P(i)) = 4.112\,02$ mbar, was to be taken from a COSMO*therm* vapor pressure prediction (Figure 20.16), and the reference pressure P was assumed to be 1 bar.

ABBREVIATIONS

BP86	approximate DFT functional, consisting of Becke's exchange functional [30] and Perdew's correlation functional [31]
COSMO	COnductor-like Screening MOdel
COSMO-RS	COnductor-like Screening MOdel for Real Solvents
DFT	density functional theory
GCM	group contribution method
MC	Monte Carlo method
MD	molecular dynamics method
MP2	second-order Møller–Plesset perturbation theory
QC	quantum chemistry/quantum chemical
rmse	root mean square error
TZVP, TZVPP	Ahlrichs' triple-zeta valence polarization basis sets [32]
vdW	van der Waals interaction

SYMBOLS

σ	screening charge density
ΔG_{fus}	free energy of fusion
ΔG_{solv}	free energy of solvation
ΔH_{vap}	heat of vaporization
E_{gas}	gas-phase energy
μ	chemical potential

REFERENCES

1. Klamt, A. and Schüürmann, G. (1993). COSMO: a new approach to dielectric screening in solvents with explicit expressions for the screening energy and its gradient. *J. Chem. Soc. Perkin Trans.* 2 799–805.
2. Klamt, A. (1995). Conductor-like screening model for real solvents: a new approach to the quantitative calculation of solvation phenomena. *J. Phys. Chem.* 99: 2224–2235.
3. Klamt, A. (2005). *COSMO-RS: From Quantum Chemistry to Fluid Phase Thermodynamics and Drug Design*. Amsterdam: Elsevier.
4. Gmehling, J. (1998). Present status of group-contribution methods for the synthesis and design of chemical processes. *Fluid Phase Equilibr.* 144: 37–47.
5. Klamt, A., Jonas, V., Bürger, T., and Lohrenz, J.C.W. (1998). Refinement and parameterization of COSMO-RS. *J. Phys. Chem. A* 102: 5074–5085.
6. Klamt, A. and Eckert, F. (2000). COSMO-RS: a novel and efficient method for the a priori prediction of thermophysical data of liquids. *Fluid Phase Equilibr.* 172: 43–72.
7. Ben-Naim, A. (1987). *Solvation Thermodynamics*. New York and London: Plenum Press.
8. Klamt, A., Eckert, F., Hornig, M. et al. (2002). Prediction of aqueous solubility of drugs and pesticides with COSMO-RS. *J. Comput. Chem.* 23: 275–281.
9. Granberg, R.A. and Rasmuson, A.C. (1999). Solubility of paracetamol in pure solvents. *J. Chem. Eng. Data* 44: 1391–1395.
10. Manzo, R.H. and Ahumada, A.A. (1990). Effects of solvent medium on solubility. V: Enthalpic and entropic contributions to the free energy changes of Di-substituted benzene derivatives in ethanol: water and ethanol: cyclohexane mixtures. *J. Pharm. Sci.* 79: 1109–1115.
11. Matcham, G., Bhatia, M., Lang, W. et al. (1999). Enzyme and reaction engineering in biocatalysis: synthesis of (S)-methoxyisopropylamine (= (S)-1-methoxypropan-2-amine). *Chimia* 53: 584–589.
12. Mota, F.L., Carneiro, A.P., Queimada, A.J. et al. (2009). Temperature and solvent effects in the solubility of some pharmaceutical compounds: measurements and modeling. *Eur. J. Pharm. Sci.* 37: 499.
13. Romero, S., Reillo, A., Escalera, B., and Bustamante, P. (1996). The behavior of paracetamol in mixtures of amphiprotic and amphiprotic-aprotic solvents. Relationship of solubility curves to specific and nonspecific interactions. *Chem. Pharm. Bull.* 44: 1061–1064.
14. Granberg, R.A. and Rasmuson, A.C. (2000). Solubility of paracetamol in binary and ternary mixtures of water + acetone + toluene. *J. Chem. Eng. Data* 45: 478–483.
15. Koch, W. and Holthausen, M.C. (2001). *A Chemist's Guide to Density Functional Theory*. Weinheim: Wiley-VCH.
16. Cramer, C.J. (2002). *Essentials of Computational Chemistry*. Chichester: Wiley.
17. TURBOMOLE V7.2 2017 (2007). A development of University of Karlsruhe and Forschungszentrum Karlsruhe GmbH, 1989–2007. TURBOMOLE GmbH, since 2007.
18. Macnab, J.I. (1981). The role of thermochemistry in chemical process hazards: catalytic nitro reduction processes. *I. Chem. E. Symp. Ser.* 68 (3/S): 1–15.
19. Abramov, Y.A., Loschen, C., and Klamt, A. (2012). Rational coformer or solvent selection for pharmaceutical cocrystallization or desolvation. *J. Pharm. Sci.* 101: 3687.
20. Karki, S., Friscic, T., Fabian, L., and Jones, W. (2010). New solid forms of artemisinin obtained through cocrystallisation. *CrystEngComm* 12: 4038–4041.
21. Corner, P.A., Harburn, J.J., Steed, J.W. et al. (2016). Stabilisation of an amorphous form of ROY through a predicted coformer interaction. *Chem. Commun.* 52: 6537–6540.
22. Abramov, Y.A. (2015). Virtual hydrate screening and coformer selection for improved relative humidity stability. *CrystEngComm* 17: 5216–5224.
23. Loschen, C. and Klamt, A. (2015). Solubility prediction, solvate and cocrystal screening as tools for rational crystal engineering. *J. Pharm. Pharmacol.* 67: 803–811.
24. am Ende, D., Klamt, A., Loschen, C. et al. (2014). *Prediction of Energetic-Energetic Cocrystal Forms and Their Competition*

with *Solvate Formation* [Internet]. Atlanta, GA, USA. https://aiche.confex.com/aiche/2014/webprogram/Paper383003.html (accessed 12 October 2018).

25. Loschen, C. and Klamt, A. (2016). Computational screening of drug solvates. *Pharm. Res.* 33: 2794.

26. Bis, J.A., Vishweshwar, P., Weyna, D., and Zaworotko, M.J. (2007). Hierarchy of supramolecular synthons: persistent hydroxylpyridine hydrogen bonds in cocrystals that contain a cyano acceptor. *Mol. Pharm.* 4: 401–416.

27. Musumeci, D., Hunter, C.A., Prohens, R. et al. (2011). Virtual cocrystal screening. *Chem. Sci.* 2: 883–890.

28. Diedenhofen, M., Eckert, F., and Klamt, A. (2003). Prediction of infinite dilution activity coefficients in ionic liquids using COSMO-RS. *J. Chem. Eng. Data* 48: 475–479.

29. Peters, M., Greiner, L., and Leonhard, K. (2008). Illustrating computational solvent screening: prediction of standard Gibbs energies of reaction in solution. *AIChE J.* 54: 2729–2734.

30. Becke, A.D. (1988). Density-functional exchange-energy approximation with correct asymptotic behaviour. *Phys. Rev. A* 38: 3098–3100.

31. Perdew, J.P. (1986). Density-functional approximation for the correlation-energy of the inhomogenous electron gas. *Phys. Rev. B* 33: 8822–8824.

32. Schäfer, A., Huber, C., and Ahlrichs, R. (1994). Fully optimized contracted Gaussian basis sets of triple zeta valence quality for atoms Li to Kr. *J. Chem. Phys.* 100: 5829–5835.

21

THERMODYNAMIC MODELING OF AQUEOUS AND MIXED SOLVENT ELECTROLYTE SYSTEMS

BENJAMIN CAUDLE, TONI E. KIRKES, CHENG-HSIU YU, AND CHAU-CHYUN CHEN

Texas Tech University, Lubbock, TX, USA

21.1 INTRODUCTION

Electrolyte systems are ubiquitous and play critical roles in a wide variety of industrial and natural processes, including basic chemical manufacturing, pharmaceutical processing, oil and gas production, flue gas desulfurization, CO_2 capture and sequestration, nuclear waste processing, electronic waste treatment, energy production and storage, etc. For example, it is estimated that half of all drug molecules used in medicinal therapy are administered as salts [1]; "produced water" from oil and gas extraction may have salinity up to 30 wt % electrolytes [2]; waste sludge resulting from nuclear materials processing contains electrolytes encompassing much of the periodic table [3]; and electrolytes are the key ingredients in electrochemical energy production and storage systems such as fuel cells, batteries, and electrochemical capacitors [4].

A core element of modern chemical engineering practice is process simulation. It has been widely accepted and routinely practiced by chemical engineers worldwide in the design, debottlenecking, and optimization of chemical processes across all industries [5]. The foundation of successful process simulation is the underlying thermodynamic model that calculates the various thermodynamic properties of a component or mixture based on thermodynamic principles and an understanding of the underlying molecular interactions. As chemical processes inevitably involve multicomponent systems and multiple unit operations such as mixing, separation, reaction, phase change, and heat and mass transport, a central requirement of a successful thermodynamic model is the capability to provide robust predictions of thermodynamic properties of multicomponent systems across process conditions with a minimum of experimental input. It cannot be overemphasized that such a comprehensive model should be extensible to multicomponent systems from thermodynamic properties of lower-order systems such as pure components, binary systems, and ternary systems. However, in the case of electrolytes, until recently thermodynamic modeling has been limited to lower-order and dilute aqueous electrolyte solutions at or near ambient conditions.

As our understanding of electrolyte systems has evolved and expanded in recent decades, advances in electrolyte thermodynamics have started to change the process simulation landscape. Accurate thermodynamic modeling and process simulation of electrolyte systems are now being pursued actively. A few recent successes include the sulfur–iodine thermal chemical cycle for hydrogen production [6], natural gas sweetening [7], CO_2 capture with amine solutions [8], sulfuric acid recovery [9], etc. This chapter describes key thermodynamic relationships for electrolyte systems and advances in electrolyte thermodynamic models, as well as three examples of how to apply these models to aqueous and mixed solvent electrolytes.

21.2 MODELING THERMODYNAMIC PROPERTIES OF ELECTROLYTE SOLUTIONS

Various thermodynamic properties of electrolyte systems, such as vapor pressure, osmotic coefficient, mean ionic activity coefficient, solution enthalpy, heat capacity, and salt

---

Chemical Engineering in the Pharmaceutical Industry: Active Pharmaceutical Ingredients, Second Edition.
Edited by David J. am Ende and Mary T. am Ende.
© 2019 John Wiley & Sons, Inc. Published 2019 by John Wiley & Sons, Inc.

solubility, are integral to process simulation of natural and industrial processes. These properties are all derived from the activity coefficients of chemical species in the system, and, therefore, accurate calculation of activity coefficients as functions of temperature and solution composition has become the focus of electrolyte thermodynamic models [10, 11].

The activity coefficient of a species in solution is most often calculated from models for excess Gibbs energy, G^{ex}. The basic relationship is

$$RT \ln \gamma_i = \left(\frac{\partial G^{ex}}{\partial n_i}\right)_{T,P,j\neq i} \quad (21.1)$$

where

γ is the activity coefficient.
n is the number of moles.
i can be a charged or uncharged chemical species.

Other important properties for electrolyte solutions are mean ionic activity coefficient ($\gamma_\pm$), activity of the solvent (a_s), and osmotic coefficient (ϕ):

$$I_x = \frac{1}{2}\sum_i x_i z_i^2 \quad (21.2)$$

$$\gamma_\pm = \left(\gamma_-^{z_-}\gamma_+^{z_+}\right)^{1/\nu} \quad (21.3)$$

$$\nu = z_- + z_+ \quad (21.4)$$

$$a_s = \gamma_s x_s \quad (21.5)$$

$$\phi = -\frac{1000}{\nu m_{MX} M_s}\ln a_s \quad (21.6)$$

$$\ln \gamma_\pm^{(m)} = \phi - 1 + 2\int_0^{m_{MX}} \frac{\phi-1}{m_{MX}^{1/2}} dm_{MX}^{1/2} \quad (21.7)$$

where

I_x is the ionic strength.
x is the mole fraction.
z is the absolute charge number.
γ is the activity coefficient.
m is molality of a solute.
M is molecular mass.

Subscripts refer to chemical species i, anion −, cation +, solvent s, or undissociated electrolyte MX. Variables with a superscript (m) are in molality scale, while all other relevant variables are in mole fraction scale [12].

21.3 ELECTROLYTE THERMODYNAMIC MODELS

Development of electrolyte thermodynamic models is a challenging task [13], as the dissociation of electrolytes into ions make it necessary to consider long-range ion–ion electrostatic interactions in addition to short-range molecule–molecule, molecule–ion, and ion–ion van der Waals interactions. All these interactions can have significant effect on the thermodynamic properties of the system.

There are numerous thermodynamic models proposed in the literature to calculate activity coefficients in electrolyte solutions, and new models and model extensions appear often. These models range from purely theoretical to purely empirical, with each type having its own advantages and drawbacks. Theoretical models are directly derived from first principles and generally apply only to very dilute electrolyte solutions. Empirical models are developed from experimental data and can be very accurate, but only when applied to the system at the conditions from which the data was gathered. Taking up the middle ground, semiempirical models correlate experimental data to derive the so-called interaction parameters within a theoretical thermodynamic framework. These models are then used to extrapolate beyond the original data range and predict the properties and behaviors of multicomponent systems. Three such models have come to widespread use in the past few decades: the Pitzer ion-interaction model, the OLI-mixed solvent electrolyte (OLI-MSE) model, and the electrolyte nonrandom two-liquid (eNRTL) model. The Pitzer model is mainly used by academia, while the latter two are extensively used in industry. The number of parameters varies from model to model and can have a great effect on the ease with which a model may be developed or expanded to suit systems of increasing size and complexity.

When dealing with thermodynamic functions that are concentration dependent, the choice of concentration scale is important. Molality concentration scale is used by the Pitzer model, while mole fraction scale is used by the other two. The outcome of activity coefficient calculation is also dependent on the choice of reference state. The Pitzer model and the OLI-MSE model use an unsymmetric reference state, where the activity coefficient of an electrolyte is assumed to be unity at aqueous phase infinite dilution extrapolated to unit concentration. The eNRTL model uses the symmetric reference state, where the activity coefficient of an electrolyte is unity when it is a hypothetical pure "molten salt" state. Given an activity coefficient in the symmetric reference state, one can always convert it to activity coefficient in the unsymmetric reference state, but the reverse is not true. When applying the eNRTL model, engineers have the choice of calculating either symmetric or unsymmetric activity coefficients. In general, unsymmetric activity coefficients are used when modeling aqueous electrolytes, while symmetric activity coefficients are required when modeling nonaqueous electrolytes.

In the following sections, this chapter summarizes the Pitzer and the eNRTL model. The OLI-MSE model can be considered as an extension of the Pitzer model and is not presented here primarily because of its complexities and an inherent double counting of ion–ion and ion–molecule interactions within the model formulation [14, 15]. Furthermore, the OLI-MSE model parameters are considered proprietary information and are not reported in the open literature.

21.3.1 The Pitzer Ion-Interaction Model

The Pitzer model is the most commonly used model for thermodynamic calculations of aqueous electrolyte solutions. Derived from considerations based on virial equations for electrolyte solutions, the Pitzer model provides an empirical mathematical form of ionic strength dependence and ion–ion binary and ion–ion–ion ternary interaction parameters suitable for describing the thermodynamic properties of aqueous electrolyte solutions up to ionic strengths of approximately 6 m [16, 17]. The model is a virial expansion in terms of solute concentration of the Debye–Hückel limiting law for electrolytes:

$$\frac{G^E}{RTw_S} = f(I_m) + \sum_i \sum_j m_i m_j \lambda_{ij} + \sum_i \sum_j \sum_k m_i m_j m_k \Lambda_{ijk} + \cdots \quad (21.8)$$

$$I_m = \frac{1}{2} \sum_i m_i z_i^2 \quad (21.9)$$

where

- w_s is the mass of solvent in kilograms.
- I_m is the molality scale ionic strength.
- m is the molality of species i, j, or k.

λ and Λ are the second- and third-order virial coefficients. Grouped from these virial coefficient terms, the ion–ion binary and ion–ion–ion ternary interaction parameters are temperature dependent, and, without guidance from the Pitzer model on the temperature dependence of the interaction parameters, up to eight temperature coefficients are often required. As a form of Debye–Hückel limiting law, $f(I_m)$ is a function of ionic strength, solvent properties, and temperature and takes the form

$$f(I_m) = -\frac{4 A_\phi I_m}{b} \ln\left(1 + b\sqrt{I_m}\right) \quad (21.10)$$

$$A_\phi = \frac{1}{3}(2\pi N_A d_s)^{1/2} \left(\frac{e^2}{4\pi\varepsilon_0 \varepsilon_S k_B T}\right)^{3/2} \quad (21.11)$$

where

- N_A is Avogadro's number.
- d_S is the molar density of the solution (mol/m^3).
- e is the electron charge.
- ε_0 is the permittivity of a vacuum.
- ε_S is the dielectric constant of the solvent.
- k_B is Boltzmann's constant.
- T is the temperature in degrees Kelvin.

The binary interaction parameters are often sufficient to describe the behavior of most dilute solutions, but it is necessary to include ternary interaction parameters to cover ionic strength up to 6 m for some highly nonideal systems. As a result, to cover both composition and temperature dependence, up to 24 adjustable parameters must be determined from experimental measurements to cover aqueous electrolyte solutions up to ionic strengths of 6 m.

The high number of adjustable parameters can easily lead to overfitting of the data, restricting the model to the temperature and concentration range in which it was developed. If multiple ions are in solution, additional terms must be regressed to capture the interactions between the different solute species [18]. While the Pitzer ion-interaction model is currently the most accurate model for binary mixtures of electrolytes in water, possessing the most comprehensive database of experimentally derived interaction parameters, its limitations make it impractical for widespread, integrated use in a process simulation setting. In summary, the Pitzer model falls short in thermodynamic modeling applications due to the very large number of adjustable interaction parameters and temperature coefficients required to cover electrolyte concentrations and temperatures of interest. Additionally, the Pitzer model is applicable to aqueous electrolytes only. It is very cumbersome to extend the Pitzer model to nonaqueous electrolyte or mixed solvent electrolyte systems.

21.3.2 The Electrolyte NRTL Model

An extension of the nonelectrolyte nonrandom two-liquid (NRTL) theory to electrolytes, the eNRTL model has had much success in industry over the past few decades. The eNRTL model integrates the Pitzer–Debye–Hückel theory for screened long-range ion–ion electrostatic interactions and the NRTL theory for molecule–molecule, molecule–ion, and ion–ion short-range interactions [19, 20]. The model further introduced the concepts of like-ion repulsion and local electroneutrality to account for the constraints on first neighbor species composition due to the long-range ion–ion electrostatic interactions. The general form is

$$G^E = G^{PDH} + G^{LC} \quad (21.12)$$

with the terms defined as

$$\frac{G^{PDH}}{nRT} = -\left(\frac{(4\pi\varepsilon_0)^3}{M_S}\right)^{1/2} \frac{4A_\phi I_x}{\rho} \ln\left(1 + \rho\sqrt{I_x}\right) \quad (21.13)$$

where

M_S is the molecular weight of the solvent.
A_ϕ is from Eq. (21.11).
ρ is the closest approach parameter.

and

$$\frac{G^{LC}}{nRT} = \sum_i x_i c_i \left(\frac{\sum_j x_j c_j G_{ji} \tau_{ji}}{\sum_k x_k c_k G_{ki}}\right) \quad (21.14)$$

$$G_{ij} = \exp(-\alpha_{ij}\tau_{ij}) \quad (21.15)$$

where

c is unity for molecules and the charge number of ions.
$\alpha_{ij}\,(=\alpha_{ji})$ is the nonrandomness factor (often set to 0.2).
$\tau_{ij}\,(\neq\tau_{ji})$ is the binary interaction parameter.

Interaction parameters are regressed for electrolyte ion pairs, rather than individual ions themselves as is the case with the Pitzer model, considering interactions between electrolytes and molecules and pairs of electrolytes that contain a common ion. So long as data is available to regress the binary interaction parameters, the eNRTL model can be easily extended to mixtures with multiple electrolytes and mixed solvents.

An advantage of the eNRTL model is its simplicity, requiring up to six parameters per solvent–electrolyte binary pair to accurately describe the short-range interactions across all concentrations from infinite dilution electrolyte solution to "molten salt" and the corresponding temperature dependence. This gives the model a strong basis for reliable extrapolation and decreases the chances of overfitting a specific data set. A major usability issue with this model is the current lack of a comprehensive parameter database, but there are many recent efforts [21–24] to expand this database.

21.3.3 Developing A Thermodynamic Model

The first stage in developing a thermodynamic model for an electrolyte system is always a thorough assessment of existing resources, including available interaction parameters and thermodynamic data. The desired model interaction parameters may already be available in commercial process simulators, and many literature sources or public databanks may contain additional parameters that have not yet been incorporated. A careful review of all sources is necessary to understand the range of conditions under which the model has acceptable accuracy. When interaction parameters are missing or poorly defined, it is necessary to use thermodynamic data to regress the missing parameters.

The literature contains a large amount of thermodynamic data that can be used to regress model parameters, although a good understanding of the data and associated uncertainty is necessary for the modeler to make proper use of the information. It is worth noting that, while electrolyte systems have been studied and measured for over a century, older data sets can be of questionable reliability since it has only recently become common practice to report data uncertainty along with the data. While reporting practices have improved in recent decades, this has unfortunately coincided with a reduction in motivation and funding for data measurement projects, and as a result, there is a dearth of reliably good data. Fortunately, computers have made it much easier to sift through the available data to find data sets that are suitable for regression, being in good agreement with each other and thermodynamically consistent. It is also important to take the range of the data into account when regressing parameters. For example, temperature-dependent parameters require a wide range – typically more than 50 K – to adequately capture the temperature dependency of the data.

Once appropriate and sufficient data has been compiled, computer regression software can be used to determine the pertinent interaction parameters. The minimization function used to determine quality of fit is an important consideration. For example, data uncertainties should be properly incorporated in the formulation of the function. Also, the minimization function should account for the negative impact of using more parameters, thus making the optimal function one that minimizes the function with the fewest adjustable parameters. Using the typical values of parameters as an initial guess reduces the risk of software converging on a local minimum that is outside the range of values consistent with the model assumptions. At this point, the model should be compared to reliable data to validate its use under desired conditions.

21.4 EXAMPLES: MODELING WITH eNRTL

The examples in this chapter use Aspen Properties software [25] to regress parameters for the eNRTL model. They cover three classes of electrolyte systems that are commonly encountered and challenging to model: the lithium nitrate + lithium chloride + water ternary system to represent aqueous electrolytes, the sodium acetate + water + acetonitrile ternary system to represent mixed solvent electrolytes, and the nitric acid system to represent reactive electrolyte systems.

21.4.1 Aqueous Electrolytes: The Lithium Nitrate + Lithium Chloride + Water Ternary System

Lithium salt solutions are widely used in absorption refrigeration systems and chemical heat pumps as alternatives to chlorofluorocarbons [26]. While they tend to be more environmentally friendly, corrosion and crystallization associated with lithium salt solutions are still challenges that need to be overcome. One of the first steps in addressing these challenges is to develop a thermodynamic model to aid in determining optimal operating conditions. This example covers the development of an eNRTL model for the lithium nitrate ($LiNO_3$) + lithium chloride (LiCl) + water (H_2O) ternary system.

The thermodynamic data used to determine the binary interaction parameters (τ_{ij}) for the binary systems $LiNO_3$ + H_2O and LiCl + H_2O includes vapor pressure, mean ionic activity coefficient, osmotic coefficient, heat capacity, and liquid enthalpy at varying temperature and composition ranges. Complete dissociation of the lithium salts into their ionic compounds is assumed, and the binary interaction parameters are regressed from the thermodynamic data using Aspen Properties. The equation for temperature dependency is

$$\tau_{ij} = C_{ij} + \frac{D_{ij}}{T} + E_{ij}\left(\frac{T_{ref}-T}{T} + \ln\frac{T}{T_{ref}}\right) \quad (21.16)$$

where

C_{ij}, D_{ij}, and E_{ij} are the temperature coefficients.
T_{ref} = 298.15 K.

Table 21.1 contains the values of τ_{ij} regressed for the binary pairs. To validate the results, the model is used to calculate vapor pressure, osmotic coefficient, and activity coefficient at various temperatures and compositions. Figures 21.1–21.2 show the results plotted alongside data from multiple sources that were used in the regression.

Because the $LiNO_3$ + LiCl binary system does not exist in a state conducive to measuring fluid properties, ternary data for an aqueous solution containing both salts is needed to regress the binary interaction parameters. Unfortunately, thermodynamic data for ternary and higher-order systems is difficult to find and can often be insufficient or inconsistent. Assuming ideal mixing of the two electrolytes, the (Li^+ NO_3^-):(Li^+ Cl^-) and (Li^+ Cl^-):(Li^+ NO_3^-) interaction parameters are set to zero. Even without reliable experimental data, using these interaction parameters from simplifying assumptions can yield a reasonable prediction. To validate the results, vapor pressure is calculated at various temperatures and compositions and compared to experimental data, shown in Figure 21.3.

To obtain a complete model, solubility data is needed to account for solid–liquid equilibrium. According to Zeng et al. [37], there are four solid crystals that precipitate out from the ternary system:

$$Li^+ + NO_3^- \rightarrow LiNO_{3(s)} \quad (21.17)$$

$$Li^+ + NO_3^- + 3H_2O \rightarrow LiNO_3 \cdot 3H_2O_{(s)} \quad (21.18)$$

$$Li^+ + Cl^- + H_2O \rightarrow LiCl \cdot H_2O_{(s)} \quad (21.19)$$

$$Li^+ + Cl^- + 2H_2O \rightarrow LiCl \cdot 2H_2O_{(s)} \quad (21.20)$$

There are several equations used to model temperature and pressure dependency of the solubility product constant, K_{sp}. An expression commonly used is given below:

$$\ln(K_{sp}) = A + \frac{B}{T} + C\ln(T) + DT + E\left(\frac{P-P_{ref}}{P_{ref}}\right) \quad (21.21)$$

where, most of the time, only A and B, and sometimes C, are necessary to accurately represent the data. For this ternary

TABLE 21.1 Regressed Values of τ_{ij} with Nonrandom Factor α_{ij} = 0.2

Component i	Component j	C_{ij}	D_{ij}	E_{ij}	τ_{ij} at 298.15 K
H_2O	(Li^+ NO_3^-)	7.273 ± 0.106	732.7 ± 31.4	−2.157 ± 0.552	9.730
(Li^+ NO_3^-)	H_2O	−3.961 ± 0.041	−300.5 ± 12.4	2.255 ± 0.166	−4.969
H_2O	(Li^+ Cl^-)	8.081 ± 0.105	970.1 ± 31.2	−8.464 ± 0.519	11.335
(Li^+ Cl^-)	H_2O	−4.192 ± 0.035	−431.9 ± 10.5	3.265 ± 0.149	−5.641
(Li^+ NO_3^-)	(Li^+ Cl^-)	0	0	0	0
(Li^+ Cl^-)	(Li^+ NO_3^-)	0	0	0	0

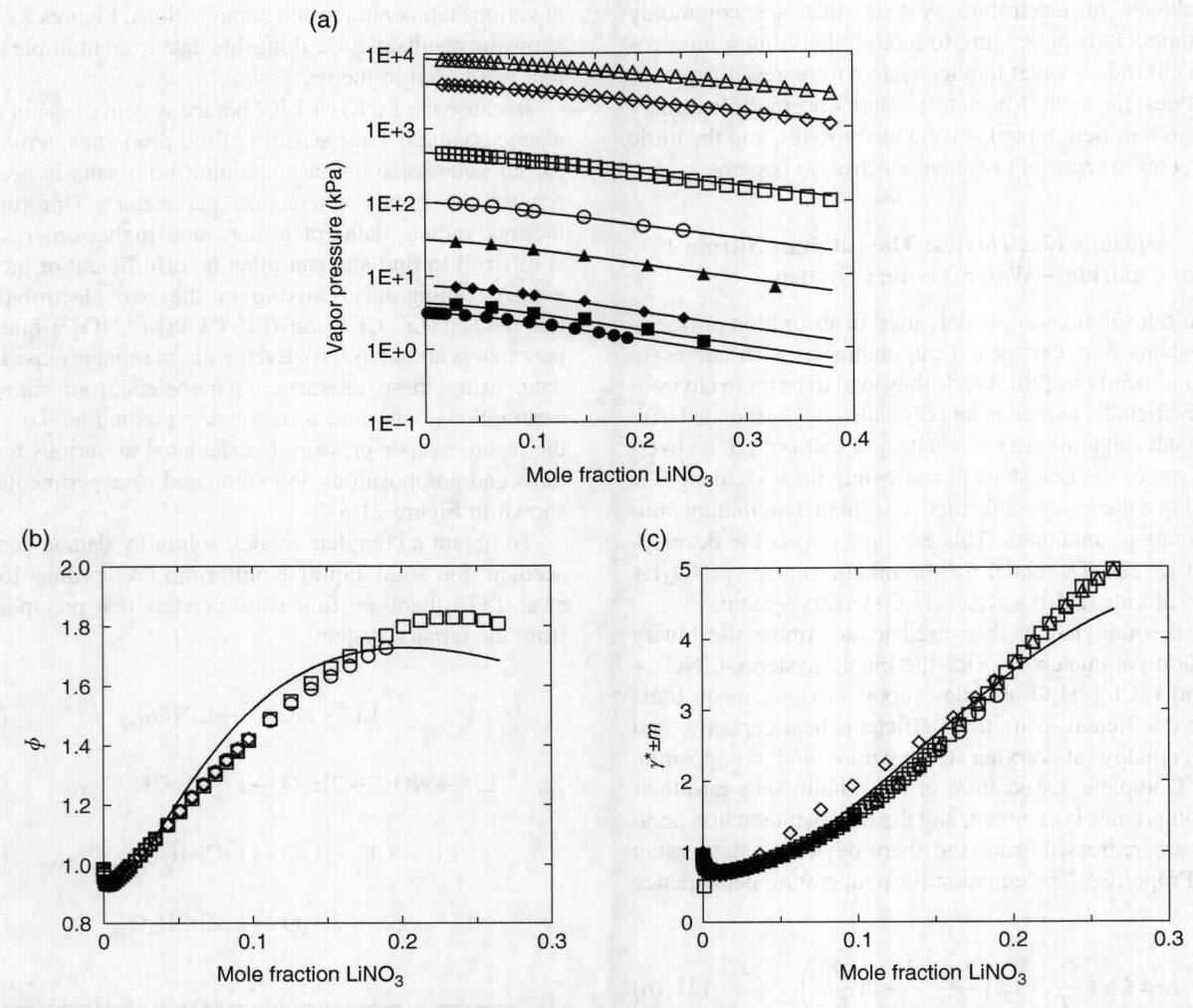

FIGURE 21.1 Model results shown as solid lines and experimental data shown as symbols for the $LiNO_3 + H_2O$ binary system. (a) Vapor pressure at (●) 298.15 K (Pearce and Nelson [27]), (■) 303.15 K (Campbell et al. [28]), (♦) 313.15 K (Patil et al. [29]), (▲) 343.15 K [28], (○) 373.15 K [29], (□) 423.15 K, (◇) 523.15 K, and (△) 573.15 K (Zaytsev and Aseyev [30]). (b) Osmotic coefficients at 298.15 K, (○) Robinson and Stokes [31], (□) Hamer and Wu [32], and (◇) Guendouzi and Marouani [33]. (c) Mean ionic activity coefficients at 298.15 K, (○) [31], (□) [30], (◇) [27], (△) [32], and (●) [33].

system temperature-dependent parameters can only be regressed for $LiNO_{3(s)}$ due to lack of temperature-dependent data for other salts. The values found are shown in Table 21.2. Figure 21.4 shows the calculated solubility of LiCl and $LiNO_3$ in the ternary system compared to experimental data that was used in the regression.

21.4.2 Mixed Solvent Electrolytes: The Sodium Acetate + Water + Acetonitrile Ternary System

Sodium acetate (NaOAc) has many applications in the textile industry, food industry, and analytical sciences. Because of this, its solubility in mixed solvents has received much attention recently [38]. Soleymani et al. reported the apparent solubility of NaOAc in binary solvents of water, acetonitrile, methanol, and n-propanol at 298.2 K without identifying the polymorphs of NaOAc in the saturated solution [39]. Among the solvents studied, the solubilities of NaOAc in pure water and pure acetonitrile (ACN) are the highest and the lowest, respectively, and the solubility of NaOAc varies over three orders of magnitude in the water + ACN binary solvent system. This example calculates the solubility of NaOAc in water + ACN binary solvent system at 298 K using the eNRTL model. This requires binary interaction parameters for water–ACN, water–(Na^+ OAc^-), and ACN–(Na^+ OAc^-) pairs as well as the solubility product constant, K_{sp}, of the NaOAc polymorphs.

NaOAc has two polymorphs, NaOAc anhydrous and NaOAc trihydrate, NaOAc•$3H_2O$. The solubility of NaOAc (anhydrous) in water is about twice that of NaOAc•$3H_2O$ at 298 K.

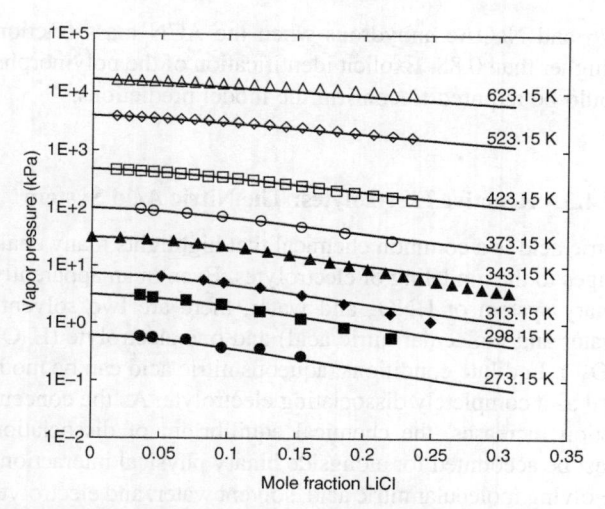

FIGURE 21.2 Vapor pressure of the LiCl + H$_2$O binary system: model results shown as solid lines and experimental data shown as symbols. (●) 273.15 K, (■) 298.15 K [30], (◆) 313.15 K [34], (▲) 343.15 K (Johnson and Molstad [35]), (○) 373.15 K, (□) 423.15 K, (◇) 523.15 K, and (△) 623.15 K [30].

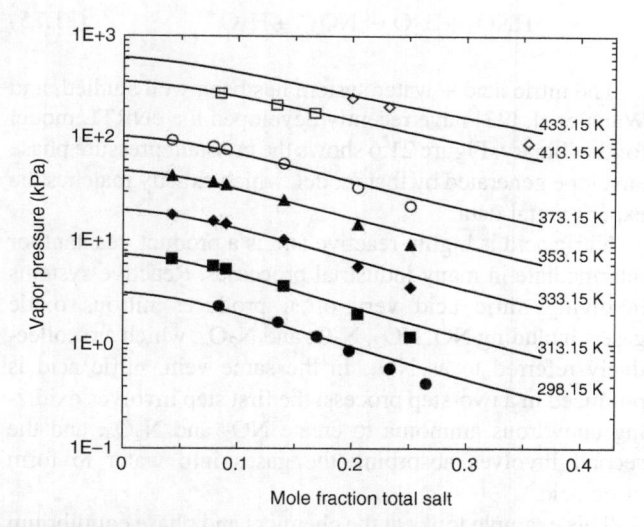

FIGURE 21.3 Vapor pressure of the LiNO$_3$ + LiCl + H$_2$O ternary system: model results shown as solid lines and experimental data shown as symbols. (●) 298.15 K (Iyoki et al. [36]), (■) 313.15 K, (◆) 333.15 K, (▲) 353.15 K, (○) 373.15 K (Patil et al. [29]), (□) 413.15 K, and (◇) 433.15 K [36]. LiCl/LiNO$_3$ salt mole ratio for Iyoki et al. is 2.8:1 and Patil et al. is 0.88:0.12.

TABLE 21.2 Regressed Coefficients for Eq. (21.21) and Values of K_{sp} for LiNO$_3$ and LiCl Salts

Solid	A	B	K_{sp} at 298.15 K
LiNO$_3$	−2.387 ± 0.518	1232 ± 162	5.73
LiNO$_3$·3H$_2$O	−3.080 ± 0.131	—	0.0460
LiCl·H$_2$O	2.476 ± 0.044	—	11.9
LiCl·2H$_2$O	1.817 ± 0.071	—	6.15

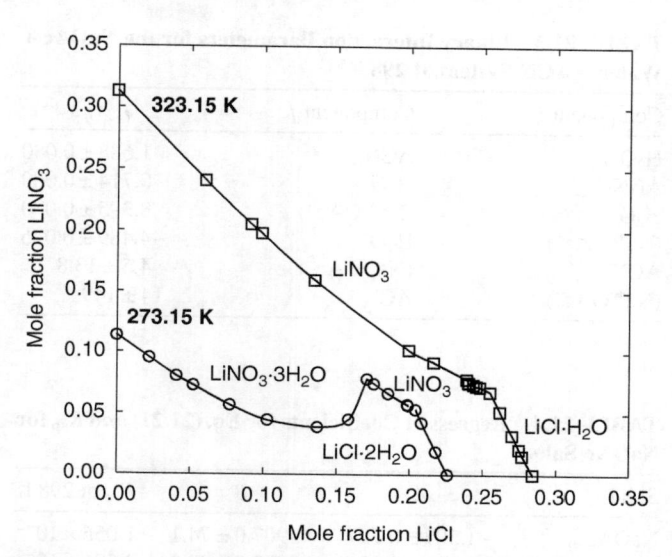

FIGURE 21.4 Solubility of LiNO$_3$ and LiCl in the LiNO$_3$ + LiCl + H$_2$O ternary system: model results shown as solid lines and experimental data shown as symbols. (○) 273.15 K and (□) 323.15 K [37].

In addition, NaOAc (anhydrous) is metastable in water in the temperature range of 273–313 K [40]. Since Soleymani et al. did not identify the NaOAc polymorphs in the saturated solution, both polymorphs are taken into consideration in developing the eNRTL model.

The solution chemistry includes the dissociation of NaOAc to ions and the formation of NaOAc (anhydrous) and NaOAc·3H$_2$O salts:

$$CH_3COONa_{(aq)} \rightarrow Na^+_{(aq)} + CH_3COO^-_{(aq)} \quad (21.22)$$

$$Na^+_{(aq)} + CH_3COO^-_{(aq)} \rightarrow CH_3COONa_{(s)} \quad (21.23)$$

$$Na^+_{(aq)} + CH_3COO^-_{(aq)} + 3H_2O \rightarrow CH_3COONa \cdot 3H_2O_{(s)} \quad (21.24)$$

The following thermodynamic data are used to identify the binary interaction parameters for the NaOAc + water + ACN system: vapor–liquid equilibrium data of water and ACN [41–43], mean ionic activity coefficient of aqueous NaOAc solution [44], and the solubility data of NaOAc in water +ACN binary solvent [39]. In addition, the solubility of NaOAc (anhydrous) in water and NaOAc·3H$_2$O in water [40] is used to fit the solubility product constants. The regressed binary interaction parameters and the solubility product constants are given in Tables 21.3 and 21.4.

Figure 21.5 plots the eNRTL model results against the experimental NaOAc solubility in the water–ACN binary solvent. In general, the eNRTL model results of NaOAc·3H$_2$O show an adequate agreement with the experimental solubility

TABLE 21.3 Binary Interaction Parameters for the NaOAc + Water + ACN System at 298 K

Component i	Component j	τ_{ij}
H_2O	ACN	1.688 ± 0.080
ACN	H_2O	0.714 ± 0.060
H_2O	$(Na^+ \, OAc^-)$	8.395 ± 0.070
$(Na^+ \, OAc^-)$	H_2O	-4.489 ± 0.026
ACN	$(Na^+ \, OAc^-)$	4.5 ± 13.8
$(Na^+ \, OAc^-)$	ACN	14 ± 149

TABLE 21.4 Regressed Coefficients for Eq. (21.21) and K_{sp} for NaOAc Salts

Solid	A	B	K_{sp} at 298 K
$NaOAc_{(s)}$	-1.507 ± 0.227	-907.0 ± 74.1	1.056×10^{-2}
$NaOAc \cdot 3H_2O_{(s)}$	2.018 ± 0.321	-2972.2 ± 99.3	3.506×10^{-4}

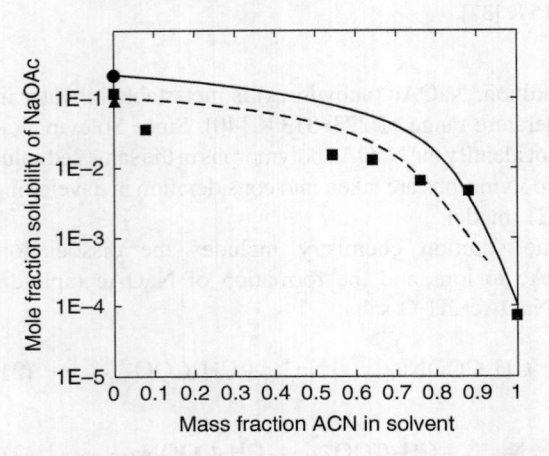

FIGURE 21.5 Mole fraction solubility of NaOAc in the binary solvent of H_2O + ACN. (—) represents the anhydrous form, while (- -) represents the trihydrate. Data from (■) [39], (●) [40], and (▲) [44].

of NaOAc when the ACN mass fraction in the binary solvent is in the range of 0–0.76. The model predicts that the NaOAc trihydrate polymorph cannot be formed when the ACN mass fraction exceeds 0.92, suggesting that water activity in the binary solvent beyond that point is insufficient for the formation of NaOAc trihydrate.

On the other hand, the eNRTL model results for NaOAc (anhydrous) show significantly higher solubility in the binary solvents than those of experimental NaOAc solubility when the ACN mass fraction in the binary solvent is in the range of 0–0.76. However, it has an excellent agreement with the experimental NaOAc solubility when the ACN mass fraction is higher than 0.88. The model results imply that the solid polymorph is $NaOAc \cdot 3H_2O$ at ACN mass fractions up to 0.76 and NaOAc anhydrous when the ACN mass fraction is higher than 0.88. Explicit identification of the polymorphs would be required to confirm the model predictions.

21.4.3 Reactive Electrolytes: The Nitric Acid System

Nitric acid is a common chemical that highlights many challenges to the modeling of electrolytes. Even in an apparently binary system of HNO_3 and water, there are two solvents (water and molecular nitric acid) and one electrolyte (H_3O^+ NO_3^-). In dilute conditions, aqueous nitric acid can be modeled as a completely dissociating electrolyte. As the concentration increases, the chemical equilibrium of dissociation must be accounted for alongside binary physical interactions involving molecular nitric acid, solvent water, and electrolyte (H_3O^+ NO_3^-). At even higher concentrations molecular nitric acid must be considered a second solvent. Furthermore, molecular nitric acid is volatile and must be included in the vapor–liquid equilibrium of the system. The dissociation of HNO_3 is modeled by the equilibrium reaction

$$HNO_3 + H_2O \leftrightarrow NO_3^- + H_3O^+ \qquad (21.25)$$

The nitric acid + water system has been well studied, and Wang et al. [23] have recently developed the eNRTL model for that binary. Figure 21.6 shows the constant pressure phase envelope generated by that model, which exactly matches the experimental data.

Nitric acid is highly reactive and is a product, reactant, or intermediate in many industrial processes. Reactive systems involving nitric acid very often produces nitrous oxide gases, including NO, NO_2, N_2O, and N_2O_4, which are collectively referred to as NO_x. In the same vein, nitric acid is produced in a two-step process: the first step involves oxidizing anhydrous ammonia to create NO_2 and N_2O_4, and the second involves absorbing the gases into water to form nitric acid.

This example looks at the chemical and phase equilibrium of a system of aqueous nitric acid and NO_x gases. Nitrous acid (HNO_2) is also included, as a reactive system with one acid present will generate appreciable quantities of the other. Nitrous acid dissociates with the reaction

$$HNO_2 + H_2O \leftrightarrow NO_2^- + H_3O^+ \qquad (21.26)$$

The combined model accounts for the equilibrium reactions involving NO_x:

$$2NO + O_2 \leftrightarrow 2NO_2 \qquad (21.27)$$

$$4NO \leftrightarrow 2N_2O + O_2 \qquad (21.28)$$

$$2NO_2 \leftrightarrow N_2O_4 \qquad (21.29)$$

$$3NO_2 + H_2O \leftrightarrow 2HNO_3 + NO \quad (21.30)$$

$$N_2O_4 + H_2O \leftrightarrow HNO_3 + HNO_2 \quad (21.31)$$

While Eq. (21.27) is a rate-limited reaction [45], it is assumed that the time scale is sufficient for equilibrium to be reached. The components N_2, O_2, NO, NO_2, N_2O, N_2O_4, and HNO_2 have associated Henry's constants for describing vapor–liquid equilibrium. The eNRTL model is used to characterize the liquid-phase components, while the Redlich–Kwong equation of state does the same for the vapor phase. The base case condition is a feed of 1 m^3/min air containing 10 000 ppm NO_2 and a feed 40 kg/min of 10 mol % nitric acid solution. Table 21.5 summarizes the "feed" compositions.

The system is allowed to reach equilibrium at 1 bar, under varying conditions. In the first case, represented in Figure 21.7, the feed conditions are held constant while the temperature is varied from 273 to 423 K. In the second case, represented in Figure 21.8, the temperature is held constant at 298 K while the composition of the liquid feed is varied from pure water to pure nitric acid. While just the overall composition is shown, it was verified that there are two phases in equilibrium (liquid and vapor) throughout the full range of conditions. Some species are not shown in the charts in order to make it easier to track individual species. Water, which is in great excess except for the high HNO_3 feed conditions, is not shown, nor is the inert gas N_2. O_2 participates in the reaction system, but is always available and not particularly interesting, so it was also excluded.

This example serves as an illustration of the power of thermodynamic modeling. The ability to track many different components, including ionic species, is invaluable in the predictive analysis of reactive systems for design and development applications.

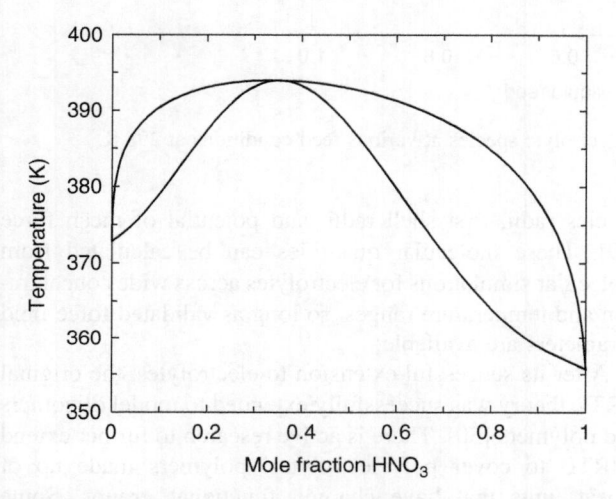

FIGURE 21.6 T-x-y diagram of the HNO_3 + H_2O binary system.

TABLE 21.5 Base Case Feed Compositions

	Mole Fraction				
Stream	N_2	O_2	NO_2	H_2O	HNO_3
Liquid	—	—	—	0.90	0.10
Vapor	0.79	0.20	0.01	—	—
Overall	0.0194	0.0049	0.0002	0.8779	0.0975

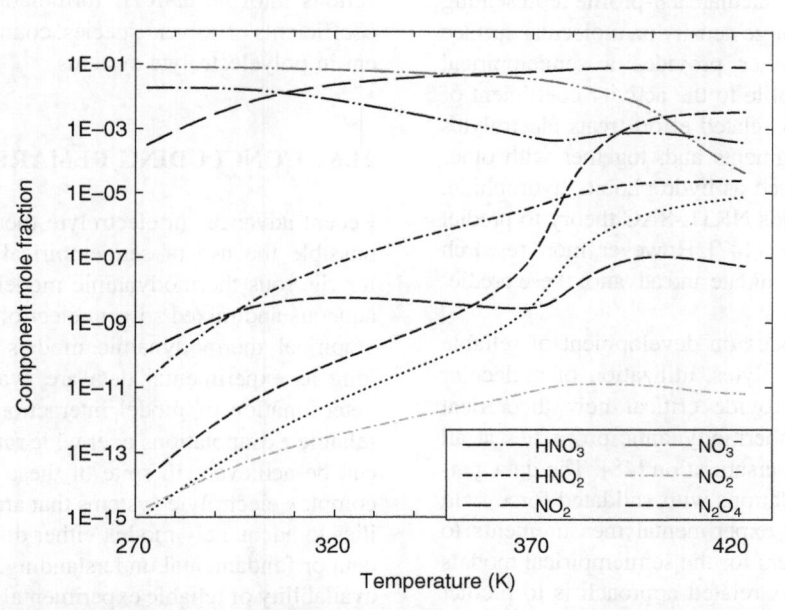

FIGURE 21.7 Overall equilibrium concentrations of NO_x and electrolyte species at various temperatures.

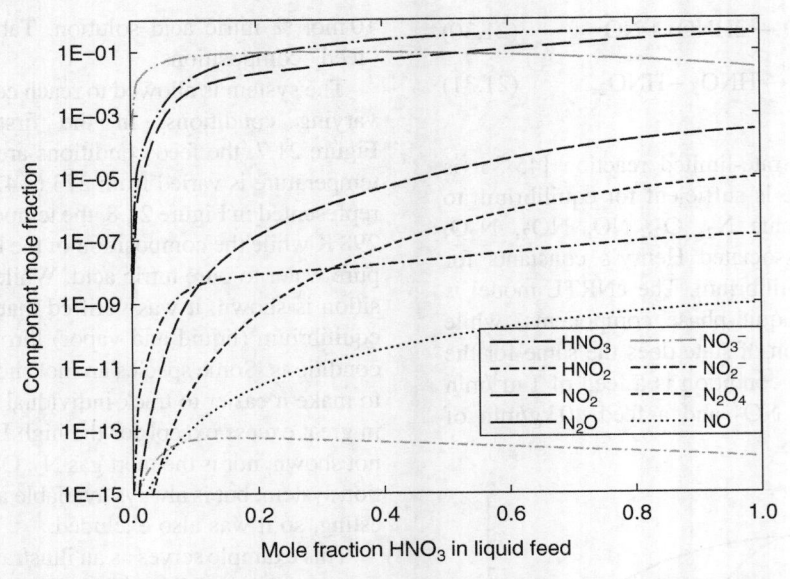

FIGURE 21.8 Overall equilibrium concentrations of NO_x and electrolyte species at various feed conditions at 298 K.

21.5 ONGOING DEVELOPMENTS

Despite recent advances in thermodynamic modeling of electrolyte systems, there is much need for further research and advances in electrolyte thermodynamics. The most prominent shortcoming of the current models is that they are correlative models and their reliance on extensive experimental data to identify model interaction parameters. Therefore, a high priority need is development and validation of robust predictive models for electrolyte thermodynamics. One such effort is COSMO-SAC (conductor-like screening model for real solvents-segment activity coefficient) [46]. This method makes use of quantum chemistry to calculate a σ-profile representing probability distribution of charge density on molecular surface segments. The method further provides a semiempirical expression relating the σ-profile to the activity coefficient of the species in the system. A related effort treats electrolytes as "charged" conceptual segments and, together with other conceptual segment types such as hydrophobic, hydrophilic, and polar for solvents, extends NRTL-SAC theory to predict mean ionic activity coefficients [47]. However, more research is needed in this direction to validate and advance these predictive models.

Other research efforts focus on development of reliable force field models for electrolytes, utilization of molecular dynamics simulations to provide critical new theoretical insights, and estimation of thermodynamic properties at all concentrations including supersaturation [48]. The data generated from molecular simulations with validated force field parameters can supplement experimental measurements to identify interaction parameters for the semiempirical models mentioned in this chapter. A related approach is to predict interaction parameters with molecular quantities such as species radii, first shell radii, and potential of mean force [49]. These molecular quantities can be calculated from molecular simulations for electrolytes across wide concentration and temperature ranges, so long as validated force field parameters are available.

After its successful extension to electrolytes, the original NRTL theory was successfully extended to model oligomers and polymers [50]. There is active research to further extend eNRTL to cover polyelectrolytes: polymers made up of repeat units that have charged functional groups. Some success has been reported by incorporating the Manning model [51] for the long-range polyion–ion electrostatic interactions into the eNRTL formulation to model the activity coefficients of solvent species, counterions, and co-ions present in polyelectrolyte systems.

21.6 CONCLUDING REMARKS

Recent advances in electrolyte thermodynamics have made possible the use of semiempirical thermodynamic models for rigorous thermodynamic modeling of a wide variety of aqueous and mixed solvent electrolyte systems. These semiempirical thermodynamic models are correlative, and, so long as experimental data are available to support proper determination of model interaction parameters, robust and reliable extrapolations over wide range of process conditions can be achieved. In spite of these successes, there are still complex electrolyte systems that are beyond current capabilities to adequately model, either due to lack of experimental data or fundamental understanding. Future advances require availability of reliable experimental data for complex electrolyte systems of high importance and new insights gained through theoretical developments and molecular simulations. Electrolyte thermodynamics is one of the most active areas of research for chemical thermodynamicists, and high-impact advances can be expected in the future. Electrolyte thermodynamics has much overlap with nonelectrolyte solution thermodynamics [12], and, while the differences between them are not trivial, it is likely that these two branches of solution thermodynamics will be unified soon.

REFERENCES

1. Stahl, P.H. and Wermuth, C.G. (eds.) (2002). *Handbook of Pharmaceutical Salts: Properties, Selection and Use.* Zürich, Switzerland: Verlag Helvetica Chimica Acta.

2. Igunnu, E.T. and Chen, G.Z. (2014). Produced water treatment technologies. *Int. J. Low Carbon Tech.* 9: 157–177.

3. Gorensek, M.B., Lambert, D.P., Edwards, T.B., and Chen, C.-C. (2014). *Smart Manufacturing: Replacing Analytical Sample Control with Model Predictive Control.* Project ID: LDRD-2014-00099, Savannah River National Laboratory.

4. Winter, M. and Brodd, R.J. (2004). What are batteries, fuel cells, and supercapacitors? *Chem. Rev.* 104: 4245–4269.

5. Chen, C.-C. and Mathias, P.M. (2002). Applied thermodynamics for process modeling. *AIChE J.* 48: 194–200.

6. Murphy, J.E. and O'Connell, J.P.A. (2010). Properties model of the $HI-I_2-H_2O-H_2$ system in the sulfur-iodine cycle for hydrogen manufacture. *Fluid Phase Equil.* 288: 99–110.

7. Austgen, D., Rochelle, G.T., Xiao, P., and Chen, C.-C. (1989). Model of vapor-liquid equilibria in the aqueous acid gas-alkanolamine system using the electrolyte NRTL equation. *Ind. Chem. Eng. Res.* 28: 1060–1073.

8. Zhang, Y. and Chen, C.-C. (2013). Modeling CO_2 absorption and desorption by aqueous monoethanolamine solution with Aspen rate-based model. *Energy Procedia* 37: 1584–1596.

9. Li, G., Asselin, E., and Li, Z. (2014). Process simulation of sulfuric acid recovery by azeotropic distillation: vapor-liquid equilibria and thermodynamic modeling. *Ind. Eng. Chem. Res.* 53: 11794–11804.

10. Saravi, S.H., Honarparvar, S., and Chen, C.-C. (2015). Modeling aqueous electrolyte systems. *Chem. Eng. Prog.* 111 (3): 65–75.

11. Wang, M., Yu, Y., and Chen, C.-C. (2016). Modeling mixed-solvent electrolytes. *Chem. Eng. Prog.* 112 (2): 34–42.

12. Prausnitz, J.M., Lichtenthaler, R.N., and de Azevedo, E.G. (1999). *Molecular Thermodynamics of Fluid-Phase Equilibria*, 3e. Upper Saddle River, NJ: Prentice Hall PTR.

13. May, P.M. and Rowland, D. (2017). Thermodynamic modeling of aqueous electrolyte systems: current status. *J. Chem. Eng. Data* 62: 2481–2495.

14. Wang, P., Anderko, A., and Young, R.D. (2002). A speciation-based model for mixed-solvent electrolyte systems. *Fluid Phase Equil.* 203: 141–176.

15. Wang, P., Springer, R.D., Anderko, A., and Young, R.D. (2004). Modeling phase equilibria and speciation in mixed-solvent electrolyte systems. *Fluid Phase Equil.* 222–223: 11–17.

16. Pitzer, K.S. (1973). Thermodynamics of electrolytes. I. Theoretical basis and general equations. *J. Phys. Chem.* 77 (2): 268–277.

17. Pitzer, K.S. and Mayorga, G. (1973). Thermodynamics of electrolytes. II. Activity and osmotic coefficients for strong electrolytes with one or both ions univalent. *J. Phys. Chem.* 77 (19): 2300–2308.

18. Voigt, W. (2011). Chemistry of salts in aqueous solutions: applications, experiments, and theory. *Pure Appl. Chem.* 83 (5): 1015–1030.

19. Chen, C.-C. and Song, Y. (2004). Generalized electrolyte-NRTL model for mixed-solvent electrolyte systems. *AIChE J.* 50 (8): 1928–1941.

20. Chen, C.-C. and Song, Y. (2009). Symmetric electrolyte nonrandom two-liquid activity coefficient model. *Ind. Eng. Chem. Res.* 48: 7788–7797.

21. Wang, M., Kaur, H., and Chen, C.-C. (2017). Thermodynamic representation of nitric acid-sulfuric acid-water ternary system. *AIChE J.* 63: 3110–3317.

22. Honarparvar, S., Saravi, S.H., Reible, D., and Chen, C.-C. (2017). Comprehensive thermodynamic modeling of saline water with electrolyte NRTL model: a study on aqueous Ba^{2+}-Na^+-Cl^--SO_4^{2-} quaternary system. *Fluid Phase Equil.* 447: 29–38.

23. Wang, M., Gorensek, M.B., and Chen, C.-C. (2016). Thermodynamic representation of aqueous sodium nitrate and nitric acid solution with electrolyte NRTL model. *Fluid Phase Equil.* 407: 105–116.

24. Que, H. and Chen, C.-C. (2011). Thermodynamic modeling of the NH_3-CO_2-H_2O system with electrolyte NRTL model. *Ind. Eng. Chem. Res.* 50: 11406–11421.

25. Aspen Properties V8.8 (2015). Burlington, MA: Aspen Technology, Inc.

26. Iyoki, S., Iwasaki, S., Kuriyama, Y., and Uemura, T. (1993). Solubilities for the two ternary systems water + lithium bromide + lithium iodide and water + lithium chloride + lithium nitrate at various temperatures. *J. Chem. Eng. Data* 38: 396–398.

27. Pearce, J.N. and Nelson, A.F. (1932). The vapor pressures of aqueous solutions of lithium nitrate and the activity coefficients of some alkali salts in solutions of high concentration at 25°C. *J. Phys. Chem.* 31: 3544–3554.

28. Campbell, A.N., Fishman, J.B., Rutherford, G. et al. (1956). Vapor pressures of aqueous solutions of silver nitrate, of ammonium nitrate, and of lithium nitrate. *Can. J. Chem.* 34: 151–159.

29. Patil, K., Chaudharl, S., and Katti, S. (1992). Thermodynamic properties of aqueous electrolyte solutions. 3. Vapor pressure of aqueous solutions of $LiNO_3$, $LiCl + LiNO_3$, and $LiBr + LiNO_3$. *J. Chem. Eng. Data* 37: 136–138.

30. Zaytsev, I. and Aseyev, G. (1992). *Properties of Aqueous Solutions of Electrolytes*. Boca Raton: CRC Press.

31. Robinson, R.A. and Stokes, R.H. (1949). Tables of osmotic and activity coefficients of electrolytes in aqueous solution at 25 °C. *Trans. Faraday Soc.* 45: 612–624.

32. Hamer, W.J. and Wu, Y.-C. (1972). Osmotic coefficients and mean activity coefficients of uni-univalent electrolytes in water at 25 °C. *J. Phys. Chem. Ref. Data Monogr.* 1: 1047–1100.

33. Guendouzi, M.E. and Marouani, M. (2003). Water activities and osmotic and activity coefficients of aqueous solutions of nitrates at 25 °C by the hygrometric method. *J. Solution Chem.* 32: 535–546.

34. Patil, K., Trlpathi, A., Pathak, G., and Katti, S. (1990). Thermodynamic properties of aqueous electrolyte solutions. 1. Vapor pressure of aqueous solutions of LiCl, LiBr, and LiI. *J. Chem. Eng. Data* 35: 166–168.

35. Johnson, E. and Molstad, M. (1951). Thermodynamic properties of lithium chloride solutions. *J. Phys. Colloid Chem.* 55: 257–281.

36. Iyoki, S., Kuriyama, Y., Tanaka, H. et al. (1993). Vapor-pressure measurements on (water + lithium chloride + lithium nitrate) at temperatures 274.15 K to 463.15 K. *J. Chem. Therm.* 25: 569–577.

37. Zeng, D., Ming, J., and Voigt, W. (2008). Thermodynamic study of the system (LiCl + LiNO$_3$ + H$_2$O). *J. Chem. Therm.* 40: 232–239.

38. Soleymani, J., Kenndler, E., Acree, W.E. Jr., and Jouyban, A. (2014). Solubility of sodium acetate in ternary mixtures of methanol, 1-propanol, acetonitrile, and water at 298.2 K. *J. Chem. Eng. Data* 59 (8): 2670–2676.

39. Soleymani, J., Zamani-Kalajahi, M., Ghasemi, B. et al. (2013). Solubility of sodium acetate in binary mixtures of methanol, 1-propanol, acetonitrile, and water at 298.2 K. *J. Chem. Eng. Data* 58 (12): 3399–3404.

40. Seidell, A. (1919). *Solubilities of Inorganic and Organic Compounds*, 2e. New York: D. Van Nostrand Company.

41. Sugi, H. and Katayama, T. (1978). Ternary liquid-liquid and miscible binary vapor-liquid equilibrium data for the two systems n-hexane ethanol acetonitrile and water acetonitrile-ethyl acetate. *J. Chem. Eng. Jpn.* 11 (3): 167–172.

42. Blackford, D.S. and York, R. (1965). Vapor-liquid equilibria of the system acrylonitrile-acetonitrile-water. *J. Chem. Eng. Data* 10 (4): 313–318.

43. Maslan, F. and Stoddard, E. Jr. (1956). Acetonitrile–water liquid–vapor equilibrium. *J. Phys. Chem.* 60 (8): 1146–1147.

44. Söhnel, O. and Novotný, P. (1985). *Densities of Aqueous Solutions of Inorganic Substances*. Amsterdam: Elsevier Publishing Company.

45. Miller, D.N. (1987). Mass transfer in nitric acid absorption. *AIChE J.* 33: 1351–1358.

46. Wang, S., Song, Y., and Chen, C.-C. (2011). Extension of COSMO-SAC solvation method for electrolytes. *Ind. Eng. Chem. Res.* 50: 176–187.

47. Song, Y. and Chen, C.-C. (2009). Symmetric nonrandom two-liquid segment activity coefficient model for electrolytes. *Ind. Eng. Chem. Res.* 48: 5522–5529.

48. Hossain, N., Ravichandran, A., Khare, R., and Chen, C.-C. (2018). Revisiting electrolyte thermodynamic models: insight from molecular simulations. *AIChE J.* 64: 3728–3734.

49. Ravichandran, A., Khare, R., and Chen, C.-C. (2018). Predicting NRTL binary interaction parameters from molecular simulations. *AIChE J.* 64: 2758–2769.

50. Chen, C.-C. (1993). A segment-based local composition model for the Gibbs energy of polymer solutions. *Fluid Phase Equil.* 83: 301–312.

51. Yu, Y., Hossain, N., and Chen, C.-C. (2018). Modeling of polyelectrolytes system with electrolyte nonrandom two-liquid model. In preparation.

22

THERMODYNAMICS AND RELATIVE SOLUBILITY PREDICTION OF POLYMORPHIC SYSTEMS

YURIY A. ABRAMOV
Global R&D, Pharmaceutical Sciences, Pfizer, Inc., Groton, CT, USA

KLIMENTINA PENCHEVA
Global R&D, Pharmaceutical Sciences, Pfizer, Inc., Sandwich, UK

22.1 INTRODUCTION

Polymorphism of the crystalline state of pharmaceutical compounds is quite a common phenomenon that has been the subject of intense investigation for more than 40 years [1]. Polymorphs may significantly differ from each other in a variety of physical properties such as melting point, enthalpy and entropy of fusion, heat capacity, density, dissolution rate, and intrinsic solubility. These differences are dictated by the differences in the free energies of the forms, which in turn determine their relative stability at specific temperatures. Two polymorphs are monotropically related to each other if their relative stability remains the same up to their melting points. Otherwise, the forms are related to each other enantiotropically and may display a solid–solid transition at a temperature below the melting point. In practice, monotropic and enantiotropic behaviors are usually differentiated by several simple rules based on the experimental heats of fusion, entropies of fusion, heat of solid–solid transition, heat capacities, and densities [2, 3].

In the pharmaceutical industry, drug polymorphism can be a critical problem and is the subject of various regulatory considerations [4, 5]. One of the principal concerns is based on an effect that polymorphism may have on a drug's bioavailability due to change of its solubility and dissolution rate [6]. A famous example of a polymorphism-induced impact is the anti-HIV drug Norvir (also known as ritonavir) [7]. Abbott Laboratories had to stop sales of the drug in 1998 due to a failure in a dissolution test that was caused by the precipitation of a more stable form II [8]. As a result, Abbott lost an estimated $250 million in the sales of Norvir in 1998 [9, 10].

A large number of studies have been focused on the polymorphism effect on solubility, many of which were summarized by Pudipeddi and Serajuddin [11]. Several-fold solubility decrease was observed for many polymorphic systems. Therefore, in pharmaceutical industry, it is quite crucial to get comprehensive experimental information on the available drug polymorphs and their relative stability and solubility. Beyond that, it is important to be able to perform an estimation of the potential impact of an unknown, more stable form on a drug's solubility. Knowledge of such an impact should be considered in a risk assessment of the API solid form nomination for commercial development.

There have been a number of studies considering the quantitative models to estimate the solubility ratio of two polymorphs based on the thermal properties of both forms [11–15]. One of the major objectives of this work is to determine the potential impact of an unknown and more stable form on the drug solubility. This is accomplished by re-evaluating those models and paying a special attention to errors that may be introduced by the most common assumptions with the hope of producing a new more accurate model. Such model should satisfy the following two conditions. It should require a smaller number of input parameters, predominately relying on the thermal properties of only one (the known) form. When applied to a pair of observed polymorphs, the accuracy of the solubility ratio prediction by this

---

Chemical Engineering in the Pharmaceutical Industry: Active Pharmaceutical Ingredients, Second Edition.
Edited by David J. am Ende and Mary T. am Ende.
© 2019 John Wiley & Sons, Inc. Published 2019 by John Wiley & Sons, Inc.

equation should be at least as accurate as any currently known model.

22.2 METHODS

Methods used in this work are based on a combination of purely theoretical considerations and statistical analysis of available experimental data. A theoretical analysis of all popular approaches for prediction of absolute and relative solubilities of crystalline forms was performed. Special attention was paid to errors that are introduced by each of the approximations. Literature reports were carefully reviewed for solubility and thermal data of the organic crystals, with focus on drug-like molecules. In order to increase the statistical significance of the analysis, a comprehensive compilation was made of available polymorph solubility ratio data. However, only low solubility data (dilute solutions) for nonsolvated polymorphs was considered.

22.3 RESULTS AND DISCUSSION

22.3.1 Solubility of a Crystalline Form

The solubility, X_{i_i}, of a crystal form i in a solution can be presented as:

$$\ln X_{i_i} = \ln\left(\frac{X_{i_i}^{id}}{\gamma_i}\right) = -\frac{\Delta G_i}{RT} - \ln \gamma_i \quad (22.1)$$

where

$X_{i_i}^{id}$ is an ideal solubility.

γ_i is an activity coefficient, which accounts for deviations from the ideal behavior in a mixture of liquid solute and solvent.

ΔG_i is a free energy difference between the liquid and solid solute at the temperature of interest, T.

R is the universal gas constant.

In case no additional phase transition takes place in the temperature range between the temperature of interest, T, and the melting point, T_m, the ΔG_i can be presented as

$$\Delta G_i = \Delta H_{fus}\left(1 - \frac{T}{T_m}\right) + \int_{T_m}^{T} \Delta C_p dT - T \int_{T_m}^{T} \frac{\Delta C_p}{T} dT \quad (22.2)$$

where

ΔH_{fus} is the heat of fusion of the polymorph i at its melting point, T_m.

ΔC_p is a difference between the heat capacities of the liquid and solid states of the form i, which is always positive.

For practical reasons, it is usually assumed that ΔC_p is constant and equal to one estimated at the T_m, ΔC_{pm}. In that case the free energy difference, ΔG_i, can be presented as:

$$\Delta G_i = \Delta H_{fus}\left(1 - \frac{T}{T_m}\right) - \Delta C_{pm}(T_m - T) + \Delta C_{pm} T \ln \frac{T_m}{T} \quad (22.3)$$

However, as a rule, the ΔC_{pm} property is not available and further approximations should be taken. The most popular assumptions that are used in the literature are $\Delta C_{pm} = 0$ (Assumption A) and $\Delta C_{pm} = \Delta S_{fus}$ (Assumption B), where ΔS_{fus} is entropy of fusion at the melting point, $\Delta S_{fus} = \Delta H_{fus}/T_m$. Equation (22.3) is simplified upon these assumptions to Eqs. (22.4) and (22.5), respectively:

$$\Delta G_i = \Delta H_{fus}\left(1 - \frac{T}{T_m}\right) \quad (22.4)$$

$$\Delta G_i = \Delta H_{fus}\frac{T}{T_m}\ln\frac{T_m}{T} = \Delta S_{fus} T \ln \frac{T_m}{T} \quad (22.5)$$

While the first assumption (A) is usually justified by negligibly low value of the ΔC_{pm} (which is not always true), the latter one (B) is based on the observation by Hildebrand and Scott that $\ln X_{i_i}^{id}$ is linearly related to $\ln T$ [16].

In order to understand the errors introduced by both assumptions, they were mathematically derived below from Eq. (22.3) based on a first-order Taylor expansion of $\ln(T_m/T) \approx (T_m/T - 1)$, which is correct only in case of T_m/T close to 1 (Table 22.1).

22.3.1.1 Assumption A
Transformation of $\ln(T_m/T)$ to $(T_m/T - 1)$ in the last term of Eq. (22.3) results in the complete cancelation of the last two terms of the equation:

$$\Delta G_i \approx \Delta H_{fus}\left(1 - \frac{T}{T_m}\right) - \Delta C_{pm}(T_m - T) + \Delta C_{pm} T\left(\frac{T_m}{T} - 1\right)$$

$$= \Delta H_{fus}\left(1 - \frac{T}{T_m}\right)$$

Thus, the applied transformation is equivalent to neglecting the ΔC_{pm} term ($\Delta C_{pm} = 0$, Eq. 22.4). Since at $T < T_m$, $(T_m/T - 1)$ is always larger than $\ln(T_m/T)$ (Table 22.1), and ΔC_{pm} is always positive. Assumption A leads to the systematic overestimation of the ΔG_i, resulting into underestimation of the solubility relative to the predictions based on Eq. (22.3). The ΔG_i error introduced by Assumption A relative to Eq. (22.3) is related to the error of the first-order Taylor series expansion and can be presented as

$$\Delta G_{i,\text{error}}^A = \Delta C_{pm} T\left[\left(\frac{T_m}{T} - 1\right) - \ln\frac{T_m}{T}\right] \quad (22.6)$$

TABLE 22.1 Relative Errors of the First-order $\ln(T_m/T)$ Expansions for Different T_m/T Values

T_m/T^a	$\ln(T_m/T)$	$T_m/T - 1^b$	Relative Error (%)^b	$2(T_m/T-1)/(T_m/T+1)^c$	Relative Error (%)^c
1.1 (330/300)	0.095	0.1	4.9	0.095	−0.1
1.2 (360/300)	0.182	0.2	9.7	0.182	−0.3
1.3 (390/300)	0.262	0.3	14.3	0.261	−0.6
1.4 (420/300)	0.337	0.4	18.9	0.333	−0.9
1.5 (450/300)	0.406	0.5	23.3	0.400	−1.3
1.6 (480/300)	0.470	0.6	27.7	0.462	−1.8

^{a}Examples of T_m and T values in K are presented in the parentheses. T is chosen to be close to the room temperature.
^{b}First-order Taylor series expansion: $\ln(T_m/T) \approx (T_m/T - 1)$.
^{c}First-order expansion adopted by Hoffman [17]: $\ln(T_m/T) \approx 2(T_m/T-1)/(T_m/T+1)$. This expansion is significantly more accurate than the first-order Taylor expansion.

This error is proportional to the ΔC_{pm} property and increases with T_m/T due to an increasing inaccuracy in the first-order Taylor expansion (Table 22.1).

22.3.1.2 Assumption B

In attempt to counterbalance the error introduced by the direct Taylor expansion transformation used in Assumption A, one may apply a reverse transformation, $(T_m/T - 1) \approx \ln(T_m/T)$, to Eq. (22.4):

$$\Delta G_i = \Delta H_{fus}\left(1 - \frac{T}{T_m}\right) = \Delta H_{fus}\frac{T}{T_m}\left(\frac{T_m}{T} - 1\right)$$

$$\approx \Delta H_{fus}\frac{T}{T_m}\ln\frac{T_m}{T} = \Delta S_{fus}T\ln\frac{T_m}{T}$$

The result is equivalent to Eq. (22.5), which was derived under assumption of $\Delta C_{pm} = \Delta S_{fus}$. Since ΔH_{fus} (and ΔS_{fus}) is always positive and the error introduced by the reverse transformation is opposite to the one introduced by the direct transformation (Assumption A), a cancelation of errors should take place. The resulting ΔG_i error introduced by Assumption B relative to Eq. (22.3) is equal to

$$\Delta G^B_{i,\text{error}} = \Delta G^A_{i,\text{error}} + \Delta S_{fus}T\left[\ln\frac{T_m}{T} - \left(\frac{T_m}{T} - 1\right)\right]$$

$$= (\Delta C_{pm} - \Delta S_{fus})T\left[\left(\frac{T_m}{T} - 1\right) - \ln\frac{T_m}{T}\right] \quad (22.7)$$

It is apparent from this equation that the error will change sign in case of $\Delta S_{fus} > \Delta C_{pm}$, resulting in underestimation of the ΔG_i and overestimation of solubility relative to the predictions based on Eq. (22.3). In the case where ΔS_{fus} is more than twice as large as ΔC_{pm}, an absolute error introduced by Assumption B will exceed the error introduced by Assumption A. This phenomenon may have resulted in the contradicting results of the relative accuracy of Assumptions A and B in the literature [18–21]. It was shown recently [22] that a relation between ΔS_{fus} and ΔC_{pm} properties is dependent on a chemical class of organic compounds. A ratio of the absolute ΔG_i errors introduced by Assumptions B (Eq. 22.7) and A (Eq. 22.6), $\Delta G^B_{i,\text{error}}/\Delta G^A_{i,\text{error}} = |\Delta C_{pm} - \Delta S_{fus}|/\Delta C_{pm}$, is presented for 68 organic compounds in Figure 22.1. Only 12 non drug-like compounds out of 68 totally displayed a relative error of more than 1, indicating that Assumption B introduces a higher absolute error than Assumption A. A majority of these compounds can be characterized by the low value of their differential heat capacities, $\Delta C_{pm} < 40$ J/mol/K (Figure 22.1). All of these considerations provide justification for the application of Assumption B over Assumption A for drug-like compounds.

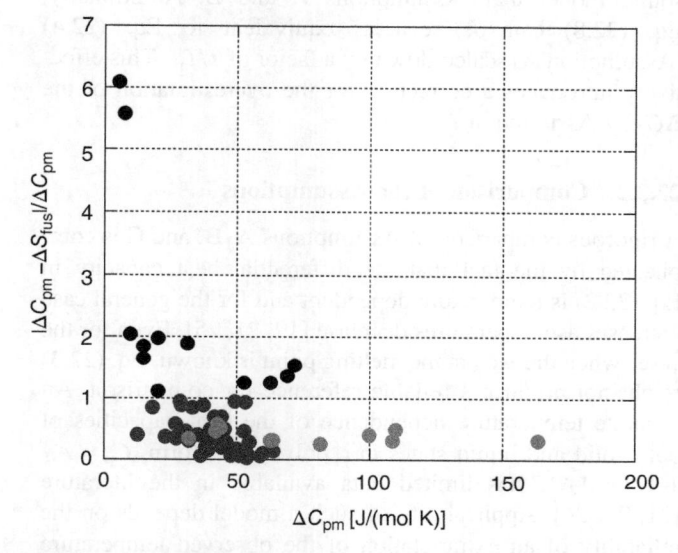

FIGURE 22.1 A ratio of the absolute ΔG_i errors introduced by Assumptions B and A relative to Eq. (22.3), $|\Delta C_{pm} - \Delta S_{fus}|/\Delta C_{pm}$, vs. differential heat capacity values ΔC_{pm} for 68 organic compounds. The closer this ratio to zero, the lower is the error introduced by Assumption B relative to Eq. (22.3). The compounds for which Assumption B introduces higher absolute error than Assumption A are are designated by the darkest bullets. The drug compounds (paracetamol [21], anisic acid [21], diethylstilbestrol [21], mannitol [21], naproxen [21], caffeine I [23], carbamazepine I [23], progesterone I [23], and acetamide [24]) are highlighted in light bullets. Source: Data from Pappa et al. [22].

22.3.1.3 Assumption C

Another valuable approximation of Eq. (22.2) was proposed by Hoffman [17] based on a significantly more accurate series expansion of $\ln(T_m/T)$ than the first-order Taylor expansion applied above (Table 22.1):

$$\Delta G_i = \Delta H_{\text{fus}}(T_m - T)\frac{T}{T_m^2} \quad (22.8)$$

The differential heat capacity, ΔC_p, is assumed to be both not negligible and independent on temperature. The lack of significant errors introduced by the $\ln(T_m/T)$ expansion, and perhaps a more justified approximation for the ΔC_p, makes Assumption C a generally more thermodynamically sound model than Assumptions A and B. Additionally, Eq. (22.8) can be seen as equivalent to Eq. (22.4) (Assumption A) scaled down by a factor of T/T_m. This effectively introduces a correction for the overestimation of the ΔG_i by Assumption A.

22.3.2 Comparison of the Assumptions

A rigorous comparison of Assumptions A, B, and C is complicated by the fact that the differential heat capacity in Eq. (22.2) is temperature dependent and for the general case increases as temperatures decrease [19, 21, 25]. Even for the cases when the ΔC_p at the melting point is known, Eq. (22.3) might not produce a reliable reference for comparison. An accurate temperature dependence of the heat capacities of both solid and liquid states in a polynomial form, $C_p = A_0 + A_1T + A_2T^2$, has limited data available in the literature [21, 23, 26]. Applicability of such a model depends on the reliability of an extrapolation of the observed temperature behavior of the heat capacities above (liquid state) and below (solid state, supercooled liquid) T_m at the temperature of interest. The difference between the coefficients (A_i) of the liquid and solid forms reflects a temperature dependence of the differential heat capacity. In such a case, the free energy difference between the liquid and solid solutes can be presented as

$$\Delta G_i = \Delta H_{\text{fus}}\left(1 - \frac{T}{T_m}\right) - \Delta A_0(T_m - T) + \Delta A_0 T \ln\frac{T_m}{T} \\ - \Delta A_1 \frac{(T_m - T)^2}{2} - \Delta A_2 \frac{2T_m^3 + T^3 - 3TT_m^2}{6} \quad (22.9)$$

where

$$\Delta A_i = A_i \text{ (liquid)} - A_i \text{ (solid)}.$$

In Table 22.2 ΔG_i predictions at room temperature using Eq. (22.3) and Assumptions A, B, and C are compared to the results based on Eq. (22.9). The differential heat capacities of all the compounds increase significantly at room temperature relative to the values at their melting points (Table 22.2). The temperature dependence of the ΔC_p leads to a decrease of the predicted ΔG_i values at the room temperature relative to the predictions based on the differential heat capacities at T_m (Eq. 22.3). A resulting mean absolute error (MAE) of Eq. (22.3) predictions is 1.0 kJ/mol (Table 22.2). This ΔG_i error corresponds to an average underestimation of the ideal solubilities at room temperature by 33%. The MAE values of the ΔG_i predictions based on Assumptions A, B, and C for the same compounds relative to results obtained by Eq. (22.9) are 3.9, 1.9, and 0.4 kJ/mol, respectively. The corresponding errors of the ideal solubility predictions at the room temperature are 79, 54, and 15%, respectively. Thus, the presented results demonstrate that the Hoffman approximation significantly outperforms Assumptions A and B. The largest error of ideal solubility prediction at ambient temperature is made using Assumption A, which introduces a large ΔG_i error.

TABLE 22.2 Absolute Errors of the ΔG_i Predictions at the Room Temperature According to Eq. (22.3) and Assumptions A (Eq. 22.4), B (Eq. 22.5), and C (Eq. 22.8) Relative to the Results Obtained Utilizing Temperature-dependent ΔC_p Values (Eq. 22.9)[a]

Name	T_m (K)	ΔH_{fus} (kJ/mol)	ΔC_{pm} (J/[mol K])	ΔC_p (T = 298.2 K) (J/[mol K])	Error Relative to Eq. (22.9) (kJ/mol)			
					Eq. (22.3)	Eq. (22.4)	Eq. (22.5)	Eq. (22.8)
Carbamazepine I [23]	463.7	26.3	109.8	164.6	0.7	4.4	2.5	1.0
Carbamazepine III [23]	452.4	27.2	111.3	184.3	1.0	4.3	2.5	1.1
Paracetamol [26]	442.2	28.1	99.6	165.8	0.6	3.3	1.6	0.3
Anisic acid [21]	455.4	27.8	81.4	150.6	0.8	3.3	1.4	0.0
Diethylstilbestrol [21]	441.8	28.8	43.8	262.3	2.0	3.2	1.5	0.2
Mannitol [21]	438.7	50.6	163.8	290.3	1.1	5.3	2.4	0.1
Naproxen [21]	428.5	31.5	108.6	220.3	0.9	3.3	1.7	0.4
MAE[b] (kJ/mol)					1.0	3.9	1.9	0.4

[a] The ΔA_2 term (Eq. 22.9) is different from zero only in the case of carbamazepine.
[b] Mean absolute error is calculated as an arithmetic average of the absolute errors of the predictions performed by the corresponding approach.

22.3.3 Application to Polymorphs Solubility Ratio

The solubility ratio of polymorphs can be presented by Eq. (22.10), and seems to be an optimal test for validation of the different ΔG_i models considered in Section 22.3.1:

$$\frac{X_i}{X_j} = \frac{X_i^{id}\gamma_j}{X_j^{id}\gamma_i} = \frac{\gamma_j}{\gamma_i}\exp\left(-\frac{\Delta G_i - \Delta G_j}{RT}\right) \equiv \frac{\gamma_j}{\gamma_i}\exp\left(-\frac{\Delta\Delta G_{ij}}{RT}\right) \quad (22.10)$$

Recently, evaluations of different models for polymorph solubility ratio prediction based on thermal properties of the polymorphs were reported [11, 12]. Pudipeddi and Serajuddin have found that for 10 polymorphic pairs of predictions based on Assumption C were "slightly closer" to the experimental data than the results obtained by Assumption A [11]. Mao et al. have considered calculations based on Assumption A [12]. A validation of this approach on nine polymorphic systems led to the conclusion that the utilization of Assumption A typically leads to an error of only 10% or less. An obvious drawback of these two studies is that the very limited data sets of the polymorph pairs were adopted for the testing of only selected assumptions. Therefore, to increase statistical significance of the results, further side-by-side verification of all three assumptions using larger experimental data sets could prove to be very important.

Two different data sets were selected for the model validation in this study, which contains 10 monotropically related (Table 22.3) and 18 enantiotropically related (Table 22.4) pairs of nonsolvated polymorphs. Each data point in these sets contains information on experimental properties such as solubility, X_i, melting point, T_m, and heats of fusion, ΔH_{fus}. The following considerations were taken into account during the data selection. There is quite a common misperception that the polymorph solubility ratio is solvent independent. However, according to Eq. (22.10), this is only true when the activity coefficients for the two polymorphs are identical to each other in any solvent [12, 47]. This takes place in the case of an infinite solubility limit (dilute solution). In such a case, each polymorph in the liquid state is not a significant part of the solvent system in which the actual solubility is measured. Thus, whenever possible, solubility data was chosen for polymorphs approximately tens of mg/ml or less. Moreover, at these low concentrations, there is no need to convert mg/ml or µg/ml units to mol fractions, in which Eq. (22.10) is presented. One drawback of the selection of very low solubility data is a higher standard deviation of the experimental polymorph solubility ratios (Appendix 22.A).

TABLE 22.3 Comparison of the Experimental and Predicted Solubility Ratios for Monotropically Related Polymorphic Pairs

Compound	T_{m1} (K)	ΔH_{fus1} (kJ/mol)	T_{m2} (K)	ΔH_{fus2} (kJ/mol)	T (K)	$S_1/S_{2, exp}$	Assumption A	Assumption B	Assumption C	Eq. (22.13)	Eq. (22.14)	Eq. (22.15)	Eq. (22.16)
Chloramphenicol palmitate (A/B) [24, 27]	362	41.9	368	64.1	303	4.2	5.95	4.93	4.19	5.14	4.74	4.10	3.60
Tolbutamide (I/III) [24, 28]	379	18.5	400	24.5	303	1.22	2.42	2.08	1.84	1.29	1.78	1.65	1.55
Ritonavir (I/II) [9]	395.2	56.4	398.2	63.3	298	2.39[a]	2.29	2.01	1.80	2.17	2.02	1.83	1.69
MK571(II/I) [29]	425.2	49.0	437.2	54.0	309	1.64	2.59	2.08	1.77	1.56	1.77	1.61	1.50
Cyclopenthiazide (II/I) [30][b]	496.2	98.42	512.5	105.5	310	1.78	6.32	3.40	2.30	1.72	2.96	2.31	1.93
E2101 (II/I) [31]	413.0	35.2	421.3	38.2	298	1.25	1.74	1.54	1.40	1.32	1.43	1.35	1.28
Indomethacin (α/γ) [15]	429.2	36.14	435.2	36.49	318	1.1	1.19	1.14	1.10	1.11	1.04	1.03	1.03
Acemetacin (II/I) [3]	423.2	48.4	423.7	50.7	293	1.67	1.36	1.29	1.23	1.52	1.34	1.27	1.22
Torasemide (II/I) [32]	430.2	29	434.7	37.2	293	2.74	3.26	2.58	2.16	2.51	3.00	2.45	2.10
Cimetidine (A/B) [33, 34]	413.5	44.03	413.7	44.08	298	1.15	1.01	1.01	1.01	1.21	1.01	1.00	1.00
MAE							**1.01**	**0.50**	**0.32**	**0.19**	**0.38**	**0.27**	**0.33**

Results of application of Eqs. (22.14)–(22.16) adopting $T_m = T_{m2}$ are presented. The explicit equations for solubility ratio predictions are presented in Appendix 22.B.
[a]Polymorph solubility data in ethyl acetate:heptanes 2 : 1 mixture is adopted for solubility ratio estimation.
[b]An enantiotropic relationship between forms I and II with a very low transition temperature was proposed in the literature [14].

TABLE 22.4 Comparison of the Experimental and Predicted Solubility Ratios for Enantiotropically Related Polymorphs

Compound	T_{m1} (K)	ΔH_{fus1} (kJ/mol)	T_{m2} (K)	ΔH_{fus2} (kJ/mol)	T (K)	$S_1/S_{2,exp}$	Assumption A	Assumption B	Assumption C	Eq. (22.14)	Eq. (22.15)	Eq. (22.16)
Axitinib (IV/VI) [35–37]	491.90	47.15	484.80	51.79	310	1.25	1.62	1.53	1.45	1.91	1.67	1.51
Axitinib (IV/XXV) [35–37]	491.90	47.15	490.40	50.43	310	1.25	1.54	1.42	1.33	1.60	1.45	1.34
Paracetamol (II/I) [24, 38]	429	26.9	442	28.1	303	1.3	1.45	1.30	1.21	1.16	1.13	1.11
Buspirone-HCL (II/I) [39]	476.8	42.24	463.0	47.45	303	1.7	1.49	1.49	1.46	2.04	1.78	1.60
Carbamazepine (I/III) [40]	462	26.4	448	29.3	299.2	1.20	1.19	1.21	1.21	1.47	1.37	1.30
F2692 [24]	453	27.17	445	29.32	303	1.8	1.15	1.16	1.15	1.31	1.25	1.20
Gepirone hydrochloride (II/I) [41]	485	41.6	453	47.1	303	2.01	0.99	1.19	1.31	2.06	1.80	1.62
Indiplon[a]	465.51	41.63	463.08	45.89	298	1.1	1.74	1.58	1.46	1.85	1.63	1.48
Nimodipine (I/II) [3]	397.2	39	389.2	46	298	2	1.52	1.49	1.46	1.94	1.78	1.66
Phenylbutazone (II/III) [24]	370	21.9	368	24.4	303	1.1	1.15	1.14	1.13	1.19	1.17	1.16
Piroxicam (I/II) [42]	475.8	36.54	472.9	37.43	310	0.99	1.06	1.07	1.06	1.13	1.10	1.08
Propranolol hydrochloride (I/II) [43]	436.2	31.35	435.0	36.62	293	1.34	1.98	1.75	1.60	2.03	1.78	1.61
Retinoic acid (II/I) [44]	456.3	36.8	456.9	37.1	310	1.32	1.05	1.04	1.03	1.04	1.03	1.03
RG12525 (II/I) [45]	431.0	43.10	427.8	46.86	304	1.26	1.41	1.35	1.31	1.54	1.43	1.36
Sulfathiazole (I/III) [24]	474.2	27.75	446.8	29.47	303	1.68	0.81	0.93	1.01	1.25	1.20	1.16
WIN63843 (I/III) [46]	337.7	28.87	334.4	31.88	296	1.04	1.04	1.04	1.05	1.15	1.14	1.13
Cimetidine (C/B) [33, 34]	417.5	43.13	413.7	44.08	298	1.23	0.99	1.01	1.03	1.11	1.09	1.08
Cimetidine (C/A) [33, 34]	417.5	43.13	413.5	44.03	298	1.07	0.98	1.01	1.02	1.11	1.09	1.08
MAE							0.34	0.28	0.25	0.29	0.24	0.22

Results of application of Eqs. (22.14)–(22.16) adopting $T_m = T_{m2}$ are presented. The explicit equations for solubility ratio predictions are presented in Appendix 22.B.
[a]Collman B, private communication.

22.3.3.1 Monotropic Case

The initial validation of the solubility ratio models was performed using monotropically related polymorphs. In the following discussions notation 1 and 2 will refer to the higher and lower soluble polymorphs. Equations used for the solubility ratio predictions in this section are explicitly listed in Appendix 22.B. Given that low solubility experimental data was selected for the test, it seems reasonable to expect that cancelation will take place not only between the activity coefficients of both polymorphs in the solution but also between the errors introduced by the ΔG_i assumptions. Results of the relative solubility predictions utilizing Assumptions A, B, and C (Appendix 22.B) for each polymorph are presented in Table 22.3. The corresponding MAE values relative to the experimental X_1/X_2 observations are 1.01, 0.50, and 0.32, respectively. These observations disagree with previous reports that Assumption A results in an error of only 10% or less [12] and that Assumption C is just slightly closer to the experimental data than the results obtained by Assumption A [11]. The obtained MAE values demonstrate that a complete cancelation of errors does not take place and, as a result, Assumption C remains significantly more accurate than the others.

According to the error analysis presented in Section 22.3.1 (Eqs. 22.6 and 22.7), the lack of error cancelation in the $\Delta\Delta G_{12}$ prediction can be accounted for by non-negligible differences of the ΔC_{pm} ($[\Delta C_{pm} - \Delta S_{fus}]$ in case of Assumption B) and/or T_m properties between the two polymorphs. For example, in case of Assumption A, the error of the $\Delta\Delta G_{12}$ prediction relative to the one based on Eq. (22.3) can be presented as a difference of $\Delta G^A_{i,\text{error}}$ errors (Eq. 22.6) between two polymorphs:

$$\Delta\Delta G_{\text{error}} = T\left\{\Delta C_{pm1}\left[\left(\frac{T_{m1}}{T}-1\right)-\ln\frac{T_{m1}}{T}\right]\right.$$
$$\left. -\Delta C_{pm2}\left[\left(\frac{T_{m2}}{T}-1\right)-\ln\frac{T_{m2}}{T}\right]\right\} \quad (22.11)$$

The following two limiting cases can be derived from Eq. (22.11). In the case of relatively insignificant variations of the ΔC_{pm} terms, the $\Delta\Delta G_{\text{error}}$ is proportional to $T\{[(T_{m1}/T - 1) - \ln(T_{m1}/T)] - [(T_{m2}/T - 1) - \ln(T_{m2}/T)]\}$. In the case where variations of ΔC_{pm} are noticeably more significant than the variations of T_m, the $\Delta\Delta G_{\text{error}}$ is proportional to $(\Delta C_{pm1} - \Delta C_{pm2})$. The ΔC_{pm} values should be replaced by the $(\Delta C_{pm} - \Delta S_{fus})$ differences for the error estimation of the $\Delta\Delta G_{12}$ prediction based on Assumption B.

It is easy to show from Eq. (22.10) that for dilute solutions, the difference between the natural logarithms of polymorph solubility ratios as predicted by Assumption A or B and Eq. (22.3) should be proportional to the $\Delta\Delta G_{\text{error}}$:

$$\ln\left(\frac{X_1}{X_2}\right)_{A,B} - \ln\left(\frac{X_1}{X_2}\right)_{Eq.(22.3)} = -\frac{\Delta\Delta G_{\text{error}}}{RT} \quad (22.12)$$

According to Eq. (22.11) in case of Assumption A, this difference will be equal to $\{\Delta C_{pm2}[(T_{m2}/T - 1) - \ln(T_{m2}/T)] - \Delta C_{pm1}[(T_{m1}/T - 1) - \ln(T_{m1}/T)]\}/R$. It is reasonable to propose that the difference between natural logarithms of polymorph solubility ratios as predicted by Assumption A or B, and those observed experimentally, may be described by the similar factors as presented in Eqs. (22.11) and (22.12). In the absence of the ΔC_{pm} values, a correlation was tested between the $\ln(X_1/X_2) - \ln(X_1/X_2)_{\text{exp}}$ predictions based on the different assumptions and $\{[(T_{m2}/T - 1) - \ln(T_{m2}/T)] - [(T_{m1}/T - 1) - \ln(T_{m1}/T)]\}$ property (Figure 22.2). High linear correlation coefficients, R^2, of 0.92 and 0.91 were found for the predictions based on Assumptions A and B, respectively (Figure 22.2a and b). This observation suggests a higher and a more systematic contribution to the $\Delta\Delta G_{\text{error}}$ by the differences in the T_m values, rather than by the differences in the ΔC_{pm} or $(\Delta C_{pm} - \Delta S_{fus})$ properties. A noticeably weaker correlation (having an R^2 of 0.72) (Figure 22.2c) was observed for the predictions based on Assumption C.

Found simple linear regressions (Figure 22.2) can be used for estimations of likely errors in the solubility ratio predictions of monotropically related polymorphs based on the different assumptions. The MAE values of the prediction (0.24, 0.22, and 0.19, respectively) are based on Assumptions A, C, and B after the errors are corrected by using simple functions of the melting points. The latter result corresponds to the best agreement with the experimental observations using the approaches presented in Table 22.3. This suggests that the polymorph solubility ratio of the monotropically related polymorphs can be best predicted through the following relationship:

$$\ln\frac{X_1}{X_2} = \ln\left(\frac{X_1}{X_2}\right)_B + 0.188 - 43.096$$
$$\left\{\left[\left(\frac{T_{m2}}{T}-1\right)-\ln\frac{T_{m2}}{T}\right] - \left[\left(\frac{T_{m1}}{T}-1\right)-\ln\frac{T_{m1}}{T}\right]\right\} \quad (22.13)$$

Based on the above observation of the high contribution to the $\Delta\Delta G_{\text{error}}$ by the differences in the T_m values, an alternative approach can be suggested. In order to better counterbalance the prediction errors, it was proposed to adopt a single T_m value for both polymorphs used in the solubility ratio predictions. In this case, an improvement of the predictions should take place through the increase of ΔG_1 ($T_m = T_{m2}$) or the decrease of ΔG_2 ($T_m = T_{m1}$). The following simplifications of the $\Delta\Delta G_{12}$ calculation based on Assumptions A, B, and C are proposed:

$$\Delta\Delta G_{12} = (\Delta H_{fus1} - \Delta H_{fus2})\left(1 - \frac{T}{T_m}\right) \quad (22.14)$$

$$\Delta\Delta G_{12} = (\Delta H_{fus1} - \Delta H_{fus2})\frac{T}{T_m}\ln\frac{T_m}{T} \quad (22.15)$$

$$\Delta\Delta G_{12} = (\Delta H_{fus1} - \Delta H_{fus2})(T_m - T)\frac{T}{T_m^2} \quad (22.16)$$

Besides a possible improvement of the polymorph solubility prediction, the proposed equations more importantly depend on only two input parameters: T_m of one of the forms and a difference between the enthalpies of fusion of the two polymorphs. This fact makes these equations useful for solving one of the major objectives of the current study – the development of a working equation for the estimation of the potential impact of an unknown and more stable form on drug solubility.

The application of Eqs. (22.14)–(22.16) in predicting the solubility ratio of monotropically related polymorphs adopting $T_m = T_{m2}$ is presented in Table 22.3. Eqs. (22.14) and (22.15) dramatically improve agreement with the experimental data. The MAE drops from 1.01 to 0.38 for Assumption A using Eq. (22.14). In the case of Assumption B, the MAE changes from 0.50 to 0.27 using Eq. (22.15). No improvement was found for Assumption C, in which the MAE value practically does not change when adopting

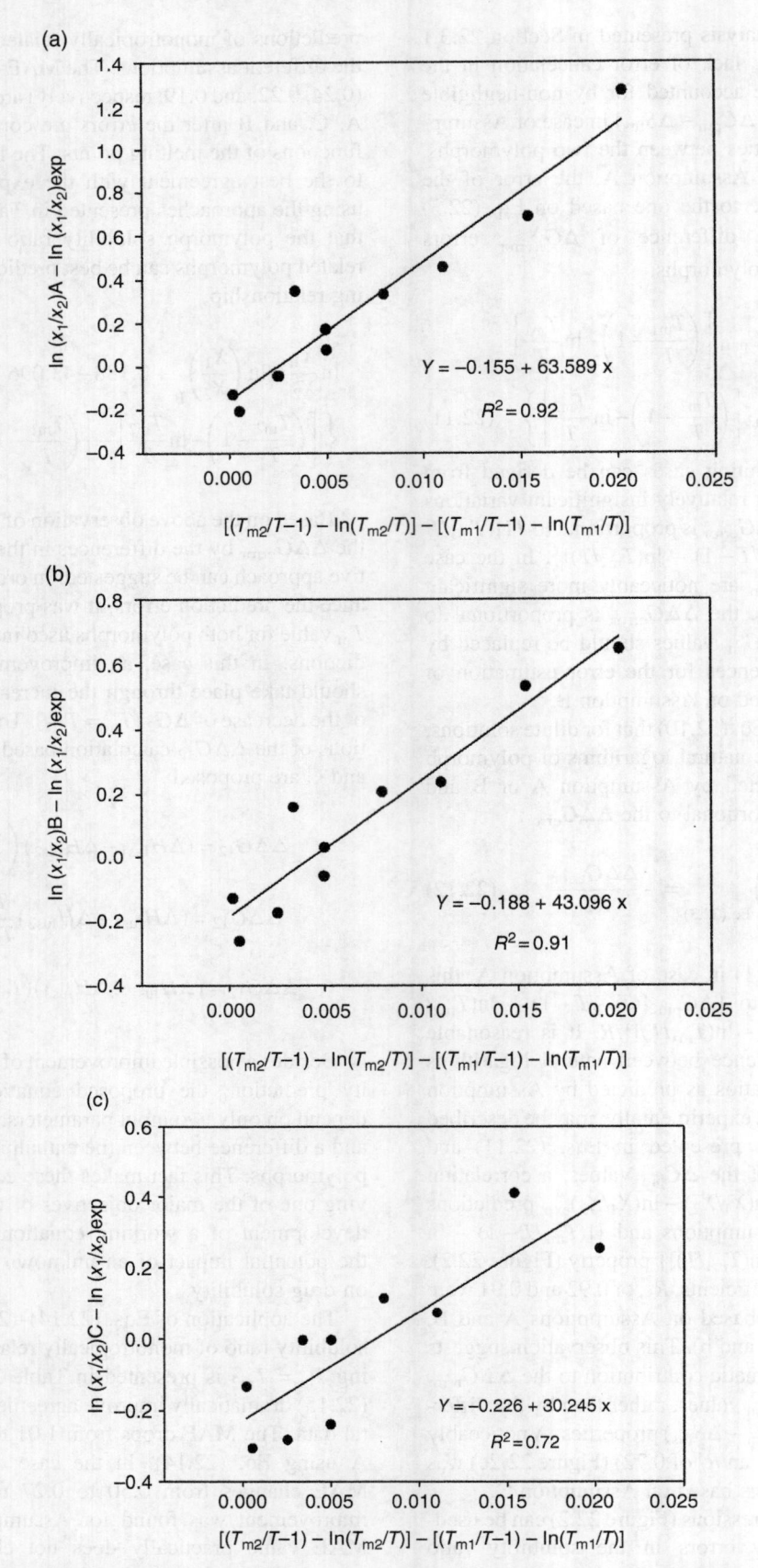

FIGURE 22.2 A correlation between the $\ln(X_1/X_2) - \ln(X_1/X_2)_{\exp}$ values based on Assumptions A (a), B (b), and C (c) and the $\{[(T_{m2}/T - 1) - \ln(T_{m2}/T)] - [(T_{m1}/T - 1) - \ln(T_{m1}/T)]\}$ property.

Eq. (22.16) with a single T_m value of T_{m2}. When $T_m = T_{m1}$, the MAE values for Eqs. (22.14)–(22.16) are 0.28, 0.29, and 0.35, respectively. This demonstrates behavior of the X_1/X_2 predictions similar to those found with $T_m = T_{m2}$.

22.3.3.2 Enantiotropic Case

A thermodynamic expression of the solubility ratio of enantiotropically related polymorphs requires knowledge of the temperature and enthalpy of the solid–solid transition [12], which is often difficult to measure accurately. For this reason, only enantiotropic systems with available melting properties for both polymorphs were included in this study. Results of the application of Assumptions A, B, and C to the predictions of the solubility ratio of the enantiotropically related polymorphs are presented in Table 22.4. An overall accuracy of the predictions is noticeably better than it was found for the monotropic system (Table 22.3). As in the monotropic case, the agreement with the experimental data is worse for the calculations based on Assumption A (MAE value is 0.34), relative to those based on Assumptions B and C (MAE values are 0.28 and 0.25, respectively).

No strong correlation was found between the $\ln(X_1/X_2) - \ln(X_1/X_2)_{exp}$ values and the $\{[(T_{m2}/T - 1) - \ln(T_{m2}/T)] - [(T_{m1}/T - 1) - \ln(T_{m1}/T)]\}$ property in case of enantiotropic system based on the different assumptions. Thus, error correction similar to that proposed by Eq. (22.13) is not applicable to the enantiotropic case. However, an improvement of the predictions based on Assumptions A, B, and C is possible by the application of Eqs. (22.14)–(22.16), respectively (Table 22.4). The best performance was found for Eq. (22.16) and Eq. (22.15), both resulting in MAE values of 0.22 and 0.24 (where $T_m = T_{m2}$ or T_{m1}), respectively.

It should be noted that Eqs. (22.14)–(22.16) cannot describe the change of the relative stability of the enantiotropically related polymorphs with temperature. To do so would result in the $\Delta\Delta G_{12}$ property having the wrong sign above the solid–solid transition, T_t ($\Delta\Delta G_{12}(T_t) = 0$). Thus, the application of these equations to enantiotropic polymorphs is limited to systems with temperatures below T_t.

From the above considerations, which are based on the analysis of the largest reported experimental data set of both monotropic and enantiotropic systems, an application of the original (Eq. 22.8) (Appendix 22.B) and modified (Eq. 22.16) Hoffman approaches as well as of Eq. (22.15) are recommended for an accurate solubility ratio prediction for both monotropic and enantiotropic polymorphic systems. Since the latter two approaches utilize the melting temperature measurements of only one form, T_{m2} or T_{m1}, they can be used in combination with the statistical analysis of the differences of the heat of fusions of polymorphs, for an estimation of the potential impact of an unknown and more stable form on drug solubilities (see Section 22.4 for more details).

EXAMPLE PROBLEM 22.1 Prediction of Ideal Solubility Ratio Between Forms II and I of Acemetacin at 293 K Based on Assumptions A, B, and C and Regression Eq. (22.13)

The thermal data for both forms of acemetacin are presented in Table 22.3. Initially the ΔG_{II} and ΔG_I properties should be calculated for each form adopting equations corresponding to Assumptions A (Eq. 22.4), B (Eq. 22.5), and C (Eq. 22.8). For example, adopting Assumption A one should get

$$\Delta G_I = \Delta H_{fusI}\left(1 - \frac{T}{T_{mI}}\right) = 50.7\left(1 - \frac{293}{423.7}\right) = 14.90\,\text{kJ/mol}$$

$$\Delta G_{II} = \Delta H_{fusII}\left(1 - \frac{T}{T_{mII}}\right) = 48.4\left(1 - \frac{293}{423.3}\right) = 15.64\,\text{kJ/mol}$$

The resulting ΔG_i values based on all three assumptions are listed in Table 22.5. At the next step differences between ΔG_{II} and ΔG_I properties should be calculated to obtain $\Delta\Delta G$ values. In order to calculate $\ln\left(X_{II}^{id}/X_I^{id}\right)$ values, the negative of the $\Delta\Delta G$ predictions should be divided by RT factor: $\ln\left(X_{II}^{id}/X_I^{id}\right) = -\Delta\Delta G/RT$. RT at 293 K is equal to: 8.314×10^{-3} (kJ/mol·K) × 293 (K) = 2.436 kJ/mol. For example, based on Assumption A, the $\Delta\Delta G$ value between two acemetacin forms is equal to $\Delta G_{II} - \Delta G_I = 15.64 - 14.90 = -0.74$ kJ/mol, which corresponds to $\ln\left(X_{II}^{id}/X_I^{id}\right)$ value of 0.30. The $\ln\left(X_{II}^{id}/X_I^{id}\right)$ value calculated based on Assumption B is used in combination with the $\{[(T_{m2}/T - 1) - \ln(T_{m2}/T)] - [(T_{m1}/T - 1) - \ln(T_{m1}/T)]\}$ property for the $\ln\left(X_{II}^{id}/X_I^{id}\right)$ prediction based on the regression Eq. (22.13). All the above steps are combined in the explicit

TABLE 22.5 Prediction of Ideal Solubility Ratio Between Forms II and I of Acemetacin at 293 K Based on Assumptions A, B, and C and the Regression Eq. (22.13)

Approach	ΔG_{II} (kJ/mol)	ΔG_I (kJ/mol)	$\Delta\Delta G$ (kJ/mol)	$\{[(T_{mI}/T - 1) - \ln(T_{mI}/T)] - [(T_{mII}/T - 1) - \ln(T_{mII}/T)]\}$	$\ln\left(X_{II}^{id}/X_I^{id}\right)$	X_{II}^{id}/X_I^{id}
Assumption A	14.90	15.64	−0.74		0.30	1.36
Assumption B	12.33	12.93	−0.61		0.25	1.28
Assumption C	10.31	10.82	−0.50		0.206	1.23
Eq. (22.13)				4.2E−4	0.42	1.52

equations presented in Table 22.B.1 in Appendix 22.B. Results of all the intermediate calculations are summarized in Table 22.5. Finally, exponent of the $\ln(X_{II}^{id}/X_I^{id})$ results gives the polymorph solubility ratio predictions based on all four methods. For this particular example the best and the worst agreements with the experimental value of 1.67 are obtained by Eq. (22.13) and Assumption C, respectively.

22.4 APPLICATION TO AN ESTIMATION OF LIKELY IMPACT ON DRUG SOLUBILITY BY UNKNOWN MORE STABLE FORM

Below we present two approaches to predict a likely change of drug solubility due the form change. The first thermal data approach is based on a combination of statistical analysis of the experimental heat of fusion differences between polymorphic pairs and the proposed in the current work Eq. (22.16) for the ideal solubility ratio prediction. The second, solubility ratio approach is based on statistical results from experimental solubility ratio observations.

22.4.1 Thermal Data Approach

This approach is based on application of the modified Hoffman Eq. (22.16) coupled with statistical analysis of experimentally determined heat of fusion differences between polymorphs. The ideal solubility ratio predictions can be carried out for a known form with available melting temperature and likely changes in heat of fusion, $\Delta\Delta H_{fus}$. In order to do that, a survey of thermal data for 101 polymorphic pairs was carried out, where most of the data where found in one literature source [24] and the rest were taken from Tables 22.3 and 22.4 of the current study. Trends in heat of fusion changes between polymorphs were presented in the form of the cumulative relative frequency distribution in Figure 22.3. The cumulative relative frequency distribution is particularly useful for describing the likelihood that a variable (heat of fusion difference) will not exceed a certain value. It was found that there is a 50% probability that the change in heat of fusion for a polymorphic pair is less than or equal to 3.0 kJ/mol (Figure 22.3). The probability of heat of fusion difference between a pair of polymorphs not exceeding values of 6.2 and 16.7 kJ/mol is, respectively, 80 and 95%. Combining these $\Delta\Delta H_{fus}$ values with Eq. (22.16) allows estimation at the different probability levels of maximum impact on ideal solubility by a new and more stable polymorph. Though the thermal data approach relies on the statistical analysis (of $\Delta\Delta H_{fus}$), it introduces some degree of the dependence on the thermal properties (T_m) of the reference form through Eq. (22.16). Therefore predictions based on this method are form specific.

22.4.2 Solubility Ratio Approach

An alternative approach is based on the statistical analysis of the polymorph solubility ratio observations. A survey of solubility changes for 153 polymorphic pairs was carried out, where most of the data were found in open literature sources [11] and some were extracted from in-house Pfizer data or provided by company-associated institutions. A statistical

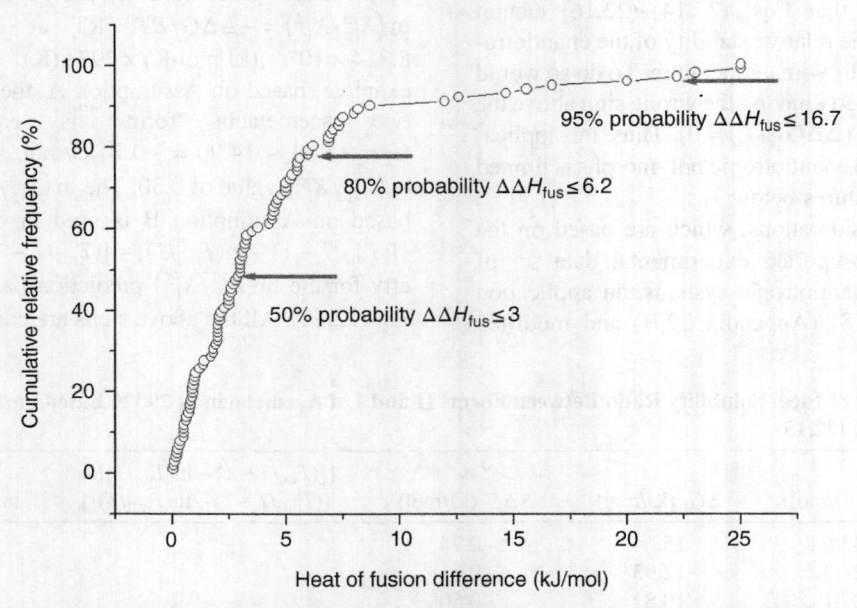

FIGURE 22.3 Cumulative relative frequency distribution of experimental differences of heats of fusion, $\Delta\Delta H_{fus}$, for 101 polymorphic pairs. Data points corresponding to 50, 80, and 95% probabilities of $\Delta\Delta H_{fus}$ not exceeding a certain threshold are indicated by arrows.

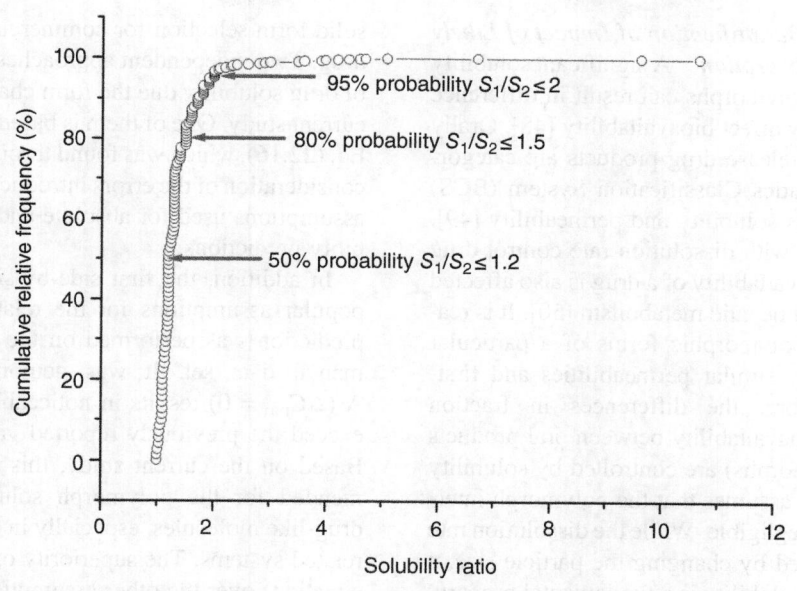

FIGURE 22.4 Cumulative relative frequency distribution of experimental solubility ratios, X_1/X_2, for 153 polymorphic pairs. Data points corresponding to 50, 80, and 95% probabilities of solubility ratio not exceeding a certain threshold are indicated by arrows.

analysis of the experimental data was performed on the basis of cumulative relative frequency distribution, presented in Figure 22.4. It was found that there is a 50% probability that the change in solubility is less than 1.2-fold for a polymorphic switch (Figure 22.3). The probability of the solubility ratio between a pair of polymorphs not exceeding value of 1.5 is 80%. It was also shown that only in 5% of the studied cases the change in solubility for polymorphic pairs would be more than twofold (Figure 22.4).

The presented trend of probabilities of the relative solubility changes is purely statistical and does not provide any direct dependence on thermal data of the current form. Therefore this approach may be considered as form nonspecific one. In addition, majority of the solubility ratio measurements are performed at different temperatures in the range of 20–40 °C [11], rather than at the room temperature. That may introduce some level of noise in the predictions based on this method.

EXAMPLE PROBLEM 22.2 Estimation of a Likely Impact on Solubility by a New Form of Ritonavir

Analysis of a possible impact on solubility by a new form should be defined by a selected probability limit of the X_1/X_2 (the solubility ratio approach) and $\Delta\Delta H_{fus}$ (the thermal data approach) changes. In this example the 80% probability was selected to provide a reasonably high level of confidence of predictions by both methods. In case more than one polymorphic form exist, a probability of further increase of the polymorphs solubility ratio as well as of the $\Delta\Delta H_{fus}$ should be estimated relative to the least stable form. In case of ritonavir the most unstable and soluble form is I. The 80% probability of heat of fusion increase according to the thermal data approach is not exceeding 6.2 kJ/mol (Figure 22.3). This $\Delta\Delta H_{fus}$ value together with the melting point of form I, $T_{mI} = 395.2$ K (Table 22.3), is used to predict a likely change of solubility of form I at the room temperature by Eq. (22.16):

$$\Delta\Delta G_{12} = \Delta\Delta H_{fus}(T_m - T)\frac{T}{T_m^2} = 6.2(395.2 - 298)\frac{298}{395.2^2}$$

$$= 1.17 \text{ kJ/mol}$$

This $\Delta\Delta G_{12}$ value corresponds to solubility ratio between two polymorphs of 1.60:

$$\frac{X_1}{X_2} = \exp\left(-\frac{\Delta\Delta G_{12}}{RT}\right) = \exp\left(-\frac{1.17}{0.008314*298}\right) = 1.60$$

Therefore the estimated impact is not exceeding 1.60-fold. The 80% probability of the change in solubility between two polymorphs according to the solubility ratio approach is less than or equal to 1.5-fold (Figure 22.4). The two methods provide similar estimated change of the ideal solubility with respect to the least stable form for a probability level of 80%. It is known that a more stable form II was discovered later for ritonavir. The observed heat of fusion difference and solubility ratio between forms II and I are, respectively, 6.9 kJ/mol and 2.39 (Table 22.3). These values correspond to, respectively, 82 and 98% probability limits of the $\Delta\Delta H_{fus}$ and X_1/X_2 changes and exceed the thresholds predicted within the probability limit of 80%. Therefore both approaches suggest a quite low probability of further impact on solubility by a hypothetical new stable form.

22.4.2.1 Qualification/Quantification of Impact of Likely Form Change on Drug Absorption

A significant solubility difference between two polymorphs can result in difference in oral absorption and may affect bioavailability [48]. Orally administrated immediate-release drug products are categorized in the Biopharmaceutics Classification System (BCS) according to their aqueous solubility and permeability [49]. These properties together with dissolution rate control drug absorption. Absolute bioavailability of a drug is also affected by first-pass intestinal and hepatic metabolism [50]. It is reasonable to assume that polymorphic forms of a particular compound should display similar permeabilities and first-pass clearances. Therefore, the differences in fraction absorbed and absolute bioavailability between oral products (based on different polymorphs) are controlled by solubility and dissolution rate. This assumes that the polymorph interaction with excipients is negligible. While the dissolution rate can be generally controlled by changing the particle size, a thermodynamic aqueous solubility is a fundamental property of the polymorphic form that cannot be modified.

The classification of drug form solubility is based on dimensionless dose number D_0, which is a function of maximum dose strength D (mg) and solubility S (mg/ml):

$$D_0 = \frac{D}{V_0 S} \quad (22.17)$$

Here V_0 is volume of water taken with the dose, which is generally set to 250 ml. Solid forms of drugs with D_0 equal to or less than 1 are considered being highly soluble [51]. According to the BCS system, such forms are characterized as classes I and III. An estimation of a likely change in solubility due to transformation to a new and more stable form allows prediction of the potential impact on D_0 that a new form could present. The potential risk associated with the late discovery of a new stable form can be accessed based on a degree of probability of solubility (and D_0) change, as discussed above, and projected change of drug absorption. A qualitative analysis of the impact of form change on absorption can be based on the BCS system; here we classify risk as associated with a potential change of the drug class from I to II or from III to IV. In addition, computational simulations (e.g. GastroPlus, Simulations Plus, Inc., Lancaster, CA) may be adopted for a (semi-)quantitative analysis of sensitivity of drug absorption to a potential form change.

22.5 CONCLUSION

One of the main purposes of this study is to develop valid methods for the estimation of a potential impact of an unknown and more stable form on drug solubility. This information has a crucial practical application in the pharmaceutical industry by supporting the risk assessment of an API solid form selection for commercial development of an oral drug. Two independent approaches to predict a likely change of drug solubility due the form change were suggested in the current study. One of them is based on the modified Hoffman Eq. (22.16), which was found through a consistent theoretical consideration of the errors introduced by the different popular assumptions used for absolute and relative polymorph solubility predictions.

In addition, the first side-by-side validation of all three popular assumptions for the relative polymorph solubility prediction was performed on the largest up-to-date experimental data set. It was demonstrated that Assumption A ($\Delta C_{pm} = 0$) results in noticeable errors that significantly exceed the previously reported values of 10% or less [12]. Based on the current study, this assumption is not recommended for the polymorph solubility ratio prediction of drug-like molecules, especially in case of the monotropically related systems. The superiority of Assumption C (Hoffman equation) over the other assumptions, and in particular over Assumption A, was found to be much stronger than was previously reported [11]. Assumption B ($\Delta C_{pm} = \Delta S_{fus}$) demonstrated an intermediate performance between Assumptions A and C.

Finally, based on the error analysis, a new model, Eq. (22.13), was proposed for the solubility ratio prediction of the monotropically related polymorphs. This model provided the best agreement with the experimental data set of 10 polymorphic pairs.

22.A PROPAGATION OF ERRORS OF THE SOLUBILITY RATIO MEASUREMENTS

Assuming independence of the solubility measurements of two polymorphs, X_1 and X_2, the standard deviation of the solubility ratio, $k = X_1/X_2$, can be expressed as [52]:

$$\sigma(\kappa) = \left\{ \left[\sigma(X_1)\frac{\partial k}{\partial X_1}\right]^2 + \left[\sigma(X_2)\frac{\partial k}{\partial X_2}\right]^2 \right\}^{1/2} \quad (22.A.1)$$

In case $\sigma(X_1) \approx \sigma(X_2) = \sigma(X)$, the following resulting equation can be obtained:

$$\sigma(\kappa) = \left\{ \left[\frac{\sigma(X)}{X_2}\right]^2 + \left[\frac{k\sigma(X)}{X_2}\right]^2 \right\}^{1/2} = \frac{\sigma(X)}{X_2}(1+k^2)^{1/2} \quad (22.A.2)$$

Equation (22.A.2) demonstrates that the error of the solubility ratio measurements increases with the increase of the k value and with the decrease of the polymorph solubility, X_2. For example, for the solubility X_2 of 0.2 µg/ml, $\sigma(X)$ equal to 0.02 µg/ml, and k value of 2, the $\sigma(k)$ is equal to 0.22.

TABLE 22.B.1 Explicit Equations Used for Predictions of Polymorphs Solubility Ratio

Based On	Equation	Comments
Assumption A	$\ln \frac{X_1}{X_2} = -\left\{ \Delta H_{fus1}\left(1-\frac{T}{T_{m1}}\right) - \Delta H_{fus2}\left(1-\frac{T}{T_{m2}}\right)\right\}/RT$	
Assumption B	$\ln \frac{X_1}{X_2} = -\left\{ \Delta H_{fus1}\frac{T}{T_{m1}}\ln\frac{T_{m1}}{T} - \Delta H_{fus2}\frac{T}{T_{m2}}\ln\frac{T_{m2}}{T}\right\}/RT$	
Assumption C	$\ln \frac{X_1}{X_2} = -\left\{ \Delta H_{fus1}(T_{m1}-T)\frac{T}{T_{m1}^2} - \Delta H_{fus2}(T_{m2}-T)\frac{T}{T_{m2}^2}\right\}/RT$	
Eq. (22.13)	$\ln \frac{X_1}{X_2} = \ln\left(\frac{X_1}{X_2}\right)_B + 0.188 - 43.096\left\{\left[\left(\frac{T_{m2}}{T}-1\right)-\ln\frac{T_{m2}}{T}\right]-\left[\left(\frac{T_{m1}}{T}-1\right)-\ln\frac{T_{m1}}{T}\right]\right\}$	Only for the monotropic system
Eq. (22.14)	$\ln \frac{X_1}{X_2} = -(\Delta H_{fus1}-\Delta H_{fus2})\left(1-\frac{T}{T_m}\right)/RT$	$T_m = T_{m2}$ or T_{m1}
Eq. (22.15)	$\ln \frac{X_1}{X_2} = -(\Delta H_{fus1}-\Delta H_{fus2})\frac{T}{T_m}\ln\frac{T_m}{T}/RT$	$T_m = T_{m2}$ or T_{m1}
Eq. (22.16)	$\ln \frac{X_1}{X_2} = -(\Delta H_{fus1}-\Delta H_{fus2})(T_m-T)\frac{T}{T_m^2}/RT$	$T_m = T_{m2}$ or T_{m1}

22.B SUMMARY OF EXPLICIT EQUATIONS USED FOR THE SOLUBILITY RATIO PREDICTIONS

Explicit equations used for predictions of solubility ratio of polymorphs are presented in Table 22.B.1.

ACKNOWLEDGMENTS

The authors would like to thank Mr. Brian Samas, Dr. Neil Feeder, Dr. Paul Meenan, Dr. Robert Docherty and Dr. Bruno Hancock for the valuable comments and discussions. The authors also wish to thank Mr. Anthony M. Campeta for providing the experimental thermal and solubility data on the axitinib polymorphs. YAA is thankful to Mr. Brian D. Bissett for a thorough review of the manuscript.

REFERENCES

1. Brittain, H.G. (ed.) (1999). *Polymorphism in Pharmaceutical Solids*. New York: Marcel Dekker.
2. Burger, A. and Ramberger, R. (1979). On the polymorphism of pharmaceuticals and other organic molecular crystals. I: theory of thermodynamic rules. *Mikrochim. Acta* II: 259–271.
3. Grunenberg, A., Henck, J.-O., and Siesler, H.W. (1996). Theoretical derivation and practical application of energy/temperature diagrams as an instrument in preformulation studies of polymorphic drug substances. *Int. J. Pharm.* 129: 147–158.
4. Byrn, S., Pfeiffer, R., Ganey, M. et al. (1995). Pharmaceutical solids: strategic approach to regulatory considerations. *Pharm. Res.* 12: 945–954.
5. Yu, L.X., Furness, M.S., Raw, A. et al. (2003). Scientific considerations of pharmaceutical solid polymorphs in abbreviated new drug applications. *Pharm. Res.* 20: 531–536.
6. Brittain, H.G. and Grant, D.J.W. (1999). Effects of polymorphism and solid-state solvation on solubility and dissolution rate. In: *Polymorphism in Pharmaceutical Solids* (ed. H.G. Brittain), 279–330. New York: Marcel Dekker.
7. Kempf, D.J., Marsh, K.C., Denissen, J.F. et al. (1995). ABT-538 is a potent inhibitor of human immunodeficiency virus protease and has high oral bioavailability in humans. *Proc. Natl. Acad. Sci. U. S. A.* 92: 2484–2488.
8. Bauer, J., Spanton, S., Henry, R. et al. (2001). Ritonavir: an extraordinary example of conformational polymorphism. *Pharm. Res.* 18: 859–866.
9. Chemburkar, S.R., Bauer, J., Deming, K. et al. (2000). Dealing with the impact of ritonavir polymorphs on the late stages of bulk drug process development. *Org. Process Res. Dev.* 4: 413–417.
10. Morissette, S.L., Soukasene, S., Levinson, D. et al. (2003). Elucidation of crystal form diversity of the HIV protease inhibitor ritonavir by high-throughput crystallization. *Proc. Natl. Acad. Sci. U. S. A.* 100: 2180–2184.
11. Pudipeddi, M. and Serajuddin, A.T.M. (2005). Trends in solubility of polymorphs. *J. Pharm. Sci.* 94: 929–939.
12. Mao, C., Pinal, R., and Morris, K.R. (2005). A quantitative model to evaluate solubility of polymorphs from their thermodynamic properties. *Pharm. Res.* 22: 1149–1157.
13. Grant, D.J.W. and Higuchi, T.T. (1990). *Solubility Behavior of Organic Compounds*. New York: Wiley.
14. Gu, C.-H. and Grant, D.J.W. (2001). Estimating the relative stability of polymorphs and hydrates from heats of solution and solubility data. *J. Pharm. Sci.* 90: 1277–1287.
15. Hancock, B.C. and Parks, M. (2000). What is the true solubility advantage for amorphous pharmaceuticals? *Pharm. Res.* 17: 397–403.
16. Hildebrand, J.H. and Scott, R.L. (1962). *Regular Solutions*. Englewood Cliffs, NJ: Prentice-Hall.
17. Hoffman, J.D. (1958). Thermodynamic driving force in nucleation and growth processes. *J. Chem. Phys.* 29: 1192–1193.
18. Mishra, D.S. and Yalkowsky, S.H. (1992). Ideal solubility of a solid solute: effect of heat capacity assumptions. *Pharm. Res.* 9: 958–959.
19. Neau, S.H. and Flynn, G.L. (1990). Solid and liquid heat capacities of *n*-alkyl para-aminobenzoates near the melting point. *Pharm. Res.* 7: 1157–1162.

20. Neau, S.H., Flynn, G.L., and Yalkowsky, S.H. (1989). The influence of heat capacity assumptions on the estimation of solubility parameters from solubility data. *Int. J. Pharm.* 49: 223–229.
21. Neau, S.H., Bhandarkar, S.V., and Hellmuth, E.W. (1997). Differential molar heat capacity to test ideal solubility estimations. *Pharm. Res.* 14: 601–605.
22. Pappa, G.D., Voutsas, E.C., Magoulas, K., and Tassios, D.P. (2005). Estimation of the differential molar heat capacities of organic compounds at their melting point. *Ind. Eng. Chem. Res.* 44: 3799–3806.
23. Defossemont, G., Randzio, S.L., and Legendre, B. (2004). Contributions of calorimetry for C_p determination and of scanning transitiometry for the study of polymorphism. *Cryst. Grow. Des.* 4: 1169–1174.
24. Yu, L. (1995). Inferring thermodynamic stability relationship of polymorphs from melting data. *J. Pharm. Sci.* 84: 966–974.
25. Gracin, S., Brinck, T., and Rasmuson, A.C. (2002). Prediction of solubility of solid organic compounds in solvents by UNIFAC. *Ind. Eng. Chem. Res.* 41: 5114–5124.
26. Hojjati, H. and Rohani, S. (2006). Measurement and prediction of solubility of paracetamol in water–isopropanol solution. Part 2. Prediction. *Org. Process Res. Dev.* 10: 1110–1118.
27. Aguiar, A.J., Krc, J. Jr., Kinkel, A.W., and Samyn, J.C. (1967). Effect of polymorphism on the absorption of chloramphenicol from chloramphenicol palmitate. *J. Pharm. Sci.* 56: 847–853.
28. Rowe, E.L. and Anderson, B.D. (1984). Thermodynamic studies of tolbutamide polymorphs. *J. Pharm. Sci.* 73: 1673–1675.
29. Ghodbane, S. and McCauley, J.A. (1990). Study of the polymorphism of 3-((3-(2-(7-chloro-2-quinolinyl)-(*E*)-ethenyl) phenyl)((3-(dimethylamino-3-oxopropyl)thio)methyl)-thio) propanoic acid (MK571) by DSC, TG, XRPD and solubility measurements. *Int. J. Pharm.* 59: 281–286.
30. Gerber, J.J., vander Watt, J.G., and Lötter, A.P. (1991). Physical characterization of solid forms of cyclopenthiazide. *Int. J. Pharm.* 73: 137–145.
31. Kushida, I. and Ashizawa, K. (2002). Solid state characterization of E2101, a novel antispastic drug. *J. Pharm. Sci.* 91: 2193–2202.
32. Rollinger, J.M., Gstrein, E.M., and Burger, A. (2002). Crystal forms of torasemide: new insights. *Eur. J. Pharm. Biopharm.* 53: 75–86.
33. Shibata, M., Kokubo, H., Morimoto, K. et al. (1983). X-ray structural studies and physicochemical properties of cimetidine polymorphs. *J. Pharm. Sci.* 72: 1436–1442.
34. Crafts, P.A. (2007). The role of solubility modelling and crystallization in the design of active pharmaceutical ingredients. In: *Chemical Product Design: Toward a Perspective Through Case Studies* (ed. K.M. Ng, R. Gani and K. Dam-Johansen), 23–85. Dordrecht, The Netherlands: Elsevier.
35. Ye, Q., Hart, R.M., Kania, R. et al. (2006). Polymorphic forms of 6-[2-((methylcarbomoyl)phenylsulfanyl]-3-E-[2-(pyridin-2-yl)ethenyl]indazole. US Patent 0094763.
36. Campeta, A.M., Chekal, B.P., McLaughlin, R.W., and Singer, R.A. (2008). Novel crystalline forms of a VEGF-R inhibitor. PCT Int. Appl. WO 2008122858.
37. Chekal, B., Campeta, A., Abramov, Y.A. et al. (2009). Facing the challenges of developing an API crystallization process for a complex polymorphic and highly-solvating system. Part I. *Org. Process Res. Dev.* 13: 1327–1337.
38. Sohn, Y.T. (1990). Study on the polymorphism of acetaminophen. *J. Korean Pharm. Sci.* 20: 97–104.
39. Sheikhzadeh, M., Rohani, S., Traffish, M., and Murad, S. (2007). Solubility analysis of buspirone hydrochloride polymorphs: measurements and prediction. *Int. J. Pharm.* 338: 55–63.
40. Behme, R.L. and Brooke, D. (1990). Heat of fusion measurement of a low melting polymorph of carbamazepine that undergoes multiple-phase changes during differential scanning calorimetry analysis. *J. Pharm. Sci.* 80: 986–990.
41. Behme, R.J., Brooke, D., Farney, R.F., and Kensler, T.T. (1985). Research article characterization of polymorphism of gepirone hydrochloride. *J. Pharm. Sci.* 74: 1041–1046.
42. Vrečer, F., Vrbinc, M., and Meden, A. (2003). Characterization of piroxicam crystal modifications. *Int. J. Pharm.* 256 (256): 3–15.
43. Bartolomei, M., Bertocchi, P., Ramusino, M.C. et al. (1999). Physico-chemical characterization of the modifications I and II of (R, S) propranolol hydrochloride: solubility and dissolution. *J. Pharm. Biomed. Anal.* 21: 299–309.
44. Caviglioli, C., Pani, M., Gatti, P. et al. (2006). Study of retinoic acid polymorphism. *J. Pharm. Sci.* 95: 2207–2221.
45. Carlton, R.A., Difeo, T.J., Powner, T.H. et al. Preparation and characterization of polymorphs for an LTD_4 antagonist, RG 12525. *J. Pharm. Sci.* 85: 461–467.
46. Rocco, W.L. and Swanson, J.R. (1995). WIN 63843 polymorphs: prediction of enantiotropy. *Int. J. Pharm.* 117: 231–236.
47. Higuchi, W.I., Lau, P.K., Higuchi, T., and Shell, J.W. (1963). Solubility relationship in the methylprednisolone system. *J. Pharm. Sci.* 52: 150–153.
48. Singhal, D. and Curatolo, W. (2004). Drug polymorphism and dosage form design: a practical perspective. *Adv. Drug Deliv. Rev.* 56: 335–347.
49. Amidon, G.L., Lennernas, H., Shah, V.P., and Crison, J.R.A. (1995). Theoretical basis for a biopharmaceutic drug classification: the correlation of *in vitro* drug product dissolution and *in vivo* bioavailability. *Pharm. Res.* 12: 413–420.
50. Varma, M.V.S., Obach, R.S., Rotter, C. et al. (2010). Physicochemical space for optimum oral bioavailability: contribution of human intestinal absorption and first-pass elimination. *J. Med. Chem.* 53: 1098–1108.
51. Kasim, N.A., Whitehouse, M., Ramachandran, C. et al. (2004). Molecular properties of WHO essential drugs and provisional biopharmaceutical classification. *Mol. Pharm.* 1: 85–96.
52. Taylor, J.R. (1982). *An Introduction to Error Analysis*. Mill Valley, CA: University Science Books.

23

TOWARD A RATIONAL SOLVENT SELECTION FOR CONFORMATIONAL POLYMORPH SCREENING

Yuriy A. Abramov, Mark Zell[†], and Joseph F. Krzyzaniak

Global R&D, Pharmaceutical Sciences, Pfizer, Inc., Groton, CT, USA

23.1 INTRODUCTION

Crystalline solids with the same chemical composition but different molecular arrangements in the crystal lattice are known as polymorphs [1]. Changes in polymorphic form during pharmaceutical development can have a negative impact on a drug's performance, i.e. solubility and bioavailability (Chapter 22), chemical and physical stability, and mechanical properties. Therefore, it is necessary to identify the stable crystal form under normal manufacturing and storage conditions to ensure that this form does not change during the life cycle of the drug product.

Polymorph screens are conducted early in drug development to identify unique crystal forms of the active pharmaceutical ingredient (API). Each crystal form discovered is characterized to identify whether the crystalline phase is an anhydrous form or a solvate. The polymorphic lattice can also consist of either the same or different molecular conformations. Conformational polymorphism describes the latter case when different conformations of the same molecule occur in different crystal forms [2]. Solid-state characterization studies are then conducted to develop an understanding of the stability relationship between all crystalline phases since the thermodynamically stable form is directly related to conditions (crystallization, environmental, and manufacturing) in which the API is exposed to during the drug development process [1].

During the preparation of the desired polymorphic form, the science of crystallization has shown to be a very complex phenomenon, which is dictated by interplay between different thermodynamic and kinetic factors. The presence of different molecular conformations in saturated solution introduces an additional degree of complexity allowing crystallization of polymorphs different not only in the packing arrangement but also in molecular geometry as seen in conformational polymorphs [1–3]. Crystallization is believed to be a multiple stage process in which molecules associate into pre-nucleation molecular clusters followed by their assembling into crystal nuclei leading to crystal growth (Figure 23.1) [3, 4]. It was assumed [5] and later demonstrated [6, 7] that a saturated phase contains clusters of molecules displaying packing of all possible polymorphs. Final growth of a specific polymorph can be achieved by altering crystallization conditions, such as degree of supersaturation, type of solvent, and additives [8–11].

From a thermodynamic viewpoint, a primary factor for conformational polymorph formation is stabilization of the conformer free energy in the crystalline environment relative to that in saturated solution. That consideration defines the type of solvent as one of the major factors in polymorphic selectivity. Solvent selection for polymorph crystallization is usually based on achieving a reasonably high API solubility [12] to facilitate crystal growth during drowning-out, evaporative, cooling, or slurry crystallization techniques. With this, it is reasonable to assume that a higher population of a specific molecular conformation is needed to feed a crystallization of a corresponding conformational polymorph. A higher conformer population should contribute to increased nucleation of the corresponding conformational polymorph, structural

[†]Current address: Takeda Oncology, Cambridge, MA, USA

Chemical Engineering in the Pharmaceutical Industry: Active Pharmaceutical Ingredients, Second Edition.
Edited by David J. am Ende and Mary T. am Ende.
© 2019 John Wiley & Sons, Inc. Published 2019 by John Wiley & Sons, Inc.

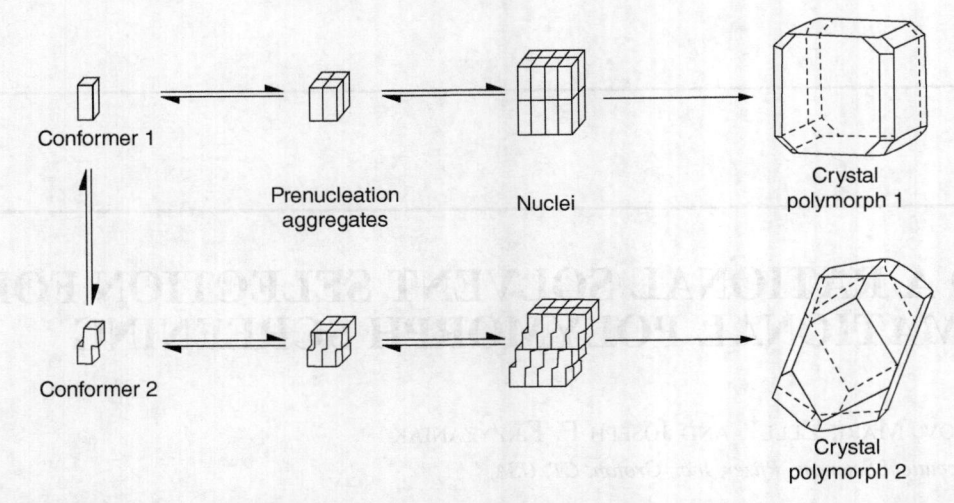

FIGURE 23.1 An illustration of the crystallization of conformationally flexible molecules. *Source*: Reproduced from Yu et al. [3] with permission. Copyright (2010) American Chemical Society.

organization of which is most readily derived from the preferred conformations in solution [13, 14]. The focus of this study is testing a computational approach for conformational population prediction in different solvent media in order to explore diversification of conformational populations in solution. A working hypothesis is that controlling the selective conformer's population should facilitate a rational solvent selection for conformational polymorph screening. In addition, prediction of conformer population in a supercooled liquid (self-media) will also be performed and should mimic stabilization of molecular conformations in an amorphous solid state. An assumption is made that while the conformer distribution in solution is important for nucleating a conformational polymorph, a high conformation population in the supercooled liquid should reflect stability of the conformation in a solid state neglecting long-order contributions. The interplay between conformer distribution in solution and self-media should reflect a driving force for crystallization of the different molecular conformations.

In addition, the proposed conformational population analysis technique will be discussed for preferable conformation selection for crystal structure prediction (CSP). CSP is an important computational tool, which is valuable in guiding polymorph screening [15–17] and for performing a polymorphic risk assessment for solid form nomination during pharmaceutical development [18]. However, the currently available CSP methods heavily rely on a correct selection of the starting molecular conformations, which are typically held rigid at least during initial crystal packing generation. That is why a rational selection of the preferred starting molecular conformations for CSP of complex flexible molecules is of great importance [19–21].

The following topics and considerations though important are out of scope of this work.

1. Due to the effect of different crystalline environments, molecular conformations may be distributed around local minima of the potential energy surface as defined in the reference media (e.g. gas phase) [22]. Molecular conformations considered in this work are defined strictly by a local minimum in the aqueous or gas media neglecting crystal packing effects. Therefore, we will be considering types of conformations representing possible conformers with slightly different geometries. This is especially true for the freely rotatable bonds, rotation of which is defined by a flat potential energy surface in the gas phase or in implicit solution, while torsion angle of such bonds may be fixed in a crystalline environment by a specific hydrogen bonding (HB) interaction (e.g. hydroxyl and carboxyl groups).

2. Relative stability of polymorphs is described by an equilibrium phase diagram and cannot be adequately determined by the methods developed in the current study. Moreover, the intention of the current work is to facilitate crystallization of a diverse set of conformational polymorphs including those that are metastable.

3. Rotational barriers, which are important for kinetics of conformational interconversion, will be not considered. Therefore, only thermodynamic factors will be used in theoretical evaluation of the conformational distributions.

4. In addition, no considerations of kinetic factors that may affect selectivity of the polymorph crystallization will be addressed. Among such factors are the degree of supersaturation, cooling and stirring rates, impurities and additives that affect the polymorphic form, morphology, and crystal growth [8–11].

23.2 METHODS

23.2.1 Theoretical

The conformational population study was performed in two steps. During the first step, a stochastic conformational search was performed using the MOE 2008.10 software package [23] with the following parameters: MMFF94x force field with distance-dependent dielectric, chiral constraints, allowing amide bond rotation and energy cutoff of 7 kcal/mol. After the conformational search was complete, the conformations generated were further optimized using aqueous media at PBE/DNP/COSMO level of theory (defined by theoretical method and a basis set) as implemented in DMol^3 [24–26]. This utilizes density functional theory (DFT) PBE approximation [27] with all-electron double-numeric-polarized basis set. The effect of bulk water is estimated by conductor-like screening model (COSMO) as implemented in DMol^3 [28]. It was demonstrated recently [29] that PBE is one of the best DFT functionals in the prediction of energies of HB systems. The *cosmo* files generated by the PBE/DNP/COSMO calculations for each conformer are further used for solvation free energy calculations, ΔG_{solv}, and for prediction of conformational distribution in different solvents adopting COSMO-RS theory [30] as implemented in COSMOtherm software [31].

An obvious advantage of the application of the PBE/DNP/COSMO calculations is that the COSMOtherm parameters are specifically optimized for this level of theory. However, the DFT is known to underestimate strong electron correlation effects (e.g. dispersion energy) [32]. Therefore, in order to assure the quality of the PBE/DNP calculations, we have also performed calculations at a combined level of theory according to the following procedure (Chapter 20). The conformer free energy in solution was presented as

$$G(i) = E_{gas}(i)^{gas} + ZPE(i) + \Delta G_{solv}(i) \quad (23.1)$$

where

$E_{gas}(i)$ is the gas-phase energy of the conformer.

$ZPE(i)$ is its zero-point vibrational energy.

$\Delta G_{solv}(i)$ was calculated at the PBE/DNP/COSMO level as described above, while the gas free energy of the conformer was calculated at the RI-MP2/TZVPP level [33] adopting TURBOMOLE software [34].

The $ZPE(i)$ contribution was estimated at the DMol^3 PBE/DND level of theory.

In the following discussions, the two theoretical approaches described above for simplicity will be referred to as PBE and RI-MP2 levels, respectively.

The equilibrium conformer population, $p(i)$, was calculated according to the following equation [30]:

$$p(i) = \frac{\varpi(i)\exp(-G(i)/RT)}{\sum_j \varpi(j)\exp(-G(j)/RT)} \quad (23.2)$$

where

$\varpi(i)$ is a multiplicity of the conformer i, which is based on geometrical degeneration factors.

R is the universal gas constant.

T is the temperature in Kelvin.

A challenge in an accurate prediction of $p(i)$ is introduced by an exponential dependence on the calculated $G(i)$ values, so that a relatively small error in conformer's free energy transforms into significant errors in conformer populations. For example, in the case of a simple system with two conformations displaying similar energies and multiplicities, the true population of each conformer is 50%. However, an error in predicted relative $G(i)$ values, $G(2)–G(1)$, of 0.4 kcal/mol would result in $p(i)$ error at room temperature of ±16.3%. The predicted conformer populations would be 66.3 and 33.7%.

23.2.2 Experimental

23.2.2.1 NMR Measurements

^{1}H, gHMBC, and G-BIRD$_{R,X}$-CPMG-HSQMBC experiments were performed on a Bruker Avance DRX 600 spectrometer equipped with a 5 mm BBO probe with z-axis gradient. All experiments were performed at a temperature of 298 K.

Samples were prepared at concentrations of 10, 100, and 300 mg/ml in acetone-d_6, acetonitrile-d_3, and methanol-d_4 with tetramethylsilane as an internal standard. All experiments were performed using Bruker standard pulse sequences, except for G-BIRD$_{R,X}$-CPMG-HSQMBC, which was written and implemented by a staff member in our laboratory.

For ^{1}H NMR analysis, typically one transient was acquired with a 1 second relaxation delay using 32 K data points. The 90° pulse was 10.5 μs and a spectral width of 10 775 Hz was used.

The gHMBC spectrum was acquired with 4096 data points for $F2$ and 128 $F1$ increments. gHMBC data was acquired with 4 K data points in $F2$, 128 increments for $F1$ (16 scans per increment) and $F2 \times F1$ spectral window of 5200 × 23 800 Hz. Data was processed with 4 K data points zero-filled to 8 K in $F2$ and 128 data points zero-filled to 1 K in $F1$.

The G-BIRD$_{R,X}$-CPMG-HSQMBC spectra were acquired in approximately 14 hours with 4 K data points in $F2$, 128 increments for $F1$ with 256 scans per increment and $F2 \times F1$ spectral window of 5200 × 23 800 Hz. Data was processed with 4 K data points zero-filled to 8 K in $F2$ and 128 data points zero-filled to 1 K in $F1$. A sine squared window function was applied to the $F1$ dimension before Fourier transformation, and no apodization was applied in the $F2$ dimension. The gradient ratios for G-BIRD$_{R,X}$-CPMG-HSQMBC were G1 : G2 : G3 : G4 : G5 = 2.5 : 2.5 : 8 : 1 : ±2. The delay for long-range polarization transfer was set to 63 ms. The delay used for delta was set to 200 μs. All measured coupling constants are believed to be within ±0.5 Hz. The values of $^3J_{CH}$ were determined from direct measurement and subsequent manual peak fitting analysis [35].

23.3 RESULTS AND DISCUSSION

23.3.1 Test of Accuracy of Conformational Population Predictions

The accuracy of conformational population prediction for flexible organic molecules was tested at different levels of theory. For this, three test cases were selected consisting of N-substituted amide series [36], N-(pyridin-2-yl)benzamide chemical series [37], and S-ibuprofen. While accurate experimental conformational populations based on the solution NMR experiments were available for the first two cases, a separate experimental work was performed for S-ibuprofen.

23.3.1.1 N-Substituted Amide Series
Yamasaki et al. [36] reported a detailed study of the cis–trans ratio of conformer population of a series of N-substituted amides (Figure 23.2) in DCM-d_2, methanol-d_4, and acetone-d_6 at 183 K. It was demonstrated that compounds 1 and 2 display cis and trans conformations in all three solvents, respectively. A very weak solvent dependence was found for compounds 6 and 7, the former being preferably in the trans conformations, while the latter displaying close to a uniform cis–trans distribution. The most remarkable result of the study was an observation of a pronounced solvent-dependent conformational switching of the phenylhydroxamic acids (compounds 3–5, Figure 23.3, Table 23.1) from the predominantly cis-conformations in DCM to preferably trans conformations in acetone.

Theoretical predictions at the DFT (PBE) and RI-MP2 levels are in agreement with the experimental observations (Table 23.1). In particular, the strong solvent dependence of conformations of compounds 3–5 was reproduced correctly, though the absolute values of cis populations in methanol and acetone are underestimated for compounds 3 and 4, especially by the DFT method (Figure 23.3). At the same time both levels of the predictions overestimated cis populations of compound 7. The overestimation is more pronounced in the case of the predictions using the RI-MP2 level of theory.

1. R_1=H, R_2=H
2. R_1=CH$_3$, R_2=H
3. R_1=OH, R_2=H
4. R_1=OH, R_2=CH$_3$
6. R_1=NH$_2$, R_2=H
7. R_1=OCH$_3$, R_2=H

FIGURE 23.2 Chemical structures of N-substituted amide series [36]. *Source*: Reproduced with permission of American Chemical Society.

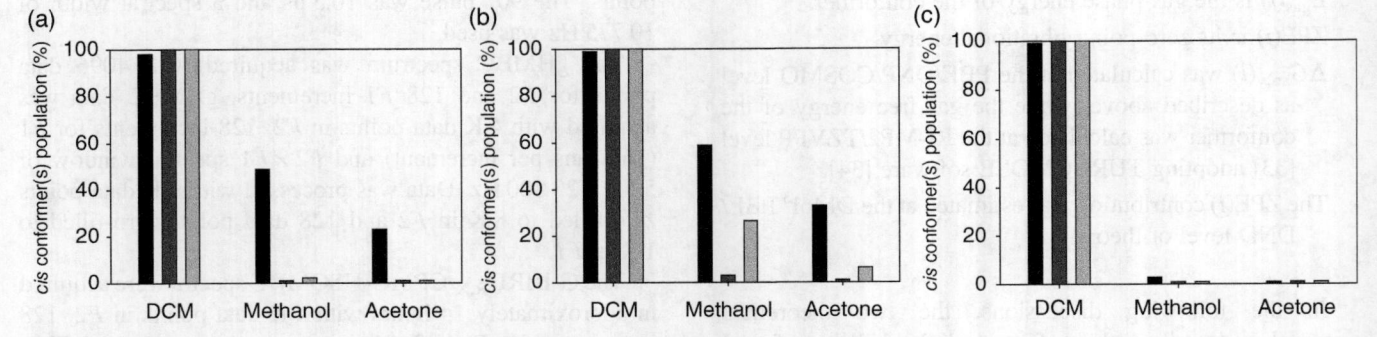

FIGURE 23.3 Cis conformer(s) populations of compounds 3–5 (a–c) of N-substituted amide series (Table 23.1, Figure 23.2) [36]. Experimental, PBE, and RI-MP2 results are represented, respectively, by black, dark gray, and gray bars. *Source*: Reproduced with permission of American Chemical Society.

TABLE 23.1 Experimental [36] and Predicted *Cis* Conformer Populations (%) of N-Substituted Amides (Figure 23.2) at −90 °C in Three Solvents

Molecule	DCM			Methanol			Acetone			Self-Media	
	Exp	PBE	RI-MP2	Exp	PBE	RI-MP2	Exp	PBE	RI-MP2	PBE	T (°C)
1	<1	<1	<1	<1	<1	<1	<1	<1	<1	<1	−90
2	98	98.8	>99	>99	95.4	>99	>99	98.8	>99	97.0	−90
3	98	>99	>99	49	<1	2.2	23	<1	<1	98.1	−90
4	>99	>99	>99	59	3.0	26.4	33	<1	6.4	98.6	−90
5	>99	>99	99.6	3	<1	<1	<1	<1	<1	79.6	−90
6	98	>99	>99	95	86.5	>99	85	60.2	>99	97.2	−90
7	50	88.1	>99	63	88.7	>99	55	52.2	>99	45.5	−90

Source: Reproduced with permission of American Chemical Society.
Details of the theoretical approaches based on PBE and RI-MP2 levels of theory are described in Section 23.2.1.

23.3.1.2 N-(Pyridin-2-yl)benzamide Series

Populations of *cis* conformations of a series of *N*-(pyridin-2-yl)benzamides (Figure 23.4, compounds 1–4) and of *N*-(2,6-dimethylphenyl)acetamide (Figure 23.4, compound 5) were studied by means of solution NMR spectroscopy in three solvents: chloroform-d_1, methanol-d_4, and acetone-d_6 at 243 K (Table 23.2) [37]. It was demonstrated that compounds 1 and 3 display *cis* and *trans* conformations in all three solvents, respectively. These observations are analogues to the results reported Yamasaki et al. [36] for the compounds 1 and 2 (Figure 23.2, Table 23.1). Solvent-dependent conformational behavior was observed for compounds 2 and 5. Theoretical predictions at the DFT PBE level are in a quite good agreement with the experimental observations (Table 23.2). In particular, the solvent dependence of compound 2 and 5 conformations was reproduced correctly, though the absolute values of *cis* populations in methanol and acetone are somewhat overestimated (Figure 23.5). The overestimation is more pronounced in case of the predictions at the RI-MP2 level of theory.

It is found for both N-substituted amide and *N*-(pyridin-2-yl)benzamide series that a combined approach with RI-MP2/TZVPP gas-phase calculations does not demonstrate an advantage over PBE/DNP/COSMO level of theory in predicting conformational distributions in different solvents. Therefore, a less demanding PBE/DNP/COSMO analysis was adopted for further calculations.

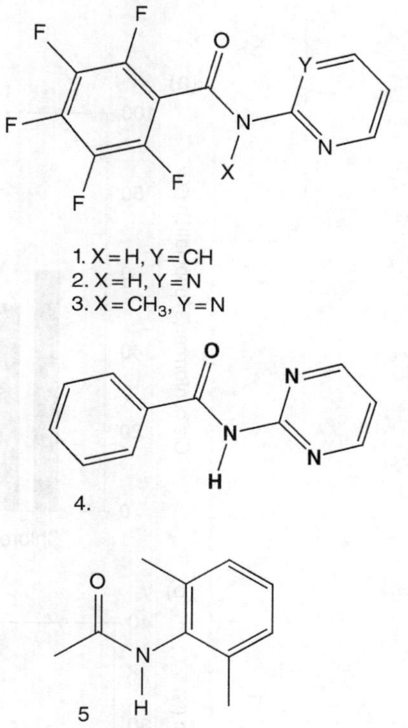

EXAMPLE PROBLEM 23.1 S-Ibuprofen Conformational Population in Methanol, Acetone, and Acetonitrile

Ibuprofen is a nonsteroidal anti-inflammatory drug whose activity is usually associated with the S (+) isomer. As it is shown in Figure 23.6, the ibuprofen molecule displays four flexible torsion angles. The initial stochastic conformational search was performed using the MOE software package [23] adopting default parameters (MMFF94x force field with distance-dependent dielectric, chiral constraint, and energy cutoff of 7 kcal/mol). The 25 lowest-energy conformations generated were further optimized in water at the PBE/DNP/COSMO level of theory. The optimized S-ibuprofen conformations are aligned and presented in Figure 23.7. The *cosmo* files generated were used for the conformational distribution study in methanol, acetone, and acetonitrile solvents at 25 °C, adopting COSMOtherm software [31]. No significant solvent effect on conformer distribution was

FIGURE 23.4 A series of *N*-(pyridin-2-yl)benzamides (1–4) and *N*-(2,6-dimethylphenyl)acetamide (5) [37]. *Source*: Reproduced with permission of American Chemical Society.

TABLE 23.2 Experimental [37] and Predicted *Cis* Conformer Populations (%) of *N*-(Pyridin-2-yl)benzamides (Figure 23.4, 1–4) and *N*-(2,6-dimethylphenyl)acetamide (Figure 23.4, 5) at −30 °C in Three Solvents

Molecule	CHCl$_3$			Methanol			Acetone			Self-Media	
	Exp	PBE	RI-MP2	Exp	PBE	RI-MP2	Exp	PBE	RI-MP2	PBE	T (°C)
1	<1	<1	<1	<1	<1	<1	—	<1	<1	<1	−30
2	60.6	51.7	66.5	7.4	26.1	56.3	50.3	78.2	96.4	80.0	−30
3	>99	>99	>99	—	>99	>99	—	>99	>99	>99	−30
4	<1	<1	8.4	—	<1	1.3	—	<1	32.4	<1	−30
5	20.6	25.4	35.9	—	7.0	7.8	2.9	3.3	4.2	8.3	−30

Source: Reproduced with permission of American Chemical Society.
Details of the theoretical approaches based on PBE and RI-MP2 levels of theory are described in Section 23.2.1.

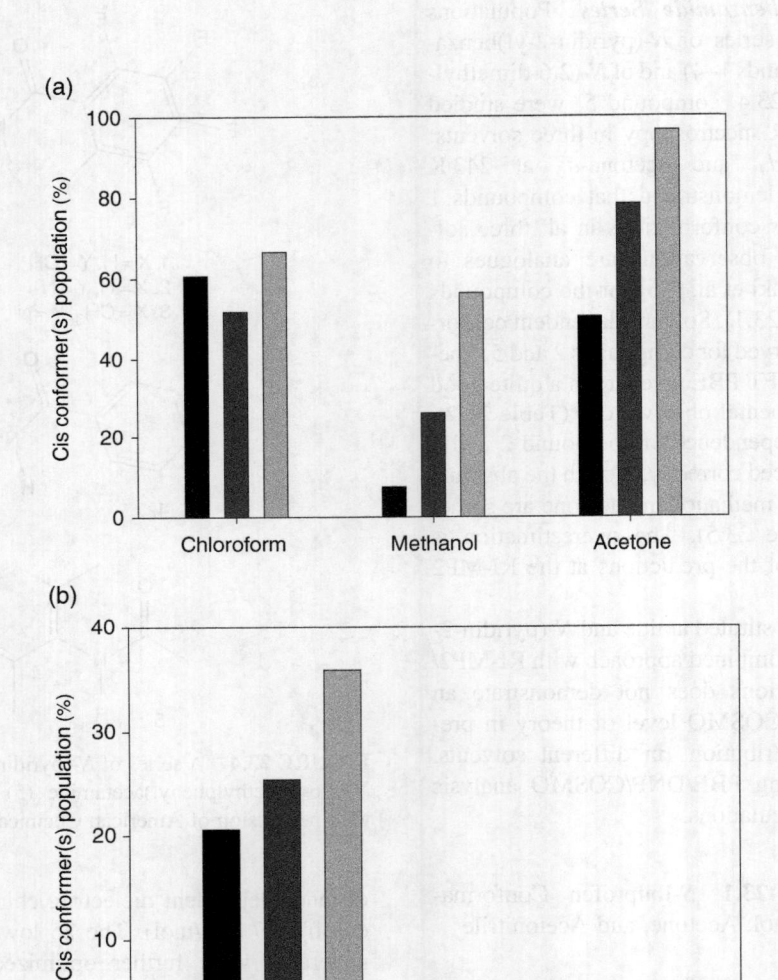

FIGURE 23.5 *Cis* conformer(s) populations of (a) 2,3,4,5,6-pentafluoro-*N*-(pyrimidine-2-yl)benzamide (Table 23.2, Figure 23.4, compound 2) and (b) *N*-(2,6-dimethylphenyl)acetamide (Table 23.2, Figure 23.4, compound 5) [37]. Experimental, PBE, and RI-MP2 results are represented, respectively, by black, dark gray, and gray bars. *Source*: Reproduced with permission of American Chemical Society.

FIGURE 23.6 Structure of ibuprofen molecule with four flexible torsion angles.

TABLE 23.3 Torsion Angles (in Degrees, Figure 23.6) of the Selected Four Conformations of S-Ibuprofen with the Highest Populations in the Three Solvents

Conformer	τ_1	τ_2	τ_3	τ_4
1	94.7	−65.6	−73.1	−63.2
2	96.0	−62.7	−105.3	−173.2
3	95.8	−64.2	103.6	−65.5
4	107.3	−65.7	71.8	−172.8

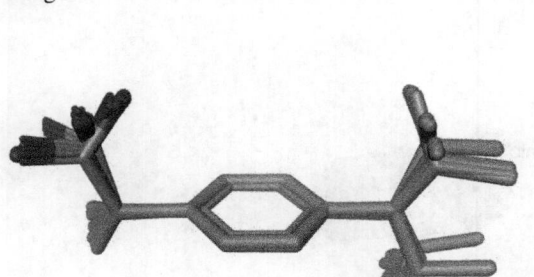

FIGURE 23.7 Aligned 25 S-ibuprofen conformations optimized in water at PBE/DNP/COSMO level of theory. All hydrogens are omitted.

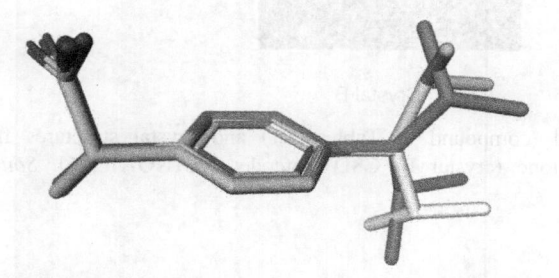

FIGURE 23.8 Aligned four S-ibuprofen conformers which displayed the highest populations in methanol, acetone, and acetonitrile solvents at 25 °C. All hydrogens are omitted.

found. Population of four conformers (Figure 23.8) was found to be in the range of 12–23% in all solvents under consideration, which significantly exceeded the population of any other conformation. The torsion angles of the preferred conformations are listed in Table 23.3. However, it should be expected that in a crystalline environment, the τ_1 angle may change (e.g. switch by ~180°) in order to accommodate intermolecular HB interaction.

The solvent and concentration dependences of S-ibuprofen conformations were investigated by G-BIRD$_{R,X}$-CPMG-HSQMBC experiment. For this the $^3J_{CH}$ heteronuclear coupling constants that define the flexible torsion angles (τ_{1-4}) in the S-ibuprofen molecule were extracted (Table 23.4). Due to the fast conformational interconversion at room temperature, the NMR observations are averaged over specific conformational distributions.

The $^3J_{CH}$ coupling constant data presented in Table 23.4 suggests that there is no change in the molecular conformation of S-ibuprofen in the solution, within the limits of the NMR measurements (>±0.5 Hz) as a function of either solvent composition or concentration. This data further supports the computational results for S-ibuprofen presented above that also suggest no change in conformational populations as a function of solvent composition. An exercise of converting the measured $^3J_{CH}$ coupling constant values into their respective torsion angles was not carried out.

23.3.2 Conformational Distribution and Polymorph Crystallization

Polymorph crystallization may be performed by utilizing different crystallization techniques. Typically a slurry experiment is considered to be the best to induce solvent-mediated transformation to the most stable form [38, 39]. The system is preferably under thermodynamic control, and the solution is saturated with respect to a metastable form and supersaturated with respect to a more stable form. It has been shown that a solvent that gives high solubility provides faster transformation to the most stable form and usually 8 mM solubility threshold has been adopted when designing the slurry experiment [39]. At the same time, cooling and evaporation crystallization in general produce a supersaturated solution with respect to all possible forms, and the system is preferably under kinetic control. In that case, polymorph crystallization typically follows Ostwald law of stages [40], and crystallization of the least stable form is expected first, followed by transformation to a more stable polymorph. The focus of the current study is to explore a correlation between conformational polymorph distributions and conformational polymorph crystallization. Unfortunately, there is very limited information available in the literature addressing the effect of solvent on both conformational distribution and crystallization. Below we will consider four examples – two slurry and two cooling crystallization experiments, for which conformational distribution in different

TABLE 23.4 $^3J_{CH}$ Values (Hz) Obtained with the G-BIRD$_{R,X}$-CPMG-HSQMBC Experiment for Ibuprofen in Acetone-d_6, Acetonitrile-d_3, and Methanol-d_4

Correlation	CD$_3$OD 300 mg/ml	CD$_3$OD 100 mg/ml	CD$_3$OD 10 mg/ml	CD$_3$CN 300 mg/ml	CD$_3$CN 100 mg/ml	CD$_3$CN 10 mg/ml	(CD$_3$)$_2$CO 300 mg/ml	(CD$_3$)$_2$CO 100 mg/ml	(CD$_3$)$_2$CO 10 mg/ml
H$_3$–C$_4$	4.6	4.6	4.5	4.4	4.5	4.5	4.6	4.7	4.7
H$_8$–C$_1$	5.1	5.0	5.1	5.1	5.0	5.1	5.1	5.1	5.1
H$_{12(13)}$–C$_{10}$	4.7	4.6	4.7	4.8	4.7	4.6	4.7	4.7	4.7
H$_{11}$–C$_7$	2.5	2.4	2.4	2.5	2.4	2.4	2.5	2.5	2.3

The numbering scheme is similar to the one presented in Figure 23.6.

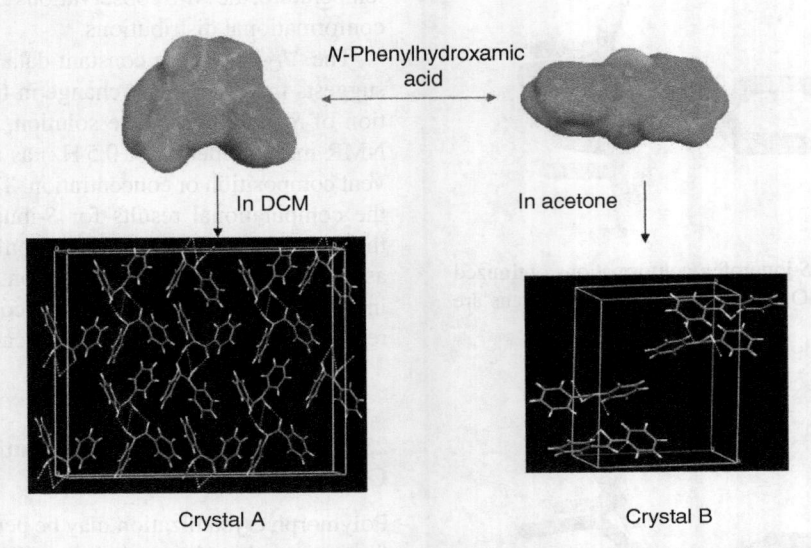

FIGURE 23.9 Conformational equilibrium of *N*-phenylhydroxamic acid (compound 3, Table 23.1) and crystal structures from recrystallization in DCM (crystal A, CSD Refcode: XERSOD) and acetone (crystal B, CSD Refcode: XETNOA) [36]. *Source*: Reproduced with permission of American Chemical Society.

solvents will be predicted by adopting theoretical approaches described above (PBE/DNP/COSMO level of theory).

23.3.2.1 N-Phenylhydroxamic Acids

Experimental and conformational distributions of *N*-phenylhydroxamic acids in three different solvents were reported by Yamasaki et al. [36] and discussed above (Table 23.1, compounds 3–5). Both NMR spectroscopy and theoretical calculations (Table 23.1, compounds 3–5) demonstrated that all *N*-phenylhydroxamic acids display switching from the *cis* conformations in dichloromethane to the *trans* conformations in methanol and especially in acetone (183 K). Yamasaki et al. recrystallized compound 3 from DCM and acetone producing polymorphs with *cis* (crystal A) and *trans* (crystal B) molecular conformations, respectively (Figure 23.9). It appears from differential scanning calorimetry (DSC) profile [36] that an enantiotropic relationship exists between A and B forms, so that at room temperature the crystal B is presumably more stable. The stability assignment is opposite to the one that may be based on the predicted population distribution in the self-media (or an amorphous solid, Table 23.1). This reflects the importance of the long-range order contributions in the *N*-phenylhydroxamic acid crystals. An important conclusion from the results of the Yamasaki et al. study [36] is that the polymorph crystallization in different solvents may follow the conformational population trend rather than the Ostwald rule of stages.

23.3.2.2 Taltirelin

Taltirelin ((4*S*)-*N*-[(2*S*)-1-[(2*S*)-2-carbamoylpyrrolidin-1-yl]-3-(3*H*-imidazol-4-yl)-1-oxopropan-2-yl]-1-methyl-2,6-dioxo-1,3-diazinane-4-carboxamide), a central nervous system activating agent, was reported to have two crystalline tetrahydrate forms, a metastable α-form and stable β-form (Figure 23.10) [41, 42]. It was found that the solvent-mediated transformation to the β-form occurring in the water slurry can be significantly promoted by adding a small amount (10 wt. %) of MeOH. It was demonstrated that

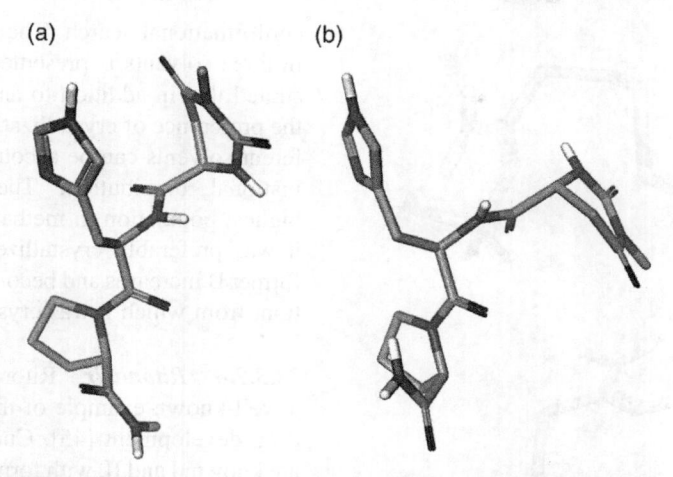

FIGURE 23.10 Conformations of taltirelin α- (a) and β- (b) forms. Only polar hydrogens are shown.

although the polymorph solubility had little affect with the added MeOH (it is actually slightly decreasing), an increase in methanol concentration causes an induction period of transformation to become shorter. In addition, Shoji Maruyama and Hiroshi Ooshima [41] demonstrated by means of nOe NMR analysis that MeOH causes the conformation change of taltirelin from the α-form to β-form conformers through the solute–MeOH interaction. That observation supports the importance that the conformer population has on driving the corresponding conformational polymorph crystallization.

Taltirelin is a very flexible molecule with eight rotatable bonds, which make a reliable conformation search a very challenging task. In order to test whether the MeOH effect on conformational populations can be predicted, the following calculations were performed. The α-form conformation was taken from the crystal structure available in the Cambridge Structural Database [43] (CSD, reference code REPLIH[1]). The β-form conformation was reconstructed from that of α-form by rotation of two single bonds as described in the literature (Figure 23.10) [41, 42]. The α- and β-form conformations were further adopted for the conformational population analysis at the PBE/DNP/COSMO level of theory. Though no significant effect was reproduced at 10 wt. % MeOH concentration, the calculations demonstrated a general qualitative increase of the β-conformer population when switching from aqueous to methanol solution (Figure 23.11).

23.3.2.3 Famotidine

Famotidine, a histamine H2-receptor antagonist, is a very flexible molecule that has two known conformational polymorphs: A and B. These two polymorphs are monotropically related with form

[1] REPLIH conformer represents a mirror image ((4R)-N-[(2R)-1-[(2R)-2-carbamoylpyrrolidin-1-yl]-3-(3H-imidazol-4-yl)-1-oxopropan-2-yl]-1-methyl-2,6-dioxo-1,3-diazinane-4-carboxamide tetrahydrate) of taltirelin α-form as it is described in Ref. [41] and [42].

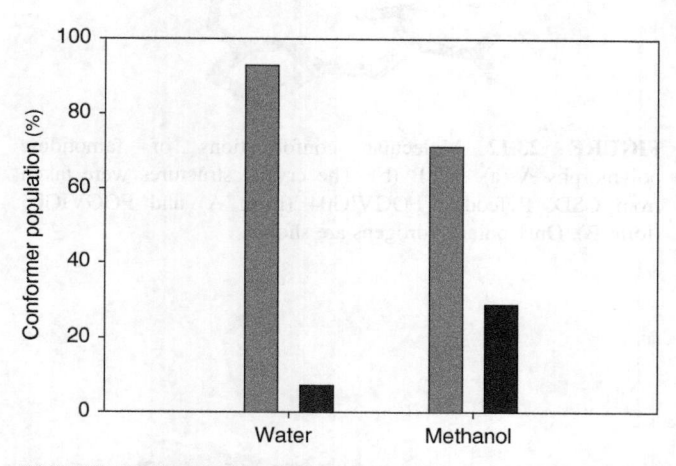

FIGURE 23.11 Predicted populations of the α- (gray) and β-form (black) taltirelin conformers in water and methanol.

A being more stable [44]. Since form B is the metastable form, it is kinetically favored and according to the Ostwald law should crystallize first when cooling a saturated solution. Selective cooling crystallization of famotidine polymorphs was reported by Lu et al. [44]. It was found that the form prepared was not only influenced by the cooling rate but also affected by the solvent of crystallization. For example, form B was crystallized from a water solution with high initial drug concentration independent of the rate of cooling. Additionally, the stable form A was crystallized from methanol and acetonitrile solutions at low initial drug concentrations. No conformational population study in solutions was reported by the authors.

Conformational populations of forms A and B were predicted at a crystallization temperature of 50 °C in three solvents at the PBE/DNP/COSMO level of theory. Due to the very high flexibility of the drug, crystallographic conformations (Figure 23.12) were adopted for the calculations with no

528 CHEMICAL ENGINEERING IN THE PHARMACEUTICAL INDUSTRY

(a)

(b)

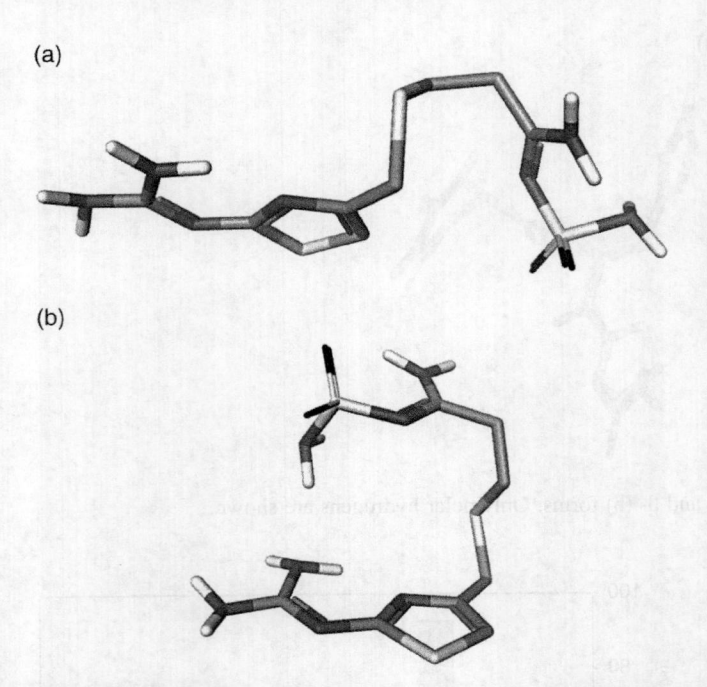

FIGURE 23.12 Molecular conformations of famotidine polymorphs A (a) and B (b). The crystal structures were taken from CSD, Refcodes: FOGVIG04 (form A) and FOGVIG05 (form B). Only polar hydrogens are shown.

conformational search. The resulting conformer population in three solvents is presented in Figure 23.13. It is demonstrated that in addition to any solubility and kinetic factors, the preference of crystallization of B or A conformers in different solvents can be accounted for by a trend in a conformational distribution. The conformer A displays the highest population in methanol and acetonitrile from which it was preferably crystallized. The population of the conformer B increases and becomes the highest in the water solution, from which it was crystallized.

23.3.2.4 Ritonavir

Ritonavir, a HIV protease inhibitor, is a well-known example of the impact polymorphism has on drug development [45]. Currently, two polymorphic forms are known, I and II, with form II being the most stable at room temperature. A solvent-mediated polymorphic conversion study was reported by Miller et al. [39]. It was demonstrated that, although the slurry crystallization is preferentially thermodynamically controlled, relatively high drug solubility (>8 mM but <200 mM) is needed to ensure solvent-mediated conversion to the most stable form. These conclusions were supported by polymorph screening in 13 solvents. Conformational populations of forms I and II in 13 solvents were predicted at room temperature at the PBE/DNP/COSMO level of theory. Due to the extremely high flexibility of the ritonavir

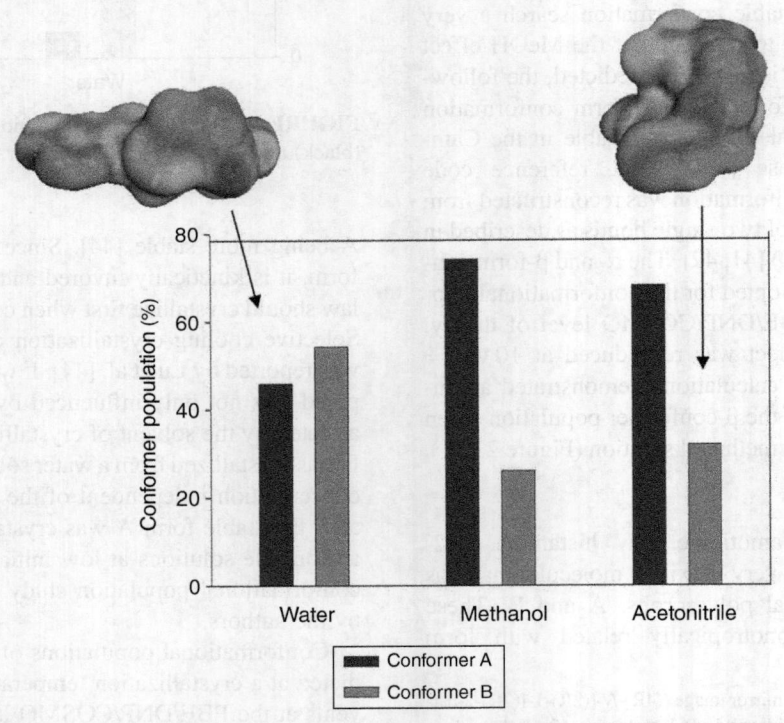

FIGURE 23.13 Histogram of predicted conformational distributions of famotidine in three solvents at 50 °C. σ-Surfaces of the two conformers are shown on the top.

drug, the crystallographic conformations were adopted (Figure 23.14) for the calculations. The conformations were taken from CSD database (reference codes are YIGPIO and YIGPIO01). The resulting conformer populations together with the polymorph screening results are presented in Table 23.5. It is demonstrated that except for MTBE the results of two-week slurry crystallization follow the trend of the preferred conformer population in the corresponding solvent.

23.3.3 Implication to Crystal Structure Prediction

CSP is becoming a useful tool conducted in parallel to polymorph screening as well as being able to assess the risk of discovering a more stable form [15–18]. A starting point for CSP is the selection of molecular conformations, which are as a rule held rigid during the generation of potential packing diagrams. Typically, the conformations are generated in the gas phase [19–21]. Since crystallization in the pharmaceutical industry is not performed from the gas phase but rather from different solvents, we recommend adopting the method of conformer distribution analysis, considered in the current study, be applied to the conformation selection for virtual polymorph screening. For this, a diverse set of solvents should be considered in order to determine whether the solvent induces conformational switching of the active pharmaceutical. An indication that a molecule may switch conformations is the presence of intramolecular HB or a noticeable variation of molecular hydrophobic and hydrophilic surfaces. We propose a small diverse set of solvents to be considered

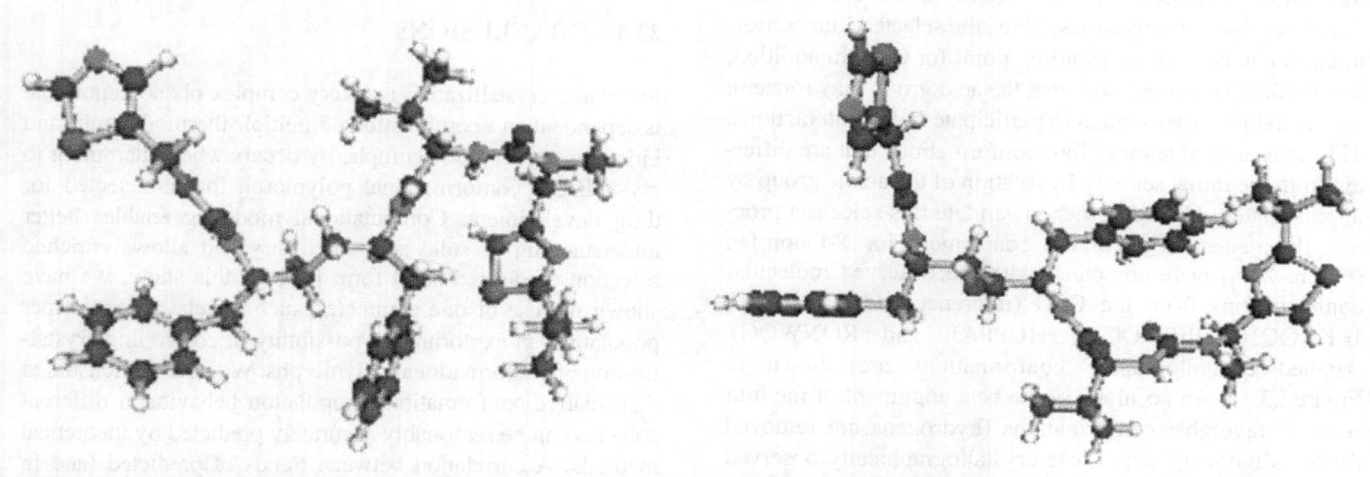

FIGURE 23.14 Crystallographic conformations of ritonavir forms I and II. The crystal structures were taken from CSD (reference codes: YIGPIO and YIGPIO01).

TABLE 23.5 Results of Polymorph Screen of Ritonavir [39] and Predicted Conformer Populations at Room Temperature

Solvent	Solid Form two Days[a]	Solid Form two Weeks[a]	Conformer I Population[b] (%)	Conformer II Population[b] (%)
Water	I	II	21	79
Hexane	I	I, II mixture	85	15
Methyl t-butyl ether	I	I	39	61
1,2-Xylene	I	I	70	30
Toluene	I	I	68	32
Nitromethane	II	II	31	69
Ethyl acetate	II	II	35	65
Acetonitrile	II	II	26	74
2-Propanol	II	II	32	68
2-Butanone	II	II	34	66
Acetone	II	II	27	73
1,2-Dimethoxyethane	II	II	28	72
Ethanol	II	II	30	70

Source: Reproduced with permission of Taylor & Francis.
[a]Results of stable polymorph screen reported by Miller et al. [39].
[b]Predictions at PBE/DNP/COSMO level of theory.

for selecting molecular conformations used in CSP: polar protic, water (both HB donor and acceptor capabilities) and diethylamine (HB donor capabilities); polar aprotic, acetone (HB acceptor capabilities); nonpolar, hexane; and self-media to mimic solid amorphous. A combined set of conformations that displayed the highest populations in any of the above solvents can be recommended for the virtual conformational polymorph screening.

In order to illustrate this point, a conformer selection in support of CSP analysis was performed using a theoretical conformational population study of the flexible S-ibuprofen molecule. This study allowed for the selection of only four preferred conformations (Figure 23.8) with the highest populations using the following three solvents: methanol, acetone, and acetonitrile. Theoretical conformational distribution analysis in water, diethylamine, hexane, and self-media at the PBE/DNP/COSMO level of theory discovers the same four favorable conformations. Thus, the selected four conformations can be used as a starting point for CSP. In addition, we should take into account that the acid group may rotate in the crystalline environment to participate in HB interactions. This adds at least another four conformations that are different from the initial set only by rotation of the acidic group by approximately 180°. In order to validate this selection process, the generated favorable conformers for S-ibuprofen (Figure 23.8) were compared with the observed molecular conformations from the CSD (reference codes IBPRAC, JEKNOC10, JEKNOC11, HUPPAJ, and RONWOG). Aligned crystallographic conformations are shown in Figure 23.15. An accuracy of the best alignment of the four selected favorable conformations (hydrogens are removed for the alignment) with those crystallographically observed is presented in Table 23.6. Quite low RMSD values demonstrate a good quality of theoretical selection of the conformations. It is interesting that while all crystallographic conformations were represented in the selected set of the favorable conformations of S-ibuprofen, one of the preferred conformations (Conformer 3 in Table 23.3; Figure 23.8) has not been yet crystallographically observed (Figure 23.15).

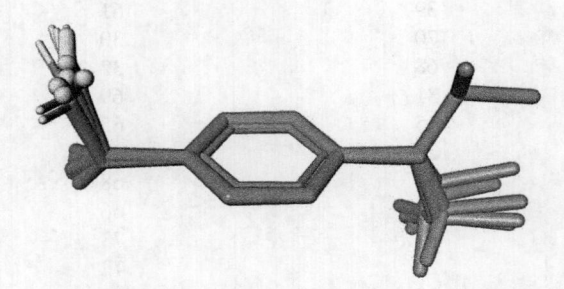

FIGURE 23.15 Alignment of the crystallographically observed S-ibuprofen conformations (CSD reference codes are IBPRAC, JEKNOC10, JEKNOC11, HUPPAJ, and RONWOG). All hydrogens are omitted.

TABLE 23.6 Best Alignments of the Calculated Favorable Conformations of S-Ibuprofen with the Similar Crystallographic Conformations from CSD

Conformer	RMSD (Å)	Reference Code
1	0.27	RONWOG
2	0.23	IBPRAC, JEKNOC10, JEKNOC11
4	0.12	HUPPAJ

The successful pharmaceutical applications of the conformational distribution analysis in the diverse set of solvents to support CSP study were reported for the oncology drug crizotinib (Xalcory®) [46] and the antibacterial drug candidate sulopenem [47].

23.4 CONCLUSIONS

Molecular crystallization is a very complex phenomenon that is dependent on a combination of multiple thermodynamic and kinetic factors. Added complexity occurs when attempting to crystallize a conformational polymorph that is selected for drug development. Computational modeling enables better understanding of solid state chemistry and allows enriched selection of desired solid form [48]. In this study, we have shown the use of one parameter, such as relative conformer population, in exploring the possibility of controlling crystallization of conformational polymorphs. We have demonstrated that relative conformational population behavior in different solvents can be reasonably accurately predicted by theoretical methods. A correlation between trends of predicted (and in some cases also observed) conformational distribution changes in different solvents and crystallization results for conformational polymorphs was found in four examples considered in the current study. Though a more extended testing is probably required, the following applications may be proposed.

1. The solvent selection strategy may focus on an attempt to diversify predicted conformer distribution in general, adopting crystallographic and/or theoretically generated conformations. This may facilitate crystallization of new form(s) that are both stable and metastable.
2. Another application of solvent selection is based on predicted change (increase or decrease) of the population(s) of the known conformer(s). This may allow tuning (promotion or inhibition) of the corresponding polymorph(s) crystallization. This approach was applied to a targeted polymorph screening of (Inlyta®) [49].

In all cases, the final selection of solvents should also satisfy the criteria of relatively high (predicted or measured) solubility of the compound under consideration. In addition, the theoretical prediction of conformer distribution in different solvents, including self-media, was proposed for the selection of preferable starting conformation(s) for virtual polymorph screening via CSP.

ACKNOWLEDGMENTS

The authors would like to thank Mr. Brian Samas and Dr. Brian Marquez for the valuable comments and discussions. YAA is thankful to Dr. Alex Goldberg for the valuable suggestions in running DMol^3 applications. YAA is grateful to Dr. Frank Eckert for the valuable consultations on COSMOtherm implementation of conformational population analysis.

REFERENCES

1. Bernstein, J. (2002). *Polymorphism in Molecular Crystals*. Oxford: Clarendon.
2. Nangia, A. (2008). Conformational polymorphism in organic crystals. *Acc. Chem. Res.* 41: 595–604.
3. Yu, L., Reutzel-Edens, S.M., and Mitchell, C.A. (2000). Crystallization and polymorphism of conformationally flexible molecules: problems, patterns, and strategies. *Org. Process Res. Dev.* 4: 396–402.
4. Weissbuch, I., Kuzmenko, I., Vaida, M. et al. (1994). Twinned crystals of enantiomorphous morphology of racemic alanine induced by optically resolved α-amino acids; a stereochemical probe for the early stages of crystal nucleation. *Chem. Mater.* 6: 1258–1268.
5. Weissbuch, I., Popoviz-Biro, R., Leiswerowitz, L., and Lahav, M. (1994). Lock-and-key processes at crystalline interfaces: relevance to the spontaneous generation of chirality. In: *The Lock-and-Key Principle* (ed. J.-P. Behr), 173–246. New York: Wiley.
6. Lee, A.Y., Lee, I., Dette, S.S. et al. (2005). Crystallization on confined engineered surfaces: a method to control crystal size and generate different polymorphs. *J. Am. Chem. Soc.* 127: 14982–14983.
7. Lee, A.Y., Lee, I.S., and Myerson, A.S. (2006). Factors affecting the polymorphic outcome of glycine crystals constrained on patterned substrates. *Chem. Eng. Technol.* 29: 281–285.
8. Weissbuch, I., Addadi, L., and Leiswerowitz, L. (1991). Molecular recognition at crystal interfaces. *Science* 253: 637–645.
9. Datta, S. and Grant, D.J.W. (2005). Effect of supersaturation on the crystallization of phenylbutazone polymorphs. *Cryst. Res. Technol.* 40: 233–242.
10. Davey, R.J., Blagden, N., Righini, S. et al. (2002). Nucleation control in solution mediated polymorphic phase transformations: the case of 2,6-dihydroxybenzoic acid. *J. Phys. Chem. B* 106: 1954–1959.
11. Davey, R.J., Blagden, N., Potts, G.D., and Docherty, R. (1997). Polymorphism in molecular crystals: stabilization of a metastable form by conformational mimicry. *J. Am. Chem. Soc.* 119: 1767–1772.
12. Mullin, J.W. (1992). *Crystallization*. Oxford: Butterworth-Heinemann.
13. Threlfall, T. (2003). Structural and thermodynamic explanations of Ostwald's rule. *Org. Process Res. Dev.* 7: 1017–1027.
14. Hursthouse, M.B., Huth, S., and Threlfall, T.L. (2009). Why do organic compounds crystallise well or badly or ever so slowly? Why is crystallization nevertheless such a good purification technique? *Org. Process Res. Dev.* 13: 1231–1240.
15. Blagden, N. and Davey, R.J. (2003). Polymorph selection: challenges for the future? *Cryst. Growth Des.* 3: 873–885.
16. Cross, W.I., Blagden, N., Davey, R.J. et al. (2003). A whole output strategy for polymorph screening: combining crystal structure prediction, graph set analysis, and targeted crystallization experiments in the case of diflunisal. *Cryst. Growth Des.* 3: 151–158.
17. Price, S.L. (2008). From crystal structure prediction to polymorph prediction: interpreting the crystal energy landscape. *Phys. Chem. Chem. Phys.* 10: 1996–2009.
18. Price, S.L. (2004). The computational prediction of pharmaceutical crystal structures and polymorphism. *Adv. Drug Deliv. Rev.* 56: 301–319.
19. Ouvrard, C. and Price, S.L. (2004). Toward crystal structure prediction for conformationally flexible molecules: the headaches illustrated by aspirin. *Cryst. Growth Des.* 4: 119–1127.
20. Day, G.M., Motherwell, W.D.S., and Jones, W. (2008). A strategy for predicting the crystal structures of flexible molecules: the polymorphism of phenobarbital. *Phys. Chem. Chem. Phys.* 9: 1693–1704.
21. Lupyan, D., Abramov, Y.A., Sherman, W., (2012). Close intramolecular sulfur–oxygen contacts: modified force field parameters for improved conformation generation. *J. Comput. Aided Mol. Des.* 26: 1195–1205.
22. Weng, Z.F., Sam, W.D., Motherwell, W.D.S. et al. (2008). Conformational variability of molecules in different crystal environments: a database study. *Acta Crystallogr.* B64: 348–362.
23. MOE2008.10. Montreal, Quebec, Canada: Chemical Computing Group, Inc.. www.chemcomp.com.
24. Delley, B. (1990). An all-electron numerical method for solving the local density functional for polyatomic molecules. *J. Chem. Phys.* 92: 508–517.
25. Delley, B. (2000). From molecules to solids with the DMol3 approach. *J. Chem. Phys.* 113: 7756–7764.
26. Accelrys Software (2008). *Material Studio 4.4, Dmol3*. San Diego: Accelrys Software Inc.
27. Perdew, J.P., Burke, K., and Ernzerhof, M. (1996). Generalized gradient approximation made simple. *Phys. Rev. Lett.* 77: 3865–3868.
28. Andzelm, J., Kölmel, C., and Klamt, A. (1995). Incorporation of solvent effects into density functional calculations of molecular energies and geometries. *J. Chem. Phys.* 103: 9312–9320.

29. Zhao, Y. and Truhlar, D.G. (2005). Benchmark databases of nonbonded interactions and their use to test density functional theory. *J. Chem. Theory Comput.* 1: 415–432.
30. Klampt, A. (2005). *COSMO-RS; From Quantum Chemistry to Fluid-Phase Thermodynamics and Drug Design*. Amsterdam: Elsevier.
31. *COSMOTherm* Version C2.1_0108. Leverkusen, Germany: COSMOLogic GmbH.
32. Lein, M., Dobson, J.F., and Gross, E.K.U. (1999). Towards the description of van der Waals interactions within density functional theory. *J. Comput. Chem.* 20: 12–22.
33. Weigend, F., Häser, M., Patzelt, H., and Ahlrichs, R. (1998). RI-MP2: optimized auxiliary basis sets and demonstration of efficiency. *Chem. Phys. Lett.* 294: 143–152.
34. TURBOMOLE V6.0 2009, a development of University of Karlsruhe and Forschungszentrum Karlsruhe GmbH, 1989–2007. TURBOMOLE GmbH, since 2007.
35. Keeler, J., Neuhaus, D., and Titman, J.J. (1988). A convenient technique for the measurement and assignment of long-range carbon-13 proton coupling constants. *Chem. Phys. Lett.* 146: 545–548.
36. Yamasaki, R., Tanatani, A., Azumaya, I. et al. (2006). Solvent-dependent conformational switching of N-phenylhydroxamic acid and its application in crystal engineering. *Cryst. Growth Des.* 9: 2007–2010.
37. Forbes, C.C., Beatty, A.M., and Smith, B.D. (2001). Using pentafluorophenyl as a lewis acid to stabilize a cis secondary amide conformation. *Org. Lett.* 3: 3595–3598.
38. Gu, C., Young, V. Jr., and Grant, D.J.W. (2001). Polymorphs screening: influence of solvents on the rate of solvent-mediated polymorphic transformation. *J. Pharm. Sci.* 90: 1878–1890.
39. Miller, J.M., Collman, B.M., Greene, L.R. et al. (2005). Identifying the stable polymorph early in the drug discovery–development process. *Pharm. Dev. Technol.* 10: 291–297.
40. Ostwald, W. (1897). Studien Uber Die Bildung und Umwandlung Fester Korper. *Z. Phys. Chem.* 22: 289–302.
41. Maruyama, S., Ooshima, H., and Kato, J. (1999). Crystal structures and solvent-mediated transformation of taltirelin polymorphs. *Chem. Eng. J.* 75: 193–200.
42. Maruyama, S. and Ooshima, H. (2001). Mechanism of the solvent-mediated transformation of taltirelin polymorphs promoted by methanol. *Chem. Eng. J.* 81: 1–7.
43. Allen, F.H., Bellard, S., Brice, M.D. et al. (1979). The Cambridge Structural Database: a quarter of a million crystal structures and rising. *Acta Crystallogr.* B35: 2331–2339.
44. Lu, J., Wang, X.-J., Yand, X., and Ching, C.-B. (2007). Characterization and selective crystallization of famotidine polymorphs. *J. Pharm. Sci.* 96: 2457–2468.
45. Chemburkar, S.R., Bauer, J., Deming, K. et al. (2000). Dealing with the impact of ritonavir polymorphs on the late stages of bulk drug process development. *Org. Process Res. Dev.* 4: 413–417.
46. Abramov, Y.A. (2013). Current computational approaches to support pharmaceutical solid form selection. *Org. Process Res. Dev.* 17: 472–485.
47. Krzyzaniak, J.F., Meenan, P.A., Doherty, C.L. et al. (2016). Integrating computational materials science tools in form and formulation design. In: *Computational Pharmaceutical Solid State Chemistry* (ed. Y.A. Abramov), 117–144. Hoboken, NJ: Wiley.
48. Abramov Y.A. (ed.) (2016). *Computational Pharmaceutical Solid State Chemistry*. Hoboken, NJ: Wiley.
49. Campeta, A.M., Chekal, B.P., Abramov, Y.A. et al. (2010). Development of a targeted polymorph screening approach for a complex polymorphic and highly solvating API. *J. Pharm. Sci.* 99: 3874–3886.

PART VII

CRYSTALLIZATION AND FINAL FORM

24

CRYSTALLIZATION DESIGN AND SCALE-UP

ROBERT MCKEOWN, LOTFI DERDOUR, AND PHILIP DELL'ORCO
Chemical Development, GlaxoSmithKline, King of Prussia, PA, USA

JAMES WERTMAN
Technical Operations, Theravance Biopharma, South San Francisco, CA, USA

24.1 INTRODUCTION

Crystallization can be defined as the formation of a solid crystalline phase of a chemical compound from a solution in which the compound is dissolved. In the synthesis of fine chemicals and pharmaceuticals, crystallization is extensively employed to achieve separation, purification, and product performance requirements. Despite its industrial relevance, an understanding of crystallization as a unit operation is often de-emphasized in academic engineering curricula and "learned on the job" in industrial settings.

In order to improve the knowledge and practice of crystallization science, several excellent volumes have been published, which provide a comprehensive treatment of the subject [1–3]. The objective of this chapter is not to repeat these comprehensive overviews, but rather to provide a concise, basic understanding of crystallization design and scale-up principles, which the reader can apply toward common industrial problems. The focus is on batch rather than continuous crystallization processes, as batch crystallization is the predominant processing method used in the pharmaceutical industry today.

The chapter begins with a discussion of crystallization design objectives and constraints on design, including a description of physical properties important to product performance. Thermodynamic principles of crystallization are then reviewed, followed by a discussion of crystallization kinetics. Common modes of crystallization design which integrate thermodynamic and kinetic principles are then presented, along with scale-up considerations for each mode. Finally, advanced topics relevant to crystallization of pharmaceutical compounds are reviewed. Throughout the chapter, industrially relevant examples are used to illustrate the concepts presented.

24.2 CRYSTALLIZATION DESIGN OBJECTIVES AND CONSTRAINTS

Crystallization is used in pharmaceutical synthesis to accomplish the following two objectives: (i) separation and purification of organic compounds and (ii) delivery of physical properties suitable for downstream processing and formulation. In achieving these objectives, a crystallization design is constrained by economic and manufacturing considerations, such as yield, throughput, environmental impact, and the ability to scale the process. An overview of these topics is presented in this section.

24.2.1 Separation and Purification

The synthesis of an active pharmaceutical ingredient (API) from raw materials involves a multistep synthetic procedure during which the raw materials undergo numerous chemical transformations and purification steps to ultimately prepare the desired molecular structure in high purity (typically >99%). An example from the literature is used to exemplify a synthetic route and its separation/purification challenges, all of which are illustrated in Figure 24.1 [4].

For this example, the API is produced in five stages from raw materials. Four of these five stages have crystallization steps to achieve purification, while the final stage uses a

---

Chemical Engineering in the Pharmaceutical Industry: Active Pharmaceutical Ingredients, Second Edition.
Edited by David J. am Ende and Mary T. am Ende.
© 2019 John Wiley & Sons, Inc. Published 2019 by John Wiley & Sons, Inc.

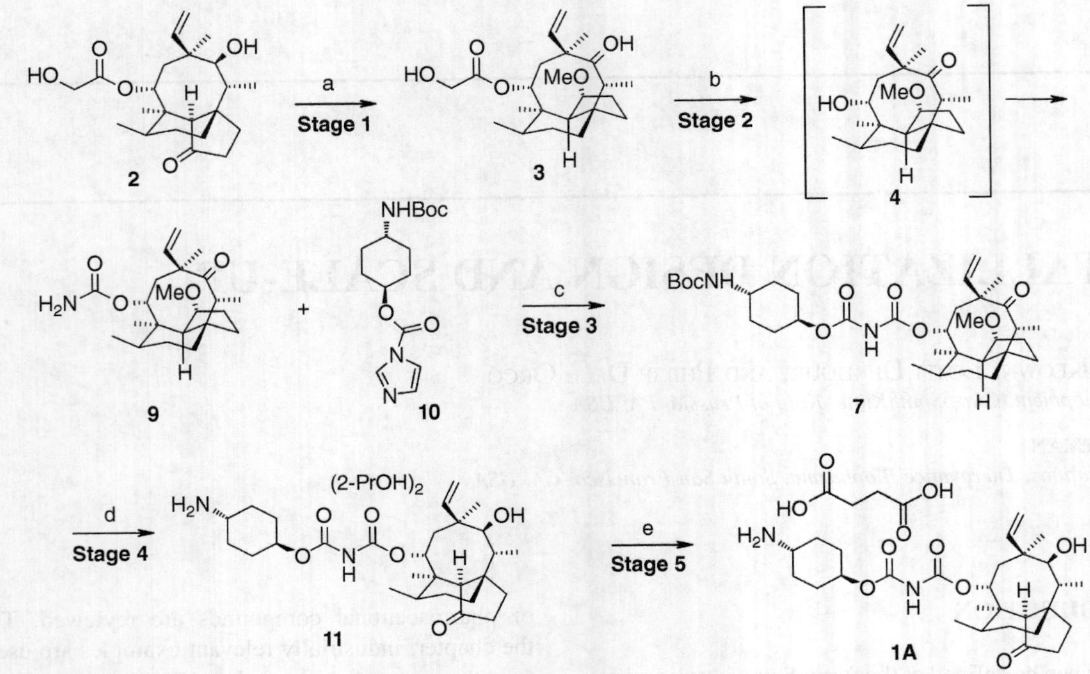

FIGURE 24.1 Schematic of a typical synthetic route to an active pharmaceutical ingredient (SB-742510) [4]. This particular route uses six chemical transformations with five crystallizations to achieve the purity required to ensure product quality. *Source*: Reprinted with permission from Andemichael et al. [4]. Copyright 2009 American Chemical Society.

crystallization to control the composition and physical properties of the final molecule. The reference describes in detail the rationale for the placement of crystallization steps. In particular, the stage 1 process was used to control key impurities in the process, as the input raw material [2] had approximately 35 impurities with the reaction producing additional impurities. This stage used a design space approach across the reaction and crystallization to ensure complete purging of raw material [2] at levels up to 3% and of an alkene impurity on the cyclohexyl ring of intermediate [3] at levels up to 4%. Crystallizations in stages 2–4 increased the organic purity of the product from approximately 97% to greater than 99% prior to stage 5 so that this step could focus solely on the formation of the desired salt and control of resultant physical properties. Moreover, the use of crystallization in this synthesis allows intermediates to be "stabilized" by forming a less reactive solid phase, preventing solution phase side reactions (e.g. racemization), and allowing for material storage.

While a primary purification concern is organic impurities related to the molecular structure of the intermediates and products, inorganic and organic reagents also require separation. These include simple salts (e.g. NaCl, K_2CO_3, NaOAc) and reagents (e.g. triethylamine, Pd(OAc)$_2$, triphenylphosphine) commonly used in pharmaceutical synthesis. In addition, crystallization is also often used to control chiral purity often through the use of chiral resolving agents.

In all cases, purification is enabled by selecting a solvent in which impurities are dissolved at the point where the desired product can be crystallized. This is discussed in further detail in the solubility Section 24.3. While thermodynamics often are the primary factor in achieving purification, kinetics may also impact the impurity content of a product by entrapment of impurities in the crystal lattice (referred to as "inclusion"). This is often induced by rapid crystallization processes that effectively trap impurities or solvents within the crystal lattice. This is discussed in additional detail later in the chapter.

24.2.2 Product Performance

The second objective of crystallization is related primarily to API rather than to intermediate production. It concerns the delivery of the appropriate material physical properties to ensure acceptable downstream processing (e.g. isolation, drying, size reduction, and formulation unit operations) as well as the in vivo/in vitro performance in the formulated product. Of particular concern are the crystalline form of the molecule, the particle size distribution of the active ingredient, and the morphology and flow properties of the product.

24.2.2.1 Crystalline Form Pharmaceutical solids are known for their ability to form multiple solid phases. A brief schematic of common solid phases is shown in Figure 24.2.

Solid phases commonly exist as either "polymorphs" or "pseudopolymorphs." Pseudopolymorphs are also referred to as solvates/hydrates. Polymorphism occurs when a single

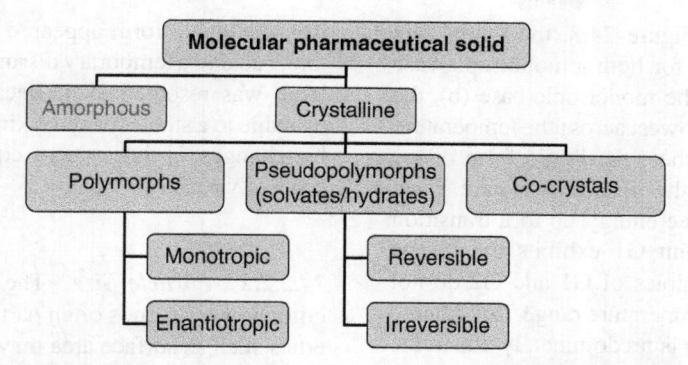

FIGURE 24.2 A map of the forms that a molecular pharmaceutical solid can exhibit. Crystalline materials are polymorphs, solvates/hydrates, and co-crystals. Co-crystals can be considered special cases of solvates, in which the "solvent" is instead an involatile compound that noncovalently bonds to the molecular solid in a regular, ordered manner. Irreversible solvates can convert either to polymorphs, different solvates, or amorphous materials upon desolvation/dehydration.

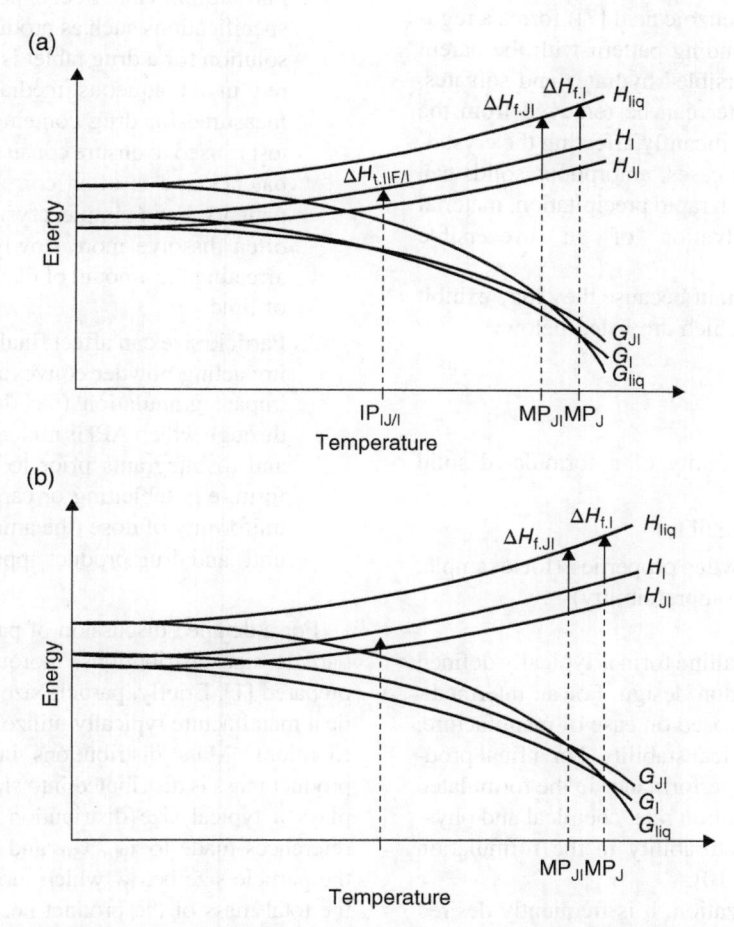

FIGURE 24.3 Thermodynamic description of polymorphism: enantiotropic (a) and monotropic (b) systems [5]. In the monotropic system, the stability order of forms is the same up to the melting point. For the enantiotropic system, a crossover temperature exists where the stability order changes. *Source*: Reprinted with permission from Henck and Kuhnert-Brandstatter [5]. Copyright (1999) Elsevier.

compound exists in two or more solid-state forms that have identical chemical structures but that have different crystal lattice structures.

Polymorphs can have either "monotropic" or "enantiotropic" relationships. This behavior is exhibited in Figure 24.3.

When two polymorphs have a monotropic relationship, they exhibit the same relative stability up to the normal melting point of both polymorphs. When an enantiotropic polymorphic relationship exists, the polymorphs change stability order at a transition temperature below the normal melting point of either polymorph. In Figure 24.3, the Gibbs free energy of polymorphs is shown for both a monotropic case and an enantiotropic case. For the monotropic case (b), the Gibbs free energy of form G1 is lowest across the temperature range until the melting point, where the liquid form of the material becomes most stable. In the enantiotropic case, Form G2 exhibits the lowest Gibbs free energy up to a transition temperature ($T_{p/st}$) at which point G1 exhibits the lowest Gibbs free energy. As the enthalpies of G1 and G2 do not change substantially up to the temperature range, this change in Gibbs free energy relationship is predominately due to the $T\Delta S$ term.

Solvates and hydrates occur when a solvent or water molecule is integrated into the crystal lattice through a repeating, non-covalent bonding arrangement with the parent molecule. Co-crystals are similar to solvates, except that a nonvolatile solute (e.g. nicotinamide [6], benzoic acid [7]) forms a regular non-covalent, non-ionic bonding pattern with the parent molecule. In the case of "reversible" hydrates and solvates, the molecule(s) of solvent/water can be removed from the pseudopolymorph without significantly affecting the crystallinity of the solid [8]. In some cases, amorphous solids can also be formed, typically through rapid precipitation, material comminution, or the desolvation of an irreversible pseudopolymorph.

Crystalline forms are important because they may exhibit different properties, some of which are listed below:

- Solubility
- Melting point
- Dissolution rate/bioavailability of a formulated solid dosage form
- Chemical and physical stability
- Habit and associated powder properties (for example, flow, bulk density, and compressibility)

As a result, the desired crystalline form is typically defined prior to initiating a crystallization design. For an intermediate, the form is often selected based on ease of manufacture, filterability, and chemical/physical stability. For a final product, it is often chosen based on performance in the formulated product (bioavailability, dissolution rate, chemical and physical stability) and on manufacturability in the formulation process (e.g. bulk density) [9, 10].

When designing a crystallization, it is frequently desired to produce the polymorph or crystalline form that is most stable at the solution composition and temperature prior to isolation. If the form being produced is not thermodynamically stable, it is possible for form conversion to occur at some point in the life cycle of the product, with the unstable form being difficult or nearly impossible to manufacture. An excellent example of this is the oft-quoted ritonavir example, where a stable form appeared during commercial manufacture, causing a temporary disruption to supply, while the form issue was resolved [11]. Because the conversion from an unstable to a stable form is a kinetic process, it can be affected by changes in impurities, equipment, concentration, and process variables.

24.2.2.2 Particle Size The second product performance criteria of concern is often particle size, although other properties such as surface area may be of equal or greater importance. The reasoning for the focus on particle size, and its frequent specification for APIs, is due to its potential impact on the performance of solid dosage forms:

1. Particle size can affect exposure to patients and in vitro specifications such as product dissolution. Product dissolution for a drug tablet is a test where a tablet is stirred in an aqueous media with the aqueous media measured for drug content as a function of time. The test is used to ensure consistency between drug product batches and is often correlated to exposure levels in patients. For example, crystals with larger particle sizes often dissolve more slowly than small particle sizes, affecting the amount of material dissolved as a function of time.
2. Particle size can affect final dosage form production by impacting powder conveyance and mixing, which can impact granulation (the drug product unit operation through which API is mixed with excipients, lubricants, and disintegrants prior to preparing the final dosage form, e.g. tabletting or capsule filling), and affect the uniformity of dose (the amount of drug in each dosage unit) and drug product appearance (color, shape, etc.).

For a detailed discussion of particle size and the nature of particle size distributions, thorough descriptions have been prepared [1]. Briefly, particle size distributions in pharmaceutical manufacture typically utilize mass- or volume-based distributions. Mass distributions indicate what percentage of product mass is distributed into size intervals. Figure 24.4 displays a typical size distribution for a pharmaceutical, with references made to x_{90}, x_{50}, and x_{10}. These values represent the particle size below which more than 90, 50, and 10% of the total mass of the product lie, respectively. Specifications for active ingredients most often contain a specification range for particle size that include one or more of x_{90}, x_{50}, and x_{10}. While the example in Figure 24.4 reflects a measurement obtained through a laser light scattering method for particle sizing, alternative methods such as sieving are still routinely employed. Example Problem 24.1 illustrates a particle size determination problem.

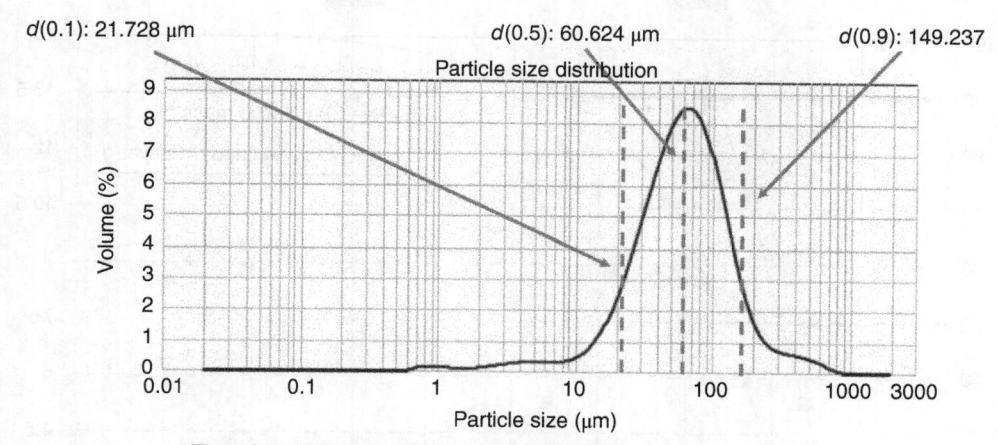

FIGURE 24.4 Particle size distribution (vol %) of a typical pharmaceutical product, measured by laser light diffraction. The $d(0.9)$ or d_{90} corresponds to the size at which 90% of the area of the curve is contained.

TABLE 24.1 Sieve Mass Fractions for an API

Sieve No.	Size Opening (μm)	Mass Retained on Sieve (g)	Sieve No.	Size Opening (μm)	Mass Retained on Sieve (g)
—	0	0.2	30	595	12.2
80	177	0.7	25	707	8.7
70	210	1.3	20	841	5.3
60	250	2.7	18	1000	3.2
50	297	5.3	16	1190	2.4
45	354	7.8	14	1410	1.2
40	420	10.9	12	1680	0
35	500	13.3	10	2000	0

EXAMPLE PROBLEM 24.1 Particle Size Distribution Calculations

Seventy-five gram of API has been sieved using a cascade of 16 sieves. The mass on top of each sieve has been weighed and is shown in Table 24.1. Estimate the x_{90}, x_{50}, and x_{10} for the material.

For sieving, the first step is to assign a particle size to each interval. The amount on top of a sieve indicates that the particle size is between the size opening of the sieve the material is retained on and the next highest sieve size. The size of a fraction is frequently estimated by averaging these two sieve sizes. Once this is done, the mass amount is tabulated as a function of sieve size. The mass amounts are then cumulatively added across sieve sizes with the amount at each interval divided by the total amount of material input to the sieve test. This gives a % of material retained as a function of particle size. The mass on each sieve and cumulative % mass are then plotted as a function of particle size to produce Figure 24.5. The x_{90} can be estimated as the particle size at which the cumulative mass = 90%. A similar approach is used for x_{50} and x_{10}, giving values of $d_{90} = 890$ μm, $d_{50} = 520$ μm, and $d_{10} = 300$ μm.

24.2.2.3 Morphology and Powder Flow Properties

A knowledge of crystal habits is important in understanding the particle size distribution and the behavior of powders and slurries, as different crystalline habits have different characteristic lengths and different flow characteristics. Flow characteristics are important because poor flowing powders can influence the ability of a powder to blend with excipients into a uniform granule and can also impact the flow of powder through primary and secondary unit operations, causing phenomena such as "rat-holing" in powder feed hoppers.

Figure 24.6 displays a summary of common habits observed in pharmaceutical crystallizations. Equant/block habits typically result in products that are easy to isolate, dry, and handle, due to their relatively low surface area/volume ratio. Acicular and thin blade crystals, which are common in pharmaceuticals, tend to pose more processing difficulties, such as long isolation times, agglomeration, and poor flow and handling properties. Despite processing

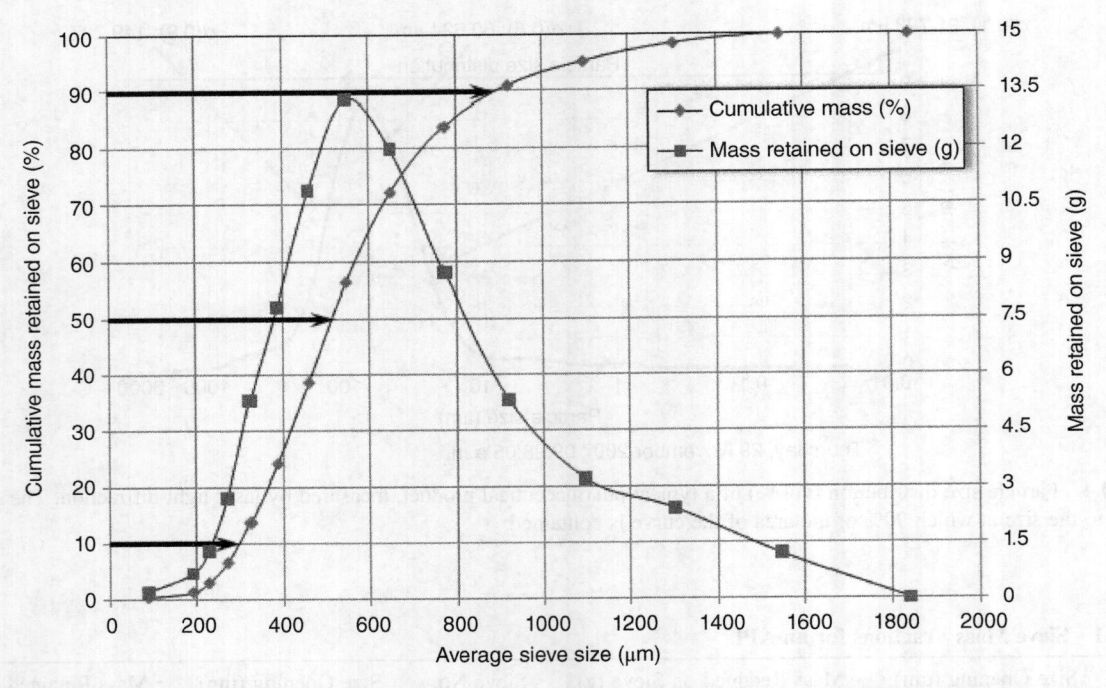

FIGURE 24.5 Size distribution estimated from sieve analysis data. Details of plot are described in Example Problem 24.1.

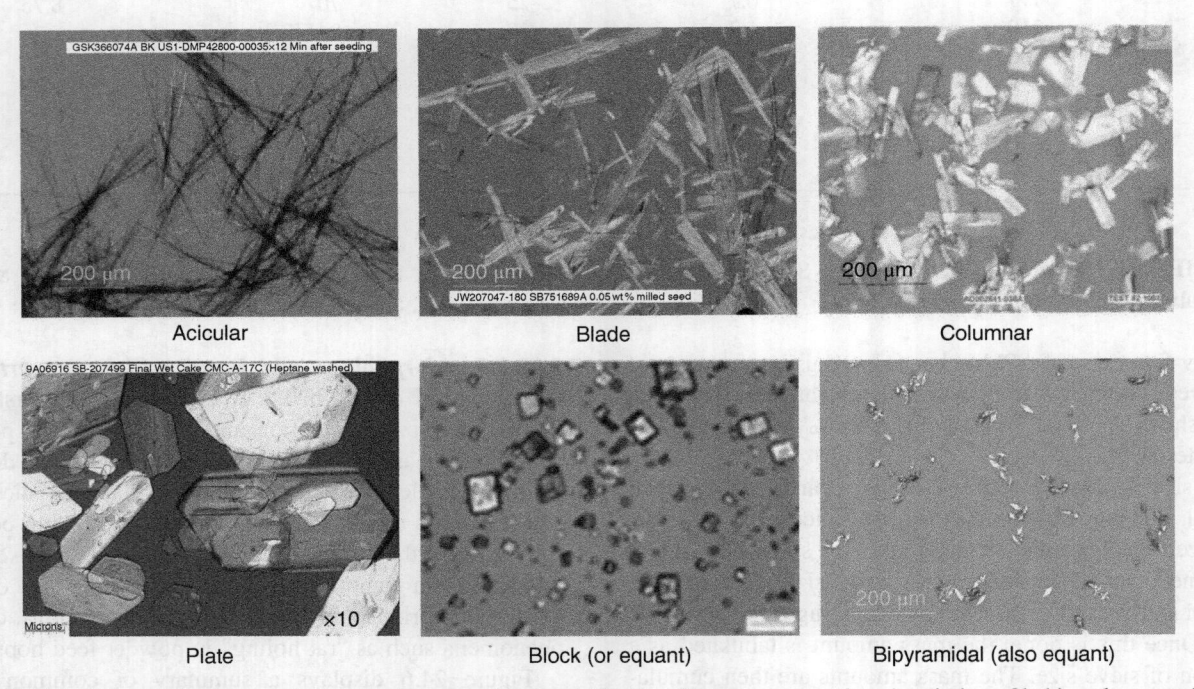

FIGURE 24.6 Commonly observed habits in pharmaceutical crystallization. For a more complete description of habits, please see Ref. [1].

difficulties, these habits may also result in high surface area materials, which can positively impact bioavailability performance in a formulation.

The flow characteristics of a powder are frequently inferred from knowledge of powder densities, commonly the bulk and tapped density. The bulk density of a powder is the density measured "as is," while the tapped density uses mechanical "taps" to facilitate further packing and settling of the powder. The ratio of the tapped to bulk density is referred to as the Hausner ratio and provides an indication of the compressibility of powders and as a result the ease of powder conveyance. Materials with a Hausner ratio of less than 1.2 are generally considered to have good flow properties, while those with a ratio greater than 1.4 are considered to have poor flow properties. The absolute values of density are also important, as they affect the level of fill for a piece of equipment. If material A has twice the bulk density of material B, the size of a batch can potentially be twice as large, resulting in throughput savings. Acicular materials routinely have Hausner ratios > 1.3 and bulk densities < 0.2, while block and equant habits often have bulk densities > 0.3 and Hausner ratios < 1.2 [12]. In addition to habit, particle size plays a strong role in affecting material bulk densities.

24.2.3 Manufacturability

Common manufacturability criteria include yield, cycle and batch times, environmental impact, and processability. Theoretical process yields are calculated with equilibrium solubility data and are described in the next section. Cycle time is affected by the time it takes for the crystallization to proceed and by the time spent in the isolation and drying unit operations. For instance, a crystallization can require a long time to maximize particle size, resulting in short isolation and drying times, or the crystallization can occur quickly, resulting in a short crystallization time with small particle generation causing long isolation times. The integration of crystallization, isolation, and drying must be considered to deliver an optimum throughput for a product. An example of batch time analysis for a crystallization and isolation process is shown in Example Problem 24.2.

EXAMPLE PROBLEM 24.2 Estimation of Optimal Cycle Time for a Crystallization, Isolation, and Drying Process

The API shown in Figure 24.7 is prepared by reactive crystallization where a base (ethylenediamine) is added to a molecular free acid. There are two potential crystallization processes that can meet product performance needs.

Process 1 is performed in *tert*-butyl methyl ether (TBME) and requires 12 hours to crystallize. Process 2 is performed in isopropyl alcohol (IPA) and requires five hours to crystallize. Both crystallizations use 10 l solvent/kg product, and the yields for both processes are the same, and both products are isolated at 20 °C. Both products are planned to use 1 m² filter area with a mass loading of 100 kg and a 1 bar filtration pressure. The filter cake resistances (α) of the products were measured through small-scale tests to be 1.1×10^{11} m/kg and 0.5×10^{11} m/kg, respectively, where the cake resistance is estimated through Eq. (24.1). In Eq. (24.1),

$$\frac{t}{m_f} = \frac{\alpha \nu c}{2A^2 P} m_f \qquad (24.1)$$

where

- t is the filtration time.
- m_f is the mass of filtrate.
- ν is the kinematic viscosity.
- c is the mass of solids per mass of filtrate.
- A is the filtration area.
- P is the filtration pressure.

Which process offers the minimum batch time through isolation and by how much?

To solve the problem, the filtration time for each material must be estimated. At 20 °C, TBME has a kinematic viscosity of 4.7×10^{-7} m²/s, while isopropanol has a kinematic viscosity of 2.9×10^{-6} m²/s. From the equation provided, the isolation times for a 100 kg batch with 10 l solvent/kg product can be estimated as 4.8 and 15.8 hours for the TBME and IPA crystallizations, respectively. So the total batch time using TBME is 12 + 4.8 = 16.8 hours and that for IPA is 5 + 15.8 hours = 20.8 hours. TBME offers a lower batch time by four hours, mainly due to the lower kinematic viscosity of TBME compared with IPA. A more detailed approach to a similar problem is provided in [5], which optimizes crystallization parameters to minimize the isolation time of a product.

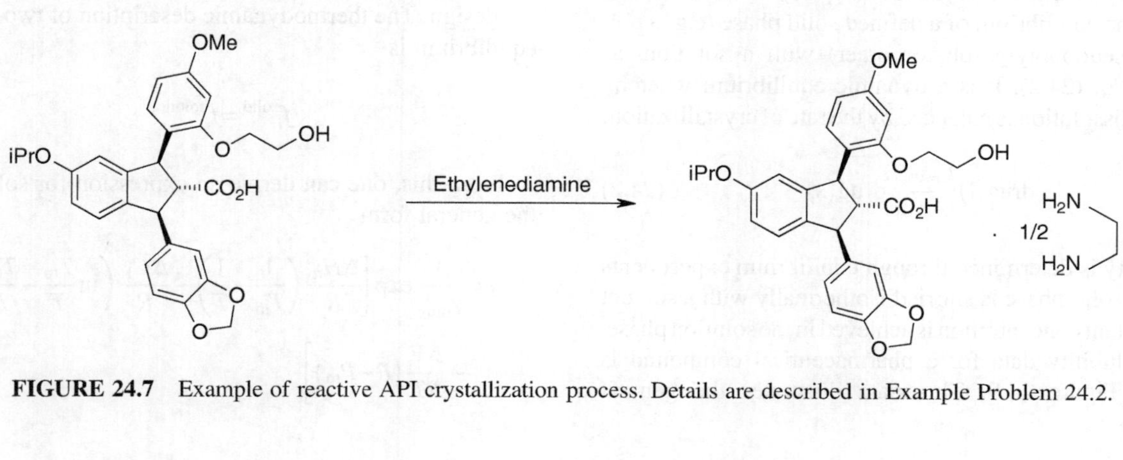

FIGURE 24.7 Example of reactive API crystallization process. Details are described in Example Problem 24.2.

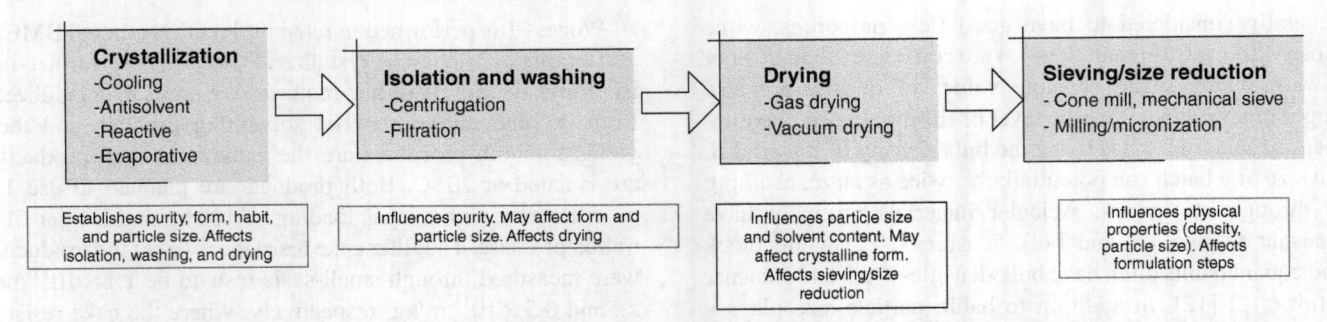

FIGURE 24.8 The relationships of crystallization with downstream processing steps. Crystallization typically needs to be studied in conjunction with downstream steps to understand the complete control of desired attributes.

Example Problem 24.2 illustrates a key aspect of crystallization design, which is shown in Figure 24.8. The crystallization in the example affects isolation performance, which in turn affects drying performance, which affects sieving and size reduction performance, which ultimately affects performance in the formulation process. As a result, it is often necessary to study crystallization in conjunction with several unit operations. As shown in Example Problem 24.2, it is straightforward to assess the dependence of the crystallization on the isolation step, and the drying step can further be integrated into structured studies. Each downstream unit operation must be investigated for its impact in ensuring the process, and not just the crystallization, achieves the desired performance objective.

24.3 SOLUBILITY ASSESSMENT AND PRELIMINARY SOLVENT SELECTION

An understanding of a compound's solubility is the normal starting point for crystallization design. The solubility of a compound determines the throughput and yield; it is often the key property in selecting a solvent system and designing the crystallization procedure. Coupling the API solubility with the solubility of impurities can assist in predicting selectivity or impurity purging.

Solubility is a thermodynamic property of a solute that describes the equilibrium of a defined solid phase (e.g. a polymorph, pseudopolymorph, or other) with a solution, as shown in Eq. (24.2). It is a dynamic equilibrium whereby the rate of dissolution is balanced by the rate of crystallization:

$$\text{drug}(l) \underset{k_{\text{dissolution}}}{\overset{k_{\text{crystallize}}}{\leftrightarrow}} \text{drug}(s) \quad (24.2)$$

Solubility is determined through equilibrium experiments in which a solid phase is slurried isothermally with a solvent until a constant concentration is achieved in the solution phase. Typical solubility data for a pharmaceutical compound is shown in Figure 24.9. This data illustrates the general approach to data measurement: solid of a known phase is added to a predefined amount of solvent at low temperature, an equilibrium concentration is achieved, and the temperature is then increased, ensuring solids are still present. At the end of the experiment, the solid phase of the material is assessed to ensure the crystalline form has not changed, as different forms have different solubility equilibria. If binary or ternary solvents are being studied, the phase composition of the mixture should also be measured at the end of the experiment to ensure no changes. Each compound possesses its own time scale in which equilibria are achieved. This typically ranges from minutes to several hours, although days are required in some circumstances. As illustrated in Figure 24.10, in-line methods of solute concentration measurement are useful in understanding time to equilibria. When in-line methods are not practical, measurement of the filtered equilibrium solution by HPLC (or gravimetric analysis if the material is relatively free from impurities) is commonly used.

While Figure 24.10 displays solubility as a function of temperature, it may also be measured as a function of solvent composition. When temperature is used to generate solubility differences, the result is a "cooling" crystallization. When solvent composition is used, the result is often called an "antisolvent" crystallization. In all cases, the preferred units of solubility are mass solute/mass solvent. Using these units simplifies subsequent engineering calculations.

Once data is collected, it can be evaluated for use in process design. The thermodynamic description of two phases in equilibrium is

$$f_i^{\text{solid}} = f_i^{\text{solution}} \quad (24.3)$$

From this, one can derive an expression for solubility of the general form

$$x_i = \frac{1}{\gamma_{\text{drug}}} \exp\left[\frac{\Delta H_{\text{tp}}}{R}\left(\frac{1}{T_{\text{tp}}} - \frac{1}{T}\right) - \frac{\Delta C_P}{R}\left(\ln\frac{T_{\text{tp}}}{T} - \frac{T_{\text{tp}}}{T} + 1\right)\right.$$
$$\left. - \frac{\Delta V}{RT}(P - P_{\text{tp}})\right] \quad (24.4)$$

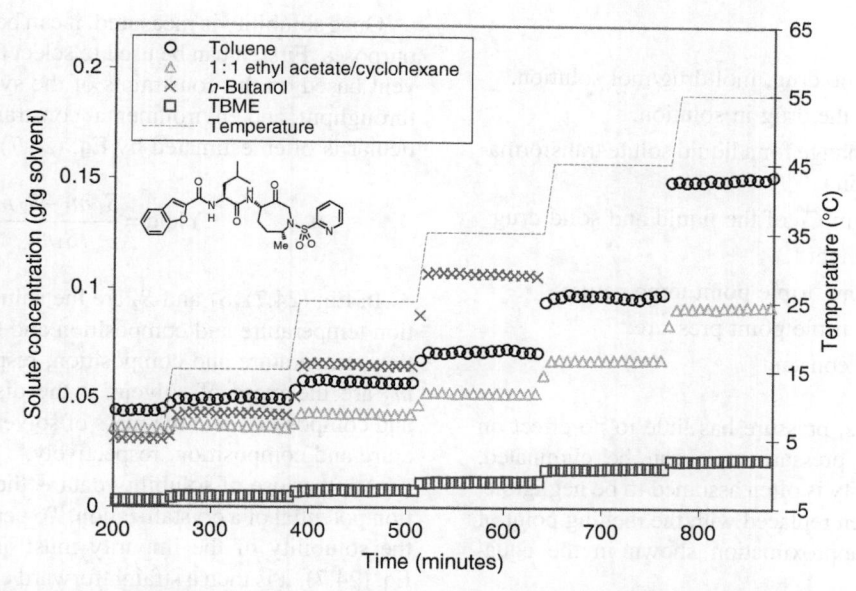

FIGURE 24.9 Results from a solubility experiment using diamond attenuated total reflectance IR spectroscopy to measure concentration *in situ* for a pharmaceutical active ingredient. At increasing temperatures, the solute achieves an equilibrium in the different solvent mixtures, indicated by the plateau in concentration upon achieving a new temperature. In this case, equilibrium is rapidly achieved.

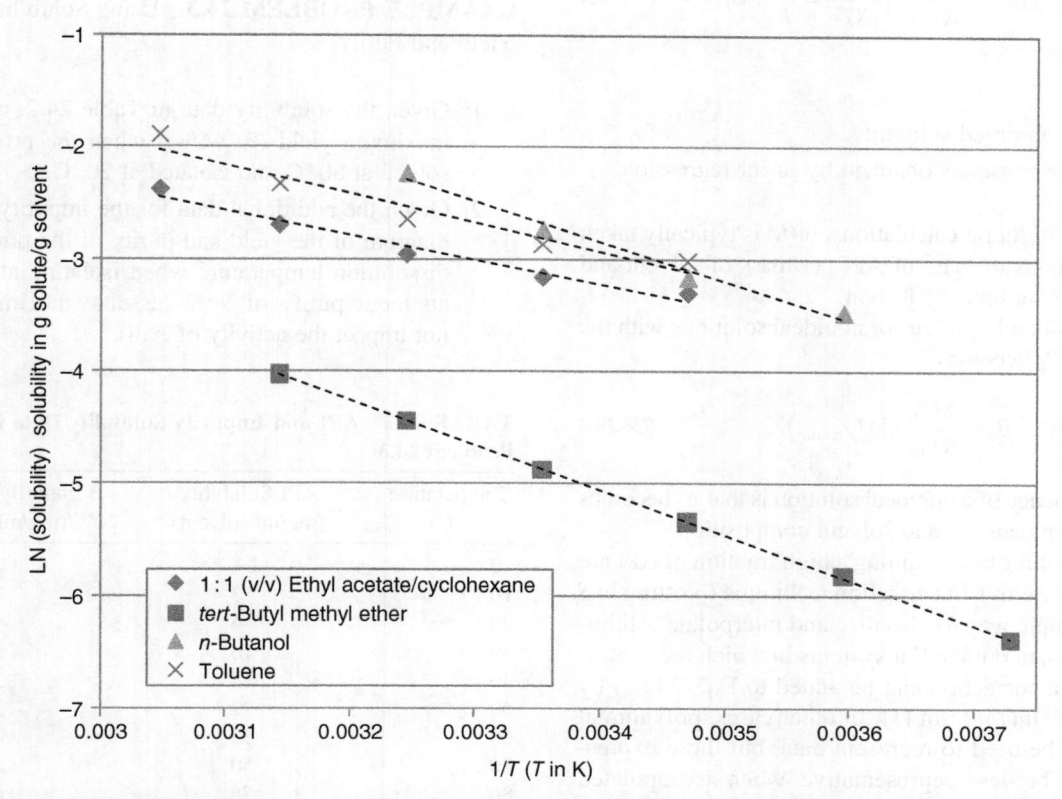

FIGURE 24.10 Solubility data from Figure 24.9 regressed using a simple van't Hoff relationship. For some solvents, the fit is very linear (e.g. *tert*-butyl methyl ether). Other solvents would benefit from an additional fitting parameter due to curvature (e.g. toluene using the $C \ln(T)$ term).

where

- x_i is the solubility of the drug, mol drug/mol solution.
- γ_{drug} is the activity of the drug in solution.
- ΔH_{tp} is the enthalpy change for a liquid solute transformation at the triple point.
- ΔC_P is the difference in C_P of the liquid and solid drug.
- ΔV is the volume change.
- T, T_{tp} is the temperature, triple point temperature.
- P, P_{tp} is the pressure, triple point pressure.
- R is the universal gas constant.

In almost all situations, pressure has little to no effect on solubility; therefore the pressure term can be eliminated. The change in heat capacity is often assumed to be negligible, and the triple point is often replaced with the melting point of the solid to yield the approximation shown in the equation below:

$$x_i = \frac{1}{\gamma_{drug}} \exp\left[\frac{\Delta H_m}{R}\left(\frac{1}{T_m} - \frac{1}{T}\right)\right] \quad (24.5)$$

For an ideal solution, this can be reduced to a van't Hoff type expression and linearized

$$\ln x_i = \ln(S_{solute}) = \frac{\Delta H_m}{RT_m} - \frac{\Delta H_m}{RT} = \frac{A_s}{T} + B_s \quad (24.6a)$$

where

- S_{solute} is the observed solubility.
- A_s and B_s are constants obtained by linear regression.

For ease of use in future calculations, $\ln(S)$ is typically taken with S having units of mass of API per mass of solvent and temperature having units of Kelvin.

The linearization is similar for nonideal solutions with the exception that B_s becomes

$$B_s = \frac{\Delta H_m}{RT_m} - \ln(\gamma_{solute}) \quad (24.6b)$$

The consequence of a nonideal solution is that B_s becomes a function of temperature and solvent composition.

Almost all solutions containing a high fraction of API are nonideal; however this linearization technique (plotting $\ln S$ vs. $1/T$) is a simple way to visualize and interpolate solubility from a few data points. For systems in which these plots are nonlinear, a correction can be added to Eq. (24.6a) by including a $[+C \ln(T)]$ term [1]. In other cases, polynomial functions may be used to represent data, but these expressions tend to be less representative when extrapolated beyond the range of temperature studied in the solubility experiment.

Once solubility is measured, it can be used for a number of purposes. First, it can be used to select the crystallization solvent based on the constraints of the system, typically yield, throughput, and environmental constraints. The yield in particular is often estimated by Eq. (24.7):

$$\text{Yield} = \frac{S_1 m_1 - S_2 m_2}{S_1 m_1} \quad (24.7)$$

In Eq. (24.7), S_1 and S_2 are the solubilities at the dissolution temperature and composition and the solubility at isolation temperature and composition, respectively, and m_1 and m_2 are the mass of solvent at the dissolution temperature and composition and the mass of solvent at isolation temperature and composition, respectively.

Another use of solubility data is the estimate of purification potential of a crystallization. To perform this assessment, the solubility of the impurity must also be known. Using Eq. (24.7), it is then a straightforward calculation to calculate the mass of impurity and desired solute out to estimate the potential for the crystallization to achieve purification goals. The actual purification can be less than the calculated values due to impurity inclusion or occlusion in the product. The following example illustrates the use of solubility data to estimate the purification potential of a solvent.

EXAMPLE PROBLEM 24.3 Using Solubility to Predict Yield and Purity

1. Given the solubility data in Table 24.2, calculate the maximum yield (% wt/wt) when the product is dissolved at 80 °C and isolated at 20 °C.
2. Given the additional data for the impurity, generate a diagram of the yield and purity of the product versus dissolution temperature when isolating at 10 °C with an input purity of 96% (assume the impurity does not impact the activity of API).

TABLE 24.2 API and Impurity Solubility Data for Example Problem 24.3

Temperature (°C)	API Solubility (mg/ml solvent)	Impurity Solubility (mg/ml solvent)
0	1.8	0.9
10	3	1.2
20	6	1.7
30	10	2.2
40	15	2.7
50	22	3.3
60	33	4.1
70	50	5.0
80	78	6.2
90	120	7.7

3. The yield can be calculated by performing a mass balance on the drug in the liquid phase.
   The formula is
   $$\text{Yield} = \frac{S_1 V_1 - S_2 V_2}{S_1 V_1} \quad (24.8)$$
   where

   S_1, S_2 is the solubility at the dissolution temperature and composition and the solubility at isolation temperature and composition, respectively.

   V_1, V_2 is the volume of solvent at the dissolution temperature and composition and the volume of solvent at isolation temperature and composition, respectively.

   A more in-depth explanation is that at dissolution all of the API is in the liquid at a concentration of 78 mg/ml. At isolation, the minimum concentration that can be in the liquid is 6 mg/ml. Therefore, the amount of solid is 78 − 6 = 72 mg/ml. The yield is then 72 mg/ml/78 mg/ml = 92.3%.

4. The data for API and impurity can be linearized and plotted using the van't Hoff equation to give better resolution (Figure 24.10); however, for this example only the points given will be used. The yield of API for each dissolution temperature can be calculated using the formula above (Figure 24.11). However, the level of impurity will always start at 4% of the initial API concentration, not the solubility limit. Therefore, the yield of impurity can be calculated as

   $$Y_{\text{impurity}} = 0.04\, S_1^{\text{API}} - \frac{S_2^{\text{impurity}}}{0.04\, S_1^{\text{API}}} \quad (24.9)$$

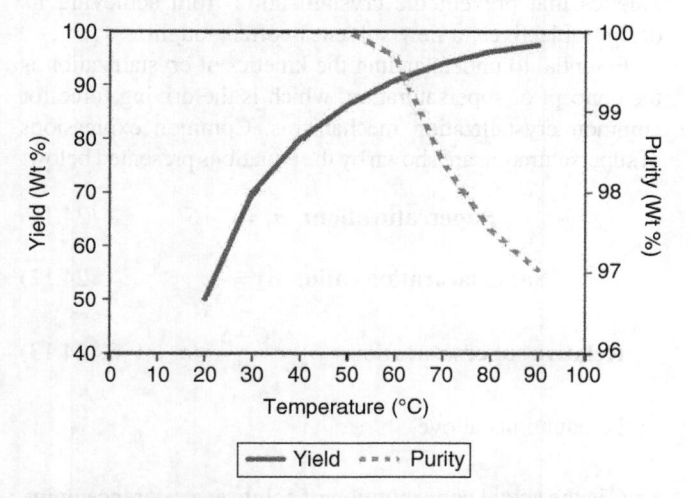

FIGURE 24.11 Illustration of the trade-off between API purity and yield. Details of the plot are described in Example Problem 24.3.

The overall purity is then

$$\text{Purity} = \frac{m_i^{\text{API}} Y_{\text{API}}}{m_i^{\text{API}} Y_{\text{API}} + m_i^{\text{impurity}} Y_{\text{impurity}}} = \frac{m_i^{\text{API}} Y_{\text{API}}}{m_i^{\text{API}} Y_{\text{API}} + 0.04\, m_i^{\text{API}} Y_{\text{impurity}}}$$

$$= \frac{Y_{\text{API}}}{Y_{\text{API}} + \left(1 - \text{Purity}_{\text{input}}\right) Y_{\text{impurity}}} \quad (24.10)$$

Finally, solubility and related experiments are essential in understanding the relative stability of crystalline forms. When slurrying solids at a constant composition and a constant temperature, the crystalline form may either stay the same or fully/partially convert to another form. If conversion occurs, the new form is the most stable form at that temperature/concentration condition. The typical path for this type of process is shown in the scheme below for the formation of a hydrate from an anhydrate form. The anhydrate will often initially dissolve, forming a dissolved and solvated solute molecule. This dissolved species then spontaneously crystallizes as the more stable form, and the newly formed crystalline form continues to grow due to the higher solubility of the anhydrate relative to the hydrate. As the processes for spontaneous crystallization and growth can be slow, these experiments often take days or even weeks to achieve equilibrium and can be accelerated by performing the slurry experiment with both forms present initially or through minor fluctuations in temperature (e.g. 1–2 °C):

$$A(s) + x\,H_2O(l) \Leftrightarrow [A \cdot x\,H_2O](l) \Leftrightarrow A \cdot x\,H_2O(s)$$

The phase diagram that can be constructed upon performing such experiments is illustrated in Figure 24.12 [13].

The solubility of a compound can be dramatically affected by the presence of impurities or residual solvents in crystallization liquors. When measuring the solubility for crystallization design purposes, it is recommended that the first measurement be made on relatively "pure" materials (>98%) in pure solvents. This gives a baseline understanding of behavior in the absence of nonidealities. Then, measurements of representative materials in representative process solvents should be taken. The values in actual systems should be used for forward design purposes; the differences between "ideal" and actual systems can often be narrowed through an adjustment of the actual system (e.g. removal of an impurity or better control of small quantities of undesired solvent).

Nonidealities caused by compositional differences can often be used advantageously. A specific instance involves the use of water as a cosolvent for poorly soluble intermediates and APIs. Water, when present as a minor component of a solvent system, often has a dramatic solubility enhancement affect that can be used advantageously, especially for compounds that exhibit poor solubility in neat solvents. This behavior is illustrated in Figure 24.8, which indicates that

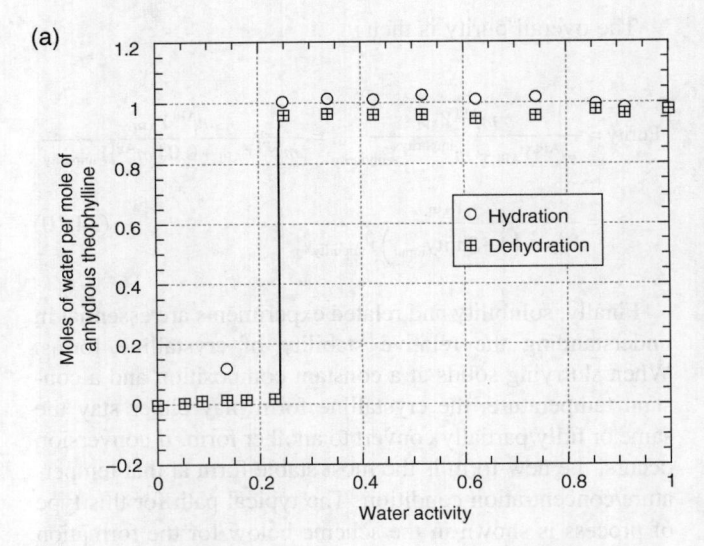

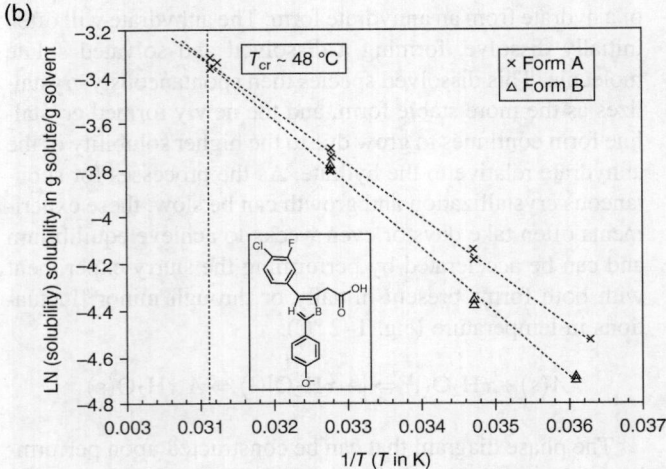

FIGURE 24.12 The use of equilibration and solubility experiments to understand form stability relationships. (a) Hydrate/anhydrate phase diagram that shows a critical water activity of approximately 0.25 needed to achieve full hydration. Below this activity, the anhydrate will be the most stable polymorph. The data was collected over five days starting from both 100% hydrate in one case (squares) and 100% anhydrate in the other case (circles). *Source*: Reprinted from Zhu et al. [13]. Copyright 1996 with permission from Elsevier. (b) An enantiotropic system of an API. The solubility of Form B is less than the solubility of Form A at temperatures up to approximately 48 °C (i.e. the crossover temperature), at which point Form A exhibits a lower solubility and becomes more stable.

the compound (carbamazepine) has essentially no solubility in water and relatively low solubility in neat solvent (ethanol) but is highly soluble at 20% water content [14].

Solvent evaluation using solubility data is essential to achieving many of the objectives for a crystallization: yield, throughput (i.e. amount of solvent required), and environmental impact (i.e. hazard of solvent and amount required). In addition, solubility data and assessment is a primary determinant of the "mode" of crystallization chosen. Without changes in solubility or the ability to increase solute concentrations above the equilibrium solubility, the material will not crystallize from solutions. Frequently used crystallization modes in pharmaceutical production are shown in Table 24.3, along with the solubility behavior that typically leads to the use of each mode. While solubility is important to the initial mode selection, an understanding of kinetics, as described in Section 24.4, is important in defining the parameters required to meet other crystallization objectives, such as particle size and crystalline form.

24.4 CRYSTALLIZATION KINETICS AND PROCESS SELECTION

Solubility, like any thermodynamic relationship, provides a starting point and an endpoint for a process. Knowledge of crystallization kinetics is critical in determining the path through which the beginning and endpoint are linked. For chemical reactions, kinetics is used to indicate the rate of change of molecular species; for crystallizations, kinetics is used to indicate the rate of solute mass transfer from solution phase to a solid phase. While solubility is often a primary control for achieving purification and separation objectives, the kinetic mechanisms of a crystallization are typically the primary determinant for physical properties. Readers are referenced to more comprehensive discussions on these topics, for example [3].

Following the preliminary selection of one or multiple solvents from solubility data, the kinetics of the system must be understood in order to choose the conditions under which the crystallization operates and to validate the choice of solvent. If the chosen solvent presents significant challenges related to kinetics that prevent the crystallization from achieving its design objectives, a new solvent is often sought.

Essential to understanding the kinetics of crystallization is the concept of supersaturation, which is the driving force for common crystallization mechanisms. Common expressions of supersaturation are shown by the equations presented below:

Supersaturation: $\quad \sigma_1 = C - S \quad$ (24.11)

Supersaturation ratio: $\quad \sigma_2 = \dfrac{C}{S} \quad$ (24.12)

Relative supersaturation: $\quad \sigma_3 = \dfrac{C-S}{S} = \sigma_2 - 1 \quad$ (24.13)

In the equations above,

C is the actual concentration of solute at a reference point.
S is the equilibrium solubility at the same point.

TABLE 24.3 Common Crystallization Modes and the Influence of Solubility Behavior on the Selection of Modes

Crystallization Mode	Description	Solubility Behavior Leading to Mode Selection
Cooling	Crystallization is achieved by cooling solvent from a high temperature to a low temperature at constant solvent composition. Temperature is used to reduce solubility	Compound is soluble in a solvent at an elevated temperature below the normal boiling point of the solvent (e.g. >100 g/g solvent) but relatively insoluble at a lower temperature (e.g. <10 g/g solvent)
Antisolvent addition	Crystallization is achieved by adding an antisolvent to a solvent in which the solute is soluble. Composition is used to reduce solubility	Cooling crystallizations cannot achieve yield constraints (e.g. >90%) at reasonable dilutions (e.g. <20 l solvent/kg compound). The addition of an equal volume of antisolvent to a solvent reduces the solubility of a compound by >50%
Reactive	Crystallization is achieved by changing the compound ionically or structurally through reaction. The reactants are often soluble with the product being insoluble. Reaction is used to change the concentration of the product above the solubility limit	The product is completely insoluble in all potential solvents, and the precursors are readily soluble
Evaporative	Crystallization is achieved by the concentration of a solute from a concentration where the material is soluble to a concentration at which it is insoluble. Concentration is used to increase the solute concentration above the solubility limit	Often used in combination with a cooling crystallization. For instance, if a cooling crystallization without evaporation can come close to meeting yield requirements, further concentration through distillation will allow additional mass to be recovered

σ_1 is the supersaturation in absolute terms (concentration).
σ_2 is referred to as the supersaturation ratio.
σ_3 is referred to as the relative supersaturation.

All three terms are frequently used in the analysis of crystallization processes.

Crystal mass is generated by either nucleation or growth. Nucleation can be described as the formation of new crystals from a solution or slurry, while growth can be defined as the deposition of solute mass on existing crystals of that solute. Nucleation can further be divided into two mechanisms: "primary" nucleation, which is the formation of new crystals from solutions devoid of crystals, and "secondary" nucleation, which is the formation of new crystals in the presence of existing crystals. Primary nucleation can occur within solutions (homogeneously) or at surfaces (heterogeneously, e.g. on crystallizer walls, agitators, etc.).

Nucleation generates small crystals, which can be useful in preparing small particle size powders. However, nucleation can also lead to significant downstream processing problems, such as long isolation times, significant agglomeration leading to poor performance in a formulation, and batch-to-batch variability. Crystal growth is used to increase the size of the product, reduce batch-to-batch variability, and overcome downstream processing and handling issues [15]. It is also used to control the crystalline form of the compound being prepared for systems in which multiple forms may nucleate. As a result of the benefits afforded by crystal growth, it is generally preferred as the dominant mechanism in crystallization design, especially for controlling physical properties of materials.

24.4.1 Nucleation Kinetics and the Metastable Limit

When solutions are supersaturated, they are thermodynamically unstable. Like chemical reactions that do not react spontaneously, a thermodynamically unstable solution does not necessarily crystallize spontaneously as an energy barrier must be overcome to form a surface, analogous to the activation energy associated with a chemical reaction. Solutions that are supersaturated but do not spontaneously crystallize are referred to as "metastable." A complete description as to the origin of metastable solutions can be found in Ref. [1]. It is quite common for pharmaceutical intermediates and APIs to form metastable solutions at supersaturation ratios between 1.0 and 1.2. Eventually (weeks to years), many "metastable" systems might nucleate, but over the time scales associated with processing (minutes to hours), nucleation typically does not occur. A solute is said to be at its metastable limit when it is at the maximum supersaturation at which primary nucleation does not spontaneously occur.

After the solubility, the metastable limit is the next critical measurement in crystallization design. It is measured by two primary methods, which are described in detail in Table 24.4.

An example of metastable limit determination and data reduction is provided by Example Problem 24.4 for a cooling crystallization and an antisolvent crystallization.

TABLE 24.4 Description of Metastable Limit Measurement Techniques

Method	Description
Cooling rate (this method is primarily applicable to cooling crystallizations but can also be applied to evaporative crystallizations)	1. Using solubility data, a saturated solution of compound in solvent is prepared at a temperature close to maximum temperature of the crystallization in a reactor equipped with a particle measurement device (e.g. turbidity, Lasentec FBRM) 2. The solution is cooled at a slow rate (~0.1 °C/min) until particles are observed by the measurement device. The temperature at which crystallization is observed is recorded 3. The experiment is repeated several times at faster cooling rates (e.g. 0.25, 0.5, 0.75, and 1 °C/min) 4. A graph is prepared plotting crystallization temperature as a function of cooling rate. This plot should be linear; a linear fit will indicate the nucleation temperature at a 0 °C/min cooling rate 5. The supersaturation at the 0 °C/min cooling rate is the metastable limit at this temperature. This limit is often expressed as a temperature difference between the saturation temperature and the temperature estimated at the 0 °C/min cooling rate 6. The experiment can be repeated at lower temperatures. It is best to start first at a temperature near the isolation condition. If there is little difference in the metastable limit in supersaturation terms relative to the high temperature point, additional data is not necessary 7. It is recommended to perform duplicate experiments in the same and different equipment to understand the potential error associated with the measurement 8. For screening purposes, the observed crystallization temperature at the lowest cooling rate (0.1–0.25 °C/min) can be approximated as a 0 °C/min cooling rate and the metastable limit estimated from a single data point
Nucleation induction time method (this method is applicable to all crystallization types)	1. Saturated solutions of compound are prepared under conditions (temperature and composition) anticipated to be near the starting point of the crystallization. A particle detection probe is inserted (e.g. turbidity or Lasentec FBRM) 2. Supersaturation is generated by one of the following methods: (i) cooling the solution as rapidly as possible to a temperature at which the solution is supersaturated, (ii) adding non-solvent to a solvent composition at which the solution is supersaturated, (iii) performing a partial reaction to generate supersaturation, and (iv) rapidly evaporating a fraction of the solvent 3. After the rapid generation of supersaturation in (2), the solution is held isothermally until a particle detection device indicates that particles have formed. The time (i.e. nucleation induction time) to particle formation is recorded 4. The experiment is performed at multiple conditions (e.g. different temperatures, different amounts of antisolvent added, different amounts of reactants used) 5. The nucleation induction time is plotted as a function of supersaturation. An asymptote will be observed. The supersaturation value at this asymptote is the metastable limit 6. Duplicate the experiments under the same conditions to understand variability

EXAMPLE PROBLEM 24.4 Estimation of Metastable Zone Width by Cooling and Nucleation Induction Time Methods

Compound A is to be crystallized through cooling crystallization in neat ethyl acetate. Compound B is to be crystallized using the addition of n-heptane (antisolvent) to a solution of tetrahydrofuran (THF, solvent). For compound A, the "cooling rate" method is applied, and for compound B, the nucleation induction time method is used. Here is the data for both studies:

Compound A: A solution of compound A is saturated in ethyl acetate at 65 °C. The solubility follows a van't Hoff relationship, with $A = -3269.2$ and $B = 8.13$ (S is in units of g solute/g solvent). The solution is heated to 70 °C and cooled at 0.25 °C/min. The crystallization temperature, recorded by turbidity, is 50.4 °C. After crystallization, the solution is reheated to 70 °C to achieve dissolution, and cooled at 0.5 °C/min, with a crystallization temperature of 48.9 °C. The procedure is repeated at 0.67 °C/min and at twice at 1 °C/min with crystallization temperatures of 47.3, 45.9, and 45.4 °C, respectively. Estimate the metastable limit of the compound. Report results as supersaturation in units of g solute/g solvent.

Compound B: A solution of compound B is saturated in THF at 20 °C. To the THF, different amounts of heptane (non-solvent) are added, and the time to crystallization is noted. Relevant data is reported in Table 24.5. Estimate the metastable limit and report as relative supersaturation.

Solution

Compound A: With the data displayed above, a plot can be made of crystallization temperature as a function of cooling rate, and this data can be linearly fit and extrapolated to a 0 °C cooling rate. This is illustrated in Figure 24.13a, with the extrapolation leading to a predicted nucleation temperature of approximately 52 °C at 0 °C/min. Estimating the solution concentration and the concentration at 52 °C with the solubility equation and taking the difference, results are in an estimated metastable limit of 0.068 g solute/g solvent for a solution saturated at 65 °C. The supersaturation ratio is approximately 1.46.

Compound B: Induction time is plotted as a function of the supersaturation ratio that is determined from the data in Table 24.5 (Figure 24.13b). An asymptote is estimated from a power law fit to the data, giving a supersaturation ratio of approximately 5. As a result, the relative supersaturation is estimated to be 4.

Of the two methods for estimating metastable zone width, the nucleation induction time method is often the most general, as it is broadly applicable to all crystallization modes. In addition, the nucleation induction time method gives an indication of the time window that is available to allow the addition and growth of seed materials and mix antisolvents/reactants with the main solvent. The nucleation induction time is also useful in understanding the scale-up implications of crystallizations and is described later in this chapter.

While the metastable limit is useful in understanding the conditions under which primary nucleation occurs, the potential for secondary nucleation must also be considered. Secondary nucleation occurs through several mechanisms, with the most common mechanism being contact nucleation. Contact nucleation is micro-attrition of crystals, resulting in small crystalline fragments (<10 μm) being present in the slurry [16]. The rate of contact nucleation is influenced by crystal–crystal, crystal–impeller, and crystal–wall collisions. A common expression for contact nucleation is indicated by Eq. (24.14):

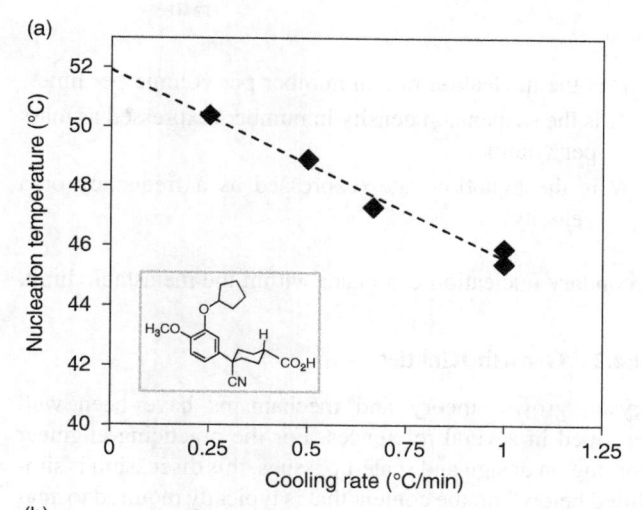

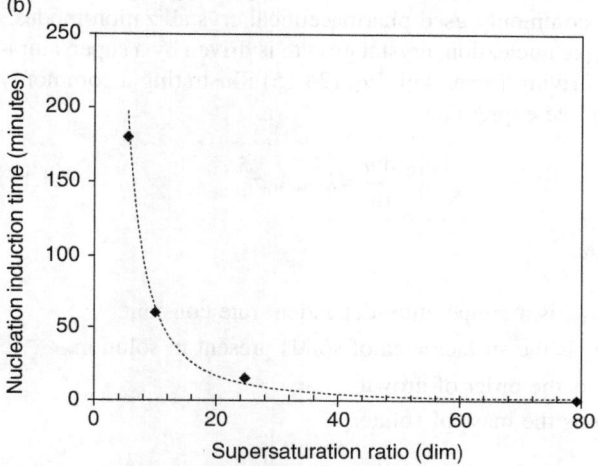

$$B = k_N M^j N^k \sigma^b \qquad (24.14)$$

FIGURE 24.13 Metastable zone width measurement using the (a) cooling method for compound A and (b) nucleation induction time method for compound B. Details are included in Example Problem 24.4.

TABLE 24.5 Metastable Limit Example – Nucleation Induction Time Data for Compound B

Heptane Volume Fraction	Induction Time (min)	Solute Concentration (g/g Solvent)	Solubility (g/g Solvent)
0.055	180	0.104	0.0186
0.11	60	0.097	0.0097
0.22	15	0.085	0.0034
0.44	1	0.069	0.0009

where

- B is the nucleation rate in number per volume per time.
- M is the suspension density in number (expressed as mass per volume).
- N is the agitation rate (expressed as a frequency or a velocity).

Secondary nucleation can occur within the metastable limit.

24.4.2 Growth Kinetics

Crystal growth theory and mechanisms have been well described in several references. For the practicing engineer working on design and scale-up issues, this discussion is simplified below into the content that is typically required to analyze commonly used pharmaceutical crystallization modes.

Like nucleation, crystal growth is driven by a supersaturation driving force, with Eq. (24.15) illustrating a commonly used rate expression:

$$\frac{dm}{dt} = k_{GM} A_c \sigma_l^\gamma \qquad (24.15)$$

where

- k_{GM} is a temperature-dependent rate constant.
- A_c is the surface area of solids present in solution.
- γ is the order of growth.
- m is the mass of solute.

Growth occurs either on material that has already been generated by nucleation or on material that has been purposefully added to the solution. Purposefully added material is referred to as "seed." Seeding is frequently employed in pharmaceutical crystallization processes to control crystalline form and physical properties, especially particle size. Seed material may be prepared from an alternative processing method, such as milling, or may be taken from one batch of material and added to a subsequent batch.

If a system exhibits growth as the primary mechanism for crystal mass formation, the amount of seed added controls the particle size of the product. A common expression used to interpret the relationship of seed amount to particle size for batch crystallization processes is expressed by the proportionality shown below:

$$\frac{m_s}{m_p} \propto \left(\frac{d_s}{d_p}\right)^n \qquad (24.16)$$

In the proportionality,

- m_s represents the mass of seed.
- m_p represents the mass of product.
- d_s and d_p represent sizes of the seed and product, respectively.

Typically, d_{50} or d_{90} values from laser light diffraction measurement or sieve measurement are used for the size terms. The exponent term n is related to the morphology of the crystal. For a perfectly spherical crystal, the exponent n would equal 3, and the proportionality would be an equality. In practice, the relationship between seed amount and particle size at the end of the crystallization is obtained by an empirical regression of the seed response curve (prepared by running the process at several seed loading [m_s/m_p] values and trending versus product size for several seed types). The preparation of seed response curves and subsequent analysis is described by Table 24.6.

After generation of the seed response curve, the amount and size of seed required to deliver a desired particle size can be interpolated from the resultant data/correlations. When

TABLE 24.6 Method for Generation of Seed Response Curves

Step Number	Description of Step
1	Two to three different types of seed are generated by the following methods: (i) taking existing crystals from the as-is crystallization; (ii) taking prepared crystals and performing a particle size reduction step, such as milling, micronization, or sonication (can also be done using a mortar and pestle or a blender); and (iii) crystallizing material in a different way (e.g. use of other solvents, other modes of crystallization). Each material is measured for particle size by an appropriate technique. The seed loading and seed sizes employed should vary by at least one order of magnitude
2	At least three experiments are performed with each seed at different seed loadings. The experiments are performed by seeding within the metastable limit, allowing the solution to fully desupersaturate. For the remainder of the crystallization, supersaturation is generated very slowly until the crystallization has reached its completion (e.g. slow cooling rate or antisolvent addition rate). Supersaturation and particle size are monitored to ensure that the mechanism is growth throughout the crystallization
3	The product is harvested and weighed, giving the m_s/m_p term. The product is then measured for particle size
4	A seed response curve plots the size of the product versus the seed loading (m_s/m_p) for each different size. The particle size as a function of seed loading can be fit through a variety of empirical functions

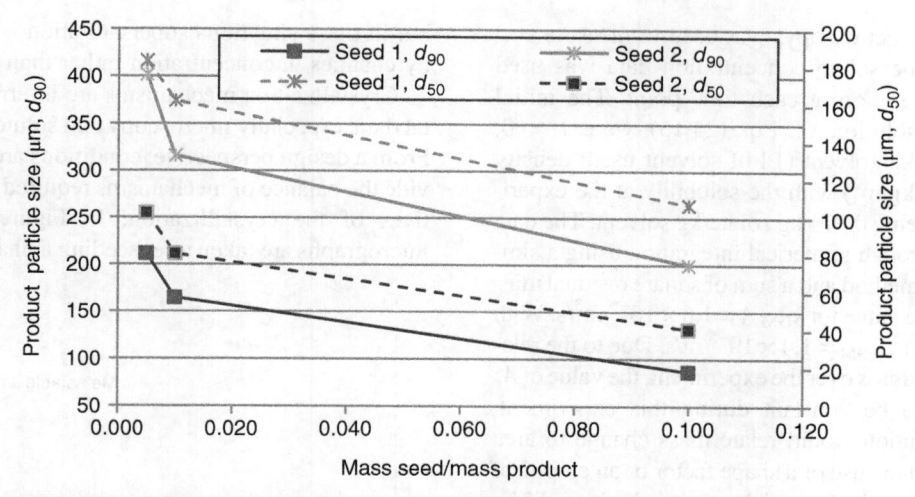

FIGURE 24.14 Typical seed response curve plot. Seed 1 in this case has a d_{90} value of 6.1 μm and a d_{50} of 2.2 μm, while Seed 2 has a d_{90} of 11.5 μm and a d_{50} of 4.4 μm. Both d_{90} and d_{50} values exhibit the anticipated response.

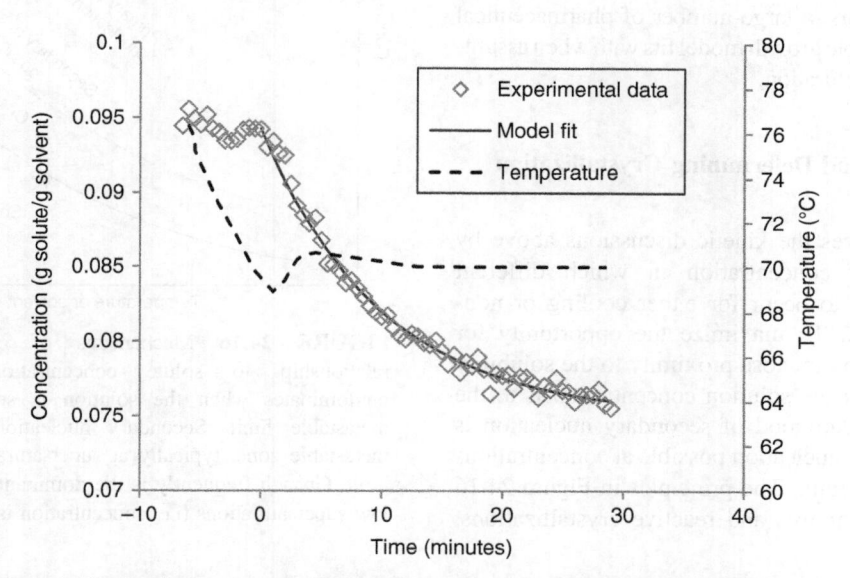

FIGURE 24.15 Crystal growth kinetics from concentration data. Crystallization seeded at time 0. Details are provided in the text. A finite differences approach was used to solve for the growth rate constant, $k_{GM} = 1.4 \times 10^{-5}$ m/s, with seed surface area $A_c = 0.116$ m^2 and solvent volume $V_{solvent} = 1$ l.

conducting these experiments, it is important to conduct the experiments at supersaturations within the metastable limit, so that growth is the predominant mechanism. An example of a seed response curve is illustrated in Figure 24.14.

In performing seed response curve experiments, the kinetics of crystal growth can also be measured. Growth kinetics are valuable in crystallization design, as they can be used to determine the required rates of cooling, antisolvent addition, distillation, and reagent addition for the corresponding crystallization mode. Many methods for the measurement of crystal growth kinetics can be used [2, 3]. One useful method for growth kinetics determination in batch systems is through the measurement of solute concentration as a function of time following the addition of seed to a supersaturated solution. Varying the seed amount changes the "area" term in Eq. (24.15), allowing the determination of the rate constant. The area term is often indicated using a "specific surface area" measurement determined through nitrogen adsorption methods or by understanding the size distribution of a material and the shape factor through which an area of a material can be related to its mass or characteristic length. Measurement of the rate constant as a function of temperature enables a correlation over the entire path of a crystallization.

Figure 24.15 displays a typical crystal growth kinetic experiment and associated data fit. For this particular case, an online method (measurement of concentration by reflectance infrared spectroscopy) was used to measure solute concentration. The solute concentration data was used to calculate supersaturation at each time point. The initial conditions for the solution of Eq. (24.15) were $t = 0$, $C = 0.094$ kg solute/kg solvent (1 l of solvent used; density of the solvent = 800 kg/m^3) with the solubility at the experimental temperature being 0.074 kg solute/kg solvent. The data was adequately fit through numerical integration using a simple finite differences method and a sum of square residual minimization to estimate a value for $k_{GM} A = 1.6 \times 10^{-6}$ m^3/s. With a seed area of 0.116 m^2, $k_{GM} = 1.4 \times 10^{-5}$ m/s. Due to the relatively small change in mass over the experiment, the value of A_c was approximated to be constant during this experiment. A more rigorous solution would relate mass change to area change either through the use of a shape factor or an empirical correlation, or alternatively an initial rates method would be used with a constant area approximation. Crystallization growth rate orders are often low, with values between 0.5 and 1. In the experience of the authors, a large number of pharmaceutical systems exhibit reasonable growth model fits with when assuming first order in supersaturation.

24.4.3 Controlling and Determining Crystallization Mechanisms

Figure 24.16 summarizes the kinetic discussions above by displaying regions of concentration in which different mechanisms are likely to occur for either cooling or nonsolvent crystallizations. To maximize the opportunity for crystal growth, operation in close proximity to the solubility is preferred. The nearer the solution concentration is to the metastable limit, the likelihood of secondary nucleation is increased, with primary nucleation possible at concentrations beyond the metastable limit. The principles in Figure 24.16 are applicable to evaporative and reactive crystallizations, but in these situations supersaturation is typically generated by changes in concentration rather than solubility.

Crystallization mechanisms are inferred from experimental data, especially microscopy and solute concentration data. From a design perspective, conditions are sought, which provide the balance of mechanisms required to deliver the objectives of the crystallization. In Figure 24.17, successive micrographs are taken after seeding a pharmaceutical product

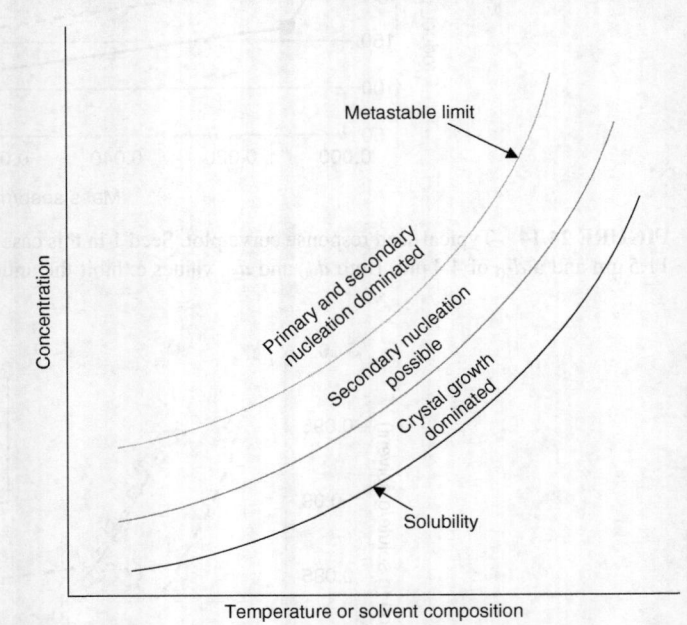

FIGURE 24.16 Mechanisms of crystallization and their relationship to solute concentration. Primary nucleation predominates when the solution is supersaturated beyond the metastable limit. Secondary nucleation can occur within the metastable zone, typically at supersaturations near the metastable limit. Growth frequently is the dominant mechanisms at relatively low supersaturations (i.e. concentration is close to the solubility).

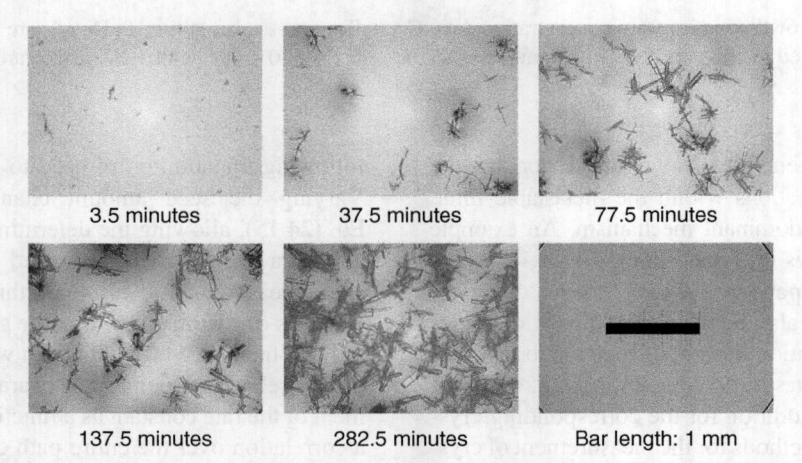

FIGURE 24.17 Illustration of crystal growth mechanism observation by optical microscopy. The processes is seeded at time 0 with the process held isothermally for 282.5 minutes. Fine particles disappear as coarse particles appear, which grow larger throughout the duration of the experiment with no evidence of fine reappearance [17]. *Source*: Reproduced with permission of Elsevier.

within the metastable zone. As observed in the figure, the seed materials grow successively larger with time with no new fine crystals are observed. This is the type of behavior representative of a growth-dominated system and indicates that the crystallization is operating sufficiently close to the solubility curve to minimize secondary nucleation.

In Figure 24.18, micrographs are overlaid with solute concentration data for a system that exhibits both growth and secondary nucleation. The solution was supersaturated beyond the metastable limit before adding seeds. Some crystal growth is observed immediately after seeding. After a growth period where seed material has clearly disappeared and been enlarged through growth, a change is seen in the slope of the solute concentration curve, indicating a mechanism change from growth to nucleation. Upon further desupersaturation, the crystals have not grown larger, and the habit has slightly changed from a columnar habit to more of a thin plate habit (the form has not changed). The change in the rate of desupersaturation combined with the change in crystal habit and the lack of crystal enlargement is clear evidence of secondary nucleation.

Of particular use in mechanism detection are process analytical instruments that detect particulate matter, such as in-line microscopy or Lasentec™ focused beam reflectance microscopy. Overviews of the utility of FBRM in crystallization mechanism inference can be found at [17].

When considering a design for a crystallization that will encourage growth, a recommended starting point for a design is to add seed at a concentration no more than midway between the solubility curve and the metastable limit and to maintain this minimum proximity to the solubility curve for the remainder of the crystallization process. Initial experiments are performed using microscopy, concentration, and online particle detection methods to ensure the selected conditions are indeed producing the desired mechanism. Conditions are then verified at larger scale to ensure that changes in mixing or equipment geometry and materials of construction have not altered the mechanistic behavior. Acicular habits, in particular, are highly prone to contact nucleation as they can easily be broken.

While crystal growth and nucleation are most frequently observed mechanisms, other important mechanisms are possible. These include oiling and aggregation or agglomeration. In the case of oiling, supersaturation is typically in great excess of the metastable limit or the solution sufficiently concentrated with nucleation inhibiting impurities that the solute forms a liquid phase consisting of a solvent–solute concentrate rather than a stable crystalline form. Oils are metastable and may crystallize spontaneously with sufficient holding times. Aggregation occurs when two crystals collide in solution and adhere to each other through surface–surface interactions. Once an aggregate forms, it can either be disrupted with the crystals regaining their individual identity (this typically occurs through fluid shear), or the crystals may fuse

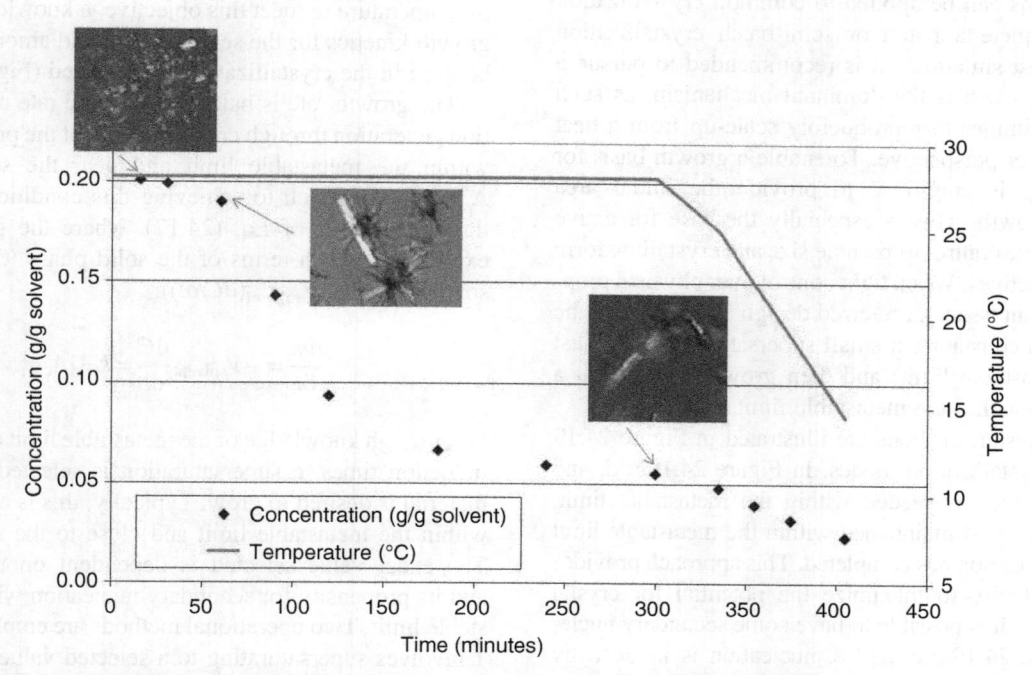

FIGURE 24.18 A crystallization experiment exhibiting both growth and secondary nucleation mechanisms. Initially, seeds grow as indicated by the presence of large crystals shortly after seeding. With additional time, a change in slope during desupersaturation is observed, indicating a change in mechanisms to secondary nucleation.

together due to growth that links the two surfaces. When crystals fuse together, the result is frequently called an agglomerate. Aggregation is often severe in processes in which nucleation occurs at high supersaturations and is common for acicular habits or for crystals with high specific surface areas. Severe aggregation is often manifested as an immobile slurry that does not mix well and under certain conditions can result in "shelves" of solid on internal reactor equipment. As aggregation and agglomeration can affect processability, physical properties, and mixing scale-up, the approach is often taken to minimize supersaturation to reduce the driving force for aggregate or agglomerate formation. In certain instances, aggregation or agglomeration is purposefully attempted to prepare a material that behaves like a small particle from a pharmaceutics perspective (e.g. rapid dissolution in a dosage form) but behaves like a larger particle from a manufacturability perspective (e.g. short filtration times). Examples of the favorable use of aggregates or agglomerates include spherical crystallizations [18]. The design of a spherical crystallization process is highly dependent on the properties of a molecule and often must be established on a trial-and-error basis.

24.5 APPLICATION OF SOLUBILITY AND KINETICS DATA TO CRYSTALLIZATION MODES

The kinetic and thermodynamic principles described in the previous sections can be applied to common crystallization modes to complete a batch or semi-batch crystallization design. For most situations, it is recommended to pursue a design where growth is the dominant mechanism, as such processes are simpler to reproducibly scale-up from a heat and mass transfer perspective. To enable a growth basis for design, seeding is employed to provide the initial area required for growth. This is especially the case for active ingredients where control of particle size and crystalline form are the key objectives. When tight control over physical properties is less of an issue, a preferred design approach may be to nucleate at a constant but small supersaturation (e.g. just outside of metastable limit) and then grow the nuclei at a supersaturation within the metastable limit.

These two design methods are illustrated in Figure 24.19 for common crystallization modes. In Figure 24.19b, d, and f, the crystallization is seeded within the metastable limit, and supersaturation is maintained within the metastable limit until the crystallization has completed. This approach provides the best opportunity to maximize the potential for crystal growth, although it is possible to have some secondary nucleation. In Figure 24.19a, c, and e, nucleation is induced by increasing the supersaturation beyond the metastable limit, but supersaturation is then controlled to ensure that the rest of the crystallization occurs close to the solubility line.

A common error in crystallization design and scale-up is the generation of supersaturation to levels where nucleation is a dominant mechanism. This is caused by a lack of understanding of the time scales over which controlling process parameters need to be changed. This section addresses simple approaches to estimating the time scales for the controlling parameters in each major crystallization mode:

- Rate of temperature change (cooling crystallizations).
- Rate of addition of antisolvent (antisolvent crystallizations).
- Rate of reaction through control of temperature or addition of reactants (reactive crystallizations).
- Rate of solvent removal (evaporative crystallizations).

Only simple methods for the estimation of process parameters are provided. There are many published examples of more rigorous approaches to solving similar problems, particularly the work of the Bratz[1] and Rawlings[2] research groups, which are recommended for further study.

24.5.1 Cooling Crystallization

Seeded cooling crystallizations represent perhaps the simplest design approach for achieving consistent crystallizations while minimizing scale-up challenges. The challenge is to balance the growth rate with the rate of supersaturation generation. In order to understand the required rate of change of temperature to meet this objective, a knowledge of crystal growth kinetics for the seed material and amount planned to be used in the crystallization are required (Figure 24.15).

The growth rate is balanced with the rate of supersaturation generation through cooling such that the process remains within the metastable limit and near the solubility line. A simple approach to achieving this condition is achieved through the use of Eq. (24.17), where the growth rate is expressed both in terms of the solid phase (dm/dt) and the solution phase ($V_{solvent}(dC/dt)$):

$$\frac{dm}{dt} = -V_{solvent}\frac{dC}{dt} = k_{GM}A_c\sigma_1^g \qquad (24.17)$$

Through knowledge of the metastable limit and nucleation induction times, a supersaturation is selected at which the material is desired to grow. Typically, this is a concentration within the metastable limit and close to the solubility line. The exact value selected is dependent on the compound and its propensity for secondary nucleation within the metastable limit. Two operational methods are employed. Method 1 involves supersaturating to a selected value, seeding, and

---

[1] Bratz [Online]. http://brahms.scs.uiuc.edu.
[2] Rawlings [Online]. http://jbrwww.che.wisc.edu.

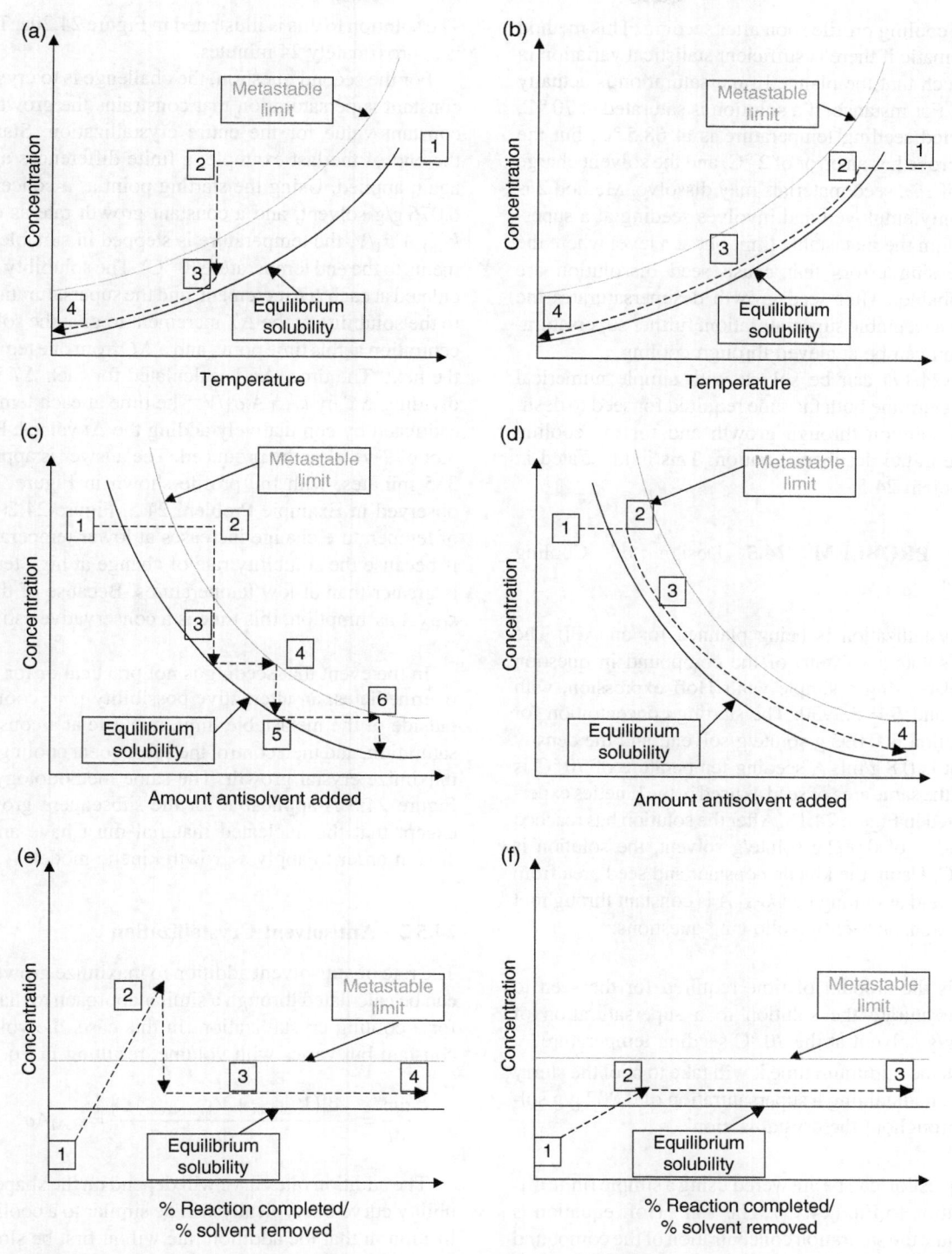

FIGURE 24.19 Frequently used crystallization design modes in pharmaceutical processes. (a) and (b) represent cooling crystallizations, (c) and (d) represent crystallizations induced by antisolvent addition, and (e) and (f) represent both reactive and evaporative crystallizations, which can be similar in terms of behavior. (a), (c), and (e) represent situations where initial crystal mass is generated by primary nucleation, with growth occurring after the nucleation event through controlled generation of supersaturation. The amount of nucleation is dependent on how much supersaturation is generated prior to crystallization. (b), (d), and (f) represent seeded crystallizations, where seed is added in the metastable zone. In each plot, (1) represents the starting point of the crystallization, where all material is dissolved. (2) represents the point at which crystallization mass is first generated, either by seeding or by nucleation. The highest number on each plot represents the point at which the crystallization is complete.

initiation of a cooling profile soon after seeding. This method can be problematic if there is sufficient statistical variation in the process such that the planned supersaturation is actually not achieved. For instance, if a solution is saturated at 70 °C, and the selected seeding temperature is at 68.5 °C, but the temperature probe has an error of 2 °C, and the solvent charge has an error of 1%, seed materials may dissolve. Method 2 is more commonly employed and involves seeding at a supersaturation within the metastable limit but at a level where the typical processing errors that cause seed dissolution are highly improbable. After seed growth desupersaturates the solution to an acceptable supersaturation, further supersaturation generation can be achieved through cooling.

Equation (24.17) can be solved with simple numerical solutions to determine both the time required for seed to desupersaturate a solution through growth and for the cooling rates after the initial desupersaturation. This is illustrated in Example Problem 24.5.

EXAMPLE PROBLEM 24.5 Design of Cooling Crystallization

A cooling crystallization is being planned for an API. The solubility (g solute/g solvent) of the compound in question can be modeled using a simple van't Hoff expression, with $A = -3773.0$ and $B = 8.3930$. The starting concentration for the crystallization is 0.095 g solute/g solvent, and the density of the solvent is 0.8 g/ml. A seeding temperature of 70 °C is selected with the same area of seed as used in the kinetics experiment illustrated in Figure 24.15. After the solution has reached a supersaturation of 0.002 g solute/g solvent, the solution is cooled to 0 °C. Using the kinetic constant and seed area from Figure 24.15, and assuming that $k_{GM} A$ is constant throughout the crystallization, answer the following questions:

1. What is the amount of time required for the seed to desupersaturate the solution to a supersaturation of 0.002 g/g solvent at the 70 °C seeding temperature?
2. What is the minimum time it will take to cool the slurry to 0 °C, maintaining a supersaturation of 0.002 g/g solvent throughout the crystallization?

The first question can be answered using a simple finite differences solution to Eq. (24.15). The van't Hoff equation is used to calculate the saturation concentration of the compound at 70 °C (0.074 g/g solvent). This allows a calculation of supersaturation at time 0, which can be multiplied by the rate constant and by an appropriately small Δt to provide a value for ΔC, which can be subtracted from the initial concentration to make a new concentration for the next time step. The new concentration is used to calculate a new supersaturation, from which a new ΔC is calculated, and so on until the solution has reached a supersaturation level of 0.002 g solute/g solvent.

The solution to this is illustrated in Figure 24.20a. The answer is approximately 24 minutes.

For the second question, the challenge is to crystallize at a constant supersaturation that constrains the growth rate to a constant value for the entire crystallization. Starting from the end of the last example, a finite differences approach is again applied. Using the starting point as a concentration of 0.076 g/g solvent, and a constant growth rate as defined by $k_{GM} A \sigma_1/V$, the temperature is stepped in suitable ΔT increments to the end temperature (0 °C). The solubility is then calculated at each ΔT increment, and the supersaturation is added to the solubility at the ΔT increment to give the solution concentration at this time point, and a ΔC from one temperature to the next. The time, Δt, is calculated for each ΔT interval by dividing ΔC by $k_{GM} A \sigma_1/V$. The time at each temperature is estimated by cumulatively adding the Δt values. From this a plot of T vs. t can be produced. The answer is approximately 355 minutes, with the profile shown in Figure 24.20b. As observed in Example Problem 24.5, Figure 24.20b, the rate of temperature change increases at lower temperatures. This is because the solubility rate of change at high temperatures is greater than at low temperatures. Because of the constant $k_{GM} A$ assumption, this time is a conservative estimate.

In the event that seeding is not practical or for the case of intermediates, an alternative possibility is to cool to a point outside of the metastable limit, nucleate at a constant supersaturation, and then control the subsequent cooling in order to maximize crystal growth. The same methodology shown in Figure 24.5 is applicable for the subsequent growth phase, except that the nucleated material must have an estimated area in order to apply a growth kinetic model.

24.5.2 Antisolvent Crystallization

The rate of antisolvent addition to maximize growth potential can be calculated through a similar approach to that described for a cooling crystallization. In this case, the volume is not constant but varies with volume, resulting in Eq. (24.18):

$$\frac{dm}{dt} = -\frac{d[(V_{solvent} + V_{antisolvent})C]}{dt} = k_{GM} A \sigma_1^g \quad (24.18)$$

The addition rate curve will depend on the shape of the solubility curve but typically will be similar to a cooling crystallization in that the addition rate will at first be slow and will ramp up near the end of the crystallization [19]. As with a cooling crystallization, the first step is to add antisolvent to generate a supersaturation within the metastable limit and then calculate the time needed to desupersaturate for a given seed load. The next step is then to calculate the rate of antisolvent addition. This can be done through a numerical integration of Eq. (24.18). It is first necessary to select a supersaturation appropriate to maintain crystal growth using metastable limit

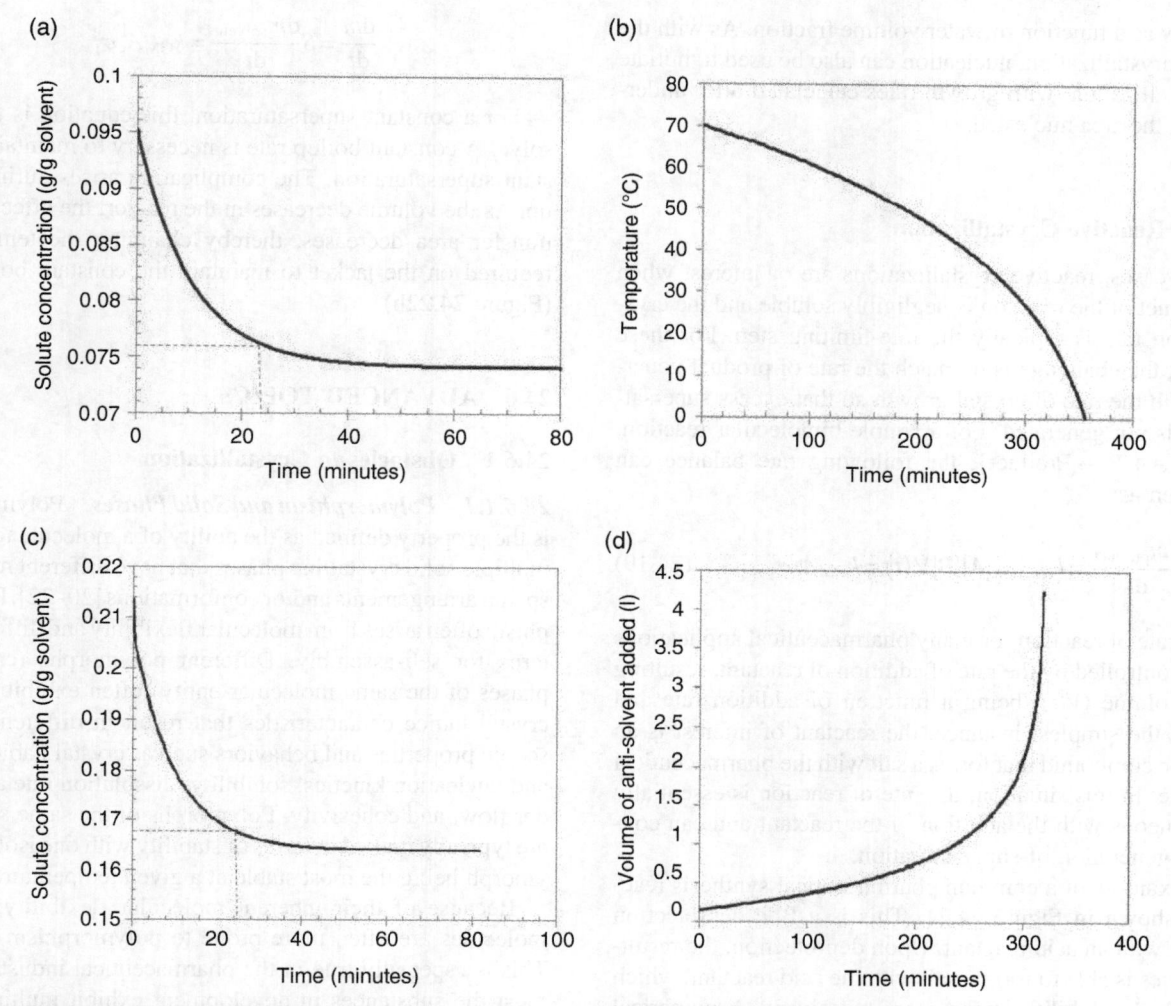

FIGURE 24.20 Calculated crystallization time scales and rate changes for a cooling crystallization (a) and (b) and an antisolvent addition (c) and (d). The cooling crystallization is described by Example Problem 24.5, and the antisolvent example is described in the text. For (a) and (b), $k_{GM} A_c$ = constant = 1.6×10^{-6} m^3/s and $A_c = 0.116$ m^2. The solubility was represented by a simple van't Hoff correlation with $A = -3773.0$ and $B = 8.390$, and the solvent volume was 1 l. Seeding was performed at 70 °C, with the criteria that a supersaturation, σ_1, of 0.002 g/g solvent be reached during isothermal seed growth (a), and with that value maintained during cooling (b). For (c) and (d), $k_{GM} A_c$ = constant = 1×10^{-5} m^3/s, with $A_c = 0.71$ m^2. The initial volume of methanol was 6 l, with 0.15 l water added to supersaturate. Seed was isothermally grown until a supersaturation of 0.004 g/g solvent was achieved (c), and then the balance of water, 4.1 l, was added isothermally while maintaining this supersaturation (d).

data. A simple finite difference approach involves starting at this supersaturation and calculating the growth rate at this supersaturation. The volume is then incremented in small intervals, with the change in solute mass calculated over each volume interval through the difference of the solubility times the mass of solvent across the volume interval. The time required for each added increment of antisolvent is calculated by dividing the change in mass over a time interval by the growth rate. The results of such a calculation are shown in Figures 24.20c and d for a system in which water is added to a methanol solution containing an active ingredient, again assuming that $k_{GM} A$ is constant over the entire crystallization

(0.6 g/min). For this particular scenario, the solvent is methanol and the antisolvent is water. The solubility of the binary system is described by an exponential function, $S = 0.22 \exp(-12.04 \phi_{water})$, where ϕ_{water} is the water volume fraction. Initially, 1 kg of material is dissolved in 6 l of methanol, and 0.15 l of water is added to generate supersaturation at 20 °C. The solution is then held until a supersaturation of 0.004 g/g solvent is achieved. Antisolvent is then added at a rate such that this supersaturation is maintained to achieve a total added volume of 4.25 l. The result is a seed hold time of approximately 24 minutes and a total addition time of approximately 317 minutes. The shape of the addition curve mirrors the solubility as a function of water volume fraction. As with the cooling crystallization, nucleation can also be used to initiate the crystallization, with growth rates calculated after understanding the area nucleated.

24.5.3 Reactive Crystallization

In most cases, reactive crystallizations are of interest when the product of the reaction is negligibly soluble and the crystallization rate is typically the rate-limiting step. For these systems, the challenge is to match the rate of product generation with the rate of crystal growth so that excess supersaturation is not generated. For a simple bimolecular reaction, where $\{A + B \rightarrow \text{Product}\}$, the following rate balance can be written as:

$$\frac{dm_{product}}{dt} = k_{reaction}[A][B]V(t) = k_{GM}A_c\sigma_1^g \quad (24.19)$$

The rate of reaction for many pharmaceutical applications can be controlled by the rate of addition of reactant, resulting in the volume ($V(t)$) being a function of addition rate and time. In the simplest instance, the reactant of interest is an ionizable compound that forms a salt with the pharmaceutical molecule. In this situation, the rate of reaction is essentially instantaneous with the addition of the reactant and can control the generation of supersaturation.

An example of a common pharmaceutical synthesis reaction is shown in Figure 24.21. This is a Boc deprotection reaction with an acid reactant. Upon deprotection, the resultant species is able to form a salt with the acid reactant, which has limited solubility in the reaction solvent. A numerical solution to Eq. (24.19) is illustrated in Figure 24.22a for this example.

24.5.4 Evaporative Crystallization

In evaporative crystallization, the solubility is constant, so the rate of supersaturation generation is proportional to the amount of solvent mass removed (dm/dV):

$$\frac{dm}{dt} = S\frac{dm_{solvent}}{dt} = k_{GM}A_c\sigma_1^g \quad (24.20)$$

For a constant supersaturation, this equation is trivial to solve. A constant boilup rate is necessary to maintain a constant supersaturation. The complication arises during scale-up. As the volume decreases in the reactor, the effective heat transfer area decreases, thereby changing the temperature required on the jacket to maintain the constant boilup rate (Figure 24.22b).

24.6 ADVANCED TOPICS

24.6.1 Obstacles to Crystallization

24.6.1.1 Polymorphism and Solid Phases Polymorphism is the property defined as the ability of a molecule to exist in multiple solid crystalline phases that have different molecular spatial arrangements and/or conformations [20–22]. Polymorphism often arises from molecular flexibility and different patterns for self-assembly. Different polymorphs (crystalline phases of the same molecular entity) often exhibit different crystal lattice characteristics that result in different macroscopic properties and behaviors such as crystal habit, growth and nucleation kinetics, solubility, dissolution rate and powder flow, and cohesivity. Polymorphs of the same substance are typically ranked in terms of stability with one isolated polymorph being the most stable at a given temperature.

Because of their inherent molecular flexibility, organic molecules are often more prone to polymorphism [23–28]. This is especially true in the pharmaceutical industry where most the substances in development exhibit multiple polymorphs [29].

From an industrial perspective, polymorphism represent two main challenges related to product quality and intellectual property (IP), respectively:

- Product quality: The majority of drug substances developed in the pharmaceutical industry are crystalline solids [30]. Because bioavailability and hence exposure to the medicine depends on solubility and dissolution, it

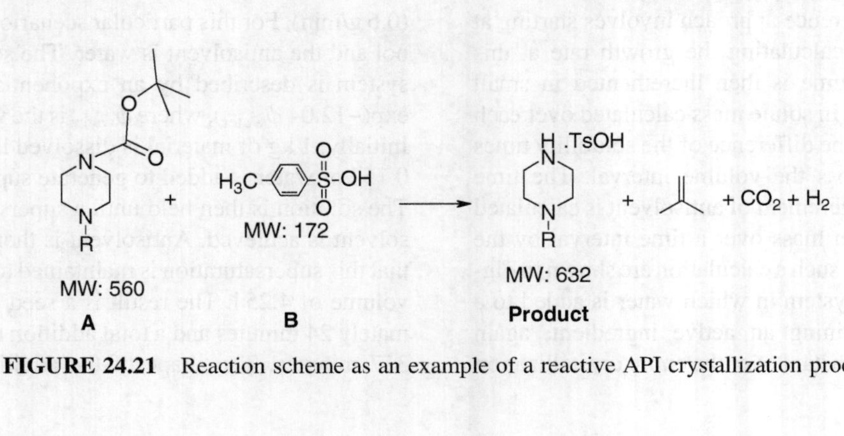

FIGURE 24.21 Reaction scheme as an example of a reactive API crystallization process.

CRYSTALLIZATION DESIGN AND SCALE-UP 559

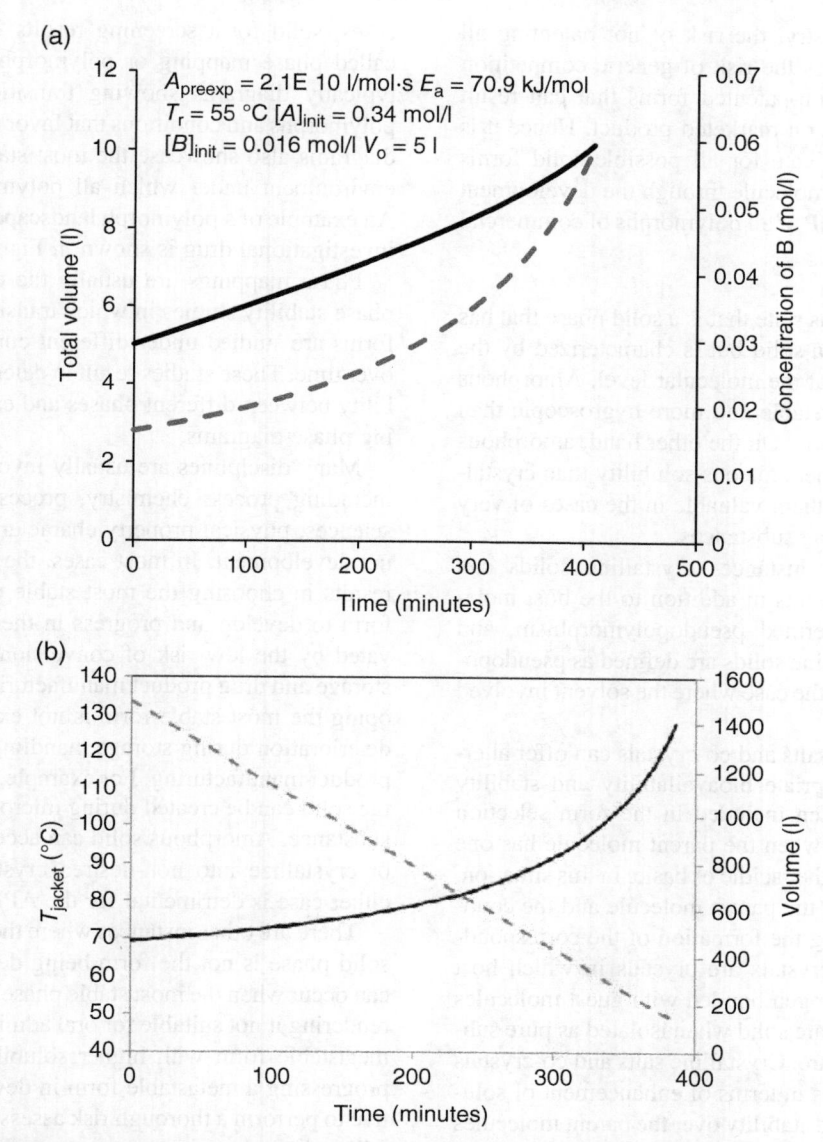

FIGURE 24.22 Examples illustrating constant supersaturation control for (a) reagent addition for a reactive crystallization and (b) jacket temperature for an evaporative crystallization. The reaction for (a) is shown in Figure 21, run isothermally at 55 °C (E_a = 70.9 kJ/mol, A_{preexp} = 2.1 × 10^{10} l/mol s). Starting amount of compound A is 1 kg in 5 l solvent. Initially, the reaction was run to approximately 5% completion, through an addition of 5 mol % compound B, and a product concentration of 0.004 g/g solvent was achieved after seeding and seed growth. The balance of B was then added in a feed solution with a concentration of 0.336 mol/l. Compound B was added such that a supersaturation of 0.004 g/g solvent was maintained. $k_{GM} A_c$ = constant = 9.8 × 10^{-6} m^3/s. For (b), the solvent was distilled from 1500 to 100 l; $k_{GM} A_c$ = constant = 2.9 × 10^{-4} m^3/s. The vessel has a linear correlation between surface area and volume A_{vessel} (m^2) = 0.003 22 $V_{solvent}$ + 0.53. The overall heat transfer coefficient was 340 W/m^2 K, the solvent was methanol, and the supersaturation upon seeding was 0.005 g/g solvent; this was maintained throughout the crystallization.

is a primary requirement that the API be present in a well-defined and pure crystalline form (i.e. polymorph). The existence of multiple polymorphs of the API can create the challenge of manufacturing the API in the pure desired polymorph. This risk of contamination with undesired solid forms is increased by the fact that differences between polymorphs' lattice energies are typical small (<3 kcal/mol, [31]) and the possibility of less stable forms crystallizing before the most stable form (Oswald's rule of stages) [32].

- IP: As mentioned above, if a candidate drug is developed as a solid, it is required that the physical (solid) form is defined and controlled. Other possible solid forms can also have value and be developed. In the pharmaceutical industry, the risk of not patenting all possible forms creates the risk of generic competition acquiring IP over non-patented forms that can result in a negative impact on marketed product. Hence it is imperative to investigate for all possible solid forms when progressing a molecule through the development milestones to insure IP of all polymorphs of commercial and scientific value.

A case is the amorphous state that is a solid phase that has mechanical properties of a solid but is characterized by the lack of short range order at the molecular level. Amorphous solids are inherently less stable and more hygroscopic than their crystalline counterparts. On the other hand, amorphous solids usually exhibit higher aqueous solubility than crystalline solids, which makes them valuable in the cases of very poorly solubility crystalline substances.

In addition, in many instances crystalline solids can include molecules of solvents in addition to the host molecule. This property is termed pseudopolymorphism, and the corresponding crystalline solids are defined as pseudopolymorphs and hydrates in the case where the solvent involved is water.

Moreover, crystalline salts and co-crystals can offer alternatives to achieve appropriate bioavailability and stability when needed and are often included in the form selection process. Salts are sought when the parent molecule has one or more ionizable sites, either acidic or basic. In this situation, a proton transfer between the parent molecule and the counter-ion is possible, making the formation of the corresponding salt achievable. Co-crystals are crystals in which host molecules (API) are hydrogen bonded with guest molecules called co-formers, which are solid when isolated as pure substances at room temperature. Crystalline salts and co-crystals can often offer advantages in terms of enhancement of solubility, dissolution rate, and stability over the parent molecules [33, 34].

Studies aimed at identifying, and isolating polymorphs is often termed polymorph or solid form screening. In most cases, solid form screening results in establishing the so-called phase mapping or polymorph landscape, which are typically diagrams showing transitions between different polymorphs and conditions that favor those transitions. These diagrams also showcase the most stable solid form and the environment under which all polymorphs can be present. An example of a polymorph landscape of a GlaxoSmithKline investigational drug is shown in Figure 24.23.

Phase mappings are usually the end result of extensive phase stability studies in which transitions between different forms are studied under different conditions and monitored over time. These studies result in determining the relative stability between different phases and establishing corresponding phase diagrams.

Many disciplines are usually involved in form selection, including process chemistry, process engineering, particle sciences, physical property characterization, and drug product development. In most cases, the form selection activity results in choosing the most stable polymorph as the solid form to develop and progress in the pipeline. This is motivated by the low risk of conversion to other forms during storage and drug product manufacturing. Nonetheless, developing the most stable form is not exempt of risk of quality deterioration during storage, handling, processing, and drug product manufacturing. For example, amorphous domains in the solid can be created during micronization of a crystalline substance. Amorphous solid can accelerate degradation and/or crystallize into non-desired crystalline domains that in either case is detrimental for the API end use.

There are other instances where the most stable crystalline solid phase is not the form being developed. This situation can occur when the most stable phase has very poor solubility rendering it not suitable for oral administration. In this case a metastable form with higher solubility is preferred. When progressing a metastable form in development, it is imperative to perform a thorough risk assessment to ensure the crystalline form does not convert to the most stable form during manufacturing and the life cycle of the substance up to patient administration to guarantee bioavailability and treatment

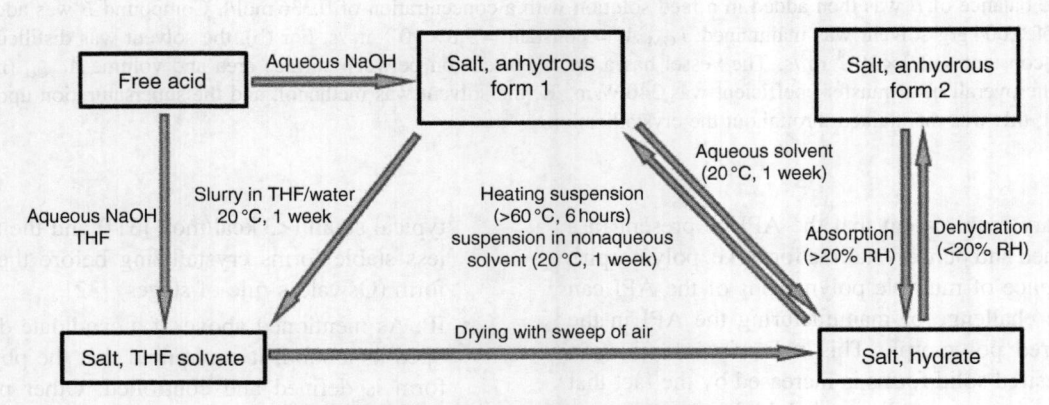

FIGURE 24.23 Example of polymorph landscape.

efficiency. Another situation where a metastable form can be selected for development can arise from challenges related to manufacturing the most stable form in pure phase, whereas the manufacturing of a less stable form offers more robustness and flexibility.

In summary, the selection of the solid form for development depends mainly on the drug substance achieving the following objectives:

- It is produced in the required quality standards including chemical and physical purities.
- It is stable through its life cycle.
- It achieves the intended exposure and patient compliance.
- It is obtained via a robust manufacturing process.

24.6.1.2 Oiling Out
Oiling out, also referred to as demixing or liquid–liquid phase separation (LLPS), is a common phenomenon that occurs during crystallization development in the pharmaceutical industry. LLPS is usually unwanted because it impedes crystallization of the drug. Nevertheless, LLPS can be advantageous for the separation of fatty acids as reported by Ref. [35]. In the pharmaceutical industry processes involving LLPS are highly undesirable because they lack sufficient engineering process controls upon scale-up and robustness. Crystallization processes that exhibit LLPS can create molten phases that stick to reactor walls and render the crystallization procedure useless by concentrating impurities in the solute-rich phase, which leads to high impurity integration in the crystals upon nucleation. Figure 24.24 shows a microscope image of a typical oiling out phenomenon.

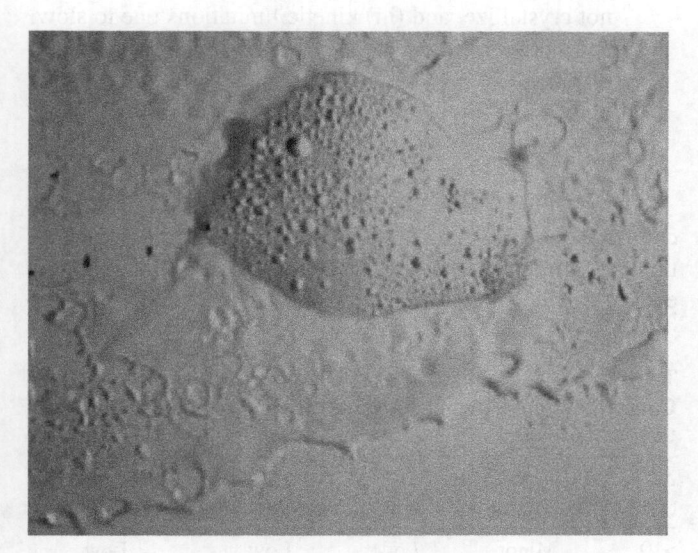

FIGURE 24.24 Microscope image of a typical LLPS. *Source*: Reprinted with permission from Derdour [36]. Copyright (2010) Elsevier.

The literature reports that there was an increase in the number of less hydrophilic and less polar drug candidates upon discovery of more potent and improved target-specific drug molecules [34, 36]. Less hydrophilic molecules have an inherent low polarity with a lack of anchoring sites, and therefore they do not easily self-assemble in an organized manner. Often, during the early development of drug candidates, seeds of crystalline forms are not available, and creation of first crystals via self-nucleation is often needed to produce a crystalline solid form. Thus, generating supersaturation is a prerequisite for obtaining a solid form. In early-stage development, limited data are available regarding the solubility of drug candidates in different solvents. Therefore, supersaturation is usually created by screening methods. Trial and error is also utilized at this stage. However, attempts to crystallize drug substances often lead to a phase separation in which the active molecule is either concentrated in one phase or distributed between many phases.

Two types of oiling out are usually found in the literature [36]:

- Liquid–liquid separation occurs where each phase contains reasonable amounts of solute. This situation can occur when a mixture of solvents is used [37].
- Liquid–liquid separation occurs where one phase contains the solvent(s) and the other phase is mainly formed by the solute in the form of a very heavy viscous oil-like phase, hence the name oiling out. This phenomenon occurs usually for high solute concentrations at moderately high temperatures.

In all cases, the determination of the phase diagram can help greatly in selecting crystallization conditions to avoid LLPS phenomenon. Once crystalline solids are isolated, oiling out can be avoided via an appropriate seeding strategy [36].

24.6.1.3 Gelation
Molecular gels are defined as 3D assemblies composed of 1D high aspect ratio structures themselves made of non-covalently linked small organic molecules. Small molecules such those most substances involved in the pharmaceutical industry (except biomolecules) that can form gels are termed low molecular-mass organic gelators (LMOGs). The 1D assemblies are thought to form due to aggregation of the small molecules upon separation from organic solutions. Because of the high aspect ratio of the 1D assemblies making up the molecular gels, the term "self-assembled fibrillar networks (SAFINs)" is usually employed to denote them [38].

Due to the non-covalent nature of the intermolecular interactions involved in SAFINs, these are often thought of as a metastable state that is readily disrupted by applying heat, dilution, shear, or other type of perturbations.

Gels are typically formed via similar phenomena as crystals: an initial aggregation is followed by nucleation, and then the nuclei align to form fibers that ultimately self-assemble into 3D SAFINs. However, it is still unclear how certain molecules tend to form gels, while others do not [38]. It is currently impossible to predict the propensity of molecule to gelify and the conditions that would favor gel formation. From the crystallization of the pharmaceutical substances perspective, gelation is an undesired phenomenon because of the following reasons:

- Gels are viscoelastic and behave like a fluid, which renders their isolation practically impossible.
- Gels are typically highly viscous and can be "non-flowing" [39], which results in increased resistance to mixing due to increased viscosity, which can result in equipment failure.
- Because of decreased ordered structure at the molecular level, gelation does not provide the impurity rejection advantage that crystal growth offers [39].

On the other hand, gels were found to be advantageous in providing enhanced means for drug delivery [40, 41] and can be used an additive in crystallization from solution to improve polymorphic selectivity [42].

24.6.1.4 Molecular Rigidity
A common characteristic of organic molecules is their ability to adopt multiple conformations due to their inherent high degree of freedom compared with inorganic substances. In theory, for solution organic chemistry, conformational change always occurs because most organic molecules contain at least one degree of freedom such as a single covalent bond that permits free rotation. Therefore, it is intuitive to expect that in most cases, organic molecules exhibit conformational changes in solution.

There is no consensus about the effect of the existence of several conformers for a given molecule on its crystallizibility, i.e. its propensity to self-assemble in a new crystalline phase (nucleation). Two general ideas can be extracted from the literature:

1. The presence of multiple conformers in solution is expected to increase the barrier to nucleation and decrease the propensity for polymorphism [25]. These authors consider that systems with higher degrees of freedom (i.e. high number of conformers) are expected to be more difficult to nucleate due to lower effective concentration of the conformer that crystallizes, which results in low "effective" supersaturation. Recently, based on ab initio calculations and NMR measurements, it was proposed this behavior is likely to occur if the energy barrier of transition between conformers is higher than approximately 10 kcal/mol [23, 24]. These authors showcased that the presence of rotomers in solution NMR spectra can indicate challenges to nucleation of the species at hand.
2. Another school of thought considers that the presence of multiple conformations in solution favors nucleation and polymorphism [26, 28, 31]. It is noted that for most organic compounds, energy difference between different conformers is of the same order of magnitude as the difference in lattice energy between polymorphs (~2 kcal). Therefore, strained conformers, i.e. conformers with higher energy, can crystallize if the resulting crystal lattice provides a better recognition pattern and/or stronger anchoring sites, resulting in low lattice energy.

With respect to crystal growth, two main possibilities are extracted from the literature with regard to the influence of the presence of multiple conformations in solution on crystal growth rate [24]:

1. In most cases, the energy barrier to conformational change is relatively low, and forces present on crystal surfaces can readily modify the conformation of the crystallizing species. Thus, it is expected that the presence of multiple conformations in solution does not affect crystal growth kinetics.
2. If the energy barrier to conformational change is relatively higher than approximately 10 kcal/mol, the effect of crystal forces on the conformation of the solute is decreased. In this case, crystal growth can be impeded due to one or the combination of the following factors: (i) lower effective supersaturation, (ii) integration site inhibition by conformers that do not crystallize, and (iii) kinetic limitations due to slow equilibration between conformers. The negative effect on crystal growth kinetics is accentuated if the crystallizing species is a high-energy conformer, i.e. the minor conformer in solution.

Table 24.7 summarizes the expected effect of multiple conformations in solution on propensity for primary nucleation, polymorphism, and crystal growth kinetics based on the energy barrier to conformational conversion.

TABLE 24.7 Expected Effect of Energy Barrier to Conformational Change on Nucleation Polymorphism and Crystal Growth

ΔE (kcal/mol)	Right Conformer	Nucleation	Polymorphism	Crystal Growth
>10	Minor	Low	Low	Low
	Major	No effect	Low	No effect
<10	NA	High	High	No effect

It is being noted that an increasing number of atropisomers (conformers with half-life higher than 1000 seconds) are being developed in the pharmaceutical industry due to enhanced activity and target specificity they may offer over more flexible molecules [43–48]. This trend may explain the increasing number of cases of molecular rigidity affecting API crystallization development and form selection [24].

24.6.2 Comprehensive Process Models and Optimization

In recent years, modeling had become increasingly employed in research and development in the pharmaceutical industry. Two main types of modeling can be extracted from the literature:

1. Theoretical and semi-theoretical modeling: The models developed are mainly based on first principles and understanding of the underlying phenomena and mechanisms involved in the processes involved. These models often require some experimental data input to identify kinetic and thermodynamic parameters. Development of theoretical models becomes increasingly challenging with complexity of the mechanisms and phenomena involved in the process (e.g. nonlinearity, large number of phenomena involved, coupling…). Often, these models require adopting some justified hypotheses to simplify the mathematical resolution. Caution should be taken to ensure the models are utilized within the ranges of parameters in which they were validated.
2. Empirical or statistical models where models developed are so-called "black box" models and where the objective is to determine an empirical relationship between outputs and inputs of the model with minimal emphasis of the phenomena and mechanisms involved. These models are usually heavily dependent on experimental data and their accuracy, and predictive power typically improves with the amount of experimental data utilized to build the models. These models have the advantage for removing the mathematical complexity. On the other hand, they present the disadvantage of providing limited improvement of process knowledge.

Modeling is increasingly becoming a core part of research and development of pharmaceutical substances and processes for their manufacture throughout the projects timelines.

24.6.3 Early-stage Development

Often, in early-stage development of pharmaceuticals, many drug candidates are transitioned from development. Material availability at this stage is limited, which results in limited experimental data being available and/or obtainable. In addition, because limited data on substance safety and efficacy is available at this stage, it is not justified to invest time and effort to obtain a full and accurate characteristic data set. This peculiarity makes predictive modeling an attractive tool to estimate key parameters that can be indicative of the substance performance and developability. For example, modeling with DFT and Gaussian can inform about molecular flexibility based only on the chemical structure. The results can inform about the crystallizibility of the molecule [24, 49]. Predictive modeling of the solubility based on packages such as NRTL-SAC and COSMO-SAC can help design focused experiments to determine accurate solubility data on only selected solvents that can save material and shorten development time. An example of predictive modeling for a marketed drug is shown in Figure 24.25 [50].

24.6.4 Late-stage Development

As the program progresses, substance under development can be withdrawn from the project for the following reasons:

- The efficacy of a pharmaceutical substance in achieving its intended physiological response is often demonstrated in earlier-stage development. That data combined with early proof of safety justify progressing the program through the development life cycle. Initial tests are usually conducted at lowest doses, and during the project progression, dosage is usually increased gradually, and more scrutiny is given to substance safety. The project removes a substance from development if negative data about its safety emerges as dosage increases.
- Often two or more drug substances are progressed in the same program with one molecule being the lead candidate. Withdrawal of one or all contingency molecules can be justified if a substance within the program displays superior efficacy and safety data, minimizing the risk of its failure.

Substances that satisfy efficacy and safety go into late-stage development. At this stage, larger quantities of the substance are needed to feed various clinical trials, material characterization, and process development efforts. At this stage, a green, efficient, and robust crystallization process is required. Identification of critical process parameters (CPPs) and process knowledge is required. The relationship between CPPs and product attributes must be established in order to establish the design space of the operation, defined as the intervals or ranges of variation of process parameters proven to consistently result in product with quality attributes within specifications. In late-stage development modeling is becoming increasingly employed and is driven by the following objectives:

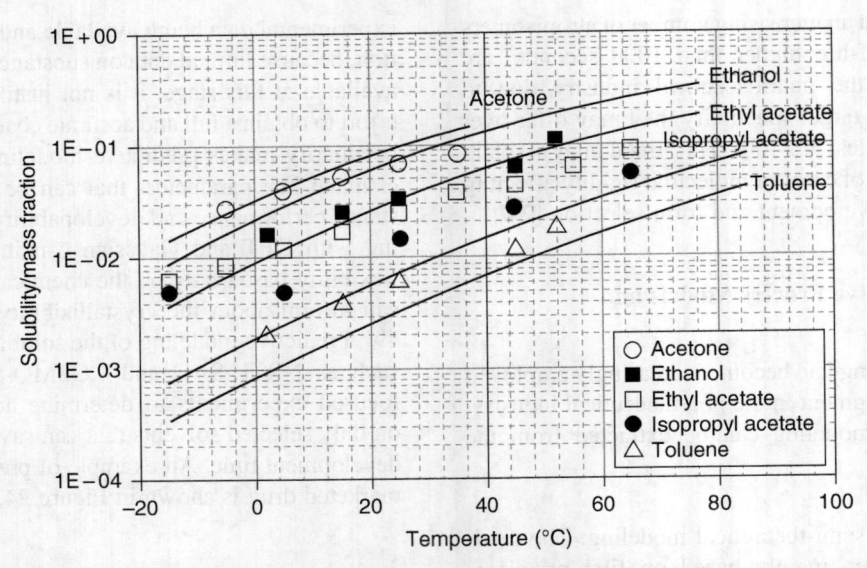

FIGURE 24.25 Predicted and measured solubility of Lovastatin in various solvents as a function of temperature. *Source*: Reprinted with permission from Tung et al. [50]. Copyright (2007) Elsevier.

- Generate process knowledge and understanding.
- Predict process behavior across and outside the design space.
- Assist with process optimization.
- Minimize experimental work.

24.6.5 Parameter Estimation

In late-stage development, theoretical and semi-theoretical modeling are preferred as they generate process knowledge and understanding as opposite to empirical models. Nevertheless, theoretical models still require the knowledge of key kinetic and thermodynamic parameters that are often determined from experiments. In the case of crystallization, several phenomena can occur during the operation. Evidently crystal growth is always present for crystallization to occur, but other phenomena such as nucleation, agglomeration, and breakage can also occur to a certain extent. In addition, these phenomena can have different mechanisms depending on the system at hand and/or the stage of the operation. Often, only one or two phenomena are predominant, which allows for modeling simplification. Hence experiments utilizing process analytical tools (PATs) are typically designed to identify kinetic parameters for main phenomena occurring during the particle-forming step. Several software packages were developed in recent years to assist scientists and engineers with the estimation of kinetic and thermodynamic parameters that can be utilized in predictive modeling. Such packages are available from companies such as Scale-up System, ASPEN, and G-prom. Examples of kinetic parameters estimation for growth-dominated crystallization can be found in Ma and Wang [51] and Derdour and Chan [52].

24.6.6 Predictive Modeling

Once kinetic and thermodynamic parameters needed to build a model are identified, they are utilized to build predictive models, and the resulting computations are compared with experimental data. This exercise is part of model validation that informs about the predictive power of the model; in other words, how accurate is the model prediction within the process parameters ranges for which it is intended to be valid.

The interval of confidence of computation depends on the criticality of the predicted response. Typically, a high interval of tolerance is required when the predicted response is a CQA.

Once a model to predict a given response is established and validated, it can provide a full landscape for the relationship response/process parameters and be utilized to ascertain the design space and be part of quality by design. In this situation, modeling can also justify the design space and provide confidence about process behavior. Besides chemical and physical quality, manufacturability is an important aspect in drug development. Powder flowability is a required attribute to enable a robust drug manufacturing process. The Carr index is often utilized as a descriptor to assess flowability and is often correlated with the aspect ratio in the case of elongated crystals. An example of predictive modeling of the aspect ratio of an investigational drug is shown in Figure 24.26. In that study [52], predictive modeling was utilized to determine aspect ratio range that would result in an acceptable powder flowability based on the predicted Carr

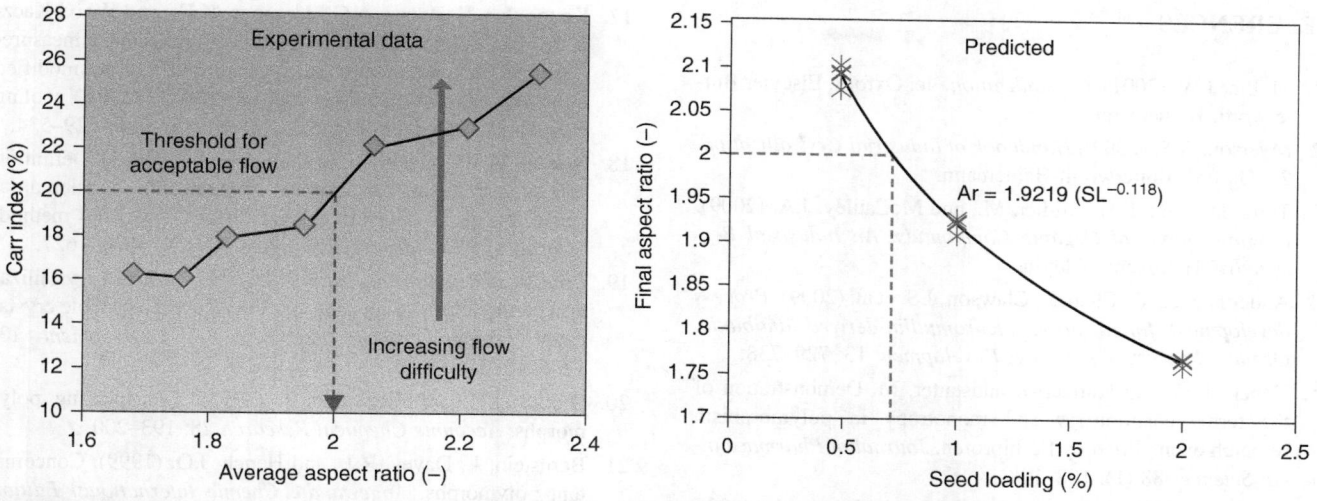

FIGURE 24.26 Predictive modeling of the aspect ratio of an investigational drug (CQA: Ar, aspect ratio: average aspect ratio of crystals [–]; CPP: SL, seed loading [–]). *Source*: Reprinted with permission from Derdour and Chan [52]. Copyright (2015) John Wiley & Sons.

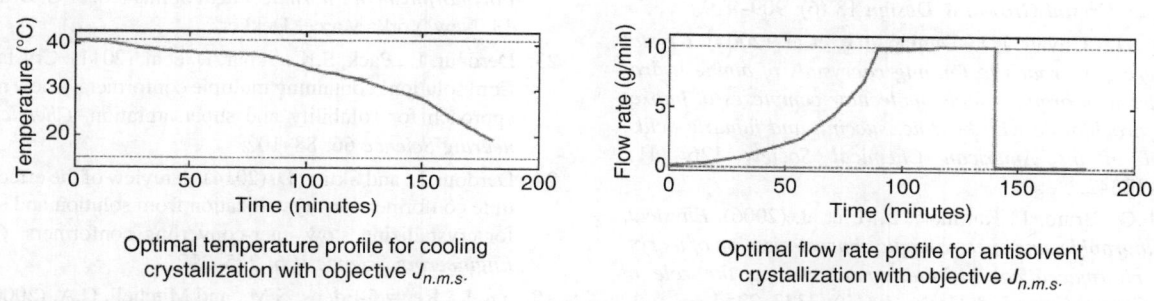

FIGURE 24.27 Optimized crystallization trajectories for Lovastatin [53]. *Source*: Reprinted with permission from Nagy et al. [53b]. Copyright (2008) Elsevier.

index. In this case, predictive modeling confirmed that the seed loading was the only CPP with respect to the aspect ratio (and powder flowability).

24.6.7 Optimization

Another application of modeling is optimization. This exercise is intended to identify operating conditions to attain the following objectives in order of priority:

- Obtain the best product quality.
- Enhance process greenness and robustness.
- Minimize operating costs.

Optimization is typically approached utilizing process parameters at their center point levels. Model predictions of a given response (typically a CQA) are then determined by varying the levels of the inputs (process parameters). The mathematical principle of optimization is usually based on solving first derivative equations for the response where the identified solutions (input parameters) lead to a minimal value of the response.[3] In other words, optimization is essentially the resolution of the following equation:

$$\frac{d(R=f(I_i))}{dt}=0 \qquad (24.21)$$

For example, for the case of minimizing the level of an impurity, resolution of the equation above identifies input parameters, I_i, which result in the minimal level of that impurity in the product (response, R). Optimal conditions are identified as the process parameters that result in the best level in terms of quality of the predicted CQA. This exercise can be repeated for other objectives, but it must not be detrimental to product quality (i.e. not deteriorating CQAs levels). For crystallization, particle size distribution can be a CQA, and this is typically controlled by a suitable seeding strategy and supersaturation control. An example of optimized crystallization trajectories for a cooling and antisolvent where the optimized CPP is supersaturation is shown in Figure 24.27 [53].

---
[3] In some cases, a maximum value is desired, for example, maximizing the yield of an operation.

REFERENCES

1. Mullin, J.W. (2001). *Crystallization*, 4e. Oxford: Elsevier Butterworth-Heinemann.
2. Myerson, A.S. (2002). *Handbook of Industrial Crystallization*, 2e. Oxford: Butterworth-Heinemann.
3. Tung, H., Paul, E.L., Midler, M., and McCauley, J.A. (2009). *Crystallizations of Organic Compounds: An Industrial Perspective*. Hoboken: WileyInc.
4. Andemichael, Y., Chen, J., Clawson, J.S. et al. (2009). Process development for a novel pleuromutilin-derived antibiotic. *Organic Process Research & Development* 13: 729–738.
5. Henck, J.-O. and Kuhnert-Brandstatter, M. Demonstration of the terms enantiotropy and monotropy in polymorphism research exemplified by flurbiprofen. *Journal of Pharmaceutical Sciences* 88 (1): 103–108.
6. Zaworotko, M.J., Kuduva, S.S., McMahon, J.A. et al. (2003). Crystal engineering of the composition of pharmaceutical phases: multiple-component crystalline solids involving carbamazepine. *Crystal Growth & Design* 13 (6): 909–919.
7. Childs, S.L., Chyall, L.J., Dunlap, J.T. et al. (2004). Crystal engineering approach to forming cocrystals of amine hydrochlorides with organic acids, molecular complexes of fluoxetine, hydrochloride with benzoic, succini, and fumaric acids. *Journal of the American Chemical Society* 126 (41): 13335–13342.
8. Vogt, F.G., Brum, J., Katrincic, L.M. et al. (2006). Physical, crystallographic, and spectroscopic characterization of a crystalline pharmaceutical hydrate: understanding the role of water. *Crystal Growth & Design* 6 (10): 2333–2354.
9. Gibson, M. (2004). *Pharmaceutical Preformulation and Formulation*, 1e. Boca Raton, FL: Interpharm/CRC.
10. Byrn, S., Pfeiffer, R., Ganey, M. et al. (1995). Pharmaceutical solids: a strategic approach to regulatory considerations. *Pharmaceutical Research* 12 (7): 945–954.
11. Chemburkar, S.R., Bauer, J., Deming, K. et al. (2000). Dealing with the impact of ritonavir polymorphs on the late stages of bulk. *Organic Process Research & Development* 4: 413–417.
12. Santomaso, A., Lazzaro, P., and Canu, P. (2003). Powder flowability and density ratios: the impact of granules packing. *Chemical Engineering Science* 58 (13): 2857–2874.
13. Zhu, H., Yuen, C., and Grant, D.J.W. (1996). Influence of water activity in organic solvent+water mixtures on the nature of the crystallizating drug phase 1-theophylline. *International Journal of Pharmaceuticals* 135: 151–160.
14. Li, Y., Chow, P.S., Tan, R.B.H., and Black, S.N. (2008). Effect of water activity on the transformation between hydrate and anhydrate carbamazepine. *Organic Process Research & Development* 12: 264–270.
15. Matthews, H.B. and Rawlings, J.B. (1998). Batch crystallization of a photochemical: modeling, control, and filtration. *AIChE Journal* 44 (5): 1119–1127.
16. Larson, M. and Khambaty, S. (1978). Crystal regeneration and growth of small crystals in contact nucleation. *Industrial & Engineering Chemistry Fundamentals* 17: 160–165.
17. Kougoulos, E., Jones, A.G., Jennings, K.H., and Wood-Kaczmar, M.W. (2005). Use of focused beam reflectance measurement (FBRM) and process video imaging (PVI) in a modified mixed suspension mixed product removal (MSMPR) cooling crystallizer. *Journal of Crystal Growth* 273 (3–4): 529–534.
18. Nocent, M., Bertocchi, L., Espitalier, F. et al. (2001). Definition of a solvent system for spherical crystallization of salbutamol sulfate by quasi-emulsion solvent diffusion (QESD) method. *Journal of Pharmaceutical Sciences* 90 (10): 1620–1627.
19. Cote, A., Zhou, G., and Stanik, M. (2009). A novel crystallization methodology to ensure isolation of the most stable crystal form. *Organic Process Research & Development* 13: 1276–1283.
20. Dunitz, J.D. and Bernstein, J. (1995). Disappearing polymorphs. *Accounts Chemical Research* 28: 193–200.
21. Bernstein, J., Davey, R.J., and Henck, J.O. (1999). Concomitant polymorphs. *Angewandte Chemie International Edition* 38: 3441–3461.
22. Grant, D.J.W. (1999). Theory and origin of polymorphism. In: *Polymorphism in Pharmaceutical Solids* (ed. H.G. Brittain), 1–33. New York: Marcel Dekker.
23. Derdour, L., Pack, S.K., Skliar, D. et al. (2011). Crystallization from solutions containing multiple conformers: a new modeling approach for solubility and supersaturation. *Chemical Engineering Science* 66: 88–102.
24. Derdour, L. and Skliar, D. (2014). A review of the effect of multiple conformers on crystallization from solution and strategies for crystallizing slow inter-converting conformers. *Chemical Engineering Science* 106: 275–292.
25. Yu, L., Reutzel-Edens, S.M., and Mitchell, C.A. (2000). Crystallization and polymorphism of conformationally flexible molecules: problems, patterns and strategies. *Organic Process Research & Development* 4: 396–402.
26. Buttar, D., Charlton, M.H., Docherty, R., and Starbuck, J. (1998). Theoretical investigations of conformational aspects of polymorphism. Part 1: o-acetamidobenzamide. *Journal of the Chemical Society, Perkin Transactions* 2: 763–772.
27. Yu, L. (2002). Color changes caused by conformational polymorphism: optical-crystallography, single-crystal spectroscopy, and computational chemistry. *Journal of Physical Chemistry A* 106: 544–550.
28. Nangia, A. (2008). Conformational polymorphism in organic molecules. *Accounts of Chemical Research* 41: 595–604.
29. Arora, K.K. and Zaworotko, M.J. (2009). Pharmaceutical co-crystals: a new opportunity in pharmaceutical science for a long-known but little-studied class of compounds. In: *Polymorphism in Pharmaceutical Solids*, 2e (ed. H.G. Brittain), 282–317. New York: Informa Healthcare.
30. Morissette, S.L., Almarsson, Ö., Peterson, M.L. et al. (2004). High-throughput crystallization: polymorphs, salts, co-crystals and solvates of pharmaceutical solids. *Advanced Drug Delivery Reviews* 56: 275–300.
31. Starbuck, S., Docherty, R., Charlton, M.H., and Butter, D. (1999). A theoretical investigation of conformational aspects of polymorphism. Part 2. Diarylamines. *Journal of the Chemical Society, Perkin Transactions* 2: 677–691.

32. Brittain, H.G. (1999). Application of the phase rule to the characterization of polymorphic systems. In: *Polymorphism in Pharmaceutical Solids*, 35–72. New York: Marcel Dekker.
33. Derdour, L., Reckamp, J.M., and Pink, C. (2017). Development of a reactive slurry salt crystallization to improve solid properties and process performance and scalability. *Chemical Engineering Research & Design* 121: 207–218.
34. Serajudin, A.T.M. and Pudipeddi, M. (2002). Salt selection strategies. In: *Handbook of Pharmaceutical Salts: Properties, Selection and Use* (ed. P.H. Stahl and C.G. Wermuth), 147–172. Zürich: Verlag Helvetica Chimica Acta.
35. Maeda, K., Nomura, Y., Fukui, K., and Hirota, S. (1997). Separation of fatty acids by crystallization using two liquid phases. *Korean Journal of Chemistry* 14: 175–178.
36. Derdour, L. (2010). A method to crystallize substances that oil out. *Chemical Engineering Research & Design* 88: 1174–1181.
37. Veesler, L., Lafferrère, L., Garcia, E., and Hoff, C. (2003). Phase transitions in supersaturated drug solution. *Organic Process Research & Development* 7: 983–989.
38. Caran, K.L., Lee, D.C., and Weiss, R.G. (2013). Molecular gels and their fibrillar networks. In: *Applications, Soft Fibrillar Materials: Fabrication and Use* (ed. X.Y. Liu and J.L. Li), 1–75. Weinheim: Wiley-VCH.
39. Yin, Y., Gao, Z., Bao, Y. et al. (2014). Gelation phenomenon during antisolvent crystallization of cefotaxime sodium. *Industrial & Engineering Chemistry Research* 53: 1286–1292.
40. Langer, R. (2000). Biomaterials in drug delivery and tissue engineering: one laboratory's experience. *Accounts of Chemical Research* 33: 94–101.
41. Xing, B., Yu, C.W., Chow, K.H. et al. (2002). Hydrophobic interaction and hydrogen bonding cooperatively confer a vancomycin hydrogel: a potential candidate for biomaterials. *Journal of the American Chemical Society* 124: 14846–14847.
42. Diao, Y., Whaley, K.E., Helgeson, M.E. et al. (2012). Gel-induced selective crystallization of polymorphs. *Journal of the American Chemical Society* 134: 673–684.
43. Laplante, S.E., Fader, L.D., Fandrick, K.R. et al. (2011). Assessing atropisomer axial chirality in drug discovery and development. *Journal of Medicinal Chemistry* 54: 7005–7022.
44. Laplante, S.E., Edwards, P.J., Fader, L.D. et al. (2011). Revealing atropisomer axial chirality in drug discovery. *ChemMedChem* 6: 505–513.
45. Bringmann, G., Price Mortimer, A.J., Keller, P.A. et al. (2005). Atroposelective synthesis of axially chiral biaryl compounds. *Angewandte Chemie International Edition* 44: 5384–5427.
46. Zhou, Y.S., Tay, L.K., Hughes, D., and Donahue, S. (2004). Simulation of the impact of atropisomer interconversion of plasma exposure of atropisomers of an endothelin receptor antagonist. *The Journal of Clinical Pharmacology* 44: 680–688.
47. Albert, J.S., Ohnmacht, C., Bernstein, P.R. et al. (2004). Structural analysis and optimization of NK1 receptor antagonists through modulation of atropisomer interconversion properties. *Journal of Medicinal Chemistry* 47: 519–529.
48. Fukuyama, Y. and Asakawa, Y. (1991). Novel neurotrophic isocuparane-type sesquiterpene dimers, mastigophorenes A, B, C and D, isolated from the liverwort mastigophora didados. *Journal of the Chemical Society, Perkin Transactions* 1 (11): 2737–2741.
49. Price, S.L. (2013). Why don't we find more polymorphs? *Acta Crystallographica Section B: Structural Science Crystal Engineering and Materials* 69: 313–328.
50. Tung, H.H., Tabora, J., Variankaval, N. et al. (2008). Prediction of pharmaceutical solubility via NRTL-SAC and COSMO-SAC. *Journal of Pharmaceutical Sciences* 97: 1813–1820.
51. Ma, C.Y. and Wang, X.Z. (2012). Model identification of crystal facet growth kinetics in morphological population balance modeling of L-glutamic acid crystallization and experimental validation. *Chemical Engineering Science* 70: 22–30.
52. Derdour, L. and Chan, E.J. (2015). A model for supersaturation and aspect ratio for growth dominated crystallization from solution. *The AIChE Journal* 61: 4456–4469.
53. (a) Nagy, Z.K., Fujiwara, M., and Braatz, R.D. (2007). Recent advances in the modelling and control of cooling and antisolvent crystallization of pharmaceuticals. *Proceedings of the 8th IFAC Symposium on Dynamics and Control of Process Systems*, Cancún, Mexico (4–8 June 2007), Vol. 2, pp. 29–38.
(b) Nagy, Z.K., Fujiwara, M., and Braatz, R.D. (2008). Modelling and control of combined cooling and antisolvent crystallization processes. *Journal of Process Control* 18 (9): 856–864.

25

INTRODUCTION TO CHIRAL CRYSTALLIZATION IN PHARMACEUTICAL DEVELOPMENT AND MANUFACTURING

Jose E. Tabora, Shawn Brueggemeier, and Jason Sweeney
Product Development, Bristol-Myers Squibb, New Brunswick, NJ, USA

Michael Lovette
Drug Substance Process Development, Amgen Inc., Thousand Oaks, CA, USA

Synthetic active pharmaceutical ingredients (API) (referred to herein as drug substances or DS) may contain one or more stereo center, resulting in an undesired enantiomer (U) and potentially multiple diastereomers of the desired product (D). For simplicity, this work considers the case of a single chiral center, resulting in a single desired and undesired enantiomer. The undesired enantiomer of a drug substance may differ in biological activity, toxicology, pharmacokinetics, and metabolism than the desired enantiomer. Therefore, the control of chiral purity is an important part of process development for synthetic DS.

As shown in Figure 25.1 in the period from 1983 to 2001, the proportion of approved chiral versus non-chiral pharmaceuticals increased from 60 to 75%. Moreover, the proportion of chiral pharmaceuticals approved as pure enantiomers (v 1:1 mixtures of both enantiomers) increased from approximately 30 to 100% [1]. By 2020 it is anticipated that 95% of the marketed synthetic pharmaceuticals will be pure enantiomers [2].

The impact of this transformation in pharmaceutical development is an increased reliance on stereo-specific chemical transformations and chiral purification. Of the latter, chiral crystallization is the preferred method of chiral purification at commercial scale.

Nguyen and coworkers [3] define separate chiral drugs into three different groups depending on the bioactivity (for a given target) relationship between the desired and undesired enantiomers. The groups defined therein are drugs with (1) one major bioactive enantiomer, (2) equally bioactive enantiomers, and (3) in vivo chiral inversion. Examples of group (1) drugs, which are shown in Figure 25.2, include the beta-blocker (S)-propranolol, which is approximately 100 times more active than (R)-propranolol, and the central-acting analgesic (R)-methadone, which is approximately 25–50 times more potent than (S)-methadone.

Group (2) drugs, those with equally bioactive enantiomers, are rare and include the example flecainide, shown in Figure 25.3.

Group (3) chiral drugs can undergo unidirectional or bidirectional (racemization) chiral inversion in vivo. This inversion complicates the distinction of the differences in the biological activity, toxicology, and pharmacokinetic properties of the drug owed to the chiral purity of DS from properties inherent to the achiral structure of the molecule [3]. Ibuprofen is the canonical example of unidirectional chiral inversion, with the biologically inactive (R)-ibuprofen converting to the active (S)-ibuprofen in vivo (shown in Figure 25.4).

Thalidomide is an example of rapid bidirectional in vivo chiral inversion (shown in Figure 25.5). This is particularly problematic as the (S)-thalidomide and its metabolites are known to be teratogenic, while (R)-thalidomide is not.

However, since the enantiomers exhibit in vivo chiral inversion, exposure to either enantiomer of thalidomide risks the teratogenic effects of the (*S*)-enantiomer [3, 4].

This is in contrast to the belief that the thalidomide tragedy could have been avoided if (*R*)-thalidomide as opposed to the racemic mixture had been used as a treatment for morning sickness.

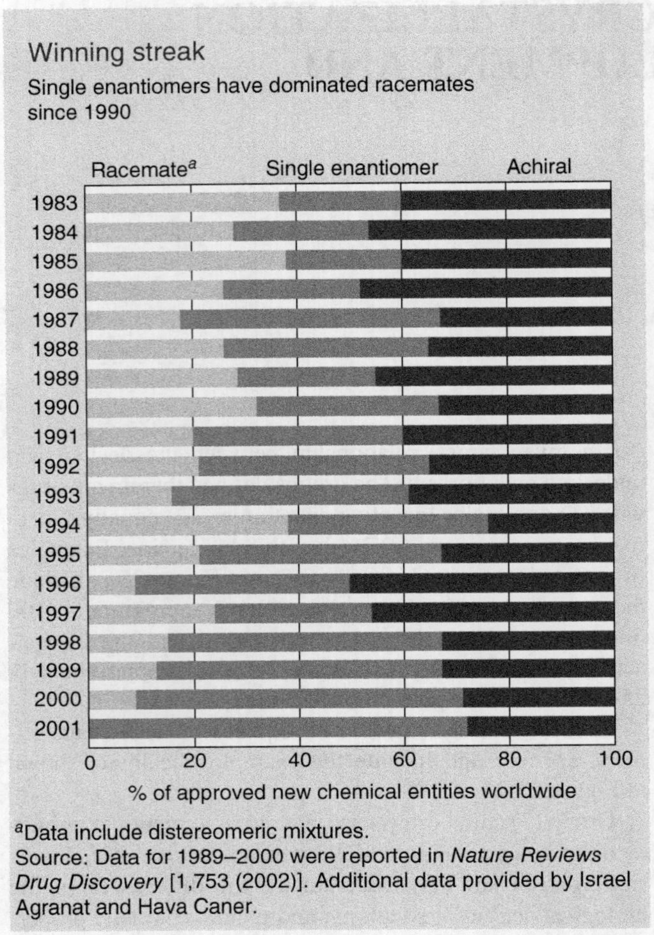
Winning streak
Single enantiomers have dominated racemates since 1990
Racemate[a]
Single enantiomer
Achiral
1983
1984
1985
1986
1987
1988
1989
1990
1991
1992
1993
1994
1995
1996
1997
1998
1999
2000
2001
0
20
40
60
80
100
% of approved new chemical entities worldwide
[a]Data include distereomeric mixtures.
Source: Data for 1989–2000 were reported in *Nature Reviews Drug Discovery* [1,753 (2002)]. Additional data provided by Israel Agranat and Hava Caner.

FIGURE 25.1 Proportion of pure enantiomers vs. achiral or racemate approved worldwide from 1983 to 2001 [1].

Depending on the specific differences in biological activity, toxicology, pharmacology, metabolic activity, manufacturability, and cost concerns, it may be beneficial to develop drugs as either racemic mixtures or single enantiomers of the same compound. However, there are several examples where racemic mixtures and single enantiomers of the same compound are marketed including ibuprofen (Advil/Motrin) and dexibuprofen (Seractil), omeprazole (Prilosec) and esomeprazole (Nexium), and zopiclone (Imovane) and eszopiclone (Lunesta). The majority of pharmaceutical compounds currently in development are being pursued as single enantiomers with the undesired enantiomer treated as a process impurity controlled to specifications guided by toxicity and regulatory expectations.

The regulatory expectations for chiral quality in synthetic DS were established in the International Council for Harmonization (ICH) Q6A guideline. The guideline requires chiral identity (i.e. the identification of chiral impurities), chiral assay (i.e. the amount of desired product), and enantiomeric impurity (i.e. the levels of the undesired enantiomers) testing for DS or "applying limits to appropriate starting materials or intermediates when (*suitably*) justified from developmental studies." The development, validation, and maintenance of analytical methods for chiral identity, chiral assay, and enantiomeric impurity are typically less cumbersome for starting

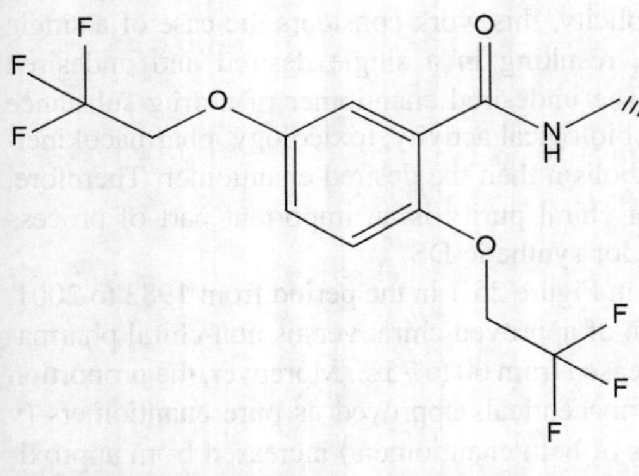
F
F
F
O
O
N
H
H
N
O
F
F
F

FIGURE 25.3 Flecainide, an example of a drug with equally bioactive enantiomers.

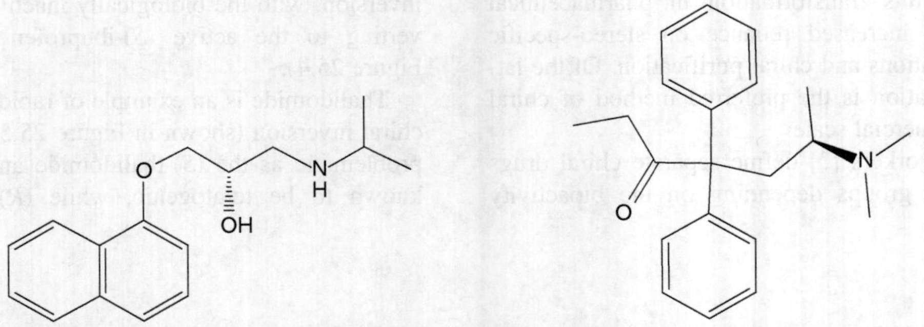
O
N
H
OH
O
N

FIGURE 25.2 (*S*)-propranolol (left) and (*R*)-methadone (right). The opposite enantiomers of these drugs exhibit significant reductions in their potencies.

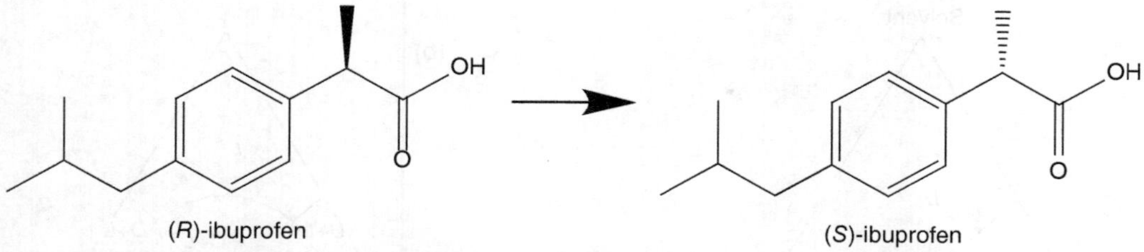

FIGURE 25.4 The biologically inactive (R)-ibuprofen converts in vivo to the active (S)-ibuprofen.

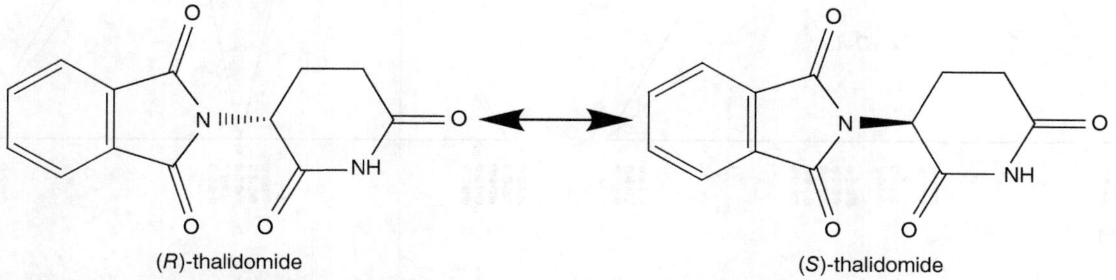

FIGURE 25.5 Teratogenic (S)-thalidomide and non-teratogenic (R)-thalidomide interconvert in vivo, resulting in toxicology concerns for exposure to either enantiomer (or a racemic mixture of both).

materials and intermediates than for DS release. It is therefore desirable to use starting materials and intermediates with the desired chiral purity. Furthermore, it is undesirable to further process batches of starting materials or intermediates that will not result in DS meeting specifications. As such, the approach of determining and justifying starting material and intermediate chiral purity limits and control strategies is often used to establish chiral quality in DS. As a result, the isolation of the smallest fragment of DS containing the chiral center in question is frequently used as a control point to establish chiral purity [5].

At present, most asymmetric synthesis techniques used to form chiral starting materials and intermediates provide some but not all of the purity required within a given synthetic step (a potential exception being biocatalysis). This is referred to as partial resolution. Chiral purity upgrades of partially resolved mixtures are often obtained through subsequent crystallization processes. Typically, the resource investment in developing crystallizations for starting materials and intermediates is significantly less than the comparable investment in final DS crystallizations where the solid-state properties of DS such as particle size and polymorphic form are established. Therefore, there is a need to develop simple process design strategies that establish chiral control for use in intermediate and starting material crystallization processes. This chapter aims to provide a background in the underlying thermodynamics, characterization techniques, and process design fundamentals required to develop simple crystallizations able to upgrade enantiomeric mixtures to meet chiral purity requirements.

25.1 TERNARY PHASE DIAGRAMS

The characterization of chiral systems and the design of crystallization strategies are often aided by the use of ternary phase diagrams (TPDs). Although the relevant design calculations may be performed without TPDs, they provide a useful visualization for the design of crystallization processes. Additionally, they are ubiquitous in their application for interpreting the underlying thermodynamics of chiral systems, i.e. the phase behavior. The reader unfamiliar with their generation and interpretation is directed to the formalism described in Ref. [6].

25.2 TERNARY PHASE DIAGRAMS OF CHIRAL SYSTEMS CONTAINING SOLIDS

For chiral crystallizations, the components of the ternary system are the desired and undesired enantiomers and the solvent or solvent mixture used in the crystallization. For pharmaceutical compounds, there are often several solid and potentially liquid phases present within the system. For simplicity, we will primarily consider cases with a single liquid phase – without solvates/hydrates, salts, or multiple polymorphs present. In this limited set of cases, the phase behavior is restricted such that there is a limited set of three solid phases possible depending on the system in question.

The three most common solid phases are pure (i) desired and (ii) undesired enantiomer solids and (iii) racemic solids containing a one-to-one ratio of desired and undesired

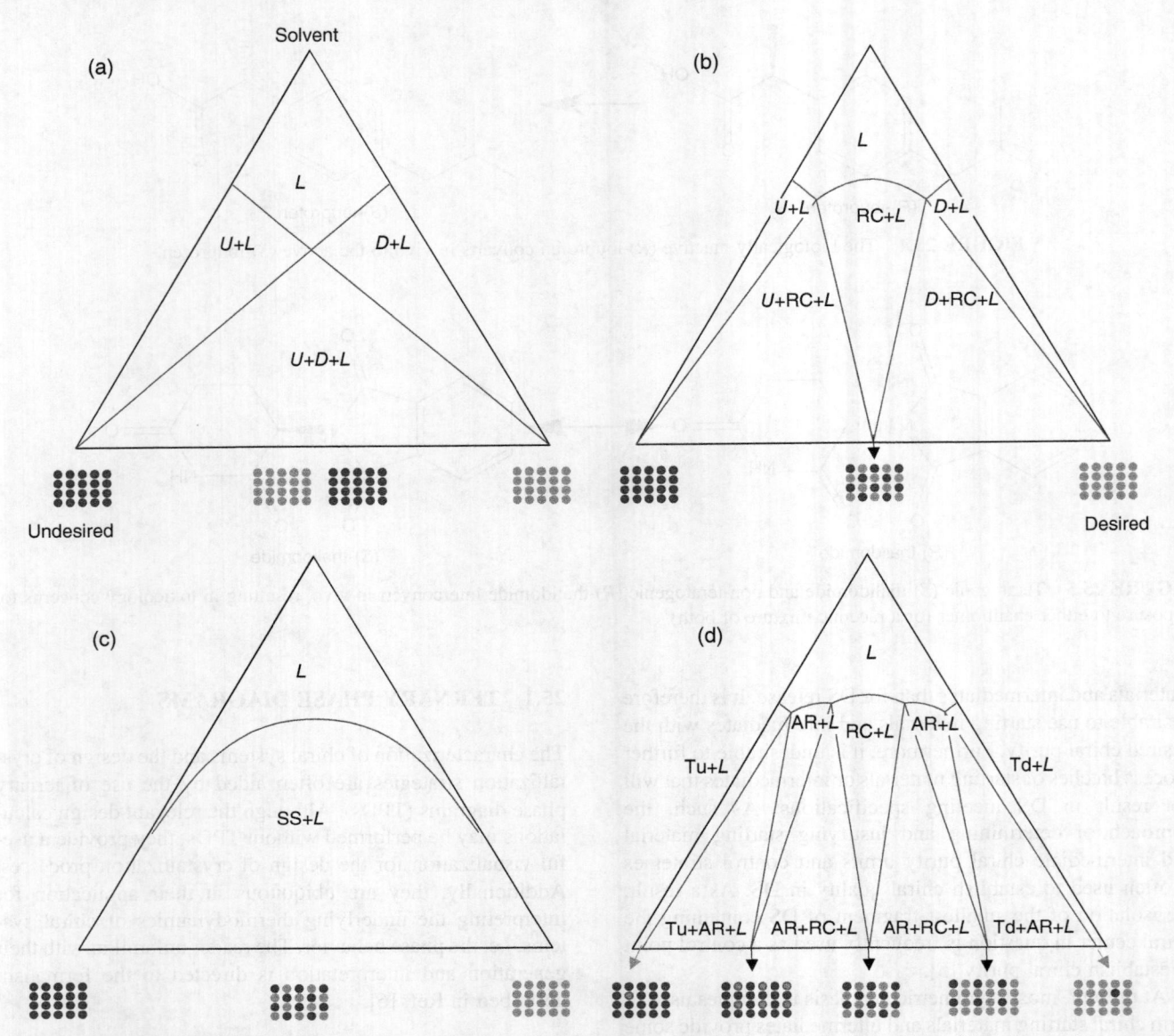

FIGURE 25.6 Schematic representations of the solids present in (a) conglomerate, (b) racemic, (c) solid solution, (d) a system with a racemic, a 3 : 1 anomalous racemate, and terminal solution. The light and dark circles indicate opposite enantiomers of the same compound. D, desired enantiomer; U, undesired enantiomer; RC, racemic compound; SS: solid solution; L, liquid; AR, anomalous racemate; Tu/Td, terminal solution for undesired and desired enantiomers, respectively.

stereoisomers. It is often estimated that approximately 90% of synthetic organic molecules form racemic solids. In conglomerate systems, a racemic solid does not form, and there are only two solid forms: (i) pure D and (ii) pure U. Other possible phases that are encountered infrequently are solid solution, terminal solid solutions, and anomalous racemates. Solid solutions occur when there is full miscibility in the solid phases, allowing an enantiomer within crystals of the opposite enantiomer. Terminal solid solutions occur when there is partial miscibility in the solid phases, allowing an enantiomer within crystals of the opposite enantiomer or racemic solid to

a limited degree. Terminal solid solutions can occur for systems that do or do not form racemic crystals. In addition, some systems form solid phases incorporating D and U in non-equimolar (1 : 1) ratio; these systems are referred to as anomalous racemates [7]. Although these last three systems are uncommon, they are important in establishing chiral control policies and are frequently not included in discussions of ternary systems, and their frequency of occurrence is not known. Representative TPDs for these classes are shown in Figure 25.6. A schematic representation of the solid-phase composition is presented in the solid (bottom) axis of the diagram to depict the solid phases that are in equilibrium in the respective regions.

Traditionally the enantiomer purity is expressed in enantiomeric excess (ee) as this value was associated with optical activity present in chiral compounds. The concentrations of desired $[D]$ and undesired $[U]$ enantiomers are related to the ee, the fractional purity of desired enantiomer $x_D = [D]/([D] + [U])$, and the ratio of the desired enantiomer to undesired enantiomer $\xi_D = [U]/[D]$ as follows:

$$ee = \frac{[D] - [U]}{[D] + [U]} \quad (25.1a)$$

$$ee = 2x_D - 1 \quad (25.1b)$$

$$ee = \frac{\xi_D - 1}{\xi_D + 1} \quad (25.1c)$$

The concentrations may be expressed in any conventional unit (mg/ml, molarity, mg/g or wt % basis); however, mg/g is typically useful in the context of crystallization design. The selection of ee, fraction, or ratio is mainly determined by the preference of the researcher, with the first two frequently expressed as a percent purity. As the molecular weights of the enantiomers are identical, any concentration unit will result in the same value of the enantiomeric purity.

Typically, the equivalent solid phases will be formed for the solids of pure D and U. These solids will have the same physical properties (e.g. density, melting point), resulting in the same solubility in achiral solvents and same eutectic point with the racemic crystal (when applicable). In these cases, the TPDs contain a mirror plane at zero enantiomeric excess. Eutectic points are equilibrium conditions between solid phases – as such the solvent-free concentrations of solid phases at eutectics do not vary from solvent to solvent. As will be shown, determining the eutectic point composition and the phase diagram is a key aspect of process design. Figure 25.7 shows a generic phase diagram for a racemic forming system. The relevant equilibrium that describes the limits of solubility in the phase diagram is represented in the figure and described below along with the corresponding thermodynamic equations [4].

The equations for the equilibrium lines are indicated in terms of the enantiomeric ratio; however, enantiomeric excess or enantiomer fraction may also be used. Their relationship along the axis line is indicated.

As depicted in the region in which only the pure D solid phase is stable (region $D + L$ in Figure 25.6), the relevant equilibrium is

$$D(s) \underset{K_P}{\longleftrightarrow} D(l) \quad (25.2a)$$

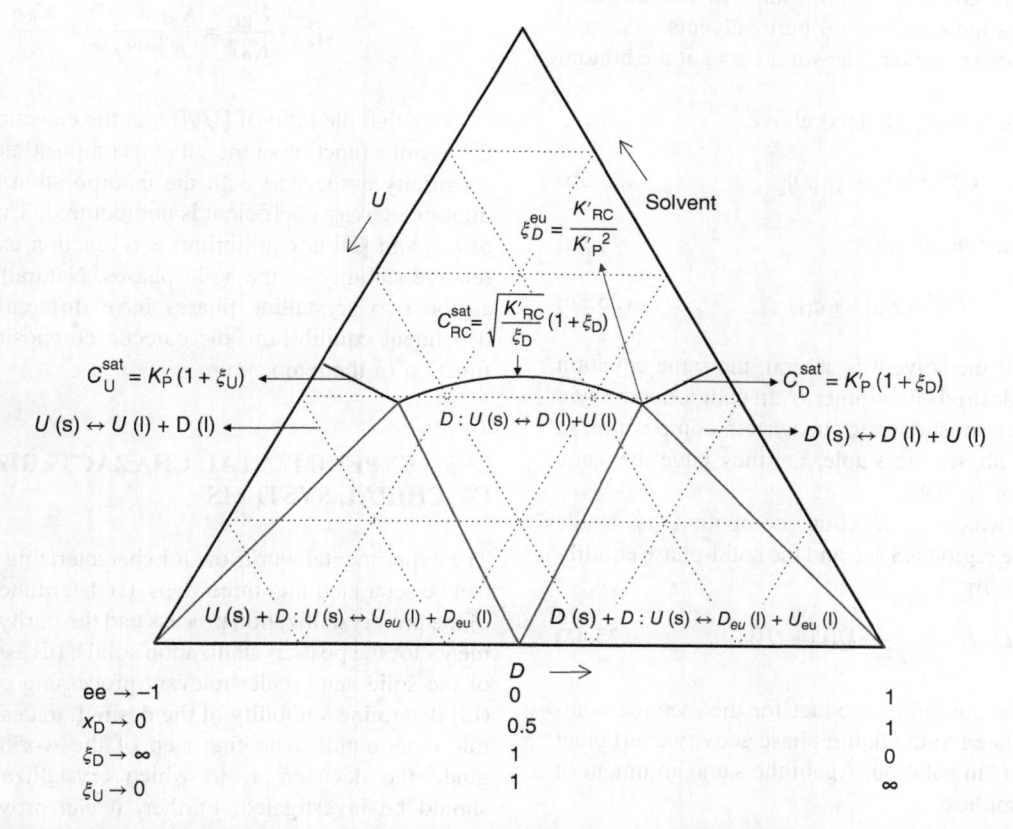

FIGURE 25.7 Generic racemic-forming system with thermodynamic equilibrium relationships. The black line corresponds to the liquidus limit; above this line the only phase is solution. Along the line the concentration $[C] = [D] + [U]$.

The solubility product is related to the liquid-phase activity coefficient of species D in solution. In pharmaceutical systems the concentrations are frequently such that the infinite dilution approximation applies:

$$K_P = a_D = \gamma [D]_e \approx \gamma^\infty [D]_e \qquad (25.2b)$$

We can define a solvent-specific solubility product K'_P:

$$K'_P = [D]_e \qquad (25.2c)$$

The black line in Figure 25.7 delimits the solution-phase concentration at the top of the TPD, and the straight line in the right-hand side triangular region (solution + pure D solids) corresponds to the overall concentration $[C] = [D] + [U]$; this solubility limit is represented by

$$C_D^{\text{sat}} = [D]_e + [U] \qquad (25.3a)$$

where $[D]_e$ correspond to the concentration of D satisfying the equilibrium in Eq. (25.2). Note that the subscript D refers only to the fact that the concentration is in equilibrium with pure D in the solid phase; the concentration is the sum of both D and U concentrations.

It is important to recognize that we use concentrations to establish the thermodynamic equilibrium. In the diagram, low concentrations increase toward pure solvents, and high concentration decreases toward the solid curve at the bottom of the TPD.

Substituting the ratio ξ_D defined above,

$$C_D^{\text{sat}} = [D]_e + \xi_D [D]_e \qquad (25.3b)$$

And invoking the equilibrium

$$C_D^{\text{sat}} = K'_P (1 + \xi_D) \qquad (25.3c)$$

By symmetry if the solvent is achiral, the same equation applies for the undesired enantiomer U. In conglomerate systems, the eutectic corresponds to the system composition in which both solid phases are stable, i.e. they have the same solubility, or when $\xi_D = 1$.

In cases in which a racemic compound is stable (Figure 25.6b), the region RS + L and the solid-phase equilibrium are described by

$$D:U(s) \underset{K_{RC}}{\longleftrightarrow} D(l) + U(l) \qquad (25.4a)$$

In this case, the solubility product for the racemic compound $D:U$ is related to the liquid phase activity coefficient of species D and U in solution. Again the same argument of infinite dilution applies:

$$K_{RC} = a_D a_U = \gamma_D [D] \gamma_U [U] \approx \gamma_D^\infty [D] \gamma_U^\infty [U] \qquad (25.4b)$$

As previously we can define a solvent-specific solubility product KRC':

$$K'_{RC} = [D][U] \qquad (25.4c)$$

The boundary line depicted in the diagram above the $L + $ RC region corresponds to the overall concentration $[C] = [D] + [U]$; therefore the solubility line corresponding to the region $D + L$ is

$$C_{RC}^{\text{sat}} = [D] + [U] \qquad (25.5a)$$

where $[D]$ and $[U]$ satisfy the equilibrium product in Eq. (25.4c). Substituting the ratio ξ_D defined above and the equilibrium restriction,

$$C_{RC}^{\text{sat}} = \sqrt{\frac{K'_{RC}}{\xi_D}(1 + \xi_D)} \qquad (25.5b)$$

In the case of racemic compound, the eutectic composition on the D-enriched side of the phase diagram corresponds to the point in which the solubility of solid D is the same as that of the racemic $D:U$ (i.e. when Eqs. 25.2c and 25.4c intersect), which occurs when the value of the enantiomeric ratio is

$$\xi_D^{\text{eu}} = \frac{K'_{RC}}{K'^2_P} = \frac{K_{RC} \gamma^\infty \gamma^\infty}{K_P^2 \gamma^\infty \gamma^\infty} = \frac{K_{RC}}{K_P^2} \qquad (25.6)$$

Note that the ratio of $[U]/[D]$ at the eutectic composition, ξ_D^{eu}, is not a function of the solvent composition as the solvent specificity associated with the incorporation of the infinite dilution activity coefficient is not required. The relative ratio of $[U]$ and $[D]$ at equilibrium is a function explicitly of the relative stability of the solid phases. Naturally, in as much as the two crystalline phases have different temperature-dependent equilibrium, the eutectic composition may be a function of the temperature.

25.3 EXPERIMENTAL CHARACTERIZATION OF CHIRAL SYSTEMS

The experimental workflow for characterizing chiral systems can be separated into three steps: (i) determine the capability of the pre-crystallization process and the purity/yield requirements for the post-crystallization solids, (ii) assess the phases of the solid state under relevant processing conditions, and (iii) determine solubility of the desired, undesired, and racemic compounds. The first step of the workflow serves to guide the decision as to which crystallization approach should be investigated. Further, it can provide insight to upstream processes where a higher initial chiral purity would be beneficial.

The second step of the workflow starts in a manner similar to a polymorphic form screen. In this screen, physical mixtures across a range of enantiomer excess (and possibly different solvents) are equilibrated. Thermal cycling may be applied during the equilibration period. The goal of the screening is to identify and isolate a racemic crystal, if possible, and to determine if that racemic crystal is stable at the equilibration temperature. Analytical techniques used to characterize the solids formed during the screen include X-ray powder diffraction (form identification), differential scanning calorimetry (DSC) (melting point and form identification), thermogravimetric analysis (solvate identification), and optical microscopy. Alternatively, melting point diagrams can be constructed from physical mixtures of solids across the range of interest to identify potential eutectic points.

Once this screen has determined if a racemic crystal is possible, a broad set of physical mixtures of solids at different enantiomeric ratios are equilibrated with varied volumes V of the intended solvent system at the isolation temperature. The pure desired enantiomer should be included in this set. If enough of the racemic solid has been formed previously, it is recommended to use the racemic solid to provide the necessary content of undesired enantiomer. After equilibration, the concentration of the desired and undesired enantiomers in the supernatant is measured using high performance liquid chromatography (HPLC) or a related technique to generate the solubility data. The resulting solids are then isolated and characterized to determine their solid phase and enantiomeric purity. A TPD is constructed by generating tie lines connecting the chiral purity of the resulting solids (along the binary axis at the bottom of the diagram), through the initial composition, and to the liquid concentration obtained through characterizing the supernatant. If a number of initial conditions result in nearly the same concentrations of desired and undesired enantiomers in the supernatant, these points likely occur near a eutectic. These points are averaged to provide a rough outline of the TPD.

After the second step of the workflow, a general understanding can be obtained regarding the feasibility of different approaches for establishing chiral purity control. For instance, if a rough outline of the TPD reveals a solid solution, further investigation of a direct crystallization as a means of establishing purity should be abandoned. However, if a conglomerate or racemic system is identified, step 3 should include a careful determination of the equilibrium concentration of the eutectic point. This is necessary because of the importance of these quantities in establishing process designs. In this step, several physical mixtures of the desired and undesired (or racemic solid if applicable) are suspended for extended periods in the intended solvent at the isolation temperature. The goal of this step is to accurately determine the solubility of the enantiomer and racemic solids under relevant process conditions. This step of the workflow can be extended to include the incorporation of other process impurities as necessary. Equations (25.3c), (25.5b), and (25.6) may be used to model the concentrations of the liquidus line in the TPD. It is important at this point to ensure that the infinite dilution assumption is valid.

25.4 CHIRAL ANALYTICAL METHODS IN PHARMACEUTICAL DEVELOPMENT

The availability and application of adequate methods to analyze the process streams (liquids and solids) associated with the workflows described earlier is an important consideration in the development of chiral crystallizations. The two primary methods employed are chiral HPLC and powder X-ray diffraction (PXRD). HPLC is the most used method and is essential in evaluating the chiral purity of the starting material, the mother liquors, and solids both to define the TPD and in developing and evaluating the process. It is important that the HPLC method can fully separate the enantiomers, and linearity is demonstrated for concentration calculations. PXRD is used to confirm the form of the pure solids and the racemic solids when present. Form can also be used to evaluate the difference between a conglomerate and racemic mixture as the racemate will generally have a different crystal structure than the pure enantiomers. Additional techniques that may be employed are solid-state NMR (ssNMR), DSC, and thermal gravimetric analysis (TGA). ssNMR is an orthogonal technique to PXRD and can provide insight into crystal structure and phase purity. DSC is used to evaluate the solid's melting point, which can give insight into the type of system and can also be used to generate the TPD [8]. TGA provides insight into the solvate presence within the final crystalline material and is used in conjunction with DSC to characterize thermal events within the DSC trace.

With respect to DSC measurements, the melting point of solids of intermediate chirality is related to the chiral purity x_D. As the chiral purity is decreased from pure enantiomer, the melting point decreases according to the Schroeder-van Laar equation (Eq. 25.7a). In cases in which a racemic compound is present, the melting point has a local maxima at the $x_D = 0.5$ and decreases as the chiral purity is increased (or decreased) according to the Prigogine–Defay equation[1] (Eq. 25.7b).

The melting curves of the racemic compound and the pure enantiomer meet at the eutectic corresponding to the composition with a lowest melting point. The eutectic composition

---

[1] The Prigogine–Defay equation as originally published predicts the liquidus composition curve for an additional compound of two enantiomers in any stoichiometry corresponding to a composition x_c, with stoichiometries v_1 and v_2 [9]. In the case of the racemic compound $v_1 = v_2 = 1$ and $x_c = 1/2$ yielding Eq. (25.7b).

$$-\ln x^{v_1}(1-x)^{v_2} + \ln x_c^{v_1}(1-x_c)^{v_2} = \frac{\Delta H_c^f}{R}\left(\frac{1}{T_c^f} - \frac{1}{T^f}\right)$$

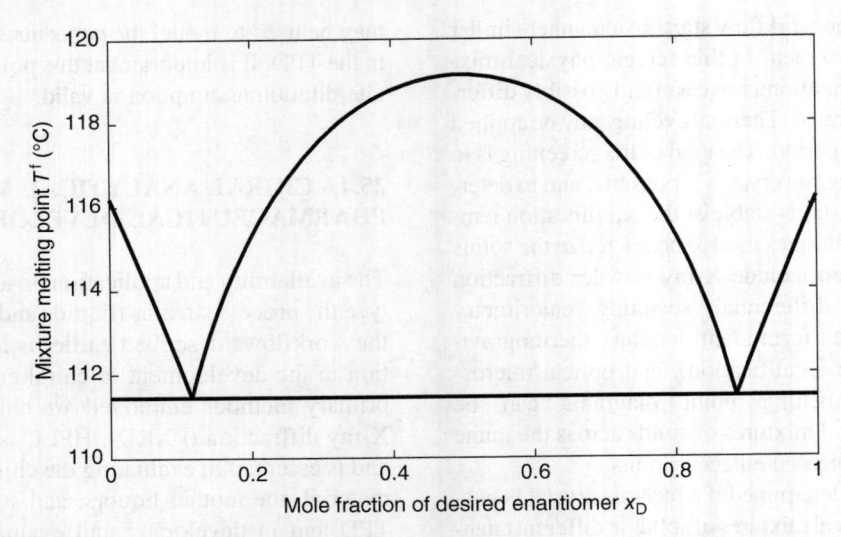

FIGURE 25.8 Melting point versus chiral purity for a chiral system forming a racemic compound.

observed from the DSC is equivalent to the one observed in the TPD. Figure 25.8 shows a schematic of the melting point dependence on chiral purity for a system forming a racemic compound.

The Schroeder-van Laar equation provides the melting point depression as a function of the chiral purity (x_D) of the solid mixture:

$$\ln x_D = \frac{\Delta H_A^f}{R}\left(\frac{1}{T_A^f} - \frac{1}{T^f}\right) \quad (25.7a)$$

$$\ln 4x_D(1-x_D)^D = \frac{2\Delta H_{RC}^f}{R}\left(\frac{1}{T_{RC}^f} - \frac{1}{T^f}\right) \quad (25.7b)$$

where

ΔH_A^f and ΔH_{RC}^f are the enthalpy of melting of the pure enantiomer and the racemic compound, respectively.

T_A^f and T_{RC}^f are the corresponding melting points.

In cases in which pure racemic and pure enantiomer are available, the DSC analysis may be used to provide an estimate of the eutectic, which, as will be seen, is a critical parameter in designing chiral crystallization.

25.5 PROCESS DESIGN OF CHIRAL CRYSTALLIZATIONS

25.5.1 Direct Crystallization of Partially Resolved Systems

Chiral separation by direct crystallization is the most straightforward approach to chiral upgrade since it relies on system thermodynamics and does not require additional additives or a complex equipment setup. Similar to a conventional crystallization, the key aspects for chiral crystallization are the polymorphic form, solubility, metastable zone, and nucleation and growth kinetics. This section assumes the polymorphic form for the desired, undesired, and racemic solids is established, and the system solubility has been determined using the relevant TPD workflow as described previously. This section will focus only on the two most common general classes of direct crystallization (racemic and conglomerate systems).

For simplicity it is assumed that the crystallization or dissolution process will be run in the same solvent system at a constant temperature. In this case the entire process can be visualized with the same ternary diagram by adjusting only the location of the process in terms of overall concentration (total solid mass/total solvent) or solvent to material (D and U) ratio. If changes in temperature or solvent composition are used, the limits of the phase diagram will change as the process evolves and multiple solubility curves in the ternary diagram would be required. Conceptually the analysis does not change, but the visualization in the TPD becomes more difficult to represent. To analyze the conditions of the system in the examples below, we will consider situations in which the crystallization design is being performed for a system of known endpoint solubility. The relevant parameters for the crystallization are described in Table 25.1.

The parameters related to purity are only a function of the relative amounts of D and U (mass, mole, %, HPCL, AP, etc.). They may be obtained for any point in the ternary diagram by projecting a curve from the vertex of the TPD to the bottom axis. Any of the three definitions of chiral purity can be used; for graphical representation we will use ee in the discussion below, although Eqs. (25.1a)–(25.1c) can be used to convert from one measure of purity to the other. The crystallization solvent ratio corresponds to the ratio of the mass of the solvent to the total mass of D and U in the crystallizer (present in the liquid and solid phases). In the diagram, the height of the point represents the solvent ratio (i.e. the lower the point, the less proportion of solvent to solids is being used in the crystallization). The yield is the total mass of material isolated (regardless of purity), and the isolation phase may be the liquid or the solid depending on the process as described below. If the recovery is from the liquid phase, the yield of interest would be $1 - Y$, as Y is strictly used for the solid yield.

For a system forming a racemic compound, the chiral purity of the eutectic at the point of isolation determines the achievable purity of the crystallized solids. Two scenarios are possible as the purity of the initial (and partially resolved) system may be higher or lower than the purity of the eutectic composition. A generic racemic system is depicted in Figure 25.9, presenting the solubility at the crystallization endpoint (C in Table 25.1) with the eutectic point marked as E and the pure enantiomer solubility as A. The two diagrams correspond to the same isolation endpoint and crystallization solvent ratio (P) for two processes, **1** (left) and **2** (right). The initial purity in process **1** is lower than that one of process **2** ($ee_i^1 < ee_i^2$). In particular, the starting purity ee_i^1 of process **1** is below the chiral purity of the eutectic composition, ee_{eu}. In both cases the crystallization solids will be a combination of D solids and racemic compound. However in process **1**, the liquid phase will be enriched over the initial chiral purity, and the isolated material will have a lower purity ($ee_f^1 < ee_i^1$), whereas the opposite will be true for process **2**, i.e. $ee_f^2 > ee_i^2$.

As mentioned above in cases in which $ee_i > ee_{eu}$, the isolated solids will be of higher chiral purity than the starting chiral purity ($ee_f > ee_i > ee_{eu}$). Multiple formalisms are available to design the crystallization [6]; the one presented here is generic and may be easily extended to other conditions.

For the first example the formalism will be shown in detail. To accomplish this we will define a system with properties as indicated in Table 25.2. The properties are not particularly realistic and are provided for illustration purposes only. The system is defined only for the relevant portion of the diagram, but it can be extended by symmetry to the other half.

The first four parameters are properties of the system; however, the concentrations at the solubility point (S_D, S_{RC}, and C_{eu}), which are experimentally determined, may be adjusted by manipulating the temperature and solvent composition or both. For this example it is assumed that

TABLE 25.1 Parameters for Chiral Crystallization Design

Parameter	Variable Name(s)
Initial chiral purity	ee^i, x_D^i, ξ_D^i
Solid chiral purity	ee^S, x_D^S, ξ_D^S
Solution (liquid) purity (eutectic)	ee^L, x_D^L, ξ_D^L
Liquid concentration	$C = [D] + [U]$
Total material (D and U)	$M_T = M_U + M_D$
Crystallization solvent ratio	$P = \dfrac{M_{solvent}}{M_T}$
Crystallization yield	$Y = \dfrac{M_{isolated}}{M_T}$

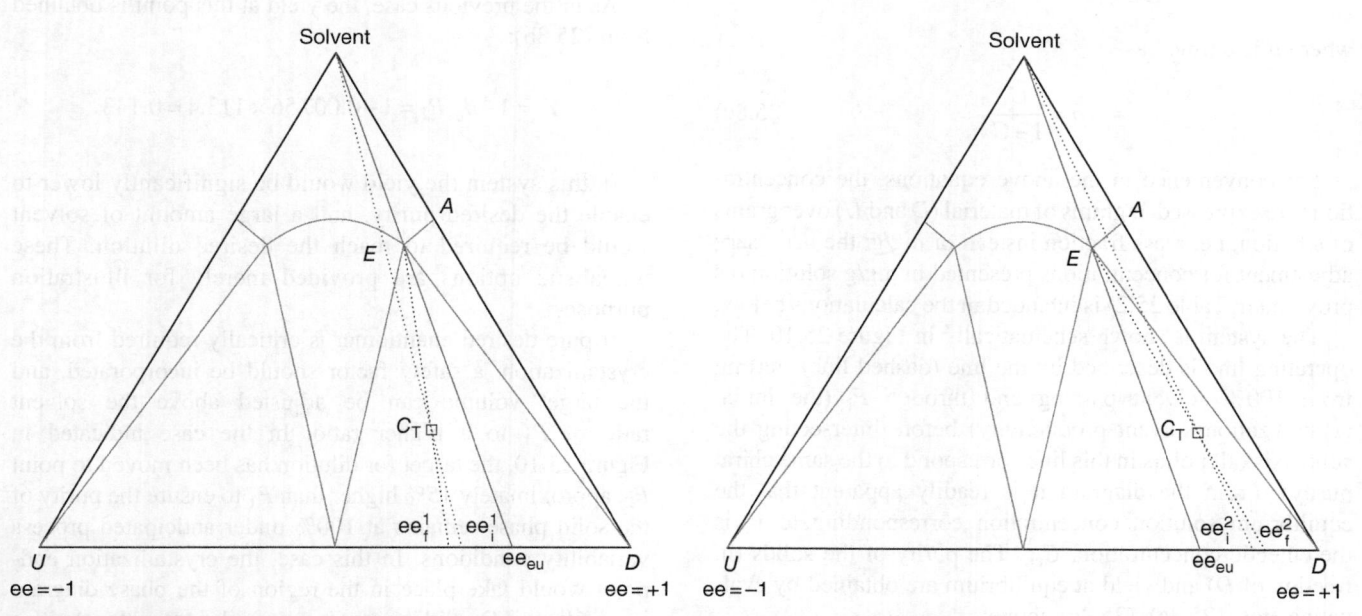

FIGURE 25.9 Ternary phase diagram demonstrating the impact of starting composition on the ability to result in a chiral upgrade in either phase. For systems starting at purity below the eutectic purity ee_{eu} (diagram on the left), enrichment will be in the mother liquors, i.e. the eutectic composition. For systems starting at purity higher than the eutectic purity ee_{eu} (diagram on the right), enrichment will be in the solid phase. A, pure enantiomer solubility; E, eutectic; D, desired enantiomer; U, undesired enantiomer.

TABLE 25.2 Properties of Ternary Phase Equilibrium for the First Quantitative Analysis Considered

Parameter	Value
Solubility of pure D	$S_D = 5.0$ mg/g
Solubility of racemic compound	$S_{RC} = C_D + C_U = 7.07$ mg/g
Eutectic chiral purity	$\xi_D^{eu} = 0.5$
	$x_D^{eu} = 0.667$
	$ee^{eu} = 0.333$
Eutectic concentration	$C_{eu} = 7.5$ mg/g
Initial chiral purity	$\xi_D^i = 0.4$
	$x_D^i = 0.714$
	$ee_i = 0.429$
Initial crystallization solvent ratio	32.3 g solvent/g D and U

the solubility point has been selected as the desired operating point. Potentially, the initial purity may also be subject to modification, but, for demonstration purposes, it is assumed that this modification is not possible. Therefore, for this study we are interested in determining the final crystallization solvent productivity that will result in adequate isolated purity (in this case the solid phase expressed as ee^S, x_D^S, or ξ_D^S) and the corresponding crystallization yield Y. However, the equations may be adjusted to address any design option.

The relevant design equations are presented below:

$$x_D^S = \frac{x_D^i - x_D^L \beta P}{1 - \beta P} \tag{25.8a}$$

$$Y = 1 - \beta P \tag{25.8b}$$

where β is defined as

$$\beta = \frac{C}{1-C}. \tag{25.8c}$$

For convenience in the above equations, the concentrations are expressed in grams of material (D and U) over grams of solution, i.e. mass fraction instead of mg/g; the necessary adjustment for concentrations presented in mg/g solution (as provided in Table 25.2) is included in the calculations below.

The system is shown schematically in Figure 25.10. The operating line is described by the line (dashed line) starting from 100% solvent passing and through P_i (the initial crystallization solvent productivity) before intersecting the solid axis (all points in this line correspond to the same chiral purity). From the diagram it is readily apparent that the equilibrium solution concentration corresponding to P_i is the eutectic concentration, C_{eu}. The purity of the solids (in fraction of D) and yield at equilibrium are obtained by evaluating Eqs. (25.8a)–(25.8c) above:

$$\beta_i = \frac{C}{1-C} = \frac{C_{eu}}{1-C_{eu}} = \beta_{eu} = \frac{7.5/1000}{1-7.5/1000} = 0.00756$$

$$Y_i = 1 - \beta P = 1 - \beta_{eu} P = 1 - 0.00756 \times 32.3 = 0.756$$

$$x_D^S = \frac{x_D^i - x_D^L \beta P}{1 - \beta P} = \frac{x_D^i - x_D^{eu} \beta_{eu} P}{1 - \beta_{eu} P}$$

$$= \frac{0.714 - 0.667 \times (0.00756 \times 32.2)}{1 - 0.00756 \times 32.2} = 0.730$$

In this case, the reasonable yield of 75.6% results in a marginal increase in chiral purity. If the system is diluted to the point P_1 or greater, the solids will be pure desired enantiomer and the mother liquor will be at the ee_{eu}. The point P_1 corresponds to maximum yield that results in pure solid material while still in equilibrium with the eutectic point. At this condition the solvent volume and yield can be calculated solving for P in Eq. (25.8a) by setting the solid-phase chiral purity to 100% ($x_D^S = 1$) with the liquid phase corresponding to the eutectic composition, i.e. $x_D^L = x_D^{eu}$ with concentration C_{eu}:

$$P_{min} = \frac{1 - x_D^i}{(1 - x_D^{eu})\beta_{eu}} \tag{25.9}$$

Applying Eq. (25.9) to the system described above, we obtain

$$P_{min} = \frac{1 - x_D^i}{(1 - x_D^{eu})\beta_{eu}} = \frac{1 - 0.714}{(1 - 0.667) \times 0.00756}$$

$$= 113.4 \text{ g solvent/g product } (D + U)$$

As in the previous case, the yield at this point is obtained from (25.8b):

$$Y_1 = 1 - \beta_{eu} P_1 = 1 - 0.00756 \times 113.4 = 0.143$$

In this system the yield would be significantly lower to enable the desired purity, and a large amount of solvent would be required to reach the desired dilution. These unrealistic options are provided merely for illustration purposes.

If pure desired enantiomer is critically required from the crystallization, a safety factor should be incorporated, and the target volume can be adjusted above the solvent ratio of P_1 to a higher ratio. In the case indicated in Figure 25.10, the target for dilution has been moved to point P_2, approximately 15% higher than P_1 to ensure the purity of the solid phase remains at 100% under anticipated process variability conditions. In this case, the crystallization endpoint would take place in the region of the phase diagram in which pure D solid is present in equilibrium with solutions of D and U. As noted from the tie line for point P_2 (dashed line), the concentration in the liquid phase will no longer be defined by C_{eu}, but rather will lie on the concentration line

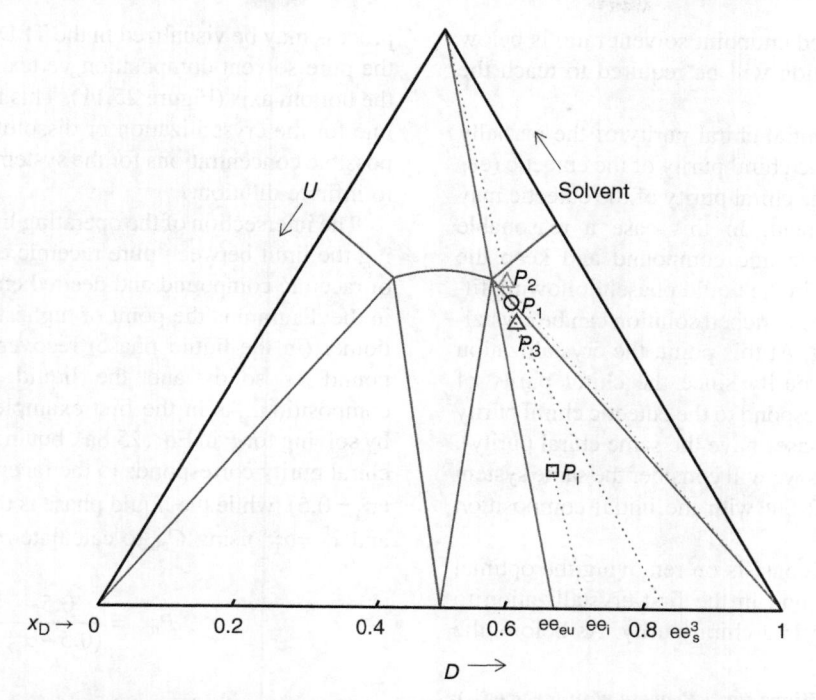

FIGURE 25.10 Ternary phase diagram showing the operating line for a case with $ee_i > ee_{eu}$. The initial solvent ratio of P_i (square) may be diluted to point P_1 (circle) that corresponds to the dilution of highest yield with pure D present in the solid phase. Point P_2 (upper triangle) corresponds to a condition with equal isolated purity but lower yield, while point P_3 (lower triangle) represents higher yield but lower purity (ee_s^3). Each system has the corresponding tie line represented by a dashed line.

defined by Eq. (25.3c) above. In this region the solution-phase chiral ratio ξ_D^L is obtained from Eq. (25.10) below:

$$\xi_D^L = \frac{1 - x_D^i}{K_P'(1+P)} \qquad (25.10)$$

where K_P' is defined in Eq. (25.2c) above, which with the solvent ratio increased by 15% over P_1 yields

$$\xi_D^L = \frac{1 - x_D^i}{K_P'(1+P)} = \frac{1 - x_D^i}{S_D(1 + 1.15 P_1)}$$

$$= \frac{1 - 0.714}{5/1000 \times (1 + 1.15 \times 113.4)} = 0.436$$

The chiral purity ratio is used to estimate the concentration in the liquid phase corresponding to the tie line from Eq. (25.3c):

$$C = K_P'(1 + \xi_D^L) = \frac{5}{1000}(1 + 0.436) = 0.00717 \text{ or } 7.17 \text{ mg/g}$$

Once the equilibrium concentration in the liquid phase has been determined, Eq. (25.8b) and (25.8c) are used to calculate the yield:

$$\beta_2 = \frac{C}{1-C} = \frac{7.17/1000}{1 - 7.17/1000} = 0.00723$$

$$Y_2 = 1 - \beta_2 P_2 = 1 - \beta_2 \, 1.15 \times P_1 = 1 - 0.00723 \times 1.15 \times 113.4$$
$$= 0.0571 \text{ or } 5.71\% \text{ yield}$$

Alternately, if the pure enantiomer is not required, the target volume may be decreased to increase total yield. For instance, if desired purity of the isolated product corresponded to an enantiomeric excess of no less than 70% ($ee_3^S \geq 0.70$), the minimum target solvent ratio would be obtained from Eq. (25.8a) by solving for P, since this equation is expressed in terms of the fractional purity we transform from enantiomeric excess:

$$x_D^S = \frac{ee_3^S + 1}{2} = \frac{0.70 + 1}{2} = 0.85$$

$$P_3^{min} = \frac{x_D^S - x_D^i}{(x_D^S - x_D^{eu})\beta_{eu}} = \frac{0.85 - 0.714}{(0.85 - 0.667) \times 0.00756}$$

$$= 97.96 \text{ g solvent/g material } (D \text{ and } U)$$

The yield at this solvent ratio is obtained from Eq. (25.8b):

$$Y_3 = 1 - \beta_{eu} P_3 = 1 - 0.00756 \times 97.96 = 0.259 \text{ or } 25.9\% \text{ yield}$$

The operating conditions to achieve the desired crystallization endpoint by dissolution can be defined as described above. If the volume is below the desired isolation volume, additional solvent will be required.

Alternately, if the desired endpoint solvent ratio is below the initial volume, distillation will be required to reach the desired final volume.

In cases in which the initial chiral purity of the partially resolved system is below the chiral purity of the eutectic ($ee_i < ee_{eu}$), it is possible that the chiral purity of the eutectic may provide sufficient enrichment. In this case a reasonable approach is to filter out racemic compound and keep the desired product in solution in the liquid phase. Following filtration of racemic solids, the enriched solution can be crystalized to isolate the product. At this point, the crystallization will not afford increased purity since the chiral purity of the isolated solids will correspond to the eutectic chiral purity (i.e. the liquid and solid phases have the same chiral purity).

For illustration purposes we will consider the same system as described in Table 25.2 but with the initial composition defined in Table 25.3.

The design in this case consists on removing the optimal amount of racemic compound in the first crystallization to afford a solution of the desired chiral purity. As before, the process may be visualized in the TPD by drawing a line from the pure solvent composition vertex through the point P_i to the bottom axis (Figure 25.11). This represents the operating line for the crystallization or dissolution process, i.e. all the possible concentrations for the system going from pure solids to infinite dilution.

The intersection of the operating line with the line demarking the limit between pure racemic compound and mixtures of racemic compound and desired enantiomer denoted as P_1 in the diagram is the point of highest yield for desired enantiomer (in the liquid phase) recovering pure racemic compound as solids and the liquid phase at the eutectic composition. As in the first example, this point is obtained by solving for P in Eq. (25.8a), but in this case the solid-phase chiral purity corresponds to the racemic compound, i.e. 50% ($x_D^S = 0.5$), while the liquid phase is the eutectic composition, and $x_D^L = x_D^{eu}$ using C_{eu} to calculate β:

$$P_{min} = \frac{0.5 - x_D^i}{(0.5 - x_D^{eu})\beta_{eu}} \quad (25.11)$$

Using the previous value of β_{eu} since we are in the same system as before,

$$P_{min} = \frac{0.5 - x_D^i}{(0.5 - x_D^{eu})\beta_{eu}} = \frac{0.5 - 0.565}{(0.5 - 0.667) \times 0.00756}$$
$$= 51.48 \text{ g solvent/g product } (D+U)$$

TABLE 25.3 Initial Conditions for a Process with ($ee_i < ee_{eu}$): The System Is Defined in Table 25.2

Initial chiral purity	$\xi_D^i = 0.770$
	$x_D^i = 0.565$
	$ee_i = 0.130$
Initial crystallization solvent ratio	32.3 g solvent/g D and U

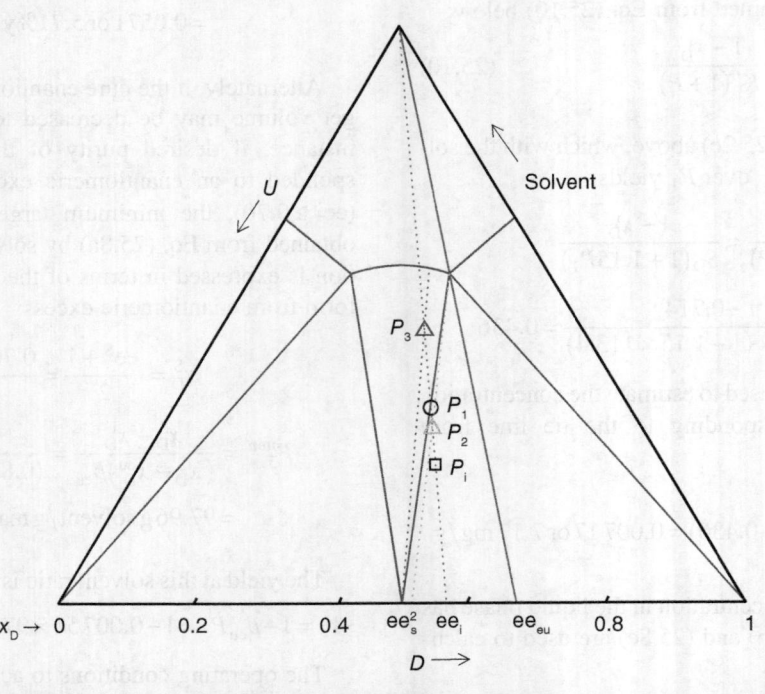

FIGURE 25.11 Ternary phase diagram showing the operating line for a starting composition of $ee_i < ee_{eu}$. Point P_1 (circle) is the point of highest yield in which the liquid phase corresponds to the eutectic and the solid phase is pure racemic compound (R). Points P_2 and P_3 correspond, respectively, to a conservative and aggressive processing conditions relative to the optimum.

As in the previous case, the yield at this point is obtained from Eq. (25.8b):

$$Y_1 = 1 - \beta P_1 = 1 - 0.00756 \times 51.48 = 0.611 \text{ or } 61.1\%$$

However, the yield in Eq. (25.8b) corresponds to the solid phase and therefore the amount of material that would be carried forward following the removal of the racemic compound, i.e. the liquid-phase yield is $1 - Y = 38.9\%$.

If there is a potential for variability and a 15% safety margin is used for the optimal solvent ratio to ensure that the liquid phase will correspond to the eutectic composition, a lower solvent to solid ratio would be used as indicated by the point P_2 in Figure 25.12. In this case the solvent ratio is given by $P_2 = 0.85 \times P_1$, and Eq. (25.8a)–(25.8c) may be used to calculate the yield and solid-phase purity. In this region of the phase diagram, the liquid-phase composition corresponds to the eutectic composition:

$$Y_2 = 1 - \beta P_2 = 1 - \beta 0.85 P_1 = 1 - 0.00756 \times 0.85 \times 51.48$$
$$= 0.669 \text{ or } 66.9\%$$

$$x_D^S = \frac{x_D^i - x_D^L \beta P}{1 - \beta P} = \frac{x_D^i - x_D^{eu} \beta_{eu} P_2}{1 - \beta_{eu} P_2}$$

$$= \frac{0.565 - 0.667 \times (0.00756 \times 0.85 \times 51.48)}{1 - 0.00756 \times 0.85 \times 51.48} = 0.515$$

As before the solution yield would be given by $1 - Y_2 = 33.1\%$. The enhanced control strategy for the process has 6% yield penalty as solid phase consists of a mixture of pure desired enantiomer in racemic compound (waste).

Finally, if the target purity for the material was above the initial purity but below the eutectic purity, a higher dilution to point P_3 could be performed to dissolve some racemic and increase the yield. For instance, if in this case a purity of 60% D in the liquid phase was sufficient, then point P_3 would correspond to a dilution in which the liquid phase has the desired composition or

$$\xi_D^L = \frac{1 - x_D^L}{x_D^L} = \frac{1 - 0.6}{0.6} = 0.667$$

In the region of the phase diagram corresponding to the location of P_3, the equilibrium is defined by Eq. (25.5b) above. The value of K'_{RC} can be obtained by evaluating Eq. (25.5b) at the solubility of the racemic compound ($\xi_D = 1$) provided experimentally in Table 25.2:

$$K'_{RC} = \left(\frac{C_{RC}^{sat}}{1 + \xi_D}\right)^2 \xi_D = \left(\frac{7.1 \text{ mg/g}}{1 + 1}\right)^2 1 = \left(\frac{7.1 \text{ mg/g}}{2}\right)^2 = 12.6 \text{ mg}^2/\text{g}^2$$

We use the value and the liquid-phase purity to evaluate the concentration corresponding to the tie line of point P_3:

$$C_3 = \sqrt{\frac{K'_{RC}}{\xi_D^L}(1 + \xi_D^L)} = \sqrt{\frac{12.6 \text{ mg}^2/\text{g}^2}{0.667}}(1 + 0.667) = 7.25 \text{ mg/g}$$

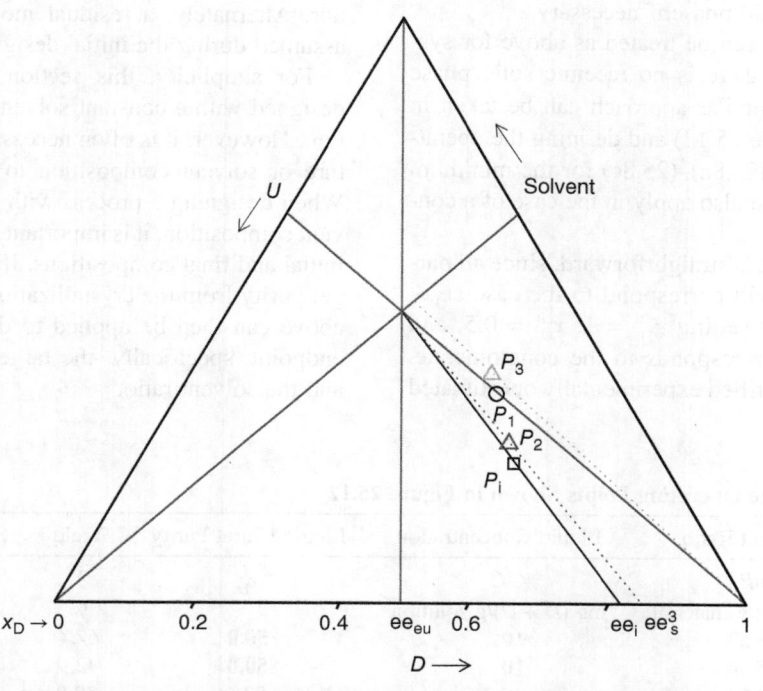

FIGURE 25.12 Ternary phase diagram for a conglomerate system showing the operating line for a starting composition at P_i. Point P_1 is the point of highest yield, while point P_2 corresponds to the conservative isolation ensuring pure desired enantiomer, and point P_3 the process targeting a lower purity (ee_s^3) with higher yield.

We now have all the information to estimate the yield at point P_3 with Eq. (25.11). First we need to calculate the value of β_3 with Eq. (25.8c):

$$\beta_3 = \frac{C_3}{1-C_3} = \frac{7.25/1000}{1-7.25/1000} = 0.00730$$

This value of β_3 is used in Eq. (25.11) to determine solvent ratio corresponding to P_3:

$$P_3^{min} = \frac{x_D^S - x_D^i}{(x_D^S - x_D^L)\beta_3} = \frac{0.5 - 0.714}{(0.5 - 0.6) \times 0.00730} = 97.96$$

As before, the yield at this is obtained from Eq. (25.8b):

$$Y_3 = 1 - \beta P_3 = 1 - 0.00730 \times 97.96 = 0.285 \text{ or } 28.5\%$$

As before the solution yield would be given by $1 - Y_3 = 71.5\%$. By relaxing the target purity of the isolation, we have increased the yield substantially (in Figure 25.11 this is appreciated by visualizing how much higher, i.e. lower solvent ratio, P_3 is in comparison with the other crystallization options). One consideration in this analysis is that since a second crystallization is required, the yield in that crystallization should be close to 100% to ensure purity is preserved, as the solid composition will be lower than the starting liquid composition. The same analysis as the one just performed may be used to design the second crystallization and adjust the target purity of the liquid phase if necessary.

A conglomerate system can be treated as above for systems with $ee_i > ee_{eu}$ since there is no racemic solid phase ($ee_{eu} = 0$ or $x_D^L = 0.5$). A similar approach can be taken in developing the TPD (Figure 25.12) and defining the operating conditions. Equations (25.8a)–(25.8c) for the minimum volume and maximum yield also apply in the case of a conglomerate-forming system.

The analysis in this case is straightforward, since all partially resolved solutions will correspond to the case $ee_i > ee_{eu}$ discussed above, but setting $\xi_D^{eu} = 1$, $x_D^{eu} = 0.5$, and $ee^{eu} = 0$ as the eutectic corresponds to the conglomerate. The solubility should be verified experimentally or estimated from Eq. (25.3c) setting $\xi_D = \xi_D^{eu} = 1$. It is left to the reader to verify that for the system properties indicated in Table 25.1 but with a eutectic concentration $C_{eu} = 10.0$ mg/g, the processing points indicated in the diagram would correspond to the values in Table 25.4. Bolded entries indicate they are defined for the entry and not the result of a calculation, and italic entries are calculated from the formalism presented above.

The much favorable location of the eutectic chiral purity results in higher attainable yield for a given purity target.

When starting from a solution, the process should first be either concentrated or adjusted in solubility to a supersaturated state and then seeded with the racemic form prior to concentrate to the desired process volume. Though a discussion of form control is outside of the scope, multiple racemic, desired, and undesired forms may exist that would alter the TPD. As with any controlled crystallization, seeding with the desired form may be critical to ensure the process consistently achieves the desired chiral purity.

Though less critical than the crystallization conditions, the cake wash conditions should also be considered when planning the crystallization process. The solubility and position within the TPD for the wash may be important to optimize the yield and purity. If the wash solvent is of the same composition and temperature as the crystallization solvent, the same TPD can be used to visualize the step. If the residual mother liquor held in the isolated cake is known, the ee can be calculated using the mother liquor concentration, anticipated weight of the solids, and the amount of residual mother liquor. Alternately, a residual mother liquor of 30% can be assumed during the initial design.

For simplicity, this section assumed the process was designed with a constant solvent composition and temperature. However, it is often necessary to adjust either temperature or solvent composition to address ease of operation. When designing a process with variable temperature or solvent composition, it is important to understand the TPD at the initial and final compositions. If the input purity and the target purity from the crystallization are known, the formalism above can then be applied to determine the crystallization endpoint, specifically, the target equilibrium concentration and the solvent ratio.

TABLE 25.4 Values for the Operating Points Shown in Figure 25.12

Processing Point	Solvent Ratio	Liquid Concentration	Liquid Chiral Purity	Yield	Solid-Phase Purity	β
Symbol	P	C	x_D^L	Y	x_D^S	
Units	g solvent/g material	mg $(D+U)$/g solution		%		
P_i	**32.3**	**10**	**50.0**	*67.4*	*0.818*	*0.0101*
P_1	*56.6*	**10**	**50.0**	*42.9*	**1.0**	*0.0101*
P_2	*65.1*	*9.33*	*53.6*	*38.8*	**1.0**	*0.00941*
P_3	*42.8*	**10**	**50.0**	*56.8*	**0.85**	*0.0101*

25.5.2 Preferential Crystallization

During direct crystallization, the output purity and process yield are limited by the input chiral purity. In cases where increased yield is desired or the earlier partial resolution steps are unable to sufficiently upgrade the chiral purity, alternative purification processes are required. Preferential crystallization may be employed to enhance the chiral purity while maximizing the potential yield [4]. This technique relies on the ability to maintain the crystallization in a supersaturated state with respect to the undesired solid forms (pure enantiomer U or racemic compound). Figure 25.13 shows the schematic of the process for a system in which the equilibrium includes the formation of a racemic compound. The metastable solid phases during the process are denoted by a dashed line, while the relevant active equilibrium is denoted by the solid line. In this strategy, crystallization is adjusted to a slightly supersaturated state, and seeds of the desired enantiomer are added (Figure 25.13a). Preferential growth of the desired enantiomer will then occur, reducing the supersaturation with respect to the "perceived" thermodynamic equilibrium. The solubility is decreased, while the systems maintain equilibrium with the solid phase of the desired enantiomer. All other phases remain supersaturated (Figure 25.13b).

This technique is limited, since reducing the solubility sufficiently to recover a large portion of desired enantiomer will increase the driving force to nucleate the other potential solid compounds. In cases in which the racemic compound crystallizes at supersaturation levels required for adequate yield (i.e. the metastable zone of the racemic compound is not wide enough), sequential batch or semi-batch configurations could be considered. One option is to isolate the desired enantiomer prior to reaching a high supersaturation for the undesired enantiomer. This leaves mother liquors enriched in undesired enantiomer. By seeding with undesired enantiomer, a slurry of undesired enantiomer can be obtained. This will leave racemic mother liquors at the original concentration and the cycle can be repeated [7, 10, 11].

Alternatively, a coupled batch process may be employed in cases where the desired and undesired enantiomers are crystallized simultaneously in crystallizers in which the mother liquors or liquid phase is in equilibrium between the two vessels. This setup reduces the operational complexity and, assuming the crystallization is occurring at similar rates, ensures the mother liquors remain as pure racemic, reducing the possibility of nucleating undesired enantiomer. Theoretically the process could be intensified by reducing the temperature or adding in additional racemic solution. An alternate approach would be to maintain coupled crystallizers with a supersaturated solution crystallizing the desired enantiomer in one vessel and a racemic slurry in the other vessel [12]. By coupling the liquid streams, the mother liquor is maintained as a racemic mixture similar to the simultaneous crystallization. The difference in this case is that the solids in the saturated slurry vessel will become enriched in undesired enantiomer as desired enantiomer is dissolved to maintain a racemic composition in the liquid. Ultimately this would yield two slurries: one pure desired enantiomer and one pure undesired enantiomer.

Though the above approaches are technically feasible, these approaches are more complex and would require a relatively wide metastable zone and low nucleation propensity of the solid phases. It would also be important to understand the growth and nucleation kinetics as well as the impact mixing has on nucleation and the ability to maintain a racemic mother liquor. Finally, the capabilities of the pilot scale and commercial facility should also be considered to ensure adequate equipment and controls are available to successfully implement the process.

25.5.3 Diastereomeric Salt Resolutions

As an alternative to direct crystallization from an enantiomeric system, if the molecule has the ability to form a salt with a chiral modifier, the enantiomeric system can be transformed into a diastereomeric system in order to provide alternative options for the purification of D [13, 14]. Typically the

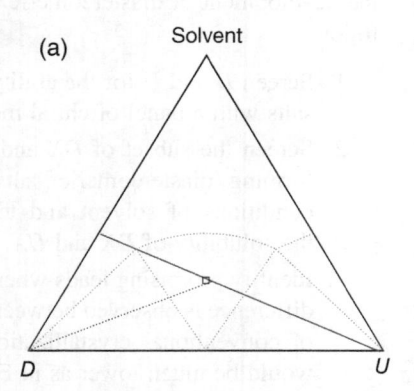

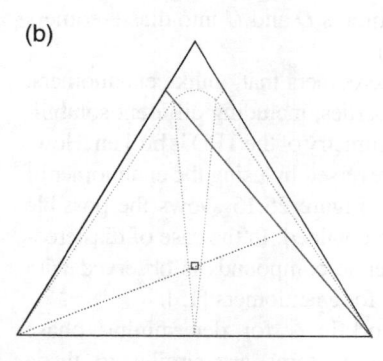

FIGURE 25.13 Schematic of a preferential resolution crystallization indicated the initial seed point (a) and the final pre-filtration point (b). The dashed equilibrium lines indicate the systems that are supersaturated during the process.

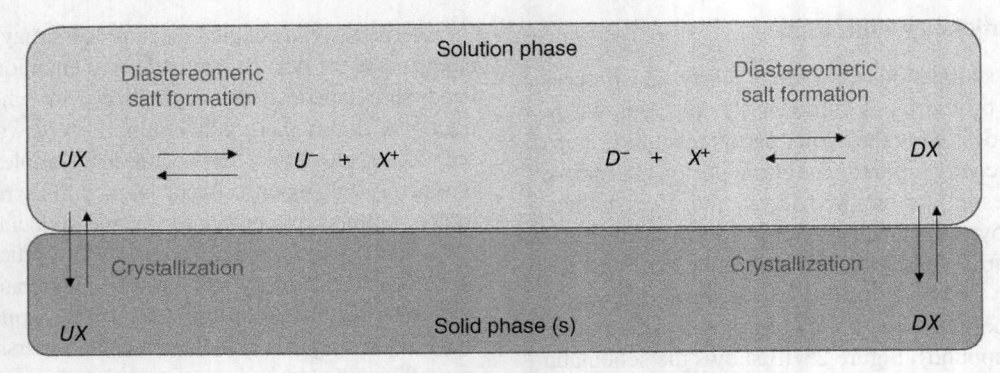

FIGURE 25.14 Transforming an enantiomeric crystallization into a diastereomeric crystallization system by the use of a chiral modifier. Example is shown for the formation of a diastereomeric salt via the addition of a chiral acid (X) to an enantiomeric mixture of free base (U and D).

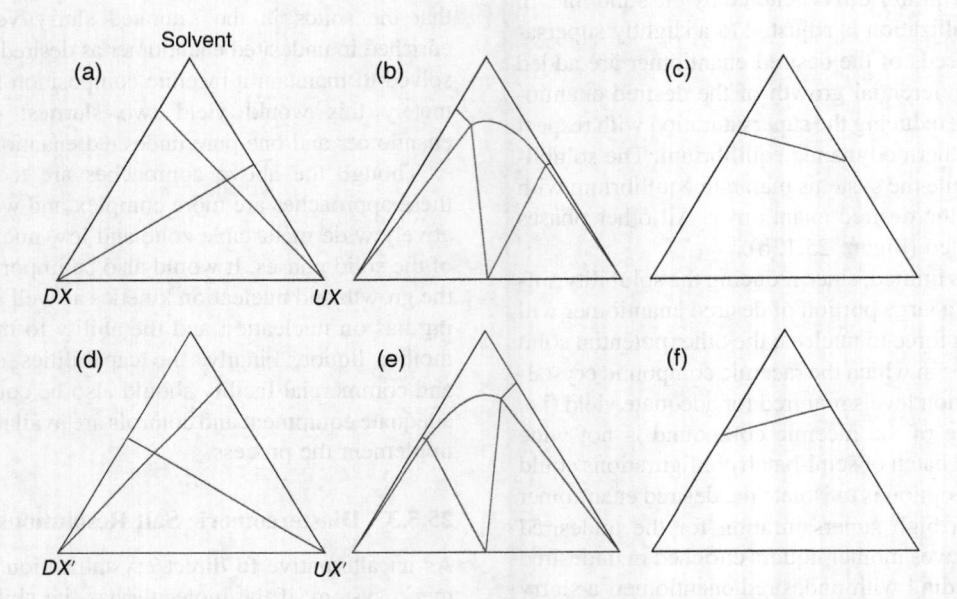

FIGURE 25.15 Typical phase diagrams for diastereomeric systems. (a and d) Simple single eutectic system. (b and e) Complex crystal phase incorporating both DX and UX (top) or (DX' and UX') (bottom) into a single crystalline lattice. (c and f) Solid solution of the diastereomers.

modifier is a chiral acid or base (X). In this case the mirror image symmetry of the enantiomeric phase diagram is broken by transforming the enantiomers D and U into diastereomers DX and UX (Figure 25.14).

As DX and UX are diastereomers that, unlike enantiomers, have different physical properties, including different solubilities in achiral solvent, the symmetry of the TPD is broken. However, the symmetry can be reversed by using the enantiomer of the chiral acid (or base) X'. Figure 25.15 shows the possible phase diagrams that may be obtained. In the case of diastereomers, the formation of racemic compounds is observed with much lower frequency than for enantiomers [15].

Experimentally, the workflows for determining phase behavior of diastereomeric systems are similar to those described above with the added complexity that the relative solubilities of DX and UX are now impacted by the achiral solvent selection and temperature. A typical workflow for the development of diastereomeric salt crystallizations is as follows:

1. Screen D and U for the ability to form diastereomeric salts with a panel of chiral modifiers X.
2. Screen the subset of DX and UX systems capable of forming diastereomeric salts against crystallization conditions of solvent and temperature to determine the solubility of DX and UX.
3. Identify promising leads where a significant solubility difference is observed between DX and UX (in the case of conventional crystallization, the solubility of DX would be much lower as in Figure 25.15a).
4. Fully characterize the identified system(s) to develop the TPD(s) and crystallization conditions.

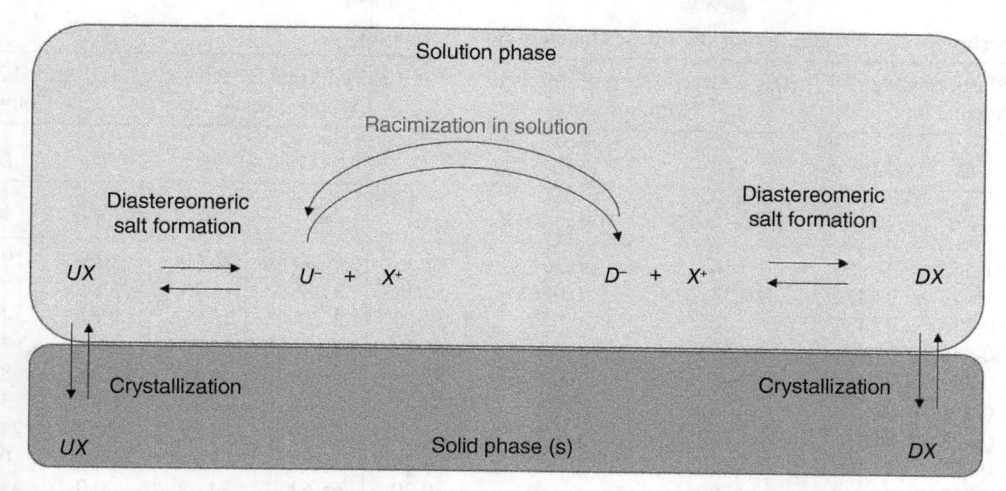

INTRODUCTION TO CHIRAL CRYSTALLIZATION IN PHARMACEUTICAL DEVELOPMENT AND MANUFACTURING 585

FIGURE 25.16 Extension of the diastereomeric crystallization to a "dynamic diastereomeric resolution" where undesired U in solution is able to racemize back to a 50/50 mixture of U and D in the liquid phase. The crystallization can then be driven to nearly pure D in the solid phase via the preferential crystallization of DX due to the asymmetry of the diastereomeric salt TPD, i.e. the solubility of DX is much lower than the solubility of UX.

5. For the identified system(s), develop a salt break procedure(s), typically an acidic or basic extraction process, to separate D from X and allow the recovery of purified D.

While the TPD asymmetry of diastereomeric salt systems can be exploited to obtain increased yields of D from direct crystallization processes as compared with symmetric enantiomeric systems, yields of direct diastereomeric salt crystallizations are still limited by the input purity of the desired molecule. The best possible outcome is high recovery of D and full purge of U. However, in certain instances it may be possible to racemize U into D in the solution phase and use the asymmetry of the DX and UX phase diagrams to drive the maximum yield of the desired product in the solid phase to the combined amount of $D_i + U_i$ at the purity of pure D. These systems are termed dynamic diastereomeric salt resolutions and described further in Figure 25.16.

An example of the direct crystallization system characterization and crystallization design described above is provided in Section 25.5.4.

25.5.4 Case Study: Development of a Direct Crystallization for the Isolation of Pure Chiral Compound A

In this case study the control strategy for a direct chiral crystallization of a partially resolved solution of compound A is presented. The workflow includes characterization of an enantiomeric system of the desired enantiomer D and the undesired enantiomer U, including the creation of a TPD. This characterization formed the foundation for an enantiomeric purity control strategy and the design of a direct crystallization process. The description that follows includes both the practical aspects of enantiomeric crystallization characterization and a description of how a relatively lean data set formed the foundation for the overall chiral purity control strategy for this API step.

The characterization of the TPD of compounds D and U was the foundation for the development of an enantiomeric purity control strategy and the design of a direct crystallization process. The description that follows includes both the practical aspects of enantiomeric crystallization characterization and a description of how a relatively lean data set formed the foundation for the overall chiral purity control strategy for this step.

The mixtures of varying chiral purity presented as slurries were prepared as shown in Table 25.5 and aged overnight at the final crystallization solvent composition and temperature. Each slurry sample was filtered, and the concentration and enantiomeric purity in the mother liquor was determined via HPLC analysis (Table 25.5). The mol % data for the total system composition and equilibrated mother liquor was then transformed by applying an arbitrary scaling factor of the enantiomer values relative to solvent in order to provide the "readable TPD" as shown in Figure 25.17. The solubility of the pure enantiomers is inconsistent (lower) with the partially resolved slurries. Although the exact cause is unclear, the lower solubility may be attributed to method of preparation of the pure enantiomers, resulting in high purity (of other contaminates in the process) and consequently lower solubility. The partially resolved mixtures were more representative of actual processing material and therefore of anticipated crystallizer composition. Therefore the partially resolved liquid-phase equilibria were used to estimate the location of the eutectic. Samples 1 and 7 are equilibrated in racemic compound region of the phase diagram. The data was fitted to Eqs. (25.3c) and (25.5b) (solid lines in Figure 25.17), and

TABLE 25.5 Experimental Determination of the Phase Diagram for Compound A

	(A) Slurry Sample Preparation			(B) Analysis of Equil. Mother Liquor		(C) Scaled Total System Composition			(D) Scaled Mother Liquor Composition		
	D	U	Solvent	D	U	D	U	Solvent	D	U	Solvent
Sample	wt %	wt %	wt %	HPLC wt %	HPLC wt %	%	%	%	%	%	%
1	1.218	1.232	97.550	0.164	0.161	41.45	41.94	16.60	19.9	19.6	60.5
2	1.525	1.003	97.472	0.377	0.066	50.58	33.26	16.17	40.1	7.0	52.9
3	1.790	0.773	97.437	0.446	0.045	58.68	25.34	15.97	45.1	4.6	50.3
4	2.009	0.526	97.464	0.459	0.078	66.47	17.41	16.12	44.4	7.5	48.1
5	4.346	0.479	95.176	0.437	0.087	81.99	9.03	8.98	42.8	8.5	48.7
6	0.937	0.294	98.769	0.430	0.030	54.33	17.04	28.63	44.9	3.1	52.0
7	0.707	0.772	98.521	0.180	0.190	35.84	39.18	24.98	20.7	21.9	57.4
8	0.891	0.000	99.109	0.340	0.000	64.26	0.00	35.74	40.6	0.0	59.4
9	0.004	0.953	99.042	0.000	0.360	0.29	65.62	34.09	0.0	41.9	58.1

(A) Slurry samples were prepared at the total system compositions shown. (B) The slurry samples were equilibrated at the final crystallization conditions and the mother liquor concentrations of the compound A and the undesired enantiomers were measured via HPLC. Using a scaling factor of 1/200 for the solvent % to enable graphical visualization, the total system composition (C) and mother liquor results (D) were transformed for TPD generation.

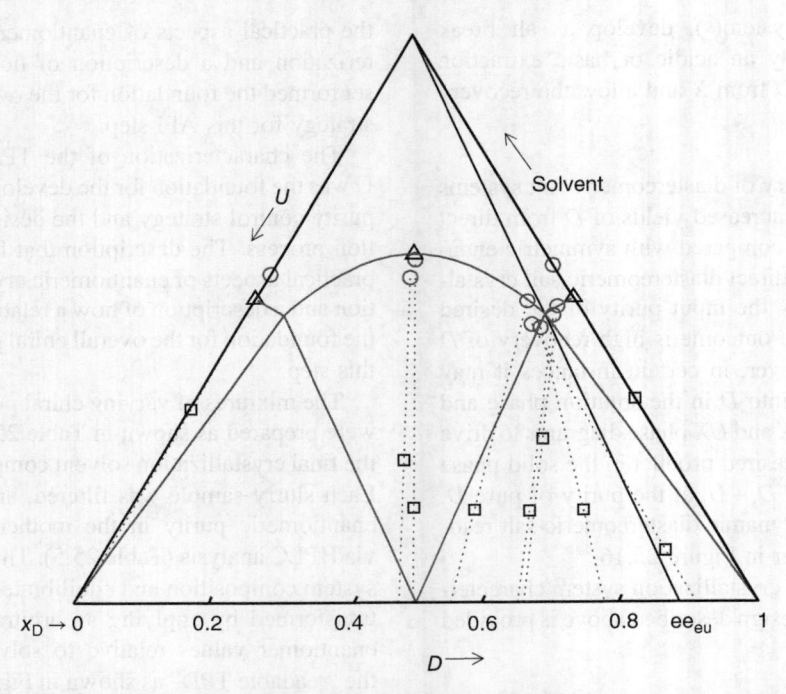

FIGURE 25.17 Ternary phase diagram for compound A. The circles correspond to the data. The squares correspond to the estimated solubility of the pure enantiomers and racemic compound used to establish the liquid equilibrium lines (note the deviation in the pure enantiomer experimental solubility). The squares represent the overall system composition. The chiral purity of the eutectic composition is projected by a solid line to the X_D axis.

it was determined that the enantiomeric pair forms a racemic solid with a eutectic point of 88.5 wt % D (ξ_D^{eu} = 0.130, x_D^{eu} = 0.885, ee^{eu} = 0.77) and equilibrium concentration C_{eu} = 0.0488 g/g solution (4.88 mg/g). Sample 1 indicated a lower solubility, so the estimate of the racemic compound solubility was slightly lower in solubility to ensure that a conservative estimate of the eutectic was used for process design.

This behavior and eutectic point composition was subsequently confirmed via isolating the solids from these slurry samples and determining the phase diagram via melting point analysis as described earlier (Figure 25.18). Equations (25.7a) and (25.7b) were evaluated using only the data for the pure solid phases (enantiomers and racemic compound). For pure phases a single melting point is

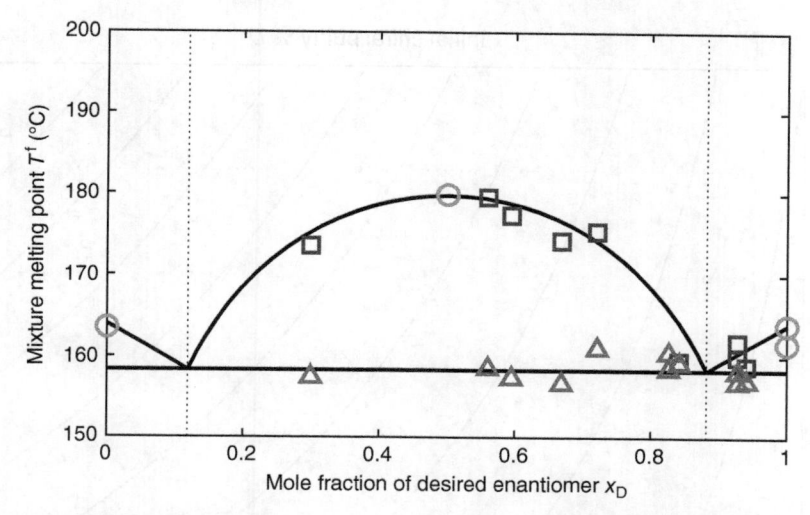

FIGURE 25.18 DSC of the isolated solids from the equilibration experiments (triangles, first melting point; squares, second melting point; circles, estimated melting points for the pure solid phases) and the corresponding Schroeder–van Laar and Prigogine–Defay equations (solid lines) obtained using the thermodynamic properties of the solid phases. The dashed lines correspond to the eutectic composition.

determined (circles in the diagram), whereas for mixtures of phases points are observed in the melting point curves; the first event is indicated with triangles, and the second with squares. The eutectic composition obtained from Eqs. (25.7a) and (25.7b) corresponds to $x_D^{eu} = 0.881\,50$ that is in good agreement with the one determined from the slurry equilibration experiments ($x_D^{eu} = 0.885$).

Having determined that the enantiomer system of compound A forms a racemic compound with a eutectic point of 88.5 wt % D, the overall chiral purity control strategy for this step could be defined. It was desired to obtain chiral purity solids from the crystallization in excess of 98% ($x_D^S > 0.98$). As a direct crystallization was selected, the required minimum chiral purity of the stream used as input to the crystallization is at least 88.5 wt % D to enable the crystallization to upgrade the solids to a chiral purity greater than 98%.

Direct crystallization requires that the purity of the input stream can be consistently provided higher than the eutectic purity. The goal of the analysis is to establish the balance between the specification that will be selected for the input chiral purity and the yield that will be obtained from the crystallization. Equations (25.8b), (25.8c), and (25.9) (with Eq. 25.8c evaluated with the eutectic concentration) can be used to generate a family of curves that relate the yield and isolated purity for solvent ratios below the optimal ratio (since pure material is not required). Figure 25.19 shows the curves collected for different input chiral purities and solvent ratios from zero to the optimal solvent ratio for the corresponding input chiral purity of the system. If the desired yield for the crystallization is required to be above 85%, then the input purity should be higher than 97% as indicated by the intersection of the red dashed lines corresponding to the desired yield (horizontal) and isolated purity (vertical).

Alternatively if the isolation purity and yield are relaxed to 97.5 and 80%, respectively, then the input specification should be higher than 95% chiral purity. As discussed before an adequate margin should be placed on the solvent to solid ratio to ensure sufficient control under any anticipated variability in the process.

Figure 25.20 demonstrates the usefulness of the TPD in visualizing two different options for the system under development. However, for quantification, graphical projections of different possible outcomes corresponding to different operating conditions allow for a more nuanced navigation of the accessible operating space.

In this case setting the input chiral purity specification at 97% allows for the crystallization to be operated with a 90% yield and produces isolated material with chiral purity of 98%.

25.6 SUMMARY AND CONCLUSIONS

Here we have presented the concepts and general workflow associated with the characterization of a chiral system for the purpose of designing an optimal crystallization to achieve a given target chiral purity. Generally, the manufacture of pharmaceutical API consists of a series of synthetic steps with multiple isolation points. Partially resolved streams are prevalent due to the almost universal implementation of enantioselective chemistry to introduce chiral centers in the product synthesis; therefore direct resolution is the typical chiral purification process. In cases in which the chirality is introduced before the final isolation, it is useful to design the chiral crystallizations as an integrated effort across all the available isolations. The step with the most favorable TPD should be used

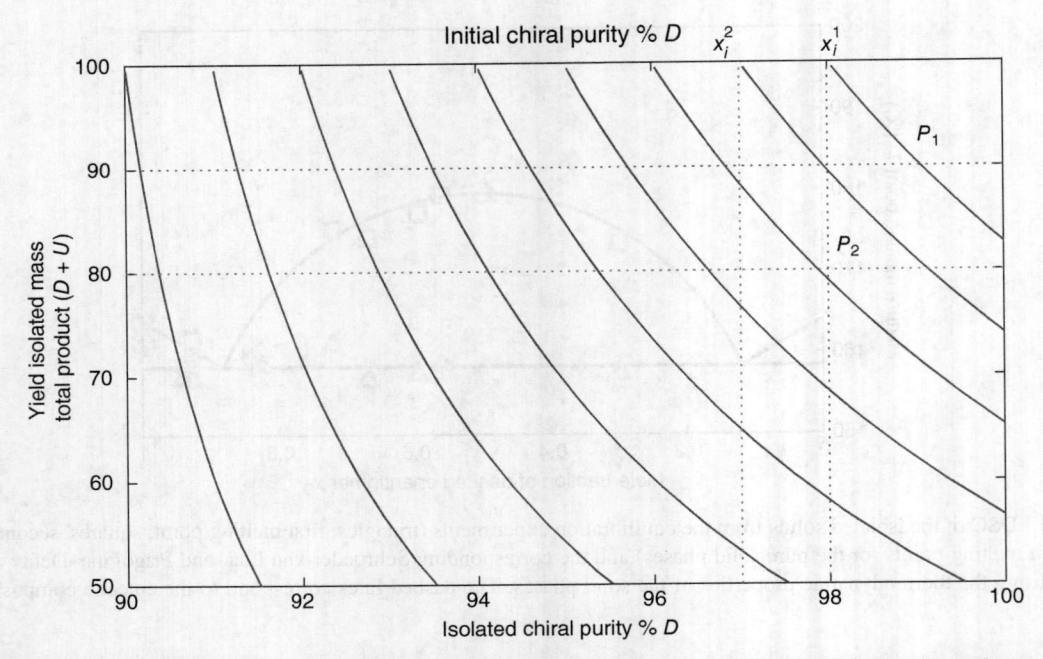

588 CHEMICAL ENGINEERING IN THE PHARMACEUTICAL INDUSTRY

FIGURE 25.19 Comparison of the impact of input enantiomeric purity on the output yield and crystallization volume during the direct crystallization of molecule A. The solid lines indicate the relationship between the isolated purity (x axis) and yield (y axis) for a given initial purity (indicated by the value at 100% yield). Point P1 and P2 represent the yield (90 and 80%) associated with a purification of 97% to 98% and 95.3 to 97% respectively.

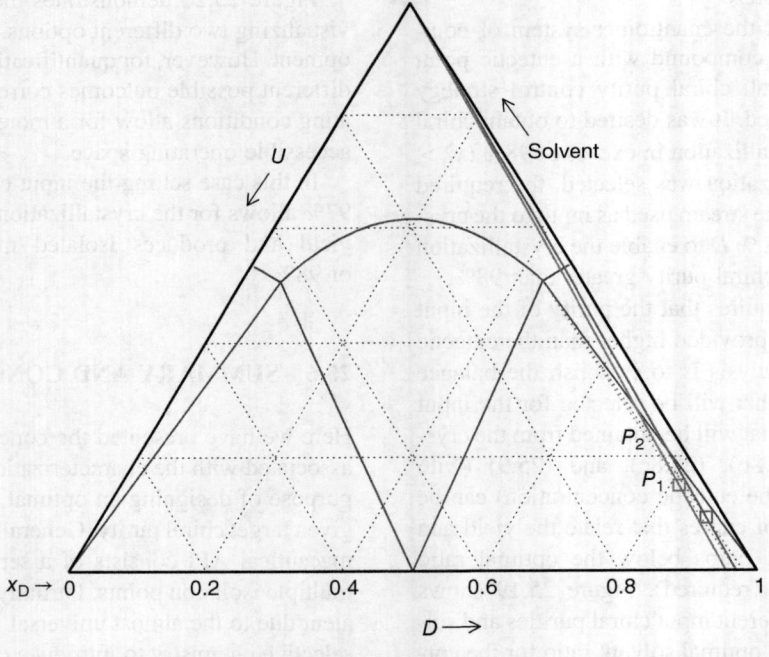

FIGURE 25.20 TPD projection of the two options for trade of initial purity requirements and isolated purity shown in Figure 25.18.

as the quality gatekeeper as it would provide the highest possible yield to reach a given chiral purity. Although the discussion has been presented mainly from the perspective of direct crystallizations, with a thorough understanding of the workflows and the underlying design and thermodynamic equations, the procedures can be readily adapted to more complex (and rare) chiral crystallizations.

ACKNOWLEDGMENT

All graphics were created with Matlab 2017b by the MathWorks. The ternplot function was used with a GNU General Public License from the source created by Sandrock and Afshari [16].

REFERENCES

1. Rouhi, A.M. (2003). Chiral business – fine chemicals companies are jockeying for position to deliver the increasingly complicated chiral small molecules of the future. *Chem. Eng. News* 81 (18): 56–61.
2. Challener, C.A. (2016). Expanding the chiral toolbox: recent chiral advances demonstrate promise for API synthesis. *Pharm. Technol.* 40 (7): 28–29.
3. Nguyen, L.A., He, H., and Pham-Huy, C. (2006). Chiral drugs: an overview. *Int. J. Biomed. Sci.* 2 (2): 85–100.
4. Smith, S. (2009). Chiral toxicology: it's the same thing...only different. *Toxicol. Sci.* 110 (1): 4–30.
5. Faul, M.M., Busacca, C.A., Eriksson, M.C. et al. (2014). Part 2: designation and justification of API starting materials: current practices across member companies of the IQ consortium. *Org. Process Res. Dev.* 18 (5): 594–600.
6. Jacques, J., Collet, A., and Wilen, S.H. (1981). *Enantiomers, Racemates, and Resolutions*. New York: Wiley.
7. Kotelnikova, E.N., Isakova, A.I., and Lorenz, H. (2017). Non equi-molar discrete compounds in binary chiral systems of organic substances. *Cryst. Eng. Comm.* 19: 1851.
8. Lorenz, H., Perlberg, A., Sapoundjiev, D. et al. (2006). Crystallization of enantiomers. *Chem. Eng. Process.* 45: 863–873.
9. Prigogine, I. and Defay, R. (1962). *Chemical Thermodynamics*, Translated by D. H. Everett. New York: Wiley.
10. Chen, A., Wang, Y., and Wenslow, R. (2006). Purification of partially resolved enantiomeric mixtures with the guidance of ternary phase diagram. *Org. Process Res. Dev.* 12: 271–281.
11. Angelov, I., Raisch, J., Elsner, M.P., and Seidel-Morgenstern, A. (2006). Optimization of initial conditions for preferential crystallization. *Ind. Eng. Chem. Res.* 45: 759–766.
12. Angelov, I., Raisch, J., Elsner, M.P., and Seidel-Morgenstern, A. (2008). Optimal operation of enantioseparation by batchwise preferential crystallization. *Chem. Eng. Sci.* 63: 1282–1292.
13. Levilain, G., Eicke, M.J., and Seidel-Morgenstern, A. (2012). Efficient resolution of enantiomers by coupling preferential crystallization and dissolution. Part 1: experimental proof of principle. *Cryst. Growth Des.* 12: 5396–5401.
14. Mitchell, A.G. (1998). Racemic drugs: racemic mixture, racemic compound, or pseudoracemate? *J. Pharm. Pharm. Sci.* 1: 8–12.
15. Mersmann, A. (1995). *Crystallization Technology Handbook*, 417–418. New York: Marcel Dekker, Inc.
16. Sandrock, C. and Afshari, S. (2016). alchemyst/ternplot: DOI version [Data set]. Zenodo. http://dx.doi.org/10.5281/zenodo.166760 (accessed March 2018).

26

MEASUREMENT OF SOLUBILITY AND ESTIMATION OF CRYSTAL NUCLEATION AND GROWTH KINETICS

NANDKISHOR K. NERE, MANISH S. KELKAR, ANN M. CZYZEWSKI, KUSHAL SINHA, AND EVELINA B. KIM

Process Research and Development, AbbVie Inc., North Chicago, IL, USA

26.1 INTRODUCTION

The performance of a crystallization process in terms of the resulting particle size distribution and impurity rejection is largely dictated by the relative rates of crystal nucleation and growth. Hence, characterization of nucleation and growth kinetics is of paramount importance to enable the efficient development of robust crystallization processes. The fundamental basis on which to understand any crystallization process is solubility; this chapter begins with a discussion of solubility measurement methodologies, best practices, and relevant examples. We then delve into a discussion of different approaches to interrogating nucleation and growth kinetics, including both experimental and computational methodologies. Lastly, relevant case studies are presented to exemplify how a fundamental understanding of nucleation and growth kinetics can be used in the development of an appropriate crystallization process and control strategy.

26.2 SOLUBILITY

The term *solubility* is defined as "analytical composition of a mixture or solution, which is saturated with one of the components of the mixture or solution, expressed in terms of the proportion of the designated component in the designated mixture or solution" [1]. While the "mixture" or "solution" could be composed of any physical state, this discussion is focused on systems in which a crystalline solid (solute) is dissolved in a liquid (solvent), as those pertain to industrial crystallizations of drug substances (DSs) and/or their chemical intermediates.

When a solid is contacted with a liquid, a dynamic process is initiated wherein molecules from the solid phase dissolve into the liquid phase and dissolved molecules re-adsorb onto the solid surface. Initially, the rate of solute molecules going into the liquid phase is higher than that of solute molecules adsorbing back on the solid surface, resulting in increased concentration of solute in the solution. Eventually, a dynamic equilibrium is reached, where the rates of dissolution and adsorption are equal, and the solute concentration reaches a steady state. This concentration is defined as the solubility of the solute in a given solvent. For any solute A, this equilibrium can be depicted as

$$A_{(S)} \leftrightarrow A_{(L)} \tag{26.1}$$

$$\mu(A,S) = \mu(A,L) \tag{26.2}$$

where

$A_{(S)}$ is solute in solid phase.
$A_{(L)}$ is solute in liquid phase.
$\mu(A, S)$ is the chemical potential of A in solid phase S.
$\mu(A, L)$ is the chemical potential of A in liquid phase L.

This solid–liquid equilibrium can be mathematically represented by the van't Hoff equation as

---

Chemical Engineering in the Pharmaceutical Industry: Active Pharmaceutical Ingredients, Second Edition.
Edited by David J. am Ende and Mary T. am Ende.
© 2019 John Wiley & Sons, Inc. Published 2019 by John Wiley & Sons, Inc.

$$\ln(\gamma x) = \frac{\Delta H_m}{R}\left[\frac{1}{T_m} - \frac{1}{T}\right] \quad (26.3)$$

where

- ΔH_m is the molar enthalpy of melting.
- R is the universal gas constant.
- T_m is the absolute melting temperature of the solute.
- T is the absolute temperature of the solution.
- x is the mole fraction of the solute.
- γ is the activity coefficient.

As solubility is significantly influenced by properties of both the solvent and solute, characterization of the solid phase is very important for accurate definition *and measurement* of solubility.

Discussed in this section are several of the more prevalent methods used to measure solubility as the first step toward the goal of developing commercializable crystallization processes for pharmaceutical compounds. Measurements near the critical region, under high pressures, or advanced methods used in bioavailability/pharmacokinetic studies and during drug discovery phase are not discussed here.

26.2.1 Shaker-Flask Method

Typically, measurement of solubility involves some variation of the shaker-flask method and includes the following steps:

1. Solid–liquid contact and attainment of equilibrium.
2. Sampling or separation of solid and liquid phases.
3. Measurement of solute concentration in liquid phase.
4. Confirmation of crystal form of the solid phase.

Some considerations for obtaining accurate solubility measurements in the context of the shaker-flask procedure are outlined in the following subsections.

26.2.1.1 Material Properties

26.2.1.1.1 Purity The presence of very small amounts of impurities in a solubility study can significantly affect the measured solubility. Impurities can not only change the activity coefficient in Eq. (26.3), thus altering the solubility of solute significantly, but in some cases, they can also alter the thermodynamic stability of the solute solid form. Impurities can also impact nucleation and growth kinetics, which can in turn affect the solubility measurement.

To ensure the most accurate and representative solubility measurements are obtained, it is important to avoid contamination with extraneous impurities and ensure the system is stable such that degradants do not form over time. It is equally important to interrogate the impact of process-related impurities on the measured solubility when relevant to the system being evaluated.

26.2.1.1.2 Effect of Solute Content and Particle Size The amount of solute present in a solubility experiment should be sufficient to ensure it will not completely dissolve over the time course of the study and have enough solids to evaluate crystal form while not using an excessive amount, making it hard to mix. In certain situations, such as in the case of indomethacin [2], the amount of suspended solid could affect the measured solubility due to an imbalance between solute crystallization and dissolution rates. It is recommended that the amount of solute present in the experiment be reported and, if necessary, the impact of solute amount on measured solubility be explored.

Studies have also shown the effect of particle size on the measured solubility through Gibbs–Thomson type of equation [3]:

$$\ln\left[\frac{c(r)}{c^*}\right] = \frac{2M\gamma}{\vartheta RT\rho r} \quad (26.4)$$

where

- $c(r)$ is solubility of solid with particle radius of r.
- c^* is the equilibrium solubility of the solid substance.
- M is the molar mass of solid in solution.
- γ is the interfacial tension of solid and liquid.
- ρ is the density of the solid.
- ϑ is the number of moles of ions generated from 1 mol of solute.

For nonionic substance, $\vartheta = 1$. Ratios $[c(r)/c^*]$ of as high as approximately 13 have been observed in the literature for some inorganic compounds [3].

26.2.1.2 Experimental Setup and Design
Important considerations for appropriate equipment setup and design include material compatibility, agitation design, and approach and attainment of equilibrium.

26.2.1.2.1 Material Compatibility The compatibility of the system (solute and solvent) with the container and sampling device needs to be considered. In general, glass vials/reactors are used on account of their reasonable chemical inertness. Vials should be tightly sealed to avoid evaporation of solvent (especially when dealing with solvent mixtures) and exposure to air/oxygen/moisture. If the study materials are photosensitive, exposure to light should be reduced by using dark/amber-colored vials or by keeping the vials in dark container.

26.2.1.2.2 Agitation
Agitation promotes intimate contact between solid and liquid phases, which is required to facilitate proper wetting of the solids and equilibration of the two-phase system. Several types of equipment (rocking table, vortex mixer, mechanical arm, etc.) can be used to keep the solubility vial in motion, thus providing mixing. Typically, this type of mixing is expected to be gentler on the crystals and is employed when solubility screening is coupled with crystal morphology screening. Alternatively, an overhead stirrer or magnetic stirrer can be used to internally agitate the solubility mixture rather than relying on the motion of the container to contact the phases. This type of mixing is usually more vigorous and can accelerate attainment of equilibrium, which is especially useful in the systems with slow dissolution/de-supersaturation kinetics. However, it should be noted that magnetic stirrers can grind the solids, thus impacting the resultant crystal morphology, particle size, and crystallinity to some extent.

26.2.1.2.3 Approach and Attainment of Equilibrium
During the equilibration period, the solubility can be approached by either (i) dissolving more solute in an undersaturated solution (saturation experiment) or (ii) removing solute from a supersaturated solution (de-supersaturation experiment). Schematically, these approaches to equilibrium solubility are represented in Figure 26.1. Since theoretically both approaches should result in the same value, it is recommended that solubility measurements be made both ways as a means of verification. A difference in measured values from these two approaches may indicate that impurities are impacting solubility or de-supersaturation kinetics. This impact can be further discerned by repeating the approach from undersaturation experiments in the presence of spiked impurities (one at a time) and measuring solubility. A difference in the measured solubility may confirm the impurity's impact on solubility, while a similar value may indicate an effect of an impurity on crystallization kinetics of the solute.

The attainment of equilibrium in a two-phase system takes time, typically on the order of hours to days of sustained intimate contact. The time needed to reach equilibrium can vary based on liquid and solid physical properties as well as experimental conditions; however, equilibrating mixtures for 24 hours before sampling for solubility measurements is recommended. Best practice is to take at least two consecutive measurements, which should measure within experimental error, to confirm equilibrium has been reached. Moreover, care must be taken to ensure equilibrium is not disturbed during the hold period by, for example, adding new solvent to the mixture or exposing the system to variable temperatures. Lastly, it is very important that the solute is chemically stable throughout the duration of the hold period so as to avoid impact of the degradants on the solubility measurements (see Section 26.2.1.1).

26.2.1.3 Equilibration Parameters
Two of the most notable parameters that can significantly impact solubility measurements in a given solvent system are temperature and pH. The impact of temperature on solubility in different systems can vary significantly; however, it has been reported that, for example, controlling temperature to within ±1 °C for an organic compound with a heat of solution of approximately 10 kcal/mol leads to variability in measured solubility of about ±10% [4]. Best practice, therefore, is to accurately control temperature to the extent possible both during the equilibration and during the sampling for solubility measurements.

The pH of a mixture can impact solubility (particularly salts) as well as the distribution of ionic species and hence the crystal form of the solid. It is notable that solubility of a salt can also be affected by the choice of acid/base through the common ion effect, particularly at extremes of the pH scale. Artificially low solubility may be observed due to salting out of the desired compound due to the presence of common ions introduced through acids or bases.

26.2.1.4 Sampling
The conventional sampling method involves removing a sample for solubility measurement. After allowing the solids to settle as much as possible, a sample of the liquid phase is removed and filtered through an

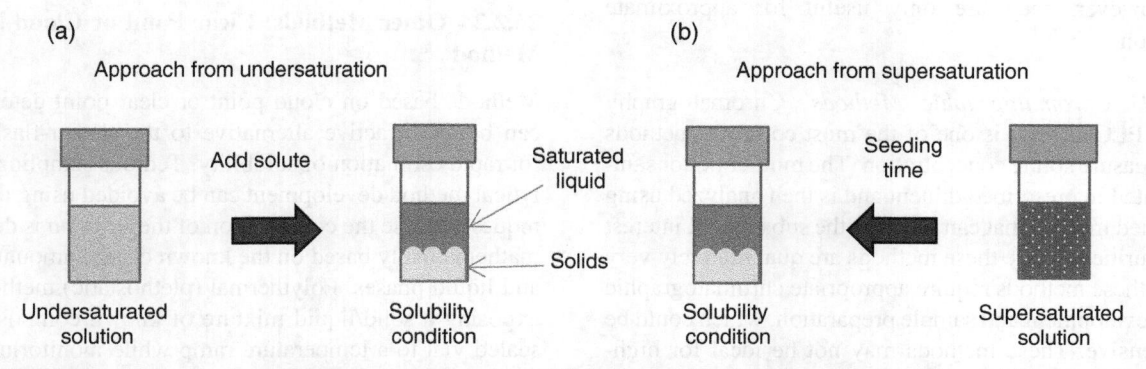

FIGURE 26.1 Schematic demonstrating an approach to equilibrium solubility from (a) undersaturation and (b) supersaturation.

appropriate microporous disposable syringe filter (0.2 or 0.45 μm). It is important to ensure that the selected filters do not adsorb the solute and are used correctly in order to not adversely impact the solubility measurement. Using this sampling method, it can be difficult to maintain equilibrium conditions throughout isolation of the sample, particularly at high and low temperatures. Typically, syringes/probes used to sample slurries/liquids are preheated/precooled to appropriate temperature to avoid disturbing the equilibrium while sampling. On the other hand, this challenge can also be overcome using automated sampling techniques discussed below.

Sampling robots or auto-sampling assemblies, such as Global FIA® FloPro® and Komplx® KS-1, can be used efficiently to sample systems with minimal disturbance to equilibration. These sampling systems can automatically filter a slurry and sample only the mother liquor, making them useful for tracking de-supersaturation. Most sampling systems can also automatically quench or dilute the samples to ensure the composition does not change over time. The Komplx® KS-1 system has the ability to dilute the sample at the probe, and FloPro® has introduced heated probes, both enhancements that can further minimize temperature effects during sampling.

In isosystic systems, where analysis is performed with *in situ* probes/process analytical tools (PAT), external sampling of slurry or mother liquor is not necessary. In such cases, disturbance of the system equilibrium is minimized. These systems are particularly useful in data intensive studies, wherein sample preparation and analysis can become burdensome.

26.2.1.5 Measurement of Solute Concentration in Liquid Phase Many different destructive and nondestructive methods can be used to analyze solute concentration in the liquid phase. Some of the most commonly used techniques used for solubility measurements are described in the subsections below. The simplest method to analyze the liquid-phase concentration is to evaporate a known quantity of well-characterized solvent and quantitate the resultant solid residue. Alternatively, solute can be added incrementally to a solution to approximate the solubility limit. These methods are inexpensive and do not need elaborate method development; however, they are only useful for approximate quantitation.

26.2.1.5.1 Chromatographic Methods Chromatography (HPLC/UPLC or GC) is one of the most common methods used to measure solute concentration. The mother liquor sample is diluted in prescribed diluent and is then analyzed using a predefined method that can separate the substrate of interest from impurities. While these methods are quantitatively very accurate, these methods require appropriate chromatographic method development and sample preparation, which could be labor intensive. These methods may not be ideal for high-throughput solubility measurements; however, autosampling robots and online HPLC systems can help circumvent some of these challenges.

26.2.1.5.2 In Situ Methods Various *in situ* measurement methods such as electrical methods (conductometry, electromotive force, polarography, pH), optical methods (colorimetry, refractometry, polarimetry), spectrophotometric methods (UV/Vis, infrared [IR], near IR [NIR], Raman), and densitometry can be employed to determine the concentration of the solute in solution. The selectivity and sensitivity of these nondestructive methods (including impact of impurities) need to be evaluated for each of the solubility measurement systems. Quantitative solubility determination requires development of a calibration curve. When applicable, these high-throughput methods are ideal for solubility screening as they are able to generate large amounts of data that can inform solubility behavior.

26.2.1.6 Confirmation of the Crystal Form of the Solid Phase As solubility is defined with respect to a specific solid crystal form, the solid form of the equilibrated solids is typically analyzed using a powder X-ray diffractometer (XRD) and compared to reference patterns. Solids can be isolated using disposable or fritted centrifuge tubes, and the wet cake prepared for X-ray powder diffraction (XRPD) analysis. Due to the potential for solid form to change during isolation and drying, it is best to analyze a wet cake rather than fully dried solids. High-throughput XRD machines are available to efficiently screen a large number of samples (e.g. Bruker's D8 Discover HTS^2 system). Other off-line techniques such as differential scanning calorimetry (DSC) or thermogravimetric analysis DSC (TGA-DSC) can also be used to verify solid forms based on known melting point, melting point shift, and/or mass loss (for solvates/hydrates).

In situ methods such as spectroscopic techniques including IR, NIR, or Raman can also be used to distinguish between solid forms. The applicability of such techniques for each system needs to be evaluated before the use, as interference from solvents in the region of interest may render these methods inappropriate.

26.2.2 Other Methods: Clear Point or Cloud Point Method

Methods based on cloud point or clear point determination can be an attractive alternative to the shaker-flask method for rapid estimation of solubility. Tedious sampling and analytical method development can be avoided using these techniques because the composition of the solution is determined mathematically based on the known charge amounts of solid and liquid phases. Polythermal (plethostatic) methods entail exposing a solid/liquid mixture of known composition in a sealed vial to a temperature ramp while monitoring the turbidity of the mixture. The clear point is the temperature at which solids dissolve. While the clear point temperature is fundamentally different than the solubility temperature, it approaches the solubility temperature when the heating rate is slow relative to the kinetics of dissolution and hence can be used to estimate solubility.

Once a clear/homogeneous solution is obtained, the solution can then be cooled until it becomes cloudy, which marks the cloud point temperature. In conjunction with the clear point temperature, the cloud point temperature measurement can generate information about the metastable zone width (MSZW). As the cooling rate is decreased, the cloud point temperature approaches the clear point temperature. These temperature variation methods are reversible such that the same sample can typically be heated and cooled multiple times to get precision on the measurement provided the crystal form of the solids does not change. These methods can also be used to learn about liquid–liquid-phase separation (LLPS) during crystallization.

Isothermal methods, on the other hand, alter the composition of the vial while holding at constant temperature until a clear solution is obtained. The composition at which the solids completely dissolve is called the clear point composition. When the solvent is added slowly enough, the clear point composition, and hence the solubility at the given temperature, can be estimated with reasonable accuracy. Unlike the temperature variation method, the solvent addition method is not reversible; fresh solvent/solute mixtures must be prepared and tested to evaluate reproducibility of the measurement.

Because these methods require less manual intervention and are amenable to automation, multi-reactor setups such as Crystal16® and Crystalline® are well suited for these measurements. While Crystalline® can execute both temperature variation and solvent addition methods, Crystal16® is not currently designed to execute solvent addition experiments readily.

There are several considerations that can factor into the applicability of clear and cloud point measurements to a particular study. One consideration is that these methods require *a priori* knowledge of the approximate solubility and polymorph landscape. The experimental conditions must be selected to ensure the clear/cloud points will be traversed over the range of interest. Moreover, because solids dissolve and are not available for crystal form testing, the crystal form landscape must be understood to confirm that (i) the desired form is stable over the experimental range of the studies and (ii) no metastable forms exist. Additionally, if the dissolution kinetics are unusually slow, errors in solubility measurement can be incurred. Measurements at different temperature ramps and solvent addition rates are hence recommended to avoid erroneous conclusions. Lastly, the clear point temperature can be underestimated if the system under study is prone to "crowning," i.e. creeping/crystallization of solid phase on the walls of the vial above the liquid level. The erroneous conclusion could also be reached if the solids tend to cream or settle readily, and the optical sensor may not detect the solids present in the experiment. This issue can be somewhat mitigated by using alkaline earth metal stir bars; however, it is a good practice to observe the vials periodically to confirm if such problems exist and rectify those in timely manner.

26.2.3 Regression of Solubility Data

Mathematical treatment of solubility data can oftentimes be insightful to garner more information from the discrete and limited measurements. During early process development, solubility data from the solvent screening can be appropriately modeled to render solubility prediction in a variety of solvents, which can significantly reduce the number of experiments. For the later stages of process development when the solvent system is already established, solubility experiments are designed to collect detailed solubility data at various process-relevant conditions. These data are typically regressed to get a mathematical representation of solubility as a continuous function of a relevant process parameter (solvent composition, temperature, pH, etc.). The resulting solubility equation is customarily the backbone upon which the crystallization process parameters are built.

Various mathematical models have been used in the literature to fit solubility data [3]; some of the more commonly used model equations are summarized below in Table 26.1.

For the limiting case of one solvent ($x_1 \rightarrow 0$), the model equations (26.8)–(26.10) for two-solvent systems equations convert to a form similar to Eq. (26.6). Models represented by Eqs. (26.8)–(26.10) are also included in the Crystallization

TABLE 26.1 Solubility Regression Models Describing Equilibrium Solubility ($c*$) for One-solvent and Two-solvent Systems

One-Solvent Systems		Two-Solvent Systems	
$\ln c^* = A + BT + CT^2$	(26.5)	$\ln c^* = A + BT + Cx_1 + \frac{Dx_1}{T}$	(26.8)
$\ln c^* = A + \frac{B}{T} + \frac{C}{T^2}$	(26.6)	$\ln c^* = Ax_1^2 + \frac{Bx_1}{T} + \frac{C}{T} + D\exp(x_1)$	(26.9)
$\ln c^* = A + \frac{B}{T} + C\ln T$	(26.7)	$\ln c^* = A + \frac{B}{T} + \frac{C}{T^2} + D\exp(x_1)$	(26.10)

T is the absolute temperature.
x_1 is the anti-solvent weight fraction.
A, B, C, and D are the constants/fit parameters.

Toolbox utility from DynoChem®. This software utility can fit the measured solubility data and choose the best fit equation for further crystallization process modeling and design.

Other models, e.g. nonrandom two-liquid segment activity coefficient (NRTL-SAC) and UNIQUAC functional group activity coefficients (UNIFAQ), are also frequently used to model and fit the solubility data. They are included in some of the commercially available software such as Aspen Properties®.

The solubility of a salt as a function of pH is typically modeled using Henderson–Hasselbalch equation. For a monoacidic salt at a pH > pK_a, this equation can be written as Eq. (26.11). Caution must be used in extrapolating this model beyond the measured pH window to account for the stability of crystal form:

$$c^* = c_0 \left[1 + 10^{(pH - pK_a)} \right] \quad (26.11)$$

Other approach-based theories of solid–liquid equilibria also exist for the prediction of solubilities of crystalline solids in liquids and are not discussed here.

26.2.4 Recommended Protocol for Solubility Measurement

A recommended protocol for measuring solubility is summarized below. Care must be taken to use consistent units (mg/ml solution, mg/g solution, mg/g solvent, wt %) while measuring, reporting, and using the solubility data. A significant error can be introduced through the density of solution (ml vs. g) or basis of measurement (solvent vs. solution):

1. Plan experiments to approach solid–liquid equilibrium from both saturation and de-supersaturation.
2. Mix the desired solid form with solvents/solutions with magnetic stir bar.
3. Equilibrate at desired pH and/or temperature for approximately 24 hours if the chemical stability is not of concern and kinetics of solubilization are not known *a priori*.
4. Confirm the presence of solids and ensure that desired temperature and/or pH along with the total mass is maintained. If not, set the system to desired state point and equilibrate for another 24 hours.
5. If desired state point is maintained, sample liquid phase without disturbing the equilibrium.
6. Analyze the liquid phase for concentration of the compound of interest using appropriate technique (i.e. HPLC, IR, etc.).
7. Stir for another approximately 24 hours.
8. Confirm that the desired state point is maintained and sample the liquid phase again.
9. Analyze the liquid phase again for concentration of the compound of interest.
10. If the two measured concentrations are within acceptable tolerance, sample the solid phase to confirm crystal form.
    a. If mixture of solvents is used, sample liquid phase (using GC, KF, etc.) for solvent composition.
11. If crystal form is consistent and solvent composition is maintained, the measured concentration is the solubility at the given state point (crystal form, temperature, solvent composition, impurity profile, pH).
12. Evaluate the effect of process-relevant impurities by comparing saturation and de-supersaturation experiments and spiking specific impurities during the solubility measurement experiments.
13. Regress the data using an appropriate mathematical model to enable solubility predictions over the desired state space.

26.2.5 Example Applications of Solubility Measurement

In this section, practical examples of how the solubility measurement recommendations and best practices have been applied to crystallization process design and development are presented. The first example describes the use of a NRTL-SAC model to develop the process. The second case illustrates the regression of solubility data using the DynoChem® Crystallization Utility. The third example showcases the use of automated lab reactors and PAT to aid in solubility measurements. The final example combines various aspects of solvent screening/detailed study, use of NRTL-SAC, regression, and influence of the solubility on the crystallization process development.

26.2.5.1 Solubility Measurements as a Function of Temperature and Antisolvent Amount with Data Fitting Using NRTL-SAC Model This example summarizes the measurement of solubility data for pharmaceutical compound A, the desired form of which is a solvate hydrate. The solubility of compound A was measured in methanol as a function of water (antisolvent) content and temperature. The data were then fitted to the NRTL-SAC model using Aspen Properties® V8.6. Figure 26.2 shows measured solubility data, and the resulting model fits for the desired crystal form (confirmed by XRPD).

These data were the basis for designing a methanol–water antisolvent crystallization that is capable of delivering the desired yield. Furthermore, this understanding of solubility variation over the solvent composition and temperature range

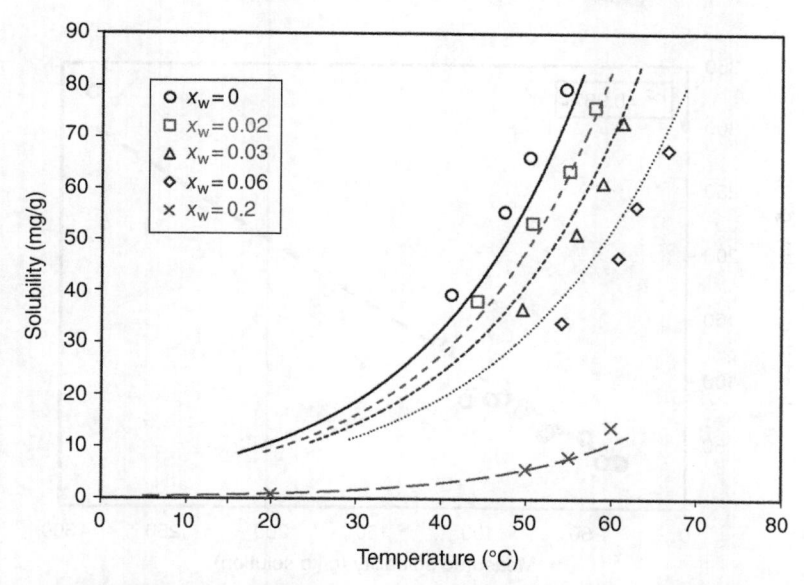

FIGURE 26.2 Raw solubility data of compound A in methanol as a function of water content (antisolvent) and temperature. x_W is the weight fraction of water in the solution.

informed the optimal seeding conditions and helped to design the procedure used to reduce fines through Ostwald ripening.

26.2.5.2 Solvent Screening and Solubility Measurements at Different Temperatures and Regression with DynoChem®

In this example, solubility of pharmaceutical compound B was screened in different solvents and/or solvent combinations to measure how much solute could be dissolved at the given process conditions. Based on the solvent screening experiments and the choice of upstream reaction solvents, 2-MeTHF–water–heptane system was selected. Water was a good solvent, 2-MeTHF was a moderate solvent, and heptane was used as an antisolvent. Detailed solubility measurements were carried out for compound B in the miscibility region of 2-MeTHF–water system with the recommended protocol (Section 26.2.4). HPLC was used to analytically measure the concentration in the solvent system. The data were then regressed using five different equations in DynoChem®, and Eq. (26.9) that yielded the best fit was selected. Figure 26.3 shows the fitted solubility as a function of water/2-MeTHF composition and temperature.

As can be noticed from Figure 26.3b, as water content decreases, the temperature dependence of solubility also diminishes. Similar data were also collected for 2-MeTHF–water–heptane system (not shown here). The data helped design the amount of water that needs to be removed and amount of heptane that needs to be added for improving the yield and performance of the crystallization. The estimated data also suggest that at lower water content, the temperature dependence of solubility is nonexistent. Hence, if temperature cycling is required for reduction of fines, it would be better to carry out the temperature cycling at higher water content to take advantage of the solubility.

26.2.5.3 Solubility Measurement with EasyMax® and Online HPLC

Solubility of compound C was measured in ethyl acetate as a function of temperature. The data were collected from the experiments conducted in a 100 ml automated reactor using an online HPLC system.

The solubility of compound C was screened in different solvents, and ethyl acetate was picked as the solvent of choice. More elaborate solubility measurements were then carried out to further understand the crystallization process and impact on the impurity rejection. Crude C was suspended in ethyl acetate such that representative impurities would be present in the system. 100 ml Mettler Toledo® EasyMax® reactor was used with overhead stirring to simulate representative process conditions. The automated reactor was fitted with an online HPLC probe (KS-1 from Komplx®). The probe is designed to sample the mother liquor (ML) of the slurry by filtering it through a Teflon filter and diluting it immediately with the prescribed diluent and then running it automatically on the HPLC.

The system was programmed to equilibrate at different temperatures from 0 °C up to 75 °C. The system was equilibrated for 4 hours at each temperature, and sampling was carried out every 30 min. Compound C was studied a priori to ensure the stability of the desired crystal form under explored temperature range in ethyl acetate. The system was also cooled back down to 0 °C while equilibrating at the same temperatures as used during ramp up. The data are summarized in Figure 26.4. As can be seen from the data, four hours was enough to reach equilibration for this system during

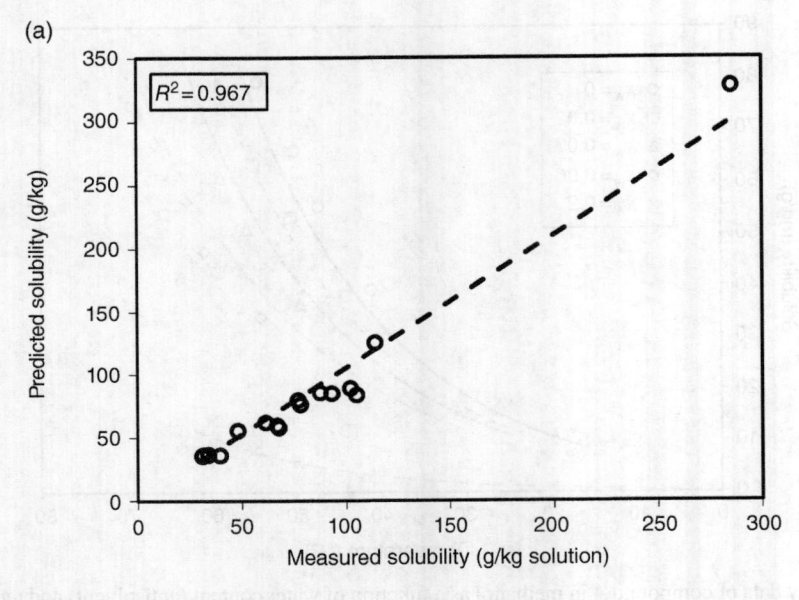

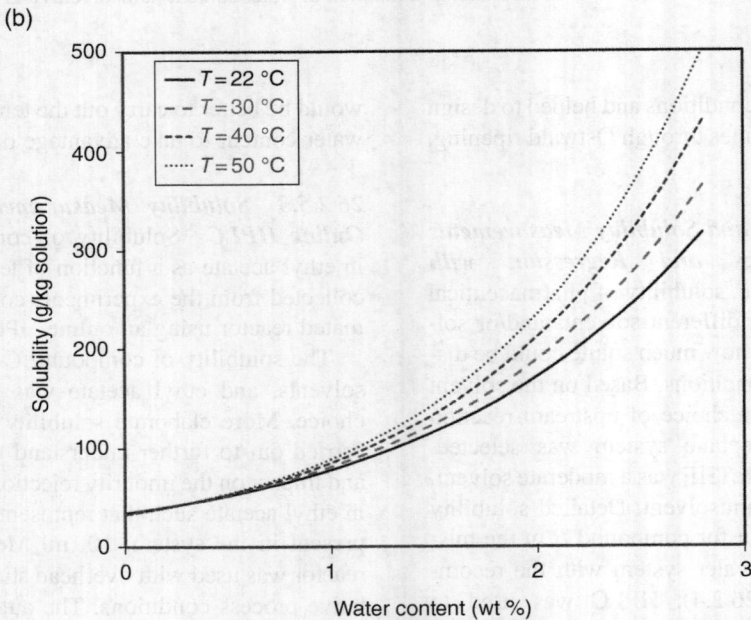

FIGURE 26.3 Estimated solubility of compound B using best-fit equation (Eq. (26.9) from Table 26.1) in DynoChem® as a function of water composition in a designated temperature range. (a) Parity plot for comparison between measured and predicted solubilities. (b) Estimated solubility as a function of water content (wt %) and temperature.

heating. It can also be seen that during cooldown, as temperature decreased, the time required to reach equilibrium increased.

Measured solubility data were useful to understand the impact of temperature and ethyl acetate charge on the yield of the process. As crude solids for compound C were used with representative impurities, online HPLC could be used to determine the purge of impurities in the ML at various temperatures. The data were also used to understand the rate of de-supersaturation at low temperature and to design the equilibration period to reduce concentration before the filtration of the crystallization slurry.

26.2.5.4 Solubility Measurement of Carbamazepine This example summarizes the solubility measurement of carbamazepine (CBZ) dihydrate. Solubility of CBZ dihydrate was measured in ethanol as a function of water (antisolvent) content at two different temperatures. Experiments were conducted in 100 ml Mettler Toledo® EasyMax® reactor with overhead stirring, and samples were analyzed using HPLC.

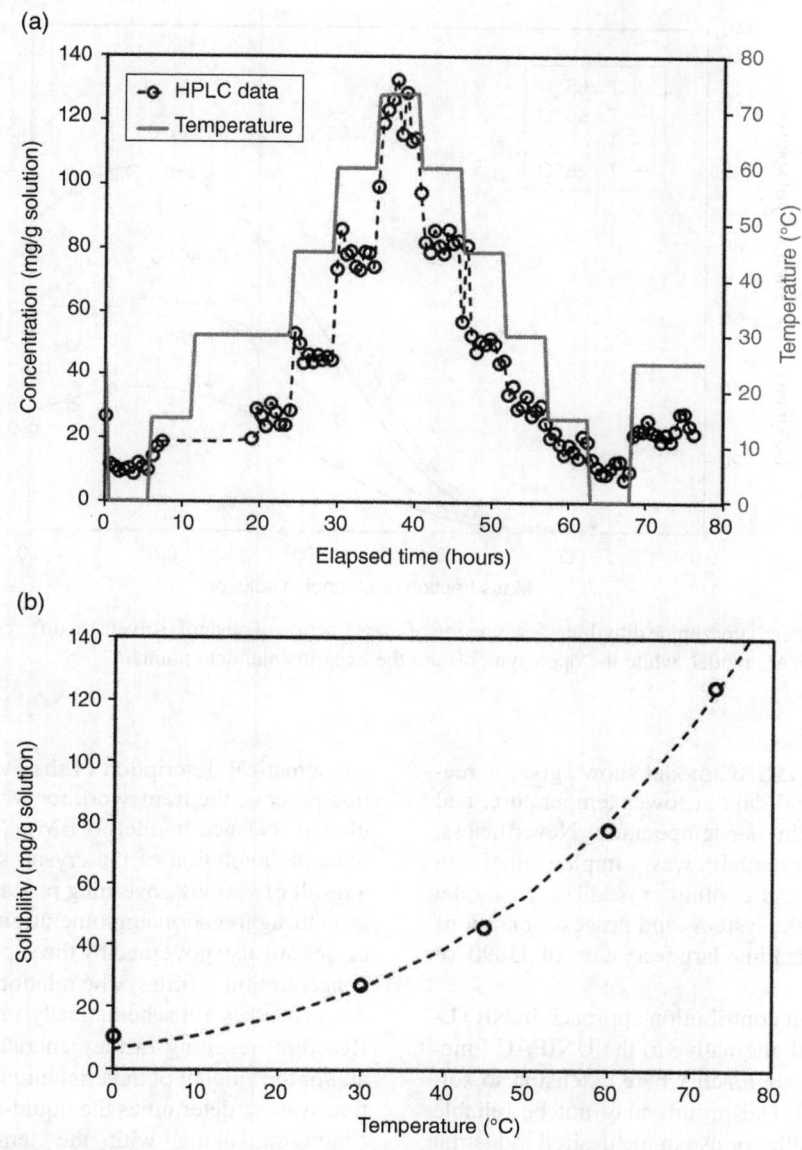

FIGURE 26.4 (a) Raw data of solubility and temperature collected in an automated reactor using online HPLC. (b) Summarized solubility data with an exponential model fit for the solubility.

CBZ is widely studied model compound for crystallization, and the solubility data reported in the literature as a function of temperature in other solvents were used to build a robust NRTL-SAC model. Aspen NRTL-SAC model treats the non-idealities in the mixture containing a complex organic molecule (solute) and small molecules (solvents) by considering pairwise interaction between three conceptual segments: hydrophobic segment (x), hydrophilic segment (z), and polar segments ($y-$ and $y+$) [5]. In practice, these conceptual segments can be viewed as molecular descriptors representing the molecular surface characteristics of each solute or solvent molecule. In order to build a robust NRTL-SAC model, solubility data should be collected for solvents with hydrophobic, hydrophilic, and polar segments, respectively. The molecular parameters for all other solvents can be determined by regression of available VLE or LLE data for binary systems of solvent and the reference molecules or their substitutes. NRTL-SAC model can also be used to model electrolyte systems.

In this case, the solubility data in ethanol, 2-propanol, and tetrahydrofuran were used to build the NRTL-SAC model. Solubility measurements in ethanol were carried out, while the data for 2-propanol and tetrahydrofuran were taken from the literature [6]. Figure 26.5 shows the solubility of CBZ dihydrate (form was confirmed using XRPD) as a function of mass fraction of ethanol at various temperatures.

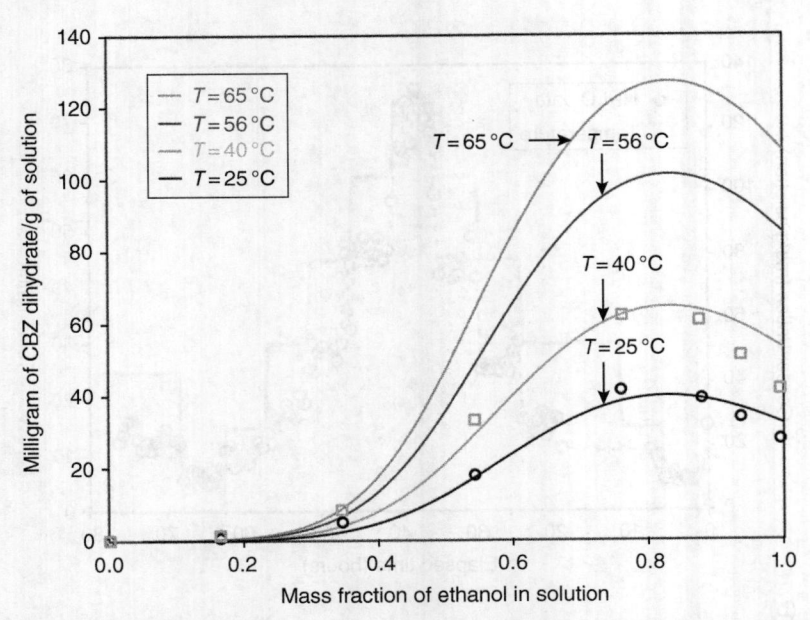

FIGURE 26.5 Solubility of carbamazepine dihydrate as a function of mass fraction of ethanol (solvent) at different temperatures. Solid curves are predictions from NRTL-SAC model, while the open symbols are the experimental data points.

As can be observed, NRTL-SAC model shows good agreement with the experimental data at lower temperature, and the deviation is larger at higher temperature. Nevertheless, NRTL-SAC solubility model was implemented in DynoChem® antisolvent and cooling crystallization model to probe a variety of solvent systems and process conditions for the objective of generating large crystals of $D_V 90$ of around 300 μm.

The conceptual segment contribution approach in NRTL-SAC represents a practical alternative to the UNIFAC functional group contribution approach where extension to solvents with different functional groups may not be reliable [5]. This approach is suitable for use in the desired industrial practice of carrying out measurements for a few selected solvents and then using NRTL-SAC to quickly predict other solvents or solvent mixtures and to generate a list of suitable solvent systems. In this case, the NRTL-SAC model not only helped in solvent selection but, in conjunction with DynoChem® models for the antisolvent and cooling crystallizations, also enabled crystallization process optimization and scale-up.

26.3 ESTIMATION OF NUCLEATION AND GROWTH KINETICS

With a rigorous understanding of equilibrium solubility, another important insight into crystallization process design is afforded by exploring nucleation and growth kinetics. Estimation of nucleation and growth kinetics requires mathematical description of the evolution of the crystallization process, the framework for which is an appropriate population balance model (PBM). The PBM describes the dynamic evolution of the crystal size distribution (CSD) as a result of various governing processes including nucleation, growth, aggregation/agglomeration, and breakage. These processes are also governed by the supersaturation and hence the concentration profiles. The relationships between these various processes are schematically represented by Figure 26.6. Reaction, resulting in the generation/depletion of a species and/or the amount of material initially present in a crystallization system, determines the liquid-phase concentration. Concentration, along with the temperature and/or solvent composition, leads to supersaturation, which induces nucleation, growth, and potentially particle aggregation/agglomeration. Growth affects concentration and the product CSD. Nucleation as well as aggregation/agglomeration also affects the resultant CSD, but their impact is oftentimes influenced by the system hydrodynamics and interactions between various solid interfaces. Hydrodynamics and solid–solid interactions often affect the extent of crystal breakage/attrition due to shear and impact, which also controls the product CSD. Boundary conditions for the solution of a PBM are described by initial concentration, solvent composition, temperature, and the seed CSD where appropriate.

Thus there exists a complex interplay between various underlying processes, and its quantification can be of significant use in the holistic design of crystallization processes. However, this section will focus on the methodologies to estimate only nucleation and growth kinetics for the sake brevity.

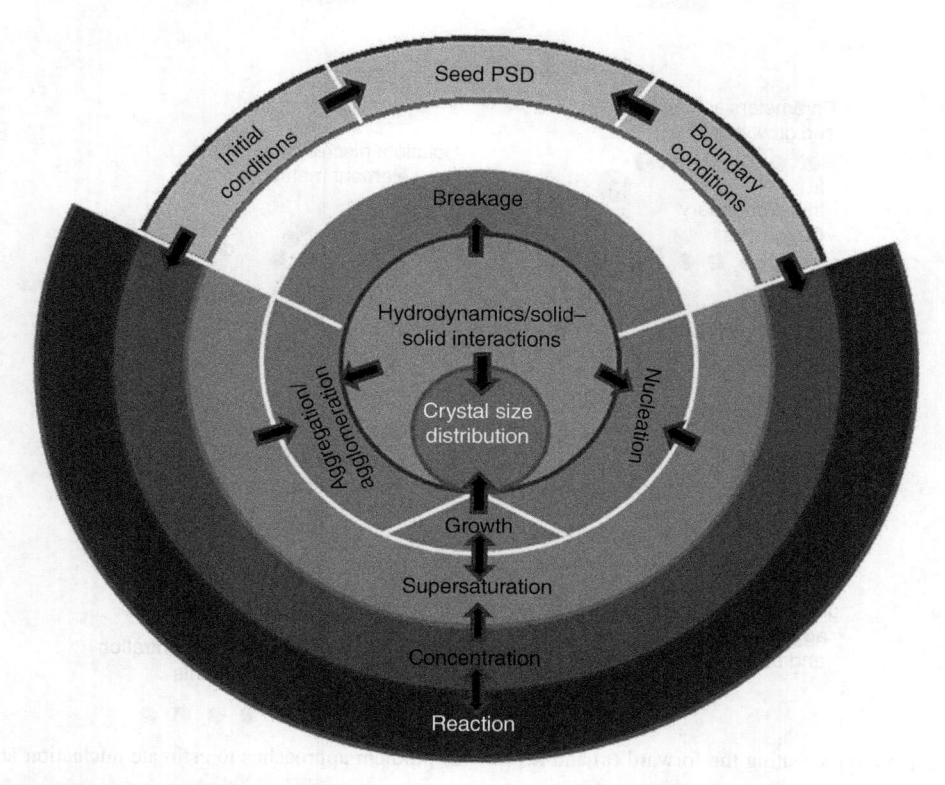

FIGURE 26.6 Interplay between various underlying processes that impact the crystallization process and resultant crystal size distribution.

26.3.1 Forward and Inverse Problem Approach-based Methodologies

Most of the methods reported in the literature to estimate the kinetics of crystal nucleation and growth can be broadly classified as using either a forward problem or inverse problem approach. A schematic depicting the essence of these two approaches to model extraction is shown in Figure 26.7. The forward problem approach shown in Figure 26.7a is based on a starting point of a presumed form of the model with correct inputs, parameter settings, initial and boundary conditions, etc., which is used to arrive at the final solution (CSD). Using this approach, the kinetics of primary nucleation is oftentimes extracted using MSZW estimation and induction time measurements. Growth kinetics are estimated by regressing its parameters so as to yield a better fit of the experimental data with the predictions due to PBM in conjunction with the concentration profiles. This approach is predicated on an *a priori* assumption of a particular form of growth kinetics, with unknown parameters yielded as an output of the regression exercise.

In the absence of known nucleation kinetics, one can also resort to regressing its parameters along the lines of growth kinetics estimation.

The inverse approach relies on starting with the solution (CSD derived from experimental data) to determine what an appropriate model is for estimating nucleation and growth rates. This approach relies on mathematical analysis of PBM and dynamic data on CSD and concentration obtained from the careful design of experiments. This approach, unlike the forward problem approach, is independent of any assumption regarding the form of the nucleation or growth models. Although obviously powerful, this approach is rarely used in industrial practice, seemingly due to the perceived complexity of the methodology. However, the power of this approach warrants further advancements and simplifications in order to promote industrial use.

Most practical approaches today are based on the simplifying assumption of having a spatially and temporally homogeneous particulate system, which is characterized by a single particle dimension. However, the convergence of highly efficient computational methodologies and novel technologies to characterize three-dimensional crystal size and shape distributions holds promise for reducing such simplifying assumptions in the future.

26.3.2 Case Studies for Estimation of Nucleation and Growth Kinetics

Three case studies employing some of the popular methodologies to estimate nucleation and growth kinetics are discussed in this section, each showcasing details regarding the approach and its implementation. The first case sheds

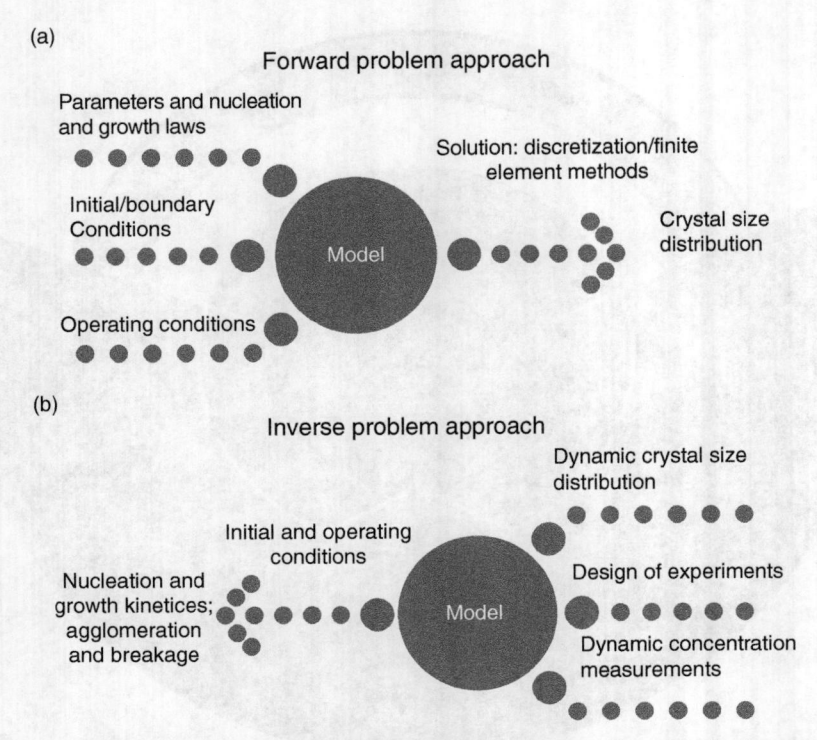

FIGURE 26.7 Schematic representing the forward (a) and inverse (b) problem approaches to estimate nucleation and growth kinetics.

light onto the inverse PBM approach-based technique, reported by Mahoney et al. [7]. It is then followed by an example in which the approach was used to estimate particle growth rates to inform the selection of appropriate cooling rate to support process scale-up. The second case study, based on the work presented by Mitchell and Frawley [8], illustrates the approach of induction time measurement via *in situ* techniques to estimate the nucleation kinetics. The final case study describes how the data obtained from experiments involving supersaturation spikes were used to determine nucleation and growth kinetics for both a cooling and an antisolvent crystallization. Taken together, this work provides practical theory coupled with industrially relevant case studies, providing an overview of options for interrogating nucleation and growth kinetics.

26.3.2.1 Case Study 1: Inverse Problem Approach to Extract Crystallization Kinetics
This section elaborates on the inverse problem approach to extract crystallization kinetics, details of which have been reported in the literature [7]. Application of a part of this methodology has been then illustrated through the case of CBZ dihydrate crystallization.

The population balance equation (PBE) describing the number density of the crystals (defined as the number of crystals per unit of "size" measure per unit volume in the crystallizers) involving only nucleation and growth can also be written as

$$\frac{\partial n(L,t)}{\partial t} + \frac{\partial}{\partial L}\{n(L,t)G[L,\sigma(t)]\} = 0 \quad (26.12)$$

with the boundary condition described by

$$n(0,t) = \frac{B[t]}{G[0,\sigma(t)]} \quad (26.13)$$

where

- $n(L, t)$ is the number density of the crystals with the internal coordinate of crystal size, L.
- $G[L, \sigma(t)]$ is the growth rate that depends on both the size, L, and the supersaturation, σ.
- B is the nucleation rate.

Note that the supersaturation can also be expressed in terms of time, t, to get a simple algebraic expression for subsequent numerical treatment.

In this approach, the PBE is solved by the method of characteristics under the suitable assumptions of the deterministic growth rate and no or minimal aggregation/breakage. This associates the population of crystals of any size at any time with a single point from the initial (i.e. existing crystals from seeding) or boundary condition (i.e. nucleation). The key to leveraging this approach is the recognition that these growth characteristics correspond to the size history of individual crystals and that they can be linked with the fixed cumulative number of the crystals. These growth characteristics are directly identified from the experimental time measurement of crystal size or chord length distributions. The variability in the size for a given set of cumulative oversize number (CON) (counts) along these characteristics provides decoupled equations that can be used to determine the growth rate.

The solution to the PBE along these growth characteristics is given by

$$n[L(t;l_0,t_0,),t] = n(l_0,t_0,)\exp\left[-\int_{t_0,}^{t}\frac{dG\{L(t';l_0,t_0,),\sigma(t')\}}{\partial L}dt'\right] \quad (26.14)$$

For no aggregation, the crystal number density depends only on the characteristic passing through the initial size. The key idea is that the characteristics can be readily determined from the data on the CSD as a function of time by identifying that there is no influence from the nuclei to the population of larger particles. Quantiles refer to the locus of constant numbers over a size cut. The cut indicative of size must move as the characteristic. Under suitable assumptions of deterministic growth and known (or absent) aggregation, characteristics correspond to a constant number oversize.

Figure 26.8 shows how the method of characteristics is used to construct the solution to the PBE.

The solution in terms of the number density along the characteristic curves of the first-order partial differential equation follows individual particle growth paths. One can readily envision that the number density can change due to growth and aggregation. The paths are solely dependent on particle growth. Thus, the characteristic curves show how the single particles develop from their initial sizes. Furthermore, they directly relate to both growth of nuclei and initial crystals. Each of the constant CON, defined as the total number of crystals with the size larger than a given size under consideration, lines would intersect the CON time profiles, and the projections of these intersections on the particle size coordinate would yield a size vs. time trajectory. These trajectories are nothing but the growth characteristics and are depicted in Figure 26.9.

The crystal growth characteristics in the plot shown in Figure 26.9 are represented by Eq. (26.15):

$$\frac{dL}{dt} = G[L,\sigma(t)] \quad (26.15)$$

The following subsection describes the estimation of growth and nucleation kinetics in further detail.

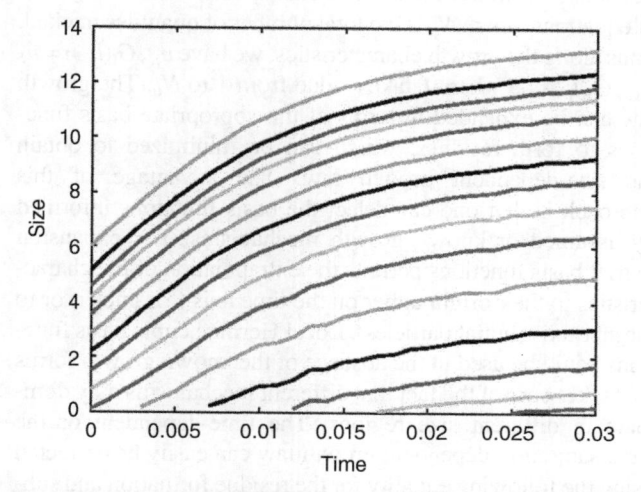

FIGURE 26.9 Growth characteristics extracted from Figure 26.8.

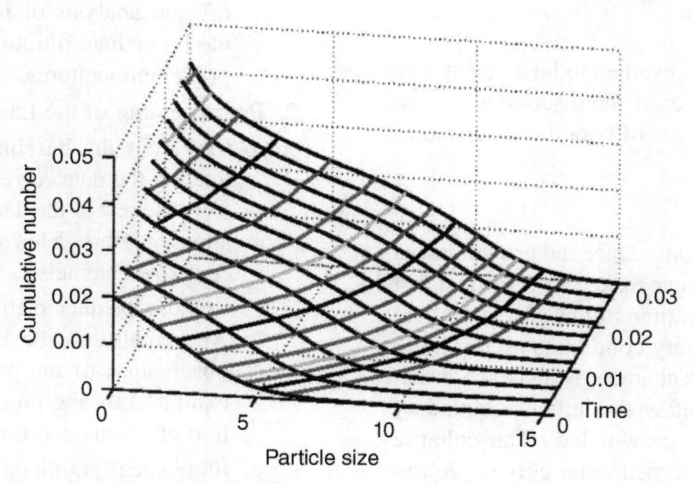

FIGURE 26.8 Plot of cumulative oversize number as a function of particle size for various times.

26.3.2.1.1 Kinetics of Crystal Growth
The assumption of treating the growth rate expression as a degenerate function of size- and supersaturation-dependent components makes the determination of the growth kinetics in more of a complete sense. The product of the crystal number density and the size dependent part of the growth expression can be readily shown to be an exact invariant along the growth characteristic. Thus one may solve the characteristic equation as above (Eq. 26.15) and notice that the right-hand side is independent of the quantile, while the left-hand side depends on it. For each quantile one can calculate the right-hand side and examine if the calculations coincide. Thus for a separable growth law (i.e. degenerate form), we have

$$G(l,\sigma) = G_l(l)G_\sigma(\sigma(t)) = G_l(l)G_t[t] \quad (26.16)$$

The variation of the number density along the characteristics is used with the conserved quantity to provide the system of equations where N_q is the total number of quantiles tracked. Thus along the growth characteristics, we have $n_{j,k}G_l(l_{j,k}) = n_{j+1,k}G_l(l_{j+1,k})$ where k takes value from 1 to N_q. The growth law can be expanded in terms of the appropriate basis functions to form residues, which can be minimized to obtain the size-dependent growth law. The advantage of this approach is that one can select the basis functions informed by assumed or known growth mechanisms. The expansion in trial basis functions permits the extrapolation of the characteristics to their origin either on the time axis (for nuclei) or to the size axis (initial particles). Local Hermite cubic basis functions could be used in the absence of the known growth forms as it takes care of the fact that different mechanisms may dominate in different size regions. The time-dependent or the supersaturation-dependent growth law can easily be extracted using the following equality for the residue formation and subsequent error minimization:

$$\int_{l_{j,k}}^{l_{j+1,k}} \frac{dl'}{G_l(l)} = \int_{t_j}^{t_{j+1}} dt' G_t[t'] \quad (26.17)$$

Thus the overall growth law presumed to be a degenerate function as discussed on the onset of the discussion can be completely determined as a function of both the supersaturation and size.

26.3.2.1.2 Kinetics of Nucleation
Once the growth law is completely determined, then one can make use of the fact that the measured CSD evolution over time from each experiment arises from the initial and boundary conditions according to the solution of the PBE along the characteristics. Each measurement can be traced back to either the initial or boundary condition using the determined growth law. The collapse of these measurements onto the same master curve is equivalent to the prediction of the data, and the dispersion of the collapse is due to experimental or model error. This allows a verification of the model assumptions, i.e. deterministic separable growth law, and also gives the initial and boundary conditions more accurately than using the experimental data nearest to those conditions. These conditions can be used subsequently in the determination of nucleation law.

While the reader is referred to the paper by Mahoney et al. [7] for full details, a practical protocol to extract the growth and nucleation kinetics is presented below.

26.3.2.1.3 Overall Protocol for the Kinetic Model Extraction
Below is a protocol to estimate nucleation and growth rates using the inverse problem approach.

1. Experimental protocol:
    a. Determine the phase diagram for the crystallization of the compound of interest using guidance from Section 26.2 on solubility measurements.
    b. Estimate the temperature and/or solvent composition profile that will keep the crystallization system in a metastable region throughout the process so that nucleation can be controlled to promote growth rate. Reduced nucleation will lead to reduced aggregation, thus enabling the growth rate to be determined with better confidence.
    c. Carry out measurement of chord length distribution (CLD) using FBRM® technique with the optimal optical settings and physical position. Check the consistency of the total FBRM® count for different concentrations of crystals to ensure appropriateness of measurements. Ideally, the depth of FBRM® laser penetration and measurement zone should remain constant to keep the observable volume constant to the extent possible.
    d. Carry out the measurement of concentration as a function of time using ATR-FTIR, online LC, or off-line analysis of mother liquors obtained from the immediate filtration of slurry samples at appropriate temperatures.

2. Preprocessing of the Lasentec® FBRM® data:
    a. Carry out the baseline correction by simply subtracting the data corresponding to the clear solution from the rest of the data. Replace negative counts (if any) by interpolated values corresponding to adjacent size channels.
    b. Smooth the data to eliminate abnormal fluctuations. Abnormality can be discerned based on knowledge/observation of the physical behavior that may be exhibited by the process. Check for the conservation of counts and the trend shown by total counts.
    c. Eliminate the outliers. This can be done by observing the evolution of the profile in the time coordinate.

3. Conversion of the CLD measured by Lasentec® FBRM® to CSD:
   a. The CLD can be either directly used as is (the growth rate will correspond to the chord length) or transformed into CSD using various algorithms [9–12] of varying complexity available in commercial software (e.g. DynoChem®).
   b. It is advisable to verify the conversion of CLD to CSD to have meaningful data analysis and further interpretations. The CLD can be used as a direct measure to assess the crystallization performance if the relationship between the chord length and the target crystal size attribute can be established.
4. Data processing to extract nucleation and growth kinetics using the inverse population balance approach:
   a. Discretize the time and size domain for CSD measurements done using an off-line analysis method. For in-line measurement, such as FBRM®, one can use appropriately spaced time points to reduce the burden on the amount of data to be processed. Data filtering should be done so as not to affect the extracted kinetics.
   b. Plot cumulative number count data.
   c. Extract crystal growth characteristics at various CONs.
   d. Select basis functions, and form a matrix equation for inversion to get the size-dependent growth and then the supersaturation (which changes as a function of time due to progression of crystallization) growth, followed by the determination of the nucleation rate.
   e. Carry out the forward simulation using the extracted growth and nucleation rates to check with the measured evolution of CLD or CSD.
   f. Fit the discrete model form obtained from the inverse problem approach to a continuous one, assuming some kind of "well-defined model" with accurate parameters.
   g. Carry out another experiment within the operating range, and measure the CSD/CLD time evolution along with the concentration measurements, and verify the same using the forward simulation with the extracted models following the earlier steps.

While the abovementioned exhaustive protocol could be followed for the complete determination of growth and nucleation kinetics, the following subsection illustrates an example of quick growth rate estimation.

26.3.2.1.4 Example Application of Inverse Problem Approach for Growth Rate Estimation The following example shows how the inverse problem approach was used to assess the impact of cooling rate on crystal growth rates for a cooling crystallization of CBZ dihydrate from an ethanol–water solution. The information gained regarding growth rates from this exercise was used to optimize a linear cooling profile (parabolic cooling was not accessible with available equipment) to improve product physical properties under constrained project timelines.

A cooling crystallization was designed based on the solubility data generated in Section 26.2.5.4, the objective of which was to produce larger and thicker CBZ dihydrate crystals. Experiments (10 g scale) were performed at three different cooling rates (6, 2, and 1 °C/h) in an effort to identify the appropriate cooling rate to achieve desired crystal growth. While reduced cooling rates would be expected to favor growth, reduction beyond what is required to achieve the objective needlessly lengthens process cycle time and is therefore undesirable. In each experiment, a seed slurry (generated using a high-shear wet mill) was added to a supersaturated solution at 48 °C, and the crystallization mixture was cooled down to 22 °C at the prescribed rate. Each experiment was monitored using Lasentec® FBRM to track the CLD over time. CLD data collected from the 1 °C/h experiment at discrete time points over the course of the crystallization is shown as an example in Figure 26.10. While the shape of the distribution is very consistent throughout the experiment, the evolution of the profile suggests growth over time. In this case, the chord length measurements were found to correlate with and be representative of the CSD, as confirmed by photomicrographs taken by laser microscopy.

The FBRM® data were then converted to CON profiles as a function of chord length. This analysis was performed on selected time points to simplify the extraction of growth characteristics in the next step. An example of the CON profile for the experiment using a cooling rate of 1 °C/h is shown in Figure 26.11.

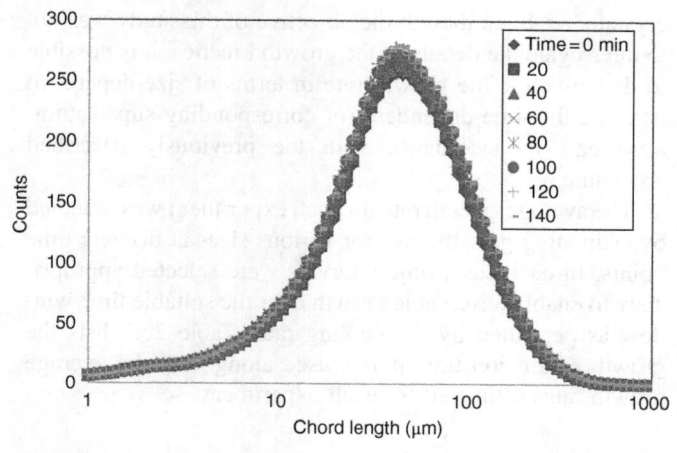

FIGURE 26.10 Raw FBRM data from the experiments carried out at the cooling rates of 1 °C/h.

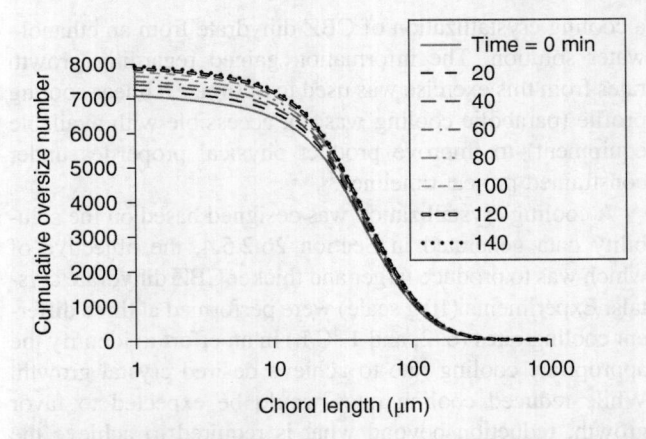

FIGURE 26.11 Cumulative oversize number as a function of chord length for selected time points for the cooling rate of 1 °C/h. Results presented as smoothed lines to improve readability.

The growth characteristics were then extracted for each experiment, as described previously, and shown in Figure 26.12. Equations describing the change in chord length over time are displayed for three selected crystal sizes (chord length at $t = 0$) for each experiment. The low chord length numbers in the 6 °C/h (Figure 26.12a) experiment compared with that of either the 2 or 1 °C/h experiments (Figure 26.12b and c, respectively) may be a manifestation of a greater extent of secondary nucleation occurring with increased cooling rates. Moreover, the growth trajectory associated with faster cooling experiments is enhanced compared with slower cooling experiments.

The growth characteristics were then used to extract growth rates for various initial crystal sizes (indicated by the intercept on the y-axis of Figure 26.12) at selected time points. A plot of the growth rate corresponding to initial crystal size for the cooling rate of 1 °C/h is shown in Figure 26.13 as an example.

Figure 26.13 shows that the growth rate is a function of the crystal size. Even though the objective of this study was not to interrogate the details of the growth kinetics, it is possible to de-convolute the growth rate in terms of size-dependent part and the time-dependent (or corresponding supersaturation) part in accordance with the previously described procedure.

The average growth rate for each experiment was obtained by estimating growth rates for various sizes at discrete time points. In each case, time intervals were selected appropriately to enable observable growth over the suitable time window as permitted by the cooling rate. Table 26.2 lists the growth estimation time points used along with the average growth rates estimated for each experiment.

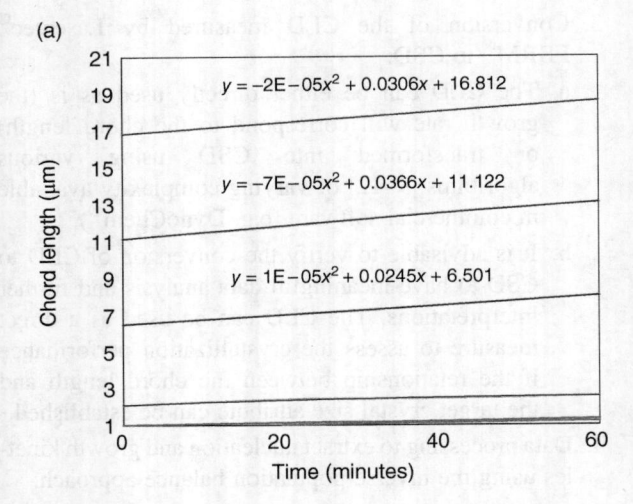

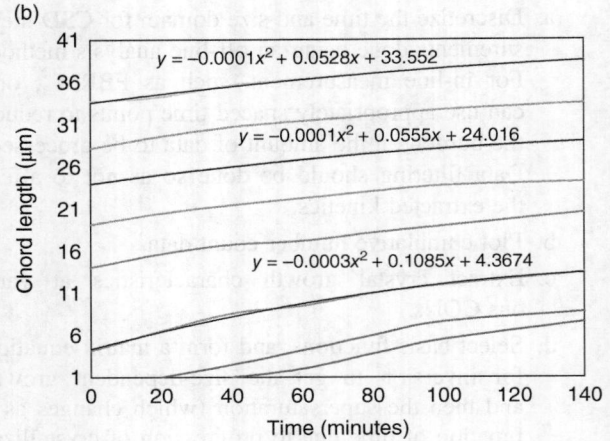

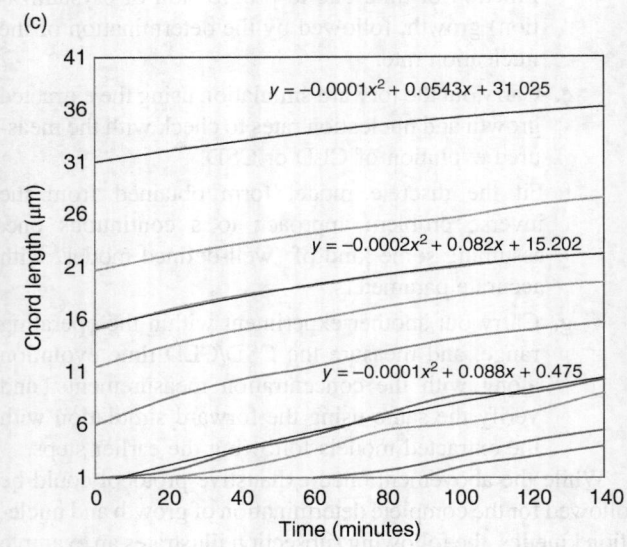

FIGURE 26.12 Growth characteristics extracted from the cumulative oversize plots (a) 6 °C/h, (b) 2 °C/h, and (c) 1 °C/h. Results presented as smoothed lines to improve readability.

Equations in Figure 26.12:

(a) 6 °C/h:
- $y = -2\text{E}-05x^2 + 0.0306x + 16.812$
- $y = -7\text{E}-05x^2 + 0.0366x + 11.122$
- $y = 1\text{E}-05x^2 + 0.0245x + 6.501$

(b) 2 °C/h:
- $y = -0.0001x^2 + 0.0528x + 33.552$
- $y = -0.0001x^2 + 0.0555x + 24.016$
- $y = -0.0003x^2 + 0.1085x + 4.3674$

(c) 1 °C/h:
- $y = -0.0001x^2 + 0.0543x + 31.025$
- $y = -0.0002x^2 + 0.082x + 15.202$
- $y = -0.0001x^2 + 0.088x + 0.475$

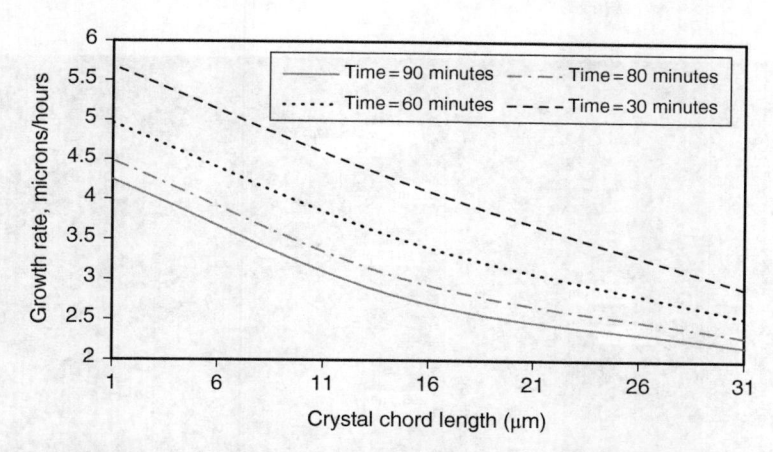

FIGURE 26.13 Growth rates extracted from the characteristics as a function of the crystal size represented as the chord length for selected time instances. Results presented as smoothed lines to improve readability.

TABLE 26.2 Average Growth Rates for Various Cooling Rates

Cooling Rate (°C/h)	Growth Estimation Time Points (min)	Average Growth Rate (μm/h)
6	5, 10, 15	1.802
2	15, 30, 45	3.336
1	30, 60, 90	3.660

The data in Table 26.2 shows the expected trend such that the growth rate increases as the rate of cooling decreases. While there is a significant increase in growth rate from 6 to 2 °C/h, the increase in growth rate from 2 to 1 °C/h is only marginal. Based on this finding, the cooling rate implemented in the 15 kg pilot-scale run was 2 °C/h. This allowed the processing to complete in the available manufacturing time with the assurance that sufficient growth of the crystal population would not be significantly compromised. The process scale-up delivered product with the desired physical properties. The photomicrographs of the crystals obtained from the pilot-scale run (2 °C/h) compared with the laboratory experiment, which was cooled at 6 °C/h, are shown in Figure 26.14a and b, respectively. The crystals using the slower cooling profile were both larger in size and exhibit a reduced, more favorable aspect ratio, as evidenced by the more platelike morphology of crystals shown in Figure 26.14a.

In summary, the presented example shows that an inverse population balance approach is useful for estimating crystal growth rates without the need to rely on complex calculations. This approach was enabled in a timely fashion by three simple lab-scale experiments and appropriate treatment of CLD data. These data provided the basis to incorporate a timesaving optimization of the crystallization process while minimizing risk to product quality.

26.3.2.2 Case Study 2: Measurement of Nucleation Rate From Induction Time Measurement In this example reported by Mitchell and Frawley, nucleation kinetics were estimated based on induction time via measurement of the MSZW [8, 13, 14]. Estimations of nucleation kinetics were presented in this work using two theoretical approaches: (i) classical nucleation theory reported by Nyvlt [15] and (ii) a modification of the Nyvlt approach in which MSZW is assumed to be dependent upon the technique used to detect particle formation as reported by Kubota [16].

The MSZW is defined as the gap between thermodynamic solubility and the supersolubility required for solid particles to form. The schematic shown in Figure 26.15 shows the solubility and supersolubility plotted as a function of concentration and temperature; at any given concentration, the temperature gap between these curves is defined as the MSZW.

Based on classical nucleation theory and appropriate assumptions, the cooling rate, R, can be expressed in terms of the MSZW (expressed as ΔT_{max}), the apparent nucleation order, m, the nucleation rate constant, k_n, and the thermodynamic solubility concentration, $c*$, as shown in Eq. (26.18):

$$\ln R = m*\ln(\Delta T_{max}) + \ln k_n + (m-1)\ln\left(\frac{dc^*}{dT}\right) \quad (26.18)$$

Using this approach put forth by Nyvlt, the log/log plot of cooling rate vs. MSZW yields a straight line with a slope commensurate with the apparent nucleation order (m) and k_n estimated from the y-intercept.

Kubota's modification of classical nucleation theory expresses the nucleation rate in terms of the number density (N_m/V) and the true nucleation order (n) as shown in Eq. (26.19). This enables estimation of a number-based nucleation rate rather than the mass-based nucleation rate afforded by Nyvlt's approach:

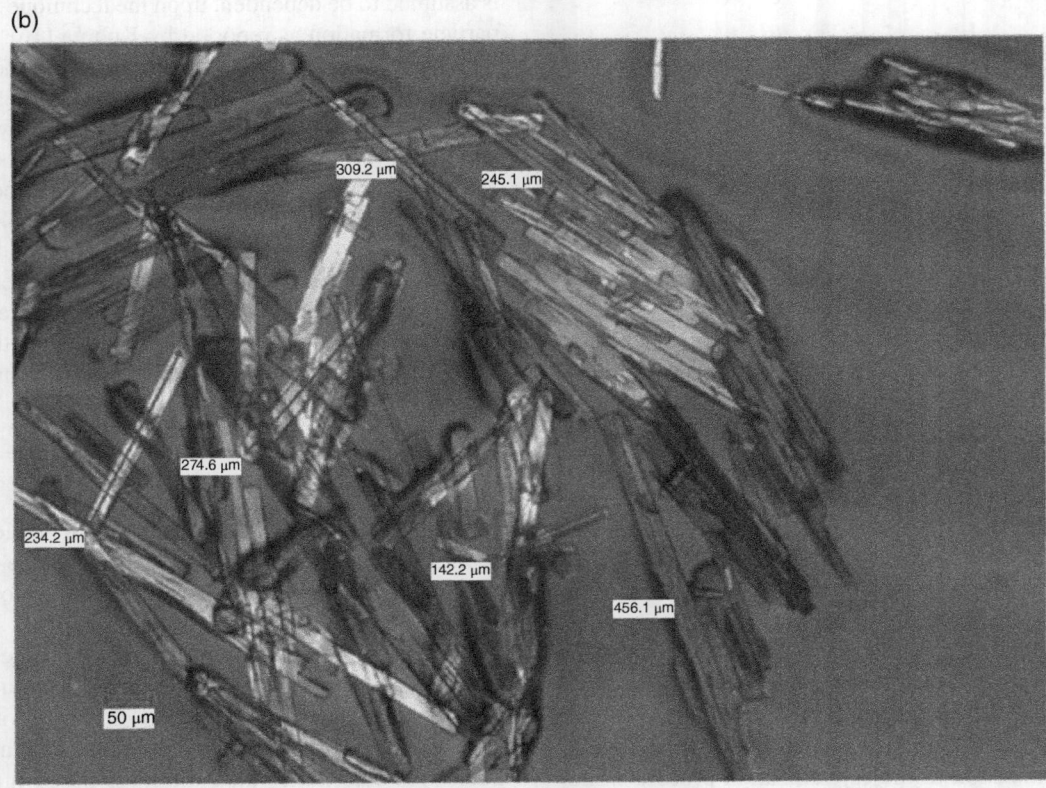

FIGURE 26.14 Photomicrographs of crystals resulting from the cooling rate of (a) 2 °C/h from the pilot plant and (b) 6 °C/h from laboratory experiment.

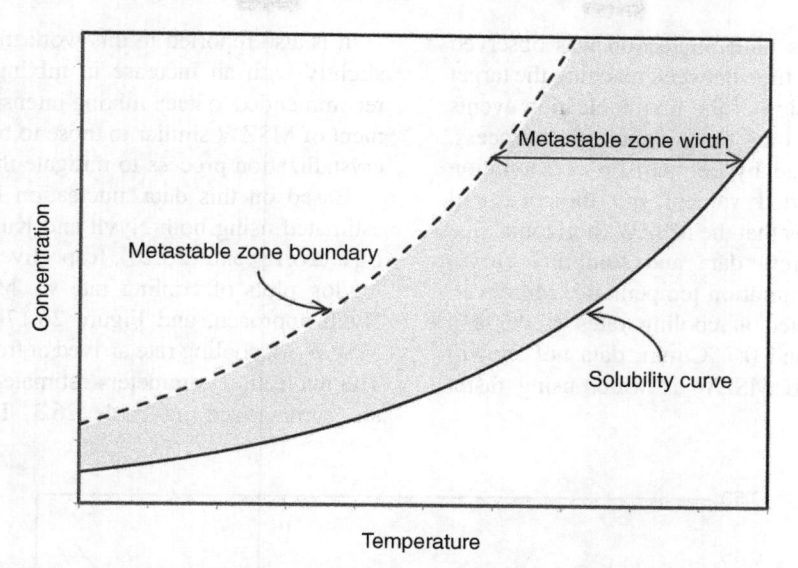

FIGURE 26.15 Schematic showing typical solubility and metastable zone width boundary curves in relation to the metastable zone width (MSZW).

TABLE 26.3 Summary of Nucleation Parameters Estimated from Nyvlt and Kubota Approaches Along with Estimated Values for Paracetamol in Ethanol System Based on Induction Time Measurement

Approach	Basis	Nucleation Order	Kinetic Nucleation Parameter
Nyvlt	Equation (26.18), Figure 26.17a	m, apparent nucleation order m = slope	k_n
	Estimated value, paracetamol in ethanol	1.68 ± 1.25	0.206 ± 0.13
Kubota	Equation (26.19), Figure 26.17b	n, true nucleation order n = slope $-$ 1	$\dfrac{N_m}{k_n V} = \dfrac{1}{(n+1)e^{\text{Intercept}}}$
	Estimated value, paracetamol in ethanol	0.76 ± 0.66	191.7 ± 464.1

$$\ln(\Delta T_{\max}) = \left(\frac{1}{n+1}\right)\ln\left[\left(\frac{N_m}{k_n V}\right)(n+1)\right] + \left(\frac{1}{n+1}\right)\ln R \quad (26.19)$$

Equation (26.19) can be readily rearranged, resulting in Eq. (26.20):

$$\ln R = (n+1)\ln(\Delta T_{\max}) + \ln\left[\frac{k_n V}{N_m(n+1)}\right] \quad (26.20)$$

Thus, using Kubota's approach, the nucleation order can be estimated from the slope of the line on a log/log plot of MSZW vs. cooling rate at a given saturation concentration. The nucleation rate constant, k_n, can be extracted from the intercept with an assumed value of the detectable number density for the detection technique used. Table 26.3 summarizes the nucleation parameters estimated from each approach and how they are derived from measurement of the MSZW.

A significant benefit of the approach put forth by Kubota is that it establishes a relationship between MSZW and nucleation induction time. The induction time is described as the time required for the number density (N_m/V) to reach a fixed value. Assuming induction time can be described by a power law expression and the ΔT is constant (experiments are conducted isothermally), the induction time (t_{ind}) can be written as a function of the number density, nucleation constant, and the degree of supercooling as shown in Eq. (26.21):

$$t_{\text{ind}} = \frac{N_m(\Delta T)^{-n}}{k_n V} \quad (26.21)$$

Both the Nyvlt and Kubota approaches were applied to characterizing the nucleation of paracetamol in ethanol as described below. Figure 26.16 shows the cloud point of paracetamol/ethanol solutions as a function of concentration and temperature. Reported methodology involved rapidly cooling solutions to the desired temperature and then holding under isothermal conditions until nucleation was observed using a turbidity probe. The time between reaching the target supercooling and the detection of the first nucleation events was taken as the induction time for the nucleation process. The MSZW was determined by comparison of saturation conditions (literature reported values) and the measured cloud point. These data show that the MSZW (horizontal distance between supersolubility data and solubility curve) increases with increasing saturation temperature. Moreover, comparison of data collected at cooling rates at 0.5 and 1 °C/min (as well as 0.2 and 0.7 °C/min, data not shown) suggests that the measured MSZW is wider using faster cooling rates.

It is also reported in this work that the MSZW decreases slightly with an increase in mixing intensity. Hence, it is recommended to keep mixing intensities during the measurement of MSZW similar to those to be used in the anticipated crystallization process to mitigate this effect.

Based on this data, nucleation kinetic parameters were estimated using both Nyvlt and Kubota's approaches using Eqs. (26.18) and (26.20), respectively. Figure 26.17a shows log/log plots of cooling rate vs. MSZW derived from the Nyvlt approach, and Figure 26.17b shows log/log plot of MSZW vs. cooling rate arrived at from the Kubota approach. The nucleation parameters estimated using both approaches are summarized in Table 26.3. It is notable that using

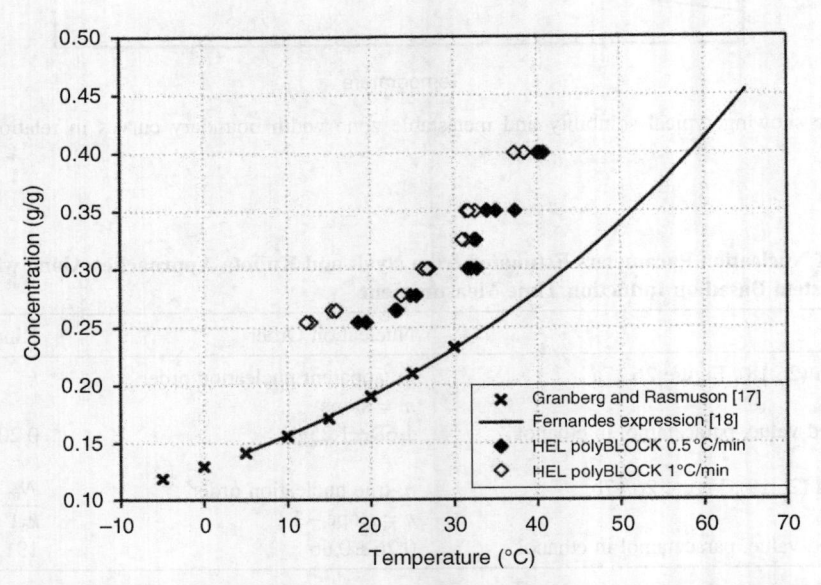

FIGURE 26.16 Measure cloud point for cooling rates of 0.5 and 1 °C/min along with the literature solubility data for paracetamol in ethanol.

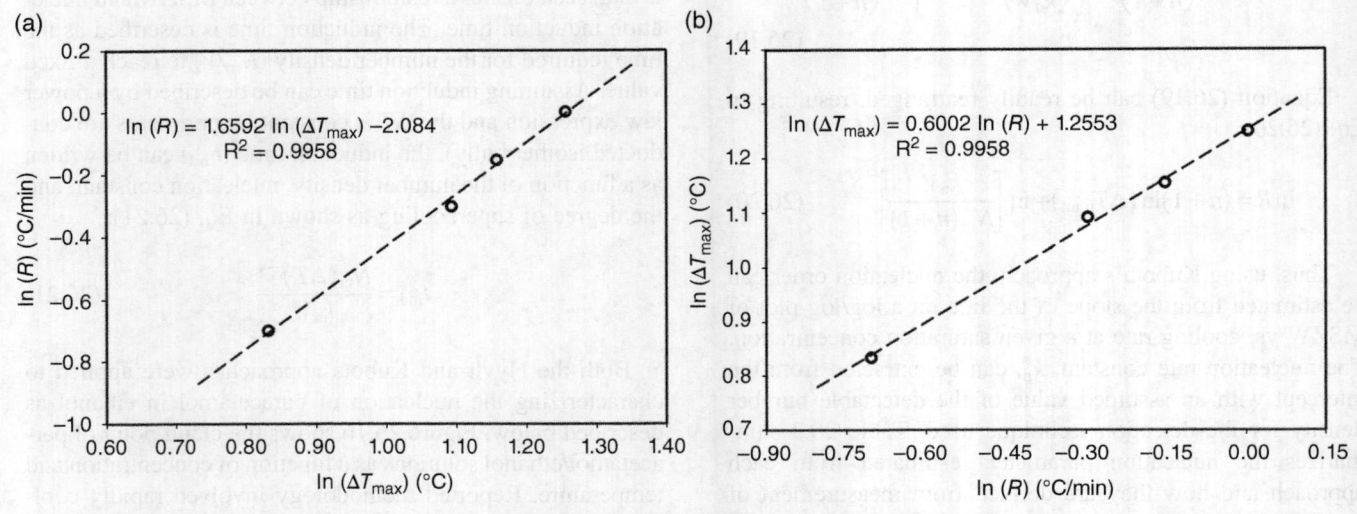

FIGURE 26.17 Nucleation order estimated from MSZW data for a saturation temperature of 44 °C using (a) Nyvlt approach derived from classical nucleation theory and (b) Kubota-modified approach that corrects for the limit of detection of the turbidity measurement.

Figure 26.17a: $\ln(R) = 1.6592 \ln(\Delta T_{max}) - 2.084$, $R^2 = 0.9958$

Figure 26.17b: $\ln(\Delta T_{max}) = 0.6002 \ln(R) + 1.2553$, $R^2 = 0.9958$

Kubota's approach, k_n can be evaluated by assuming a value of the detectable nuclei size of the cloud point detection technique used; a detectable nuclei size of 10 μm was used.

The absolute values of the constants differ significantly, and this results in a difference in the estimated nucleation rates. A plot of the nucleation rate vs. supersaturation estimated using both the Nyvlt and Kubota approaches is shown in Figure 26.18. This figure shows that the nucleation rate profile predicted using the Kubota approach is different from that obtained using the Nyvlt approach, with a crossover point between the two prediction curves.

A key insight provided by the Kubota approach is the ability to predict nucleation induction time, which was demonstrated and confirmed experimentally in this work. Figure 26.19 shows the measured induction time at two saturation temperatures (30 and 40 °C) vs. the degree of supercooling compared with the predictions using the Kubota approach. The experimental data are in close agreement with theoretical predictions.

In summary, the work of Mitchell and Frawley demonstrates how the measurement of MSZW can be used to estimate nucleation kinetic parameters using the approach of

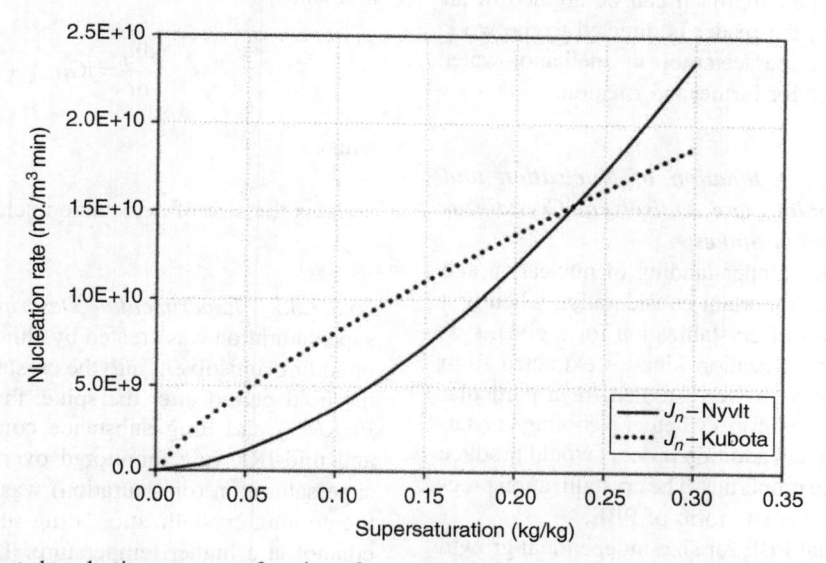

FIGURE 26.18 Estimated nucleation rates as a function of supersaturation obtained using both the Nyvlt and Kubota approaches.

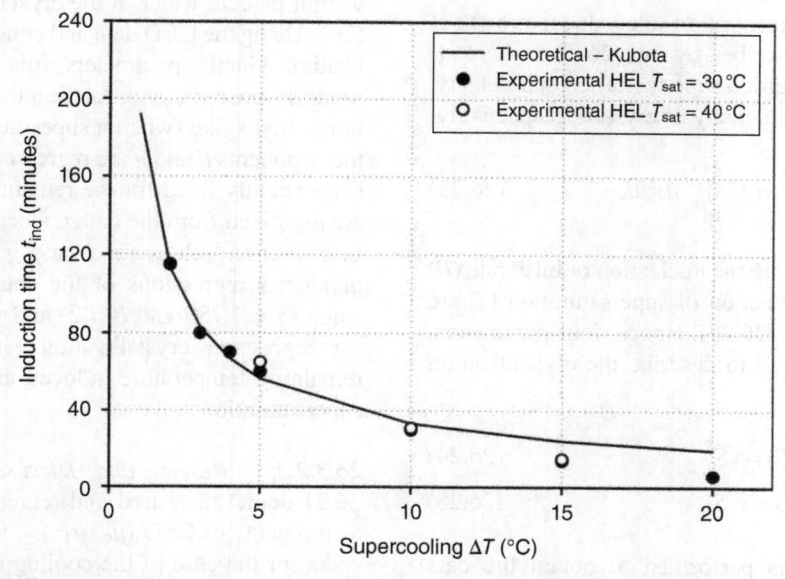

FIGURE 26.19 Experimentally measured induction time as a function of supercooling for paracetamol in ethanol mixtures vs. predicted induction times using Kubota's approach.

Nyvlt (based on classical nucleation theory) and Kubota's modified approach. Kubota's approach affords key advantages that include (i) the estimation of a number-based nucleation rate, for which the limit of detection for the cloud point determination technique factors into the calculation, and (ii) ability to relate the MSZW to the nucleation induction time for a given system. The measured induction time for the paracetamol–ethanol system studied showed close agreement with induction times predicted using Kubota's approach across the range of supersaturations evaluated. While this example demonstrates estimation of nucleation kinetics for a cooling crystallization, an analogous approach can be applied to an antisolvent crystallization; the reader is directed to the work published by authors on paracetamol in methanol–water antisolvent crystallizations for further information.

26.3.2.3 Case Study 3: Estimation of Nucleation and Growth Kinetics for Cooling and Antisolvent Crystallizations Using Supersaturation Spikes

26.3.2.3.1 Introduction Understanding of nucleation and growth kinetics can be an important consideration when first evaluating different modes of crystallization for a given system. In this example, crystallization kinetics extracted from the forward problem approach were studied for a particular drug substance of interest to inform whether a cooling crystallization process or antisolvent addition process would produce the desired drug substance properties. The crystallization process can be described by alternate form of PBE.

Let us put conventional PBE for size-independent growth rate here that is used to derive moment equations:

$$\frac{\partial n}{\partial t} + G\frac{\partial n}{\partial L} = B \quad (26.22)$$

The change in solute concentration solely due to growth of the seed crystals can be described by Eq. (26.23), where k_v, ρ_c, G, and n are the volumetric shape factor, crystal density, growth rate, and population density of crystals, respectively:

$$\frac{dc}{dt} = -3k_v\rho_c G \int_0^\infty nL^2 dL \quad (26.23)$$

The expressions describing the nucleation or birth rate (B) and growth rate (G) as a function of supersaturation (S) are shown in Eqs. (26.24) and (26.25). Hence, four parameters, k_b, b, k_g, and g, are required to describe the crystallization kinetics using this model:

$$B = k_b S^b \quad (26.24)$$

$$G = k_g S^g \quad (26.25)$$

The simple experiments performed to obtain the data needed to regress these crystallization parameters involve generating spikes of supersaturation within the crystallization system and then monitoring particle properties and drug substance concentration over an isothermal/isocratic hold period at these conditions. As described below, analysis of data from these experiments can provide valuable insight into crystallization kinetics that can inform the crystallization development process.

The methodology is predicated on the moment equations derived from PBE. For size-independent growth rate, the PBE for the system accounting for only nucleation and growth reduces to the following general equation for jth moment:

$$\frac{d\mu_j}{dt} = jG\mu_{j-1} + BL_0^j \quad (26.26)$$

where

L_0 is the size of detectable nuclei.

26.3.2.3.2 Experimental Description In these studies, supersaturation was created by either decreasing temperature or adding antisolvent into the crystallization mixture. During the hold period after the spike, the CLD (using Lasentec® FBRM®) and drug substance concentration (using HPLC and mid-IR) were monitored over time until complete desupersaturation (or saturation) was achieved. In the case of the cooling crystallization, drug substance was dissolved in ethanol at a higher temperature. Then the first temperature spike was generated at 45 °C and subsequently reduced stepwise to 40, 35, 25, and 5 °C. In the case of the antisolvent addition crystallization, water was added to an ethanol solution of drug substance in incremental spikes, making the weight percent water in the crystallization range from 34 to 59%. Using the CLD data and concentration profiles, crystallization kinetic parameters for simple nucleation/growth models were regressed. As the most reliable data are collected in the first spike (when a supersaturated solution is seeded), the exponents b and g are regressed for only that step and are subsequently fixed for the remaining spikes; only k_b and k_g are regressed from the data collected from remaining spikes. In this case, values for b and g were obtained from four parameter regressions of the data from 45 °C temperature spike ($g = 1.789$ and $b = 2$) and were fixed at these values for regression crystallization kinetic parameters for the remaining temperature-induced and all antisolvent-induced supersaturation spikes.

26.3.2.3.3 Results and Discussion Figures 26.20 and 26.21 depict measured and regressed concentration as well as moments of CSD (μ_0, μ_1, μ_2, and μ_3) for a first (seeded) spike for the case of the cooling crystallization and antisolvent addition crystallization, respectively. In general,

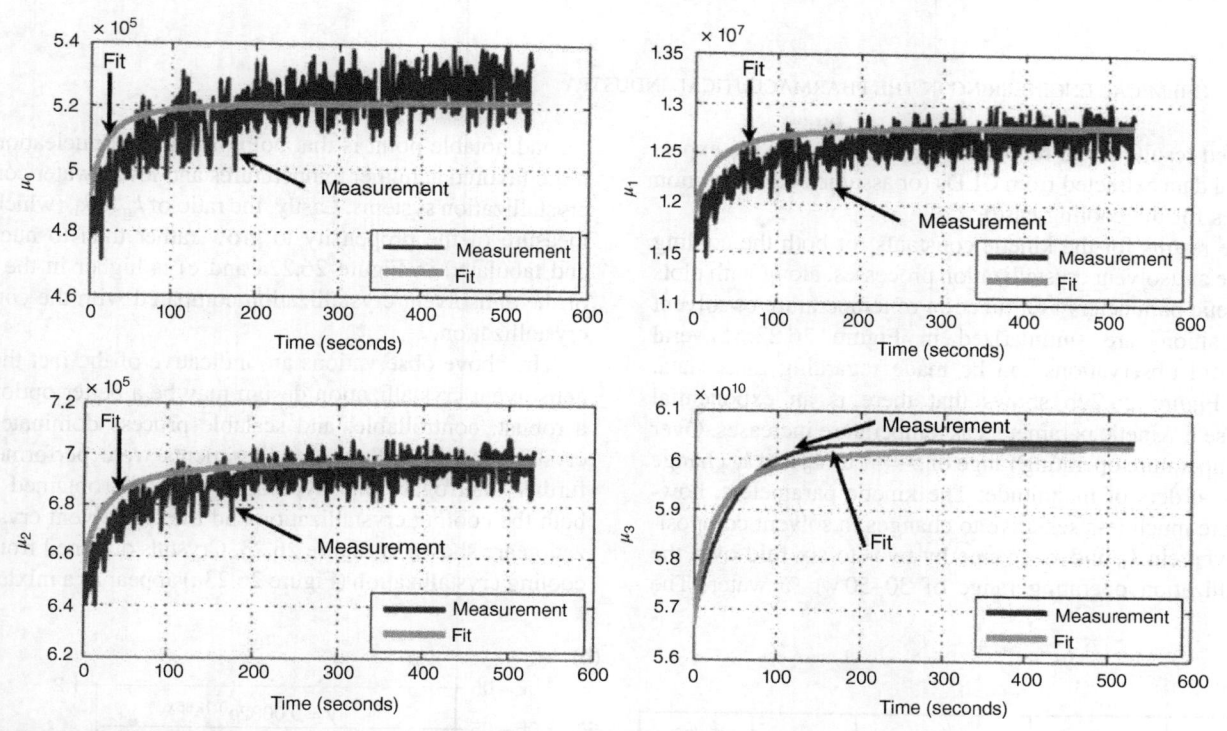

FIGURE 26.20 Modeled and measured concentration and CSD moment profiles (μ_0, μ_1, μ_2, and μ_3) at $T = 45\,°C$ for the cooling crystallization.

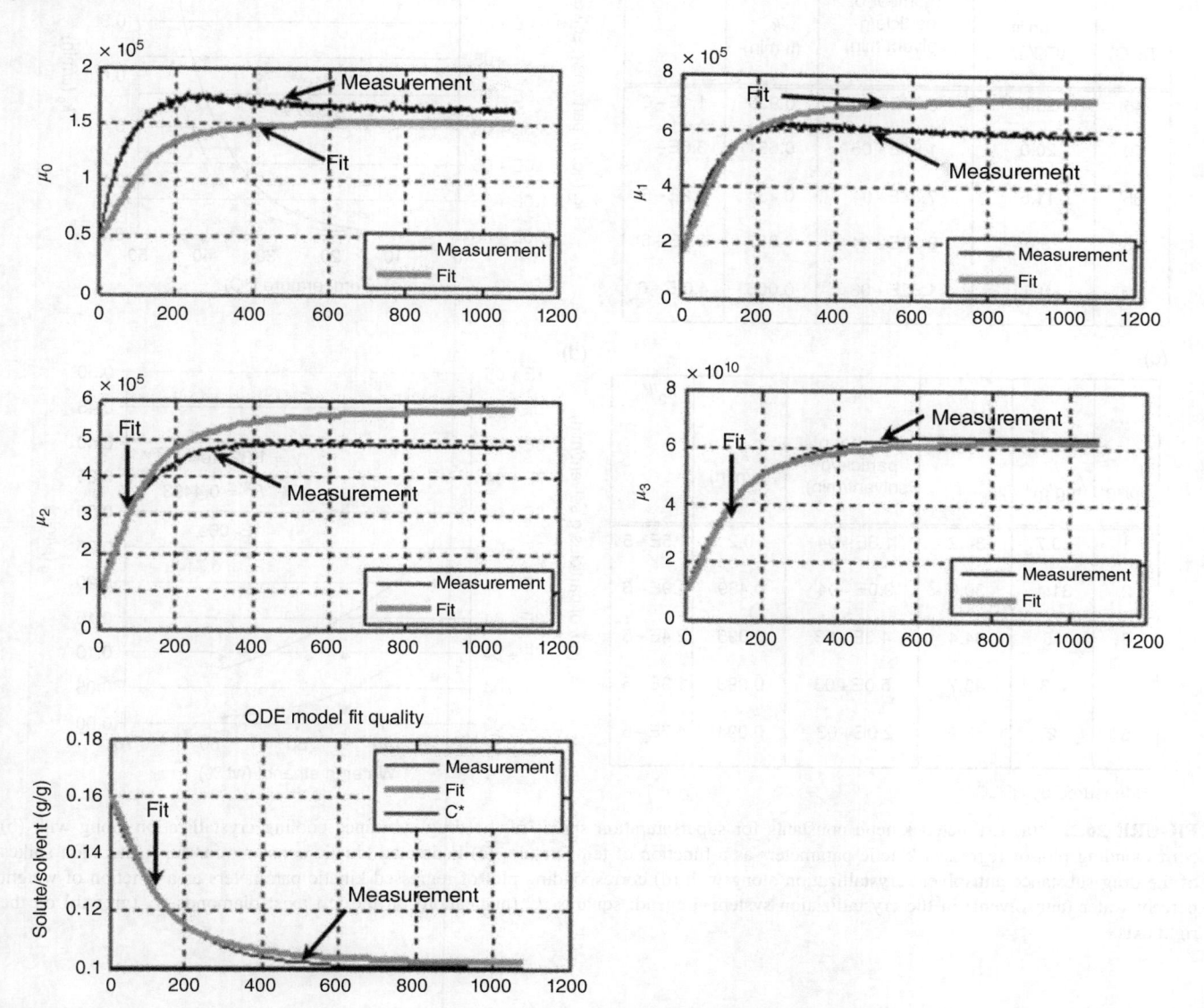

FIGURE 26.21 Modeled and measured concentration and PSD moment profiles of antisolvent crystallization at 34 wt % water and $T = 40\,°C$.

modeled results are in reasonable agreement with the experimental data extracted from CLDs (or assumed concentration profiles for the cooling case).

The results for the kinetic constants for both the cooling and the antisolvent crystallization processes, along with plots of kinetic parameters as a function of temperature or solvent composition, are summarized in Figure 26.22. Several insightful observations can be made regarding these data. First, Figure 26.22b shows that there is an exponential increase in kinetic parameters as temperature increases. Over the temperature operating range of 5–45 °C, k_b and k_g change by two orders of magnitude. The kinetic parameters, however, are much less sensitive to changes in solvent composition, wherein k_b and k_g change by two- to six fold over the crystallization operating range of 30–50 wt % water. The second notable point is that both growth and nucleation are more favored at higher temperatures and lower water content crystallization systems. Lastly, the ratio of k_g to k_b (which is a measure of the propensity to grow rather than to nucleate and tabulated in Figure 26.22a and c) is higher in the case of the antisolvent crystallization compared with the cooling crystallization.

The above observations are indicative of the fact that an antisolvent crystallization design may be a better option for a robust, controllable, and scalable process dominated by crystal growth. Follow-on experiments were performed to further interrogate this hypothesis. Crystals obtained from both the cooling crystallization and the antisolvent crystallization are shown in Figure 26.23. Crystals obtained from the cooling crystallization (Figure 26.23a) appear as a mixture of

(a)

T (°C)	Solubility (mg/g)	k_b (number of particle/g solvent/min)	k_g (μ/min)	k_g/k_b
45	26.8	8.04E+05	0.920	1.1E−6
40	20.0	1.84E+05	0.661	3.6E−6
35	11.5	7.84E+04	0.254	3.2E−6
25	5.0	2.58E+04	0.082	3.2E−6
5	0.9	1.52E+03	0.006	4.0E−6

(b)

[Plot: k_b (no. of particles/g solvent/min) and k_g (μ/min) vs Temperature (°C). Curves: $y = 0.0032e^{0.1285x}$, $R^2 = 0.9959$; $y = 644.21e^{0.1476x}$, $R^2 = 0.9798$.]

(c)

Spike	Conc. (mg/g)[a]	Water (wt %)	k_b (number of particle/g solvent/min)	kg (μ/min)	k_g/k_b
1	60.7	34.2	1.3E+04	0.2	1.5E−5
2	31.4	39.0	9.0E+04	0.439	4.9E−6
3	15	44.4	4.0E+03	0.095	2.4E−5
4	8.3	48.7	5.0E+03	0.093	1.9E−5
5	2	58.9	2.0E+03	0.094	4.7E−5

[a]Measured by HPLC.

(d)

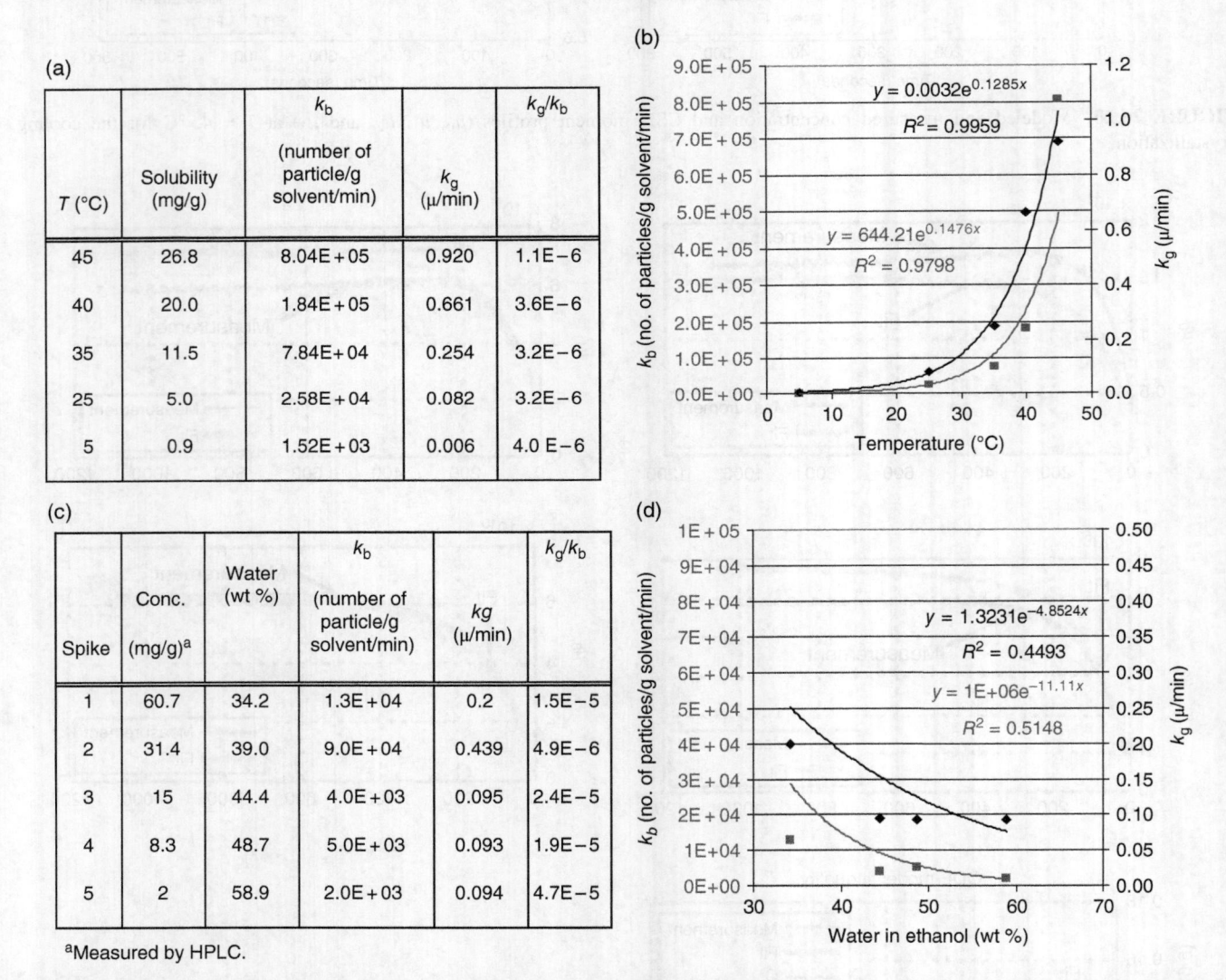

[Plot: k_b (no. of particles/g solvent/min) and k_g (μ/min) vs Water in ethanol (wt %). Curves: $y = 1.3231e^{-4.8524x}$, $R^2 = 0.4493$; $y = 1E+06e^{-11.11x}$, $R^2 = 0.5148$.]

FIGURE 26.22 (a) Estimated kinetic constants for supersaturation spikes of the drug substance cooling crystallization along with (b) corresponding plot of regressed kinetic parameters as a function of temperature. (c) Estimated kinetic constants for supersaturation spikes of the drug substance antisolvent crystallization along with (d) corresponding plot of regressed kinetic parameters as a function of weight percent water (antisolvent) in the crystallization system. Legend: squares, k_b (nucleation) on the left axis; diamonds, k_g (growth) on the right axis.

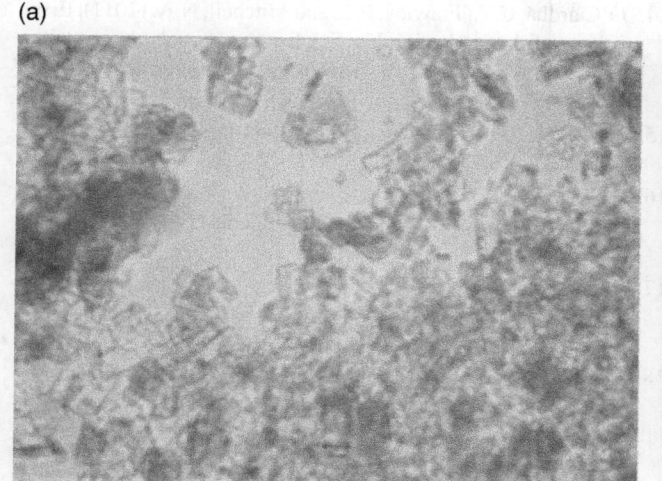

FIGURE 26.23 Polarized light microscopy images of product from (a) cooling crystallization and (b) antisolvent addition crystallization.

primary particles and agglomerates exhibiting a thin lath morphology. From the antisolvent crystallization (Figure 26.23b), thicker prismatic crystals were obtained, which proved to be more easily processable in the downstream isolation and drug product process.

The knowledge of crystallization kinetics and the MSZW was used to inform the appropriate antisolvent addition rate to sustain crystal growth throughout the crystallization. Furthermore, the seed loading was increased after initial pilot-scale batches to minimize secondary nucleation upon scale-up. This process, developed based on a solid foundation of nucleation and growth kinetics, has been successfully commercialized and shown to consistently deliver the appropriate DS physical properties.

26.4 SUMMARY

Fundamental understanding of solubility and nucleation/growth kinetics is the foundation upon which to develop and scale-up robust, commercial-ready crystallization processes. Solubility considerations, best practices, and example applications discussed herein provide insight into readily accessible means of enhancing understanding gleaned from these fundamental studies. Different approaches to understanding nucleation and growth kinetics are also presented and then discussed in the context of both literature examples and industrially relevant process development challenges. As demonstrated in this chapter, the meaningful impact that fundamental understanding can have on process development decisions and timelines is significant; the techniques and approaches discussed are appropriate for use in routine crystallization process development.

ACKNOWLEDGMENT

Authors would like to acknowledge experimental support from Bradley Greiner, Pankaj Shah, Onkar Manjrekar, and Carlos Orihuela and appreciate useful technical discussions with Shailendra Bordawekar, Samrat Mukherjee, and Ahmad Sheikh.

REFERENCES

1. Gamsjäger, H., Lorimer, J.W., Scharlin, P., and Shaw, D.G. (2008). Glossary of terms related to solubility: IUPAC goldbook. *Pure Appl. Chem.* 80 (2): 233–276.
2. Kawakami, K., Miyoshi, K., and Ida, Y. (2005). Impact of the amount of excess solids on apparent solubility. *Pharm. Res.* 22 (9): 1537–1543.
3. Mullin, J.W. (2001). *Crystallization*, 4e. Oxford: Elsevier Butterworth-Heinemann.
4. Murdande, S.B., Pikal, M.J., Shanker, R.M., and Bogner, R.H. (2011). Aqueous solubility of crystalline and amorphous drugs: challenges in measurement. *Pharm. Dev. Technol.* 16 (3): 187–200.
5. *ASPEN Properties*, Version 8.4. Burlington, MA: Aspen Technology, Inc.
6. Liu, W., Dang, L., Black, S., and Wei, H. (2008). Solubility of carbamazepine (Form III) in different solvents from (275 to 343) K. *J. Chem. Eng. Data* 53 (9): 2204–2206.
7. Mahoney, A.W., Doyle, F.J., and Ramkrishna, D. (2002). Inverse problems in population balances: growth and nucleation from dynamic data. *AIChE J.* 48 (5): 981–990.
8. Mitchell, N.A. and Frawley, P.J. (2010). Nucleation kinetics of paracetamol-ethanol solutions from metastable zone widths. *J. Cryst. Growth* 312: 2740–2746.

9. Tadayyon, A. and Rohani, B. (1996). Determination of particle size distribution by ParTec_100: modeling and experimental results. *Part. Part. Syst. Charact.* 15: 127.
10. Worlitschek, J. (2003). Monitoring, modeling and optimization of batch cooling crystallization. PhD Dissertation, Swiss Federal Institute of Technology, Zurich, Switzerland.
11. Li, M. and Wilkinson, D. (2005). Determination of non-spherical particle size distribution from chord length measurements. Part 1: theoretical analysis. *Chem. Eng. Sci.* 60: 3251.
12. Nere, N.K., Ramkrishna, D., Parker, B.E. et al. (2006). Transformation of the chord length distributions to size distributions for non spherical particles with orientation bias. *Ind. Eng. Chem. Res.* 46: 3041–3047.
13. Mitchell, N.A., Ó'Ciardhá, C.T., and Frawley, P.J. (2011). Estimation of the growth kinetics for the cooling crystallization of paracetamol and ethanol solutions. *J. Cryst. Growth* 328: 39–49.
14. Ó'Ciardhá, C.T., Frawley, P.J., and Mitchell, N.A. (2011). Estimation of the nucleation kinetics for the anti-solvent crystallization of paracetamol in methanol/water solutions. *J. Cryst. Growth* 328: 50–57.
15. Nyvlt, J. (1968). Kinetics of nucleation in solutions. *J. Cryst. Growth* 3–4: 377–383.
16. Kubota, N. (2008). A new interpretation of metastable zone widths measured for unseeded solutions. *J. Cryst. Growth* 310: 629–634.
17. Granberg, R.A., Rasmuson, A.C. (1999). Solubility of paracetamol in pure solvents. *J. Chem. Eng. Data* 44 (6): 1391–1395.
18. Fernandes, C. (1999). Effect of the nature of the solvent on the crystallization of paracetamol. *Proceedings of the 14th International Symposium on Industrial Crystallization*. Cambridge, UK.

27

CASE STUDIES ON CRYSTALLIZATION SCALE-UP

NANDKISHOR K. NERE, MOIZ DIWAN, ANN M. CZYZEWSKI, JAMES C. MAREK, AND KUSHAL SINHA
Process Research and Development, AbbVie Inc., North Chicago, IL, USA

HUAYU LI
Material and Analytical Sciences, Boehringer Ingelheim Pharmaceuticals, Inc., Ridgefield, CT, USA

27.1 INTRODUCTION

Reliable scale-up of crystallization processes requires a fundamental understanding of mixing and its impact on desired process performance attributes. The term "mixing" encompasses all aspects of hydrodynamics that dictate heat and mass transfer along with phase dispersion characteristics. In turn, hydrodynamics is governed by several equipment and process considerations, namely, (i) crystallizer geometry (size, aspect ratio), (ii) internals geometry/configuration (e.g. impeller type, size, location, number, and clearance; baffle type, size, location, and number; pump type, piping/tubing geometry, rotor–stator geometry, and configurations as applicable), (iii) physicochemical properties of fluid and solids, and (iv) operating parameters (speed of impeller, pump, and rotor–stator; fill level; feed rate/location; flow rates). Understanding and controlling hydrodynamics is a key to enabling robust process design and scale-up; for further detailed characterization of hydrodynamics, the reader is referred to Ranade [1]. The key governing parameter characterizing the hydrodynamics of a system is the turbulent kinetic energy dissipation rate, which is defined as the rate of turbulent kinetic energy loss due to viscous forces in turbulent flow. The turbulent kinetic energy dissipation rate exhibits spatiotemporal heterogeneity. As depicted in Figure 27.1, the turbulent kinetic energy dissipation rate determines the kinetics of various underlying processes (i.e. heat and mass transfer, breakage/agglomeration, phase dispersions, etc.).

In the case of crystallization processes, reliable scale-up is heavily dependent upon the reproducibility of both local and global mixing characteristics across laboratory to pilot to commercial scale. While cooling crystallization is governed by heat transfer (hence both local and global mixing), antisolvent and reactive crystallizations are most strongly influenced by local mixing and will be the focus of this discussion. The reader is referred to Chapter 24 for in-depth discussion related to cooling crystallization. The four case studies showcased in this discussion highlight how a thorough understanding of hydrodynamics and fundamentally the energy dissipation rate that is reflected through the micromixing time considerations led to reliable scale-up. The first case study demonstrates that shear imparted by the impeller across scales in a given process can play a key role in not only crystal breakage and attrition but also the extent of secondary nucleation, both of which can impact crystal size distribution (henceforth referred to as particle size distribution or PSD). The second case study details an antisolvent crystallization process designed to circumvent negative impact of variability in local mixing intensities in a large batch vessel by instituting a high-shear device outside the vessel in a recirculation loop. The third case study elaborates on the added complexity introduced when using an external mixing element in a crystallization process, namely, the need to characterize the evolution of the solvent composition of streams both in the crystallizer and in the recirculation loop. The final case study highlights the approach used to scale-up a fed-batch reactive crystallization in such a way to deliver product with consistent particle morphology. Taken together, these case studies demonstrate that fundamental understanding of mixing and associated considerations is imperative to developing scalable and robust crystallization processes.

Chemical Engineering in the Pharmaceutical Industry: Active Pharmaceutical Ingredients, Second Edition.
Edited by David J. am Ende and Mary T. am Ende.
© 2019 John Wiley & Sons, Inc. Published 2019 by John Wiley & Sons, Inc.

27.2 CASE STUDY I: DESIGNING SHEAR EXPOSURE TO ACHIEVE SIMILAR BREAKAGE/ATTRITION ACROSS SCALES

27.2.1 Background

The example described in this section demonstrates a case in which careful design to control shear exposure had significant impact on improving particle physical properties. During laboratory-scale development, a cooling crystallization process consistently delivered drug substance (DS) containing a significant amount of fines. Experiments were typically run at the 50 g scale in a 1L jacketed vessel with overhead stirring. Upon scale-up to 50 gal pilot-scale equipment, the crystallized product contained a significantly reduced amount of fines compared to laboratory batches. Scanning electron micrograph (SEM) images of the particles from both the lab scale and pilot scale demonstrate increased presence of fines in lab-scale batches as evidenced by the increased number of fine particles observed on the surfaces of larger crystals in Figure 27.2a compared with Figure 27.2b).

Smaller particle size crystals tend to cause slower and more operationally challenging isolations and can also negatively impact drug product processing. Therefore, ensuring control of the fines generated upon further scale-up would ensure a more robust process, and efforts were directed at understanding and mitigating the root cause of their formation. The parabolic cooling profile employed at a supersaturation ratio (SSR) of 1.2 showed little to no evidence of causing secondary nucleation. Therefore, it was hypothesized that fines were caused primarily due to the energy input into the process via mixing. Further efforts were made to understand what aspects of the mixing may be impacting particle secondary nucleation and attrition.

27.2.2 Mixing Characterization

Breakage and attrition of crystals is primarily governed by the frequency of particle–particle, particle–wall, and particle–impeller collision as well as intensity of the impact. Both the frequency and intensity of collisions are dictated by local shear and the global convective flow pattern. The impeller region typically exhibits an order of magnitude higher shear rate compared with the global average shear. The shear imposed is directly related to the power input per unit volume, where the power is calculated as a volumetric integration of the turbulent energy dissipation rate (ε) as shown in Eq. (27.1):

$$P = \int_0^V \rho \varepsilon \, dV \qquad (27.1)$$

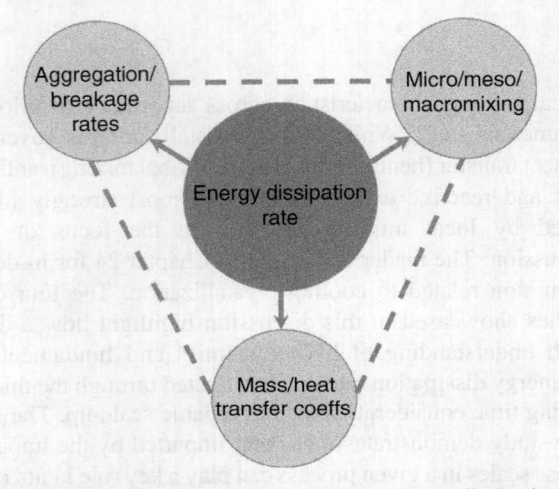

FIGURE 27.1 Scale-up triangle showing energy dissipation rate controlling various key attributes for reliable scale-up of crystallization processes.

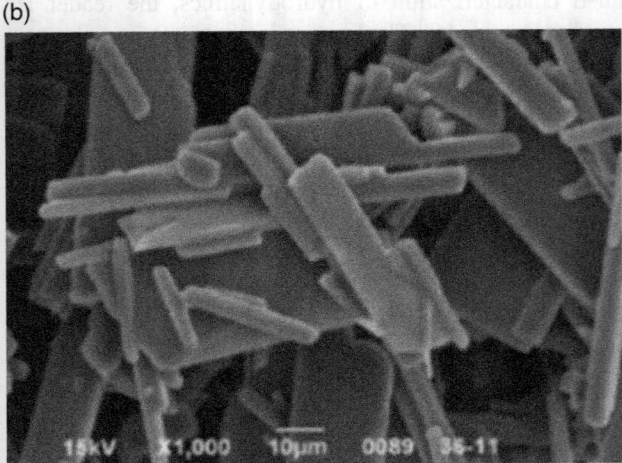

FIGURE 27.2 SEM images of crystals made from (a) lab-scale cooling crystallization and (b) pilot-scale cooling crystallization.

where

ρ is the average fluid/dispersion density.

In addition to the maximum shear, the cumulative time the crystallizer contents are passing through the impeller region also can impact particle physical properties. The average time particles spend passing through the impeller region can be quantified by calculating the number of turnovers (or number of passes, N_{pass}) the slurry makes through the impeller region. A schematic depicting the direction of flow in and out of the impeller region of an agitated vessel is shown for reference in Figure 27.3.

The number of passes through the impeller zone can be readily assessed based on parameters gleaned through computational fluid dynamics (CFD) simulations. The primary flow number (N_{qd}) represents the pumping capacity of the impeller and is calculated by determining the fluid flux out of the impeller region [2]. The primary flow number can be expressed as a function of the impeller speed (N), impeller diameter (D), and the radial or axial fluid velocity out of the impeller (v_r or v_z) as shown in Eqs. (27.2a) and (27.2b) for radial and axial flow impellers, respectively:

$$N_{qd} = \frac{1}{ND^3} \int_0^{R_I} 2\pi r v_z dr \quad (27.2a)$$

$$N_{qd} = \frac{1}{ND^3} \int_0^{h} 2\pi r v_r dh \quad (27.2b)$$

where

R_I and h are radius of the impeller for axial flow impeller and height of the blade for radial flow impeller.

The number of passes through the impeller zone (N_{pass}) can then be calculated as a function of the flow number, impeller diameter, agitation speed, and operating solution/slurry volume (V) as shown in Eq. (27.3):

$$N_{pass} = \frac{N_{qd}D^3N}{V} \quad (27.3)$$

27.2.3 Results and Discussion

To understand the impact of mixing on the generation of fines, an analysis of the mixing from both the 50 g (1 L volume) and 10 kg (50 gal volume) DS crystallization processes was performed and is summarized in Table 27.1. It is notable in this example that neither the change in impeller tip speed nor the power per unit volume is consistent with the observation that a significantly greater number of fine particles were observed at the 1 L scale compared with the 50 gal scale. Looking at the number of passes through the impeller zone, however, it is clear that particles in the 1 L equipment experience higher cumulative exposure to shear, with a more than 7 times increase in the number of passes through the impeller zone compared with the same process run at the 50 gal scale. This significant increase in

TABLE 27.1 Comparison for Mixing Characteristics in 1 L and 50 Gal Equipment

	1 L Vessel (200 rpm)	50 Gal Vessel (90 rpm)
Impeller tip speed (m/s)	0.70	1.38
Power per unit volume relative to 250 ml vessel (W/m³)	1.05	0.95
Passes through impeller zone (N_{pass}, 1/h)	1072	142

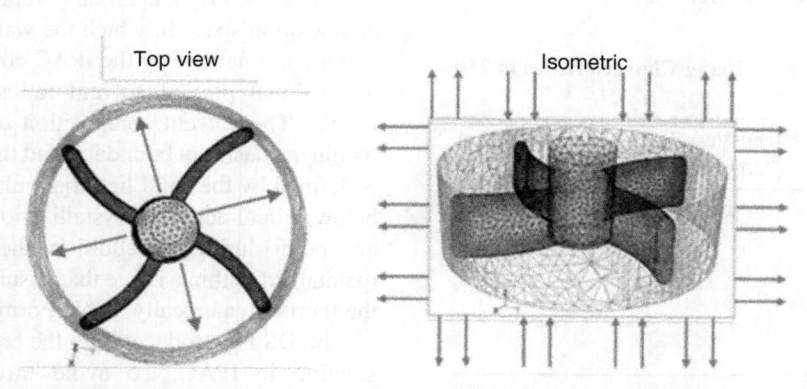

FIGURE 27.3 Top and isometric views of a rotating impeller and associated region. Arrows indicate the typical direction of fluid flow for radial flow impeller.

exposure of a crystallization mixture to the high-shear mixing zone of the vessel can result in a greater extent of secondary nucleation as well as particle attrition, both of which contribute to more fine particles being observed in the product.

The results of CFD modeling of a more extreme change in scale for the same crystallization process are summarized in Table 27.2. Here, the mixing characteristics of a 250 ml laboratory-scale vessel with a retreat curve agitator are compared with those of a 500 gal vessel equipped with a curved blade turbine impeller, similar to that shown in Figure 27.3. In this case, agitation speeds of the laboratory-scale vessel (250 rpm) and the plant scale (40 rpm) were selected to deliver equivalent power per unit volume (5.8 W/m^3).

These data show that although power per unit volume is a commonly used parameter to scale-up mixing, it may not be the only (or the most important) parameter that can significantly impact crystallization product particle size and shape distributions. In this example, while power per unit volume is relatively consistent across scales, the number of passes through the impeller zone is almost an order of magnitude greater in the 250 ml laboratory-scale vessel compared with the 500 gal plant scale.

Since many factors can impact the success of a process scale-up, it is important to reflect on multiple aspects of the hydrodynamic behavior that can impact product physical properties. The maximum shear imparted (reflected in the tip speed or power per unit volume) as well as cumulative shear exposure (quantified by the number of passes through the impeller zone) can both have significant impact on process outcomes in different situations. In this example, when the observed particle properties were not consistent with scale-up based on maximum shear, deeper insight into additional contributing factors provided a context in which to understand the scale-up outcome. In general, it is recommended that when scaling up a process for the first time, due consideration be given to potential impact of hydrodynamics in terms of both maximum and cumulative shear exposure in order to ensure robust process design.

TABLE 27.2 Comparison for Mixing Characteristics in 250 ml and 500 Gal Equipment

	250 ml Vessel (250 rpm)	500 Gal Vessel (40 rpm)
Impeller tip speed (m/s)	0.62	1.40
Power per unit volume relative to 250 ml vessel (W/m^3)	1.08	1.0
Passes through impeller zone (N_{pass}, 1/h)	1384	146

27.3 CASE STUDY II: TAILORING MIXING TO ACHIEVE DESIRED CRYSTAL FORM AND PARTICLE SIZE DISTRIBUTION

The second case study describes a more complex crystallization process wherein control and proper scale-up of mixing is required to achieve the desired DS crystal form in addition to appropriate physical properties. In this section, the complexity of the polymorph landscape coupled with the miscibility boundary of the ternary crystallization solvent mixture is described. The impact of mixing on the ability to deliver a product with a reproducible and scalable PSD is also discussed. Lastly, the process used to deliver quality product within both of these process design constraints is described followed by a summary of observed process performance across scales.

27.3.1 Polymorph Landscape and Basic Process Design

There are two polymorphs of the DS relevant to this case study: Form II is the desired thermodynamically stable form, and Form III is the undesired metastable crystal form. These polymorphs have distinct morphologies as shown by the polarized light microscopy (PLM) images shown in Figure 27.4.

The ternary crystallization solvent system consisting of isopropyl acetate (IPAC), water, and isopropyl alcohol (IPA) was designed to deliver the desired Form II of the DS. Form competition and solubility studies (discussed in detail in Chapter 18) were used to establish understanding of the polymorph landscape and to construct the ternary phase diagram shown in Figure 27.5.

The circles in the phase diagram represent the solvent compositions where Form II is thermodynamically stable, and the corresponding space is marked by the dashed line on the plot. Outside of this region, particularly in IPA-rich conditions, there is likelihood of crystallizing the undesired Form III, as indicated by the squares. The triangular markers demarcate the immiscibility boundary within which two liquid phases exist, both with and without the presence of DS.

Figure 27.6 is an alternative representation of the solvent composition space in which the water content and IPA content are normalized by the IPAC content, resulting in a two-dimensional plot of solvent ratios (IPA/IPAC vs. water/IPAC). The solvent composition design space is enclosed within the dash-dot boundary, and the immiscibility envelope is defined by the solid line/triangular markers. As described below, a dual-addition crystallization was designed so that at any point during the addition, the solvent composition is maintained within a range that ensures the generation of only the thermodynamically stable Form II.

The DS is introduced into the crystallization process as a solution in IPAC. To avoid traversing through solvent

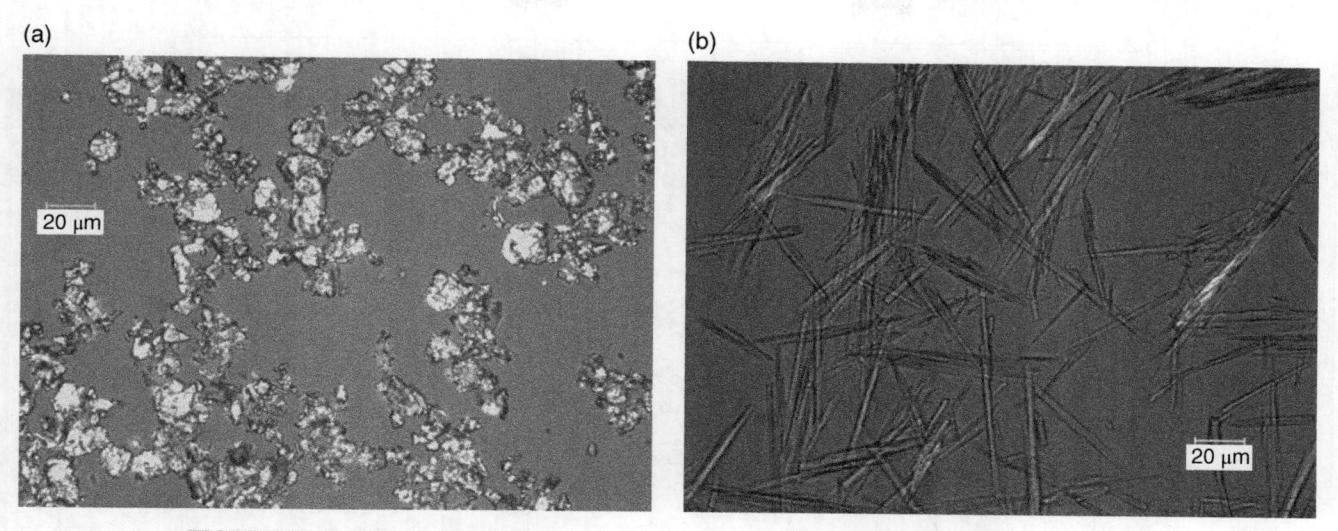

FIGURE 27.4 Polarized light microscope images of (a) Form II and (b) Form III drug substance.

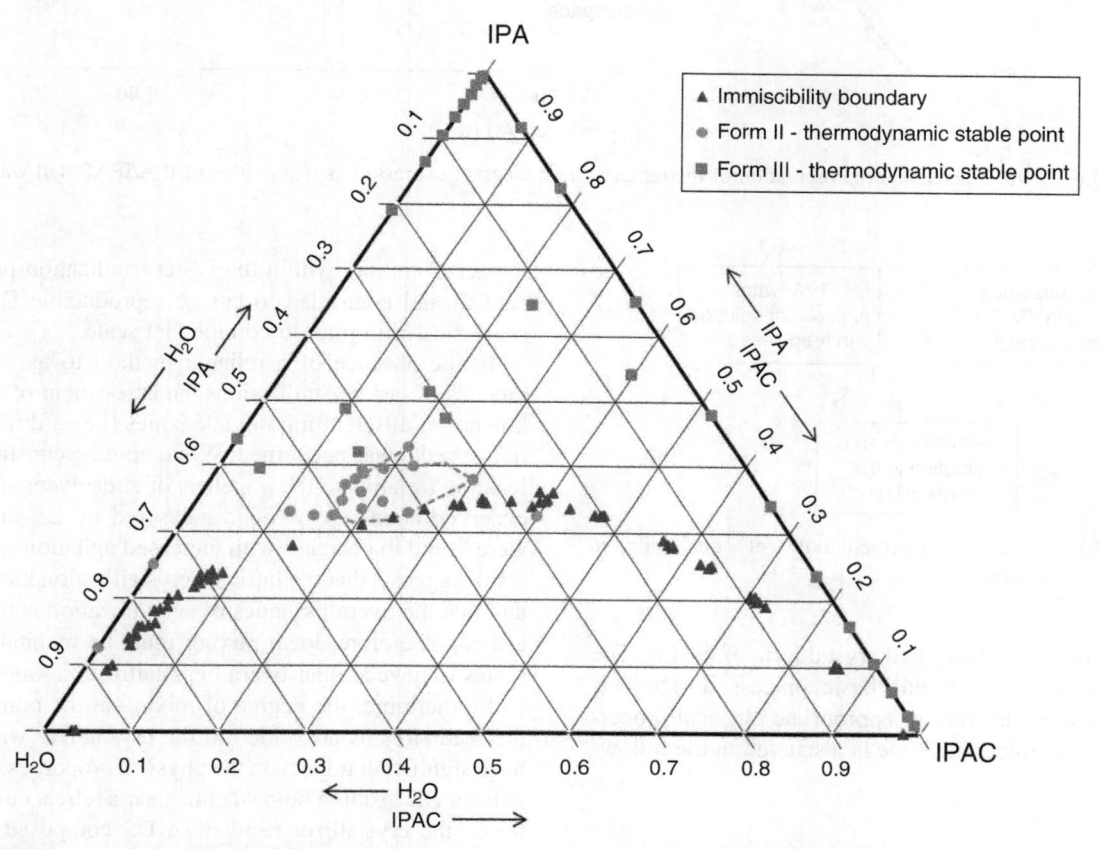

FIGURE 27.5 Ternary phase diagram (mass fraction) summarizing the thermodynamically stable crystal form at various solvent compositions relevant to the drug substance crystallization.

compositions within a Form III stable region of the phase diagram, the DS solution in IPAC and a premixed antisolvent mixture of IPA/H$_2$O are added simultaneously (and at a specific ratio of addition rates) to a premixed IPAC/IPA/H$_2$O solution in the crystallization vessel. This design, schematically represented in Figure 27.7, ensures that the entire crystallization is performed within solvent compositions in which Form II is thermodynamically stable, making the process

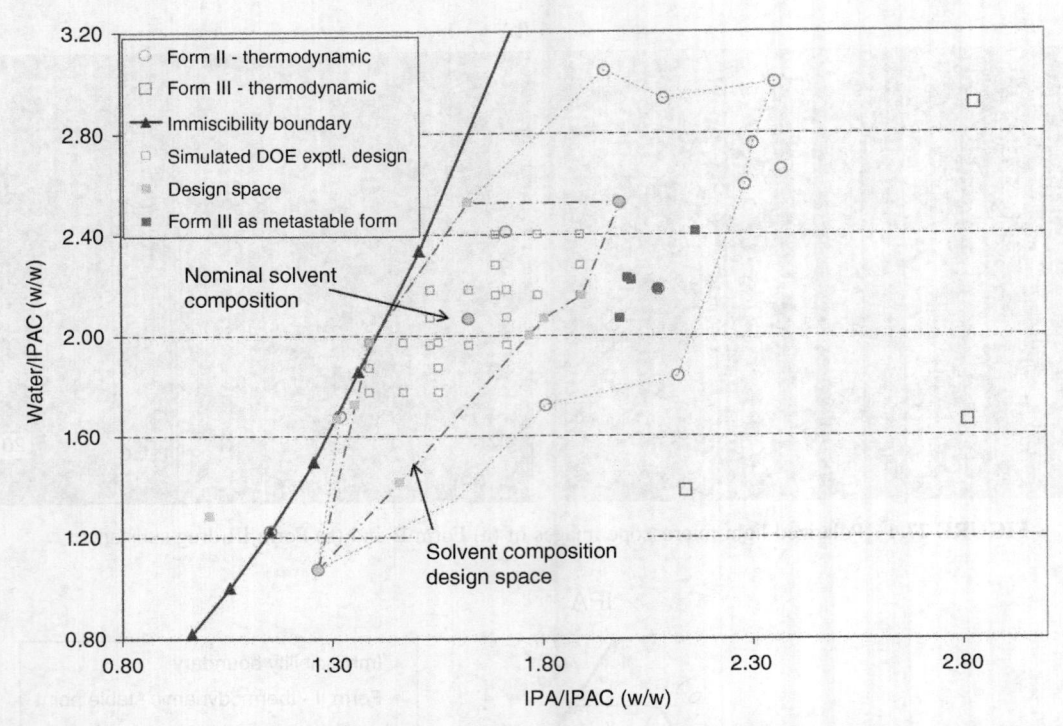

FIGURE 27.6 Two-dimensional representation of the ternary phase diagram expressed as mass ratios of IPA/IPAC and water/IPAC.

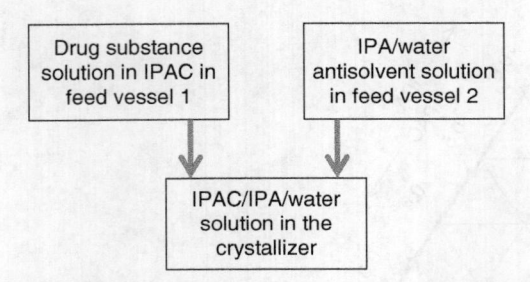

FIGURE 27.7 Schematic representation of dual-addition antisolvent crystallization.

robust for delivering the required crystal form of the DS. The modus operandi in which this basic process design was enhanced to ensure delivery of appropriate physical properties at pilot and commercial scale is discussed in the following section.

27.3.2 Control of Physical Properties

Because the formulation relevant to this case study is based on hot melt extrusion technology, tight control of the DS PSD is required to ensure successful drug product processing. Specifically, the formulation was found to be sensitive to the presence of agglomerates in the DS, due to the need to control residual DS crystallinity in the amorphous solid dispersion (ASD) dosage form. Following is a description on how the impact of mixing within the DS crystallization process was studied and controlled to ensure reproducible DS physical properties from pilot to commercial scale.

In the absence of a reliable method to assess intrinsic kinetics of fast crystallizations, an assessment of the relative kinetics at different mixing intensities (i.e. at different agitation speeds) was performed. While holding constant rate and location (relative to the impeller) of antisolvent addition, the observed counts per second measured by Lasentec FBRM were found to decrease with increased agitation speeds. This result suggests that the intrinsic crystallization kinetics is fast and that the overall kinetics of crystallization is mixing controlled. Therefore, local mixing must be maintained across scales to have similar overall crystallization kinetics.

Furthermore, the degree of mixing at the point in which the feed streams are added to the crystallizer was found to have significant impact on the physical properties of the crystallized DS. Adding both streams near a retreat curve agitator inside the crystallizer resulted in DS composed of loosely held agglomerated particles, as depicted in Figure 27.8. Due to the rapid nucleation and growth kinetics (compared to the micromixing time), the size of agglomerates generated was dependent on mixing efficiency during product solution addition. Once the agglomerates were formed, breaking them using dry milling or high-shear rotor–stator (wet milling) technologies resulted in a bimodal PSD due to variability in agglomerate strength. The bimodal PSD presented two undesirable attributes. The agglomerates promoted residual crystallinity in the drug product, and the material exhibited poor flow properties, which negatively impacted manufacturability of the downstream drug product manufacturing process.

Based on these observations, higher intensity, more consistently controlled mixing during crystallization was required across various scales to ensure suitable control of DS physical properties. Options for enhancing mixing outside of the crystallizer were considered. Typical mixing time within a recirculation loop, for example, with an assumed flow rate of 140 L/min and 1.5 in diameter tube (flow velocity of ~2 m/s) is on the order of 50 ms. This mixing may be slightly better compared with the mixing time obtained near the agitator of the stirred tank, but still would not be sufficient to minimize particle agglomeration.

To overcome mixing limitations at the point of addition of the DS feed stream, a high-shear rotor–stator mixer (wet mill) was incorporated into the process, wherein the micromixing time inside the rotating elements is estimated to be less than 10 ms. The high-shear mixer is composed of three rotor–stator pairs in series that interact with the fluid. As shown in the process flow diagram in Figure 27.9, the DS feed solution is added into a recirculation stream, followed immediately by passage of this stream through a high-shear mixer. Ensuring the micromixing time (<10 ms) is sufficient results in DS that is predominantly primary particles, as shown by the PLM image depicted in Figure 27.10. Using the high-shear mixer minimizes the local buildup of high supersaturation and results in a process that is independent of the design of the crystallizer while achieving consistent mixing across scales of interest.

The complete, final design of the DS crystallization process is shown schematically in Figure 27.9. As depicted in the

FIGURE 27.8 Polarized light microscopy image of agglomerated Form II particles crystallized from the dual addition of feed streams into the crystallizer.

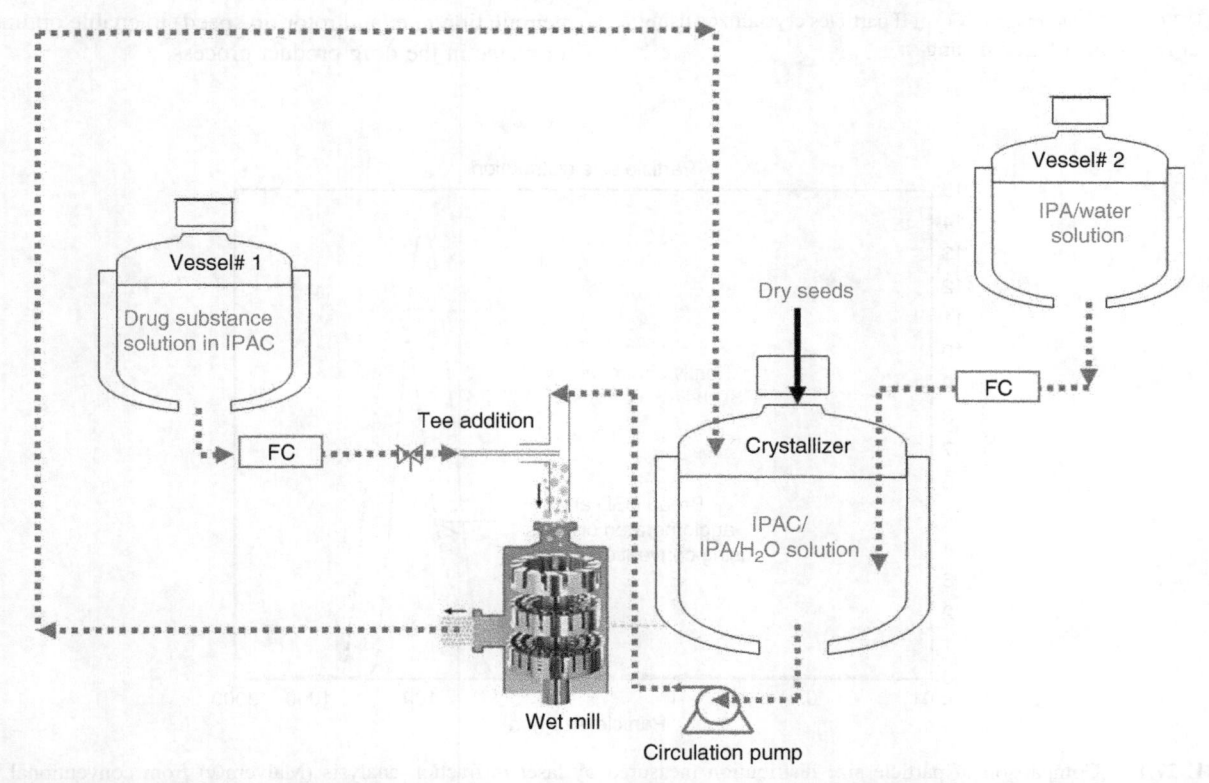

FIGURE 27.9 Drug substance crystallization process flow diagram using high-shear rotor–stator-based mixing in the recirculation loop.

process flow diagram, vessel 1 contains the DS solution in IPAC, and vessel 2 holds the IPA/water antisolvent mixture. A mixture of IPAC/IPA/water is prepared in the crystallizer. The next step is to generate a saturated DS solution that will sustain the charge of seed crystals. The saturated solution is created by simultaneously adding a portion of the vessel 1 solution to the tee at the inlet of the high-shear rotor–stator mixer and a portion of the vessel 2 solution to the crystallizer at a fixed addition rate ratio (antisolvent to DS solution). Once the solution is saturated, DS Form II seeds are charged.

The slurry is then circulated through the high-speed rotor–stator-based mixer with the simultaneous addition of a portion of vessel 1 contents and a portion of the vessel 2 contents at a fixed addition rate ratio to crystallize the product.

It is important that the ratio of the recirculation rate to the product solution addition rate is held constant during crystallization. At a given recirculation rate through high-shear mixer, slow batch addition affords good mixing of the added streams and produces predominantly primary particles. However, a study done with fast batch addition during crystallization resulted in loosely held agglomerates (due to locally high supersaturation), which broke to a greater extent upon filtration and drying and led to a bimodal PSD.

The impact of using a high-shear mixer as described on the DS PSD was investigated. As shown in Figure 27.11, using a conventional process (adding the DS solution inside the reactor) resulted in a broad PSD compared with the narrower distribution obtained using the high-shear mixer.

A summary of the PSD of various batches of DS made throughout the development and into commercial production is shown in Table 27.3. It is notable that the variability in PSD observed in the early pilot plant batches made without the use of the high-shear mixer was greatly mitigated in the final crystallization process. Implementation of the high-shear mixer delivered product of consistent particle size at both the pilot plant and commercial scales. Moreover, these data demonstrate that the PSD can be tuned by adjusting the high-shear mixer parameters (rotor–stator configuration, recirculation rate, and rotor tip speed) to enable optimal performance in the drug product process.

FIGURE 27.10 PLM image of Form II particles crystallized using high-shear rotor–stator-based mixing

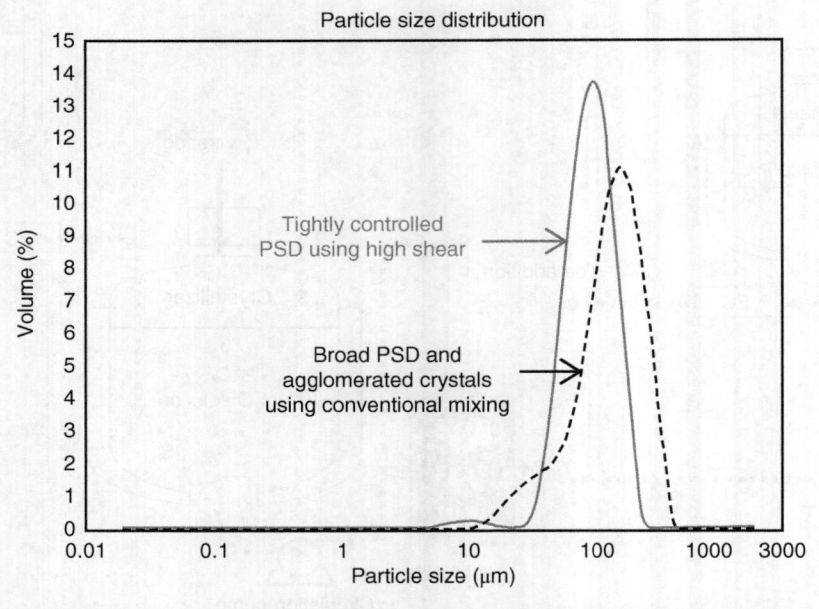

FIGURE 27.11 Comparison of particle size distribution measured by laser diffraction analysis (Malvern®) from conventional mixing (dashed lined) and high-shear rotor–stator-based mixing (continuous line).

CASE STUDIES ON CRYSTALLIZATION SCALE-UP

TABLE 27.3 Particle Size Distribution Results Using Various Mixing Configurations

Mixing Configuration	Lot ID	PSD (μm)		
		d_{10}	d_{50}	d_{90}
Conventional mixing	Early Pilot Plant Lot 1	78	184	339
Non-reproducible PSD with traditional process	Early Pilot Plant Lot 2	22	66	144
High-shear rotor–stator-based mixing in recirculation loop	Pilot Plant Lot 1	20	64	143
	Pilot Plant Lot 2	28	67	133
	Pilot Plant Lot 3	13	54	145
Consistent performance of the manufacturing process at scale using optimized mixing Ease of process transfer to commercial plant	Commercial Lot 1	11	48	158
	Commercial Lot 2	13	52	141
	Commercial Lot 3	11	46	159
High-shear mixer parameter studies	Very small particles	7	17	52
Ability to tune particle size distribution (PSD) by varying process parameters for optimal drug product performance	Large predominately primary particles	15	54	148
	Large agglomerated particles	61	165	296

27.4 CASE STUDY III: SCALE-UP CONSIDERATIONS FOR ANTISOLVENT ADDITION INTO A RECIRCULATION LOOP

As was demonstrated in the previous case study, intense and consistent mixing is achieved by addition of antisolvent into a high shear zone. Consistent and sufficient mixing can in some cases be alternatively accessed in the impeller region of the crystallizer and/or within a recirculation loop (high velocity addition into tubing/piping or into the impeller eye of the centrifugal pump). To ensure development of a robust crystallization process, mixing in these areas needs to be appropriately scaled. Higher flow velocities naturally provide a viable means of achieving better mixing; however flow velocity cannot be increased arbitrarily without considering the relevant residence time. The ratio of the average residence time (calculated by dividing the operating volume by the recirculation flow rate) to the mixing/blend time of at the least 10 is expected to approximate a well-mixed batch vessel (akin to the analogy of ideal CSTR behavior requirements; see Chapter 13). For practical purposes it is recommended to maintain a ratio significantly higher than 10; as the scale of crystallizer operating volume increases, more heterogeneity in the mixing is expected.

An additional consideration in the design of antisolvent crystallizations is that the solvent composition and solute concentration (and thus supersaturation) changes as a function of time. Therefore, in the design of a process in which antisolvent is added into a recirculation loop, it is important to ensure that the evolution of solvent composition and solute concentration in both the crystallizer and recirculation loop is consistent across scales. In this way, the rate of antisolvent addition and recirculation flow rate need to be chosen not only to deliver appropriate mixing characteristics but also to achieve similar temporal profiles of solvent composition and solute concentration in both the recirculation loop and the main crystallizer across various scales of operation. The following theoretical example focuses on the due considerations around these aspects.

27.4.1 Theoretical Analysis of Antisolvent Crystallization Added in the Recirculation Loop

Figure 27.12 depicts a schematic of an antisolvent crystallization configuration in which the antisolvent (solvent B) is added into the high-shear region of the recirculation loop. The main crystallizer contains a co-solvent (solvent A) and also contains the solute to be crystallized totaling a mass, M_T. The concentration of each solvent in the loop ($S_{A,loop}$ and $S_{B,loop}$) and solute in the loop ($S_{s,loop}$) and the corresponding concentrations in the tank ($S_{A,T}$, $S_{B,T}$, and $S_{S,T}$) are functions of time, as shown. The flow rates into and out of the vessel are shown as $\dot{Q}_{in}$ and $\dot{Q}_o$, respectively, and the antisolvent addition flow rate into the addition tee is designated as $\dot{S}_B$. The total holdup in the recirculation loop (R_T) consists of the holdup in the lines before the antisolvent addition (R_B) and after the antisolvent addition (R_A) and holdup in the process equipment (high-shear mixer and the pump, R_{eqmt}).

The objective of this type of evaluation is to understand how different the solute and solvent concentration profiles in the tank are from the respective concentrations in the recirculation loop over the course of the antisolvent addition. Any observed difference can then be assessed in the context of its impact on supersaturation and hence on the crystallization process. This evaluation begins with establishing the material balance as the basis for the calculations.

Concentrations of solvent A and solute S in the crystallizer change over time due to the dilution effect caused by the addition of antisolvent B. For example, the total solute concentration can readily be calculated from the Eq. (27.4), and a similar equation could be written for solvent A:

$$S_{S,T}(t) = \frac{S_{S,T}(0) M_T(0)}{M_T(0) - R_T + \dot{S}_B t} \quad (27.4)$$

However, concentration profile of the antisolvent B in the crystallizer can be obtained by solving the following ordinary differential equation using Matlab® or Polymath® software with the known initial conditions of the total mass in the tank $[M_T(0)]$, the antisolvent mass fraction $[S_{B,T}(0) = 0]$, and the total mass hold in the loop for a given mass flow rate of the antisolvent, $\dot{S}_B$:

$$\frac{d}{dt}\left\{S_{B,T}(t)\left[M_T(0) + \dot{S}_B t - R_T\right]\right\} = \dot{S}_B \quad (27.5)$$

where

$R_T = R_A + R_B + R_{eqmt}$ is the total holdup volume in the recirculation loop and associated equipment.

Furthermore, based on simple mass balance around the point of addition of antisolvent in the recirculation loop, one can easily infer the concentrations of solvents and solute in the loop. For example, the concentration of solvent A in the loop would be given by

$$S_{A,\,loop}(t) = \left(\frac{\dot{Q}_o}{\dot{Q}_o + \dot{S}_B}\right) S_{A,T}(t) \quad (27.6)$$

For this theoretical case study, application of the analysis described above is demonstrated in the context of a simple antisolvent addition crystallization process. The crystallization process is designed such that initially, the solute is dissolved in 0.19 kg solute/kg solution in the solvent A. Antisolvent (solvent B, 9.6 kg/kg of solute) is then added into a recirculation loop over a period of three hours to complete the crystallization. Two batch configurations are considered, including a 25 kg nominal batch size and a 175 kg batch size, representing a 7× scale-up of the process. Relevant parameters and initial conditions for two such batches are summarized in Table 27.4.

The temporal evolution of the ratio of solvent B (process antisolvent) to solvent A (process solvent) in both the crystallizer and the recirculation loop is shown in Figure 27.13a for the 25 kg batch size and Figure 27.13b for the 175 kg batch

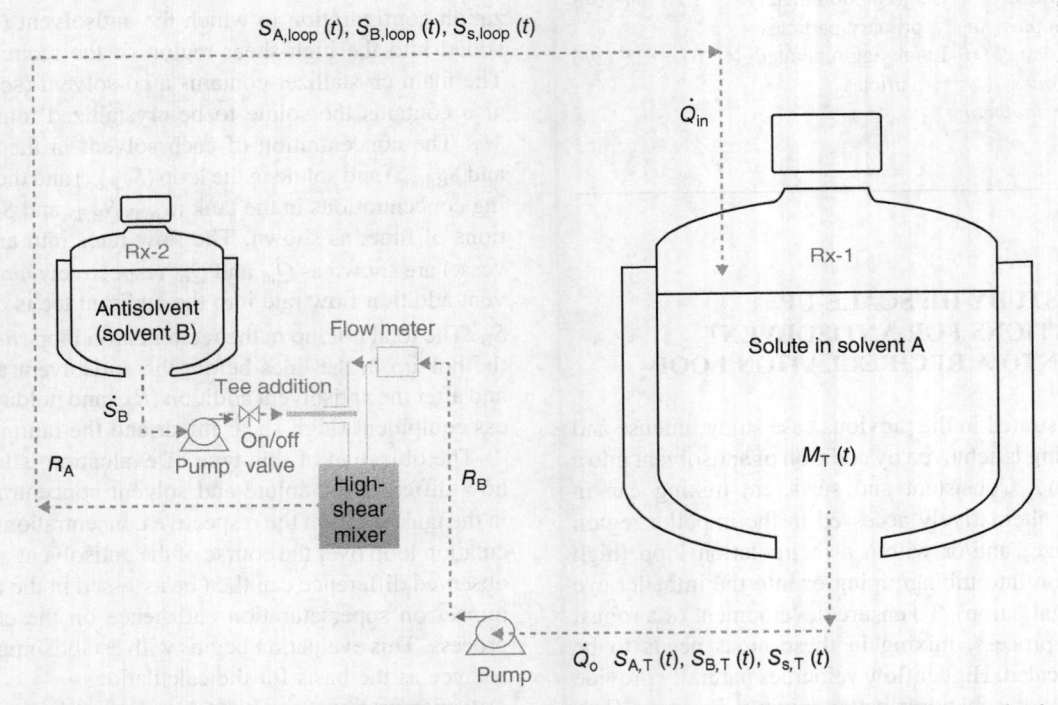

FIGURE 27.12 Schematic of the recirculation loop along with the stream compositions of solvents and solute.

TABLE 27.4 Process Parameters and Initial Conditions for a 25 and 175 kg Batch Size Antisolvent Addition Crystallization Process

Batch Size (kg)	Initial Mass in the Crystallizer, $M_T(0)$, (kg)	Mass Fraction of Solvent, $S_{A,T}(0)$ kg Solvent/kg Solution	Mass Flow Rate of Antisolvent, $\dot{S}_B$ (kg/min)	Recirculation Flow Rate (kg/min)	Antisolvent Added (kg)	Time of Antisolvent Addition (h)
25	137.5	0.81	1.14	33.14	240	3
175	962.5	0.81	7.98	115.99	1678	3

size. Notably, the ratio of antisolvent in the crystallizer and in the recirculation loop diverges from each other in both cases, and the difference becomes more pronounced at increased batch scales. Moreover, the difference in solvent ratio in the crystallizer and the recirculation loop grows over time in both cases, with the recirculation loop being enriched in antisolvent. While it may appear at first glance that these differences in crystallizer/recirculation loop solvent ratios over time and across scales are relatively minor, such a change may have significant impact on supersaturation, depending on the slope of the solubility curve and on the rate of crystallization kinetics. Hence, the systems that have the fastest crystallization kinetics and steepest solubility curves are expected to have the greatest impact on supersaturation. In such a case, one may anticipate the potential for a divergence in physical properties of crystals nucleated and/or grown in the crystallizer compared with the recycle loop over time.

By a similar analysis, the ratio of antisolvent (solvent B) to total solute was profiled over time for batches at both the 25 and 175 kg scales, as shown in Figure 27.14a and b, respectively. A similar trend is observed in this data, with a notable divergence between the crystallizer and recirculation loop concentrations of antisolvent with increased scales.

Figure 27.15 displays a profile of how the total solute concentration in both the crystallizer and recirculation loop evolves over the course of the antisolvent addition at both the 25 and 175 kg scales. The total solute concentration in the crystallizer is slightly higher than that in the recirculation loop, driving supersaturation higher in the crystallizer compared with the recirculation loop.

This analysis shows that these highly interrelated measures of solvent ratios and solute concentration can impact the SSR differently in different parts of the reactor configuration. It should also be noted that these differences *may get amplified in the systems involving more than two solvents*, and it is therefore recommended to carry out similar analyses to inform appropriate scale-up considerations.

27.4.2 Summary Considerations

The case study shown here elaborates possible concerns that need to be addressed for designing the scale-up of the process

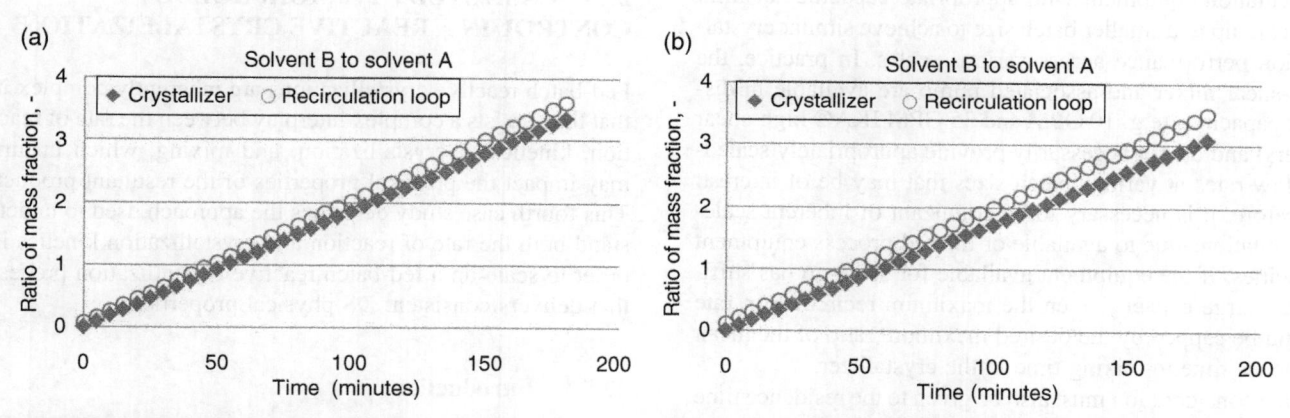

FIGURE 27.13 Ratio of solvent B (antisolvent) to solvent A (solvent) in the crystallizer and the recirculation loop over the course of a three hour antisolvent addition for the process described in Table 27.4 at (a) 25 kg scale and (b) 175 kg scale.

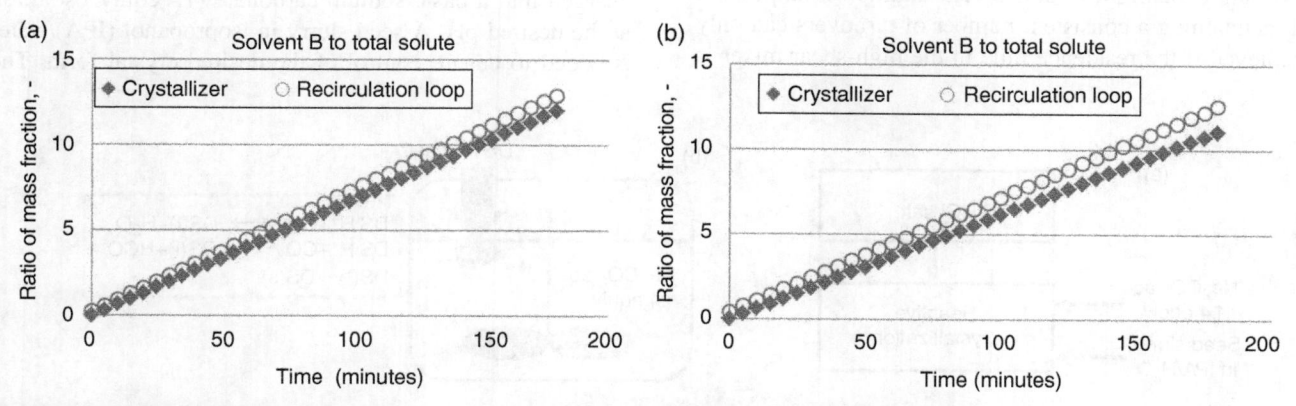

FIGURE 27.14 Ratio of solvent B (antisolvent) to total solute in the crystallizer and the recirculation loop over the course of a three hour antisolvent addition for the process described in Table 27.4 at (a) 25 kg scale and (b) 175 kg scale.

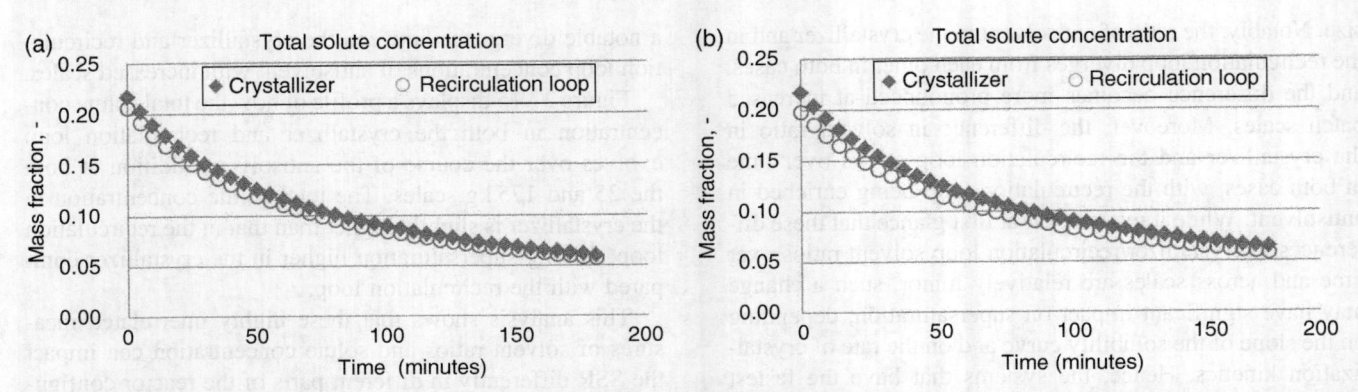

FIGURE 27.15 Total solute concentration in the crystallizer and the recirculation loop over the course of a three hour antisolvent addition for the process described in Table 27.4 at (a) 25 kg scale and (b) 175 kg scale.

involving the use of a recirculation loop meant to enhance mixing. While local mixing can be managed well in such a process design, it is also important to keep the crystallization trajectories the same in both the recirculation loop and the crystallizer to yield similar process performance. In order to ensure similar performance, it is necessary to either use recirculation equipment with appropriate capacities or limit the scale-up to a smaller batch size to achieve similar crystallization performance across various scales. In practice, the high-shear mixer and associated pump are available in discrete capacities (e.g. 10 GPM and 35 GPM IKA® high-shear mixers) and do not necessarily provide appropriately scaled-up flow rates at various batch sizes that may be of interest. Therefore, it is necessary to be cognizant of inherent scale-up limitations due to available or limited process equipment capacities. If the equipment available for scale-up has sufficiently large capacity, then the maximum recirculation rate should be capped by the desired maximum ratio of the mean residence time to mixing time in the crystallizer.

Due consideration must also be given to the residence time in the high-shear mixer, which could be an important scale-up criterion, as discussed in the first case study of this chapter. For example, if scale-up of "breakage" in a high shear zone that occurs concurrent to antisolvent addition is important, then maintaining a consistent number of turnovers can only be achieved if the residence time in the high-shear mixer is kept constant. Higher flow rates required to achieve good mixing may result in a lower residence time and hence may require an increased number of turnovers through the high-shear mixer upon scale-up.

27.5 CASE STUDY IV: MORPHOLOGY CONTROL IN A REACTIVE CRYSTALLIZATION

Fed-batch reactive crystallizations are inherently complex in that there exists a complex interplay between the rate of reaction, kinetics of crystallization, and mixing, which in turn may impact the physical properties of the resultant product. This fourth case study describes the approach used to understand both the rate of reaction and crystallization kinetics in order to scale-up a fed-batch reactive crystallization process that delivers consistent DS physical properties.

27.5.1 Introduction

The process flow diagram of a reactive crystallization used to manufacture the DS of interest in this case study is shown in Figure 27.16a. The monohydrochloride salt of the DS is charged into a basic sodium carbonate (1.4 equiv.) solution at the desired pH. A seed slurry in isopropanol (IPA)/water is added to ensure control of the desired crystal form. The

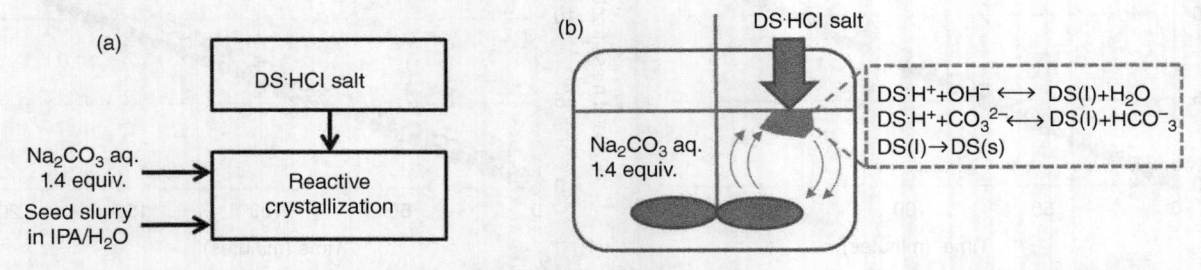

FIGURE 27.16 (a) Crystallization process flow diagram. (b) Schematic depicting impact of mixing on reaction/crystallization at the point of addition of DS solution into the basic solution.

monohydrochloride salt immediately reacts with the base, resulting in the generation of supersaturation followed by the crystallization of DS free base.

Figure 27.16b is a schematic that depicts the complex processes involved at the location where the DS solution contacts the base. It is at this point where both reaction and crystallization are occurring, the relative rates of which, along with the intensity of mixing, impact the local supersaturation. As described in this case study, understanding these phenomena was important to establishing a control strategy to reproducibly deliver the desired product morphology.

27.5.2 Crystallization Kinetics

The solubility of the DS free base was measured as a function of base concentration and isopropanol content at 20 °C and is summarized in Figure 27.17. The solubility decreases with increasing sodium carbonate content but is very low, ranging from 1.2 to <0.1 mg/g, throughout the range studied. Moreover, the isopropanol introduced through the seed slurry into the crystallization process does not impact the DS solubility significantly. This low solubility in the crystallization solvent system results in inherently high supersaturation of the DS free base at the point of contact between the hydrochloride salt and basic crystallization medium due to fast reaction.

A very rapid nucleation of crystals observed under nominal processing conditions is not suitable for interrogating crystallization kinetics due to the inability to sample and monitor the process over time. White solids were observed to precipitate upon the dropwise addition of the DS HCl solution to sodium carbonate base under nominal processing conditions. The nominal condition is indicated by a large circle in the upper right corner in the plot of DS HCl concentration as a function of sodium carbonate concentration (Figure 27.18).

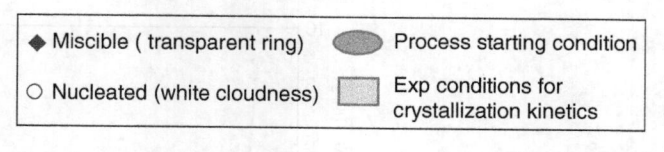

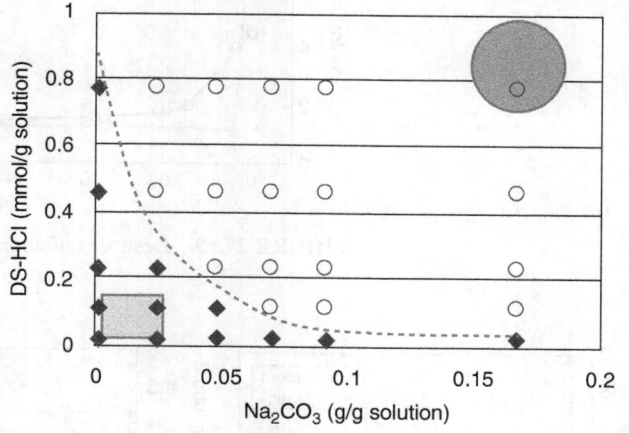

FIGURE 27.18 Precipitation domain map as a function of sodium carbonate concentration and DS HCl salt concentration at 20 °C.

Similar observations of rapid crystallization were noted at all conditions marked with open circles in Figure 27.18. At sufficiently lower base concentrations and/or DS HCl solution concentrations, phase separation was observed, followed by slow conversion to crystalline product. This behavior was observed at conditions marked with diamonds in Figure 27.18. The low base/low DS HCl concentration conditions (marked in the lower left corner of Figure 27.18) were selected as appropriate conditions for accessing crystallization kinetics.

Three experiments were performed in which a supersaturation spike was generated to a desired level and the desupersaturation profile was monitored over time. All experiments were executed under the same agitation rate (400 rpm) and under conditions with the same thermodynamic solubility (1.16 mg/g at 20 °C). Table 27.5 lists SSR (denoted by S) along with the seed loading used for each of the spike experiments.

Desupersaturation profiles for each of these spike experiments are summarized in Figure 27.19.

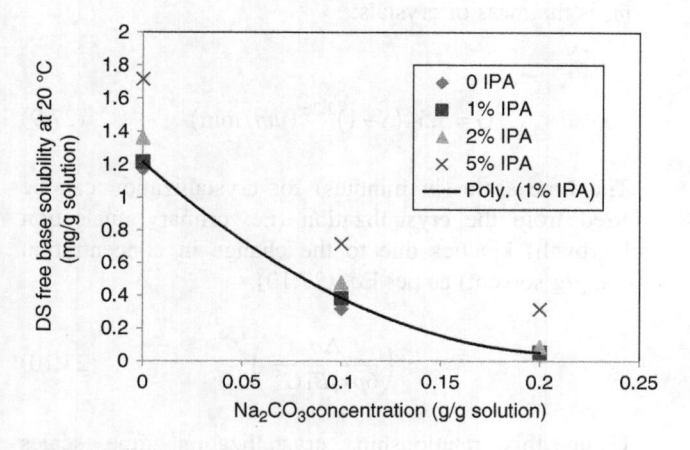

FIGURE 27.17 Solubility of the DS free base as a function of sodium carbonate concentration and isopropanol content at 20 °C.

TABLE 27.5 Supersaturation Ratios (SSR) and Seed Loading Used for Spike Experiments

Spike Number	SSR	Seed Concentration (mg/g)
1	7.6	1.28
2	4.2	0.19
3	4.2	0

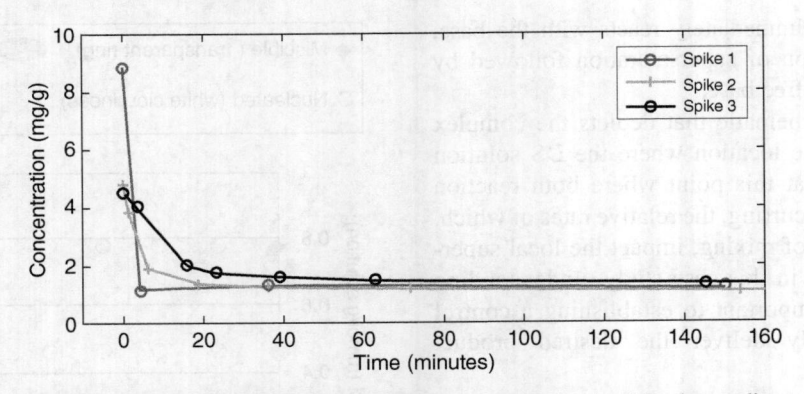

FIGURE 27.19 Desupersaturation profile for crystallization kinetics studies.

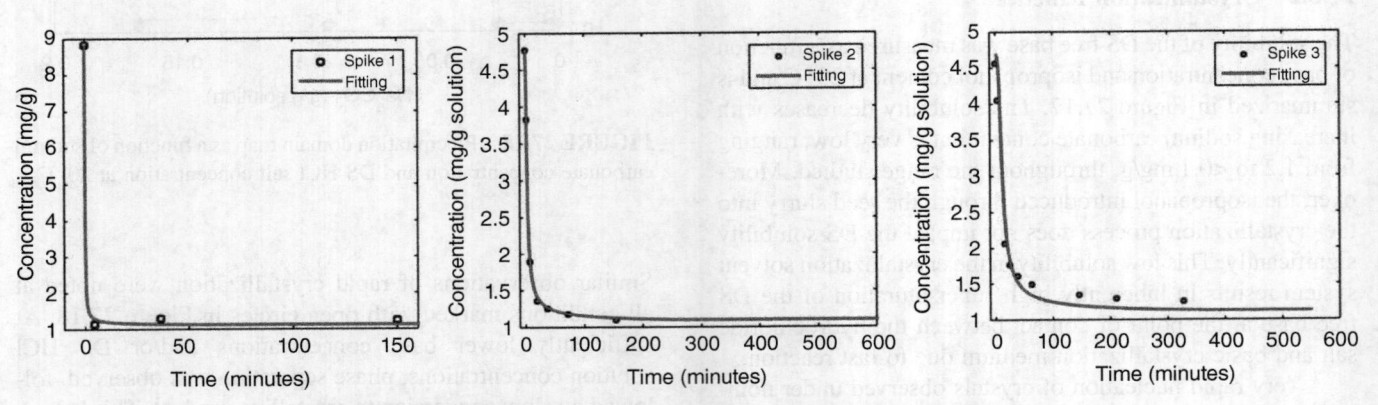

FIGURE 27.20 Parameter fitting results from spiking experiments.

These desupersaturation profiles were used in population balance modeling and parameter fitting to estimate the primary nucleation rate (B_1), secondary nucleation rate (B_2), and crystal growth rate (G). These parameters were modeled using the method of moments [3] and minimization of the sum of squared error ($\sum(c_{sim} - c_{exp})^2$) between experimental data and simulation results [4, 5]. Fitting results for the three spiking studies are shown in Figure 27.20.

Equations for the primary nucleation rate, secondary nucleation rate, and growth rate are shown in Eqs. (27.7), (27.8), and (27.9), respectively. The exponent of 22.7 in Eq. (27.7) is significantly higher than the typical value of 3–4. A high value for this exponent suggests the process is dominated by primary nucleation [6]. It is likely, however, that variability in local supersaturation contributed to estimation of this extraordinarily high result. The apparent difference between the locally high supersaturation due to pockets of inhomogeneity at the point of nucleation (exacerbated by low solubility for a fast acid–base reactive crystallization), which is responsible for underlying crystallization, and the average (lower value) supersaturation used for parameter fitting might have resulted in overestimation of the primary nucleation rate exponent.

$$B_1 = 6.31 \times 10^{-11} (S-1)^{22.7} \ (\#/\text{g solvent}/\text{min}) \quad (27.7)$$

$$B_2 = 20.0 (S-1)^{0.39} m_s^{1.58} \ (\#/\text{g solvent}/\text{min}) \quad (27.8)$$

where

m_s is the mass of crystals:

$$G = 3.54(S-1)^{2.26} \ (\mu m/\text{min}) \quad (27.9)$$

The time scale (in minutes) for crystallization can be inferred from the crystallization (i.e. primary nucleation and growth) kinetics due to the change in concentration (Δc in g/g solvent) as per Eq. (27.10):

$$t_{cryst} = \left(\frac{\Delta c}{6\rho k_v B_1 G^3}\right)^{1/4} \quad (27.10)$$

Using this relationship, crystallization time scales were calculated as a function of SSR and plotted in Figure 27.21. Over the SSR range of 5–50, the crystallization time varies from microseconds to seconds. In the event that this time scale of crystallization is similar to that of the micromixing time, it may be the case that the crystallization and mass transfer are competing phenomena. As discussed in the next section, insufficient mixing may allow DS free base to nucleate under locally high SSR.

27.5.3 Mixing Characterization

CFD simulations were performed to assess the hydrodynamics in the crystallizer and calculate the time scale of mixing. Both the micromixing time and mesomixing time (Figure 27.22a and b, respectively) were calculated throughout the fluid domain under operational conditions. While the micromixing time represents the time scale of mixing at diffusion length scales (Kolmogorov length scale), the mesomixing time is the time it takes to dissipate a feed stream into a batch by turbulent dispersion.

The mesomixing time at the liquid surface of the agitated vessel where the feed stream is introduced to the main batch during the DS mono HCl addition is of the order of hundreds of milliseconds. The relative rate of the mesomixing time compared with reaction/crystallization rate impacts physical properties of the resultant DS, as discussed in the following section.

27.5.4 Impact on Particle Morphology

As was discussed in case study II, in the absence of a reliable method to assess intrinsic kinetics of fast crystallizations, it can be informative to assess relative kinetics. In this case, experiments were performed to evaluate how variation in the reaction rate (modulated via the base concentration) and variation in the degree of mixing impact the particle morphology of the resultant product. Figure 27.23 shows scanning electron microscope (SEM) images of products obtained from (Figure 27.23a) the nominal base concentration, (Figure 27.23b) increase of the base concentration to 133% of nominal, and (Figure 27.23c) 50% of the nominal base concentration (reduced by half) while holding the mixing rate constant. At higher base concentration, the image in Figure 27.21b shows that the increased reaction rate resulted in agglomerated particles composed of fine primary particles. At reduced base concentration, the reduced reaction rate resulted in large primary particles with smooth surfaces (Figure 27.23c). The nominal base concentration resulted in a mixture of larger primary particles and fines (Figure 27.23a).

The ability to modulate the morphology by varying the base concentration is indicative of the crystallization being reaction rate limited. At reduced base concentrations, the slowed reaction rate leads to reduced DS free base concentrations and a low SSR, which leads to large primary particles with smooth crystal surfaces. On the other hand, increased base concentration leads to faster reaction rate and higher

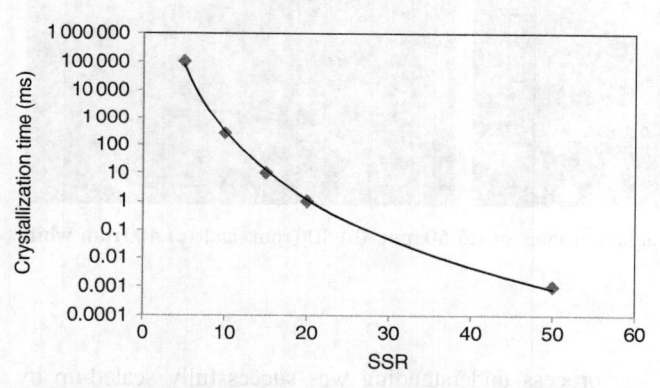

FIGURE 27.21 Calculated time scale of crystallization as a function of supersaturation ratio.

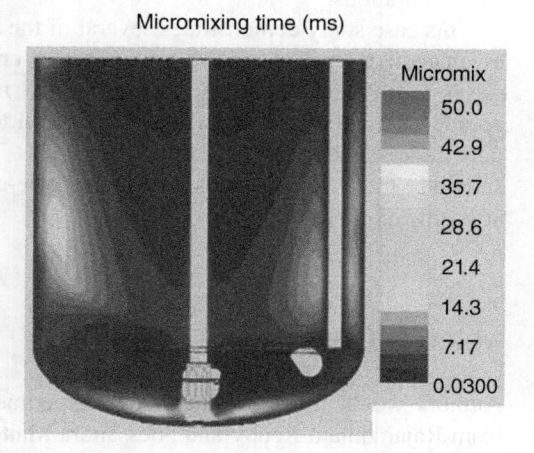

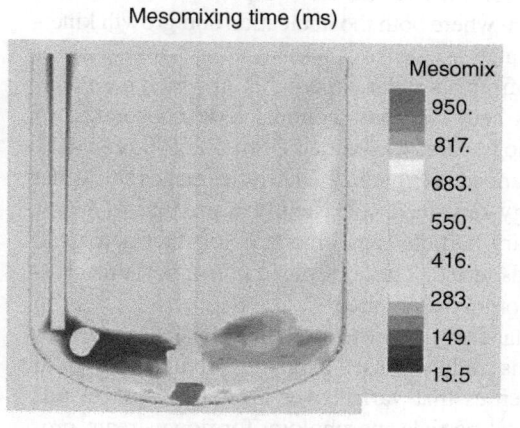

FIGURE 27.22 Contour plots of micromixing and mesomixing times calculated from CFD simulations in 500 L reaction vessel with the agitation speed of 100 rpm.

FIGURE 27.23 Scanning electron micrograph (SEM) images of products from crystallizations run at (a) nominal base concentration, (b) 133% nominal base concentration, and (c) 50% decrease from nominal base concentration while maintaining consistent agitation speed.

FIGURE 27.24 SEM images of products from crystallizations run at agitation rates of (a) 50 rpm, (b) 100 rpm, and (c) 400 rpm while maintaining nominal base concentration.

DS free base concentrations, resulting in a high SSR, causing rapid nucleation of fine particles, which agglomerate into irregular shapes. The nominal base concentration resulting in a mixture of both particle morphologies appears to be operating in conditions where the local supersaturation is in a transition region where both the nucleation and growth kinetics are prevalent.

The results from a similar experiment in which base concentration was held constant (nominal base concentration) and the agitation speed was varied from 50, 100, and 400 rpm are summarized in Figure 27.24. These images show that the morphology remained consistently a mixture of larger and fine primary particles agglomerated together across all agitation speeds tested. These results confirm that this crystallization is not mixing limited.

The understanding of how to modulate physical properties gained from this study was used to readily produce batches of DS with properties that varied over a broad range. In this case, the optimal particle morphology for downstream processing and formulation was the finer. Agglomerated particles were observed from the higher base concentrations. The process understanding was successfully scaled-up by controlling the base addition (and thus the rate of reaction) and ensuring mixing was sufficient across all scales to maintain similar overall rates of reaction and hence the local supersaturations.

This case study demonstrates several of the complexities of characterizing and scaling up a reactive crystallization. Assessment of reaction, crystallization, and mixing using appropriately designed experiments is critical to fundamentally understanding key aspects of the process. These results should be then used as a basis for reliable scale-up of reactive crystallization processes.

ACKNOWLEDGMENT

Authors would like to acknowledge experimental support from Rajarathnam Reddy and Alessandra Mattei and useful technical discussions and insights from Shailendra Bordawekar, Samrat Mukherjee, and Ahmad Sheikh.

REFERENCES

1. Ranade, V.V. (2002). *Computational Flow Modelling for Chemical Reactor Engineering*. New York: Academic Press.
2. Nere, N.K., Patwardhan, A.W., and Joshi, J.B. (2003). Liquid-phase mixing in stirred vessels: turbulent flow regime. *Industrial and Engineering Chemistry Research* 42 (12): 2661–2698.
3. Randolph, A. (1988). *Theory of Particulate Processes: Analysis and Techniques of Continuous Crystallization*, 2e. San Diego, CA: Elsevier Inc.
4. Miller, S.M. and Rawlings, J.B. (1994). Model identification and control strategies for batch cooling crystallizers. *AIChE Journal* 40 (8): 1312–1327.
5. Togkalidou, T., Tung, H., Sun, Y. et al. (2004). Parameter estimation and optimization of a loosely bound aggregating pharmaceutical crystallization using in situ infrared and laser backscattering measurements. *Industrial and Engineering Chemistry Research* 43 (19): 6168–6181.
6. Aoun, M., Plasari, E., David, R., and Villermaux, J. (1999). A simultaneous determination of nucleation and growth rates from batch spontaneous precipitation. *Chemical Engineering Science* 54 (9): 1161–1180.

28

POPULATION BALANCE-ENABLED MODEL FOR BATCH AND CONTINUOUS CRYSTALLIZATION PROCESSES

AJINKYA PANDIT AND RAHUL BHAMBURE
Chemical Engineering and Process Development Division, CSIR – National Chemical Laboratory, Pune, MH, India

VIVEK V. RANADE
School of Chemistry and Chemical Engineering, Queen's University Belfast, Belfast, UK

28.1 INTRODUCTION

Crystallization is a key unit operation in the pharmaceutical process industries that helps in the separation and purification step of the active pharmaceutical ingredients (APIs). In case of the solid dosage forms, physicochemical properties of the API crystals define the "critical quality attributes" (CQAs) like dissolution rate, stability, and in turn the overall therapeutic efficacy of the drug product [1]. Accurate and precise control of the solid dosage form CQAs can be achieved by controlling various crystal parameters such as crystal size, shape, polymorphic form, compactability, density, flowability, and solubility [2]. The crystal size distribution and the crystal shape are the critical factors that determine the CQAs for APIs. During the last decade, most of the pharmaceutical manufacturers adopted the quality by design (QbD) approach to achieve the improved operational control and compliance resulting from continuous real-time quality assurance. QbD has been defined in the ICH Q8 guideline as "a systematic approach to process development that begins with predefined objectives and emphasizes product and process understanding and process control, based on sound science and quality risk management" [3].

Despite the widespread application of the batch crystallization for manufacturing of various APIs, the control of this critical unit operation is generally based on empirical approaches involving the use of empirical kinetic expressions for predicting the batch CSD information [4]. Due to the intrinsic nonlinear dynamics associated with nonideal mixing and kinetic variations, process design and control in a crystallization is a challenging task. Recent studies have shown that operating crystallizers in a continuous mode potentially offer several benefits over the batch mode such as better product consistency, more control over product quality, and reduced operating cost and space [5, 6]. However, implementation of continuous crystallizers can be cumbersome due to the complex process dynamics involved. Recent studies that investigated the potential benefits of continuous crystallization were performed on a lab scale, and issues related to scale-up were not addressed [7]. Mathematical modeling can go a long way toward aiding in process design, identifying optimum operating protocols and process control of batch and continuous crystallization processes especially when complex process dynamics are involved. Validated mathematical models may also be used to reduce the uncertainties associated with scale-up of crystallization processes.

Modeling of crystallization processes is important from a process design and control point of view. It is critical to identify suitable operating protocols to achieve the target product CQAs. Modeling of crystallization processes can help in identifying optimum operating protocols as also in the process design and control. For the reliable prediction of crystallizer behavior, it is important to have accurate estimates for the values of key crystallization kinetic parameters that describe phenomena like growth, nucleation, and dissolution. However, the values for these parameters for even model systems such as paracetamol–ethanol or ibuprofen–ethanol are not readily available in literature ([8]). Further, there is

---

Chemical Engineering in the Pharmaceutical Industry: Active Pharmaceutical Ingredients, Second Edition.
Edited by David J. am Ende and Mary T. am Ende.
© 2019 John Wiley & Sons, Inc. Published 2019 by John Wiley & Sons, Inc.

a lack of general consensus with regard to which rate laws (and consequently the choice of parameters) best describe crystallization processes (nucleation, growth, dissolution) and how to go about estimating the parameter values for the same. Typically, these parameter values are highly system specific and need to be estimated through experimentation. Further, previous studies, which estimate the kinetic parameter values, validate those values over a limited range of operating conditions and focus primarily on seeded crystallization [8, 9].

The chapter presents an approach that integrates population balance modeling with the process analytical technology for modeling batch and continuous crystallization processes. The proposed work is motivated by the necessity of prediction of crystallization events like onset of the crystallization, nucleation, growth kinetics, etc. using the mechanistic model. The population balance framework in the form of partial differential equation defines the number of crystals in the crystallizer, which is a function of time and particle size coordinate. Figure 28.1 shows the schematic of the generalized modeling framework used for predicting the batch and continuous crystallization processes. The proposed model makes use of the popular tanks-in-series modeling framework to accurately capture the mixing behavior of different crystallizers [10]. Different crystallizers may be modeled within the same framework by varying the number of tanks to match the crystallizer mixing behavior.

28.2 POPULATION BALANCE FRAMEWORK FOR CRYSTALLIZATION

The population balance approach for crystallization involves a combination of a set of equations describing rate equations and the conservation laws for the system [6, 8, 9]. This approach helps in capturing the crystal size distribution, which is the CQA for the crystallization. The population balance equation tracks the change in the number density distribution function of the crystals in the crystallizer.

28.2.1 Model for Batch Mode

For the batch mode crystallization, a transient population balance equation needs to be solved. The transient PBE may be written as

$$\frac{\partial n(L,t)}{\partial t} + \frac{\partial G(L,t)n(L,t)}{\partial L} = B_0(t)\delta(L-L_{\text{nuc}}) \quad (28.1)$$

where

- $B_0(t)$ stands for the rate of nucleation of particles at time "t."
- L is the particle size coordinate.
- G is the crystal growth rate.
- δ is the Kronecker delta function.
- L_{nuc} is the size of the nucleating particles.
- $n(L, t)$ is the number density distribution function.

Other phenomena such as breakage and agglomeration may also be represented by adding suitable source terms on the right-hand side [9, 11, 12]. The overall approach however is similar, and for the sake of clarity, the phenomena of breakage and agglomeration were not considered here. The presented framework can be extended in a relatively straightforward manner for the cases where it is necessary to account for breakage and agglomeration.

28.2.2 Model for Continuous Mode

The general form of the population balance equation for continuous process can be written by accounting for inflow and outflow terms as

$$\frac{\partial n(L,t)}{\partial t} + \frac{\partial n(L,t)G(t)}{\partial L} = \frac{v_0 n_0(L,t) - v(t)n(L,t)}{V_R} \quad (28.2)$$

It should be noted that the outlet volumetric flow for fixed volume crystallizers will not be equal to the inlet flow rate due to the changing slurry volume on account of crystallization.

28.2.3 Solution Methods for the PBE Framework

Randolph and Larson [13] have provided the analytical solution for the steady-state population balance equation. However, it is difficult to find the analytical solutions for crystallization systems involving complex nucleation and growth expressions. Analytical solutions to a system of equations (mass, species, and energy) coupled to the PBE as also

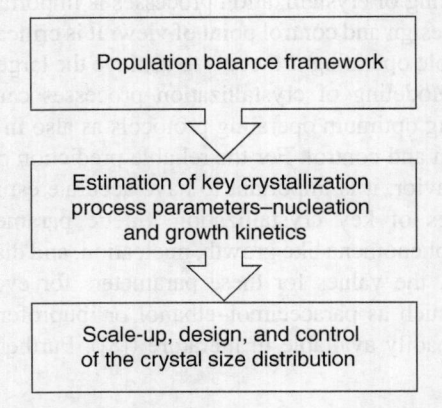

FIGURE 28.1 PBE framework scale-up, design, and control of the crystal size distribution in batch and continuous crystallization processes.

for the tanks-in-series framework are even more cumbersome to implement. Numerical methods present a reliable alternative for the quick and robust implementation of even complex population balance-based models. Numerical methods for the solution of PBEs may be broadly classified into moment-based methods and bin-based methods. A detailed review of a popular subset of the methods to solve the PBE was given by Ramakrishna [14]. Ramakrishna [14] reviewed analytical solutions, moment-based methods (standard method of moments [MOM] and quadrature MOM), discretization-based methods (fixed pivot and moving pivot), and the method of weighted residuals. Hulbert and Katz [15] first applied the standard MOM to solve the 1D and 2D PBE for continuous crystallizers. Similarly, Randolph and Larson [13] have used this method for solving the PBE under steady- and unsteady-state continuous crystallizer.

28.2.3.1 Moment-based Methods MOM for solving PBE track moments of the particle size distribution rather than tracking the complete particle size distribution. The standard MOM, however, has its limitations while accounting for complex process dynamics involving nonlinear growth rate laws, breakage, and agglomeration. For this purpose, a variant of conventional MOM, the quadrature MOM, was proposed by McGraw [16]; this method requires a relatively small number of scalar equations for tracking the moments of population with small errors. In general, the moment-based methods for solving the PBE are computationally efficient and easy to implement. Although the complete determination of the PSD is difficult to achieve, moment-based methods can help in capturing key design criteria for crystallization process like mean crystal size and key process variables like the total particle surface area or solid volume.

28.2.3.2 Bin-based Methods Bin-based methods offer the solution to the PBE by discretization along the particle size coordinate (to form bins) and transformation of the PBE into a set of ordinary differential equations (ODEs). Essentially, the number of particles in each bin is tracked with time by solving the ODEs. Discretization-based methods for the solution of the PBE enable the complete description of the PSD. However, one should be careful in implementing these methods as they are prone to numerically induced errors. The resolution of the bin-based methods depends upon the number of discretization classes, resulting in a trade-off between the accuracy and speed of solving.

28.3 A GENERALIZED MODEL FOR BATCH AND CONTINUOUS CRYSTALLIZATION PROCESSES

A schematic of the generalized modeling framework and the key process variables considered are given in Figure 28.2. The tanks-in-series framework effectively helps to capture different types of mixing behaviors (CSTR to PFR) for continuous crystallizer configurations. Equations for the total liquid mass, solid mass, dissolved solute mass, and particle number density (in terms of PBE) were solved for each tank. The present framework was formulated for the fixed volume continuous crystallizers. Hence, additional equations for the outlet flow rate from each tank need to be solved to account for the volume changes due to crystallization. Batch crystallization is a special case for the presented framework and can be solved using a single tank. However, for batch crystallization, an additional equation for the dynamic slurry volume needs to be solved to account for the volume changes.

Following the method by Randolph and Larson [13], the standard MOM solution to the PBE can be written in terms of moment equations as

$$\frac{dM_i}{dt} = 0^i B_0 + iGM_{i-1} + \frac{v_0 M_{i,0} - vM_i}{V_R} \quad (28.3)$$

In the case of variable crystallizer volume (as in the cases of batch crystallizers), equation can be rewritten as

$$\frac{dV_R M_i}{dt} = 0^i B_0 V_R + iGM_{i-1} V_R + v_0 M_{i,0} - vM_i \quad (28.4)$$

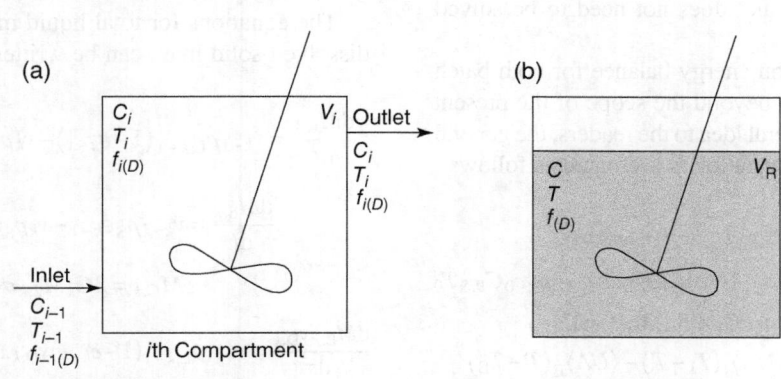

FIGURE 28.2 (a) Model schematic of continuous crystallization. (b) Model schematic for batch crystallization.

The mass balances can be written separately for the solid and the liquid phase as

$$\frac{dM_L}{dt} = v_0 \rho_{L,0}(1-\epsilon_0) - v\rho_L(1-\epsilon) - \dot{M}_C \qquad (28.5)$$

$$\frac{dM_S}{dt} = v_0 \rho_S \epsilon_0 - v\rho_S \epsilon + \dot{M}_C \qquad (28.6)$$

The equation for the rate of mass transfer to the crystals after some manipulations to the rate change equation of the third moment can be written as

$$\dot{M}_C = 3GM_2 \sigma_V \rho_S V_R \qquad (28.7)$$

The mass balance equation for the dissolved solids can be written as

$$\frac{dM_L y_D}{dt} = v_0 \rho_{L,0}(1-\epsilon_0) y_{D,0} - v\rho_L(1-\epsilon) y_D - \dot{M}_C \qquad (28.8)$$

The only thing that is unknown in the above equations is how to account for the unknown outlet velocity. An explicit formulation for the outlet velocity may be obtained by making either of two assumptions: (i) constant liquid-phase density or (ii) linearly varying liquid-phase density. Depending of the assumption, we obtain two formulations for the outlet velocity:

Constant density

$$v = v_0 - \dot{M}_C \left\{ \frac{\rho_S - \rho_L}{\rho_S \rho_L} \right\} \qquad (28.9)$$

Linearly varying density

$$\rho_L = A + By_D \qquad (28.10)$$

$$v = \left\{ v_0 \frac{\rho_{L,0}}{\rho_L}(1-\epsilon_0) + v_0 \epsilon_0 \right\} - \frac{BM_L}{\rho_L^2}\frac{dy_D}{dt} - \left\{ \frac{\dot{M}_C}{\rho_L} - \frac{\dot{M}_C}{\rho_S} \right\} \qquad (28.11)$$

$$\frac{dy_D}{dt} = \frac{1}{M_L}\{v_0 \rho_{L,0}(1-\epsilon_0)(y_{D,0}-y_D) - \dot{M}_C(1-y_D)\} \qquad (28.12)$$

It should be noted that for the constant density and linearly varying density assumptions, the outlet volumetric flow rate may be obtained explicitly and does not need to be solved simultaneously.

The detailed discussion on energy balance for both batch and continuous processes is beyond the scope of the present text. However, to give a general idea to the readers, the general form of the energy balance equation is presented as follows:

$$\frac{d(M_L C_{P,MIX} + M_S C_{P,S})T}{dt} = v_0 \rho_{L,0}(1-\epsilon_0)C_{P,MIX,0}T_0$$
$$- v\rho_L(1-\epsilon)C_{P,MIX}T + v_0 \rho_S \epsilon_0 C_{P,S}T_0$$
$$- v\rho_S \epsilon C_{P,S}T + \dot{M}_C(-\Delta H_C)$$
$$+ \langle UA \rangle_J (T_J - T) - \langle UA \rangle_E (T - T_E) \qquad (28.13)$$

where

$$C_{P,MIX} = y_D C_{P,S} + (1-y_D)C_{P,L}$$

28.3.1 Model Extension for Batch Mode

For the case of a batch reactor, instead of solving for the outlet velocity, the equation for the crystallizer volume needs to be solved. The equations for the rate of change in crystallizer volume depending upon the density assumptions are given below:

Constant density

$$\frac{dV_R}{dt} = -\frac{\dot{M}_C}{\rho_L} + \frac{\dot{M}_C}{\rho_S} \qquad (28.14)$$

Linearly varying density

$$\frac{dV_R}{dt} = -\frac{\dot{M}_C}{\rho_L} - \frac{BM_L}{\rho_L^2 \frac{dy_D}{dt}} + \frac{\dot{M}_C}{\rho_S} \qquad (28.15)$$

$$\frac{dy_D}{dt} = \frac{-\dot{M}_C(1-y_D)}{M_L} \qquad (28.16)$$

28.3.2 Equations for Tanks in Series

For multiple tanks in series, the above equations can be written as

$$\frac{dM_{i,k}}{dt} = 0^i B_{0,k} + iG_k M_{i-1,k} + \frac{v_{k-1}M_{i,k-1} - v_k M_{i,k}}{V_{R,k}} \qquad (28.17)$$

and for nonconstant volume as

$$\frac{dV_{R,k}M_{i,k}}{dt} = 0^i B_{0,k}V_{R,k} + iG_k M_{i-1,k}V_{R,k} + v_{k-1}M_{i,k-1} - v_k M_{i,k} \qquad (28.18)$$

The equations for total liquid mass, total solid mass, and dissolved solid mass can be written as

$$\frac{dM_{L,k}}{dt} = v_{k-1}\rho_{L,k-1}(1-\epsilon_{k-1}) - v_k \rho_{L,k}(1-\epsilon_k) - \dot{M}_{C,k} \qquad (28.19)$$

$$\frac{dM_{S,k}}{dt} = v_{k-1}\rho_S \epsilon_{k-1} - v_k \rho_S \epsilon_k + \dot{M}_{C,k} \qquad (28.20)$$

$$\dot{M}_{C,k} = 3G_k M_{2,k}\sigma_V \rho_S V_{R,k} \qquad (28.21)$$

$$\frac{dM_{L,k}y_{D,k}}{dt} = v_{k-1}\rho_{L,k-1}(1-\epsilon_{k-1})y_{D,k-1} - v_k \rho_{L,k}(1-\epsilon_k)y_{D,k} - \dot{M}_{C,k} \qquad (28.22)$$

The relations for the tank outlet velocities can be written as
Constant density

$$v_k = v_{k-1} - \dot{M}_{C,k} \left\{ \frac{\rho_S - \rho_L}{\rho_S \rho_L} \right\} \quad (28.23)$$

Linearly varying density

$$v_k = \left\{ v_{k-1} \frac{\rho_{L,k-1}}{\rho_{L,k}} (1 - \epsilon_{k-1}) + v_{k-1} \epsilon_{k-1} \right\}$$
$$- \frac{BM_{L,k}}{\rho_{L,k}^2} \frac{dy_{D,k}}{dt} - \left\{ \frac{\dot{M}_{C,k}}{\rho_{L,k}} - \frac{\dot{M}_{C,k}}{\rho_S} \right\} \quad (28.24)$$

$$\frac{dy_{D,k}}{dt} = \frac{1}{M_{L,k}} \left\{ v_{k-1} \rho_{L,k-1} (1 - \epsilon_{k-1})(y_{D,k-1} - y_{D,k}) - \dot{M}_{C,k}(1 - y_{D,k}) \right\}$$
$$(28.25)$$

The equation for the energy balance for each tank may be written as

$$\frac{d(M_{L,k} C_{P,MIX,k} + M_{S,k} C_{P,S}) T_k}{dt} = v_{k-1} \rho_{L,k-1} (1 - \epsilon_{k-1})$$
$$C_{P,MIX,k-1} T_{k-1} - v_k \rho_{L,k}$$
$$(1 - \epsilon_k) C_{P,MIX,k} T_k + v_{k-1} \rho_S$$
$$\epsilon_{k-1} C_{P,S} T_{k-1} - v_k \rho_S \epsilon_k C_{P,S}$$
$$T_k + \dot{M}_{C,k}(-\Delta H_C) + \langle UA \rangle_{J,k}$$
$$(T_J - T_k) - \langle UA \rangle_{E,k}(T_k - T_E)$$
$$(28.26)$$

28.3.3 Constitutive Laws

28.3.3.1 Crystal Growth
Crystal growth is typically a two-step process. First, the solute molecules are convected/diffused from the solution medium toward the crystal surface and then are integrated into the crystal matrix [17]. Typically, the surface integration step is assumed to be rate controlling, and the growth rate may then be written for each crystallizer unit by an Arrhenius-type rate expression [8] as

$$G_k = k_{g,0} \exp\left(-\frac{E_A}{RT_k}\right) \left(C_k - C_k^*(T_k)\right)^g \quad (28.27)$$

28.3.3.2 Primary Nucleation
The primary nucleation rate is typically represented by a simple power law relation as [8, 18]

$$B_{0,k,\text{pri}} = k_1 \left(C_k - C_k^*(T_k)\right)^{n_1} \quad (28.28)$$

28.3.3.3 Secondary Nucleation
There is a lot of ambiguity with regard to the modeling of the secondary nucleation rate. A detailed discussion regarding the different rate laws used to represent secondary nucleation was given in Pandit and Ranade [8]. In the present study, the rate law proposed by Pandit and Ranade [8] was used. The rate law assumes that the secondary nucleation rate is directly proportional to the total surface area (second moment) in addition to a power of the super saturation. This assumption is consistent with the mechanistic understanding that secondary nucleation occurs because of "chipping" (dislodging of small fragments) at the crystal surfaces due to crystal–crystal or crystal–impeller collision interactions:

$$B_{0,k,\text{sec}} = k_2 M_{2,k} \left(C_k - C_k^*(T_k)\right)^{n_2} \quad (28.29)$$

The total nucleation rate can be written as the sum of the primary and the secondary nucleation rates.

28.3.4 Onset of Crystallization

In cooling crystallization, the metastable zone width (MSZW) is the temperature difference between the solubility temperature and the temperature at which crystallization is first detected. The MSZW phenomenologically depends on several factors, among which the effects of solute concentration [19, 20], cooling profile [19, 20], agitation speed [20], and volume of crystallizer [20–22] have been reported. The MSZW is also known to depend upon the detection technique employed to determine the onset of crystallization [23]. Thus, the MSZW can be defined as the temperature difference between the solubility temperature and the temperature after which *observable* crystallization initiates. From this statement, we may infer that there are in fact non-observable crystallization events occurring even before the crystallization is first detected. Nothing can be said about when these non-observable events actually initiate after the solution reaches supersaturation. However, it may be hypothesized that non-detectable crystallization events start occurring just as the solution starts getting supersaturated.

Using the above hypothesis, Kubota [23] proposed a model for determining the MSZW by modeling the kinetics of the non-observable nucleation events using a simple power law model. The crystallization was said to be detectable only when a certain critical number density of the nuclei was reached. The model proposed by Kubota [23] was successful in predicting the MSZW and also relating the MSZW to the induction time using the same nucleation kinetics. Model parameter values were obtained by fitting simulation results to experimental MSZW data. Recently, a model proposed by Nagy et al. [24] used the same starting hypothesis as used by Kubota [23]. However, the kinetics of the nucleation process were solved more rigorously by solving simultaneous growth and nucleation events using the MOM solution to the population balance equation. The nucleation kinetic parameter values were obtained by fitting model predictions to experimental FBRM (particle counts) and FTIR (solute concentration) data.

The model proposed by Kubota [23] only accounted for solute concentration and a linear cooling profile. The effect of variables such as volume of crystallizer and nonlinear cooling profile was not considered. From a process fundamental's point of view, the kinetics of growth of the crystals was not considered. In the models proposed by both Kubota [23] and Nagy et al. [24], nothing was mentioned as to how the nucleation kinetic parameter values in the non-observable region related to those in the observable region. Intuitively, it may be hypothesized that these parameter values do not change due to the "artificial" transition from the non-observable to the observable region. In the present study, the onset of detectable crystallization was modeled similar to the method proposed by Braatz (2008). The onset of observable crystallization was said to have initiated when a change in the concentration profile greater than the least count of measurement was observed. After the onset of observable crystallization, the same model equations were solved with the inclusion of secondary nucleation and using the kinetic parameter values for primary nucleation and growth as obtained during the non-observable region. The kinetic parameter values were obtained by fitting model predictions to experimental data for solute concentration and final mean of the particle size distribution.

28.3.5 Model Implementation

The nondimensional form of the tanks-in-series model (Eqs. 28.17–28.25) may be solved simultaneously using an ODE solver. For the batch mode, the equation for the changing process fluid volume may be solved by considering a general case of liquid-phase density varying linearly with solute mass fraction (Eq. 28.10). For the case of continuous mode, the equation for the outlet volumetric flow rate may be calculated dynamically again by considering a case of linearly varying density (Eq. 28.24). The special case of a constant density may be obtained by just setting the value of the linear component to zero. Figure 28.3 demonstrates the overall approach for modeling crystallization processes.

In certain cases, crystallizing systems may be exothermic and endothermic to such an extent so as to affect the dynamics of crystallization. For example, in the case of the ibuprofen–ethanol system, crystallization leads to significant release of heat that increases slurry temperature and adversely affects the supersaturation profile. The additional heat released causes the system to deviate from the "ideal" super saturation trajectory, which in turn affects product quality. Thus it becomes important to estimate the rate (and quantity) of the heat release (heat removal in the case of endothermic crystallization). The

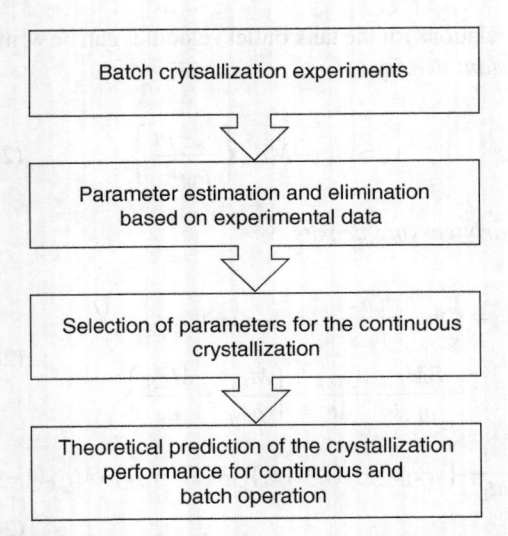

FIGURE 28.3 Overall approach for modeling crystallization processes.

heat released (heat removed) may then be compensated appropriately to maintain an "ideal" supersaturation trajectory through proper control of the jacket temperature. In such cases, the energy balance equation (Eq. 28.26) additionally needs to be solved along with the proposed model framework to account for the heat effects. The heat of crystallization and the jacket- and environmental-side heat transfer coefficients are required to completely describe the heat transfer model. These parameter values may be estimated by comparing model predictions with experimental data (for crystallizer temperature) through controlled heat transfer and crystallization experiments.

In the following Sections 28.4 and 28.5 two case studies are presented to demonstrate the capability of the developed model framework for modeling batch and continuous crystallization processes. The first case study investigates the crystallization of paracetamol–ethanol system. Unseeded batch cooling crystallization experiments were performed using batch and continuous crystallizers. The crystallization model parameters were estimated by comparing the model predictions with the corresponding experimental data for batch crystallization. The approach for the estimation of crystallization kinetic parameter values was discussed in detail and can be easily applied for other systems. Continuous crystallization simulations were performed using the kinetic parameter values obtained through batch crystallization studies. Validation of the continuous crystallizer model was performed by comparing the experimental and simulated outlet concentration profiles. The validated models were used for investigating the effect of scale for batch crystallization and residence time for continuous crystallization.

The second case study investigated the crystallization of a sodium nitrite–water system. The methodology presented in the first case study was followed for the same. Key crystallization kinetic parameters were estimated through batch studies. Continuous cooling crystallization experiments were performed in a novel continuous crystallizer assembly. Classical residence time distribution (RTD) experiments were performed to characterize the mixing behavior of the novel crystallizer. The number of tanks in series required to capture the mixing behavior was determined by comparing model-predicted RTD with the experimental RTD. The crystallization kinetic parameter values estimated through batch studies were used to simulate the continuous crystallization experiments. The continuous crystallizer model was used to investigate the effect of key operating/design parameters (number of tanks, residence time) on key process quality parameters (yield, mean particle size, variance in particle size).

28.4 CASE I: PARACETAMOL–ETHANOL SYSTEM

In the present section, the cooling crystallization of paracetamol as the solute and ethanol as the solvent was investigated. The experimental studies done by Addis et al. [25] are presented. Crystallization was brought about in the present study by cooling the solution to below the solubility temperature. Experiments were performed in both batch and continuous mode of operations. Paracetamol (98%, Acros Organics) was used as the solute and EMSURE grade ethanol (Merck) as the solvent for experimentation. The paracetamol concentration was measured using refractive index detector after calibration against standard solutions. The average particle size of the paracetamol crystals was measured after filtering out the particles and observing under a microscope. Simulations were performed to estimate kinetic parameter values through batch crystallisation studies. Estimated parameter values were then used to simulate for continuous crystallization and to investigate the effect of scale up (batch mode) and residence time (continuous mode).

28.4.1 Experimental Methods

28.4.1.1 Batch Crystallization A 300 ml ChemRxnHub (ChemGlass) was used for the batch crystallization experiments, and a Huber K-6 for was used for the temperature control (Figure 28.4). First, the solution was equilibrated at about 8 °C above the expected solubility temperature. Then the solution was cooled at cooling rates of approximately 0.6 K/min to achieve a temperature difference of 35–40 °C from initial temperature. The experimental temperature profiles were measured using a hand held Thermometer (Fisherbrand Kangaroo with stainless steel probe, ±0.1 °C) and were supplied as an input during simulations. Table 28.1 indicates the type of experiments performed in batch mode. One experiment was repeated three times to prove repeatability and then the starting concentration was varied in two further experiments to understand how the system reacts.

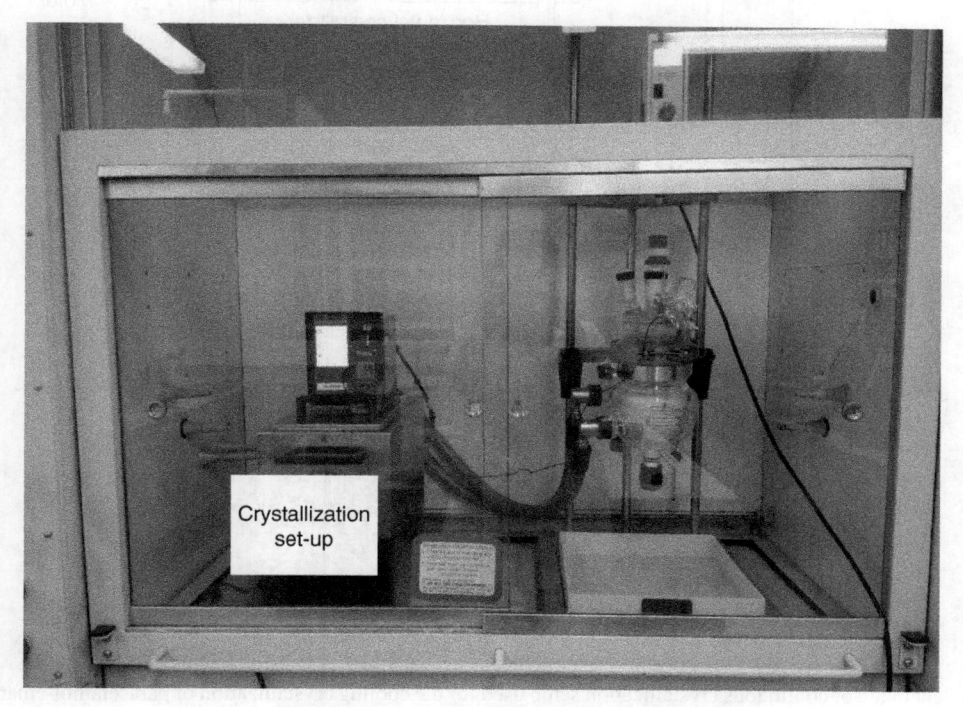

FIGURE 28.4 Batch experimental setup.

28.4.1.2 Continuous Crystallization

The continuous crystallizer, pumps, dissolution, and receiving vessels were designed to ensure the residence time within the continuous crystallizer is sufficient to allow for crystallization to occur. A schematic the continuous crystallizer setup is shown in Figure 28.5a. A novel horizontal stirred crystallizer (using a double impeller) used in the present case is shown in Figure 28.5b. A target residence time of 20 minutes was chosen. A diameter of 70 mm ID for the continuous crystallizer was identified as a size that further industrial scale-up can be based from. An H/D ratio of 5.0 is an ideal plug flow system ratio. Based on this, a continuous crystallizer of approximately 700 ml was constructed, and so holding vessels of appropriate size will allow for the Continuous Crystallizer to reach steady state and produce repeatable results over approximately several hours.

A refractive index measurement was used to measure concentration. A calibration curve of different concentrations was generated using the under saturated solutions, where under saturated is defined by concentration

$$C < C^*$$

where C^* is the theoretical solubility based on temperature.

28.4.2 Model Implementation

The code to solve the developed tanks-in-series model was written in MATLAB. A simple spreadsheet interface to provide inputs for the code was designed using Excel. The stiff ODE solver "ode23s" from the MATLAB suite of ODE solvers was used. Absolute and relative tolerances of 1×10^{-12}

TABLE 28.1 Batch Experiments Naming Nomenclature [25]

Batch Experiment No.	1–3	4	5
Starting concentration (g.paracetamol/g.ethanol)	0.26	0.22	0.28
Expected solubility (°C)	35.4	26.94	39.28

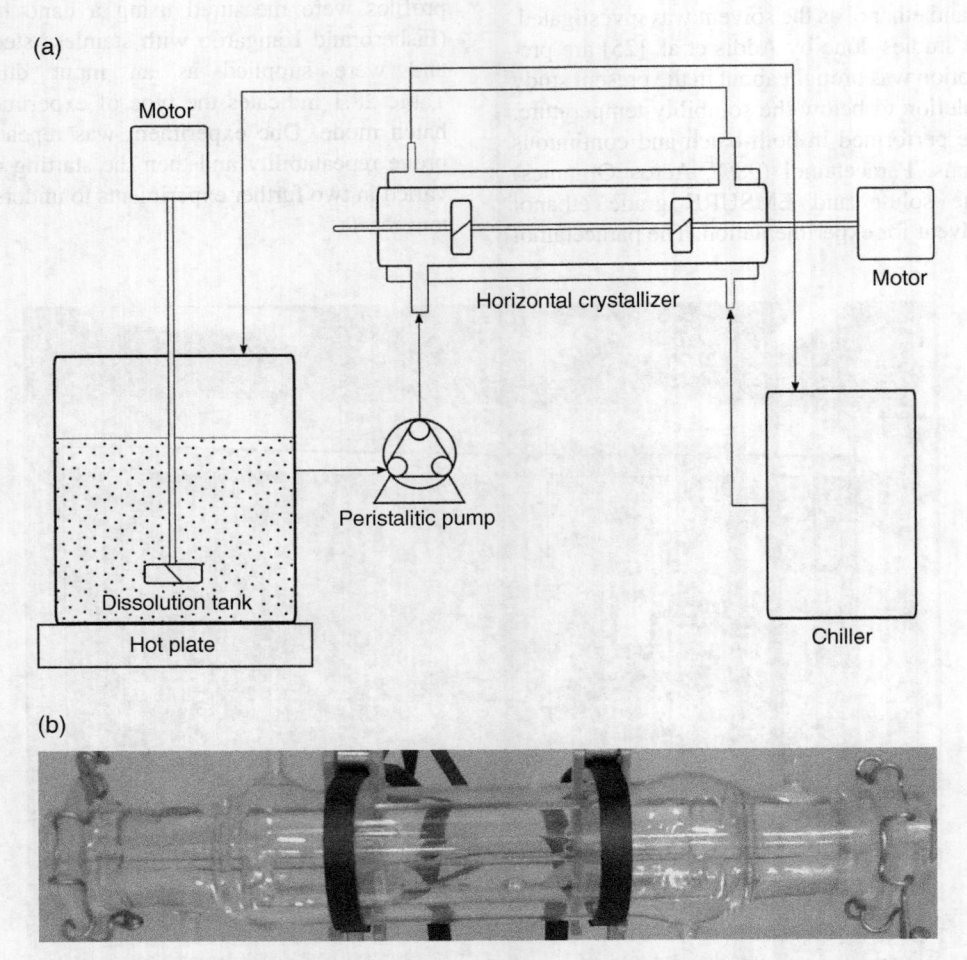

FIGURE 28.5 (a) Schematic of continuous crystallization setup used for the cooling crystallization of paracetamol-ethanol. (b) Continuously stirred horizontal tank for continuous experiments.

were used to solve the code. The model verification was performed independently for the tanks-in-series framework and the crystallization model. For continuous crystallization simulations, a single tank was considered. For a detailed discussion with regard to the verification of the solver, kindly refer to Appendix 28.A. The model was then used to model batch and continuous crystallization processes.

28.4.2.1 Parameter Estimation and Elimination for the Simulations
The group of parameters describing the crystallization can be divided into two parts:

1. Integral parameters
   a. Growth rate constant ($k_{g,0}$)
   b. Primary nucleation rate constant (k_1)
   c. Secondary nucleation rate constant (k_2)
2. Tertiary parameters
   a. Growth rate order (g)
   b. Growth rate activation energy (E_A)
   c. Primary nucleation order (n_1)
   d. Secondary nucleation order (n_2)

Simulations were performed for the case of batch crystallization using conditions of **Experiment 4** (Table 28.1). It was observed that when a tertiary parameter value was changed from the base case value, simulation results similar to the base case could be recreated by varying the values of the integral parameters. The value of the specific tertiary parameter considered in each case was changed within a span of the expected value range of the corresponding parameter known from literature [8]. The solute phase concentration, normalized 0th moment, and the average particle diameter profiles were considered for the comparison between the base case and modified case simulation results. These particular comparison points were chosen because experimental data was available for the same and the final parameter set after the entire exercise would be closest to realistic crystallization kinetics of the system.

For the case of E_A and n_1, it was observed that even though their corresponding values were varied (individually), the base case simulation results could be reproduced almost exactly by varying the integral parameter values. After varying the value of g and readjusting the values of the integral parameters, it was observed that simulation results for the modified cases agreed with the base case results within the confines of the comparison points chosen. That is, the solute phase concentration profile and the endpoint average diameter matched. However, the profiles of the average diameter, the 0th moment, and the variance of the PSD were slightly different. Similarly for the case of n_2, simulation results of the modified cases matched those of the base case within the confines of the comparison points chosen. However, the profiles of the average diameter, the 0th moment, and the variance of the PSD were even more different than observed for the case of g.

Through batch simulations, different sets of parameter values were obtained for each of the tertiary parameters corresponding to the modified and base cases (three sets per tertiary parameter including base case parameter set). Continuous crystallization simulations were performed for each of the tertiary parameters using the different parameter sets, and the results were compared with each other and against experimental data for solute phase concentration. Continuous crystallization simulations were performed considering the conditions corresponding to the experiment. It was observed that using different sets of parameter values, no significant change was observed for the cases of E_A, n_1, and g. However, for the case of n_2, significant differences were seen in the simulation profiles for the different parameter sets. However, the solute phase concentration profile was seen to not vary for different parameter sets and also reasonably matched the experimental profile.

From the above study, it was concluded that the parameters of E_A, n_1, and g can effectively be eliminated from the cumbersome exercise of estimating the values by comparing simulation results with experiments. The respective values for these parameters may be assigned from literature, or a token value may be specified within an expected value range of the specific parameters. It was seen to be not advisable to assign a token value to n_2. However, to simultaneously estimate the values for all four parameters – the integral parameters and n_2 – only the experimental data for solute phase concentration profile and the endpoint average particle diameter were not enough. The authors recommend obtaining intermediate values for the average particle diameter for narrowing down the parameter set values.

With regard to the integral set of parameters, each of the parameters affected the simulation results uniquely. Given the experimental data, it was useful to observe the effect of the integral parameters on three important criteria:

1. Onset crystallization temperature (T_{nuc}).
2. Rate of equilibration of solute phase concentration profile (RE).
3. Endpoint average particle diameter (d_{10}).

The RE criteria can loosely be defined as the rate at which the solute concentration profile reaches equilibrium *in comparison* with the experimental one. Table 28.2 lists how increasing each value of the integral parameters ($k_{g,0}$, k_1, k_2) affects the simulation results in terms of the criteria mentioned above. Decreasing the values for the integral parameters will have an inverse effect than those mentioned in Table 28.2.

TABLE 28.2 Effect of Increasing the Integral Parameters on Key Simulation Characteristics

	$k_{g,0}$	k_1	k_2
T_{NUC}	Increases	Increases	No effect
RE	Increase	Increases	Increases
d_{10}	Increases	Increases	Decreases

From simulation results it was observed that $k_{g,0}$ and k_2 affected the simulation results the most. The effect of k_1 was comparatively smaller but was observable. After fixing the values for the tertiary parameters, the values for $k_{g,0}$ and k_2 were varied so as to match the criteria of d_{10} and RE. The value of k_1 was then tweaked to fine-tune the match between the simulation results and the experimental data. The effect of the parameters on the criteria specified in Table 28.2 is useful to find out the direction in which the respective parameter values need to be changed to match simulation results to experimental data. *It was observed that for a fixed set of tertiary parameters, a unique set was obtained for the integral parameter values for the experimental data available.*

Results for the elimination of activation energy as a parameter are shown in Figures 28.6 and 28.7. As mentioned in the above discussion, the parameter value for the activation energy was varied, and one of the integral parameter values (growth rate constant) was tweaked to reproduce the base case results (set 1). The corresponding tweaked values of the integral set of parameters and the activation energy are given in Table 28.3. As can be seen from Figures 28.6 and 28.7, there is sufficient agreement between the model predictions using the three sets of parameter values. The comparative analysis was performed for both batch and continuous crystallizations. This proved that activation energy may be eliminated from the exercise of estimating parameter values and may be assigned a fixed *representative* value. Through a similar process, the parameters of growth rate exponent, primary nucleation rate exponent, and secondary nucleation rate exponent were eliminated [25].

After assigning fixed representative values to the tertiary set of parameters, the integral set of parameter values were varied such that simulation results matched the experimental results for Experiment 4 in batch crystallization. The final set of parameter values chosen for further simulations is given in Table 28.4.

28.4.3 Results and Discussions

28.4.3.1 Batch Crystallization The integral parameter values were estimated by a simultaneous comparison of model predictions to experimental results for the average end particle size (Figure 28.6a) and concentration profile (Figure 28.6b) of Experiment 4 of batch crystallization. Further validation of the chosen parameter set was performed for different experimental conditions given in Table 28.1. A detailed comparison of the experimental and simulation results for the same is given in Addis et al. [25].

28.4.3.2 Continuous Crystallization The set of crystallization kinetic parameters estimated using batch crystallization studies were used to simulate for the conditions of the continuous crystallizer. It was seen that the concentration profile in the continuous crystallizer was captured accurately using the presented model and batch crystallization kinetic data (Figure 28.7b). Simulations revealed that for the geometries considered, there was no significant difference in the average end particle size obtained through batch and continuous crystallization processes (Figures 28.6a & 28.7a).

28.4.3.3 Effect of Scale The validated model was used to investigate the effect of scale on the batch crystallization process. The conditions of Experiment 4 of the batch crystallization were chosen as the base case for comparison. Simulations were carried out for three different scales of 500 ml, 5 L, and 50 L by appropriately scaling up (subsequently multiplying by a factor of 10 each time) the initial fill conditions for Experiment 4 (corresponding to 500 ml scale). It was observed through simulation results that there was no observable impact of scaling up on the average particle size (Figure 28.8a) and outlet concentration profiles (Figure 28.8b). This was anticipated as the model assumed an ideal mixing behavior for all the scales considered. Differences in the product quality during scaling up typically occur due to the deviations in the mixing behavior across different scales (nonideal mixing).

28.4.3.4 Effect of Residence Time The developed continuous crystallizer model was used to investigate the effect of the residence time on the average outlet particle size and outlet concentration profiles. Simulations were carried out for residence times of 10, 20, and 40 minutes. It was observed that increasing the residence time (decreasing the flow rate) leads to an (anticipated) increase in the steady-state average outlet particle size (Figure 28.9a). Increasing the residence time also resulted in a decreased steady-state outlet concentration profile (Figure 28.9b) (reduced outlet supersaturation) because of the longer time available for equilibration.

28.5 CASE II: SODIUM NITRITE–WATER CRYSTALLIZATION

28.5.1 Experimental Methods

28.5.1.1 Batch Crystallization Batch crystallization experiments for sodium nitrite–water system were carried out in a setup as shown in Figure 28.10 [7]. A 250 ml OptiMAX reactor setup was used. The jacket temperature in OptiMAX setup was controlled electronically using Peltier elements. That coupled with a sophisticated control system

(a)

(b)

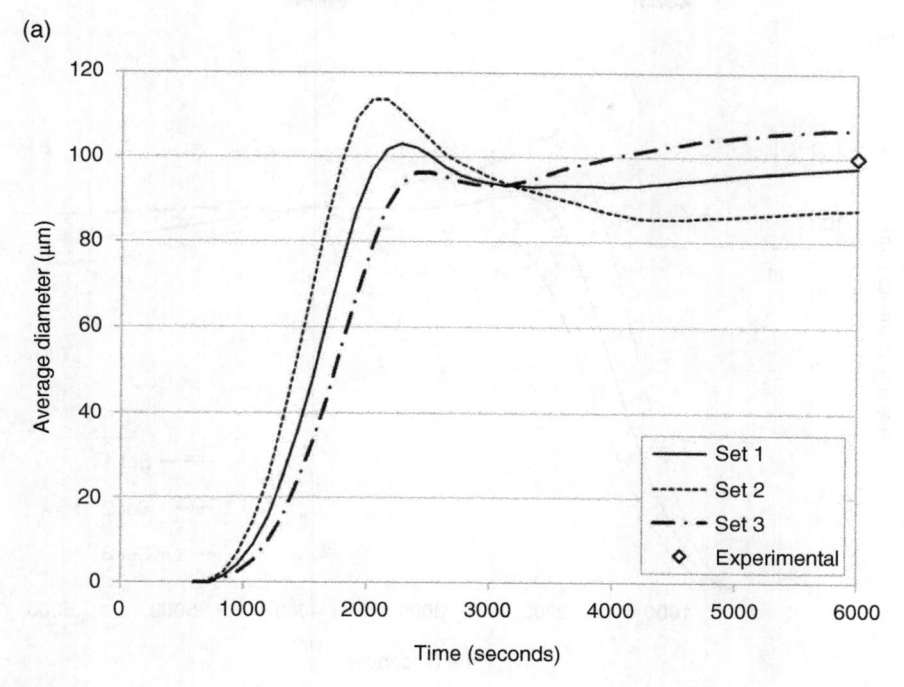

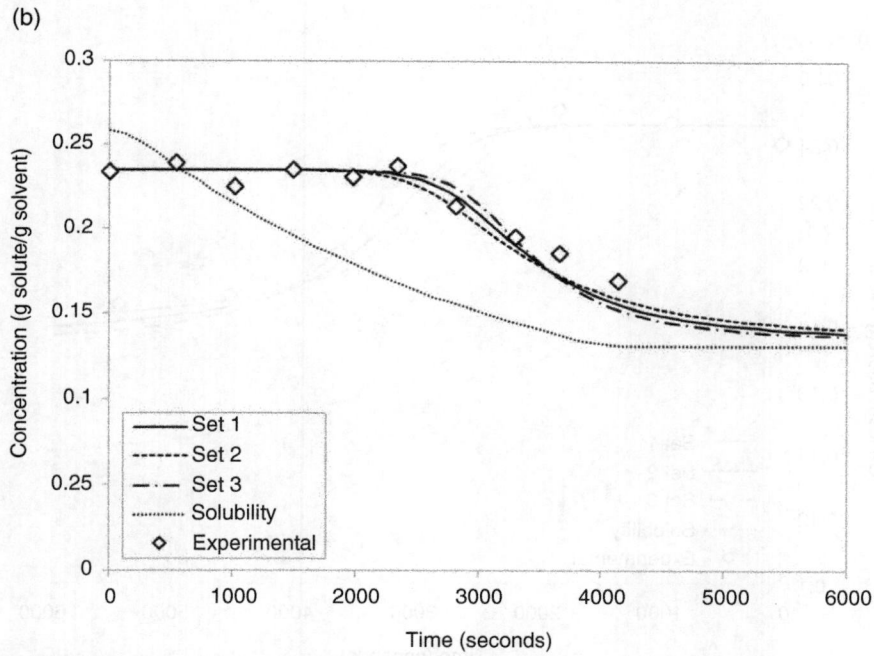

FIGURE 28.6 (a) Comparison between the simulated average diameter profiles (d_{10}) using different sets of parameter values and the final experimental average diameter for batch crystallization. (b) Comparison between the simulated average solute concentration profiles using different sets of parameter values and the experimental solute concentration profile for batch crystallization.

allows for precise control over the crystallizer temperatures. The particle counts and chord length distributions were measured using an FBRM probe. Temperature cycling was done between defined temperatures depending upon the solubility of the considered system. Equal magnitudes were employed for both the heating and cooling rates for each cycle. The effect of the heating/cooling rate on the crystallization for a system of sodium nitrite in water for values of 0.3, 0.5, and 0.7 K/min was investigated.

28.5.1.2 Continuous Crystallization Setup for Sodium Nitrite–Water System

A novel continuous crystallizer provided by Technoforce was used for continuous crystallization processes. The crystallizer was a jacketed plug flow

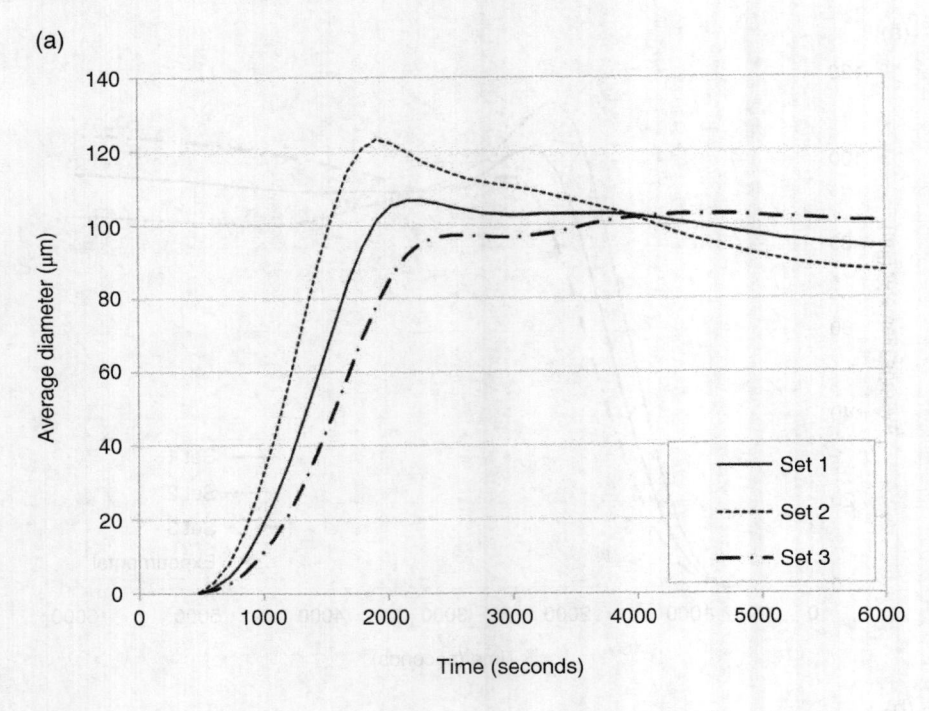

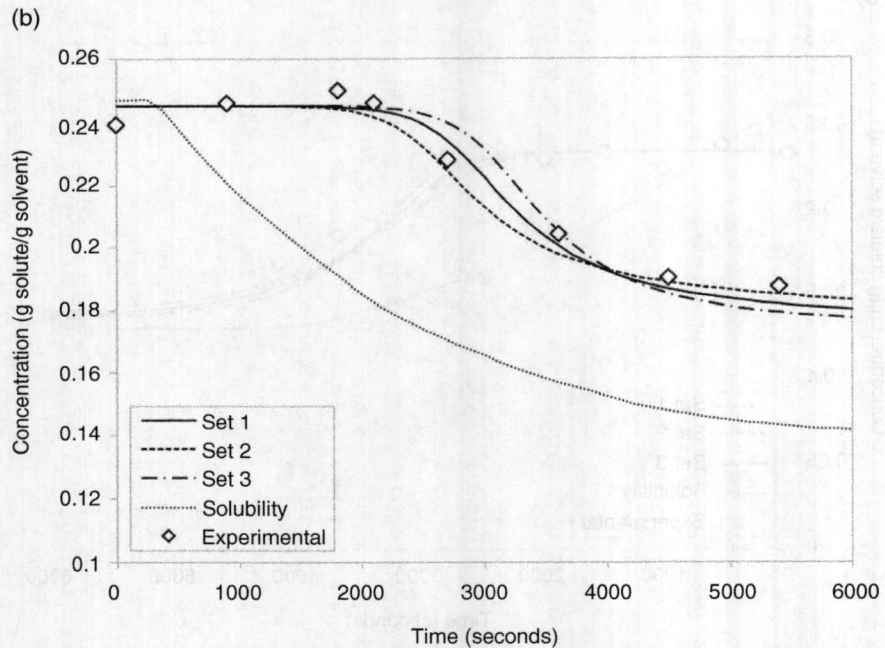

FIGURE 28.7 (a) Comparison between the simulated average diameter profiles (d_{10}) using different sets of parameter values and the final experimental average diameter for continuous crystallization. (b) Comparison between the simulated average solute concentration profiles using different sets of parameter values and the experimental solute concentration profile for continuous crystallization.

crystallizer with a novel impeller design. The crystallizer assembly and the novel impeller design are shown in Figure 28.11a. The overall schematic of the experimental setup is shown in Figure 28.11b. The impeller shaft was connected to a motor via a reduction box. The reduction box reduced the impeller speed from 1300 RPM of the motor to a maximum of 62 RPM in crystallizer. The impeller speed was controlled using a variable frequency drive (VFD) that was connected to a three-phase power supply. The impeller speed was kept at 62 RPM under all conditions. The total crystallizer liquid volume was 1.3 L. The jacket connected to a Julabo chiller assembly (FP50–HL) that pumps cooling fluid through the jacket at a high flow rate of 24 LPM. The solution was pumped into the crystallizer using a peristaltic pump in the flow rate range of (20–80 ml/min).

Classical RTD experiments were done to characterize the mixing behavior of the novel crystallizers. A saturated salt solution was used as the tracer. The tracer was injected into the inlet stream using a syringe to resemble a pulse input. The outlet concentration of the salt solution was monitored using a conductivity probe. RTD experiments were carried out at different flow rates and different crystallizer configurations. For crystallization experiments, a solution of sodium nitrite and DI water saturated at room temperature (~24 °C) was used as the feed solution.

Continuous crystallization experiments were carried out under a fixed flow rate of 20 ml/min. Different methods of conductivity measurements, absorbance measurements, and refractance measurements were evaluated for their effectiveness to measure the liquid-phase concentration. It was determined that the refractance-based measurements yielded the most accurate and repeatable results. The liquid sample for concentration measurements was collected after vacuum filtration of the outlet slurry.

28.5.2 Model Implementation

28.5.2.1 Parameter Estimation and Elimination for Crystallization Simulation
Simulations were first carried out for the batch crystallization of sodium nitrite–water system to estimate key crystallization kinetic parameters. The kinetic parameter values were estimated by simultaneously matching the simulated and experimental normalized 0th moment curves (Figure 28.12a) and the simulated and experimental Sauter mean diameter (d_{32}) curves (Figure 28.12b) for one set of experiments (0.3 K/min). The comparison was then done for the other sets of experiments (0.5 K/min, Figure 28.13; 0.7 K/min, Figure 28.14) using the parameter values estimated in the first step. It was observed that the parameter values estimated using the first experiment could very well predict the relevant profiles for the other sets of experiments. The list of parameter values obtained by the matching simulated data to experiments is shown in Table 28.5.

As was discussed previously, it is possible to have multiple solutions for the crystallization parameter values if just the normalized 0th moment curves are considered for comparison. This may be achieved, starting from a set of already "validated" parameters by increasing the constant for the growth kinetics while simultaneously reducing the constants for the nucleation kinetics and vice versa. However, the story changes if both the normalized 0th moment curve and the Sauter mean diameter curves are considered for comparison. For instance, if the value of the growth rate constant is increased and the values of the nucleation constants are reduced to match the normalized moment curves, this effectively serves in increasing the average Sauter mean diameter curve from the base case. And decreasing growth rate constant and increasing the nucleation rate constants effectively serves in decreasing the average Sauter mean diameter curve from the base case. Thus, we can say that for the present case, the solution is unique. Multiple solutions may in fact be possible for the growth rate constant and the activation energy for the estimated nucleation kinetic parameters. However, nothing can be said about this as most of the crystallization occurs very fast; it does not span a very wide temperature change. Hence, the effect of temperature on the growth rate parameters cannot be studied very well. However, from the preceding exercise we may say that the obtained total contribution of the growth rate term will not vary drastically.

TABLE 28.3 Parameter Sets Used to Demonstrate Elimination of Activation Energy as a Parameter

Symbol	Set 1	Set 2	Set 3
$k_{g,0}$	40.0	2×10^5	8×10^{-3}
E_A	4×10^7	6×10^7	2×10^7
R	8314.0	8314.0	8314.0
g	1.9	1.9	1.9
k_1	2×10^6	2×10^6	2×10^6
n_1	1.5	1.5	1.5
k_2	4×10^6	4×10^6	4×10^6
n_2	2.5	2.5	2.5

TABLE 28.4 List of Crystallization Kinetic Parameter Values to Simulate for the Batch Cooling Crystallization of Paracetamol in Ethanol

Symbol	Description	Value	Unit
$k_{g,0}$	Pre-exponential constant for growth rate	40.0	$(m/s)(kmol/m^3)^{-g}$
E_A	Activation energy	4×10^7	$J/(kmol \cdot K)$
R	Universal gas constant	8314.0	$J/kmol$
g	Growth rate exponent	1.9	—
k_1	Constant for primary nucleation rate	2×10^6	$No./([m^3 \cdot s][kmol/m^3]^{-n_1})$
n_1	Exponent for primary nucleation rate	1.5	—
k_2	Constant for secondary nucleation rate	4×10^6	$No./([m^2 \cdot s][kmol/m^3]^{-n_2})$
n_2	Exponent for secondary nucleation rate	2.5	—

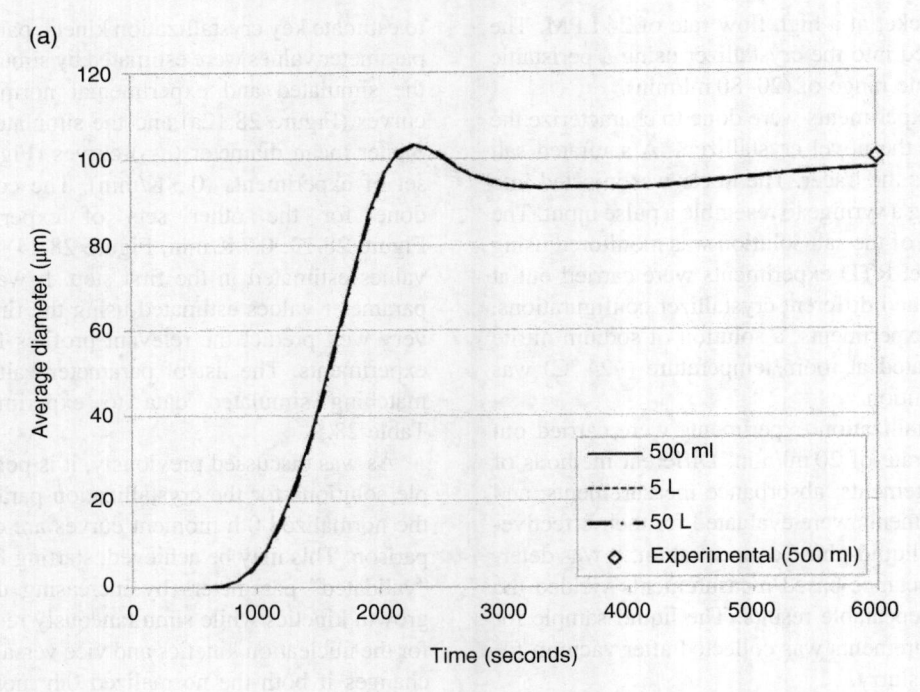

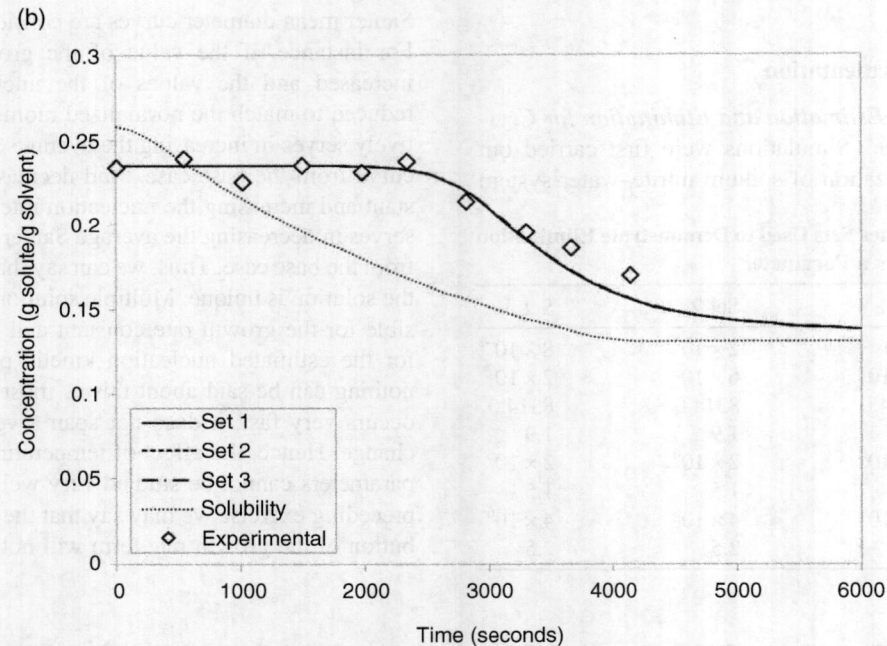

FIGURE 28.8 (a) Comparison between the average diameter profiles for simulation runs using multiple parameter values. (b) Comparison between the concentration profiles for simulation runs using multiple parameter values.

28.5.3 Results and Discussions

28.5.3.1 Continuous Crystallization Simulations were carried out by using the crystallization parameter values obtained during the batch crystallization. The experimental conditions of the *CC8* experiment reported by Pandit and Ranade [7] were simulated. The outlet flow rate was fixed at 20 ml/min. The inlet concentration of the solvent–solute mixture was taken to be equal to the initial experimental outlet concentration. Simulation profiles for the outlet normalized 0th moment and the outlet temperature are shown in Figure 28.15a. Simulation profiles for the mean of the outlet PSD and the outlet supersaturation ratio are shown in Figure 28.15b. The model was not validated due to a lack of experimental data.

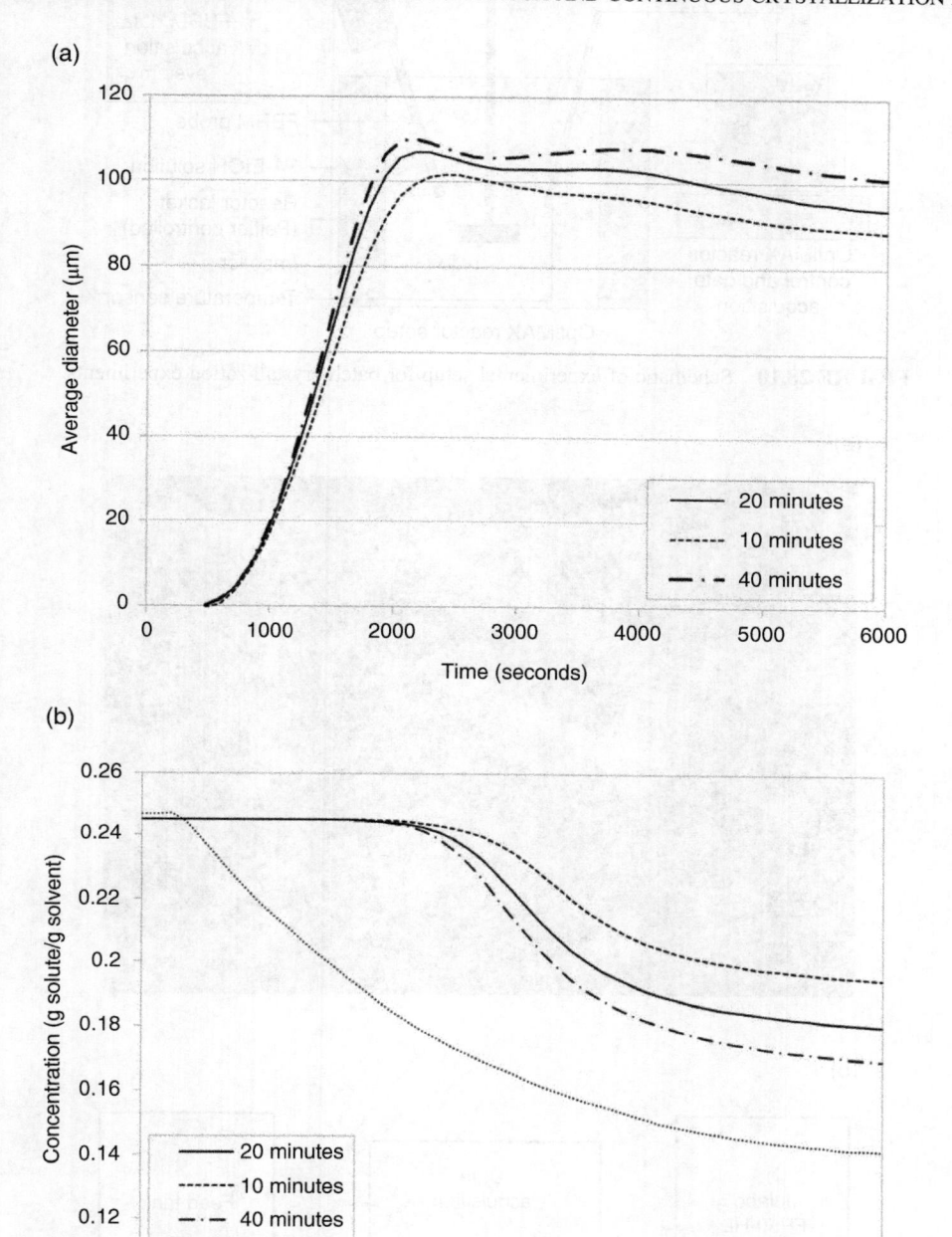

FIGURE 28.9 (a) Comparison between the average outlet diameter profiles of continuous crystallization simulations for different residence times. (b) Comparison between the outlet concentration profiles of continuous crystallization simulations for different residence times.

28.5.3.2 Influence of Operating Parameters

The developed model was used to understand the influence of key operating parameters on key product quality parameters. The product quality parameters of the mean and the variance of the PSD were considered. The mean of the PSD was defined using the moments as follows:

$$\mu = \frac{\int_0^\infty L n(L) \mathrm{d}L}{\int_0^\infty n(L) \mathrm{d}L} = \frac{M_1}{M_0} \quad (28.30)$$

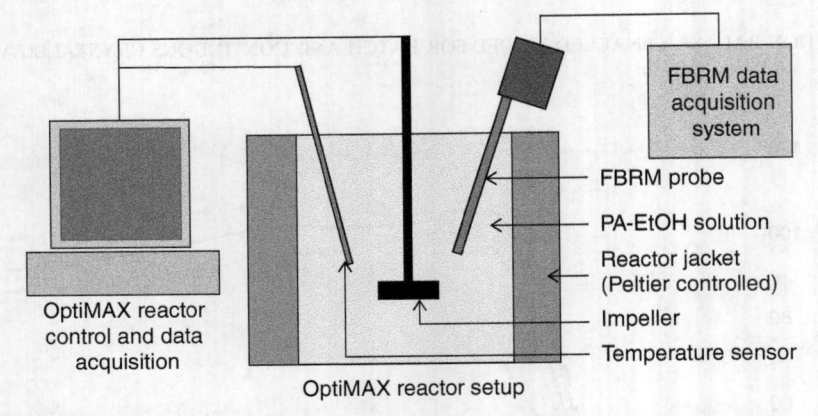

FIGURE 28.10 Schematic of experimental setup for batch crystallization experiments.

(a)

(b)

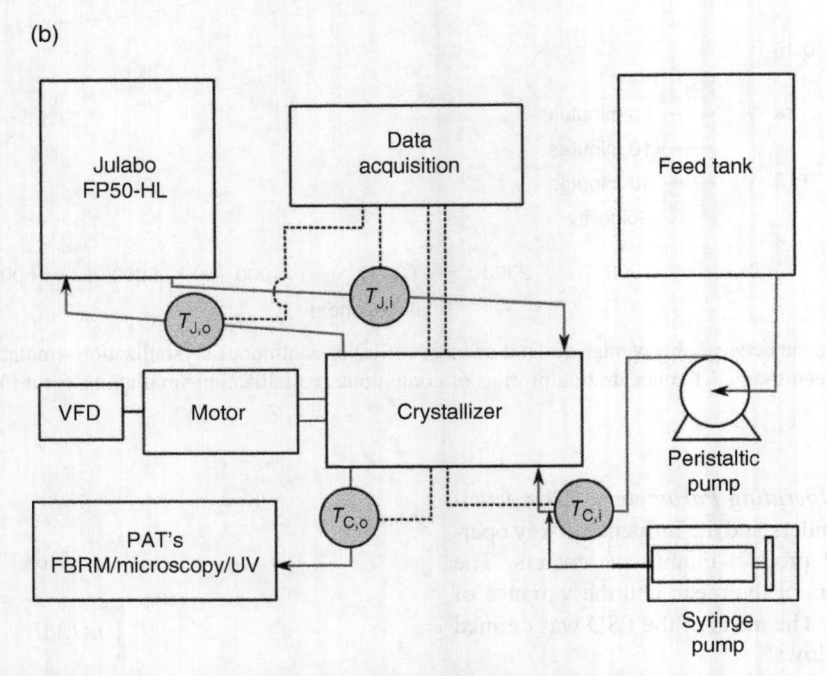

FIGURE 28.11 (a) The continuous crystallizer used in sodium nitrite crystallization experimentation. (b) Schematic of the continuous crystallizer setup used for sodium nitrite crystallization.

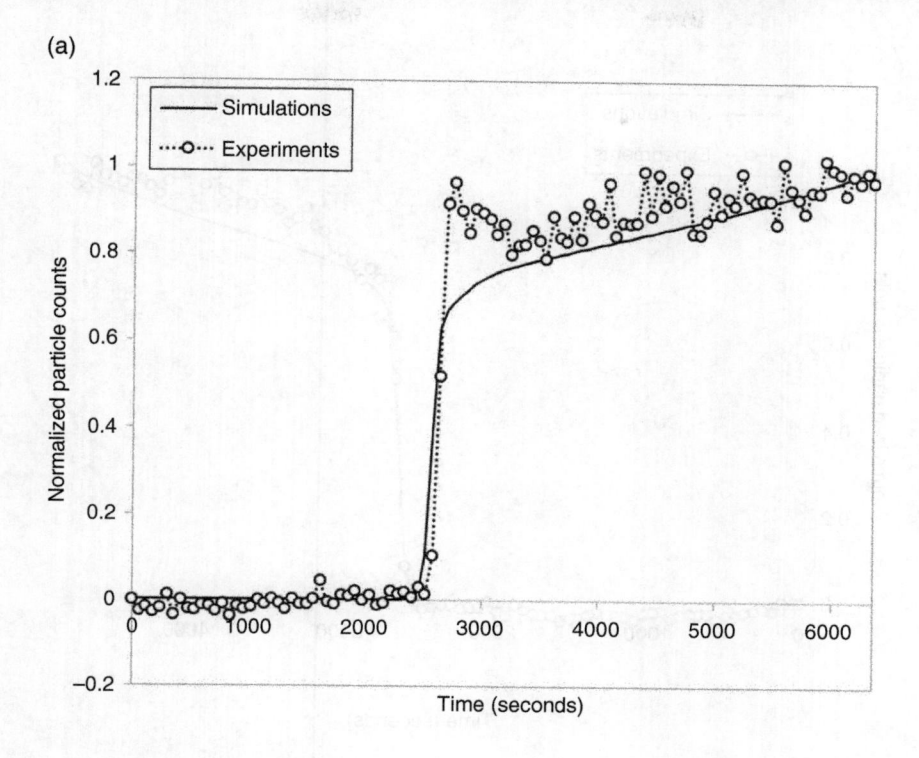

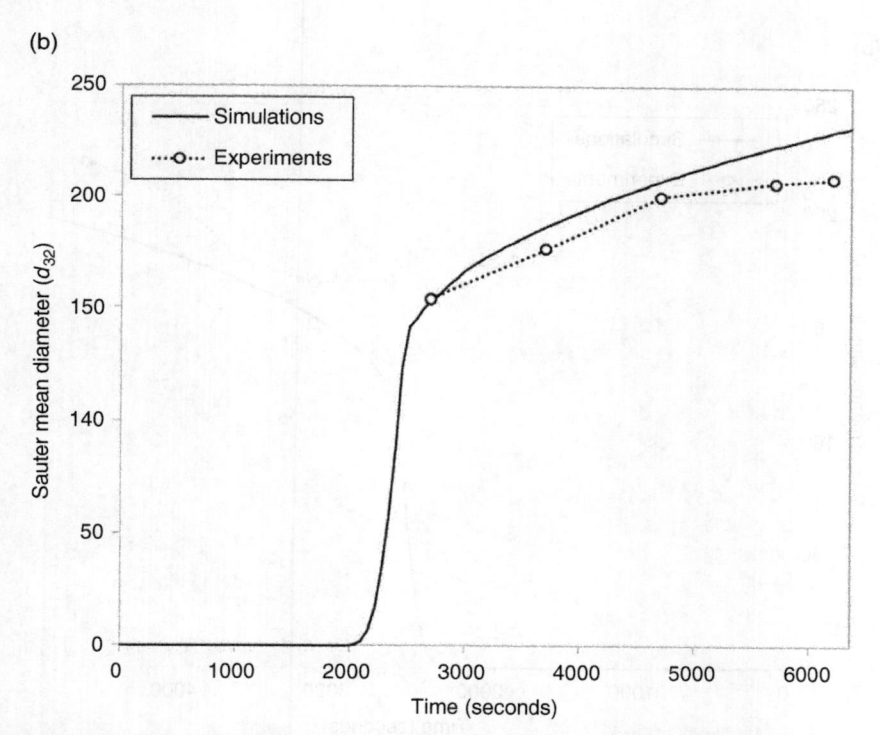

FIGURE 28.12 (a) Comparison between experimental and simulated normalized 0th moment for 0.3 K/min experiment. (b) Comparison between experimental and simulated Sauter mean diameter (d_{32}) for 0.3 K/min experiment.

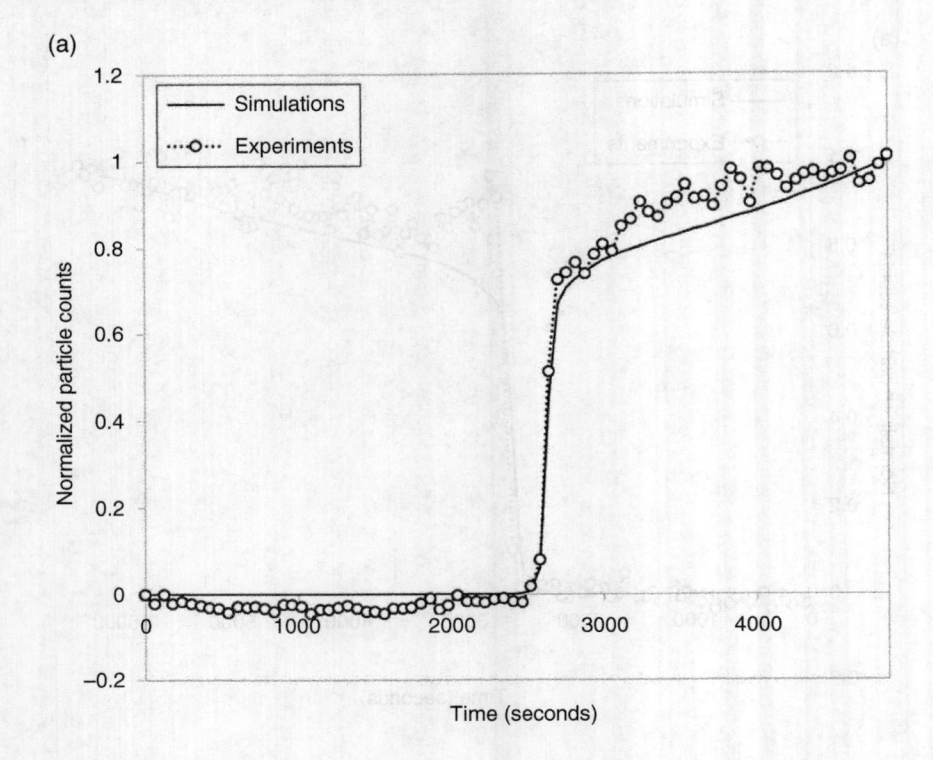

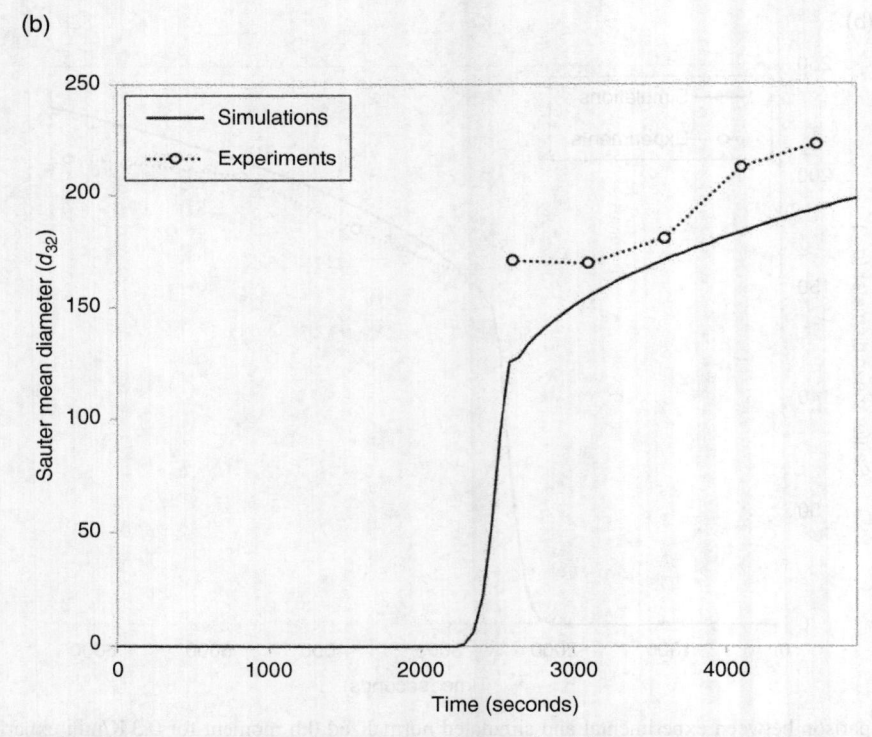

FIGURE 28.13 (a) Comparison between experimental and simulated normalized 0th moment for 0.5 K/min experiment. (b) Comparison between experimental and simulated Sauter mean diameter (d_{32}) for 0.5 K/min experiment.

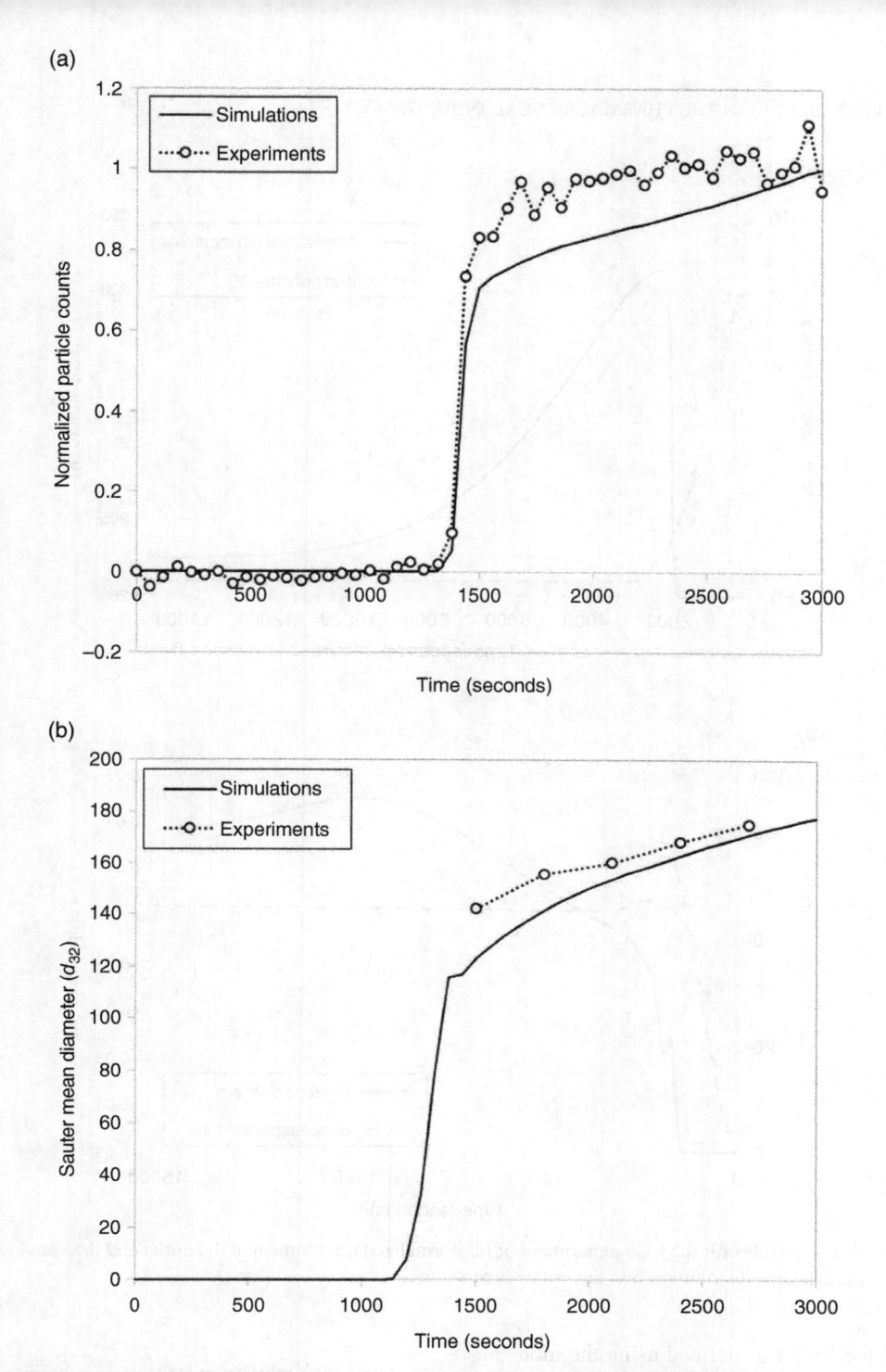

FIGURE 28.14 (a) Comparison between experimental and simulated normalized 0th moment for 0.7 K/min experiment. (b) Comparison between experimental and simulated Sauter mean diameter (d_{32}) for 0.7 K/min experiment.

TABLE 28.5 List of Crystallization Kinetic Parameter Values to Simulate for the Cooling Crystallization of Sodium Nitrite in Water

Symbol	Description	Value	Unit
$k_{g,0}$	Pre-exponential constant for growth rate	190.0	$(m/s)(kmol/m^3)^{-g}$
E_A	Activation energy	4×10^7	$J/(kmol \cdot K)$
R	Universal gas constant	8314.0	J/kmol
g	Growth rate exponent	2.0	—
k_1	Constant for primary nucleation rate	1×10^8	$No./([m^3 \cdot s][kmol/m^3]^{-n_1})$
n_1	Exponent for primary nucleation rate	2	—
k_2	Constant for secondary nucleation rate	1×10^{11}	$No./([m^2 \cdot s][kmol/m^3]^{-n_2})$
n_2	Exponent for secondary nucleation rate	4.5	—

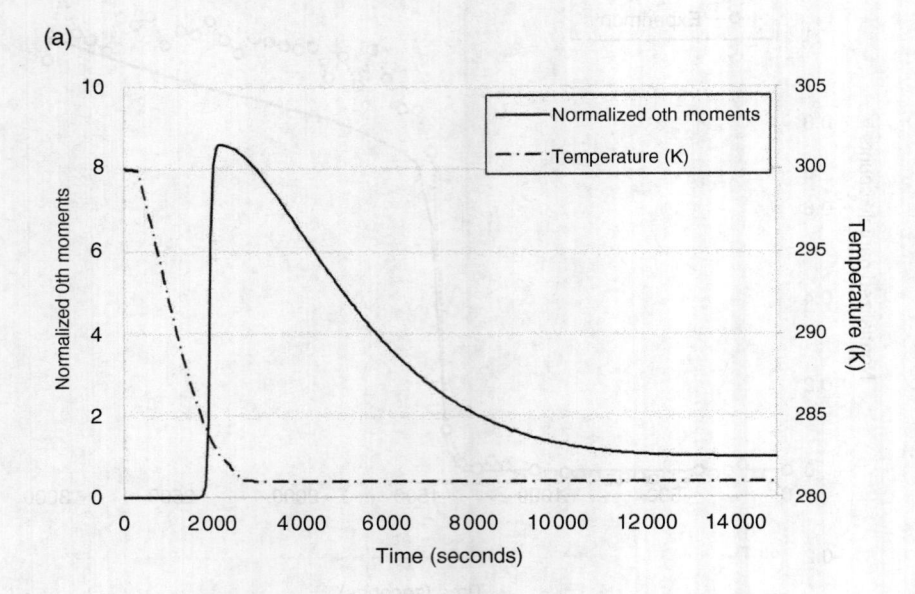

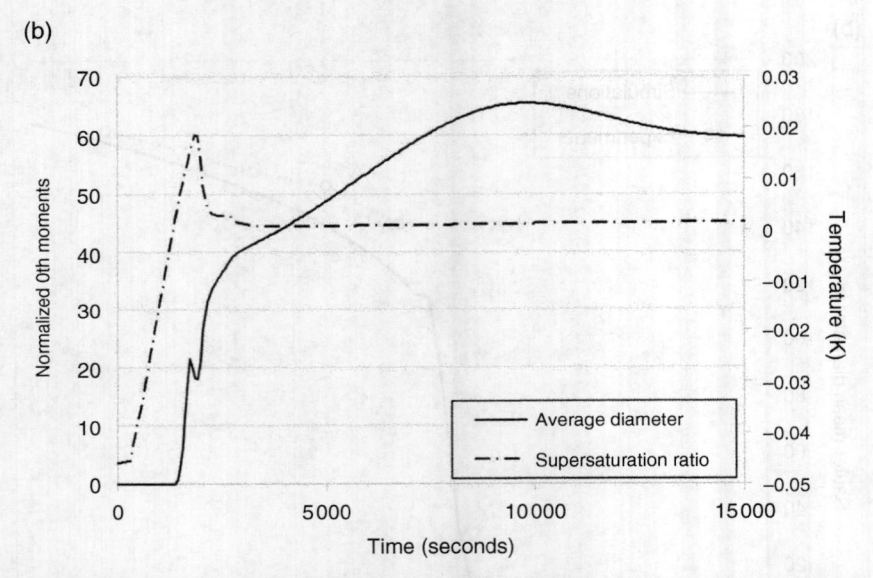

FIGURE 28.15 Simulation profiles for the CC8 experiment of (a) normalized 0th moment at the outlet and the simulated outlet temperature (b) mean of the outlet particle size distribution and the outlet supersaturation ratio.

The variance of the PSD was defined using the moments as follows:

$$\sigma^2 = \frac{\int_0^\infty (L-\mu)^2 n(L) dL}{\int_0^\infty n(L) dL} = \frac{M_2}{M_0} - \mu^2 \quad (28.31)$$

Besides these, the yield of the crystallization process and the time required to reach steady state were also investigated as key process performance parameters. The yield of the process can be defined as

$$\text{Yield}(\%) = 100 \left(\frac{C_{in} - C_{out}}{C_{in} - C^*(T_{out})} \right) \quad (28.32)$$

The time required to reach steady state is important from a process control point of view. A system that has a faster response to step changes can essentially be controlled more easily. The time required to reach steady state was inferred from the simulated normalized 0th moment, mean particle size, and outlet concentration profiles. For a particular simulation run, three steady-state time values may be obtained from the aforementioned profiles by recording the time after which the change in the simulated values was less than the specified tolerance (1×10^{-6}).

The influence of the crystallizer mixing behavior (number of tanks) and key operating parameters of residence time and jacket temperature on the product quality and process performance parameters was investigated. Simulations were carried out considering a constant jacket temperature of 280 K and a very high heat transfer coefficient such that the reactor operating temperature is maintained equal to the jacket temperature. A sample jacket temperature profiles similar to the once reported by Pandit and Ranade [7] was considered to simulate a typical case of cooling crystallization. The inlet concentration of the sodium nitrite solution was 0.81 g-solute/g-solvent. For carrying out the base case simulations, a flow rate of 20 ml/min (residence time of 65 minutes) and a mixing behavior equivalent of 2 tanks in series were considered.

As can be seen from Figure 28.16a, increasing the number of tanks in series leads to a substantial decrease in the time required to reach steady state. This can be observed for all the flow rates considered. Hence, we may conclude that crystallizers with more of a plug-flow-like mixing behavior respond faster to step changes and can be controlled easily. The time was normalized using residence time. From Figure 28.16b, it can be seen that the mean size of the outlet particle size distribution keeps on decreasing with decreasing residence time (increasing flow rates). The difference between the mean and the variance of the outlet particle size distribution, however, seems to remain unaffected by the residence time. A shift toward smaller particle sizes is also seen on increasing the number of tanks in series across all residence times. Hence, it may be concluded that to achieve smaller particle sizes, either the flow rates must be increased or the crystallizer should be designed to have a more "plug-flow-like" mixing behavior.

On increasing the number of tanks in series, the difference between the mean and the variance of the outlet particle size distribution also increased. Thus, narrower particle size distributions are expected for crystallizers with a mixing behavior approaching plug flow. With regard to yield (Figure 28.16c), it was observed that the yield decreases with decreasing residence time (increasing flow rates) regardless of the mixing behavior. Also, crystallizers with a mixing behavior approaching plug flow were seen to give significantly higher yields for the same residence time than their CSTR counterparts.

From the results discussed above, it may be concluded that crystallizers with mixing behaviors approaching plug flow *offer higher yield, narrower outlet particle size distributions, and better process control* than their CSTR counterparts. Plug flow crystallizers were also seen to offer smaller particle sizes for the same residence time than their CSTR counterparts. The RTD was also identified as a key variable to control the mean size of the outlet particle size distribution. Decreasing the residence time (increasing flow rate) was seen to provide smaller particle sizes, however, at the expense of a lower yield.

Two case studies were presented to demonstrate usefulness of the developed generalized framework for modeling cooling crystallization processes. First, the unseeded cooling crystallization of paracetamol in ethanol was considered. Crystallization kinetic model (CKM) parameter values were estimated by comparing simulation results with experimental results for batch crystallization. Key aspects of the procedure for estimating the CKM parameter values were discussed in detail. Using the batch crystallization model, it was found that there was no effect of scaling up on the crystallization dynamics. CKM parameter values obtained through batch studies were used for unseeded continuous cooling crystallization simulations. The continuous crystallization model was validated by comparing the model-predicted outlet concentration profile with the experimental profile. The continuous crystallization model was used to study the effect of residence time of the crystallization dynamics.

The second case study dealt with the unseeded cooling crystallization of sodium nitrite in water. Using a procedure similar to the one described in the first case study, key CKM parameter values were estimated through batch studies for one experiment. The batch crystallization model was validated by comparing model predictions with experiments for other experiments. Unseeded continuous crystallization studies were performed using a novel continuous crystallizer. By comparing model predictions to results of classical RTD experiments, it was found that the mixing behavior of the novel crystallizer resembled that of two tanks in series. Continuous crystallization simulations were performed using CKM parameter values estimated through batch studies. The developed continuous crystallization model was used to investigate the effect of key process design (mixing behavior) and operating (residence time) parameters on key process quality attributes (yield, mean and variance of particle size, time to reach steady state). It was concluded that moving toward plug flow mixing behavior promotes narrower particle size distributions, higher yields, and better process control. The developed model was thus seen to provide useful insights into key aspects of process design for crystallization processes.

28.6 SUMMARY AND RECOMMENDATIONS

A generalized population balance-enabled framework was presented in this chapter to model for both batch and continuous crystallization processes. The model was developed using the popular tanks-in-series framework that allowed to capture the mixing behavior specific to the continuous crystallizer considered. The present model considers fixed volume continuous crystallizers instead of fixed flow rate continuous crystallizers considered by previous

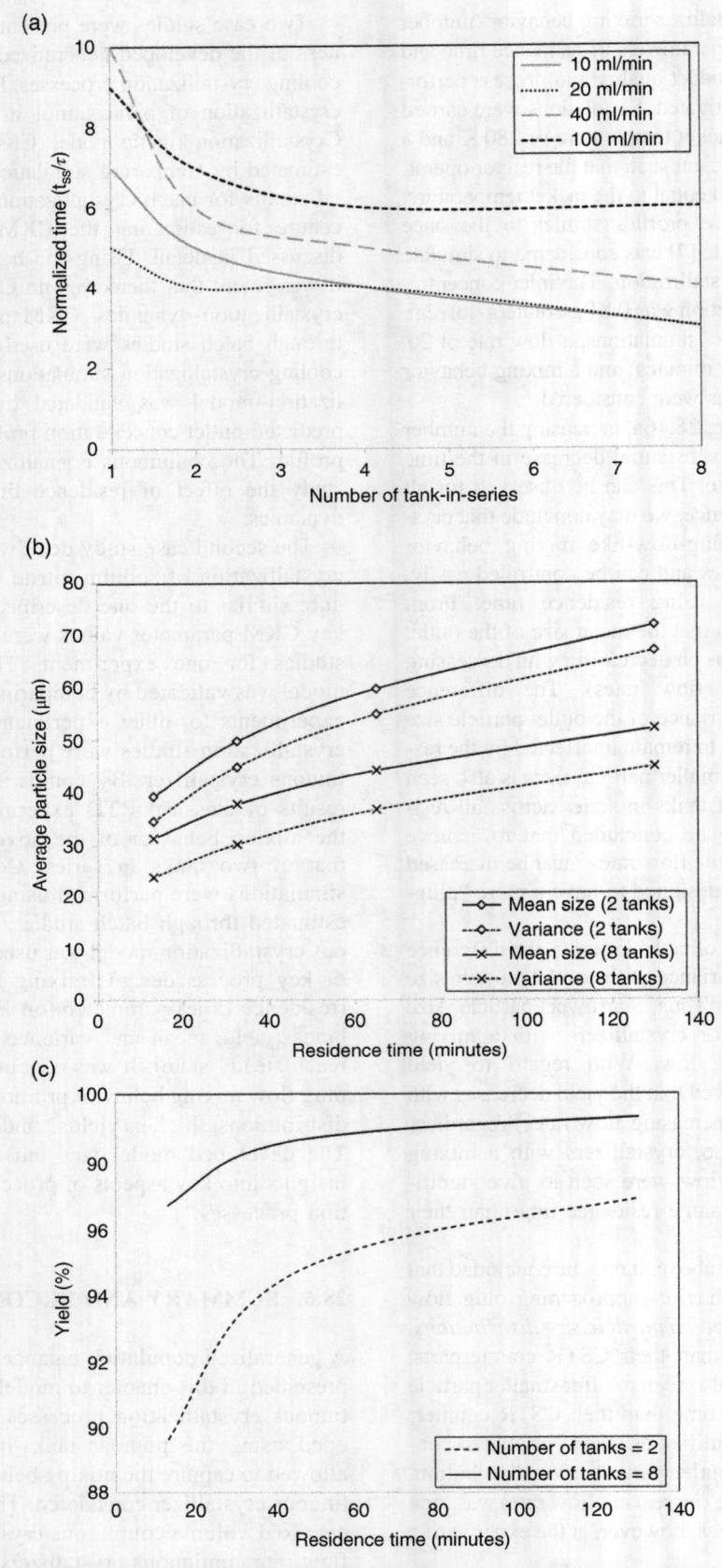

FIGURE 28.16 Comparison between the (a) normalized time to reach the steady state for different flow rates, (b) mean and variance of the particle size distribution using 2 and 8 tanks in series, and (c) yield versus residence time plots obtained using 2 and 8 tanks in series.

studies. Volume changes during crystallization were considered by writing a separate equation for change in slurry volume for the case of batch and equations for outlet flow rate for each tank in case of continuous crystallization. The standard MOM provided an efficient and convenient implementation of the presented population balance equations for the generalized framework. Nondimensionalized equations for dissolved solid mass, solid-phase mass, and the liquid-phase mass were solved simultaneously along with the moment equations for each tank. The model was implemented using MATLAB and uses an MS Excel-based interface.

Key aspects of the procedure to estimate CKM parameter values were demonstrated using a suitable case study. It was proposed that the CKM parameters can be reduced to an integral set of parameters that are sufficient to describe the crystallization system considered. The CKM parameter values were estimated through batch studies, and the same set of parameters were extended for continuous crystallization simulations. The proposed modeling framework provides a novel approach for predicting the onset of crystallization. The CKM typically used to model observable crystallization events was used to predict the onset of crystallization. It was assumed that nucleation events occurring before the detectable onset of crystallization follow the same crystallization kinetics as those after. Previous studies did not reconcile between the crystallization kinetics for detectable and non-detectable crystallization events.

The model may be effectively used to understand the impact of various critical process/design parameters (scale, residence time, mixing, temperature profile, etc.) on key process quality attributes (yield, mean and variance of particle size, process control, etc.). Effective implementation of QbD principles in API manufacturing requires a thorough mechanistic understanding of the crystallization dynamics, which is enabled by the present modeling framework. The presented modeling framework does not take into account certain crystallization events such as agglomeration or breakage that may be relevant in specific cases. Heat effects due to crystallization are critical in capturing the dynamics of certain crystallization systems. For example, the crystallization of ibuprofen in ethanol is a highly exothermic process, and the released heat significantly affects the slurry temperature (and hence the supersaturation trajectory). An additional heat balance equation needs to be solved in such cases along with the presented model.

The chapter demonstrates the effective use of experimentation integrated with modeling framework to capture key characteristics of crystallization events such as onset of crystallization, particle size distribution, concentration profiles, etc. The proposed modeling pathway is extensible and may be tailored to best suit specific nuances for various crystallization systems.

28.A APPENDIX

28.A.1 CLD to PSD

For the purposes of estimating reliable values for crystallization kinetics, it is important to obtain estimates for the average particle size distribution during the crystallization process. The model described in Pandit and Ranade [26] was used to estimate the mean particle size for the present case. The model was validated for a variety of shapes and particle systems. Hence, the authors believe that the values obtained using the model would serve as a good starting point for estimating crystallization kinetics. Kinetic parameter values may be reestimated when more reliable values for the mean particle size become available.

The PSD can be estimated by comparing the experimental normalized CLD against the model-predicted one. For this purpose, as described in Pandit and Ranade [26], both the unweighted and the square-weighted forms of the CLD were used depending upon the system considered. In most cases (regular unimodal systems), the PSDs obtained using both forms of the CLD are nearly identical. However, in certain cases, such as where the shape of the original CLD is not entirely unimodal (due to a disproportionate recording of smaller particles), different PSDs are predicted using the two CLD forms. The CLD to PSD procedure was implemented for both the forms of the CLD, and comparisons between the normalized experimental and model-predicted CLDs are shown in Figure 28.A.1. The CLD for a particular instance after crystallization had occurred in the 0.5 K/min cooling rate experiment was considered.

A limitation of the CLD to PSD model is that it can model only normal or lognormal-shaped distributions. It can be seen from Figure 28.A.1 that the experimental unweighted CLD was not properly captured using the model-predicted one ($R^2 = 0.93$). This leads to a significant difference between even the mean chord lengths predicted by the model and the experiment. On the other hand, the model-predicted normalized square-weighted CLD resembled the experimental one to a very high degree ($R^2 = 0.97$) as can be seen in Figure 28.A.1. Thus, the average particle sizes were obtained using the square-weighted form of the CLD, which was seen to give predictions consistent with experimental data.

28.A.2 Verification of Tanks-in-series Model

For the verification of the tanks-in-series model, a comparison was made between the model-predicted solution and the

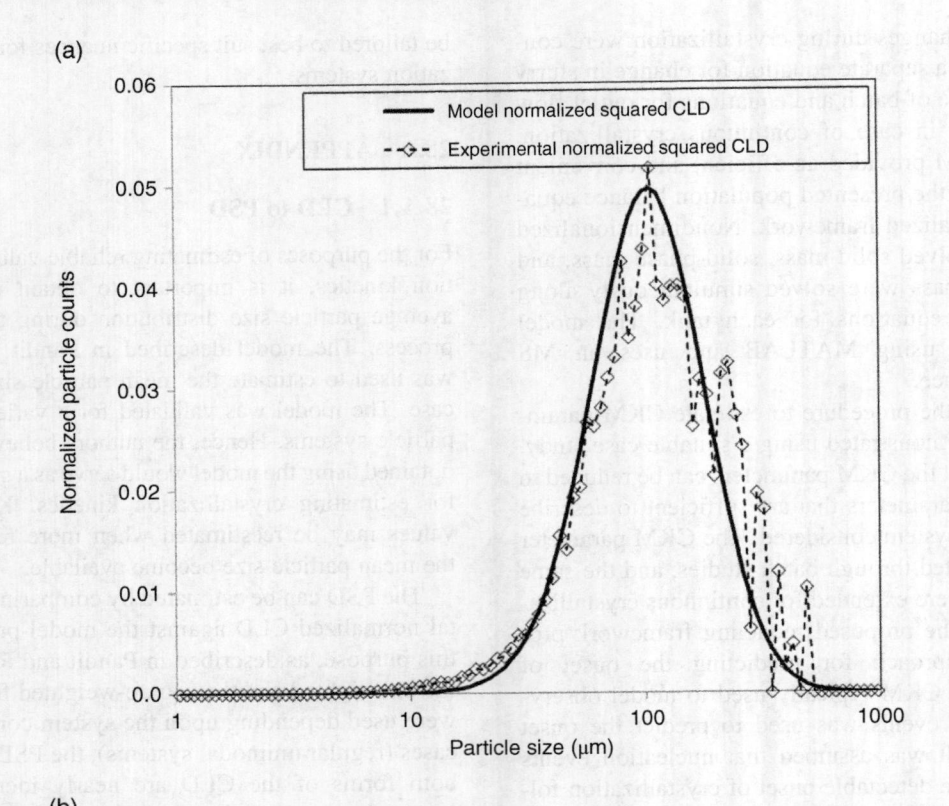

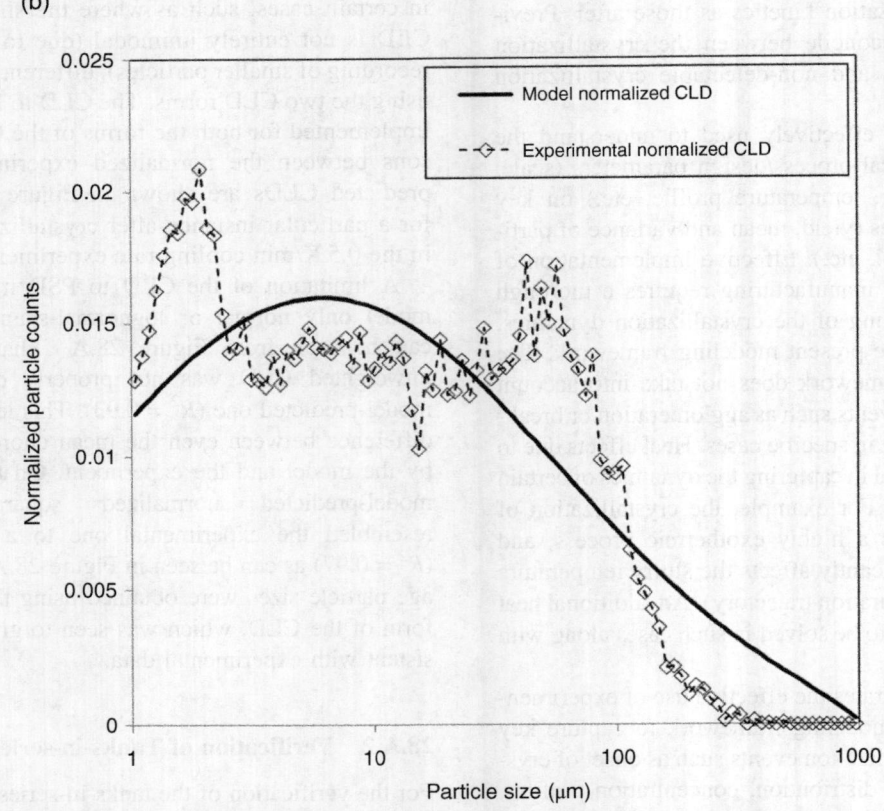

FIGURE 28.A.1 (a) Comparison of the measured squared normalized CLD to the model-predicted squared normalized CLD. (b) Comparison of normalized experimental CLD to the normalized model-predicted CLD.

solution obtained by directly solving the tanks-in-series ODEs. In a nondimensional form, the ODEs for tanks-in-series model can be written as

$$\frac{dC_1'}{dt'} = n(1 - C_1') \qquad (28.A.1)$$

$$\frac{dC_i'}{dt'} = n(C_{i-1}' - C_i') \quad \text{for } i > 1 \qquad (28.A.2)$$

It should be noted that n is the number of tanks and it appears in the equations because the reference time used for nondimensionalization was taken as the total residence time rather than the residence time for a tank. The inlet concentration was used for the nondimensionalization of concentration. Both the model equations and the above equations were solved for 8 tanks in series for a standard step response type of tracer experiment. A comparison between the model-predicted and the tanks-in-series ODE-predicted concentration profiles for the first and the fifth tanks is shown in Figure 28.A.2. As can be seen, there is excellent agreement between the two. Hence, the model was verified for the tanks-in-series model.

28.A.3 Crystallization Model

For the verification of the crystallization part of the solver, the simulation results were compared with the simulation results obtained using a previous solver. Simulations were performed for a model system of paracetamol–ethanol having a concentration of 1.37 kmol/m^3 corresponding to a solubility temperature of approximately 326 K. The solution was assumed to be completely clear. The solution temperature was assumed to be held constant at 285 K to simulate induction time. The period after the nucleation events had started was simulated. For the new solver, simulations were performed using both constant and variable liquid density assumptions. The crystallization kinetic parameter values estimated in Chapter 3 were used for the simulations.

A comparison between the simulated concentration profiles using the previous solver and the new solver (using constant and varying liquid density assumptions) is shown in Figure 28.A.3. A comparison between the simulated 0th moment profiles using the previous solver and the new solver (using constant and varying liquid density assumptions) is shown in Figure 28.A.4. A comparison between the simulated Sauter mean diameter profiles using the previous solver and the new solver (using constant and varying liquid density assumptions) is shown in Figure 28.A.5. Let us first consider the case of comparison between the previous solver and the new solver implemented using a constant density assumption. As can be seen from Figure 28.A.3, there is an excellent agreement

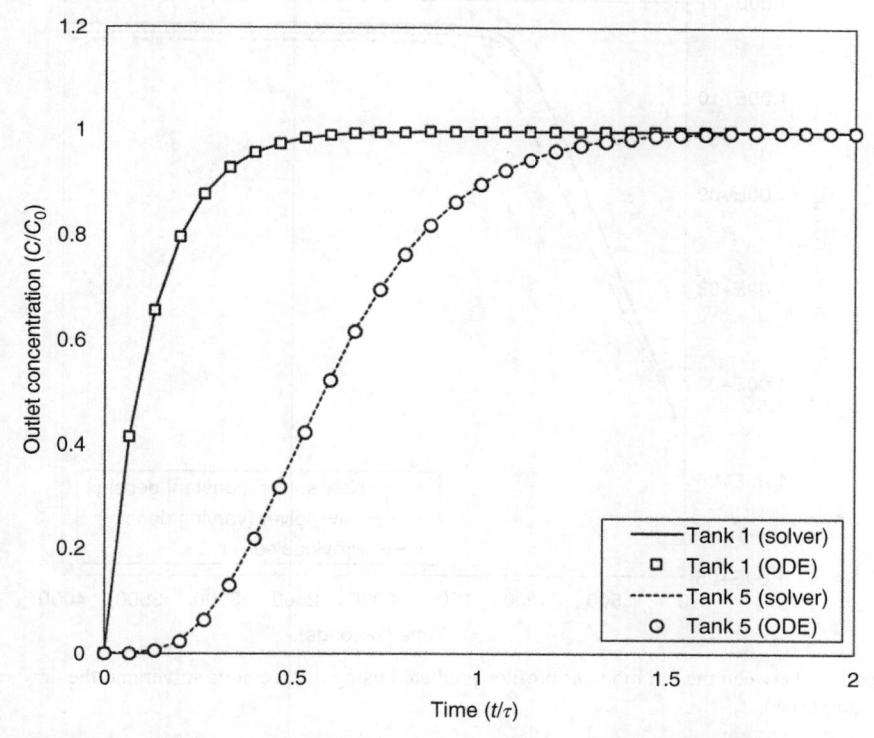

FIGURE 28.A.2 Comparison between simulated and ODE results for tanks-in-series model for a step response tracer experiment (solved for 8 tanks).

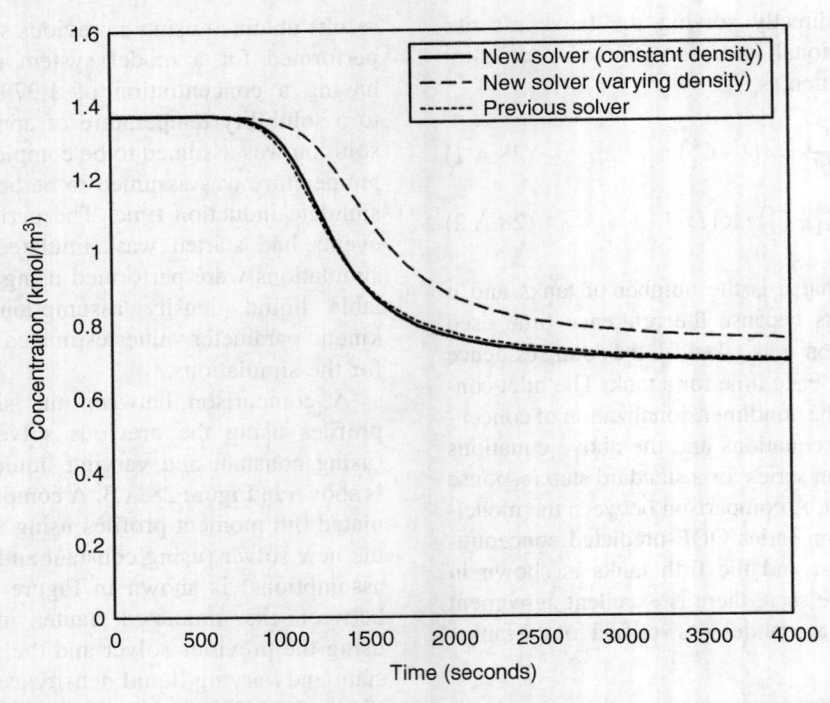

FIGURE 28.A.3 Comparison between the concentration profiles predicted using the previous solver and the new solver (using constant and variable liquid density assumptions).

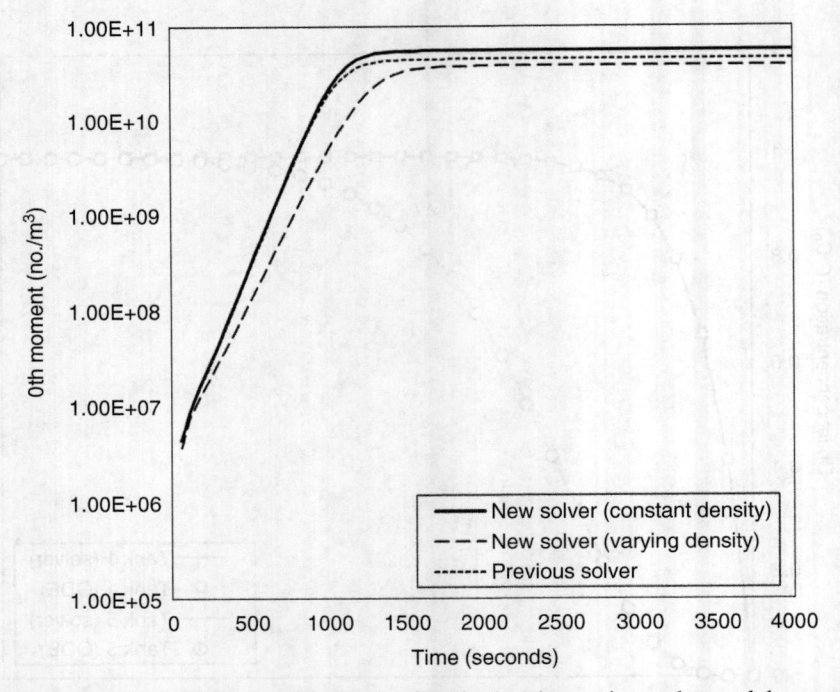

FIGURE 28.A.4 Comparison between the 0th moment profiles predicted using the previous solver and the new solver (using constant and variable liquid density assumptions).

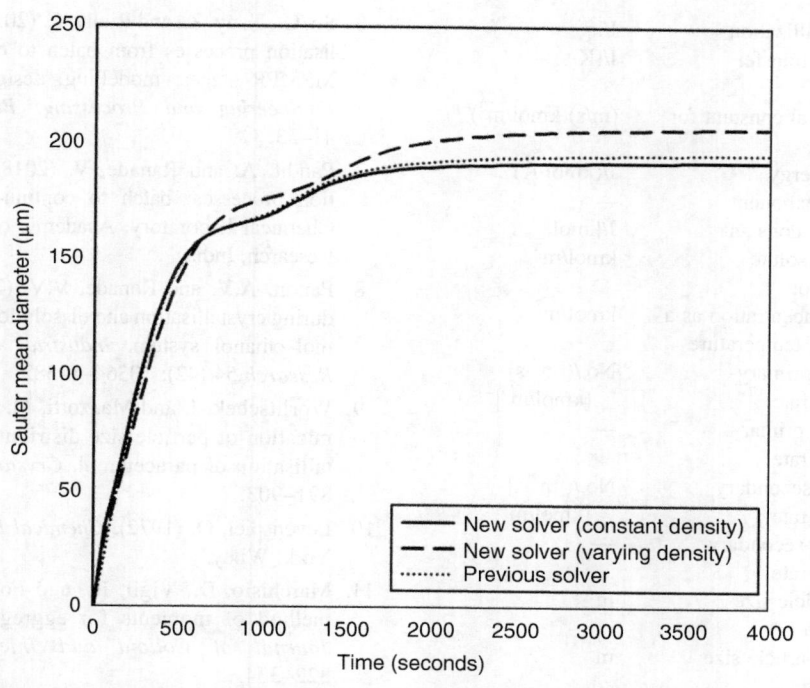

FIGURE 28.A.5 Comparison between the Sauter mean diameter profiles predicted using the previous solver and the new solver (using constant and variable liquid density assumptions).

between the concentration profiles, although the agreement is not exact. As seen from Figures 28.A.4 and 28.A.5, there is good agreement between the 0th moment and Sauter mean diameter profiles although there is some discrepancy toward the end. The discrepancy arises due to the fact that the previous solver does not consider the volume change in the liquid phase due to the precipitation of solids. As the volume of the liquid phase will decrease when this happens, the 0th moment (which is the number density of particles) will increase correspondingly.

As can be seen from Figures 28.A.3–28.A.5, there is a difference between the concentration, 0th moment, and Sauter mean diameter profiles predicted using the varying density assumptions as opposed to using the constant density assumption. However, the difference is not drastic as is indicated by the relatively close values for the Sauter mean diameters (~200 μm) for both cases. Hence values estimated for the crystallization kinetic parameters estimated using the previous solver and the new solver will not differ by order(s) of magnitude.

ACKNOWLEDGMENT

The authors would like to thank the Council of Scientific and Industrial Research for the Senior Research Fellowship to one of the authors (AVP).

NOTATIONS

Symbol	Description	Units
L	Particle size coordinate to describe the number distribution	m
t	Time	s
$n(L, t)$	Number density distribution function	No./(m^3·m)
$G(L, t)$	Crystal growth rate	m/s
$B0$	Nucleation rate	No./(m^3·s)
δ	Kronecker delta function	—
v	Volumetric flow rate	m^3/s
V_R	Volume of crystallizer	m^3
M_i	ith moment of the number density distribution function	m^i/m^3
M	Total phase mass	kg
ρ	Phase density	kg/m^3
ϵ	Solid phase hold-up	—
$\dot{M}_C$	Rate of mass transfer between solid and liquid phases	kg/s
σ_V	Particle volume shape factor	—
y_D	Mass fraction of dissolved solute in liquid phase	—
A, B	Constants for the linear liquid-phase density function	kg/m^3
C_P	Specific heat capacity	J/(kg·K)
T	Temperature	K

$(-\Delta H_C)$	Heat of crystallization	J/kg
$\langle UA \rangle$	Lumped heat transfer coefficient	J/(K·s)
$k_{g,0}$	Pre-exponential constant for growth rate	$(m/s)(kmol/m^3)^{-g}$
E_A	Activation energy	J/(kmol·K)
g	Growth rate exponent	—
R	Universal gas constant	J/kmol
C	Liquid-phase solute concentration	$kmol/m^3$
$C^*(T)$	Solubility concentration as a function of temperature	$kmol/m^3$
k_1	Constant for primary nucleation rate	No./([m^3·s] [$kmol/m^3$]$^{-n_1}$)
n_1	Exponent for primary nucleation rate	—
k_2	Constant for secondary nucleation rate	No./([m^2·s] [$kmol/m^3$]$^{-n_2}$)
n_2	Exponent for secondary nucleation rate	—
μ	Mean of particle size distribution	m
σ^2	Variance in particle size distribution	m^2

Subscripts	Description
i	Moment index
k	Tank index
o	Crystallizer inlet
L	Liquid phase
S	Solid phase
J	Jacket
E	Environment

REFERENCES

1. Yu, L., Amidon, G., Khan, M. et al. (2014). Understanding pharmaceutical quality by design. *The AAPS Journal* 16 (4): 771–783.
2. Gao, Z., Rohani, S., Gong, J., and Wang, J. (2017). Recent developments in the crystallization process: toward the pharmaceutical industry. *Engineering* 3 (3): 343–353.
3. FDA (2018). *Guidance for Industry: Q8 (R2) Pharmaceutical Development* (April 2018). Silver Spring, MD: US Department of Health and Human Services, Food and Drug Administration (FDA).
4. Nagy, Z. and Braatz, R. (2012). Advances and new directions in crystallization control. *Annual Review of Chemical and Biomolecular Engineering* 3: 55–75.
5. Nagy, Z., Fevotte, G., Kramer, H., and Simon, L. (2013). Recent advances in the monitoring, modelling and control of crystallization systems. *Chemical Engineering Research and Design* 91 (10): 1903–1922.
6. Su, Q., Nagy, Z., and Rielly, C. (2015). Pharmaceutical crystallisation processes from batch to continuous operation using MSMPR stages: modelling, design, and control. *Chemical Engineering and Processing: Process Intensification* 89: 41–53.
7. Pandit, A. and Ranade, V. (2018). Modeling of crystallisation processes: batch to continuous. PhD thesis. National Chemical Laboratory, Academic of Scientific and Innovative Research, India.
8. Pandit, A.V. and Ranade, V.V. (2015). Modeling hysteresis during crystallisation and dissolution: application to a paracetamol–ethanol system. *Industrial & Engineering Chemistry Research* 54 (42): 10364–10382.
9. Worlitschek, J. and Mazzotti, M. (2004). Model-based optimization of particle size distribution in batch-cooling crystallisation of paracetamol. *Crystal Growth & Design* 4 (5): 891–903.
10. Levenspiel, O. (1972). *Chemical Reaction Engineering*. New York: Wiley.
11. Marchisio, D., Vigil, R., and Fox, R. (2003a). Quadrature method of moments for aggregation–breakage processes. *Journal of Colloid and Interface Science* 258 (2): 322–334.
12. Marchisio, D., Pikturna, J., Fox, R. et al. (2003b). Quadrature method of moments for population-balance equations. *AIChE Journal* 49: 1266–1276.
13. Randolph, A.D. and Larson, M.A. (1988). *Theory of Particulate Processes: Analysis and Techniques of Continuous Crystallisation*. San Diego, CA: Academic Press, Inc.
14. Ramkrishna, D. (2000). *Population Balances: Theory and Applications to Particulate Systems in Engineering*, 1e. San Diego, CA: Academic Press.
15. Hulbert, H.M. and Katz, S. (1964). Some problems in particle technology. *Chemical Engineering Science* 19: 555–574.
16. McGraw, R. (1997). Description of aerosol dynamics by the quadrature method of moments. *Aerosol Science and Technology* 27 (2): 255–265.
17. Karpinski, P. (1985). Importance of the two-step crystal growth model. *Chemical Engineering Science* 40 (4): 641–646.
18. Myerson, A.S. (2002). *Handbook of Industrial Crystallization*. Boston, MA: Butterworth-Heinemann.
19. Barrett, P. and Glennon, B. (2002). Characterizing the metastable zone width and solubility curve using Lasentec FBRM and PVM. *Chemical Engineering Research and Design* 80 (7): 799–805.
20. Mitchell, N.A. and Frawley, P.J. (2010). Nucleation kinetics of paracetamol–ethanol solutions from metastable zone widths. *Journal of Crystal Growth* 312 (19): 2740–2746.
21. Kadam, S.S., Kulkarni, S.A., Ribera, R.C. et al. (2012). A new view on the metastable zone width during cooling crystallization. *Chemical Engineering Science* 72: 10–19.
22. Kubota, N. (2015). Analysis of the effect of volume on induction time and metastable zone width using a stochastic model. *Journal of Crystal Growth* 418: 15–24.

23. Kubota, N. (2008). A new interpretation of metastable zone widths measured for unseeded solutions. *Journal of Crystal Growth* 310 (3): 629–634.

24. Nagy, Z.K., Fujiwara, M., Woo, X.Y., and Braatz, R.D. (2008). Determination of the kinetic parameters for the crystallization of paracetamol from water using metastable zone width experiments. *Industrial & Engineering Chemistry Research* 47 (4): 1245–1252.

25. Addis, C., Pandit, A., and Ranade, V. (2017). Towards continuous crystallisation. Master's in engineering thesis. School of Chemistry and Chemical Engineering, Queen's University, Belfast, UK.

26. Pandit, A.V. and Ranade, V.V. (2016). Chord length distribution to particle size distribution. *AIChE Journal* 62 (12): 4215–4228.

29

SOLID FORM DEVELOPMENT FOR POORLY SOLUBLE COMPOUNDS

ALESSANDRA MATTEI, SHUANG CHEN, JIE CHEN, AND AHMAD Y. SHEIKH

Solid State Chemistry, AbbVie Inc., North Chicago, IL, USA

29.1 INTRODUCTION

In recent years the number and complexity of targets being studied for developing drugs have increased dramatically across multiple therapeutic areas. Additionally, modalities for developing therapeutics for these targets have increased well beyond typical small molecules and humanized monoclonal antibodies to include many new approaches, such as various constructs of antibody drug conjugates and cell-based therapies. Even many of the recently approved chemically synthesized small molecules are not small, with molecular masses well in excess of 500 Da. Consequently, the percentage of drug candidates with poor aqueous solubility emerging from drug discovery has steadily increased. As observed by Loftsson and Brewster in 2010 [1], "while 40% of currently marketed drugs are poorly soluble based on the definition of the biopharmaceutical classification system (BCS), nearly 90% of drug molecules in the development pipeline can be characterized as poorly soluble compounds." Figure 29.1 illustrates how the majority of the portfolio of developmental drugs falls into the two low aqueous solubility categories of the BCS (i.e. classes II and IV) [2]. An increasing trend toward poorly soluble drug candidate molecules represents a major challenge for drug development as the formulation of poorly soluble compounds can be very complex.

Many small molecule drug candidates are delivered to patients in a crystalline state, due to its more desirable properties, including stability, purity, reproducibility, and ease of manufacture. The crystalline state of a molecule is highly consequential to two important elements of the target product profile in drug development: bioavailability and shelf life. Bioavailability and shelf life are in turn related to solubility and stability, respectively. These two attributes can be in conflict, and finding an optimum between the two is an essential element of preclinical development. Higher solubility and faster dissolution rate can lead to measurable increases in bioavailability and, likely, therapeutic efficacy. A number of pharmaceutical strategies can be applied to enhance solubility, dissolution rate, and ultimately bioavailability of poorly water-soluble drug candidates. Crystal structure modifications, particle size reduction, and transformation of a crystalline active pharmaceutical ingredient (API) into a stabilized amorphous state are some of the main approaches that can be utilized to overcome drugability challenges. The crystal-structure-modification-based approach includes, but is not limited to, the use of higher solubility polymorphs, salts, and cocrystals. Micronization and nano-milling can be used for dissolution rate enhancement due to increased surface area; and amorphous solid dispersions (ASDs) offer a viable path to increase solubility while ensuring the stability of the high-energy amorphous state.

Enhancement of solubility and dissolution rate can be achieved by molecular modifications. Different crystal forms from the same organic molecule, known as polymorphs [3], exhibit unique X-ray diffraction patterns and have distinct melting points and solubility properties in addition to other well-defined physicochemical properties. Up to fivefold solubility differences between different crystal forms have been reported in the literature [4, 5]. As such, higher-energy crystal forms that offer acceptable solid-state properties and are also

---

Chemical Engineering in the Pharmaceutical Industry: Active Pharmaceutical Ingredients, Second Edition.
Edited by David J. am Ende and Mary T. am Ende.
© 2019 John Wiley & Sons, Inc. Published 2019 by John Wiley & Sons, Inc.

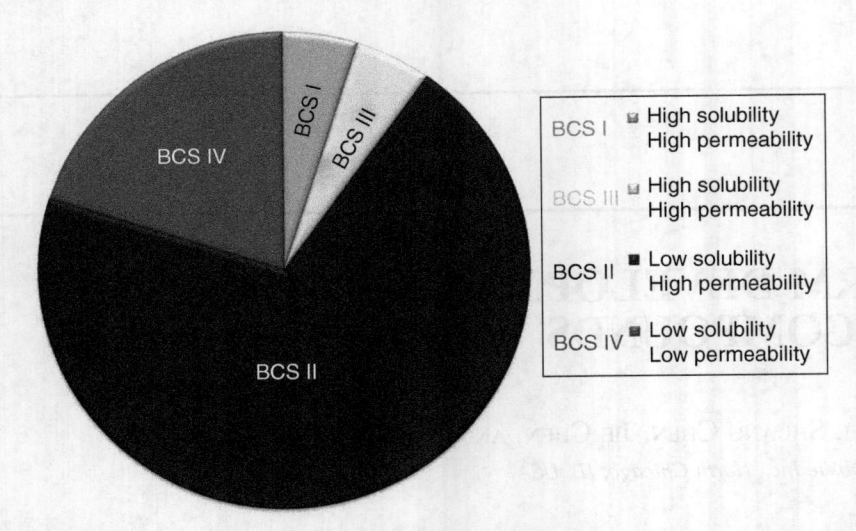

FIGURE 29.1 Trend of pipeline drugs toward low aqueous solubility. *Source*: Data from Lipp [2].

kinetically stable can be considered for development with well-defined and robust control strategies.

At a fundamental level, the solubility enhancement via salt and cocrystal formation is a consequence of the pH-solubility profile of acidic/basic drugs and the solubility profile of a cocrystal as a function of the coformer concentration, as illustrated in Figure 29.2. Sound understanding of concepts, including intrinsic solubility, K_a (ionization constant), K_{sp} (solubility product), pH_{max} (pH of maximum solubility), and critical coformer activity for cocrystals, is essential to fully leverage the opportunities offered by salts and cocrystals. As such, salts and cocrystals are routinely explored in the search of commercially viable solid forms. Pharmaceutical cocrystals can in fact be considered a natural extension of the opportunity to modify molecular properties of drugs that cannot form salts either due to the lack of ionizable moieties or insufficient pK_a differences to make a stable salt. Indeed, the stoichiometric nature of cocrystals predisposes them to large yet predictable changes in solubility and thermodynamic stability as solution conditions vary (i.e. pH and drug solubilizing agents) [7–9]. Physical and chemical stability of salts or cocrystals is also a key consideration in their development. It is important to understand the dissociation of a salt to its parent species or of a cocrystal to its individual components during processing conditions, storage, or in the presence of formulation excipients as it can lead to poor performance.

The dissolution rate and solubility of crystalline drugs can be affected by the particle size. In fact, particle size reduction is a fairly common and at times effective approach for dissolution rate enhancement of poorly soluble compounds [10, 11]. Conventional mechanical milling can reduce particle size to the 2–5 μm range, resulting in a moderate enhancement of

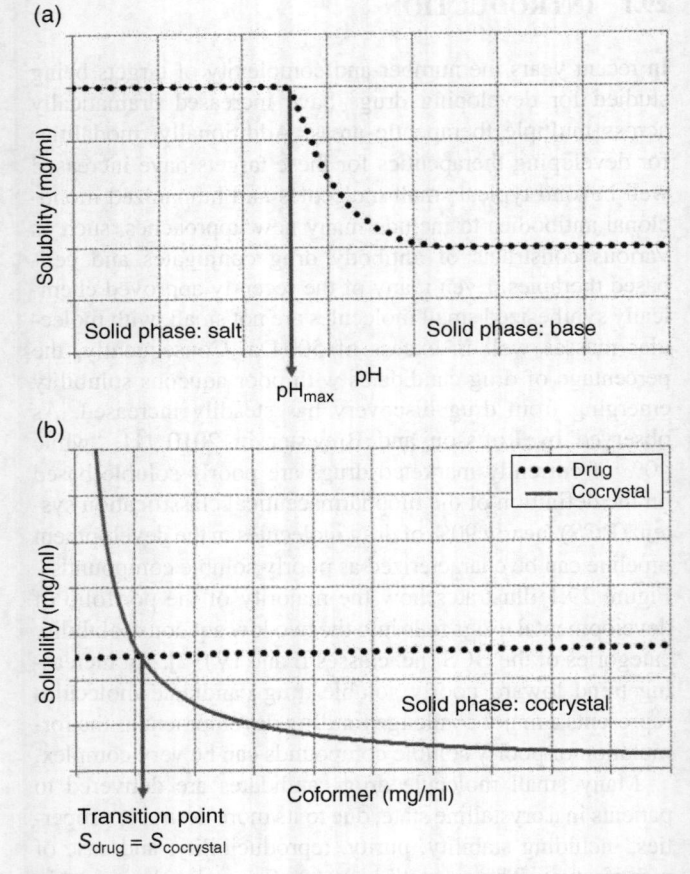

FIGURE 29.2 (a) pH-solubility profile for a weak base. (b) Solubility curve as a function of coformer concentration. *Source*: Adapted from Serajuddin [6] and Kuminek et al. [7] with permission.

surface area and, hence, dissolution rate. Nano-milling can further reduce the size to the 100 nm range. Particle size reduction into the nanometer range can generate a 100-fold increase in surface-area-to-volume ratio. This increase in surface area is often accompanied by an increase in surface energy due to the generation of structural disorder or exposure of high-energy crystal facets. The net result can be not only a profound improvement in dissolution rate but also an increase in apparent solubility. However, nano-milling is not possible if molecular organic crystals are soft and undergo elastic deformation. A careful understanding of intrinsic mechanical properties is thus a requisite for materials undergoing nano-milling as well as conventional high-impact mechanical milling. Even when material properties are conducive to nano-milling, nanometer-sized drug particles can show a tendency for agglomeration due to attractive interparticle forces, which can decrease the effective surface area, thereby negating the benefit of size reduction on dissolution enhancement. In such cases, wetting agents, such as surfactants, can be used to retain the effective surface area. Implications of high-energy surfaces for physical and chemical stability also need to be thoroughly assessed in the context of shelf life determination for the drug substance.

When crystal structure modifications or particle size reduction are insufficient to improve solubility and dissolution rate, enabling formulations that do not incorporate the drug substance in the crystalline state can be evaluated. Chief among such options is ASDs, which can provide superior bioavailability when compared with drug products containing a crystalline drug. At the most basic level, ASDs favor enhanced dissolution or increased solubility of poorly soluble crystalline compounds due to the formation of a high-energy amorphous state [12–14]. The increased solubility can also be attributed to solubilization effects of the polymer, often used in ASDs to stabilize the amorphous state during manufacturing and storage [15]. Polymers are also selected to help maintain supersaturation during in vivo dissolution by inhibiting recrystallization within the time scale relevant to absorption. It has been reported that even the flux across a membrane can be enhanced by the polymer in supersaturated solutions [16]. At a fundamental level, in vitro and in vivo stabilization of ASDs depends on both thermodynamic and kinetic factors. Maximum physical stability is achieved when the drug and excipients remain intimately mixed. Assurance of thermodynamic stability requires quantitative understanding of the solubility/miscibility of the drug in the polymer-rich excipient matrix. Recently, relatively simple but reliable experimental approaches have been developed and implemented to obtain the phase diagram of API/polymer systems and determine the maximum drug load for a fully miscible ASD [17–19]. Mathematical modeling solutions to estimate solubility/miscibility of drugs in polymers have been also reported, with the perturbed-chain statistical associating fluid theory (PC-SAFT) [20, 21] being especially noteworthy due to its origins in polymer chemistry. Designing a thermodynamically stable and molecularly dispersed ASD, wherein the drug load does not exceed the solubility/miscibility limit in the polymer matrix, offers excellent stability. However, this ASD system often translates in relatively low drug load, which can lead to pill burden challenges for high dose drug candidates. In such cases, kinetically frozen ASDs, where drug molecules are not molecularly mixed, can be considered with appropriate process controls and storage conditions.

In vivo stability of an ASD is an equally important consideration in achieving target bioperformance. For an enhanced absorption to take place, supersaturation must be obtained and maintained in the gastrointestinal environment. This can be achieved by designing ASD formulations that upon dissolution form nanosized drug-rich droplets. Implicit in the drug-rich domain formation is the benefit offered by rapid dissolution rates and equilibration with the aqueous phase, saturated with the drug at the amorphous solubility, to replenish the molecularly dissolved drug, as the drug diffuses into systemic circulation. Therefore, drug-rich droplets can serve as reservoirs to sustain the thermodynamic activity of the drug at its maximum value [22, 23]. The underlying phenomenon causing the formation of drug-rich droplets upon dissolution of an ASD is the liquid–liquid phase separation (LLPS) [24]. A schematic representation of the drug uptake of an ASD undergoing LLPS is displayed in Figure 29.3. Whether an ASD dissolves to form a solution in which LLPS occurs is likely to depend on a number of factors, including inherent drug properties (e.g. crystallization tendency), drug loading, and polymer type in the formulation [25]. The role of LLPS is the foundation for a well-designed ASD formulation; as long as LLPS can be maintained within the gastrointestinal absorption window, the formulation is close to optimal. Multiple processing options including drum drying, spray drying, and hot melt extrusion exist to manufacture ASDs. Advances in the hot melt extrusion technology have in particular accelerated industrial applications of ASDs for the delivery of poorly soluble drugs.

As part of the process of identifying and efficiently selecting the most appropriate solubility-enhancing strategy, solid form development plays an integral role in the development of new chemical entities. Solid form screening experiments can be tailored not only to the physicochemical properties of the drug molecule but also the downstream formulation options. As a consequence, solid form selection criteria can be adapted to the specific objectives for the drug substance molecule. In this chapter, we provide a general current perspective on solid form development in the pharmaceutical industry and two case studies on solid form development of poorly soluble compounds. Through the case studies we illustrate the impact of solid form on drug substance as well as on drug product manufacturing processes.

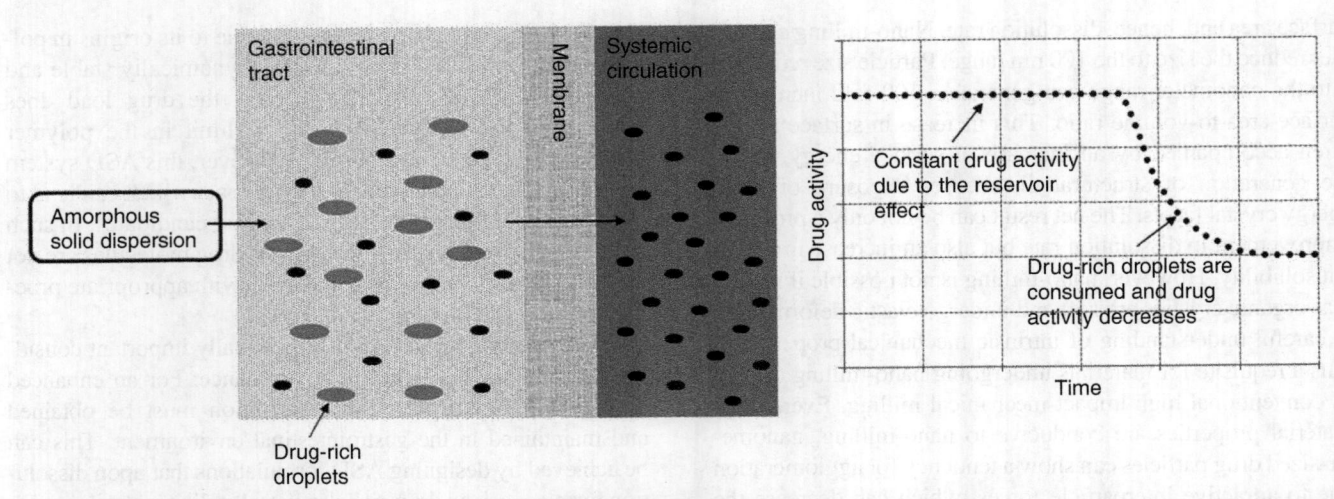

FIGURE 29.3 Schematic representation illustrating the drug uptake of an ASD formulation that undergoes LLPS. *Source*: Adapted from Raina et al. [22] with permission of Elsevier.

29.2 PERSPECTIVE ON SOLID FORM SCREENING IN DRUG DEVELOPMENT

The solid form development process starts with a discussion on strategic objectives related to dosage forms to achieve a target product profile. Determinations are made on whether chemical (salts/cocrystals) and/or physical (polymorphs) modifications are needed based on the physicochemical properties of the molecule and early predictions of the projected dose by physiologically based pharmacokinetics models. The aim is to generally identify a crystalline form that shows optimal solid-state properties and a relatively simple solid form landscape. In addition, a developable solid form is able to reject impurities during crystallization, is amenable to downstream processing, and ensures consistency in the safety and efficacy profile of the drug product throughout its shelf life. Overall, among the potentially many solid forms of an API, a crystal form that is stable, helps achieve bioperformance, and can be manufactured into the dosage form is generally progressed in pharmaceutical development. A thermodynamically most stable form is highly desirable, because a significant drop in solubility due to the discovery of a more stable form can profoundly affect bioperformance and dosage form. This is especially true for soluble compounds where solubility modification techniques to maintain supersaturation through *in vivo* dissolution and absorption are not built into the dosage form.

When salts and/or cocrystals are anticipated to address significant solubility or stability issues, experimentation begins with screening appropriate counterions and/or coformers. Some of the recently matured *in silico* tools can be used to augment experimental efforts to improve outcomes and save resources [26]. Generally, salt or cocrystal screening does not add value for molecules that are deemed to be developable only in the amorphous state via, for instance, the ASD route. Once the species (salt, cocrystal, or parent molecule) has been selected, extensive polymorph screens are conducted in stages during preclinical development to eventually identify a developable commercially viable crystal form generally well before registration supporting clinical studies. Although the solid form screen is a routine activity for any given small molecule drug candidate, the actual recipe for searching the crystal forms of a compound for which the crystal form landscape has never been explored is uniquely tied to the structure of the molecule itself [27]. In other words, polymorph screening is by no means routine [28], and the design of each solid form screen has to be customized based on (i) physicochemical properties of the molecule and (ii) the intended formulation.

Figure 29.4a illustrates some of the key considerations that drive customization of form screening experiments for commercially viable crystal form selection. The interconnectivity between crystal form complexity, formulation, manufacturing processes, and bioperformance, along with the associated need for customization of screens, becomes even more important for poorly soluble compounds. By way of example, Figures 29.4b and c illustrate an approach to customization based on solid form complexity and manufacturing process choices. The illustrations capture a scenario where wet milling is intended to be used either to induce nucleation under high-shear conditions or control particle size and hot melt extrusion for the preparation of an ASD-based dosage form. For the latter, since hydrates or low-melting but sufficiently stable solid forms are preferable over high-melting polymorphs, a strong emphasis is placed on searching for hydrates and crystallizing under high-supersaturation conditions. The use of shear can reduce the induction time and facilitate nucleation and even isolation of metastable forms.

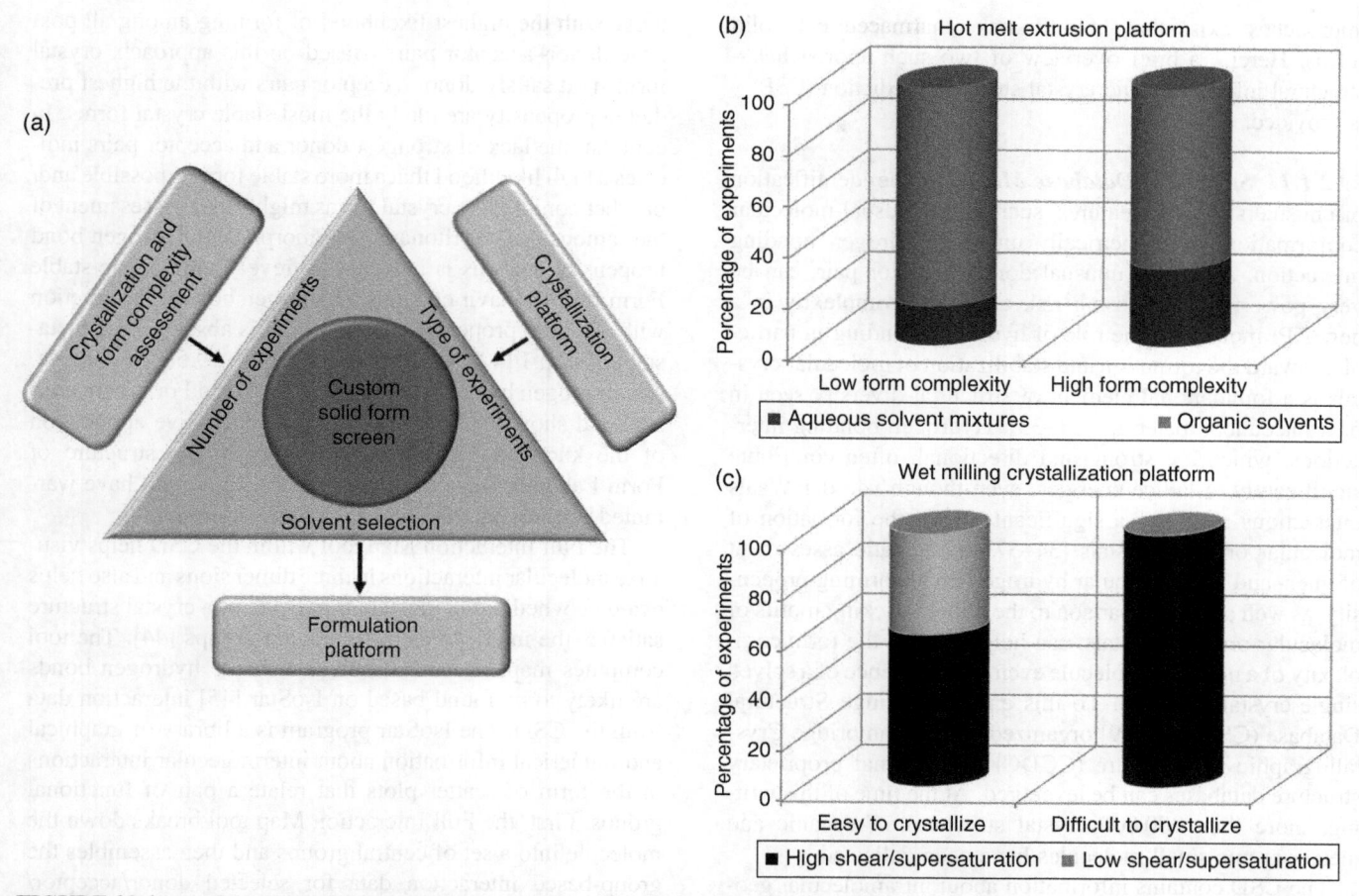

FIGURE 29.4 (a) Schematic representation of the strategy for a customized solid form screen, (b) solvent selection guidance based on the solid form complexity and formulation platform, and (c) experimental condition guidance based on the ease of crystallization and the crystallization platform.

Relatively large molecules that do not nucleate under mild supersaturation conditions afforded by the classic solution-mediated phase transformation-based polymorph screening or that have the tendency of oiling out can especially benefit from high-shear/high-supersaturation conditions. Miniaturized Taylor–Couette [29] flow system can be used for screening purposes to mimic high-shear conditions and achieve Reynold's numbers not too dissimilar to those generated in commonly used rotor–stator-based wet mills. Complexity of a crystal form landscape would necessitate focusing more experiments in the process-relevant space. In the end, the number of experiments performed is somewhat commensurate with the solid form complexity.

A comprehensive perspective on a solid form screen can provide a route map toward a fully integrated, holistic product design process and enable the journey from molecule to crystal to product performance. The framework needs to be customized for each drug molecule due to the distinct nature of crystal chemistry and different crystallization tendencies, either of which can limit the diverse experimental conditions that can be applied in the search for crystal forms. For these reasons, solid form control is, in the end, primarily a risk assessment and mitigation exercise for the selected, commercially viable crystalline form, from drug substance isolation through drug product manufacture and storage. Rapidly evolving *in silico* tools can also effectively augment the well-established crystal form risk assessment tools typically used in pharmaceutical development and help de-risk the detrimental implications of late discovery of a new crystal form.

29.2.1 *In Silico* Modeling in Pharmaceutical Solid-State Chemistry

Advances in computational solid-state chemistry are increasingly becoming important for de-risking crystal form complexity in pharmaceutical development [30]. *In silico* modeling, for example, provides unique insights into the fundamentals of crystal chemistry that often are inaccessible through traditional experimental methods. Within the broad spectrum of computational solid-state chemistry, multiple approaches exist that help de-risk pharmaceutical solid forms. Herein, a brief overview of two such approaches – structural informatics and crystal structure prediction (CSP) – is provided.

29.2.1.1 Structural Database Mining

The identification of unusual structural features, such as an unusual molecular conformation, a geometrically unusual hydrogen bonding interaction, and/or an unusual donor–acceptor pair, can be very powerful in assessing risk and form complexity in a new API molecule. The role of hydrogen bonding in particular toward the formation and stabilization of molecular crystals is a foundational element of structural diversity seen in pharmaceutical materials [31–33]. Hydrogen bonding interactions, which are strong and directional, often contribute significantly to lattice energies, even though van der Waals interactions also play a significant role in the formation of molecular organic crystals [34–37]. Systematic assessment of inter- and intramolecular hydrogen bond-forming propensity, as well as its comparison to the known packing motifs of molecular organic crystals, can help estimate the form complexity of a new API molecule even in the absence of a solved single-crystal structure. To this end, Cambridge Structural Database (CSD) [38, 39] organized by the Cambridge Crystallographic Data Centre (CCDC) and internal proprietary structure databases can be leveraged. At the time of this writing, more than 900 000 crystal structures of organic and metal-organic small molecules have been collected.

The CSD contains information about intramolecular geometries and molecular conformations as well as intermolecular interactions. Once a crystal structure of an API molecule has been solved, additional insights can be obtained through the analysis of full interaction maps within the crystal lattice and with the comparison of the known structures. A stagewise combination of hydrogen bond propensity analysis and full interaction maps can even provide insights into the relative stability of polymorphs and aid polymorph risk assessment/mitigation. The CSD offers a suite of software that provides a wide range of tools for both of these structural informatics-based approaches to readily execute the assessment at a quite modest computational cost.

Consistent with the important correlation between diversity of hydrogen bonding motifs and polymorphism, CSD analysis shows that when molecules have multiple hydrogen bond donors and acceptors, a variety of hydrogen bond pairings can be observed among polymorphs. Statistical models of experimental observations in the crystallographic database can be applied to compute the likelihood that a given hydrogen bond pairing will occur. The hydrogen bond propensity model offers a prediction for the presence or absence of a hydrogen bond between a specified donor and acceptor atom in a crystal structure, based on related known crystal structures, their chemical functionality, and their molecular environment [40, 41]. The key assumption is that actual hydrogen bonds directing the formation of a crystal structure will be those with the highest likelihood of forming among all possible donor–acceptor pairs. Based on this approach, crystal forms that satisfy donor–acceptor pairs with the highest predicted propensity are likely the most stable crystal forms. In contrast, the lack of strongest donor and acceptor pairs indicates a high likelihood that a more stable form is possible and/or other competitive crystal forms might exist. Assessment of the famous case of ritonavir polymorphs via hydrogen bond propensity analysis is indeed very revealing [5]. The stable Form II of ritonavir contains a hydrogen bonding interaction with a higher propensity to occur that is absent in the metastable Form I [42], as depicted in Figure 29.5a. In contrast, two hydrogen bonds in the initially marketed Form I are unusual and show low propensity values. Proactive application of the knowledge-based methodology to the structure of Form I at early stage of drug development would have warranted further experimental solid form screening.

The Full Interaction Map tool within the CSD helps visualize molecular interactions in three dimensions and also helps evaluate whether the molecular packing of a crystal structure satisfies the interactions of functional groups [44]. The tool computes maps around a molecule where hydrogen bonds are likely to be found based on IsoStar [45] interaction data from the CSD. The IsoStar program is a library of graphical and numerical information about intermolecular interactions in the form of scatter plots that relate a pair of functional groups. First, the Full Interaction Map tool breaks down the molecule into a set of central groups and then assembles the group-based interaction data for selected donor/acceptor/hydrophobic probes around each central group. Environmental effects, including steric factors, are taken into account to create a full three-dimensional picture of molecular interactions. A comparison of the fit of the calculated maps with the observed interactions allows the crystal structure to be assessed in terms of how well intermolecular interactions are satisfied by the existing lattice. For a polymorphic compound, the different interactions observed in each polymorph can be examined and assessed. Revisiting the case of ritonavir with full interaction maps shows that Form II has a crystal structure with hydrogen bonding consistent with the most favorable interactions, as shown in Figure 29.5b. A complementary method of comparing known crystal forms is represented by Hirshfeld fingerprint plots, which represent a two-dimensional visualization of the space (surface) occupied by a molecule in a crystal. The fingerprint plots provide a quantitative analysis of various intermolecular interactions, including close contacts in the crystal [46, 47].

29.2.1.2 Crystal Structure Prediction

As stated by Cruz-Cabeza et al. [27], targeted polymorphism remains out of reach for molecular organic crystals, such as pharmaceuticals. In fact, given the unique features of every compound, we have no means of knowing the number of crystal forms that can exist, nor the number or type of experiments that need to be carried out. Structural informatics and risk

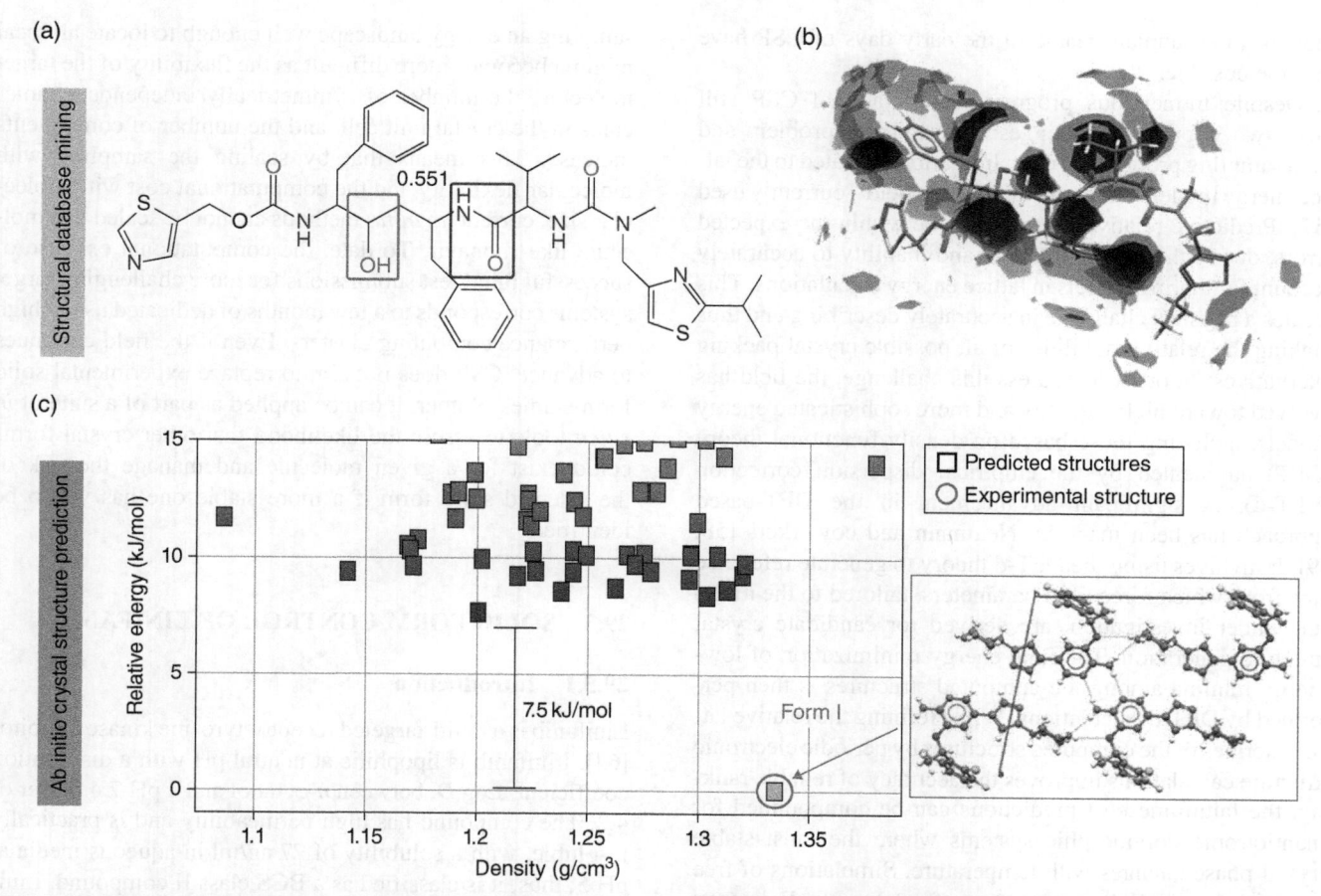

FIGURE 29.5 *In silico* tools for solid form de-risking: (a) donors and acceptors in ritonavir molecule with the highest propensity of the hydrogen bonding observed in Form II predicted between the amide donor and the hydroxyl acceptor, (b) full interaction map for ritonavir Form II, and (c) crystal energy landscape of the model pharmaceutical crizotinib *Source*: Adapted with permission from Abramov [43]. Copyright © 2013 American Chemical Society.

experimental methodologies described above can help reduce risk within the limitations of the approaches but not eliminate it. *Ab initio* CSP is in principle the most comprehensive approach to explore the complete form landscape and fully understand risks. The field has made significant strides since the gauntlet was thrown down by Dunitz in the early 1990s [48].

CSP attempts to predict all the possible crystal structures of a molecule, given only the two-dimensional chemical structure of the compound in question. The most general and commonly applied method that has been developed for CSP is global lattice energy minimization [49]. The approach involves locating and assessing the relative stabilities of all local energy minima on the lattice energy landscape. The result is an assembly of plausible crystal structures (i.e. a crystal energy landscape) that are ranked in order of their lattice energy, computed at absolute zero temperature, and separated by their density, as shown in Figure 29.5c. In this example, the lowest-energy predicted structure of the pharmaceutical molecule crizotinib corresponds to the experimental crystal structure and is approximately 7.5 kJ/mol below the next lowest-energy structure [43]. A key assumption of CSP is that the structure corresponding to the global minimum is the most likely observed structure [50]. Because the lattice energies of possible crystal structures in a given molecule are often found to be within a few kilojoules per mole [51], ranking crystal structures can be a challenge. However, CSP can be applied to identify and rank potential polymorphs of a given crystal form by lattice energy.

Many theoretical methods of lattice energy calculations have been developed. The performance of different approaches has been assessed in six blind tests hosted by the CCDC [50, 52–56], where predictions are performed in advance of the real crystal structures being made available. Genuine successes in predicting crystal structures have been achieved during these blind tests. The latest blind test [56], held in 2016, shows the continuing development and increased maturity of the computational methods with more challenging target systems included in the test, such as two large conformational flexible molecules, a salt hydrate, and a cocrystal as multicomponent systems. All experimental crystal structures of the target systems but one were predicted by at least one submission. Thus, tasks that were thought to be unmanageable in the early days of CSP have become possible.

Despite tremendous progress, computational CSP still faces two important challenges: (i) a ranking problem and (ii) a sampling problem. A main limitation is related to the lattice energy model – the atom–atom force field – currently used [57]. Predictive relative energies can be within the expected errors, due to the model potential and inability to accurately account for thermal effects in lattice energy calculations. This creates a physical challenge in accurately describing and thus ranking the relative stabilities of all possible crystal packing alternatives. In order to address this challenge, the field has evolved toward higher quality and more sophisticated energy models, including those based on density functional theory (DFT) augmented by the empirical dispersion correction (DFT-d). A significant advancement in the DFT-based approach has been made by Neumann and coworkers [58, 59]. It involves using the DFT-d theory to generate reference data from which force field parameters, tailored to the molecule under investigation, are derived for candidate crystal structure generation. The final energy minimization of low-energy minima among the computed structures is then performed by DFT-d calculations. While refining the relative lattice energies of the computed structures by periodic electronic structure calculations improves the accuracy of reliable ranking, the faithfulness of predictions can be compounded for enantiotropic polymorphic systems where the most stable crystal phase changes with temperature. Simulations of free energy tackle the finite temperature effect directly. Enhanced molecular dynamics algorithms, based on multiscale modeling, have been developed with the aim of addressing theoretically challenging questions on computed crystal energy landscape at ambient temperature with high accuracy [60, 61]. Benchmark free energy calculations of small rigid molecules, such as benzene and naphthalene, have proven to be successful for generating and thermodynamically ranking their crystal structures [62]. It is worth emphasizing that, at present, free-energy-based approaches can only handle molecules with significantly low structural complexity compared with lattice-energy-based methods.

A second challenge relies on finding all low minima locations on the multidimensional lattice energy surface, which is defined by the degrees of freedom or structural variables of a system. Navigating and exploring the complex energy surface is not a trivial task. Most of the plausible crystal structures for a molecule are generated based on the input molecular connectivity and stoichiometry and the user-specified range of space groups and number of independent molecules in the unit cell. In order to determine the relevant local minima, the vast majority of studies have utilized the random sampling approach, which has proven to be fairly effective with structures up to 20 degrees of freedom [63]. This number of degrees of freedom covers structure generation for a rigid molecule with up to two independent molecules in any space group. Nevertheless, it is known that sampling an energy landscape well enough to locate all local minima becomes more difficult as the flexibility of the target molecule, the number of symmetrically independent molecules in the crystal unit cell, and the number of components increase. This means that by scaling the sampling with molecular flexibility and the computational cost with molecular size, current *ab initio* methods cannot be scaled to a molecule like ritonavir. To date, the computational expense of successful blind test submissions for more challenging target systems corresponds to a few months of dedicated use of high performance computing clusters. Even if the field continues to advance, CSP does not aim to replace experimental solid form studies. Rather, it can be applied as part of a suite of *in silico* tools to assess the likelihood that other crystal forms could exist for a given molecule and manage the risk of the selected solid form if a more stable one has yet to be identified.

29.3 SOLID FORM CONTROL OF LINIFANIB

29.3.1 Introduction

Linifanib is a multi-targeted receptor tyrosine kinase inhibitor [64]. Linifanib is lipophilic at neutral pH with a distribution coefficient, Log D, between n-octanol and a pH 7.4 buffer of 4.2. The compound has high permeability and is practically insoluble, with a solubility of 27 ng/ml in aqueous media at pH 5; thus, it is classified as a BCS class II compound. Linifanib exhibits an exceedingly slow dissolution rate, suggesting that oral absorption through the membrane is dictated completely by dissolution (i.e. the absorption is dissolution rate limited). Simply increasing the dissolution rate via salt formation or particle size reduction may not be sufficient to achieve the desired bioavailability. An ASD-based formulation will be required for overcoming problems associated with dissolution rate absorption, as it ensures that the drug is present in a reservoir with a low barrier to permeate through the membrane. Linifanib is formulated as ASD by hot melt extrusion technology for enhancing solubility, thereby improving dissolution rate and oral bioavailability.

The molecular structure of linifanib, shown in Figure 29.6, contains hydrophobic aromatic rings as well as polar

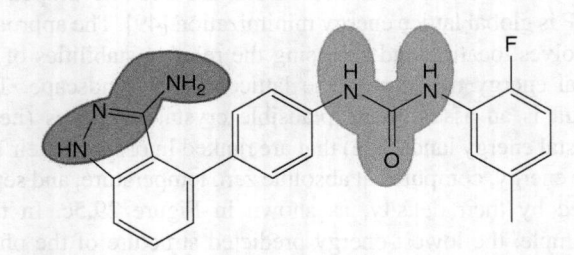

FIGURE 29.6 Molecular structure of linifanib, highlighting donor–acceptor functional groups.

substituents (i.e. urea and indazole groups), which provide hydrogen bonding donating and accepting capacity. The presence of rotatable bonds permits the molecule to exist in a number of different plausible conformations. In principle, these features of the molecular structure should create the opportunity for linifanib to form a variety of potential molecular packing motifs, giving rise to different polymorphic forms. This case study illustrates the application of systematic yet comprehensive solid form screening experiments coupled with *in silico* modeling tools for solid form risk assessment. This assessment includes crystallographic knowledge and structural informatics toward selecting a solid form that possesses desirable solid state and physicochemical properties and also enables both the drug substance quality and an optimal drug product dosage form.

29.3.2 Structural Analysis of Linifanib Ethanol Solvate

29.3.2.1 Crystal Structure
During the early stage of development, various isomorphic solvated crystalline forms, including ethanol and toluene solvates, were discovered as part of the initial solid form screening. These solid forms were characterized by single-crystal X-ray structure determination. The crystal structure of the ethanol solvate, named as Form I, is shown in Figure 29.7a. Form I crystallizes in the triclinic lattice system and with the space group P-1, wherein linifanib molecules form a layered structure. Solvent molecules are arranged in "pockets" between drug chains. The spacing between drug chains can increase in order to accommodate solvent molecules larger than ethanol, thus accounting for the ability of linifanib to form structurally similar solvates with various organic solvents.

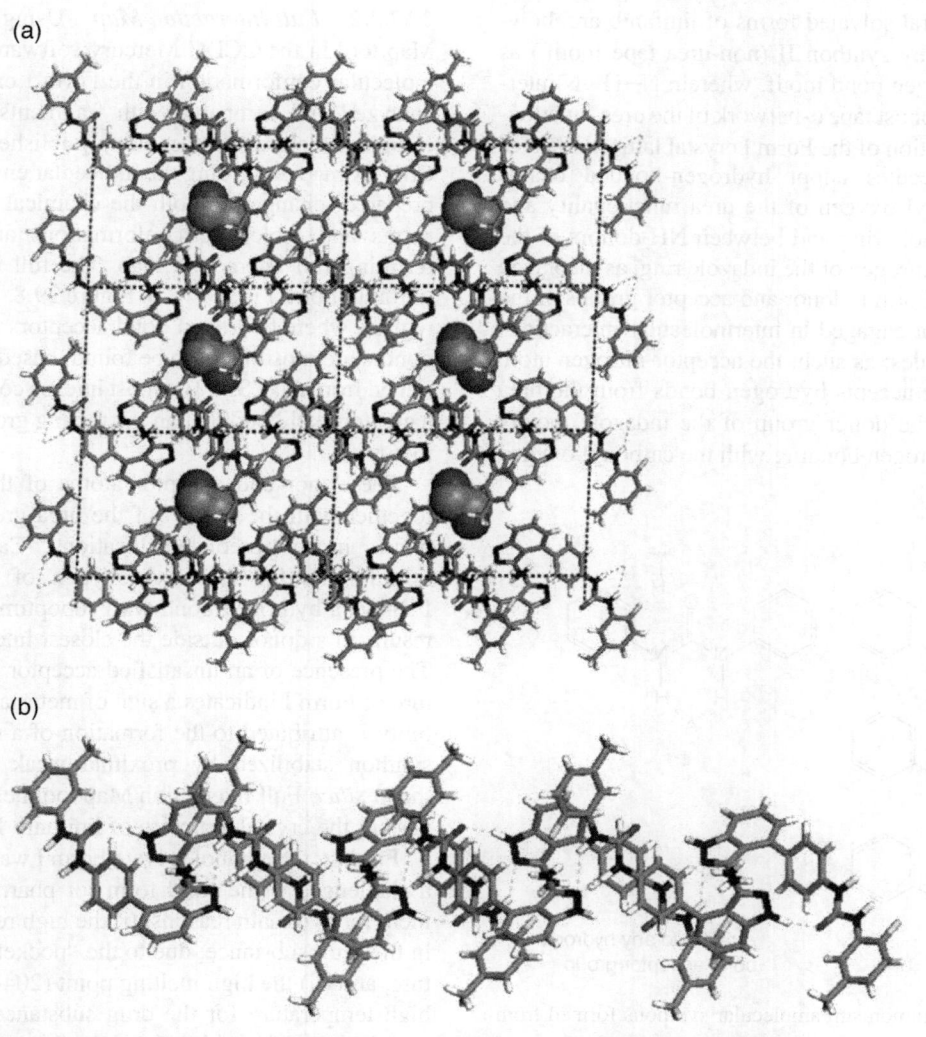

FIGURE 29.7 (a) Molecular packing and (b) intermolecular hydrogen bonding network, highlighted as dashed lines, of linifanib ethanol solvate Form I.

The analysis of the strength and directionality of intermolecular interactions in a crystal structure usually allows the classification of molecular organic crystals depending upon the type of their basic structural motif or synthon. Supramolecular synthons express the core features of a crystal structure; as such, they are considered a reasonable approximation of the entire crystal [65]. Linifanib contains a N-N'-diaryl urea moiety. Even though the urea functional group has only one acceptor atom, the carbonyl oxygen, it can form N—H···O interactions with two neighboring molecules. Crystal structures characterized by the urea functional group can be principally classified into two categories depending upon the common hydrogen bond pattern or supramolecular synthon: the urea tape α-network (synthon I) and the non-urea tape structure (synthon II), as displayed in Scheme 29.1. Typically, synthon I (i.e. urea···urea hydrogen bond motif) is known to be the strongest and most dominant motif of the urea functional group [66]. In the urea tape α-network, the acceptor oxygen atom receives hydrogen bonds from two equivalent donor groups, thus forming a cyclic eight-member ring. The isostructural solvated forms of linifanib are, however, characterized by synthon II (non-urea tape motif) as the dominant hydrogen bond motif, wherein N—H···N interactions disrupt the robust tape α-network of the urea functionality. A close inspection of the Form I crystal lattice indicates that linifanib molecules adopt hydrogen-bonded chains between the carbonyl oxygen of the urea functionality and the NH of the indazole ring and between NH donors of the urea group and the nitrogen of the indazole ring, as displayed in Figure 29.7b. In Form I, donor and acceptor groups of the indazole ring are not engaged in intermolecular interactions with solvent molecules; as such, the acceptor nitrogen atom of the indazole ring accepts hydrogen bonds from the urea NH donors, while the donor group of the indazole moiety is involved in a hydrogen bonding with the carbonyl oxygen of the urea functional group. These interactions alter the structural arrangement of linifanib molecules and result in a synthon that prevents the geometry and topology of the strong hydrogen bonding urea tape motif.

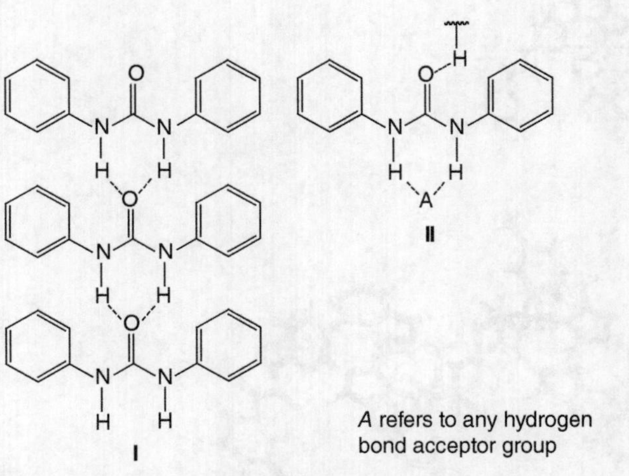

The hydrogen bond motif present in the crystal structure of Form I is thought to be related to the molecular conformation. The urea moiety in the Form I crystal structure shows the characteristic feature of adopting a near-planar conformation with N–Ph torsional angles of 176.1° and 158.3°. This conformation is further stabilized by an intermolecular interaction with the indazole ring of a neighboring drug molecule. Thus, in the Form I crystal structure, drug molecules reside in a stable conformation with approximate coplanarity between the urea group and the aryl rings, but do not generate the strong hydrogen bonding of the urea tape α-network. To go beyond the functional group-based viewpoint, intermolecular interactions of the whole molecule in the crystal structure of linifanib Form I were then evaluated using the Full Interaction Map tool.

29.3.2.2 Full Interaction Map

Using the Full Interaction Map tool in the CCDC Mercury software package [67], the molecular conformation in the Form I crystal structure was analyzed and compared with fragments exhibiting similar chemical features to the many published structures in the CSD. A map describing the molecular environment was then prepared; changes in both the chemical nature of the fragments and molecular conformations are reflected in the resulting full interaction map. The full interaction map for linifanib Form I is shown in Figure 29.8. The clouds indicate regions where hydrogen bond acceptors and hydrogen bond donors are most likely to be found, based on interaction data mined from the CSD. The most intense contour maps indicate a greater likelihood that an interacting group will be found in that region.

The donor and acceptor atoms of the indazole moiety together with the donors of the urea group form hydrogen bonds in the predicted locations. Careful investigation reveals that the carbonyl oxygen of the urea in Form I shows a hydrogen bond with suboptimal geometry, which results in a donor outside the closest interaction map cloud. The presence of an unsatisfied acceptor in the crystal structure of Form I indicates a sign of metastability. The metastability is attributed to the formation of a rare hydrogen bond synthon stabilized by proximal weak interactions. Thus, the *in silico* Full Interaction Map tool helps confirm a weakness in the crystal structure of linifanib Form I.

Further, the ethanol solvate Form I was found to represent a challenge, as the solid form for pharmaceutical development for two main reasons: (i) the high residual solvent level in the drug substance, due to the "pocket-like" crystal structure, and (ii) the high melting point (204 °C) that results in a high temperature for the drug substance dissolution in the extrudate matrix with potential high risk of processability

A refers to any hydrogen bond acceptor group

SCHEME 29.1 Common supramolecular synthons formed from urea functional groups.

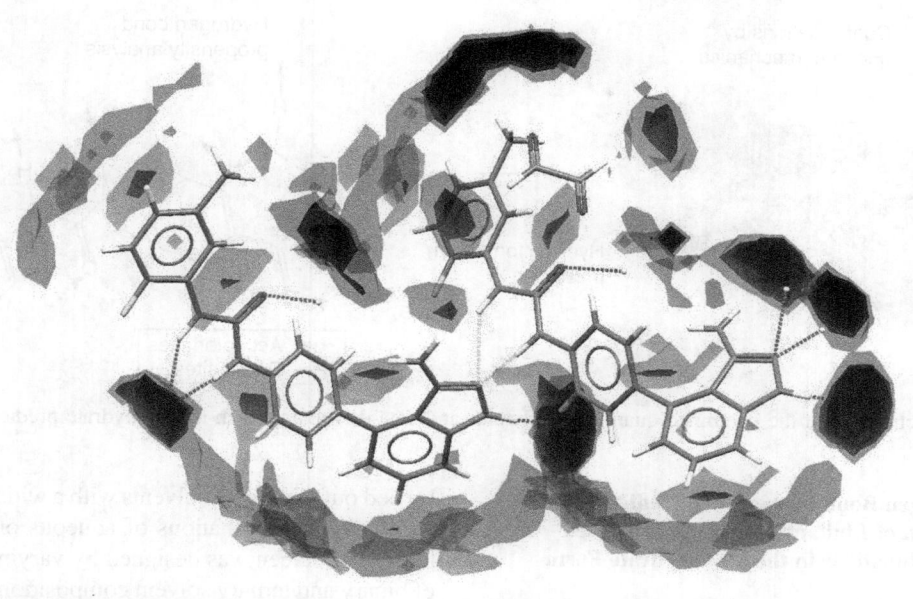

FIGURE 29.8 Full interaction map around molecular conformations of linifanib ethanol solvate Form I.

for the hot melt extrusion. Therefore, based on structural informatics principles, the statistical treatment of intermolecular interactions in the Form I crystal structure, together with its solid-state properties and drug product manufacturability, warranted the search for alternative solid forms, including low-melting anhydrates or hydrates with low dehydration temperatures. *In silico* methods for the hydrate formation prediction have proven to be of high value in guiding a more systematic experimental screening. In the following section, additional details on the approach and its application to linifanib are summarized.

29.3.3 Linifanib Hydrate Formation: Computational and Experimental Approaches

The solid form screening conducted early in drug development did not yield crystalline hydrates, despite including crystallization conditions favoring their formation. As such, computational approaches were undertaken to predict the hydration propensity of linifanib. Computational and statistical models have been used to rationalize hydrate formation of molecular organic crystals [68–70]. The free energy of mixing of the molecule of interest with water in a supercooled liquid phase has been shown to be an effective descriptor of the propensity of small molecule drug candidates to form hydrates [71]. Hydrogen bond propensity analysis has been also utilized to understand hydration in molecular organic crystals [40]. However, two main considerations are (i) the probability of low-energy geometrical distortions of a molecule from its ideal geometry and (ii) the existence of multiple conformers on the intramolecular energy surface. Building upon these concepts, a method based on molecular conformation estimation and hydrogen bond propensity has been developed for predicting hydrate formation [72]. Gauging molecular conformations representative of the solid state can guide a more accurate description of intermolecular hydrogen bonding interactions in hydrates and anhydrous crystals, because intramolecular interactions and sterically inaccessible polar regions can significantly affect whether intermolecular interactions can form in the solid state.

A combined quantum mechanics and data-driven approach, according to the schematic illustrated in Figure 29.9, has been developed and utilized to predict the hydrate formation of linifanib. Conformers were generated using quantum mechanical calculations by the conductor-like screening model for realistic solvation (COSMO-RS). This method adopts a quantum approach in combination with statistical thermodynamics [73, 74]. Donor and acceptor groups available for intermolecular interactions were analyzed by evaluating representative molecular conformations. API–water and API–API propensities for each donor/acceptor group were determined to provide insights on potential hydrate formation. Hydrogen bond propensities, modeled by a probability function using data mined from relevant CSD crystal structures, were predicted for linifanib fragments. API–water and API–API propensities were evaluated in a hierarchical manner by pairing the best donor with the best acceptor, the next best donor with the next best acceptor, and so on. The results of the analysis are presented in Table 29.1. The strongest donor–acceptor pair is between the primary amine and water, followed by water and the urea group, with propensity values of 0.71 and 0.70, respectively. The strongest API–API propensity is between the primary amine and the urea groups with a value of 0.76. Since each functional group is allowed to donate and/or accept once, another strong

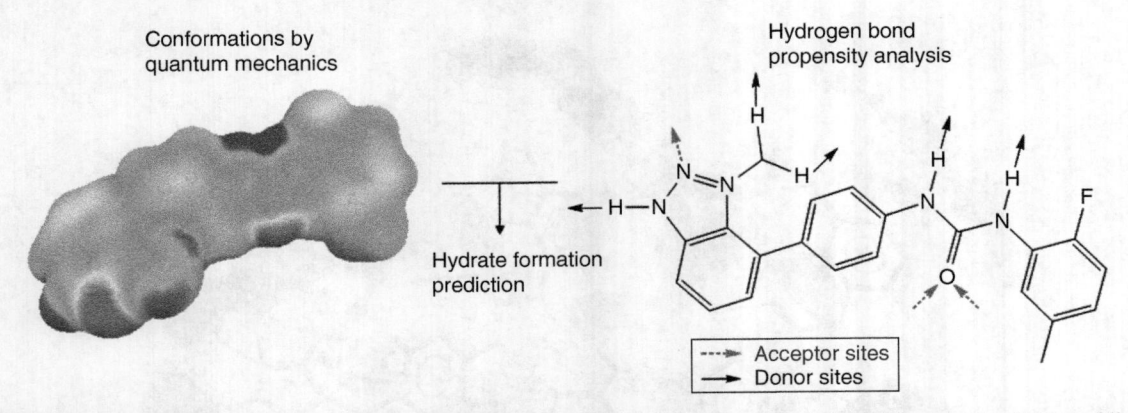

FIGURE 29.9 Schematic of the combined quantum mechanics and data-driven approach for the hydrate prediction of linifanib.

TABLE 29.1 Hydrogen Bond Propensities Predicted for the Hydrate Formation of Linifanib Molecule. Observed Donor–Acceptor Combinations in the Monohydrate Form V Are Indicated

Donor	Acceptor	Propensity	Form V
NH amine	O urea	0.76	X
NH amine	O water	0.71	√
OH water	O urea	0.70	X
NH amine	N indazole	0.67	X
OH water	O water	0.65	X
OH water	N indazole	0.60	√
NH urea	O urea	0.58	√
NH indazole	O urea	0.55	X
NH urea	O water	0.51	X
NH indazole	O water	0.48	√
NH urea	N indazole	0.46	X
NH indazole	N indazole	0.44	√

Note: √, observed; X, not observed.

API–API interaction is between urea and indazole groups. Using hydrogen bond propensities, a multi-differential hydrogen bond propensity score can be calculated, according to the equation

$$\text{Score} = \sum_{D,A} \text{HBP}_{\text{API-API}} - \text{HBP}_{\text{API-water}} \quad (29.1)$$

The overall hydrogen bond propensity score of the model is −0.08. A negative score indicates that hydrate formation is favorable for linifanib. This prompted the initiation of a more systematic solid form screen with the intent to find developable crystalline hydrates.

The solid form screening of linifanib required overcoming challenges presented by its poor aqueous solubility properties and the relative ease with which the molecule seems to crystallize as solvates. The screen comprised mainly solution-based recrystallizations and included both thermodynamic and kinetic approaches. The thermodynamic screening was carried out in organic solvents with a wide range of properties and various combinations of aqueous organic solvent mixtures. The screen was designed by varying the water content of binary and ternary solvent compositions as a means of regulating the water activity, so as to favor the formation of potential hydrates. For the kinetic screening, stressed crystallization experimental conditions were employed (e.g. elevated supersaturation ratio, high antisolvent addition rate, forward/reverse antisolvent addition, high shear rate, and/or fast cooling rate) with the aim of identifying metastable forms. These targeted experiments resulted in the isolation of a monohydrate crystal form, designated as Form V.

29.3.4 Structural Analysis of Linifanib Monohydrate

29.3.4.1 Crystal Structure In the crystal structure of the monohydrate Form V, linifanib molecules are arranged in a linear array via the urea tape α-network, as shown in Figure 29.10. Intermolecular interactions between the indazole rings contribute to hydrogen-bonded chains, while water molecules interact within the indazole hydrogen bonding network.

Molecules in the Form V crystal structure adopt a twisted conformation with N–Ph torsional angles of −132.6° and 147.6°. An overlay of Form I and Form V conformers is displayed in Figure 29.11. The rotation of the N–Ph moiety in the molecule, from a near-planar conformation (as observed in Form I) to a twisted conformation (as observed in Form V), is required for the urea tape synthon I self-assembly. In order to form the urea tape α-network, the molecule has to take an energy penalty by rotating around the N–Ph bond that is energetically compensated by forming a strong intermolecular hydrogen bond. This implies that the molecular conformation is related to the urea tape α-network and non-urea hydrogen bonding synthon. This leads to the idea of exploiting the hard and soft acids and bases (HSAB) principle [75] to evaluate intermolecular interactions and understand molecular assembly in molecular organic crystals. It has been shown that the

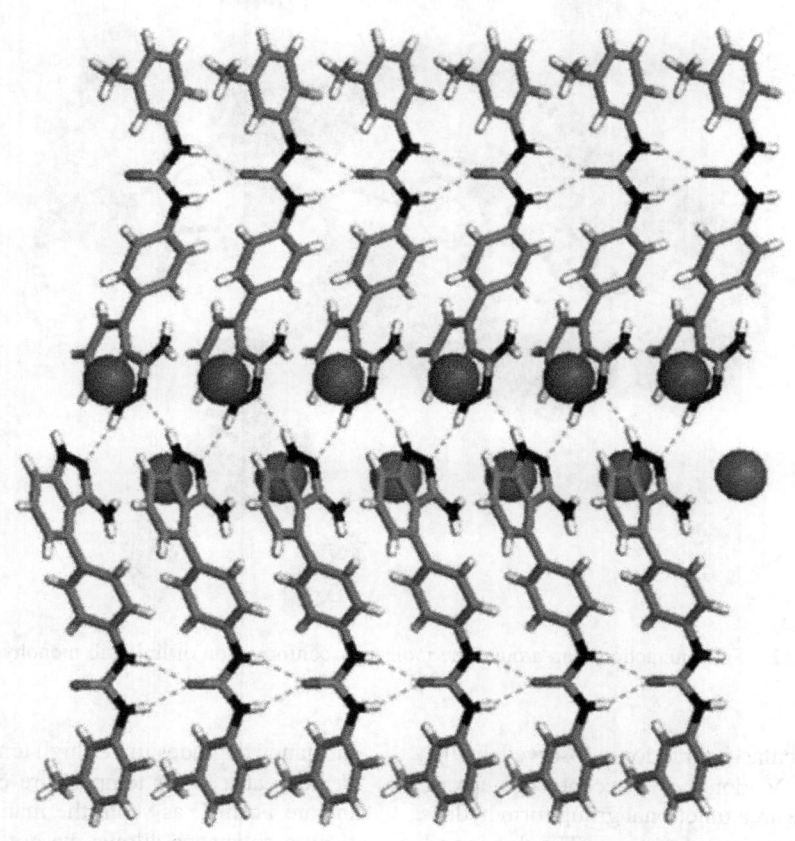

FIGURE 29.10 Molecular packing and intermolecular hydrogen bonding network, highlighted as dashed lines, of linifanib monohydrate Form V.

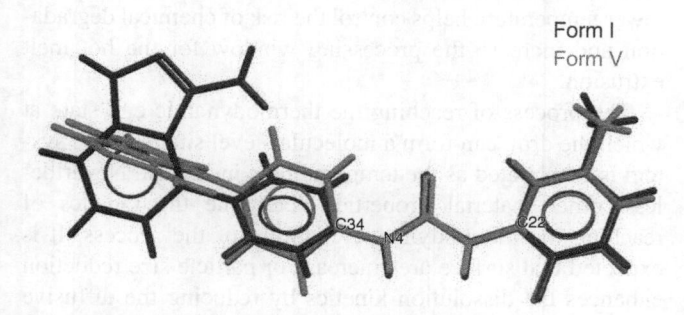

FIGURE 29.11 Overlay of Form I and Form V conformers. The three pairs of atoms used for the overlay are highlighted.

twisting of the aryl groups in flexible diaryl urea structures makes the carbonyl oxygen of the urea functional group a better hydrogen bond acceptor [76], able to interact with the strong urea NH donors and form the urea tape array. Thus, there is a key link between the intrinsic molecular structure and the crystal packing, allowing a profile of important interactions to be built up within families of compounds. To provide additional support for intermolecular packing effects, two CSD-based structural informatics approaches – hydrogen bond propensity and full interaction map – have been applied.

29.3.4.2 Hydrogen Bond Propensity As discussed previously, a predictive indicator of whether a hydrogen bonding interaction is unusual is vital in assessing solid form stability during pharmaceutical product development. Table 29.1 shows hydrogen bonding interactions in the known monohydrate Form V compared with the predictions of a set of potential donor–acceptor pairs, as an assessment of solid form stability. In the known hydrate crystal structure, the observed hydrogen bonding between the amine group of linifanib and the oxygen atom of the water molecule is ranked the second most likely donor–acceptor pair (0.71) and shows a hydrogen bond propensity higher than that predicted between water molecules. The results suggest that intermolecular interactions between linifanib and water molecules are favored compared with those between water molecules. This may also imply that the direct formation of linifanib hydrate is possible when water is used as the crystallization solvent.

29.3.4.3 Full Interaction Map A full interaction map was computed for linifanib Form V using data from fragments found in the CSD. The results are shown in Figure 29.12. The urea and indazole groups in Form V have distinct regions of interaction density located near them. By

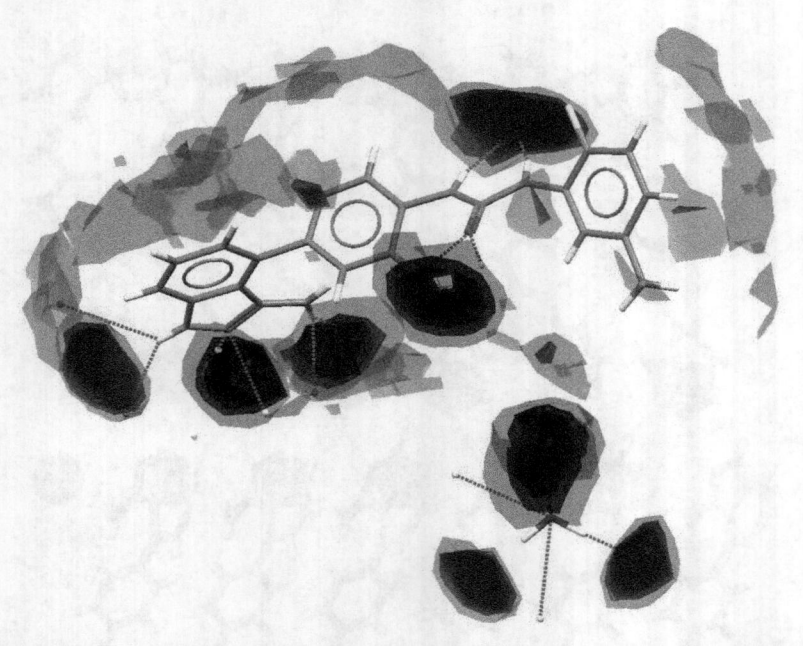

FIGURE 29.12 Full interaction map around the molecular conformation of linifanib monohydrate Form V.

overlaying the intermolecular interactions observed in the crystal structure of Form V, donor and acceptor groups of the indazole group and the urea functional group form hydrogen bonds in the exactly predicted locations. The donor and acceptor groups of the water molecule are also satisfied and fit the map well. Overall, the crystal structure of linifanib Form V matches the predicted interaction geometry. Thus, the known monohydrate crystal form exhibits a satisfactory network of intermolecular interactions. This supports the hypothesis that a different molecular packing would not likely result in a more stable hydrate form.

29.3.5 Solid Form Selection and Impact on the Downstream Processing

Linifanib monohydrate Form V was selected as the solid form for continued development based on its favorable solid state and physicochemical properties. Obtaining an API in the right solid form and with consistent physical properties is critical not only from the drug substance manufacture standpoint but also from the perspective of drug product processing, performance, and stability.

The high-energy amorphous state of the hot melt extrusion process features the dissolution of the drug substance into the polymer matrix, which is facilitated by applying shear stress and thermal energy to a powder blend. During processing, drug substances are thereby exposed to elevated temperatures for prolonged periods of time. High processing temperatures together with shear stresses and melt viscosities can induce decomposition of thermally unstable drugs. Linifanib drug substance degrades under high temperatures. Linifanib Form V melts at a lower temperature compared with the ethanol solvate Form I; as such, the manufacture of linifanib Form V drug substance allows the hot melt extrusion technology to operate at a reduced temperature in order to molecularly dissolve the drug substance into the polymer matrix and achieve a homogeneous ASD. The ability to operate at a lower temperature helps control the risk of chemical degradation and increase the processing window for the hot melt extrusion.

The process of reaching the thermodynamic end state at which the drug can form a molecular-level single-phase system is accelerated as the temperature is increased. Nevertheless, other material properties determine the kinetics of reaching the thermodynamic endpoint of the process. It is expected that surface area increase or particle size reduction enhances the dissolution kinetics by reducing the diffusive mixing required for dissolution. Mathematically, this can be described by the traditional Noyes–Whitney equation [77], wherein the dissolution rate or rate of mass transport is a function of the difference in solubility of the drug in the polymer, C_s, and the concentration of the drug dissolved in the polymer, $C(t)$; the diffusion coefficient, D; the surface area, A; and the boundary layer thickness, h, as follows:

$$\frac{dM}{dt} = \frac{DA}{h}(C_s - C(t)) \quad (29.2)$$

It becomes apparent that the dissolution rate can be enhanced by increasing the surface area of the components through effective dispersion or decreasing the boundary layer

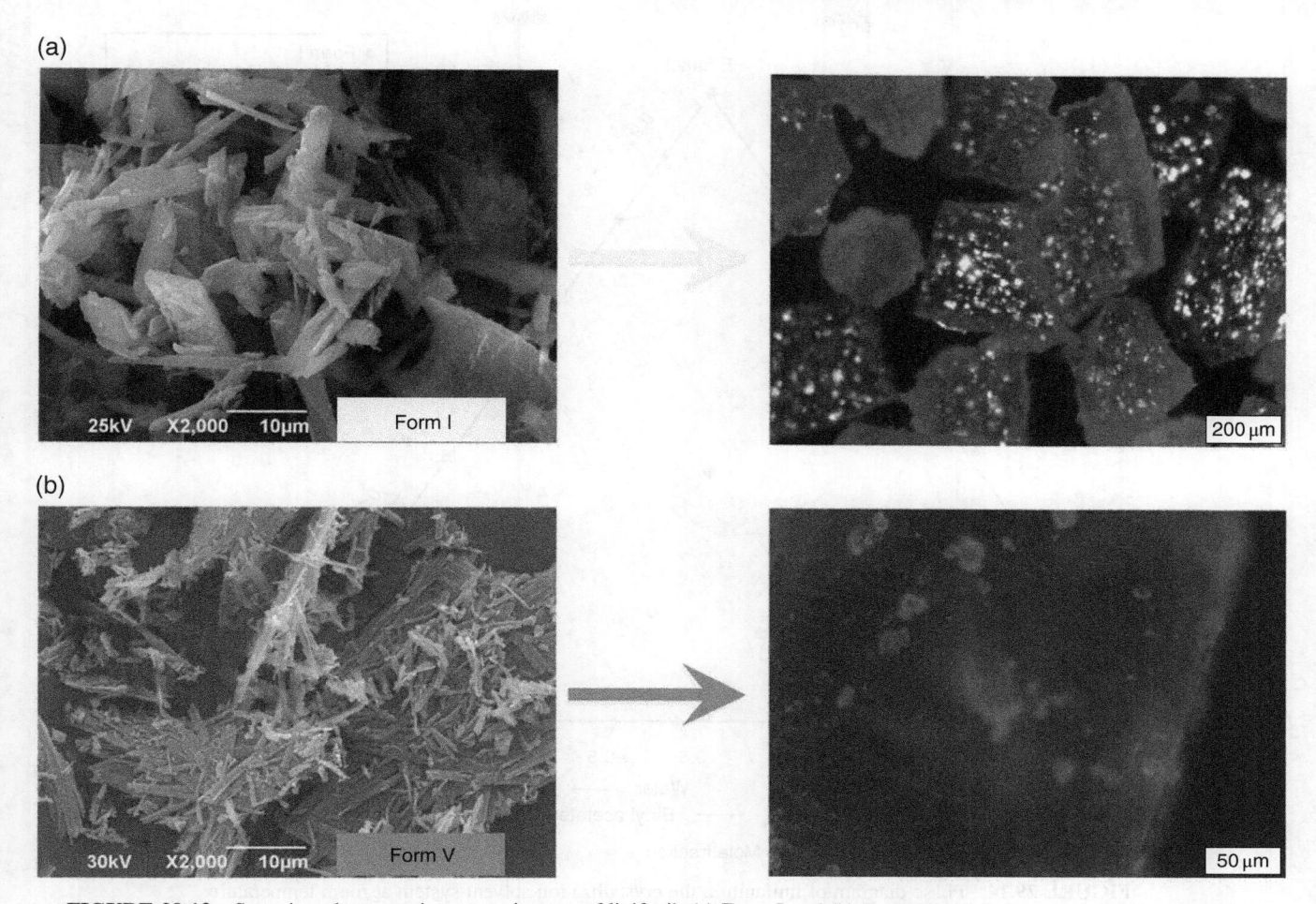

FIGURE 29.13 Scanning electron microscopy images of linifanib (a) Form I and (b) Form V and their corresponding extrudates.

thickness through shearing of the particles. Linifanib Form V crystallizes as small and thin needlelike particles, resulting in high surface area. This is beneficial for the complete dissolution of linifanib drug substance in the hot melt extrusion process. The resulting extrudates, manufactured using Form V, were crystal-free, while those manufactured using Form I were found to contain small amounts of residual crystallinity, as illustrated in Figure 29.13. The incomplete suppression of the drug crystallinity, as observed by polarized light microscopy, is attributed to the specific surface area and particle morphology of the drug substance.

29.3.6 Manufacturing Process of Linifanib Monohydrate

Based on slurry competition studies, linifanib Form V was established to be thermodynamically stable only at high water activities (>0.9). This represents a significant challenge for a poorly soluble molecule in terms of solvent selection and conceptual design of the crystallization process. Based on experiments conducted in various solvent compositions to evaluate solubility and purification potential, a crystallization solvent system comprising water, ethanol, and ethyl acetate was selected. The relative stability of process-relevant solid forms in the ternary solvent system was assessed based on the equilibrium solubility, which was determined by high performance liquid chromatography, and the solid phase, which was characterized by X-ray powder diffraction. The phase diagram built in the ternary solvent system, based on the mole fraction of solvent mixtures, shows that either an anhydrate crystal form (Form VII) or a mixture of polymorphs was obtained in most of the solvent region, as displayed in Figure 29.14. This implies that most of the solvent domain is not appropriate for the consistent isolation of Form V under thermodynamic control. Despite challenges related to the very tight region involved in Form V stability, the low solubility, and chemical instability in many organic solvents, a crystallization process was designed to selectively yield the monohydrate Form V by direct precipitation without interferences from Form I and/or Form VII. The developed crystallization process consisted of charging a solution of linifanib in ethanol/water to water for seed

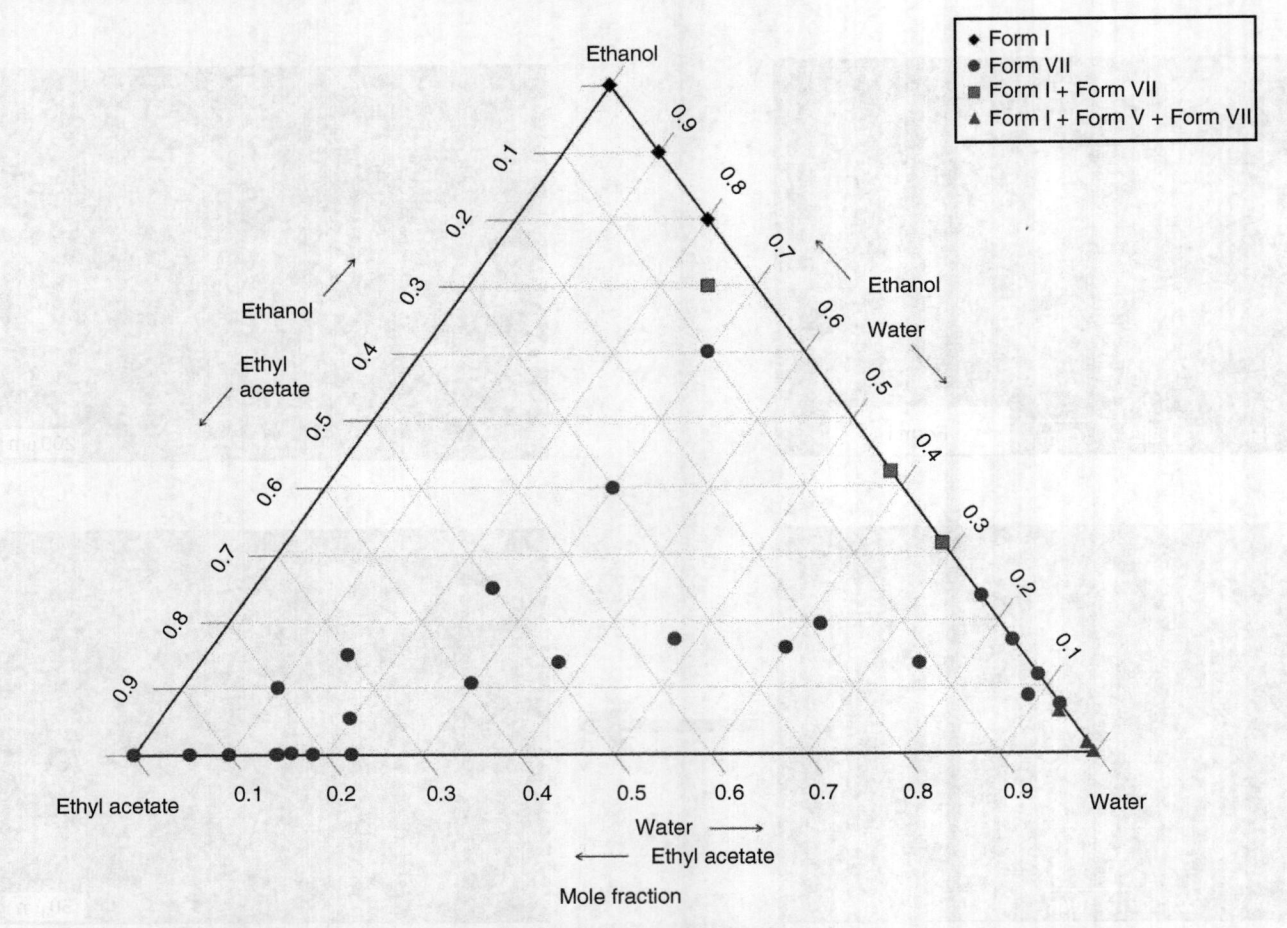

FIGURE 29.14 Phase diagram of linifanib in the crystallization solvent system at room temperature.

generation. The seed slurry was added to a concentrated solution of linifanib with an appropriate ethyl acetate/water composition. The product slurry was then charged to water while distilling to remove organic solvents and maintain a water-rich solvent composition. The product was isolated by filtration, washed with water, and then dried. During the drying process, the integrity of the hydrate state of the desired solid form was carefully studied, and appropriate controls were established. Appropriate storage conditions were also instituted to preserve the monohydrate Form V, as it is stable at a relative humidity equal to or greater than 30% at 25 °C.

29.3.7 Summary

A multidisciplinary approach, involving solid form screening experiments and *in silico* modeling tools as well as the correlation with information extracted from a given single-crystal structure, is of paramount importance to better understand the structure and function of small pharmaceutical molecules. Through tailored designed experiments and *in silico* modeling tools, a fundamental understanding of the structure, thermodynamics, and kinetics of process-relevant solid forms has been gained for the poorly soluble compound linifanib. When multiple crystalline forms are identified from solid form screens, it is the connection between crystal structure, solid-state properties, drug substance, and drug product processing and performance that ultimately determine which form advances the development of the drug product. A metastable form of linifanib was selected for development to meet the nature or requirements of the drug product, thus facilitating the enabling formulation technology. This case study thus emphasizes the importance of applying integrated evaluation of drug substance properties and drug product processing technology and their interplay to enable rational design and development of crystalline drugs with poor aqueous solubility.

29.4 SOLID FORM CONTROL OF DASABUVIR

29.4.1 Introduction

Dasabuvir is a non-nucleoside inhibitor of the hepatitis C virus (HCV) NS5B polymerase. The parent compound of dasabuvir, shown in Scheme 29.2, is lipophilic with the

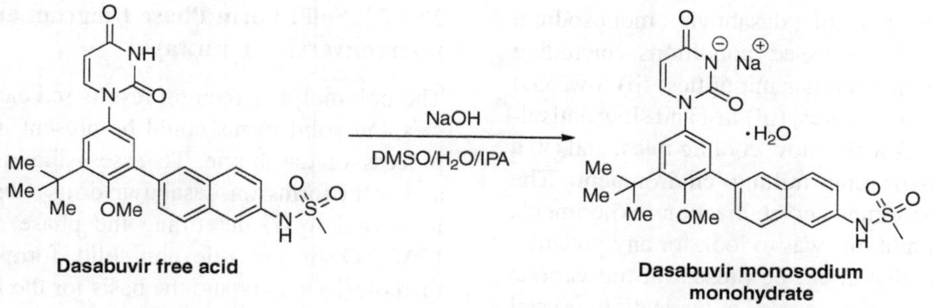

SCHEME 29.2 Preparation of dasabuvir monosodium monohydrate.

distribution coefficient, Log D, between octanol and a pH 7.4 buffer of 4.5 and exhibits low aqueous solubility (0.15 μg/ml at pH 7.4, 25 °C). Due to its low aqueous solubility, good permeability by Caco-2 assay (~40×10^{-6} cm/s), and projected human dose (400 mg per day), the free form of dasabuvir is classified as a BCS class II compound. It is a highly crystalline solid exhibiting a high melting point (>200 °C), which limits the feasibility of the hot melt extrusion technology as an enabling formulation.

Based on *in vitro* tests, the dissolution of the parent form of dasabuvir in aqueous media is extremely slow. Correspondingly the compound exhibits low bioavailability (9.3%), as determined by *in vivo* dog PK studies. It was, therefore, desirable to conduct salt screening to identify a faster dissolving solid form that enables formulating the compound into a conventional oral dosage form. The free form of dasabuvir is a weak acid drug with pK_a values of 8.2 and 9.2. The weak acidic functional groups (i.e. uracil and sulfonamide moieties) of the compound limit its salt formation to only a few pharmaceutically acceptable strong bases, including sodium hydroxide, potassium hydroxide, and choline hydroxide. As a result of the salt screening, a monohydrate form of the monosodium salt was discovered and selected as the lead solid form for development based on its favorable solid-state properties, physicochemical properties, and manufacturability. The monosodium salt shows a higher apparent water solubility (0.47 mg/ml) and a faster dissolution rate compared with the parent compound. The dissolution rate enhancement was confirmed in a dog PK study, where the aqueous suspension of the salt achieved a sevenfold increase (72.6%) in bioavailability.

29.4.2 Solid Form Screening and Salt Formation Process

The main purpose of the solid form screening was to verify that the dasabuvir monosodium salt monohydrate selected for development was thermodynamically stable under process conditions. The polymorph screen consisted of thermodynamic and kinetic approaches. Emphasis was placed on identifying anhydrates and hydrates of the sodium salt and its free acid, as well as process-relevant solvates.

In the thermodynamic solid form screening, the starting material dasabuvir monosodium salt monohydrate was suspended in various organic solvents at ambient temperature. Wet cakes were sampled at different time points up to three months. The design of the thermodynamic polymorph screening was based on solution-mediated phase transformation to facilitate the conversion from metastable solids to stable phases (more stable solid species, polymorphs, hydrates, and solvates). The guiding principles for selecting non-process-relevant solvents for the thermodynamic polymorph screen included (i) the selection of a diverse range of solvent properties within a practical number of experiments, (ii) the search for potential anhydrate and/or hydrate forms, and (iii) the consideration of the water-soluble formulation platform.

As the parent compound of dasabuvir is a weak acid, the manufacturing of its sodium salt hydrate poses a few challenges, such as the disproportionation and solid form control of the monosodium salt. In order to design the process and control salt formation to ensure consistent isolation of the desired monosodium salt monohydrate, a series of experiments were conducted to, first, select a solvent system for the conceptual design of the crystallization process and, second, identify process-relevant solid forms. As a result of comprehensive solvent screening studies, a solvent system composed of dimethyl sulfoxide (DMSO), water (H_2O), and isopropanol (IPA) was selected for the crystallization of the monosodium salt. DMSO provides most of the solubility for the API, while IPA and H_2O act as antisolvents, with the latter providing the required water activity for yielding the monohydrate. The general scheme for manufacturing dasabuvir monosodium salt monohydrate from its free acid is represented in Scheme 29.2.

Detailed analysis of solvent compositions at various steps of the process – salt formation, nucleation and growth, and isolation of the monosodium salt monohydrate – was crucial to define the solvent compositions used in the thermodynamic screening. Various solvent combinations of DMSO, IPA, and H_2O were used in the thermodynamic screen as process-relevant solvents. In the kinetic solid form screening, crystallization experiments of dasabuvir monosodium salt were executed under stressed conditions, including (i) spiking with structurally similar impurities, (ii) low seed amounts and holding temperatures, (iii) fast and slow antisolvent addition rates, (iv) fast and slow cooling rates, and (v) a process-relevant high-shearing milling environment. The main purpose of performing crystallization experiments under these stressed conditions was to look for any potential metastable solid forms that could be present at the various stages of the manufacturing process (i.e. nucleation, crystal growth, and isolation of the monosodium salt monohydrate).

As a result of thermodynamic and kinetic screens, the following process-relevant dasabuvir species and solid forms thereof were discovered: monosodium salt monohydrate, monosodium salt monoDMSO solvate, free acid triDMSO solvate, free acid anhydrate, and disodium salt DMSO/H_2O mixed solvate. Specifically, during the polymorph screening and crystallization development, it was observed that the disodium salt DMSO/H_2O mixed solvate could appear as a kinetic form during self-nucleation and/or addition of a low seed amount of monosodium salt monohydrate. The disodium salt DMSO/H_2O mixed solvate converted to the desired monosodium salt monohydrate during the holding period post-seeding addition or IPA/H_2O antisolvent addition. The monosodium salt monoDMSO could be also observed as a kinetic form, being promoted by high-shear milling, and it converted to the monosodium salt monohydrate during IPA/H_2O antisolvent addition. Finally, an uncontrolled crystallization of the monosodium salt could yield a hydrate crystalline form, solvates, or anhydrate of the free acid in organic/water solvent mixtures.

29.4.3 Solid Form Phase Diagram and Interconversion Pathways

The polymorph screening results indicate that multiple species and solid forms could be present in the crystallization process of dasabuvir. To assess the risk of these species and solid forms on dasabuvir drug substance manufacture, it is vital to (i) determine the phase diagram in DMSO/IPA/H_2O of the relevant solid forms and relationships thereof; (ii) understand the basis for the formation of various species, including the free acid and disodium salt solid forms, in the monosodium salt reaction crystallization; and (iii) evaluate how well the crystallization process controls the desired monosodium salt monohydrate.

The dasabuvir solid form phase diagram in the DMSO/IPA/H_2O ternary solvent system at ambient conditions was constructed based on the thermodynamic screening results, as shown in Figure 29.15. A schematic representation of the chemical species interconversion pathways in DMSO/IPA/H_2O solvent system is depicted in Scheme 29.3. Given the complexity of the solid form and species landscape and their interconversions, a detailed assessment is provided.

As shown in both Figure 29.15 and Scheme 29.3, dasabuvir monosodium salt undergoes disproportionation to yield the corresponding free acid and disodium salt in DMSO–H_2O-rich regions. The disproportionation of monosodium salt in a DMSO–H_2O-rich region is likely due to two factors. First, the monosodium salt and its disproportionation products (i.e. free acid and disodium salt) differ in solubility; second, water (pK_a 14) acts as proton donor and acceptor, thus

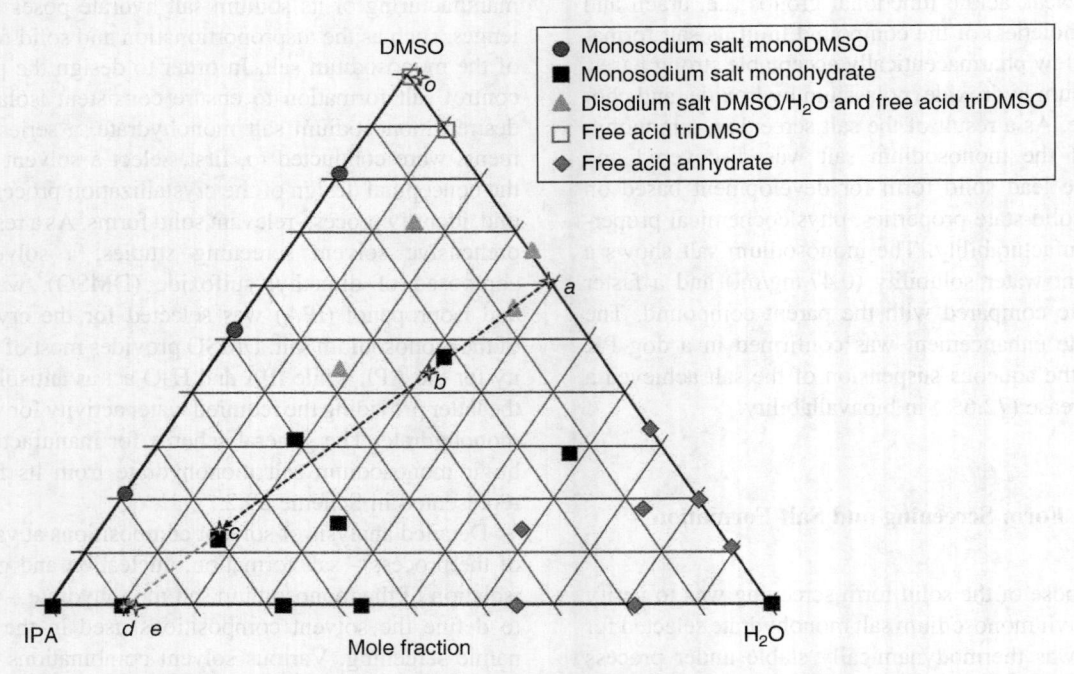

FIGURE 29.15 Phase diagram of dasabuvir in the crystallization solvent system at room temperature.

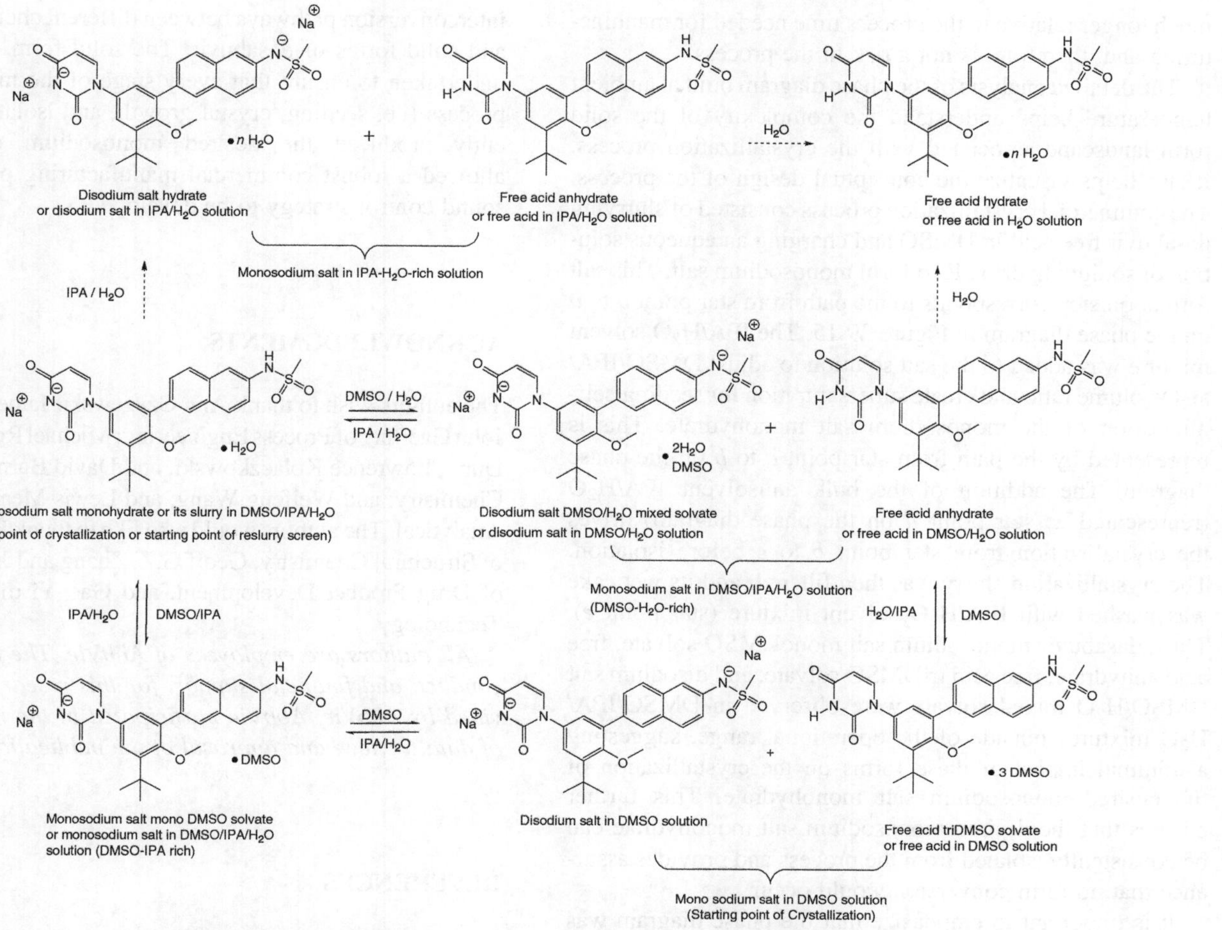

SCHEME 29.3 Interconversion pathways of dasabuvir species and solid forms.

facilitating the disproportionation process. When the DMSO/H_2O volume ratio was high enough, the free acid is converted to triDMSO solvate. Similar behavior occurred in DMSO/IPA/H_2O systems in which DMSO was the dominant component. When the DMSO content was at its extreme (i.e. 100% DMSO), the free acid triDMSO solvate became the least soluble solid form due to very high DMSO activity, and it precipitated out with or without trace amounts of water. In addition, the disodium salt, the other portion of the disproportionation product, could exist as a DMSO/H_2O mixed solvate or remained in solution in the aforementioned DMSO–H_2O-rich solvent regions. When the monosodium salt monohydrate was reslurried in DMSO–IPA-rich regions, the monosodium salt was stable as a monoDMSO solvate and did not undergo disproportionation. Besides solubility reasons, the lack of disproportionation may be due to the fact that IPA (pK_a 16.5) is a less efficient protic solvent than water and thus not able to facilitate the process.

When the monosodium salt monohydrate was reslurried in IPA–H_2O-rich regions, the monosodium salt was stable and remained as the monohydrate crystalline form. This is most likely due to the very limited solubility of the compound in such solvent compositions and to the water activity being not high enough to effectively promote disproportionation of the monosodium salt. However, when the water content was significantly high (e.g. 30% or above), the disproportionation of the monosodium salt to its free form occurred, and thus the free acid anhydrate became more stable. It is noteworthy that in neat water the monosodium salt monohydrate was found to be stable over four weeks. This further confirms that both a good solubility differential between the monosodium salt and free acid/disodium salt and an effective protic solvent are necessary to effectively promote the disproportionation of the monosodium salt. This also holds true in neat IPA where the monosodium salt monohydrate was found to be stable over 12 weeks. Yet, from a theoretical point of view, the monosodium salt should eventually undergo disproportionation in both H_2O and IPA to yield the corresponding hydrate or anhydrate/solvate of the free acid and disodium salt. Practically though, the time frame for this to happen is much longer relative to the process time needed for manufacturing and, therefore, is not a risk to the process.

The detailed analysis of the phase diagram built at ambient temperature helps understand the complexity of the solid form landscape associated with the crystallization process. It also helps visualize the conceptual design of the process. The commercial crystallization process consisted of slurrying dasabuvir free acid in DMSO and charging an aqueous solution of sodium hydroxide to form monosodium salt. This salt formation step corresponds to the path from star point o to a on the phase diagram in Figure 29.15. The IPA/H_2O solvent mixture was added to the salt solution to adjust DMSO/IPA/H_2O volume ratio and create supersaturation for seeding/self-nucleation of the monosodium salt monohydrate. This is represented by the path from star point a to b on the phase diagram. The addition of the bulk antisolvent IPA/H_2O (represented as star point d on the phase diagram) drives the crystallization from star point b to c before isolation. The crystallization slurry was then filtered, and its wet cake was washed with IPA/H_2O solvent mixture (star point e). Thus, dasabuvir monosodium salt monoDMSO solvate, free acid anhydrate, free acid triDMSO solvate, and disodium salt DMSO/H_2O mixed solvate were observed in DMSO/IPA/H_2O mixtures outside of the operational range, suggesting a minimal impact of these forms on the crystallization of the desired monosodium salt monohydrate. This further ensures that the desired monosodium salt monohydrate can be consistently isolated from the process and provides assurance that no form conversion would occur.

It is important to emphasize that the phase diagram was constructed for process-relevant solid forms at ambient temperature. This phase diagram may not necessarily provide an accurate view of the solid phase stability at elevated temperatures, particularly during the initial crystallization steps. Constructing a complete and accurate phase diagram at high temperatures is experimentally challenging given the potential solvent evaporation and the compound stability over a prolonged period of time. Nevertheless, the ambient temperature phase diagram provides highly valuable information regarding potential solid forms present at the beginning of the crystallization step, as well as guidance on how to conduct process control justification work with appropriate solid form focus.

29.4.4 Summary

The salt formation strategy was utilized to improve the bioavailability of the poorly soluble HCV polymerase inhibitor dasabuvir. In order to ensure a robust solid form control in the manufacturing process of dasabuvir monosodium salt monohydrate, comprehensive thermodynamic and kinetic solid form screens were carefully designed and executed. The results from the solid form screening led to the construction of the phase diagram and to the fundamental understanding of interconversion pathways between different chemical species and solid forms of dasabuvir. The solid form development undertaken to ensure that every stage of the manufacturing process (i.e. seeding, crystal growth, and isolation) consistently produced the desired monosodium monohydrate allowed a robust commercial manufacturing process and a sound control strategy to be developed.

ACKNOWLEDGMENTS

The authors wish to thank Ann Czyzewski, James Marek, and John Gaertner of Process Engineering; Michael Rozema, Travis Dunn, Lawrence Kolaczkowski, and David Barnes of Process Chemistry; and Weifeng Wang, and Lewis Meads of Process Analytical. The authors would also like to thank Rodger Henry of Structural Chemistry, Geoff G.Z. Zhang and Xiaochun Lou of Drug Product Development, and Gao Yi of Science and Technology.

All authors are employees of AbbVie. The design, study conduct, and financial support for this research were provided by AbbVie. AbbVie participated in the interpretation of data, review, and approval of the publication.

REFERENCES

1. Loftsson, T. and Brewster, M.E. (2010). Pharmaceutical applications of cyclodextrins: basic science and product development. *Journal of Pharmacy and Pharmacology* 62 (11): 1607–1621.
2. Lipp, R. (2013). The innovator pipeline: bioavailability challenges and advanced oral drug delivery opportunities. *American Pharmaceutical Review* 16 (3): 10–16.
3. Bernstein, J. (2002). *Polymorphism in Molecular Crystals*. New York: Oxford University Press.
4. Chemburkar, S.R., Bauer, J., Deming, K. et al. (2000). Dealing with the impact of ritonavir polymorphs on the late stages of bulk drug process development. *Organic Process Research & Development* 4 (5): 413–417.
5. Bauer, J., Spanton, S., Henry, R. et al. (2001). Ritonavir: an extraordinary example of conformational polymorphism. *Pharmaceutical Research* 18 (6): 859–866.
6. Serajuddin, A.T.M. (2007). Salt formation to improve drug solubility. *Advanced Drug Delivery Reviews* 59 (7): 603–616.
7. Kuminek, G., Cao, F., Bahia de Oliveira da Rocha, A. et al. (2016). Cocrystals to facilitate delivery of poorly soluble compounds beyond-rule-of-5. *Advanced Drug Delivery Reviews* 101: 143–166.
8. Good, D.J. and Rodríguez-Hornedo, N. (2009). Solubility advantage of pharmaceutical cocrystals. *Crystal Growth & Design* 9 (5): 2252–2264.

9. Blagden, N., Coles, S.J., and Berry, D.J. (2014). Pharmaceutical co-crystals – are we there yet? *CrystEngComm* 16 (26): 5753–5761.
10. Kawabata, Y., Wada, K., Nakatani, M. et al. (2011). Formulation design for poorly water-soluble drugs based on biopharmaceutics classification system: basic approaches and practical applications. *International Journal of Pharmaceutics* 420 (1): 1–10.
11. Merisko-Liversidge, E. and Liversidge, G.G. (2011). Nanosizing for oral and parenteral drug delivery: a perspective on formulating poorly-water soluble compounds using wet media milling technology. *Advanced Drug Delivery Reviews* 63 (6): 427–440.
12. Newman, A. (2015). *Pharmaceutical Amorphous Solid Dispersions*. Hoboken, NJ: Wiley.
13. Baghel, S., Cathcart, H., and O'Reilly, N.J. (2016). Polymeric amorphous solid dispersion: a review of amorphization, crystallization, stabilization, solid-state characterization, and aqueous solubilization of biopharmaceutical classification system class II drugs. *Journal of Pharmaceutical Sciences* 105 (9): 2527–2544.
14. DeBoyace, K. and Wildfong, P.L.D. (2018). The application of modeling and prediction to the formation and stability of amorphous solid dispersions. *Journal of Pharmaceutical Sciences* 107 (1): 57–74.
15. Law, D., Krill, S.L., Schmitt, E.A. et al. (2001). Physicochemical considerations in the preparation of amorphous ritonavir–poly(ethylene glycol) 8000 solid dispersions. *Journal of Pharmaceutical Sciences* 90 (8): 1015–1025.
16. Miller, J.M., Beig, A., Carr, R.A. et al. (2012). A win-win solution in oral delivery of lipophilic drugs: supersaturation via amorphous solid dispersions increases apparent solubility without sacrifice of intestinal membrane permeability. *Molecular Pharmaceutics* 9 (7): 2009–2016.
17. Tao, J., Sun, Y., Zhang, G.G.Z., and Yu, L. (2009). Solubility of small-molecule crystals in polymers: D-mannitol in PVP, indomethacin in PVP/VA, and nifedipine in PVP/VA. *Pharmaceutical Research* 26 (4): 855–864.
18. Sun, Y., Tao, J., Zhang, G.G.Z., and Yu, L. (2010). Solubilities of crystalline drugs in polymers: an improved analytical method and comparison of solubilities of indomethacin and nifedipine in PVP, PVP/VA, and PVAc. *Journal of Pharmaceutical Sciences* 99 (9): 4023–4031.
19. Kyeremateng, S.O., Pudlas, M., and Woehrle, G.H. (2014). A fast and reliable empirical approach for estimating solubility of crystalline drugs in polymers for hot melt extrusion formulations. *Journal of Pharmaceutical Sciences* 103 (9): 2847–2858.
20. Lehmkemper, K., Kyeremateng, S.O., Heinzerling, O. et al. (2017). Long-term physical stability of PVP- and PVPVA-amorphous solid dispersions. *Molecular Pharmaceutics* 14 (1): 157–171.
21. Lehmkemper, K., Kyeremateng, S.O., Bartels, M. et al. (2018). Physical stability of API/polymer-blend amorphous solid dispersions. *European Journal of Pharmaceutics and Biopharmaceutics* 124: 147–157.
22. Raina, S.A., Zhang, G.G.Z., Alonzo, D.E. et al. (2014). Enhancements and limits in drug membrane transport using supersaturated solutions of poorly water soluble drugs. *Journal of Pharmaceutical Sciences* 103 (9): 2736–2748.
23. Taylor, L.S. and Zhang, G.G.Z. (2016). Physical chemistry of supersaturated solutions and implications for oral absorption. *Advanced Drug Delivery Reviews* 101: 122–142.
24. Ilevbare, G.A. and Taylor, L.S. (2013). Liquid-liquid phase separation in highly supersaturated aqueous solutions of poorly water-soluble drugs: implications for solubility enhancing formulations. *Crystal Growth & Design* 13 (4): 1497–1509.
25. Jackson, M.J., Kestur, U.S., Hussain, M.A., and Taylor, L.S. (2016). Dissolution of danazol amorphous solid dispersions: supersaturation and phase behavior as a function of drug loading and polymer type. *Molecular Pharmaceutics* 13 (1): 223–231.
26. Abramov, Y.A., Loschen, C., and Klamt, A. (2012). Rational coformer or solvent selection for pharmaceutical cocrystallization or desolvation. *Journal of Pharmaceutical Sciences* 101 (10): 3687–3697.
27. Cruz-Cabeza, A.J., Reutzel-Edens, S.M., and Bernstein, J. (2015). Facts and fictions about polymorphism. *Chemical Society Reviews* 44 (23): 8619–8635.
28. Caira, M.R. (1998). Crystalline polymorphism of organic compounds. In: *Design of Organic Solids* (ed. E. Weber), 163–208. Berlin: Springer.
29. Liu, J. and Rasmuson, A.C. (2013). Influence of agitation and fluid shear on primary nucleation in solution. *Crystal Growth & Design* 13 (10): 4385–4394.
30. Abramov, Y.A. (2016). *Computational Pharmaceutical Solid State Chemistry*. Hoboken, NJ: Wiley.
31. Etter, M.C. (1991). Hydrogen bonds as design elements in organic chemistry. *The Journal of Physical Chemistry* 95 (12): 4601–4610.
32. Etter, M.C. and Reutzel-Edens, S.M. (1991). Hydrogen bond directed cocrystallization and molecular recognition properties of acyclic imides. *Journal of the American Chemical Society* 113 (7): 2586–2598.
33. Reutzel-Edens, S.M. and Etter, M.C. (1992). Evaluation of the conformational, hydrogen bonding and crystal packing preferences of acyclic imides. *Journal of Physical Organic Chemistry* 5 (1): 44–54.
34. Desiraju, G.R. (1991). The C–H⋯O hydrogen bond in crystals. What is it? *Accounts of Chemical Research* 24 (10): 290–296.
35. Aakeröy, C.B. and Seddon, K.R. (1993). The hydrogen bond and crystal engineering. *Chemical Society Reviews* 22 (6): 397–407.
36. Dunitz, J.D. and Gavezzotti, A. (2005). Molecular recognition in organic crystals: directed intermolecular bonds or nonlocalized bonding? *Angewandte Chemie International Edition* 44 (12): 1766–1787.
37. Desiraju, G.R., Vittal, J.J., and Ramanan, A. (2011). *Crystal Engineering: A Textbook*. Singapore: World Scientific Publishing Co. Pte. Ltd.
38. Allen, F.H. (2002). The Cambridge Structural Database: a quarter of a million crystal structures and rising. *Acta Crystallographica Section B* 58 (3): 380–388.

39. Groom, C.R. and Allen, F.H. (2014). The Cambridge Structural Database in retrospect and prospect. *Angewandte Chemie International Edition* 53 (3): 662–671.
40. Galek, P.T.A., Fábián, L., Motherwell, W.D.S. et al. (2007). Knowledge-based model hydrogen-bonding propensity in organic crystals. *Acta Crystallographica Section B* 63 (5): 768–782.
41. Galek, P.T.A., Fábián, L., and Allen, F.H. (2010). Truly prospective prediction: inter- and intramolecular hydrogen bonding. *CrystEngComm* 12 (7): 2091–2099.
42. Galek, P.T.A., Allen, F.H., Fábián, L., and Feeder, N. (2009). Knowledge-based H-bond prediction to aid experimental polymorph screening. *CrystEngComm* 11 (12): 2634–2639.
43. Abramov, Y.A. (2013). Current computational approaches to support pharmaceutical solid form selection. *Organic Process Research & Development* 17 (3): 472–485.
44. Wood, P.A., Olsson, T.S.G., Cole, J.C. et al. (2013). Evaluation of molecular crystal structures using Full Interaction Maps. *CrystEngComm* 15 (1): 65–72.
45. Bruno, I.J., Cole, J.C., Rowland, R.S. et al. (1997). Isostar: a library of information about nonbonded interactions. *Journal of Computer-Aided Molecular Design* 11 (6): 525–537.
46. McKinnon, J.J., Spackman, M.A., and Mitchell, A.S. (2004). Novel tools for visualizing and exploring intermolecular interactions in molecular crystals. *Acta Crystallographica Section B* 60 (6): 627–668.
47. Spackman, M.A. and Jayatilaka, D. (2009). Hirshfeld surface analysis. *CrystEngComm* 11 (1): 19–32.
48. Dunitz, J.D. (2003). Are crystal structures predictable? *Chemical Communications* 0 (5): 545–548.
49. Day, G.M. (2011). Current approaches to predicting molecular organic crystal structures. *Crystallography Reviews* 17 (1): 3–52.
50. Lommerse, J.P.M., Motherwell, W.D.S., Ammon, H.L. et al. (2000). A test of crystal structure prediction of small organic molecules. *Acta Crystallographica Section B* 56 (4): 697–714.
51. Gavezzotti, A. and Filippini, G. (1995). Polymorphic forms of organic crystals at room conditions: thermodynamic and structural implications. *Journal of the American Chemical Society* 117 (49): 12299–12305.
52. Motherwell, W.D.S., Ammon, H.L., Dunitz, J.D. et al. (2002). Crystal structure prediction of small organic molecules: a second blind test. *Acta Crystallographica Section B* 58 (4): 647–661.
53. Day, G.M., Motherwell, W.D.S., Ammon, H.L. et al. (2005). A third blind test of crystal structure prediction. *Acta Crystallographica Section B* 61 (5): 511–527.
54. Day, G.M., Cooper, T.G., Cruz-Cabeza, A.J. et al. (2009). Significant progress in predicting the crystal structures of small organic molecules – a report on the fourth blind test. *Acta Crystallographica Section B* 65 (2): 107–125.
55. Bardwell, D.A., Adjiman, C.S., Arnautova, Y.A. et al. (2011). Towards crystal structure prediction of complex organic compounds — a report on the fifth blind test. *Acta Crystallographica Section B* 67 (6): 535–551.
56. Reilly, A.M., Cooper, R.I., Adjiman, C.S. et al. (2016). Report on the sixth blind test of organic crystal structure prediction methods. *Acta Crystallographica Section B* 72 (4): 439–459.
57. Price, S.L. (2009). Computed crystal energy landscapes for understanding and predicting organic crystal structures and polymorphism. *Accounts of Chemical Research* 42 (1): 117–126.
58. Neumann, M.A. (2008). Tailor-made force fields for crystal-structure prediction. *The Journal of Physical Chemistry B* 112 (32): 9810–9829.
59. Neumann, M.A., Leusen, F.J.J., and Kendrick, J. (2008). A major advance in crystal structure prediction. *Angewandte Chemie International Edition* 47 (13): 2427–2430.
60. Rosso, L., Mináry, P., Zhu, Z., and Tuckerman, M.E. (2002). On the use of the adiabatic molecular dynamics technique in the calculation of free energy profiles. *The Journal of Chemical Physics* 116 (11): 4389–4402.
61. Chen, M., Yu, T.Q., and Tuckerman, M.E. (2015). Locating landmarks on high-dimensional free energy surfaces. *PNAS* 112 (11): 3235–3240.
62. Yu, T.Q. and Tuckerman, M.E. (2011). Temperature-accelerated method for exploring polymorphism in molecular crystals based on free energy. *Physical Review Letters* 107 (1): 015701.
63. van Eijck, B.P. and Kroon, J. (2000). Structure predictions allowing more than one molecule in the asymmetric unit. *Acta Crystallographica Section B* 56 (3): 535–542.
64. Linifanib (2010). *Drugs R&D* 10 (2): 111–122.
65. Desiraju, G.R. (1995). Supramolecular synthons in crystal engineering – a new organic synthesis. *Angewandte Chemie International Edition* 34 (21): 2311–2327.
66. Reddy, L.S., Chandran, S.K., George, S. et al. (2007). Crystal structures of N-aryl-N'-4-nitrophenyl ureas: molecular conformation and weak interactions direct the strong hydrogen bond synthon. *Crystal Growth & Design* 7 (12): 2675–2690.
67. Bruno, I.J., Cole, J.C., Edgington, P.R. et al. (2002). New software for searching the Cambridge Structural Database and visualizing crystal structures. *Acta Crystallographica Section B* 58 (3): 389–397.
68. Hulme, A.T. and Price, L.S. (2007). Toward the prediction of organic hydrate crystal structures. *Journal of Chemical Theory and Computation* 3 (4): 1597–1608.
69. Takieddin, K., Khimyak, Y.Z., and Fábián, L. (2016). Prediction of hydrate and solvate formation using statistical models. *Crystal Growth & Design* 16 (1): 70–81.
70. Bajpai, A., Scott, H.S., Pham, T. et al. (2016). Towards an understanding of the propensity for crystalline hydrate formation by molecular compounds. *International Union of Crystallography* 3 (6): 430–439.
71. Abramov, Y.A. (2015). Virtual hydrate screening and coformer selection for improved relative humidity stability. *CrystEngComm* 17 (28): 5216–5224.
72. Tilbury, C.J., Chen, J., Mattei, A. et al. (2018). Combining theoretical and data-driven approaches to predict drug substance hydrate formation. *Crystal Growth & Design* 18 (1): 57–67.

73. Klamt, A. (2011). The COSMO and COSMO-RS solvation models. *WIREs: Computational Molecular Science* 1 (5): 699–709.
74. Klamt, A. (1995). Conductor-like screening model for real solvents: a new approach to the quantitative calculation of solvation phenomena. *The Journal of Physical Chemistry* 99 (7): 2224–2235.
75. Pearson, R.G. (2005). Chemical hardness and density functional theory. *Journal of Chemical Sciences* 117 (5): 369–377.
76. Nangia, A. (2010). Hydrogen bonding and molecular packing in multi-functional crystal structures. In: *Organic Crystal Engineering Frontiers in Crystal Engineering* (ed. E.R.T. Tiekink, J. Vittal and M. Zaworotko), 151–189. Chichester, UK: Wiley.
77. Noyes, A. and Whitney, W. (1897). The rate of solution of solid substances in their own solutions. *Journal of the American Chemical Society* 19 (12): 930–934.

30

MULTISCALE ASSESSMENT OF API PHYSICAL PROPERTIES IN THE CONTEXT OF MATERIALS SCIENCE TETRAHEDRON CONCEPT

RAIMUNDO HO, YUJIN SHIN, YINSHAN CHEN, LAURA POLONI[*], SHUANG CHEN, AND AHMAD Y. SHEIKH

Solid State Chemistry, AbbVie Inc., North Chicago, IL, USA

30.1 INTRODUCTION

Materials science and engineering is an interdisciplinary field focusing on the relationships between material's structure, properties, processing, and performance. The interplay between these elements of materials science is best represented as the materials science tetrahedron (MST) (Figure 30.1) [1, 2]. The ultimate goal of materials science and engineering is to enable the design and manufacture of product for a targeted performance criterion via application of engineering principles to control material properties from fundamental understanding of the structure–property relationships. Characterization, the core of MST, provides the bridge between the four interdependent elements of the MST that form the fundamental basis for engineering materials to meet target performance needs. A thorough understanding of material properties, process, and performance is an essential part of the pharmaceutical development process. Such insights are crucial to understand (i) material structure–property relationships, (ii) sensitivity of processing on material attributes, and (iii) the influence of material attributes on product performance such as safety, bioavailability, efficacy, manufacturability, stability, etc.

In the context of small molecule active pharmaceutical ingredient (API), the properties of materials at the bulk scale are intrinsically related to the molecular structure and its packing across multiple length scales (Figure 30.2). This multiscale structure–property relationship originates from the molecular flexibility of chemical structure that gives rise to polymorphism of the solid crystalline phase. The molecular arrangement and conformation of molecules in the crystal lattice are dictated by thermodynamic and kinetic factors, which determine the crystal properties at the single particle level. The API properties measured at the bulk level and, therefore, its bulk powder behavior are a consequence of the distribution of single-crystal properties from each individual particulate that make up the bulk material as well as the interactions between single particles. Needless to say, API physical/chemical properties across scales can significantly influence formulation choice/composition, manufacturing route, and performance of the pharmaceutical product.

It remains an aspiration in the field of pharmaceutical materials science to predict the API bulk powder behavior from the fundamental structural properties of the crystalline species. The challenge lies in the complexity to link and model multilevel structure–property relationships. Although molecular modeling has been used to predict crystal properties from the crystal structure, correlating single-crystal properties to the bulk scale is not straightforward due to crystal anisotropy and potential interparticle interactions. Despite the challenge, characterization tools are becoming increasingly refined to study material properties at different scales and can be used to develop practical structure–property–performance relationships. These advanced characterization tools provide deep understanding of structural, surface, bulk,

---

[*] Current address: Department of Chemical Engineering and Applied Chemistry, University of Toronto, Ontario, Canada

Chemical Engineering in the Pharmaceutical Industry: Active Pharmaceutical Ingredients, Second Edition.
Edited by David J. am Ende and Mary T. am Ende.
© 2019 John Wiley & Sons, Inc. Published 2019 by John Wiley & Sons, Inc.

and mechanical properties of materials and enable our understanding of the correlations between structural parameters and physical/chemical properties of APIs at different length scales, and ultimately facilitate the design of drug substance and drug product (DP) processes to achieve targeted performance criterion.

In this chapter, a broad range of characterization techniques that can be used to elucidate the structure–property–performance relationships of APIs are summarized. Additionally, two case studies are presented to demonstrate the application of the MST concept in process development and optimization of APIs to meet required performance targets. The case studies are related to (i) modeling of API powder flowability from fundamental API physical properties and (ii) impact of API intrinsic mechanical properties on process-induced disorder.

30.2 STRUCTURAL, SURFACE, BULK, AND MECHANICAL CHARACTERIZATION TOOLS

API physical and chemical properties can be broadly classified into four major categories – structural, surface, bulk, and mechanical properties. These properties and the associated characterization techniques contribute to a comprehensive and in-depth understanding of the APIs for process optimization and formulation development. Table 30.1 provides an overview of characterization tools for API properties and will be further discussed in this section.

30.2.1 Understanding Crystal Structure and Molecular Interactions

API molecule can crystallize in a number of crystal forms, including hydrates, solvates, salts, and co-crystals (Figure 30.3). The crystal form and polymorph that is observed is dependent on thermodynamic and kinetic factors, including solvent composition, concentration of counterions or coformers, and temperature/supersaturation [3]. Many of the emergent properties of the powder, including surface, mechanical, and bulk properties, have their origins in the

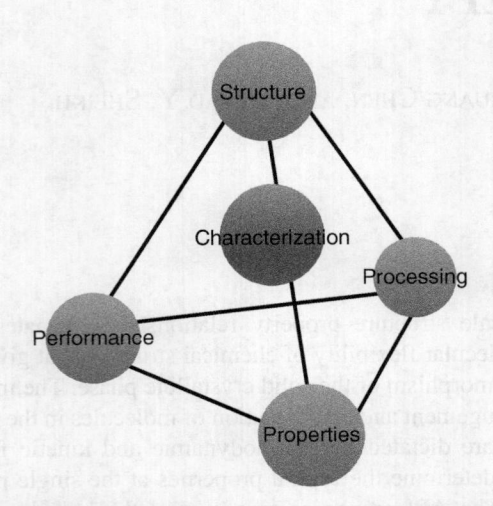

FIGURE 30.1 Materials science tetrahedron. The materials science paradigm depicts correlation of the structure of materials, their properties and processing, and subsequently performance of applications.

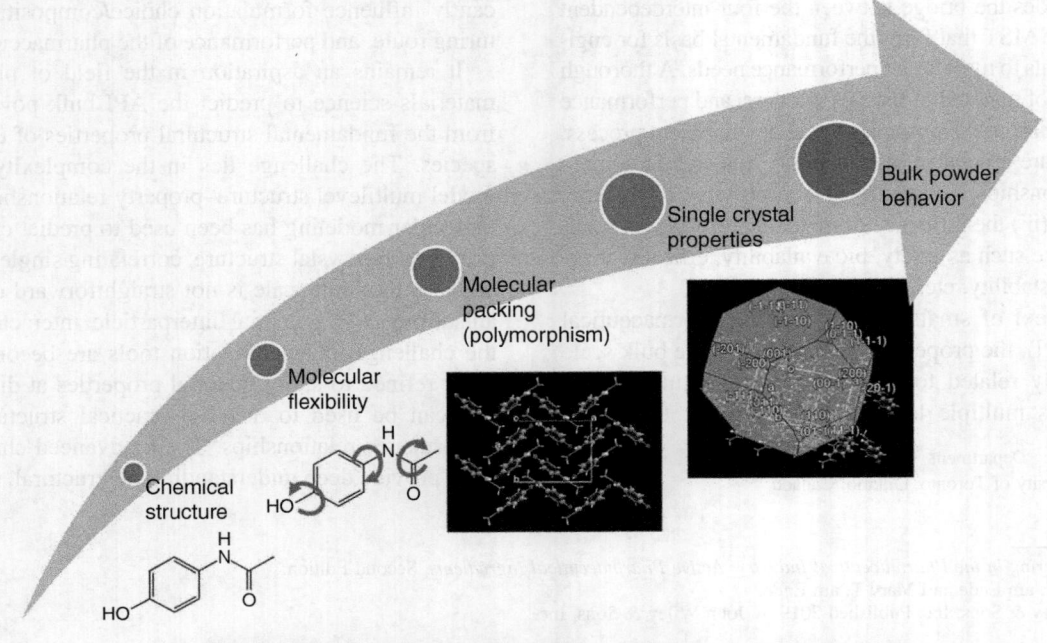

FIGURE 30.2 Multiscale relationships of API physical properties.

TABLE 30.1 API Multiscale Characterization Tools for Systematic Assessment of API Attributes

Solid-State Properties	API Physical Properties	Characterization Tools
Structure	• Crystal form • Packing • Weak interactions • Local/long-range order	• Thermal techniques (TGA/DSC) • Single-crystal and powder X-ray diffraction (SCXRD and PXRD) • Solid-state nuclear magnetic resonance (SS-NMR) • Total scattering atomic pair distribution function (TS-PDF)
Surface/interface	• Crystal morphology/shape • Surface topography • Surface area • Surface energy/chemistry	• Scanning electron microscopy (SEM) • Confocal laser light microscopy • Atomic force microscopy (AFM) • Gas adsorption: specific surface area • Inverse gas chromatography (IGC) • Dynamic vapor sorption (DVS)
Bulk	• Particle size/shape distribution • Powder flow properties (cohesion, friction, flow function) • Densities	• Laser diffraction/static and dynamic image analysis • Ring shear tester • Gas pycnometry
Mechanical properties	• Hardness (H) • Young's modulus (E) • Fracture toughness (K_{IC}) • Brittleness (BI) • Friction • Tensile strength • Agglomerate strength	• Nano-indentation (hardness, Young's modulus, fracture toughness, friction) • Texture analyzer • Compaction simulator • Hydraulic press • Atomic force microscopy (AFM)

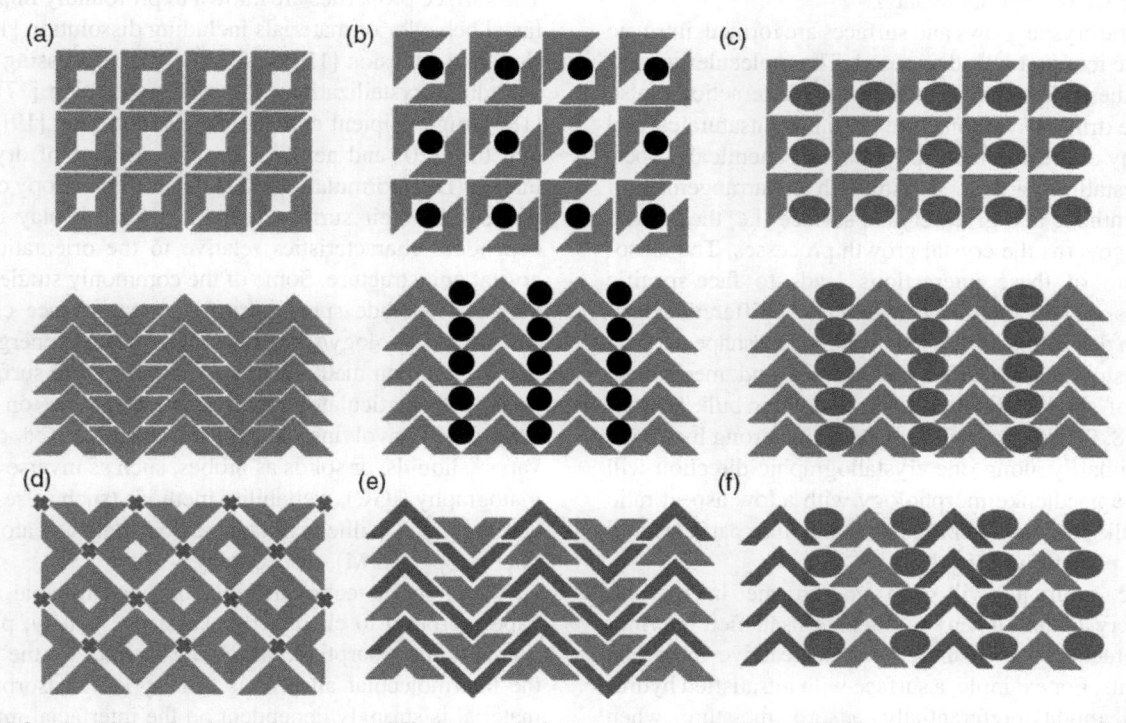

FIGURE 30.3 An API molecule can have many different crystal forms and polymorphs of these forms, including (a) anhydrous forms, (b) hydrates (water molecules shown as circles), (c) solvates (solvent molecules shown as ellipsoids), (d) salt forms (counterion shown as crosses), (e) cocrystals (conformer shown as thin arrows), and (f) hydrates and solvates of salts and co-crystals.

packing of molecules within the crystalline lattice. A thorough understanding of structural features such as hydrogen bonding network, van der Waals interactions, planar chemistry, and slip planes can therefore help understand and rationalize behavior and performance of drug substance.

The packing of molecules in the solid state is a result of intermolecular interactions between constituents comprising a crystal, also known as synthonic interactions or intrinsic synthons, and includes ionic interactions, π–π stacking, hydrogen bonding, and van der Waals interactions. The strength and three-dimensional arrangement of these interactions determine the bulk properties of the crystalline material. Intermolecular interactions in molecular crystals are typically anisotropic, leading to anisotropic properties of single crystals, e.g. direction-dependent mechanical properties resulting from preferred slip planes for deformation and propensity for cleavage and fracture [4–6]. In a review by Reddy et al. on the structural–mechanical property correlations in molecular crystals, it was noted that polymorph with a layered structure is subjected to slipping and shearing, commonly due to the weakness of the interlayer interaction, which leads to bending and twining of the crystals [7]. Molecular crystals that show bending phenomenon occur when the packing is anisotropic in a way that strong and weak interaction patterns are in nearly perpendicular directions, and they tend to be highly plastic and ductile. In contrast, hard and brittle crystals that fracture and break down into pieces on application of stress are generally due to isotropic interactions in all three dimensions of the structure, where the interactions can be strong (H-bonding) or weak (van der Waals).

As a single crystal grows and surfaces are formed, intrinsic synthons are terminated and exposed. The molecules associated with these "broken" intermolecular interactions, also known as extrinsic synthons, are partially unsaturated and govern many of the interfacial physical and chemical properties of a crystalline particle. The strength and arrangement of extrinsic synthons at a given crystal surface (i.e. the surface chemistry) governs the crystal growth processes. The anisotropic nature of these interactions leads to face-specific growth rates, and the relative growth rates of different crystal faces in turn determine the crystal morphology/shape. Crystal shape can significantly affect the surface and mechanical properties of single crystals and thereby their bulk powder properties [8, 9]. For example, crystals with strong hydrogen bonding primarily along one crystallographic direction will often form a needlelike morphology with a low aspect ratio (AR). A bulk powder composed of needlelike particles will likely have poor flowability [10].

Extrinsic synthons will also govern the interactions between a crystal and its environment. Unsatisfied intermolecular interactions on a surface can be reactive in certain environments. For example, a surface with unsatisfied hydrogen bonds might preferentially adsorb moisture when exposed to air, which could be detrimental to the stability and performance of an API powder. Extrinsic synthons can affect crystal–crystal interfacial interactions in an API powder by mediating the processes that can lead to cohesion and agglomeration of particles. In a study on the sticking propensity of ibuprofen, unsatisfied hydrogen bonds due to exposed surface carboxylic acid group were found to dominate the specific surface energy and stickiness of the ibuprofen [11]. It should also be noted that the extrinsic synthons play a role in the dissolution of a given crystal surface, akin to their role in growth phenomena at the same surface [12]. Depending on the formulation of the final DP, API wettability and dissolution plays an important role in the design of the DP for drug delivery or may impact the bioavailability of the API *in vivo*. Extrinsic synthons significantly influence the wettability of API crystal [13]. As the hydrophobicity of extrinsic synthons increases the contact angle, it may compromise the dissolution performance of the API [9].

Overall, by understanding packing of molecules within the crystalline lattice and the molecular interactions at the surface, it can lead to rational design and production of crystalline particles with desirable shape and size distributions – properties that stem from the crystal structure of the API – for consistent physical properties of the API powder and subsequent streamlined downstream processing to manufacture the DP.

30.2.2 Surface/Interfacial Interactions and Characterization

The surface properties are known to profoundly impact interfacial behavior of materials including dissolution [14], cohesion, and adhesion [15] and a variety of processing behavior including crystallization [16], wet granulation [17], milling [18], drug-excipient compatibility and mixing [19], coating, tableting [20] and aerolization performance of dry powder inhaler (DPI) formulations [21]. Due to anisotropy of crystalline solids, their surface properties also display direction-dependent characteristics relative to the orientation of the crystal unit structure. Some of the commonly studied surface attributes include specific surface area, surface chemistry, surface morphology/topography, and surface energy of particles. Common methods of characterizing the surface properties of particulate pharmaceuticals rely on indirect approaches involving the use of characterized or *known* vapors, liquids, or solids as probes, such as inverse gas chromatography (IGC), wettability methods (such as sessile drop contact angle, Wilhelmy plate method, etc.), and atomic force microscopy (AFM), respectively.

Vapor probe techniques rely on two principal modes of vapor sorption to characterize surfaces, namely, physisorption and chemisorption, which are distinct by the nature of the intermolecular attractions. The type of adsorption on a material is strongly dependent on the interfacial intermolecular interactions, vapor pressure, and temperature. Surface adsorption models, such as Langmuir, Brunauer–Emmett–Teller (BET), and Guggenheim–Anderson–de Boer (GAB), were developed for monolayer and multilayer gas adsorption from which enthalpy of adsorption and monolayer capacity (and therefore specific surface area) can be determined. Other methods such as Kelvin's equation, Barrett–Joyner–Halenda (BJH), and Dollimore–Heal (DH) are used to model capillary condensation, where micro/meso porosity and pore size distribution can be determined.

IGC, which is a vapor probe technique, is an evolving technology in the field of pharmaceutical R&D, and its utility spans a variety of applications including batch-to-batch variability, solid–solid transitions, physical stability, interfacial behavior in powder processing, and more. Figure 30.4a shows the schematic setup of IGC. In IGC, probe molecules in the vapor phase (adsorbate with known properties) are passed over a sample solid (stationary phase adsorbent of interest), which is packed in a column of inert surface, via a carrier gas (mobile phase). The partition coefficient, K_R, of the adsorbate between the mobile and stationary phase at a particular temperature and pressure is directly related to the net retention volume, V_N, which is the volume of carrier gas required to elute the injected adsorbate through the column. Both K_R and V_N are indications of the interaction strength between the adsorbate and the solid sample of interest. Together, they form the fundamental bases for the derivations of surface-specific thermodynamic parameters such as surface energies, surface glass transition temperature, free energy of adsorption, enthalpy of adsorption, electron-donating or electron-accepting properties, etc. When the concentration of adsorbate is increased to a level such that adsorption goes beyond Henry's law region, IGC can be used

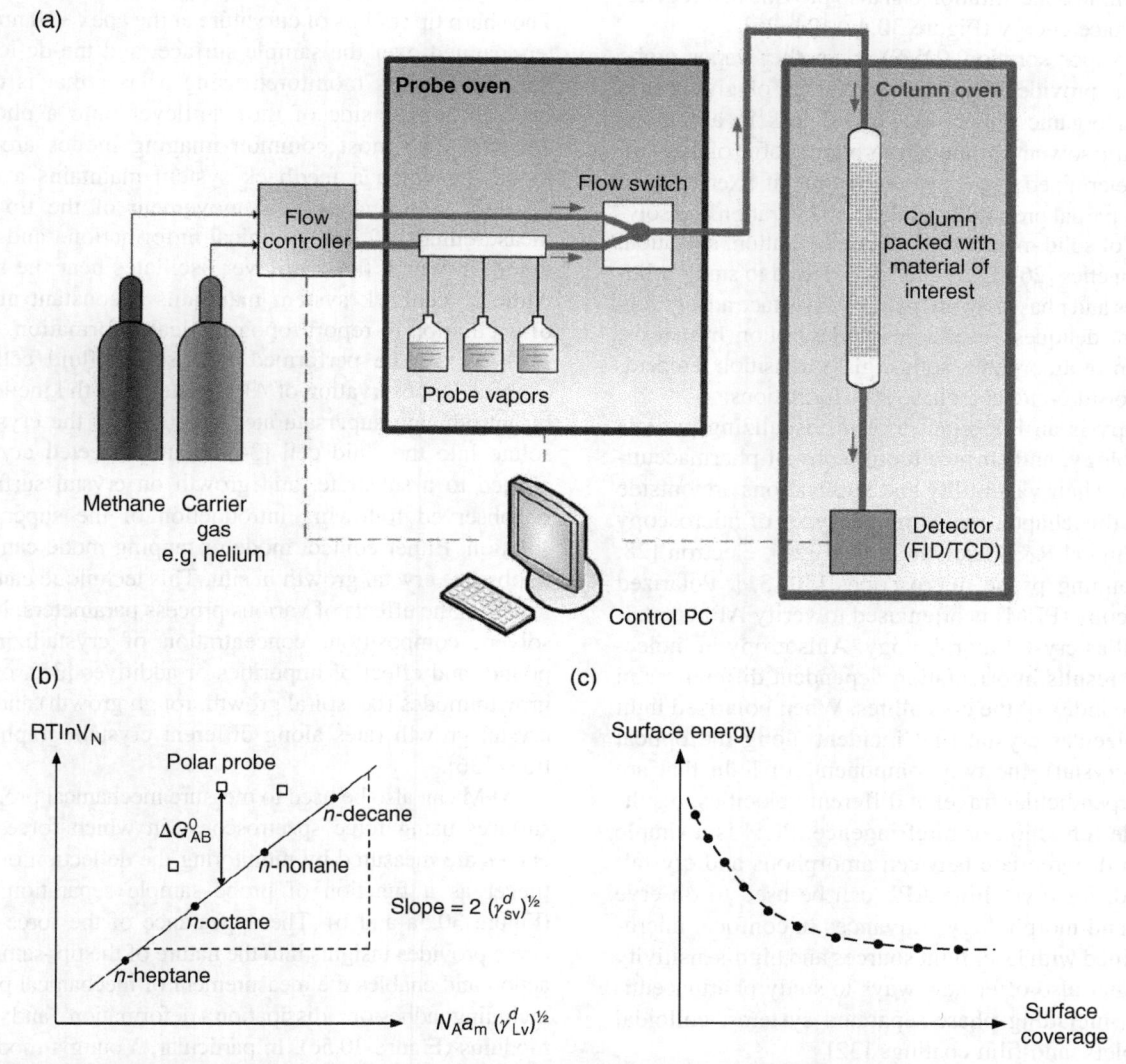

FIGURE 30.4 (a) Schematic of inverse gas chromatography. (b) Schultz approach for determination of the dispersive surface energy, γ^d, and specific free energy of the material (ΔG^0_{AB}). (c) Surface energy heterogeneity profile of the material as a function of probe surface coverage.

to characterize surface energy/chemistry heterogeneity, specific surface area, and bulk properties such as solubility parameters (e.g. Hansen) and interaction parameters (e.g. Flory–Huggins). An extensive review of IGC and its utility in pharmaceutical development can be found elsewhere [22].

For surface energetics, a common methodology employs the Schultz approach by calculating the dispersive and specific components using a series of n-alkane and polar probes, respectively (Figure 30.4b). For n-alkane probes, the dispersive component of the surface energy is derived from the slope of the linear regression (i.e. alkane line) of the retention volumes as a function of the dispersive liquid surface tension of the probes. The vertical distance between the polar probe points and the alkane straight line represents specific free energy (ΔG_{AB}^0), which can be used to further derive the specific or acid–base surface energy component and surface chemistry [22, 23]. In addition to surface energy determination, IGC at finite concentration can also provide heterogeneity of the surface energy (Figure 30.4c) [24, 25].

Dynamic vapor sorption (DVS) is another vapor probe technique that provides rapid measurements of gravimetric moisture and organic vapor uptake and loss in a sample. The gravimetric solvent sorption/desorption profile of the sample can be determined as a function of time at fixed relative humidities or partial pressures. Aside from hydration and solvation studies of solid materials including hydration, solvation, and drying kinetics [26], DVS can be also used to study solid-state reactions and phase transitions such as surface adsorption, bulk sorption, deliquescence, sorption/desorption hysteresis, crystallization of amorphous solids, glass transition temperatures, and moisture-induced phase transformations.

Microscopy is an important tool in visualizing particle size, morphology, and surface topography of pharmaceutical materials. Their vast utility and applications are outside the scope of this chapter, but common types of microscopy in pharmaceutical R&D include optical [27], electron [28, 29], and scanning probe microscopes [30, 31]. Polarized light microscopy (PLM) is often used to verify API crystallinity as well as crystal morphology. Anisotropy in molecular crystals results in orientation-dependent differences in the refractive index of the crystallites. When polarized light enters a molecular crystal (not incident along the optical axis of the crystal), the two components of light that are mutually perpendicular travel at different velocities, resulting in double refraction or birefringence. PLM is a simple technique to differentiate between amorphous and crystalline API and, for crystalline API, can be used to observe crystal size and morphology. Advances in confocal microscope combined with laser light sources and high-sensitivity photomultiplier also offer new ways to study pharmaceutical systems, including phase-separated systems, colloidal systems, tablets, and film coatings [32].

Electron microscopy is a powerful technique to study organic crystalline materials [33]. Among the many types of electron microscopy, scanning electron microscopy (SEM) is the most widely used in the pharmaceutical industry. SEM uses a focused electron beam that is raster-scanned across the specimen, and high-energy backscattered electrons are measured using a secondary electron detector to produce images. SEM is commonly used to observe crystal and particle morphology and size and can be combined with elemental analysis. Conventional SEM is performed under high vacuum and with high voltage, and so molecular crystalline samples typically need to be sputter coated with a conductive coating to prevent charging of the sample. Metal coatings are typically used because they are good secondary electron emitters.

AFM is a versatile technique that can be used to explore many properties of pharmaceutical materials. AFM is a type of scanning probe microscopy (SPM) and measures the interaction forces between the surface of interest (i.e. an API crystal surface) and a sharp tip mounted on a cantilever. The sharp tip (radius of curvature at the apex ~10 nm) is raster-scanned over the sample surface, and the deflection of the cantilever is monitored using a laser that is reflected from the upper side of the cantilever onto a photodiode detector. The most common imaging modes are contact mode, in which a feedback system maintains a constant tip deflection and vertical movement of the tip enables measurement of topographical information, and tapping mode, in which the cantilever oscillates near the substrate while a feedback system maintains a constant amplitude of oscillation to report topographical information.

AFM can be performed in a sealed fluid cell, which enables the observation of API crystal growth kinetics in situ by introducing supersaturated solutions of the crystallizing solute into the fluid cell [34]. Mature faceted crystals are affixed to a substrate, and growth on crystal surfaces can be observed following introduction of the supersaturated solution. Either contact mode or tapping mode can be used to observe crystal growth in situ. This technique can be used to assess the effects of various process parameters, including solvent composition, concentration of crystallizing compound, and effect of impurities or additives [35], on crystal growth modes (i.e. spiral growth, rough growth) and relative crystal growth rates along different crystallographic directions [36].

AFM can also be used to measure mechanical properties of surfaces using force spectroscopy, in which force–distance curves are measured by monitoring the deflection of the cantilever as a function of probe–sample separation distance (Figure 30.5a and b). The appearance of the force–distance curve provides insights into the nature of the tip–sample interaction and enables the measurement of mechanical properties including adhesion, dissipation, deformation, and Young's modulus (Figure 30.5c). In particular, Young's modulus can be obtained by fitting the retraction curve (Figure 30.5c) with Derjaguin–Muller–Toporov (DMT) model by [37]

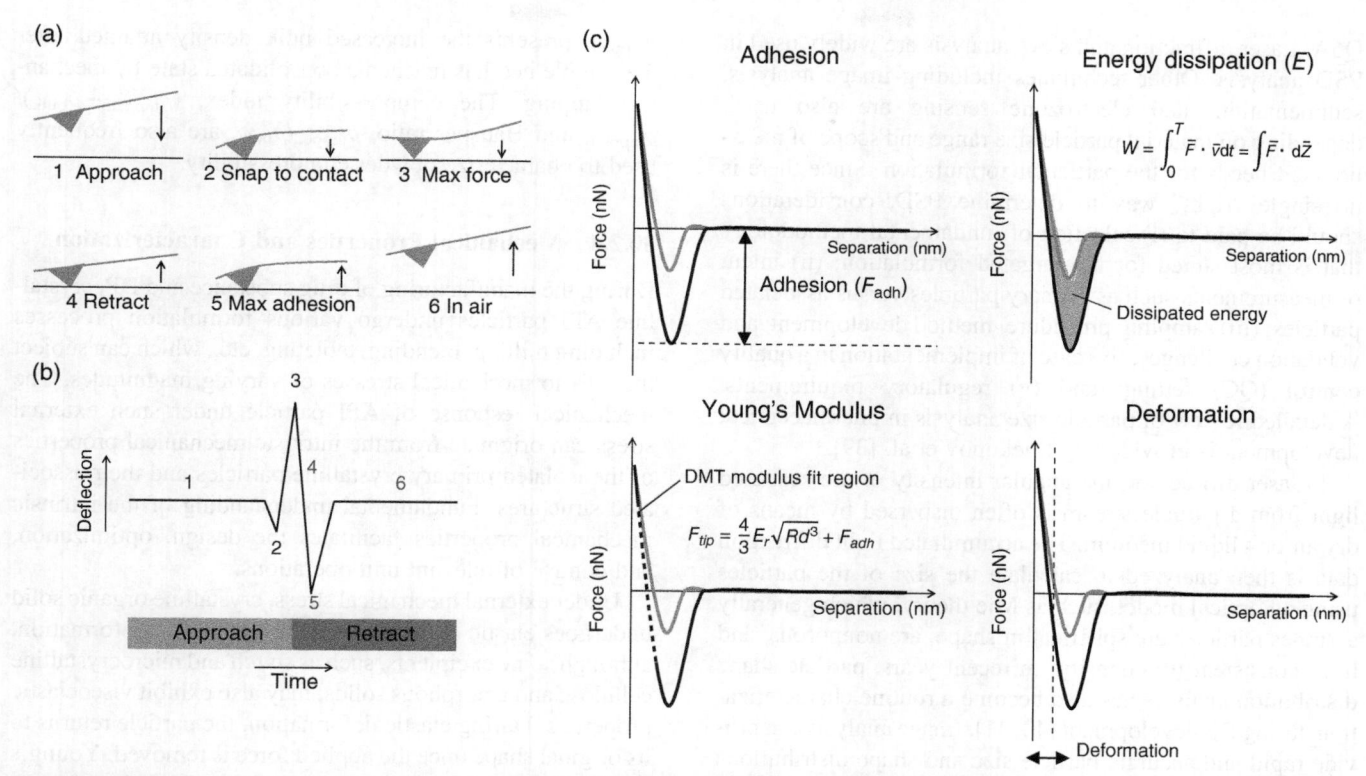

FIGURE 30.5 Force vs. distance curves obtained upon approach (gray line) and retraction (black line) of a cantilever from a surface. (a) Movement of the cantilever and (b) corresponding measured deflection, which is converted into a force vs. separation curve (c) from which mechanical properties can be determined.

$$E_r = \frac{3(F_{tip} - F_{adh})}{4\sqrt{Rd^3}}$$

where

F_{tip} is the force on the cantilever.
F_{adh} is the adhesion force between the AFM tip and sample.
R is the AFM tip radius.
d is the deformation of the sample.
E_r is the reduced Young's modulus.

Young's modulus of the sample can then be further determined from the reduced modulus. The recent introduction of the PeakForce imaging mode has increased the ease with which force curves can be obtained and analyzed, enabling real-time simultaneous mapping of several mechanical properties with high spatial resolution.

Chemical force microscopy (CFM) is an application of force spectroscopy in which gold-coated AFM tips are functionalized with alkanethiols to investigate interactions between specific functional groups and crystal surfaces [38]. For example, AFM tips can be coated in alkanethiols with terminal COO^-, NH_3^+, OH, and CH_3 groups, and tip–crystal adhesion forces can be measured on crystallographically unique crystal surfaces. These interactions can provide insights to crystal–crystal aggregation and the effects of crystal morphology on crystal agglomeration.

30.2.3 Bulk Properties and Characterization

Crystalline API particles undergo various formulation processes including but not limited to mixing, granulation, drying, blending, tableting, and coating. In many of these unit operations, API bulk physical properties are key determinants of the process performance. Therefore, API process optimization is traditionally focused on API bulk property improvements. The bulk physical properties are also key elements of the API physical property control strategy. As a consequence both from quality and regulatory perspectives, it is important to understand and monitor consistency of API bulk physical properties to ensure robust DP manufacturing based on alignment of drug substance and DP control strategies.

Among the bulk properties, particle size distributions (PSD) are considered most pertinent to ensure the final DP quality. The ICH guidelines explicitly state that particle size is one of the physical properties that influences DP processability and performance. Further regulatory guidance on setting particle size acceptance criterion is described in ICH

Q6A. Laser diffraction and sieve analysis are widely used in PSD analysis. Other techniques including image analysis, sedimentation, and electrozone sensing are also used, depending on material particle size range and scope of measurement needs for the particular formulation. Since there is no single "right" way to determine PSD, considerations should be paid to (i) selection of fundamental methodology that is most suited for the targeted formulation; (ii) intent of measurements such as primary particles versus associated particles; (iii) sampling procedure, method development, and validation challenges; (iv) ease of implementation in a quality control (QC) setting; and (v) regulatory requirements. A detailed review of particle size analysis in pharmaceutical development is provided by Shekunov et al. [39].

In laser diffraction, the angular intensity of the scattered light from a particle stream is often dispersed by means of dry air or a liquid medium. The accumulated light diffraction data is then analyzed to calculate the size of the particles using an optical model, such as Mie theory, which generally assumes particles are spherical in shape, are nonporous, and have consistent true density. In recent years, particle shape distribution analysis has also become a routine characterization during DS development [40, 41]. Image analysis can provide rapid and accurate particle size and shape distributions for dry powders and wet dispersions. Image analysis can be categorized into static and dynamic image analysis (DIA). Static techniques generally utilized microscopy equipped with motorized stage that moves stepwise to capture images of dispersed particles. DIA uses a pulse light source to illuminate dry or wet dispersions that are streamed continuously through the measuring zone and captured by high-speed camera. Based on the two-dimensional (2D) images of particles, a number of particle size and shape factors including circular equivalent diameter, Feret diameters, circularity, solidarity, area, aspect ratio, etc. can be quantified for nonspherical-shaped particles most commonly encountered for the APIs. In the static techniques, the preferred orientation (e.g. the longest dimensions) can be easily captured because particles are typically deposited on the microscope stage, whereas dynamic techniques allow more random particle orientations and a statistically larger number of particles can be analyzed quickly. In many cases, particle size and shape distributions can also indirectly capture bulk powder attributes, such as flowability. This topic is the focus of the first case study.

Other bulk properties that are routinely characterized for APIs are true/apparent density, bulk density, and tapped density. True density is the density of solid material excluding any open and closed pores and can be calculated from the lattice parameters for a given crystal form. It can be measured to high accuracy by gas pycnometry. When the materials contain inaccessible pores, the measured density can be referred as apparent density. The bulk density, ρ_{bulk}, is the ratio of the untapped powder sample and its volume including contribution of interparticulate void volume. The tapped density, ρ_{tapped}, presents the increased bulk density attained after the sample bed has reached a consolidated state by mechanical tapping. The compressibility index, $(\rho_{tapped} - \rho_{bulk})/\rho_{tapped}$, and Hausner ratio, $\rho_{tapped}/\rho_{bulk}$, are also frequently used to characterize powder compressibility.

30.2.4 Mechanical Properties and Characterization

During the manufacturing of drug substance and DP, crystalline API particles undergo various formulation processes including milling, blending, tableting, etc., which can subject the API to mechanical stresses of varying magnitudes. The mechanical response of API particle under such external stress can originate from the intrinsic mechanical properties of the isolated primary crystalline particles and their associated structures. Fundamental understanding of the intrinsic mechanical properties facilitates the design, optimization, and control of relevant unit operations.

Under external mechanical stress, crystalline organic solid undergoes elastic deformation and then plastic deformation, although some excipients, such as starch and microcrystalline cellulose and amorphous solids, may also exhibit viscoelastic properties. During elastic deformation, the particle returns to its original shape once the applied force is removed. Young's modulus (E) is a measure of the resistance to elastic deformation. During plastic deformation, the particle experiences a permanent deformation by gliding, twining, and kinking of molecular layers. Hardness (H) is the measure of resistance to plastic deformation. When the strain is high enough, fracture intervenes, and a material experiences a failure into pieces. Fracture toughness (K_{IC}) is a parameter to measure the resistance to propagation of a fracture. This parameter is related to the brittleness of a material. Brittleness index (BI), which is defined as H/K_{IC}, is used to describe the overall brittleness of a material. In the full strain–stress curve, tensile strength is typically defined as the stress at which a material begins to deform plastically or develops macroscopic damage. In addition to nano/microscale elastic and plastic deformation, irreversible defects/imperfections at a molecular level can also occur under mechanical stress. Total diffraction analysis, where both the Bragg and diffuse scattering elements of the data are thoroughly assessed, provides an avenue to assess these molecular level imperfections in bulk solids at angstrom-level resolution. The latter can be further improved by using diffraction data from low wavelength synchrotron-based radiation sources to better understand the impact of processing on local-, short- and long-range order in pharmaceutical solids. The influence of structural and mechanical properties of APIs in process-induced defects is the subject of the second case study.

30.2.4.1 Micro/Macroscale Mechanical Assessment
Micro-indentation and macro-indentation are used to characterize the E, H, K_{IC}, and brittleness of macroscopic particles,

such as agglomerates and granules. Such characterization is important in understanding the processing and the influence on consolidation, extrusion, and attrition of particles [42].

The strength of a granule can be interpreted as its resistance to permanent deformation and fracture during a stressing event, and it is normal to consider granule strength as the maximum stress value of the granule before fracture occurs [43]. Granules are not continuum solid particles; instead they are clusters of small particles held together by interparticle bonds. Stress applied at a local point may cause the granule to shear apart at the point of application before the load is transmitted to the rest of the solid volume. The strength of a granule is, therefore, determined by the nature and concentration of its internal bonds and granule microstructure. Apart from the tensile strength, mechanical properties of granular materials can also be described by Young's modulus, Poisson's ratio, tensile yield stress, shear strength, fracture toughness, and hardness [43, 44], which can be measured by various techniques such as confined compression tests, indentation, shearing tests, three-point bending test, single particle compression, and wall impact tests depending on the choice of the mechanical parameter one wish to obtain.

Texture analyzer is an emerging technique in studying mechanical properties at macroscale that are suitable for both crystals and granular materials. It achieves multifunctional applications by taking advantages of the versatility of the indenters. In a typical measurement, as the indenter approaches and withdraws from the specimen, the force is monitored against travel distance in either compression or tensile testings. Similar to three-point bending method, when a tablet is compressed, the breaking force measured can be used to characterize the tensile strength of APIs from the compactibility plot (i.e. tensile strength at zero porosity) [45]. Under confined uniaxial compression, mechanical strength of agglomerates can be characterized. When agglomerated powder is compressed in the uniaxial bed, the force–displacement data are measured. According to Kawakita's model, agglomerate strength can be extrapolated from the fitting of volume–pressure relationship [46]. If the single agglomerate strength needs to be assessed, Adam's model could be applied to extrapolate the agglomerate strength based on the fitting of pressure–strain relationship [47]. Measured strength reflects the initial cracking of agglomerates. Together with the macroscopic mechanical properties mentioned above, agglomerate strength can be used to predict the tableting behavior and impact of agglomerated particles on downstream processes and product performance [42].

Additional dynamic testing of API can be done using hydraulic pressing, single station tablet press, compaction simulator, etc. From the compression profile of solid fraction-compaction pressure, yield strength of material under the dynamic condition can be determined as a measure of deformability [48]. Other measured information like compatibility, tabletability, and compressibility are significantly applied in understanding and process design of API and DP.

30.2.4.2 Nanoscale Mechanical Assessment Quasistatic nano-indentation has emerged as an effective material sparing technique to characterize the mechanical properties of molecular crystals. In the measurement, the applied load and the corresponding depth of the penetration of the indenter are continuously monitored. When the indenter is withdrawn, the unloading displacement is also monitored until the zero load is reached and a final penetration depth is measured. From the loading–unloading curve, E and H can be easily calculated. Nano-indentation has also been applied to measure K_{IC} of molecular crystals. The method is based on the radial cracking when brittle material is indented with a sharp indenter. K_{IC} can be calculated from a semiempirical relation of peak load, E, H, and the length of the radial cracks. These intrinsic properties measured by nano-indentation have been widely applied in the study of milling, and various practical performance-indicating models have been developed. In Vogel and Peukert's model of breakage probability, the rate of particle breakage during milling is described as a function of E, crack propagation velocity, fracture energy (K_{IC}), elastic energy wave propagation, particle size, and flaw size [49–51]. A similar model derived by de Vegt et al. shows that the breakage rate is a function of E, H, and K_{IC} [52]. These models have been applied and validated by the quantitative nano-indentation measurement of E, H, K_{IC}, and H/K_{IC} in Meier et al.'s experimental study [53]. In Taylor et al.'s study, an elegant empirical correlation has been derived between milling effectiveness (% size reduction ratio) and BI (H/K_{IC}) measured with nano-indentation by studying five model drug systems [54]. Nano-indentation is also used to study the mechanical anisotropic properties and long-range molecular/layer migration (e.g. slip plane) of organic crystals [55, 56]. By measuring the slip plane, the fracture behavior and plasticity of the crystal can be predicted. The information is used to investigate the tableting performance of the crystal particles [57]. Finally, nano-indentation equipped with a lateral force transducer can also be extended to characterize friction and wear. Besides, cutting-edge techniques, such as the AFM, can also greatly broaden the scope in the assessment of material intrinsic mechanical properties. For example, AFM can achieve mapping of mechanical properties across the surface in nanoscale. It is possible to study the heterogeneity of mechanical properties, such as stiffness and elastic modulus, across a surface at previously unattainable resolutions. When choosing between nano-indentation equipment for intrinsic mechanical property characterization, factors such as the nature of material (i.e. hardness and its response to force), dynamic force range, load and displacement sensitivity, and the need for high resolution imaging should be considered for the desired application. For example, specialized nano-indentation device with a maximum force load in the mN range may be more suitable for pharmaceutical organic crystals, whereas the AFM nano-indentation, due to its tip size (usually <10 nm) and force range, may be more applicable to mechanical characterization of surface layers and for conducting high resolution imaging.

30.2.4.3 Molecular-Level Assessment of Process-Induced Defects
From a biopharmaceutical perspective, smaller particle size is desirable, because it leads to an increase in specific surface area and surface to volume ratio, thereby improving dissolution rate. Improvements in dissolution rates can improve bioavailability when the overall rates are not permeability limited. Particle size reduction processes are therefore often employed during API production, such as wet milling, screen milling, and impact milling. The milling processes provide small PSD but can also generate surface- or structural-level defects in the particles. These process-induced imperfections can manifest as increased chemical reactivity, loss of crystallinity/physical stability, increase in hygroscopicity, loss of mechanical integrity, and increased propensity for aggregation and agglomeration. Many of these consequences negatively impact downstream processing performance. The defects generated by milling processes can be localized and are not easy to detect with conventional characterization techniques.

A high-energy powder X-ray diffraction (PXRD) method known as total scattering (TS) analysis is a powerful tool that utilizes Bragg and diffuse scattering content of diffraction data to establish a relationship between the real-space arrangement of atoms and X-ray diffraction intensity through the atomic pair distribution function (PDF). When the data is collected at low wavelength over wide diffraction angles, atomic distance resolutions of less than 0.3 Å can be achieved. Such resolution is ideally suited to study existence and emergence of local, short, and long order/disorder in pharmaceutical solids. This structural information is obtained by mathematically treating the TS data to obtain PDF. The pair distribution function, $G(r)$, provides information on local molecular packing arrangement by giving the number of atoms in a spherical shell of unit thickness at a distance r from any reference atom. An advantage of PDF $G(r)$ analysis is that the amplitude of the oscillations in PDF plot gives a direct measure of the structural coherence of the sample. In crystalline samples with some degree of disorder, the amplitude of $G(r)$ falls off faster than dictated by the resolution of scattered wave factor Q ($Q = 4\pi \sin(\theta)/\lambda$). As the positions of the PDF peaks give the separations of pairs of atoms in the structure directly and the width and shape of the PDF peaks contain information about the real atomic probability distribution, the combined assessment of amplitude and peak position/shape therefore provides a complete understanding of local, short- and long-range order/disorder in materials.

30.3 MODELING POWDER FLOWABILITY FROM FUNDAMENTAL PHYSICAL PROPERTIES

30.3.1 Introduction

The powder flow properties of API are very important for high drug loading formulations because they affect downstream processing, such as mixing/blending, capsule and die filling, hopper transfer, fluidization, and coating [58]. Optimization of bulk powder flowability therefore becomes a fairly routine objective during the development of API crystallization and isolation processes. In the practical industrial application, powder flowability is determined by the driving force relative to the flow constraints [59]. The driving force is provided by the unit operation arising from agitation, pressure, shear, gravitational forces, etc. The flow constraints include both extrinsic constraints, which are defined by physical volumes and stresses (e.g. hopper dimension, consolidation stress due to weight of materials, etc.), and intrinsic constraints, which are related to the fundamental material characteristics and properties. Processing challenges related to powder flow can often be overcome by increasing the driving force or by equipment design and selection to relieve extrinsic constraints. However, by understanding the intrinsic material properties that affect powder flowability, crystal engineering solutions can be chosen and tailored to maximize target flow property performance criterion that is dominated by intrinsic constraints of powder flow.

Many examples of modeling powder flowability from particle size and shape distributions have been reported [60–63]. In our case study, we present a rigorous modeling approach based on partial least squares regression (PLS) to predict powder flowability from comprehensive particle physical properties covering both bulk powder properties and surface properties. The latter addresses very important aspects related to interparticle cohesion. The surface properties are important contributors to powder flowability in the context of nongranular materials such as API, as they often exhibit higher surface area to volume ratios due to their relatively smaller particle size compared to commonly used pharmaceutical excipients. The main purpose of this study is to establish and gain fundamental insights into the key material characteristics that would give rise to better powder flowability for more efficient API process development.

30.3.2 Physical Properties that Impact Powder Flowability

Eight APIs and seven excipients with various particle size and shape distributions were used in this study. The flow behavior of each individual material was quantified using the flow function coefficient ff_c (preconsolidated at 1 kPa) obtained from shear cell tests (ring shear tester, Dietmar Schulze, Germany). The ff_c, which is the ratio of consolidation stress to unconfined yield strength, is a commonly used flowability performance indicator in drug process development [58]. Materials with better flow performance are associated with higher ff_c. The ff_c of the model materials were modeled by PLS regression with their bulk and surface properties as tabulated in Table 30.2. The particle size and shape distributions were obtained using DIA (QICPIC, Sympatec, Germany) operated with a wet dispersion system. In DIA, the particle size is characterized by the circular equivalent diameter (CED or D), which is calculated by taking the 2D projected area of the particle image and then calculating the diameter of a circle with the same area. The aspect ratio (AR) is calculated by dividing the minimum Feret's diameter by the maximum Feret's diameter taken perpendicular to the minimum. The circularity (C) is calculated from the ratio of the CED perimeter to the perimeter of the projected area. The solidarity (S), which also describes the macroscale roughness of the particle, is calculated by dividing the projected area by the area of the particle created by smoothing edges and filling in voids. The volume-based (v) size and shape factors were used in this case study. The surface properties of the model compounds were characterized by their dispersive (van der Waals) surface energies, γ^d, and acid–base surface energies, γ^{ab}, using IGC (Surface Energy Analyzer, Surface Measurement Systems, United Kingdom). The total surface energy, γ^{TOTAL}, is the summation of both dispersive and acid–base components. The dispersive surface energies were measured following the Schultz method using decane, nonane, octane, and heptane as probes, and the acid–base surface energies were calculated using monopolar probes, ethyl acetate and dichloromethane [22]. The surface energies were measured at a probe surface coverage of 5%, a background RH of 32%, and a temperature of 30 °C. All samples were stored at approximately 32% RH in a controlled humidity chamber for at least seven days before characterization.

30.3.3 Relationships Between Powder Flowability and Physical Properties

The model APIs and excipients exhibited a broad range of particle size and shape distributions as shown in Figure 30.6, with Dv50 (median volume-based CED) in the range of 21–301 μm. The morphologies of the model materials include acicular, columnar, blade, plate, irregular, and granular morphologies with ARv50 (median volume-based aspect ratio) in the range of 0.21–0.75. The total surface energies varied from 42.5 mJ/m² (low-energy materials) to 77.6 mJ/m² (representing highly energetic and cohesive organic solids), with γ^d ranging between 33.1 and 61.9 mJ/m² and γ^{ab} between 3.8 and

TABLE 30.2 Physical Properties That Impact Powder Flowability

Material Physical Properties			
Parameter	Definition	Equation	Illustration
CED (Dv)	Diameter of a circle with equivalent projected area		
Aspect ratio (ARv)	Feret min divided by Feret max	$\dfrac{F_{min}}{F_{max}}$	
Circularity (Cv)	CED perimeter divided by real perimeter	$\dfrac{2 \times \sqrt{\pi \times Area}}{P}$	
Solidarity (macro-roughness) (Sv)	The object area divided by the area enclosed by the convex hull	$\dfrac{Area\,B}{Area\,A + Area\,B}$	
Surface energies	Surface van der Waals and acid–base interactions	$\gamma^{TOTAL} = \gamma^d + \gamma^{ab}$	
Flowability Performance Indicator			
Parameter	Definition	Equation	
Flow function coefficient (ff_c)	Ratio of major principal stress and unconfined yield strength/stress at a defined normal load	$\dfrac{\sigma_1}{\sigma_c}$	

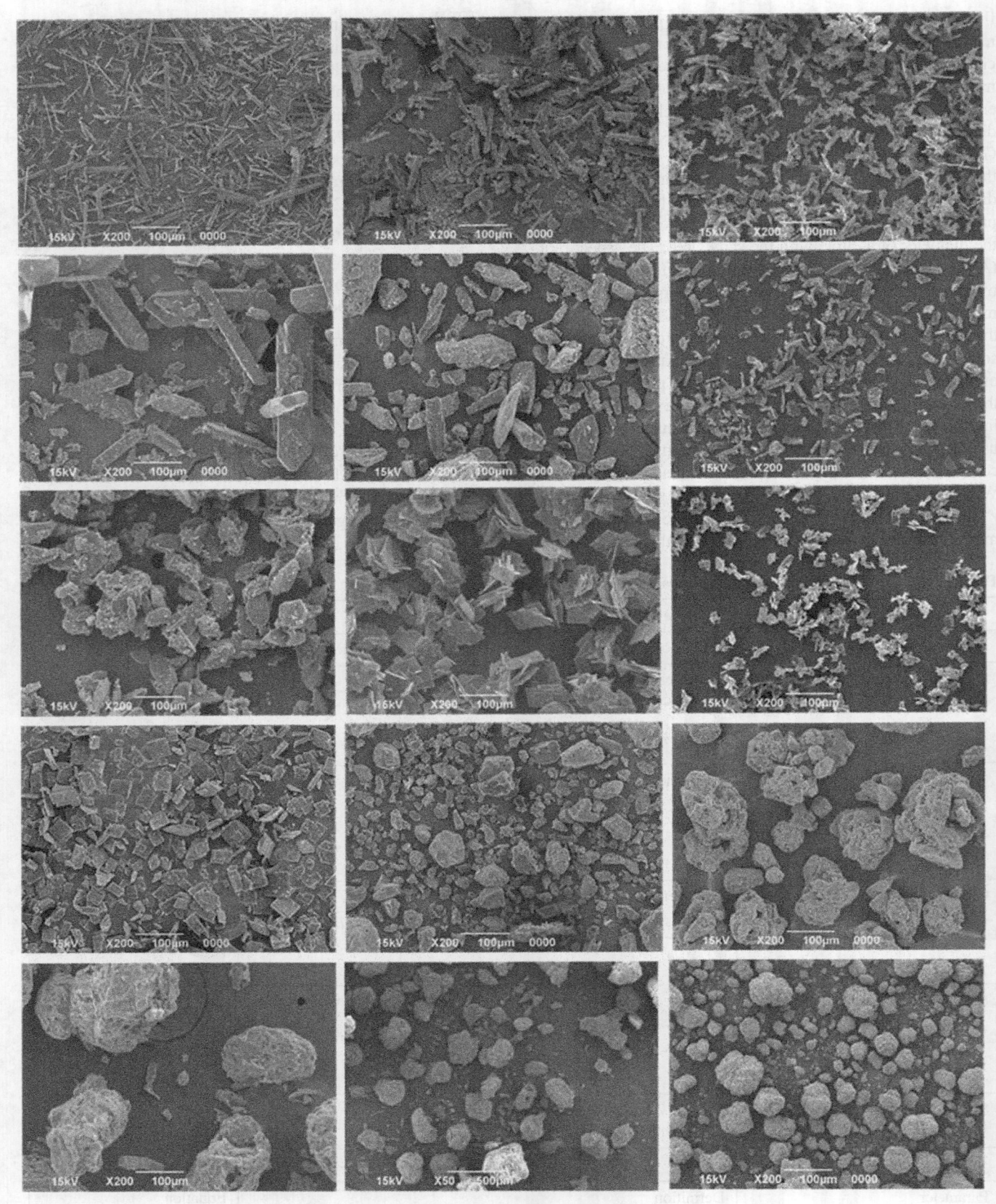

FIGURE 30.6 SEM images of model APIs and excipients.

15.7 mJ/m². The range of ff_c varied between 1.4 (representing very cohesive, poor flowing material) and 12 (representing free flowing material).

The relationships between particle size, shape, surface energies, and ff_c of the compounds are shown in Figure 30.7. Several observations are noted as follows:

- As the particle size (CED, Dv50) increased, ff_c improved.
- At a fixed particle size, increasing particle shape factors (ARv50, Cv50, and Sv50) greatly enhanced ff_c. As can be seen from the changes in the contour lines, the improvement in ff_c became more substantial above a threshold shape factor. For instance, when the ARv50 increased beyond 0.60, the ff_c improved significantly.
- Reducing total surface energy had minor influence on ff_c, unless the surface energies were substantially low.

Through general inspection of the raw data in Figure 30.7, it is therefore clear that higher ff_c are correlated to increase in particle size, shape factors, and reduction in surface energies.

30.3.4 Partial Least Squares Modeling

In order to further establish the influence of particle size, shape, and surface energies on powder flowability, a PLS model was constructed to link the input variables X (material properties) to the model response Y (ff_c). PLS regression can effectively reduce high-dimensional data matrices into low-dimensional subspace and provide good estimates for model response. In order to capture the particle size and shape distributions as input variables, the 10th, 50th, and 90th percentiles of the CED, AR, C, and S were selected together with the two surface energy components, γ^d and γ^{ab}. The input variables and responses were first centered and scaled to have mean of 0 and standard deviation of 1 to prevent data bias due to range of the magnitude of their numerical values. The model compounds were divided into two subsets of data: a model training set with a total of 11 compounds and a validation data set with a total of 4 compounds. The PLS model was constructed using data from the training set and then validated with the validation data set. The optimum number of latent factors was chosen based on cross-validation method

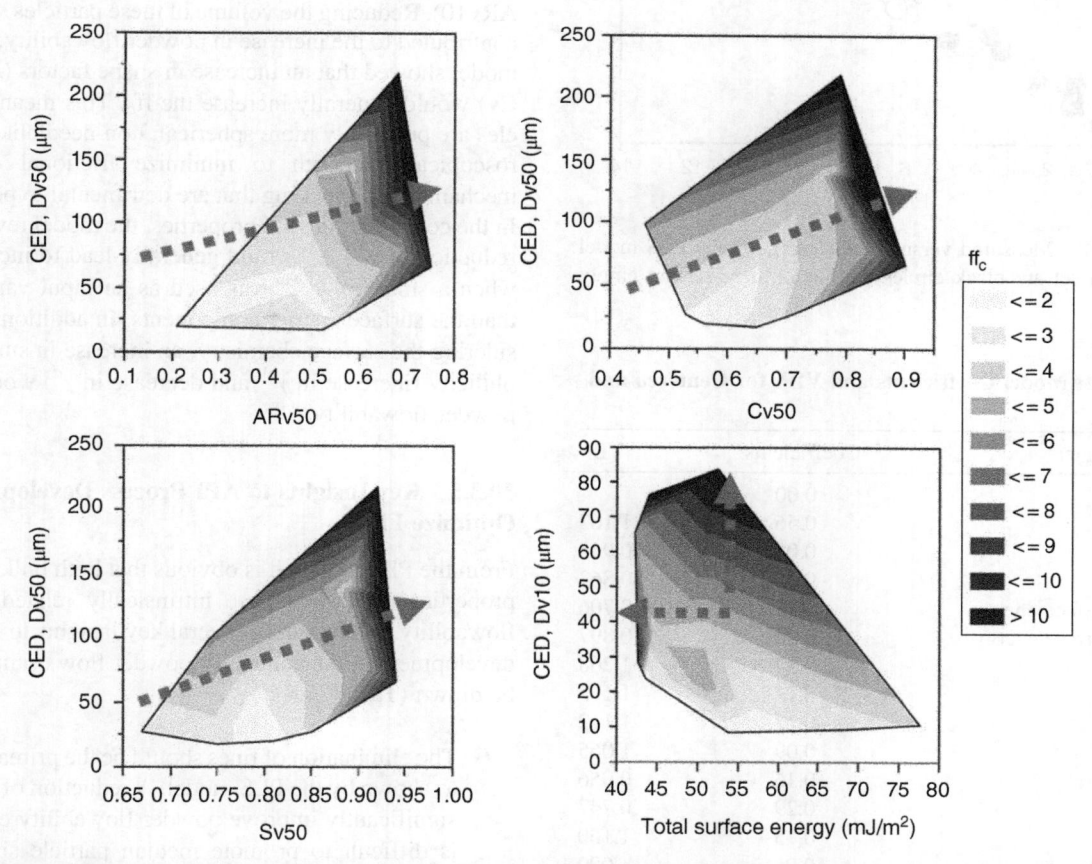

FIGURE 30.7 Contour plots showing relationships between median particle size, median shape factors, total surface energies, and ff_c. Arrow indicates the direction of significant ff_c improvement.

by minimizing root mean predicted residual sum of squares (PRESS) while maximizing percent variation explained for X and Y. The final PLS showed both excellent X and Y variations explained (94.7 and 95.0%, respectively) and low root mean PRESS (1.31).

Figure 30.8 shows the measured ff_c versus predicted ff_c from the PLS model covering both the training and validation data set. It can be seen that the final PLS model had an excellent predictability on ff_c across the range of compounds studied. The model coefficient and variable importance in projection (VIP) score for each input variable X are tabulated in Table 30.3. The VIP score is a measure of a variable's importance for projecting the data to the X–Y scores. If a variable has a small coefficient and a VIP with a threshold value of less than 0.8, then it can be considered as a candidate for removal from the model. It can be seen from Table 30.3 that most variables have VIP scores larger than the threshold value except for γ^d, Cv90, and Sv90. Their relatively low VIP scores were likely due to the fact that the variation in these variables for the model compounds was substantially lower than other variables. For instance, organic crystalline solids have very similar dispersive surface energy component because of their similar London, Debye, and Keesom interactions but widely different acid–base surface energy due to vast differences in the surface propensity for hydrogen bonding, coulombic, and other acid–base interactions. According to the model coefficients, the most significant contributor to ff_c was Dv10, followed by ARv10, γ^{ab}, and Dv90 (excluding those variables with lower than threshold VIP scores). It is known from previous studies that increasing particle size would increase ff_c [60–63], but our model also showed that reducing volume percent of small particles or fines (increase in Dv10) and a reduction of the span of the PSD (as defined by difference of Dv90 and Dv10 divided by Dv50) would also promote an increase in powder flowability. In terms of shape factors, the most significant contributor to ff_c came from particles that were most acicular or needlelike (low ARv10). Reducing the volume of these particles significantly contributed to the increase in powder flowability. In fact, our model showed that an increase in shape factors (ARv, Sv, or Cv) would generally increase the ff_c. This means that particles are preferably more spherical, non-needlelike, and macroscopically smooth to minimize frictional effects and mechanical interlocking that are detrimental to powder flow. In the context of surface properties, the model revealed that a reduction in γ^{TOTAL} would generally lead to increase in ff_c, when a single γ^{TOTAL} was used as an input variable rather than the surface energy components. In addition, when considering the surface chemistry, an increase in surface hydrophilicity (increase in γ^{ab} and decrease in γ^d) would enhance powder flowability.

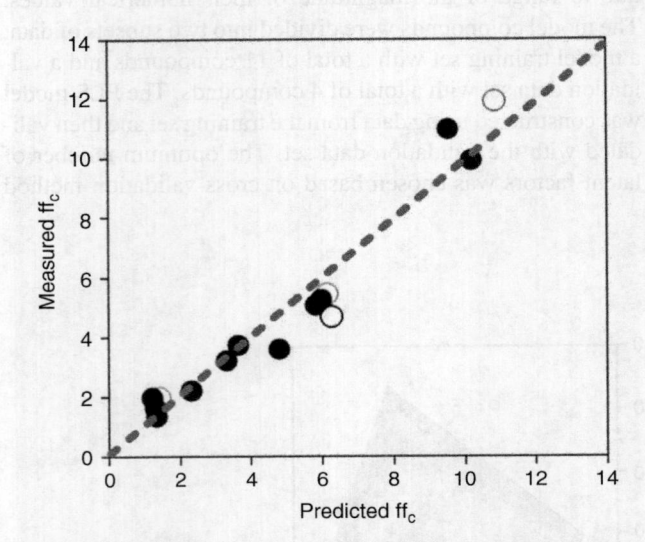

FIGURE 30.8 Measured versus predicted ff_c by the PLS model (training data set are black circles; validation data set are empty circles).

TABLE 30.3 Model Coefficients and VIPs for Centered and Scaled Data

Model Parameters	Coefficients	VIP
Intercept	0.00	
CED, Dv10	0.56	1.163
CED, Dv50	0.05	0.968
CED, Dv90	−0.25	0.863
Dispersive surface energy	−0.13	0.706
Acid–base surface energy	0.26	0.897
ARv10	0.37	1.268
ARv50	0.11	1.165
ARv90	−0.22	1.112
Cv10	0.09	1.065
Cv50	−0.16	0.956
Cv90	0.29	0.747
Sv10	0.15	1.160
Sv50	0.06	1.009
Sv90	0.09	0.703

30.3.5 Key Insights to API Process Development to Optimize Flow

From the PLS model, it is obvious that both bulk and surface properties of particles are intrinsically related to powder flowability. In our study, several key insights to API process development in the context of powder flow optimization can be drawn (Table 30.4):

- The elimination of fines should be the primary objective as shown by the PLS model. A reduction of fines would significantly improve powder flowability even when it is difficult to promote median particle size increase. This could be achieved, for instance, by adding heat–cool cycles at the appropriate stage in the crystallization

TABLE 30.4 Key Physical Insights from the PLS Model of API Physical Properties and ff$_c$

Model Prediction	Direction	Physical Meaning	Effect
Dv10	Increase	Reduction in fines	Significant
Dv10, Dv90	Increase and decrease, respectively	Reduction in PSD span (narrower PSD)	Intermediate
Shape factors (ARv, Cv, and Sv)	Generally increase	Favored cubical, spherical, and macroscopically smooth particles	Intermediate
Total surface energies	Decrease	Reduction in surface cohesive forces	Intermediate
Surface chemistry	Increase in acid–base and decrease in dispersive	Increased hydrophilicity	Intermediate

process to promote Oswald ripening for systems which display temperature-dependent solubility. Careful selection of the milling equipment and design of operation based on the breakage behavior of the API particles can also be used to reduce fines generation and negative consequences for flow.

- While increasing particle size can improve flowability, consideration should also be given to the shape of the PSD with the objective of achieving more uniform PSD. The use of wet milling in conjunction with heat–cool cycles described above is generally applicable solutions to tighten the span of size distribution. Continuous crystallization offers a distinct avenue toward tightening size distribution and achieving consistency.
- Spherical, non-acicular, and macroscopically smooth particles exhibit better flowability, but reducing the amount of the longest needles would also enhance flow. Understanding the impact of solvent and supersaturation on crystal shape can help design crystallization processes that intrinsically provide optimum crystal shape for the selected polymorphic form. A combination of in silico and *in vitro* screening experiments along with direct mechanistic understanding of relative growth rates and mechanism, for instance, via in situ AFM, can be used to build high level of confidence in the scalability and consistency of the shape control. Process-focused optimization techniques mentioned above, including in situ wet milling combined with heat–cool cycles during crystallization, can help further optimize crystal morphology.
- Both surface energies and surface chemistry were shown to play a role in powder flowability. API becomes more cohesive and exhibits higher surface energies primarily due to exposed surface functional groups (as dictated by its shape), surface defect sites, and disorders. From the PLS model, the surface chemistry effect on powder flow was favored by more hydrophilic surfaces such that hydrophobic cohesive interactions were reduced. While attempts can be made to focus the morphology optimization efforts described above toward increasing the relative surface area of hydrophilic surfaces, such attempt may not always be practical when all the other constraints for process design are considered. In such cases, it is of paramount importance that the multiscale understanding of the API physical properties feeds into the formulation selection process, so that appropriate composition and/or processing choices are made to reduce the risk to overall drug substance and DP processing and performance.

30.3.6 Conclusions

In this study, the concept of MST was applied to extract property–performance relationships to provide critical insights for the development of API isolation processes in the context of powder flow optimization. This case study demonstrated that fundamental physical characteristics of API for better powder flowability are limited not necessarily only to larger PSD and more ideal shape factors but also amount of fines, shape of the PSD, and the surface properties such as surface intermolecular interactions and chemistry.

30.4 IMPACT OF API INTRINSIC MECHANICAL PROPERTIES ON PROCESS-INDUCED DISORDER

30.4.1 Introduction

During the manufacturing of DP containing crystalline APIs, drug crystals are often subjected to unit operations such as milling. It is well documented that this pharmaceutical process can generate various types of defects in API crystals [64, 65]. These crystal defects represent regions of higher disorder and higher energy relative to the average overall energy of the crystalline material [66]. These high-energy regions can ultimately affect a number of important pharmaceutical properties of APIs including dissolution rate [67–69], chemical stability [70–72], mechanical properties [73], and moisture sorption [74]. Historically, a range of "bulk" analytical techniques have been used to understand the extent of disorders in pharmaceutical solids including X-ray powder diffraction [75], density [76], heat of solution [77], infrared spectroscopy [78], dissolution rate [79], Raman spectroscopy [80],

SCHEME 30.1 Chemical structures of compound 1 and 2.

solid-state NMR [81], dynamic mechanical analysis [82], differential scanning calorimetry [83], and water vapor sorption. The disorders generated during milling are, however, more local in nature, and are not always easily detectable with the aforementioned methods. It is, therefore, important for pharmaceutical industry to investigate and develop analytical methods that are capable of accurately detecting and assessing the extents of subtle changes in API local structures that can be linked with product stability and performance.

In this case study, a collection of advanced complimentary characterization and analysis tools including synchrotron-based total scattering pair distribution function (TS-PDF), nano-indentation, in situ SPM, and single-crystal X-ray diffraction were applied to holistically characterize and correlate the mechanical properties and crystal structural orders of two API crystals, PARP inhibitor compound 1, 2-([R]-2-methylpyrrolidin-2-yl)-1H-benzimidazole-4-carboxamide, and S1P1 antagonist compound 2, ((1R,3S)-1-amino-3-(4-(7-methoxyheptyl) phenyl)cyclopentyl)methanol(S)-mandelate, following milling under various process conditions [84]. The chemical structures of the two compounds are depicted in Scheme 30.1.

30.4.2 Mechanical Properties

Nano-indentation (Triboindenter TI900, Hysitron, Minneapolis, MN) equipped with a three-sided pyramidal cube corner probe was used to differentiate the intrinsic mechanical properties of 1 and 2. Figure 30.9 shows the characteristic force–displacement profiles of the two compounds subjected to maximum load force (P_{max}) of up to 9 000 µN. Compound 1 required much larger force to reach the same indent depth compared with compound 2, suggesting that compound 1 is an intrinsically harder material. In situ SPM images (Triboindenter TI900, Hysitron, Minneapolis, MN) obtained immediately after nano-indentation for compounds 1 and 2 are displayed in Figure 30.10. For compound 1, cracking lines and material "sink-in" around the indents, especially at high load force, are clearly visible. The cracking of the crystal and the collapse of material inward are indicative of the brittle nature of the material [54]. Compared with compound 1, indents of compound 2 were clear triangular indents. Around the indents, the presence of material "pileup" suggests that this material is much softer, and deformation of this material can occur via a plastic flow mechanism [85, 86]. The shallower slopes of the unloading curves for compound 2 also suggest that this material is less stiff compared with compound 1.

Figure 30.11 shows a comparison of particle reduced elastic modulus and hardness as a function of indentation depth for compounds 1 and 2. The hardness and reduced elastic modulus of the crystals were determined using the method described by Oliver and Pharr [87]. Young's modulus, E, was determined, with Poisson's ratio, v, of 0.30 via

$$E = E_r(1 - v^2)$$

Although the H and E_r are not constant within the indentation depth range explored, compound 1 displays much larger H and E_r compared with compound 2. At constant indent depth, Young's modulus (E) and hardness (H) of compound 1 are almost four times greater than those of compound 2 (Table 30.5). The calculated mechanical properties are

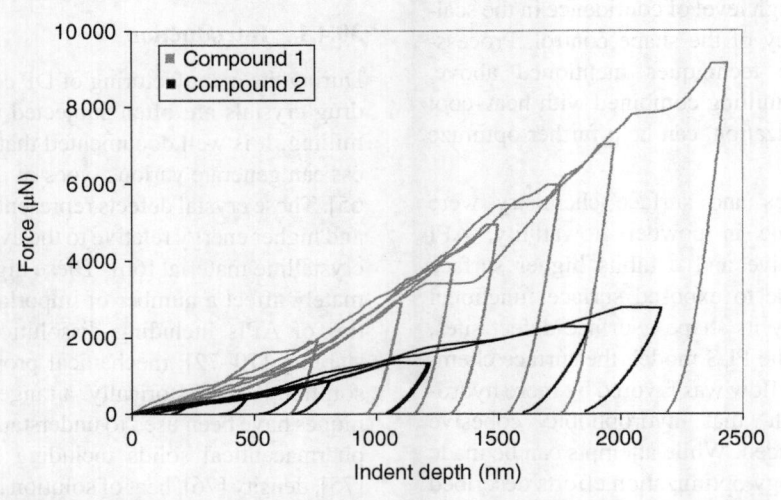

FIGURE 30.9 Representative force–displacement curves of different P_{max} for compounds 1 and 2.

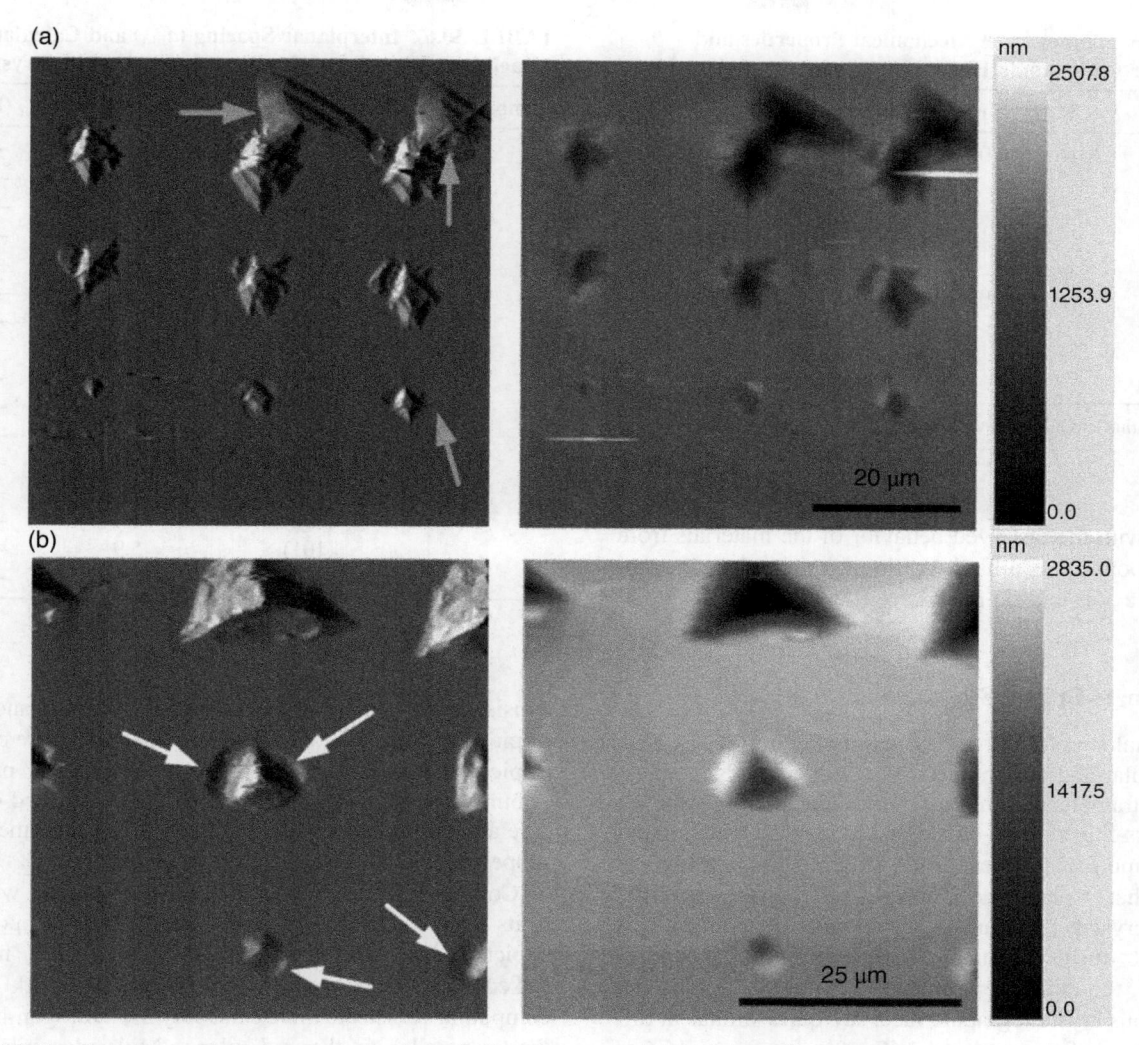

FIGURE 30.10 Scanning probe microscopy gradient (left) and topography (right) profiles of (a) compound 1 and (b) compound 2 immediately post-nano-indentation. Crystal cracking and material "sink-in" (as indicated by arrows) are evident in compound 1, whereas material "pileup" (as indicated by arrows) is evident in compound 2.

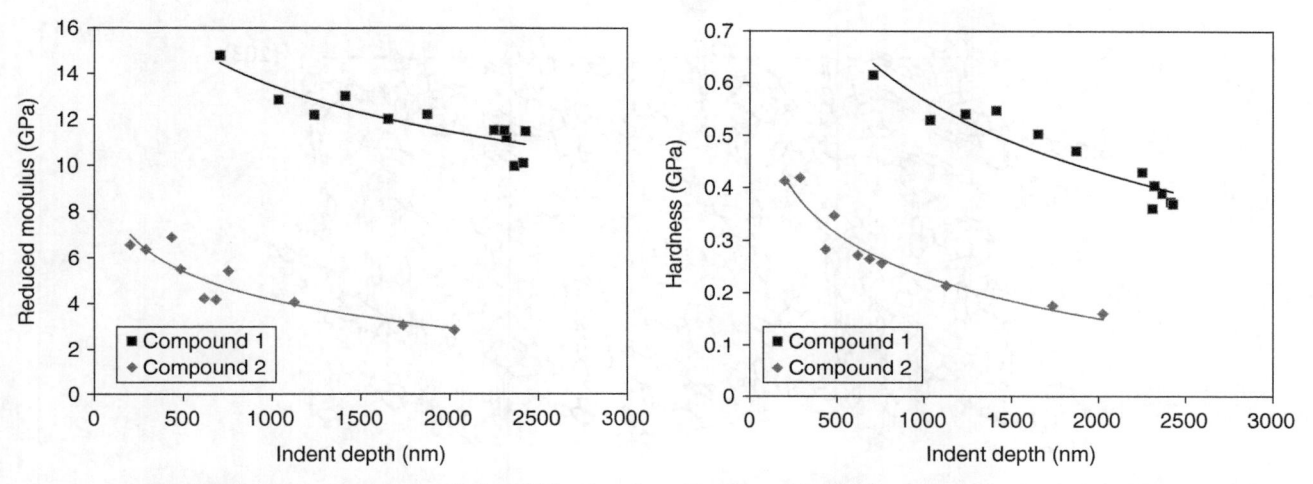

FIGURE 30.11 Reduced modulus and hardness as a function of indent depth for compounds 1 and 2.

TABLE 30.5 Summary of Mechanical Properties and Behavior for Compounds 1 and 2

Compound	E (GPa)	H (GPa)	Observed Events
1	10.4	0.43	Cracking and material "sink-in" around indents show that material exhibits fracture upon indentation (brittleness)
2	2.7	0.15	"Pileup" events indicate that material undergoes plastic flow upon deformation (softness)

Young's modulus and hardness values were calculated at an indent depth of 2000 nm.

TABLE 30.6 Interplanar Spacing (d_{hkl}) and Calculated Attachment Energies of the Compounds 1 and 2 Crystals

Compound	Lattice Plane (hkl)	d_{hkl} (Å)	E_{att} (kcal/mol)
1	(101)	8.0	−16.6
	(102)	7.4	−17.9
	(004)	9.0	−19.6
	(110)	5.8	−19.9
	(103)	6.8	−20.5
	(111)	5.7	−21.4
	(112)	5.5	−22.3
	(104)	6.0	−22.7
2	(001)	23.9	−10.7
	(010)	9.0	−18.6
	(011)	8.3	−18.6
	(110)	5.6	−21.8
	(111)	5.6	−22.3
	(012)	7.1	−22.7
	(101)	5.9	−23.4
	(100)	6.0	−26.0

consistent with the observed behavior of the materials from SPM: compound 1 is a hard, brittle material, whereas compound 2 is a soft, plastic material.

30.4.3 Single-Crystal Structures

In the crystal structure of compound 1 (Figure 30.12), both intramolecular and intermolecular hydrogen bonds are present. The intramolecular and intermolecular hydrogen bonds form a cross-linked three-dimensional network with comparable intermolecular interactions in the three orthogonal directions that is a structure feature known for hard and brittle molecular crystals [7]. Nano-indentation of compound 1 was carried out on its dominant (103) face. As shown in Table 30.6, the attachment energy and interplanar spacing of the various crystallographic faces are quite similar in different directions for compound 1 (E_{att} ranging from −16.6 to −22.7 kcal/mol and d-spacing from 5.5 to 9.0 Å). This is consistent to the bipyramidal crystal morphology and the isotropic structural feature of the compound. Because of its isotropic lattice packing, the intrinsic mechanical properties (Young's modulus, E, and hardness, H) measured on (103) face are considered a good indicator of the bulk mechanical properties of compound 1 crystals [88, 89].

Compound 2 exhibits a thin-plate morphology with (001) as its dominant face. The crystal structure of compound 2 is depicted in Figure 30.13. Unlike compound 1, no cross-linked three-dimensional hydrogen-bond network exists in compound 2. Instead, there exist only 2D hydrogen-bond networks parallel to the a–b plane. Molecular bilayers are formed based on the 2D hydrogen-bond networks. The

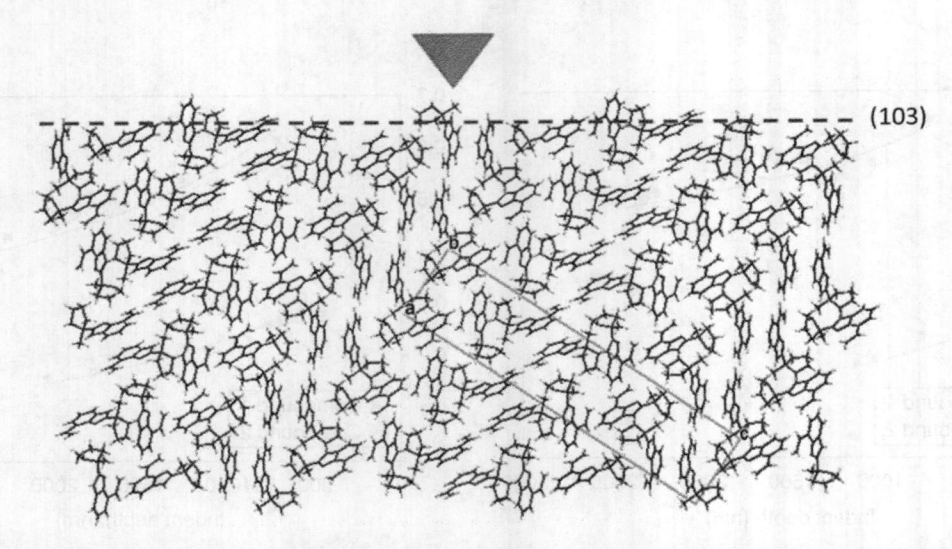

FIGURE 30.12 Single-crystal structure of compound 1 viewed along b-axis. The triangle represents the indentation direction.

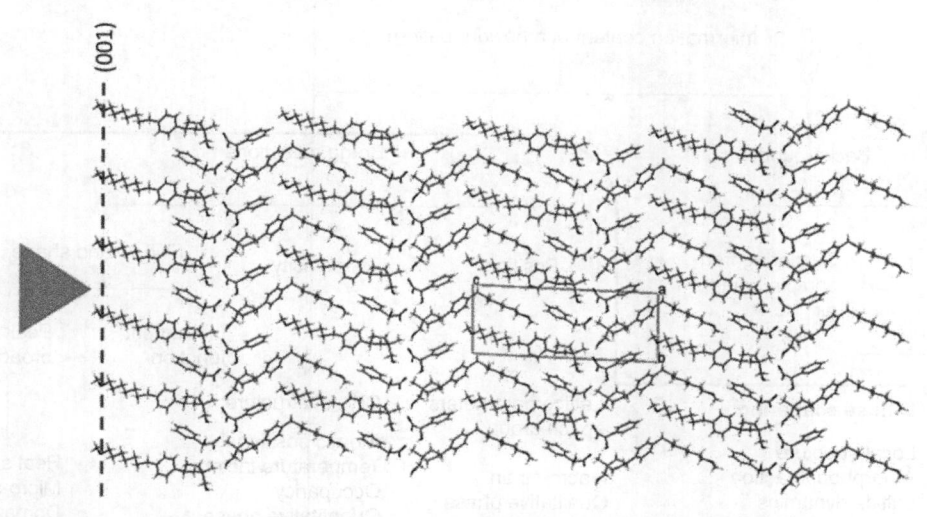

FIGURE 30.13 Single-crystal structure of compound 2 viewed along a-axis. The triangle represents the indentation direction.

packing of compound 2 lattice is anisotropic in such a way that strong and weak interaction patterns are present in nearly orthogonal directions, a structural characteristic typically present in soft and plastic organic crystals [7]. As shown in Table 30.6, the attachment energies of the different faces of compound 2 and their corresponding interplanar distances vary quite significantly in different directions. The (001) face has lowest absolute value of attachment energy (−10.7 kcal/mol), and its corresponding interplanar distance is 23.9 Å. Nano-indentation of compound 2 was carried out on its dominant (001) face that coincides with the primary slip plane of the crystal. Because of the significant anisotropy of compound 2 crystals (thin plate), it was only possible to indent the crystals in a direction normal to (001) face. Nevertheless, the low Young's modulus, E, and hardness, H, obtained on the (001) face are still considered indicative of the overall soft and plastic nature of the compound.

30.4.4 Pair Distribution Function Analysis

Traditionally, powder X-ray diffraction analysis has been focused on the assessment of position and intensity of Bragg reflection peaks to provide information about the long-range order (or the average structure) of a given crystalline phase. The method of TS analysis, on the other hand, uses both the Bragg reflections and diffuse scattering on an equal basis. Assessment of the diffuse scattering component of the powder data provides quantifiable information regarding deviations from the average lattice properties by revealing differences in the local and short-range structure of the material (Figure 30.14) [90]. This structural information is obtained by mathematically treating the TS data to obtain PDF. The pair distribution function, $G(r)$, provides information on local molecular packing arrangement by giving the number of atoms in a spherical shell of unit thickness at a distance r from any reference atom. The crystalline samples were filled into 1 mm diameter Kapton capillary tubes. High-energy X-ray (58 keV, $\lambda = 0.2128$ Å) scattering measurements were performed at beamline 11-ID-B of the Advanced Photon Source (APS) located at Argonne National Laboratory (ANL). A PerkinElmer amorphous Si 2D image plate detector was placed perpendicular to the high-energy X-ray beam, 170 mm behind the capillary samples. Scattering patterns for the samples were collected as a function of the magnitude of the scattered wave vector Q, which is given by $Q = 4\pi \sin(\theta)/\lambda$. To obtain high accuracy and adequate real-space resolution of PDF, the scattering data were measured at a scattered angle, 2θ, up to 55°. The corresponding Q_{max} was greater than 27 Å^{-1} (compared with a value of ~8 Å^{-1}, equivalent to 160° 2θ, typically achieved with a lab Cu-based power X-ray diffraction). To achieve good statistic counts, each sample was collected for 600 seconds.

Compounds 1 and 2 crystalline materials were milled under four different milling conditions with increasing intensity and time: low-speed 9 000 rpm pin mill (9 000 rpm pm), high-speed 36 000 rpm pin mill (36 000 rpm pm), short-time (1 minute) cryomill (1 min cm), and long-time (7 minutes) cryomill (7 min cm). For pin milling, the residence time of sample powder in the mill was less than one minute. The TS-PDFs from compound 1 samples obtained under different milling conditions are shown in Figure 30.15. The PDF of unmilled compound 1 was added for comparison. The PDFs of compound 1 milling samples, with the exception of the seven minute cryomilling sample, were visually identical to that of the unmilled, suggesting the hard crystals of compound 1 exhibit very little or essentially no disorders under typical pin milling (9 000–36 000 rpm) and short-time (1 minute) cryomilling conditions. Only when subjected to extensive milling, such as long-time (seven minute) cryomilling, crystals of compound 1 began

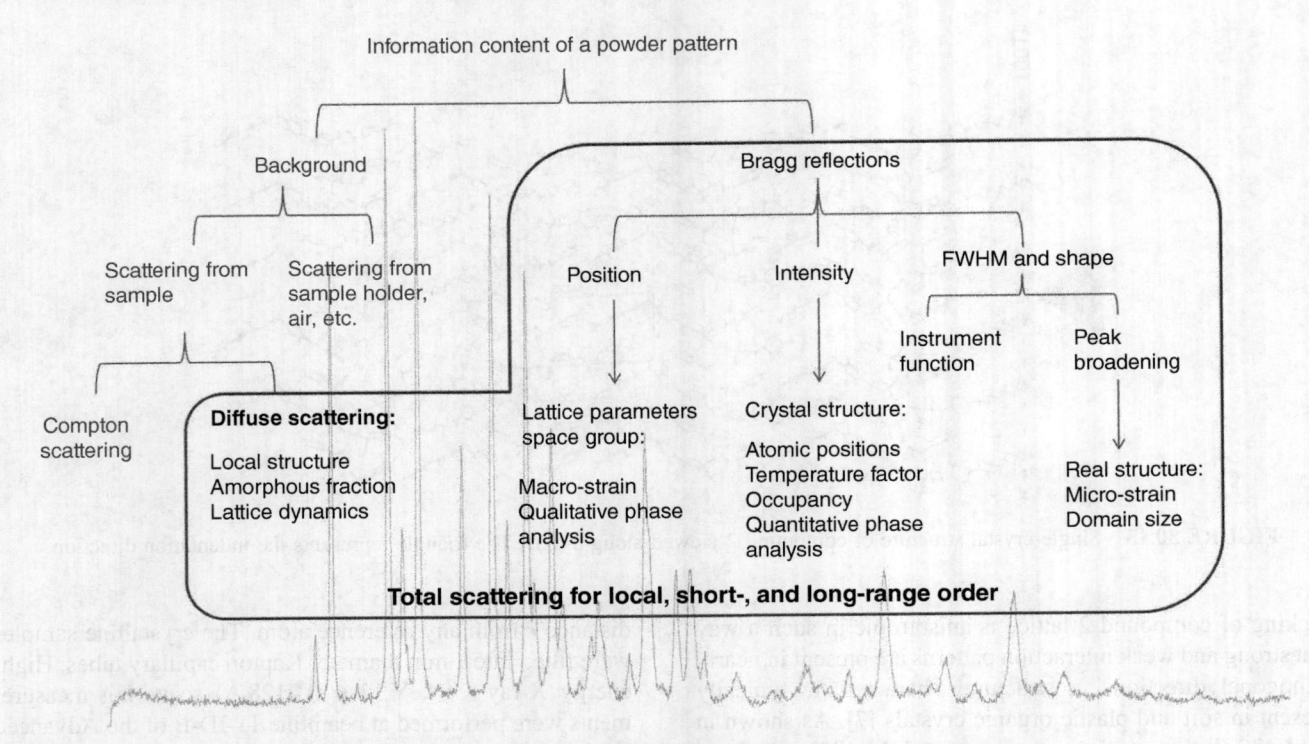

FIGURE 30.14 Information content of a powder X-ray diffraction pattern.

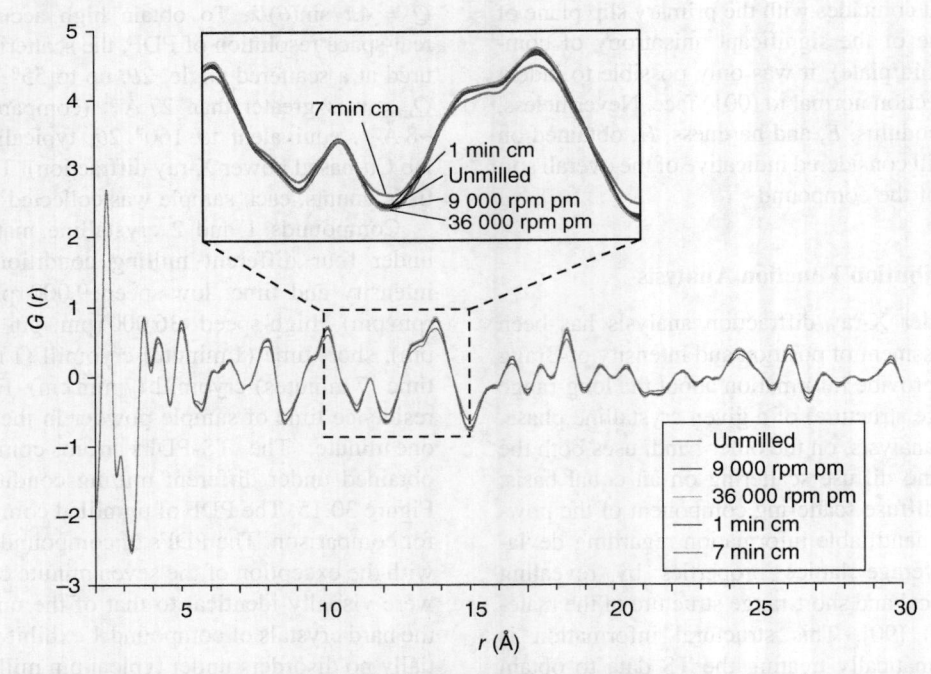

FIGURE 30.15 Comparison of total scattering pair distribution functions from samples of compound 1 prepared under different milling conditions: unmilled, 9 000 rpm pin milling, 36 000 rpm pin milling, 1 minute cryomilling, and 7 minutes cryomilling.

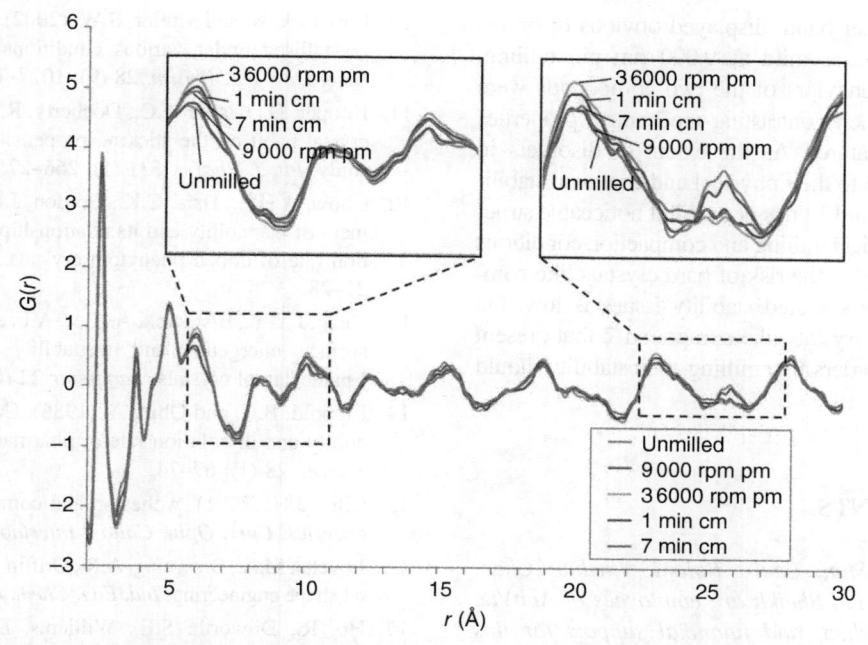

FIGURE 30.16 Comparison of total scattering pair distribution functions from samples of compound 2 prepared under different milling conditions: unmilled, 9 000 rpm pin milling, 36 000 rpm pin milling, 1 minute cryomilling, and 7 minutes cryomilling.

to show discernible changes in its PDF plot. The PDF plots of compound 1 milling samples indicate that while the particle size of compound 1 was reduced with increasing milling intensity and duration, its internal structure was left largely intact.

A similar analysis was also carried out for compound 2. The TS-PDFs of compound 2 milling samples are shown in Figure 30.16. Compared with the unmilled, the PDF plots of milling samples showed not only obvious changes in peak amplitudes but also small and noticeable differences in peak positions and shape, especially at radial distances r of about 6, 8, 10, 23, and 26 Å. An advantage of PDF $G(r)$ analysis is that the amplitude of the oscillations in PDF plot gives a direct measure of the structural coherence of the sample. In crystalline samples with some degree of disorder, the amplitude of $G(r)$ falls off faster than dictated by the resolution of scattered wave factor Q ($Q = 4\pi \sin(\theta)/\lambda$). As the positions of the PDF peaks give the separations of pairs of atoms in the structure directly and the width and shape of the PDF peaks contain information about the real atomic probability distribution [89], the PDF peak position and shape changes observed in compound 2 milling samples indicate substantial changes in local structural orders. Moreover, as the longest intramolecular atom–atom distance in compound 2 crystals is about 17 Å and the shortest non-H-bonding intermolecular distance is about 4 Å, the PDF difference seen above between the unmilled and milled samples suggests that there are likely structural changes in both the molecular configuration and the intermolecular stacking.

The different milling behaviors of the two compounds, as detected through the PDF analysis, are consistent with their crystal structure features and the different mechanical properties measured by nano-indentation. For the hard and brittle material of compound 1, the molecules are interlocked through the cross-linked three-dimensional hydrogen-bond network, resulting in a very rigid isotropic crystal lattice. Compound 1 is, therefore, more resistant to the destructive milling force. On the other hand, for the soft and plastic material of compound 2, the 2D hydrogen-bond networks hold the molecules together in bilayers, and the long methoxyheptyl tails among the stacking bilayers lack strong interactions with each other, particularly along the c-axis. Therefore the compound has a higher propensity to change its molecular configuration and stacking structure upon milling.

30.4.5 Conclusions

In this case study, nano-indentation was successfully used to determine the intrinsic mechanical properties, hardness, and reduced modulus of two API compounds 1 and 2. Synchrotron TS and pair distribution function analysis was successfully applied to assess the effects of different milling conditions on the structural disorders of the two drug crystals. The study showed that the hard compound 1 crystals exhibited very little to essentially no structural disorders under typical pharmaceutical milling (pin mill speed of 9 000–36 000 rpm) conditions. To create discernible disorders in hard crystals of compound 1, long-time cryomilling had to be used. The soft crystals of compound 2, on the other hand, displayed obvious disorders under milling conditions as mild as 9 000 rpm pin milling. The different milling behaviors of the two compounds were found consistent with their contrasting mechanical properties and crystal structure features. As the extent of disorders in API crystals is often tied to their physical and chemical stabilities, the fact that compound 1 does not exhibit noticeable structural disorders under typical milling and compaction conditions allows one to conclude that the risk of hard crystals like compound 1 showing process-related stability issues is low. On the other hand, for soft crystals like compound 2 that present noticeable structural disorders after milling, their stability should be monitored closely.

ACKNOWLEDGMENTS

Raimundo Ho, Yujin Shin, Laura Poloni, Yinshan Chen, Shuang Chen, and Ahmad Sheikh are employees of AbbVie. The design, study conduct, and financial support for this research was provided by AbbVie. AbbVie participated in the interpretation of data, review, and approval of the publication.

REFERENCES

1. Sun, C.C. (2009). Materials science tetrahedron—a useful tool for pharmaceutical Research and Development. *J. Pharm. Sci.* 98 (5): 1671–1687.
2. Yang, P. and Tarascon, J.-M. (2012). Towards systems materials engineering. *Nat. Mater.* 11: 560.
3. Brittain, H.G. (2009). *Polymorphism in Pharmaceutical Solids*, 2e. New York: Informa Healthcare USA, Inc.
4. Beyer, T., Day, G.M., and Price, S.L. (2001). The prediction, morphology, and mechanical properties of the polymorphs of paracetamol. *J. Am. Chem. Soc.* 123 (21): 5086–5094.
5. Khomane, K.S. and Bansal, A.K. (2013). Weak hydrogen bonding interactions influence slip system activity and compaction behavior of pharmaceutical powders. *J. Pharm. Sci.* 102 (12): 4242–4245.
6. Yadav, J.A., Khomane, K.S., Modi, S.R. et al. (2017). Correlating single crystal structure, nanomechanical, and bulk compaction behavior of Febuxostat polymorphs. *Mol. Pharm.* 14 (3): 866–874.
7. Reddy, C.M., Padmanabhan, K.A., and Desiraju, G.R. (2006). Structure–property correlations in bending and brittle organic crystals. *Cryst. Growth Des.* 6 (12): 2720–2731.
8. Bag, P.P., Chen, M., Sun, C.C., and Reddy, C.M. (2012). Direct correlation among crystal structure, mechanical behaviour and tabletability in a trimorphic molecular compound. *CrystEngComm* 14 (11): 3865–3867.
9. York, P. (1983). Solid-state properties of powders in the formulation and processing of solid dosage forms. *Int. J. Pharm.* 14 (1): 1–28.
10. Rasenack, N. and Muller, B.W. (2002). Properties of ibuprofen crystallised under various conditions: a comparative study. *Drug Dev. Ind. Pharm.* 28 (9): 1077–1089.
11. Hooper, D., Clarke, F.C., Docherty, R. et al. (2017). Effects of crystal habit on the sticking propensity of ibuprofen—a case study. *Int. J. Pharm.* 531 (1): 266–275.
12. Chow, A.H.L., Hsia, C.K., Gordon, J.D. et al. (1995). Assessment of wettability and its relationship to the intrinsic dissolution rate of doped phenytoin crystals. *Int. J. Pharm.* 126 (1): 21–28.
13. Heng, J.Y.Y., Bismarck, A., Lee, A.F. et al. (2006). Anisotropic surface energetics and wettability of macroscopic form I paracetamol crystals. *Langmuir* 22 (6): 2760–2769.
14. Lippold, B.C. and Ohm, A. (1986). Correlation between wettability and dissolution rate of pharmaceutical powders. *Int. J. Pharm.* 28 (1): 67–74.
15. Clint, J.H. (2001). Adhesion and components of solid surface energies. *Curr. Opin. Colloid Interface Sci.* 6 (1): 28–33.
16. Lovette, M.A., Browning, A.R., Griffin, D.W. et al. (2008). Crystal shape engineering. *Ind. Eng. Chem. Res.* 47 (24): 9812–9833.
17. Ho, R., Dilworth, S.E., Williams, D.R., and Heng, J.Y.Y. (2011). Role of surface chemistry and energetics in high shear wet granulation. *Ind. Eng. Chem. Res.* 50 (16): 9642–9649.
18. Newell, H.E., Buckton, G., Butler, D.A. et al. (2001). The use of inverse phase gas chromatography to measure the surface energy of crystalline, amorphous, and recently milled lactose. *Pharm. Res.* 18 (5): 662–666.
19. Rumondor, A.C.F., Marsac, P.J., Stanford, L.A., and Taylor, L.S. (2009). Phase behavior of poly(vinylpyrrolidone) containing amorphous solid dispersions in the presence of moisture. *Mol. Pharm.* 6 (5): 1492–1505.
20. Fichtner, F., Mahlin, D., Welch, K. et al. (2008). Effect of surface energy on powder compactibility. *Pharm. Res.* 25 (12): 2750–2759.
21. Chow, A.H.L., Tong, H.H.Y., Chattopadhyay, P., and Shekunov, B.Y. (2007). Particle engineering for pulmonary drug delivery. *Pharm. Res.* 24 (3): 411–437.
22. Ho, R. and Heng, J.Y.Y. (2013). A review of inverse gas chromatography and its development as a tool to characterize anisotropic surface properties of pharmaceutical solids. *KONA Powder Part. J.* 30: 164–180.
23. Das, S.C., Larson, I., Morton, D.A.V., and Stewart, P.J. (2011). Determination of the polar and total surface energy distributions of particulates by inverse gas chromatography. *Langmuir* 27 (2): 521–523.
24. Ho, R., Heng, J.Y.Y., Dilworth, S.E. et al. (2009). Determination of surface heterogeneity of D-mannitol by sessile drop contact angle and inverse gas chromatography. *Int. J. Pharm.* 387 (1–2): 79–86.
25. Smith, R.R., Williams, D.R., Burnett, D.J., and Heng, J.Y.Y. (2014). A new method to determine dispersive surface energy site distributions by inverse gas chromatography. *Langmuir* 30 (27): 8029–8035.
26. Khoo, J.Y., Williams, D.R., and Heng, J.Y.Y. (2010). Dehydration kinetics of pharmaceutical hydrate: effects of environmental conditions and crystal forms. *Dry. Technol.* 28 (10): 1164–1169.

27. Vitez, I.M., Newman, A.W., Davidovich, M., and Kiesnowski, C. (1998). The evolution of hot-stage microscopy to aid solid-state characterizations of pharmaceutical solids. *Thermochim. Acta* 324 (1): 187–196.

28. Eddleston, M.D., Bithell, E.G., and Jones, W. (2010). Transmission electron microscopy of pharmaceutical materials. *J. Pharm. Sci.* 99 (9): 4072–4083.

29. Klang, V., Valenta, C., and Matsko, N.B. (2013). Electron microscopy of pharmaceutical systems. *Micron* 44 (Supplement C): 45–74.

30. Liao, X. and Wiedmann, T.S. (2004). Characterization of pharmaceutical solids by scanning probe microscopy. *J. Pharm. Sci.* 93 (9): 2250–2258.

31. Shur, J. and Price, R. (2012). Advanced microscopy techniques to assess solid-state properties of inhalation medicines. *Adv. Drug Deliv. Rev.* 64 (4): 369–382.

32. Pygall, S.R., Whetstone, J., Timmins, P., and Melia, C.D. (2007). Pharmaceutical applications of confocal laser scanning microscopy: the physical characterisation of pharmaceutical systems. *Adv. Drug Deliv. Rev.* 59 (14): 1434–1452.

33. Martin, D.C., Chen, J., Yang, J. et al. (2005). High resolution electron microscopy of ordered polymers and organic molecular crystals: recent developments and future possibilities. *J. Polym. Sci. B Polym. Phys.* 43 (14): 1749–1778.

34. Poloni, L.N., Zhong, X., Ward, M.D., and Mandal, T. (2017). Best practices for real-time in situ atomic force and chemical force microscopy of crystals. *Chem. Mater.* 29 (1): 331–345.

35. Poloni, L.N., Zhu, Z., Garcia-Vázquez, N. et al. (2017). Role of molecular recognition in l-cystine crystal growth inhibition. *Cryst. Growth Des.* 17 (5): 2767–2781.

36. Poloni, L.N., Ford, A.P., and Ward, M.D. (2016). Site discrimination and anisotropic growth inhibition by molecular imposters on highly dissymmetric crystal surfaces. *Cryst. Growth Des.* 16 (9): 5525–5541.

37. Maugis, D. (2000). *Contact, Adhesion and Rupture of Elastic Solids*. Berlin, Heidelberg: Springer-Verlag.

38. Mandal, T. and Ward, M.D. (2013). Determination of specific binding interactions at l-cystine crystal surfaces with chemical force microscopy. *J. Am. Chem. Soc.* 135 (15): 5525–5528.

39. Shekunov, B.Y., Chattopadhyay, P., Tong, H.H.Y., and Chow, A.H.L. (2007). Particle size analysis in pharmaceutics: principles, methods and applications. *Pharm. Res.* 24 (2): 203–227.

40. Nalluri, V.R., Schirg, P., Gao, X. et al. (2010). Different modes of dynamic image analysis in monitoring of pharmaceutical dry milling process. *Int. J. Pharm.* 391 (1): 107–114.

41. Yu, W. and Hancock, B.C. (2008). Evaluation of dynamic image analysis for characterizing pharmaceutical excipient particles. *Int. J. Pharm.* 361 (1–2): 150–157.

42. Bika, D.G., Gentzler, M., and Michaels, J.N. (2001). Mechanical properties of agglomerates. *Powder Technol.* 117 (1): 98–112.

43. Reynolds, G.K., Fu, J.S., Cheong, Y.S. et al. (2005). Breakage in granulation: a review. *Chem. Eng. Sci.* 60 (14): 3969–3992.

44. Bika, D.G., Gentzler, M., and Michaels, J.N. (2001). Mechanical properties of agglomerates. *Powder Technol.* 117 (1–2): 98–112.

45. Hancock, B.C., Clas, S.D., and Christensen, K. (2000). Microscale measurement of the mechanical properties of compressed pharmaceutical powders. 1: The elasticity and fracture behavior of microcrystalline cellulose. *Int. J. Pharm.* 209 (1–2): 27–35.

46. Kawakita, K. and Lüdde, K.-H. (1971). Some considerations on powder compression equations. *Powder Technol.* 4 (2): 61–68.

47. Adams, M.J., Mullier, M.A., and Seville, J.P.K. (1994). Agglomerate strength measurement using a uniaxial confined compression test. *Powder Technol.* 78 (1): 5–13.

48. Amidon, G.E., Secreast, P.J., and Mudie, D. (2009). Particle, powder, and compact characterization. In: *Developing Solid Oral Dosage Forms* (ed. Y. Qiu, Y. Chen, G.G.Z. Zhang, et al.), 163–186. San Diego: Academic Press.

49. Vogel, L. and Peukert, W. (2002). Characterisation of grinding-relevant particle properties by inverting a population balance model. *Part. Part. Syst. Charact.* 19: 149–157.

50. Vogel, L. and Peukert, W. (2003). Breakage behaviour of different materials—construction of a mastercurve for the breakage probability. *Powder Technol.* 129 (1): 101–110.

51. Vogel, L. and Peukert, W. (2005). From single particle impact behaviour to modelling of impact mills. *Chem. Eng. Sci.* 60 (18): 5164–5176.

52. de Vegt, O., Vromans, H., Faassen, F., and van der Voort Maarschalk, K. (2005). Milling of organic solids in a jet mill. Part 1: Determination of the selection function and related mechanical material properties. *Part. Part. Syst. Charact.* 22 (2): 133–140.

53. Meier, M., John, E., Wieckhusen, D. et al. (2009). Influence of mechanical properties on impact fracture: prediction of the milling behaviour of pharmaceutical powders by nanoindentation. *Powder Technol.* 188 (3): 301–313.

54. Taylor, L.J., Papadopoulos, D.G., Dunn, P.J. et al. (2004). Predictive milling of pharmaceutical materials using nanoindentation of single crystals. *Org. Process. Res. Dev.* 8 (4): 674–679.

55. Bhatt, P.M., Ravindra, N.V., Banerjee, R., and Desiraju, G.R. (2005). Saccharin as a salt former. Enhanced solubilities of saccharinates of active pharmaceutical ingredients. *Chem. Commun.* (8): 1073–1075. https://doi.org/10.1039/B416137H.

56. Varughese, S., Kiran, M.S.R.N., Ramamurty, U., and Desiraju, G.R. (2013). Nanoindentation in crystal engineering: quantifying mechanical properties of molecular crystals. *Angew. Chem. Int. Ed.* 52 (10): 2701–2712.

57. Feng, Y., Grant, D.J.W., and Sun, C.C. (2007). Influence of crystal structure on the tableting properties of n-alkyl 4-hydroxybenzoate esters (parabens). *J. Pharm. Sci.* 96 (12): 3324–3333.

58. Jager, P.D., Bramante, T., and Luner, P.E. (2015). Assessment of pharmaceutical powder flowability using shear cell-based methods and application of Jenike's methodology. *J. Pharm. Sci.* 104 (11): 3804–3813.

59. Behringer, R., Louge, M., McElwaine, J. et al. (2017). *Report of the IFPRI Powder Flow Working Group SAR30–08*.

60. Capece, M., Ho, R., Strong, J., and Gao, P. (2015). Prediction of powder flow performance using a multi-component granular bond number. *Powder Technol.* 286 (Supplement C): 561–571.

61. Fu, X., Huck, D., Makein, L. et al. (2012). Effect of particle shape and size on flow properties of lactose powders. *Particuology* 10 (2): 203–208.
62. Sandler, N. and Wilson, D. (2010). Prediction of granule packing and flow behavior based on particle size and shape analysis. *J. Pharm. Sci.* 99 (2): 958–968.
63. Yu, W., Muteki, K., Zhang, L., and Kim, G. (2011). Prediction of bulk powder flow performance using comprehensive particle size and particle shape distributions. *J. Pharm. Sci.* 100 (1): 284–293.
64. Dialer, V. and Kuessner, K. (1973). High surface energy absorption in vibration milling and influence in conditions of fine grinding. *Kolloid-Z.U.Z. Poly.* 251: 710–715.
65. Saleki-Gerhardt, A., Ahlneck, C., and Zografi, G. (1994). Assessment of disorder in crystalline solids. *Int. J. Pharm.* 101 (3): 237–247.
66. Zhang, J., Ebbens, S., Chen, X. et al. (2006). Determination of the surface free energy of crystalline and amorphous lactose by atomic force microscopy adhesion measurement. *Pharm. Res.* 23 (2): 401–407.
67. Burt, H.M. and Mitchell, A.G. (1981). Crystal defects and dissolution. *Int. J. Pharm.* 9 (2): 137–152.
68. Tawashi, R. (1968). Dissolution rates of crystalline drugs. *J Mondial de Pharmacie* 11: 371–379.
69. Ward, G.H. and Schultz, R.K. (1995). Process-induced crystallinity changes in albuterol sulfate and its effect on powder physical stability. *Pharm. Res.* 12 (5): 773–779.
70. Byrn, S.R., Pfeiffer, R.R., Stephenson, G. et al. (1994). Solid-state pharmaceutical chemistry. *Chem. Mater.* 6 (8): 1148–1158.
71. Byrn, S.R., Xu, W., and Newman, A.W. (2001). Chemical reactivity in solid-state pharmaceuticals: formulation implications. *Adv. Drug Deliv. Rev.* 48 (1): 115–136.
72. Shalaev, E., Shalaeva, M., and Zografi, G. (2002). The effect of disorder on the chemical reactivity of an organic solid, tetraglycine methyl ester: change of the reaction mechanism. *J. Pharm. Sci.* 91 (2): 584–593.
73. Wildfong, P.L.D., Hancock, B.C., Moore, M.D., and Morris, K.R. (2006). Towards an understanding of the structurally based potential for mechanically activated disordering of small molecule organic crystals. *J. Pharm. Sci.* 95 (12): 2645–2656.
74. Ahlneck, C. and Zografi, G. (1990). The molecular basis of moisture effects on the physical and chemical stability of drugs in the solid state. *Int. J. Pharm.* 62 (2): 87–95.
75. Klung, H. and Alexander, L. (1974). *X-Ray Diffraction Procedures for Polycrystalline and Amorphous Materials*. New York: Wiley.
76. Duncan-Hewitt, W.C. and Grant, D.J.W. (1986). True density and thermal expansivity of pharmaceutical solids: comparison of methods and assessment of crystallinity. *Int. J. Pharm.* 28 (1): 75–84.
77. Pikal, M.J., Lukes, A.L., Lang, J.E., and Gaines, K. (1978). Quantitative crystallinity determinations for β-lactam antibiotics by solution calorimetry: correlations with stability. *J. Pharm. Sci.* 67 (6): 767–773.
78. Black, D.B. and Lovering, E.G. (1977). Estimation of the degree of crystallinity in digoxin by X-ray and infrared methods. *J. Pharm. Pharmacol.* 29 (1): 684–687.
79. Hendriksen, B.A. (1990). Characterization of calcium fenoprofen 1. Powder dissolution rate and degree of crystallinity. *Int. J. Pharm.* 60 (3): 243–252.
80. Strachan, C.J., Rades, T., Gordon, K.C., and Rantanen, J. (2007). Raman spectroscopy for quantitative analysis of pharmaceutical solids. *J. Pharm. Pharmacol.* 59 (2): 179–192.
81. Offerdahl, T.J., Salsbury, J.S., Dong, Z. et al. (2005). Quantitation of crystalline and amorphous forms of anhydrous neotame using 13C CPMAS NMR spectroscopy. *J. Pharm. Sci.* 94 (12): 2591–2605.
82. Royall, P.G., Huang, C.-Y., Tang, S.-W.J. et al. (2005). The development of DMA for the detection of amorphous content in pharmaceutical powdered materials. *Int. J. Pharm.* 301 (1): 181–191.
83. Zimper, U., Aaltonen, J., McGoverin, M.C. et al. (2010). Quantification of process induced disorder in milled samples using different analytical techniques. *Pharmaceutics* 2 (1): 30–49.
84. Chen, S., Sheikh, A.Y., and Ho, R. (2014). Evaluation of effects of pharmaceutical processing on structural disorders of active pharmaceutical ingredient crystals using nanoindentation and high-resolution total scattering pair distribution function analysis. *J. Pharm. Sci.* 103 (12): 3879–3890.
85. Cao, X., Morganti, M., Hancock, B.C., and Masterson, V.M. (2010). Correlating particle hardness with powder compaction performance. *J. Pharm. Sci.* 99 (10): 4307–4316.
86. Ramos, K.J. and Bahr, D.F. (2011). Mechanical behavior assessment of sucrose using nanoindentation. *J. Mater. Res.* 22 (7): 2037–2045.
87. Oliver, W.C. and Pharr, G.M. (1992). An improved technique for determining hardness and elastic modulus using load and displacement sensing indentation experiments. *J. Mater. Res.* 7 (6): 1564–1583.
88. Egart, M., Ilić, I., Janković, B. et al. (2014). Compaction properties of crystalline pharmaceutical ingredients according to the Walker model and nanomechanical attributes. *Int. J. Pharm.* 472 (1): 347–355.
89. Egart, M., Janković, B., Lah, N. et al. (2015). Nanomechanical properties of selected single pharmaceutical crystals as a predictor of their bulk behaviour. *Pharm. Res.* 32 (2): 469–481.
90. Billinge, S. (2008). Local structure from total scattering and atomic pair distribution function (PDF) analysis. In: *Powder Diffraction: Theory and Practice* (ed. R. Dinnebier and S. Billinge), 464–493. Cambridge: The Royal Society of Chemistry.

PART VIII

SEPARATIONS, FILTRATION, DRYING AND MILLING

31

THE DESIGN AND ECONOMICS OF LARGE-SCALE CHROMATOGRAPHIC SEPARATIONS

FIROZ D. ANTIA

Antisense Oligonucleotide Process Development and Manufacturing, Biogen Inc., Cambridge, MA, USA

31.1 INTRODUCTION

Chromatography ("color writing") was the fanciful name given in 1906 by the botanist Mikhail Tswett to the adsorptive separation method he employed to separate plant pigments on columns of calcium carbonate using petroleum ether as the eluting solvent [1]. Chromatography is a differential migration process: components of a mixture that distribute differently between a particulate adsorbent and a fluid separate because they move at different rates through the fluid-perfused[1] adsorbent bed. If conditions are chosen correctly, the individual species in the mixture emerge at the outlet of the chromatographic column in pure bands. The broad array of sorbent and fluid combinations available today make chromatography a versatile and widely applied analytical and preparative purification tool at the laboratory, pilot, or industrial scale for virtually any class of pharmaceutical compound, particularly when gentle, but nonetheless highly selective, separation conditions are required. Chromatography is an enabling technology in biotechnology – practically all industrial biopharmaceutical purification processes contain one or more chromatography steps. While employed less frequently in small molecule drug (API) manufacturing processes, it is nevertheless heavily used in early stages of API development and does find important industrial applications in natural product and chiral separations. In addition, it is indispensable for the large-scale purification of synthetic peptides and oligonucleotides. In this chapter, the focus is on chromatography as it is practiced in the purification of chemically derived compounds.

Chromatography is considered by many to be an expensive step, a useful tool to obtain from milligrams to a few hundred grams of intermediates or drug substances for deliveries early in development, but ultimately a step that must be superseded in favor of more cost-effective methods (extractions and crystallizations or improved synthesis routes) before commercialization of the process. However, done properly, chromatography is economically competitive; indeed, with modern equipment, method optimization, and solvent-sparing technologies – including solvent recycling and continuous multicolumn chromatography (MCC) techniques – chromatography can be both effective and efficient for industrial-scale purifications.

This chapter will begin with an outline of the key design elements in chromatography and a discussion on fundamental chromatographic relationships such as retention and selectivity. The various available chromatographic chemistries will then be discussed. This will be followed by brief sections on chromatographic operating parameters, the choice of the mode of operation including multicolumn systems, choice of equipment, scale-up, a short section on parametric design space, and a discussion on chromatographic economics.

---

[1]The fluid may be a gas, a supercritical fluid, or a liquid; for this chapter, the discussion will be confined to liquid chromatography.

31.2 KEY DESIGN ELEMENTS

Choice of the combination of the adsorbent (also known as chromatographic media or the stationary phase) and the fluid (or mobile phase) is the first design element in chromatography. This choice sets the chromatographic "chemistry", dictating the type of physicochemical interactions that take place within the system as well as the means to manipulate their intensity to achieve the desired separation goals. Many types of adsorbents with a wide array of surface chemistries are available today, and several are prepared with large-scale applications in mind, i.e. they are manufactured under controlled conditions in large batch sizes. The mobile phase, an aqueous or solvent-based mixture that could also contain other components such as acids, bases, buffers, or salts, is selected to be compatible with the adsorbent and acts to mediate adsorption and release of the separating species from the stationary phase. Manipulation of the mobile phase composition is the primary means to control retentivity (a measure of the strength of adsorption of the separating species on the adsorbent) and obtain selectivity (a measure of the difference in retentivities of separating species, which is the key to affecting the desired separation). Solubility of the separating species in the mobile phase is also an important consideration; high solubility is desirable to achieve high productivity. A classification of the various adsorbents and compatible mobile phases based on the underlying physicochemical factors governing retention is provided in Section 31.4.

Another important design element is the choice of operating mode. Most chromatographic separations are carried out in the elution mode, where the separating species move through the system in the presence of all components of the mobile phase. However, there are other means to carry out chromatography. In the displacement mode, a mobile phase component known as the displacer (introduced after the feed) binds tightly to the adsorbent, swamping available binding sites so that the separating species move ahead of it. Continuous chromatographic techniques are operating modes in their own right; simulated moving bed (SMB) chromatography and the multicolumn solvent gradient process (MCSGP) have become more popular in recent years as applications for large-scale chromatographic processes increase. These are discussed in more detail in Section 31.5.

Once the choice of media and operating mode is determined, the chromatographic process is defined by its operating parameters such as the column dimensions, the adsorbent particle size, mobile phase flow rate, the operating temperature, and other factors pertinent to the operating mode (e.g. cycle and column switching times in SMB systems). The outcome of varying these operating parameters forms the basis for process and economic optimization as well as the regulatory design space for a chromatographic process. Operating parameters and design space are also discussed later in this chapter.

31.3 FUNDAMENTAL CHROMATOGRAPHIC RELATIONSHIPS

31.3.1 Chromatographic Velocity

Chromatographic operations take place in a column packed with particles. The volume fraction of the interstitial space between the particles in a randomly packed bed, or the interstitial porosity, ε_e, typically has a value of about 0.4. The totally porous sorbent particles commonly used in chromatographic applications often have an internal void fraction, ε_i, of approximately 50%, so that the packed bed has a total porosity, ε_T ($= \varepsilon_e + (1 - \varepsilon_e)\varepsilon_i$), of roughly 0.7. For a fluid flow rate F in a bed of cross-sectional area A, three flow velocities[2] may be defined and are related as follows:

$$\frac{F}{A} = u_s = \varepsilon_e u_e = \varepsilon_T u_0 \tag{31.1}$$

where

- u_s is the superficial velocity.
- u_e is the interstitial velocity.
- u_0, the average velocity of an unretained molecule that explores the entire void space, is the chromatographic velocity.

For modeling purposes, it is assumed that the mobile phase migrates through the system at the chromatographic velocity.[3]

31.3.2 Operating Pressure

Chromatographic operations are mostly carried out at a fixed velocity or flow rate. The pressure drop, ΔP, across a packed column is related to the chromatographic velocity, u_0; the column length, L; the adsorbent particle diameter, d_p; and the mobile phase viscosity, η, via Darcy's law as

$$\Delta P = \frac{L u_0 \eta \varphi}{d_p^2} \tag{31.2}$$

where the factor φ is a proportionality factor related to porosity given by the Kozeny–Carman equation

$$\varphi = 180 \frac{(1-\varepsilon_e)^2}{\varepsilon_e^2} \frac{\varepsilon_T}{\varepsilon_e} \tag{31.3}$$

---

[2] In this analysis, we assume one-dimensional axial flow in a cylindrical column; radial flow systems exist but will not be considered here.
[3] Strictly speaking, in the surface layer near the sorbent, the composition of a multicomponent mobile phase is often different from that in the bulk. Changes in mobile phase composition can lead to differences in migration rates of the solvent components themselves, but this subtlety can be ignored in practice.

For the typical values of the porosities given above, this factor has a numerical value of approximately 710, while in practice it may vary from 600 to 1000 depending on the exact porosities, as well as the regularity and roughness of the particles. In the equation above the variables can be expressed in any consistent units; the equation is also consistent without the need to add correcting factors if the pressure drop is expressed in bars, the velocity in centimeters per second, the column length in centimeters, the particle size in microns, and the viscosity in centipoise.

31.3.3 Mass Balance Equation

A differential one-dimensional mass balance for a retained species in a chromatographic system can be written in simplified form as

$$\frac{\partial c}{\partial t} + \phi \frac{\partial q}{\partial t} + u_0 \frac{\partial c}{\partial x} = D_e \frac{\partial^2 c}{\partial x^2} \quad (31.4)$$

where

- the parameters t and x are time and distance along the column, respectively.
- c is the concentration of the species in the mobile phase.
- q is its concentration in the stationary phase.
- ϕ is the ratio of stationary to mobile phase volumes (i.e. $(1-\varepsilon_T)/\varepsilon_T$).

In this simplified view, all dispersive effects including molecular diffusion, transport of the species to and within the particle, and any dispersive influence of slow adsorption kinetics are lumped into the effective dispersion coefficient, D_e.

31.3.4 Retention and Retention Factor

A system of mass balance equations (one for each migrating species), with suitable initial and boundary conditions, serves as an adequate model for chromatography [2]. The adsorbed concentration q is related to the mobile phase concentration c via an equilibrium relationship known as the adsorption isotherm. In general (and particularly in the case of preparative chromatography that is carried out at high concentrations of the migrating species), the adsorption of each species depends not only on its own concentration in the mobile phase but also on that of all the other species present. Consequently, migration through the system is concentration dependent; the system of equations is nonlinear and must be solved numerically. On the other hand, at low concentrations it can be assumed that the distribution coefficient ($K = q/c$) for each species is an independent constant. In this circumstance, in an ideal system without dispersion, the mass balance equation reduces to

$$\frac{\partial c}{\partial t} + \frac{u_0}{(1+\phi K)} \frac{\partial c}{\partial x} = 0 \quad (31.5)$$

This is a one-dimensional wave equation with propagation velocity – or the migration velocity of a species through the system – of $u = u_0/(1+\phi K)$. The product ϕK is known as the retention factor, k'. Despite being defined only at low concentrations in limited circumstances, k' is a key factor in the understanding and characterization of chromatographic behavior.

Conditions where the distribution coefficient K of each migrating species is independent of concentration (the Henry's law region) are known as "linear" (or "analytical" conditions, as these are conditions under which chromatographic analyses are carried out). A system in which the mobile phase composition is kept constant over time is termed "isocratic." In a linear isocratic system, if the mixture to be separated is injected to approximate a δ-function, the retention factor k' of each separating species can be determined from its retention time, t_R (the time of elution of the center of gravity of the migrating component), and the dwell time, t_0 (the time of elution of an unretained component), as follows [3]:

$$k' = \phi K = \frac{t_R - t_0}{t_0} \left(= \frac{V_R - V_0}{V_0} \right) \quad (31.6)$$

(V_R and V_0 are the corresponding elution and dwell volumes: $V_R = Ft_R$, where F is the mobile phase flow rate, and $V_0 = Ft_0 = \varepsilon_T V_C$, where V_C is the total column volume (CV).)

The distribution coefficient K is a thermodynamic property, related to the free energy of adsorption, and as a result k' is a function of temperature via the relationship [3]

$$\ln k' = -\frac{\Delta H}{RT} + \frac{\Delta S}{R} + \ln \phi \quad (31.7)$$

where

- ΔH and ΔS are the enthalpy and entropy of adsorption.
- T is the temperature in degrees Kelvin.
- R is the universal gas constant.

Values of k' can increase or decrease with increasing temperature depending on the sign of the enthalpy of adsorption; in most instances, the latter is true as adsorption is often enthalpically favored (i.e. ΔH is negative), but there are important exceptions, such as in hydrophobic interaction chromatography (HIC) of proteins and in other individual cases, where the opposite can hold.

31.3.5 Selectivity

The power of a chromatographic system to discriminate between two species is quantified by the ratio of their retention factors, termed the "selectivity," α ($= k'_1/k'_2$, where the subscripts refer to the two species and $k'_1 > k'_2$, so that $\alpha > 1$). The type of system chemistry, the mobile phase composition, and the temperature are major factors that influence selectivity. In industrial applications, concentrations are usually high, so Henry's law no longer applies, and distribution coefficients as well as selectivities are nonlinear functions of the concentration of all the locally present separating species. Nonetheless, the first step in designing a chromatographic separation is to operate in Henry's law regime to find conditions (i.e. the right stationary and mobile phase composition) that maximize selectivity. To achieve this, one must understand how retention and selectivity can be manipulated in the context of the various chromatographic chemistries. A discussion on some of the more popular chemistries is given in the next section.

31.3.6 Efficiency

During chromatography, dispersive effects counteract the effectiveness of the separation; chromatographic efficiency thus increases when dispersion is decreased. Contributions to dispersion arise from flow anastomosis, molecular dispersion, transport in and out of the particle, and slow adsorption kinetics. Efficiency is characterized by the so-called plate number, N, a concept that arises from a model that treats the chromatography column as a series of stirred cells containing equal amounts of stationary phase. Based on this model applied in Henry's law adsorption regime, the plate number is related to the width of a chromatographic peak[4] by the relationship [3]

$$N = 5.54 \left(\frac{t_R}{w_{1/2}}\right)^2 \qquad (31.8)$$

The plate number is related to the lumped dispersion coefficient in Eq. (31.4) as follows [4]:

$$N = \frac{u_0 L}{2 D_e} \qquad (31.9)$$

The larger the number of plates, the more closely the system approximates plug flow, and the more efficient it is.

The height equivalent of a theoretical plate, H ($=L/N$), is a related measure of the efficiency. The plate height is a function of the particle size and flow velocity, as elucidated first by van Deemter and his colleagues [5]:

$$H = A d_p + \frac{B D_m}{u_0} + \frac{C d_p^2}{D_m} u_0 \qquad (31.10)$$

Here A, B, and C are constants, with typical numerical values of about 1.5, 0.8, and 0.3, respectively [6], and D_m is the molecular diffusivity of the separating species in the mobile phase. The first (or A) term of this equation is independent of flow rate and arises from flow nonuniformity within the packed bed. The second (or B) term is a result of molecular dispersion and rapidly becomes insignificant as flow velocity increases. The third (or C) term is a consequence of diffusion of the separating molecule in and out of the stagnant fluid within the particle pore space. It increases linearly with flow rate and is usually the most significant factor related to bandspreading. Contributing factors to the plate height, such as diffusion through the boundary layer around the particle and the effect of slow adsorption kinetics, are usually of less import and have thus been ignored in the simplified equation presented above.

The van Deemter equation can be written in dimensionless form by introducing into it the reduced plate height, h ($=H/d_p$), and the reduced velocity, ν ($=u_0 d_p/D_m$), so that

$$h = A + \frac{B}{\nu} + C\nu \qquad (31.11)$$

Figure 31.1 shows the van Deemter curve and indicates the approximate practical range for the reduced plate height for small molecules (of 15–80, assuming diffusivities in the 2–5×10^{-6} cm^2/s range, flow velocities from 0.1 to 0.3 cm/s, and a 10 μm particle diameter). With similar flow velocities and particle size, peptides and oligonucleotides have a higher reduced velocity range because of their lower molecular diffusivities. In practice, therefore, the B term reduces to zero, and bandspreading is dominated by the effect of diffusion through the sorbent particle.[5]

With diffusion through the particle as the major contributing factor to the plate height (and thus to N), another useful scaling factor related to efficiency, termed the Lightfoot number, Li, can be constructed by taking the ratio of the characteristic diffusion time across a particle ($t_{diff} = d_p^2/D_m$) and the dwell time in the system ($t_0 = L/u_0$) so that

$$\text{Li} \equiv \frac{t_0}{t_{diff}} = \frac{L D_m}{u_0 d_p^2} \qquad (31.12)$$

31.3.7 Operation at High Concentration

As mentioned earlier, at high concentration Henry's law no longer applies, and the adsorption isotherm of each species is generally a nonlinear function of the concentration of all

---

[4]The peak is theoretically the output of a δ-function input to the system; in practice, a small analytical size injection is employed. The peak approximates a Gaussian distribution. The width at half height, $w_{1/2}$, is more convenient to use than the width at the baseline.

[5]Think of it this way: the distance between a molecule in the flowing stream outside a particle and another diffusing within a particle would increase as flow velocity increases, and this would widen the peak.

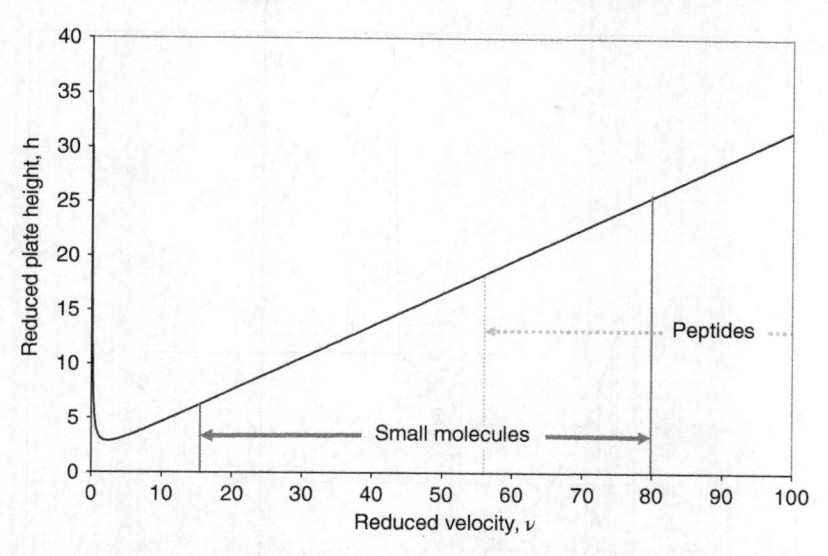

FIGURE 31.1 Van Deemter plot of reduced plate height *versus* reduced velocity showing typical operating regimes for small molecules and peptides as described in the text.

the species present. Migration of the components under these conditions is concentration dependent.

While there are many possible forms of the adsorption isotherm depending on the molecular properties of the adsorbing species, the mobile phase and the sorbent, in many cases the sorbent has a finite maximum capacity for the adsorbing species. In this situation, compounds vie for a limited number of sites on the sorbent and adsorption is competitive. At elevated concentrations then, relatively fewer sites are available, and the slope of the adsorption isotherm decreases, implying that higher concentrations migrate faster. In this circumstance, the low concentration at the leading edge of a concentration front moves more slowly than the high concentration of its trailing edge; this leads to a self-sharpening effect, countering the dilutive effect of dispersion and leading to a sharp shock wave. The opposite is true for the rear, producing a long dilutive tail. Peaks that at low concentration appear symmetrical become triangle shaped at higher concentrations. This effect is illustrated in Figure 31.2. In multicomponent systems, which are of more interest in practice, strongly bound compounds displace and thus suppress the binding of weakly adsorbed species. Under the right conditions in a chromatographic system, this results in weakly bound components being pushed ahead of strongly bound ones; as a result some concentration of the early eluting species can occur. This "displacement effect" is beneficial if an early eluting component is the target of the purification. Unfortunately, competitive adsorption also leads to a perverse "tag-along" effect: weakly bound components act to suppress the binding of strongly bound components, accelerating their motion and dragging them ahead, reducing the extent of separation. The effect is shown for two components eluting individually and together in Figure 31.3.

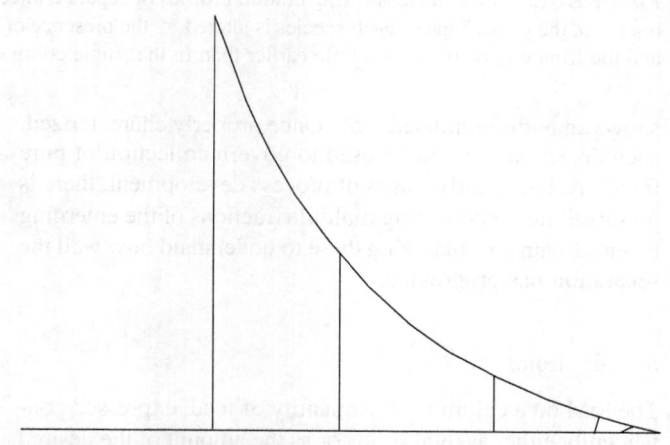

FIGURE 31.2 Schematic showing overlaid elution profiles resulting from injections of increasing concentration of a single species when adsorption is competitive. The elution profiles appear as a series of nested near-triangular shapes, with sharp fronts and extended tails. The end of each tail coincides with the retention time of the species at infinite dilution.

It is worth noting that not all systems display competitive binding and the displacement and tag-along effects are by no means universal. Some systems display cooperative binding, the chromatographic effects of which are the inverse of those discussed above. Some systems show a combination of these types of binding, and the resulting peak shapes in preparative chromatography can be quite complex.

One practical consequence of operating at high concentration is that the neat symmetric separated peaks seen in analytical separations disappear. Nonspecific detection at the end of the column (usually by UV light absorption) often

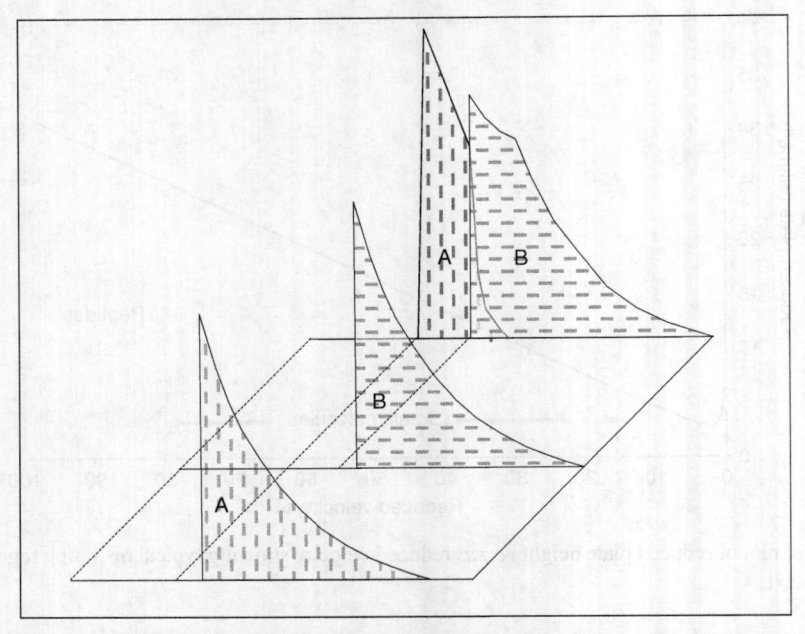

FIGURE 31.3 Schematic showing elution profiles of separate injections of two species A and B and of a mixture of A and B. In the mixed injection, the peak shape of each species is altered by the presence of the other: A is compressed and concentrated by the displacement effect, and the front end of B elutes a little earlier than in the single component case because of the tag-along effect.

shows an undifferentiated blob. Once properly characterized, such detection can still be used to govern collection of pure fractions, but in early stages of process development, there is no substitute for collecting multiple fractions of the emerging eluent stream and analyzing these to understand how well the separation has progressed.

31.3.8 Load

The load on a column is the quantity of feed, expressed conveniently either as total solids or as the amount of the desired product in the feed mixture, introduced into the system per injection. A dimensionless load parameter, Γ, can be defined as the column load divided by the mass of sorbent, and this can be used as a scaling factor in conjunction with the plate number, as discussed later.

31.4 CHROMATOGRAPHIC ADSORBENT CHEMISTRIES AND BASIS OF RETENTION

31.4.1 Normal-Phase Chromatography

"Normal-phase" chromatography implies use of a polar stationary phase, usually unmodified porous macroreticular silica, with an organic solvent blend as the mobile phase. Mobile-phase solvents covering the spectrum from the very nonpolar (heptane) to the very polar (methanol) are used. The silica surface is populated with silanol groups with a distribution of activities that depend on their structure [7], and water can bind strongly at the most polar sites. As a consequence, minor fluctuations in the water content of nominally dry solvents can have a profound effect on retention. Deliberate addition of some water (usually about 0.5% by volume, although higher amounts up to 7% have also been used) [8, 9] to the mobile phase can provide uniformity and quell variations in retention.

Retention is manipulated by changing the mobile-phase blend; addition of polar solvents generally decreases retentivity. Good guides to selection of solvents to maximize selectivity are available [10]. Solvents used have been grouped according to their electron donor, electron acceptor, and dipole properties, and blends of solvents with broadly different properties often lead to the best selectivity. Selection of the appropriate solvent blend can be assisted by the use of thin-layer chromatography (TLC) on silica-coated plates, which are in common use in organic chemistry labs. Translation from TLC to column chromatography may require some reduction in the proportion of the most polar component of the blend to increase retentivity.[6]

In normal-phase chromatography with bare silica as the stationary phase, the layer of solvent closest to the silica surface has a different composition from the bulk. Equilibration

---

[6]In TLC, practitioners like to elute components so that they migrate to about half the length of the plate, so that the "retardation factor" R_F (distance migrated/plate length) is approximately 0.5. Retention factor and retardation factors are related ($k' = 1/R_F - 1$), so the corresponding k' is approximately 1. For column chromatography $k' > 3$ is often desirable, hence the need to increase retention while translating from TLC to LC.

of the surface layer upon changes in the bulk takes time, presumably because of the heterogeneity of active sites on the silica surface. As a result, most normal-phase separations on silica are carried out isocratically (i.e. without changing the solvent composition during the chromatography).

Since many molecules of interest are very soluble in organic solvents and because silica is relatively inexpensive and has high adsorptive capacity, normal-phase chromatographic processes can be run at high concentrations, resulting in productive and cost-effective separations. On the other hand, the surface energy of bare silica can be influenced by many factors (trace metals, for instance), and a normal-phase system can suffer from variability for a variety of reasons, including variations in silica manufacturing or fluctuations in the crude feedstock that may contain components that bind practically irreversibly to the silica and can slowly change the character of the surface. Bonded phases with more homogenous surfaces are available that reduce or eliminate these sensitivities, and these can also be used in a gradient mode (i.e. where the composition of the mobile phase is changed during the chromatographic run). Phases with diol and amino functions are available that have niche applications (for instance, for separation of molecules containing polyols) with selectivities that differ considerably from bare silica. Bonding chemistry adds to the cost of the adsorbent, but this may be offset by a longer column lifetime under the right conditions.

31.4.2 Reversed-Phase Chromatography

"Reversed-phase" chromatography is carried out with nonpolar stationary phases using a mixture of water with a miscible solvent as the mobile phase (the polarity of the system is reversed compared with the "normal" phase described above). Molecules separate from each other based on their hydrophobicity; more hydrophobic species are more highly retained. Porous macroreticular silica-based stationary phases whose surfaces have been modified by bonding an alkyl group, such as octadecyl(C18), octyl(C8), or butyl(C4), are most commonly used, although a wide variety of others, including macroreticular polymer-based adsorbents, are available. Commonly employed water-miscible solvents used in the mobile phase are acetonitrile, methanol, ethanol, isopropanol, and, more rarely, tetrahydrofuran. Mobile phases are often buffered or acidified to influence retention of ionizable species. Organic or inorganic acids (e.g. acetic, trifluoroacetic, methanesulfonic, phosphoric acids) can interact with basic groups on the separating species, forming ion pairs and altering their hydrophobicity, and simultaneously suppress ionization of residual (unbonded) silanols on the stationary phase surface, reducing ionic interactions. For separation of ionizable species, manipulation of the nature of the ion-pairing agent and the pH can have a profound effect on selectivity.

For a given pH and ion-pairing agent, the retention factor in reversed-phase chromatography decreases with the increasing volume percent of miscible organic solvent in the mobile phase, ψ. In such a system, the relationship between k' and ψ can be adequately described over a broad range of ψ by the expression

$$\ln k' = \ln k'_0 - S\psi \qquad (31.13)$$

where

k'_0 is the retention factor extrapolated to zero organic content.

The slope of the plot, S, correlates roughly with molecular weight, large molecular weight species having larger S values, although there is often considerable variation for species with similar molecular weight. The relationship breaks down at high organic content, where retention can sometimes increase with increasing ψ. Figure 31.4 shows the k' vs. ψ relationship for several compounds in a hypothetical mixture.

Reversed phase is the most widely used form of chromatography on an analytical scale and is also used extensively in industrial applications. Practically every peptide manufacturing process, for instance, includes a reversed-phase chromatography step. Reversed-phase chromatography has also been used successfully for oligonucleotide separations [11]. Here polymer adsorbents find favor over silica-based bonded phases, since the latter are unstable at high pH that can be encountered in feed streams or in the eluents employed.

31.4.3 Chiral Chromatography

Perhaps the most important development in the field of chromatography over the past 25 years has been the introduction of stationary phases with chiral selectivity. Such phases have seen increasing use on an industrial scale for the separation of enantiomers, providing an alternative to stereoselective synthesis or classical resolution techniques. The phases contain a chiral selector (for instance, substituted cellulose or amylose) bonded or coated onto a chromatographic particle (usually porous macroreticular silica). A broad variety of chiral stationary phases (CSPs) are available utilizing different selectors, and these can be operated in either the reversed-phase or normal-phase modes, depending on their design and solvent compatibility. Choice of the best phase for an application is typically a process of screening a broad range of CSPs under analytical conditions for the ones with the best selectivity. Since selective adsorption capacity can be limited on CSPs, it is worth choosing a few different CSPs to see which one works best under the higher concentration conditions intended for the application.

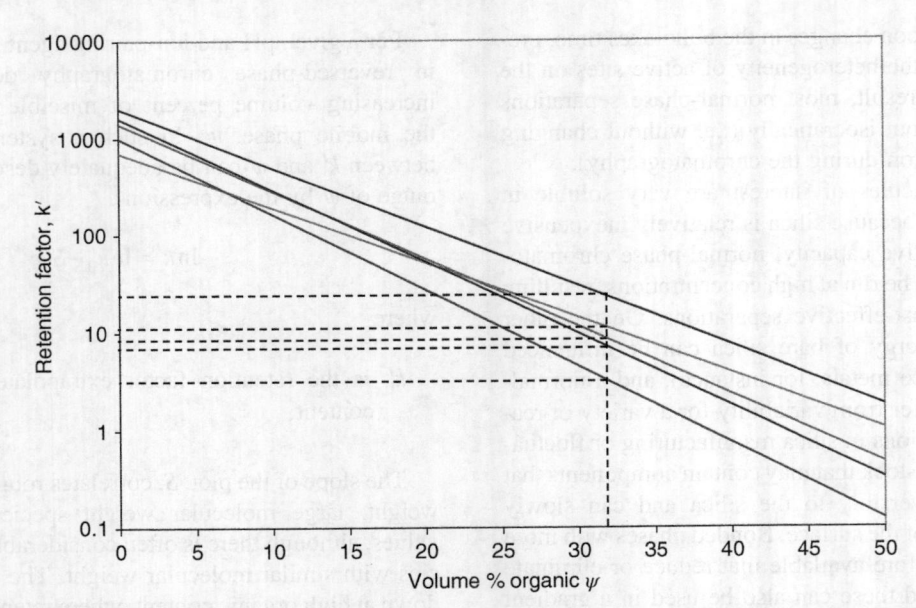

FIGURE 31.4 Plots of retention factor of various components of a mixture *versus* the organic solvent volume percentage in the mobile phase in a reversed-phase chromatography system, where there is typically a straight-line relationship between log k' and ψ.

CSPs are expensive relative to normal- or reversed-phase sorbents, and separation design efforts often focus on minimizing the amount of stationary phase. In chiral separation applications, the intent usually is to separate enantiomers; the presence of other impurities is incidental. For this reason, SMB chromatography and related multicolumn technologies, which are ideally suited to binary separations, have seen increasing use in the pharmaceutical industry.

31.4.4 Other Chemistries

While most applications of large-scale chromatography for separation of small molecules employ the types of phases discussed above, other types of chemistries are also sometimes employed. Notable among these are phases containing weak or strong anionic or cationic ionized groups bonded on macroreticular silica or polymeric particles, commonly referred to as ion-exchange or electrostatic interaction phases. These are usually operated with aqueous or hydro-organic mobile phases containing salt and buffers for pH control. Salt screens electrostatic interactions, so retention on these phases decreases with increasing mobile-phase salt content. pH affects ionization of the separating molecules as well as the ionizable groups on the stationary phase surface and is thus a useful means to manipulate retention. Ion-exchange chromatography is used widely for separation of large molecules like proteins and antibodies but is also used for synthetic peptide and oligonucleotide purifications.

Hydrophilic interaction chromatography (HILIC) uses polar stationary phases and water-miscible solvents; retention increases with increasing polarity, and increases in the water content of the mobile phase result in a decrease in retention.

Other types of chemistries, such as those used for hydrophobic interaction chromatography (HIC), where weakly hydrophobic surfaces and high mobile phase salt concentrations are employed, or for size exclusion chromatography (SEC), where molecules are not retained but separate based on molecular size, are largely confined to large molecule separations. However, HIC has recently found use in separation of synthetic oligonucleotides [12].

31.5 OPERATIONAL ASPECTS

31.5.1 Chromatographic Applications

Chromatographic systems can be used in several ways in a chemical process. One simple application is for removal of impurities – a process stream is run through a chromatographic column, and one or more impurities are adsorbed on the bed. Once the bed is saturated, it is either discarded or regenerated, re-equilibrated, and returned to use. This requires conditions where the impurities are tightly bound and the product is minimally affected by the sorbent. Conditions for this simple operation can often be found if the product and the impurities are not very closely related, separating neutral impurities from a charged product, for instance.

Another application is the inverse operation; a dilute process stream is run through a chromatographic column to capture the product. The product is then desorbed in a concentrated form with a suitable eluent. For example, a clarified aqueous fermentation broth may be run through a reversed-phase system to capture the product, which is then eluted in concentrated form with methanol.

A chromatographic system is a convenient tool for carrying out certain solvent switches. For example, in order to change the product solvent from aqueous acetonitrile to methanol, the product could be captured on a reversed-phase column. This may require dilution of the stream with water to create strong binding conditions for the product. The column could then be washed with dilute aqueous methanol (maintaining binding conditions) to remove acetonitrile, and the product can then be eluted in methanol. A similar product capture/wash/elute approach can also be used to desalt a process stream. Yet another variation is the use of a reversed-phase system to affect a counterion switch, an approach often used after the purification of peptides. For example, a peptide purified by reversed-phase chromatography using trifluoroacetic acid (TFA) as the ion-pairing agent (often a good choice to obtain selectivity for reversed-phase separation, but usually not desired in the final product) can be switched to another counterion form, say, acetate, by a) capturing it (after suitable aqueous dilution) on a reversed-phase column, b) washing under binding conditions with streams containing first ammonium acetate to remove the TFA and then acetic acid to wash away excess ammonium ion, and finally c) eluting the peptide by increasing the organic solvent content of the acetic acid-containing mobile phase.

By far the most useful application of chromatography, of course, is the purification of a product from closely related impurities; for instance, a desired peptide from a dozen or more synthesis impurities, some with differences as subtle as, say, a β-aspartic acid substituted for aspartic acid in the original sequence, resulting in an isomer with an extra methylene group in the peptide backbone. The following operational modes are employed for such applications.

31.5.2 Elution Chromatography

Elution is the most familiar mode of operation for chromatography: species migrate through the column and separate in the presence of a mobile phase that directly mediates retention. When the mobile phase composition is held unchanged, this is known as "isocratic" elution. When the mobile phase composition is varied during the migration, this is known as "gradient" elution.

The motivation for changing the mobile phase composition in a gradient elution process lies in the potentially wide range of retentivities of all the components in the mixture to be separated. An increase in the strength of the mobile phase over time ensures that the most strongly retained component will elute off the column within a practicable time period. The difference between operating under isocratic and gradient conditions in a reversed-phase system can be understood by examining the $\ln k'$ vs. ψ plot in Figure 31.4. For a successful isocratic operation, a single solvent strength must be selected that affords selectivity between the desired compound and the other components of the mixture and also enables elution within a reasonable time frame. For instance, operating at 32 v% organic, the first and last components to elute would have k' of about 3 and 23, respectively. A gradient in solvent strength would hasten the elution of the more strongly absorbed compounds.

Gradients are most commonly either linear or stepwise, and they are usually formed by blending two streams, changing the blend composition appropriately over time. Some equipment allows formation of ternary gradients or gradients with convex or concave shapes, but this is not recommended if transfer between different brands of equipment is anticipated.

When the mobile phase strength increases over time, the front and rear of a migrating peak are often in different solvent environments, so that retention is lower at the rear of the peak than the front. This results in a peak "compression" that acts counter to and somewhat mitigates the dilution that is always observed under isocratic conditions. Reducing dilution can have significant impact on solvent volumes and processing of the collected purified material.

31.5.3 Displacement Chromatography

Under conditions of competitive binding and at high column loads, the displacement effect mentioned earlier leads to concentration of early eluting, weakly bound species displaced by more strongly bound components. Displacement chromatography, introduced in 1940s by Tiselius [13] and developed further in more recent years by the academic groups of Horváth [14] and Cramer [15], among others, exploits this effect to the fullest extent possible. A mixture introduced into the column under strong binding conditions (e.g. low salt content in an ion-exchange system, where k' is high) is pushed through the column not by the customary means of changing the eluent strength, but by introducing a solution containing a species that is more strongly bound than any of the mixture components. This component is called the displacer. Velocity of the displacer front can be manipulated by adjusting its concentration. The displacer swamps the binding sites on the sorbent, acting almost like a piston to drive the other species ahead of it. The mixture species, concentrated and moved forward by the action of the displacer, then act to displace each other in order of binding strength, until – after development of the displacement train over a sufficient column length – they separate into contiguous concentrated bands of individual components. The displacer does not mix with the separating components except by dispersion at the end of the separation train. The displacer must be removed from the column before it can be reused. This can be accomplished by changing the eluent strength or other conditions such as the pH to facilitate washing the displacer from the column.

This mode of separation is ideally suited for economical preparative separations, since it perforce runs at high concentration and makes efficient use of the capacity of the stationary phase. However, relatively few displacement separations have been implemented. One reason may be that development of a displacement separation is often time consuming and can require expert attention. Some headway in widening the use of this technique is being made, particularly in ion-exchange chromatography applications [16].

31.5.4 Multicolumn Chromatography

It has long been recognized by chemical engineers that the most effective mass transfer between two phases takes place when they are contacted in a continuous, countercurrent manner. In a chromatographic system with a particulate solid sorbent, actual movement of the solid would disrupt the packing structure, negating any potential efficiency gains. Nevertheless, many of the benefits of countercurrent solid motion can be achieved by simulating the solid motion by appropriately switching inlet and outlet positions in a multicolumn system. When the inlet and outlet ports are simultaneously switched in the direction of fluid flow, as shown in Figure 31.5, the sorbent appears to flow in the opposite direction. SMB units of this kind have been used in multiton industrial operations such as the purification of p-xylene from C8 fractions since the early 1960s, with installed capacity exceeding 10^7 tons/year [17]. More recently, SMB and related MCC technologies have been increasingly employed in pharmaceutical applications.

The configuration shown in Figure 31.5 shows two inlet streams (one for the feed mixture, the other for the eluent) and two outlet streams (one for the separated early eluting compound(s), known as the raffinate, and the other for the separated late eluting compound(s), known as the extract). The system has four zones, two separation zones on either side of the feed inlet, a solvent recycle zone that strips the early eluting compound from the solvent, sending it toward the raffinate, and a sorbent recycle zone that strips the late eluting compound from the sorbent, sending it toward the extract. The presence of two outlet streams makes the technique naturally suited to binary separations, and unsurprisingly in the pharmaceutical arena, SMB technology is used most widely for enantiomer purification. There are three major operating advantages over conventional chromatography: (i) the stationary phase is used to maximum capacity, minimizing sorbent requirements; (ii) operation can be adjusted to obtain close to quantitative yields at the targeted purity, minimizing loss of precious product; and (iii) internal recycle of the solvent minimizes the solvent consumption [18].

Various efficiency modifications have been introduced into the repertoire of multicolumn operations. A process employing asynchronous column switching, known as "Varicol," enables the assignment of a fractional part of a physical column to a given separation zone, reducing the number of columns required and thus the column hardware needed. Appropriately timed flow rate changes within a switching period (termed "power feed") have also led to significant improvement in MCC efficiency [19, 20]. A "multicolumn countercurrent solvent gradient purification (MCSGP)" process [21] has also been introduced that can carry out multicomponent separations employing mobile phase gradients. This process can employ as few as two columns in a semicontinuous countercurrent operation that in principle could be adapted to work for any existing gradient elution separation. The major advantage of the process is that the semicontinuous operation enables near quantitative yield of the desired compound at the target purity [22]. Unlike the more traditional four-zone binary systems, the MCGSP does not afford significant solvent savings. Nevertheless, yield considerations can be a sufficient driver to implement such a scheme.

Many of the advantages of the SMB can also be realized using a single or dual column multi-injection steady recycle process invented by Charles Grill in the mid-1990s [23]. This process has now been commercialized [24].

31.6 EQUIPMENT

31.6.1 Columns

Until the introduction of large-scale high performance chromatographic equipment in the 1980s, industrial chromatographic separations were carried out in large columns packed with metric ton quantities of relatively large sorbent particles (c. 100 μ diameter or larger). One consequence of large particle size was the need for significant column lengths (several meters) to achieve separations. Poor distribution in such columns created inlet and outlet flow development zones, creating serious inefficiencies and making column aspect ratio an important factor in scale-up. Modern high performance columns, now available from several vendors, have excellent flow distributors, so that performance is virtually independent of column diameter. Sorbent particle diameters used today are usually between 10 and 20 μm. While these provide high efficiency so that column lengths of less than 1 m (and usually ≪50 cm) are sufficient for many separations, they demand proper packing techniques and relatively high operating pressure of up to about 100 bar. One technology that has solved these issues and so has dominated the large-scale column market is the so-called dynamic axial compression (DAC) column. A hydraulic piston inside the column is used initially to compress a slurry of sorbent particles into a firmly packed bed, and pressure is maintained on the bed during column operation, ensuring that any holes or pockets that may arise from bed subsidence over time are eliminated, ensuring efficient long-term operation. It is not unusual with for a DAC column of 1 m diameter to show the same efficiency as an analytical column packed with the same sorbent. This

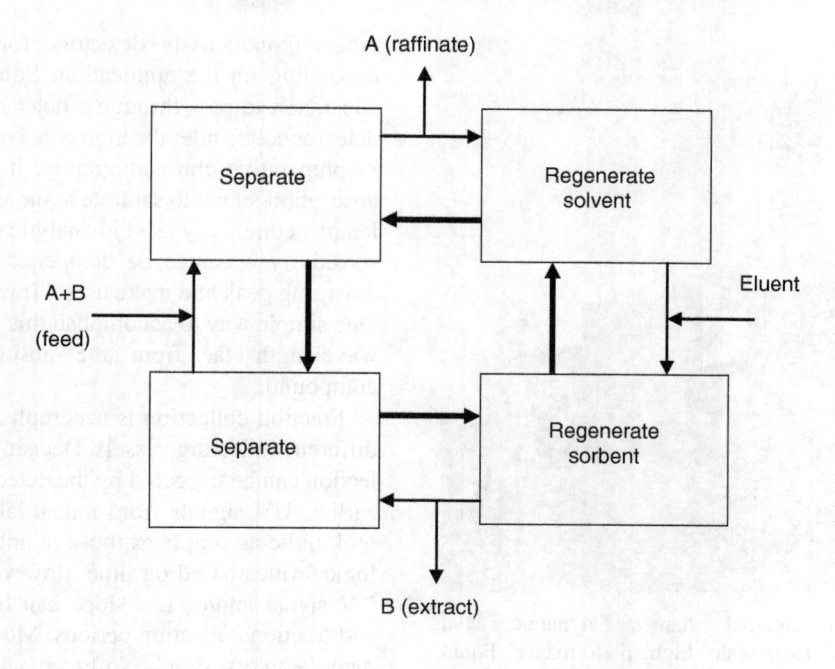

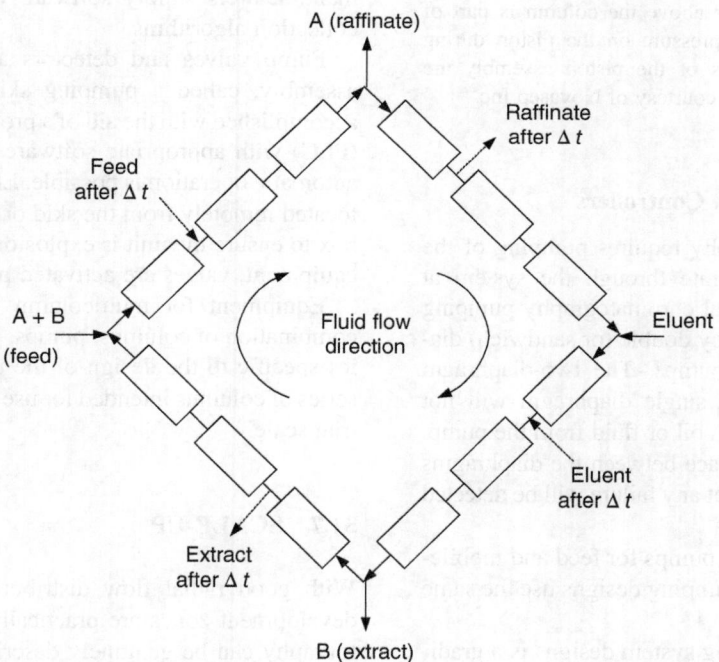

FIGURE 31.5 The four-zone simulated moving bed (SMB) chromatograph. The top drawing shows a schematic of the four-zone moving bed – dark arrows show the desired direction of solid flow. The bottom drawing show how simulation of the solid movement is carried out in a 12-column system with 3 columns in each zone. The position of each of the external flow switches by one column after each switching period Δt. Flow within the column ring is driven by a pump.

makes scale-up facile, as the only scaling parameter is then the cross-sectional area of the column. Figure 31.6 shows an example of a large-scale DAC column. The tall gray assembly is part of the hydraulic system that transmits pressure to the column piston. For unpacking, the bottom flange of the column is opened and the piston is driven downward to force the sorbent out. Column packing and unpacking of the few tens of kilos of sorbent is thus straightforward, taking only a few hours (including all preparation time) in an industrial setting.

FIGURE 31.6 A 1 m internal diameter dynamic axial compression column for large-scale high performance liquid chromatography. The gray cylinder above the column is part of the hydraulic system to maintain pressure on the piston during packing an operation. Components of the piston assembly are seen in the foreground. Photograph courtesy of Novasep Inc.

31.6.2 Pumps, Detectors, and Controllers

High performance chromatography requires pumping of the mobile phase at a fixed flow rate through the system at pressures up to 100 bar. Industrial chromatography pumping systems almost universally employ double (or sandwich) diaphragm positive displacement pumps. The two-diaphragm design assures that failure of a single diaphragm will not contaminate the process side with oil or fluid from the pump. Continuous monitoring of the space between the diaphragms for rupture provides assurance that any failure will be detected immediately.

It is advisable to use separate pumps for feed and mobile-phase streams, although some pumping designs use the same pump for both.

A common feature in pumping system designs is a gradient forming system that enables blending of two (or more) inlet streams to vary mobile phase composition for gradient elution operation. Valves and mass flow meters along with electronic controllers are often used to achieve the gradient. Gradient composition monitoring and control using PAT systems, such as near-IR probes with feedback control, are available commercially.

The effluent stream is usually monitored by a detector; signals from the detector can be used to decide when to collect fractions. The most commonly employed detector is the variable wavelength UV detector, although a plethora of others (conductivity detectors, for instance) can be used depending on the application. Some systems take a small slipstream to pass through a detector; others use a full-flow detector cell. Under the high concentration conditions typical in preparative chromatography, it is common for the UV absorption signal to saturate at the wavelengths and cell path lengths commonly used in analytical chromatography. Often sensitivity needs to be dampened to properly capture the emerging peak and make useful fraction collection decisions. One simple way to accomplish this in practice is to use a UV wavelength far from the absorption maxima of the compound.

Fraction collection is accomplished by directing flow to different collecting vessels. Decision making on fraction collection can be triggered by the detector signal. As mentioned earlier, UV signals from industrial chromatographs do not look quite as simple as those in analytical systems, so some logical rules based on time, flow volume, and thresholds of UV signal and/or UV slope can be employed to start and end fraction collection periods. Most commercial equipment manufacturers supply software with sophisticated fraction collection algorithms.

Pump valves and detectors are often mounted in one assembly, called a pumping skid. Control of the skid is accomplished with the aid of a programmable logic controller (PLC) with appropriate software running from a PC. Fully automatic operation is possible. Electronic controls are either located remotely from the skid or in an appropriately purged box to ensure the unit is explosion proof. In explosion proof equipment, valves are activated pneumatically.

Equipment for multicolumn units likewise involves a combination of columns, pumps, valves, and process analytics specific to the design of the units. Figure 31.7 shows a series of columns intended for use in a Varicol MCC at industrial scale.

31.7 SCALE-UP

With good radial flow distribution, inlet and outlet flow development zones are practically eliminated, and chromatography can be genuinely described by a one-dimensional (axial) mass balance such as that in Eq. (31.4). The simplest possible scale-up paradigm under these circumstances is to maintain the same sorbent particle size, column length, and chromatographic velocity across scales, setting all other parameters proportional to the column cross-sectional area. This will ensure practically identical pressure drop, column efficiency (plate number), separation quality, and specific production rate (production rate normalized to sorbent mass) across scales. An example showing such a scale-up from a standard 0.46 cm internal diameter (ID) column to a 45 cm ID column is shown in Table 31.1.

FIGURE 31.7 A six-column simulated moving bed unit. Each column has 1 m internal diameter. The unit was designed to operate in a Varicol process. Photograph courtesy of Novasep Inc.

TABLE 31.1 Example of Scale-Up Keeping Particle Size and Column Length Fixed

Parameters	Small Scale	Scale Factor	Large Scale	Comments
Key scale-up parameters				
Column diameter	0.46 cm		45.00 cm	
Column cross section	0.17 cm^2	9570	1590.43 cm^2	Basis for scale-up 9569.94
Parameters that must be fixed to enable scale-up based on cross-sectional area				
Particle size	10.00 μm		10.00 μm	
Column length	25.00 cm		25.00 cm	
Parameters that scale with cross-sectional area				
Mass sorbent in column	2.53 g	9570	24.25 kg	Packed density of 0.61 g/cc
Flow rate	1.50 ml/min	9570	14.35 L/min	
Feed injected per run	25.00 mg	9570	239.25 g	Load factor Γ held constant
Important parameters that remain unchanged in this scale-up paradigm				
Chromatographic velocity	0.215 cm/s		0.215 cm/s	
Pressure drop across column[a]	40 bar		40 bar	Equation (31.2): $\varphi = 710$; $\eta = 1.05$ cP
Feed composition	—		—	
Mobile phase composition	—		—	
Run time	25 min		25 min	
Gradient program	—		—	
Detector settings	—		—	
Temperature	30 °C		30 °C	
Performance attributes				
Yield	85%		85%	
Production rate	1.22 g/day		11.71 kg/day	Feed injected/day × yield
Specific production rate	0.48		0.48	kkd = kg produced/kg sorbent/day
Mobile phase used per day	2.16 L		20 671 L	
Specific solvent consumption 1	1.76 L/g		1.76 L/g	No recycle
Specific solvent consumption 2	N/A		176 L/kg	90% recycle

[a]There will be additional pressure drop caused by flow in piping, not accounted for here.

TABLE 31.2 Productivity Comparison for Columns Packed with 10 and 60 μ Particles

	Case 1	Case 2	Comments/Calculations
Scaling calculations			
Particle size (μ)	10	60	
Pressure drop (bar)	40	3	3 bar typical of low pressure equipment
Column length (cm)	25	246	Calculated based on scaling parameter
$(L/d_p^2)^2/\Delta P$ (dyne^{-1})	15 625 000	15 625 000	Scaling parameter
Productivity calculations: columns packed with 1 kg of sorbent in both cases			
Linear flow velocity, u_0 (cm/s)	0.215	0.059	Equation (31.2): $\varphi = 710$; $\eta = 105$ cP
Dwell time, t_0 (min)	1.94	69.89	$t_0 = L/u_0$
Run time, t_R (min)	25	900	Case 2 based on ratio of dwell times
Column ID that contains 1 kg sorbent (cm)	9.14	2.91	Packed density of 0.61 g/cc
Feed injected per run (g)	9.9	9.9	From Table 31.1, keeping feed to sorbent mass constant
Yield (%)	85	85	Based on Table 31.1
Production rate (g/day)	484.70	13.46	A 36-fold difference
Column dimensions and sorbent quantities required to obtain the same productivity			
Column diameter (cm)	45	86	
Column volume (L)	40	1431	Must have 36X volume to maintain equal productivity
Sorbent mass to pack column (kg)	24.25	873.15	
Feed injected per run (g)	239.25	8613	
Number of injections/ day	57.6	1.6	# = $24 \times 60/t_R$
Production rate (kg/day)	11.71	11.71	= # × feed injected per inj × yield
Specific productivity	0.48	0.013	kg produced/kg sorbent/day

If identical particle sizes are not employed, similar quality of separation can be obtained by keeping the dimensionless number Li and the normalized feed load constant across scales. Rearranging Eq. (31.2) to solve for the chromatographic velocity u_0 and substituting the result into the expression of Li from Eq. (31.12), one obtains

$$\text{Li} = \frac{\left(L/d_p^2\right)^2}{\Delta P}\left(\frac{\eta\varphi}{D_m}\right) \quad (31.14)$$

The viscosity η and diffusivity D_m are constants across scales as long as temperature is maintained constant. The bed permeability φ should remain constant across scales if similar sorbent particle geometry (e.g. spherical particles) is maintained. Thus, the pertinent invariant scale factor when sorbent particle size is not held constant is $(L/d_p^2)^2/\Delta P$.

Table 31.2 shows an example of two systems with different particle sizes, 10 and 60 μm, where the factor $(L/d_p^2)^2/\Delta P$ is held constant. Separation quality (not quantified in the table) and important performance characteristics such as solvent consumption are expected to be identical in both cases, but specific production rates are dramatically different. The smaller particle size sorbent has an increased specific productivity proportional to the square of the ratio of particle sizes (a factor of 36 in the example). If the same production rate was required in the two cases, the quantity of 60 μm sorbent required would be 36-fold that in the 10 μm system. Dimensions of identically performing columns packed with the two different particles are also given in Table 31.2. This example illustrates the incentive for using smaller particles and higher pressures for chromatographic operations.

31.7.1 Scaling of Gradients

Solvent gradients should be expressed in scalable terms. For example, in reversed-phase chromatography, varying solvent composition from X to $Y\%$ in a straight-line manner over several CV of flow (rather than expressing it as $A\%$ of solvent A to $B\%$ of solvent B over 20 minutes) enables seamless scale-up.

31.8 DESIGN SPACE

The design space is the window within the operating parameter space in which acceptable process performance is achieved. The goal of this section is to provide the reader with some appreciation for the various operating parameters and the sensitivity of conventional chromatographic operations to these parameters. Multicolumn systems are not discussed. The manner of defining the design space for regulatory agencies – carrying out design of experiments – is not within the scope of this discussion.

Figure 31.8 shows a view of a chromatographic process illustrating the various process parameters as well as salient product quality and process performance attributes. Design issues associated with some of the parameters, for instance, sorbent selection, column dimensions, and efficiency, have been discussed earlier in this chapter. Some of the other parameters are discussed below.

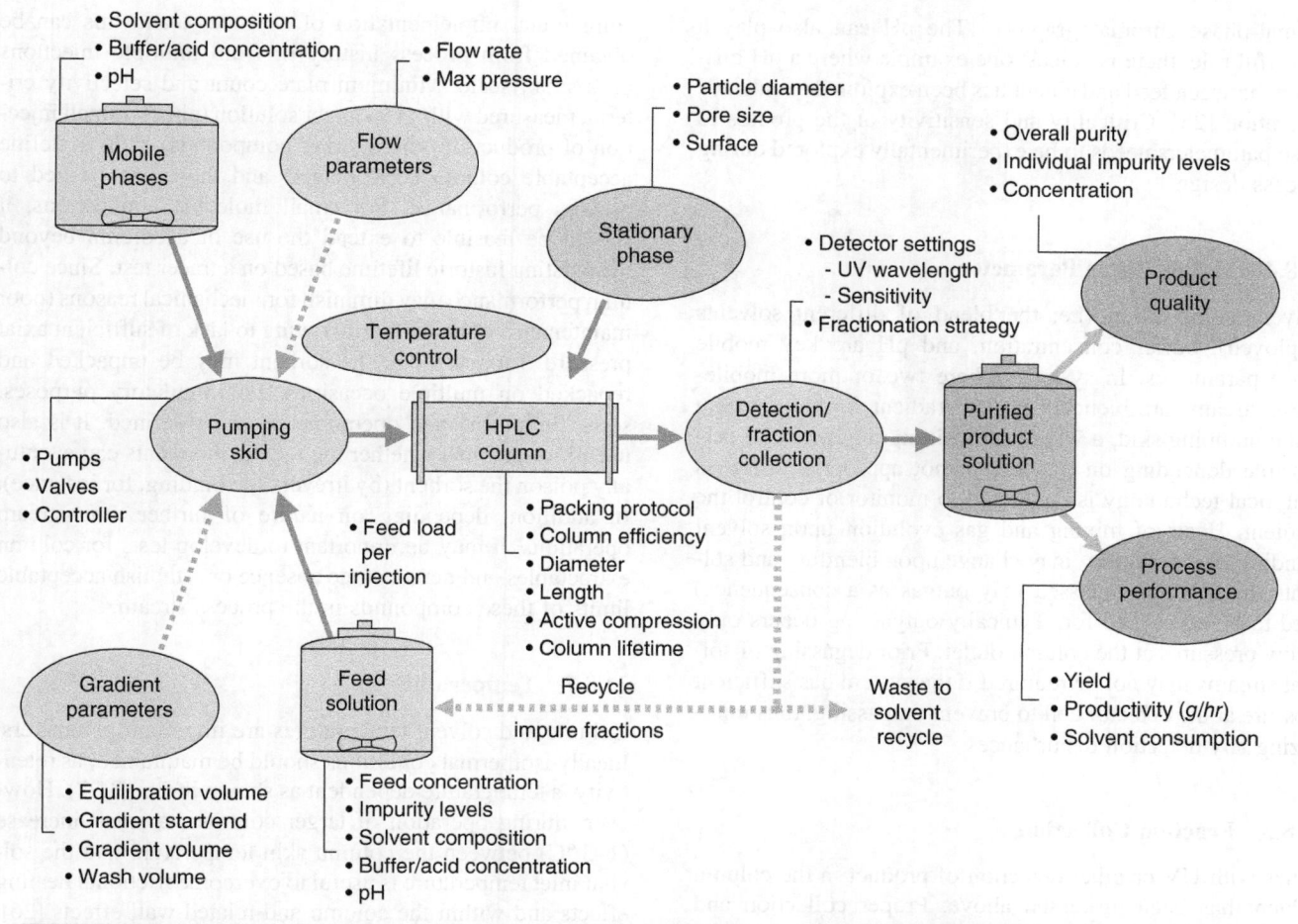

FIGURE 31.8 Overview of chromatographic operating parameters, product quality attributes, and process performance measures.

31.8.1 Process Quality Attributes

Critical quality attributes of the product stream from a chromatographic process are product purity, which must be above a minimum target (e.g. >98.5%), with key individual impurities maintained below pre-specified maximum limits. The concentration in the product stream can be a quality attribute, but a wide range is usually acceptable, and thus it is not usually critical.

31.8.2 Process Performance Attributes

Key process performance (and indeed process economic) measures are (i) yield, calculated as the moles of desired product recovered at or above the target quality as a percentage of the moles introduced in the crude feed; (ii) process productivity, which is simply a mass rate of production, expressed in g or kg/h (a specific production rate, normalized to the mass of sorbent, is sometimes a useful measure); and (iii) solvent consumption, most usefully expressed in terms of each solvent species consumed per unit mass of product. These parameters are discussed further in the Section 31.9.

31.8.3 Feed Parameters

As shown in Figure 31.8, feed concentration, impurity levels, and the feed solution characteristics including solvent composition, buffer or counterion concentration, and pH are important factors. The rule of thumb is that for the same feed load, the higher the feed concentration, the better the results. Nonetheless, feed concentration is usually more critical in isocratic elution than in gradient elution chromatography. This is because in gradient elution, feed introduction is usually under relatively strong binding conditions, enabling concentration of the feed at the column inlet regardless of its concentration. There is usually no such opportunity under isocratic conditions, unless the feed solvent composition is manipulated to afford somewhat stronger binding conditions than the eluent itself. (In reversed-phase chromatography, this would imply using lower organic modifier in the feed than in the eluent.) Feed solubility is of course a strong limiting factor influencing concentration; some practitioners using isocratic elution chromatography manipulate feed solvent composition to provide more solubility at the expense of strong binding (e.g. raising the methanol content in normal-phase chromatography).[7] The pH can also play a powerful role; there is at least one example where a pH mismatch between feed and eluent has been exploited to enhance separation [25]. Criticality and sensitivity of the process to these parameters needs to be experimentally explored during process design.

31.8.4 Mobile Phase Parameters

Solvent composition (i.e. the blend of different solvents employed), buffer concentration, and pH are key mobile phase parameters. In systems where two or more mobile-phase streams are blended by the gradient forming system in the pumping skid, a wide composition range may be permissible depending on whether or not appropriate process analytical technology is employed to monitor or control the gradient. Heats of mixing and gas evolution upon solvent blending (gas solubility may change upon blending and solvents that are not degassed may outgas as a consequence) need to be accounted for. Typically outgassing occurs only at low pressures at the column outlet. Prior degassing of solvent streams may not be required if the system has sufficient pressure at the detector cell to prevent outgassing, thus minimizing any detection disturbances.

31.8.5 Fraction Collection

Issues with UV or other detection of product in the column effluent has been discussed above. Proper collection and combining of fractions containing purified material is critical to the success of the chromatographic process. Ideally, a process should be designed to collect one purified fraction and perhaps one or two lower purity fractions that could be recycled. The remainder of the process stream can be sent to solvent recovery units. Collection of a single pure fraction is possible once the chromatographic operation has been fully understood and a fail-safe rich cut collection algorithm is devised that can handle anticipated variation in feedstock purity and minor drifts in retention caused by operating within parameter ranges. Until sufficient process history is available, it is prudent to collect multiple fractions; combination of *contiguous* fractions at different purity levels *in their entirety* to achieve a rich cut at the desired purity is operationally the same as taking a single rich fraction, and thus does not carry the regulatory stigma associated with blending of poor and high quality materials.

31.8.6 Column Lifetime

An important design and economic parameter is column lifetime. Some minimum lifetime should be specified before commercial implementation of a process, and this can be obtained from process history or from multiple injections on a small scale. Minimum plate count and selectivity criteria, measured with a test tracer solution (either a small injection of product or other marker compounds), help to define acceptable column performance, and these can be tied to process performance. For small molecule applications, it should be feasible to extend the use of a column beyond the existing historic lifetime based on a tracer test. Since column performance may diminish for mechanical reasons (poor maintenance of bed integrity owing to lack of sufficient axial pressure, for instance), the sorbent may be unpacked and repacked on multiple occasions. For regulatory purposes, some limits for such operations must be defined. It is also useful to establish whether the feed components can eventually poison the sorbent (by irreversible binding, for instance). In addition, depending on nature of further downstream operations, it may be important to develop tests for column extractables and demonstrate absence or establish acceptable limits of these compounds in the process stream.

31.8.7 Temperature

Column and solvent temperatures are important parameters. Ideally isothermal conditions should be maintained, as retentivity is temperature dependent as shown in Eq. (31.7). However, during operation of larger columns, a small increase (1–3 °C) between the column skin temperature and the solvent inlet temperature is useful to overcome frictional heating effects and within the column and related wall effects [26]. During gradient operations, heat of mixing of the solvent streams can affect the temperature and so should be suitably accounted for.

31.9 ECONOMICS

Given the many parameters associated with design of a chromatographic process, rigorous economic assessment and optimization of chromatography is a complex task. Key economic parameters, such as return on capital, Lang factors (the ratio of total installation cost to cost of equipment), etc., vary from company to company, so that outcomes of net present value (NPV) assessments of the same separation problem may lead to different conclusions at different locations. Nevertheless, cost drivers for chromatographic operations are common to most other processes; in an industrial manufacturing operation, the overall cost, expressed per kg or ton of product purified, is the sum of costs of amortized capital, labor, consumables, waste, and product loss. In addition, there are intangibles, such as development opportunity costs, that are not usually taken into account because they are hard to quantify. For instance, a chiral separation is typically easier and quicker to implement and scale-up than an asymmetric

---

[7]See Ref. [9].

synthesis. If the unwanted enantiomer can be racemized, the major objection to a separation approach (50% yield loss associated with the unwanted enantiomer) may be mitigated. With commensurate investments in optimization and engineering, cost differences between the chiral chromatography and a more elegant asymmetric synthesis may not ultimately be significant, but the time that talented chemists would have spent on the synthesis development could have been used more fruitfully elsewhere.

Capital and labor costs are usually specific to the site of operation. In this author's opinion, cost for large-scale high performance chromatographic equipment and installation is not prohibitive compared with new installation of other process equipment. There is sufficient competition among vendors and important equipment designs are now off patent. The key economic concerns for the chromatography development engineer should be the minimization of cost of consumables and product loss. Several good literature references for quantitative optimization are available [27, 28, Vol. 1].

31.9.1 Stationary Phase

During development, cost of the stationary phase can be high. Sorbent usually is dedicated to a particular product, and, since product failure rates are high, there is limited opportunity to spread the cost of the phase over multiple batches. On the other hand, during commercial operation, sorbent costs, while significant, are not usually a limiting factor since column lifetime can be quite long. For instance, CSP lifetimes of several years have been reported. Unpublished reports suggest that stationary phase costs in such applications can be held below $10/kg purified (Olivier Dapremont, personal communication).

31.9.2 Solvent Consumption

Solvent volumes employed in chromatography are indeed high. Operating a 45 cm ID column at 10 L a minute involves handling 4000 gal a day. However, measures can be taken to drastically limit total solvent consumption and costs by implementing solvent recycle procedures. At full production, it is reasonable to expect upward of 90% solvent recycle on the organic component. (Water is rarely recycled; in some instances this may give organic solvent-rich mobile phases the economic edge, since disposal costs for aqueous waste must be considered.)

For separations where multicolumn techniques, such as SMB technology or multi-injection steady-state recycle technology, can be employed (mostly for chiral separations), solvent consumption can be significantly reduced compared to column chromatography. Coupled with solvent recycle, the use of SMB technology may consume less solvent than a conventional three-step classical chiral resolution by crystallization with a tartrate salt, and the total cost for an SMB operation can be competitive with or lower than such a process [29]. New CSPs with high adsorption capacity are becoming available in the market; the use of these phases is expected to improve the economics of chromatographic separation of enantiomers even further.

31.9.3 Product Loss

For high value products, yield loss can have a significant economic impact. In conventional chromatography (e.g. a gradient elution separation of peptides), yield losses can be minimized by recycling impure fractions. Regulatory agencies expect defined criteria for choosing fractions for recycling, and limits on the number of times that recycle is allowed. Multicolumn techniques have a distinct advantage since high yield is an inherent attribute of continuous processes. An example showing yield increase from 85% in a conventional gradient peptide purification to greater than 95% in an MCSGP process, with a concomitant 25-fold productivity increase, has been published [29].

31.10 CONCLUSIONS

With modern high performance equipment and improvements in operating strategies, including the increasing use of multicolumn technologies, robust and economic large-scale chromatographic processes can be designed for purification of a wide variety of pharmaceutical compounds.

ACKNOWLEDGMENTS

I am grateful to Dr. Olivier Dapremont, of Ampac Fine Chemicals, for insightful discussion on sorbent costs and economics of simulated moving bed chromatography operations. I also thank Dr. Henri Colin, Drew Katti, and the late Professor Georges Guiochon for useful discussions. Also, I am grateful to Novasep Inc. for providing the photographs of their equipment used in Figures 31.6 and 31.7.

REFERENCES

1. Twett, M.S. (1906). *Ber Dtsch Botan Ges* 24: 316–323. and 384–393.
2. Antia, F.D. and Horváth, C. (1989). *Ber Bunsenges. Phys. Chem.* 93: 961–968.
3. Horváth, C. and Melander, W.R. (1983). *Chromatography: Fundamentals and Applications of Chromatographic and Electrophoretic Methods, Part A* (ed. E. Heftmann), 27–135. Amsterdam: Elsevier.
4. Antia, F.D. and Horváth, C. (1989). *J. Chromatogr.* 484: 1–27.

5. van Deemter, J.J., Zuiderweg, F.J., and Klinkenberg, A. (1956). *Chem. Eng. Sci.* 5: 271–289.
6. Antia, F.D. and Horváth, C. (1988). *J. Chromatogr.* 435: 1–15.
7. Unger, K.K. (1978). *Porous Silica*. Amsterdam: Elsevier.
8. Roush, D.J., Antia, F.D., and Göklen, K.E. (1998). *J. Chromatogr. A* 827: 373–389.
9. Nti-Gyabaah, J., Antia, F.D., Dahlgren, M.E., and Göklen, K.E. (2006). *Biotechnol. Prog.* 22: 538–546.
10. Meyer, V. (2010). *Practical High-Performance Liquid Chromatography*, 5e. New York: Wiley.
11. Capaldi, D.C. and Scozzari, A.N. (2008). *Antisense Drug Technology: Principles, Strategies and Applications*, 2e (ed. S.T. Crooke), 401–434. Boca Raton: CRC Press.
12. Gronke, R. (2017). New purification process for antisense oligonucleotides. *Webinar presented at KNect365 TIDES Digital Week*, (18–21 September 2017). https://event.on24.com/wcc/r/1481597/CAEFB5DFCDA434F834AFA5191E99592A/150584 (accessed 22 October 2018).
13. Tiselius, A. (1943). *Kolloid Z.* 105: 101.
14. Frenz, J. and Horváth, C. (1985). *AIChE J.* 31: 400–409.
15. Tugcu, N., Bae, S., Moore, J.A., and Cramer, S.M. (2002). *J. Chromatogr. A* 954: 127–135.
16. Liu, J., Hilton, Z., and Cramer, S.M. (2008). *Anal. Chem.* 80 (9): 3357–3364.
17. Gattuso, M.J. (1995). *Chim. Oggi. (Chem. Today)* 13: 18–22.
18. Antia, F.D. (2002). *Scale-Up and Optimization in Preparative Chromatography: Principles and Biopharmaceutical Applications* (ed. A.S. Rathore and A. Velayudhan), 173–201. New York: Marcel Dekker.
19. Ludemann-Hombourger, O. and Nicoud, R.M. (2000). *Sep. Sci. Technol.* 35: 1829–1862.
20. Zhanga, Z., Mazzotti, M., and Morbidelli, M. (2003). *J. Chromatogr. A* 1006: 87–99.
21. Aumann, L. and Morbidelli, M. (2007). *Biotechnol. Bioeng.* 98: 1043–1055.
22. Müller-Späth, T. and Morbidelli, M. (2017). In: *Integrated Continuous Biomanufacturing III* (ed. S. Farid, C. Goudar, P. Alves, and V. Warikoo), ECI Symposium Series. Hotel Cascais Miragem, Cascais. http://dc.engconfintl.org/biomanufact_iii/23 (accessed 22 October 2018).
23. Grill, C.M. (1997). US Patent 5,630,943, 30 November 1995, Granted date 20 May 1997.
24. Valery, E., Ludemann-Hombourger, O., Boni, J., and Hauck, W. Novasep publication. https://www.novasep.com/media/-articles-and-publications/50c%201%20-%20cyclojet%20a%20powerful%20tool%20for%20binary%20separations.pdf (accessed 22 October 2018).
25. Vailaya, A., Sajonz, P., Sudah, O. et al. (2005). *J. Chromatogr. A* 1079: 85–91.
26. Dapremont, O., Cox, G.B., Martin, M. et al. (1988). *J. Chromatogr. A* 796: 81–99.
27. Guiochon, G., Shirazi, D.G., Felinger, A., and Katti, A.M. (2002). *Fundamentals of Preparative Chromatography*, 2e. Amsterdam: Elsevier Chapter 18.
28. Katti, A. (2000). *Handbook of Analytical Separations*, vol. 1 (ed. K. Valkó), 213–291. Amsterdam, Oxford: Elsevier.
29. Auman, L., Stroehlein, G., Schenkel, B., and Morbidelli, M. (2009). *Biopharm. Int.* 22 (1): 46–53.

32

MEMBRANE SYSTEMS FOR PHARMACEUTICAL APPLICATIONS

DIMITRIOS ZARKADAS
Merck & Co., Inc., Rahway, NJ, USA

KAMALESH K. SIRKAR
New Jersey Institute of Technology, Newark, NJ, USA

32.1 INTRODUCTION

Membrane separation technologies are being rapidly incorporated in a number of industries. There are a number of reasons; they are often cheaper, modular, athermal and can achieve separations difficult to achieve otherwise. In specific industries, e.g. desalination/water treatment, they are becoming the dominant technology. In biopharmaceutical industry, the processes of dialysis for buffer adjustment, microfiltration for clarification, ultrafiltration, and membrane chromatography are widely used. An earlier brief review of applications of membrane technologies in pharmaceutical industries is available in Sirkar [1]. There have been however limited applications of membrane technologies in the pharmaceutical industry during active pharmaceutical ingredient (API) processing in the presence of organic solvents. The membrane technologies that are being used and/or explored in a more than cursory fashion in pharmaceutical processing are pervaporation and organic solvent nanofiltration (OSN). To that extent our focus in this chapter will be on pervaporation first and then on OSN. At the end we will briefly focus on membrane solvent extraction.

32.2 PERVAPORATION IN THE PHARMACEUTICAL INDUSTRY

32.2.1 Introduction

Pervaporation is a process in which a feed liquid mixture at atmospheric or higher pressure is brought into contact with a membrane, which allows the selective removal of one or more components of the feed stream into a gaseous/vapor stream on the other side of the membrane (permeate side). Separation is achieved by maintaining a difference between the species partial pressure in equilibrium with the feed liquid and the permeate side partial pressure of the species in the feed to be removed. The partial pressure differential is commonly established by applying vacuum at the permeate side, flowing an inert gas, or a combination of the above techniques. When the feed is a vapor stream, the process is called vapor permeation. Some researchers treat the two processes as different. However, the operating principles and the membranes used in pervaporation or vapor permeation are similar, and the two processes will be treated here as variations of the same technique.

The term pervaporation is a composite of the words permeation and evaporation. The evaporation heat required to transfer the permeating component(s) from the liquid phase in the feed to the vapor phase in the permeate side is supplied by the sensible heat of the feed stream. Separation in pervaporation processes is the outcome of a sequence of three steps [2]:

1. Preferential sorption of one or more components into the feed side of the membrane.
2. Selective diffusion through the membrane.
3. Desorption to the vapor phase at the permeate side.

It is apparent that complex mass and heat transfer phenomena occur during pervaporation. The membrane acts

Chemical Engineering in the Pharmaceutical Industry: Active Pharmaceutical Ingredients, Second Edition.
Edited by David J. am Ende and Mary T. am Ende.
© 2019 John Wiley & Sons, Inc. Published 2019 by John Wiley & Sons, Inc.

in two ways: first as a physical barrier between the liquid and vapor phases and second as an additional component in the system, which alters its thermodynamics and allows the separation of its components. The last point is significant in understanding why pervaporation has been successfully used to break azeotropes (i.e. ethanol–water), for which conventional distillation is unsuccessful. Separation efficiency in distillation is governed by the vapor–liquid equilibria (VLE) of the system, which for an azeotrope cannot change. In pervaporation separation is driven by differences in solubility and diffusivity of the components in the feed stream through the membrane used.

Pervaporation has found industrial applications in various fields including the dehydration of organic solvents [3, 4], the separation of organic–organic mixtures [5–7], the concentration/extraction of aroma compounds from water solutions in the food industry [8], the removal of Volatile Organic Compounds (VOCs) from aqueous waste streams [9], and the enhancement of reaction conversion/rate by removal of water during condensation or esterification reactions [10]. In the above applications pervaporation is applied either as a stand-alone technique or in a hybrid process combined with distillation. When the feed stream contains chemicals harmful to the membrane or solids, which would foul the membrane and reduce performance, vapor permeation can be applied. In this case, the stream is evaporated via distillation and while at the vapor state passed through the pervaporation module. Vapor permeation has found extensive application in distillation–pervaporation hybrid units [11].

32.2.2 Process Description and Theory

A schematic of the process is shown in Figure 32.1. A liquid feed containing components 1 and 2 is entering the membrane device at a temperature $T_{f,in}$ and pressure $P_{f,in}$. A reduced pressure P_p is applied at the permeate side of the membrane. The membrane preferentially permeates component 1 over 2. Under these conditions component 1 permeates through the membrane and appears in the vapor phase on the permeate side. The net outcome is the removal of component 1 from the feed stream. The heat of evaporation for the permeating component(s) is supplied by the sensible heat of the feed. The latter is therefore cooled to a temperature $T_{f,out}$ at the outlet of the membrane.

Figure 32.2 shows a schematic of the two operating configurations most commonly used in pervaporation applications.

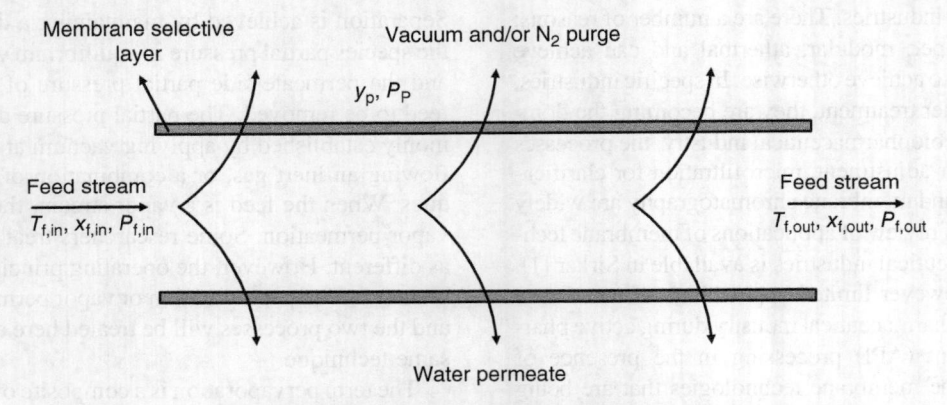

FIGURE 32.1 Operating principle of pervaporation.

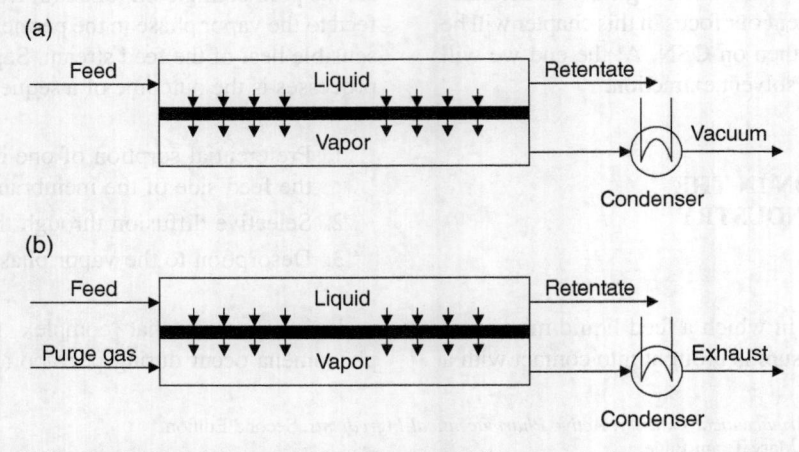

FIGURE 32.2 Schematic of different pervaporation operating schemes. (a) Vacuum on the permeate side and (b) inert carrier gas on the permeate side.

Both of them utilize a pump to circulate the feed solution through the membrane module and optionally a heat exchanger to preheat the feed stream to the appropriate temperature, although an effort is always made to use heat available from upstream processing to minimize operating costs. At the permeate side a condenser is available to collect the permeating species. The two configurations shown differ only in the way the driving force across the membrane is established: in Figure 32.2a, a vacuum pump is used, while in Figure 32.2b, a carrier gas is used to reduce the mole fraction y_i of the permeating species and hence its partial pressure.

The quality of the separation is commonly expressed in terms of the separation factor, α, which for a binary system of species 1 and 2 is given by

$$\alpha = \frac{y_1 x_2}{x_1 y_2} \quad (32.1)$$

where

y_i: mole fraction of component i in the permeate
x_i: mole fraction of component i in the feed stream

The flux of component i through the membrane is given by the following expression:

$$J_i = \frac{Q_i}{l} \Delta p_i = \frac{Q_i}{l} (p_{i,f} - p_{i,p}) \quad (32.2)$$

where

J_i: flux of component i through the membrane, kg/m²/s
Q_i: permeability of component i through the membrane, kg/m/s/Pa
l: effective membrane thickness, m
$p_{i,f}$: partial pressure of component i in a gas stream in equilibrium with the feed liquid stream, Pa
$p_{i,p}$: partial pressure of component i in the permeate, Pa

The amount of component i that can be removed in time t from the feed stream is a function of the flux and the membrane area used to achieve the separation according to the following equation:

$$m_i = J_{i,ave} A t \quad (32.3)$$

where

m_i: amount of component i removed, kg.
A: membrane surface area, m².
t: operating time, s.

and an average value is used for the flux of component i to account for the fact that the flux declines as component i is removed from the feed stream to the permeate.

The permeability coefficient is dependent on the membrane material and varies with temperature and composition of the liquid feed stream. It is considered to be the product of the solubility and diffusivity of component i through the membrane:

$$Q_i = S_i(T,C) D_i(T,C) \quad (32.4)$$

where

S_i: solubility of component i, kg/m³/Pa
D_i: diffusivity of component i, m²/s

Although thermodynamic and transport models can be applied to calculate, respectively, the solubility and diffusivity of a species through a membrane, in almost all practical applications, the permeability coefficient is estimated based on experimental data.

The partial pressure difference across the membrane is given by

$$\Delta p_i = x_i \gamma_i p_i^{sat} - y_i P_p \quad (32.5)$$

where

γ_i: activity coefficient of component i in the feed stream
p_i^{sat}: vapor pressure of component i at the feed temperature, typically described by the Antoine equation

Combining Eqs. (32.2) through (32.5), a final expression for the flux of component i is obtained:

$$J_i = \frac{S_i D_i}{l} (x_i \gamma_i p_i^{sat} - y_i P_p) \quad (32.6)$$

Equation (32.6) can provide an insight as to how the flux of the permeating species can be increased and hence the operation time needed to achieve a specified separation can be minimized. Four different cases and their combinations can be identified:

1. Increase the permeability coefficient. Since the latter is dependent primarily on the membrane material, an increase of the flux can be achieved by selecting a membrane with a more open structure at the expense, however, of lower selectivity values. The allowable limits for this trade-off between flux and selectivity will generally depend on the intended application and can be decided by the process engineer during the process development stage.

2. Decrease the effective membrane thickness. This is also a membrane property, which has to be considered during the process development stage. Membranes with thin selective layers and open support layers to minimize diffusion limitations are the best choices.
3. Increase the temperature of the feed. This action maximizes the vapor pressure of the permeating component and hence the partial pressure driving force across the membrane according to Eq. (32.5). Upper temperature limits are dictated by the boiling point of the liquid feed and the temperature stability of the API. The second limitation is more severe in terms of process design. Boiling point limitations, which would result in cavitation of the feed pump and reduced process performance, can be partially overcome by pressurizing the feed. Such an action raises the boiling point according to the Clausius–Clapeyron equation. This is the most effective and easy to implement modification to increase the flux of the permeating species due to the exponential dependence of vapor pressure on temperature.
4. Decrease the partial pressure of component i in the permeate side. This can be achieved by reducing the permeate absolute pressure at the expense, however, of higher pumping costs. Alternatively, an inert gas can be pumped through the permeate side, which will reduce the mole fraction y_i to low levels. A combined approach, namely, starting by operating only the vacuum pump at the beginning of the process and then introducing a purge stream of an inert gas (i.e. N_2) toward the end of the process, will be more effective since the partial pressure difference across the membrane is reduced throughout the process and becomes very small when most of the separation has been performed. This approach is similar to well-established guidelines for conventional drying and is sometimes called the Combo Mode.

Equation (32.5) can be used to identify appropriate operating conditions with respect to the permeate pressure for a given separation. A positive flux and hence a separation are obtained if $\Delta p_i > 0$. For a given final concentration x_i in the feed stream, Eq. (32.7) then yields the maximum allowable operating pressure in the permeate side:

$$P_{p,\max} = \frac{x_i \gamma_i p_i^{sat}}{y_i} \quad (32.7)$$

Li et al. [12] used the above expression to calculate the permeate pressure for benzene dehydration at 70 °C and select a vacuum pump appropriate for the desired levels of water removal. Such calculations require the estimation of the activity coefficient γ_i, which can be performed by group contribution or semi-empirical models such as UNIFAC and NRTL. Activity coefficient calculations are easy to perform for solvent–water systems but become more complicated for solvent–water–API/intermediate systems. The API or intermediate will alter the activity coefficient of water and will reduce or increase the water activity coefficient, depending on the nature of its interaction with water.

The flux of component i can also be expressed in terms of an overall mass transfer coefficient and concentrations:

$$J_i = K_i (C_{i,f} - C_{i,p}) \quad (32.8)$$

where

K_i: overall mass transfer coefficient of component i through the membrane, m/s
$C_{i,f}$: concentration of component i in the feed stream, kg/m^3
$C_{i,p}$: concentration of component i in the permeate, kg/m^3

The overall mass transfer coefficient can be expressed based on film theory as the sum of three resistances in series, feed side, and membrane and permeate side:

$$\frac{1}{K_i} = \frac{1}{k_{i,f}} + R_m + \frac{1}{k_{i,p}} \quad (32.9)$$

where

$k_{i,f}$: mass transfer coefficient of component i at the feed side, m/s
R_m: membrane mass transfer resistance, s/m
$k_{i,p}$: mass transfer coefficient of component i at the feed side, m/s

Equations (32.8) and (32.9) can be used to calculate fluxes based on film theory. In principle, mass transfer correlations can be used to calculate the feed and permeate side mass transfer coefficients with reasonable accuracy, i.e. see Ortiz et al. [13]. Experimentation is needed to determine the membrane resistance. In practice, experimentation is performed, and the overall coefficient is expressed as a function of the feed side mass transfer coefficient, since the membrane and permeate resistances are usually small compared with the feed side mass transfer resistance. The combined membrane and permeate resistance can then be obtained as the intercept in a Wilson plot [13].

The dependence of the flux of component i on temperature is usually expressed via an Arrhenius-type relationship [14]:

$$J = J_0 \exp\left(-\frac{E_a}{RT}\right) \quad (32.10)$$

where

E_a: activation energy for permeation, J/mol
R: universal gas constant, J/mol
T: temperature, K

The activation energy E_a in Eq. (32.10) is a compounded parameter accounting for the variation of permeation flux with both the membrane permeability and the permeation driving force Δp_i, as pointed out by Feng and Huang [14]. Both the pre-exponential term and the activation energy can be expressed as functions of temperature, feed, and permeate mole fractions of component i by curve fitting laboratory experimental data obtained for the intended application. These expressions can then be substituted in Eq. (32.10) and used to perform scale-up calculations, as detailed in Section 32.2.4.2 [15].

32.2.3 Pervaporation Membranes

The nature of the pharmaceutical industry poses certain limitations to the selection of membranes for manufacturing API or intermediates. First, the membrane must be compatible with the pharmaceutical stream to be processed. The use of harsh solvents (i.e. DMF (dimethylformamide), THF (tetrahydrofuran)) is still widespread in the industry, although a turn toward greener chemistry has been experienced in the last few years. In addition, many times the active ingredient will make the stream acidic or basic, or it might increase the interaction between the stream and the membrane. The question of leachables into the pharmaceutical stream and its impact on the final drug substance (DS) quality must then be addressed appropriately. Membranes with increased chemical and thermal stability are preferable. A second characteristic of the pharmaceutical industry is the generally low production requirement compared with the chemical industry and the fact that only one in nine new chemical entities entering phase I clinical trials reaches commercialization [16]. Many times the intended application will have to be abandoned. Therefore, a membrane that can be used for a variety of processes/streams will be clearly preferable. Membranes tailored to a specific process would make sense only if the intended application enters commercial production. In summary, the three desirable characteristics for a pervaporation membrane in the pharmaceutical industry are the following:

1. Good chemical stability.
2. Good thermal stability.
3. Ability to handle a variety of streams and/or process conditions without significant loss in performance.

Three types of membranes are available for pervaporation: polymeric, inorganic, and mixed matrix[1] membranes [17]. Only the first two types of membranes are available commercially. Both of them are asymmetric membranes, namely, they consist of a porous support layer and a thin dense selective layer coated onto the support layer. The selective layer is the one in contact with the process stream and hence must have good chemical and thermal stability. It is also the layer that will start leaching first to the process stream.

Commercial polymeric hydrophilic membranes have a polyvinyl alcohol (PVA)-selective layer. Sulzer Chemtech is the leader in the industry with probably more than 90% of market share. Spiral wound or hollow fiber modules are available. PVA membranes can tolerate temperatures of up to about 90 °C and mild to relatively strong acidic and basic conditions. Fluxes reported for typical dehydration applications (water < 10 wt %) range between 1 and 2 kg/m^2/h [17, 18]. PVA membranes have a proven track record in a number of industrial dehydration applications. Their operating temperature limitations and stability issues when exposed to certain organic solvents should be taken into consideration. The above limitations make it unlikely that PVA membranes could be used as a general tool for dehydration problems in the pharmaceutical industry. However, they can still be useful in specialized applications. Recently, polymeric membranes with a fluoropolymer-selective skin have been developed by Compact Membrane Systems (DE, USA) [19]. The selective layer can withstand temperatures of up to 200 °C and a variety of different chemical conditions. It seems that this membrane can satisfy the performance criteria listed above and potentially be very useful for applications in the pharmaceutical industry. However, a suitable substrate must be selected for high temperature applications to avoid membrane damage due to thermal stresses.

Inorganic membranes can overcome the temperature limitations of polymeric membranes; certain types of inorganic membranes are resistant to a variety of chemical conditions also. Table 32.1 summarizes the commercially available inorganic membranes for dehydration applications. Two types of inorganic membranes exist: zeolite and ceramic. Zeolite membranes exhibit excellent selectivity and good fluxes [20–23], but they can only be used for a limited pH range, between 6 and 8 [1, 21]. These properties make zeolite membranes an excellent choice for solvent dehydration; however, in most cases zeolite membranes will be not able to withstand the presence of an API or an intermediate. Microporous silica membranes on the other hand can tolerate a much broader pH range, which makes them ideal for the dehydration of most pharmaceutical streams. The membranes manufactured by Pervatech BV can tolerate a pH down to 2–3, as has been confirmed with an actual pharmaceutical stream [24]. Silica membranes exhibit inferior separation factors compared to

---

[1] Mixed matrix membranes consist of a polymeric base membrane impregnated with inorganic material.

zeolite membranes, but their flux is higher [20] and comparable if not better than polymeric membranes.

All commercially available inorganic membranes are of tubular geometry. Figures 32.3–32.5 show pictures of the membrane modules described in Table 32.1. The zeolite membrane modules from Mitsui have a typical shell-and-tube configuration [25]. The feed is introduced on the shell side, and the baffles present force it to a path perpendicular to the tube length leading to higher feed side mass transfer coefficients. The permeate is collected in the tube side. The configuration of the Pervap® SMS module by Sulzer Chemtech is shown in Figure 32.4. The membrane tubes are placed inside the tubes of a heat exchanger to achieve isothermal operation and increase permeation rate. The feed flows in the annular space between the heat exchanger and the membrane tubes, while the permeate is collected inside the tubes. The membrane tubes can either be connected in series or in parallel. Both configurations yield acceptable pressure drops [26]. The modules by Pervatech BV have also a shell-and-tube configuration. They are made of two identical parts consisting of 54 tubes, 50 cm long, stacked one upon the other. The two parts are connected by a plate with machined channels connecting the individual tubes. The feed is introduced in the tube side and the permeate is collected at the shell side. The tubes are connected in series, and the manufacturer recommends a linear velocity of 2 m/s through the tubes to minimize concentration polarization effects. This requirement results in a pressure drop of about 4 bar as measured

TABLE 32.1 Commercially Available Inorganic Pervaporation Membranes

Membrane Type	Selective Layer	Support Layer	ID/OD (mm)	Selective Layer Location	Manufacturer
Zeolite	Zeolite (A-, T-, Y-)	Alumina	9/12	Outside of tube	Mitsui Engineering and Shipbuilding Ltd.
Ceramic	Amorphous silica	Alumina, titania	8/14	Outside of tube	Sulzer Chemtech
Ceramic	Amorphous silica	Alumina, titania	7/10	Inside of tube	Pervatech BV

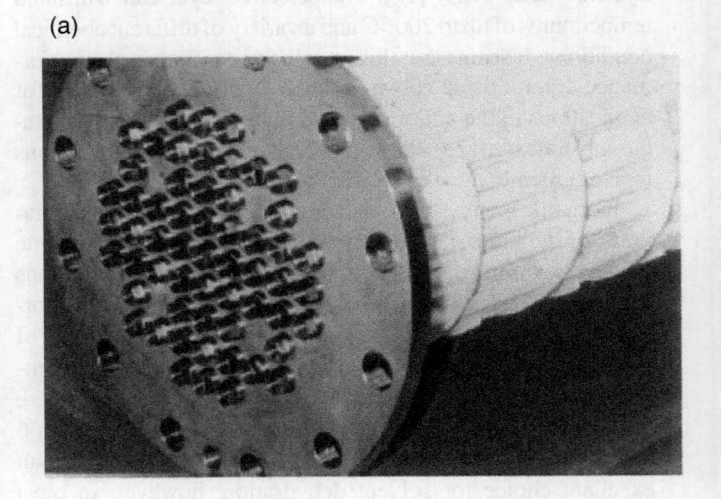

FIGURE 32.3 Zeolite NaA membranes from Mitsui Engineering and Shipbuilding Ltd. (a) Module configuration and (b) module layout. *Source*: From Moriagami et al. [25], with permission.

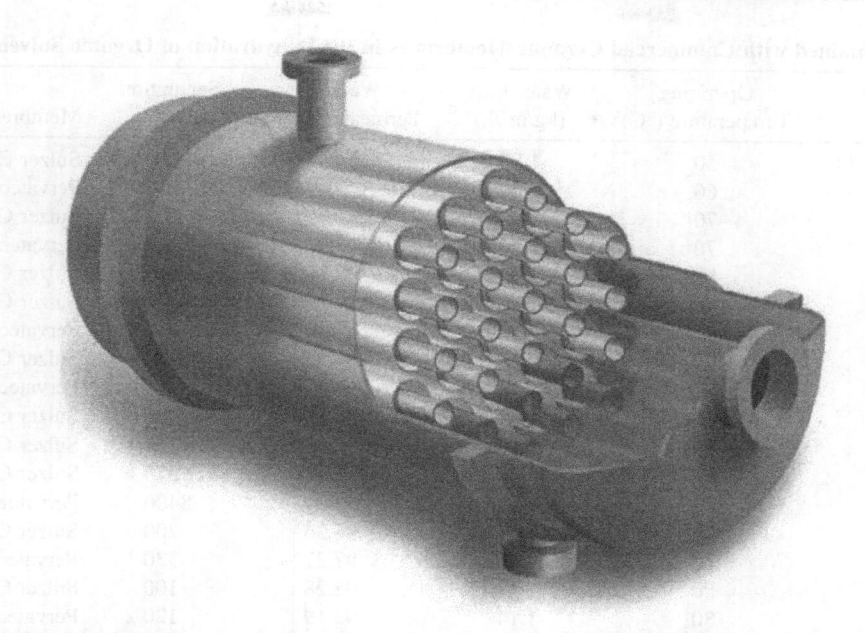

FIGURE 32.4 Pervap SMS silica membrane modules from Sulzer Chemtech. *Source*: From Sommer et al. [26], with permission.

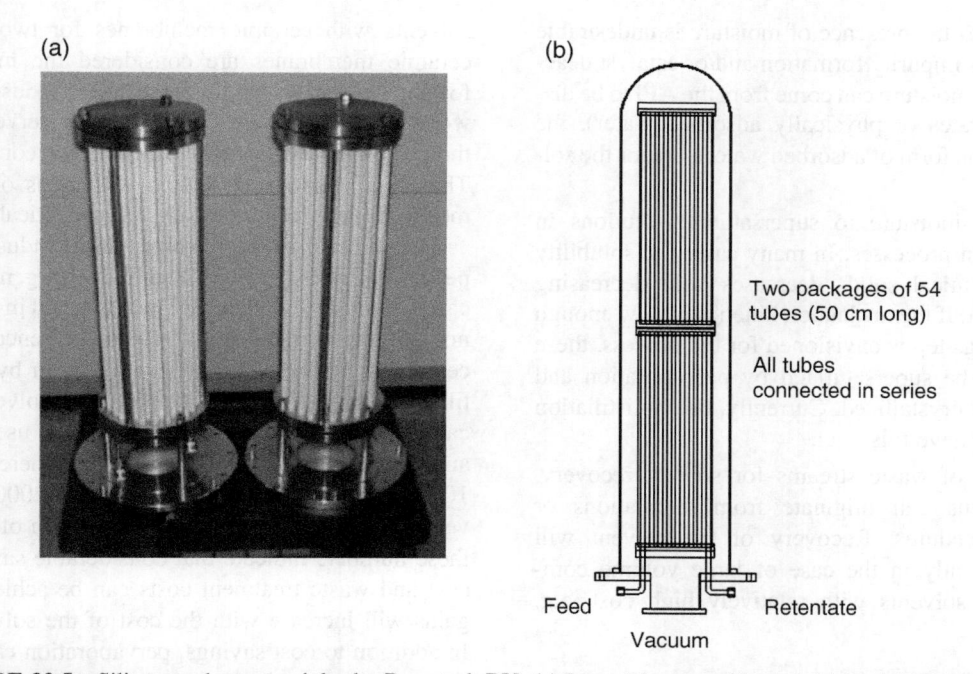

FIGURE 32.5 Silica membrane modules by Pervatech BV. (a) Inner tube assembly and (b) module configuration.

for water by the manufacturer (F. Velterop, personal communication with D. Zarkadas).

32.2.4 Pervaporation Applications in the Pharmaceutical Industry

The majority of potential pervaporation applications in the pharmaceutical industry are related to dehydration of pharmaceutical streams. Pharmaceutical streams need to be dehydrated in the following cases:

1. Removal of moisture during reactions. Certain types of reactions, i.e. esterification or condensation reactions, require the removal of water to increase conversion and yield. Such applications have already been demonstrated with nanofiltration (NF) membranes [27, 28].

TABLE 32.2 Results Obtained with Commercial Ceramic Membranes in the Dehydration of Organic Solvents

Solvent	Feed Water Content (wt %)	Operating Temperature (°C)	Water Flux (kg/m^2/h)	Water in Permeate (wt %)	Separation Factor	Membrane	Reference
Methanol	10.4	60	1.87	58.84	10	Sulzer Chemtech[a]	Ref. [20]
	10.5	60	0.39	71.69	20	Pervatech	
Ethanol	10.3	70	2.33	86.37	60	Sulzer Chemtech	Ref. [20]
	11.0	70	2.00	95.26	160	Pervatech	
IPA	4.5	80	1.86	98.10	1150	Sulzer Chemtech	Ref. [30]
	10.2	75	2.76	91.06	90	Sulzer Chemtech	Ref. [20]
	9.8	75	2.55	95.33	190	Pervatech	Ref. [20]
Acetone	10.0	70	0.52	>99.5	n/a	Sulzer Chemtech	Ref. [31]
	10	70	2.72	>99.5	n/a	Pervatech	Ref. [31]
Acetic acid	10.4	80	1.91	86.80	60	Sulzer Chemtech	Ref. [20]
Ethyl acetate	2.0	70	3.16	93.71	750	Sulzer Chemtech	Ref. [20]
THF	11.8	60	3.47	96.53	210	Sulzer Chemtech	Ref. [20]
	11.5	60	3.30	99.91	8400	Pervatech	
Acetonitrile	11.9	70	2.73	96.36	200	Sulzer Chemtech	Ref. [20]
	9.7	70	3.90	97.27	330	Pervatech	
DMF	10.2	80	1.53	92.28	100	Sulzer Chemtech	Ref. [20]
	9.1	80	1.14	92.19	120	Pervatech	

[a]In the original text the membrane is quoted as ECN. This membrane has been commercialized by Sulzer Chemtech and is referenced by this name here.

In many cases the presence of moisture is undesirable due to excess impurity formation and/or catalyst deactivation. The moisture can come from the API to be dissolved (hydrates or physically adsorbed water), the catalyst (in the form of adsorbed water), and/or the solvent itself.

2. Removal of moisture to supersaturate solutions in crystallization processes. In many cases the solubility of pharmaceutical solids decreases with decreasing water content. If the nucleation kinetics is slow enough and a seeding step is envisioned for the process, these streams can be supersaturated by pervaporation and subsequently crystallized. Currently, batch distillation is used to achieve this goal.

3. Dehydration of waste streams for solvent recovery. These streams can originate from distillations or workup procedures. Recovery of the solvent will make sense only in the case of large volume compounds and solvents with relatively high cost, e.g. 2-MeTHF.

Another potential application of pervaporation is the removal of VOCs from pharmaceutical waste streams [29]. Since this application does not involve streams used in the production of an API, it will not be further explored here. The remainder of this section will explore in more detail the dehydration applications listed above.

32.2.4.1 Dehydration Applications of Pervaporation

This section will examine in more detail the work that has been performed up to now in the dehydration of organic solvents with ceramic membranes for two reasons. First, ceramic membranes are considered the most appropriate for applications in the pharmaceutical industry. Second, the work performed in pure solvents can serve as a basis for the design of applications with streams containing an API. The remainder of the section will focus on the operation modes that can be adopted in pharmaceutical manufacturing.

Dehydration in the pharmaceutical industry is primarily performed by distillation. Other means, e.g. molecular sieves, can find only specialized applications and in many cases will not perform satisfactorily due to the presence of the pharmaceutical ingredient. The removal of water by azeotropic distillation can result in the intense use of solvents, even in the case where continuous[2] distillation is used. One of the authors is aware of several examples where the removal of 100–200 kg of water requires the use of 2000–3000 L of solvent per batch. For an annual production of 20–50 batches, these numbers indicate that considerable savings in solvent cost and waste treatment costs can be achieved. The fiscal gains will increase with the cost of the solvent to be used. In addition to cost savings, pervaporation can lead to much greener solutions. In almost all cases, the cost savings from reduced solvent usage will be enough to repay the capital investment required to purchase the membrane modules.

Table 32.2 summarizes the results that have been obtained with commercial silica membranes during the dehydration of organic solvents by various researchers. Only solvents

---

[2]The term continuous is rather misleading; distillation is performed by continuously adding solvent and removing only distillate but not the batch, which is the heavy fraction.

relevant to pharmaceutical applications are reported here; additional information can be found in the original references. The quoted fluxes can serve as a basis for initial design according to Eq. (32.3). A sample calculation is given in Example Problem 32.1. As mentioned in Section 32.2.2, the presence of a pharmaceutical ingredient in the stream will alter its thermodynamics and especially the activity coefficient of water. According to Eq. (32.6) this will increase or decrease the water flux depending on the nature of the interaction of the pharmaceutical ingredient with the rest of the stream components. The pressure required in the permeate side to effect a certain separation will also change accordingly. The presence of a pharmaceutical ingredient in a stream will also change its viscosity and hence the hydrodynamic conditions in the feed side. The latter can often negatively impact the flux through the membrane due to concentration polarization [26, 32]. Pervatech recommends a feed linear velocity of 2 m/s through the membrane to avoid concentration polarization (F. Velterop, personal communication with D. Zarkadas). From the above discussion, it follows that laboratory experimentation should be performed prior to scale-up to determine the actual flux and optimal operating conditions to adjust initial estimates of membrane surface area. As a rule of thumb, the flow should always be kept in the turbulent or at least in the transitional regime to avoid concentration polarization effects.

EXAMPLE PROBLEM 32.1

A pharmaceutical stream of 5000 L contains an API, 6 wt % water and ethyl acetate. Stability concerns dictate that the removal of water must be performed in 24 hours. Calculate the membrane surface area that is needed for the separation.

Solution

$$m_i = J_{i,\text{ave}} A t$$

Rearrangement of Eq. (32.3) for area results in the following:

$$A = \frac{m_i}{J_{i,\text{ave}} t}$$

A water flux from ethyl acetate can be obtained from Table 32.2. Based on a value of 3.16 kg/m^2/h and $t = 24$ hours and substituting values yields a membrane area of 3.56 m^2:

$$A = \frac{m_i}{J_{i,\text{ave}} t} = \frac{(5000 \text{ l})(0.9 \text{ kg/l})(0.06 \text{ wt \%})}{(3.16 \text{ kg/m}^2\text{h})(24 \text{ hours})}$$

$$= \frac{270 \text{ kg}}{75.84 \text{ kg/m}^2} = 3.56 \text{ m}^2$$

This will give an initial estimate of capital cost to purchase the membrane modules. Additional laboratory experimentation with the actual pharmaceutical stream should be performed to obtain a more accurate estimate of water flux and hence membrane surface area.

Figure 32.6 shows a schematic for a typical pervaporation setup in the pharmaceutical industry. This setup will be useful when the pharmaceutical stream contains no solids. The batch is recycled between the reactor and the pervaporation module by means of a pump. A heat exchanger can be

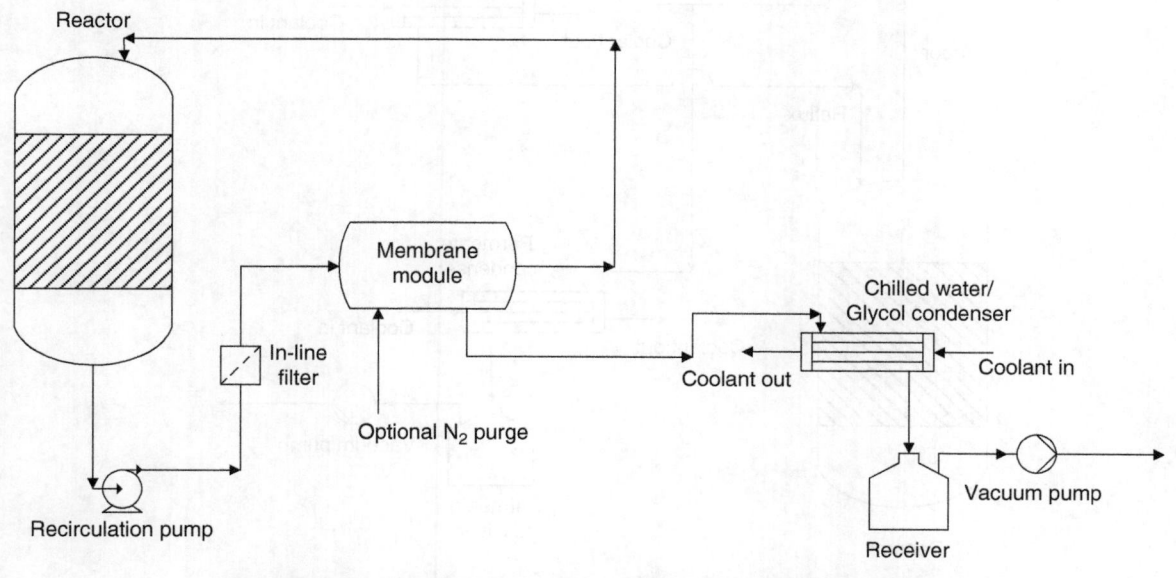

FIGURE 32.6 Typical pervaporation setup in the pharmaceutical industry.

optionally used to ensure that the feed inlet temperature is maintained at appropriate levels. Alternatively, the reactor can be pressurized and its temperature increased to levels that will compensate any heat losses between the reactor and the pervaporation module. A typical cartridge filter is installed prior to the pervaporation module to minimize membrane fouling and potential flux decline with time. The permeate side of the membrane module is connected to a condenser and a vacuum pump. In most cases, it will be practical to use chilled water (available at a temperature of about 6 °C) for the condenser; however, glycol or Syltherm® condensers can also be used. It will usually be better to use a dedicated vacuum pump to ensure that the appropriate vacuum levels are maintained in the permeate side during operation. However, house vacuum can also be considered if the intended application does not require permeate pressures lower than 30–40 mmHg. An option to use a nitrogen purge stream is also depicted in Figure 32.6; its use can be decided based on laboratory process development. A portable skid arrangement for the setup depicted in Figure 32.6 will be preferable, since it gives the flexibility to use the unit in multiple applications. This is important in an industry where the majority of the products scaled up will not reach commercial-scale production.

Figure 32.7 shows a typical setup for vapor permeation. This mode of operation will be useful when solids are present in the pharmaceutical stream (i.e. heterogeneous reactions, insoluble catalyst), which would foul the membrane and render it useless. This mode of operation is not as energy efficient as the one depicted in Figure 32.6; the heat of vaporization needs to be supplied to effect the separation. However, this is of relatively small concern for the pharmaceutical industry, whose end products are of high value. The membrane module is positioned between the reactor and the condenser. Water is removed from the vapor phase and the rest of the vapor is condensed and returned as reflux to the reactor. Concentration polarization will be of smaller concern with this operation mode since the mass transfer resistance in the vapor phase will be considerably smaller compared with the liquid phase.

32.2.4.2 Process Modeling As Figures 32.6 and 32.7 illustrate, pervaporation processes in the pharmaceutical industry operate batchwise. During operation, part of the pharmaceutical stream is removed through the membrane, and the stream mass decreases by the amount permeated. The temperature of the stream entering the membrane module in such applications can be considered constant, since it is regulated in the reactor containing the batch and the amount of stream present inside the module and the recirculation loop is small compared to the total batch volume. However, a temperature decrease is experienced inside the module, since the heat required for the removal of water is supplied by the sensible heat of the stream. Therefore, the temperature of the stream will vary with axial position inside the module. From the above discussion it is apparent that both the mole fraction of the permeating species and the stream temperature are functions of the axial position inside the module. According to Eqs. (32.6) and (32.10), this implies that the permeation flux also changes along the membrane module.

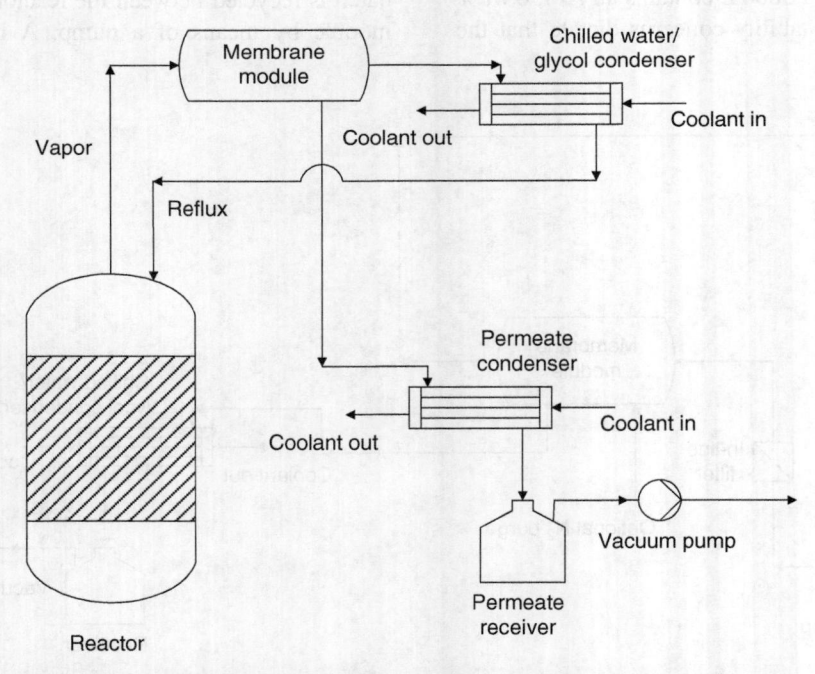

FIGURE 32.7 Typical vapor permeation setup in the pharmaceutical industry.

For a membrane module consisting of n tubes, the following governing equations can be written [15]:

Overall System

Overall mass balance

$$dM = -J_t A_T dt \qquad (32.11)$$

Mass balance for water

$$d(Mx_F) = -J_t A_T y_M dt \qquad (32.12)$$

Combining Eqs. (32.11) and (32.12) the following expression can be obtained:

$$dx_F = \frac{(x_F - y_M) J_t A_T}{M} dt \qquad (32.13)$$

Membrane Module

Overall mass balance

$$dF = -Jn\pi D dz \qquad (32.14)$$

Mass balance for water

$$d(Fx) = -Jyn\pi D dz \qquad (32.15)$$

Energy balance

$$d(FH_f) = -J\Delta H_v n\pi D dz \qquad (32.16)$$

Equations (32.14) and (32.15) can be combined as follows:

$$dx = \frac{(x-y)}{F} Jn\pi D dz \qquad (32.17)$$

Equation (32.16) may be rewritten as

$$FC_p dT = -J\Delta H_v n\pi D dz \qquad (32.18)$$

The following notation is used in the above equations:

M: mass of batch at time t, kg.

J_t: average flux through the membrane module for the time interval dt, kg/m^2/s.

A_T: total membrane area, m^2.

x_F: water mole fraction in feed tank, dimensionless.

y_M: average water fraction in permeate for the time interval dt, dimensionless.

F: mass flow rate through membrane module at time t, kg/s.

z: axial displacement inside the membrane module from entrance, m.

J: local flux through a differential length dz of a membrane tube, kg/m^2/s.

D: membrane tube diameter (inside or outside), m.

H_f: enthalpy of the stream entering the membrane module, J/kg.

ΔH_v: heat of vaporization of water, J/kg.

C_p: specific heat capacity of the stream entering the module, J/kg/K.

The model represented by Eqs. (32.11), (32.13), (32.14), (32.17), and (32.18) neglects any heat losses due to vaporization of organic solvent. Its accuracy is not expected to deteriorate significantly due to this fact for the majority of organic solvents used with the exception of methanol, ethanol, and acetic acid, which as shown in Table 32.2 can permeate through the membrane to an appreciable extent. The system of Eqs. (32.11), (32.13), (32.14), (32.17), and (32.18) can be solved by applying a finite difference scheme. Discretization is performed in both time and axial position inside the module domains. A prerequisite for this task is to have expressions of the flux J and the permeate water mole fraction as functions of feed water mole fraction and temperature:

$$J = f(x, T) \qquad (32.19)$$

$$y_M = g(x, T) \qquad (32.20)$$

These expressions can only be obtained through laboratory experimentation. The latter should include permeation runs at different feed compositions and temperatures. Flux and permeate composition data will then need to be curve fitted against temperature and composition data. Equation (32.10) can be used for the flux, where the flux and/or activation energies are expressed as functions of feed composition; any expression for the water permeate fraction can be applied [15]. Based on the above discussion, the initial and boundary conditions to solve Eqs. (32.11), (32.13), (32.14), (32.17), and (32.18) are the following:

$$\begin{aligned} M(0) &= M_0 \\ x_F(0,0) &= x_0 \\ x_F(t,0) &= x_i \\ J(0,0) &= f(x_0, T_F) \\ J(t,0) &= f(x_i, T_F) \\ y(0,0) &= g(x_0, T_F) \\ y(t,0) &= g(x_i, T_F) \\ F(0,0) &= F(t,0) = F_0 \\ T(0,0) &= T(t,0) = T_F \end{aligned} \qquad (32.21)$$

A simpler but less accurate model can be obtained if the variation of flux inside the module is neglected. In this case only Eqs. (32.11) and (32.13) need to be solved with the aid of Eqs. (32.19) and (32.20). This is an initial value problem that can be solved by numerical integration (e.g. fourth order Runge–Kutta scheme). The initial conditions are summarized below:

$$M(0) = M_0$$
$$x_F(0) = x_0$$
$$J(0) = f(x_0, T_F) \quad (32.22)$$
$$y(0) = g(x_0, T_F)$$
$$T_F = \text{constant}$$

This model will probably be adequate for scale-up calculations without the need to face the complexity of the more rigorous model presented previously. The approximation will be better for modules where the tubes are arranged in parallel. When tubes are connected in series, the variation of composition and temperature along the tube length will be significant, and the simple model will tend to overpredict the flux and hence underpredict batch processing time.

32.3 OSN IN PHARMACEUTICAL INDUSTRY

The API solutes of interest in pharmaceutical synthesis vary in molecular weight between 200 and 1000 Da. NF membranes originally developed for treatment of aqueous solutions are particularly suited for rejecting solutes in such a size range and passing water through [33]; the solute molecular weight range is said to vary between 150 and 1000 Da. OSN is directed toward achieving the same goal with organic solvents.

32.3.1 Process Description and Principles

In NF, the feed solution is brought under pressure of as much as 35 bars to one side of the membrane (Figure 32.8). The solvent flows through the membrane to the other (permeate) side at a lower pressure, generally atmospheric. The membrane is supposed to prevent the transmission of solutes of MW > 150~200 Da. The extent of solute retention by the membrane is described by solute rejection, R_i:

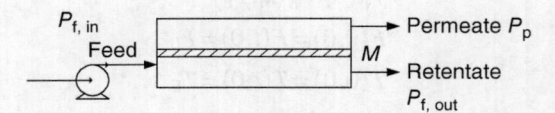

FIGURE 32.8 Schematic of a nanofiltration membrane unit for continuous operation.

$$R_i = 1 - \frac{C_{ip}}{C_{if}} \quad (32.23)$$

where

C_{ip}: solute concentration in permeate side, kg/m^3
C_{if}: solute concentration in the feed side, kg/m^3

If the NF membrane retains the solute completely, then $C_{ip} = 0$ and $R_i = 1$. There are a variety of membranes; some membranes will reject higher molecular weight solutes/species/catalysts in the range from 200 to 1000 Da but not in the lower molecular weight range. The molecular weight cut-off (MWCO) of the membrane is defined to be the solute molecular weight which will yield a value of $R_i = 0.95$; a smaller molecular weight will pass through the membrane more readily and have a lower R_i. An additional quantity of great interest is the solvent flux through the membrane N_i in gmol/cm^2·s. It is reported often as a volume flux J_v in L/m^2·h (LMH); typical values at, say, 30 bars and 30 °C may be between 10 and 150 LMH.

There are two general models for species transport through OSN membranes: (i) solution–diffusion model and (ii) pore–flow model. The molar flux expression J_i suggested [34] for solution–diffusion of a species through a nonswollen selective layer of the membrane (analogous to that for reverse osmosis by Wijmans and Baker [35]) is

$$J_i = Q_{i,m}\left[x_i - \frac{\gamma_{ip}}{\gamma_i} y_i \exp\left(-\frac{\bar{V}_i(P_f - P_p)}{RT}\right)\right] \; (\text{mol/m}^2\cdot\text{h})$$
(32.24)

where

$Q_{i,m}$: membrane permeability, mol/m^2·h
γ_{ip}: activity coefficient of component i in the permeate at pressure P_p
x_i: mole fraction of component i in the feed
y_i: mole fraction of component i in permeate
$\bar{V}_i$: partial molar volume of species i
P_f: feed pressure, Pa
P_p: permeate pressure, Pa

According to this model, each component i gets dissolved in the membrane at the feed interface, then diffuses through the membrane, and is desorbed at the other interface without any consideration of other species being transported through the membrane.

In the pore–flow model of transport for cylindrical pore models, the solvent flux (or volume flux) expression is

$$J_v = -\frac{\varepsilon\, d_{\text{pore}}^2}{32\mu\tau}\nabla P \quad (\text{m}^3/\text{m}^2/\text{s}) \tag{32.25}$$

where

d_{pore}: diameter of the pore, m
τ: pore tortuosity dimensionless
μ: viscosity of the solvent, kg/m/s
ε: porosity of the microporous membrane, dimensionless
∇P: pressure gradient, Pa/m

If the pore structure can not be modeled as consisting of effectively cylindrical capillaries of diameter d_{pore}, the solvent volumetric flux is characterized as

$$J_v = -\frac{Q_{\text{sm}}}{\eta}\nabla P \tag{32.26}$$

where

Q_{sm} is the solvent permeability through the membrane.

The molar solute flux in the pore–flow model is described via

$$N_i = -\alpha'_i C_{im} J_v \tag{32.27}$$

where

C_{im} is the molar concentration of solute i in the membrane.
α'_i is a viscous flow characterization parameter [34].

Generally, membrane–solvent interactions occur and they would influence the parameters to be used in pore–flow models. The behavior of solute rejection R_i in such a model as a function of permeate volume flux (which increases essentially linearly with $-\nabla P$) has been illustrated for MPF-60 membrane in Whu et al. [27]. How to obtain various parameters for both models has been illustrated in Silva et al. [34] who have described the earlier literature. The local permeate solute concentration C_{ip} for component i may be described via

$$\frac{C_{ip}}{C_{\text{sp}}} = \frac{N_i}{N_s} \tag{32.28}$$

where

N_s is the molar solvent flux.
C_{sp} is the solvent concentration in the permeate.

We could rewrite Eq. (32.28) for a dilute solution of component i as

$$C_{ip} = \frac{N_i}{J_v/\bar{V}_s} C_{\text{sp}} \cong \frac{N_i}{J_v} \tag{32.29}$$

where the partial molar volume of solvent is $\bar{V}_s$ yielding

$$R_i \cong 1 - \frac{N_i}{J_v C_{if}} \tag{32.30}$$

Most performance data in OSN reported in literature were obtained with small membrane samples. In larger units, there will be variation of feed concentration along the membrane. There is very little information in the literature on analysis of such a situation. However, in any situation, with small or larger membrane sample, one has to account for concentration polarization. Since OSN is used often to reject a solute, which is an API, while the solvent is going through, the rejected solute concentration will increase on the membrane surface to C_{iw}, which is larger than the feed concentration C_{if}. The extent of this increase will depend on the solute mass transfer coefficient in the feed solution k_{if} and the volume flux J_v as is observed in the processes of reverse osmosis and ultrafiltration:

$$\frac{J_v}{k_{if}} = \ln\left(\frac{C_{iw} - C_{ip}}{C_{if} - C_{ip}}\right) \tag{32.31}$$

Therefore when using the two models of membrane transport, C_{iw} should be used instead of C_{if} unless the ratio (J_v/k_{if}) is very small.

32.3.2 OSN Membranes

OSN membranes may be ceramic or polymeric in nature. There are no practical ceramic membranes whose MWCO value is less than 1000. Therefore we will focus here on polymeric OSN membranes. Most polymeric membranes used in a variety of industries are prepared via the phase inversion technique in the first step of which the polymer is dissolved in a solvent. Therefore most of those polymers are not suitable for OSN since the membrane lacks solvent stability. Among those few that may be suitable, polyimide (PI) and polyacrylonitrile (PAN) stand out. A brief description of various aspects of the materials of OSN membranes is provided in Silva et al. [36].

The polymeric membranes of these materials may be of the asymmetric type or a composite membrane. In asymmetric membranes, a very thin skin at the top of the membrane is the solute-selective layer with the rest of the membrane providing a low transport resistance mechanical support; however, the whole membrane is of the same material and these are often described as integrally skinned. In composite membranes, the top selective skin is of a different material compared to the material of the porous support. This selective layer is cross-linked to impart substantial solvent stability. Cross-linked elastomeric barrier layers are often made out of polydimethylsiloxane on a PAN support.

Integrally skinned polyimide membranes of the STARMEM® type are available from W.R. Grace, Columbia, MD, USA, and Membrane Extraction Technology Ltd., UK. The composite polymeric membranes with silicone top layers are identified as MPF types and are available from Koch Membrane Systems, Wilmington, MA, USA, in small sizes as well as spiral wound modules. These membranes are good with many organic solvents such as toluene, methanol, ethyl acetate, etc. However, polar aprotic solvents such as methylene chloride, DMF, n-methyl pyrrolidone (NMP), and THF are demanding; successful OSN has been reported with these solvents using integrally skinned Lenzing P84 polyimide membranes chemically cross-linked with aliphatic diamines [37].

32.3.3 Potential Applications of OSN in Pharmaceutical Industry

Pharmaceutical synthesis of small molecules generally involves 4–20 reaction steps. There are many separation steps involved in between where one could use OSN. It may involve separation of the catalyst and its reuse/recycle, removal of solvent and its exchange with a different solvent, concentration of a product/intermediate molecule rejected completely by the membrane, removal of smaller molecules/impurities from the reaction mixture along with the solvent through the membrane, recycling of resolving agents in chiral resolution processes, etc. These steps will most likely be implemented in a NF membrane unit external to the vessel/reactor where synthesis is being implemented (Figure 32.9) in a batch/semi-batch fashion with concentrate recycle.

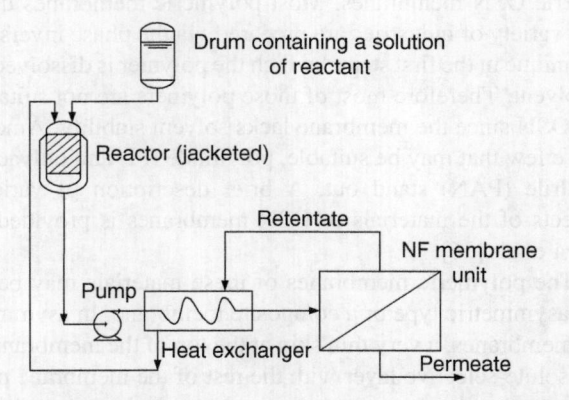

FIGURE 32.9 Schematic of a batch/semi-batch reactor coupled externally with an OSN unit. *Source*: Whu et al. [27] with permission.

There are two basic properties of successful membranes in OSN: small molecular weight solvent and solutes pass through easily and the solvent flux is acceptable; molecules larger than the MWCO size of the membrane are essentially retained completely. The membrane MWCO may be 220/250, 400, 700 Da, implying that species with molecular weight larger than the membrane MWCO is very likely to have a solute rejection (R_i) value greater than or equal to 0.95. A basic expectation is also that the membrane performance will not change much with time either with respect to flux or solute rejection; therefore membrane swelling in the solvents should be very low.

Scarpello et al. [38] demonstrated very high rejections of the following catalysts, Jacobsen catalyst (622 Da), Wilkinson catalyst (925 Da), and Pd-BINAP (849 Da) using STARMEM™ membranes of different kinds, STARMEM™ 122 (MWCO, 220 Da), STARMEM™ 120 (MWCO, 200 Da), and STARMEM™ 240 (MWCO, 400 Da) in the presence of different solvents, such as ethyl acetate, and THF. Usually STARMEM™ 240 showed higher rejection values, while STARMEM™ 120/122 showed very high rejection values greater than 0.95 bordering onto 1. For smaller Pd-based catalysts used in Suzuki coupling, the leakage of catalyst through the STARMEM™ 122 membrane led to Pd levels in the product at a higher than the desired level [39]; Pd content was brought down to acceptable levels (<10 mg Pd kg/product) by using adsorbents on the permeate from OSN employed on the post-reaction solution [40]. Adsorbents used alone for treating the post-reaction solution were unsuccessful in achieving the desired level unless a very large amount was employed. Homogeneous Heck catalysts were recycled from post-reaction mixtures using OSN [41]. Recovery and reuse of ionic liquids used as solvents was also demonstrated during these studies with OSN for pharmaceutical synthesis processes [42]. Phase-transfer catalysts such as tetraoctylammonium bromide were separated and successfully reused via OSN [43, 44]. Recycle and reuse of organic acid resolving agents such as di-p-toluoyl-1-tartaric acid (DTTA) used for resolution of chiral bases such as racemic amines was demonstrated via OSN [45].

Solvent exchange is an important step in pharmaceutical synthesis. It is usually performed by batch distillation, which results in an intensive use of energy and solvent. In addition, the separation achieved by conventional distillation is not very satisfactory. The room temperature exchange of the solvent ethyl acetate with methanol in a solution containing the solute erythromycin (734 Da) was demonstrated via discontinuous and continuous diafiltration [46, 47] using MPF-50 and MPF-60 membranes (Figure 32.10). In Figure 32.10 E Conc represents the concentration of erythromycin in the permeate or retentate. This process avoids distillation processes where thermally sensitive APIs are likely to be affected. Toluene as the solvent for solutes such as tetraoctylammonium bromide (547 Da) was exchanged with methanol

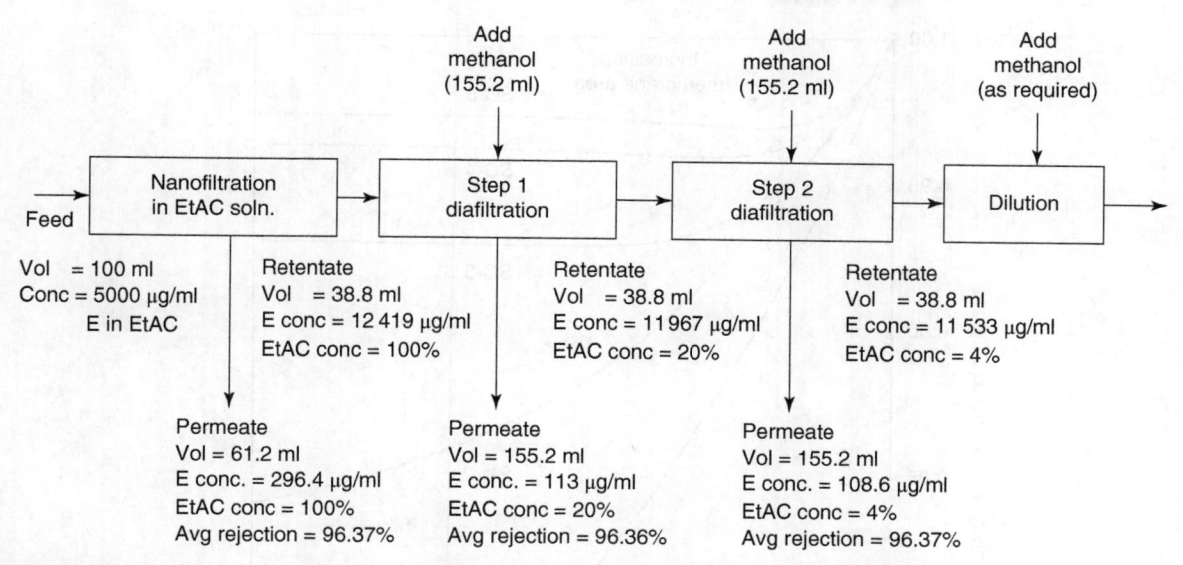

FIGURE 32.10 Mass balance during the pre-concentration and discontinuous DF steps for the MPF-60 membrane in solvent exchange for erythromycin from ethyl acetate to methanol. *Source*: From Sheth et al. [47] with permission.

in batch distillation using STARMEM™ 122 membrane with the solute retention exceeding 99% [28] compared with average solute rejection of around 96.37% in Sheth et al. [46, 47]. A continuous countercurrent cascade of three NF membrane cells for the feed solution and the exchange solvent entering at two ends of the cascade was demonstrated for a test solute such as tetraoctylammonium bromide (547 Da) and the solvents methanol and toluene [48].

Membrane-assisted organic synthesis wherein NF membranes are utilized to remove undesirable byproducts/intermediates from the reaction mixture as the reaction proceeds and concentrate the reaction product has not received as much attention. Whu et al. [27] pointed out via modeling how OSN may facilitate organic synthesis. Consider the reaction

$$\underset{(MW \sim 400)}{A} + \underset{(MW \sim 50)}{B} = \underset{(MW \sim 400)}{C} + \underset{(MW \sim 50)}{D} \quad (32.32)$$

where the product D participates in a side reaction, which consumes the reactant A to produce an undesired by-product E via

$$A + D = E \quad (32.33)$$

If we carry out OSN in an integrated fashion with a batch/semi-batch reactor, a few benefits are apparent. If Reaction (32.32) is equilibrium limited, removal of D via OSN will shift the equilibrium to the right. Although reactant B is also removed through the membrane, one could continuously add reactant B from a drum in a solvent to replenish it. Removal of D reduces the extent of loss of A via side Reaction (32.33). The concentration of C in the final reaction mixture can be considerably enhanced. Further, the conversion time may be significantly reduced.

Figure 32.11 from Whu et al. [27] illustrates the enhancement of the selectivity with respect to species C as a function of dimensionless time for five modes of operation: BU-1 represents a batch reactor uncoupled from an OSN unit; SU-2 represents a semi-batch reactor uncoupled from an OSN unit; and SC-3, SC-4, and SC-5 represent a semi-batch reactor coupled externally with an OSN unit with increasing membrane area and increasing concentration of species B in the drum and the volumetric rate of addition. This figure demonstrates that a much higher selectivity with respect to species C is achieved when the semi-batch reactor is coupled externally with an OSN unit to remove the solvent and the product D. Further, this enhanced selectivity is achieved in much less time.

32.4 NONDISPERSIVE MEMBRANE SOLVENT EXTRACTION

In pharmaceutical synthesis, sometimes one needs to extract a solute/API from an organic phase to an aqueous phase or vice versa. Much less frequently, one encounters extraction from a highly polar aqueous phase to a nonpolar immiscible organic phase or vice versa. Conventionally these extractions are carried out in a mixer–settler type of extraction device by dispersing one phase as drops in another phase. Alternately, centrifugal extractors (Podbielniak) are employed. However,

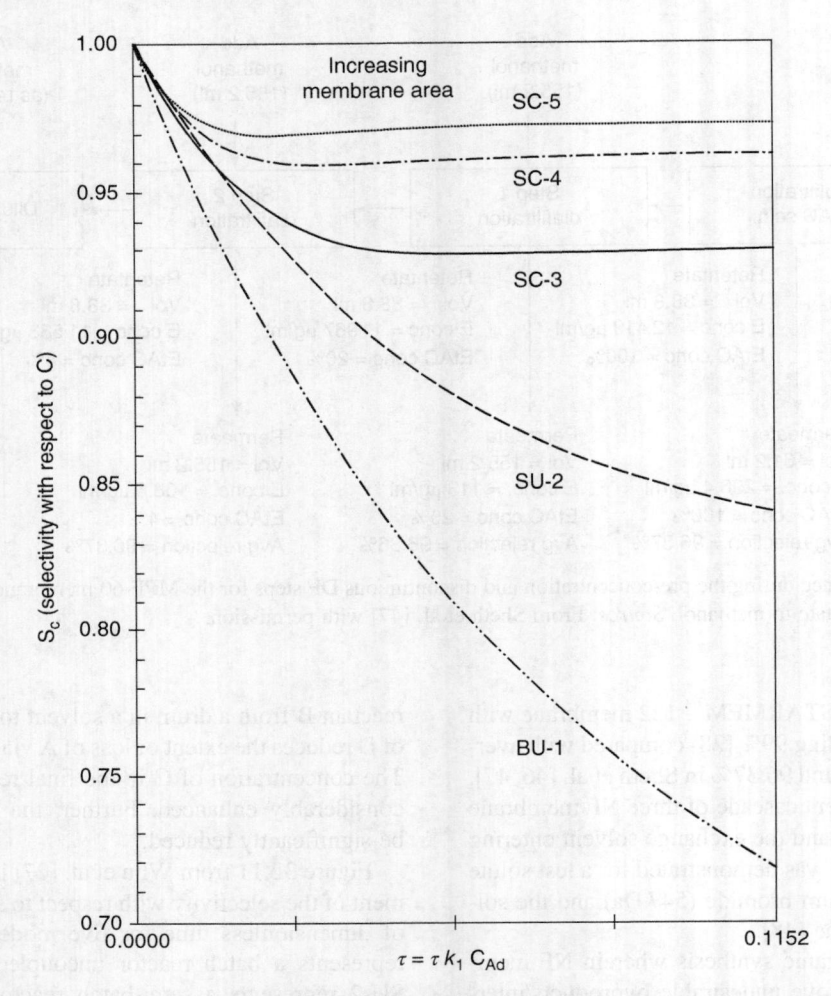

FIGURE 32.11 Selectivity with respect to the desired product C (S_c) as a function of time. See text for more details. *Source*: From Whu et al. [27] with permission.

the possibility of a stable emulsion is a major problem in such processing. Nondispersive solvent extraction using microporous membranes has been developed to avoid such problems [49]. In this technique, the organic phase flows on one side of a porous hollow fiber membrane, and the aqueous phase flows on the other side. One of the phases preferentially wets the membrane pores. Hydrophobic membrane pores are wetted by the organic phase. Hydrophilic membrane pores are wetted by the aqueous phase. However, the phase not wetting the pores is maintained at a pressure equal to or higher than the pressure of the phase present in the pores. There is no dispersion. One can have a wide range of phase flow rates on the two sides of the membrane. The aqueous–organic or organic–organic phase interfaces at the pore mouths are stable as long as the required relative phase pressure conditions are maintained.

Examples of commercial exploitation of such a technique are provided in Sirkar [50] for aqueous–organic systems. The commonly used porous hollow fiber membranes to this end are made of polypropylene with the tube sheet made out of polyolefin resins. Therefore only some of the more common and less aggressive solvents can be used at lower temperatures. Large modules are commercially available (Celgard/Membrana, Charlotte, NC). High performance ceramic membranes having a fluoropolymer coating are available from Kühni AG (Allschwil, Switzerland; Stanley, NC (Kühni, USA)). These modules can be used up to a temperature of 150 °C and can be steam sterilized.

In such membrane solvent extraction techniques, porous hydrophobic membranes provide mass transfer advantages when extracting solutes/APIs from an aqueous into a solute-preferred organic phase. On the other hand, when the solute/API in an organic phase prefers the aqueous phase and an aqueous wash is desirable, a porous hydrophilic membrane is desirable. Porous nylon-based membranes are quite suitable for this goal [51].

32.5 CONCLUDING REMARKS

Pervaporation for removing water from organic reaction medium is an attractive opportunity in pharmaceutical processing. Large membrane modules are available. OSN can provide multiple processing opportunities of great relevance; these include catalyst recovery/recycle, solvent exchange, product concentration, enhancement of reaction conversion/selectivity, and reduction of reaction time. Large modules for nondispersive membrane solvent extraction can facilitate pharmaceutical processing as long as the membrane module has acceptable solvent resistance.

REFERENCES

1. Sirkar, K.K. (2000). Application of membrane technologies in the pharmaceutical industry. *Current Opinion in Drug Discovery and Development* 3 (6): 714–722.
2. Mulder, M. (1996). *Basic Principles of Membrane Technology*. Dordrecht: Kluwer Academic Publishers.
3. Sander, U. and Jannsen, H. (1991). Industrial application of vapor permeation. *Journal of Membrane Science* 61: 113–129.
4. Jonquiéres, A., Clément, R., Lochon, P. et al. (2002). Industrial state-of-the-art of pervaporation and vapor permeation in the western countries. *Journal of Membrane Science* 206: 87–117.
5. Luo, G.S., Niang, M., and Schaetzel, P. (1997). Separation of ethyl tert-butyl ether-ethanol by combined pervaporation and distillation. *Chemical Engineering Journal* 68: 139–143.
6. Hommerich, U. and Rautenbach, R. (1998). Design and optimization of combined pervaporation/distillation processes for the production of MTBE. *Journal of Membrane Science* 146: 53–64.
7. Maus, E. and Brüscke, H.E.A. (2002). Separation of methanol from methyl ethers by vapor permeation: experiences of industrial applications. *Desalination* 149: 315–319.
8. Lipinzki, F., Olson, J., and Trägård, G. (2002). Scale-up of pervaporation for recovery of natural aroma compounds in the food industry. *Journal of Food Engineering* 54: 197–205.
9. Abou-Nemeh, I., Majumdar, S., Saraf, A. et al. (2001). Demonstration of pilot-scale pervaporation systems for volatile organic compound removal from a surfactant enhanced aquifer remediation fluid. II: Hollow fiber membrane modules. *Environmental Progress* 20 (1): 64–73.
10. Lim, S.Y., Park, B., Hung, F. et al. (2002). Design issues of pervaporation membrane reactors for esterification. *Chemical Engineering Science* 57 (22–23): 4933–4946.
11. Wynn, N. (2001). Pervaporation comes of age. *Chemical Engineering Progress* 96: 66–72.
12. Li, J., Chen, C., Han, B. et al. (2002). Laboratory and pilot-scale study on dehydration of benzene by pervaporation. *Journal of Membrane Science* 203: 127–136.
13. Ortiz, I., Urtiaga, A., Ibanez, R. et al. (2006). Laboratory and pilot plant scale study on dehydration of cyclohexane by pervaporation. *Journal of Chemical Technology and Biotechnology* 81: 48–57.
14. Feng, X. and Huang, R.Y.M. (1997). Liquid separation by membrane pervaporation: a review. *Industrial and Engineering Chemistry* 36: 1048–1066.
15. Baig, F.U. (2008). Pervaporation. In: *Advanced Membrane Technology and Applications* (ed. N.N. Li, A.G. Fane, W.S. Winston Ho and T. Matsuura), 469–488. Hoboken, NJ: Wiley.
16. Kennedy, T. (ed.) (2008). Strategic project management at the project level. In: *Drugs and the Pharmaceutical Sciences, Vol 182: Pharmaceutical Project Management*, 2e. New York: Informa Healthcare.
17. Chapman, P.D., Oliveira, T., Livingston, A.C., and Li, K. (2008). Membranes for the dehydration of solvents by pervaporation. *Journal of Membrane Science* 318: 5–37.
18. Kuzawski, W. (2000). Application of pervaporation and vapor permeation in environmental protection. *Polish Journal of Environmental Studies* 9 (1): 13–26.
19. Nemser, S.M., Majumdar, S., and Pennisi, K.J. (2008). Removal of water and methanol from fluids. US Patent Application 2008/0099400A1.
20. Sommer, S. and Melin, T. (2005). Performance evaluation of microporous inorganic membranes in the dehydration of industrial solvents. *Chemical Engineering and Processing* 44: 1138–1156.
21. Urtiaga, A., Gorri, E.D., Casado, C., and Ortiz, I. (2003). Pervaporative dehydration of industrial solvents using a zeolite NaA commercial membrane. *Separation and Purification Technology* 32: 207–213.
22. Van Hoof, V., Dotremont, C., and Buekenhoudt, A. (2006). Performance of Mitsui Na type zeolite membranes for the dehydration of organic solvents in comparison with commercial polymeric pervaporation membranes. *Separation and Purification Technology* 48: 304–309.
23. Caro, J., Noack, M., and Kölsch, P. (2005). Zeolite membranes: from the laboratory scale to technical applications. *Adsorption* 11: 215–227.
24. Zarkadas, D.M. and Lekhal, A. (2009). Unpublished data for a solution of an investigational drug substance in 2-methyl THF.
25. Morigami, Y., Kondo, M., Abe, J. et al. (2001). The first large-scale pervaporation plant using tubular-type module with zeolite NaA membrane. *Separation and Purification Technology* 25: 251–260.
26. Sommer, S., Klinkhammer, B., Schleger, M., and Melin, T. (2005). Performance efficiency of tubular inorganic membrane modules for pervaporation. *AIChE Journal* 51 (1): 162–177.
27. Whu, J.A., Baltzis, B.C., and Sirkar, K.K. (1999). Modeling of nanofiltration-assisted organic synthesis. *Journal of Membrane Science* 163: 319–331.
28. Livingston, A., Peeva, L., Han, S. et al. (2003). Membrane separation in green chemical processing. Solvent nanofiltration in liquid phase organic synthesis reactions. *Annals of the New York Academy of Sciences* 984: 123–141.

29. Shah, D., Bhattacharyya, D., Magnum, W., and Ghorpade, A. (1999). Pervaporation of pharmaceutical waste streams and synthetic mixtures using water selective membranes. *Environmental Progress* 18: 21–29.
30. Van Veen, H.M., Van Delft, Y.C., Engelen, C.W.R., and Pex, P.P.A.C. (2001). Dewatering of organics by pervaporation with silica membranes. *Separation and Purification Technology* 22–23: 361–366.
31. Casado, C., Urtiaga, A., Gorri, D., and Ortiz, I. (2005). Pervaporative dehydration of organic mixtures using a commercial scale silica membrane. Determination of kinetic parameters. *Separation and Purification Technology* 42: 39–45.
32. Cuperus, P.F. and Van Gemert, R.W. (2002). Dehydration using ceramic silica pervaporation membranes-the influence of hydrodynamic conditions. *Separation and Purification Technology* 27: 225–229.
33. Schäfer, A.I., Fane, A.G., and Waite, T.D. (2005). *Nanofiltration-Principles and Applications*. Oxford: Elsevier.
34. Silva, P., Han, S., and Livingston, A.G. (2005). Solvent transport in organic solvent nanofiltration membranes. *Journal of Membrane Science* 262: 49–59.
35. Wijmans, J.G. and Baker, R.W. (1995). The solution diffusion model: a review. *Journal of Membrane Science* 107: 1–21.
36. Silva, P., Peeva, L.G., and Livingston, A.G. (2008). Nanofiltration in organic solvents. In: *Advanced Membrane Technology and Applications* (ed. N.N. Li, A.G. Fane, W.S. Winston Ho and T. Matsura), 451–467. Hoboken, NJ: Wiley.
37. Toh, Y.H.S., Lim, F.W., and Livingston, A.G. (2007). Polymeric membranes for nanofiltration in polar aprotic solvents. *Journal of Membrane Science* 301: 3–10.
38. Scarpello, J.T., Nair, D., Freitas dos Santos, L.M. et al. (2002). The separation of homogeneous organometallic catalysts using solvent resistant nanofiltration. *Journal of Membrane Science* 203: 71–85.
39. Wong, H.T., Pink, C.J., Ferreira, F.C., and Livingston, A.G. (2006a). Recovery and reuse of ionic liquids and palladium catalyst for Suzuki reactions using organic solvent nanofiltration. *Green Chemistry* 8: 373–399.
40. Pink, C.J., Wong, H.T., Ferreira, F.C., and Livingston, A.G. (2008). Organic solvent nanofiltration and adsorbents: a hybrid approach to achieve ultra low palladium contamination of post coupling reaction products. *Organic Process Research and Development* 12: 589–595.
41. Nair, D., Scarpello, J.T., Vankelecom, I.F.J. et al. (2002). Increased catalytic productivity for nanofiltration-coupled Heck reactions using highly stable catalyst systems. *Green Chemistry* 4: 319–324.
42. Wong, H.T., Toh, Y.H.S., Ferreira, F.C. et al. (2006b). Organic solvent nanofiltration in asymmetric hydrogenation: enhancement of enantioselectivity and catalytic stability by ionic liquids. *Chemical Communications* 42: 2063–2065.
43. Luthra, S.S., Yang, X., Freitas dos Santos, L. et al. (2001). Phase-transfer catalyst separation and re-use by solvent resistant nanofiltration membranes. *Chemical Communications* 37: 1468–1469.
44. Luthra, S.S., Yang, X., Freitas dos Santos, L. et al. (2002). Homogeneous phase transfer catalyst recovery and re-use using solvent resistant membranes. *Journal of Membrane Science* 201: 65–75.
45. Ferreira, F.C., Macedo, H., Cocehini, U., and Livingston, A.G. (2006). Development of a liquid-phase process for recycling resolving agents within diastereomeric resolutions. *Organic Process Research and Development* 10: 784–793.
46. Sheth, J.C., Qin, Y., Baltzis, B.C., and Sirkar, K.K. (2000). Nanofiltration-based diafiltration process for solvent exchange in pharmaceutical manufacturing. *Proceedings of the 11th Annual Meeting of the North American Membrane Society*, Boulder, CO (27 May).
47. Sheth, J.C., Qin, Y., Sirkar, K.K., and Baltzis, B.C. (2003). Nanofiltration-based diafiltration process for solvent exchange in pharmaceutical manufacturing. *Journal of Membrane Science* 211: 251–261.
48. Lin, J.C.-T. and Livingston, A.G. (2007). Nanofiltration membrane cascade for continuous solvent exchange. *Chemical Engineering Science* 62: 2728–2736.
49. Prasad, R. and Sirkar, K.K. (2001). Membrane-based solvent extraction. In: *Membrane Handbook* (ed. W.S. Winston Ho and K.K. Sirkar) Chap 41, 727–763. Boston: Kluwer Academic.
50. Sirkar, K.K. (2008). Membranes, phase interfaces and separations: novel techniques and membranes-an overview. *Industrial and Engineering Chemistry Research* 47: 5250–5266.
51. Kosaraju, P. and Sirkar, K.K. (2007). Novel solvent-resistant hydrophilic hollow fiber membranes for efficient membrane solvent back extraction. *Journal of Membrane Science* 288 (1–2): 41–50.

33

DESIGN OF DISTILLATION AND EXTRACTION OPERATIONS

Eric M. Cordi
Chemical R&D, Pfizer, Inc., Groton, CT, USA

33.1 INTRODUCTION TO SEPARATION DESIGN BY DISTILLATION AND EXTRACTION

Distillation and extraction are two important but often neglected unit operations in pharmaceutical process design and scale-up. Other than filtration following crystallization, these two processes represent the primary means of separating the important products of a complex manufacturing recipe from unwanted by-products, solvents, and other wastes. Successful design and execution of distillation and extraction operations have a direct impact on the purity and efficiency of downstream processing. These unit operations are material and cycle time intensive – often representing the point of maximum dilution and longest processing times in a synthetic manufacturing step. The cumulative effect of numerous distillation and extraction systems in active pharmaceutical ingredient (API) manufacture can be a determining factor in work center productivity in terms of kilograms of product per cubic meter of vessel space per week of operation. Beyond direct productivity, the overdesign of either unit operation leads to impacts on inventory and waste treatment systems – from thermal incinerators to solvent recovery to wastewater treatment facilities. Significant capital investments are often required to transport, store, recover, or treat liquids separated from products. The decision to waste or recover these liquids will have a direct impact on common green chemistry metrics such as the E-factor, which is defined as kilograms of waste per kilogram of product. Therefore, it is vital for process development chemists and their chemical engineering counterparts to jointly design the most efficient use of time and materials outside of traditional interests of bond-forming chemistry and product isolation.

33.1.1 Phase Equilibrium Thermodynamics

33.1.1.1 Phase Rule Distillation and extraction design is based on the fundamentals of phase equilibrium thermodynamics. Two phases in equilibrium must exist at the same temperature and pressure, and each species distributed between phases must possess identical partial molar Gibbs free energies in each phase.

$$T_\alpha = T_\beta$$
$$P_\alpha = P_\beta$$
$$G_i(T_\alpha, P_\alpha) = G_i(T_\beta, P_\beta)$$

The Gibbs phase rule for nonreactive components states that the number of degrees of freedom or the number of intensive state variables, available to fully specify a closed system at equilibrium is

$$F = C + 2 - P$$

where

C is the number of distinct components.
P is the number of phases in equilibrium.

Chemical Engineering in the Pharmaceutical Industry: Active Pharmaceutical Ingredients, Second Edition.
Edited by David J. am Ende and Mary T. am Ende.
© 2019 John Wiley & Sons, Inc. Published 2019 by John Wiley & Sons, Inc.

Intensive variables are those that are independent of the size of a system as opposed to extensive variables that are a function of size. Mass and volume are examples of extensive variables, while temperature, pressure, and mass or mole fractions are intensive. For example, a binary system in vapor–liquid equilibrium (VLE) consists of two components and two phases in equilibrium, which results in two degrees of freedom ($F = 2 + 2 - 2 = 2$). Therefore, exactly two independent, intensive state variables (temperature, pressure, or mass/mole fraction of one species) may be specified to define the system at equilibrium.

33.1.1.2 Fugacity/Partial Fugacity

As shown previously, partial molar Gibbs free energies are equivalent at equilibrium. Partial molar Gibbs free energy is also known as the chemical potential, μ_i, in phase equilibrium thermodynamics; therefore, the chemical potentials of each component in each phase are equivalent at system equilibrium. Since chemical potential cannot be directly quantified, it can be conveniently redefined as a pseudo-pressure known as fugacity by the relationship

$$f_i = C \exp\left(\frac{\mu_i}{RT}\right)$$

where

C is a temperature-dependent constant [1].

Partial molar fugacities of each component in each phase are therefore equivalent at equilibrium.

$$f_{i\alpha} = f_{i\beta}$$

In the vapor phase under ideal gas conditions, partial molar fugacity is equal to partial pressure (p_i) or

$$p_i = \frac{P}{y_i}$$

where

P is total pressure.
y_i is the mole fraction of component i.

For ideal gas conditions or systems in which molecular interactions are absent in the vapor phase, it can be assumed at moderate temperatures and at a pressure of less than 10 bar, which holds true of most systems in pharmaceutical processing. Activity is defined as the partial molar fugacity of a component relative to its fugacity at some standard state. If the standard state is chosen as the fugacity of the pure component at the same temperature and pressure, then activity is

$$a_i = \frac{f_i}{f_o} \text{ or } f_i = a_i f_o$$

Partial molar activity, a_i, can be further defined as the product of an activity coefficient, γ_i, and the mole fraction of the component, or for the liquid phase

$$a_i = \gamma_i x_i$$

where

x_i is the mole fraction of a component in the liquid phase.

Combining terms, the partial molar fugacity of a component in the liquid phase is therefore

$$f_i = \gamma_i x_i f_o$$

The standard state fugacity for a pure component at a given temperature is known as the vapor pressure, or $P^*(T)$.

33.1.1.3 Raoult's Law

As stated previously, in VLE, the partial molar fugacities are equal, which form the basis for Raoult's law for VLE

$$f_{(i,v)} = f_{(i,l)}, \text{ or } y_i P = \gamma_i x_i P_i^*(T)$$

For an ideal solution of components, the activity coefficient would be unity for all components across all possible compositions. Few multicomponent solutions exhibit ideal behavior upon mixing; therefore, activity coefficient data are an essential part of phase equilibrium calculations. For non-condensable species with a mole fraction in solution approaching zero, Raoult's law can be restated by combining the activity coefficient and vapor pressure terms into what is known as a Henry's constant, H_i, and therefore Henry's law.

$$y_i P = x_i H_i$$

Liquid–liquid equilibrium (LLE) follows the same thermodynamic principles; therefore, equilibrium between phases can be described in a similar fashion and may include equilibrium with a vapor phase as well. That is, the partial molar fugacities of each component of each phase must be equal, or

$$f_{i\alpha} = f_{i\beta} = f_{i\gamma}$$

where

α is the vapor phase.
β is the first liquid phase.
γ is the second liquid phase.

Substituting known values for partial molar fugacities for each phase and assuming ideal gas behavior for the vapor phase, we have

$$y_i P = \gamma_{i\beta} x_{i\beta} P_i^*(T) = \gamma_{i\gamma} x_{i\gamma} P_i^*(T)$$

Since the pure component vapor pressure is independent of phase composition, this term can be eliminated from the equality to define vapor–liquid–liquid (VLL) equilibrium as

$$\frac{y_i P}{P_i^*(T)} = \gamma_{i\beta} x_{i\beta} = \gamma_{i\gamma} x_{i\gamma}$$

For non-condensable vapor, Henry's law for equilibrium between VLL phases would be

$$y_i P = H_{i\beta} x_{i\beta} = H_{i\gamma} x_{i\gamma}$$

33.1.1.4 Extended Antoine Equation Pure component vapor pressure data is well correlated as a function of temperature by an empirical relation known as the Antoine equation. In its extended form, the Antoine equation is

$$\ln P_i^*(T) = C_{1i} + \frac{C_{2i}}{(T + C_{3i})} + C_{4i} T + C_{5i} \ln(T) + C_{6i} T^{C_{7i}}$$

where

$P_i^*(T)$ is vapor pressure of component i.
T is temperature.

Parameter values are widely available in online databanks and within commercial properties and process simulation packages such as Aspen Properties and Aspen Plus.

33.1.1.5 K Values Phase equilibrium is also represented frequently by ratios known as K values. In VLE, this is simply the ratio of vapor to liquid mole fractions of a single component.

$$K_i = \frac{y_i}{x_i}$$

In relation to Raoult's law, this equilibrium constant can be restated as

$$K_i = \frac{\gamma_i P_i^*(T)}{P}$$

In the LLE case, the K value is a ratio of mole fractions in each liquid phase and is known as a distribution coefficient,

$$K_{iD} = \frac{x_{i\alpha}}{x_{i\beta}}$$

or through the phase equilibrium definition for the liquid–liquid system:

$$K_{iD} = \frac{\gamma_{i\alpha}}{\gamma_{i\beta}}$$

Ratios of VLE constants for a pair of components, i and j, are known as relative volatility, α, while the ratio of LLE constants for a pair of components, i and j, is referred to as relative selectivity, β:

$$\alpha = \frac{K_i}{K_j} = \gamma_i \frac{P_i^*(T)}{\gamma_j P_j^*(T)}$$

$$\beta = \frac{K_{iD}}{K_{jD}} = \frac{\gamma_{i\beta} \gamma_{j\alpha}}{\gamma_{i\alpha} \gamma_{j\beta}}$$

These ratios provide an alternative means of describing equilibrium between components though they remain temperature and composition dependent. Further, the ratios often provide a basis of comparison when selecting components in separation design. For instance, relative volatility may be used to determine the value of constant-level distillation as a means of separating a pair of components. Also, relative selectivity is useful in evaluating liquid–liquid extraction systems, where the choice of solvent is important for efficient enrichment of the extract phase in the solute without undesirable carryover of feed solvent to the same.

33.1.2 Process Design from Lab to Manufacturing Plant

Accurate and efficient design of equilibrium stage separations requires fundamental physical property data for individual components and binary pairings as well as computational tools to store properties in databanks, to predict phase equilibrium, and to design new equipment or specify the operation of an existing plant. The range of activities leading to process design begins with component screening, phase diagramming, and single-stage equilibrium modeling. Most often in pharmaceutical manufacturing operations, a process design must fit the constraints of existing plant equipment, which is usually limited to relatively simple vessels with heat transfer utilities for establishing phase equilibrium, and in the case of VLE an overhead condenser to separate vapor as distillate from the equilibrium vessel. Distillate is typically collected in a similar type of vessel, and this by-product is sometimes sent to a separate facility on site for solvent recovery through further distillation. Liquid–liquid extractions are performed in similar equipment except that the receiver is used to collect the lower phase once equilibrium is established at the desired temperature and sufficient time is allowed for phases to separate. The separated phases in their respective vessels are then further processed with additional extractions using fresh solvent to achieve the desired design endpoint. For the reasons outlined here, the focus of subsequent sections will be single-stage equilibrium and separation design, though many of the early-stage activities detailed here are applicable to selection and evaluation of solvents for relative volatility and selectivity as a basis for comparison of systems.

33.2 DESIGN OF DISTILLATION OPERATIONS

33.2.1 Vapor–Liquid Equilibrium Modeling

The graphical presentation of VLE requires a collection of vapor pressure and activity coefficient data to calculate the mole fraction of each component in each equilibrium phase. Primary data from the lab are preferred as a starting point, but another valuable source of data is that contained within well-recognized databanks such as those of NIST, DIPPR, and DETHERM. In the absence of laboratory or a databank source, pure component and binary properties may be estimated with the aid of property evaluation tools such as COSMOthermX or Aspen Properties. These tools use quantum chemical calculation of screening charge density or functional group contribution methods to provide pure component and binary data estimates.

33.2.1.1 Pure Component Vapor Pressure Data Fugacity of the liquid phase in Raoult's law is the product of an activity coefficient, liquid mole fraction, and the pure component vapor pressure. Vapor pressure is a function of temperature, and the data is often stored in tabular form or regressed into parameters of the extended Antoine equation for convenience and for use in modeling software. Figure 33.1 shows a collection of vapor pressure data plotted in semi-log fashion for a set of common solvents.

33.2.1.2 Binary Activity Coefficient Data Activity coefficient data are vital to ensuring accurate nonideal solution VLE modeling. Whether obtained from laboratory or estimated from equilibrium data, activity coefficients from binary pairs are typically regressed into NRTL binary interaction parameters (BIPs), which may or may not be temperature dependent depending on the level of accuracy. The NRTL model is preferred for its ability to provide activity coefficient values for vapor–liquid and systems that display liquid–liquid immiscibility and predict maxima and minima in activity coefficients, where other models like the Wilson equation are limited to vapor–liquid systems. The utility of this data is not limited to binary system, however, as the NRTL equation uses BIPs of possible parings of interest in a given system to calculate the liquid-phase activity coefficients for binary, ternary, and higher-order systems. Figure 33.2 shows a set of binary activity coefficient plots for solvent pairs exhibiting a variety of behavior in solution. In ideal solutions, the activity coefficients of all components would be equal to unity. Strongly interacting solvents exhibit negative deviations from unity, while repulsive interactions are represented by positive deviations. In binary systems, the activity coefficient of a solvent is equal to unity at infinite dilution of all other components, which is shown in all plots at a mole fraction of unity.

33.2.1.3 Binary Phase Diagrams Simple binary phase diagrams contain a wealth of information about phase equilibrium behavior for a pair of solvents. Figure 33.3 shows examples of diagrams that combine vapor pressure and activity coefficient data of binary mixtures into a graphical representation of phase equilibrium for vapor–liquid and vapor–liquid–liquid systems. Generally, phase diagrams

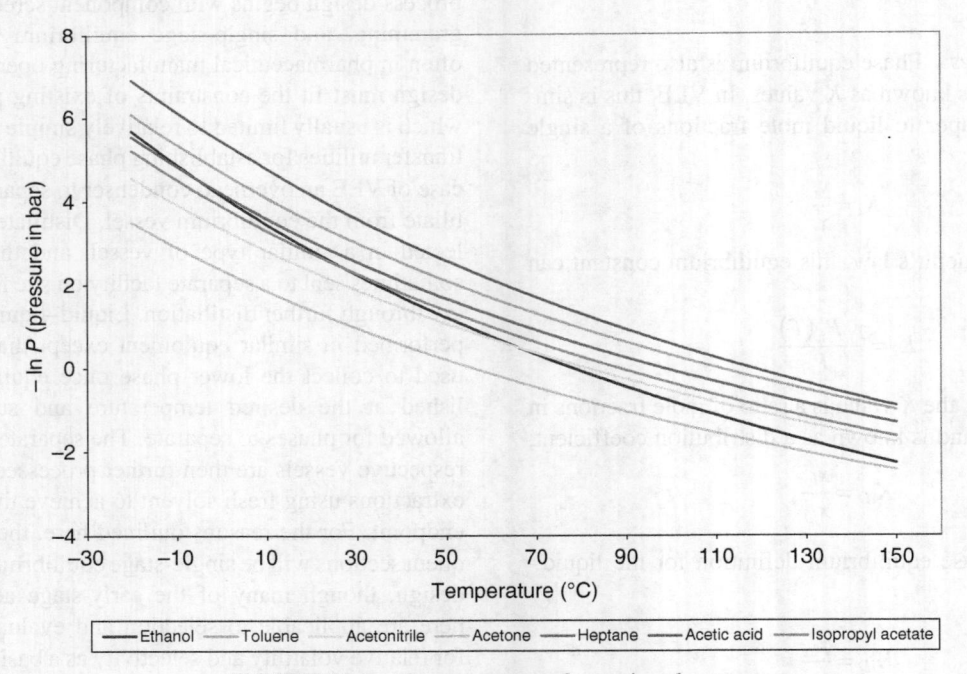

FIGURE 33.1 Vapor pressure of organic solvents.

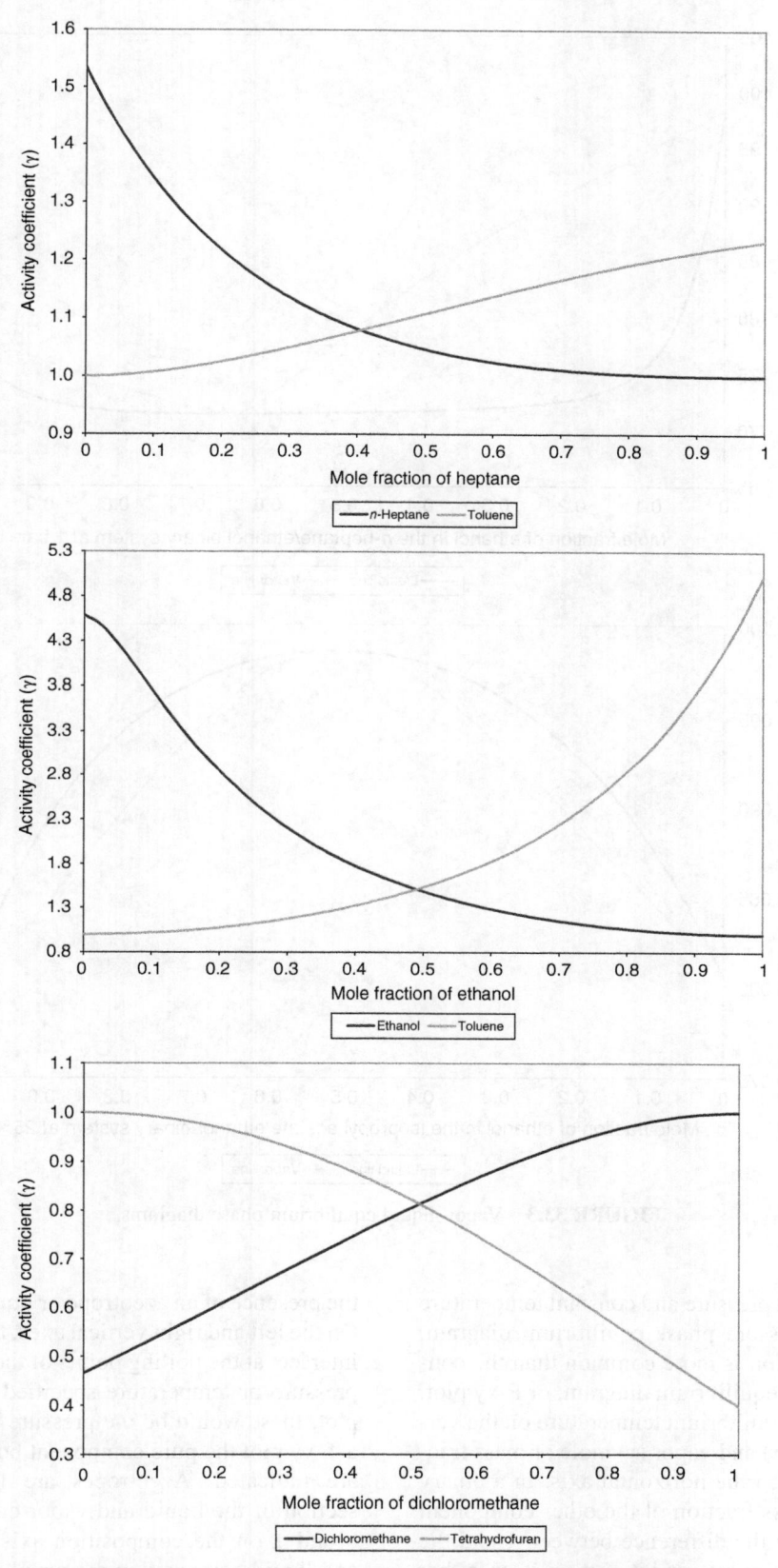

FIGURE 33.2 Liquid activity coefficients for the binary systems.

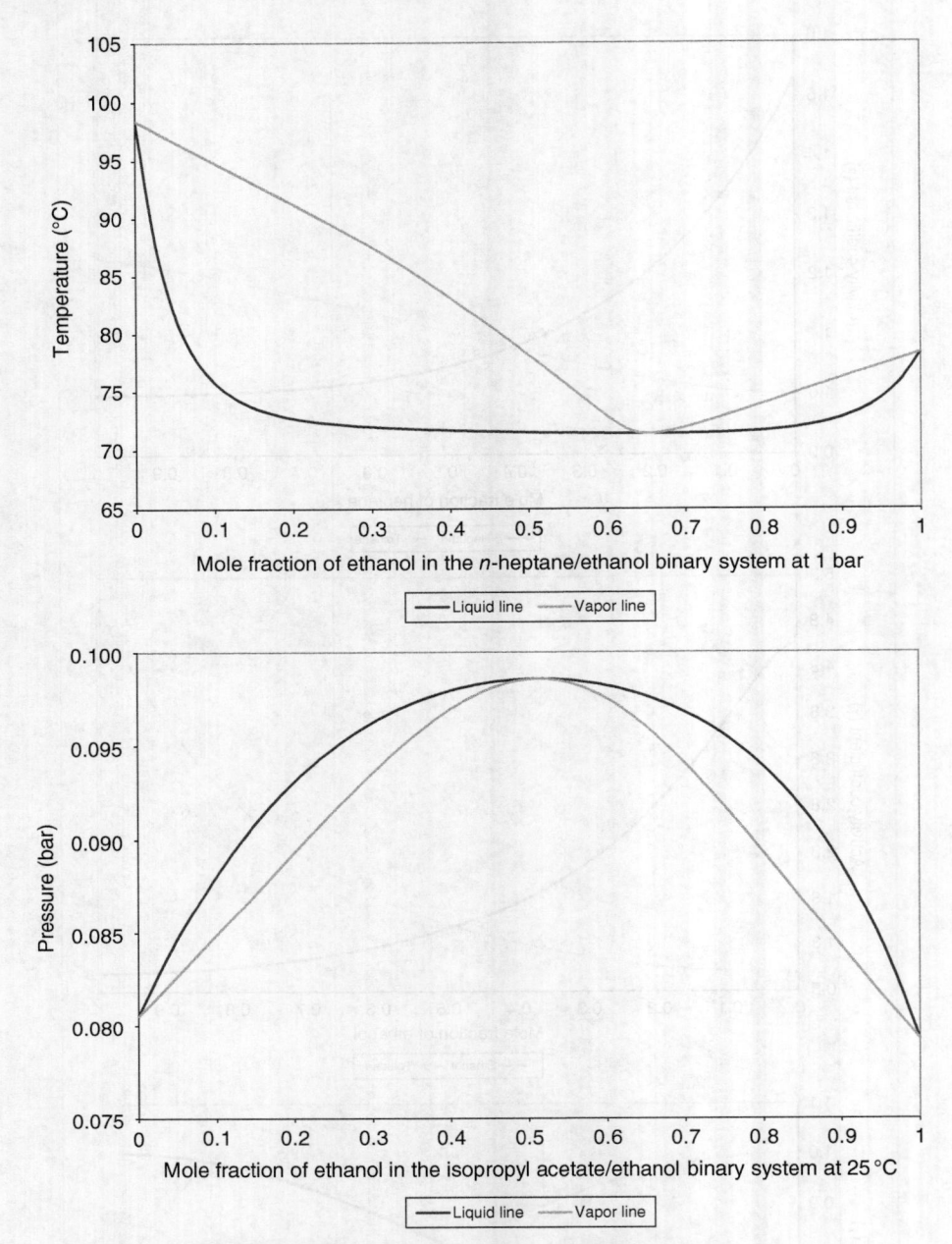

FIGURE 33.3 Vapor–liquid equilibrium phase diagrams.

are divided into constant pressure and constant temperature plots. The constant pressure phase equilibrium diagram, also known as a T-xy plot, is more common than the constant temperature phase equilibrium diagram, or P-xy plot. The T-xy plot displays equilibrium temperature on the vertical axis versus liquid (x) and vapor (y) mole or mass fraction of one component on the horizontal axis. In a binary system, the mole or mass fraction of the other component in equilibrium is simply the difference between 1 and the plotted fraction. Several points of interest exist on either type of binary phase diagram including those that indicate the presence of an azeotrope or liquid–liquid immiscibility. On the left and right vertical axes, the liquid and vapor lines intersect at the boiling points of the pure components at the pressure or temperature specified by the plot. On a P-xy plot, these would be the pressure boiling points, while on a T-xy plot the pure component boiling point temperatures are indicated. Azeotropes are indicated by the intersection of the liquid and vapor curves anywhere between 0 and 1 on the composition axis. In other words, vapor and liquid composition are equal at this point and are further defined by a temperature or pressure based on vertical position on the plot. On the T-xy plot, if this intersection occurs at a temperature less than the boiling temperature of either pure component (i.e. below the intersection of the liquid and vapor lines at either end of the composition axis), this is a minimum boiling azeotrope. If this temperature is greater than the boiling temperature of either pure component, this is a maximum boiling azeotrope. The reverse is true for less commonly used P-xy diagrams since pressure is plotted on the vertical axis. That is, minimum boiling azeotropes appear above the boiling point pressures on either vertical axis. Azeotropes are divided into two classes: heterogeneous and homogeneous. Homogeneous azeotropes are those that form in a single liquid phase, while heterogeneous azeotropes form in mixtures with liquid–liquid immiscibility. Further, heterogeneous azeotropes are always minimum boiling azeotropes since a significant positive deviation in activity must occur for liquid-phase splitting to occur. On binary phase diagrams, liquid-phase immiscibility is indicated by a horizontal line for some portion of the composition range. A heterogeneous azeotrope occurs at the intersection of the vapor curve with this horizontal line. Heterogeneous liquid systems boil at a constant temperature indicated by the horizontal line position on the vertical axis of the T-xy plot. The composition of the liquid phases in equilibrium for these systems is indicated by the endpoints of the horizontal line. Constant temperature boiling continues until the system becomes homogeneous in the liquid phase, at which point the pot boiling point increases as the more volatile component evaporates from the mixture.

EXAMPLE PROBLEM 33.1

Distillation of a *n*-heptane and ethanol binary mixture – a 100 mol mixture of *n*-heptane (50 mol%) and ethanol (50 mol%) will be distilled to displace *n*-heptane from ethanol at ambient pressure. What is the initial boiling point of this mixture? As the mixture is distilled, which component will be enriched in the pot? If ethanol is not enriched in the pot, how much ethanol must be added to start at a composition that will distill to enrich the pot in ethanol? If sufficient ethanol is added to a point of 85 mol% ethanol, what is the initial boiling point of the mixture, and what is the vapor composition in equilibrium at the start of distillation? What is the relative volatility at the initial mixture?

Solution
The T-xy phase diagram in Figure 33.3 shows that the initial boiling point of the mixture is about 72 °C, and as the mixture boils, the vapor composition is approximately that of the azeotrope (65 mol% ethanol, 35 mol% *n*-heptane), which will enrich the pot in *n*-heptane as the boiling point increases to

that of 98 °C. To begin at a point that will enrich the pot in ethanol, additional ethanol must be added to achieve a composition beyond the minimum boiling azeotrope. To reach an initial mixture composition of 85 mol% ethanol and 15 mol% *n*-heptane, an ethanol material balance is used to determine the amount of extra ethanol to add:

$$100 \text{ mol} \times (50 \text{ mol\% ethanol}) + A \text{ mol ethanol} = B \text{ mol} \\ \times (85 \text{ mol\% ethanol})$$

$$100 \text{ mol} + A \text{ mol} = B \text{ mol}$$

Substituting the total mole balance for B into the first equation,

$$50 \text{ mol ethanol} + A \text{ mol ethanol} = (100 \text{ mol} + A \text{ mol}) \\ \times 85 \text{ mol\% ethanol})$$

$$A = 233 \text{ mol of ethanol added}$$

$$B = 333 \text{ mol total moles in new mixture}$$

The molar relative volatility of *n*-heptane to ethanol in the new mixture is

$$\alpha = \frac{y_1/x_1}{y_2/x_2}$$

where

Component 1 is *n*-heptane.
Component 2 is ethanol.

$$= \frac{(0.30/0.15)}{(0.70/0.85)} = 2.4$$

The relative volatility of *n*-heptane over ethanol at ambient pressure is fairly low, which often results in inefficient batch strip-and-replace distillation. An alternative process in the form of constant-level distillation is a way to improve the time and material efficiency of batch distillation. Constant-level distillation will be discussed in detail in an upcoming section of the chapter.

33.2.1.4 Ternary Phase Diagrams
Systems containing three components in VLE are graphically represented in ternary phase diagrams, which are sometimes referred to residue plots. Unlike binary phase diagrams, these phase diagrams contain mostly information about the liquid-phase composition – hence the name residue plot – or in the case of vapor–liquid–liquid systems, the liquid–liquid phase envelope with associated tie line endpoints defining the composition of each liquid phase. Examples of ternary phase diagrams are shown in Figure 33.4 for a homogeneous azeotrope system and in Figure 33.5 for a system with a heterogeneous azeotrope. Binary azeotropes are represented by points on the triangle edges labeled with the associated boiling point. Boiling points of individual components are included at the vertices of the triangle. When binary azeotropes are present, the ternary phase diagram is divided into distillation regions

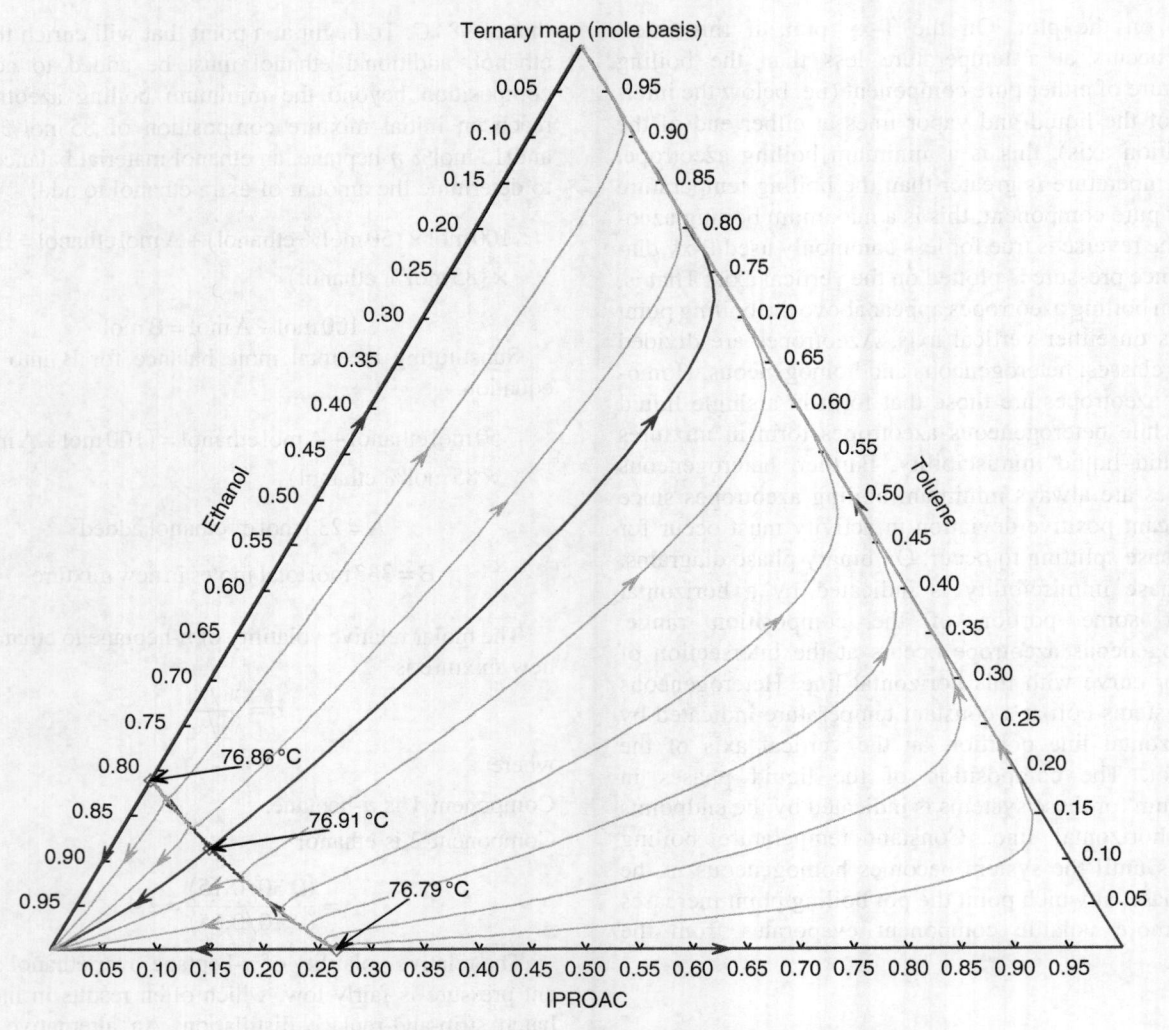

FIGURE 33.4 Ternary phase diagram with homogeneous azeotrope.

separated by distillation boundaries. In the case of three or more binary azeotropes, ternary azeotropes may also form at the intersection of distillation boundaries. Pot composition trajectories are represented on the plot by residue curves that follow increasing temperature, either from one pure component vertex to another or between an azeotrope and a pure component. Analysis and design of distillation systems using ternary phase diagrams can become complex with the number of possible scenarios posed by multiple binary and ternary azeotropes with and without LLE phase envelopes, and there are many references available to consider for such in-depth analysis. As with binary diagrams, it is important for the chemical engineer to have access to the property databanks and plotting tools that allow quick access to system behavior in graphical form, which leads to a level of understanding for the purpose of process modeling and design that would be much more difficult and time-consuming with the limitation of pure component boiling and perhaps binary azeotrope data at ambient pressure. These are the limitations imposed by tabulated data that can easily be overcome by computational tools such as Aspen Plus that provide flexibility in terms of depth of component choices and process conditions.

EXAMPLE PROBLEM 33.2

Distillation of ethanol and water from toluene – a 100 mol mixture of ethanol (63 mol%) and water (37 mol%) will be displaced with toluene at ambient pressure. Using the ternary VLE diagram in Figure 33.5, determine the moles of toluene that are added to reach the saddle ternary azeotrope near the center of the diagram. What is the significance of the mixture at this point? How much additional toluene is added to reach a starting distillation composition of 45 mol% toluene, 35 mol% ethanol, and 20 mol% water? Is this composition

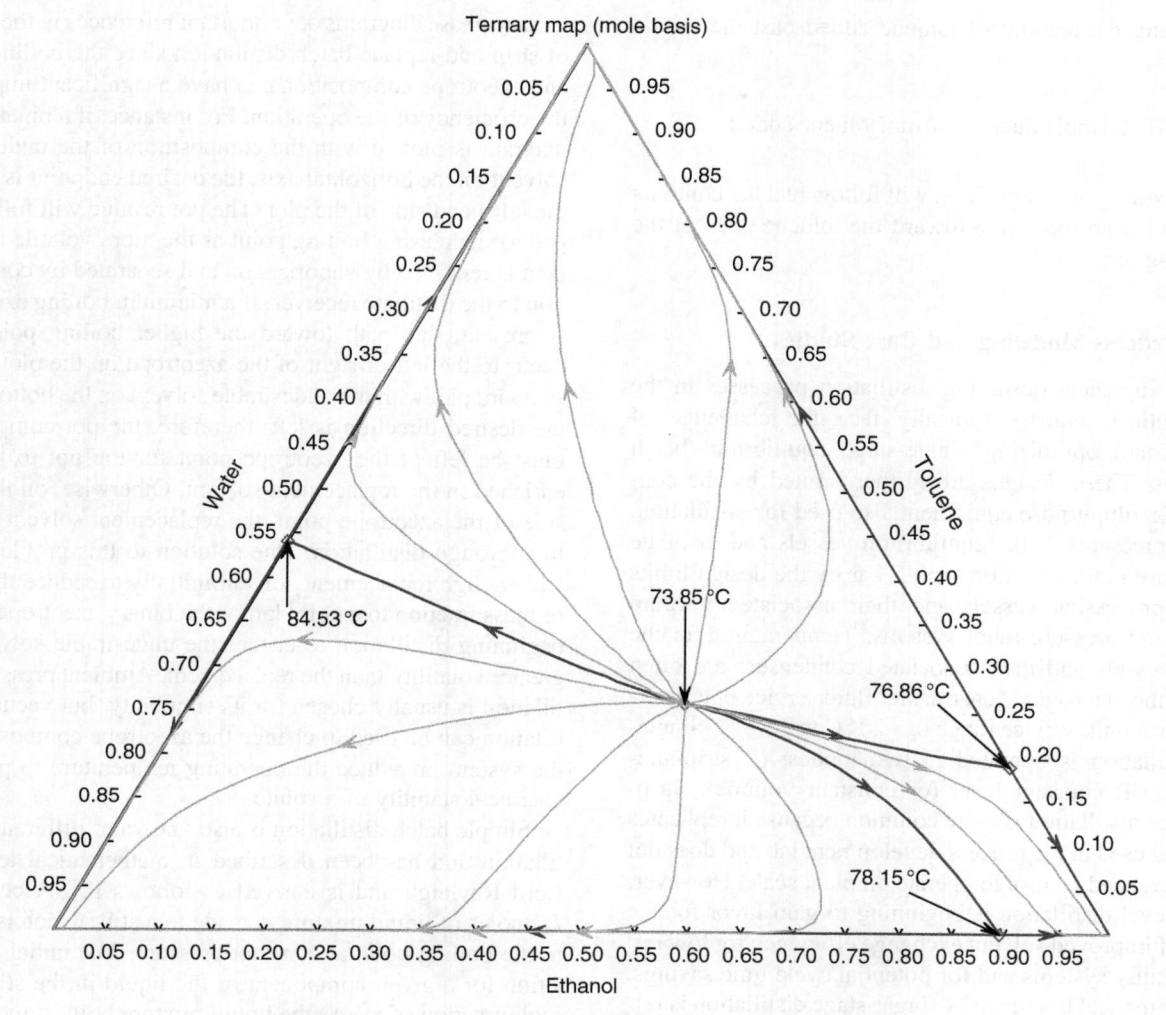

FIGURE 33.5 Ternary phase diagram with heterogeneous azeotrope.

sufficiently high in toluene to ensure that distillation enriches the pot with toluene as distillate, which is collected from the condensed vapor?

Solution
A straight material balance line is drawn from the starting composition of 63 mol% ethanol and 37 mol% water to the toluene apex of the triangle. This line passes through the saddle azeotrope with a composition of 26 mol% toluene, 47 mol% ethanol, and 27 mol% water. A material balance on ethanol can be used to determine the total moles in the mixture at the saddle azeotrope:

$$100 \text{ mol } (63 \text{ mol\% ethanol}) = A \text{ mol } (47 \text{ mol\% ethanol})$$

$$A = 134 \text{ mol in total mixture at saddle azeotrope}$$

A material balance on toluene determines the amount of toluene added to reach the saddle azeotrope.

$$134 \text{ mol } (26 \text{ mol\% toluene}) = 34 \text{ mol of toluene added}$$

The saddle azeotrope represents the ternary point of constant boiling composition at ambient pressure. The distillate and pot compositions will be equivalent during the entire distillation process. This can only be overcome by adding additional solvent. In this case, additional toluene is added to achieve a composition within the distillation boundary that favors ethanol and water volatility over that of toluene.

The material balance line for toluene lies on the starting point of interest for the distillation. A material balance on ethanol between the toluene-free mixture and the desired mixture distillation starting point will determine the amount of toluene added:

$$100 \text{ mol } (63 \text{ mol\% ethanol}) = B \text{ mol } (35 \text{ mol\% ethanol})$$

$$B = 180 \text{ mol of mixture at distillation starting point}$$

The amount of toluene in this mixture is

$$180 - 100 \text{ mol} = 80 \text{ mol of toluene}$$

Therefore, the amount of toluene added past the saddle azeotrope is

$$80 - 34 \, \text{mol toluene} = 46 \, \text{mol toluene added}$$

The mixture pot composition will follow residue contours interpolated from the figure toward the toluene apex of the ternary diagram.

33.2.2 Process Modeling and Case Studies

Chemical engineers designing distillation processes in the pharmaceutical industry typically face the challenge of designing and optimizing single-stage equilibrium batch separations. These designs are further limited by the constraints of multipurpose equipment also used for distillation. Absolute pressures in the equilibrium vessels and distillate receivers are limited to 0.05 bar – 3 bar by the design limits of batch processing vessels and their associated vacuum pumping and pressure relief systems. Temperatures for the jacketed vessels and their associated condensers are often limited to the plant-wide heat transfer fluid service or temperature control unit, service range of −20 °C to 150 °C. Single-stage distillation is operated in two modes: (i) strip-and-replace or (ii) constant level (or constant volume). Strip-and-replace distillation is more common because it replicates the method used in the process development lab and does not require advanced control to operate on plant scale. However, constant-level distillation is beginning to gain favor for the benefits of improved solvent exchange efficiency for low relative volatility systems and for potential cycle time savings. The mathematical treatment of single-stage distillation is relatively easy for either mode, but speed of design is enhanced by the use of dynamic process simulators such as Aspen Batch Modeler, which is an adaptation of the more general dynamic modeler known as Aspen Custom Modeler. The scope of Aspen Batch Modeler goes beyond single-stage distillation, if needed, and also employs process control simulation, which is useful for implementing constant-level distillation models.

33.2.2.1 Strip-and-Replace Batch Distillation

The more common mode of distillation in a pharmaceutical manufacturing facility is that of strip-and-replace distillation under ambient or vacuum conditions. This is achieved by stripping an initial mixture often containing a nonvolatile solute to a desired concentration or volume and then adding a replacement solvent to the equilibrium still or pot. Vapor is condensed into liquid distillate by an overhead condenser. Distillate is collected by gravity flow into a receiver that is usually similar in construction to the equilibrium still. The strip-and-replace procedure is repeated until a desired endpoint of composition and volume is obtained. Binary and ternary phase diagrams are important references in the design of strip-and-replace batch distillation since the boiling point and azeotrope composition can have a significant impact on the efficiency of the operation. For instance, if a binary T-xy diagram is plotted with the composition of the undesirable solvent on the horizontal axis, the desired endpoint is toward the left-hand side of the plot. The pot residue will follow the path of increasing boiling point as the more volatile component is removed by vaporization and separated by condensation to the distillate receiver. If a minimum boiling azeotrope is present, that path toward the higher boiling point may occur to the left or right of the azeotrope on the plot. In the standard plot with the undesirable solvent on the bottom axis, the desired direction is left; therefore, the pot composition must be left of the azeotrope point for the pot to become enriched in the replacement solvent. Otherwise, on the right side of the azeotrope point, the replacement solvent will be lost through distillation. One solution to this problem is to add enough replacement solvent initially to reduce the mole or mass fraction to a point left of the binary azeotrope before beginning distillation to ensure the undesirable solvent has greater volatility than the replacement. Ambient pressure distillation is usually chosen for its simplicity, but vacuum distillation can be used to change the azeotrope composition of the system or reduce the operating temperature to preserve chemical stability of a solute.

Simple batch distillation is also known as differential distillation and has been described in mathematical terms by Lord Rayleigh and is derived as follows [2]. A charge of L_0 moles of liquid mixture is made to a still, which is heated to the boiling point at constant pressure. The initial composition for a given component of the liquid in the still has a mole fraction of x_1. As the liquid mixture boils at mole fraction x, the vapor V in equilibrium with the pot liquid is y, and this vapor is completely condensed to $y_d = x_d$. A material balance can be performed on this dynamic system for total moles.

$$\frac{dL}{dt} = -V \quad \text{(total material balance)}$$

Or for a given component, the material balance is

$$\frac{d(Lx)}{dt} = -Vy \quad \text{(component material balance)}$$

$$L\frac{dx}{dt} + x\frac{dL}{dt} = -Vy$$

If the above equation is combined with the total material balance,

$$L\frac{dx}{dt} + x\frac{dL}{dt} = y\frac{dL}{dt}$$

Multiplying through by dt to eliminate time dependence,

$$L dx + x dL = y dL$$

Rearranging L and mole fractions to either side,

$$L dx = (y - x) dL$$

Or finally, the Rayleigh equation

$$\frac{dL}{L} = \frac{dx}{(y-x)}$$

A more useful form of the Rayleigh equation for numerical solution of time-independent batch distillation is

$$\frac{dx}{dL} = \frac{(y-x)}{L}$$

The values of y and x depend on knowledge of VLE for a given system at temperature T and total pressure P.

This derivation describes composition trajectory of a solution during concentration in the batch still. If fresh replacement solvent is charged to the system, the new initial concentration can be determined by material balance, and a composition new trajectory is then calculated again using the Rayleigh relationship. When a solute is present, one or more simplifying assumptions can be made when the nature of the solute in terms of mole fraction or binary interactions is unknown. The calculations may be performed ignoring the solute mole fraction, and the results will be stated in terms of solvent fractions. This ignores the contribution of the solute as a mole fraction and presumes that the solute is nonvolatile. If solute is likely to impact boiling point temperature by the nature of its mole fraction in the solution, it may be included while still assuming the solute is nonvolatile, but this still ignores any binary interactions with solvent in the solution. Use of either assumption is reasonable for feasibility case studies or for initial design of the distillation process. The impact of a solute on VLE could be assessed by estimating BIPs for the NRTL model using COSMOthermX or group contribution methods such as UNIFAC or ultimately measuring VLE for the system in the lab with a range of compositions and concentrations. The latter is the most time-consuming and should only be used in the final stages of design of a commercial process.

EXAMPLE PROBLEM 33.3

Batch strip-and-replace distillation with ethanol and n-heptane – a 100 L mixture of ethanol (65 wt%) and n-heptane (35 wt%) is added to a single-stage batch distillation vessel. Using a batch distillation simulator, determine the amount of ethanol that must be used in strip-and-replace distillation to achieve less than 0.1 wt% n-heptane at ambient pressure. The endpoint must be 100 L of final mixture, and the volume cannot exceed 200 L at any point in the distillation.

Solution

Using Aspen Batch Modeler, the simulation case is constructed for ambient pressure distillation of a single-stage system. The pot is filled with 100 L of 65 wt% ethanol and 35 wt% n-heptane. Next, 100 L of ethanol were added, and distillation commences until a pot volume of 100 L is achieved at 77 °C. An additional 100 L of fresh ethanol were added, and distillation resumes until a pot volume of 100 L is achieved. After consuming 200 L of fresh ethanol in this operation, the pot contains 0.07 wt% n-heptane, which meets the requirement of achieving less than 0.1 wt% n-heptane by the end of distillation at a temperature of 78 °C. In fact, the 0.1 wt% n-heptane specification was achieved at the 107 L mark. An additional 7 L of volume were distilled from the pot to achieve the final volume requirement of 100 L. Note: the volumes are approximate due to thermal expansion of solvents under ambient pressure distillation conditions.

33.2.2.2 Constant-Level Batch Distillation
A variation of traditional batch distillation is constant-level distillation, which is sometimes referred to as constant volume in this mode of operation. In this scenario, batch distillation takes place with a feed of fresh replacement solvent that matches the distillate rate in a fashion that maintains a constant liquid level in the equilibrium still. It is usually most efficient to concentrate the original solution to the greatest degree before adding replacement solvent, which minimizes the overall consumption of the added solvent to achieve the same endpoint. This effect is achieved my minimizing the amount of replacement solvent that is evaporated during the exchange. Analysis shows that systems with low relative volatility, $\alpha < 5$, enable replacement solvent savings of greater than 50% [3]. Constant-level distillation is an advance from traditional single-stage batch distillation efficiency when column rectification with reflux is unavailable. Distillation with partial reflux requires significant capital investment and results in increased batch cycle time and energy use. However, constant-level distillation may suffer inefficiencies without optimization of operating conditions. For instance, traditional batch strip-and-replace distillation uses a much larger fraction of the available heat transfer surface area in a vessel than constant-level distillation would under suboptimal conditions [4]. To maximize heat transfer area, the initially concentrated solution could be transferred a suitably sized constant-level distillation vessel for batch processing in a full vessel. In a development or kilogram laboratory, constant-level distillation can be controlled by manual adjustment of replacement solvent flow that matches distillate collection. When volumes of solvent are small, the

installation of advanced controls does not provide significant cost advantage. However, installation of level control would provide efficiencies in a pilot plant or commercial manufacturing facility under routine processing conditions. Automation enables programming of a defined endpoint in terms of replacement solvent volume charged. Level control can be maintained by a variety of inputs such as the still or distillate vessel level. Less direct due to differences in solvent density is the still or distillate mass measured by strain gauge as a control input. Beyond solvent exchange efficiency are the benefits of a reduced temperature profile over time experienced by the batch and reduced energy use by reduced solvent loading per batch. The reduced temperature profile may preserve chemical stability of solute leading to potentially higher yield and lower impurity loading in the isolated product.

EXAMPLE PROBLEM 33.4

Constant-level distillation with ethanol and n-heptane – a 100 L mixture of ethanol (65 wt%) and n-heptane (35 wt%) is added to a single-stage batch distillation vessel. Using a constant-level distillation simulator, determine the amount of ethanol that must be used in constant-level distillation to achieve less than 0.1 wt% n-heptane at ambient pressure. The endpoint must be 100 L, which is identical to the starting mixture. Compare the amount of ethanol consumed in this example with that of the batch strip-and-replace distillation in Example Problem 33.3, and determine which distillation method is more material efficient in this scenario.

Solution

Using Aspen Batch Modeler, the simulation case is constructed for ambient pressure distillation of a single-stage system. The pot is filled with 100 L of 65 wt% ethanol and 35 wt% n-heptane. The batch is heated to its initial boiling point of 72 °C, and as distillate begins to collect, a level controller regulates the feed of fresh ethanol to the pot to maintain a 100 L volume. Distillation continues until 0.1 wt% n-heptane remains in the pot at a temperature of 78 °C. The simulation shows that 150 L of fresh ethanol were consumed to achieve the n-heptane endpoint specification. The constant-level distillation processing scenario consumed 50 L less ethanol than the batch strip-and-replace distillation specified in Example Problem 33.4. In the batch strip-and-replace scenario, 200 L of ethanol were consumed to achieve the 0.1 wt% n-heptane endpoint specification. Therefore, the constant-level distillation operation provided a savings of 25% of fresh ethanol used for strip-and-replace distillation. This level of savings is typical for scenarios in which the relative volatility of components is low.

33.2.3 Laboratory Investigation

33.2.3.1 Vapor–Liquid Equilibrium Data Collection

As the number of possible binary combinations is quite large, a process developer or modeler is faced with screening and design without the benefit of BIPs for solvent pairs that are necessary to correlate activity coefficient for a given system as a function of composition and temperature. For instance, a selection of 100 solvents relevant to the pharmaceutical industry would provide 4950 binary combinations ($100 \cdot 99/2$) among the set. Though the presence of a solute is often assumed to have negligible vapor pressure and minimal impact on the solution boiling point or VLE of the system due to binary interactions, these assumptions may be ultimately tested in the laboratory. Accurate prediction requires careful measurement of important parameters in a laboratory equilibrium still. The experimentalist may choose to study binary systems for regression of collected data into BIPs, such as those for the NRTL model, which are temperature and composition dependent. Relying on Gibbs phase rule, the appropriate number of degrees of freedom is predetermined, and then the full condition of the system at equilibrium is recorded. Typically, in a binary system the pressure and total composition are predetermined, and the batch is heated to its boiling point. At equilibrium, representative samples of vapor and liquid are extracted for analysis, and system temperature is recorded. This data, converted to mole or mass fraction versus temperature, is then regressed within a convenient platform such as Aspen Properties into the a_{ij}, a_{ji}, b_{ij}, b_{ji}, and c_{ij} parameters (as sub-parameters of α_{ij} and τ_{ij}) of the NRTL model for later use in system phase diagramming and process modeling. Applications designed for regression of such data also conveniently perform thermodynamic consistency testing with the ability to exclude data points that are inconsistent for the system. Thermodynamic consistency testing is derived from the Gibbs–Duhem equation [5]:

$$\sum x_i d(\ln \gamma_i) = 0$$

There are two derived forms for thermodynamic consistency testing based on the Gibbs–Duhem equation. First, the integral test is [6]

$$\int \ln \frac{\gamma_i}{\gamma_j} dx_i = 0$$

Graphically, the net area under a plot of $\ln(\gamma_i/\gamma_j)$ vs. x_i must equal zero, or typically to some tolerance such as 10%.

Alternately, a differential or point test that tests consistency at a single composition may be constructed from the same equation and subjected to a tolerance similar to the integral test [7].

$$\frac{(\mathrm{d}\ln)\gamma_j^\circ}{\mathrm{d}x_j} = -\frac{x_i}{1-x_i} \cdot \frac{(\mathrm{d}\ln)\gamma_i^\circ}{\mathrm{d}x_j}$$

Graphically, the slopes of a combined plot of $\ln \gamma_i^\circ$ and $\ln \gamma_j^\circ$ versus x_i at a given composition must satisfy the equation to establish thermodynamic consistency. The results of consistency tests speak nothing about the accuracy of the data, only that it fits within the bounds imposed by partial molar Gibbs free energy of components or chemical potential, within a system at equilibrium.

33.2.3.2 Batch Distillation Simulation in the Laboratory

Performing batch distillations in the lab as part of process development is not a trivial exercise if reliable scale-up to manufacturing conditions is the goal. Lab-scale distillations should be studied in equipment representative of a single stage still on scale, with the ability to collect samples of off-line analysis or insert analytical probes. The equipment must be adequately insulated from ambient conditions to prevent condensation and liquid holdup in places other than the condenser and distillate receiver, respectively. In either strip-and-replace or constant-level distillation, an accurate material balance and associated parameters should be used to record and describe the procedure. The total amount of replacement solvent consumed until the endpoint of the distillation is important as well as the liquid volumes and compositions in the still and the distillate receiver. A temperature profile should be noted as it may have a direct impact on product quality since distillation time on scale will be longer than in the laboratory. Vacuum distillations should be simulated by first applying vacuum to a desired set point that matches plant capacity and then slowly applying heat to reach the solution boiling point to mimic common practice in manufacturing facilities. This practice also reduces the likelihood of bumping or boiling over of pot contents into the receiver. Unintended refluxing before the condenser will result in a distillation with higher efficiency than achievable in a single stage on scale. During distillation development, other effects such as foaming or solute oiling should be noted. Uncontrolled crystallization of the solute may occur at any pressure during distillation, which may impact the quality of the isolated product by incorporating impurities or producing an undesired polymorph. Finally, an important simulation for batch distillation is that of extended time at a given temperature, whether it is the initial or final boiling point. The final boiling point will be higher than the start, so this will represent the most stressed condition in terms of chemical stability in the still. Stability at final reflux temperature is often tested for at least 24 hours and sometimes longer to check for degradation that may occur on scale due to extended processing time or unexpected delays. The heat transfer efficiency of the plant will also determine the time to heat up and cool down the solution before and after distillation, and this temperature history must be incorporated into the design. Similarly, there may also be heat and cool cycles associated with requirements around minimum temperature for sampling or solvent charging that should be considered.

33.2.4 Process Scale-Up

33.2.4.1 Equipment Selection

The selection of equipment for batch distillation is normally limited in the pharmaceutical industry due to the design of synthesis plants around the strategy of maximum flexibility. Therefore, equipment selection is about choosing single-stage vessels appropriate for the task, which is to efficiently concentrate or exchange solvents without impacting the quality of pharmaceutical intermediates or active ingredients carried through. Efficiency is about choosing the volume of vessel that provides maximum heat transfer area per unit volume and with a minimum stir volume that meets the needs of the process. Further, a system capable of vacuum or pressure should be chosen carefully when necessary to achieve a distillation target. The vacuum and pressure limits should be well understood, and the heat removal performance of the condenser in terms of operating temperature should be noted. If a constant-level distillation is planned, the replacement solvent should be added to the still in a fashion that allows rapid mixing with the batch. In other words, avoid adding the replacement solvent in such a way that it runs down the heated wall of the vessel and has opportunity to vaporize before mixing with the contents of the still.

33.2.4.2 Process Control and Unit Operation

In operation the batch still requires temperature and perhaps vacuum control. In the mode of constant-level distillation, the replacement solvent feed and distillate rates must be matched with appropriate automation. For temperature control, the most beneficial mode of operation is temperature differential across the jacket to maintain vessel integrity and to preserve solute chemical stability by keeping the temperature at the vessel wall below a reasonable specification. A typical temperature differential set point is less than 30 °C to achieve a desirable boilup rate at a given pressure. In this regime, the temperature of the jacket will increase with the boiling point to maintain the specified temperature differential through the distillation endpoint.

33.2.4.3 Endpoint Determination

The desired endpoint distillation is usually based upon a desired solution composition or complete displacement of one solvent from the system. Detection of this endpoint is available through a variety of direct or indirect measurements that can be made online or off-line. Sampling of the still contents for off-line solvent analysis is the most direct and most likely option in most pilot- and manufacturing-scale facilities. However, the turnaround time for results may be on the order of hours, and sampling may require cooling of the batch to a safe temperature. Sometimes the batch temperature is a sufficient endpoint when a complete exchange of solvents is desired and the downstream process is insensitive to the presence of small amounts of the replaced solvent or water. In that case, the operator would look to continue distillation until the pot temperature reaches the boiling point of the pure solvent or that of the solution tested in the lab since the presence of a solute can raise the boiling point by one or more degrees Celsius. If the distillation is run under vacuum, it is useful to provide boiling point data at the desired system pressure as a guide to distillation endpoint. Note that heterogeneous mixtures will exhibit a constant temperature boiling point until the mixture becomes homogeneous. From that point, there is often a rapid increase in temperature to the replacement solvent endpoint.

33.3 DESIGN OF EXTRACTION OPERATIONS

33.3.1 Liquid–Liquid Equilibrium Modeling

Equilibrium phase models for the liquid–liquid case requires demonstration of equivalent fugacity in each phase for each component, and this case may be extended further by the inclusion of vapor-phase equilibrium. As with VLE, lab-measured data is the preferred source of equilibrium data, and even without the facility to determine LLE parameters directly, one may rely on well-populated databanks of NIST, DIPPR, and DETHERM for raw data. Or, as with VLE, the LLE data may be estimated by computational tools such as COSMOthermX, and activity coefficient data may be extracted for regression into nonideal solution models such as NRTL. With BIPs in hand, one may begin to plot liquid–liquid phase diagrams that are descriptive for binary or ternary systems. Higher-order systems may be modeled easily with the BIPs even though graphical representation of equilibrium between four and more components is not as helpful.

33.3.1.1 Binary Activity Coefficient Data
Focusing here on the liquid–liquid case alone, the important temperature- and composition-dependent element of equilibrium is the activity coefficient of each component of each phase. As discussed previously, a variety of nonideal solution models may be used to correlate activity coefficients for components as a function of temperature and composition, and in the case of LLE, the NRTL model is preferred because of the flexibility in the description of liquid-phase splitting as well as VLE. Equilibrium phase splitting is only possible in nonideal solutions with significant deviations in the activity coefficient for each component away from unity.

33.3.1.2 Phase Diagrams
The primary method of graphical representation of LLE between three components is a ternary phase diagram. There are several methods for plotting three-component liquid equilibrium, and the most common method is a triangular plot as shown in Figure 33.6. The plot may be formatted as an equilateral or isosceles triangle with the sides of either format divided as mole or mass fraction of each of the three components. Within the bounds of the triangle edges, a boundary line is drawn to divide single- and two-phase regions for a mixture of three components. Tie lines within the two-phase envelope determine the equilibrium mole or mass fractions of each component in each liquid phase at a given temperature. As LLE is also temperature dependent, the size and shape of the two-phase region will be affected by changes in temperature. The point at which the tie lines become infinitely small is known as the plait point, where both liquid phases have identical compositions at their limit of immiscibility.

EXAMPLE PROBLEM 33.5

Extraction of acetone from water into toluene – a 100 mol mixture of acetone (70 mol%) and water (30 mol%) is added to an extraction vessel. Determine the minimum number of moles of toluene that must be added to the system to obtain a liquid–liquid system in equilibrium. Further, how much additional toluene must be added to the system to obtain a liquid–liquid system with the light phase containing 45 mol% acetone. What is the mole composition of the light and heavy phase in equilibrium at 20 °C? Finally, determine the molar distribution coefficient of acetone and the selectivity of toluene for acetone relative to water in the final mixture.

Solution
Use the ternary phase diagram in Figure 33.6 for the toluene, water, and acetone system to determine the behavior of a ternary mixture of these solvents. A straight line drawn from the 70 mol% acetone and 30 mol% water intersection to the 100% toluene apex of the triangle determines the composition path for the addition of pure toluene to the system. The straight line intersects the phase boundary at 23 mol% toluene, 54 mol% acetone, and 23 mol% water. Performing a mole balance on acetone, we have

$100 \text{ mol} \times (70 \text{ mol\% acetone}) - A \text{ mol} \times (54 \text{ mol\% acetone}) = 0;$

$A = 130$ mol of mixture after toluene is added to the point of phase separation.

A balance on toluene on the mixture shows

$130 \text{ mol} \times (23 \text{ mol\% toluene}) = 30$ mol of toluene added to reach the beginning of phase separation at 20 °C.

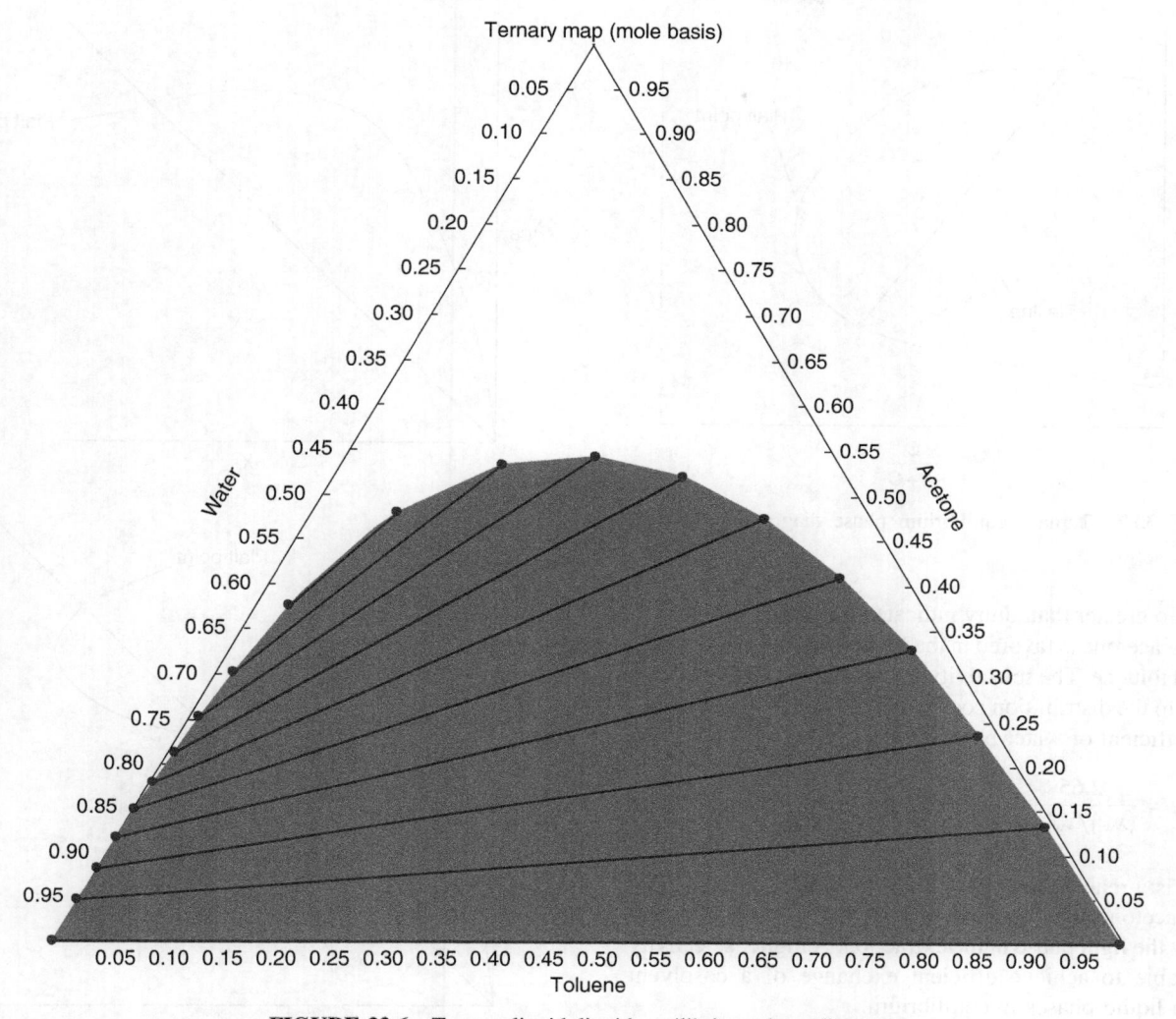

FIGURE 33.6 Ternary liquid–liquid equilibrium phase diagram.

A tie line is constructed at the intersection of 45 mol% acetone, 46 mol% toluene, and 9 mol% water to the other side of the phase boundary at 17 mol% acetone and 83 mol% water. This tie line intersects the composition path line at 40 mol% toluene, 42 mol% acetone, and 18 mol% water. Performing another acetone mole balance, we find

$100 \, \text{mol} \times (70 \, \text{mol\% acetone}) - B \, \text{mol} \times (42 \, \text{mol\% acetone}) = 0; B = 167 \, \text{mol total mixture}$

The total amount of toluene in this second mixture is

$167 \, \text{mol} \, (40 \, \text{mol\% toluene}) = 67 \, \text{mol toluene}$

The amount of extra toluene added beyond the initial phase split is

$67 - 28 \, \text{mol} = 39 \, \text{mol of toluene added in total}$

The compositions of the two phases in equilibrium based on the endpoints of the phase envelope tie line are

45 mol% acetone, 46 mol% toluene, and 9 mol% water (light phase)

17 mol% acetone, 83 mol% water, and trace toluene (heavy phase)

The molar distribution coefficient of acetone in terms of light-to-heavy phase compositions is the ratio of the mol% of acetone in the toluene phase divided by the mol% in the aqueous phase:

$$\frac{x_{a,l}}{x_{a,h}} = \frac{45 \, \text{mol\%}}{17 \, \text{mol\%}} = 2.65$$

where

Component 1 is acetone in the light phase.
Component 2 is acetone in the heavy phase.

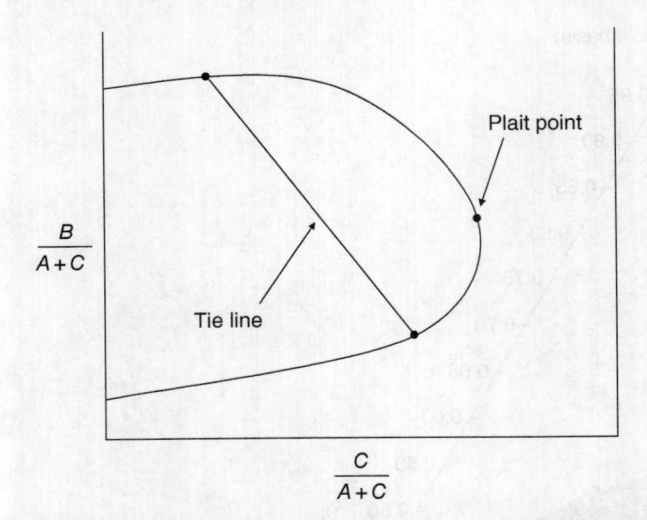

FIGURE 33.7 Ternary equilibrium phase diagram in Janecke format.

A ratio greater than unity indicates the equilibrium distribution of acetone is favored in the light phase, which contains the most toluene. The selectivity of the light phase for acetone is equal to the distribution coefficient divided by the distribution coefficient of water between phases:

$$\beta = \frac{2.65}{(x_{w,l}/x_{w,h})} = \frac{2.65}{(9 \text{ mol}\%/83 \text{ mol}\%)} = 24.4$$

This is a relative measure of the propensity of toluene to extract acetone while accounting for the amount of water taken by the light phase in the extraction. A higher selectivity is desirable to achieve efficient exchange of a cosolvent between liquid phases in equilibrium.

An alternative to triangular form for the ternary diagram is a Janecke plot on rectangular coordinates, as demonstrated in Figure 33.7. The rectangular plot contains the same amount of information as the triangular plot though not as direct since each axis is a ratio of component fractions rather than pure component fractions. Another feature of interest on a ternary phase diagram is the tie line that represents an equilibrium mixture in which each phase is of equal density, which may therefore present challenges in phase separation. This is referred to as the isopycnic line and is usually represented by a dashed line on the phase diagram.

Distribution and selectivity of solute between the extract and raffinate phases are of interest in the selection of solvents and design of extractive systems. Figure 33.8 is a demonstration of distribution plots of a solute between two liquid phases. The 45° line represents the boundary of unity in the distribution coefficient. The distribution curve ends at the plait point for the system. In some systems, the distribution curves may cross the 45° line before ending at the plait point. Such a system is referred to as solutropic because the

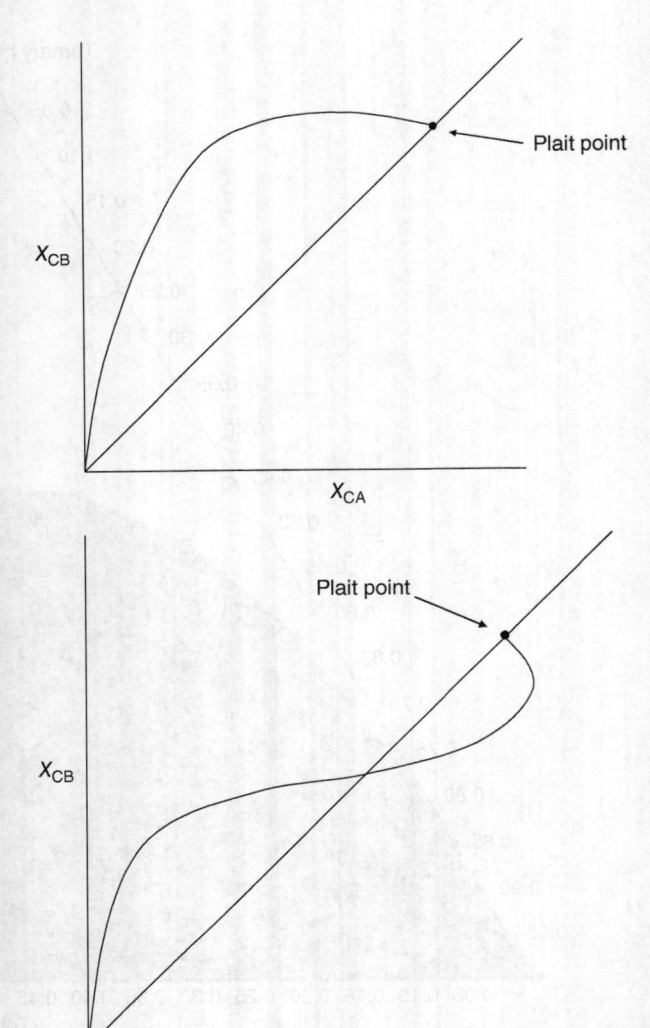

FIGURE 33.8 Liquid–liquid distribution curves.

distribution passes through unity. On a triangular ternary plot, the slope of tie lines changes sign in solutropic systems, which may also pose challenges in the design of a liquid–liquid extraction system.

33.3.2 Process Modeling and Case Studies

Similar to the case of batch distillation, liquid–liquid extraction processes in the pharmaceutical industry involving solutes of moderate manufacturing volume and high value are typically operated as single-stage operations in multipurpose manufacturing vessels. Multistage extractors present an opportunity for efficient separation of a solute from a given phase when the distribution coefficient with a given solvent is small. However, significant investment in processing equipment and control systems may be required for multistage systems, whereas a single-stage extraction is easy to operate in flexible plant equipment at hand.

33.3.2.1 Single-Stage Extraction
A single-stage liquid–liquid extraction may be performed in the same bulk vessel used for most other operations in the batch pharmaceutical plant. The temperature range of −15 °C to 150 °C in these jacketed vessels provides plenty of flexibility in the design of liquid–liquid extraction. Phase equilibrium operations such as this are easily modeled in a process simulator such as Aspen Plus using a decanter (LLE) or three-phase flash (VLLE) model. Temperature and overall composition are required inputs, and results are posted in terms of mass or mole fraction plus total mass and volume for each resulting liquid phase. If liquid-phase splitting will not occur under the specified conditions, only a single liquid stream flow and composition is reported. Case studies are easily generated by creating sensitivity plots or by running optimization routines for parameters of interest such as temperature or solvent volumes. These models rely on accurate prediction of activity coefficients through a nonideal solution model such as NRTL, with temperature-dependent parameters stored in model databanks. Aspen Plus may also be used to generate triangular ternary phase plots for LLE systems as a function of temperature. Multiple single-stage contactors can be linked in series with fresh solvent input to each block to model multiple extraction operations. The results of multiple single-stage operations will differ from the countercurrent operation known as multistage extraction.

EXAMPLE PROBLEM 33.6

Multiple single-stage extraction operations – using the starting mixture from Example Problem 33.5 of acetone (70 mol%) and water (30 mol%), determine the final composition of the light phase if the initial mixture and two intermediate heavy phases are each contacted with 67 mol of fresh toluene in a series of single-stage extractions at 20 °C.

Solution
Either the ternary LLE phase diagram for the toluene–acetone–water system shown in Figure 33.6 or a process simulator such as Aspen Plus may be used to determine the outcome of this series of single-stage extractions. In Aspen Plus, a series of three DECANT blocks would be linked by the heavy phase streams and each DECANT block. Equal amounts of fresh toluene are fed to each DECANT block. The first-stage extraction mixes 100 mol of acetone (70 mol%) and water (30 mol%) with 67 mol of fresh toluene. At equilibrium, the first-stage phases contain:

Light phase: 146 mol (45.9 mol% toluene, 45.5 mol% acetone, 8.6 mol% water)

Heavy phase: 21 mol (0.4 mol% toluene, 17 mol% acetone, 82.6 mol% water)

The second stage receives the heavy phase from the first stage and 67 mol of fresh toluene. At equilibrium, the second stage phases contain:

Light phase: 71 mol (95.0 mol% toluene, 4.7 mol% acetone, 0.3 mol% water)

Heavy phase: 18 mol (trace toluene, 1.7 mol% acetone, 98.3 mol% water)

Finally, the third stage receives the heavy phase from the second stage and 67 mol of fresh toluene. At equilibrium, the third stage phases contain:

Light phase: 67 mol (99.4 mol% toluene, 0.4 mol% acetone, 0.02 mol% water)

Heavy phase: 18 mol (trace toluene, 0.14 mol% acetone, 99.86 mol% water)

In three single-stage extractions with a total of 201 mol of fresh toluene, the acetone in the water was reduced to 0.1 mol%. The heavy, aqueous phase contains only 0.002 mol of toluene, but the combined light, organic phases contains a total of 12.8 mol of water, which is 42% of the original 30 mol of water in the initial feed. Though toluene extracts acetone very well from water, this system is not very selective for acetone because a significant amount of water is extracted into the light, organic phase as well.

33.3.2.2 Multistage Countercurrent Extraction
The use of multiple-stage countercurrent extraction to improve the efficiency of liquid–liquid extractions is easily modeled by most process simulators. In multistage countercurrent extraction, feed and solvent streams enter opposite ends of the column. The product streams exiting the column are the extract and the raffinate. The extract stream contains the solvent with the desired component transferred between phases, whereas the raffinate is the stream remaining after the desired component is transferred to the extract. The internals of a multistage countercurrent extraction column can be quite varied. Stages are formed by zones of intense mixing between independent liquid phases, and this is accomplished by column packing in a variety of configurations (structured or random), by mechanical agitators, such as those seen in Scheibel, Karr, or rotating disk columns, or by liquid dispersion with static sieve trays. Regardless of the liquid-phase contacting technology, the important characteristic for a given column operating under specific temperature and flow rate conditions is the number of equivalent equilibrium stages the equipment achieves. If the equilibrium behavior of component distribution between phases is well understood and mapped with the aid of a ternary diagram or expressed in fundamental terms of activity coefficient model binary parameter terms, a process

model may be used to explore operating conditions. The inputs for the model block are identical to that single-stage operation, except that the total number of equilibrium stages must also be specified. The results obtained from this countercurrent phase contacting scheme are reported in the same fashion as the single-stage model. Example Problem 33.7 demonstrates the difference in efficiency based on choice of countercurrent multistage extraction in place of the multiple single-stage operations shown in Example Problem 33.6.

EXAMPLE PROBLEM 33.7

Multistage countercurrent extraction – using an acetone (70 mol%) and water (30 mol%) feed equivalent to that of Example Problem 33.6 (100 mol/h), determine how many moles of fresh toluene must be fed to a three-stage countercurrent extraction column at 20 °C to achieve 0.14 mol% acetone in the heavy, aqueous phase at the outlet of the extraction column.

Solution

An Aspen Plus simulation is a convenient manner of solving this type of multistage extraction problem, though graphical methods using an equilibrium and operating line approach for staged operations may also be used to solve this type of problem. In Aspen Plus, the EXTRACT column model is used to simulate the countercurrent contact of the heavy feed stream entering the top of the column with a feed of fresh toluene entering the bottom of the column. Three equilibrium stages and adiabatic operation are specified for the model. The feed to the top of the column is 100 mol/h in total containing 70 mol% acetone and 30 mol% water at 20 °C. The objective of the simulation is to add only enough fresh toluene to achieve 0.14 mol% acetone in the aqueous raffinate stream at the exit of the bottom of the column. This can be achieved by trial-and-error manipulation of the fresh toluene feed while monitoring the raffinate composition, or in Aspen Plus a design specification algorithm may be specified to manipulate the toluene feed rate with the objective set at the specified acetone composition in the raffinate. If design specification is used to solve the problem, the solution is to feed 48.5 mol/h of toluene to the column to achieve 0.14 mol% acetone in the aqueous raffinate. In this countercurrent extraction example, 20.5 mol/h of water are extracted by the fresh toluene compared with 12.8 mol of water in the previous example. This represents a loss of 68% of the total water fed to the column. The raffinate contains 0.001 mol/h of toluene, however, which is as low as the loss of toluene observed in Example Problem 33.6. The most striking result, however, is that the countercurrent column requires only 48.5 mol/h of toluene compared with the multiple single-stage operation that requires 201 mol of fresh toluene. This represents 24% of the toluene requirement compared to the multiple single-stage operation.

33.3.3 Laboratory Investigation

The design of liquid–liquid extraction operations in the development laboratory may be divided generally into equilibrium data collection and operation stressing. Both require effective lab-scale equipment to provide robust data and scale-up assessment on the way to the design of an efficient design, whether single or multistage extraction is the objective.

33.3.3.1 Equilibrium Data Collection
LLE data is often collected in a single-stage cell in which conditions temperature and total composition are altered to allow measurement of phase equilibrium across a wide operating space. An accurate representation of equilibrium data relies on accurate analytical measurement of individual phase compositions. The cell is loaded with components and agitated sufficiently to increase the surface area between the dispersed and continuous phases. The surface area between phases is the mass transfer area across which components distribute between phases to an equilibrium condition. Sufficient time must be given to ensure an equilibrium condition is met, and a time series of phase samples should be analyzed to ensure a constant value is obtained in the absence of some interfering form of component degradation. After equilibrium is achieved, agitation is stopped and the resulting phases are allowed to split into individual layers. In some cases only a single homogeneous solution may result, or sometimes relatively stable emulsion may form between phases. Both of these conditions signify a combination of components that is unsuitable for use in liquid–liquid extraction since the goal of separation is not achievable at a given temperature and total composition. Data collected in the lab with the single-stage equilibrium cell is used to provide the basis for a ternary phase diagram or for regression into BIPs of activity coefficient model such as NRTL. An early indication of unit operation behavior in terms of relative difficulty of phase separation by settling, the accumulation of interfacial contaminants, or potential component degradation can be observed and recorded during equilibrium data collection. That is, the experience obtained in equilibrium data collection my bring insight into the behavior of the system upon scale-up to manufacturing.

33.3.3.2 Extraction Lab Simulation
Single- or multistage lab models may be easily constructed to further assess behavior of system components under the conditions of interest and to refine those conditions to optimize the desired separation. Single-stage lab models are easiest to construct since they are closest to the equilibrium cell described previously.

In fact, there is no reason that a single-stage model could not be used for the purpose of gathering equilibrium data. The single-stage model incorporates several elements in terms of geometry and agitation conditions that mimic that which will be encountered on scale for the purpose of providing a close scale-down representation of manufacturing technology. In the pharmaceutical industry, this is typically the same batch vessel used for other unit operations such as reaction, distillation, and crystallization. In addition to providing equilibrium data, the single-stage model should be used to examine and understand the dynamic approach to the equilibrium state. One can examine the impact of agitator choice and the power applied at the blade tip, where the highest-shear forces are found, to determine the propensity for the system to irreversibly emulsify. The position of the agitator relative to vessel fill height is also an important factor in the extraction efficiency and approach to equilibrium. A system with a large difference in densities between phases may make efficient mixing difficult to achieve with an agitator set low in the vessel as is often the case in pharmaceutical manufacturing facilities for the purpose of enabling low stirring volumes. Multiple-stage models are scaled-down versions of the manufacturing scale in terms of the intended phase mixing technology. A direct scale-down of the contacting technology is necessary to ensure that the dynamic and equilibrium behavior of the components in the system behave in a desirable manner. These scaled-down systems are typically constructed of glass-walled columns to allow easy visibility of conditions within the operating space. This allows the user to observe the presence of emulsions or the accumulation of solid particulate or oily components within the equilibrium stages that could impact extraction efficiency or the mechanical operation of the system over an extended operating time.

33.3.3.3 Phase Split Efficiency The rate of phase separation is important to ensure an efficient operation of the extraction process. Since it is difficult or impossible to see into manufacturing-scale equipment to assess the quality of the liquid-phase splitting, the predicted time to phase separation on manufacturing scale must be known or inferred from laboratory behavior.

33.3.3.4 Impact of Missed Phase Splits and Desired Fate of Phase Interface The process designer must specify the desired fate of the phase interface to ensure proper operation of liquid–liquid extraction as a separation technique. In other words, the designer must state explicitly which phase must be disposed of entirely while keeping in mind that a small amount of one phase will always be carried with the other as the result of a phase split. Even after specifying the phase split strategy with regard to the interface after a batch liquid–liquid extraction operation, it is important to test the outcome of the reversed condition, or in other words a missed phase split on subsequent extractions, downstream operations, or perhaps process yield. Use the resulting product phase containing a small amount of the waste phase and any associated rag material in downstream operations, and note any impact on the process. If the impact is significant, then the phase split strategy or other specification may be enhanced to limit the extent of potential downstream problems.

33.3.3.5 Difficult Separations In some cases of liquid–liquid extraction, ease of dispersion is less of a problem than separating the resulting liquid phases after components are distributed in an equilibrium state. In the worst case, the dispersed phase remains stable as an emulsion, and this is a terminal condition to avoid. There are generally three physical properties of components and their mixtures that determine the relative ease of separation after equilibrium extraction: (i) liquid-phase density differences, (ii) viscosity, and (iii) interfacial tension. A large density difference between resulting phases enhances the rate of phase settling, while a low viscosity of the continuous phase also promotes separation. Moderate interfacial tension between the dispersed and continuous phase allows relatively easy separation to take place by coalescence of droplets, but equally it does not hinder the dispersion of one liquid phase into another.

Occasionally an emulsion cannot be avoided due to the nature of the system or due to batch variability in manufacturing. During process design or in a manufacturing environment, there are several methods of handling emulsified phases to recover the phase separation. Sometimes an interfacial particulate or rag layer may be responsible for stabilizing an emulsion, and its influence on the system may be tested by filtering the mixture to remove the material. The phases may separate as expected in the absence of this material. The other technique is to alter the physical properties of phases mentioned previously in a manner that benefits the rate of separation. For instance, density and interfacial tension of an aqueous layer can be increased by the addition of inorganic salts. These salts will also serve to reduce the solubility of water in the organic phase that is a common contributor to emulsification. An increase in temperature will typically reduce the continuous phase viscosity to further improve separation. Finally, a mechanical means of increasing the gravitational force, such as a centrifugal separator, can be used to increase the rate of sedimentation of the dispersed phase. Increasing the volume of the dispersed phase relative to the continuous phase may also accelerate coalescence of droplets to aid separation.

33.3.3.6 Process Scale-Up The selection of scale-up equipment and process controls should be considered as laboratory process development activities near completion. The primary objectives of the liquid extraction scale-up design should be to achieve time and material efficient separation and purification of components.

33.3.3.7 Equipment Selection

Two factors for consideration are the agitator type and fill level of the vessel for efficient equilibrium mixing. Most general purpose vessels have agitators placed relatively low to the maximum fill volume. If the densities of liquid phases are significantly different and the vessel is nearly full, the time to equilibrium during agitation may be extended. High-shear agitator types and vessel baffles make this operation more efficient under these conditions.

33.3.3.8 Process Control

The most common parameter under adjustment and control in liquid–liquid extraction other than temperature is pH, which is often adjusted with aqueous acid or base. An adjustment in pH is used to convert the component of interest into a free acid, free base, or salt to enable it to distribute preferentially into a desirable liquid phase for further processing. The acid or base is added to the agitated solution, and the pH is either measured in place by a probe installed in the tank or in a recirculation loop. Alternatively, a sample of the mixture or aqueous phase may be taken from the vessel for off-line analysis. The pH specification for a process should be carefully defined with regard to the state of agitation. The pH measurement of an aqueous and organic liquid-phase mixture results in a different pH value than measurement of the aqueous phase alone. Temperature also impacts solution pH, and this effect must be accounted for when taking a sample for off-line analysis.

33.3.3.9 Unit Operation

Maximum agitation is most often a benefit to liquid–liquid extraction operation since impeller shear creates interfacial surface area between phases. Greater surface area between phases means increased mass transfer rates toward the equilibrium state. On a manufacturing scale in a batch pharmaceutical plant, the agitator is typically set very low to achieve stirring at very low volumes. This can hinder liquid–liquid extraction when the vessel is more than half full since the upper portions of the fluid are less turbulent and coalescence between dispersed droplets is more likely. Using impeller types with greater shear characteristics may assist the operation by ensuring minimum droplet size near where the impeller is formed. Where one impeller would be damaging in a crystallization process, the same impeller would typically improve liquid–liquid extraction performance. Once equilibrium is achieved, agitation is stopped, and the phases are allowed to settle into top and bottom layers.

33.3.3.10 Phase Split Determination

Separation of phases from a single-stage extractor is usually achieved by draining the lower layer from the vessel through a bottom valve. The endpoint of the phase split is normally determined by visual inspection through a sight glass near the bottom valve. The sight glass allows detection of the interface between phases. When there is color similarity between phases, a conductivity probe may also be used in conjunction with the sight glass to detect the step change in conductance between phases as the split is nearing completion. The aqueous phase normally has greater conductivity than the organic phase, and change between phases with the sensor is either indicated by the change in a signal light or displayed by digital indication of the conductivity measurement.

33.A GUIDE TO GENERATION OF BINARY INTERACTION PARAMETERS FOR SOLVENT PAIRS

33.A.1 Software

COSMOthermX v.C30_1701 and Aspen Properties V9 (or Aspen Plus V9)

33.A.2 Purpose

This guide provides instruction on the generation of NRTL BIPs for liquid compounds pairs using COSMOthermX and Aspen Properties. Once the appropriate data are in Aspen Properties, they can be used as the basis for distillation or liquid–liquid extraction modeling. Estimation of BIPs through COSMOthermX is only required if the NRTL BIPs are not predefined in the Aspen Properties databank. This guide assumes the user has basic skills in operating COSMOthermX and Aspen software packages.

33.A.3 Procedure

The flowchart in Figure 33.A.1 describes the workflow for generating BIPs using COSMOthermX and Aspen software interfaces. The workflow demonstrates three paths to estimate for BIPs, and each is ranked by the accuracy of results based on the complexity of inputs used for estimation. The rankings for NRTL parameter estimation are summarized here:

- *Good*: Provides a set of *infinite dilution* activity coefficients for a pair of components at a *single temperature*. The NRTL parameters are estimated within Aspen.
- *Better*: Provides NRTL parameters directly from COSMOthermX, but the *parameters are insensitive to temperature* and are generally applicable to the pressure or isothermal conditions chosen in COSMOthermX.
- *Best*: Regression of activity coefficients from COSMOthermX is performed within Aspen to provide *temperature-dependent* NRTL parameters within Aspen.

Screenshots from both applications are shown after the flowchart as further guidance on the process. After determining the desired path through the workflow, the user should proceed to the matching workflow section of the guide for step-by-step instructions. A methanol–water system is used throughout the examples. A comparison of results for three binary solvent systems is also provided at the end of the section.

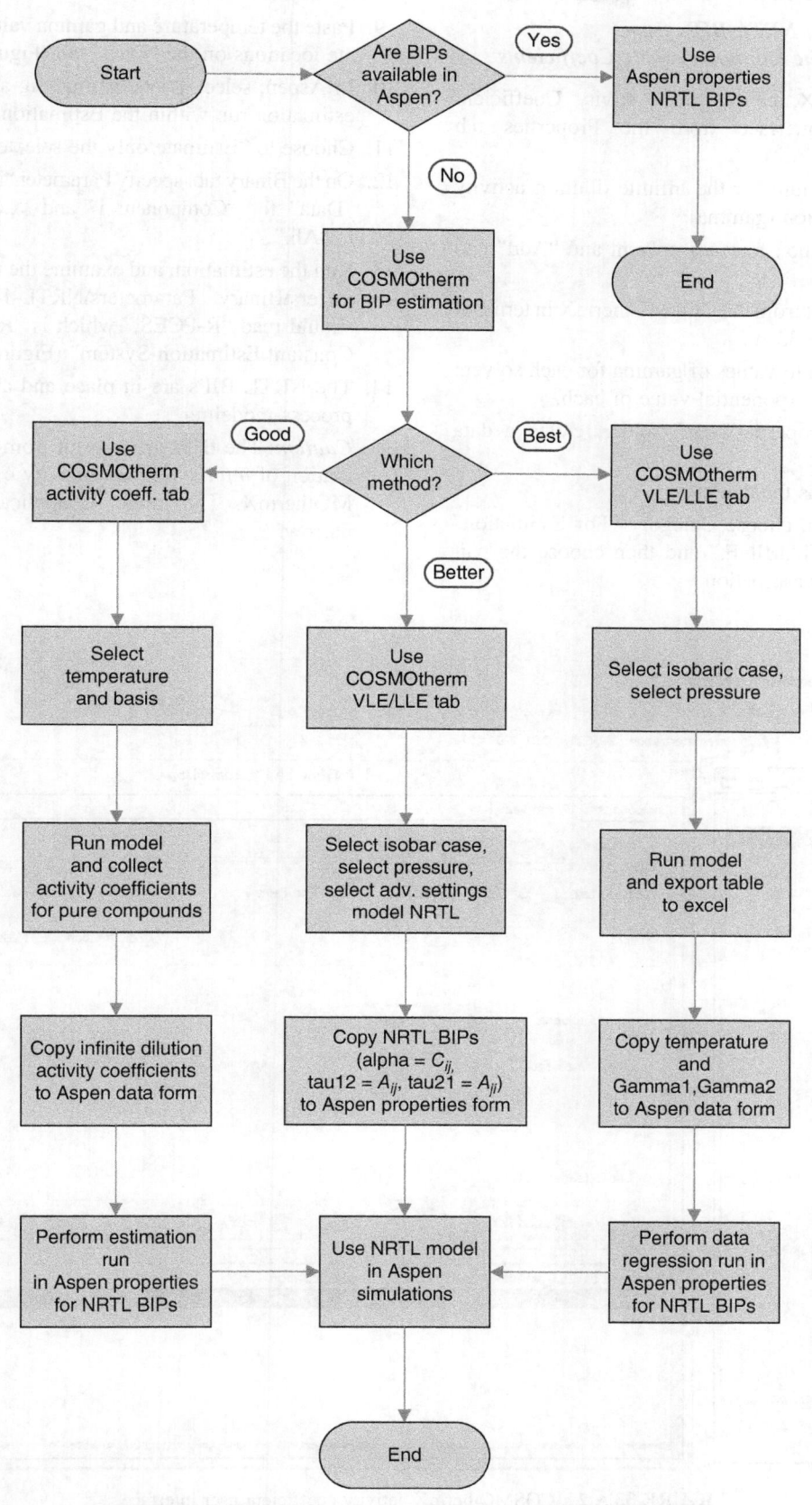

FIGURE 33.A.1 Binary interaction parameter estimation workflow.

33.A.3.1 Estimation of NRTL BIPs from COSMOthermX Infinite Dilution Activity Coefficients

1. In COSMOthermX, choose the Activity Coefficient button in the interface from the Properties tab section (Figure 33.A.2).
2. Select the temperature for the infinite dilution activity coefficient estimation (gamma).
3. Check the "Pure" box for each solvent and "Add" and "Run" the cases.
4. Examine the output from the COSMOthermX in terms of ln(gamma) (Figure 33.A.3).
5. Convert the results to values of gamma for each solvent by calculating the exponential value of each.
6. In the Aspen "Properties\Data" folder, create a data entry case.
7. Select "Mixture" as the data type.
8. In the "Setup" tab, choose Category "For Estimation" and Data Type "GAMINF," and then choose the pair of components for estimation.
9. Paste the temperature and gamma values in the appropriate locations on the "Data" tab (Figure 33.A.4).
10. In Aspen, select Tools/Estimation, and then request an estimation run within the Estimation folder.
11. Choose to "Estimate only the selected parameters."
12. On the Binary tab, specify Parameter "NRTL" and Method "Data" for "Component i" and "Component j" equal to "All."
13. Run the estimation, and examine the results in the "Parameters\Binary Parameters\NRTL-1" folder. Source should read "R-PCES," which is "Regression Property Constant Estimation System" (Figure 33.A.5).
14. The NRTL BIPs are in place and are ready for use in process modeling.
15. *Caution*: The BIPs are a result from an *isothermal* estimation of *infinite dilution* activity coefficients in COSMOthermX. The range of applicability is relatively narrow.

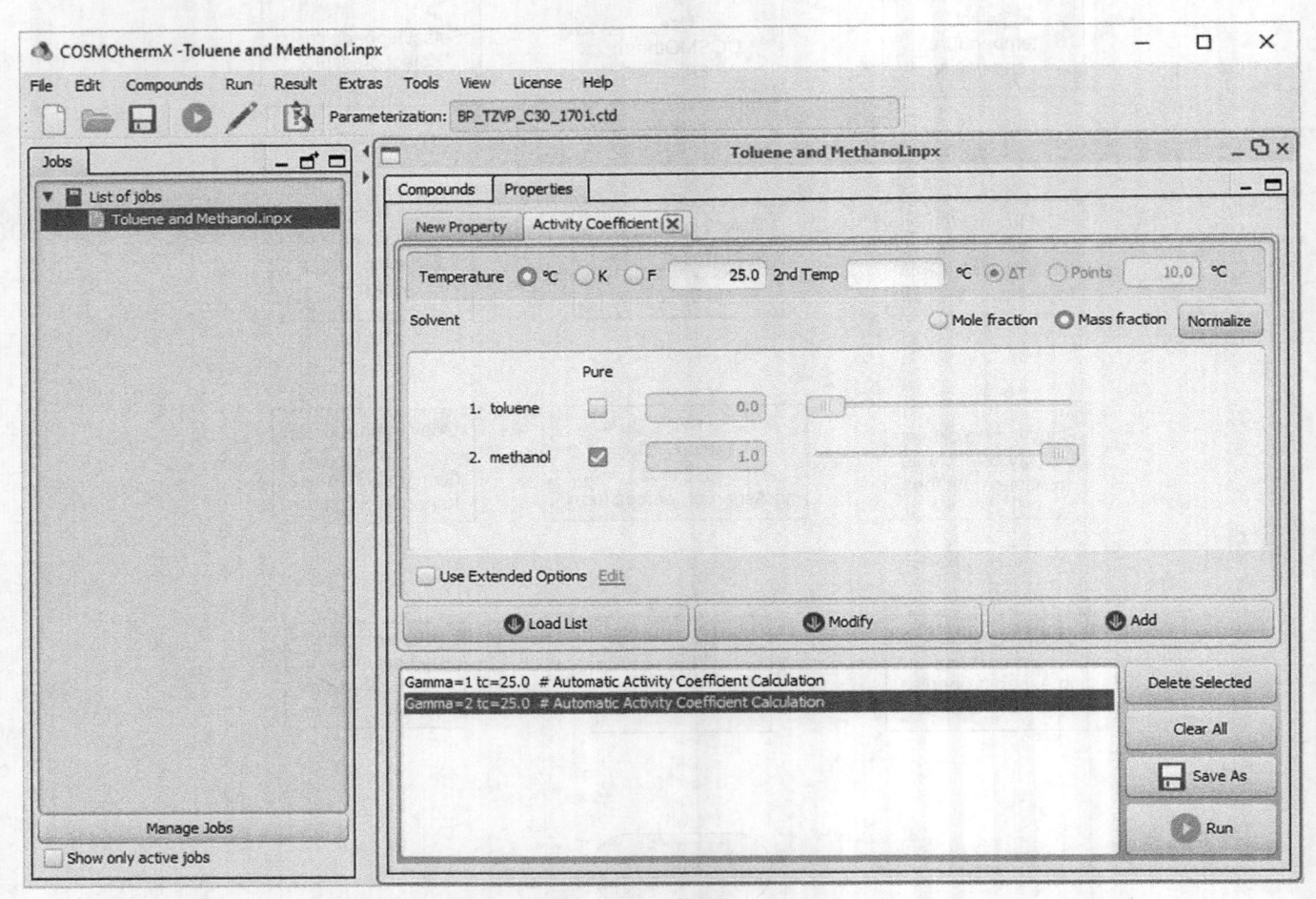

FIGURE 33.A.2 COSMOthermX activity coefficient user interface.

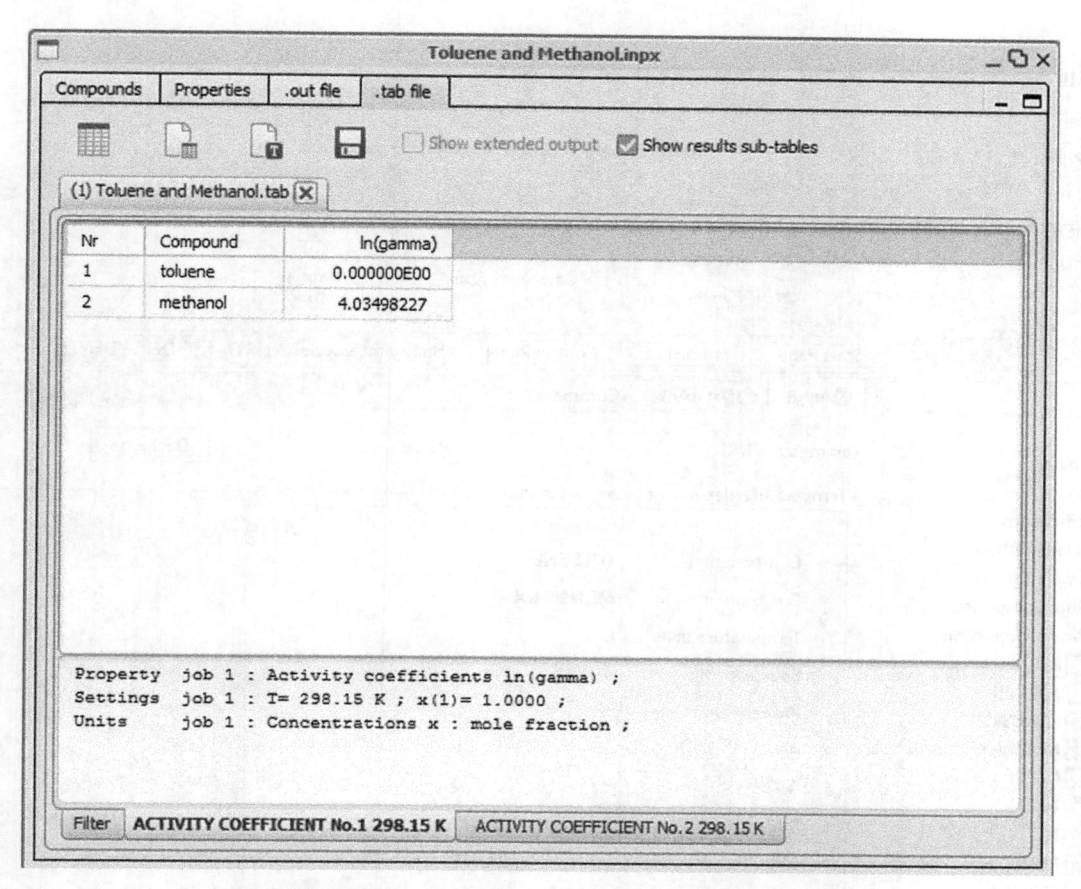

FIGURE 33.A.3 COSMOthermX activity coefficient output.

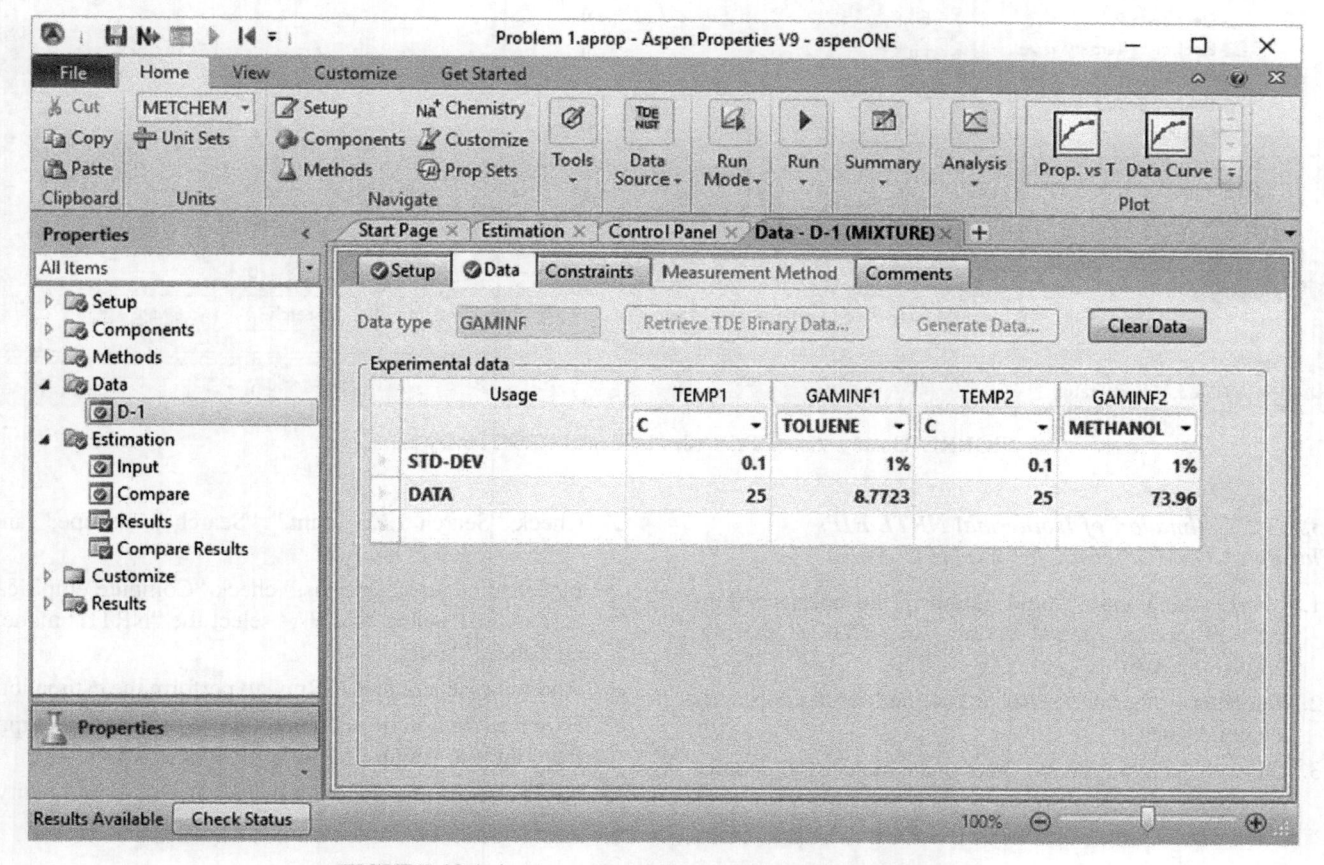

FIGURE 33.A.4 Aspen Properties activity coefficient data input.

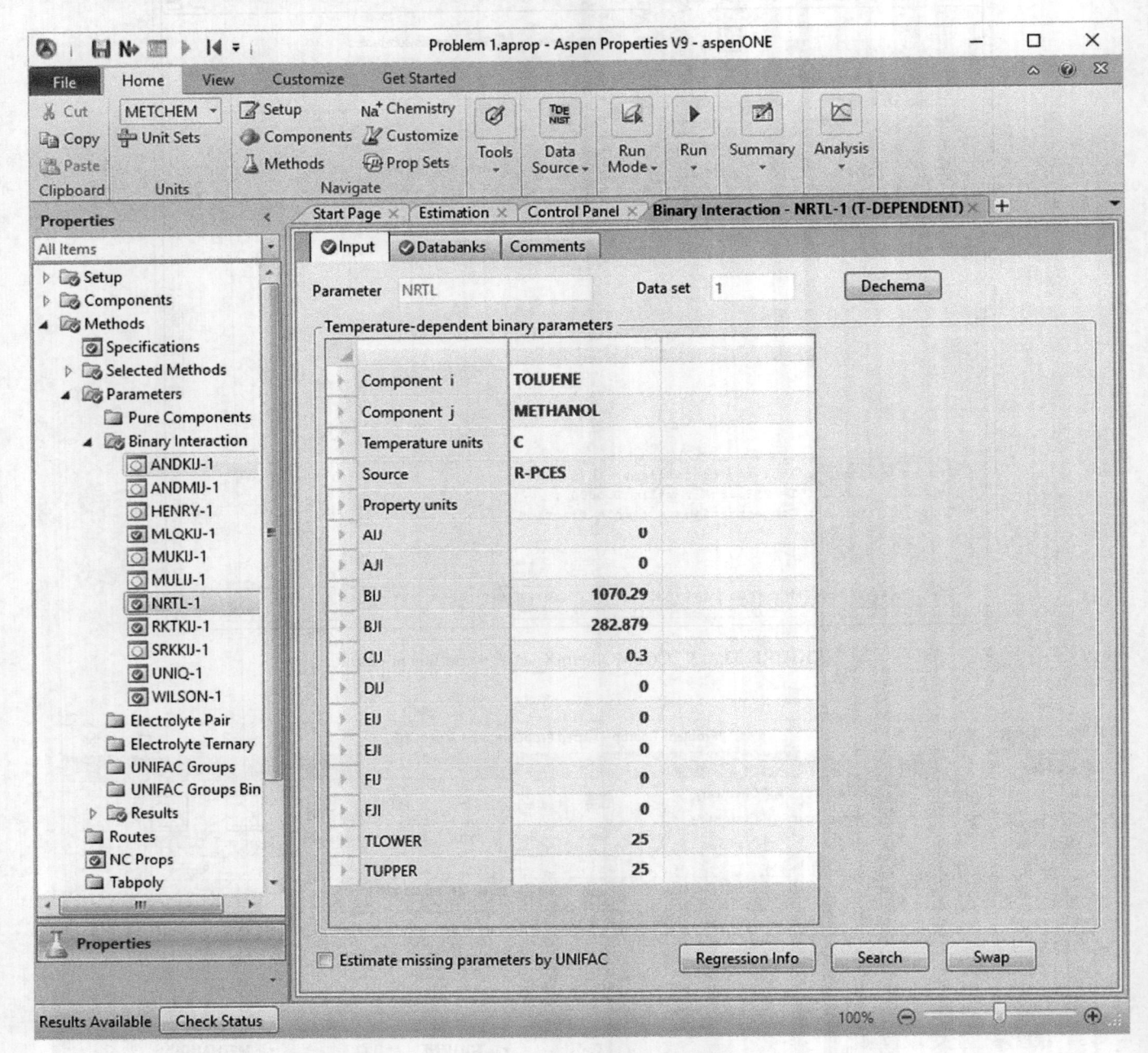

FIGURE 33.A.5 Aspen Properties estimated NRTL binary parameters.

33.A.3.2 Estimation of Isothermal NRTL BIPs Through COSMOthermX for Export to Aspen

1. Choose the Vapor–Liquid button in the interface from the Properties tab section in COSMOthermX (Figure 33.A.6).
2. Select an isothermal or isobar case and set the appropriate conditions.
3. Choose the components and make note of the order of entry.
4. Check "Search LLE point," "Search azeotrope," and "Use Extended Options."
5. Under "Extended Options," check "Compute empirical activity coefficient models," select the "NRTL" model, and click "Apply."
6. Add the case and press "Run" to perform the estimation.
7. Examine the bottom section of the ".out file" text output from the run (Figure 33.A.7).
8. NRTL rms is an indication of parameter fitting quality.

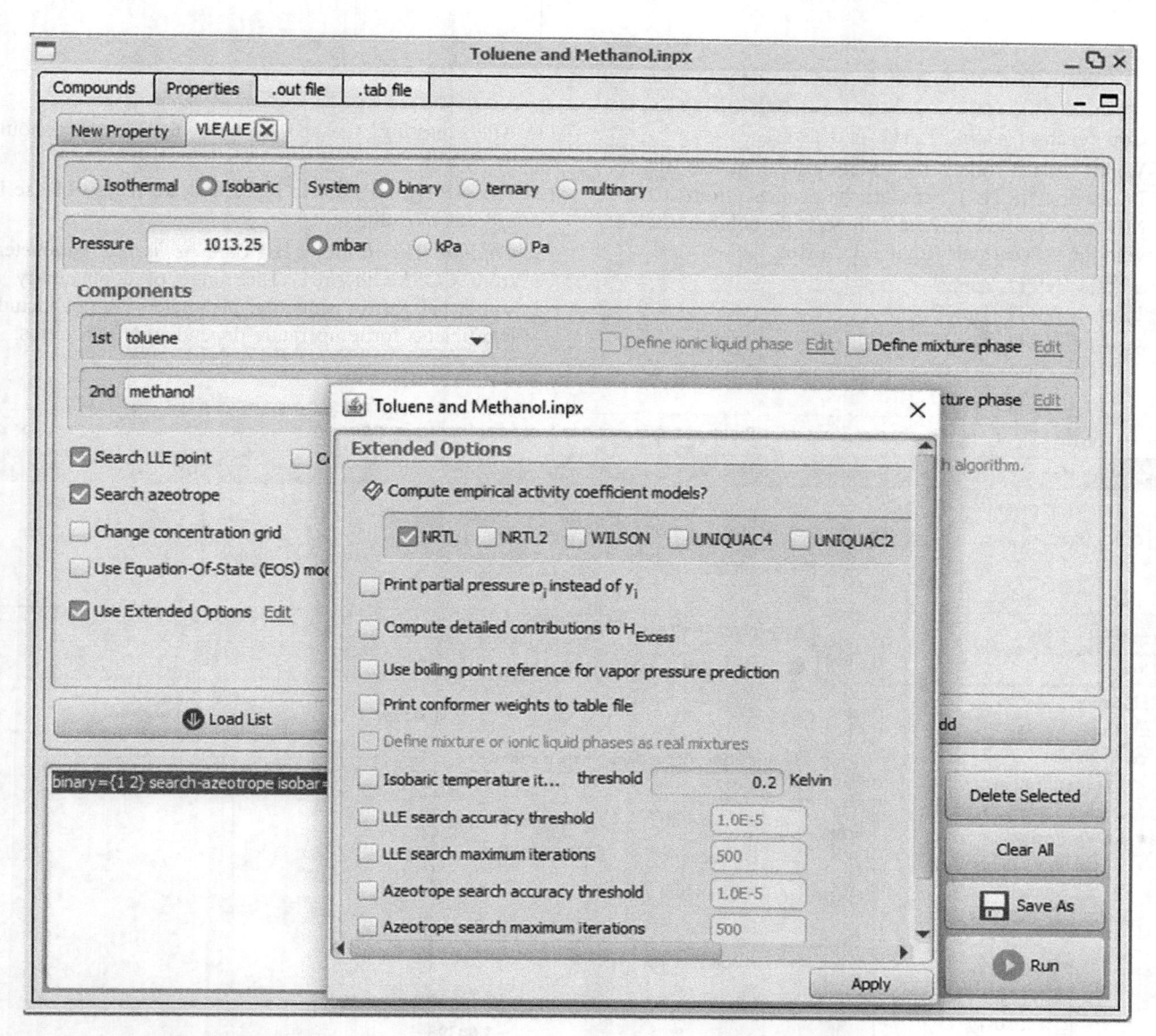

FIGURE 33.A.6 COSMOthermX VLE/LLE user interface for NRTL binary parameter estimation.

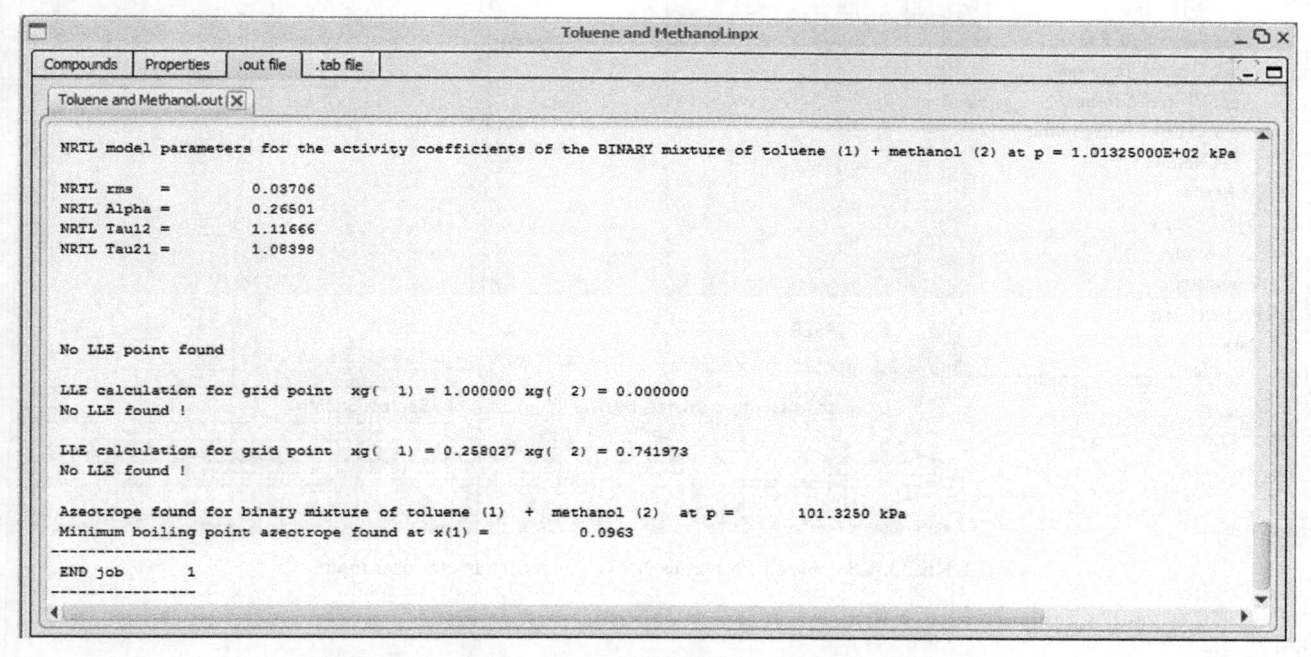

FIGURE 33.A.7 COSMOthermX NRTL binary parameter output.

9. Copy each of Alpha, Tau12, and Tau21.
10. Within Aspen, under the "Methods\Parameters\Binary Interaction\NRTL-1," paste the values from COSMOthermX NRTL BIPs as follows in the column labeled with the solvent pair (Figure 33.A.8):
    a. C_{ij} = NRTL Alpha
    b. A_{ij} = NRTL Tau12
    c. A_{ji} = NRTL Tau21
11. After pasting the parameters, the source should be "USER."
12. The NRTL BIPs are in place and are ready for use in process modeling.
13. *Caution*: The resulting BIPs are *isothermal* parameters from COSMOthermX. The range of applicability is somewhat narrow since the parameters do not include coefficients for temperature-dependent NRTL terms.

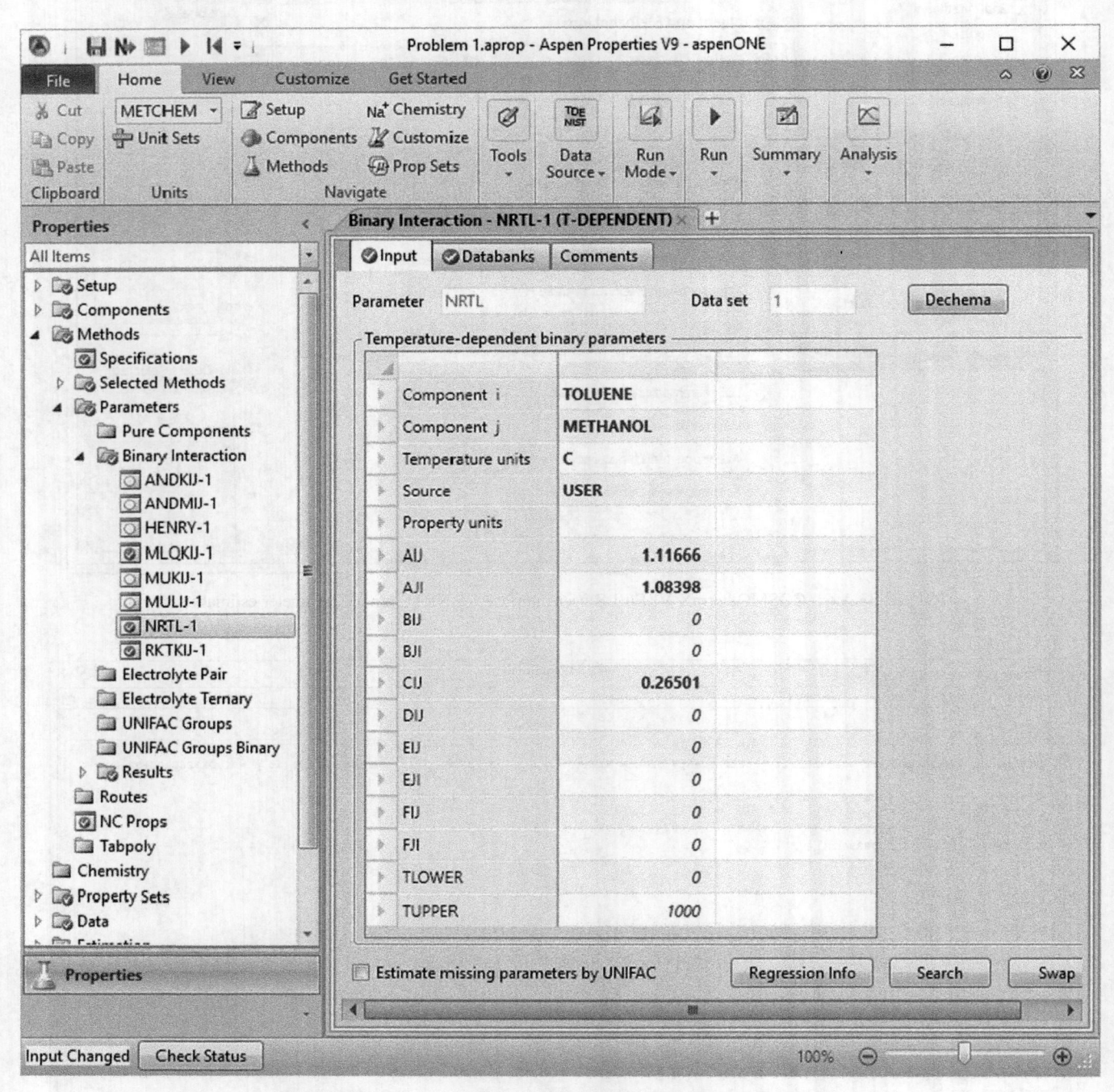

FIGURE 33.A.8 Aspen Properties NRTL binary parameter user input.

33.A.3.3 Estimation of NRTL BIPs from COSMOthermX Liquid Activity Coefficients

1. Choose the Vapor–Liquid button in the interface from the Properties tab section in COSMOthermX (Figure 33.A.9).
2. Select an isothermal or isobar case and set the appropriate conditions.
3. Choose the components and make note of the order of entry.
4. Select "Search LLE point" and "Search azeotrope."
5. Add the case and press Run to perform the estimation.
6. Examine the tabular output (".tab file" section) from the run (Figure 33.A.10).
7. Choose "Open as .xls" button, and save the file to a convenient location with a desirable filename.
8. In the Excel spreadsheet, insert columns after $\ln(\gamma_1)$ and $\ln(\gamma_2)$ as shown below.
9. In the new columns, calculate "γ_1" and "γ_2" as $\exp(\ln(\gamma))$ for each (Figure 33.A.11).
10. In the Aspen "Data" folder, create a data entry case for a mixture.
11. In the "Setup" tab, choose Category "Thermodynamic" and Data Type "GAMMA," and choose the binary pair for estimation. Set the pressure at the bottom to that of the system.
12. Paste the temperatures and gamma values in the appropriate locations (Figure 33.A.12).

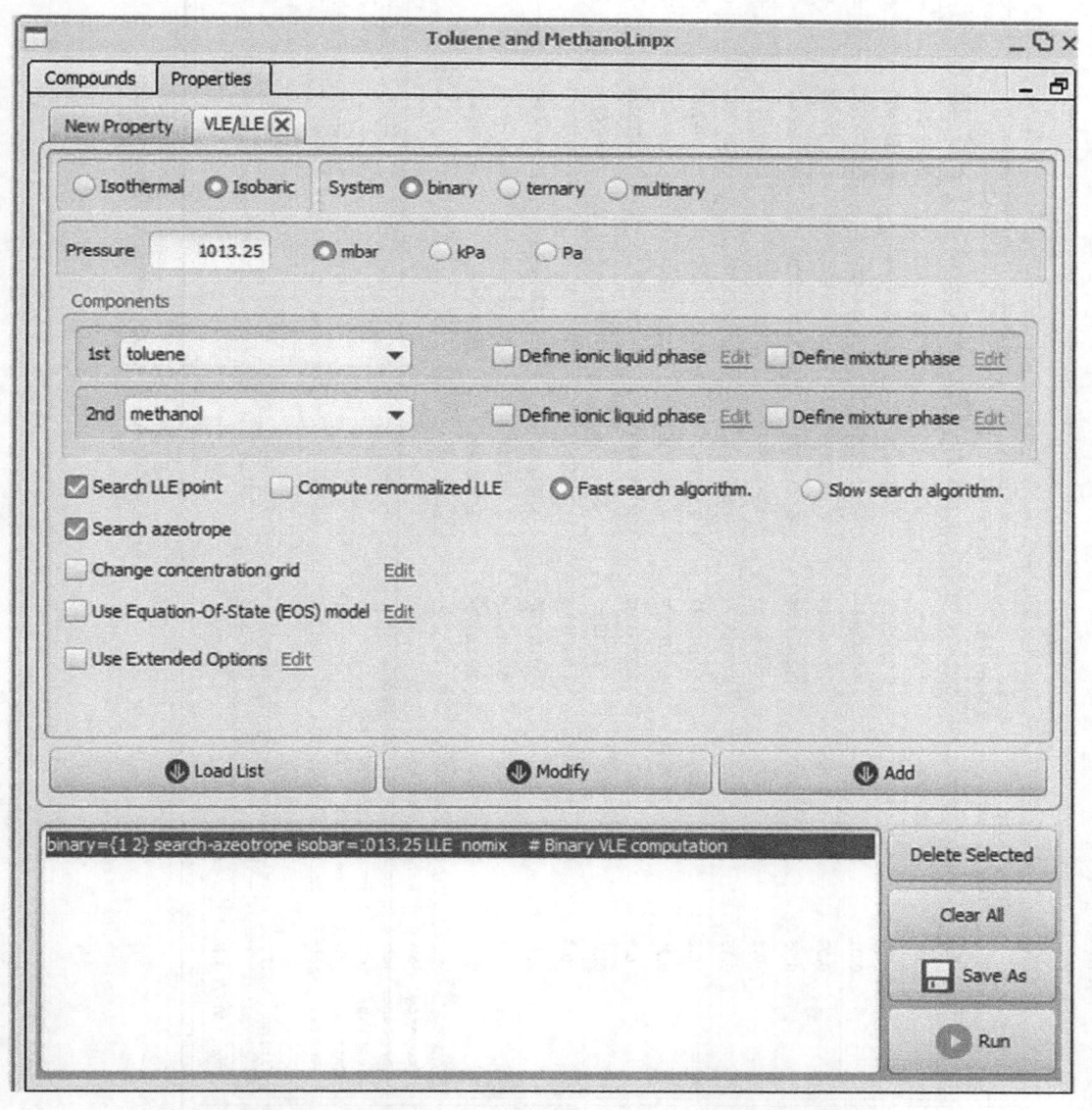

FIGURE 33.A.9 COSMOthermX VLE/LLE user interface for data estimation.

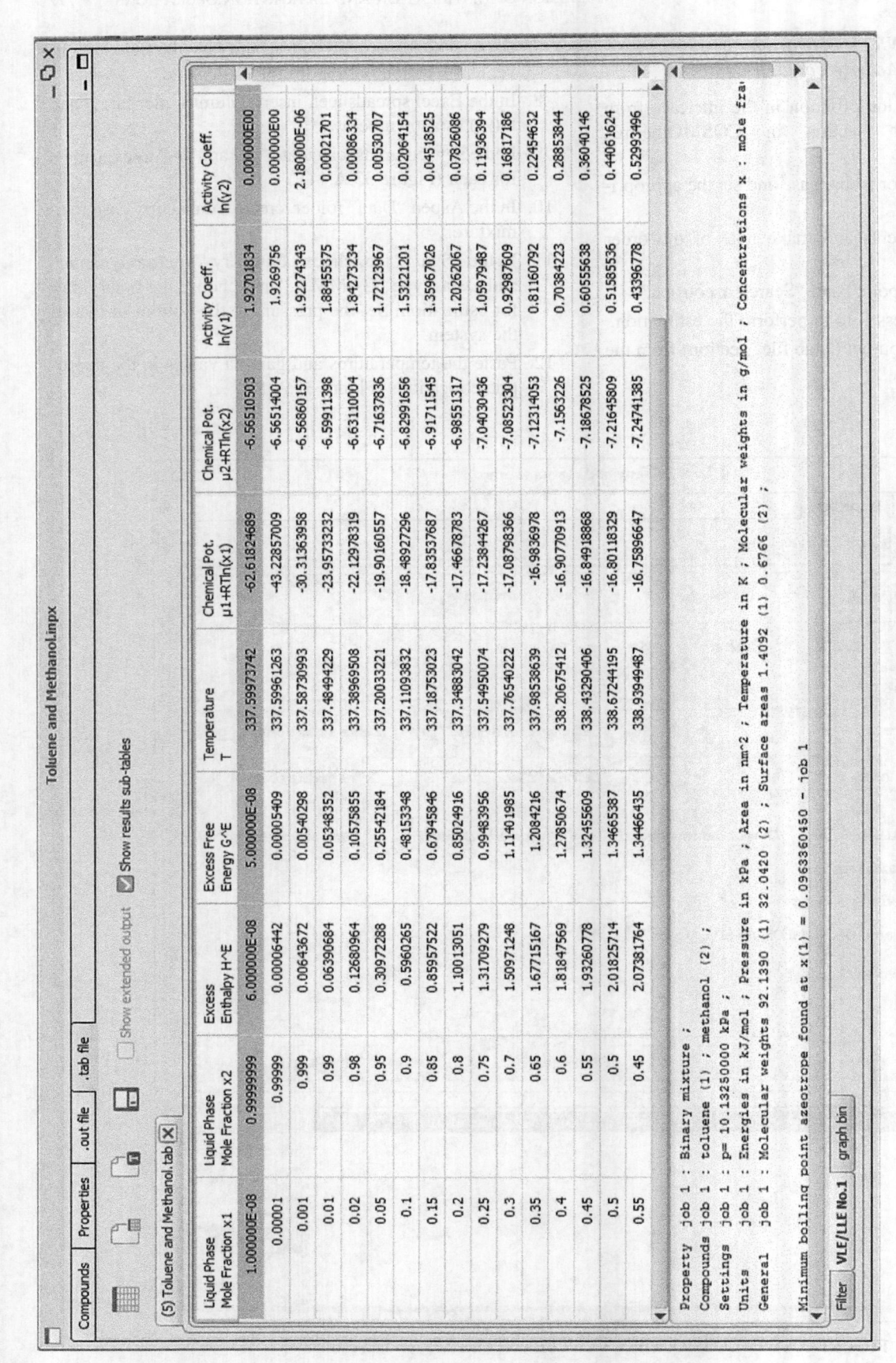

FIGURE 33.A.10 COSMOthermX VLE data output.

property job 1 : binary mixture ; compounds job 1 : toluene (1) ; methanol (2) ; settings job 1 : p= 10.13250000 kpa ; units job 1 : energies in kj/mol ; pressure in kpa ; area in nm^2 ; temperature in k ; molecular weights in g/mol ; concentrations x : mole fraction ; general job 1 : molecular weights 92.1390 (1) 32.0420 (2) ; surface areas 1.4092 (1) 0.6766 (2) ;

Liquid Phase Mole Fraction x1	Liquid Phase Mole Fraction x2	Excess Enthalpy H^E	Excess Free Energy G^E	Temperature T	Chemical Pot. μ1+RTln(x1)	Chemical Pot. μ2+RTln(x2)	Activity Coeff. ln(y1)	gamma1	Activity Coeff. ln(y2)	gamma2	Gas Phase Mole Fraction y1	Gas Phase Mole Fraction y2
0.00000001	0.99999999	0.00000006	0.00000005	337.59973742	-62.61824689	-6.56510503	1.92701834	6.869	0.0	1.000	0.00000002	0.9999999
0.00001	0.99999	0.00006442	0.00005409	337.59961263	-43.22857009	-6.56514004	1.9269756	6.869	0.0	1.000	0.00001494	0.99998506
0.001	0.999	0.00643672	0.00540298	337.58730993	-30.31363958	-6.56860157	1.92274343	6.840	0.00000218	1.000	0.00148668	0.99851332
0.01	0.99	0.00639684	0.05348352	337.48494229	-23.95733232	-6.59911398	1.88455375	6.583	0.00021701	1.000	0.01425353	0.98574647
0.02	0.98	0.12680964	0.10575855	337.38969508	-22.12978319	-6.63110004	1.84273234	6.314	0.00066334	1.001	0.02723947	0.97276053
0.05	0.95	0.30972288	0.25542184	337.20033221	-19.90160557	-6.71637836	1.72123967	5.591	0.00530707	1.005	0.05986995	0.94013006
0.1	0.9	0.5960265	0.48153348	337.11093832	-18.48927296	-6.82991656	1.53221201	4.628	0.02064154	1.021	0.09877545	0.90122456
0.15	0.85	0.8957522	0.67945846	337.18753023	-17.83537687	-6.91711545	1.35967026	3.895	0.04518525	1.046	0.12505133	0.87494867
0.2	0.8	1.10013051	0.85024916	337.34883042	-17.46678783	-6.98551317	1.20262067	3.329	0.07826086	1.081	0.14338998	0.85661002
0.25	0.75	1.31709279	0.99483956	337.54950074	-17.23844267	-7.04030436	1.05977487	2.886	0.11936394	1.127	0.15658569	0.84341432
0.3	0.7	1.50971248	1.11401985	337.76540222	-17.08798366	-7.08523304	0.92987609	2.534	0.16817186	1.183	0.16638382	0.83361618
0.35	0.65	1.67715167	1.2084216	337.98536639	-16.9836978	-7.12314053	0.81160792	2.252	0.22454632	1.252	0.17392337	0.82607663
0.4	0.6	1.81847569	1.27850674	338.20675412	-16.90770913	-7.1563226	0.70384223	2.022	0.28853844	1.334	0.17998173	0.82001827
0.45	0.55	1.93260778	1.32455609	338.43290406	-16.84918868	-7.18678525	0.60555638	1.832	0.36040146	1.434	0.18511864	0.81488136
0.5	0.5	2.01825714	1.34665387	338.67244195	-16.80118329	-7.21645809	0.51585536	1.675	0.44061624	1.554	0.1897671	0.8102329
0.55	0.45	2.07381764	1.34466435	338.93949487	-16.7589647	-7.24741385	0.43396778	1.543	0.52993496	1.699	0.19429803	0.80570198
0.6	0.4	2.09722298	1.31819578	339.25537991	-16.71909954	-7.28215274	0.35924045	1.432	0.62945214	1.877	0.19907585	0.80092415
0.65	0.35	2.08572756	1.26654409	339.65233089	-16.67878297	-7.32405057	0.29113516	1.338	0.74071721	2.097	0.20452214	0.79547786
0.7	0.3	2.03555021	1.18860326	340.18110084	-16.63543621	-7.3781954	0.22923106	1.258	0.86591238	2.377	0.21121296	0.78878704
0.75	0.25	1.94124473	1.0287814	340.92704672	-16.58607959	-7.4530184	0.17323814	1.189	1.00813272	2.740	0.22006821	0.77993179
0.8	0.2	1.794468	0.94643139	342.04755367	-16.5264594	-7.5645426	0.12303402	1.131	1.17180754	3.228	0.23279649	0.76720351
0.85	0.15	1.58122307	0.7760202	343.87300092	-16.44892232	-7.7472022	0.07875793	1.082	1.3631678	3.909	0.25317183	0.74682817
0.9	0.1	1.27440175	0.5650117	347.2476515	-16.33575616	-8.10127604	0.04107882	1.042	1.58930192	4.900	0.29184432	0.70815568
0.95	0.05	0.80734723	0.30577716	355.17553678	-16.12877393	-9.05106198	0.01219243	1.012	1.83923819	6.292	0.3945319	0.60504681
0.98	0.02	0.37643417	0.1234982	367.38164096	-15.86512752	-11.12194467	0.00196194	1.002	1.92503821	6.855	0.60136468	0.39863532
0.99	0.01	0.1950807	0.06071726	374.49866826	-15.72440551	-13.14578489	0.00046236	1.000	1.90419568	6.714	0.75502565	0.24497435
0.999	0.001	0.0197851	0.00589776	382.75782249	-15.5682095	-20.48767108	0.00000416	1.000	1.84906847	6.354	0.97017902	0.02982098
0.99999	0.00001	0.0001979	0.00005874	383.77506418	-15.54944592	-35.21021299	0.0	1.000	1.84079928	6.302	0.99969519	0.00030482
0.99999999	0.00000001	0.0000002	0.00000006	383.78543499	-15.54925477	-57.25333209	0.0	1.000	1.84071369	6.301	0.9999997	0.00000031

Property job 1 : Binary mixture ;
Compounds job 1 : toluene (1) ; methanol (2) ;
Settings job 1 : p= 10.13250000 kPa ;
Units job 1 : Energies in kJ/mol ; Pressure in kPa ; Area in nm^2 ; Temperature in K ; Molecular weights in g/mol ; Concentrations x : mole fraction ;
General job 1 : Molecular weights 92.1390 (1) 32.0420 (2) ; Surface areas 1.4092 (1) 0.6766 (2) ;

FIGURE 33.A.11 COSMOthermX VLE data exported to Microsoft Excel.

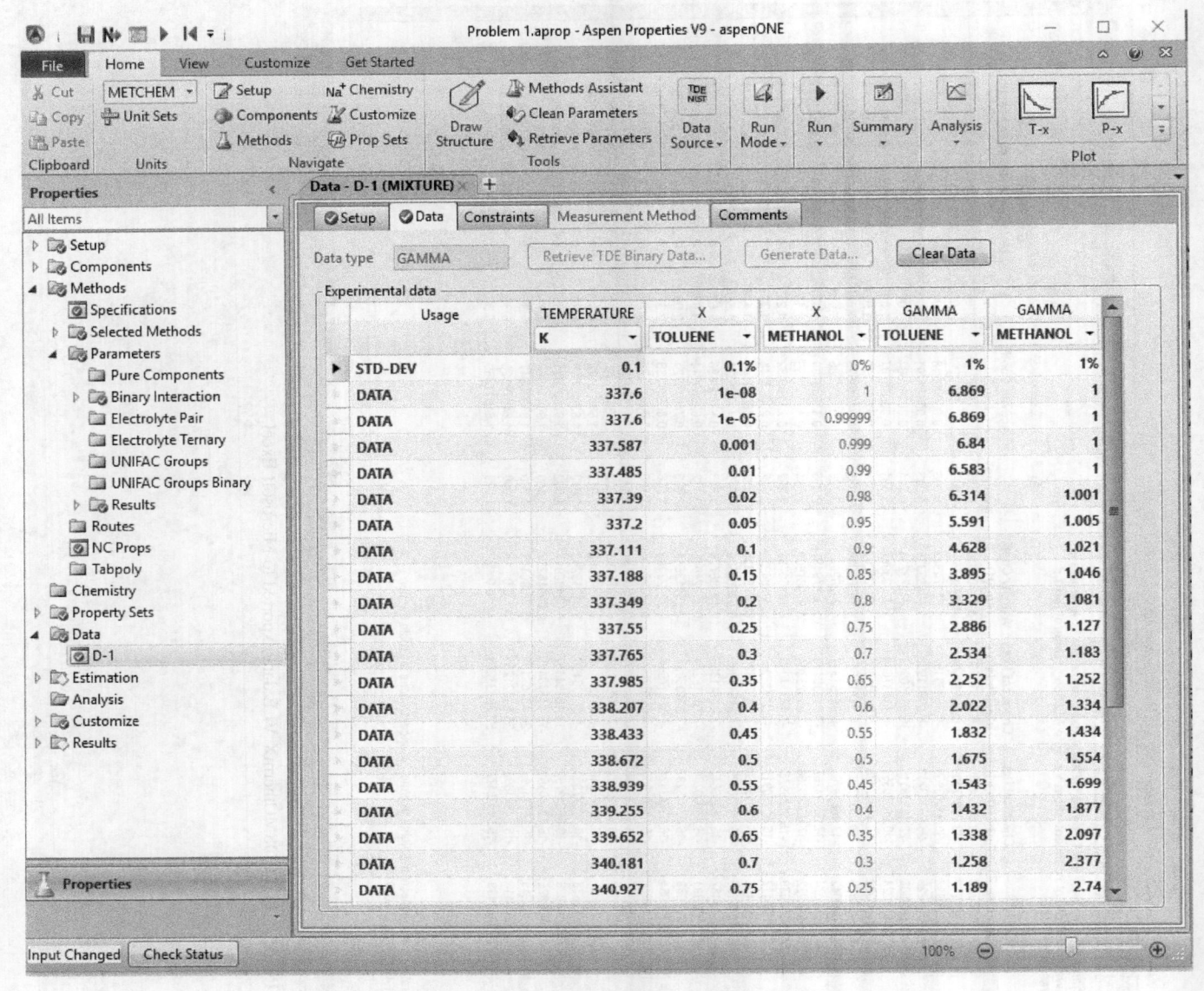

FIGURE 33.A.12 Aspen Properties user interface for VLE data input.

13. In the Aspen toolbar, choose Regression, and then start a new case in the Regression folder. Choose the name of data set in the "Setup" section of Regression.
14. Under the Parameters tab, choose "Binary parameter" for type, select the Name as NRTL, and set Elements 1, 2, and 3 as shown below (Figure 33.A.13).
15. For components, select the pair used in the COSMOthermX estimation.
16. Repeat the component pair for the same element but in reverse order, except for Element 3, which is symmetric and only has one value. Optionally, "fix" the value of Element 3 under the "Usage" option.
17. Run the regression and examine the results under the Regression folder for closeness of fit. Predicted vs. entered data can be plotted in the same area.
18. Examine the results in the "Parameters\Binary Parameters \NRTL-1" folder. Source should read "R-DR-#," which means a result from regression case (Figure 33.A.14).
19. The NRTL BIPs are in place and are ready for use in process modeling.
20. *Caution*: The BIPs are a result of a regression across the boiling point temperature range of the liquid pair (i.e. an isobar case), which defines the range of applicability. To obtain NRTL BIPs at a different pressure, either *extend*

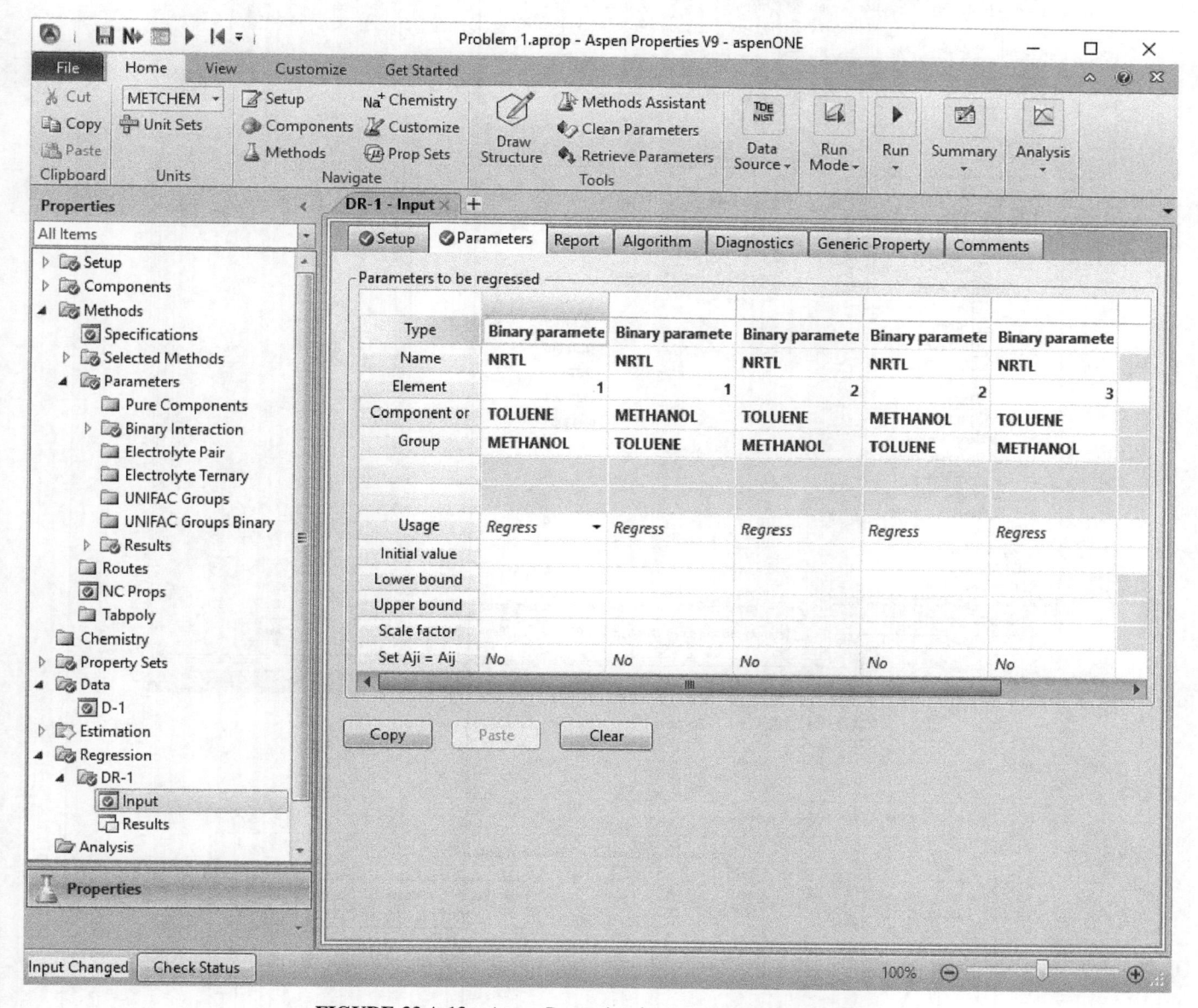

FIGURE 33.A.13 Aspen Properties data regression user interface.

the first data set by running a new case in COSMOthermX at a different isobaric condition *or create a separate regression case* in Aspen for the additional COSMOthermX isobaric data.

33.A.3.4 Examining the Quality of Liquid Activity Coefficients after Estimation or Regression

1. In Aspen, view the NRTL BIPs in the "Methods\Binary Interaction\NRTL-1" folder and select the appropriate "source."
2. In the Aspen toolbar, select Binary and choose Analysis type Txy or Pxy, set other conditions to match those of the regressed or estimated BIPs, and then run the analysis.
3. After seeing the resulting phase diagram, close the plot, and choose the "Activity Coeff." plot from the toolbar above the input form for the binary case.
4. An example of the resulting plot of liquid activity coefficients for each component is shown (Figure 33.A.15).

33.A.3.5 Comparison of T-xy Diagrams and Liquid Activity Coefficients by Source

A graphical comparison of VLE and liquid activity coefficient data retrieved from various Aspen Properties databanks and generated through COSMOthermX is shown in Figures 33.A.16–33.A.21. Three pairs of binary systems are considered in these examples: (i) the zeotropic system of

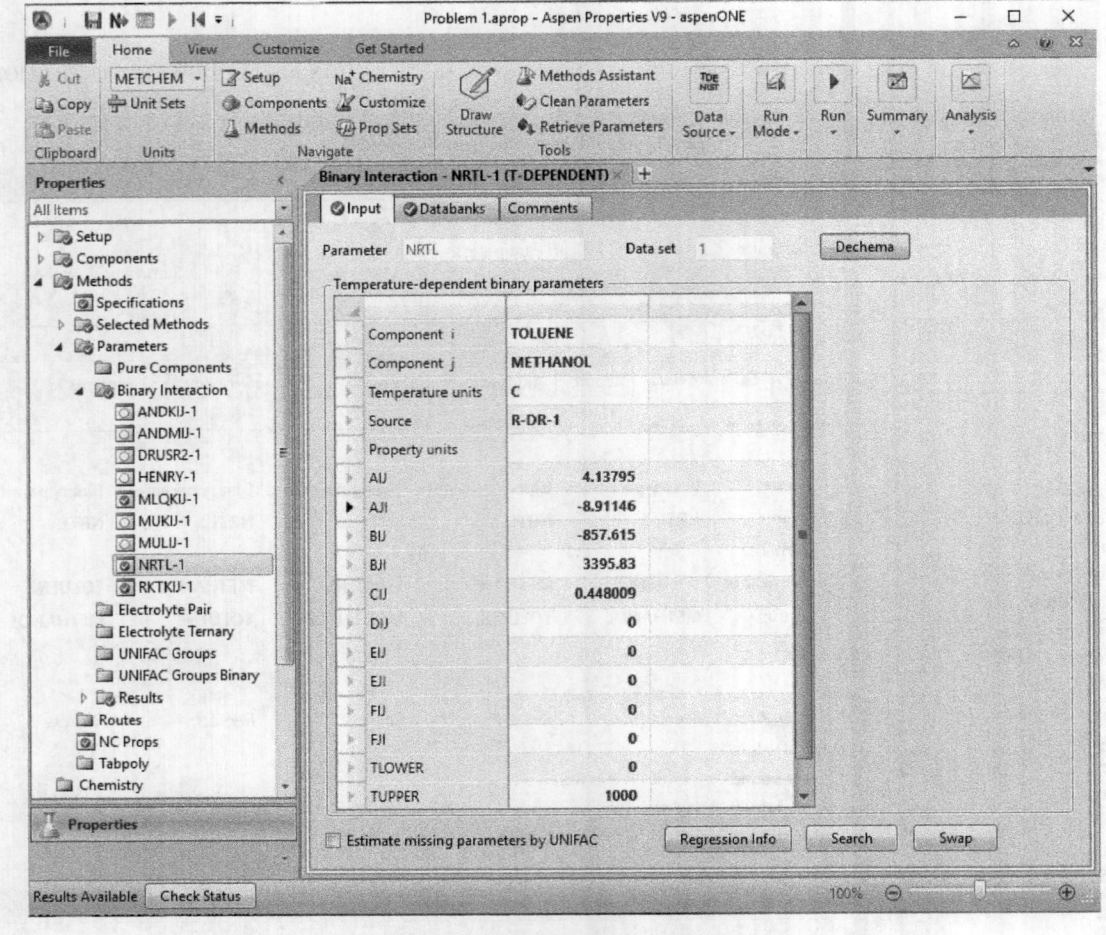

FIGURE 33.A.14 Aspen Properties VLE data regression results.

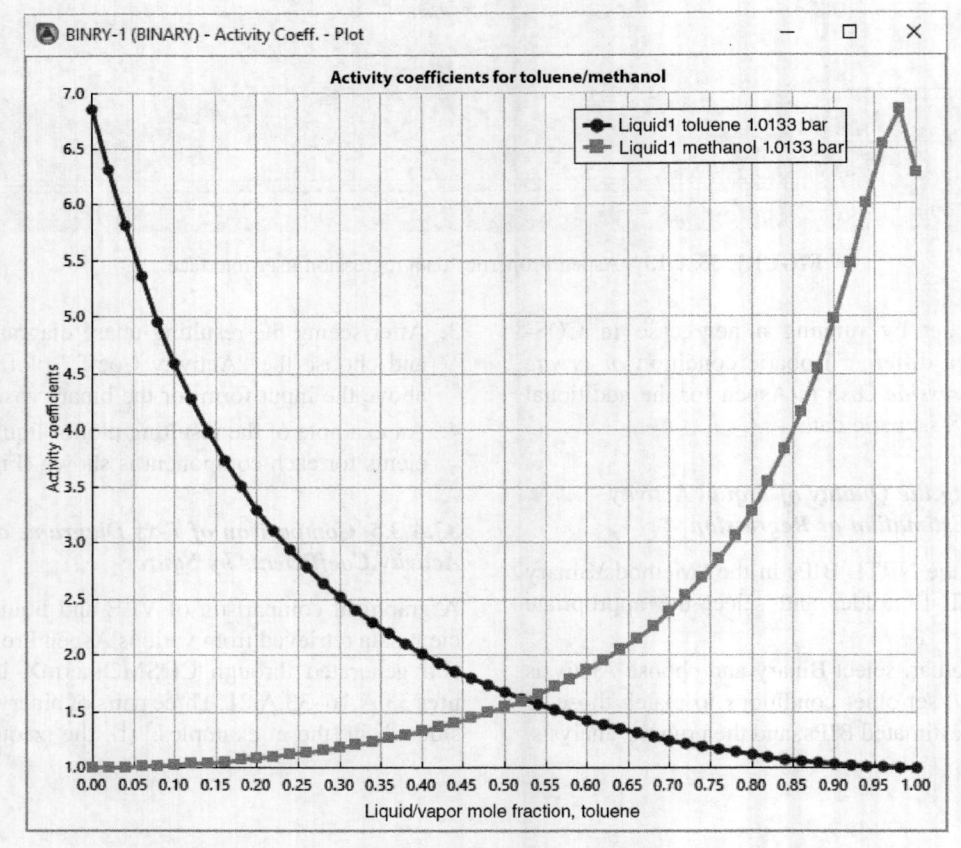

FIGURE 33.A.15 Aspen Properties plot of binary activity coefficients.

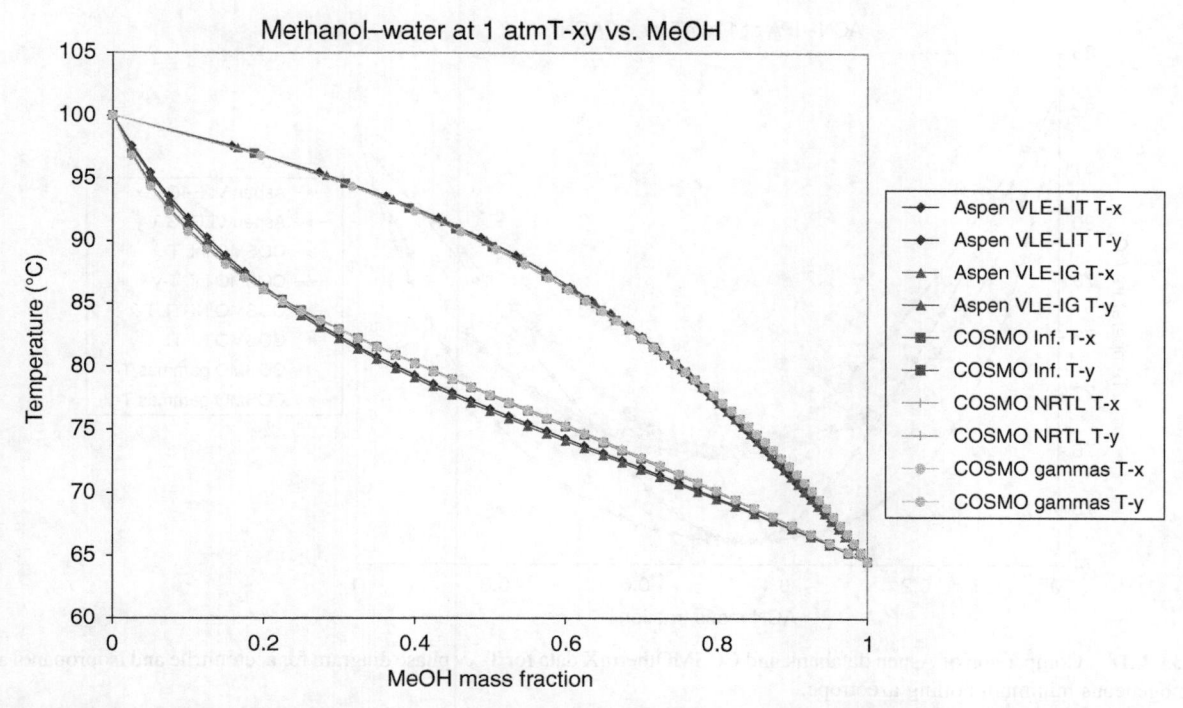

FIGURE 33.A.16 Comparison of Aspen databank and COSMOthermX data for T-xy phase diagram for methanol and water at 1 atm.

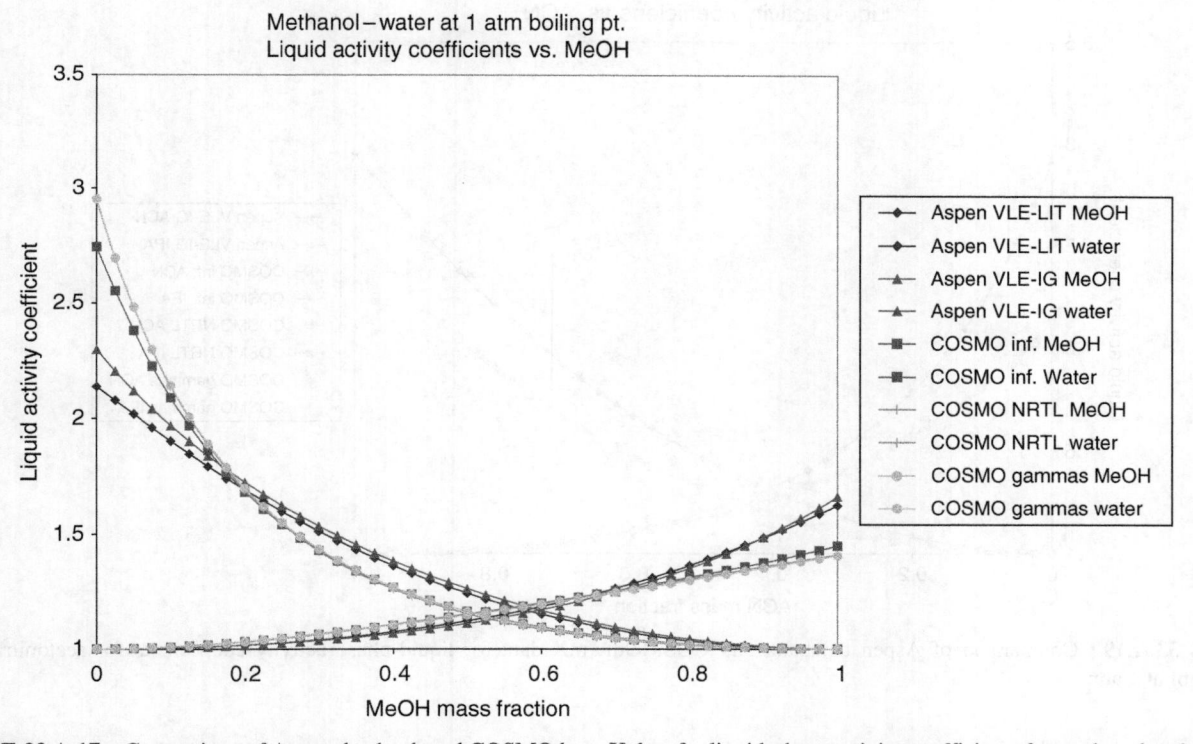

FIGURE 33.A.17 Comparison of Aspen databank and COSMOthermX data for liquid-phase activity coefficients for methanol and water at 1 atm.

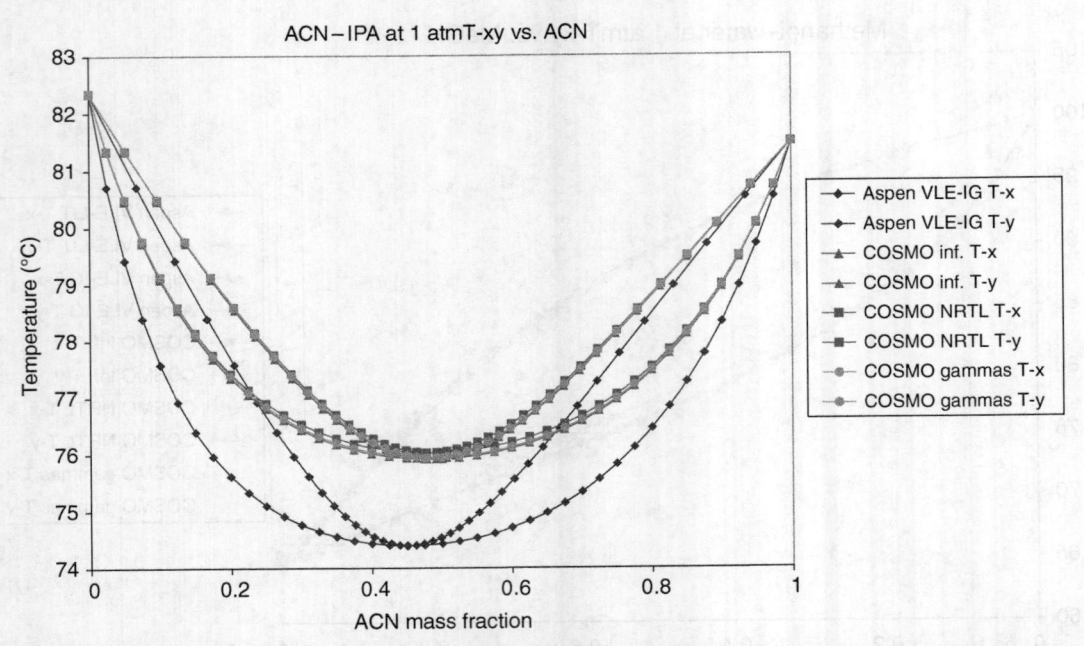

FIGURE 33.A.18 Comparison of Aspen databank and COSMOthermX data for T-xy phase diagram for acetonitrile and isopropanol at 1 atm with a homogeneous minimum boiling azeotrope.

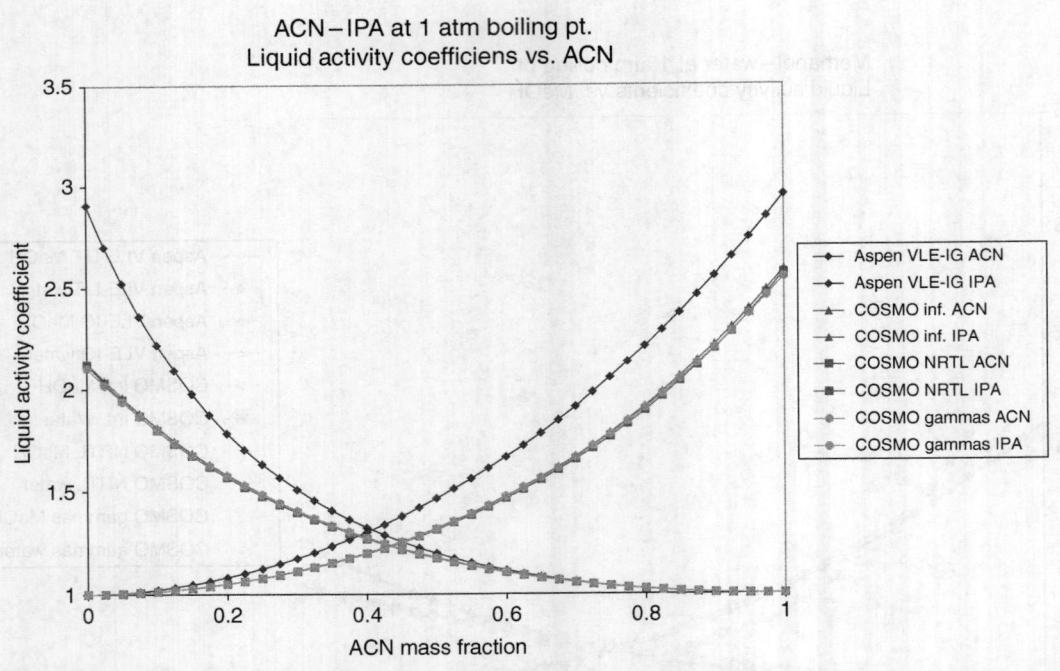

FIGURE 33.A.19 Comparison of Aspen databank and COSMOthermX data for liquid-phase activity coefficients for acetonitrile and isopropanol at 1 atm.

methanol and water (Figures 33.A.16 and 33.A.17); (ii) the minimum boiling, homogeneous azeotropic system of acetonitrile and isopropanol (Figures 33.A.18 and 33.A.19); and (iii) the minimum boiling, heterogeneous azeotropic system of n-butanol and water (Figures 33.A.20 and 33.A.21). In each case, the T-xy phase diagrams and liquid-phase activity coefficients plots were generated with the aid of COSMOthermX using the infinite dilution, NRTL parameter,

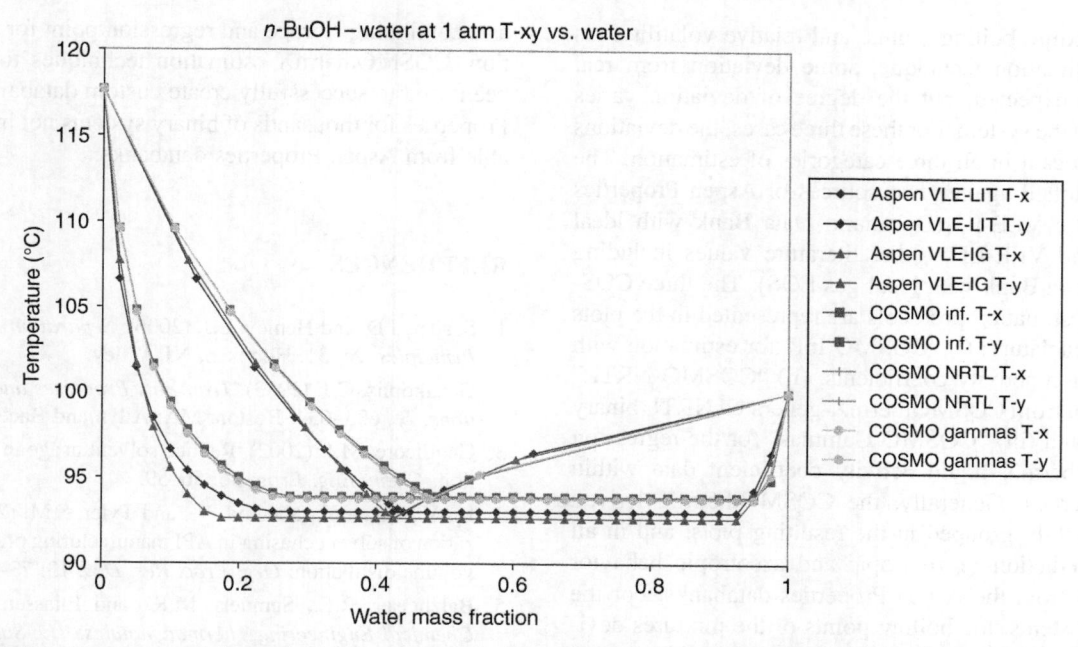

FIGURE 33.A.20 Comparison of Aspen databank and COSMOthermX data for T-xy phase diagram for *n*-butanol and water at 1 atm with a heterogeneous minimum boiling azeotrope.

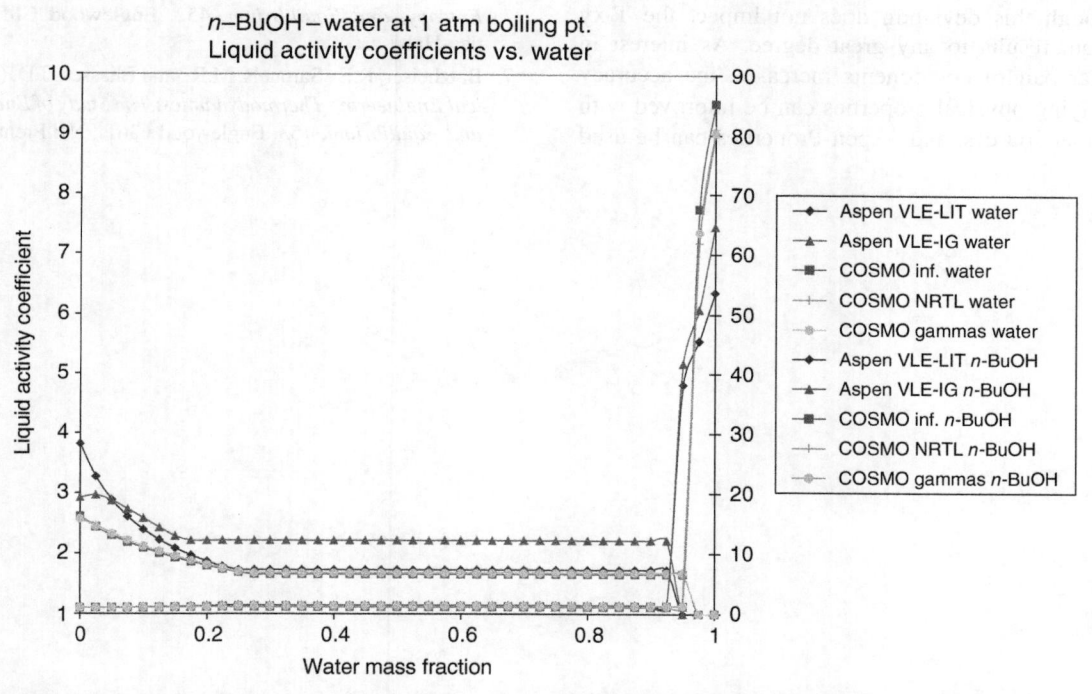

FIGURE 33.A.21 Comparison of Aspen databank and COSMOthermX data for liquid-phase activity coefficients for methanol and water at 1 atm.

and gamma regression protocols described in this guide. These three cases demonstrate that COSMOthermX is an exceptional method of obtaining estimated VLE data and associated BIP data, in the absence of laboratory or literature data. Aspen Properties is a convenient platform for the collection and regression of equilibrium data for further use in process modeling. The COSMOthermX estimation method provides immediate answers about the potential of azeotropic behavior, mixture boiling points, and relative volatility. As with any estimation technique, some deviation from real properties is expected, but the degree of deviation varies depending on the system. For these three cases, the deviations are not significant in all three categories of estimation. The results are plotted against two sources of Aspen Properties databank data: VLE-IG (Dortmund Data Bank with ideal gas EOS) and VLE-LIT (other literature values including Dortmund Data Bank with ideal gas EOS). The three COSMOthermX estimation protocols are represented in the plots by this nomenclature: (i) "COSMO Inf" for estimation with infinite dilution activity coefficients, (ii) "COSMO NRTL" for estimation from COSMOthermX-generated NRTL binary parameters, and (iii) "COSMO Gammas" for the regression of COSMOthermX liquid activity coefficient data within Aspen Properties. Generally, the COSMOthermX-derived results are tightly grouped in the resulting plots, and in all cases the prediction of zeotropic and azeotropic behavior matches that from the Aspen Properties databanks. For the azeotropic systems, the boiling points of the mixtures deviates by less than 2 °C and the composition of the binary azeotrope is offset by less than 0.1 mass fraction units. The activity coefficient plots show the greatest deviation from the databank values at the endpoints approaching infinite dilution, though this deviation does not impact the T-xy phase diagram results to any great degree. As interest in any particular pair of components increases, the accuracy of the underlying physical properties can be improved with data from other sources, and Aspen Properties can be used as a continuing storage and regression point for the information. COSMOthermX estimation techniques for BIPs have been used to successfully create custom databanks in Aspen Properties for thousands of binary systems not initially available from Aspen Properties databanks.

REFERENCES

1. Seader, J.D. and Henley, E.J. (2006). *Separation Processes and Principles*, 2e, 31. Hoboken, NJ: Wiley.
2. Geankoplis, C.J. (1983). *Transport Processes and Unit Operations*, 2e, 633–634. Boston, MA: Allyn and Bacon, Inc.
3. Gentilcore, M.J. (2002). Reduce solvent usage in batch distillation. *Chem. Eng. Prog.* 56: 56–59.
4. Li, Y., Yang, Y., Kalthod, V., and Tyler, S.M. (2009). Optimization of solvent chasing in API manufacturing process: constant volume distillation. *Org. Proc. Res. Dev.* 13: 73–77.
5. Balzheiser, R.E., Samuels, M.R., and Eliassen, J.D. (1972). *Chemical Engineering Thermodynamics: The Study of Energy, Entropy, and Equilibrium*, 387. Englewood Cliffs, NJ: Prentice-Hall.
6. Balzheiser, R.E., Samuels, M.R., and Eliassen, J.D. (1972). *Chemical Engineering Thermodynamics: The Study of Energy, Entropy, and Equilibrium*, 452. Englewood Cliffs, NJ: Prentice-Hall.
7. Balzheiser, R.E., Samuels, M.R., and Eliassen, J.D. (1972). *Chemical Engineering Thermodynamics: The Study of Energy, Entropy, and Equilibrium*, 453. Englewood Cliffs, NJ: Prentice-Hall.

34

CASE STUDIES ON THE USE OF DISTILLATION IN THE PHARMACEUTICAL INDUSTRY

Laurie Mlinar, Kushal Sinha, Elie Chaaya, and Subramanya Nayak
Process Research and Development, AbbVie Inc., North Chicago, IL, USA

Andrew Cosbie
Drug Substance Technologies and Engineering, Amgen Inc., Thousand Oaks, CA, USA

34.1 INTRODUCTION

Within the pharmaceutical industry, distillation is generally utilized to efficiently remove one or more solvents or to exchange solvents in preparation for isolation of an active ingredient or intermediate. Distillation modeling can also be a beneficial *in silico* tool for appropriate solvent selection and can significantly reduce the number of laboratory experiments required to optimize distillation parameters. Calculating and incorporating the heat transfer coefficients for each reactor in the plant strengthen the utility of a distillation model. In this chapter, the advantages and disadvantages of various modes of distillation along with practical considerations for implementing and optimizing distillation scale-up will be discussed. Several examples of how distillation models have both provided deeper insight into a distillation process and enabled process design and scale-up will be provided. While distillation modeling can be used to fundamentally understand and design a robust process, the limitations of these models and processes must also be considered. In the following case studies, examples of using UNIFAC- vs. nonrandom two liquid (NRTL)-based models, enhancing distillation models with laboratory data, selecting distillation solvents to optimize distillation time and solvent usage, designing space optimization to control distillation time, and using distillation as a means of crystallization will be discussed.

In addition to the case studies presented within this chapter, a detailed review of key chemistry and chemical engineering concepts as they relate to distillation has been presented in Chapter 33.

34.2 INTRODUCTION TO DISTILLATION

Distillations are largely performed in the pharmaceutical industry for concentration, solvent exchange, azeotropic drying, and solvent recovery. In this section we will only target single-stage batch distillation without reflux.

34.2.1 Equipment Used for Distillation

Laboratory distillations are generally performed under vacuum using a rotary evaporator, also known as a roto-vap, due to its ease of operation. The roto-vap consists of a heating bath, roto-vap ball or round-bottom flask, condenser, receiver, and vacuum pump. This assembly helps to evaporate solvent and condense it into the receiving vessel. In a manufacturing plant, due to large volume requirements, distillations are performed using processing vessels equipped with an agitator, condenser, and heating/cooling and vacuum capabilities that can accommodate both vacuum and atmospheric distillations.

Chemical Engineering in the Pharmaceutical Industry: Active Pharmaceutical Ingredients, Second Edition.
Edited by David J. am Ende and Mary T. am Ende.
© 2019 John Wiley & Sons, Inc. Published 2019 by John Wiley & Sons, Inc.

34.2.2 Modes of Distillation

The first decision to be made when performing a distillation is to determine whether it will be carried out under vacuum or atmospheric pressure. Both modes of distillation have their pros and cons as shown in Table 34.1.

Generally, in the laboratory setting, the use of distillation to solvent exchange or azeotropically dry the product to remove water is performed by distilling the product to "dryness." In this case, the solvent is completely stripped and the product precipitates as solids in the roto-vap. The new solvent will then be added to the precipitated product and redissolved. In the plant setting, distilling to "dryness" is operationally challenging; therefore, these distillations are performed using either constant volume distillation (CVD) or put-and-take. In a constant volume distillation, the batch is concentrated to a desired volume (V_{conc}), and the volume is maintained throughout the unit operation at V_{conc} by continuously adding fresh solvent at the same rate of solvent removal, until the desired residual solvent levels are achieved. In a put-and-take distillation, the batch is concentrated to a desired volume (V_{conc}), and then a shot of fresh solvent (V_{add}) is added, which will increase the volume. The batch is again distilled to V_{conc}. This put-and-take process is repeated until the desired target level of residual solvent is achieved. The value of V_{conc} depends on factors such as the minimum stir volume, solubility of the product, and heat transfer area. The most common methods to call the end of distillation are done via gas chromatography or Karl Fischer analysis (KF, for residual water). The advantages of CVD over traditional put-and-take distillation are lower amounts of solvent needed, smaller vessels, and potentially less time. However, CVD requires extra automation and controls such as a level indicator in the distillation vessel and a flow controller on the solvent charging line that can communicate with each other. The performance of CVD vs. put-and-take is a function of several factors, namely, jacket temperature, pressure, V_{conc}, vessel heat transfer, and V_{add}, for put-and-take distillation. While designing and optimizing a distillation process, each of these factors must be considered.

34.2.3 Distillation Startup

Below, a typical procedure for startup of a distillation vessel is described.

> Verify that the condenser is on and at the desired temperature. The desired condenser temperature should be significantly lower than the boiling point and above the freezing point of the solvents to avoid freezing the solvent and clogging the condenser lines. For example, when distilling water, the condenser temperature should be set to approximately 2 °C.
>
> Set up the distillation tank and receiver for distillation but leave the vapor line closed on the distillation tank at this time.
>
> Set up a means to charge fresh solvent to the *dip tube* of the distillation tank. Charging fresh solvent subsurface through the dip tube reduces flashing of the solvent prior to mixing with the main batch, which minimizes wastage.
>
> At the operating pressure, make sure that the distillation tank is cooled below the boiling point of the solvent to avoid yield loss due to excessive vaporization of the product solution.
>
> Apply cooling to the receiver; if possible use the same temperature as the condenser.
>
> Lower the receiver pressure to the desired vacuum level. When the receiver reaches the desired vacuum, slowly open the vapor line on the distillation tank.
>
> When the desired pressure is achieved in the distillation tank, start ramping the jacket temperature up to the target value until the desired distillation rate is achieved.
>
> *NOTE:* This startup procedure is to control foaming of the batch during startup. If foaming occurs and is getting close to the top of the distillation tank, slowly apply N_2 to keep it from entering the vapor line. This may take several minutes until no more N_2 is needed.
>
> Distill the batch to the desired volume and charge fresh solvent to the distillation vessel until desired residual solvent endpoint is achieved. This can be done via put-and-take or constant volume distillation.

34.2.4 Practical Considerations for Scale-Up

1. Know the heat transfer on the distillation vessel (mixing can enhance heat transfer).
2. Know the condenser capacity.
3. Know the vapor–liquid equilibrium (VLE) for the desired solvent system.
4. Create a checklist for the setup.

TABLE 34.1 Comparing Distillation Modes

Atmospheric	Vacuum
Used when high temperature is needed during distillation mostly to avoid precipitation	Lower temperature distillation can be carried out for temperature sensitive compounds
Higher temperature on the condenser can be used (water cooled, no need for heating and cooling skids)	Lower jacket temperature on distillation tank can be used (no need for high boiling heat transfer fluid on jacket)
Requires less energy input when distilling low boiling solvents	Higher solvent removal rate but with a higher chance of foaming and yield loss due to bumping
Less capital and operating cost (no need for vacuum skids)	Energy efficient when distilling at lower jacket temperatures due to less heat loss to surrounding

5. Know the product's solubility to avoid precipitation of solids during distillation, unless it is desired or is not a concern.
6. Use a secondary condenser when distilling low boiling compounds to avoid having vapors in the vacuum lines and thus reduce emission of vapors to atmosphere.

As a general consideration, the maximum jacket temperature should be set below the temperature where the product is deemed stable.

34.3 DISTILLATION MODELING

VLE is a key thermodynamic element in the design and modeling of distillations. By understanding the phase compositions and temperature vs. pressure relationships, operations can be designed and optimized to yield the desired endpoint composition while minimizing solvent usage and processing time. The VLE of a solvent system can be experimentally determined but requires a significant amount of experimentation and sample analysis to sufficiently map out the system across a range of pressures and compositions. Alternatively, VLE models can be used to explore a design space and simulate distillations. These models employ the use of activity coefficients to describe the nonideality of the liquid mixture. There are several thermodynamic models that can be used to calculate VLE, including Wilson [1], UNIQUAC [2], Van Laar, and Margules, but the more commonly used methods for modeling VLE are UNIFAC [3] and NRTL [4].

34.3.1 Comparing UNIFAC vs. NRTL

The UNIFAC method is semi-empirical and is based on functional group contributions. An advantage of using this method is that it can be applied to multicomponent systems where no experimental data is available. One disadvantage is that the functional group contributions in the model may lack accuracy, particularly at the edges of the VLE diagrams or when the molecules of interest have a large number of functional groups. Also, UNIFAC VLE models may not accurately predict immiscibility of components or the existence of azeotropes.

The NRTL model is empirically derived and uses binary interaction parameters (BIPs) to calculate activity coefficients. These BIPs are regressed directly from experimental data sets for the components of interest, typically resulting in a more accurate model compared to UNIFAC. A large library of experimental data and regressed BIPs exists in various sources including DECHEMA and DIPPR. The downside is that BIPs or experimental data for the specific system of interest are needed, though not always readily available. For process design, NRTL should be used if data is available. In both cases (NRTL and UNIFAC) the accuracy of the model should be verified in the lab.

To demonstrate the difference between the activity coefficient models, the VLE for a solvent system consisting of a mixture of tetrahydrofuran (THF) and water are plotted as binary T-xy diagrams in Figures 34.1 and 34.2. The VLE predicted by these two models show significant differences, particularly between 80 and 100 wt % THF. At 1 bar, the NRTL model predicts a minimum boiling point azeotrope at 95.3 wt % THF, where the UNIFAC model does not predict the existence of an azeotrope at all. The effect of this difference between the models can have a significant impact on the accuracy of a distillation design or simulation.

34.3.2 *In Silico* Distillation Design

Advancements in *in silico* distillation modeling have made it possible to design and scale-up a distillation process even in the absence of laboratory data. Over the last few decades,

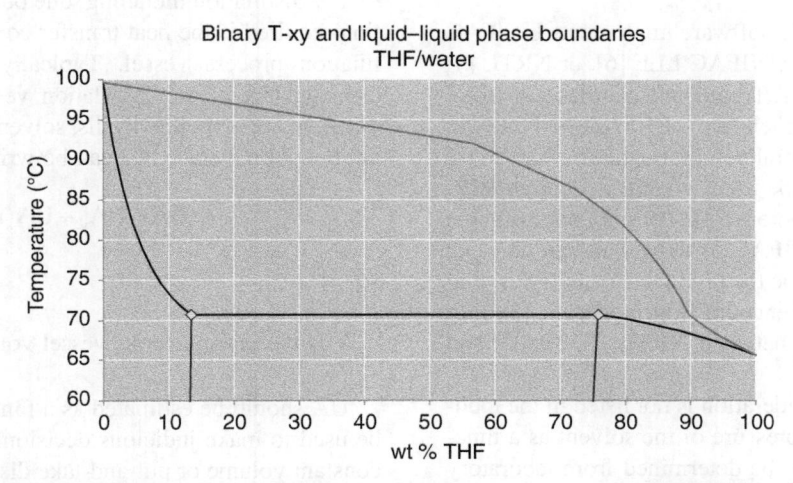

FIGURE 34.1 VLE of THF/water at 1000 mbar, UNIFAC model. *Source*: From DynoChem VLLE utility.

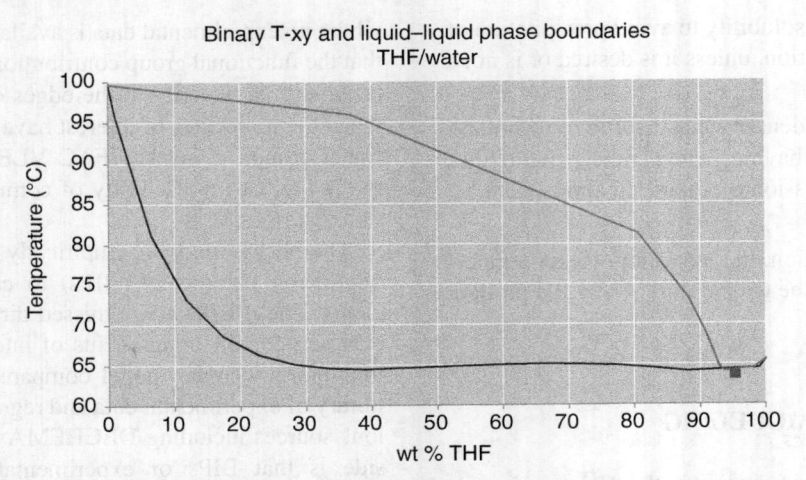

FIGURE 34.2 VLE of THF/water at 1000 mbar, NRTL model with square marker at location of azeotrope. *Source*: From DynoChem VLLE utility.

multiple software tools have aided in the process such as ASPEN, COSMOthermX, gPROMS, and DynoChem, among others. As outlined in Chapter 33, the following key parameters should be obtained for distillation design: (i) vapor pressure vs. temperature data and/or VLE data, (ii) solvent properties, and (iii) the heat transfer coefficient, UA, of the distillation vessel as a function of volume or details of vessel geometry. VLE data can be obtained, for example, via COSMOthermX or ASPEN as outlined in the Appendix 3.A, from DynoChem, or by internally developed tools. Most software provide a vapor–liquid and liquid–liquid equilibrium tool (VLLE tool) which contains the data for most commonly used solvents in the pharmaceutical industry. These tools are designed to generate thermodynamic phase diagrams for binary and ternary component systems. They calculate both VLE and liquid–liquid equilibrium (LLE) using standard activity coefficient methods to account for liquid-phase nonideality. Most of these models treat the gas phase as an ideal gas as the components are common solvents at low pressure.

Most commonly available software implements UNIFAC [3], modified UNIFAC [5], UNIFAC LLE [6], or NRTL [4] for the liquid-phase behavior. To model a distillation, two of the most popular VLE models are UNIFAC and NRTL. NRTL predictions are generally more accurate than those from UNIFAC and should be used when the required BIPs are available. In the case where NRTL BIPs are not available for solvents of interest, UNIFAC or its variant can be used; however, one should examine the predicted VLE T-X,Y diagram carefully. Readers are encouraged to see Refs. [3–6] to understand the differences between NRTL, UNIFAC, and other UNIFAC variants.

If the solvent under consideration is not listed in the modeling tool, saturated vapor pressure of the solvent as a function of temperature needs to be determined from laboratory experiments. Alternatively, vapor pressure versus temperature data can be obtained from the DIPPR or DECHEMA database. Vapor pressure data is used to regress Antoine coefficients (A, B, and C; see Section 33.1.1.4), which are required to calculate NRTL BIPs. With the knowledge of Antoine coefficients and VLE data obtained from a database or experiments, NRTL BIPs can be fitted. Other relevant solvent properties such as density and heat capacity can be estimated from the DIPPR database as well. These solvent parameters can be taken back to the distillation model to design and optimize distillation processes.

In any VLLE tool used, special care should be taken to inquire for the presence of azeotropes. Azeotropes are not distillation boundaries for solvent swap distillation as fresh solvent is added to the solution. Continual solvent addition aids in pushing through the azeotrope boundary. However, when distillation is used to concentrate a solution and fresh solvent is not added, the distillation cannot move beyond the azeotrope composition, if one is present.

For distillation modeling, one other key piece of information required is the heat transfer coefficient (UA) of the distillation process vessel. Typically, models available can estimate UA of the distillation vessel if geometric details of the vessel, impeller details, solvents used, and service fluid are known. Overall, UA can be written as

$$UA = UA_0 + UA(v) \times V, \qquad (34.1)$$

where

V is the instantaneous vessel volume.

UA should be estimated as a function of volume that can be used to make judicious decisions on holding volumes in constant volume or put-and-take distillations. To get a better estimate of UA, an engineering run can be performed using the plant distillation vessel with process-relevant solvent. At any specified fill volume, the solution temperature and jacket temperature need to be recorded as a function of time; adequate mixing must be achieved in the reactor. The same experiment is repeated at different fill volumes. The experimental data should be fit to Eq. (34.1) to estimate UA as a function of volume.

EXAMPLE PROBLEM 34.1 Solvent Swap from Tetrahydrofuran (THF) to Isopropyl Alcohol (IPA)

Use a modeling tool to simulate a solvent swap from 889 kg (1000 L) of THF to IPA such that the final solution contains less than 1 wt % of THF. An upper limit on the operating temperature is set to 45 °C to keep impurity generation below a specified limit. (a) Determine the operating pressure and then predict the quantity of fresh solvent required and distillation time for CVD vs. put-and-take distillation modes. Also investigate the impact of jacket temperature. (b) Investigate distillation performance as a function of the concentrated volume, V_{conc}, and put-and-take volume, V_{add}, for CVD and put-and-take processes.

Solution
In this example, DynoChem 4.1 was used to model the solvent swap distillation. Open DynoChem VLLE utility and choose IPA and THF as solvents. Notice that for the IPA/THF system in DynoChem, NRTL BIPs are not available. The procedure outlined in Section 34.3.2 can be used to obtain NRTL BIPs either through experiments or through models. In Section 34.4.1, a case study is presented where experiments were a preferred route to obtain NRTL BIPs. For the current example, the modified UNIFAC model was selected. Operating pressure was adjusted to maintain a liquid-phase temperature below the maximum operating temperature of 45 °C during the entire operation; this yields a pressure of 180 mbar as shown in Figure 34.3. In Figure 34.3 and subsequent VLE plots, top, and bottom curves are for gas and liquid phases, respectively. In plant operations, pressure can fluctuate. Thus, an operating pressure below the maximum pressure is chosen keeping engineering safety margins during operation. It should be noted that the boiling point of a solution mixture will decrease at lower operating pressure. This will lead to a faster distillation rate at a constant jacket temperature. An example demonstrating this effect has been presented in the case study discussed in Section 34.4.3. In this example, we will select an operating pressure of 140 mbar to demonstrate the impact of different jacket temperatures. The jacket temperature chosen should be above the mixture boiling point. At 140 mbar operating pressure, the highest mixture boiling point is 39.7 °C, which allows for study of jacket temperature between 40 and 45 °C.

After obtaining VLE behavior and operating pressure, the DynoChem solvent swap model for two miscible liquids is used to design the distillation. For partially miscible and immiscible solvents, ensure that the model takes care of deviations from ideality. In the solvent swap model, keep the operating pressure at 140 mbar and the feed tank at 20 °C. Ensure the feed tank IPA amount is greater than what is needed for the distillation to avoid a negative value. The distillation vessel UA can either be calculated through a DynoChem utility or through an engineering run and is entered in the worksheet. For the current example, use UA of 60 W/K and UA(v) of 0.78 W/L·K as determined using DynoChem utility for UA estimation. Concentrated volume

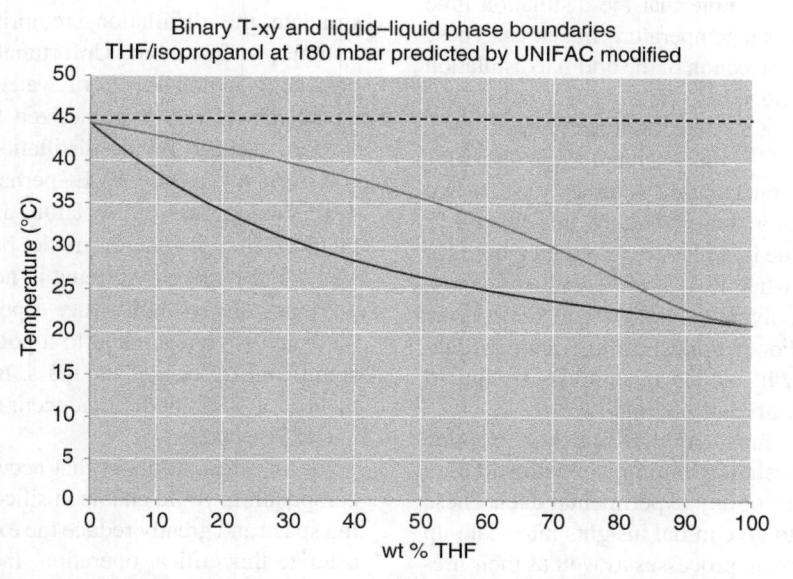

FIGURE 34.3 VLE of isopropanol/THF at 180 mbar using the modified UNIFAC model. *Source*: From DynoChem VLLE utility.

TABLE 34.2 Comparison of Distillation Time and Fresh Feed Required for Different Distillation Modes and Jacket Temperatures

Scenario	Distillation Mode	Jacket Temperature (°C)	V_{add} (L)	Total Distillation Time (h)	No. of Put-and-Takes	Volume of Fresh IPA Added (L)
1.	CVD	40	—	44	—	135
2.	Put-and-take	40	50	48	4	177
3.	Put-and-take	40	100	56	2	207
4.	CVD	45	—	21	—	135
5.	Put-and-take	45	50	23	4	205
6.	Put-and-take	45	100	23	2	220

TABLE 34.3 Comparing the Impact of Concentrated Volume with Different Distillation Modes

Scenario	Distillation Mode	V_{conc} (L)	V_{add} (L)	Total Distillation Time (h)	No. of Put-and-Takes	Volume of Fresh IPA Added (L)
1.	CVD	150	—	24	—	273
2.	Put-and-take	150	50	23	6	328
3.	Put-and-take	150	100	24	4	391
4.	CVD	300	—	25	—	542
5.	Put-and-take	300	50	24	12	589
6.	Put-and-take	300	100	25	6	620

of 75 L is chosen for CVD and put-and-take, and two different V_{add} volumes are explored in Table 34.2. Table 34.2 shows the results for the two distillation modes and various jacket temperatures.

As we can see from Table 34.2, put-and-take distillation mode requires a larger quantity of fresh solvent as well as time to achieve the same distillation endpoint. However, as noted in Section 34.2.2, CVD requires automation and additional manufacturing controls. The difference in performance between CVD and put-and-take will be minimal as put-and-take-added volume V_{add} is reduced. Also, note that the distillation time reduces significantly as jacket temperature increases. These studies can be used to guide selection of the optimal distillation mode and jacket temperature.

For part (b), the DynoChem model built earlier is utilized. Jacket is maintained at 45 °C in this example. New scenarios are explored for different concentrated volumes V_{conc} in two different distillation modes and are tabulated in Table 34.3.

Table 34.3 shows that the fresh IPA required increased for both modes of distillation when V_{conc} is increased. In general, distillation time and fresh solvent required will increase as the level of V_{conc} increases. For put-and-take operation, reduction in the volume V_{add} will reduce fresh solvent required and will lead to higher number of put-and-takes.

Distillation modeling can be used to quickly examine numerous scenarios and design and optimize the distillation process even in the absence of any experimental data. These models can be employed to give initial insights into scale-up and tech transfer of distillation processes as well as their life-cycle management.

34.4 CASE STUDIES ON THE USE OF DISTILLATION MODELS

34.4.1 Using Lab Data to Enhance a Distillation Model

A distillation was designed to reduce the concentration of water in a solution of THF and a dissolved substrate from 10 to less than 2 wt %. This reduction in water content can be achieved by concentrating the batch, followed by diluting with fresh solvent. The NRTL model was used to design and simulate the distillation, requiring 4 put-and-takes of 7 volumes at 200 mbar. Unfortunately, the simulations were predicting higher levels of water than what was observed in the lab. It was hypothesized that at high concentration (toward the end of the distillation) the dissolved substrate was influencing the VLE, perhaps due to an affinity for THF. Many times, nonvolatile, dissolved substrates are not included in a distillation model because their impact on the VLE of the solvent systems is negligible. This assumption does not always hold true, especially in processes where the batch is concentrated to a point where the substrate is a significant fraction of the mass. In these cases, the development of a VLE model that accurately predicts the system at hand is desirable.

In this case, a model that accurately predicts the solvent composition would enable in silico exploration of the operating space and greatly reduce the experimental burden to characterize this critical operation. In order to develop a model more representative of our specific system, a single batch distillation experiment was conducted at laboratory bench scale with a representative mixture of the substrate dissolved in wet THF (85 wt % THF, 15 wt % water on a solvent basis). The batch was heated to reflux and sampled, both batch and distillate, for solvent composition and the batch temperature recorded. The batch was concentrated intermittently to change the composition by directing distillate to a receiver and then sampling after re-equilibration. This cycle was repeated until the batch composition reached 99 wt % THF on a solvent basis. It is important to collect batch and distillate stream samples at the same time, as these pairs of data points are needed to construct the T-xy diagram. If automated sampling tools are available, this can enable sampling the two streams simultaneously. For distillations performed at atmospheric pressure, distillate samples can be taken directly from the condenser outlet. For distillations performed under reduced pressure, the distillate stream can be directed through a series of valves between the condenser and distillate receiver, allowing for isolation of a small amount of the distillate stream without breaking vacuum on the entire system. Distillate stream samples should not be taken directly from the distillate receiver, as they are not representative of the vapor phase at the given time point, but an accumulation of the distillate stream.

The plot in Figure 34.4 shows the data captured in this experiment. This data was then used to regress NRTL binary interaction parameters that could be used in further simulations of the distillation. These values are captured in Table 34.4 and compared with the values included in the DIPPR database.

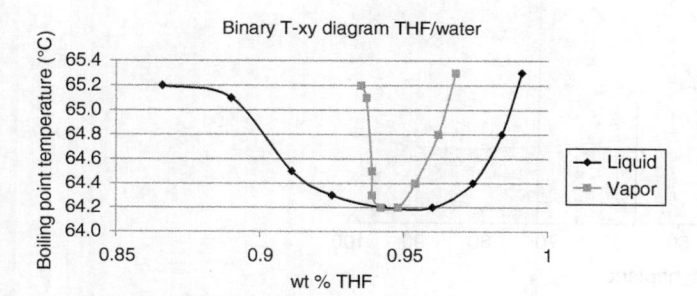

FIGURE 34.4 VLE plot at 1000 mbar for THF/water with dissolved substrate present, from experimental data.

TABLE 34.4 Binary Interaction Parameters for THF/Water Solvent System

NRTL Model BIPs	Δg_{12} (cal/mol)	Δg_{21} (cal/mol)	Alpha
From database	915.7	1725	0.4522
Regressed from experimental data	916.0	1869	0.4472

The distillation was modeled using both the BIPs from the database and again with the experimentally fit BIPs. Using the database values, the model predicts that the batch will have 97.6 wt % water remaining under the previous outlined conditions, failing the specification. Using the experimentally fit BIPs, the model predicts the batch will have 98.7 wt % water remaining under the same process conditions, meeting the specification and aligning with experimental results. While that difference may not seem very significant, small amounts of water carried forward in a process can have significant impact on solubility or reactivity in downstream operations and may need to be carefully controlled. Refinement of the VLE model using experimental data enabled more accurate model predictions to be made and utilization of the model for in silico experimentation.

34.4.2 Selection of Solvent Pairs with Azeotropes to Improve Efficiency of Solvent Removal

A distillation process was being used to remove isobutanol and isopropanol from a solution of substrate dissolved in THF. Since residual isobutanol and isopropanol contributed to the formation of impurities in the subsequent reaction, each solvent needed to be reduced to very low levels (≤2 wt % total).

In this situation, the batch was concentrated by distillation to a minimum agitation volume, while the boiling point temperature was within the thermal stability range of the substrate. This was followed by addition of fresh THF and repeated until the desired isobutanol and isopropanol levels in the batch were obtained. In practice, 10 put-and-takes with THF were required to reach the desired levels.

From examination of the T-xy diagram for THF and isobutanol at 50 mbar pressure (Figure 34.5), it becomes apparent that removing isobutanol from a solution of THF is inefficient. THF is much more volatile than isobutanol, so the vapor phase that is removed and condensed during a batch distillation is enriched in THF (as opposed to the preferred isobutanol). This mismatch in relative volatility of the desired solvent and undesired solvent is why 10 put-and-takes are necessary to arrive at the desired endpoint.

In order to optimize the process, the use of an alternate solvent that forms an azeotrope with isobutanol was explored. The existence of a minimum boiling point azeotrope between isobutanol and another solvent (n-heptane in this case) facilitates the efficient removal of the less volatile component. When the batch composition lies on the right-hand side of the azeotrope (see Figure 34.6), the vapor phase is enriched in isobutanol and results in greater removal of isobutanol compared to the THF/isobutanol system. This phenomenon can be observed between many organic solvents and water, allowing for efficient drying of organic solvent systems by batch distillation, even though the organic solvent is oftentimes the more volatile component. For the process discussed in this case study, the addition of n-heptane reduced the

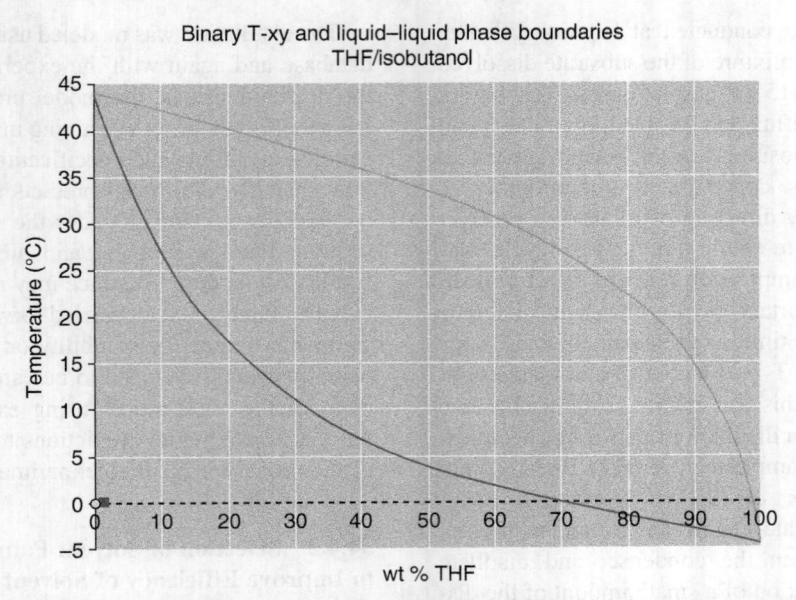

FIGURE 34.5 VLE for THF/isobutanol at 50 mabr. *Source*: From DynoChem VLLE utility.

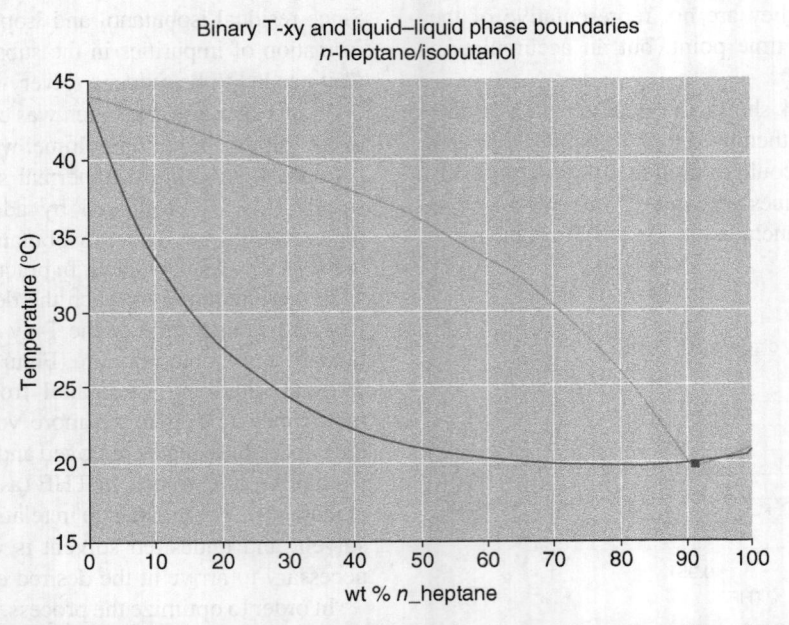

FIGURE 34.6 VLE for *n*-heptane/isobutanol at 50 mabr. *Source*: From DynoChem VLLE utility.

required number of put-and-takes from ten to two and resulted in lower levels of isobutanol and isopropanol than the original process. It should be noted that *n*-heptane was determined to be a suitable cosolvent to THF in the subsequent reaction, allowing for its use in this process.

34.4.3 Optimization of Distillation Parameters to Enable Efficient Solvent Removal

In this case study, a solution of water-saturated 2-methyl tetrahydrofuran (2-me THF) was being distilled to efficiently remove water from the system, which included a dissolved substrate. The presence of water combined with sufficiently high temperatures and long distillation times could lead to the formation of impurities. More broadly, undesired residual solvents present during isolation of an active pharmaceutical ingredient or intermediate can oftentimes, if present in sufficient quantities, lead to the formation of an undesired crystal form, impact particle size distribution following isolation, result in a difference in purification characteristics, or even form a new impurity. In this case study, distillation modeling was conducted to understand the impact of distillation parameters such as temperature and pressure on distillation time to achieve acceptable residual water content. Using this approach, the design space for the distillation was identified.

Figure 34.7 shows the results of distillation modeling performed using DynoChem. In this example, a range of pressures and temperatures were investigated and the resulting time to achieve the desired water content was calculated. Given a maximum distillation time of 30 hours, the distillation model helps define the optimal temperature and pressure and predicts the distillation time for a given temperature and pressure. In order to model the distillation time, it is necessary to calculate UA for the operating vessel (see Section 34.3.2 for more details). In this case study, a put-and-take distillation method was employed with a maximum solution temperature of 55 °C. The lowest obtainable pressure for the system under consideration was 100 mbar.

It is well known that a lower operating pressure and a higher temperature during a batch distillation will result in a shorter distillation time when removing a trace solvent via a chase distillation. Distillation modeling in this case study, however, provided bounds for the pressure and temperature needed in a given manufacturing vessel, which would otherwise only be achievable by testing these parameters at scale. Using the predicted bounds, minimal experiments could be done to verify the boundaries of the design space where the process would be routinely operated. Equipment capabilities, such as the required temperature and pressure as well as the distillation mode, were also identified via the model.

34.4.4 Distillation as a Means of Crystallization (Evaporative Crystallization)

When tightly controlled crystallization is not a requirement for isolation, as is often the case with starting materials or early intermediates, crystallization via distillation can result in cycle time reduction. It is common to see both distillation and crystallization unit operations in the isolation process because the solvent used in the reaction may not be the desired solvent for the crystallization. To crystallize while distilling, also known as evaporative crystallization, solvent must be continuously removed until the solution is supersaturated. By carefully controlling the supersaturation ratio (Eq. 34.2) during distillation, crystallization kinetics can be controlled (to a limited extent), and particle growth can be achieved. It is important, however, to first understand the solubility of the solid mass throughout the solvent compositions encountered during the distillation so that the supersaturation ratio at any point in time can be calculated. The distillation rate can therefore be established, which results in the desired purity and physical properties attributes.

$$\text{SSR} = \frac{\text{Concentration of solid in solvent}}{\text{Solubility of solid in solvent}} \quad (34.2)$$

Supersaturation ratio (SSR)

Evaporative crystallization can be modeled in DynoChem to provide additional insight into the crystallization process, optimize distillation parameters, and enable the design of a robust distillation and crystallization process. An example output from a DynoChem model for distillation as a means of crystallization is provided in Figure 34.8. The parameters chosen are those of a typical solid particle crystallizing from a mixture of 2-me THF and water. The results shown are for a put-and-take simulation where water is being chase-distilled with an excess of 2-me THF. In this example, the solute was supersaturated at the beginning of the distillation process. Only a portion of the distillation is shown to provide an example of the output from an evaporative crystallization

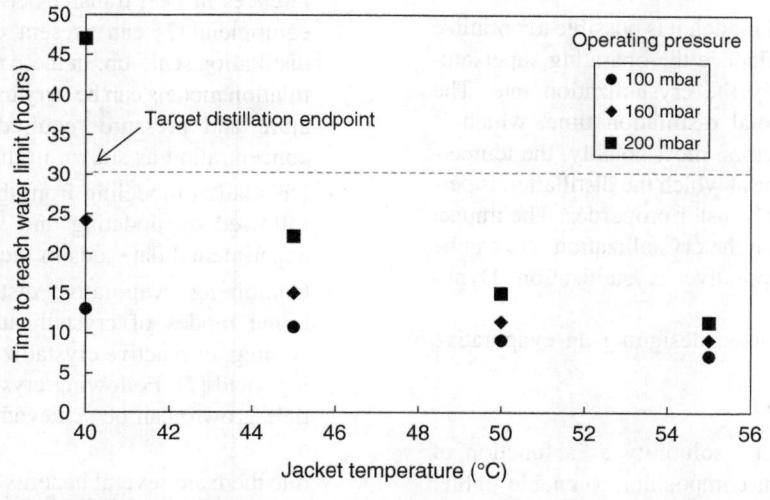

FIGURE 34.7 Distillation modeling to understand the temperature and pressure ranges that achieved a desired water limit in the user-defined time frame. *Source*: Model built in DynoChem.

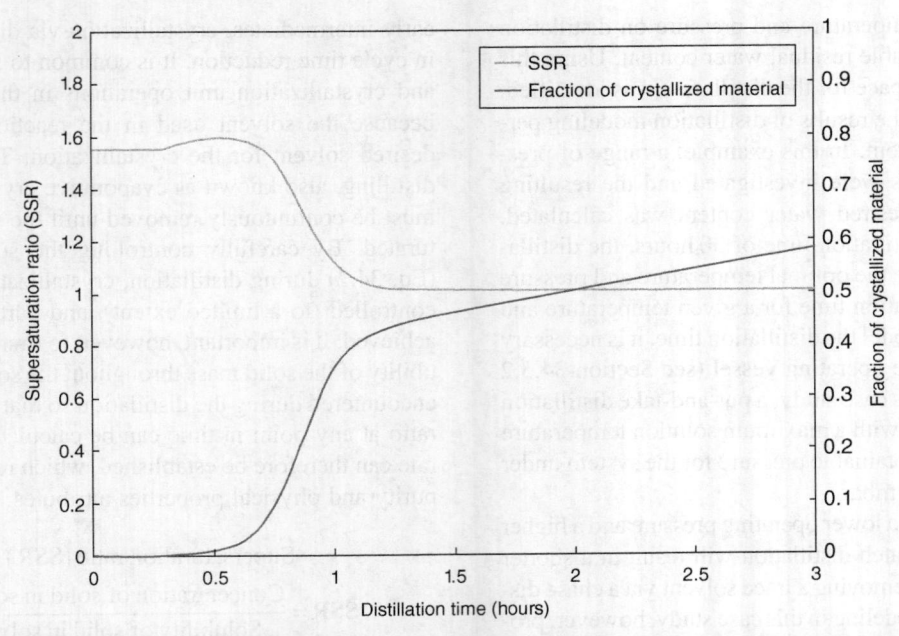

FIGURE 34.8 Results from a DynoChem simulation of evaporative distillation of a model compound.

model. To model an evaporative crystallization process, solubility at numerous solvent compositions and temperatures should be obtained at the outset. The resulting solubility equation can be incorporated into the evaporative crystallization model. While multiple outputs can be used to assess the progression of the crystallization, two examples are shown in Figure 34.8. Monitoring the supersaturation ratio (SSR, calculated by applying Eq. (34.2) to the output data) with time *in silico* provides a means to understand the crystallization driving force. Modeling the fraction of solute crystallized provides a measurement of the expected yield. The SSR at any given time and the resulting fraction of solute that has been crystallized can be changed by varying the parameters discussed below.

After creation of the initial model, it is possible to optimize the distillation process to influence the operating supersaturation ratio and consequently the crystallization rate. The model also calculates the total distillation time, which is impacted by a number of factors, most notably, the temperature, the pressure, the volume at which the distillation is performed, and the vessel's heat transfer properties. The impact of each of these parameters on the crystallization rate can be predicted through the evaporative crystallization DynoChem model.

Practical considerations when designing an evaporative crystallization process:

- Understand the product's solubility as a function of temperature and solvent composition to enable tighter control of the crystallization process.

- Carefully design the distillation/crystallization vessel, agitator, and agitation protocol considering the change in volume expected in the reactor during the course of the distillation. Understanding the impact of shear on particle attrition can aid in selecting the appropriate agitator design and agitation rate.

- The ability to make small changes to temperature and pressure in the distillation vessel enables control of the distillation through highly supersaturated regimes. Scale-up from laboratory to the plant is nontrivial for evaporative distillation processes as the mechanisms of temperature and pressure control between the laboratory and plant, in particular, are distinctly different. Differences in heat transfer between laboratory and plant equipment [7] can present challenges in evaporative distillation scale-up. In these scenarios, evaporative distillation models can be very useful to predict the temperature and pressure required to achieve the desired concentration as shown in this case study.

- Distillation modeling from the onset of process design followed by updating and verifying the model with experimental data aids in the utility of the model.

- Combining evaporative distillation with other traditional modes of crystallization, such as antisolvent, cooling, or reactive crystallization, can aid in maximizing yield [7]. Following crystallization, additional particle growth can be achieved via thermal ripening.

While there are several benefits to evaporative distillation, it is important to note that there may be challenges to this approach. The inherent inability to tightly control the distillation and therefore crystallization parameters could lead to uncontrolled crystallization. Controlling the nucleation rate is difficult due to the inhomogeneity of supersaturation as a result of temperature gradients throughout the vessel [7]. Due to this effect combined with volume reduction as a result of distillation, particles can form a crust on the reactor wall at the solid–liquid boundary as the volume of liquid in the crystallizer vessel is decreasing. This material is subject to high temperatures for an extended period of time and can result in unstable material that can decompose or result in impurity formation if the jacket temperature is not appropriately chosen [7]. Given the challenges that can be encountered to control the nucleation rate as well as particle encrustation in the reactor, attaining consistent physical properties from batch-to-batch can be difficult. If tight control of the product's physical properties is required, evaporative distillation may not be a good option. If the resulting isolated material following evaporative crystallization can meet the physical properties' requirements and leads to the desired yield and purity, combining distillation and crystallization can simplify the isolation process and reduce the isolation cycle time.

ACKNOWLEDGMENTS

We would like to thank Shailendra Bordawekar and Nandkishor Nere for the opportunity to author the chapter, Samrat Mukherjee and Moiz Diwan for critically reviewing sections of the chapter, Mathew Mulhern for discussions on distillation parameter selection, and the AbbVie API Pilot Plant for the exceptional execution and development of best practices for distillations.

Laurie Mlinar, Kushal Sinha, Elie Chaaya, and Subramanya Nayak are employees of AbbVie. The design, study conduct, and financial support for this research were provided by AbbVie. AbbVie participated in the interpretation of data, review, and approval of the publication.

REFERENCES

1. Wilson, G.M. (1964). Vapor-liquid equilibrium. XI. A new expression for the excess free energy of mixing. *Journal of the American Chemical Society* 86 (2): 127–130.
2. Abrams, D.S. and Prausnitz, J.M. (1975). Statistical thermodynamics of liquid mixtures: a new expression for the excess Gibbs energy of partly or completely miscible systems. *AIChE Journal* 21 (1): 116–128.
3. Fredenslund, A., Jones, R.L., and Prausnitz, J.M. (1975). Group-contribution estimation of activity coefficients in nonideal liquid mixtures. *AIChE Journal* 21: 1086.
4. Renon, H. and Prausnitz, J.M. (1968). Local compositions in thermodynamic excess functions for liquid mixtures. *AIChE Journal* 14 (1): 135–144.
5. Magnussen, T., Rasmussen, P., and Fredenslund, A. (1981). UNIFAC parameter table for prediction of liquid-liquid equilibriums. *Industrial and Engineering Chemistry Process Design and Development* 20 (2): 331–339.
6. Gmehling, J., Li, J., and Schiller, M. (1993). A modified UNIFAC model. 2. Present parameter matrix and results for different thermodynamic properties. *Industrial and Engineering Chemistry Research* 32 (1): 178–193.
7. Tung, H.-H., Paul, E.L., Midler, M., and McCauley, J.A. (2009). Evaporative crystallization. In: *Crystallization of Organic Compounds: An Industrial Perspective*. Hoboken, NJ: Wiley.

35

DESIGN OF FILTRATION AND DRYING OPERATIONS

PRAVEEN K. SHARMA
Chemical Development, Tetraphase Pharmaceuticals, Inc., Watertown, MA, USA

SARAVANABABU MURUGESAN* AND JOSE E. TABORA
Product Development, Bristol-Myers Squibb, New Brunswick, NJ, USA

35.1 INTRODUCTION

In this chapter we consider two important unit operations in the pharmaceutical industry, filtration and drying. As part of the manufacturing process of active pharmaceutical ingredients (APIs), these two operations follow naturally from the main mechanism of isolation and purification used in this industry, i.e. crystallization. It is in fact the dominance of crystallization as an isolation and purification technology that dictates the corresponding use of filtration and drying. As unit operations, both filtration and drying have wide applicability in the broad chemical and pharmaceutical industry; however, for brevity and clarity we will limit our discussion here to situations in the pharmaceutical industry in which these unit operations are a continuation of an isolation train that is preceded by a crystallization. In this context the slurry resulting from the crystallization is a mixture of the suspended solids (valuable product) and the supernatant (waste solvent or mother liquor).

The goal of the filtration process is to separate the solids from the supernatant. In many cases, filtration operations are carried out to remove undesirable solid matter when the valuable material is present in the liquid phase. Waste filtration will not be discussed in this chapter, and we will concentrate on filtration as a unit operation to isolate the solid API from the solvent in which it was crystallized.

The process of filtration results in the separation of the solids as a wet cake that requires subsequent washing

*Current address: Product Engineering, Becton Dickinson, Franklin Lakes, NJ, USA

Chemical Engineering in the Pharmaceutical Industry: Active Pharmaceutical Ingredients, Second Edition.
Edited by David J. am Ende and Mary T. am Ende.
© 2019 John Wiley & Sons, Inc. Published 2019 by John Wiley & Sons, Inc.

followed by drying to yield a product of high purity. This is subsequently packaged or processed further (e.g. milling) as a dry powder to complete the manufacturing of the drug substance. As discussed in previous chapters, the chemical manufacturing of an API proceeds via a number of sequential chemical conversions during which some of the intermediates are isolated in the solid form. Although the isolation of all the intermediates may require the implementation of a filtration and drying sequence, these two unit operations acquire extreme importance in the final step when the API is isolated. Both filtration and drying operations play a significant role in the quality and the physical properties of the API.

35.2 FILTRATION

35.2.1 Background and Principles

In the pharmaceutical industry, filtration is the process of separating solids from a supernatant by exposing the slurry to a filter medium (filter cloth, screen, etc.) and applying a pressure gradient across the medium. The medium retains all or a large portion of the solids, and the supernatant flows through the medium producing the separation. For the purpose of this discussion, it is assumed that we are interested in filtering a slurry stream generated from a crystallization, to recover the solids (product) in high purity from the supernatant (waste, often called mother liquor). Typical methods to provide a pressure gradient are applied static pressure, applied vacuum downstream from the separating medium, and centrifugal force. The nature of the filtered slurry and the operating parameters of the isolation equipment dictate the filtration rate, and thereby the cycle time of this unit operation during the process and the properties of the resulting wet cake.

35.2.1.1 Solid–Liquid Separation
The typical solid–liquid separation performed in the pharmaceutical industry can be depicted as a cake filtration or filtration through porous medium.

Cake filtration is performed by forcing slurry against a filter medium whose pore sizes are smaller than the average particle size of the solid particles present in the slurry so that the liquid phase passes through the medium, while the solid phase is retained by the medium. During this process, cake is built up over time on the filter medium (Figure 35.1).

35.2.1.1.1 Filtration Principles
In cake filtration, the resistance to flow increases with time as the cake builds on the filter medium. Initially the only resistance to flow is provided by the filter medium; however, as the cake builds, the cake itself becomes a resistance that must be overcome by the fluid being removed. If the filtration is performed at a constant pressure, the flow rate will decrease monotonically during the filtration. This is called a *constant pressure filtration*. In a less common *constant rate filtration*, the pressure drop is increased gradually to afford a constant flow rate.

In general, the rate of filtrate flow in a pressure-driven filtration can be depicted as shown in Figure 35.2. The rate of filtrate flow (q) is directly proportional to the driving force, pressure differential ($\Delta P = P_1 - P_2$) and inversely proportional to the length (L) of the cake (which in turn is a function of the amount of cake filtered).

For the same amount of filtered slurry, an increase in the filtration area (A) would provide more surface area for the filtrate flow and a reduction in the cake height L. An increase in the filtrate viscosity would adversely affect the filtration rate. Hence, the filtrate flow can be given as follows:

$$q = \frac{dV}{dt} \propto \frac{\Delta P_{Cake} A}{\mu L} \qquad (35.1)$$

where

$q = dV/dt$ = filtration flow rate (m³/s)
ΔP_{Cake} = pressure drop across the cake (Pa)
A = filtration area (m²)
μ = filtrate viscosity (Pa·s)
L = filtration length or cake height (m)

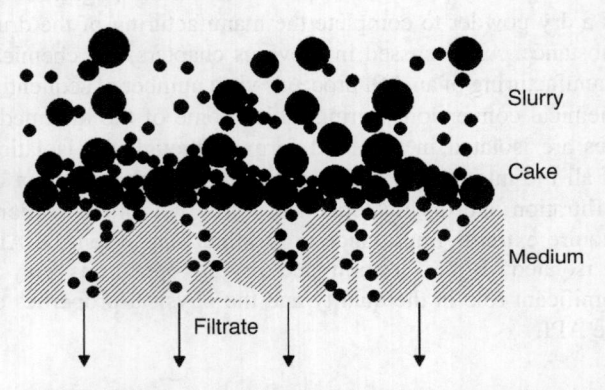

FIGURE 35.1 Cake filtration through porous medium.

Equation (35.1) is called Darcy's law [1] with a proportionality constant k defined as the cake permeability:

$$\frac{dt}{dV} = \frac{\mu}{\Delta P_{Cake} A} \left(\frac{L}{k}\right) \qquad (35.2)$$

where

k = permeability of the cake (m²)

The term (L/k) is equivalent to the cake resistance (R_c) faced by the fluid flow during filtration:

$$\frac{dt}{dV} = \frac{\mu}{\Delta P_{Cake} A} (R_c) \qquad (35.3)$$

where

R_c = cake resistance (m⁻¹)

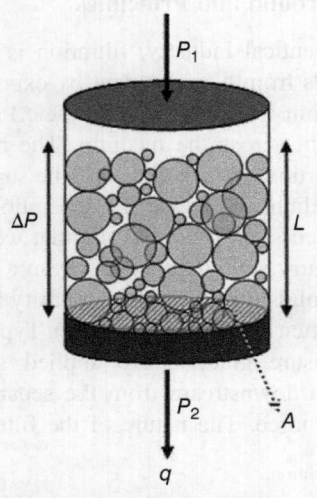

FIGURE 35.2 Schematic of solid–liquid separation.

However, in addition to the cake resistance, the flow also faces the medium resistance ($R_m = L_m/k_m$). These resistances can be presented as resistances in series in determining the rate of filtration (rate at which volume of filtrate is collected) [2]

$$\frac{dV}{dt} = \frac{\Delta P A}{\mu} \frac{1}{(R_c + R_m)} \quad (35.4)$$

where

ΔP is the combined pressure drop (Pa) across the cake and the filter medium.

Specific cake resistance (α) is then given by

$$\alpha = \frac{R_c A}{CV} \quad (35.5)$$

where

C = concentration of slurry (kg/m^3)
V = volume of filtrate collected (m^3)

Specific cake resistance (m/kg) is an intrinsic property of the material being filtered and is a function of the particle size, cake porosity, particle density, and shape [3].

At time zero, the term ($\alpha CV/A$) is zero as the filtration has not been initiated ($V = 0$), and there is no cake built up, and hence there is no resistance due to the cake ($R_c = 0$). The total resistance at this time is due to only the medium resistance (R_m). As the filtration proceeds ($V > 0$), the cake starts to build up, and the value of R_c in Eq. (35.4) increases, thereby resulting in a corresponding increase in the overall resistance and further decreasing the rate of filtration. After a certain point of filtration, R_c dominates and R_m becomes negligible. In general terms, R_c originates from the packing of the particles as the cake forms, restricting the space available for the fluid to flow through. The fluid must then flow through the spaces between the particles with increasing friction, which in turn, for the same pressure drop, results in a decrease in the filtrate flow rate.

Combining Eqs. (35.4) and (35.5) and rearranging,

$$dt = \frac{\mu}{\Delta P A} \left(\frac{\alpha CV}{A} + R_m \right) dV \quad (35.6)$$

Integrating Eq. (35.6) from the start of filtration ($t = 0$, $V = 0$) to an arbitrary time t at which a volume V of filtrate has been collected, we obtain

$$t = \frac{\mu}{\Delta P A} \left(\frac{\alpha CV^2}{2A} + R_m V \right) \quad (35.7)$$

Equation (35.7) can be further rearranged as

$$\frac{V}{At} = \frac{\Delta P}{\mu} \left(\frac{1}{\frac{\alpha CV}{2A} + R_m} \right) \quad (35.8)$$

The term V/At in Eq. (35.8) corresponds to the volume of the filtrate filtered through unit area in unit time and is denoted as the average filtration flux or simply the filtrate flux. Note that the average filtration flux is an inverse function of the amount of the cake collected ($CV = M$).

In some cases, the flow itself influences the packing of the solid particles that may result in an increased resistance. In a filtration process this will be experienced as a change in the specific cake resistance with increased pressure (flow), and the cake is then said to be *compressible*. Specific cake resistance in the case of compressible cakes may be represented by an average-specific cake resistance $\bar{\alpha}$ in the following equation:

$$\bar{\alpha} = \alpha_0 \Delta P^n \quad (35.9)$$

where

α_0 and n are empirical constants.
n is called the compressibility index of the cake.

For incompressible cakes, α is independent of ΔP, and the compressibility index n is 0. If the cake is compressible (dependent on pressure), then n is greater than 0. Typical pharmaceutical cakes have a compressibility index in the range of 0.2–1.

The average filtrate flux profiles of compressible and incompressible cakes are illustrated in Figure 35.3 as a function of the pressure drop across the cake. As seen in the figure, in the case of incompressible cakes ($n = 0$), an increase in the pressure drop results in a corresponding increased filtrate flow, while in moderately compressible cakes ($0 < n < 1$), the filtrate flow monotonically increases but with decreasing slope, and in highly compressible cakes ($n > 1$), the flow reaches a maximum and starts to decline as the pressure is increased further. This is mainly because of the fact that in

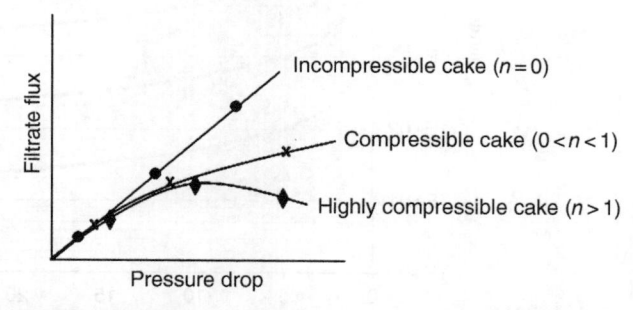

FIGURE 35.3 Filtrate flux vs. pressure drop as a function of the compressibility index. *Source*: Reprinted with permission from Ref. [4]. Copyright 1997, Elsevier.

the case of highly compressible cakes, after a certain point an increase in pressure compresses the cake, filling the interstitial spaces between the particles, thereby reducing the space available for the filtrate to flow through the cake.

Derivation of Eqs. (35.4) through (35.8) assumed that solid particles are deposited on the filter cake and there is no migration of particles across the surface of the cake.

Figure 35.4 shows the profile of a typical pressure drop seen across the cake and medium from the initiation ($t = t_0$) to the completion of filtration ($t = t_f$).

At $t = t_0$, there is no cake, and hence the overall pressure drop is only due to the medium resistance, and at $t = t_{1/2}$, the overall pressure drop is due to the combination of cake and medium resistances with cake resistance contributing to the majority of the pressure drop. At the end of the filtration, the overall pressure drop is mainly due to the cake resistance with minimal contribution from the medium. The linearity and the precise slopes of the profiles of the pressure drops depend on the compressibility of the cake.

From Figure 35.4, at

$$t = t_f, \Delta P = P_a - P_f = (P_a - P_c) + (P_c - P_f) = \Delta P_{cake} + \Delta P_{medium} \quad (35.10)$$

As a result of the increasing resistance over time, the filtration flux decreases over time. The profile of the average filtration flux as a function of total cake formation (mass of cake over unit area of filtration, M/A) and specific cake resistance can be estimated for a given applied pressure differential and viscosity by using Eq. (35.8) as shown in Figure 35.5. A significant reduction in the flux occurs during the initial period of filtration as the cake resistance begins to dominate over the medium resistance. The profiles in Figure 35.5 are estimated for filtration at pressure 20 psi, viscosity of 1.2 cP, and R_m 1×10^6 m^{-1} as an example case.

It is also evident from the figure that reporting the average filtration flux to compare the filtration performances of two different slurries could be misleading, as those two streams could result in the same filtration flux but for different levels of cake formation. A comparison of just the filtration fluxes (e.g. from Buchner funnel filtration) can be performed as long as the filtration pressure and the M/As are similar. Typical ranges for the ratio M/A seen in the laboratory (left-most gray region) and pilot plant (right-most gray region) are also given

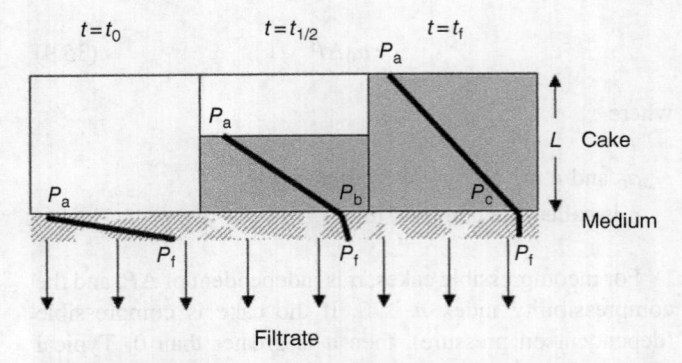

FIGURE 35.4 Pressure drop across the cake during filtration.

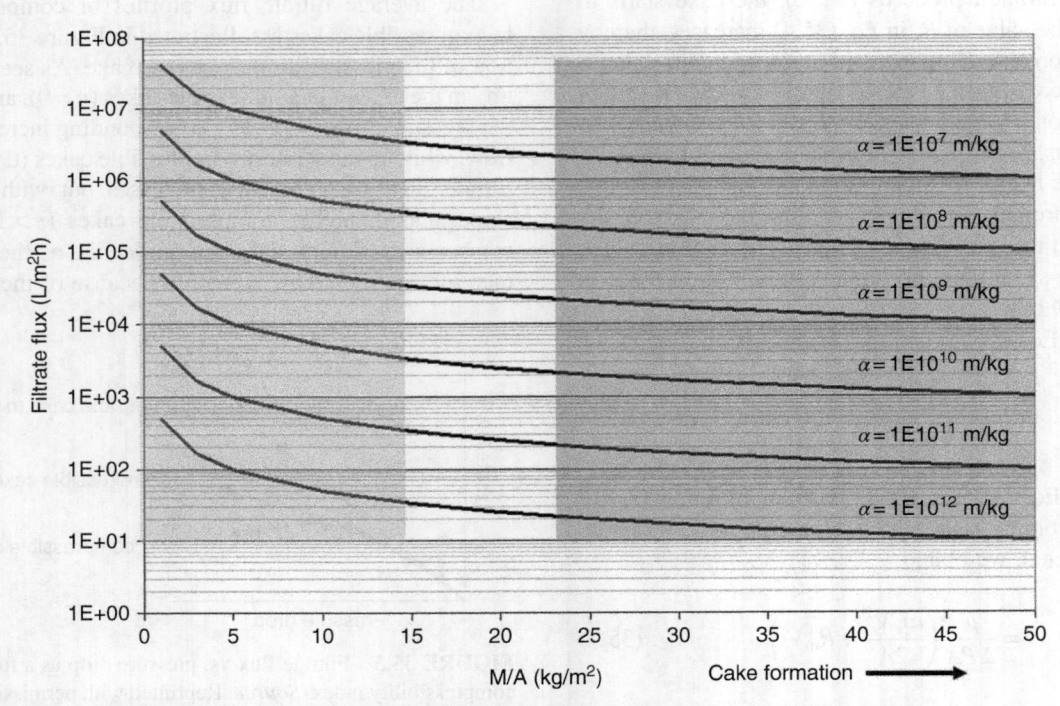

FIGURE 35.5 Filtrate flux profile vs. M/A as a function of α (note the logarithmic scale on the y-axis).

in Figure 35.5 for better understanding. Table 35.1 gives a general guidance of the filtration performance in terms of specific cake resistance.

EXAMPLE PROBLEM 35.1

A pilot plant filtration needs to be performed at 12 psi to separate 20 kg of API from 200 L of solvent. The area of the filter dryer is 0.5 m². Assume $\alpha = 1 \times 10^{10}$ m/kg for this API and $R_m = 1 \times 10^6$ m^{-1} for the filter cloth. Viscosity of the filtrate is 1.62 cP. Using Eq. (35.8), (a) generate a plot of filtration flux profile as a function of M/A (similar to Figure 35.5) and (b) calculate and compare the intrinsic cycle times to complete this filtration when the filtration is performed in (a) 1 load, (b) 3 loads, and (c) 5 loads.

TABLE 35.1 Alpha vs. Practical Filtration Performance in Conventional Equipment

Alpha (m/kg)	Filtration Performance
1×10^7 to 1×10^8	Fast filtering
1×10^8 to 1×10^9	Moderately fast filtering
1×10^9 to 1×10^{10}	Slow filtering
$>1 \times 10^{10}$	Very slow filtering

Solution

The following are the M/A values:

a. M/A = 40 kg/m²
b. M/A = 13 kg/m²
c. M/A = 8 kg/m²

a. Using $\Delta P = 12$ psi, $\alpha = 1 \times 10^{10}$ m/kg, $\mu = 1.62$ cP, and $R_m = 1 \times 10^6$ m^{-1}, a profile (Figure 35.6) of filtration flux could be generated as a function of M/A using Eq. (35.8) as shown below (similar to Figure 35.5).
b. Plotting the above M/A values against the flux curve gives the average filtration flux for each case.

The corresponding filtration fluxes are approximately

a. 920 L/m² h
b. 2800 L/m² h
c. 4600 L/m² h

These filtration flux values are then used to calculate the cycle time to filter 200 L of solvent by using the following equation:

$$\text{Cycle time} = \frac{\text{volume of filtrate}}{\text{filtration area} \times \text{filtration flux}} \times \text{number of loads}$$

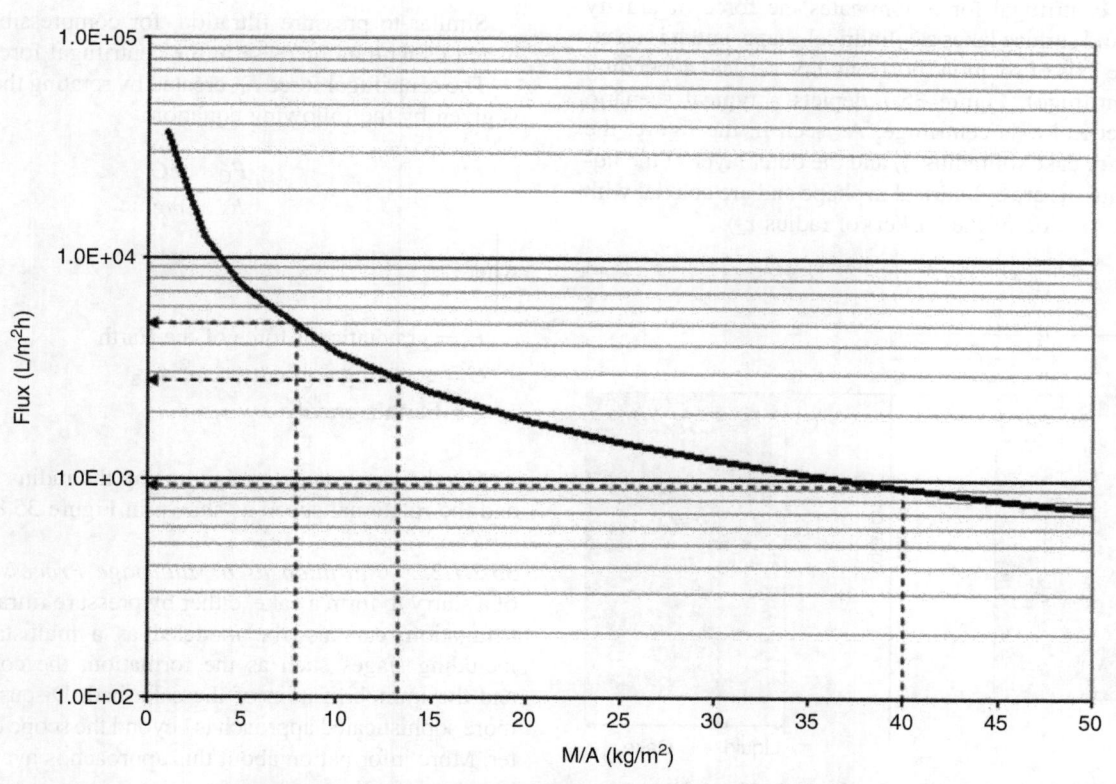

FIGURE 35.6 Filtrate flux profile vs. M/A at $\alpha = 1 \times 10^{10}$ m/kg.

The calculated cycle times are

a. 0.435 hours
b. 0.143 hours
c. 0.087 hours

It is worth noting that these cycle times do not include washing and discharging after the completion of filtering each load. Hence, in this case, splitting the batch into multiple loads to reduce the cake height (and to increase the filtration flux) may not reduce the *overall* cycle time for filtration.

35.2.1.2 Centrifugation

Centrifugation is another type of filtration used to separate solids from liquids (or slurry) where the pressure drop across the filter medium is generated by using centrifugal force. Centrifugation is more common for larger batch sizes compared with pressure filtration, as it can be performed in a relatively continuous fashion. In centrifugation, the filter medium supports the solid particles as they settle due to the centrifugal force (F_G) that can be orders of magnitude higher than standard gravitational force (F_g), and the liquid passes through the cake and then through the medium.

35.2.1.2.1 Centrifugation Principles

In a small-scale centrifuge with a lower rotational speed, the liquid surface usually takes the shape of a paraboloid [5]. However, in large-scale industrial centrifuges, due to high rotational speed, the centrifugal force dominates the force of gravity and the liquid surface takes a cylindrical shape, and it is coaxial with the axis of rotation (horizontal or vertical depending on the centrifuge). Figure 35.7 depicts a typical scenario inside a vertical axis centrifuge. As seen in the figure, the outer layer of cake (of radius r_i) and the outer layer of the liquid (of radius r_1) are cylindrical in shape and are coaxial with the axis of rotation of the basket (of radius r_2).

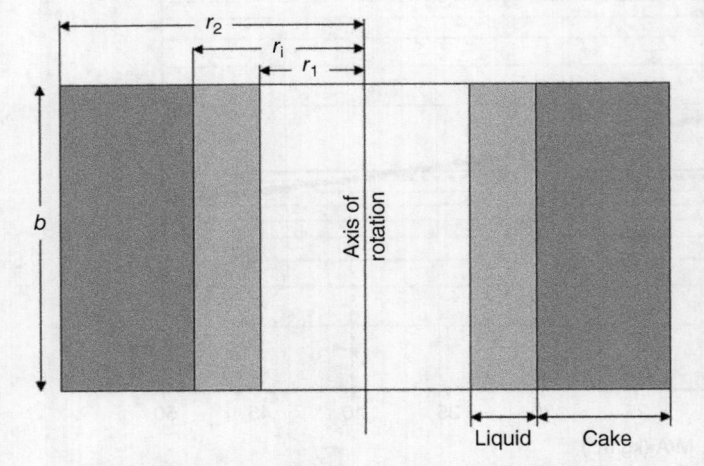

FIGURE 35.7 Cake filtration in a centrifugal filter.

The pressure drop across the cake and liquid layer is [3]

$$\Delta P = P_2 - P_1 = \frac{\omega^2 \rho (r_2^2 - r_1^2)}{2} \quad (35.11)$$

where

ω = angular velocity of the basket (rad/s)
ρ = slurry density (kg/m³)

Combining Eqs. (35.6) and (35.11) and substituting for ΔP,

$$\frac{dt}{dV} = \frac{2\mu}{\omega^2 \rho (r_2^2 - r_1^2) A} \left(\frac{\alpha C V}{A} + R_m \right) \quad (35.12)$$

Equation (35.12) has been derived with the following assumptions:

1. The filtration area does not change with the cake formation.
2. The liquid layer is on top of the cake layer at all times.
3. Gravitational force is negligible when compared with the generated centrifugal force.
4. Medium resistance is constant throughout the filtration.

Similar to pressure filtration, for compressible cakes, α increases with an increase in the centrifugal force.

The centrifugal force F_G created by rotating the centrifuge is given by the following equation:

$$\frac{F_G}{F_g} = \frac{mG}{mg} \quad (35.13)$$

where

F_g = gravitational force of the Earth
G = centrifugal gravity = $\omega^2 \times r_2$
g = Earth's gravity

G values can be determined using the radius of the bowl and the rotational speed as shown in Figure 35.8.

35.2.1.2.2 Filtration as a Multistage Process

Filtration of a slurry to form a cake, either by pressure filtration or centrifugation, can also be modeled as a multistage process including stages such as the formation, the consolidation, and the immobilization of the cake bed. Discussion of this more sophisticated approach is beyond the scope of this chapter. More information about this approach is available in the literature [6, 7].

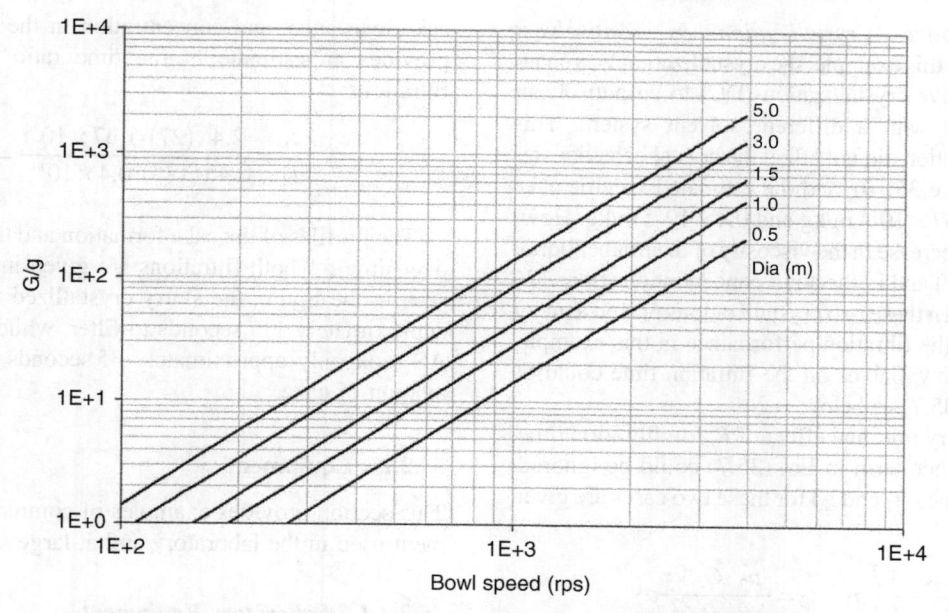

FIGURE 35.8 Centrifugal gravity versus bowl speed as a function of basket radius.

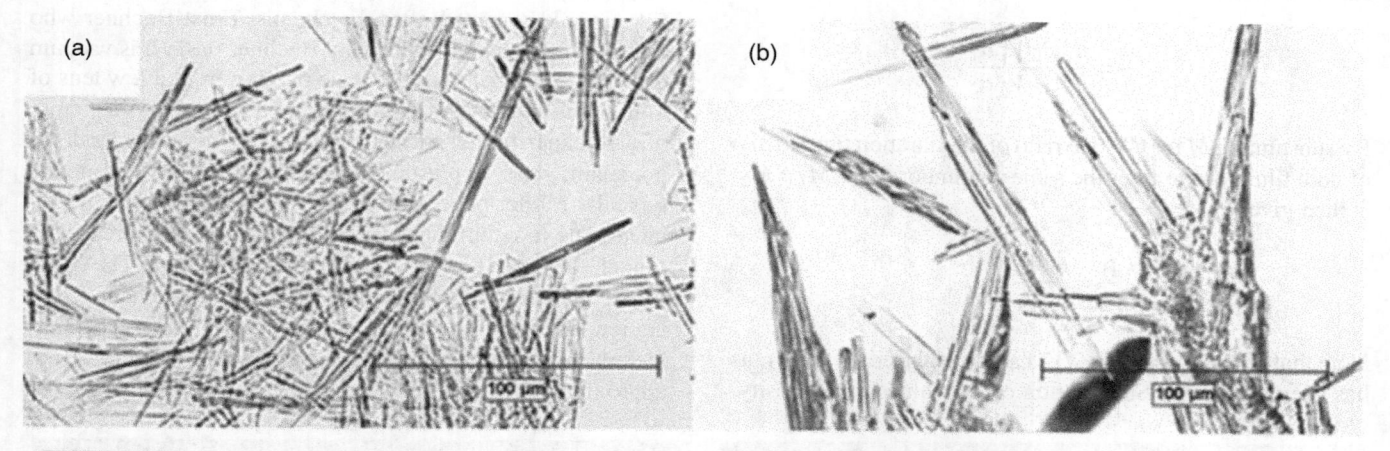

FIGURE 35.9 Optical microscopy images of the crystals obtained through (a) spontaneous nucleation and (b) seeded crystallization.

35.2.1.3 Effect of Slurry Properties
As mentioned earlier in this chapter, the nature of the slurry as defined by particle size, concentration and morphology heavily influences *filtration design* and *filtration performance*. *Filtration design* includes the choice of the equipment, wash solvents, and filtration conditions, while *filtration performance* includes the purity/impurity profile of the resulting cake, its wetness, and cycle time for filtration. Hence, in several occasions, filtration performance may be improved by adjusting the crystallization conditions to yield a slurry with good filtering properties. Related crystallization concepts are described in detail in Chapter 24.

Here, we provide example cases in which minor adjustments in crystallization procedure resulted in significant improvements in filtration performance.

35.2.1.3.1 Spontaneous Nucleation vs. Seeded Crystallization
Figure 35.9 shows the comparison of two crystals produced from the same pharmaceutical intermediate but through different crystallization conditions. The spontaneously nucleated batch (Figure 35.9a) has a significant portion of fines with a needlelike morphology and a high aspect ratio, while the seeded batch (Figure 35.9b) has a thicker rodlike morphology and comparatively low aspect ratio. In this case, the modification in crystallization reduced the filtration time significantly. The specific cake resistance measured for the needles (Figure 35.9a) was 6.9×10^{10} m/kg, while the specific cake resistance for the rods (Figure 35.9b) manufactured with an improved crystallization was 9.0×10^9 m/kg.

35.2.1.3.2 Distillative Crystallization vs. Antisolvent Crystallization

In this example, the crystallization was changed from a distillative crystallization (DC) to an antisolvent crystallization (AS) with a different solvent system. This change in crystallization did not affect the crystal behavior significantly (see Figure 35.10), and the estimated specific cake resistances were 1.37×10^{11} m/kg and 0.4×10^{11} m/kg. However, a significant decrease in the viscosity of the mother liquor (from 2.4 to 0.45 cP) and in the solid concentration (from 126 to 27 mg/ml) between these two crystallizations played a major role in influencing the filtration performance in this example. The impact of these variables on the filtration time could be explained by Eq. (35.7) as below.

Since there is very minimal effect of R_m on filtration time, the medium resistance term in Eq. (35.7) could be ignored, and the filtration times (t_1 and t_2) for these two cases are given as follows:

$$t_1 = \frac{\mu_1}{\Delta P_1} \frac{C_1 V_1^2}{2 A_1^2} \alpha_1, \quad t_2 = \frac{\mu_2}{\Delta P_2} \frac{C_2 V_2^2}{2 A_2^2} \alpha_2$$

Since the values of ΔP and A were the same for both the cases, the ratio of t_1 and t_2 is

$$\frac{t_1}{t_2} = \frac{\mu_1}{\mu_2} \frac{C_1}{C_2} \frac{V_1^2 \alpha_1}{V_2^2 \alpha_2}$$

By substituting M for CV, the ratio of the filtration times (t_1/t_2) of both filtrations to filter the same amount of cake ($M_1 = M_2$) is then given by

$$\frac{t_1}{t_2} = \frac{\mu_1}{\mu_2} \frac{C_2}{C_1} \frac{\alpha_1}{\alpha_2}$$

Note that when $M_1 = M_2$, V_1/V_2 can be replaced by C_2/C_1, as has been done here. Substitution of the values for viscosity, cake resistance, and concentration in the above expression provides an estimate of the time ratio between the two filtrations:

$$\frac{t_1}{t_2} = \frac{2.4}{0.45} \frac{(27)}{(126)} \frac{1.37 \times 10^{11}}{0.4 \times 10^{11}} = 3.9$$

The profiles of the cake formation and the filtrate received over time for both filtrations are given in Figure 35.11. As seen in the figure, the slurry crystallized from the DC took approximately 117 seconds to filter, while the one from the AS took only approximately 35 seconds to filter a similar amount of cake.

35.2.2 Equipment

This section provides examples of common isolation equipment used in the laboratory and at large scales.

35.2.2.1 Laboratory Equipment

35.2.2.1.1 Pressure Filtration

35.2.2.1.1.1 Buchner Funnels Buchner funnel filtration is one of the simplest and most widely used forms of laboratory filtration. It is named after the chemist Ernst Buchner who invented the Buchner funnel and Buchner flask. It is vacuum driven and can be used in a scale ranging from a few tens of milligrams to a few kilograms. Figure 35.12a shows a picture of a Buchner funnel filtration setup. The funnel is made of porcelain, glass, or plastic (Figure 35.12b). The filter media, typically a filter paper or a filter cloth, is kept on the perforated plate to retain the solid. In the case of a sintered glass funnel, the fritted surface acts as the filter medium. The usage of the Buchner funnel is mainly driven to implement a separation and they are rarely used to collect scale-up parameters; however, with some care, they can provide a first approximation to the filtration properties of the cake. After

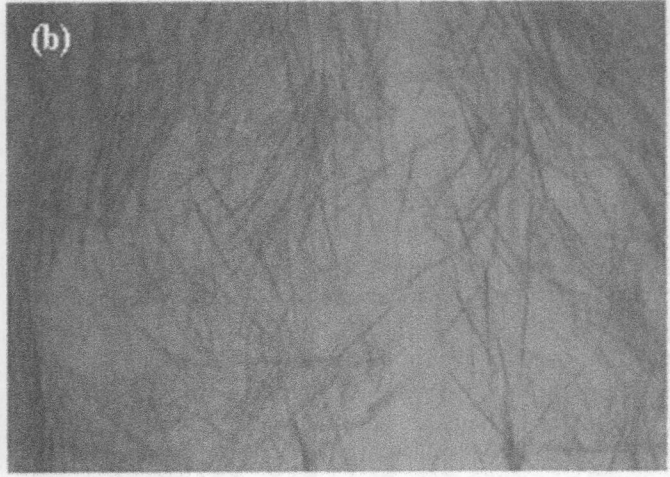

FIGURE 35.10 Optical microscopy images of the crystals through (a) distillative crystallization and (b) antisolvent crystallization.

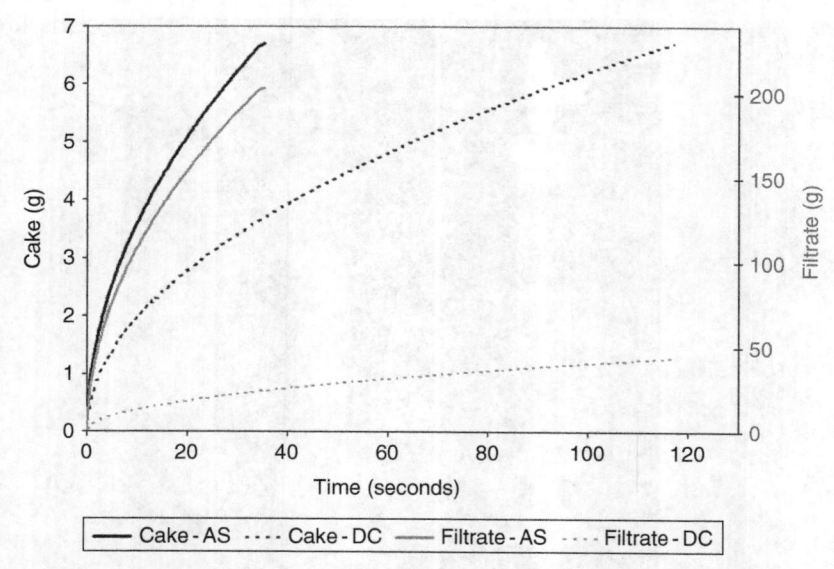

FIGURE 35.11 Profiles of cake formation and filtrate collection as a function of time for the two crystallizations.

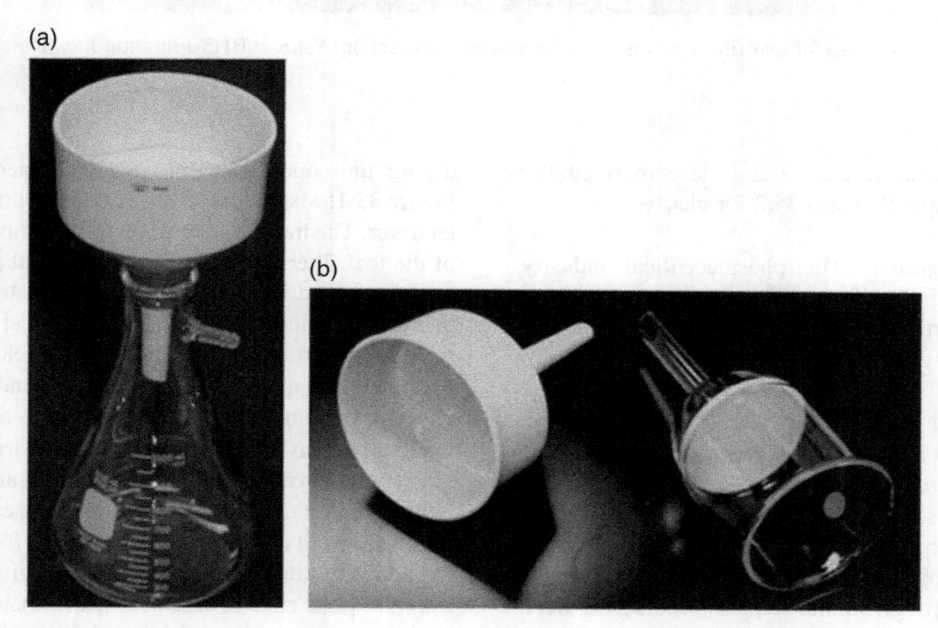

FIGURE 35.12 Pictures of Buchner funnel in (a) a filtration setup and (b) ceramics with perforated plate and glass with fritted filtration surface. *Source*: ChemGlass, Inc.

completing the isolation, cake washes can be performed by adding the solvent on the wet cake and starting the filtration.

35.2.2.1.1.2 *Leaf Filters* Leaf filters are relatively larger-scale filters compared with Buchner funnel filters. Their use is driven by the need to characterize slurry filtration properties and requires detailed measurement of the filtration profile, i.e. rate of filtrate collection. Figure 35.13 shows a few examples of leaf filters of different materials of construction. Leaf filters consist of three parts: a bottom portion with flow valve, a middle stem portion, and a top lid portion. The bottom portion has the provision to place the filter medium, the middle stem portion holds the slurry, and the top portion closes the stem portion. After connecting the bottom portion (with filter medium of desired pore size) to the stem portion, the slurry is poured into the stem. The top lid portion is then connected to the stem. The chosen pressure set point is attained usually through compressed nitrogen/air. Cake washes can also be performed in a similar fashion. Filtration is performed at multiple pressures to fully characterize the cake properties

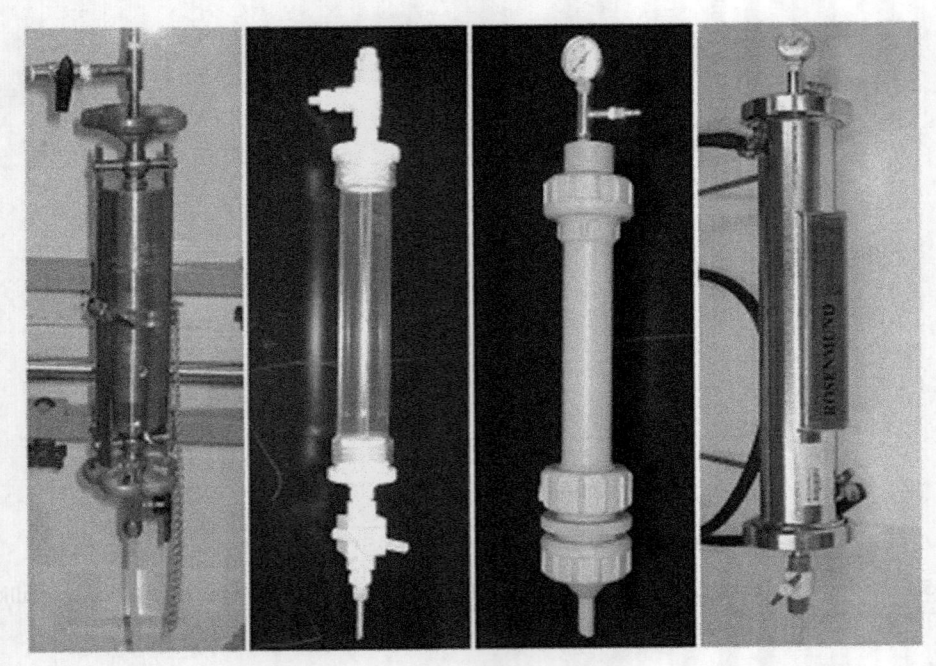

FIGURE 35.13 Examples of leaf filters of various materials of construction. *Source*: BHS-Filtration Inc., Rosenmund, Inc.

such as the specific cake resistance and the compressibility index. Refer to Example Problem 35.2 for clarity.

35.2.2.1.2 Centrifugation The pharmaceutical industry uses various types of centrifuges. This section provides an example of a vertical axis, batch mode centrifuge (Figure 35.14a). It uses a perforated basket (Figure 35.14b) rotating on a vertical axis to produce a centrifugal force on the slurry that is fed to the basket through the feed pipe. The perforated basket is lined with a filter bag of desired pore size. This filter bag retains the solid particles as wet cake, while the filtrate passes through the filter medium. The filtrate flows out of the basket through the perforations and gets collected outside through the outlet valve. Cake washes can then be performed through similar steps by feeding the wash solvent to the basket.

35.2.2.2 Large-Scale Filtration Equipment The large-scale isolation equipment used in pilot and manufacturing plants have the same working principle as the laboratory filtration equipment. The operating conditions of the large-scale equipment, however, differ significantly from those of the laboratory equipment. Selection of appropriate isolation equipment for a specific process depends on various factors, and this issue is discussed in detail in the next section.

35.2.2.2.1 Filter Dryer Filter dryers are one of the most robust isolation equipment used in the pharmaceutical industry. As the name implies, it encompasses the filtration and the drying unit operations into a single piece of equipment. Figure 35.15 shows the cross-sectional representation of a filter dryer. The mode of operation for filtration is similar to that of the leaf filter. The required differential pressure is generated by introducing either pressure upstream of the filter medium or vacuum downstream. Typical operation would include equipment setup (fitting a filter cloth of appropriate pore size and material of construction and performing leak checks), loading the slurry, deliquoring the mother liquor, loading the wash, deliquoring the wash, drying, and discharging. Filter dryers are usually fitted with an agitator that can be lowered or raised as needed. The presence of an agitator makes filtering and cake washing more efficient as the agitator aids in mixing the cake uniformly. Agitation also helps in smoothing the cake in case of crack formation that otherwise could cause channeling of the wash solvents resulting in inefficient cake washing. Filter dryers are also fitted with a jacket to facilitate heating in the drying step. Provision of a jacket also allows for performing filtrations at different temperatures. After completing the filtration and washing steps, the cake can be dried in the same equipment, and the dry cake can be discharged through the discharge port with the help of the agitator. There is typically an offset between the lowest position of the agitator and the filter cloth in order to avoid lifting or rupturing the cloth. This offset often results in residual cake (denoted as "heel") left on the cloth after the discharge. The heel could potentially slow down the filtration of the next batch. Heel removal is discussed in a later Section 35.2.3.2.1.

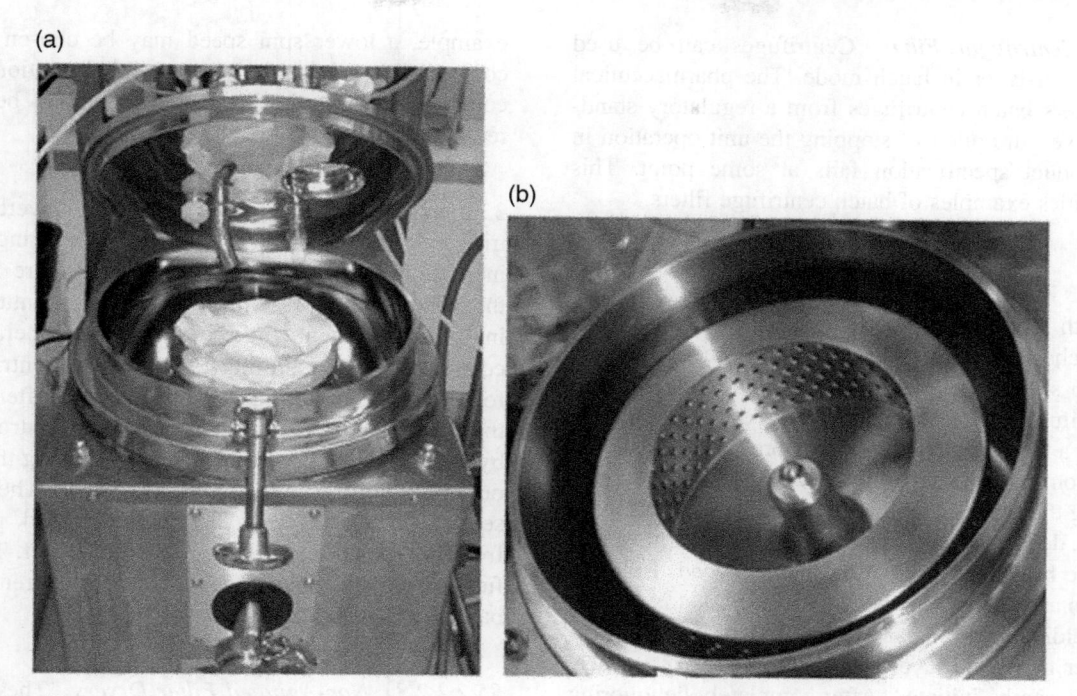

FIGURE 35.14 Pictures of (a) a vertical axis laboratory centrifuge with filter bag installed and (b) centrifuge basket. *Source*: Rousselet Robatel.

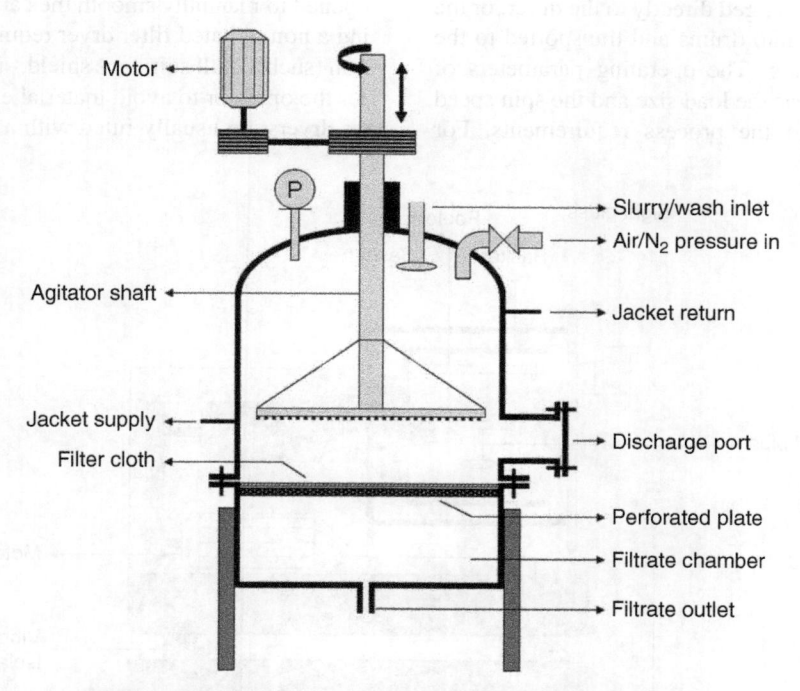

FIGURE 35.15 Cross-sectional representation of a filter dryer.

35.2.2.2.2 Centrifugal Filter
Centrifuges can be used either continuously or in batch mode. The pharmaceutical industry prefers batch centrifuges from a regulatory standpoint as it gives an option of stopping the unit operation in case the product specification fails at some point. This section provides examples of batch centrifuge filters.

35.2.2.2.2.1 Peeler Centrifuge Peeler centrifuges are available with a horizontal or vertical axis of rotation. The mode of discharge is through a mechanical peeling action. Figure 35.16 shows the cross-sectional representation of a peeler centrifuge. Typical operation involves equipment setup (fitting a filter bag of appropriate pore size and material of construction, performing leak checks), initiating the spinning, loading the slurry, deliquoring the mother liquor, loading the wash, deliquoring the wash, peeling, discharging, and removing the heel before initiating the next load. Isolation starts as soon as the slurry is loaded, as the basket is already spinning, yielding the necessary centrifugal force for the separation. After the liquid layer goes below the cake surface, cake washing can be initiated. After completely deliquoring the wash solvents, the cake is peeled off using a hydraulically activated peeler. The peeled solids are then discharged through a chute for the next unit operation. Often, there is an offset between the peeler's closest position and the cloth, resulting in a residual heel. Depending on the available equipment train, either the discharge chute could be connected to a dryer, and the material discharged directly to the dryer, or the cake could be discharged into drums and transported to the dryer for further processing. The operating parameters of the peeler centrifuge such as the load size and the spin speed are adjusted according to the process requirements. For example, a lower spin speed may be chosen to isolate a compressible cake and to prevent the formation of a highly compressed cake, the heel of which might be difficult to remove.

35.2.2.2.2.2 Inverting Bag Centrifuge Inverting bag centrifuges are horizontally or vertically rotating centrifuges incorporating an automatic discharge. Figure 35.17 shows the schematic representation of a horizontally rotating, inverting bag centrifuge. The order of operation in this centrifuge is very similar to the peeler centrifuge except for the mode of discharge. In this case, after completing the filtration and washing, the filter bag is stroked forward by a hydraulically controlled piston, turning the bag inside out, allowing the solids to be discharged. The filter bag is secured to the basket wall at the front end, preventing it from completely detaching from the basket. These centrifuges offer an efficient and faster cake removal as the bag is turned inside out.

35.2.2.2.3 Non-Agitated Filter Dryers
These are similar to filter dryers except that they are not fitted with agitators. Figure 35.18 shows a schematic representation of a non-agitated filter dryer in closed and cross-sectional views. The absence of an agitator may necessitate manual mixing of the cake during washing and drying, for content uniformity. In case of crack formation in the cake, the lid needs to be opened to manually smooth the cake. For this reason, operating a non-agitated filter dryer requires more personal protection (such as full suit, face shield, supplied breathing air, etc.) for the operator to avoid material exposure. Non-agitated filter dryers are usually fitted with a jacket.

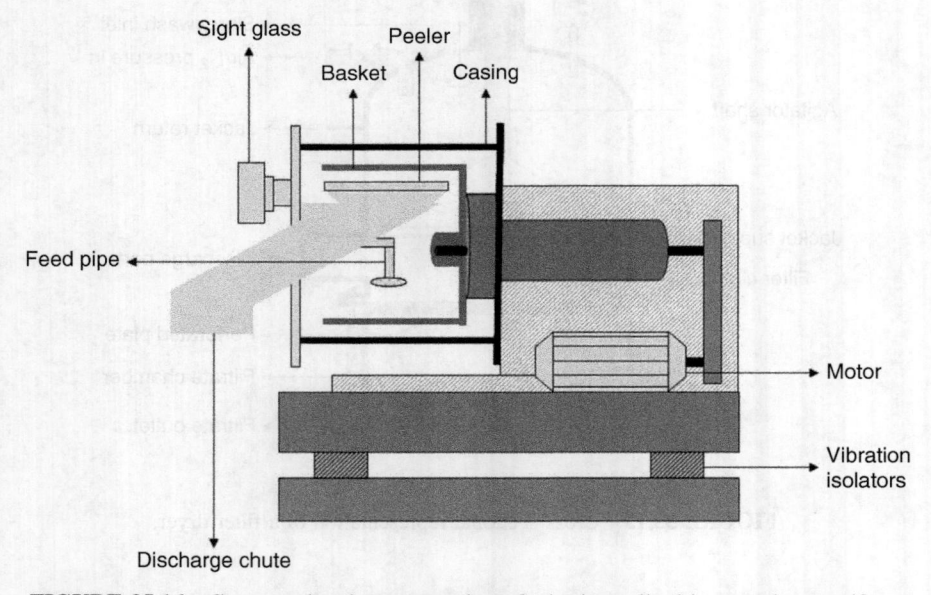

FIGURE 35.16 Cross-sectional representation of a horizontally driven peeler centrifuge.

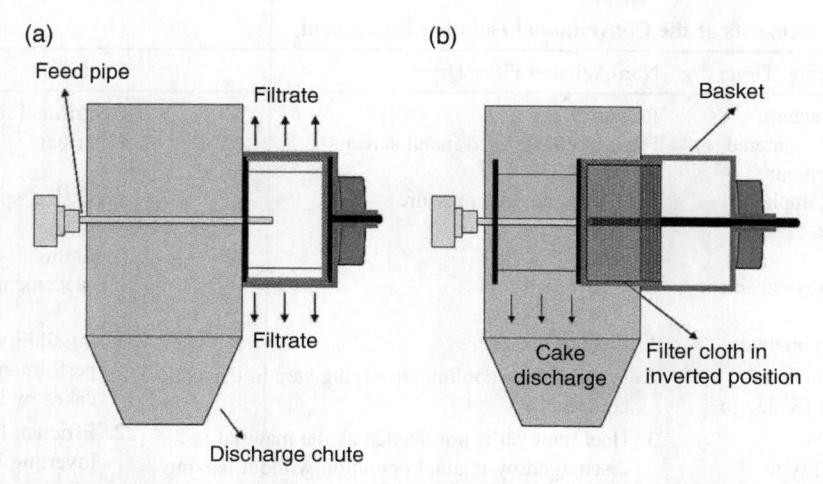

FIGURE 35.17 Schematic representation of an inverting bag centrifuge during (a) deliquoring phase and (b) discharge phase. *Source*: Derived from a model of Henkel, Inc.

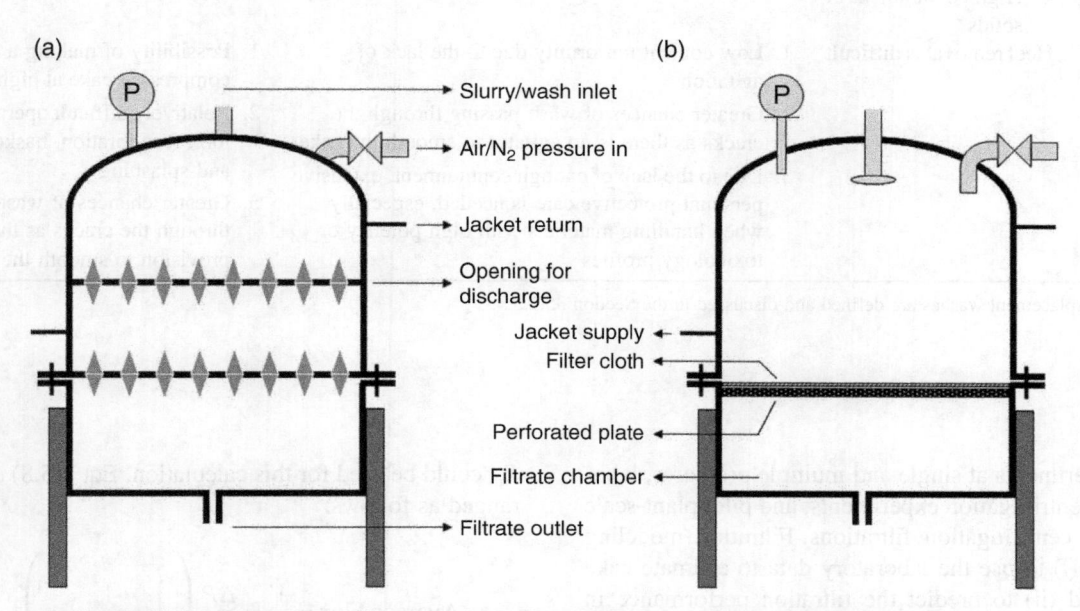

FIGURE 35.18 Non-agitated filter dryer in (a) closed position and (b) cross-sectional representation.

35.2.2.3 Equipment Selection
The specific equipment is selected for isolating a material based mainly on the nature of the slurry. The following are the factors that could affect the decision on the equipment selection:

1. Compressibility of the cake.
2. Susceptibility of the cake to agglomerate upon agitation.
3. Susceptibility of the cake to attrite upon agitation.
4. Nature of the wet cake to crack during isolation/washing.
5. Nature and availability of the equipment train.
6. Environmental health and safety impact of the exposure of solids.

Table 35.2 compares the merits and demerits of the equipment discussed above.

35.2.3 Design

35.2.3.1 Data Collection and Modeling of Filtration
Several data could be collected from laboratory filtration experiments to assess the potential performance of filtration on scale. These include data from Buchner funnel filtration, leaf

TABLE 35.2 Merits and Demerits of the Conventional Isolation Equipment

Filter	Agitated Filter Dryer	Non-Agitated Filter Dryer	Centrifuge
Driving force	Pressure/vacuum	Pressure/vacuum	Centrifugal force
Filter medium	Filter cloth, sintered metal screens	Filter cloth, sintered metal screens	Filter bag
Operating parameters	Load size, applied pressure	Load size, applied pressure	Load size, spin speed, feed rate
Robustness	High	Medium	Medium
Washes possible[a]	Reslurry, displacement	Displacement	Displacement
Advantages	1. Easy to operate 2. High content uniformity due to agitation 3. Possibility to continue to drying step in the same equipment 4. High containment of solids	1. Easy to operate 2. Possibility to continue to drying step in the same equipment 3. Heel removal is not needed as the material is discharged by manual operation without leaving any heel	1. Possibility to enhance the filtration performance for incompressible cakes by increasing the spin speed 2. Efficient heel removal in case of inverting bag centrifuge
Disadvantages	Heel removal is difficult	1. Low content uniformity due to the lack of agitation 2. Greater chances of wash passing through the cracks as there is no agitator to smooth the cake 3. Due to the lack of enough containment, extensive personal protective care is needed, especially when handling materials with high potency or toxicology profiles	1. Possibility of making a highly compressed cake at high spin speeds 2. Relatively difficult operation, need to look for vibration, basket balance and splashing 3. Greater chances of wash passing through the cracks as there is no provision to smooth the cake

[a]Reslurry and displacement washes are defined and discussed in the Section 35.2.3.3.

filtration experiments at single and multiple pressures, laboratory-scale centrifugation experiments, and pilot plant-scale pressure and centrifugation filtrations. Filtration modeling is performed (i) to use the laboratory data to estimate cake properties and (ii) to predict the filtration performance in the plant by using the estimated cake properties.

35.2.3.1.1 Buchner Funnel Filtration: First-Order Approximation As mentioned earlier, Buchner funnel filtration is performed when there is no requirement for extensive data collection. It is possible only to record the total amount of filtrate collected (V) and the total time (t) to complete the filtration at a single differential pressure (vacuum driven either through a vacuum pump or house vacuum). An average filtrate flux could be calculated from V, t, and the filtration area. The average filtration flux (L/m²h) obtained from a Buchner funnel filtration is sufficient to estimate the average-specific cake resistance ($\bar{\alpha}$) by using Eq. (35.8) as a first-order approximation. An approximate R_m could be used for this calculation. Eq. (35.8) can be rearranged as follows:

$$\text{Average Flux} = \frac{V}{At} = \frac{\Delta P}{\mu} \left(\frac{1}{\left(\dfrac{\bar{\alpha} M}{2 A}\right) + R_m} \right) \quad (35.14)$$

where

$M = CV$ = mass of dry solids (kg)

The pilot plant filtration performance (filtration flux and cycle time) can then be calculated by using the estimated $\bar{\alpha}$ and the pilot plant operating parameters such as differential pressure (ΔP), filtration area (A), and batch size (M). Since, in typical pharmaceutical separations, the medium resistance is negligible when compared with the cake resistance, estimation of specific cake resistance through this approach by using an approximate low R_m is fairly reasonable. It should be noted that $\bar{\alpha}$ has been estimated from a single data point obtained by conducting filtration at a single pressure. Hence, it is not possible to calculate the compressibility index of the cake through this approach.

35.2.3.1.2 Leaf Filtration: Usage of t/V at a Single Pressure With the help of balances equipped with automated data collection, the instantaneous filtrate volume at a particular time interval could be collected during leaf filtration experiments at multiple pressures. However, in cases of limited material availability, data from filtration at a single pressure may be used to calculate the cake properties more accurately when compared with Buchner funnel filtration. Equation (35.7) can be rearranged as follows:

$$\frac{t}{V} = \frac{\mu \alpha C}{2A^2 \Delta P} V + \frac{\mu R_m}{A \Delta P} \qquad (35.15)$$

From the t vs. V data, t/V can be plotted against V to obtain a straight line, the slope of which is $\mu \alpha C / 2A^2 \Delta P$ and the intercept is $\mu R_m / A \Delta P$. With all the other parameters such as μ, C, A, and ΔP known, α and R_m can be calculated. Pilot plant filtration performance can then be estimated using the similar approach described earlier. Compressibility index cannot be calculated through this set of data.

35.2.3.1.3 Leaf Filtration: Usage of t/V at Multiple Pressures Inverse filtration rate (t/V) data at multiple pressures is the most complete data set required to calculate the cake properties including the compressibility index. Equation (35.9) can be modified as

$$\ln(\alpha) = \ln(\alpha_o) + (n)\ln(\Delta P) \qquad (35.16)$$

After calculating α and R_m for the individual differential pressures as explained above, $\ln(\alpha)$ can be plotted against $\ln(\Delta P)$ to obtain a straight line. The slope of this straight line is the compressibility index (n) and the intercept is $\ln(\alpha_o)$.

Even though this approach is more complete and reliable, the previous two approaches can also be beneficial in cases of limited data.

35.2.3.1.4 Leaf Filtration: Filtration over Heel/Cake Washes Cake washes are typically performed when the cake is already deposited and the clean solvent is filtered through the deposited cake to displace the mother liquor. In often cases, crystal slurry is also filtered over the heel from the previous batch. Pilot plant performances of these cases can also be predicted by using similar equations. Equation (35.8) can be given as

$$\text{Average Flux} = \frac{V}{At} = \frac{\Delta P}{\mu} \left(\frac{1}{\left(\dfrac{\bar{\alpha} M}{2A}\right) + R_m^{obs}} \right) \qquad (35.17)$$

where

$R_m^{obs} = R_m$, in the case of slurry filtration over a clean filter cloth.

$R_m^{obs} = R_m + \dfrac{\alpha_{heel} m_{heel}}{A}$, in the case of slurry filtration over a heel.

$R_m^{obs} = R_m + \dfrac{\alpha_{cake} m_{cake}}{A}$, in the case of a cake wash.

In the case of a cake wash, the term M ($=CV$) in Eq. (35.17) is zero as the concentration of API in wash solvent is zero.

35.2.3.1.5 Laboratory and Pilot Plant Centrifuge Filtration A similar set of data is collected from laboratory and pilot plant centrifuges as the pressure filtration. Spin speed (rpm) and feed rate (kg/h) are the only two parameters required in addition to the data set of t vs. V collected from a leaf filter at multiple pressures. Spin speed can be collected through the motor of the centrifuge, and the feed rate can be collected through a mass flow meter, radar, or load cell.

t vs. V data from leaf filtration is sufficient for estimating the specific cake resistance as discussed previously. In addition to α, for modeling a centrifuge operation in the pilot plant, the slurry feed rate and the centrifuge spin speed need to be optimized for the following constraints: (i) to minimize centrifuge vibration, (ii) to maintain uninterrupted filtration for completing a load, (iii) to evenly distribute the cake, (iv) to minimize compression of the cake (in the case of compressible cakes), and (v) to prevent the formation of the standard wave. In simplified terms, Eq. (35.12) can be used for this purpose in a similar fashion as described for pressure filtration.

EXAMPLE PROBLEM 35.2

In an attempt to optimize the crystallization of an API, various crystallization parameters were studied that resulted in three different slurries. The three streams were then filtered separately in a leaf filter to assess the filtration performance of the streams. The following table provides the t vs. V data for the streams filtered at 20 psi. The diameter of the leaf filter used was 5 cm. The viscosity of the mother liquor was 1.2 cP. The concentration of all the streams was 50 mg/ml. Calculate the specific cake resistances and the medium resistances for the three cakes.

Stream 1		Stream 2		Stream 3	
t (s)	V (ml)	t (s)	V (ml)	t (s)	V (ml)
0	0	0	0	0	0
3	22.48	9	29.69	3	27.06
6	40.88	18	47.70	6	49.32
9	55.72	27	61.15	9	67.06
12	68.44	36	72.48	12	82.34
15	79.74	45	82.50	15	95.92
18	90.03	54	91.62	18	108.27
21	99.52	63	100.05	21	119.65
24	108.38	72	107.95	24	130.28
27	116.72	81	115.58	27	140.28
30	125.13	90	122.68	30	149.76
33	132.63	99	129.47	33	159.37
36	139.64	108	135.95	36	167.95
48	166.57	117	142.03	–	–
–	–	126	148.07	–	–
–	–	135	153.89	–	–
–	–	144	159.48	–	–
–	–	155	166.12	–	–

Solution

From the given t vs. V data, the following table was prepared with V vs. t/V in SI units.

Stream 1		Stream 2		Stream 3	
V (m^3)	t/V (s/m^3)	V (m^3)	t/V (s/m^3)	V (m^3)	t/V (s/m^3)
0	–	0	–	0	–
2.25E−05	1.33E+05	2.97E−05	3.03E+05	2.71E−05	1.11E+05
4.09E−05	1.47E+05	4.77E−05	3.77E+05	4.93E−05	1.22E+05
5.57E−05	1.62E+05	6.12E−05	4.42E+05	6.71E−05	1.34E+05
6.84E−05	1.75E+05	7.25E−05	4.97E+05	8.23E−05	1.46E+05
7.97E−05	1.88E+05	8.25E−05	5.45E+05	9.59E−05	1.56E+05
9.00E−05	2.00E+05	9.16E−05	5.89E+05	1.08E−04	1.66E+05
9.95E−05	2.11E+05	1.00E−04	6.30E+05	1.20E−04	1.76E+05
1.08E−04	2.21E+05	1.08E−04	6.67E+05	1.30E−04	1.84E+05
1.17E−04	2.31E+05	1.16E−04	7.01E+05	1.40E−04	1.92E+05
1.25E−04	2.40E+05	1.23E−04	7.34E+05	1.50E−04	2.00E+05
1.33E−04	2.49E+05	1.29E−04	7.65E+05	1.59E−04	2.07E+05
1.40E−04	2.58E+05	1.36E−04	7.94E+05	1.68E−04	2.14E+05
1.67E−04	2.88E+05	1.42E−04	8.24E+05	–	–
–	–	1.48E−04	8.51E+05	–	–
–	–	1.54E−04	8.77E+05	–	–
–	–	1.59E−04	9.03E+05	–	–
–	–	1.66E−04	9.33E+05	–	–

By plotting t/V vs. V, the following graph (Figure 35.19) is obtained for all the three streams.

The slopes and the intercepts of the straight lines are obtained from the plot and are given below:

Streams	Slope	Intercept
Stream 1	1E+09	102 365
Stream 2	5E+09	160 315
Stream 3	8E+08	85 283

Calculation of α and R_m is then performed by using the Eq. (35.15).

The slope $= \dfrac{\mu \alpha C}{2A^2 \Delta P}$ and the intercept $= \dfrac{\mu R_m}{A \Delta P}$.

By introducing the values for μ (1.2 cP, 0.0012 Pa·s), A ($\pi \times (0.05/2)^2$, 1.96 E-03 m^2), ΔP (20 psi, 137 894 Pa), and C (50 mg/ml, 50 kg/m^3), the α and R_m values are calculated as follows:

For example, in the case of stream 1:

$$\alpha = \frac{(1\text{E}+09)(2)(1.96\text{E}-03)^2(137\,894)}{(0.0012)(50)} = 1.77\text{E}+10 \text{ m/kg}$$

$$R_m = \frac{(102\,365)(1.96\text{E}-03)(137\,894)}{(0.0012)} = 2.31\text{E}+10 \text{ m}^{-1}$$

The α and R_m values are calculated for all the three streams in a similar fashion.

Streams	α (m/kg)	R_m (m^{-1})
Stream 1	1.77E+10	2.31E+10
Stream 2	8.83E+10	3.61E+10
Stream 3	1.41E+10	1.92E+10

By comparing the values of α, it seems that Stream 2 is the one with the highest specific cake resistance and Stream 3 with the lowest specific cake resistance. Thus, the filtration performance of Stream 3 is expected to be much better than that of Stream 2 in the pilot plant

Not surprisingly, R_m values are relatively comparable for all the streams as R_m depends mainly on the filter cloth and not on the cake properties.

35.2.3.2 Troubleshooting

35.2.3.2.1 Heel Removal The formation of a heel may be an unavoidable problem in cake filtration in most of the equipment, except in cases such as inverting bag centrifuges. As discussed earlier in this chapter, the offset between the agitator or peeler and the filter medium results in the formation of a heel. The next load/batch could be filtered over the heel from the previous load/batch as long as it does not adversely affect the filtration performance, but the heel would become the primary filtration medium for the next load. For compounds with low toxicology profiles, a manual removal of the heel is a viable option. However, for compounds with high potency/toxicology profiles, removal of a heel in filter dryers/centrifuges needs an additional unit operation involving the dissolution of the cake in a solvent and passing the solution through the filter cloth. In modern-day centrifuges, there are provisions such as blowing compressed nitrogen from behind the cloth to remove the heel. Another possible solution in a centrifuge is to loosen the heel by rewetting it with wash solvent and discharging it.

35.2.3.2.2 Filter Medium Blinding The filter medium may play an important role in determining the filtration

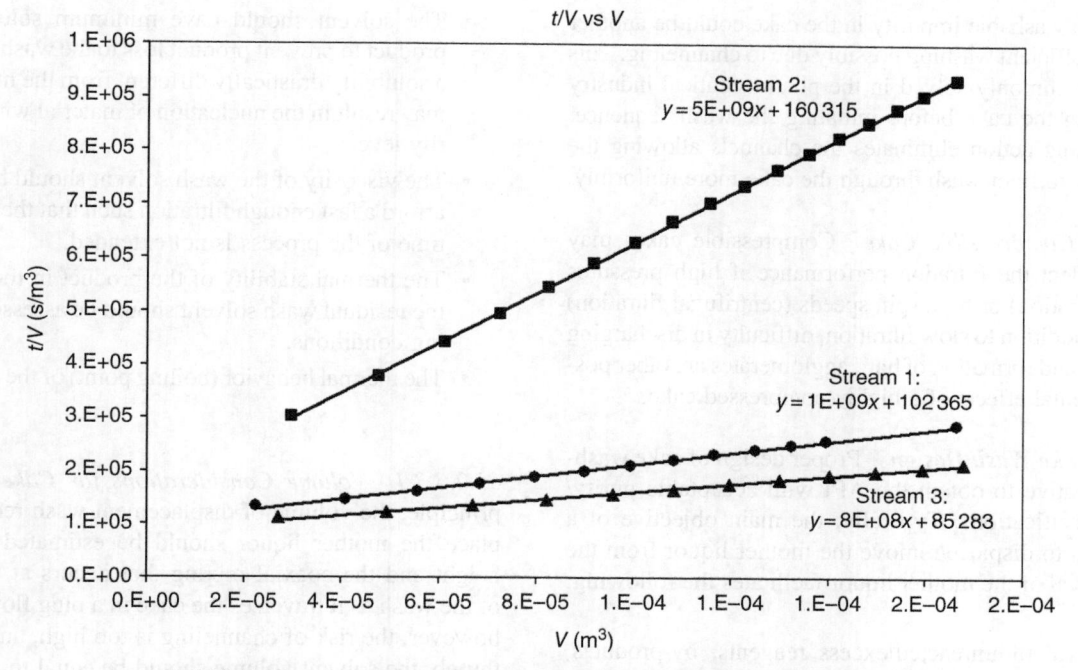

FIGURE 35.19 t/V versus V profiles of all the three crystallization streams.

performance. At the beginning of the filtration process, the filter medium retains the solid particles while allowing the filtrate to pass through. The cake that is formed on the medium as a result then does most of the filtration. However, in some cases it has been observed that the filtration performance on a used filter medium is significantly different from that on a new one. This behavior could be observed even in the absence of heel over the medium. This is called "filter medium blinding" [4]. It results from clogging of the pores by the solid particles, thereby preventing the free movement of the filtrate through the pores as originally observed in a new medium. Due to this blinding, the initial medium resistance for the next batch/load might rise. Figure 35.20 shows the effect of filter cloth blinding on the medium resistance. From this figure, it is clear that the filter medium resistance increased gradually over the number of filtration cycles.

It is generally not possible to remove blinding through mechanical agitation or by passing compressed gas (nitrogen or air) through the medium, as it could result in permanently damaging the medium, rendering it unfit for use for the next batch/load. A typical solution for this problem is to use a solvent to dissolve off the entrained solid particles.

35.2.3.2.3 Cake Cracking In many cases, mostly after the deliquoring step, cake cracking is observed. In cases where a complete removal of the mother liquor (along with the unreacted starting material, reagents and by-products) is relied upon by washing the wet cake, cake cracking may pose a detrimental effect on the quality of the batch. Cracking of the cake results in the formation of various channels. When

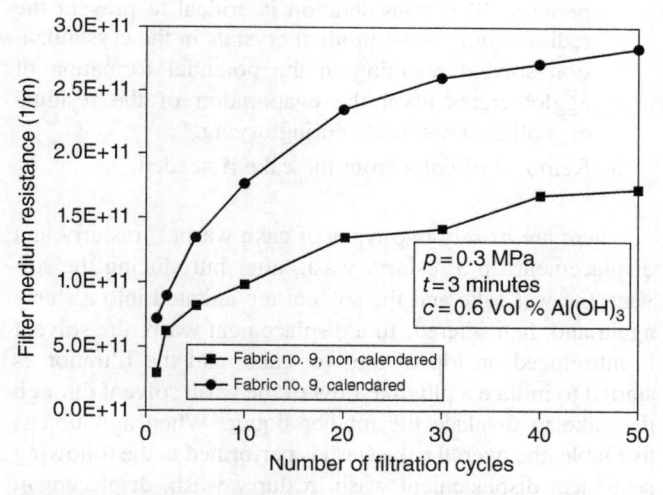

FIGURE 35.20 The effect of filter cloth blinding on the filter medium resistance. *Source*: Reprinted with permission from Weighert and Ripperger [4]. Copyright 1997, with permission from Elsevier.

washed with solvents without reslurrying the cake, the wash solvents tend to pass through these channels (paths of least resistance), leaving many parts of the cake unwashed. Cake cracking becomes more frequent when the average porosity of the cake is high [8], resulting in the formation of channels. The presence of cracks could be identified by (i) visual inspection through the sight glass, (ii) monitoring the filtrate flow (an abnormally fast deliquoring could be a sign of cracking), and (iii) measuring the purity of the cake, and the presence of a washable impurity in the cake could be another sign of an inefficient washing possibly due to channeling. This problem is commonly solved in the pharmaceutical industry by smoothing the cake before initiating the wash sequence. This smoothing action eliminates the channels allowing the wash solvent to filter/wash through the cake more uniformly.

35.2.3.2.4 Compressible Cake Compressible cakes may adversely affect the filtration performance if high pressures (pressure filtration) or high spin speeds (centrifugal filtration) are used. In addition to slow filtration, difficulty in discharging the wet cake and formation of hard agglomerates are other possible detrimental effects of a highly compressed cake.

35.2.3.3 Cake Wash Design Proper design of cake washing is imperative to obtain the API with acceptable purity/impurity specifications. Typically, the main objective of a cake wash is to displace/remove the mother liquor from the cake. Removal of the mother liquor facilitates the following:

1. Removal of unreacted/excess reagents, by-products, and impurities rejected by crystallization.
2. Removal of the crystallization solvent in case of significant product solubility at the anticipated drying temperature. This consideration is critical to prevent the redissolution of the product crystals in the crystallization solvent resulting in the potential formation of agglomerates upon the evaporation of the residual crystallization solvent during drying.
3. Removal of color from the cake if needed.

There are mainly two types of cake washes: reslurry and displacement. In a reslurry wash, after introducing the solvent, the wet cake and the solvent are agitated into a slurry again and then filtered. In a displacement wash, the solvent is introduced on top of the wet cake, and the filtration is started to initiate a pluglike flow of the wash solvent through the cake to displace the mother liquor. When agitation is available, the overall cake wash is performed in the following sequence: displacement wash, reslurry wash, displacement wash (DRD).

There are various factors that need to be considered to design and develop an efficient cake wash in order to attain the abovementioned objectives. In typical cases, the crystallization solvent is the preferred solvent for cake washing and is generally used to displace the mother liquors even when a different solvent is implemented for other reasons. The following are considerations for choosing a solvent for the cake wash:

- Identification of a suitable solvent that will have maximum solubility of the unwanted impurities. A low solubility might crystallize those impurities along with the already crystallized product.

- The solvent should have minimum solubility of the product to prevent product loss to the washes. However, a solubility drastically different from the mother liquors may result in the nucleation of material with high impurity levels.
- The viscosity of the wash solvent should be minimal to afford a fast enough filtration such that the overall cycle time of the process is not extended.
- The thermal stability of the product in the presence of the residual wash solvent should be assessed under drying conditions.
- The thermal behavior (boiling point) of the wash solvent.

35.2.3.3.1 Volume Considerations for Cake Washes In principle, the volume of displacement wash required to displace the mother liquor should be estimated by the cake height and the coaxial mixing that occurs at the front end of the wash as it traverses the cake in a plug flow. Generally, however, the risk of channeling is too high, and as a rule of thumb, the solvent volume should be equal to at least thrice the volume of wet cake. In general terms, a reslurry wash requires more solvent than a displacement wash. If the isolation is performed in multiple loads, the cake wash volumes should be adjusted accordingly. This is true for both pressure filtration and centrifugation.

35.2.3.3.2 Uniform Washing To facilitate an efficient cake wash, care should be taken to ensure a uniform wash of the entire cake bed. Even though there are various approaches (such as spray balls, mists, or weirs) available to spray the solvent evenly across the surface of the cake, care should be taken to ensure that the cake surface is adequately smoothed to prevent (i) more solvent stagnating on troughs and (ii) channeling of the wash solvent through cracks formed across the bed. These two factors might cause uneven washing of the cake, increasing the likelihood of inefficient washing of some parts of the cake.

In-process controls could be put in place to ensure the complete removal of the impurities and mother liquor. Possible controls include parameters such as purity, color, and pH. However, the accuracy of these methods is directly related to the uniformity of the wash.

35.2.4 Summary

A thorough understanding and proper design of the filtration unit operation is imperative in the pharmaceutical industry to yield a high quality product. Optimizing crystallization and filtration parameters to achieve an efficient filtration might also help in reducing the cycle time for the overall manufacturing process. Use of laboratory filtration data in the filtration equations derived from Darcy's law provides a prediction of the filtration performance upon scale-up in pilot and manufacturing plants. There is an array of isolation equipment available with various sizes and operating parameters intended for laboratory, pilot plant, and manufacturing scales. An efficient cake wash design also aids in enhancing the product quality and the process yield.

35.3 DRYING

35.3.1 Background and Principles

Drying is an integral part of the isolation process and is important for the production of consistent and stable product. It is especially important in the case of the API where product properties can have a direct impact on product performance and the levels of residual solvent are considered a regulatory specification. Drying is basically a complex distillation carried out in a heterogeneous system involving solid, liquid, and vapor. Distillation involves vaporization of solvents from the liquid phase, the same phase transformation that in drying removes the solvents from "wetted" solids that were retained following the filtration process. One of the primary objectives of a drying operation is the removal of solvent to meet specifications in the product. In the case of API, the acceptable levels of residual solvent may be set according to safety (solvent toxicity), regulatory, or stability requirements (see Chapter 44). For earlier pharmaceutical intermediates, these specifications may be set according to the tolerance of solvents in chemistry steps that are immediately downstream of the drying operation. For APIs, another key objective of drying is to either achieve or maintain a specified crystallographic form. In most cases, the required form may have been produced during crystallization of the product, and it is important not to effect a change in form during the drying operation. In some cases, the required form may not be readily accessible through direct crystallization and is more practically obtained during the drying operation.

In pharmaceutical processing, the drying operation is almost always preceded by filtration. The output from filtration is referred to as wet cake since it contains solvent (typically 5–50%), and the output from a successful drying operation is referred to as dry cake since it usually contains only trace levels of residual solvent. A typical drying operation consists of the following general steps. The wet cake is charged/loaded into the dryer, which is then heated in order to vaporize the solvent. The dryer is often equipped with a vacuum pump connected to the vessel's vent line, and the vaporized solvent is removed through this line for downstream recovery, which is typically done by means of a condenser. Alternatively, and sometimes in combination with the vacuum operation, a continuous stream of inert carrier gas is passed through the dryer to sweep the vaporized solvent through the vent line. When the material is dry, it is discharged/unloaded from the dryer and packaged.

35.3.1.1 Phase Equilibria Since drying involves the vaporization of solvents, the vapor pressure, enthalpy of vaporization, and in some instances molecular diffusivity of the solvent are fundamental properties defining the drying operation. For some solvents, the boiling point will be beyond the operational range that can be achieved in a plant dryer, so the pressure in the dryer is reduced to effect a decrease in the boiling point. Often, pharmaceutical compounds are unstable at elevated temperatures and must be dried at lower temperatures, which may also require reduced pressures. Hence, it is useful to understand the relationship between vapor pressure and temperature. The Antoine equation (shown below) provides a reasonable estimation of the vapor pressure of pure solvents:

$$\ln P^{\text{sat}} = A - \frac{B}{T+C} \qquad (35.18)$$

where

P^{sat} is the vapor pressure.
T is the temperature.

The constants A, B, and C are empirically determined constants that are readily available for many solvents.

This relationship is used to estimate the pressure required to bring the drying operation within the limits of the dryer's capabilities (or the product's stability) for the particular solvent that is being removed. For more accurate representation of vapor pressure data over a wide temperature range, an equation of greater complexity such as the Riedel equation could be used [9]. However for most applications in the pharmaceutical industry, the Antoine equation provides a satisfactory approximation.

Figure 35.21 shows some examples of the vapor pressure versus temperature relationship for common solvents used in the pharmaceutical industry. Solvents with an equilibrium curve to the right of *n*-heptane are generally harder to remove in a drying operation. When the temperature at which drying is performed is above the boiling point of the solvent, the drying rate is typically increased significantly until mass transfer of the solvent becomes the rate-limiting step.

When a solid material is held in an environment with constant temperature and humidity, it will reach steady-state equilibrium with respect to its moisture content. The *equilibrium moisture content* is the limit of the water content that the material can retain under specific conditions of temperature and humidity. Materials encountered in drying operations can either be hygroscopic (e.g. porous materials) or nonhygroscopic (e.g. nonporous materials); however, typical pharmaceutical compounds are nonporous and do not exhibit significant moisture (or other solvent) movement unless they exist as hydrates (or solvates).

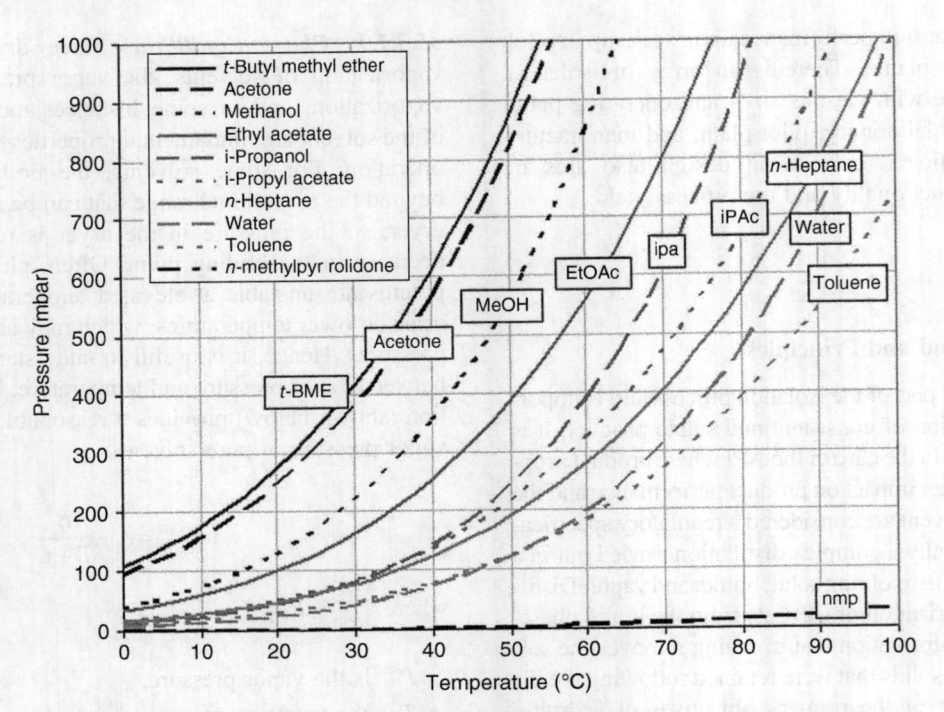

FIGURE 35.21 Example vapor pressure chart generated using Antoine's equation.

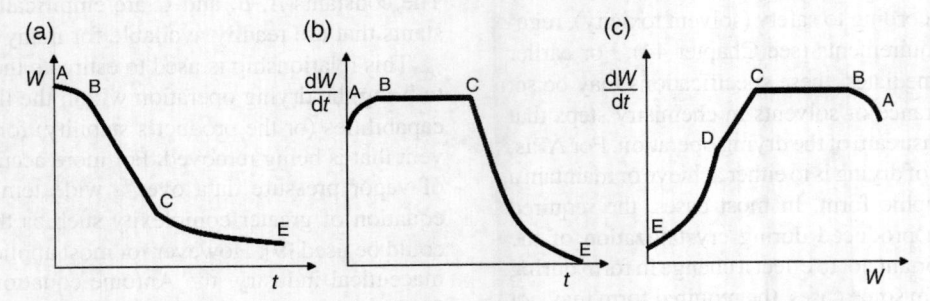

FIGURE 35.22 Typical drying curves. *Source*: Adapted from R. H. Perry [9].

35.3.1.2 Drying Mechanisms/Periods

Several mechanisms are involved during a drying operation, and the drying process can be divided into different stages that reflect the mechanism that dominates the rate of solvent removal. Initially, drying is generally slow during the warm-up period as the system reaches steady state. Following this stage, drying may be faster during the heat transfer limited period, after which the rate of solvent removal may become considerably slower as the system becomes mass transfer limited near the end of the drying process. These mechanisms are described in more detail below.

The drying process can be monitored by following the solvent content versus time. The data can then be presented in several ways to determine when different drying mechanisms are dominant. These plots are called drying curves, and an example schematic is shown in Figure 35.22. The first curve (Figure 35.22a) shows a typical drying curve for solvent content, W versus t. In the second curve (Figure 35.22b), the derivative of W with respect to t (i.e. the change in solvent content versus time or the drying rate) is plotted against time. In the third curve (Figure 35.22c), the drying rate is plotted against the solvent content. From each of these curves, distinct stages or periods can be identified and are denoted by points A through E in Figure 35.22 [9].

35.3.1.2.1 Jacket Warm-Up Period The period between A and B is called the warm-up period and accounts for the time taken for the system to heat up toward the target set point of the jacket temperature controller. At this point, drying will proceed as solvent saturating the surface of the solids is vaporized. Under reduced pressure, this vapor-phase solvent would be continuously removed through the dryer vent. In the case where reduced pressure is not used, a continuous gas flow (i.e. sweep) through the dryer is utilized to ensure that vaporized solvent is removed through the dryer vent with the inert carrier gas (e.g. nitrogen). Under vacuum conditions, the cake will often cool down to below its original temperature due to the energy transfer associated with the solvent enthalpy of vaporization. In this case, the cake temperature will increase back to the jacket set point temperature when drying is essentially complete and all the solvent has been vaporized.

35.3.1.2.2 Constant Rate Period The period between B and C is typically characterized by a constant drying rate (as can be seen from the second and third drying curves in Figure 35.22). The drying mechanism during this period is dominated by heat transfer, and the drying rate is usually limited by the rate of heat transfer to the system. During this period, solvent that has saturated the surface of the solids is rapidly vaporized. The driving force during this constant rate period is the temperature difference between the dryer jacket set point and the temperature of the solvent being removed (at the system pressure). The drying rate can hence be increased by either increasing the jacket temperature set point or decreasing the pressure (i.e. to decrease the boiling point). It is important to note that the rate of solvent removal is also dependent on the size of the vacuum pump. If the pressure setting is too low, the rate of solvent vaporization may exceed the capacity of the vacuum pump to remove solvent from the system. In this case, the system settles at a higher pressure than the desired set point, and the drying rate is determined by the vacuum pump flow rate rather than the heat transfer rate. For the drying rate to be controlled by the heat transfer rate, the vacuum pump must be appropriately sized for the system so that it imposes no limitations on solvent removal from the dryer. As would be expected, the drying rate is also affected significantly by the surface area available for heat transfer. As the cake dries, this area is reduced; however, agitation of the cake exposes "fresh" solvent to the heated surface and increases the area for heat transfer. Hence, agitation of the cake allows for an increase in the area available for heat transfer.

35.3.1.2.3 Falling Rate Period At the point C, solvent is no longer abundant on the solid surface: i.e. the surface becomes unsaturated, and the drying rate begins to fall. This point is typically called the *critical moisture content*. Note that the term "moisture" derives from historical convention when early drying studies were carried out with water, but the concept of a critical transition point applies to any solvent system. The critical moisture content is a property of the material being dried that also depends on the thickness of the material and the relative magnitudes of the internal and external resistances to heat and mass transfer. In some cases, it could also be a function of the total surface area of the solid phase and hence the particle size [3]. It is important to take into account possible variations in the value of the critical moisture content during scale-up of a drying process.

During the period from C to E, heat transfer no longer becomes the limiting mechanism for drying. Instead, mass transfer becomes the dominant mechanism limiting the rate of solvent removal from the heated surface. As a result, the drying rate is typically observed to fall as depicted in the second and third drying curves of Figure 35.22. In some cases, the decrease in the rate of drying can become significant and lead to extremely long drying times. At some point during the falling rate period, the solvent present on the surface is depleted, and solvent present inside the internal structure of the solids remains (either as a solvate or an occluded solvent that may not be removable). The drying rate is then determined by the movement of this internal solvent to the heated surface for vaporization. This transition is given at point D and can sometimes be seen by plotting the drying rate versus the solvent content as shown in the third drying curve of Figure 35.22. It is important to note that this solvent migration is extremely difficult to model and may not be a feature clearly visible on many drying curves.

35.3.1.3 Effect of Wet Cake Properties
For any process operation, it is important to understand the properties of the input to the operation, and drying is not an exception. Careful analysis of the wet cake used in the drying process can provide insight into common problems faced during drying. There are properties of the wet cake that can have significant impact on the drying operation, such as the particle size distribution, the particle morphology/shape, and the solvent content. For example, the particle size distribution and the particle morphology can have a direct impact on the surface area available for heat transfer, while the initial solvent content will clearly affect the overall drying cycle time. Furthermore, the solvent(s) that the cake is wet with can impact both the drying cycle time and the final particle properties. A higher initial solvent content can result from inadequate deliquoring (removal of solvent), practical equipment limitations during filtration, or a propensity of the material to retain solvent even in extreme deliquoring conditions. The interactions between solvent and solid phases can have an impact on the degree of particle attrition and/or agglomeration and hence the final particle size distribution. In addition, the composition of the solvents and the resulting vapor–liquid–solid equilibria can have an impact on the final particle properties (see Section 35.3.3.4).

The final particle properties of an API product are important characteristics that must be carefully controlled. The mechanical impact of the drying operation is exceedingly difficult to model and predict, but recent work using torque measurements to calculate the total work imparted on the product can generate correlations to predict the final particle size of a given API [10]. Crystallization is a much better understood unit operation, and control of particle size distribution and morphology is typically achieved through careful design and execution of this unit procedure. The typical goal and general expectation for a drying process is therefore to have little to no impact on these particle properties and to attempt to maintain the same properties as the wet cake input.

35.3.2 Equipment

Typical dryers can be classified as either *convective* or *conductive*. Convective dryers operate by passing heated air or gas over or through the material to be dried. Some common advantages of these types of dryers are that they tend to keep the product temperature relatively low and they can be scaled up to large sizes fairly easily. Some disadvantages are that they tend to have poor energy efficiency and require elaborate dust collection/capture units on the gas effluent side [11]. In pharmaceutical plants, conductive (or *contact*) dryers are by far the most commonly used type. In this type of dryer, the product is in direct contact with a heated surface. Contact dryers are mostly used in batch mode and are often more energy efficient than convective dryers [11]. Some common types include vacuum tray dryers, filter dryers, rotary cone dryers, and tumble dryers. The discussion here on large-scale equipment will focus on these types of dryers.

35.3.2.1 Laboratory Drying Equipment
It is often difficult to accurately simulate the drying process in the laboratory; however, there are some common types of lab-scale equipment that can be helpful in providing guidance toward the selection of plant equipment, providing estimates of drying cycle time on scale, and determining equipment settings necessary to achieve design objectives.

Figure 35.23 shows an oven balance that can be used to determine a drying curve for a particular material. The setup consists of a temperature-resistant balance, placed inside a vacuum oven. The balance is connected to a computer outside of the oven that captures the mass of the product as it is dried. The solvent content at any time can be calculated from the product mass. The oven balance is therefore useful for generating drying curves such as those shown in Figure 35.22 and hence determining the critical moisture content. One advantage of using an oven balance is that it is not an *invasive* technique: that is to say that significant and discrete samples of the product are not required; hence data can be collected on a very small scale. One disadvantage is that drying in an oven is a purely static process and provides no information on the effect of agitation on the material.

The effect of agitation can sometimes be estimated using a pan dryer in the lab. A pan dryer is basically a closed jacketed vessel with an overhead motorized agitator with impeller blades. The jacket provides a heated surface to contact the product, and the impeller blades allow the cake to be agitated (or turned over). The product and jacket temperatures can be recorded using thermocouples that provide the temperature differential (the driving mechanism for solvent evaporation in the constant rate period), and the agitator power can be recorded using a torque meter. The shape, type, and size of the impellers in a pan dryer can be adjusted so that it is geometrically similar to the large-scale plant dryers; however, the shear force exerted on the cake in a lab setup is typically an order of magnitude lower than that in large-scale units used in plants. In order to simulate the shear forces in large-scale plant dryers, weights can be placed on top of the cake in the lab scale to generate the same hydraulic

FIGURE 35.23 An oven balance used to generate drying curves.

pressure or normal stresses experienced in the large-scale unit [10]. The drying process is typically faster in a pan dryer than an oven due to the larger relative surface area in contact with the material from the effect of agitation. The use of a pan dryer is often helpful in determining the effect of agitation on powder properties of the product (e.g. particle size). The impact of different modes of agitation (i.e. turning over the cake at intermittent intervals or continuously agitating the cake throughout the process) can also be examined in a qualitative manner.

A typical setup for using a pan dryer in the lab is shown in Figure 35.24. Here, the solvent vapor removed from the product is condensed outside the dryer and collected in a vessel located on a balance. This allows an estimate of the solvent content to be determined (from a mass balance) at any time during the process without the need to take discrete samples, especially if the balance is connected directly to a computer or data recorder. Figure 35.25 shows a modified setup incorporating the use of weights to simulate the level of shear the cake would experience in a large-scale dryer [10, 12].

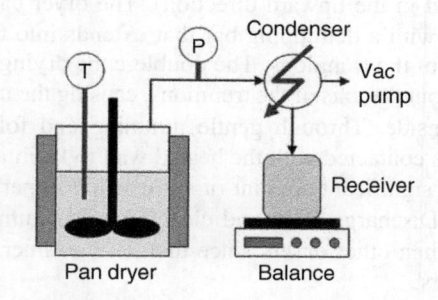

FIGURE 35.24 Typical laboratory setup.

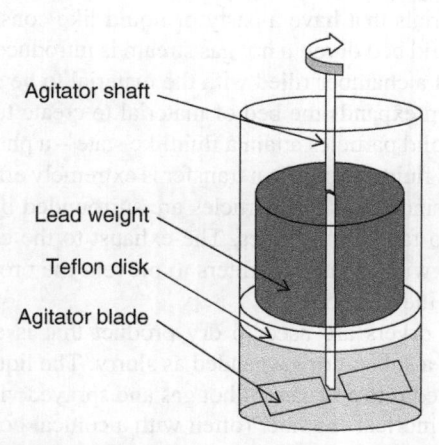

FIGURE 35.25 Lab pan dryer with weight to simulate normal stresses typical of large-scale dryer. *Source*: Reprinted with permission from Remy et al. [10]. Copyright 2015, American Institute of Chemical Engineers.

35.3.2.2 Large-Scale Drying Equipment
The four most common types of dryers used for pharmaceutical products are vacuum tray dryers, filter dryers, rotary cone dryers (or conical dryers), and tumble dryers. These units are shown in Figures 35.26–35.29, respectively.

A tray dryer (see Figure 35.26) is basically a large oven connected to a vacuum pump and/or an inert gas supply. Product is loaded onto metal trays that are placed onto shelves in the dryer. Samples can be taken only by opening the oven. When the product is dry, the final dry cake is unloaded manually from the trays into product containers/drums.

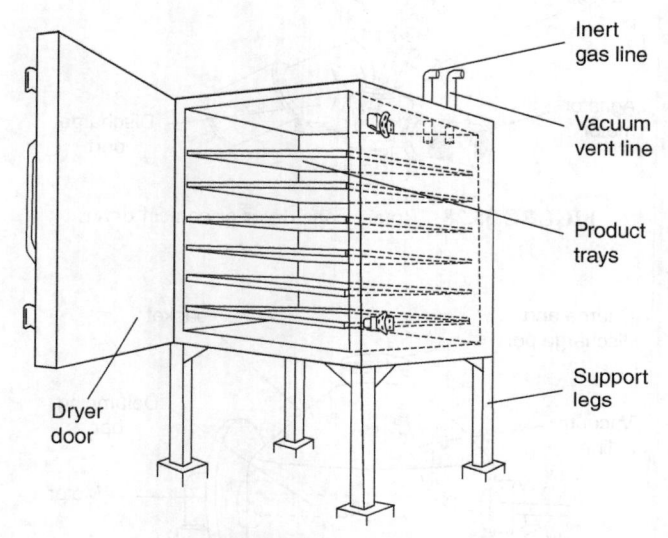

FIGURE 35.26 Tray dryer.

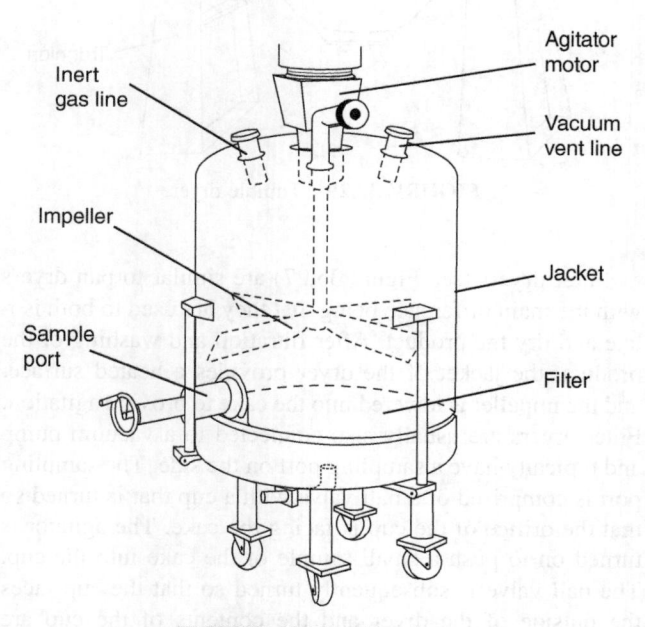

FIGURE 35.27 Filter dryer.

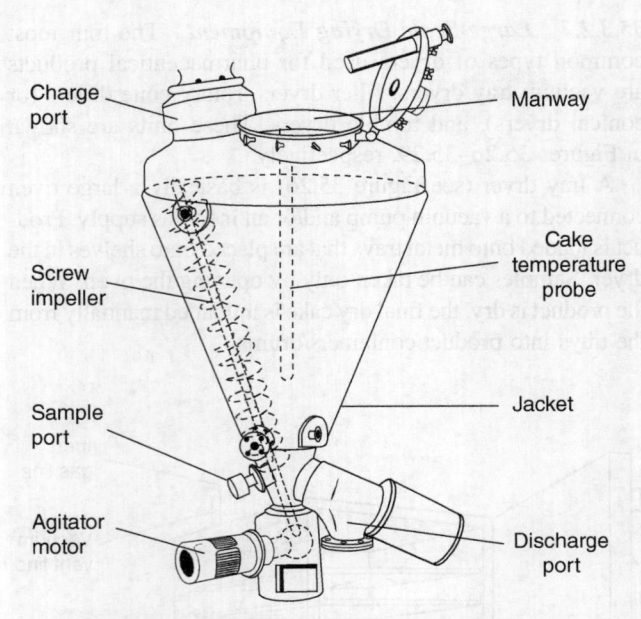

FIGURE 35.28 Rotary cone dryer or conical dryer.

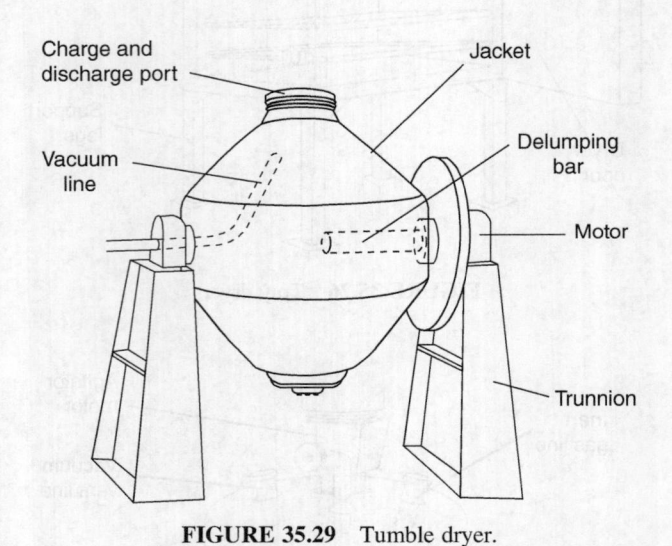

FIGURE 35.29 Tumble dryer.

Filter dryers (see Figure 35.27) are similar to pan dryers with the main difference being that they are used to both isolate and dry the product. After filtration and washing of the product, the jacket of the dryer provides a heated surface, and the impeller is lowered into the cake to provide agitation. Filter dryers are usually also connected to a vacuum pump and typically have a sampling port on the side. The sampling port is comprised of a ball valve with a cup that is turned so that the orifice of the cup is facing the cake. The agitator is turned on to push a small sample of the cake into the cup. The ball valve is subsequently turned so that the cup faces the outside of the dryer and the contents of the cup are removed. Obtaining a significantly sized and representative sample of the cake can sometimes be challenging; however, this method provides a lower risk of exposure to the operator than attempting to access the cake directly.

A rotary cone dryer (also known as a conical dryer) is a jacketed dryer with an internal agitator (see Figure 35.28). The shape of the rotary cone dryer provides a larger surface area to volume ratio for contact drying than a filter dryer. A shaft that is permanently embedded within the cake provides the agitation and in some instances acts as a heat source as well. As the shaft turns, a screw built onto the shaft lifts the cake from the bottom of the cone to the top. In addition, the shaft itself orbits the vessel so that all parts of the cake can be turned over. Samples are typically obtained from the conical dryer via a sampling port similar to the one described for the filter dryer, and the temperature of the cake is measured by a temperature probe lance that extends down into the cake. Wet cake is charged into the top of the dryer, and dry product is discharged from a special port built into the bottom of the cone.

A tumble dryer (see Figure 35.29) consists of a rotating or tumbling vessel and is most commonly found in a double-cone shape supported by two stationary trunnions. The vessel is surrounded by a heated jacket, and a small vacuum line is installed within one of the trunnions and extends into the vessel (angled in the upward direction). The dryer can also be equipped with a delumping bar that extends into the vessel from one of the trunnions. The double-cone drying chamber rotates about the axis of the trunnions, causing the material to cascade inside. Through gentle tumbling and folding, the material is contacted with the heated wall to facilitate drying without a significant amount of shear being imparted to the material. Discharge from and cleaning of the tumble dryer is easier than other dryers since there are no internal shafts or agitators.

Other types of dryers that are available for use include fluid bed dryers and spray dryers. Fluid bed dryers are suitable for drying granular crystalline materials but not for very wet materials that have a pasty or liquid-like consistency.

In a fluid bed dryer, a hot gas stream is introduced into the bottom of a chamber filled with the material to be dried. The gas stream expands the bed of material to create turbulence, and the solid particles attain a fluidlike state – a phenomenon known as fluidization. Heat transfer is extremely efficient and uniform since the solid particles are surrounded by hot gas, leading to fast drying times. The exhaust to the chamber is equipped with particulate filters to prevent the product from escaping the chamber.

Spray dryers are used to dry product that is either dissolved in a solvent or suspended as slurry. The liquid stream is dispersed into a stream of hot gas and sprayed via a nozzle into a cylindrical chamber (often with a conical bottom) as a mist of fine droplets. Solvent is vaporized rapidly to leave a residue of dry solid product particles in the chamber. It is important to ensure that particles of product are not wet with solvent when they touch the walls of the chamber and hence

DESIGN OF FILTRATION AND DRYING OPERATIONS 823

TABLE 35.3 Advantages and Disadvantages of Typical Dryers

Dryer Type	Advantages	Disadvantages
Tray dryer	Simple operation Easier to scale-up Easier to sample Easier to clean	Labor intensive Long drying times Poor uniformity Operator exposure
Filter dryer	No product loss after isolation Various agitation modes Lower risk of exposure	Difficult to sample Agglomerate formation Particle attrition Difficult to scale up
Conical dryer	Active agitation Good homogeneity Lower risk of exposure	Difficult to sample High particle attrition Difficult to scale up
Tumble dryer	Simple operation Easier to clean Lower capital cost Suitable for shear-sensitive materials	Lower efficiency Material must be free flowing Long drying times
Fluid bed dryer	Good homogeneity Lower risk for agglomeration Short drying times	Unsuitable for pasty or liquid materials Product recovery may be required from exhaust High volumetric consumption of inert gas
Spray dryer	Suitable for heat-sensitive materials Good product uniformity Combines crystallization and drying	Liquid or slurry feed required Difficult to control bulk density High volumetric consumption of inert gas

spray drying chambers tend to have large diameters. The heating period is very short; hence functional damage to the product is usually not an issue.

35.3.2.3 Equipment Selection Selection of the appropriate type of dryer to be used is usually based on the objectives of the drying operation. Some typical objectives include minimization of drying time, achieving or maintaining powder properties (i.e. prevention of particle agglomeration), and maintaining crystallographic form. Table 35.3 shows some advantages and disadvantages for each of the dryer types discussed that may be used for equipment selection.

EXAMPLE PROBLEM 35.3

Select and recommend appropriate drying equipment for the following three products (note there may be several dryers that are suitable for these systems):

a. Crystals of product A are needles that tend to break easily, and it is desirable to maintain the integrity of both their particle size distribution and the bulk density of the material while achieving a uniform powder. There are no exposure issues associated with product A.

b. Product B is a potent compound that may require careful control of exposure hazards. A short cycle time for the drying operation is of critical importance, and the formation of agglomerates causes issues in the downstream formulation process.

c. Product C is in the early stages of development and requires rapid scale-up. The crystallization and isolation procedures have not been finalized yet, and the material is unstable at higher temperatures.

Solution

a. The agitation within a filter dryer or conical dryer may be too strong for the particles, and it could be difficult to maintain control over the bulk density of the product if a spray dryer were used. Due to the shear-sensitive nature of the product and the fact that it has no exposure issues, a tray dryer would be a better option. However, this could lead to poor product uniformity. So, assuming that long drying times are not of major concern, a *tumble dryer* may be the best drying equipment to use for product A.

b. Since a low cycle time is critical, a dryer with agitation may be preferable to increase the heat transfer rate and reduce the overall drying time. Agglomerate formation can often be more significant in a filter dryer, so a *conical dryer* would be preferable for this material. A conical dryer would also provide the necessary containment to handle the exposure issues associated with a potent compound. Alternatively, a *fluid bed dryer* would also provide a short drying time and lower risk for agglomeration; however, recovery of material from the exhaust may be required and could lead to containment issues.

c. A *spray dryer* would provide a combined crystallization and drying operation and would alleviate the temperature sensitivity issue. However, rapid and simple scale-up may not be possible as the operation would require more detailed development in order to implement successfully. For relatively fast, simple, and low cost scale-up, a *tray dryer* or *tumble dryer* could be used, especially since the isolation has not been finalized and there may be an opportunity to influence which solvent the product input will be "wetted" with. The operating pressure would need to be designed in conjunction with this to ensure that the product temperature does not exceed the stability limit.

35.3.3 Design

The two primary concerns during scale-up are powder properties and drying cycle time. Powder property control is important for APIs since particle size, particle morphology and solvate form can all be impacted by the drying operation. For intermediates, scale-up is generally only a concern if there is a tendency to form hard macroscopic lumps during drying.

Scale-up of drying operations is not well understood and similar performance between lab and plant equipment is often difficult to achieve, making the selection of the correct scale-up parameters challenging. Data collection and modeling are important aspects to designing a drying operation and are discussed below.

35.3.3.1 Data Collection The drying process is typically monitored using temperature and pressure since most dryers already have existing product temperature probes and pressure sensors installed. In the case of temperature, correct placement of the temperature probe in the drying cake is critical to obtaining reliable information from the data. It should be noted that during static drying, a temperature probe may not provide a good representation of the average temperature of the cake. Agitation of the cake may be required to get a representative measure of the bulk temperature. Alternatively, it can be helpful to have multiple temperature probes situated at different locations within the cake. This can be useful for both static and agitated drying.

Pressure and an estimate of the solvent flow from the dryer can help to monitor the stage of the drying operation and determine the endpoint. One possible way to identify when solvent is being vaporized is to monitor the position of the valve controlling the pressure of the dryer. As solvent is vaporized, it causes an increase in pressure in the dryer, and in order to maintain vacuum, the vacuum control valve in the dryer vent opens. So while solvent is being removed,

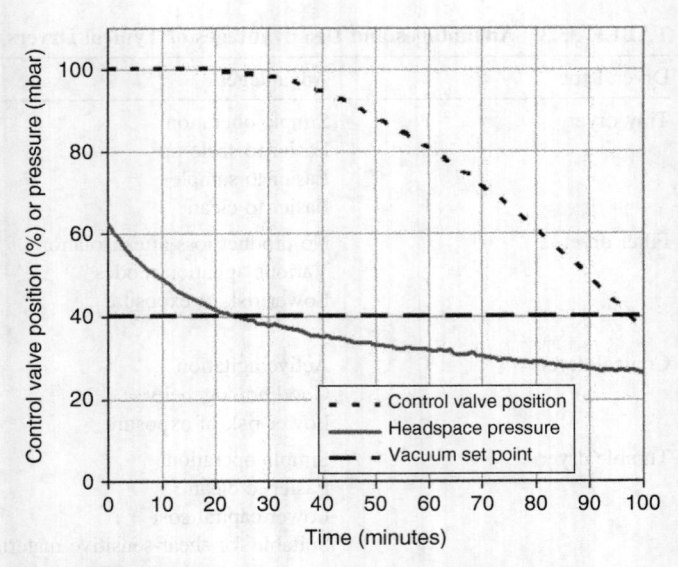

FIGURE 35.30 Using pressure and vacuum control valve position to determine the endpoint of drying.

the vacuum control valve remains open. When there is no longer a significant amount of solvent to create enough pressure in the dryer, the vacuum control valve begins to close until it reaches a position corresponding to the leak rate of the dryer setup. This movement of the vacuum control valve can be used as an indication of the end of the heat transfer limited period of the drying process. Figure 35.30 shows an example of how the vacuum control valve position starts to decrease from fully open (a value of 100%) as the pressure in the dryer approaches the vacuum set point. In this case, the position corresponding to the leak is approximately 35%.

Drying is also typically monitored by estimating the residual solvent content of the drying solids and is accomplished by either a direct measurement or by inference from surrogate process parameters. Direct measurement methods include loss on drying (LOD), gas chromatography (GC), and Karl Fischer (KF) coulometric or volumetric titration.

An LOD instrument rapidly heats up a sample of the solids to a temperature high enough to vaporize most solvents at atmospheric pressure (typically 120 °C is used). The sample is then held at this temperature for approximately 30 seconds to ensure that any solvent is removed before cooling back to ambient temperature. The mass of the sample before and after heating is recorded and the change in mass is expressed as a fraction of the starting mass to obtain the final LOD result.

GC also directly measures the content of volatile materials in a sample. A priori calibration of the GC measurements with solvent standards is typically required.

KF titration is a technique that is used specifically to determine the water content of a sample. Residual water down to the parts per million (ppm) range can often be detected using a KF instrument.

Estimating the solvent content using offline methods such as LOD, GC, and KF can have potential drawbacks. Firstly, these methods are "obtrusive" as they may require significantly sized samples to be taken to obtain a representative and reproducible result; this may be a problem when carrying out drying studies in the lab in a small-scale dryer. Furthermore, samples taken from a plant dryer may not be representative of the bulk material, depending on the location and size of the sampler. Another important disadvantage of taking samples is the risk of exposure to operators and analysts. This is especially the case when the product being dried is toxic, as can be the case for many pharmaceutical intermediates. It is therefore desirable to utilize online monitoring techniques to minimize or even eliminate the need to take samples from a dryer. Online techniques can also reduce the analysis time associated with many offline methods used to determine drying endpoint. Brief introductions to some of these online methods are given below.

Mass spectrometry (MS) has gained popularity recently and is particularly useful when different solvents are being removed from the cake. The vent of the dryer is sampled for the vapor content, and molecular masses of ionized species in the vapor are determined. The intensity of identified peaks obtained from a mass spectrum enables the determination of solvent ratios, and through the use of standards and appropriate calibration, solvent amounts can also be determined.

Spectroscopic methods including infrared spectroscopy (IR), near-infrared spectroscopy (NIR), and Raman spectroscopy depend on quantification of the solvent content relative to the solid phase (for contact measurements, i.e. direct contact of the probe with the cake) or quantification of the vapor stream leaving the dryer vent. These methods have been successfully implemented in many pharmaceutical processes to monitor drying operations [13].

Raman spectroscopy can also be used to determine if the solids have undergone a form transformation during the drying operation. This technique uses vibrational information to provide a fingerprint for the chemical environment of a molecule that is sensitive to its crystallographic arrangement.

The removal of water during drying can be monitored using a dew-point hygrometer that is typically installed in the vent line between the dryer and the vacuum pump. The partial pressure of water in the vapor from the dryer is calculated through the use of dew-point/frost-point tables. In cases where it is important to control the hydrate form of the product, the partial pressure of water can then be used to calculate the relative humidity of the vapor stream [14].

An off-line technique frequently used to characterize the crystal structure and detect changes in morphology or hydrate/solvate forms is X-ray powder diffraction (XRPD). Peaks associated with different crystallographic faces can be identified in the pattern from an XRPD measurement. The existence of different polymorphic forms will result in the appearance of new or shifted peaks in the pattern and a fingerprint pattern can be established for each particular form. As with other off-line techniques, the disadvantage of XRPD is that it is an analytical technique that must be carried out on samples that have been taken of the drying cake that may not be representative of the bulk material in the dryer.

35.3.3.2 Modeling of Drying
Modeling the rate at which solvent is removed from the drying cake can be useful to determine optimal parameter settings for the dryer or to choose the appropriate equipment. To model the drying process, mass and energy balances must be carried out on the system. By coupling these two balances, it is possible to calculate the solvent content of the cake at any time and hence the drying rate. A simple static drying model [3] is presented here that should enable simulations of the constant rate period and falling rate period to be run, given some basic information about the system. For more complex drying models, the reader is referred to more detailed treatments proposed in the literature [15, 16].

Consider a bed of product wet cake in a static dryer where the cake is in contact with a wall heated by a jacket service. The heat transfer rate from the jacket to the cake ($\dot{q}$) is given by the heat transfer coefficient (U), the area available for heat transfer (A_{ht}), and the difference between the dryer jacket temperature (T_j) and the vaporized solvent temperature (T), which is related to the operating pressure by Antoine's equation:

$$\dot{q} = UA_{ht}(T_j - T) \quad (35.19)$$

At steady state, the heat transfer rate from the wall into the cake is completely consumed by solvent vaporization, giving the following equation for the mass transfer rate of solvent during the constant rate period ($\dot{m}$):

$$\dot{m} = \frac{\dot{q}}{\Delta H_{vap}} \quad (35.20)$$

where

ΔH_{vap} is the heat of vaporization of the solvent.

The mass transfer rate of the vapor allows calculation of the rate of change of solvent content in the cake versus time.

$$-\frac{dX}{dt} = \frac{\dot{m}}{m_{drycake}} \quad (35.21)$$

In Eq. (35.21), X is the solvent content in the cake per unit mass of dry cake, and $m_{drycake}$ is the final mass of dry product cake obtained. Substituting Eqs. (35.19) and (35.20) into

Eq. (35.21) and integrating allows calculation of the drying time during the constant rate period (t_C):

$$t_C = \frac{m_{drycake}\Delta H_{vap}}{A_{ht}} \frac{(X_1 - X_C)}{U(T_j - T)} \quad (35.22)$$

where

X_1 is the initial solvent content.

X_C is the critical moisture content.

Below the critical moisture content, the model must be modified to take into account the fact that the drying rate is no longer controlled by the heat transfer rate. As an approximation, the mass transfer rate of the solvent can be assumed to be proportional to the solvent content in the falling rate period [3]:

$$\dot{m} = aX \quad (35.23)$$

where

a is a proportionality constant that may be related to the molecular diffusivity of the solvent.

By substituting Eq. (35.23) into Eq. (35.21) and integrating, the drying time during the falling rate period (t_F) can be estimated:

$$t_F = \frac{m_{drycake}}{a} \ln\left(\frac{X_C}{X_2}\right) \quad (35.24)$$

where

X_2 is the final solvent content at the end of drying.

Equations (35.22) and (35.24) can be combined to calculate the total drying time (t_T) for the process:

$$t_T = t_C + t_F \quad (35.25)$$

Given the pressure set point, jacket temperature, dryer heat transfer coefficient, enthalpy of vaporization, and critical moisture content, the equations can be used to generate solvent content profiles as a drying operation proceeds. Alternatively, experimental data on the solvent content profile versus time can be used to determine the critical moisture content and/or the value of the constant a through a data fitting approach. Example simulations based on these equations are shown in Figure 35.31, where the DynoChem® software application was used to solve the equations in conjunction with complete material and energy balances. The simulations were carried out under identical operating conditions ($T_j = 60\,°C$, $T_{initial} = 20\,°C$, $P = 150$ mbar, $UA = 10$ W/K, $X_1 = 1$ kg/kg, $X_C = 0.15$ kg/kg, $X_2 = 0.0101$ kg/kg, $m_{drycake} = 5$ kg, and $a = 0.1$ min^{-1}) for five typical solvents. The results illustrate the impact of solvent properties on drying performance and hence wash solvent choice during upstream isolation.

The charts in Figure 35.31a–c are typical drying curves that can be obtained from the model equations, and the transition from the constant rate period to the falling rate period is clearly visible. The deviation from expected behavior seen for some of the simulations at the start and near the critical moisture content is related to the choice of numerical integration parameters in DynoChem. Comparison of the profiles for each of the solvents shows that more volatile solvents such as acetone and ethyl acetate are removed very rapidly relative to less volatile solvents such as water that can take a prohibitively longer time to dry (in fact the water simulation does not reach the critical moisture content until ~84 hours). The chart in Figure 35.31d shows that the cake temperature reaches a steady-state value corresponding to the boiling point during the constant rate period. At the transition point, the temperature starts to increase toward the jacket set point (in this case 60 °C). When the boiling point of a volatile solvent is below the initial system temperature, the cake will be cooler than the initial temperature for the duration of the constant rate period. According to Antoine's equation, this behavior should be expected when the system pressure is set low enough that the boiling point of the solvent is below the initial conditions. In this case, the saturation temperature of acetone is just below 10 °C at a pressure of 150 mbar, so the cake is seen to cool down from an initial temperature of 20 °C, until the end of the constant rate period.

An important point to note is that the model does not take into account the effect of agitation that would lead to the redistribution of "wet" and "dry" portions of the cake effectively changing the area available for heat transfer (above the critical moisture content). An approach to describe the effect of agitation on heat transfer during drying has been proposed in the literature [17]. Random particle motion during agitation is represented by an empirical mixing parameter known as the mixing number. This parameter essentially represents how many revolutions of the agitator are necessary to achieve full turnover or redistribution of the wet and dry zones in the cake so that they can no longer be distinguished as separate zones on either side of a drying front moving through the cake. The larger the value of the mixing number, the less efficient the mixing provided by the agitator. The drying rate was shown to be sensitive to the mixing number in the case of finely grained particles, and the reader is referred to the original reference for a more detailed description of how to use this approach for agitated systems [17].

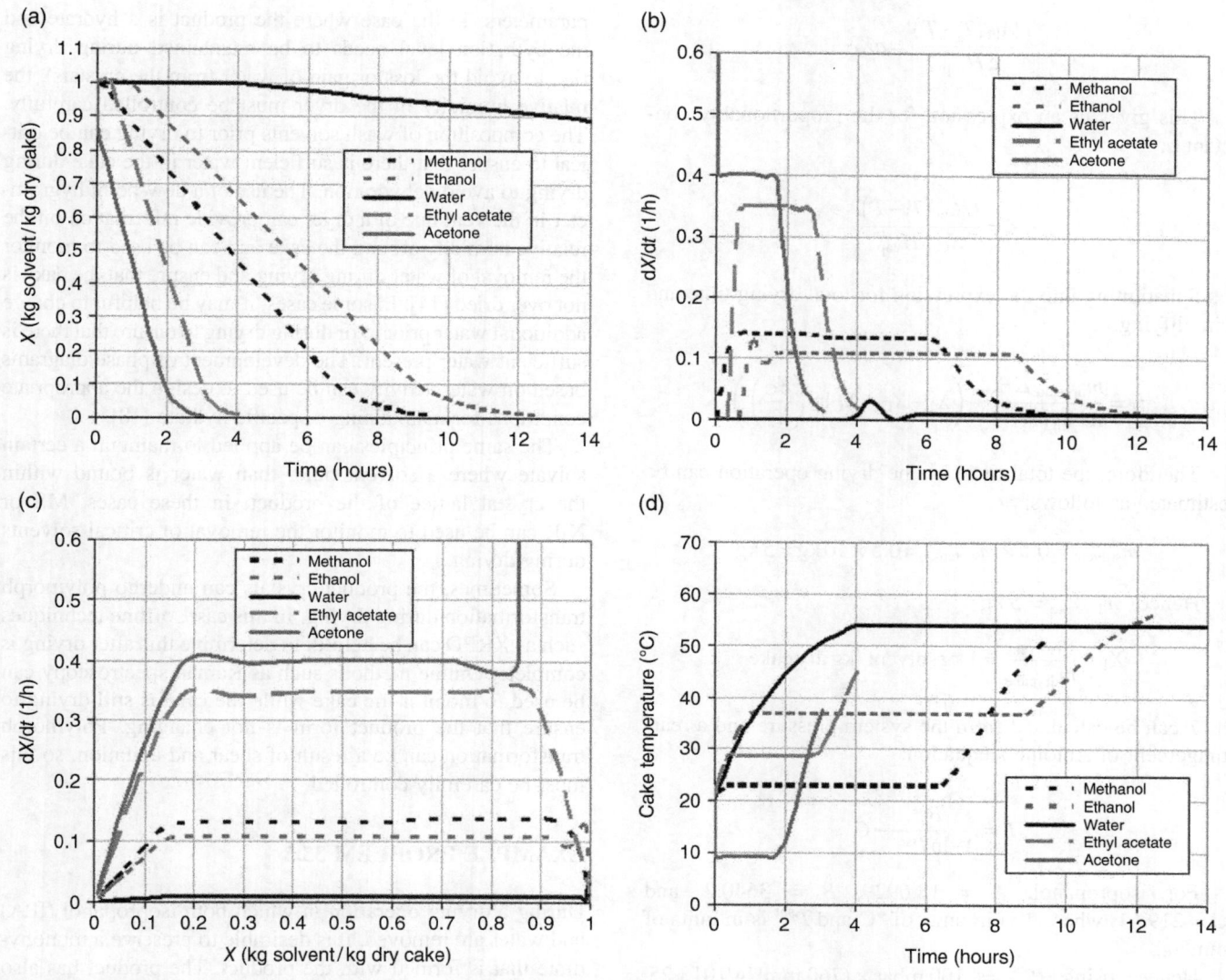

FIGURE 35.31 DynoChem® simulations of the model equations for drying. (a) Solvent content vs. time. (b) Drying rate vs. time. (c) Drying rate vs. Solvent content. (d) Cake temperature vs. time.

EXAMPLE PROBLEM 35.4

A total of 10 kg of wet cake that is initially 50 wt % wet with isopropanol is to be dried at 100 mbar pressure and 60 °C jacket temperature in a dryer that has 0.125 m² surface area available for heat transfer and a heat transfer coefficient of 40 W/m²/K. The enthalpy of vaporization of isopropanol is 662.789 kJ/kg, and the Antoine coefficients are $A = 18.6929$, $B = 3640.2$, and $C = 219.61$, where T is in units of °C and P^{sat} is in units of mmHg. If the critical moisture content is expected to be 0.1 kg solvent/kg of dry cake and the target solvent content is 0.02 kg solvent/kg dry cake, provide a first-order estimate for the total time it will take to complete the drying operation.

Solution

The portion of the total time spent in the constant rate period is given by Eq. (35.22), the portion of the total time spent in the falling rate period is given by Eq. (35.24), and the total drying time is the sum of these as given by Eq. (35.25). Hence,

$$t_T = t_C + t_F = \frac{m_{drycake} \Delta H_{vap}}{A_{ht}} \frac{(X_1 - X_C)}{U(T_j - T)} + \frac{m_{drycake}}{a} \ln\left(\frac{X_C}{X_2}\right)$$

At the critical moisture content, X_C, the final mass transfer rate of solvent from the constant rate period must be equal to the initial mass transfer rate of solvent from the falling rate period. Hence, from Eqs. (35.19), (35.20), and (35.23), we can write

$$\frac{UA_{\text{ht}}(T_j - T)}{\Delta H_{\text{vap}}} = aX_C$$

This gives us an expression for the proportionality constant a:

$$a = \frac{UA_{\text{ht}}(T_j - T)}{X_C \Delta H_{\text{vap}}}$$

Substituting into the expression for total drying time and simplifying,

$$t_T = \frac{m_{\text{drycake}} \Delta H_{\text{vap}}}{UA_{\text{ht}}(T_j - T)} \left(X_1 - X_C + X_C \ln\left(\frac{X_C}{X_2}\right) \right)$$

Therefore, the total time for the drying operation can be estimated as follows:

$$m_{\text{solvent}} = 0.5 \times m_{\text{wetcake}} = 0.5 \times 10\,\text{kg} = 5\,\text{kg}$$

Hence, $m_{\text{drycake}} = 5\,\text{kg}$

$$X_1 = \frac{m_{\text{solvent}}}{m_{\text{drycake}}} = 1\,\text{kg solvent/kg dry cake}$$

T can be calculated from the system pressure and a rearrangement of Antoine's equation:

$$T = \frac{B}{A - \ln P^{\text{sat}}} - C$$

For isopropanol, $A = 18.6929$, $B = 3640.2$, and $C = 219.61$, where T is in units of °C and P^{sat} is in units of mmHg.

Hence, using $P^{\text{sat}} = 100\,\text{mbar} \times (760\,\text{mmHg}/1013.25\,\text{mbar}) = 75.0\,\text{mmHg}$ in the equation,

$$T = 33.6\,°\text{C}.$$

Now, substituting this into the expression, we obtain the following total drying time:

$$t_T = \frac{(5)(663 \times 1000)}{(40)(0.125)(60 - 34)} \left((1) - (0.1) + (0.1)\ln\left(\frac{(0.1)}{(0.02)}\right) \right)$$

$$= 27\,100\,\text{seconds}$$

So, the estimated total drying time for this operation is seven to eight hours. *Note that ΔH_{vap} and T were approximated for the calculation since the objective was simply a first-order estimate of the drying time.*

35.3.3.3 Form Control Hydrates, solvates, or other forms may require careful control of humidity or other drying parameters. In the case where the product is a hydrate and the hydration level needs to be maintained during drying (i.e. to avoid the loss or gain of water from the crystals), the relative humidity in the dryer must be controlled carefully. The composition of wash solvents prior to drying can be critical to ensure that there is sufficient water in the cake during drying to avoid dehydration. The use of a dew-point hygrometer in the vent line of a dryer can provide information on the relative humidity around the cake and can be used to monitor the removal of water during drying and ensure that the cake is not over dried [14]. In some cases, it may be helpful to charge additional water prior to or during drying to ensure that there is sufficient water present. The development of phase diagrams based on water activity can be used to design the appropriate conditions for maintaining a specific hydrate [18].

The same principles can be applied to maintain a certain solvate where a solvent other than water is bound within the crystal lattice of the product. In these cases, MS or NIR can be used to monitor the removal of critical solvents during drying.

Sometimes, the product crystals can undergo polymorph transformation during drying. In this case, offline techniques such as XRPD can be helpful to determine this after drying is complete. Online methods such as Raman spectroscopy can be used to monitor the cake while the cake is still drying to ensure that the product form is not changing. Polymorph transformation can be a result of shear and agitation, so this must be carefully controlled.

EXAMPLE PROBLEM 35.5

During a drying operation in which both isopropanol (IPA) and water are removed, it is desirable to preserve a monohydrate that is formed with the product. The product has also been shown to form an undesirable solvate with IPA that must be avoided. Use the data given in Figure 35.32 to suggest a strategy for maintaining the desired monohydrate while avoiding the undesired IPA solvate. Given that the composition of the azeotrope formed between IPA and water is 16 wt % water at 1 mmHg, would the strategy be successful at any operating pressure?

Solution
From the solubility plot it is clear that the solubility of the IPA solvate and monohydrate intersect at approximately 8 wt % water. At any given condition, the polymorph with the lower solubility will be the thermodynamically stable crystal form, and hence the monohydrate will be favored over the IPA solvate if the water content is maintained above 8 wt %.

In order to form and preserve the monohydrate, it is necessary to maintain a solvent composition in the drying cake of more than 8 wt % water. From the xy diagram, there is a minimum boiling azeotrope between IPA and water at 12 wt %

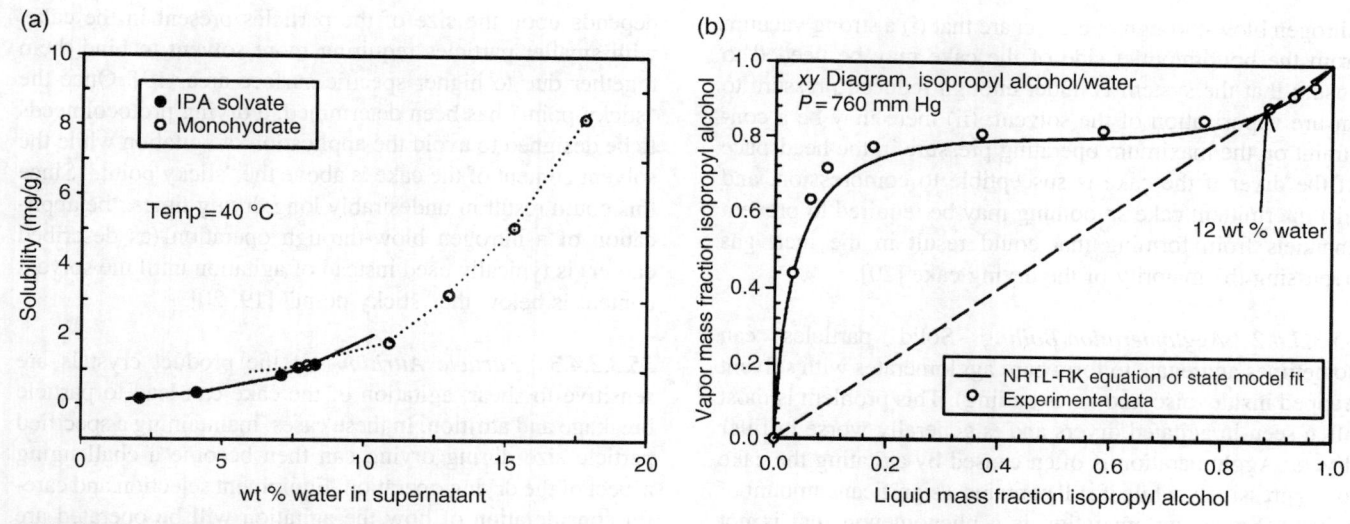

FIGURE 35.32 Solubility plot (on the left) showing the relative solubility of the IPA solvate and the monohydrate at different water contents (although the data was collected at 40 °C, the results are applicable at all temperatures – see monotropism) and an x–y-diagram (on the right) for IPA/water showing the existence of a minimum boiling azeotrope at 12 wt % water. *Source*: Reprinted with permission from Cypes et al. [14]. Copyright 2004, American Chemical Society.

water. If the final cake wash (i.e. the starting point for the drying operation) contains less than 12 wt % water, the liquid phase will shift to lower water concentrations as evaporation takes place, and it will be possible to reach a point during drying where the solvent composition in the cake is less than 8 wt % water. In this case, the monohydrate will not be stable, and the IPA solvate could be formed instead. If the final cake wash composition is greater than 12 wt % water, the solvent composition in the dryer will never be below the solvent composition transition point due to the presence of the azeotrope, and the monohydrate will remain the more stable form over the drying process. A suggested strategy is therefore to use a final wash composition of, say, 20 wt % water to ensure that the solvent composition always remains above 8 wt % water during the drying operation.

Using any final wash composition containing more than 16 wt % water should result in a successful strategy for obtaining the monohydrate, down to an operating pressure of 1 mmHg.

35.3.3.4 Troubleshooting
There are some common challenges faced during drying that we discuss below.

35.3.3.4.1 Long Drying Times
Long drying times can cause significant bottlenecks in the overall process and is most often a result of a prolonged period under the mass transfer limited mechanism. Drying times can be improved by ensuring that most of the drying operation is carried out under heat transfer-limited conditions and that both the driving force and the heat transfer area available are maximized as much as possible (within the constraints of product stability and equipment limitations). From Eq. (35.19), it can be seen that the difference between the jacket temperature and the product temperature provides the driving force for heat transfer and the heat transfer coefficient is equally as important. Static dryers such as a vacuum oven often do not provide enough surface area for heat transfer, whereas drying in agitated dryers can provide the benefits of increased surface area for heat transfer. Note that certain solvents such as water can be difficult to remove regardless of the equipment used, leading to longer drying times. Estimation of the critical moisture content in the lab and simulation of the drying process with a model can be helpful to ensure that appropriate operating conditions are used in the plant to achieve optimal drying cycle times.

An approach that is often used in manufacturing to reduce drying times is to flow an inert gas such as nitrogen through the drying cake. This strategy utilizes convection to remove residual solvent from the wet solids as the inert carrier gas passes through the cake. A strategy for scaling up this approach was proposed by Birch and Marziano [19], using a simplification of the Carman–Kozeny equation for flow through particulate beds [19]. In their work, they showed that the pressure drop across the cake is proportional to the specific gas flow rate (i.e. per unit cross-sectional area) multiplied by the depth of the cake. The proportionality constant is a function of the void fraction of the solid bed, the particle diameter of the solids, and the viscosity of the solvent being removed. It can be determined by lab experimentation and assumed to have the same value on a larger scale if the physical properties of the solid being isolated and the wash solvent remain the same. The pressure drop needed to maintain a specific gas flow rate on scale can then be calculated for the larger-scale bed depth to achieve the same rate of solvent removal [19]. Some important considerations when using a nitrogen blow-through operation are that (i) a strong vacuum from the bottom/outlet side of the cake may be needed to ensure that the system is under enough reduced pressure to ensure vaporization of the solvent, (ii) there may be a constraint on the maximum operating pressure in the headspace of the dryer if the cake is susceptible to compression, and (iii) intermittent cake smoothing may be required to prevent channels from forming that could result in the inert gas bypassing the majority of the drying cake [20].

35.3.3.4.2 Agglomeration/Balling Solid particles can sometimes aggregate to form hard agglomerates with solvent trapped inside (also known as balling). This problem is most often seen in agitated dryers and is generally worse in filter dryers. Agglomeration is often caused by agitating the cake too aggressively while it still contains a significant amount of solvent. Since agglomeration is a phenomenon that is not well understood and there are no specific guidelines on how to avoid the issue during drying, a general approach is to avoid agitation in the initial stages of the drying. One factor that may be important for avoiding the formation of agglomerates is the solubility of the product in the individual solvents that are being removed. Sometimes the cake has been washed with a mixture of solvents prior to drying and if there is enough solubility in one of these solvents by itself, the product could become partially dissolved that could lead to fusion of the crystals by the formation of solid bridges between them as the solvent is removed. One approach to avoiding agglomeration is therefore to adjust the solvent composition of the final cake wash to avoid the potential enrichment of highly soluble solvents during the drying operation. Agglomeration has been found to occur when the drying rate is high and the shear rate is relatively low [21, 22], indicating that there is a higher risk for agglomeration if agitation is applied during the initial stages of drying when the process is heat transfer controlled as opposed to mass transfer controlled. Indeed, there is often a critical amount of solvent, known as the "sticky point," at which the tendency to form agglomerates may be maximal [12]. This is due to the presence of solvent capillary forces strong enough to cause particles to agglomerate together if agitation is applied, in a similar fashion to how particles stick together in a wet granulation process. Birch and Marziano ([19]) showed that the "sticky point" can be determined experimentally using mixer torque rheometry, since the torque required to maintain a constant rotational speed during agitation will increase as binder solvent is added to the solids until it is at a maximum value when the agglomerates are strongest (i.e. at the "sticky point") [19]. The magnitude of the torque at this point is also an indication of the hardness of the agglomerates that are formed. However, a potentially more reliable estimate of agglomerate hardness (along with the extent of agglomeration) can be obtained by sieving samples of the solids [19]. Birch and Marziano [19] also showed that the "sticky point" depends upon the size of the particles present in the cake, with smaller particles requiring more solvent to bind them together due to higher specific surface area [19]. Once the "sticky point" has been determined, a drying protocol needs to be designed to avoid the application of agitation while the solvent content of the cake is above the "sticky point." Since this could result in undesirably long drying times, the application of a nitrogen blow-through operation (as described earlier) is typically used instead of agitation until the solvent content is below the "sticky point" [19, 20].

35.3.3.4.3 Particle Attrition If the product crystals are sensitive to shear, agitation of the cake can lead to particle breakage and attrition. In these cases, maintaining a specified particle size during drying can then become a challenging aspect of the drying operation. Equipment selection and careful consideration of how the agitation will be operated are critical for ensuring particle size integrity. Based on the geometry of the equipment and how the agitator interacts with the cake, the force exerted on the particles should be considered. Particles are likely to experience less force in a tumble dryer as opposed to a filter dryer or conical dryer. When the use of a tumble dryer is not an option, lab studies to determine the maximum shear that the particles can endure before breakage becomes a problem can be carried out. The plant agitation protocol could then be scaled up based on maintaining a lower tip speed than the critical value. However, it is important to note that the ranges of tip speeds achievable in the lab often do not overlap with those achievable in the plant. Particle attrition has been shown to occur when the drying rate is low and the shear rate is high [21, 22], indicating that there is a higher risk for particle attrition if continuous agitation is applied during the later stages of drying when the drying rate is limited by mass transfer.

Remy et al. [10] correlated the impact of agitated drying on powder properties with the amount of work done by blades on the cake, which is determined by the normal stresses (hydrostatic pressure) and the bulk friction coefficient of the cake (affected by the degree of wetness). The publication demonstrated an experimental setup that allowed for the determination of the cake's bulk friction coefficient that in turn could be used to estimate the work exerted on the cake in different large-scale equipment. From a laboratory setup, the impact on the powder properties at scale-relevant conditions could then be evaluated.

35.3.4 Summary

Proper control of the drying process is important especially for API compounds whose solvent content, particle size, and final form need to meet specifications. Depending on the solvent system and the equipment being used, a drying operation can often have a long cycle time and become a bottleneck in the overall process. Agitation is helpful to avoid this issue but can lead to other problems such as agglomeration or attrition. It is therefore important to select the appropriate equipment and carefully design the operating conditions. The scale-up of drying processes can be challenging due to the lack of understanding associated with some of the issues encountered; however, there are many tools and methods available for simulation, monitoring, and data collection that can be applied for design purposes.

REFERENCES

1. Darcy, H. (1856). *Les Fontaines Publiques de la Ville de Dijon ("the Public Fountains of the Town of Dijon")*. Paris: Dalmont.
2. Ruth, B.F., Montillo, G.H., and Montonna, R.E. (1933). Studies in filtration: I. Critical analysis of filtration theory; II Fundamentals of constant pressure filtration. *Industrial and Engineering Chemistry* 25: 76–82. 153–161.
3. McCabe, W.L., Smith, J.C., and Harriott, P. (2005). *Unit Operations of Chemical Engineering*, 7e. New York: McGraw-Hill.
4. Weigert, T. and Ripperger, S. (1997). Effect of filter fabric blinding on cake filtration. *Filtration and Separation* 34 (5): 507–510.
5. Bird, R.B., Stewart, W.E., and Lightfoot, E.N. (2001). *Transport Phenomena*, 2e. Hoboken, NJ: Wiley.
6. Barr, J.D. and White, L.R. (2006). Centrifugal drum filtration: I. A compression rheology model of cake formation. *AIChE Journal* 52: 545–556.
7. Landman, K.A., White, L.R., and Eberl, M. (1995). Pressure filtration of flocculated suspensions. *AIChE Journal* 41: 1687–1700.
8. Rideal, G. (2006). Roundup. *Filtration and Separation* 43 (1): 36–37.
9. Perry, R.H. (1984). *Perry's Chemical Engineer's Handbook*, 6e. New York: McGraw-Hill.
10. Remy, B., Kightlinger, W., Saurer, E.M. et al. (2015). Scale-up of agitated drying: effect of shear stress and hydrostatic pressure on active pharmaceutical ingredient powder properties. *AIChE Journal* 61: 407–418.
11. McConville, F.X. (2004). *The Pilot Plant Real Book: A Unique Handbook for the Chemical Process Industry*. Worcester, MA: FXM Engineering and Design.
12. Papadakis, S.E. and Bahu, R.E. (1992). The sticky issues of drying. *Drying Technology* 10 (4): 817–837.
13. Burgbacher, J. and Wiss, J. (2008). Industrial applications of online monitoring of drying processes of drug substances using NIR. *Organic Process Research and Development* 12: 235–242.
14. Cypes, S.H., Wenslow, R.M., Thomas, S.M. et al. (2004). Drying an organic monohydrate: crystal form instabilities and a factory-scale drying scheme to ensure monohydrate preservation. *Organic Process Research and Development* 8: 576–582.
15. Kohout, M., Collier, A.P., and Štěpánek, F. (2006). Mathematical modeling of solvent drying from a static particle bed. *Chemical Engineering Science* 61: 3674–3685.
16. Michaud, A., Peczalski, R., and Andrieu, J. (2008). Modeling of vacuum contact drying of crystalline powder packed beds. *Chemical Engineering and Processing: Process Intensification* 47 (4): 722–730.
17. Schlünder, E.-U. and Mollekopf, N. (1984). Vacuum contact drying of free flowing mechanically agitated particulate material. *Chemical Engineering and Processing* 18: 93–111.
18. Variankaval, N., Lee, C., Xu, J. et al. (2007). Water activity-mediated control of crystalline phases of an active pharmaceutical ingredient. *Organic Process Research and Development* 11: 229–236.
19. Birch, M. and Marziano, I. (2013). Understanding and avoidance of agglomeration during drying processes: a case study. *Organic Process Research and Development* 17: 1359–1366.
20. Nere, N.K., Allen, K.C., Marek, J.C., and Bordawekar, S.V. (2012). Drying process optimization for an API solvate using heat transfer model of an agitated filter dryer. *Pharmaceutical Technology* 101 (10): 3886–3895.
21. Lekhal, A., Girard, K.P., Brown, M.A. et al. (2003). Impact of agitated drying on crystal morphology: KCl-water system. *Powder Technology* 132: 119–130.
22. Lekhal, A., Girard, K.P., Brown, M.A. et al. (2004). The effect of agitated drying on the morphology of L-threonine (needle-like) crystals. *International Journal of Pharmaceutics* 270: 263–277.

36

FILTRATION CASE STUDIES

SETH HUGGINS AND ANDREW COSBIE
Drug Substance Technologies and Engineering, Amgen Inc., Thousand Oaks, CA, USA

JOHN GAERTNER
Process Research and Development, AbbVie Inc., North Chicago, IL, USA

36.1 INTRODUCTION

This chapter presents various approaches used to apply filtration fundamentals to develop efficient and robust processes, which consistently produce API meeting the physicochemical requirements for small molecule compounds. In these case studies, an appropriate level of understanding is developed to ensure confident determination of the design space and a scalable process. This can be challenging at times due to the complexity of the filtration processes and the limited availability of appropriate scale-down equipment and diagnostic tools. It can be further complicated by the wide variety of equipment used for filtration.

36.2 FILTRATION DECISION TREE

Complementary to Chapter 35, this section will highlight several case studies of how filtration characteristics are used to influence process parameters and equipment selection to arrive at optimal filtration conditions. The primary purpose of characterizing filtration at the lab scale is to ensure that the performance across scales, equipment, and operating conditions does not produce any unexpected response in rate or efficiency. In the case studies that follow, the experimental lab procedure and equipment were chosen to enable rapid filtration data collection and assessment in parallel to the standard commercial process development workflows.

To perform the bench-scale filtration characterization, any set of equipment may be used provided the filter is pressurated across a meaningful range, and the filtration mass (or volume) may be measured at regular intervals. As outlined in the Chapter 35, the bench-scale procedure may be used either at a single pressure to collect cake and media resistance or at several pressures to enable regression of the compressibility term. The equipment and procedures typically used in Amgen's Synthetic Technologies and Engineering labs are as follows:

- Ertel Alsop® 4T (250 ml) and/or Ertel Alsop® 10T (1.3 l) filters.
- Filter media to match the anticipated pilot- and commercial-scale application:
  – Typically PTFE, PP, nylon, or metal mesh with a porosity ranging from 10 to 50 μm.
- Mettler Toledo® PG or XS series scales.
- Mettler Toledo® balance link software.

The procedure utilizes a multistage pressure modulation over the course of a single experiment to regress the filtration characteristics from Eq. (36.1) [1]. Characterization is typically performed with 100 mL to 1 L of slurry. Filtrate mass is collected at one to five second intervals, where pressure is increased in a stepwise manner such that approximately one-third of the filtration is performed at a lower initial pressure (typically 5–14 psi), a third at an intermediate pressure (10–40 psi), and the remainder at a higher pressure (30–90 psi). This process is repeated over several lots or when there are any notable changes to the solid-state properties of the

---

Chemical Engineering in the Pharmaceutical Industry: Active Pharmaceutical Ingredients, Second Edition.
Edited by David J. am Ende and Mary T. am Ende.
© 2019 John Wiley & Sons, Inc. Published 2019 by John Wiley & Sons, Inc.

material to aid in understanding of the inherent variability and potential relationship to particle properties.

In addition to pressure (ΔP) and filtrate rate (m/t), data is collected for mass fraction of the solids (w), mass of filtrate (m), filter surface area (A), filtrate density (ρ), and filtrate viscosity (μ). This data is used along with the modified Darcy's law to regress the average cake resistance (α_{avg}), media resistance (R_m), and compressibility index (n) Figure 36.1:

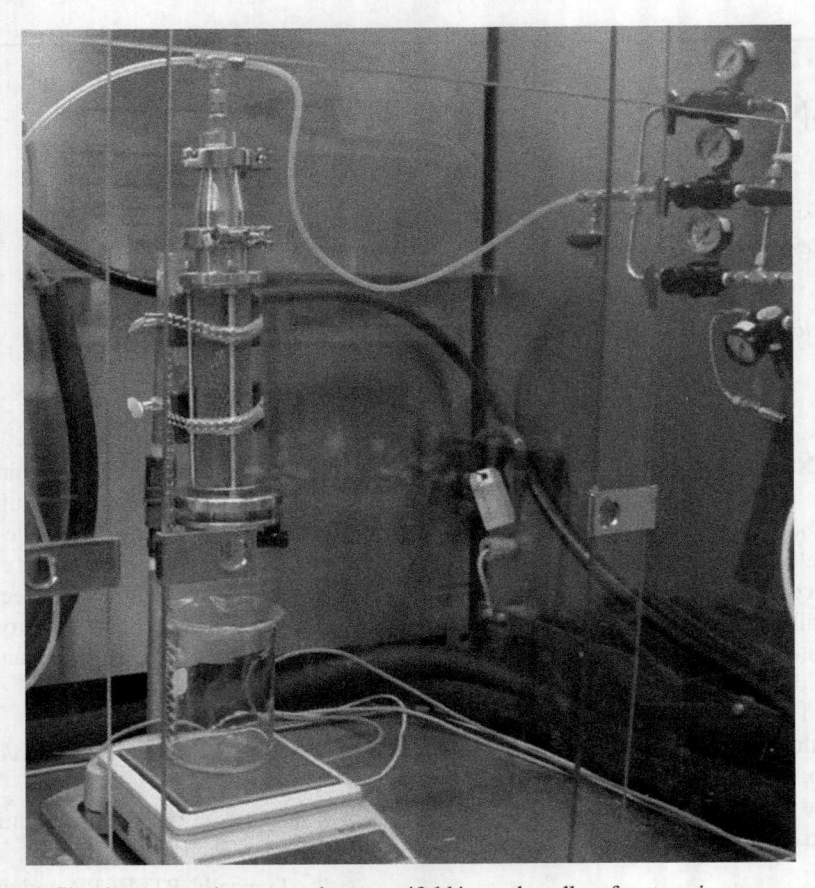

FIGURE 36.1 Standard bench filtration setup where a regulator manifold is used to allow for stepwise pressure modulation during filtration.

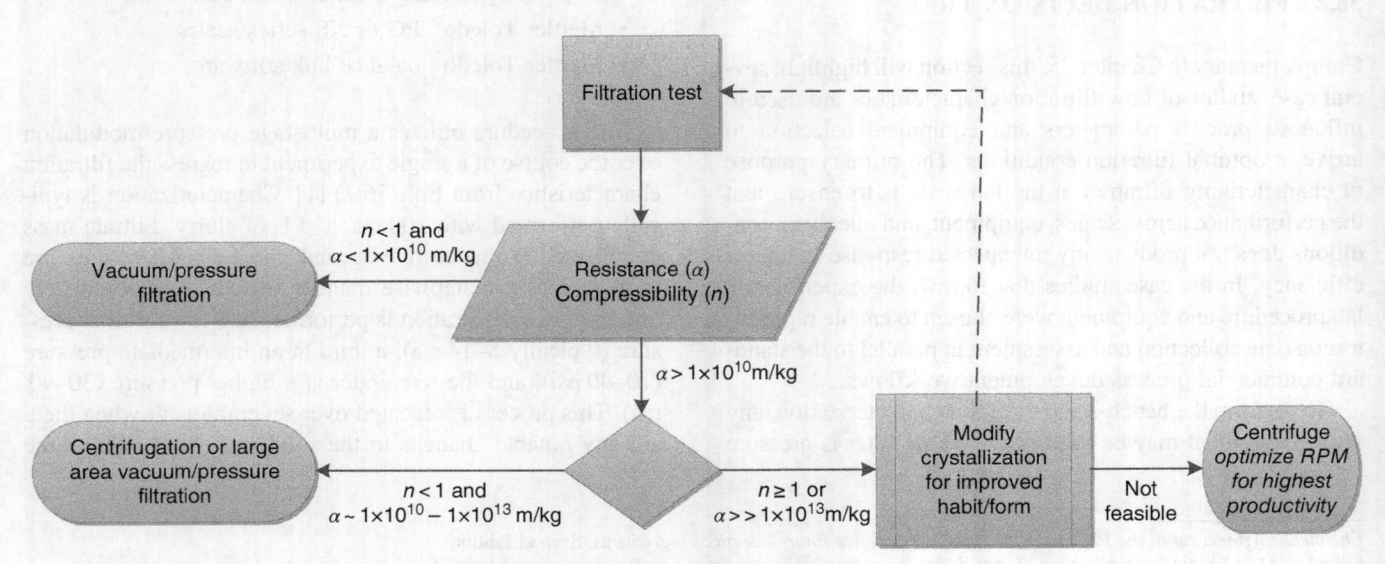

FIGURE 36.2 Guide for utilization of filtration information.

$$\left(\frac{t}{m}\right) = \left(\frac{\mu R_m}{\rho A \Delta P}\right) + \left(\frac{\mu \alpha_{avg} \dfrac{w}{1-w}}{\rho 2 A^2 \Delta P}\right) m \quad \text{where} \quad \alpha_{avg} = \alpha_o \Delta P^n$$

(36.1)

The regression is best accomplished through a fitting software either commercially available or with "home-grown" code. One such platform widely used across the pharma industry is DynoChem® software from Scale-up Systems Ltd. Both the regression and simulation capabilities for this application are available in DynoChem® and were used in the majority of filtration case studies that follow. Regardless of the regression platform, it is important that the number of data points be approximately equal across the different experimental pressures employed to ensure that the regression fit is not biased toward the experimental conditions with the largest number of data points. Alternatively, a proportional weighting could be applied to balance the data sets.

A decision tree (Figure 36.2) was developed to use the filtration characteristics to guide the development strategy and to inform productivity impact from equipment selection during scale-up or tech transfer. This guide, in conjunction with known logistical and process constraints, is intended to be used as part of the strategy that determines the equipment and processing options.

The case studies that follow begin with the most ideal state, incompressible cake with low resistance, and increase in complexity while highlighting the rationale of the above guidance.

36.3 LOW TO MODERATE CAKE RESISTANCE

In the cases where the resistance is low, average cake resistance (α_{avg}) < 1×10^{10} m/kg, it is often suitable to move directly to utilization of the characteristic information to simulate the performance of the filtration in the available equipment (Figure 36.3). However, care should be given to ensure the properties of the solid state will not change appreciably upon scaling. The following case demonstrates the scenario when filtration resistance is low and outlines the typical flow of experimentation, characterization, and utilization of the filter cake parameters to simulate filtrations at larger scales to ensure proper selection of equipment and operating parameters.

Compound A is crystallized as rod-shaped particles on the order of 50–100 μm in length and an aspect ratio of about 3 : 10 (Figure 36.4). This slurry is observed to filter rapidly and the particles are not prone to attrition while in the slurry state. From bench-scale filtration, filtrate mass vs. time data was used to regress a specific cake resistance, media resistance, and compressibility index. The results are shown in Table 36.1, indicating the slurry was suitable for filtration in standard pressure filtration equipment such as an agitated filter dryer (AFD) or an Aurora A-20 as shown in Figure 36.5. These values were subsequently used to simulate filtration performance, prior to executing in the kilo lab. The case modeled was 1.6 kg of compound A in 28 kg of solvent with a viscosity of 0.777 cP filtered on a filter surface area of 0.05 m² with a driving force of 14 psi. The simulation predicted a filtration time of 11 minutes, indicating the operation was suitable for the available equipment. This scenario was executed in a kilo lab, and the filtration was completed in 13 minutes, aligning well with the model prediction.

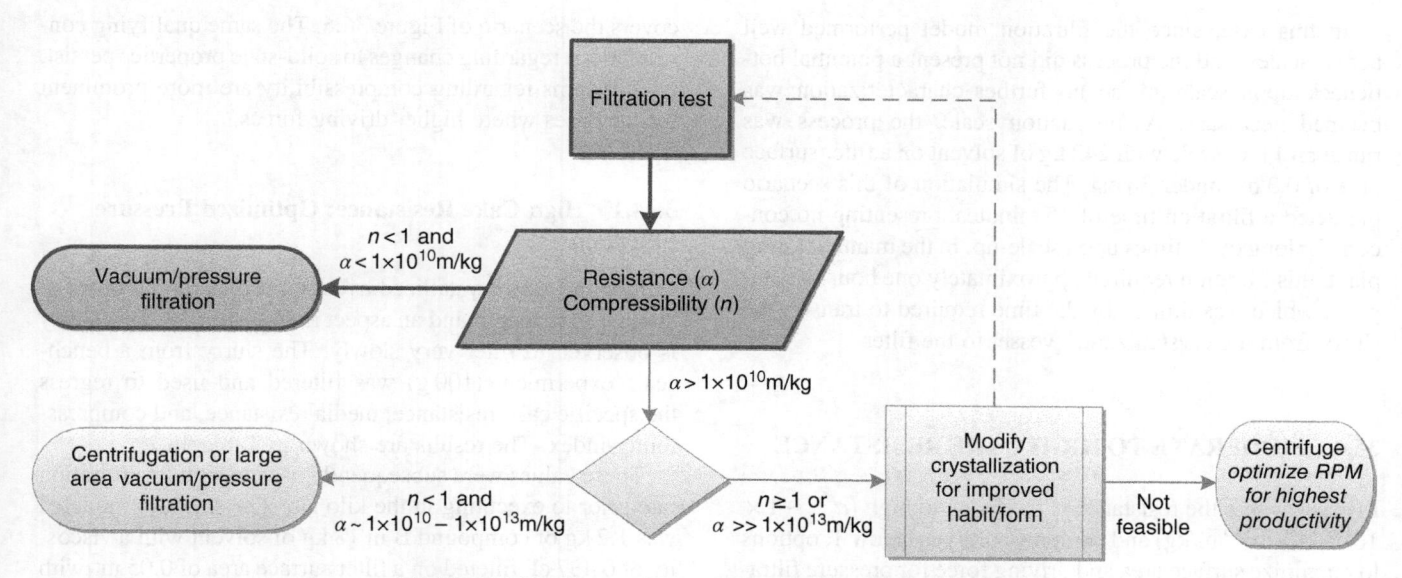

FIGURE 36.3 Low resistance, low compressibility scenario.

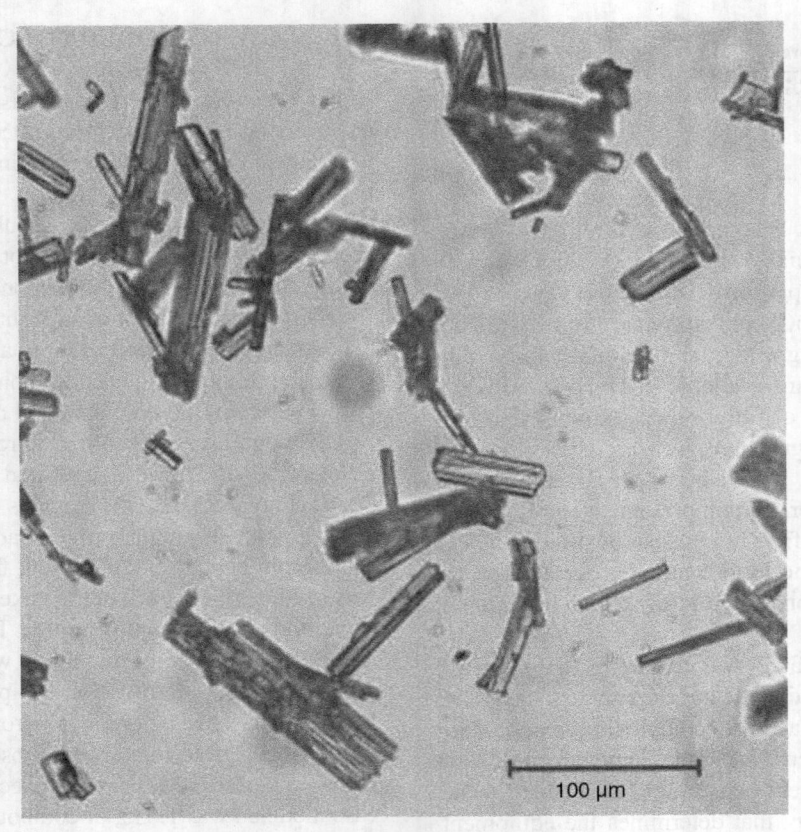

FIGURE 36.4 Microscope image of compound A crystals in slurry.

TABLE 36.1 Filtration Parameters for Compound A

Scenario	Specific Cake Resistance (α)	Media Resistance (R_m)	Compressibility Index (n)
Bench-scale data	1.03×10^{10} m/kg	1.81×10^9 1/m	0.6323

In this case, since the filtration model performed well across scales, and the process did not present a potential bottleneck upon scale-up, so no further characterization was deemed necessary. At production scale, the process was run at an 11 kg scale with 242 kg of solvent on a filter surface area of 0.3 m² under 30 psi. The simulation of this scenario predicted a filtration time of 15 minutes, presenting no concern for long cycle times upon scale-up. In the manufacturing plant, this filtration required approximately one hour to complete, which was limited by the time required to transfer the slurry from the crystallization vessel to the filter.

36.4 MODERATE TO HIGH CAKE RESISTANCE

In cases where the resistance is moderate to high ($\alpha_{avg} \sim 1 \times 10^{10} - 1 \times 10^{13}$ m/kg) and compressibility is below 1, options to maximize surface area and driving force for pressure filtration or to employ the use of a centrifuge are evaluated. This covers the scenario of Figure 36.6. The same qualifying considerations regarding changes to solid-state properties persist, but concerns regarding compressibility are more prominent for the cases where higher driving forces.

36.4.1 High Cake Resistance: Optimized Pressure Filtration

Compound B is crystallized as long needles on the order of 20–50 μm in length and an aspect ratio of 10 : 30. This slurry is observed to filter very slowly. The slurry from a bench-scale experiment (100 g) was filtered and used to regress the specific cake resistance, media resistance, and compressibility index. The results are shown in Table 36.2.

These values were subsequently used to simulate filtration rate prior to executing in the kilo lab. The scenario modeled was 1.2 kg of compound B in 18 kg of solvent with a viscosity of 0.437 cP filtered on a filter surface area of 0.05 m² with a driving force of 14 psi. The simulation predicted a filtration

FIGURE 36.5 Aurora filter used in kilo lab.

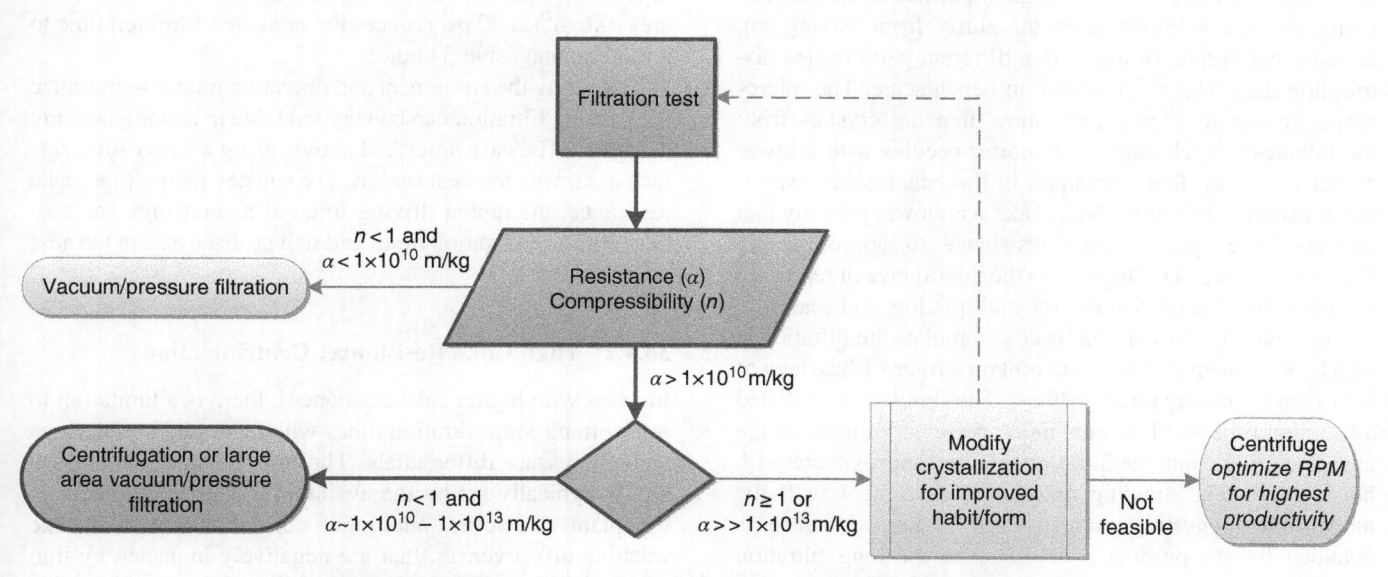

FIGURE 36.6 Moderate to high resistance, low compressibility scenario.

TABLE 36.2 Filtration Parameters for Compound B

Scenario	Specific Cake Resistance (α)	Media Resistance (R_m)	Compressibility Index (n)
From bench-scale data	2.44×10^{11} m/kg	1.54×10^{10} 1/m	0.8875

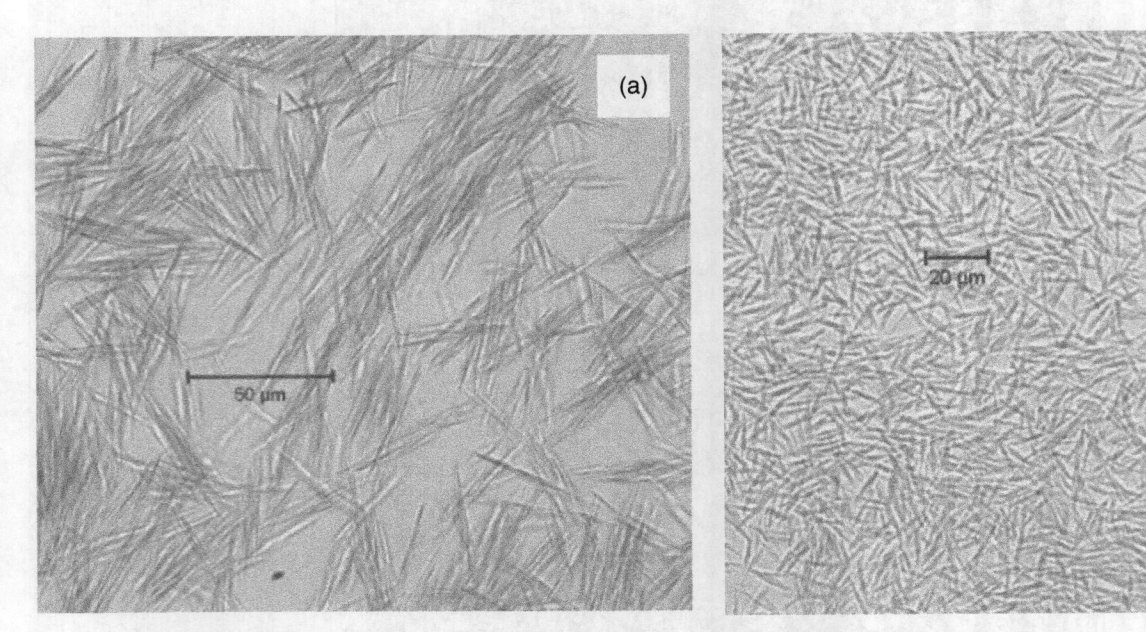

FIGURE 36.7 Microscope image of compound B crystals in slurry from lab-scale experiment (a) and kilo-scale batch (b).

time of 71 minutes. This scenario was executed in a kilo lab and the filtration required 137 minutes to complete.

While the regression of filtration parameters for the bench-scale data resulted in a good fit, that does not mean that the experiment is necessarily representative of performance on larger scale. This particular crystallization is nucleation dominant and can thus result in a range of particle sizes. Furthermore, the forces imparted on the slurry from mixing can increase with scale, resulting in a different particle size distribution than what is observed on bench scale. The microscope images in Figure 36.7 show that the crystals from the kilo-scale batch consist of shorter needles with a lower aspect ratio than those obtained in the bench-scale experiment, resulting in a more dense cake with lower porosity that increased the specific cake resistance to approximately 4.81×10^{11} m/kg. This highlights the importance of repeating the filtration characterization for multiple lots and scales.

For scale-up, this data was used to simulate the filtration at a 16 kg scale with 260 kg of solvent on a filter surface area of 0.3 m^2 and a driving force of 30 psi. The simulation predicted a filtration time of 11.4 hours under these conditions. At the manufacturing plant, the filtration required approximately 14 hours per batch, showing reasonable agreement with the model but suggesting that an even higher resistance is obtained for the product slurry. Due to the long filtration time, higher pressures and greater surface areas were explored in order to understand these parameter's sensitivities with respect to cycle time. Due to the high compressibility index, continuing to increase pressure will afford diminishing returns in reducing filtration time. Doubling the driving force pressure to 60 psi (maximum for this vessel) only reduces predicted filtration time to 10.5 hours (8% reduction). Alternatively, using a filter with twice the surface area (0.6 m^2) at 30 psi reduces the predicted filtration time to a more manageable 3 hours.

Exploring the equipment and operating ranges to optimize the product filtration can be very valuable in debottlenecking a process. The case described above, using a larger filter surface area, was the best option. For slurries with a high cake resistance, the higher driving force of a centrifuge can also help optimize filtration times and will be discussed in the next case study.

36.4.2 High Cake Resistance: Centrifugation

In cases with higher cake resistances, there is a limitation to overcoming long filtration times with increased surface area and/or pressure differentials. The upper end to the surface area is typically set by the availability of the equipment in the plant. However, additional considerations, like cake washing effectiveness, that are negatively impacted by thin cake heights should also be considered when evaluating this option. The limit to the pressure differential for cakes with compressibilities lower than one is typically set by the equipment specification or the utility capabilities of the production facility. While some Nutsche-type filters are rated for pressures in the tens of bar, it is more common for pressure filters to be operated with a pressure differential up to 6 bars due to equipment rating and/or available pressure from nitrogen utilities [2]. This driving force is significantly lower than the driving forces that can be applied through the use of a centrifuge. In addition to increased filtrate flux, there are many other considerations that should be included during the equipment selection strategy, which may include deliquoring efficiency, washing efficiency, containment, scheduling, and others.

The following case study exemplifies a typical centrifuge operation for moderate to high cake resistance, where the compressibility is less than one. In this case, an API intermediate was being isolated at a contract manufacturing facility where little characterization of the filtration had been performed. Upon the initial scaling, a long filtration time had been observed. As a response, the vendor split the 45–50 kg batches across two filters to double the filter area in efforts to minimize the filtration time. This was successful to a limited extent, and the resulting filtration time averaged approximately at 16 hours. With a shift in material demand, the batch size needed to be increased to 115 kg. To accommodate this increase in scale, and to improve productivity, a centrifuge was evaluated. To ensure that equipment choice would improve performance, the cake resistance and compressibility were characterized, and equipment information was collected to model and simulate the centrifugation performance. The resistance and compressibility were determined to be 1.2×10^{11} m/kg and 0.9, respectively. The details for the centrifuge surface area, diameter, and rotation speeds were collected and used to simulate the process as seen in Figure 36.8. The results from the simulation suggested that the primary filtration should be complete at approximately 115 minutes, indicating the equipment and operating parameters for the filtration were suitable for the scale increase. The process was subsequently scaled and filtered. The observed filtration time was 125 minutes, in agreement with the model and further demonstrating the value of collecting and using the characteristic data to aid in equipment selection and setting of operation parameters.

36.4.3 High Cake Resistance: Centrifugation and Compressibility

All of the previous examples have had compressibilities less than one. By examination of the modified Darcy's law (36.1), it can be seen that when a material is incompressible ($n = 0$),

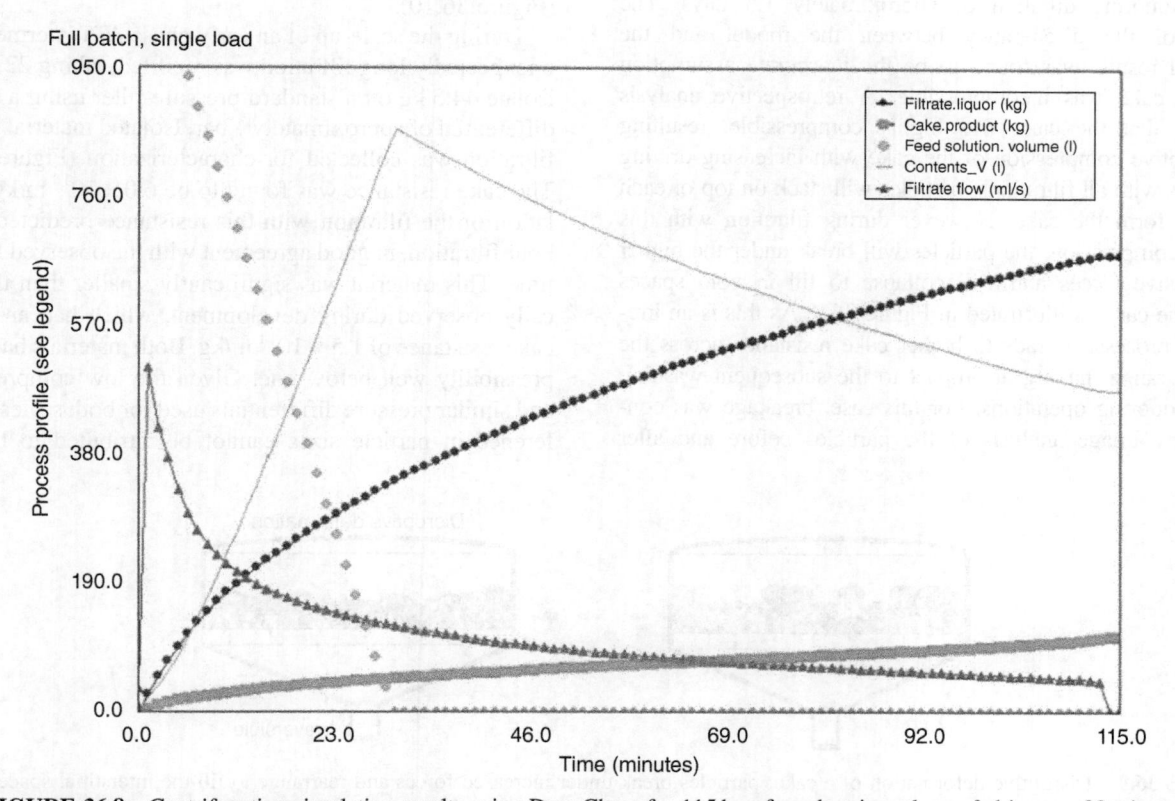

FIGURE 36.8 Centrifugation simulation results using DynoChem for 115 kg of product in a slurry, fed in over 30 minutes.

the rate of filtration is linearly proportional to the pressure differential. However, most of the synthetic materials that have been characterized within Amgen's Synthetic Technologies and Engineering labs have ranged from 0.2 to 0.9, in agreement with other commonly reported ranges [3]. Within this range, filtration rate will increase with an increasing driving force, but the proportionality of this relationship diminishes as the compressibility increases. As most materials fall into this low to moderate compressibility range, it is not unreasonable to make this assumption in cases where time or material availability limits the characterization opportunities. This assumption was made for an API intermediate that needed to be rapidly scaled to 3 kg in a kilo lab facility. Due to the limited time and material, cake resistance was measured at a single pressure and determined to be 1.0×10^{13} m/kg. The material was assumed to be incompressible, and the filtration was simulated with various equipment to identify a suitable configuration that minimized the filtration time. Using a standard filter in the kilo lab with an area of $0.2\ m^2$ predicted a filtration time of approximately 1690 minutes. As this was not a desirable duration, a centrifuge with a bowl height of 28 cm and a diameter of 40 cm was simulated at 2000 rpm, predicting a filtration time of approximately 85 minutes. The operation was later carried out in a centrifuge with an observed filtration time of 480 minutes, indicating the model was not accurate. The issue causing the slower rate carried into the washing, giving a total filtration and washing and deliquoring duration of approximately 1.5 days. The source of the discrepancy between the model and the observed results was found to be the inaccurate assumption that the cake was incompressible. A retrospective analysis revealed that the cake was highly compressible, resulting in disruptive compression of the cake with increasing driving force. As with all filtrations, particles will stack on top of each other to form the cake. However during filtration with this type of compression, the particles will break under the higher compressive forces and will collapse to fill in void spaces within the cake as illustrated in Figure 36.9. As this is an irreversible process, it leads to higher cake resistance across the pressure range, having an impact to the subsequent washing and deliquoring operations. For this case, breakage was confirmed by image analysis of the particles before and after centrifugation. Analysis of filtration performance for the material revealed that cake has a compressibility greater than or equal to 1.15. In the regime of compressibility greater than or equal to 1.0, it is possible to increase the driving force to an extent where the filtration rate starts to decrease as pressure is further increased. In this example, if the compressibility of 1.15 had been used to simulate the centrifugation, it would have resulted in a predicted filtration time of 500 minutes. Further, the model would have shown a more optimal rotation speed of 600 rpm would have further decreased the filtration time to approximately 270 minutes. In addition to the value of incorporating compressibility into the filtration characterization, this case also highlights the need to ensure that filtration characterization is completed over a pressure range that covers the driving forces that will be used during production.

36.5 REDESIGN OF SOLID-STATE PROPERTIES FOR IMPROVED FILTRATION

In cases where the resistance or the compressibility of the cake is high, it may not be feasible to achieve a desirable filtration rate via equipment selection or modifications to the operating parameters. When presented with this scenario, it is advantageous to evaluate modifications to the process that alter the solid-state properties of the material being isolated (Figure 36.10).

During the scale-up of an early phase API intermediate an unexpectedly long filtration was reported, taking 22 hours to isolate 44.5 kg on a standard pressure filter using a pressure differential of approximately 1 bar. Isolated material from the filtration was collected for characterization (Figure 36.11). The cake resistance was found to be 6.0×10^{14} m/kg. Simulation of the filtration with this resistance predicted a 23.5 hour filtration, in good agreement with the observed filtration time. This material was significantly smaller than that typically observed during development, which had an average cake resistance of 1.5×10^{11} m/kg. Both materials had a compressibility well below one. Given the low compressibility and similar pressure differentials used for both scales, the difference in particle sizes cannot be attributed to breakage

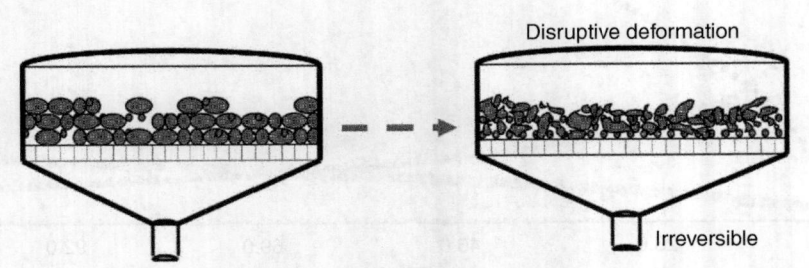

FIGURE 36.9 Disruptive deformation of a cake: particles break under increased forces and rearrange to fill the interstitial space between particles, resulting in a smaller void fraction and observed increase in cake resistance.

FILTRATION CASE STUDIES   841

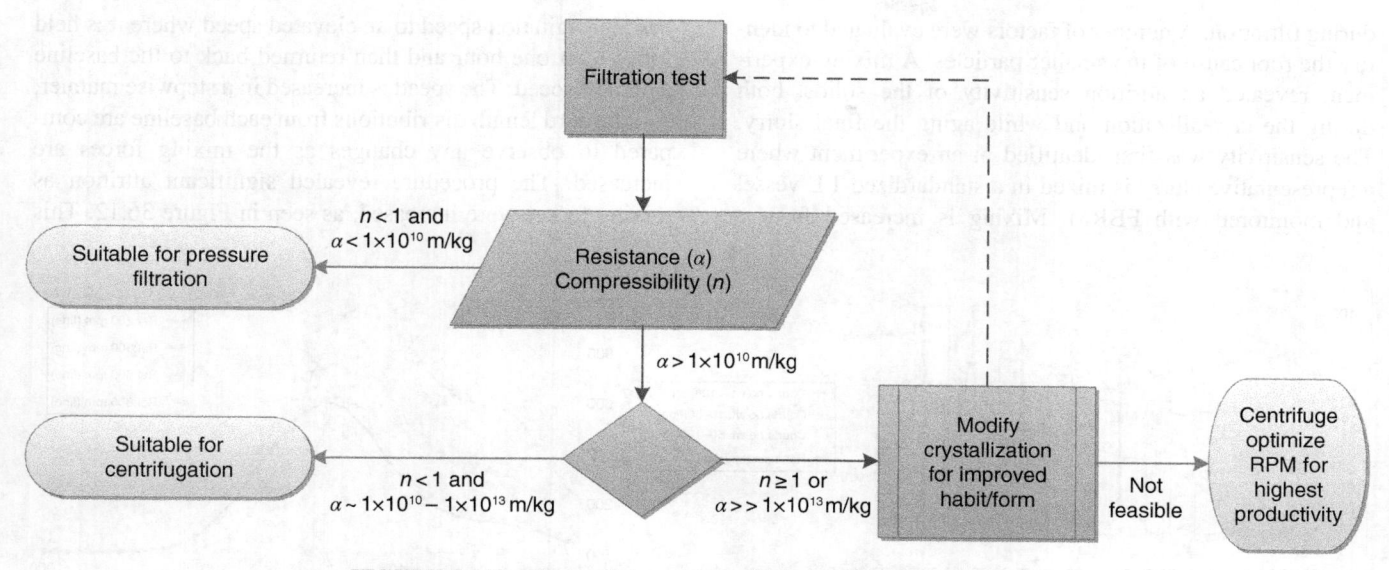

FIGURE 36.10 High resistance and/or high compressibility scenario.

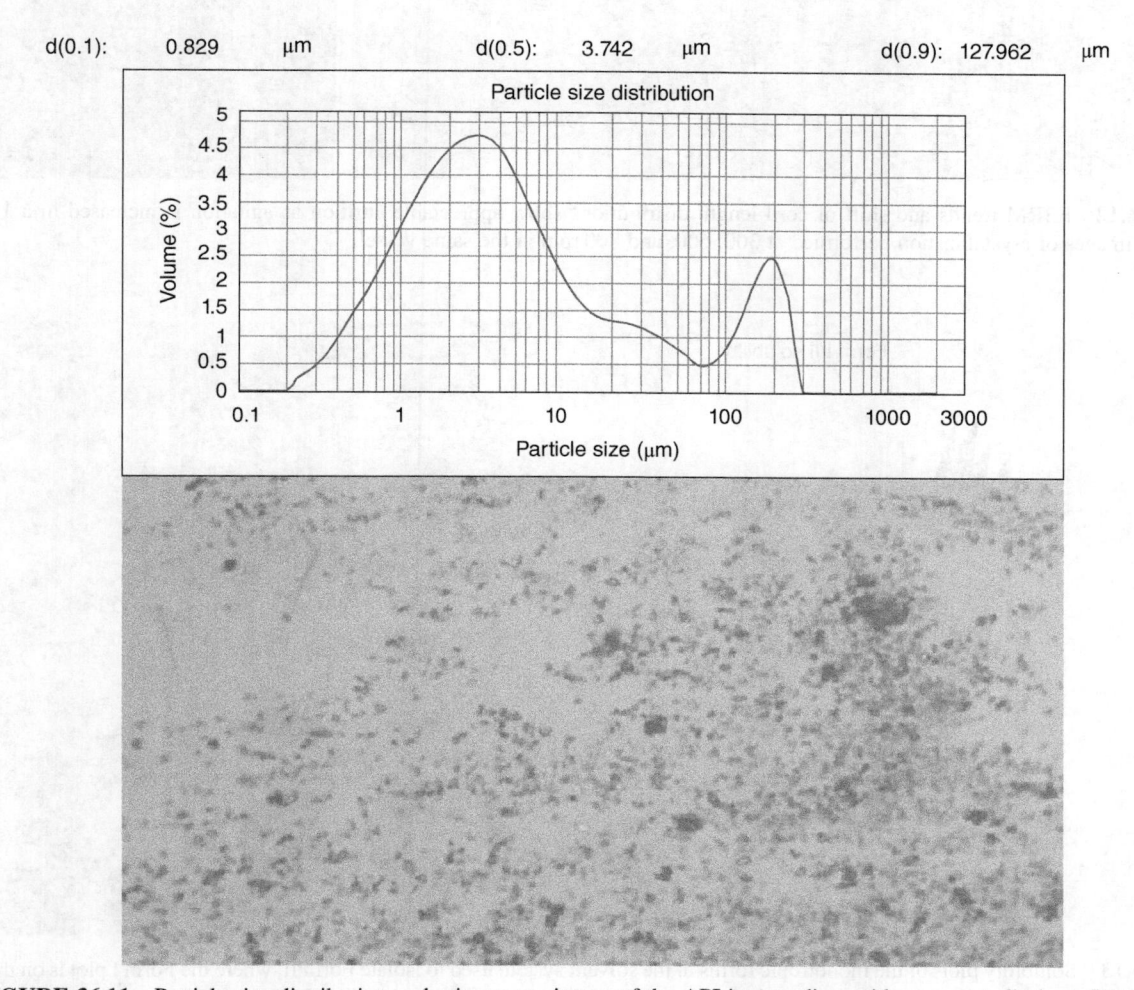

d(0.1): 0.829 μm   d(0.5): 3.742 μm   d(0.9): 127.962 μm

FIGURE 36.11 Particle size distribution and microscope image of the API intermediate with unexpectedly long filtration.

during filtration. A number of factors were evaluated to identify the root cause of the smaller particles. A mixing experiment revealed an attrition sensitivity of the solids, both during the crystallization and while aging the final slurry. The sensitivity was first identified in an experiment where a representative slurry is mixed in a standardized 1 L vessel and monitored with FBRM. Mixing is increased from a baseline agitation speed to an elevated speed where it is held for at least one hour and then returned back to the baseline agitation speed. The speed is increased in a stepwise manner, and the cord length distributions from each baseline are compared to observe any changes as the mixing forces are increased. The procedure revealed significant attrition as mixing forces were increased, as seen in Figure 36.12. This

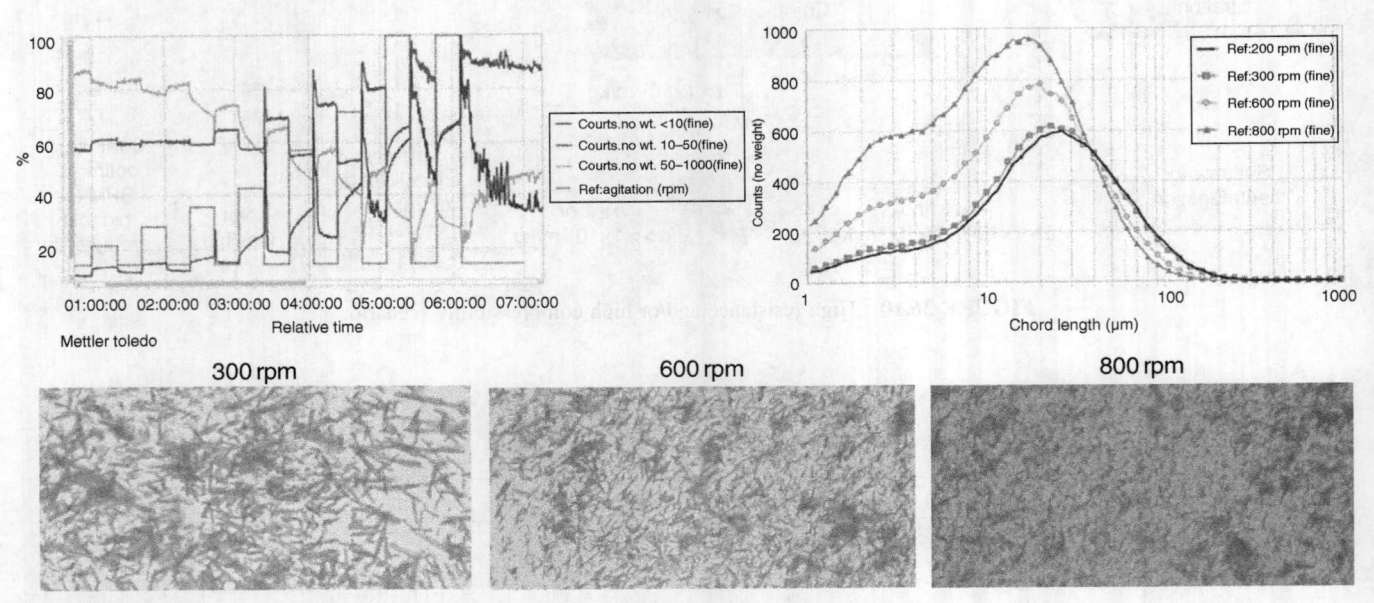

FIGURE 36.12 FBRM trends and shift of cord length distributions show appreciable attrition as agitation is increased in a 1 L vessel. Microscope images of crystallization performed at 300, 600, and 800 rpm in the same vessel.

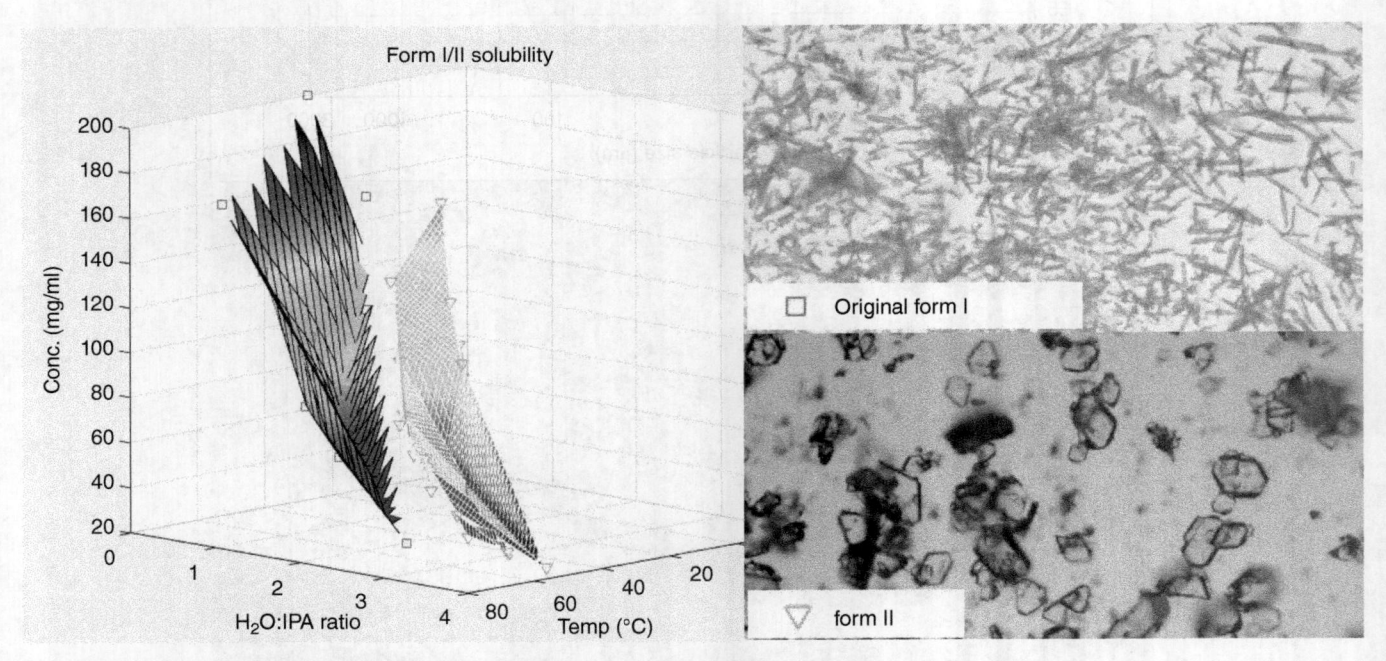

FIGURE 36.13 Solubility plots of the monotropic forms in the solvent system used to isolate Form II, where the Form I plot is on the left and Form II on the right. Form I has a melting point of 150 °C and Form II has a melting point of 170 °C.

highlights the importance of completing studies of this nature to ensure that the materials being used to characterize the filtration are truly representative of the *range* of particles that will be achieved through the crystallization and scaling.

As the attrition of the particles leads to higher cake resistance and mixing forces are expected to increase with scale, prolonged filtration may not be easily resolved with changes to filter type, size, or driving force. Instead, to mitigate the issue, efforts were applied to change the solid-state properties of the material. The investigation focused on solvents known to be advantageous to the upstream reaction, where the solvent type and ratios were varied to investigate the impact to habit or form. While slightly different sizes could be obtained from some of the combinations, all materials of this form exhibited similar propensity for attrition. However, an alternative polymorph (Form II) was identified in a water, isopropyl alcohol solvent system that exhibited a different habit with improved filtration and resistance to attrition. A comparison of the habit and solubility of the two forms is shown in Figure 36.13. Form II had a cake resistance of 3×10^9 m/kg and no notable propensity for attrition under relevant mixing conditions, alleviating the prolonged filtrations.

Ideally, problematic characteristics of the cake are identified early in the process development cycle such that there is ample time to investigate modification to the material's solid-state properties to improve the filtration. In the previous example, a change to the crystal form was the means to modify the solid-state properties. Other common practices include changes to the supersaturation profile during crystallization to achieve larger or more uniform PSDs, heat cycling to remove fines, modifications of solvent/cosolvent system to promote a different form or habit, or changes to purity profiles that impact the particle aspect ratios.

36.6 CUMULATIVE PLUGGING OF AGITATED FILTER DRYER PLATE

During process development of a compound with prismatic habit, lab pressure filtrations of a crystal slurry were performed to measure the flow resistance of the incompressible wet cake. A low resistance of approximately 10^6 m/kg was measured indicating a rapidly filtering cake. Filtrations and cake washes performed in the pilot plant using AFDs and Nutsche filters were also fairly rapid. Between batches, the AFD was cleaned per standard pilot plant practice. The Nutsche filters, shown in Figure 36.14, have a perforated bottom support plate and were lined with polymeric filter cloth. The polymeric filter cloth used in the Nutsche filter is removed and discarded after each batch and replaced with a new filter cloth.

Due to the good filtration properties of the compound – including absence of cake cracking – observed during pilot plant batches, filtration received modest attention during process scale-up and transfer to the production site. The initial sets of batches at the production site consisted of three campaigns of 2, 2, and 5 batches. Production batches of a given compound are frequently campaigned for efficiency to minimize equipment cleaning time. Inter-batch cleaning within a campaign consisted of cleaning the reactors but not the AFD. Inter-batch cleaning of the AFD would reduce the yield of each batch. A thorough cleaning of the entire

FIGURE 36.14 Side and top views of an open Nutsche filter.

TABLE 36.3 Production-Scale AFD Filtration Rates

	Campaign 1		Campaign 2		Campaign 3				
	Batch 1 Rate (cm/min)	Batch 2 Rate (cm/min)	Batch 1 Rate (cm/min)	Batch 2 Rate (cm/min)	Batch 1 Rate (cm/min)	Batch 2 Rate (cm/min)	Batch 3 Rate (cm/min)	Batch 4 Rate (cm/min)	Batch 5 Rate (cm/min)
Filtration of crystal slurry	15.6	7.0	12.9	3.0	13.8	9.1	6.4	1.9	1.7
Filtration of Wash #1	7.2	0.9	13.4	3.4	12.8	8.7	7.1	0.7	2.2
Filtration of Wash #2	7.8	1.3	6.2	2.7	7.0	4.1	3.7	1.8	1.5

Expressed as superficial linear velocity of the filtrate.

equipment train (including AFD) is performed at the end of each campaign.

When an AFD is unloaded, nearly all of the dried compound is removed from the AFD by the action of the blade; in some cases a rake-like instrument may be used to manually remove additional dried solids. However, a thin heel of dry solids typically remains on the AFD filter plate below the bottom of the blade. This thin layer may be partially compressed or attrited – resulting in additional flow resistance in subsequent batches of the campaign. In addition, cumulative plugging of the AFD filter plate can occur during the campaign; this results from penetration of fines into the filter plate with each batch.

After the crystal slurry is filtered in the AFD, the cake is displacement washed with two equal portions of wash. This is followed by a two hours nitrogen blow to deliquor the cake and then vacuum drying. The production site filtration receiver (unlike the pilot plant receiver) had a level sensor enabling accurate measurement of filtration and cake wash rates. Filtration rates for the crystal slurry filtration and two displacement cake washes appear in Table 36.3 for the nine batches, which were of the same batch size. The filtration rate decreases over the course of each campaign for all three campaigns. For example, in campaign 2, the batch 2 filtration rate was 70% less than the batch 1 rate. This reduction in filtration rate increased the overall processing time.

The cumulative filter plate plugging impacted downstream processing. After completion of cake washing, pressurized nitrogen is applied to the AFD headspace for two hours to fully deliquor the cake by flowing the nitrogen through the cake and out the bottom AFD discharge port. The bottom AFD outlet is vented at atmospheric pressure. This deliquoring reduces the wet cake residual solvent content from 30 to ~10 wt %. Regions of wet cake above partially plugged portions of the filter plate were less effectively deliquored by nitrogen flow. This increased the burden on the subsequent heated vacuum drying step.

In the vacuum drying step, the AFD jacket is heated, and full vacuum is applied to the bottom AFD outlet port. A slight flow of N_2 is applied to the AFD headspace for convective transport of solvent vapor through the cake. The transition from the positive pressure of the N_2 blowdown to vacuum for heated drying in the upper AFD chamber typically takes about five minutes as shown in the process trend in Figure 36.15. This was the case for the first batch of the campaign. However, due to the cumulative partial plugging of the filter plate, it took three hours in the fourth batch (in campaign 3) to remove the N_2 in the AFD headspace through the filter plate to the fully evacuated bottom chamber of the AFD.

Over the course of the heated vacuum drying, the cake in the AFD was sampled for residual solvent analysis. The residual solvent concentration of the first sample of cake (taken at the start of drying when the dryer reached temperature) increased over the course of the campaign (see Figure 36.16). Similarly, the required drying time increased during the campaign. The fifth batch took twice as long as the first batch to meet the residual solvent specification.

Increased operational efficiency can be achieved by monitoring cumulative filter plate plugging. During a campaign of batches, one approach to monitor cumulative AFD plate plugging is to measure the combined resistance of the heel layer and filter plate in the emptied AFD using a stream of nitrogen. This can be performed during the pressure check of the AFD that precedes each batch. The combined resistance should be no more than about 25% of the expected cake resistance. This corresponds to a combined plate–heel resistance of approximately 1E+11 1/m (for the case of a 30 cm tall

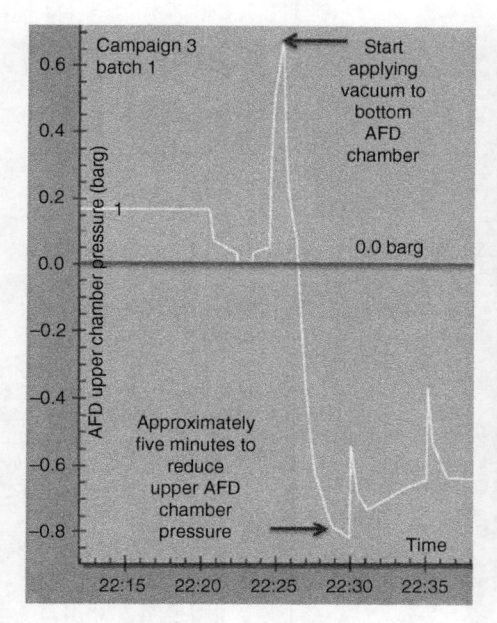

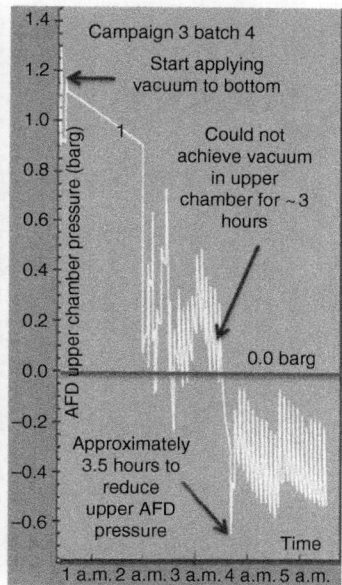

FIGURE 36.15 Trends of the upper AFD chamber pressure (barg) for batches 1 and 4 during the switch from the positive pressure of the N_2 blowdown to vacuum heated drying. The two trends have different time scales on the x-axis.

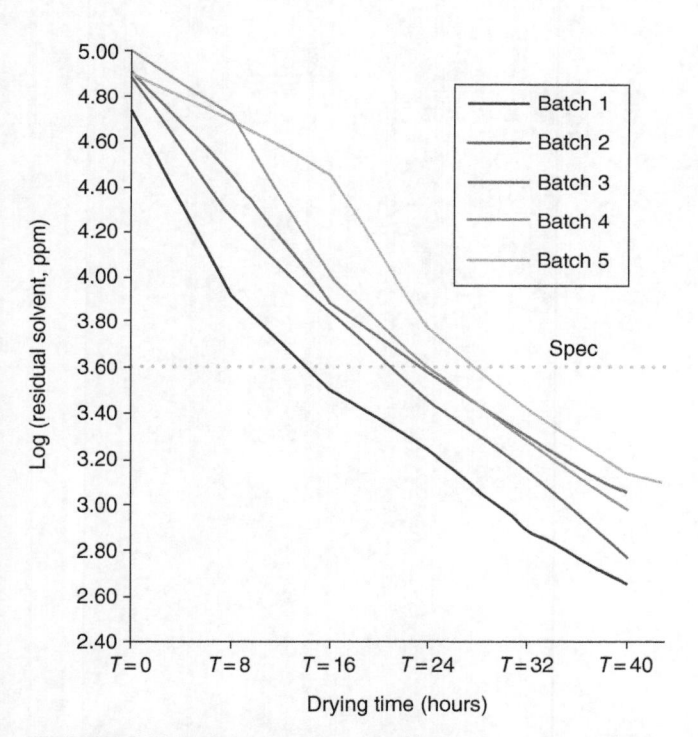

FIGURE 36.16 Concentration of residual solvent (log scale) in the wet cake during the course of drying.

cake bed). For example, when 100 SCFH of N_2 is applied to an AFD with a plate area of 1.0 m² (with discharge at atmospheric pressure), the pressure drop should not exceed 0.15 psi. If substantial filter plate plugging/heel resistance is detected, then a cleaning should be performed to remove the heel as well as solids embedded in the filter plate. If the filter cake becomes compressed or the filter plate becomes plugged during the processing of a batch, then one option to accelerate drying is to pull vacuum from the AFD headspace during the drying.

REFERENCES

1. Murugesan, S., Hallow, D.M., Vernille, J.P. et al. (2012). Lean filtration: approaches for the estimation of cake properties. *Org. Process Res. Dev.* 16: 42–48.
2. Palosaari, S., Louhi-Kultanen, M., and Sha, Z. (2015). Industrial crystallization. In: *Handbook of Industrial Drying*, 4e (ed. A.S. Mujumdar), 1286–1287. Boca Raton, FL: CRC Press.
3. Murugesan, S., Sharma, P.K., and Tabora, J.E. (2010). Design of filtration and drying operations. In: *Chemical Engineering in the Pharmaceutical Industry: R&D to Manufacturing* (ed. D. am Ende), 317. Hoboken, NJ: Wiley.

37

DRYING CASE STUDIES

John Gaertner, Nandkishor K. Nere, James C. Marek, Shailendra Bordawekar, Laurie Mlinar, and Moiz Diwan
Process Research and Development, AbbVie Inc., North Chicago, IL, USA

Lei Cao
Operations Science and Technology, AbbVie Inc., North Chicago, IL, USA

37.1 INTRODUCTION

This chapter presents various approaches used to apply the drying fundamentals covered in Chapter 35 to develop efficient and robust processes that consistently produce API, meeting the physicochemical requirements for small molecule compounds. In these case studies, an appropriate level of understanding is developed to ensure confident determination of the design space and a scalable process. This can be challenging at times due to the complexity of drying processes and the limited availability of appropriate scale-down equipment and diagnostic tools. It can be further complicated by the wide variety of equipment used for drying.

The manufacture of an API involves many synthetic steps, producing a large number of intermediates. Many of these intermediates are crystallized, filtered/washed, and dried. In the drying of an intermediate, the levels of residual solvents must be reduced sufficiently to produce a stable compound; these solvents should not interfere with processing in the subsequent step. The drying of an API must generate solids with appropriate particle size distribution and crystal form and with residual solvent levels conforming to regulatory guidelines. The API drying step may need to maintain the crystal form produced by the crystallization or transform the crystal to the desired form. The properties of the dried API are particularly important since they may impact processing in the formulation steps.

Despite the common use of drying, this unit operation is not well understood. Heat and mass transfer in the solid phase can be complicated, requiring measurement of a large number of material-dependent properties. There is complex interplay between material properties and liquid-phase and vapor-phase processes, as well as equipment-dependent parameters. Typically, a limited amount of material is available in early development for use in drying studies; due to the unique properties of materials, the use of a surrogate is often unrealistic. Due to limitations in fundamental understanding of drying mechanisms and rates and inadequate knowledge of material properties, scale-up often involves the use of empirical correlations with limited applicability and limited supporting data. Scale-up correlations developed for one dryer type may be inappropriate for application to other dryer types due to the unique aspects of each dryer.

An objective of drying development is to develop an appropriate level of understanding to enable identification and addressing of potential issues that may arise during scale-up or technology transfer to an alternate equipment train. The drying case studies presented here demonstrate this approach. These case studies include development of a computational model to simulate the drying process and application of this model to optimize drying parameters to limit drying cycle time. Also described are process transfer to alternate equipment, achieving the desired crystal form either through desolvation or humidification, and the application of relevant process analytical technology (PAT). Challenges encountered in pharmaceutical processing are also described.

A large number of different types of dryers are used in the pharma industry. Agitated filter dryers (AFDs) are commonly

---
Chemical Engineering in the Pharmaceutical Industry: Active Pharmaceutical Ingredients, Second Edition.
Edited by David J. am Ende and Mary T. am Ende.
© 2019 John Wiley & Sons, Inc. Published 2019 by John Wiley & Sons, Inc.

used due to their dual functionality (filtration and drying) and ability to safely process potent compounds. Other commonly used dryers include conical screw dryers (CSDs), spherical dryers, tumble (or biconical) dryers, and tray dryers. The first drying case study describes the development and scale-up of a process using an AFD that was subsequently transferred to an equipment train employing a centrifuge and CSD.

37.2 PROCESS TRANSFER TO ALTERNATE EQUIPMENT TRAIN

A goal of process design and equipment selection in the manufacture of a compound is to achieve process efficiency and robustness while ensuring product quality attributes are met. In achieving efficient commercial production, processing may be constrained by the type of equipment in available equipment trains. The processing route and operating conditions may be modified to fit the equipment.

37.2.1 Development and Scale-Up in Agitated Filter Dryers

The API compound is a channel solvate crystallized from a dual solvent solution. The crystal slurry is filtered to isolate the compound. The filter cake resistance and compressibility as measured by a leaf filter were low, indicating good pressure filtration (and centrifugation) characteristics. The crystal slurry filtration and cake wash are followed by a two hour nitrogen blow to deliquor and partially dry the wet cake. This reduces the wet cake solvent content to approximately 8–12 wt %.

TGA measurements of wet cake with an initial solvent content of 27 wt % showed that the drying rate starts falling when the solvent content drops to 11–12 wt % at which point the remaining solvent is largely internal to the crystals. The solvent content of the fully solvated crystal is 19 wt %. Based on TGA measurements, a minimum temperature of 80 °C is required to achieve a reasonable desolvation rate; these results were confirmed by means of drying rate measurements performed in a lab vacuum tray dryer at several temperatures. Based on these results, the final drying temperature was increased from 50 to 80 °C. The jacket temperature is ramped from 20 to 80 °C over six hours to limit the rate of solvent vapor generation, avoiding condensation in the dryer and vacuum line.

The propensity for particle attrition and agglomeration was measured in a lab-scale AFD over a solvent content range of 28–0 wt %. There was no agglomeration or snowballing; however, attrition occurred over the entire range and was particularly high at high solvent concentrations. In scaling up to the pilot plant and production AFDs, attrition was managed by reducing the solvent content (by nitrogen cake blow) prior to initiating AFD blade rotation and by minimizing the number of AFD blade rotations. Agitation was initiated at low cake solvent content in which the particles are less sensitive to stress. The AFD blade is typically rotated 40 times at the lowest speed every two hours, achieving a total of approximately 900 revolutions during drying. Attrition was also measured in pilot-scale AFDs. One batch that was exposed to a large amount of agitation (2200 revolutions) exhibited a approximately 50% reduction in particle size (see Table 37.1); this was greater attrition than a similar batch agitated for 610 revolutions at a slightly lower tip speed.

37.2.2 Production-Scale Equipment Change

After successfully completing many batches at the production site, a switch from an AFD to a centrifuge and CSD was evaluated in order to increase throughput and free up the AFD equipment train for another compound. The modified process was developed at lab and pilot scales and then scaled up and demonstrated at production scale.

37.2.3 Switch from Pressure to Centrifugal Filtration

Centrifugation was demonstrated using a lab-scale 12 cm diameter vertical bowl centrifuge with perforated bowl (Figure 37.1). This centrifuge is capable of generating a g-force of approximately 600 g, which is less than 1000–1500 g at large scale. An additional limitation of the lab-scale centrifuge is the low liquid head pressure (~15 psi) that is generated; this is less than the 100–200 psi at large scale. The decision to proceed to pilot scale was supported by polarized light microscopy (PLM) and particle size analyses that showed

TABLE 37.1 Particle Size Distribution (PSD) of Crystal Slurry and Dried Crystals Filtered and Dried in Agitated Filter Dryers

				Particle Size Distribution (µm)		
AFD	AFD Blade Revolutions	Blade Tip Speed (m/s)	Processing Point	D10	D50	D90
0.6 m² AFD			Crystal slurry	30	74	142
	610	0.23	Dried crystals	12	41	93
0.3 m² AFD			Crystal slurry	13	38	112
	2200	0.31	Dried crystals	6	21	55

PSD measured by laser diffraction.

FIGURE 37.1 Tabletop vertical bowl centrifuge with perforated bowl.

TABLE 37.2 Effect of Wash Solution Composition on Lab-Scale Centrifuge Wet Cake Residual Solvent Composition

Cake Wash Composition (EtAc:DCM, w/w)	Wet Cake Residual Solvent Composition (EtAc:DCM, w/w)
0.6	0.6
1.3	0.9
2.7	1.5
6.7	1.4

Cake LOD ~20 wt % after spin-down.

no significant fracture of crystals in centrifuged wet cake and wet cake from high pressure filtrations.

Centrifugation was evaluated using two pilot-scale centrifuges. The solvent content of washed centrifuge wet cake (~20 wt %) is higher than that of deliquored AFD wet cake (~8–12%) since the two hour nitrogen blow of the AFD wet cake is effective in evaporating a large fraction of the more volatile solvent (dichloromethane (DCM)), but less effective in removing the higher boiling ethyl acetate (EtAc) antisolvent. Consequently, the deliquored AFD wet cake contains a higher concentration of EtAc internal to the channel solvate crystals than centrifuge cake. In order to achieve similar solid-state drying behavior in the two different equipment trains, similar wet cake solvent compositions prior to the start of heated drying were targeted. One approach to accomplish this is to sweep nitrogen through the washed wet cake in the centrifuge bowl; however, the available production-scale centrifuge lacked this capability. An alternative approach consisting of modifying the composition of the centrifuge cake wash solution was pursued.

The primary purpose of the cake wash is to remove the mother liquor containing dissolved impurities from the cake. In order to ensure effective impurity removal, a plan to apply two sequential washes to the centrifuge cake was considered. The composition of the first wash was the standard 0.6 : 1 (w/w) EtAc: DCM; this composition approximates the mother liquor composition. This wash was followed by a wash solution with high EtAc composition. EtAc is an antisolvent, so the second wash may be less effective in removing impurities. This approach was rejected because it required dual solvent wash tanks for multiple centrifuge loads; however, just a single wash tank was available at the production site.

A single wash with high EtAc content was evaluated in lab-scale centrifugation experiments (see Table 37.2). EtAc contents from 1.3 : 1 to 6.7 : 1 (w/w) EtAc : DCM were tested. Application of a wash of 2.7 : 1 EtAc : DCM increased the composition of the residual solvent in the centrifuge cake to 1.5 : 1; a further increase in the EtAc content of the applied wash did not further increase the EtAc content in the wet cake. The wash amount (6.4 kg/kg dry solids) used for the centrifugation matched the total wash amount used for the pressure filtration in the AFD. However, the wash solution has a short contact time with the centrifuge wet cake; this provides limited opportunity for DCM in isolated pockets of the wet cake to diffuse to the main flow channels in the cake. DCM present in the internal channels of the solvate crystals is more difficult to remove by the cake wash solution. The single wash with high EtAc content was shown to achieve acceptable removal of impurities present in the mother liquor, and the wash ratio of 2.7 : 1 was implemented.

X-ray powder diffraction (XRPD) analyses of lab-dried wet cake showed that a residual solvent EtAc–DCM ratio of 1.0 : 1 (w/w) or greater at the start of heated drying generates the desired solid structure when the wet cake is dried under appropriate conditions. So a residual solvent composition of 1.0 : 1 or greater was targeted for the centrifuge wet cake.

Residual solvent compositions exceeding 1.0 : 1 EtAc : DCM (w/w) were achieved in spun-down centrifuge wet cake

TABLE 37.3 Effect of Wash Solution Composition on Pilot-Scale Centrifuge Wet Cake Residual Solvent Composition

Pilot-Scale Centrifuge	Scale (Total Dry Solids kg for All Drops of the Batch)	Cake Wash Composition (EtAc:DCM, w/w)	Wet Cake Residual Solvent Composition (EtAc:DCM, w/w)	Wet Cake Residual Solvent Content (wt %)
Vertical bowl	15	0.6	0.5	22
Horizontal peeler	80	0.6	0.6	18
Vertical bowl	41	0.6	0.5	23
Vertical bowl	9	2.7	1.3	22
Horizontal peeler	33	2.7	2.6	31

in pilot plant batches using a wash solution composition of 2.7 : 1 EtAc : DCM (Table 37.3) for vertical bowl and horizontal peeler centrifuges (Figure 37.2).

Operating conditions and process data for representative centrifugations are summarized in Table 37.4. The g-forces used for slurry loading/cake washing and for final spin down in the pilot batches with the largest pilot plant centrifuge (horizontal peeler) were similar to those implemented at production scale.

The wet cake produced using the pilot-scale centrifuge with highest g-force exhibited an acceptable degree of attrition relative to the crystal slurry feed (Table 37.5).

The process was transferred to the production-scale train, which employs a horizontal inverting bag centrifuge. Application of a cake wash with a 2.7 : 1 (w/w) ratio of EtAc : DCM generated a final wet cake residual solvent composition of 1.1 : 1 EtAC : DCM; this met the target of greater than or equal to 1.0 : 1. In the new production train, the centrifuge wet cake is transferred to a CSD for the subsequent drying operation.

37.2.4 Development and Scale-Up in Conical Screw Dryer

Advantages of CSDs include high heat and mass transfer coefficients resulting from agitation that exposes fresh wet solids to the heated surfaces and headspace. Screw rotation achieves mixing in the radial and axial directions, while screw orbiting mixes in the angular dimension. The gentle agitation introduces relatively little stress, resulting in less particle attrition for many compounds.

Lab- and pilot-scale drying studies were performed to develop and scale up the conical screw drying to a 3000 L production-scale dryer. CSD equipment information is summarized in Table 37.6. Large CSDs have significantly less heat transfer area per volume of cake than a lab-scale CSD. All of the dryers have a tapered screw that achieves better mixing and conveying of solids near the top of the dryer than a cylindrical screw. The screws for the lab- and pilot-scale CSDs have closed flights that provide more area for vertical solid conveyance, while the production-scale dryer

FIGURE 37.2 Vertical bowl (top) and horizontal peeler (bottom) pilot-scale centrifuges.

screw has open flights (lacking recesses) that facilitate screw cleaning.

Since the drying rate of the lab 1.5 L CSD (Figure 37.3) is not readily scalable to large dryers, the lab experiments focused on the impact of drying process parameters and

TABLE 37.4 Centrifuge Operating Conditions at Lab, Pilot, and Production Scales

Scale	Centrifuge	Bowl ID (cm)		Speed (rpm)	G-Force	LOD of Spun Cake (wt %)	Wet Cake per Drop (kg)
Lab	Vertical bowl	12	Slurry loading; cake washing	1000	67		0.1
			Final spin	3000	605	23	
Pilot	Vertical bowl	40	Slurry loading; cake washing	1000	224		9
			Final spin	2000	894	22	
Pilot	Horizontal peeler	63	Slurry loading; cake washing	900	286		19
			Final spin	1800	1143	20	
Production	Horizontal inverted bag	80	Slurry loading; cake washing	900	363		35
			Final spin	1600	1147	20	

TABLE 37.5 Particle Attrition Resulting from High G-Force Centrifugation

	Particle Size Distribution (μm)		
	D10	D50	D90
Crystal slurry	19	45	94
Spun filter cake	16	40	86

TABLE 37.6 Equipment Parameters for Lab-, Pilot-, and Production-Scale CSDs

Scale	Lab	Pilot	Plant
Maximum working volume (L)	1.5	30	3000
Heated wall area/vol (1/dm)	3.8	1.4	0.3
Screw Flight design	Tapered Closed flight	Tapered Closed flight	Tapered Pharma open flight
Max tip speed (m/s)	0.41	0.89	1.63

mechanical stress on particle physical properties (e.g. attrition and agglomeration). In one experiment, drying with constant agitation (at 40% of maximum speed) was continued for 17 hours after drying completion; this extended period of agitation of the dried particles resulted in essentially no particle attrition. Similarly, agitation at maximum tip speed did not attrite the particles. The ease of heel removal from the wall during dryer unloading was also assessed. A mass spectrometer was employed to monitor drying rate, facilitate the decision on drying completion, and help detect an improper mechanical seal in the dryer. Wet cake from the pilot-scale centrifuge was used for these studies.

Pilot-scale studies in the 30 L CSD using centrifuge wet cake were performed to measure the effect of different profiles of agitation, vacuum pressure, and jacket temperature on drying rate and dried particle properties. Heat and mass transfer rates of the 30 L dryer can be used for scale-up

FIGURE 37.3 1.5 L conical screw dryer.

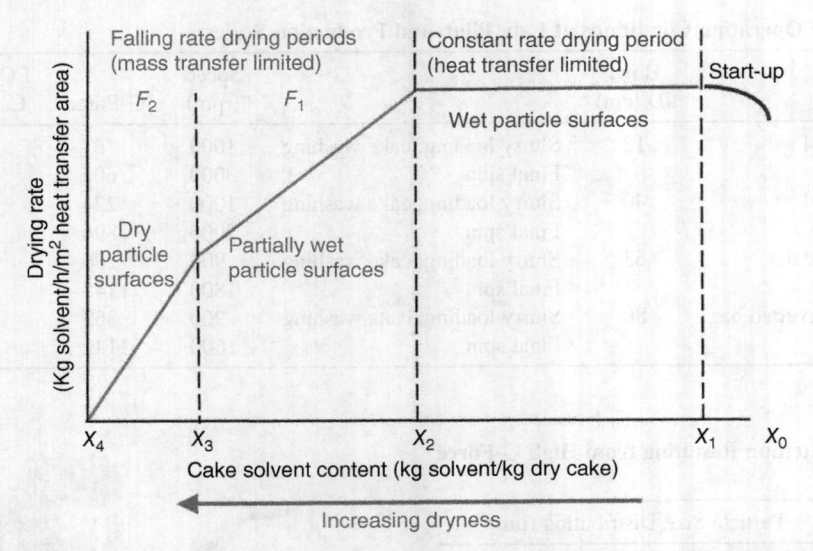

FIGURE 37.4 Drying rate curve showing different regions used for scale-up calculations.

(within certain constraints); this is typically done by measuring the drying rate of a pilot-scale batch employing the design temperature, pressure, and agitation profiles. The drying rate is expressed as a function of the residual solvent content (see Figure 37.4). The resulting drying rate curve is typically divided into three or more regions that correspond to different rate-controlling mechanisms.

The drying rate for a particular region of the curve is similar across scales. The drying time for each region is calculated separately using Eqs. (37.1) and (37.2):

$$\text{Constant rate region}: \quad t_C = \frac{m_{dry}}{A} \frac{(X_1 - X_2)}{\dot{m}_C} \quad (37.1)$$

$$\text{Falling rate regions}: \quad t_{F_1} = \frac{m_{dry}}{A} \frac{(X_2 - X_3)}{(\dot{m}_2 - \dot{m}_3)} \ln \frac{\dot{m}_2}{\dot{m}_3} \quad (37.2)$$

where

t_C is the constant rate drying time (hours).
t_{F_1} is the falling rate region 1 drying time (hours).
m_{dry} is the dry mass (kg).
X is the solvent content of cake (kg solvent/kg dry cake).
A is the heat transfer area (m^2).
$\dot{m}_C$ is the drying rate of constant rate region (kg/m^2/h).

The total drying time is the sum of the individual drying times: $t_{total} = t_C + t_{F_1} + t_{F_2}$. Scale-up using the drying rate curve approach assumes equivalency across scales of heat and mass transfer coefficients, temperature driving forces, vacuum pressure, nitrogen sweep, mixing, dryer geometry, and cake characteristics.

The mixing turnover time is calculated using equations 37.3 and 37.4 and is maintained across scales:

$$\text{Mixing turnover time} = \frac{\text{total solids volume}}{\text{screw conveyance rate}} \quad (37.3)$$

$$\text{Screw conveyance rate} = \frac{\pi}{4}(D^2 - d^2)nP \quad (37.4)$$

where

D is the screw outer diameter (cm).
d is the screw shaft diameter (cm).
n is the screw rotation speed (rpm).
P is the screw pitch (cm).

For open flight screws, the screw shaft diameter is replaced with an equivalent diameter selected to reflect the reduced flight area available for conveyance. For the case at hand, a mixing turnover time of three minutes is achieved in the pilot- and production-scale CSDs (filled to capacity) using screw rotation speeds of 11 and 45 rpm, respectively. The ratio of screw rotation speed (rpm) to orbiting speed (rpm) is maintained at 50–100 to ensure mixing in all three spatial dimensions.

Solvent vapor is removed through a vacuum port on the top of the CSD. The upward transport of solids by the screw brings fresh solids to the surface, resulting in a high vapor mass transfer coefficient. Nitrogen may be applied to the bottom of the CSD to further improve mass transfer by providing a convective sweep resulting from the flow of N_2 from the bottom to the top of the bed. In pilot studies, it was found that substantial particle attrition resulted from a high N_2 gas velocity; similarly high attrition also resulted from a leak in the bottom discharge valve of the CSD that led to airflow into the bottom of the evacuated CSD. For this reason, N_2 was not applied to the bottom of the CSD, and the importance

TABLE 37.7 Reduction in Particle Size of Crystals (i.e. Attrition) During Drying in AFD and CSD

Dryer Type	Scale	Agitation Frequency	Tip Speed (m/s)	Particle Size Reduction (%)		
				D10	D50	D90
AFD	Plant	Intermittent, 5 min every 2 h	0.34	85	61	50
CSD	Lab	Continuous	0.14	21	23	18
CSD	Pilot	Continuous	0.18	29	32	22
CSD	Plant	Continuous	1.05	25	26	25

Expressed as drop in particle size relative to size of particles in initial wet cake.

of a pre-batch vacuum leak test of the CSD was emphasized. For other less stress-sensitive products, application of a slow N_2 flow to the CSD bottom may be appropriate; in this case, the additional flow area provided by a sintered N_2 distribution ring at the bottom of the CSD may result in reduced N_2 gas velocity and less particle attrition in comparison with an N_2 distributor consisting of just a few drilled holes that provide a small N_2 injection area.

A low level of attrition of crystals during agitated drying was measured in lab- and pilot-scale CSD tests; consistent results were obtained in the plant CSD. Table 37.7 compares the size of the crystals composing the centrifuge wet cake charged to the CSD with the size of the dried crystals. The degree of attrition in the CSD is less than half of the attrition measured in the AFD – despite the use of continuous agitation in the CSD. For both dryer types, agitation was initiated soon after the dryer jacket reached its final temperature of 80 °C.

The information generated by the lab- and pilot-scale studies enabled a successful scale-up and demonstration batch in the production-scale CSD. The desired crystal structure was achieved by desolvating the crystal solvate at high temperature by rapidly heating the dryer jacket to 80 °C with no agitation and limited vacuum. The vacuum level in the initial drying phase was sufficient to avoid vapor condensation in the CSD and vacuum exhaust line; it also limited the amount of vaporization and resulting cooling. Agitation and high vacuum were applied after the jacket temperature reached 80 °C; the agitation distributed the heat imparted by the jacket. Residual solvents were reduced to passing levels within 36 hours of drying after high vacuum was achieved.

37.3 OPTIMIZATION OF SOLVATE DRYING THROUGH HEAT TRANSFER MODELING

There is no universal protocol for drying using an AFD that is applicable to all pharmaceutical compounds; this is due to differences in physicochemical properties of the wet cake and the desired dried product attributes. Optimization of AFD performance for a given compound requires the proper determination of operating parameters including the headspace pressure, nitrogen sweep rate, level of applied vacuum, duration and intensity of mixing, and jacket and mixing blade temperature. In the prior case study, the API solids dried sufficiently rapidly that it was unnecessary to reduce the cycle time through an intensive computational analysis of drying. In other cases, drying can be a significant bottleneck and warrants deeper study. This was the case for the drying of an early-stage API candidate that required excessively long cycle times in a pilot plant AFD. Modeling was shown to be of great utility in determining proper operating conditions specific to the compound. Insights from modeling reduced experimental efforts (saving expensive materials), improved understanding of the drying process at bench scale, and ensured predictable robust performance on scale-up. The key to increasing the drying rate was to minimize mass transfer limitations and to ensure sufficient heat transfer by use of an appropriate mixing protocol. A composite model was developed that incorporates models for heat transfer and desolvation kinetics. The heat transfer model accounts for the effect of a gas gap between the dryer wall and adjacent API solids on the overall heat transfer coefficient. It also accounts for the effect of headspace pressure on the mean free path length of the inert gas and therefore on the heat transfer between the vessel wall and the first layer of solids. The computational methodology was developed using a modified model framework within commercially available simulation software. An AFD operational protocol was designed based on the desolvation kinetics, thermal and crystal form stability, and the understanding gained through model simulations. Implementation of this protocol reduced the drying time fourfold.

37.3.1 Desolvation Kinetics and Form/Thermal Stability

The intrinsic desolvation kinetics were calculated by measuring the solvent (ethanol) content as a function of time in dynamic vapor sorption (DVS) experiments carried out at three different temperatures (40, 60, and 70 °C) under nitrogen sweep. A thin layer of solids and high N_2 sweep rate were used to minimize the mass transfer resistance and ensure measurement of intrinsic desolvation rate. The net desolvation rate constant (k_D; units of s^{-1}) for each temperature was calculated by fitting the desolvation data to first-order kinetics. The resulting rate constants were graphed in an Arrhenius plot to quantify the temperature (T in K) dependence:

$$\ln k_D = -0.0002 \frac{1}{T} + 0.0026 \quad (37.5)$$

Drying results of early pilot plant batches were consistent with the findings of the desolvation kinetics study. In early pilot plant batches, the dryer was operated at a nominal jacket temperature of 50 °C, resulting in a maximum internal temperature of about 40 °C; the drying process took nearly a week at this temperature to reduce the ethanol content to an acceptable level (≤0.1 wt %).

Thermal stability and form stability studies were performed to determine the maximum acceptable drying temperature. At temperatures above 80 °C, wet cakes began to discolor – indicating thermal degradation. For the case of properly deliquored wet cake, the desolvate was shown by XRPD to be the stable crystal form at temperatures below 80 °C, which is well below the API melting temperature of approximately 150 °C. Amorphization of the desolvate crystal was not observed under these conditions.

37.3.2 Composite Drying Model

Drying of this compound consists of two main phases. The first phase is relatively short and consists of rapid evaporation of unbound solvent accomplished by the nitrogen cake blow that follows proper cake deliquoring. Desolvation is insignificant during this phase due to its high energy barrier.

The second phase consists of removal of bound solvent (primarily as solvent internal to the crystals) in three steps:

1. *Desolvation*: The first step is the desolvation of the ethanol solvate. The net desolvation rate (R_D) is governed by temperature. The desolvation results in the separation of solvent in liquid form:

$$\text{API solvate} \xrightarrow[R_D]{\text{Heat}} \text{Solvent(l)} + \text{Desolvated API}$$

2. *Evaporation*: The second step is the evaporation of the free liquid solvent after desolvation:

$$\text{Solvent(l)} \xrightarrow{\text{Heat}} \text{Solvent(v)}$$

3. *Convection*: In the third step, the evaporated solvent is convected away by the nitrogen sweep. The nitrogen is expected to flow around the particles at a sufficiently high velocity to alleviate diffusion limitations in this third step.

The desolvation and evaporation rates depend on the local temperature and heat transfer rate, which are governed by the physical properties of the cake, jacket fluid temperature, heat transfer area, and the mixing provided by the agitator. By applying a sufficient nitrogen gas flow rate for rapid convection of solvent vapor, desolvation or evaporation becomes the rate-determining step. The effective rate of solvent loss is the smaller of the desolvation rate and the solvent evaporation rate. At a given temperature T, the desolvation rate is readily calculated using the equation $R_D = k_D(T)C_{s,b}$, where $C_{s,b}$ is the ratio of the mass of the bound solvent in the form of solvate to the mass of desolvated solids.

The heat transfer model used in the simulation is based on the penetration theory proposed by Tsotsas and Schlunder [1] with a few modifications. The heat transfer mechanism is dominated by thermal conduction. The AFD is divided into three regions in the heat transfer analysis: (i) the vessel wall, (ii) the gap between the outer layer of the dry solid bed and the vessel wall (this is assumed to be composed of the first layer of solids and gas gap), and (iii) the dry solids. Further description of the mathematical modeling is provided in Ref. [2].

In the absence of agitation during drying, a gas gap is created by preferential flow of N_2 along the vessel wall, generating additional resistance to heat transfer. The contribution of this gas gap to the overall heat transfer resistance was found to be 10–50% depending on the fraction of wet solids; this justifies its incorporation in the heat transfer model. The model also incorporates an equation for the contact heat transfer coefficient (h_{ws}) by Hoekstra et al. [3] for the first layer of solids adjacent to the dryer wall; h_{ws} is inversely dependent on the mean free path length for N_2 gas molecules. Use of this equation enables simulation of the effect of dynamic pressure in the filter dryer. The pressure variation may be significant due to the N_2 gas flow rate, the vacuum drawn at the AFD bottom, and dynamics of pressure drop across the cake. A headspace pressure increase from 50 to 1000 mbar absolute increases h_{ws} by 2.5x.

37.3.3 Regression of Model Parameters Through Simulations

All simulations were carried out using the modified model framework within a commercially available simulation software. The contact time for mixing was calculated from the mixing number, N_{mix}, which is empirically related to the Froude number. Two constants in the N_{mix} correlation that are characteristic of the agitator type were estimated by forcing a reasonable fit of the model results with the temporal profiles of the measured solvent content and solids temperature of an early pilot plant batch (Figures 37.5 and 37.6).

The initial solvent composition of 5.4 wt % in Figure 37.5 represents the first available measurement. This solvent content is below the bis-solvate solvent composition of 8.8 wt %, indicating that the crystals are partially desolvated when heated drying is initiated after the completion of the nitrogen blow of the cake. Note in Figure 37.6 the initial phase where the solids temperature increases (from ~19 °C used for the N_2 blow) before attaining a somewhat steady value. The time at which the internal temperature achieves a somewhat steady value corresponds to mono-solvate composition based on Figures 37.5 and 37.6. The first solvent molecule of the bis-solvate likely has a lower energy barrier for desolvation than the second solvent molecule. During removal of the first solvent molecule, evaporation of the desolvated solvent may be the rate-determining step, thereby increasing the solids temperature. The thermocouple tip is flush with the AFD wall surface and is in contact with the gas flowing through the gap between the dryer wall and the solids. Consequently, the measured temperature does not exactly match the dry solids temperature predicted by the model. This is supported by the observed jumps in the measured temperature during the mixing period as shown in Figure 37.6.

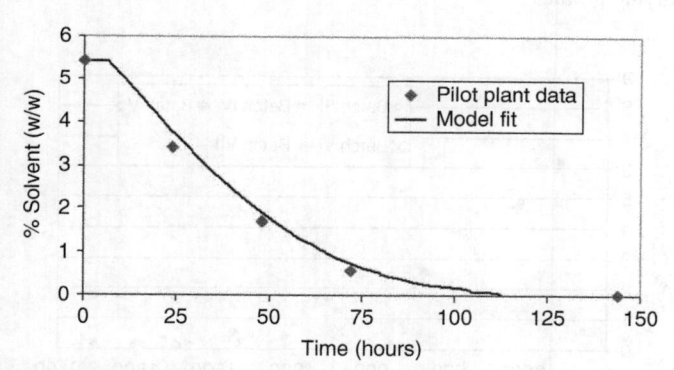

FIGURE 37.5 Model fit of the measured cake solvent content of early pilot plant batch.

37.3.4 Improved Drying Through Model Simulations

Simulations were performed using the computational model to better understand the dynamics of agitated drying. The heat transfer rate is dependent on the jacket temperature and the agitation speed and time. A jacket temperature upper limit of 70 °C was imposed to assure thermal stability of the product and acceptable crystal form, which in this case demanded a gradual increase in the drying temperature. It was seen that the agitation frequency, duration, and speed play a significant role in enhancing the drying rate. Continuous agitation results in the fastest drying at a given jacket temperature. Since agitation can also result in significant attrition of product crystals, intermittent agitation is commonly used in practice. The mixing period and intensity that produce acceptable particle attrition form a practical basis for the design of the agitator operating protocol. Model simulations were run using various headspace pressures and agitation profiles and yielded the improved drying protocol in Figure 37.7.

Prior to the start of heated contact drying, the wet cake is blown with nitrogen for three hours. The jacket temperature is then gradually increased over 12 hours to avoid excessive heating of wet cake with unbound solvent as it may result in the formation of a dry coating on the jacketed wall, reducing the heat transfer rate. No agitation is performed at high cake solvent content to avoid agglomeration induced by mixing that can generate hard lumps upon heating – potentially reducing the drying rate. The original and improved drying protocols are compared in Table 37.8.

In the improved drying protocol, the agitation speed is reduced to reduce particle breakage and attrition, while the total mixing time is increased. This protocol was implemented in a pilot plant batch. Figure 37.8 compares the model predictions and observed drying performance, demonstrating that the model simulations are in close agreement with the measured solvent content throughout the drying process.

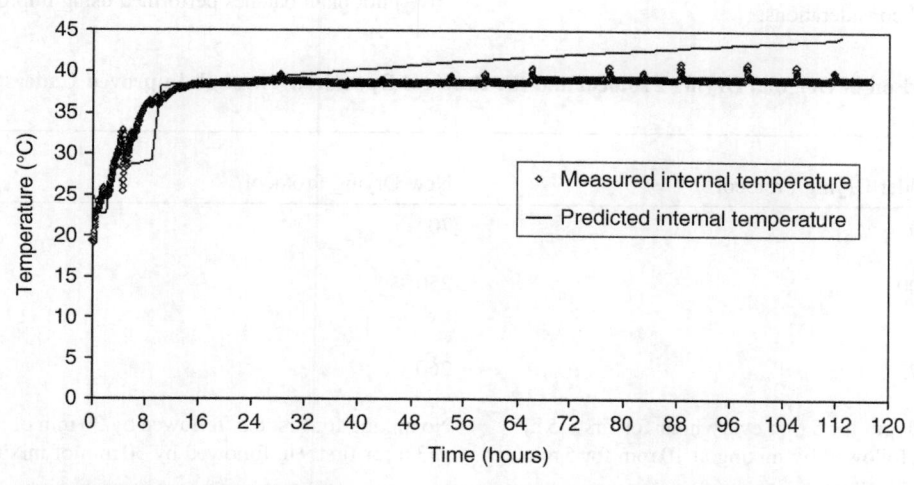

FIGURE 37.6 Model fit of the measured internal temperature of early pilot plant batch.

The drying time to reduce the residual solvent content to 0.1 wt % (from 8.8 wt %) was approximately 35 hours. This constitutes a fourfold reduction from the 140 hours required in the early pilot plant batch. Four additional API pilot plant batches were dried using the improved operating conditions. Figure 37.9 depicts the temporal profiles of measured solvent content for all five batches dried using the improved protocol. All batches showed similar drying performance, thereby verifying the computational model description of the drying process. The model could not be verified beyond one set of operating parameters at scale due to constraints on the use of pilot plant equipment and the high API cost.

37.3.5 Guidelines for AFD Operation

On the basis of the model studies, pilot-scale demonstration, and practical considerations, the following guidelines are recommended to achieve efficient AFD operation. In practice, the drying protocol will need to be tailored to address the specific issues related to the physicochemical characteristics of the particular product:

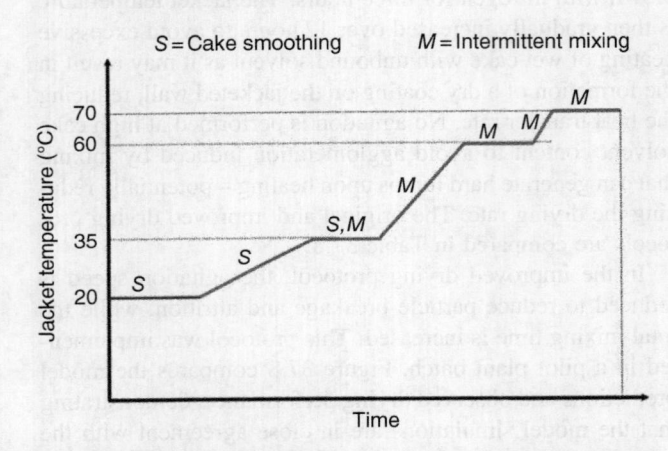

FIGURE 37.7 Target jacket temperature, intermittent mixing (*M*), and cake smoothing (*S*) profiles determined through model simulations and practical considerations.

1. Efficient N_2 flow through the entire cake is important; the headspace pressure should not be allowed to increase beyond the point that causes cake compression, decreased inert gas flow rate, or increased mass transfer resistance. Intermittent cake smoothing must

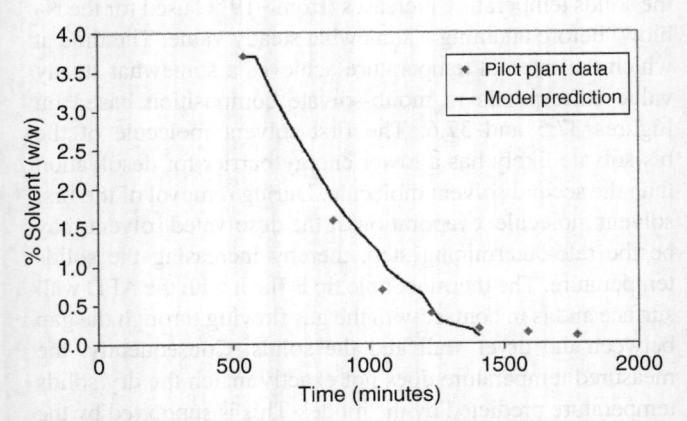

FIGURE 37.8 Comparison of model predictions of wet cake solvent content with pilot plant measurements for improved drying profile.

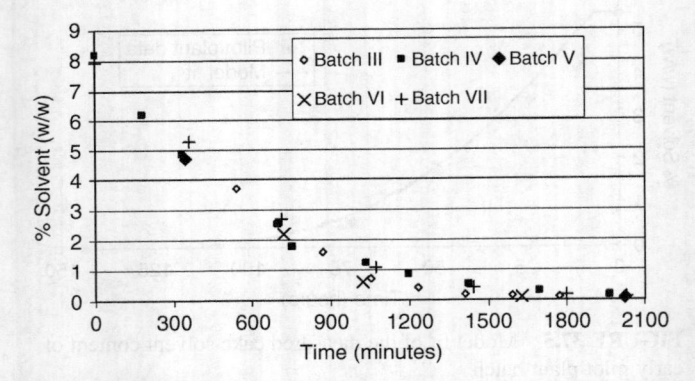

FIGURE 37.9 Measured wet cake solvent temporal profiles for five pilot plant batches performed using improved drying protocol.

TABLE 37.8 Comparison of Original Drying Protocol and the Protocol Optimized Through Improved Understanding from Model Simulations

Operational Parameter	Older Drying Protocol	New Drying Protocol
Jacket temperature (°C)	50	70
Average headspace pressure (mm Hg)	300	250
Total mixing time (min)	175	360
Mixing rpm and frequency	10 rpm for 5 min every hour for first 15 h, followed by mixing at 10 rpm for 5 min every 6 h	No mixing for first 6 h, followed by 30 min of mixing at 2 rpm every 3 h for first 9 h, followed by 30 min of mixing at 3 rpm every 2 h

be performed to prevent preferential inert gas channeling or bypassing. This will alleviate mass transfer limitations relating to the movement of solvent vapors.

2. Cake that is substantially wet should not be mixed in order to avoid granulation. The wet granules can form hard lumps upon heating that can be difficult to dry, resulting in increased drying time.

3. Heat the wet cake gradually after ensuring deliquoring to the fullest practical extent. Heating wet cake with unbound solvent in the proximity of vessel wall may result in formation of a dry coat on the jacketed wall, causing fouling and a reduced heat transfer rate. An excessively high vaporization rate can also result in condensation within the dryer, resulting in formation of liquid droplets that may promote dissolution or granulation.

37.4 DRYING WITH HUMIDIFIED NITROGEN

In the prior agitated filter drying case study, elevated temperature was used to desolvate the API to achieve the desired desolvated crystal form. A significant number of APIs are hydrates; many of these require maintenance of appropriate humidity during drying to produce the desired crystal form. This is true of the following case study, which represents an example where humidification during drying can be employed to accelerate desolvation of the crystal form originating from the crystallization.

This case study describes a readily encountered case of crystal form control during API drying. The crystal form generated by the crystallization and isolated by filtration is a mixed solvate (form B) containing both water and methanol. During drying the methanol level must be reduced below the ICH limit, and the crystal must be transformed to the desired hydrate (form A) using humidified nitrogen.

Crystal forms A and B are isostructural, exhibiting similar XRPD patterns. Drying studies employing variable relative humidity (RH) indicated that complete desolvation of the mixed solvate results in a desolvate crystal form (form C), which can be converted to the desired form A upon exposure to appropriate RH. Based on a fundamental understanding of crystal form conversions, the drying to produce the desired crystal form can be achieved via two different approaches. The first approach is a two-stage serial drying process involving removal of methanol using vacuum drying assisted by dry nitrogen sweep followed by hydration using humidified nitrogen. The second approach is a single-stage parallel drying process involving concomitant removal of methanol with hydration using vacuum drying assisted by humidified nitrogen.

In the two-stage serial drying process (Figure 37.10), the wet cake is first dried using dry nitrogen to reduce the residual solvent level below the ICH limit. Appropriate vacuum drying can be done at the elevated temperature dictated by thermal stability considerations to accelerate drying kinetics. This results in crystal form transformation from the mixed solvate to desolvate form C. In the second stage, the crystals are hydrated using humidified nitrogen to obtain the desired hydrate crystal form as judged by conformity testing of the XRPD pattern against the reference form A crystal pattern. The minimum RH required for crystal form conversion from desolvate form C to hydrate form A at a given temperature can be determined based on the DVS isotherm confirmed by variable RH XRPD studies. Figure 37.11 depicts the DVS isotherms at two different temperatures. The hydration stage can be split in two parts. In the first part, higher RH (e.g. 50%) and temperature are used to increase the driving force and diffusivity to accelerate hydration of the crystals to the target water content corresponding to the hydrate. In the second part, a lower RH (e.g. 30%) is used at the dryer offload temperature (which is less than the drying temperature). Since the equilibrium water content of the crystals at a given RH decreases with temperature, this water equilibration should be carried out at the dryer offload temperature.

Alternatively, the wet cake can be dried with humidified nitrogen throughout the drying process as shown schematically in Figure 37.12. The removal of methanol below the ICH limit is achieved while preserving the crystal hydrate structure. The kinetics of methanol removal from the crystal lattice is facilitated by favorable activity gradients. Humidified nitrogen provides higher activity of water in the gas phase as compared with the solid phase; the opposite applies for the methanol. The resulting activity gradients accelerate the diffusion of water molecules into the crystal lattice while providing the way for methanol molecules to diffuse out of the crystal lattice. At a given temperature, the equilibrium water content increases as the RH is increased from the minimum RH required to stabilize the hydrate to the upper practical limit. The upper practical RH limit is designed in light of the following considerations:

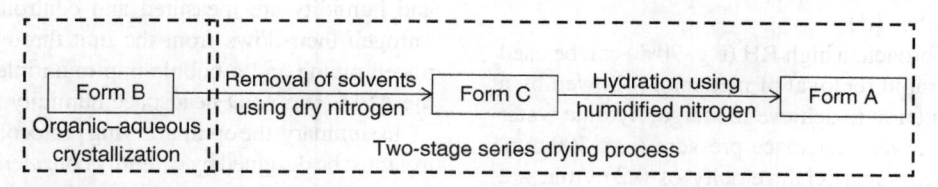

FIGURE 37.10 Schematic of API crystal form conversion and two-stage serial drying process.

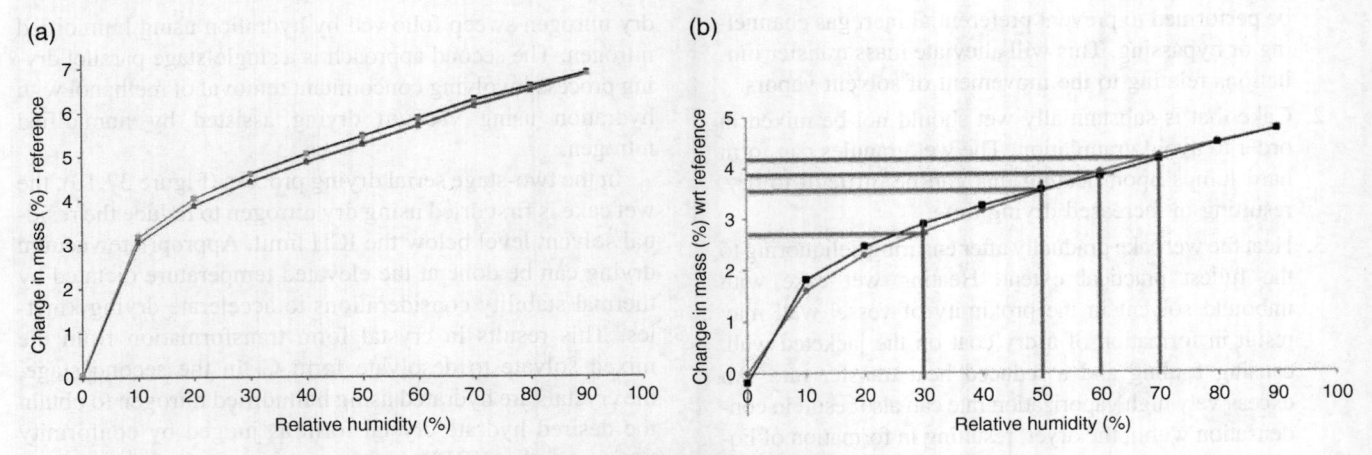

FIGURE 37.11 Dynamic vapor sorption isotherms at (a) 25 and (b) 60 °C.

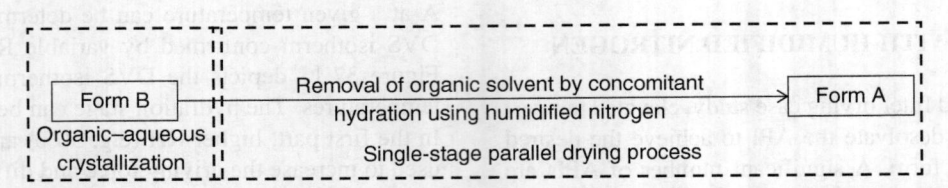

FIGURE 37.12 Schematic of the API crystal form conversion in single-stage parallel drying process.

1. The dew point temperature of the humidified nitrogen should always be lower than any surface temperature that comes in contact with the humidified nitrogen stream. Note that the wet cake temperature in an AFD can be significantly lower than the jacket temperature.

2. Apart from thermal stability considerations, the upper temperature of the humidified nitrogen stream is also governed by the capability of the humidifier and the control of water vapor condensation on the product. Higher temperature along with the higher RH will need higher water vaporization from the humidifier. It will also exhibit higher dew point leading to higher probability of condensation of water that, in turn, can affect the product physical properties adversely.

While higher temperature and RH accelerate removal of methanol, the equilibrium water content can also drop at higher temperature; this will require humidification at lower temperature and higher RH.

In the parallel approach, a high RH (e.g. 90%) can be used initially to achieve rapid removal of methanol followed by a subsequent RH reduction to achieve the target hydrate water level. A significant dryer headspace pressure (e.g. 300 mm HgA) and a nitrogen superficial velocity of 0.2 m/min are effective.

Drying by the series approach can be made time efficient as compared with the parallel drying process. However, the impact of each of the processes on crystal surface should be considered. Crystal form transformation often results in surface defects that make the particles weaker and more prone to breakage or attrition in an AFD. The time-efficient serial drying process can cause such damage. Figure 37.13 shows high magnification laser microscopic images of particles produced by the two different drying approaches. It clearly depicts the surface damage and defects resulting from form conversion to the desolvate followed by conversion to the hydrate as compared with the smoother particle surfaces produced by humidified drying wherein the isostructural crystal lattice is preserved.

Dryer humidification can be accomplished using a commercially available automated humidification unit. Pressurized water supplied to the unit is evaporated. Nitrogen is also supplied to the unit and flows through a heater; the heated N_2 is combined with the water vapor, and the stream's temperature and humidity are measured and controlled. The humidified nitrogen then flows from the unit through insulated or heat traced piping and a bubble trap to an inlet port on the top of the AFD. The AFD headspace humidity is also measured.

In summary, the optimal drying protocol for crystal form control must be designed by carefully considering multiple interacting parameters and the desired product physical properties.

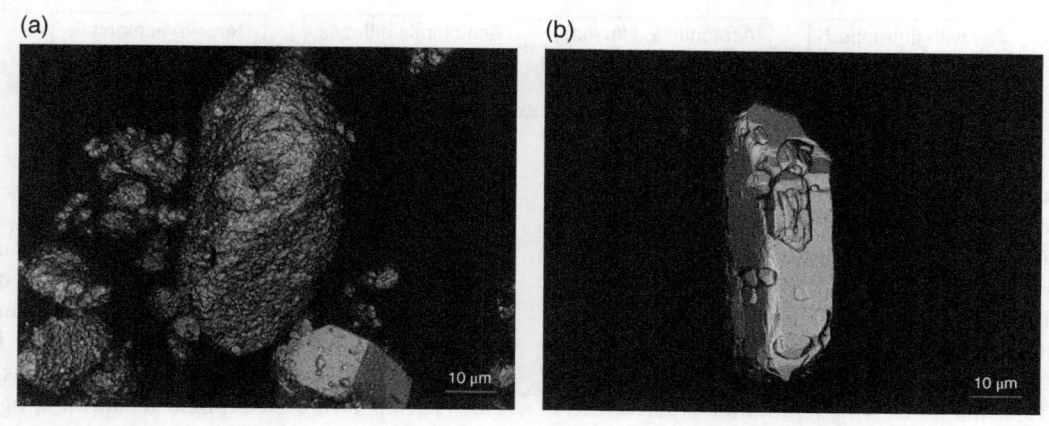

FIGURE 37.13 Differences in crystal surfaces resulting from different drying approaches. (a) Two-stage series drying process. (b) Single-stage parallel drying process.

37.5 DISTILLATIVE DRYING IN AN AGITATED FILTER DRYER

In an antisolvent-based crystallization, the compound has poor solubility in the antisolvent but may have moderate to high solubility in other crystallization solvents. The presence of these solvents during the drying process can lead to local regions of high solubility, which, upon agitation or heated drying, leads to bridging of particles to form agglomerates. Multiple antisolvent washes (displacement and slurry) are usually applied to efficiently remove these crystallization solvents. However, these washes are only able to remove the unbound surface solvents and not the solvents that are part of the crystal lattice (e.g. hydrates and solvates). The following case study presents an example of the use of slurry distillation to remove volatile unbound and bound solvents from the AFD and hence significantly reduce the extent of agglomerate formation during the later stages of drying. Particle agglomeration is a commonly encountered problem in APIs isolated from AFDs; agglomerates can form via a multitude of mechanisms, for instance, induced by shear and agitation of wet cake or heat.

In this case study, the crystallization solvent system yields API with an acetonitrile solvate crystal form. The dual solvent system consists of acetonitrile, which is a crystallization solvent, and dibutyl ether (DBE), which is the antisolvent. Following multiple antisolvent washes of the wet cake in the AFD, residual acetonitrile could not be removed to a level less than the critical solvent composition for formation of the mono-solvate. The use of heated, prolonged slurry, or antisolvent washes in large excess was unsuccessful at removing acetonitrile from the crystal lattice. In the nominal process, after washing the wet cake with the antisolvent, an extended blowdown of the filter cake was completed to remove free solvent. Undesirable agglomerates formed upon cake agitation if the level of acetonitrile was not sufficiently reduced by the blowdown (Figure 37.14). The time required to blow the wet cake to remove acetonitrile to a sufficiently low level to minimize or avoid agglomerate formation significantly increased the process cycle time.

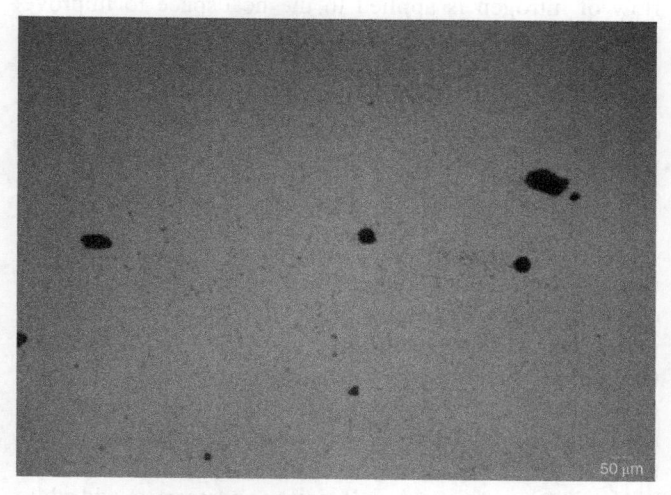

FIGURE 37.14 Typical agglomerates formed in the absence of the distillation; primary particles were isolated through implementation of distillation in the AFD.

In order to remove acetonitrile from the wet cake, a chase distillation was employed in the AFD. Acetonitrile has a sufficiently lower boiling point than DBE, and therefore, DBE was utilized to chase-distill the contents of the AFD. Implementation of the chase distillation reduced the amount of acetonitrile by greater than 90% versus the API. The removal of acetonitrile reduced potential agglomeration of the API during drying.

A schematic showing the proposed mechanism for solvent removal is detailed in Figure 37.15. Following slurry and displacement washes, acetonitrile remains as it is entrapped in the API crystal lattice. In the presence of an excess of the antisolvent and heat, acetonitrile in the bulk or liquors is removed via distillation. As acetonitrile is removed from the bulk,

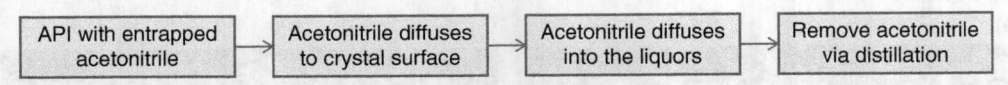

API with entrapped acetonitrile → Acetonitrile diffuses to crystal surface → Acetonitrile diffuses into the liquors → Remove acetonitrile via distillation

FIGURE 37.15 Schematic of the process for removing solvent within crystals.

additional acetonitrile that has diffused within the crystal to the crystal surface moves into the bulk. Continuous agitation of the slurry aids mass transfer from the crystal surface into the bulk as well as through the slurry bed height to facilitate distillation. This process continues until the level of acetonitrile is significantly reduced.

The distillation is performed by charging approximately 3 L DBE/kg API to the AFD. The distillation is performed using a jacket temperature of 65 °C and a headspace pressure of approximately 0.1 barA and with continuous agitation in which the blade is moved up and down. A slow flow of nitrogen is applied to the headspace to improve convection of vapor from the AFD to the overhead condenser/vacuum pump. Additional DBE (~5 volumes) at 50 °C is charged to the AFD in shots to maintain slurry consistency. Distillation is terminated after an in-process sample taken without breaking vacuum using a special slurry sampler indicates an acceptable level of acetonitrile relative to API.

Several parameters play a key role in implementing distillation in the AFD. The most important distillation parameter to optimize is agitation of the slurry. As the process is mass transfer limited by the diffusion of acetonitrile through the boundary layer of the crystal to the bulk as well as the movement of acetonitrile throughout the bed height, continuous and elevated agitation decreases the process cycle time. It is important to ensure the agitation applied does not lead to particle attrition. It may be necessary to add additional antisolvent to chase-distill acetonitrile, in which case the consistency of the slurry must be sufficiently liquid-like both to enable mass transfer of acetonitrile through the bed and to avoid particle agglomeration. Preheating the antisolvent prior to addition and adding the antisolvent without breaking vacuum both aid in optimizing the process on scale.

In summary, when traditional methods for removal of crystallization solvents, such as displacement or slurry washes (with or without the use of heat), do not sufficiently remove the crystallization solvent from the wet cake, a slurry distillation can be implemented in the AFD. In this case study, distillation of the acetonitrile crystallization solvent successfully led to isolation of non-agglomerated API.

37.6 CONCLUDING REMARKS

These drying case studies illustrate the importance of developing an appropriate level of understanding of the drying process to fit the purpose. Early-stage compounds with relatively straightforward drying behavior require less development than complicated late-stage compounds that exhibit slow drying or are susceptible to attrition, agglomeration, or form change. Analysis performed during early stages of development may be limited by availability of a sufficient quantity of material for testing. Later-stage studies are facilitated by increased availability of compound resulting from the increase in process equipment size. Appropriate lab- and pilot-scale studies should be performed to identify the acceptable design space during drying development and process transfer to alternate equipment. This can be constrained by incomplete understanding of material properties, challenges in the calculation of temperature and solvent gradients within a dryer, and the limited capability of analytical tools used to monitor drying.

ACKNOWLEDGMENTS

John Gaertner, Nandkishor Nere, James Marek, Shailendra Bordawekar, Laurie Mlinar, Moiz Diwan, and Lei Cao are employees of AbbVie. The design, study, and financial support were provided by AbbVie. AbbVie participated in the interpretation of data, review, and approval of the publication.

REFERENCES

1. Tsotsas, E. and Schlunder, E.U. (1987). Vacuum contact drying of mechanically agitated beds: the influence of hygroscopic behavior on drying rate curve. *Chem Eng Process* 21: 199–208.
2. Nere, N.K., Allen, K.C., Marek, J.C., and Bordawekar, S.V. (2012). Drying process optimization for an API solvate using heat transfer model of an agitated filter dryer. *J Pharm Sci* 101: 3886–3895.
3. Hoekstra, L., Vonk, P., and Hulshof, L.A. (2006). Modeling the scale-up of contact drying processes. *Org Process Res Dev* 10: 409–416.

38

MILLING OPERATIONS IN THE PHARMACEUTICAL INDUSTRY

KEVIN D. SEIBERT, PAUL C. COLLINS, AND CARLA V. LUCIANI
Eli Lilly and Company, Indianapolis, IN, USA

ELIZABETH S. FISHER
Merck & Co., Inc., Rahway, NJ, USA

38.1 INTRODUCTION

Two aspects of a successful formulation are that it produces consistent results in vivo and it can be manufactured reproducibly. One of an engineer's main goals in pharmaceutical development is to design a process that results in a high level of consistency and control of final product performance from research through manufacturing. Particle size of ingredients in a formulation, especially the active pharmaceutical ingredient (API), greatly impacts bioperformance and process capability, and as such it is an important parameter to understand and control. Milling, or mechanical size reduction of solids, is frequently used to achieve API or formulated intermediate particle size control and can also be used for reagent and excipient size control [1]. Size reduction can be performed with enough energy to break individual particles or with less energy to break granules or agglomerates during formulation; it can be performed on dry solids, partially wet solids, or in slurry mode; it can be performed on reagents and excipients as well as final API. While there are other processing options that can be used to achieve these goals, milling is a powerful tool that is frequently chosen as a relatively straightforward way not only to evaluate whether particle size control will provide the desired performance and consistency but also to effect the desired process or product performance control from early development through commercial-scale manufacturing.

Milling can increase the surface area of solids and their dissolution rate. This can be extremely important for improving bioavailability of a formulated drug product. The Biopharmaceutics Classification System [2–4] is one way of characterizing whether increasing API dissolution rate may improve its bioperformance; in this system, compounds are classified as BCS II or IV when their solubility is low based on a set of standard dissolution test conditions. While clearly particle size does not affect compound equilibrium solubility, dissolution rate is impacted and is directly proportional to particle size (see Noyes–Whitney equation, Fick's first law).

Control of dissolution rate can also be important for facilitating or improving chemical reaction performance when a reagent is charged in solid form. Increasing surface area of the API can also increase its effectiveness as seed for crystallization. As discussed in previous chapters, providing enough available surface of the correct crystal form is important for controlling API crystal size and form uniformity.

Size reduction of API and/or excipients can also be important to ensure consistency of formulation processing. This processing frequently involves dry blending of API with one or more excipients, and the mean size and particle size distribution (PSD) of the API relative to the excipients can influence the tendency of the API to either segregate or not mix uniformly with the other ingredients. Low dose formulations can be especially sensitive to blend uniformity problems linked to variable API particle size and disparity between

---

Chemical Engineering in the Pharmaceutical Industry: Active Pharmaceutical Ingredients, Second Edition.
Edited by David J. am Ende and Mary T. am Ende.
© 2019 John Wiley & Sons, Inc. Published 2019 by John Wiley & Sons, Inc.

the API particle size and that of the excipients. Milling can produce tight PSDs and small API mean size that may be necessary for drug product quality attributes [5]. Suspension or controlled release formulations may also have tight requirements on mean and top (largest) size in the API PSD. Additionally, API that is used in dry powder or metered dose inhalers must have sizes smaller than 5–6 μm and larger than 0.5–1 μm to reach deep lung and not be exhaled [6, 7].

In a milling operation, mechanical energy imparts stress to particles, and they are strained and deformed. When strained to the point of failure, cracks are formed and can propagate through the particle and result in breakage. Crystalline materials tend to break along crystal planes, while amorphous materials break randomly, and flawed particles will be easier to break than those with fewer internal weaknesses. When sufficient force is rapidly applied normal to a particle surface and directed toward its center, mass fracture or breakage into a few large fragments usually occurs. When force is applied more slowly, compression will be the main cause of particle breakage. If force is applied parallel to the surface of the solid, over time the particle can break into many fine particles, which is usually described as attrition [1]. Mill design and operation play a large role in determining which mode of breakage occurs.

There are several techniques available to characterize the distribution of particle and granule sizes produced by milling. Optical microscopy or image analysis is useful when little material is available, and observations of particle/granule morphology can be invaluable in troubleshooting milling problems or interpreting other particle size analysis results and powder flow behaviors. Laser light scattering particle size analysis requires more material but is commonly used to evaluate API PSDs. Solids to be analyzed can be dispersed in a carrier gas or liquid, and some force (air pressure or sonication) is applied to disrupt agglomerates and permit analysis of individual or primary particles (Figure 38.1). This technique can also be used online as milling takes place and therefore permit additional process understanding and control [8–10]. Sieve analysis, or determining the amount of material that passes through or is retained on standard mesh screens, is also used to assess size distribution of granules in a formulation process. *Perry's Chemical Engineers' Handbook* [11] summarizes several size analysis techniques and provides additional references on this topic.

It is important to note that in certain situations milling can be challenging and alternative size reduction techniques should be considered. If the milling operation results in material that contains a high proportion of fines (particles much smaller than the average size), processing problems can occur in subsequent unit operations. Large proportions of fine particles can increase cohesive behavior of the bulk solid, resulting in poorly flowing powder as the API is handled during formulation [12]. When solids are milled in a slurry and then need to be isolated from the liquors, fines can pass through the filter media and reduce overall yield. Fines can also reduce cake and filter media permeability and reduce filtration productivity. At pilot and manufacturing scale, formulation usually involves mechanical solids feeding, which requires that the solids flow in a predictable manner. This "flowability" can be assessed in different ways [13–15], and there is no standard industry practice for this testing in pharmaceutical manufacturing. However, it is clear that solids that tend to demonstrate cohesive behaviors and clump or stick to equipment surfaces can be challenging to process reproducibly and efficiently in formulation without additional effort and safeguards.

Applying mechanical energy via milling does more than break crystals. As the applied energy increases to be sufficient to make individual crystals smaller, the force and stress

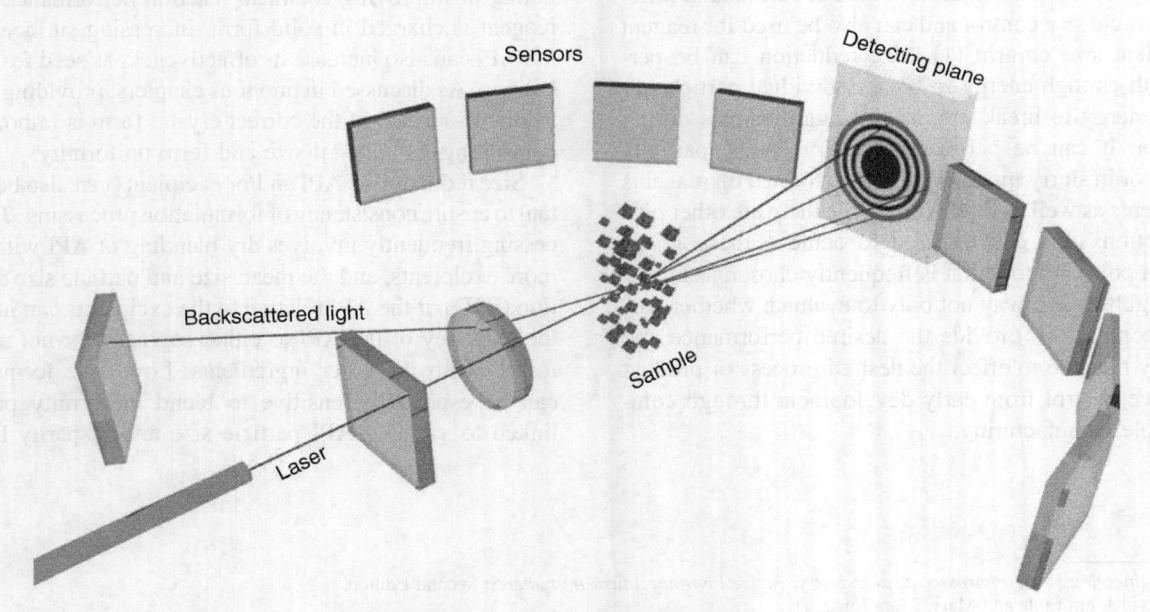

FIGURE 38.1 Laser light scattering device.

that cause fracture can also induce changes in the crystal. These changes may manifest themselves in a variety of ways, from increasing numbers of surface cracks and flaws that may increase specific surface area and surface roughness to changes in crystal form, including conversion to an amorphous form. All of these types of changes can affect performance in formulation/in vivo and can affect physical and chemical stabilities. This observation is more common for dry milling operations that break primary particles than it is for delumping/granule breaking operations or for wet milling. While milling-induced disorder likely occurs during a wet milling process, contact of the crystal with a liquid medium facilitates surface annealing via dissolution and subsequent redeposition, and the liquid present can transfer heat more readily than a gas stream. As understanding of solid surface chemistry and crystal structure has increased, and analytical detection and quantification capabilities have improved, these consequences of milling have become more frequently reported and studied. Ongoing work in this area continues to show that while milling can be a desirable way to control particle size, it is important to fully characterize the quality of the milled product. Some analytical techniques to assess amorphous content or crystal form include X-ray powder diffraction and differential scanning calorimetry, and techniques to assess surface energy include atomic force microscopy and inverse gas chromatography [16–18].

Each type of milling equipment is designed to impart different amounts of energy to the solids and therefore can produce different mean sizes and size distributions. Product requirements and process limitations will inform what target size is required and/or largest size is allowed. The following sections provide some guidance on how to use that information to select milling equipment and the key parameters that can influence its performance. The resulting mean particle size and PSD depend not only on the mill but also on physical properties of the feed solids such as brittleness and hardness [19], as well as its initial size distribution. Generally speaking, the most efficient way to assess milling feasibility is to perform trial experiments that can be combined with mechanistic tools to describe particle breakage. Even when combined with a size classification operation, it is not possible to use a mill to precisely "dial in" a mean size or PSD with any reliability without the use of feedback control from inline particle size analysis equipment. However, engineering understanding of mill operation and key process parameters permits effective scale-up of milling, so that the particle size to be achieved at large scale can be predicted accurately from carefully conducted gram-scale bench experiments.

38.2 SAFETY AND QUALITY CONCERNS

Chemical engineers must proactively address process safety concerns and quality risks associated with wet and dry milling processes. While these operations are not unique to the pharmaceutical industry, performing these operations on potent organic molecules can pose some unique challenges.

The main quality concern for both wet and dry milling is to minimize introduction of extraneous matter into the milled material. The most common material of construction of milling equipment is stainless steel, although ceramic, Hastelloy C, and various types of Teflon are also frequently used depending on the material to be milled. Abrasion of the mill surfaces by API or excipient particles can result in erosion of the mill material, and levels of residual metals such as Cr and Ni must be minimized relative to the human dose of API [20]. Assessing the hardness of the API during milling process development, either empirically through small-scale milling experiments or fundamentally through experiments such as indentation testing [19], will help prevent unexpected outcomes on scale-up.

The most obvious safety risk associated with dry milling is personnel exposure to highly potent airborne dust. This is especially important to consider during drug development, before full API safety testing has been completed, and a conservative equipment setup is recommended at early development stages. While most mill equipment is or can be enclosed during operation, minimizing this exposure risk, solids recovery/packaging operations and equipment cleaning should not be overlooked as opportunities for exposure. Typical methods to address this risk include clean-in-place systems and performing the entire milling operation in a negative pressure isolator with HEPA filtered exhaust. Multiple levels of exposure control, beginning with these engineering controls and continuing through personal protective equipment (PPE), are often required to ensure adequate operator protection.

Perhaps less apparent but no less serious is the risk of dust explosion during dry milling or post-milling activities [21, 22]. There have been numerous reported incidents of explosions occurring in the chemical, food, and pharmaceutical industries, in situations where fine powders are dispersed in air and a spark may be present. The most straightforward way to address this risk in development is to mill in an inert atmosphere (no/limited oxygen), using nitrogen or argon, and maintaining a scrupulously clean processing area. When sufficient API is available (tens of grams), the explosive properties of the dust should be quantified to better understand dust explosion risks. Typical measurements include determining the minimum ignition energy (MIE, smallest amount of energy required to ignite a dust cloud of optimal concentration) and minimum ignition temperature (minimum temperature at which a dust cloud will ignite) [23, 24]. If an explosive dust (MIE < 20 mJ) will be dry milled at pilot or manufacturing scale, engineering controls must be employed to minimize risk of explosion [25]. The force of a possible explosive event can be quantified by measuring K_{st} (expression of the burning dust's rate of pressure increase), and this parameter can be used in equipment/facility design.

Engineering or design solutions include inertion of the process train and the use of rupture disks or automatic shutoff valves that quickly close when pressure spikes are detected.

The most common risk associated with wet milling is that of accumulating a static charge in a recirculating loop, especially when a nonconductive solvent such as hexane, heptane, or toluene is used. Electrostatic charge accumulation results from frictional contact between flowing liquids and solids in recirculating loops and the surfaces of equipment (pipes, vessels, agitators). The resulting potential difference between batch and processing surfaces could lead to a static discharge, which could ignite a fire or explosion, especially while processing with flammable organic solvents. Conductivity of organic solvents can range from 10^7 to <1 pS/m. Another way to interpret these values is to estimate the time required for an accumulated charge to decay (a multiple of the relaxation time). For 99% charge decay this may range from milliseconds for water to several minutes for n-hexane [26]. Slurry conductivities of less than 100 pS/m are considered at high risk for electrostatic discharge if a mechanism for charge formation is present (such as a recycle loop or wet mill). Charges can be encouraged to dissipate by the use of conductive equipment (stainless steel in place of Teflon-lined pipe) and grounding and bonding the equipment. Additional precautions include operating under nitrogen and keeping linear velocity through the recirculation loop as low as is practical (while maintaining adequate flow rates to ensure solids do not settle within the recirculation piping and low points in the loop).

38.3 TYPES OF MILLING AND MILL EQUIPMENT

The two primary means of reducing particle size of a solid product are wet and dry milling as defined by the media in which the engineer chooses to carry out the milling activity. Selecting a milling strategy depends on a variety of factors including particle physical and chemical properties, chemical stability, and safety-related issues.

API is most often isolated by crystallization, followed by filtration of the solids and a terminal drying step. Based on the thermodynamic and kinetic properties of the system, the physical properties of the solids of interest, and the dynamics of the environment within the crystallizer, several outcomes are possible. Crystallizations that end in final isolated particles at the desired PSD specification require no further processing to be carried into the downstream processing train and are often the result of targeted particle engineering efforts [27–29]. Particles below the target PSD can either be tested for processability in the drug product manufacturing, or the size can be increased through modified crystallization strategies to yield larger particles of the desired PSD [30]. Most often, and generally by design, the isolated crystals are larger than the target PSD as established through formulation development or bioavailability testing. In these situations the resulting material must be milled to reduce the PSD to the target.

Deliberately growing and isolating particles larger than the target PSD may offer several advantages such as cycle time savings from reduced filtration time during isolation. While the filtration time for small particles can be reduced through increased pressure, equipment limitations and nonlinearity associated with compressible solids may result in diminishing returns with respect to cycle time savings. As a result, it is often advantageous to attempt to grow, isolate, and dry larger particles in the final processing steps of an API synthesis. In situations with solution instability, this may be the only suitable processing option as the time associated with milling a slurry could negatively impact API purity in these cases. Therefore, in considering the overall manufacturing efficiency associated with making API, the purposeful generation of large particles, which are later reduced in size, is a viable option to consider. In these scenarios, dry milling is the preferred technology for micronization.

Dry milling may also be preferred over wet milling due to incompatibility between the processing solvent and the available wet milling equipment, such as when the process stream is corrosive to stainless steel. Additionally, dry milling may be a desirable unit operation if an isolation lends itself to some process advantage. Motivating factors driving an isolation prior to reducing particle size may include impurity removal or rejection resulting from a selective crystallization and rapid removal of the liquors.

While growing of oversized crystals followed by terminal dry milling offers a unique set of advantages with respect to overall cycle time, there are other significant disadvantages when compared with particle size reduction using wet milling techniques. Oftentimes dry milling equipment and facilities are difficult and time consuming to clean when compared with in-line wetted pipes that can easily be cleaned with circulating cleaning solutions. For highly potent or toxic compounds, operator exposure is generally considered less complex to achieve in wet milling infrastructure. Lastly, the risk for dust-related events is nearly eliminated when choosing a wet milling approach.

Wet milling offers an advantage in that milling can be combined with terminal isolation such as selective crystallization of the desired product. When impurity removal is still of principal concern, techniques such as decantation, crossflow microfiltration with diafiltration, or filtration with reslurry to remove crystallization mother liquors may be employed as a way to remove rejected impurities or selectively switch to a desired solvent system.

An additional process design benefit is potentially realized with wet milling, as it allows for greater flexibility in terms of how a crystallization process is developed. As an example, as crystallization progresses the mean size of the crystals can be simultaneously reduced by milling during the crystallization process. This is time sparing and also presents crystallization benefits by continually exposing new surface area for use in the ongoing crystallization. Wet mills are also preferred if the material being milled exhibits undesirable physical properties or phase changes at higher temperatures. The increased heat capacity of the liquid carrier media allows for smaller temperature fluctuations during milling. This can be especially important for materials that have either a low melting point or are susceptible to crystal form conversion at lower temperature.

Wet milling may offer significant operating advantages over dry milling when the solvent system used for particle size reduction is directly compatible with the downstream drug product processing operation. Drug product unit operations including spray coating and wet granulation require the blend of the milled drug substance with liquid and may therefore be amenable to wet milling followed by direct use of the slurry stream in the formulation processing.

38.3.1 Dry Milling Setup

A general dry milling setup is shown in Figure 38.2. Unmilled solids are charged to the feed hopper via several methods, generally designed to contain any particle dusting during the charge operation. Solids are fed from the feed hopper through a feed device that will allow a constant mass flow to the milling unit. They are conveyed by either gravitational flow through the milling apparatus, through screw conveyance, or by conveyance with a carrier gas such as nitrogen. Typically, rotary valves and screw feeders are used to ensure constant feed rates and prevent backflow of solids due to the slightly positive gas pressures in the mill itself. Feed hoppers are generally designed with sloped sides to enable proper solids flow into the rotary valve and screw feeder system. For poorly flowing solids, "bridge breakers" or small agitators designed to facilitate solids flow from the hopper into the screw feeder are also installed.

Gas is fed to the mill along with the solids. Depending on the type of mill being used, some of the gas may be introduced to the mill separately from the entering solids, such as in the design of a loop or spiral jet mill, or the gas may be used as a carrier fluid to convey the solids to the mill as is the case in a hammer or pin mill design. While air may be used to convey and mill the solids, nitrogen is most often used to ensure adequate inertion of the environment around the solid phase. For solids with a low melting point, or a propensity to undergo a pressure-induced phase transition to an amorphous solid, liquid nitrogen may be supplied via an orifice upstream of the milling operation to maintain cryogenic temperatures within the mill internals.

Depending on the mechanism of grinding within the mill, the residence time of the solids may vary from one to several seconds of average residence time. Once milled, particles will exit the grinding chamber and be carried to product collection. When tighter particle size control is desired, a classifier may be installed in-line that allows for larger particles to be returned to the feed hopper while particles that have been adequately reduced will be routed to the collection area. Product collection from the gas stream will frequently use a cyclone to separate most of the larger solids from the gas stream and then send the gas-containing finer particles to a bag filter and HEPA filter for final dust removal.

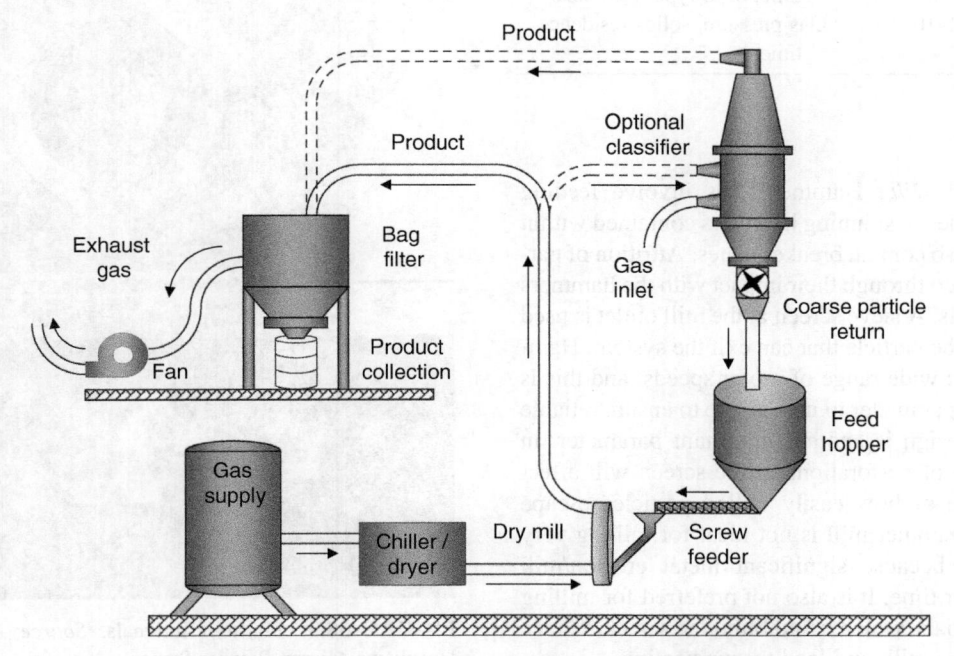

FIGURE 38.2 General setup for dry milling operations.

In most size reduction milling equipment, residence time in the mill is a key parameter determining outlet particle size. As a result, the feed rate becomes a significant part of the control strategy and must be carefully considered on scale-up.

38.3.2 Dry Milling Equipment

Dry milling may be accomplished by grinding or high force collisions of particles with a moving pin or hammer. Alternatively, it may be accomplished by high-energy particle–particle and particle–wall collisions as is the case with jet mills. The following sections briefly describe different types of dry mills. Note that for reduction of particle size to submicron, dry milling strategies are not usually adequate, and therefore wet milling strategies are typically employed. The table below summarizes some different types of dry mills, their key scale-up parameters, and typical minimum particle sizes they can produce. As noted above, solids residence time in the mill is always an important scale-up parameter for dry milling. This can be easily measured experimentally and can be controlled by limiting solid and gas feed rates to match mill efficiency.

Mill Type	Typical Minimum Milled Size (µm)	Key Parameters
Hammer mill	20–60	Mill speed, solids residence time, hammer type, screen size/type, feed size
Universal/pin mill	15–30	Mill speed, solids residence time, head type, feed size
Jet mill	2–10	Gas pressure, solids residence time

38.3.2.1 Hammer Mill
Hammer mills involve feeding solids through a series of spinning hammers contained within a casing that may also contain breaker plates. Attrition of particles is accomplished through their impact with the hammers and the mill internals. A sieve screen at the mill outlet is used to limit the size of the particle that can exit the system. Hammer mills come in a wide range of motor speeds, and this is the most important parameter to investigate to ensure reliable scale-up. Screen design is another important parameter, in that size and shape of perforations in the screen will affect mill residence time or how easily milled particles escape the chamber. The hammer mill is not ideal for milling very abrasive materials because significant metal erosion/mill wear can result over time. It is also not preferred for milling highly elastic materials, which could blind the screen, reduce gas flow through the mill, and lead to overheating.

38.3.2.2 Universal/Pin Mill
The term "universal" mill usually refers to a mill configuration where multiple milling heads can be used. These mills often can be fitted with pin, turbo rotor, and hammer-type heads. Pin mills operate similarly to hammer mills but with typically faster rotor speeds and small clearances between rotating and stationary pins. Solids are fed to a milling chamber in a conveyance gas stream. The milling chamber contains a high-speed rotor/stator configuration of pins, which impact the particles as solids are directed from the center of the pin disk out through all the rows of intermeshing pins (Figure 38.3). Control parameters to adjust and vary the output PSD include pin gap or pin spacing, the rotational speed of the rotor, solids feed rate, size of the mill, and velocity of the carrier gas used to convey the solids out of the mill.

38.3.2.3 Jet Mill
Jet mills are an alternative to hammer or pin milling, where the primary mode of action is mechanical impact of the mill with the particle. With jet milling, micronization is accomplished mainly through particle–particle and particle-wall collisions caused by high velocity gas streams. Spiral jet mills, loop jet mills, and fluidized bed jet mills are examples of jet or fluid energy mills. These jet mills use the same general operating principle. Through the introduction of high pressure air or other carrier gas (i.e. nitrogen) into

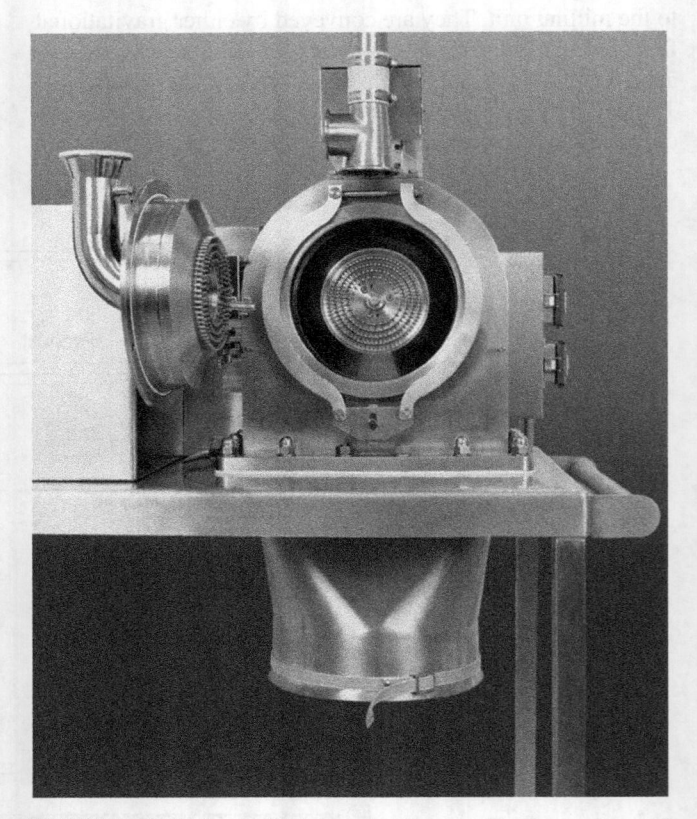

FIGURE 38.3 Pin mill internals. *Source*: Photo courtesy of Hosokawa Micron Powder Systems.

specially designed nozzles, the potential energy of the compressed gas is converted into a grinding stream at sonic or supersonic velocities.

Differences in the various jet mills are predominantly in the geometry of the grinding chamber itself. Spiral mills create a high velocity helix of gas that rotates around the center of the circular jet mill. Solids are introduced via a venturi feed eductor (Figure 38.4) and become entrained in the turbulent helical flow. The resulting high-energy collisions between particles as well as between the particles and the mill internals fracture particles to micron and submicron size. For loop jet mills, air or carrier gas is injected into an oval grinding loop or "race track" through specially designed nozzles (Figure 38.5). Solids are introduced in a manner similar to the spiral jet mill. In both designs, particles will classify based on their relative inertia toward the outlet of the mill, resulting in larger particles remaining in the grinding chamber longer while smaller particles are carried out of the mill with the overall gas exhaust to the collection cyclone or chamber. This permits some internal classification and makes the spiral or loop mill performances somewhat less dependent on feed particle size, compared with hammer and pin mills. A derivation is available for spiral jet mill cut size as a function of micronization settings, gas properties, and other empirical constants for the material and mill [31]. Key scale-up parameters are the grinder nozzle pressure and the mill residence time, typically determined as particle density in the mill (ratio of solid mass flow to volumetric gas flow through the mill).

In fluidized bed jet mills, the grinding chamber is oriented as a fluidized bed, with specially designed nozzles introducing the grinding gas at the bottom of the mill chamber, creating high intensity collisions between particles (Figure 38.6). Net vertical gas flow out of the mill fluidizes the milled material in the grinding chamber. Fluidized bed jet mills are usually fitted with a classification wheel. Based on the rotational speed of the classifier wheel, large particles gain radial momentum and are returned to the grinding zone of the fluidized bed. Lighter particles can escape the mill through vanes in the classifier wheel, carried by the main gas exhaust. This type of arrangement can lead to much narrower PSDs than other types of jet milling equipment. In addition to grinder pressure, classifier speed is clearly an important scale-up parameter. To control the solids residence time in the mill, nozzle diameter, total gas flow rate, and mill chamber

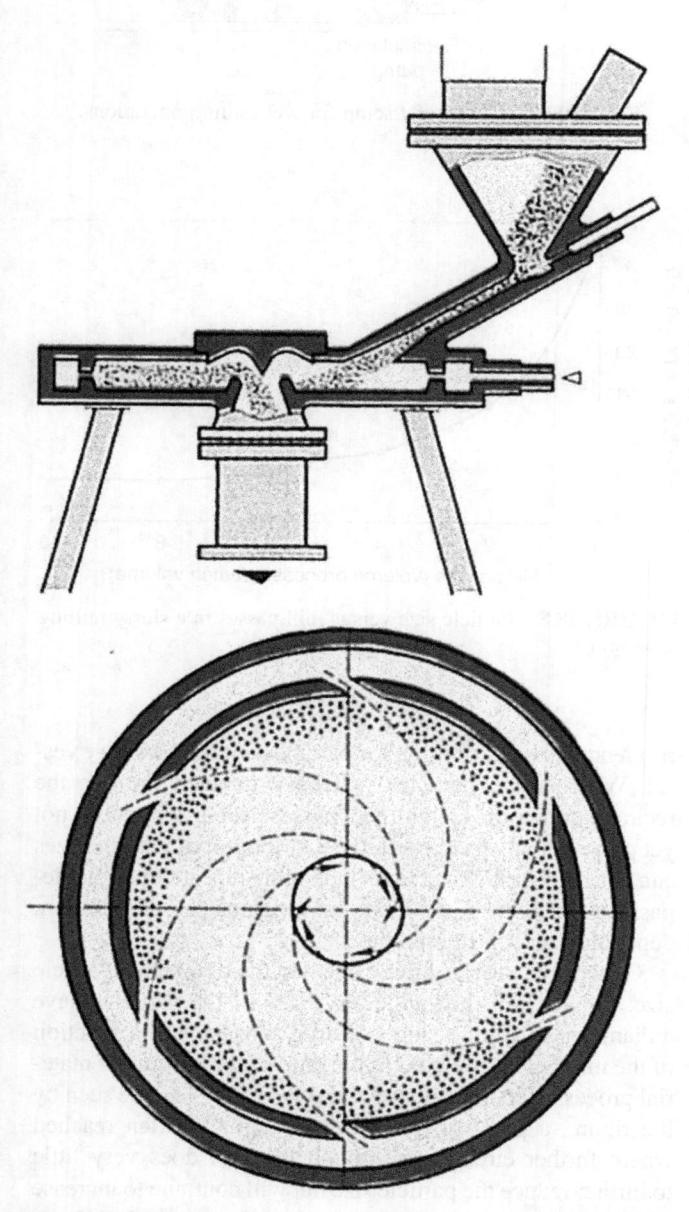

FIGURE 38.4 Spiral jet or "pancake" mill. *Source*: Photo courtesy of Hosokawa Micron Powder Systems.

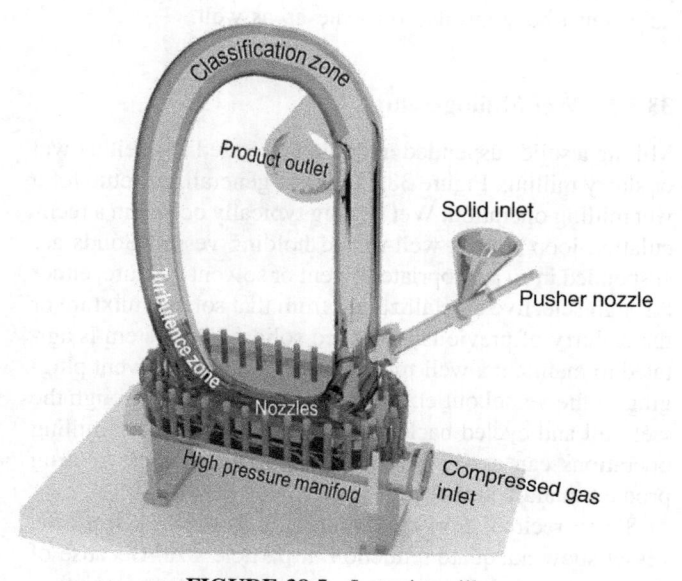

FIGURE 38.5 Loop jet mill.

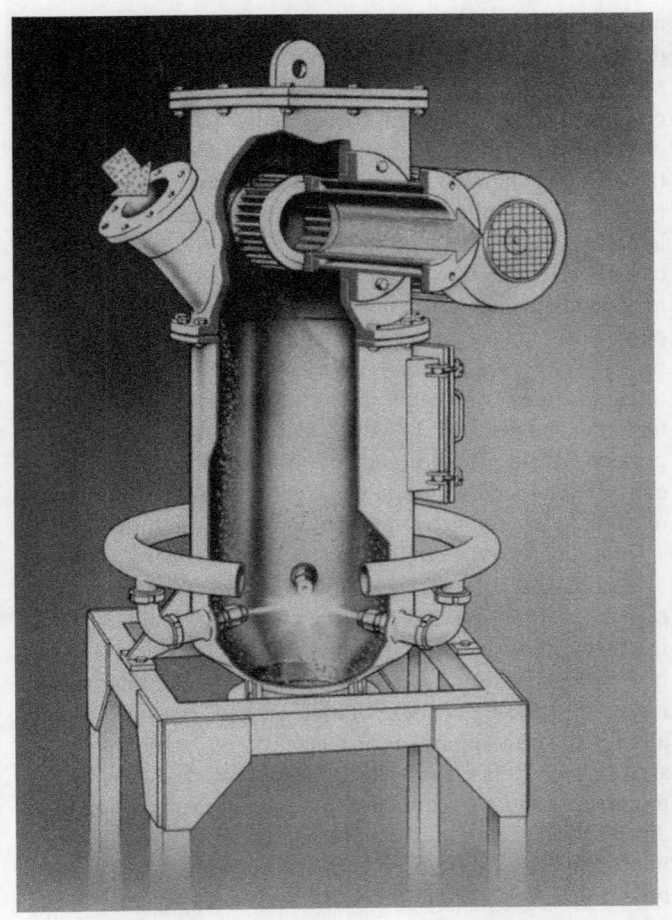

FIGURE 38.6 Fluidized bed jet mill with classification. *Source*: Photo courtesy of Hosokawa Micron Powder Systems.

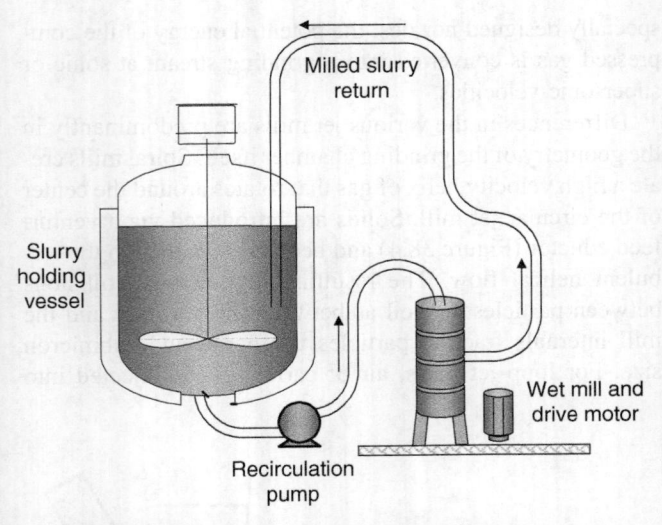

FIGURE 38.7 General setup for wet milling operations.

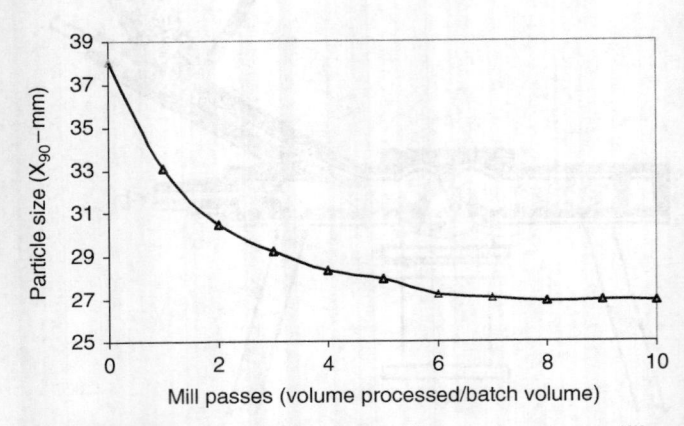

FIGURE 38.8 Particle size versus mill passes in a slurry milling operation.

pressure can be varied with solids feed rate, and these variables must be controlled on scale-up as well.

38.3.3 Wet Milling Setup

Milling a solid suspended in liquid is referred to as either wet or slurry milling. Figure 38.7 shows a generalized setup for a wet milling operation. Wet milling typically occurs in a recirculation loop from a well-mixed holding vessel. Solids are suspended in an appropriate solvent or solvent mixture, either through selective crystallization from that solvent mixture or the reslurry of previously isolated solids. The system is agitated to maintain a well-mixed suspension and prevent plugging of the vessel outlet. The slurry is circulated through the wet mill and cycled back to the holding vessel. Wet milling operations can easily be part of a continuous API or drug product manufacturing process.

Slurry recirculation is continued until samples from the vessel show adequate reduction in particle size. Because of the nature of the recirculation operation, there exists a residence time distribution for particles within the slurry vessel. Where some particles will have traveled through the recirculation loop in multiple passes, some will have not yet passed at all. As a result, the distribution of particle sizes can tend to broaden over time. Operating repeatedly in single-pass or "once-through" mode may tighten the distribution, depending on solid properties.

Generally, a lower limit exists for the obtainable particle size. Figure 38.8 shows the X_{90} (90% of the particles have a diameter $\leq X_{90}$) of a slurry of drug substance as a function of the mill passes (defined as the ratio of the volume of material processed to the volume of the slurry). As can be seen by the figure, a point of diminishing returns is often reached where further circulation through the mill does very little to further reduce the particle size but will continue to increase the levels of fines produced by chipping rather than mass fracture.

38.3.4 Wet Milling Equipment

Wet milling offers many advantages to dry milling, though there are some limitations. One significant advantage is the ability to eliminate a separate unit operation of dry milling. Wet milling can be carried out as part of the final crystallization–isolation sequence. This approach eliminates the cycle time and cost associated with an extra unit operation, which can be particularly significant if special containment is necessary (for personnel protection). Wet mills also provide the ability to protect the product from the heat input from some dry milling equipment.

Multiple types of wet mills are available. The three most commonly used in pharmaceutical manufacturing are toothed rotor–stator mills, high pressure homogenizers, and media mills. The table below summarizes typical minimum particle sizes and main scale-up parameters for these mills. If the mills are operated in recycle mode but not allowed to reach "steady-state" milled particle size, feed particle size will also be an important scale-up parameter.

Mill Type	Typical Minimum Milled Size (μm)	Key Parameters
Toothed rotor–stator mill	20–30	Mill speed, tooth spacing, rotor–stator design, milling time (number of passes)
High pressure homogenizers	<1–10	Gap size, pressure, number of homogenization cycles
Media mill	<1–10	Mill speed, slurry composition, amount and size of media in mill

38.3.4.1 Toothed Rotor–Stator Mills

Rotor–stator mills (Figure 38.9) consist of a rotating shaft (rotor), with an axially fixed concentric stator. Toothed rotor–stator mills have one or more rows of intermeshing teeth on both the rotor and the stator with a small gap between the rotor and stator. Variations in the number of teeth, teeth spacing, angle of incidence, etc. all impact the milling efficiency of toothed rotor–stator mills [32–34].

The differential speed between the rotor and the stator imparts extremely high-shear and turbulent energy in the gap between the rotor and stator. In this configuration, particle size is reduced by both the high shear created in the annular region between the teeth and the collisions of particles on the leading edge of the teeth (Figure 38.10). Size is affected by selecting rotor–stator pairs with different gap thickness or by operating at different rotational rates (or tip speeds) of the rotor. Tip speed is a very important factor when considering the amount of shear input into the product. Tip speed (S_r) is determined according to the following equation:

$$S_r = \pi D \omega \quad (38.1)$$

where

D is the rotor diameter

ω is the rotation rate (rev/min)

The average shear created between the rotor and stator is a function of the tip speed and the gap thickness. The average shear rate $\dot{\gamma}_{av}$ is given by

$$\dot{\gamma}_{av} = \frac{S_r}{\delta} \quad (38.2)$$

where

δ is horizontal gap distance between the rotor and stator.

Another important factor is the number of occurrences when rotor and stator openings mesh. This is known as the shear frequency (f_s) and is calculated as

FIGURE 38.9 Wet mill internals. *Source*: Photo courtesy of IKA Works Inc.

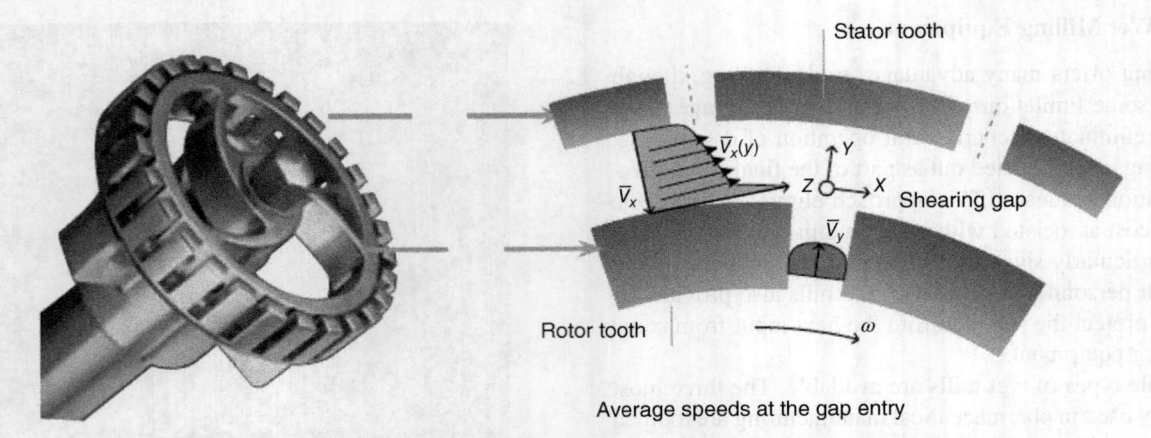

FIGURE 38.10 Rotor–stator working principle. *Source*: Photo courtesy of IKA Works Inc.

$$f_s = N_r N_s \omega \tag{38.3}$$

where

N_r and N_s are number of teeth on the rotor and stator, respectively.

The shear number (N_{Sh}) is given by

$$N_{Sh} = f_s \dot{\gamma}_{av} \tag{38.4}$$

Depending upon the rotor–stator design, the number of rows should be accounted for when calculating the shear number. Scale-up of toothed rotor–stator mills is accomplished by determining for the material in question whether shear frequency or tip speed has greater control over milled size and then working with rotor–stator design options and mill speed to maintain the key parameter across scales. If milling will be stopped before steady state/minimum size is achieved, the process must also be scaled on batch turnovers or passes through the mill.

In addition to the scale-up guidelines mentioned above, integral scale-up factors have been proposed. The term "integral" is used here to indicate that the milling time is also taken into account [35]. One approach calculates the normalized energy imparted to particles during milling (E^*) and introduces a new scale-up parameter [35]:

$$E^* = \varphi_r E_r \tag{38.5}$$

$$E_r = \frac{S_r^2}{S_{rb}^2} \tag{38.6}$$

$$\varphi_r = \frac{N_r N_s \omega}{N_{rb} N_{sb} \omega_b} \frac{N_{bt} N_g Q_{mill}}{N_{btb} N_{gb} Q_{mill,b}} \frac{W_r H_r}{W_{rb} H_{rb}} \tag{38.7}$$

where

Q_{mill} is the volumetric flow rate.

W_r and H_r are the rotor slot width and height, respectively.

Subscript b indicates a reference experiment.

N_{bt} is the total number of batch turnovers.

N_g is the number of generators (rotor/stator pairs) in the mill.

Another approach defines a number of slot events (N_{se}) to correct the conventional number of batch turnovers as an integral scale-up factor [36]. In this approach, the same tip speed is used at different scales but incorporates geometrical considerations of slots as well as of the bypass region to correct the elapsed milling time as follows [36]:

$$f_{se} = N_{bt} N_r \tag{38.8}$$

$$P_{se} = \frac{H_s W_s N_r}{H_s W_s N_r + \pi D_r \ell} \tag{38.9}$$

$$N_{se} = f_{se} P_{se} \tag{38.10}$$

where

W_s and H_s are the stator slot width and height, respectively.

D_r is the rotor diameter.

ℓ is the vertical gap between rotor and stator.

38.3.4.2 High Pressure Homogenizer High pressure homogenization is a high-energy process in which size reduction of API particles is achieved by repeatedly cycling a slurry through a gap or orifice at high velocity, around 500 m/s, and pressure, 1000–1500 bars (Figure 38.11) [37]. When the slurry is forced through a gap at a high flow rate, the downstream pressure on the liquid falls below the vapor pressure of the liquid at the system temperature, and gas bubbles are formed. Bubbles collapse when the liquid exits from the gap. The cavitation forces arising from the formation and

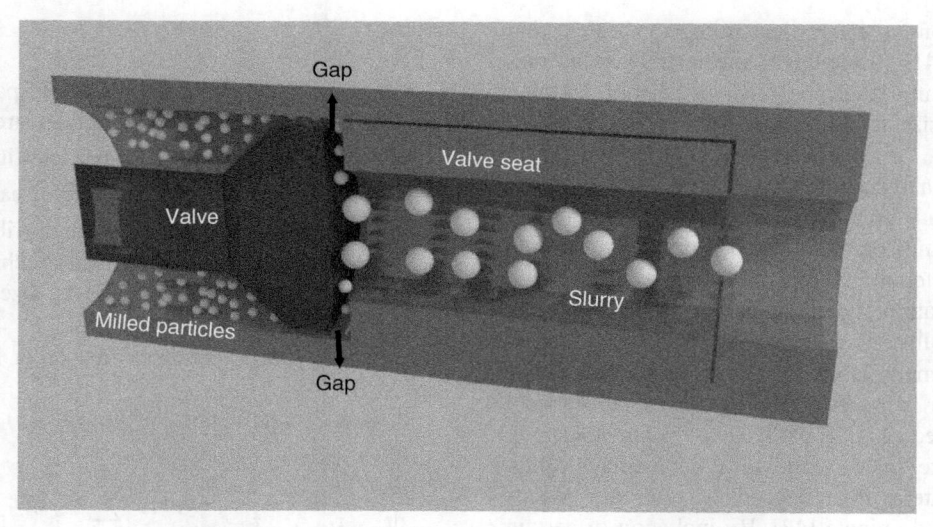

FIGURE 38.11 Schematic diagram of a high pressure homogenizer.

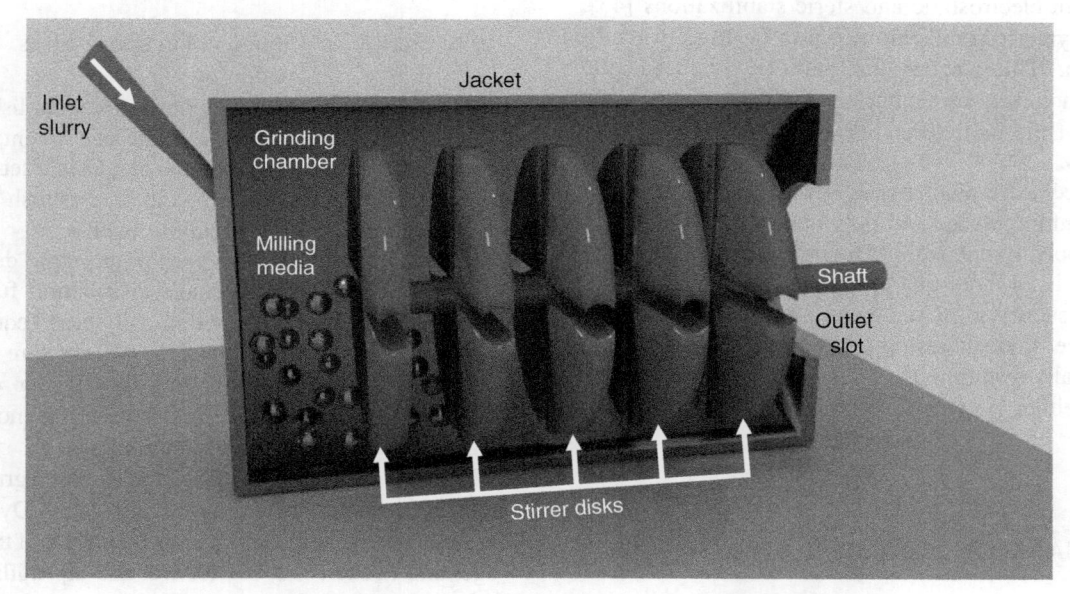

FIGURE 38.12 Schematic diagram of a media mill.

collapse of the gas bubbles, coupled with very high shear stresses, acceleration, impact, and intense turbulent flow in the liquid, result in effective particle breakage. The highly disruptive energy generated by high pressure homogenizers can produce extremely fine suspensions with particle sizes of less than 1 μm.

High pressure homogenization is widely used in the pharmaceutical and food industries [38, 39].

High pressure homogenizers expose drug particles to a power density greater than 1000 W/m^3. The high energy activates the drug particles, potentially leading to partial or complete amorphization of the drug [38, 40].

38.3.4.3 Media Mills Media mills, also referred to as pearl or bead mills, are much different in operation than a rotor–stator mill. The mill is composed of a milling chamber, milling shaft, and product recirculation chamber (Figure 38.12). The milling shaft extends the length of the chamber. A shaft can have either radial protrusions or fingers extending into the milling chamber, a series of disks located along the length of the chamber, or a relatively thin annular gap between the shaft and mill chamber. The chamber is filled with spherical milling media usually less than 2 mm in diameter and typically 1 mm in diameter or less. Media are retained in the mill by a screen located at the exit of the mill. The rotation of the shaft causes the protrusions to move the milling media, creating high shear forces [41]. Scale-up of media mills is accomplished by maintaining residence time in the mill, keeping the milling media size constant, and holding energy input constant.

The high energy and shear that result from the movement of the milling media is imparted to the particles as the material is held in the mill or recirculated through the milling chamber by an external recirculation loop. Particles are ground by a combination of particle–media, particle–particle, and particle–wall collisions. These mills can produce submicron particles. Thermally labile material is easily handled as the milling chamber is jacketed. By utilizing smaller media (<100 μm), nanosized (20 nm) particles are achievable.

At the small size scales achievable by media milling, particle–particle interactions caused by van der Waals forces can begin to dominate [42]. By inclusion of certain additives to the dispersion fluid, the possible agglomeration and resulting reduced efficiency and reduced effectiveness of the mill can be mitigated. Surfactants can inhibit agglomeration by both electrostatic and steric stabilizations [43]. Similarly, polymeric stabilizers can also be used to retard agglomeration. The smaller the particle produced, the greater the amount of surfactant needed since the specific surface area of the particle increases proportionately with particle radius.

Milling media are available as glass, metals, ceramics such as zirconium oxide, and polymers such as a highly cross-linked polystyrene resin. The proper selection of the milling media is an important criterion in the milling process. Materials such as glass and metals are typically not used for APIs since some abrasion of the milling media can occur, potentially resulting in extraneous material concerns. As a result, either polymeric or ceramic media are usually used.

38.4 MECHANISTIC MODELS

One of the most common ways to mathematically describe milling processes involves the use of population balance equations (PBEs). One of the key benefits of using PBEs is the ability to predict evolution of PSDs when multiple phenomena are involved such as primary nucleation, agglomeration, dissolution, and growth.

For a continuous milling process where nucleation, dissolution, agglomeration, and growth are neglected, the PBE can be simply expressed as follows [44]:

$$\frac{\partial m(x,t)}{\partial t} = \frac{1}{\tau(x)} m_0(x,t) - \frac{1}{\tau(x)} q m(x,t) - S(x) m(x,t) + \int_x^\infty b(x,y) S(y) m(y,t) dy \quad (38.11)$$

where

$m(x,t)$ and $m_0(x,t)$ are the masses of particles of size x at time t at the mill outlet and inlet streams, respectively.
$\tau(x)$ is the residence time of particles in the mill of size x.
$S(x)$ is the specific breakage rate of particles of size x.
$b(x,y)$ is the mass-based breakage distribution function that specifies the mass of fragments of child particle of size x formed when larger particles of size y are broken.

Equation (38.11) assumes first-order kinetics for breakage [45].

A size-discrete version of Eq. (38.11) is as follows:

$$\frac{dm_i}{dt} = \frac{1}{\tau_i} m_{i,0} - \frac{1}{\tau_i} m_i - S_i m_i + \sum_{j=1}^{i-1} b_{i,j} S_j m_j \quad \forall \ i = 1, 2, \ldots, N$$

(38.12)

where

subscripts i and j represent bin sizes/classes.

The most commonly used convention establishes that bin 1 contains the coarsest particle, while size N contains the finest particle. The size discretization results in a set of ordinary differential equations, where S_i and $b_{i,j}$ establish both breakage rate and breakage distribution function.

A variety of models have been proposed to describe specific breakage rate and breakage distribution functions for milling processes including empirically based equations [46, 47], statistically based models [48, 49], fracture mechanics-based expressions [50–54], and data-driven expressions [55, 56]. Table 38.1 summarizes some of the most common equations used to describe grinding processes.

Commercial implementations of PBE-based grinding models are available for practitioners. For instance, DynoChem by Scale-up Systems provides an implementation of the model by Vogel and Peukert [52, 53] for wet and dry milling applications. The mill model in gPROMS FormulatedProducts by Process Systems Enterprise Ltd. accounts for most of the specific breakage rates and breakage distribution function models listed in Table 38.1. Moreover, gPROMs Formulated Products allows milling unit operations to be integrated with drug substance and drug product unit operations in a single flowsheet.

38.5 CASE STUDIES

Throughout the development cycle of a drug substance, different PSDs may be studied to determine the optimum size range for an API or excipient. Milling process parameters that achieve these profiles have often been determined in an empirical fashion through the use of test milling – portions of a material lot are

TABLE 38.1 Examples of Common Equations Used to Describe Grinding Processes

Approach	Equations	Uses	Reference
Empirical	$B_{i,j} = \varphi\left(\frac{x_i}{x_j}\right)^\gamma + (1-\varphi)\left(\frac{x_i}{x_j}\right)^\beta$ $b_{i,j} = B_{i,j} - B_{i+1,j}$ $S_i = S_1\left(\frac{x_i}{x_1}\right)^\alpha$	Ball and rod mill, ball-and-race pulverizer, shredder-cutter mill, hammer mill, conical mill, and wet mill. Model has been tested for a variety of materials	Refs. [46, 47]
Statistical	*Product function with a power law form:* $b_n(v,w) = \frac{pv^m(w-v)^{m+(m+1)(p-2)}[m+(m+1)(p-1)]!}{w^{pm+p-1}m![m+(m+1)(p-2)]!}$ *Summation function with a power law form:* $b_n(v,w) = \frac{m+1}{w^{m+1}}[v^m + (v-w)^m]$ for $p=2$ $b_n(v,w) = \frac{A_{m,p}}{w^{m+p-1}}\left[v^m(w-v)^{p-2}\prod_{i=0}^{p-3}\frac{1}{i+1} + (w-v)^{m+p-2}\left(\frac{p-1}{A_{m,p-1}}\right)\right]$ with $A_{m,2} = m+1$ and the following recursive relationship for $A_{m,p>2}$ $\frac{p}{A_{m,p}} = \frac{(p-2)!m!}{(m+p-1)!}\prod_{i=0}^{p-3}\frac{1}{i+1} + \left(\frac{p-1}{A_{m,p-1}}\right)\frac{1}{m+p-1}$ *Bivariate kernels* have been also developed	Stirred vessels and various mills. Model has been tested for sugar crystals	Refs. [48, 49]
Fracture mechanics (energy) based	$P_B = 1 - \exp[-f_{Mat} \times k(W_{m,kin} - W_{m,min})]$ $B_{i,j} = \left(\frac{x_i}{x_j}\right)^n \frac{1}{2}\left[1 + \tanh\left(\frac{x_i - y'}{y'}\right)\right]$	Hammer mill and air classifier mill. The model performance has been tested for variety of materials	Refs. [52, 53]
	$S_i = c\frac{(E_{kin}/V)(E_{fract}/V_i)\sqrt{P_y/\rho}}{H\sqrt{x_i}K_{1c}}\left(\frac{1}{x_i}\right)$ $b_{i,j} = \frac{S_{i-1} - S_i}{S_j}$	Jet mill. It was tested with several organic compounds and APIs	Refs. [50, 51, 54]
Data driven	$\beta_{i,j} = (1-k_c)\sigma_j$ for $i = j+1$ $\beta_{i,j} = k_c\sigma_j$ for $i =$ smallest bin $\beta_{i,j} = 0$ for any other bin $\sigma_i = \frac{1}{1+\exp(-n_i)}$ $n_i = w_0 + w_{GP}GP + w_{PP}PP + w_{FR}FR + w_{size}x_i + w_{GPPP}GP \cdot PP$ $k_c = \frac{1}{2}\cos(\theta) + \frac{1}{2}$ $\theta = a_0 + a_{GP}GP + a_{PP}PP + a_{FR}FR$	Spiral air jet mil. The model was tested with sodium bicarbonate, α-lactose monohydrate, and sucrose	Refs. [55, 56]

subjected to different milling conditions, and the resultant output is tested to determine PSD. While useful in establishing milling ranges for one particular lot of material, test mill runs do not easily allow extrapolation to different milling input conditions. As quality-by-design expectations continue to increase within the pharmaceutical processing world, an understanding of milling process parameters, as well as input PSD and their effect on final API attributes, is desirable.

In what follows, two examples of potential approaches to describe milling processes are presented: (i) statistical model and (ii) mechanistic model.

38.5.1 Milling: Statistical Model

As an example, a simple model for particle breakage, like the one shown in Eq. (38.13), would likely predict an output parameter, such as the X_{50} of the PSD, as a function of one or more input parameters:

$$X_{50,\text{milled}} = \text{function (input size, mill energy, residence time)} \quad (38.13)$$

To utilize this type of model, the input properties must be chosen, and parameters simulating residence time and energy input must be determined. For example, consider the case of dry milling of a material that has been crystallized and then dried in an agitated filter dryer. As a first approach, a static parameter such as the unmilled X_{50} of the slurry could be chosen as the input size, the mass feed rate (m) to the mill chosen as a surrogate for residence time, and rotor tip speed (ν) chosen for energy input:

$$X_{50,\text{milled}} = K(X_{50,\text{unmilled}})^\varphi m^\alpha \nu^\beta \quad (38.14)$$

Figure 38.13 shows the relationship between predicted and actual milled values using such a model following a calibration exercise to fit the exponents and the constant K.

While a rough trend may be observed between predicted and actual, the accuracy of this model is not sufficient for

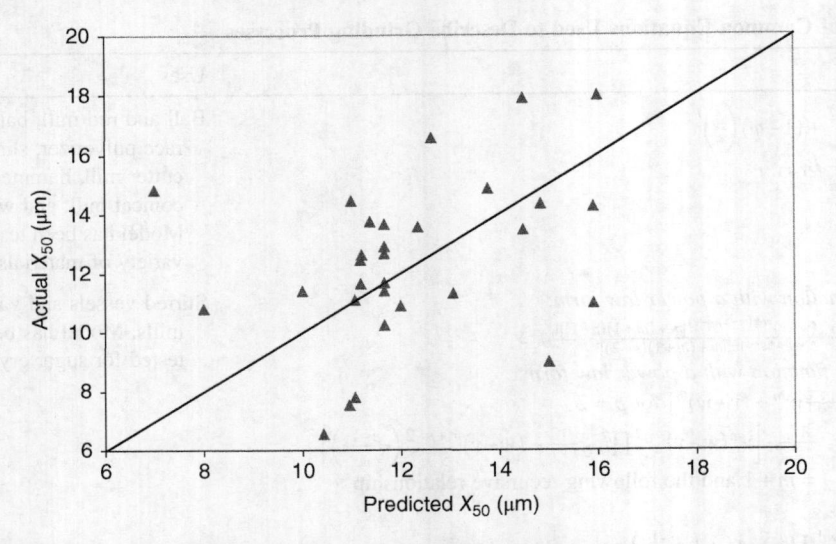

FIGURE 38.13 Predictive ability of a milling model utilizing unmilled slurry X_{50} as an input parameter.

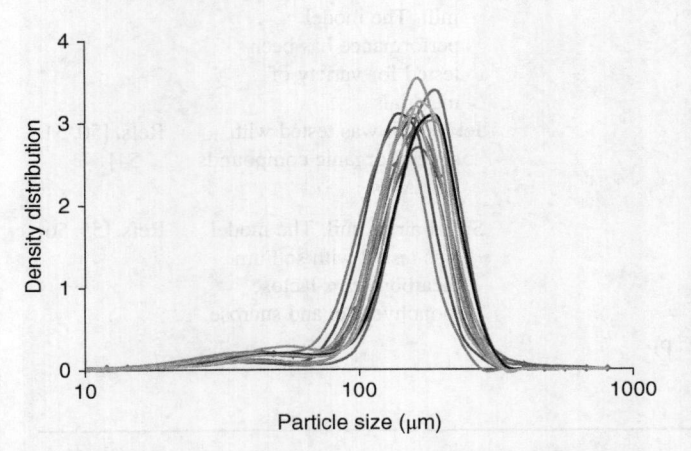

FIGURE 38.14 Particle size of the API coming out of the crystallizer.

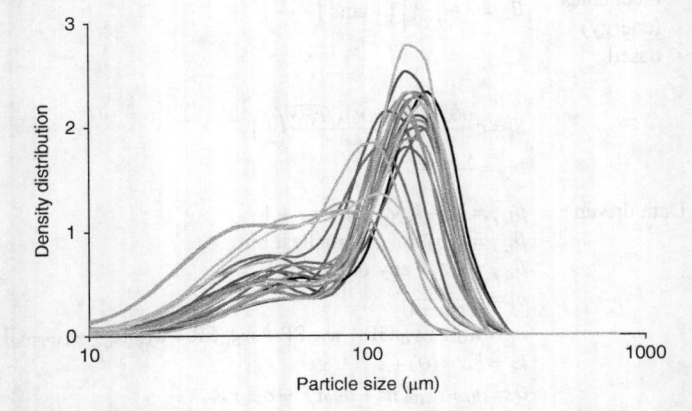

FIGURE 38.15 Particle size of API following agitated drying.

predicting actual output PSD. Further examining the choice of input parameters for the model, one can observe that while the X_{50} of an in-process sample from the crystallizer may represent the slurry PSD adequately prior to drying (Figure 38.14), this parameter may be insufficient to capture variations in actual mill input material following the final drying unit operation (Figure 38.15). The relatively unimodal peaks from the slurry samples clearly change to a wide variety of average sizes and modalities after drying in an agitated filter dryer. It is this variability in size entering the milling step that makes prediction difficult for a simple model such as that of Eq. (38.14).

To account for the variability seen in the dried material prior to milling, a simple summary statistic cannot be used. Rather, a measure of the entire PSD should be utilized to inform the model of the distribution of particle sizes moving into the breakage operation. Most particle size analyzers generate histograms with multiple channels of data describing the overall distribution of particle sizes in the sample. Chemometric methods such as principal component analysis (PCA) allow the full shape of the PSD to be reduced from a very large number of histogram components to 2–3 principal components describing the breadth, shape, and skewedness of the distribution. Using PCA in this manner maintains model inputs to a minimum number without significant loss in ability to characterize the full PSD spectrum.

After adding the principal components to the model to accurately represent the shape of the distribution, our equation becomes

$$X_{50,\text{milled}} = K(\text{PCA1})^{\varphi}(\text{PCA2})^{\mu}(\text{PCA3})^{\gamma}m^{\alpha}v^{\beta} \quad (38.15)$$

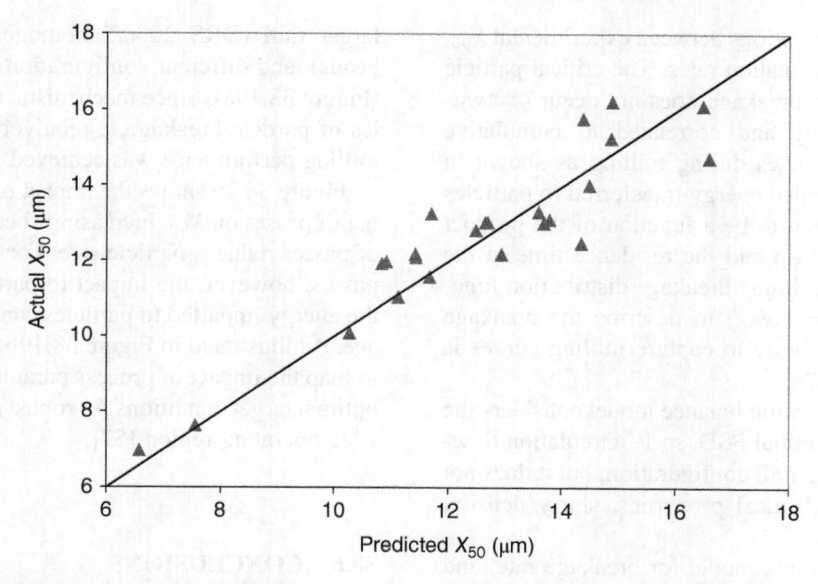

FIGURE 38.16 Predictive ability of a milling model incorporating full PSD spectrum as an input.

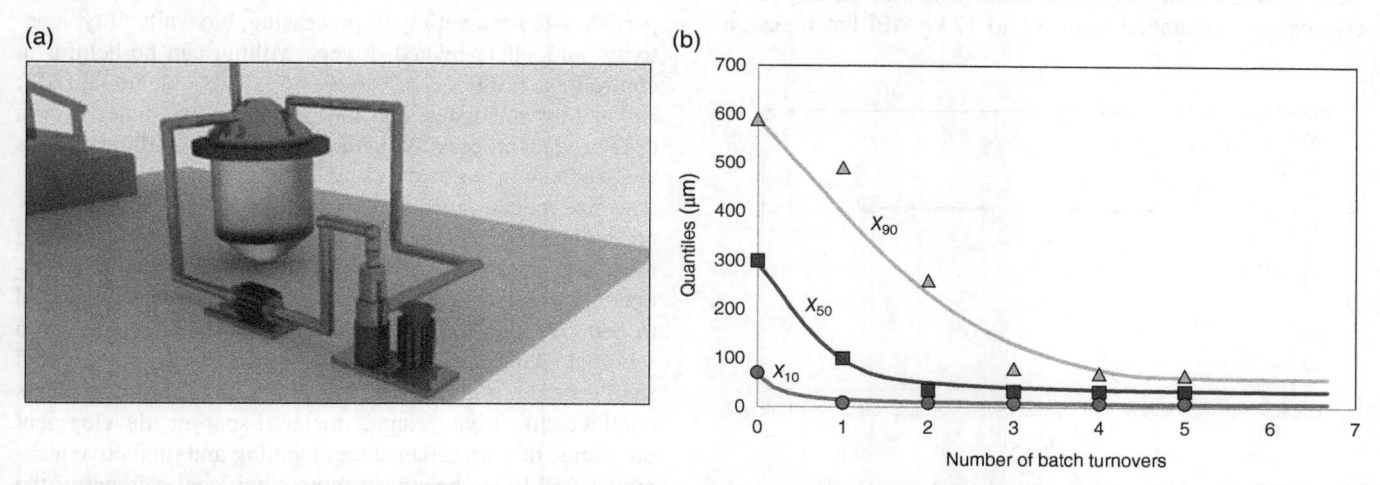

FIGURE 38.17 (a) Investigated recirculation loop (small-scale) wet mill configuration. (b) Experimental and simulated milling curve (small-scale) (experiments, symbols; simulation, traces).

where

PCA1, PCA2, and PCA3 represent the three principal components of the PSD.

This enhancement to the model, following a short number of calibration runs, allows much better prediction of output X_{50}, as shown in Figure 38.16.

The final predictive model is still limited to the API that was used to generate the data. However, the model allows for prediction as needed, and the process of constructing the model may provide valuable information regarding the API, including the relative importance of inclusion of the entire PSD spectrum as well as the impact that filtration and drying technique may have upon milling performance.

38.5.2 Wet Milling: Mechanistic Model Approach

Mechanistic models describe the physics of the particle breakage process, and therefore they can be extrapolated as far as the underlying assumptions are met. Mechanistic models can be used not only for scale purposes but also for risk assessment and definition of proven acceptable ranges as well as design space [57].

As an example of the use of mechanistic models to scale up a wet milling process, small-scale experiments were performed at different mill rotation rates using an IKA Magic Lab homogenizer (three fine heads) using a recirculation loop configuration (Figure 38.17a). A PBE was used following the same equations as in Luciani et al. [44] Parameters to define specific breakage rate were correlated with the shear number (i.e. $S_i = S_c(x_i/x_c)^\alpha$, with $S_c = AN_{Sh} + B$) by minimizing the sum of relative squared deviations between experimental X_{90}, X_{50}, and X_{10} for different rotation rates. The critical particle size below which particle breakage does not occur (x_c) was determined experimentally and correlated to cumulative energy transferred to particles during milling as shown in Figure 38.18. The cumulative energy transferred to particles during milling was adopted to be a function of the product between shear number (N_{Sh}) and the residence time in the mill (τ). For simplicity, a simple breakage distribution function was adopted ($B_{i,j} = (x_i/x_j)$) to describe the breakage mechanism. The model ability to capture milling curves is illustrated in Figure 38.17b.

As described, the population balance model considers the impact of mill geometry, initial PSD, scale, circulation flowrate, mill rotation rate, and mill configuration, but it does not account for particle mechanical properties, slurry density, and suspension medium.

Equation (38.12), Austin's model for breakage rate, and the aforementioned simplified breakage distribution function were used to define target conditions when the milling process scale was changed from 0.1 to 12 kg API batch size, a larger mill (DRS 2000/5 homogenizer with 3 super fine heads), and different configuration (single pass), were used (Figure 38.19a). Since mechanistic models capture the physics of particle breakage, a relatively good prediction of the milling performance was achieved in this case.

Figure 38.20 shows the impact of shear number and number of passes on X_{90}. Increasing shear number and/or number of passes reduces particle size. For large N_{Sh} and number of passes, however, the impact to particle size is limited since the energy imparted to particles cannot induce further breakage. As illustrated in Figure 38.19b, models can also be used to map the impact of process parameters on X_{90}, determining optimal target conditions for robust milling as well as the feasible operating region [57].

38.6 CONCLUSIONS

Milling is a powerful unit operation to aid in the control of particle size for a variety of processing, bioavailability, reactivity, and safety-related drivers. Milling can be helpful in controlling PSD for a variety of solids, including API, isolated intermediates, and formulated intermediates from a drug product process. While both wet and dry milling options are available to engineers as potential unit operations for particle size reduction, each infrastructure offers unique advantages and disadvantages that can be exploited to ensure the most robust, cost-effective, and safe process can be developed. A variety of modeling techniques have been employed to better predict milling performance at large scale from a minimal data set of particle physical properties and smaller-scale milling trials. As time and cost pressures continue to drive leaner, material-sparing development programs, heavier reliance on modeling and predictive techniques will likely become a more significant element of the engineer's efforts in process optimization.

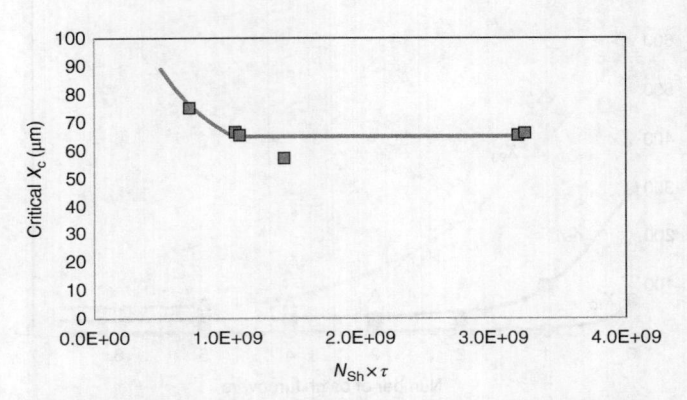

FIGURE 38.18 Impact of shear number (N_{Sh}) and residence time in mill (τ) on diameter below which breakage does not take place.

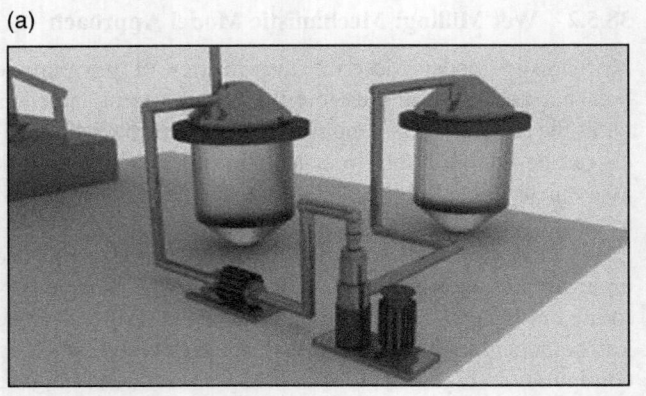

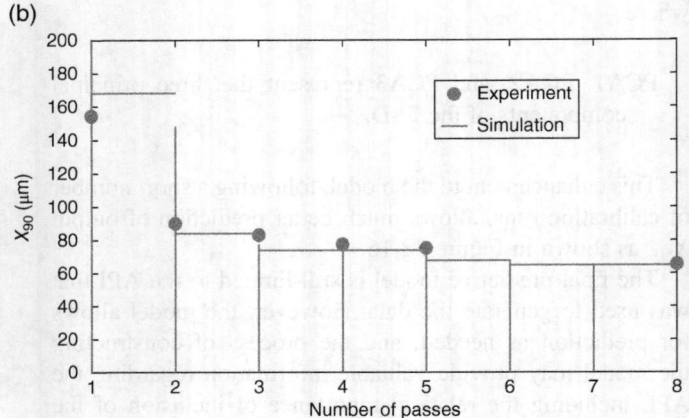

FIGURE 38.19 (a) Investigated large scale wet mill single pass configuration. (b) Comparison of experimental and simulated milling curves (large-scale).

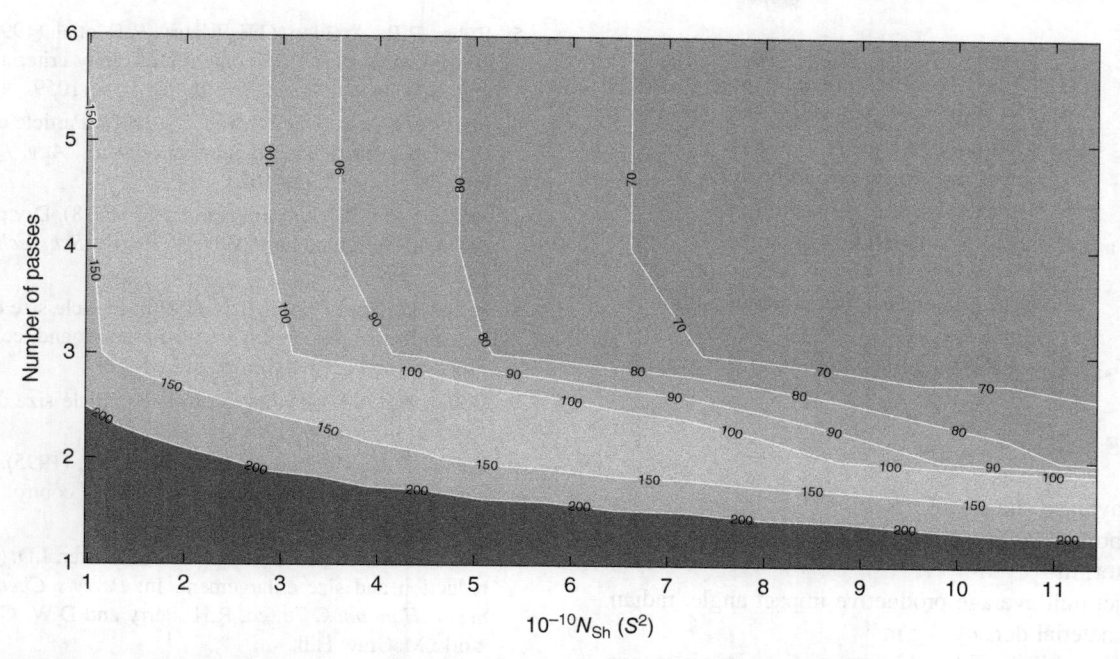

FIGURE 38.20 Contour plot of X_{90} (in μm) as a function of shear number and number of passes for single-pass configuration wet mill.

38.7 NOMENCLATURE

a_{GP}, a_{PP}, and a_{FR}	jet mill-dependent parameter
a_0	jet mill material-dependent parameter
$b(x,y)$	distribution of fragments of child particle of size x formed when larger particles of size are broken, dimensionless
$b_{i,j}$	discrete distribution of fragments of child particle of size i formed when larger particles of size j are broken, dimensionless
$B_{i,j}$	discrete cumulative distribution of fragments of child particle of size i formed when larger particles of size j are broken, dimensionless
c	energy-based model constant
D	wet mill rotor diameter, m
E_{fract}	fracture energy, J/m^3
E_{kin}	kinetic energy of the particles, J
f_{Mat}	mass-based material strength
FR	jet mill feed flow rate, kg/s
f_S	wet mill shear frequency, 1/s
GP	jet mill grinding pressure, Pa
H	material hardness, Pa
H_r	wet mill rotor slot height, m
H_s	wet mill stator slot height, m
k	number of impacts, dimensionless
k_c	conditional probability of chipping upon breakage, dimensionless
K_{1c}	stress intensity factor, Pa·m$^{1/2}$
l/x_i	relative length of the initial cracks, dimensionless
m	positive integer use in statistical-based breakage model, dimensionless
$m(x,t)$	mass of particles of size x at time t at the mill outlet, kg
$m_0(x,t)$	mass of particles of size x at time t at the mill inlet, kg
n	power law coefficient, dimensionless
N_{bt}	number of batch turnovers, dimensionless
N_r	number of teeth on wet mill rotor, dimensionless
N_s	number of teeth on wet mill stator, dimensionless
N_{se}	number of slot events, dimensionless
N_{Sh}	shear number, 1/s^2
p	number of child particles formed upon breakage
P_B	probability of breakage upon impact, dimensionless
PP	jet mill pusher pressure, Pa
P_y	yield pressure, Pa
Q_{mill}	wet mill volumetric flow rate, m^3/s
$S(x)$	specific breakage rate of particles of size x, 1/s
S_i	discrete specific breakage rate, 1/s
S_r	wet mill rotor tip speed, m/s
V	volume of mill chamber, m^3
V_i	volume of the particle, m^3

$W_{m,kin}$	mass-specific impact energy
$W_{m,min}$	mass-specific threshold energy (mass-specific energy that a particle can absorb without fracture)
W_r	wet mill rotor slot width, m
W_s	wet mill stator slot width, m
$w_0, w_{GP}, w_{PP}, w_{FR},$ and w_{GPPP}	jet mill-dependent parameters
w_{size}	jet mill material-dependent parameter
x, y, y'	particle sizes, m
X_{10}, X_{50}, X_{90}	PSD quantiles, μm

SYMBOLS

$\dot{\gamma}_{av}$	average shear rate, 1/s
δ	horizontal gap distance between the rotor and stator, m
θ	jet mill average productive impact angle, radian
ρ	material density, kg/m^3
σ_j	probability of breaking a particle in given residence time, dimensionless
$\tau(x)$	residence time of particles of size x, s
φ, γ, β	Austin's model parameters, dimensionless
ω	wet mill rotation rate, 1/s
ℓ	vertical gap distance between the rotor and stator, m
$\sqrt{P_y/\rho}$	crack propagation velocity, m/s

SUBSCRIPTS

1	coarsest particle size
i, j	bin classes

REFERENCES

1. Parrott, E.L. (1974). Milling of pharmaceutical solids. *J. Pharm. Sci.* 63 (6): 813–829.
2. Amidon, G.L., Lennernas, H., Shah, V.P., and Crison, J.R. (1995). A theoretical basis for a biopharmaceutic drug classification: the correlation of in vitro drug product dissolution and in vivo bioavailability. *Pharm. Res.* 12: 413–420.
3. Oh, D.M., Curl, R., and Amidon, G. (1993). Estimating the fraction dose absorbed from suspensions of poorly soluble compounds in humans: a mathematical model. *Pharm. Res.* 10: 264–270.
4. FDA (2015). Waiver of in vivo bioavailability and bioequivalence studies for immediate-release solid oral dosage forms based on a biopharmaceutics classification system. U.S. Department of Health and Human Services Food and Drug Administration Center for Drug Evaluation and Research (CDER) May 2015, Revision 1.
5. Rohrs, B.R., Amidon, G.E., Meury, R.H. et al. (1996). Particle size limits to meet USP content uniformity criteria for tablets and capsules. *J. Pharm. Sci.* 95 (5): 1049–1059.
6. Shoyele, S.A. and Cawthorne, S. (2006). Particle engineering techniques for inhaled biopharmaceuticals. *Adv. Drug Deliv. Rev.* 58 (9–10): 1009–1029.
7. Malcolmson, R.J. and Embleton, J.K. (1998). Dry powder formulations for pulmonary delivery. *Pharm. Sci. Technol. Today* 1: 394–398.
8. Holve, D.J. and Harvill, T.L. (1996). Particle size distribution measurements for in-process monitoring and control. *Adv. Powder Metall. Part. Mater.* 1: 4.81–4.93.
9. Collin, A. (1997). Online control of particle size distribution. *Spectra Analyse* 26 (196): 31–33.
10. Harvill, T.L., Hoog, J.H., and Holve, D.J. (1995). In-process particle size distribution measurements and control. *Part. Part. Syst. Charact.* 12 (6): 309–313.
11. Snow, R.H., Allen, T., Ennis, B.J., and Litster, J.D. (1997). Size reduction and size enlargement. In: *Perry's Chemical Engineers' Handbook*, 7e (ed. R.H. Perry and D.W. Green). New York: McGraw Hill.
12. Popov, K.I., Krstic, S.B., Obradovic, M.C. et al. (2003). The effect of the particle shape and structure on the flowability of electrolytic copper powder. I. Modeling of a representative powder particle. *J. Serb. Chem. Soc.* 68 (10): 771–777.
13. Brittain, H.G., Bogdanowich, S.J., Bugay, D.E. et al. (1991). Review: physical characterization of pharmaceutical solids. *Pharm. Res.* 8: 963–973.
14. Gabaude, C.M.D., Gautier, J.C., Saudemon, P., and Chulia, D. (2001). Validation of a new pertinent packing coefficient to estimate flow properties of pharmaceutical powders at a very early development stage, by comparison with mercury intrusion and classical flowability methods. *J. Mater. Sci.* 36: 1763–1773.
15. Lindberg, N.O., Palsson, M., Pihl, A.C. et al. (2004). Flowability measurements of pharmaceutical powder mixtures with poor flow using five different techniques. *Drug Dev. Ind. Pharm.* 30: 785–791.
16. Ward, G.H. and Shultz, R.K. (1995). Process-induced crystallinity changes in albuterol sulfate and its effect on powder physical stability. *Pharm. Res.* 12: 773–779.
17. Buckton, G. (1997). Characterization of small changes in the physical properties of powders of significance for dry powder inhaler formulations. *Adv. Drug Deliv. Rev.* 26: 17–27.
18. Price, R. and Young, P.M. (2005). On the physical transformations of processed pharmaceutical solids. *Micron* 36: 519–524.
19. Ghadiri, M., Kwan, C., and Ding, Y. (2007). Analysis of milling and the role of feed properties. In: *Handbook of Powder Technology*, vol. 12. Elsevier B.V.
20. ICH Harmonized Guideline (2014). Guideline for elemental impurities Q3D, 16 December 2014.
21. Cashdollar, K. (1996). Coal dust explosivity. *J. Loss Prev. Process Ind.* 9 (1): 65–76.
22. Eckhoff, R.K. (2003). *Dust Explosions in the Process Industries*, 3e. Burlington, MA: Gulf Professional Publishing.

23. Eckhoff, R.K. (1991). *Dust Explosions in the Process Industries*. Oxford: Butterworth-Heinemann Ltd.
24. Jaeger, N. and Siwek, R. (1999). Prevent explosions of combustible dusts. *Chem. Eng. Prog.* June: 25–37.
25. Stevenson, B. (2001). Preventing disaster: analyzing your plant's dust explosion risks. *Powder Bulk Eng.* 15: 19–27.
26. Britton, L.G. (1999). *Avoiding Static Ignition Hazards in Chemical Operations*. New York: Center for Chemical Process Safety of the AIChE.
27. Liotta, V. and Sabesan, V. (2004). Monitoring and feedback control of supersaturation using ATR-FTIR to produce and active pharmaceutical ingredient of a desired crystal size. *Org. Process. Res. Dev.* 8: 488–494.
28. Dennehy, R.D. (2003). Particle engineering using power ultrasound. *Org. Process. Res. Dev.* 7: 1002–1006.
29. Kim, S., Lotz, B., Lindrud, M. et al. (2005). Control of particle properties of a drug substance by crystallization engineering and the effect on drug product formulation. *Org. Process. Res. Dev.* 9: 894–901.
30. Tung, H.H., Paul, E.L., Midler, M., and McCauley, J.A. (2009). *Crystallization of Organic Compounds: An Industrial Perspective*. New York: Wiley.
31. MacDonald, R., Rowe, D., Martin, E., and Gorringe, L. (2016). The spiral jet mill cut size equation. *Powder Technol.* 299: 26–40.
32. Urban, K., Wagner, G., Schaffner, D. et al. (2006). Rotor-stator and disc systems for emulsification processes. *Chem. Eng. Technol.* 29 (1): 24–31.
33. Lee, I., Variankaval, N., Lindemann, C., and Starbuck, C. (2004). Rotor-stator milling of APIs: empirical scale up parameters and theoretical relationships between the morphology and breakage of crystals. *Am. Pharm. Rev.* 7 (5): 120–123.
34. Atiemo-Obeng, V. and Calabrese, R. (2003). Rotor stator mixing devices. In: *Handbook of Industrial Mixing*, 1e (ed. E.L. Paul, V. Atiemo-Obeng and S. Kresta). New York: Wiley.
35. Engstrom, J., Wang, C., Lai, C., and Sweeney, J. (2013). Introduction of a new scaling approach for particle size reduction in toothed rotor-stator wet mills. *Int. J. Pharm.* 456: 261–268.
36. Harter, A., Schenck, L., Lee, I., and Cote, A. (2013). High-shear rotor–stator wet milling for drug substances: expanding capability with improved scalability. *Org. Process. Res. Dev.* 17: 1335–1344.
37. Loh, Z.H., Samanta, A.K., and Heng, P.W.S. (2015). Overview of milling techniques for improving the solubility of poorly soluble drugs. *Asian J. Pharm Sci* 10: 255–274.
38. Mohr, K.H. (1987). High-pressure homogenization. Part I. Liquid-liquid dispersion in turbulence fields of high energy density. *J. Food Eng.* 6: 177–186.
39. Pandolfe, W.D. (1981). Effect of dispersed and continuous phase viscosity on droplet size of emulsions generated by homogenization. *J. Dispers. Sci. Technol.* 2: 459–474.
40. Müller, R.H. and Peters, K. (1998). Nanosuspensions for the formulation of poorly soluble drugs. I. Preparation by a size-reduction technique. *Int. J. Pharm.* 160: 229–237.
41. Kwade, A. and Schwedes, J. (2007). Wet grinding in stirred media mills. In: *Handbook of Powder Technology*, vol. 12 (ed. A.D. Salman, M. Ghadiri and M.J. Hounslow), 229–249. Amsterdam: Elsevier.
42. Russel, W.B., Saville, D.A., and Schowalter, W.R. (1991). *Colloidal Dispersions*. New York: Cambridge University Press.
43. Bilgili, E., Hamey, R., and Scarlett, B. (2006). Nano-milling of pigment agglomerates using a wet stirred media mill: elucidation of the kinetics and breakage mechanisms. *Chem. Eng. Sci.* 61 (1): 149–157.
44. Luciani, C., Conder, E., and Seibert, K. (2015). Modeling-aided scale-up of high-shear rotor–stator wet milling for pharmaceutical applications. *Org. Process. Res. Dev.* 19 (5): 582–589.
45. Bilgili, E., Yepes, J., and Scarlett, B. (2006). Formulation of a non-linear framework for population balance modeling of batch grinding: beyond first-order kinetics. *Chem. Eng. Sci.* 61: 33–44.
46. Austin, L., Shoji, K., Bhatia, V. et al. (1976). Some results on the description of size reduction as a rate process in various mills. *Ind. Eng. Chem. Process. Des. Dev.* 15: 187–196.
47. Austin, L., Klimpel, R., and Luckie, P. (1984). *Process Engineering of Size Reduction: Ball Mill*. New York: Society of Mining Engineers of the American Institute of Mining, Metallurgical, and Petroleum Engineers.
48. Hill, P. and Ng, K. (1996). Statistics of multiple particle breakage. *AIChE J.* 42: 1600–1611.
49. Hill, P. (2004). Statistics of multiple particle breakage accounting for particle shape. *AIChE J.* 50: 937–952.
50. de Vegt, O., Vromans, H., Faasen, F., and van der Voort Maarschalk, K. (2005). Milling of organic solids in a jet mill. Part 1: Determination of the selection function and related mechanical material properties. *Part. Part. Syst. Charact.* 22: 133–140.
51. de Vegt, O., Vromans, H., Faassen, F., and van der Voort Maarschalk, K. (2005). Milling of organic solids in a jet mill. Part 2: Checking the validity of the predicted rate of breakage function. *Part. Part. Syst. Charact.* 22: 261–267.
52. Vogel, L. and Peukert, W. (2003). Breakage behaviour of different materials: construction of a mastercurve for the breakage probability. *Powder Technol.* 129: 101–110.
53. Vogel, L. and Peukert, W. (2005). From single impact behaviour to modelling of impact mills. *Chem. Eng. Sci.* 60: 5164–5176.
54. Berthiaux, H. and Varinot, C. (1996). Approximate calculation of breakage parameters from batch grinding tests. *Chem. Eng. Sci.* 51: 4509–4516.
55. Starkey, D., Taylor, C., Morgan, N. et al. (2014). Modeling of continuous self-classifying spiral jet mills. Part 1: Model structure and validation using mill experiments. *AIChE J.* 60: 4086–4095.
56. Starkey, D., Taylor, C., Siddabathuni, S. et al. (2014). Modeling of continuous self-classifying spiral jet mills. Part 2: Powder-dependent parameters from characterization experiments. *AIChE J.* 60: 4096–4103.
57. García-Muñoz, S., Luciani, C.V., Vaidyaraman, S., and Seibert, K. (2015). Definition of design spaces using mechanistic models and geometric projections of probability maps. *Org. Process. Res. Dev.* 19 (8): 1012–1023.

PART IX

STATISTICAL MODELS, PAT, AND PROCESS MODELING APPLICATIONS

PART IX

STATISTICAL MODELS, PAT, AND PROCESS MODELING APPLICATIONS

39

EXPERIMENTAL DESIGN FOR PHARMACEUTICAL DEVELOPMENT

GREGORY S. STEENO

Worldwide Research and Development, Pfizer, Inc., Groton, CT, USA

39.1 INTRODUCTION

In the pharmaceutical industry and in today's regulatory environment, process understanding, in terms of characterization or optimization, is critical in developing and manufacturing new medicines that ensure patient safety and drug efficacy. Experimentation is the key component in building that knowledge base, whether those activities are physical experiments or trials run *in silico*. And this understanding comes from relating variation in observed response(s) back to changes, both intended and observational, in factors. If the factor changes are deliberate, then the scientist is testing hypotheses of factor effects and qualifying or quantifying their impact. What can be often neglected, however, is how important the *quality* of the experimental plan is and how that connects to making inferences from those data. This chapter focuses on both of these aspects: the experimental design and the data modeling.

For engineers characterizing a chemical reaction, experimental inputs span catalyst load, reaction concentration, jacket temperature, reagent amount, and other factors that are continuous in nature, as well as types of solvents, bases, catalysts, and other factors that are discrete in nature. These are examples of *controllable factors* and are represented as $x_1, x_2, \ldots, x_p$. Additional elements that can influence reaction outputs, such as analysts, instruments, lab humidity, etc., are examples of *uncontrollable factors* and are represented as $z_1, z_2, \ldots, z_q$. Figure 39.1, as shown by Montgomery [1], depicts a general process where both factor types impact the output.

The set of trials to develop relationships between factors and outputs, and the structure of how those trials are executed, is the *experimental design*. What follows are strategies for sound statistical design and analysis.

As a motivating example, consider a process where the output is yield (grams) and is expected to be a function of two controllable factors, reaction time (minutes) and reaction temperature (°C). That is, yield = f(time, temperature). As a first step in understanding and optimizing this process, experiments were executed by fixing time at 30 minutes and observing yield across a range of reaction temperatures. Figure 39.2 shows the data, and it is concluded that 35 °C is a reasonable choice as an optimal temperature. For the second step, experiments were executed by now fixing temperature at 35 °C and observing yield across a range of reaction times. Figure 39.3 shows the data, and it is concluded that the optimal time is about 40 minutes. The combined information from the totality of experiments produces an optimal setting of (temperature, time) = (35 °C, 40 minutes) with a predicted yield around 70 g.

This experiment was conducted using a *one-factor-at-a-time* (OFAAT) approach. And while easy to implement and instinctively sensible, there are a few shortcomings when compared with a statistically designed experiment. In general, OFAAT studies (i) are not as precise in estimating individual factor effects; (ii) cannot estimate multivariate factor effects, such as linear × linear interactions; and (iii) as a by-product are not as efficient at locating an optimum. Figure 39.4 illustrates the experimental path (●······●), the

---

Chemical Engineering in the Pharmaceutical Industry: Active Pharmaceutical Ingredients, Second Edition.
Edited by David J. am Ende and Mary T. am Ende.
© 2019 John Wiley & Sons, Inc. Published 2019 by John Wiley & Sons, Inc.

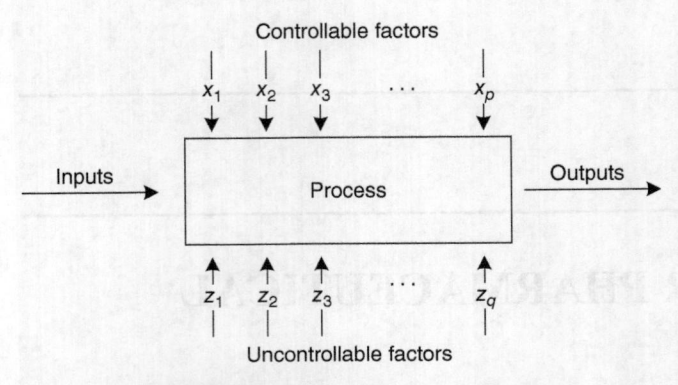

FIGURE 39.1 General process diagram.

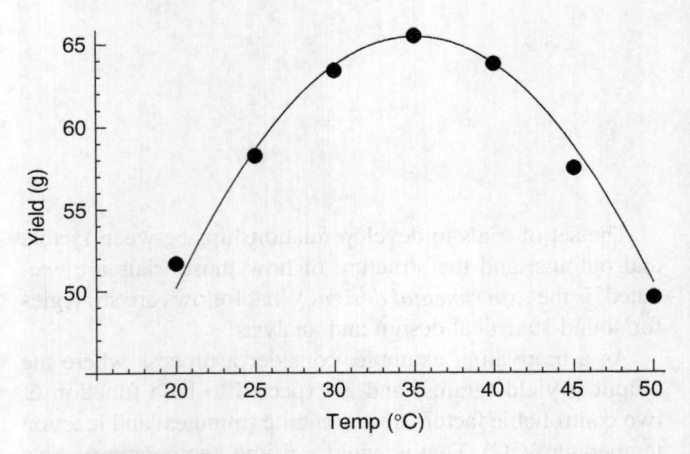

FIGURE 39.2 Scatterplot of yield against temperature with empirical fit.

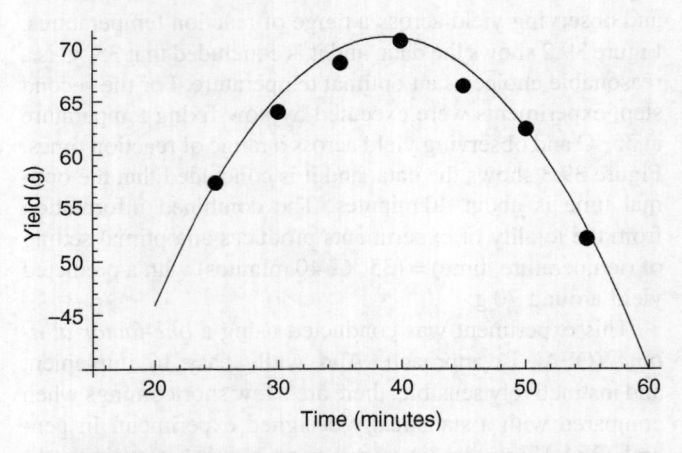

FIGURE 39.3 Scatterplot of yield versus time with empirical fit.

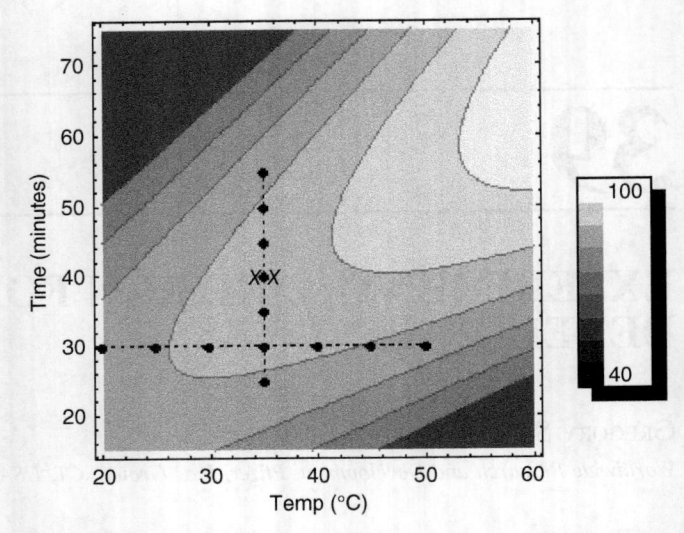

FIGURE 39.4 Contour plot of underlying mechanistic relationship of temperature and time on yield, along with experimental path (●·····●) and chosen optimum (XX).

chosen optimum (XX), and the actual underlying relationship between time and temperature on yield. Since the joint relationship between time and temperature was not fully investigated, the true optimum around (temp, time) ≈ (60 °C, 60 minutes) was missed.

This example illustrates that even with only two variables, the underlying mechanistic relationship between the factors and response(s) can be complex enough to easily misjudge. This is especially true of many pharmaceutical processes where functional relationships are dynamic and nonlinear. Due to this complexity, experiments that produce the most complete information in the least amount of resources (time, material, cost) are vital. This is a major aspect of *statistical experimental design*, which is an efficient method for changing process inputs in a systematic and multivariate way. The integration of experimental design and model building is generally known as *response surface methodology*, first introduced by Box and Wilson [2]. The data resulting from such a structured experimental plan coupled with regression analysis easily lend to deeper understanding of where process sensitivities exist, as well as how to improve process performance in terms of speed, quality, or optimality.

The experimental design and analysis procedure is straightforward and intuitive but is described below for completeness to ensure that all information is collected and effectively used:

1. Formulate a research plan with purpose and scope.
2. Brainstorm explanatory factors denoted $X_1, X_2, X_3, \ldots$, that could impact the response(s). Discuss ranges and omit factors that have little scientific value.
3. Determine the responses to be measured, denoted $Y_1, Y_2, Y_3, \ldots$, and consider resource implications.
4. Select an appropriate experimental design in conjunction with purpose and scope. Consider the randomization sequence. Hypothesize process models.
5. Execute the experiment. Measurement systems should be accurate and precise. The randomization sequence is

key to balancing effects of random influences and propagation of error.
6. Appropriately analyze the data.
7. Draw inference and formulate next steps.

The statistical experimental designs discussed in this chapter are used to help estimate approximating response functions for chemical process modeling. To begin, the functional relationship between the response, y, and input variables, $\xi_1, \xi_2, \ldots, \xi_k$, is expressed as

$$y = f(\xi_1, \xi_2, \ldots, \xi_k) + \varepsilon$$

which is unknown and potentially intricate. The inputs, $\xi_1, \xi_2, \ldots, \xi_k$, are called the natural variables, as they represent the actual values and units of each input factor. The ε term represents the variability not explicitly accounted for in the model, which could include the analytical component, the lab environment, and other natural sources of noise. For mathematical convenience, the natural variables are centered and scaled so that coded variables, $x_1, x_2, \ldots, x_k$, have mean zero and standard deviation one. This does not change the response function, but it is now expressed as

$$y = f(x_1, x_2, \ldots, x_k) + \varepsilon$$

If the experimental region is small enough, $f(\cdot)$ can be empirically estimated by lower-order polynomials. The motivation comes from Taylor's theorem that asserts any sufficiently smooth function can locally be approximated by polynomials. In particular, first-order and second-order polynomials are heavily utilized in response modeling from designed experiments.

A first-order polynomial is referred to as a *main effects model*, due to containing only the primary factors in the model. A two-factor main effects model is expressed as

$$y = \beta_0 + \beta_1 x_1 + \beta_2 x_2 + \varepsilon$$

where

β_1 and β_2 are coefficients for each factor.
β_0 is the overall intercept.

This function represents a plane through the (x_1, x_2) space. As an example, consider an estimated model:

$$\hat{y} = 100 - 10x_1 + 5x_2$$

Figure 39.5 shows a 3D view of that planar response function, also called a surface plot. Figure 39.6 represents the 2D analogue called a contour plot. The contour plots are often easier to read and interpret since the response function height is projected down onto the (x_1, x_2) space. If there is an interaction between the factors, it is easily added to the model as follows:

$$y = \beta_0 + \beta_1 x_1 + \beta_2 x_2 + \beta_{12} x_1 x_2 + \varepsilon$$

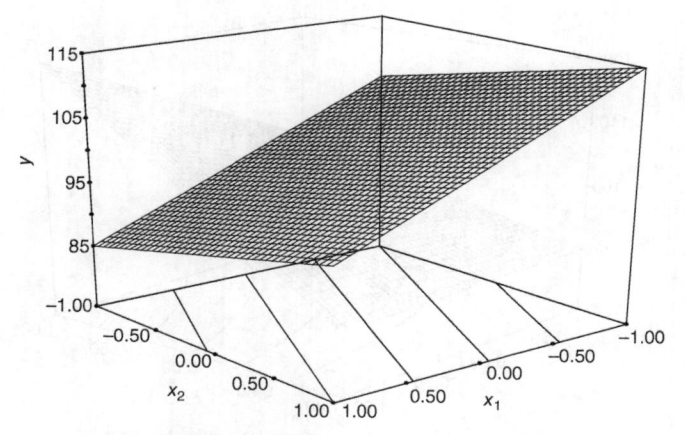

FIGURE 39.5 Surface plot of $\hat{y} = 100 - 10x_1 + 5x_2$.

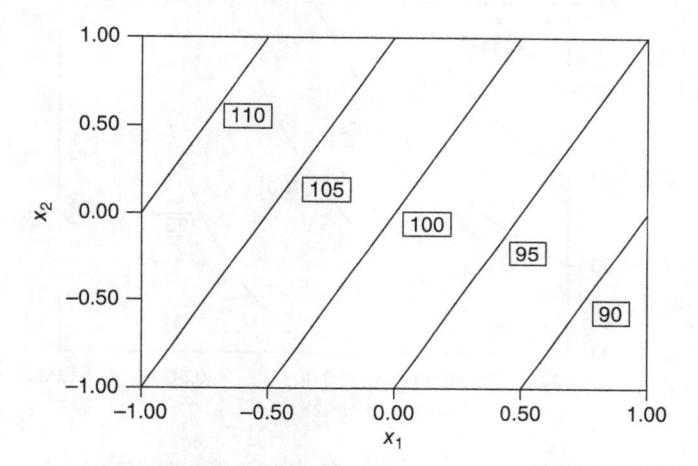

FIGURE 39.6 Contour plot of $\hat{y} = 100 - 10x_1 + 5x_2$.

This is called a first-order model with interaction. To continue with the example, let the estimated model be

$$\hat{y} = 100 - 10x_1 + 5x_2 - 5x_1 x_2$$

The additional term $-5x_1 x_2$ introduces curvature in the response function, which is displayed on the surface plot in Figure 39.7 and the corresponding contour plot in Figure 39.8. Occasionally, the curvature in the true underlying response function is strong enough that a first-order plus interaction model is inadequate for prediction. In this case, a second-order (quadratic) model would be useful to approximate $f(\cdot)$ and takes the form

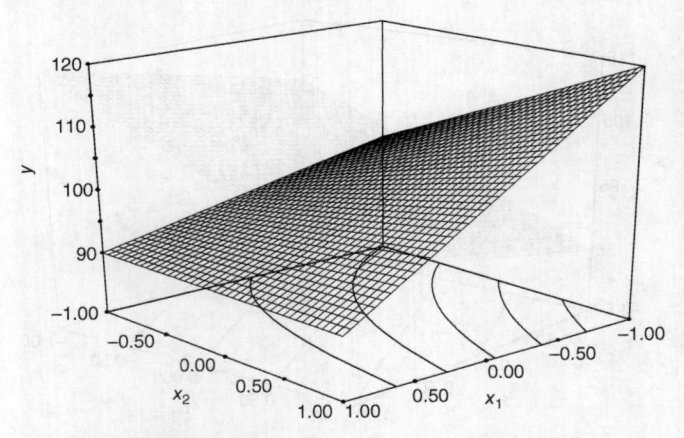

FIGURE 39.7 Surface plot of $\hat{y} = 100 - 10x_1 + 5x_2 - 5x_1x_2$.

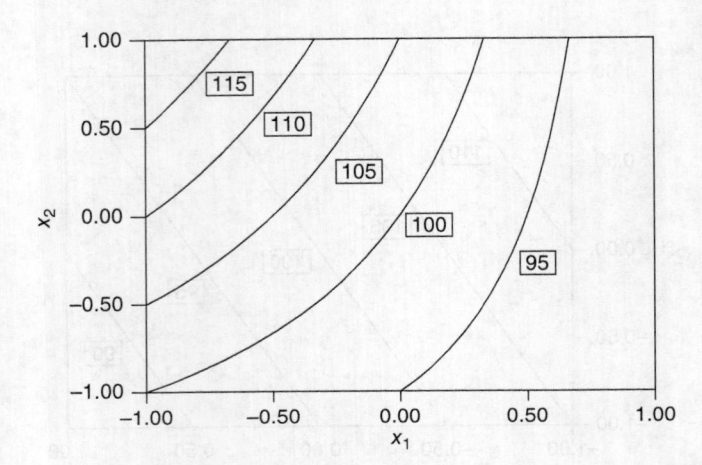

FIGURE 39.8 Contour plot of $\hat{y} = 100 - 10x_1 + 5x_2 - 5x_1x_2$.

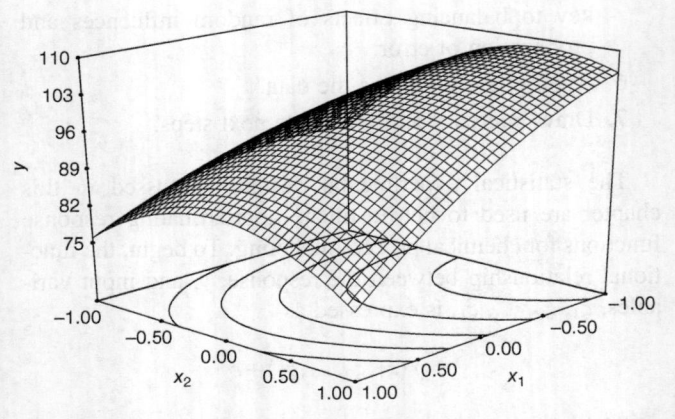

FIGURE 39.9 Surface plot of $\hat{y} = 100 - 10x_1 + 5x_2 - 5x_1x_2 - 4x_1^2 - 10x_2^2$.

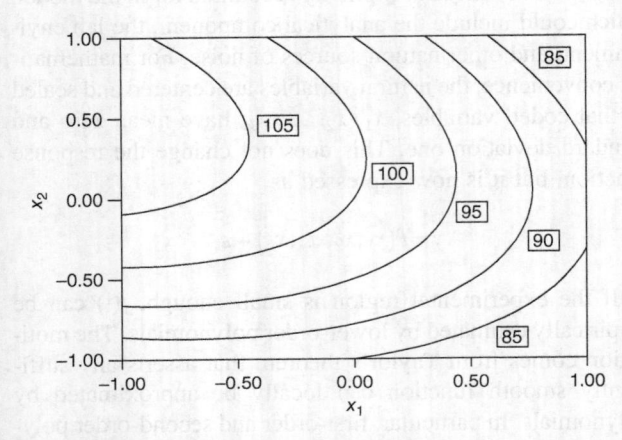

FIGURE 39.10 Contour plot of $\hat{y} = 100 - 10x_1 + 5x_2 - 5x_1x_2 - 4x_1^2 - 10x_2^2$.

$$y = \beta_0 + \beta_1 x_1 + \beta_2 x_2 + \beta_{12} x_1 x_2 + \beta_{11} x_1^2 + \beta_{22} x_2^2 + \varepsilon$$

To finish the example, let the estimated model be

$$\hat{y} = 100 - 10x_1 + 5x_2 - 5x_1x_2 - 4x_1^2 - 10x_2^2$$

Figure 39.9 shows a parabolic relationship between y and x_1, x_2, while Figure 39.10 displays the typical elliptical contours generated by this model.

There is an iterative, sequential nature to understanding and optimizing the performance of chemical processes. If the goal is to first identify the most important factors for further study, a screening design may be carried out. This is sometimes referred to as *phase zero* of the study. Once this is complete, the next objective is to determine if the optimum lies within current experimental region or if the factors need adjustment to locate a more desirable one, say, by using methods of steepest ascent/descent. This is referred to as *phase one* of the study, also known as *region seeking*.

Finally, once in the region of desirable response is established, the goal becomes to precisely model that area and establish optimal factor settings. In this case, higher-order models are employed to capture likely curvature about the optimum point.

39.2 THE TWO-LEVEL FACTORIAL DESIGN

Factorial designs are experimental plans that consist of all possible combinations of factor settings. As an example, a factorial design with three different catalysts, two different solvents, and four different temperatures produces a design with $3 \times 2 \times 4 = 24$ unique experimental conditions. The advantage of these designs is that all joint effects of factors can be investigated. The disadvantage is that these designs become prohibitively large and impractical when factors contain more than just a few levels or the number of factors under investigation is large.

The simplest and most widely used factorial designs for industrial experiments are those that contain two levels per factor, called 2^k factorial designs, where k is the number of factors under investigation. The two levels for each factor are usually chosen to span a practical range to investigate. These designs can be augmented into fuller designs and are very effective in terms of time, resource, and interpretability. The class of 2^k factorial designs can be used as building blocks in process modeling by:

- Screening the most important variables from a set of many.
- Fitting a first-order equation used for steepest ascent/descent.
- Identifying synergistic/antagonistic multifactor effects.
- Forming a base for an optimization design, such as a central composite.

Factors can either be continuous in nature or discrete. For the rest of the chapter, it is assumed that the factors are continuous. This allows for predictive model building and regression analysis that includes the linear and interaction terms and subsequently the quadratic/second-order terms.

To illustrate a two-level factorial design, consider the previous case where there are $k = 2$ factors, temperature (factor A) and time (factor B), each having an initial range under investigation. In coded units, the low level of the range for each factor is scaled to -1, and the high level of the range is scaled to $+1$. The 2^2 experimental design with all four treatment combinations is shown in Table 39.1 and graphically depicted via (○) in Figure 39.11. Notice that each treatment condition occurs at a vertex of the experimental space. For notation, the four treatment combinations are usually represented by lowercase letters. Specifically, a represents the combination of factor levels with A at the high level and B at the low level, b represents A at the low level and B at the high level, and ab represents both factors being run at the high level. By convention, (1) is used to denote A and B, each run at the low level.

Two-level designs are used to estimate two types of effects, main effects and interaction effects, and these are estimated by a single degree of freedom *contrast* that partitions the design points into two groups: the low level (-1) and the high level ($+1$). The contrast coefficients are shown in Table 39.1 for each of the main factors of temperature and time, as well as the temperature × time interaction obtained through pairwise multiplication of the main factor contrast coefficients.

A main effect of a factor is defined as the average change in response over the range of that factor and is calculated from the average difference between data collected at the high level ($+1$) and data collected at the low level (-1). For the 2^2 design above and using the contrast coefficients in Table 39.1, the temperature (A) main effect is estimated as

TABLE 39.1   Example 2^2 Experimental Design with $n = 2$ Replicates per Design Point

Treatment	Design			Response	
	Temperature (A)	Time (B)	Temp × Time (AB)	Rep 1	Rep 2
(1)	-1	-1	1	y_{11}	y_{12}
a	1	-1	-1	y_{21}	y_{22}
b	-1	1	-1	y_{31}	y_{32}
ab	1	1	1	y_{41}	y_{42}

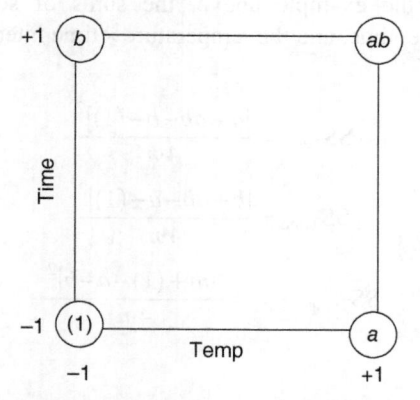

FIGURE 39.11   Factor space of 2^2 experimental design.

$$A = \bar{y}_{\text{Temp}+} - \bar{y}_{\text{Temp}-} = \frac{ab+a}{2n} - \frac{b+(1)}{2n} = \frac{1}{2n}[ab+a-b-(1)]$$

where

(1), a, b, and ab are the respective sum total of responses across the n replicates at each design point ($n = 2$ in Table 39.1).

Geometrically this is a comparison of data on average from the right side to the left side of the experimental space in Figure 39.11. If the estimated effect is positive, the interpretation is that the response increases as the factor level increases. Similarly, the time (B) main effect is estimated as

$$B = \bar{y}_{\text{Time}+} - \bar{y}_{\text{Time}-} = \frac{ab+b}{2n} - \frac{a+(1)}{2n} = \frac{1}{2n}[ab+b-a-(1)]$$

which is a comparison of data on average from the top side to the bottom side in Figure 39.11.

An interaction between factors implies that the individual factor effects are not additive and that the effect of one factor depends on the level of another factor(s). As with the main effects, the interaction is estimated by partitioning the data into two groups and comparing the average difference. The contrast coefficients in Table 39.1 show that the temperature × time (AB) interaction effect is estimated as

$$AB = \bar{y}_{\text{Temp} \times \text{Time}+} - \bar{y}_{\text{Temp} \times \text{Time}-} = \frac{ab+(1)}{2n} - \frac{a+b}{2n}$$

$$= \frac{1}{2n}[ab+(1)-a-b]$$

which is a comparison on average of data on the right diagonal against the left diagonal in Figure 39.11.

The sum of squares for each effect is mathematically related to their corresponding contrast. Specifically, the sum of squares for an effect is calculated by the squared contrast divided by the total number of observations in that contrast. For the example above, the sums of squares for temperature, time, and the temperature × time interaction are

$$SS_{\text{Temp}} = \frac{[a+ab-b-(1)]^2}{4 \cdot n}$$

$$SS_{\text{Time}} = \frac{[b+ab-a-(1)]^2}{4 \cdot n}$$

$$SS_{\text{Temp} \times \text{Time}} = \frac{[ab+(1)-a-b]^2}{4 \cdot n}$$

In 2^k designs, the contrasts are orthogonal and thus additive. The total sum of squares, SS_T, is the usual sum of squared deviations of each observation from the overall mean of the data set. Because the contrasts are orthogonal, the error sum of squares, denoted SS_E, can be calculated as the difference between the total sum of squares, SS_T, and the sums of squares of all effects. For the 2^2 example, $SS_E = SS_T - SS_{\text{Temp}} - SS_{\text{Time}} - SS_{\text{Temp} \times \text{Time}}$. With this information, the analysis of variance (ANOVA) table is constructed, as shown in Table 39.2.

The ANOVA table contains all numerical information in determining which factor effects are important in modeling the response. The hypothesis test on individual factor effects is conducted through the F-ratio of MS_{Factor} against MS_{Error}. If this ratio of "signal" against "noise" is large, this implies that the factor explains some of the observed variation in response across the design region and should be included in the process model. If the ratio is not large, then the inference is that the factor is unimportant and should be deleted from the model. All statistical evidence of model inclusion comes via the p-value. Large F-ratios imply low p-values, and a common cutoff for model inclusion of a factor is $p \leq 0.05$, although this should be appropriately tailored with experimental objectives, such as factor screening where the critical p-value is normally bit higher.

It is important to identify all influential factors for modeling variation in response back to change in factor level. An *underspecified* model, one that does not contain all the important variables, could lead to bias in regression coefficients and bias in prediction. One approach to mitigate this issue is to not model edit but rather incorporate all factor terms in the process model, including those that contribute very little or nothing of value in predicting. However, an *overspecified* model, one that contains insignificant terms, produces results that lead to higher variances in coefficients and in prediction. Thus a proper model will be a compromise of the two. This can be completed manually, say, by investigating the full model ANOVA and then deleting insignificant effects one at a time, all while updating the ANOVA after every step. Yet for models that could contain many effects, this exercise becomes cumbersome. There are several variable selection procedures that can aid in helping identify smaller-sized candidate models. The most common algorithms used in standard software packages entail either sequentially bringing in significant factors to build the model up (called forward selection), sequentially eliminating regressors from a full model (called backward elimination), or a hybrid of the two (called stepwise regression). The procedures typically involve defining a critical p-value for factor inclusion/exclusion in the model-building process. Once a term enters or leaves, factor significance is recalculated, and the process is repeated for the next step. The engineer should use these tools not as a panacea to the model-building process, but rather as an exercise to see how various models perform. For more information on the process and issues of model selection, the reader is instructed to see Myers [3].

Once the significant factors are selected, the estimated regression coefficients in the linear predictive model are functionally derived from the factor's effect size. To estimate the regression coefficient β_i for factor i, its effect is divided by

TABLE 39.2 ANOVA Table for Completely Randomized 2^2 Design with n Replicates per Design Point

Source	SS	DF	MS	F	p-Value
Temp	SS_{Temp}	1	MS_{Temp}	MS_{Temp}/MS_E	p_{Temp}
Time	SS_{Time}	1	MS_{Time}	MS_{Time}/MS_E	p_{Time}
Temp × Time	$SS_{\text{Temp} \times \text{Time}}$	1	$MS_{\text{Temp} \times \text{Time}}$	$MS_{\text{Temp} \times \text{Time}}/MS_E$	$p_{\text{Temp} \times \text{Time}}$
Error	SS_E	$4(n-1)$	MS_E		
Total	SS_T	$4n-1$			

two. The rationale is that by definition the regression coefficient represents the change in y per unit change in x. Since each factor effect is calculated as a change in response over a span of two coded units (−1 to 1), division by two is needed to obtain the per-unit basis. Finally, the model intercept, β_0, is calculated as the grand average of all the data.

EXAMPLE PROBLEM 39.1 2^3 Factorial Design

A factorial experiment is carried out to investigate three factors on percent reaction conversion: catalyst load (factor A), ligand load (factor B), and temperature (factor C). Each experimental condition is completely randomized and independently replicated ($n = 2$). The design in coded units, full model, and data are listed in Table 39.3 and depicted in Figure 39.12. Note that the run sequence in Table 39.3 is in *standard order*, as opposed to a randomized order for the actual experiment.

As previously remarked, the estimated effects and associated sums of squares are functions of their respective contrasts. Using catalyst load (A) as an example, the contrast coefficients shown in Table 39.3 represent the comparison between data on the right side of the cube (+) in Figure 39.12 to the data on the left side of the cube (−). The data where catalyst load is high sum to

$$a + ab + ac + abc = 76.4 + 74.9 + 76.3 + 76.0 + 77.6 + 80.4 + 79.0 + 77.2 = 617.8$$

while the data where catalyst load is low sum to

$$b + c + bc + (1) = 66.5 + 63.7 + 78.7 + 77.7 + 78.0 + 81.3 + 63.8 + 64.1 = 573.8$$

The estimated effect of catalyst load is calculated as

$$A = \bar{y}_{\text{Cat}+} - \bar{y}_{\text{Cat}-} = \frac{a+ab+ac+abc}{4 \cdot n} - \frac{b+c+bc+(1)}{4 \cdot n}$$
$$= \frac{617.8}{4 \cdot 2} - \frac{573.8}{4 \cdot 2} = 5.5$$

The interpretation is that the average yield increases by 5.5% as the catalyst load increases from low to high. The corresponding sum of squares for the catalyst load effect is calculated as

$$\text{SS}_A = \frac{[a+ab+ac+abc-b-c-bc-(1)]^2}{8 \cdot n}$$
$$= \frac{(617.8-573.8)^2}{8 \cdot 2} = 121$$

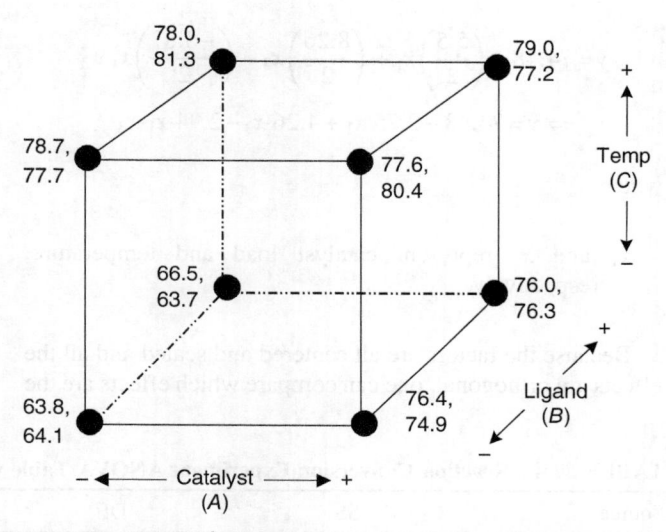

FIGURE 39.12 2^3 reaction conversion experimental space and corresponding data.

TABLE 39.3 2^3 Reaction Conversion Experimental Design, Model, and Data

	Design							Data Conversion	
Treatment	Catalyst A	Ligand B	Temp C	AB	AC	BC	ABC	Rep 1	Rep 2
(1)	−1	−1	−1	1	1	1	−1	63.8	64.1
a	1	−1	−1	−1	−1	1	1	76.4	74.9
b	−1	1	−1	−1	1	−1	1	66.5	63.7
ab	1	1	−1	1	−1	−1	−1	76	76.3
c	−1	−1	1	1	−1	−1	1	78.7	77.7
ac	1	−1	1	−1	1	−1	−1	77.6	80.4
bc	−1	1	1	−1	−1	1	−1	78	81.3
abc	1	1	1	1	1	1	1	79	77.2

The other estimated factor effects and sums of squares follow the same logic as above and are trivial to calculate. The full model ANOVA table is shown in Table 39.4. Based on the information, there are three highly significant effects ($p < 0.0001$): catalyst load, temperature, and their corresponding interaction. All other effects are insignificant across the study range. The final model and ANOVA table after sequential model editing are shown in Table 39.5.

This model accounts for approximately 96.3% of the observed variability in reaction completion, as determined by the coefficient of determination, R^2:

$$R^2 = \frac{SS_{Model}}{SS_{Total}} = \frac{SS_A + SS_C + SS_{AC}}{SS_{Total}} = \frac{549.76}{570.87} = 0.9630$$

The final regression model expressed in coded units is estimated as

$$\hat{y} = 74.48 + \left(\frac{5.5}{2}\right)x_1 + \left(\frac{8.26}{2}\right)x_3 + \left(\frac{-5.87}{2}\right)x_1 \cdot x_3$$

$$\Rightarrow \hat{y} = 74.48 + 2.75 \cdot x_1 + 4.26 \cdot x_3 - 2.94 \cdot x_1 \cdot x_3$$

where

x_1 and x_3 represent catalyst load and temperature, respectively.

Because the factors are all centered and scaled and all the effects are orthogonal, one can compare which effects are the most dominant by the size of the coefficient. Using the reaction conversion model, temperature has the largest effect, followed by the catalyst × temperature interaction and then catalyst. However, the estimated main effects of temperature and catalyst have lost direct interpretation due to the presence of their interaction. Specifically, the conclusion from the temperature effect is that for every coded unit change in temperature, the conversion increases 4.26% via the coefficient on the temperature term. But this estimate is a pooled average over the other factors. That is, it smoothes over the significant joint effect between temperature and catalyst. The information contained in the interaction will need to be visually explored in greater detail.

After obtaining the final equation, residual analysis and other model diagnostics are carried out. This is critical step in assessing the process model and having trust in its ability to accurately predict over the experimental region. Standard residual analyses consist of inspecting the normality assumption, checking for constant variance, identifying outliers relative to the model, and observing patterns in residuals over time. There are many flavors of model diagnostic information, both numerical and graphical. With the aid of Design-Expert [4] software package, highlighted below are two that, in this author's view, capture a significant snapshot of model performance. First, the normal probability plot is an effective graphical tool to verify the normality assumption of the errors as well as for outlier detection. Figure 39.13 shows this plot for the reaction conversion residuals using the final regression model. If the residuals fall along a straight line, then the normality assumption is valid. Any large errors or outliers from the fitted

TABLE 39.4 Reaction Conversion Experiment ANOVA Table with Full Model

Source	SS	DF	MS	F	p-Value
A-catalyst	121.00	1	121.00	58.24	<0.0001
B-ligand	1.21	1	1.21	0.58	0.4673
C-temp	290.70	1	290.70	139.93	<0.0001
AB	2.25	1	2.25	1.08	0.3284
AC	138.06	1	138.06	66.46	<0.0001
BC	0.30	1	0.30	0.15	0.7127
ABC	0.72	1	0.72	0.35	0.5717
Error	16.62	8	2.08		
Total	570.87	15			

TABLE 39.5 Reaction Conversion Experiment ANOVA Table After Model Editing

Source	SS	DF	MS	F	p-value
A-catalyst	121.00	1	121.00	68.80	<0.0001
C-temp	290.70	1	290.70	165.29	<0.0001
AC	138.06	1	138.06	78.50	<0.0001
Error	21.10	12	1.76		
Total	570.87	15			

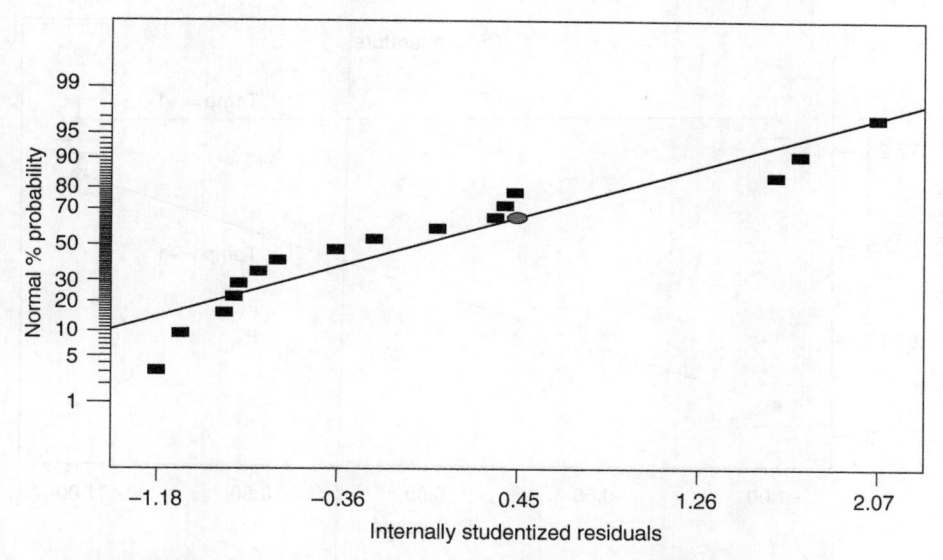

FIGURE 39.13 Normal probability plot of residuals from reaction conversion model.

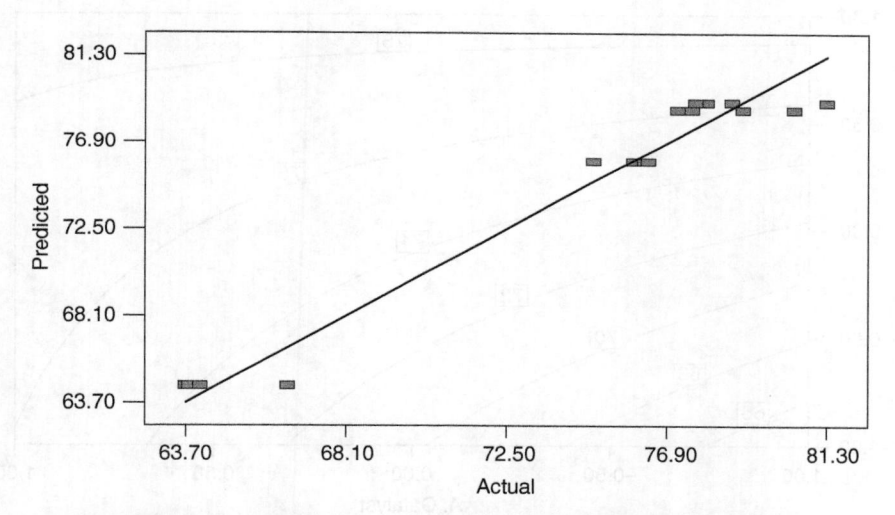

FIGURE 39.14 Predicted vs. actual plot using reaction conversion example.

model would be visually apparent by significantly falling off the line. The interpretation of Figure 39.13 is that there is no severe problem with the normality assumption.

Second, another very effective plot is model-based prediction against the observed data, often referred to as "Predicted vs. Actual." This reveals how well the model predicts back the original data and is a graphical depiction of the calculated R^2 value. Additionally, it will aid in identifying (sets of) data that are not well captured by the model, as well as indicate any trends of nonconstant variance across the prediction range. Figure 39.14 shows an example using the reaction conversion data and final model. The interpretation from this graph is that the regression model is performing well and the spread of about 45° line is relatively constant across the range.

Once the engineer is satisfied with the diagnostics information, visualizing the change in response across the subspace of significant factors is the next step. This is best accomplished by main effect and interaction plots. If factors are continuous, then contour and surface plots are very informative and more descriptively illustrate bivariate factor effects on response. As previously shown with the reaction conversion example, the interaction between catalyst and temperature is significant, and therefore the synergistic relationship between these two factors needs further inspection. Figure 39.15 displays an interaction plot for these two factors. Notice that when temperature is at the high (+1) level, there is no observed effect of catalyst load on the reaction conversion. That is, it is robust to changes in catalyst.

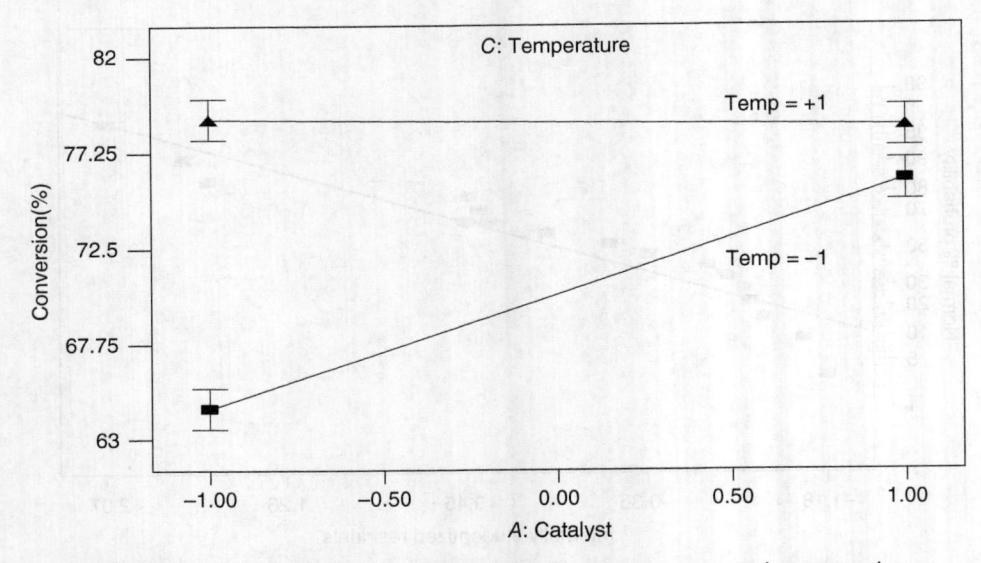

FIGURE 39.15 Interaction plot of catalyst and temperature on reaction conversion.

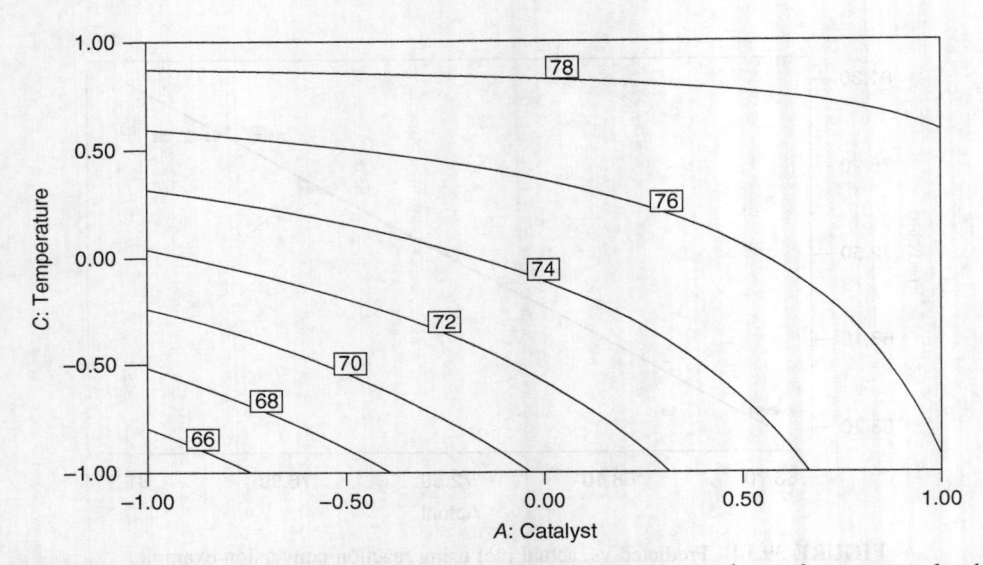

FIGURE 39.16 Contour plot of predicted reaction conversion across catalyst and temperature levels.

However, when temperature is at the low (−1) level, reaction conversion is now a function of catalyst load, and higher load leads to higher predicted conversion. Figure 39.16 shows the corresponding contour plot of predicted change in reaction conversion over changes in temperature and catalyst. This also illustrates the predicted change in conversion across the temperature and catalyst levels but takes advantage of the continuous nature of the factors. Regardless, the inference is the same: at high temperatures, the prediction is virtually constant across catalyst (~78% conversion), while at lower temperatures, the catalyst effect is present.

Two-level designs are very intuitive, comprehensive, and powerful in identifying main and, in particular, interaction effects on responses of interest. However, even at two levels, they can become impractically large as the number of factors increases. For example, if $k = 6$ factors are under investigation, then a full factorial experiment with no replication would consist of $2^6 = 64$ runs, which is not a cost-effective experiment. If this design were executed, the full regression model with all possible main effects and interactions would leave zero degrees of freedom to estimate variability for statistical inferences on factor effects. There are strategies and tools that combat the issues of unreplicated designs and variance estimation. One such strategy takes advantage of the *sparsity-of-effects principle* (or similarly, the Pareto principle), which states that a process is usually dominated by only the vital few effects, such as main effects and two-factor interactions, from the trivially many. More specifically, observing effects from higher-order terms such as three-factor interactions and beyond is rare in practice. If it is reasonable to assume that effects from higher-order interactions are negligible, then those terms can be pooled to estimate variability through the mean squared error. Another approach is to use a normal probability plot or half-normal probability plot to determine the significant effects. Proposed by Daniel [5], these are effective graphical tools that use the estimated effects to highlight which are important in explaining response variation relative to those that do not. Negligible effects are assumed to be normally distributed with mean zero and variance σ^2, while significant effects are assumed normally distributed at their true effect size different from zero with variance σ^2. Normal and half-normal effect plots are standard in most statistical software packages.

39.3 BLOCKING

There are situations that call for the design to be run in groups or clusters of experiments. This often occurs when the number of factors is large, in which case the design would be broken down into smaller *blocks* of more homogeneous experimental units. These are referred to as incomplete blocks, as not all treatment combinations occur in these smaller sets. This situation also occurs with equipment setup or limitations, where groups of experiments are executed in "batches" at the same time. As one example, consider a replicated 2^2 design that investigates temperature and solvent concentration on reaction completion. The equipment chosen for the experiment is a conventional process chemistry workstation that has four independent reactors. A natural and appropriate way to execute this design is to run each replicate of the 2^2 design on the workstation with random assignment of the reactor vessels to each treatment combination. By doing this, any block-to-block variability is accounted for and does not influence the analysis on factor effects. As another example, consider a 2^3 design using the four-reactor workstation that investigates temperature, solvent concentration, and catalyst loading on reaction completion. To execute this study, the design needs to be split into two groups of four, since it is not possible to execute all experiments in one block.

Both of these examples result in observations within the same group to be more homogeneous than those in another group. For situations where the experiment is subdivided, appropriate blocking can help in optimally constructing a design based on the assumption or knowledge that certain (higher-order) interactions are negligible. This design technique is called *confounding* or *aliasing*, where information on treatment effects is indistinguishable from information on block effects. The number of blocks for the two-level design is usually a multiple of two, implying designs are run in blocks of two, four, eight, etc.

To illustrate, consider the previous 2^3 design with temperature (*A*), solvent concentration (*B*), and catalyst loading (*C*) as the factors, with the design executed in two blocks of four on the workstation. The design and full model with associated contrast coefficients are shown in Table 39.6.

One candidate design partition into two blocks of four is to group all combinations where *ABC* is at the low level (−1) into one block and all combinations where *ABC* is at the high level (+1) into the other block. Schematically, each block would appear as in Figure 39.17.

Recall the contrast to estimate the effect of *A* (temperature) is written as

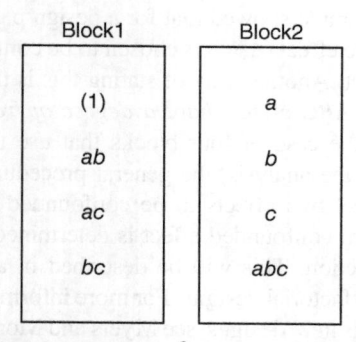

FIGURE 39.17 Schematic of 2^3 factorial divided into two blocks of size four.

TABLE 39.6 2^3 **Design Full Model Contrast Coefficients**

Treatment	A	B	C	AB	AC	BC	ABC
(1)	−1	−1	−1	1	1	1	−1
a	1	−1	−1	−1	−1	1	1
b	−1	1	−1	−1	1	−1	1
ab	1	1	−1	1	−1	−1	−1
c	−1	−1	1	1	−1	−1	1
ac	1	−1	1	−1	1	−1	−1
bc	−1	1	1	−1	−1	1	−1
abc	1	1	1	1	1	1	1

$$A = \frac{1}{4}[a + ab + ac + abc - (1) - b - c - bc]$$

Any block effect between experiments conducted in Block 1 and those conducted in Block 2 is canceled out in this contrast, as the overall effect of A is actually a pooled sum of the *simple within-block effects* of A. That is, let A_1 be the comparison of A at the high level versus A at the low level within Block 1 and A_2 be the corresponding comparison within Block 2. The overall effect of A is calculated as

$$A_1 = [ab + ac - (1) - bc]$$
$$A_2 = (a + abc - b - c)$$
$$\Rightarrow A = \frac{1}{4}(A_1 + A_2)$$

This holds for all other effects except the ABC effect, where its corresponding contrast from Table 39.6 is

$$ABC = \frac{1}{4}[a + b + c + abc - (1) - ab - ac - bc]$$

By design, the difference in those treatment combinations corresponds exactly to the design partition into Blocks 1 and 2. That is, the effect of ABC is not estimable; it is confounded with blocks. Assuming the ABC effect is negligible, this is an optimally constructed design, as the ABC effect was intentionally confounded to preserve inferences on lower-order effects in the presence of any block-to-block variation.

The last example showed that for a design partitioned into two blocks, one effect (ABC) is chosen to be confounded with the block effect. Another way of stating this is that the block effect and the ABC effect *share a degree of freedom* in the analysis. For the case of four blocks that use three degrees of freedom in the analysis, the general procedure is to independently select two effects to be confounded with blocks, and then a third confounded effect is determined by the generalized interaction. This will be described in a more detail with fractional factorial designs. For more information on constructing blocks in 2^k designs, see Myers and Montgomery [6].

39.4 FRACTIONAL FACTORIALS

Consider an unreplicated 2^6 design that contains 64 unique treatment combinations of the 6 factors and therefore 63 degrees of freedom for effects. Of those 63 degrees of freedom, only 6 are used for main effects (factors A, B, ..., F) and 15 for two-factor interactions (AB, AC, ..., EF). Assuming the sparsity-of-effects principle holds, only a subset of the total degrees of freedom are used for vital few effects that should adequately model the process. This discrepancy gets bigger as k gets larger, making full factorial designs an inefficient choice for experimentation. As with blocking, it is possible to optimally construct designs based on the assumption or knowledge that higher-order interactions are negligible, which are smaller in size yet preserve critical information about likely effects of interest. These are called *fractional factorial* designs, and they are widely utilized for any study involving, say, five or more factors. In particular, these are highly effective plans for *factor screening*, the exercise to whittle down to only the crucial process factors to be subsequently studied in greater detail. As with the full factorial designs, the fractional factorials are balanced, and the estimated effects are orthogonal. Two-level fractional factorial designs are denoted 2^{k-p}, where k still represents the number factors in the study and p represents fraction level. A 2^{k-1} design is called a one-half fraction of the 2^k, a 2^{k-2} design is called a one-quarter fraction of the 2^k, and so on. A 2^{k-p} design is a study in k factors, but executed with 2^{k-p} unique treatment combinations.

Consider the 2^3 design, but due to limited resources, only four of the eight treatment combinations can be studied. The candidate design is a one-half fraction of a 2^3 factorial, denoted 2^{3-1}. The primary questions in design construction are similar to those encountered in blocking and center on exactly how the four treatment combinations should be chosen and what information is contained in those experiments. Under the sparsity-of-effects principle, the ABC effect is negligible. Thus one choice of design is to choose those treatment combinations that are all positive in the ABC contrast coefficients. Equivalently one could choose the set that are all negative in ABC. Table 39.7 shows the experimental design for those coefficients positive in ABC.

It should be clear to the reader from the contrast coefficients in Table 39.7 that (i) all information for the ABC term is sacrificed in creating this design and (ii) the contrast coefficients that estimate one effect exactly match those for another. For instance, the contrast to estimate the A effect simultaneously estimates the BC effect. That is,

$$\frac{1}{2}(a + abc - b - c) = A + BC$$

As previously discussed with the blocks, the effect of A is said to be confounded (aliased) with BC. Similarly, the B effect is confounded with the AC effect, and the C effect is confounded with the AB effect. There is no way to individually estimate those effects, only their linear combinations.

TABLE 39.7 One-Half Fraction of the 2^3 Full Factorial Design

Treatment	A	B	C	AB	AC	BC	ABC
a	1	−1	−1	−1	−1	1	1
b	−1	1	−1	−1	1	−1	1
c	−1	−1	1	1	−1	−1	1
abc	1	1	1	1	1	1	1

This pooling of effect information is a by-product of fractional factorial designs.

In the previous example, the *ABC* term was assumed the least important effect and used as the basis for constructing the 2^{3-1} experimental design. Formally, *ABC* is called the *design generator* and is algebraically expressed in the relation

$$I = ABC$$

This is known as the defining relation, where *I* stands for identity, and implies that the *ABC* effect is confounded with the overall mean. Knowing this relationship helps determine details about the alias structure. This is accomplished by multiplying each side of the defining relation by an effect of interest and deleting any letter raised to the power 2 (i.e. via modulo 2 arithmetic). Any effect multiplied by *I* gives the effect back. Below demonstrates how to determine which effects are confounded with the main effect of *A*:

$$A \cdot I = A \cdot ABC$$
$$\Rightarrow A = A^2 BC$$
$$\Rightarrow A = BC$$

The interpretation is that the estimated *A* effect is confounded with the *BC* effect ($A = BC$), which was previously observed via the contrast coefficients in Table 39.7. Likewise it is trivial to show $B = AC$ and $C = AB$.

The defining relation for the chosen fraction above is more descriptively expressed as $I = +ABC$, since all contrast coefficients were positive in *ABC*. This is called the principle fraction and will always contain the treatment combination with all levels at their high setting. Alternatively, the complementary fraction could have been selected such that the defining relation would be expressed as $I = -ABC$. As a consequence, the linear contrasts would estimate $A - BC$, $B - AC$, and $C - AB$. Irrespective of sign, both fractions are statistically equivalent as main effects are confounded with two-factor interactions, although there may be a practical difference between the two.

Appropriately, one-half fraction designs are always constructed with the highest-order interaction in the defining relation. For instance, a 2^{5-1} experiment uses $I = ABCDE$ to create the fraction and to investigate the alias structure, as this interaction is assumed the least likely effect to significantly explaining response variation. However, the one-half fraction may still be too large to feasibly execute, and therefore fractions of higher degrees should be considered. The quarter-fraction design, denoted 2^{k-2}, is the next highest degree fraction from the half-factorials and comprises a fourth of the original 2^k factorial runs. These designs require two defining relations, call them $I = E_1$ and $I = E_2$, where the first designates the half fraction based on the "+" or "−" sign on the E_1 interaction and the second divides it further into a quarter fraction based on the "+" or "−" sign on the E_2 interaction. Note that all four possible fractions using $\pm E_1$ and $\pm E_2$ are statistically equivalent, with the principle fraction corresponding to choosing $I = +E_1$ and $I = +E_2$ in the defining relation. In addition, the generalized interaction $E_3 = E_1 \cdot E_2$ using modulo 2 arithmetic is also included. Thus to investigate the alias structure for a 2^{k-2} fractional factorial design, the complete defining relation is written as $I = E_1 = E_2 = E_3$. These interactions need to be chosen carefully to obtain a reasonable alias structure.

As an example, consider the 2^{6-2} design for factors *A* through *F*, and let $E_1 = ABCE$ and $E_2 = BCDF$. The generalized interaction, E_3, is computed by multiplying the two interactions together and deleting any letter with a power of 2. Thus

$$E_3 = (ABCE) \cdot (BCDF) = AB^2 C^2 DEF = ADEF$$

and hence the complete defining relation is expressed as

$$I = ABCE = BCDF = ADEF$$

It is easy to show for this design that main effects are confounded with three-factor interactions and higher (e.g. $A = BCE = ABCDF = DEF$) and that two-factor interactions are confounded with two-factor interactions and higher ($AB = CE = ACDF = BDEF$). Again, individual effects are not estimable, only the linear combinations.

To succinctly describe the alias structure of fractional factorials, *design resolution* is introduced. The resolution of a fractional factorial design is characterized by the length of the shortest effect (often referred to as the *shortest word*) in the defining relation and is represented by a Roman numeral subscript. The one-half fraction of a 2^5 factorial with defining relation $I = ABCDE$ is called a resolution V design and is formally denoted 2^{5-1}_V. Similarly, the one-quarter fraction of a 2^6 factorial with defining relation $I = ABCE = BCDF = ADEF$ is a resolution IV design and is denoted 2^{6-2}_{IV}. The design resolutions of greatest interest are described below.

- *Resolution III*: There exist main effects aliased with two-factor interactions. These designs are primarily used for screening many factors to identify which are the most influential in process modeling.
- *Resolution IV*: Main effects are confounded with three-factor interactions, and two-factor interactions are confounded with each other. Assuming the sparsity-of-effects principle, main effects are said to be estimated free and clear, since three-factor interactions and higher are considered negligible.
- *Resolution V*: Main effects are confounded with four-factor interaction, and two-factor interactions are

confounded with three-factor interactions. Assuming the sparsity-of-effects principle, main effects and two-factor interactions are said to be estimated free and clear, since three-factor interactions and higher are considered negligible.

39.5 DESIGN PROJECTION

One of the major benefits to using the two-level factorials and fractional factorials is to take advantage of the *design projection property*. This states that factorial and fractional factorial designs can be projected into stronger designs in a subset of the significant factors. In the case of unreplicated full factorials, those designs project into full factorials with replicates. For example, by disregarding one insignificant factor from a 2^4 full factorial, the design becomes a full $2^{4-1} = 2^3$ factorial with 2^1 replicates at each point. In the case of fractional factorial designs of resolution R, those designs project into full factorials in any of the $R - 1$ factors, conceivably with replicates. It may be possible to project to a fuller design with more parameters than the $R - 1$ rule dictates, but this is not guaranteed.

As a specific example, consider the 2_{III}^{3-1} fractional factorial design with defining relation $I = +ABC$ that investigates catalyst equivalents (A), ligand equivalents (B), and solvent volume (C). The four treatment combinations are depicted on the cube in the middle of Figure 39.18. As this is a resolution *III* design, it projects into a full factorial in any two of the three factors, also displayed in Figure 39.18. The projection property of two-level designs is very important, simply due to its usefulness in obtaining full modeling information on a subset of factors and its implicit use in sequential experimentation.

39.6 STEEPEST ASCENT

The content so far has focused on employing experimental designs with the sole purpose of identifying magnitude of effects plus two-parameter synergies. Often though, the

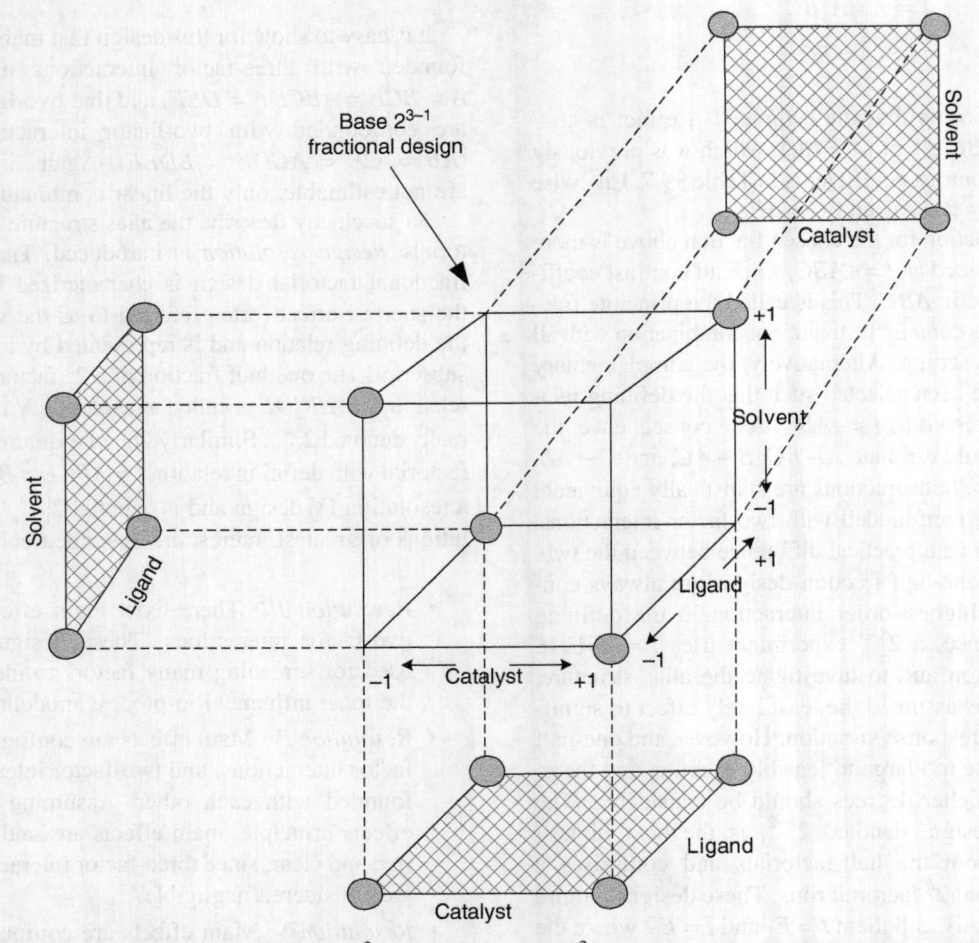

FIGURE 39.18 Illustration of projecting a 2^3 fraction onto a full 2^2 factorial in a subspace of the experimental region.

process models are used for optimization or improvement. The combination of experimental design, model building, and *sequential experimentation* used in searching for a region of improved response is called the method of steepest ascent (or steepest descent if explicitly minimizing). The goal is to effectively and efficiently move from one region in the factor space to another. Model simplicity and design economy are very important. The general algorithm consists of:

- Fitting a first-order (main effects) model with an efficient two-level design.
- Computing the path of steepest ascent (descent), where there is an expected maximum increase (decrease) in response.
- Conduct experiments along the path. Eventually response improvement will slow or start to decline.
- Carry out another factorial/fractional design.
- Recompute new path, or augment into optimization design.

Constructing the path of steepest ascent is straightforward. Consider all points that are a fixed distance from the design region center (i.e. radius r) with the desire to seek the parameter combination that maximizes the response. Mathematically, one uses the method of Lagrange multipliers to find where the maximum response lies, constrained to the radius r. Intuitively, the path of steepest ascent is proportional to the size and sign of the coefficients for the first-order model in coded units. For example, let fitted equation be $2 + 3x_1 - 1.5x_2$. As shown in Figure 39.19, the path of steepest ascent will have x_1 moving in a positive direction and x_2 in a negative direction. More specifically, the path is such that for every 3.0 units of increase in x_1, there will correspondingly be 1.5 units of decrease in x_2. For steepest decent, the path is chosen using the opposite sign of the coefficients.

The success of the steepest ascent method rests on whether the region where the path is constructed is main effect driven. Steepest ascent should still be successful in the presence of curvature (interaction or quadratic), as long as it is small relative to size of the main effects. If curvature is large, then this exercise is self-defeating. In addition, the success of the path is dependent on the overall process model. Models that are poor and have high uncertainty lead to paths with high uncertainty. Finally, modifying the steepest ascent path with linear constraints is both mathematically and practically easy to incorporate. For more information on process improvement with steepest ascent, see Myers and Montgomery.

39.7 CENTER RUNS

One of the model assumptions in using a two-level design is linearity across the experimental region. If the region is small enough, this is a fair assumption. If the region spans a somewhat bolder space and/or the region contains the optimal process condition, then it would not be surprising if nonlinearity exists. Unfortunately, two-level designs by themselves cannot even detect any curvilinear relationship across the design region, much less model it. A cost-effective strategy to initially identify curvature and also have an independent estimate of variance is to add *center runs* to the experimental design. This second point is critical as in practice most 2^k designs are unreplicated. Using the standard ±1 scaling of factor levels, center runs are replicated n_c times at the design point $x_i = 0$, $i = 1, 2, \ldots, k$. Note that adding center runs produces no impact on the usual effect estimates in the 2^k design. The pure error variance is estimated with $n_c - 1$ degrees of freedom, and the test for nonlinearity is via a single degree of freedom contrast that compares the average response at the center to the average response from the factorial points. If nonlinearity is nonexistent across the design region, these averages should be comparable. Specifically, let $\bar{y}_c$ be the average of the n_c center points, and let $\bar{y}_f$ be the average of the n_f factorial points. The formal hypothesis test of nonlinearity is conducted by comparing the sum of squares for curvature

$$SS_C = \frac{n_f n_c (\bar{y}_f - \bar{y}_c)^2}{n_f + n_c}$$

against the mean squared error. This test does not give any information on which factors contribute the sources of curvature, only whether curvature exists or not. If $\bar{y}_f - \bar{y}_c$ is large,

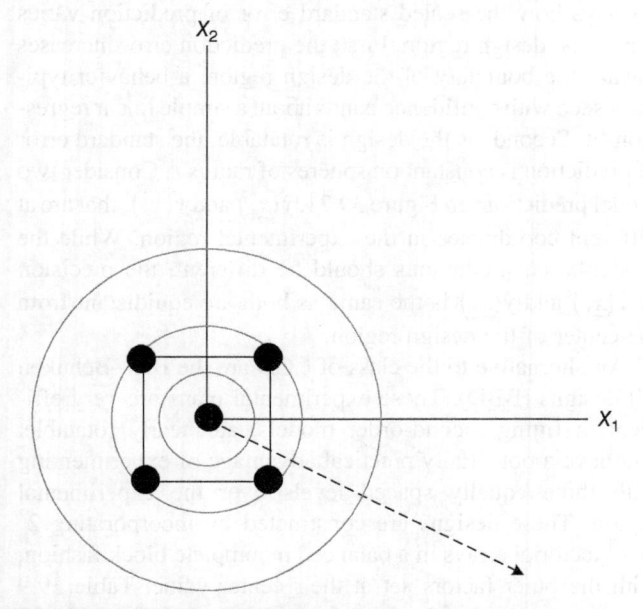

FIGURE 39.19 Path of steepest ascent for the model $\hat{y} = 2 + 3x_1 - 1.5x_2$.

then curvature is present across the design region. The implication is that the linear model with main effects and interactions is inadequate for prediction and additional design points or a more advanced design is necessary to identify which specific factors are contributing to the nonlinearity in order to accurately predict across the experimental region.

39.8 RESPONSE SURFACE DESIGNS

Previous discussion has centered on fitting first-order and first-order plus interaction models. However, a higher-order model is necessary when in the neighborhood of optimal response. In this case, second-order models are very good approximations to the true underlying functional relationship when curvature exists. These take the form

$$y = \beta_0 + \beta_1 x_1 + \beta_2 x_2 + \cdots + \beta_k x_k$$
$$+ \beta_{12} x_1 x_2 + \beta_{13} x_1 x_3 + \cdots + \beta_{k-1,k} x_{k-1} x_k$$
$$+ \beta_{11} x_1^2 + \beta_{22} x_2^2 + \cdots + \beta_{kk} x_k^2 + \varepsilon$$

To estimate the second-order model, each factor must have at least three levels. Implicitly there has to be as many unique design points as model terms. Many efficient designs are available that accommodate the above model. The most common is the central composite design (CCD) (Box and Wilson). The k-factor CCD is comprised of three components: a full 2^k factorial or resolution V fraction, center runs, and $2 \cdot k$ axial points. One compelling feature of CCD is that the axial points are a natural augmentation to the standard 2^k or 2_V^{k-p} plus center run designs.

As the name suggests, the axial points lie on the axes of each factor in the experimental space. In coded units they are set at a distance $\pm \alpha$ from the center of the design region. The axial value, α, can take on any value, which speaks to the flexibility of these designs. In practice they are usually taken at either $\alpha = 1$ for a face-centered design, $\alpha = \sqrt{k}$ for a spherical design, or $\alpha = f^{1/4}$ for a rotatable design, where f is the size of factorial or fractional used in the CCD. Table 39.8 shows an example of a two-factor CCD, and Figure 39.20 displays the experimental region.

Referring to Figure 39.20, if the axial value is set at $\alpha = 1$, then all of the experimental conditions except the center runs lie on the surface of the cube. Similarly, if the axial value is set at $\alpha = \sqrt{k}$, all of the experimental conditions except the center runs lie on the surface of a sphere. For rotatable designs, the precision on the model prediction is a function of only the distance from the design center and the error variance, σ^2. This is illustrated through the two-factor CCD in Table 39.8. The number of factorial points is $n_f = 4$, which yields $\alpha = 4^{1/4} = \sqrt{2}$ as the axial value for rotatability. Assuming the design contains $n_c = 3$ center runs, Figure 39.21

TABLE 39.8 General Two-Factor Central Composite Design

X_1	X_2	
−1	−1	
1	−1	} Factorial runs
−1	1	
1	1	
0	0	} Center runs (≥1)
−α	0	
α	0	} Axial runs
0	−α	
0	α	

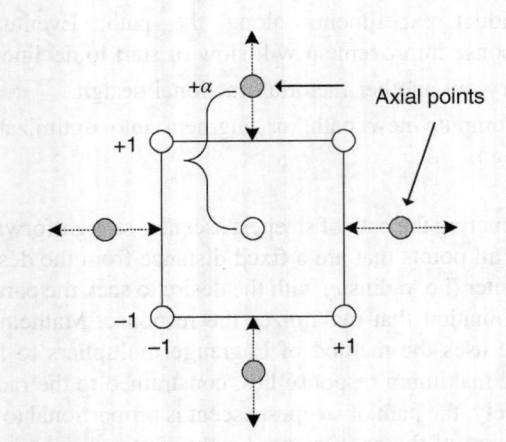

FIGURE 39.20 Illustration of general two-factor central composite design.

displays how the scaled standard error of prediction varies across the design region. First, the prediction error increases toward the boundary of the design region, a behavior typically seen with confidence bands about a simple linear regression fit. Second, as the design is rotatable, the standard error of prediction is constant on spheres of radius r. Consider two model predictions in Figure 39.21, $\hat{y}(\underline{x}_1)$ and $\hat{y}(\underline{x}_2)$, that are at different coordinates in the experimental region. While the model-based predictions should be different, the precision of $\hat{y}(\underline{x}_1)$ and $\hat{y}(\underline{x}_2)$ is the same as both are equidistant from the center of the design region.

An alternative to the class of CCD are the Box–Behnken [7] designs (BBD). These experimental plans are very efficient in fitting second-order models, are nearly rotatable, and have a potentially practical advantage of experimenting with three equally spaced levels over the experimental region. These designs are constructed by incorporating 2^2 or 2^3 factorial arrays in a balanced incomplete block fashion, with the other factors set at their center value. Table 39.9 shows an example of a three-factor BBD, and Figure 39.22 displays the design in graphical form. The BBDs are

EXPERIMENTAL DESIGN FOR PHARMACEUTICAL DEVELOPMENT 899

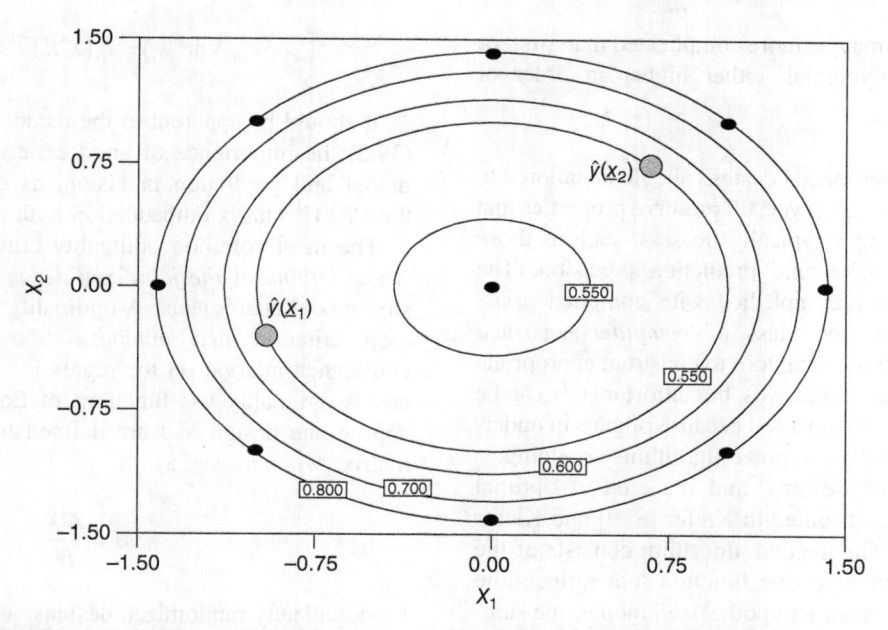

FIGURE 39.21 Illustration of rotatability using the standard error of prediction across factor space.

TABLE 39.9 Three-Factor Box–Behnken Design

X_1	X_2	X_3
−1	−1	0
1	−1	0
−1	1	0
1	1	0
−1	0	−1
1	0	−1
−1	0	1
1	0	1
0	−1	−1
0	1	−1
0	−1	1
0	1	1
0	0	0

spherical designs, and there are no factorial "corner points" or face points. For the $k = 3$ design shown in Figure 39.22, all conditions except the center runs are at $\sqrt{2}$ distance from the center. This should not deter the engineer from using this design, especially if predicting at the corners is not of interest, potentially due to cost, impracticality, feasibility, or other issues.

39.9 COMPUTER-GENERATED DESIGNS

Classical experimental designs such as those discussed so far may not be appropriate for a practical situation due to one or more constraints. These could include:

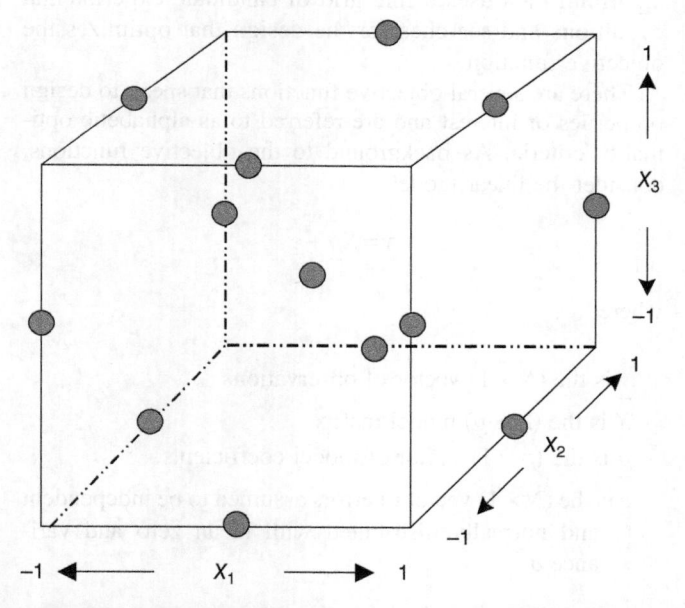

FIGURE 39.22 Illustration of three-factor Box–Behnken design.

- Sample size limitations due to time, budget, or material, which may yield a nonstandard number of runs, say, 11 or 13.
- Nonviable factor settings, therefore impacting the experimental region geometry. Examples include solubility limitations, safety concerns, and small-scale mixing sensitivities.
- Desired factor levels are nonstandard, say, 4 or 6.
- Factors are both qualitative and quantitative.

- Proposed model may be more complicated than first- or second-order polynomial, either higher in order or nonlinear.

For such cases, experimental designs should be tailored to accommodate any constraints yet still preserve properties that the more classical designs typically possess, such as those based on model precision and prediction precision. The design construction is accomplished with computer assistance and falls under the class of *computer-generated designs*. The computer is a vital tool to construct appropriate designs that meet certain objectives, but unfortunately can be viewed as a black box and misused because of gaps in understanding exactly what the computer algorithms are doing.

Computer-generated designs, and the area of optimal design theory, can be attributed to Kiefer [8, 9] and Kiefer and Wolfowitz [10]. The general algorithm consists of the engineer providing an objective function that reflects the design property of interest, a hypothesized model, the sample size for the study, and other design elements potentially corresponding to blocks, center runs, and lack of fit. The algorithm then uses a fine grid of candidate experimental conditions and searches for the design that optimizes the objective function.

There are several objective functions that speak to design properties of interest and are referred to as alphabetic optimality criteria. As background to the objective functions, consider the linear model

$$y = X\beta + \varepsilon$$

where

- y is the $(N \times 1)$ vector of observations.
- X is the $(N \times p)$ model matrix.
- β is the $(p \times 1)$ vector of model coefficients.
- ε is the $(N \times 1)$ vector of errors assumed to be independent and normally distributed with mean zero and variance σ^2.

The ordinary least squares estimate of model coefficients is

$$\hat{\beta} = (X'X)^{-1}X'y$$

and the covariance matrix of those estimates is given by

$$\mathrm{Var}(\hat{\beta}) = (X'X)^{-1}\sigma^2 \qquad (39.1)$$

In addition, the variance of a predicted mean response, $\hat{y}_0$, at coordinates x_0 is given by

$$\mathrm{Var}(\hat{y}_0) = x_0'(X'X)^{-1}x_0\sigma^2 \qquad (39.2)$$

It should be apparent to the reader from Eqs. (39.1) and (39.2) the importance of good experimental design on the model and prediction precision, as demonstrated through the $(X'X)^{-1}$ matrix embedded in both of those quantities.

The most common optimality criterion is D-optimality, which *minimizes the joint confidence region* on the regression model coefficients. A-optimality is a similar and common criterion that *minimizes the average size* of a confidence interval on the regression coefficients. Both D- and A-optimality use functions of Eq. (39.1) to obtain the appropriate design and are defined by the scaled moment matrix, M, expressed as

$$M = \frac{X'X}{N}$$

for completely randomized designs. The scaling takes away any dependence on σ^2, a constant independent of the design, and the sample size, N, which allows for comparisons across designs of different size. The algorithm finds that set of design points that maximizes the determinant of M for D-optimality and minimizes the trace of M^{-1} for A-optimality.

Two other criteria are G-optimality and IV-optimality, which use functions of the prediction variance in Eq. (39.2) to obtain a candidate design. A G-optimal design minimizes *the maximum size* of a confidence interval on a prediction over the entire experimental region, whereas IV-optimality minimizes *the average size* of a confidence interval on a prediction over the entire experimental region. Similar to the scaling done with the D-criterion above, these criteria are defined by the scaled prediction variance given by $v(x) = Nx'(X'X)^{-1}x$.

Research and software application in the area of optimal design theory has grown immensely over the recent past. Clearly the flexibility is appealing and at times invaluable. It allows the engineer to generate an experimental design for any sample size, number of factors (both discrete and continuous), type of model (linear or nonlinear), and randomization restrictions. Here are some additional notes and cautions regarding computer-generated designs:

- They are *model-dependent* optimal. Occasionally, the engineer will propose a mechanistic model that represents the true relationship between y and x. Often though, empirical models are proposed and inevitably edited after collecting data. The edited designs are suboptimal with respect to the criterion, and process modeling performance could deteriorate. There are strategies and graphical tools available to help generate and assess model robustness, such as those proposed by Heredia-Langner [11].

- The optimal design for one criterion is usually robust/near optimal across other criteria [12]. This is not overly surprising due to the importance of the $(X'X)^{-1}$ matrix. However, this is not guaranteed as it is possible to generate a D-optimal design that has poor prediction variance.
- Slight variations in algorithms and software packages could lead to generated designs that are statistically equivalent but different experimentally. As a matter of good scientific practice, the engineer needs to scrutinize the design for merit and practicality.
- Two-level designs for main effects only and main effects plus two-factor interaction models are all A-, D-, G-, and IV-optimal.
- Classical CCD and BBD for second-order models are near optimal. That is, they are highly efficient relative to the most optimal designs.

39.10 MULTIPLE RESPONSES

Up until now any discussion involving effect identification and regression analysis has focused on a single response, and the process model for that response can be used to hone in on a region of *desirability* in the factor space, as defined by the engineer. However, rarely in practice are experiments conducted when only a single response is collected. With chemical reaction trials, natural outputs could include various impurity levels, completion time, and yield, just to name a few. Standard analysis practice would involve (i) modeling each of those outcomes separately, (ii) defining their respective region of acceptable response over the factor space, and (iii) identifying the intersection of individual regions where *all* responses are deemed acceptable. Some software packages include this feature of *overlaying* contour plots, which is an effective approach in locating an optimal process operating region. However, when the number of responses and/or factors gets somewhat large, this exercise can become quite cumbersome. In addition, it is not surprising to have competing responses, meaning the optimal region for one response is suboptimal for other responses. This commonly occurs with crystallization processes, where maximum impurity purge is often at the sacrifice of higher yield, due to similar solubility properties of the chemical species. The question then becomes how to effectively merge process model information to identify conditions that are optimally balanced across multiple criteria.

The idea of desirability functions introduced by Derringer and Suich [13] addresses this problem. This is a formula scaled between [0, 1] inclusively, where the researchers own priorities and requirements are built into the optimization procedure. To illustrate, consider minimizing the total impurity level (%) from a chemical reaction and assume that any reaction that produces less than 0.5% is highly desirable. On the other hand, assume a reaction with more than 3% is unacceptable. The desirability function, d, for that response is expressed as

$$d = \begin{cases} 1, & \hat{y} \leq 0.5 \\ \left(\dfrac{3.0-\hat{y}}{3.0-0.5}\right)^S, & 0.5 < \hat{y} < 3.0 \\ 0, & \hat{y} \geq 3.0 \end{cases}$$

Model predictions less than 0.5% get the highest desirability score of one, whereas model predictions higher than 3% get the lowest desirability score of zero. For cases in between those levels, there exists a gradient of desirability scores that are a function of both the model prediction and a weight, S. This weight is chosen by the engineer, and it determines the severity of not achieving the most desirable goal, which in this case is 0.5% or less. Figure 39.23 gives a visual of how that desirability function behaves for various values of S.

For a given factor setting, each of the m responses has its own desirability score. That is, response i gets desirability, d_i, $i = 1, 2, \ldots, m$. To obtain the overall desirability across m responses at any experimental condition, the overall desirability score, D, is calculated as the geometric mean of each individual d_i, expressed as $D = \{d_1 \cdot d_2 \cdot, \ldots, \cdot d_m\}^{1/m}$. This overall score is easily modified when responses vary in importance. For example, impurity responses that affect drug product quality (and therefore affect the patient) are considered more important versus, say, yield that affects a sponsor's bottom line. The final objective is to locate parameter conditions that make D largest. This is normally accomplished via response surface modeling of D across experimental space and/or numerical techniques. Many software packages include this functionality as part of optimization. Note that any identified conditions should be confirmed for acceptability.

EXAMPLE PROBLEM 39.2 Three-Factor Central Composite Design

A three-factor face-centered central composite design is carried out to identify factor combinations that simultaneously minimize total impurities (%) and maximize yield (g). The three factors are catalyst (factor A), concentration (factor B), and temperature (factor C). Each experimental condition is completely randomized, and the center runs were replicated three times ($n_c = 3$). Total size of the study is 17 runs. The randomized design in coded units and data are listed in Table 39.10 and depicted in Figure 39.24. A full second-order model was fit to each response. The final ANOVA, fit statistics, predicted model equation, and relevant plots are presented below.

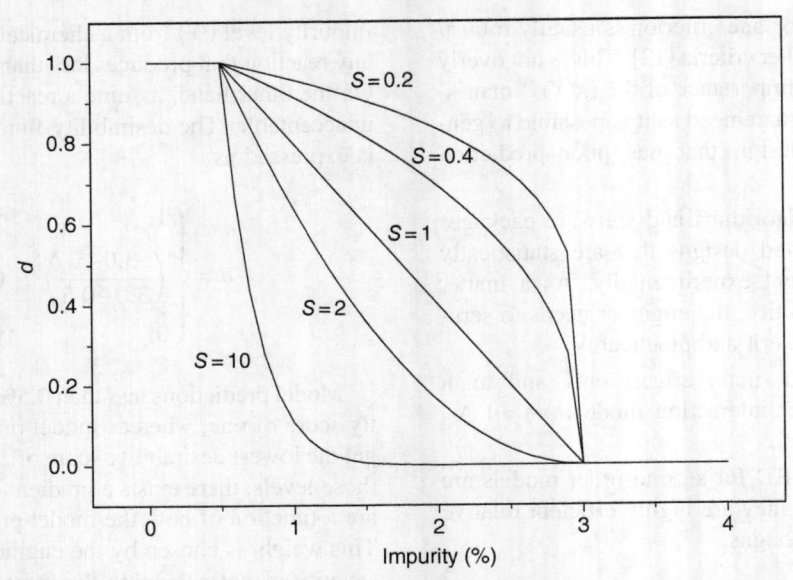

FIGURE 39.23 Example total impurity desirability function across various weights of S.

TABLE 39.10 Three-Factor CCD Design to Optimize Total Impurities and Yield

Design			Data	
Catalyst	Concentration	Temperature	Total Impurities	Yield
−1	1	−1	23.54	33.95
−1	−1	−1	10.57	15.27
0	0	0	12.20	35.10
1	1	−1	19.74	15.43
−1	1	1	5.20	18.68
1	−1	−1	8.61	2.50
0	0	0	11.10	37.20
1	−1	1	0.79	24.00
0	0	1	8.63	33.40
−1	−1	1	2.59	8.30
0	0	0	10.10	42.20
1	0	0	11.96	32.16
1	1	1	4.40	42.10
0	−1	0	5.47	31.90
0	0	−1	15.98	21.60
0	1	0	17.07	47.80
−1	0	0	12.82	31.17

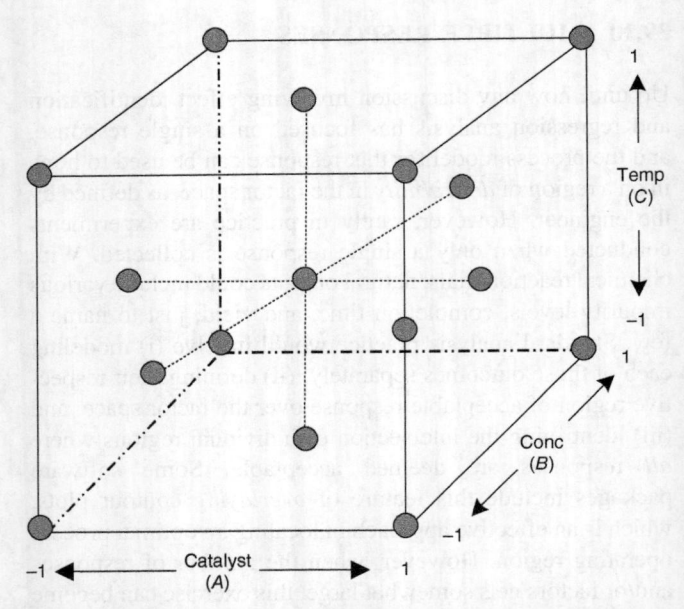

FIGURE 39.24 Illustration of three-factor face-centered CCD design.

Figure 39.25 gives an example snapshot of all relevant information in modeling total impurities. The table shows that total impurities are jointly impacted by concentration and temperature, as shown by the significant p-value on the interaction term, and independent of catalyst level. The model explains approximately 90.7% of the variation in the data as calculated by the R^2 value, and the final predicted model equation in coded units is embedded as well. Two previously highlighted diagnostic plots are included in this snapshot, although there could be others of interest that depict aspects of the data/model. The normal probability plot of residuals shows no deviation from the normality assumption, and the predicted vs. actual plot indicates that the model is performing well with constant variability across the range of prediction. Finally, a model-based contour plot of predicted total impurities across concentration and temperature is shown along with its 3D analogue surface plot. From this, total impurities are minimized at the

Source	SS	DF	MS	F	p-value
B - conc	175.7286	1	175.73	41.46	< 0.0001
C - temperature	322.9649	1	322.96	76.19	< 0.0001
BC	39.9618	1	39.96	9.43	0.0089
Error	55.10706	13	4.24		
Total	593.7624	16			

$R^2 = 90.7\%$ $\hat{y}_{imp} = 10.63 + 4.19(\text{Conc}) - 5.68(\text{Temp}) - 2.24(\text{Conc})(\text{Temp})$

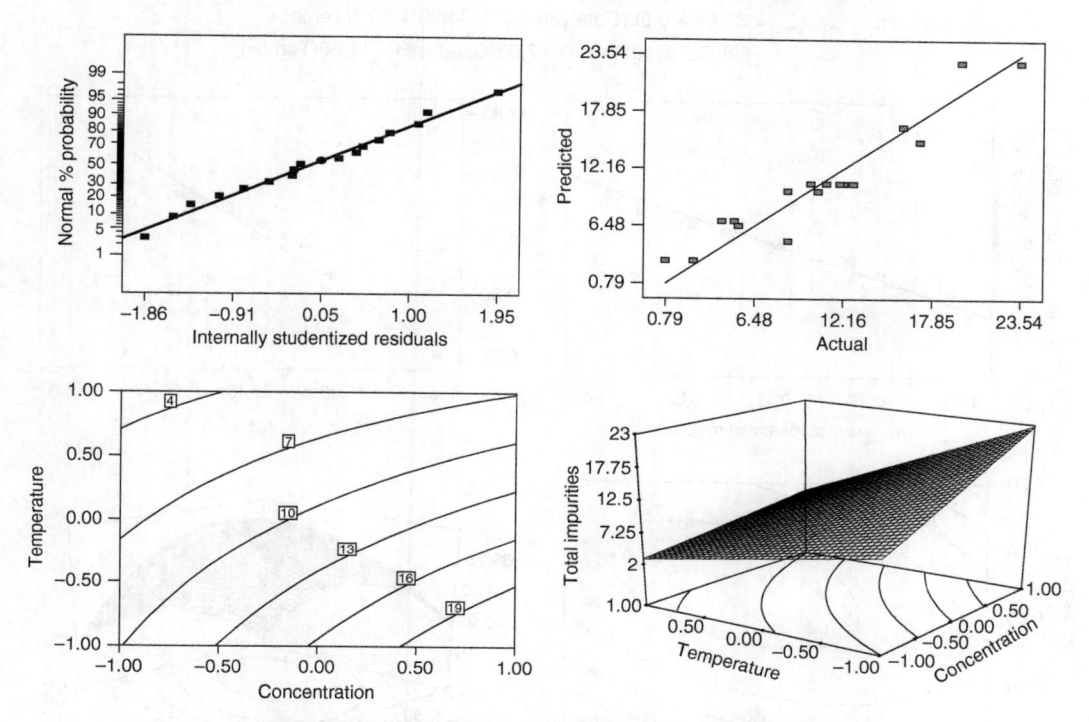

FIGURE 39.25 Summary of total impurity modeling.

combination lower concentration and higher temperature levels, independent of catalyst. That is, (concentration, temperature) = (−1, 1).

Figure 39.26 shows all relevant information in modeling yield. The embedded table shows that yield is impacted by all three factors, including second-order catalyst and temperature effects. Notice also that the main effect of catalyst is included even though the p-value is insignificant. This is to preserve *model hierarchy* as catalyst does explain some of the variation in response, but in conjunction with higher-order terms (either interactions and/or as a second-order effect). The model explains approximately 97.4% of the variation in the data as calculated by the R^2 value, and the final predicted model equation in coded units is shown. The normal probability plot of residuals demonstrates that the residuals can be assumed normally distributed, and the predicted vs. actual plot indicates that the model is performing quite well with no departures from the constant variability assumption. A model-based contour plot of predicted yield across catalyst and temperature levels at the high concentration is displayed along with its corresponding surface plot. The concentration level is set to high as the main effect is positive; higher concentration predicts higher yield. From this information, yield is maximized at (catalyst, concentration, temperature) ≈ (0.25, 1.0, 0.25) in coded units.

In identifying a candidate process operating region that is optimal over both responses, criteria that determine acceptability are established. For this example, assume that the process is acceptable if the predicted total impurities are below 10% and the yield is above 40 g. Figure 39.27 displays at the high (+1) concentration level the subspace (in white) of catalyst–temperature combinations where simultaneously both responses meet the specified process performance criteria. Similarly, one solution from using the desirability function predicts a candidate optimal setting at (catalyst, concentration, temperature) ≈ (0.67, 1, 1). This is denoted in the upper part of the white area by (●). Of course, this condition should be verified experimentally.

Source	SS	DF	MS	F	p-value
A - catalyst	7.78	1	7.8	1.2	0.3047
B - conc	577.45	1	577.4	86.9	< 0.0001
C - temp	142.36	1	142.4	21.4	0.0009
AC	619.70	1	619.7	93.2	< 0.0001
A^2	162.79	1	162.8	24.5	0.0006
C^2	400.29	1	400.3	60.2	< 0.0001
Error	66.46	10	6.6		
Total	2536.27	16			

$R^2 = 97.4\%$ $\hat{y}_{yield} = 38.88 + 0.88(\text{Catalyst}) + 7.6(\text{Conc}) + 3.77(\text{Temp})$
$+ 8.8(\text{Catalyst})(\text{Temp}) - 7.33(\text{Catalyst}^2) - 11.50(\text{Temp}^2)$

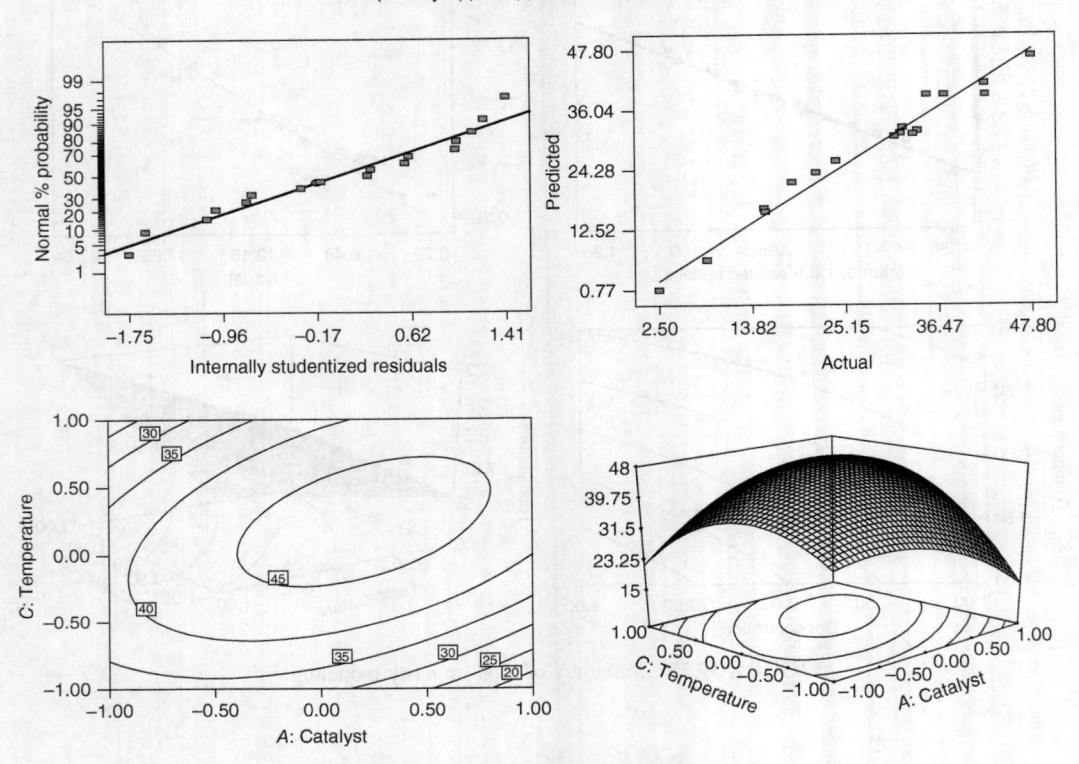

FIGURE 39.26 Summary of yield modeling.

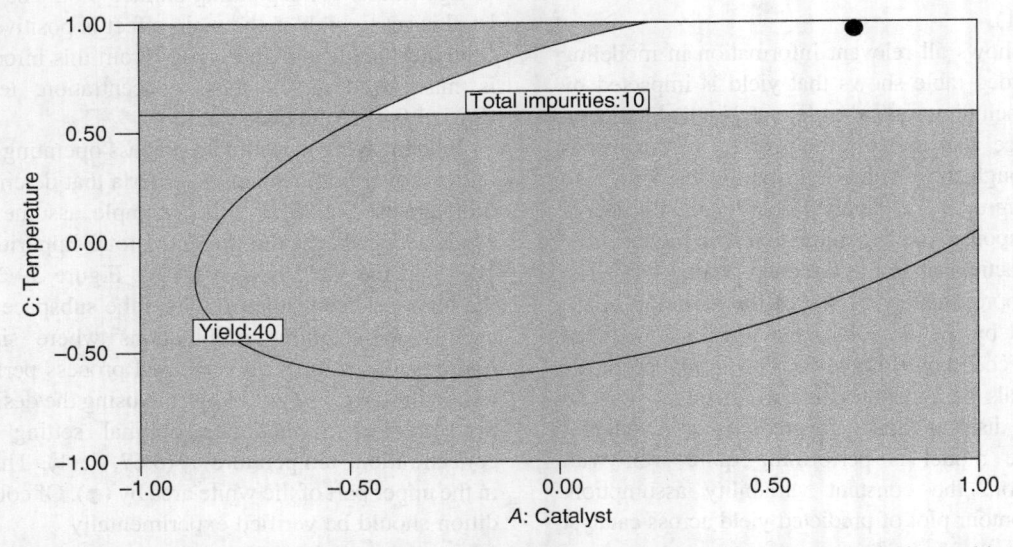

FIGURE 39.27 Overlay plot of catalyst–temperature combinations that meet specified performance criteria for total impurities and yield, with candidate optimal setting (●).

39.11 ADVANCED TOPICS

39.11.1 Industrial Split-Plot Designs

One assumption made throughout the chapter is that all experimental designs have complete randomization of the treatment combinations, which are generally called *completely randomized designs*. However, for experiments run at larger scale and/or with equipment limitations, complete randomization of the experiments is arduous to execute. This happens with factors that are difficult to independently change for each design run or impractical when certain factors can and should be held constant for some duration of the experiment due to resource or budget constraints. Temperature and pressure are examples that immediately come to mind of hard-to-change factors. When this situation occurs, part of the design is executed in "batch mode." That is, certain treatment combinations are fixed across a sequence of experiments, without resetting the actual treatment combination. This type of execution results in nested sources of variation and is called a *split-plot design*. These designs were originally developed for agronomic experiments, but its applicability easily spans all fields of science, even as the agricultural naming conventions have endured.

The basic split-plot experiment can be viewed as two experiments that are superimposed on each other. The first corresponds to a randomization of hard-to-change factors to experimental units called *whole plots*, while the second corresponds to a separate randomization of the easy-to-change factors within each whole plot, called *subplots*. This is different than completely randomized designs, which have unrestricted randomization factor combinations across all experimental units. The blocking of the hard-to-change factors along with the two separate randomization sequences creates a correlation structure among data collected within the same whole plot. Inferentially, the associated ANOVA needs to reflect that the experiment was executed in a split-plot fashion. If the data were analyzed as if the experiments were completely randomized, then it is possible to erroneously conclude significance of hard-to-change effects when they are not, while conclude insignificance of easy-to-change effects when they are. A good discussion on classical split-plot designs can be found in Hinkelmann and Kempthorne [14], as well as in Box and Jones [15] regarding response surface methodology.

In the recent past considerable attention has been given to constructing and evaluating optimal split-plot designs, especially with the computational horsepower of today's computers. Topics span from algorithms for D-optimal split-plot designs [16, 17] to comparing the performance between classical response surface designs in a split-plot structure [18] to graphical techniques for comparing competing split-plot designs [19]. This is an important area of research as many, if not most, industrial experiments are conducted in a split-plot fashion, whether deliberately planned or not. For those cases when it is planned, more software tools are becoming widely available so that the split-plot experiment are both powered sufficiently and analyzed appropriately.

39.11.2 Nonstandard Conditions

Model diagnostics and the analysis of residuals are a crucial part of any model-building exercise and are standard with any software package with a regression component. The information can numerically or visually identify any conditions from the standard assumption of normal and independent residuals with mean zero and common variance σ^2. Examples of nonstandard conditions are heterogeneous variance, data transformations, outliers, and non-normal errors.

In practice, the assumption of homogeneous variance is most likely violated. For many scientific applications, it is natural that variability increases with either response or regressor. From a theoretical perspective, this is simple to accommodate as the ordinary least squares estimate of regression coefficients, $\hat{\beta}$, is slightly altered to a weighted least squares estimate, where the weights are the inverse of the variances. The implication is that residuals with large variances (small weights) should not count as much in the model fit as those with small variances (large weights). For a given matrix of weights, W, the expression for the weighted least squares estimate of β is given by

$$\hat{\underline{\beta}}_{WLS} = (X'WX)^{-1}X'W\underline{y}$$

The immediate practical issue is how to obtain the weights. One approach is to collect multiple data at each experimental condition and use the corresponding estimate of variance. However, this is problematic if the sample variances are based on a small number of data and therefore potentially unreliable as a solution. A loose rule of thumb is that any estimated weights should be based on no less than a sample of nine [20]. Less than that can result in very poor performance of the weighted least squares solution, and ignoring the weights would be the better course of action.

One effective approach concerning the issue of increasing variance relative to increasing response is through a response transformation. The most common data transformation is the log transformation ($y^* = \ln(y)$ or $y^* = \log_{10}(y)$), where the mechanism of change in y across $\underline{x}$ is exponential. Other less common transformations include the square root ($y^* = \sqrt{y}$) and reciprocal ($y^* = y^{-1}$). All of these examples fall under the class of power transformations, where $y^* = y^\lambda$. The Box–Cox method [21] is one particular and powerful technique that aids in properly transforming y into y^* and takes the form

$$y^* = \begin{cases} \dfrac{y^\lambda - 1}{\lambda y^{\lambda-1}} & \lambda \neq 0 \\ y\ln(y) & \lambda = 0 \end{cases}$$

where

y is the geometric mean of the original data.

The assignment of $\ln(y)$ when $\lambda = 0$ comes from the limit of $(y^\lambda - 1)/\lambda$ as λ approaches zero, which allows for continuity in λ. The use of y^* rescales the response to the same units so that the error sum of squares can be comparable across different values of λ. As mentioned, common power transformation values of λ are -1, 0, 0.5, and 1, the last corresponding to no response transformation. In practice, after an appropriate model is fit to the data, an estimate of λ that minimizes the error sum of squares is estimated. If this value is significantly different than 1.0, a transformation on the data is recommended to produce a better fit, with potentially a different model. Most software packages that contain the Box–Cox procedure as part of model diagnostics provide a confidence interval on λ that simultaneously tests the hypothesis of no response transformation needed and puts a bound on a recommended one.

39.11.3 Designs for Nonlinear Models

Except for a brief mention in the section on computer-generated designs, all the discussion and examples in this chapter have only considered models that are linear in their parameters, which are then (successfully) used to approximate underlying nonlinear behavior over a small region. By definition, a linear model is one where the partial derivatives with respect to each model parameter are only a function of the factor/regressor. A simple example is the two-factor interaction model

$$y = \beta_0 + \beta_1 x_1 + \beta_2 x_2 + \beta_{12} x_1 x_2$$

It is trivial to show that $\partial y/\partial \beta_i$, ($i = 0, 1, 2$, and 12), does not depend on β_i. Conversely, a nonlinear model is defined as one where at least one of the partial derivatives is a function of a model parameter. An example of a nonlinear model would be

$$y = \beta_0 e^{-\beta_1 x}$$

where

both $\partial y/\partial \beta_0$ and $\partial y/\partial \beta_1$ are still functions of the unknown β-parameters.

Of course, chemical engineers are fully aware that the majority of kinetic models are nonlinear and not even closed form but expressed as differentials.

As opposed to linear models, in most cases classical experimental designs are not appropriate or optimal for nonlinear models. (The notable exception is in some applications of generalized linear models, as discussed by Myers et al [22].) Thus practitioners typically rely on computer-generated designs that are optimal in some respect, such as by D- or G-optimality criteria previously described. However, this is problematic and circular in terms of design construction, since *the optimal design for a nonlinear model is a function of the unknown parameters*. To circumvent that issue, initial parameter estimates are used that are often the results of previous studies and/or scientific knowledge. These designs are called *locally optimal*, and Box and Lucas [23] discuss locally D-optimal designs for nonlinear models. Intuitively, if the initial model parameter estimates are poor, then the locally optimal design will suffer in performance. One avenue that gets away from parameter point estimates is to use Bayesian approach to experimental design [24], where the scientist would postulate a *prior distribution* on the parameters plus specify a utility function, which is similar in spirit to the objective functions discussed in classical alphabetic optimal criteria. In any event, optimal designs for nonlinear models are inherently sequential, which may be an obstacle in adoption. In addition, commonly available technology has not caught up to the contemporary thinking in this field, adding another very tangible barrier. Nevertheless, independent of model type, it should be apparent that the process model fit is only as good as the design behind the data.

REFERENCES

1. Montgomery, D.C. (2005). *Design and Analysis of Experiments*, 6e. New York: Wiley.
2. Box, G.E.P. and Wilson, K.B. (1951). On the experimental attainment of optimum conditions. *Journal of the Royal Statistical Society, Series B* 13: 1–45.
3. Myers, R.H. (1990)). *Classical and Modern Regression with Applications*, 2e. Belmont, CA: Duxbury Press.
4. Design Expert version 7.1.6 (2007). Stat-Ease, Inc., Minneapolis, MN.
5. Daniel, C. (1959). Use of half-normal plots in interpreting factorial two-level experiments. *Technometrics* 1: 311–342.
6. Myers, R.H. and Montgomery, D.C. (2002). *Response Surface Methodology: Process and Product Optimization Using Designed Experiments*, 2e. New York: Wiley.
7. Box, G.E.P. and Behnken, D.W. (1960). Some new three-level designs for the study of quantitative variables. *Technometrics* 2: 455–475.
8. Kiefer, J. (1959). Optimum experimental designs. *Journal of the Royal Statistical Society, Series B* 21: 272–304.

9. Kiefer, J. (1961). Optimum designs in regression problems. *Annals of Mathematical Statistics* 32: 298–325.
10. Kiefer, J. and Wolfowitz, J. (1959). Optimum designs in regression problems. *Annals of Mathematical Statistics* 30: 271–294.
11. Heredia-Langner, A., Montgomery, D.C., Carlyle, W.M., and Borror, C.M. (2004). Model robust designs: a genetic algorithm approach. *Journal of Quality Technology* 36: 263–279.
12. Cornell, J.A. (2002). *Experiments with Mixtures: Designs, Models, and the Analysis of Mixture Data*, 3e. New York: Wiley.
13. Derringer, G. and Suich, R. (1980). Simultaneous optimization of several response variables. *Journal of Quality Technology* 12: 214–219.
14. Hinkelmann, K. and Kempthorne, O. (1994). *Design and Analysis of Experiments Volume I: Introduction to Experimental Design*. New York: Wiley.
15. Box, G.E.P. and Jones, S. (1992). Split-plot designs for robust product experimentation. *Journal of Applied Statistics* 19: 3–26.
16. Goos, P. and Vandebroek, M. (2001). Optimal split plot designs. *Journal of Quality Technology* 33: 436–450.
17. Goos, P. and Vandebroek, M. (2003). D-Optimal split plot designs with given numbers and sizes of whole plots. *Technometrics* 45: 235–245.
18. Letsinger, J.D., Myers, R.H., and Lentner, M. (1996). Response surface methods for bi-randomization structure. *Journal of Quality Technology* 28: 381–397.
19. Liang, L., Anderson-Cook, C.M., Robinson, T.J., and Myers, R.H. (2006). Three dimensional variance dispersion graphs for split-plot designs. *Journal of Computational and Graphical Statistics* 15: 757–778.
20. Deaton, M.L., Reynolds, M.R. Jr., and Myers, R.H. (1983). Estimation and hypothesis testing in regression in the presence of nonhomogeneous error variances. *Communications in Statistics, B* 12 (1): 45–66.
21. Box, G.E.P. and Cox, D.R. (1964). An analysis of transformations (with discussion). *Journal of the Royal Statistical Society, Series B* 26: 211–246.
22. Myers, R.H., Montgomery, D.C., and Vining, G.G. (2002). *Generalized Linear Models With Applications in Engineering and the Sciences*. New York: Wiley.
23. Box, G.E.P. and Lucas, H.L. (1959). Design of experiments in non-linear situations. *Biometrika* 46: 77–90.
24. Chaloner, K. and Verdinelli, I. (1995). Bayesian experimental design: a review. *Statistical Science* 10: 273–304.

40

MULTIVARIATE ANALYSIS IN API DEVELOPMENT

James C. Marek

Process Research and Development, AbbVie Inc., North Chicago, IL, USA

40.1 INTRODUCTION

When more than one variable is involved to determine a response, the mathematical space under consideration is multivariate. Chemical and biological processes are inherently multivariate, ensuring that the synthetic pharmaceutical and biopharmaceutical processes that rely on them are also multivariate in nature. Multivariate analysis (MVA) speaks specifically to the techniques within the statistical sciences that involve two or more variables. This chapter serves several purposes. As an introduction to MVA, it will give the reader a behind-the-scene understanding of the complex mathematical treatments. It is also intended to give the reader sufficient breadth and depth of the subject to convey the potential to benefit the process development effort and even support manufacturing controls.

Why is MVA important? MVA is recognized in ICH Q11 [1] as an important and integral part of the enhanced approach to process development. Specifically, the approach entails multivariate design of experiments, simulations, and modeling to establish control strategies for appropriate material specifications and process parameter ranges. Further, the US FDA provided guidance on process analytical technology (PAT) as early as 2004 [2]. PAT involves in situ measurements during operation of batch, semi-batch, or continuous processes. PAT collects data in real time, affording several benefits such as (i) speedy acquisition of the desired information; (ii) the need for in-process control samples in the manufacturing environment that may be replaced by PAT; (iii) process development data that may be collected essentially continuously throughout the duration of the experiment, without the need of human intervention or automatic samplers; (iv) the ability to track a process where traditional sampling is complicated by safety or sample stability considerations; and (v) acquisition of data to control the process in a feedback or feedforward control loop.

PAT may range from simple sensors to complex spectrometers. Thus, PAT includes providing simple measurements of parameters like density, conductivity, optical refractive index, single wavelength UV absorbance, etc. to very complex spectral data from infrared (IR) spectrometers, near-IR spectrometers, Raman spectrometers, and laser diffraction analyzers. Analysis of data from simple sensors may be univariate; however the analysis of spectral data will rely on chemometrics. Chemometrics is defined as the statistical analysis of chemical data, and more often than not relies on multivariate statistical methods to incorporate the absorbance from more than one wavelength, Raman shift, particle chord length, etc. Spectra may be collected at frequent time intervals throughout the course of the process or experiment, allowing determination of reaction kinetics with multiple PAT instruments [3]. Each spectrum may contain hundreds, perhaps even thousands, of data points, consisting of the measured absorbances over a range of several hundred wave numbers.

40.2 APPROACHES TO MODEL DEVELOPMENT

The goal of quantitative MVA is often to develop a predictive model of the concentration of species of interest. This can be

---

Chemical Engineering in the Pharmaceutical Industry: Active Pharmaceutical Ingredients, Second Edition.
Edited by David J. am Ende and Mary T. am Ende.
© 2019 John Wiley & Sons, Inc. Published 2019 by John Wiley & Sons, Inc.

accomplished by either principal component regression (PCR) or the more commonly used technique of partial least squares (PLS). These techniques reduce the dimensionality of the data sets and involve factor spaces and eigenvectors, approaches that some researchers may find as somewhat complex and abstract. Therefore, it is useful to first consider some simpler, more traditional approaches for model development to assist understanding the more complex techniques.

40.2.1 Absorption Theory: Beer–Lambert Law

The absorption from a single species i at wavelength λ, A_λ, is proportional to the product of the absorption coefficient at the wavelength λ, α_λ; the path length, L; and the species concentration, C_i:

$$A_\lambda = \alpha_\lambda L C_i \quad (40.1)$$

In theory the concentration of species i could be determined by collecting spectra and using a single wavelength for the calibration. Assuming that compound i absorbs at multiple wavelengths, these absorptions could be combined to develop a multiple regression fitting of C_i with n wavelengths:

$$A_{\lambda 1} = \alpha_{\lambda 1} L C_i \quad (40.2)$$

$$A_{\lambda 2} = \alpha_{\lambda 2} L C_i \quad (40.3)$$

$$A_{\lambda n} = \alpha_{\lambda n} L C_i \quad (40.4)$$

$$C_i = \left(\frac{1}{nL}\right)\left(\frac{A_{\lambda 1}}{\alpha_{\lambda 1}} + \frac{A_{\lambda 2}}{\alpha_{\lambda 2}} + \frac{A_{\lambda 3}}{\alpha_{\lambda 3}} + \ldots + \frac{A_{\lambda n}}{\alpha_{\lambda n}}\right) \quad (40.5)$$

Combining terms for the constants to generate coefficients $b_1, b_2, b_3, \ldots, b_n$,

$$b_n = \frac{1}{nL\alpha_{\lambda n}} \quad (40.6)$$

Then a multiple regression equation for C_i results:

$$C_i = (b_1 A_{\lambda 1} + b_2 A_{\lambda 2} + b_3 A_{\lambda 3} + \ldots + b_n A_{\lambda n}) \quad (40.7)$$

Figure 40.1 shows data from laboratory experiments for disappearance of a starting material (SM) (reactant) in a batch reactor from grab samples taken as a function of reaction time. The FTIR absorption noted for the SM was clear of overlap from other species in the system at a wavelength of 1397.32 cm^{-1}. Hence, simple linear regression of the first derivatives at 1397.32 cm^{-1} was tested as an option to build a calibration model for the residual SM. The model, built using 72 data points collected in four experiments, yielded

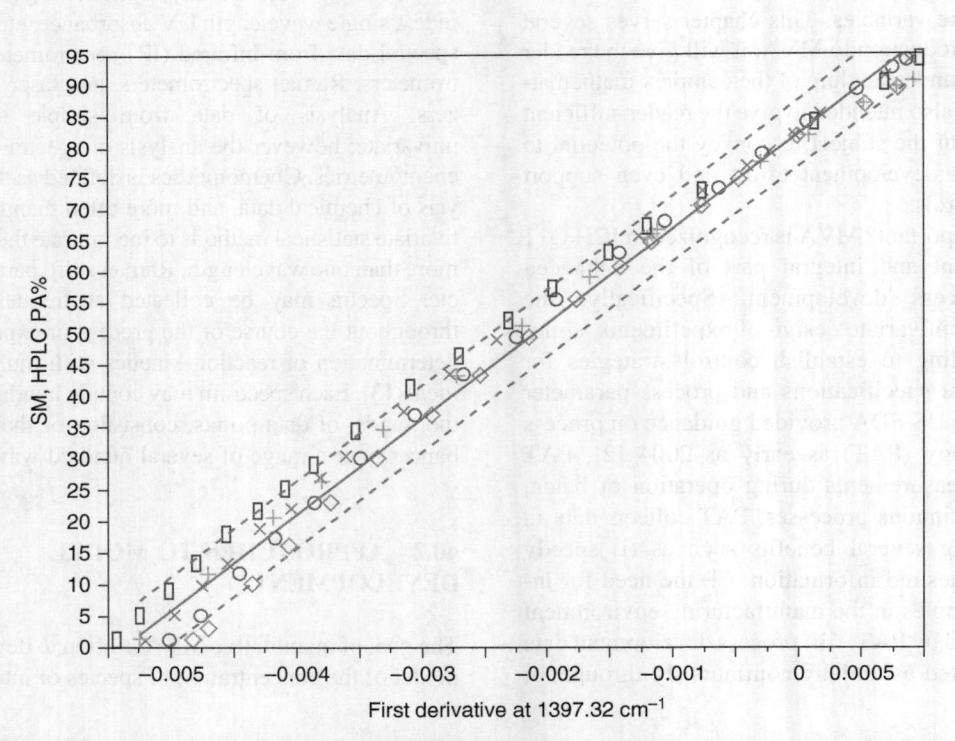

FIGURE 40.1 Residual SM from a reaction measured by HPLC PA% is plotted versus first derivative of mid-IR absorption at 1397.32 cm^{-1}. The five different symbols represent data from five separate laboratory reactions, the solid line is the regression fit, and the dashed line the 95% confidence interval for predicting an individual value.

unsatisfactory predictions for the residual SM using spectra collected in a fifth experiment (Figure 40.1, hollow squares). Next, a multiple regression model was created to enable fitting the residual SM concentration (as determined by off-line HPLC analysis of grab samples, PA%) to the mid-IR absorbances collected in four of the experiments. The model was built using absorbances at four wave numbers and included three statistically significant two-factor interactions. The model was then applied to the absorbance data from the fifth batch (hollow squares) to predict the SM concentration profile and compared to the corresponding HPLC data (Figure 40.2).

The data suggest that the fit of the model to the calibration data from the four trials was excellent, but the model was not capable of accurately predicting the behavior of the fifth batch, particularly in the lower PA% region, which is the region of main interest to determine reaction completion in manufacturing batches. These examples from a complex reaction system illustrate cases where simple calibration models are unsatisfactory. Calibrations using a single wavelength, or multiple regression of absorbances from a few wavelengths, can be appropriate in some cases. For example, simple models may be suitable for analysis of solutions of high purity drug prepared for solubility determination.

In most systems of interest in pharmaceutical development, other species (reactants, solvents, related-substance impurities, etc.) may be present that absorb at the same wavelength as the molecule of interest. Hence, we need to consider the contributions from more than one component to the observed absorption, in addition to more than one wavelength:

$$A_1 = \alpha_{11}LC_1 + \alpha_{12}LC_2 + \ldots + \alpha_{1i}LC_i \qquad (40.8)$$

$$A_2 = \alpha_{21}LC_1 + \alpha_{22}LC_2 + \ldots + \alpha_{2i}LC_i \qquad (40.9)$$

$$A_n = \alpha_{n1}LC_1 + \alpha_{n2}LC_2 + \ldots + \alpha_{ni}LC_i \qquad (40.10)$$

where

A_n is the absorbance at the nth wavelength.

α_{ni} is the absorbance coefficient at the nth wavelength for the ith component.

L is the path length.

C_i is the concentration of the ith component.

i is the number of components.

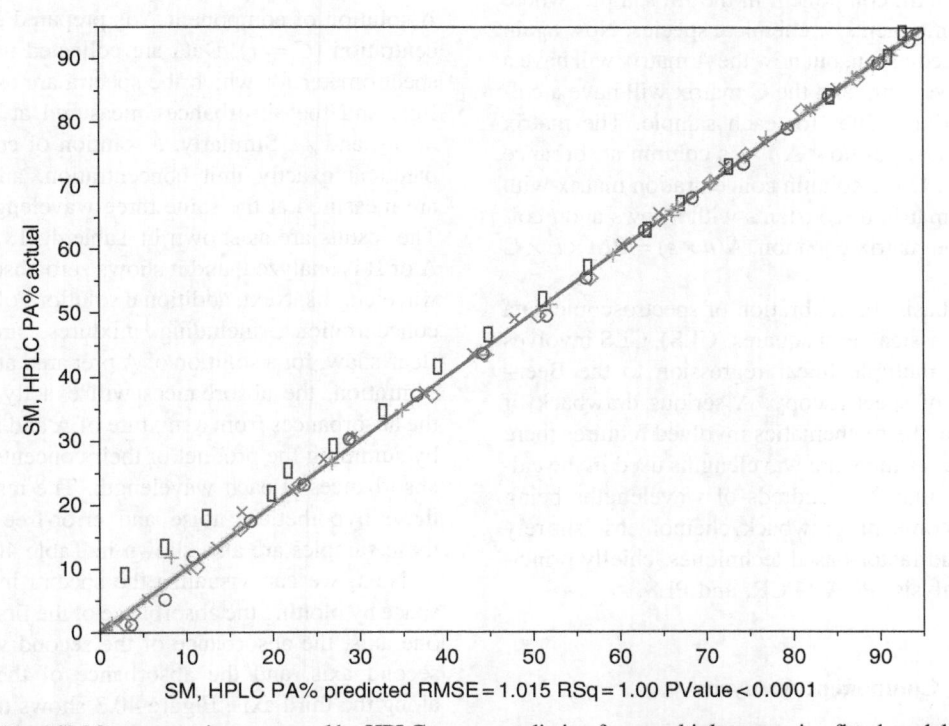

FIGURE 40.2 Residual SM from a reaction measured by HPLC versus prediction from multiple regression fitted model for mid-IR data. The five different symbols represent data from five separate laboratory reactions, the solid line is the regression fit, and the dashed line the 95% confidence interval.

This framework now has the absorbance at a given wavelength properly determined by the sum of the absorbances at that wavelength for each of the i components. The equation may be expressed in a modified form, taking b as the product of the absorption coefficient and the path length, wherein it becomes

$$A_n = \sum_{i=1}^{i} b_{ni} C_i \qquad (40.11)$$

This expression further simplifies to the matrix equation:

$$\mathbf{A} = \mathbf{BC} \qquad (40.12)$$

Here $\mathbf{A}$ is a single-column absorbance matrix with n rows, ($n \times 1$). $\mathbf{B}$ is a matrix of constants with n rows and i columns, and $\mathbf{C}$ is a single-column concentration matrix with i rows, ($i \times 1$). Hence, $\mathbf{A}(n \times 1) = \mathbf{B}(n \times i) \times \mathbf{C}(i \times 1)$. The $\mathbf{B}$ matrix is essentially a matrix of the absorbances produced at the n wavelengths with each of the i components at unit concentration.

It is a simple matter to extend the situation to the collection of spectra from more than one sample, each spectrum being generated from the unique concentrations of $C_1, C_2, C_3, \ldots, C_i$ in the sample.

Let A_{ns} be the absorbance at the nth wavelength for sample s, b_{ni} be the absorbance coefficient for at the nth wavelength times the path length for the ith component, and C_{is} be the concentration of the ith component in the sth sample, where there are n total components or chemical species. Now again we have the matrix equation, but now the $\mathbf{A}$ matrix will have a column for each spectrum, and the $\mathbf{C}$ matrix will have a column of concentration values for each sample. The matrix equation above holds, but now $\mathbf{A}$ is a s column absorbance matrix with n rows, $\mathbf{C}$ is s column concentration matrix with i rows, and $\mathbf{B}$ is a matrix of constants with n rows and i columns. Hence, in the matrix equation, $\mathbf{A}(n \times s) = \mathbf{B}(n \times i) \times \mathbf{C}(i \times s)$.

This forms the basis for calibration of spectroscopic data by the method of classical least squares (CLS). CLS involves the application of multiple linear regression to the Beer–Lambert equation of spectroscopy. A serious drawback of this approach is that the mathematics involved requires there be as many samples as there are wavelengths used in the calibration, and there may be hundreds of wavelengths being measured. To overcome this drawback, chemometricians rely on factor spaces and factor-based techniques, chiefly principal component analysis (PCA), PCR, and PLS.

40.2.2 Principal Component Analysis

PCA is an orthogonal mathematical transformation to convert a set of observations that may be correlated into a smaller set of linearly uncorrelated variables termed the principal components, which are eigenvectors. In doing so, the dimensionality of the data set is reduced. For example, an IR spectrum with absorbance values measured at several hundred wave numbers may be reduced to considerably fewer, perhaps a handful of, principal components after PCA. PCA itself does not perform a regression; rather it operates on a set of data. Hence, p variables will end up with a set of p principal components [4].

The first principal component, PC1, is selected by finding a vector describing the maximum variation in the data set. To find the second principal component, PC2, a vector orthogonal to the first is identified that explains the greatest amount of the remaining variation in the data set. Additional principal components are found in this manner until the variation in the data set is explained to within the expected variation due to random error or noise. The determination of the appropriate numbers of principal components to include can be done by a variety of numerical methods. This is essential to avoid overfitting the data and creating inaccurate predictions by including too many factors that describe random error and noise. PCA by itself does not directly yield quantitative values for prediction of concentrations in samples, but is extremely useful to compare data collected under different processing conditions, such as monitoring reactions or crystallizations performed at different temperatures, reagent concentrations, etc.

To help visualize how principal components work, consider a two-component system, a mixture of A and B. A solution of component A is prepared at exactly unit concentration ($C = 1$). Data are collected using a hypothetical spectrometer for which the spectra are assumed to be error-free, and the absorbances measured at three wavelengths, λ_1, λ_2, and λ_3. Similarly, a solution of component B is prepared at exactly unit concentration, and the absorbances are measured at the same three wavelengths, λ_1, λ_2, and λ_3. The results are as shown in Table 40.1. A solution with no A or B is analyzed, and it shows zero absorbance at the three wavelengths. Next, additional solutions of A and B at various concentrations, including mixtures, are prepared. From Beer's law, for a solution of A prepared at twice the unit concentration, the absorbances will exactly double. Similarly, the absorbances from a mixture of A and B can be calculated by summing the product of their concentrations and the unit absorbances at each wavelength. The results for a series of these hypothetical noise and error-free measurements on these samples are also shown in Table 40.1.

Next, we can visualize the spectra in three-dimensional space by plotting the absorbance of the first wavelength along one axis, the absorbance of the second wavelength along a second axis, and the absorbance of the third wavelength along the third axis. Figure 40.3 shows the plot for the data in Table 40.1.

Note that the spectra, although measured at three wavelengths, lie in a two-dimensional plane. Regardless of how

TABLE 40.1 Absorbances from Error-Free Measurements of A and B

Conc. of A	Conc. of B	A_{λ_1}	A_{λ_2}	A_{λ_3}
0	0	0	0	0
1	0	1	0.5	0.25
2	0	2	1	0.5
3	0	3	1.5	0.75
0	1	0.25	0.5	1
0	2	0.5	1	2
0	3	0.75	1.5	3
1	1	1.25	1	1.25
1	2	1.5	1.5	2.25
1	3	1.75	2	3.25
1	4	2	2.5	4.25
1	5	2.25	3	5.25
1	6	2.5	3.5	6.25
1	7	2.75	4	7.25
2	1	2.25	1.5	1.5
2	2	2.5	2	2.5
2	3	2.75	2.5	3.5
2	4	3	3	4.5
2	5	3.25	3.5	5.5
2	6	3.5	4	6.5
2	7	3.75	4.5	7.5
3	1	3.25	2	1.75
3	2	3.5	2.5	2.75
3	3	3.75	3	3.75
3	4	4	3.5	4.75
3	5	4.25	4	5.75
3	6	4.5	4.5	6.75
3	7	4.75	5	7.75
4	1	4.25	2.5	2
4	2	4.5	3	3
4	3	4.75	3.5	4
4	4	5	4	5
4	5	5.25	4.5	6
4	6	5.5	5	7
4	7	5.75	5.5	8
5	1	5.25	3	2.25
5	2	5.5	3.5	3.25
5	3	5.75	4	4.25
5	4	6	4.5	5.25
5	5	6.25	5	6.25
5	6	6.5	5.5	7.25
5	7	6.75	6	8.25
6	1	6.25	3.5	2.5
6	2	6.5	4	3.5
6	3	6.75	4.5	4.5
6	4	7	5	5.5
6	5	7.25	5.5	6.5
6	6	7.5	6	7.5
6	7	7.75	6.5	8.5

many wavelengths are used in the measurements, even cases where several hundred wavelengths are used, the data for two components will lie in a two-dimensional plane. The principal components are then found by finding the eigenvector in

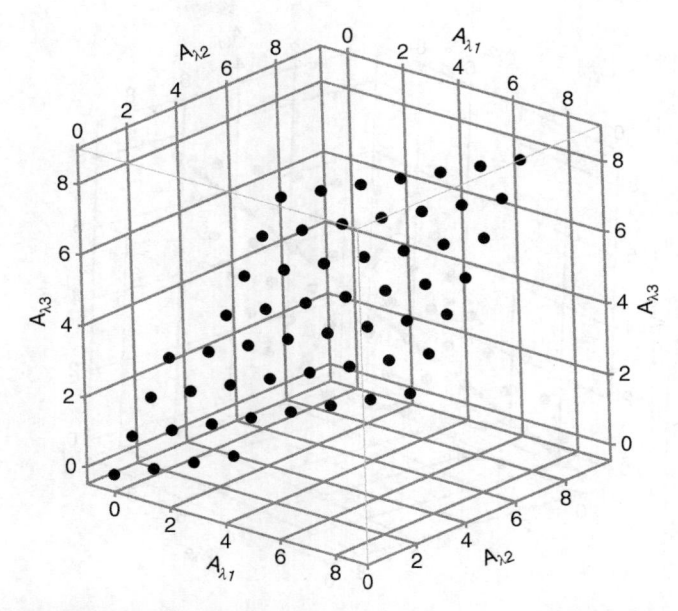

FIGURE 40.3 Absorption data from A and B mixtures from Table 40.1.

the direction of maximum variation of the data set and then a second eigenvector orthogonal to the first in a direction that accounts for the most of the remaining variation in the data set. The first principal component must lie in the plane of the data points to span the maximum possible variation of the data. Similarly, the second principal component will be orthogonal to the first but also lies in the plane of the data set. The principal components for this system are also shown in Figure 40.4. Note that the first principal component explains greater than 87% of the variation in the data set and the second explains a little less than 13%.

It is apparent that we could define a pair of axes in this plane that form a new coordinate system to plot the data and then describe each spectrum on the basis of the distance from the two new axes. This would reduce the dimensionality from 3 to 2. Similarly, had we used three chemical species, A, B, and C, and made similar measurements, the PCA would reduce the dimensionality to a data set described by three new axes.

For each of the data points, the distance of the data point from the PC1 or PC2 vector, often described as the projection of the data point onto the PC1 or PC2 vector, is called a score value. Scores are computed for each data point with respect to PC1 and then again for each data point with respect to a different principal component, typically PC2. Thus a score plot is created, PC1 versus PC2, which allows visualization of how different samples are related to one another. The eigenvalue of the eigenvector is the sum of squares of the projections of the data onto the eigenvector, and the projections are of course the distance along the vector of each data point. Principal component plots, termed loading plots, allow identification of how different variables are related to each other.

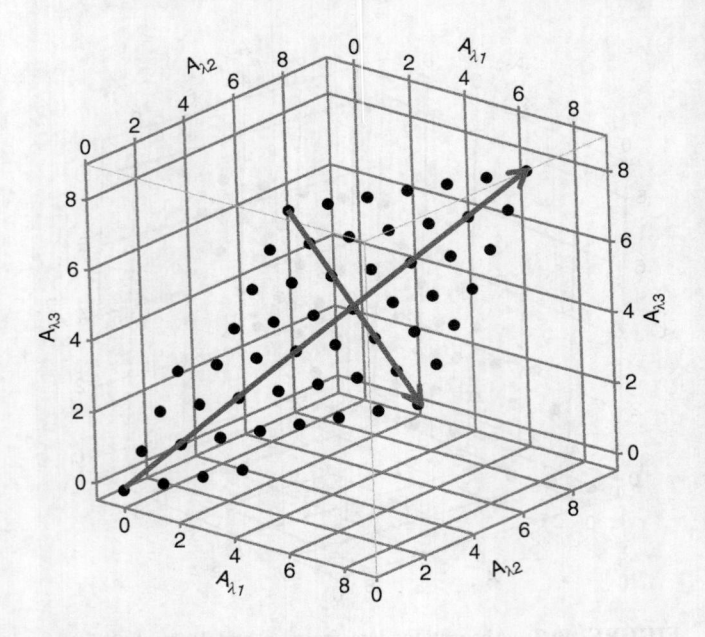

FIGURE 40.4 Principal components for A and B absorptions at A_{λ_1}, A_{λ_2}, and A_{λ_3}.

Principal components

Number	Eigenvalue	Percent	Cum percent
1	2.6185	87.283	87.283
2	0.3815	12.717	100.000

Working with scaled variables is preferred, since the PCs and scores then become dimensionless variables.

Most commercial software packages output the eigenvalues of the principal component, as well as the relative amount of the variation in the data that is explained by each principal component, enabling interpretation as to the number of vectors needed to reduce the dimensionality of the data set.

40.2.3 Principal Component Regression (PCR)

For spectra collected on a set of samples of known concentration, the principal components of both the absorbances, X, and the known concentrations, Y_n, can be calculated. For the set of observations, the matrix of regression coefficients, B, is computed:

$$B = Y_n X^{-1} \quad (40.13)$$

This is termed building a calibration model. As additional spectra on samples of unknown concentration are collected in the future, it is possible to estimate the concentration of the unknowns, Y_u, from the calibration model and the measured absorbance, X_u:

$$Y_u = BX \quad (40.14)$$

40.2.4 Partial Least Squares

PLS is an extension of PCR where again both the X and Y data are considered. However, the PLS equation or calibration is based on decomposing both the X and Y data into a set of scores and loadings, similar to PCR, but with one significant difference: the scores for both the X and Y data are not selected based solely on the direction of maximum variation, but instead are selected in order to maximize the correlation between the scores for both the X and Y variables. The result is termed a set of latent vectors, or basis vectors, or sometimes simply factors. The selection of the appropriate number of factors is important to properly fit the model, and the software packages have built-in capability to recommend the number of factors to keep based on the predicted residual error sum of squares (PRESS residuals). Most commercial software packages [5] come with the ability to do PCA, PCR, and PLS.

PLS decomposition of X and Y into scores and loadings [5, 6] is given as

$$X = TP^T + E \quad (40.15)$$

$$Y = UQ^T + F \quad (40.16)$$

T and U are the score matrices for X and Y and P and Q are the loading matrices. PLS does not calculate the scores independently, but rather it seeks optimal congruence of the u and t vectors by using the u vector as the starting point for calculating the t vectors. This self-consistent approach allows for a set of scores and loadings that represent the variation in the Y data set. Therefore, the scores and loadings are much better than PCR scores [4, 5] and loadings for quantitative prediction; hence PLS regression is the preferred technique for building calibration models and unknown prediction. The algorithm proceeds by mean centering the data and then finding the first loading spectrum and first component scores. The prediction of a PLS method is summarized in the regression vector or coefficient, B. The predictions are related to the X sample data by equation ($Y = BX$).

40.2.5 Data Preprocessing

It is important to note that while chemometric techniques may be applied to raw absorbance values, this is not usually the case. Before performing the various multivariate calibration and prediction techniques, some form of data transformation or preprocessing, while not mandatory, is nearly always warranted. There can be bias in the estimates if the predictors and the response variables are significantly different in magnitude, resulting in unequal leverage. Hence, each variable may be scaled by its mean and standard deviation to place the predictors and responses on more equal footing. Scaling the data at each wavelength as well as scaling of the response data used to derive the calibration is common. In fact, mean centering and scaling is so commonplace that some investigators take it for granted and do not report that they performed it prior to the analysis.

Derivative processing is common with mid-IR and NIR spectra. Taking the first derivative of the spectra corrects for baseline shift during the data acquisition. Taking the second derivative of the spectra corrects for baseline drift. There are several methods for calculation of the spectral derivatives, with the most common being the method of Savitzky–Golay [7]. Examples of the raw absorbance spectrum and the resulting first derivatives of the spectra are shown in Figures 40.5 and 40.6.

Depending on the sample type, particulate containing slurries and solutions may show light scattering effects with mid-IR or NIR, and Raman spectra may exhibit background scattering. Hence, another technique applied to the spectra depending on the amount of diffuse reflectance is the multiplicative scattering correction (MSC). The pretreatments can be applied to the spectra or to the response data (y-data).

40.3 BUILDING THE CALIBRATION MODEL

A more thorough discussion of methodologies in multivariate calibration model building is available in Ref. [8]. Building a chemometrics model would appear to be straightforward but can be deceptively tricky. A simple approach would be to prepare a set of samples containing the chemical species of interest dissolved in a relevant solvent at multiple, known concentrations, collect the spectra, and compute the calibration using PLS or PCR.

However, a calibration model built in this manner may prove deficient when applied to spectra collected from realistic chemical processing conditions, for example, monitoring the disappearance of a reactant, or the crystallization of an API from mother liquors. In both cases the spectral absorbances from the chemical species of interest may be mixed with absorbances from species in the system, such as reagents, solvents, impurities, and by-products. Hence, although it requires more effort, it is usually more useful to build a robust calibration model by collecting spectra and simultaneously analyzing the solution for the concentration

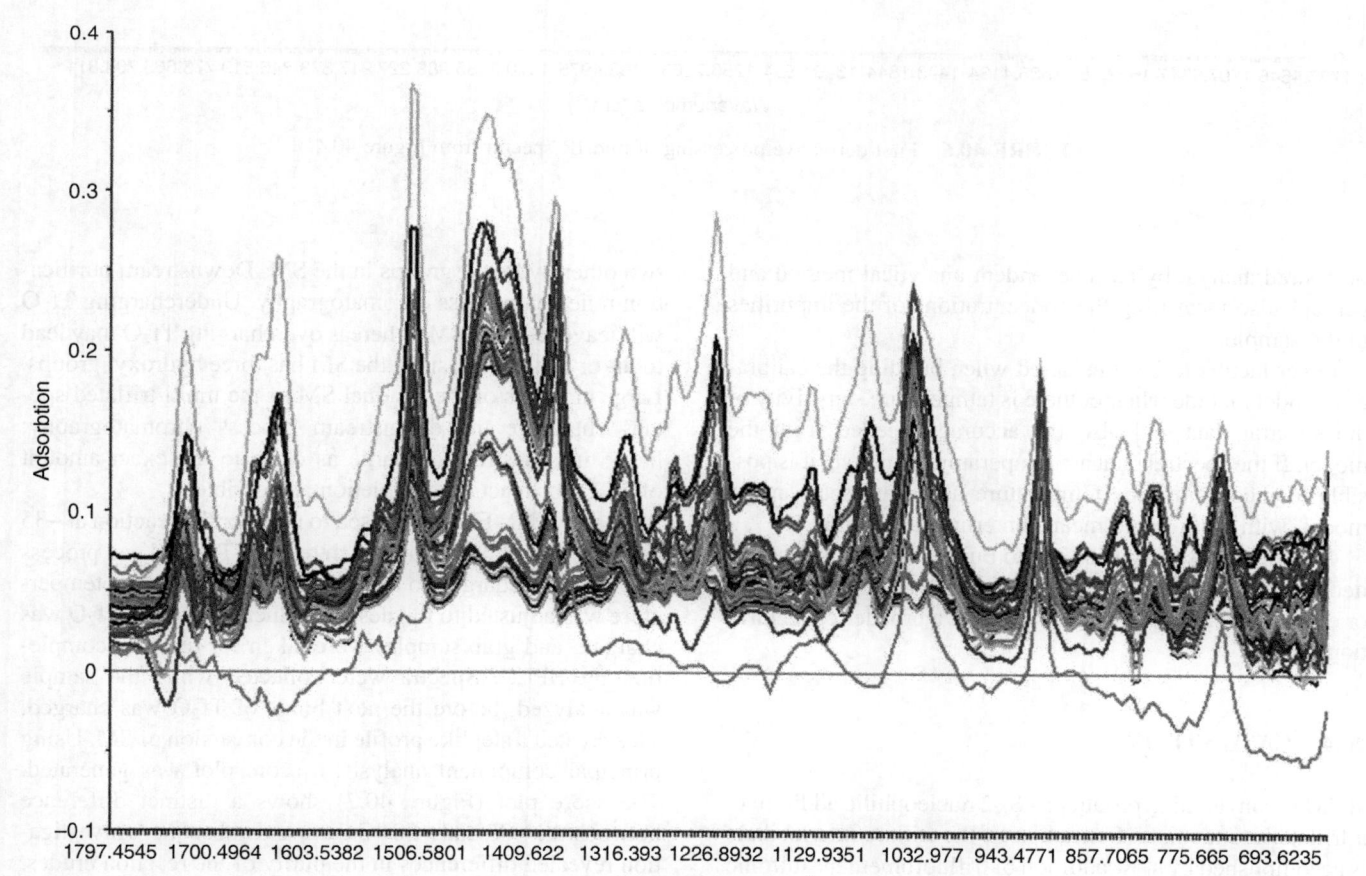

FIGURE 40.5 Series of mid-IR spectra absorbances from 600 to 1797 cm^{-1}. The raw data are not centered or scaled.

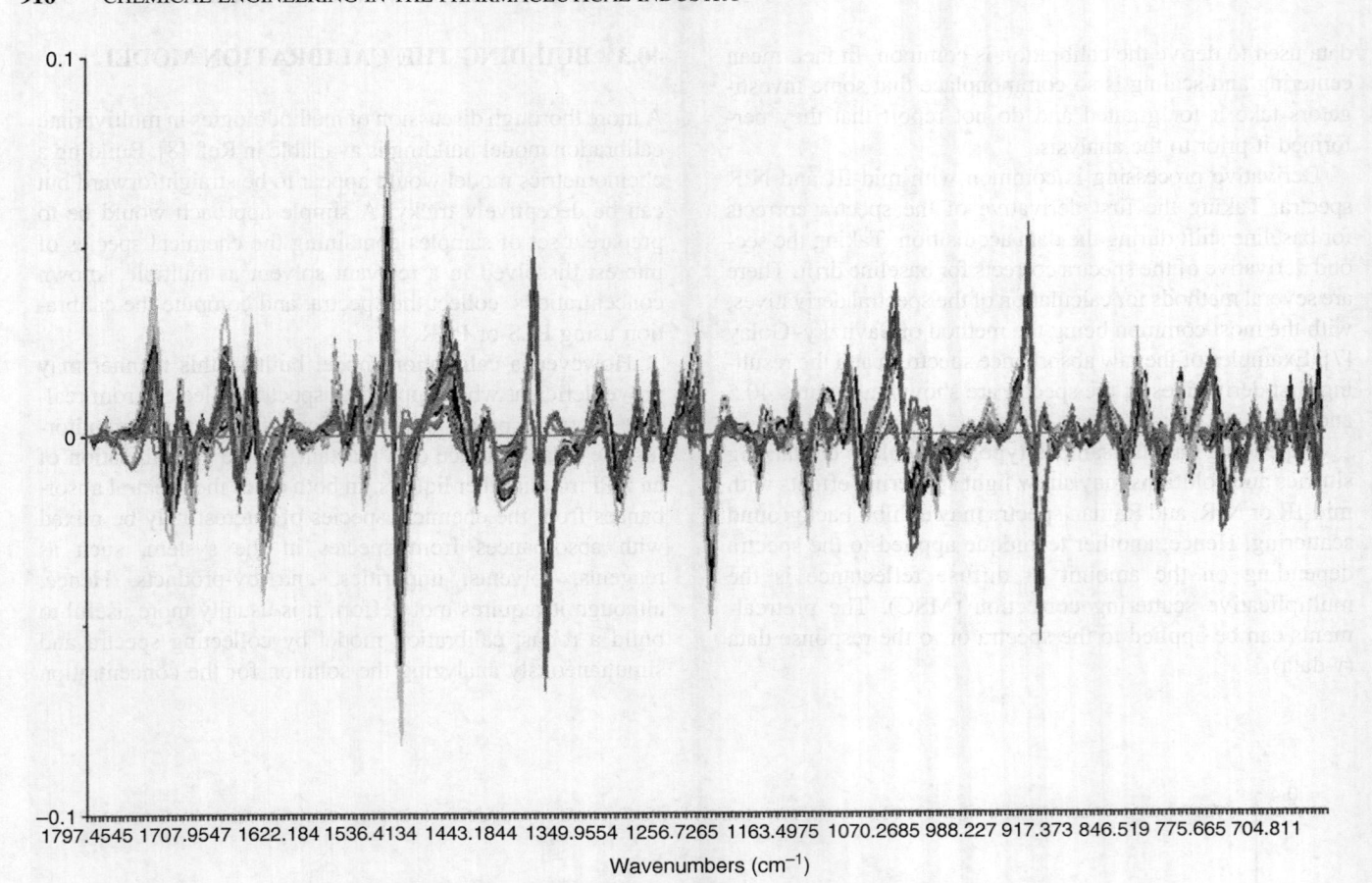

FIGURE 40.6 First derivative processing of mid-IR spectra from Figure 40.4.

of desired analyte by an independent analytical method and perhaps also measuring the concentrations of the impurities in the sample.

Other factors to be considered when building the calibration model include whether there is temperature sensitivity of the spectral data and also the accuracy needed from the model. If the spectral data are temperature sensitive, it is possible to incorporate the temperature as a parameter in the model, with appropriate mean centering and scaling.

In many cases it is necessary to build a model over a limited range of concentrations to achieve the required accuracy or even to use multiple models to cover a broader concentration range.

40.4 CASE STUDY

A SM is converted to product by $S_N 2$ nucleophilic addition to a triflate intermediate. Formation of the triflate intermediate is accomplished by slow addition of trifluoromethanesulfonic anhydride (Tf_2O), which reacts favorably at a hydroxyl site on the SM. The triflation is complicated by the presence of two other hydroxyl groups in the SM. Downstream purification relies on process chromatography. Undercharging Tf_2O will leave residual SM, whereas overcharging Tf_2O may lead to di- or tri-triflates, since the SM has three hydroxyl groups. Large amounts of the residual SM or the multi-triflated species challenge the downstream process chromatography; hence it is desired to charge as close to the exact amount of Tf_2O for reaction completion as possible.

In situ ATR–FTIR was used to monitor the reaction at −35 and −50 °C in a pilot plant batch reactor. The SM and process solvent were charged to the reactor, and the internal temperature was adjusted to the desired value. A bolus of Tf_2O was charged, and grab samples were taken for reaction completion by HPLC. Spectra were collected while the sample was analyzed, before the next bolus of Tf_2O was charged. This created a steplike profile in the conversion of SM. Using principal component analysis, a score plot was generated. The score plot (Figure 40.7) shows a distinct difference between the −35 and −50 °C data, and subsequent investigation revealed differences in the purity of the reaction crudes. Ultimately, the reaction temperature set point was fixed at −50 °C, and all future batches were conducted at that

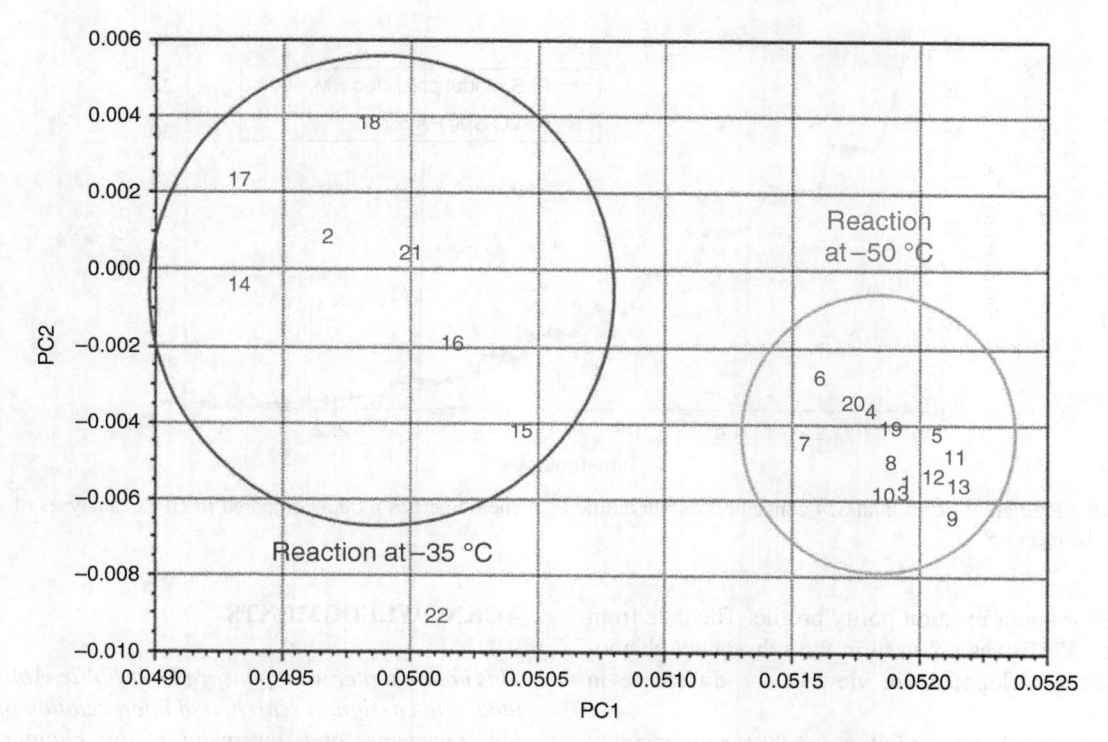

FIGURE 40.7 Score plot for triflation reaction at −35 and −50 °C.

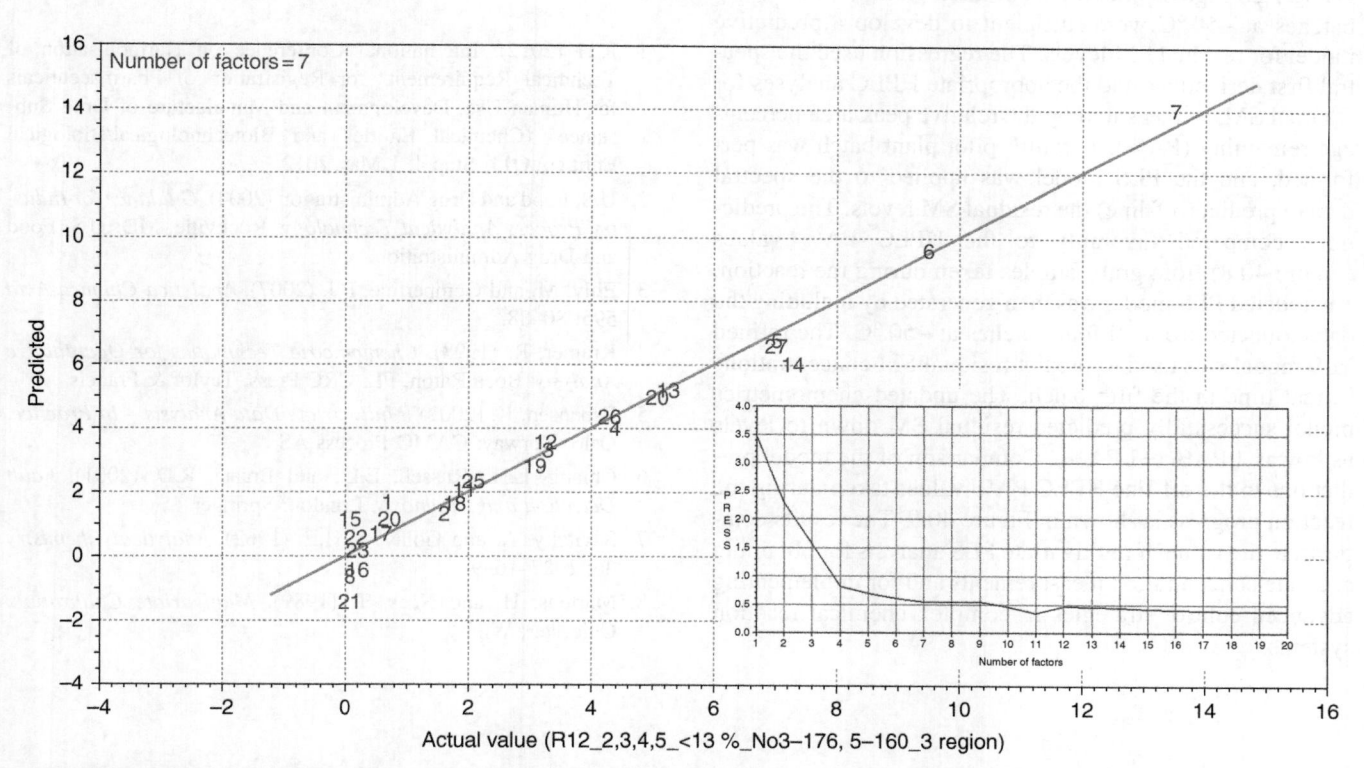

FIGURE 40.8 Training set comparison for PLS model built with data from three batches using seven factors to predict the fourth batch offline. Also shown in the inset is the predicted residual error sum of squares (PRESS) as a function of number of factors.

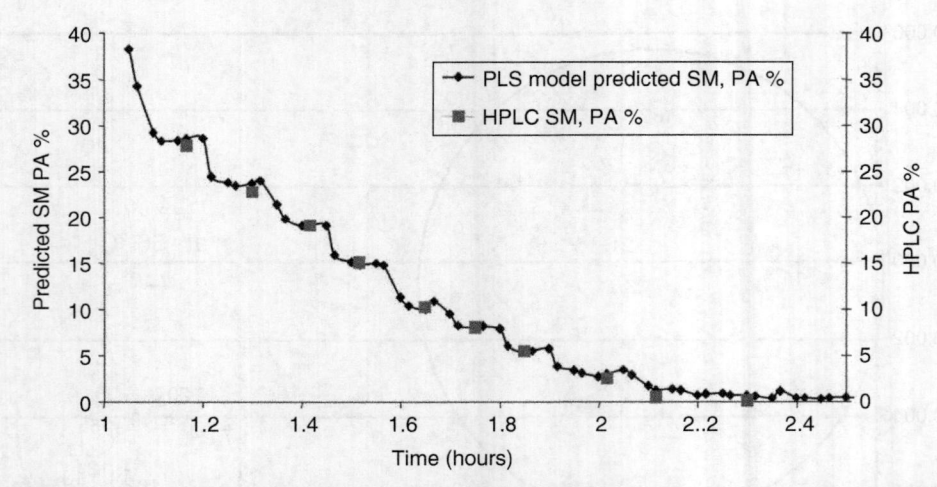

FIGURE 40.9 Prediction of residual SM concentration via online PLS chemometrics model compared to HPLC analyses of grab samples taken during the reaction.

temperature for better reaction purity profile. The data from the batch at −35 °C was not used in the subsequent chemometrics model development, in view of the difference in the score plot.

Attempts to develop a predictive model for the residual SM concentration from the spectra using both simple and multiple linear regressions were described earlier in this chapter, with neither approach yielding a robust prediction. Using PLS regression, data from three pilot plant triflation batches at −50 °C were sufficient to develop a predictive model for residual SM levels. The regression used the spectral first derivatives and the appropriate HPLC analyses for residual SM, as measured by the relative peak area percentage remaining (PA%). A fourth pilot plant batch was performed, and the PLS model was applied to the spectral data to predict (off-line) the residual SM levels. The predictions compared favorably to the HPLC PA% values (Figure 40.8) from grab samples taken during the reaction. An updated PLS model was then generated by including the data collected from all four batches at −50 °C. The refined PLS model was used to predict residual SM concentrations in real time in the fifth batch. The updated chemometrics model successfully predicted residual SM down to levels as low as 1 PA% (±1 PA%). Comparison of the model predictions to the off-line HPLC PA% values taken during the reaction progress is shown in Figure 40.9. The results demonstrate the value of multivariate PLS analysis for obtaining accurate concentration measurements and for implementing advanced control strategies in complex chemical reaction systems.

ACKNOWLEDGMENTS

This book chapter was sponsored by AbbVie. AbbVie contributed to the design, research, and interpretation of data, writing, reviewing, and approving of this chapter. James C. Marek is an employee of AbbVie.

REFERENCES

1. ICH (2012). International Conference on Harmonisation of Technical Requirements for Registration of Pharmaceuticals for Human Use, Development and Manufacture of Drug Substances (Chemical Entities and Biotechnological/Biological Entities) Q11, Step 4, 1 May 2012.
2. U.S. Food and Drug Administration (2004). *Guidance for Industry Process Analytical Technology*. Rockville, MD: U.S. Food and Drug Administration.
3. Ehly, M. and Gemperline, P.J. (2007). *Analytica Chimica Acta* 595: 80–88.
4. Kramer, R. (1998). *Chemometric Techniques for Quantitative Analysis*. Boca Raton, FL: CRC Press, Taylor & Francis.
5. Esbensen, K. (2002). *Multivariate Data Analysis – In Practice*. Oslo, Norway: CAMO Process AS.
6. Chiang, L.H., Russell, E.L., and Braatz, R.D. (2001). *Fault Detection and Diagnosis*. London: Springer.
7. Savitzky, A. and Golay, M.H.E. (1962). *Analytical Chemistry* 36: 1627–1639.
8. Martens, H. and Naes, T. (1989). *Multivariate Calibration*. Chicester: Wiley.

41

PROBABILISTIC MODELS FOR FORECASTING PROCESS ROBUSTNESS

JOSE E. TABORA, JACOB ALBRECHT, AND BRENDAN MACK
Product Development, Bristol-Myers Squibb, New Brunswick, NJ, USA

41.1 INTRODUCTION

The key goal of pharmaceutical process development is to design processes that enable the safe, reliable, and cost-effective manufacture of the active pharmaceutical ingredient (drug substance (DS)). As these materials are intended for therapeutic treatment, both the product and the processes are highly regulated. In particular, the quality of the product must meet specific targets, which can be measured and quantified, such as purity, particle size distribution, crystallinity, etc. Control of these attributes will ensure adequate product performance while minimizing the risk to the patient. The quality-by-design (QbD) paradigm championed by health authorities is intended to ensure that the quality of the product is designed into the process. In a broad sense, this means that any unit operation in the manufacturing train achieves its intended purpose. Generally this expectation is defined during process development by ensuring that the process operates reliably at different values of the process inputs (materials or process parameters). Once development is completed and the process is transitioned to a manufacturing environment, it is anticipated that the process will be operated under tighter ranges of the process parameters than the one explored during development. Ultimately the goal is to provide consistent and reliable DS following implementation of the manufacturing process in a commercialization facility intended for long-term production.

Section 41.2 describes some of the statistical tools that are used to quantify process robustness. Section 41.3 provides a background on the regulatory expectations driving the process design and characterization of process robustness, and Sections 41.4 and 41.5 provide a probability-based mathematical formalism to model and estimate robustness. Sections 41.6 and 41.7 describe approaches to define and calculate these robustness metrics. Finally, a case study is presented to incorporate the concepts described in this chapter.

41.2 MEASURING PROCESS ROBUSTNESS

41.2.1 Statistical Process Control

Variability is intrinsic to any manufacturing process due to both unavoidable, random fluctuations in process parameters and inevitable problems with equipment performance, analytical method drift, and input material attribute changes. Under routine conditions, process outputs will fluctuate in a statistically normal fashion due to common causes of process variation, which are unmeasurable and random fluctuations in process conditions [1]. Under other circumstances, process variation is due to assignable causes, which are differences in process conditions that can be identified and potentially corrected. Statistical process control (SPC) is the practice of using standardized, and statistically relevant, methods to compare current process information with historical information to determine if fluctuations in the process are coming from common or assignable causes. SPC involves tracking the fluctuating process responses and using the measured fluctuations to make inferences about how well the process parameters are controlling the process outputs.

Chemical Engineering in the Pharmaceutical Industry: Active Pharmaceutical Ingredients, Second Edition.
Edited by David J. am Ende and Mary T. am Ende.
© 2019 John Wiley & Sons, Inc. Published 2019 by John Wiley & Sons, Inc.

For a typical SPC project, individual measurements are compared against statistically determined expected process variation to determine whether or not the process is in control. If the measurement falls outside of those limits, an investigation is initiated to determine whether the variability is due to a common cause or assignable cause. If an assignable cause is identified, future variability can be reduced by placing stricter controls around that process parameter or input attribute.

41.2.2 Control Charts: Uni/Multivariate

The most common statistical tool to monitor process performance over many batches is the control chart. A control chart plots the performance of multiple batches in a series fashion so that any individual batch can be compared against the expected range of behavior of previous batches. A control chart has a defined upper control limit (UCL) and lower control limit (LCL), which are statistically defined limits that help identify batches that may have variability due to an assignable cause [2]. An example control chart tracking the mean response for an impurity can be found in Figure 41.1. The mean (μ) is clearly delimited by the solid horizontal line, with the red lines showing the control limits. The out of control batch is easily identified at a glance by its position over the UCL.

Control charts are classified by the type of data they contain. In addition to the mean control chart in Figure 41.1, some different types of control charts are summarized in Table 41.1 [3]. The type of control chart that you select depends on the application and the parameter of interest from a quality perspective. There are many resources available for choosing an appropriate control chart, but the overarching goal of any chart is to measure a quality-relevant parameter and track it across multiple manufacturing batches to determine when a process is performing unacceptably and should be fixed.

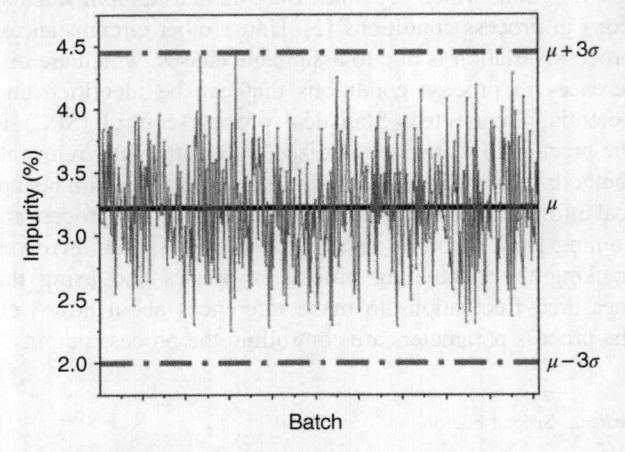

FIGURE 41.1 Example of mean control chart.

TABLE 41.1 Control Chart Types

Chart Type	Typical Use
Mean ($\bar{X}$)	Most common type of chart that tracks the mean variable response for a given batch, best used when the central tendency of the process is the desired response
Range (R)	Monitors the variability of the process over time and is best used when the dispersion of samples is the process metric of greatest interest
Proportion (P)	Measures the proportion of failures within a given batch, best used when the response of interest is a binary failure/pass and the ratio of failures is the metric of interest
Conformity (C)	Provides a count of nonconforming samples, used in similar situations to P chart, but when the total sample size is unclear
Hotelling's T^2	Multivariate extension of $\bar{X}$ chart, used when multiple correlated variables are tracked and individual control charts may not provide an accurate view of process control

Control limits are typically calculated from the mean and the standard deviation (σ) of all samples. Typically, limits are bounded by 3σ, as shown in Eqs. (41.1) and (41.2):

$$UCL = \mu + 3\sigma \qquad (41.1)$$

$$LCL = \mu - 3\sigma \qquad (41.2)$$

This technique assumes that the data are normally distributed and control limits encompassing 99.7% of the expected random variability are sufficient. In general, this is an acceptable estimate that will only identify true outliers but could break down if the distribution of data is highly non-normal.

When process responses outside of the limits, this indicates a process that is "out of control," and an investigation to ascertain a cause is justified. If responses fall within the control limits, it is not necessarily true that a process is in control since other trends in the data might indicate a poorly performing process. A process that has started to routinely show responses higher than the mean may warrant investigation, since it is highly unlikely to observe a long string of such responses if the data were normally distributed. A gradually increasing or decreasing response may indicate that the process is slowly shifting to a new mean, a phenomenon that should be investigated, understood, and possibly controlled.

When a single variable is not an accurate predictor of pharmaceutical process performance, it is appropriate to use a control chart of a multivariate statistic to monitor the process [4, 5]. Multivariate control charts plot a scalar that is calculated using information from several variables. Common multivariate charts plot the Hotelling's T^2 statistic or the

Mahalanobis distance [3]. These control charts are particularly useful when following multiple, correlated process parameters and responses that are not fully informative on their own but present a picture of process performance when combined in a multivariate statistic. When a multivariate statistic is found to be out of control, the scientist typically must reanalyze univariate data to assign a cause due to the highly correlated nature of the multivariate statistics.

41.2.3 Process Capability

The capability of a process to pass process specifications (predetermined ranges of acceptable process quality or performance) is independent of the control chart statistics. For instance, a process may be statistically "in-control" and routinely failing specifications. Alternately, a process may be "out of control" but has such wide specification range that no failures are observed at all. Process capability is a measurement of how frequently a process, operating with expected common cause variability, is able to routinely meet a quality or performance specification.

Measuring process capability involves comparing the control range with the range of acceptable values defined by the specifications. Similar to control limits, specifications can be defined by an upper specification limit (USL) and lower specification limit (LSL). Five commonly computed process capability metrics are shown in Table 41.2. The choice of process capability measurement depends on the circumstance under which you use it.

In general, a process capability less than 1 describes an unacceptable system, where normal common cause variability will routinely result in failed specifications. The process is considered not capable of meeting that specification. A process with measured capability greater than 1 is considered capable of routinely meeting specifications, but in practice a process capability greater than 1.33 is often desirable so that small changes in the state of the process will not result in the process becoming incapable of achieving specifications.

41.2.4 Process Robustness During Development

During process development, it is desirable to understand process robustness even though the number of batches will be small and most batches will be run on a scale inappropriate to manufacturing. Metrics of robustness during development tend to be experimentally defined, with projection of process variability being based on expected performance given controlled process ranges. The key objective during development is to answer the question: "what range of known process parameters is acceptable to achieve product quality?"

Due to insufficient historical data of a process, models for the behavior of the new process are developed on a small laboratory scale, and inferences from those models are generalized to larger scales. Verification of those models comes only during relatively rare scale-up campaigns to deliver DS for clinical supplies and/or drug product development. Variability is assumed by understanding plant capabilities and getting feedback from commercial manufacturing sites.

41.3 REGULATORY REQUIREMENTS FOR PROCESS ROBUSTNESS

41.3.1 Guidelines for Process Development and Validation

The International Conference on Harmonization (ICH) development and validation guidelines [ICH Q7–Q11] describe various means to help companies improve the transition from development to commercial and assure that the drug product will be manufactured fit for its intended use. In the ICH quality framework, quality, safety, and efficacy are designed into the product, and each step of a manufacturing process is controlled to assure that the finished product meets all specified quality attributes [6].

A common approach to develop a well-understood process, designed to integrate quality, is to identify critical quality attributes (CQAs) associated with desired product quality and create predictive mathematical models that quantify the relationship between the various process parameters and the product CQAs. The models are subsequently used to define a design space: a range of process parameters that ensures the quality attributes will meet desired quantitative criteria (typically defined as ranges of acceptability, limit values, or specifications). This goal of QbD requires more fundamental understanding and modeling compared with the more

TABLE 41.2 Process Capability Metrics

	Formula	Conditions for Use
C_p	$\dfrac{USL-LSL}{6\sigma}$	If the process mean is roughly centered between lower and upper specifications
C_{pk}	$\min\left(\dfrac{USL-\mu}{3\sigma}, \dfrac{\mu-LSL}{3\sigma}\right)$	If lower and upper specifications are set, but only the distance from the closest limit is of relevance
C_{pm}	$\dfrac{USL-LSL}{6\sqrt{\sigma^2+(\mu-T)^2}}$	If a target value (T) is the desired mean operating state, C_{pm} adjusts the mean value to reflect the deviation from target
C_{pu}	$\dfrac{USL-\mu}{3\sigma}$	If only a USL is required
C_{pl}	$\dfrac{\mu-LSL}{3\sigma}$	If only a LSL is required

conventional quality-by-control approach that is taken by control charts.

In parallel to this quantification, precise analytical methods to measure the CQA and other important attributes are also developed. Careful selection of process parameters during development leads to the largest possible sets of tolerable process conditions to produce high quality materials, fit for the intended purposes. There is confidence (based on experience or scale testing) that the final process based on this design space will perform consistently in the commercial manufacturing facility. Furthermore, specifications put in place for an appropriate CQA ensure that the product will have the predicted quality. All this process performance is managed in manufacturing by the Monitoring and Control Plan defined for the process.

Processes designed in the rigorous way described by the ICH guidelines usually transfer well into a manufacturing environment and deliver the expected target behavior. However, a precise estimation of the anticipated variability under manufacturing conditions is not directly accessible from development work for several reasons. Primarily, the limited development experience does not provide sufficient samples to understand the true process variability. In addition, the differences in process control are not well understood between development and manufacturing equipment trains, and there is likely more variability in manufacturing process control. Finally, the manufacturing analytical measurements often have more variability than in the development environment. This process variability is not predicted in the design space because the design space analysis and resulting unit operation models are designed to predict average process behavior. Not understanding the expected process variability, rather than just the average process behavior, may lead a development scientist to conclude that a process is robust when, in fact, there is a risk of process failure within the design space.

In order to best incorporate variability into the assessment of process robustness, we turn to a branch of statistics known as Bayesian statistical modeling. In contrast to frequentist statistics that are more commonly used in process control literature, Bayesian approaches utilize prior knowledge and limited data to estimate the plausible behavior of a process. To address the ICH guidance on design space, a Bayesian approach to understanding uncertainty has been used [7–9] not only to describe limits of failures for parameter ranges but also to more accurately map failure *probability* given a model and data. An advantage of the Bayesian approach over frequentist approaches is that Bayesian analysis more intuitively incorporates past observations when estimating the likelihood of unknown events and is therefore more aligned with the ICH guidance. By applying this flexible approach to determining manufacturing process capability, we aim to develop more informed and accurate models for variability while avoiding many assumptions used in standard control charts.

41.4 MODELING PROCESS ROBUSTNESS

41.4.1 Robustness of Pharmaceutical Processes

The manufacturing process for the DS may be abstracted as a mathematical relationship between the process inputs (parameters and quality descriptors of the input materials) and the process outputs, i.e. the yield and qualitative properties of the DS. This abstraction may be represented by Figure 41.2 as a mathematical function that converts inputs such as process or equipment settings into quality attribute outputs. The manufacturing process could consist of a series of unit operations incorporating chemical transformations, extractions, and isolations in the DS manufacturing process.

Typically, it is impossible to operate a process in separate instances (batch) or over time (continuous processing) under identical and/or invariant settings. Under any realistic control scheme, the input parameters will fluctuate around a target set point.

In this context, the robustness of a process may be defined as the lack of sensitivity of the process outputs, such as quality attributes, to fluctuations in the process inputs. Figure 41.3 shows a schematic of a process operated at three different target locations of the single process parameter with the same parameter variability (locations A, B, and C, in the x-axis). The resulting distribution has a lower measure of variability in the location B, since in this region the function is less sensitive to variation in the input X. In this context the process is said to be more robust for condition B than for A or C. The input and output distributions use arbitrary units for probability density but show how identical variabilities in the inputs map to different output distributions, depending on the nature of the underlying process model. A sharp output distribution indicates that the process is robust at that input condition. In this example, operating at condition B appears to be much better than at A or C.

In the example above, the source of variability (and corresponding process robustness) was restricted to variability in the input parameters. However, a given measure of the quality of the pharmaceutical product (potency, purity, etc.) will also be subjected to the intrinsic process equipment variability as well as measurement or analytical variability.

Process parameters
$x_1, x_2, x_3, \ldots, x_N$
$x_{Set\ point}$
X

Unit operation process experiment

Quality attributes
$y_1, y_2, y_3, \ldots, y_N$
y_{Actual}
Y

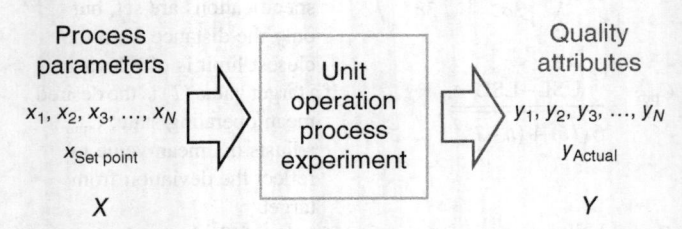

FIGURE 41.2 Abstract representation of a pharmaceutical process (lab: experiment or manufacturing process) where y denotes all the quality attributes of the drug substance and x the input parameters.

These additional sources need to be incorporated to determine a comprehensive analysis of the process variability for a given outcome and the corresponding relevant process robustness. Figure 41.4 shows a process with the same input parameter sensitivity but with an additional measure of process variability using random error with a relative standard deviation of 5%. The output distribution is now affected by both the input distribution on the model and the random process variability.

In comparing Figures 41.3 and 41.4, it is evident that the inputs that lead to optimal robustness are different than if we only considered the mean response from the model. The extra noise incorporated into the robustness assessment means that input condition A now has a narrower output distribution than condition B or C. The general approach to characterize process robustness is to measure the process outcome as a function of the process inputs and build a model that estimates the mean of the observed outcome, i.e. a causal model. Subsequently the probability of a given observation deviating from the model is estimated by constructing a model for the process that takes this uncertainty into account.

The established view of process variability (the random distribution around the mean value) is that it is generated from random fluctuations of variables that are not necessarily accounted for in the process description. To the extent that a source of variability can be understood, quantified, and included in the causal model as an input variable, the robustness of the process can be improved by adding control over that variability source (a causal input).

The next sections will discuss general procedures to determine, evaluate, and control process variability in the context of process robustness optimization.

41.5 PROCESS UNCERTAINTY

41.5.1 A Manufacturing Process as a Random Number Generator

The general practice in chemical engineering pharmaceutical development is to represent relevant unit operations involved in the manufacturing as mathematical models that provide a quantitative relationship between the control variables and the quality attributes, which may be represented by Eq. (41.3):

$$\mathbf{y} = f(\mathbf{x}, \boldsymbol{\theta}) \qquad (41.3)$$

where, in this case,

$f(\ldots)$ represents the structure of the model.

$\mathbf{x}$ and $\boldsymbol{\theta}$ are the process inputs of the control variables and model parameters.

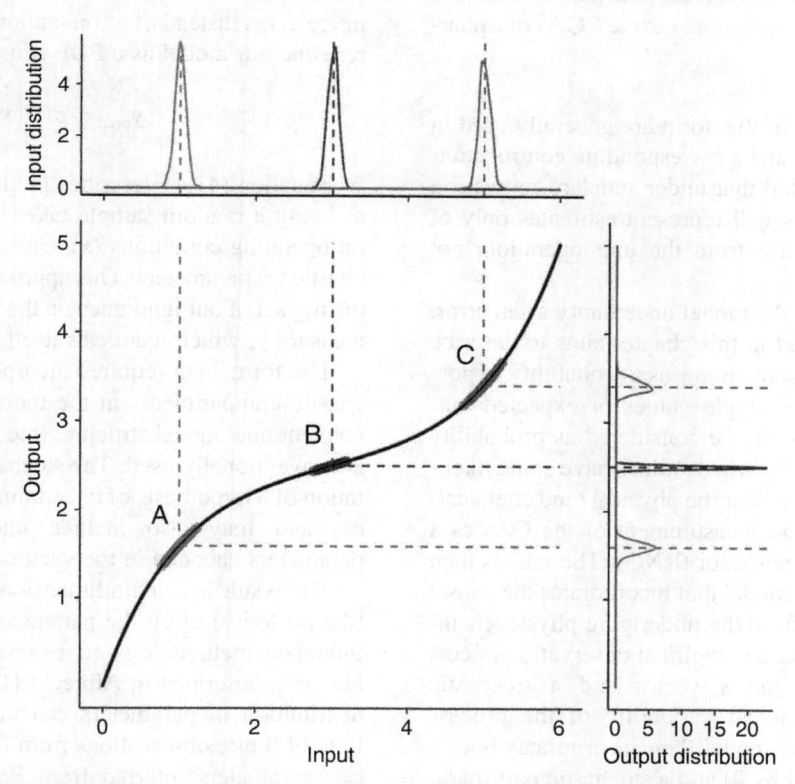

FIGURE 41.3 Sensitivity of the output (y) to a given variability in the input (x) at different locations of the process input parameters.

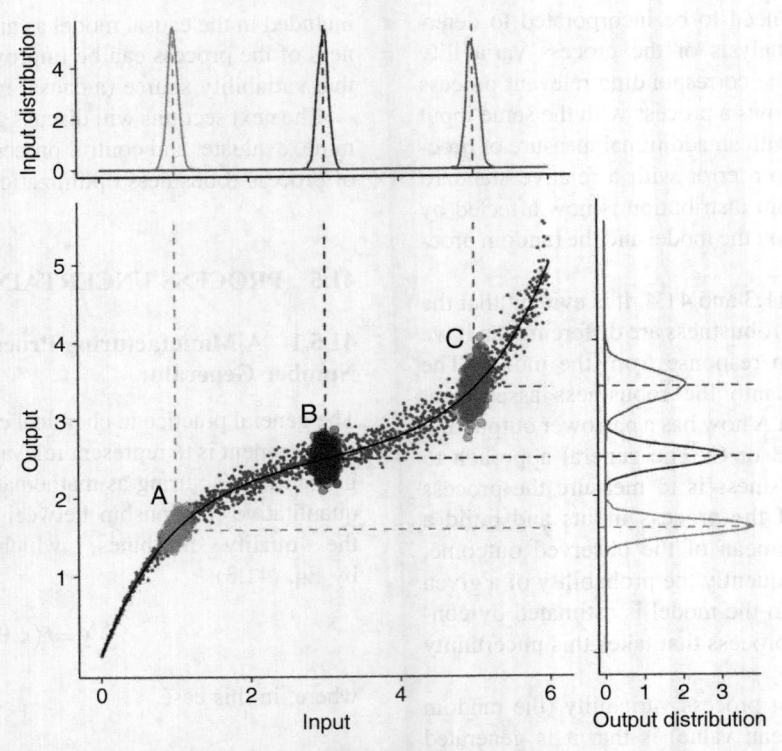

FIGURE 41.4 Output variability in a process with random variation in addition to input variability.

y represents the quality-relevant attributes that are impacted by the unit operation, i.e. the CQAs of a pharmaceutical process.

Although relationships of this form are generally used in developing a design space and a corresponding control strategy, it should be recognized that under standard regression techniques, the predictions will represent estimates only of the expected mean response from the unit operation, not the risk of deviations.

Rather than expressing the model uncertainty as an error, the formalism implemented in this chapter aims to describe the estimate of the process outcome **y** as a probability of possible outcomes. Instead of single values or expected outcomes, inputs and outputs can be considered as probability distribution functions (PDFs) where values have some likelihood of occurrence. We consider the physical (and chemical) process of manufacture and measurement of the QAs as a physical random number generator (RNG). The goal is then to generate a probabilistic model that incorporates the causal relationships determined from the underlying physicochemical nature of the process or an empirical observation of correlation between **y** and the **x** vector and a stochastic contribution from the inherent variability of the process and the measurement. The model then incorporates both a causal part (f parametrized by θ) and a stochastic part (parametrized by τ). In practice, we wish to define a RNG whose distribution closely simulates the observable quantity of interest, i.e. instead of a conventional use of a model, we will redefine our model as a PDF (Eq. 41.4):

$$\mathbf{y}_{pred} \sim p(\mathbf{y} \mid \mathbf{x}, \theta, \tau) \qquad (41.4)$$

Equation (41.4) describes predicted quality attributes (**y**) as being a random sample taken from a PDF that depends on operating conditions (**x**), causal (θ) parameters, and stochastic (τ) parameters. This approach is an attempt to capture the impact of our ignorance on the exact ability to predict and measure **y**, which manifests itself as uncertainty.

The formalism requires incorporation of causal and stochastic compartments in the model. The causal part is the conventional model structure (mechanistic or empirical) that is conventionally used. The stochastic part requires incorporation of a hypothesis of uncertainty propagation in the process and may also include uncertainty in the control parameters encoded in the vector **x**.

The result is a multidimensional probability distribution (the posterior) of all the parameters in the model. Bayesian inference methodology to estimate the model parameters has been described in Albrecht [10] and Gelman [11]. This distribution of parameters can then be used to obtain the PDF of future observations from Eq. (41.3), using the posterior parameters inferred from Eq. (41.4). Specifically the methodology results in the definition of an RNG whose distribution best describes the data that was sampled from the physical RNG that we are attempting to understand.

Therefore, the process of estimating the model (causal and stochastic) parameters provides a framework to estimate the PDF of future observations (y_{pred}). Ultimately, we estimate the probability of **y** given **x**, conditional on prior observations of the relationship of **y** to **x** (the data) under a postulated mathematical relationship (the model) that is a function of model parameters θ and observational noise τ. Each of these three groups of values has their own types of uncertainty.

Later in this chapter, we describe a methodology to build RNGs for pharmaceutical development applications by incorporating Bayesian framework. A model is developed to predict the quality attribute of interest. The process variability is simulated by incorporating the variability of the input process parameters, model uncertainty, and analytical (measurement) variability. These models were incorporated and simulated using Stan [12] from the RStan package available in R [13]. This approach allows great flexibility in proposing and testing new model structures, both to best develop models not only for the process and to describe the sources and distributions of process, model, and measurement errors. Hierarchical models to describe the common and individual errors could therefore be used to describe inter- and intra-lot measurement variability. Once the RNG is developed based on available experimental data, it is subsequently used to sample the distribution of potential outcomes in the commercial manufacturing environment.

41.5.2 Variability from Process Parameters and Inputs/Control Parameter Variability

During pharmaceutical manufacturing, there are many potential causes of variability due to the relative complexity and uniqueness of each processing step and the difficulties in perfect control during batch processing. Most commonly, variability arises from the failure to perfectly achieve target parameter states. Much of this variability is expected and should be accommodated by robust manufacturing processes. Material charges, such as reagents, solvents, and catalysts, may not be exactly as expected due to errors in weighing or material hang up during the charging procedure. Temperature may not be controlled exactly as specified due to drifts in thermocouple calibrations, inability for reactor jackets to cool against exothermal events, controller oscillation, or many other reasons. These variations are usually mild and do not cause process failure, but the result is a process that displays significant common cause variability during normal manufacturing conditions. Drifts in manufacturing equipment may also cause process variability, due to slow but steady breakdown in the operating condition of any particular equipment train. This equipment is periodically adjusted through calibration or scheduled maintenance, causing variability in the processing conditions but not outright failure.

Equipment may also be replaced, or the process may be shifted to new equipment, causing many slight differences in processing conditions that may not be easy to identify when taken in total with all common cause variability.

Input material quality fluctuations can cause process variability if the quality attributes of the product depend on the quality attributes of the input material. Sometimes, impurities from inputs are directly present in output product, causing a chain of process variability that cascades down the synthetic sequence. Alternately, attributes of the input material like water content, residual solvent, residual acid/base, or inorganic salt content can affect the process responses of an output in a way that increases process variability. For example, an organic solvent like acetone may have variable water content depending on the source or batch used on a particular day. This water variability (potentially measured in ppm) may cause a water-sensitive reaction to behave slightly differently. This difference in behavior may lead to a slight increase in impurity level, so the variability in measured impurity level becomes dependent on the source and batch of acetone used in the reaction, which may change over time.

41.5.3 Model Uncertainty/Epistemic Uncertainty

In quantifying uncertainty for process models, the term epistemic uncertainty describes uncertainty that is due to insufficient knowledge of a system. Process input variability was described in the previous section and described the contribution of variability in the value of **x** from Eq. (41.4). Epistemic uncertainty stems from a lack of knowledge of some "ground truth," such as a regressed model parameter θ in Eq. (41.4). By increasing the amount of data and analysis, the uncertainty of these parameters will shrink until only the true value can be considered possible. Examples of epistemic uncertainty include confidence intervals for regression, as well as the relative confidence in the ability of different models to describe a set of data. Often, the assumption that the model structure is appropriate must be examined. This can be performed by assessing the performance of a family of potential models though approaches such as stepwise regression [14] or by performing experiments that best discriminate among models [15].

41.5.4 Observational Variability/Aleatory Variability

Aleatory variability is defined as an uncontrollable random effect, such as a game of chance that produces a random outcome where the degree of risk can be known in advance. The overall probability distribution of aleatory variability can be knowable, but any single observation of a system that is subject to aleatory variability cannot be predicted exactly. Even with a large number of identically prepared samples, the observed values will vary. Aleatory variability occurs at each step in the sampling collection, analysis, and reporting. For example, even on a calibrated analytical instrument, a reference standard will have some observed variability in measurement due to the inherent precision of the machine. The resulting signal over many repeated measurements will average to the true reference value, despite the noise in the individual measurements. Analytical equipment used in the pharmaceutical industry typically has a high signal-to-noise ratio to reduce this source of aleatory variability, but there are other sources. A common problem arises in sampling from a heterogeneous mixture. Such samples will have variability in their analysis, but as more samples are taken, the average of the samples will become closer to the true value. While aleatory variability can only be reduced with changes to sample collection and analysis, its contribution to the overall uncertainty in y_{pred} can be estimated from data. For simple linear models, approaches such as analysis of variance (ANOVA) [14] are commonly used to estimate the contribution of τ in Eq. (41.4). For nonlinear models that are relevant to the fundamental mechanics of a system, Bayesian approaches are recommended to most accurately separate epistemic uncertainty from parametric and aleatory variability.

41.6 BRIEF OVERVIEW OF PROBABILISTIC GRAPHICAL MODELS

41.6.1 Background and Core Concepts

Probabilistic graphical models (PGMs) are a powerful framework to define probabilistic analysis problems [16]. This formalism allows for definition of a large variety of problem statements while also enabling rigorous solution methods for that problem. They are a simple visual tool to define problem statements according to some basic rules, but a problem that can be defined by a PGM can be solved using a suite of specialized mathematical algorithms such as those outlined in the next section. PGMs represent uncertain probabilities as oval nodes connected by arrows that represent causal influence. Many problems can be formulated by a subset of as PGMs known as directed acyclic graphs (DAGs). In a series of nodes connected by directional arrows, the acyclic constraint of DAGs forbids a loop around a group of nodes. Arrows between nodes define the direction of causality of how one node influences the value of another. While the nature of that influence is not explicitly labeled in the graph, it must be accounted for when formulating and computing the model. Of the many solution methods that exist, Markov chain Monte Carlo (MCMC) (Section 41.6.3) will be presented as the preferred solution algorithm for the types of problems experienced during process development.

In this chapter, uncertainty is considered in the probabilistic sense of a value existing within a probability distribution.

In using models for process robustness, there are two phases: first quantifying the uncertainty of a process given any available data or fundamental knowledge and then applying the degree of uncertainty to make future projections of robustness. The previous Sections 41.5.3 and 41.5.4 described the sources and nature of epistemic and aleatory variability. Both types of uncertainties will be considered in the modeling of process robustness using PGMs.

41.6.2 Hierarchical Models

The term hierarchical models, while not formally defined, typically refers to model with multiple layers of complexity relating inputs to outputs. For process models, the hierarchy starts with uncertain process inputs followed by uncertain model predictions, followed by observational errors. Variability in CQA measurements can be likewise described as contributions stemming from analyst, equipment, and intra-lot and inter-lot variability components. Figure 41.5 shows an example of measuring the composition of a heterogeneous batch, factoring in variability in collecting samples and measurement error.

In Figure 41.5, the observed composition for any sample depends on the true sample composition that in turn depends on the true batch composition. Noise associated with batch sampling variability and analysis variability together influence any measurements of composition. While the batch and sample compositions have a single ground truth value, the aleatory variability introduced by the process of gathering a sample and the sample analysis will cause the observed composition to vary. With only composition observation from multiple analyses, it is possible to determine the true value of the sample and batch composition (within a range of epistemic uncertainty). The variability in the observations can be separated into the contributions from the aleatory variability of the sampling and analysis process to identify the remaining epistemic uncertainty of the batch composition. This degree of confidence in our knowledge of the underlying truth is needed to provide a proper accounting of robustness and risks.

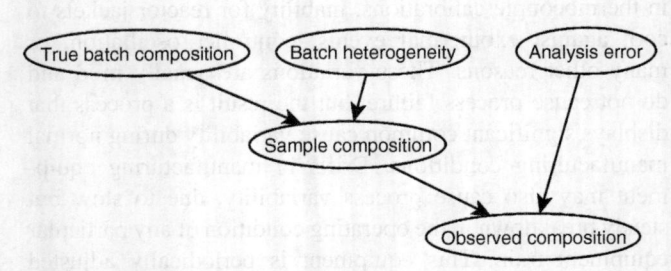

FIGURE 41.5 Hierarchical graphical model example for analyzing a heterogeneous batch for content uniformity.

41.6.3 Capturing Uncertainty with Markov Chain Monte Carlo

With MCMC, any computational model with uncertain parameters can have the probability distributions of those parameters be fully estimated. MCMC can be conceptualized as a series of random jumps or hops (called a Markov chain) between possible parameter values. Each hop represents a set of model parameter values that are randomly selected from a distribution. Using these parameters, the model is used to make predictions of the observations. These predictions are compared with the observed values and the goodness of fit is assessed. As with regular Monte Carlo, samples are drawn from a random distribution, but with MCMC, the performance of each set of samples is compared with the performance from the previous iteration. Combinations of model parameters that better describe the observations are identified, and the MCMC algorithm is more likely to use similar values for future hops. As the number of hops between states becomes very large, the resulting distribution of parameter values, called the posterior distribution, begins to approximate the true parameter value distribution. MCMC approaches, while computationally expensive, are essential tools to estimate parameter uncertainty ranges for models with non-normally distributed errors or with nonlinear models [10].

The output of an MCMC simulation is a list of typically thousands of posterior samples of the uncertain parameters. Each sample contains a self-consistent set of parameters that is credible given the data and underlying model. Combinations of parameters that do a better job of predicting the observations will appear more frequently in this list, and less likely combinations of parameters will appear less often. By plotting histograms of the individual parameters, summary statistics such as mean and standard deviation of model parameters can be calculated. By plotting 2D histograms of two parameters, joint probabilities can be visualized. This approach can be used to quickly understand correlations between modeled parameters. The samples from the parameter posterior distributions represent all credible combinations of regressed parameter values; any estimation of model robustness must use these samples to properly account for uncertainty in predicted outcomes.

For a simple example of MCMC, consider a simple nonlinear equation in Eq. (41.5), where a result, y, is a function of input x, determined by parameters a and b:

$$y = a + x^b \tag{41.5}$$

Data is collected for y with $a = 1$ and $b = 0.5$, with normally distributed error with a standard deviation of 0.1 added, shown in Table 41.3.

In the observed data, underlying true model and MCMC results are plotted in Figure 41.6.

TABLE 41.3 Example Data for Parameter Regression

x	y True	y Observed
1	2.00	1.90
2	2.41	2.49
3	2.73	2.67
4	3.00	3.20
5	3.24	3.34
6	3.45	3.55
7	3.65	3.65
8	3.83	3.90
9	4.00	4.06
10	4.16	4.25

In Figure 41.6, the MCMC results are plotted as family of faint prediction trends, one trend from each of the posterior samples of a and b. In aggregate, the prediction trends describe the data well and are in agreement with the ground truth model. The distribution of these parameters and their correlation are plotted in Figure 41.7. From MCMC, the most likely values for a and b (1.01 and 0.51, respectively) are very close to their true values. This pairwise plot shows the distribution of a and b along the diagonal, the correlation coefficient (upper right), and joint probability contours (lower left) of a vs. b.

The negative correlation between a and b is important to carry forward when using the model to make robustness predictions. Focusing only on the individual distributions of the regressed parameters makes the mistake of implying that they are independent, contrary to what the MCMC regression has shown. If one parameter is at the upper end of its distribution, the other must be near the low end of its distribution to make credible predictions. While conventional nonlinear regression approaches using a frequentist approach to statistics [17] can calculate parameter covariance, there is the underlying assumption of linear behavior near the optimal parameter values. Only Bayesian approaches such as MCMC can determine the full joint probability distributions needed for an accurate assessment of robustness and the likelihood of rare events.

41.6.4 Utilizing Uncertainty for Process Robustness

Once the epistemic uncertainty of the model parameters is characterized by MCMC, the model can be applied to any scenario where the process has some degree of aleatory variability. This involves utilizing each sample of the MCMC posterior with a simple Monte Carlo simulation. The resulting values combine the process variability with the epistemic uncertainty of the process model. From these values, forecastable metrics such as failure rate and C_{pk} can be calculated. This two-step process of regressing model parameters using

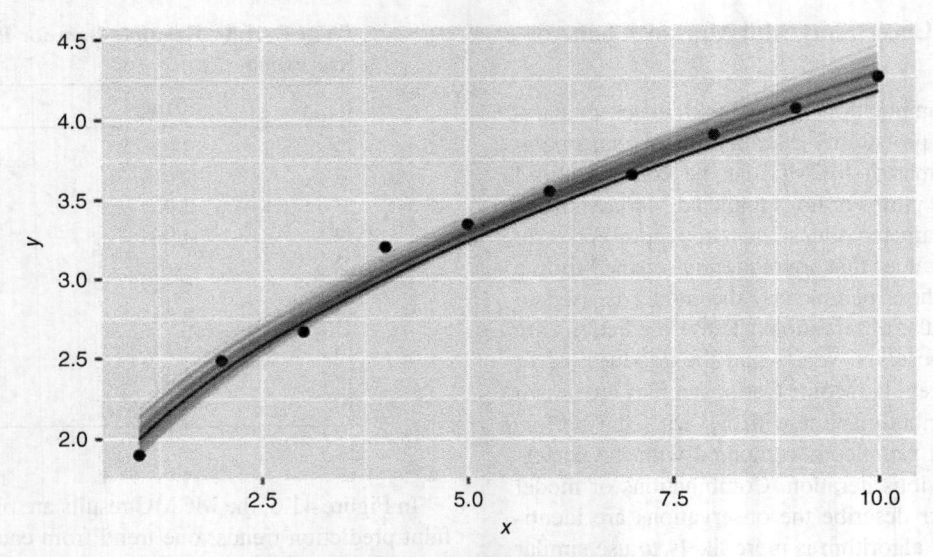

FIGURE 41.6 Regression using MCMC.

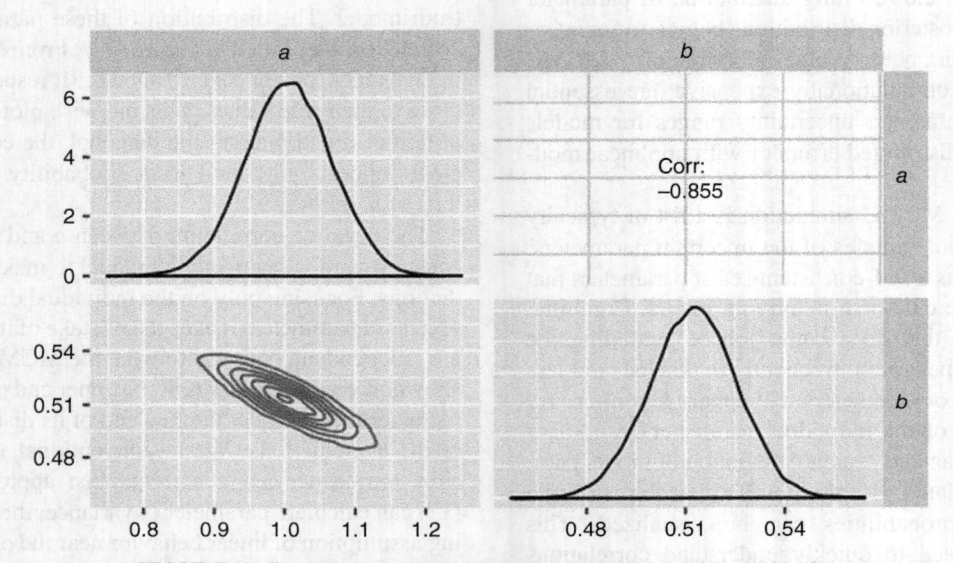

FIGURE 41.7 Correlation between parameters regressed by MCMC.

MCMC followed by using the parameter uncertainty to predict a range of probable future outcomes using Monte Carlo is illustrated in Figure 41.8.

41.7 SIMULATION TOOLS FOR BAYESIAN STATISTICS

Modeling uncertainty is a persistent challenge, especially for nonlinear models. Typical approaches such as confidence intervals cannot be accurately applied to nonlinear models and to models where the sources of noise are not uniform. Fortunately, Bayesian statistical inference is now easier than ever, thanks to modern computational approaches [11]. With the growing advances, it is now possible to compute even large hierarchical models with hundreds or thousands of variables. MCMC is an essential tool in estimating uncertainty for complex systems. MCMC approaches have been recently recognized within the chemical engineering community as being well suited to pharmaceutical process modeling as they can incorporate expert judgment with sparse data sets to yield estimates of risk in a manner consistent with regulatory guidance [7, 18]. The interpretability and communication of the results becomes more straightforward, as the confidence intervals from frequentist statistics unnaturally describe the range of conclusions expected over many replicated studies. By contrast, Bayesian credible intervals more intuitively describe the likelihood that a parameter truly has a given

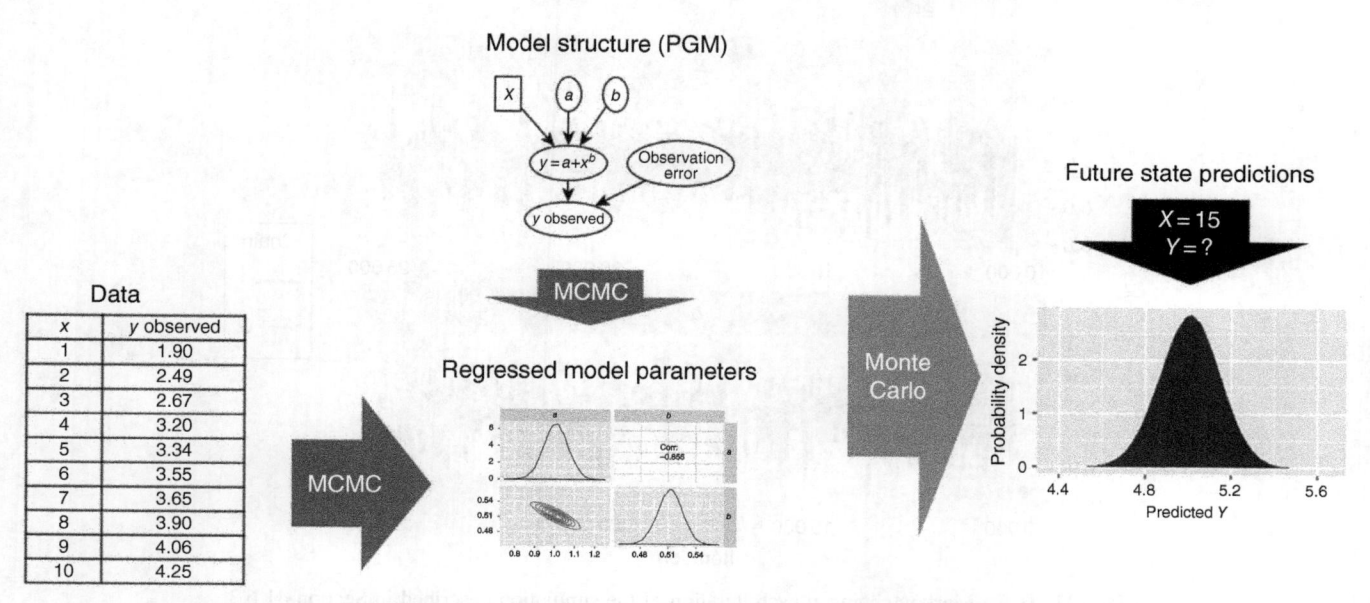

FIGURE 41.8 Workflow to convert a model and data to a prediction range while preserving uncertainty.

value. By avoiding the some of the assumptions used in frequentist statistics, this approach can more accurately estimate the intervals of parameters regressed using nonlinear regression [10, 18].

41.7.1 Overview of Available Software Tools

Basic versions of the MCMC algorithm are straightforward to program [19]; therefore, implementations can and do exist in many programming languages. However, as more sophisticated and computationally efficient approaches are developed, using a MCMC library maintained by a third party enables more rapid setup and execution of a simulation. A recent (2017) list of these packages used by the statistics and modeling communities includes Stan [12], WinBUGS [20], PyMC3 [21], and Emcee [22].

Because the selection of software is highly dependent on the end user's circumstances and need, a deeper assessment of each of these tools is left to the individual user. Attributes to keep in mind when selecting a software package include setup time and cost, available documentation, and available algorithms. In this section and the following section, Stan was selected due to its availability as an open-source program, refined interface with R via RStan [13], and portability of Stan models between other programming languages.

41.7.2 Model Diagnostics

Assessing the quality of the modeling is an important check against implementation errors and poorly specified models. MCMC simulations in particular have two main challenges to performance [23]: convergence and autocorrelation.

Checks of these performance measures should be done to ensure that the MCMC results are valid and can be used for predicting process robustness.

41.7.2.1 Convergence
A common problem with MCMC is a lack of convergence of the Markov chains, resulting in an incomplete coverage of parameter space. This is equivalent to local minima in an optimization problem. Fortunately, there are many approaches to prevent and identify when a MCMC chain does not converge. The most common approach is to run multiple chains simultaneously in parallel and to ensure the chains reach a consensus distribution. A warm-up (sometimes called burn-in) period of several thousand iterations is also used to ensure that the chains have enough time to move from their random starting points to their final distribution and that they are in agreement. This consensus can be visualized by plotting the distribution of chains over time or assessed statistically with metrics such as potential scale reduction factor R [23]. For a basic visual check, make sure that the multiple chains overlap (mix) and that there is no overall drift in the distribution as the simulation is run. Figure 41.9 shows the trace of the Markov chains from the example in Section 41.6.3. The chains show no long-term trends and are in good agreement, indicating convergence. The R values for each parameter are equal to 1, indicating ideal performance.

41.7.2.2 Autocorrelation
Other issues related to MCMC are due to autoregression in the sampling. Autoregression occurs when the sampling is inefficient, causing the Markov chain to slowly converge to the final posterior distribution. This causes the individual parameter samples to remain

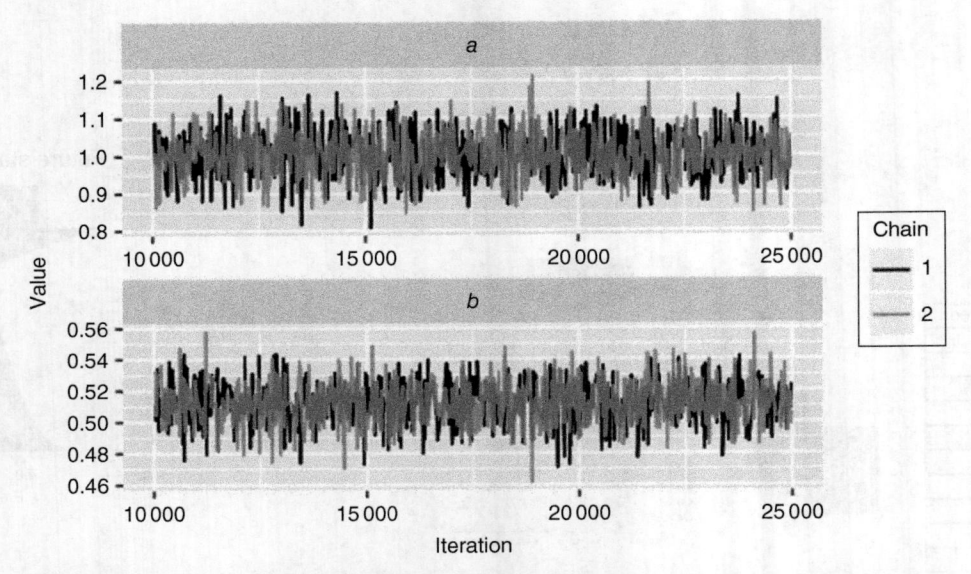

FIGURE 41.9 Markov chains at each iteration of the simulation described in Section 41.6.3.

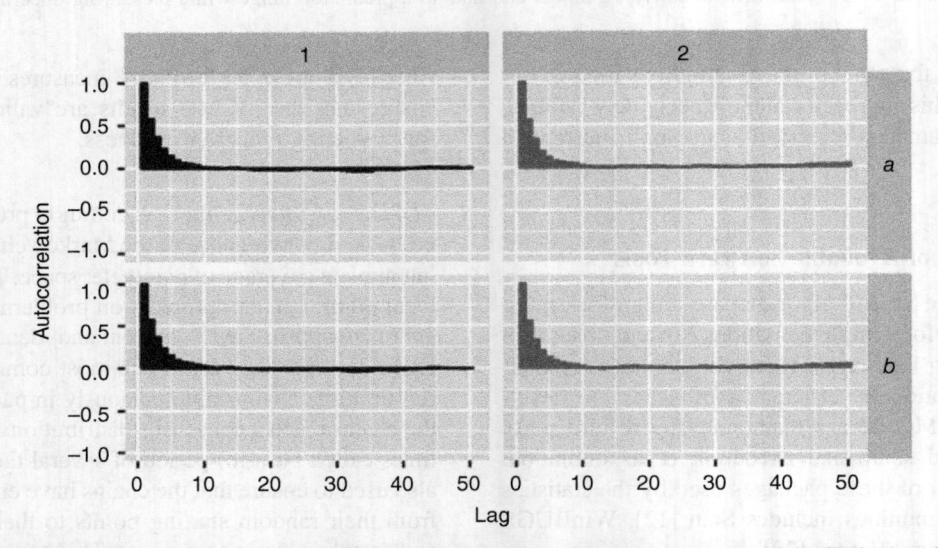

FIGURE 41.10 Plots of autocorrelation for parameter regression for parameters a and b in Chain 1 and Chain 2.

unchanged over several Monte Carlo iterations, distorting the posterior distribution. Sample thinning is an approach that reduces the overall number of samples by keeping only a subset (such as every 10th or every 50th sample) of the posterior samples. The degree of thinning increases the accuracy of the posterior distribution, but because many iterations are discarded from the analysis, there is a trade-off of wasted computation time. Figure 41.10 shows the autocorrelation of the Markov chains from the example in Section 41.6.3. For each bar plot panel, the autocorrelation at a lag 0 is always 1 (a sample is perfectly correlated with itself). After one random sampling iteration, the sampled parameter value is approximately 50% correlated with the previous sample for both parameters a and b in each chain. This correlation across samples introduces biases in any subsequent analysis. As the chain is iterated, the autocorrelation eventually drops to near 0. In Figure 41.10, the autocorrelation after a lag of 10 iterations is roughly zero. This means that the chain should be thinned by only keeping every 10th sample to ensure that each sample is independent. Plotting the chains after this thinning step would show a plot with a value of 1 at lag 0 but nearly zero correlation for all other values of lag.

41.8 EXAMPLE OF MODELING PROCESS ROBUSTNESS

An example calculation of process robustness will be presented in this section. The goal of this example is to present scenarios similar to real situations in which MCMC has been applied to inform decision makers during pharmaceutical process development. The goal in the example is to assess process robustness of a crystallization operation at a larger scale, accounting for the uncertain model and uncertain inputs. The example will use R and Stan [13] to analyze the raw data, regress the model, and determine the uncertainty of the outcome under different conditions.

41.8.1 Problem Statement

In this example, a crystallization process follows an empirical model under uncertain operating conditions. Here, the batch process will be run at a facility that has a known distribution of process control. The final crystals must have a particle size (D90) of 100 ± 30 μm. Some experiments have been conducted on a different scale to collect data to build a model. Table 41.4 shows a set of DoE data collected at the lab scale to develop a crystallization model. At each condition, three physical samples were each analyzed twice to understand the sources of sample collection heterogeneity and analytical error. These steps introduce errors that confound the regression of the model parameters.

41.8.2 Process Model Definition

The desired empirical model describes a linear relationship between the particle size (expected D90), seed temperature T, and cooling rate X:

$$\text{Expected D90} = A + BT + CX \quad (41.6)$$

With the model in Eq. (41.6), the goal is to determine the likelihood that the variability in process inputs will cause the product to be out of specification. This will in turn determine the distribution of C_{pk} that will be expected at a manufacturing scale. The model for expected D90 is purely deterministic, so no variability is expected between batches with constant operating conditions; however, we know that there will be some random variability in sampling and in sample analysis. Because the batch may have some intrinsic heterogeneity, Eq. (41.7) models the sampling variability as a normally distributed multiple of the expected D90:

$$\text{Sample D90} = \text{Expected D90} \cdot N(1, \sigma_{\text{samp}}) \quad (41.7)$$

Likewise, repeated analysis of the same sample is expected to result in some variability of the measurement. Equation (41.8) adds a similar error source as in Eq. (41.7) to the batch sample:

$$\text{Observed D90} = \text{Sample D90} \cdot N(1, \sigma_{\text{meas}}) \quad (41.8)$$

To add more complexity to the problem of estimating the model parameters, the degree of sampling and measurement variability σ_{samp} and σ_{meas} are assumed to be unknown in the fitting exercise.

The model definition and errors due to sampling and analysis can be summarized by the PGM in Figure 41.11.

In the PGM in Figure 41.11, the boxes represent a stacking of separate nodes for each of the replicates; in this case there will be nine nodes each for expected batch D90, T, and X to represent the individual experiments. There will be 27 (3×9) sample D90 nodes for each sample and experiment and 54 ($2 \times 3 \times 9$) observed D90 nodes for each analysis of each sample at each experimental condition.

41.8.3 Estimating Uncertainty

The data in Table 41.4 can be fit simultaneously to Eq. (41.8) using Stan while capturing hierarchical sources of error stemming from measurement and prediction errors. Here the MCMC routine uses 15 000 iterations with two chains, for a total of 30 000 samples. Convergence is assessed by manually examining the chains as well as the conversion statistics R. In this case, convergence and autocorrelation measures confirmed that the simulation was successful.

The distribution of the regressed parameters is shown in Figure 41.12, showing that there is some covariance between the A and B model parameters but that the other pairs of parameters are independent distributions.

TABLE 41.4 Crystallization Data

Scale	T	X	D90 Sample 1 Analysis 1	D90 Sample 2 Analysis 1	D90 Sample 3 Analysis 1	D90 Sample 1 Analysis 2	D90 Sample 2 Analysis 2	D90 Sample 3 Analysis 2
Lab	40.0	5.0	93.2	100.1	94.1	91.7	93.9	93.7
Lab	40.0	10.0	85.2	75.4	76.1	81.9	76.0	69.2
Lab	40.0	15.0	69.6	73.2	69.8	66.1	77.4	73.9
Lab	50.0	5.0	131.9	117.5	1.2	134.1	120.5	126.4
Lab	50.0	10.0	117.0	98.3	103.1	116.4	96.9	98.0
Lab	50.0	15.0	104.7	87.9	90.9	97.6	82.1	86.5
Lab	60.0	5.0	143.8	153.6	137.4	150.8	155.0	4.0
Lab	60.0	10.0	127.2	128.4	125.9	135.0	122.0	122.4
Lab	60.0	15.0	116.4	108.0	105.8	116.9	116.5	119.0

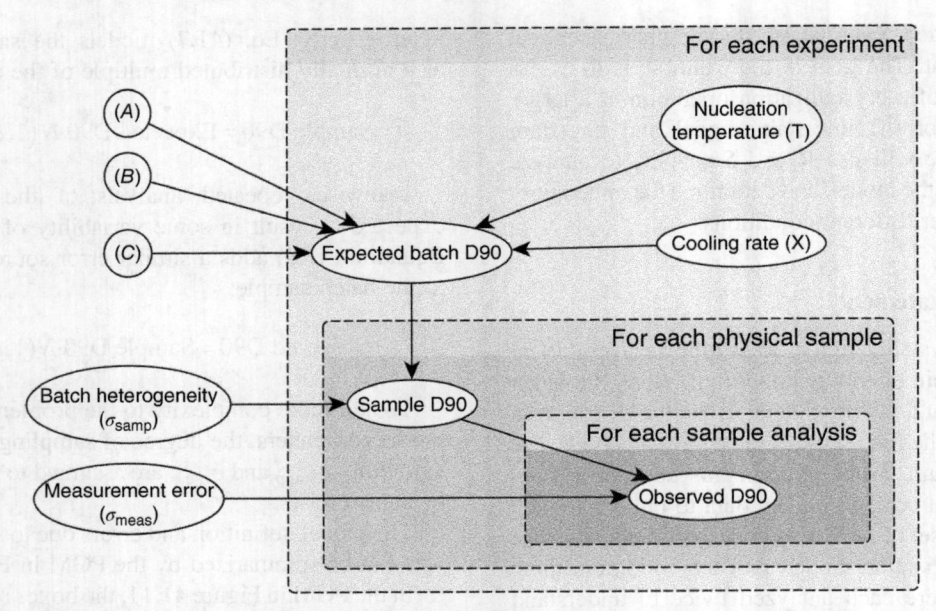

FIGURE 41.11 Crystallization robustness described as a probabilistic graphical model.

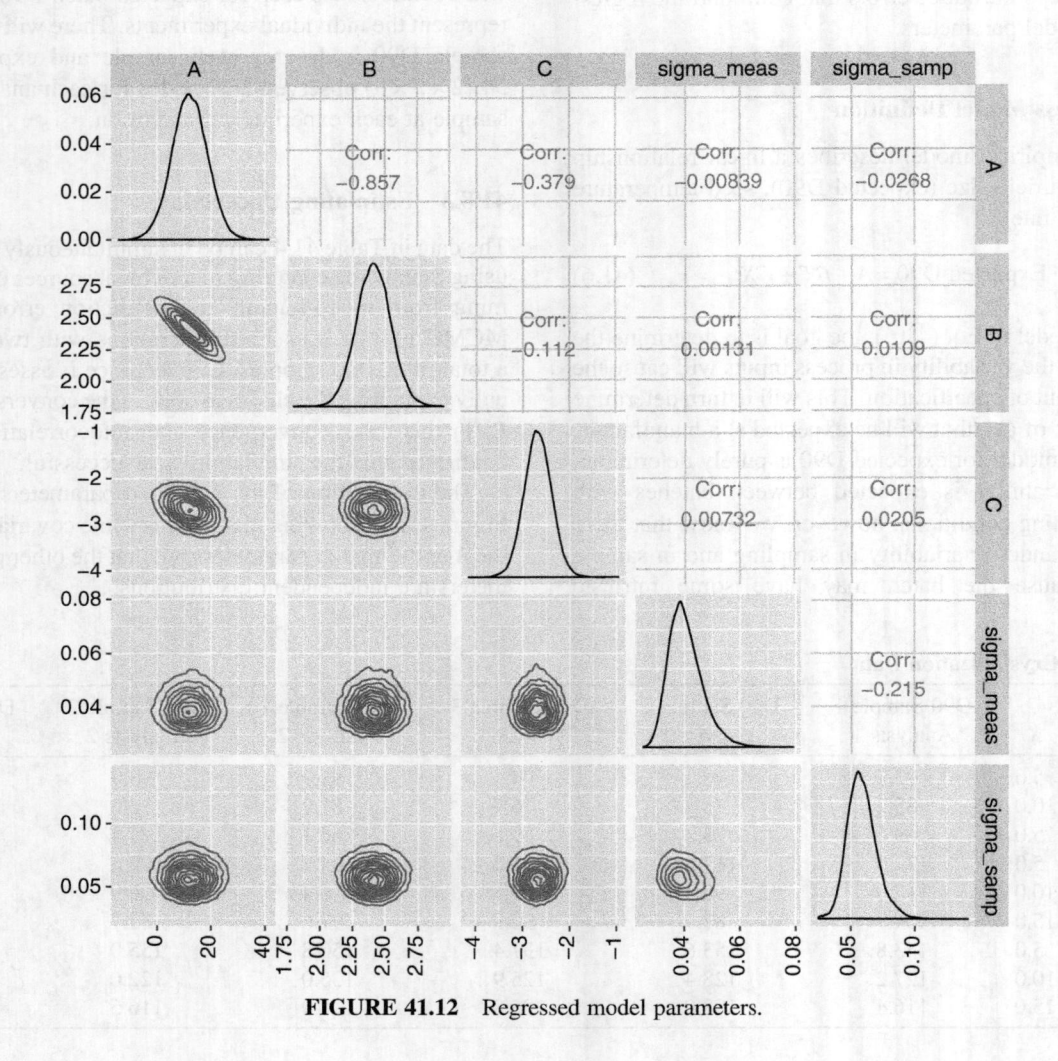

FIGURE 41.12 Regressed model parameters.

TABLE 41.5 Parameter Values Estimated from MCMC

Parameter	Parameter Mean	Parameter Standard Deviation	True Value
A	12.17	6.57	20
B	2.41	0.13	2.2
C	−2.68	0.29	−2.5
σ_{samp}	0.06	0.01	0.05
σ_{meas}	0.04	0.01	0.05

The MCMC routine works well to separate and quantify the sources of the uncertainty in observed D90. Table 41.5 shows the parameter values estimated by MCMC. The true value of the parameter (used to generate the data in Table 41.4) is compared to the mean and standard deviation to confirm that the MCMC model can recapture the true model parameters within the 95% interval of its estimates. In the case of this model, the larger CV for the model parameters A, B, and C relative to the magnitude of the sampling and measurement variability indicates that much of the uncertainty is due to epistemic uncertainty; therefore more experimental data will increase the model confidence and therefore prediction accuracy.

The next step is to determine the optimal operating conditions to minimize variability. Because the model equation is simple, an algebraic approach can be used here. Using distributions of the parameters A, B, and C, the optimal values for T and X to meet the target D90 of 100 μm can be determined to satisfy Eq. (41.9):

$$T = \frac{100 - A - CX}{B} \quad (41.9)$$

Scanning across a range of X, the corresponding values of T can be calculated from the regressed parameters, and the model credibility can be estimated from the posterior samples. Figure 41.13 plots the combination of X and T values that are expected to result in a D90 of 100 μm. While any point along the line is expected to meet this target, the uncertainty of the model can also be incorporated to find conditions where the model is more confident. The optimal value of X to minimize variability due to epistemic uncertainty (thinnest region of the uncertainty band in Figure 41.13) of the model parameters corresponds to approximately $X = 10$, $T = 48$. These are the best conditions to use in order to maximize manufacturing robustness.

41.8.4 Robustness Predictions

With the regressed model and posterior samples, the next step is forecasting the risk of generating batches that exceed the particle size specification. Using the terminology from Section 41.5.3, the posterior distribution from MCMC contains the epistemic uncertainty of the parameters; resampling

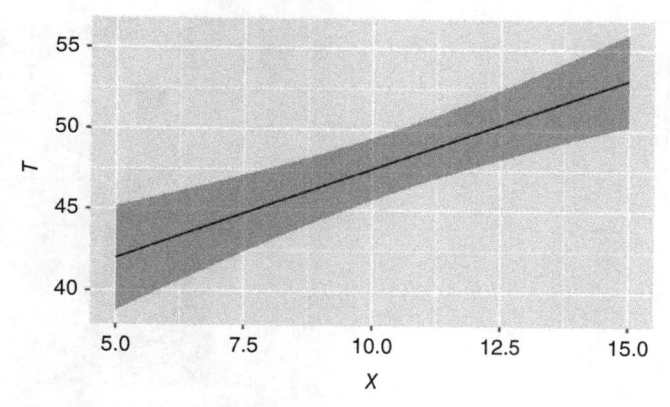

FIGURE 41.13 Combinations of inputs X and T to meet product specifications, with a range of uncertainty.

from these results and using the aleatory variability in a manufacturing facility will forecast the CQA variability. This approach allows for an arbitrary posterior distribution and a more accurate representation of the true model error under different circumstances. In this example, we suppose that an independent study determines that the nucleation temperature T and cooling rate X in manufacturing is controlled to be within 5 °C and 0.5 °C/h, respectively, 99% of the time, following a normal distribution. The analytical variability σ_{meas} is also independently determined to be 5%. Now with these sources of aleatory variability, the risk of exceeding the particle size specification can be calculated from the posterior distribution.

Using these ranges, the model described in Figure 41.11 is run using MCMC to calculate the distributions of the model parameters. These posterior distributions for the parameters are then resampled using Monte Carlo to calculate the range of future observed D90 values and the corresponding values for the C_{pk}. Random samples for X and T are taken from their distributions and used; next, random noise is added. These model outputs are plotted as histograms in Figure 41.14. Here, the C_{pk} we expect to observe in manufacturing may be between 0.75 and 1.25 but is most likely to be near 1.

Additionally from the distribution of D90, the empirical cumulative distribution function (ECDF) can be used to determine the likelihood of observing a D90 value below a certain specification. The ECDF is simply the cumulative integral of the probability distribution calculated by Monte Carlo and is used to determine the proportion of outcomes that are below a given value. Table 41.6 highlights how the likelihood of the process meeting a specification changes as the USL is increased. For example, a specification of 100 μm would be near the average expected D90, so the chance of meeting that specification would be approximately 50%.

This type of analysis also allows for teams to quickly asses the quality manufacturing predictions as new batches are run. For example, if a batch had an observed D90 of 90 μm, this

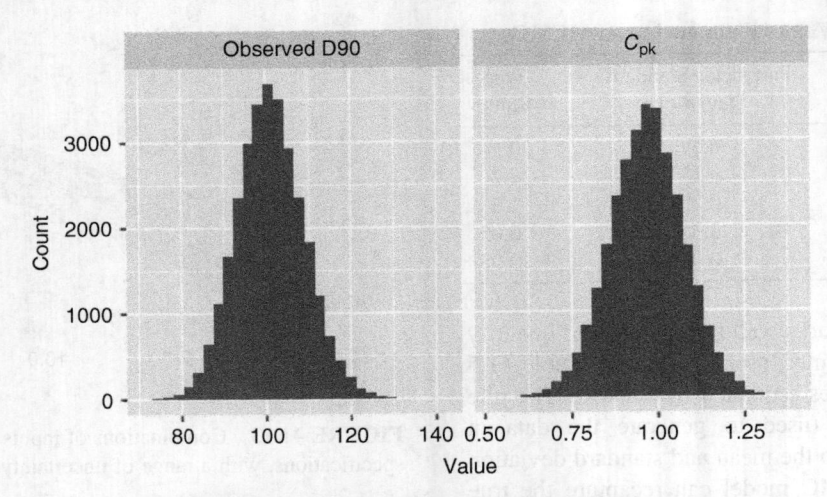

FIGURE 41.14 Future predictions of D90 at manufacturing.

TABLE 41.6 Empirical Cumulative Distribution Function Values from Expected Range of D90

D90 Upper Specification Limit	Expected Chance of Observing D90 Below Specification Limit (%)
80	0.4
90	8.0
100	45.4
110	86.9
120	98.8

difference from the expected D90 could be within normal variability, as there is a predicted 8% (or 1 in 12) chance of this occurring. Such a difference may not warrant immediate intervention. However, if a batch has an observed D90 of 80 μm, the much lower likelihood (1 in 250) of this occurrence would strongly suggest to the team that process is not performing consistently with the model built from the lab experiments and should be investigated further to identify a cause.

41.9 SUMMARY

In summary, pharmaceutical process development is uniquely sensitive to variability due to the common use of batch processing, relatively sparse process knowledge, time constraints, and high process complexity. Models for process robustness are becoming an essential tool for process development thanks to the triple convergence of business needs for faster development times, computing hardware, and publicly available software to lower the barrier to apply these approaches. Such approaches now enable the estimation of the process capability to expect in any manufacturing facility, greatly facilitating technology transfer activities. While the recent and ongoing advances in robustness assessment enable more quantitative assessments of pharmaceutical quality prediction, there is no substitute for an engineer's expert judgment and critical thinking. To maximize the process robustness under a given set of constraints, the control strategy, process sensitivity, process variability, and selection of quality limits are all important considerations. This chapter demonstrates that by embracing a probability mind-set, powerful statistical tools can enable a rational, quantitative assessment of risk.

SUPPORTING INFORMATION

The source code used to create the example data, probabilistic models, and figures are available as a Jupyter notebook (jupyter.org) at DOI: 10.6084/m9.figshare.7607354. The supporting information also includes version numbers and installation instructions for the programming language R and associated function libraries.

REFERENCES

1. Anderson, D.R., Sweeney, D.J., Williams, T.A. et al. (2010). *Statistics for Business and Economics*. Independence, KY: South-Western CENGAGE Learning.
2. Reid, R.D. and Sanders, N.R. (2011). *Operations Management: An Integrated Approach*, 4e. Hoboken, NJ: Wiley.
3. NIST/SEMATECH (2013). NIST/SEMATECH e-Handbook of Statistical Methods. http://www.itl.nist.gov/div898/handbook (accessed 4 April 2018).
4. Nomikos, P. and MacGregor, J.F. (1995). *Multivariate SPC charts for monitoring batch processes. Technometrics* 37 (1): 41–59.

5. Anderson, T.W. (2003). *An Introduction to Multivariate Statistical Analysis.* New York: Wiley.
6. ICH (2011). ICH-endorsed guide for ICH Q8/Q9/Q10 implementation. ICH Quality Implementation Work Group Points to Consider (R2) (accessed 4 April 2018).
7. Peterson, J.J. (2008). *A Bayesian approach to the ICH Q8 definition of design space. Journal of Biopharmaceutical Statistics* 18 (5): 959–975.
8. Peterson, J.J. (2010). *Design space and QbD: It's all about the (stochastic) distributions.* AIChE Annual Meeting, Conference Proceedings, Salt Lake City, UT (7–12 November 2010).
9. Tabora, J.E. and Domagalski, N. (2017). *Multivariate analysis and statistics in pharmaceutical process research and development. Annual Review of Chemical and Biomolecular Engineering* 8: 403–426.
10. Albrecht, J. (2013). *Estimating reaction model parameter uncertainty with Markov Chain Monte Carlo. Computers and Chemical Engineering* 48: 14–28.
11. Gelman, A., Carlin, J.B., Stern, H.S. et al. (2013). *Bayesian Data Analysis*, 3e. Boca Raton, FL: Taylor & Francis.
12. Stan Development Team (2016). Stan modeling language users guide and reference manual, version 2.14.0.
13. Stan Development Team (2016). RStan: the R interface to Stan, R package version 2.14.1.
14. Kutner, M.H., Nachtsheim, C.J., and Neter, J. (2003). *Applied Linear Regression Models.* New York: McGraw-Hill.
15. Box, G.E.P. and Hill, W.J. (1967). Discrimination among mechanistic models. *Technometrics* 9 (1): 57–71.
16. Koller, D. and Friedman, N. (2009). *Probabilistic Graphical Models: Principles and Techniques.* Cambridge, MA: MIT Press.
17. Seber, G.A.F. and Wild, C.J. (eds.) (2005). Statistical inference. In: *Nonlinear Regression*, 191–269. Hoboken: Wiley.
18. Blau, G., Lasinski, M., Orcun, S. et al. (2008). *High fidelity mathematical model building with experimental data: a Bayesian approach. Computers and Chemical Engineering* 32 (4–5): 971–989.
19. Beers, K.J. (2007). *Numerical Methods for Chemical Engineering.* New York: Cambridge University Press.
20. Lunn, D.J., Thomas, A., Best, N., and Spiegelhalter, D. (2000). *WinBUGS: A Bayesian modelling framework: concepts, structure, and extensibility. Statistics and Computing* 10 (4): 325–337.
21. Salvatier, J., Wiecki, T.V., and Fonnesbeck, C. (2016). *Probabilistic programming in Python using PyMC3. PeerJ Computer Science* 2: e55.
22. Daniel, F.-M., Hogg, D.W., Lang, D., and Goodman, J. (2013). emcee: The MCMC hammer. *Publications of the Astronomical Society of the Pacific* 125 (925): 306.
23. Gelman, A., Carlin, J.B., Stern, H.S., and Rubin, D.B. (2003). *Bayesian Data Analysis*, 2e. Boca Raton, FL: Taylor & Francis.

42

USE OF PROCESS ANALYTICAL TECHNOLOGY (PAT) IN SMALL MOLECULE DRUG SUBSTANCE REACTION DEVELOPMENT

DIMITRI SKLIAR, JEFFREY NYE, AND ANTONIO RAMIREZ
Product Development, Bristol-Myers Squibb, New Brunswick, NJ, USA

42.1 INTRODUCTION

Reaction engineering is an essential component of quality risk management in pharmaceutical drug substance manufacturing. Identification of critical impurities, elucidation of reaction mechanisms leading to them, and characterization of reaction parameter space are key elements of robust and manufacturing-ready chemical processes. Historically, chromatographic methods have been the analytical workhorse for monitoring chemical reactions throughout the pharmaceutical industry, as they provide unmatched sensitivity and specificity for a variety of organic compounds. However, one inherent shortcoming is the "off-line" nature of the technique – samples must be collected and prepared prior to analysis, which may change the chemical nature of the analyte of interest. And while UPLC has dramatically shortened the time required for chromatographic separation, an inherent time delay exists between sampling and analysis.

Real-time analytical tools provide an attractive complement to traditional chromatography by delivering a continuous stream of data from the direct interface with the reaction stream. Such tools are referred to as process analytical technology (PAT) analyzers, which can include simple univariate measuring devices such as temperature and pressure sensors but more commonly involve "in-line" analyzers such as mid-infrared (IR), near-infrared (NIR), and Raman spectrometers. PAT analyzers provide real-time information of chemical reactions by tracking the spectra of the molecules involved in the transformation.

Though the FDA [1] and EMEA definitions of PAT analyzers pertain to a manufacturing setting, these tools offer a number of benefits in laboratory development of chemical reactions as well. The exquisite time resolution associated to the capacity of collecting data points on the time scale of seconds allows for detailed kinetic analysis of complex reactions. Moreover, since these tools directly interface with the reaction stream, they enable the detection of reactive or short-lived intermediates, which cannot be accurately identified via chromatographic methods. The in-line nature also allows for analysis of hazardous systems, eliminating the requirement that an operator collects samples. In this manner, PAT analyzers provide improved safety when designing or optimizing processes involving highly toxic chemicals, reactions under pressure, or reactions at extreme temperatures. Finally, the fast acquisition time of PAT analyzers allows for timely response of processing conditions to reaction progress. This can enable feedback control when implementing fully developed processes and is a crucial component of flow reactions [2]. With these benefits in mind, significantly improved and user-friendly experimental PAT units have become available for laboratory use, which have enabled facile experimentation on even very small scales, and have streamlined transfer of knowledge and methods to pilot and manufacturing scale. This chapter presents an overview

---

Chemical Engineering in the Pharmaceutical Industry: Active Pharmaceutical Ingredients, Second Edition.
Edited by David J. am Ende and Mary T. am Ende.
© 2019 John Wiley & Sons, Inc. Published 2019 by John Wiley & Sons, Inc.

TABLE 42.1 Spectroscopic Techniques Based on Electromagnetic Radiation

Spectroscopy Type	Wavelength Range	Wavelength Range (cm^{-1})	Type of Quantum Transition
Gamma-ray emission	0.005–1.4 Å	—	Nuclear
X-ray absorption, emission, fluorescence, and diffraction	0.1–100 Å	—	Inner electron
Vacuum ultraviolet absorption	10–180 nm	1×10^6 to 5×10^4	Bonding electrons
Ultraviolet–visible absorption, emission, and fluorescence	180–780 nm	5×10^4 to 1.3×10^4	Bonding electrons
Near-infrared and mid-infrared absorption and Raman scattering	0.78–300 μm	1.3×10^4 to 33	Rotation and vibration of molecules
Microwave absorption	0.75–3.75 mm	13–27	Rotation of molecules
Electron spin resonance	3 cm	0.33	Spin of electrons in a magnetic field
Nuclear magnetic resonance	0.6–10 m	5×10^{-2} to 1.3×10^{-3}	Spin of nuclei in a magnetic field

of standard spectroscopy-based PAT analyzers and their application to pharmaceutically relevant transformations. A number of case studies are chosen to illustrate both the enabling aspects of PAT in laboratory reaction development as well as their implementation in a manufacturing setting as part of a control strategy.

42.2 PAT METHODS

While the physics underlying the various spectroscopic techniques discussed in this chapter varies, the general principles of implementing PAT are common to all of them. Light is directed from a source within the spectrometer through a conduit into a probe that is immersed in the reaction stream of interest. Analyzed light, be it an absorption spectrum or scattered photons, is then directed back to a detector, processed, and presented as a spectrum that changes with time as the reaction progresses. Spectral data can be processed and interpreted in a variety of ways, both quantitative and qualitative. Table 42.1 lists the wavelength and frequency ranges for common analytical techniques based on electromagnetic radiation [3]. Ultraviolet–visible, NIR, mid-IR, and Raman spectroscopies are the four most common methods used for reaction monitoring and analysis. These techniques cover the wavelength range of the electromagnetic spectrum between 180 nm and 300 μm (see highlighted rows in Table 42.1).

A typical in-line instrument setup for use in laboratory consists of a probe that can be inserted into a reaction vessel, a spectrometer, and a computer interface where data can be processed (see Figure 42.1). Standard probe diameters allow for utilization of vessel sizes ranging from several milliliter vials all the way to multi-liter scale reactors (see Figure 42.2a and b). For improved and automated analysis, a variety of instruments are available that combine spectroscopic collection with process parameter monitoring (such as temperature, pressure, agitation speed, etc.) in one package (see Figure 42.2c).

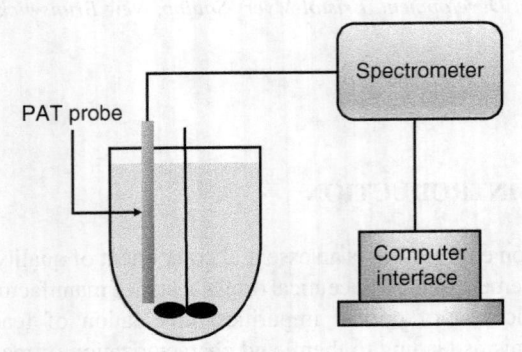

FIGURE 42.1 Schematic of laboratory setup for in-line spectroscopic monitoring.

Several probe configurations are suitable for scale-up implementation of PAT. The reaction mixture can be analyzed by inserting a probe through a reactor port, a bottom valve, or a recirculation loop (Figure 42.3). Commercially available instruments for the spectroscopy types discussed in this chapter can be utilized on scale in these configurations.

42.2.1 Mid-infrared Spectroscopy

IR spectroscopy examines molecular vibrations based on the absorption of light in the mid-IR region of the electromagnetic spectrum (~400–4000 cm^{-1}). As the IR absorption bands are often unique to the vibrational modes of chemical functional groups, IR spectroscopy provides a high degree of structural specificity in the analysis of liquids, solids, and gases and has long been used by analytical chemists in structural determination of unknown materials. The enabling technology underlying most IR spectrometers is the Michaelson interferometer, a common optical assembly that uses light interference to measure distances in wavelength units. IR light is transmitted from a source through a beam splitter

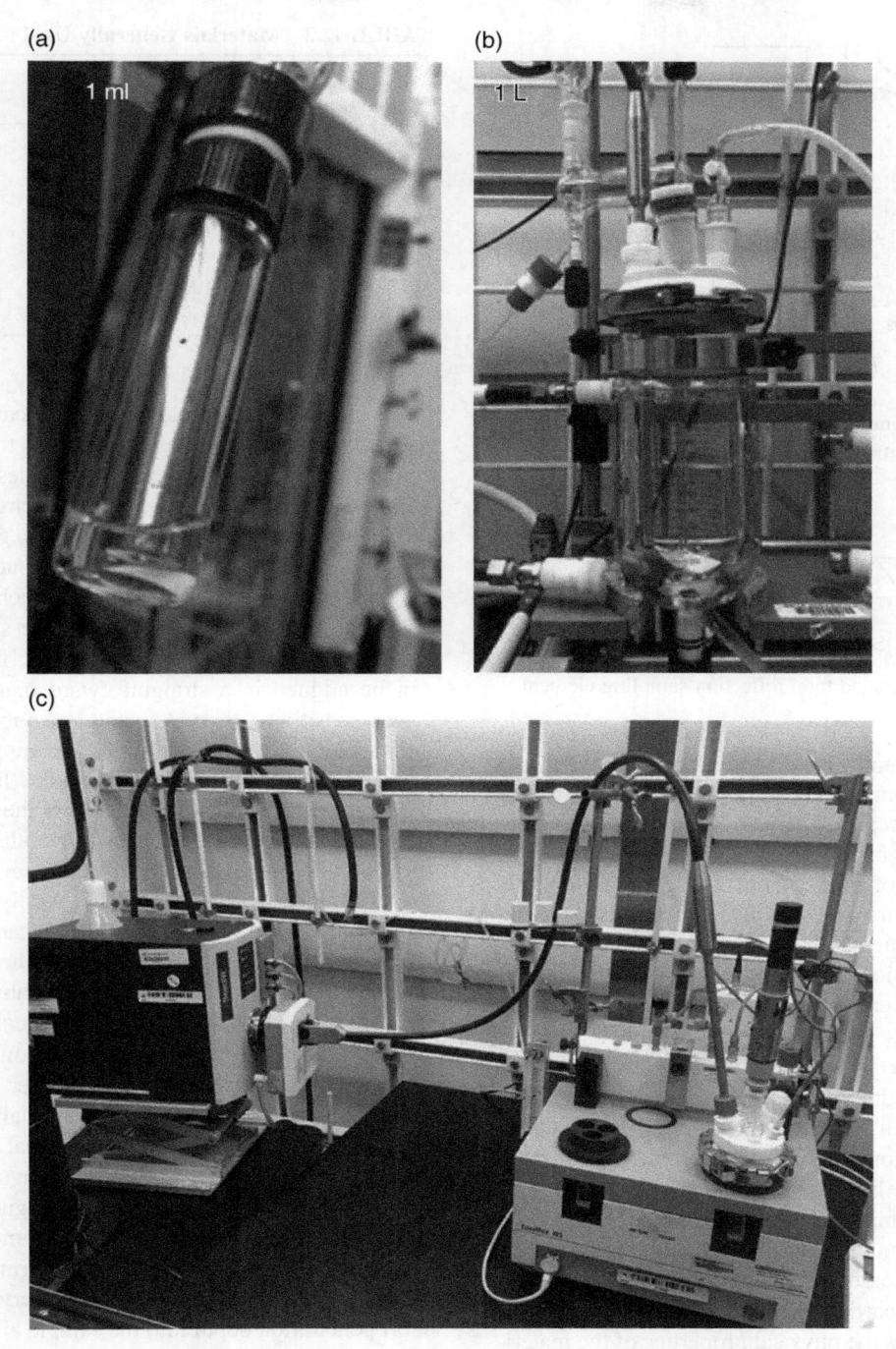

FIGURE 42.2 (a) Spectroscopic probe inserted into 4 ml vial. (b) Spectroscopic probe inserted into 1 L reactor. (c) Mettler Toledo ReactIR 45 M instrument coupled with EasyMax reactor setup.

and movable mirror; after interaction with the sample, an interferogram is detected and then converted to a true absorption spectrum by Fourier transform. Hence the technique is commonly known as Fourier transform infrared spectroscopy (FTIR).

A key difference in FTIR spectrometers used for real-time analysis relative to standard benchtop units is the configuration of the optical element that transmits light from the spectrometer to the interface with the reaction stream. Traditional benchtop instruments measure IR absorption through a fixed length of sample in a transmission mode. In contrast, online measurements primarily employ a sampling element based on attenuated total reflectance (ATR) (Figure 42.4). In ATR measurements, light is transmitted through a solid material and reflected on the surface in contact with the sample before being sent back to the detector. This reflection

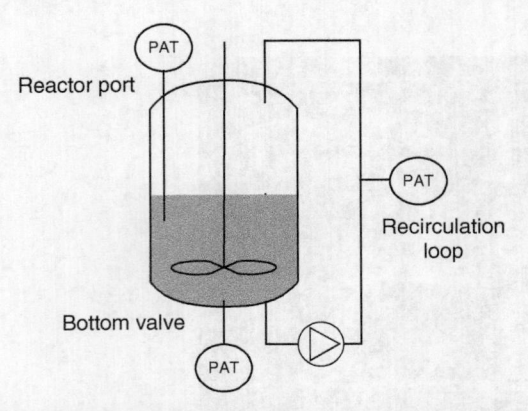

FIGURE 42.3 Schematic of different possible PAT configurations for implementation on scale.

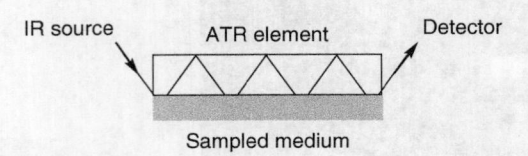

FIGURE 42.4 Attenuated total reflection sampling element.

TABLE 42.2 Materials Generally Used in ATR Elements

Material	Spectral Range (cm^{-1})	Refractive Index	Depth of Penetration (μm)
Diamond	4500–2500 1667–33	2.4	1.66
Si	8300–1500 360–70	2.37	1.73
Ge	5500	4	0.65
KRS-5	20 000–400	2.37	1.73
ZnSe	20 000–650	2.4	1.66

establishes an evanescent wave that penetrates less than 2 μm into the sampled medium. Thus the ATR element enables in situ monitoring of the reaction being analyzed with zero sample preparation. ATR materials are chosen for low absorbance and high index of refraction, but as they are directly immersed in reaction streams, chemical inertness is a requirement. Diamond is frequently the material of choice, as it exhibits good IR transmission and is chemically inert to the wide range of chemistries employed in chemical development. Moreover, diamond is very hard, resists abrasion by solid particles, and is stable across a wide range of temperatures and pressures. One drawback is that diamond strongly absorbs IR radiation in the 1900–2300 cm^{-1} region, which overlaps with vibrational modes corresponding to triple and cumulated double bonds. Silicon is an alternative ATR element that provides high sensitivity across the IR spectrum and is thus ideal for monitoring reactions involving azides, acetylenes, carbon dioxide, etc. However, silicon is less chemically resistant, particularly to strong aqueous bases. Table 42.2 summarizes the physical properties of the materials generally used in ATR elements.

An important challenge of PAT tools that incorporate ATR elements is the difficulty of transmitting light from the spectrometer to the sampled medium. Housing the ATR element in a metal probe, usually Hastelloy for its chemical inertness, facilitates light transmission to the reaction medium as well as the adequate ruggedness to withstand the breadth of chemical environments used in chemical development (strong acids and bases, very high or low temperatures, high pressure, etc.). In early FTIR instruments, the probe and the spectrometer were connected by a hollow light path with a series of adjustable mirrors to focus the IR laser beam to the probe. This system provides high performance and low attenuation across the spectrum, but achieving proper alignment of the probe requires a good level of expertise. A more robust, user-friendly technology is the use of polycrystalline silver halide fibers, which provide a flexible connection between the probe and the spectrometer. Compared with probes with mirrored ends, fiber-optic conduits can be aligned in a straightforward manner and display a lower sensitivity to movement and vibration. Since light attenuation through fiber optics increases with length and can be affected by conduit curvature, limitations exist on the physical size and configuration of the fibers, with typical fiber lengths being less than 3 m. One other consideration for the use of FTIR spectroscopy is that water strongly absorbs in the IR region. While this does not totally preclude the technique's used in aqueous systems [4], it can make analysis difficult as many important regions of the spectrum may be obscured. In general, for solvents with absorptions that overlap with the analyte bands, adequately scaled solvent subtraction techniques [5] may facilitate reaction monitoring. When solvent subtraction is not suitable, other chemometric treatments such as principal component analysis (PCA), partial least squares (PLS) regression, principal component regression (PCR), or multiple linear regression (MLR) may be used to deconvolute reaction mixtures [6]. Figure 42.5 displays the IR transmission bands for solvents commonly used in chemical reactions. In addition to monitoring reactions, FTIR probes are routinely used to monitor crystallizations as the short penetration depth into the sample allows for supersaturation studies [7].

42.2.2 Raman Spectroscopy

Raman spectroscopy probes molecular vibrations and rotations in the 100–3000 cm^{-1} frequency range. It is considered a complimentary technique to mid-IR spectroscopy that analyzes the wavelength shifts associated to Raman scattering. When a sample is excited with monochromatic light, a small portion of the scattered radiation undergoes quantized

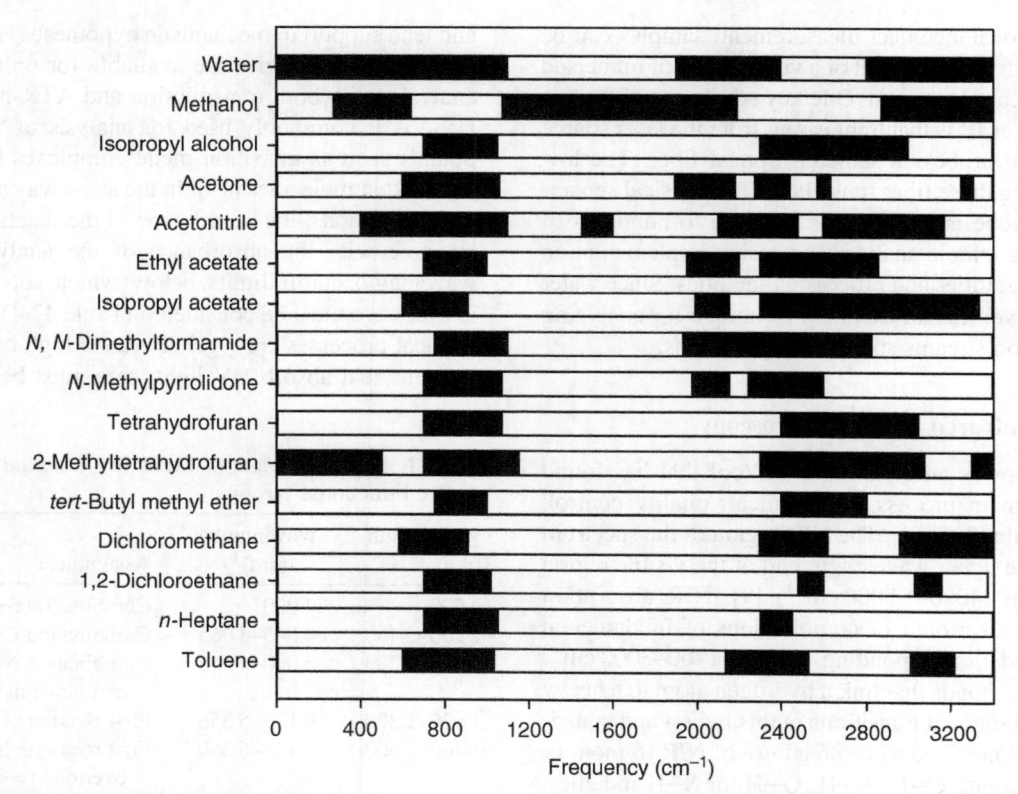

FIGURE 42.5 Infrared transmission for typical solvents used in chemical reactions. Dark regions with less than 50% transmission represent bands where analyte monitoring may require solvent subtraction.

changes in frequency that result in vibrational and rotational transitions. The Raman effect accounts for the new frequencies in the spectrum of the monochromatic light and is reported in "Raman shift" away from the monochromatic light source: $\Delta \nu = \nu_o - (\nu_o \pm \nu_1) = \pm \nu_1$, which is independent on the frequency of the monochromatic light, ν_o.

Whereas IR absorption only occurs when the excited vibrational mode induces a change in dipole moment of the molecule, Raman scattering relies on a change in the molecule's dipole polarizability – the temporary distortion of the electron density around bonds. Hence, a given vibrational mode may be "Raman active," "IR active," or both, and Raman spectra and IR spectra can be very different for a given molecule. For example, Raman spectroscopy can detect symmetric vibrations in diatomic molecules such as chlorine, oxygen, etc., whereas IR spectroscopy cannot. In a standard Raman spectrometer, monochromatic light from a laser is directed to a probe immersed in a reaction stream, where it interacts with the analyte. The scattered photons are then returned to the spectrometer and passed through a series of optical filters selected to exclude the wavelength of the laser itself – the majority of the light returning to the spectrometer will be Rayleigh scattered and thus be the same as the excitation wavelength. The strength of the Raman effect depends on the excitation wavelength and is dramatically increased at high excitation frequencies. Ideally, a lower wavelength light source will provide a stronger Raman spectrum. In practice, however, many samples will fluoresce when irradiated with low wavelength light. This fluorescence overwhelms the Raman scattered photons, making spectral interpretation difficult or impossible. Most commercial laboratory Raman instruments use one of three readily available laser lines, 532, 785, or 1000 nm. Among these, the 785 nm wavelength is the most common, as it provides a good balance between Raman intensity and low fluorescence in reaction systems relevant to pharmaceutical development. It should be noted that ambient light collected by the probe will also be detected and interferes with Raman spectra, so quantitative analysis requires that samples be protected from stray light sources.

Like FTIR probes, Raman probes generally consist of an optical window at the end of a Hastelloy pipe. The optical elements of a Raman probe are not ATR based, but instead rather focused into the reaction sample a few millimeters away from the probe tip. Many materials can be used for the window, including diamond, quartz, fused silica, and sapphire. The latter is the most common material because it offers a suitable arrangement of chemical inertness, strength, and availability. Since glass is Raman inactive, Raman probes can be fit with long working distance optics that enable indirect or noncontact measurement; samples can be analyzed through the glass wall of a vial, instead of immersed directly in the reaction stream. One key advantage of Raman spectroscopy over IR is that light is sent from the laser source to the immersed probe via standard optical fiber. The low attenuation along these fiber runs allows for physical separation between probe and spectrometer of up to hundreds of meters, enabling remote analysis and easing implementation in crowded laboratories and processing facilities. Since water is Raman inactive, it is an excellent technique for analyzing biotransformation streams such as fermentations.

42.2.3 Near-infrared (NIR) Spectroscopy

NIR spectroscopy is an effective PAT tool that has found wide application in process development, quality control, and raw material testing [8]. The NIR region of the spectrum extends from the upper wavelength end of the visible region at 770–2500 nm (4000–13 000 cm^{-1}) [9]. NIR absorption bands represent overtones or combinations of fundamental vibrational bands corresponding to the 1700–3000 cm^{-1} region. Chemical bonds that link a hydrogen atom to a heavy atom display vibrational transitions in this region and constitute the foundation of the widespread use of NIR to monitor molecules containing C—H, N—H, O—H, or S—H moieties. Due to the overtone nature of the vibrations, NIR spectroscopy is not suited for structural elucidation or identification of compounds, but is well qualified to perform quantitative analysis aided by chemometric techniques. Nevertheless, despite the complexity of NIR spectra, wavelength–structure correlations that can be used for guidance in qualitative analysis have been tabulated (Table 42.3) [10]. In addition to in situ monitoring of chemical processes, NIR spectroscopy has been used in a variety of fields including medical diagnostics, food quality control, and atmospheric chemistry. Typical instruments available for reaction monitoring are both benchtop off-line units, capable of analyzing solids or liquids, and in-line fiber-coupled probes for direct analysis in a reactor setting.

42.2.4 Ultraviolet–Visible Spectroscopy

Ultraviolet–visible (UV-Vis) spectroscopy analyzers measure the absorption of a sample in the mid-ultraviolet to visible region of the electromagnetic spectrum that arise from electronic transitions (~200–700 nm, 50 000–14 000 cm^{-1}). While lacking the chemical specificity of the techniques discussed above, it is commonly used to monitor liquid phase reactions due to its simplicity, sensitivity, and time resolution [11]. Despite spectral overlap, absorbance changes occurring during the reaction can be tracked and assigned to specific components with the aid of calibrated standards, chemometric techniques, or both. Moreover, analysis of the isosbestic points can help elucidate the number of absorbing species

and lend support to mechanistic hypotheses [12]. A variety of sampling technologies are available for online and off-line analysis, with both transmission and ATR-based interfaces. UV-Vis is commonly used for analysis of "colorful" compounds such as transition metal complexes [13] and highly conjugated molecules [14]. In the same way as other spectroscopic methods, the absorbance of the reaction solvents can interfere with the absorbance of the analytes, and lower wavelength cutoff limits below which solvent absorbance is excessive must be considered (Table 42.4). Since pharmaceutical processes are typically run at high concentrations in solvents that absorb UV light, care must be taken to avoid

TABLE 42.3 Wavelength–Structure Correlations for NIR Active Functional Groups

Wavelength (nm)	Wavenumber (cm^{-1})	Assignment
2 500	4 000	Combination S—H stretch
2 200–2 460	4 545–4 065	Combination C—H stretch
2 000–2 200	5 000–4 545	Combination N—H stretch; combination O—H stretch
1 620–1 800	6 173–5 556	First overtone C—H stretch
1 400–1 600	7 143–6 250	First overtone N—H stretch; first overtone O—H stretch
1 300–1 420	7 692–7 042	Combination C—H stretch
1 100–1 225	9 091–8 163	Second overtone C—H stretch
1 020–1 060	9 804–9 434	Combination S=O stretch
950–1 100	10 526–9 091	Second overtone N—H stretch; second overtone O—H stretch
850–950	11 765–10 526	Third overtone C—H stretch
775–850	12 903–11 765	Third overtone N—H stretch
600–700	16 667–14 286	Combination C—S stretch
450–550	22 222–18 182	Combination S—S stretch

TABLE 42.4 Lower Cutoff Wavelength Limits for Solvents Commonly Used in Chemical Reactions

Solvent	UV Cutoff (nm)
Water	191
Methanol	203
Isopropyl alcohol	205
Acetone	330
Acetonitrile	190
Ethyl acetate	256
N,N-Dimethylformamide	268
N-Methylpyrrolidone	285
Tetrahydrofuran	215
tert-Butyl methyl ether	210
Dichloromethane	232
1,2-Dichloroethane	228
n-Heptane	200
Toluene	284

quantitative analysis in high-absorbing regions of the spectrum by reducing path length in transmission mode or focusing on longer wavelengths.

42.2.5 Spectroscopy Interpretation (IR, Raman)

Complementing the observation of spectral differences detected during the course of a reaction with sound structural characterization can provide insight into reaction mechanisms. Spectral interpretation handbooks and databases are available in a number of online and printed resources [15],[1] and most contemporary software packages provide approximate functional group assignments for the data collected.[2] In general, this information should be helpful for a scientist to be able to assign characteristic peaks to a species one expects to observe in a particular reaction. For more accurate predictions of vibrational modes for a compound of interest, quantum mechanical packages such as Gaussian can be utilized [16].

42.3 PAT APPLICATION IN PHARMACEUTICAL REACTION ENGINEERING

42.3.1 General Reaction Types

Table 42.5 provides a summary of literature reports for common reaction types with their associated functional group transformations in the mid-IR region. The abundance and diversity of the reactions explored with in situ spectroscopy methods speaks to the value of PAT to advance the design, implementation, and characterization of chemical processes.

42.3.2 PAT Application in Complex Reaction Systems

The analysis of transformations involving reactive intermediates or multiphase media often presents a challenge for conventional off-line analytical techniques (e.g. HPLC, GC, NMR). During process development and plant execution of such processes, utilization of in situ spectroscopic methods offers a significant advantage from accuracy, time of analysis, and progress monitoring perspectives.

Off-line techniques are often destructive for reactive intermediates and may not be able to detect the high-energy species due to their thermal or chemical instability (Figure 42.6a). While the addition of a derivatization step may facilitate conventional analysis, reproduction of the intermediate concentrations demands an innocuous sampling and a derivatization reaction that is viable, quantitative, and significantly faster than the reaction being monitored.

[1] http://webbook.nist.gov/chemistry/.
[2] iC IR™, Mettler Toledo.

Similarly, representative sampling can be difficult for heterogeneous reactions where multiphasic mixtures (e.g. solid–liquid) are present (Figure 42.6b). A common off-line approach to measure analyte concentration in slurries involves sampling the slurry, filtering out the solids, and then performing the analysis using chromatographic or spectroscopic methods. Immersion probes can simplify this measurement by using direct detection since the ATR element is in close contact with the liquid phase and the interference from the solids is generally inconsequential [44].

For pressurized reactions the use of in-line spectroscopy can be crucial, as it eliminates the disturbance to the system associated with conventional sampling (Figure 42.6c). For example, sampling a catalytic hydrogenation under pressure often leads to modification of the reaction conditions due to exposure of the reaction mixture to non-inert conditions, changes in composition, or variations in the partial pressure of hydrogen upon reaction restart. The ability to perform real-time analysis using in-line FTIR or at-line NIR has been exploited to control selectivities in hydrogenations that afford a mixture of products [45]. Table 42.6 provides a brief summary of the benefits offered by PAT tools relative to conventional off-line analysis methods.

42.4 CASE STUDIES

The case studies that follow highlight the use of PAT tools in complex reaction systems. They cover examples of the application of PAT to establish in-process controls (IPCs), the description of method development, and the method implementation in a production setting.

42.4.1 Case Study 1: Control of Ammonia Content in Solution Using NIR Spectroscopy

Consider the chemical transformation depicted in Figure 42.7, where a pharmaceutical intermediate is synthesized via a reductive amination that involves the use of gaseous reagents under pressure [46]. The reaction is performed in two steps: (i) ketone ammonolysis by reaction with ammonia to form an intermediate ketimine and (ii) heterogeneous catalytic hydrogenation of the ketimine intermediate to give primary amine.

In general, for reactions conducted under pressure where dissolved gas is present in significant amounts, it is important to design a robust purging protocol. Ensuring the solution is free of gas in downstream processing is necessary to guarantee safe handling and emissions controls. In this particular case, residual ammonia from the ammonolysis part of the process negatively impacted the hydrogenation step due to its correlation with increased levels of impurity and ammonia off-gassing that complicated the charge of Pd/Al_2O_3 catalyst. During the first kilo-scale implementation of the process,

TABLE 42.5 Spectroscopically Active Reaction Types Relevant to Pharmaceutical Reaction Development

Reaction	Starting Material Characteristic Functional Group and Corresponding Vibrational Frequency (cm^{-1})	Product Characteristic Functional Group and Corresponding Vibrational Frequency (cm^{-1})
Nitroaromatic hydrogenation [17]	Ar—NO$_2$ (1530)^a	[Ar—NH—OH (3310)]^b
		Ar—NH$_2$ (3380)
Methyl nicotinate hydrogenation [18]	Pyr-3-CO$_2$Me (1305, 1720)	Pip-3-CO$_2$Me (1660)
Pyrazine hydrogenation [19]	Pyrazine (1675)	[Dihydropyrazine (1660)]
		Piperazine (1600)
α,β-Unsaturated carboxylic acid [20] hydrogenation	RR'C=CH—CO$_2$H·TMG (1625)	RR'CHCH$_2$—CO$_2$H·TMG (1625)
Grignard formation [21]	Ar—I	Ar—Mg (1035, 880)
3-Bromoquinoline Br-metal exchange [22]	Bu$_3$MgLi (700, 735)	Ar$_3$MgLi (1270, 1335)
		ArMgBr (1260)
		Ar$_2$Mg (1075)
		ArLi (1065, 1255)
ortho-Lithiation [23]	Ar—H (1520)	ArLi (1420)
Lithiation of 3,5-dichloropyridine [24]	3,5-diCl-Pyr-H (1110, 1010, 885, 810, 695)	3,5-diCl-Pyr-Li (753, 1007, 1140)
Aryl carbamate *ortho*-lithiation [25]	R$_2$NCO$_2$Ar (1720)	R$_2$NCO$_2$ArLi (1660)
Grignard quench with CO$_2$ [26]	CO$_2$ (2340)	[RC=C—COMgCl (1650–1500)]
Acetylide addition to aldimine [27]	R—C(H)=N—R' (1670)	[R—C(H)=N—R'—BF$_3$] (1713)
Acetylide addition to Weinreb amide [28]	RCON(OMe)Me (1681 in pentane; 1663 in THF)	RCOR' (1670, after quench)
Benzaldimine formation with LiHMDS [29]	ArCOH (1700)	ArC=N—TMS (1655)
Ester enolization [30]	R$_2$CHCO$_2$t-Bu (1730)	R$_2$C—CO$_2$(Li)t-Bu (1645–1665)
Oxazolidinone enolization [31]	R—CH$_2$CO—Ox (1800)	R—CHCO(K)—Ox (1740–1715)
Sultam enolization [32]	RCH$_2$CONRSO$_2$R* (1695)	RCHLiCONRSO$_2$R* (1620)
Formamide deprotonation [33]	HCONH$_2$ (1710)	HCONHNa (1580)
Benzoic acid methylation [34]	PhCO$_2$H (1700)	PhCO$_2$Me (1725) [35], CH$_2$N$_2$ (2100)
Carboxylic acid to acid chloride [36]	R—CO$_2$H (1710)	R—COCl (1805)
Carboxylic acid to acid chloride [37]	Ar—CO$_2$H (1690)	Ar—COCl (1745, 1770)
Acid chloride to ketene [38]	RR'CHCOCl (1790)	RR'C=C=O (2200)
Acid chloride to ketene [39]	ArCH$_2$COCl (1800)	ArHC=C=O (2330)
Modified Arndt–Eistert reaction [34]	ArCOCl (1775), CH$_2$N$_2$ (2100)	[ArCOCHN$_2$ (2090)], CH$_2$N$_2$ (2100)
Amide bond formation with CDI [40]	ArCOImidazole (1790)	—
Curtius rearrangement to carbamate [18]	—	[RCON$_3$ (1710)]
		RNCO$_2$R (1725)
Aldol reaction [41]	H$_2$C(OR)=CH(OR)OSiMe$_3$ (1670)	[H$_2$C(OR)=CH(OR)OBu$_4$N (1625)]
		[H$_2$C(OR)=CH(OR)OCuL$_n$ (1690, 1550)]
		R'CHOXC(OR)=CH(OR)OBu$_4$N (1730)
Azide–alkyne cycloaddition [42]	Pyrimidinone (1680)	C=C triazole (1625)
		C=N, N=N triazole (1235)
Baeyer–Villiger oxidation [43]	RCOR	[ArCO$_3$H (1165)], [peroxide Int (950)]

^{a}Frequency numbers rounded to the closest frequency within 5 cm^{-1} intervals.
^{b}Structures in brackets represent reaction intermediates.

ammonia was removed from solution through a sequence of nitrogen purges. Due to inability to continuously subsurface sparge with nitrogen in the reaction vessel and high ammonia solubility in MeTHF (>1 mol/L), a large number of purge cycles needed to be implemented. Each purge cycle consisted of the following steps: venting of headspace to atmospheric pressure with agitation turned off, nitrogen introduction into the headspace to 1–2 psig, and agitation of solution for several minutes until pressure stabilization (i.e. a new ammonia liquid–vapor equilibrium was established), which was then followed by stopping of the agitator and venting of the headspace gas to a scrubber. This operation helped reduce headspace pressure close to atmospheric and enabled the charge of solid Pd/Al$_2$O$_3$ catalyst for the subsequent hydrogenation. Figure 42.8 depicts the pressure profile during the purge of residual ammonia for a representative reaction

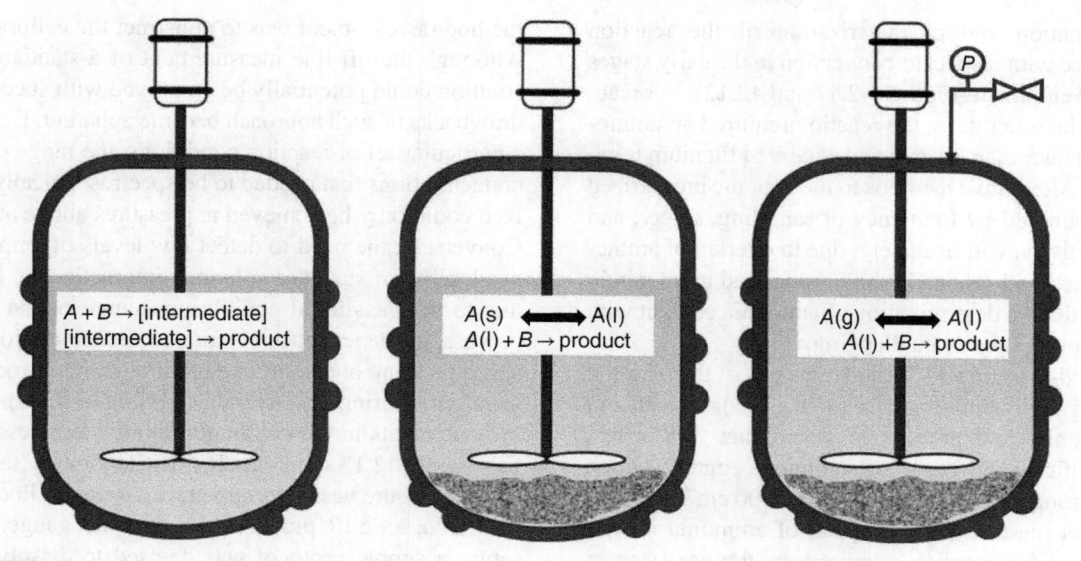

FIGURE 42.6 Schematic illustration of reactions where PAT offers significant advantage over conventional techniques.

TABLE 42.6 Benefits of PAT in Complex Reaction Systems

"Standard" Analytical Techniques	Spectroscopy-based Techniques
Reaction can be disturbed due to physical sampling	No physical sampling necessary
Unstable intermediates could be undetectable	Enable continuous monitoring of reactive intermediates
Long analysis time	Short analysis time that enables real-time control

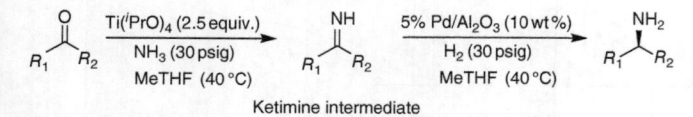

FIGURE 42.7 Reductive amination process.

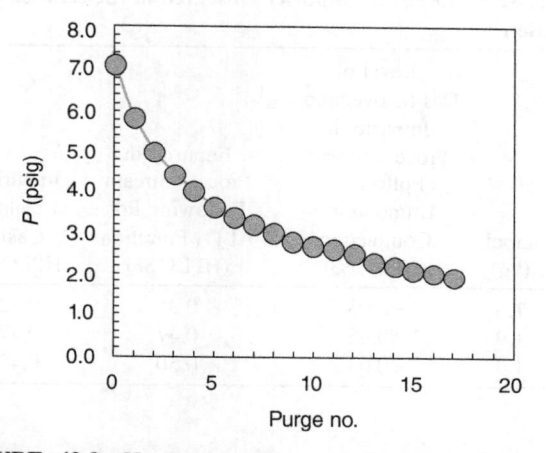

FIGURE 42.8 Headspace pressure during ammonia removal through nitrogen purges.

executed in a 20 L reaction vessel. Table 42.7 summarizes the impurity levels observed at different process stages in the three batches executed during the scale-up campaign.

The impurity formed during filtration of the Pd/Al$_2$O$_3$ catalyst, following hydrogenation completion. Moreover, further investigations demonstrated that impurity formation rates were accelerated by increasing levels of residual ammonia. Interestingly, such high levels of impurity were never observed in laboratory experiments prior to scale-up. Whereas filtration times are on the order of minutes in the laboratory, typical filtration times for plant-scale hydrogenations (~1000 L) are on the order of hours. Therefore, implementing a control strategy that established acceptable ammonia levels in post-ammonolysis stream to prolong post-hydrogenation stream stability had to be considered. Evaluation of multiple spectroscopic techniques for detection of key reaction components is a good general practice when facing a new chemical transformation in the context of PAT development. In-line or at-line testing of reaction components in the solvent system of interest or reaction matrix allows to quickly assess if a particular technique has specificity for the compounds of interest. Figures 42.9 and 42.10 show, respectively, FTIR and NIR spectra of the reaction components. Inspection of the FTIR spectra in the 1500–1800 cm^{-1} region indicated that the carbonyl band of the starting material at approximately 1700 cm^{-1} is well separated from bands of other species and suggested that FTIR was well suited for monitoring the ammonolysis. Indeed, in situ FTIR monitoring enabled the rapid collection of kinetic information and characterization of the reaction parameter space with respect to conversion in the early stages of reaction development (Figures 42.11 and 42.12). For reaction times on the order days, the reaction required an ammonia pressure greater than 30 psig and excess of titanium tetraisopropoxide. Alternative methods to monitor the pressurized reaction were limited by frequency of sampling, safety, and moisture sensitivity. Unfortunately, due to overlap of amine-containing functional groups in the product and ammonia in solution, quantitative determination of ammonia content with FTIR at low levels proved challenging.

As presented in Figure 42.10, examination of the overlaid NIR spectra of starting material, imine intermediate with and without ammonia, and product indicated that NIR offers excellent specificity with respect to ammonia content, which appears as a strong peak at approximately 5000 cm^{-1}. Having established that the characteristic peak of ammonia is well separated from other reaction components, the next step in method development was to construct the calibration curve. Although the off-line measurement of a standard ammonia solution could potentially be employed with success, several drawbacks of such approach became apparent. Depending on a particular set of reaction conditions, the range of ammonia concentrations that needed to be spectroscopically characterized could only be achieved at pressures above atmospheric. Conversely, the need to detect low levels of ammonia made availability of standard solutions impracticable. In addition, due to the gas–liquid equilibrium nature of the dissolution process, the dependence of ammonia concentrations on headspace pressure and solution handling could introduce significant error during the calibration. Using in-line spectroscopic measurements in the calibration eliminated these uncertainties. Figure 42.13 shows a schematic laboratory setup that utilizes a pressure vessel equipped with a supply line of gaseous ammonia, an NIR probe, and a pressure gauge. Using this setup, a simple protocol was devised to dissolve a known amount of gas in the reaction solvent. First, the reactor headspace was pressurized to a known value P_o while the agitator was turned off. Second, the gas supply was turned off, and the agitator activated until an equilibrium pressure P_f at which a constant gauge reading was achieved. Using ideal gas law, namely,

$$\Delta n(NH_3)_{headspace} = (P_o - P_f) \cdot \frac{V_{headspace}}{RT} = \Delta n(NH_3)_{solution},$$

allows to determine the number of moles of ammonia dissolved in solution for each such operation. Repeating this sequence several times (Figure 42.14) affords the calibration curve shown in Figure 42.15, where ammonia concentration is proportional to ammonia NIR peak height. In this particular case, the ammonia peak at approximately 5000 cm^{-1} was normalized relative to the MeTHF solvent peak. Development of the analytical method for in situ ammonia detection enabled

TABLE 42.7 Levels of Impurity Observed in the Reductive Amination

Batch	Input (kg)	Level of Des-halogenation Impurity in Process Stream Following Ammonolysis Completion (HPLC %)	Impurity in Process Stream Following Pd/Al$_2$O$_3$ Filtration (HPLC %)	Impurity in Isolated Cake (HPLC %)
1	1.8	<0.05	0.36	0.32
2	1.9	<0.05	0.47	0.47
3	1.9	<0.05	0.50	0.43

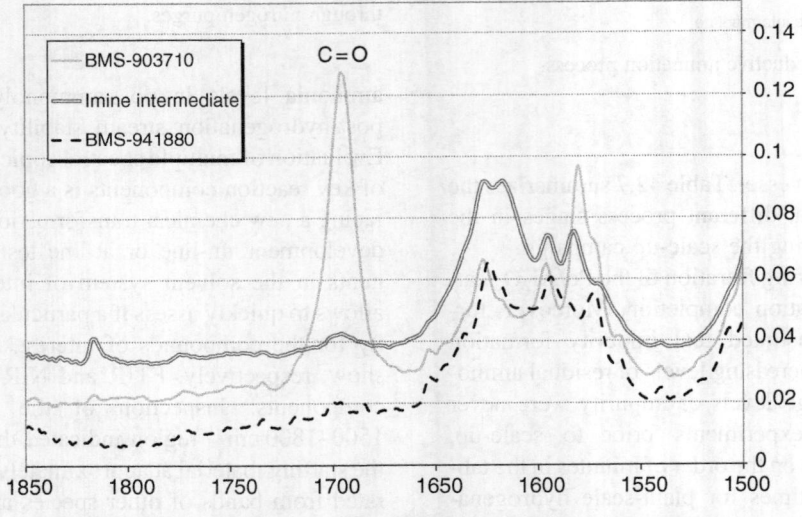

FIGURE 42.9 FTIR spectra of various mixtures of reaction components.

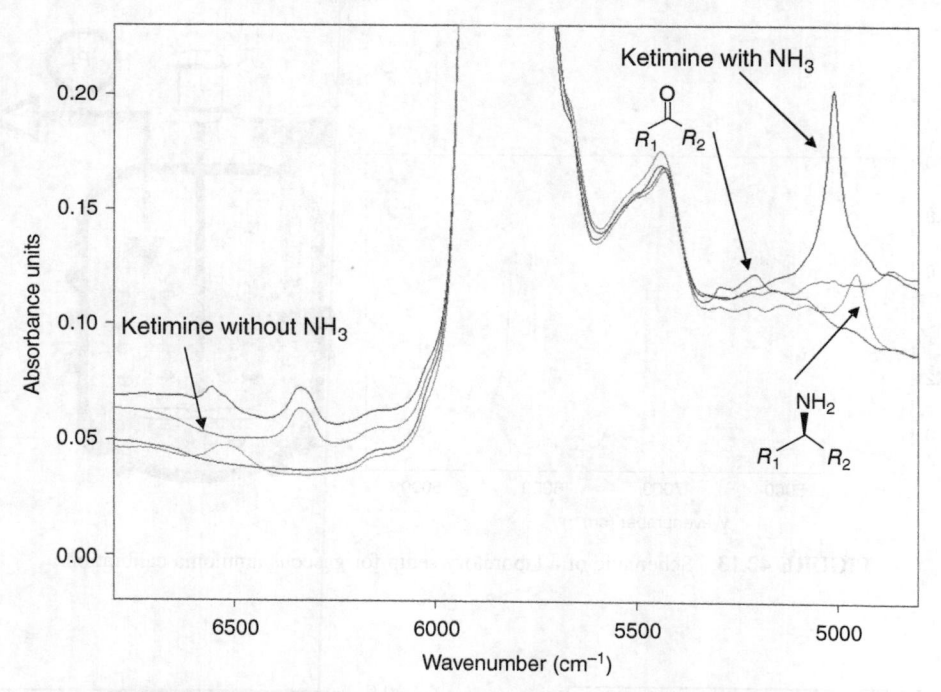

FIGURE 42.10 NIR spectra of various mixtures of reaction components.

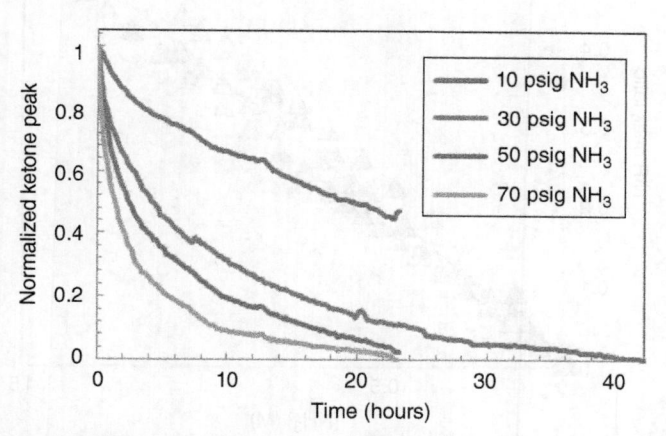

FIGURE 42.11 Ammonolysis reaction progress versus ammonia pressure.

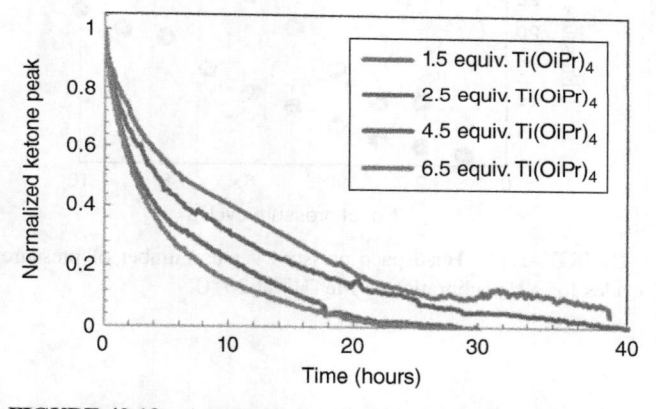

FIGURE 42.12 Ammonolysis reaction progress versus Ti(OiPr)$_4$ equivalents.

conduction of laboratory stability studies for post-hydrogenation streams of variable ammonia content. It was established that ammonia concentrations lower than 0.1 M afforded extended multi-day stream stability. NIR IPC for residual ammonia concentrations lower than 0.1 M was set in place at the end of the ammonolysis reaction. Implementation of this spectroscopy IPC during continuous subsurface nitrogen sparge in scale-up batches afforded significant reduction of the impurity levels as shown in Table 42.8.

42.4.2 Case Study 2: FTIR Spectroscopy Enabled Control Strategy in Manufacturing of Brivanib Alaninate

The last step of the chemical sequence for the preparation of anticancer drug brivanib alaninate is shown in Figure 42.16 [47]. The overall transformation involves conversion of the parent drug into a prodrug by formation of its alaninate ester. Two reactions are telescoped in a single step without isolation of the intermediate species: (i) esterification of the parent

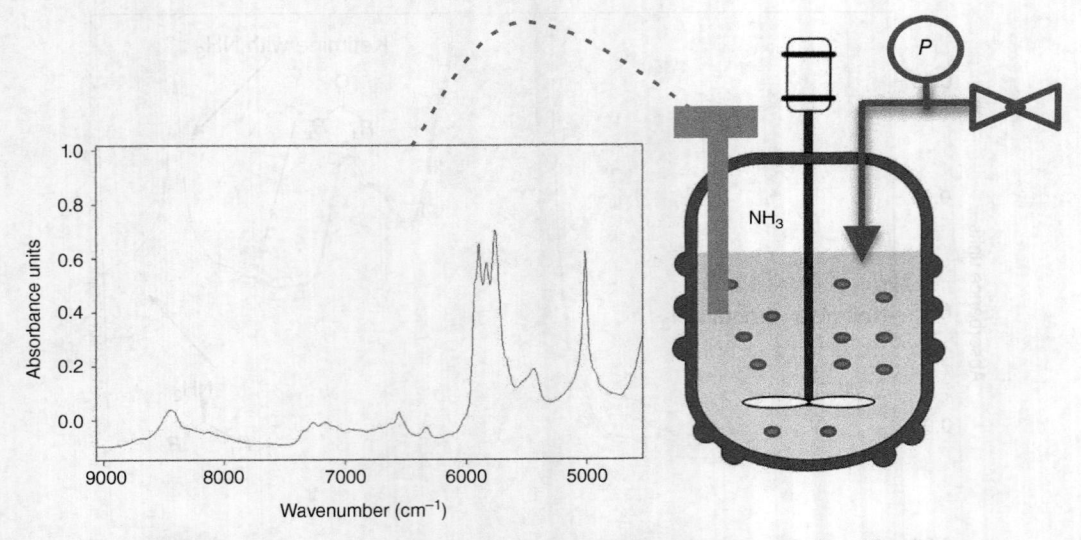

FIGURE 42.13 Schematic of a laboratory setup for gaseous ammonia calibration.

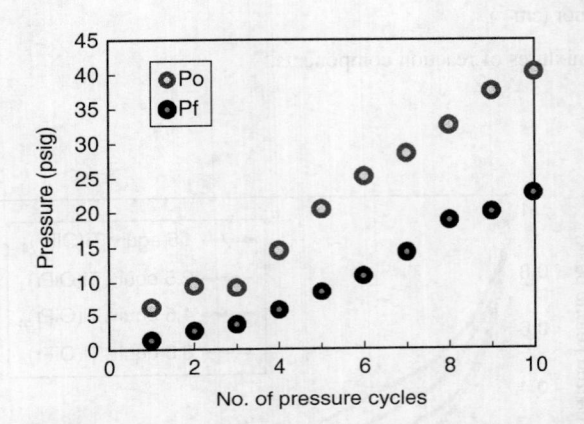

FIGURE 42.14 Headspace pressure versus number of pressure cycles for NH_3 calibration in MeTHF at 20 °C.

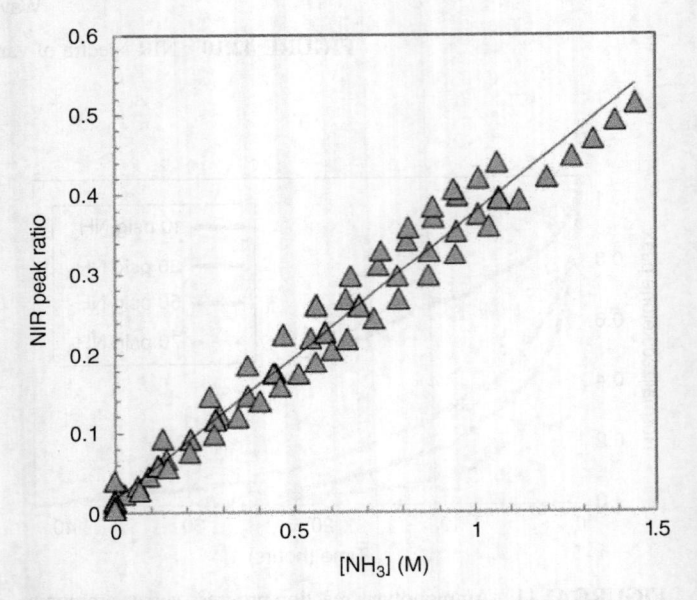

FIGURE 42.15 NIR calibration curve for NH_3 concentration in MeTHF at 20 °C.

drug with Cbz-protected alanine and (ii) removal of the alanine Cbz protecting group via hydrogenolysis. Prolonged aging of the reaction mixture during the hydrogenolysis could result in elevated levels of the impurities shown in Figure 42.17.

During laboratory development, it was established that if hydrogen supply is insufficient or interrupted at early stages of the reaction, catalyst poisoning by CO_2 formed during the hydrogenolysis could occur and result in significant reaction rate inhibition. To address this risk and assure manufacturing robustness, a control strategy that implemented two PAT-based IPCs was developed: (i) reaction progress IPC-1 at 20 minutes to protect against stalling associated with catalyst poisoning and (ii) end-of-reaction IPC-2 to protect against over-aging. An additional IPC-3 was established to measure

TABLE 42.8 Levels of Impurity Observed During Scale-up Implementation

Batch	Input (kg)	Impurity in Process Stream Following Pd/Al_2O_3 Filtration (HPLC %)	Impurity in Isolated Cake (HPLC %)
1	37	0.25	0.23
2	50	0.19	0.14
3	56	0.21	0.17

FIGURE 42.16 Final step for the manufacturing of brivanib alaninate.

FIGURE 42.17 Brivanib alaninate hydrogenolysis impurities.

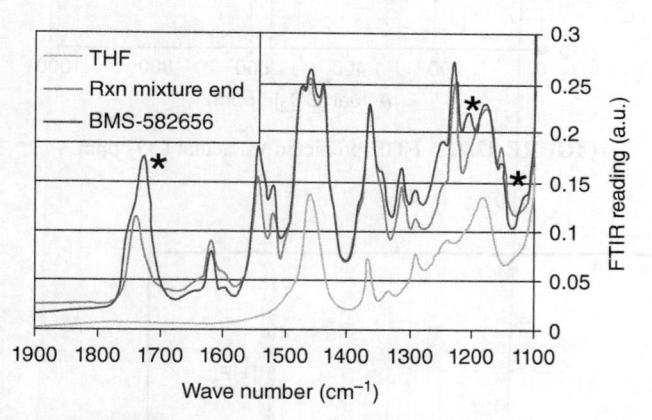

FIGURE 42.18 FTIR spectra of brivanib alaninate hydrogenolysis components.

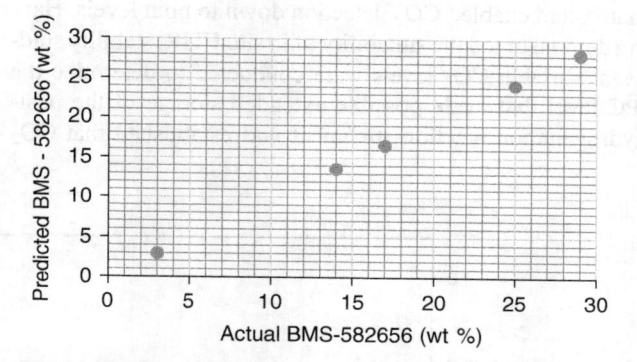

FIGURE 42.19 FTIR predicted vs. actual BMS-582656 wt %.

residual levels of CO_2 at hydrogenation completion due to the accelerating effect of CO_2 on the hydrolysis reaction rate that affords parent drug as an undesired impurity.

In order to establish suitability of FTIR spectroscopy for both kinetics tracking and CO_2 concentration, reaction components were evaluated with in situ measurements. Figure 42.18 shows spectra of the THF solvent, starting material (BMS-582656 in THF), and end-of-reaction mixture containing brivanib alaninate product in the 1100–1900 cm^{-1} region. Three peaks corresponding to starting material at ~1710 cm^{-1}, ~1200 cm^{-1}, and ~1120 cm^{-1} could be monitored during reaction progress. A PLS regression model fit to a series of calibration solutions spectra correlates integration of the above IR peaks with concentrations of starting material during the reaction. In order to improve model accuracy, the training sets covered a range of reaction parameters such as temperature (20–35 °C) and catalyst loading (5–17 wt %). Figure 42.19 illustrates the predictive power of the model obtained. Spectroscopic sensitivity is a key issue worth highlighting in the context of starting material concentration determination. Due to stringent conversion requirements encountered in pharmaceutically relevant reactions, most spectroscopic methods will not be able to detect residual starting material at levels of less than or equal to 1

HPLC %. The FTIR detection limit for the brivanib alaninate hydrogenolysis was approximately 2.5 wt %, which is significantly higher than the desired conversion of 99%. In order to overcome this limitation, a prediction-based end-of-reaction IPC was developed. The idea is to evaluate the reaction rate at conversion levels close to the limit of sensitivity and predict reaction completion time in accordance with a kinetic law. A first-order rate law

$$\frac{d[BMS-582656]}{dt} = -k[BMS-582656]$$

was tested and proved effective at predicting conversion in the concentration regime below FTIR sensitivity. In terms of algorithm design, several data points were measured before the starting material concentration reaches 3–4 wt %, and the first-order equation was fit to them, thus providing as an output a time when greater than 99% conversion would be achieved.

Figure 42.20 shows the IR spectra of reaction components in the expanded approximately 700–4000 cm^{-1} region. The CO_2 peak at approximately 2300 cm^{-1} was well resolved from the rest of reaction components, and a calibration curve could be easily established in a fashion similar to the one described in Case Study 42.1 (Figure 42.21). In situ FTIR was highly sensitive with respect to CO_2 in the reaction matrix and enabled CO_2 detection down to ppm levels. Having demonstrated the suitability of in situ FTIR, stability studies at variable CO_2 levels were conducted to determine the IPC level that made possible extended storage of the post-hydrogenation reaction stream. It was established that CO_2 levels of less than or equal to 250 ppm were acceptable and appropriate as the IPC-3 limit.

Following extensive laboratory studies, FTIR-based IPCs were implemented during a pilot plant campaign in 1000 L hydrogenator. For this first phase of IPC implementation in a scale-up setting, calculations were performed manually on plant floor using a computer program and logged into quality assurance data systems, following standard GMP practices. Table 42.9 summarizes results from three batches executed during the campaign. All of the IPCs performed well and proved to be an effective element of the overall control strategy to assure API quality.

Following successful implementation of spectroscopy-based IPCs in the pilot plant, the process transitioned into the manufacturing validation campaign. To eliminate the need to perform manual calculations and automate data

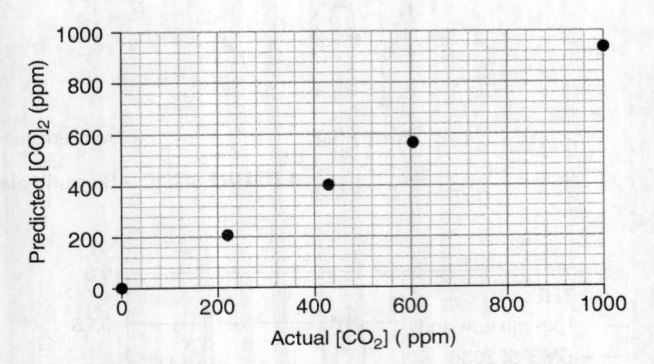

FIGURE 42.21 FTIR predicted vs. actual CO_2 ppm.

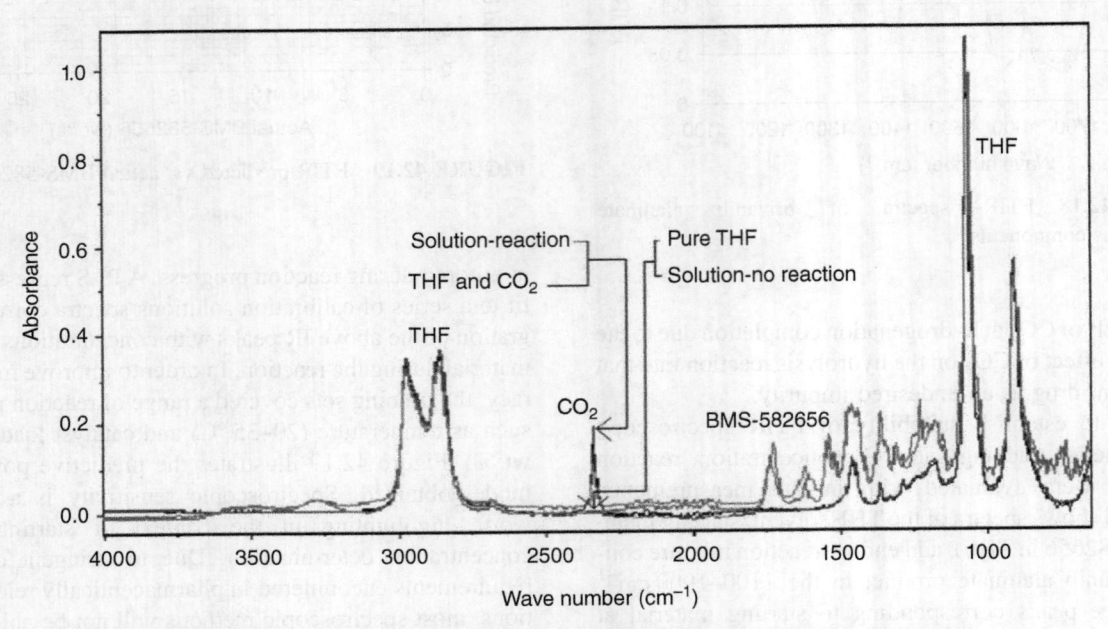

FIGURE 42.20 FTIR spectra of brivanib alaninate hydrogenolysis reaction components.

acquisition and processing in the manufacturing setting, a graphic operator interface was developed. Figure 42.22 displays a screenshot of the graphic interface during the execution of a batch showing trends for BMS-582656 wt % (measured and extrapolated), CO_2 levels, and the status of each IPC. Both analytical method and instrumentation demonstrated robustness in all of twenty validation campaign hydrogenation loads. Minor probe realignment and algorithm parameter adjustments to accommodate lower signal-to-noise ratios were required only once.

A summary of hydrogenation reaction times to reach greater than 99% conversion as determined by IPC-2 is shown in Figure 42.23. Reaction time spread is evident, confirming inherent variability associated with the process. Moreover, IPC-1 failure in one hydrogenation load allowed identification of an incipient leak in the agitator nitrogen seal as the hydrogen partial pressure was not preserved (Figure 42.24). Addition of catalyst kicker charge as prescribed by IPC-1 logic allowed to drive reaction to completion despite the leak and ensured successful recovery of product. It also informed plant maintenance personnel to investigate the agitator integrity and replace the seal prior to initiation of another batch. Overall, implementation of PAT offered significant advantages over standard at-line methods to mitigate risks associated with reaction robustness and ensure impurity control.

TABLE 42.9 FTIR-Based IPC Results During Pilot Plant Implementation

	Batch 1	Batch 2	Batch 3
BMS-582656 wt % at $t = 0$ min	23.2	20.9	22.4
IPC-1: BMS-582656 wt % at $t = 20$ min	3.5	3.1	4.3
IPC-1 conversion (%)	85.0	85.2	80.7
IPC-2: predicted time to purge initiation (at 3 wt %) (min)	67	79	117
IPC-2: predicted time to purge initiation (at 4 wt %) (min)	51	72	78
IPC-3 result by HPLC, HPLC% BMS-582656	ND	0.2	0.2
Time to purge CO_2 (min)	247	248	262

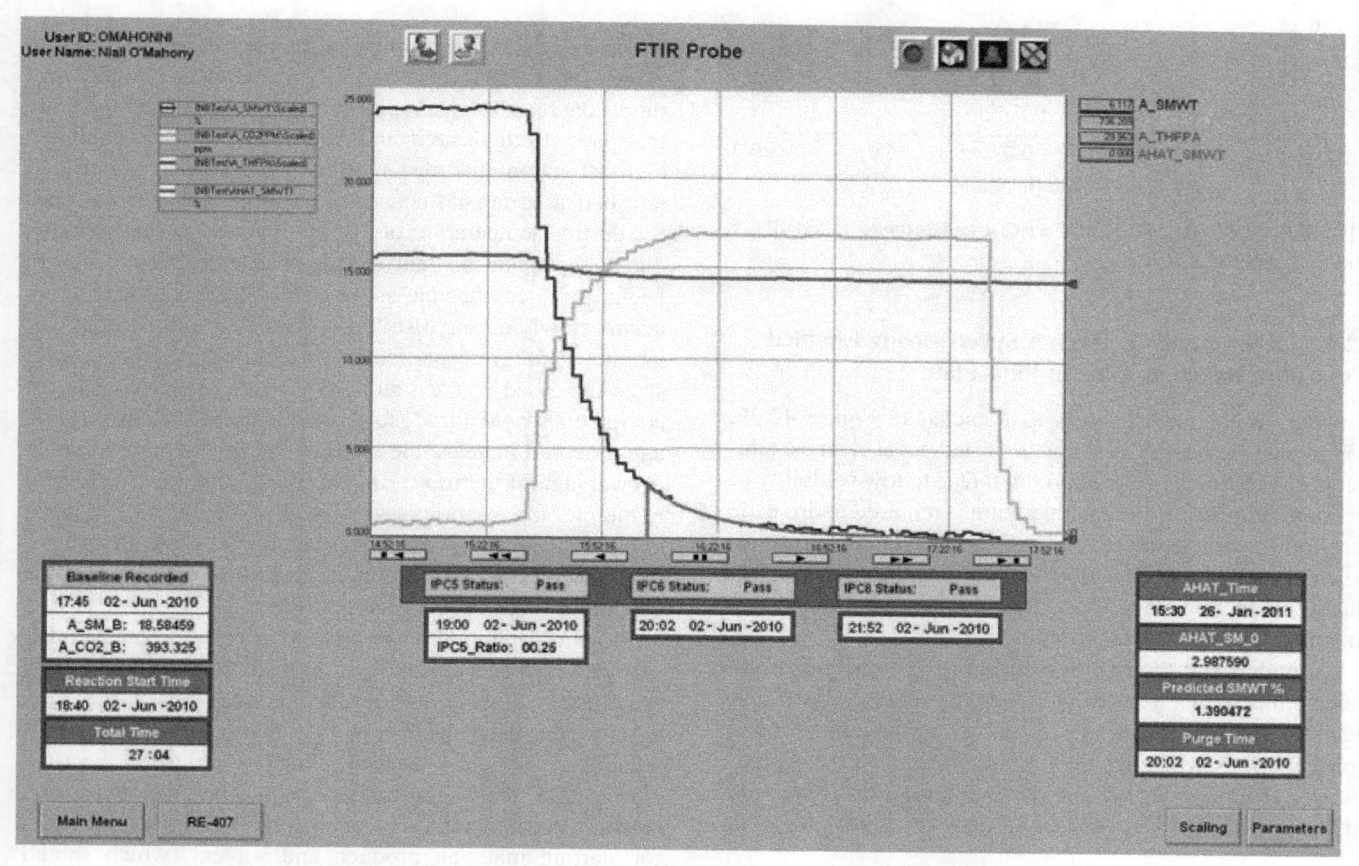

FIGURE 42.22 Plant operator interface showing real-time concentrations of starting material and CO_2.

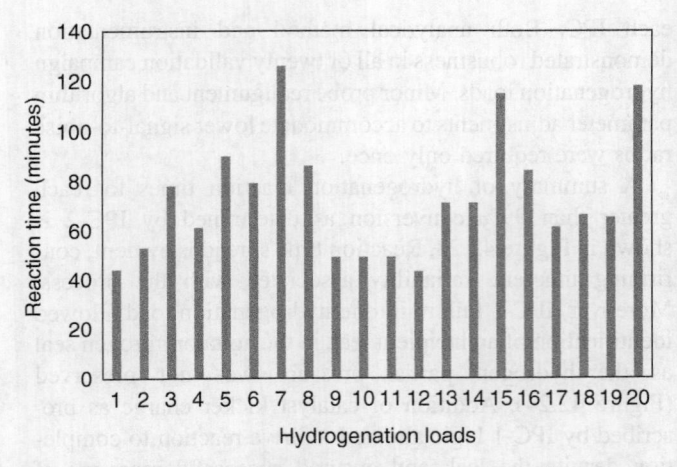

FIGURE 42.23 Hydrogenation reaction time to achieve greater than 99% conversion as predicted by IPC-2.

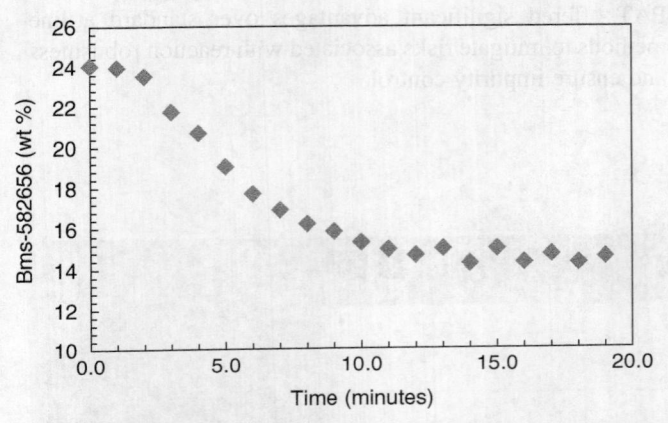

FIGURE 42.24 BMS-582656 wt % trend prior to failed IPC-1 due to agitator seal leak.

42.4.3 Case Study 3: Raman Spectroscopy Enabled Control of Reaction Time in Pilot Plant

Consider a chemical process as depicted in Figure 42.25, wherein an elimination reaction is carried out at elevated temperature to form the desired product. Due to low solubility of the starting material, high temperature is required to drive the reaction to completion, but extended holds at elevated temperature lead to product degradation and the formation of quality-impacting impurities. Thus, control of reaction age time is critical to successful implementation.

To understand the risk of over-aging the reaction at elevated temperature, a series of kinetic experiments were conducted using HPLC analysis under varying temperatures and concentrations. Modeling the reaction system in DynoChem[3] showed that both desired and undesired reactions were well fit by unimolecular, first-order rate expressions with

[3]Dynochem Software, Scale-up Systems Ltd., Dublin, Ireland.

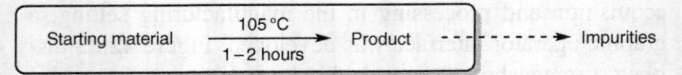

FIGURE 42.25 Elimination reaction process.

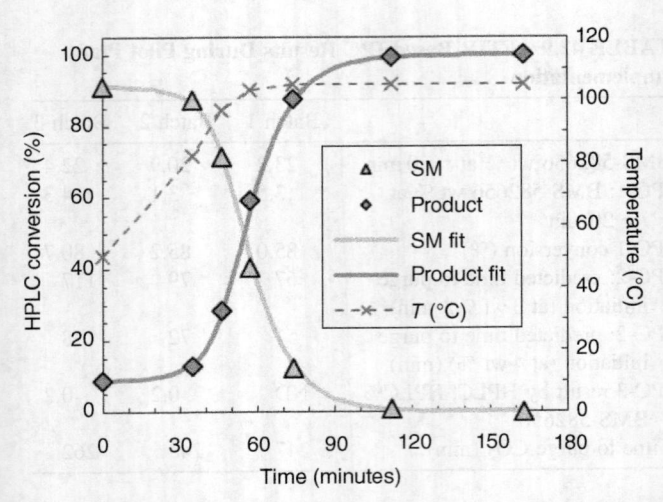

FIGURE 42.26 Kinetics of the desired reaction (HPLC data and fit).

Arrhenius-style temperature dependence. Figure 42.26 shows an example data set for conversion as a function of time – the reaction rate rapidly accelerates as the temperature increases. Both desired and undesired reactions exhibited high activation energies, and thus variability in the heating rate had a strong influence on the optimal age time once the desired temperature of 105 °C is reached. Furthermore, due to upstream solvent exchange, the reaction does not begin at 0% conversion, as the thermal elimination reaction occurs slowly during distillation, introducing additional variability. This grew particularly concerning as the process scale increased and the reduced heat transfer rate in larger equipment would introduce variability to the optimal reaction age time and increase the risk of impurity formation due to over-aging. Furthermore, off-line HPLC analysis is sufficient to monitor the reaction in small-scale experiments, but is not viable as a control strategy on scale due to the long analysis time and the processing delays associated with cooling the batch to a safe sampling temperature.

Reaction endpoint determination via in-line Raman analysis provides an elegant solution to these problems. Using spectroscopy to monitor reaction conversion in real time ensures that each batch is held for the optimal age time, regardless of the thermal "history" during heat-up. Figure 42.27 shows the Raman spectra over the course of a reaction. A chemometric model was built using distinct peaks for starting material, product, and solvent (which should remain constant throughout) and was demonstrated to be

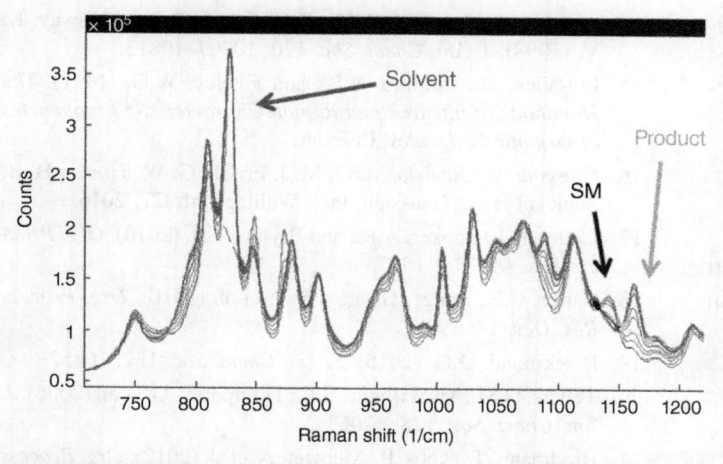

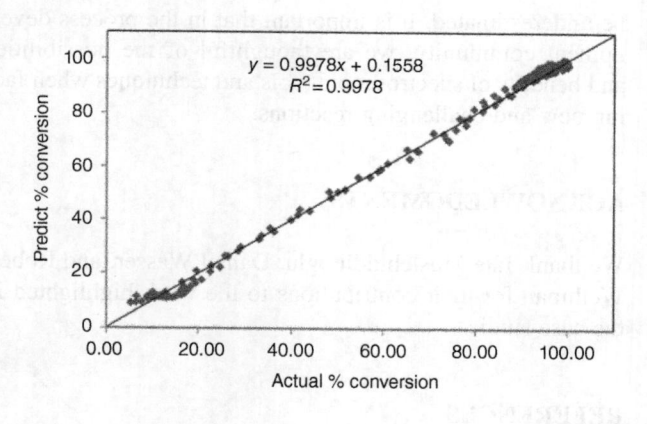

FIGURE 42.27 Raman spectra during elimination reaction progress (left) and linearity of HPLC results versus Raman prediction of reaction completion for calibration set spectra (right).

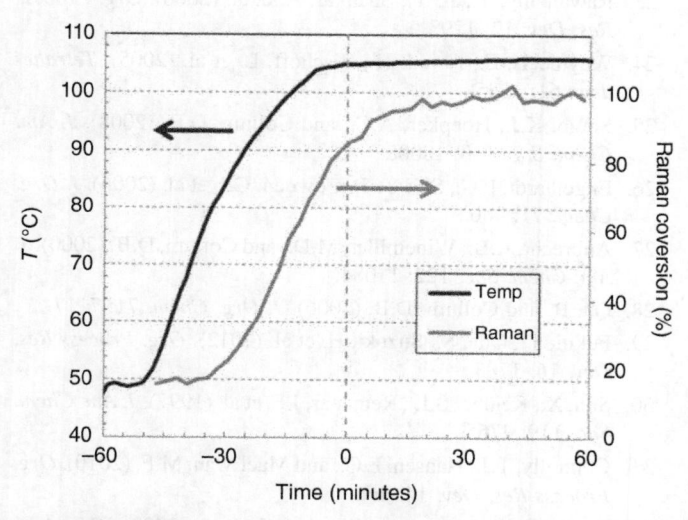

FIGURE 42.28 On scale temperature and Raman product reaction profiles.

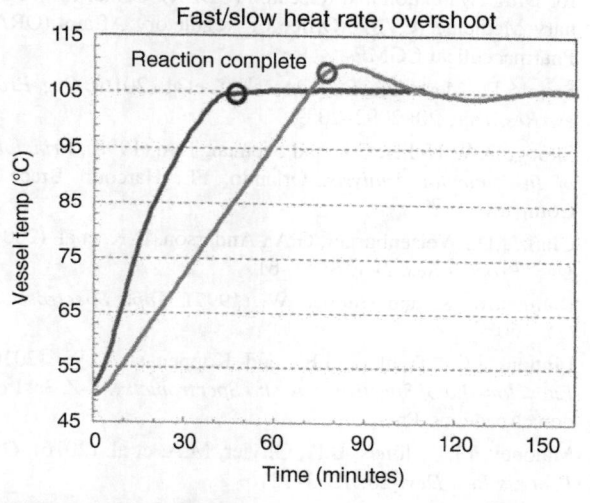

FIGURE 42.29 Raman predicted reaction completion time (circles) for two different heating protocols on scale.

sufficiently quantitative throughout the reaction and within the temperature range of interest. An example trace of reaction rate as monitored by PAT is shown in Figure 42.28, in which reaction conversion via Raman tracks with reactor temperature.

This in-line Raman method allowed for facile technology transfer, as the reaction hold time was determined independently of equipment parameters. Figure 42.29 shows reaction progression in two different, but similarly sized, pilot-scale reactors. The Raman method reduced risk to quality by ensuring timely calling of the reaction endpoint, even in the case with slower heat-up and overshoot, and was successfully demonstrated in twelve pilot-scale batches to convert up to 80 kg of starting material.

42.5 SUMMARY

PAT-based techniques are an important component in the characterization, development, and implementation of pharmaceutically relevant chemical processes. They provide a wealth of information complementary to standard off-line methods, as illustrated by case studies described in this chapter. These benefits can be realized across all stages of development, from early reaction identification and optimization on laboratory scale all the way to process characterization and control strategy design on manufacturing scale. With the stringent quality and robustness requirements of pharmaceutical products, and the increasing complexity of the chemical processes required to make them, the role of PAT cannot be underestimated. It is important that in the process development community, we are thoughtful of the possibilities and benefits of spectroscopic tools and techniques when facing new and challenging reactions.

ACKNOWLEDGMENTS

We thank Jale Muslehiddinoglu, Daniel Wasser, and Robert Wethman for their contributions to the work highlighted in the case studies.

REFERENCES

1. Guidance for Industry PAT — A Framework for Innovative Pharmaceutical Development, Manufacturing, and Quality Assurance. (September 2004). U.S. Department of Health and Human Services, Food and Drug Administration, Center for Drug Evaluation and Research (CDER), Center for Veterinary Medicine (CVM), Office of Regulatory Affairs (ORA), Pharmaceutical CGMPs.
2. Šahnić, D., Mestrovic, E., Jednačak, T. et al. (2016). *Org. Process Res. Dev.* 20: 2092–2099.
3. Skoog, D.A., Holler, F.J., and Nieman, T.A. (1998). *Principles of Instrumental Analysis*. Orlando, FL: Harcourt Brace & Company.
4. Clark, J.D., Weisenburger, G.A., Anderson, D.K. et al. (2004). *Org. Process Res. Dev.* 8: 51–61.
5. Rahmelow, K. and Hubner, W. (1997). *Appl. Spectrosc.* 51 (2): 160.
6. Lindon, J.C., Tranter, G.E., and Koppenaal, D.W. (2016). *Encyclopedia of Spectroscopy and Spectrometry: S-Z*, 3e. London: Academic Press.
7. Maloney, M.T., Jones, B.P., Olivier, M.A. et al. (2016). *Org. Process Res. Dev.* 20: 1203–1216.
8. (a) Bordawekar, S., Chanda, A., Daly, A.M. et al. (2015). *Org. Process Res. Dev.* 19: 1174–1185.
   (b) Zhou, G., Grosser, S., Sun, L. et al. (2016). *Org. Process Res. Dev.* 20: 653–660.
9. Burns, D.R. and Ciurczak, E.W. (eds.) (1992). *Handbook of Near-Infrared Analysis*. New York: Marcel Dekker.
10. Stuart, B.H. (ed.) (2004). Organic molecules. In: *Infrared Spectroscopy: Fundamentals and Applications*. Hoboken, NJ: Wiley.
11. (a) Ben Said, M., Baramov, T., Herrmann, T. et al. (2017). *Org. Process Res. Dev.* 21: 705–714.
    (b) Hart, R., Herring, A., Howell, G.P. et al. (2015). *Org. Process Res. Dev.* 19: 537–542.
    (c) Grabow, K. and Bentrup, U. (2014). *ACS Catal.* 4: 2153–2164.
12. Croce, A.E. (2008). *Can. J. Chem.* 86: 918–924.
13. Mas-Ballesté, R. and Que, L. Jr. (2007). *J. Am. Chem. Soc.* 129: 15964–15972.
14. Abbotto, A., Leung, S.S.-W., Streitwieser, A., and Kilway, K.V. (1998). *J. Am. Chem. Soc.* 120: 10807–10813.
15. Lin-Vien, D., Colthup, N.B., and Fateley, W.G. (1991). *The Handbook of Infrared and Raman Characteristic Frequencies of Organic Molecules*. Elsevier.
16. *Gaussian 09, Revision A.02*, M. J. Frisch, G. W. Trucks, H. B. Schlegel et al., Gaussian, Inc., Wallingford, CT, 2016.
17. Littler, B.J., Looker, A.R., and Blythe, T.A. (2010). *Org. Process Res. Dev.* 14: 1512.
18. Carter, C.F., Lange, H., Ley, S.V. et al. (2010). *Org. Process Res. Dev.* 14: 393.
19. Blackmond, D.G. (2015). *J. Am. Chem. Soc.* 137: 10852.
20. Tellers, D.M., McWilliams, J.C., Humphrey, G. et al. (2006). *J. Am. Chem. Soc.* 128: 17063.
21. Brodmann, T., Koos, P., Metzger, A. et al. (2012). *Org. Process Res. Dev.* 16: 1102.
22. Dumouchel, S., Mongin, F., Trécourt, F., and Quéguiner, G. (2003). *Tetrahedron* 59: 8629.
23. Rawalpally, T., Ji, Y., Shankar, A. et al. (2008). *Org. Process Res. Dev.* 12: 1293.
24. Weymeels, E., Awad, H., Bischoff, L. et al. (2005). *Tetrahedron* 61: 3245.
25. Singh, K.J., Hoepker, A.C., and Collum, D.B. (2008). *J. Am. Chem. Soc.* 130: 18008.
26. Engelhardt, F.C., Shi, Y.-J., Cowden, C.J. et al. (2006). *J. Org. Chem.* 71: 480.
27. Aubrecht, K.B., Winemiller, M.D., and Collum, D.B. (2000). *J. Am. Chem. Soc.* 122: 11084.
28. Qu, B. and Collum, D.B. (2006). *J. Org. Chem.* 71: 7117.
29. Fukui, Y., Oda, S., Suzuki, H. et al. (2012). *Org. Process Res. Dev.* 16: 1783.
30. Sun, X., Kenkre, S.L., Remenar, J.F. et al. (1997). *J. Am. Chem. Soc.* 119: 4765.
31. Connolly, T.J., Hansen, E.C., and MacEwan, M.F. (2010). *Org. Process Res. Dev.* 14: 466.
32. Koch, G., Loiseleur, O., Fuentes, F. et al. (2002). *Org. Lett.* 22: 3811.
33. Ramirez, A., Mudryk, B., Rossano, L., and Tummala, S. (2012). *J. Org. Chem.* 77: 775.
34. Dallinger, D., Pinho, V.D., Gutmann, B., and Kappe, C.O. (2016). *J. Org. Chem.* 81: 5814.
35. Smith, B.C. (ed.) (1998). Functional groups containing the CO bond. In: *Infrared Spectral Interpretation: A Systematic Approach*. Boca Raton, FL: CRC Press.
36. Dozeman, G.J., Fiore, P.J., Puls, T.P., and Walker, J.C. (1997). *Org. Process Res. Dev.* 1: 137.
37. Agbodjan, A.A., Cooley, B.E., Copley, R.C.B. et al. (2008). *J. Org. Chem.* 73: 3094.
38. France, S., Wack, H., Taggi, A.E. et al. (2004). *J. Am. Chem. Soc.* 126: 4245.
39. Liang, J.T., Mani, N.S., and Jones, T.K. (2007). *J. Org. Chem.* 72: 8243.

40. Mennen, S.M., Mak-Jurkauskas, M.L., Bio, M.M. et al. (2015). *Org. Process Res. Dev.* 19: 884.
41. Pagenkopf, B.L., Krüger, J., Stojanovic, A., and Carreira, E.M. (1998). *Angew. Chem. Int. Ed.* 37: 3124.
42. Stefani, H., Silva, N.C.S., Manarin, F. et al. (2012). *Tetrahedron Lett.* 53: 1742.
43. Lan, H.-Y., Zhou, X.-T., and Ji, H.-B. (2013). *Tetrahedron* 69: 4241.
44. Togkalidou, T., Fujiwara, M., Patel, S., and Braatz, R.D. (2001). *J. Cryst. Growth* 231: 534.
45. De Smet, K., van Dun, J., Stokbroekx, B. et al. (2005). *Org. Process Res. Dev.* 9: 344.
46. Skliar, D., Nye, J., Hickey, M. et al. (2015). Spectroscopy assisted process design for complex multiphase organic reactions. 2015 *AIChE Annual Meeting*.
47. Lobben, P. et al. (2015). *Org. Process Res. Dev.* 19: 900–907.

43

PROCESS MODELING APPLICATIONS TOWARD ENABLING DEVELOPMENT AND SCALE-UP: CHEMICAL REACTIONS

ANUJ A. VERMA, STEVEN RICHTER, BRIAN KOTECKI, AND MOIZ DIWAN

Process Research and Development, AbbVie Inc., North Chicago, IL, USA

43.1 INTRODUCTION

With an increase in the number of pipeline compounds in CMC development among pharmaceutical companies that need to be delivered using limited pool of resources, efficient tools for process understanding need to be established to enable predictive scale-up of chemical reactions in the pilot plant and in manufacturing. Process modeling is an example of such efficient tools available among the chemical engineering community. Through an appropriate application of this tool, a process engineer can obtain multivariate understanding of the various factors affecting the scale-up of a chemical reaction with limited data. Process models can further assist in simulating the impact of various process parameters on quality target product profile. As a result, an appreciable number of pharmaceutical companies have dedicated resources to modeling chemical reaction processes toward enabling development and scale-up.

Process modeling of chemical reactions uses mathematical principles of chemical reaction engineering to prepare mechanistic multivariate models that agree with experiment. A typical roadmap of such an analysis is shown in Figure 43.1.

An aim for a process modeler is to assess the suitability or robustness of a contacting device that conducts the chemistry of interest with regard to three major areas of investigation: kinetics and thermodynamics, heat and mass transfer, and flow patterns. Having a multivariate understanding of these factors enables understanding of the fate of chemical reaction as the reactant is converted to product, by-products, and impurities. A brief description of these areas of investigation is provided below:

- **Kinetics and thermodynamics.** The analysis of intrinsic reaction kinetics and chemical reaction equilibria enables a process modeler to assess the forward and net rate of conversion of a starting material to product/impurity in the absence of mass transfer limitations. For example, in a simplistic example of bimolecular second-order reactions, kinetics provide a measure of the probability of a successful collision that forms the product/impurity out of all the collisions between the two sets of molecules. These probabilities change with concentration and reaction temperature and are generally described adequately by the Arrhenius equation coupled with chemical reaction equilibria.

- **Transport phenomena.** Analysis of transport phenomena enables a process modeler to assess the total number of molecular collisions (irrespective of number of successful collisions leading to a chemical reaction). This is typically calculated as the rate at which the contacting device can bring the reacting molecules to collide and is a function of mixing intensity in a single phase and interphase transport rate in a multiphase reaction. This rate is always greater than or equal to the rate of chemical reaction involving bimolecular collisions. Mixing

---

Chemical Engineering in the Pharmaceutical Industry: Active Pharmaceutical Ingredients, Second Edition.
Edited by David J. am Ende and Mary T. am Ende.
© 2019 John Wiley & Sons, Inc. Published 2019 by John Wiley & Sons, Inc.

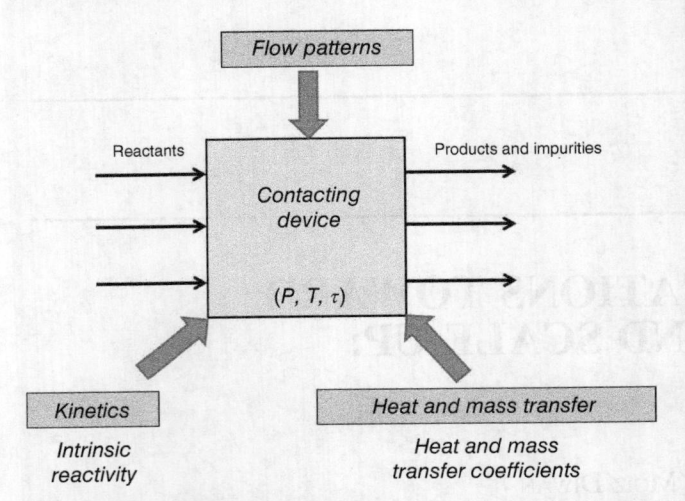

FIGURE 43.1 Typical roadmap for process modeling in chemical reaction engineering.

$$R-OH + MsCl \xrightarrow{Et_3N} R-OMs + Et_3\overset{\oplus}{N}HCl$$
$$MsCl + Solvent \longrightarrow Impurity$$

SCHEME 43.1 Reaction of an alcohol with methanesulfonyl chloride (MsCl). The solvent has an alternate reaction with MsCl.

contacting devices in order to enable predictive scale-up of business-relevant chemical transformations.

43.2 PREDICTIVE SCALE-UP OF A HIGHLY EXOTHERMIC REACTION

43.2.1 Introduction

In the pharmaceutical industry, the focus of process development shifts from obtaining an adequate yield and quality in a chemical reaction in early stage (before phase I clinical readout) to a robust process with a proper understanding of critical quality attributes (CQAs) on reaction performance and product quality in later stages (phase II clinical readout and beyond). For late-stage development, chemical reactions can be challenging to scale-up due to a lack of fundamental understanding of the impact of transport phenomena in impurity generation and overall robustness of the reaction process. Herein, we apply the reaction process modeling roadmap to an example of a mixing sensitive mesylation reaction in a late-stage drug substance process. The reaction pathway is shown in Scheme 43.1.

43.2.2 Problems Faced in Reaction Scale-up

Methanesulfonyl chloride (MsCl) undergoes a fast reaction with starting material alcohol in the presence of triethylamine to form the desired product and triethylammonium hydrochloride salt. This reaction is very exothermic with a 250 kJ/mol heat of reaction and an adiabatic temperature rise (ATR) of 120 °C in that solvent. On the other hand, methanesulfonyl chloride also reacts with the solvent to form an impurity (Scheme 43.1). A fed-batch process was developed with slow dosing of MsCl to keep the overall reaction temperature at less than 10 °C (Figure 43.2). Due to the side reaction with the solvent, early process development work recommended the use of 1.3 equiv of MsCl. Further lowering the MsCl equivalents led to incomplete conversion, while higher equivalents led to additional impurity generation. It was desired that the reaction endpoint would have no more than 2% starting material alcohol remaining in the reaction mixture (>98% conversion). A higher level of remaining starting material alcohol and an increased impurity burden were separately shown to have an inadequate purge in downstream processing.

Preliminary reaction scale-up in fed-batch reactors of increasing sizes (1–1000 L) revealed undesirable scale-up

also affects the heat distribution locally in an exothermic or endothermic chemical reaction. This is shown mathematically by similar dimensionless equations of heat and mass transport locally in a differential fluid volume [1]. Moreover, this analysis has a direct impact on the kinetics of reaction due to its ability to map out the variation of concentration of chemical entities and temperature in the contacting device. Under a large number of practical scenarios, computational fluid dynamics (CFD) becomes an important tool for understanding transport phenomena.

- **Flow patterns.** Most analysis of kinetics and transport phenomena locally depends on the reaction fluid following a variation of an ideal flow pattern, i.e. plug flow (in a plug flow reactor [PFR]) or perfectly back-mixed flow (in a CSTR). In real contacting devices, however, the analysis of flow patterns is paramount to the performance of the same because these flow patterns and other system nonidealities (like dead volumes and recirculation zones) affect the generation of impurities and by-products. Often, based on the kinetics of impurity generation, or from an operational viewpoint (slurry processing, for example), a particular flow pattern can be desired. As a result, a systematic analysis of global flow patterns in the contacting device helps to ensure robustness of the reaction process with respect to impurity/by-product generation. In flow reactors, a combination of kinetics and transport phenomena can help in guiding the desirable flow patterns in a continuous reactor that maximize yield and minimize impurities.

This chapter provides three detailed examples where the chemical reaction process modeling roadmap has been utilized to design and understand the robustness of respective

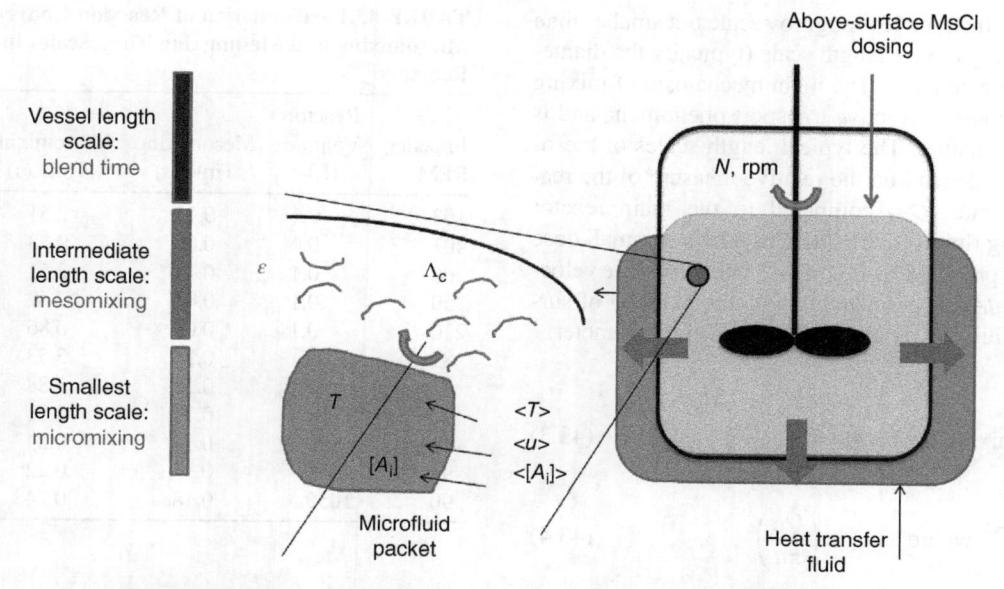

FIGURE 43.2 Process diagram for the mesylation reaction in a fed-batch reactor.

effects. There were instances of reaction stalling at suboptimal conversion of the alcohol. These batches would be subjected to additional charges of MsCl, leading to increased impurity levels. Due to these variabilities, there was risk of failing reaction endpoint specification. This was concerning as the reaction was a part of a late-stage compound and this observation would hamper efficient technology transfer to the manufacturing site. As a result, the reaction process modeling roadmap was applied to provide reaction robustness and enable efficient technology transfer to manufacturing site.

43.2.3 Analysis of Transport Phenomena

Early process development work pointed toward the reaction being mixing limited. In phenomenological terms, the observed rate of reaction was equal to the rate of mixing locally in the contacting device. This can be characterized mathematically by using the reaction Damköhler number defined by Eq. (43.1). In other words, the reaction Damköhler number was greater than one:

$$\text{Damkohler number} = \frac{\text{Intrinsic rate of reaction}}{\text{Rate of mixing}} \quad (43.1)$$

If proper mixing protocols are not followed at different scales of operation, there could be concentration and temperature gradients (due to high exothermicity of the reaction). The presence of local hotspots was more concerning due to possibility of preferential impurity generation and loss of MsCl toward the main reaction. This problem was expected to become worse with increasing reaction scale. In order for the reaction to be robust upon scale-up, a quantitative *fit-for-purpose* analysis of the mixing in the contacting device (stirred reactor) was needed. This exercise would provide for a reasonable *scale-independent* measure of mixing time to ensure a passing reaction endpoint.

The quantification of mixing was assessed by mixing time scales during chemical reaction (Figure 43.2). In general, three mixing time scales can be quantified in a stirred vessel. Its quantification is provided by the Baldyga and Bourne mixing model, and the full details of this model are provided in Ref. [2]. A short description of the different mixing time scales is provided below:

- **Micromixing engulfment time.** The micromixing engulfment time scale is the inverse of the rate of mixing in fluid packets that have a length scale smaller than the Kolmogorov length [1]. Below this length scale, the main mode of mixing is laminar. Large-scale eddies break down to below the Kolmogorov scale where mixing continues and is governed by laws of laminar mixing. The micromixing time scale defined by Bourne and Baldyga is shown below:

$$\text{Micromixing time scale} = 17\left(\frac{\nu}{\varepsilon}\right)^{0.5} \quad (43.2)$$

The rate of mixing, which is the inverse of this micromixing time scale, depends on the kinematic viscosity (ν) and the dissipation rate of turbulent kinetic energy (per unit mass, ε) in those fluid packets. This rate of mixing affects both the mass and heat transfer.

- **Mesomixing time.** The mesomixing time scale is the inverse of rate of mixing in fluid packets having a length scale larger than the Kolmogorov scale but smaller than the contacting device length scale (typically the diameter of the stirred tank). The main mechanism of mixing occurs through convective transport phenomena and is turbulent in nature. The typical length scales of mesomixing (Λ_c) depend on the relative measure of the reagent dose rate (Q_B) compared to the main reactor surrounding fluid velocity (u). For relatively small dose rates, Λ_c is provided by Eq. (43.4), while for dose velocity ($u = Q_B/A_{pipe}$) comparable with the velocity of surrounding fluid, Λ_c is close to the feed pipe diameter:

$$\text{Mesomixing time scale} = 2\left(\frac{\Lambda_c^2}{\varepsilon}\right)^{0.33} \quad (43.3)$$

$$\text{where} \quad \Lambda_c = \left(\frac{Q_B}{\pi u}\right)^{0.5} \quad (43.4)$$

Hence these time scales depend mainly on the dissipation rate of turbulent kinetic energy and an intermediate measure of length. There is a possibility that upon scale-up, a mixing-limited reaction can change their mechanism from being micromixing limited in the laboratory to being mesomixing limited in the plant. Hence, one must always analyze micromixing and mesomixing time scales together and be observant on possible changes in rate-limiting steps.

- **Blend time.** It is the time taken by the fluid to be homogeneous in the length scale of the entire vessel. This time scale is usually higher than either the micromixing or the mesomixing time scales.

For mixing-limited reactions with Da > 1, the time scale of the reaction is usually compared and ranked with either micromixing time, mesomixing time, or blend time. A controlling mechanism is inferred, and appropriate scale-up guidance is provided. For our reaction of interest, the relative measures of these time scales are provided in Table 43.1.

The analysis of fluid mixing rates also provides for a direct correlation with the temperature inhomogeneity at appropriate reaction scales as shown in Figure 43.3. This is due to the similarity of the dimensionless equations of local heat and mass transfer with respect to fluid mixing [1]. Suppose a fluid packet is at temperature T with the surrounding fluid at temperature $<T>$ ($\neq T$). A lower rate of mixing leads to a suboptimal rate of swirling of the surrounding fluid into the test fluid packet to equalize the temperature. This leads to lower heat transfer rates and temperature inhomogeneity. For our example, temperature "hotspots," in addition to mass transfer inhomogeneity, could potentially lead to MsCl being consumed in unproductive pathways.

TABLE 43.1 Tabulation of Reaction Conversion with Micromixing and Mesomixing Time Scales in Different-Sized Reactors

Impeller RPM	Reactor Volume (L)	Mesomixing Time (s)	Micromixing Time (s)	Conversion (Peak Area %)
103	0.1	0.17	2.51	99.05
401	0.1	0.03	0.33	99.86
100	0.1	0.17	2.51	99.49
50	0.1	0.44	7.42	98.4
210	0.1	0.07	0.86	99.73
82	0.1	0.25	3.53	98.47
77	0.1	0.28	3.88	99.22
50	0.5	0.71	8.01	97.47
90	1050	0.18	0.24	99
94	1050	0.17	0.22	99.8
90	1050	0.18	0.24	99.3

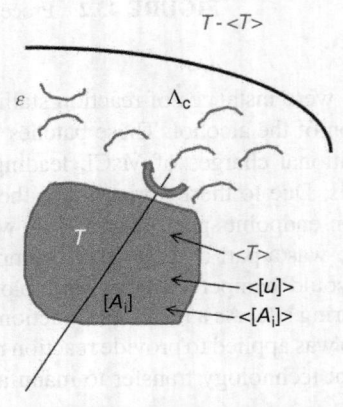

FIGURE 43.3 Use of mixing time scale to correlate with the local temperature excursion in the exothermic reaction.

Returning to Figure 43.3, an analysis of mixing time scales was to be done in the zone of chemical reaction within the stirred reactor. A best-case scenario for fluid mixing is to dose the MsCl subsurface very close to the impeller. This configuration effectively utilizes the highest rate of mixing due to very high levels of turbulence generated close to the impeller. Manufacturing limitations, however, prevented this configuration and restricted the MsCl dosage to be above surface. One main reason for this restriction was that substantial efforts would be required to retrofit the subsurface dip tube with a nozzle of appropriate diameter to effectively prevent the reactive mixture to diffuse into the tube. Applying the Baldyga and Bourne mixing model [2], three key requirements for mixing analysis are the kinematic viscosity, mesomixing length scale (Λ_c), and turbulent dissipation rate at the location on the liquid surface where the drop impinges. The kinematic viscosity of the endpoint reaction mixture was experimentally obtained from a rheometer in the laboratory at process-relevant temperatures. The dissipation rate of turbulent kinetic energy (ε) was calculated by the following relation:

$$\varepsilon(\text{reaction zone}) = \varepsilon_{\text{impeller}} + \varepsilon_{\text{feed jetting}} + \varepsilon_{\text{buoyancy}} \quad (43.5)$$

where

- $\varepsilon_{\text{impeller}}$ was the turbulence generated by the action of the impeller at the feed location on the liquid surface and was calculated by CFD applied to reactors of various sizes of interest.
- $\varepsilon_{\text{feed jetting}}$ was the turbulence generated by the accelerating above-surface drop in a gravitational field, whose value is provided by Newton's laws.
- $\varepsilon_{\text{buoyancy}}$ was the turbulence generated by the difference in buoyancy between the droplet and the reaction liquid. This was once again provided by Newton's laws.

It should be emphasized that $\varepsilon_{\text{feed jetting}} + \varepsilon_{\text{buoyancy}}$ was less than 3% contribution to the total turbulence intensity and was negligible in comparison with the turbulence provided by the impeller.

The mesomixing length scale was calculated by estimating the surrounding fluid velocity in the reaction zone via CFD and using Eq. (43.4) for a known dose volumetric flow rate of MsCl.

These values were subsequently input in a *DynoChem* mixing utility to calculate the micro- and mesomixing time scales for various reactor configurations and various intensities of agitation during the experiment. The three volume scales that were experimentally evaluated were 0.1, 0.5, and 1000 L.

43.2.4 Results

Table 43.1 and Figure 43.4 display the results from various experiments conducted in the laboratory with different impeller speeds (to simulate different mixing time scales) and pilot-scale campaigns.

From Figure 43.4, an increase in mesomixing time caused a linear decrease in reaction conversion. This was consistent with the expectation that longer mixing times lead to more inhomogeneity in temperature and concentration across the reactor, leading to progressively more suboptimal usage of MsCl. To demonstrate that this measure of reaction mixing time scale was an appropriate scale-independent parameter, a representative experiment was performed at a higher scale in the laboratory (0.5 versus 0.1 L), and the reaction performance scaled appropriately (as shown in Figure 43.4). Based on these recommendations, three clinical trial supply batches executed in the pilot plant at 1050 L scale also provided passing results consistently.

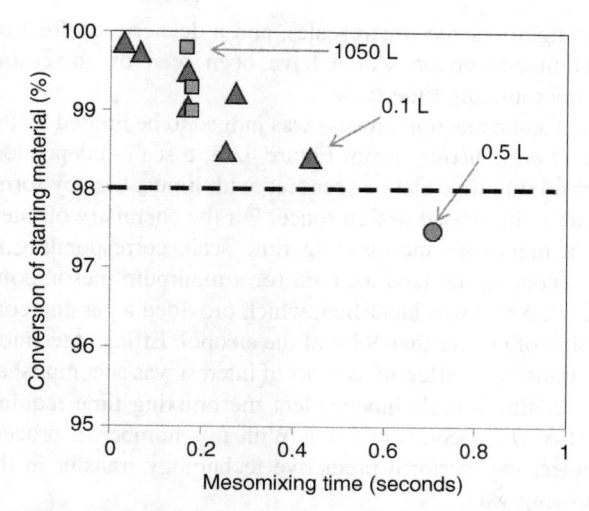

FIGURE 43.4 Variation of the conversion of alcohol with mesomixing time for different reactor scales.

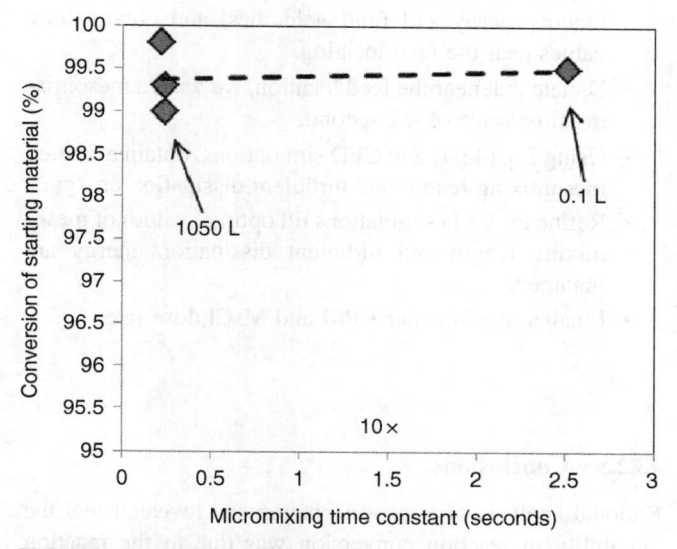

FIGURE 43.5 Variation of conversion of alcohol versus the micromixing time scale. These experiments were performed at the same mesomixing time scale (0.17 s).

Separately, an analysis of micromixing time scales in the same experiments revealed that the reaction is not limited by micromixing. Figure 43.5 shows the variation of the reaction conversion with the change in micromixing time scale.

A 10-fold change in micromixing rate led to almost no change in reaction conversion. This was consistent with the reaction Da $\gg$ 1 as a result of which the reaction volume was differentially small and localized even at 1000 L scale. Due to this localization, bulk of the heat and mass transfer was to be accomplished when the MsCl droplet enters the mesoscale and completes the reaction in those scales. Otherwise, MsCl droplet would have continued to react below the

Kolmogorov scale (microscale), and a deleterious effect on reaction conversion would have been seen by increasing the micromixing time scale.

Thus, the reaction process was judged to be limited by the *rate of mesomixing*. From Figure 43.5, a scale-independent mesomixing time can be specified with limits, thereby forming a mixing-based design space. For the chemistry of interest, a maximum mesomixing time scale corresponding to 0.4 second at the feed location (or a minimum mesomixing rate = $2.5 \, s^{-1}$) was identified, which provided a passing conversion of greater than 98% of the alcohol. Efficient technology transfer to different reactors of interest was accomplished by dictating a scale-independent mesomixing time requirement of 0.2 second (or less). With this number, a process engineer can perform predictive technology transfer in the following way:

- Obtain complete reactor geometry and perform CFD simulations to map out the dissipation rate of turbulent kinetic energy and fluid velocities, and obtain those values near the feed location.
- Dictate that near the feed location, we want a mesomixing time scale of 0.2 second.
- Using Eq. (43.4) and CFD simulations, obtain a desired mesomixing length and turbulent dissipation energy.
- Refine the CFD simulations till optimal values of mesomixing length and turbulent dissipation energy are obtained.
- Finalize the impeller RPM and MsCl dose rate.

43.2.5 Conclusions

Rational analysis of transport phenomena revealed that the variability in reaction conversion was due to the reaction being mesomixing limited. As the scale of operation increased, there is increased variability in temperature and mass inhomogeneity, leading to suboptimal usage of MsCl, which decreased conversion and increased impurity generation. Using the approach highlighted in this example, proper scale-up guidance was provided to enable successful technology transfer to the manufacturing site with consistent reaction performance.

43.3 EFFICIENT SCALE-UP OF A MULTIPHASE EXOTHERMIC NITRATION REACTION

43.3.1 Introduction

The nitro derivatives of aromatic compounds are frequently used as intermediates to form amines in the pharmaceutical industry. These reactions, however, are very energetic, producing a powerful exotherm that can be hard to control upon reaction scale-up. Moreover, there is potential for over-nitration. This leads to a safety concern because dinitro and trinitro compounds are often labeled as explosives by the safety laboratory. As a result, the level of over-nitrated compounds has to be controlled via a strict control of nitration stoichiometry along with a preference of performing flow chemistry for such reactions [3–5]. The advantages of flow chemistry for nitration include high rates of heat and mass transfer caused by higher rates and more homogeneous mixing compared with batch reactors. Among flow chemistry contacting devices, many academic groups preferred the use of microreactors [4, 6, 7]; however, more industrial nitration reactors are composed of coiled metallic pipes and CSTRs [3, 8].

Although there have been various successful demonstrations of continuous flow nitration [3–8], a rational mathematical framework to evaluate the scale-up of such reactions in the pharmaceutical setting has been missing. In this example, nitration of 4-trifluoromethyl toluene using KNO_3/H_2SO_4 mixture has been investigated with intent to scale up to multi-kilograms production using flow chemistry. With this aim, this example provides the mathematics needed for such an exercise using the process modeling roadmap described in the beginning of the chapter.

43.3.2 Problems Faced in Scale-up of Nitration Reaction

The nitration reaction of interest is shown in Scheme 43.2. A mixture of KNO_3 in 98% H_2SO_4 undergoes a biphasic reaction with 4-trifluoromethyl toluene to form the nitrated products. The reaction is solvent-less and composed of two liquid phases. The dinitrated product (a potential explosive) is one potential impurity formed in the reaction. Among the by-products, $KHSO_4$ is formed due to the reaction of sulfuric acid with KNO_3. Potassium bisulfate precipitates in the reaction mixture, thereby making it an L-L-S reaction type.

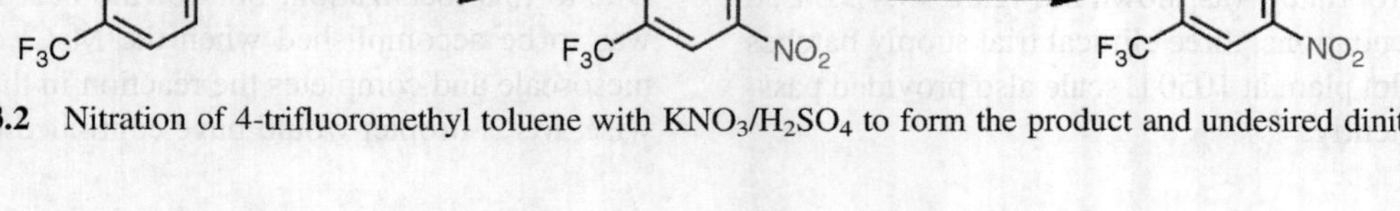

SCHEME 43.2 Nitration of 4-trifluoromethyl toluene with KNO_3/H_2SO_4 to form the product and undesired dinitro impurity.

The reaction is very exothermic with an ATR of 175 °C. In addition, due to the solventless nature of the chemistry, this reaction mixture is not amenable to good liquid–liquid (L-L) dispersion (a requirement for this chemistry). This is due to the density difference ($\Delta\rho$) being approximately 1 g/ml. Early-stage reaction scale-up involved a fed-batch addition of the nitrating mixture (KNO_3/H_2SO_4) into a stirred vessel of 4-trifluoromethyl toluene. Scale-up of the fed-batch process from 1 g to 1 kg product led to the dinitro impurity increase from 0.6% to almost 4%. This scale-up was considered problematic from a safety viewpoint due to the risk of enhanced impurity formation in the pilot plant upon further scale-up. As a result, the process modeling roadmap was applied to this reaction to ensure robust scale-up to the pilot plant and manufacturing.

43.3.3 Kinetic Analysis

Small-scale batch experiments were performed in 5 ml stirred vials to assess the reaction kinetics. The nitrating mixture was prepared beforehand and added as a bolus into the starting material. An exotherm was observed and kinetic data points were collected. It was observed that identical experiments at small scale (from a macro viewpoint) provided for significantly different results with respect to reaction conversion and dinitro impurity levels. It was judged that the reaction is indeed sensitive to mixing and efficient interphase contact was critical for reaction propagation. Later experiments in larger reactors confirmed this hypothesis (Figure 43.6). As a result, even though the rate constant was elusive, the mixing limitations were revealed even at a small scale. A measure of the rate constant was obtained from the literature by Esteves et al. [9] wherein the pseudo-first-order rate constant for toluene nitration (a surrogate reaction) was approximately 0.05 s^{-1} with activation energy is equal to 30 kJ/mol.

43.3.4 Analysis of Transport Phenomena

Successful scale-up of the nitration reaction required a rigorous understanding of the interphase transport between the aqueous and organic phases. The rate of transport phenomena is dependent on the reactor design. As a result, a series of experiments were conducted in a laboratory CSTR with different impeller stir rates under otherwise identical process configuration. If the reaction is sensitive to mixing, then different impeller power inputs should produce different levels of conversion. This variation is presented in Figure 43.6 and shows that the reaction is indeed mixing limited.

The main driver for this reaction was identified to be the rate of interphase transport. It was desired to model this behavior with the application of concepts from mass transfer with chemical reaction to reveal the multivariate interactions that was played by the fluid mechanics to the rate of this chemical reaction. Any learning from the analysis of this multivariate interaction would be important for rational scale-up of this reaction process.

Based on similar work performed by Baptista et al. [10], a film theory model was used to quantitatively match the experimental profile from the CSTR. The pictorial depiction of the mass transfer of the organic phase component into the aqueous phase is shown in Figure 43.7. Each "phase" is described by an appropriate mass balance equation. The model equations governing these mass balances are also shown in Figure 43.7. The reactive component is present in a well-mixed organic phase, where there is a constant rate of replenishment from the feed flow in a flow process. That component traverses through the mass transfer resistance film and can participate in the chemical reaction in the film itself as it traverses. Finally, the remaining organic reaction component then enters the well-mixed aqueous phase wherein the rest of the chemical reaction occurs. The thickness of the mass transfer resistance film is directly related to the dissipation rate of turbulence provided by the mixing

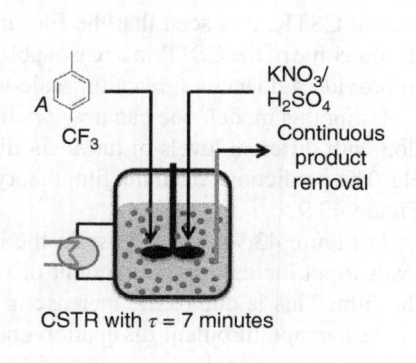

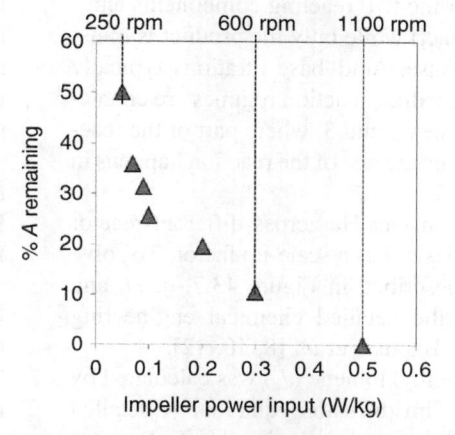

FIGURE 43.6 Continuous nitration in a CSTR. Figure on the right shows the amount of 4-trifluoromethyl toluene remaining with an increase in the impeller power per unit mass imparted.

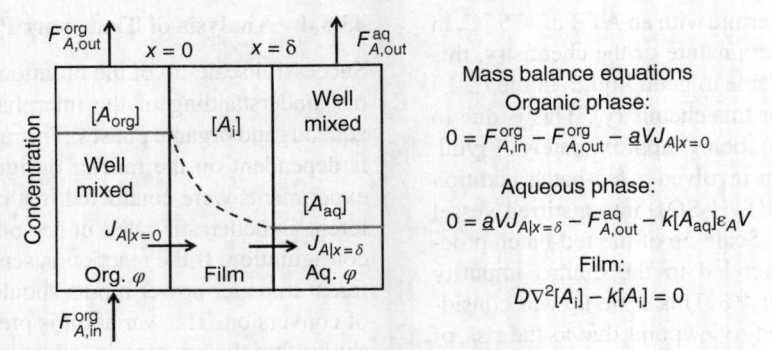

FIGURE 43.7 Film theory model for interphase transport with chemical reaction.

Mass balance equations
Organic phase:
$$0 = F_{A,in}^{org} - F_{A,out}^{org} - \underline{a}VJ_{A|x=0}$$

Aqueous phase:
$$0 = \underline{a}VJ_{A|x=\delta} - F_{A,out}^{aq} - k[A_{aq}]\varepsilon_A V$$

Film:
$$D\nabla^2[A_i] - k[A_i] = 0$$

element in the contacting device. For example, for a CSTR, the agitator system provides the turbulent dissipation and governs that film thickness.

For simple systems, this set of equations can be solved analytically. The main dimensionless number that comes out from such analysis is the Hatta number (Ha). It is defined as the extent of reaction that occurs in the film compared with the total reaction extent and for a pseudo-first-order reaction is mathematically defined by

$$\text{Ha} = \sqrt{\frac{\text{Reaction rate in film}}{\text{Diffusion rate in film}}} = \frac{\sqrt{kD}}{k_L} \quad (43.6)$$

The Hatta number is an important dimensionless number that characterizes the nature of mixing limitations in multiphase mixing-sensitive reactions. Ha divides multiphase reactions into various regimes. A few of them are described here. A comprehensive description of Ha and its implications in multiphase chemical reaction engineering are described by Doraiswamy and Sharma [11]. For example, Ha < 0.3 means that the bulk phase is saturated and the reaction occurs at the solubility in the bulk phase, as is the case for various gas–liquid (G-L) or L-L reactions. On the other hand, Ha > 3 signifies an instantaneous reaction where the entire reaction is completed in the film (or the two reacting components annihilate each other on contact) while only the product is transported across the interphase. Acid–base titrations typically follow this regime. Intermediate reaction regimes are characterized by Ha being between 1 and 3, where part of the reaction occurs in the film while the rest of the reaction happens in the bulk.

For our example, a constant Ha across different reactor scales of operation was desired as a scale-up factor. To solve the film theory model described in Figure 43.7, $\underline{a}$, D, and k_L were estimated from the detailed chemical engineering correlations provided by Baptista et al. [8, 10, 12].

The flux of trifluoromethyl toluene (J_a) was calculated by analytically solving the film diffusion equation. A detailed solution of these equations can be obtained from Baptista et al. [12]. The flux at $x = 0$ and δ was estimated. Using the global mass balance equations, the molar flux of

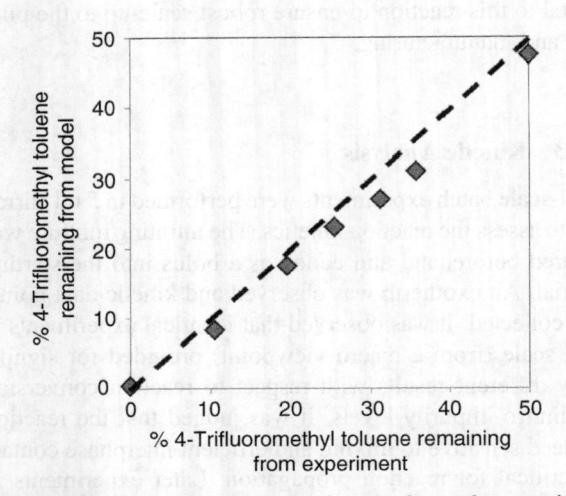

FIGURE 43.8 Agreement between the experimental conversion of 4-trifluoromethyl toluene and the modeled conversion as a function of different mixing intensities (quantified by Ha).

4-trifluoromethyl toluene was then evaluated and correlated with the experimental data. The fit of the model with the experimental data is shown in Figure 43.8.

It should be emphasized that each data point on Figure 43.8 represents a different Ha due to different mixing intensities imparted during each steady state in the experimental CSTR. It is seen that the film theory model predicts the outcome of the CSTR in a reasonable way and can be used to provide guidance on reaction scale-up.

Using this model, one can now predict the extent on reaction with different levels of turbulent dissipation energy and Ha. The predictions from the film theory model are shown in Figure 43.9.

In Figure 43.9a, an increase in the impeller power input leads to an increase in the amount of reaction completed in the film. This is due to the increase in Ha with an increase in the average turbulent dissipation energy. Probing deeper into the phenomenological aspect of the impact of Ha to reaction reveals that higher Ha provides for a higher interfacial area for reaction. The increase in interfacial area is obtained by smaller dispersed phase droplets as the impeller pumps in

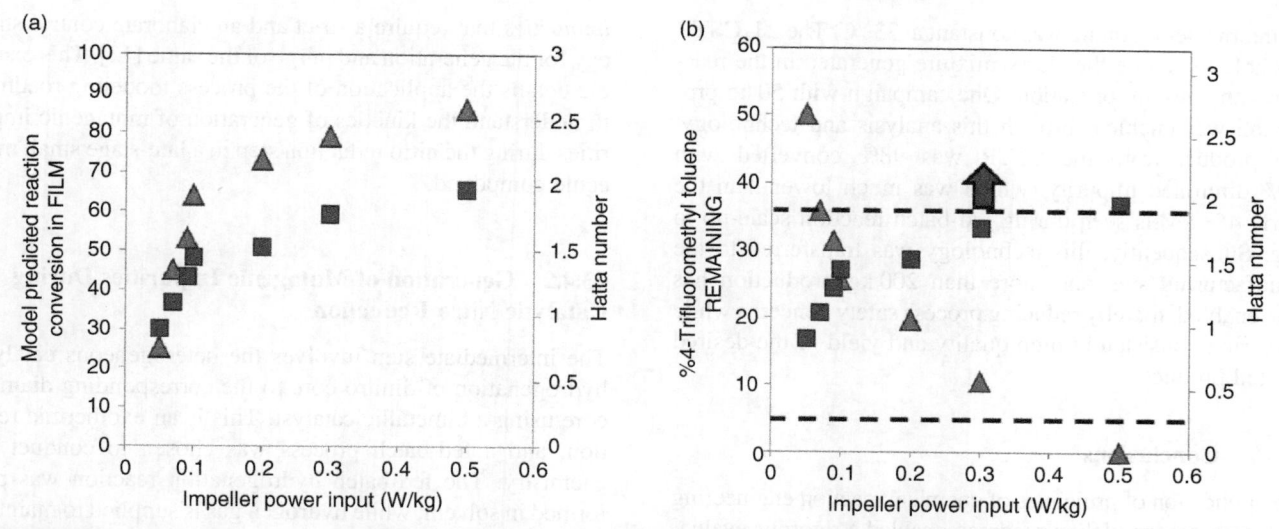

FIGURE 43.9 (a) Variation of the model-predicted % reaction conversion in the film as a function of Ha and impeller power input. (b) Requirement of minimum Ha to achieve the desired reaction endpoint of no more than 5% 4-trifluoromethyl toluene remaining.

more energy per unit mass into the contacting device. However, from Figure 43.9a, the analysis tells us that initially increasing the impeller stir rate decreases the droplet sizes easily, but it becomes more difficult to reduce them below a certain size due to surface tension effects manifested by the Weber number.

From a manufacturing viewpoint, a favorable reaction endpoint was judged to be less than 5% 4-trifluoromethyl toluene remaining, due to relative ease of downstream separation versus added impurity generation. In Figure 43.9b, the desired endpoint of the reaction could be achieved if Ha > 2. The requirement of Ha > 2 in processing could be stated as the requirement that almost 80% of reaction should be done in the interphase film (Figure 43.9a). This is achieved by increasing the interfacial area as a result of more intense agitation provided by the impeller or other mixing devices. In summary, from a process parameter viewpoint, the droplet size distribution has to remain identical with reactor scale-up. This requirement ensures linear scalability of the interfacial area and constant Ha with increasing reaction scale of operation.

43.3.5 Analysis of Flow Patterns

Two flow patterns were evaluated for the nitration reaction scale-up. Static mixers were used to evaluate PFR geometry, while a stirred reactor setup in CSTR mode was used to evaluate back-mixed flow. Assuming heat transfer to not be problematic, the reaction behavior was identical in both if a constant interfacial area criterion was met, manifested in mixing intensity. From a heat transfer viewpoint, the highly exothermic chemistry could favor a tubular flow reactor due to higher heat transfer area per unit volume compared with a stirred reactor. The tubular flow reactor geometry, however, was considered more risky due to potential for solids plugging during a prolonged operation of the L-L-S reaction. As a result, a 2 L Hastelloy CSTR with a high L/D (2.5 : 1) ratio was chosen for production.

It should be emphasized that even though a decision was to go ahead with the CSTR, from a fundamental viewpoint, there was a greater risk of spatial variation of Ha in a 2 L CSTR compared with a jacketed static mixer due to a higher volume of reacting fluid with the same heat transfer area. This would potentially result in reaction zones that are not as well mixed (with a lower Ha) and could result in non-predictive scale-up. One strategy would be to use CFD to simulate the multiphase flow in a 2 L CSTR, map out the droplet size distribution, and recommend an impeller RPM on the requirement of an acceptable dispersed phase droplet size distribution. However, timelines prevented us from doing this analysis. As a result, a laboratory glass CSTR was already available with geometric similarity to the plant CSTR, and a representative experiment was performed at half the production flow rate in the laboratory. A separate film theory analysis on the glass reactor dictated that the impeller RPM should be set at 700 RPM. This experiment provided passing results for reaction endpoint (99% reaction conversion and 0.3% dinitro impurity) and validated the appropriate scale-up parameter to be a constant Ha.

43.3.6 Production-scale CSTR System

Success from using the 2 L glass CSTR in the laboratory provided confidence that we could use the CSTR of the same size in the pilot plant. Completion of the reaction campaign in a reasonable time required removal of 1.5 kW of heat from the CSTR during steady-state operation. As a result, appropriate chilling capabilities were enabled such that during steady-state operation, the jacket was kept at 15 °C while the internal temperature was constant at 33 °C. The 2 L CSTR was able to handle the slurry mixture generated in the reaction with constant operation. One campaign with 50 kg production was enabled through this analysis and technology. The product from the CSTR was 98% converted with 0.5% dinitrated impurity, which was much lower than the nearly 4% levels seen during fed-batch reaction scale-up to 1 kg. Subsequently, this technology was transferred to the manufacturing site, and more than 200 kg production has been enabled, thereby reducing process safety concerns while delivering consistently high quality and yield of the desired nitrated product.

43.3.7 Conclusions

The application of principles of chemical reaction engineering via the process modeling roadmap enabled a rigorous evaluation of transport phenomena in a multiphase mixing-limited nitration reaction. Modeling enabled identification of appropriate scale-up parameters (Hatta number) and ensured that the reaction was scaled up in a safe manner with respect to the exotherm and minimization of the dinitro impurity. This enabled an easier technology transfer to the manufacturing site.

43.4 KINETIC JUSTIFICATION-BASED CONTROL FOR POTENTIAL MUTAGENIC IMPURITIES

43.4.1 Introduction

Catalytic heterogeneous hydrogenations are one of the most important classes of chemical reactions scaled up in the pharmaceutical industry. These reactions are typically of G-L-S type, and a keen knowledge of kinetics and mass transfer is required to ensure proper scale-up. An added complexity in the specific case of nitro reduction is the high exotherm of such reactions and the potential to form *mutagenic impurities* that require a strict and an elaborate control strategy for the generation and purge of the same [13]. This example details the application of the process modeling roadmap to understand the kinetics of generation of mutagenic impurities during the nitro reduction step in a late-stage small molecule compound.

43.4.2 Generation of Mutagenic Impurities During Catalytic Nitro Reduction

The intermediate step involves the heterogeneous catalytic hydrogenation of dinitro core to the corresponding diamino core using a bimetallic catalyst. This is an exothermic reaction, and a fed-batch process was chosen to conduct the chemistry. The fed-batch hydrogenation reaction was performed in solvent, while hydrogen gas is supplied to maintain a constant overall hydrogen pressure. This ensures a constant solubility of hydrogen in the reacting solvent. Moreover, earlier studies have shown that the rate of dosing at scale ensured that the reaction had a Hatta number (Ha) < 0.3, which meant the hydrogenation intrinsic kinetics was rate limiting.

Aryl nitro hydrogenations proceed via two monomeric intermediates: an aryl nitroso compound and an aryl hydroxylamine. The presence of the two nitro groups in dinitro pro-core means that at least eight intermediates are possible along the reaction pathway as shown in Figure 43.10.

In silico evaluation identified the aryl nitro, aryl nitroso, and aryl hydroxylamine substructures as alerts for potential mutagenicity. Since many of the reaction intermediates are unstable, analytical standards could not be prepared, and the mutagenicity of the reaction intermediates could not be determined. A kinetic justification strategy was required to ensure that each hydrogenation intermediate will not be present in diamino core above 1 ppm at the end of hydrogenation reaction process. This was a conservative criterion more than 100-fold below the Threshold of Toxicological Concern (TTC) in the API.

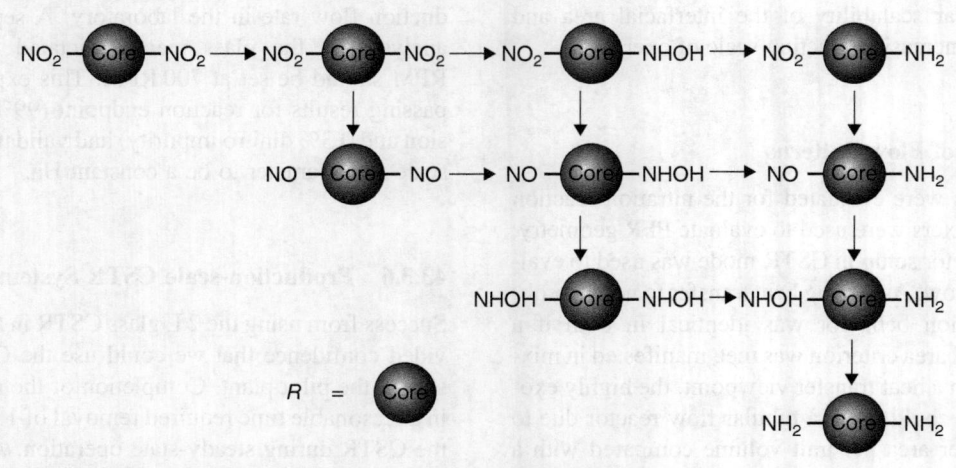

FIGURE 43.10 Dinitro pro-core reduction pathway during catalytic hydrogenation.

As a result, a multistep kinetic analysis was required to understand the multivariate interactions, leading to the prediction of such mutagenic impurities with change in process parameters.

To allow for the observation of transient reaction intermediates to develop the kinetic justification, semi-batch hydrogenation experiments were performed under multivariate conditions that significantly retard the rate of hydrogenation in comparison to the commercial manufacturing process. Further experiments were conducted to prove that the reaction rate was sufficiently slowed by running at a lower temperature and hydrogen pressure, using less catalyst, and running more dilute using additional solvent. These conditions are collectively referred to as the "lowest rate process." A comparison between the process parameters used in "nominal process" and the "lowest rate process" is shown in Table 43.2.

Under the "lowest rate process" conditions, the dinitro core is quickly consumed, and three out of the eight possible intermediates (Figure 43.10) along the hydrogenation pathway were detected by anaerobic sampling of the reaction and HPLC analysis. Of these three intermediates, the "NO—NH_2" species was fleetingly observed at trace levels before becoming undetectable. The remaining two detectable intermediates, "NHOH—NHOH" and "NHOH—NH_2," and the diamino core ("NH_2—NH_2") were used to develop an operational kinetic model and determine kinetic rate constants (k's) using *DynoChem* software. The reactions included in the kinetic model are listed in Table 43.3.

The "*" in the reaction steps correspond to a catalytic site. The number of catalytic sites was assumed to be proportional to the total weight loading of catalyst as determined by the vendor. Step 4 in Table 43.3 was added to the model to account for possible catalyst inhibition by the diamino core ("NH_2—NH_2"). To test for catalyst inhibition, experiments were conducted at the "lowest rate process" conditions with the diamino core charged into the reactor before beginning the reaction. Figure 43.11 shows that as the charged equivalents of the diamino core increased from 0 to 1, the conversion rate of NHOH—NH_2 decreased. In other words, the time required to achieve NMT 1 ppm of the reaction intermediates is dependent on the diamino core concentration.

The concentration vs. time profiles for "NHOH—NHOH," "NHOH—NH_2," and "NH_2—NH_2" are shown in Figure 43.12 (no additional "NH_2—NH_2") and Figure 43.13 (1 equiv "NH_2—NH_2"). The dinitro core is the most quickly consumed reaction species (k_1 is greater than k_2 and k_3) and is not shown in Figures 43.12 and 43.13 for clarity.

The kinetic model accounts for the inhibition caused by the diamino core and predicts the experimental data well. Based upon the kinetic model, the predicted amount of time required for the reaction operating under the "lowest rate process" conditions to reach NMT 1 ppm of potentially mutagenic intermediates is 6.1 hours.

TABLE 43.2 Comparison Between the Nominal Hydrogenation Conditions (Labeled "Nominal Process") and the "Lowest Rate Process"

Parameter	Nominal Process	Lowest Rate Process
Temperature (°C)	70	60
H_2 pressure (bar absolute)	2.9	2.0
Catalyst loading (kg cat/kg dinitro core)	0.063	0.042
kg solvent 1/kg dinitro core	8.5	10.2

TABLE 43.3 Data Fitting for the Most Prevalent Intermediates Under the "Lowest Rate Process" Conditions

Step	Reaction	Parameter[a]
1	NO_2—core—NO_2 + * → NHOH—coreNHOH + *	$k_1 = 2.56 \pm 0.23$ L/mol s
2	NHOH—core—NHOH + * → NHOH—core—NH_2 + *	$k_2 = 1.57 \pm 0.05$ L/mol s
3	NHOH—core—NH_2 + * → NH_2—core—NH_2 + *	$k_3 = 1.06 \pm 0.03$ L/mol s
4	NH_2—core—NH_2 + * ↔ NH_2—core—NH_2*	$k_4 = 1 \times 10^4$ L/mol s[b] $k_4 = 15.31 \pm 0.88$ L/mol

[a]The rate constants are reported at 60 °C. The ± values represent the standard error obtained from the DynoChem fitting of the data to the kinetic model (95% confidence interval can be obtained by multiplying the standard error values by 1.96).

[b]The rate of adsorption of the diamino core to the catalyst surface (k_4) was set to 1×10^4 L/mol s based upon the assumption that adsorption to catalyst surface is very fast. This assumption is implicitly utilized in Eqs. (43.1)–(43.3) as reactive species must absorb to the surface prior to reaction.

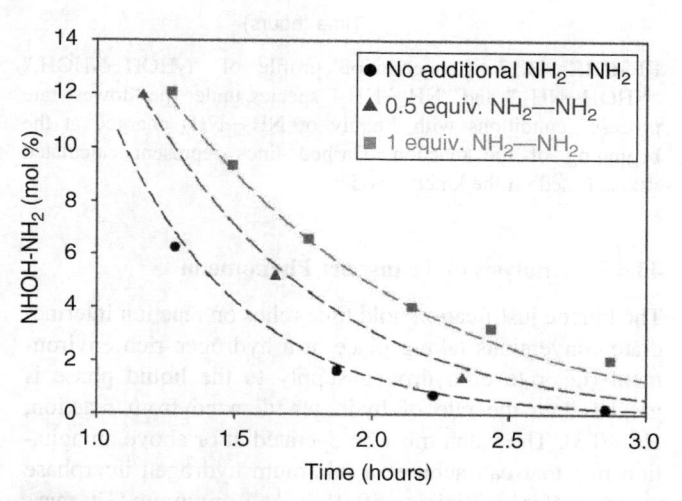

FIGURE 43.11 Comparison of the concentration profile of NHOH—NH_2 as a function of the equivalents of the diamino core, NH_2—NH_2, charged into the reactor prior to start of the reaction. Dashed lines represent calculated values based on the kinetic model.

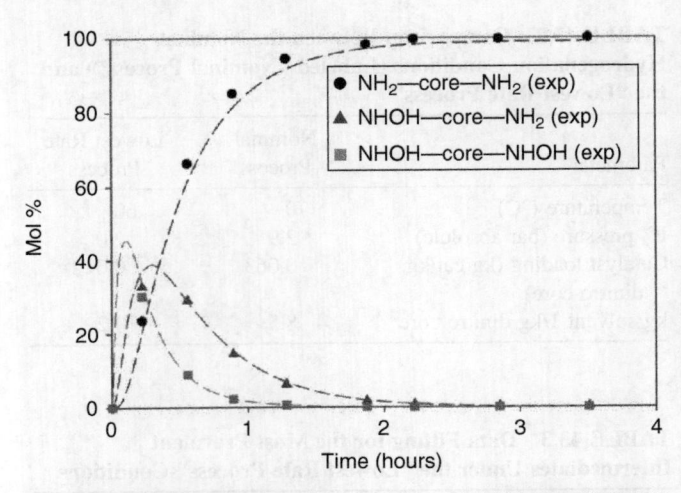

FIGURE 43.12 Concentration profile of "NHOH—NHOH," "NHOH—NH$_2$," and "NH$_2$—NH$_2$" species under the "lowest rate process" conditions. Dashed lines represent calculated values based on the kinetic model.

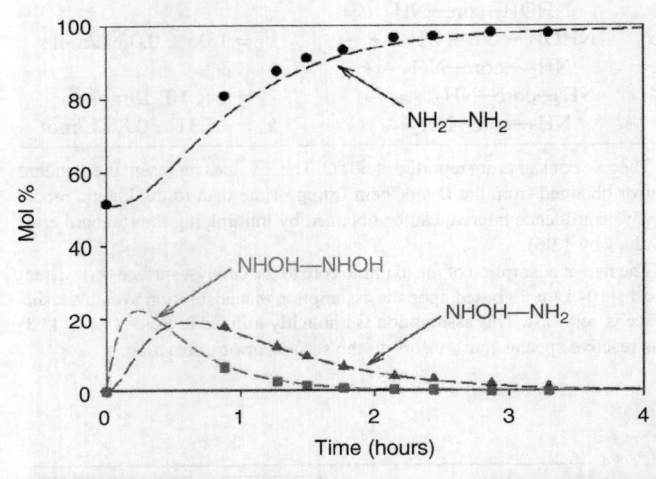

FIGURE 43.13 Concentration profile of "NHOH—NHOH," "NHOH—NH$_2$," and "NH$_2$—NH$_2$" species under the "lowest rate process" conditions with 1 equiv of NH$_2$—NH$_2$ charged at the beginning of the reaction. Dashed lines represent calculated values based on the kinetic model.

43.4.3 Analysis of Transport Phenomena

The kinetic justification hold time relies on reaction intermediate conversions taking place in a hydrogen-rich environment (i.e. rate of hydrogen supply to the liquid phase is greater than the rate of hydrogen demand from reaction, Ha < 0.3). The batch must be operated at or above an agitation rate that can achieve a minimum hydrogen interphase mass transfer coefficient of 0.011 s^{-1}. A minimum G-L mass transfer coefficient (k_La) is specified for nominal hydrogenation process based on the conservative assumptions:

- The overall reaction rate is always limited by G-L mass transfer (i.e. the intrinsic chemical conversion rate would be significantly higher if not restricted by slow delivery of gas into solution), and, therefore, the rate of reaction is equal to the G-L mass transfer rate.
- Delivery of stoichiometric amount of hydrogen into solution is accomplished within a 2-hour period.

Based upon these two boundary conditions and assumptions, 0.011 s^{-1} was the calculated minimum requirement for k_La for the nominal hydrogenation process. The minimum k_La was determined in this manner:

$$R = k_La([A^*] - [A_t]) \qquad (43.7)$$

where

- $[A^*]$ is the equilibrium solubility of H$_2$ in 2-methyltetrahydrofuran (2-MeTHF) determined at 2 barg H$_2$ and 70 °C = **0.007 25** mol H$_2$/L (from DynoChem – H$_2$ solubility and heat of reaction utility)
- $[A_t]$ is the concentration of H$_2$ on catalyst surface = 0 mol H$_2$/L (mass transfer limited)
- From DynoChem simulations, R required = 7.839 × 10^{-5} mol/(L s):

$$k_La = \frac{R}{[A^*]} = 7.839 \times 10^{-5}\,\text{mol}/(\text{L s})/0.007\,25\,\text{mol/L} = \mathbf{0.011\,s^{-1}}$$

The k_La measurement was carried out in the commercial equipment using the primary process solvent at a volume equal to the final batch volume.

As an additional level of assurance that the process is not G-L mass transfer limited, a mass transfer hold of 4 hours, twice the 2-hour period stipulated above, was used to ensure that potentially mutagenic intermediates are consumed even when operating at the lower limit of the k_La. This mass transfer hold time under hydrogenation conditions is followed by the reaction kinetics control hold time of 12 hours. Thus, the total reaction hold time is 16 hours, which is the sum of the mass transfer and reaction kinetics hold times. A 16-hour hold time under the nominal hydrogenation conditions provides multiple layers of assurance that the reaction intermediates will be NMT 1 ppm in the hydrogenation.

43.4.4 Conclusions

An understanding of the kinetics of the reaction pathway observed during catalytic hydrogenation was required to ensure process robustness with respect to the purge of the mutagenic impurities that can potentially appear in the API. This was accomplished by building a multivariate kinetic model coupled with an analysis of the G-L mass transfer of hydrogen under worst-case scenarios. The model recommendations were supplemented with mass transfer experiments performed at the manufacturing-scale reactors.

This multivariate understanding of reaction kinetics and the transport phenomena provided a control strategy to limit the amount of mutagenic impurity at the end of the hydrogenation process for a late-stage compound.

ACKNOWLEDGMENTS

This work was sponsored by AbbVie. AbbVie contributed to the design, research, and interpretation of data, writing, reviewing, and approving the publication. Anuj Verma, Steven Richter, Brian Kotecki, and Moiz Diwan are employees of AbbVie. The authors would like to thank Niharika Chauhan, Samrat Mukherjee, Kaid Harper, Benoit-Cardinal David, Shailendra Bordawekar, and Nandkishor Nere for providing experimental results and helpful comments.

LIST OF SYMBOLS

P	total pressure
T	temperature
τ	space time
Da	Damköhler number
ν	kinematic viscosity
ε	dissipation rate of turbulent kinetic energy
Λ_c	mesomixing length scale
Q_B	reagent dose rate
A_{pipe}	cross-sectional area of feed pipe
u	velocity of surrounding fluid in reactor
$<T>$	average temperature of the surrounding fluid
$[A_i]$	concentration in test volume/within the film
$<[A_i]>$	average concentration of active species in the surrounding fluid
$<u>$	average velocity of the surrounding fluid in reactor
$\Delta\rho$	density difference
δ	film thickness
$F_{A,in}^{org}$	molar flow rate of component A entering the organic phase
$F_{A,out}^{org}$	molar flow rate of component A exiting the organic phase
$[A_{org}]$	concentration of A in the well-mixed organic phase
$J_{A|x=0}$	flux of A at $x = 0$
$J_{A|x=\delta}$	flux of A at $x = \delta$
$F_{A,out}^{aq}$	molar flow rate of component A exiting the aqueous phase
a	interfacial area
V	volume of phase
ε_a	holdup fraction
k	pseudo-first-order rate constant
D	diffusivity
Ha	Hatta number
k_L	mass transfer coefficient
$[A^*]$	solubility/equilibrium concentration

REFERENCES

1. Bird, R.B., Stewart, W.E., and Lightfoot, E.N. (2002). *Transport Phenomena*. Hoboken, NJ: Wiley.
2. Bourne, J.R. (2003). Mixing and the selectivity of chemical reactions. *Organic Process Research & Development* 7 (4): 471–508.
3. Pelleter, J. and Renaud, F. (2009). Facile, fast and safe process development of nitration and bromination reactions using continuous flow reactors. *Organic Process Research & Development* 13 (4): 698–705.
4. Burns, J.R. and Ramshaw, C. (2002). A microreactor for the nitration of benzene and toluene. *Chemical Engineering Communications* 189 (12): 1611–1628.
5. Brocklehurst, C.E., Lehmann, H., and La Vecchia, L. (2011). Nitration chemistry in continuous flow using fuming nitric acid in a commercially available flow reactor. *Organic Process Research & Development* 15 (6): 1447–1453.
6. Hartman, R.L., McMullen, J.P., and Jensen, K.F. (2011). Deciding whether to go with the flow: evaluating the merits of flow reactors for synthesis. *Angewandte Chemie International Edition* 50 (33): 7502–7519.
7. Kulkarni, A.A., Kalyani, V.S., Joshi, R.A., and Joshi, R.R. (2009). Continuous flow nitration of benzaldehyde. *Organic Process Research & Development* 13 (5): 999–1002.
8. Quadros, P.A., Oliveira, N.M.C., and Baptista, C.M.S.G. (2005). Continuous adiabatic industrial benzene nitration with mixed acid at a pilot plant scale. *Chemical Engineering Journal* 108 (1–2): 1–11.
9. Esteves, P.M., de M. Carneiro, J.W., Cardoso, S.P. et al. (2003). Unified mechanistic concept of electrophilic aromatic nitration: convergence of computational results and experimental data. *Journal of the American Chemical Society* 125 (16): 4836–4849.
10. Quadros, P.A., Reis, M.S., and Baptista, C.M.S.G. (2005). Different modeling approaches for a heterogeneous liquid–liquid reaction process. *Industrial & Engineering Chemistry Research* 44 (25): 9414–9421.
11. Doraiswamy, L.K. and Sharma, M.M. (1984). *Heterogeneous Reactions: Analysis, Examples, and Reactor Design. 2. Fluid-fluid-solid Reactions*. New York: Wiley.
12. Quadros, P.A., Oliveira, N.M.C., and Baptista, C.M.S.G. (2004). Benzene nitration: validation of heterogeneous reaction models. *Chemical Engineering Science* 59 (22–23): 5449–5454.
13. Betori, R.C., Kallemeyn, J.M., and Welch, D.S. (2015). A kinetics-based approach for the assignment of reactivity purge factors. *Organic Process Research & Development* 19 (11): 1517–1523.

PART X

MANUFACTURING

44

PROCESS SCALE-UP AND ASSESSMENT

ALAN BRAEM, JASON SWEENEY, AND JEAN TOM
Product Development, Bristol-Myers Squibb, New Brunswick, NJ, USA

44.1 INTRODUCTION

One of the key roles for chemical engineers in drug substance process development is transforming an active pharmaceutical ingredient (API) synthesis route into a scalable commercial process. In this chapter, we present an approach to process assessment and scale-up focused on risks to safety, quality, and business that takes into consideration the product's stage of development and the magnitude of scale-up. Emphasis is placed on understanding common unit operations' scale-up factors and use of this knowledge in the assessment and definition of a development strategy.

The initial chemical synthesis of a small molecule API is typically developed by an exploratory or discovery chemistry group. The goal of the initial synthesis is to quickly enable production of milligrams to grams of the API to support exploratory studies and to confirm the biological activity of the molecule. This initial route is designed to be divergent and allow access to a variety of targets and is not designed for further scale-up to kilogram scale, much less to a manufacturing process. It is the role of process chemists to design a synthesis that can be developed into a scalable process to deliver sufficient API quantity and quality to support clinical toxicology assessment and downstream formulations. It is a collaboration of process chemical engineers with the process chemists that shapes the synthesis from a procedure to a plant-scale process.

44.1.1 Phases of Development

The focus of the chemical engineer in transforming a synthesis into a scalable process changes as the compound goes through the various stages of development. The evolution of the synthesis route is tied to the clinical timeline. Table 44.1 provides a simplified overview of the evolution of the synthesis and product requirements as the stage of development progresses.

In reality, the progression of development is likely not as clearly defined, and overlap of these categories will occur, depending on the compound's potency, the synthesis and molecular complexity, the therapeutic class of the compound, and the infrastructure and organization of the involved research and development groups. However, the process development goals can be generally delineated within these milestones in terms of the magnitude of scale-up and the process safety, business, and quality risks. This, in turn, guides the allocation of engineering resources and determines the level of risk assessment needed.

44.1.1.1 Early Development In early development, the synthesis milestone is the selection of an appropriate route for the initial scale-up. The key considerations are process safety; chemical hygiene; number of synthetic steps; availability of reagents, raw materials, and intermediates; and ability of the synthesis to address API quality. The top process assessment priority is to identify hazardous reactions and reagents and to evaluate safe operating limits and material exposure limits. Altering the reagents and conditions and developing engineering controls to ensure safe operation are the main engineering foci. The secondary process assessment priority is to meet targeted process development goals. These goals would be (i) to obtain sufficient process knowledge to support at-scale operations (first-order-scale effects,

---

Chemical Engineering in the Pharmaceutical Industry: Active Pharmaceutical Ingredients, Second Edition.
Edited by David J. am Ende and Mary T. am Ende.
© 2019 John Wiley & Sons, Inc. Published 2019 by John Wiley & Sons, Inc.

TABLE 44.1 Process Research and Development Requirements During Stages of Development [1, 2]

Stage of Development	Discovery	Early Development	Full Development	Launch
Clinical stage	IND toxicology	Phase I	Phase II	Phase III/launch
Type of synthesis	Expedient	Practical	Efficient	Optimal
Synthesis milestone	Enabling synthesis	Route (intermediates) selected	Sequence of unit operations finalized	Process parameter ranges finalized
Amount of compound	10 mg–10 g	100 g–10 kg	10–100 kg	>100 kg
Site for preparation	Laboratory	Kilo laboratory	Pilot plant	Manufacturing plant
Number of batches	1–5	1–10	10–100	10–1000
Probability of success to next stage of development [3–6]	40–60% Preclinical to clinical	40–60% Phase I to phase II	40–60% Phase II to phase III	80–100% Phase III to NDA filing and launch

stability over duration of unit operations, heat/mass transfer, and hold points) and (ii) understanding but not optimization of the process conditions (stoichiometry, temperatures, concentration, filtration) to enable appropriate equipment usage. Process knowledge may be limited to single point information rather than a design range. The API product quality must meet an initial set of specifications, which may include form, purity, stability, and impurities.

At this stage of development, the probability of the molecule achieving success for a New Drug Application (NDA) is still low (<10%), and there is a high likelihood that the chemistry will not be used beyond this stage of development. Thus, fewer engineering resources will generally be allocated.

44.1.1.2 Full Development In this stage of development, the compound has demonstrated some key human safety milestone (phase I) and/or some evidence of clinical response or efficacy. A clinical timeline can be projected for the compound that will lead to an NDA.

By this stage, the process chemists will have evaluated alternate synthetic routes and will have determined the desired sequence of intermediates. The development will focus on finalizing the chemistry and reagents and defining the API crystal form and powder attributes. The process scale-up assessment will continue to have strong process safety and quality focus. In addition, an evaluation of the business risks will guide the efforts to develop a robust, efficient, and economical manufacturing process. Key aspect to evaluate for this assessment includes yield, cycle time, equipment usage, waste output, and the need for analytical support such as in-process assays and process analytical technology.

At this phase of development, the probability of success for the molecule to be filed for approval has increased significantly. The optimized process must consistently meet the quality requirements at the pilot plant scale before the next stage of development.

44.1.1.3 Launch This stage of development will focus on the final process optimization and providing a full understanding of the chemistry, manufacturing, and controls (CMC) for the NDA. Detailed information on process parameter ranges and fundamental understanding of the key unit operations (e.g. reactions, crystallizations) will be required to support the process validation at the manufacturing site and the submission of the NDA.

44.1.2 Process Safety and Risk Assessments

Risk is often defined as the combination of the probability and the severity of a harmful occurrence. Regulatory guidance to the pharmaceutical industry on evaluating risk is that the protection of the patients is of prime importance [7]. Protection of workers and maximizing business objectives are also key goals in assessing risk. Thus, the assessment of a chemical process must cover three key aspects: safety, product quality, and business parameters. Safety and quality are inherent in all phases of development, while the emphasis on business risk assessment increases in the later stages of development. Process safety is tied to standards and regulations governed by government safety agencies, and acceptable and appropriate product quality is guided by the various ICH guidelines for registration of pharmaceuticals for human use [8].

44.1.2.1 Process Safety Assessment Process safety assessment is one of the main chemical engineering concerns throughout process development. As Chapters 3–11 cover this topic in more depth, this chapter will briefly review some of the key elements of such an assessment and its impact on process scale-up.

For any scale-up, the first step is a review of process safety to (i) examine the reaction issues such as impact of chemical mixing, rate, sequence and mode of charges, parameters for self-heat, potential gas evolution, and corrosivity; (ii) evaluate the thermal and chemical stability; and

(iii) evaluate the electrostatic and dust hazards. Issues raised in this review require identification of safe limits by understanding the mechanisms of the hazards and an engineering evaluation to provide appropriate controls to meet such safe limits. This review is typically followed by a safety assessment against the scale-up implementation plan. An example would be a process hazard analysis (PHA) [9] that uses a standard methodology to evaluate potential process and equipment hazards that could cause the catastrophic release of hazardous materials or other significant safety impacts and ensures that appropriate safeguards are in place to prevent, detect, or mitigate these occurrences.

As an example, for a highly exothermic reaction, chemical engineers can (i) evaluate the equipment capabilities (i.e. heat transfer or mass transfer rates), (ii) optimize the chemistry via reactants and reaction conditions (i.e. changing the mode, sequence, or rate of a reagent charge, which changes the kinetic stoichiometry), and (iii) adjust the process setup to mitigate the hazard (i.e. emergency quench vessel, addition of external heat exchanger to increase cooling).

44.1.2.2 Process Risk Assessment There are various goals in a process risk assessment, which may take many forms. Regulatory guidance sets the primary goal of a risk assessment as identification of issues, resulting in drug product quality that adversely impacts the patient's health. For API synthesis, this means identifying the process parameters that can cause the drug substance to fail its critical quality attributes (CQAs). Such CQAs may include product potency, crystal form, impurity levels, and physical properties such as particle size distribution. Process issues that impact overall productivity, capacity, or process greenness are considered business risks, because they have little or no impact on the quality of the drug substance.

Some risk assessment goals are short term such as an evaluation to ensure the chemistry is scalable in the proposed equipment. This usually involves having sufficient prior experience to maximize the probability of success to safely produce material of desired quality. There are also longer-term risk assessment goals such as evaluating the process design and synthesis against the combination of desired safety, quality, and business criteria.

The risk management process may be informal, using relatively simple empirical tools such as flow charts, check sheets, questionnaires, process mapping, and cause-and-effect diagrams to organize the data and facilitate decisions. Risk evaluation may also utilize formal processes and methodologies. Two recognized tools are:

1. Failure mode evaluation and analysis (FMEA) [7] to score and quantify risks by identifying potential failure modes and the impact on product quality.
2. Kepner–Tregoe analysis [10, 11] to provide a quantitative assessment of the synthesis, taking into consideration of both quality and business risks.

Other tools, adapted from safety risk analysis, such as PHA, hazard operability (HAZOP) analysis, and hazard analysis and critical control points (HACCP) [7], apply a failure analysis to meet set criteria of safety, product quality, and/or business deliverables. These approaches are particularly valuable when performing a systematic review of the processibility and scalability of the chemistry for commercial manufacturing. A more detailed discussion on quality risk assessments is covered in Chapter 49 of this book.

44.1.3 Manufacturing Considerations

The short-term focus of process scale-up and assessment is on glass plant or pilot plant processing; however, as a project moves through development, the focus will shift to address manufacturing-scale concerns. There are a number of differences to consider when assessing the process for either the pilot plant or manufacturing scale-up. Some of these differences are generalized and tabulated in Table 44.2.

These differences can have significant implications for the process scale-up assessment. The flexibility and technical support in a pilot plant environment may allow equipment setups and tighter control of process parameters than is typical or possible in a manufacturing plant. These factors must be understood to design a robust commercial process.

TABLE 44.2 Pilot Plant vs. Manufacturing Differences

Consideration	Pilot Plant Scale	Manufacturing Scale
Equipment size	50–4000 L	400–12 000 L
Operating hours	5 d (≤24 h/d)	7 d (usually 24 h/d)
Technical support	Process engineer/chemist	Plant engineer
Automation	Manual to fully automated	More likely automated
Analytical support	Short turnaround (<1 h)	Long turnaround
Equipment setup	• More likely to utilize unique equipment with flexible setup • Mostly sequential operations	• More likely to utilize only standardized equipment • More likely to have parallel operations

44.2 DRIVERS FOR DEVELOPMENT/RISK ASSESSMENT

The process development strategy is defined by risk management across three main areas: safety, quality, and business. Both safety and quality are necessary attributes of a scalable synthesis at any phase of development. Therefore, at a minimum, there must be sufficient development to manage the risk to safety and quality. In the absence of safety and/or quality concerns, business considerations define the development strategy. The challenge is defining the level and timing of development work as projects transition from early to full development. Throughout the stages of development, it will be impossible to understand and eliminate all the risks. The key is to eliminate sufficient risk to ensure that the differences upon scale-up do not impact the development goals.

By understanding the goal of the scale-up and performing an assessment against the goal, the development needs can be prioritized. In early development, the goal will include safety and quality aspects. While the business factors for process optimization are not key drivers, there are still scenarios to consider such optimization. These drivers will be based on the complexity of the synthesis route, the long-term synthesis strategy, the project timeline, and the facilities constraints. Some examples of such drivers are outlined in Table 44.3. As the project moves forward through development, business goals will become a higher priority.

The initial process scale-up assessment is typically performed as a paper exercise guided by prior experience. Then, a laboratory assessment is usually necessary to evaluate unknown risks. Evaluation of such unknown risk can usually be performed in a few well-planned experiments. For example, the impact of processing time is typically unknown in early development and is highly likely to change as the process is scaled. During laboratory runs of the process, sampling at key points and aging the samples at the processing conditions will provide sufficient information about the process stability over the plant-scale time frame. Example Problem 44.1 is given to illustrate an initial process assessment for an early development project. The impact of each unit operation on safety, quality, and business is evaluated (Table 44.4).

EXAMPLE PROBLEM 44.1   Process Description

1. Compound **1** is mixed with solvent and base, heated to 40 °C, and aged for one hour to activate the amide hydrogen (1) (Scheme 44.1).
2. The solution is then cooled to 0 °C and excess acid chloride (2) is added, initiating the coupling reaction to form amide (3). The reaction is exothermic and the temperature must be maintained at less than or equal to 5 °C.
3. The reaction is then quenched at less than or equal to 20 °C by addition of isopropyl alcohol. The quench is exothermic. Several process impurities are formed during the reaction and quench, with impurity A found to be a suspected carcinogenic compound.
4. Compound **3** is then crystallized at 20 °C by addition of an antisolvent. Impurity A co-crystallizes with compound **3**.
5. The slurry is then filtered, the wet cake is washed to remove impurity A, and solid compound **3** is dried under vacuum at 50 °C.

In this assessment (Table 44.4), the initial activation is shown as low priority, and the crystallization and stability are both shown as medium priority. In early development, the absence of information regarding the activation is not considered an issue since it is likely that the process will not deviate significantly from the lab procedure. Similarly, the crystallization and stability should be manageable by keeping as close to the lab procedure as possible. A medium risk was assigned since mixing during the crystallization and time are key scale factors that could result in differences between the lab and glass plant procedures. If the same process was assessed for later-stage development, all of the unit operations would receive a medium-to-high priority since the process knowledge is limited.

TABLE 44.3   Drivers for Development Outside of Safety/Quality Issues in Early Development Programs

Issue	Driver for Development	Potential Development Activities
Length of synthesis	Availability of plant time to prepare the chemistry to meet API needs	Alternate route development
Low yield in synthesis		Increased throughput by • Higher yields • Smaller maximum volume (Vmax) • Decreased cycle time
Sourcing of reagents	Key reagent not available in time for scale-up activities	Alternate route development Replacement of reagent
Intellectual property	Some key element of synthesis under patent protection	Alternate route development to replace this element of synthesis

TABLE 44.4 Step Assessment for Example Problem 44.1

Unit Operation	Risk to			Priority
	Safety	Quality	Business	
Activation	Unknown – no apparent exotherm	Unknown – no observed degradation	Incomplete activation leading to low yield	Low
Reaction and quench	Highly exothermic, HCl liberated	Key impurities form at higher temperature	None	High – impurity formation and exotherm must be understood and controlled
Crystallization	Highly acidic	Key impurity rejection	None	Medium – key impurity rejection is sufficient but should be understood
Isolation	None	Impurity A is removed during the cake wash	None	High – process impurity A must be controlled
Drying	None	None	None	Low
Impact of time	None	Unknown		Medium – effect of age time should be understood

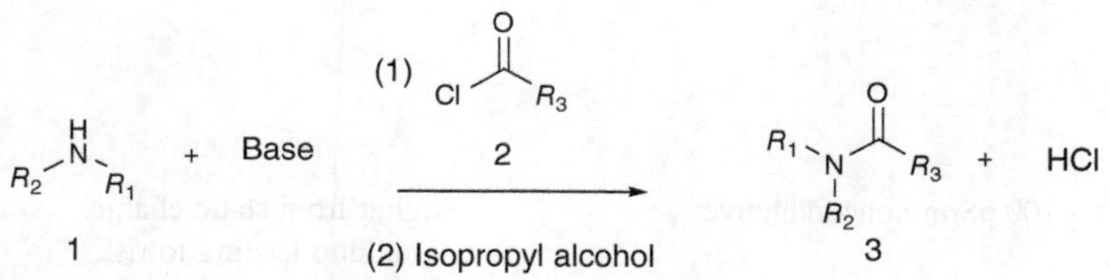

SCHEME 44.1 Example coupling reaction.

The subsequent sections will discuss in more detail areas that should be considered when evaluating the development needs for a program. It is important to keep in mind that these factors are not independent, and therefore a compilation of factors may in itself be a driver for development.

44.2.1 Process Safety

In general, safety is a risk to be understood and then managed. For most risks, mitigation strategies can be developed, once the risk is understood and acceptable risks have been defined. The definition of acceptable risk will likely change as the scale of operation increases, and this could drive process development as the project moves forward along its timelines. For example, the safety analysis for a less than or equal to 20 L scale-up may be limited to a paper review when no sign of exotherm has been observed, whereas at larger-scale thermal and corrosion testing would be required.

44.2.1.1 Personnel Safety The safety issues related to personnel may include exposure to highly potent and toxic compounds (i.e. teratogens, mutagens), sensitizers, and genotoxic and cytotoxic intermediates. Highly potent and toxic compounds are usually characterized by having an exposure limit less than 1 µg/m^3. The majority of pharmaceutical intermediates have exposure limits in the 10–100 µg/m^3 range. Compounds with exposure limits of 1–10 µg/m^3 are considered to have medium to high potency/toxicity [12]. Personnel risk can typically be managed by a combination of engineering controls such as closed isolation and handling equipment and personnel protective equipment (e.g. breathing apparatus and chemically resistant clothing, gloves, and face/eye protection) and by administrative controls such as restricted access and specialized training. With advances in containment technology, exposure levels down to less than 1.0 µg/m^3 can be achieved. This level is typical of exposure guidelines for compounds considered to be highly potent and toxic. However, the cost associated with purchasing and maintaining the appropriate high containment equipment for highly potent and toxic compounds can drive the decision to look for an alternative chemistry route when possible.

44.2.1.2 Exceptional Process Hazards The assessment of a process should include investigating exceptional process hazards. In general, pilot scale and manufacturing facilities are set up to handle a "typical process" so variables such as high or low temperature (−20 to 110 °C), solids charging, and slight exotherms ($\Delta T_{ad} < 5$ °C) would not be considered unusual when performing a hazard assessment. However, many processes have one or more steps that include additional hazards such as gas evolution, dust explosion, and static and/or significant exotherms (Table 44.5). Such hazards need to be addressed with further development or by appropriate equipment selection, additional engineering controls, and personnel training.

TABLE 44.5 Exceptional Process Hazard Examples

Hazard	Safety Limit	Risk	Mitigation Strategy
Gas evolution	Dependent on equipment vent capacity and gas properties (i.e. minimum ignition energy, flammability)	Equipment overpressure, hazardous, or combustible gas release	Understand gas formation mechanism to develop control strategy
High exotherm [13]	$\Delta T_{ad} > 50\,°C$ $TMR_{ad} \leq 24\,h$ Note: proximity of the operating temperature to the initiation temperature for secondary decomposition exotherms should also be considered	Runaway reaction and potential thermal explosion	Control reaction by addition method and cooling; maintain reaction sufficiently; maintain distance from decomposition exotherm. Training and awareness
Dust explosivity [14]	$K_{ST} \geq 1\,bar\,m/s$	Potential explosion	Based on explosion risk, consider not isolating alternative form or alternative intermediate. Inert handling, containment, explosion suppression, blowout panel. Training and awareness
Static [15]	<100 pS/m nonconductive	Arcing from static charge buildup leading to risk to personnel, equipment damage (glass or Teflon), fire and/or explosion	Bonding and grounding, inert handling, antistatic additives, solvent changes, conductive components (pumps, antistatic bags), appropriate hold times to match relaxation time for solvent. Training and awareness
Hydrogenation [16]	Concentration <4% or >75%	Potential fire and/or explosion, hydrogen embrittlement	Pressure testing, inert handling, grounding, explosion protection systems, hydrogen/LEL monitoring, equipment selection (motor and equipment rating), H_2 rated flash arrestors, open to air venting handling procedures
Flammable liquids [16]	Flash point <60.5 °C closed cup	Potential fire and/or explosion	Inert handling, bonding, and grounding

In the case of gas evolution, understanding the chemistry and particularly the source and identity of the gas is important. Generation of CO_2 as a by-product in a reaction may not pose a significant risk, if the total gas generated is too low to result in a significant pressure increase for the processing equipment or if the rate of CO_2 evolution can be controlled by adjusting reaction rates or reagent addition times. However, if the amount, gas composition, and generation rate are not known, further development is necessary to ensure the gas evolution does not pose a significant hazard. Release of a flammable gas such as hydrogen poses an additional challenge since it has a wide flammability range (4–75% in air) and low minimum ignition energy [16].

44.2.1.3 Material Compatibility Material compatibility refers to the ability of the materials in a given equipment train to withstand exposure to the process streams (i.e. to maintain mechanical integrity at the temperature and time scale for the process). Most lab development occurs in glassware using Teflon accessories (agitators, seals, etc.), which ensures compatibility with the exception of hydrogen fluoride. In the pilot plant and manufacturing facilities, the range of physical equipment (glass lined, stainless steel, Hastelloy C, tantalum) will likely provide the flexibility to ensure a compatible fit for any process. However, an initial assessment is usually necessary to ensure the equipment is properly selected. This understanding of materials and process stream compatibility is required to ensure that the right set of equipment is selected for any scale-up. It is also important to be aware not just of the raw materials (solvents, reagents), the intermediates, and the products but also of by-products in the process. Table 44.6 describes some common equipment material of construction and materials to avoid. Though some plastics and elastomers are compatible with many solvents, leaching must be considered since it would be difficult to find an impurity leached from the polymer in the final API. In certain cases, compatibility issues may conflict. For example, the use of heptane in a highly acidic environment may present a glass (static) and Hastelloy (corrosion) concern. Also, material concerns should be extended to include interactions of the process stream with jacket and condenser fluids. This is of particular concern with water-sensitive process steams.

Material compatibility will likely not play a significant role in driving process development for early- and mid-stage processes unless the equipment available is limited. Even then, a development effort based on incompatibility is not necessary until the program progresses to full development. When transferring a process to manufacturing, it becomes highly desirable to reduce or eliminate material compatibility issues to allow easy movement of a process between facilities.

TABLE 44.6 Examples of Incompatible Materials

Equipment Material of Construction	Incompatible Material
Carbon (condensers)	Bromine, NMP
Glass	Hydrogen fluoride, inorganic base at high temperature
Stainless steel	Acids, acid salts, chlorinating reagents
Polypropylene (filter media)	Some solvents (i.e. methylene chloride, heptane, toluene)
Hastelloy B	Ferric and cupric salts
Elastomers (seals)[a]	Some solvents
EPDM	Organic chlorides, cyclohexane
Neoprene	Ethers, acetates, acids
Viton	Acetone, amines, ammonia, acetates, ethers, ketones, caustics

[a]Cole-Parmer Chemical Compatibility Charts. https://www.coleparmer.com/chemical-resistance.

44.2.1.4 Hazardous Reagent Handling Highly hazardous reagents are materials that warrant special consideration as the general safety hazards are well known throughout the chemical industry. In general, these materials should be limited when developing a commercial process since in the complexity associated with risk mitigation can be costly and difficult to manage. However, most highly hazardous materials have been well studied, and methods to mitigate the risk have been developed. Despite the added cost, it is not uncommon to use a hazardous material in early development when the scale-up is still limited and the risks are well known and can be mitigated. The key is risk awareness and communication to ensure that all parties involved are aware of the hazards and the necessary controls used to address them. Table 44.7 lists some examples of hazardous reagents used in API syntheses.

44.2.2 API Quality

Prior to use in human clinical studies, impurities and other foreign contaminants within the API must be controlled at a level specified by regulatory authorities. Quality in pharmaceuticals refers to adhering to these regulatory rules as well as understanding the impact of process variations. Similar to safety, product quality is a key development driver. In this section we focus on discussion of key aspects of API quality: (i) CQAs, (ii) genotoxic risk, and (iii) process robustness.

44.2.2.1 Critical Quality Attributes CQAs are quantifiable properties of an intermediate or final product that are considered critical for establishing the intended purity, efficacy, and safety of the product. For API, CQAs must be met prior to release for formulation and eventual use in clinical trials or commercial production. These may include overall purity, levels of impurities, form, color, metals, solvent content, and powder properties (Tables 44.8–44.12). The selection of CQAs will be based on ensuring the API does not pose a significant risk to patients.

For impurities, the target levels are conservatively set assuming a high dose of 1 g without consideration of the actual clinical trial dosage or therapy duration. It is therefore possible that if the risk is sufficiently understood and can be managed, these target properties can be adjusted to less conservative values for compound used at low dosage and/or in a short duration clinical study. The inability to meet a purity CQA will drive additional development. At early stages of development, this may mean developing a rework strategy for the API and, subsequently in later development, may entail obtaining a detailed and complete understanding of the mechanism of formation for a key impurity.

The powder properties of the API, such as particle size and morphology, can impact the design of the formulation as well as the formulation process performance and need to be addressed in concert with drug product development. Particle

TABLE 44.7 Examples of Hazardous Reagents

Chemical	Hazard	Risk Mitigation
t-Butyl lithium [16]	Pyrophoric, air sensitive	1. Proper handling and inert handling to prevent exposure to air
Azides (e.g. sodium azide) [16]	1. Under acidic conditions, sodium azide forms highly toxic and explosive hydrazoic acid gas 2. Sodium azide and hydrazoic acid are extremely toxic and can be absorbed through the skin 3. Heavy metal azides are shock sensitive	1. Proper use and waste handling techniques. Avoid contact with water or acid 2. When handling impervious protective clothing and proper PPE 3. Plastic materials are preferred for handling or non-sparking metals
Hydroxybenzotriazole (HOBt, anhydrous) [17]	Explosive, flammable solid	Proper handling and storage. Avoid heat and mechanical shock. Consider using the HOBT hydrate due to reduced explosivity risk
Alkali and alkaline hydrides (e.g. sodium hydride) [16]	Sodium hydride reacts violently with water. The heat given off is sufficient to ignite the hydrogen decomposition product. Sodium hydride is spontaneously flammable in air	Proper handling and storage avoiding contact with water
Alkali metal cyanides (e.g. sodium cyanide) [16]	Highly toxic	Use only in well-ventilated area. Note: hydrogen cyanide can be absorbed through the skin Avoid contact with water or acid Communicate the hazards prior to use
Hydrazine [16]	Highly toxic Highly explosive (limits in air 4.7–100%, flash point 52 °C)	Proper PPE and communication of hazards prior to use. Maintain inert environment and avoid contact with oxidants and catalyst
Chlorinating agents (e.g. oxalyl chloride, $SOCl_2$, $POCl_3$, PCl_3) [18]	Highly toxic. React violently with water and release HCl gas. In addition thionyl chloride ($SOCl_2$) and oxalyl chloride (($COCl)_2$) release SO_2 and CO gas, respectively	Proper use and waste handling techniques. Avoid contact with water or acid When CO and SO_2 gas release are anticipated, ensure proper ventilation and consider monitoring
Metal and supported metal catalyst (e.g. Pd or Pt on carbon, Raney nickel) [18]	Catalysts such as Raney nickel or activated heterogeneous catalysts such as supported Pt and Pd are pyrophoric. The risk of fire is especially high in the presence of flammable liquids	Proper use and waste handling techniques. Catalyst in contact with flammables should be kept under nitrogen inertion. When kept wet with water, Pt- and Pd-supported catalysts can be handled in air
Grignard reagents [19, 20]	Grignard reagent formation and reactions are highly exothermic. The exotherm is particularly hazardous since the reaction initiation can be delayed. Grignard reagents react exothermically with air water and CO_2	Proper transportation, handling, and storage. Avoid contact with air and water. Evaluate the reaction exotherm and the equipment cooling capacity to ensure proper control can be achieved. Charge no more than 20% of the reagent prior to reaction initiation. Consider online techniques such as FTIR to monitor the reaction progress. Evaluate the potential for water content including jacket services. Care should be taken when disposing of residual magnesium turnings
Halogenating agents (Cl_2, Br_2, I_2) [16]	Highly toxic vapors. Water reactive	Proper handling and storage to address inhalation hazards
Hydrogen fluoride [18]	Highly toxic. Fluoride ion can penetrate the skin and bind with ions such as calcium, potassium, and magnesium in the body	Select proper PPE to prevent inhalation or direct exposure. HF is incompatible with glass and most metals[a] [16]
Phosgene [16]	Highly toxic gas reacts in lungs, reducing lung function leading to suffocation	Thorough plant controls to prevent emission Implementation of leak detection devices and personnel monitors, and availablity of escape breathing mask or respirator

[a]Cole-Parmer Chemical Compatibility Charts. www.coleparmer.com/techinfo.

TABLE 44.8 Typical API Properties Analyzed Prior to Releasing the Material for Drug Product Formulation [21]

Property	Purpose
Purity/impurity profile	The weight percentage can be measured by HPLC or titration. An HPLC purity profile can also show the impurity concentration and indicate whether there is a significant amount of unknown present. The purity of an API is regulated by ICH guidelines (Table 44.9)
Chiral purity	Chiral purity for single chiral left compounds is defined by the enantiomeric excess (EE) and is derived from HPLC using chiral columns. EE is defined as $(R-S)/(R+S)$ where R, S are the fractions of the enantiomers and $R+S=1$. Chiral purity for diastereomers (multiple chiral lefts) is also derived from HPLC
Crystal form (polymorph, solvate, salt)	Form is usually verified by X-ray powder diffraction, solid-state NMR, or spectroscopic methods (i.e. Raman). The appropriate form ensures that the compound has good physical and chemical stability as well as pharmaceutical properties
Color	Color can be an indicator of an unidentified impurity or degradate. It is also important for ensuring a uniform tablet color. Color can be assessed visually or quantitatively with UV
Inorganic impurities (including metals)	Inorganic impurities can be quantified by residue on ignition or atomic adsorption spectroscopy. A generic heavy metal test is performed by ICPMS. Additionally, individual metals known to be present in the process streams such as Pd and Pt are monitored and need to be controlled based on dose. A typical target for heavy metals is ≤10 ppm [22]
Solvent content (including water)	A GC analysis of the final API is performed specifically looking for any solvent present within the final two API synthesis steps. Solvents have differing level of toxicity and therefore different target limits (Tables 44.10, 44.11, and 44.12). Residual water is important for compounds that are hygroscopic, degradable by moisture, or known hydrates. Standard methods include Karl Fisher titration or loss on drying
Powder properties	Typically, particle size measured by laser diffraction, crystal habit assessed by microscopy, and form measured by XRD or Raman are of primary concern. Other measures such as surface area and density may also be appropriate. Powder properties can have a significant impact on the formulation process and the API's bioavailability
Microbial limits	Such assays include total count of aerobic microorganism, yeast, or molds and the absence of specific bacteria. The need for such testing is based on the nature of drug substance and intended use of drug product (e.g. endotoxin testing for drug substance to be formulated into injectable drug product)

TABLE 44.9 International Conference on Harmonization (ICH) Reporting Guidelines for Impurities Present in an API

Maximum Daily Dose[a]	Reporting Threshold[b]	Identification Threshold[c]	Qualification Threshold[d]
≤2 g/d	0.05%	0.10% or 1.0 mg/d intake (whichever is lower)	0.15% or 1.0 mg/d intake (whichever is lower)
>2 g/d	0.03%	0.05%	0.05%

Source: Adapted from ICH Q3A [8].
[a]The amount of drug substance administered per day.
[b]The reporting threshold is a limit above (>) which an impurity should be reported. Higher reporting thresholds should be scientifically justified.
[c]Identification threshold is a limit above (>) which an impurity should be structurally identified.
[d]Qualification threshold is a limit above (>) which an impurity should be qualified in clinical studies. Lower thresholds can be appropriate if the impurity is unusually toxic.

TABLE 44.10 ICH Class 1 Solvents Should Be Avoided for Use in Drug Substance Synthesis

Solvent	Concern
Benzene	Carcinogen
Carbon tetrachloride	Toxic and environmental hazard
1,2-Dichloroethane	Toxic
1,1-Dichloroethene	Toxic
1,1,1-Trichloroethane	Environmental hazard

Source: Data from ICHQ3A [8].

TABLE 44.11 ICH Class 2 Solvents Should Be Limited Because of Their Inherent Toxicity

Solvent	PDE (mg/d)	Concentration Limit (ppm)
Acetonitrile	4.1	410
N,N-Dimethylacetamide	10.9	1090
N,N-Dimethylformamide	8.8	880
Methanol	30	3000
N-Methylpyrrolidone	5.3	530
Tetrahydrofuran	7.2	720
Toluene	8.9	890

Source: Data from ICHQ3A [8].

TABLE 44.12 ICH Class 3 Solvents May Be Regarded as Less Toxic and of Lower Risk to Human Health

Solvent	PDE (mg/d) Concentration Limit (ppm)
Acetic acid	Class 3 solvents may be regarded as less toxic to human health and need to be controlled to <0.5% or 50 mg/d. This specification does not require specific testing as long as the product loss on drying test is <0.5 wt %
Acetone	
Ethanol	
Ethyl acetate	
Heptane	
Isopropyl acetate	
Methyl ethyl ketone	
Methyl isobutyl ketone	
Isopropanol	

Source: Data from ICHQ3A [8].

TABLE 44.13 Biopharmaceutical Classification System[a]

Class I	High permeability[b], high solubility[c]
Class II	High permeability, low solubility
Class III	Low permeability, high solubility
Class IV	Low permeability, low solubility

[a]https://www.fda.gov/aboutfda/centersoffices/officeofmedicalproductsandtobacco/cder/ucm128219.htm.
[b]A drug substance is considered HIGHLY PERMEABLE when the extent of absorption in humans is determined to be greater than 90% of an administered dose, based on mass balance or in comparison with an intravenous reference dose.
[c]A drug substance is considered HIGHLY SOLUBLE when the highest dose strength is soluble in less than 250 ml water over a pH range of 1–7.5.

TABLE 44.14 Acceptable Intakes for an Individual Impurity from FDA Guidance on DNA-Reactive Impurities [23]

Duration of treatment	<1 mo	>1–12 mo	>1–10 yr	>10 yr to lifetime
Daily intake	120 µg	20 µg	10 µg	1.5 µg

size becomes more of an issue depending on the Biopharmaceutics Classification System (BCS) since the higher the solubility and permeability, the less likely that a change in the particle size will have an impact (Table 44.13). The crystal form of the API will impact the compound's chemical and physical stability as well as its pharmaceutical properties (solubility, permeability). Most APIs are crystalline and can exist as different polymorphs, solvates, or salts or as co-crystals with other organic compounds. An optimal crystal form needs to be selected based on its stability and pharmaceutical properties. Identification of the various forms and selection of the most appropriate form are primary objectives in early development of the compound.

Changes to CQAs of the API that significantly impact the drug product could impact the program's clinical timeline, if additional clinical studies are needed to demonstrate equivalency of the drug products.

44.2.2.2 Genotoxic Impurity Risk Genotoxic compounds have the potential to impact cells in a mutagenic or carcinogenic manner or considered DNA reactive. All intermediates, solvents, reagents, and catalysts and known impurities present in the API need to be analyzed to assess their genotoxicity. This is typically done first by simulation against a database of known genotoxic structural moieties (*in silico*) and then followed up with tests on bacteria (Ames test) to verify positive results. Genotoxic impurities (GTI) present a significant challenge for drug development since they must be controlled to levels much lower than the standard HPLC detectability limit. The limits for these impurities are set based on daily intake and duration the patient will receive the drug (Table 44.14). Guidance for the industry from International Council for Harmonisation (ICH) M7 has aligned the efforts and limits using a Threshold of Toxicological Concern (TTC) concept tied to a numerical cancer risk value [23]. Therefore at high drug load given for extended duration, the limit may be so low the impurity cannot be detected with standard analytical methods.

As with API quality attributes, the primary risk mitigation for genotoxic compounds is sufficient removal or prevention of its formation. The added cost associated with the development of a control strategy and appropriate analytical testing methods makes the presence of genotoxic compounds a formidable development challenge [24]. The need to develop a control strategy for GTI will often force a reexamination of the synthetic route to either eliminate the formation of the compound from the synthesis or move its formation earlier in the synthetic sequence, allowing the subsequent reaction, workup, and isolation steps to more effectively remove the impurity prior to the API step. If an intermediate is genotoxic, a new synthetic route that avoids the intermediate may be designed. In the case of an impurity, a detailed understanding of how the impurity forms may afford a method to limit or prevent the formation. Typical control strategies for the impurity might include reaction conditions and/or extraction and crystallization design. Where a potential risk has been identified for an impurity, an appropriate control strategy utilizing process understanding and/or analytical controls will need to be developed to ensure the genotoxic impurity is at or below the acceptable cancer risk level. Exemptions to this approach could be considered for drugs intended for cancer patients.

44.2.2.3 Process Robustness The ability of a process to demonstrate acceptable quality and performance while tolerating variability in inputs is referred to as robustness [25]. Robustness is a function of both the process design (synthesis route selected, the equipment capabilities and settings and environmental conditions) and the process inputs (quality of raw materials). Process robustness and therefore process understanding is of critical importance to enabling commercialization of a drug. The use of in-process controls and assays ensures that processing activities produce API with the required quality. Understanding of the process variability is critical to ensuring that the API quality will be consistently achieved.

As a project moves into full development and toward commercialization, increased emphasis is placed on understanding which step and process parameters have the potential to impact an API CQA. A design range for these parameters can then be defined to ensure the API quality is consistently achieved. Within the design range, a target set point is selected, and the process and equipment capability are then used to define a normal operating range. Critical process parameters (CPPs) are parameters when varied beyond a limited range have a direct and significant influence on a CQA. Failure to operate within the defined range leads to a high likelihood of failing a CQA specification. Numerous approaches have been presented on defining this range to distinguish a CPP from non-CPPs. One approach is to evaluate whether material of acceptable quality can be made within 6σ of the normal operating range, where σ is the equipment-specific operational variability [26].

Building process knowledge is typically a significant undertaking in the later stages of development since each unit operation may have up to 10 process parameters, and it is unlikely that each of these process variables has been studied. For example, for a given reaction, process parameters could include reaction temperatures, time to ramp up to the temperature, reaction time at temperature, agitation, variability of charge equivalents, sequence of charges, hold times between charges, and concentrations. An earlier stage assessment would likely focus the development effort only on a subset of these parameters with the largest impact on process robustness. In addition, the potential for multivariable interaction such as time and temperature must be evaluated at later stages.

An important component of process robustness for a given step is the understanding of process impurity generation and rejection. It is then necessary to determine the "fate" of the impurity in later processing steps; more specifically, is it inert or transformed into other process impurities, and is the original impurity or new impurity removed during an extraction or crystallization? By carrying this analysis forward through the API step, the impurity "tolerance" or limit can be established. Target purity profiles for each intermediate can then be defined through similar analysis with all process impurities. This assessment is often completed to support the establishment of the appropriate CQAs for the drug substance.

44.2.3 Business Optimization

In the absence of quality or safety issues, process development is driven by optimization of parameters to improve the business of manufacturing drug substance (e.g. productivity, flexibility, or throughput). Often, the majority of this development effort can be deferred until there is a high probability the compound will be commercialized. A key milestone for any compound is the achievement of proof of concept in the therapeutic hypothesis, typically successful completion of phase IIA clinical trials. However, the complexity of the process and the duration of the overall clinical program will play a role in assessing the timing of process optimization.

There are two key business drivers for API process development: (i) meeting the project timeline and (ii) reducing the cost of manufacturing. The first driver applies to both early- and late-stage products. The second priority, reducing the cost to manufacture, is not usually considered in the early stage of development unless there are specific issues that will impact scale-up to generate the required quantity of API for the program's development. Reducing the cost of manufacturing involves both synthesis design to reduce the material cost and number of steps and process optimization to address productivity and capacity through improvements in yield and volume efficiency.

As part of the business drivers for process development, we examine the impact of project timelines, process fit and ease of manufacturing, and process greenness. Evaluation of a process' productivity and fit into a manufacturing plant through time cycles, yield, and mass balance will also provide insight into setting the direction for process development. The use of process metrics is an important tool that will enable a common platform to evaluate the evolution of a process.

44.2.3.1 Project Timeline Hierarchically, the project timeline does not drive development but rather the development strategy. Generally, this timeline will define both immediate and long-term API needs. The immediate needs are driven by API required for the clinical studies, the drug safety studies, and drug product development. All of these activities are on the critical path to bringing a drug to market. The highest business priority is delivering sufficient API to meet the clinical and drug safety study timelines. The short-term needs may drive changes that will be covered in the next few sections. Long term, the project timeline will drive development decisions. With an extended project timeline due to long clinical trials (10–14 years), deferring process development focused on optimization allows resources to be used on programs with shorter timelines. Alternatively, an accelerated program (five to seven years) leaves little time for process development and may drive parallel development efforts to meet short-term needs as well as to develop the manufacturing process.

44.2.3.2 Process Cycle Time At commercial scale, an optimized cycle time is critical to control cost and manufacturing capacity utilization. The process cycle time can be thought of at multiple levels including the time to complete (i) the entire synthetic sequence, (ii) one isolated intermediate or process step, and (iii) an individual unit operation. In early development, optimization of the cycle time is less critical. Time cycle optimization would only be considered in the rare case the program timeline cannot be met. Even then, the first choice would be to use alternative equipment such as larger vessels or filters to accelerate the timeline. At the transition to full development, a significant effort will be placed on improving the overall process cycle time. This timing will vary depending on the severity of the bottleneck and the potential for changes to impact the API quality. As a project moves through development, emphasis will shift from individual step and unit operation optimization to debottlenecking of the whole synthetic sequence. In development, individual batches are typically run in sequence, so a reduction in time anywhere will lead to a shorter overall delivery time. At commercial scale, the process unit operations and process steps will likely be run in parallel, so resources should be more strategically placed on true bottlenecks. These bottlenecks will not likely be known until the commercial process fit is identified since multiple unit operations may be planned for the same equipment. This is most often the case with isolation and drying where filtering on a centrifuge and drying in a conical dryer can process in parallel, whereas filtration and drying on a filter dryer must occur in series.

As an example a simple process timeline involving a reaction, aqueous workup, solvent swap, crystallization, isolation, and drying steps is depicted in Figure 44.1. For ease of discussion, each of these steps is assumed to have an eight-hour cycle time. The first timeline illustrated is a process fit with two vessels and a filter dryer. During the isolation step, both the crystallizer and the filter dryer are active, which extends the time cycle in the crystallizer. In the second timeline, the fit is the same, but the equipment downtime has been minimized by running processing steps in parallel. This provides a significant improvement in time cycle, allowing the third batch to be completed in a similar time frame as the second development batch. The bottleneck however in this process is vessel 2, as it requires 24 hours vs. the 16 hours for all other equipment. If you add an additional vessel to hold the slurry during isolation, as shown in the third timeline, the parallel processing timeline is reduced by eight hours. An alternative approach to debottleneck is a focused development effort on reducing the total time cycle to complete the solvent swap, crystallization, and isolation. It is therefore important to evaluate the process as a whole. In this case, the only equipment change that would optimize the process time cycle is debottlenecking vessel 2. Similarly, the process as a whole should be considered when evaluating where to focus the development effort. In this example, all steps were assumed to take the same amount of time, whereas in reality the processing time for each unit operation will vary widely and should be considered in the evaluation.

44.2.3.3 Process Fit and Ease of Manufacture

The process design can have a significant impact on the manufacturing cost and flexibility as well as the process portability. Ideally a process is flexible enough to fit into any facility, regardless of the equipment available. However, there are often process constraints such as volume requirements as well as specialized processing equipment needs such as hydrogenators, cryogenic reactors, or continuous processing that impact the selection of manufacturing equipment. In early development, process fit is of little concern since the equipment flexibility is built into glass plant and pilot-scale facilities. However, as a project moves toward full development, the flexibility of the process becomes a significant development driver. In general, development would focus on improving the process flexibility and reducing the process complexity. Therefore as a project moves toward full development, the desire to reduce the equipment cost and operational complexity may mean minimizing the use of specialized technology.

The first step in improving process fit and ease of manufacture is minimizing the number of unit operations by reducing the need for solvent exchanges, extractions, and isolations. In early development, the process may not be well understood, and process steps are added to ensure a quality material is achieved. An increase in process knowledge around impurity formation and control will allow optimized solvent usage and may allow for the elimination of extractions and isolations. Once the process steps are defined, a high priority is placed on reducing the maximum process volume as well as providing a wider volume range. For a given process train, the maximum volume will dictate the maximum batch size and therefore manufacturing efficiency. Outside of volume reduction, the goal is to crystallize product that can be isolated and dried in any either a filter dryer or in a centrifuge followed by a dryer. Flexibility in crystallization, isolation, and drying is typically linked. Regarding reactions, flexibility in scale-up is typically associated with mixing in heterogeneous systems. It is often not possible to eliminate heterogeneous reactions, so the focus is placed more on understanding and minimizing the impact at scale. Similarly, while hydrogenations can impact the process fit, it is often a very efficient chemical transformation and in the case of asymmetric hydrogenations, there can be a significant increase in the overall process yield.

44.2.3.4 Process Greenness

Process greenness is typically considered as part of the overall development strategy to select the final synthetic route and is rarely the main driver for process development. This is especially true in early development where the waste treatment cost and environmental impact are minimal. There are many aspects to

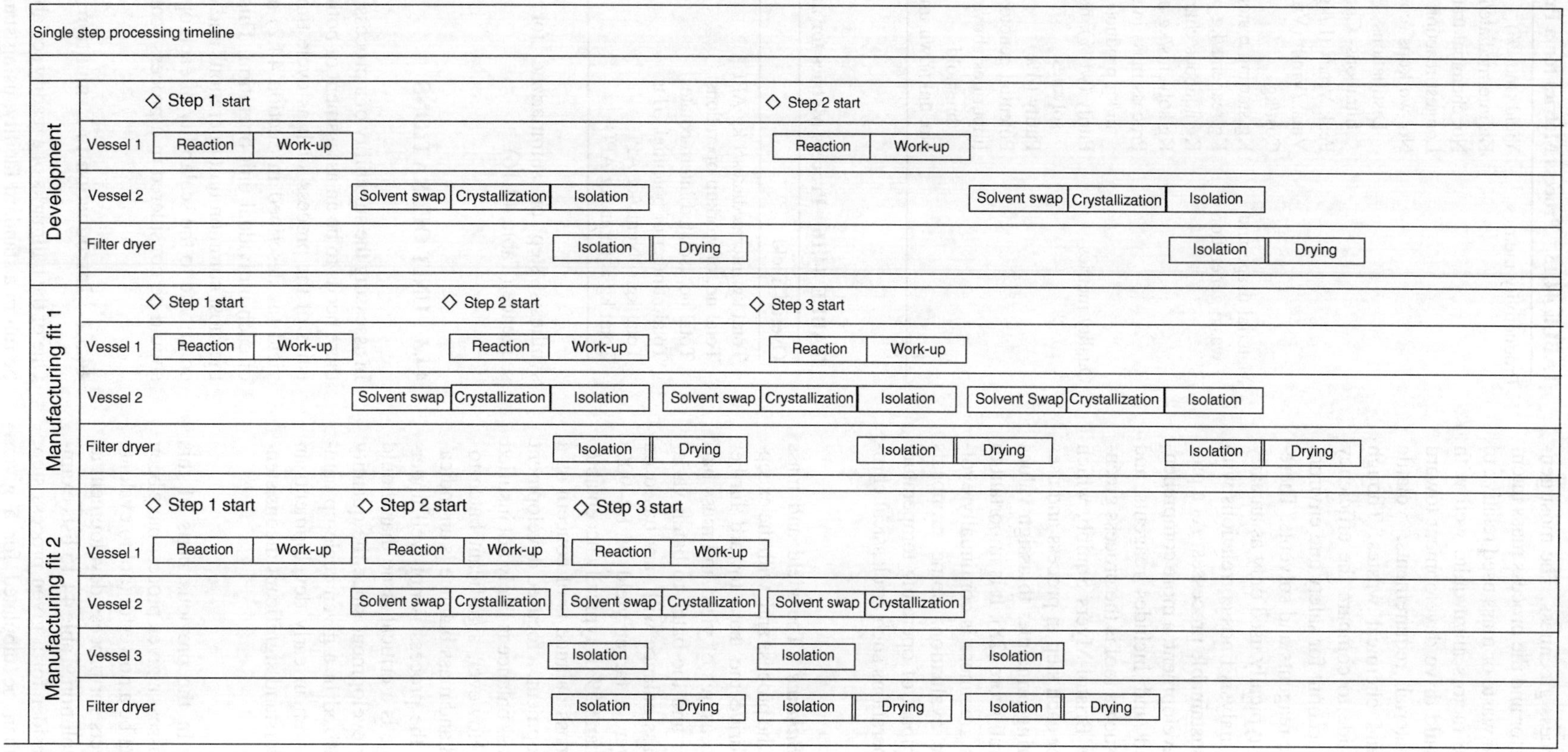

FIGURE 44.1 Single step processing timeline.

consider when discussing process greenness. The most general methods such as the E-factor and the process mass intensity (PMI) account for the total waste or mass used relative to the product mass [27]. These factors align quite well with business priorities since it would drive development toward lower cost by reducing material requirements, volatile organic carbon emissions, and chemical wastes. Though these factors give a quick guide to compare the efficiency of materials used, they fail to account for safety and environmental risks posed by specific reagents and solvents. Therefore the E-factor and PMI are typically used only as an early guide, although more recent analysis finds correlations with lower PMI values and more sustainable processes. As a program moves through to full development, a more comprehensive evaluation is performed and includes reagents and solvent risks. An example of such a tool is the process greenness scorecard, developed by Bristol-Myers Squibb, which tracks about 15 parameters for each step in process and uses green chemistry and engineering principles to assign values that are weighted into an overall score [28]. It is important to note the definition of process greenness is continually evolving toward a more holistic evaluation. Some proposals include factors such as the impact of operating temperature and certain inefficient unit operations such as classical chromatography [29, 30].

44.2.3.5 Yield and Mass Balance

The yield and mass balance are key indicators for the process and, with the exception of early development, drive the team toward further development. The two key measures of yield and mass balance are the absolute number and the batch-to-batch variation. The target yield and mass balance will vary based on the step complexity; however, a target yield of 80–90% and mass balance of greater than 95% is typically acceptable. Though a low yield and/or mass balance is of concern as a process moves through development, a focused development effort to improve yield and mass balance is likely not justified if the process is consistent. However, significant batch-to-batch variability in both yield and mass balance is an indication that a key parameter in the process is not well understood. This lack of knowledge is a critical issue that should be considered even in early development since the quantity or quality of the API synthesized in a given scale-up campaign is at risk. Therefore, even in early development, an effort should be made to understand significant inconsistencies in yield or mass balance.

44.2.3.6 Process Metrics

In the previous parts of this chapter, numerous factors to assess a given process have been discussed. Process metrics can be a powerful tool to evaluate the quality and business drivers for process development as well as to track the process evolution. Table 44.15 lists example of process metrics to consider for a given process step. Such process metrics can then be tabulated for a given

TABLE 44.15 Process Metrics for a Process Step

Productivity metrics	Yield (mol %)
	Kg intermediate/kg API
	No. chemical transformations
	Longest reaction time (h)
	No. workups (count no. of below total)- Distillations- Extractions- Waste filtrations- Chromatography
	Peak Vmax (L/kg)
	Vmax/Vmin (Vmax swings)
	Cost/kg
Material usage and waste generation	Kg starting material/kg product
	Kg reagents/kg product
	Kg aqueous charges/kg product
	Kg solvents/kg product
	Process mass intensity (total kg material in/kg product)
Quality metrics	Purity (wt %), normalize for salts and solvates
	Purity ($A\%$)
	Potential genotoxic impurities (GTIs)
	Impurities above ICH identification threshold
	No. unknown impurities

TABLE 44.16 Process Metrics for an Overall Synthesis

Overall Yield
Total kg intermediates/1 kg API
Total no. of workup operations
Total no. isolated intermediates
Total number of potential GTIs
Total kg solvents/kg API
Total kg aqueous/kg API

synthesis step or summarized for the entire synthetic sequence (Table 44.16).

44.3 UNIT OPERATIONS

In assessing the suitability of a process to run at a given scale, there needs to be an assessment of both the overall characteristics of the process such as cycle time, cost of goods, and yield (as described in Section 44.2) and the characteristics of each individual unit operation. This section will examine the most common unit operations and enumerate factors that contribute to the scalability of each operation. These factors should be considered in a process scale-up assessment.

44.3.1 Introduction to Evaluation

One of the hallmarks of a readily scalable process is that it can be run in a standard facility, using standard equipment, with an ordinary degree of control over the process parameters. Therefore it is critically important for an engineer to understand how processes are generally run on pilot and manufacturing scale.

The vast majority of pharmaceutical processes are run as a batch operation rather than as a continuous or semicontinuous operation. The process train typically consists of multiple stirred vessels, pumps and lines for liquid charges/transfers, waste receiver vessels for distillate, mother liquors, waste streams, product isolation equipment (pressure filters, centrifuges, or filter dryers), and dryers (tray, conical, rotary, filter dryers). The batch reactors are generally equipped with ports for charges/feeds/probes, a bottom valve for discharge, a fixed agitator type and fixed baffling configuration, and overhead piping system for providing venting, vacuum, and emergency pressure relief, typically with a condenser on the main vent path. Flexible lines and a manifold system are commonly used to allow transfers from vessel to vessel or from vessels to the isolation equipment. Common instrumentation on the equipment includes the temperature and pressure of the equipment's contents, the temperature of the equipment's jacket, and product stream's density.

Given the standardized nature of the equipment, standard unit operations are preferred to achieve the process goals. A typical sequence of unit operations includes solution preparation, reaction, separation (extraction, distillation), crystallization, isolation, and drying.

44.3.1.1 Selection of Unit Operations Before the sequence of unit operations for a given step can be determined, an understanding of the objectives for the step is needed. The objectives of each step in the synthetic sequence should be considered collectively, since there are likely trade-offs between steps in the sequence with respect to yield, quality, process cycle time, and the need for specialized equipment. Key to assessing these trade-offs are well-established API quality requirements (including powder property requirements) and knowledge of the material value for a given step (e.g. what is the value of an additional 5% yield). The intermediate quality requirements can then be defined after considering trade-offs between the steps. For example, one may tighten the quality specification in an early step at the expense of step yield, in exchange for eliminating the need for difficult or costly purification downstream.

Once the objectives for the overall step are established, the objectives of each individual unit operation should be understood. An optimized process will involve no additional operations (or more complicated operations) than needed to safely, reliably, and robustly meet the process objectives. Prior practice at smaller scales may dictate the initial choice of unit operations, but as the process is optimized, the number and type of operations is expected to change. For example, a prior iteration of a process may involve multiple liquid–liquid extractions designed to remove a key process impurity. If subsequent improvement to the reaction conditions reduces the number or extent of side reactions, fewer or no extractions may be needed. The process optimization to reduce the number and complexity of unit operations is a key process development objective.

44.3.1.2 Process Fit Another core process engineering activity is understanding the process fit. Engineers are frequently tasked with fitting a process in an existing facility in such a way to minimize capital expenditure (modifications to existing equipment or purchase of new equipment) and to minimize the deployment of shared resources (portable equipment). To accomplish this task, the engineer must clearly understand the capabilities and limitations of the plant. Specifically, vessel configurations (minimum and maximum volumes, baffles, number and type of agitators), vacuum and temperature control capabilities, heat and mass transfer coefficients, filtration capabilities (e.g. centrifuge vs. pressure filter, filter area, filter porosity), and drying capabilities (e.g. agitated vs. non-agitated, heat transfer, vacuum control) will need to be considered. The engineer will then be able to assess if the process as designed can operate in the plant without modification and, if necessary, modify the process to fit existing equipment.

44.3.1.3 Common Scale-up Factors There are many scale-up factors that are not specific to any one particular unit operation. Time is a particularly important example. Nearly every activity requires more time to accomplish at manufacturing scale compared with the laboratory scale. The ramifications of this will be discussed in the individual unit operations section. One concern that is common to all the unit operations is stability. The stability of the reaction mixture with respect to undesired side reactions (degradation) must be assessed for each unit operation on time scales relevant to the plant scale.

Another issue common to most unit operations is the potential for residual material in process lines and dead legs to interact with material being charged or discharged. For example, a single charge line may be reused for multiple reagent charges, with a solvent flush in between each charge. If the flush is inadequate (or not done), and the materials are not compatible with each other, a deleterious reaction may occur. It is critically important for both safety and quality reasons for the engineer to be cognizant of what lines are being used for what purpose and to systematically consider what residues may be left behind as process fluids are transferred throughout the equipment train.

For all unit operations where heat transfer is important, the surface area to volume ratio will be a common issue. As scale (vessel size) increases, the surface area to volume ratio decreases. Since the rate of heat flow is proportional to the heat transfer area, and the overall heat capacity of the system is proportional to the mass of the batch (and thus the volume), heat transfer will be significantly slower as a process is scaled up.

For the reaction, extraction, and crystallization unit operations, mixing is a common scale-up factor. Generally speaking, the mixing power is much greater in the plant than in the lab. It can be challenging to simulate the mixing behavior that will be obtained on scale in the lab, since there are many variables one can choose to hold constant between the experiment and the plant run. These variables include power, power per volume, tip speed, rotational speed, flow per volume, torque per volume, Reynolds number, blend time, and geometric similarity (ratio of impeller diameter to vessel diameter). Different phenomena scale with different variables, and it is not always well understood which variable is the best choice for scale-up and scale-down. Some case studies and rules of thumb are available in the literature [31, 32].

A final consideration that applies to several unit operations is the issue of dip tube depth. For any operation that involves sampling, the engineer must consider whether or not the dip tube is below the liquid level to allow a sample to be taken. In this case, the minimum volume for a unit operation may need to be increased.

The following sections discuss the common individual unit operations. Detailed treatment of the chemical engineering theory of heat transfer, mass transfer, thermodynamics, chemical kinetics, etc. and its application to batch reactors is available elsewhere and is outside the scope of this chapter. Instead, a brief discussion of the factors an engineer needs to consider is presented.

44.3.2 Reaction

The objective of the reaction unit operation is to convert a starting material or materials into the desired product, with maximum yield and minimum degree of by-product (impurity) formation. In the laboratory, reagent selection, solvent selection, stoichiometry, sequence of addition, and temperature are generally established. This list of process variables is unlikely to change upon scale-up, since, as Caygill et al. [33] state, "chemical rate constants are scale independent, whereas physical parameters are not." The many physical parameters that play a role in the outcome of the reaction that are scale dependent (see Table 44.17) are the main cause of scale-up problems.

A key consideration is whether the reaction is homogeneous (single phase) or heterogeneous (multiphase). Generally speaking, a standard batch reactor is configured such that reagents in a homogeneous reaction can be sufficiently well mixed to avoid the need for detailed consideration of mixing and mass transfer. There are, of course, exceptions, such as highly exothermic reactions, where temporary hot spots can cause a high level of impurity formation before reagents are well mixed. In contrast to homogeneous reactions, heterogeneous reactions (reactions with separate liquid–liquid, liquid–solid, and liquid–gas phases that participate in the reaction) are likely to be highly dependent on mass transfer considerations. Table 44.17 enumerates several scale-up factors for reactions.

44.3.3 Separation

44.3.3.1 Extraction
The objective of an extraction unit operation is to remove undesired components (organic impurities, inorganic salts) from the product solution and in some cases to quench the reaction. This is achieved by adding a liquid that is immiscible with the reaction mixture. Typically the reaction mixture is organic, and the added liquid is water or an aqueous salt solution, but the reverse situation is possible. In the laboratory, the liquid is added to the vessel and stirred, or the liquids are combined in a separatory funnel and shaken together. The agitation is stopped and the phases are allowed to settle, followed by separation. The relative densities of the phases are a key parameter in determining how quickly the phases will settle. Table 44.18 enumerates several scale-up factors for extraction.

44.3.3.1.1 Emulsions Several differences between the lab and the plant scale can contribute to emulsion formation. One is the mixing power per volume. Most often the plant-scale agitation is high power per volume, and thus there may be a greater tendency to form emulsions. Another factor is the likelihood to precipitate either product or salts during the extraction. In the lab, the midpoints of acceptable temperature and solvent composition (distillation endpoints, charge ranges) are often studied, whereas in the plant the parameters may be near the upper or lower part of the range. If one of the phases is near the solubility limit for a component, tiny particles that have the potential to stabilize an emulsion may form. Also, at scale the reagents may introduce tiny particulates or impurities that affect solubility. A final consideration is the position of the agitator blade relative to the phase boundary. This can influence which phase is dispersed in which, potentially affecting the stability of the dispersed phase. These factors may be proactively investigated in the lab to determine if an emulsion is likely.

If an emulsion is formed, methods to break the emulsion should be studied. If the emulsion is seen for the first time in the plant, such a study may be undertaken with a batch sample. Typical means of breaking an emulsion include addition of either solvent or water to change the composition, heating, filtration (to remove stabilizing entities such as tiny particles), pH adjustment, salt addition, and, in rare cases, addition of a demulsifier. Some case studies and rules of thumb are available in the literature [36].

44.3.3.2 Distillation
Generally, the objective of a distillation operation is to change the solvent composition of the system to facilitate downstream processing. This is generally

TABLE 44.17 Scale-up Factors for Reactions

Factor	In Lab	At Scale	Impact	Means to Evaluate
Time to charge reagents	One minute or less	Between 5 and 60 min	Different stoichiometry profiles with time may impact reaction kinetics	Simulate longer additions at lab scale
Charge method	Pouring, pump, addition funnel	Pump from drum, pressure from vessel, vacuum from drum	Choice of charge method may impact rate. Vacuum charges may cause volatilization of components	Simulate charge method
Charge port	Generally above surface	Above surface, subsurface, spray ball	Backmixing may occur during subsurface charges. Use of above surface ports may leave material on the vessel walls. Use of spray ball can help rinse solids from sides of the vessel	Backmixing calculation from engineering correlations [34]
Sequence of addition	Based purely on chemistry/convenience	Limited number of lines and ports may necessitate different orders of addition (e.g. to avoid incompatibles in the same line). Also, order of solids versus liquids may differ in the plant based on considerations such as inert handling	Can affect the kinetics of main and side reactions	Test different orders of addition in lab experiments
Mixing time	Can vary over wide range	Varies, max agitation likely affords longer mixing time than lab maximum	If reaction time is fast compared with mixing time, undesired reactions may occur	Experiments to determine reaction kinetics + blend time calculation. Can evaluate Damköhler number. If large, mixing is an issue
Solid suspension	Typically not an issue	May be an issue	Insufficient suspension equals lower effective surface area of solids	Njs (agitator speed to just suspend) calculation [35]
Mass transfer	k_{La} = 0.02 to 2 s^{-1}	k_{La} = 0.02 to 0.2 s^{-1} (batch reactor) k_{La} = 1 to 3 s^{-1} (Buss loop)	Either mass transfer or chemical kinetics may be rate limiting at different k_{La}. This will impact reaction profile	Gas uptake experiments in lab and at scale to determine k_{La}. k_{La} predictions by engineering correlations
Heat transfer	Excellent, high area/volume	Lower area/volume as scale increases	Safety (runaway reaction), excursion from acceptable temperature range	UA evaluation at scale (mock batch/solvent trial heating trend data may be used) and in the lab

performed in a semi-batch mode either by continually adding the new solvent at a constant volume or by sequentially adding the new solvent and then distilling down to the original volume, sometimes repeatedly (put/take). More rarely, reactive distillation may be used in cases where a volatile component must be removed to drive the reaction to completion. Distillation is also occasionally used to change the solvent composition to drive crystallization of the product (distillative crystallization).

A good first step in understanding a distillation operation is to obtain thermodynamic vapor–liquid equilibrium (VLE) data for the solvent system in question. The effect of pressure, the presence or absence of azeotropes, and the difference in vapor compositions across the liquid composition space are all easily visualized (for two solvent systems) with an x-y or T-x-y diagram (or several diagrams for different pressures). Several software packages (e.g. DynoChem™, Aspen™) are available to perform VLE calculations and distillation

TABLE 44.18 Scale-up Factors for Extractions

Factor	In Lab	At Scale	Impact	Means to Evaluate
Tendency to form emulsions	May be seen	Additional factors cause emulsions to be seen even if not seen in lab (e.g. increased power)	Stable emulsions must be broken before processing can continue	Test extremes of composition. Maximize mixing to stress. See text for means to break emulsions
Settling time (settling velocity)	Variable	Variable	Settling *velocity* should be similar between lab and plant. Settling time therefore will be much longer in a plant vessel	Measure settling time and height of phase boundary in the lab. Estimate time on plant scale using constant velocity
Mixing	Shaking in sep funnel, stir bar, and overhead stirring. Generally easy to mix the phases but low power	Various agitator, vessel, and baffle configurations. Much greater power. For fixed equipment operated within normal volume ranges, mixing is typically good	Adequate mixing needed to properly mix the phases and equilibrate composition. Excessive agitation may promote stable emulsion formation	Determine at-scale blend time to ensure phases will be adequately mixed. Stress the process by mixing as vigorously as possible in the lab
Ability to catch the split	Generally easy	May be difficult to see through small sight glass. Use of mass meters to detect density differences to determine split	Missing the phase boundary can cause repeated operations (waste of time) or loss of yield	Note when the phase boundary is more difficult to see (e.g. phases have similar phase color and opacity) and inform the plant staff of the need for caution. Before plant run, calculate expected phase volumes, so the rate of discharge can be adjusted depending on how close the phase boundary is
Rag layer	Minimal or not seen	Rag layer of significant volume may be present	Must determine disposition of rag layer in advance – inclusion can affect purity, and exclusion can affect yield	Difficult to assess in lab. Establish rag layer disposition in advance of processing based on quality/yield requirements. Typically rag layer is kept with the product phase, except for the final phase split

simulations. Typically, calculations and simulations based on pure solvents (ignoring the presence of the product or starting material) provide sufficiently accurate estimates. Table 44.19 enumerates several scale-up factors for distillations.

44.3.3.3 Color/Metal Removal The objective of a color or metal removal unit operation is to purify the process stream with respect to color bodies or metals. Typically this is accomplished through the use of an adsorbent material. Common examples include activated carbon, functionalized silica, or functionalized polymeric materials. Use of this unit operation at scale is not desirable since color and metal removal requires special materials and often special equipment. If other means of meeting product specifications are available, they should be considered.

There are two typical ways that an adsorption step is scaled up: (i) slurry of loose adsorbent followed by filtration and (ii) filtering the process stream through a cartridge or a filtration equipment (sparkler, Nutsche) containing the adsorbent. If the cartridge option is available, it is preferred,

TABLE 44.19 Scale-up Factors for Distillation

Factor	In Lab	At Scale	Impact	Means to Evaluate
Time	Can vary from very short to very long, depending on ΔT	Typically longer than in the lab; hours	Instability of the stream at elevated temperature can cause a quality issue	Establish stability by refluxing the process stream at the highest anticipated pressure and at both extremes of composition
Vacuum control	Generally excellent	Depends on equipment. May be poor. May be difficult to achieve pressures lower than 50 mmHg	Fluctuations in vacuum = fluctuations in boiling point. Some process streams have a tendency to foam or to "bump."	Carefully test the effect of decreased pressure on distillation
Jacket temperature	Choice of ΔT between jacket and boiling point drives distillation. Lower jacket temperatures might be adequate in the lab to achieve a reasonable distillation rate	Due to lower heat transfer area per volume, greater jacket temperatures may be desirable	Higher jacket temperatures may result in decomposition of any solids that are deposited onto the vessel walls during distillation	Assess the stability of the product at high temperature to see if high jacket temperature is an issue
Minimum volume	Often very low in terms of liters per kilogram of input (ability to agitate small volumes) or to use a rotary evaporator	Usually larger in terms of liters per kilogram of input. Some conical bottom vessels can have low minimum agitable volumes. There may be safety issues associated with highly concentrated process streams on scale	More solvent is needed to achieve the endpoint as the minimum volume increases	Use the plant's minimum volume (in L/kg) in the lab study. Perform safety evaluation of concentrated process streams
Endpoint – volume	Easy to mark a volume endpoint on glassware	May be difficult to see volume landmarks in the vessel; radar-level sensor may be present, but may not be very accurate (±5% typical, sometimes ±10%). Foaming can result in inaccurate radar measurement	If no in-process control is established (e.g. quantitation of the product concentration), the concentration going into the next unit operation may be off target	Assess the sensitivity of downstream unit operations to variations in the concentration representing both under- and over-distillation

(continued)

TABLE 44.19 (Continued)

Factor	In Lab	At Scale	Impact	Means to Evaluate
Endpoint – composition	In early development, the process stream may be rotovaped to an oil followed by dissolution in the next solvent. In later development, a put/take or constant volume distillation is done, and the process stream is analyzed for solvent composition by GC	Can choose between sampling the product stream for solvent composition, analyzing the distillate composition, PAT monitoring of the distillate, use product temperature as a guide, or fixed distillation protocol with no in-process control	If no in-process control is established (e.g. GC on the process stream), the composition going into the next unit operation may be off target	Assess the sensitivity of downstream unit operations to variations in the composition representing both under- and over-distillation. Develop reliable control scheme as needed
Constant volume distillation	Usually requires full-time supervision	Usually requires full-time supervision	Constant volume distillation is usually much more solvent efficient than put/take distillation; however the need for supervision may make it a less desirable choice	Evaluate both constant volume and put/take distillation modes
Sampling	Can draw a sample through a septum	Sampling often requires the distillation be temporarily stopped to draw the sample due to limitations on the sample temperature or the possible need to pressurize the vessel to draw a sample	Sampling can increase cycle time	To minimize the number of samples, establish vessel landmarks, endpoint temperature (as a function of pressure), or, if the distillation is a critical operation, PAT monitoring

since the loose materials are often challenging to filter from the process stream and are difficult to clean from process equipment.

In any investigation of absorbents, there are two key criteria for absorbent selection: (i) degree of removal of the color or metal (as a function of % loading of the adsorbent) and (ii) loss of product to the adsorbent. Secondary considerations include cost of the adsorbent and availability (lead time) of the adsorbent. As a general rule, activated carbons are cheaper and more readily available compared with functionalized materials.

A typical protocol for studying the adsorption unit operation is described in Example Problem 44.2

EXAMPLE PROBLEM 44.2

The final intermediate in the synthesis of an API is received from a vendor and found to have a dark brown color. The intermediate (designated compound A) is used in a lab run to produce API, which is found to also have a brown color. The specification for the API is off-white, so color will need to be removed. This could be done either via rework of the intermediate or as a processing step in the API step.

At the beginning of the API step, compound A is dissolved in 20 L of methanol per kg of compound A. The project team decides to pursue color removal by carbon filtration after this dissolution step.

The first step is to screen various potential adsorbents. The team has five common carbons available for scale-up. One hundred milligram of each carbon is placed in a vial along with 4 ml of compound A solution in methanol. Since the solvent quantity in the solution is 20 L/kg (or 20 ml/g), the 4 ml solution contains 200 mg of compound A. Thus the loading of carbon in the screening experiment is 50% (100 mg of carbon to 200 mg of compound A). The samples are placed in a shaker block for 60 minutes and then filtered. The color is inspected visually, and the concentration of the filtrate is analyzed by HPLC for wt %. The recovery of compound A is calculated based upon the HPLC quantitation, and results are shown in Table 44.20.

Carbon 3 and carbon 5 are the only adsorbents that afford color removal. Carbon 5 results in the best color; however too much of the desired compound is lost to the carbon. The team decides to use carbon 3 for scale-up.

TABLE 44.20 Example Problem 44.2: Screening Results

Carbon Type	Color (By Visual)	Recovery of Compound A
Carbon 1	Brown	97%
Carbon 2	Brown	95%
Carbon 3	Very light yellow	93%
Carbon 4	Brown	98%
Carbon 5	Clear	82%

Since the pilot plant will use carbon cartridges, a breakthrough study is performed in the lab to simulate the plant operation and to determine what carbon area is needed per liter of process stream to be decolorized. A 47 mm carbon disk is set up in a filter housing. This disk is known from vendor literature to have an effective carbon surface area of 0.0135 ft^2. A fluid reservoir is connected to a pump, which is subsequently connected to the carbon disk. Downstream from the carbon disk is a filter and a UV/Vis detector. First, methanol is flushed through the system for 20 minutes at 5 ml/min. Then, the feed is switched to a reservoir of 300 ml of compound A solution. UV/Vis monitoring is started and continues until all the solution is passed through the pad. The solution that has passed through the pad is collected in 5 ml fractions. A plot of the absorbance at 310 nm versus time is presented in Figure 44.2. The color begins to break through at about 30.5 minutes, and after 35 minutes the color breakthrough is increasing rapidly. Judging the breakthrough point to be 35 minutes, the team pools all the fractions from 20 to 35 minutes and proceeds with the API chemistry. The resulting material is found to be white, so 35 minutes is verified to be an acceptable breakthrough point.

Given the 20 minutes of flush and the 5 ml/min flow rate, the breakthrough point in milliliter is calculated to be 75 ml (35 minutes – 20 minutes = 15 minutes × 5 ml/min = 75 ml). Given the 0.0135 ft^2 carbon area of the 47 mm pad, the carbon "life" is 5.56 L/ft^2. The flux, or flow per area, was 0.37 L/min/ft^2.

For the scale-up to 5 kg API batch, the batch volume will be 100 L at the point of dissolution. Given the life of 5.56 L/ft^2, the needed carbon area to remove color in this batch is 18 ft^2 (100 L/5.56 L/ft^2 = 18 ft^2). Based on the flux of the experiment (0.37 L/min/ft^2), a total minimum time of 15 minutes is required for the operation (100 L/0.37 L/min/ft^2/18 ft^2). This could also be expressed as a flow rate of 6.7 L/min.

44.3.4 Crystallization

The objective of the crystallization operation is to isolate the product as a solid, purify by leaving impurities in the liquid phase, and create particles of the correct form and desired physical properties (i.e. size distribution, density, surface area). As a brief review of fundamentals, crystallization consists of several physical phenomena, the most important of which are nucleation and growth (others include attrition and aggregation). Nucleation refers to the formation of very tiny crystals from the solution, and growth refers to the increase in size of the nuclei by transfer of product from the solution to the crystal faces. The balance of the rate of nucleation and growth is a key determinant of the particle size distribution. If the nucleation rate is dominant throughout the crystallization, small particles with a nonuniform distribution

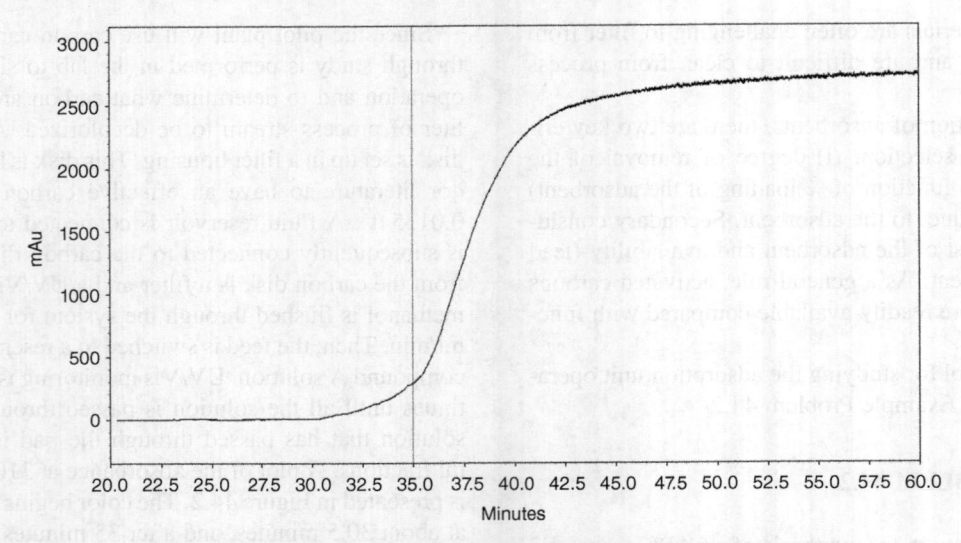

FIGURE 44.2 Carbon breakthrough curve for Example Problem 44.2.

will form. If growth is dominant throughout the crystallization, large particles with a more monodisperse distribution will form.

Supersaturation (i.e. the state where the product concentration is above the equilibrium solubility) is required for nucleation and growth. Supersaturation is often induced by the addition of antisolvent or by lowering the batch temperature. Generally, very high supersaturation favors nucleation over growth, and low supersaturation favors growth over nucleation. In a system with no nuclei present (added seeds or foreign matter that can act as nuclei), spontaneous nucleation does not happen immediately at the onset of supersaturation. The region in the parameter space (concentration and solvent composition or concentration and temperature) in which the solubility is exceeded but spontaneous nucleation does not occur is referred to as the metastable zone. The width of the metastable zone depends not only on the inherent characteristics of a given system but also on physical parameters such as agitation, rate of cooling or rate of antisolvent addition, and the presence of other nuclei (foreign matter or seeds). For this reason, metastable zone width depends on scale. Table 44.21 enumerates several scale-up factors for crystallizations.

44.3.5 Isolation

The objective of the isolation operation is to separate the solids (product or waste) from the mother liquors as rapidly as possible and efficiently wash non-desired components (organic impurities, inorganic salts, solvents, or product) from the isolated material. The primary aspect is the filtration, and this is covered in greater detail in Chapter 35. Table 44.22 enumerates several scale-up factors for isolations.

44.3.6 Drying

The objective of the drying operation is to remove solvents to achieve a final product solvent specification and to maintain or create desired powder properties. A typical drying target is set to remove solvent below a maximum allowable concentration. When drying a solvate crystalline form, minimum and maximum solvent content criteria will be set. Chapter 35 provides more detailed discussion of this unit operation. Table 44.23 enumerates several scale-up factors for drying.

44.3.7 Particle Size Reduction (Milling)

An API typically has a specification related to the powder properties. Particle size control may also be critical for process intermediate seeds to ensure sufficient impurity rejection or to improve filterability. The most common specification is related to final particle size distribution and often given as a single number that characterizes the particle size distribution. Example specifications include the mean (volume or mass based) or a D "number" (i.e. D50, D90, D97), which refers to a value on the distribution such that the "number" % (by mass) of the particles have a diameter of this value or less. Different moments of the particle size distribution as well as surface area and bulk density may also be chosen as a specification.

Development scientists can attempt to address the powder property requirement by several means, including crystallization engineering, wet milling, and dry milling. Each of these technologies is addressed in more detail elsewhere, and crystallization scale-up factors are discussed above. Making amorphous API will also bring in additional technologies such as spray drying to consider. Issues related to scale-up of milling processes are dependent on (i) the equipment for

TABLE 44.21 Scale-up Factors for Crystallization

Factor	In Lab	At Scale	Impact	Means to Evaluate
Metastable zone width (see text for explanation)	Generally broader	Generally narrower	Narrower metastable zone width may result in spontaneous nucleation prior to seeding and thus uncontrolled crystallization (lack of polymorph control, lack of control of particle size distribution)	Test metastable zone width in the lab as a function of agitation, cooling rate (if applicable), and antisolvent addition rate (if applicable). Generally, to grow large particles of a desired polymorph, it is best to seed the batch under conditions of very low supersaturation [37]
Seeded/unseeded	Either is possible – easy	Either is possible – for some equipment, solids charge is difficult, and the seeds are transferred as a slurry. Seed charges can be small in quantity, and it may be challenging to ensure the seed charge reaches the batch	Polymorph control and particle size distribution control	Evaluate effect of amount of seeds (loading), seeding point, and dry vs. slurry transfer
Mixing	Stir bar or overhead stirring. Generally easy to achieve good mixing	Various agitator, vessel, and baffle configurations. For fixed equipment operated within normal volume ranges, mixing is typically good	Inadequate mixing can result in hot spots of high supersaturation and thus uncontrolled nucleation. High agitation may reduce metastable zone width	Evaluate the impact of mixing in scale-down experiment One means of eliminating scale dependence is by performing the mixing outside the vessel, e.g. with a T-mixer or a jet mixer
Attrition	Variable	Variable	Some particles are easily broken. This process may be sensitive to agitation parameters (including time). This may affect particle size distribution, filterability, and ultimate powder properties	Explore the particle size distribution as a function of agitation using overhead stirring, or stress the process by using a high-shear mixer
Measurement of temperature	By thermocouple, not typically a problem	Temperature probes can sometimes be coated with crystals and provide false information	If jacket temperature control is used, this may not be a processing issue, but accurate temperature data will not be collected. If operating in batch temperature control mode, the control system may adjust the jacket temperature and thus move the batch temperature outside the normal operating range	Observe the tendency for coating during the crystallization. Specify jacket temperature control rather than batch temperature control

TABLE 44.22 Scale-up Factors for Isolation

Factor	In Lab	At Scale	Impact	Means to Evaluate
Filtration flux	Up to 10× greater, depending on lab vs. plant cake thickness and cake compression	Often up to 10× slower than lab or longer	Longer cycle time	Measure filtration flux as a function of cake height or mass of cake. Use engineering correlations to predict at-scale performance (see Chapter 35)
Filter media	Typically done with filter paper, 6–25 μm	Limited choices of pore sizes and material of construction	Potential for filter media to blind or pass through of product	Evaluate plant-scale filter media in lab-scale experiments
Compressibility	Low ΔP compared with the plant, so effect of compressibility is less of a factor	Higher ΔP, so effect of compressibility is more of a factor	Can greatly slow down the filtration at high ΔP or high centrifugation spin speeds	Evaluate compressibility with leaf filter studies (pressure filtration measurements of rate versus ΔP)
Cake wash	Able to smooth cracks in the wet cake	Sometimes not able to smooth cracks in the wet cake (this can be done in a filter dryer)	Channeling of cake wash through cracks results in poor washing of the cake, affecting impurity profile and solvent content	Evaluate propensity to crack by allowing cake to deliquor completely between filtration and each wash
Extent of deliquoring	Typically easy to achieve low solvent content	Solvent content after isolation may be much greater	Greater solvent content impacts cake wash efficiency and drying operations. Stability of the product may be an issue	Study stability of wet cake under very wet conditions (e.g. 50% wash solvent)
Discharge	Easy – by scooping wet cake from Buchner funnel into a drying dish	May be challenging depending on equipment. Safety considerations such as electrostatic buildup from nonconductive washes may dictate the need to delay (for relaxation of charge)	Wet cake properties needed to select appropriate parameters on equipment (i.e. peeler centrifuges, LOD, wet cake density) Longer discharge requires additional product stability under wet conditions	Study stability of wet cake under very wet conditions (e.g. 50% wash solvent)

TABLE 44.23 Scale-up Factors for Drying

Factor	In Lab	At Scale	Impact	Means to Evaluate
Agitated drying	Not always evaluated	Agitated filter dryers, rotary tumble, and conical dryers are most common drying methods. The LOD in which agitation begins is important parameter for determining powder properties	Agitation can promote lump/ball/boulder formation in cohesive powders. Agitation can influence all of the key final powder properties through breakage or attrition of particles – bulk density, particle size distribution, flowability, electrostatics	Laboratory-scale agitated dryer units are available. Scale-down of agitation drying experiments is not straightforward, so laboratory data may only provide trends or insights into tendencies of the system, not quantitative prediction of scale behavior. Recent work utilizing torque and power measurements in scale-down system provides some quantitative prediction [38]
Bulk density	Easy to adapt to low bulk density	Bulk density dictates needed dryer size	Can greatly affect the choice of equipment or number of dryer loads. May affect formulation performance	See above
Sampling	Scoop/spatula	Sampling configurations differ between different dryers. The operation may be difficult, and samples may not be fully representative. Multiple samples generally taken	Too frequent sampling adversely affects cycle time. Nonrepresentative samples can result in false passing results from in-process controls	Establish tolerance for solvents/water in downstream processing (or API release). PAT methods for monitoring drying may sometimes be implemented if drying is a critical operation. PAT may be especially useful if attempting to maintain a solvate (to prevent over-drying)
Discharge	Scoop/spatula	Depends on dryer – discharge may occur through a small port and may not be trivial. Often, a significant heel is left behind after discharge	Poorly flowing powders can be very difficult to discharge and may require excessive time/operator intervention	Measure the flow characteristics of the powder after agitated drying

the specific milling technology and (ii) the physical properties of the compound (bulk density, flowability, morphology, tendency for compaction, fragility). The parameters to consider for scale-up will vary with milling technology since the mechanism for attrition is different. For dry milling scale-down, laboratory-sized units are available for experiments in the 10–100 g scale. While scale factors and empirical rules are used to determine the initial parameters for scaling up milling operations, a small test batch is often run to verify the physical properties (PSD) prior to milling the entire batch. PAT monitoring of the particle size distribution through online particle size analysis (Insitec, FBRM) is a prudent means of ensuring the correct particle size is achieved. Scaling-down of wet milling is often tied to crystallization as it could impact the fundamental processes (nucleation, growth, attrition, and agglomeration) occurring in the crystallization. Some of the key advantage and parameters/issues to consider for the various milling technologies are described in Table 44.24. More detailed information about milling is provided in Chapter 38.

TABLE 44.24 Summary of Milling Technologies

Milling Technology	Key Advantages[a]	Key Parameters and Issues for Evaluation and Scale-up
Air attrition jet milling	• Capable of attrition down to D97 of 2–10 µm • No heat generation – ideal for heat-sensitive compounds • Easy maintenance – no moving parts • Inert milling	• Pressures (pusher and grinding) • Mass solids/gas flow rate ratio • Tendency of material to compact and stick to raceway surface • Type of jet mill (loop raceway or spiral) • Number of nozzles in spiral configuration
Fluidized bed air attrition mills w/classifiers	• Capable of attrition down to D97 of 2–10 µm • No heat generation – ideal for heat-sensitive compounds • Steeper particle size distributions are achievable • Inert milling	• Pressure • Classifier speed • Nitrogen flow to achieve fluidization • Product feed rate/product removal rate
Impact milling (hammer, pin)	• Capable of attrition down to D97 of 30–50 µm • Large industrial scale units for very high throughput	• Pin or hammer speed • Product feed rate • Sensitivity of compound to temperature
High shear rotor–stator wet milling [39, 40]	• Capable of attrition down to 10–30 µm as mean • Technique can be set up as a recycle of the crystallized slurry • No exposure to dry powders • More suited for "needle" morphologies to reach lower end of attrition	• Rotor–stator configuration (no. of teeth, tooth geometry, no. of generators) • Tip speed • Shear frequency • Slurry concentration • Batch turnovers • Flow rate through the mill • Point of wet milling initiation during the crystallization time cycle • Product filterability • Normalized energy ($E*$) (composite of multiple inputs in list) [41]
Media and ball milling	• Capable of attrition down to D97 <1 µm • Technique can be set up as a recycle of the crystallized slurry • No exposure to dry powders	• Media size • Media material compatibility • Duration of the milling run • Product filterability

[a] www.hmicronpowder.com.

44.4 SUMMARY

Understanding process scale-up and assessment is a core activity for process chemical engineers in the pharmaceutical industry. It enables transformation of a chemical synthesis to a scalable pilot plant process and then to a robust manufacturing process. The numerous factors to consider in the scale-up and assessment encompass addressing the specific risks to safety, quality, and manufacturing productivity as well as the more general strategic risks in managing a portfolio of projects that span different stages of development. In this chapter, we have discussed the many drivers for development, including the requisite process and personnel safety, product quality, and business optimization. Understanding these drivers is the key to both efficiently prioritizing development activities for a given project's stage of development and ensuring resources are appropriately prioritized across the portfolio. We have also discussed the unit operations that constitute a typical process. Understanding of the process fit and scale-up factors for these unit operations is critical to defining and executing a process development strategy. By applying these concepts, along with more detailed insights from the other chapters in this book, the process engineer will be well prepared to meet the challenges of API process development.

REFERENCES

1. Anderson, N. (2000). *Practical Process Research & Development*. San Diego, CA: Academic Press.
2. Rubin, E., Tummala, S., Both, D. et al. (2006). Emerging technologies supporting chemical process R&D and their increasing impact on productivity in the pharmaceutical industry. *Chemical Reviews* 106 (7): 2794–2810.
3. DiMasi, J.A., Hansen, R.W., Grabowski, H.G., and Lasagna, L. (1995). Research and development costs for new drugs by therapeutic category. *Pharmaco Economics* 7 (2): 152–169.
4. Pisano, G.P. (1997). *The Development Factory*, 99. Boston, MA: Harvard Business School Press.
5. Gilbert, J., Henske, P., and Singh, A. (2003). Rebuilding big pharma's business model. *In Vivo Business and Medicine Report* 21 (10) (November).
6. Adams, C.P. and Brantner, V.V. (2003) *New Drug Development: Estimating Entry From Human Clinical Trials* (7 July 2003). Federal Trade Commission. http://www.ftc.gov/be/workpapers/wp262.pdf (accessed 9 February 2009).
7. US Department of HHS, FDA, CDER and CBER (2006). *Guidance for Industry: Q9 Quality Risk Management* (June 2006). Rockville, MD: FDA.
8. International Conference on Harmonization of Technical Requirements for Registration of Pharmaceuticals for Human Use. Q3A – Impurities in New Drug Substances, Q3C – Impurities – Residual Solvents, Q8 – Pharmaceutical Development.
9. Noren, A., Naik, N., Wieczorek, G., and Basu, P. (1999). Keep pharmaceutical pilot plants safe. *Chemical Engineering Progress* 95 (3): 39–43.
10. Parker, J.S. and Moseley, J.D. (2008). Kepner-Tregoe decision analysis as a tool to aid route selection. Part 1. *Organic Process Research & Development* 12: 1041–1043.
11. Moseley, J., Brown, D., Firkin, C.R. et al. (2008). Kepner-Tregoe decision analysis as a tool to aid route selection. Part 2. Application to AZD7545, a PDK inhibitor. *Organic Process Research & Development* 12: 1044–1059.
12. Naumann, B.D., Sargent, E.V., Starkman, B.S. et al. (1996). Performance-based exposure control limits for pharmaceutical active ingredients. *American Industrial Hygiene Association Journal* 57: 33–42.
13. Stoessel, F. (1993). What is your thermal risk? *Chemical Engineering Progress* 89 (10): 68–75.
14. Ebadant, V. and Pilkington, G. (1995). Assessing dust explosion hazards in powder handling operations. *Chemical Processing* (September): 74–80.
15. Gritton, L.G. (1999). *Avoiding Static Ignition Hazards in Chemical Operations*. New York: CCPS-AICHE.
16. V. Pilz, H. Bender, M. Müller et al. (2000). Plant and process safety. doi: https://doi.org/10.1002/14356007.b08_311.
17. Wehrstedt, K.D., Wandrey, P.A., and Heitkamp, D. (2005). Explosive properties of 1-hydroxybenzotriazoles. *Journal of Hazardous Materials* 126 (1–3): 1–7.
18. SIGMA-ALDRICH Sigma-Aldrich MSDS Database. https://www.sigmaaldrich.com/safety-center.html.
19. Rakita, P.E., Aultman, J.F., and Stapleton, L. (1990). Handling commercial Grignard reagents. *Chemical Engineering* 97 (3): 110–113.
20. Silverman, G.S. and Rakita, P.E. (1996). *Handbook of Grignard Reagents*. New York: Marcel Dekker.
21. EMEA (2000). *International Conference on Harmonization Topic Q6A, Specifications: Test Procedures and Acceptance Criteria for New Drug Substances and New Drug Products: Chemical Substances* (CPMP/ICH/367/96, May 2000). London: EMEA.
22. EMEA, Committee for Human Medicinal Products (2007). Guideline on the specifications limits for residues of metal catalyst [Draft] (January 2007). London: EMEA.
23. U.S. Department of Health and Human Services, FDA, CDER, CBER (2018). M7 assessment and control of DNA reactive (mutagenic) impurities in pharmaceuticals to limit potential carcinogenic risk, guidance for industry. https://www.fda.gov/downloads/Drugs/GuidanceComplianceRegulatoryInformation/Guidances/UCM073385.pdf (May 2015).
24. Pierson, D.A., Olsen, B.A., Robbins, D.K. et al. (2009). Approaches to assessment, testing decisions, and analytical determination of genotoxic impurities in drug substances. *Organic Process Research & Development* 12: 285–291.
25. Glodek, M., Liebowitz, S., McCarthy, R. et al. (2006). Process robustness – A PQRI white paper. *Pharmaceutical Engineering* 25: 1–11.

26. Seibert, K., Sethuraman, S., Mitchell, J. et al. (2008). The use of routine process capability for the determination of process parameter criticality in small-molecule API synthesis. *Journal of Pharmaceutical Innovation* 2: 105–112.
27. Jimenez-Gonzalez, C., Ponder, C., Broxterman, Q.B., and Manley, J. (2011). Using the right green yardstick: why process mass intensity is used in the pharmaceutical industry to drive more sustainable processes. *Organic Process Research & Development* 15: 912–917.
28. Thayer, A. (2009). Sustainable synthesis. *C&E News* 87 (23): 12–22.
29. Van Aken, K., Strekowski, L., and Patiny, L. (2006). EcoScale, a semi-quantitative tool to select an organic preparation based on economical and ecological parameters. *Beilstein Journal of Organic Chemistry* 2 (3): https://doi.org/10.1186/1860-5397-2-3.
30. Ritter, S.K. (2008). Chemists and chemical engineers will be providing the thousands of technologies needed to achieve a more sustainable world. *C&E News* 86 (33): 59–68.
31. Paul, E.L., Atiemo-Obeng, V., and Kresta, S. (2004). *Handbook of Industrial Mixing*. Hoboken, NJ: Wiley.
32. Oldshue, J.Y. and Herbst, N.R. (1990). *A Guide to Fluid Mixing*. Rochester, NY: Mixing Equipment Co.
33. Caygill, G., Zanfir, M., and Gavriilidis, A. (2006). Scalable reactor design for pharmaceuticals and fine chemicals production. 1. Potential scale-up obstacles. *Organic Process Research & Development* 10: 539–552.
34. Fasano, J., Penney, R., and Xu, B. (1992). Feedpipe backmixing in agitated vessels. *AIChE Symposium Series* 293: 1–7.
35. Zweitering, T.N. (1958). Suspension of solid particles in liquid by agitators. *Chemical Engineering Science* 8: 244–253.
36. Flynn, D. (ed.) (2018). Emulsion treatment. In: *The Nalco Water Handbook*, 4e. McGraw-Hill.
37. Singh, U.K., Pietz, M.A., and Kopach, M.E. (2009). Identifying scale sensitivity for API crystallizations from desupersaturation measurements. *Organic Process Research & Development* 13: 276–279.
38. Remy, B., Kightlinger, W., Saurer, E. et al. (2014). Scale-up of agitated drying: effect of shear stress and hydrostatic pressure on active pharmaceutical ingredient powder properties. *AIChE Journal* 61 (2): 408–418.
39. Kamahara, T., Takasuga, M., Tung, H.H. et al. (2007). Generation of fine pharmaceutical particles via controlled secondary nucleation under high shear environment during crystallization – process development and scale-up. *Organic Process Research & Development* 11: 699–703.
40. Harter, A., Schenck, L., Lee, I., and Cote, A. (2013). High-shear rotor–stator wet milling for drug substances: expanding capability with improved scalability. *Organic Process Research and Development* 17: 1335–1344.
41. Engstrom, J., Wang, C., Lai, C., and Sweeney, J. (2013). Introduction of new scaling approach for particle-size reduction in toothed rotor-stator wet mills. *International Journal of Pharmaceutics* 456: 261–268.

45

SCALE-UP DO'S AND DON'TS

FRANCIS X. MCCONVILLE

Impact Technology Development, Lincoln, MA, USA

45.1 INTRODUCTION

One of the chemical engineer's primary responsibilities in the pharmaceutical industry is to assist in scaling up laboratory or development-stage processes to commercialization. An unfortunate fact is that so much of the really practical information that can make scale-up more efficient and ultimately more successful is generally acquired in the workplace only through years of on-the-job training. Most university engineering curricula are simply not geared toward teaching students about the many real-world issues that can complicate scale-up or lead to unexpected results, unsuccessful campaigns, and potentially dangerous situations.

45.2 LEARNING THE HARD WAY

Many of the most important things I know about technology transfer and process scale-up were learned throughout my career, on the floor of the pilot plant long after I left school. And suffice it to say that some of these lessons were hard won, sometimes at the cost of out-of-spec batches, close calls, and unnecessary delays. Lucky is the new graduate in a position to be mentored by someone who has learned about scale-up through experience by putting time in "in the trenches."

It sometimes seems that scale-up is simply a lesson in learning to expect the unexpected. But over time, one realizes that many of the surprises encountered during scale-up could have been anticipated and often prevented by paying attention to the appropriate details, conducting some relatively simple laboratory studies, or collecting the right quantitative data during early process development. That is why I have always been a strong proponent of appropriate process engineering studies early on in new process design. There are countless simple laboratory measurements that allow the process chemist or engineer to characterize and quantify the behavior of the reactants and other materials used in the process, and this can go a long way toward streamlining scale-up.

45.3 TYPICAL SCALE-UP ISSUES

One of the most common effects of scale-up is a change in reaction selectivity, especially in semi-batch reactions. This results mainly from differences in mixing between the laboratory or kilo lab scale and the commercial scale. This is discussed in more detail in Chapter 12. Changes in selectivity can lead to lower yields and higher levels of impurities in the final product, or changes in the impurity profile in turn can alter the physical form or polymorph of crystalline products. The appearance of previously unseen polymorphs of pharmaceutical solids upon scale-up is all too common, and this change in impurity profile is one of the major underlying reasons.

Product isolation can also lead to unexpected results upon scale-up. Despite the industry's best efforts to design better and more efficient filters and centrifuges for the recovery of solid products, the fact remains that the removal of impurities in the cake washing step is often not as complete or as efficient at large scale as in the lab. This is in part simply due to the difficulty of ensuring even distribution of the cake and cake wash at large scale.

Chemical Engineering in the Pharmaceutical Industry: Active Pharmaceutical Ingredients, Second Edition.
Edited by David J. am Ende and Mary T. am Ende.
© 2019 John Wiley & Sons, Inc. Published 2019 by John Wiley & Sons, Inc.

Other unexpected consequences of scale-up result from the very long times required to complete many processing steps at large scale. Operations as fundamental as charging raw materials can take many hours, likewise for transfers, distillations, and product isolations. It is critical to conduct the necessary laboratory stability studies to ensure that the various process streams do not undergo degradation during these prolonged processing times.

Sometimes problems result from poor communication at the tech transfer stage. One of my more uncomfortable memories involves an API process being transferred to the 8000 L scale at a foreign CMO. It was only upon arrival at the manufacturing site that our team learned that the designated reactor train was not equipped for vacuum distillation – all stripping operations were to be conducted at atmospheric pressure. However, all previous development work for this process had utilized vacuum distillation. We somewhat foolishly proceeded with the campaign based on the results of a quick lab test and wound up cooking our final product stream until it was dark brown and failed spec, because of the very elevated temperature and the many hours that the distillation took at that scale.

45.4 PURPOSE OF THIS CHAPTER

Throughout my years as a process engineer and working in kilo labs and pilot plants, I came to develop a list of things that I felt we were doing right, things that made pilot operations more efficient and safer, and a list of things that should be avoided, or operations that could not be scaled up effectively. I sometimes worked with scientists with little scale-up experience and often used this list to help educate them about the types of procedures that might work well in a plant setting and those that would probably not. This list evolved into a set of scale-up "do's and don'ts" for professionals involved in technology transfer, operating pilot plants, or laboratory personnel developing processes for eventual scale-up that I first published in 2002 [1] and that I expand upon here. This is a very subjective list, representing my own personal take on scale-up. Many of the activities or approaches I recommend are considered *de rigor* for companies involved in the GMP manufacture of pharmaceutical materials to comply with ICH guidelines, but even for small companies and start-ups, these practices often just make good sense and can be beneficial for long-term safety and success.

45.5 THINGS TO DO DURING SCALE-UP

45.5.1 Make It a Team Effort

One thing that has helped me greatly in my career is establishing strong channels of communication with the many process development chemists with whom I have worked. I feel strongly about the important role that process engineering plays in process development. There have been those few inexperienced chemists who tended to practice "over the wall" chemistry – that is, they would develop a new process and then toss it "over the wall" to the engineering/pilot plant group and expect to never see it again.

We know this cannot happen in the real world. It is only through close cooperation between the various disciplines that a successful, scalable process can be obtained. The best process chemists understand this. While the chemist has a deeper understanding of the effects of the various process variables on product quality, the engineer may have a better appreciation for the physical limitations of pilot equipment, or, in short, what operations can or cannot be conducted safely in a pilot or commercial plant. Each has his or her own set of priorities, and they must be communicated clearly to each other, and the earlier in the development cycle, the better.

For example, the engineer can inform the chemist that his choice of solvents will not be acceptable in the plant or that certain laboratory operations, such as evaporating to dryness, cannot be scaled up. Likewise, the chemist can communicate the need for tight temperature control during a certain critical step or the need for a certain degree of agitation.

Engineers bring many valuable skills to the table to assess and improve the scalability of novel processes. Below is a list of just some of the many areas in which process engineers can make important contributions to the development effort:

- Identify and determine limits for critical process parameters.
- Identify process hazards, and conduct the necessary hazard analysis, including isothermal and adiabatic calorimetry, explosivity studies, etc.
- Compare projected commercial costs of alternate routes (COGS).
- Complete mixing and heat transfer calculations for scale-up, and assist with "scale-down" experimental design.
- Conduct reactor design calculations.
- Size and specify equipment.
- Complete material and energy balances.
- Investigate opportunities for continuous or other alternate processing.

However, as mentioned above, there are many simple experiments and measurements that fall under the category of process engineering that any laboratory scientist can carry out easily, such as solids drying studies, distillation stability studies, and simply measuring and recording operating pressures and temperatures and the densities and other physical properties of the various process streams. This type of information will be a great aid in speeding scale-up.

45.5.2 Develop an Operating Philosophy

No matter how large or small your facility, it is very important to lay down some ground rules for the safe transfer and scale-up of laboratory processes to the kilo lab or pilot plant. For example, at one company we made a key decision to not operate the new kilo lab under cGMP. This eliminated the need for strict compliance to regulations imposed by outside agencies, streamlined scale-up, and allowed more operating flexibility. This did not mean, however, that we threw caution to the wind and worked haphazardly.

We established clear, strict requirements for all processes to be transferred to the kilo lab. With management support, we were able to adhere to these requirements even in the face of pressure to meet aggressive timelines. For example, we required a written batch record for all processes and that the batch record be proven by using it to run a minimum of three laboratory-scale batches (see below). One of the three lab runs would typically serve as the raw material use test, conducted using the actual kilo-scale raw material lots.

We also called for strict cleaning and rinse-test protocols for all equipment to minimize the possibility of cross contamination between batches. Another important requirement was the completion of a HAZOP study, conducted by a team of at least three individuals representing the kilo lab staff, process chemistry, and when possible a company safety officer.

Although it may appear that these requirements would hinder efficiency, that was not the case. Once it was clear that the rules would be strictly enforced, everyone on the development team adjusted their thinking accordingly with the end result that over many years of operation, that kilo lab experienced no serious accidents and virtually no failed batches.

45.5.3 Establish Use-and-Maintenance Files

Again, good engineering practice, and certainly required in a cGMP environment, but also just good common sense. No matter how small your operation, developing a sound record-keeping system to capture how each piece of equipment in your kilo lab or plant has been used and maintained, starting from the moment of installation, can pay big dividends in extending the life of the equipment, minimizing life-cycle cost, and providing traceability for important clinical or preclinical materials.

Established companies will have detailed Installation Qualification and Operational Qualification (IQ-OQ) protocols and preventive maintenance (PM) systems. Smaller companies should at a minimum establish a logbook (a lab notebook works well) in which they can record manufacturer data and wiring and engineering diagrams, in addition to information about each batch of material processed, the results of any cleaning operations or rinse tests, any maintenance performed, any calibrations, or other measurements made such as heat transfer coefficients, volume calibrations, etc. Such records should be kept over the life of each reactor or mixing vessel, each filter or centrifuge, and even reusable transfer hoses and other portable components.

45.5.4 Establish a Sample Database

Another record-keeping practice that will prove invaluable is maintaining a sample database in your kilo lab or pilot plant. Keep a permanent record of each and every sample of any process stream that is collected for analysis or observation during your operations. That includes any in-process batch samples, distillates, wet cakes, final products, and waste streams. Every sample should be given a unique sample number and the batch, time, date, amount, and reason for its collection, and any other important observations should be recorded. Larger companies may have a fully integrated digital laboratory information management system (LIMS), but for kilo-scale operations, any kind of logbook can work. Provide columns to make sure that all the pertinent data is recorded.

In one kilo lab, we used a lab notebook for the purpose and simply assigned unique sequential numbers to all process and retain samples. We frequently needed to go back and reconfirm the identity of past samples for research or regulatory purposes, for conducting mass balances, and the like, and the information we needed was readily available.

45.5.5 Collect Retain Samples

Along with a sample database, it is important to institute a retain sample system. Samples of all dried isolated intermediates and final products should be kept in appropriate sealed containers stored in a cool, dry dark place, well organized and readily accessible should the need arise. The exact sample size, storage container, and conditions will of course depend on the specifics of the material being stored, but the important things are that the containers be clearly and permanently labeled and that the storage system be carefully thought out and enable easy retrieval of samples when needed.

45.5.6 Fix the Process Before Scaling

This point goes hand in hand with developing a consistent operating philosophy. Last-minute changes to a process being scaled up can lead to serious unexpected consequences and possibly unsafe situations. It is important to minimize last-minute changes by fixing the process well in advance of scale-up and ensure that laboratory demonstration runs of the actual final process have been conducted prior to scale-up.

The cases of processes that have failed due to on-the-fly changes in the plant are legion. One example that comes to mind is a failed selective diastereomer crystallization that was "bumped" into a 12 000 L stainless steel reactor because the usual glass-lined reactor was occupied by another process. The expected enantiomeric enrichment did not occur, and the batch failed miserably, most likely due to material surface effects or differences in the nature of mixing in the two reactors. This was a harmless enough, albeit tremendously expensive lesson, but last-minute changes can and often have led to the creation of very serious hazards.

45.5.7 Conduct a HAZOP Review

There are two terms commonly heard in the chemical process industries, "HAZAN" (for hazard analysis) and "HAZOP" (for hazard and operability study). The former is a detailed examination of a particular hazard, such as a potential decomposition reaction, in which one might conduct calorimetry to determine onset temperature, time to maximum rate, maximum adiabatic temperature rise, etc. The HAZOP on the other hand is a more general study, conducted by a team, in an attempt to identify all the potential hazards of a process prior to scale-up.

Most companies insist on some level of HAZOP study before scaling up new processes, even to the kilo scale, and of course a detailed study should be absolutely required for larger-scale operations. The first step is assembling the team, which should consist of chemists and process and safety experts from within the company who are most familiar with the process and the plant. This team approach eliminates potential oversights by individuals working in isolation. The next step is the preparatory work and assembling the necessary documentation (batch record, plant P&ID's, process flow diagrams, etc.) to facilitate the study. And then finally there are the HAZOP meetings themselves, wherein the team leader should encourage free expression of ideas and uninhibited "what-if" thinking in an effort to develop a list of potential dangers and "maloperations" for each and every process step. Clearly documenting the findings and proceedings of the meetings is also a very important part of the exercise.

There is much excellent information on including HAZOPs as part of a safety management program, and the reader is encouraged to take advantage of these and other resources [2, 3].

45.5.8 Quantify Reaction Energetics

Underestimating or failing to recognize the potential hazards of exothermic reactions is perhaps the single most common cause of harmful and destructive process industry accidents. One of the reasons for this is a failure to understand the very limited heat transfer (cooling) available in large chemical reactors, a consequence of the low surface area per unit volume.

Thus an important part of safe reaction scale-up should be calorimetry or similar studies designed to quantify the exotherm, identify the potential maximum rate of reaction, predict the adiabatic temperature rise, etc. For companies that cannot conduct such studies in house, many safety labs offer these services on a contract basis. A number of innovative reaction classification systems have been proposed to help categorize the potential hazards of chemical reactions based on calorimetric parameters and to better ensure safe scale-up [4, 5].

The same can be said for determining the explosion potential of dusts and powders, along with minimum ignition energy (MIE), limiting oxygen concentration (LOC) for ignition, maximum attainable pressure, shock sensitivity, etc. Standard test methods exist for conducting and interpreting all of these characteristics, and the information from these studies can help design safer procedures to prevent explosions and to engineer improved mitigative equipment [6].

Of course, engineering judgment and chemical wisdom must be applied to interpretation and use of the data obtained from such studies, as the data often does not tell the whole story. Gustin [7] reports the case of a nitration reaction that counterintuitively exploded after the reaction was completed and the reactor was being cooled. The reason appears to have been the crystallization of a highly nitrated shock-sensitive sodium salt that came out of solution upon cooling. A piece of this material may have collected on the impeller shaft, broken free, and impacted the impeller or baffle, resulting in the explosion. Calorimetry or other studies might not have predicted this event, but an examination of the chemistry by someone experienced with these compounds might have identified the potential for this compound to form. Then more exacting safety studies could have been conducted to determine the risk and potential consequences of this happening.

45.5.9 Create a Written Batch Record

Variously called a batch log sheet, batch ticket, or batch record, this is an approved document (either paper or electronic) to be filled out by the operating staff as the batch is conducted. The batch record is based on a detailed process description prepared by the project chemist or process engineer and formally approved by his or her direct supervisors or others as necessary. It is a step-by-step recipe sheet, if you will, with spaces for recording pertinent data such as raw material charges, processing times and temperatures, etc. and with spaces for the initials or signatures of the individuals completing each task and checkers where necessary.

The batch record should be a controlled document. There is no need to point out the importance of ensuring that most current version of the record is in use and that it is as free of errors as possible. The tremendous convenience and speed of modern word processing has led to more than one pilot plant mishap due to careless cut-and-paste errors and poor proofreading. That is why a laboratory shakedown run is so important in ensuring the correctness of the batch record.

45.5.10 Understand Your Raw Material Grades

Many common raw materials and chemical reagents are available from a variety of sources and in a variety of qualities and purities or grades. By convention some terminology has arisen around these various grades, such as reagent grade and technical grade, but the precise meaning of these terms is anything but consistent from manufacturer to manufacturer, and the global supply chain creates even more confusion.

What one company may call spectrophotometric grade, another may call reagent grade and vice versa. Many supply companies have devised their own systems of nomenclature and provide various proprietary grades or purity levels that are impossible to directly compare with those of other suppliers. That is why it is important to understand the critical quality attributes of the raw materials used in a process and ensure that the chosen supplier can consistently provide the quality needed. If there is a particular impurity in a raw material that can adversely affect a reaction, then its effects must be quantified, and the specification for that impurity must be set based on the resulting data.

Much laboratory work uses reagent grade solvents and materials, but beware that the commercial grades available at large scale may not match the purity of many of these substances. It is wise to work with commercial grade materials as early in development as practical to better anticipate the results upon scale-up. This is also why it is so critical to conduct the lab-scale raw material use test prior to the first scale-up batch (see below).

Chemical purity is not the only characteristic to be concerned about. The physical form, such as particle size in the case of solids, can have a major effect on results of the process. A well-known example of this is the use of K_2CO_3 as an acid-sequestering agent in many alkylations and other organic reactions. Moseley [8] tells a tale of woe about their experiences in scaling up just such a process, due to the fact that a granular form of the carbonate with relatively low specific surface area was used during some large-scale runs. Their problems were exacerbated by the fact that the mixing conditions in the large-scale vessel were insufficient to suspend these granular solids, and so the carbonate lay unmoving and inaccessible on the bottom of the reactor, a very common phenomenon in large vessels.

45.5.11 Conduct a Raw Material Use Test

It should be considered an absolute requirement that a laboratory-scale experiment be conducted, ideally following the written batch record, that uses the very same raw material lots or batches of intermediates that will be used in the scale-up campaign. This demonstration batch should be carried to completion, and the product should be isolated and analyzed in full just as the pilot batch would be. The pilot batch should not proceed until all analytical results are reported, and it has been demonstrated that the batch passes all specifications. This way, if and when there are quality or processing issues in the scale-up batch, the raw materials can for the most part be eliminated as a cause. This can save a great amount of work and head-scratching and can prevent an investigation of a problem from being focused in the wrong area.

45.5.12 Make the Most of the Opportunity

A tremendous amount of time, effort, and money is expended in preparing for and conducting pilot-scale batches. These batches can consume alarming amounts of precious raw materials and labor. Consequently, in most cases only a limited number of pilot-scale batches can be conducted. That is why it is important to try to learn as much as possible from each batch. For example, a well-planned sampling and analytical plan will allow you to complete a mass balance, identify unexpected side products, and otherwise troubleshoot batches that have not gone as expected. Basically every process stream, including wasted streams, should be weighed and sampled. There may never be another opportunity to collect many of the samples generated in a pilot batch.

All observations should be carefully noted, and retain samples of isolated intermediates and final products should be saved for future reference if necessary. In a nutshell, one should make the best use of the opportunity to gather as much scale-up data as possible and clearly document the results of the batch in a comprehensive campaign report.

45.6 THINGS TO AVOID DURING SCALE-UP

45.6.1 Avoid Complexity

We have all no doubt heard of the famous "KISS" principle. There is certainly much to be said for keeping it simple, particularly in chemical process development and scale-up. The less complexity, the less opportunity for processing errors, operator slips, and unforeseen complications.

Commercial processes are carried out by a chemical operations staff, well trained, but often not educated as chemists or engineers, and certainly not as familiar with the idiosyncrasies and hazards of new processes as their developers are. These operators will carry out their jobs only as well as their training, experience, and operating instructions allow them. A clearly written batch record is supremely important, and this is much more difficult to achieve if a process is overly complex. A simple example would be making an extra effort in development to find a single solvent or mixture of solvents that can be used throughout reaction, workup, and crystallization steps, so as to eliminate the need for time-consuming and wasteful solvent switches between each operation.

Of course, most development chemists understand the importance of simplicity not only for improved safety and efficiency but also in minimizing processing time, minimizing waste, etc. However, it is not always possible to keep things simple when the process involves chemistry that requires sophisticated controls or specialized equipment such as hydrogenation reactions, nitrations, aminations, or other types of reactions that could potentially be hazardous. It is always an option to contract out these particular steps to manufacturers who have expertise in those types of reaction and the equipment to carry them out.

45.6.2 Avoid "All-In-and-Heat" Reactions

One of the most dangerous practices in chemical processing, and one that is frowned upon at all scales, is to charge all of the reactants to a batch vessel and then begin heating it up. The danger is that once the mixture reaches the onset temperature of reaction and the reaction starts, there will be no way of controlling it. Many reactions are highly exothermic, and once they "kick off," they will continue to heat themselves to higher and higher temperatures, possibly exceeding the boiling point of the mixture and erupting. The mixture could also begin to decompose at higher temperatures. Many times, the decomposition itself is self-accelerating and more exothermic than the process reaction. Explosive gases can be evolved and a highly energetic explosion could ensue.

Of course, when the chemistry is well understood and known to be safe, this type of all-in operation may be acceptable. But when scaling up new processes for the first time, it should be forbidden.

A number of factors make carrying out exothermic reactions at large scale much more dangerous than at laboratory scale. Of course, the consequences of a large explosion are more devastating than a small one, but the key difference at large scale is the limited heat transfer area characteristic of large reactors and the long response time if cooling is suddenly required. It can take many hours to cool a very large chemical reactor, by which point it may be too late.

The recommended approach for scaling up exothermic reactions is to maintain some degree of control. The most common approach for this is to use a "controlled addition" scheme in which the nonreactive components of the batch are charged to the reactor, and then the reactive (controlling) reagent is slowly added with agitation, allowing the reaction to proceed at a controllable pace. In the event of a temperature excursion, addition of the reagent can quickly be stopped, halting the reaction.

It is important in this method to ensure that the reaction is in fact proceeding and that the reactive agent is being consumed as it is added. If for some reason this is not happening, the reaction "stalls," for instance, then the reagent will accumulate in the reactor and may suddenly react all at once, putting us back where we started. Many methods are available for monitoring the progress of a reaction, but the simplest is to monitor the batch temperature and ensure that the anticipated rise is observed.

45.6.3 Do Not Apply Heat Without Agitation

A former colleague of mine would attest to the practical nature of this advice. He was performing a toluene/aqueous extraction experiment at about 70 °C in a round-bottom flask with a heating mantle. He stopped the agitator briefly to let the phases separate, collected his sample, and then restarted the agitator. The entire contents of the flask instantly erupted out the top of the reflux condenser. Luckily no one was hurt. We surmised that while the agitator was stopped, the glass surface must have exceeded the 85 °C boiling point of the toluene–water azeotrope. As soon as the agitator was restarted, the mixture boiled violently.

In my opinion it is never acceptable to apply heat to a reactor without agitation. Heat transfer is severely limited in large reactors to begin with, and what little heat transfer does occur is highly dependent on the degree of agitation in the vessel and the convection that it creates.

In addition to the incidents such as the one described above, countless other undesirable effects can result. Without agitation there can be no accurate reading of the internal reactor temperature. Because the reactor wall temperature is quite often much hotter than the bulk batch temperature, product can easily become overheated and "baked" against the side of the vessel, which can lower yield and lead to dangerous degradation reactions. Temperature gradients created by insufficient agitation can result in poor reaction selectivity and out-of-spec products.

Countless cases of violent reactions and explosions have been attributed to "agitation issues" of one kind or another. For example, one incident occurred during a highly exothermic reaction between sulfuric acid and an organic amine. This biphasic reaction was normally carried out by slow addition of the amine to the hot acid with vigorous agitation. One fateful day, at shift change, the agitator was inadvertently left off when amine addition was started. The amine pooled in the bottom of the reactor but did not react. Much later the second shift noted that the agitator was off and proceeded to turn it on, at which point the reactor exploded as all of the materials reacted instantaneously. A neighborhood in Frankfurt, Germany, was dusted with a yellow coating of o-nitroanisole in a similar incident that eventually resulted in the company going out of business.

45.6.4 Do Not Ignore Potential Decomposition Reactions

This recommendation goes hand in hand with Section 45.5.8. Not only must the necessary calorimetry be conducted on exothermic reactions, but also the possibility of self-heating decomposition reactions must be considered. This may require additional testing, such as adiabatic reaction calorimetry (ARC). Such testing should be considered for all process streams if it is believed that the particular chemistry involved has the potential to create unstable decomposition products.

One of the difficulties is that these self-heating decomposition reactions may happen very slowly at first, so slowly that they may not be identified in routine testing. Even below the onset temperature, exothermic reactions are still happening at some finite rate. In one case, a reactor exploded many hours after the reaction was completed, the services shut off, and the reactor left to cool on its own. A previously unknown decomposition reaction was occurring so slowly that no one noticed the temperature rising in the reactor. Eventually the temperature reached the onset temperature, and the self-accelerating reaction kicked in, resulting in an explosion (it is generally safer to cool a reactor to a safe temperature using the jacket rather than allow it to simply cool down on its own). Waste stills have been known to explode days after being charged with a waste stream that underwent a slow decomposition and unanticipated climb in temperature.

45.6.5 Avoid Adding Solid Reactants to Reacting Mixtures

Another common laboratory technique that is not easy to scale (without some advance planning) is the portion-wise addition of solid reagents to a reacting mixture. Lab scientists routinely use this technique to avoid the "all-in-and-heat" approach, but in the lab it is a simple matter to remove a glass stopper from the flask in a fume hood and add a small spatula-full of the solid and close it back up. This is repeated until the addition is complete.

There are a number of reasons that this becomes much more difficult at scale. First, it is most inadvisable to open the manway of a large reactor containing flammable solvents because an ignitable atmosphere may form as vapors exit the vessel or air enters it. This is especially true if the contents of the reactor are being heated. Second, it is quite possible that the reagent may react quickly, causing the eruption or ejection of material out of the manway, endangering personnel (sadly, operators have died because of this very thing). Therefore, the solids addition must somehow be accomplished with the reactor sealed.

One possibility is to reconsider the order of addition. It is always much easier to charge solids to a reactor first and then the solvents and liquid ingredients. Of course, changing the order of the addition may change the selectivity of the reaction or, worse, may remove an important component of exotherm control. Another possibility is to make a solution of the solid, which can be conveniently charged by pump or other methods or even a slurry, although charging slurries at a consistent rate is somewhat more difficult than charging a solution.

Where modifying the process is not possible, a number of solids charging apparatus do exist that enable the controlled addition of solids to a reacting mixture. An excellent review of available options was recently published that discusses the advantages and weaknesses of each approach for a number of given situations [9]. Many of these devices are also very helpful for improving the dissolution of hard-to-wet solids or solids that tend to float or form large lumps, which can become major processing issues in commercial-scale operations.

45.6.6 Avoid Evaporating to Dryness

Generally speaking, the common laboratory technique of evaporating a process stream to dryness by removing all solvent on a rotary evaporator or similar piece of equipment simply will not fly in pilot-scale equipment. First, toward the end of distillation, the liquid level will fall below the agitator (minimum stir volume) in most standard reactors. As discussed above, it is inadvisable to continue applying heat without adequate agitation.

One of the consequences of heating without agitation, or stripping to dryness, could be the decomposition of the product in contact with the hot reactor walls. This presents not only quality but also safety issues. When complete removal of the solvent is necessary so that a different solvent can be introduced for the next processing step, the standard approach is to conduct a "solvent exchange." In this operation, the solvent is first partially distilled down to a safe (but mixable) level; the second solvent is then added, and distillation is continued. This process is repeated until the concentration of the first solvent is as low as required. Of course, the specifics of the operation and its success depend on the relative volatilities of the solvents and whether or not they form an azeotrope. It is also well known that so-called constant volume solvent exchanges are much more efficient than the add-and-distill approach [10, 11].

45.6.7 Do Not Underestimate Processing Times at Scale

I have often said that one of the biggest surprises that an R&D chemist experiences when bringing a new process to the pilot plant for the very first time is how long everything takes. My first scale-up campaign at a CMO was no exception.

I remember arriving early on that first day eager to get the process underway. Step 1, charging a major raw material, a solid, took the entire 8 hour first shift. This included time for completing the release paperwork, transporting the material from the warehouse, the operators suiting up, the laborious act of charging the material manually, coffee breaks, etc. Later in the process, a distillation step that we routinely accomplished in about 30 minutes in our kilo lab took over 12 hours, including heating up and cooling down the reactor. The final product isolation, which used a product centrifuge, required 7 or 8 separate centrifuge loads; the entire recovery operation took over 24 hours. Again, this was something I was used to completing in under an hour, even at the pilot scale, where one filter load could accommodate the entire batch.

Extended processing times are a fact of life in plant-scale operations, but the unprepared will find that they are in for a real test of their patience. More importantly, serious quality and safety issues might arise if the question of process stream stability has not been considered prior to the scale-up. As a case in point, Dunn [12] describes a deprotection step involving trifluoroacetic acid (TFA) that was subsequently distilled off. This technique was quite successful in the lab, but in the very first scale-up batch, that distillation step took many hours, and when it was completed only about 5% of the product remained! No one had recognized that the product was not stable in the presence of TFA at elevated temperature because the stripping only took a few minutes in a laboratory rotovap.

A simple stability experiment in which the product stream is cooked at distillation temperature for a time would have given the researchers a heads-up that there could be a problem. Every step of a new process should be considered from this perspective since, even under the best of conditions, notwithstanding equipment failures or scheduling delays, every operation will take much longer in the plant than in the laboratory or kilo lab.

45.6.8 Do Not Ignore Plant-Scale Solvent Issues

Chemists often develop processes using their favorite solvents, those that provide solubility for a wide range of substances or those that are most easily removed in distillation and drying. Unfortunately, some of these very solvents may need to be avoided in pilot or commercial operations due to safety concerns or environmental issues.

Hexane comes to mind as an excellent organic solvent for running many types of reactions and for crystallizing a broad range of organic compounds. However, with a flash point of only −23 °C, a relatively low enthalpy of vaporization, and very low conductivity (making it prone to electrostatic buildup), many processing plants will simply not allow it as a production solvent. It is also toxic, as are a number of other common organic solvents.

Ethyl ether, methylene chloride, chloroform, methyl isobutyl ketone (MIBK), methyl ethyl ketone (MEK), and N-methylpyrrolidone (NMP) are but a few of the solvents that have been used widely in industry for decades, but suffer from either flammability and safety issues, high water solubility and the accompanying environmental concerns, toxicity, or other reasons that make their large-scale use unfavorable, or in many cases prevented by law. There are also those solvents that, because of health concerns, are not allowed by the FDA and other regulatory bodies to be present in the final formulation of human drug products, and this list of solvents is constantly evolving. Finally there may be waste disposal concerns for certain solvents in particular locales.

Thus it is important for those working in process development to understand as early as possible what limitations on solvents they will have to deal with in their companies or the plants or countries they will be operating in. Sometimes it is not easy to find replacement solvents with a better safety profile, and the process may need to be entirely redesigned, but the earlier this is recognized, the better.

45.6.9 Avoid Hot Filtration Operations

A common processing step is the "polish filtration," in which a final product solution is filtered through a small-pore cartridge filter or a filter coated with celite or other filter aid to remove any particulates or undissolved contaminants as a final polishing step. This is usually carried out just prior to isolation of a final product by crystallization.

At laboratory scale this step is often ignored, but plant operators know that small amounts of dust and dirt and other undissolved solids often wind up in a product batch after multiple processing steps and that this material needs to be removed prior to crystallization. Unfortunately, the way many crystallization processes are designed, the product solution is supersaturated to help maximize crystal yield. Thus the polish filtration step must be carried out at elevated temperatures to ensure that no product crystallizes out in the filter or the pipes connecting it.

This can present some difficulties and potentially unsafe situations on scale-up. In order to prevent material crystallizing in the filter, the filter and all lines leading to it must be heated, perhaps by steam tracing, which complicates the operation. If the temperature of the pipes falls too low, the lines and filter may become plugged with solids. Also the handling and transferring of heated flammable solvents is not a particularly safe practice.

For these reasons it is best to avoid filtering heated supersaturated solutions. The most obvious way to do this is to dilute the product stream so that it is not supersaturated at ambient temperature and then distill off the solvent prior to crystallization. This, of course, can be time consuming and could potentially affect product quality in other ways, illustrating again that process scale-up always involves compromises and trade-offs and that processing decisions must be made while keeping the whole process in mind.

45.6.10 Do Not Underestimate the Quench/Extraction Step

Most reactions carried out in organic solvents are conveniently "quenched" by adding an aqueous solution of an acid, base, buffer, or other quenching agent. This rapidly stops the reaction by neutralizing the reactive species and, via the phase separation step that follows, provides a convenient way to extract and remove unreacted starting materials, side products, and other impurities from the reaction mixture. However, all too often this important step is taken for granted and simply tacked on at the end of a clever new synthesis with the attitude of "workup as usual." This is unfortunate because the quench and extraction (or "workup") step is a source of countless problems during scale-up and as much care should go into the design of this step as goes into the chemistry that precedes it.

For one thing, the quench usually represents the highest volume step, and to maximize volumetric productivity, the workup should be designed to minimize the use of extract phase while still accomplishing the necessary goals. This also minimizes waste disposal. Also, the settling and phase separation part of the operation can take much longer than in the lab due to a number of factors, including finer dispersions and more of a tendency to emulsify due to the higher tip speeds and higher shear associated with commercial-scale impellers. Inexperienced operators may operate agitators at speeds much higher than necessary during extraction steps and create emulsions that can take many hours to separate.

There is also the issue of phase continuity. Every dispersion consists of a continuous phase and a dispersed phase. Depending on the relative ratios of the solvents, their surface tensions and densities, and other factors, the aqueous phase may be dispersed in the organic, or the organic phase may be dispersed in the aqueous. These two systems can often show drastically different behaviors, sometimes creating an intractable emulsion in one case and readily separating into two phases in the other. These differences should be studied in the lab in order to minimize difficulties on scale-up. This phenomenon is described in better detail in Atherton [13].

Unfortunately, emulsions are sometimes unavoidable in batch chemical scale-up, but an awareness of how easily a mixture can become emulsified and a familiarity with the ways to minimize or deal with emulsions will make the scale-up much less problematic.

45.6.11 Avoid Routine Reliance on Flash Chromatography

Chromatographic separation techniques are important tools in process development, and in certain sectors of the industry, biomolecule and protein production, for example, they play a major role in commercial production. Certain specialized types of chromatography such as simulated moving bed (SMB) are also used in full-scale production of small molecule products for the separation of chemical enantiomers.

But then there is so-called "flash" chromatography, the purification of a product stream by means of a single pass through a packed silica gel column and then eluting out the various bands with solvents. This is a favorite technique of many laboratory chemists, especially early in development where the goal is to simply isolate a small amount of product for analysis and characterization. Unfortunately, this technique is not very amenable to scale up for commercialization.

This type of chromatography can use very large amounts of solvent. Solvents are generally responsible for the majority of the environmental impact of chemical processes in the pharma industry, and chromatography utilizes a disproportionately large amount of solvent per unit product. This solvent needs to be disposed of or recovered and purified for reuse. It is also difficult to design, manufacture, and pack very large chromatography columns so that they will operate without shortcutting and back mixing. There will be much higher pressure drops through larger columns, necessitating larger pumps, and the temperature and flow control becomes more difficult for very large columns.

For these reasons, it is best for the process developer to recognize that flash chromatography is generally a method for the lab only and that a reliable, scalable purification process, such as crystallization, will be much better for commercialization. There are numerous ways to approach this, either by crystallizing the product directly from a solvent or mixture of solvents or if necessary by forming a crystalline salt form of the molecule with an appropriate anion or cation.

45.6.12 Play It Safe

While safety always has to be the number one priority, what I specifically refer to here is minimizing the risk of losing all of your valuable raw materials or intermediates in a single batch, especially when scaling up for the first time. Avoid the temptation to go for the home run, and convert all of that hard-earned feedstock to final product at once. No matter how careful you are, and how much time you spend going over the details of the process, there will always be unexpected occurrences the first time a new process is scaled up.

An instruction might be misinterpreted by an operator, processing steps will take longer than expected, possibly resulting in product degradation, and a key piece of equipment may not operate as anticipated. Better to play it safe by running two or more smaller batches to ensure that the program is not stopped in its tracks because all of the key intermediate or custom raw material is used up.

Running smaller batches has other advantages. There is improved heat transfer area per unit volume, and there will be fewer issues with mixing and chemical transfers if the scale-up factor is smaller. One group described scaling up a hydrogen-generating reaction and their plans to dilute the hydrogen in the off-gas with nitrogen to keep it below its lower flammability limit. Halfway through the first batch, an unexpected alarm went off, indicating that the liquid nitrogen tank that supplied the whole building was empty. They had completely drained it; to complete the campaign they split the remaining work into a number of smaller batches to eliminate this from recurring.

45.7 CONCLUSIONS AND FINAL THOUGHTS

It should be clear from the above that experience plays a major role in the speed and success of any scale-up campaign. There is also a tremendous amount of valuable published information about scale-up and process safety available, and the process engineer should certainly make an effort to access it and learn from it. Although it is impossible to anticipate every possible mishap during first-time scale-ups, I hope that the above list of do's and don'ts will provide food for thought and a better appreciation for the fact that making cross-disciplinary communication and cooperation high priorities (i.e. keeping it a team effort) will maximize your chances for success.

REFERENCES

1. McConville, F. (2007). *The Pilot Plant Real Book*, 2e. FXM Engineering and Design: Worcester, MA.
2. Kletz, T. (1999). *Hazop and Hazan*, 4e. Philadelphia, PA: Taylor & Francis.
3. Kletz, T. (2007). How we changed the safety culture. *Organic Process Research and Development* 11 (6): 1091.
4. Bollyn, M. (2006). DMSO can be more than a solvent: thermal analysis of its chemical interactions with certain chemicals at different process stages. *Organic Process Research and Development* 10 (6): 1299. November–December 2006 reaction classification for safety.
5. Saraf, S.R., Rogers, W.J., and Mannan, M.S. (2004). Classifying reactive chemicals. *Chemical Engineering Progress* 100 (3): 34.
6. Zalosh, R., Grossel, S.S., Kahn, R., and Sliva, D.E. (2005). Safely handle powdered solids. *Chemical Engineering Progress* 101: 23.
7. Gustin, J. (1998). Runaway reaction hazards in processing organic nitro compounds. *Organic Process Research and Development* 2 (1): 27.
8. Moseley, J.D., Bansal, P., Bowden, S.A. et al. (2006). Trouble with potassium carbonate and centrifuges: mass transfer and scale-up effects in the manufacture of ZD9331 POM quinacetate. *Organic Process Research and Development* 10: 153.
9. Glor, M. (2007). Prevent explosions during transfer of powders into flammable solvents. *Chemical Engineering* 114 (10): 88.
10. Chung, J. (1996). Solvent exchange in bulk pharmaceutical manufacturing. *Pharmaceutical Technology* 20: 38–48.
11. Li, Y.E., Yang, Y., Kalthod, V., and Tyler, S.M. (2009). Optimization of solvent chasing in API manufacturing process: constant volume distillation. *Organic Process Research and Development* 13 (1): 73.
12. Dunn, P.J., Hughes, M.L., Searle, P.M., and Wood, A.S. (2003). The chemical development and scale-up of sampatrilat. *Organic Process Research and Development* 7: 244–253.
13. Atherton, J.H. and Carpenter, K.J. (1999). *Process Development: Physicochemical Concepts*. Oxford: Oxford University Press.

46

KILO LAB AND PILOT PLANT MANUFACTURING

MATTHEW CASEY[1], JASON HAMM, MELANIE MILLER, TOM RAMSEY[2], RICHARD SCHILD[3], ANDREW STEWART[4], AND JEAN TOM

Product Development, Bristol-Myers Squibb, New Brunswick, NJ, USA

46.1 INTRODUCTION

The pharmaceutical industry has traditionally divided its scientific efforts to bring a drug to market into three stages: drug discovery, process development, and manufacturing. Process development is the link between the worlds of the laboratory and the commercial manufacturing plant. To accomplish this mission, chemical engineers in process development groups must be knowledgeable and capable in both arenas – laboratory experimentation and plant scale-up.

The two primary deliverables from the development efforts are the supply of drug substance or active pharmaceutical ingredient (API) and the process knowledge generated. The API fuels several key activities including clinical studies, toxicological studies, and formulation development. The process knowledge forms the basis of a sustainable commercial manufacturing process. Figure 46.1 shows the work process and the organizational structure metaphorically represented as a bridge. In this case the bridge is process development physically linking drug discovery to manufacturing. At the foundation of the bridge is knowledge, both process knowledge and regulatory knowledge as required by agencies such as the FDA. Also shown are several steps of the bridge representing the progression and deliverables of the development process (toxicological study supplies, clinical supplies, and process optimization). Two important tools available to process development groups to carry out their mission are kilo labs and pilot plants.

There can be significant differences across the pharmaceutical industry in what is meant by the terms kilo lab and pilot plant. Some companies differentiate kilo labs from pilot plants by equipment size and may refer to kilo labs as glass plants because of the extensive use of glass equipment typically found in these facilities. Other companies designate all of their scale-up facilities as pilot plants, making no distinction between equipment size and scale of the facility.

Kilo labs are typically utilized for the first scale-up to produce sufficient quantities for initial drug testing, i.e. toxicological studies or phase I clinical trials. Typical chemistry used is less developed than at later stages and is often a modification of the longer, less efficient discovery chemistry route. Kilo labs are designed to yield quantities ranging from 100 g to under 10 kg of materials with typical batch sizes of 2–3 kg (hence the historic reference to kilo lab). Reactor sizes usually range from 20 to 100 L where the material of construction is often glass. The kilo lab represents the first step out of the lab, and the equipment design is typically closer to a laboratory (larger glassware) than a manufacturing plant (industrial equipment).

Kilo labs typically have a high degree of flexibility to accommodate a range of complex chemistries. Portable equipment setups are configured to meet specific process needs, and disposable components may be used to control product cross contamination or decrease turnaround time for process areas. Ideally, the kilo lab infrastructure is sufficient to support "mini-piloting" or scale-down studies to explore process changes prior to scale-up in the pilot plant. Equipment similar in design to the pilot plant equipment (e.g. filter dryers) is often

[1]*Biogen Inc., Durham, NC, USA*
[2]*Janssen, Raritan, NJ, USA*
[3]*TG Therapeutics, Inc. New York City, NY, USA*
[4]*Allergan, Irvine, CA, USA*

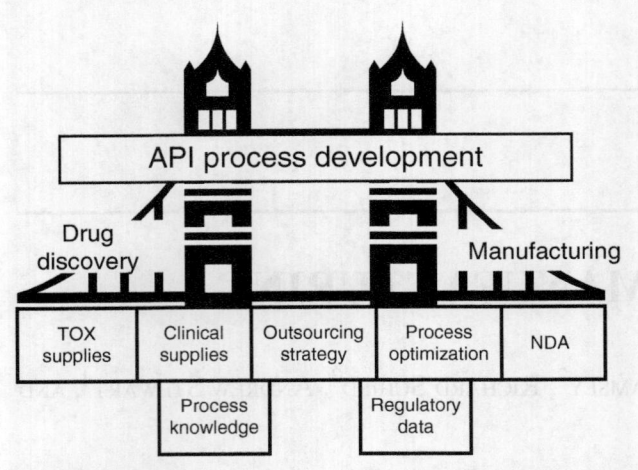

FIGURE 46.1 Process development links drug discovery and manufacturing.

available to accelerate identification of scalability issues. While many kilo labs are likely manually operated, kilo labs with robust data collection systems can provide key process information to further development, making the kilo lab more than a place to generate kilogram deliveries.

Pilot plants in contrast are typically larger than kilo labs with designs that more closely resemble the commercial manufacturing plant. Supplies generated in pilot plants often support phase II and later clinical or chronic toxicology studies. At the pilot plant stage, demonstration of potential commercial routes and generation of process knowledge through data collection can be key objectives. Though design standards and size differ among companies, pilot plant reactors typically range in scale from 200 to 4000 L.

The pilot plant, like the kilo lab, must operate with a degree of flexibility and absorb process and schedule uncertainties. However, pilot plants typically have more substantial infrastructure and regulatory requirements for quality, safety, and environmental concerns. Pilot plant operations often run one to three shifts per day and five to seven days per week as needed to manage process chemistry requirements and changes in campaign objectives (e.g. timing or quantity). Portable equipment will also supplement the fixed equipment trains to support process-specific needs. Examples of such equipment include portable tanks, filters, pumps, wet mills, chromatography equipment (i.e. columns, pumping skids), and process analytical technology (PAT) instruments (Figure 46.2).

One element that differentiates facilities in the pharmaceutical industry from those in the chemical process industry is the requirement to follow current good manufacturing practices or cGMPs when preparing clinical supplies. Good manufacturing practices are governed by regulatory health authorities such as Food and Drug Administration (FDA) and European Medicines Agency (EMA) and are defined within six quality systems: quality, production, facilities and equipment, laboratory controls, materials, and packaging and labeling. More information is provided in Section 46.3.

Figure 46.3 shows a graphical representation that matches the facilities and scale to the phase of clinical development. Many pharmaceutical companies have laboratories, kilo labs, and pilot plant facilities available as an internal capability. Other companies leverage vendor facilities to supplement or replace internal capabilities at each scale of development. The individual company strategy may vary to outsource several of the early steps in the synthesis to vendors while executing only the last several steps internally or outsourcing of full synthetic route depending upon business and regulatory drivers. Each company develops its strategy through assessment of the cost and benefit of maintaining internal capabilities and to what scale and capacity.

The basis of an operating philosophy for a kilo lab or pilot plant is safe production of high quality product while meeting all necessary regulatory requirements. Process safety is constant throughout process development and a primary consideration for process design and execution decisions. Appropriate systems and procedures regarding plant operation and material flow following cGMP provide the basis for the operating philosophy to support generation of high quality supplies for clinical studies. Processing activities in the facility are the source of process data and knowledge, and the plant's capability to generate such information is a key part of the operating philosophy for the facility. Business objectives for cost- and time-efficient operation of the facilities must also be met. As a result, there is a complex balance of timely and cost-effective manufacture of intermediates and APIs while obtaining sufficient processing experience to advance knowledge and development of the process.

46.1.1 Operating Staff

The personnel overseeing and working in the facility are key components of a kilo lab or pilot plant facility's ability to operate in a manner demanded by safety and regulatory requirements. They play a critical role in the workflows to operate the facility as well as compliance with the regulatory drivers. The size and makeup of the facility staff will also determine how the facility is run.

In the kilo lab, operating staff may include chemists or chemical engineers. The kilo lab scientist supplements the process development scientists and brings an understanding of scalability issues and process safety risks as well as strong process troubleshooting skills. In some kilo lab facilities, a lab manager may oversee the facility and its maintenance while coordinating the project scientists who execute process activities in the facility.

In a pilot plant, many of the named roles are similar to those in a manufacturing plant: plant manager, process engineer, chemical operator, and facility support staff, such as maintenance, process automation, and supply chain. The process safety group can be part of the facility staff, but in some companies it is part of the process development groups.

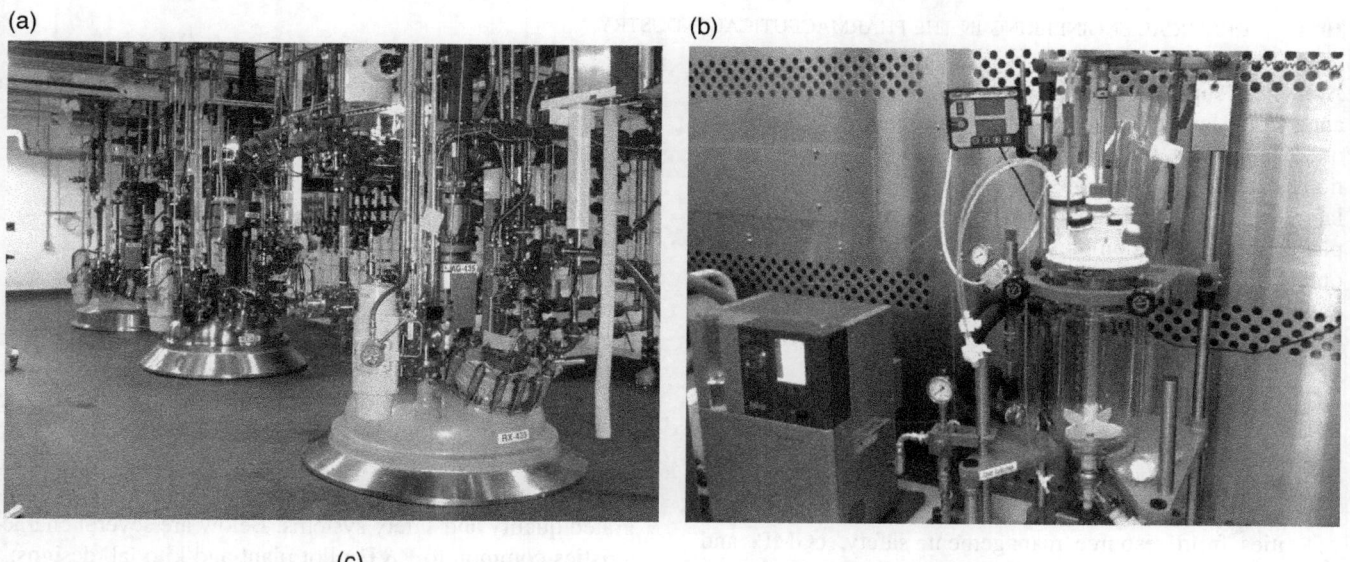

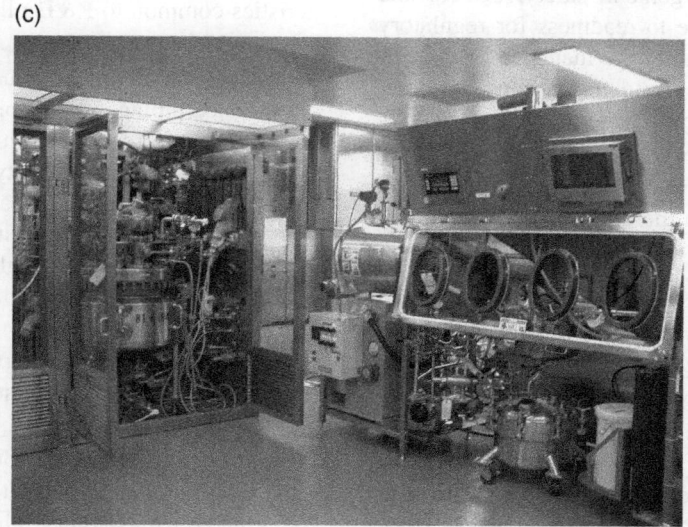

FIGURE 46.2 Examples of pilot plants and kilo labs. (a) Pilot plant process area showing three reactors (1000–4000 L) and associated overhead piping. (b) A typical kilo lab with small glass reactor (10 L). (c) Kilo lab process area showing larger reactors (left) in downflow booth and Nutsche filter and tray dryer inside isolator (right) with filtrate receiver underneath.

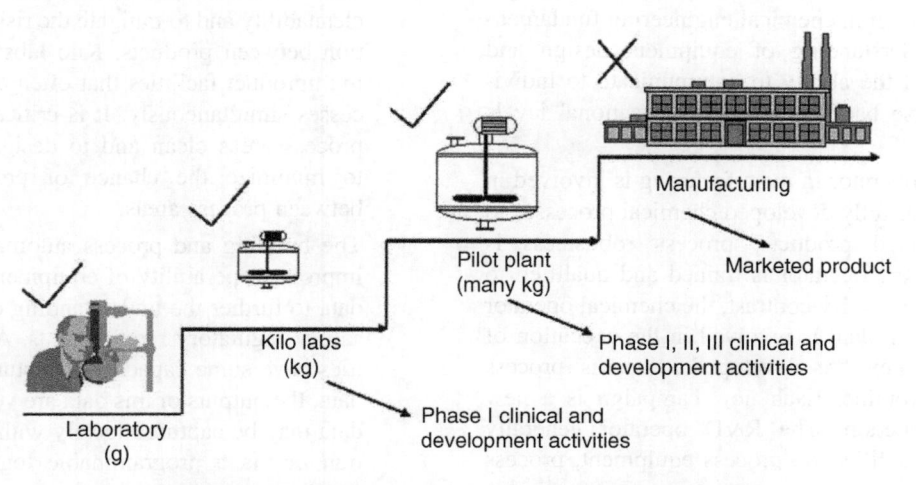

FIGURE 46.3 Progression of scale-up of chemical processes.

Analytical support for in-process assays and quality assurance groups are important links to ensuring overall operation of the pilot plant, but typically are not part of the plant operations staff. The number of staff in each role is determined in large part by the planned capacity utilization of the pilot plant, the size and complexity of the infrastructure, and the operational cost base. For the plant manager, the process engineer, and the chemical operator, there are significant differences between the skill sets required between R&D and manufacturing:

- The plant manager of a pilot plant is at the intersection of process development activities, campaign execution, and facility support with a broad range of responsibilities from resource management, safety, cGMP, and environmental compliance to readiness for regulatory inspection. It is essential that the manager have sufficient knowledge to understand the various customer needs and manage the dynamic environment of new process implementations.
- The process engineer serves an important function in transferring/implementing the defined process into the facility. In partnership with the process development team, the process engineer identifies the process fit, drives equipment decisions, and supports modeling decisions to help define equipment operating parameters. Examples include agitation rates for solids suspension, reactor heat and mass transfer determinations, distillation modeling, and drying times. During execution, the process engineer may work as a team with the process supervisor and chemical operator to ensure successful implementation and proper mitigation of risks. Upon completion of a batch and/or campaign, the process engineer also supports data analysis activities to capture key learnings and implement improvements for future campaigns. In addition to process data, the process engineer will summarize all material accountability and yield data with the complete data package. A solid foundation in chemical engineering fundamentals, good understanding of equipment design and operations, and the ability to communicate to individuals of diverse backgrounds and educational levels is important.
- The chemical operator in manufacturing is involved in the execution of fully developed chemical processes to prepare marketed products; process robustness is expected, and the operator is trained and qualified to execute the process. By contrast, the chemical operator in an R&D pilot plant is involved in the execution of processes as they are developed, whereas process variability is routine. Each new campaign is a new process introduction. The R&D operator generally trains and is qualified on process equipment, process troubleshooting, and unit operations independent of a specific process.

46.2 KILO LAB AND PILOT PLANT FACILITY DESIGN AND UNIT OPERATIONS

For both the kilo lab and the pilot plant, the key objective is to provide for as much flexibility and agility as possible to manage a dynamic portfolio of potential products while meeting environmental, health, and safety (EHS), product quality, and business requirements. Though they may appear separate, a well-designed facility acknowledges the benefits of integrated quality and safety systems. Below are several characteristics common to R&D pilot plant and kilo lab designs:

- The materials of construction of all equipment and associated piping systems provide corrosion resistance to a range of process conditions. This reduces concerns of material incompatibility between equipment components and process streams and allows a wide variety of chemistries to be run. The most common material choices are glass or glass-lined equipment, Teflon®, stainless steel, and high-end corrosion-resistant alloys such as Hastelloy®.
- The equipment is designed to cover broad temperature and pressure ranges to support safe execution of all typical unit operations.
- Pilot plant facilities should incorporate a blend of fixed and portable equipment in a complementary fashion to increase the diversity of available equipment configurations while managing the amount of time required to change over from one process to the next. The equipment can be set up relatively fast and inexpensively through the use of standard equipment designs and utilization of quick connect or single-use hoses.
- The equipment and facility should be designed for cleanability and to mitigate the risk of cross contamination between products. Kilo labs and pilot plants are multiproduct facilities that often execute multiple processes simultaneously. It is critical to keep individual process areas clean and to design building air flows to minimize the chance of product contamination between process areas.
- The building and process automation systems enable improved operability of equipment and the capture of data to further the understanding of the process and to satisfy regulatory requirements. Although many facilities have some capacity to capture and retain process data, the outputs of this data are very different. Process data may be captured locally within batch documentation or via a programmable logic controller (PLC).

More recently designed facilities could have distributed control systems (DCS) that have the availability to retain data in process data historians on local, enterprise, or cloud-based servers. Scientists and process engineers will benefit from advanced process control systems that interface with common engineering analysis/modeling tools to summarize and analyze data in a semi-automated method.

- The facility and equipment should include industrial hygiene engineering controls (i.e. barrier technology) as the primary defense against operator exposure to chemicals, drug candidates, and their intermediates. Many advances have been made with both fixed and disposable isolator technology (see Section 46.2.5) that should be considered and implemented to improve the containment/exposure design of the facility. Personal protective equipment (PPE), although a required safety measure, should not be used as the primary control to mitigate the risk of worker exposure.
- The facility should be able to execute products at both early and late stages of the development cycle. A kilo lab may be called upon to execute a low volume product at a late stage in development, while a pilot plant may be the right venue for early stage products with high volume requirements.
- The facility must support the ability to safely produce clinical material being consumed by humans with the proper employment of cGMP for the stage of development. Adherence of key aspects of the facility to cGMP needs to be considered. These include raw material and personnel flow; heating/ventilation/air conditioning (HVAC) design and air balance; maintainability; and cleanability.
- The facility must meet all relevant safety codes as required by regional, federal, and state regulations.

Each chemical process can be broken down into a series of unit operations and may be conducted in a batch, semi-batch, or continuous mode of operation. The most common unit operations for a typical process are described below:

- Equipment preparation: A critical aspect of all process implementation is the proper preparation of the equipment from both safety and quality perspectives. When the process is ready to start, all equipment must undergo an integrity check to ensure no leaks are present and residual oxygen has been removed. This preparation is best completed by performing a pressure and/or vacuum leak check and inertion cycles with nitrogen. Once the integrity of the equipment is confirmed, preprocess rinses, with supplemental in-process control samples, may be required to ensure the equipment is in the proper state (i.e. free of water) prior to proceeding with the process.
- Charging components: The starting component is typically a solid and is charged to the reactor through an open system, such as scooping material in through the reactor manway or through a closed system. Closed charging systems protect the worker and the material from contamination and include split butterfly valves, powder transfer systems, and glove box technology. Liquids are charged through reactor nozzles by pump, vacuum, or pressure transfers.
- Reaction: The reaction is where the molecule is being synthesized within each batch step progressing to the preparation of the API. It normally involves heating, cooling, controlled addition rates of reagents, and agitation. Thermal chemistry studies and energy balances should be completed prior to execution to ensure safe processing.
- Workup: A workup is typically completed post-reaction and can be composed of multiple unit operations including reaction quench, extractions, adsorption (color or impurity removal), polish filtrations, and distillations.
- Crystallization: Following the workup, the crystallization is performed to precipitate the solid product from the solution by modifying the solubility of the solution via cooling, antisolvent addition, volume reduction, and/or seeding. The crystallization is a critical aspect of the process typically requiring thorough understanding of the crystallization process and product solubility to control impurity purging, desired material form/morphology, and/or powder properties. The crystallization is one of the most complex and important steps in controlling product quality and downstream processing characteristics.
- Filtration: The crystallized solids are separated from the carrying liquids, also known as mother liquors, using filtration equipment. The filtered cake is washed with appropriate solvent(s) to facilitate impurity and residual solvent removal. Filter dryers, pressure filters (Nutsche or plate), and centrifuges are common filtration devices found in pilot plants and kilo labs.
- Drying: The solids are dried typically using heat and vacuum to remove residual solvent. The dryer may be controlled to produce the desired polymorph or form of the product. Commonly found contact dryers include tray dryers, agitated dryers, and tumble dryers (see Section 46.2.4.3).
- Dry powder finishing: Milling or micronization (for particle size reduction) as needed based on the powder property attributes of the API and the requirements for successful formulation into the drug product.

46.2.1 Kilo Lab Design

Kilo labs are typically installed within a laboratory building and are generally not found as stand-alone facilities. Kilo labs in close proximity to process development and analytical laboratories encourage interactions between scientists, allowing for greater synergy between laboratory and scale-up operations. Kilo lab workflow as an extension of laboratory practices can improve efficiency and decrease the time needed to execute a process. However, the shared infrastructure can present challenges to the control of material and personnel workflow. Requirements for controlled personnel access and separate physical areas to manage segregated storage of raw materials and material subdivision apart from laboratory operations will increase the complexity of the area management.

Laboratory buildings typically employ central corridors for transport of materials and personnel. Air locks are often employed as physical separation of the kilo lab process areas from the rest of the building and reduce unnecessary personnel flow into the areas. The air locks also help to maintain proper air balance and pressure differentiation between the process areas and the corridor (see Section 46.2.3.1).

Most of the processing for a single intermediate or API is performed in one room or area. This includes all the material charging, reactions, workup, isolation, and filtration. The most common exceptions are drying and milling, which may be performed in separate locations specifically designed for handling dry powders.

Potential for cross contamination between products should be considered at each phase of facility and process design. Contamination occurs primarily through airborne particulates and/or insufficient cleaning of product contact surfaces. Many approaches designed to minimize contamination also have benefit in protecting the kilo lab scientist from exposure to the compound. Downflow booths, fume hoods, barrier technology, and other engineering controls may be used to achieve closed processing are good examples (see Section 46.2.5). Placement of reactor systems into down flow booths or fume hoods, as shown in Figure 46.4, can increase process segregation for safety and quality. The same approach can be employed to segregate isolation and drying equipment. Where kilo labs are an extension of laboratory operations, specific laboratories can be designated for preparation of supplies destined for clinical use to further segregate process operations.

Equipment cleanability should be a consideration in the initial design of the equipment as well as in the design of the process equipment configuration for a specific product. Cleaning procedures should be developed to address a wide range of chemistries and process conditions, recognizing that little is known about the cleanability of the product at the early stages of development.

46.2.2 Pilot Plant Design

Unlike kilo labs, pilot plants are commonly designed as stand-alone facilities. As the scale of the equipment

FIGURE 46.4 Typical kilo lab reactor shown in a downflow booth.

increases, the set of equipment, often called the equipment train, designed to work together for producing a single intermediate or product will expand into multiple rooms. Since the 1990s, it has been common for new pilot plants to be constructed with a "gravity" design where the process will flow from upper floors to lower floors in the direction of gravity (Figure 46.5). In such a design, smaller equipment is typically located in the upper floors, and the larger vessels and filtration and drying equipment are located in the lower floors.

Another design aspect is the use of closed space for a single train of equipment vs. an open floor space without walls or physical barriers between multiple equipment trains. The latter is common to designs of commercial pilot plant facilities where product segregation is supported by procedural controls rather than physical separation between the individual equipment trains for each product.

In stand-alone pilot plants, loading dock areas, proximity of elevators, corridor layout, storage space, air locks, utilities, and processing areas can be addressed in a complimentary design. Material flow, personnel flow, and HVAC design can be engineered soundly with good anticipation of the end user's needs. Dryer discharge areas and milling areas may employ temperature and humidity monitoring and/or control. Like kilo labs, pressure differentials are utilized to mitigate risk of cross contamination from one processing area to another.

Pilot plants utilize a vast array of different surface finishes designed to promote compliance with cGMP requirements and a cost-effective building maintenance life cycle. There is substantial emphasis on corrosion-resistant finishes in wet chemistry areas where surfaces are subject to heavy traffic and potential chemical spills. Isolation and drying areas

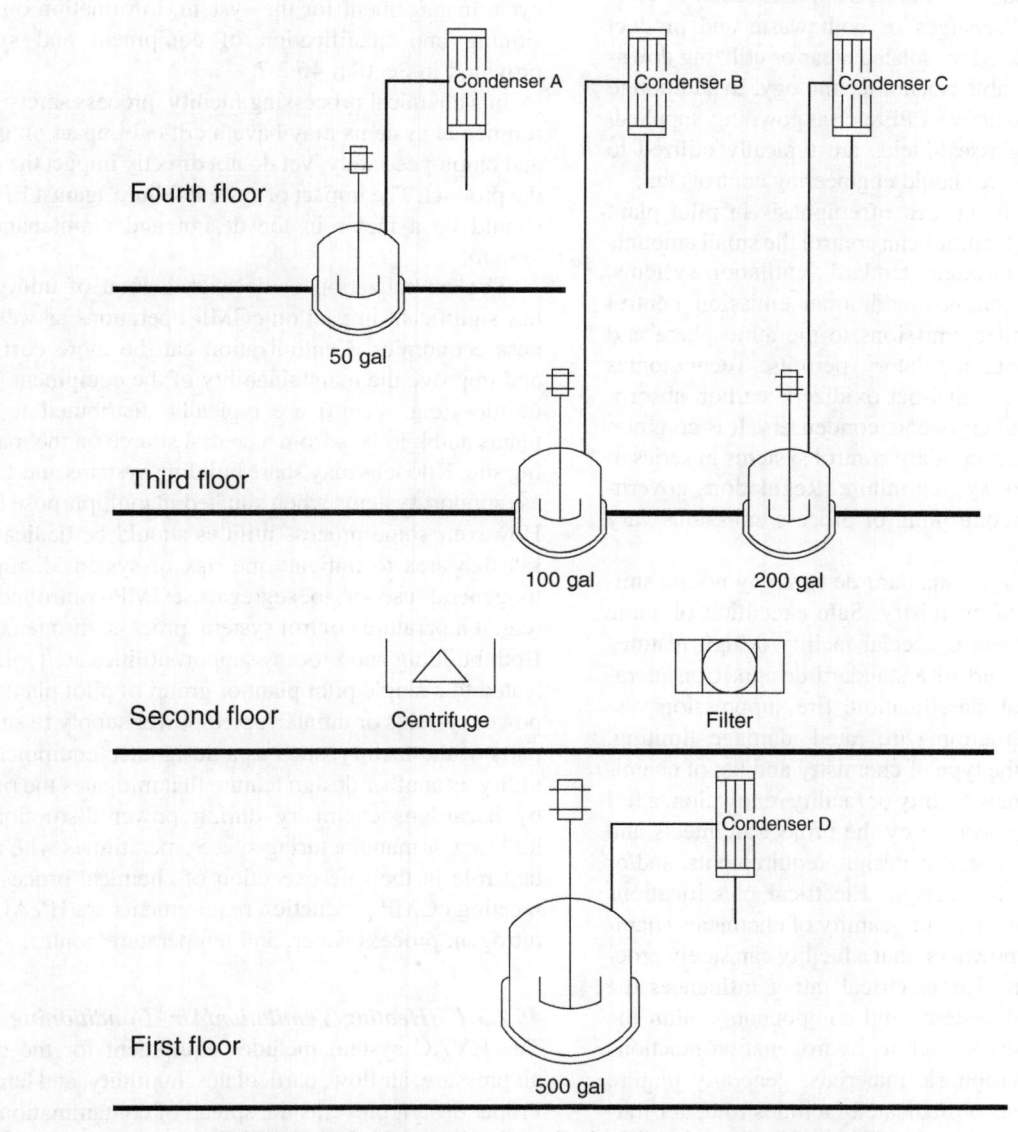

FIGURE 46.5 Schematic of gravity feed flow in a pilot plant facility.

are designed with less concern about corrosive chemical exposure and greater emphasis on finishes that are easy to clean, including floor, ceiling, and wall surface areas.

A critical aspect of chemical processing is to control and mitigate the risk of operator exposure to solvents, reagents, and active ingredients with the use of engineering containment technologies. The scale and size of the equipment in a kilo lab scale may allow for extended use of isolator technology; however larger-scale pilot plant facilities require more use of unit operation-based containment technology. Areas and/or unit operations where materials are introduced to or exiting from the process train require the highest level of engineering controls to prevent operator exposure. Material charges to the equipment may be controlled using a single-use isolator technology or a contained charging area to prevent facility contamination. Sampling of potent or hazardous compounds may include an isolated sample location or local equipment isolator. Discharges of both waste and product streams may be completed in isolated areas or utilizing downflow booths and disposable isolator technology. In addition to the engineering controls above, PPE such as gowning, supplied/filter air, and safety glasses/shields are typically utilized to prevent operator exposure should engineering controls fail.

Emissions control is an area of emphasis in pilot plant design. While kilo lab facilities can control the small amounts of process emissions through standard ventilation systems, pilot plants typically employ additional emission control technologies to minimize emissions to the atmosphere and ensure compliance with regulatory permits. Technologies used include condensers, thermal oxidizers, carbon absorption, gas scrubbers, and cryogenic condensers. It is common to employ primary and secondary control systems in series to minimize the impact of system failure. Regulations governing the reporting and controlling of process emissions vary by country or region.

Pilot plants following a standard design may not be suitable to run all types of chemistry. Safe execution of some chemical processes requires special facility design features that are not generally part of a standard design. Considerations include electrical classification, fire suppression systems, and wall construction (fire rated, damage limiting, blast resistant). Once the type of chemistry and list of chemicals is identified for a new facility or facility renovation, a full code review can be performed by the project architects and engineers to identify specific design requirements and/or gaps against the existing design. Electrical classifications or ratings determine the list and quantity of chemicals (flammable liquids and dry powders) that a facility can safely process within that design. The electrical rating influences the design of all electrical systems and components within the facility. Some chemistries, such as hydrogenation reactions or those involving pyrophoric materials, generally require specially designed areas or dedicated facilities to better mitigate the process safety risks.

46.2.3 Utility Systems

Utility systems generally fall into two groups: building systems and process support systems. Building utilities include electrical systems, potable water systems, utility steam/condensate systems, compressed air systems, fuel oil/natural gas systems, and control systems for process emissions (e.g. thermal oxidizer). Building utility systems do not come into direct contact with the product and generally do not impact its quality. Process support utilities include process gases (e.g. nitrogen, hydrogen), process solvents, process water, and temperature control fluids (e.g. glycol, thermal oils). These utilities influence the process environment through either direct process contact or their effect on process conditions. The impact of the utility on the product quality and the ability to meet cGMP production requirements determines the commissioning, qualification, and life cycle management for the system. Information on commissioning and qualification of equipment and systems is provided in Section 46.2.7.

In a chemical processing facility, process safety and environmental systems may have a critical impact on operations and employee safety, yet do not directly impact the quality of the product. The impact of system failure against EHS criteria should be a factor in the design and maintenance of the system.

The centralization or decentralization of utility systems has significant impact on cGMP operations as well as business economics. Centralization can be more cost effective and improve the maintainability of the equipment [1]. Some utilities (e.g. steam) are typically distributed to the pilot plants and kilo labs from a central source on the manufacturing site. Kilo labs may share building systems and some process support systems when situated in multipurpose buildings. However, some process utilities should be dedicated to the kilo lab area to mitigate the risk of system disruption due to general use or to segregate cGMP-controlled systems (e.g. temperature control system, process nitrogen, vacuum). Both building and process support utilities are typically dedicated to a single pilot plant or group of pilot plants. Backup power systems or uninterrupted power supply to support key parts of the facility, such as a designated equipment train or utility, is another design feature that mitigates the risks posed by hazardous chemistry during power disruptions in the building or manufacturing site. Some utilities with an important role in the safe execution of chemical processes while meeting cGMP production requirements are HVAC, process nitrogen, process water, and temperature control systems.

46.2.3.1 Heating/Ventilation/Air Conditioning (HVAC)

The HVAC system includes equipment for the control of air pressure, air flow, particulates, humidity, and temperature. Proper design prevents the spread of contamination throughout a facility, which in turn provides a layer of protection to

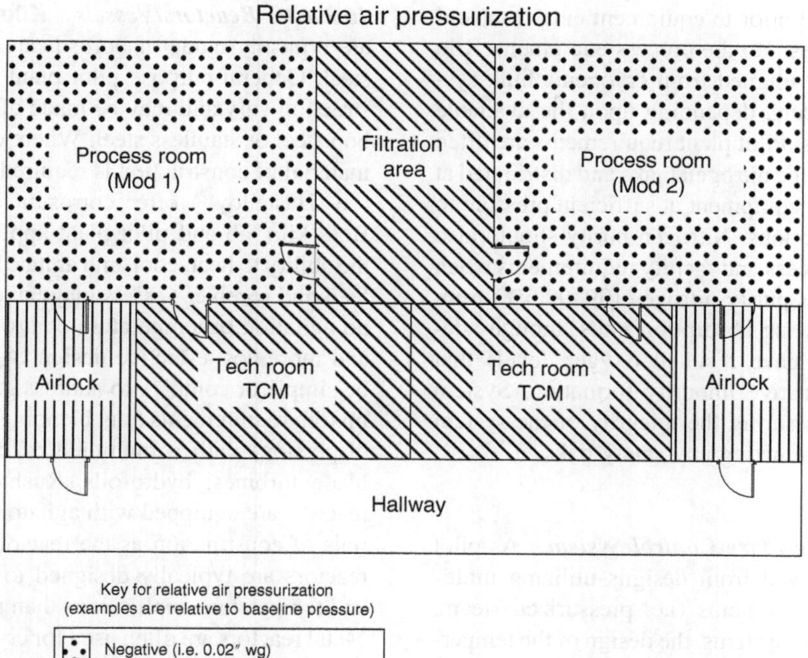

FIGURE 46.6 Example of design pressure in a pilot plant facility.

the worker from airborne compound exposure as well as protection of the product from airborne contamination of other products (cross contamination). The location of the facility and cGMP requirements are important considerations in determination of the needed HVAC capacity and control strategies. Air change requirements are typically 10–15 air changes per hour, requiring large air movement equipment (air handlers and exhaust fans). Considerations for design include the requirements for air balancing within and between manufacturing spaces, directional airflow, environmental control in process areas, and the potential for compound exposure to the environment. In general, the process area will have a negative pressure differential relative to the air lock, the air lock is typically positive relative to both the process area and the corridor, and the corridor is a relatively neutral area (Figure 46.6). Air locks may also be designed to be negative relative to the corridor depending on the room function and company design practices.

46.2.3.2 Process Water The design of the process water system is driven by the quality requirements for the water produced, which are based on its intended use. Several factors influence those requirements including where the water is used in the synthesis (intermediate or API, early or late) and any microbial specifications for the API (sterile or low endotoxin). Water quality standards for pharmaceutical processing range from potable water, to endotoxin reduced, to water for injection (sterile API). There are several guidelines available to guide design of pharmaceutical water systems [2].[1] Some considerations include the water quality and temperature, routine sampling and monitoring requirements, system sanitization and cleaning, and maintainability. Process water systems should be designed for high reliability and availability; system efficiency and cost effectiveness are other important considerations.

46.2.3.3 Process Nitrogen Gaseous process nitrogen has multiple uses in chemical processing. As a critical process safety system, nitrogen is used to inert process equipment used with flammable solvents and explosive dry powders, effectively reducing the potential for fires by displacing air as a source of oxygen from the system (one leg of the fire triangle) [3]. As a process support utility, nitrogen is used to assist several process operations including pressurizing equipment to perform liquid transfers and liquid sparging to remove unwanted dissolved gases. Liquid nitrogen can also be used on the jackets of equipment to achieve low reaction temperatures (i.e. <−15 to −90 °C).

---

[1] United States Pharmacopeia Monograph: Purified Water.

The nitrogen is filtered prior to equipment entry to avoid introduction of particulates into the process equipment. Kilo lab requirements may be sourced from cryogenic bulk systems or individual cylinders, depending upon the available source and capacity needs. Pilot plant requirements are often sourced from bulk cryogenic nitrogen tanks and distributed at high purity to the process equipment at sufficient pressure to support pressurized operations. Nitrogen purity should also be considered against process and facility requirements when selecting a source. Lower purity nitrogen (99–99.5%) from membrane or pressure swing absorption units, though adequate for inertion, may have residual oxygen levels that can increase risk of oxidative impurity formation. System design should consider assuring the integrity of the system and equipment while minimizing any safety risks.

46.2.3.4 Process Temperature Control System As pilot plant facilities have evolved from designs utilizing multi-fluid heating and cooling systems (i.e. pressurized steam, water, brine) to single fluid systems, the design of the temperature control system has become more complex. The temperature control module (TCM) has become an integral component of equipment heat transfer as a means to ensure robust process temperature control. TCM is a term given to a collection of piping, instrumentation, valves, pumps, and heat exchangers designed to act as a unit to provide control of temperature to process equipment, temperature control within one or two degrees from a desired set point. TCM operating ranges vary across the industry based on the heat transfer fluid selected and are chosen based on their range of heat transfer capability, cost, and maintainability. Two frequently used heat transfer fluids are ethylene glycol–water and silicone fluids such as Syltherm XLT™.

For TCMs to function properly, a facility must have a cold loop as part of its infrastructure and an ability to heat (usually via heat exchanger) using electricity or steam. When the equipment requires heating, the TCM pump recirculates the heat transfer fluid through a heat exchanger or electric heater in a closed loop until the desired temperature is reached. When the equipment requires cooling, the system bleeds in fresh fluid from the facility cold loop until the target temperature is reached. Cooling performance is adequate with these types of systems; however, heating performance can be slow, particularly when using electric heaters.

46.2.4 Equipment Design

The design of equipment differs for kilo lab and pilot plant applications, yet there are several common key attributes. As engineering control technologies evolve, the design for kilo lab equipment to address containment has become more similar to pilot plant equipment designs.

46.2.4.1 Reactors/Vessels Kilo lab reactors and vessels are typically made of borosilicate glass or glass on carbon steel. General-purpose pilot plant reactor materials of construction vary and may be made of borosilicate glass on carbon steel or stainless steel. Where chemically resistant metal material of construction is required, high performance alloys (i.e. Hastelloy®) offer corrosion and oxidative resistance. Both kilo lab and pilot plant applications generally require the ability to heat, cool, and agitate vessel contents. However, feed and receiver vessels may have limited capabilities (e.g. no agitation or temperature control) and can offer a simpler and more cost-effective design. Agitator system design can be simple or complex to address general-purpose or specific mixing requirements (e.g. blending, solids suspension, or gas dispersion) with a multitude of options available (pitched blade turbines, hydrofoils, Rushtons, and anchors). Most reactors are equipped with agitators made of the same materials of construction as the reactor they support. Pilot plant reactors are typically designed to achieve higher pressures and temperature applications than their kilo lab counterparts. Metal reactors are often used for cryogenic temperature application (e.g. -30 to $-90\,^{\circ}C$); however, glass or stainless steel vessels can also be used for milder cryogenic applications. Some pilot plants will have reactor systems designed to specifically support gas–liquid reactions, such as hydrogenations, utilizing specific gas mixing or hollow shaft agitator system configurations and pressure ratings.

An often used strategy to increase process configuration flexibility is to standardize reactor nozzle piping designs (i.e. number, type, and size of connections) within a facility or plant. If implemented correctly, the approach will significantly reduce customization of process equipment trains for each new process introduction. A thorough review of the common nozzle requirements for charging, material transfers, sampling, workup (extraction, distillation, concentration), recirculation loops, and PAT instruments should be performed. This information can be incorporated into a standard design that meets the general requirements of most processes.

46.2.4.2 Portable/Miscellaneous Equipment The process design may require the implementation of portable equipment based either on the scale, process requirements, or technology requirements. Portable equipment selection varies by manufacturer and may include, but is not limited to, vessels, filters, wet mills, cone mills, jet mills, PAT (i.e. NIR, Raman, FBRM, etc.), isolators, chromatography equipment, etc. Below is a description of the equipment detailed above and information on their setup and operation:

- Filters: Portable filtration equipment is used to isolate product, insoluble salts or by-products, adsorbents, or contaminants/foreign matter. The filter material, size, and porosity varies based on the processing needs such as the expected amount of material to be removed from the process stream or particle size.

- Wet mills: High-shear rotor–stator mixers or wet mills are designed to reduce the size of larger particle size materials in a heterogeneous (solid/liquid) solution. They are typically employed in a pump-around loop. Common uses of a wet mill may include particle size reduction, crystallization control, improved polymorphic form transformation kinetics, process intensification (reactions), or particle size reduction.
- De-agglomeration/delumping: Mechanical mills such as Co-mills® and mechanical sieve mills are typically used to mechanically break large agglomerates formed during drying following the discharge from drying equipment. Larger material agglomerates may impact the drug product quality and are typically more difficult to process during downstream operations in the drug product manufacturing process and therefore require delumping and/or sieving.
- Micronization: Micronization utilizes mechanical interaction or gas (typically nitrogen) to trigger particle–particle or particle–equipment collisions to produce small particle size materials. Sub 50 μm particle size material may be required to improve drug solubility and dissolution characteristics for poorly soluble drug substance. Examples of micronization mills include pin mills, loop mills, or spiral jet mills.
- PAT: In-line or at-line analytical equipment is used to gain process understanding real time via the use of analytical equipment that is installed "in-line" in the process equipment or process stream (via recirculation loop) or "at-line" close to the sample generation point. The PAT instrument, once validated, can be used as an analytical method to determine in-process unit operation endpoints (i.e. reaction, distillation, drying, etc.).
- Chromatography: Chromatography equipment is utilized to complete large-scale, continuous process purification or separation with packed columns. The high pressure and low process flow rate required for this setup can be challenging and requires experienced technical personnel for the implementation.

46.2.4.3 Isolation Equipment (Filtration and Drying)

As described in Section 46.1, the manufacture of an API proceeds through a number of chemical steps during which some intermediates are isolated in solid forms to obtain purified forms of the intermediate. While the filtration and drying sequence occurs in each case, the operations have increased importance in the final isolation of the API. Both the filtration and drying equipment design and operation play a significant role in the quality and physical properties of the isolated API (see Chapter 35). Filtration and drying operations can have a substantial impact on overall plant throughput and manufacturing costs. Slow filtrations or drying cycles can become the rate-limiting steps for a process, increasing cycle time and reducing availability of the equipment for subsequent batches.

The filtration and drying operations are dependent upon the characteristics of the crystal slurry and are best developed as a coordinated effort with the crystallization process. Kilo lab equipment design is often an extension of the laboratory and should facilitate process troubleshooting and manage a broad range of slurry characteristics. Pilot plant-scale equipment is closer to that found in commercial manufacturing facilities, providing the ability to demonstrate the performance of the operations with different equipment designs.

46.2.4.3.1 Filters Considerations for filter selection include compressibility of the solid cake, susceptibility of the solid cake to agglomeration or attrition upon agitation, compound containment requirements, filter media compatibility, occupational health concerns, the design of the equipment train, and development objectives to inform equipment selection for manufacturing. Commonly used filters include filter dryers, centrifuges, and Nutsche filters:

- Filter dryer: The filter dryer is found in both the kilo lab and the pilot plant and combines filtration and drying unit operations within a single unit. As a single plate filter, it is fitted with an agitator, which can be raised and lowered as needed, and a jacket to facilitate the drying operation. The filter media may be a polymeric cloth (e.g. polypropylene), single layer screen, or sintered metal. In some designs, the agitator and filter base can also be heated, increasing the heated contact surface area. The agitator is used to mix, reslurry, and smooth the cake during the filtration and washing protocol, minimizing crack formation and increasing the efficiency of the filtration and wash operations. During drying, the agitator is used to increase the uniformity of the cake temperature and efficiency of the solvent removal. The agitator also assists in the removal of the cake during discharge through the discharge port. Engineering controls can be added to the discharge configuration to avoid worker exposure to the product. The closed system design from slurry entry through discharge has made this a frequent choice for potent and highly potent compounds.
- Centrifuge: Two batch centrifuge designs commonly found in pilot plants are the peeler centrifuge and the inverting bag centrifuge. Both have the same sequence of operations with the exception of the discharge. The peeler centrifuge is defined by its mechanical peeling action during discharge. The inverting bag centrifuge incorporates an automatic discharge achieved by

inverting the bag (turning the bag inside out), allowing the solids to discharge. The inverting bag centrifuge offers a more efficient and faster heel removal than the peeler centrifuge. Centrifuges are commonly found in the pilot plant but are used less frequently in the kilo lab. The most often used centrifuge in the kilo lab is a basket centrifuge. As the name implies, the slurry is fed into the top of the "basket" and manually discharged.

- Nutsche filters: Nutsche filters are single plate filters that may include agitation and/or temperature control and are commonly used in both pilot plant and kilo lab operations. The absence of the agitator may require the operator to manually mix and smooth the cake during filtration. The potential for operator exposure to dust and flammable solvent during use is increased, and additional engineering controls and/or PPE are needed to provide adequate worker protection. Older designs also require manual removal of the cake upon discharge. More recent designs include engineering controls to reduce the risk of operator exposure during filtration and discharge (Figure 46.7).

46.2.4.3.2 Dryers Drying equipment used in pilot plants are generally contact dryers (solid in contact with heated surface) and operated in batch mode. Considerations for dryer selection include the acceptable drying time, the impact on powder properties (i.e. prevention of particle agglomeration, particle attrition, or crystallographic form change), and the explosibility of the dry powder. Commonly found types of dryers include tray dryers, agitated dryers, tumble dryers, etc. Drying equipment used in kilo labs is generally limited to tray dryers and filter dryers. Pilot plants typically standardize on a few different types of dryers, harmonizing where possible with commercial manufacturing. PAT (Raman, IR, mass spectroscopy) can also be incorporated to the dryer to monitor the off-gas composition or solids (e.g. polymorphic form) to monitor the drying process and determine drying endpoint:

- Tray dryer: A tray dryer is a large oven connected to a vacuum pump or inert gas supply. Product is loaded onto trays (metal or glass), which are placed onto shelves in the dryer. Sampling is achieved by interrupting the drying cycle and removing material from the trays. The dry product is manually unloaded from the trays into product containers. The tray dryer design offers little containment, and additional engineering controls are needed to provide worker protection. Like Nutsche filters, tray dryers are typically not used for other than laboratory amounts of potent compounds unless additional containment is included in the equipment design.
- Agitated dryer: The agitated dryer is also usually connected to a vacuum pump during the drying operation and typically has a sampling port on the side of the filter wall that facilitates contained sampling operations. Typical agitated dryer designs include conical dryers,

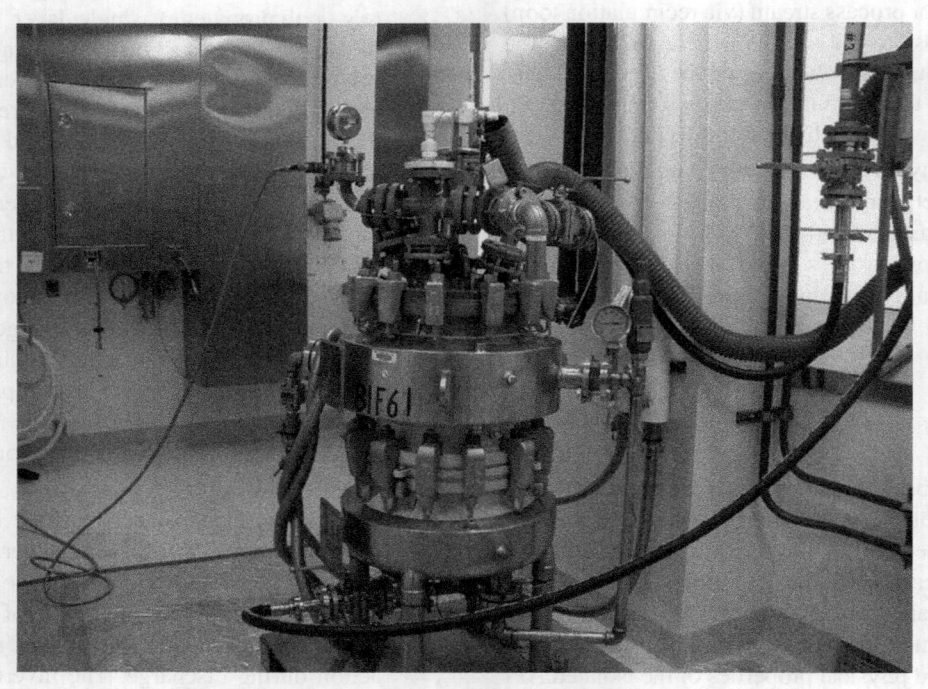

FIGURE 46.7 Nutsche filter for kilo lab or pilot plant use.

spherical dryers, and pan dryers with screw or blade agitations offering different shear profiles and attrition risk. A standard agitation protocol (intermittent vs. continuous, agitation rate) can be deployed and then varied as needed based on the project need to achieve the desired drying endpoint and powder properties. Multiple types of sampling ports are available to allow for monitoring of the drying progress.

- Rotary dryer: The rotary or double-cone dryer, similar to the tray and filter dryer, is connected to a vacuum and heating source. In place of agitation, the rotary dryer rotates axially to gently mix the product during the drying unit operation.

Product recovery yield from drying equipment can vary based on the size and configuration of the dryer. Due to the possibility of large losses to the dryer (10–20% yield), product recovery may be an option to ensure campaign deliverables are met. The recovery can be performed through the dissolution of the remaining product left within the equipment. This is often referred to as heel dissolution. The equipment design may be inclusive of the ability to recover dissolved product solutions for further processing including sprayball charge lines and liquid-sealed discharge capabilities.

46.2.4.4 Equipment Cleaning Design Equipment, although designed for the needs of the process, also incorporate post-process cleaning and contamination removal. Since drug substance manufacturing typically is performed in multiproduct equipment and facilities, process equipment must have the ability to prevent cross contamination from past processes. The equipment design may include the use of fixed or dynamic spray devices, or sprayballs, to ensure total wetting of the interior surfaces, inclusive of the reactor riser and condenser. In cases where sprayballs are not incorporated into the design, alternatives such as solvent reflux or liquid filling of the equipment may be necessary to rinse and wet all surfaces. Additionally, equipment may require some disassembly to complete a thorough hand cleaning and visual inspection.

46.2.5 Engineering Control Equipment

Engineering control equipment provides the primary level of protection for the worker from the hazards associated with chemical or pharmaceutical exposure during process operations. The engineering controls should provide protection for the specific operations where the operator exposure levels (OEL) are above a threshold of concern for an individual chemical, intermediate, or API (see Section 46.3). Typically the operations with the greatest potential for worker exposure are charging, discharging, sampling, milling, and cleaning operations. The standard designs for equipment used in these operations do not provide sufficient containment on their own, and another layer of protection is needed. Common types of engineering controls found within pilot plants and kilo labs are flow hoods, barrier isolation technology, solids transfer technology, and sampling technology.

46.2.5.1 Flow Hoods Flow hoods employ various technologies to achieve their containment target. Common types of flow hoods found in the pharmaceutical industry are downflow hoods and chemical fume hoods. Downflow hoods provide a high level of solids containment by directing the flow of particulates in the air away from the worker and into the hood exhaust system and particulate filtration system. Chemical fume hoods provide a constant flow of air from the front of the hoods that is exhausted to the atmosphere directly or via a filter. Fume hoods are effective for reducing exposure levels from chemical vapors and tend to be less effective for solids containment compared with downflow hoods. The small scale of equipment used in kilo labs often enables installation of the reactor vessels within the fume hood or downflow hood, allowing for sufficient worker protection during operations presenting the greatest risk of exposure. Figure 46.8 shows an open filter design often found in kilo labs. The filter design enables manual manipulation of the wet cake and facilitates troubleshooting during filtration yet offers no protection to the worker from the product. The filter must be placed in a downflow booth or used with other engineering controls (including PPE) to ensure adequate worker protection.

46.2.5.2 Isolator Technology Barrier isolation technology directly separates the worker from the exposure hazards via a temporary or permanent wall. The barrier isolator may resemble a box and enables the worker to perform all of the

FIGURE 46.8 Open filter often found in kilo labs.

necessary standard operations (i.e. charging, discharging, sampling) through the use of glove ports and pass through ports through the walls of the box. This technology may be incorporated into standard types of equipment such as reactors, filters, and mills at both the pilot plant and kilo lab scales to perform operations involving potent or highly potent compounds (see Figure 46.2c). Additionally, isolator technology may be used to handle hazardous solids that require inerted conditions. Further information is provided in Section 46.3. Additionally, disposable isolators may be utilized in place of the fixed isolator design that allows for a lower capital investment and ease of cleaning (see Figure 46.9). Barrier isolators contain the particulates from the outside environment through the use of high efficiency particulate air (HEPA) filters on the exhaust. Where needed, the degree of containment achieved can be increased by maintaining the internal environment in the isolator under a negative pressure differential. Isolator technology is commonly used today where primary equipment design is insufficient to provide sufficient worker protection during operation.

46.2.5.3 Solids Transfer Technology Solids transfer technology significantly reduces the worker exposure levels resulting from the charging/discharging operation. The solids are pre-charged/transferred into the charging bags in a contained location prior to processing within the chemical processing facility. The technology allows material charging and discharging operations to be performed in a closed system preventing direct worker exposure to the compound as well as the room contamination that results from open handling. Some common solids charging and discharging technologies used in the pharmaceutical industry are barrier isolators, split butterfly valve technology, powder transfer systems, disposable containment systems, and continuous liners. The barrier isolator can be installed as a hard wall or soft wall isolator at the equipment charge or discharge ports to provide the desired level of containment. Split butterfly valve technology supports low operator exposure through the use of active and passive valves. The active and passive valves are mounted separately to the charge/discharge vessel nozzle and the equipment as shown in Figures 46.10 and 46.11. The solids transfer operation can proceed only when the active and passive valves are joined or docked together. Continuous liners made of pliable material are used to increase the flexibility of solids containment. A continuous liner mounted as the receiving vessel can be crimped to isolate an amount of solids and then cut to form a separate bag containing the solids without breaking containment.

46.2.5.4 Sampling Engineering controls used during sampling operations are designed to minimize operator exposure potential during the direct sampling of reactors via pressure or caused by sample spills. This is of increased importance when the intermediates and/or APIs are highly potent, have substantial toxicological concerns, or are at high temperature or pressure conditions. To minimize operator exposure during the sampling operation, barrier technology,

FIGURE 46.9 Disposable isolator technology for a mill.

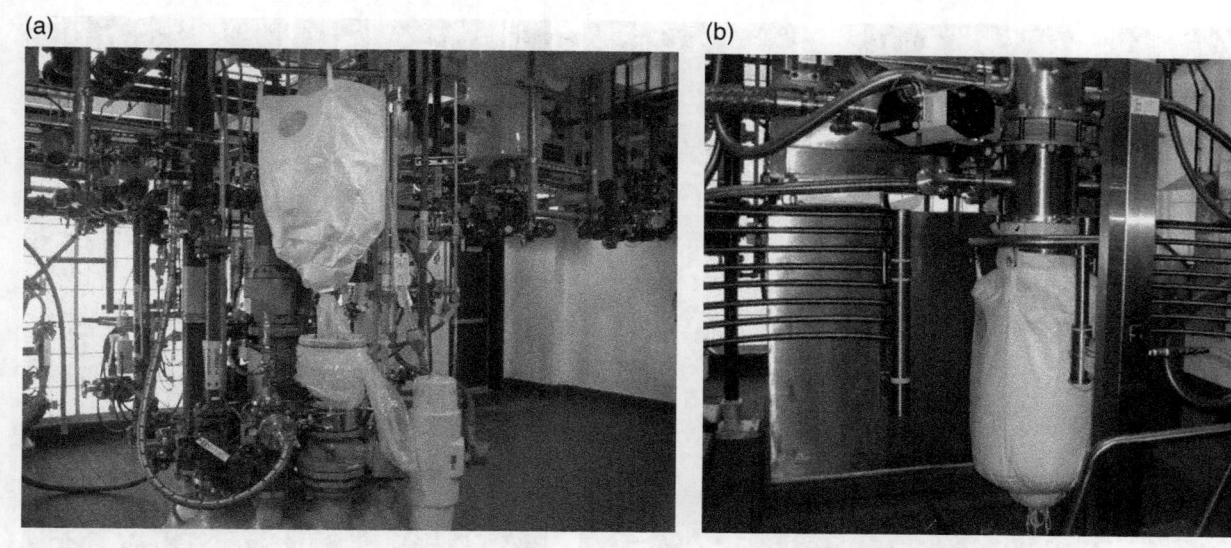

FIGURE 46.10 Solids handling using split butterfly valve technology. (a) Material charge into a reactor using a disposable isolator. (b) Discharging the wet cake from a centrifuge into a flexible liner.

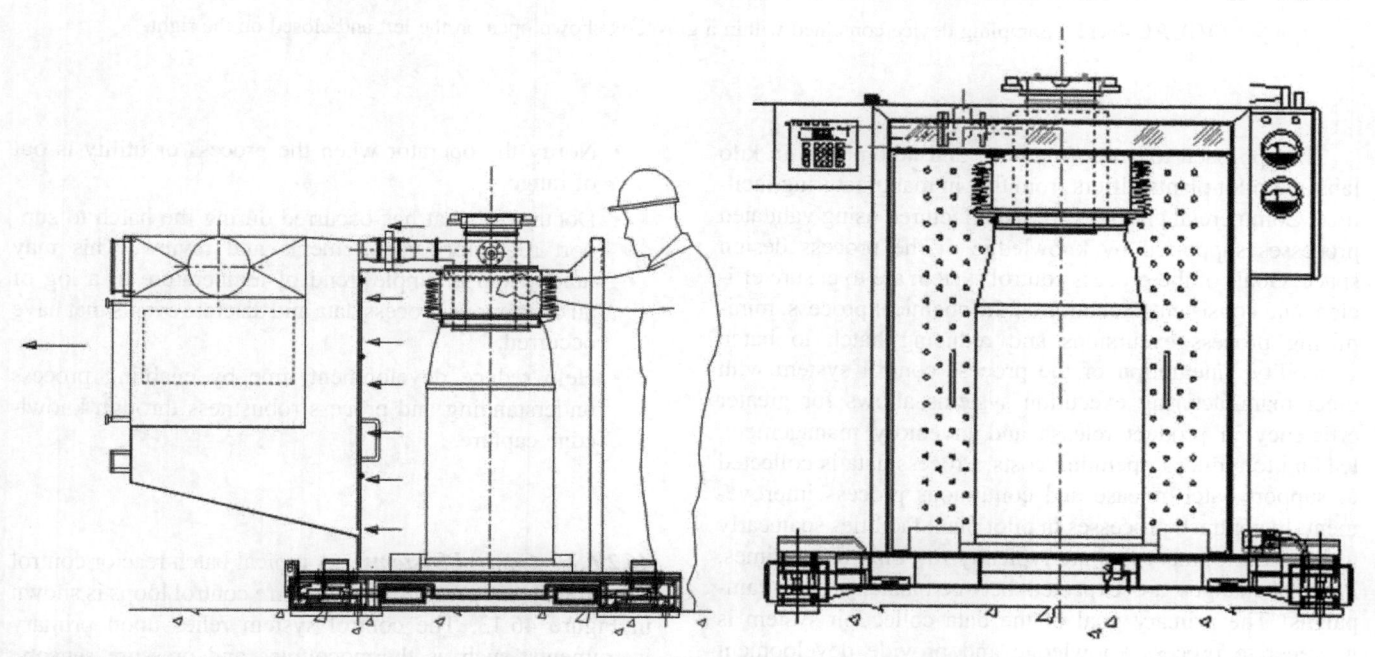

FIGURE 46.11 Schematic of closed discharge system consisting of discharge drum in laminar flow booth connected by continuous liner.

closed vent sampling devices, or similar technologies may be installed within the scale-up facilities as shown in Figure 46.12.

46.2.6 Process Automation

Today, most pilot plant facilities have modern process control systems that provide basic control of primary process parameters such as temperature and pressure. Less common to pilot plant facilities are the recipe-driven process automation systems (recipe control) found in manufacturing, which may feature advanced process control and business systems interfaces. Most pilot plants will also have data historians to capture process data (i.e. pressure, temperature) as well as events (i.e. valve opening, changed set points). Kilo labs that are an extension of laboratory systems may have basic control for reactor temperature and pressure, while others deploy process control systems for regulatory control. Some kilo labs have the extensive process control systems found in pilot plant facilities.

FIGURE 46.12 Sampling device contained within a glovebox shown open on the left and closed on the right.

The approach to process control and automation in kilo labs and pilot plants differs from that in manufacturing facilities. Commercial products are manufactured using validated processes supported by knowledge of the process design space. Goals of the process control system are to ensure efficient and consistent execution of the chemical process, minimizing process excursions and reducing batch to batch variability. Integration of the process control system with other manufacturing execution systems allows for greater efficiency in product release and inventory management, leading to reduced operating costs. Process data is collected to support batch release and continuous process improvement. In contrast, processes in pilot plant facilities span early to late development and are typically run only a few times. Process changes are expected between batches and campaigns! The primary goal of the data collection system is to generate process knowledge and provide development scientists access to process data. The process control system should be designed for flexibility and robustness to accommodate changing requirements from each process, to facilitate process troubleshooting, and to enable chemical operators to focus on the process.

Kilo lab and pilot plant process control system have the following general goals:

- Provide control of process and utility parameters to maintain safe operation and reduce variability of the process. This will reduce the risks encountered when scaling process chemistry from one scale to the next.
- Notify the operator when the process or utility is out of range.
- Document what has occurred during the batch to support regulatory requirements and review. This may range from a simple trend of temperature to a log of all continuous process data and discrete events that have occurred.
- Help reduce development time by enabling process understanding and process robustness through knowledge capture.

46.2.6.1 Control Systems A typical batch reactor control system with temperature and pressure control loops is shown in Figure 46.13. The control system relies upon primary instruments, such as thermocouples and pressure sensors, and control elements, such as control valves, which are wired back to stand-alone controllers or to a computerized process control system such as a DCS where hundreds of control loops are executed. These sensors also facilitate capture of continuous data (e.g. temperature) and discrete-event data (e.g. when does a valve open and close during a charge).

The basic regulatory control process, PID (proportional–integral–derivative) controller, is still used in most pilot plants though there are more advanced approaches such as model predictive control. In modern pilot plants, the flexibility inherent in these digital systems not only allows for plug and play of various unit operations equipment but also allows

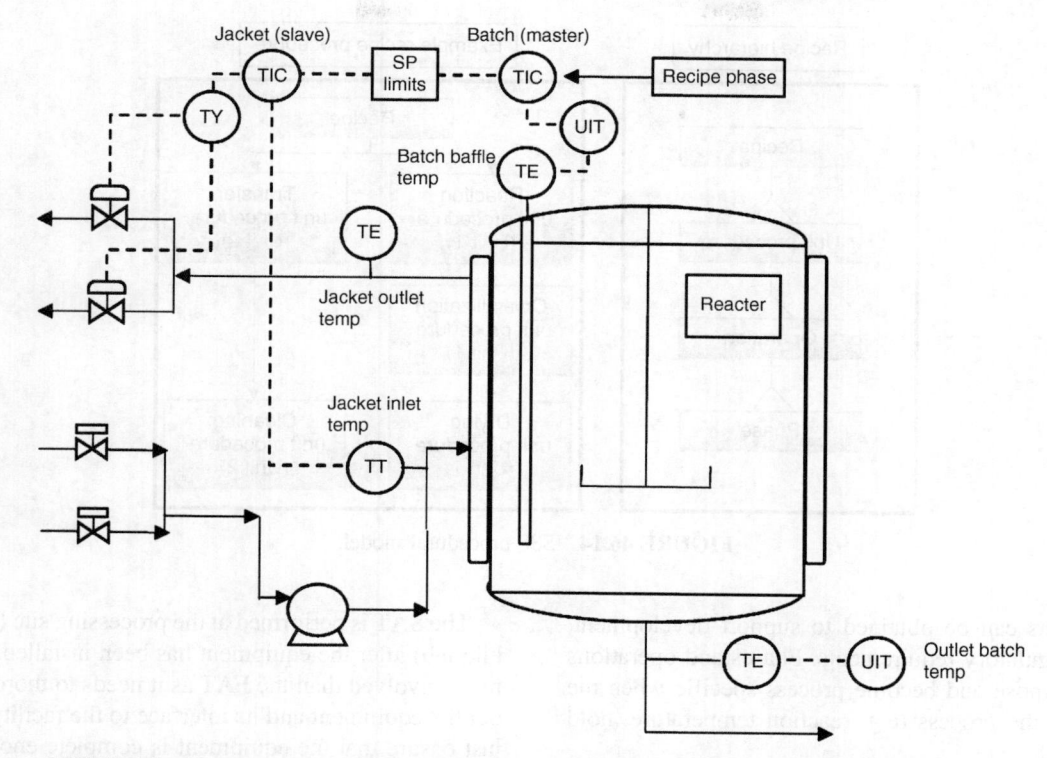

FIGURE 46.13 Batch reactor with typical split range configuration to control batch temperature.

the control strategies to be easily changed to meet the needs of each product, facilitates regulatory compliance, and supports generation of process knowledge.

46.2.6.2 Recipe-Driven Batch Control
A "batch" consists of a sequence of unit operations to produce a product. For example, to carry out a reaction, a typical sequence of events is shown in Table 46.1.

The instructions or recipe to execute the sequence is written in the control software and includes the parameters specific to the process. The recipe will contain set points and alarm limits for each step. It can also select the desired control strategy, such as jacket temperature or batch temperature control.

In 1995, the ISA-S88 batch manufacturing standard was issued describing the definition and control of batch processes.[2] At the core of the standard is the separation of the product knowledge from equipment capability, the "S88" model.

"S88" provides a logical, consistent structure for building batch recipes as shown in Figure 46.14, which combines these elements. The "recipe" sequences the different "unit procedures" such as those described in Table 46.1. Each unit

[2]ANSI/ISA88.00.xx (IEC61512); the Instrument, Systems and Automation society sponsored the creation of the S88 standard and supports the expansion and development of industrial instrumentation and control expertise for process and other industries.

TABLE 46.1 Typical Sequence of Events

Sequence	Event
1	Inert the vessel with nitrogen
2	Charge solvents, substrates, and reagents to vessel
3	Turn on agitator
4	Heat to reaction temperature
5	Hold until reaction is complete as determined by sampling
6	Sample by pressurizing vessel, withdrawing sample, and depressurizing
7	Cool the vessel and crystallize product
8	Transfer the product to filter
9	Dry the product

procedure consists of a series of "operations" (e.g. heat, hold, sample, cool) that contain parameters (e.g. set points and limits) specific to that step. Each operation then consists of a series of "phases" that instruct the equipment-specific elements to carry out that operation (open valve X-101 on reactor R-101). During the execution of the recipe, the control system captures what was done (operations and phases), when it was done, who did it (by electronic signatures), and how it was done (process data capture of continuous, temperature and discrete data, weight of a drum charge).

With a control system using an S88 recipe approach and a process and event data historian, a complete record of the

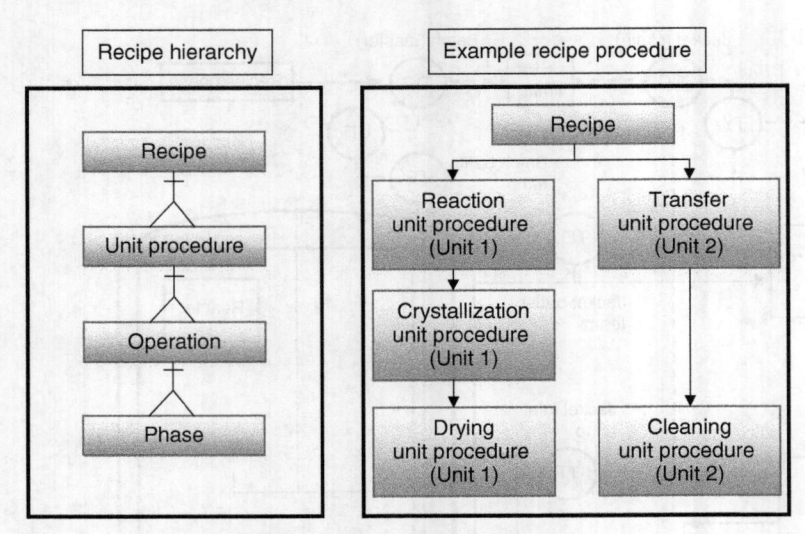

FIGURE 46.14 S88 procedural model.

batch operations can be obtained to support development, quality, and regulatory requirements. Phases and operations are general purpose and become process specific when the parameters for the process (e.g. reaction temperature, hold time) are added.

46.2.7 Equipment Commissioning and Startup

When starting up a new or renovated facility or new equipment in an existing facility, the building systems and process equipment must be tested to ensure that the design specifications are met. This is required for adherence to regulations (cGMP and EHS) and good engineering practices (GEP). The standard processes to ensure that a system or piece of equipment is started up, qualified, and rendered ready for use in a pharmaceutical environment consist of commissioning, qualification, and validation (CQV). The CQV plan includes acceptance test criteria and an impact analysis to evaluate the impact of the system or equipment on the quality of the product produced (direct, indirect, or no impact).[3] The outcome of the impact assessment determines the extent of testing required for an individual system or equipment.

The Factory Acceptance Test (FAT) entails sending a team to the equipment manufacturer's location to execute an approved document geared toward assessing that the equipment can perform as specified. The goal is to ensure that the equipment is suitable to be shipped and installed.

Commissioning is an engineering approach of bringing a facility and/or equipment into an operational state. For qualified systems, commissioning activities can overflow into the Site Acceptance Test (SAT). For a nonqualified system, the commissioning is usually more extensive since there is no subsequent activity to ensure operational readiness.

The SAT is performed at the processing site (pilot plant or kilo lab) after the equipment has been installed. It is usually more involved than the FAT as it needs to thoroughly check out the equipment and its interface to the facility, rather than just ensure that the equipment is complete enough to allow for shipment. While it is possible to leverage the work at the FAT to reduce the SAT activities, careful consideration must be given to the installation work and how the final facility conditions differ from the factory (vendor) conditions.

The next phase of the process is the qualification, which is captured in a series of protocols to confirm the installation, operation, and performance of the equipment. Installation qualification (IQ) confirms that the correct equipment has been installed as specified in the design documentation. It is a documentation review that compares nameplate data (serial number, model number, pressure rating, etc.) with the appropriate specification to ensure compliance with the design. Additional information is also collected for analysis and future use including materials of construction, construction techniques, testing, and similar information. The operational qualification (OQ) is a protocol to systematically test the equipment to ensure that it can perform according to the user requirements for the equipment. Equipment classes tend to have similar check out requirements for individual equipment within that class. Examples of reactor class testing include pressure, temperature control, and agitation testing.

The performance qualification (PQ) is perhaps the most detailed manner of check out due to the inclusion of a test specification. Water and nitrogen system are the two most common systems that require PQs in the pilot plant and kilo lab. This is to ensure that materials charged to the batch (i.e. water) or in contact with the batch (i.e. nitrogen for inerting) meet the required quality specifications.

Successful startup and qualification of equipment and facility systems requires a systematic approach. The

---

[3]ISPE Baseline Guide for Commissioning, Qualification, and Validation.

employment of FAT, SAT, IQ, OQ, and PQs provides a consistent, methodical means to prove facility integrity. Subsequent changes to the facility are managed using a change management program to ensure that facility and equipment integrity is maintained.

46.3 OPERATING PRINCIPLES AND REGULATORY DRIVERS

The design and operation of scale-up facilities in the pharmaceutical industry is influenced by a matrix of regulatory guidelines. Current good manufacturing practices (cGMPs), established to ensure the quality of the drug delivered to the patient, overlay EHS regulations at the local, state, and federal level established to protect the worker, the community, and the environment. cGMPs must be followed when producing API for clinical studies to ensure that the quality and purity of the drug substance is adequately controlled. Guidelines for cGMPs are governed by the regulatory health authorities, such as FDA and the EMA. The guidelines of each agency may differ, and substantial effort has gone into establishing a consistent set of practices across agencies through the International Conference on Harmonization (ICH).[4] Scale-up facilities located in the United States are governed by EHS agencies that include the Environmental Protection Agency (EPA)[5], Occupational Safety Health Authority (OSHA)[6], and equivalent state agencies. Equivalent agencies govern chemical process operations outside of the United States. In many cases, the design of a compliant EHS programs can be structured around the scale of operations. Kilo labs are commonly classified as laboratories, which enables these facilities to fall under the local, state, and federal regulations for laboratories. In most cases, pilot plants operate at a scale that precludes laboratory classification and must comply with regulations established for plant scale of operation and purpose. Additional EHS regulations, such as the Toxic Catastrophe Prevention Act or TCPA in the United States, apply where the quantity of listed hazardous chemicals exceeds a threshold value.

The expectations set by cGMP and EHS guidelines are shared in many areas. Most companies develop an integrated approach to operating procedures, work practices, and staffing to more effectively meet safety and quality requirements. Once established, robust training, documentation, and monitoring programs ensure compliance against regulations.

46.3.1 Process Safety

Prior to the implementation of an intermediate or API process step into a kilo lab or pilot plant, the chemistry and process details must undergo a process safety evaluation. The safety evaluation should provide data to determine the thermal hazards of the solids, key process streams, and any relevant off-gassing data. The safety evaluation should also examine the physical properties, toxicity hazards, fire and explosion hazards, and hazardous interactions between different chemicals to determine the safety risks associated with running the process in a scale-up facility. For projects that are further in development, a safety evaluation on the powder properties of any isolated intermediates and the API should be completed. The complete evaluation should be used to determine the intrinsic safety of the process design and whether a process can be safely implemented at the desired scale.

During the transfer of process knowledge from the process development group to the plant operations group, an assessment of the process hazards should be completed following a standard methodology to identify the process safety risks of execution into the specific scale-up facility. Some common methodologies used for the hazard assessment are a what-if analysis, checklist of hazards, process hazard analysis (PHA), hazard and operability study (HAZOP), failure mode and effect analysis (FMEA), or fault tree analysis. The data from the safety evaluation, plant equipment details, plant procedures, and the intended process batch instructions should be used to develop the safety assessment. The intent of the assessment is to address hazards from the process, consequences of failure from engineering and administrative controls within the plant, and potential human errors. Any process changes proposed after the initial hazard assessment is completed should be examined prior to introduction to the scale-up facility using the facility management of change policy to ensure that all safety precautions are taken. Further information is available in Chapter 3 and related.[7,8]

46.3.2 Current Good Manufacturing Practices

The Quality Management section of ICH Q7A[9] specifically states that "quality should be the responsibility of all persons involved in manufacturing" and that each individual manufacturer is responsible for establishing and documenting effective quality systems to manage quality of products produced both internally and through external partners. cGMP can be classified into six quality systems: facilities and equipment, production, packaging and labeling, material systems,

---

[4]International Conference on Harmonization of Technical Requirements for Registration of Pharmaceuticals for Human Use, ICH Q7A.
[5]www.epa.gov.
[6]www.osha.gov.
[7]Guidelines for Process Safety in Batch Reaction Systems, AICHE Center for Chemical Process Safety.
[8]Guidelines for Process Safety Fundamentals in Plant Operations, AICHE Center for Chemical Process Safety.
[9]International Conference on Harmonization of Technical Requirements for Registration of Pharmaceuticals for Human Use, ICH Q7A.

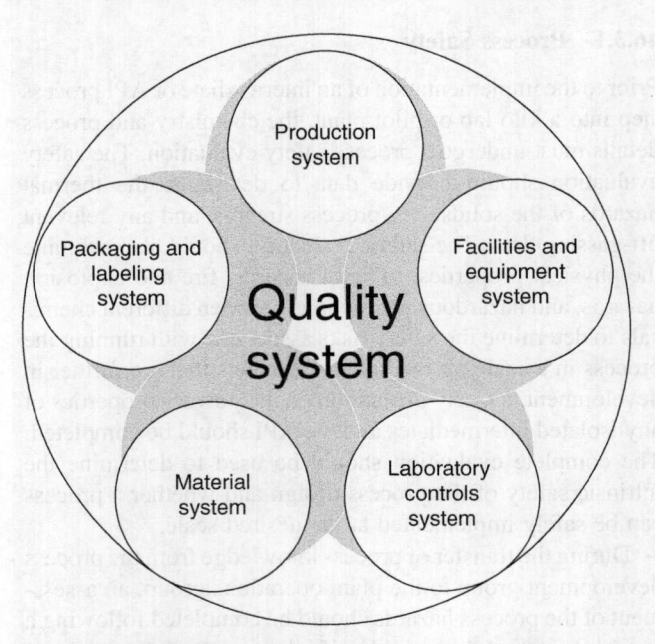

FIGURE 46.15 The six system approach to quality.

quality, and laboratory controls. Each requirement listed in ICH Q7A will map to one of the six systems (Figure 46.15).

These systems address review of completed manufacturing records for critical process steps prior to the release of API, making sure that equipment used for manufacturing is maintained, calibrated, and fit for its intended use and that the proper level of documentation for all activities is reviewed, complete, and accurate. An internal audit program designed to ensure that a state of compliance is maintained must be established, including a defined audit frequency with documented audit findings and corrective and preventative actions (CAPAs) to address identified quality risks.

A sound personnel training and qualification program is an important element of the cGMPs. The roles and responsibilities for employees associated with pharmaceutical manufacturing facilities and operations should be defined, documented, and maintained. Requirements for education, experience, general training, job function training, proper hygiene and sanitation habits, and suitable clothing worn during manufacturing or lab operations are included. A comprehensive and well-documented employee qualification program assists operations managers in assigning the right employee with the right skill set to the right task in compliance with cGMP.

cGMP considerations for scale-up facilities and equipment are described in Section 46.2 and include design, construction, and startup activities. Facility and equipment design and construction should facilitate maintenance and cleaning, minimize contamination of the product, and provide adequate control systems. Materials of construction for product contact equipment are also an important consideration. Materials that are incompatible with process streams can contribute to quality issues, potentially rendering the product unusable. Process safety issues may also result where the introduction of contaminants into the process stream causes an unexpected reaction. CQV activities are completed to demonstrate that the facilities, equipment, and systems were constructed as designed and are fit for their intended use in the manufacture of APIs and chemical intermediates. Additional facility considerations include material management such as material receipt, identification, sampling, storage, and separation of released and unreleased input materials, intermediates, and APIs.

Proper facility and equipment qualification should be maintained through a change management program for systems and equipment that impact the quality of the API and/or chemical intermediates. Written procedures for facility and equipment maintenance and cleaning operations should be designed to promote consistent execution over time independent of an individual operator.

Documentation and records is one of the most visible cGMP requirements. This is the primary evidence of "what, when, who, and how" as it relates to an operation or task. Documentation requirements vary by task and may include what task was performed, who performed it, when it was performed, how it was performed, and reconciliation of any deviation from the expected procedure. The documentation is the link to ensure that an operation or task was completed as intended and that the impact of any deviation on the quality of the outcome is evaluated and understood. cGMP documentation includes equipment cleaning records, laboratory (analytical) controls, training records, and batch records, which document the execution of the manufacturing process. Batch records contain input material information, a description of executed unit operations, analytical control results, and process data. They should also contain an adequate explanation of any process deviations that occurred during processing. Any changes to batch records during execution should follow a change management program that includes appropriate comment to reconcile the change with the original entry.

46.3.3 Health and Safety

Kilo lab and pilot plant operations include the use of hazardous chemicals that may be toxic, flammable, corrosive, and/or explosive. The health and safety of the employees must be the highest priority, and worker protection programs should be in place before hazardous chemicals are introduced into the facilities. Regulations governing worker and workplace safety can vary substantially across the world, and many companies have established a minimum standard for all of their facilities to supplement country, regional, and local regulations. In the Unites States, OSHA sets standards to ensure worker and workplace safety; the standards can be found

TABLE 46.2 Some US Health and Safety Regulations for Kilo Labs and Pilot Plants

Topic	Regulation Standard (29 CFR)
Emergency action plans	1910.38
Occupational noise exposure	1910.95
Process safety management of highly hazardous chemicals	1910.119
Hazardous waste operations and emergency response	1910.120
Permit-required confined spaces	1910.146
The control of hazardous energy (lockout/tagout)	1910.147
Hazardous (classified) locations	1910.307
Air contaminants	1910.1000
Hazards communication	1910.1200
Occupation exposure to hazardous chemicals in laboratories	1910.1450
Hazardous materials	1910 Subpart H
Personal protective equipment	1910 Subpart I
Fire protection	1910 Subpart L

under the Regulation Standard 29 CFR 1910 Occupational Safety and Health Standards.[10] Table 46.2 lists some of the regulations under 29 CFR 1910. A chemical hygiene plan (CHP) or similar program is typically developed for kilo lab and pilot plant facilities to ensure compliance with OSHA standards. The plan or program consists of policies and procedures around specific topics that define the operational safety framework, including hazard communication, occupational exposure, PPE, and the procurement.

46.3.3.1 Hazards Communications

A well-defined hazard communication procedure is essential to understanding, accessing, and communicating the hazards associated with chemicals used in the kilo labs and pilot plants. Required elements include a hazard determination process, a procedure for management and use of material safety data sheets (MSDS), proper container labeling requirements, and effective information and training on hazardous chemicals.

The hazard determination process should involve the evaluation of chemicals purchased for use in and produced by the kilo lab and pilot plant facilities to determine if they are hazardous. Hazard information for chemicals that are purchased is contained in the MSDS provided by the vendor. For compounds produced as part of the research and development process, the company producing the compound is responsible for generating the MSDS. MSDS information for all hazardous chemicals used in the scale-up facilities should be made available to the operating staff.

An MSDS contains the chemical name of the compound as found on the label and other common names in addition to any available hazard information. If the material is a mixture, the MSDS will list the major components and nominal composition. The types of hazard information commonly found are related to: the properties of the material, such as toxicological information, stability and reactivity, exposure control, environmental impact (i.e. ecological concerns and disposal procedures), and procedures to handle the material (i.e. storage, accidental release measures, and transportation information). There are also additional sources for hazard information beyond that found on the MSDS [4].[11]

Proper labeling of hazardous chemical containers is another vehicle for hazard communications. The identity of the chemical, the appropriate hazard warning, and the name of the chemical manufacturer or other responsible party must be listed on the container. The hazard warnings may take the form of words, pictures, symbols, or a combination. It is important to note that the requirements of the hazard communication regulation do not supersede the labeling requirements of other US government regulations (i.e. DOT, EPA).

The hazard information for hazardous chemicals and appropriate training must be made available to all employees and contractors who have the potential to come in to contact with the chemical. The information and training should include the details of the hazard communication program, the physical and health hazard of chemicals in the labs or plant, and the methods of protection to prevent chemical exposure.

46.3.3.2 Occupational Exposure for Pharmaceutical Compounds

Companies are responsible for developing the hazard information for research and development compounds. There is often limited information on the compound early in development. Therefore, a variety of methods, including toxicology studies and chemical structure-based assessments, are often used to develop the information needed to evaluate the potential hazards to employees within the laboratories and scale-up facilities. An evaluation of the available data is typically performed by an industrial hygienist, who assigns an exposure control limit (ECL) to the compound if sufficient data is available. It is often not feasible to assign a single number ECL early in pharmaceutical development, and most companies use a performance-based exposure control band strategy, assigning the compound into a control band. The ECL for each control band are set with consideration of engineering control technologies, PPE, procedural controls, and other available tools to mitigate the risks of worker exposure as required [5]. The data generated from the toxicological evaluation and testing is typically added to the MSDS for the API or intermediates.

---

[10] 29 CFR 1920 Occupational Safety and Health Standards.

[11] 29 CFR 1920 Occupational Safety and Health Standards.

TABLE 46.3 NIOSH Control Banding

Band No.	Target Range of Exposure Concentration	Hazard Group	Control
1	>1–10 mg/m^3 dust >50–500 ppm vapor	Skin and eye irritants	Use good industrial hygiene practice and general ventilation
2	>0.1–1 mg/m^3 dust >5–50 ppm vapor	Harmful on single exposure	Use local exhaust ventilation
3	>0.01–0.1 mg/m^3 dust >0.5–5 ppm vapor	Severely irritating and corrosive	Enclose the process
4	<0.01 mg/m^3 dust <0.5 ppm vapor	Very toxic on single exposure, reproductive hazard, sensitizer	Seek expert advice

Although, there is not one standard control band classification used across the pharmaceutical industry, the National Institute for Occupational Safety and Health (NIOSH) defines targeted control bands[12] as detailed in Table 46.3. Based on the control band assignment, the handling procedures and industrial hygiene sampling (facility surface testing), for a compound in the laboratory, kilo lab, and pilot plant, are designed to ensure that the potential worker exposure is controlled or contained within acceptable limits. Engineering controls should be used as the primary containment strategy (Section 46.2.5) supplemented by PPE as necessary to provide redundancy. PPE should not be used as the primary control.

Employee exposure – controlled with engineering controls, procedures, and PPE – is also prevented by employing sound industrial hygiene sampling following the completion of a campaign. For example, following a campaign using highly potent compounds, equipment exterior and facility surfaces are decontaminated and then sampled to ensure potent compounds have been removed to prevent worker exposure. The sampling protocol and limits are defined by the facility and based on operator exposure guidelines.

46.3.3.3 Personal Protective Equipment (PPE) The proper selection and use of PPE is important to worker safety in the kilo lab and pilot plant. Various types of PPE are available and should be used as appropriate to mitigate risk of chemical exposure. A standard policy defining the minimum

[12] www.cdc.gov.

TABLE 46.4 Common Minimum PPE Requirements for Kilo Labs and Pilot Plants

Type of PPE	Common Requirements
Eye protection	Safety glasses with side shields. Protection should meet the ANSI Z87 standard for occupational and education eye and face protection
Foot protection	Steel-toed shoes that are static dissipating
Protective apparel	Fire-retardant uniform or laboratory coat
	Chemical-resistant gloves and clothing
Head protection (if applicable to facility)	Protection should meet the ANSI S89 standard for protective headwear for industrial workers requirements

requirements for all workers and visitors is typically developed for each facility based on facility operations and construction. For kilo labs and pilot plants, minimum requirements generally consist of the items found in Table 46.4.

There may be additional requirements for PPE for use when handling specific chemicals or performing certain operations where the PPE is applicable to the hazards involved. The selection criteria for protective face shields, glove type, respiratory protection, and outer safety garments should include potential hazards as well as known hazards.

Gloves provide a barrier between the worker and chemical or physical hazard. When choosing a glove, the kilo lab or pilot plant operator should select one that provides the type of protection necessary for the operation being performed and also provides the dexterity necessary to complete the task. There is no single type of glove that provides protection against all chemicals, and it is important to carefully select the appropriate glove for the operation. The respiratory protection used in a kilo lab or pilot plant must provide protection against gas, vapors, mists, and solid particulates. The type of respiratory protection selected must be able to maintain the employee's exposure levels below permissible levels. The types of respirators currently used in pharmaceutical scale-up facilities include air-purifying respirators, powered air-purifying respirators, and supplied air respirators. The MSDSs for the chemicals to be used and additional literature sources should be consulted to determine the appropriate glove and respirator selection.

46.3.3.4 Management of Hazardous Chemicals Procedures should be established for the procurement, distribution, and storage of hazardous chemicals that provide for adequate control over chemical inventory to mitigate the risk of safety-related issues. Consideration of interactions and incompatibilities between different classes of chemicals should be included at all phases of the management process. The design and operation of chemical storage and staging areas must comply with regulations for segregation of certain hazardous chemicals.

46.3.4 Environmental

Pharmaceutical kilo labs and pilot plants handle larger quantities of hazardous and toxic chemicals than standard laboratories and have a responsibility to prevent accidental release of substances into the environment. Local, state, and federal governments have established laws and regulations to prevent the accidental release of hazardous substances into the environment. These laws and regulations typically require that the facilities have procedures and safeguards for air pollution and controls, site management of toxic substances, and waste disposal management. Examples of US environmental regulations are the Clean Air Act, Resource Conservation and Recovery Act (RCRA), and Toxic Substances Control Act.

The government regulations for air pollution are set to provide protections for the public health against hazardous pollutants. The regulations aim to reduce the overall emissions by setting limits on certain pollutants, setting performance levels for pollutant controls, managing environmental permit programs, and providing enforcement powers. Facilities, such as kilo labs and pilot plants, which produce above certain quantities of pollutants, as defined by local, state, or federal regulations, are required to obtain operating permits that limit air pollutants. The permit will typically require the facilities to report the actual air pollutants generated through the use of theoretical models. A typical emissions report provides the actual emissions produced categorized by type of pollutant and documents that the emissions do not exceed the allowable permitted quantity for each type of pollutant. Scale-up facilities will typically reduce or limit air pollutants with air emission controls, such as condensers, cryogenic condensers, cold traps, gas scrubbers, and thermal oxidizers connected to the facility or equipment train. When properly engineered, these devices can significantly reduce or eliminate specific pollutants.

46.3.4.1 Waste Disposal
Scale-up facilities must have a waste management program for waste that is generated during processing. There are local, state, and federal government laws and regulations for the management of hazardous waste and nonhazardous waste. Hazardous waste is considered as any waste with properties that makes it dangerous or capable of having a harmful effect on human health or the environment. Nonhazardous solid waste is not based on any physical form (i.e. solid, liquid, gaseous) but is any waste that is generated by industrial process or non-process sources that does not meet the classification of hazardous waste. The regulations surrounding hazardous waste requires control from "cradle to grave." Procedures for the generation, storage, treatment, transportation, and disposal of hazardous waste must be developed to ensure compliance with regulations.

46.3.4.2 Emergency Response for Spills and Accidents
An emergency response plan should be established for kilo labs and pilot plants to enable rapid and appropriate response. The response plan should include EHS hazards associated with hazardous material spills, fires, and explosion emergencies, proper notification details, the containment and control of the hazard, cleanup and restoration of the area, a list of the emergency response groups, the evacuation plan, and reporting requirements. Some companies train operating staff in kilo labs and pilot plants, who are familiar with the hazards associated with the chemicals, to provide the first response to spill cleanup. Emergency response procedures should be used when there are properties of the hazardous substances, circumstances of the release, and other factors in the work area that require more than an incidental cleanup.

46.3.5 Programs and Procedures

Procedures and programs developed for kilo labs and pilot plants should be structured to promote compliance with cGMP, health and safety, and environmental regulations while enabling process development. Operating procedures, employee training, equipment and facility cleaning, housekeeping, maintenance, and inspection programs provide a foundation for operation of the kilo lab and pilot plant.

46.3.5.1 Operating Procedures
Operating procedures and other written instructions for kilo labs and pilot plants should provide process operators with clear instructions on how to safely operate equipment and conduct routine plant or kilo lab operations. Equipment instructions should describe startup operations, normal operations, shutdown operations, and emergency operations for the specified equipment. Additional information about equipment operating limits and an interlock or safety system list may be included. Other topics for written instructions include industrial hygiene testing, hazardous and/or toxic materials handling, hazardous waste handling, and energy control procedures required to manage the multiple sources of hazardous energy during maintenance and operations (see Section 46.3.5.3).

46.3.5.2 Employee Training
A well-defined and documented employee training program is essential to ensuring compliance with procedures and program expectations. The training plan is role based and assigned to an individual based on his or her role within the kilo lab or pilot plant operations. The training objective may range from awareness of the topic to hands-on application within daily operations. Awareness training will usually apply to policies or procedures where the knowledge of the general context is

important but the individual is not responsible for performing the operations. Application training is used when the knowledge is necessary to perform an assigned task. Between the two is comprehension-based training, which applies to employees that need to understand the specifics of the procedures due to their role in the organization. An essential element of a good training program is the requirement for refresher training at a specified time interval to ensure that operating staff maintain a competent skill level and knowledge needed to be compliant with regulatory requirements.

46.3.5.3 Equipment/Facility Cleaning Equipment and facility cleaning programs are designed to reduce the risk of product cross contamination between processes. Although many facilities will have unique cleaning approaches and procedures, generally a defined plan will be in place to ensure appropriate identification of residual product in the equipment train consistent with ICH Q7A. The cleaning criteria are based on the toxicity or acceptable daily intake (ADI) of the previous product that is carried over into the next product. Since the carryover is variable based on the product dose, this parameter is also used to define cleaning limits. After the cleaning limits have been established, the post-process cleaning is performed based on the facility procedures and equipment design. A combination of organic solvents, with acceptable solubility of the product, and aqueous detergents may be used to execute the cleaning. Once the cleaning process has been completed, equipment samples will be analyzed, and a visual inspection of the equipment will be completed to ensure process residues have been removed from the equipment train.

46.3.5.4 Housekeeping, Maintenance, and Inspection Program Preventative programs are designed to proactively address workplace safety and ensure cGMP compliance of the facility. A housekeeping policy, preventative maintenance (PM) program, and inspection practices allow for early identification of issues and corrective actions before problems arise. A housekeeping policy for kilo labs or pilot plants will include keeping process areas clear of obstructions, ensuring clean and sanitary work areas, and the proper storage of hazardous chemicals.

PM and mechanical integrity programs help to ensure that the systems installed within the facility are functioning properly and within the design specifications. A well-designed PM program enables determination of whether equipment replacement or calibration is necessary and allows the tasks to be scheduled with the least disruption to processing activities. The program includes examination of instruments, such as pressure gauges, transmitters, and relief valves; evaluation of wear on equipment train components such as agitator seals, pumps, and valves; and testing of facility-related equipment such as safety showers, fire extinguishers, and engineering controls (i.e. laminar flow hoods, isolators).

A "lockout–tagout" procedure for the control of hazardous energy in a kilo lab or pilot plant is necessary when performing maintenance activities on equipment with multiple energy sources (i.e. electricity, pressure, mechanical). Maintenance on the equipment is defined as any activity performed on the equipment, including any setup, adjustment, or inspections. The procedure is necessary to prevent the unexpected energization of equipment while the maintenance is being performed causing serious injury to the employees. The basis of the procedure should contain the identification of the energy sources, proper training of the affected personnel, and periodic inspections of the energy sources. To control the hazardous energy, lockout or tagout devices should be installed on the energy sources to prevent the accidental energization on the equipment or machine. The use of locks is the preferred method for any energy control procedures as tags are only a warning against hazardous conditions. It is critical for the individual or groups of individuals who are performing the maintenance operations to confirm the de-energization of all equipment being serviced.

A safety inspection program acts to alert the staff to any safety risks associated with inadequate housekeeping and deficiencies within the PM program. It can also provide a real-time check on compliance with safety procedures such as labeling, hazardous chemical handling, and improper use of PPE. Completing the integration of safety and quality, the inspection program can also supplement the cGMP internal audit through identification of potential quality risks.

46.4 SUMMARY

Kilo labs and pilot plants are important tools supporting process development in the pharmaceutical industry. Designed for flexibility and responsiveness to change, they are an extension of the scientist's laboratory supporting process knowledge generation and successful introduction of new chemistries at all stages of process development. Governed by cGMP and EHS regulations, they are designed and operated in compliance with a multitude of regulations. Protection of the patient, the worker, and the environment are top priorities. Joined together, the kilo lab and the pilot plant can provide a competitive edge within process development in the pharmaceutical industry.

46.5 EXERCISE

cGMP regulations require investigation of unplanned events that occur within scale-up facilities to determine the effect on the quality of the product manufactured. Regulations define certain elements that must be included within a compliant "quality events program"; however, individual company programs will vary. At the core of a quality events program is the investigation of the unplanned event to enable identification of the root cause and determination of the impact to the quality of the product. Corrective actions are then implemented that address the root cause mitigating the risk of a future occurrence. Facility operations at the time of the event, the state of the equipment, input materials, and the chemical process are some factors to consider during the investigation. Investigations in kilo labs and pilot plants can be further complicated where there are limitations on the process knowledge available for the product. A commonly used methodology for investigation of such events is the Kepner Tregoe process (www.kepner-tregoe.com). A situational analysis is typically performed to determine if an investigation is needed based on the potential for impact to the quality of the product. As with any investigation, constructing a timeline and collecting "forensic evidence" are important first steps in determining if the product quality is affected.

You have been asked to lead the situational appraisal for an unplanned event that occurred in the manufacture of an API. Below is a summary of information provided to you.

Where: Pilot Plant

What: An API slurry was transferred from a crystallizer into a filter dryer using nitrogen pressure. The mother liquor was collected in a receiver downstream of the filter. After the slurry transfer was complete, the API was washed with fresh solvent charged into the filter from solvent drums to remove any residual mother liquor from the cake. The process flow diagram (PFD) for the operation in depicted in Figure 46.16. During the washing step, black particulates were observed by the operator on the top of the wet cake in the filter dryer.

Background:

- The API is designated for use in a clinical phase II study.
- The crystallizer and filter dryer are used for processing of both intermediates and API.
- After each campaign, the equipment is cleaned and inspected.
- The filter dryer has been used in the manufacture of multiple products since the last PM, both intermediates and API.
- The crystallizer has not been used since the last PM was completed. It was verified as clean prior to use in this batch.
- The process stream and the crystallization antisolvent were filtered through a 1 μm filter before they were charged to the crystallizer.

A. Construct the timeline for the event.
B. What is your next step? Should an investigation be conducted? Explain your answer.

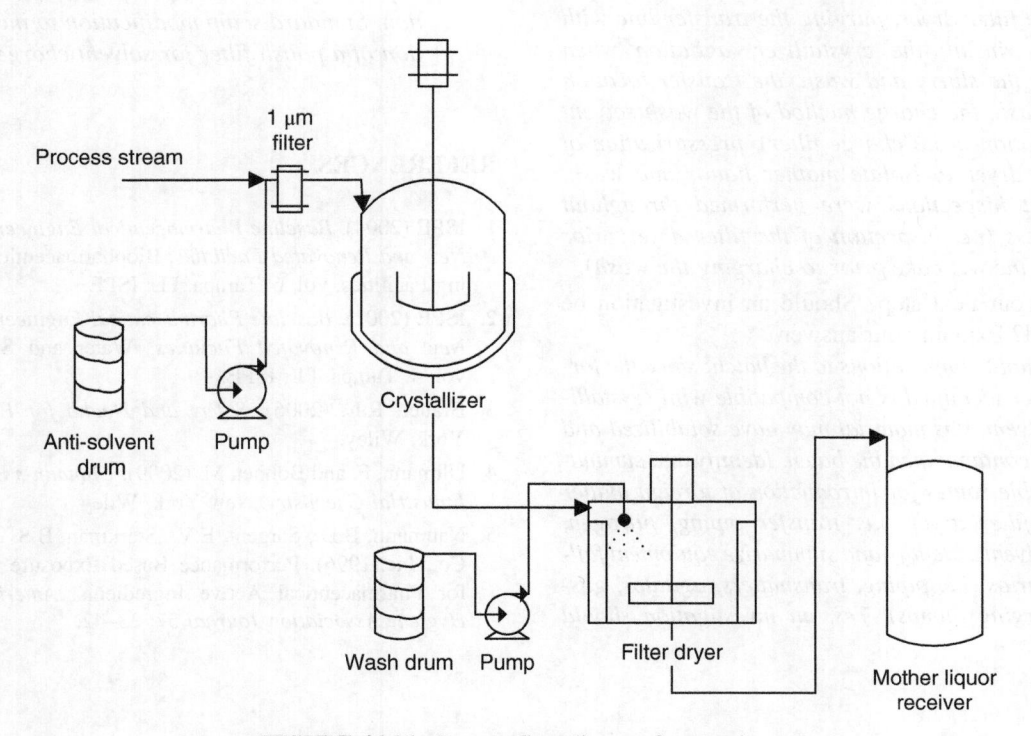

FIGURE 46.16 Process flow diagram for exercise.

C. Would your conclusion change if the particulates were found before the manufacture of the processing began?

Additional data:
- The black particulates were eventually determined to be a polymer frequently found in o-rings, which are used to seal equipment to seal against the escape of gas or fluid.
- There are several o-rings in the filter dryer assembly that come into contact with the process stream.
- The polymer is listed as incompatible with the crystallization solvent in the available literature and is known to swell after a short period of time when immersed in the solvent.

D. Consider the additional data and review the PFD. What is the likely source of the black particulates observed on the filter cake?

E. Is there another obvious potential source?

F. Should the investigation be expanded to include other products? Why or why not?

G. What actions would you take to prevent a reoccurrence?

ACCEPTABLE RESPONSES:

A. Construct the timeline for the event.
Include all operations (every detail) leading up to the identification of the foreign matter. Key things to highlight may include the transfer from the crystallizer to the filter dryer, purging the transfer line with nitrogen, rinsing the crystallizer, agitation when receiving the slurry and wash, the transfer location for the wash, the charge method of the wash solvent (identification of no charge filter), pressurization of the filter dryer to isolate mother liquor and wash, and what inspections were performed throughout the process (i.e. inspection of the filter dryer prior to batch, the wet cake prior to charging the wash).

B. What is your next step? Should an investigation be conducted? Explain your answer.
Identify quality implications to the batch; since the foreign matter identified is not compatible with crystallization solvent, this material may have solubilized and therefore contaminated the batch. Identify and summarize possible routes for introduction of foreign matter into the filter dryer (i.e. transfer piping, nitrogen, slurry, solvent), identify and summarize equipment failure scenarios (i.e. piping, transmitters, agitator, gaskets, charging pumps). Yes, an investigation should be conducted to identify the source, root cause, and corrective actions and to determine path forward for the use of the batch.

C. Would your conclusion change if the particulates were found before the manufacture of the processing began?
No, investigation should still be conducted to understand source, root cause, corrective actions, and possible implication to previous batches.

D. Consider the additional data and review the PFD. What is the likely source of the black particulates observed on the filter cake?
Since the particulates were not observed prior to introduction of the wash, and a charging filter was not used, it is likely that the material was introduced from an equipment failure on the charging pump or charging lines. This should be confirmed with an inspection of the pump.

E. Is there another obvious potential source?
The foreign matter could have also been from the solvent container, if the solvent container was contaminated.

F. Should the investigation be expanded to include other products? Why or why not?
Yes, the investigation team would need to understand where the pump was used and when the last maintenance was performed (or last good state of the pump).

G. What actions would you take to prevent a reoccurrence?
Improved frequency of pump maintenance or inspection. Standard setup modification to include installation of a polish filter for solvent charge lines.

REFERENCES

1. ISPE (2004). *Baseline Pharmaceutical Engineering Guides for New and Renovated Facilities*, Biopharmaceutical Manufacturing Facilities, vol. 6. Tampa, FL: ISPE.
2. ISPE (2001). *Baseline Pharmaceutical Engineering Guides for New and Renovated Facilities*, Water and Steam Systems, vol. 4. Tampa, FL: ISPE.
3. Brauer, R.L. (2006). *Safety and Health for Engineers*. New York: Wiley.
4. Ullmann, F. and Bohnet, M. (2009). *Ullmann's Encyclopedia of Industrial Chemistry*. New York: Wiley.
5. Naumann, B.D., Sargent, E.V., Starkman, B.S. et al., Merck & Co., Inc.(1996). Performance Based Exposure Control Limits for Pharmaceutical Active Ingredients. *American Industrial Hygiene Association Journal* 57: 33–42.

47

THE ROLE OF SIMULATION AND SCHEDULING TOOLS IN THE DEVELOPMENT AND MANUFACTURING OF ACTIVE PHARMACEUTICAL INGREDIENTS

DEMETRI PETRIDES, DOUG CARMICHAEL, CHARLES SILETTI, AND DIMITRIS VARDALIS
Intelligen, Inc., Scotch Plains, NJ, USA

ALEXANDROS KOULOURIS
Alexander Technological Education Institute of Thessaloniki, Thessaloniki, Greece

PERICLES LAGONIKOS
Merck & Co. Inc., Singapore, Singapore

47.1 INTRODUCTION

The global competition in the pharmaceutical industry and the increasing demands by governments and citizens for affordable medicines have driven the industry's attention toward manufacturing efficiency. In this new era, improvements in process and product development approaches and streamlining of manufacturing operations can have a profound impact on the bottom line. It has been estimated that improvements in supply chain management alone could provide $65 billion in benefits to the pharmaceutical industry [1]. Computational tools such as process simulation and scheduling applications can play an important role in these endeavors. The role of such tools in the development and operation of processes for the manufacture of pharmaceutical products has been reviewed previously [2]. This chapter focuses on the role of these tools in the process development and manufacturing of active pharmaceutical ingredients (APIs) and specifically on small molecule APIs that are produced through organic synthesis. Information on the role of such tools in the development and manufacturing of biologics is also available in the literature [3].

Process simulation and scheduling tools serve a variety of purposes throughout the life cycle of product development and commercialization in the pharmaceutical industry [3–6]. During process development, process simulators are used to facilitate the following tasks:

- Represent the entire process on the computer.
- Perform material and energy balances.
- Estimate the size of equipment.
- Calculate demand for labor and utilities as a function of time.
- Estimate the cycle time of the process.
- Perform cost analysis.
- Assess the environmental impact.

The availability of a good computer-based model improves the understanding of the entire process by the team members and facilitates communication. What-if and sensitivity analyses are greatly facilitated by such tools. The objective of such studies is to evaluate the impact of critical parameters on various key performance indicators (KPIs), such as production cost, cycle times, and plant throughput.

Chemical Engineering in the Pharmaceutical Industry: Active Pharmaceutical Ingredients, Second Edition.
Edited by David J. am Ende and Mary T. am Ende.
© 2019 John Wiley & Sons, Inc. Published 2019 by John Wiley & Sons, Inc.

If there is uncertainty in some of the input parameters, sensitivity analysis can be supplemented with Monte Carlo simulation to quantify the impact of the uncertainty. Cost analysis, especially capital cost estimation, facilitates decisions related to in-house manufacturing versus outsourcing. Estimation of the cost of goods identifies the expensive processing steps, and the information generated is used to guide R&D work in a judicious way.

When a process is ready to move from development to manufacturing, process simulation facilitates technology transfer and process fitting. A detailed computer model provides a thorough description of a process in a way that can be readily understood and adjusted by the recipients. Process adjustments are commonly required when a new process is moved into an existing facility whose equipment is not ideally sized for the new process. The simulation model is then used to adjust batch sizes, figure out cycling of certain steps (for equipment that cannot handle a batch in one cycle), estimate recipe cycle times, etc.

Production scheduling tools play an important role in manufacturing (large scale as well as clinical). They are used to generate production schedules on an ongoing basis in a way that does not violate constraints related to the limited availability of equipment, labor resources, utilities, inventories of materials, etc. Production scheduling tools close the gap between enterprise resource planning (ERP)/manufacturing resource planning (MRP) II tools and the plant floor [7, 8]. Production schedules generated by ERP and MRP II tools are typically based on coarse process representations and approximate plant capacities, and, as a result, solutions generated by these tools may not be feasible, especially for multiproduct facilities that operate at high capacity utilization. That often leads to late orders that require expediting and/or to large inventories in order to maintain customer responsiveness. "Lean manufacturing" principles, such as just-in-time (JIT) production, low work in progress (WIP), and low product inventories, cannot be implemented without good production scheduling tools that can accurately estimate capacity.

47.2 COMMERCIALLY AVAILABLE SIMULATION AND SCHEDULING TOOLS

Computer-aided process design and simulation tools have been used in the chemical and petrochemical industries since the early 1960s. Simulators for these industries have been designed to model continuous processes and their transient behavior for process control purposes. Although there is ongoing research into continuous manufacturing of pharmaceuticals, there are substantial complications and challenges to implementing this type of process in the pharmaceutical industry, and therefore most APIs and pharmaceutical products are currently manufactured in batch or semicontinuous modes [9, 10]. Such processes are best modeled with batch process simulators that account for time dependency and sequencing of events. *Batches* from Batch Process Technologies, Inc. (West Lafayette, IN) was the first simulator specific to batch processing. It was commercialized in the mid-1980s. All of its operation models are dynamic, and simulation always involves integration of differential equations over a period of time. In the mid-1990s, Aspen Technology (Burlington, MA) introduced *Batch Plus*, a recipe-driven simulator that targeted batch pharmaceutical processes. Around the same time, Intelligen, Inc. (Scotch Plains, NJ) introduced *SuperPro Designer*. The initial focus of SuperPro was on bioprocessing. Over the years, its scope has been expanded to include modeling of small molecule API and secondary pharmaceutical manufacturing processes.

Discrete-event simulators have also found applications in the pharmaceutical industry, especially in the modeling of secondary pharmaceutical manufacturing processes. Established tools of this type include *ProModel* from ProModel Corporation (Orem, UT); Arena and Witness from Rockwell Automation, Inc. (Milwaukee, WI); and Extend from Imagine That, Inc. (San Jose, CA). The focus of models developed with such tools is usually on the minute-by-minute time dependency of events and the animation of the process. Material balances, equipment sizing, and cost analysis tasks are usually out of the scope of such models. Some of these tools are quite customizable, and third-party companies occasionally use them as platforms to create industry-specific modules.

Microsoft Excel is another common platform for creating models for pharmaceutical processes that focus on material balances, equipment sizing, and cost analysis. Some companies have even developed models in Excel that capture the time dependency of batch processes. This is typically done by writing extensive code (in the form of macros and subroutines) in Visual Basic for Applications (VBA) that comes with Excel. K-TOPS from Biokinetics, which is now part of Amec Foster Wheeler (Philadelphia, PA), belongs to this category.

In terms of production scheduling, established tools include *Optiflex* from i2 Technologies, Inc. (Irving, TX); *SAP APO* from SAP AG (Walldorf, Germany); *ILOG* Plant PowerOps from ILOG SA (Gentilly, France); *Aspen SCM* (formerly Aspen MIMI) from Aspen Technology, Inc. (Burlington, MA); etc. Their success in the pharmaceutical industry, however, has been rather limited thus far. This is partly due to the complexity and time-consuming nature of scheduling based on mathematical optimization, as well as the primary focus on discrete manufacturing (as opposed to batch chemical manufacturing) for many supply chain management tools. In contrast, *SchedulePro* from Intelligen, Inc. (Scotch Plains, NJ) is a finite capacity scheduling tool that focuses on scheduling of batch and semicontinuous pharmaceutical and related processes. It is a recipe-driven tool with emphasis on generation of feasible solutions that can be readily improved by the user in an interactive manner.

47.3 MODELING AND ANALYSIS OF AN API MANUFACTURING PROCESS

The steps involved during the development of a model will be illustrated with a simple process that represents the manufacturing of an active compound for skin care applications.

The first step in building a simulation model is always the collection of information about the process. Engineers rely on draft versions of process descriptions, block flow diagrams, and batch sheets from past runs, which contain information on material inputs and operating conditions, among others. Reasonable assumptions are then made for missing data.

The steps of building a batch process model are generally the same for all batch process simulation tools. The best practice is to build the model step by step, gradually checking the functionality of its parts. The registration of materials (pure components and mixtures) is usually the first step. Next, the flow diagram (see Figure 47.1) is developed by putting together the required unit procedures and joining them with material flow streams. Operations are then added to unit procedures (see next paragraph for explanation), and their operating conditions and performance parameters are specified.

In SuperPro Designer, the representation of a batch process model is loosely based on the ISA S-88 standards for batch recipe representation [11]. A batch process model is in essence a batch recipe that describes how to make a certain quantity of a specific product. The set of operations that comprise a processing step is called a "unit procedure" (as opposed to a unit operation, which is a term used for continuous processes). The individual tasks contained in a procedure are called "operations." A unit procedure is represented on the screen with a single equipment-looking icon. Figure 47.2 displays the dialog through which operations are added to a vessel unit procedure. On the left-hand side of that dialog, the program displays the operations that are available in the context of a vessel procedure; on the right-hand side, it displays the registered operations (Charge Quinaldine, Charge Chlorine, Charge Na_2CO_3, Agitate, etc.). The two-level representation of operations in the context of unit procedures enables users to describe and model batch processes in detail.

For every operation within a unit procedure, the simulator includes a mathematical model that performs material and energy balance calculations. Based on the material balances, it performs equipment-sizing calculations. If multiple operations within a unit procedure dictate different sizes for a certain piece of equipment, the software reconciles the different demands and selects an equipment size that is appropriate for all operations. The equipment is sized so that it is large enough and, hence, not overfilled during any operation, but it is not larger than necessary (in order to minimize capital costs). If the equipment size is specified by the user, the simulator checks to make sure that the vessel is not overfilled. In addition, the tool checks to ensure that the vessel contents do not fall below a user-specified minimum volume (e.g. a minimum stirring volume) for applicable operations.

In addition to material balances, equipment sizing, and cycle time analysis, the simulator can be used to carry out cost-of-goods analysis and project economic evaluation. The Sections 47.3.1 through 47.6.1 that follow provide illustrative examples of the above.

Having developed a good model using a process simulator, the user may begin experimenting on the simulator with alternative process setups and operating conditions. This has the potential of reducing the costly and time-consuming laboratory and pilot plant effort. Of course, the garbage-in, garbage-out (GIGO) principle applies to all computer models. If critical assumptions and input data are incorrect, so will be the outcome of the simulation.

When modeling an existing process, input data required by the model can be extracted from the data recorded by the actual process. A communication channel must, therefore, be established between the modeler and the operations department. The application of some data mining technique is usually required to transform the process data to the form required by the model. When designing a new plant, experience from similar projects can be used to fill in the information gaps. In all cases, a certain level of model verification is necessary after the model is developed. In its simplest form, a review of the results by an experienced engineer can play the role of verification. Running a sensitivity analysis on key input variables can reveal the parameters with the greatest impact on the model's most important outputs. These parameters would then constitute the focal points in the data acquisition effort in an attempt to estimate their values and uncertainty limits with the best possible accuracy.

47.3.1 Design Basis and Process Description

A simple batch process is used to illustrate the steps involved in building a model with SuperPro Designer. It is assumed that the process has been developed at the pilot plant and it is ready to be moved to large-scale manufacturing. Based on input from the marketing department, the objective is to produce at least 27 000 kg of active ingredient per year at a cost of no more than $330/kg. A production suite can be dedicated to this process that includes two 3800 L reactors (R-101 and R-102), one 2.5 m^2 Nutsche filter (NFD-101), and a 10 m^2 tray dryer (TDR-101).

The entire flow sheet of the batch process is shown in Figure 47.1. It is divided into four sections: (i) product synthesis, (ii) isolation and purification, (iii) final purification, and (iv) crystallization and drying. A flowsheet section in SuperPro Designer is simply a group of unit procedures (processing steps). The unit procedures of each section are marked by distinct colors.

The formation of the final product in this example involves 12 unit procedures. The first reaction step

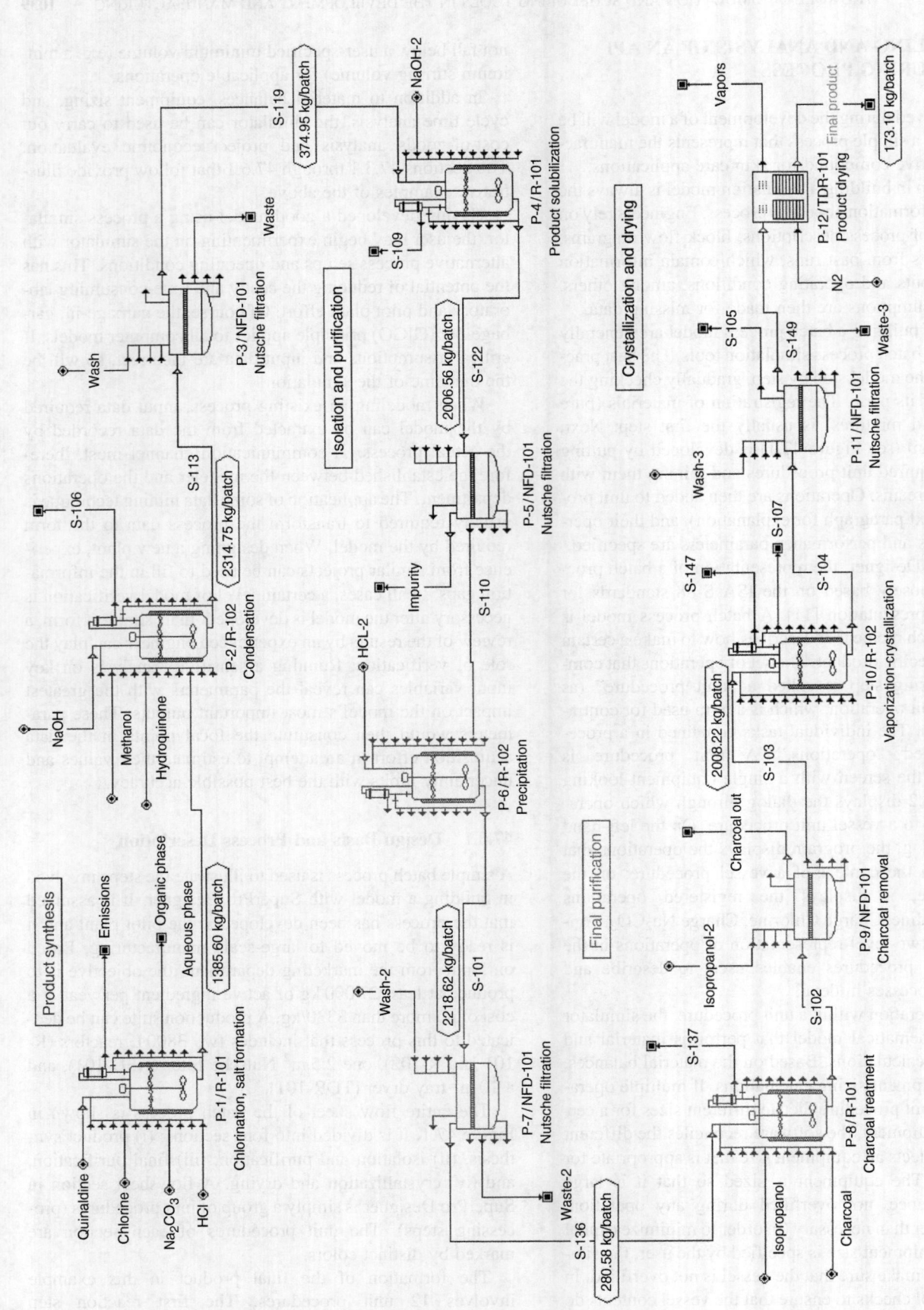

FIGURE 47.1 Flow diagram of the API process.

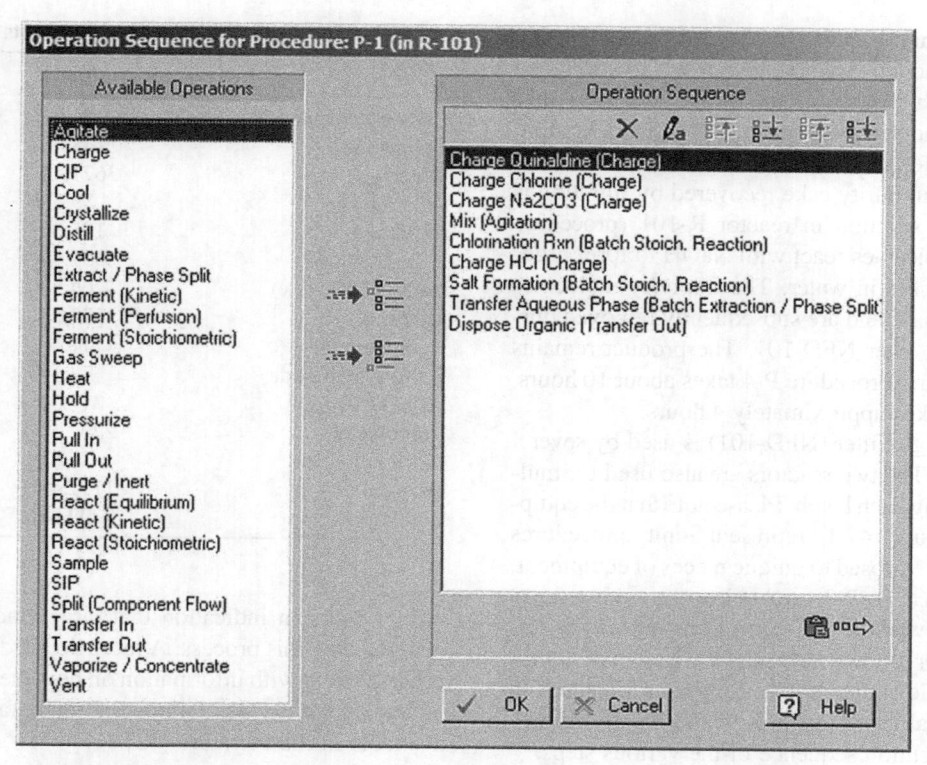

FIGURE 47.2 The operations associated with the first unit procedure of Figure 47.1.

(procedure P-1) involves the chlorination of quinaldine. Quinaldine is dissolved in carbon tetrachloride (CCl_4) and reacts with gaseous Cl_2 to form chloroquinaldine.[1] The conversion of the reaction is around 98% (based on amount of quinaldine fed). The generated HCl is then neutralized using Na_2CO_3. The stoichiometry of these reactions follows:

$$Quinaldine + Cl_2 \rightarrow Chloroquinaldine + HCl$$

$$Na_2CO_3 + HCl \rightarrow NaHCO_3 + NaCl$$

$$NaHCO_3 + HCl \rightarrow NaCl + H_2O + CO_2$$

The small amounts of unreacted Cl_2, generated CO_2, and volatilized CCl_4 are vented. The above three reactions occur sequentially in the first reactor vessel (R-101). Next, HCl is added in order to produce chloroquinaldine-HCl. The HCl first neutralizes the remaining $NaHCO_3$ and then reacts with chloroquinaldine to form its salt, according to the following stoichiometries:

$$NaHCO_3 + HCl \rightarrow NaCl + H_2O + CO_2$$

$$Chloroquinaldine + HCl \rightarrow Chloroquinaldine.HCl$$

[1] Note that carbon tetrachloride is an ideal solvent for this specific reaction from a chemistry perspective but that from an environmental, health, and safety perspective, this solvent is considered highly undesirable.

The small amounts of generated CO_2 and volatilized CCl_4 are vented. The presence of water (added with HCl as hydrochloric acid solution) and CCl_4 leads to the formation of two liquid phases. Then the small amounts of unreacted quinaldine and chloroquinaldine are removed with the organic phase. The chloroquinaldine-HCl remains in the aqueous phase. This sequence of operations (including all charges and transfers) requires about 14.5 hours.

After removal of the unreacted quinaldine, the condensation of chloroquinaldine and hydroquinone takes place in reactor R-102 (procedure P-2). First, the salt chloroquinaldine-HCl is converted back to chloroquinaldine using NaOH. Then, hydroquinone reacts with NaOH and yields hydroquinone-Na. Finally, chloroquinaldine and hydroquinone-Na react and yield the desired intermediate product. Along with product formation, roughly 2% of chloroquinaldine dimerizes and forms an undesirable by-product impurity. This series of reactions and transfers takes roughly 13.3 hours. The stoichiometry of these reactions follows:

$$Chloroquinaldine.HCl + NaOH \rightarrow NaCl + H_2O + Chloroquinaldine$$

$$2 Chloroquinaldine + 2 NaOH \rightarrow 2 H_2O + 2 NaCl + Impurity$$

$$Hydroquinone + NaOH \rightarrow H_2O + Hydroquinone.Na$$

$$Chloroquinaldine + Hydroquinone.Na \rightarrow Product + NaCl$$

Both the product and impurity molecules formed during the condensation reaction precipitate out of solution and are recovered using a Nutsche filter (procedure P-3, filter NFD-101). The product recovery yield is 90%. The filtration, wash, and cake transfer time is 6.4 hours.

Next, the product/impurity cake recovered by filtration is added into a NaOH solution in reactor R-101 (procedure P-4). The product molecules react with NaOH to form product-Na, which is soluble in water. The impurity molecules remain in the solid phase and are subsequently removed during procedure P-5 in filter NFD-101. The product remains dissolved in the liquors. Procedure P-4 takes about 10 hours, and procedure P-5 takes approximately 4 hours.

Notice that the single filter (NFD-101) is used by several different procedures. The two reactors are also used for multiple procedures during each batch. Please note that the equipment icons in Figure 47.1 represent unit procedures (processing steps), as opposed to unique pieces of equipment. The procedure names (P-1, P-3, etc.) below the icons refer to the unit procedures, whereas the equipment tag names (R-101, R-102, etc.) refer to the actual physical pieces of equipment. The process flow diagram in SuperPro Designer is essentially a graphical representation of the batch "recipe" that displays the execution sequence of the various steps.

After the filtration in procedure P-5, the excess NaOH is neutralized using HCl, and the product-Na salt is converted back to product in reactor R-102 (procedure P-6). Since the product is insoluble in water, it precipitates out of solution. The product is then recovered using another filtration step in procedure P-7. The product recovery yield is 90%. The precipitation procedure takes roughly 10.7 hours, and the filtration takes about 5.7 hours. The recovered product cake is then dissolved in isopropanol and treated with charcoal to remove coloration. This takes place in reactor R-101 under procedure P-8. After charcoal treatment, the solid carbon particles are removed using another filtration step in procedure P-9. The times required for charcoal treatment and filtration are 15.9 and 5 hours, respectively.

In the next step (procedure P-10), the solvent is distilled off until the solution is half its original volume. The product is then crystallized in the same vessel with a yield of 97%. The crystalline product is recovered with a 90% yield using a final filtration step (procedure P-11). The distillation and crystallization steps take approximately 18.3 hours, and the filtration requires roughly 3.3 hours. The recovered product crystals are then dried in a tray dryer (procedure P-12, TDR-101). This takes an additional 15.6 hours. The amount of purified product generated per batch is 173.1 kg.

Table 47.1 displays the raw material requirements in kg per batch and per kg of main product (MP = purified product) that correspond to the maximum batch size achievable with the available equipment. Note that around 54.3 kg of raw materials (solvents, reagents, etc.) is used per kg of main product produced. Thus the product to raw material ratio is only 1.84%, an indication that large amounts of waste are generated by this process. A more detailed description of this process along with information on how the pilot plant process is transferred to the large-scale manufacturing facility is available in the literature [12].

TABLE 47.1 Raw Material Requirements

Material	kg/Batch	kg/kg MP
Carbon tetrachloride	497.31	2.87
Quinaldine	148.63	0.86
Water	3621.44	20.92
Chlorine	89.52	0.52
Na_2CO_3	105.06	0.61
HCl (20% w/w)	357.44	2.07
NaOH (50% w/w)	204.52	1.18
Methanol	553.26	3.20
Hydroquinone	171.45	0.99
Sodium hydroxide	74.16	0.43
HCl (37% w/w)	217.57	1.26
Isopropanol	2232.14	12.90
Charcoal	15.85	0.09
Nitrogen	1111.49	6.42
Total	**9399.84**	**54.30**

47.3.2 Process Scheduling and Cycle Time Reduction

Figure 47.3 displays the equipment occupancy chart for three consecutive batches (each color represents a different batch). The process batch time is approximately 92 hours. This is the total time between the start of the first step of a batch and the end of the last step of that batch. However, since most of the equipment items are utilized for shorter periods within a batch, a new batch can be initiated every 62 hours, which is known as the minimum cycle time of the process. Multiple bars on the same line (e.g. for R-101, R-102, and NFD-101) represent reuse (sharing) of equipment by multiple procedures. If the cycle times of procedures that share the same equipment overlap, scheduling with the assumed equipment designation is infeasible. White space between bars represents idle time. The equipment with the least idle time between consecutive batches is the *time (or scheduling) bottleneck* (R-102 in this case) that determines the maximum number of batches per year. Its occupancy time (~62 hours) is the minimum possible time between consecutive batches.

Scheduling in the context of a simulator is fully process driven, and the impact of process changes can be analyzed in a matter of seconds. For instance, the impact of an increase in batch size (that affects the duration of charge, transfer, filtration, distillation, and other scale-dependent operations) on the plant batch time and the maximum number of batches can be seen instantly. Due to the many interacting factors involved with even a relatively simple process, simulation tools that allow users to describe their processes in detail

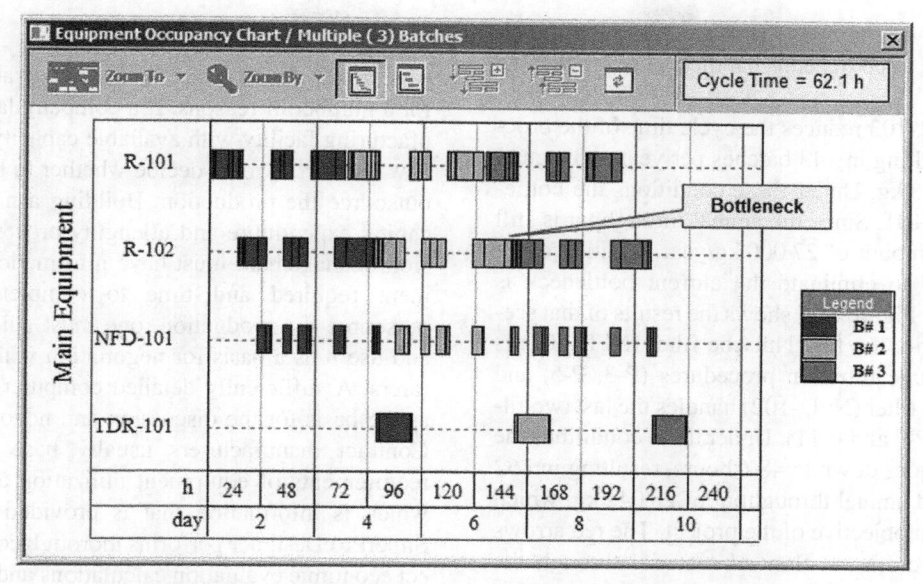

FIGURE 47.3 Equipment occupancy chart for three consecutive batches.

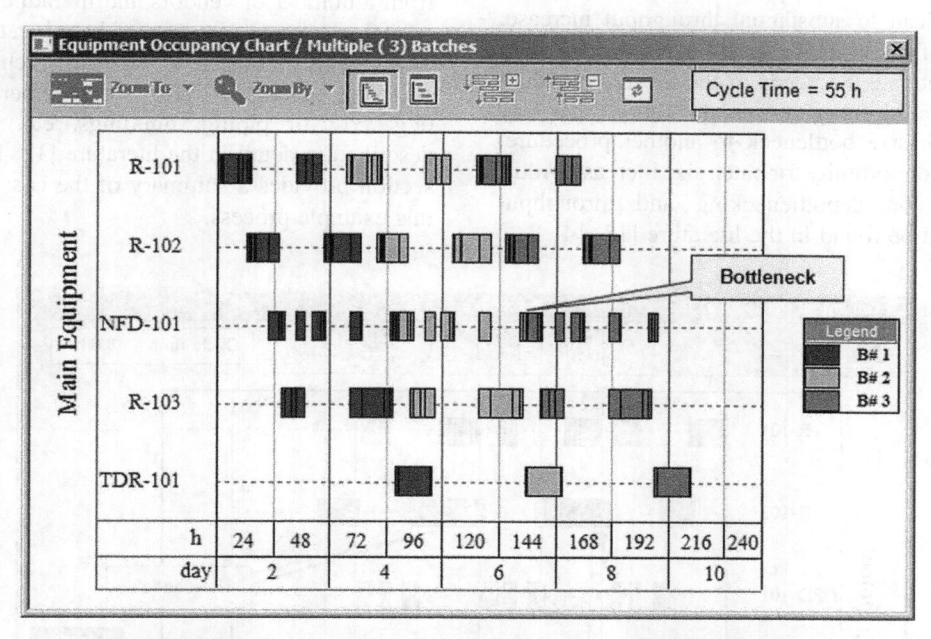

FIGURE 47.4 Equipment occupancy chart for the case with three reactors.

and to quickly perform what-if analyses can be extremely useful.

If this production line operated around the clock for 330 days a year (7920 hours) with its minimum cycle time of 62 hours, its maximum annual number of batches would be 126, leading to an annual production of 21 810 kg of API (126 batches × 173.1 kg/batch), which is less than the project's objective of 27 000 kg. And since the process operates at its maximum possible batch size, the only way to increase production is by reducing the process cycle time and thus increasing the number of batches per year. The cycle time can be reduced through process changes or by addition of extra equipment. However, major process changes in GMP manufacturing usually require regulatory approval and are avoided in practice. Addition of extra equipment is the practical way for cycle time reduction. Since R-102 is the current bottleneck, addition of an extra reactor can shift the bottleneck to another unit. Figure 47.4 displays the effect of the addition of an extra reactor (R-103). Please note that under the new conditions each reactor handles two procedures instead of three.

The addition of R-103 reduces the cycle time of the process to 55 hours, resulting in 143 batches per year and annual throughput of 24 753 kg. Under these conditions the bottleneck shifts to NFD-101. Since the annual throughput is still below the desired amount of 27 000 kg/year, addition of an extra Nutsche filter to eliminate the current bottleneck is the next logical step. Figure 47.5 shows the results of that scenario. In this scenario, the first Nutsche filter (NFD-101) is used for the first three filtration procedures (P-3, P-5, and P-7), and the second filter (NFD-102) handles the last two filtration procedures (P-9 and P-11). Under these conditions the process cycle time goes down to 48.6 hours, resulting in 162 batches per year and annual throughput of 28 042 kg, which meets the production objective of the project. The red arrows of Figure 47.5 represent the flow of material through the equipment for the first batch.

Debottlenecking projects that involve installation of additional equipment provide an opportunity for batch size increases that can lead to substantial throughput increase. More specifically, if the size of the new reactor (R-103) is selected to accommodate the needs of the most demanding vessel procedure (based on volumetric utilization) in a way that shifts the batch size bottleneck to another procedure, then, that creates an opportunity for batch size increase. Additional information on debottlenecking and throughput increase options can be found in the literature [13, 14].

47.3.3 Cost Analysis

Cost analysis and project economic evaluation is important for a number of reasons. If a company lacks a suitable manufacturing facility with available capacity to accommodate a new product, it must decide whether to build a new plant or outsource the production. Building a new plant is a major capital expenditure and a lengthy process. To make a decision, management must have information on capital investment required and time to complete the facility. To outsource the production, one must still do a cost analysis and use it as a basis for negotiation with contract manufacturers. A sufficiently detailed computer model can be used as the basis for the discussion and negotiation of the terms. Contract manufacturers usually base their estimates on requirements of equipment utilization and labor per batch, which is information that is provided by a good model. SuperPro Designer performs thorough cost analysis and project economic evaluation calculations and estimates capital as well as operating costs. The cost of equipment is estimated using built-in cost correlations that are based on data derived from a number of vendors and literature sources. The fixed capital investment is estimated based on total equipment cost using various multipliers, some of which are equipment specific (e.g. installation cost), while others are plant specific (e.g. cost of piping, buildings, etc.). The approach is described in detail in the literature [12, 15]. The rest of this section provides a summary of the cost analysis results for this example process.

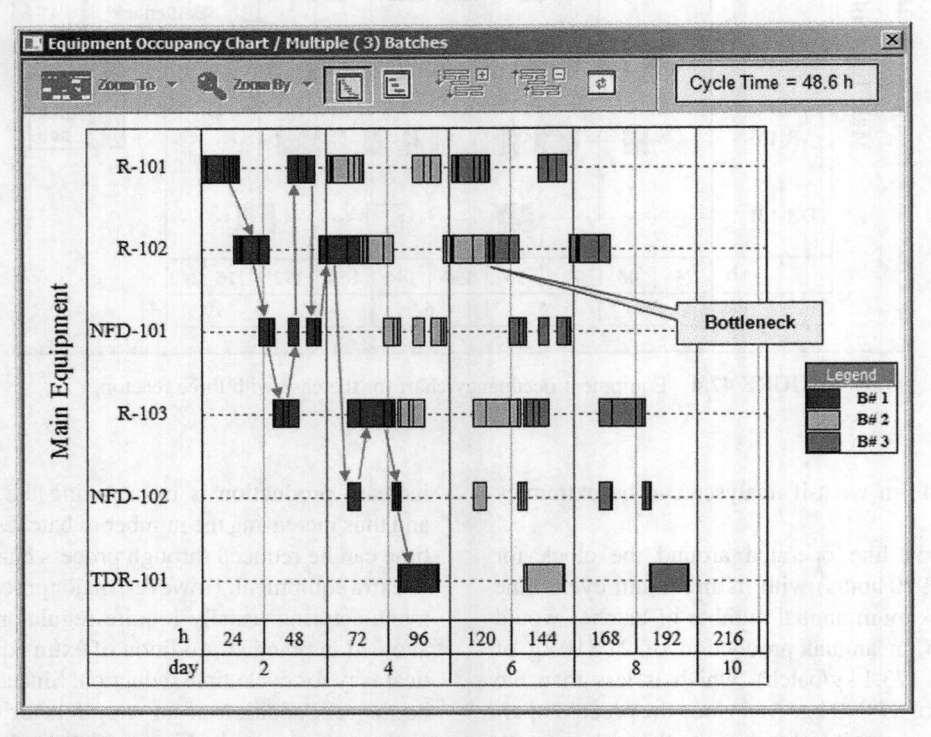

FIGURE 47.5 Equipment occupancy chart for the case with three reactors and two filters.

Table 47.2 shows the key economic evaluation results for this project. Key assumptions for the economic evaluations include (i) a new plant will be built and dedicated to the manufacturing of this product, (ii) the entire direct fixed capital is depreciated linearly over a period of 12 years, (iii) the project lifetime is 15 years, and (iv) 27 000 kg of final product will be produced per year.

For a plant of this capacity, the total capital investment is around $19.5 million. The unit production cost is $318/kg of product, which satisfies the project's objective for a unit cost of under $330/kg. Assuming a selling price of $450/kg, the project yields an after-tax internal rate of return (IRR) of 14% and a net present value (NPV) of $8.5 million (assuming a discount interest of 7%).

Figure 47.6 breaks down the manufacturing cost. The facility-dependent cost, which primarily accounts for the depreciation and maintenance of the plant, is the most important item accounting for 35.74% of the overall cost. This is common for high-value products that are produced in small facilities. This cost can be reduced by manufacturing the product at a facility whose equipment has already been depreciated. Raw materials are the second most important cost item accounting for 32.12% of the total manufacturing cost. Furthermore, if we look more closely at the raw material cost breakdown, it becomes evident that quinaldine, hydroquinone, and isopropanol make up more than 80% of this cost (see Table 47.3). If a lower-priced quinaldine vendor could be found, the overall manufacturing cost would be reduced significantly.

Labor is the third important cost item accounting for 18.8% of the overall cost. The program estimates that 12 operators are required to run the plant around the clock supported by 3 QC/QA scientists. This cost can be reduced by increasing automation or by locating the facility in a region of low labor cost.

47.4 UNCERTAINTY AND VARIABILITY ANALYSIS

Process simulation tools typically used for batch process design, debottlenecking, and cost estimation employ deterministic models. They model the "average" or "expected" situation commonly referred to as the base case or most likely scenario. Modeling a variety of cases can help determine the range of performance with respect to key process parameters. However, such an approach does not account for the relative likelihood of the various cases. Monte Carlo simulation is a practical means of quantifying the risk associated with uncertainty in process parameters [16]. In a Monte Carlo simulation, uncertain input variables are represented with probability distributions. A simulation calculates numerous scenarios of a model by repeatedly picking values from a user-defined probability distribution for the uncertain variables. It then uses those values in the model to calculate and analyze the outputs in a statistical way in order to quantify risk. The outcome of this analysis is the estimation of the confidence by which desired values of KPIs can be achieved. Inversely, the analysis can help identify the input parameters with the greatest effect on the bottom line and the input value ranges that minimize output uncertainty.

TABLE 47.2 Key Economic Evaluation Results

Total capital investment	$19.5 million
Plant throughput	27 000 kg/year
Manufacturing cost	$8.6 million/year
Unit production cost	$318/kg
Selling price	$450/kg
Revenues	$12.2 million/year
Gross margin	29.3%
Taxes (40%)	$1.1 million/year
IRR (after taxes)	14.0%
NPV (for 7% discount interest)	$8.5 million

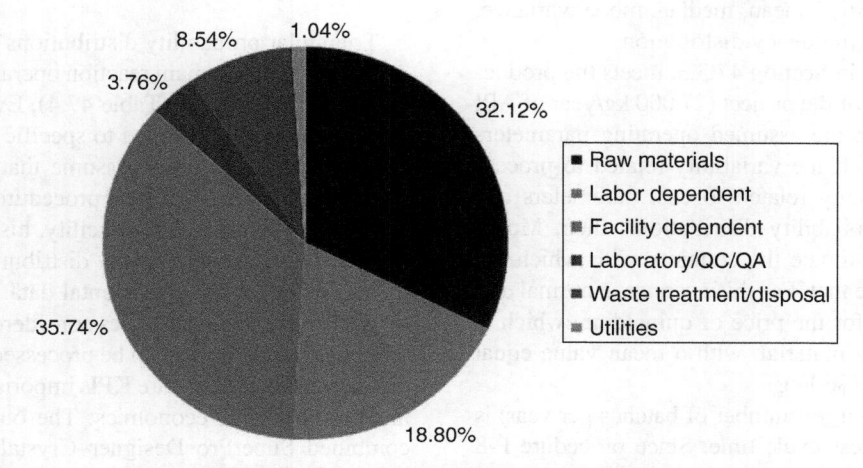

FIGURE 47.6 Manufacturing cost breakdown.

TABLE 47.3 Raw Material Requirements and Costs

Bulk Material	Unit Cost ($/kg)	Annual Amount (kg)	Annual Cost ($)	%
Carbon tetrachloride	0.80	77 581	62 065	2.25
Quinaldine	60.00	23 187	1 391 215	50.40
Water	0.10	564 944	56 494	2.05
Chlorine	3.30	13 965	46 083	1.67
Na_2CO_3	6.50	16 389	106 528	3.86
NaOH (50% w/w)	0.15	31 905	4 786	0.17
Methanol	0.24	86 308	20 714	0.75
Hydroquinone	18.00	26 746	481 427	17.44
Sodium hydroxide	2.00	11 569	23 138	0.84
HCl (37% w/w)	0.17	33 942	5 770	0.21
Isopropanol	1.10	348 214	383 035	13.88
Charcoal	2.20	2 473	5 440	0.20
Nitrogen	1.00	173 393	173 393	6.28
Total		1 466 376	2 760 088	100.00

In batch pharmaceutical processing uncertainty can emerge in operation- or market-related parameters. Process times, equipment sizes, material purchasing, and product selling prices are common uncertain variables. Performing a stochastic analysis early on in the design phase increases the model's robustness and minimizes the risk of encountering unpleasant surprises later on.

For models developed in SuperPro Designer, Monte Carlo simulation can be performed by combining SuperPro Designer with *Crystal Ball* from Oracle Corporation (Redwood Shores, CA). Crystal Ball is an Excel add-in application that facilitates Monte Carlo simulation. It enables the user to designate the uncertain input variables, specify their probability distributions, and select the output (decision) variables whose values are recorded and analyzed during the simulation. For each simulation trial (scenario), Crystal Ball generates random values for the uncertain input variables selected in frequency dictated by their probability distributions using the Monte Carlo method. Crystal Ball also calculates the uncertainty involved in the outputs in terms of their statistical properties, mean, median, mode, variance, standard deviation, and frequency distribution.

Scenario 3, analyzed in Section 47.3.3, meets the production and cost objectives of the project (27 000 kg/year of API for <$330/kg) based on the assumed operating parameters and material unit costs. If the variability related to process parameters and uncertainty related to cost parameters can be represented with probability distributions, then, Monte Carlo simulation can estimate the certainty with which the project objectives can be met. For this exercise, a normal distribution was assumed for the price of quinaldine, which is the most expensive raw material, with a mean value equal to that of the base case ($60/kg).

The annual throughput (or number of batches per year) is determined by the process cycle time. Since procedure P-8 that utilizes vessel R-102 is the time bottleneck, any variability in the completion of P-8 leads to uncertainty in the annual throughput. Variability in the completion of P-8 can be caused by variability in the operations of P-8 as well as by variability in the operations of procedures upstream of P-8. Common sources of process time variability in chemical manufacturing include:

1. Fouling of heat transfer areas that affect duration of heating and reaction operations.
2. Fouling of filters that affect duration of filtration operations.
3. Presence of impurities in raw materials that affect reaction rates.
4. Off-spec materials that require rework.
5. Random power outages and equipment or utility failures.
6. Differences in skills of operators that affect setup and operation of equipment.
7. Availability of operators.

Triangular probability distributions were assumed for the duration of the two main reaction operations and the filtration steps that precede P-8 (Table 47.4). Even though variability distributions were assigned to specific operations, it may be deemed more accurate to assume that they account for the composite variability of their procedures. If this type of analysis is done for an existing facility, historical data should be used to derive the probability distributions. Crystal Ball has the capability to fit experimental data.

The two decision variables considered in this study are the number of batches that can be processed per year and the unit production cost. These are KPIs important for production planning and project economics. The output variables, of the combined SuperPro Designer–Crystal Ball simulation, are quantified in terms of their mean, median, mode, variance,

TABLE 47.4 The Input Parameters Used for the Monte Carlo Simulation and Their Variation

Variable	Base Case Value	Distribution	Variation and Range
Quinaldine cost	60 ($/kg)	Normal	SD = 10 [30–90]
Chlorination reaction time (in P-1)	6 h	Triangular	[4–8]
Condensation reaction time (in P-2)	6 h	Triangular	[4–8]
Cloth filtration flux in P-3, P-5, P-7, and P-9	200 (L/m^2·h)	Triangular	[150–250]

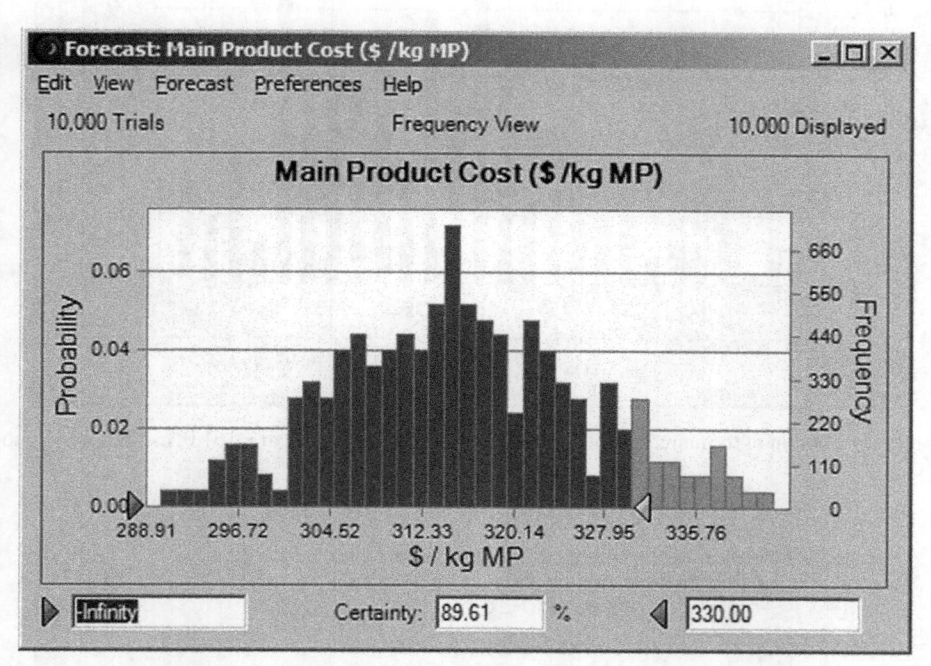

FIGURE 47.7 Probability distribution of the unit production cost (10 000 trials) (mean = 315.52, median = 315.03, SD = 10.49, range = 290.19–342.25).

and standard deviation. These results are shown in Figures 47.7 and 47.8 for the "unit production cost" and the "number of batches," respectively. Based on the assumptions for the variation of the input variables, we note that average values (mean/median/mode) calculated for the decision variables satisfy the objective. The certainty analysis reveals that we can meet the unit production cost goal (unit cost of under $330/kg) with a certainty of 89.6% (Figure 47.7). The certainty of meeting the production volume goal (of 27 000 kg or 156 batches) is only 82.4% (Figure 47.8). Such findings constitute a quantification of the risk associated with a process and can assist the management of a company in making decisions on whether to proceed or not with a project idea.

The dynamic sensitivity charts provide useful insight for understanding the variation of the process. They illustrate the impact of the input parameters on the variance (with respect to the base case) of the final process output, when these parameters are perturbed simultaneously. This allows us to identify which process parameters have the greatest contribution to the variance of the process and thus focus on them for process improvement. The sensitivity analysis for the *maximum number of batches per year* is displayed in Figure 47.9. The flux of the filtration operations has the greatest impact on the number of batches and consequently the annual throughput. If the management of the company is seriously committed to the annual production target, it would be wise to allocate R&D resources to the optimization of the filtration operations.

47.5 PRODUCTION SCHEDULING

After the process is developed and transferred to a manufacturing facility for clinical or commercial production, it becomes the job of the scheduler to ensure that all the activities are correctly sequenced and that the necessary labor, materials, and equipment are available when needed. The short-term schedule includes the upcoming production campaigns and may span from a week to several months. The general workflow begins with the long-term plan, which describes how much of each product should be made over the planning period. The long-term plan, which is described

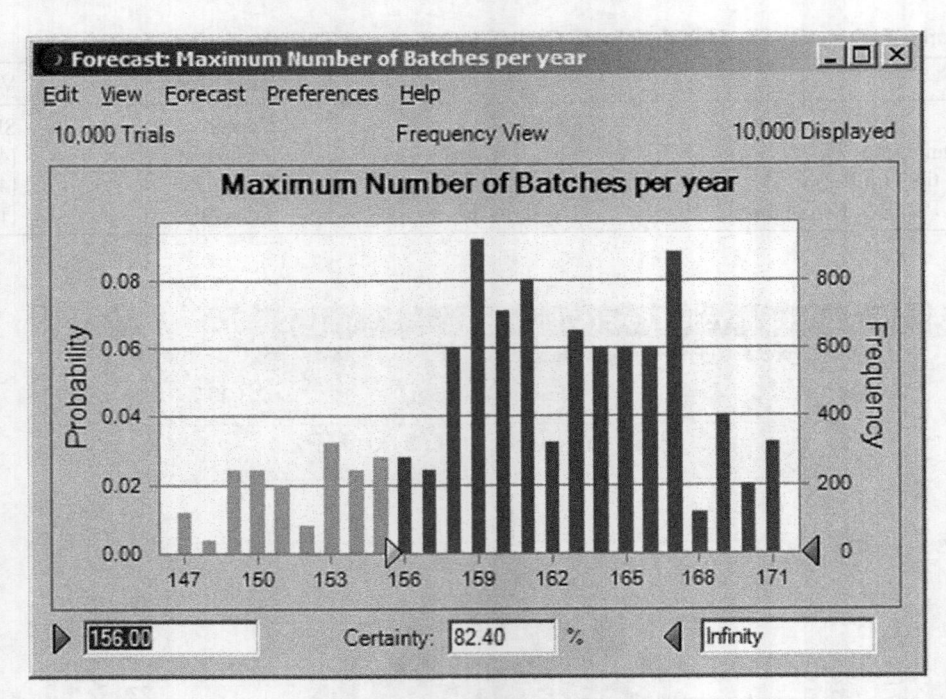

FIGURE 47.8 Probability distribution of the annual number of batches (10 000 trials) (mean = 161.0, median = 161, mode = 159, SD = 5.72, range = 147–171).

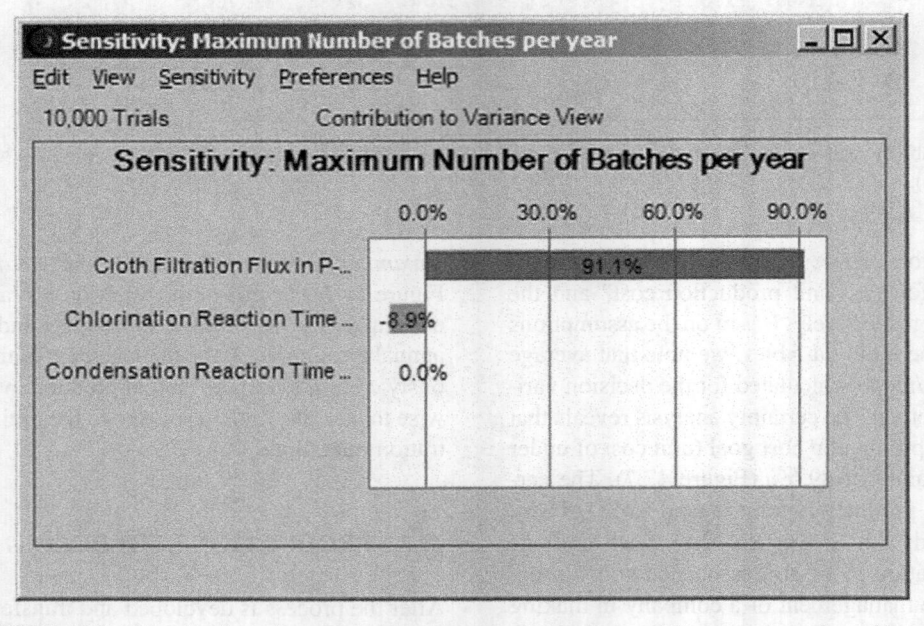

FIGURE 47.9 Contribution of uncertain parameters to the variance of the annual number of batches.

in Section 47.6, does not include details about process activities. The scheduler uses the plan and knowledge about the process and available equipment and resources to generate a detailed production plan, i.e. the short-term schedule, and communicate it to the appropriate staff. As the schedule is executed, there may be deviations between the schedule and the actual process execution. For example, tests may need to be redone, operations may take longer than assumed, or equipment may fail. The scheduler must recalculate the production schedule to reflect changes in resource availability and then notify the appropriate staff.

In addition to planning-and-scheduling tools, pharmaceutical companies use a variety of other plant systems to enable efficient and reliable manufacturing. ERP/MRP systems keep track of the quantity of resources, such as materials or labor. Manufacturing execution systems (MES) ensure that the process proceeds according to precise specifications. Process control systems interface with the equipment and sensors to carry out steps and to maintain the process parameters according to specification [17]. Short-term scheduling is often managed manually or with stand-alone systems, but it could potentially interface with ERP/MRP and even MES programs.

The following examples utilize SchedulePro as a scheduling tool. SchedulePro does not model the detailed material and energy balances of a process; it is mainly concerned with the time and resources that tasks consume. If a user is interested in detailed process modeling as well as scheduling, he/she can generate the process model in SuperPro Designer, perform the material and energy balances there, and then export the model as a recipe to SchedulePro for a thorough capacity planning or scheduling analysis in the context of a multiproduct facility. Relative to SuperPro, SchedulePro provides a variety of scheduling enhancements:

- The rigidity of recipe execution is relaxed with the introduction of "flexible shifts," which may delay the start of an operation, and "interruptibility," which may break the execution of an operation (if the resources it requires are not available).
- For every procedure, an equipment pool can be defined to represent the list of alternative equipment units that could potentially host that procedure.
- Auxiliary equipment (including pools of equipment) can be assigned to operations.
- Materials supplied or generated through operations can be linked to supply, intermediate, or receiving storage units.

The inclusion of this additional information in the production model is motivated mainly by the needs of the pharmaceutical/biotech industry where the manufacturing bottlenecks are often related to the use of auxiliary equipment (e.g. CIP skids, transfer panels) or to support activities (e.g. cleaning, buffer preparation), which tend to have flexible execution.

With the product recipes, facilities, and resources defined, simulation of a production plan in SchedulePro can proceed through the definition and scheduling of campaigns. A *campaign* is defined as a series of batches of a given recipe, leading to the production of a given quantity of product. A series of campaigns organized in a prioritized list constitute the production plan. As a finite capacity tool, SchedulePro attempts to schedule production of campaigns while respecting capacity constraints stemming from resource unavailability (e.g. facility or equipment outages) or availability limitations (e.g. equipment can only be used by one procedure at a time). Conflicts (i.e. violations of constraints) can be resolved by utilizing alternative resources declared as candidates within equipment pools, introducing delays or breaks if this flexibility has been declared in the corresponding operations, or moving the start of a campaign or batch to a time where the required resources are available. The automatically generated schedule can subsequently be modified by the user through local or global interventions in every scheduling decision. Through a mix of automated and manual scheduling, users can formulate a production plan that is feasible and satisfies their production objectives.

EXAMPLE PROBLEM 47.1 Campaign Planning for a Four-Step Process

In this example, a pharmaceutical product is synthesized in four distinct steps. Each step of the process includes a chemical reaction, as well as purification and solvent removal procedures. Based on a detailed analysis done in a process simulator like SuperPro Designer, the amounts of raw materials consumed and intermediate/product produced in each step have been calculated. The raw materials and intermediates that are consumed in order to produce the final product may be displayed in a synthesis tree, as shown in Figure 47.10. In this figure, "RM" prefixes denote raw materials, "Int" prefixes denote intermediates, and the final product is simply called "API." This diagram shows the amounts of the key materials that are required to produce a single 456 kg batch of AA-API. Reading from the bottom of the diagram to the top, it is apparent that 115 kg of RM-1 and 138 kg of RM-2 are required to produce 223 kg of AA-Int-1, 115 kg of RM-3 is required (along with the 223 kg of AA-Int-1) to produce 287 kg of AA-Int-2, etc. Note that Figure 47.10 has been simplified to *only* include the *key* input molecules and product of a given step; inexpensive raw materials, by-products, and solvents are ignored. Furthermore, this diagram is not "balanced" from a mass balance perspective due to the exclusion of by-products, waste, and inexpensive raw materials, as well as the fact that certain reagents are fed in stoichiometric excess, reactions do not necessarily proceed to 100% completion, and there is lost yield during purifications and transfers.

Also note that in this example (and in many batch industrial processes), there is not necessarily a 1-to-1 correspondence between batches of different intermediates. For instance, in this case only about 47% (i.e. 0.47 batches) of the AA-Int-1 intermediate that is produced in Step 1 is needed in order to eventually produce a single 456 kg batch of AA-API. In other words, Figure 47.10 shows the *relative* amounts of the intermediates and some key raw materials required for this process, and it therefore represents a simplified "bill of materials" (BOM) for a single batch of the final API.

The creation of a synthesis tree like the one shown in Figure 47.10 is facilitated by the use of stock keeping units

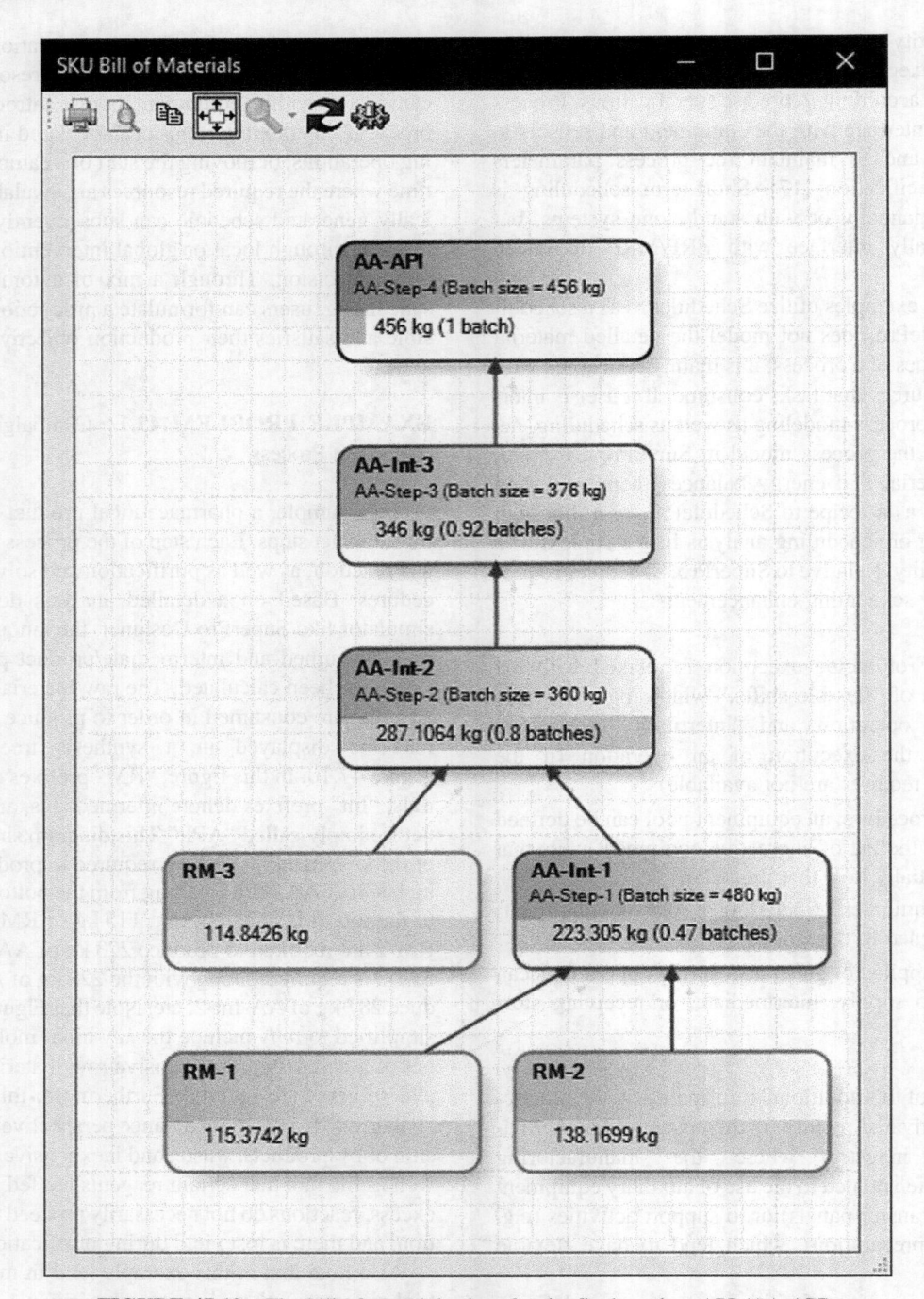

FIGURE 47.10 The bill of materials chart for the final product API (AA-API).

(SKUs), which represent each of the intermediates and the final product. The intermediate and product SKUs in this model were each assigned a BOM, which defines the relationship between a given SKU and other SKUs or materials needed to produce it. For instance, the BOM for the AA-Int-1 SKU specifies that 248 kg of RM-1 and 297 kg of RM-2 are required in order to produce a single (full-size) 480 kg batch of AA-Int-1. Note that the amounts shown in Figure 47.10 correspond to the quantities needed to produce *one batch of the final API* and that one batch of AA-API only requires about half a batch (223.3 kg) of AA-Int-1.

The use of SKUs is also beneficial because their BOM relationships allow the scheduling tool to automatically determine the number of batches that will be required to generate a specific amount of final product. As mentioned previously, there is not necessarily a 1-to-1 correspondence

TABLE 47.5 Batch Size and Demand for Each Step of the Process

Step No.	Batch Size (kg)	Number of Batches
1	480	6 (6 × 480 = 2880 kg of Int-1)
2	360	10 (10 × 360 = 3600 kg of Int-2)
3	376	11 (11 × 376 = 4136 kg of Int-3)
4	456	11 (11 × 456 = 5016 kg of AA-API)

between batch sizes of different intermediates. As a result, different numbers of batches will need to be performed for each step of the process. For instance, if the demand for AA-API is 5000 kg, the number of batches of each recipe step that need to be executed is shown in Table 47.5.

The batch totals in Table 47.5 are determined by calculating the minimum number of integer batches required for each step of the process in order to produce at least 5000 kg of AA-API product using full-size batches at each step. Although it may be possible to run batches below their full size, it is usually desirable to run them at full size in order to maximize equipment capacity utilization and labor resources. Note that these batch calculations assume there is no intermediate or final production inventory associated with these steps prior to their production campaigns. In addition, since full batches are produced at each step of the process, some excess inventory will remain at the end of the production run. This excess inventory can be used for the next set of campaigns for this drug. (Alternatively, batch sizes could be scaled so that all material from each step is consumed in the next step's campaign. This would be the preferred approach if the drug is made infrequently and any excess intermediate might expire before it can be used.)

The scheduling model for this example uses separate recipes for each step of the four-step process. These recipes capture the timing and duration of each individual operation, as well as the use of resources such as equipment, materials, utilities, labor, etc. A portion of the recipe block diagram for the first process step is shown in Figure 47.11. This recipe has eleven "unit procedures" (reaction, quench solution, product phase storage, etc.), although not all of them are displayed in this figure. Each unit procedure has one or more operations within it. The timing links between operations and procedures are displayed as arrows on the right side of the diagram (i.e. once the charge materials operation in the first procedure is completed, the react operation begins; after the react operation is completed, the transfer in quench operation begins; etc.). This recipe block diagram also displays the main equipment units (R-101, T-101, etc.) assigned to each procedure. Note that this recipe has been simplified to only include the key process operations; other operations, such as setup, equipment rinses, etc., have been excluded. However, additional details could be added if the desired scheduling and analysis activities that will be performed justify the additional

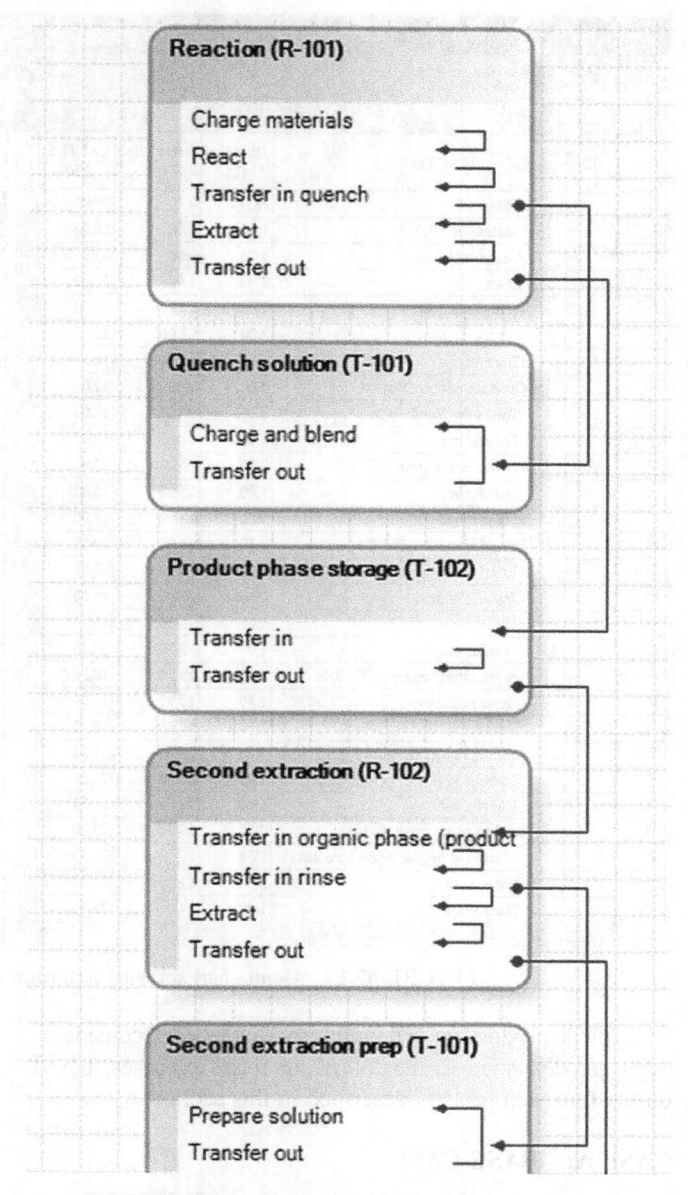

FIGURE 47.11 The recipe block diagram for Step 1.

effort required to model the use of plant resources more precisely.

A more detailed view of the scheduling specifications for the Step 1 recipe is provided by the recipe Gantt chart shown in Figure 47.12. This chart displays the durations as well as the start and end times (relative to the batch start) for each operation in the first seven procedures of the recipe. Each of the procedures is shown in dark color, while the operations within a procedure are shown in a lighter color underneath each procedure. This type of chart provides an excellent way to check a recipe to ensure that the durations and timing links between operations have been specified correctly.

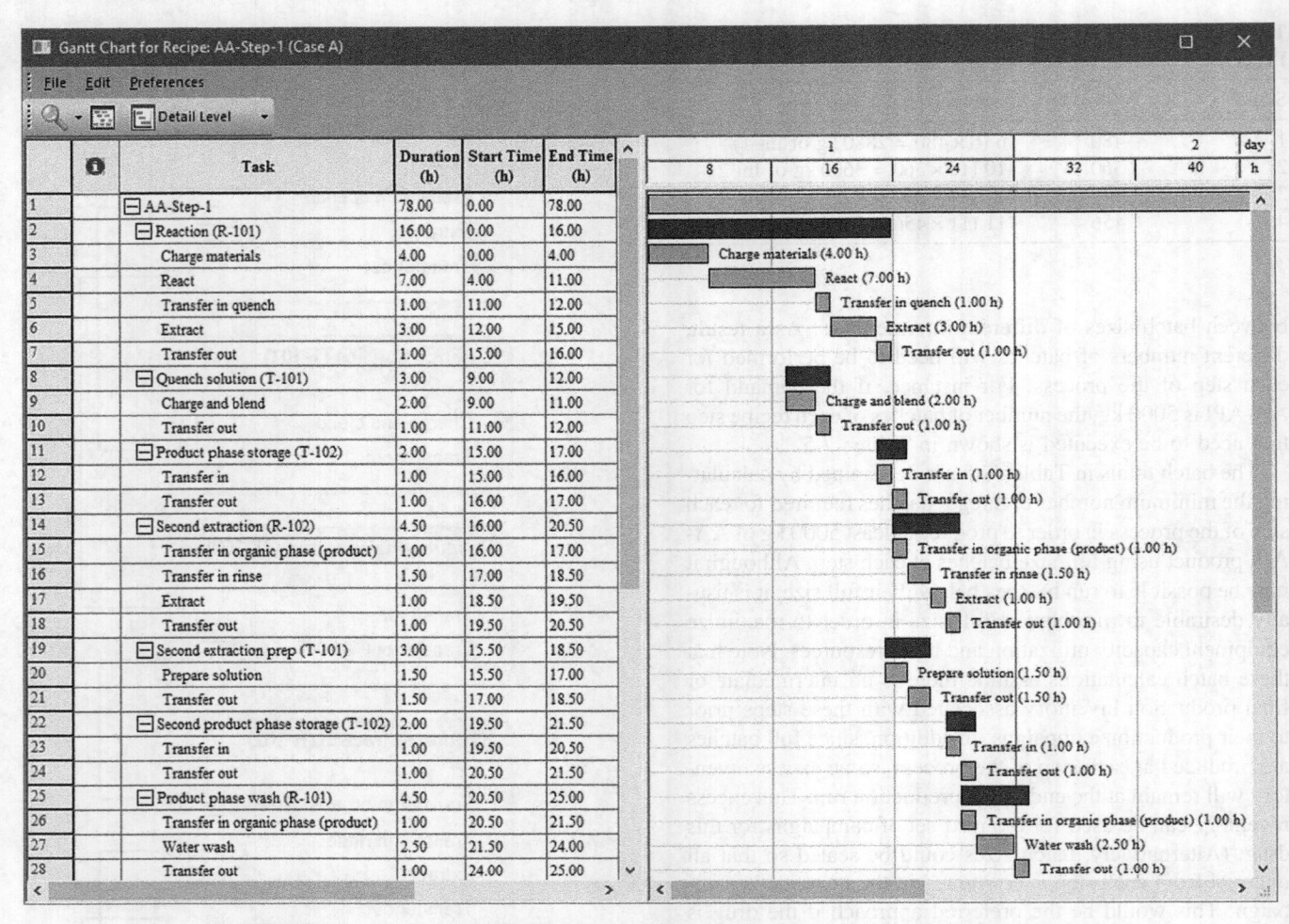

		Task	Duration (h)	Start Time (h)	End Time (h)
1		⊟ AA-Step-1	78.00	0.00	78.00
2		⊟ Reaction (R-101)	16.00	0.00	16.00
3		Charge materials	4.00	0.00	4.00
4		React	7.00	4.00	11.00
5		Transfer in quench	1.00	11.00	12.00
6		Extract	3.00	12.00	15.00
7		Transfer out	1.00	15.00	16.00
8		⊟ Quench solution (T-101)	3.00	9.00	12.00
9		Charge and blend	2.00	9.00	11.00
10		Transfer out	1.00	11.00	12.00
11		⊟ Product phase storage (T-102)	2.00	15.00	17.00
12		Transfer in	1.00	15.00	16.00
13		Transfer out	1.00	16.00	17.00
14		⊟ Second extraction (R-102)	4.50	16.00	20.50
15		Transfer in organic phase (product)	1.00	16.00	17.00
16		Transfer in rinse	1.50	17.00	18.50
17		Extract	1.00	18.50	19.50
18		Transfer out	1.00	19.50	20.50
19		⊟ Second extraction prep (T-101)	3.00	15.50	18.50
20		Prepare solution	1.50	15.50	17.00
21		Transfer out	1.50	17.00	18.50
22		⊟ Second product phase storage (T-102)	2.00	19.50	21.50
23		Transfer in	1.00	19.50	20.50
24		Transfer out	1.00	20.50	21.50
25		⊟ Product phase wash (R-101)	4.50	20.50	25.00
26		Transfer in organic phase (product)	1.00	20.50	21.50
27		Water wash	2.50	21.50	24.00
28		Transfer out	1.00	24.00	25.00

FIGURE 47.12 Gantt chart showing a portion of procedures and operations in the Step 1 recipe.

Typical production scheduling activities and considerations are described in the following series of cases, based on the four-step process described in this section.

CASE A: BASE CASE

Problem Statement

Assess whether 5000 kg of the final product can be produced in the next 3 months using a single production suite.

Assumptions

- The production suite contains all necessary equipment to run each step of the process (i.e. two reactors, two blending tanks, a chromatography column, a filter, and a dryer).
- The production plan will be fulfilled using four campaigns in which all batches of the first intermediate are run first, all batches of the second intermediate are run next, etc.
- The approximate batch-to-batch cycle times for the four steps are 27, 29.5, 36, and 39 hours, respectively. Each campaign will run batches back-to-back as quickly as possible, with batch start times based on the cycle time bottleneck for each recipe (i.e. the equipment with the least idle time between consecutive batches).
- A three day changeover is required between campaigns of each step, to account for extended equipment cleaning and other changeover activities.

Methodology

The plant scheduler must create a schedule that meets the demand for product while also respecting the resource limitations of the facility. In this case, the demand forecast requires 6, 10, 11, and 11 batches of Steps 1, 2, 3, and 4, respectively (as shown previously in Table 47.5). Therefore, campaigns with an appropriate number of batches were defined for each processing step. The four-campaign production plan was then scheduled.

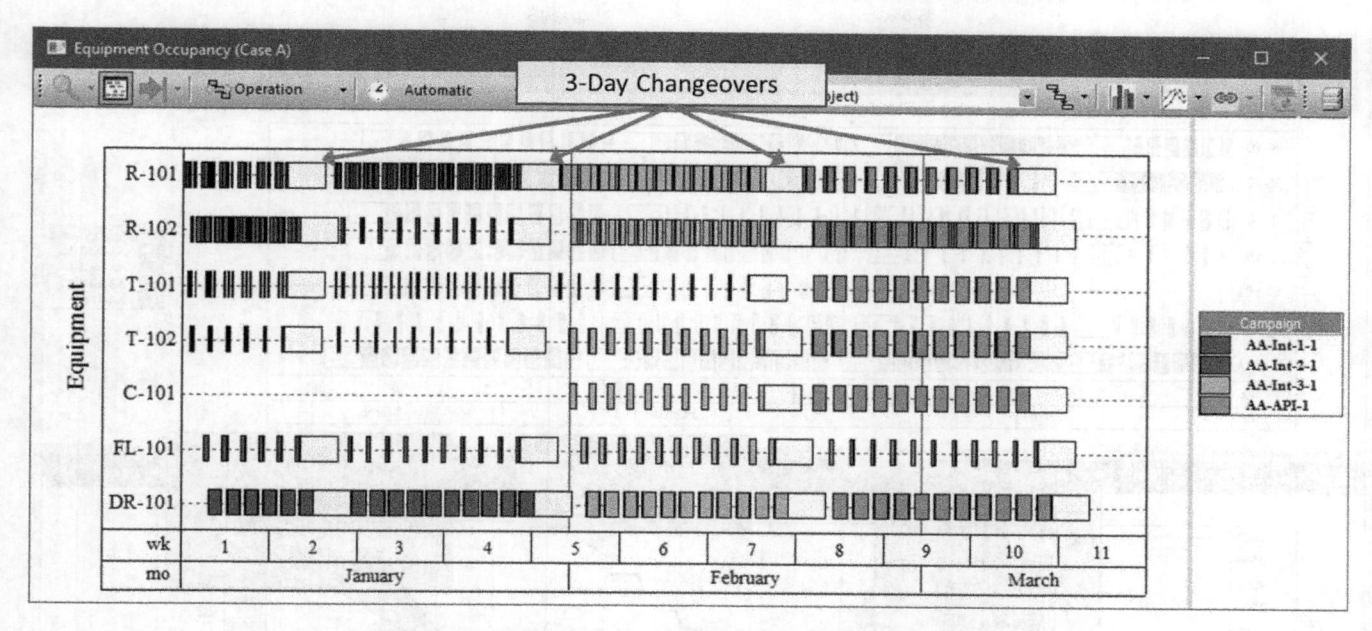

FIGURE 47.13 Equipment occupancy chart for Case A (base case).

Results and Conclusions

The equipment occupancy results for all four campaigns are displayed in Figure 47.13. Each campaign is shown as a separate color in the chart, and the three day changeovers between campaigns are shown in white. This production plan easily achieves the objective of producing 5000 kg of AA-API within 3 months.

CASE B: MULTIPLE SMALL CAMPAIGNS OF EACH STEP

Problem Statement

Long campaigns of a given step, such as the ones in Case A, are often advantageous because they utilize plant equipment relatively efficiently. However, in Case A, no AA-API product will be ready for delivery to customers until the first Step 4 batches are completed during week 8. This may be unacceptable if customers need the product sooner. Another issue with the Case A production plan is that large amounts of inventory must be stored in the warehouse. This can be seen in Figure 47.14, which shows that over 4000 kg of inventory space must be available for Int-3 at the end of week 7. Thousands of kg of inventory space is also required for Int-1 and Int-2 earlier in the schedule. A more effective supply chain management strategy is to minimize working capital associated with inventory [18]. In Case B, the goal is to determine whether the production requirements from Case A (i.e. 5000 kg of the AA-API product manufactured within the next 3 months using a single production suite) may be met using much smaller campaigns.

Assumptions

- The original long campaigns of each recipe will be broken into multiple mini-campaigns of 2 to 4 batches, each of which is staggered over the course of the production plan. The total number of batches of each step of the process is the same as in Case A (i.e. there are a total of 6 Step 1 batches, 10 Step 2 batches, etc.).
- As before, there is a three day changeover between campaigns.

Methodology

In the scheduling tool, a new set of multiple mini-campaigns was created and scheduled.

Results and Conclusions

The equipment occupancy results for this case are shown in Figure 47.15. As before, each step is shown in a different color on the figure. In this case, the first three batches of the final product are run in week 4, which is much sooner than in Case A. This provides a shorter time to market. Another advantage of the Case B "small-campaign" strategy is that the in-process inventory levels are greatly reduced. This can be seen in the equipment occupancy chart and the linked Int-3 inventory chart in Figure 47.16. In this case,

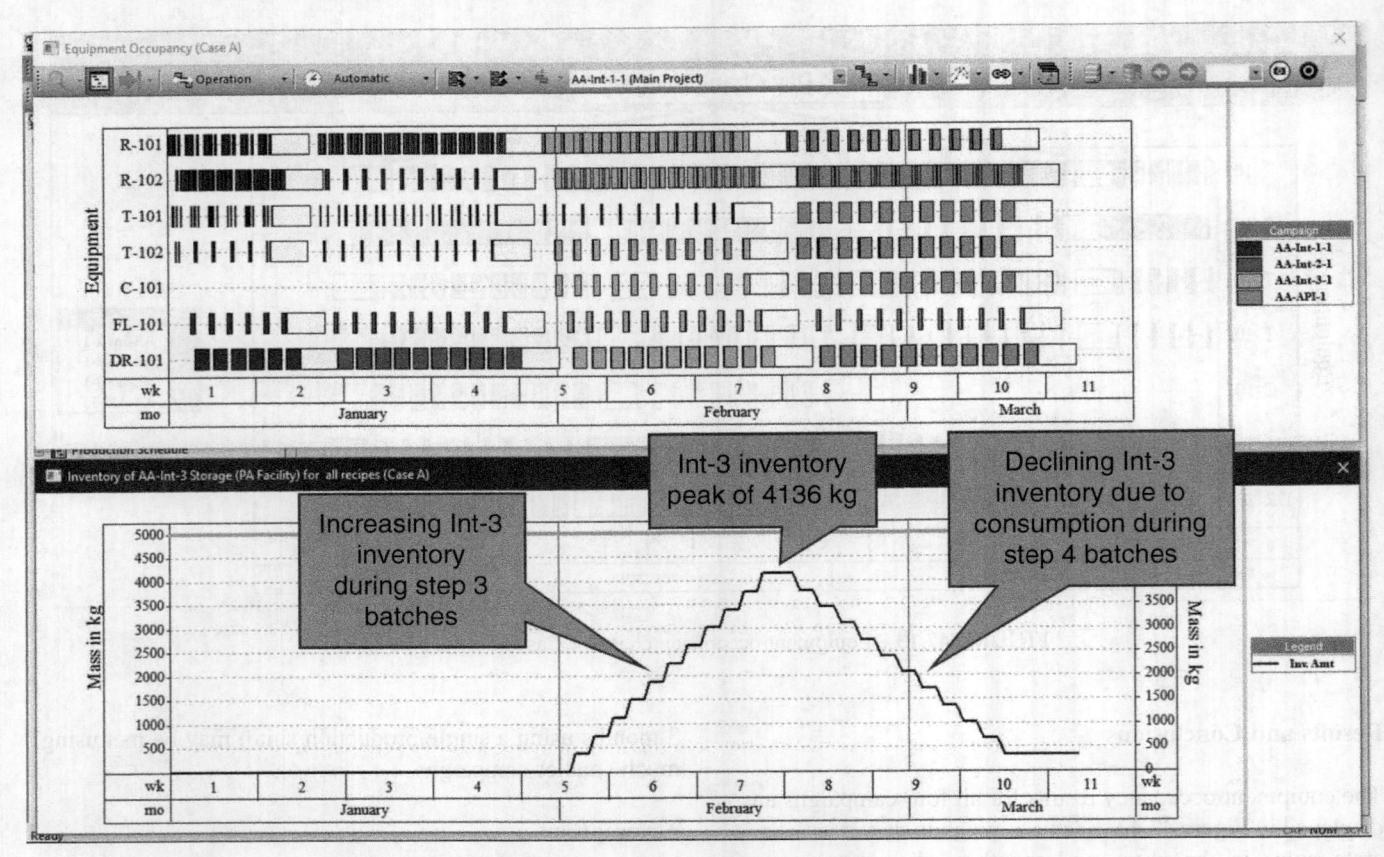

FIGURE 47.14 Int-3 inventory level during the Case A production plan.

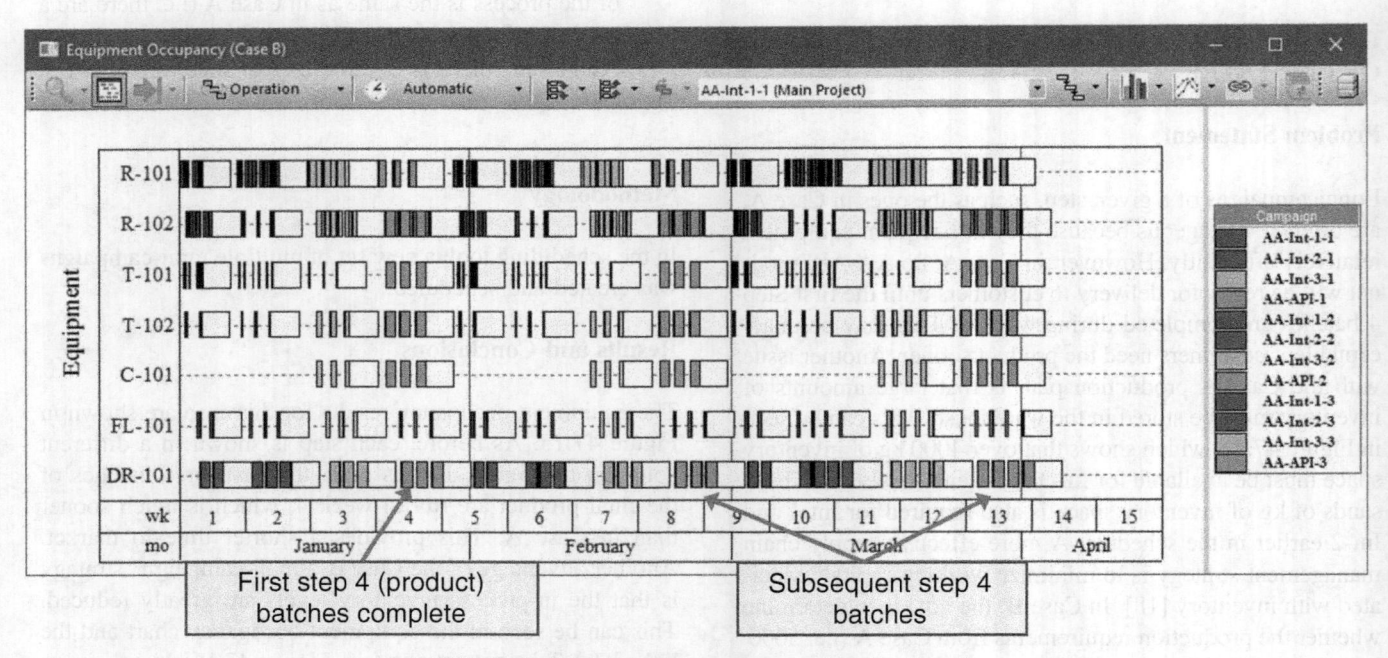

FIGURE 47.15 Equipment occupancy chart for Case B.

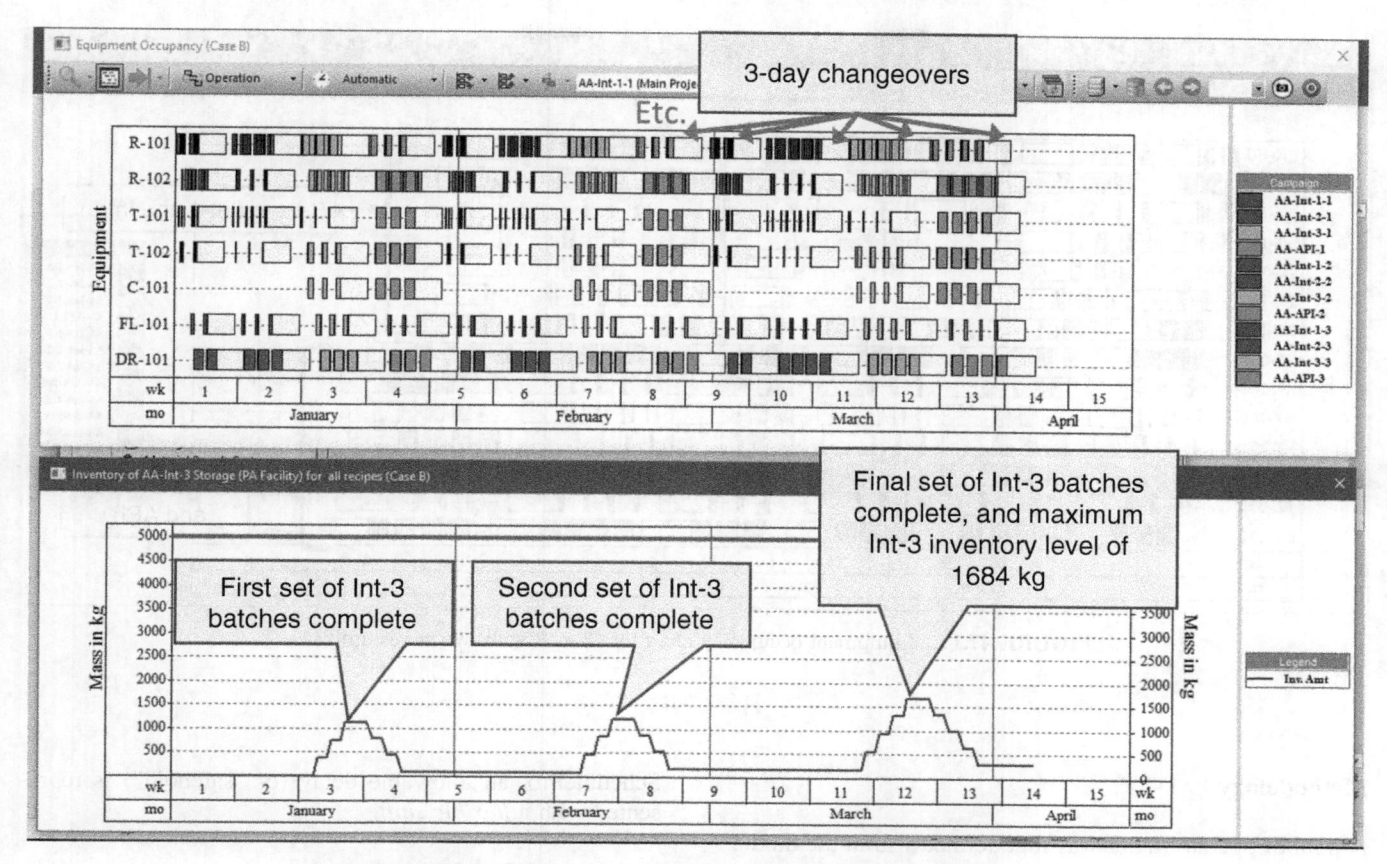

FIGURE 47.16 Int-3 inventory level during the Case B production plan.

the in-process inventory for Int-3 never goes above 1684 kg. The in-process inventory limit for the other two intermediates is similarly reduced (compared to their levels in Case A). The use of multiple small campaigns rather than single large campaigns represents a scenario that is closer to a JIT manufacturing plan. However, Case B also results in a significantly less efficient overall schedule due to the higher number of time-consuming changeovers (shown in white) between these short campaigns. As a result, the total duration (or "makespan") for the production plan increases to over 13 weeks (from under 11 weeks previously). This fails to meet the objective of producing 5000 kg of AA-API product within 3 months. Even though the production plan would come close to meeting that objective, if there were delays in any of the batches (which is a common occurrence in industry), delivery of the product might be seriously delinquent.

Note that at this point, the scheduler could choose to lengthen each campaign in order to find a solution that meets the 3 month production objective while also producing final product batches early in the schedule. Alternatively, the schedule could be condensed by running campaigns in multiple production suites simultaneously. This scenario is displayed in Figure 47.17. The use of two production suites (denoted by equipment with a 1XX identifier or a 2XX identifier) reduces the makespan for these campaigns to less than nine weeks.

Although Figure 47.17 shows that it would be possible to meet the three month production deadline with multiple suites, for the sake of the remaining cases in this section, we will use the "single long campaigns of each step within a single suite" strategy that was demonstrated in Case A.

CASE C: LABOR RESTRICTIONS

Problem Statement

The scheduler must determine whether the required production level described in Case A can be achieved with a crew size of two operators assigned to this production suite.

Assumptions

- The operators work in shifts, and the plant runs 24 hours per day and 7 days per week. All shifts have the same two-operator limit in the suite.

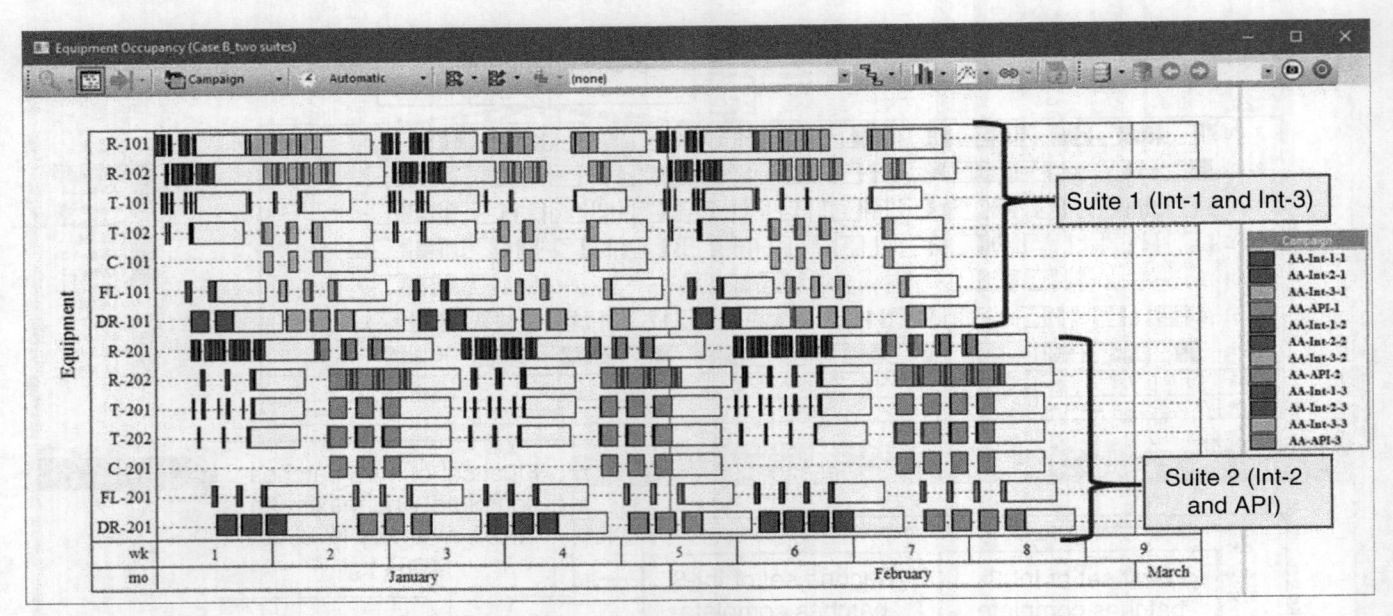

FIGURE 47.17 Equipment occupancy chart for Case B with two suites utilized.

Methodology

A pool of operator labor was first defined within the model. Then specific operator labor levels were assigned to various operations in the recipes. A partial list of the labor requirements associated with the Step 1 recipe is shown in Figure 47.18. Note that the labor level for some operations remains at zero operators since there are some operations that are part of two different procedures (such as a "transfer out" from one vessel and its associated "transfer in" to another vessel) and assigning labor to both operations would result in double-counted labor. Also note that fractional operators (e.g. 0.5 operators) were assigned to operations that do not require an operator to be constantly working on them (such as a long reaction or drying procedure, which may only require occasional checkups).

Next, a small amount of flexibility was added to the recipes so that conflicts related to resources such as labor could be automatically resolved by allowing the scheduling tool to shift the start times of certain operations (e.g. delay an unload filter operation until sufficient resources are available or perform an operation such as a setup operation earlier than its default start time). Note that pharmaceutical process scheduling is unlike scheduling of assembly manufacturing because delays or stoppages in various steps of a pharmaceutical process may be unacceptable due to chemical stability limitations. In this example, most tasks must progress one after the other without significant delay or interruption when the product is dissolved in solution, but the product may be held for up to six hours when it is in a solid form. Within SchedulePro, an allowable delay (or "safehold") is represented with a *flexible shift*.

Results and Conclusions

After the labor assignments were completed for all recipes, the production plan was scheduled, and the equipment occupancy chart was generated (see Figure 47.19). This chart shows that after the production plan begins, the number of operators required at any given time varies from 0 to 2.5. Since the goal is to use no more than two operators at any time, this initial schedule does not meet our objective (i.e. the two operators would sometimes have too many activities that they need to perform or monitor simultaneously). However, by activating a constraint on the labor level, the production plan can be automatically rescheduled to conform to the two-operator limit. The updated schedule is shown in Figure 47.20. This production plan "load levels" the labor in order to keep it at or below the two-operator limit. It does this by shifting certain operations or even shifting entire batches in order to ensure the labor limit is not exceeded. Notice that by enforcing the labor limit, the batches in the Step 3 campaign require a small delay between consecutive batch starts. This results in a delay of several days (relative to the unconstrained labor scenario in Figure 47.19) for the completion of the production plan.

Short-term schedulers often need to account for fluctuations in labor availability as well. For instance, there may be fewer (or zero) operators available on weeknights, weekends, and holidays. Variability in labor limits may be

Labor Assignments for Recipe AA-Step-1

Procedure	Operation	Operator (in PA Facility)
Reaction (R-101)	Charge materials	1.00
	React	0.50
	Transfer in quench	0.00
	Extract	0.50
	Transfer out	0.50
Quench solution (T-101)	Charge and blend	1.00
	Transfer out	0.50
Product phase storage (T-102)	Transfer in	0.00
	Transfer out	0.50
Second extraction (R-102)	Transfer in organic phase (product)	0.00
	Transfer in rinse	0.00
	Extract	0.50
	Transfer out	0.50
Second extraction prep (T-101)	Prepare solution	1.00
	Transfer out	0.50
Second product phase storage (T-102)	Transfer in	0.00
	Transfer out	0.50
Product phase wash (R-101)	Transfer in organic phase (product)	0.00
	Water wash	0.50
	Transfer out	0.50
Third product phase storage (T-101)	Transfer in	0.00
	Transfer out	0.50
Distillation - Precipitation (R-102)	Transfer in organic phase (product)	0.00
	Solvent switch (distillation)	0.50
	Distillation	0.50
	Pricipitation	0.50
	Transfer out	0.50

FIGURE 47.18 Labor requirements for the Step 1 recipe.

accounted for by setting different limits for different shifts and/or dates.

Maintaining the Schedule

During execution of the schedule in the plant, the timing of activities may deviate from the original plan. Some of the causes of these timing differences include planned maintenance (e.g. preventive maintenance (PM) of equipment), unplanned outages (e.g. equipment failures), operational delays due to a lack of resources or process variability, etc. Delays such as these, as well as campaign modifications and new additions, require schedulers to update the production schedule regularly with the latest information. They must also decide how to resolve any future resource conflicts that are created by the schedule changes. Accounting for changes to a complex schedule is difficult and error prone in traditional tools such as Excel. In contrast, modern constraint-based scheduling applications allow a conflict-free production plan to be quickly and easily scheduled. In the subsequent pages, SchedulePro is used to demonstrate several typical types of scheduling modifications. A web-based application is then used to demonstrate how the updated production plan may be communicated to various stakeholders within an organization.

CASE D: ACCOUNTING FOR EQUIPMENT OUTAGES

Problem Statement

Reconciling production activities with equipment maintenance activities is one of the scheduler's routine tasks. From

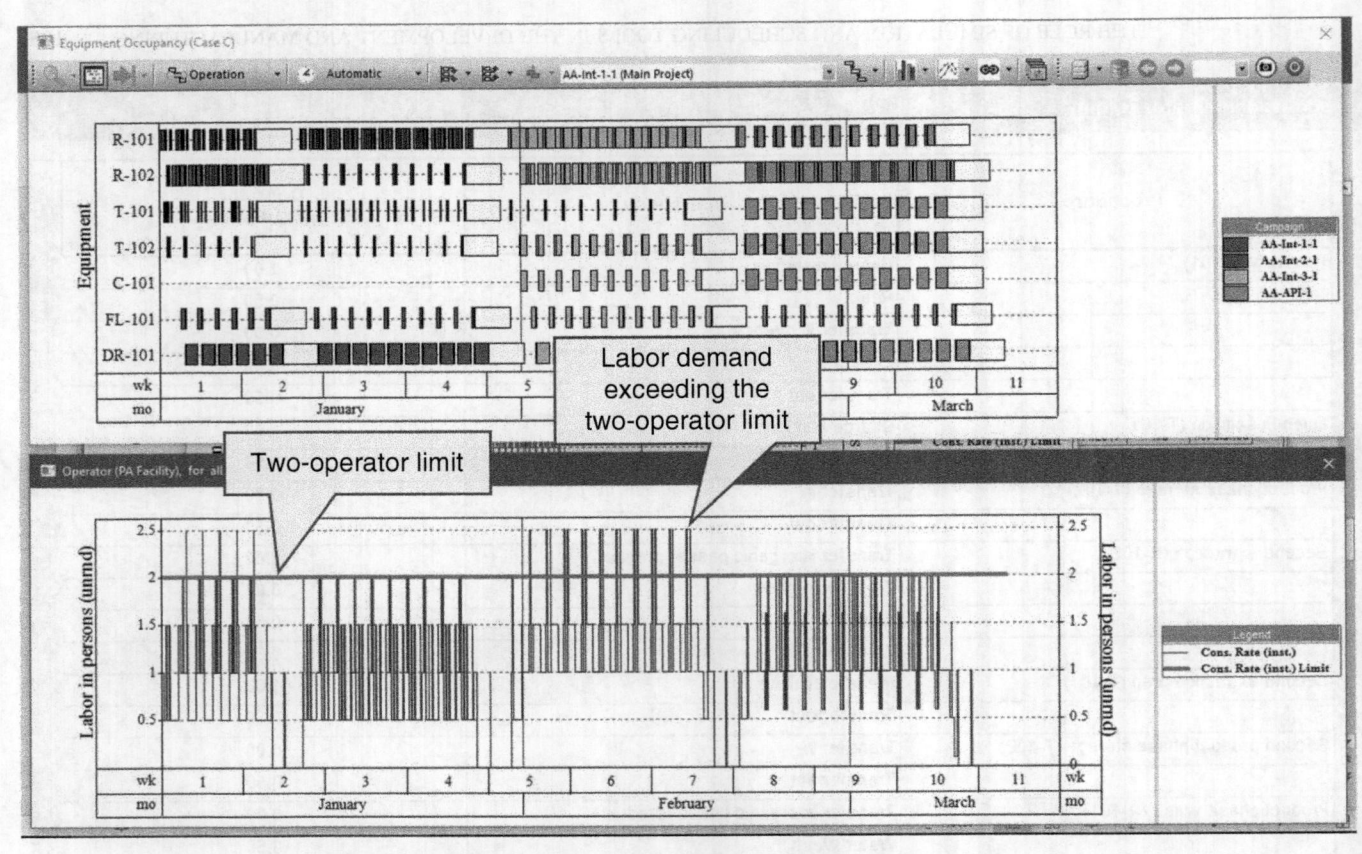

FIGURE 47.19 Production plan for Case C with unenforced labor constraint.

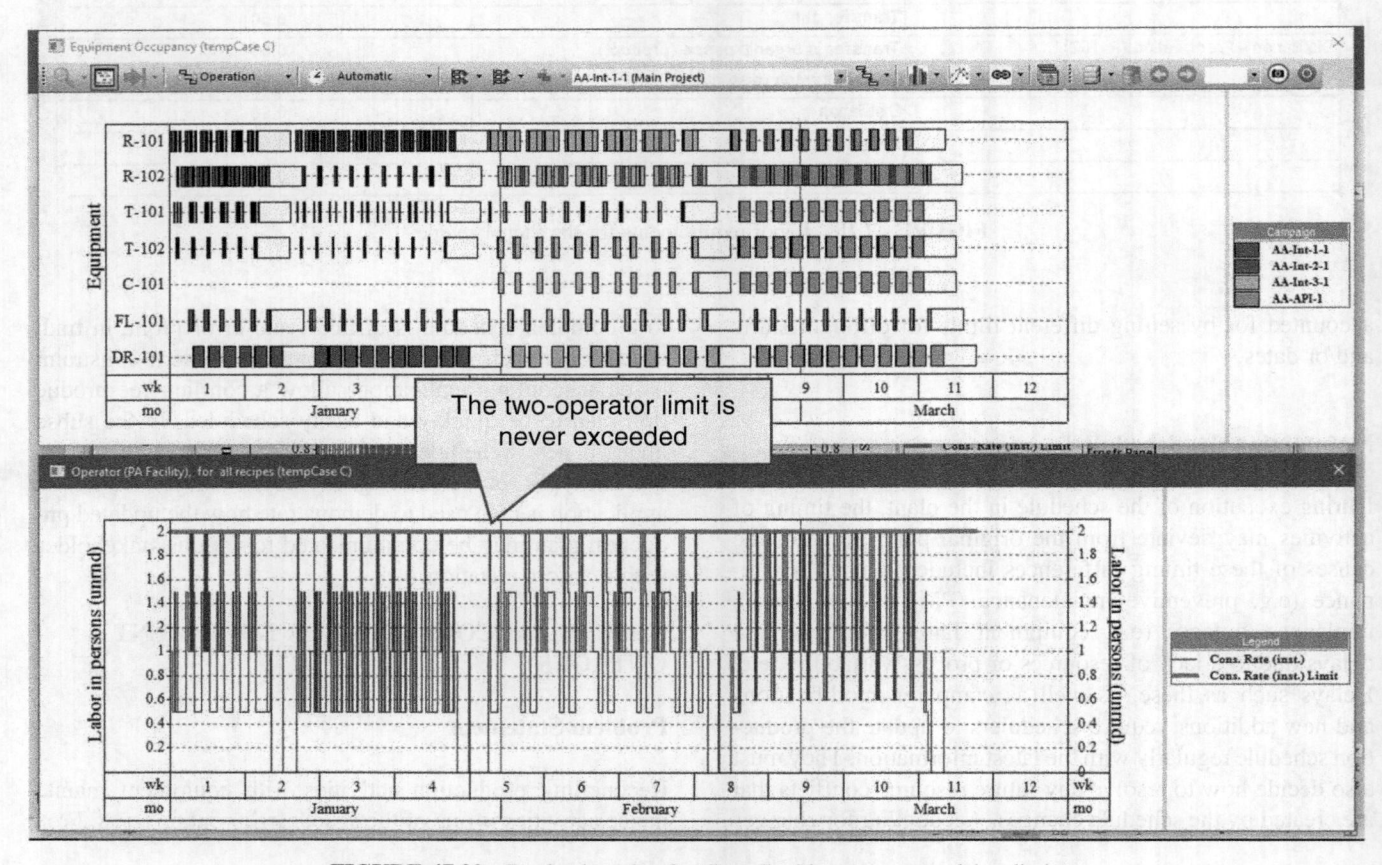

FIGURE 47.20 Production plan for Case C with two-operator labor limit.

a scheduling standpoint, maintenance creates periods of unavailability. Maintenance may be fixed to a particular date and time, or it may be allowed to "float" within a certain time frame.

Assumptions

In this case, the production schedule has progressed to the end of week 6. At this point, the motor for the agitator in tank T-102 begins to fail and needs to be replaced as soon as possible. Therefore, critical maintenance work must be scheduled for T-102. This work is expected to take five full days, beginning on 12 February (i.e. shortly after the current procedure in T-102 is complete). The scheduler needs to determine the impact of this maintenance on the schedule. An eight hour PM activity also needs to be performed on filter FL-101, although this maintenance can be done at any time prior to the end of the month.

Methodology

The fixed maintenance associated with T-102 was added as a hard constraint associated with that equipment unit's schedule. This outage will take priority over processing activities, so it may cause delays in the schedule. In contrast, to accommodate the PM for filter FL-101, the scheduler creates a "batch" of maintenance for that unit and allows it to occur during the first convenient window of opportunity prior to the end of February. This PM batch is given a lower priority than the processing batches, and therefore the brief PM activity will be allowed to "float" to a convenient time when that filter is not needed for the process.

Results and Conclusions

Figure 47.21 illustrates each type of maintenance. Tank T-102 has a firm maintenance outage for five days at the beginning of week 7 (the outage is shown with a gray box). This causes conflicts with subsequent batches that need to use T-102 (the conflicting activities are outlined in red on the chart). In contrast, the PM activity in FL-101 is automatically scheduled to occur at a time when it will *not* cause a usage conflict. Since there is not enough time to complete the FL-101 PM activity prior to the start of the next production procedure that utilizes the filter, the PM activity is instead scheduled to begin immediately after the next production procedure is completed. Note that a vertical "current timeline" has also been included on this chart in order to differentiate between past activities and future activities and to facilitate monitoring production progress. The current timeline results in the division of activities into three categories: completed (displayed by a crosshatched pattern), in progress (diagonal hatch), and not started (filled pattern).

To eliminate the conflict caused by the emergency maintenance in T-102, the scheduling tool "plans around" the outage by delaying the batch that uses this tank (as well as subsequent batches). This is shown in Figure 47.22.

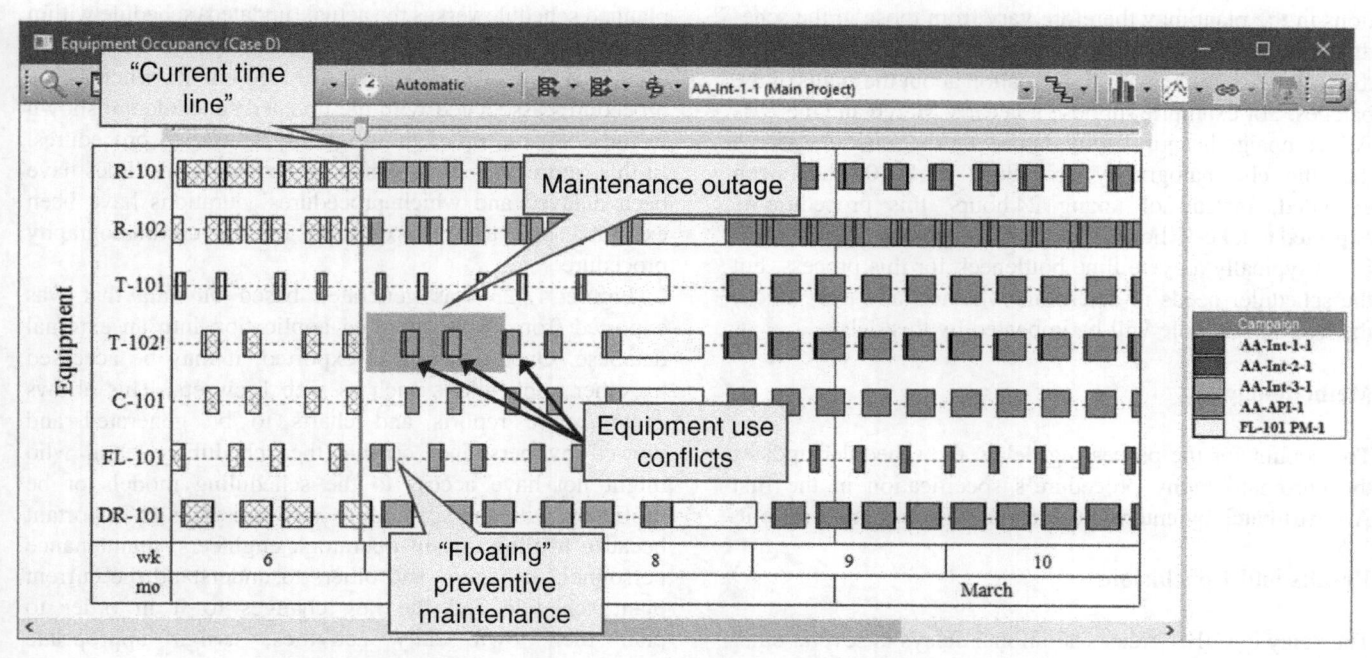

FIGURE 47.21 Equipment occupancy chart for Case D after maintenance is added.

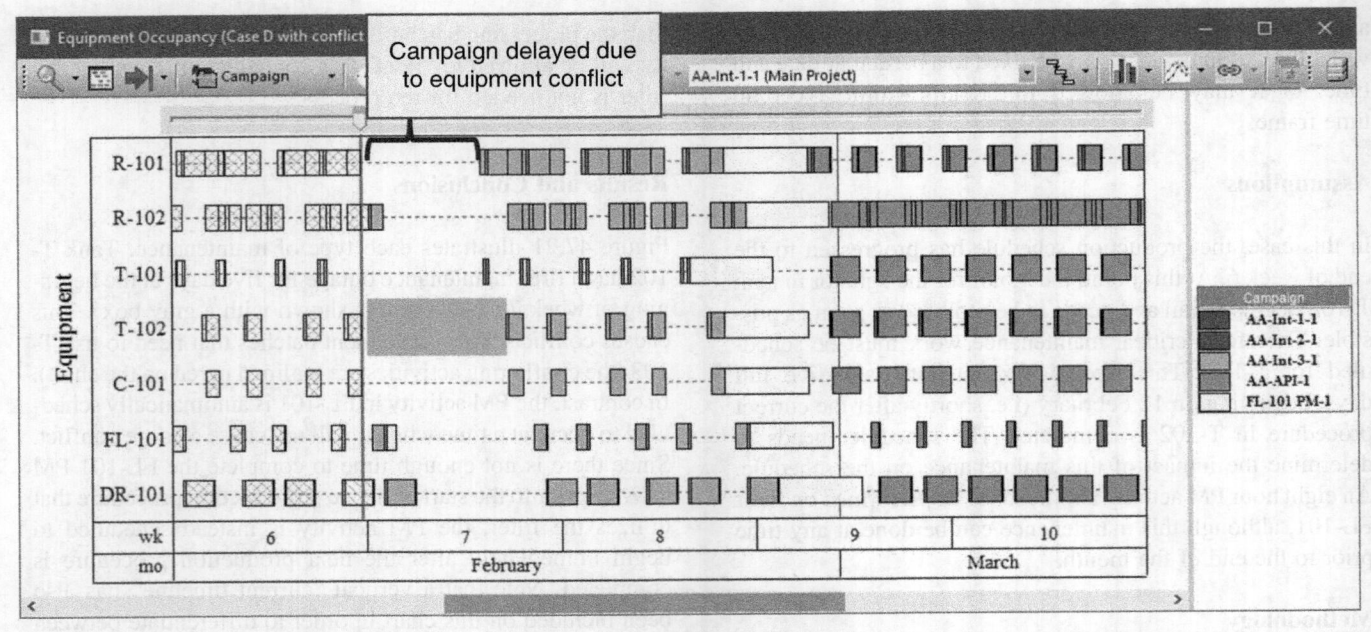

FIGURE 47.22 Equipment occupancy chart for Case D after T-102 conflict is resolved.

CASE E: ACCOUNTING FOR AN OPERATIONAL DELAY

Problem Statement

During pharmaceutical processing, the completion of an operation is not always determined based on a set duration. The concentration of a key component may, for example, be the primary specification. The actual durations of operations in the plant may therefore vary from those in the scheduling recipe. As a result, the scheduler must regularly update the schedule based on new information about the status of the batches. For example, suppose it is early March and the AA-API campaign has just begun. However, the scheduler learns that the chromatography procedure in C-101 has been extended. Instead of taking 24 hours, this procedure is expected to take 36 hours. The chromatography column itself is not typically a cycle time bottleneck for this process, but the scheduler needs to determine whether (and how much) the overall schedule will be impacted by this delay.

Methodology

To account for the processing delay, the scheduler updates the chromatography procedure's specification in the first AA-API batch by entering the actual 36 hour duration for it.

Results and Conclusions

The delay in C-101 creates additional delays in several other units. This includes the operations feeding the column from tanks T-101 and T-102, as well as the eluent receipt operation in R-102. At this point, the scheduler may manually resolve the conflicts, or he/she may allow the scheduling tool to attempt to resolve conflicts automatically. The conflict resolution in this case results in a half-day delay for the remaining batches in the campaign, relative to their initially planned batch start times. The timing of the procedures associated with the first batch of the AA-API recipe is displayed in Figure 47.23. This figure shows a representation of the planned schedule versus the actual (updated) schedule within a web page. The procedures associated with the original (planned) schedule are shown in faded colors, whereas the procedures associated with the updated schedule are shown in full color on top of the originally scheduled procedures. In this figure, it is clear which procedures' start times have been delayed and which procedures' durations have been extended due to the extension of the chromatography procedure.

Figure 47.23 was generated based on data that was exported from the scheduling application into an external database. Once the data is exported, it may be accessed by other applications such as web browsers. This allows customizable reports and charts to be generated and viewed by personnel outside the scheduling team, who might not have access to the scheduling models or be proficient with the scheduling software. This is important because it allows plant operators, engineers, maintenance personnel, managers, and others to understand the current plant schedule and the key changes to it in order to plan their own daily activities, assign appropriate resources, etc.

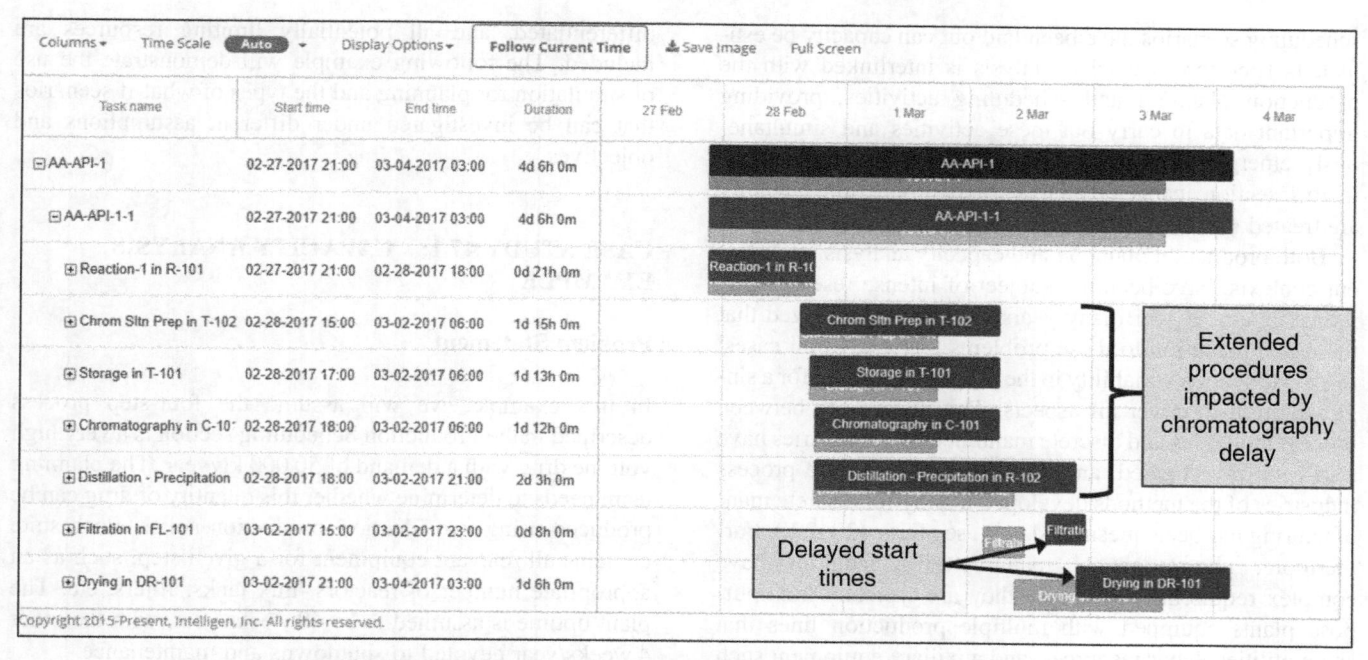

FIGURE 47.23 "Planned-versus-actual" schedule information viewed in a web page.

47.6 CAPACITY ANALYSIS AND PRODUCTION PLANNING

The short-term scheduler usually starts with a rough long-term production plan that is based on product demand, estimated plant capacity, and/or inventory constraints. Since these long-term plans generally do not require all the resource and timing details associated with short-term scheduling, production planners can create their long-term plans in systems with less scheduling functionality, such as an ERP tool. Alternatively, a scheduling tool with simple recipes can be an effective planning system. Regardless of the specific planning system used, an automated link between planning and scheduling can streamline both the planning process and the scheduling process. For example, the scheduler may download campaign information with estimated dates from the planning system. As the schedule progresses, the scheduler uploads revised date and production information to the planning system. This type of integration saves time and reduces the potential for human error during the planning-and-scheduling cycle.

In order to generate the initial production plan, it is important to consider the approximate plant capacity. Capacity is a measure of how much product a manufacturing system can make. The amount of product manufactured in a given time period (hour, day, week, etc.) or the time required to produce a given quantity of product are the most intuitive and commonly used measures of capacity. The capacity of a manufacturing system should, on average, exceed demand. However, excess capacity is costly and undesirable [19]. Increasing capacity to meet demand might require capital investments in equipment and buildings or an extension of the manufacturing uptime (through labor overtime or additional shifts). *Effective capacity* is the actual capacity achieved in practice. The effective capacity is usually less than the nominal plant capacity due to equipment maintenance, unexpected breakdowns, scheduling inefficiencies, labor unavailability, and other factors.

The need for an estimate of the plant capacity arises in different supply chain management activities, including aggregate planning (i.e. generating feasible long-range or medium-range production plans that can satisfy expected lumped demand for a range of aggregate products), inventory management, batch sizing, and operation scheduling. Depending on the complexity of the production system, the range of different products produced, and the diversification of their routings (recipes), the level of difficulty in estimating a plant's capacity can vary from trivial to formidable. The capacity of a single-product batch plant depends only on the batch size, the cycle time, and the allocation of production time. If greater capacity is required, then either the production time should be extended or the cycle time reduced by removing bottlenecks. However, in multiproduct or multipurpose facilities with complex material flows, multiple equipment used in parallel, shared resources, and sequence-dependent changeover and cleaning times, the estimation of the capacity is far from trivial. In fact, in these cases, capacity estimates emerge through the same activities that capacity analysis is supposed to serve, i.e. planning and scheduling. In other words, only after specific production planning and scheduling scenarios have been laid out can capacity be estimated. Therefore, capacity analysis is interlinked with the production planning and scheduling activities, providing important data to carry out these activities and simultaneously emerging as their outcome. This is the reason why, in this section, capacity analysis and production planning are treated simultaneously.

Both production planning and capacity analysis, in different contexts, have been the subject of intense research and industrial activity for many years. It is now recognized that there is no solution to these problems that can fit all cases; there is too much variability in the problem structure for a single solution to cover all aspects. The differences between process industries and discrete manufacturing industries have also been investigated, and the applicability in the process industries of the methods developed mainly for discrete manufacturing has been questioned (e.g., see Refs. [20–22]). Furthermore, pharmaceutical manufacturing facilities have complex requirements because they are typically multipurpose plants equipped with multiple production lines that share utilities, labor resources, and auxiliary equipment such as CIP skids, transfer panels, delivery lines, and occasionally main equipment. In addition, production is typically campaigned, and considerable changeover time is often required between campaigns of different products. API synthesis, in particular, is characterized by complex material flows and the need to handle and store a variety of required intermediates.

Simulation is an appropriate tool to cope with the complexity of production planning and capacity analysis in pharmaceutical manufacturing. Rather than attempting to formalize a single model and come up with a single solution (as optimization-based methods do), simulation allows the planner to formulate and analyze different scenarios and select the one that best fits the objectives and constraints of the problem. Such "what-if" analyses can generate feasible production plans utilizing the available capacity or provide justifications for facility expansions and/or outsourcing of production. The types of capacity analysis questions that can be answered using simulation will be demonstrated in this section with the use of the software tool SchedulePro.

47.6.1 Simulating the Production Plan

Production planning involves assigning key facility resources to processing tasks. A simulation-based approach can be used to support both planning/capacity analysis and scheduling activities. The level of detail included in the simulation model is the only difference between the two. In planning, the recipe representations are coarse, products may be lumped in aggregates with similar production recipes, and only the most basic resources are considered. As a result, a high-level planning model may be created very quickly. In scheduling, recipes are expanded to their fullest detail, products are differentiated, and all potentially limiting resources are included. The following example will demonstrate the use of simulation for planning and the types of what-if scenarios that can be investigated under different assumptions and objectives.

CASE STUDY 47.1: CAPACITY ANALYSIS EXAMPLE

Problem Statement

In this example, we will assume the four-step process described in the Production Scheduling section is a very high volume drug with a demand of 50 000 kg/year. The planning team needs to determine whether this quantity of drug can be produced using two identical production suites. Each suite contains all relevant equipment for a given step, such as an appropriate number of reactors, mix tanks, filters, etc. The plant uptime is assumed to be 48 weeks/year, with the other 4 weeks/year devoted to shutdowns and maintenance.

Methodology

Unlike the detailed models described earlier in Section 47.5, models for long-term planning do not require all resources to be accounted for or each individual operation to be defined. Therefore, instead of using recipes comprised of groups of procedures that each contain multiple operations, the four recipes for this case have each been represented as a *single procedure containing a single long operation*. Each recipe is assigned to one of the available production suites. In other words, the entire recipe is abstracted to a single processing task, and all required resources (equipment, labor, etc.) are represented through a single resource corresponding to a particular suite. Assuming that the capacity of each suite to execute each detailed step's recipe has been checked, the above simplification comes at no loss of generality. The advantages of faster implementation and production plan development easily exceed the effects of possible inaccuracies (such as end effects in the planning horizon's beginning or end) caused by this simplified representation.

In this representation, each recipe's procedure is assigned a duration based on the corresponding process step's cycle time, which was calculated earlier. It should be noted that the cycle times used for this long-term plan were extended a bit compared with the minimum (optimal) cycle times calculated during the detailed analysis performed previously in Section 47.5. Furthermore, the changeover times between campaigns of different steps were extended to five days. Slightly extending the batch cycle times and changeover times enables the schedule to absorb any delays (due to variability in the plant and occasional equipment failures) without large deviations from the original plan.

Results and Conclusions

Several campaign configurations were analyzed. The first was based on single long campaigns of each step. In this case, the recipes for AA-Int-1 and AA-Int-3 were assigned to run in Suite 1, whereas the recipes for AA-Int-2 and AA-API were assigned to run in Suite 2. With this setup, the production objective can be achieved within the 48 week annual plant uptime. This is shown in Figure 47.24. In this chart, each individual colored bar represents a single batch of a given step. The number of batches of each step is dependent upon the batch size of the step, the ratios between the intermediates and the product, and the required amount of final product. Given the BOM shown earlier (in Figure 47.10), 110 batches of Step 4 are needed in order to produce at least 50 000 kg of AA-API (50 000 kg ÷ 456 kg batch size for AA-API = 110). Furthermore, 102 batches of Step 3 are needed to support the input requirement for Step 4 since Figure 47.10 indicates there are roughly 0.92 batches of AA-Int-3 needed per batch of AA-API (110 × 0.92 = 101.2, which is rounded up to 102 full batches). The number of batches of Steps 1 and 2 can be calculated in a similar way.

Although the production objective is met with the configuration shown in Figure 47.24, the production between the two suites is not "balanced," i.e. the total processing duration in Suite 2 is much longer than the total processing duration in Suite 1. To improve the capacity utilization of the plant and reduce the makespan of the production plan, recall the initial assumption that both suites are identical. Therefore, they may be used interchangeably for every step. As a result, the second campaign configuration allows AA-API batches to be produced in Suite 1 *and* Suite 2 (in parallel) after the AA-Int-3 campaign in Suite 1 is completed. This increases the utilization for Suite 1 and reduces the total makespan of the production plan from approximately 46 to 41 weeks (see Figure 47.25). However, note that both this configuration and the previous configuration will result in high intermediate inventory capacity requirements.

The final campaign configuration for this example is displayed in Figure 47.26. In this configuration, multiple

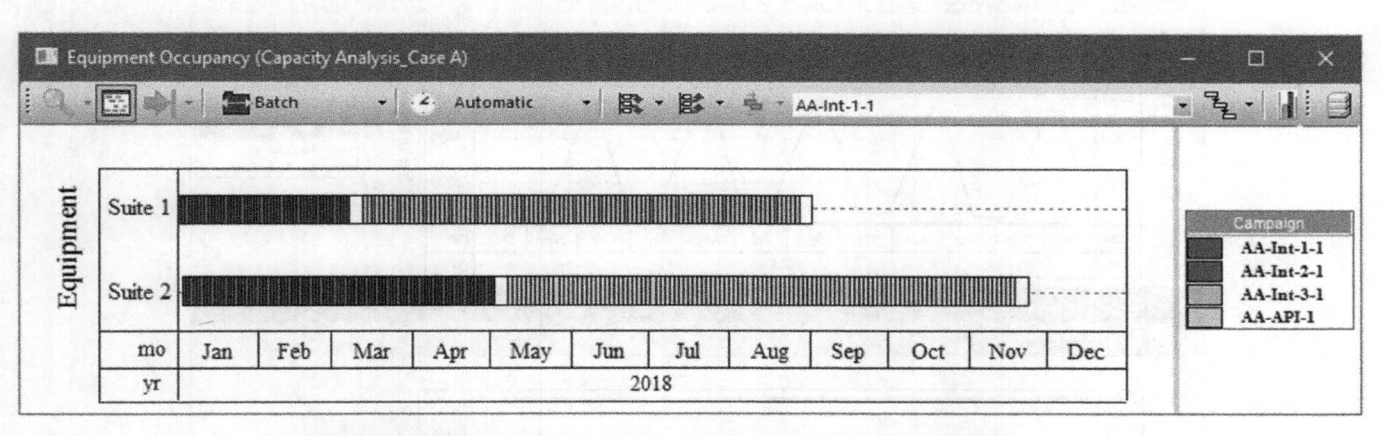

FIGURE 47.24 Equipment occupancy chart for the base production scenario.

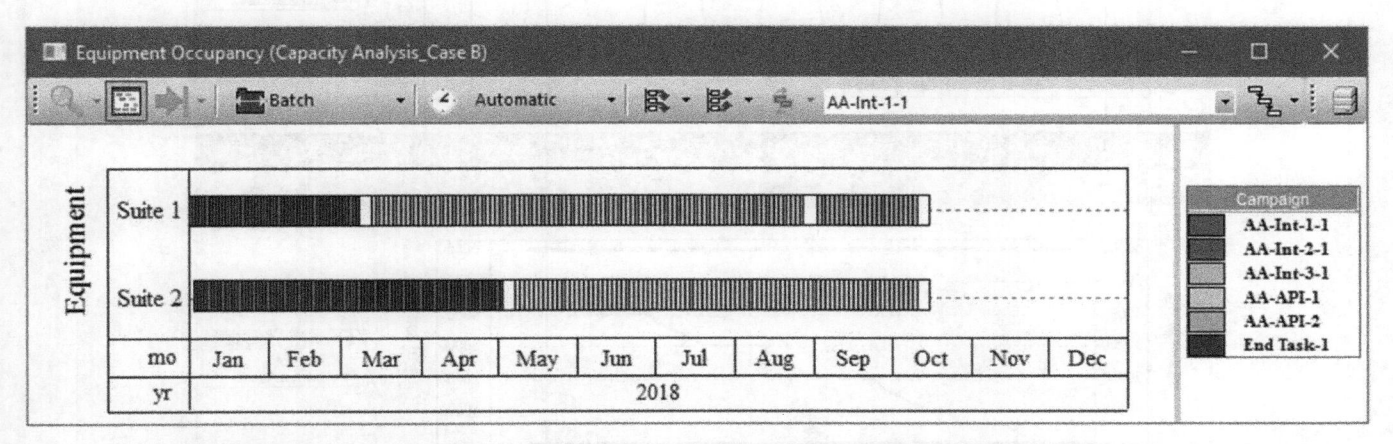

FIGURE 47.25 Equipment occupancy chart with AA-API batches following AA-Int-3 batches in Suite 1.

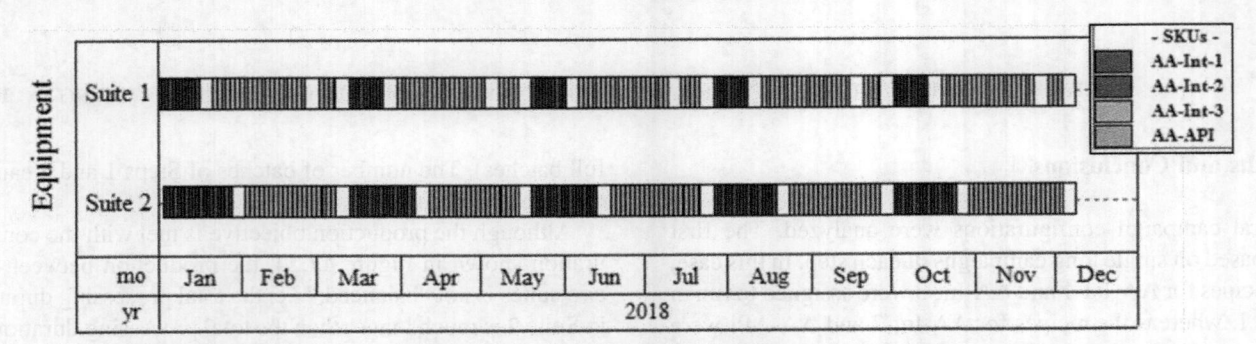

FIGURE 47.26 Equipment occupancy chart with shorter campaigns and AA-API production in Suites 1 and 2.

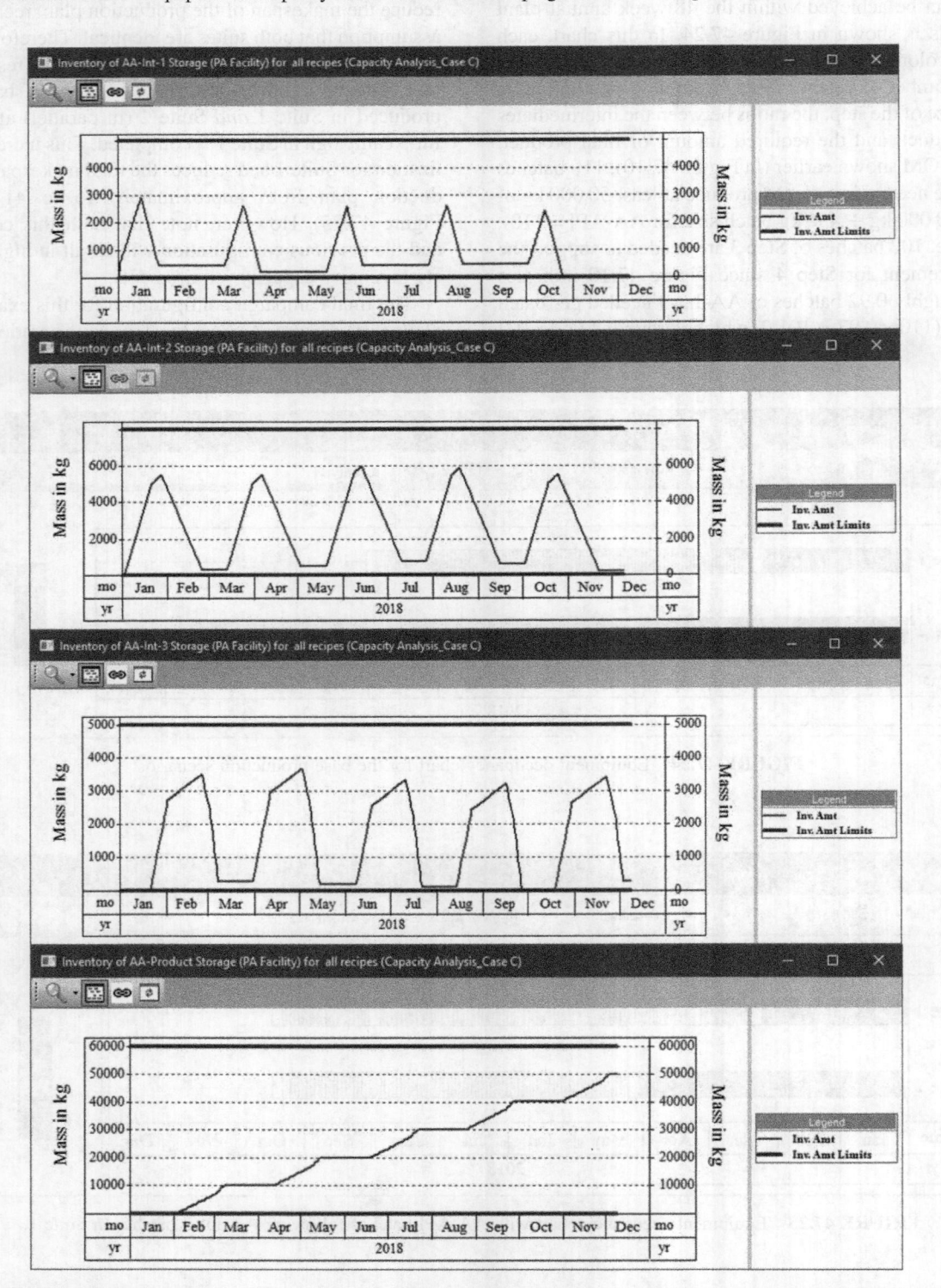

FIGURE 47.27 Inventories of Int-1, Int-2, Int-3, and AA-API, based on the production scenario in Figure 47.26.

campaigns of each step are scheduled, and Step 4 may be produced in either Suite 1 or Suite 2. Although the reduced campaign sizes in this case increase the makespan to approximately 48 weeks, they also reduce the in-process inventory relative to the previous cases and allow some batches of AA-API final product to be produced as soon as early February (rather than starting the AA-API production in early May, as was done in the previous cases). The inventory levels for each intermediate and the final product are shown in Figure 47.27.

Although the 48 week production objective has been met in the previous case, there is significant inefficiency in this production plan due to the use of some very short campaigns that are followed by extended changeovers. This issue could be addressed by creating campaigns of roughly equal durations for each intermediate and by running the AA-API batches in Suite 1 as longer campaigns that are performed once or twice per year. One could think of numerous other variations for modeling this process under different assumptions and objectives. Furthermore, as with the variability analysis of an individual process described in Section 47.4, variability analyses of multiple recipes may be performed in order to understand the range of potential outcomes that may be associated with a production plan. This could involve specification of probability distributions for specific operations' durations, product yields from a given batch, etc. The execution of a possible production plan generated based on these probability distributions could then be visualized and analyzed. Alternatively, many simulations could be performed to determine the best-case scenario, worst-case scenario, average-case scenario, etc. As long as the capacity and planning constraints can be intuitively captured, formulating and developing feasible and satisfactory solutions under variable assumptions and objectives can be easily performed in a simulation environment.

47.7 SUMMARY

Process simulation and production scheduling tools play an important role throughout the life cycle of product development and commercialization. In process development, process simulation tools are becoming increasingly useful as a means to analyze, improve, communicate, and document processes. During the transition from development to manufacturing, they facilitate technology transfer and process fitting. In manufacturing, production scheduling tools play a valuable role by generating production schedules based on an accurate estimation of plant capacity, thus minimizing late orders and reducing inventories. Such tools also facilitate capacity analysis and debottlenecking tasks. The pharmaceutical industry has begun making significant use of process simulation and scheduling tools. Increasingly, universities are incorporating the use of such tools in their curricula.

In the future, we can expect to see increased use of these technologies and tighter integration with other enabling IT technologies, such as supply chain tools, MES, batch process control systems, process analytics tools (PAT), and so on. The result will be more robust processes and efficient manufacturing, leading to more affordable medicines.

REFERENCES

1. Lainez, J.M., Schaefer, E., and Reklaitis, G.V. (2012). Challenges and opportunities in enterprise-wide optimization in the pharmaceutical industry. *Computers and Chemical Engineering* 47: 19–28.
2. Papavasileiou, V., Koulouris, A., Siletti, C.A., and Petrides, D. (2007). Optimize manufacturing of pharmaceutical products with process simulation and production scheduling tools. *Chemical Engineering Research and Design* 85 (A7): 1–12.
3. Petrides, D.P., Calandranis, J., and Cooney, C.L. (1996). Bioprocess optimization via CAPD and simulation for product commercialization. *Genetic Engineering News* 16 (16): 24–40.
4. Petrides, D.P., Koulouris, A., and Lagonikos, P.T. (2002). The role of process simulation in pharmaceutical process development and product commercialization. *Pharmaceutical Engineering* 22 (1): 1.
5. Hwang, F. (1997). Batch pharmaceutical process design and simulation. *Pharmaceutical Engineering* 17: 28–43.
6. Thomas, C.J. (2003). A design approach to biotech process simulations. *BioProcess International* 1 (10): 2–9.
7. Petrides, D.P. and Siletti, C.A. (2004). The role of process simulation and scheduling tools in the development and manufacturing of biopharmaceutical. In: *Proceedings of the 2004 Winter Simulation Conference* (ed. R.G. Ingalls, M.D. Rossetti, J.S. Smith and B.A. Peters), 2046–2051. Piscataway, NJ: IEEE Service Center.
8. Plenert, G. and Kirchmier, B. (2000). *Finite Capacity Scheduling: Management, Selection, and Implementation*. New York: Wiley.
9. Lakerveld, R., Heider, P.L., Jensen, K.D. et al. (2017). End to end continuous manufacturing: integration of unit operations. In: *Continuous Manufacturing of Pharmaceuticals* (ed. P. Kleinebudde, J. Khinast and J. Rantanen). Hoboken, NJ: Wiley.
10. Gerogiorgis, D.I. and Jolliffe, H.G. (2015). Continuous pharmaceutical process engineering and economics: investigating technical efficiency, environmental impact and economic viability. *Chimica Oggi* 33 (6): 29–32.
11. Parshall, J. and Lamb, L. (2000). *Applying S88: Batch Control from a User's Perspective*. Research Triangle Park: ISA.
12. Intelligen (2007). Chapter 2 of the User's Guide of SuperPro Designer. http://www.intelligen.com/demo (accessed 22 October 2018).
13. Petrides, D., Koulouris, A., and Siletti, C. (2002). Throughput analysis and debottlenecking of biomanufacturing facilities, a job for process simulators. *BioPharm International* 15: 28–34.

14. Tan, J., Foo, D.C.Y., Kumaresan, S., and Aziz, R.A. (2006). Debottlenecking of a batch pharmaceutical cream production. *Pharmaceutical Engineering* 26: 72–82.
15. Harrison, R.G., Todd, P., Rudge, S.R., and Petrides, D.P. (2003). *Bioseparations Science and Engineering*. New York: Oxford University Press.
16. Achilleos, E.C., Calandranis, J.C., and Petrides, D.P. (2006). Quantifying the impact of uncertain parameters in the batch manufacturing of active pharmaceutical ingredients. *Pharmaceutical Engineering* 26: 34–40.
17. Abel, J. (2008). Enterprise knowledge management for operational excellence. *Pharmaceutical Engineering* 28 (2): 1–6.
18. Lainez, J.M., Reklaitis, G.V., and Puigjaner, L. (2010). Linking marketing and supply chain models for improved business strategic decision support. *Computers and Chemical Engineering* 34: 2107–2117.
19. Sipper, D. and Bulfin, R. (1997). *Production: Planning, Control and Integration*. Singapore: McGraw-Hill.
20. Crama, Y., Pochet, Y., and Wera, Y. (2001). *A Discussion of Production Planning Approaches in the Process Industry*, CORE Discussion Papers, Center for Operations Research and Econometrics, vol. #2001042. Louvain-la-Neuve: Université Catholique de Louvain.
21. Kallrath, J. (2002). Planning and Scheduling in the Process Industry. *OR Spectrum* 24: 219–250.
22. Schuster, E.W. and Allen, S.J. (1995). A New Framework for Production Planning in the Process Industries. *APICS Conference Proceedings* 38: 31–35.

PART XI

QUALITY BY DESIGN AND REGULATORY

48

SCIENTIFIC OPPORTUNITIES THROUGH QUALITY BY DESIGN

Timothy J. Watson and Roger Nosal

Global CMC, Pfizer, Inc., Groton, CT, USA

Quality by design (QbD) is about understanding the relationships between the patient's needs and the desired product attributes by ensuring that all process attributes and parameters that are functionally related to safety and efficacy are consistently met. The value of proactively gaining enhanced process knowledge and understanding significantly minimizes patient risk while balancing continued product improvements, innovation, business needs, and regulatory oversight. It becomes the foundation that links research and development, manufacturing, and regulatory agencies through a fundamental common language that is based on a science- and risk-based philosophy.

For decades, much of the activity in quality and quality management was placed on compliance rather than utilizing an understanding of the science. Business practices adapted to procedures and focused on minimizing regulatory risk. The implementation of new technology was not typically part of a strategy because oversight for novel technology was not precedented. Extremely risky and high attrition rates of research programs, unlike any other industry, coupled with the lack of global harmonization between agencies, significantly challenged investing in new technologies and using modern methodologies for development and in submissions. In the 1990s the use of PAT started to gain interest in pharmaceutical manufacturing; however, it was primarily used for business purposes and not seriously considered for regulatory purposes. Describing to regulatory authorities a comprehensive view of process understanding was avoided in fear of being held accountable and overregulated.

In August 2002, the FDA launched their GMPs for the twenty-first-century initiative in partial response to academics and consultants, indicating the pharmaceutical industry was not manufacturing to very high standards. Companies were encouraged to use risk-based assessments, in particular when identifying product quality attributes, and adopt integrated quality systems throughout the life cycle of a product. This movement toward science-based regulations has not been limited to the United States as seen by the guidance provided by the International Conference on Harmonization (ICH). Thus, QbD for the pharmaceutical industry evolved from a vision to obtain a mutually efficient, agile, flexible sector that reliably produces high quality drug products without extensive oversight.[1] Guidance for QbD was crafted through the ICH process to what is now considered the "QbD trio": ICH Q8, Q9, and Q10 (ICH Q11 for drug substance in draft as of 2009). This movement away from prescriptive development programs has become an exciting and empowering platform for chemists, scientists, formulators, and engineers. While many elements associated with QbD, i.e. risk assessments, design of experiments (DoE), operational control strategies, etc., have been employed well before the adoption of the ICH guidelines, application was frequently not systematic, concerted, or prospective, but rather retrospective in response to issues or problems encountered during development or after commercial launch. Consequently, companies were reluctant to pursue a QbD

---
[1] Janet Woodcock, MD, Pharmaceutical Quality Assessment Workshop, 5 October 2005.

Chemical Engineering in the Pharmaceutical Industry: Active Pharmaceutical Ingredients, Second Edition.
Edited by David J. am Ende and Mary T. am Ende.
© 2019 John Wiley & Sons, Inc. Published 2019 by John Wiley & Sons, Inc.

approach or introduce supplemental studies on process capability for fear of unnecessarily increasing regulatory "burden" and potentially delaying regulatory approvals.

QbD begins with a prospective vision that accepts and builds upon a science- and risk-based platform with a commitment to never lose sight of customer focus. It starts with the quality target product profile (QTPP) and works backward through the drug product and drug substance platforms to establish a holistic understanding of what attributes are linked to patients' needs and how these attributes are functionally related through the entire process. Ensuring patient safety is not about "what we apply" to maintain the QTPP, but it is about "how we apply" process understanding, and it is about "what we need and what do we have" to ensure that quality is consistently met. It is about identifying the relationship between each attribute and its manufacturing origin and consistently controlling these relationships [1].

A drug product or drug substance control strategy should describe the functional relationship between drug product/drug substance critical quality attributes and material attributes and process parameters that govern the manufacture and effectively demonstrate control of quality. It is generally recognized that the fundamental concept of QbD is control strategy, and ICH was clear to state that every CQA has a control strategy, where concepts like design space could be (but not mandatory) an element for the control chosen for any given CQA. The development of potential design spaces as an element of the control strategy is a _symbiotic relationship_ that encompasses all of the concepts contained within the chapters of this book. In adopting a QbD approach and applying the science- and risk-based principles to assess quality attributes and process parameters, design space can be created to describe the boundaries within which unit operations of a manufacturing process may operate. In essence, design space can demonstrate control of variables that may impact a critical quality attribute, and a control strategy can be established to accommodate design space. For example, a combination of well-defined design space boundaries and real-time release testing can effectively demonstrate and confirm control and serve as the basis for release of the product without the need for specific end-product testing. Therefore, where the risk is understood and the severity and probability of impact are controllable, the demonstration of process control through the creation of design space could conceivably reduce the need to perform in-process testing as well. Continuous formal verification to demonstrate process capability in accordance with well-grounded design space criteria could serve as the basis for product release to a specification derived largely from critical quality attributes (Figure 48.1).

Scientist that embraces an enhanced approach to development should consider these types of questions:

- How is prior knowledge substantiated, and how can internal and external knowledge be used to leverage more accurate risk assessments?
- What level of detail is required to justify risk assessments?
- How should design space be presented and conveyed to demonstrate quality assurance?
- How can modeling be used to justify commercial manufacturing process changes?
- How should the control strategy connect drug product and drug substance quality attributes to process parameters and material attributes?

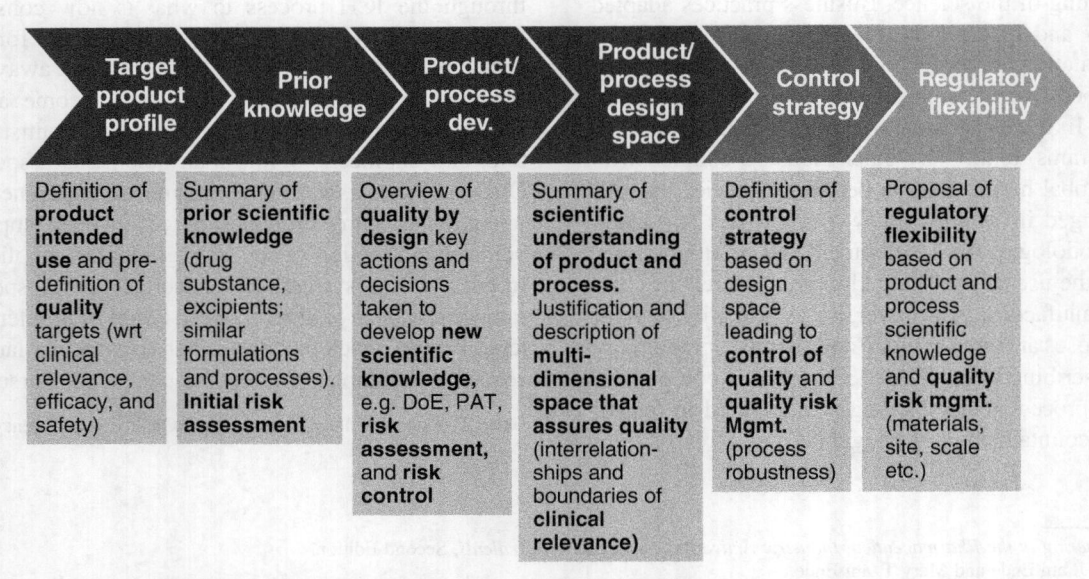

Target product profile	Prior knowledge	Product/process dev.	Product/process design space	Control strategy	Regulatory flexibility
Definition of **product intended use** and pre-definition of **quality** targets (wrt clinical relevance, efficacy, and safety)	Summary of **prior scientific knowledge** (drug substance, excipients; similar formulations and processes). **Initial risk assessment**	Overview of **quality by design** key actions and decisions taken to develop **new scientific knowledge**, e.g. DoE, PAT, **risk assessment,** and **risk control**	Summary of **scientific understanding of product and process.** Justification and description of **multi-dimensional space that assures quality** (interrelationships and boundaries of **clinical relevance)**	Definition of **control strategy** based on design space leading to **control of quality** and **quality risk Mgmt.** (process robustness)	Proposal of **regulatory flexibility** based on product and process scientific knowledge and **quality risk mgmt.** (materials, site, scale etc.)

FIGURE 48.1 General outline of approach to application of quality by design.

Work flow for the drug substance development of the illustrative example

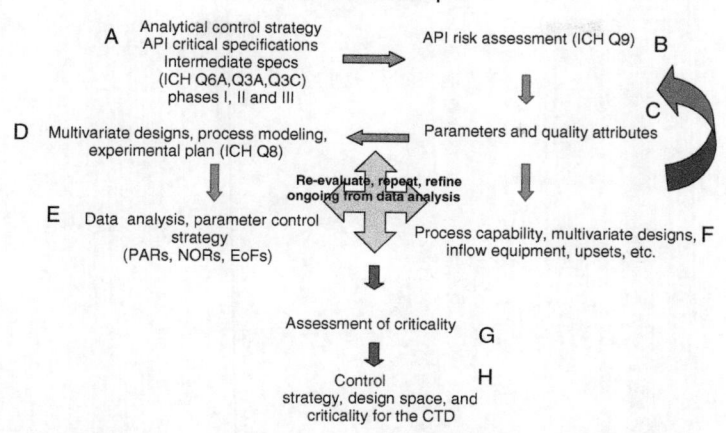

A — Analytical control strategy, API critical specifications, Intermediate specs (ICH Q6A, Q3A, Q3C) phases I, II and III
B — API risk assessment (ICH Q9)
C — Parameters and quality attributes
D — Multivariate designs, process modeling, experimental plan (ICH Q8)
E — Data analysis, parameter control strategy (PARs, NORs, EoFs)
F — Process capability, multivariate designs, inflow equipment, upsets, etc.
G — Assessment of criticality
H — Control strategy, design space, and criticality for the CTD

Re-evaluate, repeat, refine ongoing from data analysis

FIGURE 48.2 Small example of drug substance workflow for QbD.

- Is there an attenuation of regulatory latitude for post-approval optimization and continual improvement?

In addition, there are many general processes that can be adapted to sketch out a general procedure for any team to adapt a science- and risk-based approach. One example is given in Figure 48.2. Common themes to any of these processes are to constantly perform risk assessments on the attributes/parameters and their connectivity to the QTPP, use an iterative approach to risk and experimental data evaluation, be creative to understand parameter interaction in a multivariate process, present a solid design space and control strategy to ensure safety and quality are top priority, and be open and transparent to ensure the data generated can stand peer review and "pass the red face test." Far from suppressing progress, these, among many other questions, stimulated regulatory authorities and industry to pursue clarification. A fair measure of subsequent progress has improved the consistent application and value of these concepts.

The intrinsic advantages of investing in process understanding increase confidence and assurance of product quality. Tangible benefits, reductions in manufacturing costs associated with improved efficiencies and expeditious innovations, and reduction in manufacturing recalls, failures, or extraneous investigations attributed to uncertainty are largely realized over the long term, as the life cycle of a product matures.

The fundamental scientific premise as a result of QbD that attracts scientific support from every discipline across this industry is driven by a common passion to develop improved process understanding and product knowledge. The movement away from prescriptive development programs has become an exciting and empowering platform for chemists, scientists, formulators, and engineers. QbD has also played an instrumental role in establishing the value and importance of cross-functional, scientific relationships in pharmaceutical development through proactively developing and understanding processes and formulations. Perhaps, most importantly, the application of a QbD approach and investment in robust pharmaceutical quality systems are expected to reduce unexpected variability in manufacturing processes and unanticipated failures in product quality, thereby improving quality assurance of products.

ACKNOWLEDGMENT

The authors would like to thank Robert Baum from Pfizer for sharing his knowledge and thoughts on the historical perspective on QbD.

REFERENCE

1. Watson, T.J.N., Nosal, R., am Ende, D. et al. (2007). API quality by design example from the torcetrapib. *Journal of Pharmaceutical Innovation* 2 (3–4): 71.

49

APPLICATIONS OF QUALITY RISK ASSESSMENT IN QUALITY BY DESIGN (QbD) DRUG SUBSTANCE PROCESS DEVELOPMENT

ALAN BRAEM
Product Development, Bristol-Myers Squibb, New Brunswick, NJ, USA

GILLIAN TURNER
Product Development and Supply, GlaxoSmithKline, Stevenage, UK

49.1 WHY RISK ASSESSMENT IS USED IN THE PHARMACEUTICAL INDUSTRY

Pharmaceutical manufacturers and the global regulatory authorities acknowledge that the manufacture of drug substance (the bulk active ingredient) and the drug product (the finished dosage form that the patient uses) both entail some degree of risk to quality. The International Conference on Harmonization of Technical Requirements for Registration of Pharmaceuticals for Human Use (ICH) establishes regulatory guidance[1] for the development and manufacture of pharmaceutical drug substances and drug products that incorporates this understanding of risk. The ICH quality guidelines ICH Q8 [1], Q9 [2], and Q11 [3] describe a quality-by-design (QbD) and quality risk management (QRM) approach to pharmaceutical development and manufacturing that has quality risk assessment (QRA) as a core element of the development process. Identifying and managing risks to product quality is crucial to ensuring the safety and efficacy of pharmaceutical products. Chemical engineers often have a key role in conducting or contributing to these QRAs.

QbD is a systematic approach to development that begins with predefined objectives and emphasizes product and process understanding and process control, based on sound science and QRM. The goals of QbD typically include:

[1] http://www.ich.org/products/guidelines/quality/article/quality-guidelines.html

- The definition of meaningful product quality specifications based on clinical performance.
- Improved process capability and reduction of product variability and defects through enhancing product and process design, understanding, and control.
- Improved development and manufacturing efficiency.
- Enhanced capability for root cause analysis and post-approval change management.

In particular, QbD is dependent on effective use of risk assessment and management tools (ICH Q9 Quality Risk Management [2]) and needs to be accompanied by good development science (ICH Q8(R2) Pharmaceutical Development [1] and ICH Q11 Development and Manufacture of Drug Substance [3]) and at an appropriate stage integrated into a suitable quality system (ICH Q10 Pharmaceutical Quality System [4]).

Prior to assessing risks to quality, the quality requirements for the drug product and drug substance need to be understood. ICH Q8 [1] indicates that this can be done by "Defining the quality target product profile (QTPP) as it relates to quality, safety and efficacy, considering e.g., the route of administration, dosage form, bioavailability, strength, and stability." Delineating the QTPP allows definition of potential drug product critical quality attributes (CQAs). Some of the drug product CQAs will be linked only to aspects of the drug product process, but typically several are linked back to drug substance and excipient CQAs, for example, purity and

Chemical Engineering in the Pharmaceutical Industry: Active Pharmaceutical Ingredients, Second Edition.
Edited by David J. am Ende and Mary T. am Ende.
© 2019 John Wiley & Sons, Inc. Published 2019 by John Wiley & Sons, Inc.

powder properties. Given that the drug substance manufacturing process typically entails a multistep synthetic process, one can in turn identify the quality attributes of synthetic intermediates that have potential to impact the drug substance. Each of these drug product, excipient, drug substance, and intermediate CQAs needs to be controlled within defined limits to ensure product quality.

With these quality targets in hand, a control strategy must be developed to ensure targets are met. This control strategy can consist of:

- Procedural controls – definition of procedural steps and their order.
- Technical requirements – defined limits on facilities and equipment used.
- Parametric controls – limits on process parameters.
- In-process controls (IPCs) – analysis of process streams against control limits, with an accompanying failure pathway (procedure) that addresses failure modes.
- Analytical test and specifications on both input materials and isolated products.

In order to understand what aspects of the process require control, the process must undergo, as described in ICH Q8 [1], "a systematic evaluation… including

- Identifying (through, e.g., prior knowledge, experimentation, and risk assessment) the material attributes and process parameters that can have an effect on product CQAs
- Determining the functional relationships that link material attributes and process parameters to product CQAs"

Risk assessment is used to distinguish between those process and input material aspects that pose low risk to quality from those that pose moderate or high risk to quality. Thus, risk assessment can be used to focus control strategy development on only those aspects with significant potential to impact quality.

A general framework for QRA is discussed in the ICH Q9 QRM guidance [2]. Q9 defines risk as "the combination of the probability of occurrence of harm and the severity of that harm" and risk assessment as "the identification of hazards and the analysis and evaluation of risks associated with exposure to those hazards."

The fundamental questions that relate to risk assessment therefore include the following:

- What may go wrong?
- What is the likelihood that something will go wrong?
- What are the consequences?

These three questions are addressed by QRA processes through the identification of failure modes that may cause harm, the estimation of probability of failure mode occurrence, and an analysis of the severity of the resulting outcome. The tools used, the level of effort, and the formality and extent of documentation of the process can all vary with the stage of development but should be commensurate with the level of risk. Different companies can use different tools and strategies, but there are many approaches that are commonplace throughout the industry.

In the sections that follow, we will describe common risk assessment approaches and how they are applied at various stages of drug substance development. We will discuss how risk assessments can be used early in the development cycle for risk-based prioritization of development efforts and identification of potential CQAs and process parameters. For later stages, we will describe how risk assessments enable identification of a robust control strategy and how they are typically used for formal identification of criticality, and we will touch on the associated regulatory implications. We will also cover the assessment of residual risk after control strategies are implemented, risk review, and operational risk assessment. The examples shared are focused on small molecule drug substances rather than biologics that, due to their complexity, have additional challenges to address in the risk management process.

It is important to appreciate that risk assessment is not a single event but a continual process of identification, assessment, mitigation, and reassessment that is driven by increasing knowledge and experience. The effectiveness of the risk management process depends on having the right people contributing sufficient data and knowledge of the process under review and a well-defined risk assessment process.

49.2 OVERVIEW OF RISK ASSESSMENT PROCESS

The risk assessment process is a structured approach that enables development of a robust control strategy, thereby ensuring manufacturing operations remain in control. It is composed of several steps (see Figure 49.1) including risk identification, risk analysis and evaluation, risk reduction and acceptance, and risk communication and review, and here approaches to each of these steps will be discussed in more detail.

Risk assessment processes are used throughout the development of the API from early synthetic route and process selection through to definition of a commercial control strategy. Risk assessment can also be used in life-cycle management of commercial processes to assess changes to the process or manufacturing facility. Before working through the risk process, it is important to define the scope of the activity to ensure that the discussions remain focused and

Risk assessment step	Risk questions
Risk identification	What failures can impact the process or the quality of the product?
Risk analysis and evaluation	What is the risk of these failures? What should I study more?
Risk reduction and acceptance	What controls or process changes do I need to address risks? Is a risk low enough to accept as is?
Risk communication and risk review	Are all relevant stakeholders informed of the risks? Is my assessment correct in light of additional information?

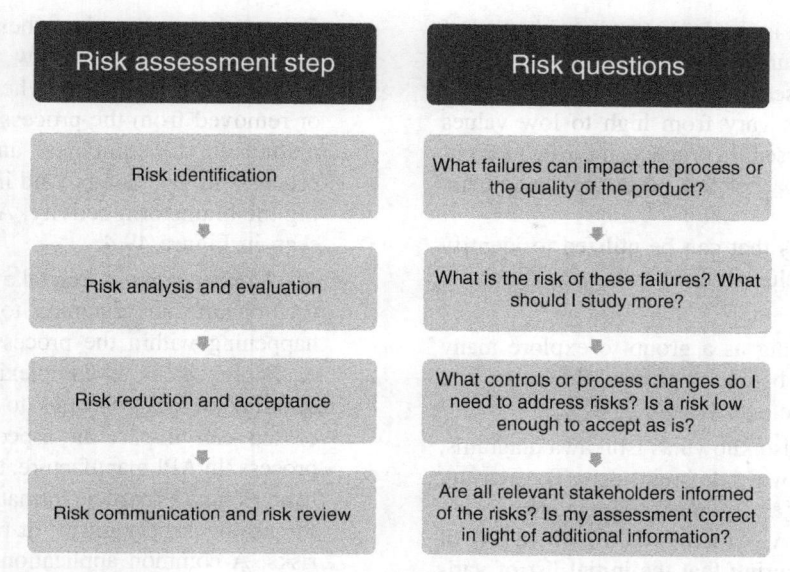

FIGURE 49.1 Risk questions that arise during development and associate risk assessment step.

the output is aligned with business needs. The majority of risk assessments focus on the quality of the output API through linkages to the API CQAs, though some will also include elements related to manufacturability too. In particular, risks associated with yield are often included and evaluated though these are not necessarily directly linked to the quality of the API. For more information of the various types of risk assessment, see Section 49.3.

49.2.1 Risk Identification

In early risk assessment, risk identification activities focus on the ability of the potential routes and process to meet both the QTPP and any constraints or opportunities provided by the use of novel and developing technologies. Examples of such technologies are flow and biocatalysis, safety and environmental considerations, and manufacturability aspects including cost, cycle time, and operability. A common approach taken here is structured brainstorms guided by criteria derived from the QTPP in addition to manufacturing considerations.

When route and process have been defined, the later risk identification activities can commence. At this stage, the identification of potential CQAs is usually possible, and this provides one aspect of the scope of the activity. The aim of this step is to identify what could potentially go wrong during the operation of a process and then outline the possible causes and consequences. There are a number of different ways that this can be achieved, and they fall into two main categories, the "top-down" approach and the "bottom-up" approach. Ideally each of these will deliver the same end result but the journey will look different. The top-down approach focuses the initial effort on identification of failure modes that have the potential to impact the quality of the API. This activity can be informed by several different inputs, and those most commonly used include historical data and experience with the specified process, in addition to expert opinions based on information from similar processes and equipment. Once credible failure modes have been identified, the consequences need to be fully explored as the direct impacts can themselves have a secondary impact further through the process. An example of such a situation would be variations in reaction conditions that increase the level of impurity A, and the impurity is not removed in the subsequent downstream operations, resulting in higher levels of the impurity in the next stage of the process. In addition, it may be found that higher levels of impurity A entering the next stage result in a change to the reaction rate or the performance of the crystallization and give rise to elevated levels of a second impurity, impurity B. Increased levels of both impurity A and impurity B are then consequences of the failure mode.

Once a failure mode has been identified, the risk assessment process can then explore the potential causes of the failure. For APIs, these are often variation in parameter settings and levels of impurities, though they can also be due to equipment selection. The detail to which the potential causes are identified can vary depending on the type of risk assessment being carried out and the desired outputs. In some situations, identification of the parameter(s) directly causing the failure modes is sufficient; in other cases, the root cause(s) behind the parameter variation is also recorded. The potential for two or more parameters and attributes to interact and jointly influence a particular failure mode should also be considered. Interactions commonly occur within the same unit operation though they may also include parameters and attributes that are introduced in different parts of the process.

The bottom-up approach is similar but starts with a list of the parameters and other sources of variation and uses either data or expert opinion to assess what the likely failure modes could be as the parameters vary from high to low values within the range being assessed. In practice, it is often a combination of these two approaches that produces a pragmatic list of the potential risks.

A number of approaches that can be utilized to identify risks and common ones include the following:

- *Brainstorming.* Working as a group to explore many "what-if" scenarios before focusing down to the likely/credible/realistic ones.
- *Fishbone diagrams*, also known as Ishikawa diagrams, provide a structure on which to brainstorm. They ensure that possible causes of a particular failure mode, or variables impacting a CQA, are explored from a variety of perspectives, thus ensuring that the initial list of variables (parameters and attributes) included in the risk assessment is comprehensive. An example of this can be seen in Figure 49.2.
- *Input process output (IPO) diagrams.* These are visual representations that focus on identifying the inputs and outputs of a process. For a risk assessment, the inputs are typically the starting materials, reagents, solvents, excipients, etc., and the outputs are the CQAs. The parameters, both those that can be controlled and that contribute to the noise, are then added. This is another way to encourage teams to think more broadly than just the process aspects and help to link the causes of failure to the effects. An example of this type of diagram can be seen in Figure 49.3.
- *Gemba* literally means "the real place" and refers to the place where value is created [5]. The risk identification activity happens as the process operation is observed and provides an excellent way of engaging with the practicalities. It has the advantage of highlighting potential failure modes that may be less obvious from written instructions or equipment diagrams.
- *Process mapping* describes the individual steps within a process and how they link together and includes information on where substances are added or removed from the process train. Additional information on the equipment and physical state of the process can be added to aid in risk identification during the brainstorm activity. A simple example can be seen in Figure 49.4.
- *Mechanism mapping* can take a number of formats and uses pictures and diagrams to look in detail at what is happening within the process. It is ideally suited to explaining areas of complexity where a deep understanding of how changes to the material being processed can impact an aspect of the output of the process. In API manufacture, this is typically employed in areas such as particle formation, where understanding the physical mechanisms at play is key to identifying risks. A common application of mechanism mapping for API is in the formation and fate of impurities; visual representations that describe how impurities form and transform, including any mechanistic understanding, and how and where they purge, are used to direct the brainstorming activities.

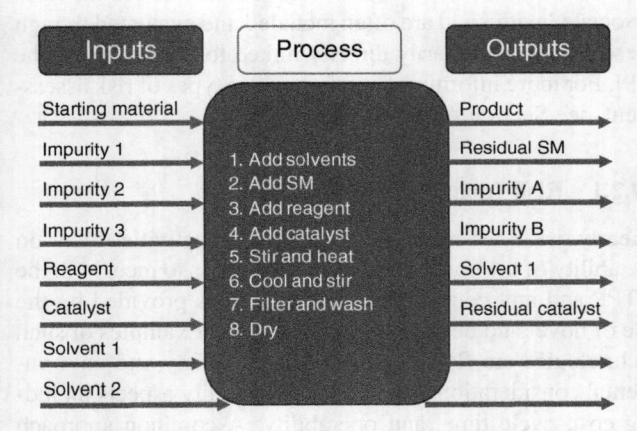

FIGURE 49.3 Input process output diagram.

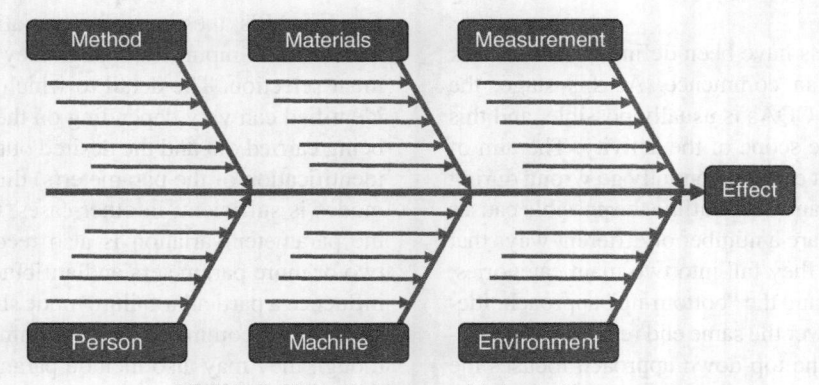

FIGURE 49.2 Fishbone or Ishikawa diagram.

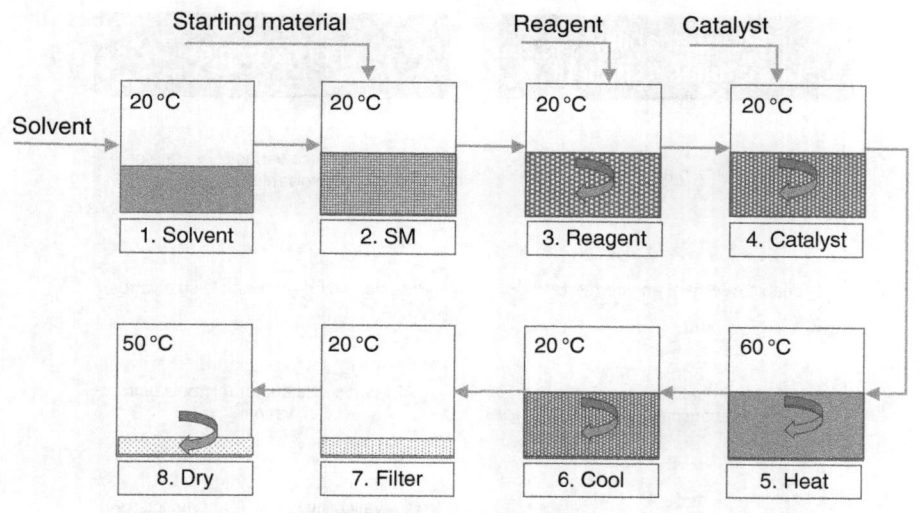

FIGURE 49.4 Process map.

As with most activities, the quality of the input has a bearing on the quality of the output; the same principles can be applied to the risk assessment process. Having a range of expertise available to work as a team to complete the various stages of the risk assessment will result in a more rigorous assessment and hence ensure that no significant risks are omitted. When assessing an API process, the range of technical experts may typically include representation from engineering, chemistry, analysis, formulation, production, statistics, quality, and regulatory affairs who work together with a facilitator. The value of team members who have practical experience of the process under evaluation cannot be overemphasized; they are in an excellent position to identify any practical constraints that may lead to failure and will have awareness of any previous deviations that have given rise to unanticipated results.

49.2.2 Risk Analysis and Evaluation

Having identified the risks, the next step is to provide an assessment of the adverse impact of the risk and its likelihood of occurrence, in either a quantitative or qualitative manner. In many processes, there is also a consideration of the ability to detect the risk, and in some cases, a knowledge rating may also be included either in the evaluation of the risk or to provide context to the scoring that has been applied. Following these steps, the risks are then ordered, or a particular set of criteria applied, to identify any actions required during risk mitigation.

In the early phases of risk assessment, and in instances where knowledge may be limited, the quantification of risk may not be possible, and in these cases a qualitative system is often used, such as high/low or a traffic light system where red, amber, and green indicate the degree of risk posed by each failure mode. The scoring is often reached through discussion within an expert group, and it is acknowledged that as more information becomes available, the scoring may require reevaluation. It is important to draw the distinction between a lack of knowledge and high risk; not all knowledge gaps result in high risks when data does become available; in fact, it is often the case that those areas where little knowledge exists ultimately prove to be low risk, but this cannot be guaranteed.

As development of the process proceeds and data is generated to support the risk assessment, a more detailed scoring of the risk becomes possible, and at this phase quantitative systems are more prevalent. Typically, tools such as failure mode, effects, and criticality analysis (FMECA) and failure mode effect analysis (FMEA) are used to capture the scoring; see Section 49.4.3 for further information.

Scoring for severity or impact typically considers the link between either the failure mode or the causes of the failure and the ultimate impact that has on the patient. This may be considered as a direct link, by assessment of the impact of the effect of the failure on the patient, for example, or it could be linked through a series of steps as shown in Figure 49.5.

The scoring system may be based on a generic assessment of classes of parameters or attributes or may require an individual assessment of each parameter or material attribute. Inputs to consider when determining the severity score in the latter way include any univariate or multifactorial data and the range of the parameter or material attribute that has been investigated. As is the case with risk identification, risk scoring also requires the right technical input to interpret the data and produce a well-reasoned score. This is particularly important as once a high level of process understanding

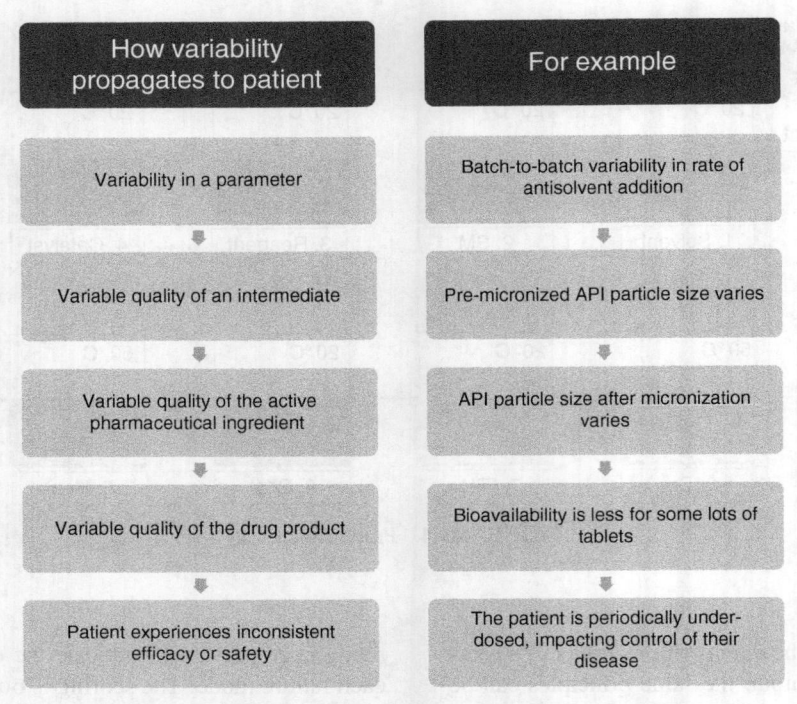

```
How variability                    For example
propagates to patient

Variability in a parameter         Batch-to-batch variability in rate of
                                   antisolvent addition
            ↓                                    ↓
Variable quality of an intermediate   Pre-micronized API particle size varies
            ↓                                    ↓
Variable quality of the active     API particle size after micronization
pharmaceutical ingredient          varies
            ↓                                    ↓
Variable quality of the drug product   Bioavailability is less for some lots of
                                       tablets
            ↓                                    ↓
Patient experiences inconsistent   The patient is periodically under-
efficacy or safety                 dosed, impacting control of their
                                   disease
```

FIGURE 49.5 How upstream variability links to patient impact.

has been achieved, the individual severity scores should not be reduced as a means of reducing the risk score.

Scoring the probability or occurrence of the failure mode requires an assessment of how likely it is that the failure mode will occur, either through a direct scoring of the probability of the failure mode occurring or linked to the underlying causes such as parameters varying from their target value or range. This is typically assigned a numerical probability based on the projected frequency of occurrence and may also take into consideration capability of the process or equipment to deliver within a specified range. For example, it could be assumed that the variability in process parameters fits a normal distribution, and plant capability, or normal operating range (NOR), could be defined as one standard deviation; the probability will then be scored directly based on the capability value calculated. More qualitative ways of scoring probability may take account of the type of controls in place and any historical knowledge or data of how these typically perform.

The third element of risk scoring represents detection, and this links directly to the control strategy that is being employed. There are many methods of detection, and common ones can include analysis of starting material, reagents, and solvents to determine if a batch is of the appropriate quality, the use of PAT and in-line or off-line analysis during processing to monitor physical and chemical changes, and analysis of the isolated material. This score may reflect either where in a process any detection of the failure mode takes place or how capable the detection method is. Some systems will incorporate both of these elements into the scoring, and emphasis is often placed on whether the detection method can enable an avoiding action to be employed or whether it is just observing, and preventing, onward processing of a batch of insufficient quality. In some cases, where the link between the cause of failure and the failure mode is well established, it may be appropriate to consider detection of the cause in addition to the failure mode as this may offer more opportunity for developing a preventative control strategy.

Following the scoring of the individual components of each risk, the information is combined to provide a risk evaluation. The output of this provides a ranking of the identified risks. The evaluation of the risks often takes one of two common approaches; if quantitative scoring has been applied in the analysis phase, then application of an appropriate formula will result in a numerical scoring of each risk. Where risks have been assessed using qualitative descriptors such as "high," "medium," or "low," an alternative is to make use of a risk matrix. Once the numerical scores have been calculated for each risk, predetermined cutoff values may be applied, or risk ranking tools such as Pareto charts can be used. These enable identification of a specific proportion of the highest scoring risks. The risk matrix approach applies predefined criteria to each combination of scoring for severity, probability, and detection and results in a categorization being applied that feeds into the risk reduction and acceptance phase.

49.2.3 Risk Reduction and Acceptance

Having evaluated the risk, decisions then needs to be made to determine whether the risk will be accepted or whether actions will be required to reduce the risk to an acceptable level. A number of factors may contribute to the decision-making process including the size of the risk, whether it is practical to implement the changes required to reduce the risk, the cost benefit analysis of making any improvements to the controls, and the potential impact that any risk may have on the patient. For those risks identified as requiring mitigating actions, the choices are the following:

- Reduce the severity of the risk; this can rarely be achieved without introducing a modification to the process as the severity often describes a fundamental relationship.
- Reduce the probability of occurrence of the risk; this is often targeted as a way of reducing risk and relies on improvements either to ways of working or to instrument or equipment capability.
- Improve the detection of the failure mode, typically by the addition of analytical controls or in some cases by moving, or improving, existing controls.

Whenever changes to the control strategy are introduced, it is important to reevaluate the risk assessment to determine if new risks have been created and if the scoring of existing risks is impacted by the change.

Some risks will be of low significance, and the decision may be to accept these risks without making further attempts to reduce them. Elimination of risk entirely is not possible, and there will come a point where it may be impractical to reduce the risk further without committing a disproportionate amount of resource.

49.2.4 Risk Communication and Review

Following a risk assessment exercise, the conclusions concerning the risks and those identified as posing the greatest risk are communicated to the stakeholders. Mitigation plans to reduce the risk and actions to manage any residual risk once the improved controls have been implemented should also be included.

Risk management plays a fundamental part in ensuring the quality of a product or process and needs to be regularly utilized during development to assess any changes and their impact on the control strategy. As the product moves from development into its product life cycle, the risk management process will continue to be used to assess the impact of both planned and unplanned changes.

49.3 RISK ASSESSMENT TYPES

As depicted in Figure 49.6, different risk assessment types and tools are used throughout the development cycle to achieve different purposes. In this section, we will discuss different types of risk assessment, and in Section 49.4 we will discuss risk assessment tools used.

49.3.1 Attribute Risk Assessment (Defining CQAs Based on Product/Patient Risk)

As mentioned in the preceding sections, it is necessary to establish which attributes of the API are critical. ICH Q8 [1] defines CQAs as follows: "A CQA is a physical, chemical, biological, or microbiological property or characteristic that should be within an appropriate limit, range, or distribution to ensure the desired product quality." In the case of solid oral dosage forms, CQAs are related to those attributes that impact "product purity, strength, drug release, and stability." For other dosage forms, additional drug product attributes may be identified, and the starting point for the drug substance risk assessment is to understand the API CQAs that are known to link to these DP CQAs.

Purity is the most straightforward – if impurities are present in the API, they will be present in the drug product and are potentially impactful to the patient. In general, impurities observed in the drug substance are deemed to be CQAs of the API. More specifically, for regions following ICH guidance, non-mutagenic impurities that are typically found in the API above the ICH Q3A reporting limits [6] are considered to impact the impurity profile of the drug substance and may be deemed as critical impurities. Health authorities require detailed knowledge of impurity formation and purge throughout the synthetic process to make the API. In the case where impurities are related to those formed in synthetic steps prior to the final API step, the precursor impurities may also be designated as critical impurities if they are not completely purged and have the potential to result in impurities in drug substance. Genotoxic impurities and known, or suspected, human carcinogens are special classes of impurities [7] that are typically always deemed critical if they can be present in the API. Typically, a limit may be applied to this, and an example would be that genotoxic impurities potentially present at, or above, 30% of the ICHM7 Threshold of Toxicological Concern are defined as CQAs. Solvents and elemental impurities used in the manufacturing process may also be defined as API CQAs. Assessment of the potential impurities should consider not only the input materials and the process but also the storage of the API. New or raised levels of impurities seen on stability should be included as API CQAs. Water present in API, if shown to increase bioburden or impact the chemical stability, may also be defined as a CQA. The strength or dose of a drug product has two

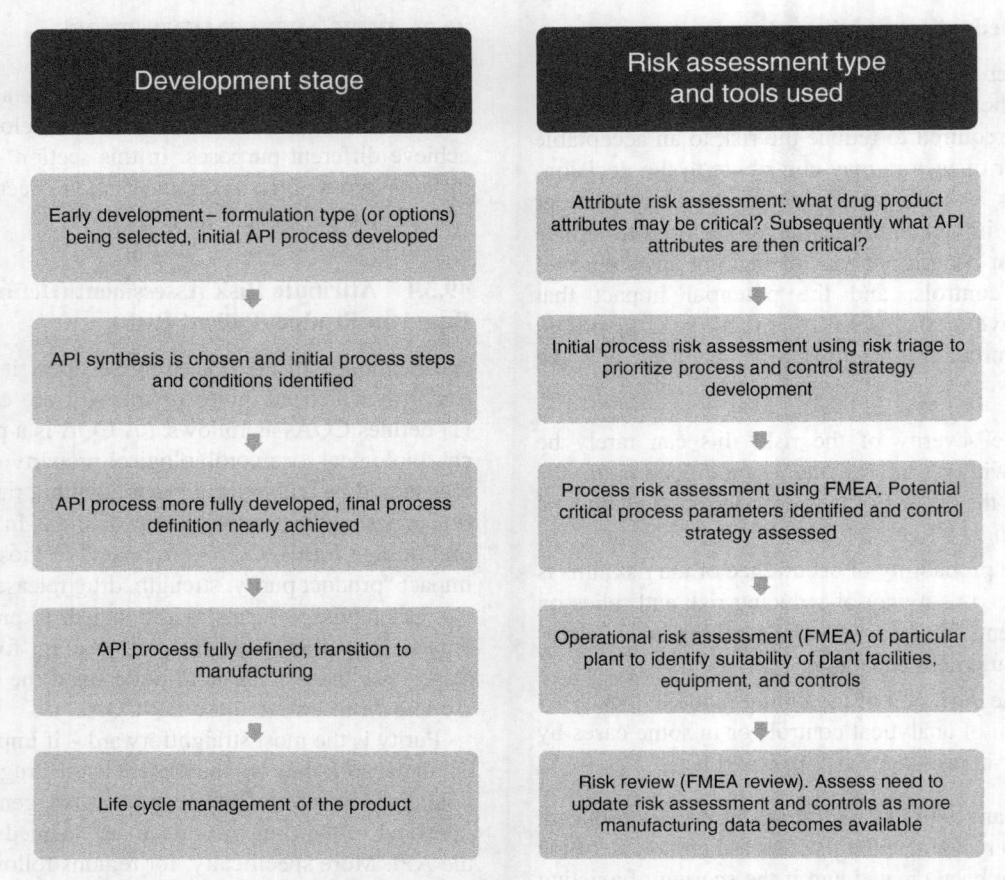

FIGURE 49.6 Risk assessment types and tools used through the development cycle.

Development stage	Risk assessment type and tools used
Early development – formulation type (or options) being selected, initial API process developed	Attribute risk assessment: what drug product attributes may be critical? Subsequently what API attributes are then critical?
API synthesis is chosen and initial process steps and conditions identified	Initial process risk assessment using risk triage to prioritize process and control strategy development
API process more fully developed, final process definition nearly achieved	Process risk assessment using FMEA. Potential critical process parameters identified and control strategy assessed
API process fully defined, transition to manufacturing	Operational risk assessment (FMEA) of particular plant to identify suitability of plant facilities, equipment, and controls
Life cycle management of the product	Risk review (FMEA review). Assess need to update risk assessment and controls as more manufacturing data becomes available

aspects – the average dose and the dose uniformity (also called content uniformity). Each of these is required to be maintained within a range to ensure safe dosing of the patient. The criticality of dose and content uniform may depend on the therapeutic window of the drug (dose range that is effective and safe) and the consequences of chronic or acute under- or overdosing. In turn, API properties can impact the effective tablet dose (e.g. API assay) and content uniformity (e.g. particle size distribution). For example, for a low dose drug product, if the particle size distribution includes very large particles, the presence of a few large particles in a tablet can have a large influence on the percent assay of active ingredient in the tablet. In such cases, the particle size distribution can be deemed critical.

Bioavailability of the drug and dissolution performance of the drug product are other typical CQAs of the drug product. API crystal form (which impacts solubility and potentially dissolution) and particle size distribution (which impacts dissolution in most cases) may be designated as critical if they impact bioavailability and dissolution.

An example of how these and a few other aspects are used to generate an attribute risk assessment is given in Table 49.1. In this case, the risk is a simple qualitative risk assessment using expert knowledge of the drug product and drug substance synthesis to arrive at a rationale for criticality or noncriticality of the attributes. In some instances, the link between drug product CQAs and drug substance CQAs needs to be assessed through the generation of data to establish a link; an example of where this is particularly important is in the determination of API physical property CQAs.

49.3.2 Input Material Quality Risk Assessment

The parameters, the elements of the process that are typically described in the process definition and explored during experimentation, are often the initial focus of the risk assessment in building a comprehensive control strategy. However, consideration of parameters alone will lead to an incomplete control strategy without also considering the quality of the input materials. In an API process the main aspect of input quality comes from the impurities in the starting materials, intermediates, reagents, and solvents. When developing a drug product process or, in some instances, the latter stages of an API process, the impurities and material properties of excipients also need to be included in the risk assessment.

TABLE 49.1 Example of an API Attribute Risk Assessment

Attribute	Critical, Noncritical, or Not Applicable	Rationale
Identification	Critical	The patient must receive the correct active ingredient to ensure safety and efficacy
Purity/impurity profile	Critical	Impurities impact patient safety
Assay	Critical	The drug product has a narrow therapeutic window, and the correct dose is critical
Chiral purity	Not applicable	This drug is achiral
GTI/potential carcinogen	Critical	There are two Ames-positive impurities that are potentially present in the API
Metals	Critical	Catalytic palladium is used in the final step of the API synthesis
Residual solvents	Noncritical	Only less toxic (ICH Class III [8]) solvents are used in the final synthetic steps of the API, and all solvents are reduced to a level where there is no impact to patient safety
KF/water content	Noncritical	Water does not impact the form or stability of the API and drug product
Form	Critical	Multiple polymorphs of the API have been identified and can potentially be produced in the API process
Particle size distribution	Critical	The drug product is very low dose (0.5 mg), and particle size can impact content uniformity
Color	Noncritical	The drug product is a film-coated tablet, and API color does not impact product appearance

To enable a rigorous assessment of the impurities entering the process, data from multiple representative batches of all inputs should be assembled and then reviewed against any existing specifications. Where possible, knowledge of the synthetic route used to generate the starting materials and more complex intermediates should be obtained for the risk assessment. This data provides the starting point for assessing the probability of encountering these impurities at a level that could impact the quality of the final drug substance. In addition, an understanding of the fate of these impurities in the stage in which they are initially added to the process, and all downstream stages until they are fully purged, is required to determine the severity of impact that they may impose on the process. The ability to detect these impurities is also important to the assessment. If analytical methods are not capable, or are not in use, the risk to the patient may be increased.

With the information outlined above, it is possible to include the quality inputs directly into the same assessment as the parameters. Where there is significant complexity (i.e. numerous potential impurities), it is recommended to complete an initial triage of the impurities using risk-based tools before combining with the other elements of the process.

Consideration of the input quality should not be limited to impurities alone. Several of the more common material attributes that also require consideration are listed below:

- *Physical properties* – Variation of physical properties in the inputs can play a significant role in the performance of the process and needs to be carefully considered in the risk assessment. Particle size and the polymorphic form of the input material in particular can impact the rate of dissolution that in turn can affect the rate of the reaction and relative stoichiometry, giving rise to multiple potential failure modes. For heterogeneous reactions, the particle surface area, in addition to PSD, can also impact reaction rate. For catalysts, there is also an influence of surface area and surface activity that can impact both the reaction rate and the reaction mechanism, giving rise to new impurities and varying levels of known impurities. When considering seeded crystallizations, for example, the characteristics of the seed including PSD, form, and even the mode of seed generation should be included in the risk assessment as all have the potential to influence the isolated material.
- *Color* – A significant number of inputs into reactions either contain colored impurities or are themselves colored when isolated. Depending on the formulation of the drug product selected, there is potential for any color introduced into the process to carry forward with the API into the drug product. In some situations, the color can be controlled through specific impurity control. In other cases, when the link is difficult to establish, the risk assessment will need to draw on data and process experience to assess the risk posed by this input material attribute.
- *Bioburden* – Bioburden is defined as the combination of microbiological and endotoxin contamination. In processes where bioburden in the drug substance poses a risk to patients, the potential to introduce bioburden in the input material needs to be included in the risk assessment.

49.3.3 Process Risk Assessment/Technical Risk Assessment

Once the CQAs of the drug substance are established via the attribute risk assessment, an assessment of the process or technical risks can occur. In these risk assessments, aspects of the process that may have variability (e.g. process parameters, input material quality) are examined to understand the risk posed by that variability on quality. It is typically taken as a given that the procedural steps that constitute the process will be followed per instructions and that a facility will be used that is suitable for drug substance or intermediate manufacture. The GMP systems that safeguard these aspects can be subject to their own, separate risk assessment and do not need to be included in a risk assessment of a particular process.

In preparation for a risk assessment, the background information is typically compiled in advance of a risk assessment session. This may include the synthetic scheme, a process description, a parameter table and/or Ishikawa diagram, a list of IPCs including procedures to be followed in case of failure to meet control limits, prior performance information, and information on impurities and other quality attributes such as residual solvents, metals, or crystal form. This information is used along with an understanding of the equipment and capabilities of a typical pharmaceutical manufacturing plant to perform the risk assessment. The knowledge of attribute criticality developed in the attribute risk assessment may be used to limit the risk assessment scope to only those items that impact critical attributes.

Either the top-down or bottom-up approach (or both), discussed in Section 49.2, may then be used to perform the risk assessment. In the top-down approach, the individual quality attributes (either all attributes or only the critical ones) are listed, and potential causes are considered. The methods discussed in Section 49.2 (brainstorming, fishbone diagrams, etc.) may be used at this point. An example of top-down risk identification and enumeration of potential causes is given in Table 49.2. The bottom-up approach can be particularly effective later in the development cycle when deviations in most or all of the process parameters have been studied, and thus a highly comprehensive examination of failure modes around each is possible. An example of bottom-up risk assessment is given in Table 49.3.

Once the failure types and potential causes are identified, the risk associated with each can be assessed. The manner of assessment depends on the risk assessment tool being used. These tools and examples using them will be provided in Section 49.4.

A final step in the risk assessment is to describe any risk mitigations that have been, or will be, put into place to mitigate the risk and to evaluate their effectiveness in risk reduction. These controls, which can be of the various types discussed in Section 49.1, collectively can be referred to as the control strategy. The level of risk that exists after controls are put in place is known as the residual risk. In general, mitigations are put into place until the residual risk falls either into the "low risk" category or a "moderate risk, but as low as reasonably practicable (ALARP)." The ALARP concept acknowledges that with infinite resources, a risk could be driven very low; however it is not appropriate to spend excessive resources to affect a minor reduction in risk. In instances where the residual risk after controls are put in place is high, many companies will deem the risk as unacceptable and look

TABLE 49.2 Identified Risks and Potential Causes Using the Top-Down Approach

Failure Mode	Effect	Causes
Degradation in the reaction	Impurity A present in the API	High reaction temperature. Long reaction time
Dimerization in the reaction	Impurity B present in the API	Too much reagent A charged
Insufficient drying	Residual solvent present in the API	Short drying time, poor agitation regimen
Seed dissolution during crystallization	Incorrect API crystal form	Amount of seed crystals used in crystallization was too low. Excessive amount of solvent B present
Insufficient generation of nucleation sites	Particle size distribution too large	Antisolvent addition rate too slow

TABLE 49.3 Identified Failures and Associated Risks of a Reaction Step Using the Bottom-Up Approach

Potential Failure	Risk
Too little solvent used	Reduced stability of the reaction stream resulting in formation of impurity A
Too much solvent used	No impact
Too little reagent A used	Competitive side reaction occurs resulting in formation of impurity B
Too much reagent A used	No impact
Low reaction temperature	Incomplete reaction resulting in residual starting material present in the downstream steps
High reaction temperature	Decomposition of product to impurity A
Reaction time too short	Incomplete reaction resulting in residual starting material present in the downstream steps
Reaction time too long	None, no additional side reactions anticipated

to avoid the risk altogether, for example, by redesign of the process or use of different process chemistry or technology.

Effective application of risk management requires that the risk assessment process is regularly updated throughout development. As a greater understanding of the process is formed, and knowledge gaps are filled, the assessment of severity can be refined. Increasing experience in operation of the process leads to a better estimate of the probability of a failure mode occurring, and with evolving analytical controls, and enhanced monitoring of the process, changes to the detection scoring may also be required.

49.3.4 Manufacturing/Operational Risk Assessment

Manufacturing or operational risk assessments differ from process risk assessments in that they are generally focused on a specific plant or equipment train. The risk assessment may still be a step-by-step examination of the process; however the probability of failure, failure detection systems, and controls will be specific to the facility. As examples, the precision of the weigh scales used, accuracy of temperature probes and control capability of the heat transfer system, and specific alarms and interlocks could all influence the probability of failure and ability to detect and control the risk.

Often, the general process or technical risk assessment will be taken as the input to the operational risk assessment. The risk rating (if applicable, the probability and detectability ratings) can be determined for the particular plant implementation. A brief example of operational risk assessment is provided in Table 49.4.

49.4 RISK ASSESSMENT TOOLS

Different tools can be applied at the various stages of risk assessment, depending on the objective, the amount of

TABLE 49.4 Example of an Operational Risk Assessment

Failure Mode (From Process Risk Assessment)	Risk (From Process Risk Assessment)	Probability (From Process Risk Assessment)	Operational Controls	Operational Risk Assessment of Probability	Operational Risk Assessment of Detectability	Residual Risk
Temperature too high	Decomposition of product to form impurity A	High	1. Automated control system 2. Temperature alarm limits with failure set points 3. Continuously manned/monitored operation	Medium	High detectability	Medium (ALARP)
Amount of reagent A too low	Side reaction occurs to form impurity B	Medium	1. Pre-weighing of material in warehouse 2. Sensitive weighing equipment on production floor 3. Frequent calibration and maintenance 4. Weight check after material charge 5. Additional material available to charge	Low	Moderate detectability	Low

information available prior to risk assessment, and the stage of development. Two relatively simple qualitative tools are risk triage and risk relationship matrices. In later-stage development, when more development information is available, failure mode and effects analysis (FMEA) is very commonly used for thoroughly assessing process risk and determining critical process parameters (CPPs) and control strategy. Each of these tools is discussed in the following section.

49.4.1 Risk Triage/Three-Level Qualitative Risk Assessment

The simplest method of risk assessment may be the risk triage. This qualitative risk assessment approach has the advantage of being expedient, and it results in concise, easily digestible output. The main disadvantage is that it is less thorough and nuanced than tools such as FMEA.

A typical risk triage can consider the impact or severity of the risk and the likelihood or probability. These together can be used to define an overall risk level. A scoring guide is typically used to ensure consistent risk rating between assessors. An example scoring system is given in Table 49.5.

Using this guide and the example failure information from Table 49.3, Table 49.6 is an example qualitative risk assessment. In this example, four failures with moderate or high risk were identified, which necessitate the implementation of controls.

49.4.2 Risk Relationship Matrices

Relationship matrices can be used throughout the risk management process and provide a visual way to convey the areas of potential risk that need to be considered. They are structured to enable an assessment of risk, or a known relationship, to be recorded and typically have a list of CQAs on one axis. The other axis can be populated at various levels depending on the information available and may include, for example, each stage of a process, or each unit operation, or each parameter as shown in the example in Table 49.7. The strength of relationship, or size of the risk, can be represented by a categorical or numeric input. These can be particularly useful as a quick way to brainstorm where there may be risks that need further analysis and which parts of a process present lower risk.

Similar matrices can also be used to represent the relationships between API CQAs and drug product CQAs.

49.4.3 Failure Mode and Effects Analysis (FMEA)

Failure mode and effects analysis is a tool that was developed in the late 1950s to study potential malfunctions of military systems. It is a highly structured and systematic method of failure analysis that is well suited to detailed assessment of pharmaceutical (and other industries') manufacturing processes. There are several types of FMEA analysis that are used in different contexts, but in this chapter, we will focus on process FMEAs. Process FMEAs are ubiquitous in pharmaceutical process development and manufacturing.

In later-stage development, failure mode and effect analysis (FMEA) is recommended as a risk management tool given its thorough and detailed approach and broad acceptance by global health authorities. FMEA provides for a numerical evaluation of various potential failure modes and their likely effect on product performance and safety. Failure modes are identified and scored according to the mode severity, probability of occurrence, and detectability. Risk is then calculated as a product (Risk Priority Number or RPN) by

$$RPN = severity \times probability \times detectability$$

(Detectability may not always be employed as a separate step; see Section 49.4.3.3.)

49.4.3.1 Severity Severity (S) is defined as the measure of the possible consequence of the hazard. Two different example systems for assessing severity are provided in Tables 49.8 and 49.9; one links to patient impact, and the other reflects the relationship (or lack thereof) and sensitivity to variability between the cause and the resultant effect on an API CQA. An example of this would be to determine the strength of the relationship between a process parameter, such as temperature, and the formation of an impurity. In addition, the part played by a parameter in an interaction should also be considered in the assessment of severity. Some parameters can show little or no impact when evaluated alone but have a more significant impact as part of an interaction with other parameters or attributes.

49.4.3.2 Probability Probability (P) is defined as the likelihood of the occurrence of the failure mode. In some instances, the occurrence of the harm is distinct from the occurrence of the failure mode and is also taken into account.

TABLE 49.5 Example of a Qualitative Risk Assessment (Risk Triage) Scoring Guide

Risk Rating	Impact Guide	Probability Guide
Low	Non-detectable impact or detectable but not significant impact to CQAs	Not likely (occurrence rate <1 in 100 batches)
Medium	Significant impact to CQAs – potential to impact patient safety	Moderately likely (occurs in 1 in 10 batches to 1 in 100 batches)
High	Impact to CQAs with potential for severe harm to patient	Likely (more frequent than 1 in 10 batches)

TABLE 49.6 Example Qualitative Risk Assessment

Potential Failure	Risk	Impact Rating	Probability Rating	Overall Risk Rating
Too little solvent used	Reduced stability of the reaction stream resulting in formation of impurity A	High	Low	Moderate
Too much solvent used	No impact	Low	Low	Low
Too little reagent A used	Competitive side reaction occurs resulting in formation of impurity B	Moderate	Moderate	Moderate
Too much reagent A used	No impact	Low	Moderate	Low
Low reaction temperature	Incomplete reaction resulting in residual starting material present in the downstream steps	Moderate	Moderate	Moderate
High reaction temperature	Decomposition of product to impurity A	High	High	High
Reaction time too short	Incomplete reaction resulting in residual starting material present in the downstream steps	Moderate	Low	Low
Reaction time too long	None, no additional side reactions anticipated	Low	Low	Low

TABLE 49.7 Example of a Risk Relationship Matrix

	Impurity A	Genotoxic Impurity B	Form	Particle Size Distribution
Reaction temperature	Strong relationship	Moderate relationship	No known relationship	No known relationship
Amount of reagent	No known relationship	Strong relationship	No known relationship	No known relationship
Crystallization temperature	Moderate relationship	No known relationship	No known relationship	Moderate relationship
Crystallization seed amount	No known relationship	No known relationship	Moderate relationship	Moderate relationship

TABLE 49.8 Example Severity Levels and Terminology Based on Patient Safety Impact

Severity Level (Score)	Potential Impact to Patient Safety
Disastrous (5)	Will cause permanent impairment or damage of a body structure or function. Could lead to patient death
Critical (4)	Could cause permanent impairment or damage to a body structure or function, but is not fatal or life threatening. Likely to impact product efficacy
Moderate (3)	May cause significant temporary unintended impairment of a body function. May impact product efficacy
Minor (2)	May cause transient, self-limiting, unintended impact to a body function. May cause dissatisfaction to the patient and customer complaint
Irrelevant (1)	No performance impact to patient. May have cosmetic defect that is unlikely to cause dissatisfaction to the patient

TABLE 49.9 Example of a Different Severity Scoring System Based on Sensitivity of CQAs to the Failure Mode

Score	Description	Criteria
1	Not severe	The attribute or process parameter has no impact on CQAs
4	Slightly severe	Large change of this attribute or process parameter in combination with other factors has a significant impact on a CQA
7	Moderately severe	Large change of this attribute or process parameter or a small change in this attribute or parameter in combination with other factors has a significant impact on a CQA
10	Extremely severe	Small to moderate change of this attribute or process parameter has a significant impact on a CQA

Different risk assessments need to account for different magnitudes of failure frequency: drug substance and some drug product risk assessments consider failure modes that would cause an entire batch to fail, whereas device risk assessments and some other drug product risk assessments consider the failure rate of individual units in a production run. The former case has a context where only 5–20 batches may be run per year, whereas in the latter case hundreds of thousands or millions of units may be manufactured per batch. In light of these different contexts, it is possible that multiple probability scales can be defined, e.g. one for

TABLE 49.10 Example Probability Scoring

Probability Term	Score	Probability (For Batch Occurrence)	Probability (For Unit Occurrence)
Improbable	1	≤1 in 1000	≤1 in 100 000
Remote	2	>1 in 1000	>1 in 100 000
Occasional	3	>1 in 100	>1 in 10 000
Probable	4	>1 in 10	>1 in 1 000
Frequent	5	>1 in 5	>1 in 100

occurrence rates impacting the batch and one for rates impacting individual unit occurrence. An example of such scoring is given in Table 49.10.

Where insufficient data exists to enable any numerical estimate of probability, it may be more pragmatic to consider other approaches such as comparison of NORs with proven acceptable ranges (PARs) or design space ranges or a generic assessment of individual operations to determine their inherent variability based on, for example, the level of automated versus manual operations. In these cases, the probability of occurrence may not score the same for each end of a range, and an understanding of the cause and mechanism of the failure being assessed is key to defining an appropriate probability score.

49.4.3.3 Detectability

Detectability (D) is the ability to discover or determine the existence, presence, or fact of a failure mode and provide a control action on the failure mode and/or effect. An example detectable scoring system is given in Table 49.11.

The use of a separate detectability element is not always mandated. For example, detectability may not be employed in early-stage (design) assessments where detection systems have not been developed or where inherent risk is required to be assessed before corrective measures are put in place.

Where a Risk Prioritization Number (RPN) is to be calculated and detectability is not utilized, a score of 5 should be assumed for D, i.e. to reflect no separate detection measures.

49.4.3.4 RPN

The composite risk score for each risk assessed is called a Risk Priority Number (RPN) that is the product of its three individual component ratings: severity, probability, and detectability. This composite risk score can be reduced in two ways, mitigation (see Section 49.4.3) or detection.

Table 49.12 shows an example risk matrix with acceptability criteria. A "not-acceptable" (dark shaded) composite score indicates that the anticipated failure mode is likely to produce an unacceptably severe effect on the patient or the manufacturing site. Further development to reduce the risk by redesign, mitigation, or detection is required. An "ALARP" (lightly shaded) composite score indicates that all reasonable efforts should be made to reduce the risk by mitigation (reduction of probability or severity) or detection. An "acceptable" (unshaded) composite score indicates that the risk has been reduced to acceptable levels and no further development actions are needed.

An alternative way to identify the areas of greatest risk is through the use of a Pareto chart. This displays the risks in size order, and then a suitable RPN cutoff can be defined that accounts for a proportion of the total risk. For example, a cutoff figure of 80% is shown in Figure 49.7. Risks that are above the cutoff are further examined to determine any additional controls that can be reasonably employed to reduce the risk. Treating risk in this way ensures that 80% of the cumulative risk is being evaluated for further action. Typical examples of mitigating actions for the detection score include additional analysis of materials prior to use, additional parametric monitoring, and the introduction of improved analytical methods. Opportunities to reduce the probability score typically come from modification in the way the process is executed, introduction of equipment with enhanced control, and modification of the parameter settings.

One additional consideration in risk assessment scoring is that at earlier stages of development, knowledge of the failure modes may be incomplete. This can be handled either by incorporating an explicit knowledge score or by defaulting the severity and/or the probability to a high level until further knowledge is obtained. The risk assessment can be repeated and finalized later in the development cycle when better information is available to assess the true risk.

49.4.3.5 FMEA Assessment Tool

With such scoring system in place, the FMEA exercise is undertaken by considering each failure mode, the effect of the failure mode, the associated severity, probability, and detectability, and the RPN number. Often an additional action plan is documented along with the risk assessment. An example FMEA risk assessment, following the risks previously given in Table 49.3, is presented in Table 49.13.

In this example, it becomes clear where additional development efforts are required to address the unacceptable or ALARP items. Another outcome of the risk assessment can be the designation of CPPs. In the above example, reagent A quantity and reaction temperature could be designated as CPPs. When taking an enhanced QbD approach to product development, identification of any CPPs in the process is required by the health authorities. For all batches manufactured, the manufacturer must track whether the CPPs have been maintained with a range that was proven to result in an acceptable outcome (PAR). Deviation of a CPP from a PAR in a given batch triggers a deviation investigation by the quality unit, and in some instances a notification to some health authorities must be provided.

TABLE 49.11 Example Detectability Scoring System

Detectability (D)	Ranking Score	Explanation
High degree of detectability	1	Controls will almost certainly detect potential failure modes to enable a control action, e.g. validated automated detection system that is a direct measure of failure. Two or more manual operated validated detection systems, direct or indirect (e.g. monitored control range and IPC)
Good detectability	2	Controls will have a good chance of detecting potential failure modes to enable a control action, e.g. single non-automated, validated detection system that is a direct measure of failure (e.g. IPC of failure, validated PAT)
Likely to detect	3	Controls may detect the existence of failure to enable a control action, e.g. single manually operated validated detection system that is not a direct measure of failure (e.g. PAT measurements or IPCs not directly linked to failure)
Fair detectability	4	Controls may not detect failure modes, e.g. non-validated (manual or automated) detection such as a visual level check, visual inspection)
Low or no detectability	5	Controls will very likely not detect the existence of potential failure mode, e.g. no detection system of any kind in use

TABLE 49.12 RPN Matrix Example

RPN Scores

	Severity (Severity of Failure Increases from Left to Right)				
Probability	Irrelevant (1)	Minor (2)	Moderate (3)	Critical (4)	Disastrous (5)
Frequent (5)	25	50	75	100	125
Probable (4)	20	40	60	80	100
Occasional (3)	15	30	45	60	75
Remote (2)	10	20	30	40	50
Improbable (1)	5	10	15	20	25

A case where detectability = 5.

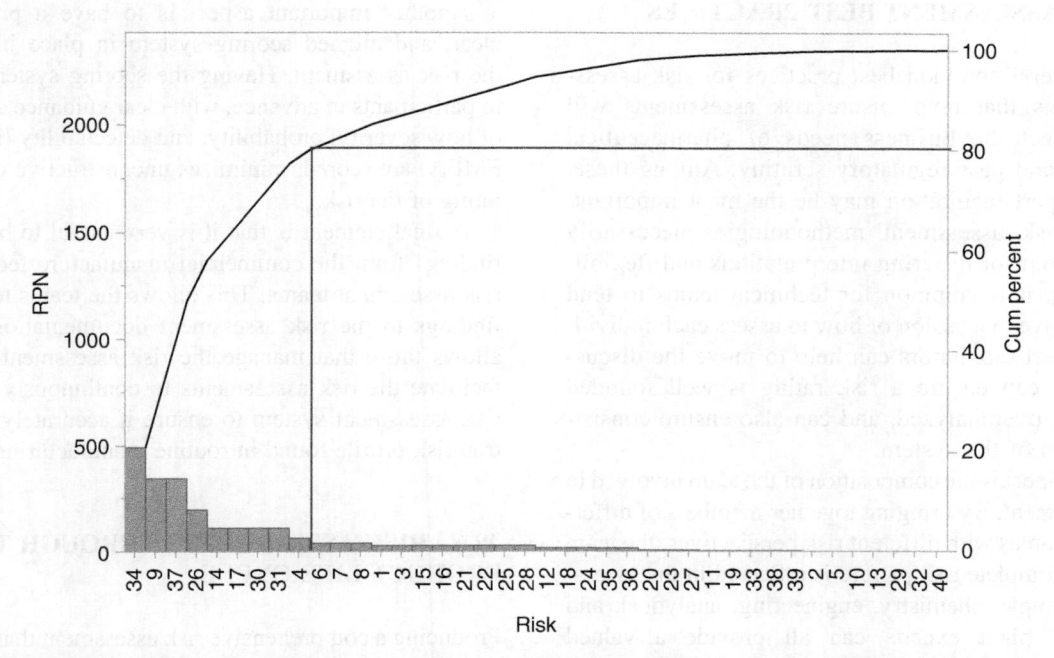

FIGURE 49.7 Example Pareto chart of the relative risk (RPN) and cumulative risk (cum. percent) against risk number ordered by risk size.

TABLE 49.13 Example FMEA Risk Assessment

Failure Mode	Effect	Detection/Mitigation Available	Severity	Probability	Detectability	RPN	Action Plan
Too little solvent used	Reduced stability of the reaction stream resulting in formation of impurity A	None	4	1	5	20	Determine if additional controls can be developed
Too much solvent used	No impact	None	1	1	5	5	
Too little reagent A used	Competitive side reaction occurs resulting in formation of impurity B	None	3	3	5	45	Additional in-process control to be developed
Too much reagent A used	No impact	None	1	3	5	15	
Low reaction temperature	Incomplete reaction resulting in residual starting material present in the downstream steps	In-process control point (analytical test and control actions)	3	3	1	9	
High reaction temperature	Decomposition of product to impurity A	Temperature alarm with control action	4	4	3	48	Additional process development needed
Reaction time too short	Incomplete reaction resulting in residual starting material present in the downstream steps	In-process control point (analytical test and control actions)	3	2	1	6	
Reaction time too long	None, no additional side reactions anticipated	None	1	1	5	5	

49.5 RISK ASSESSMENT BEST PRACTICES

There are several common best practices for risk assessment programs that help ensure risk assessments will effectively meet the business needs of pharmaceutical development and pass regulatory scrutiny. Among these, the use of expert facilitation may be the most important. Given that risk assessment methodologies necessarily have some room for differing interpretations and flexibility in scoring, it is common for technical teams to tend toward excessive discussion of how to assess each individual item. Expert facilitators can help to move the discussion forward, can ensure a risk rating is well founded without being overanalyzed, and can also ensure consistent application of the system.

A second aspect is the composition of the team involved in the risk assessment. By bringing together members of different functional areas with different risk perspectives, the team can ensure a complete risk picture is reflected in the assessment. For example, chemistry, engineering, analytical, and manufacturing plant experts can all provide a valued perspective.

Another important aspect is to have a predetermined, clear, and aligned scoring system in place in advance of the risk assessment. Having the scoring system distributed to participants in advance, with clear guidance and examples of how severity, probability, and detectability (in the case of FMEA) are scored, minimizes unconstructive debate on the rating of the risk.

A final element is that it is very useful to have data and findings from the commercial manufacture fed back to the risk assessment teams. This allows the teams to add missed findings to the risk assessment documentation. Further, it allows those that manage the risk assessment process and facilitate the risk assessments to continuously improve the risk assessment system to ensure it accurately captures the true risk profile found in routine manufacturing.

49.6 RISK ASSESSMENTS THROUGH THE PRODUCT LIFE CYCLE

Producing a comprehensive risk assessment that supports the transfer of the process from R&D to manufacturing, in addition to the regulatory submissions, is not the end of the process. During the life cycle of the product, a regular assessment of risk and the performance of the control strategy are highly recommended. This activity should consider a review of recent batch performance, any additional laboratory data that has been generated, equipment changes, and experience of additional batches of starting material, reagent, and solvent supplies. A more comprehensive reassessment is required if post-approval changes to the process are planned such as changing the site of manufacture or the introduction of a new processing step or different technology. If any of these changes result in the identification of a new CQA, then the whole process must be reassessed. It is also possible that additional experience gained in commercial manufacturing results in the definition of new CPPs when a rigorous risk management process is applied.

49.7 CONCLUDING REMARKS

In this chapter, we have outlined the role of QRA in QbD development, various risk assessment approaches and tools, and examples of how these tools may be applied during the development cycle. In our exemplification, we have focused on technical risks to quality, leading to a robust control strategy based on QbD principles. Many pharmaceutical companies use these or similar approaches in a structured or formal way throughout product development, and they can provide an excellent framework for in-depth scientific discussion in addition to managing risk. A final aspect to acknowledge is that as well as such codified work practices, less formal risk assessment also has application as part of sound process development. Individual scientists, or teams, can make use of informal risk assessment to prioritize their efforts and convey risk to their management as a justification of resource spend or development timelines. Chemical engineers are often called upon to lead or perform these risk assessments, particularly in later stages of development. Thus, an understanding of risk assessment principles, terminology, and practices is fundamental to any process development engineer in the pharmaceutical industry.

REFERENCES

1. ICH (2009). Q8(R2): Pharmaceutical development. *International Conference on Harmonization of Technical Requirements for Registration of Pharmaceuticals for Human Use*. https://www.ich.org/fileadmin/Public_Web_Site/ICH_Products/Guidelines/Quality/Q8_R1/Step4/Q8_R2_Guideline.pdf.
2. ICH (2005). Q9: Quality risk management. *International Conference on Harmonization of Technical Requirements for Registration of Pharmaceuticals for Human Use*. https://www.ich.org/fileadmin/Public_Web_Site/ICH_Products/Guidelines/Quality/Q9/Step4/Q9_Guideline.pdf.
3. ICH (2012). Q11: Development and manufacture of drug substances. *International Conference on Harmonization of Technical Requirements for Registration of Pharmaceuticals for Human Use*. https://www.ich.org/fileadmin/Public_Web_Site/ICH_Products/Guidelines/Quality/Q11/Q11_Step_4.pdf.
4. ICH (2008). Q10: Pharmaceutical quality system. *International Conference on Harmonization of Technical Requirements for Registration of Pharmaceuticals for Human Use*. https://www.ich.org/fileadmin/Public_Web_Site/ICH_Products/Guidelines/Quality/Q10/Step4/Q10_Guideline.pdf.
5. Womack, J. (2011). *Gemba Walks*, 348. Lean Enterprise Institute, Inc.
6. ICH (2006). Q3A(R2): Impurities in New Drug Substances. International Conference on Harmonization of Technical Requirements for Registration of Pharmaceuticals for Human Use. https://www.ich.org/fileadmin/Public_Web_Site/ICH_Products/Guidelines/Quality/Q3A_R2/Step4/Q3A_R2__Guideline.pdf.
7. ICH (2017). M7: Assessment and control of DNA reactive (mutagenic) impurities in pharmaceuticals to limit potential carcinogenic risk. *International Conference on Harmonization of Technical Requirements for Registration of Pharmaceuticals for Human Use*. https://www.ich.org/fileadmin/Public_Web_Site/ICH_Products/Guidelines/Multidisciplinary/M7/M7_R1_Addendum_Step_4_2017_0331.pdf.
8. ICH (2018). Q3C(R7): Impurities: guideline for residual solvents. *International Conference on Harmonization of Technical Requirements for Registration of Pharmaceuticals for Human Use*. https://www.ich.org/fileadmin/Public_Web_Site/ICH_Products/Guidelines/Quality/Q3C/Q3C-R7_Document_Guideline_2018_1015.pdf.

50

DEVELOPMENT OF DESIGN SPACE FOR REACTION STEPS: APPROACHES AND CASE STUDIES FOR IMPURITY CONTROL

SRINIVAS TUMMALA, ANTONIO RAMIREZ, AND SUSHIL SRIVASTAVA
Bristol-Myers Squibb, New Brunswick, NJ, USA

DANIEL M. HALLOW
Noramco, Athens, GA, USA

50.1 INTRODUCTION

The objective of pharmaceutical development is to advance a commercial manufacturing process to consistently deliver drug substance and drug product of the intended quality through the life cycle of the product [1]. The scientific knowledge gained from pharmaceutical development and manufacturing experience provides understanding to enable process optimization and support the establishment of chemistry and manufacturing controls that assure product quality. Elements of the control strategy can include drug substance specifications, procedural controls, controls in material attributes (MAs), and parameter controls. In the case of parameter controls, the selection of ranges that provide material meeting quality criteria can be presented in terms of univariate intervals that keep other parameters at their intended target value. Alternatively, the parameter range selection can be built upon multivariate experimentation to define a design space, "the multidimensional combination and interaction of input variables and process parameters that has been demonstrated to provide assurance of quality" [2].

Although there is no regulatory requirement for the establishment of a design space to define a successful control strategy, design space selection is associated with an enhanced approach to pharmaceutical development that can provide a more robust process and lead to regulatory flexibility. For example, working within an approved design space is not considered an operational change and could facilitate continuous improvement throughout the product life cycle. A design space is proposed by the applicant as part of the regulatory submission and subject to the approval of the regulatory agencies.

This chapter examines current approaches to establish a design space for reaction steps. The review does not intend to give an exhaustive description but rather distill key concepts from existing ICH guidance and illustrate them with case studies. Certainly, the selection of these examples is biased by the author's interests and experience. Efforts to maintain a uniform terminology throughout the chapter were made for the sake of clarity but may have not captured subtle modifications due to varying approaches specific to different companies, processes, or products.

50.2 ELEMENTS OF PHARMACEUTICAL DEVELOPMENT

50.2.1 Quality Target Product Profile

The first step toward achieving the goals of pharmaceutical development is to outline a summary of the characteristics of the drug product that will ideally be achieved to ensure the desired quality from a safety and efficacy perspective. These characteristics are commonly referred to as the quality target product profile (QTPP). The mode of delivery, dosage strength, factors that impact release of the therapeutic moiety,

---

Chemical Engineering in the Pharmaceutical Industry: Active Pharmaceutical Ingredients, Second Edition.
Edited by David J. am Ende and Mary T. am Ende.
© 2019 John Wiley & Sons, Inc. Published 2019 by John Wiley & Sons, Inc.

and the drug product quality attributes (QAs) (e.g. sterility, purity, and stability) must be considered when assessing the QTPP. In order to guarantee product safety and efficacy for the patient, only characteristics pertinent to the patient should be included in the QTPP. For example, if particle size is critical to the dissolution of an oral solid product, the QTPP should refer to dissolution instead of particle size [3]. A typical example of QTPP for an oral solid dosage product is described in Table 50.1.

50.2.2 Potential Critical Quality Attributes

The potential critical quality attributes (pCQAs) of the drug product and, subsequently, those of the drug substance are identified based on the QTPP and initial process knowledge. ICH Q8 defines a CQA as "a physical, chemical, biological, or microbiological property or characteristic that should be within an appropriate limit, range, or distribution to ensure the desired product quality." CQAs of an oral solid dosage product typically include characteristics affecting product purity, strength, drug release, and stability. For a drug substance, the pCQAs include those properties that could potentially affect drug product CQAs. The first column of Table 50.2 illustrates common pCQAs for a drug substance.

Impurities are a fundamental class of drug substance pCQAs due to their ability to impact the safety and efficacy of the drug product. For chemical entities, impurities can involve organic impurities – including potentially genotoxic impurities (GTIs) [4], inorganic impurities (e.g. residual metals), and residual solvents. Understanding the key steps where the impurities form and purge in the process and establishing controls to ensure that these impurities do not affect the quality of the drug substance is a key focus of drug substance development.

50.2.3 Quality Risk Assessment

Quality risk assessment is a tool used to (i) evaluate if a pCQA is likely to impact quality, (ii) review which steps of the synthesis are likely to affect these CQAs, and (iii) identify and rank MAs and process parameters that may influence the QAs of the step's output and ultimately impact drug substance quality. The characteristics of an input material that should be within an appropriate limit, range, or distribution to ensure the desired product quality are commonly referred to as MAs [5]. For a discussion of quality risk assessment in drug substance process development, see Chapter 49.

TABLE 50.1 Example of QTPP for an Oral Solid Dosage Product

QTPP	Target	Justification
Dosage form	Capsule	Proven effective dosage form
Dosage design	Immediate release soft gelatin capsule	Efficacy of the drug product
Dosage strength	100 mg	Efficacy of the drug product
Route of administration	Oral	Safe and effective route
Identity	Drug substance is "name"	Safety of the drug product
Assay	95–105%	Efficacy of the drug product
Impurity profile	Control of potential impurities in drug substance	Safety of the drug product

TABLE 50.2 Example of pCQAs for an Oral Solid Drug Substance

Quality Attribute	Tests	Criticality	Rationale
Identity	IR/Raman/HPLC	Yes	ICH Q6A
Assay	HPLC	Yes	Impacts drug product potency
Impurity			
Related substances	HPLC	Yes	Impurity x in the Final Intermediate has potential to impact drug substance quality and is controlled to ≤0.20 RAP
Genotoxic impurities	LC-MS	Yes	Residual y and p-TsOEt are GTIs. These impurities are controlled below the TTC limits, i.e. ≤3.7 ppm each for y and p-TsOEt
Diastereomers	HPLC	Yes	ICH Q6A Flowchart #5 "z" is an optically active substance
Inorganic impurities (including heavy metals)	ICP-MS	No	No metals used in the synthesis of the drug substance. The level of inorganic impurities is controlled at ≤20 ppm
Class 2 residual solvents	GC	Yes	Class 2 solvents are used in the manufacturing process. These solvents are controlled below limits defined in ICH Q3 (R6)
Particle size distribution	LLS	No	Drug substance is dissolved during drug product manufacture
Solid-state crystal form	PXRD	No	Drug substance is dissolved during drug product manufacture

The TTC acronym stands for "Threshold of Toxicological Concern" for genotoxic impurities. See Section 50.7.

A criticality assessment – a yes or no evaluation primarily based on severity of harm – is employed to determine which of the potential CQAs of the drug substance are likely to impact the drug product CQAs. This assessment does not change as a result of risk management [6] and is iterative in nature because criticality of the pCQAs may evolve during process development: as process knowledge increases, the evaluation of the impact of a pCQA on the quality of the drug product may change. The last two columns of Table 50.2 illustrate a criticality assessment and its rationale for pCQAs.

Once the synthesis for the commercial process is established, a collective process risk assessment is carried out to identify the steps in the synthetic sequence whose QAs have the potential to impact drug substance quality. These steps can be designated as "critical steps." For most early steps, the assessment focuses on evaluating if impurities from these steps impact the drug substance impurity pCQAs. Typical levels of impurities in a given step are evaluated for their fate and purge in the subsequent unit operations. Criticality of a process-related impurity can only be established when the functional relationship between the impurity and the drug substance pCQA is mapped and understood. For the final steps, additional properties are assessed beside impurity profile, including physical properties of the drug substance pCQAs such as crystal form, particle size, etc. Finally, an individual risk assessment of each step is performed to assess MAs and process parameters that could impact the QAs of that step. MAs and process parameters are then prioritized for additional investigation.

50.2.4 Control Strategy

A central tenet of the quality-by-design (QbD) paradigm for process development is to understand the relationship between MAs and process parameters and the QAs for the various steps in the synthesis. Subsequent sections of this chapter will outline two methodologies to establish this link, namely, the mechanistic approach and the empirical approach. If cause–effect analysis demonstrates that there is a link between the MAs, process parameters, or both and the QAs, then a control strategy must be developed to ensure that consistent quality can be achieved. The control strategy for the impurity CQA can entail one or several of the following elements: (i) specifications for the input materials; (ii) specifications for impurities in the starting materials and intermediates of the chemical synthesis based on fate and purge during the process; (iii) process design to mitigate impurity formation, purge of the impurity, or both; and (iv) ranges for MAs and process parameter controls to reduce impurity formation or ensure its purge in subsequent processing steps [7]. A representative flow diagram of elements of pharmaceutical development in drug substance manufacture is shown in Figure 50.1.

50.2.4.1 Proven Acceptable Ranges and Design Space
The ranges for MAs and process parameters in which quality is assured can be expressed either as a series of proven acceptable ranges (PARs) or as a multivariate design space. PARs for a given parameter denote the univariate range that the parameter could vary within and still assure quality either because (i) it has minimal or no interaction with other MAs and process parameters on how it impacts the quality attribute or (ii) it assumes that all the other attributes and parameters are at or close to their target values. When parameters have significant multivariate interactions, the ranges can be expressed as a design space. The final control strategy may include a mixture of PARs, design spaces, or both across the unit operations and reaction steps. Choosing the approach for each operation should consider the risk and impact of multivariate interactions on product quality. The normal operating range (NOR) of a parameter corresponds to the manufacturing range inside the PAR that is defined by the target value and extent in which the parameter is controlled (Figure 50.2) [8].

50.2.4.2 Unit Operation Design Space
Impurities with potential impact on the quality of the drug substance can (i) originate from input raw materials, (ii) form in earlier steps of the synthesis, or (iii) be generated in the drug substance step. During the assessment of the impact of MAs and process parameters on the QAs of a step, dividing the step into unit operations helps to decide which operations should be selected for further evaluation. Let us consider an intermediate step that generates an impurity with potential to impact the QAs (Eqs. 50.1 and 50.2, Table 50.3). The reaction that forms the impurity and the crystallization that purges it can be chosen for additional analysis. Establishing limits for the impurity levels that must be met at the end of the reaction and after the crystallization facilitates the independent study of the operations – i.e. understanding the impact of the process parameters and defining process parameter ranges or design spaces that assure quality. The following sections of this chapter focus on reaction steps:

$$\text{Input} + \text{Reagent} \rightarrow \text{Product} \quad (50.1)$$

$$\text{Product} + \text{Reagent} \rightarrow \text{Impurity} \quad (50.2)$$

50.3 REACTION DESIGN SPACE DEVELOPMENT

50.3.1 Establishing Functional Relationships

While the development of the steps in the synthesis of a drug substance must consider all the unit operations, the reaction step assumes significance because it often is the origin of the impurities that could be quality impacting. As discussed previously, an individual risk assessment of the reaction step is

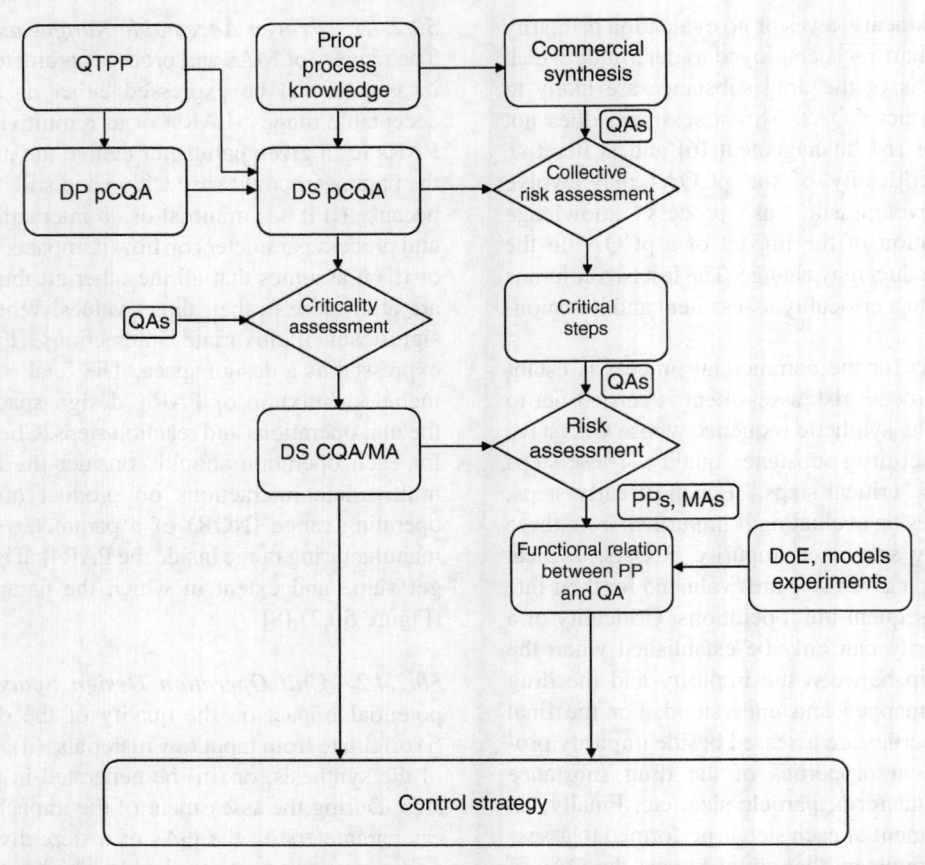

FIGURE 50.1 Elements of pharmaceutical development in drug substance manufacture. DP, drug product; DS, drug substance; QAs, quality attributes of a given step; PPs, process parameters; MAs, material attributes.

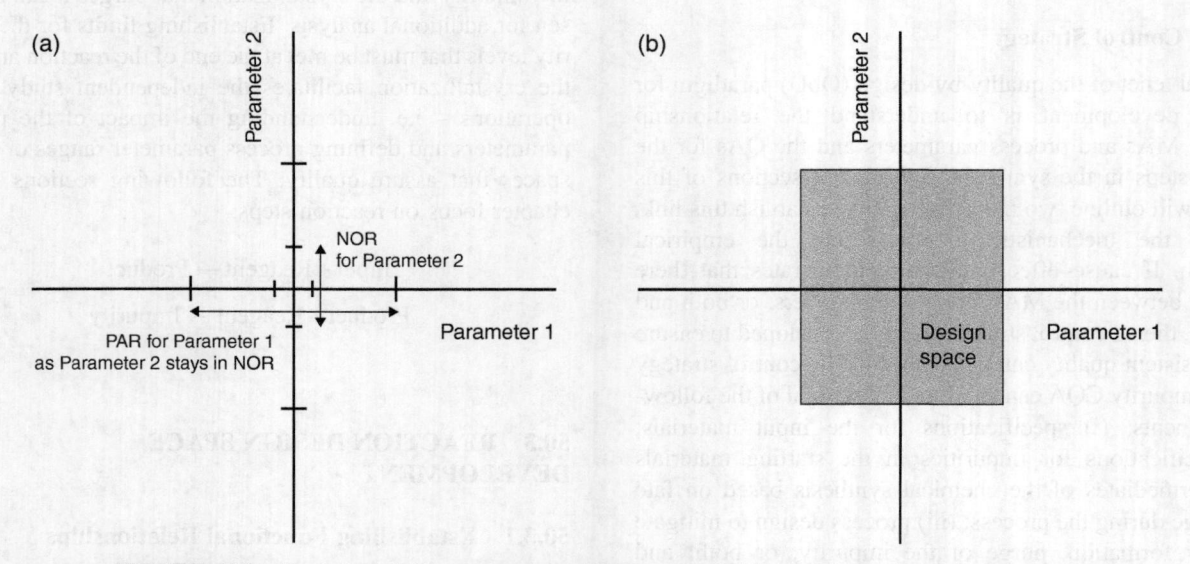

FIGURE 50.2 Depiction of (a) proven acceptable ranges and (b) a two-dimensional design space.

TABLE 50.3 Representative Assessment of Unit Operations

Unit Operation	Assessment	Potential to Impact QAs
Reaction	Impurity formation step	High
Quench, extraction, washes	Minimal impact on Impurity level	Low
Distillation	Minimal impact on Impurity level	Low
Crystallization	Impacts Impurity purge level	Medium
Drying	Minimal impact on Impurity level	Low

carried out to identify the MAs and process parameters that can impact impurity levels. Subsequently, the functional relationship between these impurities, MAs, and process parameters needs to be established. This relation can be determined using a univariate analysis in which the effect of a single variable on a given quality attribute is measured while keeping the other variables at their "target" value. Alternatively, the relationship can be established using a multivariate analysis that considers the effect of multiple variables at a time.

Two strategies based on the elaboration of predictive models can be adopted to define a multivariate functional relationship: the mechanistic and the empirical approach. Mechanistic models entail the elucidation of kinetic and thermodynamic expressions that define the formation of product and relevant impurities. In contrast, empirical models utilize design of experiment (DOE) [9] and statistical analysis to regress polynomial expressions that help establish the desired relationship. From an ideal standpoint, the mechanistic approach would be the preferred strategy since it is underpinned by fundamental knowledge. However, in complex systems where developing a full understanding is not viable, the empirical approach may be employed [10].

50.3.2 Development of the Mechanistic Model

Mechanistic models for a reaction step correlate concentrations of the species involved in the transformation with the underlying processes and variables deemed to affect them, according to physical organic principles. The models aim at describing the time-course fate of input materials, intermediates, and products and require an understanding of the elementary steps that represent the reaction, including kinetic and thermodynamic variables. The development of a mechanistic model involves four basic stages: (i) description of the elementary steps that define the transformations of interest, (ii) mechanistic studies to define model variables, (iii) regression of model variables using an experimental building data set, and (iv) model evaluation using an experimental verification data set.

50.3.2.1 Description of Elementary Steps
Model development starts by defining the sequence of elementary steps that represents the overall reaction. Literature precedence, control experiments, or a combination of both can be used to lay out the adequate network of discrete processes that includes chemical transformations and mass transfer events. Description of the elementary chemical transformations entails the extraction of structural information to specify balanced equations for starting materials and products. Similarly, characterization of mass transfer events (e.g. solid–liquid, liquid–liquid, gas–liquid) exploits evidence for physical transport to define balanced equations that influence the concentrations of the chemical species. Methods to characterize chemical transformations and mass transfer events are routinely reported in the chemical literature, and a revision of the techniques involved is beyond the scope of this chapter. However, note that in situ spectroscopic methods offer structural and kinetic information that can precisely link starting materials with products – or reactive intermediates – of an elementary reaction [11]. They can also shed light into the nature of mass transfer events and help evaluate the driving forces behind them.[1]

The elementary steps constitute the foundation upon which the mechanistic model is built. Elementary steps are statistically independent but linked by the concentration of the components they have in common [13]. Consequently, an imprecise definition of the step sequence may limit its predictive capability, undermining subsequent kinetic, thermodynamic, and numerical analyses required to complete the model. For example, impurities that do not affect the quality of the product and are generated within the purge capability of the process should be incorporated in the model when their formation influences the concentrations of the main reaction components. While elementary steps are represented by simple balanced equations, recent practice has adopted the use of process schemes as a generic and visual depiction that contains the phases and rates involved in the reaction step (Figure 50.3) [14].

50.3.2.2 Mechanistic Studies
The next stage to build the model entails gathering mechanistic information for the kinetic and thermodynamic variables associated with the elementary steps. The goal of this stage is feeding the model with experimentally measured variables (e.g. reactant orders, rate and equilibrium constants, activation energies, solubility) to simplify and guide the subsequent numerical regression. For the elementary steps shown in Table 50.4, kinetic studies would afford reactant orders and initial values for rate

---
[1] For leading references, see Ref. [12].

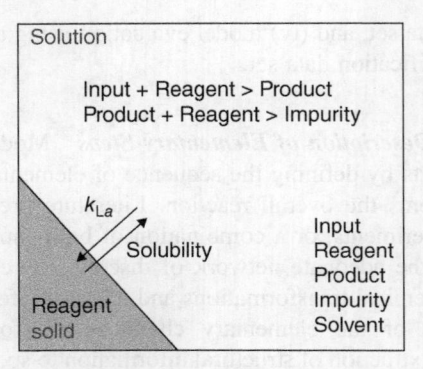

FIGURE 50.3 Representative process scheme for a solid–liquid reaction.

TABLE 50.4 Rate Equations and Reaction Orders of Elementary Steps

Elementary Step	Rate Equation	Reaction Order
$A \rightarrow P$	Rate = $k[A]$	1st
$A + B \rightarrow P$	Rate = $k[A][B]$	1st in A, 1st in B, 2nd overall
$2A \rightarrow P$	Rate = $2k[A]^2$	2nd
$2A + B \rightarrow P$	Rate = $2k[A]^2[B]$	2nd in A, 1st in B, 3rd overall

TABLE 50.5 Equilibrium Reaction with Forward and Reverse Rates

Equilibrium	Requirement	Forward Rate	Reverse Rate
$2A \leftrightarrow P$	$K_{eq} = [P]/[A]^2$	Rate = $k_a[A]^{1.35}$	Rate = $k_b[P]/[A]^{0.65}$

constants (k), while temperature dependencies of the rate constants would provide activation energies (E_a) in agreement with the Arrhenius empirical equation ($k = A \exp[-E_a/RT]$). Similarly, for reversible reactions, understanding the equilibrium expression and its position would afford the equilibrium constant (K_{eq}) along with forward and reverse rate equations in thermodynamic consistency with equilibrium requirements (Table 50.5). Examination of the temperature dependence of the equilibrium constants would provide activation energies (E_a) in agreement with the van't Hoff equation ($K_{eq} = \exp^{-\Delta rHo/RT} \exp^{\Delta rSo/R}$).

Mechanistic studies and elementary steps should be self-consistent: e.g. an experimental rate law should concur with the proposed sequence of steps, and a kinetic isotope effect ought to match with the proposed rate-limiting step in the sequence. In complex reactions, varying concentration–time curves due to changes in mechanism should be traceable to a common set of steps. Convoluted scenarios such as water-sensitive transformations, autocatalysis or autoinhibition [15], catalyst deactivation [16], or mass transfer-limited reactions [17] should be captured by the model, and their significance under different conditions ought to reflect the diverse kinetic profiles. Take, for example, the solid–liquid reaction shown in Figure 50.4. Simulation of mass transfer-limited and non-limited regimes in reagent (1.25 equiv. relative to input) maintaining k_1 and k_2 constant ($k_1/k_2 = 50$, Eqs. (50.4) and (50.5)) affords distinct profiles that differ in the nature of the rate-limiting step. Note that a simple 10-fold increase in the mass transfer coefficient of the solid reagent shifts the rate-limiting step from a mass transfer event (Eq. 50.3) to a chemical transformation (Eq. 50.4) and modifies conversion rates and curvatures for all components:

$$\text{Reagent(s)} \overset{K}{\rightleftharpoons} \text{Reagent(sat)} \quad (50.3)$$

$$\text{Input} + \text{Reagent} \overset{k_1}{\rightarrow} \text{Product} \quad (50.4)$$

$$\text{Product} + \text{Reagent} \overset{k_2}{\rightarrow} \text{Impurity} \quad (50.5)$$

Since an account of the approaches used to conduct mechanistic studies falls beyond the scope of this chapter, secondary sources are selected in the references [18]. Traditionally perceived as resource intensive, mechanistic studies are currently facilitated by advancements of process analytical technologies (PAT) [12], parallel experimentation, and laboratory automation [19]. Collection of kinetic and thermodynamic knowledge can build on literature precedent and does not require recreation of the conditions used in the actual reaction step. For example, fast reactions can be slowed down at lower temperatures or higher dilutions to enable accurate rate measurements. Similarly, product degradation pathways leading to the formation of impurities can be studied in isolation from their direct precursors to maximize impurity levels and minimize experimental and analytical error.

50.3.2.3 Regression of Model Variables Elementary steps and initial variables describe a preliminary model to which model-building experiments run under representative process conditions will be regressed. The objective of this stage is to refine kinetic and thermodynamic variables characterized in the mechanistic studies and to regress the variables that were not experimentally measured. A DOE is completed to ensure that the model-building data set covers the parameter space of interest. Parameters with potential impact on the QAs are chosen and their ranges selected to explore possible edges of failure based on existing knowledge. Typically, reaction aliquots are taken at intervals of adequate conversion and analyzed off-line by HPLC to monitor the reactant and product concentrations, including impurities formed at low levels. The concentration–time data for each component are then introduced into the modeling

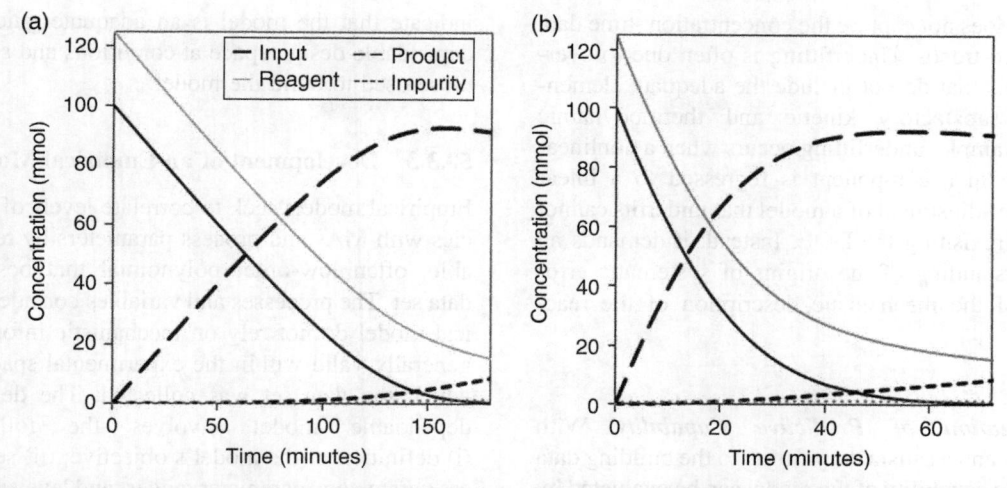

FIGURE 50.4 Concentration–time curves for a solid–liquid reaction under (a) mass transfer-limited and (b) non-limited regimes in reagent following 10-fold increase in the mass transfer coefficient (see Figure 50.3 for the corresponding process scheme).

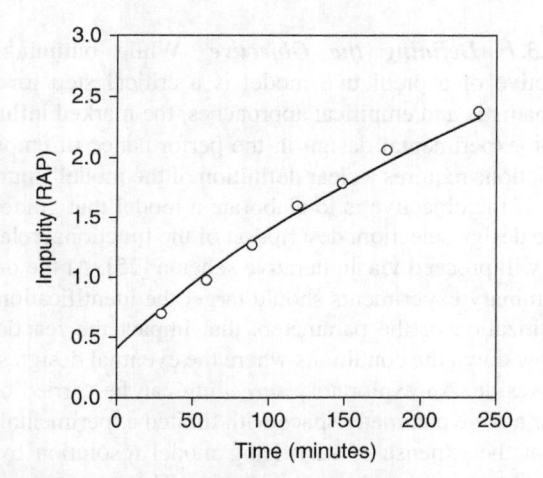

FIGURE 50.5 Data and model fitting for impurity formation in representative model-building experiment.

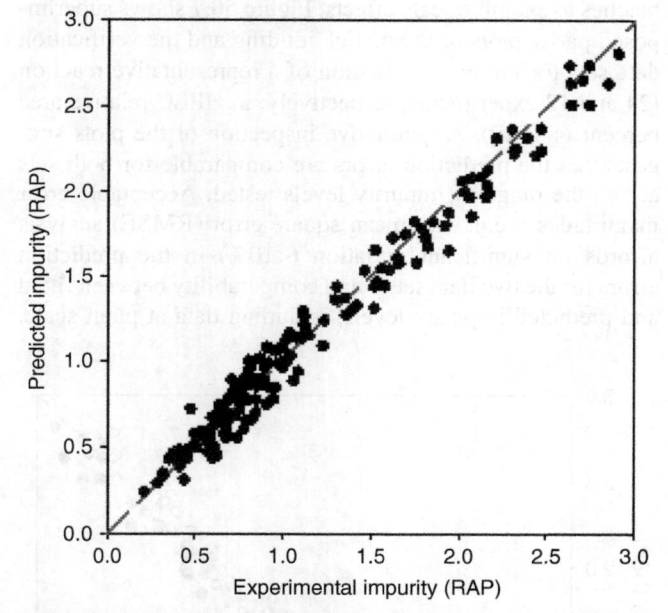

FIGURE 50.6 Parity plots for impurity formation in the model-building experiments. *Source*: From Ref. [23]. Reprinted with Permission. Copyright (2011) Springer Nature.

software (e.g. DynoChem®)[2] to numerically fit the experimental values until adequate convergence criteria are met (Figure 50.5).

Acceptable fitting of experimental data to a model alone does not validate the predictive capability of the model. For example, overly complex models or models that regress an excessive number of variables relative to the size of the building data set may lead to overfitting and poor performance. A model overfits the building data set when it fits well this data set but has limited predictive power on an independent verification data set. Overfitting can be diagnosed via cross-validation tests that split the building data set in subgroups for building and verification purposes. If overfitting occurs, iterative steps based on mechanistic information must

[2]Scale-up Systems Ltd., Dublin, Eire.

be undertaken to mitigate the disconnection between model building and model verification. Poor performance due to model complexity can be addressed following simplification strategies such as network reduction [20], setting stoichiometric constraints [21], or applying the quasi-equilibrium approximation when appropriate [22]. The regression of a disproportionate number of variables can be tackled by fixing them at values or ranges consistent with the mechanistic studies, and an inadequate sample size can be resolved by redefining the DOE. Alternatively, a model underfits the building data set when it does not capture the concentration–time data and shows biased trends. Underfitting is often due to oversimplified models that do not include the adequate elementary steps or satisfactory kinetic and thermodynamic variables. For example, underfitting occurs when a nonlinear rate dependence in a component is regressed to a linear dependence. The adjustment of a model that underfits cannot be achieved by revisiting the DOE. Instead, it demands an improved understanding of the origins of systematic error and a review of the mechanistic description of the reaction step.

50.3.2.4 Evaluation of Predictive Capability With refined variables and a satisfactory fitting to the building data set, the predictive capability of the model can be evaluated by comparing its predictions against the experimental values measured in an independent verification data set. At this stage, laboratory experiments can be combined with plant batches to examine scale effects. Figure 50.7 shows superimposed parity plots of the model building and the verification data sets for impurity formation of a representative reaction (24 and 12 experiments, respectively, in HPLC relative area percent or RAP). A qualitative inspection of the plots suggests that the prediction errors are comparable for both sets across the range of impurity levels tested. Acceptable error magnitudes – e.g. root mean square error (RMSE) analysis affords no significant variation (<10%) in the prediction errors for the two data sets – and comparability between fitted and predicted impurity levels, including data at plant scale,

indicate that the model is an adequate guide to develop a dependable design space at conditions and scales other than those used to build the model.

50.3.3 Development of an Empirical Model

Empirical models seek to correlate levels of significant species with MAs and process parameters by regressing a suitable, often low-order polynomial that best describes the data set. The processes and variables considered in an empirical model do not rely on mechanistic information and are generally valid within the experimental space of interest in which the data set was collected. The development of a dependable model involves the following stages: (i) definition of the model's objective; (ii) selection of process parameters, parameter ranges, and levels; (iii) selection of the response variable; (iv) choice of experimental design; (v) performing the experiment; and (vi) statistical analysis of the data.

50.3.3.1 Defining the Objective While outlining the objective of a predictive model is a critical step for both mechanistic and empirical approaches, the marked influence of the experimental design in the performance of empirical predictions requires a clear definition of the model's purpose [24]. If the objective is to elaborate a model that guides the space design selection, description of the functional relationship will proceed via an iterative scheme [25]. At the outset, preliminary experiments should target the identification and prioritization of the parameters that impact the reaction to narrow down the conditions where the eventual design space may reside. An exploratory *screening* can be carried out to cover a large parameter space with limited experimental burden at the expense of sacrificing model resolution by, for example, assuming linear responses. The screening could then be followed by *optimization* experiments to regress an improved model that adequately correlates the responses with process parameters. Graphical visualization of the functional relationship using three-dimensional surface plots or two-dimensional contour plots may prompt further revision of the empirical model or conclude that an appropriate relationship has been achieved. In contrast, if the goal is to explore process robustness within predetermined conditions (e.g. PARs), full iterative development of an empirical model is unnecessary, and the study of the functional relationship is limited to verifying a minimal or lack of response to parameter changes.

50.3.3.2 Selection of Process Parameters, Ranges, and Levels Selection of the process parameters under scrutiny requires careful balance between the inclusion of a prohibitive number of parameters that may impact the reaction and neglecting parameters that may be critical to develop the functional relationship. An adequate choice relies on

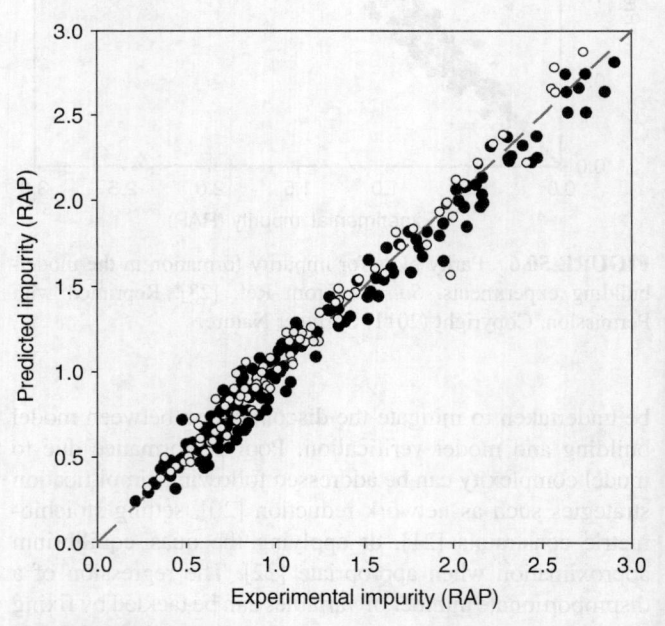

FIGURE 50.7 Parity plot of impurity formation for model-building (○) and verification (●) data sets. *Source*: From Ref. [23]. Reprinted with Permission. Copyright (2011) Springer Nature.

experimental information and scientific understanding. Risk assessment techniques used to prioritize the influence of process parameters often classify them into categories such as known to be impactful, suspected, possible, and unlikely. Reactant stoichiometries, concentration, and temperature are parameters examined routinely in reaction steps. Parameters not chosen for study can be held at a constant value or may be allowed to vary. For example, if extended reaction times lead to high levels of impurity, the experimental design can hold the reaction time constant at the highest value accepted for a commercial manufacturing setting, thereby representing the worst-case scenario. In contrast, factors whose impact is very small within an anticipated range can be allowed to vary. For example, if multiple batches of input material have slightly different impurity profiles that are not expected to influence the results, a decision can be made to ignore batch variability and rely on randomization to balance out the effects that this difference could have.

Once the design parameters are selected, parameter variation ranges and specific levels at which the experiments will be carried out must be determined. Prior knowledge is often a good starting point. Early in model development, when the objective of experimentation is parameter screening, it is preferable to limit the number of levels to high and low. In general, the knowledge space is larger than the eventual design space, and, therefore, the range of parameter variation in this stage is broad. During the optimization phase, the two-level design can be augmented to make the data set more suitable for model fitting. The ranges chosen for optimization must be relevant, achievable, and practical both from an experimental design and a commercial manufacturing perspective. Consideration of the plant capabilities where the process will be carried out in the long term is therefore an important aspect of parameter selection.

50.3.3.3 Selection of the Response Variable Typical response variables selected for a reaction are yield, conversion, and level of the impurities that may impact the QAs of the step. Although several responses can be studied in a single DOE, it is important to search for unexpected responses. For example, if unknown impurities or unanticipated levels of known impurities result, further investigation is warranted to assess the impact of response variability upon quality.

50.3.3.4 Choice of Experimental Design The distinct interconnection between model performance and experimental data demands a careful choice of the experimental design, which must consider data sets aligned with the objective of the model [26]. The design must provide a reasonable distribution of data points throughout the space of interest to ensure that the appropriate information can be collected and analyzed by statistical methods. It should allow for investigation of model adequacy, such as lack of fit, and provide a good profile of the prediction variance. Moreover, the design ought to facilitate the inclusion of follow-up experiments needed to refine the model. For example, a two-level factorial design adequate at the screening stage may need augmentation with center point replicates to estimate the experimental error. Similarly, a first-order model that exhibits a lack of fit can be augmented by adding axial points to create a central composite design [27] or by including incomplete block designs [28] that lead to higher-order terms in the fitting equation. Multiple types of experimental designs can be applied to the study of reactions, including computer-generated optimal designs that are pertinent to special situations (e.g. irregular spaces of interest, nonstandard models, or costly experimentation). A detailed description of these designs is extensively discussed elsewhere [29].

50.3.3.5 Performing the Experiment The execution planning of the actual experiments is often underestimated during the design stage. Performing a large number of experiments while pursuing accuracy and reproducibility to obtain meaningful results can be labor intensive. Modern high-throughput and parallel automation tools can simplify the experimental setup and reduce material requirements by carrying out the reactions at milligram or gram scale. However, the implications of scaling down from commercial manufacturing conditions must be considered during model development. Furthermore, prior to conducting the experiment, a series of "familiarization" runs are recommended to probe reproducibility (e.g. triplicate experiments at the center point), obtain information about the consistency of the inputs and the measurement system, and provide an opportunity to practice the overall experimental technique.

50.3.3.6 Statistical Analysis of the Data The results of the DOE are analyzed using statistical methods. Several software packages that can also be used in the experimental design are available to carry out the study. Initially, a combination of analysis of variance (ANOVA), post hoc tests, and graphical visualization will help identify the main effects of the significant variables and the interactions thereof. The experimental results are then regressed to fit an empirical model, that is, an equation derived from the data that expresses the relationship between the response and the important design parameters. Since the response variables (e.g. conversion, levels of impurity) follow an Arrhenius relationship, using a low-order polynomial to fit the data may not be adequate. In these instances, variable transformation (e.g. using logarithm of time or inverse of temperature) can be employed to facilitate the polynomial fitting. Two techniques commonly used to help select the most relevant equation terms and improve predictive power are the forward stepwise regression and, more recently, genetic algorithms [11]. Forward step regression chooses the terms that offer the best response prediction and, sequentially, includes additional terms whose model coefficients are statistically significant. A genetic algorithm is a heuristic search that follows a defined set of rules to iteratively identify the subset terms that offer the best predictive power. The result of employing a genetic algorithm is a series of models that need further evaluation.

Completion of residual analysis and model adequacy checking helps to validate the appropriate mathematical model and provides guidance for additional follow-up experimentation that could be required. A typical ANOVA and residual analysis using JMP® statistical software is shown in Figure 50.8. In this analysis report, an actual versus predicted plot (parity plot) provides a visual for the goodness of fit and the confidence curves for the model applied. Next, a summary of the model fit includes R^2 (a measure of the variability of the response attributable to the regressor variables), R^2_{adj} (an adjusted form of R^2 that helps guard against the inclusion of extra nonsignificant terms), and the RMSE (an estimate of the standard deviation of the random error). Following the summary of fit, the results of the ANOVA review the comparison of the fitted model against a model in which all the predicted values equal the mean response. The calculations of the sources of variance culminate in the F-ratio (which analyzes the variance for the significance of regression) and the p-value (which provides evidence that there is at least one significant effect in the model). The final sections display the parameter estimates for the regressed model (including p-values for the individual parameters), a residual plot, and the final predictive equation (more appropriately described in Eq. (50.6)). Although the statistical analysis facilitates the initial validation of the model, its predictive power must be verified by testing the predictions for an independent data set not used in the regression of the model variables.

Equation (50.6) – where A equals the initial KF values, B is the equivalents of NaOEt, T corresponds to the reaction temperature, and C equals the term $(T - 49.8)$ – represents the analysis output that can be used to predict Impurity levels within the ranges and constraints considered in the building set:

$$\text{Impurity} = 0.0847 + 0.9498A - 0.3674B + 0.0087T \\ + 0.0003C^2 + 1.9801(B - 1.2053)^2 \quad (50.6)$$

50.4 DEFINING THE DESIGN SPACE

Translation of the process model, mechanistic or empirical, into a design space requires consideration of an approach that (i) can ensure drug substance quality, (ii) can be applied in routine manufacturing, and (iii) can be adequately presented to regulatory agencies for acceptance. This section outlines some of the most important considerations in developing and applying a model toward the definition of a design space.

50.4.1 Response Surfaces: Model Predictions and Impurity Level Contours

In the QbD paradigm, the terms knowledge space, design space, and operating space are used to describe the important processing regions where every parameter is a distinct dimension of a conceptual space (Figure 50.9). The knowledge space represents the region where the process is well understood by way of experiments, batch history, and empirical or mechanistic modeling, that is, operating at any combination of multiple parameters within this space has a predictable outcome. The design space consists of the region that has been demonstrated to assure quality product, and the operating space is the region targeted by the manufacturer for routine processing.

Once the model and the knowledge space where the model is valid are established, the goal of the practitioner is to define the design space, an inclusive subspace that considers the edges of failure and affords quality product at very high frequency. Additional laboratory experiments, pilot- or manufacturing-scale batches, are often required to test the high risk regions and provide verification of quality product throughout the design space. Ultimately, the operating space is established by considering factory control capabilities, balance of efficiency (e.g. yield or cycle time), and high risk regions within the design space.

The most common approach to presenting and evaluating a model output is to compute response surfaces from the model equations with respect to two parameters. Figure 50.10 provides an example of a response surface, and Table 50.6 lists common parameters and responses evaluated in a reaction model. It is important to identify parameters that could be scale dependent, especially for multiphase reactions, to target a design that can be applied across scales. A response surface is also generally preferred by the regulatory agencies for presentation of the output of the model and justification of the design space. Several examples using response surfaces for presenting a design space are provided in the ICH Q8 (R2) Appendix [2].

It is worth considering the enhancement in process knowledge gained through modeling compared with the traditional approach that defines PARs by testing one factor at a time. Figure 50.11 illustrates how PARs may fail to capture multivariate interactions between parameters in the context of response surfaces. While the operating conditions represented by point A are located within the PARs of Parameter 1 and Parameter 2, they lie beyond the edge of failure. Quality can be compromised if parameter interactions are not well understood and the operating conditions change for more than one parameter, even within their PARs.

Response impurity whole model actual by predicted plot

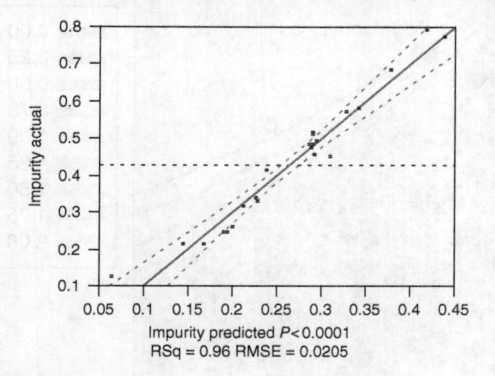

[Plot: Impurity actual vs Impurity predicted, P<0.0001, RSq = 0.96, RMSE = 0.0205]

Summary of fit

RSquare	0.957 125
RSquare adj	0.945 215
Root mean square error	0.020 523
Mean of response	0.264 102
Observations (or sum wgts)	24

Analysis of variance

Source	DF	Sum of squares	Mean square	F Ratio
Model	5	0.169 239 38	0.033 848	80.3646
Error	18	0.007 581 22	0.000 421	Prob > F
C. total	23	0.176 820 60		0.0001*

Parameter estimates

Term	Estimate	Std error	t Ratio	Prob>|t|
Intercept	0.084 710 7	0.043 356	1.95	0.0664
Pre-dehydration KF (wt %)	0.949 795 9	0.091 787	10.35	<0.0001*
T (°C)	0.008 680 3	0.000 79	10.98	<0.0001*
(T (°C)–49.7083)*(T (°C)–49.7083)	0.000 356 7	0.000 148	2.41	0.0269*
Net equiv NaOMe	−0.367 396	0.028 483	−12.90	<0.0001*
(Net equiv NaOMe-1.20535)*(Net equiv NaOMe-1.20535)	1.980 060 4	0.280 905	7.05	<0.0001*

Residual by predicted plot

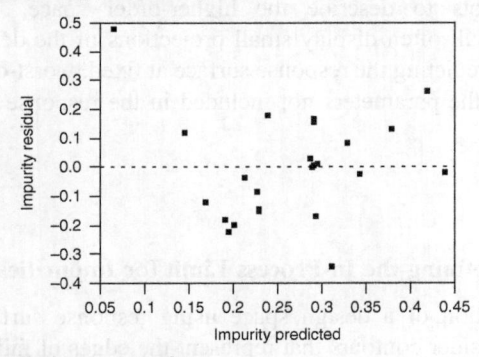

[Plot: Impurity residual vs Impurity predicted]

Prediction expression

0.084 710 7 + 0.949 795 9* Pre-dehydration KF (wt %) + 0.008 680 3* T (°C) −0.367 396* Net equiv NaOMe + 0.000 356 7*(T (°C)–49.708 3)*(T (°C)–49.708 3) + 1.980 060 4*(Net equiv NaOMe-1.205 35)*(Net equiv NaOMe−1.205 35)

FIGURE 50.8 Representative example of ANOVA and residual analysis using JMP® statistical software.

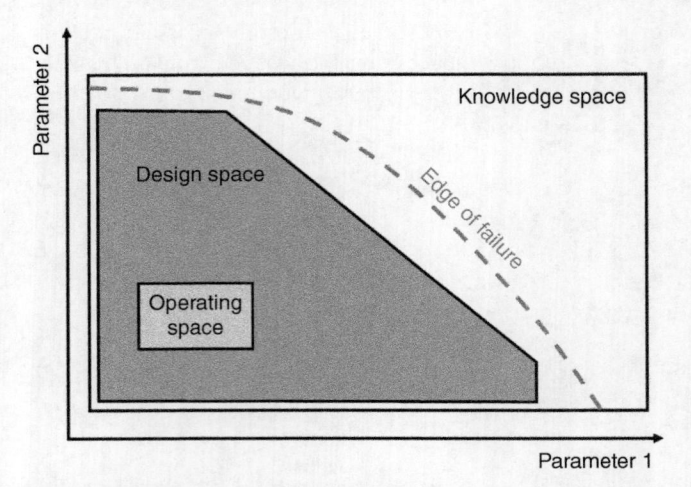

FIGURE 50.9 Illustration of a conventional knowledge, design, and operating space.

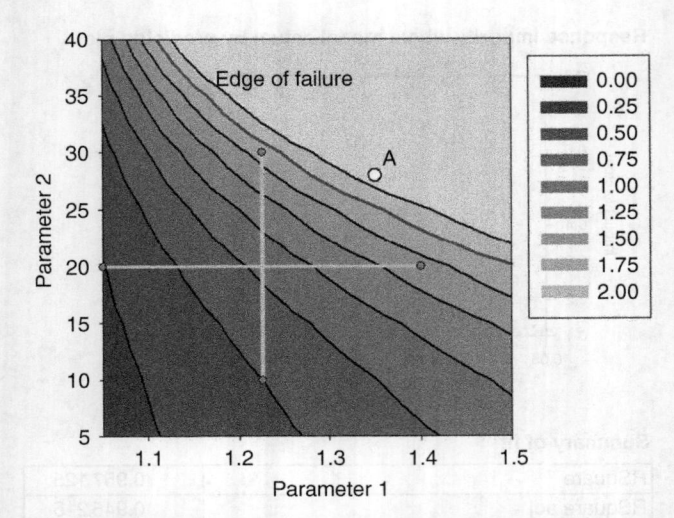

FIGURE 50.11 Representation of an edge of failure in the context of PARs.

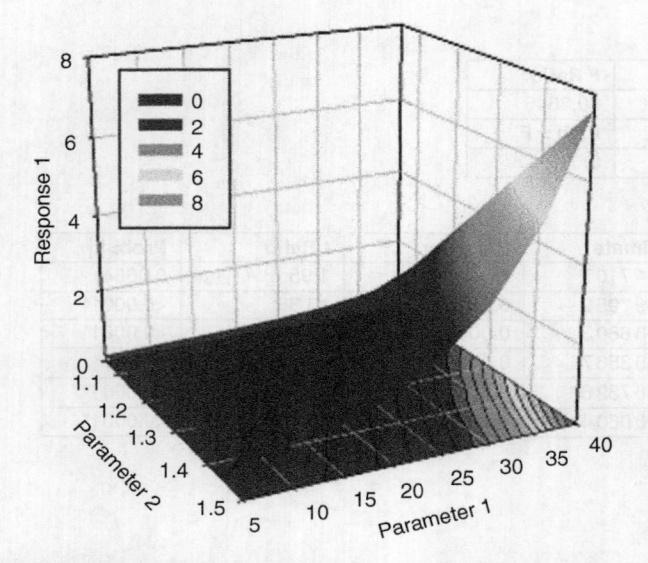

FIGURE 50.10 Example of a response surface with respect to two parameters.

Multiple responses can be displayed together (e.g. yield, impurities, cycle time) by selectively shading or displaying specific contours. Visualization of the multiple responses affords a rich picture of the risks and benefits associated to operating at different points within the knowledge space. A limitation of response surfaces is that only two parameters can be displayed versus the response, whereas often the process requires to consider a conceptual space of higher order, i.e. n dimensions formed by n parameters – often termed as a hypercube for any space greater than three dimensions. The restriction may be overcome by either (i) reducing the discussion to the two parameters of highest impact in view of the practical variability or (ii) displaying multiple response surface plots to describe the higher-order space. This approach will often display small projections of the design space by predicting the response surface at fixed, worst-case, values for the parameters not included in the response surface graph.

50.4.2 Defining the In-Process Limit for Impurities

The definition of a design space using response surfaces should consider contours that represent the edges of failure, that is, the limit for the response that ensures quality product. For a reaction model, each contour would commonly represent the in-process limit of a given impurity. In-process limits are determined from the drug substance specifications and fate and purge studies, which in turn may encompass several steps of the synthesis. Fate and purge studies often operate with intermediate limits and can include additional models to describe other unit operations.

TABLE 50.6 Common Parameters and Responses Evaluated in a Reaction Model

Common Reaction Parameters	Common Responses
Temperature profile	Impurity levels
Amount of starting material, reagents, product	Input conversion
Reaction volume	Yield
Reaction age	
Dosing rate	
$k_L a$	
Agitation	

50.4.3 Specifying and Presenting a Design Space in a Regulatory Submission

With a response surface and in-process limits, a design space can now be defined. There are two general strategies to present a design space: (i) employing the entire mathematical model with n independent variables as a virtual n-dimensional design space and *dynamically* determining, for each batch, the ranges of MAs and parameters that ensure quality and (ii) using the model predictions to select a subset projection of the space to define an *invariant* design space that can be represented by ranges of MAs and process parameters or a combination thereof [21]. To illustrate the two strategies, let us consider that all potential combinations of MAs and parameters defined by a mathematical model afford acceptable impurity levels in a tridimensional space characterized by the trapezoidal prism in Figure 50.12. A dynamic, model-defined, approach could be chosen that provides access to all the parameter combinations within the prism. In contrast, an invariant approach would, for example, fix parameter z at its most conservative level, z_1, and select the rectangular space represented by the small base of the prism. Whereas operating within the xy ranges denoted by this space would ensure success for *all* production batches, the invariant strategy provides access to only a *subset* of the parameter combinations. Allowing parameter z to change from z_1 to z_2 would expand the original projection to the rectangle represented by the large base of the prism. It is important to note that while in this example the invariant design space is shown as a two-dimensional rectangle, there are several ways to express it. Design spaces could be specified through simple ranges that would form a square for two dimensions, a cube for three dimensions, or a hypercube for $n > 3$ dimensions. Alternatively, the parameters could be described in functions that afford a design space with irregular shape. Assigning a function to describe the edge of a design space can provide a wider operational space or enable access to a desired region. An example of a design space with a non-cubic shape is provided in Section 50.5.2.

Although the dynamic strategy affords more flexibility, practical considerations have made the use of an invariant design space the method of choice for reaction steps. An essential requirement for a design space established through either approach is to verify that operating within the parametric region indeed ensures quality. This entails assessment against experimental data and verification across scales for application in a commercial process. A dynamic design space requires significant resources for rigorous model characterization and verification, as well as a long-term commitment for model maintenance and revision. The importance of a model at ensuring product quality has implications on the level of detail in a regulatory submission, the degree to which the model should be assessed for fidelity, and its life-cycle management. ICH provides a framework to categorize a model in low, medium, or high impact, depending on its importance to ensure quality [30]. Models used for design space selection are generally categorized as medium impact models. These models are expected to be described in detail with information pertaining to inputs, outputs, assumptions, and relevant equations. In the invariant approach case, a risk-based, less resource-intensive strategy can be used since the design space is a more constrained subset defined through parameter ranges. Furthermore, from an operational convenience perspective, a clear definition of ranges within which the production batches must always be carried out removes the need to dynamically calculate acceptable ranges for each batch by the manufacturing personnel.

50.5 VERIFICATION OF THE DESIGN SPACE

In a joined communication issued by the FDA and EMA, design space verification is defined as the demonstration that the proposed combination of input process parameters and MAs is capable of manufacturing quality product *at commercial scale* [31]. Verification is an experimental measure of performance directed to confirm that potential scale-up effects or model assumptions are controlled within the design space limits and do not threaten the expected product quality. While it is possible to execute verification studies during process validation, design space verification should not be confused with process validation. Design space verification demonstrates *adequate performance within the area of design space* in relation to operations for which the space has been proposed. In contrast, process validation demonstrates *consistency of the process at NORs* and covers all the operations of the manufacturing process. Verification may occur over the life cycle of the product if changes within the design space pose unknown risks due to scale-up effects.

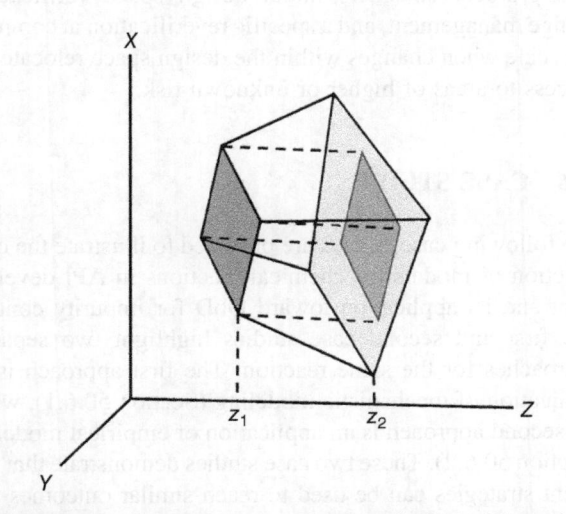

FIGURE 50.12 Representation of a hypothetical tridimensional trapezoidal prism operating space.

50.5.1 Risk-Based Approach to Design Space Verification

A risk-based approach to design space verification for the final process prioritizes examination of the reaction step under worst-case conditions. Since the limits of the design space are defined by specific ranges, not by model predictions, the model is simply used to guide the space selection, and the focus of the verification is the design space denoted by the parameter ranges rather than the model itself (Figure 50.13). Select experiments including replicates are run along the predicted edges of failure to independently explore the impact of process parameters and MAs at their high risk values. The experiments must be performed at a representative scale, and the reaction stream can be carried through the downstream operations chosen to isolate the product. For the verification to indicate that material of acceptable quality is generated throughout the design space, the isolated product has to meet the established specifications for *all* impurities and not just the impurity that may have been the focus of the design space development.

50.5.2 Relationship of Design Space to Scale and Equipment

A risk-based approach can be applied to complete the design space verification prompted by changes on scale, equipment, or manufacturing site, since verification at commercial scale may not be possible. When a drug substance manufacturer demonstrates that a design space is scale independent, additional steps are not required for verification. However, changes within an unverified design space area that may pose risks due to scale-up effects demand that these risks are understood and evaluated utilizing an appropriate control strategy. Scale independence can be justified based on scientific knowledge – including first principles, models, and equipment scale-up factors [30, 32]. Ultimately, changes on scale or equipment require proper risk assessment and confirmation that the control strategy in place can deliver drug substance with acceptable quality.

For homogeneous reactions that have not shown scale or equipment limitations during development, scale and equipment independence can be confirmed by stressing multivariate experiments under worst-case conditions of the proposed commercial equipment and demonstrating that the outcome matches design space predictions. Exploration of worst-case conditions can entail the expansion of temperature ranges, simulation of heating or cooling profiles according to plant capabilities, or extension of the gap between laboratory and manufacturing reaction times. For heterogeneous or multiphasic reactions *affected by mass transfer events*, execution of laboratory simulations under worst-case conditions must take place to demonstrate the lack of scale-up effects on quality. For example, the anticipated role of scale and equipment upon mixing rates in a solid–liquid reaction can be deemed acceptable if product quality is not affected by particle size, agitation speed, or temporary loss of agitation under stressed experiments.

It is important to note that regulatory expectations about design space verification at commercial scale are not necessarily harmonized across health authorities [8] and that design space verification on commercial scale may be deemed indispensable to establish the commercial batch size. A protocol for design space verification based on a structured risk assessment has been reported in instances where adequate on-scale data may not be available during filing [33]. This protocol considers initial design space verification, change management, and a specific reverification at commercial scale when changes within the design space relocate the process to areas of higher or unknown risk.

50.6 CASE STUDIES

The following case studies are provided to illustrate the construction of models for chemical reactions in API development and its application toward QbD for impurity control. The first and second case studies highlight two separate approaches for the same reaction. The first approach is an application of mechanistic modeling (Section 50.6.1), while the second approach is an application of empirical modeling (Section 50.6.2). These two case studies demonstrate that different strategies can be used to reach similar outcomes and allow a comparison of the advantages and disadvantages associated to each approach. The third case study is an

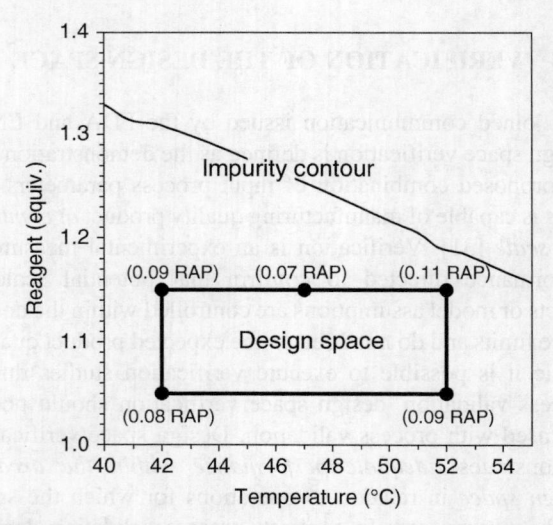

FIGURE 50.13 Representative design space verification. Numbers in parentheses represent impurity level (RAP) in isolated drug substance. The drug substance CQA limit for this impurity is less than or equal to 0.15%. *Source*: From Ref. [23]. Reprinted with Permission. Copyright (2011) Springer Nature.

application of mechanistic modeling to a much more complicated reaction (Section 50.6.3). It discusses a characteristic scenario including product degradation and the application of PAT to increase process robustness.

50.6.1 Mechanistic Model to Control a Critical Impurity in the Penultimate Step of a Drug Substance

This case study shows the development of a mechanistic model to control a critical impurity in the penultimate step of a drug substance [34]. The model contributes to the process design and control strategy by assisting in the selection and verification of the design space. This work was performed during the late-stage process development of a drug substance and was used to provide an enhanced regulatory submission [35]. An impurity in the drug substance, Impurity B, was deemed critical after process development and risk assessment because it could impact the quality of the drug substance. Impurity B is formed from Impurity A, generated in the penultimate step reaction, and, therefore, also deemed critical (Figure 50.14). The final control strategy for Impurity B included the selection of a design space around the penultimate step reaction to control Impurity A to an acceptable in-process level at reaction completion, which was established through fate and purge studies. The other major element of the control strategy was the specification for Impurity A in Final Intermediate (Table 50.7).

The reactions of the penultimate step include the desired lactamization and an undesired lactam ethanolysis (Scheme 50.1 and Table 50.8). The lactamization of the Input material was divided in two steps: (i) a reversible deprotonation of the Input promoted by sodium ethoxide (NaOEt) and (ii) a first-order rate-limiting ring closure. The undesired ethanolysis occurs when NaOEt reacts with the product to open the lactam and afford the corresponding ethyl ester. During this reaction, the ethyl ester impurity exists as a salt that is later quenched with acetic acid to produce Impurity A. These transformations constitute a series of parallel reactions as NaOEt reacts with the Final Intermediate to generate the undesired impurity.

Figure 50.15 shows the reaction profile of a representative experiment including the consumption of Input and the formation of Impurity A (Figure 50.15a and b, respectively) for one of the model-building experiments (i.e. 1.25 equiv. NaOEt and 26°C). The reaction is typically completed in less than 1 hour, but aging past completion leads to additional growth of Impurity A. Extended reaction times are common for batch processes because additional time is taken to allow for a measurement of the reaction (e.g. HPLC analysis) to ensure completion before proceeding to the next unit

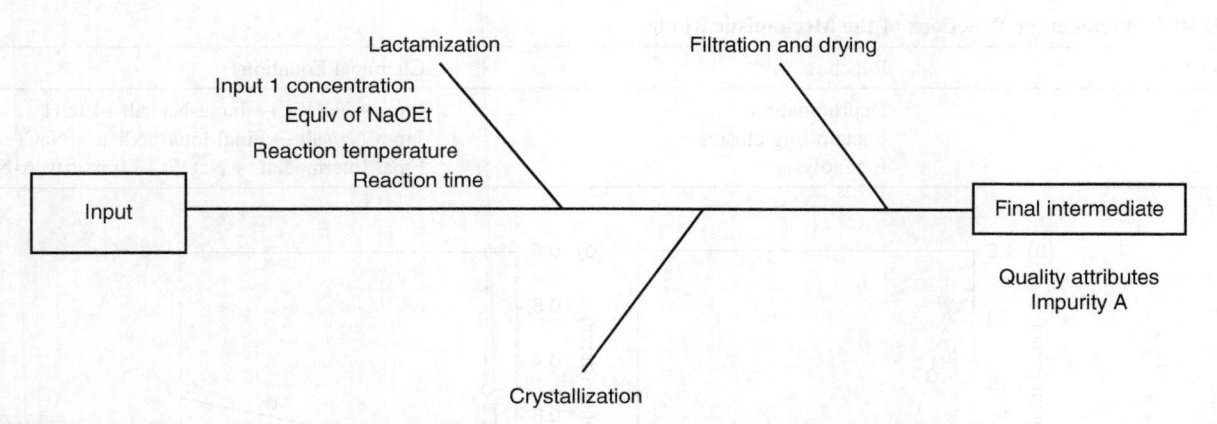

FIGURE 50.14 Ishikawa diagram for the Final Intermediate step. *Source*: From Ref. [34]. Reprinted with Permission. Copyright (2010) Springer.

TABLE 50.7 Elements of Control Strategy for a Drug Substance

Drug Substance Specifications	Type of Control		
	Procedural Controls	Material Attributes	Parameter Controls
Impurity B NMT 0.15%	Crystallization of Final Intermediate and drug substance shown to purge Impurities A and B, respectively	In-process controls: none Intermediate testing: Impurity A, NMT 1.0% in Final Intermediate	Design space of the penultimate reaction to control Impurity A to NMT 1.76% post-reaction

SCHEME 50.1 Formation of Final Intermediate and Impurity A. *Source*: From Ref. [34]. Reprinted with Permission. Copyright (2010) Springer.

TABLE 50.8 Elementary Reactions of the Mechanistic Model

Equation No.	Reaction	Chemical Equation
50.7	Deprotonation	Input + NaOEt ↔ Input-Na salt + EtOH
50.8	Lactam ring closure	Input-Na salt → Final Intermediate + NaCl
50.9	Ethanolysis	Final Intermediate + NaOEt → Impurity A-Na salt

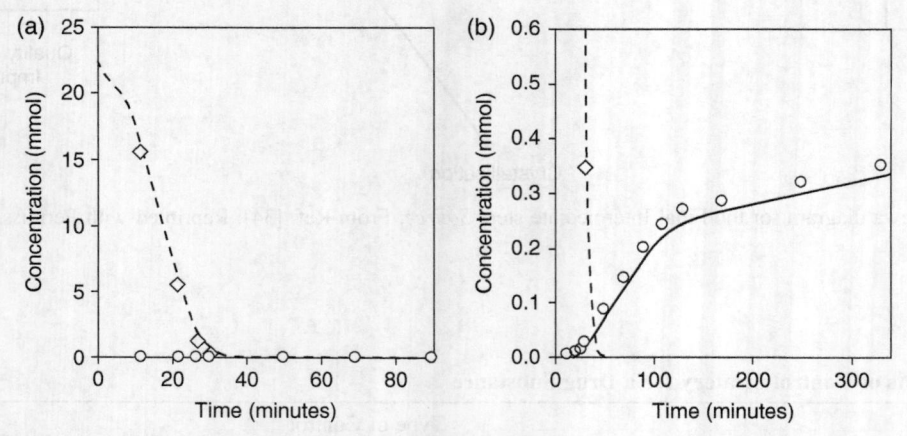

FIGURE 50.15 (a) Concentration–time curves and model fitting for the consumption of Input (◊) and formation of Impurity A (o). (b) Rescaled version of Figure 50.15a to display Impurity A. *Source*: From Ref. [34]. Reprinted with Permission. Copyright (2010) Springer.

operation. The inflection in the formation of Impurity A results from a procedural decrease in the reaction temperature implemented to help suppress the formation of Impurity A during the extended reaction time.

Preliminary studies indicated that temperature and excess NaOEt were the parameters with the highest impact to the formation of Impurity A. Consequently, a central composite design with 10 experiments was conducted across the space of interest to generate the model-building data set for parameter regression (Figure 50.16). It is worth mentioning that NaOEt charges were made in an ethanol solution and that changes in excess NaOEt inflicted variations in other parameters known to impact the reaction, such as concentration and solvent composition.

The reaction constants, activation energies, and equilibrium constant for the deprotonation of the Input were regressed using DynoChem® modeling software [20] (Table 50.9). Figure 50.17 shows concentration–time curves for Impurity A and associated model fitting for 12 experiments of the model-building set, which included changes in the excess NaOEt (1.05–1.50 equiv.) and temperature (12–40 °C). Note that an imposed temperature profile consisting of the reaction temperature and a cooling ramp to suppress reaction progress during analytical testing was applied to mimic the intended manufacturing procedure. Profiles represented by solid curves result from fitting the data to the mechanistic model under the corresponding reaction conditions. The differential equations based on the rate expression and the regressed variables provided a model that could be used to predict the consumption of Input and the formation of Impurity A with respect to temperature profile, reaction time, concentration of Input, and equivalents of NaOEt.

The predictive capability of the model was assessed by comparing the predicted versus experimental levels of Impurity A for model-building and verification data sets in a parity plot and by quantitative error analysis for each data set (Figure 50.18 and Table 50.10, respectively). The assessment consisted of nine lab-scale experiments and five manufacturing-scale batches that were not included in the model regression. The strong agreement between the model-building set and the verification set demonstrates the high fidelity of the model and the independence of scale, which was expected for this liquid-phase homogeneous reaction.

Despite the high accuracy of the model, the elementary equations in Table 50.8 represent a simplified system that does not capture all of the chemical interactions. For instance, it was found that the ethanol to N-methylpyrrolidone solvent ratio significantly impacts the ethanolysis reaction. This parameter was considered intractable for inclusion in the model but was captured by determining acceptable ranges and holding it at a constant worst-case value in relation to impurity formation. This approach simplified model development while accounting for the influence of solvent composition.

To determine the parameter ranges that would constitute the design space, response surfaces were generated for consumption of Input at the time of reaction sampling and levels of Impurity A upon reaction quench with respect to temperature and excess NaOEt (Figure 50.19). The response surfaces represent the computational prediction of the concentration of Input and Impurity A across the parameter ranges and initial conditions simulated in the virtual experiments. Moreover, it was necessary to define values for other parameters that affected the levels of Impurity A - (Table 50.11). These parameters were varied in the model-building experiments and shown to influence model prediction but were less impactful than temperature and excess NaOEt. The values of these less impactful parameters were chosen within predefined ranges that would afford the highest level of Impurity A, thereby providing the most conservative projection of the edges of failure.

The response surfaces were displayed as contour plots to visualize the desired and undesired regions of operation.

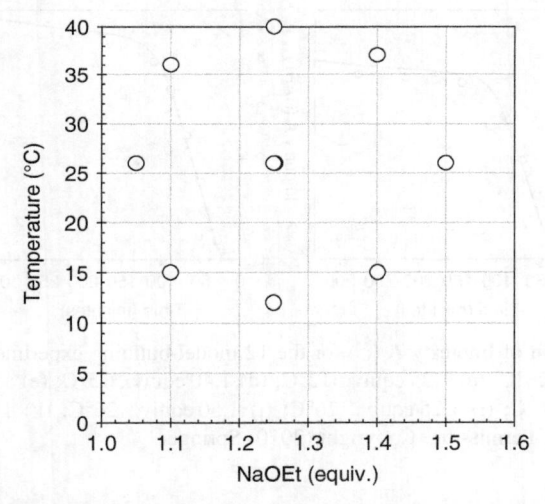

FIGURE 50.16 Graphical representation of the experimental conditions for the model-building set. *Source*: From Ref. [34]. Reprinted with Permission. Copyright (2010) Springer.

TABLE 50.9 Rate Expressions and Regressed Model Variables

Equation No.	Reaction	Rate Expression	k^a (l/mol·min)	E_a (kJ/mol)	$K_{eq}{}^a$
50.7	Deprotonation	$r_d = k_d$ [Input][NaOEt]	161.3 (19%)[b]	26.6 (41%)[b]	1.54 (65%)[b]
50.8	Lactam ring closure	$r_1 = k_1$ [Input-Na salt]	0.0194[c] (38%)[b]	87.0 (13%)[b]	—
50.9	Ethanolysis	$r_e = k_e$ [Final Intermediate][NaOEt]	0.0027 (2.9%)[b]	81.9 (4%)[b]	—

[a] $T_{ref} = 25\,°C$.
[b] 95% confidence interval.
[c] k value in 1/s units.

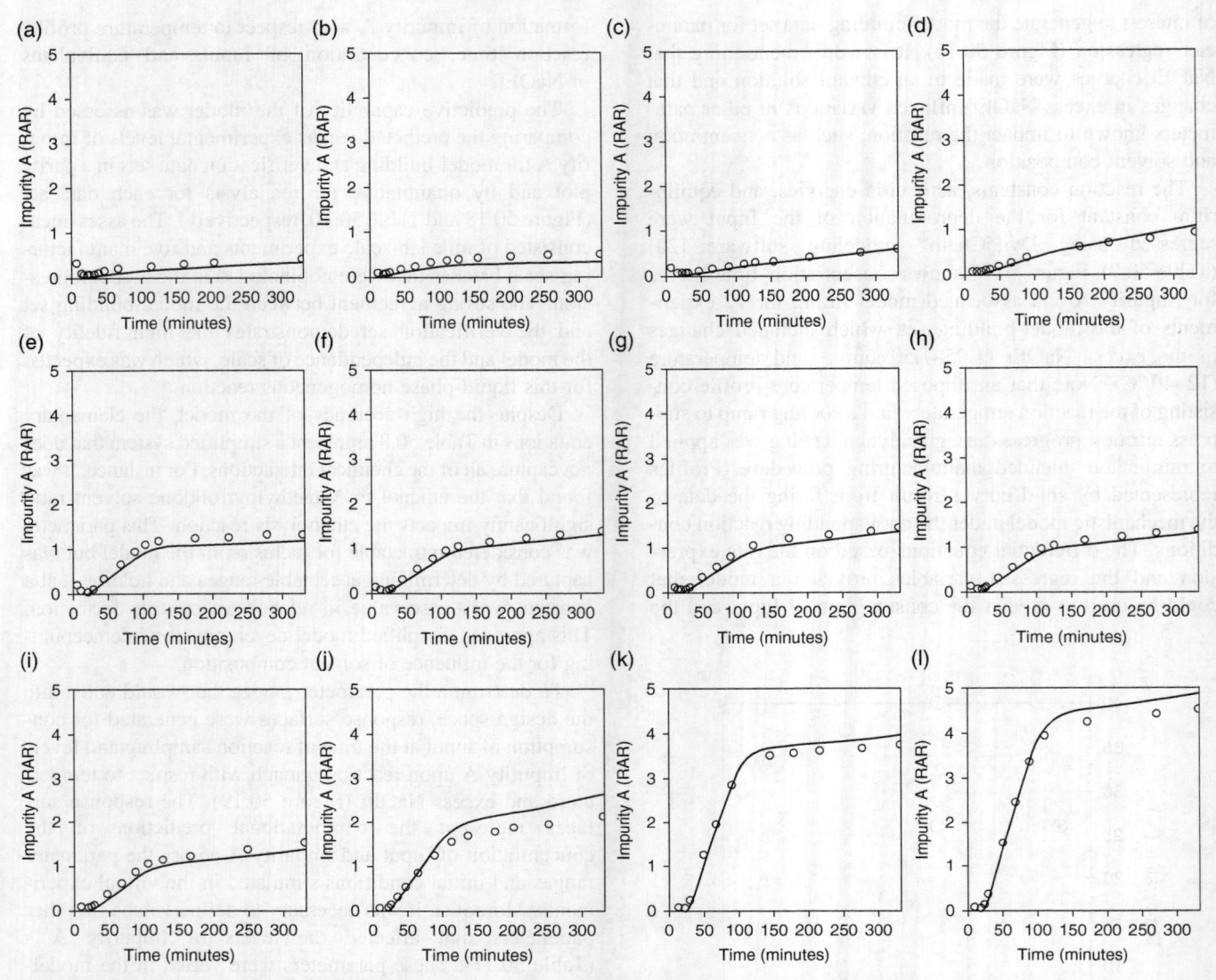

FIGURE 50.17 Concentration–time curves and model fitting for the formation of Impurity A (○) for the 12 model-building experiments varying NaOEt equiv. and temperature: (a) 1.10 equiv., 15°C; (b) 1.05 equiv., 26°C; (c) 1.25 equiv., 12°C; (d) 1.40 equiv., 15°C; (e) 1.10 equiv., 36°C; (f) 1.25 equiv., 26°C; (g) 1.25 equiv., 26°C; (h) 1.25 equiv., 26°C; (i) 1.25 equiv., 26°C; (j) 1.50 equiv., 26°C; (k) 1.25 equiv., 40°C; and (l) 1.40 equiv., 37°C. *Source*: From Ref. [34]. Reprinted with Permission. Copyright (2010) Springer.

Figure 50.20 shows how low temperatures should be avoided to ensure reaction completion and minimize residual Input plots at the time of reaction sampling, while high temperatures and high equivalents of NaOEt should be avoided to minimize the levels of Impurity A upon reaction quench. The acceptable operating region was resolved by combining Figure 50.20a and b and displaying the contours of interest, that is, the acceptable in-process levels for Input and Impurity A (Figure 50.21).

Guided by these conservative response surfaces, the final design space was selected as a trapezoidal region between the edges of failure for residual input and Impurity A (Figure 50.22). This design space was described by a temperature range of 15–30 °C, a NaOEt range of 1.1–1.4 equiv., and an equation for the edge of the trapezoid, $T = -70B + 113$, where T is the temperature in °C and B is the excess NaOEt in equivalents. Design space verification was carried out by conducting a set of experiments at the extreme points of the trapezoid and taking the reaction to the isolated Final Intermediate for analysis using the release method. Table 50.12 displays the results of these experiments, including predicted and experimental in-process levels of Impurity A, and the value of Impurity A in isolated cake. The model showed strong performance for the prediction of Impurity A at the same time that the Final Intermediate met the specification for Impurity A (NMT 1.0%). Based on

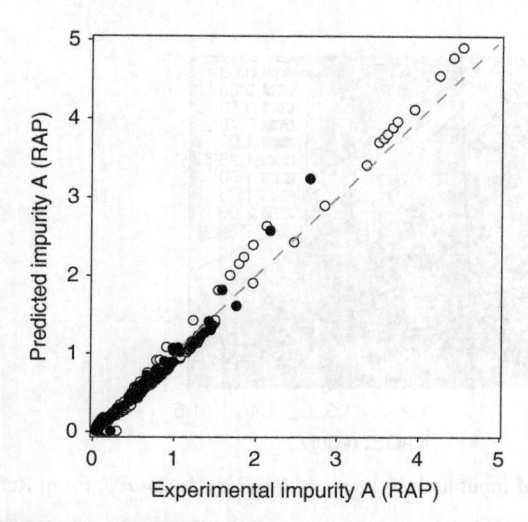

FIGURE 50.18 Parity plot of Impurity A formation for the model-building (○) and verification (●) data sets. *Source*: From Ref. [34]. Reprinted with Permission. Copyright (2010) Springer.

these results, there was high confidence that the design space selected would provide a robustness element in the overall control strategy for Impurity A.

Given that any model is a simplified representation of the actual system and that the experimental measurements and the regression have inherent uncertainty, the predicted edges of failure had some uncertainty. Rather than attempting to calculate the uncertainty about the edge of failure, the variability around the design space with respect to failure was assessed. For this analysis, multiple experiments were conducted in the highest risk region, and the variability around this point was determined. Table 50.12 shows the results of experiments conducted at 26 °C and 1.25 equiv. of NaOEt, and Table 50.13 shows the calculation of the variability at this point. The distribution of in-process and isolated cake impurity levels showed that the mean in-process level and dry cake level were 1.45 and 0.81, respectively. Furthermore, the upper tolerance limit (95% confidence and 90% proportion) for the in-process and isolated cake distribution was 1.67 and 0.99. The tolerance limit for both in-process and

TABLE 50.10 Quantitative Error Analysis for Model-Building and Model Verification Data Sets

	Model-Building Set	Model Verification Set
RMSE (RAP)	0.16	0.13
Mean relative error (%)	11.2	10.6
Maximum absolute error (RAP)	0.51	0.40

Source: From Ref. [34].

TABLE 50.11 Conservative Values of the Process Parameters for the Lactamization

Process Parameter	Conservative Values	
	Minimum	Maximum
Substrate concentration	8.9 wt %	
Reaction age time		60 min
Total reaction age time above hold temperature		180 min
Reaction hold temperature		10 °C

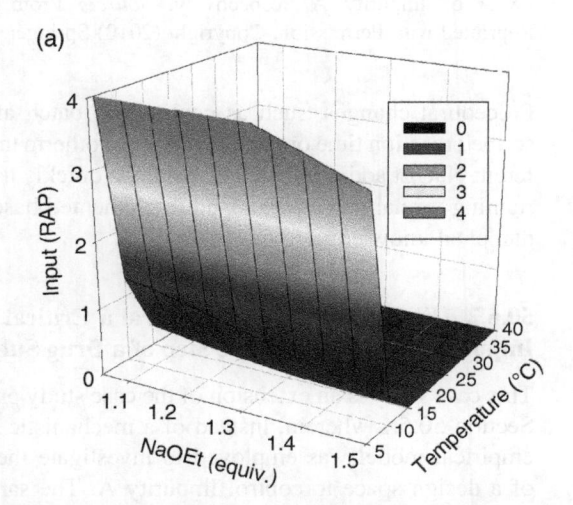

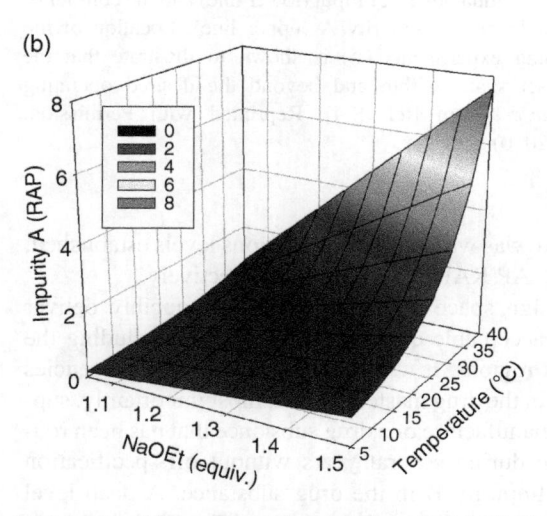

FIGURE 50.19 Response surfaces for (a) residual levels of input and (b) levels of Impurity A. *Source*: From Ref. [34]. Reprinted with Permission. Copyright (2010) Springer.

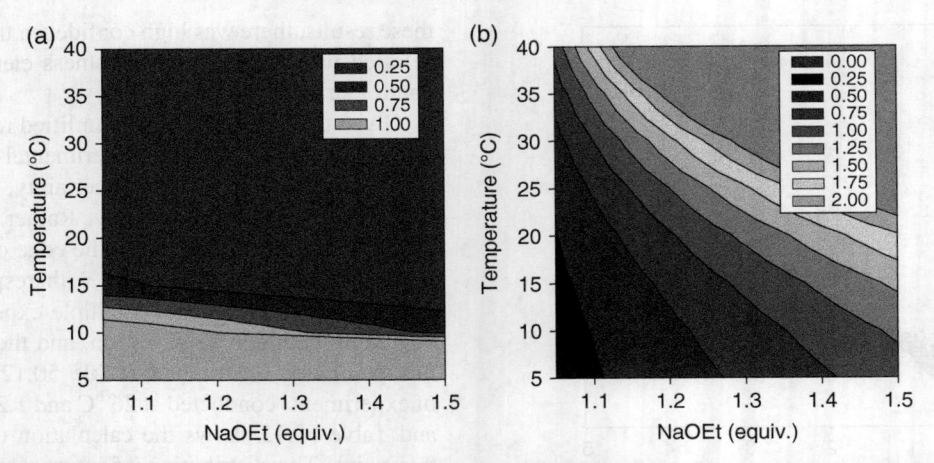

FIGURE 50.20 Contour plots of the response surface for (a) residual levels of Input and (b) levels of Impurity A. *Source*: From Ref. [34]. Reprinted with Permission. Copyright (2010) Springer.

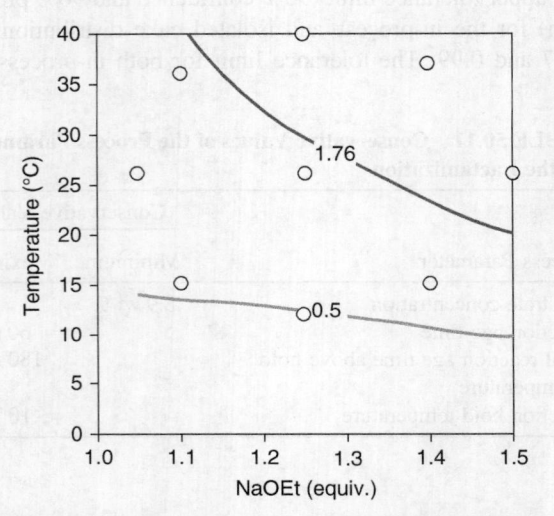

FIGURE 50.21 Combined contour plots including the contour of 1.76 RAP for residual levels of Input (lower line) and the contour of 0.5 RAP for levels of Impurity A (upper line). Location of the model-building experiments (○) is shown to illustrate that the calibration set spans within and beyond the desired operating region. *Source*: From Ref. [34]. Reprinted with Permission. Copyright (2010) Springer.

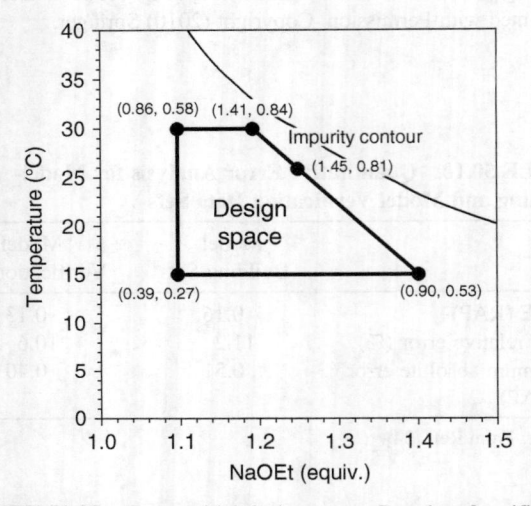

FIGURE 50.22 Trapezoidal design space. Results of verification experiments are given in parenthesis for in-process and isolated cake levels of Impurity A, respectively. *Source*: From Ref. [34]. Reprinted with Permission. Copyright (2010) Springer.

isolated cake was within the specifications levels established, that is, 1.76 AP RAP and 1.0 RAP, respectively.

This design space was verified to reproducibly deliver product of acceptable quality at all locations, including the highest risk regions. It was accepted by regulatory agencies as reported in the drug master file and, most importantly, supported the manufacture of a drug substance that has been routinely made during several years without off-specification results for Impurity B in the drug substance. A deep level of knowledge was obtained that shaped the final design of the process through building the mechanistic model.

Procedural changes, such as cooling the batch after a prescribed reaction time or controlling the exotherm through different NaOEt addition profiles, could be quickly assessed by running virtual experiments and implemented based on fundamental knowledge.

50.6.2 Empirical Model to Control a Critical Impurity in the Penultimate Step of a Drug Substance

This case study is an extension of the case study presented in Section 50.5.1 wherein, instead of a mechanistic model, an empirical model was employed to investigate the selection of a design space to control Impurity A. The same overall control strategy (Table 50.7) and reactions (Scheme 50.1) are relevant to building the empirical model, but rather than

TABLE 50.12 Results of Verification Experiments for In-Process and Isolated Cake Levels of Impurity A

NaOEt (equiv.)	Temperature (°C)	Impurity A In-Process Model Prediction (RAP)	Impurity A In-Process Experimental (RAP)	Impurity A Isolated Cake Value (RAP)
1.10	15	0.32	0.39	0.27
1.10	30	0.80	0.93	0.58
1.40	15	1.08	0.91	0.53
1.19	30	1.41	1.43	0.84
1.25	26	1.41	1.50	0.85
1.25	26	1.33	1.47	0.83
1.25	26	1.36	1.44	0.78
1.25	26	1.32	1.38	0.76

TABLE 50.13 Variability of Experiments Conducted at 26 °C and 1.25 equiv. of NaOEt

	Impurity A In-Process RAP	Impurity A Isolated Cake RAP
Mean	1.45	0.81
Standard deviation	0.054	0.044
Upper tolerance limit (1-alpha = 0.95, proportion = 0.90)	1.67	0.99

determining the mechanism and rate expression for the reaction, a polynomial expression of the statically significant factors was used to predict the edge of failure. In addition, the reaction procedure and target completion value were identical as described in Section 50.5.1. This study demonstrates that the empirical approach can achieve a satisfactory outcome and highlights some of its advantages and disadvantages relative to the mechanistic approach.

A central composite experimental design was conducted with the intention to build either a mechanistic or empirical model (Figure 50.16). The DOE consisted of a full-factorial, two-factor, and two-level design with four center points and four additional axial points to facilitate the accurate prediction of the curvature of the response. The two factors considered were reaction temperature and equivalents of NaOEt. The response measured in the DOE was the final level of Impurity A prior to quenching the reaction, which includes the extended reaction time required for in-process control analysis of reaction completion. The analysis of the DOE data was performed using JMP® statistical software package, including the generation of a model that considered first- and second-order effects as well as potential interactions between temperature and the equivalents of NaOEt. The JMP® output provided (i) the ANOVA, which tests the hypothesis that parameter terms are zero with exception of the intercept; (ii) the lack of fit, which assesses whether the model fits the data well; and (iii) parameter estimates for the significant factors that define the empirical model. Equation (50.11), where B equals the equivalents of NaOEt and T equals the temperature of the reaction, is the final output of this analysis and can be applied to predict levels of Impurity A within the ranges of the data set and the procedural constraints applied in the model-building set. Analysis tables of primary interest shown in Figure 50.23 indicate that all the parameter estimates are statistically significant and support a good model fit as shown by the high R^2 value and low RMSE:

$$\text{Impurity A} = -7.43 + 4.70B + 0.11T + 0.0043T^2 + 0.386B \cdot T \tag{50.11}$$

Similar to the mechanistic model analysis, a parity plot can provide a qualitative assessment of the model fit (Figure 50.24).

The empirical model was used to predict the contour of the 1.76 RAP in-process specification for Impurity A. Figure 50.25 compares contours predicted with both the empirical and the mechanistic models. Interestingly, the mechanistic and empirical models show nearly identical predictions for the edge of failure, indicating that both approaches can guide the selection of the design space below the edge of failure and ultimately define the trapezoidal space shown in Figure 50.22.

As expected, the empirical approach was more efficient than the mechanistic approach from an experimental and computational perspective. Overall, the empirical approach involved a more simplistic tactic where no consideration of the reaction mechanism was required. Moreover, the model provided an equation that was easily described and presented. However, while the mechanistic model required much more consideration of the underlying physical organic principles and additional experiments to confirm the mechanistic hypothesis, this effort provided its primary benefit and in-depth understanding of the reaction. In many cases fundamental knowledge can be leveraged to improve the process and provide a more efficient or robust design. For instance, in this case study, the effect of cooling the reaction after a defined reaction age to reduce impurity formation during analytical testing could be easily explored with the mechanistic model. In contrast, the empirical model was valid for a very specific procedure and could not predict deviations regarding the timing of the operations or alternative profiles of variant parameters such as temperature. This could be a limitation as the scale of the reaction increases. For example, while the dosing profile of NaOEt produces an exotherm that results in a minor increase in temperature at the laboratory scale, if the manufacturing scale provided a different degree of heat removal or the dosing profile was altered slightly, the

Analysis of variance

Source	DF	Sum of squares	Mean square	F Ratio
Model	4	17.656707	4.41418	72.7830
Error	10	0.606485	0.06065	Prob > F
C. total	14	18.263192		<.0001*

Lack of fit

Source	DF	Sum of squares	Mean square	F Ratio
Lack of fit	4	0.29094116	0.072735	1.3830
Pure error	6	0.31554371	0.052591	Prob > F
Total error	10	0.60648488		0.3435
				Max RSq

Parameter estimates

Term	Estimate	Std error	t Ratio	Prob>|t|
Intercept	−7.426384	0.739003	−10.05	<0.0001*
Base	4.6988523	0.561602	8.37	<0.0001*
Temp	0.1092044	0.008425	12.96	<0.0001*
(Base-1.25333)*(Temp-25.9333)	0.3869077	0.076439	5.06	0.0005*
(Temp-25.9333)*(Temp-25.9333)	0.004291	0.000861	4.98	0.0006*

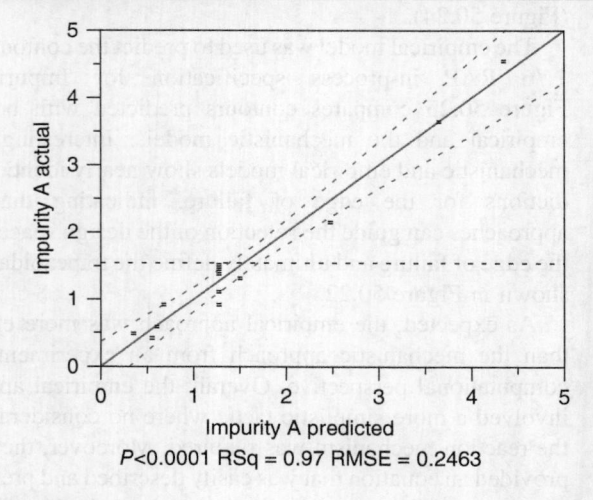

$P<0.0001$ RSq = 0.97 RMSE = 0.2463

Summary of fit

RSquare	0.966792
RSquare adj	0.953509
Root mean square error	0.246269
Mean of response	1.5434
Observations (or sum wgts)	15

FIGURE 50.23 Representative ANOVA and residual analysis using JMP® statistical software.

temperature profile could be altered. Although the design space could still be valid, the empirical model would not be able to provide an accurate prediction.

The practitioner must first weigh the complexity of building a mechanistic model and the importance of fundamental understanding against the efficiency and expedience gained through an empirical approach. Careful consideration of the procedural constraints and the variant parameters must be considered in either approach but especially if an empirical approach is taken.

50.6.3 Mechanistic Model and Role of PAT in the Penultimate Step of a Drug Substance

Selection of the design space for the Final Intermediate step of a drug substance was guided by a mechanistic model [36].

Preparation of the Final Intermediate entailed the condensation of a heteroaryl halide (Input 1) with a highly functionalized starting material (Input 2). The transformation involved (i) deprotonation of Input 2 with potassium *tert*-butoxide (*t*-BuOK) to form a tripotassium salt (Compound 1) that was not isolated, (ii) condensation between the heteroaryl halide and the alkoxide group of Compound 1 to form Compound 2 as its salt, and (iii) aqueous quench upon reaction completion. The Final Intermediate would be obtained following workup, crystallization, isolation, and drying operations.

A collective risk assessment of the drug substance CQAs and a process risk assessment of the parameters affecting the step identified the Final Intermediate QAs as assay, identity, and impurities. Six process-related impurities were studied (i.e. impurities A–F), but only two were critical to quality and yield (impurities A and B, respectively). Impurity B is the major impurity generated in the reaction, but it purges to a high degree during the crystallization. In contrast, Impurity A is of high interest due to its lower purging in the subsequent operations and its capability to lead to an impurity that may impact drug substance quality. The levels of impurities C–F remained within the purging capability of the process. The simplified Ishikawa diagram shown in Figure 50.26 depicts the step with regard to the QAs of the Final Intermediate.

Reaction conditions that would enable the control of impurities A and B within acceptable limits were evaluated by means of a mechanistic model. Construction of the model began with kinetic studies to understand Final Intermediate formation, consumption of input materials, and the generation of impurities. The knowledge thus gathered was used to articulate the elementary reactions described in Table 50.14. As shown in Scheme 50.2, condensation of Input 1 (the electrophile) and Compound 1 (the nucleophilic tripotassium salt of Input 2) affords Compound 2, the dipotassium salt that forms Final Intermediate upon aqueous quench. Deprotonation of Input 2 by addition of greater than 3 equiv. of *t*-BuOK was demonstrated to be complete and instantaneous under studied temperatures, and, consequently, the model approximates the initial concentration of Compound 1 with the concentration of its precursor Input 2.

Impurity A is generated by addition of unreacted Compound 1 to a transient species derived from Compound 2,

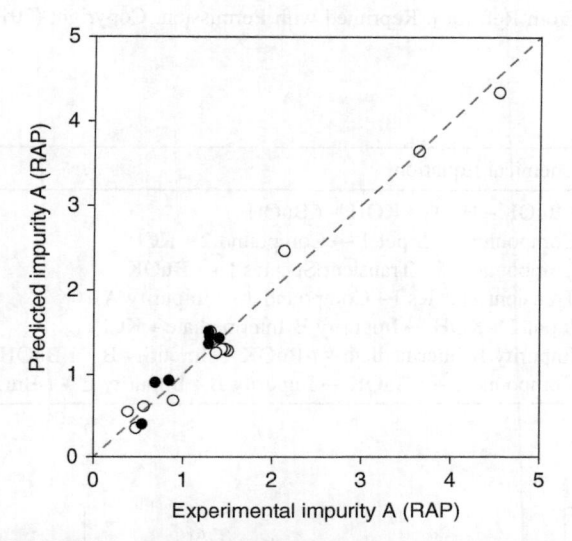

FIGURE 50.24 Parity plot of Impurity A formation for the model-building (○) and verification (●) data sets. *Source*: From Ref. [34]. Reprinted with Permission. Copyright (2010) Springer.

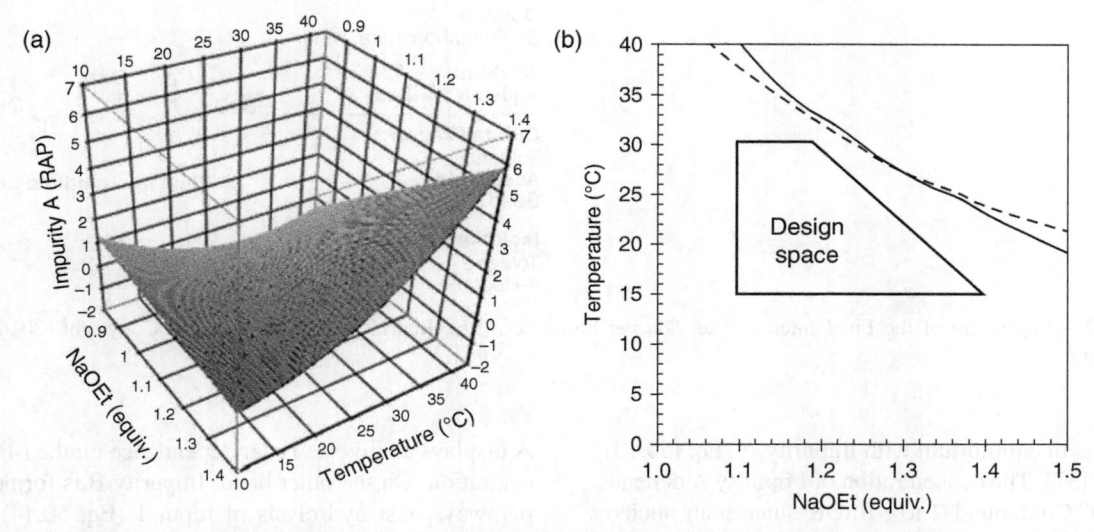

FIGURE 50.25 (a) Response surface for levels of Impurity A predicted by the empirical model. (b) Empirical (---) and mechanistic (—) model predictions for the edge of failure. *Source*: From Ref. [34]. Reprinted with Permission. Copyright (2010) Springer.

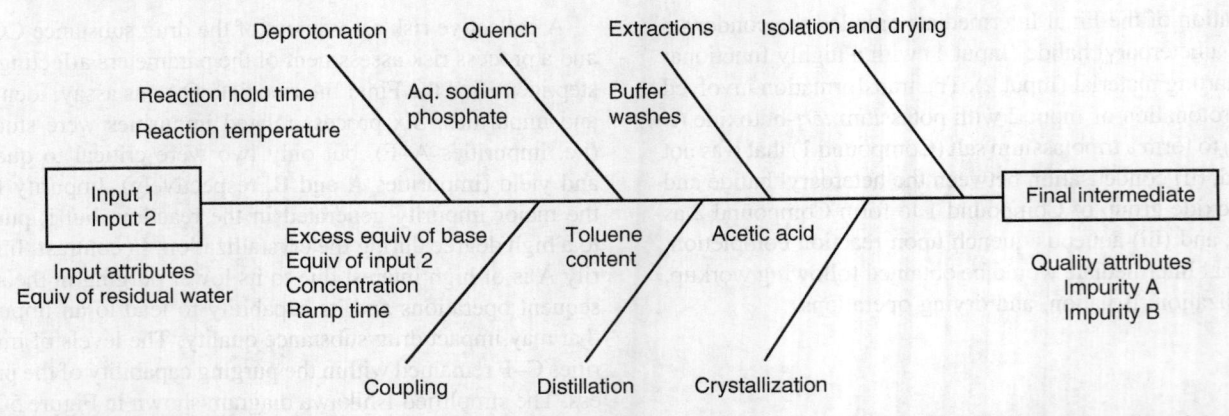

FIGURE 50.26 Ishikawa diagram for the Final Intermediate step. *Source*: From Ref. [36]. Reprinted with Permission. Copyright (2016) American Chemical Society.

TABLE 50.14 Elementary Reactions of the Mechanistic Model

Equation No.	Reaction	Chemical Equation
50.17	Reaction of base with water	$t\text{-BuOK} + H_2O \rightarrow KOH + t\text{-BuOH}$
50.12	Formation of Final Intermediate	Compound 1 + Input 1 $\rightarrow$ Compound 2 + KCl
50.18	Final Intermediate equilibrium	Compound 2 $\leftrightarrow$ Transient Species I + t-BuOK
50.13	Formation of Impurity A	Transient Species I + Compound 1 $\rightarrow$ Impurity A
50.14	Formation of Impurity B (hydrolysis)	Input 1 + KOH $\rightarrow$ Impurity B Intermediate + KCl
50.15	Formation of Impurity B (hydrolysis)	Impurity B Intermediate + t-BuOK $\rightarrow$ Impurity B + t-BuOH
50.16	Formation of Impurities B and C (β-elimination)	Compound 2 + t-BuOK $\rightarrow$ Impurity B + Impurity C + t-BuOH

Source: From Ref. [34].

SCHEME 50.2 Preparation of the Final Intermediate. *Source*: From Ref. [36]. Reprinted with Permission. Copyright (2016) American Chemical Society.

which in turn is in equilibrium with Impurity A (Eq. (50.13), Scheme 50.3) [37]. The concentration of Impurity A depends on the ratio of Compound 2 to t-BuOK since both nucleophiles compete for the transient intermediate. In agreement with this observation, the reaction that forms Impurity A displays an inverse-order dependence on the t-BuOK concentration. On the other hand, Impurity B is formed via two pathways, fast hydrolysis of Input 1 (Eq. 50.14) and slow degradation of the desired product through β-elimination promoted by t-BuOK (Eq. 50.15). Accordingly, the growth of

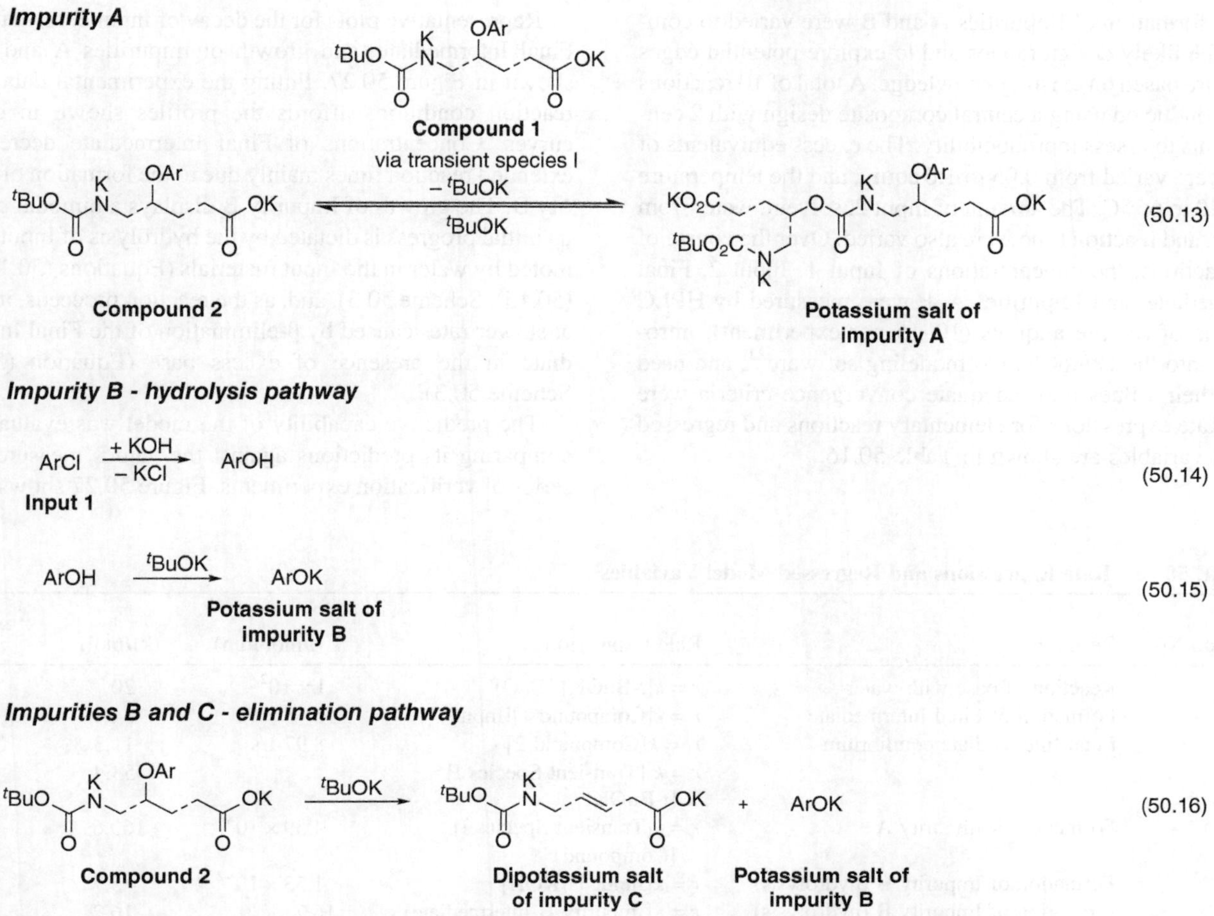

SCHEME 50.3 Formation of Impurities A and B. Equations 50.17 and 50.18 in Table 50.14 are not shown for simplification. *Source*: From Ref. [36]. Reprinted with Permission. Copyright (2016) American Chemical Society.

impurity B shows a bimodal kinetic profile with a rapid onset followed by a slow progression. Superimposition of the inhibitory effect of *t*-BuOK concentrations upon the levels of Impurity A and the promoting effect of *t*-BuOK concentrations upon the levels of Impurity B explains why controlling the charge of base is crucial. Low excess base affords higher levels of Impurity A, whereas high excess base leads to decomposition of Compound 2 by β-elimination, resulting in higher levels of Impurity B.

Although Impurities C–F were always generated within the purging capability of the process and did not affect the quality of the Final Intermediate, they were incorporated in the mechanistic model because their formation influences the concentrations of the main reaction components, including impurities A and B. For example, Impurity C is formed during the β-elimination pathway that generates Impurity B and contributes to the consumption of *t*-BuOK (Eq. 50.16).

Since the reaction mixture formed a slurry comprising Compounds 1 and 2, the solubility of the two salts was investigated. The solubility of Compound 1 was nearly constant within the studied temperature range, whereas the solubility of Compound 2 showed a marked dependence on temperature. The van't Hoff equation [38] was used to correlate the experimental solubility of Compound 2 with temperature, and the solubility equilibria for Compounds 1 and 2 were incorporated into the mechanistic model (Table 50.15).

Model variables for the elementary reactions were regressed using a series of model-building experiments. In these experiments, process parameters with a high impact

TABLE 50.15 Solubility of Compounds 1 and 2

Species	Solubility	ln(As)	Bs (kJ/mol)
Compound 1	0.06 mol/l	—	—
Compound 2	Solubility (mg/ml) = As·e$^{(Bs/RT)}$	10.53	17.80

The van't Hoff equation defines the logarithm of solute concentration as a linear function of the reciprocal of absolute temperature; variables A and B were calculated using a least squares analysis.

on the formation of Impurities A and B were varied to comprise the likely design ranges and to explore potential edges of failure based on existing knowledge. A total of 10 reactions were conducted using a central composite design with 2 center points to assess reproducibility. The excess equivalents of base were varied from 0.05 to 1.0 equiv. and the temperature from 40 to 65 °C. The amount of Input 2, solvent, water from Inputs, and reaction time were also varied. Over the course of the reactions, the concentrations of Input 1, Input 2, Final Intermediate, and Impurities A–F were measured by HPLC analysis of sample aliquots (10–15 per experiment), introduced into the DynoChem® modeling software[21], and used to fit their values until adequate convergence criteria were met. Rate expressions for elementary reactions and regressed model variables are shown in Table 50.16.

Representative plots for the decay of Input 2, formation of Final Intermediate, and growth of Impurities A and B are shown in Figure 50.27. Fitting the experimental data to the reaction conditions affords the profiles shown in dashed curves. Concentrations of Final Intermediate decrease at extended reaction times mainly due to the formation of Impurity B. The growth of Impurity B displays a bimodal profile: its initial progress is dictated by the hydrolysis of Input 1 promoted by water in the Input materials (Equations (50.14) and (50.15), Scheme 50.3), and, as the reaction proceeds, it forms at slower rates caused by β-elimination of the Final Intermediate in the presence of excess base (Equation (50.16), Scheme 50.3).

The predictive capability of the model was evaluated by comparing its predictions against the values measured in a series of verification experiments. Figure 50.27 shows parity

TABLE 50.16 Rate Expressions and Regressed Model Variables

Equation No.	Reaction	Rate Expression	k (l/mol·min)	E_a (kJ/mol)	$K_{eq} \times 10^3$ (1/s)
50.17	Reaction of base with water	$r = k[t\text{-BuOK}][H_2O]$	1×10^2	20	—
50.12	Formation of Final Intermediate	$r = k[\text{Compound 1}][\text{Input 1}]$	6.48×10^{-3}	84.6	—
50.18	Final Intermediate equilibrium	$r_f = k[\text{Compound 2}]$	8.97 1/s	11.3	0.54
		$r = k[\text{Transient Species I}][t\text{-BuOK}]$		53.4	
50.13	Formation of Impurity A	$r = k[\text{Transient Species I}][\text{Compound 1}]$	4.59×10^{-4}	160.6	—
50.14	Formation of Impurity B (hydrolysis)	$r = k[\text{Input 1}][\text{KOH}]$	1.53×10^{-2}	5.7	—
50.15	Formation of Impurity B (hydrolysis)	$r = k[\text{Impurity B Intermediate}][t\text{-BuOK}]$	16.7	16.2	—
50.16	Formation of Impurities B and C (β-elimination)	$r = k[\text{Compound 2}][t\text{-BuOK}]$	6.21×10^{-5}	97.0	—

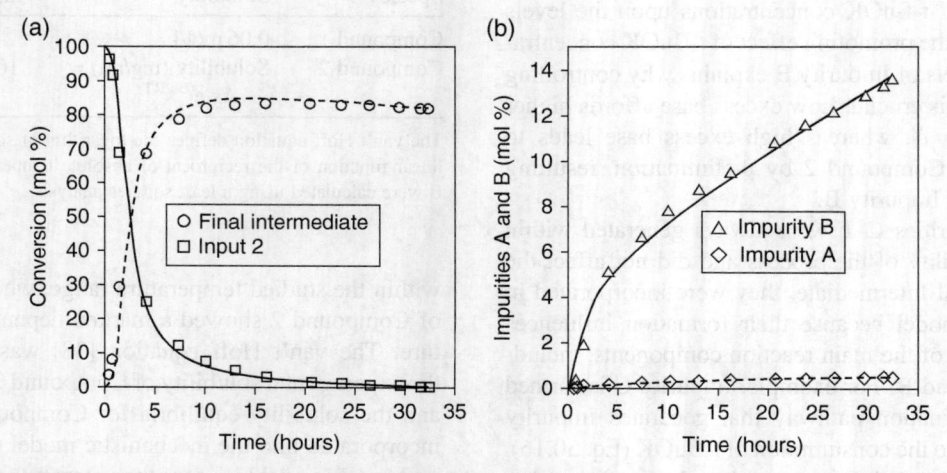

FIGURE 50.27 Concentration–time curves and model fitting for (a) consumption of Input 2 and formation of Final Intermediate and (b) growth of impurities A and B for the same reaction. The experimental data corresponds to a reaction performed at 45 °C with 0.80 equiv. of excess t-BuOK. *Source*: From Ref. [36]. Reprinted with Permission. Copyright (2016) American Chemical Society.

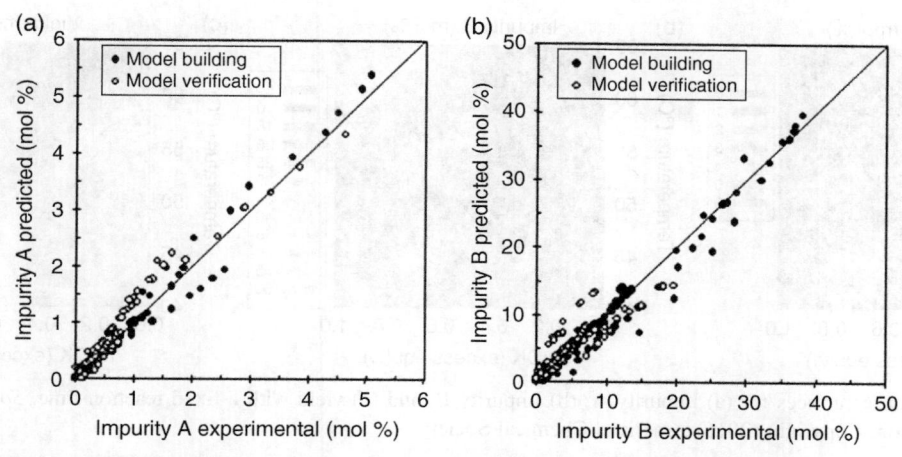

FIGURE 50.28 Parity plots of the model verification and building data sets for (a) Impurity A and (b) Impurity B. *Source*: From Ref. [36]. Reprinted with Permission. Copyright (2016) American Chemical Society.

plots of the model-building data set and the verification data set for Impurities A and B. The verification data set consisted of six reactions at laboratory scale and eight batches at pilot plant scale. A RMSE analysis of the parity plots shows that the prediction errors in the model-building data set and the verification data set are comparable across the range of impurity levels tested (RMSE < 10%). Acceptable error magnitudes and comparability between fitted and predicted impurity levels indicated that the model is an adequate guide to develop a dependable design space.

Specifications for impurities A and B in the isolated Final Intermediate needed to be first established and then correlated with in-process levels to achieve a conservative prediction of the multidimensional space that affords compliance with drug substance CQAs. The specifications for impurities A and B in the isolated Final Intermediate were established at 0.30 and 0.20 wt %, respectively, based on fate and purge studies in the drug substance manufacturing process. Next, the criteria were correlated with in-process levels according to impurity purging during workup and crystallization. Impurity A, which purges in both the workup and the crystallization, was assigned an in-process limit of 3 mol % to meet the less than 0.30 wt % target in the isolated Final Intermediate. Impurity B, which purges in the crystallization to a 100% of its solubility limit, was assigned an in-process limit of 14 mol % to meet the less than 0.20 wt % target in isolated Final Intermediate. Although residual Input 2 purges in the workup, an in-process limit of 4 mol % was established to minimize yield losses.

The design space for the preparation of the Final Intermediate should consider all process variables that could affect the level of impurities A and B. Excess equivalents of base, coupling reaction temperature, equivalents of residual water, and deprotonation reaction hold time had the highest impact on the formation of the critical impurities. However, considering ranges for every single parameter would lead to a

TABLE 50.17 Fixed Values of Process Parameters Used in the Development of the Design Space

Process Parameter	Design Space Range
Reaction time	22–30 h
Equivalents of Input 2	1.05–1.15 equiv.
Solvent volume	17–19 l/kg
Water content	Maximum from all Input materials ≤0.2 equiv.

multidimensional space of great complexity. For this reason, the multidimensional space was reduced to the two parameters that have the largest impact on the impurity levels – the excess equivalents of base and the reaction temperature – while the remaining parameters were held at fixed values within proposed ranges that could lead to the formation of the highest impurity levels (Table 50.17).

Response surfaces were generated for potential levels of impurities and yields using the DynoChem® design space exploration tool[21] (Figure 50.29). Fixed reaction time scenarios were simulated within the range of 0–1.0 excess base equivalents at intervals of 0.1 equiv. as well as temperatures within the range of 40–65 °C at intervals of 2.5 °C. As expected, the model predicted that decreasing the excess base favors the formation of Impurity A, whereas increasing the excess base favors the formation of Impurity B. Raising the temperature promotes the formation of both impurities, and the highest yields can be achieved using low excess base (0.1–0.3 equiv.) and low temperatures (45–55 °C). Projection of the response surfaces provided contour plots that depict combinations of excess base and temperatures predicted to afford the in-process limits for impurities A and B along with residual Input 2. Thus, maintaining fixed parameters at their limit values afforded contours that enclosed the acceptable operating space shown in Figure 50.30.

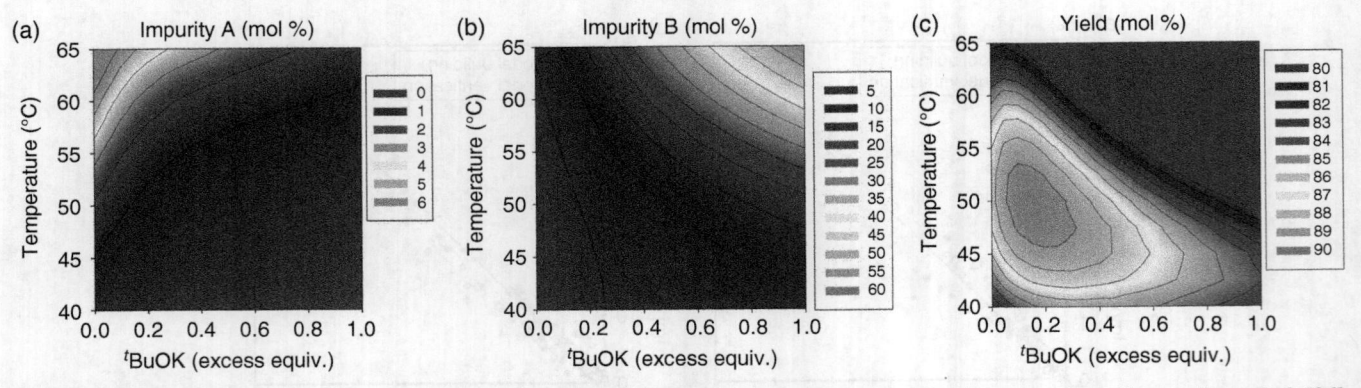

FIGURE 50.29 Response surfaces for (a) Impurity A, (b) Impurity B, and (c) yield with a fixed reaction time. *Source*: From Ref. [36]. Reprinted with Permission. Copyright (2016) American Chemical Society.

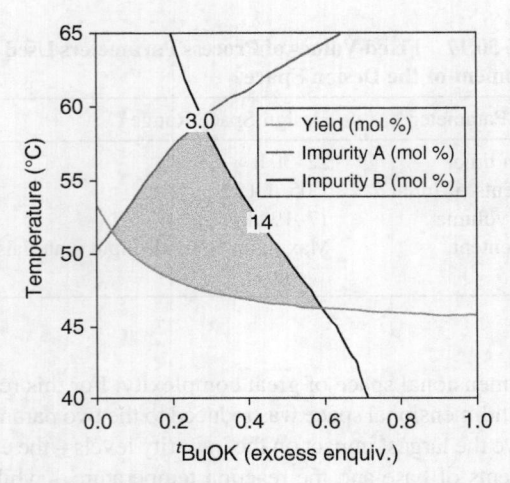

FIGURE 50.30 Combined contour plots of interest for the response surfaces of Impurity A, Impurity B, and yield. *Source*: From Ref. [36]. Reprinted with Permission. Copyright (2016) American Chemical Society.

The fixed reaction time approach enabled control of the impurities within a wide range of reaction times (22–30 hours). To reduce this range, reaction times were partitioned into the sum of a variable reaction time and a post-completion reaction hold time. The variable reaction time could be assessed by continuously monitoring the reaction for completion. To that end, an in situ ATR Raman spectroscopy method was developed by correlating Raman and HPLC responses for 98 data point collected in 8 experiments. The mechanistic model was then used to simulate multiple scenarios within the ranges shown in Table 50.17 with exception of the fixed reaction time, since continuous monitoring ensured completion. Variable reaction time simulations presented trends for impurity formation analogous to those predicted by the fixed reaction time approach (cf., Figures 50.29a and 50.31a, as well as Figures 50.29b and 50.31b). Most importantly, removal of the fixed reaction time expanded the space associated to response surfaces that furnished maximum yields (cf., Figures 50.29c and 50.31c). The impurity profile was always within the in-process limits at the instance of reaction completion (Figure 50.32a and b) and only exceeded the limits when the reaction was held for an extended post-completion time.

Since the post-completion reaction hold time could become a possible failure mode, response surfaces were predicted to estimate the time between reaction completion and exceeding the in-process limits. Excess base and reaction temperature simulations were carried out for each impurity while conservative values were assigned to the remaining parameters (worst-case scenario for each impurity, Figure 50.33a and b). For the major part of the parameter space, the hold time could be extended beyond 30 hours without exceeding the in-process limits. However, because at the high risk edges the acceptable hold time becomes shorter, the post-completion hold time was conservatively established at 5 hours. The combined contour plots of the response surface for impurities A and B represent the most conservative projection of the acceptable operating space and primarily characterize a region of excess base that balances the formation of the two impurities (Figure 50.33c).

The variable reaction time approach using continuous monitoring led to the expanded design space shown in Figure 50.34: a rectangle contained by an excess base range of 0.10–0.55 equiv. and a reaction temperature range of 40–60 °C. The reaction was continuously monitored using Raman or HPLC analysis and cooled to less than or equal to 25 °C within less than or equal to 5 hours after considered complete to halt impurity formation (Figure 50.35). Most importantly, this approach enabled the development of a robust process and made possible the definition of a control strategy for the Final Intermediate (Table 50.18).

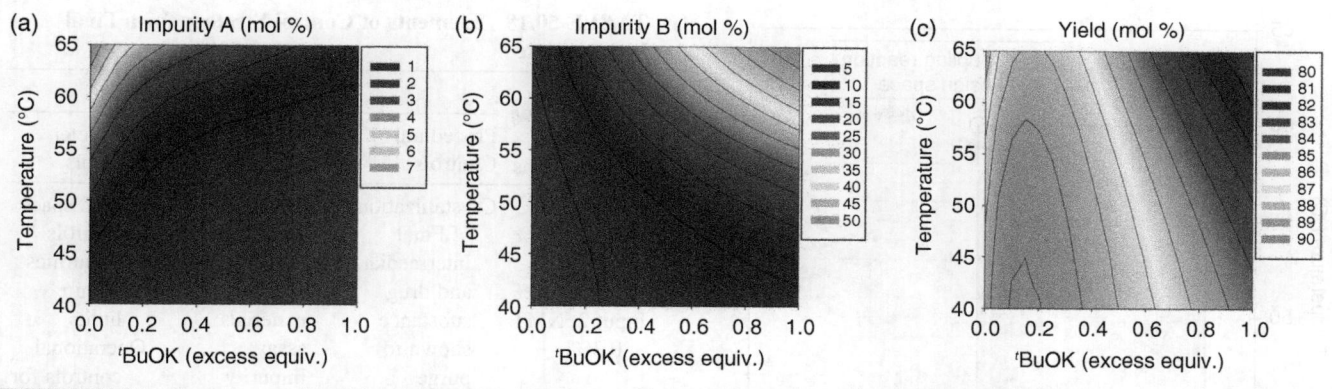

FIGURE 50.31 Response surfaces for (a) Impurity A, (b) Impurity B, and (c) yield with a variable reaction time. *Source*: From Ref. [36]. Reprinted with Permission. Copyright (2016) American Chemical Society.

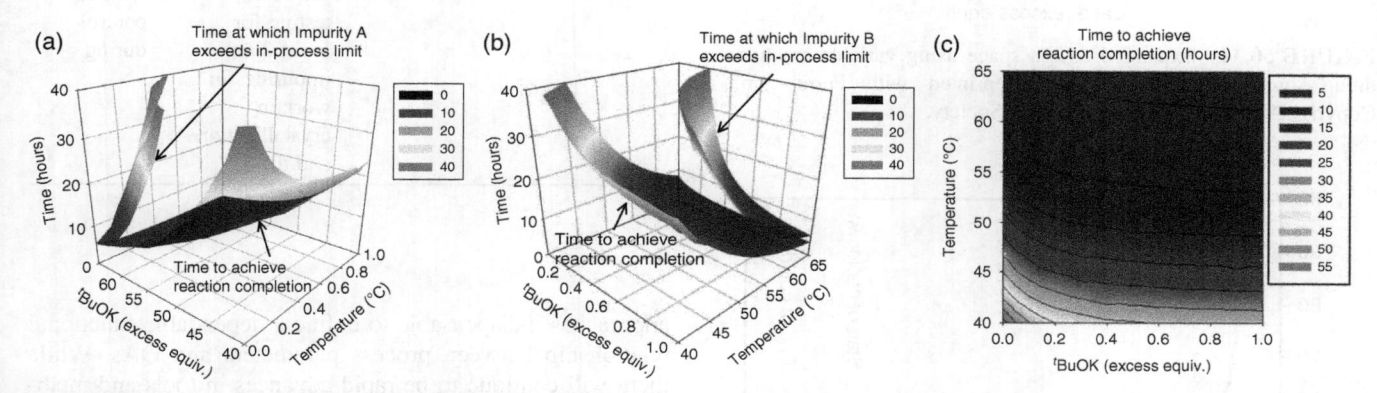

FIGURE 50.32 Response surfaces comparing the time to achieve reaction completion and time to exceed in-process limits for (a) Impurity A and (b) Impurity B. (c) Response surface for reaction time required to achieve reaction completion. *Source*: From Ref. [36]. Reprinted with Permission. Copyright (2016) American Chemical Society.

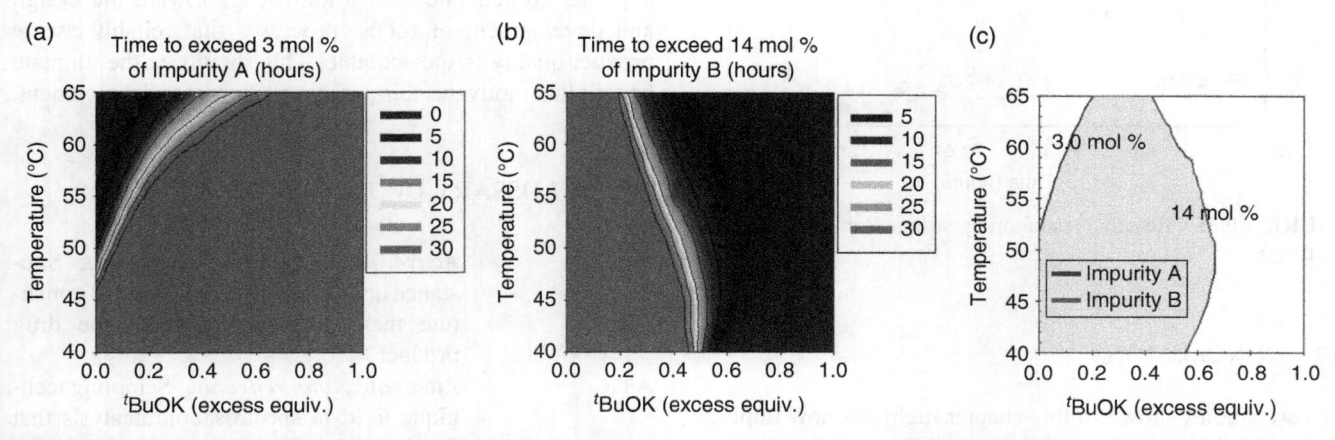

FIGURE 50.33 Response surfaces for time post-reaction completion to exceed the in-process limit for (a) Impurity A and (b) Impurity B. (c) Combined contour plots of the response surface for impurities A and B. *Source*: From Ref. [36]. Reprinted with Permission. Copyright (2016) American Chemical Society.

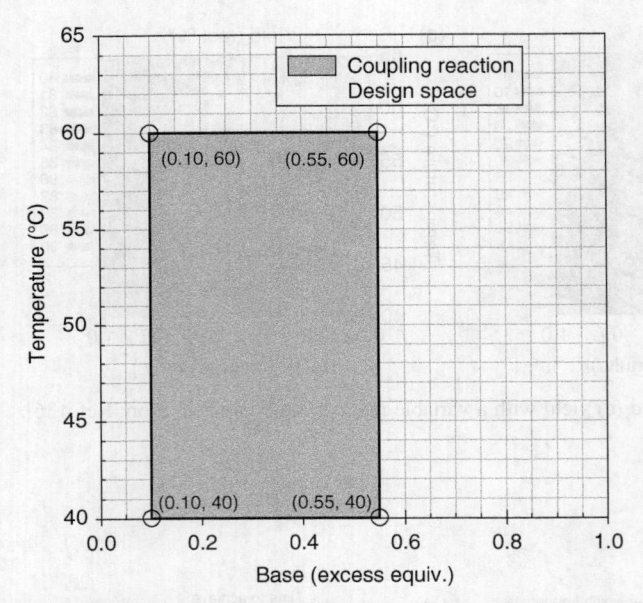

FIGURE 50.34 Expanded design space using variable reaction time. *Source*: From Ref. [34]. Reprinted with Permission. Copyright (2016) American Chemical Society.

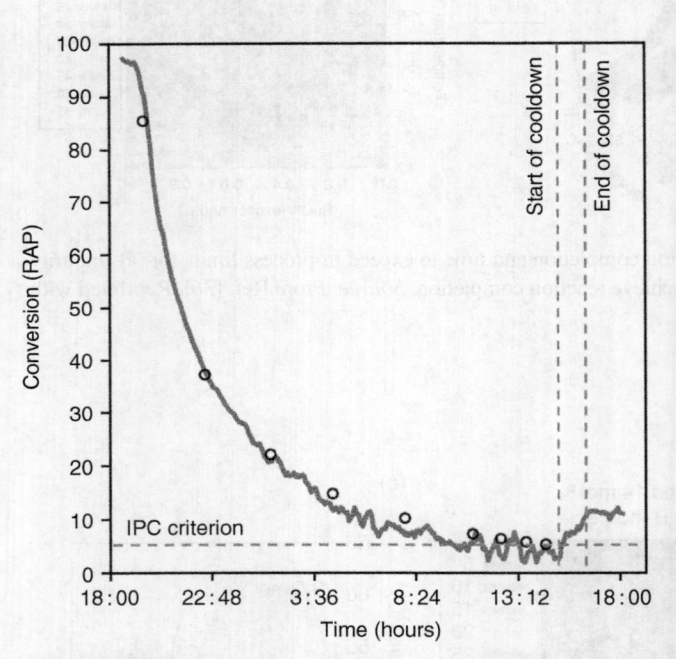

FIGURE 50.35 Reaction monitoring using Raman (–) and HPLC (○).

50.7 CONCLUSIONS

The case studies shown in this chapter highlight how impurity formation in reaction steps is often influenced by a combination of process parameters and MAs. The complexity of these scenarios makes the adoption of multivariate approaches indispensable to define a dependable functional relationship between process parameters and QAs. While there will continue to be rapid advances in tools and methodologies to simplify the collection and analysis of data, some of the core principles outlined in the chapter will remain relevant for the foreseeable future. The value proposition of an enhanced strategy that presents the design space in a regulatory submission continues to evolve with different cost–benefit perspectives. However, the importance of this approach to generate critical knowledge toward the design and development of robust processes that reliably assure product quality is indisputable. This should be the ultimate benefit that motivates this paradigm of process development.

TABLE 50.18 Elements of Control Strategy for a Final Intermediate

Final Intermediate Specifications	Type of Control		
	Procedural Controls	Material Attributes	Parameter Controls
Impurity A, NMT 0.3% Impurity B, NMT 0.2% Input 2, NMT 0.3%	Crystallization of Final Intermediate and drug substance shown to purge Impurities A and B, respectively	Attribute specification control on input materials: assay; impurity specifications; water content and alcohol content In-process controls: testing for Input 2 and impurities in workup crystallization and drying	Design space controls Impurities within limits Operational controls for residual water, alcohol, and oxygen in nitrogen Parameter control during

50.8 GLOSSARY OF TERMS

API
: *Active pharmaceutical ingredient*. Substance or mixture of substances that constitute the active ingredient of the drug product.

ATR
: *Attenuated total reflection*. Sampling technique used in spectroscopic analysis that facilitates the direct study of solid, liquid, or gases with little or no sample preparation.

Center point	An experiment with all numerical factor levels set at their midpoint value.
Central composite design	A design for response surface methods that is composed of a core two-level factorial plus axial points and center points.
Contour plot	Topographical map drawn from a mathematical model, usually in conjunction with response surface methods for experimental design. Each contour represents a continuous response at a fixed value.
DOE	*Design of experiment*. Methodical approach to determine and understand how the parameters of a process affect the performance of that process.
ICH	*International Council for Harmonization of Technical Requirements for Pharmaceuticals for Human Use*. ICH's mission is to achieve greater harmonization worldwide to ensure that safe, effective, and high quality medicines are developed and registered in the most resource-efficient manner.
ICP-MS	Inductively coupled plasma mass spectrometry.
IR	*Infrared spectroscopy*.
GC	Gas chromatography. Chromatographic technique in which the mobile phase is a gas used to separate compounds that can be vaporized.
GTIs	*Genotoxic impurities*. Compounds that have been demonstrated to induce genetic mutations and have the potential to cause cancer in humans.
KF	*Karl Fischer titration*. Volumetric or coulometric method for the determination of moisture content in a sample.
$k_L a$	*Volumetric mass transfer coefficient*. Experimental measure in reciprocal time that represents the rate of mass transfer in a given reactor and mode of operation.
LC-MS	*Liquid chromatography–mass spectrometry*. Analytical technique that combines liquid chromatography with mass spectrometry.
LLS	*Laser light scattering*. Laser diffraction technique used to analyze particle size distribution in the nm-to-mm range.
NMT	*No more than*.
p-TsOEt	*Ethyl p-toluenesulfonate*. Alkylating agent and potentially genotoxic impurity that requires a control strategy in process development.
PAT	*Process analytical technologies*. Tools for designing, analyzing, and controlling manufacturing through timely measurements of critical quality and performance attributes of raw and in-process materials. See Chapter 42.
PXRD	*Powder X-ray diffraction*. Analytical technique primarily used for phase identification of a crystalline material. The analyzed material is finely ground and homogenized, and average bulk composition is determined.
QbD	*Quality by design*. A systematic approach to development that begins with predefined objectives and emphasizes product and process understanding and process control, based on sound science and quality risk management.
RAP	*Relative area percent*. Estimation of percent composition that divides the area of each peak by the total area of all peaks and multiplies by 100%. Assumes that all the mixture components cause the same response in the detector.
RMSE	*Root mean square error*. The square root of the residual mean squared error. It estimates the standard deviation associated with experimental error.
TTC	*Threshold of Toxicological Concern for genotoxic impurities*. Refers to a threshold exposure level to compounds that does not pose a significant risk for carcinogenicity or other toxic effects.

REFERENCES

1. ICH Q11 (2011). *Development and Manufacture of Drug Substances (Chemical Entities and Biotechnological/Biological Entities)*. Rockville, MD: U.S. Department of Health and Human Services, Food and Drug Administration, Center for Drug Evaluation and Research (CDER).

2. ICH (2009). Harmonised tripartite guideline: pharmaceutical development Q8 (R2). *International Conference on Harmonisation of Technical Requirements for Registration of Pharmaceuticals for Human Use*, Geneva (August 2009).

3. Yu, L.X., Lionberger, R., Olson, M.C. et al. (2009). *Pharmaceutical Technology* 33 (10): 122–127.

4. Lee, H. (ed.) (2014). *Pharmaceutical Industry Practices on Genotoxic Impurities*. Boca Raton, FL: CRC Press, Taylor & Francis.

5. Sakuramill Sakuramil S2 mock of ICH Q11 Guideline regarding development and manufacture of drug substance. http://www.nihs.go.jp/drug/section3/H23SakuramillMock(Eng).pdf (accessed 13 November 2018).

6. ICH (2009). ICH Harmonised Tripartite Guideline: Q8 (R2) Pharmaceutical Development. https://www.ich.org/fileadmin/

Public_Web_Site/ICH_Products/Guidelines/Quality/Q8_R1/Step4/Q8_R2_Guideline.pdf (accessed 13 November 2018).

7. Thomson, N.M., Singer, R., Seibert, K.D. et al. (2015). *Organic Process Research and Development* 19: 935–948.
8. Glodek, M., Liebowitz, S., McCarthy, R. et al. (2006). *Pharmaceutical Engineering* 26: 1–11.
9. Weissman, S.A. and Anderson, N.G. (2015). *Organic Process Research and Development* 19: 1605–1633.
10. Domagalski, N.R., Mack, B.C., and Tabora, J.E. (2015). *Organic Process Research and Development* 19: 1667–1682.
11. Bakeev, K.A. (2010). *Process Analytical Technology: Spectroscopic Tools and Implementation Strategies for the Chemical and Pharmaceutical Industries*, 2e. Chichester: Wiley.
12. Zaborenko, N., Linder, R.J., Braden, T.M. et al. (2015). *Organic Process Research and Development* 19: 1231–1243.
13. Helfferich, F. (2004). *Kinetics of Multistep Reactions*, 2e, 7–16. Amsterdam: Elsevier.
14. Hannon, J. (2011). Characterization and first principles prediction of API reaction systems. In: *Chemical Engineering in the Pharmaceutical Industry. R&D to Manufacturing* (ed. D.J. am Ende). Hoboken: Wiley.
15. Boudart, M. (1991). *Kinetics of Chemical Processes*. London: Butterworth-Heinemann, Reed Publishing.
16. Murzin, D.Y. and Salmi, T. (2016). *Catalytic Kinetics: Chemistry and Engineering*, 2e. Cambridge: Elsevier.
17. Cussler, E.L. (1997). *Diffusion: Mass Transfer in Fluid Systems*, 2e. London: Cambridge University Press.
18. (a) Carpenter, B.K. (1984). *Determination of Organic Reaction Mechanisms*. New York: Wiley.
    (b) Espenson, J.H. (1995). *Chemical Kinetics and Reaction Mechanisms*, 2e. New York: McGraw-Hill.
    (c) Houston, P.L. (2006). *Chemical Kinetics and Reaction Dynamics*. New York: Dover Publications, Inc.
19. Caron, S. and Thomson, N.M. (2015). *The Journal of Organic Chemistry* 80: 2943–2958.
20. Madelaine, G., Lhoussaine, C., and Niehren, J. (2015). Structural simplification of chemical reaction networks preserving deterministic semantics. In: *Computational Methods in Systems Biology. CMSB 2015*, Lecture Notes in Computer Science, vol. 9308 (ed. O. Roux and J. Bourdon). Cham: Springer.
21. Klamt, S. and Stelling, J. (2006). Stoichiometric and constraint-based modeling. In: *Systems Modeling in Cellular Biology* (ed. Z. Szallasi, J. Stelling and V. Periwal). Cambridge, MA: MIT Press.
22. Turányi, T. and Tomlin, A.S. (2014). *Analysis of Kinetic Reaction Mechanisms*. Berlin, Heidelberg: Springer–Verlag.
23. Burt, J.L., Braem, A.D., Ramirez, A. et al. (2011). *The International Journal of Pharmaceutics* 6: 181–192.
24. Thomas, B.G. and Brimacombe, J.K. (1997). Process modeling. In: *Advanced Physical Chemistry for Process Metallurgy* (ed. N. Sano, W.-K. Lu, P.V. Riboud and M. Maeda). San Diego, CA: Academic Press.
25. Ekins, S. (2006). *Computer Applications in Pharmaceutical Research and Development*, 2e. Hoboken, NJ: Wiley.
26. Montgomery, D.C. (2013). *Design and Analysis of Experiments*, 8e. Danvers, MA: Wiley.
27. Oehlert, G.W. (2000). *A First Course in Design and Analysis of Experiments*. New York: W. H. Freeman.
28. Robinson, T.J. (2014). Box-Behnken Designs. Wiley Stats Ref: Statistics Reference Online.
29. (a) Dean, A.M. and Voss, D. (1999). *Design and Analysis of Experiments*. New York: Springer-Verlag.
    (b) Klimberg, R. and McCullough, B.D. (2016). *Fundamentals of Predictive Analytics with JMP®*, 2e. Cary, NC: SAS Institute.
30. U.S. Food and Drug Administration (2012). Guidance for Industry Q8, Q9, & Q10 Questions and Answers – Appendix Q&As from Training Sessions. U.S. Department of Health and Human Services Food and Drug Administration Center for Drug Evaluation and Research (CDER) Center for Biologics Evaluation and Research (CBER) July 2012 ICH.
31. European Medicines Agency (2013). Questions and Answers on Design Space Verification. Retrieved 2 May 2017. https://www.ema.europa.eu/documents/other/questions-answers-design-space-verification_en.pdf (accessed 13 November 2018).
32. Garcia, T., McCurdy, V., Watson, T.N. et al. (2012). *The Journal of Pharmaceutical Innovation* 7: 13–18.
33. Watson, T.J., Bonsignore, H., Callaghan-Manning, E.A. et al. (2013). *The Journal of Pharmaceutical Innovation* 8: 67–71.
34. Hallow, D.M., Mudryk, B.M., Braem, A.D. et al. (2010). *The Journal of Pharmaceutical Innovation* 5: 193–203.
35. ICH (2011). ICH quality implementation working group points to (R2) ICH–Endorsed Guide for ICH Q8/Q9/Q10 Implementation. Document date: 6 December 2011.
36. Ramirez, A., Hallow, D.M., Fenster, M.D.B. et al. (2016). *Organic Process Research and Development* 20: 1781–1791.
37. Tom, N.J., Simon, W.M., Frost, H.N., and Ewing, M. (2004). *Tetrahedron Letters* 45: 905–906.
38. (a) Prausnitz, J.M., Lichtenthaler, R.N., and Azevedo, E.G. (1986). *Molecular Thermodynamics of Fluid-Phase Equilibria*. Englewood Cliffs, NJ: Prentice Hall.
    (b) Grant, D.J.W., Mehdizadeh, M., Chow, A.H.L., and Fairbrother, J.E. (1984). *The International Journal of Pharmaceutics* 18: 25–38.

INDEX

Ab initio, 184, 480, 486, 562, 671–672
Absorbance, 118, 403, 647, 909, 911–915, 940, 942, 947–948, 950, 993
Accelerating Rate Calorimetry (ARC), 10, 48, 62, 64, 68, 70, 72–76, 79, 85, 89, 94–97, 102–104, 110–114, 164, 219, 225, 344, 382, 1007
Acetaminophen, 472–478, 518
acid–base surface energy, 694, 702
Activity coefficient, 442–443, 446, 463, 470–471, 479, 491, 493–495, 497–499, 502–504, 506, 509–510, 574, 592, 596, 735–736, 741, 744, 752, 754–755, 762, 764, 767–768, 770–774, 777, 781–786, 789–790, 797
Adhesion, 692, 694–695, 710–712
Adiabatic, 30–31, 33–35, 45, 47–52, 54–56, 64–65, 70–71, 73–74, 89, 92, 94, 96–97, 103–104, 111–113, 162, 334, 342, 344–345, 362, 379, 686, 768, 958, 969, 1002, 1004, 1007
Adiabatic calorimetry, 47, 49, 51, 54, 64, 70, 1002
Agglomerate strength, 622, 691, 697, 711
Agglomeration, 158, 204, 236, 253, 294, 301–302, 539, 547, 553–554, 564, 600–602, 617, 623, 636–637, 657, 667, 692, 695, 698, 819, 823–824, 830–831, 848, 851, 855, 859–860, 872, 998, 1021–1022
Agitated drying, 824, 830–831, 853, 855, 874, 997, 1000
Aleatory Variability, 925–927, 933
Amorphous, 293, 423, 430–431, 437–438, 443, 465, 485, 490, 517, 520, 526, 530, 537–538, 560, 615, 622, 665, 667–668, 678, 685, 694, 696, 707–708, 710, 712, 738, 862–863, 865, 994
Amorphous solid dispersion (ASD), 5–6, 423, 433, 622, 665, 667–668, 672, 678, 685, 710
Analysis of Variance (ANOVA), 888, 890, 901, 905, 926, 1099–1101, 1111–1112
Angular velocity, 399, 804
Anisotropic, 445, 692, 697, 707, 710–711

Anomalous racemate, 572
Antisolvent, 293, 298–301, 316, 424, 426–427, 439, 542, 547–551, 554–557, 565, 567, 596–598, 600, 602, 612–615, 617, 621–622, 624–628, 676, 681–682, 684, 796, 806, 849, 859–860, 976, 994–995, 1015, 1035, 1078, 1082
Antisolvent addition, 547, 550–551, 555–557, 612, 615, 622, 625–628, 676, 682, 994–995, 1015, 1078, 1082
Antisolvent crystallization, 298–301, 316, 547, 554, 556, 565, 567, 596, 602, 612–615, 617, 622, 625, 806
Antoine equation, 735, 753–754, 817
Aqueous electrolytes, 495, 497
Arrhenius plot, 107, 348, 854
ARSST, 62, 70, 74–75, 89
Artemisinin, 484–485, 490
Aspect ratio, 261, 391, 395, 410, 561, 564–565, 567, 607, 617, 692, 696, 699, 724, 805, 835–836, 838, 843
Aspen Properties, 496–497, 503, 596, 615, 753–754, 762, 770–771, 773–774, 776, 780–782, 785–786
Asymmetric hydrogenation, 161, 191, 193–194, 200–201, 369–370, 385, 750, 984
Asymmetric hydrogenation of Geraniol, 193–194
Atomic force microscopy, 691–692, 712, 863
Attachment energy, 706–707
Attrition, 4, 6, 247, 253, 292, 338, 549, 600, 617–618, 620, 697, 796, 819, 823, 830–831, 835, 842–843, 848, 850–853, 855, 858, 860, 862, 866, 993, 995, 997–998, 1021–1023, 1069
Austin's model, 876, 878
Autocatalysis, 56, 1096
Autocorrelation, 929–931
Axial dispersion, 162, 221, 327, 348, 350–351, 355, 361–362, 368, 371–372, 374–375
Axial flow, 251–253, 315, 395, 619, 716
Azeotrope, 734, 756–760, 774, 777, 784–786, 789–790, 793, 828–829, 989, 1006–1007

Chemical Engineering in the Pharmaceutical Industry: Active Pharmaceutical Ingredients, Second Edition.
Edited by David J. am Ende and Mary T. am Ende.
© 2019 John Wiley & Sons, Inc. Published 2019 by John Wiley & Sons, Inc.

INDEX

Baffles, 243, 245–246, 252–253, 273, 276, 305–307, 317, 396, 738, 770, 987
Balling, 830, 848
Batch process simulation, 1039
Bayesian, 183, 226, 906–907, 922, 924–928, 935
Beer–Lambert (Beer), 119, 910, 912, 935
Bicalutamide, 486
BINAP-Ru(II), 193
binary interaction parameters (BIPs), 435–436, 495–500, 504, 754, 761–762, 764, 768, 770–772, 774, 776–777, 780–781, 786, 789–791, 793
Biologics, vii, xi–xii, xv, 4, 8–10, 13–15, 384, 387, 389–392, 394, 396, 398, 400, 402, 404, 406, 408–410, 412, 414–416, 1037, 1074, 1122
Biopharmaceutical Classification System (BCS), 516, 665–666, 672, 681, 685, 861, 982
Bioreactor, 230, 308, 389, 391–399, 409–411, 414–416
Biphasic, 207, 225, 255–256, 262, 322, 333, 354, 424, 431–432, 962, 1006
Biphasic reactions, mixing in, 255
Birefringence, 694
Blend time, 241, 246–248, 250–251, 254, 257, 259, 625, 959–960, 988–990
BLEVE, 91–92, 98, 109
Borane–THF (BTHF), v, 91–112, 114
Bourne reaction, 159
Breakage, 204, 236–237, 262, 269–270, 279, 292, 294, 303–304, 312, 315, 564, 600–602, 617–618, 628, 636–637, 657, 662, 697, 703, 711, 830, 840, 855, 858, 862–863, 871–877, 879, 997
Breakage distribution function, 872, 876
Breakage rate, 618, 697, 872, 875–877
Brittleness, 691, 696, 706, 863
Buchner funnel, 802, 806–807, 811–813, 996
Bulk density, 538, 540–541, 696, 823, 994, 997–998

Caffeine, 457–459, 464, 466, 485, 507
Cake cracking, 815, 843
Cake Filtration, 800, 804, 814, 831
Cake permeability, 800
Cake resistance, 541, 800–803, 805–806, 808, 812–814, 834–836, 838–840, 843–844, 848
Cake smoothing, 830, 856
Cake wash, 582, 807–808, 810, 813, 816–817, 829–830, 838, 843–844, 848–851, 977, 996, 1001
Cake wash design, 816–817
Calibration, 33, 36–37, 39, 45, 113, 221, 347, 594, 641–642, 824–825, 873, 875, 910–912, 914–916, 918, 925, 946, 948–950, 953, 1003, 1034, 1083, 1110
Calorimetry, 32–35, 40, 42, 47, 49, 51, 54, 60, 62, 64–65, 70, 79, 89, 94, 163–164, 171–173, 219, 344–345, 348, 360, 365, 390, 420, 437, 472, 518, 526, 575, 594, 704, 712, 863, 1002, 1004, 1007
Cambridge Crystallographic Data Centre (CCDC), 670–671, 674
Cambridge Structural Database (CSD), 293, 299, 301, 526–530, 532, 600–601, 603–605, 612–613, 635, 670, 674–675, 677, 685–686, 848, 850–853
Campaign planning, 1049
Capacity analysis, 1061–1062, 1065
Carbonate, 75–77, 79–81, 235–239, 398, 410, 628–629, 715, 873, 1005, 1010
Carbon dioxide (CO_2), xxiii, 36, 57, 62, 68, 76, 79–81, 136, 157, 160–161, 164, 173–174, 182, 221, 227, 230, 391–392, 395–396, 398, 409–412, 416, 493, 503, 541, 549, 558, 940, 944, 948–951, 978, 980, 1041
Catalysis, 56, 184, 188, 191, 200–201, 262, 490, 571, 1075, 1096
Catalytic hydrogenation, vii, 163, 191–192, 194, 196–198, 200–202, 241, 275, 943, 966, 968
CCPS (Center for Chemical Process Safety), 60, 71, 89, 114, 879, 999, 1029
C-curve, 374
Cell Culture, 17, 389, 391–400, 405–410, 415–416
Cell selection, 65
Centrifugal filter, 804, 810
Centrifugal force, 399, 799, 804, 808, 810, 812
Centrifugation, 393, 399, 542, 804, 808, 812, 816, 834–835, 837–841, 848–851, 996
Centrifuge, 26, 358, 399–400, 594, 804, 808–814, 834–841, 848–851, 853, 984, 987, 996, 1001, 1003, 1008, 1010, 1015, 1017, 1021–1022, 1025
CFD *see* Computational fluid dynamics (CFD)
Chemical force microscopy, 695, 711
Chemical reaction, viii, 33, 48, 56, 60, 146, 153–156, 171, 183–185, 188, 192, 203–209, 213–214, 220, 226, 232–234, 239, 241, 246–247, 250–251, 254, 256, 259, 262, 279, 310, 314–315, 365–368, 382, 385, 389, 394, 440–441, 444, 447, 449–450, 478, 546–547, 662, 861, 883, 901, 918, 937, 940–942, 957–960, 962–964, 966, 968–970, 1004, 1049, 1104, 1122
Chemical Safety Board (CSB), 61
Chemometrics, 909, 915, 918
Chirality, 531, 567, 575, 587
Chiral systems, 571, 574, 589
Cinchonidine-modified Pt, 193–194
Circular equivalent diameter, 696, 699
Circularity, 696, 699
Clear Point, 594–595
Cloud Point, 594–595, 609–612
Cocrystal, vii, 419, 427–430, 437–441, 444, 447, 449–450, 452–455, 458, 460–463, 465–468, 470, 472, 474, 476, 478, 480, 482–488, 490–492, 537–538, 560, 566, 665–666, 668, 671, 684–685, 690–691, 976, 982
Cohesion, 691–692, 698
Compressibility, 538, 540, 696–697, 801–802, 808, 811, 813, 833–841, 848, 996, 1021
Compressible cake, 801–802, 804, 810, 812–813, 816, 835
Computational fluid dynamics (CFD), 22, 221, 243, 249, 254, 257–259, 261, 263, 267, 270, 272–276, 279–282, 284, 290, 292, 294–296, 298–299, 303, 306, 308, 311, 314–317, 402–403, 416, 619–620, 631, 958, 961–962, 965
Computer-aided process design, 1038
Conductivity, xviii, xx, 37, 59, 98, 118, 249–250, 259, 348, 403, 647, 726, 770, 864, 909, 1008
Conductor-like screening model for realistic solvation (COSMO-RS), 467–472, 475–476, 478, 480–481, 483–487, 489–491, 521, 532, 675, 687
Conical dryer, 821–823, 830, 984, 997, 1022
Constant-level distillation, 753, 757, 760–763

Constant pressure filtration, 800, 831
Constant rate filtration, 800
Constant rate period, 819–820, 825–827
Continuous reactor, vii, 59, 154–155, 161–162, 183, 219–220, 321–324, 326–330, 332, 334, 336, 338, 340, 367–370, 372, 374, 376–378, 380–384, 386, 958
Continuous stirred tank reactor (CSTR), 128, 130, 159, 219, 226, 250–251, 259, 285, 326–329, 338, 350–351, 353, 355–356, 367–373, 375, 377–383, 385, 625, 637, 655, 958, 962–966
Contour plots, 363, 631, 701, 885, 901, 1098, 1107, 1110, 1117–1119
Control Chart, 920–922
Control Limit, 413, 920–921, 999, 1031, 1036, 1074, 1082
Control Strategy, 4, 179, 186, 225, 351, 372, 423, 426, 581, 585, 587, 591, 629, 684, 695, 866, 924, 934, 938, 945, 947–948, 950, 952–953, 966, 969, 978, 982, 1027, 1070–1071, 1074, 1078–1080, 1082, 1084, 1089, 1091, 1093–1094, 1104–1105, 1109–1110, 1118, 1120–1121
Corrosion, 32, 36, 65–66, 264, 382, 497, 977, 979, 1014, 1017, 1020
COSMO, xii, xiv, 467–487, 489–491, 502, 504, 521, 523, 525–532, 563, 567, 675, 687, 754, 761, 764, 770–781, 783–786, 790
COSMO-RS, 467–472, 475–476, 478, 480–481, 483–487, 489–491, 521, 532, 675, 687
COSMOthermX, 473–474, 476, 486, 754, 761, 764, 770, 772–781, 783–786, 790
Cost analysis, 1037–1038, 1044
Cowles high-shear impeller (Cowles), 253, 302
Critical impurity, 1105, 1110
Criticality class, 29–31
Critical moisture content, 819–820, 826–827, 829
Critical particle size, 876
Critical Quality Attributes (CQA), 186, 345, 413, 564–565, 635–636, 729, 921–922, 924, 926, 933, 958, 975, 979, 982–983, 1005, 1070, 1073–1076, 1079–1080, 1082, 1084–1085, 1089, 1092–1094, 1104, 1113, 1117
Crystal Ball, 1046
Crystallization Kinetics, 293–294, 535, 546, 593, 602, 612, 615, 622, 627–630, 643, 657, 795
Crystal Morphology, 593, 691–692, 694–695, 703, 706, 831
Crystal structure prediction (CSP), 472, 520, 529–531, 670–672, 686, 721–722, 731
CSTR *see* Continuous stirred tank reactor (CSTR)
Cumulative oversize number, 603, 606
Cycle time, 15–16, 154, 218, 322, 541, 605, 751, 760–761, 795, 797, 800, 803–805, 812, 816, 819–820, 823–824, 829–830, 836, 838, 847, 853, 859–860, 864, 869, 974, 976, 983–984, 986–987, 992, 996–997, 1021, 1037–1039, 1042–1044, 1046, 1052, 1060–1062, 1075, 1100, 1102
Cycle time analysis, 1039

Damkohler number, 256–257, 259, 959
Danckwerts boundary condition, 375
Darcy's law, 716, 800, 816, 834, 839
Dasabuvir, 423, 680–684
Data preprocessing, 914
debottlenecking, 367, 493, 838, 984, 1044–1045, 1065–1066
Debye–Hückel limiting law, 495

Design space, ix, 3, 164, 176, 178–180, 186, 204–207, 209, 214, 216–218, 220, 222, 224–226, 317, 343, 351, 353, 365, 536, 563–564, 620, 622, 715–716, 728, 789, 795, 833, 847, 860, 875, 879, 921–922, 924, 935, 962, 1026, 1070–1071, 1086, 1091, 1093–1094, 1096, 1098–1100, 1102–1112, 1114, 1116–1118, 1120, 1122
Desolvation, 419–422, 490, 537–538, 685, 847–848, 853–855, 857
Detonation, 83
Dew-point hygrometer, 825, 828
Diastereomeric salt, 583–585
DIERS (Design Institute for Emergency Relief Systems), 60, 75, 89
Dimensionless length, 374
Dimensionless time, 374, 747
Directed Acyclic Graphs, 926
Dispersion-corrected density functional theory (DFT-d), 672
Dispersive surface energy, 693, 702, 710
Displacement wash, 812, 816, 844, 859
Dissolution rate, 230, 306, 419–420, 437, 505, 516–517, 538, 558, 560, 592, 635, 665–667, 672, 678, 681, 698, 703, 710, 712, 861
Distillative crystallization, 806, 989
Distributed control system (DCS), 344, 379, 397, 1015, 1026
Distribution coefficient, 672, 681, 717–718, 753, 764–766
DMSO (dimethyl sulfoxide), 64, 71–73, 75, 83–84, 89–90, 175, 235, 475, 558, 681–684, 1010
Double-cone dryer, 1023
Drop-weight, 83
Drugability, 665
Drug product manufacturing, 450, 560, 623, 667, 864, 868, 1021
Drug product processing, 618, 622, 678, 680, 865
Drum, 36, 47, 57, 76–78, 81, 91, 98, 667, 746–747, 810, 821, 831, 989, 1025, 1027, 1035
Dryer, 26, 803, 808–812, 814, 817–831, 835, 843–844, 847–848, 850–855, 857–860, 865, 873–874, 984–985, 987, 996–997, 1011, 1013, 1015, 1017, 1021–1023, 1035–1036, 1039, 1042, 1052
Drying, vii–viii, xv, xxi–xxii, xxv, 24, 26, 203–204, 337, 390, 415, 420, 422–423, 426, 430–431, 438, 536, 541–542, 560, 594, 624, 667, 680, 694–695, 713, 736, 787, 793, 799, 802, 804, 806, 808, 810, 812, 814, 816–832, 844–845, 847–860, 864, 874–875, 977, 981–982, 984–985, 987, 994, 996–997, 1000, 1002, 1008, 1014–1017, 1021–1023, 1028, 1039–1040, 1056, 1082, 1095, 1105, 1113–1114, 1120
Drying curve, 818–820, 826
Drying mechanisms, 818, 847
Drying modeling, 825
Drying periods, 852
Drying rate, 817–819, 825–827, 830, 848, 850–853, 855, 860
Drying rate curve, 852, 860
Drying time, 541, 819, 822–823, 826–830, 844–845, 852–853, 856–857, 1014, 1022, 1082
Dry Milling, 236, 622, 711, 863–866, 869, 872, 876, 994, 998
DSC (differential scanning calorimetry), 19, 32, 44–46, 51, 53–56, 62, 64–68, 71–76, 82–90, 111, 344–345, 365, 382, 390, 420–422, 431–432, 447, 472, 493, 518, 526, 531, 560, 564, 575–576, 587, 594–595, 620, 668–669, 671–672, 682, 684, 686, 691, 704, 863
DSC, 100-Degree rule, 46, 65
Dynamic image analysis, 691, 696, 711
Dynamic vapor sorption (DVS), 433, 435, 691, 694, 853, 857–858

Dynochem, 127, 225–226, 361–362, 385, 596–598, 600, 605, 789–792, 794–796, 826–827, 835, 839, 872, 952, 961, 967–968, 989, 1097, 1107, 1116–1117

EasyMax, 79, 131, 939
Economic evaluation, 1039, 1044–1045
Eddy, 120, 243, 259, 265–266, 293, 401, 632, 686, 692, 710
Edges of failure, 216, 1096, 1100, 1102, 1104, 1107–1109, 1116
Effective hydrogen pressure, 200
Eigenvalues, 914
Eigenvectors, 910, 912
Elastic deformation, 667, 696
Electrolyte NRTL model, 495, 503
Electrolyte thermodynamic models, 493–494, 504
Electrolyte thermodynamics, 493, 502–503
Electron microscopy, 679, 691, 694, 711
Elementary steps, 169, 172, 1095–1096, 1098
Empirical model, 85, 177, 225, 405, 409, 442, 448, 451, 453–455, 458, 486, 494, 502, 564, 736, 900, 931, 1095, 1098–1099, 1104, 1110–1113
Emulsion, 220–221, 261–262, 416, 566, 748, 768–769, 879, 988, 990, 1000, 1009
Enantiomeric excess (EE), 256, 573, 579, 981
Enantiotropic, 420–421, 505, 509–510, 513, 526, 537–538, 546, 672
Enantiotropic systems, 513
Energy balance, v, xv, 27, 64, 99, 107, 110, 163, 330, 335, 343, 345, 353, 359, 379, 409, 638–640, 743, 825–826, 1002, 1015, 1037, 1039, 1049
Energy dissipation, 221, 243–244, 246–247, 251, 254, 259, 265, 267, 270, 273, 279–280, 297, 312–313, 315, 617–618, 695
Enthalpy of vaporization, 817, 819, 826–827, 1008
Epistemic Uncertainty, 925–927, 933
Equilibrium constants, 182, 410, 441, 467, 479, 1095–1096
Equilibrium data, 471, 499, 504, 754, 762, 764, 768–769, 785
Equilibrium diagram, 427, 756
Equilibrium moisture content, 817
Equipment selection, 289, 353, 763, 770, 811, 823, 830, 833, 835, 839–840, 848, 977–978, 1020–1021, 1075
Ethyl pyruvate, hydrogenation of, 193–194, 201
Evaporation, 47, 52, 65, 70–71, 124, 143, 262, 293, 344, 353–354, 427, 431, 525, 547, 592, 684, 733–734, 816, 820, 829, 854–855
Excess enthalpy, 483–486
Exclusivity, 7–9, 14–15
Exothermic, xxvii, 29–30, 32–33, 35–36, 41, 44–45, 47–49, 51–52, 55, 57–58, 61–62, 65–66, 71–76, 80, 82, 85, 87, 89–90, 109, 113, 161, 165, 206, 211, 226, 294, 322, 325–326, 329, 332, 334, 342–343, 367–368, 378–379, 383, 481, 640, 657, 958–960, 962–963, 965–966, 975–977, 980, 988, 1004, 1006–1007
Exotherm initiation temperature, 64, 75, 89
Experimental design, viii, 176–178, 181, 183, 186, 206, 209, 236, 406, 883–890, 892, 894–900, 902, 904–908, 1002, 1098–1099, 1111, 1121
Explosion, v, 30–31, 34, 48, 51, 59, 61, 83, 89, 91–92, 94, 96–98, 100, 102–104, 106, 108–112, 114, 227, 369, 726, 863–864, 878–879, 977–978, 999, 1004, 1006–1007, 1010, 1029, 1033
Extraction, vii, 119, 203–204, 219, 221, 228, 241, 262, 321, 337, 353, 357–358, 365–366, 493, 585, 601, 604–605, 715, 733–734, 746–751, 753–754, 756, 758, 760, 762, 764, 766–770, 772, 778, 782, 784, 786, 922, 982–984, 986–988, 990, 1006, 1009, 1015, 1020, 1095, 1114
Extrinsic synthons, 692

Falling rate period, 819, 825–827
Fate and purge, 1093, 1102, 1105, 1117
F-curve transition, 373
Fed batch, 211, 219
Feed rate, 293, 333, 377, 617, 768, 812–813, 865–866, 868, 873, 998
Fenofibrate, 451–452, 461, 464, 466
Feret's diameter, 699
Fill fraction, 68, 383
Filter bag, 808–810, 812
Filter cloth, 799, 803, 806, 808–809, 811–815, 843
filter dryer, 26, 803, 808–812, 814, 820–823, 830–831, 835, 843, 847–848, 854, 859–860, 873–874, 984–985, 987, 996–997, 1011, 1015, 1021–1023, 1035–1036
Filter medium, 407, 799–801, 804, 806–808, 812, 814–815
Filter medium blinding, 814–815
Filter paper, 806, 996
Filter plate, 844–845
filtrate flux, 405, 407, 801–803, 812, 839
Filtration decision tree, 833
Filtration flux, 801–804, 812, 996, 1047
filtration modeling, 812
Filtration principles, 800
Finite capacity scheduling, 1038, 1065
Flowability, 564–566, 635, 690, 692, 696, 698–699, 701–703, 711, 862, 878, 997–998
Flow function coefficient, 698–699
Flow Number, 259, 292, 313, 619
Flow patterns, 51, 250–252, 281, 308, 957–958, 965
Fluid bed dryer, 822–823
FMEA (failure models and effects analysis), 61, 975, 1029, 1077, 1080, 1084, 1086, 1088
Fracture toughness, 691, 696–697
Freezing point, xviii, 71–72, 75, 156, 382, 788
Friction, 246, 393, 403, 691, 697, 702, 730, 801, 830, 864
FTIR (Fourier-transform infrared), xvi, 79–80, 333, 604, 640, 879, 910, 916, 939–941, 943, 945–947, 949–951, 980
Fugacity, 752, 754, 764
Full interaction map, 670–671, 674–675, 677–678, 686

γ-geraniol, 193
Gantt chart, 1051–1052
Gas chromatography, 118, 424, 433, 691–693, 710, 788, 824, 863, 1121
Gas generation, 29, 31, 57, 62–65, 69–73, 75–81, 87–88
Gas-liquid mass transfer coefficient (kLa), 160–161, 187, 192–200, 207–210, 220, 229–234, 238, 370, 392, 397, 410, 968, 989, 1096, 1102, 1121
Gas pycnometry, 691, 696
Gelation, 561–562, 567
geometric similarity, 242–243, 246, 248, 254, 308, 392, 965, 988
Gibbs–Duhem equation, 762
Gibbs Free Energy of Mixing, 434–436, 483
gPROMS custom modeler, 375
Grignard, 154, 159, 162, 165, 225, 358–359, 362–364, 366, 369, 377–379, 383, 385, 944, 980, 999

INDEX

Grignard formation reaction, 369, 377–378, 383
Growth characteristics, 602–606
Growth kinetics, vii, 299, 550–551, 554, 562, 567, 576, 591–592, 594, 596, 598, 600–602, 604–606, 608, 610, 612, 615–616, 622, 632, 636, 647, 694
Growth rate, 4, 7, 293–294, 299–300, 392–394, 551–552, 554, 556–558, 562, 601–607, 612, 630, 633, 636–637, 639, 643–644, 647, 653, 661–662, 692, 694, 703
gSAFT, 450–452, 454–455, 457–459, 461–463, 465

Hammer Mill, 866, 873
Hardness, 687, 691, 696–697, 704–707, 709, 712, 830, 863, 877
Hausner ratio, 540–541, 696
HAZOP, 60–61, 71, 75, 975, 1003–1004, 1010, 1029
Heat exchangers, 323–325, 332–337, 368, 1020
Heat Flow, xxviii–xxix, 33, 65–67, 72–73, 76, 79, 82–84, 86–87, 163–164, 173, 204, 208, 210, 213–215, 218–219, 224, 344–345, 987
Heat of decomposition, 82–83
Heat of reaction, xxvii, 32, 63, 70, 107–108, 111, 113, 162–163, 183, 206, 224, 322, 330, 335–336, 344, 382, 467, 479–481, 487, 958, 968
Heat of vaporization, 68, 467, 479–482, 487, 490, 742–743, 825
Heat removal rate, 379
Heat transfer coefficient, xviii, 33, 37–42, 64, 99, 101, 162–163, 183, 219, 224, 249–250, 259, 330, 335–336, 342–343, 346, 348, 354–355, 362, 559, 640, 655, 662, 787, 790, 825–827, 829, 853–854, 1003
Heat-wait-search, 49, 52, 70, 73–75, 79, 103
Heel, 381, 808, 810, 812–815, 844–845, 851, 867, 997, 1022–1023, 1038
Heel removal, 808, 812, 814, 851, 1022
Helical anchor, 252
Henry's law, xxii–xxiii, 68, 161, 207, 224, 693, 717–718, 752–753
Herschel–Bulkley, 249
Heterogeneous azeotrope, 757, 759
Heterogeneous hydrogenation, 966
Hierarchical Model, 925–926, 928
High Pressure Homogenizer, 869–871
Homogeneous, 22, 39, 155, 171, 174, 182, 191–192, 205–214, 218, 220, 226, 241, 250, 254, 294, 296, 298–299, 304, 316, 321, 328, 354, 368–369, 381–382, 395, 431–432, 547, 595, 601, 678, 746, 750, 757–758, 764, 768, 784, 893, 905, 907, 960, 962, 988, 1104, 1107
Homogeneous azeotrope, 757–758
Homogeneous catalyst, 191, 369
Homogeneous hydrogenation, 171
HPLC (high pressure/performance liquid chromatography), 116–129, 132, 144–145, 165, 169–170, 173, 181, 183, 204–205, 215–216, 225, 236, 361, 375, 430, 542, 575, 585–586, 594, 596–599, 612, 614, 729, 910–911, 916, 918, 943, 946, 948, 950–953, 967, 981–982, 993, 1092, 1096, 1098, 1105, 1116, 1118, 1120
Humidification, 847, 857–858
Humidified nitrogen, 857–858
Humidity, xxv, 32, 307, 426, 485, 490, 680, 686, 699, 817, 825, 828, 857–858, 883, 1017–1018
Hydrate, 26, 44, 68, 390, 400, 420, 423–424, 439–441, 444, 447, 449–450, 454–459, 462–463, 466, 485, 490, 498, 500, 517, 526–527, 536–538, 545–546, 560, 566, 571, 594, 596, 598–600, 602, 605, 668, 671, 675–684, 686, 690–691, 710, 739–740, 817, 825, 828–829, 831, 857–859, 873, 980–981

Hydrate prediction, 676
Hydrodynamics, 280–281, 294, 313, 315–316, 600–601, 617, 620, 631
Hydrogenation, vii, xxii, 32–34, 57, 64–65, 121, 160–161, 163–164, 171–172, 182, 191–202, 207–211, 220, 225–226, 228, 230, 241, 262, 275, 345, 367, 369–370, 375, 383, 385, 481, 750, 943–945, 947, 949–952, 966–969, 978, 984, 1006, 1018, 1020
Hydrogen bonding interactions, 445, 469, 670, 675, 677, 710
Hydrogen bond propensity, 670, 675–677
Hydrogen-rich conditions, 968
Hydrogen-starved conditions, 193
Hydrophilicity, 433, 702–703

Ibuprofen, 445, 448, 451–457, 462, 464–466, 522–523, 525–526, 530, 569–571, 635, 640, 657, 692, 710
Ideal PFR, 327, 350–351, 371–372, 375–377
Image analysis, 691, 696, 711, 840, 862
Immiscibility, 424–426, 431–432, 620–622, 754, 756–757, 764, 789
Impeller, xxiii–xxiv, 158, 197, 209, 220–221, 224, 226, 230–232, 234, 239, 241–249, 251–254, 257–259, 261, 263, 266–270, 272–279, 281–282, 284–286, 290–299, 301–302, 306–317, 355–356, 391–392, 395–397, 549, 617–620, 622, 625, 639, 642, 646, 650, 770, 790, 820–822, 960–965, 988, 1004, 1009
Impinging Jet, 299, 316
Incompressible cake, 801, 812, 835
Indent, 691, 696–698, 704–707, 709, 711–712, 863
Indentation, 691, 696–698, 704–707, 709, 711–712, 863
inertion, xxvi, 227, 368, 382, 864–865, 980, 1015, 1020
In-process limit, 1102–1103, 1117–1119
Interfacial, 161, 220, 229–231, 234, 236, 239, 256–257, 262, 265, 279, 281, 288–289, 306, 308, 310, 314, 326, 357, 397, 433, 592, 692–693, 768–770, 964–965, 969
Interfacial tension, 279, 289, 592, 769
Intermolecular interactions, 440, 442, 463, 486, 561, 670, 674–678, 686, 692, 703, 706
International Conference on Harmonization (ICH) Quality Guidelines, 4, 19, 22, 60, 91, 114, 187, 189–190, 225–226, 314–317, 339, 367, 409, 416, 423, 467, 569–570, 635, 695, 857, 878, 909, 918, 921–922, 935, 957, 974, 981–982, 986, 999, 1002, 1011, 1029–1030, 1034, 1037, 1069–1071, 1073–1074, 1079, 1081, 1089, 1091–1092, 1100, 1103, 1121–1122
Interplanar spacing, 706
Intrinsic synthons, 692
Inventory management, 1026, 1061
Inverse gas chromatography, 433, 691–693, 710, 863
Inverting bag centrifuge, 810–812, 814, 850, 1021–1022
Ishikawa diagram, 306–307, 1076, 1082, 1105, 1113–1114
Isothermal age, 67–68, 75–76, 85

Jet Mill, 711, 865–868, 873, 877–879, 998, 1020–1021

Karl Fischer titration, 424, 1121
Ketoprofen, 453–455, 462, 464, 466
Kilo lab, viii, 10, 97, 203, 219, 333, 835–838, 840, 974, 1001–1003, 1008, 1011–1018, 1020–1026, 1028–1036
Kinetic model, 56, 105–106, 109, 111, 117, 122, 124, 127–128, 130–131, 154, 157, 169, 173, 176, 180–182, 184, 186, 201, 205, 207, 209, 214, 219, 225, 329, 335–336, 348, 351, 361, 556, 604, 655, 906, 967–968

1128 INDEX

Knowledge space, 1099–1100, 1102
Kolmogorov length scale, 243, 267, 313, 631

Laboratory drying equipment, 820
Laminar flow, 159, 183, 287, 290, 316, 323, 326–327, 336, 349, 368, 1025, 1034
Laminar regime, 247, 292, 316, 348, 372, 374
Laser diffraction, 624, 691, 696, 848, 909, 981, 1121
Lattice Energy, 562, 671–672
Leaf filter, 807–808, 813, 848, 996
Lean Manufacturing, 15–16, 18, 1038
Light scattering, 118, 538, 862, 915, 1121
Linifanib, 672–680, 686
Lipitor, 7, 9–10, 12, 14
Liquid–Liquid Extraction, 366
Liquid–liquid phase separation (LLPS), 561, 595, 667–668, 685
Long-term production (long-term planning), 919, 1062
LOPA (layer of protection analysis), 75, 89, 279–280, 315, 563, 668, 676
Loss on drying, xxv, 824, 981–982

Markov Chain Monte Carlo (MCMC), 926–931, 933, 935
Mass spectrometry, 165–166, 825, 1121
Mass Transfer, vii, xxi–xxii, xxix, 22–23, 80, 120, 153–161, 163, 170, 172–175, 185, 187–188, 191–197, 199–201, 204, 207–210, 212–213, 218, 220–221, 224, 226–240, 243, 250, 256, 262, 264–265, 269, 272–273, 276, 279, 294, 299, 306, 308, 310, 314, 321–323, 325–329, 332, 337, 342, 346–348, 351, 353–354, 357–358, 365, 370–371, 379, 383, 389, 392, 395–398, 400–403, 409–410, 412, 416, 504, 546, 554, 617, 631, 638, 661, 724, 736, 738, 742, 745, 748, 768, 770, 817–819, 825–827, 829–830, 847, 850–853, 856–857, 860, 957–958, 960–963, 966, 968–969, 974–975, 987–989, 1010, 1014, 1095–1097, 1104, 1121–1122
Material balance equation, 293, 370–371
Material of construction (MOC), 65, 72–73, 75, 348, 406, 808, 810, 863, 979, 996, 1011, 1020
Materials science, vii, 406, 532, 689–690, 692, 694, 696, 698, 700, 702, 704, 706, 708, 710, 712
Materials science tetrahedron, vii, 689–690, 710
Maximum allowable working pressure (MAWP), 63, 70, 73–74, 87
Maximum boiling azeotrope, 757
mCPBA (3-chloroperbenzoic acid), 85, 87–88, 90
Mean ionic activity coefficient, 493–494, 497–499, 502
Mean residence time (t), 250–251, 259, 322, 327, 368, 371, 374, 377, 628
Mechanical properties, 438, 519, 560, 667, 690–692, 694–697, 703–704, 706, 709–712, 876
Mechanistic, 56, 105, 153–156, 165–166, 169–170, 172, 176–179, 186, 206, 209, 213–214, 217, 225, 345, 348, 359, 361–362, 409, 553, 636, 639, 657, 703, 863, 872–873, 875–876, 879, 884, 900, 924, 935, 942, 957, 969, 1076, 1093, 1095–1098, 1100, 1104–1107, 1110–1115, 1118
Mechanistic model, 179, 186, 206, 209, 213–214, 217, 361, 636, 872–873, 875–876, 879, 900, 935, 1095, 1100, 1104–1107, 1110–1115, 1118
Media mill, 685, 869, 871–872, 879
Medium resistance, 800–802, 804, 806, 812–813, 815

Melting, xix, 46, 68, 262, 419–423, 437, 442–444, 447–449, 451–452, 457–458, 460–463, 472–473, 505–506, 508–509, 511, 513–515, 517–518, 537–538, 544, 573, 575–576, 586–587, 592, 594, 665, 668, 674–675, 681, 842, 854, 865
Melting point, 46, 68, 419, 421–422, 443–444, 447, 451–452, 457, 460–462, 472–473, 505–506, 508–509, 511, 515, 517–518, 537–538, 544, 573, 575–576, 586–587, 594, 665, 674, 681, 842, 865
Mesomixing, 211–214, 293, 310, 316, 631, 959–962, 969
Metastable zone width (MSZW), 548–549, 595, 601, 607, 609–612, 615–616, 639, 662–663, 994–995
Metzner–Otto, 247, 259
Micromixers, 325–327, 365–366
Microreactor, 58, 162, 175–176, 183, 226, 322, 325–330, 332–333, 338–339, 352, 365, 368, 384, 962, 969
Microscopy, 237, 420, 422, 431, 437, 552–553, 575, 605, 615, 620, 623, 650, 679, 691–692, 694–696, 705, 711–712, 805–806, 848, 862–863, 981
Mill configuration, 866, 875–876
Milling process, 698, 711, 863, 872–873, 875–876, 994
Minimum boiling azeotrope, 757, 760, 784–785, 828–829
Mixed solvent electrolytes, 493, 497–498
Mixers, 159, 221, 226, 250–251, 254, 257, 279, 290, 292, 316, 323, 325–327, 338–339, 346, 352, 354, 363, 365–366, 628, 965, 1021
Mixing-limited reactions, 246, 255–256, 960
mixing number, 826, 854
Mixing turnover time, 852
Model-free kinetics, 84
moisture content, xxv, 433, 436, 817, 819–820, 826–827, 829, 1121
Molecular Crystals, 419, 517, 531, 670, 684, 686, 692, 694, 697, 706, 711
Molecular diffusivity, 718, 817, 826
Molecular packing, 419, 670, 673, 677–678, 687, 690, 698, 707
Monotropic, 505, 509–511, 513, 516–517, 527, 537–538, 842
Monte Carlo simulation, 927, 1038, 1045–1047
Morphology, 438, 520, 531, 536, 539, 550, 593, 607, 615, 617, 628–629, 631–632, 679, 691–692, 694–695, 703, 706, 710, 805, 819, 824–825, 831, 862, 879, 979, 998, 1015
MSZW see Metastable zone width (MSZW)
Multiphase mixing, 281, 964, 966
Multiple module calorimeter (MMC), 62, 69, 71, 76–81, 85, 87–89
Multivariate, viii, 179, 185, 409, 412–414, 416, 883–884, 909, 912, 914–916, 918, 920–921, 934–935, 957, 963, 967–969, 1071, 1091, 1093, 1095, 1100, 1104, 1120

Nano-milling, 665, 667, 879
Near-infrared spectroscopy (NIR), 333, 594, 825, 828, 831, 915, 937–938, 942–943, 945–948, 1020
Newtonian fluids, 249, 251, 286, 292
Newton's Law of Cooling, 99
Nitration, 32, 58, 154, 322, 332, 338, 342, 365, 384–385, 962–963, 965–966, 969, 1004, 1006
Nitrogen blow, xxi, 830, 844, 848–849, 855
Nitrogen sweep, 852–854, 857
Non-agitated filter dryer, 810–812
Non-Newtonian fluids, 249, 251, 286
Non-Random Two-Liquid (NRTL), 68, 207, 494–504, 563, 567, 596, 599–600, 736, 754, 761–762, 764, 767–768, 770–772, 774–777, 780–781, 783–787, 789–793, 829

Normal operating range, 402, 983, 995, 1078, 1093
Noyes–Whitney equation, 229, 678, 861
NRTL binary interaction parameters, 504, 754, 793
NRTL-SAC solubility model, 600
Nucleation rate, 188, 293, 299–300, 550, 602, 605, 607, 609, 611–612, 630, 639, 643–644, 647, 653, 661–662, 797, 993
Numerical diffusion, 287, 375
Nusselt number, xviii–xix, 250, 259, 389
Nutsche filter, 843, 1013, 1021–1022, 1039, 1042, 1044

Oiling, xviii, 31, 38, 40, 48, 52, 61, 65, 68, 71, 73, 75, 91, 119, 155–156, 230, 239, 382, 439, 465, 547, 553, 561, 669, 736, 756–764, 780, 783–786, 788–789, 791, 793, 816–817, 819, 826, 828–829, 849, 859, 991, 1006
Online mass spectrometry, 166
Onset temperature, 45–46, 49, 51–52, 54–55, 64–66, 68, 71–73, 82–83, 85, 89, 103–104, 113, 1004, 1006–1007
Operating space, 22–23, 170, 342, 587, 768–769, 792, 1100, 1102–1103, 1117–1118
Osmotic coefficient, 493–494, 497–498, 503–504
Oven balance, 820

Packed catalyst bed, 369, 382–383
Pair distribution function, 691, 698, 704, 707–709, 712
Pan dryer, 820–822, 1023
Parameter estimation, 106, 180–181, 344–345, 348–349, 457, 564, 633, 640, 643, 647, 770–771, 775
Parameter selection, 797, 1099
Parecoxib, 419–423, 437
Paritaprevir, 423–426
Parity plot, 598, 1097–1098, 1100, 1107, 1109, 1111, 1113, 1117
Partial Least Squares, 412, 698, 701, 910, 914, 940
Partial least squares regression, 698
Particle attrition, 620, 796, 819, 823, 830, 848, 850–853, 855, 860, 1022
Particle breakage, 236–237, 292, 697, 830, 855, 862–863, 871, 873, 875–876, 879
Particle size distribution, 24, 211, 231, 236–237, 241, 266, 281, 292–293, 302–303, 357, 536, 538–539, 565, 591, 616–617, 620, 624–625, 637, 640, 654–657, 662–663, 695, 794, 819, 823, 838, 841, 847–848, 851, 861, 878, 919, 975, 993–995, 997–998, 1080–1082, 1085, 1092, 1121
Pd/C Catalyst, 197–198, 200
Peeler centrifuge, 810, 850, 996, 1021–1022
Peritectic Melting, 422
Permanent gas, 47, 51, 62–63, 68–70, 73–74, 85
Personalized medicine, 16, 18
PFR (Plug flow reactor), 218, 221, 250, 285, 323, 325–330, 332–338, 350–351, 353–356, 362–363, 367–377, 380–383, 637, 958, 965
pH, 76, 79–81, 121, 164, 255, 257, 390–398, 410–412, 439, 450, 453–457, 462–463, 466, 593–596, 628, 632, 666, 672, 681, 721–723, 729–730, 737, 770, 816, 982, 988
Phase diagram, 420, 422, 424–427, 431–433, 438, 453–455, 457–459, 461–463, 466, 471, 520, 545–546, 560–561, 571, 573–574, 576–581, 584–586, 589, 604, 620–622, 667, 679–680, 682, 684, 753–754, 756–760, 762, 764–768, 781, 783–786, 790, 828
Phase equilibria, 450, 456, 465, 503, 585, 817, 1122

Phase equilibrium, 207, 427–430, 440–442, 444, 457, 463, 465–466, 500, 574, 578, 751–754, 756, 764, 767–768
Phase rule, 567, 751, 762
Phi Factor, 64, 89, 94, 99, 101, 113
pH-solubility, 666
Physical properties, vii, 33, 39–40, 42, 138, 185, 230, 295, 348, 358, 407, 419, 437, 440, 450, 466, 505, 535–536, 542, 546–547, 550, 554, 573, 584, 593, 605, 607, 615, 618–620, 622–623, 627–628, 631–632, 678, 689–692, 694–696, 698–700, 702–704, 706, 708, 710, 712, 769, 786, 795, 797, 799, 829, 851, 854, 858, 863–865, 876, 878, 940, 975, 993, 998, 1021, 1029, 1081, 1093
Pilot plant, viii, 10, 17, 64, 91–92, 97, 120, 165, 186, 203, 210, 219, 227–228, 239, 259, 281, 334, 336–337, 341, 363, 608, 624–625, 749, 762, 797, 802–803, 812–814, 817, 831, 843–844, 848, 850–851, 853–856, 916, 918, 950–952, 957, 961, 963, 965, 969, 974–975, 979, 993, 999, 1001–1003, 1005, 1007, 1010–1026, 1028–1036, 1039, 1042, 1117
Pin Mill, 707–710, 865–867, 1021
Pitzer ion-interaction model, 494–495
pKa, vii, xii, 76, 121, 189, 261–262, 267–269, 271–278, 284–286, 294–295, 308–312, 314–317, 441, 457, 462, 465, 596, 666, 681–683
Plait point, 764, 766
Plastic deformation, 696
Plug Flow, 154, 162, 218–219, 221, 226, 250, 281, 285, 323–324, 327–329, 331, 333, 335–336, 350–351, 354–355, 362, 367–369, 371–372, 374, 376, 385, 642, 645, 655, 718, 816, 958
Plug flow reactor (PFR), 162, 218, 221, 226, 250, 285, 323, 325–338, 350–351, 353–356, 362–363, 367–377, 380–383, 385, 637, 958, 965
plug flow with dispersion reactor (PFDR), 368, 372–374
Point of failure, 65, 71, 73, 862
Poisson's ratio, 697, 704
Polymorph, vii, 254, 298–299, 419, 422, 427, 437, 439, 441, 444, 457–458, 460, 462–463, 465–466, 484, 486, 498–500, 505–506, 508–520, 522, 524–532, 536–538, 542, 546, 558–560, 562, 566–567, 571, 575–576, 595, 620, 635, 665, 668–673, 679, 681–682, 684–686, 689–692, 703, 710, 763, 825, 828, 843, 981–982, 995, 1001, 1015, 1021–1022, 1081
Polymorph control, 298, 995
Polymorphism, 419, 437, 465, 486, 505, 517–519, 528, 531, 536–537, 558, 560, 562, 566–567, 670, 684–686, 689–690, 710
Population Balance Equation, 279, 602, 636, 640, 657, 872
Pore size distribution, 693
Powder X-ray diffraction, 422, 424, 575, 691, 698, 707–708, 1121
Power intensity, 243
Power number, xxiv, 244–245, 259, 272–273, 278–279, 306, 312–313, 389, 392
Power per unit volume, 262, 299, 392, 619–620
Power/volume (P/V), xix, 242–243, 245–248, 301, 392, 437, 685
Prandtl number, xix, 265, 274–275, 313
Pressure consequences, 61, 72–73, 75, 85
Pressure drop, 51, 75, 160, 195–196, 324, 328, 348, 354, 357, 369, 374, 383, 393, 402–403, 405, 407, 716–717, 726–728, 738, 800–802, 804, 829, 845, 854, 1009
Pressure filtration, 800, 804, 806, 813, 816, 831, 834–837, 841, 843, 848–849, 996
Prigogine–Defay equation, 575, 587
Principal Component Analysis (PCA), 175, 185–186, 412–414, 874–875, 912–914, 916, 940

Principal Component Regression, 910, 914, 940
Principal components, 185, 413, 874–875, 912–914
Probabilistic Graphical Models, 926, 935
Probability, 24, 156, 183–184, 186, 222–224, 315, 350, 392, 484–485, 502, 514–516, 675, 697–698, 709, 711, 858, 877–879, 890–891, 893, 902–904, 919, 922–927, 929, 933–934, 957, 974–975, 983, 1045–1048, 1065, 1070, 1074, 1078–1079, 1081, 1083–1088
Probability distribution, 183–184, 222, 350, 484, 502, 698, 709, 924–927, 933, 1045–1048, 1065
Process analytical technologies, 343, 1096, 1121
Process analytical technology (PAT), viii, xv, 3, 16–18, 23, 164–166, 169, 176, 186, 189, 236, 311, 333, 339, 343, 344, 347, 367, 382, 389, 416, 426, 564, 594, 596, 636, 650, 726, 730, 847, 881, 909, 918, 937–938, 940, 942–946, 948, 950–954, 956, 974, 992, 997–998, 1012, 1020–1022, 1065, 1069–1070, 1078, 1087, 1096, 1105, 1112, 1121–1122
Process Capability, 861, 921–922, 934, 1000, 1070–1071, 1073
process-induced defect (process-induced disorder), 690, 696, 698
Process Robustness, viii, 13, 153, 157, 186, 351, 365, 919, 921–924, 926–932, 934, 936, 968, 979, 982–983, 999, 1014, 1026, 1070, 1098, 1105
Process safety, v, xvi, 29, 31–32, 34, 36, 38, 40, 42–46, 48, 50, 52, 54, 56, 58–62, 64, 66, 68, 71, 85, 89–91, 114, 153, 156, 161, 164, 199, 204, 218, 225, 322, 332, 363, 385, 863, 879, 966, 973–974, 977, 999, 1010, 1012, 1018–1019, 1029–1031
Process scheduling, 1042, 1056
Process scheme, 205, 207, 210–211, 1095–1097
process simulation, 343, 351, 353, 365, 493–495, 503, 753, 1037–1039, 1045, 1065
Production scheduling, 1038, 1047, 1052, 1062, 1065
Proven acceptable ranges (PAR), 217, 875, 1086, 1093–1094
p-toluenesulfonic acid, 79

Qr, 41, 204, 206, 208, 210, 213, 215, 344–345, 413, 479, 489, 999, 1073–1074, 1089
Quality attributes, 177, 179, 186, 345, 393, 401, 413, 419, 563, 635, 655, 657, 729, 848, 862, 921–925, 958, 975, 979, 982, 1005, 1069–1071, 1073–1074, 1082, 1092, 1094, 1105, 1114
Quality by design (QbD), ix, xvi, 3–4, 15, 18, 22–23, 179, 203–204, 217, 219, 222, 224–226, 317, 351, 367, 409, 416, 564, 635, 657, 662, 873, 919, 921, 935, 1067, 1069–1073, 1076, 1078, 1080, 1082, 1084, 1086, 1088–1090, 1093, 1100, 1104, 1121
Quality target product profile, 957, 1070, 1073, 1091
Quantum mechanics, 676

Racemic, 193, 531, 570–578, 580–587, 589, 746
Radial flow, 251–253, 619, 716, 726
Raffinate, 724–725, 766–768
Raman spectroscopy, 166, 169, 204, 365, 703, 712, 825, 828, 940–942, 952, 1118
Raoult's law, xxi–xxii, 68, 752–754
Rate constant, xxvii, 107, 127, 138, 146, 155–156, 159, 161, 163, 165, 171, 176, 181, 204–206, 209, 212–213, 215–216, 224, 233, 239, 256, 329, 344–345, 361–362, 370, 372, 379, 383, 394, 550–551, 556, 607, 609, 631, 643–644, 647, 854, 963, 967, 969, 988, 1096
Rayleigh equation, 761
Reaction calorimetry, 33–34, 42, 60, 163–164, 172–173, 360, 365, 1007

Reaction kinetics, vii, xv, 22–23, 53, 107, 146, 151, 153–166, 168–172, 174–178, 180–182, 184, 186–188, 190, 192–193, 199–200, 203, 212, 226, 232–234, 256, 281, 310, 326, 329, 333, 336, 365, 369, 385, 394, 909, 957, 963, 968–969, 989
Reaction rate, xxvii, 31, 35, 46–48, 52–53, 61, 83, 107, 122, 127–128, 154–156, 159–163, 165, 169, 171–174, 183–185, 187, 193, 206, 220, 228, 233–236, 239, 241, 256–257, 282, 294, 310, 327, 329–330, 368, 370–371, 380, 382–383, 631, 948–950, 952–953, 959, 964, 967–968, 978, 1046, 1075, 1081, 1107, 1116
Reaction thermodynamics, vii, 467, 470, 472, 474, 476, 478, 480, 482, 484, 486, 488, 490, 492
Reactive crystallization, 369, 541, 547, 552, 554, 558–559, 617, 628, 630, 632, 796
Reactive electrolytes, 500
Recipe block diagram, 1051
Recirculation Loop, 257, 617, 623, 625–628, 742, 770, 864, 868, 872, 875, 938, 940, 1020–1021
Relative volatility, 118, 753, 757, 760–762, 786, 793
Residence time, 46, 145, 161, 175–176, 213, 218, 221, 250–251, 259, 281, 284–286, 289, 315, 322–324, 326–330, 332–334, 336–337, 339, 350, 354, 358, 360, 362–363, 366, 368–369, 371–372, 374–375, 377, 379–381, 383, 401, 625, 628, 640–642, 644, 649, 655–657, 659, 707, 865–868, 872–873, 876, 878
Residence time distribution (RTD), 250, 259, 281, 284–286, 294, 298, 308, 326–328, 339, 350–351, 354, 363, 368–369, 372–374, 401, 641, 647, 655, 868
Reslurry wash, 816
Response surface(s), 176–178, 186, 217–218, 221, 223, 884, 898, 901, 905–907, 1100, 1102–1103, 1107–1110, 1113, 1117–1119, 1121
Retention volume, 693–694
Revenue, 4–7, 9, 14, 19–22, 25–26, 341, 1045
Reynolds number, xxiii–xxiv, 157–158, 232, 239, 245, 247–250, 259, 290–292, 295, 312–313, 336, 354, 365, 372, 374, 389, 395–396, 988
Rheology, 249, 259, 263, 288, 299, 316, 831
Ring shear tester, 691, 698
Risk assessment, ix, 26, 31, 60, 89, 179, 189, 505, 516, 520, 560, 669–670, 673, 875, 973–976, 1069–1071, 1073–1086, 1088–1090, 1092–1094, 1099, 1104–1105, 1113
Ritonavir, 505, 509, 515, 517, 528–529, 532, 538, 566, 670–672, 684–685
Rotary cone dryer, 820–822
Rotational speed, 213, 224, 243, 245, 247, 259, 273, 277, 294–299, 301–302, 308, 313, 804, 830, 866–867, 988
Rotor–stator, 259, 879, 1000
Runaway, 29–30, 34–35, 47–52, 54–55, 58–59, 61–62, 73, 91, 94, 96–97, 103–104, 108–109, 111, 154–155, 206, 281–282, 315, 322, 332, 342, 365, 367, 378, 978, 989, 1010
Runaway reaction, 29–30, 34–35, 47–48, 50–52, 54–55, 58–59, 61–62, 91, 103, 109, 155, 281–282, 315, 322, 332, 342, 365, 978, 989, 1010

SADT (self-accelerating decomposition temperature), 68, 85–86, 92, 96–97, 101–103, 109–111 see also Accelerating Rate Calorimetry (ARC)
Safety concerns, 31, 34, 71, 89, 322, 369, 863, 899, 966, 1008
SAFT, 443–448, 450–465, 667

Salt, 44–45, 64, 75–76, 79, 118, 121, 149, 160, 173, 178, 239, 403, 439–441, 444, 447, 449–450, 457, 462–463, 486, 493–494, 496–500, 503, 536, 558, 560, 566–567, 571, 583–585, 593, 596, 628–629, 647, 665–666, 668, 671–672, 681–684, 690–691, 711, 716, 722–723, 731, 769–770, 925, 935, 958, 979, 981–982, 986, 988, 994, 1004, 1009, 1020, 1040–1042, 1105–1107, 1113, 1115
Sample selection, 64
Scale dependence, 64, 203–204, 995
Scale independence, 203, 209, 1104
Scanning probe microscopy, 694, 705, 711
Scanning technique, 70
Scan rate, 66, 84, 86, 422
SchedulePro, 1038, 1049, 1056–1057, 1062
Schroeder-van Laar equation, 575–576
Scores, 185, 414, 702, 901, 913–914, 1078, 1087
Seed, 292, 393, 395–396, 426, 429, 437, 540, 549–557, 559, 561, 565, 582–583, 593, 597, 600–602, 605, 612, 615–616, 623–624, 628–629, 636, 640, 655, 663, 679–680, 682, 684, 740, 805, 861, 931, 994–995, 1015, 1081–1082, 1085, 1114
Seeded crystallization, 555, 636, 805, 1081
Selectivity, 143, 155–156, 159–161, 163, 176, 184–186, 193–194, 197, 199–201, 221, 254–257, 259, 294, 321–322, 327, 329, 351, 356, 359, 367, 369, 381–383, 400, 519–520, 542, 562, 594, 715–718, 720–721, 723, 730, 735, 737, 747–750, 753, 764, 766, 969, 1001, 1006–1007
Self-reactive, 85, 89, 97, 101, 103, 111
Semenov, 101–103, 109–110
Semi-batch, xxvi, 33, 35, 48, 130, 166, 211, 251, 254, 256, 258, 316, 337, 353, 365, 380, 554, 583, 746–747, 909, 967, 989, 1001, 1015
Sensitivity, 43–45, 51, 53, 64–65, 67, 83, 90, 118, 134, 136, 140, 142–143, 153–154, 159, 165, 169, 180–181, 186, 227–228, 233, 236, 280, 307, 314, 345, 348–350, 353, 362, 516, 594, 689, 694, 697, 726, 728–730, 767, 824, 842, 916, 922–923, 934, 937, 940, 942, 946, 949–950, 991–992, 998, 1000, 1004, 1037–1039, 1047, 1084–1085
Sensitivity analysis, 348, 353, 362, 1038–1039, 1047
Shaker-flask method, 592, 594
Shear frequency, 869–870, 877, 998
Shear number, 870, 875–877
Shear rate, 242–243, 246–249, 259, 262, 294, 298, 304, 618, 676, 830, 869, 878
Shear-thinning fluids, 249
Shelf life, 57, 68, 665, 667–668
Shock sensitivity, 43–44, 1004
σ-profile, 469–471, 473, 502
Similarity, 118, 141, 204, 207, 209, 214, 242–243, 246, 248, 254, 281, 308, 392, 421, 429, 770, 960, 965, 988
Size reduction, 304, 536, 542, 550, 665–667, 672, 678, 697–698, 853, 861–862, 864–866, 870, 876, 878–879, 994, 1000, 1015, 1021
Slurry properties, 805
Sodium Borohydride, 91–92, 104–106, 111
Solidarity, 696, 699
Solid form screening, 560, 667–668, 670, 673, 675–676, 680–681, 684
Solubility prediction, vii, 185, 445, 447, 467, 471, 473, 476, 486, 490, 505, 508, 510–512, 514, 516, 518, 595–596
Solubility product constant, 497–499
Solution hydrogen concentration, 192
Solution-mediated phase transformation, 437, 681

Solvate, 65, 68, 420–424, 438–441, 444, 447, 449, 463, 485, 490–491, 506, 509, 519, 536–538, 545, 560, 566, 571, 575, 594, 596, 673–676, 678, 681–684, 686, 690–691, 817, 819, 824–825, 828–829, 831, 848–849, 853–855, 857–860, 981–982, 986, 994, 997
Specific cake resistance, 801–803, 805–806, 808, 812–814, 835–836, 838
Specific surface area, 236, 346, 551, 554, 679, 691–694, 698, 830, 863, 872, 1005
Spin speed, 810, 812–813, 816, 996
Spontaneous nucleation, 805, 994–995
Spray dryer, 822–824
Square-cube effect, 64
Stabilizer, 91–92, 97, 103–105, 108–109, 111, 114, 279, 415, 872
Static drying model, 825
Static mixer, 221, 250–251, 316, 323–324, 327, 329, 333–339, 354–355, 363, 965
Statistical model, viii, xv, 176, 226, 351, 412, 563, 670, 675, 686, 873, 881, 922
Statistical Process Control, 416, 919
Steady State, 175, 232–233, 264, 326–327, 334, 360, 368–369, 371–372, 374, 376–377, 379–381, 383, 396, 591, 642, 654–656, 818, 825, 870, 964
Stepwise isothermal, 69–70
Sticky point, 830
Stochastic, 521, 523, 662, 924–925, 935, 1046
Strip-and-replace distillation (Solvent swap), 757, 760–762, 985
Structural informatics, 670, 673, 675, 677
Substrate to catalyst ratio (S:C), 369
SuperCRC, 79
SuperPro Designer, 1039, 1042, 1044, 1046, 1049, 1065
Supersaturation, 292–294, 298–300, 427, 429–430, 502, 519–520, 531, 546–559, 561–562, 565–567, 583, 593–594, 596, 598, 600–602, 604–606, 611–612, 614, 618, 623–625, 627, 629–632, 639–640, 644, 648, 654, 657, 667–669, 676, 684–685, 690, 703, 795–797, 843, 879, 940, 994–995, 1000
Supramolecular synthons, 491, 674, 686
Surface chemistry, 692, 694, 702–703, 710, 863
Surface energies, 693, 699, 701, 703, 710
Surface tension, xxiv, 357, 694, 965, 1009
Surface topography, 691, 694
Swept volume, 243, 310
Synchrotron, 696, 704, 709
Synthesis, 17, 23, 26, 29–30, 32–33, 42–44, 48, 51, 85, 91, 120, 154, 186, 191, 200, 203–204, 218–219, 225, 321, 338–339, 345, 358, 360, 365–366, 384–385, 390, 393–395, 439, 490, 535–536, 558, 567, 571, 587, 589, 686, 715, 721, 723, 731, 744, 746–747, 749, 763, 864, 969, 973–976, 981–983, 986, 993, 999–1000, 1009, 1012, 1019, 1037, 1039–1040, 1049, 1062, 1080–1081, 1092–1094, 1102

Taylor longitudinal dispersion coefficient, 371, 374
Technology transfer, 420, 847, 934, 953, 959, 962, 966, 1001–1002, 1038, 1065
Tensile strength, 691, 696–697
Ternary Phase diagrams, 438, 571, 757–758, 760
Tetrasubstituted enone, 369
Texture analyzer, 691, 697
Thermal conductivity, xviii, xx, 37, 98, 249–250, 259

Thermal Runaway, 91, 94, 97, 103–104, 108–109, 111, 154, 206, 322, 367
Thermodynamic consistency, 762–763, 1096
Thermodynamic modeling, vii, 438, 466, 486, 493, 495–496, 498, 500–504
Thermodynamic properties, 421, 446, 463, 466–467, 469–470, 493–495, 502–504, 517, 587, 797
Thermodynamics, vii, xv–xvi, 111, 188, 200, 262, 306, 382, 417, 419, 421–422, 424, 426, 428, 430, 432, 434, 436, 438, 440, 463, 465–467, 469–472, 474, 476, 478, 480, 482, 484, 486, 488, 490, 492–493, 502–503, 505, 508, 510, 512, 514, 516, 518, 532, 536, 571, 576, 589, 675, 680, 734, 741, 751–752, 786, 797, 957, 988, 1122
Thermokinetic barrier, 67
Throughput analysis, 1065
Tip speed, 241–243, 246–248, 251, 253–254, 259, 262, 392, 411–412, 619–620, 624, 830, 848, 851, 853, 869–870, 873, 877, 988, 998, 1009
TMR (time to maximum rate), 30–31, 46, 48–56, 68, 85–86, 89, 95–96, 104, 113, 978, 1004
TMR24 (TD24), 68, 85, 89, 96–97, 103
Top-selling drugs, 10, 12, 14
Torque, 245–246, 258–259, 302, 819–820, 830, 988, 997
Total diffraction, 696
Total surface energy, 699, 701, 710
Translation equations, 247–248
Transportation, 15, 23, 44, 59, 78, 90, 97, 101, 103, 110, 402, 980, 1031, 1033
Tray dryer, 820–821, 823–824, 848, 1013, 1015, 1022, 1039, 1042
Troubleshooting, 392, 402, 412, 414, 814, 829, 862, 1012, 1014, 1021, 1023, 1026
True density, 421, 696, 712
Tumble dryer, 820–824, 830, 1015, 1022
Turbulent flow, xxiv, 232, 245–246, 249, 314, 349, 354, 391, 396, 405, 617, 633, 871

Uncertainty, 6, 23, 101, 143, 158, 180–181, 221–222, 224, 232, 259, 273, 431, 476, 496, 897, 922–929, 931, 933, 935, 1038–1039, 1045–1046, 1071, 1109
Undesired reaction, v, 61–62, 64, 66, 68, 70–76, 78, 80, 82, 84, 86, 88, 90, 125–127, 153, 206, 322, 329, 952, 989
Uniform washing, 816
Unit Operations, vii, 10, 22–24, 203–204, 206, 208, 210, 212, 214, 216, 218, 220, 222, 224, 226, 228, 293, 321–322, 337, 342–343, 353, 369, 389, 393, 399, 439, 493, 536, 539, 541–542, 695–696, 703, 751, 769, 786, 795, 799, 808, 831, 862, 865, 872, 876, 922–923, 973–974, 976, 984, 986–988, 991–992, 999, 1014–1015, 1018, 1021, 1026–1027, 1030, 1065, 1070, 1093, 1095, 1102
Uptake, 8, 160, 163–164, 184, 195–199, 209–210, 220, 230, 314, 392, 394, 397, 433–437, 667–668, 694, 989

Vaccine, 10–11, 14, 20, 23, 25, 415
Vacuum control valve, 824
Vapor pressure, xxi–xxii, xxv, 47, 51–52, 62–63, 66, 68–70, 85, 88, 122, 160, 195, 208, 367, 435, 446, 450, 471, 479–482, 489, 493, 497–499, 503–504, 692, 735–736, 752–754, 762, 790, 817–818, 870
Variability analysis, 1045, 1065
Variable importance in projection, 702
Vent sizing, 42, 47–48, 51, 59, 64, 70, 74–75
Verification, 180, 205, 217, 221–222, 344, 348, 350–351, 353, 365, 422, 509, 593, 604, 643, 657, 659, 921, 1039, 1070, 1095, 1097–1098, 1100, 1103–1105, 1107–1111, 1113, 1116–1117, 1122
Vessel vent capacity, 79
Vibration, 166, 478–482, 488–489, 521, 712, 810, 812–813, 825, 938, 940–944
Viscosity, xx–xxi, xxiii–xxiv, 33, 37, 39–40, 158–159, 191–192, 230–231, 239, 243, 245–246, 249, 252, 259, 265, 267–268, 279–280, 286, 289, 292, 294, 299, 301–302, 305, 313, 315–316, 374, 396, 399, 433–434, 436–437, 541, 562, 716–717, 728, 741, 745, 769, 800, 802–803, 806, 813, 816, 829, 834–836, 879, 959–960, 969
VisiMix®, 243–244, 259

Waste streams, 56–57, 734, 740, 750, 987, 1003
Weber number, 289, 965
Wet cake properties, 819, 996
Wet milling (Wet-milling), 622, 668–669, 698, 703, 863–866, 868–869, 875, 879, 994, 998, 1000
Wettability, 299, 307, 692, 710
Wurtz coupled impurity, 378

X-ray powder diffraction (XRPD), 518, 575, 594, 596, 599, 679, 703, 825, 828, 849, 854, 857, 863, 981

Yield stress, 249, 254, 259, 305, 316, 697
Young's Modulus, 691, 694–697, 704, 706–707